Saturn from Cassini-Huygens

Michele K. Dougherty • Larry W. Esposito
Stamatios M. Krimigis
Editors

Saturn from Cassini-Huygens

Springer

Editors

Michele K. Dougherty
Imperial College
The Blackett Laboratory
Prince Consort Road
London
United Kingdom SW7 2AZ
m.dougherty@imperial.ac.uk

Larry W. Esposito
University of Colorado
Boulder
Boulder CO 80309-0449
USA
Larry.Esposito@lasp.colorado.edu

Stamatios M. Krimigis
Johns Hopkins University
Applied Physics Lab.
11100 John Hopkins Rd.
Laurel MD 20723
USA
and
Center for Space Research and Technology
Academy of Athens
Soranou Efesiou 4
Athens 11527
Greece
tom.krimigis@jhuapl.edu

Front cover figure: This false-color image of Saturn was constructed from three infrared images taken at 2, 3, and 5 microns wavelength respectively cast to red, green and blue.
Credit: University of Arizona and NASA JPL, by Virginia Pasek and Dyer Lytle (Cassini VIMS)

Back cover figure: A dynamic, rotating ring current around Saturn
Credit: Johns Hopkins University Applied Physics Laboratory - Academy of Athens - NASA (Krimigis, S. M., N. Sergis, D. G. Mitchell, D. C. Hamilton and N. Krupp, Nature, 450, 1050, 2007)

ISBN 978-1-4020-9216-9   e-ISBN 978-1-4020-9217-6
DOI 10.1007/978-1-4020-9217-6
Springer Dordrecht Heidelberg London New York

Library of Congress Control Number: 2009935175

©Springer Science+Business Media B.V. 2009
No part of this work may be reproduced, stored in a retrieval system, or transmitted in any form or by any means, electronic, mechanical, photocopying, microfilming, recording or otherwise, without written permission from the Publisher, with the exception of any material supplied specifically for the purpose of being entered and executed on a computer system, for exclusive use by the purchaser of the work.

Printed on acid-free paper

Springer is part of Springer Science+Business Media (www.springer.com)

Preface

This book is one of two volumes meant to capture, to the extent practical, the scientific legacy of the Cassini-Huygens prime mission, a landmark in the history of planetary exploration. As the most ambitious and interdisciplinary planetary exploration mission flown to date, it has extended our knowledge of the Saturn system to levels of detail at least an order of magnitude beyond that gained from all previous missions to Saturn.

Nestled in the brilliant light of the new and deep understanding of the Saturn planetary system is the shiny nugget that is the spectacularly successful collaboration of individuals, organizations and governments in the achievement of Cassini-Huygens. In some ways the partnerships formed and lessons learned may be the most enduring legacy of Cassini-Huygens. The broad, international coalition that is Cassini-Huygens is now conducting the Cassini Equinox Mission and planning the Cassini Solstice Mission, and in a major expansion of those fruitful efforts, has extended the collaboration to the study of new flagship missions to both Jupiter and Saturn. Such ventures have and will continue to enrich us all, and evoke a very optimistic vision of the future of international collaboration in planetary exploration.

The two volumes in the series *Saturn from Cassini-Huygens* and *Titan from Cassini-Huygens* are the direct products of the efforts of over 200 authors and co-authors. Though each book has a different set of three editors, the group of six editors for the two volumes has worked together through every step of the process to ensure that these two volumes are a set. The books are scholarly works accessible at a graduate-student level that capture the approximate state of knowledge of the Saturn system after the first 4 years of Cassini's tenure in Saturn orbit. The topics covered in each volume range from the state of knowledge of Saturn and Titan before Cassini-Huygens to the ongoing planning for a return to the system with vastly more capable spacecraft.

In something of a departure from the norm for works such as these, we have included an appendix in each of the books featuring the people of Cassini-Huygens who are truly responsible for its success – the people behind the scientific scenes who ensure that everything works as flawlessly as it has. We dedicate the Cassini-Huygens volumes to them and to those who started the journey with us but could not finish it. We hope that all who read the books will share in the new knowledge and gain a deeper appreciation for the tireless efforts of those who made possible its attainment.

Bob Brown, Michele Dougherty, Larry Esposito, Stamatios Krimigis,
Jean-Pierre Lebreton and J. Hunter Waite

Contents

1. **Overview** 1
   Michele K. Dougherty, Larry W. Esposito, and Stamatios M. Krimigis

2. **Review of Knowledge Prior to the Cassini-Huygens Mission and Concurrent Research** 9
   Glenn S. Orton, Kevin H. Baines, Dale Cruikshank, Jeffrey N. Cuzzi, Stamatios M. Krimigis, Steve Miller, and Emmanuel Lellouch

3. **Origin of the Saturn System** 55
   Torrence V. Johnson and Paul R. Estrada

4. **The Interior of Saturn** 75
   William B. Hubbard, Michele K. Dougherty, Daniel Gautier, and Robert Jacobson

5. **Saturn: Composition and Chemistry** 83
   Thierry Fouchet, Julianne I. Moses, and Barney J. Conrath

6. **Saturn Atmospheric Structure and Dynamics** 113
   Anthony D. Del Genio, Richard K. Achterberg, Kevin H. Baines, F. Michael Flasar, Peter L. Read, Agustín Sánchez-Lavega, and Adam P. Showman

7. **Clouds and Aerosols in Saturn's Atmosphere** 161
   R.A. West, K.H. Baines, E. Karkoschka, and A. Sánchez-Lavega

8. **Upper Atmosphere and Ionosphere of Saturn** 181
   Andrew F. Nagy, Arvydas J. Kliore, Michael Mendillo, Steve Miller, Luke Moore, Julianne I. Moses, Ingo Müller-Wodarg, and Don Shemansky

9. **Saturn's Magnetospheric Configuration** 203
   Tamas I. Gombosi, Thomas P. Armstrong, Christopher S. Arridge, Krishan K. Khurana, Stamatios M. Krimigis, Norbert Krupp, Ann M. Persoon, and Michelle F. Thomsen

10. **The Dynamics of Saturn's Magnetosphere** 257
    D.G. Mitchell, J.F. Carbary, S.W.H. Cowley, T.W. Hill, and P. Zarka

11. **Fundamental Plasma Processes in Saturn's Magnetosphere** 281
    B.H. Mauk, D.C. Hamilton, T.W. Hill, G.B. Hospodarsky, R.E. Johnson, C. Paranicas, E. Roussos, C.T. Russell, D.E. Shemansky, E.C. Sittler Jr., and R.M. Thorne

12 **Auroral Processes** 333
   W.S. Kurth, E.J. Bunce, J.T. Clarke, F.J. Crary, D.C. Grodent, A.P. Ingersoll,
   U.A. Dyudina, L. Lamy, D.G. Mitchell, A.M. Persoon, W.R. Pryor, J. Saur,
   and T. Stallard

13 **The Structure of Saturn's Rings** 375
   J.E. Colwell, P.D. Nicholson, M.S. Tiscareno, C.D. Murray, R.G. French,
   and E.A. Marouf

14 **Dynamics of Saturn's Dense Rings** 413
   Jürgen Schmidt, Keiji Ohtsuki, Nicole Rappaport, Heikki Salo, and Frank Spahn

15 **Ring Particle Composition and Size Distribution** 459
   Jeff Cuzzi, Roger Clark, Gianrico Filacchione, Richard French, Robert Johnson,
   Essam Marouf, and Linda Spilker

16 **Diffuse Rings** 511
   M. Horányi, J.A. Burns, M.M. Hedman, G.H. Jones, and S. Kempf

17 **Origin and Evolution of Saturn's Ring System** 537
   Sébastien Charnoz, Luke Dones, Larry W. Esposito, Paul R. Estrada,
   and Matthew M. Hedman

18 **The Thermal Evolution and Internal Structure of Saturn's Mid-Sized Icy Satellites** 577
   Dennis L. Matson, Julie C. Castillo-Rogez, Gerald Schubert, Christophe Sotin,
   and William B. McKinnon

19 **Icy Satellites of Saturn: Impact Cratering and Age Determination** .. 613
   Luke Dones, Clark R. Chapman, William B. McKinnon, H. Jay Melosh,
   Michelle R. Kirchoff, Gerhard Neukum, and Kevin J. Zahnle

20 **Icy Satellites: Geological Evolution and Surface Processes** 637
   Ralf Jaumann, Roger N. Clark, Francis Nimmo, Amanda R. Hendrix,
   Bonnie J. Buratti, Tilmann Denk, Jeffrey M. Moore, Paul M. Schenk,
   Steve J. Ostro, and Ralf Srama

21 **Enceladus: An Active Cryovolcanic Satellite** 683
   John R. Spencer, Amy C. Barr, Larry W. Esposito, Paul Helfenstein,
   Andrew P. Ingersoll, Ralf Jaumann, Christopher P. McKay, Francis Nimmo,
   and J. Hunter Waite

22 **The Cassini Extended Mission** 725
   David A. Seal and Brent B. Buffington

23 **Saturn's Exploration Beyond Cassini-Huygens** 745
   Tristan Guillot, Sushil Atreya, Sébastien Charnoz, Michele K. Dougherty,
   and Peter Read

24 **Cartographic Mapping of the Icy Satellites Using ISS and VIMS Data** 763
   Th. Roatsch, R. Jaumann, K. Stephan, and P.C. Thomas

Appendix: The Cassini Orbiter, Behind the Scenes 783

Index 795

Chapter 1
Overview

Michele K. Dougherty, Larry W. Esposito, and Stamatios M. Krimigis

1.1 Introduction

This book is one of two volumes whose aim is to capture the main scientific results of the Cassini-Huygens prime mission from orbit around the Saturn system, covering observations from the first four years of orbital tour and incorporating data from July 2004 to June 2008. The second book, *Titan from Cassini-Huygens*, contains the material pertinent to Saturn's largest moon Titan; its surface, atmosphere and interaction with Saturn's magnetosphere. This book, *Saturn from Cassini-Huygens*, focuses on the new results from Saturn, its satellites (excluding Titan), rings and magnetosphere.

Details of the Cassini orbiter spacecraft and the Huygens probe, and their respective science instruments and investigations, as well as the overall design of the Cassini-Huygens mission can be found in a 3-volume series of books written prior to the arrival of Cassini-Huygens at the Saturn system. We do not attempt to reproduce any of that material here; instead the reader is referred to the relevant volumes (The Cassini-Huygens Mission 2002, 2004a, b). In this book we detail in the various chapters, information concerning the orbital tour of Cassini-Huygens, as executed during the four year prime mission, the extended Cassini Equinox mission and plans for a further extension after that, the Cassini Solstice mission.

1.2 Organization of This Volume

This volume is divided into seven main sections. The first section consists of introductory material, including this chapter and a chapter describing our knowledge of the Saturn system prior to Cassini-Huygens. The second section deals in two separate chapters with the origin and evolution of the Saturn system and Saturn's interior. Section three focuses on the atmosphere of Saturn, with four chapters describing its composition and chemistry, its structure and dynamics, clouds and aerosols and the upper atmosphere and ionosphere respectively. The fourth section is devoted to Saturn's magnetosphere and contains four chapters relating to the configuration of the magnetosphere, its dynamics, its plasma physics and the auroral processes which arise there. Section five covers the rings of Saturn and features five chapters which study their structure, dynamics, composition, the diffuse rings, as well as the origin and evolution of the ring system. The sixth section deals with Saturn's satellites, excluding Titan which is detailed in the second book *Titan from Cassini-Huygens*. There are four chapters which study the thermal evolution and internal structure of the mid-sized icy satellites, impact cratering and age determination, the geological evolution and surface processes of the satellites and then a focus on Enceladus, as an active cryovolcanic satellite. The last section of the book is dedicated to a description of the continued exploratory plan for Saturn. It contains two chapters, one detailing plans for Cassini's continued exploration of the Saturn system, and the other discussing proposals for returning to the Saturn system in the future with new and more capable spacecraft. The final chapter in the book details cartographic mapping of the satellites which has been possible to date, and an appendix features the people of Cassini-Huygens who were responsible for the success of the mission.

1.3 Synopsis of the Main Results for the Saturn System

Our understanding of the Saturn system prior to orbital insertion of Cassini-Huygens was based on decades of analysis

M.K. Dougherty (✉)
Space and Atmospheric Physics, The Blackett Laboratory, Imperial College London, SW7 2AZ, UK

L.W. Esposito
Laboratory for Atmospheric and Space Physics, University of Colorado, Boulder, CO, 80309-0449, USA

S.M. Krimigis
Applied Physics Laboratory, Johns Hopkins University, Laurel, MD, 20723, USA
and
Center for Space Research and Technology, Academy of Athens, Athens, 11527, Greece

from ground-based observations, in-situ observations from the three spacecraft flybys of Saturn in the late 1970s and early 80s, as well as observations from orbiting telescopes, such as ISO, IUE and HST. As this book will reveal, many new discoveries have been made and our understanding of the system in greatly improved. What follows, is a summary of the main results which are covered in much greater detail in the chapters of this book.

1.3.1 Origin and Interior

Cassini-Huygens results from the prime mission have aided in constraining and improving our understanding of the formation and evolution of the Saturn system, including its rings and satellites (Chapters 3 and 4). For example, the strong enrichment in carbon over solar abundances, now confirmed by Cassini, progresses the issues arising between gas instability and core accretion formation models. Implications of some recent observations are that Saturn's ring system may be more massive and longer-lived than previously thought; this is centering attention on primordial models for their formation. Cassini observations have also enabled very accurate measurements of the densities of Saturn's satellites to be made which reveal significant variations in their compositions. This is in contrast to the more systematic variation which has been observed at the Jovian Galilean satellites.

An understanding of the interior of Saturn will continue to be uncertain whilst we remain unable to attribute a unique rotation rate to its deep interior. However, there is now little uncertainty that the planet has a large mass core, the equation of state of dense hydrogen–helium mixtures is also less in doubt than previously, and fundamental measurements related to the interior are now known with improved precision.

1.3.2 Saturn's Atmosphere

In order to probe the chemical composition and chemistry of Saturn's atmosphere, remote sensing measurements via spectroscopy are required (Chapters 5–8). Our understanding of this area has greatly improved as a result of the combination of Earth-based observations, modeling work and Cassini measurements; although a great deal more remains to be done. Highlights of Cassini observations to date, include measurements of the carbon abundance and isotopic ratio. Below the ammonia cloud level; phosphine, germane and arsine have been observed to move at a more rapid rate to the shallow atmosphere than the speed at which chemical reactions can eliminate them; thus revealing pointers to the vertical mixing timescales and the composition of the deep atmosphere. In the upper troposphere, the meridional change in the abundance of phosphine has been measured, thus enabling the dynamics above the cloud levels to be better constrained.

Our current understanding of the thermal structure and circulation of Saturn's atmosphere, reveals the occurrence of a reversal in the meridional temperature gradient between the water cloud level and the upper troposphere. As a result, near the water cloud base, cyclonic regions are cool and anticyclonic regions warm; whereas near and above the visible clouds the opposite holds. Between the levels of the temperature gradient reversal and that of water condensation, zonal winds are expected to increase with height and meridional circulation reverses sign at a higher level than that of condensation of ammonium hydrosulfide.

Our present understanding of the behavior of clouds and aerosols is based on analyses of both ground and space-based observations, of which the latest Cassini observations are providing new views. This picture reveals many components of the clouds and haze over a wide range of temporal and spatial scales. Small scale convective activity is seen at some latitudes within the deepest clouds, and observed morphological differences with latitude could provide indications of the dynamics of the zonal jets, as well as of the smaller scale phenomena. Both seasonal and non-seasonal variations are observed and the hemispheric asymmetry is also clearly changing. Three latitude bands are clearly defined on regional scales and in the vertical domain two clear regions are revealed, that of the upper troposphere and the stratosphere. In addition, auroral processes seem to be important for polar stratospheric haze, similar to the processes known to occur at Jupiter. Spectral evidence for a hydrocarbon haze component in the stratosphere has also recently been obtained.

The quality of the existing information concerning the upper atmosphere and ionosphere of Saturn has greatly improved with Cassini observations, as well as the temporal and spatial coverage of these regions. The main conclusion which can be drawn, is that this upper region of the atmosphere is much more variable than had previously been thought. Preliminary Cassini results imply thermospheric temperatures much colder than inferred from Voyager data, atmospheric mixing seems to be relatively weak although vertical winds probably play a key role in controlling the methane profile in the homopause region. Recent electron density measurements of the upper atmosphere have resulted in good latitudinal coverage, with diurnal variations similar to those from Voyager times. H_3^+ appears to be the dominant low altitude ion with H^+ dominating at higher altitudes. Large variability in electron density profiles at similar latitudes and times suggests that dynamical processes play an important role in controlling the structure of the ionosphere.

1.3.3 Magnetosphere

It could be argued that the magnetosphere represents the most dynamic environment in the Saturn system (Chapters 9–12). Not only does its size change due to the interplay of the solar wind pressure externally and the plasma pressure internally, but also these variations are coupled into Saturn's auroral kilometric radiation (SKR), the aurora itself, and particle acceleration processes and plasma loss occurring near the equatorial plane on both the day and night-side of the planet. There has been a quantum leap in both the plethora of novel observational phenomena and theoretical understanding and modeling, as will become evident to the reader from the chapters on the magnetosphere. A detailed list of Cassini discoveries is contained in Chapter 9 and there is no need to repeat them here. It is, however, important that a lot of the often puzzling questions posed by observations from the previous flybys of Pioneer 11 and Voyagers 1 and 2, such as the insufficient source strength of plasma sources to supply the apparent losses, were answered with the single most important discovery of the Enceladus plasma source first inferred by the observations of the magnetic deflection of the planetary field near that satellite. Many other unexplained observations immediately fell into place, such as the richness of water group ions spanning the energy range from eV to hundreds of keV, and the relative scarcity of nitrogen ions thought previously to be a strong plasma source emanating principally from the upper atmosphere of Titan.

The variability in the period of recurring phenomena manifested in the plasma environment is, by itself, a most fascinating and still evolving story. We recall that following the Voyager encounters, the SKR period was thought to represent the rotation period of the planet itself. Ulysses observations in the nineties provided the first hint that the period may not be constant. The Cassini observations established that fact not only in SKR, but also in energetic electrons, ions, plasma, and magnetic field observations. The remotely sensed periodicity of the SKR has now been complemented by magnetospheric particle oscillations observed by the novel technique of energetic neutral atom (ENA) imaging that enables measurements of both temporal and spatial fluctuations throughout the system. Such is the power of this technique that it has become possible to observe synergistically the acceleration events near the equatorial plane, the simultaneous brightening of the UV aurora, and the triggering of intense SKR. These observations, as well as other phenomena, are described in detail and interpreted in Chapter 10, on the dynamics of the magnetosphere, and Chapter 12 on auroral processes.

It is now clear that we have extensive information on three planetary magnetospheres, the solar wind-dominated one of Earth, the rotationally dominated magnetosphere of Jupiter and the presumed hybrid one of Saturn. We have studied the dominant processes in each one and can, therefore, ask whether there exist fundamental physical mechanisms that operate in each magnetosphere, perhaps in differing parametric space. This is the topic addressed by Chapter 11, where plasma–material interactions, transport, and energy conversion processes are described, with emphasis on Saturn but with an eye to similarities and differences with Earth and Jupiter. Saturn, having the most neutral gas-dominated environment of the three makes an excellent case study of processes such as the centrifugal interchange instability, charge exchange loss of energetic ions, sputtering of material from both rings and satellite surfaces, and co-rotation lag due to plasma loading at the equator and coupling to the planetary ionosphere, to mention just a few. Plasma injection events have been studied in detail at Saturn and appear to be similar in all three planets, albeit with drastically different time scales. Comparisons of dayside reconnection show that process to be quite common at Earth but yet to be seen at Saturn. Although "substorms" could be construed to occur at all three planets, the details do not seem to work out well, in that at Earth the solar wind is a big influence while at Jupiter and Saturn internal dynamics seem to play a dominant role. Finally, energy conversion processes seem to have much in common at all three magnetospheres, including particle acceleration, current generation, wave–particle interactions, and tail reconnection. It is certainly satisfying to find that many basic physical processes are common to all three planets and that we may have enough understanding to develop predictive capability in the not too distant future.

1.3.4 Saturn's Rings

Saturn's rings are very dynamic and continually evolving with the rings changing on timescales ranging from days to 10–100 million years (Chapters 13–17). Each advance in observation of the rings reveals new structure. We find that the ring particles are mainly aggregates of smaller particles arranged into transient elongated clumps 10 m in size, with different ring regions possibly having different origins. Small moons near and within the rings are intimately involved in creating ring structure, an example of which is the density waves excited by resonances with moons making up the majority of features in the A ring. Embedded moons create satellite wakes and perturb the edges of gaps cleared by small moons. Propeller shaped structures are an intermediate stage, where the embedded object is not large enough to hold open a complete gap.

Dynamical studies allow us to understand the underlying physical processes that create the myriad structures. The self gravity of the particles is important, and the particles continually form temporary gravitationally bound clumps. The self gravity wakes show the equilibrium established between the gravitational attraction of the particles and the Kepler shear of their orbits around Saturn. Dense packing leads to strong non-local contributions to the pressure and momentum transport and this gives rise to viscous overstability, yielding axisymmetric waves of 100 m wavelength. Both accretion and fragmentation are important in the evolution of the ring system and Saturn's F ring is itself a showcase of accretion.

The question of the origin and evolution of the ring system is still unsolved. Three proposed models are that: the rings are remnants of the Saturn nebula; the debris from a destroyed satellite; or the remnants of a split comet. The rings could be ancient if they are continually renewed, and hence a solution to the puzzles of ring origin is therefore the possibility of recycling primordial material. In this case, the rings are likely much more massive than suggested by analysis of Voyager data, which calculated that the rings contain as much mass as Saturn's small moon, Mimas. If the rings were created by the destruction of a small moon during the period of the late heavy bombardment (that also created the great lunar basins), recycling would still be necessary for the rings to survive to the present time. Saturn's E ring is created by geologic activity on Enceladus and the G ring is fed by moonlets embedded within it.

The ring composition evolves with time: photons, charged particles and interplanetary meteoroids strike the rings; they may chemically evolve under the influence of oxygen in the ring environment. The ring system is not a homogeneous slab, but a two-phase system of gaps and dense clumps. Ring particles are primarily water ice, quite pure and dominantly crystalline; although the reddish color of the rings shows a non-icy component with tholins, PAH's and nanohematite being proposed as an explanation. The C ring and the Cassini Division are more contaminated, with the ring composition showing both primordial and extrinsic contributions. The lack of silicates indicates a parent body lacking a core, or sequestration of the core material keeping it from becoming mixed into the rubble of the rings.

Diffuse planetary rings are an excellent laboratory to study dusty plasma processes where much of the dust grains are likely collisional debris. Diffuse rings are associated with the tiny moons (Janus, Epimetheus, Pallene, Methone, and Anthe), reminiscent of Jupiter's ring. A population of source bodies has been found in the G ring, the dusty rings are shown to be influenced by solar radiation and sculpted by periodic forces, and the D ring material spirals into the planet as has been suggested for the Uranian rings. Ring systems also provide a local analog for more distant flattened systems like galaxies and proto-planetary disks and in the future we may find rings surrounding extra-solar planets.

1.3.5 Icy Satellites

Properties of the mid-sized icy satellites of Saturn can yield important information concerning their formation and evolution; and the new data returned to date by Cassini provides much needed geophysical constraints for relevant models (Chapters 18–21). A robust case can be made that suggests these satellites are very old objects, having formed around 4.5 Gyr ago; thereby constraining the age of Saturn and its formation period. Phoebe's density and orbital properties suggest it is a captured object formed within the solar nebula. Iapetus, which has the shape of a hydrostatic body with a 16 h rotation period, has since despun to its current synchronous rotation rate, which in turn constrains the early heating input which is required. Observations from Rhea are as yet unable to differentiate between a hydrostatic or non-hydrostatic gravitational field and both Dione and Tethys are probably linked to past tidal heating. However, the geology of Tethys is difficult to explain due to its low rock abundance and the lack of eccentricity excitation of its orbit.

Observations of impact cratering from Saturn's satellites, the regular, irregular and ring moons enable insights to be gained into the origin and history of the impacting bodies themselves. Pre-Cassini, two impactor populations were postulated: the first, from cometary objects, formed the majority of the larger older craters; the second, resulting from orbiting ejecta from satellite impacts, produced smaller and younger craters. There are presently two groups of work resulting from the analysis of Cassini observations, one of which attributes the cratering to be from an outer Solar system heliocentric source, whereas the other suggests the main-belt asteroids to be the primary impactors. What does seem clear, is that planetocentric populations have played an important role in the evolution of the Saturn system.

The icy satellites of Saturn are extremely diverse as regards size and varying geological features. Major Cassini discoveries linked to the evolution and surface processes of the Saturnian icy satellites include volcanic activity on Enceladus, the sponge-like appearance of Hyperion and the dominant equatorial ridge on Iapetus. Dark material on the surface of Iapetus has been revealed to be made up of organics and metallic iron and possibly ammonia, and is probably simply a thin surface coating. Several spectral features of this dark material match that seen on a number of the other satellites, as well as in the rings and hence probably exists in the entire Saturn system. On Iapetus in particular, this dark covering is uninterrupted by cratering, implying that recent deposition of this material has taken place. On Rhea, a bright ray system on its surface is probably the result of a recent impact, whereas wispy streaks are likely of tectonic origin, as is also observed on Dione. The icy satellites reveal old surfaces with craters and early formed tectonic structures but there is also evidence for younger processes such as crustal stress, formation of plains and surface coating by particles from the E-ring.

At Enceladus, Cassini observations have revealed a dynamic atmosphere made up primarily of water based ions which was then followed by detection of active plumes emanating, near the South Pole, from warm fractures on its surface. Hence, Enceladus is the only known icy body in our solar system to be geologically active. Plume activity is focused along four fractures which are radiating heat of up to at least 167 K, and there are multiple plume sources which are ejecting material along these fractures. The plume material constitutes primarily of water vapor along with significant hydrocarbons. In addition the moon is now confirmed to be the source of the E-ring material as well as the torus of neutral and ionized material which fills the middle magnetosphere; therefore playing a similar role to that of Io within the Jovian magnetosphere. In addition, it is clear that the level of plume activity changes between different spacecraft flybys. The implications of this geological activity as regards the interior of Enceladus are significant and our understanding of tidal heating and the orbital evolution of the icy satellites is being tested as a result. It is also becoming clear that the source of the plumes involves liquid water which may be in contact with a sub-surface ocean which, in turn, is in contact with a silicate core. This together with the energy likely to be supplied by tidal heating, as well as the rich chemical make-up of the plume, provides a promising environment for life.

1.4 Open Questions and Future Work

As will be revealed from the chapters within this book, our understanding of the Saturn system has greatly improved from the first four years of orbital tour by Cassini. There are, as always from such discovery missions, outstanding questions which yet require resolution as well as new questions which have been revealed following some unexpected discoveries.

1.4.1 Origin and Interior

The compositional variations of Saturn's satellite system in conjunction with the very ice-rich nature of the rings and ring-related moons raises complex issues for satellite formations models which require resolution. This is further complicated by the cometary D/H values observed by Cassini in the plumes of Enceladus, possibly implying the assimilation of cold icy material from the outer solar nebula. For Enceladus, its high density can potentially provide radiogenic heating but this is not sufficient to yield the power levels observed; hence tidal heating is being put forward as a probable cause which in turn has implications for its initial thermal and interior state which still have need of clarification.

For Saturn, we have yet to obtain strong gravitational or magnetic constraints on its internal structure and in particular we have been unable to resolve a unique rotation rate of the deep interior. However the possibility of high-inclination and low altitude orbits at the end of the Cassini Solstice mission will enable constraints to be derived of its interior properties, as well as comparisons to be made with similar observations at Jupiter (resulting from the planned NASA Juno mission). Resolving a unique rotation rate will lead to the ability to better constrain Saturn interior models; however an understanding of the multiplicity of rotation periods, arising from zonal flows at different depths, is also required, as well as an improved knowledge of Saturn's interior temperature profile.

1.4.2 Saturn's Atmosphere

Continued observations of Saturn's atmosphere requires remote sensing measurements from both the Earth and the Cassini orbiter in conjunction with further improvement of chemical models. Refinements which Cassini observations are yet to reveal, include elemental composition in nitrogen and sulphur as well as improved D/H ratios. Further measurements of the meridional variation in the abundance of recently measured molecules above the cloud levels will aid in constraining dynamical models of the atmosphere. In addition, tropospheric photochemistry needs to be better understood with reaction rate coefficients for the various phosphorus and nitrogen reactions required.

As regards atmospheric structure and dynamics there remain many unanswered questions, including how the nature of the zonal winds evolve down to and beneath the still unknown water cloud level. The way in which the inferred atmospheric circulation relates to the near-infrared continuum and the 5 μm cloud patterns also remains unresolved, as does the reason why the wind-albedo correlation is different to that at Jupiter. The mechanisms which drive the observed latitudinal variation of cloud level circulation is still uncertain, as well as the eddy processes driving upper tropospheric circulation. It is very clear that seasonal variability is in evidence, but observations from the Cassini Equinox and Solstice missions are necessary for hemispheric and seasonal shifts to be resolved. The existence of the prograde equatorial jet, as well as its strength and variability, remains to be elucidated.

For Saturn's clouds and aerosols, a key remaining unknown is the lack of spectral evidence for ammonia ice which was expected, based on thermochemical equilibrium models, to dominate in the high troposphere. Another expected stratospheric haze component which is yet to be observed is that of diphosphine. As the Cassini spacecraft continues to take observations in the coming years, a much better understanding of seasonal changes will result and observations at numerous phase angles and latitudes will enable radiation budget models to further improve. In addition, more detailed profiles of haze within the stratosphere as well as its compositional information will help shed light on the important chemical and physical processes arising within these atmospheric regions.

In the upper atmosphere of Saturn, the thermospheric neutral temperatures which have been observed are much higher than expected, if the assumption is purely that of absorption of solar extreme ultraviolet radiation. Modeling work needs to focus on providing insights into what the dominant heating mechanisms are, in order to resolve the thermospheric heat sources. The overall variability which has been observed in ultraviolet occultation measurements needs to be better understood and continued observations and analysis of this data will help to better constrain the circulation processes taking place within the atmosphere. A disparity between photochemical models and observations requires resolution, thereby enabling better constraints to be placed on the vertical transport and chemistry of the upper atmosphere. Ion composition of the ionosphere requires better insight, such as the importance of vibrationally excited H_2 versus water inflow or gravity waves for removal of H^+.

1.4.3 Saturn's Magnetosphere

It is not surprising to discover that all the new findings identified during the prime mission have generated a host of new questions, as is always the case for flagship-type missions of the caliber of Cassini-Huygens. Fortunately, with the extended Equinox and Solstice missions, it will be possible to pursue at least some of these. The first and foremost question relates to the key discovery of the Enceladus plumes. The temporal variability of the neutral/plasma input into the Saturn system over periods of years represents an important constraint on source and loss rates in the study of the magnetosphere. The Solstice mission, extending through 2017, will offer the opportunity to study the magnetosphere over a full solar cycle, from 2004 to 2017. Thus the influence of the evolving solar wind should help us elucidate its importance to magnetospheric dynamics when compared to internal/rotationally driven mechanisms.

Some of the observations in the magnetotail suggest that the night-side plasma sheet is relatively thin compared to the dayside and that its extent and plasma population are quite variable. The trajectory design of the Solstice mission will contain a 2-month long tail segment at low latitudes and relatively large (>30 Rs) distances. Such geometry will enable the study of tail dynamics including such questions as the frequency of occurrence of reconnection events, the loss of plasma down the tail and the extent of field line closure via studies of pitch angle evolution of energetic electrons.

Finally, it will be possible to study in substantial detail the Cassini-identified radiation belt inside the D-ring during the final months of the Solstice mission. The spacecraft trajectory, being at high inclination, will mimic a Juno-like orbit and enable study of the coupling of the radiation belts to the ionosphere, as well as search for an ionosphere associated with the rings, as indicated during Saturn Orbit Insertion back in July 2004. During this part of the trajectory, it may also be possible to identify the close coupling of the ionosphere to the magnetosphere, since it appears possible to impose a periodicity on several of the observables (SKR, ENA, plasma, magnetic field) from close to the planet to at least the planetary magnetopause.

1.4.4 Saturn's Rings

Our understanding of the structure of Saturn's rings has improved dramatically over the course of Cassini's initial four year mission. In large part this is due to the combination of observations from multiple instruments taken from a wide range of geometries. This has enabled, for the first time, a detailed exploration of non-axisymmetric structures in the rings as well as the vertical structure of the main rings. The temporal baseline of these observations, which may extend beyond a decade with the proposed Solstice mission, also provides a unique opportunity to study the changes in ring structure due to changes in gravitational forcing as the orbits of nearby ring moons shift on timescales of years. The macroscopic structure of Saturn's rings is virtually identical to that which was observed by Voyager. The most notable exception in the main rings covered here is the change in morphology of the F ring, which has continued to evolve underneath Cassini's multi-wavelength eyes. Nevertheless, as we move toward understanding the properties of the rings on the collisional scale of clumps of particles, there are several avenues for further research.

Cassini will measure the meteoroid mass flux and the ring mass. During the Equinox mission (2008–2010) the spacecraft will fly by Rhea closely to measure the mass flux indirectly, sampling the ejected mass filling its Hill sphere. The geometry of the flyby will make it possible to distinguish this ejecta from whatever equatorial debris might or might not be responsible for the charged particle absorptions observed by fields and particles instruments. At the end of Cassini's mission, it is planned that a number of orbits will be implemented with the periapse inwards of the D ring. In these close orbits, it is anticipated that a ring mass comparable to Mimas (the post-Voyager consensus) can be detected to a few percent accuracy. A primordial ring compatible with current estimates of mass flux would need to be 5–10 times more massive and would be easily detected. Until the time that these fundamental measurements can be made, the question of the ring exposure age to pollution will not be resolved.

Many of our advances in understanding the rings were not anticipated prior to Cassini. A key aspect of these advances has been the combination of observations made at different geometries and at different times, as well as the different perspective offered by multiple instruments. We can thus safely anticipate new discoveries from the remainder of Cassini's Equinox mission and final Solstice mission as the time baseline and geometric diversity of observations is extended. Improved understanding of ring microphysics enabled by modeling, in combination with Cassini observations of microstructure, may reveal the source of the formation of plateaus or the bimodal structure in the B ring. As the time baseline of measurements of the dynamic F ring grows longer, we gain a better understanding of the interaction between moon and ring and the growth and destruction of agglomerates within the F ring region. The relative spatial precision of Cassini occultation measurements will ultimately enable highly accurate kinematic models of ring features that are linked to moons. The combination of these future developments should shed light on the question of the origin of the rings as well as how they reached their current configuration.

1.4.5 The Satellite System

In order to reveal and model the evolution and internal structure of the mid-sized icy satellites of Saturn, the material properties of the satellites must be known to high accuracy. Although recent Cassini observations have increased our understanding of the properties of some of these satellites, a great deal more needs to be determined. Continued observations from the Cassini instruments on repeated flybys past numerous of the icy satellites will aid in further constraining their properties. Laboratory measurements of the behavior of icy materials at the relevant temperatures, pressures and frequencies will also allow an assessment to be made of the behavior of these icy materials during the satellite histories. Improvements in modeling work and numerical techniques will enable a far better understanding to be gained of the evolution and history of these satellites and the implications this in turn will have for the evolution of the Saturn system itself.

Analysis of the cratering record of the Saturnian satellites as updated by Cassini observations is still in its infancy, with two independent results being consistent at least in a cumulative sense. However further work needs to be carried out on analyzing existing and new observations before an interpretation will be converged on as regards the structure of the various size-frequency distributions being interpreted. In addition, in the absence of independent radiometric dates, cratering records are unable to resolve the validity of various model histories. However, continued work on crater counting, modeling of dynamical processes, simulations of impacts, crater modification and saturation processes will further our understanding of the origin and time histories of the impacting populations.

There is a great deal of variety in the extent and timing of tectonic activity of the icy satellites and our understanding of the tectonics and cryovolcanism is in its early stages. Continued Cassini observations into the Equinox and Solstice missions will enable further characterization and mapping of the tectonic features to be carried out, in order that their origins and the satellite histories may be better characterized. Further resolution of the composition of the ubiquitous dark material within the system also requires analysis in order to better understand its origin, although an external origin has been postulated for the dark material on Phoboe, Iapetus and Dione. The uniqueness and unexpected diversity of the satellite system points to a variety of geological processes and interactions which will require a continued focus in order to better understand their role within the complex planetary system.

Cassini observations at Enceladus thus far, have revealed it to be one of the most extraordinary bodies in the solar system; with geological activity, possible sub-surface liquid water and a water vapor, hydrocarbon filled plume which feeds the E-ring and the middle magnetosphere. There remain many unanswered questions, some of which will be addressed by upcoming close flybys by the Cassini spacecraft in the extended missions, but some of which will need to wait until future missions which will return to this remarkable moon. The most pressing of these questions, includes whether the high level of activity being observed by Cassini instruments is typical, and if this is so, how can the tidal heating which is driving the processes and the subsequent mass loss be sustained over long time periods. What generates the plumes at the South Pole and the mass being lost from them? Is the subsurface liquid water present as a global or regional body, or simply locally at the plume source; what is the detailed chemical make-up of these liquid reservoirs and are they able to sustain life? There are clearly extreme variations in the surface age of Enceladus as well as a symmetry in the geology about the spin axis and the direction to Saturn; interpreting such information will allow a much better understanding to be gained of the geological evolution of this moon and of the mechanisms which have produced the resulting plentiful and varied tectonic features.

1.5 Continued Saturn System Exploration

The tremendous success of the first four years of data from the Cassini-Huygens prime mission at the Saturn system, as evidenced by the results described in this book and its sister book *Titan from Cassini-Huygens,* has led to a confirmed extension by NASA of the mission for a further two years (the Cassini Equinox mission) and planning for the Cassini Solstice mission onto 2017, where an end of mission scenario will incorporate Juno type orbits of Saturn inside of the D-ring (Chapters 22 and 23). Answers to many of the remaining questions detailed above and in the rest of this book will be studied during these extended mission periods, however there will remain many unanswered questions which the Cassini observations and analyses will be unable to resolve; due both to the planned orbital tours and the spacecraft's instrumentation suite. Several recent NASA and ESA studies have considered future return missions to the Saturn system. One of which, described in Chapter 18 of the *Titan from Cassini-Huygens* book, will focus on Titan and Enceladus; this mission or one very similar may follow on from the Europa Jupiter System Mission as the next but one outer solar system Flagship/L-class mission. Studies of a Saturn probe which would measure the atmosphere of Saturn and allow important gains in our understanding of its meteorology and dynamics have also been carried out by both European and US scientists. Continued monitoring of the Saturn system by the Cassini orbiter and ground-based and Earth-orbiting instruments will allow us to continue to study this fascinating planetary system and future focused spacecraft missions will continue to fill gaps in our understanding, as well as produce new and surprising results with new questions which beg resolution.

References

The Cassini-Huygens Mission: Overview, Objectives and Huygens Instrumentarium, Vol. 1, 2002, Springer, Dordrecht, ed. C. T. Russell.

The Cassini-Huygens Mission: Orbiter In Situ Investigations, Vol. 2, 2004a, Springer, Dordrecht, ed. C. T. Russell.

The Cassini-Huygens Mission: Orbiter Remote Sensing Investigations, Vol. 3, 2004b, Springer, Dordrecht, ed. C. T. Russell.

Chapter 2
Review of Knowledge Prior to the Cassini-Huygens Mission and Concurrent Research

Glenn S. Orton, Kevin H. Baines, Dale Cruikshank, Jeffrey N. Cuzzi, Stamatios M. Krimigis, Steve Miller, and Emmanuel Lellouch

Abstract The scientific achievements of the Cassini-Huygens mission have been based on previous decades of investigations from ground-based observations, an intensive 2-year time span of exploration by three spacecraft (Pioneer Saturn, Voyagers 1 and 2) in 1979–1981, and observations by the Infrared Space Observatory, the International Ultraviolet Explorer, and Hubble Space Telescope. We review both these and research concurrent with the nominal mission which often provided directly supporting results. Saturn's "bulk" composition remains uncertain. Its few discrete cloud features include a hexagon near the sorth pole, and a major episodic storm at the equator. A heterogeneous cloud field at depth was uncovered at 5 μm. Temperatures are enhanced at Saturn's south pole both from seasonal variations of sunlight and from dynamical forcing. Zonal winds peak near Saturn's equator. Saturn possesses a well-defined ionosphere, but with significant structure and variability. The magnetic field is aligned to within a degree of Saturn's rotation axis. An equatorial ring current of $\sim 10^7$ A has been inferred, with inner and outer radii of ~ 8 and $\sim 16\,R_S$. Radiation belts at Saturn are well-established up to the edge of the outer rings, and the higherenergy (>1 meV) proton component is readily absorbed by the inner satellites. The magnetosphere includes both a well-developed plasma sheet and a magnetotail. Radio emissions and plasma waves exist throughout, and the auroral kilometric radiation (SKR), modulated at 10 h 39.4 min, has been widely adopted as a measure of the internal rotation. The plasma population consists principally of protons, but with a heavier component close to the equatorial plane widely assumed to be nitrogen or oxygen. Saturn's ring system is the most accessible in the solar system and consists of three primary components, the A, B and C rings. The A ring contains a large number of spiral density wages generated by gravitational interactions with Saturn's satellites. Other rings include the tenuous D, E, and G rings, and the narrow F ring. Voyager imaging detected "spokes" in the rings and, through occultation studies, an intricate detailed radial structure. Spectroscopic information on the icy satellites reveals the presence of crystalline water ice, mixed with a non-ice surface component with strong UV absorption. Detailed information on the geology of Saturn's eight largest icy satellites before the Cassini arrival was based entirely on Voyager observations. They were found to be surprisingly heterogeneous, with implied internal activity in Enceladus and the dichotomous albedo of Iapetus being two of the biggest mysteries. Among the smaller satellites are those embedded in Saturn's rings and irregular captured satellites outside the orbits of the co-planar satellites.

2.1 Introduction

The formation of the wide array of investigations performed by the Cassini-Huygens mission was based on years of data derived from a variety of sources. Ground-based observations were supplemented in the last few decades by

---

G.S. Orton
MS 169-237, Jet Propulsion Laboratory, California Institute of Technology, 4800 Oak Grove Drive, Pasadena, California 91109, USA

K.H. Baines
MS 183-601, Jet Propulsion Laboratory, California Institute of Technology, 4800 Oak Grove Drive, Pasadena, California 91109, USA

D. Cruikshank
245-6, NASA Ames Research Center, Moffet Field, California 94035, USA

J.N. Cuzzi
245-3, NASA Ames Research Center, Moffet Field, California 94035, USA

S.M. Krimigis
Johns Hopkins University Applied Physics Laboratory, 11100 Johns Hopkins Road, Laurel, Maryland 20723, USA
and
Academy of Athens, Athens, 28 Panepistimiou Avenue 106 79 Athens, Greece

S. Miller
Imperial College London, Gower Street, London WC1E 6BT, UK

E. Lellouch
Observatoire de Paris, Place Jules Jansen, 92195 Meudon-Cedex, France

data from Earth-orbiting spacecraft – principally the International Ultraviolet Explorer (IUE), Hubble Space Telescope (HST) and the Infrared Space Observatory (ISO). During a 2-year "golden era", Saturn was visited by Pioneer 11, and the Voyager 1 and 2 spacecraft. The data returned by these spacecraft launched the significant study of Saturn's electromagnetic environment, its detailed ring system and the morphology and history of its icy satellites.

We intend this chapter to summarize what we knew (or thought we knew) before Cassini's arrival in 2005, together with significant non-Cassini investigations during the primary mission. We review atmospheric and ionospheric properties first, not only Pioneer and Voyager spacecraft results but also the considerable body of work by Earth-orbiting and ground-based observations which followed, some concurrent with Cassini's primary mission operations. Then we "go outward" in somewhat shorter sections which concentrate on spacecraft results, reviewing Saturn's magnetosphere, followed by its rings and icy satellites. We exclude discussion of Titan, which is the subject of an independent book. This chapter should provide a good starting point for the specialized chapters which follow, especially for the reader unfamiliar with the history of exploration of the Saturn system prior to the Cassini mission with additional material on parallel exploration from Earth-based observations.

2.2 Saturn's Composition

2.2.1 Brief Historical Overview

Measurements of Saturn's composition were first obtained by Slipher (1905), who found strong absorption bands in photographic spectra of Saturn and Jupiter. It was not until 1931–1932, however, that these bands were identified as methane (CH_4) and ammonia (NH_3) absorptions (Wildt 1932). Molecular hydrogen (H_2) and helium (He), Saturn's dominant components, were detected much later. The direct detection of H_2 was achieved in 1962 by Spinrad et al. (1962), while that of He had to await the observation of its far-IR spectrum by the Voyager-1 IRIS experiment in 1980 (Hanel et al. 1981), following its detection at Jupiter in 1979 (Hanel et al. 1979). In the meantime, the development of infrared astronomy in the 1970s permitted the detection of several photochemical products of methane (ethane (C_2H_6) in 1975, acetylene (C_2H_2) in 1980 – actually discovered initially in the ultraviolet in 1979 – in the 10-μm window, of phosphine (PH_3) in 1975 at 10 μm and later at 5 μm and in the UV, and numerous studies of the CH_4 and NH_3 abundances in the near-infrared range, as well as the detection of methane isotope $^{13}CH_4$ (see the review by Prinn et al. 1984). Voyager IRIS spectra also contained information on the spatial variability of several molecules, including ammonia, hydrocarbons, and the *ortho-para* ratio of H_2. Ground-based exploration of Saturn's 5-μm spectrum in 1986–1989, a region of low opacity probing deep (2–5 bar) in Saturn's atmosphere allowed the first detection of germane (GeH_4), arsine (AsH_3) and carbon monoxide (CO). A decade later, the Infrared Space Observatory (ISO) detected H_2O in the deep atmosphere in the same spectral region. Saturn's centimeter spectrum probes even deeper down, to the $\sim$20-bar pressure level. Starting in the late 1970s, measurements in this range from ground-based radio-telescopes and interferometers determined the vertical and latitudinal distribution of NH_3 as well as constraining other species such as H_2S, although not always unambiguously. In parallel, constraints on the upper atmosphere composition and its latitudinal variability (hydrocarbons, water) were obtained from ultraviolet space-borne facilities (IUE and later HST). In the 1990s, increasing mid-infrared detector sensitivity allowed new stratospheric molecules to be measured, including methylacetylene (CH_3C_2H), diacetylene (C_4H_2), benzene (C_6H_6), water (H_2O), carbon dioxide (CO_2) and the methyl radical (CH_3) by ISO, as well as ethylene (C_2H_4) and propane (C_3H_8) from the ground. Long-slit spectrometers allowed studies of latitudinal variability of some species, particularly hydrocarbons. Finally, the exploration of the near-infrared spectrum over 2–4 μm led to the discovery of H_3^+ (Geballe et al. 1993), previously detected on Jupiter and Uranus, and the recent detailed characterization of the 3-μm window (Kim et al. 2006).

Having set the chronological stage of discovery, we now review systematically the pre-Cassini knowledge of Saturn's composition by families. Complementary and often more detailed information can be found in Chapter 5.

2.2.2 Bulk Composition: H_2, He/CH_4

Just as for the other Giant Planets, hydrogen, helium and methane constitute the bulk of Saturn's atmosphere, being uniformly mixed up to the homopause and representing altogether more than 99.9% of the atmosphere by volume. Their relative abundances have critical implications for models of Saturn's formation, evolution and interior.

Although unambiguously detected by Voyager (Hanel et al. 1980), the mixing ratio of helium relative to hydrogen remains relatively uncertain. Two methods can be applied (Conrath et al. 1984). One combines Voyager thermal infrared (IRIS) and radio-occultation (RSS) measurements. For uniformly mixed gases, the latter constrains the T(p)/m profile, where m, the mean molecular weight, depends primarily on the He/H_2 mixing ratio. Thus, the He abundance can be determined from the condition that the associated

T(p) profile matches the 20–50 μm spectrum. However, because the opacity in this range, due to H_2-He-CH_4 collisions, is spectrally dependent, the IRIS spectrum contains *in itself* information on the He/H_2 ratio, providing a second method of determination. Conrath et al. (1984) found that the first method was more accurate and inferred a helium mixing ratio of 0.034 ± 0.024, i.e., a He/H_2 mixing ratio $Y = 0.06 \pm 0.05$ by mass. This very low value, essentially three times less than Jupiter's ($Y = 0.18 \pm 0.04$, from the same method), was interpreted as due to strong helium precipitation in Saturn's interior. However, *Galileo* HAD measurements showed that the Jupiter He/H_2 mass mixing ratio is in fact $Y = 0.234 \pm 0.005$ (von Zahn et al. 1998), inconsistent with the Voyager value, and raising concerns about the accuracy of the published Voyager radio-occultation profiles. Building on their previous work, Conrath and Gautier (2000) developed an inversion algorithm for the simultaneous retrieval of the temperature, the *para* H_2 fraction (see discussion in Section 2.5), and the helium abundance from the IRIS spectra alone. They inferred a He/H_2 volume mixing ratio of 0.11–0.16 at Saturn, i.e., $Y = 0.18$–0.25. While still rather uncertain, the revised Saturn value is consistent with Jupiter's and the Sun's (0.2377); thus, there is no evidence for helium precipitation in either Jupiter or Saturn. Further progress may come from the analysis of the Cassini CIRS spectra, but ultimately, a precise helium determination in Saturn must await an *in situ* probe.

Saturn's methane abundance has been measured both in visible and near-infrared bands (0.63, 0.725–1.01, 1.1, 1.7, 3.3 μm) in the solar reflected component – mostly from the ground and in its mid-infrared ν_4 band near 7.7 μm, from Voyager and later from ISO/SWS (Fig. 2.1). These measurements and the associated values are tabulated in Fletcher et al. (2008) and described in more detail in Chapter 5. Each of the techniques has its own complications. In reflected sunlight, the methane determination requires accurate modelling of haze-scattering and the proper separation of methane lines from other potential absorbers (H_2, NH_3); it is also affected by uncertainties in methane absorption properties at low temperatures. In the thermal infrared, the main difficulty is that the contrasts of the methane features (which can appear either in emission or absorption, depending on the local opacity) depend strongly on temperature, so that the upper troposphere to mid-stratosphere temperature must be determined independently. Different methods have been devised, e.g., using the Voyager radio-occultation profile (Courtin et al. 1984), or using micro-windows at 10–11 μm probing the upper troposphere temperatures (Lellouch et al. 2001). This complexity has led to significant dispersion and uncertainties in the results. Overall, CH_4 mixing ratios in the $(2$–$5) \times 10^{-3}$ range have been inferred from the visible/near-infrared bands, and of typically $(3$–$6) \times 10^{-3}$ from the 8-μm region. A nominal mixing ratio of 4×10^{-3} corresponds

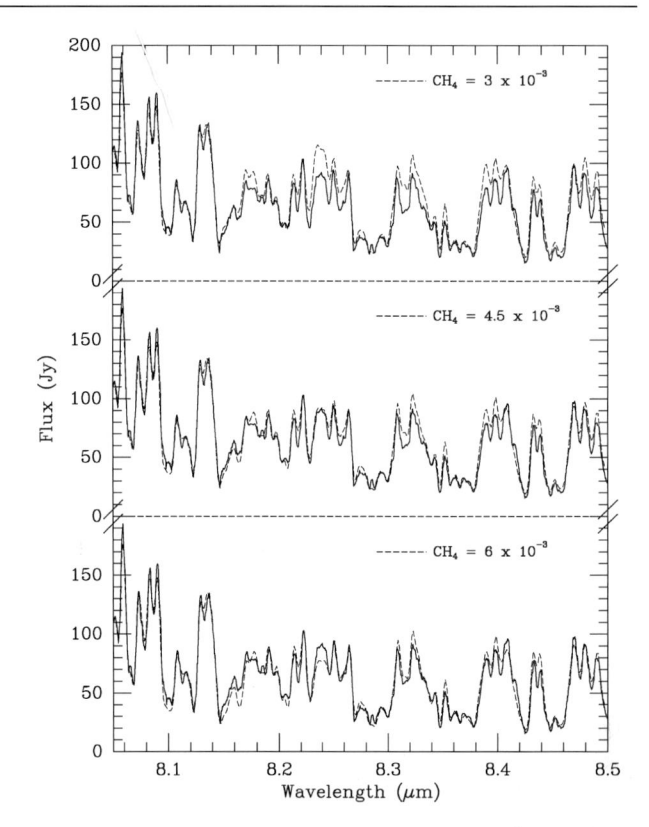

Fig. 2.1 Fit of the ISO/SWS spectrum in the 8.1–8.5 μm region with various CH_4 abundances (Lellouch et al. 2001). Reproduced with permission. © ESO

to an enrichment of the C/H ratio by a factor ∼8 over the solar value, based on the recent Grevesse et al. (2007) compilation.[1] The "weakly forbidden" lines of CH_4 in the 60–150 cm^{-1} range were first detected by the ISO/LWS instrument (Oldham et al. 1997), although an anomalously low methane abundance of $(0.7$–$1.5) \times 10^{-3}$ was inferred. These lines were later optimally exploited by Cassini CIRS to derive a more refined methane abundance (see Chapter 5). Note finally that all the above values pertain to the well-mixed atmosphere; methane is severely depleted above the homopause, which occurs near the 10^{-5} bar region (Moses et al. 2000a).

2.2.3 Tropospheric Composition

2.2.3.1 Ammonia, Water, and Hydrogen Sulfide

Methane, present throughout Saturn's atmosphere, represents the stable form of carbon in a reducing environment. The same is true of NH_3, H_2O, and presumably H_2S, which are the dominant N, O and S-bearing carriers at Saturn's

---
[1] We use hereafter this modern reference for all the solar abundances.

Fig. 2.2 Fit of millimeter-to-decimeter brightness temperatures with composition models of Saturn deep atmosphere. From van der Tak et al. (1999)

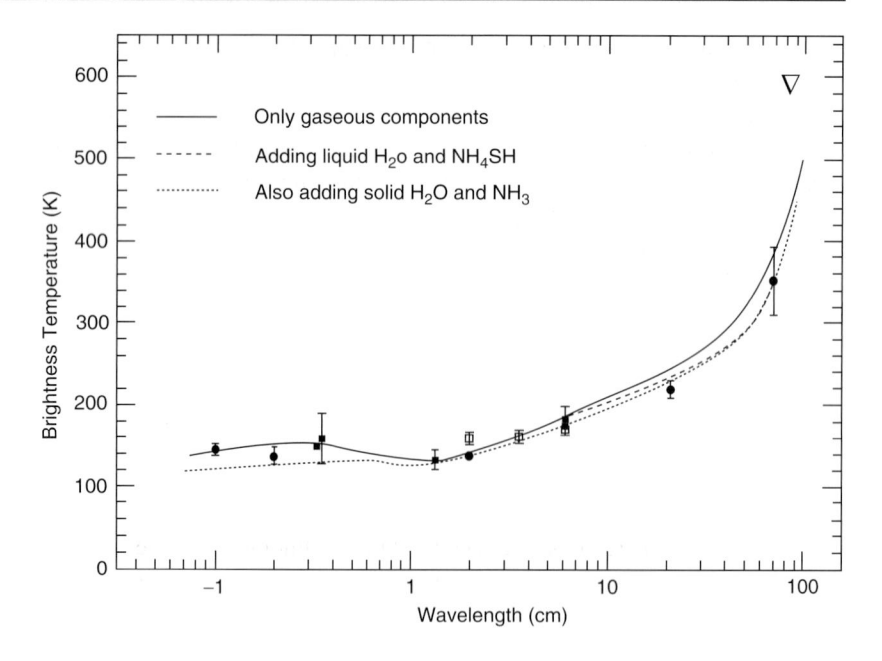

tropospheric temperatures. However, unlike methane, these species condense in or below the observable troposphere which makes determining their bulk abundances and associated elemental ratios very difficult. In fact, most observations of NH_3 and H_2O (H_2S has not been detected directly yet) probe regions located above their expected condensation level, and the measured abundances usually trace meteorological phenomena.

Originally discovered in the visible, ammonia has been studied in the thermal infrared at 5 μm (2 v_2 and v_4 bands), 10 μm (v_2), in various rotational lines over 40–500 μm (observed from Voyager and ISO/LWS), and finally at centimeter (2–6 cm) and decimeter (70 cm) wavelengths where ammonia affects the flux levels through absorption in linewings of the 1.2-cm inversion band. Generally speaking, observations at 10 μm and in pure rotational lines probe pressures around the 0.1–0.6 bar region, while 5-μm spectra sound the 2–5 bar region. The centimetric and decimetric measurements are sensitive to ammonia well below the expected NH_3/NH_4SH clouds, but their interpretation has not been straightforward, because they rely on modelling a spectral continuum to which several absorbers may contribute simultaneously (Fig. 2.2). The theoretical line shapes are uncertain, and the models are sensitive to the adopted thermal profiles. Nonetheless, the most recent and reliable of these studies (Briggs and Sackett 1989; van der Tak et al. 1999) indicated a deep (25-bar) NH_3 abundance of $(5 \pm 1) \times 10^{-4}$ (i.e., ~4 times solar), decreasing to ~10^{-4} at the ~2 bar level – consistent with inferences from the 5-μm spectrum (Fink et al. 1983; Courtin et al. 1984; Noll and Larson 1991; de Graauw et al. 1997). At higher levels, NH_3 is further depleted, with 10-μm and pure rotational features, yielding typical mixing ratios of 10^{-5} at 1 bar and 10^{-7} at 0.5 bar. This is consistent with NH_3 condensation and about 50% under-saturation at 2 bar and above.

Although water vapor was discovered in Jupiter's troposphere in 1975 (Larson et al. 1975) from 5-μm observations, its detection in Saturn was much more difficult because of its lower thermal environment and condensation at deeper levels; it was detected only by ISO (de Graauw et al. 1997). These observations indicated a H_2O mixing ratio of ~3×10^{-7} below the 3-bar level. Even more than for ammonia, this value corresponds to under-saturation, a situation similar to the case of the Jovian hot spot measured by Galileo (Niemann et al. 1996), the 5-μm ISO disk-average radiation presumably dominated by the hotter and drier regions.

Sulfur species have not been directly detected in Saturn's atmosphere, but the undersaturation of ammonia in the 2–4 bar region has been interpreted as evidence for a large H_2S abundance, implying an ammonia sink through formation of an NH_4SH cloud (Briggs and Sackett. 1989). The estimated H_2S mole fraction, 4×10^{-4}, would correspond to S/H ~15 times the solar ratio.

Images of Saturn in the centimetre-wavelength range (de Pater and Dickel 1982, 1991; Grossman et al. 1989; van der Tak et al. 1999; Fig. 2.2) indicate both latitudinal and temporal variability, presumably caused by variations of temperature and NH_4SH and NH_3-ice cloud humidity. Cassini 2–14 cm radio-science (RSS) measurements with improved calibration may shed new light in the interpretation of these measurements. In the upper atmosphere, Cassini CIRS data should ultimately provide ammonia maps, similar to those obtained from Voyager IRIS at Jupiter (Conrath and Gierasch 1986).

2.2.3.2 Disequilibrium Species: PH$_3$, AsH$_3$, GeH$_4$, CO

The detection of phosphine (PH$_3$) in Jupiter and Saturn (Gillet and Forrest 1974; Bregman et al. 1975) marked the discovery of a new class of molecules. Phosphine is very different from ammonia; under thermodynamical equilibrium conditions, it should have a negligible abundance at observable levels (Prinn and Barshay 1977; Barshay and Lewis 1978; Lewis and Fegley 1984). The dominant P-bearing species in the upper troposphere should be P$_4$O$_6$, itself ultimately converted to solid NH$_4$H$_2$PO$_4$. Its large abundance in Saturn near $\sim$2 bar, typically $(7 \pm 3) \times 10^{-6}$ inferred from the 5-μm observations (Larson et al. 1980; Bézard et al. 1989; Noll and Larson 1991; de Graauw et al. 1997) is attributed to rapid upward transport from the deep atmosphere where it is chemically stable. Although chemical reactions would tend to destroy it in the upper troposphere, they are kinetically "frozen", and the observed abundance is representative of the "quenching level", where the chemical conversion and convective timescales are equal. Knowing the chemical reactions in play (i.e., essentially the limiting chemical step of the destruction process), it is possible to estimate the quenching level, which occurs at $\sim$1,200 K for phosphine in Saturn (Fegley and Prinn 1985; Visscher et al. 2006). Because PH$_3$ is stable at this level, its abundance measured near two bars reflects the deep-atmosphere value, allowing one to derive a bulk P/H ratio $\sim$15 times solar. Note finally that PH$_3$ dominates Saturn's solar-reflected 3-μm spectrum in addition to the 5-μm window, so that Cassini VIMS experiment should be able to map its mid- and upper-troposphere abundance.

Ground-based observations of the 5-μm window (and later its re-observation by ISO) permitted the detection and measurement of three other disequilibrium species: carbon monoxide (CO, Noll et al. 1986), germane (GeH$_4$, Noll et al. 1988), and arsine (Noll et al. 1989, 1990; Bézard et al. 1989). The stable chemical forms of germanium and arsenic in the upper troposphere are GeS (and GeSe) and AsF$_3$, but – similar to phosphorus – reactions destroying GeH$_4$ and AsH$_3$ are quenched at $\sim$870 K and 1,420 K, respectively. The $\sim$2 bar abundances of GeH$_4$ and AsH$_3$ inferred from the ground-based observations (Fig. 2.3) are $\sim$4 $\times 10^{-10}$ and $\sim 2.5 \times 10^{-9}$ (although de Graauw et al. 1997 quoted a GeH$_4$ abundance about four times higher). This formally indicates Ge/H and As/H ratios of $\sim$0.05 (0.2) and $\sim$6 times the solar value, respectively. Although the As/H ratio must represent the deep value, the Ge/H does not, because GeS remains the dominant Ge-bearing species at the quench level for GeH$_4$.

Phosphine, germane and arsine are not uniformly mixed vertically. This was quickly realized in the wealth of PH$_3$ observations: in addition to the 5-μm window, phosphine shows signatures at 10 μm (Courtin et al. 1984; Lellouch et al. 2001), in rotational lines longwards of 50 μm (Weisstein and Serabyn 1996, Davis et al. 1996; Orton et al. 2000, 2001), in the reflected component at 3 μm (Bjoraker et al. 1981; Kim et al. 2006) and in the ultraviolet (Winkelstein et al. 1983; Edgington 1997). These revealed that the phosphine abundance decreases rapidly with altitude above the $\sim$600-mbar level, being depleted by a factor of $\sim$20 at 150 mbar, a behavior attributed to photolytic destruction. The same trend was shown to hold for germane and arsine. Although these species are observable only at 5 μm, radiation from Saturn in this window actually consists of the combination of thermal radiation originating from 2–5 bar with sunlight reflected near the 300-mbar level, allowing one to determine

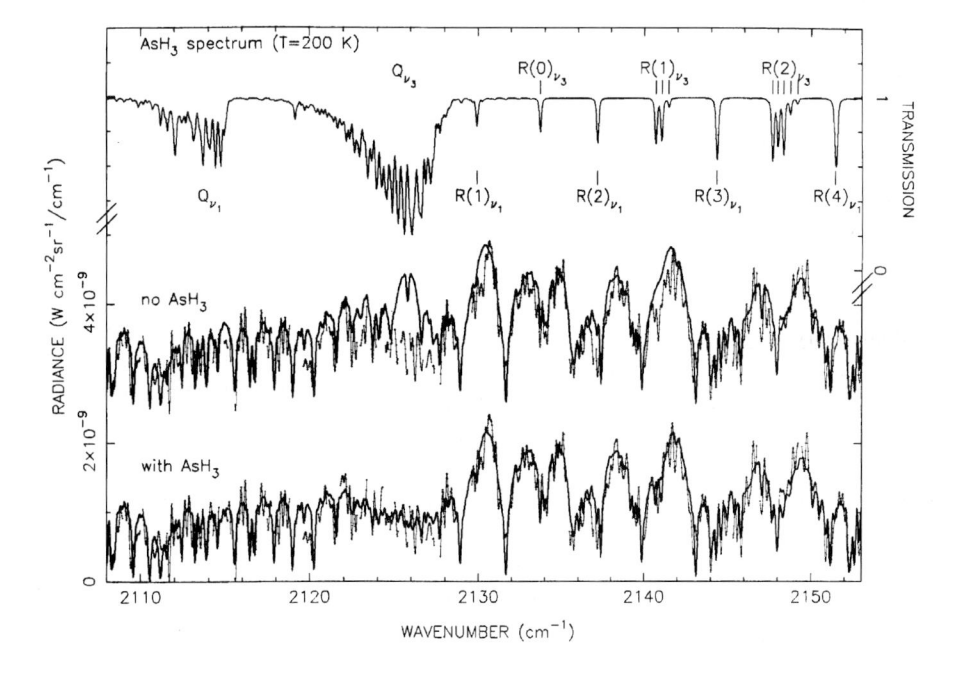

Fig. 2.3 Detection of arsine in Saturn from CFHT spectra. From Bézard et al. (1989)

the abundance of GeH$_4$ and AsH$_3$ separately at two altitudes (Bézard et al. 1989), showing that they are depleted by at least factors of 2–4 in the solar-reflected component from their deep value.

In the case of CO, first detected by Noll et al. (1986), the 5-μm spectra have not established its location in the atmosphere. Noll and Larson (1991) showed that their data could be fit either by a uniform CO mixing ratio of $(1 \pm 0.3) \times 10^{-9}$ or an abundance of 2.5×10^{-8} restricted to the stratosphere. As CO is not subject to photolysis and could also have an external origin, the two scenarios have not so far been discriminated, in spite of efforts by Rosenqvist et al. (1992) and Cavalié et al. (2008) to use millimeter/submillimeter observations to resolve the rotational line profiles and establish its vertical distribution. If CO is internal (i.e., uniformly mixed), its $\sim$2 bar abundance exceeds its thermochemical equilibrium value by a factor $\sim$40. Visscher and Fegley (2005) showed that this would imply a H$_2$O deep abundance of at least $\sim 2 \times 10^{-3}$. Combined with constraints derived from the PH$_3$ abundance and upper limits on SiH$_4$, they derived a Saturn O/H ratio in the range of 3.2–6.4. Just as for Jupiter, the deep H$_2$O abundance and hence the bulk O/H ratio has therefore not been measured yet and may need to await the deployment of a direct atmospheric probe or a Juno-like close microwave sensor experiment.

Overall, disequilibrium species have provided invaluable opportunities to (i) demonstrate the role of upward convection and estimate its strength (ii) to measure new elemental ratios. In general, Saturn appears more enriched in heavy elements than Jupiter, although there remain large variations in the enrichment factors of the various elements (from about 4 to 15 times solar), which remains to be understood.

2.2.4 Stratospheric Composition

2.2.4.1 Hydrocarbons and Photochemistry

Long expected in the Giant Planets (Strobel 1969), evidence for methane photochemistry was obtained at Saturn with the detection of C$_2$H$_6$ at 12 μm (Tokunaga et al. 1975) and C$_2$H$_2$ in the ultraviolet with IUE (Moos and Clarke 1979). Since then, a number of molecules were progressively discovered, enriching our chemical inventory and providing further insight into essential chemical pathways.

The first reliable abundances of ethane and acetylene were derived from Voyager IRIS (Courtin et al. 1984), indicating typical abundances of 3×10^{-6} and 2×10^{-7}, respectively. Although subsequent ground-based and ISO studies (Noll et al. 1986; Sada et al. 2005) led to slightly larger abundances (by factors of 1.5–3), this result confirmed that ethane was the dominant photochemical product. Using ISO data, Moses et al. (2000a) demonstrated that C$_2$H$_2$ increases with altitude, similar to Jupiter (Bézard et al. 1995; Fouchet et al. 2000), as expected for a species produced in the upper stratosphere and transported downward by eddy diffusion to the troposphere where it undergoes thermal decomposition.

The enhanced 8–20 μm sensitivity of ISO permitted the discovery of four new hydrocarbons: methylacetylene (CH$_3$C$_2$H) and diacetylene (C$_4$H$_2$) (de Graauw et al. 1997), the methyl radical (CH$_3$) (Bézard et al. 1998) and benzene (C$_6$H$_6$) (Bézard et al. 2001a). For all these species, the emission is optically thin, so that only column densities could be inferred: about 1×10^{15} cm^{-2}, 1×10^{14} cm^{-2}, 5×10^{13} cm^{-2}, and 4×10^{13} cm^{-2}, respectively. The detection of CH$_3$ was particularly significant, because it is the first radical produced in the photolysis of methane. In later ground-based observations ethylene (C$_2$H$_4$) was discovered (Bézard et al. 2001b), with a column density of $\sim 2.5 \times 10^{15}$ cm^{-2}, and then propane (C$_3$H$_8$) (Greathouse et al. 2006), with a mixing ratio of $\sim 2.5 \times 10^{-8}$ at 5 mbar.

These new detections led to a generation of photochemical models, the most extensive of which is probably that of Moses et al. (2000a). This 1-D model, primarily constrained by ISO observations, was generally successful in reproducing the observed abundances. It also demonstrated the role of (i) long-lived species such as C$_2$H$_6$ to trace vertical transport (parameterized by the eddy diffusion coefficient), and (ii) short-lived species such as CH$_3$CCH and C$_4$H$_2$ in illustrating the main chemical pathways in atmospheric chemistry. It also served to motivate a search for new species (ethylene and propane), that were ultimately discovered later. The presence of benzene, however, simultaneously detected in Jupiter, Saturn, and Titan by ISO (Bézard et al. 2001a; Coustenis et al. 2003) is not obvious to explain. Although the propergyl radical (C$_3$H$_3$) can self-add to produce a C$_6$H$_6$ isomer, a large isomerization rate is then required to produce benzene. More details can be found in Chapter 6.

Latitudinal variations of C$_2$H$_2$ and C$_2$H$_6$ were studied using ground-based IRTF/TEXES observations (Greathouse et al. 2005). Acquired near southern solstice, these data indicated that C$_2$H$_2$ tends to decrease from equator to pole, while the opposite trend was marginally observed for C$_2$H$_6$ (Fig. 2.4). The behavior of C$_2$H$_2$ could be understood because its chemical lifetime (2–7 years) is somewhat shorter than a Saturnian year; therefore C$_2$H$_2$ is expected to track the annually averaged solar insolation. In contrast, the C$_2$H$_6$ distribution could not be reproduced by 1-D seasonal models (Moses and Greathouse 2005); given its very long chemical scale (1,000 years), C$_2$H$_6$ is probably controlled by the large-scale circulation (see more details in Chapter 6). Greathouse et al. (2006) also studied the latitudinal distribution of C$_3$H$_8$ and found no significant variations.

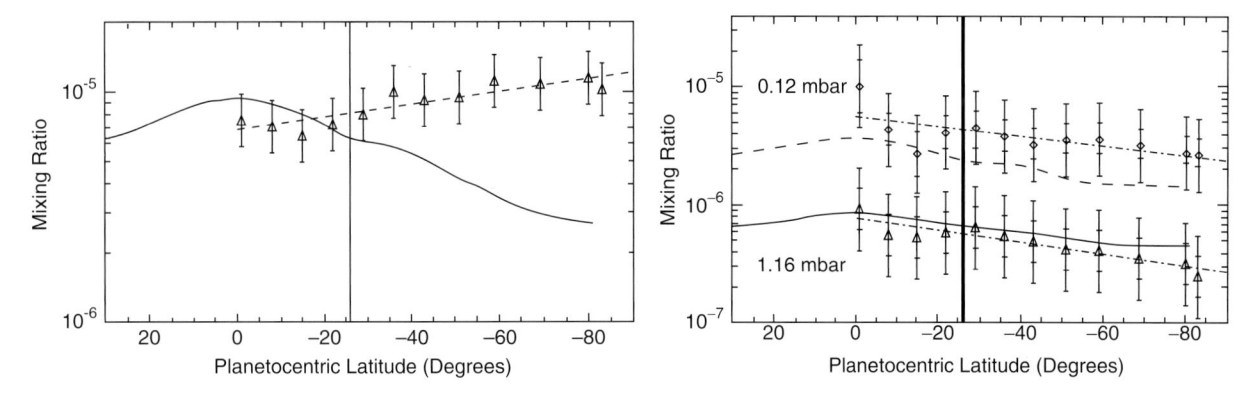

Fig. 2.4 Latitudinal variation of C_2H_6 at 2.3 mbar (*left*) and C_2H_2 at two pressures (*right*). Solid and dashed lines indicate 1-D seasonal model predictions (from Greathouse et al. 2005)

Table 2.1 CH_3D/CH_4 abundance measurements

Derived value	Source	Comment
$7 + 7 \times 10^{-5-3}$	Owen et al. (1986); de Bergh et al. (1986)	Reflected sunlight: $3\nu_2$ band, 1.55 μm
5.3×10^{-5a}	Kim et al. (2006)	Reflected sunlight: $\nu_3 + \nu_2$ band, 2.87–3.10 μm
$7.0 \pm 3.7 \times 10^{-5a}$	Noll and Larson (1991)	Thermal emission (mostly) from deep atmosphere: 4.7 μm
$8.5 \pm 5.5 \times 10^{-5a}$	Courtin et al. (1984)	Voyager IRIS: ν_6 band, 8.6 μm
$7.2 + 2.3 \times 10^{-5a-1.9}$	Lellouch et al. (2001)	ISO/SWS ν_6 band, 8.6 μm

[a]Reported originally as a volume mixing ratio for CH_3D, and converted here to a ratio using a volume mixing ratio for CH_4 of $4.7 \pm 0.2 \times 10^{-3}$ (Fletcher et al. 2008).

2.2.4.2 External Supply of Oxygen

Winkelstein et al. (1983) first suggested an external flux of oxygen to Saturn's upper atmosphere in fitting their 150–300 nm IUE spectrum which required water vapor in addition to the acetylene opacity in their models. Prangé et al. (2006) noted that Winkelstein et al. had assumed a vertically uniform mixing ratio for acetylene, requiring them to use an unrealistically large column abundance of water; a vertical distribution more consistent with photochemical models and infrared observations allowed only upper limits to the water abundance from ultraviolet spectroscopy. It was in the infrared range that the ISO Short Wavelength Spectrometer (SWS) unambiguously detected rotational water lines between 30.90 and 45.11 μm (Feuchtgruber et al. 1997), which were best fit using a column density of $1.4 \pm 0.4 \times 10^{15}$ cm^{-2} (Moses et al. 2000b). de Graauw et al. (1997) also used ISO SWS spectra to detect CO_2 for the first time using emission via the 14.98-μm band, modelled best with a column abundance of $6.3 \pm 1.0 \times 10^{14}$ cm^{-2} (Moses et al. 2000b).

The presence of H_2O and CO_2 in Saturn's stratosphere was the definite proof of a flux of external oxygen into the planet. Through the help of a photochemical model, Moses et al. 2000b performed the most detailed analysis of the ISO measurements. They concluded that (i) the data imply a total oxygen flux of (4 ± 2) O atoms cm^{-2}s^{-1} (ii) although CO_2 can in principle be formed from the reaction between OH and internal CO, a flux of H_2O alone does not satisfy the observations, and the infalling oxygen-bearing material must contain both H_2O and a C–O bond species (e.g., CO or CO_2). In contrast, they could not shed any new light on the internal vs external origin of CO, so this issue remains open.

2.2.4.3 Isotopic Ratios

The deuterium to hydrogen (D/H) ratio in Saturn's atmosphere can be determined from the HD/H_2 ratio directly or by the fraction of deuterium in hydrogen-bearing molecules. Because HD lines lie in the far infrared and are not immediately accessible to terrestrial observers, the first estimates were derived from the CH_3D/CH_4 ratio. Table 2.1 shows various values for this quantity derived using both reflected sunlight and thermal emission observations.

A conversion to D/H for H_2 also involves accounting for the existence of an isotopic exchange reaction between H_2 and CH_4, called the fractionation coefficient, $f = (D/H)_{CH4}/(D/H)_{H2} = 1.34 \pm 0.19$, estimated by Lecluse et al. (1996) and Smith et al. (1996). Accounting also for the 1 in 4 substitution, the conversion from the values of the ratio which cluster around 7–8×10^{-5}, the D/H ratio is about 1.3–1.5×10^{-5}.

Direct measurements of D/H were made by Griffin et al. (1996) using ISO Long Wavelength Spectrometer LWS measurements of the HD R(1) rotational line at 56 μm, deriving a value of $2.3^{+1.2}_{-0.8} \times 10^{-5}$. Lellouch et al. (2001) derived a D/H ratio from the ISO SWS measurements of the HD R(2) and R(3) rotational lines and derived a value for D/H for H_2 of $1.85^{+0.85}_{-0.60} \times 10^{-5}$. They combined a CH_3D and HD values to conclude that $(D/H) H_2 = 1.70^{+0.75}_{-0.45} \times 10^{-5}$. This value is compatible with the D/H ratio in Jupiter and the local interstellar medium (LISM), but the accuracy of this measurement is insufficient to distinguish it from either and thus provide meaningful constraints on models for Saturn's evolution and interior.

The few existing values for $^{12}C/^{13}C$ (Combes et al. 1977; corrected by Brault et al. 1981; Sada et al. 1996) are compatible with the terrestrial and Jovian ratios, but their high uncertainties make them very unconstraining.

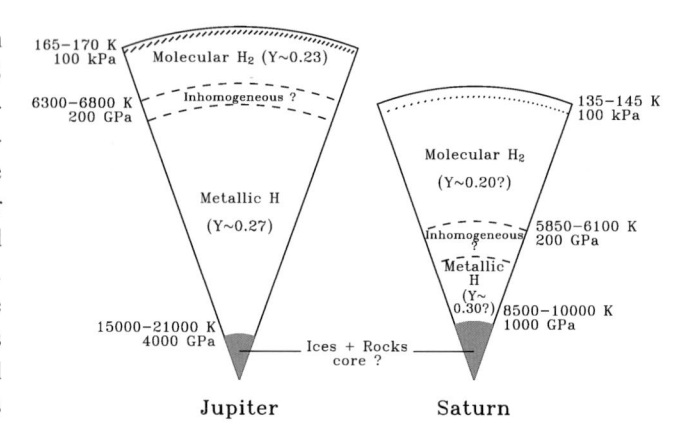

Fig. 2.5 Schematic representation of the interiors of Jupiter and Saturn. Temperature ranges are estimated from homogeneous models. Helium mass ratios, Y, are also noted. Central rock and ice core sizes are extremely uncertain. From Guillot (1999a)

2.3 Saturn's Interior

Just after the Voyager encounter, Hubbard and Stevenson (1984) pointed out that interior models were critically dependent on knowledge about the outer temperature boundary conditions, heat flow and rotation rates, which Voyager observations strongly constrained. The important zonal harmonics of the gravity field were provided by the Pioneer 11 close flyby (Null et al. 1981). The internal heat source, together with the low conductivity and viscosity of hydrogen, required that convection be the primary means of heat transport. The external gravity field and the general properties of the hydrogen equation of state required the presence of a massive, dense core – although its mass and composition were uncertain. The observed zonal winds might extend deep within the interior, producing a detectable influence on the external gravity field.

Guillot (2005), reviewing more recent work on interior structure of Saturn and other giant planets, emphasized the difficulty of calculating an appropriate equation of state (EOS) for hydrogen in regions appropriate to gas-giant interiors because of the need to account for interactions between co-existing molecules, atoms and ions of different elements. Nonetheless, it is clear that the interiors of hydrogen–helium gas giants are fluid, no matter what their age. Guillot et al. (2004) noted that the probable presence of sodium and potassium absorption in the visible spectrum in outer-planet atmospheres would wipe out a radiative layer in his original predictions (Guillot et al. 1994).

The simplest models of the interiors of Jupiter and Saturn assume the presence of three main layers (Fig. 2.5): (a) an outer hydrogen–helium envelope, whose global composition is that of the deep atmosphere (except for condensates above the 20-bar pressure level), (ii) an inner hydrogen–helium envelope, enriched in helium because the whole planet must fit the H_2–He protosolar value, and (iii) a central, dense core. Most of the uncertainty in these models arises from the fact that different hydrogen EOSs are possible.

For Saturn, the solutions depend less on the hydrogen EOS than is the case for Jupiter because the Mbar pressure region is smaller, and the total amount of heavy elements present in the planet can therefore be estimated with better accuracy. Solutions restrict the core mass to be less than 22 Earth masses; solutions are possible with no core mass (Guillot 1999b). Those solutions depend on the phase separation of an abundant species, such as water, and require the release of considerable gravitational energy. Fortney and Hubbard (2003) point out another possibility, the formation of an almost pure helium shell around a central core, which would lower the core masses derived by Saumon and Guillot (2004) by as much as 7 earth masses.

2.4 Saturn's Clouds and Aerosols

2.4.1 Pioneer/Voyager Era Measurements of Aerosol Structure

The first Saturn flybys in 1979–1981 by Pioneer 11 and the Voyager spacecraft provided observations at geometries and wavelengths unavailable from ground-based imaging. These spacecraft determined the optical properties and sizes as well as on the vertical and spatial distribution of aerosols. Measurements near phase angles of 90° in red light (0.64 μm) by the imaging photopolarimeter (IPP) on Pioneer Saturn revealed the presence of a thin layer of highly positively polarizing, small (∼0.1 μm radii) aerosols in the stratosphere (Tomasko and Doose 1984). In both blue and red light, both Pioneer Saturn and Voyager found forward-scattering particles which, unlike Jovian aerosols, exhibited a relatively flat single-scattering phase function at angles greater than 70° (Tomasko et al. 1980; West et al. 1983; Tomasko and Doose 1984), similar to ammonia ice (Tomasko and Stahl 1982; Tomasko et al. 1984) Analysis of Voyager photopolarimeter experiment maps at 0.264 μm showed that high-altitude particles are particularly UV-absorbing at the pole (West et al. 1983).

During this era of Saturn flybys, ground-based observations at wavelengths sensitive to a range of atmospheric absorption features constrained the pressure and opacity of aerosol layers. Tomasko and Doose (1984) and West et al. (1982) analyzed center-to-limb behavior of extinction to derive families of solutions for several latitudes, confirming previous work (Owen 1969; Macy 1977; Tejfel 1977a, b) that the equatorial region is particularly opaque in the visible at high altitudes. Analysis of the Pioneer-11 Infrared Radiometer 45-μm limb darkening also suggested a significant large-particle component of this cloud (Orton 1983) Ground-based methane band images clearly showed hemispherical asymmetry (West et al. 1982), indicating that aerosols in the southern hemisphere are at somewhat lower altitude than in the north (Fig. 2.6).

At the conclusion of the Pioneer and Voyager encounters, knowledge of Saturn's aerosol structure was summarized by Tomasko et al. (1984). In the stratosphere, strongly UV-absorbing 0.1-μm particles with a 0.264-um opacity of ∼0.4 exist at 30–70 mbars at low latitudes and at pressures equal to or less than ∼20 mbar at high latitudes which are likely produced by photochemical processes or by precipitating auroral ions. Significant tropospheric visual aerosol opacity occurs at higher altitudes near the equator than elsewhere. This enhancement in higher-altitude aerosols at equatorial latitudes may indicate increased vertical transport there, perhaps involving convection of ammonia ice condensates from below. The base of the NH_3 cloud is expected at its condensation level, than estimated at near 1.4 bar (Weidenschilling and Lewis 1973). As implied by the behavior of weak methane and hydrogen quadrupole lines, the visual opacity at many latitudes is on the order of 10–20 at red and near-infrared wavelengths. At other latitudes, the opacity may be ∼5, with a clearer region at pressures greater than 500 mbar.

2.4.2 From Pioneer/Voyager to Cassini: Earth-Based Observations of Aerosol Structure

In the near quarter century between the Pioneer/Voyager flybys and the arrival of Cassini-Huygens, Saturn was often scrutinized by Earth-based telescopes, resulting in a number of comprehensive and detailed analyses of haze and cloud properties (Karkoschka and Tomasko 1992, 1993; Ortiz et al. 1996; Muñoz et al. 2004; Pérez-Hoyos et al. 2005; Karksochka and Tomasko 2005). All of these analyzed individual latitudinal bands, revealing that Saturn's aerosols structure varies significantly in both latitude and time, with noticeable seasonal effects. In all cases, haze layers were modelled for both the stratosphere and troposphere, down to at least the ∼800 mbar level. Typically, for all wavelengths <0.73 μm, the net extinction of the hazes and/or overlying gaseous atmosphere reaches opacities above the 1-bar level, sufficient to prevent clear visual views of a deeper ammonia condensation cloud layer.

These analyses typically derived a stratospheric haze layer located between 1 and 50 ± 40 mbar. When modelled as Mie (spherical) scatterers, these ultraviolet-absorbing (single-scattering albedo of 0.59 at 0.3 μm) aerosols are typically found to have a mean radius of 0.15–0.2μm with a.0.3 μm opacity of ∼0.5 decreasing to ∼0.1 at 0.7 μm. In the troposphere, a blue-absorbing haze layer typically begins just underneath the tropopause level near 100 mbar and extends down to 500–1,200 mbar, depending on the region. This thicker layer has a typical opacity of 8–12 near 0.5 μm, but can change by a factor of 2 over time. Mie-scattering particles are typically 1.5–2.5 μm in mean radius. Underneath

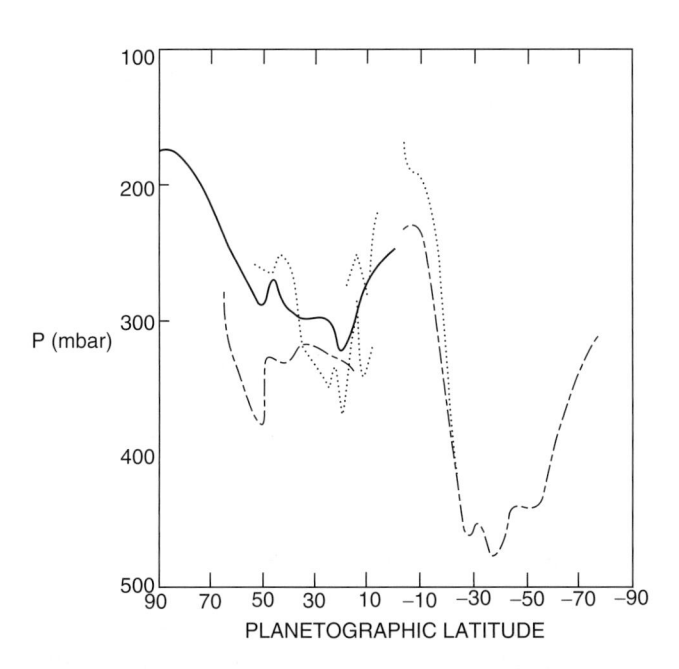

Fig. 2.6 Estimates of cloud-top pressure vs latitude from measured 0.264-μm intensities (*solid curve*), methane band to continuum radio (*dot-dashed curve*), and Pioneer polarimetry (*dotted curve*) using various simple models. Note the greater depths of clouds in the southern hemisphere. From Tomasko et al. (1984)

this haze, an optically thick ammonia cloud layer is typically modelled with a base near the 1.5–1.8-bar level expected for ammonia condensation, (depending on the ammonia mixing ratio, expected to be 1–5 times that predicted by the solar N/H abundance; Atreya et al. 1999). Its top is sometimes observed and modelled in near-infrared reflected light, especially when it can be seen rising to the ~1.2-bar level.

Major departures from this model are found at high and low latitudes. In the polar troposphere, hazes are darker than elsewhere, especially near 0.44 μm. In the polar stratosphere, aerosols are more plentiful (opacity four times that found typically equatorward of 65° latitude) and, as previously found by Voyager (West et al. 1983), unusually dark in the ultraviolet. Specifically, at 0.3 μm, particles high in the stratosphere above the 10-mbar level exhibit values for the imaginary index of refraction, n_i, and single-scattering albedo of 0.1 and 0.90, respectively (e.g., Karkoschka and Tomasko 2005). The reflectivity of these particles also exhibits a steep wavelength dependence, brightening to a single-scattering albedo of 0.99 at 0.45 μm (corresponding to a power law exponent for n_i of −6.5; Karkoschka and Tomasko 2005). The wavelength-dependent reflectivity in the deep methane absorption bands from the visible to the near-infrared show that these particles are smaller than elsewhere, with a mean radius of 0.1–0.15 μm (Karkoschka and Tomasko 1993, 2005; Pérez-Hoyos and Sánchez-Lavega 2006). The unusually dark ultraviolet single-scattering albedo, high opacity, and high altitude of these polar haze particles indicate that they are of a different composition and produced by a different formation mechanism than elsewhere. A prime candidate is ion-chemistry-generated long-chained hydrocarbon-based condensates produced by the ionization of atmospheric gas constituents – particularly hydrogen and methane – by the precipitation of energetic ions in polar aurorae.

At the equator, a significantly thicker haze layer exists at higher altitudes than elsewhere, typically extending continuously from the ~50 mbar level in the stratosphere down to ~500 mbars. Between 1979 and 2004, this layer changed substantially (Pérez-Hoyos et al. 2005), varying both its base and bottom pressures substantially (by ~25%), and sometimes increasing its opacity by a factor of 4. A major increase in opacity and haze-top altitude was associated with the 1990 outburst of the Great White Spot (Ortiz et al. 1996; Acarreta and Sanchez-Lavega 1999). Contemporaneous changes in the equatorial zonal winds (e.g., Sánchez-Lavega et al. 2003, 2004; Pérez-Hoyos and Sánchez-Lavega 2006) appear to be correlated with this outburst and the increased altitudes of the cloud features tracked.

Seasonal variations in aerosol structure have also been well-documented, especially during ring-plane crossing when the insolation and energy deposition changed significantly (Karkoschka and Tomasko 2005; Fig. 2.7). The most visible change has been in the tropospheric haze opacity

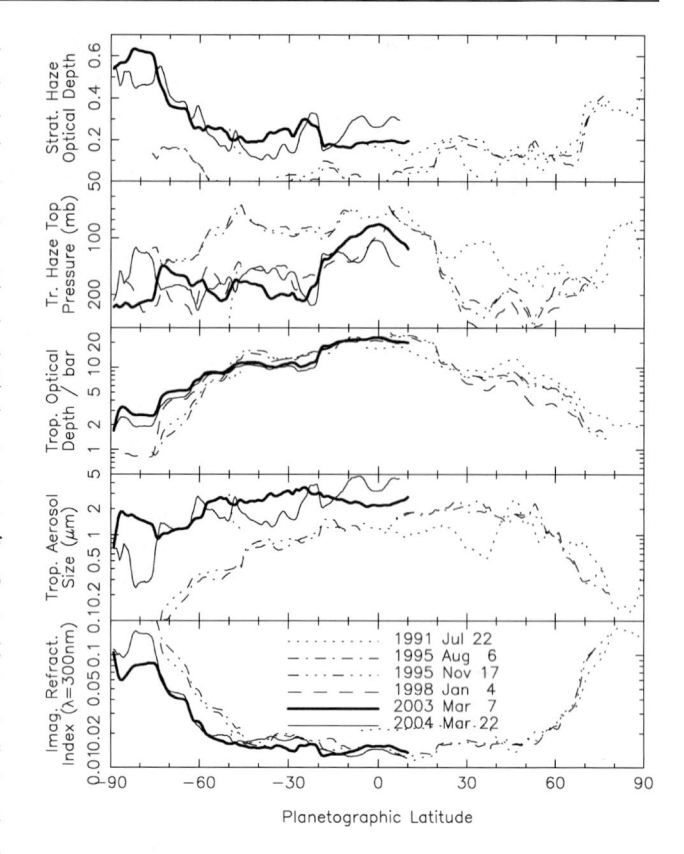

Fig. 2.7 Meridional variations of the model parameters for several observation dates examined by Karkoschka and Tomasko. (2005). The uncertainty from the data is 5–10% of the displayed range in each panel. The uncertainty due to model assumptions, such as the shape of aerosols, may be much larger

(Pérez-Hoyos et al. 2005; Karkoschka and Tomasko 2005), which reaches a minimum near maximum insolation (Pérez-Hoyos et al. 2005). The poles exhibit a factor of ~2 seasonal variability of their stratospheric opacity – the largest seasonal change in stratospheric properties observed on the planet, one which arises from changes in the column abundance of particles.

More rapid variability in aerosol microphysical properties is also observed. Most particularly near the equator or the poles, short-term brightness variations in typically 3° wide latitudinal bands have been noted by Pérez-Hoyos and Sánchez-Lavega (2006) which appear to arise from variations in the single-scattering albedo of the tropospheric haze aerosols, indicating variations of particle size or composition. The most dramatic change observed thus far on Saturn was the eruption of the Great White Spot at 5°N latitude in Saturn's Equatorial Region in September, 1990, which produced increases in both the altitude (by ~1.3 scale height) and single-scattering albedo of the cloud tops (Sánchez-Lavega et al. 1991, 1993a; Westphal et al. 1992; Barnet et al. 1992; Beebe et al. 1992; Acarreta and Sanchez-Lavega 1999).

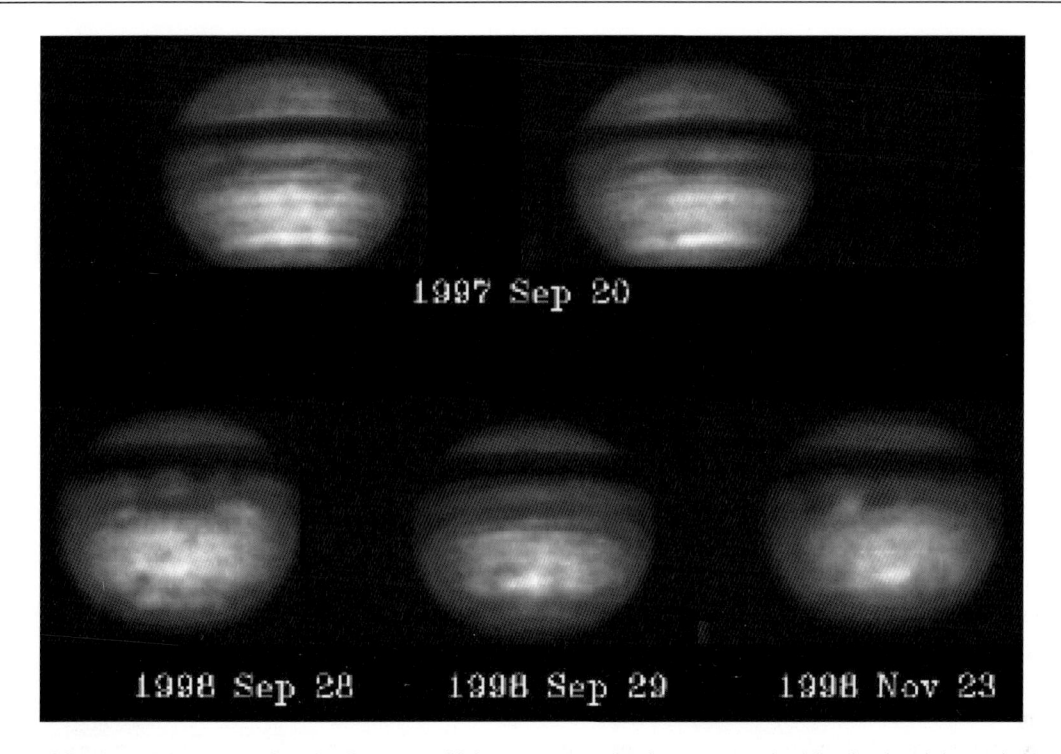

Fig. 2.8 Images of Saturn at 5.2 μm, showing the discovery of inhomogeneous cloud structure at the 1-bar level and deeper in the troposphere. From Yanamandra-Fisher et al. (2001)

Cassini-Huygens appears particularly well-suited to address the nature of clouds, meteorology, and circulation in the deep atmosphere underneath the physically- and optically thick tropospheric (putative) ammonia layer using close-up 5-μm images by VIMS, capitalizing on the revelation of structure in ground-based 5-μm images (Yanamandra-Fisher et al. 2001; Fig. 2.8). Detailed images by Cassini promise to provide insight into the nature of the well-hidden equatorial region, and the nature of the mysterious polar hexagon feature (Godfrey 1988) at depth during the dark mid-winter on Saturn. Another outstanding question is the lack of spectral identification of the aerosols observed on Saturn. Ammonia ice, which thermochemical considerations identify as the most likely primary condensable cloud in the upper atmosphere has yet to be spectroscopically identified. As suggested for Jupiter (Baines et al. 2002; Atreya et al. 2005) and recently verified in laboratory observations (Kalogerakis et al. 2008), older (>1 week) ammonia particles may be coated with impurities which sufficiently mask spectral features to prevent identification. As Galileo and New Horizons have done for Jupiter (Baines et al. 2002; Reuter et al. 2007), Cassini could identify ammonia ice spectroscopically in young, discrete clouds, particularly those forming in regions of significant vertical uplift. Thus, discovering and mapping such features may yield new insight into Saturn's tropospheric meteorology, dynamics and chemistry.

2.5 Saturn's Temperatures

The earliest models for Saturn's vertical temperature profile were obtained from radiative-convective equilibrium models (Trafton 1967). The presence of emission features of CH_4 and C_2H_6 in middle-infrared spectra (Gillett and Forrest 1974) required a stratosphere that was distinctly warmer than the temperatures at a well-defined temperature minimum. This was confirmed by observations of limb brightening of stratospheric C_2H_6 emission (Gillett and Orton 1975; Rieke 1975). Revised radiative-convective equilibrium models (Caldwell 1977; Tokunaga and Cess 1977; Appleby 1984) showed that the warmer temperatures could be explained by the absorption of sunlight by near-infrared and visible bands of CH_4, together with potential absorption of solar ultraviolet radiation by small particulates (Axel 1972; Podolak and Danielson 1977).

Ohring (1975) attempted to derive the mean temperature structure of the stratosphere using Gillett and Forrest's (1974) spectrum of its 7-μm CH_4 emission. As is also the case for H_2, CH_4 is well mixed throughout the troposphere and the stratosphere, and thus variability of emission can be directly attributed to variability of temperature. Drift-scan observations across Saturn's equator at 17.8 and 22.7 μm (Caldwell et al. 1978), sensitive to temperatures in Saturn's upper troposphere, provided the first ground-based observation of emission from a part of the spectrum sensitive to the H$_2$ collision-induced continuum.

A scan across Saturn's central meridian by Gillett and Orton (1975) showed a substantial enhancement of stratospheric C$_2$H$_6$ emission over the south polar region at 12.5 μm. They concluded that this was most likely the result of seasonal variability of stratospheric temperatures around Saturn's south pole. Tokunaga et al. (1978, 1979) confirmed this polar enhancement and demonstrated it was also true at 20 μm, implying that radiatively enhanced seasonal heating extended down to the ∼100-mbar tropopause. They also detected latitudinal variability over the rest of the planet. In 1979 the Pioneer-11 Infrared Radiometer (IRR) experiment, with broadband filters centered at 20 and 45 μm, provided an improved estimate of Saturn's effective temperature as 96.5 ± 2.5 K (Orton and Ingersoll 1980; Ingersoll et al. 1980) and mapped temperatures in the 60–160 mbar pressure range. The derived temperature profile was consistent with a broad temperature-minimum region, and temperatures were lower near ±5° latitude than at midlatitudes.

The Pioneer measurement of the phase delay of the spacecraft radio signal as it was occulted by Saturn was used to determine the refractivity of the atmosphere as a function of altitude, from which the temperature profile was derived in the 1–200 mbar range. These results (Kliore et al. 1980a, b) for the unobstructed egress ray path provided the first detailed evidence for the structure of the "inverted" stratosphere, whose temperature ranged from a high of 127 K near 20 mbar to 89 K near 100 mbar. The structure also showed evidence for vertical thermal waves with amplitudes of a few degrees.

A significant advance in measuring Saturn's temperatures was made by Voyager-1 and -2 experiments. The infrared experiment, IRIS, mapped spatially resolved spectra covering 200–700 cm^{-1} (14–50 μm) across nearly the entire disk using the collision-induced H$_2$ rotational absorption to derive 80–700 mbar temperatures. The same spectra had sufficient signal-to-noise ratio in the CH$_4$ 7-μm band to derive stratospheric temperatures near the 1-mbar level. (Hanel et al. 1981, 1982). In addition, three detailed vertical temperature profiles were derived from the Voyager-1 ingress radio occultation at high southern latitudes, and from the Voyager-2 ingress and egress occultations at 36.5°N and 31°S, respectively (Tyler et al. 1981).

The vertical resolution of IRIS-derived 100–700 mbar profiles, such as the one shown in Fig. 2.9, is approximately one half of an atmospheric scale height, a result of the pressure-squared dependence of the H$_2$ collision-induced opacity, with the largest source of uncertainty being the unknown influence of clouds on the infrared opacity. At levels above the temperature minimum, IRIS data constrain temperatures near the 1-mbar level from measurements of radiance in the CH$_4$ 7-μm band. At 10–80 mbar, the retrieved

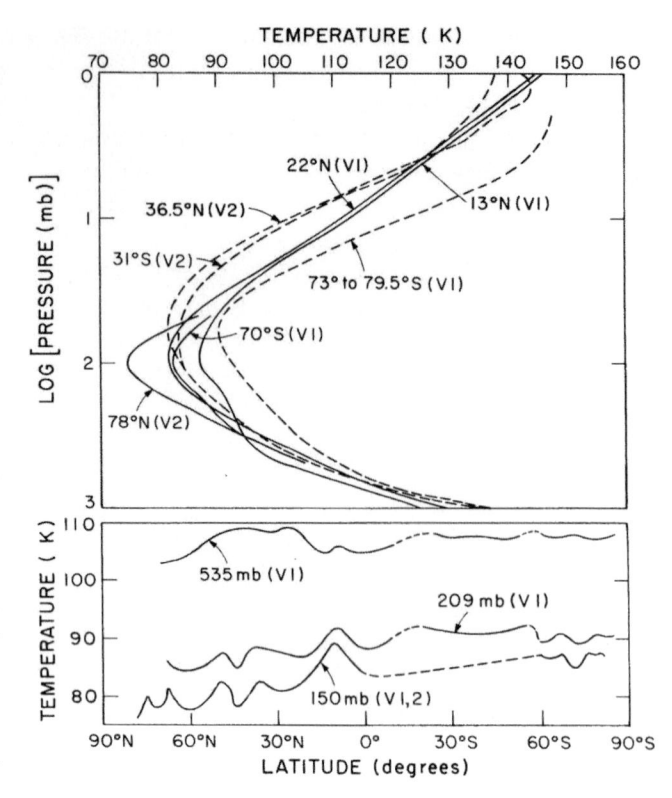

Fig. 2.9 Spatial variations of Saturn's temperature determined from Voyager 1 (V1) and Voyager 2 (V2) observations. Vertical profiles are from Voyager IRIS (*solid lines*) and radio occultations (*dashed lines*). Horizontal profiles in the lower graphs are from Voyager IRIS. Figure 1 of Prinn et al. (1984)

temperatures are essentially an interpolation between 100–700 mbar temperatures derived from emission from spectral regions dominated by H$_2$ collision-induced absorption and temperatures derived near 1 mbar from the CH$_4$ 7-μm band. The correspondence between the IRIS and the radio occultation profiles shown in Fig. 2.9 is very good in the 100–700 mbar range. Their divergence in the lower stratosphere reflects the absence of meaningful constraints on the IRIS retrievals, as noted above.

A further assumption made in deriving temperature profiles from Voyager IRIS data was that the ratio of *para*-H$_2$ to *ortho*-H$_2$ (atomic spins anti-parallel or parallel, respectively) was at its equilibrium value for the local temperature. This influences the relative strengths of the S(0) and S(1) rotational line populations, as all S(0) transitions arise from *para*-H$_2$ and all S(1) transitions from *ortho*-H$_2$. Variations from equilibrium could result from the long equilibration time associated with *ortho-para* conversion, so that hydrogen uplifted from depth faster than the conversion time scale would reflect the 1:3 para vs ortho ratio which is the high-temperature asymptotic value (Massie and Hunten 1982), so-called "normal" hydrogen. More rapid *para-ortho* H$_2$ conversion may occur as the gas encounters surfaces where catalytic processes may take place, resulting in a H$_2$ state between equilibrium and normal. Gierasch (1983) also suggested that the lapse rate could be modified by release of the latest energy of *ortho-para* conversion.

Later work by Conrath et al. (1998) combined retrievals of temperature and the *ortho-para* H_2 ratio as a function of altitude. Zonal averaged results showed significant deviations from equilibrium, implying significant influence by dynamics. Both the *para* H_2 fraction and temperature at 200 mbar is a minimum near 60°S latitude, consistent with upward motion, and a maximum near 15–20°S, consistent with downward motion. Seasonal effects produce cooler temperatures in the northern hemisphere and a higher para fraction, consistent with downward displacement in the hemisphere which just emerged from winter.

The variability of zonally averaged tropospheric temperatures was explored first by Conrath and Pirraglia (1983). Although their values were later updated by Conrath et al. (1998), there were few significant changes in the temperatures retrieved or in their dynamical implications. The lower panel of Fig. 2.14 summarizes their results, showing the latitudinal dependence of zonally averaged temperatures retrieved from Voyager IRIS data between 150 and 535 mbar pressure.

Figure 2.9 illustrates that the upper tropospheric temperatures are warmer in the northern compared with the southern hemisphere, consistent with radiative-convective models with seasonal-dependent solar radiant heating. Bézard et al. (1984) extended earlier such models for the stratosphere (Cess and Caldwell 1979; Carlson et al. 1980) to the troposphere, finding. that modulation of incident solar radiance by the ring system and emission from the rings themselves influences zonal-mean temperatures only to 1.5 K or less. Overall, the large-scale latitudinal thermal structure observed near the 150-mbar pressure level is satisfactorily explained by the radiative-convective model, although an optimal fit to the absolute values of the derived temperatures required additional opacity, presumed to be an aerosol layer with unit optical depth in the visible near 250 mbar in the equatorial region and near 350 mbar elsewhere.

Estimates of Saturn's radiative relaxation time, based on the calculations of Gierasch and Goody (1969) for hydrogen atmospheres, show that the first-order phase lag, is about 1/6 of a Saturn year (five terrestrial years). Because the sub-solar point crossed from northern to southern latitudes 6 months before the Voyager-1 encounter and 15 months before the Voyager-2 encounter, the northern hemisphere should be cooler than the southern by 6% for pressures around 200 mbar (Ingersoll et al. 1984). At greater pressures, the radiative relaxation time becomes so large that the seasonal response is much reduced, consistent with the smaller hemispherical temperature differences observed at larger pressures (Fig. 2.14, lower panel). Seasonal variability is greater in the stratosphere, where Cess and Caldwell (1979) and Carlson et al. (1980) first modeled the seasonal temperature dependence. 7 Bézard and Gautier (1985) produced a more physically sophisticated model, incorporating a complete non-gray treatment of radiative energy transfer. Their stratospheric model also accounts for variability of solar radiant flux as a result of season, including orbital eccentricity, as well as ring obscuration and planetary oblateness. Because the Voyager IRIS observations of 7-μm CH_4 emission were too noisy to derive stratospheric temperatures, Bézard and Gautier constrained their model using the Voyager 2 radio occultation profiles derived at 36.5°N and 31°S (Tyler et al. 1981) and the Voyager 1 radio occultation ingress profile at 76°S (Tyler et al. 1982). Although they matched these temperature profiles with their models by varying the assumed CH_4 abundance, they were unable to fit all three profiles with the same abundance. They suggested that the abundances of stratospheric hydrocarbons responsible for radiative cooling (ethane and acetylene) might well be variable as a function of latitude. Even more sophisticated radiative-convective-dynamical models were produced by Conrath et al. (1990). Thermal infrared images of Saturn in the 10-μm region by Ollivier et al. (2000) verified the presence of seasonal effects.

The detailed meridional variability of tropospheric temperatures is most likely the result of detailed dynamical influences. Conrath and Pirraglia (1983) derived the vertical gradient of zonal winds from the meridional gradient of temperatures. Comparisons with cloud-top winds derived from tracking features in Voyager imaging suggests that the peak-to-peak amplitude of the jet system decreases with height. Ingersoll et al. (1984) suggest that we are observing the thermal response of a statically stable, dissipative upper troposphere forced from below by an imposed zonal jet system. Deceleration on the jets arising from horizontal eddy viscosity would largely be balanced by Coriolis acceleration of meridional flow. Accompanying vertical motion would produce adiabatic heating in downward legs and cooling in upward legs, balanced by radiative sources and sinks. Alternatively, the observed upper tropospheric structure could result from meridional variations of the amount of infrared upwelling radiation or in the amount of sunlight absorbed. That there is some relationship between temperatures and aerosols is supported by the apparent correlation between meridional variations of 150-mbar temperatures and 2,460 Å aerosol absorption measured by the Voyager 2 photopolarimeter (see the review by West et al. 1983).

Among the spatially resolved observations of temperatures published since the Voyager encounters, Gezari et al. (1991), imaging 7.8-μm (stratospheric CH_4) and 11.6- and 12.4-μm (stratospheric C_2H_6) emission, detected a strong peak of radiance at the north pole, consistent with time-dependent radiative-convective models. Furthermore, they noted a meridional variability very different from tropospheric temperatures. A broad local maximum was identifiable near 30–40°N, particularly in C_2H_6 emission, and a local maximum at the equator was detected in CH_4 emission. They also detected a time-variable "clumpy" structure extending northward from the equatorial belt to about 30°N latitude which was not present at all longitudes.

Orton and Yanamandra-Fisher (2005) minimized the diffraction-limited spatial blurring of previous observations by using the 10-meter W. M. Keck Observatory, achieving a spatial resolution on the order of 3,000 km, better than Voyager IRIS, although sampling only temperatures near 3 mbar in the stratosphere and 100 mbar at the tropopause. Measuring just after Saturn's northern winter solstice, they found stratospheric temperatures increasing with higher southern latitudes, peaking near the south pole, consistent with seasonal forcing. They also observed a warm region within 15° latitude of the equator, just as seen by Gezari et al. (1991). A warmer banding near 30–40°N was also present but more subtle than in the 1989 data; similar subtle local temperature maxima were also detected at 50° and 60°N. A substantial positive meridional gradient in stratospheric temperatures was detected near 70°N, with a second gradient near 88°N surrounding peak observed temperatures at Saturn's south pole (Fig. 2.10). Their derived temperatures at 100 mbar were consistent with the zonally averaged meridional structure measured by Voyager IRIS. They also noted that, from 5° to 7°S, the 3-mbar wind shear had the opposite sign of the 100-mbar wind shear, implying that the prograde wind speed gradient increases with altitude near the 3-mbar level. Elsewhere, the wind shear at 3 mbar is lower than at 100 mbar, consistent with diminishing winds with altitude. The morphology of the polar region at 100 mbar shows a positive meridional temperature gradient near 70°S and a maximum gradient near 88°S, producing the appearance of a "hot spot" at the south pole (Fig. 2.10), a morphology similar to the stratosphere. Orton and Yanamandra-Fisher (2005) hypothesized that the warm region poleward of 70° latitude, a boundary coincident with a zonal jet, could be a warm version of cold polar vortices found in other planetary atmospheres, sustained by 15 years of continuous solar warming. On the other hand, the compact warm region at the pole itself was considered to be dynamical in origin.

Achterberg and Flasar (1996) noted the presence of zonal waves between latitudes 20° and 40°N for temperatures at 130 mbar, characterized by zonal wavenumber 2, and quasi-stationary in a reference frame associated with the deep atmosphere. Ray-tracing models suggested that these are quasi-stationary Rossby waves, confined meridionally by variations of the zonal mean winds and vertically by variations of the static stability of Saturn's atmosphere. Quasi-stationary zonal temperature waves are prominent in Jupiter's troposphere at mid-latitudes (Magalhâes et al. 1990; Orton et al. 1994) as well as its stratosphere (Orton et al. 1991), but they are characterized by higher wavenumbers. Orton and Yanamandra-Fisher (2005) noted both tropospheric and stratospheric zonal thermal waves, with the most prominent (± 1 Kelvin) stratospheric waves appearing at 29°S, tropospheric waves appearing at 32°S, and waves of lower amplitude appearing elsewhere. All are dominated by wavenumbers 9–10, and may be different from the wavenumber-2 oscillations detected by Achterberg and Flasar (1996). Unlike Jupiter, the amplitude of these waves is time variable; they often cannot be detected (see the discussion of systematic errors in the Supplemental Information for Orton et al. 2008).

A 22-year survey of stratospheric temperatures by Orton et al. (2008) revealed a semi-annual (14.8 ± 1.2 Earth year) oscillation in low-latitude and equatorial temperatures, similar to the quasi-biennial oscillation (QBO) in the Earth's atmosphere (see Baldwin et al. 2001) and the quasi-quadrennial oscillation in Jupiter's stratosphere (Leovy et al. 1991; Friedson 1991). Figure 2.11 illustrates that the phase of this oscillation also appeared to be tied to Saturn's equinox, and identified it as the manifestation of the vertical waves detected by the CIRS experiment in Saturn's stratosphere (Fouchet et al. 2008).

The radiative balance of Saturn was investigated by Hanel et al. (1983) from Voyager IRIS data. The solar input was derived from combining a wavelength-integrated geometric albedo with phase functions derived from Pioneer 11, and the thermal output from their suite of infrared spectra. They derived an effective temperature of 95.0 ± 0.4 K, and a Bond albedo of 0.342 ± 0.030, leading to the ratio of emitted to absorbed radiation of 1.78 ± 0.09. This is a major constraint on models for Saturn's interior structure.

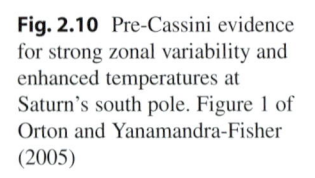

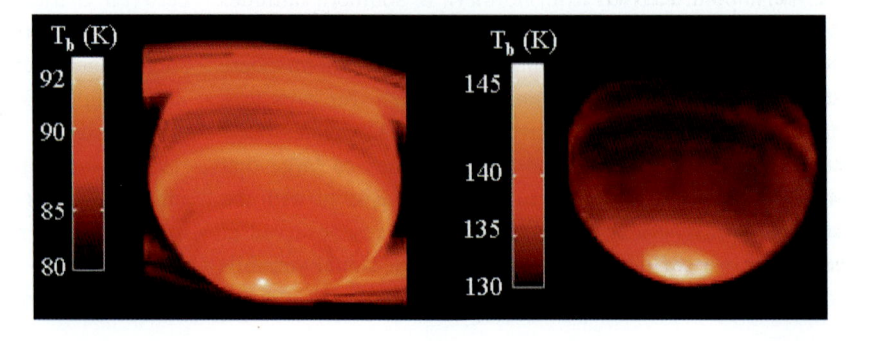

Fig. 2.10 Pre-Cassini evidence for strong zonal variability and enhanced temperatures at Saturn's south pole. Figure 1 of Orton and Yanamandra-Fisher (2005)

Fig. 2.11 Difference of zonally averaged brightness temperatures at 3.6° and 15.5°. Differences in stratospheric methane emission brightness temperatures at 7.8 μm for the northern hemisphere are given by blue symbols and for the southern hemisphere by green symbols. Differences in ethane emission brightness temperatures at 12.2 μm for the northern hemisphere are given by the orange symbols and for the southern hemisphere by the red symbols. Solid line represents a best-fit sinusoid to the individual data and dotted line a best fit to an annual average. From Orton et al. (2008), with corrections

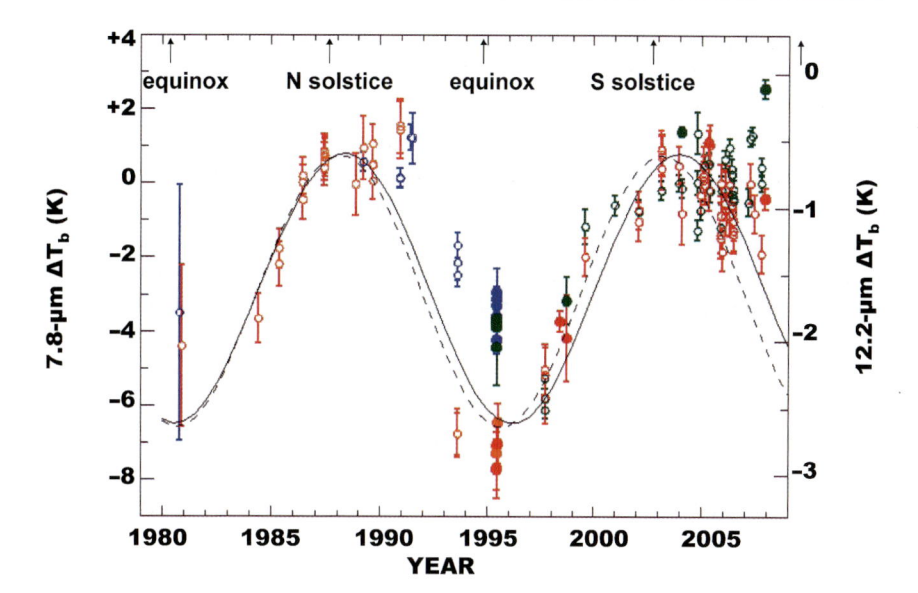

2.6 Saturn's Atmospheric Dynamics

Only a few features in the visible clouds of Saturn were tracked before Voyager. Sánchez-Lavega (1982) updated some of the earliest work by Alexander (1962), including observations up to the Voyager epoch. Ingersoll et al. (1984) noted how little the general features of Saturn's zonal wind profile have changed over a century. After 1990, quantitative information about Saturn's winds was derived from the Hubble Space Telescope (HST). It was widely assumed that Saturn's highly modulated kilometric radiation (SKR), resembling Jupiter's decametric radiation, with its 10 h 39.4 min period, could be taken as a rotation rate for the magnetic field and, thus, the interior (Desch and Rucker 1981). Voyager wind speeds were measured against this baseline.

Adding to the information originally reported by Smith et al. (1981, 1982), Ingersoll et al. (1984) analyzed about 1,000 features with measurable displacements. This profile was both remeasured and extended by Sánchez-Lavega et al. (2000). The resolution was usually about 50 km per picture element-pair in the northern hemisphere and 150–200 km per picture element-pair in the southern hemisphere. Results are shown in Fig. 2.12, together with later HST observations. Errors from uncertainties in feature identification and image navigation are significant in determining the meridional derivative of the zonal mean velocity $\bar{u}$ and the mean correlation between fluctuations of the zonal and meridional velocities u' and v' because meridional and eddy velocities are small.

Interesting aspects of these results are: the strength of the equatorial jet, the symmetry around the equator, and the dominance of eastward flow between 6° and 10°N latitude. The small offset in the profile at 30° latitude appears in both hemispheres, and it resembles a feature in Jupiter's mean zonal velocity profile at 13° latitude in both hemispheres. Ingersoll et al. (1984) pointed out a further correspondence between westward jets in Saturn at 40°, 55° and 70° and

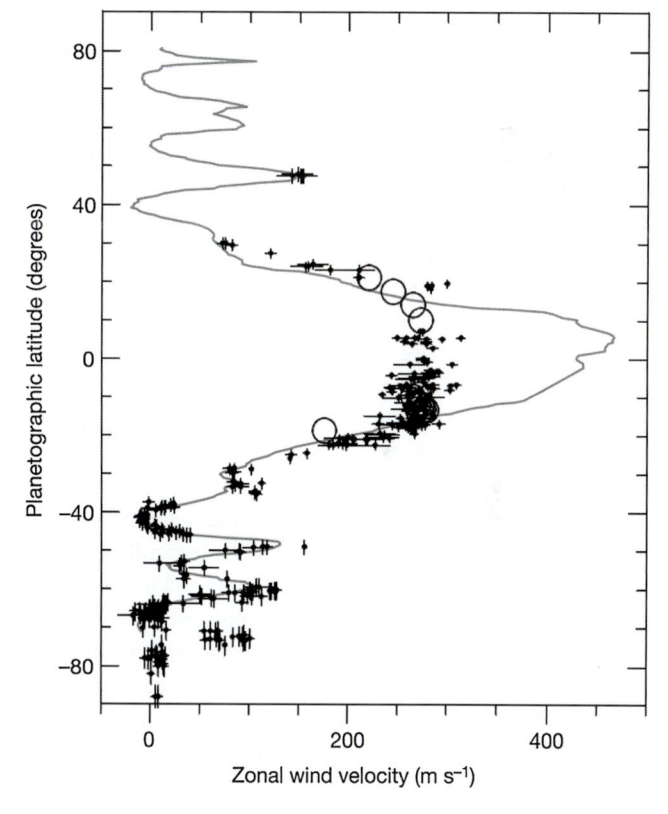

Fig. 2.12 Saturn's zonal (east–west) wind velocity profile at the cloud level as a function of latitude. The solid line shows an average of Voyager 1 and 2 results for 1980 to 1981. The open circles show results from tracking large features from ground-based telescopes from 195 to 1997. The black dots are velocities derived from HST images from 1996 to 2002. From Sánchez-Lavega et al. (2003)

those in Jupiter at 18°, 32° and 39° latitude. The mid-latitude eastward jets on Saturn are faster than those in Jupiter, but they are all dwarfed by the equatorial jet which is much larger than the corresponding jet in Jupiter. The biggest surprise of the Voyager results is the preponderance of eastward flow, which would have to be "balanced" by changing the baseline rotational period to 10 h, 30 min. Long-term HST-based imaging studies (Sánchez-Lavega et al. 2003, 2004) noted a significant decrease in the speed of the equatorial jet, although it is uncertain whether this represents a true variability of wind speed or of changes in the altitude of the clouds being tracked (Pérez-Hoyos and Sánchez-Lavega 2006).

Ingersoll et al. (1984) computed the eddy momentum transport $\overline{u'v'}$ _bar and $d\bar{u}/dy$, which varied together as a function of latitude in Jupiter and whose product in a global integral, $\{K'K\_bar\}$, was positive. For Saturn, the value is not significantly different from zero, a result which is not be too surprising, given the greater mass per area of Saturn's cloud layer and the smaller value of Saturn's radiative power per area. The smaller eddy velocities δu and δv in Saturn compared with Jupiter are consistent with this view. The greater width and speed of Saturn's equatorial jet compared with Jupiter's could also be related to the greater depth of its atmosphere.

They also noted that that the lifetime of small-scale eddies may be controlled by the shear. This view treats the eddies as a phenomenon separate from the shears of the mean zonal flow. Thus, the eddies may arise from convection, with buoyancy associated with vertical heat in the cloud zone, or, they could be associated with the shear itself. Neither view is entirely consistent with the observation of eddies at both zonal wind minima and maxima and both maxima and minima of $|\bar{u}|$.

Voyager observed few large-scale discrete features at visible wavelengths in Saturn's atmosphere, but several isolated spots were identified. These included wispy cloud feature near Saturn's equator. A large spot, prominent in reflected ultraviolet radiation, was present in both Voyager encounters near 27°N, although the area around it had changed color and albedo considerably. Three anticyclonic brown spots were seen at 42°N. A complex interaction was viewed between these and anticyclonic white spots at nearly the same latitude. In the southern hemisphere, a 3,000- by 5,000-km spot, with a color similar to Jupiter's Great Red Spot, was tracked at 55°S through both Voyager encounters. Convective clouds were also detected near the westward jet at 39°N. Between the Voyager encounters this region had changed it visible appearance substantially and displayed several different cloud systems. Unstable cloud systems were also detected by Smith et al. (1981, 1982) at latitudes where the zonal flow is close to the interior rotation rate, including a westward-moving dark spot that appeared to be shedding a several vortices eastward.

A ribbon-like wave feature at 46°N was observed traveling eastward with the 140 m s^{-1} zonal flow, peaking at intervals of 5,000 km (Sromovsky et al. 1983). Anticyclonic vortices were found south of each crest, spiraling inward with clockwise motion. The ribbon appeared in HST images over a decade later (Sánchez-Lavega et al. 2002). A hexagonal wave, embedded in an eastward jet, centered at 80°N latitude was discovered by Godfrey (1988), which was modeled by Allison et al. (1990) as a Rossby wave which is forced by a large cyclone, the North Polar Spot (NPS), found near one of the southern edges of the hexagon (Fig. 2.13). Both features were still present 10 years later (Caldwell et al. 1993; Sánchez-Lavega et al. 1993b). No southern counterpart to either of these features was detected (Sánchez-Lavega et al. 2002).

Regions of convective activity (Fig. 2.14) include a dramatic outburst known as the Great White Spot (GWS) in September of 1990 at Saturn's equator (Sánchez-Lavega et al. 1991). Based on past observations (Sánchez-Lavega et al. 1991; Barnet et al. 1992), an outburst of this sort was expected sometime in the 1990s (Sánchez-Lavega and Battaner 1986; Sánchez-Lavega 1989). It was studied intensively by HST at several wavelengths using multiple filters (Fig. 2.14). Proposed models posited that the GWS was a huge moist convective storm formed of cumulus-cloud-like structures which were higher than the surrounding clouds (Sánchez-Lavega and Battaner 1986; Hueso and Sánchez-Lavega 2004). Residuals of this outburst were observed (Sánchez-Lavega et al. 1993a) up through the appearance of a similar outburst in 1994 (Sánchez-Lavega et al. 1996;

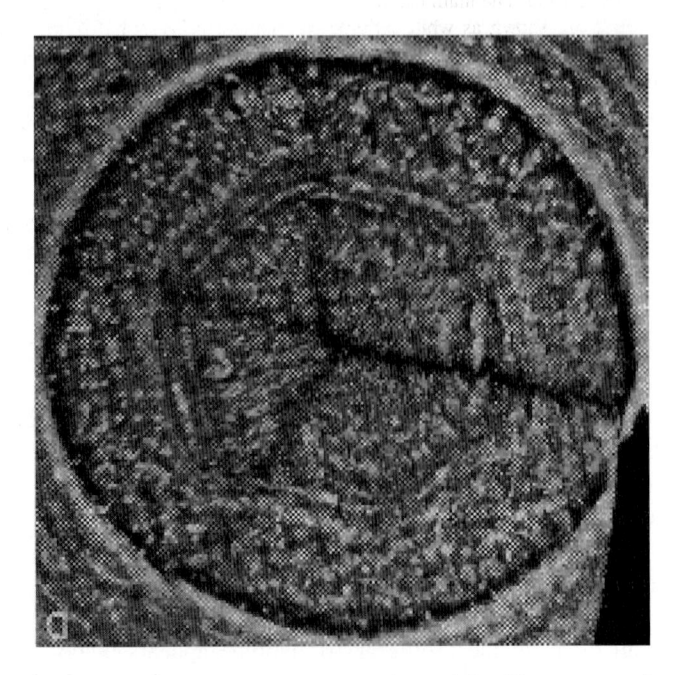

Fig. 2.13 Hexagonal wave structure observed by Voyager around Saturn's north pole. From Godfrey (1988)

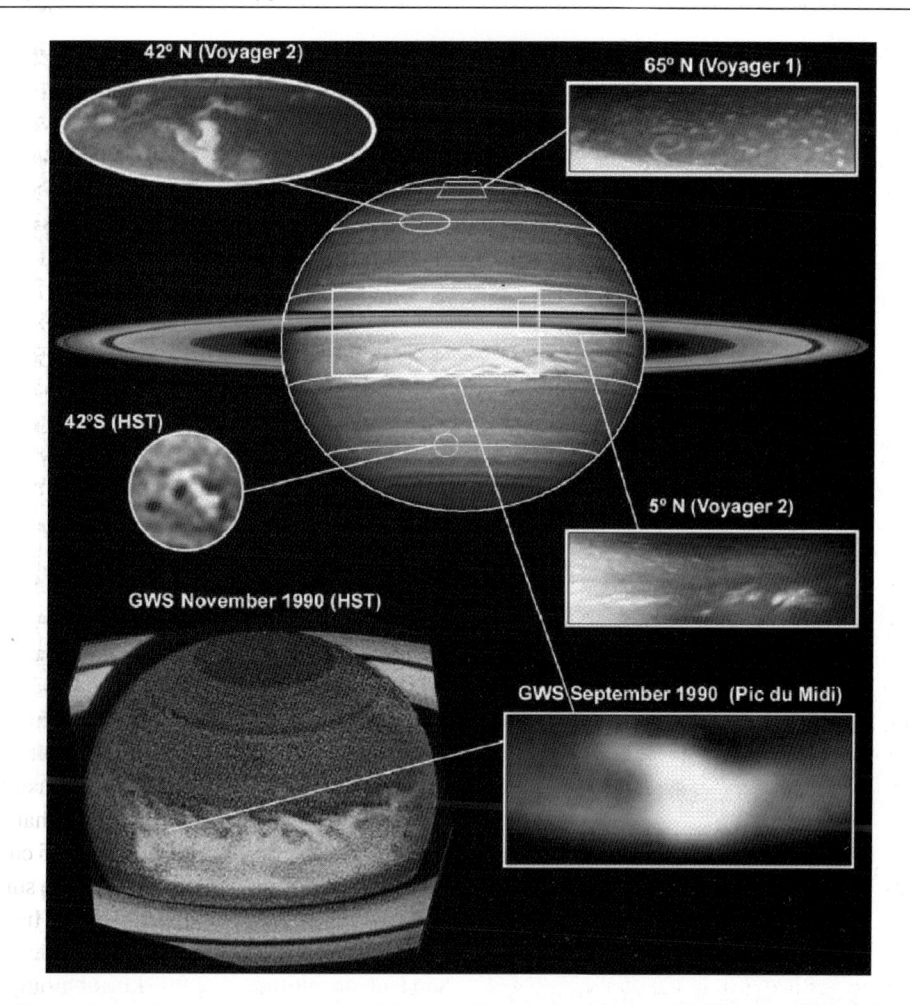

Fig. 2.14 Overview of suspected moist convective storm activity in Saturn. The upper central background images is an HST image taken in October 1966. The main latitudes where convective activity has been observed are shown as white latitude parallels, namely 65°N, 42°N, 5°N, and 42°S; the region between ±20° is highlighted as the location of the several large storms occurring in Saturn in the 1990s. For every region, an amplified image is shown of the typical storms observed there and marked in the real comparative size over the background image. Upper right: 65°N, a cluster of convective clouds imaged by Voyager 1 with typical cell sizes ∼300 km. Upper left image 24°N, strong and large-scale storms also imaged by Voyager 1. Middle right panel: 5°N weak but extended plumes (∼5,000 km) image by Voyager 2. Middle left panel 42°S, strong storm observed by HST in 2003 (size ∼3,000 km) Bottom right panel: The 1990 GWS as imaged within 1 week of the initial discovery (size ∼20,000 km). Bottom left panel. HST image of the 1990 GWS mature phase 2 months and a half after the outbreak. The scale of each enlarged storm is drawn over the background image. From Hueso and Sánchez-Lavega (2004)

Fig. 2.22). The next few years were marked by continuous formation and disappearance of bright equatorial spots (Sánchez-Lavega et al. 1999). Hueso and Sánchez-Lavega (2004) argued for the origin of these storms as instabilities in the atmosphere powered by latent heat in H_2O clouds.

Ingersoll et al. (1984) remarked that the depth of Saturn's circulation must be on the order of 10^3 to 10^4 bars or more, because Saturn's strong eastward winds cannot fall to zero in a thin layer immediately below the clouds: the thickness of the layer between constant pressures surfaces at (i) the level of the visible clouds and (ii) the level of zero motion must decrease by two scale heights from equator to pole. This argument assumes that the thickness of a gaseous layer between constant pressure surfaces in hydrostatic equilibrium is proportional to the absolute temperature of the layers. If this layer were 50 scale heights thick, then its thickness would only decrease by only about 4%, a value consistent with the observed horizontal variations of temperature at 730 mbar by the Voyager IRIS experiment. Allison and Stone (1983) point out that, if the internal rotation period were 6–8 min shorter than the SKR period, a thin convection layer model would be consistent with the data. Furthermore, important temperature changes could take place at pressures below 730 mbar, the deepest layer accessible to the IRIS experiment. This would be possible if Saturn contained 10 times more water than solar composition (Stevenson 1982). Finally, Gierasch (1983) noted that variations in the relative abundances of *ortho-* and *para*-H_2 could account for large temperature variations at depth with no motions deeper than several hundred bars of pressure.

2.7 Saturn's Upper Atmosphere and Auroral Emission

2.7.1 Ultraviolet Auroral Observations

The quest to detect auroral activity on the giant planets date back to the late 1950s, with unsuccessful attempts to measure "non-thermal" radio emission (McClain 1959) and hydrogen Hα emission (Smith et al. 1963; Dulk and Eddy 1966). In 1971, the IMP-6 spacecraft measured persistent non-thermal radio emission from Jupiter at 1 megahertz, very similar in nature to that associated with auroral activity on Earth, prompting suggestions that this could be due to Jovian aurorae (Brown 1974). The same spacecraft found similar emission from Saturn – although twice as weak, occurring during a series of storms (Brown 1975). Saturn at least had periodical aurorae. The Voyager spacecraft, equipped with instruments to measure radio emissions across a wide spectral range (Warwick et al. 1977) found kilometric radiation SKR) with an emission source region at high (>60°) latitudes (Kaiser et al. 1981). Some indication of the variability of these emissions was delivered by the 70% from of SKR intensity from Voyager 1 to Voyager 2 (Warwick et al. 1982).

In 1975, a spectrometer borne on a sounding rocket detected H Lyman −α emission from Saturn, at a level of ∼0.7 kiloRayleighs (Weiser et al. 1977), later confirmed by the ultraviolet photometer onboard Pioneer 11 (Judge et al. 1980) and short wavelength spectrograph on the Earth-orbiting International Ultraviolet Explorer (IUE) (Clarke et al. 1981). The long wavelength channel of the Pioneer-11 ultraviolet spectrometer showed variations in the disc emission with latitude. This variation was consistent with either limb brightening or auroral emissions, and in retrospect may very well have been due at least in part to polar auroral emissions. The IUE observations, reported shortly before the Voyager-1 flyby, were able to map emissions across Saturn at resolution sufficient to separate the polar regions from lower latitudes. IUE H Ly α observations confirmed an ambient level of 0.7 kR for the disk, but also reported enhancements up to 0.9 kR higher. They showed peaks at both poles which were variable with time, indicating auroral processes (Clarke et al. 1981). Further IUE observations over many years post-Voyager showed continuing variable increases in the polar H Ly α emissions (McGrath and Clarke 1992).

Voyager 1 detected polar UV H Ly α and H_2 band emissions from Saturn's upper atmosphere presenting the first unambiguous evidence for the existence of Saturnian aurorae (Broadfoot et al. 1981; Sandel and Broadfoot 1981). Broadfoot and Co-workers (1981) reported emission from a narrow latitudinal range (78°–81.5°) in both hemispheres, implying a mapping to the middle to outer magnetosphere, with a peak auroral brightness of 10–15 kR in the H_2 Lyman and Werner bands, and ∼10 kR in H Ly α. They estimated that approximately 2×10^{11} Watts of energy in the form of precipitating electrons were required to produce auroral emission corresponding to a sustained 5 kR, 10% or less of the energy powering the corresponding jovian aurorae. But Sandel and Broadfoot (1981) calculated that the bursts seen by IUE (Clarke et al. 1981) would correspond to enhancements of between 13 kR and 40 kR if localized at the Voyager auroral regions, indicating that Saturn's aurorae could brighten considerably more and for short periods of time. Voyager 2 confirmed the results of its predecessor but also found auroral bursts up to 100 kR (Sandel et al. 1982).

Modeling of Saturn's upper atmosphere indicated that the observed EUV emission could be produced using a flux of precipitating 2 keV electrons of $1\,mW\,m^{-2}$. (Note that this flux would give a total precipitation energy of 2×10^{11} W for an auroral oval ∼3,000 km wide, centered on a colatitude of 10°.) This produced peak ion densities of ∼$2 \times 10^{12}\,m^{-3}$ for H^+ and ∼$2 \times 10^9\,m^{-3}$ for H_3^+, (Gérard and Singh 1982). Analysis of the auroral spectrum showed little or no absorption from hydrocarbons, implying that the precipitating particles did not penetrate deeply into the atmosphere and thus had energies less than about 10 keV.

Occultation measurements of the ionosphere by the Voyager-1 spacecraft ingress path was the closest (at 73°S along the dusk terminator) to latitudes where strong EUV emission had been noted. The peak electron density, measured at an altitude ∼2,000 km above the 1-bar pressure level, was just over $2 \times 10^{10}\,m^{-3}$ more than twice the amount measured during egress at 1°S (Tyler et al. 1981). Voyager 2 measurements at much lower latitudes, 36.5°N near dusk for ingress and 31°S pre-dawn for egress, yielded peak electron densities of $1.6 \times 10^{10}\,m^{-3}$ and $0.6 \times 10^{10}\,m^{-3}$ respectively (Tyler et al. 1982). Barbosa (1990) examined whether a flux of particles as large as required by the EUV auroral observations and Gérard and Singh's (1982) model was available, concluding that the Voyager particles and fields data indicated that electrons with energies between 1 keV and 10 keV could only contribute ∼5×10^{10} W of the 2×10^{11} W required to power the observed aurora, with H^+ and N^+ ions contributing another 2.5×10^{10} W.

More detailed analysis of the variation of the auroral emissions for several hours during one mapping sequence showed a factor of 5 increase with a peak 1–2 h, or 50° in longitude, before the maximum probability for detection of the Saturn Kilometric Radiation (SKR) (Sandel et al. 1981), which was known to be modulated with the rotation of the planet (Gurnett et al. 1981; Warwick et al. 1981; Kaiser et al. 1981). The SKR was found to statistically be most intense when longitude $\lambda_{SLS} = 110°$ was directed toward the Sun, implying an asymmetry in the excitation or beaming process and interaction with the solar flux or solar wind.

Fig. 2.15 Image of Saturn's polar ultraviolet auroral emission obtained with the Hubble Space Telescope. NASA image 108480main_Hubble-Image-516-508.jpg

Images of Saturn's polar ultraviolet auroral emissions were obtained with the Faint Object Camera (FOC) on the Hubble Space Telescope (HST) (Fig. 2.15). The limiting sensitivity of the FOC showed the brightest of the far-ultraviolet auroral emissions, while longer wavelengths of reflected sunlight revealed a polar hood of UV-absorbing particles in Saturn's upper atmosphere (Gérard et al. 1995). Near-UV images of the polar regions near 220 nm showed the extent of the absorbing haze, presumably a result of enhanced photochemistry driven by the bright ultraviolet auroral emissions (ben Jaffel et al. 1995).

Far-UV images of Saturn with the HST Wide Field Planetary Camera 2 (WFPC 2) showed the auroral emissions with a factor of 4–5 higher sensitivity than FOC, revealing more of the full distribution of the auroral emissions and the altitude extent of the auroral curtain (Trauger et al. 1998). Modeling of the distribution of emissions including the altitude extent of the auroral curtain indicated that the auroral emissions were distributed over 74–79° latitude, and that the oval was preferentially bright between 6–10 h local time. The total radiated power was 4×10^{10} Watts, far less than the large power radiated in Jupiter's aurora (several times 10^{12} W) but larger than at the Earth (of the order of 10^9 Watts).

Far-ultraviolet images of Saturn's aurora with the Space Telescope Imaging Spectrograph (STIS) on HST provided a sensitivity which was an order of magnitude higher than WFPC 2, giving a sensitivity as low as 1–2 kilo-Rayleighs in relatively short exposures to avoid rotational blurring of the emissions. These images showed a complete auroral oval at times when the emissions were relatively bright, an overall variation of the emission brightness, and a considerable range in latitude of the auroral oval from 70–80° latitude (Gérard et al. 2006). Observations of Saturn with STIS beginning in Jan. 2004 included the approach of Cassini to Saturn, which made possible a direct comparison with solar wind conditions, and these data began the Cassini era of auroral studies (discussed later).

One issue from the EUV observations, the particles and fields data and the modeling was whether energy could be transported from the auroral/polar regions to lower latitudes, to help to explain the high thermospheric temperatures that were being deduced from the Voyager occultations at low latitudes, variously determined as 400 K (Smith et al. 1983) or 800 K (Festou and Atreya 1982).

2.7.2 Infrared Auroral Observations

The H_3^+ molecular ion was first detected in Jupiter's auroral regions in 1988 (Drossart et al. 1989), and its initial detection from Saturn was made by Geballe et al. (1993) in 1992. They found that emission was very weak, even when compared with Uranus. Spectra at the northern and southern limbs had roughly the same intensities, and they were not able to measure H_3^+ emission at the equator. However, their initial measurements indicated that, for Saturn, the line intensity fell off more slowly from the limbs to the equator than was the case for Jupiter. This indicated that the morphology was somewhat intermediate between Jupiter, with its emission strongly concentrated around the auroral/polar regions (Baron et al. 1991), and Uranus, for which a planet-wide H_3^+ glow seemed more likely (Trafton et al. 1993). Geballe et al. (1993) determined the best-fit temperature for their spectra to be ~800 K, lower than for Jupiter, for which temperatures between 900 K and 1,100 K were found in the auroral regions (Drossart et al. 1989; Lam et al. 1997), and the column density to be $\sim 1.0 \times 10^{15}$ m^{-2}, about 10–50 times less than for Jupiter's auroral zones. The temperature derived by Geballe et al. (1993) was at the highest end of those that had been derived from Voyager data (Festou and Atreya 1982). A much lower temperature, ~400 K, had been proposed by Smith et al. (1983) from the analysis of the same Voyager data.

A more accurate determination of the morphology of the H_3^+ emission was obtained by Stallard et al. (1999) in 1998. Measurements of the $Q(1, 0^-)$ line of the ν_2 band of H_3^+ at 3.953 microns peaked strongly at the poles, strongly supporting the primarily auroral nature of this emission. In 1998, the whole of the southern auroral/polar region was visible from the Earth: the H_3^+ emission showed some indication of an auroral oval, similar to contemporary UV images from the

Hubble Space Telescope (Trauger et al. 1998). But, just as for Jupiter, the H$_3^+$ emission extended completely across the auroral/polar region, with intensities much higher at the pole itself than was observed in the UV. (See Stallard et al. 2001.) Stallard et al. (1999) calculated that the total H$_3^+$ emission from Saturn might be as high as $1.5 \pm 0.3 \times 10^{11}$ Watts, if the temperature ($\sim$800 K) derived by Geballe et al. (1993) was correct. Although $\sim$50 times less than Jupiter's H$_3^+$ emission, it was still high compared with Saturn's UV emission. The associated ion wind velocities showed that, a whole, ions within the southern auroral region sub-rotated at $\sim$1/3 the rotation rate of the planet to a significantly higher sub-latitude than the location of the main auroral oval. This suggested that the source of the main oval was associated with the solar wind, rather than internal sources as seen at Jupiter.

Miller et al. (2000) challenged this high temperature for Saturn's exosphere; their analysis of later (1999) data indicated that 600 K was more likely, in between the 800 K derived by Festou and Atreya (1982) and 400 K derived by Smith et al. (1983). But it is now clear that the best-fit temperature for Saturn's upper atmosphere, at the level of the peak H$_3^+$ emission in the auroral/polar zones is even lower. Melin et al. (2007) reanalyzed the spectra taken in 1999, and others obtained in 2004, finding a best-fit temperature of 380($\pm$70)K for the 1999 data, and 420($\pm$70)K for 2004. Averaging gave 400($\pm$50)K for the best-fit thermospheric temperature. What emerged clearly was the variability in the ion column densities, – assuming a constant temperature of 400 K, with 2.1×10^{16} m^{-2} and 2.9×10^{16} m^{-2} for the 1999 and 2005 data, respectively, but 20.0×10^{16} m^{-2} for 2004. The implications of such large variations for the energy balance in Saturn's upper atmosphere are discussed below.

H$_3^+$ emission from the auroral/polar regions of emerges from a rather symmetric oval, centered on a co-latitude $\sim$15°, with lower-intensity emission coming from the polar cap region. This is reasonably consistent with the proposal by Cowley et al. (2003) that the main oval on Saturn is generated along the ionospheric footprint of the boundary between open and closed magnetic field lines. But, just as for the UV emission (e.g., Clarke et al. 2005; Esposito et al. 2005; Gérard et al. 2006), the situation is both spatially and temporally more complicated. Further information on the morphology of the H$_3^+$ emission has come both from spectral line profiles and – much more recently – from images, the first of which is shown in Fig. 2.16. Using the IRTF CSHELL spectrometer, Stallard and co-workers (2007a, b; 2008a) showed that the auroral oval can be displaced both dawnwards and duskwards, that there are dawn-dusk asymmetries, and that there are significant changes in overall intensity as a function of time. They have also demonstrated that there is a "Jovian-

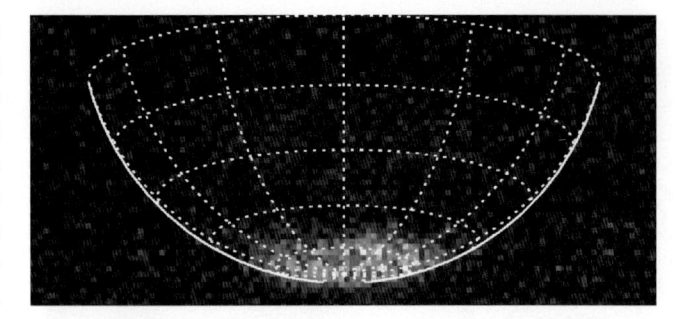

Fig. 2.16 First H$_3^+$ auroral image from the Earth (Stallard et al. 2008b)

like" aurora at a colatitude around 25°, due to the breakdown of corotation in the equatorial plasma sheet distinct from the main oval (Stallard et al. 2008c). (See Hill 1979, 2001; Cowley and Bunce 2001, for a discussion of this mechanism for Jupiter). Using images taken by the VIMS instrument on Cassini at wavelengths sensitive to H$_3^+$, Stallard et al. (2008d) have now shown that the H$_3^+$ emission morphology can have a spiral configuration, as well as the other features previously noted, and that the auroral morphology is highly time dependent.

Recently, considerable progress has been made in understanding the upper atmosphere dynamics and energetics of Saturn as a result of H$_3^+$ observations and modeling. Solar wind pressure had been found to be correlated with Saturn's kilometric radiation (Desch and Rucker 1983), suggesting that Saturn's auroral behavior might be driven by external forces, as in the case of Earth, rather than by internal factors, as in the case of Jupiter. Based on HST UV images, Cowley et al. (2003) first proposed that the main oval resulted from the interaction between the solar wind and Saturn's magnetosphere. Although other mechanisms (Hill 2005; Sittler et al. 2006) have not been entirely ruled out, this seems to be a promising proposal. Cowley et al.'s (2003) model of plasma dynamics in the ionosphere included flows from the Dungey (1961) and Vasilyunas (1983) cycles, and a general lag to corotation with the planet across the entire auroral/polar region due to interactions with the solar wind, a mechanism first put forward by Isbell et al. (1984). This mechanism gives:

$$\Omega_{\text{ion}} = \Omega_{\text{Saturn}} \left[\mu_0 {\sum}_{\text{P}}^{*} V_{\text{SW}} \right] \Big/ \left[1 + \mu_0 {\sum}_{\text{P}}^{*} V_{\text{SW}} \right]$$

where Ω_{ion} is the angular velocity of the auroral/polar ionosphere, Ω_{Saturn} the angular velocity of Saturn, μ_0 the permeability of free space, Σ_{P}^* the effective (height-integrated) Pedersen conductivity of the ionosphere, and V_{SW} is the velocity of the solar wind.

Measurements of the Doppler shifting of the H_3^+ emission line at $3.953\,\mu m$ generally supported the picture of a general lag to corotation across the auroral polar regions (Stallard et al. 2004). For a solar-wind velocity of $500\,km\,s^{-1}$, Cowley et al. (2003) derived a value of $\Omega_{ion}/\Omega_{Saturn} = 0.24$, for an effective ionospheric Pedersen conductivity of 0.5 mho, and Stallard and co-workers (2004) measured $\Omega_{ion}/\Omega_{Saturn} = 0.34$, which suggested a value of $\Sigma_P^* = 0.82$ mho.

The consequences of this lag to corotation for energy balance in the Saturnian upper atmosphere could have a bearing on the long-term issue of why the exospheric temperature of Saturn, and those of the other giant planets, are much hotter than can be explained by solar inputs alone (see Strobel and Smith 1973; Yelle and Miller 2004). The auroral/polar ionosphere is produced mainly by charged particle precipitation. For Jupiter, this amounts to a few times 10^{12} Watts planet-wide which is pretty well balanced by emission from the H_3^+ ions resulting from the ionization of the upper atmosphere (Miller et al. 1994, 2006) – the so-called H_3^+ thermostat. However, the energy generated by Joule heating and ion drag, resulting from equatorward currents across the auroral oval, and the westward winds produced by Hall drift are much more important energetically (see Smith et al. 2005; Miller et al. 2006).

For Saturn, Melin et al. (2007) showed that the H_3^+ thermostat is much less effective than it is for Jupiter (and probably for Uranus). Particle precipitation into the auroral/polar regions probably generates a few times 10^{11} W, but Joule heating and ion drag inputs are 10 times this amount (Cowley et al. 2004a, b; Miller et al. 2005). Smith et al. (2005) model the input of this energy into the upper atmosphere of Saturn, using the STIM global circulation model of Müller-Wodarg et al. (2006), finding that a total input of $\sim 8 \times 10^{12}$ W at 60 nbar pressure would produce the measured auroral/polar temperature (Melin et al. 2007) and Voyager occultation measurements derived by Smith et al. (1983) of ~ 400 K (Melin et al. 2007). However, later modeling, which included the *dynamic* impact of ion drag, found that it was not possible to distribute the Joule heating and ion drag energy towards the equator. Instead, the Coriolis forces generated by the westward ion and neutral winds turned the atmospheric circulation poleward, and heating was directed downward into the mesosphere, rather than horizontally to lower latitudes (Smith et al. 2005).

This indicates that the distribution of auroral/polar energy to lower latitudes cannot account for Saturn's long-term high exospheric temperature. Gravity waves or some other planetwide energy source may be required (Müller-Wodarg et al. 2006). To date, however, it has not been possible to model Saturn's auroral/polar ionosphere in a self-consistent way. Estimates of the Joule heating and ion drag inputs have assumed that the effective Pedersen conductivity is ~ 0.5 to 1.0 mho; the heating generated is directly proportional to Σ_P^*, which is roughly proportional to the H_3^+ column density. Melin et al. (2007) have shown that this can vary by an order of magnitude over longer time scales, which could give rise to values of Σ_P^* of 5 mho or more.

2.8 Planetary Magnetic Field and Magnetosphere

2.8.1 Magnetic Field

Pioneer 11 discovered that the magnetic axis of Saturn was aligned with its rotation axis (Smith et al. 1980a, b; Acuna and Ness 1980), a big surprise since the magnetic fields of the Sun, other stars and the other planets are all tilted relative to their rotation axes. In addition, this result was contrary to a fundamental anti-dynamo theorem introduced long ago by T. G. Cowling that showed that such an aligned axisymmetric field could not be sustained. That implied that the field might be decaying away, a possibility that could be tested by Cassini after the delay of 20 years. An alternative explanation, proposed by D. J. Stevenson (1980), was that the rotating non-axi-symmetric fields generated by the dynamo induced Eddy currents in the conducting atmosphere above the fluid core and screened them out so that they were not observable outside the planet. The planetary dipole moment (D) was determined to be 0.2 gauss R_S^3 yielding a field at the surface similar in strength to Earth's field. A significant offset of the dipole from the center of Saturn of $0.06\,R_S$ was also inferred. An offset is equivalent to a Quadrupole moment that was inferred from spherical harmonic analysis of the field measurements ($Q/D = 10\%$) along with the Octupole moment ($O/D = 11\%$). No higher order moments could be derived because of the limitations of the flyby trajectory especially the restriction in planetary latitudes. The subsequent Voyager flybys confirmed these earlier results including the absence of a dipole tilt and lead to values of the magnetic moments in agreement with Pioneer 11. A refined model was possible using data from all three encounters (the Saturn Pioneer Voyager or SPV model, Davis and Smith 1990). It was anticipated that Cassini would reexamine all three moments looking for changes with time and try to extend the field description or magnetic spectrum to higher moments. Determination of these high degree moments is important because of the information they provide regarding the internal structure of Saturn, such as the depth to the fluid core, and basic features of planetary dynamos, such as the dependence of the magnetic spectrum on degree (the magnetic spectrum of Earth is independent of degree).

The three spacecraft encounters are illustrated schematically in Fig. 2.17 (Behannon et al. 1983), and show the variability of the magnetospheric boundaries as determined by the full complement of particles and fields instrument on each. The average sunward magnetopause position (MP) is found most often outside the orbit of Titan at $\sim$22 R_S (top panel), while the bow shock (BS) in the same direction is closer to $\sim$26 R_S. Corresponding locations outbound at local morning ranged from $\sim$30 to $\sim$70 R_S (MP) to $\sim$49 to $\sim$88 R_S (BS). Clearly, the size of Saturn's magnetosphere is highly variable and is modulated by solar wind pressure as balanced by both, the planetary magnetic field and the internal plasma pressure (Krimigis et al. 1983), as will be discussed later.

The Pioneer-11 results also included information about fields generated external to Saturn in the magnetosphere. The field contributed by the ring current or magnetodisk was also inferred from spherical harmonic analysis and was found to depend on local time, specifically, a difference between dusk and dawn. A quantitative (axi-symmetric) model of the ring current was developed and applied to the Voyager data to infer the ring current properties such as the current distribution with latitude or thickness, 5 RS, radial distance, 8 to 16 RS and total current of 10 million amperes (Connerney et al. 1983 as shown in Fig.2.18. Several other refinements of the Voyager ring current model were subsequently developed in preparation for the arrival of Cassini at Saturn (Bunce and Cowley 2003; Giampieri and Dougherty 2004a,b).

Furthermore, careful reanalysis of the Pioneer and Voyager data revealed a magnetic field periodicity having essentially the same period as the Saturn Kilometric Radiation, SKR (Espinosa et al. 2003). The clearest examples of the periodicity occurred in the outer magnetosphere and were circularly polarized. Since the periodicity was too weak to be attributed to a tilt of the dipole, it was suggested that the origin was a pressure anomaly of some sort, possibly a magnetic anomaly, associated with Saturn (the so-called Cam model, Espinosa et al. 2003). This rotating anomaly was thought to create a pressure wave that propagated into the

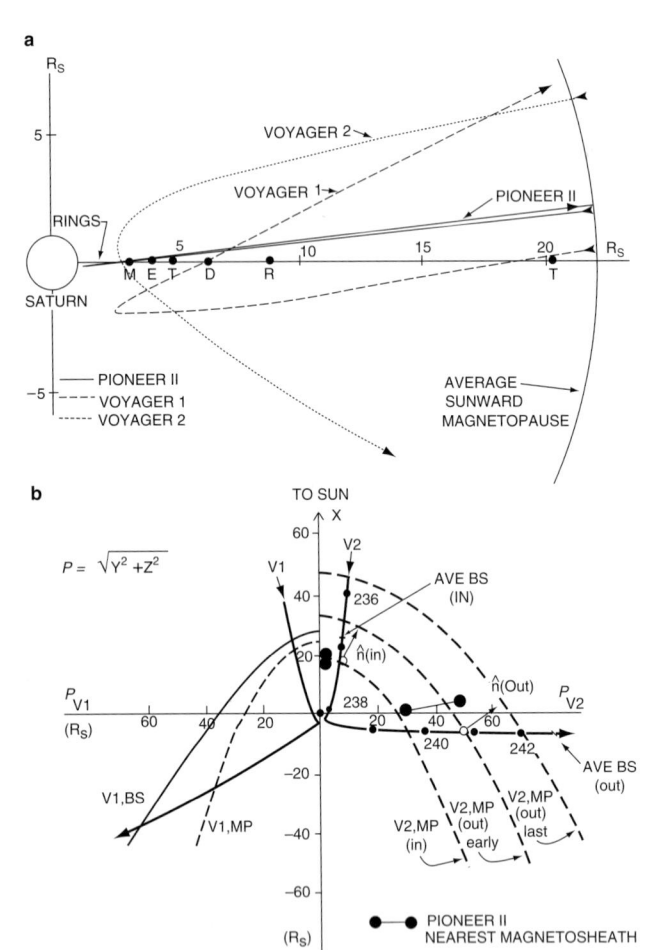

Fig. 2.17 (a) Meridional positions of the Pioneer 11, Voyager 1 and Voyager 2 trajectories (b) Saturn encounter trajectories in cylindrical coordinates. This representation gives the spacecraft position in the plane through the spacecraft and the Saturn-Sun line (x axis) with the distance from that line placed to the left for Voyager 1 and to the right for Voyager 2. For Voyager 1, the model bow shock and magnetopause boundaries are given; Pioneer 11 followed very similar trajectory. For Voyager 2, observed average inbound and outbound shock locations plus model magnetopause shapes are shown (Ness et al. 1982). The outbound "early" positions are based on the average of the first 5 outbound crossings; the outbound "last" curve is based on the last crossing (Bridge et al. 1982) and preserved the earlier shape of the magnetopause (from Behannon et al. 1983)

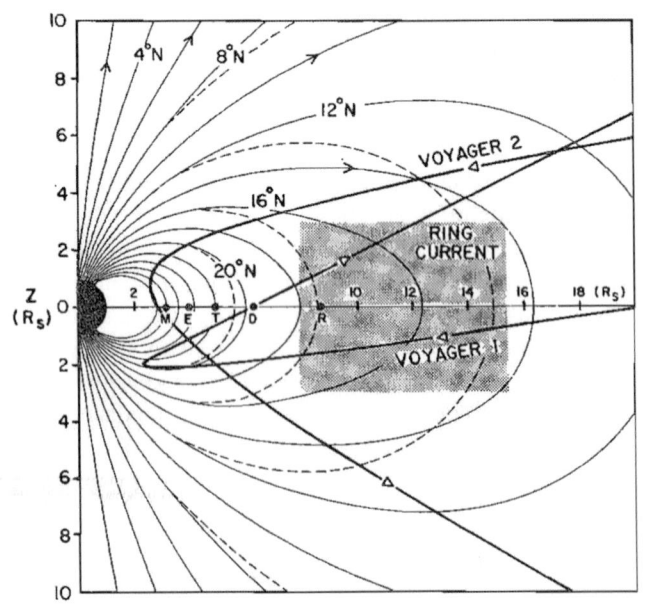

Fig. 2.18 Meridian line projections of magnetosphere field lines for a dipolar field model (*dashed*) and a model containing a dipole and a distributed ring current (*solid*). Field lines are drawn for two increments in latitude. Positions of the major satellites Mimas, Enceladus, Tethys, Dione, and Rhea are shown schematically by first letter (from Connerney et al. 1983)

magnetosphere with little change in amplitude. The existence of the periodicity, its properties and relation to SKR and to similar periodicities in magnetospheric plasma and energetic particles has proven to be a fruitful area of investigation on Cassini.

2.8.2 Magnetospheric Plasma Environment

The plasma population of Saturn's magnetosphere shows sharp density gradients at both, Enceladus on the inner edge and Rhea at the outer edge, as shown in Fig. 2.19. Here the Pioneer 11 ion density profile (Frank et al. 1980) is compared with the electron densities measured by the two Voyager spacecraft (Bridge et al. 1982), and the deduced density profile inferred from the ring current modeled from the magnetic field data (Connerney et al. 1983). In general, there is reasonable consistency in the data sets and a suggestion that there may be a torus of plasma inward of the orbit of Rhea of unknown origin.

The observations by the two Voyagers, with the Voyager-1 trajectory being close to the equator while that of Voyager 2 being at higher latitudes, helped to elucidate further the nature of the Saturnian plasma. This is illustrated in Fig. 2.20 where the evolution of the spectrum vs. mass-per-charge during part of the two trajectories is displayed (Bridge et al. 1982). Figure 2.20 (left) shows a clear H^+ and a heavier ion peak, attributed to N^+ since a source of nitrogen from Titan was expected in the outer magnetosphere. The heavier ion peak is absent from Fig. 2.20 (right) because the Voyager-2 trajectory was at higher latitudes and the heavier ions are concentrated close to the equatorial plane.

Richardson and Sittler (1990) were able to model the Voyager data by determining a spacecraft potential that showed the two principal ion components in the inner ($<12\,R_S$) magnetosphere to be H^+ and O^+. Their results are shown in Fig. 2.21, where contours of intensity for the two species plus those of electrons are included. It is evident that the scale height of oxygen ions is substantially smaller than protons closer in ($\leq 6\,R_S$) to the planet due to their large temperature anisotropy and low thermal velocity parallel to the magnetic field. Protons extend to much higher latitudes, but oxygen densities are higher than protons at the equator. Diffusive modeling by the authors shows large losses of both oxygen and protons inward of $\sim 4\,R_S$ with lifetimes of a few weeks near the equator. Their model of densities has withstood the test of time and shown to be generally consistent

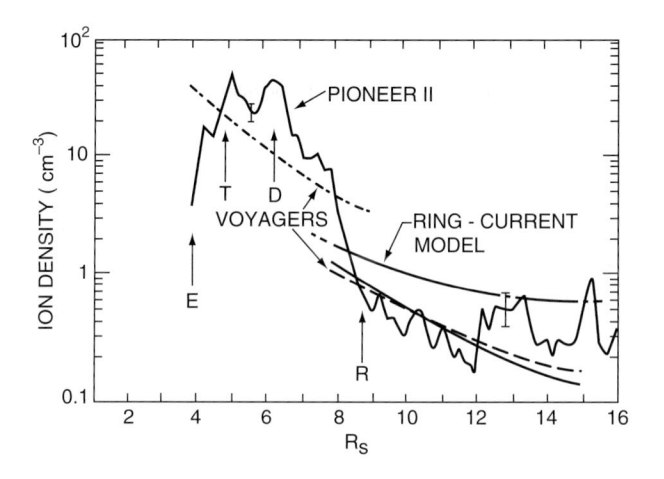

Fig. 2.19 Models of the electron density in Saturn's magnetosphere inferred from Voyagers 1 and 2 (Bridge et al. 1982) compared with the ion densities derived from Pioneer 11 plasma ion data (Frank et al. 1980). Also shown is the ion density computed from the ring-current model of Saturn's magnetic field (Connerney et al. 1983). The locations of the orbits of Enceladus, Tethys, Dione and Rhea are indicated by arrows labeled E, D, and R (from Connerney et al. 1983)

Fig. 2.20 Relative plasma distribution functions between 14 and 18 R_s observed during the inbound passes of Voyager 1 and 2 at Kronian latitudes of $-2°$ and $+18°$, respectively. Peaks appearing at a low energy per charge (10–100 V) are attributed to H^+ and those at high values (>800 V) are attributed to N^+. Note the high temperature of the H^+ ions at the latitude of Voyager 2 (18°) and the near absence of N^+ ions (from Bridge et al. 1982)

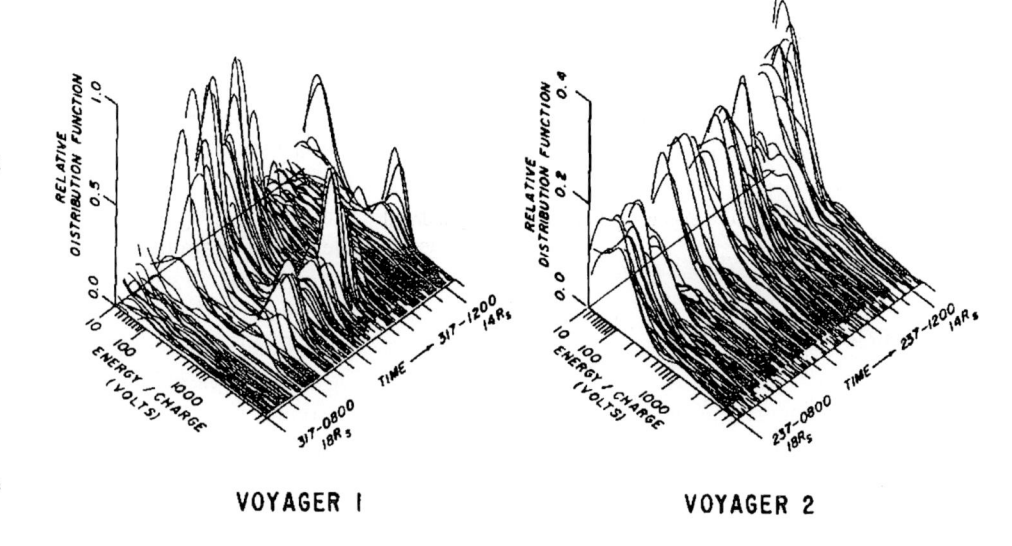

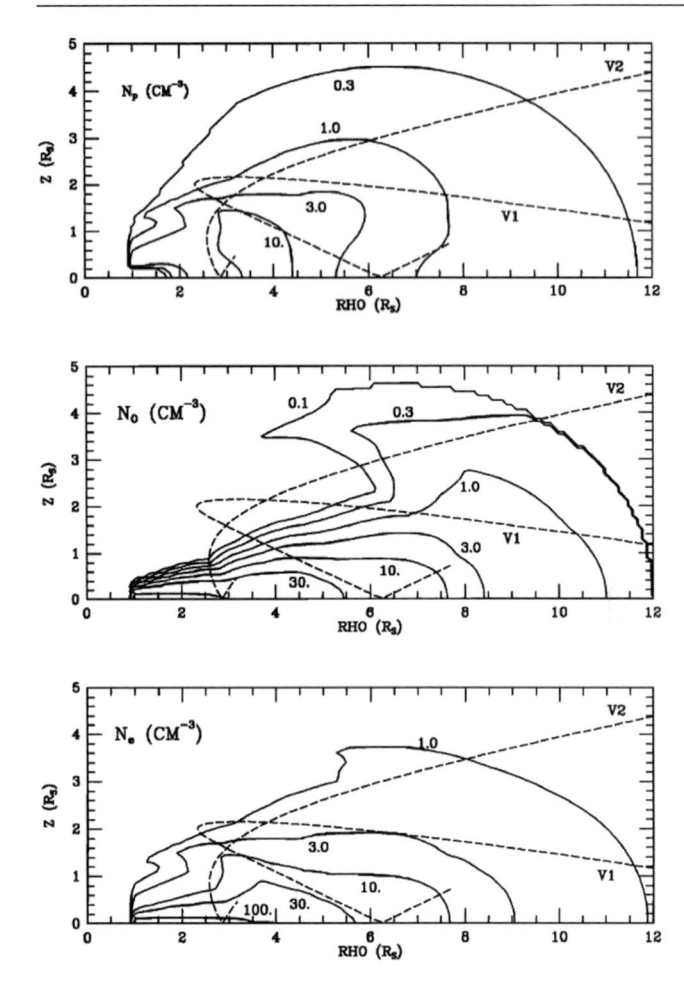

Fig. 2.21 Density contours of plasma at Saturn deduced from modeling of Voyager 1 and 2 flybys of the planet in 1980 and 1980, respectively. Scale height of protons (*top panel*) is larger than that for oxygen ions (*middle panel*). Electron contours (*bottom panel*) resemble those of protons (from Richardson and Sittler 1990)

with current results from Cassini (see Chapter 9). An important consequence of the concentration of heavy ions at the equator is mass loading of the magnetosphere that results in slowdown of co-rotation at larger radial distances from the planet. This phenomenon is similar to that observed at Jupiter, where heavy ions originating from Io cause the magnetosphere azimuthal velocity to be lower than co-rotation at larger distances.

2.8.3 Energetic Particles

In addition to low-energy plasma, Saturn's magnetosphere is populated by large fluxes of energetic ions and electrons; a comprehensive summary may be found in several chapters of the earlier Saturn review book (Van Allen 1984; Scarf et al. 1984; Connerney et al. 1984; Kaiser et al. 1984). The principal features of these populations are summarized in Fig. 2.22 (Krimigis et al. 1983) that displays the flyby of Voyager 2 over a 4-day period in 1983. The intensity of lower energy ions (upper panel) and electrons (lower panel) increases sharply at the crossing of the bow shock, typical for bow-shock crossings of planetary magnetospheres. A more pronounced increase takes place at the magnetopause at energies well over an MeV. The ions remain relatively constant to the orbit of Rhea, but then decrease substantially at the orbit of Dione before recovering inside the orbit of Enceladus, especially at the higher energies. A notable feature in the profile of ions is discrete injection events (small, sharp enhancements) appearing from just inside the magnetopause to about the orbit of Rhea, signifying the dynamical nature of ion acceleration in this part of the dayside magnetosphere. These foreshadowed the many such events detailed through the Cassini observations (Chapters 9–12 of this book).

The electrons display significantly different behavior from the ions in that their intensities continue to increase toward periapsis at basically all energies. Discrete injections can be seen, similar to those of the ions, outside the orbit of Rhea. Both ion and electron intensities decrease as the spacecraft traversed the local morning region (Fig. 2.17) of the magnetosphere, with several crossings of the magnetopause near 50 R_S, as shown.

The large intensity increase near periapsis is due to penetrating high energy protons that were seen by Pioneer 11 and again by both Voyagers (Fillius et al. 1980; Simpson et al. 1980; Van Allen et al. 1980a; Vogt et al. 1982; Krimigis et al. 1982; Krimigis and Armstrong 1982). These extend in energy to well over 80 meV and are due to neutron albedo decay from cosmic rays impinging on Saturn's upper atmosphere.

The energy spectra of the ions have been fit by a kappa-function which behaves as a Maxwellian distribution of temperature kT_H at low energies and as a power law at higher energies of the form $j = j_o (E/kT_H)^{-(k+1)}$, where E is the particle energy (Krimigis et al. 1983). It is noted that T_H is different from that of the cold plasma discussed earlier in connection with Figs. 2.18 through 2.20, which account for most of the density in the magnetosphere. Figure 2.23 shows there is a general peak in the temperature of the hot plasma outside the orbit of Dione, constituting a hot plasma torus; the value of gamma (γ) remains relatively constant to the orbit of Dione, inward of which there is rapid loss of higher energy ions. The density (lower panel) is only a small fraction of that of the cold plasma (Fig. 2.21).

An example of the decisive influence of icy moons on energetic particles (and vice-versa) is shown in Fig. 2.24 (Fillius et al. 1980). The top two traces represent the intensity of energetic electronic and the bottom curve high energy protons, inside ~3.3 R_S. The intensity of electrons, both

Fig. 2.22 Spectrograms of energetic ions (*upper panel*) and electrons (*lower panel*) measured during the Voyager 2 flyby in the indicated time internal. The radial distance and crossing of satellite L-shells are marked above the upper panel. The lower (<100 keV) energies near periapsis are dominated by penetrating high-energy protons and electrons (from Krimigis et al. 1983)

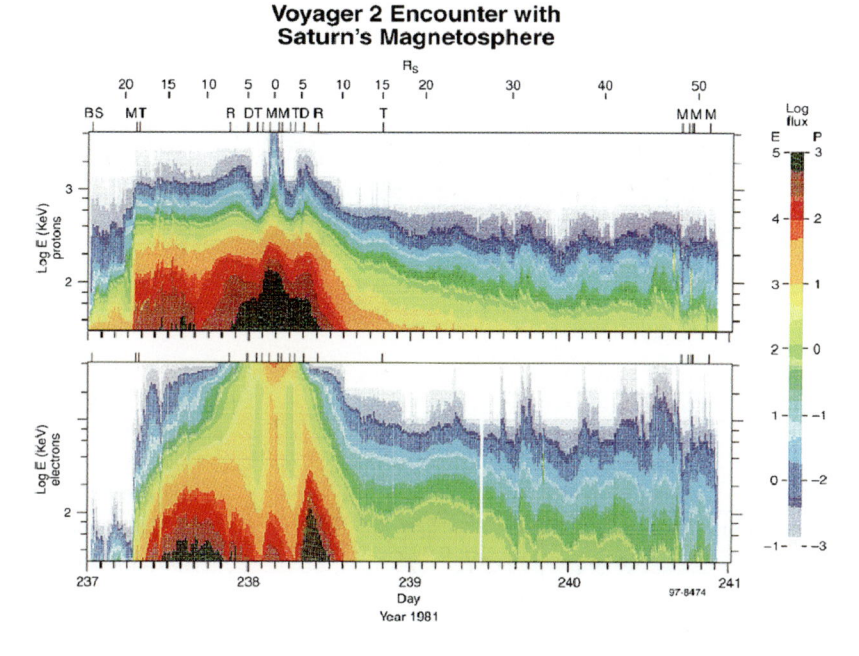

Fig. 2.23 Approximate values of temperature kT$_H$ and the spectral exponent k, derived from fits to data over the indicated energy intervals. The temperature kT$_H$ is determined by fitting the lowest three ion channels to a Maxwellian distribution and k + 1 is determined by fitting two higher energy ion channels to a power law ($\gamma = k + 1$). The bottom panel shows the ion density computed from integrating the energy spectra using 15-min averages of the data. The vertical bars indicate crossing of satellite L shells (from Krimigis et al. 1983)

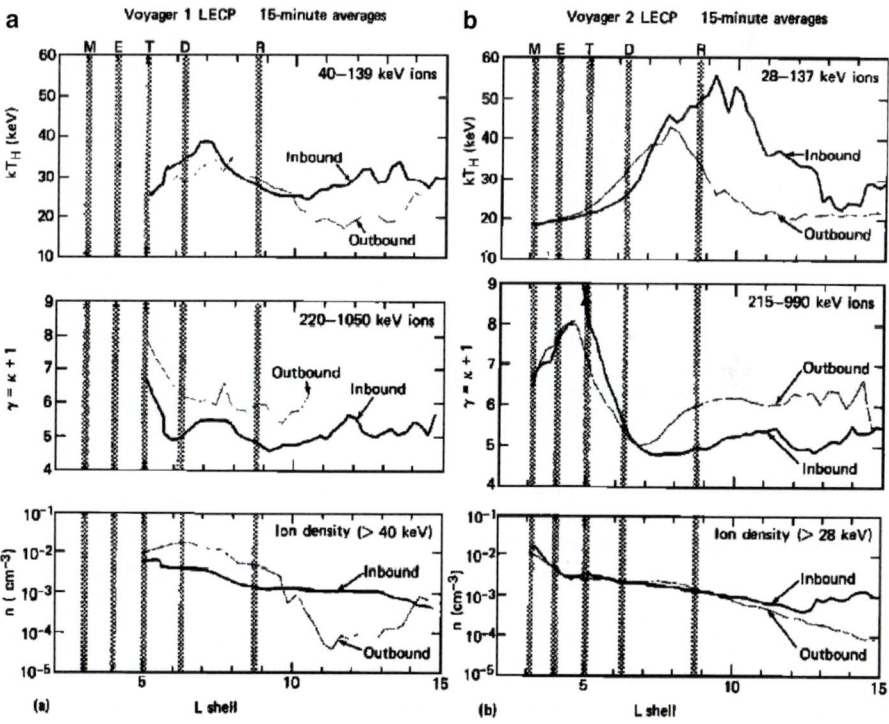

parallel and perpendicular to the planetary magnetic field is essentially continuous, with some absorption at the F-ring, but with total absorption at the outer edge of the A-ring. By contrast, protons are absorbed at the orbits of Mimas and Janus (1797-S2) and disappear at the edge of the F-ring. These effects are expected due to the gyroradii of protons in this region that are comparable to satellite diameters (e.g., Van Allen et al. 1980b). Electrons have small gyroradii and diffuse inward with relative ease.

Electrons are absorbed locally, however, when a magnetic flux tube has swept past the satellite recently, but not enough time has passed to refill the particular flux tube. Such "microsignatures" are indicators of orbiting material that could be a small moon or co-orbiting material in the vicinity of some satellites. Several such cases were seen by Voyager 1 and 2, and Pioneer 11. An example of such microsignatures is shown in Fig. 2.25 (Carbary et al. 1983) during the passage of Voyager 2 through the L shell of Dionne.

Fig. 2.24 Radial dependences of the intensities of electrons $E_E > 0.45$ meV and protons $E_P > 80$ meV measured by Pioneer 11 (Fillius et al. 1980)

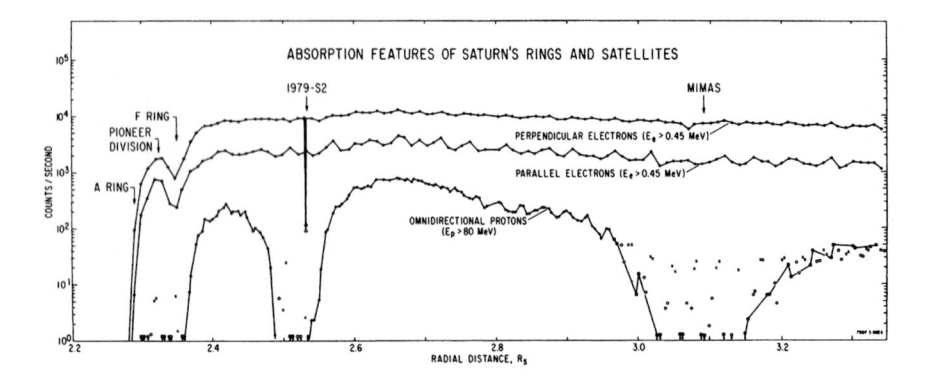

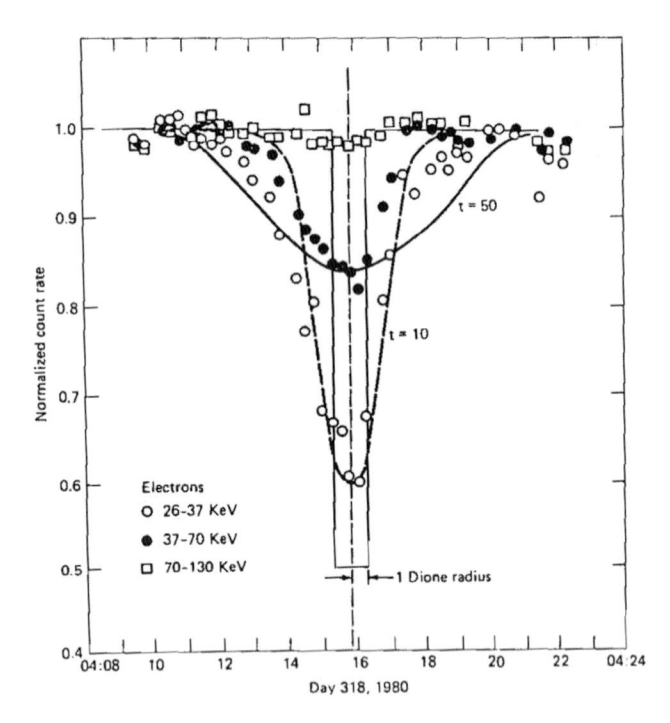

Fig. 2.25 Normalized electron counting rates at Dione. The Voyager-1 data were normalized by taking ratios of observe rates to those obtained by linearly extrapolating across the microsignature. The two theoretical fits to the data were taken from Carbary et al. (1983)

The spacecraft was located ~25° ahead of Dione in the wake region (i.e., Voyager was ahead of Dione in the co-rotation direction) and the depletion is deepest at the lowest energies but has already been filled in at the higher energies. The minimum age of the signature is ~1 h and a one-dimensional diffusion fit is consistent with a coefficient ~10^{-8} R_S^2 s^{-1}. Such microsignatures also serve to validate models of the magnetic field, in that computation of the magnetic L shell parameter for a satellite is used to predict the timing of the signature with respect to the spacecraft trajectory. If the depletion does not occur at the estimated time, it may mean that either the magnetic field model used is not sufficiently accurate, or that the observed microsignature is due to some other cause, such as co-orbiting material near the orbit of a particular moon (Carbary et al. 1983).

The overall transport of energetic particles in a magnetosphere is assessed by using observations of energy spectra and pitch angle distributions to compute phase space densities at constant first and second adiabatic invariants. An example in the case of Saturn is shown in Fig. 2.26 (Armstrong et al. 1983) for energetic ions observed during the Voyager 2 flyby. Notable features on the left panel are the flatness of the distribution at $L \geq 6 R_S$, suggesting that there is a particle source in the outer magnetosphere, and the rapid decrease at lower L values showing that ions are lost in large numbers inside the orbit of Dione. The panel on the right, at larger values of the first adiabatic invariant, also suggests a source in the outer magnetosphere but the losses appear in the vicinity of, and inward of the orbit of Rhea. Note that there are inbound-outbound asymmetries, suggesting that there exist important local time effects that have not been taken into account. The results for electrons are generally similar.

The summary of magnetospheric properties needs to consider the issue of overall energy and stress balance. There was sufficient instrumentation on the Voyagers to assess, to first order, the relative contributions of the magnetic field, the cold plasma and the hot plasma (McNutt 1984; Krimigis et al. 1983; Mauk et al. 1985). The energetic particle measurements showed that the hot plasma was a significant, and at times dominant, contributor to the inferred ring current, with beta (particle pressure/magnetic pressure) often exceeding ~1. Mauk et al. 1985, combined these measurements together with stresses due to the cold plasma (McNutt 1984) to deduce the equatorial field stresses due to the two sources, as shown in Fig. 2.27. The hot plasma stresses dominate at L < 15, while the cold plasma becomes the principal contributor at larger values; the equatorial currents (middle panel) reflect that picture. Thus, Saturn joined Jupiter as a magnetosphere where the hot plasma is a key player in the dynamics, even though MHD simulations are not able to account for this complex behavior of the plasma.

Fig. 2.26 Phase-space density of ions (assumed to be protons) at low values of (*left panel*) and high values of (*right panel*). Note the non-closure of inbound/outbound phase-space densities especially at low and high phase angles (from Armstrong et al. 1983)

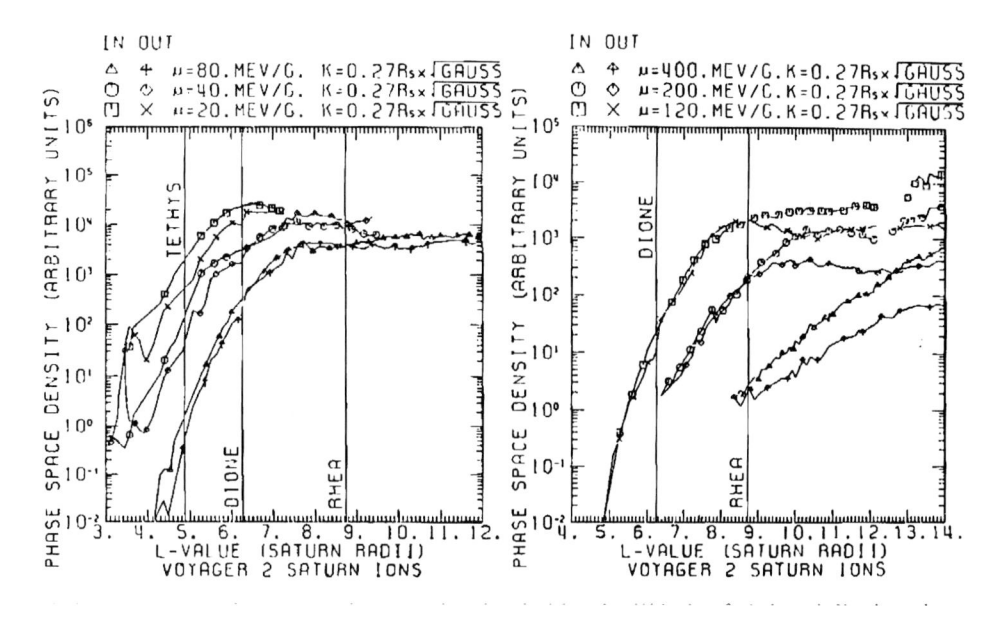

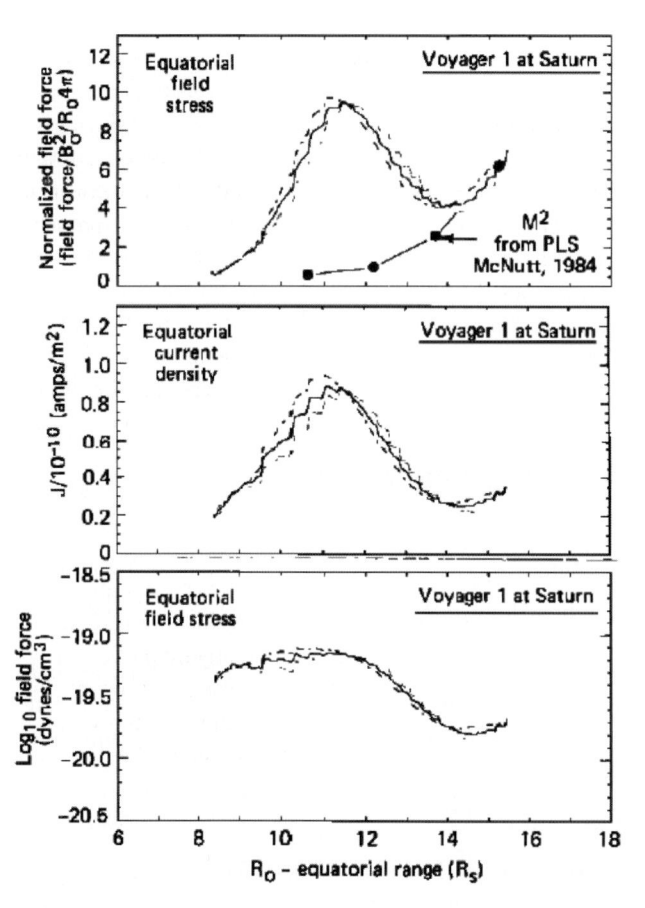

Fig. 2.27 (*Top*) Normalized equatorial field stress calculations. The normalization results in a dimensionless parameter that equals the Alfvenic Mach number squared (M^2) if the cold particle centrifugal stresses dominate the particle stresses. The solid circles are centrifugal stress measurement given by McNutt (1984). (*Middle*) Equatorial current densities corresponding to the field stresses shown in the bottom panel. (*Bottom*) Equatorial field stresses calculated using the field model developed in this study (from Mauk et al. 1985)

2.8.4 Plasma Waves

Plasma waves are intimately related to both the structure and especially the dynamics of the magnetosphere including the aurora and magnetic storms and substorms and provide diagnostic capabilities such as the measurement of electron densities. The distribution and changes at these high frequencies provide a view of the Saturn magnetosphere that complements those provided by the magnetic field, plasma and energetic particle measurements.

There is a plethora of plasma waves and radio emissions in the electromagnetic environment of Saturn (Scarf et al. 1984). Kilometric radiation (SKR) associated with Saturn's aurora is discussed in an earlier part of this chapter. A sample of plasma waves observed close to the planet during the Voyager 2 flyby is shown in Fig. 2.28.

Plasma wave emissions from the 10 s of Hz to 10 s of kHz are evident, and can be used to deduce the electron plasma frequency f_p and thus obtain an independent measurement of electron density from that of the plasma instrument. The spike in emissions during ring plane crossing is due to ring particle hits on the spacecraft that subsequently become vaporized and the antennas of the receiver sense the resulting plasma cloud.

A startling discovery by the Voyager radio receivers was the presence of intense spikes of noise at 10 s of MHz. An example is shown in Fig. 2.29 (Warwick et al. 1981) where a dynamic spectrum over a 20-min period displays several short, vertical spikes labeled as Saturn Electrostatic Discharges (SED). These have been interpreted as the manifestation of lightning strikes in Saturn's atmosphere that are coupled through the ionosphere to the magnetosphere. The lower panel shows the frequency of occurrence of SEDs

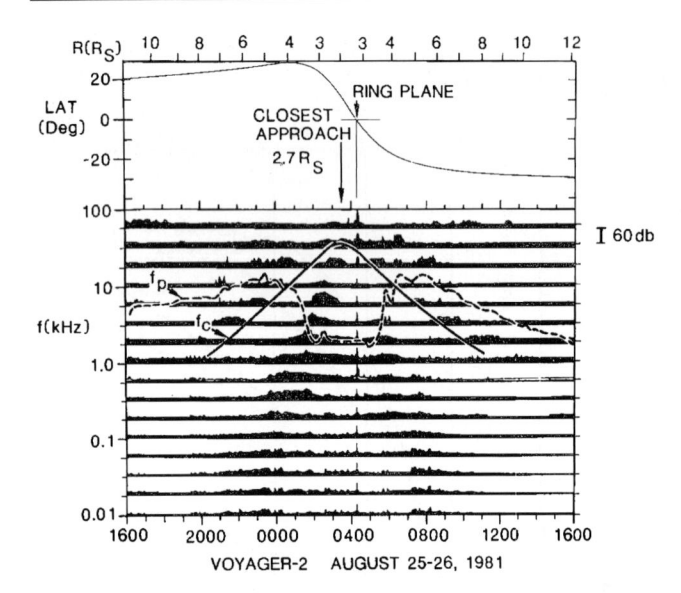

Fig. 2.28 Voyager-2 observations of the 16-channel spectrum analyzer for the 24-h interval centered about closest approach. The f_c curve is derived from magnetometer data and the f_p cure, derived from the plasma electron measurements, may represent a lower bound (from Scarf et al. 1984)

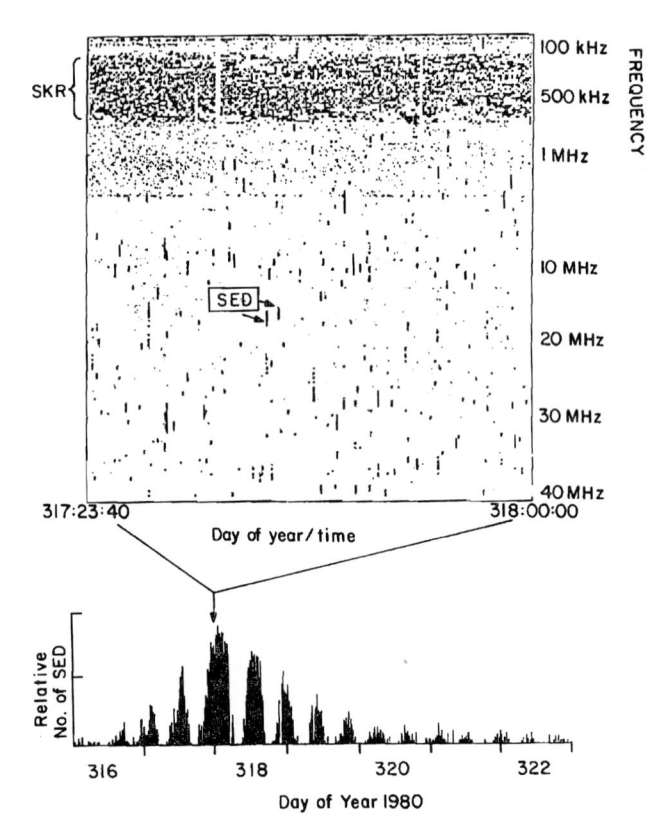

Fig. 2.29 The upper panel (from Warwick et al. 1981) shows a 40-min-long dynamic spectrum at the time of Voyager 1 closest approach. Both the PRA low and high bands are shown SED are the short, vertical streaks throughout the panel. SKR is the dark band near 500 kHz. The lower panel shows the overall occurrence of SED for the Voyager 1 encounter period as determined by a computerized detection scheme

during the Voyager 1 passage; it is evidently a common phenomenon accompanying atmospheric dynamical activity at Saturn.

2.8.5 Summary of Magnetosphere

The complex and dynamic magnetosphere of Saturn that resulted from the Voyager measurements and the earlier Pioneer 11 flyby has been synthesized in an artist's concept, as shown in Fig. 2.30 (Krimigis et al. 2004). Aside from the depiction of a magnetospheric topology that in many ways resembles that of Earth, emphasis is given to the interaction of the plasma with particulate matter orbiting the planet that seems to govern both, the source and loss of charged particles from the system. Of particular note is the expected presence of energetic neutral atoms discovered by Voyager (Kirsch et al. 1981), which held the promise of monitoring globally the dynamics of the magnetosphere, now largely fulfilled by Cassini (Krimigis et al. 2004).

Overall, the Saturn flybys nearly 30 years ago provided a remarkably clear glimpse of the planet's magnetospheric environment, and a solid foundation on which to design and build the instruments for the orbital mission that was to follow. The resulting knowledge from the Cassini measurements have provided a remarkable set of answers to vital questions but, inevitably, raised more complex issues that we need to grapple with, as detailed in Chapters 9–12 of this volume.

2.9 Saturn's Ring System

Saturn's ring system is the most prominent in the solar system. Before the twentieth century, telescopic observations discerned 3 major components of the ring system. The outer A ring comprises most ring material outside a dark division discovered by Cassini in 1675. The bright B ring lies interior to this division, and the C ring is interior to the B ring. Closer observations by Pioneer 11, and Voyager 1 and 2 provided more detailed views, showing the ring system to be much more complex, with the major rings characterized by many smaller rings. A small number of gaps are present, as well as narrow and opaque ringlets. The outermost A ring spans ~14,500 km radially, with an optical depth ranging from 0.4 to more than 2. It also contains the largest number of gravitational waves; these waves are driven by resonance interactions with nearby satellites which both confine and perturb edges of rings and create waves and wakes in them (Fig. 2.31). These waves propagate through the rings

Fig. 2.30 Artist's concept of Saturn's magnetosphere viewed from the dusk quadrant, with the solar wind indicated on the left. The equatorial plasma sheet is seen to extend to the magnetotail and the ring current is depicted inward of the orbit of Titan. Here the interaction of the hot plasma with the extended E-ring is depicted as a continuous source of Energetic Neutral Atoms (ENA), as is the interaction with the upper atmosphere of Titan (from Krimigis et al. 2004)

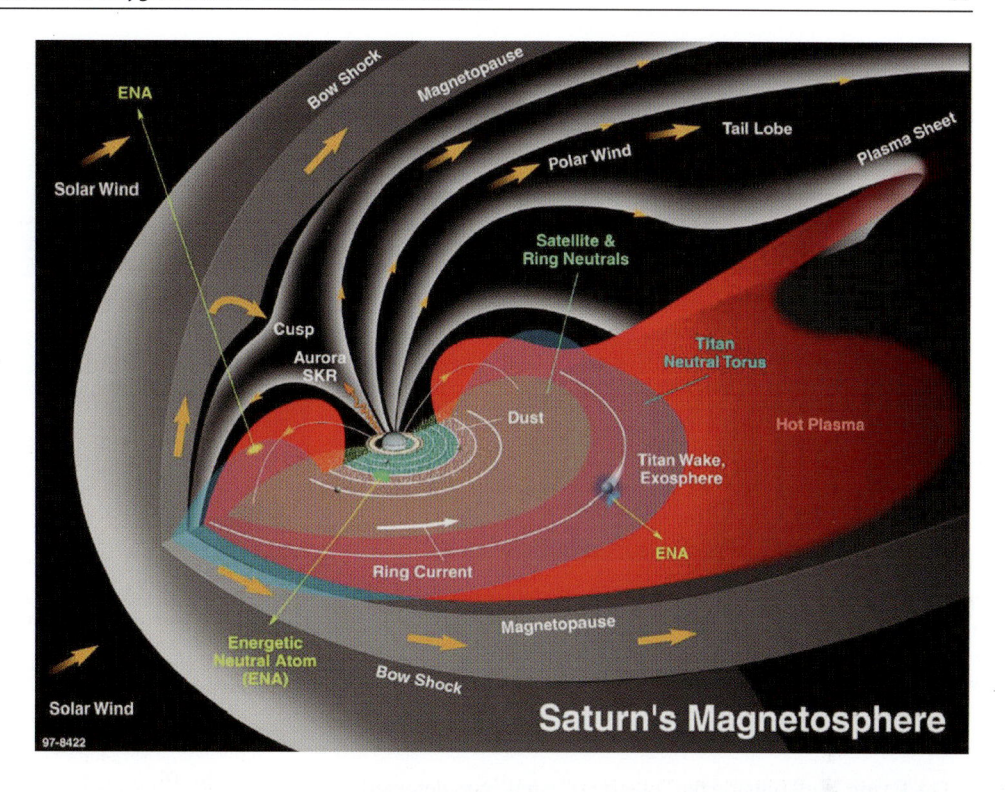

until they disperse as a result of inter-particle collisions. The A ring is asymmetric in the azimuthal variation of its brightness, a property attributed to small-scale wake patterns created by larger, embedded ring particles. The small Enke and Keeler gaps are present in the A ring. The Enke gap contains a small satellite, Pan, which keeps it open and creates one or more incomplete, clumpy ringlets (Showalter et al. 1991). Pan generates wakes in the ring material on either side of the gap which propagate and change in wavelength as it moves in its orbit (Showalter et al. 1986; Borderies et al. 1989; Stewart 1991). Pan is expected to have accreted outside the Roche limit and evolved through some process to its current location.

The Cassini Division, separating the A and B rings, is about 4,000 km wide and has a typical optical depth of 0.1, much lower than the A and B rings. The Cassini Division contains several gaps, possibly created and maintained by other small, embedded satellites (Marouf and Tyler 1986; Flynn and Cuzzi 1989).

The B ring, spanning 25,000 km in radius, contains most of the mass of the ring system. Its optical depths were found to range from 0.4 to nearly 2. Its substantial radial structure is irregular, and meteoroid ejecta may provide a source for this structure (Durisen et al. 1989). Narrow, small-scale radial and azimuthal brightness variations in the B ring (Fig. 2.32) first observed by Voyager 1 were called "spokes" (Grün et al. 1984; Doyle and Grün 1990; Tagger et al. 1991). Most of these were observed in the B-ring just after emerging from Saturn's shadow. In backscattered light, the spokes

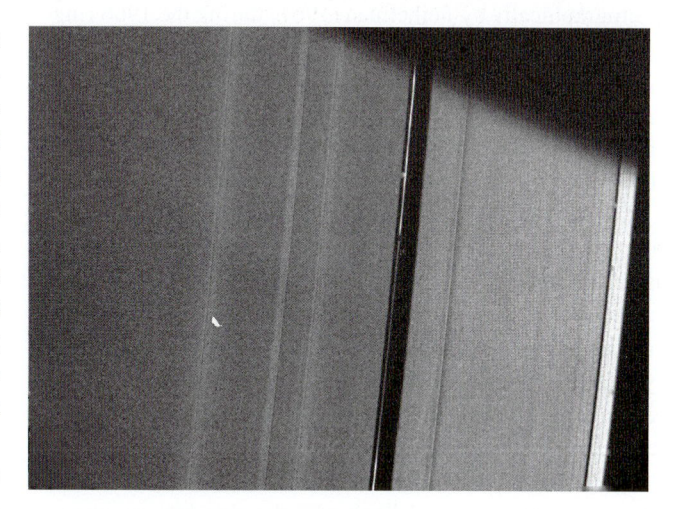

Fig. 2.31 Voyager image of a portion of the rings showing the most prominent spiral density and bending waves (together with a ringlet within the Encke gap). JPL Photo PIA01952

are dark relative to the underlying ring material, but in forward scattered light, the opposite is true, indicating a population of micron-sized particles. Dark radial bands observed by ground-based observers as early as the late nineteenth century may have been spokes.

The innermost "main" ring, the C ring, is 17,500 km wide radially and has an optical depth similar to the Cassini Division. This ring contains the bulk of radial gaps which contain narrow, opaque ringlets, and several broad 100- to 1,000-km regions of slightly greater optical depth.

Fig. 2.32 Voyager image of the B ring showing its ephemeral spoke structure. The vertex of the wedge-shaped spoke in the upper left portion of the picture is near the synchronous orbit. Rotation is anticlockwise with the leading edge being the slanted one (from B. A. Smith et al. 1982)

Fig. 2.33 The narrow F ring and two "shepherding" satellites. JPL photo P23911

The E ring, well outside the "classical" rings, was detected photographically by Feibelman (1967) during the 1966 ring-plane crossing. It is broad and diffuse, extending from some 3.5 Saturn radii (R_S = 60,330 km) to 8 R_S. It is concentrated near the orbit of Enceladus (3.95 R_S) and increases in vertical thickness away from Enceladus. Showalter et al. (1991) summarize Voyager measurements of this ring.

Guerin (1970) may have first photographed the faint D ring, just inside the C ring. Voyager images characterized it as 8,000 km wide with an optical depth much less than 0.01.

Pioneer 11 originally discovered the narrow F ring, several thousand kilometers outside the edge of the classical ring system (Gehrels et al. 1980). Its width varies from a few to several tens of kilometers, and its optical depth ranges from less than 0.1 to greater than 2 in its core. Two tiny satellites, Pandora and Prometheus, orbit on either side of the F ring and may confine ("shepherd") the ring but are just as likely to induce chaos in ring particle orbits (see Chapters 14 and 15). These two and possibly one or more embedded satellites interact gravitationally and distort the F ring by producing longitudinally and temporally varying kinks and twists in it (Kolvoord et al. 1990). Voyager images first revealed the non-Keplerian behavior of the F ring (Smith et al. 1981, 1982). Based on depletions in the Pioneer-11 measurements of charged particles, Cuzzi and Burns (1988) infer a 1,000-km wide satellite belt around the F ring in the region between the two shepherding satellites. The F ring may simply be the product of the disruption of one of these satellites (Fig. 2.33).

The tenuous G ring, lying between the F ring and E ring, lacks the fine-scale structure of the main rings. It was initially discovered by Pioneer-11 charged particle experiments (Van Allen et al. 1980a, b) and subsequently verified by Voyager images (Smith et al. 1981). Flying extremely close to its outer edge, Voyager 2 showed it to be featureless, symmetric, optically thin and several thousand kilometers wide. It is the faintest ring observed in the Saturn system (optical depth $\sim 10^{-6}$).

The reader is directed to the many detailed reviews of research on the ring system, beginning with the University of Arizona books: "Saturn" (Gehrels and Matthews 1984) and "Planetary Rings" (Greenberg and Brahic 1984). General review articles have been authored by Cuzzi (1983) and Brahic (1984), followed by reviews by Borderies (1989), Porco (1990), Nicholson and Dones (1991), Araki (1991), Esposito (1993), Colwell (1994), Cuzzi (1995), Porco (1995), Horn (1997), Dones (1998), Burns et al. (2001), and Cuzzi et al. (2002).

2.9.1 Particle Size Distribution

Estimates of the size of the particles comprising the main rings have had an interesting history (Pollack 1975 gives a thorough review). More recent reviews include Cuzzi et al. 1984; Esposito et al. 1984; Nicholson and Dones 1991; Porco 1995; Cuzzi 1995; Dones 1998; Cuzzi et al. 2002. Throughout the 1960s and early 1970s, the ring particles were thought primarily to be primarily tiny grains – much smaller than a cm. This was because early microwave observations found no trace of their expected 100 K thermal emission, leading to the conclusion that the particles were much smaller than microwave wavelengths and inefficient emitters (Berge and Read 1968; Berge and Muhleman 1973). This changed overnight when strong radar backscattering was detected by the main rings at the very same wavelengths (Goldstein and Morris 1973, Goldstein et al. 1977, Ostro et al. 1980, 1982). Such strong radar reflection proved that the particles must be quite efficient at scattering microwaves, and thus at least as large as the 4–13 cm wavelengths involved. The solution to the puzzle was first offered by Pollack et al. (1973): relatively pure water ice particles could provide both strong scattered radar reflections and weak thermal emission if their sizes were comparable to the microwave wavelengths – not much smaller and not much larger. Subsequent modeling by Cuzzi and Pollack (1979) and Cuzzi et al. (1980) showed this would apply even if the particles were realistically irregular in shape. They found that the radar and microwave brightness results, taken together, suggested a particular power-law distribution with an upper cutoff around several meters radius. Other microwave brightness measurements over the next decade, including several at shorter wavelengths (Schloerb et al. 1980; Epstein et al. 1984; Roellig et al. 1988; Grossman et al. 1989; Van der Tak et al. 1998) were consistent with this result and established an upper limit on the non-icy (i.e., silicate, carbonaceous, etc) abundance of between 1–10%.

Voyager 1 and 2 clarified the situation dramatically. A direct measurement of ring microwave transmission was obtained from the two-wavelength radio occultation of the rings by Voyager 1 in 1981 (Marouf et al. 1983, 1986; Tyler et al. 1983). Detailed analysis of these data in several regions clearly implied a power-law size distribution with a sharp upper cutoff between 1 and 5 meters radius which varied somewhat with location; the power-law distribution $n(r,dr) = n_0 r^{-q} dr$ also varied slightly, with q in the range 2.5–3.0 (Marouf et al. 1983, 1986; Zebker et al. 1985). Because of the low elevation angle of the rings, most of the A and B rings were inaccessible to this detailed size distribution analysis, but comparison between the transmissions at the two Voyager-1 radio wavelengths, together with the visual wavelength stellar occultation provided by Voyager 2 (Lane et al. 1982; Esposito et al. 1983), suggested that most of the main ring particles could obey this same kind of size distribution. A post-Voyager review of particle size constraints was presented by Esposito et al. (1984).

Other estimates of ring particle size can be obtained from measurements of the surface mass density of the rings, combined with the observed optical depth and assumed water ice composition. The Voyager stellar occultation data revealed about a dozen resolvable spiral density and bending wave-trains (Holberg et al. 1982; Esposito et al. 1983; see Fig. 2.31) which could be analyzed to give the local surface mass density (see Cuzzi et al. 1984 for a post-Voyager review

and summary). Subsequently, Cooper et al. (1985) analyzed the abundance of high-energy electron, proton, and gamma-ray data taken by Pioneer 11, which had preceded Voyager to Saturn and had pierced deeply into the region immediately above the entire main ring system. They concluded that the surface mass density of the rings was in the range of 60–100 g cm^{-2}, in general agreement with the density wave values and with the several-meter upper particle size radius observed at selected places by Voyager, thus extending the inference of a several-meter radius cutoff to the entire main ring system (at low radial resolution). This result is, however, is somewhat dependent on the assumed local structure of the rings as the source for the observed high-energy charged particles and gamma rays (see Chapters 13 and 17).

Nearly a decade after the Voyager-2 stellar occultation, Showalter and Nicholson (1990) identified a non-random variance in the stellar occultation data that could be used as an independent estimate of the size of the largest particle and produced a high-radial-resolution profile of "upper size cutoff" (which was inapplicable in most of the B ring because of its high opacity), on the order of 1-to-tens of meter radius. Some of these results were discrepant at the factor-of-several level from the radio occultation values; hindsight suggests this might be due to the presence of very fine-scale ring microstructure in the rings (see Chapters 13–15). French and Nicholson (2000) used a ground-based stellar occultation by the very bright star 28 Sagitarii to estimate the *lower* cutoff in the size distribution to be on the order of a cm in most places, as expected from comparing the Voyager radio and stellar occultations, but as large as tens of cm in the B ring. Several of these observational techniques have been greatly refined and amplified upon by Cassini, and repeated dozens of times at different ring opening angles and longitudes as well; the results are described in Chapter 15.

2.9.2 Ring Particle Composition

The dominance of water ice in Saturn's main rings was discovered by Kuiper et al. (1970) and Lebovsky et al. (1970), from its strong near-infrared absorption bands. As reviewed in detail by Pollack (1975), water ice is both cosmogonically abundant and stable against evaporation for the age of the solar system at Saturnian temperatures. More recent reviews include Cuzzi et al. 1984; Esposito et al. 1984; Nicholson and Dones 1991; Porco 1995; Cuzzi 1995; Dones 1998; and Cuzzi et al. 2002. Additional high-quality spectrophotometry by Clark and McCord (1980), in comparison with laboratory measurements of different frosts and ices, indicated that the ice was quite pure, with an intermediate grain size. The case for water ice being more than a trivialsurface coating was strengthened by its unique ability to reconcile the seemingly paradoxical low radio emissivity and high radar reflectivity of the rings, a result of its very low absorption coefficient at radio wavelengths, and by comparison of measured optical depths and surface mass densities at several locations. Various workers have estimated the "sizes" of the water ice "particles" responsible for the diagnostic near-infrared absorption bands and found them to be in the tens of microns range (Pollack et al. 1973; Clark and McCord 1980; Doyle et al. 1989); this was readily understood to represent some grain size in an icy regolith rather than the size of any independent orbiting "ring particle". Similar interpretations apply to the rolloff of brightness temperature seen from mid-infrared to microwave wavelengths (see Esposito et al. 1984; Spilker et al. 2005). Model approaches used to make this sort of grain size inference (e.g., Hapke 1981; Cuzzi and Estrada 1998; Shkuratov et al. 1999) are open to some systematic uncertainty because of their differing fundamental assumptions (Poulet et al. 2002), and some caution is needed in using them to draw quantitative conclusions regarding grain sizes or absolute mixing ratios, as discussed further in Chapter 15.

However, the rings are not pure water ice. The distinctly red color of the A and B rings suggests that water ice cannot be their sole constituent, as first pointed out by Lebovsky et al. (1970) who compared the visual color of the main rings to Jupiter's sulfurous satellite Io. Some hints were found of a possible absorption feature at 0.85 microns (Clark and McCord 1980; Karkoschka 1994), possibly due to silicates, but the reality of the feature was in question. Two independent Voyager datasets found the C Ring and Cassini Division particles to be significantly darker than their A and B ring counterparts: analysis of imaging data (Smith et al. 1981, 1982) found the C and Cassini particles to have lower visual wavelength albedos (see Fig. 2.34), and Hanel et al. (1981) found them to have higher temperatures in the thermal infrared. Cuzzi and Estrada (1998) and Poulet et al. (2003) modeled the ring composition as a spatially variable combination of an intrinsic mixture of water ice with small amounts of extrinsic reddish carbon-rich material; the very strong and very red absorption of organics ("tholins") allows them to create the observed visual redness with a small enough mixing ratio to satisfy the microwave requirements for a 90–99% water ice composition overall. Unfortunately, candidate organics have no identifiable absorption features in observable spectral regions, so their presence remains to be determined unambiguously.

The rings are enveloped in a tenuous atmosphere. Observations in the 1980s and 1990s suggested that this atmosphere was composed of water products like the OH radical, not surprisingly given the massive source of water ice and the likely ongoing vaporization by meteoroids (Ip 1997). However, Cassini in-situ observations (see Chapter 16) suggest that this is not correct.

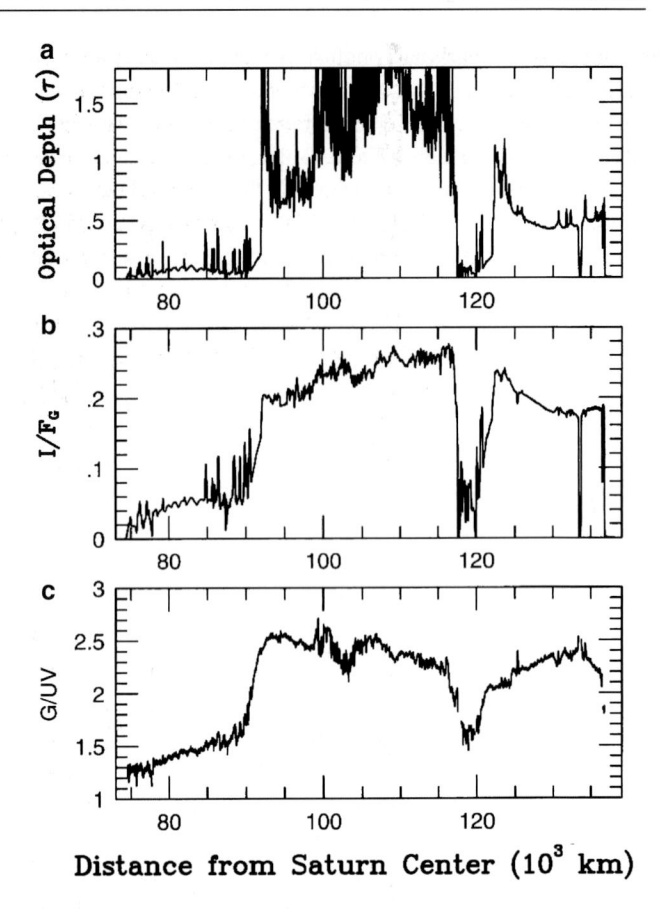

Fig. 2.34 Optical and color properties of rings as a function of radius: (**a**) ring optical depth from stellar occultation, (**b**) reflectivity in the Voyager imaging green filter, (**c**) ratio of green to ultraviolet reflectivity. From Estrada and Cuzzi (1996); see Estrada et al. (2003) for an update

2.9.3 Origin and Evolution

A number of hypotheses on the origin and evolution of the rings have been created over the last several centuries (see Pollack 1975 for a historical review, Pollack et al. 1973 for a "primary" formation scenario, and Harris 1984 for a more recent review and a more modern "secondary" formation scenario). Most workers today believe that the variety and vigor of several different kinds of evolutionary process preclude the present-day rings from being some primordial, never-accreted debris, and that instead they are the "secondary" remnant of some previously formed object that was disrupted in place or nearby, either collisionally or tidally. The timing of this formation event remains controversial, with some feeling that it might have been contemporaneous with the formation of the Saturn system, and others feeling it must be much more recent. A recent formation is, however, statistically harder to explain than a primordial formation (see Chapter 17). The apparently strange composition of the rings – somewhat different from that modeled for the surfaces of most of Saturn's classical icy moons – is clearly of great importance for questions of ring origin (see Chapter 15). However, because evolutionary effects are probably significant, it must be acknowledged that the current ring composition might not be the same as its primordial composition. For instance, Cuzzi and Estrada (1998) found that the albedo difference between the C Ring and Cassini Division particles, and the nearby A and B ring particles, could be due to surface-mass-density-dependent evolutionary effects of meteoroid bombardment (see Chapter 17). Moreover, the role of a persistent oxygen atmosphere has never been explored.

2.10 Icy Satellites

2.10.1 Introduction

When the Cassini-Huygens spacecraft entered its orbit around Saturn on June 30, 2004, the number of known satellites of Saturn was 36. On the eve of the Cassini-Huygens investigation three basic classes of satellites were recognized; the large bodies (Mimas, Enceladus, Tethys, Dione, Rhea, Titan, Hyperion, Iapetus and Phoebe), a number of coplanar, prograde small bodies a few km in size, including some within the main ring system, and the small irregular satellites with long periods in inclined orbits, some of which are retrograde.

All of the physical characteristics of a planetary satellite are interrelated. Surface composition and density are indicative of bulk composition and structure, while geological history and surface geomorphology are related to composition and internal structure. Surface microphysical properties are related to the composition and history of surface structures, their age and exposure to the space environment. It is therefore most efficient to consider the eight large, airless satellites one by one, and then to look for similarities and differences among them.

2.10.2 Sources of Information

2.10.2.1 Dimensions, Densities, and Rotational Properties

Voyagers 1 and 2 imaged all of the nine large satellites with sufficient resolution to establish their dimensions with greater precision than possible from ground-based telescopic imaging or other measurements. Mass determinations of Titan, Dione, Rhea, and Iapetus from the Voyager trajectories thus yielded their mean densities (Titan $1.88\,\mathrm{g\,cm^{-3}}$, Dione $1.43\,\mathrm{g\,cm^{-3}}$, Rhea $1.24\,\mathrm{g\,cm^{-3}}$, Iapetus $1.18\,\mathrm{g\,cm^{-3}}$). These values were determined with sufficient accuracy to demonstrate that the three airless bodies are significantly lower in density than Titan ($1.88\,\mathrm{g\,cm^{-3}}$) and to show a general agreement with previous measurements of both dimensions and masses from ground-based investigations. Voyager diameters, when combined with ground-based mass determinations, gave a general picture of the Saturn satellite density trend consistent with models of 60–80% ice, with the remainder represented as rocky material with chondritic composition (Morrison et al. 1986). It is noteworthy that the density of Enceladus was estimated as $1.1\,\mathrm{g\,cm^{-3}}$, but with an uncertainty of $0.5\,\mathrm{g\,cm^{-3}}$ (Smith et al. 1982). (The Cassini value is $0.57\,\mathrm{g\,cm^{-3}}$).

In the absence of resolved details on satellite surfaces, conjecture about their rotation periods, and in particular the matter of the synchronism of the rotation with orbital revolution about the planet, depends on photometric observations that may, or may not, show variations around the orbit. Although the airless satellites were suspected to be locked (e.g., Noland et al. 1974), Voyager images showed conclusively that six of the icy satellites (Mimas, Enceladus, Tethys, Dione, Rhea, and Iapetus are indeed in locked synchronous rotation). In the case of Iapetus it was known from the anticorrelation of the visual photometric lightcurve with the rotational modulation of the thermal radiation detected from Earth that the rotational and orbital periods are locked (Morrison et al. 1975).

2.10.2.2 Surface Compositions and Surface Optical Properties

Near-infrared spectroscopy ($\sim$1–4 μm) from ground-based telescopes and ultraviolet spectroscopy (0.2–0.45 μm) from the Hubble Space Telescope have been the principal sources of compositional information for planetary satellites prior to the mapping spectrometers on Galileo (NIMS) Cassini (UVIS and VIMS). The intermediate "visual" spectral region (0.3–1.0 μm) is also important because the frequently steep slope across this wavelength interval defines the color of the surface. Because ices are neutral in reflectance, color is an important diagnostic in establishing the identity of non-ice components. We summarize the compositional information available from reflectance spectroscopy for each of the eight large satellites (excluding Titan) on the eve of the Cassini investigation. Although the actual discoveries of H_2O ice, the principal surface component of all of the large satellites, were reported in various publications as early as the 1976 paper by Fink et al., high-quality, near-infrared spectra with spectral resolution up to 1,000 were obtained in the last decade before Cassini-Huygens entered its orbit,

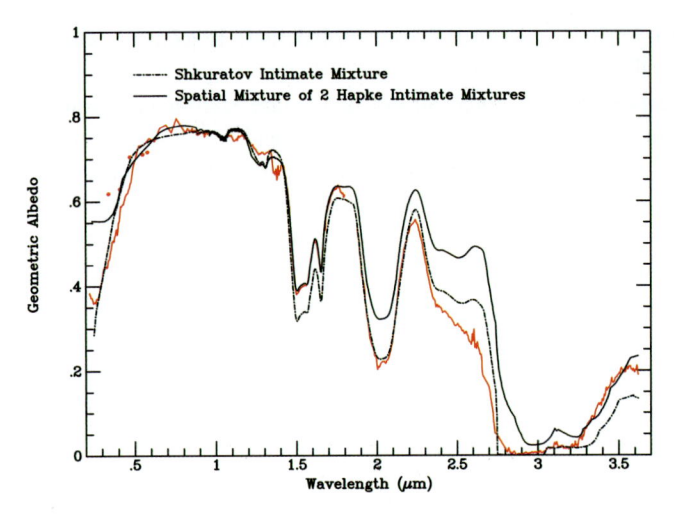

Fig. 2.35 The reflectance spectrum of Rhea, 0.2–3.6 μm, from ground-based and Hubble Space Telescope observations (*red line*). The steep slope to the short wavelength region represents absorption that increases toward the ultraviolet because of a non-ice surface component. The strong and characteristic absorption bands of H_2O ice are centered at ∼1.04, 1.25, 1.52, 2.02, and 3 μm, while the sharp absorption at 1.65 μm is indicative of the crystalline phase. The geometric albedo scale is calibrated from ground-based and Voyager photometric observations. Also shown are two scattering models calculated with the Shkuratov and Hapke radiative transfer codes. Each model reproduces the overall energy distribution and individual spectral features to some degree, but each fails to match the spectrum with a complete and satisfactory fit. From Cruikshank et al. (2004), where details and references are given

and were reported by Cruikshank et al. (2004) and Emery et al. (2005). The summary of surface compositions in this chapter is extracted from those most recent pre-Cassini papers. Figure 2.35 shows an example for Rhea.

The cameras (ISS) and infrared spectrometer (IRIS) on Voyagers 1 and 2 gave the first opportunities to observe the satellites at large phase angles – up to 175° in some cases, thus enabling the calculation of the phase integral and Bond albedo, as well as the bolometric Bond albedo (Cruikshank et al. (1984). These quantities are needed in the calculation of the temperature distribution on the satellite surfaces and are also related to parameters in the radiative transfer scattering calculations of spectral reflectance.

2.10.2.3 Geology

The pre-Cassini information on the geology of Saturn's satellites was derived entirely from images obtained by Voyagers 1 and 2 in November, 1980, and August, 1981, respectively (examples are shown in Fig. 2.36). Morrison et al. (1986) reviewed the geological investigations up to the time their paper was submitted, and some of the following description is summarized from that resource.

2.10.3 The Eight Large, Airless Satellites

2.10.3.1 Mimas

Mimas orbits just outside the A ring and within the E ring. Water ice is clearly evident from the strong and broad absorption bands centered at 1.52 and 2.02 μm, while the strong 1.65-μm absorption band of H_2O indicates the dominance of the crystalline (rather than amorphous) phase and at a low temperature consistent with the overall high albedo. Emery et al. (2005) found that the leading and trailing hemispheres have very similar spectra, but the 1.25-μm H_2O ice band is stronger on the leading hemisphere, and a possible unidentified band at 1.78 μm occurs on the trailing hemisphere. At the hemispheric spatial resolution of ground-based observations, no other absorption bands are seen in the spectral region 0.8–2.4 μm, thus eliminating NH_3 (and its hydrates), CO_2, and CH_4 with upper limits noted by Cruikshank et al. (2004) and Emery et al. (2005). Spectra at longer wavelengths would further reduce these upper limits because all the molecules mentioned (and many others) have stronger absorption bands longward of 2.5 μm.

The geometric albedo throughout the visible spectral range is high (∼0.7–0.8) and there is very little absorption toward the violet, perhaps because of a continuous source of micrometer-size particles of H_2O deposited on the surface from the E-ring.

Voyager 1 gave the best images of Mimas, with good coverage in the southern hemisphere at resolution ∼2 km. The surface is very heavily cratered and shows no clear evidence of modification by internal forces or extruded materials. Herschel, the largest crater on Mimas, has a diameter of 130 km with walls ∼5 km high, with no evidence for modification by crustal relaxation or erosion (Morrison et al. 1986). Plescia and Boyce (1982) proposed from crater counts that the surface shows a mixing of two crater populations, suggesting that an ancient modification of the original surface occurred after the initial heavy bombardment, although the surface optical properties (albedo and color) are quite uniform (Buratti et al. 1990; Verbiscer and Veverka 1992).

2.10.3.2 Enceladus

Enceladus is the next satellite outward from Saturn, orbiting at the brightest portion of the E-ring. Its composition is clearly dominated by H_2O ice from the strong absorption bands (noted above) in the near-infrared. The exceptionally high geometric albedo of Enceladus (exceeding 1.0 at 0.8 μm wavelength) results in a colder surface than the other satellites, with $T_{subsolar}$ ∼75 K. The great strength of the 1.65-μm H_2O band is consistent with this low temperature. No absorption bands other than those of H_2O are reliably detected

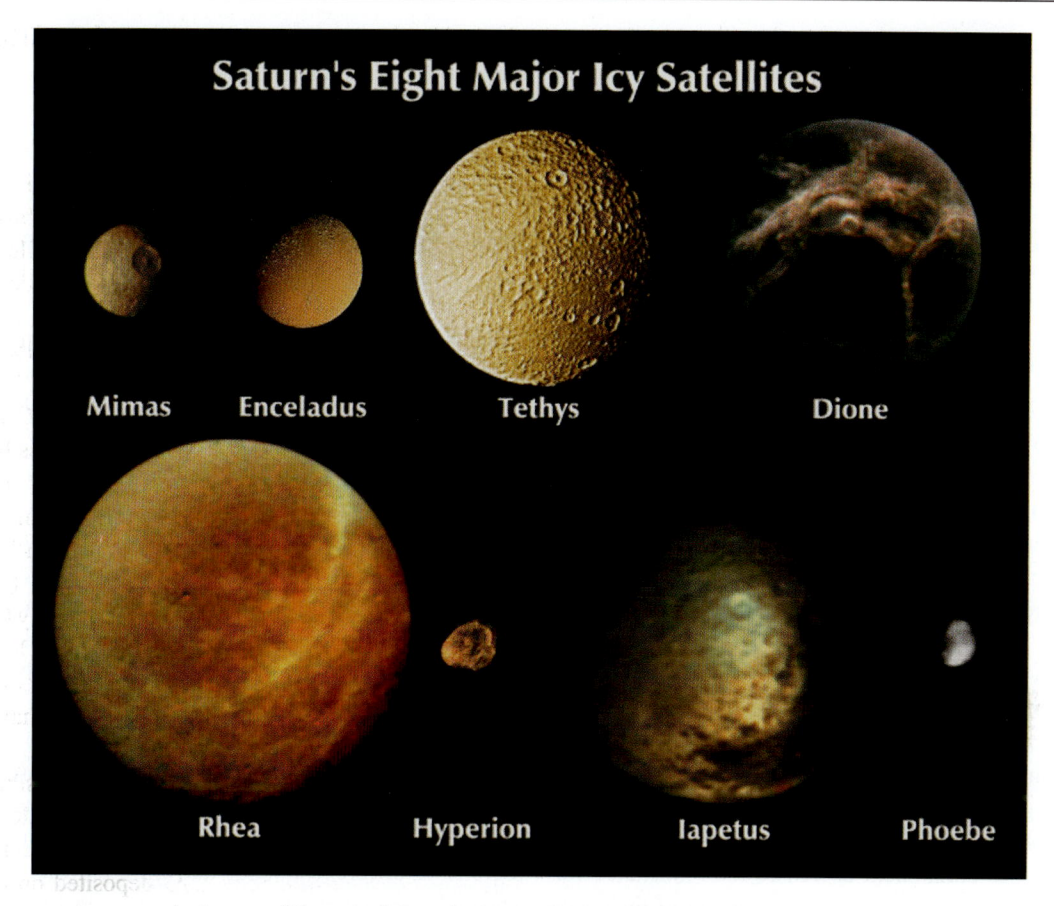

Fig. 2.36 Voyager color composite images of Saturn's eight major icy satellites. JPL Photo 46654Bc

(Cruikshank et al. 2004; Emery et al. 2005). The causal relationship of Enceladus to Saturn's E-ring, suspected from Voyager studies and clarified by Cassini, appears to affect the surface composition, albedo, and microstructure by serving as a source for the continuous deposition of very fine H_2O ice particles (with possible minor contaminants) in the present epoch. The unusually high albedo and scattering properties determined in part from Voyager observations, as well as their uniformity across the surface led Buratti et al. (1983) to propose that a relatively young surface layer covers Enceladus. Emery et al. (2005) suspected a weak absorption band near but not coincident with bands of NH_3 an ammonium hydrate, but this feature has not been confirmed in subsequent data.

Voyager-2 images of a portion of the northern hemisphere at 2 km resolution revealed a complex geomorphology of Enceladus indicative of relatively recent resurfacing by endogenic forces. Ridges, sharp and relaxed craters, craterless plains, with structures ranging in height from a few hundred meters (ridges) to 10 km (crater rim-to-bottom relief) are seen in complex patterns that define at least five distinct terrains (Morrison et al. 1986). These terrains represent a vast range in relative ages, with apparently ancient cratered regions adjacent to altered regions with crater densities indicating ages of 10^8 y or less. Morrison et al. (1986) and others suggested that an "unexpectedly large internal heat source" must be responsible for the extraordinary nature of Enceladus' surface, and while an excess of radionuclides seemed unlikely (on the basis of the poorly determined density), arguments and counterarguments for tidal heating left the matter in an ambiguous and unsettled state on the eve of the Cassini-Huygens investigation.

Soon after the Voyager flybys, investigators commented rather on the relationship between the orbit of Enceladus and its location within the E ring. Haff et al. (1983) argued that the E ring should have a lifetime of only a few 1,000 years, and the suggested meteoroidal impact ejection and geysering as mechanisms for the continuous supply of matter to the E ring. Pang et al. (1984) cited tectonic evidence to support the hypothesis that the E ring is continuously supplied by material emitted by volcanic eruptions from Enceladus. Thus, the direct evidence of geysering by several Cassini experiments was not altogether unexpected.

2.10.3.3 Tethys

Only H_2O ice has been detected on Tethys' surface with certainty, with deeper spectral bands on the leading hemisphere than on the trailing side (except the 1.04-μm band, which is stronger on the trailing side). Emery et al. (2005) extended the spectral coverage to 4.1 μm and detected the 3.12-μm Fresnel peak attributed to crystalline H_2O ice, but no other spectral bands. The surface of Tethys has a high albedo (~0.8 at 1.1 μm), but its reflectance steeply decreases for λ<0.5 μm, indicative of a non-ice, violet-absorbing material.

Tethys is densely cratered, and seen at Voyager resolution between 2 and 15 km, has two major topographic features. The giant crater (diameter 400 km) Odysseus is the largest crater imaged by Voyager among all the satellites of Saturn. Its prominent central peak stands on the crater floor, which has isostatically adjusted to the overall curvature of the satellite. A very large trough, Ithaca Chasma, extends about 3/4 of the circumference of Tethys, or about 2,500 km, in a complex system pattern of parallel fractures. Plescia and Boyce (1983) noted that there are terrains of different ages, with units of differing crater density. Ithaca Chasma may be the youngest major structure on Tethys (Morrison et al. 1984).

2.10.3.4 Dione

Among the airless icy satellites, Dione has the highest bulk density (1.43 g cm^{-3}). It is similar in size to Tethys, but its geometric albedo is significantly lower (maximum ~0.6). A non-ice surface component decreases the reflectance for λ < 0.5 μm, as with Tethys, but the remainder of the spectrum in the visible and near-infrared is entirely characterized by strong H_2O ice bands, with no other components detected in the ground-based spectra. Noll et al. (1997) found an absorption band in the ultraviolet (λ = 0.26 μm) in HST spectra and identified it as O_3 in the surface ice. Ozone is made by incident sunlight on H_2O ice, but is also rapidly destroyed. As with the other large satellites, Voyager data have been used to calculate the photometric phase integral, but the bolometric Bond albedo was not determined. The leading and trailing hemispheres show a large difference in reflectance, with the leading hemisphere having the highest albedo.

The surface of Dione has large regions of distinctly different crater density, indicating that significant internal activity occurred after the body accreted and endured the early bombardment. Heavily cratered regions contrast with lightly cratered, smooth plains units. A characteristic of the trailing hemisphere that appears unique in the Saturn system is a pattern of high-albedo deposits in approximately linear arrangements called "wispy terrain". The highest resolution Voyager images show that at least some of the wisps follow an apparently global system of fractures (horst and graben structures) induced by tectonic forces, suggesting that gases escaping from the interior along these cracks has condensed to form local and highly reflective deposits, presumably of ice. The higher density of Dione is broadly consistent with an internal radioactivity-generated heat source greater than the other satellites, perhaps driving a longer period of tectonism.

2.10.3.5 Rhea

Rhea is the largest of the airless icy satellites (radius = 764 km), although less dense than Dione. The high albedo surface is characterized by the usual strong absorption bands of H_2O ice, including the 1.65-μm band and the 3.12-μm Fresnel reflection peak that indicate the crystalline phase. A non-ice component absorbs increasingly toward the violet and ultraviolet region. As with Dione, Noll et al. (1997) found the 0.26-μm band of O_3 (see discussion above). As shown in Fig. 2.35, the models of Cruikshank et al. (2004) do not fit the spectrum uniformly well, but they do incorporate the complex organic solid tholin as a coloring agent that absorbs in the violet and is representative of materials that appear to be widespread in the outer Solar System (Cruikshank et al. 2005). This coloring component is not uniformly distributed across Rhea's surface The leading hemisphere is higher in albedo than the trailing.

Voyager obtained thermal spectrum measurements of Rhea in and out of eclipse by Saturn's shadow (Hanel et al. 1981), with results consistent with a surface having regions of different albedos and different thermal inertias. The high thermal inertia component is probably ice in blocks with a characteristic size ~10 cm, while the low thermal inertia material is probably a thin, pulverized layer of small grain size.

A heavily cratered surface with wisps of brighter material (similar to Dione) characterizes the surface structures of Rhea. Craters of small size are not uniformly distributed, and Plescia and Boyce (1982) propose that regions near the equator are mantled with a deposit some 2 km in thickness that buries most craters less than 10 km in size. Very large craters are (>30 km diameter) are lacking in the polar regions in some longitude zones, suggesting that some resurfacing of portions of Rhea's surface occurred after the supply of large impactors was depleted, but before the overall heavy bombardment ceased.

2.10.3.6 Hyperion

Voyager images reinforced conclusions reached from photometric studies that Hyperion is both irregular in shape and has a variable rotation period, with the axis oriented near the orbital plane. Its spectrum shows crystalline H_2O ice bands, although its albedo is lower than the other icy satellites (very roughly 0.3, Cruikshank et al. 1984) and there must therefore be a significant non-ice component of the optical surface. As with some of the other satellites, this component imparts a red color to portions of the surface. Voyager images showed albedo variations of 10–20% across the resolved surface of Hyperion.

At least one crater with diameter ~120 km and ~10 km vertical relief is visible in the Voyager 2 images at resolution ~10 km, even though the body itself is only ~350 km across in its maximum dimension. The highly irregular shape and surface characteristics suggest that is a collisional fragment of a larger body, and that its surface has been modified by impacts subsequent to the fragmenting event.

2.10.3.7 Iapetus

The second largest icy satellite Iapetus has the remarkable property that its leading and trailing hemispheres represent a dichotomy in surface albedo amounting to a factor of nearly 10. The hemisphere centered on the apex of Iapetus' orbital motion around Saturn has an albedo of 0.08, while the albedo of the trailing hemisphere is about 0.7. The low-albedo hemisphere shows weak H_2O ice bands and strong absorption through the visible toward the violet spectral region. Owen et al. (2001) successfully modeled the entire spectrum from 0.3 to 4.1 μm with H_2O ice and nitrogen-rich tholins. The high-albedo hemisphere shows strong bands of crystalline H_2O, similar in every regard to the other high-albedo Saturn satellites.

Images of limited regions of Iapetus by Voyager confirmed the general geographical disposition of the ice and the low-albedo material derived from photometry and ground-based radiometric observations, and showed heavily cratered terrain in the high-albedo region. Limitations of the image quality gave no information about craters in the low-albedo regions, but the dark-light boundary showed a number of dark-floored craters in the brighter material. However, there were no bright-rimed circular features detected in the low-albedo regions.

The low mean density of Iapetus $(1.16\,\mathrm{g\,cm^{-3}})$, similar to the other icy satellites of Saturn, indicated a bulk composition of ice, giving support to the nature of the low-albedo material as a surface deposit on an underlying bedrock of H_2O ice. The pre-Cassini epoch of Iapetus investigations closed with the problems of the chemistry and origin of the low-albedo material unsolved, although a preponderance of circumstantial evidence indicating a carbonaceous composition, perhaps similar to the insoluble organic matter in carbonaceous meteorites.

2.10.3.8 Phoebe

Saturn's most distant "classical" satellite is Phoebe, a body of irregular shape (average radius 110 km) in an inclined, retrograde orbit. Spectral bands of H_2O of moderate strength can be detected from Earth-based observations. Voyager detected albedo markings in the range 0.046–0.060 (normal reflectance), although the geometric albedo measured from Earth at small phase angles is 0.07, indicating a significant photometric opposition brightness surge (Cruikshank et al. 1984). Voyager images verified the supposition that Phoebe's rotation is not synchronous with its long orbital period. Phoebe's color is more neutral than that of the low-albedo material on Iapetus, although some dynamical models suggest that material ejected from Phoebe could be swept up by Iapetus as the dust particles migrated toward Saturn. The color difference has been sited as a detriment to this model.

While the irregular shape of Phoebe could be discerned in Voyager images, the spatial resolution was insufficient to reveal any craters. Although Phoebe's geological history was entirely unknown on the eve of the *Cassini* investigation, its size, dark surface, and orbital characteristics support the contention that it is a captured body.

2.10.4 The Small Satellites

2.10.4.1 Satellites Near the Rings and Associated with Larger Satellites

Small satellites near the ring system, embedded within the rings, co-orbital with other satellites, and occurring in the Lagrangian points of other satellites were mostly found from Earth-based observations during times when the Earth passed through the plane of the rings (e.g., in 1966, 1979–80, 1995–96), but some were found in the Voyager 1 and 2 images. Here we simply note the physical properties of the nine inner satellites confirmed at the time of *Cassini's* Saturn orbit insertion.

The four largest of these objects are Janus, Prometheus, Epimetheus, and Pandora, all irregular in shape and with maximum dimensions >100 km. They have high albedos (0.5–0.9) consistent with icy surfaces. The remaining five have maximum dimensions a few tens of km and similarly high albedos.

2.10.4.2 Irregular Satellites

Exterior to the system of coplanar satellites lie a number of small bodies less than ~10 km in size and occupying orbits that are eccentric and inclined. Groupings of the satellites by orbital elements suggest that many of them originated as fragments of larger bodies that were captured by Saturn and disrupted either by collision or tidal stress. In contrast to the inner satellites, these small, exterior bodies appear to have low albedos (~0.06) comparable to comets, the irregular satellite of Jupiter, and some classes of transneptunian objects.

Acknowledgements The authors would like to thank Aharon Eviatar, John Clarke, Leigh Fletcher, John Spencer, Linda Spilker, and Thomas Stallard for insights and comments on this chapter during its preparation. Portions of the work of this review were carried out at the Jet Propulsion Laboratory, California Institute of Technology, and the Applied Physics Laboratory, Johns Hopkins University, under contract to the National Aeronautics and Space Administration, as well as NASA's Ames Research Center. Funding was also provided by the UK Science and Technology Facilities Council, and the European Commission Framework 6 EuroPlaNet project.

References

Acarreta, J. R. and Sanchez-Lavega, A. (1999) Vertical cloud structure in Saturn's 1990 equatorial storm. Icarus 137, 24–33.

Alexander, A. F. O'D. (1962) *The Planet Saturn: A History of Observation, Theory and Discovery*, London, Faber & Faber.

Allison, M. and Stone, P. H. (1983) Saturn meteorology – A diagnostic assessment of thin-layer configurations for the zonal flow. Icarus 54, 296–308.

Allison, M., Godfrey, D. A., and Beebe, R. F. (1990) A wave dynamical interpretation of Saturn's polar hexagon. Science 247, 1061–1063.

Appleby, J. F. (1984) Equilibrium models of Jupiter and Saturn. Icarus 59, 336–366.

Araki, S. (1991) Dynamics of planetary rings. Am. Sci. 79, 44–59.

Armstrong, T. P., Paonessa, M. T., Bell, E. V., and Krimigis, S. M. (1983) Voyager observations of Saturnian ion and electron phase space densities. J. Geophys. Res. 88, 8893–8904.

Atreya, S. K., Wong, M. H., Owen, T. C., Mahaffy, P. R., Niemann, H. B., de Pater, I., Drossart, P., and Encrenaz, T. (1999) A comparison of the atmospheres of Jupiter and Saturn: Deep atmospheric composition, cloud structure, vertical mixing and origin. Planet. Space Sci. 47, 1243–1262.

Atreya, S. K., Wong, A. S., Baines, K. H., Wong, M. H., and Owen, T. C. (2005) Jupiter's ammonia clouds-localized or ubiquitous? Planet Space Sci. 53, 498–507.

Axel, L. (1972) Inhomogeneous models of the atmosphere of Jupiter. Astrophys. J. 173, 451–468.

Baines, K. H., Carlson, R. W., and Kamp, L. W. (2002) Fresh ammonia ice clouds in Jupiter. I. Spectroscopic identification, spatial distribution, and dynamical implications. Icarus 159, 74–94.

Baldwin, M. et al. (2001) The quasi-biennial oscillation. Rev. Geophys. 39, 179–230.

Barbosa D. D. (1990) Auroral precipitation flux of ions and electrons in Saturn's outer magnetosphere. Planet. Space. Sci. 38, 1295–1304.

Barnet, C. D., Westphal, J. A., Beebe, R. F., and Huber, L. F. (1992) Hubble Space Telescope observations of the 1990 equatorial disturbance on Saturn – Zonal winds and central meridian albedos. Icarus 100, 499–511.

Baron R. L., Joseph, R. D., Owen, T., Tennyson, J., Miller, S., and Ballester, G. E. (1991) Infrared imaging of H_3+ in the ionosphere of Jupiter. Nature 353, 539–542.

Barshay, S. S. and Lewis, J. S. (1978) Chemical structure of the deep atmosphere of Jupiter. Icarus 33, 593–611.

Beebe, R. F., Barnet, C., Sada, P. V., and Murrell, A. S. (1992) The onset and growth of the 1990 equatorial disturbance on Saturn. Icarus 95, 163–172.

Behannon K. W., Lepping, R. P., and Ness, N. F. (1983) Structure and dynamics of Saturn's outer magnetosphere and boundary regions, J. Geophys. Res. 88, 8791–8800.

ben Jaffel, L., Leers, V., and Sandel, B. (1995) Dark auroral oval on Saturn discovered in HST UV images. Science 269, 951–953.

Berge, G. L. and Muhleman, D. O. (1973) High-angular-resolution observations of Saturn at 21.1 cm wavelength. Astrophys. J. 183, 373–381.

Berge, G. L. and Read, R. B. (1968) The microwave emission of Saturn. Astrophys. J. 152, 755–764.

Bézard, B., Gautier, D., and Conrath, B. (1984) A seasonal model of the Saturnian upper troposphere: Comparison with Voyager infrared measurements. Icarus 60, 274–288.

Bézard, B. and Gautier, D. (1985) A seasonal climate model of the atmospheres of the giant planets at the Voyager encounter time. I – Saturn's stratosphere. Icarus 61, 296–310.

Bézard, B., Drossart, P., Lellouch, E., Tarrago, G., and Maillard, J. P. (1989) Detection of arsine in Saturn. Astrophys. J. 346, 509–513.

Bézard, B., Griffith, C., Lacy, J. and Owen, T. (1995) Non-detection of hydrogen cyanide on Jupiter. Icarus 118, 384, 391.

Bézard, B., Feuchtgruber, H., Moses, J. I., and Encrenaz, T. (1998) Detection of methyl radicals (CH_3) on Saturn. Astron. Astrophys. 334, L41–L44.

Bézard, B., Drossart, P., Encrenaz, T., and Feuchtgruber, H. (2001a) Benzene on the giant llanets. Icarus 154, 492–500.

Bézard, B., Moses, J., I., Lacy, J., Greathouse, T., Richter, M., and Griffith, C. (2001b) Detection of ethylene (C_2H_4) on Jupiter and Saturn in non-auroral regions. Bull. Amer. Astron. Soc. 33, 1079; Icarus 154, 492–500.

Bjoraker, G. L., Larson, H. P., Fink, U., and Smith, H. A. (1981) A study of ethane on Saturn in the 3 μm region. Astrophys. J. 248, 856–862.

Borderies, N. (1989) Ring dynamics. Celest. Mechan. 46, 207–230.

Borderies, N., Goldreich, P., and Tremaine. S. (1989) The formation of sharp edges in planetary rings by nearby satellites. Icarus 80, 344–360.

Brahic, A. (1984) Planetary rings. *Proceedings of IAU Colloquium 75, Cepadues-Editions*, Toulouse.

Brault, J. W., Fox, K., Jennings, D. E., and Margolis, J. S. (1981) Anomalous $^{12}CH_4$, $^{13}CH_4$ strengths in $3\nu_3$. Astrophys. J. 247, L101–L104.

Bregman, J., Lester, D. F., and Rank, D. M. (1975) Observations of the ν_2 band of PH_3 in the atmosphere of Saturn. Astrophys. J. 202, L55–L56.

Bridge, H. S., Bagenal, F., Belcher, J. W., Lazarus, A. J., McNutt, R. L., Sullivan, J. D., Gazis, P. R., Hartle, R. E., Ogilvie, K. W., Scudder, J. D., Sittier, E. C., Eviatar, A., Siscoe, G. L., Goertz, C. K., and Vasyliunas, V. M. (1982) Plasma observations near Saturn: Initial results from Voyager 2. Science 215, 563–570.

Briggs, F. H. and Sackett, P. D. (1989) Radio observations of Saturn as a probe of its atmosphere and cloud structure. Icarus 80, 77–103.

Broadfoot et al. (1981) Extreme ultraviolet observations from Voyager 1 encounter with Saturn. Science 212, 206–211.

Brown, L. W. (1974) Spectral behavior of Jupiter near 1 MHz. Astrophys. J. 194, L159–L162.

Brown, L. W. (1975) Saturn radio emission near 1 MHz. Astrophys. J. 198, L89–L92.

Bunce, E. J. and Cowley, S. W. H. (2003) A note on the ring current in Saturn's magnetosphere: Comparison of magnetic field data obtained during the Pioneer-11 and Voyager-1 and -2 flybys. Ann. Geophys. 21, 661–669.

Burrati, B., Veverka, J. and Thomas, P. (1983) Enceladus: Implications of its unusual photometric properties. Bull. Amer. Astron. Soc. 14, 737.

Buratti, B. J., Mosher, J. A., and Johnson, T. V. (1990) Albedo and color maps of the Saturnian satellites. Icarus 87, 339–357.

Burns, J. A., Hamilton, D. P., and Showalter., M. R. (2001) Dusty rings and circumplanetary dust, observations and simple physics. In *Interplanetary Dust*, Springer, Berlin, pp. 641–725.

Caldwell, J. (1977) The atmosphere of Saturn: An infrared perspective. Icarus 30, 493–510.

Caldwell, J., Gillett, F. C., Nolt, I. G., Tokunaga, A. (1978) Spatially resolved infrared observations of Saturn. I – Equatorial limb scans at 20 microns. Icarus 35, 308–312.

Caldwell, J., Tokunaga, A. T., and Orton G. S. (1983) Further observations of 8-μm polar brightenings of Jupiter. Icarus 53, 133–140.

Caldwell, J., Hua, X.-M., Turgeon, B., Westphal, J. A., and Barnet, C. D. (1993) The drift of Saturn's north polar spot observed by the Hubble Space Telescope. Science 260, 326–329.

Carbary, J. F., Krimigis, S. M., and Ip, W. H. (1983) Energetic particle microsignatures of Saturn's satellites, J. Geophys. Res. 88, 8947–8958.

Carlson, B. E., Caldwell, J., and Cess, R. D. (1980) A model of Saturn's seasonal stratosphere at the time of the Voyager encounters. J. Atmos. Sci. 37, 1883–1885.

Carr, T. D., Brown, G. W., Smith, A. G., Higgins, C. S., Bollhagen, H., May, J., and Levy, J. (1964) Spectral distribution of the decametric radiation from Jupiter in 1961. Astrophys. J. 140, 778–795.

Cavalié, T., Billebaud, F., Fouchet, T., Lellouch, E., Brillet, J, Dobrijevic, M. et al. (2008) Observations of CO on Saturn and Uranus at millimeter wavelengths: New upper limit determinations. Astron. Astrophys. 484, 555–561.

Cess, R. D. and Caldwell, J. (1979) A Saturnian stratospheric seasonal climate model. Icarus 38, 349–357.

Clark, R. N. and McCord, T. B. (1980) The rings of Saturn – New infrared reflectance measurements and a 0.326–4.08 micron summary. Icarus 43, 161–168.

Clarke J. T., Moos, H. W., Atreya, S. K., and Lane, A. L. (1981) IUE detection of bursts of H Ly α emission from Saturn. Nature 290, 226–227.

Clarke, J. T. and 12 co-authors (2005) Morphological differences between Saturn's ultraviolet aurorae and those of Earth and Jupiter. Nature 433, 717–719.

Colwell, J. (1994) The disruption of planetary satellites and the creation of planetary rings. Planet Space Sci. 42, 1139–1149.

Combes, M., Maillard, J. P., and de Bergh, C. (1977) Evidence for a telluric value of the $^{12}C/^{13}C$ ratio in the atmospheres of Jupiter and Saturn. Astron. Astrophys. 61, 531–537.

Connerney, J. E. P. Acuna, M. H., and Ness, N. F. (1983) Currents in Saturn's magnetosphere. J. Geophys. Res. 88, 8779–8789.

Connerney, J. E. P., Davis, L. Jr., and Chenette, D. L. (1984) Magnetic field models. In *Saturn* (T. Gehrels, M. Matthews, Eds.), University of Arizona Press, Tucson, pp. 354–377.

Conrath, B. J. and Pirraglia, J. A. (1983) Thermal structure of Saturn from Voyager infrared measurements: Implications for atmospheric dynamics. Icarus 53, 286–291.

Conrath, B. J., Gautier, D., Hanel, R. A., and Hornstein, J. S. (1984) The helium abundance of Saturn from Voyager measurements. Astrophys. J. 282, 807–815.

Conrath, B. J. and Gautier, D. (2000) Saturn helium abundance: A reanalysis of Voyager measurements. Icarus 144, 124–134.

Conrath, B. J. and Gierasch, P. J. (1986) Retrieval of ammonia abundances and cloud opacities on Jupiter from Voyager IRIS spectra. Icarus 67, 444–455.

Conrath, B. J., Gierasch, P. J., and Leroy, S. S. (1990) Temperature and circulation in the stratosphere of the outer planets. Icarus 83, 255–281.

Conrath, B. J., Gierasch, P. J., and Ustinov, E. A. (1998) Thermal structure and para hydrogen fraction on the outer planets from Voyager IRIS measurements. Icarus 135, 501–517.

Cooper, J. F., Eraker, J. H., and Simpson, J. A. (1985) The secondary radiation under Saturn's A-B-C rings produced by cosmic ray interactions. J. Geophys. Res. 90, 3415–3427.

Courtin, R., Gautier, D., Marten, A., Bézard, B., and Hanel, R. (1984) The composition of Saturn's atmosphere at northern temperate latitudes from Voyager IRIS spectra: NH_3, PH_3, C_2H_2, C_2H_6, CH_3D, CH_4 and the Saturnian D/H isotopic ratio. Astrophys. J. 287, 899–916.

Coustenis, A., Salama, A., Schulz, B., Ott, S., Lellouch, E., Encrenaz, T., Gautier, D., and Feuchtgruber, H. (2003) Titan's atmosphere from ISO mid-infrared spectroscopy. Icarus 161, 383–403.

Cowley S. W. H. and Bunce, E. J. (2001) Origin of the main auroral oval in Jupiter's coupled magnetosphere-ionosphere system. Planet. Space Sci. 49, 1067–1088.

Cowley, S. W. H., Bunce, E. J., and Prangé, R. (2004a) Saturn's polar ionospheric flows and their relation to the main auroral oval. Ann. Geophys. 22, 1379–1394.

Cowley, S. W. H., Bunce, E. J., and O'Rourke, J. M. (2004b) A simple quantitative model of plasma flows and currents in Saturn's polar ionosphere. J. Geophys. Res. 109, 5212.

Cruikshank, D. P., Veverka, J., and Lebofsky, L. A. (1984) Satellites of Saturn: Optical properties. In *Saturn* (T. Gehrels, M. S. Matthews, Eds.), University of Arizona Press, Tucson, pp. 640–667.

Cruikshank, D. P., Owen, T. C., Dalle Ore, C., Geballe, T. R., Roush, T. L., de Bergh, C., Sandford, S. A., Poulet, F., Benedix, G. K., and Emery, J. P. (2004) A spectroscopic study of the surfaces of Saturn's large satellites: H_2O ice, tholins, and minor constituents. Icarus 175, 268–283.

Cruikshank, D. P., Imanaka, H., and Dalle Ore, C. M. (2005) Tholins as coloring agents on outer Solar System bodies. Adv. Space Res. 36, 178–183.

Cuzzi, J. N. and Pollack, J. B. (1978) Saturn's rings: Particle composition and size distribution as constrained by microwave observations. I – Radar observations. Icarus 33, 233–262.

Cuzzi, J. N., Pollack, J. B., and Summers, A. L. (1980) Saturn's rings – Particle composition and size distribution as constrained by observations at microwave wavelengths. II – Radio interferometric observations. Icarus 44, 683–705.

Cuzzi, J. N. (1983) Planetary ring systems. Rev. Geophys. Space Phys. 21, 173–186.

Cuzzi, J. N., Lissauer, J. J, Esposito, L. W., Holberg, J. B., Marouf, E. A., Tyler, G. L., and Boischot, A. (1984) Saturn's rings: Properties and Processes, In *Planetary Rings* (R. Greenberg, A. Brahic, Eds.), University of Arizona Press, Tuscon, pp. 73–200.

Cuzzi, J. N. and Burns, J. A. (1988) Charged particle depletion surrounding Saturn's F ring – Evidence for a moonlet belt? Icarus 74, 284–324.

Cuzzi, J. N. (1995) Evolution of planetary ring-moon systems. In *Comparative Planetology*, also, Earth, Moon Planet. Kluwer, Dordrecht 67, 179–208.

Cuzzi, J. N. and Estrada, P. R. (1998) Compositional evolution of Saturn's rings due to meteoroid bombardment. Icarus 132, 1–35.

Cuzzi, J. N. and 10 co-authors (2002) Saturn's Rings: Pre-Cassini status and mission goals. Space Sci. Rev. 104, 209–251.

Davis, G. R. and 26 co-authors (1996) ISO LWS measurement of the far-infrared spectrum of Saturn. Astron. Astrophys. 315, L393–L396.

Davis, L., Jr. and Smith, E. J. (1990) A model of Saturn's magneitc field based on all available data. J. Geophys. Rev. 96, 15, 257–15,261.

de Bergh, C., Lutz, B. L., Owen, T., Brault, J., and Chauville, J. (1986) Monodeuterated methane in the outer solar system. II Its detection on Uranus at 1.6 microns. Astrophys. J. 311, 501–510.

de Graauw, T. and 18 co-authors (1997) First results of ISO-SWS observations of Saturn: Detection of CO_2, CH_3H_2H, C_4H_2 and tropospheric H_2O. Astron. Astrophys. 321, L13–L16.

de Pater, I. and Dickel, J. R. (1982) New information on Saturn and its rings from VLA multifrequency data. *Proceedings of Planetary Rings/Anneaux des Planètes Conference*, Toulouse, France, August 1982 (abstract).

de Pater, I. and Dickel, J. R. (1991) Multifrequency radio observations of Saturn at ring inclination angles between 5 and 26 degrees. Icarus 94, 474–492.

Desch M. D. and Rucker, H. O. (1983) The relationship between Saturn kilometric radiation and the solar wind. J. Geophys. Res. 88, 8999–9006.

Dollfus, A. (1963) Mouvements dans l'atmosphère de Saturne en 1960. Observations coordonées par l'Union Astronomique Internationale. Icarus 2, 109–114.

Dones, L. (1998) The rings of the Giant Planets. In *Solar System Ices* (B. Schmitt, C. de Bergh, M. Festou, Eds.), Kluwer, Dordrecht, pp. 711–734.

Doyle, L., Dones, R. L., and Cuzzi, J. N. (1989) Radiative transfer modeling of Saturn's outer B ring. Icarus 80, 104–135

Doyle, L. R. and Grün, E. (1990) Radiative transfer modeling constraints on the size of the spoke particles in Saturn's rings. Icarus 85, 168–190.

Drossart P., J-P. Maillard, J. Caldwell, S. J. Kim, J. K. G. Watson, W. A. Majewski, J. Tennyson, S. Miller, S. K. Atreya, J. T. Clarke, J. H. Waite Jr., and R. Wagener (1989) Detection of H_3^+ on Jupiter, Nature 240, 539–541.

Dulk, G. A. and Eddy, J. A. (1966) A new search for visual aurora on Jupiter. Astron. J. 71, 160.

Dungey, J. W. (1961) The steady state of the Chapman-Ferraro problem in two dimensions. J. Geophys. Res. 66, 1043.

Durisen, R. H., Cramer, N. L., Murphy, B. W., Cuzzi, J. N. Mulikin, T. L., and Cederbloom, S. E. (1989) Ballistic transport in planetary ring systems due to particle erosion mechanisms. I – Theory, numerical methods, and illustrative examples. Icarus 80, 136–166.

Edgington, S. (1997) Latitude variations of the abundances of ammonia, acetylene, and phosphine and vertical mixing in the atmospheres of Jupiter and Saturn. PhD Thesis. Univ. of Michigan. 243 pp..

Emery, J. P., Burr, D. M., Cruikshank, D. P., Brown, R. H., and Dalton, J. B. (2005) Near-infrared (0.8–4.0 μm) spectroscopy of Mimas, Enceladus, Tethys, and Rhea. Astron. Astrophys. 435, 353–362.

Epstein, E. E., Janssen, M. A., and Cuzzi, J. N. (1984) Saturn's rings – 3-mm low-inclination observations and derived properties. Icarus 58, 403–411.

Espinosa, S. A., Southwood, D. J., and Dougherty, M. K. (2003) How can Saturn impose its rotation period in a noncorotating magnetoshere? J. Geophys. Res. 108, NO. A2.

Esposito, L. W., O' Callaghan, M., and West, R. A. (1983) The structure of Saturn's rings – Implications from the Voyager stellar occultation. Icarus 56, 439–452.

Esposito, L. W., Cuzzi, J. N., Holberg, J. E., Marouf, E. A., Tyler, G. L., and Porco, C. C. (1984) Saturn's rings: Structure, dynamics, and particle properties. In *Saturn* (T. Gehrels, M. Matthews, Eds.), University of Arizona Press, Tuscon, pp. 463–545.

Esposito, L. W. (1993) Understanding planetary rings. Ann. Rev. Earth Planet. Sci. 21, 487–523.

Esposito, L. W. and 15 co-authors. (2005) Ultraviolet imaging spectroscopy shows an active Saturnian system. Science 307, 1251–1255.

Estrada, P. R. and Cuzzi, J. N. (1996) Voyager observations of the color of Saturn's rings. Icarus 122, 251–272.

Estrada, P. R., Cuzzi, J. N., and Showalter, M. R. (2003) Voyager color photometry of Saturn's main rings: a correction. Icarus 166, 212–222.

Fegley, B. and Prinn, R. G. (1985) Equilibrium and nonequilibrium chemistry of Saturn's atmosphere – Implications for the observability of PH_3, N_2, CO, and GeH_4. Astrophys. J. 299, 1067–1078.

Feibelman, W. A. (1967) Concerning the "D" ring of Saturn. Nature 214, 793–794.

Festou, M. C. and Atreya, S. K. (1982) Voyager ultraviolet stellar occultation measurements of the composition and thermal profiles of the Saturnian upper atmosphere. Geophys. Res. Lett. 9, 1147–1152.

Feuchtgruber, H., Lellouch, E., de Graauw, T., Bézard, B., Encrenaz, T., and Griffin, M. (1997) External supply of oxygen to the atmospheres of the giant planets. Nature 389, 159–162.

Fillius, W., Ip, W. J., McIlwain, C. E. (1980) Trapped radiation belts of Saturn: First look. Science 207, 425–431.

Fink, U., Larson, H. P., Bjoraker, G. L., and Johnson, J. R. (1983) The NH_3 spectrum in Saturn's 5-μm window. Astrophys. J. 268, 880–888.

Fletcher, L. N., and 13 co-authors (2008) Temperature and composition of Saturn's polar hot spots and hexagon. Science 319, 79–81.

Fortney, J. and Hubbard, W. B. (2003) Phase separation in giant planets: Inhomogeneous evolution of Saturn. Icarus 164, 228–243.

Fouchet, T., Lellouch, E., Bézard, B., Feuchtgruber, H., Drossart, P., and Encrenaz, T. (2000) Jupiter's hydrocarbons observed with ISO-SWS: Vertical profiles of C_2H_6 and C_2H_2, detection of CH_3C_2H. Astron. Astrophys. 355, L13–L17.

Fouchet, T., Guerlet, S., Strobel, D. F., Simon-Miller, A. A. Bézard, B., and Flasar, F. M. (2008) An equatorial oscillation in Saturn's middle atmosphere. Nature 453, 200–202.

Flynn, B. C. and Cuzzi, J. N. (1989) Regular structure in the inner Cassini Division of Saturn's rings. Icarus 82, 180–199.

Frank, L., Burek, B., Ackerson, K., Wolfe, J., and Mihalov, J. (1980) Plasmas in Saturn's Magnetosphere, J. Geophys. Res. 85 (A11), 5695–5708.

French, R. C. and Nicholson P. D. (2000) Saturn's rings II. Particle sizes inferred from stellar occultation data. Icarus 145, 502–523.

Friedson, A. J. (1999) New observations and modeling of a QBO-like oscillation in Jupiter's stratosphere. Icarus 137, 34–55.

Geballe, T. R., Jagod, M. -F., and Oka, T. (1993) Detection of H_3^+ infrared emission lines in Saturn. Astrophys. J. 408, L109–L112.

Gehrels, T. and 22 co-authors (1980) Imaging Photopolarimeter on Pioneer Saturn. Science 207, 434–439.

Gehrels, T. and Matthews, M. S., Eds. (1984) Saturn, University of Arizona Press, Tucson.

Gérard J. -C. and Singh, V. (1982) A model of energy deposition of energetic electrons and EUV emission in the Jovian and Saturnian atmospheres and implications. J. Geophys. Res. 87, 4525–4532.

Gérard, J. -C. et al. (1995) Simultaneous observations of the Saturnian aurora and polar haze with the HST FOC. Geophys. Res. Lett. 22, 2685–2688.

Gérard, J. -C. and 10 co-authors. (2006) Saturn's auroral morphology and activity during quiet magnetospheric conditions. J. Geophys. Res. 111, A12210.

Gezari, D. Y., Mumma, M. J., Espenak, F., Deming, D., Bjoraker, G., Woods, L., and Folz, W. (1991) New features in Saturn's atmosphere reveled by high-resolution thermal infrared images. Nature 342, 777–780.

Giampieri, G. and Dougherty, M. K. (2004a) Rotation rate of Saturn's interior from magnetic field observations. Geophys. Res. Lett. 31, 16701.

Giampieri, G. and Dougherty, M. K. (2004b) Modelling of the ring current in Saturn's magnetosphere Ann. Geophys., 22, 653–659.

Gierasch, P. J. (1983) Dynamical consequences of orthohydrogen-parahydrogen disequilibrium in Jupiter and Saturn. Science 219, 847–849.

Gierasch, P. J. and Goody, R. M. (1969) Radiative time constants in the atmosphere of Jupiter. J. Atmos. Sci. 26, 979–980.

Gillett, F. C and Forrest, W. J. (1974) The 7.5- to 13.5 micron spectrum of Saturn. Astrophys. J. 187, L37–L38.

Gillett, F. C. and Orton, G. S. (1975) Center-to-limb observations of Saturn in the thermal infrared. Astrophys. J. 195, L47–L49.

Godfrey, D. A. (1988) A hexagonal feature around Saturn's North Pole. Icarus 76, 335–356.

Goldstein, R. M. and Morris, G. A. (1973) Radar observations of the rings of Saturn. Icarus 20, 249–283.

Goldstein, R. M., Green, R. R., Pettengill, G. H., and Campbell, D. B. 1977. The rings of Saturn: Two-frequency radar observations. Icarus 30, 104–110.

Greathouse, T. K., Lacy, J. H., Bezard, B., Moses, J. I. Griffith, C. A., and Richter, M. J. (2005) Meridional variations of temperature, C_2H_2 and C_2H_6 abundances in Saturn's stratosphere at southern summer solstice. Icarus 177, 18–31.

Greathouse, T. K., Lacy, J. H., Bezard, B., Moses, J. I., Richter, M. J., and Knez, C. (2006) First detection of propane on Saturn. Icarus 181, 266–271.

Greenberg, R. and Brahic, A., Eds. (1984) Planetary Rings. University of Arizona Press, Tucson.

Grevesse, N., Asplund, M., and Sauval, A. J. (2007) The solar chemical composition. Space Sci. Rev. 130, 104–114.

Griffin, M. J. and 27 co-authors (1996) First detection of the 56-μm rotational line of HD in Saturn's atmosphere. Astron. Astrophys. 315, L389–L392.

Grossman, A. W., Muhleman, D. O., and Berge, G. L. (1989) High-resolution microwave images of Saturn. Science 245, 1211–1215.

Grün, E., Garneau, G. W., Terrile, R. J., Johnson, and T. V. Morfill, G. E. (1984) Kinematics of Saturn's spokes. Adv. Space Res. 4, 143–148.

Guerin, P. (1970) Sur la mise en evidence d'un quartrième anneau et d'une nouvelle division obscure dans le système des anneaux de Saturne. Comtes Res. Serie. B. Sci. Phys. 270, 125–128.

Guillot, T., Gautier, D., Chabrier, G., and Mosser, B (1994) Are the giant planets fully convective? Icarus 12, 337–353.

Guillot, T. (1999a) A comparison of the interiors of Jupiter and Saturn. Planet. Space Sci. 47, 1183–1200.

Guillot, T. (1999b) Interior of giant planets inside and outside the solar system. Science 286, 72–77.

Guillot, T., Stevenson, D. J., Hubbard, W. B., and Saumon, D. (2004) The interior of Jupiter. In *Jupiter, The Planet, Satellites, and Magnetosphere* (F. Bagenal, W. McKinnon, T. Dowling, Eds.), Cambridge Univ. Press, Cambridge.

Guillot, T. (2005) The interiors of giant planets. Ann. Rev. Earth Planet. Sci. 33, 493–530.

Gurnett, D. A., Kurth, W. S., Scarf, F. L. (1981) Narrowband electromagnetic emission from Saturn's magnetosphere. Nature 292, 733–737.

Haff, P. K., Siscoe, G. L., and Eviatar, A. (1983) Ring and plasma – The enigmae of Enceladus. Icarus 56, 425–438.

Hanel, R. A. and 12 co-authors (1979) Infrared observations of the Jovian system from Voyager 1. Science 204, 972–976.

Hanel, R. A. and 12 co-authors (1980) Infrared observations of the Jovian system from Voyager 2. Science 206, 962–956.

Hanel, R. A. and 15 co-authors (1981) Infrared observations of the Saturnian system from Voyager 1. Science 212, 192–200.

Hanel, R. A. and 12 coauthors (1982) Infrared observations of the Saturnian system from Voyager 2. Science 215, 544–548.

Hanel, R. A., Conrath, B. J, Herath, L., Kunde, V. G., and Pirraglia, J. A. (1983) Albedo, internal heat flux and energy balance of Saturn. Icarus 53, 262–285.

Hapke, B. (1981) Bidirectional reflectance spectroscopy. 1. Theory. J. Geophys. Res. 86, 3039–3054.

Harris, A. W. (1984) The origin and evolution of planetary rings. In *Planetary Rings* (R. Greenberg, A. Brahic, Eds.), University of Arizona Press, Tuscon.

Hill, T. W. (1979.) Inertial limit on corotation. J. Geophys. Res. 84, 6554–6558.

Hill, T. W. (2001) The jovian auroral oval. J. Geophys. Res. 106, 8101–8108.

Hill, T. W. (2005) Rotationally-driven dynamics in the magnetospheres of Jupiter and Saturn. In *Magnetospheres of the Outer Planets meeting*, Leicester, August 12, 2005.

Holberg, J. B., Forrester, W. T., and Lissauer, J. J. (1982) Identification of resonance features within the rings of Saturn. Nature 297, 115–120.

Hubbard, W. B. and Stevenson, D. J. 1984. Interior structure of Saturn. In *Saturn* (T. Gehrels, M. Matthews, Eds.), University of Arizona Press, Tucson, pp. 47–87.

Horn, L. J. (1997) Planetary rings. In *Encyclopedia of Planetary Sciences* (J. H. Shirley, R. W. Fairbridge, Eds.), Chapman & Hall, London, pp. 602–607.

Hueso, R. and Sánchez-Lavega, A. (2004) A three-dimensional model of moist convection for the giant planets II. Saturn's water and ammonia moist convective storms. Icarus 172, 255–271.

Ingersoll, A. P., Orton, G. S., Münch, G., Neugebauer, G., and Chase, S. C. (1980) Science 207, 439–443.

Ingersoll, A. P., Beebe, R., F., Conrath, B. J., and Hunt, G. E. (1984) Structure and dynamics of Saturn's atmosphere. In *Saturn* (T. Gehrels, M. S. Matthews, Eds.), pp 195–238.

Ingersoll A. P., Vasavada, A. R., Little, B., Anger, C. D., Bolton, S. J. Alexander, C., Klaasen, K. P., and Tobiksa, W. K. (1998) Imaging Jupiter's aurora at visible wavelengths. Icarus 135, 251–264.

Ip, W.-H. (1997) On the neutral cloud distribution in the Saturnian magnetosphere. Icarus 126, 42–57.

Isbell J., Dessler, A. J., and Waite, J. H. (1984) Magnetospheric energization by interaction between planetary spin and the solar wind. J. Geophys. Res. 89, 10716–10722.

Judge, D. L., Wu, F.-M., and Carlson, R. W. (1980) Ultraviolet spectrometer observations of the Saturnian system. Science 207, 431–434.

Kaiser, M. L., Desch, M. D., and Lecacheux, A. (1981) Saturn kilometric radiation: statistical properties and beam geometry. Nature 292, 731–733.

Kaiser, M. L., Desch, M. D., Burth, W. S., Lecacheux, A., Genova, F., Pedersen, B. M., and Evans D. R. (1984) Saturn as a radio source. In *Saturn* (T. Gehrels, M. Matthews, Eds.), University of Arizona Press, Tucson, pp. 378–415.

Kalogerakis, K. S., Marschall, J., Oza, A. U., Engel, P. A., Meharchand, R. T., and Wong, M. H. (2008) The coating hypothesis for ammonia ice particles in Jupiter: Laboratory experiments and optical modeling. Icarus 196, 202–215.

Karkoschka, E. and Tomasko, M. G. (1992) Saturn's upper troposphere 1986–1989. Icarus 97, 161–181.

Karkoschka, E. and Tomasko, M. G. (1993) Saturn's upper atmospheric hazesobserved by the Hubble Space Telescope. Icarus 106, 421–441.

Karkoschka, E. (1994) Spectrophotometry of the Jovian planets and Titan at 300–1000 nm wavelength; The methane spectrum. Icarus 111, 174–192.

Karkoschka, E. and Tomasko, M. (2005) Saturn's vertical and latitudinal cloud structure 1991–2004 from HST imaging in 30 filters. Icarus 179, 195–221.

Kim, J. H., Kim, S. J., Geballe, T. R., Kim, S. S., and Brown, L. R. (2006) High-resolution spectroscopy of Saturn at 3 microns: CH_4, CH_3D, C_2H_2, C_2H_6, PH_3, clouds and haze. Icarus 185, 476–486.

Kirsch, E., Krimigis, S. M., Ip, W. -H., and Gloeckler, G. (1981) X-ray and energetic neutral particle emission from Saturn's magnetosphere. Nature 292, 718–721.

Kliore, A. J., Lindal, G. F., Patel, I. R., Sweetnam, D. N., Hotz, H. B., and McDonough. T. R. (1980a) Vertical structure of the ionosphere and the upper neutral atmosphere of Saturn from the Pioneer radio occultation. Science 207, 446–449.

Kliore, A. J, Patel, I. R., Lindal, G. F., Sweetnam, D. N, Hotz, H. B., Waite, J. H., and McDonough, T. R. (1980b) Structure of the ionosphere and atmosphere of Saturn from Pioneer 11 Saturn radio occultation. J. Geophys. Res. 85, 5857–5870.

Kolvoord, R. A., Burns, J. A., and Showalter, M. R. (1990) Periodic features in Saturn's F ring – Evidence for nearby moonlets. Nature 345, 695–697.

Krimigis, S. M. and Armstrong, T. P., (1982) Two-component proton spectra in the inner Saturnian magnetosphere, Geophys. Res. Lett., 9, 1143–1146.

Krimigis, S. M., Carbary, J. F., Keath, E. P., Armstrong, T. P., Lanzerotti, L. J., and Gloeckler, G. (1983) General characteristics of hot plasma and energetic particles in the Saturnian magnetosphere: Results from the Voyager spacecraft, J. Geophys. Res., 88, 8871–8892.

Krimigis, S. M., Mitchell, D. G., Hamilton, D. C., Livi, S., Dandouras, J., Jaskulek, S., Armstrong, T. P., Cheng, A. F., Gloeckler, G., Hsieh, K. C., Ip, W. -H., Keath, E. P., Kirsch, E., Krupp, N., Lanzerotti, L. J., Mauk, R. B. H., McEntire, W., Roelof, E. C., Tossman, B. E., Wilken, B., and Williams, D. J. (2004) Magnetosphere Imaging Instrument (MIMI) on the Cassini Mission to Saturn/Titan, Space Sci. Rev., 114, 233–329.

Kuiper, G. P., Cruikshank, D. P., and Fink, U. (1970) Sky & Telescope 39, 80.

Lam, H. A., Achilleos, N., Miller, S., Tennyson, J., Trafton, L. M., Geballe, T. R., and Ballester, G. (1997) A baseline spectroscopic study of the infrared aurorae of Jupiter. Icarus, 127, 379–393.

Lane, A. L., Hord, C. W., West, R. A., Esposito, L. W., Coffeen, D. L., Sato, M., Simmons, K. E., Pomphrey, R. B., and Morris, R. B. (1982) Photopolarimetry from Voyager 2 – Preliminary results on Saturn, Titan, and the rings. Science 215, 537–543.

Larson, H. P., Fink, U., Treffers, R., and Gautier, T. N. (1975) Detection of water vapor on Jupiter. Astrophys. J. 197, L137–L140.

Larson, H. P., Fink, U., Smith, H. A., and Davis, D. S. (1980) The middle-infrared spectrum of Saturn – Evidence for phosphine and upper limits to other trace atmospheric constituents. Astrophys. J. 240, 327–337.

Lebovsky, L. A., Johnson, T.V., and McCord, T. B. (1970) Saturn's rings: Spectral reflectivity and compositional implications. Icarus 13, 226–230.

Lecluse, C., Robert, F., Gautier, D., and Guiraud, M. (1996) Deuterium enrichment in giant planets. Planet. Space Sci. 44, 1579–1592.

Lellouch, E., Bézard, B., Fouchet, T., Feuchtgruber, H., Encrenaz, T., and de Graauw, T. (2001) The deuterium abundance in Jupiter and Saturn from ISO-SWS observations. Astron. & Astrophys. 370, 610–622.

Leovy, C. B., Friedson, A. J., and Orton, G. S. (1991) The quasiquadrennial oscillation of Jupiter's equatorial stratosphere. Nature 354, 380–382.

Lewis, J. S. and Fegley, M. B. (1984) Vertical distribution of disequilibrium species in Jupiter's troposphere. Space Sci. Rev. 39, 163–192.

Macy, W. (1977) Inhomogeneous models of the atmosphere of Saturn. Icarus 32, 328–347.

Magalhães, J. A., Weir, A. L., Conrath B. J., Gierasch, P. J., and Leroy, S. S. (1990) Zonal motion and structure in Jupiter's upper troposphere from Voyager infrared and imaging observations. Icarus 88, 39–72.

Marouf, E. A., Tyler, G. L., Zebker, H. A., Simpson, R. A., and Eshleman, V. R. (1983) Particle size distributions in Saturn's rings from Voyager 1 radio occultation. Icarus 54, 189–211.

Marouf, E. A., Tyler, G. L., and Rosen, P. A. (1986) Profiling Saturn's rings by radio occultation. Icarus 68, 120–166.

Marouf, E. A. and Tyler, G. L. (1986) Detection of two satellites in the Cassini division of Saturn's rings. Nature 323, 31–35.

Massie, S. T. and Hunten, D. M. (1982) Conversion of para and ortho hydrogen in the giant planets. Icarus 49, 213–226.

Mauk, B. H., Krimigis, S. M., and Lepping, R. P. (1985) Particle and field stress balance within a planetary magnetosphere. J. Geophys. Res. 90, 8253–8264.

McClain, E. F. (1959) A test for non-thermal radiation from Jupiter at a wave length of 21cm. Astron. J. 64, 339–340.

McGrath, M. A. and Clarke, J. T. (1992) H I Lyman alpha emission from Saturn (1980–1990) J. Geophys. Res. 97, 13691–13703.

McNutt, R. L., Jr. (1984) Force balance in the outer planet magnetospheres. In *Proceedings of the 1982–4 Symposia on the Physics of Space Plasmas, SPI Conf. Proc. Reprint Ser.*, vol. 5 (H. S. Bridge, J. W. Belcher, T. S. Chang, B. Coppi, J. R. Jasperse, Eds.), Scientific Publishers, Cambridge, MA.

Melin, H., Miller, S. Stallard, T. Trafton L. M., and Geballe T. R. (2007) Variability in the H_3^+ emission from Saturn: Consequences for ionization rates and temperature. Icarus 186, 234–241.

Miller, S., Lam, H. A., and Tennyson, J. (1994) What astronomy has learned from observations of H_3^+. Can. J. Phys. 72, 760–771.

Miller S. and 10 co-authors (2000) The Role of H_3^+ in planetary atmospheres. Phil. Trans. Roy. Soc. 358, 2485–2502.

Miller, S., Aylward, A., and Millward, G. (2005) Giant planet ionospheres and thermospheres: The importance of ion-neutral coupling. Space Sci. Rev. 116, 319–343.

Miller, S., Stallard, T., Smith, C., Milward, G., Melin, H., Lystrup, M., and Aylward, A. (2006) H_3^+: the driver of planetary atmospheres. Phil. Trans. Roy. Soc. A. 364, 3121–3137.

Moos, H. W. and Clarke, J. T. (1979) Detection of acetylene in the Saturnian atmosphere, using the IUE satellite. Astrophys. J. 229, L107–L108.

Morrison, D., Jones, T., Cruikshank, and Murphy, R. (1975) The two faces of Iapetus. Icarus 24, 157–171.

Morrison, D., Johnson, T. V., Shoemaker, E. M., Soderblom, L. A., Thomas, P., Veverka, J., and Smith, B. A. (1984) Satellites of Saturn: Geological perspective. In *Saturn* (T. Gehrels, M. S. Matthews, Eds.), University of Arizona Press, Tucson, 609–639.

Morrison, D., Owen, T., and Soderblom, L. A. (1986) The satellites of Saturn. In *Satellites* (J. A. Burns, M. S. Matthews, Eds.), University of Arizona Press, Tucson, 764–801.

Moses, J. I., Bézard, B. Lellouch, E., Gladstone, G. R., Feuchtgruber, H., Allen, and M. (2000a) Photochemistry of Saturn's atmosphere: I. Hydrocarbon chemistry and comparisons with ISO observations. Icarus 143, 244–298.

Moses, J. I., Lellouch, E., Bézard, B., Gladstone, G. R., Feuchtgruber, H., and Allen, M. (2000b) Photochemistry of Saturn's atmosphere. II. Effects of an influx of external oxygen. Icarus 145, 166–202.

Moses, J. I. and Greathouse, T. K. (2005) Latitudinal and seasonal models of stratospheric photochemistry on Saturn: Comparison with infrared data from IRTF/TEXES. J. Geophys. Res. 110, E09007.

Müller-Wodarg, I. C. F., Mendillo, M., Yelle, R. V., Aylward, A. D. (2006) A global circulation model of Saturn's thermosphere. Icarus 180, 147–160.

Muñoz, O., Moreno, F., Molina, A., Grodent, D., Gerard, J. C., and Dols, V. (2004) Study of the vertical structure of Saturn's atmosphere using HST/WFPC2 images. Icarus 169, 413–428.

Ness, N. F., Acuna, M. H., Behannon, K. W., Burlaga, L. F., Connerney, J. E. P., Lepping, R. P., Neubauer, F. M. (1982) Magnetic field studies by Voyager 2 – Preliminary results at Saturn. Science, 215, 558–563.

Nicholson, P. D. and Dones, L. (1991) Planetary rings. Rev. Geophys. Supp. 29, 313–327.

Niemann, H. B. and 12 co-authors (1996) The Galileo Probe Mass Spectrometer: Composition of Jupiter's atmosphere. Science, 272, 846–849.

Noll, K. S., Knacke, R. F., Geballe, T. R., and Tokunaga, A. T. (1986) Detection of carbon monoxide in Saturn. Astrophys. J. 309, L91–L94.

Noll, K. S., Knacke, R. F., Geballe, T. R., and Tokunaga. A. T. (1988) Evidence for germane in Saturn. Icarus 85, 409–422.

Noll, K. S., Geballe, R. T., and Knacke, R. F. (1989) Arsine in Saturn and Jupiter. Astrophys. J. 338, L71–L74.

Noll, K. S., Larson, H. P., and Geballe, T. R. (1990) The abundance of AsH₃ in Jupiter. Icarus 83, 494–499.

Noll, K. S. and Larson, H. P. (1991) The spectrum of Saturn from 1990 to 2230/cm – Abundances of AsH₃, CH₃D, CO, GeH₄, NH₃ and PH₃. Icarus 89, 168–189.

Noll, K. S., Roush, T. L., Cruikshank, D. P., Johnson, R. E., and Pendleton, Y. J. (1997) Detection of ozone on Saturn's satellites Rhea and Dione. Nature 388, 45–47.

Noland, M., Veverka, J., Morrison, D., Cruikshank, D. P., Lazarewicz, A., Morrison, N., Elliot, Goguen, J., and Burns, J. (1974) Six-color photometry of Iapetus, Titan, Rhea, Dione, and Tethys. Icarus 23, 334–354.

Null, G. W., Lao, E. L., Biller, E. D., and Anderson, J. D. (1981) Saturn gravity results obtained fro Pioneer 11 tracking data and earth-based Saturn satellite data. Astron. J. 84, 456–468.

Ohring, G. (1975) The temperature profile in the upper atmosphere of Saturn from inversion of thermal emission observations. Astrophys. J. 195, 223–225.

Oldham, P. G., Griffin, M. J., Davis, G. R., Encrenaz, T., de Graauw, T., Irwin, P. J. Swinyard, B. M., Naylor, D. A., and Burgdorf, M. (1997) ISO LWS Far-infrared observations of Jupiter and Saturn. In *The Far Infrared and Submillmetre Universe* (A. Wilson, Ed.), ESA, Noordwijk, The Netherlands, p. 325.

Ollivier, J. L., Billebaud, F., Drossart, P., Dobrijévic, M., Roos-Serote, M., August-Bernex, T., and Vauglin, I. (2000) Seasonal effects in the thermal structure of Saturn's stratosphere from infrared imaging at 10 microns. Astron. Astrophys. 356, 347–356.

Ortiz, J. L., Moreno, F., and Molina, A. (1996) Saturn 1991–1993: Clouds and hazes. Icarus 119, 53–66.

Orton, G. S. and Ingersoll, A. P. (1980) Saturn's atmospheric temperature structure and heat budget. J. Geophys. Res. 85, 5871–5881.

Orton, G. S. (1983) Thermal infrared constraints on ammonia ice particles as candidates for clouds in the atmosphere of Saturn. Icarus 53, 293–300.

Orton, G. S. and 14 co-authors. (1991) Thermal maps of Jupiter: Spatial organization and time dependence of stratospheric temperatures, 1980 to 1990. Science 252, 537–542.

Orton, G. S. and 19 co-authors. (1994) Thermal maps of Jupiter: Spatial organization and time dependence of tropospheric temperatures, 1980–1993. Science 265, 625–631.

Orton, G. S., Serabyn, E., and Lee, Y. T. (2000) Vertical distribution of PH₃ in Saturn from observations of Its 1–0 and 3–2 rotational lines. Icarus 146, 48–59.

Orton, G. S., Serabyn, E., and Lee, Y. T. (2001) Erratum: Vertical distribution of PH₃ in Saturn from observations of its 1–0 and 3–2 rotational lines. Icarus 149, 489–490.

Orton, G. S. and Yanamandra-Fisher, P. A. (2005) Saturn's temperature field from high-resolution middle-infrared imaging. Science, 696–701.

Orton, G. S. and 27 co-authors. (2008) Semi-annual oscillations in Saturn's low-latitude stratospheric temperatures. Nature 453, 196–199.

Ostro, S. J., Pettengill, G. H., and Campbell, D. P. (1980) Radar observations of Saturn's rings at intermediate tilt angles. Icarus 41, 381–388.

Ostro, S. J., Pettengill, G. H., Campbell, D. P., and Goldstein, R. M. (1982) Delay-doppler radar observations of Saturn's rings. Icarus 49, 367–381.

Owen, T. C. (1969) The spectra of Jupiter and Saturn in the photographic infrared. Icarus 10, 355.

Owen, T., Lutz, B. L., and de Bergh, C. (1986) Deuterium in the outer solar system – Evidence for two distinct reservoirs. Nature 320, 244–246.

Owen, T. C., Cruikshank, D. P., Dalle Ore, C. M., Geballe, T. R., Roush, de Bergh, C., Pendleton, Y. J., and Khare, B. N. (2001) Decoding the domino: The dark side of Iapetus. Icarus 149, 160–172.

Pang, K., Voge, C. C., Rhoads, J. W., and Ajello, J. M. (1984) The E ring of Saturn and satellite Enceladus. J. Geophys. Res. 89, 9459–9470.

Pérez-Hoyos, S., Sánchez-Lavega, A., French, R. G., and Rolas, J. F. (2005) Saturn's cloud structure and temporal evolution from ten years of Hubble Space Telescope images (1994–2003) Icarus 176, 155–174.

Pérez-Hoyos, S. and Sánchez-Lavega, A. (2006) On the vertical wind shearof Saturn's equatorial jet at cloud level. Icarus 180, 161–175.

Plescia, J. B. and Boyce, J. M. (1982) Crater densities and geological histories of Rhea, Dione, Mimas and Tethys. Nature 295, 285–290.

Plescia, J. B., and Boyce, J. M. (1983) Crater numbers and geological histories of Iapetus, Enceladus, Tethys and Hyperion. Nature 301, 666–670.

Podolak, M. and Danielson, R. E. (1977) Axel dust on Saturn and Titan. Icarus 30, 479–492.

Pollack, J. B., Summers, A. L., and Baldwin, B. J. (1973) Estimates of the size of the particles in Saturn's rings and their cosmogonic implications. Icarus 20, 263–278.

Pollack, J. B. (1975) The rings of Saturn. Space Sci. Rev. 18, 3–93.

Porco, C. C. (1990) Narrow rings: Observation and theory. Adv. Space Res. 10, 221–229.

Porco, C. C. (1995) Highlights in planetary rings. Revs. Geophys. Sp. Phys. Supp., 497–504.

Poulet, F., Cuzzi, J. N., Cruikshank, D. P., Roush, T., and Dalle Ore, C. (2002) Comparison between the Shkuratov and Hapke scattering theories for solid planetary surfaces: Application to the surface composition of two centaurs. Icarus 160, 315–324.

Poulet, F., Cuzzi, J. N., Cruikshank, D. P., Roush, T. L., and French, R. C. (2003) Compositions of Saturn's rings A, B, and C from high resolution near-infrared spectroscopic observations. Astron. Astrophys. 412, 305–316.

Prangé, R., Fouchet, T., Courtin, R., Connerney, J. E. P., and McConnell, J. C. (2006) Latitudinal variation of Saturn photochemistry deduced from spatially resolved ultraviolet spectra. Icarus 180, 379–392.

Prinn, R. G. and Barshay, S. S. (1977) Carbon monoxide on Jupiter and implications for atmospheric convection. Science 198, 1031–1034.

Prinn, R. G., Larson, H. P., Caldwell, J. J., and Gautier, D. (1984) Composition and chemistry of Saturn's atmosphere. In *Saturn* (T. Gehrels, M. Matthews, Eds.), Tucson, pp. 88–149.

Reuter, D. C., Simon-Miller, A. A., Lunsford, A., Baines, K. H., Cheng, K. H., Jennings, D. E., Olkin, C. B., Spencer, J. R., Stern, S. A., Weaver, H. A., and Young, L. A. (2007) Jupiter cloud composition, stratification, convection, and wave motion: A view from New Horizons. Science 318, 223–225.

Richardson, J. D. and Sittler, E. C. (1990) A plasma density model for Saturn based on Voyager observations. J. Geophys. Res. 95, 12019–12031.

Rieke, G. H. (1975) The thermal radiation of Saturn and its rings. Icarus 26, 37–44.

Roellig, T. W., Werner, M. W. and Becklin, E. J. (1988) Thermal emission from Saturn's rings at 380 microns. Icarus 73, 574–583.

Rosenqvist, J., Lellouch, E., Romani, P. N, Paubert, G., and Encrenaz, T. (1992) Millimeter-wave observations of Saturn, Uranus, and Neptune–CO and HCN on Neptune. Astrophys. J. 392, L99–L102.

Sada, P. V., McCabe., G. H., Bjoraker, G. L., Jennings, D. E., Reuter, D. C. (1996) Astrophys. J. 472, 903.

Sada, P. V., Bjoraker, G. L., Jennings, D. E., Romani, P. N., and McCabe, G. H. (2005) Observations of C_2H_6 and C_2H_2 in the stratosphere of Saturn. Icarus 173, 499–507.

Sánchez-Lavega. A. (1982) Motions in Saturn's atmosphere – Observations before Voyager encounters. Icarus 49, 1–16.

Sánchez-Lavega, A., Colas, F., Lecacheux, J., Laques, P., Parker, D., and Miyazaki, I. (1991) The Great White Spot and disturbances in Saturn's equatorial atmosphere during 1990. Nature 33, 397–401.

Sánchez-Lavega, A. and Battaner, E. (1986) Long-term changes in Saturn's atmospheric belts and zones. Astron. Astrophys. Supp. Ser. 64, 287–301.

Sánchez-Lavega, A. (1989) Saturn's Great White Spots. Sky Telescope 78, 141–142.

Sánchez-Lavega, A., Colas, F., Lecacheux, J., Laques, P., Miyazaki, I., and Parker, D. (1991) The Great White Spot and disturbances in Saturn's equatorial atmosphere during 1990. Nature 353, 397–401.

Sánchez-Lavega, A., Lecacheux, J. Colas, F and Laques P. (1993a) Temporal behavior of cloud morphologies and motions in Saturn's atmosphere. J. Geophys. Res. 98, 18857–18872.

Sánchez-Lavega, A., Lecacheux, J., Colas, F., and Laques, P. (1993b) Ground-based observations of Saturn's north polar spot and hexagon. Science 260, 329–332.

Sánchez-Lavega, A., Lecacheux, J., Gomez. J. M., Colas, F., Laques, P., Noll, K., Gilmore, D., Miyazaki, I., and Parker, D. (1996) Larger-scale storms in Saturn's atmosphere during 1994. Science 271, 631–634.

Sánchez-Lavega, A., Lecacheux, J., Colas, F., Rojas, J. F., and Gomez, J. M. (1999) Discrete cloud activity in Saturn's equator during 1995, 1996 and 1997. Planet. Space Sci. 47, 10–11.

Sánchez-Lavega, A., Rojas, J. F., and Sada, P. V. (2000) Saturn's zonal winds at cloud level. Icarus 147, 405–420.

Sánchez-Lavega, A., Pérez-Hoyos, S., Acarreta, J. R., and French, R. G. (2002) No hexagonal waves around Saturn's southern pole. Icarus 160, 216–219.

Sánchez-Lavega, A., Pérez-Hoyos, S., Rojas, J. F., Hueso, R., and French, R. G. (2003) A strong decrease in Saturn's equatorial jet at cloud level. Nature 423, 623–625.

Sánchez-Lavega, A., Hueso, R., Pérez-Hoyos, S., Rojas, J. F., and French, R. G. (2004) Saturn's cloud morphology and zonal winds before the Cassini encounter. Icarus 170, 519–523.

Sánchez-Lavega, A. (2006) Solar flux in Saturn's atmosphere: Penetration and heating rates in the aerosol and cloud layers. Icarus 180, 368–378.

Sandel B. R. and Broadfoot, A. L. (1981) Morphology of Saturn's aurora. Nature 292, 679–682.

Sandel B. R. et al. (1982) Extreme ultraviolet observations from the Voyager 2 encounter with Saturn. Science 215, 248–253.

Saumon, D., and Guilot, T. (2004) Shock compression of deuterium and the interiors of Jupiter and Saturn. Astrophys. J. 609, 1170–1180.

Scarf, F. L., Frank, L. A., Gurnett, D. A., Lanzerotti, L. J., Lazarus, A., and Sittler, E. C. (1984) Measurements of plasma, plasma waves, and suprathermal charged particles in Saturn's inner magnetosphere. In *Saturn* (T. Gehrels, M. Matthews, Eds.), University of Arizona Press, Tuscon, pp. 318–353.

Schloerb, F. P., Muhleman D. O., and Berge G. L. (1980) Interferometry of Saturn and its rings at 1.30cm wavelength. Icarus 42, 125–135.

Showalter, M. R., Cuzzi, J. N. Marouf, E. A., and Esposito, L. W. (1986) Satellite 'wakes' and the orbit of the Encke Gap moonlet. Icarus 66, 297–323.

Showalter, M. R. and Nicholson, P. D. (1990) Saturn's rings through a microscope: Particle size constraints from the Voyager PPS scan. Icarus 87, 285–306.

Showalter, M. R. (1991) Visual detection of 1981S13, Saturn's eighteenth satellite, and its role in the Encke gap. Nature 351, 709–713.

Showalter, M. R., Cuzzi, J. N., and Larson, S. M. (1991) Structure and particle properties of Saturn's E ring. Icarus 94, 451–473.

Shkuratov, Y., Starukhina, L., Hoffmann, H., and Arnold, G. (1999) A model of spectral albedo of particulate surfaces: Implications for optical properties of the Moon. Icarus 137, 235–246.

Simpson, J. A., Bastian, T. S., Chenette, D. L., Lentz, G. A., McKibben, R. B., Pyle, K. R., and Tuzzolino, A. J. (1980) Saturnian trapped radiation and its absorption by satellites and rings: The first results from Pioneer 11. Science 207: 411–415.

Sittler, E. C., Blanc, M. F., and Richardson, J. D. (2006) Proposed model for Saturn's auroral response to the solar wind: Centrifugal instability model. J. Geophys. Res. 111, A06208.

Slipher, V. M. (1905) A photographic study of the spectrum of Saturn. Astrophys. J. 26, 59–62.

Smith, E. J., Davis, L., Jones, D. E., Coleman, P. J., Colburn, D. S., Dyal, P., and Sonett, C. P. (1980a) Saturn's magnetosphere and its interaction with the solar wind. J. Geophys. Res. 85, 5655.

Smith, E. J., Davis., L., Jones, D. E., Coleman, P. J., Colburn, D. S., Dyal, P., and Sonett, C. P. (1980b) Saturn's magnetic field and magnetosphere. Science 207, 407–410.

Smith, B. A., and 25 co-authors. (1981) Encounter with Saturn: Voyager 1 imaging results. Science 212, 163–191.

Smith, B. A. and 28 coauthors (1982) A new look at the Saturn system: The Voyager 2 images. Science 215, 505–537.

Smith, H. J., Rodman, J. P., and Sloan, W. A. (1963) On jovian H-α auroral activity. Astron. J. 68, 79.

Smith G. R., Shemansky, D. E., Holberg, J. B., Broadfoot, A. L., Sandel, B. R., and McConnell, J. C. (1983) Saturn's upper atmosphere from the Voyager 2 EUV solar and stellar occultations. J. Geophys. Res. 88, 8667–8678.

Smith, M. D., Conrath, B. J., and Gautier, D. (1996) Dynamical influence on the isotopic enrichment of CH3D in the outer planets. Icarus 124, 598–607.

Smith, C. G. A., Aylward, A. D., Miller, S., Müller-Wodarg, I. C. F. (2005) Polar heating in Saturn's thermosphere. Ann. Geophys. 23, 2465–2477.

Spilker, L. J., Pilorz, S. H., Edgington, S. G., Wallis, B. D., Brooks, S. M., Pearl, J. C., and Flasar, F. M. (2005) Cassini CIRS Observations of a roll-off in Saturn ring spectra at submillimeter wavelengths. Earth, Moon, and Planets 96, 149–163.

Spinrad, H, Münch, G., and Trafton, L. M. (1962) Recent spectroscopic investigations of Jupiter and Saturn. Astron. J. 67, 587.

Sromovsky, L. A., Revercomb, H. E., Krauss, R. J., and Suomi, V. E. (1983) Voyager 2 observations of Saturn's northern mid latitude cloud features: Morphology, motions and evolution. J. Geophys. Res. 88, 8650–8666.

Stallard T., Miller, S., Ballester, G. E., Rego, D., Joseph, R. D., and Trafton, L. M. (1999) The H_3^+ latitudinal profile of Saturn. Astrophys. J. Lett. 521, L149–L152.

Stallard T., Miller, S., Millward, G., and Joseph, R.D. (2001) On the dynamics of the jovian ionosphere and thermosphere I: the measurement of ion winds. Icarus 154, 475–491.

Stallard, T., Miller, S., Trafton, L. M., Geballe, T. R., and Joseph. R. D. (2004) Ion winds in Saturn's southern auroral/polar region. Icarus 167, 204–211.

Stallard T., Miller, S., Melin, H., Lystrup, M., Dougherty, M., and Achilleos, N. (2007a) Saturn's auroral/polar H_3^+ infrared emission I: General morphology and ion velocity structure. Icarus 189, 1–13.

Stallard T., Miller, S., Melin, H., Lystrup, M., Dougherty, M., and Achilleos, N. (2007b) Saturn's auroral/polar H_3^+ infrared emission II: An indicator of magnetospheric origin. Icarus 191, 678–680.

Stallard T., Miller, S., Lystrup, M., Achilleos, N., Arridge, C., and Dougherty, M. (2008a) Dusk brightening event in Saturn's H_3^+ aurora. Astrophys. J. Lett. 673, L203–L205.

Stallard T., Lystrup, M., Miller, S. (2008b) Emission line imaging of Saturn's H_3^+ aurora. Astrophys. J. Lett. 675, L117–L120.

Stallard T., Miller, S., Melin, H., Lystrup, M., Cowley, S., Bunce, E., Achilleos, N., and Dougherty, M. (2008c) Jovian-like aurorae on Saturn. Nature 453, 1083–1085.

Stallard, T. and 16 co-authors, Complex structure within Saturn's infrared aurora (2008d) Nature 456, 214–217.

Stevenson, D. J. (1980) Saturn's luminosity and magnetism. Science 208, 746–748.

Stevenson, D. J. (1982) Interiors of the giant planets. Ann. Rev. Earth and Planetary Sci. 10, 257–295.

Stewart, G. R. (1991) Nonlinear satellite wakes in planetary rings. I – Phase-space kinematics. Icarus 94, 436–450.

Strobel, D. F. (1969) Photochemistry of methane in the Jovian atmosphere. J. Atmos. Sci. 26, 906–911.

Strobel, D. F. and Smith, G. R. (1973) On the temperature of the Jovian thermosphere. J. Atmos. Sci. 30, 718–725.

Tagger, M., Henriksen, R. N., and Pellat, R. (1991) On the nature of the spokes in Saturn's rings. Icarus 91, 297–314.

Tejfel, V. G. (1977a) Latitudinal differences in the structure of Saturn's cloud cover. Solar System Res. 11, 10–18.

Tejfel, V. G. (1977b) On the determination of the two-layer planetary atmosphere model parameters from the absorption bands observations. Astron. Zh. 54, 178–189. (Russian).

Tokunaga, A., Knacke, R. F., and Owen, T. (1975) The detection of ethane on Saturn. Astrophys. J. 197, L77–L78.

Tokunaga, A. and Cess. R. (1977) A model for the temperature inversion within the atmosphere of Saturn. Icarus 32, 231–327.

Tokunaga, A. T., Caldwell, J., Gillett, F. C., and Nolt, I. G. (1978) Spatially resolved infrared observations of Saturn. II – The temperature enhancement at the south pole. Icarus 36, 216–222.

Tokunaga, A. T., Caldwell, J., Gillett, F. C., and Nolt, I. G. (1979) Spatially resolved infrared observations of Saturn. II – 10 and 20 micron disk scans at B' at −11.8°. Icarus 39, 46–53.

Tomasko, M., McMillan, R. S., Doose, L. R., Castillo, N. D., and Dilley J. P. (1980) Photometry of Saturn at large phase angles. J. Geophys. Res. 85, 8167–8186.

Tomasko, M. G. and Stahl, H. P. (1982) Measurements of the single-scattering properties of water and ammonia ice crystals. *Proceedings "Fourth Annual Meeting of Planetary Atmospheres Principal Investigators"*: Ann Arbor, MI, April 1982 (abstract).

Tomasko, M. G. and Doose, L. R. (1984) Polarimetry and photometry of Saturn from Pioneer 11: Observations and constraints on the distribution and properties of cloud aerosol particles. Icarus 58, 1–34.

Tomasko, M. G., West, R. A., Orton, G. S., and Tejfel, V. G. (1984) Clouds and aerosols in Saturn's atmosphere. In *Saturn* (T. Gehrels, M. Matthews, Eds.), University of Arizona Press, Tucson, pp. 150–194.

Trafton, L. M. (1967) Model atmospheres of the major planets. Astrophys. J. 147, 765–781.

Trafton L. M., Geballe, T. R., Miller, S., Tennyson, J., and Ballester, G. E. (1993) Detection of H_3^+ from Uranus. Astrophys. J., 405, 761–766.

Trauger J. T., Griffiths, R. E., Hester, J. J., Hoessel, J. G., Holtzman, J. A., Krist, J. E., Mould, J. R., Sahai, R., Scowen, P. A., Stapelfeldt, K. R., and Watson, A. M. (1998) Saturn's hydrogen aurora: Widefield camera 2 imaging from the Hubble Space Telescope. J. Geophys. Res. 103, 20237–20244.

Tyler, G. L., Eshleman, V. R., Anderson, J. D., Levy, G. S, Lindal, G. F., Wood, G. E., and Croft, T. A. (1981) Radio science investigations of the Saturn system with Voyager 1: Preliminary results. Science 212, 201–206.

Tyler, G. L., Eshleman, V. R., Anderson, J. D., Levy, G. S., Lindal, G. F., Wood, G. E., and Croft. T. A. (1982) Radio science with Voyager 2 at Saturn: Atmosphere and ionosphere and the masses of Mimas, Tethys and Iapetus. Science, 215, 553–558.

Tyler, G. L., Marouf, E. A., Simpson, R. A., Zebker, H. A., and Eshleman, V. R. (1983) The microwave opacity of Saturn's rings at wavelengths of 3.6 and 13 cm from Voyager 1 radio occultation. Icarus 54, 160–188.

Van Allen, J. A., Thomsen, M. F., Randall, B. A., Rairden, R. L., and Grosskreutz, C. L. (1980a) Saturn's magnetosphere, rings, and inner satellites. Science 207, 415–421.

Van Allen, J. A., Thomsen, M. F., and Randall, B. A. (1980b) The energetic charged particle absorption signature of Mimas, J. Geophys. Res., 85, 5709–5718.

Van Allen, J. A. (1984) Energetic particles in the inner magnetosphere of Saturn. In *Saturn* (T. Gehrels, M. Matthews, Eds.), University of Arizona Press, Tuscon, pp. 281–318.

Van der Tak, F., de Pater, I., Silva, A., and Millan, R. (1998) Time variability in the radio brightness distribution of Saturn; Icarus 142, 125–147.

Van der Tak, F., de Pater, I., Silva, A., and Millan, R. (1999) Time variability in the radio brightness distribution of Saturn. Icarus 142, 125–147.

Vasyliunas, V. M. (1983) Plasma distribution and flow. In *Physics of the Jovian Magnetosphere* (A. J. Dessler, Ed.), Cambridge University Press, Cambridge, pp. 395–453.

Verbiscer, A. J. and Veverka, J. (1992) Mimas-photometric roughness and albedo map. Icarus 99, 63–69.

Visscher, C. and Fegley, B. (2005) Chemical constraints on the water and total oxygen abundances in the deep atmosphere of Saturn. Astrophys. J. 623, 1221–1227.

Visscher, C., Lodders, K., and Fegley, B. (2006) Atmospheric chemistry in giant planets, brown dwarfs, and low-mass dwarf stars. II. Sulfur and phosphorus. Astrophys. J. 648, 1181–1195.

Vogt, R. E., Chenette, D. L., Cummings, A. C., Garrard, T. L., Stone, E. C., Schardt, A. W., Trainor, J. H., Lal, N., and McDonald, F.B. (1982) Energetic charged particles in Saturn's magnetosphere–Voyager 2 results. Science 215, 577–582.

von Zahn, U., Hunten, D. M., and Lehmacher, G. (1998) Helium in Jupiter's atmosphere: Results from the Galileo probe helium interferometer experiment. J. Geophys. Res. 103, 22815–22830.

Warwick J. W., Pearce, J. B., Peltzer, R. G., and Riddle, A. C. (1977) Planetary radio astronomy experiment for Voyager missions. Space Sci. Rev. 21, 309–327.

Warwick, J. W. and 12 co-authors. (1981) Planetary radio astronomy observations from Voyager 1 near Saturn. Science, 212, 239–243.

Warwick, J. W., Evans, D. R., Romig, J. H., Alexander, J. K., Desch, M. D., Kaiser, M. L., Aubier, M., Leblanc, Y., Lecacheux, A., and Pedersen, B. M. (1982) Planetary radio astronomy observations from Voyager 2 near Saturn. Science 215, 582–587.

Weidenschilling, S. J. and J. S Lewis (1973) Atmospheric and cloud structures of the jovian planets. Icarus 20, 465–476.

Weiser, H., Vitz, R. C., and Moos, H. W. (1977) Detection of Lyman-alpha emission from the Saturnian disk from the rings system.

Weisstein, E. W. and Serabyn, E. (1996) Submillimeter line search in Jupiter and Saturn. Icarus 123, 23–36.

West, R. A., Hord, C. W., Simmons, I. E., Hart, H., Esposito, L, W., Lane, A. L., Pomphrey, R. B, Morris, R. B., Sato, M., and Coffeen, D. L. (1983) Voyager photopolarimeter observations of Saturn and Titan. Adv. Space Res. 3, 45–48.

West, R. A., Tomasko, M. G., Wijesinghe, M. P. Doose, L. R. Reitsema, H. J., and Larson, S. M. (1982) Spatially resolved methane band photometry of Saturn. I. Absolute reflectivity and center-to-limb variations in the 6190-, 7250-, ad 8900-A bands. Icarus 51, 5–64.

West, R. A., Sato, M., Hart, H., Lane, A. L., Hord, C. W., Simmons, K. E., Esposito, L. W., Coffeen, D. L., and Omphey, R. B. (1983) Photometry and polarimetry of Saturn at 2640 and 7500 Å. J. Geophys. Res. 88, 8679–8697.

Westphal, J. A., Baum, W.A., Ingersoll, A. P., Barnet, C. D., De Jong, E. M., Danielson, G. E., and Caldwell, J. (1992) Hubble Space Telescope observations of the 1990 equatorial disturbance on Saturn: Images, albedos, and limb darkening. Icarus 100, 485–498.

Wildt, R. (1932) The atmospheres of the giant planets. Nature 134, 418.

Winkelstein, P., Caldwell, J., Owen, T. Combes, M., Encrenaz, T., Hunt, G., and Moore, V. (1983) A determination of the composition of the Saturnian stratosphere using IUE. Icarus 54, 309–318.

Yanamandra-Fisher, P. A., Orton, G. S., Fisher, B., and Sánchez-Lavega, A. (2001) Saturn's 5.2-μm cold spots: Unexpected cloud variability. Icarus 150, 189–193.

Yelle R. V. and Miller, S. (2004) *Jupiter's Upper Atmosphere, in Jupiter: Planet, Satellites and Magnetosphere* (F. Bagenal, W. McKinnon, T. Dowling, Eds.), (Cambridge Univ. Press), pp. 185–218.

Zebker, H. A., Marouf, E. A., and Tyler, G. L. (1985) Saturn's rings; particle size distribution for thin layer models. Icarus 64, 531–548.

Chapter 3
Origin of the Saturn System

Torrence V. Johnson and Paul R. Estrada

Abstract Cassini mission results are providing new insights into the origin of the Saturn system and giant planet satellite systems generally. The chapter discusses current models for the formation of giant planets and their satellites and reviews major Cassini findings which help advance our understanding of the system's formation and evolution to its current state.

3.1 Introduction

The results of the Cassini/Huygens mission must be interpreted in the context of what we know about the formation and history of the Saturn system. These results in turn provide many constraints on and clues to the conditions and processes which shaped the system. The purpose of this chapter is not to provide a general tutorial or textbook on the theory of giant planet and satellite formation. This is currently a very active field of planetary and astrophysical research, with new concepts being developed continually in response to the flood of new data from planetary missions and astrophysical observations of star- and planet-forming regions from ground and space-based telescopes. A full treatment of this topic is beyond the scope of the current book, let alone a single chapter. The reader is referred to a number of recent publications and reviews of the current state of research in this field (Canup and Ward 2009; Davis 2004; Estrada et al. 2009; Reipurth et al. 2007). What we hope to achieve in this chapter is to review the key concepts and current issues in planetary formation and discuss Cassini/Huygens results that relate to the problem of the origin of the system.

Historically, Saturn's system was the next most readily available after Jupiter's for early telescopic observations. By the end of the seventeenth century, less than 100 years following Galileo's first astronomical observations with the telescope, the general structure of Saturn's ring and satellite system was known, encompassing its spectacular rings, giant Titan and four of the seven medium sized 'icy satellites' of the regular satellite system (satellites in nearly equatorial, circular orbits). During the subsequent three centuries of telescopic observations many details of the ring structure were revealed and new satellites discovered, including Phoebe, the first of the irregular outer satellites to be discovered. The Pioneer 11 and Voyager 1 and 2 flybys in 1979, and 1980/81 provided the first reconnaissance of the system by spacecraft, discovering and characterizing its magnetosphere and exploring the moons and rings from close range for the first time, setting the stage for the Cassini/Huygens mission.

In the past, the Jupiter and Saturn systems were frequently referred to as archetypical examples of 'miniature solar systems' and it was suggested that their ring and satellite systems were possibly formed in an analogous manner to the planets about the Sun. Current models of star and planetary system formation suggest a more complex picture, with a range of processes and timescales leading to the formation of the sun from interstellar material, an early gas and dust solar nebula, and finally planetary formation. Cassini/Huygens observations have provided important new information and constraints on many characteristics related to the system's origin, including the composition of Saturn, the composition of the satellites and rings, the current dynamical state of the satellites (spin, orbital eccentricity, resonances), and satellites' internal structure and geological history.

In this chapter we will first review briefly current models for the formation of gas giant planets such as Jupiter and Saturn and the formation of their satellite systems. The second portion of the chapter discusses Cassini results related to the question of origins, emphasizing the chemical conditions for condensation of material in the outer solar system, and new information about the densities and structures of Saturn's satellites.

T.V. Johnson
Jet Propulsion Laboratory, California Institute of Technology,
MS 301-345E, 4800 Oak Grove Dr., Pasadena, CA 91109
e-mail: Torrence.V.Johnson@jpl.nasa.gov

P.R. Estrada
Carl Sagan Center, SETI Institute, 515 N. Whisman Rd., Mountain View, CA 94043
e-mail: Paul.R.Estrada@nasa.gov

3.2 Planet and Satellite Formation

3.2.1 Big Bang to the Solar Nebula

The events and conditions leading eventually to planetary formation in our solar system can be traced back to the Big Bang, about 13.7 Ga, based on current measurements of cosmological constants (Spergel et al. 2007). Galactic formation began within a few Gyr or less. Current estimates for the age of our Milky Way Galaxy are ∼10–13 Ga with some stars in associated globular clusters dating back even earlier. The material out of which our solar system formed was the result of billions of years of stellar evolution and processing, yielding the mix of elements in what is usually called 'cosmic' or 'solar' abundance.

It is generally agreed that the Sun and its associated circum-stellar disk formed from the collapse of a large interstellar molecular cloud, very probably in association with a massive-star-forming region similar to that seen in the Orion nebula (Boss 2007). Two basic mechanisms for the collapse have been proposed, gravitational instability and collapse triggered by a shockwave from a supernova or other nearby stellar event. Both mechanisms appear to be viable, but have different timescales and implications for the early solar nebula, with gravitational collapse taking ∼10 Myr while triggered collapse would be more rapid, ∼1 Myr. The presence of short lived radioactive isotopes in the early solar system materials has favored the supernova shock-triggered model (e.g., ^{26}Al, and particularly ^{60}Fe, which can only be made in a supernova), with other shock sources, such as outflow from massive AGB stars, regarded as less probable based on stellar evolution models (Boss 2007; McKeegan and Davies 2007). Presolar grains and the daughter products of the now extinct short lived isotopes found in primitive meteorites are the only remaining physical evidence of this early phase of the solar system's formation.

As the Sun formed from the collapsing molecular cloud, surrounded first by an envelope of gas and then an accretion disk, the earliest condensable materials appeared. These materials, preserved in refractory fragments in primitive meteorites, are known as Calcium Aluminum Inclusions (CAIs), and provide our first direct ties to the timeline and conditions which led to planet formation. CAIs have been radiometrically dated using lead isotopes, yielding an age of $4.567.2 \pm 0.6$ Ga (Amelin et al. 2002).

3.2.2 The Solar Nebula to Planets

3.2.2.1 The Inner Solar System

How rapidly the planets formed from the gas and dust of the early solar nebula is the subject of much current research and theoretical analysis. Isotopic dating of terrestrial materials, meteorites, lunar samples and Martian meteorites provides constraints on the formation times for meteorite parent bodies and the terrestrial planets. These data can be correlated using both long-lived (e.g., the U-Pb system) and short-lived chronometers (e.g., ^{26}Al/^{27}Al and ^{53}Mn/^{54}Mn) and show that the formation of chondrules and the parent bodies of igneous meteorites (Eucrites) began within a few million years following CAI condensation (McKeegan and Davies 2007). The age of the Earth and Moon can be inferred from studies of the oldest rocks on these objects and isotopic data constraining the time of core formation on the Earth. There are still significant uncertainties in these interpretations which are beyond the scope of this chapter to explore. A recent review of relevant studies up to ∼2006 has the accretion of the Earth and core formation complete by ∼4.46 Ga, with the Moon forming from a giant impact during the late stages of Earth's accretion ∼4.52 Ga (Halliday 2007).

A dated event potentially related to the final stages of planet formation is the Late Heavy Bombardment. This is based on a strong clustering of lunar impact breccia ages at ∼3.9 Ga and the lack of breccias with significantly older ages. These data have been interpreted either as the 'tail-off' of accretion or as a real 'spike' in the impact cratering rate in the inner solar system at 3.9 Ga (see Warren 2007). Recent studies of the dynamical evolution of the early solar system have provided support for the 'spike' or 'lunar cataclysm' explanation (see discussion of migration and the 'Nice' model below).

3.2.2.2 The Outer Solar System

For the outer solar system, we do not yet have sample-derived dates for the giant planets or their satellites, and so the formation times must be constrained by astrophysical observations of other star systems and theoretical studies of giant planet formation. Two major advances in astrophysical research have recently led to a greatly improved understanding of planetary formation around stars similar to our Sun: the explosive rate of discovery of extrasolar planets in the last decade and the study of protoplanetary disks, dust disks and debris disks around other stars with powerful ground and space-based telescopes (such as the Keck telescope, Hubble Space Telescope and the Spitzer mission). Two important results from this rapidly growing area of research can be summarized as: 1. Planetary formation is frequently associated with star formation and 2. Giant, gas-rich planets form very quickly – in less than 10 million years for most systems and in many cases within a few million years. In particular, evidence for gas loss in the disks around young stars suggest that giant planets (which must form before most of the gas is removed from a planetary disk) form in less than 10^7 year (Meyer et al. 2007) and gap clearing in dust disks, interpreted as evidence for planet formation, is seen in many young systems, the youngest a system only ~10^6 year old (Espaillat et al. 2007).

3.2.2.3 Giant Planet Formation

Discussions of giant planet formation have been dominated by two competing concepts for initiating the formation and capture of essentially solar composition gas from the surrounding solar nebula. These models are generally referred to as the *gas instability* model and the *core nucleated accretion* model. There are proponents of both approaches and debate continues on the pro's and con's of each. Currently, the core-accretion approach is more heavily favored by researchers in the field, although there is continuing development of both theories in response to new observations and constraints on the origin of both our solar system and exoplanetary systems.

Gas Instability Models

The gas instability (GI) approach to giant planet formation involves the formation of dense clumps in the solar nebular gas disk due to non-linear instabilities. In this model these clumps are the self gravitating precursors of the giant planets (Cameron 1978; Durisen et al. 2007; Kuiper 1951). Recent work on GI models for both solar system and exoplanet formation is associated with Alan Boss and his collaborators. The theoretical development and numerical modeling of GIs in pre-stellar gas disks is complex and there is still considerable debate about whether GIs could lead to gas giant formation in realistic models of the solar nebula (Bodenheimer et al. 2000; Durisen et al. 2007; Lissauer and Stevenson 2007). An advantage of the GI models is that they naturally lead to formation of giant planets very early in a system's lifetime, in agreement with mounting evidence that at least some stellar systems have gas giants forming within a few million years. The GI models do not do as well at explaining the enrichment of heavy elements over solar abundances in Jupiter and Saturn, since the pure instability models should produce planets reflecting the composition of the local nebula. GI models also do not address directly the formation of the terrestrial planets, although they are not intrinsically incompatible with solid planet formation by planetesimal accretion.

Core Nucleated Accretion Models

Core nucleated accretion models derive from the theories developed by Safronov (Safronov 1967, 1969, 1991; Safronov and Ruskol 1994) and others to explain the accretion of the terrestrial planets from the collision of planetesimals in a planetary accretion disk. The extension of this approach to giant planets envisions the growth of cores of rock and ice which then accrete gas from the local nebula when they become massive enough to hold substantial atmospheres (typically a few $M_\oplus$). Continued gas accumulation dominates the final stage of planet growth until the gas is depleted in the planet's vicinity, either through tidally opening a gap in the gas disk, or the nebula gas dissipating, upon which planetary growth is halted – e.g. (Estrada et al. 2009; Lissauer and Stevenson 2007).

Core nucleated models with cores enriched in rock and ice are qualitatively better able to explain the observed non-solar giant planet compositions than GI models (Estrada et al. 2009). The differences between Jupiter and Saturn and the observed enrichments in noble gases in Jupiter's atmosphere, however, are not yet fully explained by any one theory.

A difficulty with the traditional treatment of core accretion models has been the time scale for accretion of the requisite mass in the core. Extensions of Safronov's theory to the problem suggested time scales of 10^8 to 10^9 year or longer (Safronov 1969; Wetherill 1980) would be required for late-stage, high velocity growth in the outer solar system. Given the requirement that, by definition, the gas giants must form before the solar nebula gas dissipates, and the growing evidence noted earlier for formation of extrasolar giant planets in $<10^7$ years, mechanisms for much more rapid core growth are needed.

Lissauer has suggested that this implies that runaway/oligarchic growth (Kokubo and Ida 1998) must be capable of growing large cores at the distance of Jupiter, which may in turn require higher surface mass densities than given by calculations of the 'minimum mass solar nebula' (Estrada et al. 2009; Lissauer 1987, 2001; Lissauer and Stevenson 2007). In models incorporating these concepts, the time scale of giant planet formation is strongly dependent on the mass of the solid core, the rate of energy input from continued accretion of solids, and the opacity of the gaseous envelope. There are uncertainties in the appropriate values for all of these quantities, but models for reasonable ranges are capable of forming giant planets in ~1–12 Myr (Hubickyj et al. 2005; Lissauer et al. 2009). Figure 3.1 illustrates the evolution of a giant protoplanet from models of Jupiter's growth (Lissauer et al. 2009).

3.2.2.4 Planetary Migration and the Nice Model

Earlier generations of planetary formation models started with the assumption that the major planets formed at or very close to their present distance from the sun, although the effects of orbital resonances and orbit changes due to close encounters with the giant planets were recognized for their roles in the current structure of the asteroid belt (e.g., the

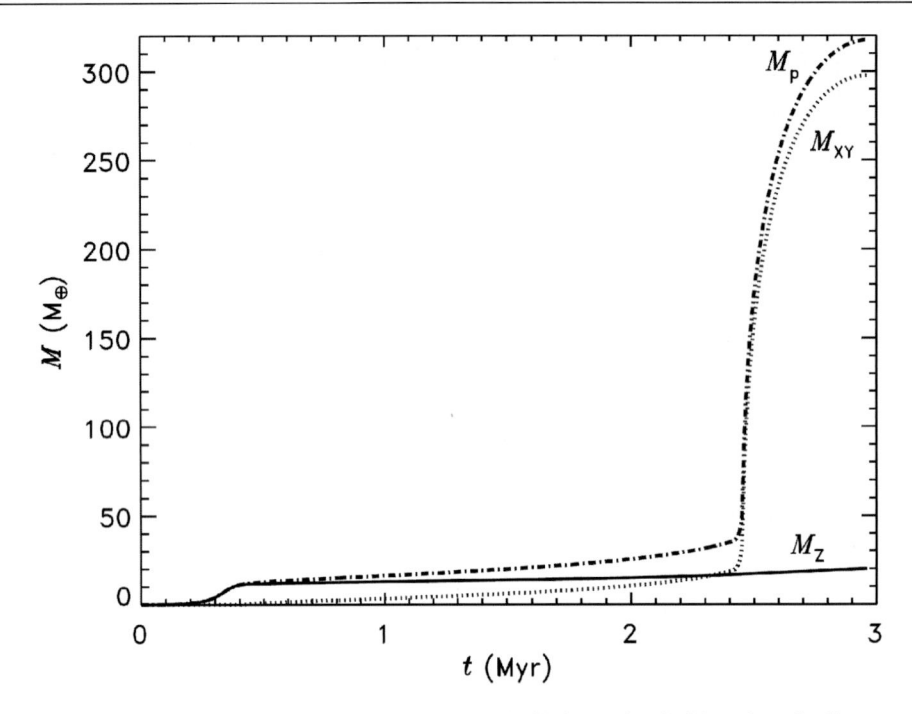

Fig. 3.1 Evolution of a proto-Jupiter: The mass of solids in the planet (*solid line*), gas in the planet (*dotted line*) and the total mass of the planet (*dot-dashed line*) are shown as functions of time. Note the slow, gradually increasing, buildup of gas, leading to a rapid growth spurt, followed by a slow tail off in accretion. Figure from Lissauer et al. (2009)

Kirkwood gaps) and comet orbits (e.g., the Oort cloud). However, planets will interact gravitationally with the disk of material from which they are forming and the gravitational torque between a planet and the disk should lead to migration of the planet's orbit (Goldreich and Tremaine 1980; Ward 1986, 1997). Many of the recently discovered exoplanet systems have giant planets orbiting extremely close to their primary star (inside 1 AU), strongly suggesting that they must have migrated inward from their formation regions in the more distant parts of the circumstellar disk. Current modeling of the early solar system addresses the possibility that significant rearrangement of planetary orbits may have occurred in our solar system as well.

How a planet migrates during growth in a massive gas disk depends on its location and whether it is massive enough to clear a gap in the disk. For lower mass objects, migration is dominated by what is known as Type I migration, due to the difference in torques exerted on material outside of and inward of their orbits. Planets which clear a gap then are subject to Type II migration, whereby they are in effect dragged along with the viscously evolving disk. Type I migration pose problems for the growth of large cores in the core nucleated accretion scenario and Type II migration creates potential difficulties for both core nucleated and gas instability models.

Once the giant planets have formed, a further complication is that in a system with multiple giant planets, the planets will create instabilities in the planetesimal disk through orbital resonances, which increase the eccentricities of planetesimals and can also affect the orbits of the other major planets, creating planet-induced migration. As the planets migrate, these resonances will 'sweep' though the system, creating the possibility of periods of rapid rearrangement of planetary orbits. Investigation of these effects with large scale computer numerical simulations is currently a rapidly growing and evolving discipline. For recent reviews, see (Levison et al. 2007; Lissauer and Stevenson 2007; Papaloizou et al. 2007).

The most comprehensive model to date for the dynamical evolution of the outer solar system through the interactions of the giant planets and the planetesimal disk has been developed by researchers at the Nice Observatory in France and their collaborators. Known as the 'Nice model' and presented in a series of papers (Gomes et al. 2005; Levison et al. 2008; Morbidelli et al. 2005; Tsiganis et al. 2005), the model attempts to explain the current location and orbital characteristics of the outer planets as well as the structure of the Kuiper Belt though the combined effects of orbital migration and orbital resonances. Numerous possible initial configurations of the outer planets were considered. The version most closely matching the observed characteristics of the current outer solar system starts with a more compact configuration of planets and the relative positions of Neptune and Uranus reversed from their final locations. In this scenario the planetary orbits evolve relatively slowly initially until Jupiter and Saturn reach a 1:2 mean motion resonance condition, at which point the orbits of all the outer planets as well as the planetesimal disk of the proto-Kuiper Belt are strongly perturbed, most of the planetesimal disk is dispersed and the outer planets

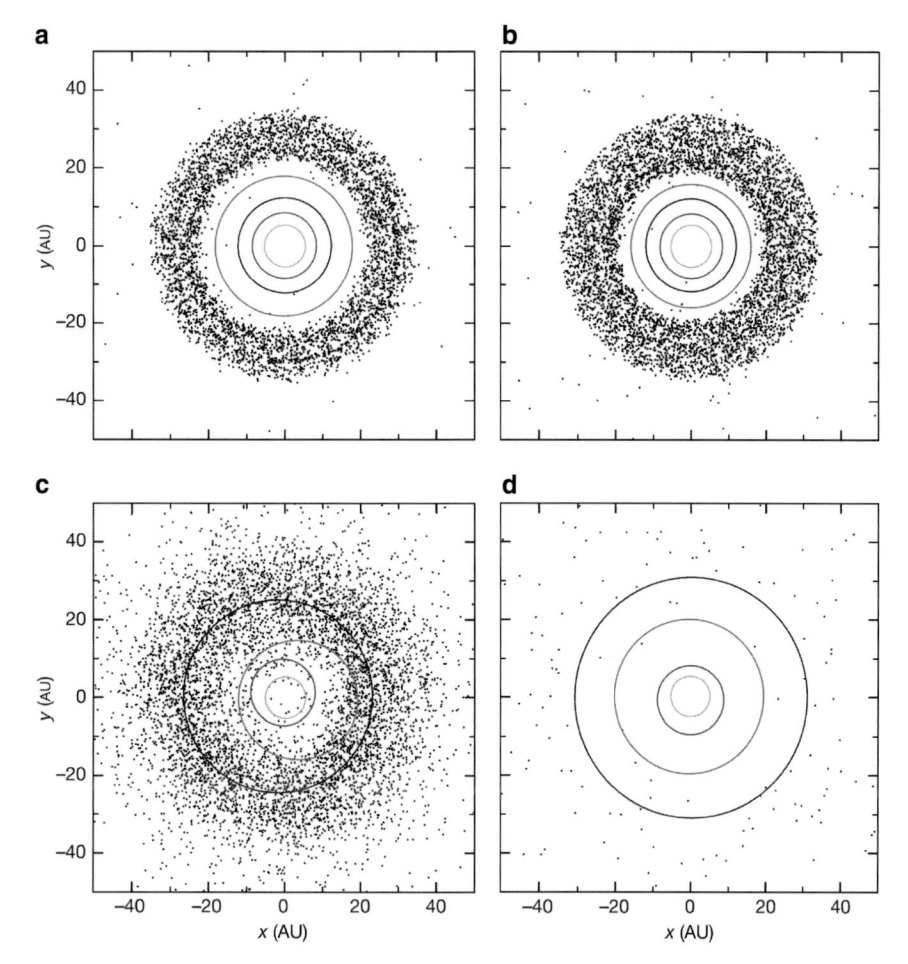

Fig. 3.2 The planetary orbits and the positions of the disk particles, projected on the initial mean orbital plane: The four panels correspond to four different snapshots taken from our reference simulation. In this run, the four giant planets were initially on nearly circular, co-planar orbits with semimajor axes of 5.45, 8.18, 11.5 and 14.2 AU. The dynamically cold planetesimal disk was 35M_E with an inner edge at 15.5 AU and an outer edge at 34 AU. Each panel represents the state of the planetary system at four different epochs: **a**, the beginning of planetary migration (100 Myr); **b**, just before the beginning of LHB (879 Myr); **c**, just after the LHB has started (882 Myr); and **d**, 200 Myr later, when only 3% of the initial mass of the disk is left and the planets have achieved their final orbits. From Gomes et al. (2005)

settle into their current more extended orbital configuration (Gomes et al. 2005). Figure 3.2 illustrates the history of the planetary orbits for this case.

In addition to rearranging the giant planets' orbital characteristics, the Nice scenario also may explain the existence and nature of the extensive population of irregular satellites at Jupiter and Saturn. Several dynamical studies have suggested that capture of scattered planetesimals can produce the observed characteristics of these irregular systems (Nesvorny et al. 2007; Turrini et al. 2008, 2009).

An interesting consequence of this model for the early history of the Saturn system is the timing of the resonance-driven rearrangement (Gomes et al. 2005). The conditions for this version of the Nice model were chosen to require the Jupiter/Saturn 1:2 resonance to occur at the time of the so-called Lunar Late Heavy Bombardment (LHB) at ∼3.9 Ga. The Nice model provides a plausible scenario for creating such a spike, and also implies that the LHB should have been a solar system-wide event. While the impacting objects on the Moon in this scenario may originate both from the gravitational scattering of bodies from the asteroid belt into the inner solar system and outer solar system planetesimals, the disruption of the planetesimal disk in the outer solar system would have created a spike in large impacts throughout the outer solar system at the same time. The large impact basins on the oldest surfaces on Saturn's icy satellites, such as Iapetus, may be evidence for this event (Johnson et al. 2007).

3.2.3 Circumplanetary Disks to Satellites

The satellites of the giant planets are characterized by prograde, low inclination orbits that are relatively compact in their spacing compared to the radial extent of their planet's Hill sphere. This indicates that they formed within a circumplanetary disk around their parent planet. Given that all the giant planets of our solar system possess regular satellite systems, it suggests that satellite formation may be a natural consequence of giant planet formation. Thus in order to account for the formation of their satellite systems, models of giant planet accretion must include the formation of their associated circumplanetary disk.

3.2.3.1 Formation of the Subnebula

The circumplanetary gas disk, or subnebula, is a by-product of the giant planet's later stages of accretion and forms over a period in which the giant planet transitions from runaway gas accretion to its eventual isolation from the circumstellar disk. This isolation, which effectively signals the end of the giant planet's formation, occurs because the giant planet successfully opens a deep, well-formed gas gap in the nebula, or the nebula gas dissipates. The process of gap-opening by the giant planet is relevant to subnebula formation because the radial extent of the circumplanetary gas disk is determined by the specific angular momentum of gas that enters the giant planet's gravitational sphere of influence.

The contraction of the giant planet envelope down to a few planetary radii occurs relatively quickly compared to the runaway gas accretion phase (see Lissauer et al. 2009), meaning that the subnebula likely begins to form relatively early on in the planetary formation process when the protoplanet is not sufficiently massive enough to open a substantial gas gap in the surrounding nebula. Prior to opening a deep, well formed gap, the planet accretes low specific angular momentum gas from its vicinity, which results in the formation of a rotationally supported compact disk component. This disk size is consistent with the observed radial extent of the giant planet satellite systems (Mosqueira and Estrada 2003a). Furthermore, since the protoplanet may still need to accrete the bulk of its gas mass after envelope collapse, this suggests that the gas mass deposited over time in the relatively compact subnebula could be substantial compared to the mass of the satellites, a picture that may be consistent with the view that the planet and disk may receive similar amounts of angular momentum (Mosqueira and Estrada 2003a, b; Stevenson et al. 1986). One potential caveat regarding Saturn is that its lower mass may mean that the time between envelope collapse and its final mass could be considerably shorter than the Jupiter-mass case.

As the giant protoplanet grows more massive and the gap becomes deeper, the continued gas inflow through this gap can significantly alter the properties of the subnebula. In particular, as gas in the protoplanet's vicinity is depleted, the inflow begins to be dominated by gas with specific angular momentum which is much higher than what previously accreted. This is because the gas must now come from increasingly farther away in heliocentric distance. A more extended, less massive disk component forms as a result. Recent simulations of gas accretion onto a giant planet embedded in a circumstellar disk that are able to resolve structure on the scale of the regular satellite systems indicate that the size of the disk formed by the inflow through the gap likely extended as much as ~ 5 times the size of the observed regular satellite systems of Jupiter and Saturn (D'Angelo et al. 2003) (see also Ayliffe and Bate 2009, who specifically treat both Jupiter- and Saturn-sized planets). The low and high specific angular momentum gas accretion phases along with the observed mass distribution of the regular satellites of Saturn (and Jupiter) thus argue in favor of a two-component circumplanetary disk: a compact, relatively massive disk that forms over the period prior to the opening of a deep, well formed gap; and, a much more extended, less massive outer disk that forms from gas flowing through the gap and at a lower inflow rate (Bryden et al. 1999; D'Angelo et al. 2003). Current models for disk and satellite formation for the Jupiter system are discussed more extensively in Estrada et al. (Estrada et al. 2009; See also Canup and Ward 2009).

3.2.3.2 Conditions for Satellite Accretion

Satellite formation is expected to begin at the tail end of planetary formation. Therefore, the character of the gas inflow through the gap during the waning stage of the giant planet's accretion is a key issue in determining the conditions under which the regular satellites form. A consequence of persistent gas inflow through the gap is that the subnebula will continue to evolve due to the turbulent viscosity generated as a result of gas accretion onto the circumplanetary disk. If the subnebula is turbulent, it can affect satellite formation in two major ways. The first is that (even weak) turbulence can pose a problem for satellitesimal formation (the subnebula equivalent of planetesimals). Depending on its strength, nebula turbulence presents a problem for particle growth because the relative velocities between growing particles can quickly become large enough to lead to fragmentation rather than growth (e.g., Dominik et al. 2007). This problem is exacerbated in the circumplanetary disk environment where gas densities, temperatures, pressures, and dynamical times are considerably different than their circumsolar analog (Estrada et al. 2009). Second, even weak, ongoing inflow through the gap can generate a substantial amount of viscous heating which would generally result in a circumplanetary disk that is too hot for ice to condense and satellites to form and survive (Coradini et al. 1989; Klahr and Kley 2006; Makalkin et al. 1999).

Despite these difficulties several models in the literature attempt to explain satellite formation under turbulent conditions, where satellite survival hinges on the turbulent dissipation of the gas disk (e.g., Alibert et al. 2005; Canup and Ward 2002; Makalkin et al. 1999; Mousis and Gautier 2004). However, given the complications associated with satellite accretion in even weak turbulence generated by the inflow, Mosqueira and Estrada (Mosqueira and Estrada 2003a, b) assumed that satellite formation did not begin until the gas inflow through the gap wanes, at which point turbulence in the subnebula may decay. Once turbulence decays, the circumplanetary gas disk becomes quiescent, so that one can not rely on turbulence to dissipate the gas disk. They assumed this because there are currently no intrinsic mechanisms that have been identified that can sustain turbulence in a relatively dense, mostly isothermal subnebula (Estrada et al. 2009). Instead, Mosqueira and Estrada (2003b) showed that the largest satellites (e.g., Titan, Ganymede), and not disk dissipation due to turbulence, may be responsible for giant planet satellite survival. This represents a key difference then between these authors' model and those listed above.

Another key issue is the source of solids which is intimately tied to the explanation for the observed mass and angular momentum of the regular satellite systems. There are several ways in which solids can be delivered to the circumplanetary disk environment, which basically fall into two categories: solid delivery via planetesimals (e.g., Estrada and Mosqueira 2006; Mosqueira and Estrada 2003a, b); and the delivery of solids via dust coupled to the gas inflow during planet accretion (e.g., Canup and Ward 2002, 2009; Makalkin and Dorofeeva 2006).

Planetesimal delivery mechanisms offer a straightforward way to explain the mass and angular momentum of the regular satellites, and help to explain the enhancement of heavy elements in the atmospheres of the giant planets (Estrada et al. 2009). At the time of planet formation, most of the mass in solids are in planetesimals with sizes larger than a kilometer (Charnoz and Morbidelli 2003; Kenyon and Luu 1999; Wetherill and Stewart 1993). As the planet approaches its final mass, planetesimals undergo an intense period of collisional grinding where a significant fraction of material is fragmented into smaller objects (Charnoz and Morbidelli 2003; Stern and Weissman 2001), but are still too large to be coupled to the gas. Much of the mass in these planetesimal fragments may then pass through the circumplanetary gas disk where, depending on the gas surface density, they may ablate, melt, vaporize, and/or be captured (Estrada et al. 2009; Mosqueira and Estrada 2003a, b). In the decaying turbulence model of Mosqueira and Estrada, gas surface densities are sufficiently high that enough material may be deposited in this way to explain the mass and angular momentum of the regular satellites. Furthermore, significant increases in the ice/rock ratio can occur because the largest first-generation planetesimal bodies (which contain most of the mass) can differentiate if a significant amount of short-lived radionuclides such as ^{26}Al are present (Jewitt et al. 2007; Merk and Prialnik 2003). Subsequent break up of these planetesimals may then lead to fractionation. A lower density outer circumplanetary gas disk might then preferentially ablate much more ice relative to rock (although much less total mass overall) compared to the denser, inner regions. Thus, for example, Iapetus' low density (compared with Callisto at Jupiter) would suggest that the Saturnian outer gas disk was less dense relative to that of Jupiter's.

Other formation scenarios which consider satellite accretion in a gas-starved disk under turbulent conditions do not include planetesimal delivery mechanisms, and rely on dust coupled to the gas inflow as the primary source of solids for generations of satellites, most of which are lost due to migration into the planet (Canup and Ward 2002). Canup and Ward (2009) note that the ablation of planetesimals cannot be a significant source of solids in their gas-starved model, presumably because their gas disk density is too low. On the other hand, it has been argued that dust inflow cannot be the dominant source of material for the satellites for a variety of reasons, and does not explain their angular momentum (see Estrada et al. 2009, and references therein). However, dust inflow may play a role in providing supplement material for the outermost satellites. For example, Mosqueira and Estrada (2003a) considered that dust inflow through the gap may augment the mass of Callisto (and Iapetus), but pointed out that too much mass delivered this way would lead to the formation of satellites in very cold regions of the disk (where they are not observed) with their full complement of ices.

In an effort to circumvent the difficulties with coagulation of dust into satellitesimals in high orbital frequency gas disks, an alternative view is that an *intrinsic* source of turbulence acts to remove most of the gas disk in a timescale shorter than the satellite formation timescale (Estrada and Mosqueira 2006). Although the residual gas density is left unspecified in this model, it is assumed to be low enough that the angular momentum of the satellite system is largely determined by circumsolar planetesimal dynamics and not gas dynamics (i.e., dust may be entrained in the gas, but satellitesimals and embryos remain largely unaffected). In this case, planetesimal material (mostly in the form of fragments) is captured and deposited in the circumplanetary disk through inelastic and gravitational collisions between planetesimals within the giant planet's Hill sphere rather than through ablation. Intermediate gas densities (where ablation would still remain unimportant) would actually make the collisional capture of planetesimals easier. One consequence of a slightly higher gas density is that, since most of the mass captured is in the form of large fragments (which sweep up the dust content in the disk), the planetesimal capture model can by-pass the problem of coagulative growth in a turbulent environment faced by models that rely only on dust inflow. However, the planetesimal capture model would still face many of the same problems associated with the gas-starved disk. For example, a similar problem faced by both Estrada and Mosqueira (2006) and Canup and Ward (2002, 2009) is that they cannot explain the origin of Iapetus.

Explaining the current state of the satellite systems is complicated by many unknowns; yet a great deal of progress continues to be made thanks to Cassini/Huygens as well as past spacecraft missions. A thorough treatment of the various possibilities and stages of satellite formation for the Jupiter system in which planetesimal delivery mechanisms provide the source of solids has been done by Estrada et al. (Estrada et al. 2009), and applies in its basic aspects to the Saturn system as well. One perhaps key difference between the Jovian and Saturnian systems is the longer planetesimal ejection times at Saturn relative to Jupiter (Goldreich et al. 2004). Although both Jupiter and Saturn would have experienced the possibility of disruptive impacts deep within their respective potential wells, longer satellite accretion times coupled with longer planetesimal 'exposure' times at Saturn relative to Jupiter may have made it more likely for satellites deep within Saturn's potential well to be disrupted by impacts helping to explain their present day configuration (Estrada and Mosqueira 2006; Mosqueira and Estrada 2003a).

Finally, in a scenario in which the giant planets of the Solar System start out much closer to each other and subsequently migrate outwards as in the Nice model (Tsiganis et al. 2005) the two planets jointly open a gap (if they grow almost simultaneously; see, e.g, Bryden et al. (2000)). In such a scenario, Saturn's local influence would dominate over that of turbulence in a moderately viscous circumsolar nebula (Morbidelli and Crida 2007). As a result, gap formation could be delayed for both planets which may influence the amount of high specific angular momentum gas they accreted. Indeed, the amount of gas left around after the opening of a combined gap may be a key factor in explaining the differences between the giant planet satellite systems, particularly their outermost satellites. In the next section we will explore how some of the results from the Cassini/Huygens mission bear on these issues and possibly provide better constraints for current models.

3.3 Cassini Results and Discussion

3.3.1 Saturn and Rings

Saturn's composition and structure provide clues to its origin and evolution. Cassini data on the atmosphere and interior of Saturn are covered elsewhere in this book (Chapters 2, 4 and 5). With respect to the issues connected with gas instability vs core accretion formation models, the strong enrichment in carbon over solar abundances, known from ground-based and Voyager observations, has been confirmed by Cassini and explored in greater detail. The existence of massive storm systems associated with evidence for lightning argues, as for Jupiter, for significant amounts of water in a condensable layer at depth, but do not provide a strong quantitative constraint on its relative abundance (Dyudina et al. 2007). Likewise, the absence of measurements of noble gases prevents a direct comparison with Galileo Probe results.

To date, no new strong gravitational constraints on interior structure have been obtained. However the possibility of a high-inclination, low altitude phase for the Cassini orbiter at the end of its extended operations may ultimately greatly improve this situation and allow comparison with similar measurements expected from the planned Juno mission to Jupiter.

How Saturn's rings are linked to the origin of the system depends on their own origin and evolution. Generally, two classes of model have been discussed in the literature: 1. Primordial models where the material in the rings is regarded as material left over from the circumplanetary disk that never accreted into satellites due to tidal forces (i.e., inside the Roche limit), and 2. Satellite disruption models, where it is hypothesized that a satellite's orbit decayed due to tidal forces or drag, taking it inside the Roche limit where it was subsequently disrupted by tidal forces, or, alternatively where a satellite was strong enough to exist inside the Roche zone but was disrupted by a large impact (either very early or recently depending on the version) and the resulting debris disk could not re-accrete into a new satellite.

Following the discovery of striking gravitationally produced wave structures in the ring system by Voyager, analyses of the angular momentum transfer involved in their production suggested that the current ring system could not survive for the age of the solar system, strongly favoring the disruption type models of recent origin. Cassini has made many detailed observations of the rings using multiple techniques. These observations and results are described in detail in the relevant chapter in this book (Chapters 13–17). Some of these results imply that the ring system is more massive and longer-lived than thought from Voyager results, focusing interest again on primordial models, possibly involving continual 'renewal' of locally ephemeral structures in the rings (Chapter 17).

3.3.2 Satellite Composition

3.3.2.1 Satellite Bulk Densities

The most important single constraint on the bulk composition of the satellites comes from measurements of their average densities. These can be used to infer the approximate proportions of condensed volatiles (primarily water ice) and 'rock' (silicates plus metal content) in the body and compare

3 Origin of the Saturn System

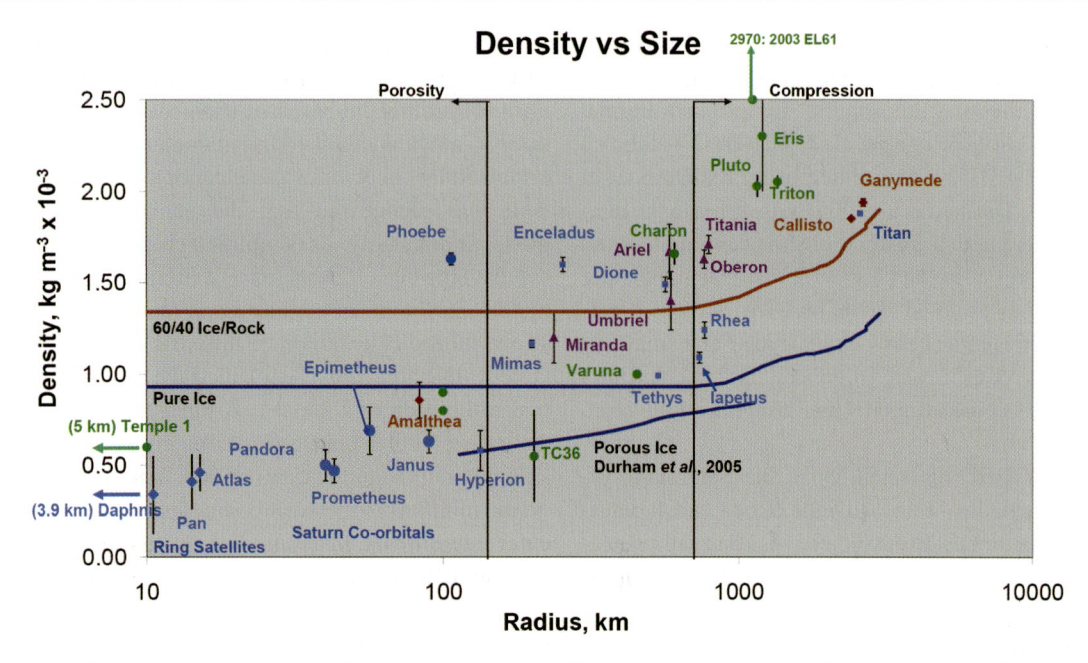

Fig. 3.3 Density of icy outer solar system bodies as a function of radius: Most of the planetary satellite values are from spacecraft observations of radius (volume) and mass from radio tracking and gravity science analyses of flybys. Data sources for the Ganymede and Callisto (orange diamonds), Uranus satellites (purple triangles), and Triton (green circle) and Titan (blue square) are found in (Yoder 1995). Amalthea data are from (Anderson et al. 2005). Mid-sized and small Saturn satellite data (radii $\sim$10–1,000 km, blue squares) are from (Jacobson 2004; Jacobson et al. 2006; Porco et al. 2005a, b, 2007; Thomas et al. 2007). Pluto and Charon and KBO's (green circles) data are, for Pluto/Charon from (Buie et al. 2006), Eris (Brown and Schaller 2007), Varuna (Jewitt and Sheppard 2002), 2003EL61 (Rabinowitz et al. 2006), 1999TC36 (Stansberry et al. 2006). The curves labeled "60/40 ice/rock" (by mass) and "Pure Ice" are theoretical calculations of densities of a sphere with a rock-ice ('rock' density $= 3,662\,\mathrm{kg\,m^{-3}}$) mixture and pure water ice composition, taking into account the presence of high density phases of water ice in the interior of a larger body (Lupo and Lewis 1979). The curve labeled "Porous Ice" represents the density of an ice sphere with a porosity throughout equal to that allowed at the central pressure of the object, based on laboratory compression experiments with cold ice compaction (Durham et al. 2005)

these with models of material composition expected from different formation models and circumplanetary nebula conditions. Figure 3.3 shows measured or derived densities for icy bodies in the outer solar system including some Kuiper Belt Objects and comets.

There are two major factors affecting the interpretation of bulk density in icy satellites, high density ice phases and porosity. The pressures in the interiors of even the largest satellites are too low for high pressure silicate phases to be significant (Titan with a radius of $\sim$2,575 km has a central pressure of $\sim$3.3 GPa). However, water ice has several high density phases that can be present at pressures of only a few hundred MPa. The temperature and pressure conditions for these phases (particularly Ice-II and Ice III) to exist can be reached in the interiors and icy crusts of satellites with radii larger than about 10^3 km, depending on their internal structure and temperatures. For these objects estimates of the uncompressed density from interior modeling are needed to constrain their composition (Lupo and Lewis 1979; Schubert et al. 1986, 2004). Two example models from Lupo and Lewis, one for a pure water ice sphere and the other for a rock fraction of 0.40 are shown in Fig. 3.3 for comparison. Of the Saturnian satellites only Titan (very similar to Ganymede and Callisto in bulk properties) is expected to have significant amounts high pressure ice in its interior resulting in model values for its uncompressed density of $\sim$1,500 kg m^{-3} (or about 0.50 rock fraction) (Schubert et al. 1986).

At the other end of the size spectrum, very small bodies with low internal pressures may sustain a high degree of bulk porosity if the internal stresses do not exceed the material strength of their constituents. Such small bodies may (and in fact, do, generally) also have arbitrarily irregular shapes, from similar considerations (Johnson and McGetchin 1973). For icy satellites, the transition from essentially spherical, non-porous objects to irregular, porous bodies occurs at about a radius of 200 km (central pressures of 1–10 MPa). Densities measured for all icy objects with radii smaller than this transition radius, except Phoebe, have densities well below 1,000 kg m^{-3}, indicative of high porosity even if they were made of pure water ice. Phoebe is also likely porous but with a significantly higher rock fraction, (Johnson and Lunine 2005).

Laboratory measurements of low temperature ice porosity confirm this general picture (Durham et al. 2005). These data show that ice even at pressures of $\sim$140 MPa can sustain significant porosity ($\sim$0.1) and at lower pressures, below 20 MPa, may have porosities of 0.3–0.5. Note that this means that even though it may not affect their bulk density significantly, large satellites may have significant porosity in their outer icy layers, affecting lithospheric strength and thermal properties (e.g., Castillo-Rogez et al. 2007). The line labeled 'Porous ice' in Fig 3.3 was calculated using the Durham et al. measurements (Durham et al. 2005), converted to the density of an ice sphere with porosity set by the central pressure. Given the lower pressures (and higher porosities) in the outer layers of a real body as noted above, pure ice, porous objects could have densities at or below these levels. The small icy Saturnian satellite densities are in general agreement with a projection from this pure ice model and must have little or no rock fraction.

The six 'mid-sized' icy satellites in the Saturnian system (Mimas, Enceladus, Tethys, Dione, Rhea and Iapetus) fall between the extremes discussed above. Their bulk densities are not strongly affected by either porosity or compression effects, and so should reflect fairly closely the density and compositional mix of their constituents. With accurate Cassini measurements of their radii and masses, the errors in satellite densities are extremely small, and the wide range of density values, from Enceladus $(1{,}600\,\text{kg}\,\text{m}^{-3})$ to Tethys $(991\,\text{km}\,\text{m}^{-3})$ must reflect real and significant variations in their compositions as reflected in rock/ice ratios. This pattern contrasts sharply with the systematic density variation of the large Galilean satellites, where the most distant, Callisto and Ganymede, are ice rich and the inner two satellites, Europa and Io, are essentially rocky objects, although tiny Amalthea, interior to Io, has a low density indicative of an ice-rich and porous small body (Anderson et al. 2005).

These compositional variations, as well as the apparent very ice-rich nature of the Rings and the inner co-orbital and ring-related moons, pose difficult questions for satellite formation models, particularly for earlier models which depended on early giant planet luminosity to explain the radial density gradient in the Galilean satellites.

3.3.2.2 Equilibrium Condensation and Solar Composition

Background

Since the 1970s, most discussions of the composition of the planets and satellites have been in the context of what is generally referred to as 'equilibrium condensation' from a solar nebula with a 'solar composition'. The foundation for this approach was laid by John Lewis and his colleagues using theoretical calculations of pressure and temperature in the early solar nebula to model the expected composition of solid material as a function of heliocentric distance (Grossman 1972; Grossman and Larimer 1974; Lewis 1971–1973). The underlying assumption in this work was that gas and solids at any point in the solar nebula could be treated as an equilibrium assemblage based on elements being present in the same as abundances as found in the Sun and primitive meteorites ('solar' or 'cosmic' abundance).

It was recognized that the equilibrium condensation model undoubtedly oversimplified processes in the real solar nebula. However, it provided an extremely powerful working hypothesis for understanding major features of what was then known of the distribution of different materials in the solar system. In particular, it explained the major dichotomy between the rock and metal rich inner terrestrial planets and the gas rich outer planets, the apparent spectral and compositional gradient in the asteroid belt, and the existence of hydrated minerals in meteoritic samples from (presumably) the outer asteroid belt. In addition a key feature of the model was that beyond the asteroid belt the composition of condensed materials should shift abruptly from being rock/metal dominated to being a roughly equal mixture of rock and water ice, producing objects with densities of $\sim 2{,}000\,\text{kg/m}^{-3}$. Lewis further suggested in his early papers that heating from the rock portion of such bodies might produce melting, differentiation and volcanic activity, aided possibly by small amounts of ammonia hydrate lowering the melting point.

At the time, few satellite densities were known accurately and none had been studied by spacecraft. By 1989, following the initial reconnaissance of the outer planets by the Pioneer and Voyager missions it was clear that ice-rich satellites were the norm, with the three largest having almost exactly the rock/ice fractions suggested by Lewis, and many relatively small icy bodies showed evidence of geologic activity. More than 30 years after Lewis' initial papers on the topic, both 'solar composition' ice/rock satellites and cryovolcanism remain at the center of discussions of satellite system formation.

Complications with the Equilibrium Model

Subsequent work has elaborated on the basic equilibrium condensation concept and modified it significantly in some respects. Although the predicted compositions of planetary materials from equilibrium condensation seem to be in general agreement with the bulk properties of planetary objects, at smaller scales there is considerable evidence for strikingly disequilibrium conditions in the formation of planetary materials. In meteorites, the existence of pre-solar grains and refractory materials such as CAIs in otherwise 'primitive' samples provides clear evidence that the nebular gas and refractory solids cannot have remained in complete equilibrium at all times (Grossman 1972). Recently returned samples from Comet Wild 2 by the Stardust mission also contain refractory grains that must have been formed under much hotter conditions close to the early Sun than the bulk of the volatile-rich cometary matter (Joswiak et al. 2008; Westphal et al. 2008).

Recent work modeling the solar nebula has also modified the earlier concepts. In current models, gas and particle interactions are modeled in more detail, resulting in the potential transport of material within the nebula by drag and turbulent processes. This may lead for instance to local enhancement in water and other volatiles due to evaporation of inwardly migrating grains at the 'snow line' (Cuzzi et al. 1993; Cuzzi and Zahnle 2004; Stevenson and Lunine 1988). Planetesimal migration to the planet forming regions can also produce volatile enhancement in the Jupiter and Saturn systems (see Estrada et al. 2009). Cassini observations of cometary H/D values in the gases in Enceladus' plumes support the suggestion that the Saturnian satellites incorporated cold icy material from the outer solar nebula (Waite et al. 2009). This observation may imply that low-temperature volatiles that condensed in the outer nebula were incorporated into migrating material, and eventually planetesimals that found their way into the feeding zones of the giant planets. Furthermore, it would tend to support the idea of planetesimal delivery mechanisms (see Section 3.2.3.2) providing the mass for the regular satellites. Whether these or other processes can fully explain compositional constraints from the Galileo Probe Jovian measurements and the new Titan and Enceladus data is still to be determined.

One of the most significant modifications of the 'pure' equilibrium condensation scheme was the recognition that solar nebula conditions in some regions might not permit equilibrium to be achieved in the age of the solar system. Studies of the kinetics of reactions in refined models of the solar nebula suggested that, in the outer portions of the solar nebula, CO and N_2 might be the primary C and N bearing species, as opposed to CH_4 and NH_3, in the warmer, denser regions (Lewis and Prinn 1980). In the satellite forming zones around the giant planets, calculations suggest that the reduced species should still predominate and produce equilibrium rock/ice ratios (Prinn and Fegley 1981, 1989).

Despite the issues and complications noted above, equilibrium condensation remains the framework for most discussions of outer solar system bodies, particularly with respect to expected rock/ice fractions in the materials forming the satellites. In this modified view of solar abundance equilibrium condensation, the mix of materials available for forming satellites depends very heavily on the amount of oxygen which is available in the form of water ice. This in turn depends critically on the partition of carbon between CO, CH_4, and solid carbon (organics and graphite). If significant carbon is tied up in CO, which does not form solids under the conditions around the giant planets, then a significant fraction of the nebular O is not available to form water

ice in these regions, and condensates would have a higher rock/ice fraction. These effects are quantitatively described in the following sections, along with their dependence on the assumed solar C and O abundances.

Effects of Carbon in the Gas Phase

Estimates of protosolar C and O abundances have varied significantly over the last 30 years due to refinements in the interpretation of solar spectra and structure. These are discussed in detail in Wong et al. (2008). The most recent values are those of Grevesse and colleagues (Asplund et al. 2006; Grevesse et al. 2007). The resultant densities of condensed material for these values are compared with earlier solar estimates in Fig. 3.4 (Wong et al. 2008), which shows the expected material density of condensates as a function of the fraction of total carbon in the form of CO in the nebular gas. Also noted on the plot are the uncompressed densities of various outer planet satellites and Pluto and Triton (as examples of far outer solar system bodies).

Effects of Carbon in the Solid Phase

The effects of solid carbon on expected condensate and satellite density depend on the gas phase chemistry discussed in the previous section. The densities of potential solid carbon species range from kerogens ($\sim$1,100 kg m^{-3}) to graphite ($\sim$2,000 kg m^{-3}). Since these densities are not greatly different from the equilibrium (CO-poor) condensate, there is little change in the expected density of materials formed in reducing conditions even if essentially all the carbon is in solid form. In contrast, for CO-rich conditions, the primary effect of solid carbon is to reduce the amount of oxygen tied up in CO, resulting in more water ice production and a decrease of the condensate density. In the extreme case where all the carbon is in solid form, gas phase carbon is irrelevant and the resultant condensate would be a mixture of rock, ice and solid carbon in roughly 40/40/20 proportions with a density ranging from $\sim$1,400 to $\sim$1,500 kg m^{-3}, depending on the solid carbon density. Figure 3.5 (Wong et al. 2008) illustrates the effects of the fraction of carbon in the solid form on condensate density for a solid carbon density of 1,700 kg m^{-3}.

Discussion of Saturn Satellite Densities

Saturn's satellites exhibit a significant range of measured densities (Fig. 3.3), implying a similar variation in their bulk composition in terms of rock/ice fraction. The only irregular satellite with an accurately determined density is Phoebe. Several lines of evidence suggest that Phoebe and the other

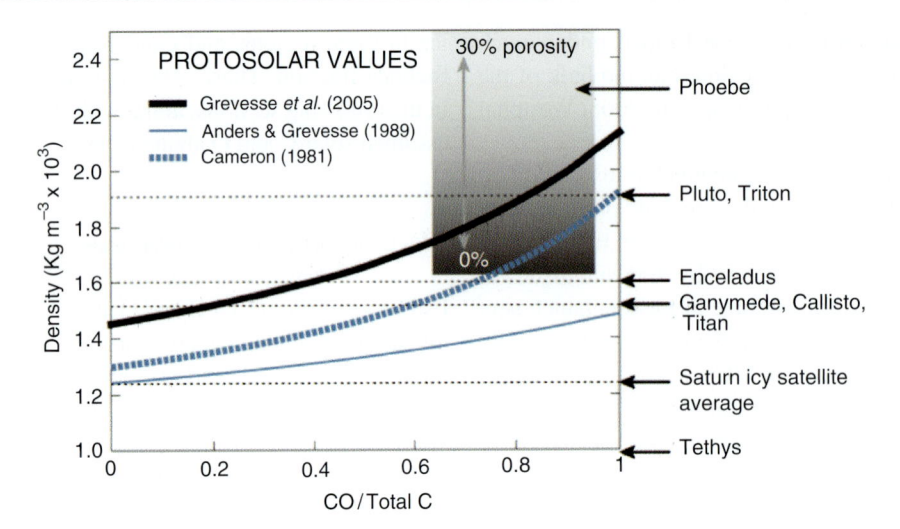

Fig. 3.4 Calculated uncompressed density of material condensed from a gas of solar composition as a function of the carbon oxidation state: Calculations for the resultant density for historical estimates of solar carbon and oxygen abundances (Anders and Grevesse 1989; Cameron 1981) and current values (Grevesse et al. 2007). Estimates of the uncompressed densities of the outer planet icy bodies are indicated by arrows. From Wong et al. (2008)

irregular satellites are captured bodies which originated in the outer solar nebula, but not within the circumplanetary subnebula of Saturn (Nicholson et al. 2008; Turrini et al. 2008, 2009). Phoebe's small size suggests that it could be highly porous, which means that its measured density of ~1,600 kg m^{-3} is probably an underestimate of the density of its constituents. Its density is in general agreement with expectations for condensates from a CO-rich outer solar nebula, as described above (Johnson and Lunine 2005). Other larger bodies in the outer solar system, Pluto, Charon, Triton, and the KBO Eris, share similarly high densities and estimated rock/ice fractions when compression is accounted for. Lower density KBOs have also been reported, however, suggesting ice-rich compositions and/or high porosities (Stansberry et al. 2006). Thus the general picture of a cold outer solar nebula where achievement of full chemical equilibrium is inhibited, producing rock-ice condensates, is consistent with the available data. However, the sample of objects in this region with accurately measured densities is still small, and the existence of low density KBOs suggests, at the least, that processing and fractionation of ice and rock in proto-KBO bodies may have played a role in the currently observed compositions.

The regular satellite system of Saturn is dominated by one massive satellite, Titan, in contrast to the Jupiter system with its four large Galilean moons. The rest of the icy satellites taken together comprise only about 4.5% of the mass of Titan. Since Titan's uncompressed density is ~1,500 kg m^{-3}, this means that most of the condensed material in the regular satellite system is consistent with near equilibrium condensation in a reducing circumplanetary nebula with a CO fraction below 0.2 (see Fig. 3.4).

The smaller 'icy' satellites present a more complex problem for any simple satellite formation model in a subnebula with temperature decreasing with distance from the planet. As illustrated in Fig. 3.3, these satellites have significant differences in their densities. In addition, taken together they have significantly lower rock/ice fractions than Titan, with a mass-weighted composite density of ~1,220 kg m^{-3}. This is lower than expected from a solar composition mixture, no matter what the reduction state or phase of carbon in the subnebula (Figs. 3.4 and 3.5). The small, very low density, high porosity satellites closer to the ring system and the very ice rich composition of the Rings themselves (Chapter 15) are even more difficult to reconcile with a solar composition condensate.

Several possibilities to explain the low icy satellite densities and the large variations among the Saturnian mid-sized satellites have been discussed (Wong et al. 2008). One possibility is that subsequent to the formation of Titan, the local Saturn environment inside of Titan's orbit became enriched in oxygen or water (Mosqueira and Estrada 2003a). These authors argue that the satellites inside of Titan do not form and have a chance to survive until the inner regions become optically thin and cool. The planetary cooling time for Saturn is ~10^5 years (Pollack et al. 1976), which is considerably shorter than the time for disk dissipation, e.g., ~10^6–10^7 years by photoevaporation (Shu et al. 1993). Condensed water may not have been available in the inner regions for this long, but as the disk cooled enough for water condensation, the inner regions would have been water-enriched due to the sublimation of inward drifting bodies, and subsequent preferential loss of silicates. A similar enrichment may not have occurred, or have been as relevant for Jupiter because the

3 Origin of the Saturn System

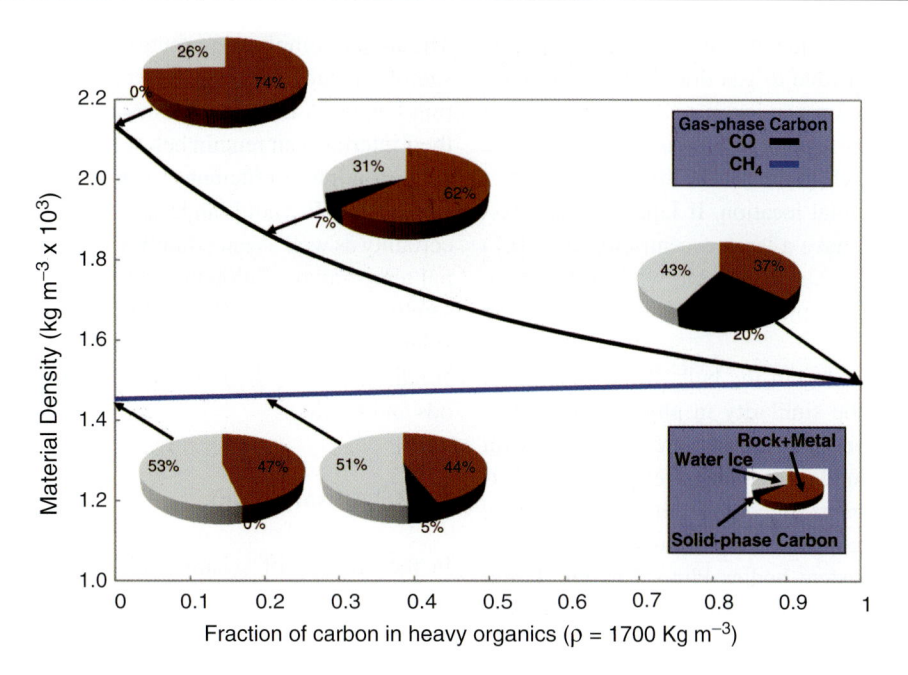

Fig. 3.5 Protosolar condensate density versus fraction of carbon in the form of refractory organics: Organic material assumed to have density of 1,700 kg m^{-3}. The upper line is the case for the rest of the carbon being in the form of CO gas, the lower line shows the result of the gaseous carbon being in the form of CH$_3$ gas. Models for intermediate oxidation states are in the shaded zone between the lines. Example condensate compositions (fraction of ice, rock, and solid carbon) for no solid carbon, 20% solid carbon and all carbon in the solid state are illustrated by the inserted pie diagrams. From Wong et al. (2008)

planetary cooling time is much longer (Pollack et al. 1976). Once these satellites form, their survival hinges on their migration times being longer than the disk dissipation time (see Mosqueira and Estrada (2003b) for a discussion of survival mechanisms). Another possibility is that the disk was enhanced in water through the ablation of large, differentiated planetesimals, and fractionated (icy) planetesimal fragments that pass through the circumplanetary gas disk (see Section 3.2.3.2).

In either case, subsequent processes are still required to create the wide range in rock/ice fractions found in the satellites (possibly including distant Iapetus). Thus the role that collisions may have played in the history of the satellites' evolution should be considered. As discussed earlier (see Section 3.2.3.2), satellites deep within the potential well of Saturn may have been more subject to disruption or fragmentation due to collisions with heliocentric interlopers than the Jovian satellites. Hyperbolic collisions might conceivably remove volatiles from the mantle of a differentiated satellite and place them on neighboring ones (analogous to the impact that may have stripped Mercury's mantle, Benz et al. (1988)).

Smaller satellites may have been disrupted and then reaccreted. Such a scenario may explain the near total absence of rock from the innermost satellites (and possibly the rings) if the collisional process leaves behind the remnants of icy shells of composite bodies whose silicate cores have since disappeared (see also Chapter 17). In a similar vein, if the Late Heavy Bombard at ~3.9 Ga was in fact a solar system-wide event as suggested by the Nice model, there would also be a significant probability that smaller satellites in the inner system might have been completely disrupted (Charnoz et al. 2009) further modifying the distribution of materials among the satellites. However, by this time the subnebula would long have dissipated, meaning there would be no gas around to assist in migration or circularization of their orbits. While intriguing, none of the above suggestions have yet received a detailed treatment explaining the current distribution of satellite density and the effects on the dynamical histories of the satellites, and the collisional mechanisms remain to be quantitatively evaluated.

On the other hand, the explanation for Iapetus' low density likely depends on whether it formed *in-situ* near its current location, or was also a product of collisional processes. The most straightforward explanation is that Iapetus formed near its present location in a low density, outer Saturnian subnebula that preferentially captured icy material versus rock through the ablation of disk-crossing planetesimal fragments (see Section 3.2.3.2). An alternative scenario which is dynamically feasible, but low probability, is the impact of a Triton-sized differentiated interloper with Titan (Mosqueira and Estrada 2005). Such a collision may produce a volatile-rich debris disk, some of which may be reaccreted by Titan, or drift inward (due to gas drag) while still leaving sufficient material to form an Iapetus-sized object. These authors estimated that its formation time is shorter than the timescale for a collision with Titan, or the timescale to scatter Iapetus to its present orbital location. If Iapetus is successfully scattered, it would have a large eccentricity (e ~ 0.7) requiring gas drag to circularize its orbit, while subsequent tidal circularization could explain Titan's anomalously large primordial eccentricity.

The low rock/ice fractions of the bulk of the material in Saturn's satellites and the similarity in physical properties between Titan, Ganymede and Callisto are consistent with roughly solar system abundance condensates in reducing circumplanetary environments. In contrast, the higher density, higher rock fraction composition of captured Phoebe and the larger outer solar system bodies Pluto, Charon, Triton and several KBOs are indicative of the more oxidizing conditions in outer solar nebula suggested by Lewis and Prinn (Lewis and Prinn 1980). However, the existence of smaller icy satellites with large differences in their rock/ice fractions and the extreme ice-rich innermost satellites, as well as evidence for ice-rich KBOs suggest that processes subsequent to formation, including collisional redistribution of material, may have altered their compositions. Therefore the small satellite densities and compositions may not provide direct clues to the composition and state of the early circumplanetary nebula.

3.3.3 Satellite Structures

3.3.3.1 Background

Following formation, the two factors which dominate a satellite's subsequent evolution and thermal history are its composition (i.e., its rock/ice fraction) and size. The amount of heat produced is simply a function of the rock fraction in the satellite, which carries the radioactive elements, while the degree to which the satellite is heated is proportional to its radius (i.e., heat production α volume, while cooling α to surface area). Lewis noted in his early papers on outer planet ice-rock satellites that they would be easier to melt and differentiate than their silicate counterparts simply because their large ice component would melt at a much lower temperature. If minor constituents such as ammonia or salts are present, he pointed out that even lower temperature melting might be achieved.

Simple thermal models for heat conduction in icy satellites confirmed that heating from long-lived radioactive isotopes of U, Th, and K could produce temperatures near the melting point of water at relatively shallow depth in large satellites (e.g., Ganymede, Callisto, and Titan) (Consolmagno and Lewis 1977, 1978; Fanale et al. 1977).

The major complicating factor for these satellites is the issue of conduction versus convection. If sub-solidus convection begins as the ice viscosity decreases with temperature, their interiors can remain below the melting point, since convection is more efficient at removing internal heat than is conduction (Reynolds and Cassen 1979). Because of this uncertainty as well as uncertainties in the satellites' initial internal temperatures following accretion, thermal models of icy satellites prior to Galileo and Cassini/Huygens usually considered both homogenous and differentiated cases as initial conditions (see Schubert et al. (1986) for a review of methods and results).

3.3.3.2 Observations

In the absence of seismic data from the surface, the best first-order indication of the internal structure of a satellite is its axial moment-of-inertia, which constrains the degree of concentration of mass towards the center. Satellites orbiting giant planets undergo distortions in their shape and gravity field due to both spin and tidal forces with the amount of distortion depending on the moment of inertia. Hubbard and Anderson (Hubbard and Anderson 1978) showed that spacecraft tracking measurements of the J_2 term of the spherical harmonic expansion of a satellite's gravity field would enable a useful determination of the moment of inertia. The major assumption required is that the body be in hydrostatic equilibrium, which fixes the relation between J_2 and C_{22} (the zonal harmonic). Although believed to be a reasonable assumption for large ice-rich objects with warm interiors, it may not be valid for small and/or very cold satellites (McKinnon 1997; Nimmo and Matsuyama 2007).

The Celestial Mechanics radio science team on Galileo successfully applied this technique to the four Galilean moons during multiple flybys and determined that Io, Europa and Ganymede all had moments of inertia indicative of significant differentiation of heavy material (rock and metal) from water in their interiors, with the number of distinct layers and their densities being less well constrained (Anderson et al. 1996a, b, 1997a, b, 1998a, b, 2001a, b; Schubert et al. 2004). The existence of a dynamo magnetic field at Ganymede is also strong evidence for differentiation of this satellite (Kivelson et al. 1996, 1997). Induction magnetic fields produced by interaction with the time variable Jovian field have also been interpreted as evidence for salty, electrically conducting subsurface oceans at Europa, Ganymede and Callisto (Khurana et al. 1998; Kivelson et al. 1999, 2000). However, Callisto, while showing a clear magnetic induction signature, has an inferred moment of inertia suggestive of only partial differentiation. Thus, while its outer layers may have been heated and melted, the deep interior has apparently not completely differentiated (Schubert et al. 2004).

At Saturn, one of Cassini/Huygens goals is to determine the interior structures of the satellites where possible. This has proved a challenging task for several reasons. The design of the Cassini Orbiter spacecraft makes it very difficult to obtain precision radio tracking data near a satellite simultaneously with remote sensing observations. This has limited dedicated radio tracking for gravity purposes to a handful of flybys of Titan and Rhea. In addition, as noted above, the validity of the hydrostatic assumption has been called into question, particularly for the smaller, very cold satellites, which may have sufficiently rigid crusts to prevent relaxation to the hydrostatic state. Therefore, several different lines of evidence must be brought to bear on the problem of internal structure and degree of differentiation. The companion volume, *Titan from Cassini-Huygens*, and Chapter 18 discuss the Cassini results in detail. We provide a brief summary of their findings below.

Titan

Based on its similar size and density to Ganymede and Callisto, Titan is expected to be at least partially differentiated (e.g., Schubert et al. 2004). Radio tracking data from Cassini show that Titan's quadrupole gravity field appears to be non-hydrostatic, making it difficult to infer internal structure from the simple hydrostatic approximation [Radio science team paper in preparation]. There is other evidence of internal heating and a subsurface liquid layer, however. This includes theoretical estimates the loss of methane from Titan's atmosphere by photodissociation and hydrogen escape (Lunine and Atreya 2008), which implies outgassing of significant quantities of methane during its history (Tobie et al. 2006). The geologically young, eroded surface of the moon also suggests internal heating and activity. Finally, cartographic comparisons of features observed by the Cassini Radar investigation show that Titan's ice crust is not rotating synchronously. This result has been interpreted as due to drag forces from atmospheric winds, requiring a liquid layer to decouple the crust from the interior (Lorenz et al. 2008). Thus the weight of the evidence suggests that Titan's interior has been warm enough for at least partial differentiation and melting. [see Titan after Cassini/Huygens for complete discussion]

Enceladus

Voyager images revealed that tiny Enceladus (R = 250 km, smaller than the largest asteroids) was a geologically active moon at some point in its history, with heavily cratered regions adjacent to smooth areas evidently resurfaced by some process. Its location close to the densest part of the diffuse E-ring and the short estimated lifetime of E-ring particles suggested that geologic activity might be on-going and responsible for replenishing the E-ring. A number of efforts were made to construct thermal models which might explain this putative activity by raising the internal temperatures to close to the water melting point, possibly with the help of 'anti-freeze' such as small amounts of NH_3 (Squyres et al. 1983). A major difficulty with all these models is the relatively low amount tidal energy available from Enceladus' current eccentricity, forced by a orbital resonance with Dione. In addition, prior to Cassini Huygens, Enceladus' mass, and therefore density and rock fraction were very poorly constrained, with most estimates suggesting a low density. Thus the amount of radiogenic heating available was highly uncertain.

Cassini discoveries have spectacularly confirmed current, geyser-like activity on Enceladus and its role supplying material to the E-ring (Dougherty et al. 2006; Hansen et al. 2006; Porco et al. 2006; Spencer et al. 2006; Waite et al. 2006). The temperature and heat flow in the South Polar Region has been measured from infrared observation, with estimates of 3–7 GW of power being emitted from the fracture system dubbed 'tiger stripes' from which the plumes originate (Spencer et al. 2006) (Chapter 21). The mass and density of Enceladus is now measured accurately from Cassini tracking and radio science data, yielding a high density of $1,600 \, kg \, m^{-3}$, providing potentially higher levels of radiogenic heating from the rock fraction. However, long-lived radiogenic heating cannot currently produce the observed power levels. Tidal heating at some point in Enceladus' history is therefore very likely and this complicates understanding of its initial thermal and interior state.

Although the active plumes in the South Polar Region and the measured high heat flow values are commonly taken as *prima facie* evidence that there is sufficient energy available to produce differentiation, there are not yet any measurements of gravity structure against which to test this assumption. Thermal models using the new density values show that early differentiation is possible due to heating from long lived radioactive isotopes, and highly probable if Saturn formed early enough for heating for short lived radioactive isotopes to be significant (Matson et al. 2007; Schubert et al. 2007). However, there is not yet a satisfactory theory for ultimate source of the total power required to explain the heat flow from the SPR either directly from global tidal dissipation (Porco et al. 2006) or from friction along the fractures in an ice shell over a liquid layer (Nimmo et al. 2007) (Chapter 21).

Rhea

Rhea's surface is relatively bright with regions of heavy tectonic fracturing exposing bright ice surface. This implies some degree of separation of ice and dark rock material in at least the upper surface layers (many meters to perhaps kilometers). Gravitational data from the radio science experiment however are consistent with either no, or very limited differentiation, although other interpretations of the single flyby radio data are possible (Anderson and Schubert 2007; Iess et al. 2007; Mackenzie et al. 2008). For a full discussion see Chapter 18. Rhea's density is close to that for Enceladus and also implies significant radiogenic heating, so a thermal history which prevents full differentiation may be required, similar to the problem of explaining Callisto's apparent liquid layer combined with a derived moment of inertia indicated limited differentiation (Anderson et al. 2001b).

Iapetus

Iapetus' low density ($1,200\,\text{kg}\,\text{m}^{-3}$) and low rock fraction, make it intrinsically difficult to differentiate from radiogenic heating. It also has an extremely non-equilibrium shape for its synchronously locked spin period of almost 80 days, implying a thick rigid lithosphere at the current time. Thermal models for Iapetus which allow enough dissipation to de-spin the satellite while cooling rapidly enough retain a 'fossil', oblate shape from an earlier more rapid spin state have been developed by Castillo-Rogez et al. (2007). These models require that Iapetus, and therefore the Saturn system, was formed about 2–5 Myr following the condensation of the first solids in the solar nebula (CAIs), consistent with general discussion of rapid giant planet formation in earlier sections. This time frame is also broadly consistent with that necessary to explain Callisto's apparent lack of complete differentiation (Barr and Canup 2008). The heating from short lived radioactive isotopes, primarily ^{26}Al, in these models is sufficient to warm the deep interior but not to trigger significant differentiation in the bulk of the satellite.

Solar wind magnetic field deflections in the vicinity of Iapetus have been interpreted (Leisner et al. 2008) as evidence for either significant crustal magnetization or a conducting layer, such as a liquid ocean with salt content (similar to those inferred from induction magnetic field signatures in the icy Galilean satellites). Based on the above discussion, a liquid layer appears highly unlikely, both from basic energy considerations and from the observed non-equilibrium shape, implying very low levels of subsurface heating in the last three to four Gyr.

3.3.4 Satellite Geological History

3.3.4.1 Crater ages

Determining absolute ages for outer planet satellite surfaces is greatly complicated by the on-going debate on the nature of the impactor population and the appropriate impact flux as a function of time and impactor size (Porco et al. 2005a; Zahnle et al. 2003). In the absence of isotopic age determinations of samples, such as are available to calibrate the lunar crater record, satellites ages are dependent on large extrapolations from current observations and uncertainties in the history of the impact flux. Very heavily cratered surfaces, at or near 'crater saturation', are generally agreed to be extremely ancient, dating back 3–4 Ga or more when essentially all models predict much higher fluxes of impacting objects than in the current solar system. Likewise, uncratered surfaces or very low crater densities certainly imply geologically young surfaces (years to millions of years old). Iapetus, discussed in more detail in the next section, is the best example of an ancient, highly cratered surface in the system, while Enceladus' South Polar Region, with its on-going plume activity is an example of an archetypically geologically young region, as is most of Titan's very sparsely cratered surface, which shows ample evidence for recent fluvial, Aeolian, tectonic, and possibly cryovolcanic modification. For most of the other icy satellites, while relative chronologies can be established based on crater statistics, estimates of absolute ages for different assumptions of impact flux can vary by factors of 2–10 (see Chapter 19 for a full discussion).

3.3.4.2 Iapetus' Formation Time

The Nice model for early solar system dynamical evolution discussed earlier provides a possible path to linking the sample based chronology of the inner solar system, including the Moon, to at least some portions of the history of the Saturn system. If the multiple large impact basins on Iapetus formed contemporaneously with the lunar basins at 3.9 Ga as a result of a solar system wide event associated with the LHB, it provides a tie point for evaluating other age estimates in the system. Such an age is basically in agreement with the interpretation of crater statistics for either of the two contending models within uncertainties at present. It is also consistent with the chronology proposed by Castillo-Rogez et al. for Iapetus, whose thermal models imply a rapidly cooling lithosphere rigid enough to retain both the satellite's oblate shape and the record of large impact basins by the time of the LHB (Johnson et al. 2007). Figures 3.6 and 3.7 illustrate

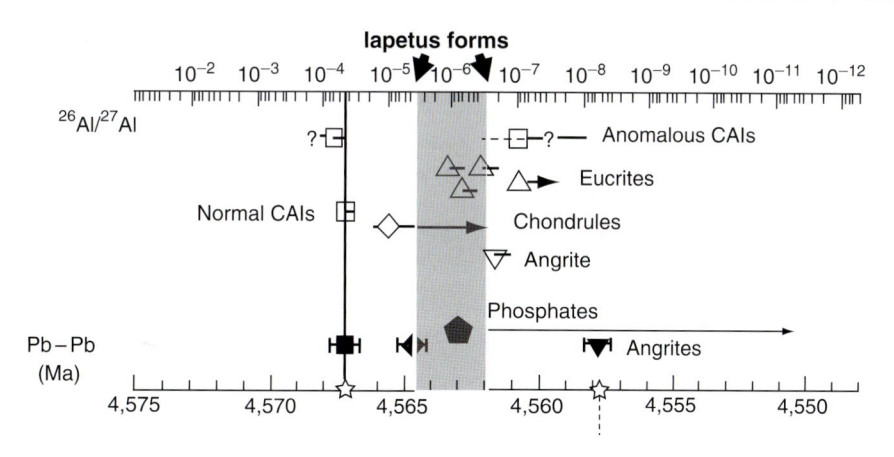

Fig. 3.6 Short lived radioisotope chronometry compared with absolute ages from Pb-Pb dating for various early chondritic and igneous (Eurcrite) meteorite samples: Modified after (McKeegan and Davies 2007), reprinted by permission from Elsevier, and compared with the estimated time of formation of Iapetus from Castillo-Rogez et al. (2007). The upper scale gives the inferred initial $^{26}Al/^{27}Al$ value tied to absolute Pb/Pb ages (*bottom scale*) using the dating of CAI formation of $4.567.2 \pm 0.6$ Ga (Amelin et al. 2002)

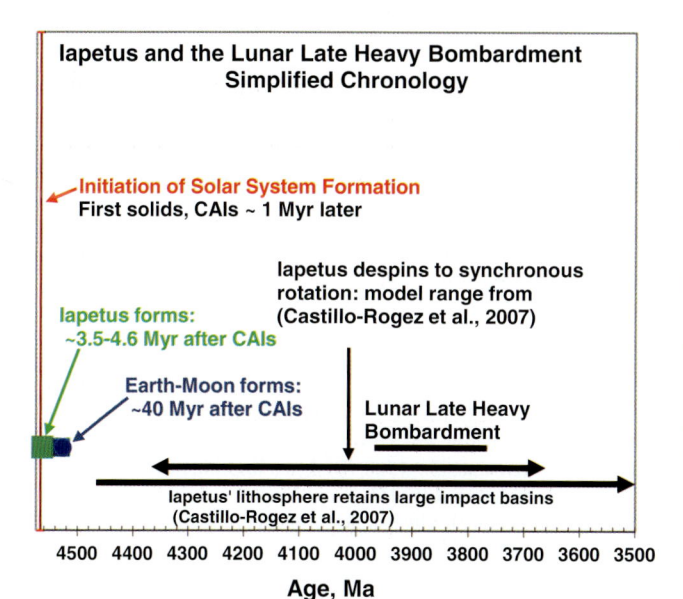

Fig. 3.7 Solar system chronology: Shows the time ranges for possible de-spinning Iapetus and for Iapetus' lithosphere to become rigid enough to retain the record of large impact basins and compares this to the time of the LHB as dated from lunar samples. From (Castillo-Rogez et al. 2007)

how this proposed chronology relates Iapetus' formation and early thermal evolution to the formation of CAIs, meteorite parent bodies and the lunar record.

3.4 Summary and Issues for the Future

This chapter summarizes how the Cassini Huygens results have constrained and added to our understanding of the formation and evolution of the Saturn system, its rings and satellites. As the Cassini orbiter continues to make new observations in the extended phase of its mission, new discoveries are likely in the future related to Saturn's internal structure, the structure of the icy satellites and Titan, the history of Titan's surface and atmosphere, the evolution of the rings, and the on-going plume activity on intriguing Enceladus. NASA and ESA are currently in the process of developing plans for future outer planet exploration which will undoubtedly include following up on many of Cassini Huygens exciting discoveries with future missions to the Saturn system.

Acknowledgements The authors wish to acknowledge Jack J. Lissauer (NASA Ames Research Center) for many useful discussions and comments on an earlier draft of this work.

A portion of this work (TVJ) has been conducted at the Jet Propulsion Laboratory, California Institute of Technology, under a contract with the National Aeronautics and Space Administration.

References

Alibert, Y., et al., 2005. Modeling the Jovian subnebula – I. Thermodynamic conditions and migration of proto-satellites. Astronomy & Astrophysics. 439, 1205–1213.10.1051/0004-6361:20052841.

Amelin, Y., et al., 2002. Lead isotopic ages of chondrules and calcium-aluminum-rich inclusions. Science. 297, 1678–1683.

Anders, E., Grevesse, N., 1989. Abundances of the elements – meteoritic and solar. Geochimica Et Cosmochimica Acta. 53, 197–214.

Anderson, J. D., et al., 2001a. Io's gravity field and interior structure. Journal of Geophysical Research-Planets. 106, 32963–32969.

Anderson, J. D., et al., 2001b. Shape, mean radius, gravity field, and interior structure of Callisto. Icarus. 153, 157–161.

Anderson, J. D., et al., 2005. Amalthea's density is less than that of water. Science. 308, 1291–1293.

Anderson, J. D., et al., 1996a. Gravitational constraints on the internal structure of Ganymede. Nature. 384, 541–543.

Anderson, J. D., et al., 1996b. Galileo gravity results and the internal structure of Io. Science. 272, 709–712.

Anderson, J. D., et al., 1997a. Europa's differentiated internal structure: Inferences from two Galileo encounters. Science. 276, 1236–1239.

Anderson, J. D., et al., 1997b. Gravitational evidence for an undifferentiated Callisto. Nature. 387, 264–266.

Anderson, J. D., Schubert, G., 2007. Saturn's satellite Rhea is a homogeneous mix of rock and ice. Geophysical Research Letters. 34, L02202–L02202.

Anderson, J. D., et al., 1998a. Distribution of rock, metals, and ices in Callisto. Science. 280, 1573–1576.

Anderson, J. D., et al., 1998b. Europa's differentiated internal structure: Inferences from four Galileo encounters. Science. 281, 2019–2022.

Asplund, M., et al., 2006. The solar chemical composition. Nuclear Physics A. 777, 1–4.10.1016/j.nuclphysa.2005.06.010.

Ayliffe, B. A., Bate, M. R., 2009. Circumplanetary disc properties obtained from radiation hydrodynamical simulations of gas accretion by protoplanets. Monthly Notices of the Royal Astronomical Society. 397, 657–665.

Barr, A. C., Canup, R. M., 2008. Constraints on gas giant satellite formation from the interior states of partially differentiated satellites. Icarus. 198, 163–177.10.1016/j.icarus.2008.07.004.

Benz, W., et al., 1988. Collisional stripping of mercurys mantle. Icarus. 74, 516–528.

Bodenheimer, P., et al., Models of the in situ formation of detected extrasolar giant planets. 2000, pp. 2–14.

Boss, A. P., 2007. The solar nebula. In: A. M. Davis, (Ed.), *Treatise on Geochemistry*: Vol. 1, *Meteorites, Comets and Planets*. Elsevier Pergamon.doi:10.1016/B0–08–043751–6/01061–6.

Brown, M. E., Schaller, E. L., 2007. The mass of dwarf planet Eris. Science. 316, 1585–1585.10.1126/science.1139415.

Bryden, G., et al., 1999. Tidally induced gap formation in protostellar disks: Gap clearing and suppression of protoplanetary growth. Astrophysical Journal. 514, 344–367.

Bryden, G., et al., 2000. Protoplanetary formation. I. Neptune. Astrophysical Journal. 544, 481–495.

Buie, M. W., et al., 2006. Orbits and photometry of Pluto's satellites: Charon, S/2005 P1, and S/2005 P2. Astronomical Journal. 132, 290–298.

Cameron, A. G. W., 1978. Physics of primitive solar accretion disk. Moon and the Planets. 18, 5–40.

Cameron, A. G. W., 1981. Elementary and nuclidic abundances in the solar system. In: C. A. Barns, et al., (Eds.), *Essays in Nuclear Astrophysics*. Cambridge University Press, New York.

Canup, R. M., Ward, W. R., 2002. Formation of the Galilean satellites: Conditions of accretion. The Astronomical Journal. 124, 3404–3423.

Canup, R. M., Ward, W. R., 2009. Origin of Europa and the Galilean satellites. In: W. McKinnon, et al., (Eds.), *Europa*. University of Arizona Press, Tucson.

Castillo-Rogez, J. C., et al., 2007. Iapetus' geophysics: Rotation rate, shape, and equatorial ridge. Icarus. 190, 179–202.10.1016/j.icarus.2007.02.018.

Charnoz, S., Morbidelli, A., 2003. Coupling dynamical and collisional evolution of small bodies: An application to the early ejection of planetesimals from the Jupiter-Saturn region. Icarus. 166, 141–156.10.1016/s0019–1035(03)00213–6.

Charnoz, S., et al., 2009. Did Saturn's rings form during the Late Heavy Bombardment? Icarus. 199, 413–428.10.1016/j.icarus.2008.10.019.

Consolmagno, G. J., Lewis, J. S., 1977. Preliminary thermal history models of icy satellites. In: J. A. Burns, (Ed.), *Planetary Satellites*. University of Arizona Press, Tucson, 492–500.

Consolmagno, G. J., Lewis, J. S., 1978. Evolution of icy satellite interiors and surfaces. Icarus. 34, 280–293.

Coradini, A., et al., 1989. Formation of the satellites of the outer solar system – Sources of their atmospheres. In: S. Atreya, et al., (Eds.), *Origin and Evolution of Planetary and Satellite Atmospheres*. University of Arizona Press, Tucson, pp. 723–762.

Cuzzi, J. N., et al., 1993. Particle gas-dynamics in the midplane of a protoplanetary nebula. Icarus. 106, 102–134.

Cuzzi, J. N., Zahnle, K. J., 2004. Material enhancement in protoplanetary nebulae by particle drift through evaporation fronts. Astrophysical Journal. 614, 490–496.

D'Angelo, G., et al., 2003. Thermohydrodynamics of circumstellar disks with high-mass planets. Astrophysical Journal. 599, 548–576.

Davis, A. M. (Ed.), 2004. *Treatise on Geochemistry*: Vol 1. *Meteorites, Comets, and Planets*. Elsevier, Pergamon, Amsterdam-Boston-Heidelberg-London-New York-Oxford-Paris-San Diego-San Francisco-Singapore-Sydney-Tokyo.

Dominik, C., et al., 2007. Growth of dust as the initial step toward planet formation. In: B. Reipurth, et al., (Eds.), *Protostars and Planets V*. University of Arizona Press, Tucson, pp. 783–800.

Dougherty, M. K., et al., 2006. Identification of a dynamic atmosphere at Enceladus with the Cassini magnetometer. Science. 311, 1406–1409.

Durham, W. B., et al., 2005. Cold compaction of water ice. Geophysical Research Letters. 32.L18202, 10.1029/2005gl023484.

Durisen, R. H., et al., 2007. Gravitational instabilities in gaseous protoplanetary disks and implications for giant planet formation. In: B. Reipurth, et al., (Eds.), *Protostars and Planets V*. University of Arizona Press, Tucson, pp. 607–622.

Dyudina, U. A., et al., 2007. Lightning storms on Saturn observed by Cassini ISS and RPWS during 2004–2006. Icarus. 190, 545–555.10.1016/j.icarus.2007.03.035.

Espaillat, C., et al., 2007. On the diversity of the Taurus transitional disks: UX Tauri A and LkCa 15. Astrophysical Journal. 670, L135–L138.

Estrada, P. R., Mosqueira, I., 2006. A gas-poor planetesimal capture model for the formation of giant planet satellite systems. Icarus. 181, 486–509.10.1016/j.icarus.2005.11.006.

Estrada, P. R., et al., 2009. Formation of Jupiter and conditions for accretion of the Galilean satellites. In: W. McKinnon, et al., (Eds.), *Europa*. University of Arizona Press, Tucson.

Fanale, F. P., et al., 1977. Io's surface and the histories of the Galilean satellites. In: J. A. Burns, (Ed.), *Planetary Satellites*. University of Arizona Press, Tucson, pp. 379–405.

Goldreich, P., Tremaine, S., 1980. Disk-satellite interactions. Astrophysical Journal. 241, 425–441.

Goldreich, P., et al., 2004. Final stages of planet formation. Astrophysical Journal. 614, 497–507.

Gomes, R., et al., 2005. Origin of the cataclysmic Late Heavy Bombardment period of the terrestrial planets. Nature. 435, 466–469.

Grevesse, N., et al., 2007. The solar chemical composition. Space Science Reviews. 130, 105–114.10.1007/s11214–007–9173–7.

Grossman, L., 1972. Condensation in primitive solar nebula. Geochimica Et Cosmochimica Acta. 36, 597–619.

Grossman, L., Larimer, J. W., 1974. Early chemical history of solar-system. Reviews of Geophysics. 12, 71–101.

Halliday, A. N., 2007. The origin and earliest history of the Earth. In: A. M. Davis, (Ed.), *Treatise on Geochemistry*: Vol. 1. *Meteorites, Comets, and Planets*. Elsevier, Pergamon. doi:10.1016/B0–08–043751–6/01070–7.

Hansen, C. J., et al., 2006. Enceladus' water vapor plume. Science. 311, 1422–5.

Hubbard, W. B., Anderson, J. D., 1978. Possible flyby measurements of Galilean satellite interior structure. Icarus. 33, 336–341.

Hubickyj, O., et al., 2005. Accretion of the gaseous envelope of Jupiter around a 5–10 Earth-mass core. Icarus. 179, 415–431.10.1016/j.icarus.2005.06.021.

Iess, L., et al., 2007. Gravity field and interior of Rhea from Cassini data analysis. Icarus. 190, 585–593.

Jacobson, R. A., 2004. The orbits of the major Saturnian satellites and the gravity field of Saturn from spacecraft and earth-based observations. Astronomical Journal. 18, 492–501.

Jacobson, R. A., et al., 2006. The GM values of Mimas and Tethys and the liberation of methane. Astronomical Journal. 132, 711–713.

Jewitt, D. C., Sheppard, S. S., 2002. Physical properties of trans-Neptunian object (20000) Varuna. Astronomical Journal. 123, 2110–2120.

Jewitt, D., et al., 2007. *Protostars and Planets V*. University of Arizona Press, Tucson, pp. 863–878.

Johnson, T. V., McGetchin, T. R., 1973. Topography on satellite surfaces and the shape of asteroids. Icarus. 18, 612–620

Johnson, T. V., Lunine, J. I., 2005. Saturn's moon Phoebe as a captured body from the outer Solar System. Nature. 435, 69–71.

Johnson, T. V., et al., 2007 Thermal and dynamical histories of Saturn's satellites: Evidence for the presence of short lived radioactive isotopes. In: R. Guandalini, et al., (Eds.), *The Ninth Torino Workshop on Evolution and Nucleosynthesis in AGB Stars and The Second Perugia Workshop on Nuclear Astrophysics*, Vol. 1001. American Institute of Physics, Perugia, Italy, pp. 262–268.

Joswiak, D. J., et al., 2008. Mineralogical origins of Wild 2 comet particles collected by the Stardust spacecraft. Geochimica Et Cosmochimica Acta. 72, A441–A441.

Kenyon, S. J., Luu, J. X., 1999. Accretion in the early outer solar system. Astrophysical Journal. 526, 465–470.

Khurana, K. K., et al., 1998. Induced magnetic fields as evidence for subsurface oceans in Europa and Callisto. Nature. 395, 777–780.

Kivelson, M. G., et al., 1997. The magnetic field and magnetosphere of Ganymede. Geophysical Research Letters. 24, 2155–2158.

Kivelson, M. G., et al., 2000. Galileo magnetometer measurements: A stronger case for a subsurface ocean at Europa. Science. 289, 1340–1343.

Kivelson, M. G., et al., 1996. Discovery of Ganymede's magnetic field by the Galileo spacecraft. Nature. 384, 537–541.

Kivelson, M. G., et al., 1999. Europa and Callisto: Induced or intrinsic fields in a periodically varying plasma environment. Journal of Geophysical Research-Space Physics. 104, 4609–4625.

Klahr, H., Kley, W., 2006. 3D-radiation hydro simulations of disk-planet interactions – I. Numerical algorithm and test cases. Astronomy & Astrophysics. 445, 747–758.10.1051/0004-6361:20053238.

Kokubo, E., Ida, S., 1998. Oligarchic growth of protoplanets. Icarus. 131, 171–178.

Kuiper, G. P., 1951. In: J. A. Hynek, (Ed.), *Proceedings of a Topical Symposium*. McGraw-Hill, New York, pp. 357–424.

Leisner, J. S., et al., 2008. The interior of Iapetus: Constraints provided by the solar wind interaction. Eos Tans. AGU. 89 (53), Fall Meet. Suppl., Abstract P31C-08.

Levison, H. F., et al., 2007. Planet migration in planetesimal disks. In: B. Reipurth, et al., (Eds.), *Protostars and Planets V*. University of Arizona Press, Tucson, pp. 669–684.

Levison, H. F., et al., 2008. Origin of the structure of the Kuiper belt during a dynamical instability in the orbits of Uranus and Neptune. Icarus. 196, 258–273.10.1016/j.icarus.2007.11.035.

Lewis, J. S., 1971. Satellites of outer planets – their physical and chemical nature. Icarus. 15, 174–185.

Lewis, J. S., 1972. Low-temperature condensation from solar nebula. Icarus. 16, 241–252.

Lewis, J. S., 1973. Chemistry of outer solar system. Space Science Reviews. 14, 401–411.

Lewis, J. S., Prinn, R. G., 1980. Kinetic inhibition of Co and N-2 reduction in the solar nebula. Astrophysical Journal. 238, 357–364.

Lissauer, J. J., 1987. Timescales for planetary accretion and the structure of the protoplanetary disk. Icarus. 69, 249–265.

Lissauer, J. J., 2001. Time for gas planets to grow. Nature. 409, 23–24.

Lissauer, J. J., Stevenson, D. J., 2007. Formation of giant planets. In: B. Reipurth, et al., (Eds.), *Protostars and Planets V*. University of Arizona Press, Tucson, pp. 591–606.

Lissauer, J. J., et al., 2009. Models of Jupiter's growth incorporating thermal and hydrodynamic constraints. Icarus. 199, 338–350.doi:10.1016/j.icarus.2008.10.004.

Lorenz, R. D., et al., 2008. Titan's rotation reveals an internal ocean and changing zonal winds. Science. 319, 1649–1651.

Lunine, J. I., Atreya, S. K., 2008. The methane cycle on Titan. Nature Geoscience. 1, 159–164.

Lupo, M. J., Lewis, J. S., 1979. Mass-radius relationships in icy satellites. Icarus. 40, 157–170.

Mackenzie, R. A., et al., 2008. A non-hydrostatic Rhea. Geophysical Research Letters. 35, L05204–L05204.

Makalkin, A. B., Dorofeeva, V. A., 2006. Models of the protosatellite disk of Saturn: Conditions for Titan's formation. Solar System Research. 40, 441–455.10.1134/s0038094606060013.

Makalkin, A. B., et al., 1999. Modeling the protosatellite circum-Jovian accretion disk: An estimate of the basic parameters. Solar System Research. 33, 456.

Matson, D. L., et al., 2007. Enceladus' plume: Compositional evidence for a hot interior. Icarus. 187, 569–73.

McKeegan, K. D., Davies, A. M., 1.16 Early solar system chronology. In: A. Davis, (Ed.), *Treatise on Geochemistry*: Vol. 1. *Meteorites, Comets, and Planets*. Elsevier, 2007.doi:10.1016/B0–08–043751–6/01147–6.

McKinnon, W. B., 1997. Mystery of Callisto: Is it undifferentiated? Icarus. 130, 540–543.

Merk, R., Prialnik, D., 2003. Early thermal and structural evolution of small bodies in the trans-Neptunian zone. Earth Moon and Planets. 92, 359–374.

Meyer, M. R., et al., 2007. Evolution of circumstellar disks around normal stars: Placing our solar system in context. In: B. Reipurth, et al., (Eds.), *Protostars and Planets V*. University of Arizona Press, Tucson, pp. 573–588.

Morbidelli, A., Crida, A., 2007. The dynamics of Jupiter and Saturn in the gaseous protoplanetary disk. Icarus. 191, 158–171.10.1016/j.icarus.2007.04.001.

Morbidelli, A., et al., 2005. Chaotic capture of Jupiter's Trojan asteroids in the early solar system. Nature. 435, 462–465.

Mosqueira, I., Estrada, P. R., 2003a. Formation of the regular satellites of giant planets in an extended gaseous nebula I: Subnebula model and accretion of satellites. Icarus. 163, 198–231.

Mosqueira, I., Estrada, P. R., 2003b. Formation of the regular satellites of giant planets in an extended gaseous nebula II: Satellite migration and survival. Icarus. 163, 232–255.

Mosqueira, I., Estrada, P. R., 2005. On the origin of the Saturnian satellite system: Did Iapetus form in-situ? Lunar and Planetary Science XXXVI, Lunar and Planetary Institute, Houston. Abstract No. 1951.

Mousis, O., Gautier, D., 2004. Constraints on the presence of volatiles in Ganymede and Callisto from an evolutionary turbulent model of the Jovian subnebula. Planetary and Space Science. 52, 361–370.10.1016/j.pss.2003.06.004.

Nesvorny, D., et al., 2007. Capture of irregular satellites during planetary encounters. Astronomical Journal. 133, 1962–1976.

Nicholson, P. D., et al., 2008. Irregular satellites of the giant planets. In: M. A. Barucci, et al., (Eds.), *The Solar System Beyond Neptune*. University of Arizona Press with Lunar and Planetary Institute, Tucson, pp. 411–424.

Nimmo, F., Matsuyama, I., 2007. Reorientation of icy satellites by impact basins. Geophysical Research Letters. 34.L19203, 10.1029/2007gl030798.

Nimmo, F., et al., 2007. Shear heating as the origin of the plumes and heat flux on Enceladus. Nature. 447, 289–291.

Papaloizou, J. C. B., et al., 2007. Disk-planet ineteractions during planet formation. In: B. Reipurth, et al., (Eds.), *Protostars and Planets V*. University of Arizona Press, Tucson, pp. 655–668.

Pollack, J. B., et al., 1976. Formation of Saturn's satellites and rings, as influenced by Saturn's contraction history. Icarus. 29, 35–48.

Porco, C. C., et al., 2005a. Cassini imaging science: Initial results on Phoebe and Iapetus. Science. 307, 1237–1242.

Porco, C. C., et al., 2005b. Cassini imaging science: Initial results on Saturn's rings and small satellites. Science. 307, 1226–1236.

Porco, C. C., et al., 2006. Cassini observes the active south pole of Enceladus. Science. 311, 1393–1401.

Porco, C. C., et al., 2007. Saturn's small inner satellites: Clues to their origins. Science. 318, 1602–1607.10.1126/science.1143977.

Prinn, R. G., Fegley, B., 1981. Kinetic inhibition of Co and N-2 reduction in circumplanetary nebulae – implications for satellite composition. Astrophysical Journal. 249, 308–317.

Prinn, R. G., Fegley, B., 1989. Solar nebula chemistry: Origin of planetary, satellite, and cometary volatiles. In: S. Atreya, (Ed.), *Origin and Evolution of Planetary and Satellite Atmospheres*. University of Arizona Press, Tucson, Arizona, pp. 78–136.

Rabinowitz, D. L., et al., 2006. Photometric observations constraining the size, shape, and albedo of 2003 EL61, a rapidly rotating, pluto-sized object in the Kuiper Belt. Astrophysical Journal. 639, 1238–1251.

Reipurth, B., et al. (Eds.), 2007. *Protostars and Planets V*. University of Arizona Press, Tucson.

Reynolds, R. T., Cassen, P. M., 1979. Internal structure of the major satellites of the outer planets. Geophysical Research Letters. 6, 121–124.

Safronov, V. S., 1967. Protoplanetary cloud and its evolution. Soviet Astronomy AJ USSR. 10, 650–658.

Safronov, V. S., 1969. *Evolution of the Protoplanetary Cloud and Formation of the Earth and Planets* (Translated in 1972 as NASA TTF-667). Nauka, Moscow.

Safronov, V. S., 1991. Kuiper prize lecture – some problems in the formation of the planets. Icarus. 94, 260–271.

Safronov, V. S., Ruskol, E. L., 1994. Formation and evolution of planets. Astrophysics and Space Science. 212, 13–22.

Schubert, G., et al., 1986. Thermal histories, compositions, and internal structures of the moons of the solar system. In: J. A. Burns, M. S. Matthews, (Eds.), *Satellites*. University of Arizona Press, Tucson, pp. 224–292.

Schubert, G., et al., 2004. Interior composition, structure and dynamics of the Galilean satellites. In: F. Bagenal, et al., (Eds.), *Jupiter: The Planet, Satellites and Magnetosphere*. Cambridge University Press, Cambridge, pp. 281–306.

Schubert, G., et al., 2007. Enceladus: Present internal structure and differentiation by early and long-term radiogenic heating. Icarus. 188, 345–355.

Shu, F. H., et al., 1993. Photoevaporation of the solar nebula and the formation of the giant planets. Icarus. 106, 92–101.

Spencer, J. R., et al., 2006. Cassini encounters Enceladus: Background and the discovery of a south polar hot spot. Science. 311, 1401–1405.

Spergel, D. N., et al., 2007. Three-year Wilkinson Microwave Anisotropy Probe (WMAP) observations: Implications for cosmology. Astrophysical Journal Supplement Series. 170, 377–408.

Squyres, S. W., et al., 1983. The evolution of Enceladus. Icarus. 53, 319–331.

Stansberry, J. A., et al., 2006. The albedo, size, and density of binary Kuiper Belt object (47171) 1999 TC36. Astrophysical Journal. 643, 556–566.

Stern, S. A., Weissman, P. R., 2001. Rapid collisional evolution of comets during the formation of the Oort cloud. Nature. 409, 589–591.

Stevenson, D. J., Lunine, J. I., 1988. Rapid formation of Jupiter by diffusive redistribution of water-vapor in the solar nebula. Icarus. 75, 146–155.

Stevenson, D. J., et al., 1986. Origins of satellites. In: J. A. Burns, M. S. Matthews, (Eds.), *Satellites*. University of Arizona Press, Tucson.

Thomas, P. C., et al., 2007. Shapes of the Saturnian icy satellites and their significance. Icarus. 190, 573–584.

Tobie, G., et al., 2006. Episodic outgassing as the origin of atmospheric methane on Titan. Nature. 440, 61–64.

Tsiganis, K., et al., 2005. Origin of the orbital architecture of the giant planets of the solar system. Nature. 435, 459–461.

Turrini, D., et al., 2008. A new perspective on the irregular satellites of Saturn – I. Dynamical and collisional history. Monthly Notices of the Royal Astronomical Society. 391, 1029–1051.10.1111/j.1365-2966.2008.13909.x.

Turrini, D., et al., 2009. A new perspective on the irregular satellites of Saturn – II. Dynamical and physical origin. Monthly Notices of the Royal Astronomical Society. 392, 455–474.10.1111/j.1365-2966.2008.14100.x.

Waite, J. H., Jr., et al., 2006. Cassini Ion and Neutral Mass Spectrometer: Enceladus plume composition and structure. Science. 311, 1419–1422.

Waite Jr, J. H., et al., 2009. Liquid water on Enceladus from observations of ammonia and ^{40}Ar in the plume. Nature. 460, 487–490.doi:10.1038/nature08153.

Ward, W. R., 1986. Density waves in the solar nebula – Differential lindblad torque. Icarus. 67, 164–180.

Ward, W. R., 1997. Protoplanet migration by nebula tides. Icarus. 126, 261–281.

Warren, P. H., 2007. The moon. In: A. M. Davis, (Ed.), *Treatise on Geochemistry*: Vol. *Meteorites, Comets, and Planets*. Elsevier, Pergamon, 10.1016/B0–08–043751–6/01149-X.

Westphal, A. J., et al., 2008. Stardust interstellar preliminary examination – First results. Meteoritics & Planetary Science. 43, A169–A169.

Wetherill, G. W., 1980. Formation of the terrestrial planets. Annual Review of Astronomy and Astrophysics. 18, 77–113.

Wetherill, G. W., Stewart, G. R., 1993. Formation of planetary embryos – effects of fragmentation, low relative velocity, and independent variation of eccentricity and inclination. Icarus. 106, 190–209.

Wong, M. H., et al., 2008. Oxygen and other volatiles in the giant planets and their satellites. In: G. J. MacPherson, (Ed.), *Oxygen in the Solar System*. Mineralogical Society of America, Chantilly, VA, pp. 241–246.

Yoder, C. F., 1995. Astrometric and geodetic properties of earth and the solar system. In: T. J. Ahrens, (Ed.), *AGU Reference Shelf1: Global Earth Physics, A Handbook of Physical Constants*. American Geophysical Union, Washington D.C., pp. 1–31.

Zahnle, K., et al., 2003. Cratering rates in the outer solar system. Icarus. 163, 263–289.

Chapter 4
The Interior of Saturn

William B. Hubbard, Michele K. Dougherty, Daniel Gautier, and Robert Jacobson

Abstract A source of uncertainty in Saturn interior models is the lack of a unique rotation rate to be ascribed to the deep (metallic-hydrogen) interior. As a result, models are not uniquely constrained by measured gravitational multiple coefficients. Further uncertainty is associated with the effect of a multiplicity of rotation periods due to zonal flows of unknown magnitude and depth (and therefore unknown mass). Nevertheless, the inference that Saturn has a large core of mass 15–20 M_E (Earth masses) is robust. The equation of state of dense hydrogen–helium mixtures is one area where uncertainty has been much reduced, thanks to new first-principles simulations. However, because there is still uncertainty in Saturn's interior temperature profile, a variety of mantle metallicities and core masses could still fit the constraints, and the question of interior helium separation is still unsettled.

Keywords Saturn interior · Saturn atmosphere · Saturn rotation · Jupiter interior

4.1 Diagnostics of Interior Structure and Dynamics

4.1.1 Gravity Field and Shape

The most direct constraints on Saturn's interior mass distribution come from measurements of the highly-oblate planet's size and mass and its response to rotation, as determined by its overall shape, and gravitational multipole moments. As a result of further measurements related to the Cassini mission and other work during the ∼ two decades between the Voyager encounters and the Cassini mission, some fundamental parameters are known with improved precision, while the rotation period(s) has (have) increased uncertainty. Table 4.1 summarizes the current situation. The observed parameters listed are Saturn's mass M, equatorial radius at 1-bar pressure a, polar radius at 1-bar pressure b, deep (solid-body) rotation period P_S, and first three even zonal harmonics J_2, J_4, and J_6. Values for the zonal harmonics were derived using the methods of Jacobson et al. (2006), but are updated in Table 4.1 to be current as of late 2008.

The linear response of the second-degree gravity potential of a liquid body to a uniform rotation rate $\Omega = 2\pi/P_S$ can be written

$$J_2 = \Lambda_2 q, \quad (4.1)$$

where $q = \Omega^2 a^3/GM$ (G = gravitational constant) and Λ_2 is a dimensionless response coefficient that contains information about Saturn's degree of central concentration. For example, a planet of infinite central concentration would have $\Lambda_2 = 0$ and a planet of uniform density (Maclaurin spheroid) would have $\Lambda_2 = 5/4$. As can be seen from Table 4.1, Saturn's $\Lambda_2 \approx 0.11$ is distinctly smaller than Jupiter's $\Lambda_2 = 0.165$, direct proof that Saturn's mass distribution is more centrally condensed.

By symmetry, all odd terms J_3, J_5, ... should be absent in the external potential of a uniformly rotating liquid body, and there is no evidence so far that any such terms are detectable in Saturn's gravity field. According to the theory of nonlinear response to uniform rotation, the leading term in J_4 should go as

$$J_4 \approx -\Lambda_4 q^2, \quad (4.2)$$

with positive Λ_4.

Saturn's axial moment of inertia C is not uniquely constrained by Λ_2, and the Radau–Darwin relation

$$\frac{C}{Ma^2} \cong \frac{2}{3}\left[1 - \frac{2}{5}\sqrt{1 + \left(\frac{5}{2}\frac{q}{e} - 2\right)}\right], \quad (4.3)$$

W.B. Hubbard
Lunar and Planetary Laboratory, University of Arizona, Tucson, AZ, USA

M.K. Dougherty
Physics Department, Imperial College, London, UK

D. Gautier
Observatoire Paris – Site de Meudon, France

R. Jacobson
Jet Propulsion Laboratory, California Institute of Technology, CA, USA

Table 4.1 Parameters constraining Saturn interior structure. Error bars on M/M_E are about equally determined by uncertainties in Saturn's mass and Earth's mass (as given at http://ssd.jpl.nasa.gov) and are not currently an important limitation for interior modeling purposes. Note the inconsistency of the Voyager-era versus Cassini-era value of P_S

Parameter	As of Voyager	As of Cassini	Reference
M (M_E)		95.16 ± 0.02	Jacobson et al. (2006)
a (km)	60268 ± 4		Lindal (1992)
b (km)	54364 ± 10		Lindal (1992)
P_s (s)	38364 ± 7	38745 ± 36	Cecconi and Zarka (2005)
$q = (2\pi/P_s)^2 a^3/GM$	0.15476 ± 0.00009	0.15173 ± 0.00031	
$J_2 \times 10^6$ (observed)		16324.19 ± 0.11	Jacobson et al. (2006)
$\Lambda_2 = J_2/q$	0.1055	0.1076	
C/Ma^2	0.2178	0.2197	
Core mass (M_E)	19.24	18.65	
$J_4 \times 10^6$ (observed)		−939.32 ± 0.98	Jacobson et al. (2006)
$\Lambda_4 = J_4/q^2$	0.039	0.041	
$J_4 \times 10^6$ (theory)	−985	−971	
$J_6 \times 10^6$ (observed)		91 ± 5	Jacobson et al. (2006)
$J_8 \times 10^6$ (assumed)		−10	Jacobson et al. (2006)

where $e = (a − b)/a$ and

$$e/q = (3\Lambda_2 + 1)/2 \quad (4.4)$$

is a very poor approximation for Saturn, overestimating its moment of inertia by almost 50%. Nevertheless, the Radau–Darwin relation is useful for estimating the impact of a change in Saturn's rotation rate on the inferred C/Ma^2.

Post-Voyager measurements of Saturn's magnetic-field rotation period, presumably coupled to the conducting metallic-hydrogen envelope, give values longer by about 381 s, or about 1% (see Table 4.1). This discrepancy is related to difficulties in measuring the rotation rate of Saturn's virtually axisymmetric magnetic field, and may not be reducible by further measurements. Thus, it is fair to ask how robust are the inferences of Saturn's C/Ma^2.

Values of C/Ma^2 given in Table 4.1 for the two different proposed rotation periods are computed for representative models using a theory valid to order q^3 and realistic equations of state for a hydrogen–helium–ice mixture (for Saturn's envelope) and olivine (for Saturn's core). (These models are based on older equations of state and do not represent the current state of the art; they are for demonstration purposes only.) The model values of C/Ma^2 differ by 0.0019, in good agreement with the shift predicted by Radau–Darwin. Figure 4.1 shows a profile of mass density ρ as a function of the average radius s of a level surface (s_0 is the average radius of the 1-bar level surface). This profile is computed for a model fitted to observed values of M, a, J_2, and the post-Voyager value of P_S; the profile for a model fitted to the Voyager-era value of P_S is very similar.

Also given in Table 4.1 is the inferred mass of the rock (olivine) core for the two different rotation rates. They differ by only 0.6 M_E. We thus conclude that a massive Saturn core, mass ∼ one Neptune mass, is a robust result from Saturn modeling and is unlikely to change in the face of continued uncertainty in P_S.

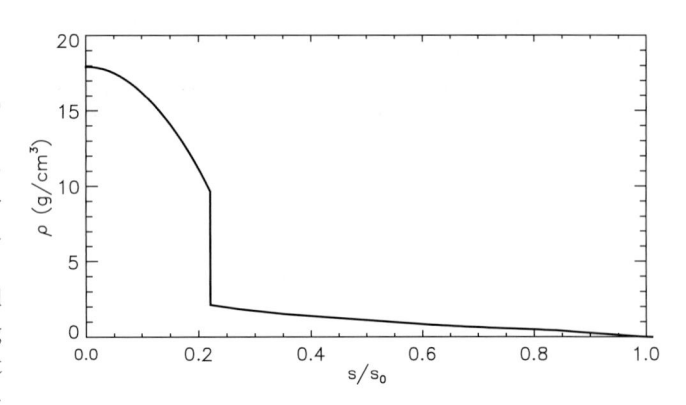

Fig. 4.1 Profile of a typical Saturn model with an envelope of solar-composition hydrogen, helium, and hydrides of C, N, and O. The massive olivine core extends to more than 20% of the radius. The envelope equation of state is obtained from the theory of Saumon, Chabrier, and Van Horn (1995)

4.1.2 Differential Rotation and Equations of State

The rotation state of Saturn's deep interior plays a role in the external gravity and the surface shape of Saturn. In principle, if sufficient mass is involved in differential motions, external gravity coefficients can be affected. Thus, a mismatch between model predictions and observed gravity coefficients could be attributed either to errors in the pressure–density relation, or to errors in the assumed rotation rate(s), or both. This problem has recently appeared in a new investigation of the interior of Jupiter using first-principles thermodynamic relations for hydrogen–helium mixtures (Militzer et al. 2008). The Jupiter model has constant entropy fixed to the measured entropy at 1 bar, with only the core mass and (constant) mantle metallicity as a adjustable parameters. This model fits Jupiter's M, a, and J_2, but the improved, more precise measurement of J_4 is not fitted within the error bar. As we see in Table 4.1, a similar situation is possibly now emerging with Saturn. The error bar on Saturn's observed J_4 is remarkably small, only $\sim 0.1\%$, and the discrepancy with the simple interior models (fitted to J_2, as was done with Jupiter) is far larger. Moreover, the sign of the discrepancy is the same for both planets.

In principle, one might resolve the discrepancy for Saturn (and for Jupiter) by assuming that the planet rotates as a solid body but with a different rotation period P_S than either of the values presented in Table 4.1. In this approach, we fix the response coefficients Λ_2 and Λ_4 and adjust the value of q to match J_2 and J_4. As demonstrated in Table 4.1, this procedure would lead to a value of P_S that differs substantially from any directly measured value, and would therefore be essentially ad hoc.

On the other hand, we have ample evidence for large and possibly variable zonal winds in Saturn's visible atmosphere, and if these winds are deep, they would involve enough mass to affect gravitational coefficients. The surface shape of Saturn can be measured with enough precision, via occultation techniques, to shed some light on this matter. Figure 4.2 shows the shape of the 100-mbar surface as measured by several spacecraft occultations and one stellar occultation.

The shape surfaces in Fig. 4.2 are referenced to a constant-potential surface defined by Saturn's $J_2, J_4, J_6, \ldots$ and solid-body rotation with the Voyager-era period P_S (Table 4.1)

passing through Saturn's equatorial 100-mbar atmosphere. These surfaces are computed by integrating the equations

$$\frac{1}{\rho}\frac{\partial P}{\partial \ell} = \frac{\partial V}{\partial \ell} + \Omega^2 \ell,$$

$$\frac{1}{\rho}\frac{\partial P}{\partial z} = \frac{\partial V}{\partial z}, \qquad (4.5)$$

where P is the pressure, V is the external gravitational potential, and ℓ and z are respectively coordinates perpendicular and parallel to the rotation axis. Eq. 4.5 gives the alignment of isobars as determined by the inertial rotation rate Ω and by itself gives no information about the depth of zonal flows. On the other hand, occultation measurements are directly sensitive to ρ and not P, so the fact that such measurements yield an overall shape surface matching isobars as determined through Eq. 4.5 means that on a planetary scale, Saturn's 100-mbar surface is also an isopycnic surface. This result implies that the Poincaré–Wavre theorem (Tassoul 1978) applies, meaning that $\Omega = \Omega(\ell)$ only (rotation on cylinders), suggesting that the large equatorial uprise (amplitude ~ 100 km when referenced to the Voyager-era period P_S) extending over a broad range of latitudes, is indeed deep-seated and would necessarily involve significant mass.

As Fig. 4.2 makes evident, we have spacecraft- and stellar-occultation measurements of Saturn's shape over planetocentric latitudes ranging from $\sim 65°$ north to $\sim 70°$ south, together with a high density of occultation data points near the equator (Fig. 4.3). Data in our Figs. 4.2 and 4.3 are exhibited slightly differently from the corresponding Fig. 4.9 of Lindal et al. (1985). In our Figs. 4.2 and 4.3, we reference

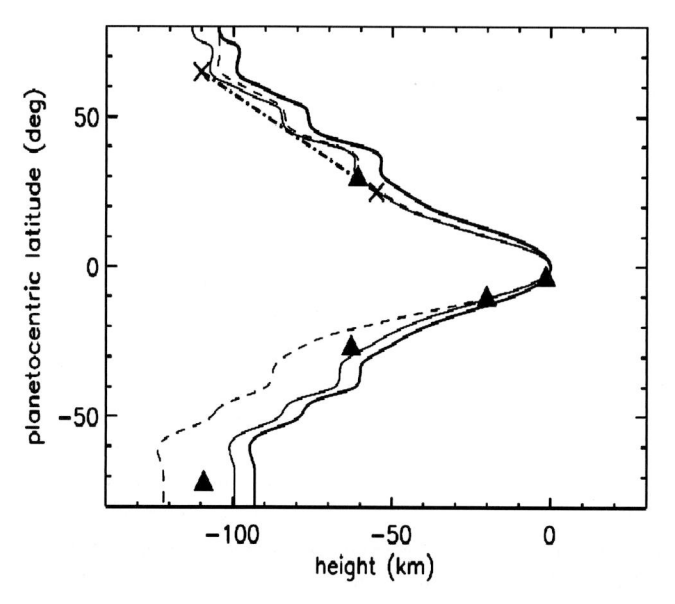

Fig. 4.2 Shape of Saturn's 100-mbar surface as predicted by measured winds (*thin solid and dashed curves* are from Voyager-era measurements of windspeeds, while the *heavy solid curve* is based on 2005-era measurements of reduced equatorial windspeeds), compared with radio-occultation data points (*triangles*; Lindal et al. 1985) and a measurement of a 1989 stellar-occultation central flash (*dashed line* between *crosses*; Nicholson et al. 1995)

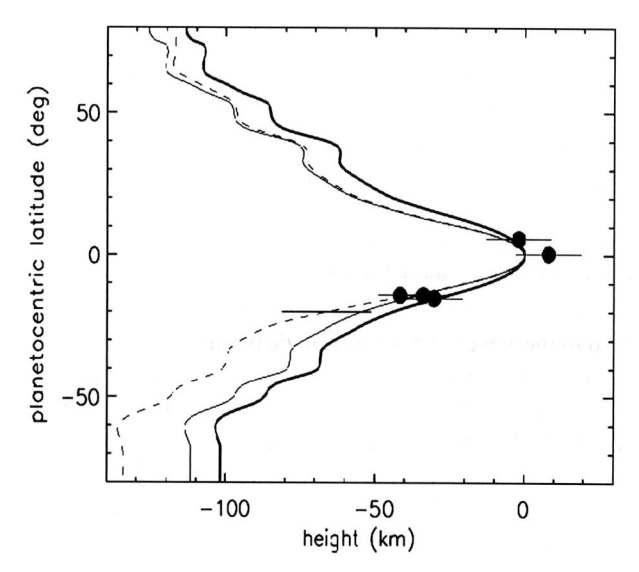

Fig. 4.3 Same as Fig. 4.2, but for the 2-µbar surface. Data points are from the July 1989 occultation of 28 Sgr by Saturn (Hubbard et al. 1997)

the atmospheric distortions to an atmosphere in uniform rotation with Voyager-era period P_S, and we further reference those distortions to the (maximum) equatorial value so as to clearly exhibit the curvature of the equatorial uplift as a function of the zonal windspeed model. The key point, as shown in Fig. 4.9 of Lindal et al. (1985) and in Figs. 4.5 and 4.6 of Hubbard et al. (1997) is that the equatorial curvature of Saturn's atmosphere, as well as the 1-bar values a and b, are well determined from multiple, consistent spacecraft and ground-based observations spanning the 10 year interval 1979–1989. Anderson and Schubert (2007) chose to relax any observational constraint on P_S and sought instead to fit uniformly-rotating interior models to Saturn's a, b and gravity field by varying the parameter q of Table 4.1, leading to $P_S = 37955 \pm 13$ s, a value considerably smaller than either value in Table 4.1, and leading to virtual disappearance of the equatorial excess bulge. Helled et al. (2009) recently investigated Saturn interior structure via models with no interior differential rotation allowed and with P_S treated as a free parameter.

Figure 4.2 suggests that the reported 2005 decrease in equatorial windspeeds (Sanchez-Lavega 2005) would not be consistent with 1979–1989 occultation data on Saturn's equatorial curvature. More evidence bearing on this matter is presented in Fig. 4.3, which shows the corresponding 2-μbar surface with stellar-occultation data points (Hubbard et al. 1997). The importance of Fig. 4.3 is that the shape of Saturn's high atmosphere in 1989 was consistent with the high-speed equatorial winds modeled by Lindal et al. (1985) based on 1980 and 1981 Voyager data and the earlier Pioneer 11 data. The heavy curve based on the much slower equatorial winds observed by Sanchez-Lavega et al. (2005) is not consistent with the 1989 occultation data. In this connection, we note that Choi et al. (2009) report recent Cassini observations showing that slower equatorial winds are underlain by faster winds consistent with the 1979–1989 data.

We conclude that (a) Saturn's excess equatorial bulge was consistently present when observed over a baseline of approximately a decade; (b) the fact that the bulge is present as nearly constant-density surfaces over many scale heights implies that it is deep rooted and may have a gravitational signature; and (c) measurements of Saturn's atmospheric shape at the present epoch could help to elucidate whether the bulge is in fact time-variable.

4.2 Evolution of Saturn

There is a long-standing problem related to Saturn's intrinsic luminosity: over the 4.5 Gyr lifetime of Saturn, not enough thermal energy can be stored to account for the planet's observed luminosity at present. Stevenson (1975) and Stevenson and Salpeter (1977a, b) proposed the standard solution for this problem. The solution involves a proposed phase diagram for binary mixtures of hydrogen and helium, causing separation of the fluid H–He mixture into a He-depleted phase which rises, a He-enriched phase which sinks, and consequent gradual conversion of gravitational potential energy into heat. An initial solar mixture would have a sufficiently large abundance of He relative to H to provide an adequate energy source to prolong Saturn's cooling by the required amount. The problem is that the Stevenson–Salpeter binary phase diagram, which is based on extrapolation of a model of fully pressure-ionized H and He to the relatively low pressures in Saturn (∼10 Mbar), turns out to not have the right behavior to jointly explain the evolution of Saturn and Jupiter.

Figure 4.4 (Fortney and Hubbard 2003) illustrates possible phase diagrams for dense hydrogen with helium impurities.

In Fig. 4.4, the upper boundary of the hatched region marked "HD" delineates where, according to the model of Hubbard and DeWitt (1985), a solar mixture of H and He first phase-separates; the HD model is equivalent to the Stevenson and Salpeter (1977a) model. The hatched region marked "Pfaffenzeller" shows where, according to Pfaffenzeller et al. (1995), phase separation occurs. The hatched region lying between the Jupiter and Saturn adiabats shows a phase-separation region modeled by Fortney and Hubbard (2003) which successfully prolonged Saturn's cooling with no prolongation of Jupiter's cooling. Figure 4.5 (from Fortney and Hubbard 2003) shows results of two cooling/phase-separation models.

The Fortney and Hubbard models predict a present-day helium abundance for Saturn which could be investigated with Cassini data. The predictions are self-consistent in the sense that they use reasonable input thermodynamics and are fitted to the observed present-day Saturn luminosity, but they are based upon an ad-hoc model for the H–He phase diagram. With these caveats, the predicted Saturn atmosphere He abundance is $Y \approx 0.185$ to 0.200, where Y is the helium mass fraction. This number may be compared with the result derived by Conrath and Gautier (2000) from a reanalysis of Voyager data: $Y = 0.18$ to 0.25.

Given that the thermodynamics of dense H–He mixtures can now be calculated from first-principles simulations, the Fortney and Hubbard scenario needs to be updated. Initial investigations of jovian-planet structure based on the new simulations are just beginning to be published, and numerous discrepancies based on modeling discrepancies are beginning to appear.

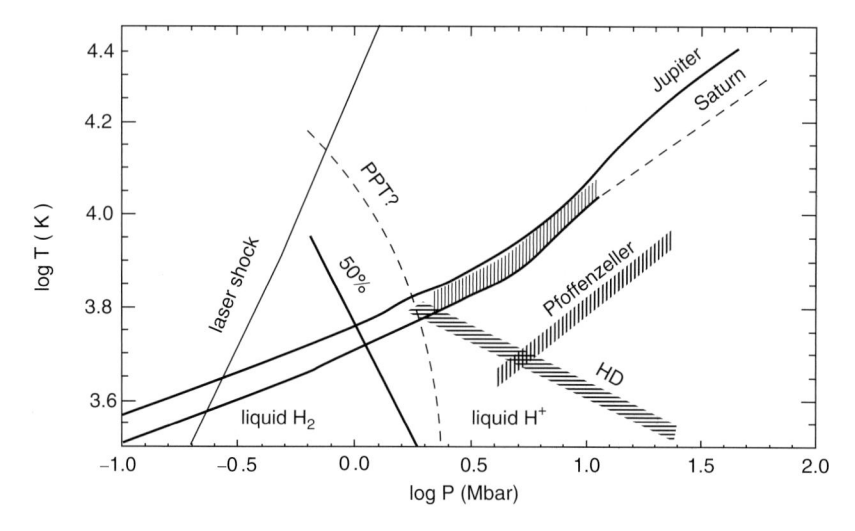

Fig. 4.4 Interior adiabats for present-day Jupiter and Saturn (*heavy curves*), together with possible regions of He–H phase separation (hatched regions). The trajectory marked "laser shock" shows an experimentally-accessible regime. The dashed curve marked "PPT" shows a putative "plasma phase transition", a first-order transition from undissociated hydrogen to ionized hydrogen. The curve marked "50%" shows where, according to one model, molecular hydrogen is 50% pressure-dissociated

Fig. 4.5 Saturn's effective temperature T_{eff} (total luminosity is proportional to T_{eff}^4) versus age of Saturn, for three different cooling models. Circles denote homogeneous (no phase separation) cooling, while models 8 and 9 have He separation with different He-solubility constants (model 8 is more realistic). The horizontal dashed line shows Saturn's observed effective temperature

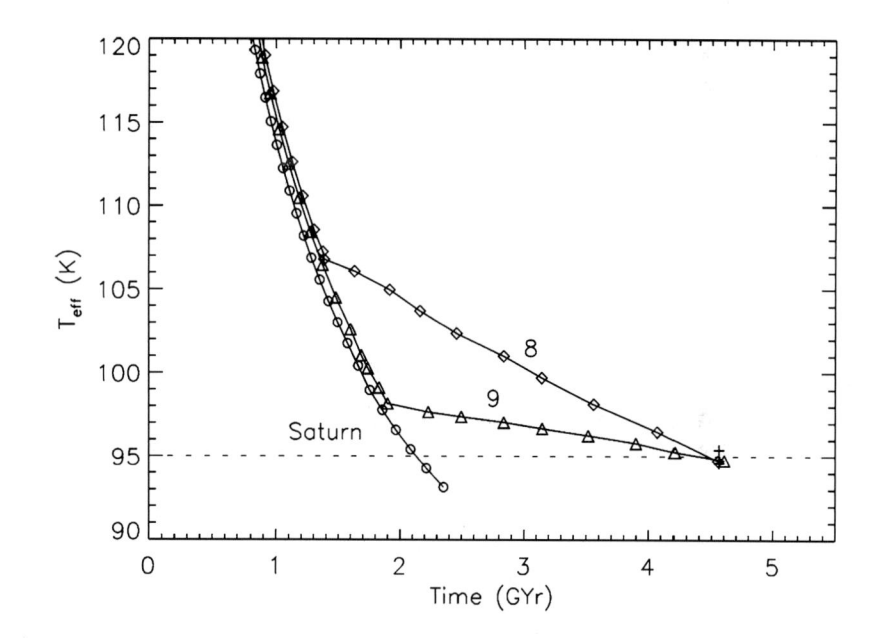

4.3 Coupling of Detailed Evolutionary Models for Saturn to the Helium Partitioning Problem, and Comparison with Jupiter

Jupiter and Saturn are the two giant planets predominantly of hydrogen and helium, so a proper theoretical synthesis of their interior structures must fit numerous simultaneous observational and theoretical constraints, beginning with a consistent thermodynamic model of hydrogen–helium mixtures. This is a daunting agenda, which is only beginning to be addressed. In 2008, the first papers based on realistic simulations of dense H–He appeared (Nettelmann et al. 2008; Militzer et al. 2008), and the discrepancies which emerged point out a number of critical issues. Both of these initial papers were devoted to Jupiter, but Saturn models should appear in the near future.

The Jupiter model published by Militzer et al. (2008) indicates that the so-called PPT sketched in Fig. 4.4 does not exist. Instead, hydrogen gradually metallizes over a range of pressures in the megabar range. Militzer's hydrogen simulations include He impurities at approximately solar concentration. The effect of the helium is to moderate a noticeable depression of the adiabatic temperature gradient in pressure range corresponding to gradual metallization, but the temperature increase in the Jupiter interior model is still much slower than depicted in Fig. 4.4. An interesting consequence of the lower interior temperatures is that Militzer et al. infer a Jupiter core mass similar to the Saturn core masses given in Table 4.1.

Since the present-day Jupiter model of Militzer et al. (2008) is substantially colder than the model depicted in Fig. 4.4, it is possible that a new Jupiter cooling model will require an additional heat source such as He separation. A consistent treatment of Saturn evolution based on the same approach will require further high-precision mapping of the phase diagram, at even lower temperatures. We note that so far such a first-principles phase diagram has not yet been incorporated in Saturn and Jupiter models, and generating such a diagram is a difficult problem, for it requires sufficient accuracy and precision in the simulation data points to accurately calculate second derivatives of thermodynamic variables. We do have a clue: The Fortney and Hubbard investigation demonstrates that the phase-separation locus in temperature–pressure space must have a maximum temperature limit that increases with pressure, which rules out the behavior indicated by the Stevenson–Salpeter (or Hubbard–DeWitt) model.

4.4 Summary

The study of Saturn's interior has reached a new phase in both theory and observation. Old paradigms that have been overturned include (a) a precise Saturn deep-rotation period with a claimed uncertainty of ± 7 s; (b) the original Stevenson He-immiscibility diagram with He miscibility increasing as a function of pressure; (c) hydrogen phase diagrams with a first-order transition associated with hydrogen metallization. The consequence of (a) is not so severe, because we still know Saturn's primary rotational disturbance with enough precision to infer the existence of a massive core. The revision to (b) can in principle be verified by high-precision many-body simulations, a process that is ongoing. The disappearance of (c) does not necessarily rule out discontinuities in the hydrogen-rich outer layers of Jupiter or Saturn, for a He-immiscibility boundary could produce such a discontinuity. The concentration of He and C, N, O hydrides on either side of such a boundary should properly follow from consistent thermodynamics rather than from ad hoc assumptions. It is premature to quote theoretical predictions for such concentrations for Saturn, since Saturn has not yet been modeled with the improved, first-principles hydrogen–helium equations of state. When such modeling has been reported, it will be possible to compare the results with determinations of atmospheric abundances from Cassini observations, such as, e.g., Hersant et al. (2008), and to update predictions such as Mousis et al. (2006).

The question of whether Saturn's higher-order gravity harmonics can be used to constrain interior structure, or whether they will be primarily sensitive to envelope dynamics, is so far unresolved. Relevant data should come from the 2016 Jupiter orbiter mission, Juno. Hubbard (1999) argued that Jupiter gravity harmonics J_n with n greater than about 8 should be dominated by the dynamics of Jupiter's outer layers. Militzer et al. (2008) argued that dynamical effects may even enter starting with $n = 4$. We have argued in this chapter that a similar situation may apply to Saturn, implying that Juno-like gravity measurements at Saturn will be most illuminating.

Acknowledgments We thank M. Podolak and another referee for helpful comments. Figures 4.4 and 4.5 are reprinted from Icarus 164, J. J. Fortney and W. B. Hubbard, "Phase separation in giant planets: inhomogeneous evolution of Saturn", pp. 228–243, Copyright 2003, with permission from Elsevier.

References

Anderson, J. D., and Schubert, G. 2007, Science 317, 1384
Cecconi, B., and Zarka, P. 2005, J. Geophys. Res. 110, A12203
Choi, D. S., Showman, A. P., and Brown, R. H. 2009, J. Geophys. Res., 114, Issue E4, CiteID E04007.
Conrath, B. J., and Gautier, D. 2000, Icarus 144, 124–134
Fortney, J. J., and Hubbard, W. B. 2003, Icarus 164, 228–243
Helled, R., Schubert, G., and Anderson, J. D. 2009, Icarus 199, 368–377.
Hersant, F., Gautier, D., Tobie, G., and Lunine, J. I. 2008. Planetary Space Sci. 56, 1103–1111
Hubbard, W. B., Porco, C. C., Hunten, D. M., Rieke, G. H., Rieke, M. J., McCarthy, D. W., Haemmerle, V., Haller, J., McLeod, B., Lebofsky, L.A., Marcialis, R., Holberg, J. B., Landau, R., Carrasco, L., Elias, J., Buie, M. W., Dunham, E. W., Persson, S. E., Boroson, T., West, S., French, R. G., Harrington, J., Elliot, J. L., Forrest, W. J., Pipher, J. L., Stover, R. J., Brahic, A., and Grenier, I. 1997, Icarus 130, 404–425
Hubbard, W. B., and DeWitt, H. E. 1985, Astrophys. J. 290, 388–393
Hubbard, W. B. 1999, Icarus 137, 357–359
Jacobson, R. A., Antreasian, P. G., Bordi, J. J., Criddle, K. E., Ionasescu, R., Jones, J. B., Mackenzie, R. A., Meek, M. C., Parcher, D., Pelletier, F. J., Owen, W. M., Jr., Roth, D. C., Roundhill, I. M., and Stauch, J. R. 2006, Astron. J. 132, 2520–2526.
Lindal, G. F., Sweetnam, D. N., and Eshleman, V. R. 1985, Astron. J. 90, 1136–1146
Lindal, G. F. 1992, Astron. J. 103, 967–982
Militzer, B., Hubbard, W. B., Vorberger, J., Tamblyn, I., and Bonev, S. A. 2008, Astrophys. J. Lett. 688, L45-L48
Mousis, O., Alibert, Y., and Benz, W. 2006, Astron. Astrophys. 449, 411–415.
Nettelmann, N., Holst, B., Kietzmann, A., French, M., Redmer, R., and Blaschke, D. 2008, Astrophys. J. 683, 1217–1228

Nicholson, P., McGhee, C. A., and French, R. G. 1995, Icarus 113, 57–83

Pfaffenzeller, O., Hohl, D., and Ballone, P. 1995, Phys. Rev. Lett. 74, 2599–2602

Sanchez-Lavega, A. 2005, Science 307, 1223–1224

Saumon, D., Chabrier, G., and Van Horn, H. M. 1995, Astrophys. J. Suppl. 99, 713–741

Stevenson, D. J. 1975, Phys. Lett. A 58, 282–284

Stevenson, D. J., and Salpeter, E.E. 1977a, Astrophys. J. Suppl. 35, 221–237

Stevenson, D. J., and Salpeter, E. E. 1977b, Astrophys. J. Suppl. 35, 239–261

Tassoul, J.-L., Theory of Rotating Stars, Princeton University Press, 1978.

Chapter 5
Saturn: Composition and Chemistry

Thierry Fouchet, Julianne I. Moses, and Barney J. Conrath

Abstract The chapter reviews our current knowledge of the molecular, elemental, and isotopic composition and atmospheric chemistry in Saturn's shallow atmosphere, i.e., between the cloud levels and the homopause. We do not restrict the review to Cassini's results, as past and current ground-based or Earth-based observations are still fundamental to draw a complete picture of the planet. We address the global composition and its importance in studying the origin of the planet, and the meridional and vertical gradients in composition, stressing their insights into Saturn's dynamics. We present the current 1D thermochemical and photochemical models, how these models fare to reproduce the observed composition, and the first attempts to design 2D chemical models. We present some directions to improve our knowledge of Saturn's composition both from observations, modelling and laboratory experiments.

5.1 Introduction

Atmospheric composition and chemistry directly determine Saturn's horizontal and vertical structure, dynamics and visible appearance. Composition affects the thermal structure through the radiative balance between solar energy deposition and thermal infrared emission. Condensation of volatiles and photochemistry determine the cloud structure, hence the visible appearance of the planet. Chemistry, either photochemistry or thermochemistry, strongly shapes the molecular composition, and vertical, meridional and seasonal gradients in composition are observed. In turn, gradients in chemical constituent abundances and latent heat release by condensibles contribute to thermal and density gradients that power atmospheric dynamics, which feeds back to affect

T. Fouchet
LESIA, Observatoire de Paris, Meudon, France

J.I. Moses
Lunar and Planetary Institute, Houston, TX, USA

B.J. Conrath
Department of Astronomy, Cornell University, Ithaca, NY, USA

the composition and thermal structure. Hence, observations of the atmospheric composition provide constraints on the present-day chemistry, dynamics, energy balance, appearance and other phenomena, whilst also providing a window into the past, to the conditions at the time of formation and evolution of the gas giants. Indeed, the elemental abundances of carbon, oxygen, nitrogen, and sulfur are key parameters needed for planetary formation scenarios. In this chapter, we address the chemical composition and the chemistry of the atmosphere from the homopause level down to below the cloud base. The composition and the structure of the interior of the planet are reviewed in Chapter 4, while the composition above the homopause level is covered in Chapter 8.

Without any *in situ* measurements to probe Saturn's composition, we have to rely exclusively on remote sensing, and more specifically on spectroscopy which provides the most powerful way to quantitatively measure gaseous composition. The full electromagnetic spectrum, from the ultraviolet to radio wavelengths, is used in order to cover the greatest possible vertical range. The early spectrometric studies of Saturn's atmosphere are reviewed in Chapter 2. However, the impact of Cassini on our knowledge of Saturn's chemical composition is still relatively immature except in some specific areas. For this reason, we do not restrict our review to Cassini results, but extend our discussion to ground-based and space-based observations up to 20 years ago, when relevant, in order to present a comprehensive and consistent picture of Saturn's composition. Ground-based observations concomitant with the Cassini mission can also prove very useful. In parallel with observations, modelers have been developing chemical models, involving either thermochemistry or photochemistry, in order to interpret and to understand the reasons for the observed composition. Many of these models were put together prior to Cassini's arrival at Saturn and need refinements in light of the newest Cassini and ground-based studies.

The chapter is divided into two large sections. Section 5.2 reviews the observed composition, starting with the problem of the global composition in helium (Section 5.2.1.1). Then, we gather the different results on the carbon (Section 5.2.1.2), nitrogen and sulfur (Section 5.2.1.3) compositions, while the isotopic composition is discussed in Section 5.2.1.4. We stress that Saturn's global oxygen content still cannot be determined from remote sensing studies at present. Section 5.2.2 reviews our current measurements of species that are not in thermodynamic equilibrium at the cloud level, but are rather rapidly transported from the deep interior. In Section 5.2.3 we discuss how photochemistry affects the observed composition. First we address the upper troposphere, then we present the family of detected hydrocarbons in the middle atmosphere and how their abundances vary horizontally and vertically. Section 5.2.4 reviews the oxygen-bearing species observed to be present in the stratosphere, which implies an external oxygen flux to Saturn. Section 5.3 presents the current status of modelling the observed composition, starting with the tropospheric chemistry, either due to thermochemical reactions (Section 5.3.1.1) or photochemical reactions (Section 5.3.1.2). In Section 5.3.2 we discuss the hydrocarbon stratospheric photochemistry. We discuss the important chemical reactions and how one-dimensional models fare in terms of reproducing the observed composition. We then describe the one-dimensional seasonal models and the first attempts to design two-dimensional models and their difficulties in fitting the observed meridional gradient in hydrocarbons abundances. Finally, Section 5.3.4 reviews our current understanding of the oxygen chemistry induced in the stratosphere by the external flux.

5.2 Observed Composition

5.2.1 Major Gases or Bulk Composition

The enrichment of Saturn in heavy elements relative to the Sun constitutes an essential constraint for formation scenarios. If our views on Saturn's composition have been improving over recent years and during the Cassini prime mission, our knowledge of the solar elemental composition has also evolved significantly. In the literature, different authors have compared Saturn's composition with different solar values, without explicitly stating their reference, making intercomparisons difficult. In addition, authors have also used the term *mixing ratio* both for volume mixing ratio and for the mixing ratio relative to H_2, different by about 20%. To avoid such a caveat in this Chapter, we quote all the measurements in Saturn in terms of mole fraction, or the equivalent volume mixing ratio. The mole fractions are then converted in terms of enrichment relative to the Sun, using the solar composition proposed by Grevesse et al. (2007). Hence, our values for Saturn enrichment may not correspond to the values quoted in the cited references, which used an earlier solar composition.

Note, however, that some caveats on the solar enrichment value still exist because of the necessity of making assumptions about the partitioning of the elements into different molecular species (e.g., some oxygen may be tied up in non-H_2O condensates in Saturn's deep atmosphere so that water may not represent the full complement of Kronian oxygen, Visscher and Fegley 2005, and other similar assumptions for the other elements come into play).

5.2.1.1 Helium Abundance

Knowledge of the helium abundance in the giant planets is important for studies of the evolutionary history of these bodies. During their formation, hydrogen and helium were acquired from the primitive solar nebula. Consequently, the volume mixing ratio of helium to molecular hydrogen $[He]/[H_2]$ is believed to have been initially uniform and equal to the protosolar value. During the evolution of Jupiter and Saturn, fractionation processes have modified this distribution. In particular, the immiscibility of helium in metallic hydrogen with resulting gravitational separation in the interiors of these planets is expected to have resulted in enhancement of helium in their deep interiors with depletion in the outer, molecular envelope. Measurements of $[He]/[H_2]$ in the observable atmosphere can strongly constrain evolutionary theories of these bodies.

A combination of Voyager infrared spectrometer (IRIS) and radio occultation (RSS) measurements were used to determine the helium abundance in the upper tropospheres of all four giant planets. Analysis of the RSS data yields a profile of T/m, the ratio of the temperature to the mean molecular mass. For an assumed atmospheric composition, the mean molecular mass and microwave refractivity coefficient can be calculated, and a profile of temperature versus barometric pressure $T(p)$ can be obtained. A radiative transfer code is used with this profile to calculate a theoretical thermal emission spectrum that is compared with measured spectra acquired near the occultation point on the planet. With this approach, a helium-to-hydrogen mixing ratio $[He]/[H_2]$ of 0.110 ± 0.032, corresponding to a mass fraction $Y = 0.18 \pm 0.04$, was obtained for Jupiter (Gautier et al. 1981; Conrath et al. 1984), representing a modest depletion with respect to the protosolar value $Y = 0.28$ (Profitt 1994), while a remarkably low value of 0.034 ± 0.024 ($Y = 0.06 \pm 0.05$) was found for Saturn (Conrath et al. 1984). Subsequent *in situ* measurements of He from the Galileo probe gave a value of $[He]/[H_2] = 0.157 \pm 0.003$ on Jupiter (von Zahn et al. 1998; Niemann et al. 1998). The Voyager Jupiter result can be made consistent with the Galileo result only if the nominal radio occultation profile (Lindal 1992) is made cooler at all atmospheric levels by ~ 2 K (Conrath and Gautier 2000). This suggests the presence of systematic errors in the Voyager measurements for Jupiter and raises the possibility of similar errors in the helium determination for Saturn.

The detailed shape of the thermal emission spectrum in the far infrared ($v < 600$ cm^{-1}) provides an additional constraint on the helium abundance. The differential spectral dependence of the S(0) and S(1) collision-induced absorption lines and translational continuum of H$_2$ is a function of the relative contribution of H$_2$–H$_2$ and H$_2$–He interactions. Conrath and Gautier (2000) used this dependence to obtain a new determination of the Saturn He abundance using Voyager IRIS measurements only. In addition to helium, this portion of the spectrum is also sensitive to the vertical thermal structure and the molecular hydrogen ortho/para ratio, and it is necessary to simultaneously retrieve the temperature profile, and the para H$_2$ profile along with the He abundance. Separation of these parameters is dependent on relatively subtle effects of each on the shape of the collision-induced H$_2$ spectrum. As a consequence, the retrieval problem is highly "ill-posed" and requires the use of *a priori* constraints in the inversion algorithm. These take the form of low-pass filtering of the temperature and para H$_2$ profiles. Because of the weak separability of the parameters, the results are strongly dependent on these constraints, and only rather broad limits can be set on the He abundance. Using this approach, Conrath and Gautier (2000) obtained values of [He]/[H$_2$] in the range 0.11–0.16 (corresponding to a helium mass fraction range 0.17–0.24), significantly larger than the value obtained with the IRIS/RSS approach. The IRIS/RSS method is strongly sensitive to systematic errors in the RSS and/or the IRIS measurements, while the IRIS-only inversion approach is sensitive to any factors that can affect the differential spectral shape such as spectrally dependent tropospheric aerosol opacities that are poorly constrained. Given these uncertainties in the Voyager results, a new determination of the Saturn helium abundance became a major Cassini science objective. The primary approach used in the Cassini analysis makes use of Composite Infrared Spectrometer (CIRS) measurements along with Cassini RSS results in a manner similar to the Voyager IRIS/RSS analysis. The range of latitudes and observational geometries covered by the Cassini radio occultations can lead to a better understanding of the limitations and possible error sources associated with this method. As an additional constraint, direct inversions of CIRS spectra are also being pursued. The extended spectral range to lower wavenumbers, along with the large spatial and temporal coverage available from Cassini, permits better separation of the parameters affecting the shape of the CIRS spectrum, which may lead to better constraints on the helium abundance with the CIRS-only approach than was possible with IRIS.

During the course of the Cassini prime mission, Saturn atmospheric radio occultations have been acquired over a range of latitudes and occultation geometries, along with near-simultaneous CIRS spectral measurements. Preliminary results based on eleven low-latitude occultation points, all within Saturn's equatorial jet, give a mean value of [He]/[H$_2$] = 0.08, corresponding to a He mass fraction of 0.13 or a mole fraction of 0.07 (Gautier et al. 2006). An example of a spectral fit is shown in Fig. 5.1. In this analysis, it was assumed that only hydrogen, helium, and methane contribute significantly to the mean molecular mass and mean refractivity coefficient, and a methane-to-hydrogen mixing ratio [CH$_4$]/[H$_2$] = 4.86×10^{-3} was used. Initial results from the CIRS-only approach of direct spectral inversion yield somewhat higher values, consistent with the Voyager

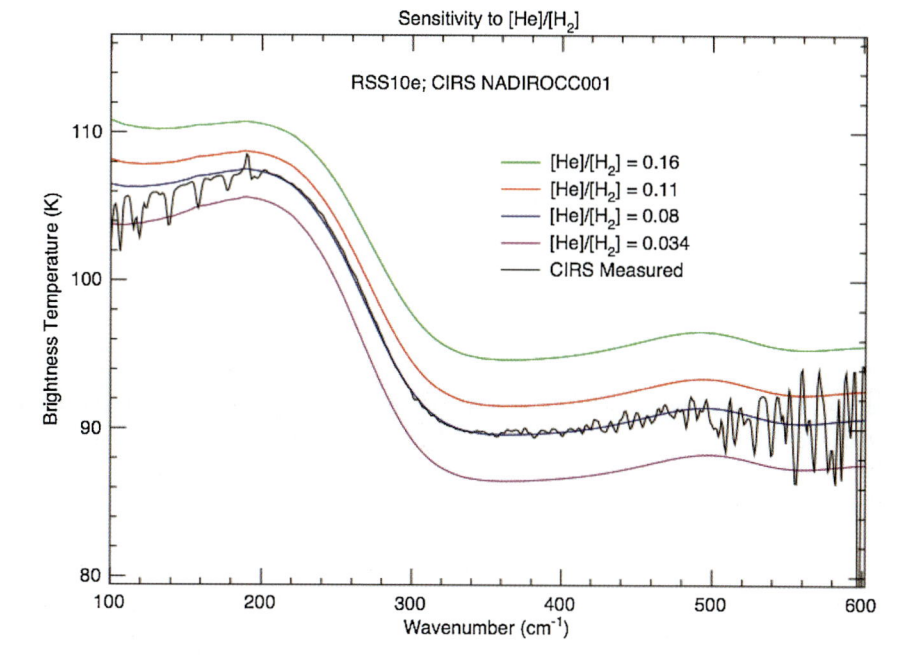

Fig. 5.1 Spectra calculated from an RSS T/m profile at 7.4°S, illustrating the sensitivity of the CIRS/RSS method to helium. The value of [He]/[H$_2$] assumed in each case is indicated. An average of 450 CIRS spectra acquired near the occultation point is shown for comparison. Only the spectral region between 220 and 500 cm^{-1} is used in the analysis

IRIS-only results of Conrath and Gautier (2000). However, the apparent discrepancy between the CIRS/RSS results and the CIRS-only results once again suggests the possibility of systematic errors. A detailed error analysis is in progress. Error sources considered include uncertainties in the absolute calibration of CIRS, gaseous absorption coefficient errors, effects of additional opacity sources such as clouds and hazes, uncertainties in radio frequency refractivity coefficients, and uncertainties in the gravitational potential surfaces used in the RSS analyses. Of particular concern is the sensitivity of the retrieved radio occultation T/m profiles to perturbations in the gravitational potential surfaces associated with the wind field. In a preliminary study of this sensitivity, two different wind models have been used. In one model, the shape of the potential surfaces is based on the Voyager cloud-tracked winds, assumed to be uniform parallel to the planetary rotation axis (barotropic). The second model assumes no winds and yields a temperature profile $\sim$2 K warmer for a given molecular mass, resulting in a value of $[He]/[H_2]$ that is $\sim$0.02 smaller than that obtained with the barotropic model (Flasar et al. 2008). Further investigations into the effects of the wind field are being carried out using radio occultations and CIRS spectra acquired at higher latitudes away from the equatorial jet. Although major uncertainties remain to be resolved, the preliminary values from both the CIRS/RSS and CIRS-only analyses fall within the He mass fraction range 0.11–0.25 suggested by evolutionary modeling results (see for example Hubbard et al. 1999; Guillot 1999), and imply at least some depletion from the proto solar value.

Some constraints on the helium abundance can also be supplied by Cassini UVIS occultation data. Preliminary UVIS results are presented in Chapter 8.

5.2.1.2 The Para Fraction of Molecular Hydrogen

Molecular hydrogen comes in two states, the ortho state, corresponding to odd rotational quantum numbers, and the para state corresponding to the even quantum rotational numbers. At deeper atmospheric levels, the ortho-para ratio approaches its 3:1 high temperature limit. In the shallow atmosphere, molecular hydrogen can depart from its thermodynamic equilibrium ortho-para ratio because of the slow conversion between the ortho and para states. An upwelling air parcel tends to retain its initial para hydrogen fraction (f_p) corresponding to the warmer temperatures of the deeper levels. As a consequence, its value of f_p can be smaller than the equilibrium value ($f_{p\,e}$) associated with the cooler temperatures of the upper troposphere. Hence, the hydrogen para fraction is of considerable interest to trace the dynamics in Saturn's upper troposphere. Since the ortho-para conversion rate can be shortened through surface catalytic processes, f_p can also be influenced by the distribution of atmospheric aerosols. The para hydrogen fraction can be measured by remote sensing using the collision-induced H_2 continuum. Conrath et al. (1998) and Fletcher et al. (2007a) both retrieved Saturn's f_p in the upper troposphere, between 400 and 100 mbar, respectively from Voyager/IRIS and Cassini/CIRS. The different patterns are not easy to interpret as the difference $f_{p\,e} - f_p$ changes sign with altitude and latitude. Conrath et al. (1998) found a f_p minimum near 60°S and a f_p maximum near 15°S, consistent with upward motion near 60°S and downwelling in the tropics. The northern hemisphere, just emerging from winter, had higher f_p. In contrast, Fletcher et al. (2007a) found the lowest f_p at the equator, a summer Southern Hemisphere with $f_{p\,e} - f_p > 0$ and a winter Northern Hemisphere with $f_{p\,e} - f_p < 0$. Since the seasons probed by Voyager and Cassini are similar, closer agreement might be expected. Because of the low upper tropospheric lapse rate on Saturn, especially in the summer hemisphere, the spectral information content on f_p is limited. As a consequence, the retrievals are strongly dependent on the *a priori* constraints assumed in the inversion algorithms, which may account for these differences, at least in part. This topic remains to be investigated in detail. As Cassini CIRS data continue to be acquired, hemispheric asymmetries associated with changing seasonal conditions will be better characterized.

5.2.1.3 Carbon Elemental Composition

The elemental composition of carbon is determined through the measurements of the methane mole fraction, since methane is well-mixed, does not condense, and is expected to be the thermodynamically stable form of carbon throughout the atmosphere. Before the review by Prinn et al. (1984), all the methane measurements in Saturn had been obtained in the solar reflected spectral range. Although all the measurements concluded that carbon was enriched with respect to the solar composition, the magnitude of the enrichment varied widely from one study to another. Since 1984, the quest to measure Saturn's C/H ratio in the visible and in the near-infrared spectral regions has continued. The principle remains the same: one needs to observe simultaneously an H_2 and a CH_4 absorption of similar strength and, if possible, close to each other in the spectral range. With this technique, the two gaseous absorptions probe similar pressure levels and can be modeled with the same cloud structure.

Recent studies have taken advantage of the improved knowledge of methane absorption coefficients, and of new and more efficient methods in multiple scattering numerical modeling. However, the large dispersion, and the incompatibility of the different measurements within their respective error bars, has not disappeared, as shown in Table 5.1. This large dispersion in methane mole fraction inferred from solar

Table 5.1 CH$_4$ volume mixing ratios and C/H enrichments relative to the solar composition. The C/H enrichments are calculated using the solar composition of Grevesse et al. (2007) and a H$_2$ mole fraction of 0.92. The measurements in the thermal infrared are the most reliable, the value measured by Fletcher et al. (2009a) being the most accurate

CH$_4$ vmr	Enrichment	Wavelength	Reference
$(3.9 \pm 0.4) \times 10^{-3}$	8.6 ± 0.9	0.725–1.01 μm	Trafton (1985)
$(2.1^{+0.8}_{-0.2}) \times 10^{-3}$	$4.6^{+1.8}_{-0.4}$	0.6 μm	Killen (1988)
$(3.0 \pm 0.6) \times 10^{-3}$	6.6 ± 1.3	0.46–0.94 μm	Karkoschka and Tomasko (1992)
$(2.0 \pm 0.5) \times 10^{-3}$	4.4 ± 1.1	1.7 μm and 3.3 μm	Kerola et al. (1997)
$(4.3^{+2.3}_{-1.3}) \times 10^{-3}$	$9.5^{+5.1}_{-2.9}$	7.7 μm	Courtin et al. (1984)
$(4.5^{+1.1}_{-1.3}) \times 10^{-3}$	$10.0^{+2.4}_{-2.9}$	8–11 μm	Lellouch et al. (2001)
$(4.5 \pm 0.9) \times 10^{-3}$	10.0 ± 2.0	96–136 μm	Flasar et al. (2005)
$(4.7 \pm 0.2) \times 10^{-3}$	10.4 ± 0.4	96–136 μm	Fletcher et al. (2009a)

Fig. 5.2 Four rotational lines of CH$_4$ observed by CIRS are compared with synthetic spectra computed for volume ming ratios of 3.0×10^{-3}, 4.5×10^{-3} and 8.0×10^{-3}. Adapted from Fletcher et al. (2009a)

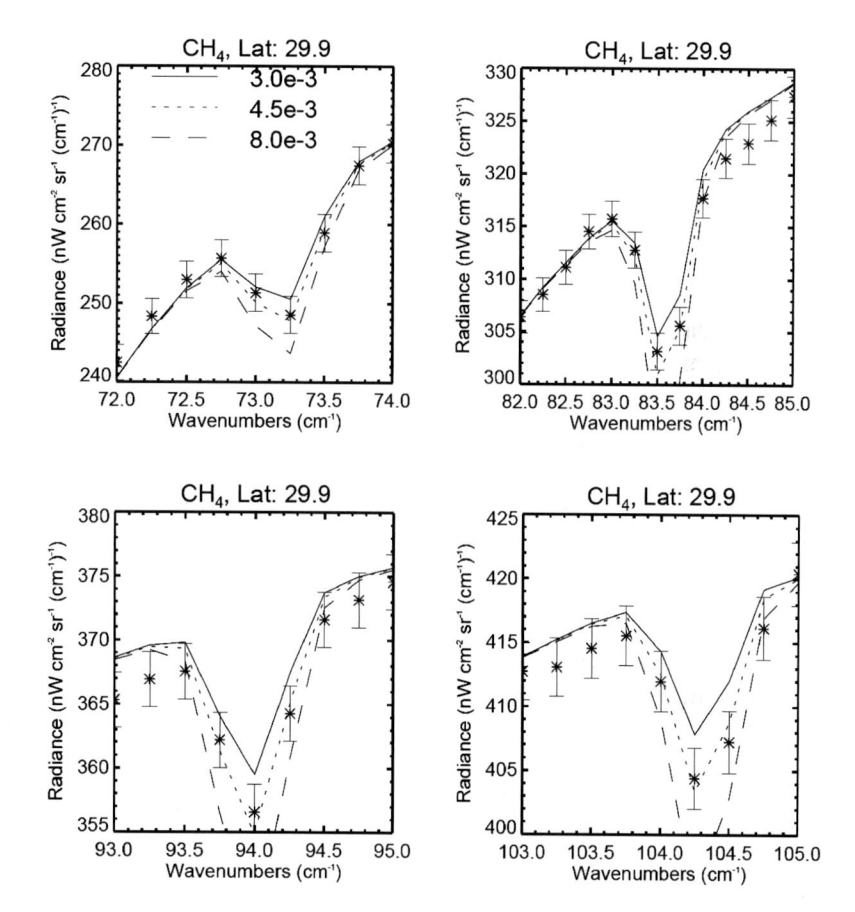

reflected observations can be attributed to the difficulty in modeling accurately the complex saturnian cloud structure, in the blending with other molecular absorptions, and also in the still large uncertainties in methane absorption coefficients at low temperatures.

In contrast, studies in the thermal infrared range have clustered their results around a methane mole fraction of 4.5×10^{-3}. Analysis in this spectral range requires first the determination of the vertical temperature profile, then, assuming this profile, the adjustment of the volume mixing ratio until the observed methane thermal emission is reproduced. Courtin et al. (1984) were the first to carry out such a measurement. They used the temperature profile determined from the Voyager radio occultation experiment along with the stratospheric CH$_4$ ν_4 emission observed by IRIS to infer a volume mixing ratio of $(4.3^{+2.3}_{-1.3}) \times 10^{-3}$. Fifteen years later, the Short Wavelength Spectrometer (SWS) aboard the Infrared Space Telescope (ISO) observed the 8–11 μm region. Although, this spectral region is dominated by PH$_3$ absorptions, micro-windows between 10 and 11 μm allowed Lellouch et al. (2001) to measure the tropospheric temperature, while the ν_4 wings in the 8.15–8.5 μm range probed the CH$_4$ tropospheric mixing ratio. Now, Cassini/CIRS observes for the first time the methane rotational lines (Fig. 5.2). These lines, formed on the H$_2$ collision-induced continuum that probes the tropospheric temperature, are well isolated from other gaseous absorption. Hence, they allow a very precise measurement. Fletcher et al. (2009a) obtain the most accurate measurement, $(4.7 \pm 0.2) \times 10^{-3}$, averaging over a larger number of CIRS observations.

This mole fraction corresponds to a carbon enrichment of 10.4 with respect to the solar abundance, assuming that a negligible amount of carbon is tied up in constituents other than CH_4. Cassini hence definitely demonstrates that Saturn lies in between Jupiter, and Uranus and Neptune, in terms of carbon elemental enrichment. This has important implications for the giant planets formation scenarios reviewed in the Formation Chapter.

5.2.1.4 Nitrogen, Sulfur and Oxygen

About ten astronomical units from the Sun, Saturn is a cold world. For this reason, the clouds form at large pressures (about 20 bars for the water cloud, a few bars for the chemical NH_4SH cloud, and around the 2-bar level for NH_3 cloud). In this situation, it is extremely difficult for a remote observer to measure the deep, global abundance of the condensibles below their associated clouds. Only observations in the centimeter range, between 6 and 70 cm, are able to probe molecular species that deep. Unfortunately, the interpretation of such observations is difficult for several different reasons. First, the absolute calibration uncertainties are large in this wavelength range, not better than 20%. Second, the lineshapes of the gaseous absorbers are not very well known, although progress has been made recently. Finally, the various absorbers, gas, liquids, or solids, have broad, relatively featureless spectra; hence the measurements in the centimeter range do not contain independent information on both temperature and composition.

For all these reasons, the published values must be taken with caution. The most convincing results have been obtained by Briggs and Sackett (1989). Their Very Large Array (VLA) and Arecibo observations were best fit by a deep ammonia volume mixing ratio of $(4.8 \pm 1) \times 10^{-4}$, hence 4.5 ± 1.0 times the solar abundance. Moreover, Briggs and Sackett (1989), (see also Killen and Flasar 1996; van der Tak et al. 1999), found that the brightness temperature in the 2–6 cm range points toward a low ammonia relative humidity between 2 and 4 bars. They interpreted this depletion of NH_3 as an evidence for a large H_2S abundance, equal to a volume mixing ratio of 4.1×10^{-4}, that served as a sink of NH_3 through the formation of a dense NH_4SH cloud. Such an abundance corresponds to 17 times the solar composition. It is hoped that measurements from the Cassini Radio Science Subsystem will improve the absolute calibration of Saturn's observation in the 2–14 cm range. Combined with the CIRS temperature profiles, a more accurate determination of the N/H and S/H ratios could be obtained.

Although tropospheric water has been detected by ISO in the 5-μm band (de Graauw et al. 1997), the inferred abundance is highly subsolar: ISO observations probe only pressure levels (2–4 bars) where water vapor condenses out. This situation precludes any measurement of the O/H ratio in Saturn, which constitutes an important missing piece of information for constraining the delivery of volatiles to the planet. Unfortunately, we cannot expect any progress on this issue from Cassini and will have to wait for a deep entry probe.

5.2.1.5 Isotopic Composition

The isotopic composition, especially the D/H ratio, is important for constraining the fraction of solids relative to gases that were accreted to form Saturn. The majority of D/H determinations in Saturn's atmosphere have been derived from the comparative abundance of methane and deuterated methane (CH_3D). Besides the uncertainty on CH_4 itself, the interpretation of the CH_3D measurements in terms of the bulk deuterium abundance (i.e., in H_2) is complicated by the existence of an isotopic exchange reaction between H_2 and CH_4 ($HD + CH_4 \rightleftharpoons CH_3D + H_2$) whose thermodynamics, kinetics and coupling with atmospheric dynamics are uncertain. According to different studies (Lecluse et al. 1996; Smith et al. 1996), the fractionation coefficient f:

$$f = \frac{(D/H)_{CH_4}}{(D/H)_{H_2}}$$

lies in the range $f = 1.34 \pm 0.19$.

Deuterated methane was first detected by Fink and Larson (1978) in its ν_2 band at 4.7 μm. Owen et al. (1986) and de Bergh et al. (1986) were the first to publish an abundance of CH_3D, using high spectral resolution observations of the $3\nu_2$ band at 1.55 μm (see Table 5.2 for the values). Another determination in the solar reflected range was obtained by Kim et al. (2006), between 2.87 and 3.10 μm (CH_3D $\nu_3 + \nu_2$ band), with a low abundance of 2.5×10^{-7}. Noll and Larson (1991) used the 4.7-μm spectral region, that probes deep in Saturn's troposphere, to determine a volume mixing ratio of $(3.3 \pm 1.5) \times 10^{-7}$. This spectral region, difficult to analyse due to some strong PH_3 absorption, yielded a similar result from ISO/SWS observations. CH_3D abundance can also be measured in the 8.6-μm region using the ν_6 thermal emission. Three studies focused in this spectral region with medium resolution observation (between 4.3 and

Table 5.2 D/H ratios in H_2

$(D/H)_{H_2}$	Method	Reference
$(1.6^{+1.3}_{-1.2}) \times 10^{-5}$	CH_3D at 8 μm	Courtin et al. (1985)
$(1.7^{+1.7}_{-0.8}) \times 10^{-5}$	CH_3D at 1.6 μm	de Bergh et al. (1986)
$(1.1 \pm 1.3) \times 10^{-5}$	CH_3D at 4.7 μm	Noll and Larson (1991)
$(2.3^{+1.2}_{-0.8}) \times 10^{-5}$	HD R(1)	Griffin et al. (1996)
$(1.85^{+0.85}_{-0.60}) \times 10^{-5}$	HD R(2) & R(3)	Lellouch et al. (2001)
$(2.0^{1.4}_{-0.7}) \times 10^{-5}$	CH_3D at 8 μm	Lellouch et al. (2001)
$(1.6 \pm 0.2) \times 10^{-5}$	CH_3D at 8 μm	Fletcher et al. (2009a)

0.5 cm^{-1}): Courtin et al., 1984 using Voyager/IRIS data, Lellouch et al. (2001) with ISO/SWS data, and Fletcher et al. (2009a) with Cassini/CIRS spectra. Overall, the different measurements cluster around a value for the D/H ratio in methane of 1.8×10^{-5}. Given the above estimate of the fractionation coefficient f, this gives a low estimate for the D/H in H$_2$: about 1.5×10^{-5}. Ground-based, high-resolution spectroscopic observations that resolved the stratospheric cores of the ν_6 CH$_3$D lines would be of great interest to improve the Cassini determination.

The first, and to date only, direct measurement of the D/H ratio in molecular hydrogen was obtained with the ISO telescope. Griffin et al. (1996) observed the R(1) rotational line at 56 μm with the Long Wavelength Spectrometer (LWS) aboard ISO. However, their measurement (Table 5.2) must be taken with caution, as the ISO/LWS data were difficult to calibrate for the very large flux of the giant planets. Lellouch et al. (2001) used ISO/SWS to observe the HD R(2) and R(3) rotational lines. From this, they derive a D/H ratio of $(1.85^{+0.85}_{-0.60}) \times 10^{-5}$. Combining HD and CH$_3$D measurements, Lellouch et al. (2001) concluded to $(D/H)_{H_2} = (1.70^{+0.75}_{-0.45}) \times 10^{-5}$. This value is not accurate enough to distinguish Saturn from Jupiter ($(D/H)_{H_2} = (2.25 \pm 0.35) \times 10^{-5}$) and from the local interstellar medium ($(D/H)_{H_2} = (1.5 \pm 0.1) \times 10^{-5}$). In this respect, Saturn does not seem to obey the trend of increasing D/H from Jupiter to Neptune. Improvements on this determination should come in the near future from the Hershell satellite when it observes the R(0) line at 2680 GHz, and from better calibrated Cassini/CIRS R(1) line observations.

Very few authors have focused on the ^{12}C/^{13}C ratio in Saturn. Combes et al. (1977) observed ^{13}CH$_4$ and ^{12}CH$_4$ lines of the $3\nu_3$ band at 1.1 μm to determined a ^{12}C/^{13}C ratio of 89^{+25}_{-18}. This ratio was later corrected to 71^{+25}_{-18} by Brault et al. (1981), who revised the ^{13}CH$_4$:^{12}CH$_4$ intensity ratio in the $3\nu_3$ band based on laboratory spectra of methane. Sada et al. (1996) measured recently a ^{12}C/^{13}C ratio of 99^{+43}_{-23} in ethane. Although C$_2$H$_6$ is not the major carbon reservoir in Saturn, we do not expect significant isotopic fraction for this species.

Hence, before the Cassini arrival, the Kronian ^{12}C/^{13}C ratio was compatible with the terrestrial value ($89.9^{+2.6}_{-2.4}$), but the measurements were not very constraining. The high quality of the CIRS dataset over the prime Cassini mission has allowed Fletcher et al. (2009a) to obtain a much more precise value. They use the ν_4 band thermal stratospheric emission to infer a ^{12}C/^{13}C ratio of $91.8^{+5.5}_{-5.3}$, still compatible with the terrestrial and the jovian values, but different from the Titan value (Niemann et al. 2005). This difference shows that methane on Titan has suffered from isotopic fractionation either at its formation or through its evolution.

5.2.2 Thermochemical Products

Some chemical species have been detected at shallow atmospheric levels with abundances that largely exceed the abundances predicted by thermodynamical equilibrium. As explained in detail in Section 5.3.1.1, the explanation for the observed abundances involves rapid convective transport from the deep, hot, atmosphere where the equilibrium abundances of these species are relatively large. The observed mixing ratios are then representative of *quenching* levels (300–500 bars) where the time constants of the conversion reactions become equal to or greater than the timescale for convective mixing. Below this level, thermochemical equilibrium prevails, whereas above, chemical destruction is inhibited so that the mole fractions remain constant in an uplifted air parcel. The distribution of thermochemical products is an important tracer of dynamical processes in the troposphere, and has been given a lot of attention. Hereafter, we describe the measurements of such disequilibrium species: phosphine, germane, and arsine. (The specific case of carbon monoxide, with two different possible origins, is detailed in Section 5.2.4.) All of these measurements were obtained in the 5-μm window of Saturn's spectrum (Fig. 5.3). The low opacity of the major gases in this spectral region offers the opportunity to probe pressure levels of several bars. This long path length allows the detection of trace constituents, while at these pressure levels the measured mixing ratios are unaffected by photochemical destruction, hence revealing the abundances of the observed species at the quench level.

The first spectrum of Saturn in the 5-μm window was obtained by Larson et al. (1980) using the Fourier

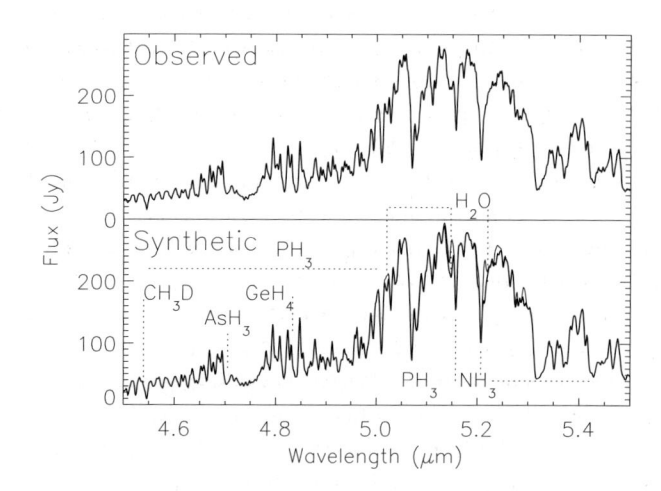

Fig. 5.3 Observed ISO-SWS spectrum (*upper curve*) and synthetic spectrum of Saturn (*lower curve*) in the 5μm region. Spectral absorptions are due to NH$_3$, PH$_3$, AsH$_3$, GeH$_4$, CH$_3$D and H$_2$O. In the lower curve, the narrow line corresponds to a calculation without H$_2$O. Adapted from de Graauw et al. (1997)

Transform Spectrometer on the Kuiper Airborne Observatory (KAO/FTS). The absorption features of phosphine were evident in this moderate spectral resolution spectrum, but a more quantitative analysis had to await higher spectral resolution observations. Bézard et al. (1989) and Larson and Noll (1991) both demonstrated that the abundance of PH_3 in the lower troposphere of Saturn, with a volume mixing ratio in the range $(4–10) \times 10^{-6}$, was larger than the value in Jupiter. ISO observations confirmed this value, with de Graauw et al. (1997) deriving a global volume mixing ratio of 4.4×10^{-6}. Saturn's PH_3 global abundance corresponds to an elemental enrichment of a factor of 10 in comparison with the solar abundance (assuming all phosphorous is tied up in PH_3 on Saturn). It would be interesting to know whether this deep mixing ratio varies with latitude, as it would indicate a meridional difference in the strength of convection from the interior. From CIRS observations, Fletcher et al. (2007b) are unable to disentangle deep variations from variations above the radiative-convective boundary. The Visual and Infrared-Mapping Spectrometer (VIMS) aboard Cassini is in position to study such a meridional variation in the lower troposphere, but no results have been released yet.

The first evidence for germane in Saturn came from observations near 4.7 μm obtained at the United Kingdom Infrared Telescope (UKIRT). Noll et al. (1988) used these data to determine a GeH_4 volume mixing ratio of $(4 \pm 2) \times 10^{-10}$. However, the analysis of spectra in the 4.7-μm region is complicated by the presence of strong PH_3 absorptions and by an equal balance between solar reflection and thermal emission. This situation has led Noll and Larson (1991) to conclude that their 4.7-μm Infrared Telescope Facility (IRTF) observations favored the presence of GeH_4 in Saturn, but that they could not exclude its absence. The confirmation of the presence of germane in Saturn hence came recently, from ISO observations. Although of moderate spectral resolution, ISO spectra achieved signal-to-noise ratios in the 5-μm region much larger than that of previous observations. Using these data, de Graauw et al. (1997) measured a deep germane volume mixing ratio of 2×10^{-9} (corresponding to 0.2 times the solar abundance), and found evidence for a decreasing mixing ratio above the 1-bar pressure level, possibly due to photolysis in the upper troposphere.

Arsine was detected independently by two different teams: Bézard et al. (1989) used the Fourier Transform Spectrometer mounted on the Canada-France-Hawaii Telescope (FTS/CFHT), while Noll et al. (1989) used the UKIRT. Noll and Larson (1991) at the IRTF and more recently de Graauw et al. (1997) using ISO data confirmed the detection. Two AsH_3 vibration bands are observed in the 4.7-μm region, the ν_1 and ν_3. The two bands allow the authors to probe the arsine vertical profile between ~4 bars and 0.2 bar. In the deep atmosphere, the published volume mixing ratios clustered around $(2.5 \pm 1) \times 10^{-9}$, while Bézard et al. (1989) measured a mixing ratio of $(3.9^{+2.1}_{-1.3}) \times 10^{-10}$ in the upper troposphere. The deep mixing ratio corresponds to a global As elemental abundance equal to five times the solar abundance. In the upper troposphere, AsH_3 may be subject to photochemical destruction as is the case for PH_3 and GeH_4. VIMS is also in position to derive a meridional distribution of the deep AsH_3 abundance.

Not all elements experience enhancements in the upper troposphere, revealing different chemistries. For example, Teanby et al. (2006) used far infrared spectra (10–600 cm^{-1}) from the Cassini Composite InfraRed Spectrometer (CIRS) to determine the best-to-date upper limits of hydrogen halides HF, HCl, HBr, and HI. These authors obtained 3-σ upper limits on HF, HCl, HBr, and HI mole fractions of 8.0×10^{-12}, 6.7×10^{-11}, 1.3×10^{-10}, and 1.4×10^{-9}, respectively, at the 500-mbar pressure level. These upper limits confirm sub-solar abundances of halide species for HF, HCl, and HBr in Saturn's upper atmosphere, consistently with predictions from thermochemical models. We note that the upper limit for HCl derived by Teanby et al. (2006) is 16 times lower than the tentative detection at 1.1×10^{-9} reported by Weisstein and Serabyn (1996).

5.2.3 Photochemical Products

5.2.3.1 Evidence for Photochemistry in the Upper Troposphere

Several molecules are known to be affected by photolysis in the upper troposphere. Weisstein and Serabyn (1994) were the first to demonstrate, from high spectral resolution of the 1–0 rotational line observations, that a cutoff in phosphine abundance exists at a pressure between 13 and 140 mbar. Orton et al. (2000, 2001), also using rotational line observations, showed that the PH_3 abundance dropped from 7.4×10^{-6} at 645 mbar down to 4.3×10^{-7} at 150 mbar. Such a behaviour was also consistent with ISO/SWS observations, as Lellouch et al. (2001) measured a profile with mole fractions of 6×10^{-6} up to the 600-mbar level, decreasing to 4×10^{-6} at 250 mbar, and 3×10^{-7} at 150 mbar. Since phosphine does not condense at the tropopause of Saturn, such a decrease with altitude can only be explained by photolysis. However, none of the products of this phosphine photochemistry has been detected to date. In this context, the monitoring of the meridional phosphine gradient is a powerful tracer of differences in vertical mixing, although it should be mentioned that aerosol opacity in the ultraviolet also strongly affects the phosphine gradient and caution should be used when trying to disentangle the relative effects of aerosol shielding and vertical mixing. In fact, phosphine photolysis may itself constitute a source of haze.

5 Saturn: Composition and Chemistry

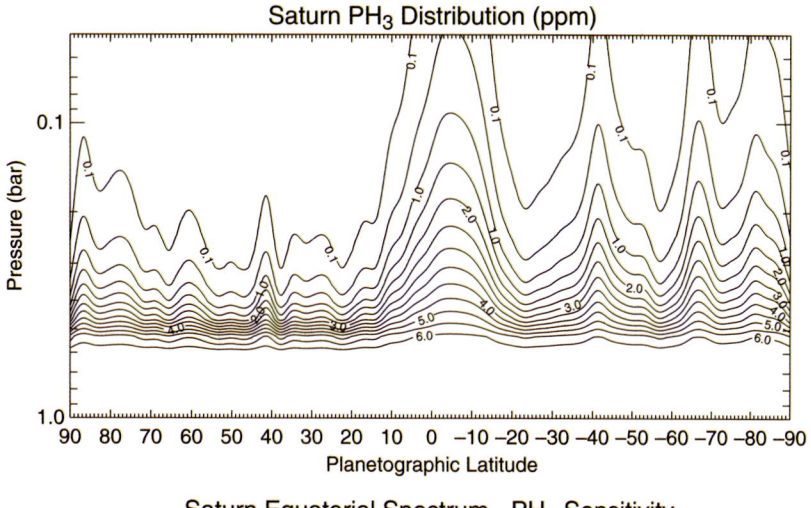

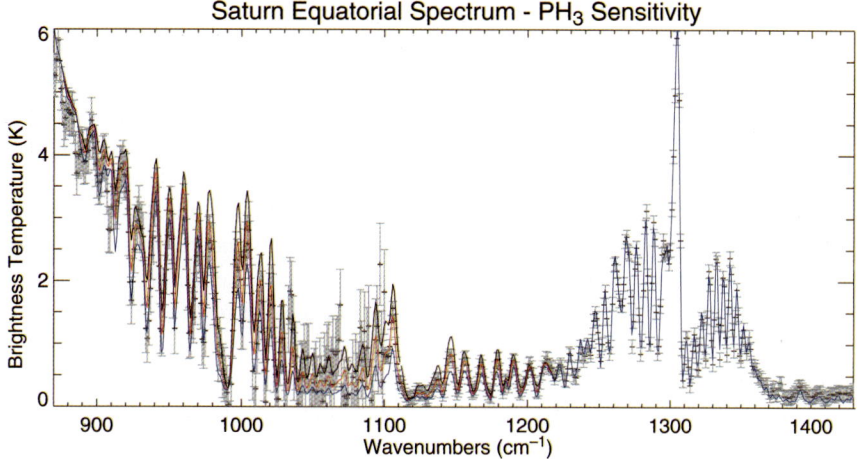

Fig. 5.4 *Top panel*: The phosphine mole fraction as a function of latitude and pressure as retrieved from Cassini/CIRS data. *Bottom panel*: CIRS spectrum of the equatorial region in the range 900–140 cm^{-1}, compared with synthetic spectra showing the sensitivity the phosphine mole fraction at 500 mbar. Adapted from Fletcher et al. (2009)

Fletcher et al. (2007b, 2008, 2009b) use the Cassini/CIRS spectra (Fig. 5.4, bottom panel) to measure the phosphine abundance as a function of latitude. Baines et al. (2005) also publish preliminary PH$_3$ meridional variations above the cloud top as retrieved from VIMS data. At 150 mbar their results are in line with previous studies, but at 250 mbar their abundance is lower than that of Lellouch et al. (2001) anywhere on the planet. Reasons for this disagreement have not been investigated. The Cassini results are summarized as follows (see also Table 5.3 and Fig. 5.4, Top panel).

1. The equatorial region presents the highest PH$_3$ abundance in the upper troposphere, which is consistent with the elevated equatorial haze, the cold temperature and the sub-equilibrium H$_2$ para fraction measured in this region.
2. Tropical latitudes (±23°) present the lowest PH$_3$ abundance, consistent with upwelling at the equator and downwelling in the neighbouring equatorial belts.
3. Mid-Latitudes at ±42° presents local maximum followed by a local minimum at ±57°.
4. Polar regions are enhanced in phosphine relative to tropical and mid-latitudes, but a sharp phosphine depletion occurs within ±2-3° of the poles, suggestive of subsidence over the high-temperature polar cyclones.
5. The northern hemisphere shows lower PH$_3$ abundance than the southern hemisphere. This is in opposition with a simple prediction based on the higher photolysis rate

Table 5.3 PH$_3$ vertical profiles measured from Cassini/CIRS mid-IR spectra, averaged in latitude regions (from Fletcher et al. (2007b)

Latitude	q_{PH_3} at 250 mbar (in ppb)	q_{PH_3} at 150 mbar (in ppb)
North Polar 65–85 °N	820 ± 380	220 ± 200
North Temperate 15–65 °N	390 ± 100	62 ± 34
Equatorial 15°S–15 °N	1700 ± 200	470 ± 90
South Temperate 65–15 °S	660 ± 80	84 ± 27
South Polar 85–65 °S	990 ± 130	160 ± 40

in the summer hemisphere compared to that in the winter hemisphere, but maybe explained by a enhanced shielding by aerosols in the southern hemisphere.

In the upper troposphere the ammonia volume mixing ratio decreases with altitude due to the combined effect of condensation and photolysis. The NH$_3$ vertical profile is slightly subsaturated around the 700-mbar pressure level (Courtin et al. 1984; Burgdorf et al. 2004), and highly subsaturated above that, hence showing evidence for photolysis at lower pressures, reaching an abundance of 10^{-9}–10^{-8} around 300–400 mbar (Kerola et al. 1997; Kim et al. 2006). In the upper troposphere, evidence for photolysis of arsine and germane has been found by Bézard et al. (1989) and de Graauw et al. (1997).

5.2.3.2 Stratospheric Products

In the upper stratosphere, methane is photodissociated by ultraviolet solar emission to produce CH, CH$_2$, and CH$_3$ radicals. Subsequent chemical reactions between the photolysis products and other atmospheric molecules yield several heavy hydrocarbons: to date, acetylene (C$_2$H$_2$), ethylene (C$_2$H$_4$), ethane (C$_2$H$_6$), propane (C$_3$H$_8$), methylacetylene (CH$_3$C$_2$H), diacetylene (C$_4$H$_2$), and benzene (C$_6$H$_6$) have been detected; the lighter radical CH$_3$ has also been observed.

C$_2$H$_6$ and C$_2$H$_2$ are the two most abundant species and, not surprisingly, were the first to be detected: acetylene in the UV range (Moos and Clarke 1979), and ethane in the thermal infrared (Tokunaga et al. 1975). Courtin et al. (1984), using Voyager/IRIS observations, were the first to determine reliable abundances: $(3.0 \pm 1.1) \times 10^{-6}$ from the ethane ν_9 band at 821 cm^{-1}, and $(2.0 \pm 1.4) \times 10^{-7}$ from the C$_2$H$_2$ ν_5 band at 729 cm^{-1}. From ground-based observations in the thermal infrared, Noll et al. (1986b) and Sada et al. (2005) found slightly larger values (the different measurements are summarized in Fig. 5.5). Using ISO/SWS observations, de Graauw et al. (1997) and Moses et al. (2000a) were the first to demonstrate that the C$_2$H$_2$ volume mixing ratio actually increases with altitude, as expected from photochemical models. Indeed, the emissions from the $\nu_4 + \nu_5 - \nu_4$ hot band detected by ISO probed higher altitudes than the ν_5 band. Moses et al. (2000a) inferred a mixing ratio of $(1.2^{+0.9}_{-0.8}) \times 10^{-6}$ at 0.3 mbar and $(2.7 \pm 0.8) \times 10^{-7}$ at 1.4 mbar. For ethane, as all lines have approximately the same strength, vertical profile information cannot be obtained; from ISO/SWS data, a volume mixing ratio at 0.5 mbar alone was determined: $(9 \pm 2.5) \times 10^{-6}$. Ethane and acetylene spectral features have also been unambiguously detected in the reflected infrared around 3 μm by Bjoraker et al. (1981) and Kim et al. (2006) (respectively the $\nu_3 + \nu_9 + \nu_{11}$ of ethane and the ν_3 of acetylene). These studies suggested a cut-off in the vertical distributions at about 10–15 mbar, but the uncertainties on the band strengths in this spectral region make the inferred mole fractions less reliable than from the thermal infrared. Kim et al. (2006) also detected ethane emissions in its ν_7 band originating from μbar levels. However, since collision relaxation parameters are not known with enough accuracy for this band, the observations could only be crudely analyzed yielding no reliable estimates of the ethane abundance.

As products of methane photolysis, ethane and acetylene abundances might be expected to exhibit meridional

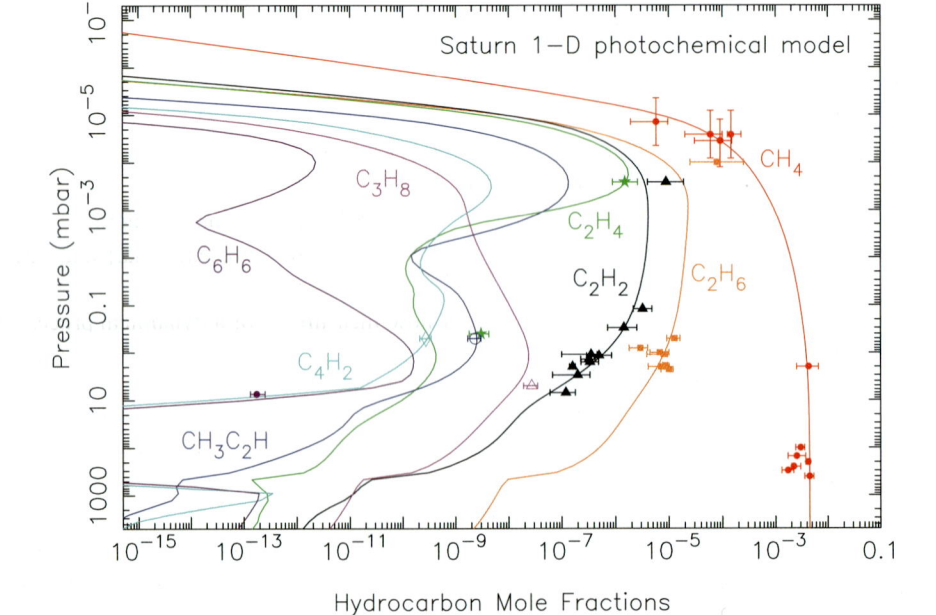

Fig. 5.5 Hydrocarbon mole fractions as a function of pressure in Saturn's upper atmosphere, as derived from the "Model C" 1-D steady-state photochemical model of Moses et al. (2005). The solid curves represent the model profiles for the individual hydrocarbons (as labeled), and the symbols with associated error bars represent various infrared and ultraviolet observations

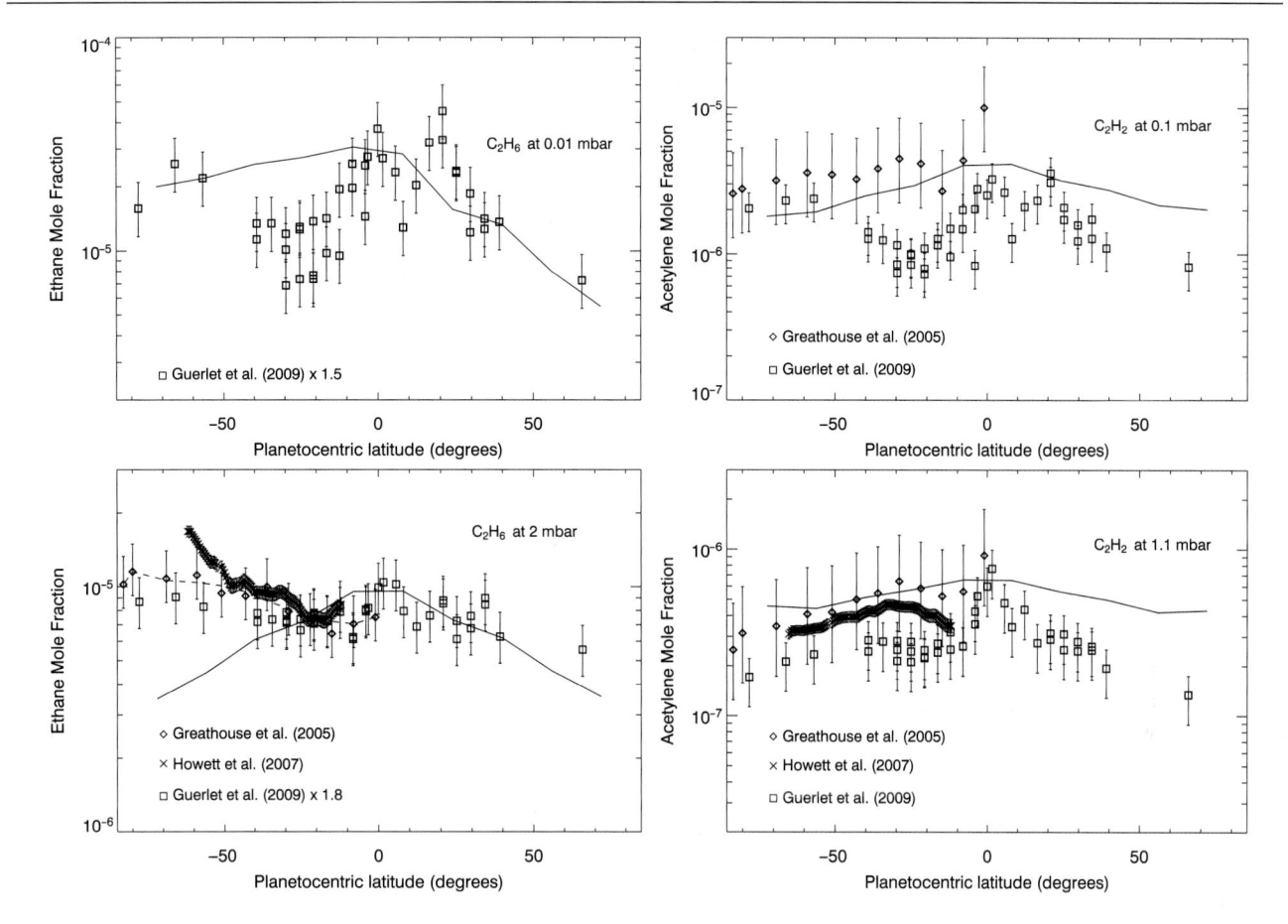

Fig. 5.6 The latitude variation of the volume mixing ratios of C_2H_6 (*left*) and C_2H_2 (*right*) at $L_s = 270°$ (southern summer solstice) at different pressure levels, as labeled. The solid lines represent the theoretical model profiles from the 1-D seasonal model of Moses and Greathouse (2005). The diamonds represent the IRTF/TEXES observations of Greathouse et al. (2005), the crosses represent the Cassini/CIRS nadir observations of Howett et al. (2007), and the squares represent the Cassini/CIRS limb observations of Guerlet et al. (2009). Note that the model reproduces the meridional behavior of C_2H_2 but not C_2H_6

and seasonal variations. The observations mentioned above were unable to access this geographical dimension as they were either disk-averaged, or targeted to a specific position. The work by Greathouse et al. (2005) hence opened the era of meridional hydrocarbon study. Using Texas Echelon Cross Echelle Spectrograph (TEXES) observations in the thermal infrared, they showed that the C_2H_2 volume mixing ratio at 1.16 and 0.12 mbar decreased by a factor of 2.5 from the equator towards the south pole (Fig. 5.6). In contrast, ethane was found to be stable within error bars, with 2.3-mbar mixing ratios of $(7.5^{+2.3}_{-1.7}) \times 10^{-7}$ at the equator and $(1.0^{+0.3}_{-0.2}) \times 10^{-5}$ at 83°S. Cassini/CIRS now measures both the C_2H_2 and C_2H_6 abundance from nadir observations (Howett et al. 2007), and from limb soundings (Fouchet et al. 2008; Guerlet et al. 2009). CIRS also extends the meridional coverage towards the northern hemisphere that was hidden by the rings to Greathouse et al. (2005). Cassini results confirm that the C_2H_6 mole fraction remains latitudinally constant at 1 mbar (although Howett et al. find a suspicious increase southward of 50°S), and that C_2H_2 decreases towards both poles (Fig. 5.6). This behaviour can be attributed to the different chemical timescales of the two molecules, C_2H_6 being a long-lived species compared to C_2H_2. Hence, the former can be horizontally transported and homogenized while the latter reflects the mean meridional solar irradiation. However, this interpretation is not validated by chemical models (see Section 5.3.2.4).

The Cassini/CIRS limb soundings also reveal some dynamical features in the hydrocarbon meridional profiles. At 1 mbar, Guerlet et al. (2009) find that the C_2H_2 profiles present a sharp and narrow maximum centered at the equator, not shown by C_2H_6. They interpret this feature as a signature of a dynamical process, the descent of air between 0.1 and 1 mbar associated with the equatorial oscillation. This descent transports C_2H_2 more efficiently than C_2H_6, as the former molecule shows a stronger vertical gradient than the latter. At lower pressures ($p < 0.1$ mbar), that can only be probed by limb soundings or occultations, Guerlet et al. (2009) find that the C_2H_2 and C_2H_6 mole fractions are the largest at mid-northern latitudes, and the lowest at mid-southern latitudes. This situation is in complete opposition with photochemical model predictions of an abundance following the solar irradiance at these pressure levels. The observations thus suggest the existence of a strong meridional circulation, transporting molecules from the summer hemisphere across the equator towards the winter hemisphere.

In the upper stratosphere, ethane, ethylene, and acetylene have been detected from spacecraft ultraviolet observations. The Ultraviolet Spectrometer (UVS) on Voyager 1 and 2 recorded six solar and stellar occultations (Broadfoot et al. 1981; Sandel et al. 1982; Festou and Atreya 1982; Smith et al. 1983; Vervack and Moses 2009). The Vervack and Moses (2009) study is the only one to analyze all six occultations. Vertical profiles for C_2H_2, C_2H_4, and C_2H_6 have been derived from this analysis, and Vervack and Moses (2009) confirm that the mixing ratios for all three species have a peak at high altitudes, with a decreasing mixing ratio with decreasing altitude away from this high-altitude peak. The variability of the results from the six UVS occultations also demonstrates that the pressure level of the peak mixing ratio varies with latitude and/or time on Saturn, which has interesting implications with respect to atmospheric transport. To date, the Cassini Ultraviolet Imaging Spectrograph (UVIS) has also performed 15 stellar occultations and 7 solar occultations (see Shemansky and Liu 2009, and Chapter 8). The results from some of these occultations are presented by Shemansky and Liu (2009), who derive C_2H_2 and C_2H_4 profiles and further confirm the distinct mixing-ratio peak and the meridional variability. The ultraviolet occultation results are discussed in more detail in Chapter 8.

The ISO/SWS instrument, due to its unprecedented sensitivity in the thermal infrared, has allowed the detection of several complex hydrocarbons. de Graauw et al. (1997) reported the detection of CH_3C_2H and C_4H_2 in the range 615–640 cm^{-1}. The emissions are optically thin, and only column densities can be derived: $(1.1 \pm 0.3) \times 10^{15}$ molecules.cm^{-2} for CH_3C_2H, and $(1.2 \pm 0.3) \times 10^{14}$ molecules.cm^{-2} for C_4H_2 were reported from the ISO/SWS analysis of Moses et al. (2000a). Bézard et al. (2001a) reported the detection of benzene, from emission at 674 cm^{-1}, with a column density of $(4.7^{+2.1}_{-1.1}) \times 10^{13}$ molecules.cm^{-2}. Cassini/CIRS is in position to derive meridional distributions for these species, but no results have been published in the refereed literature yet. ISO/SWS also unveiled the ν_2 emission from CH_3 at 16.5 μm. Bézard et al. (1998) derived a CH_3 column density in the range 1.5–7.5×10^{13} molecules.cm^{-2}, taking into account uncertainties in the data calibration, the CH_3 band strength, and the temperature and methyl vertical profiles. Given recent updates in the measured line strengths of the of ν_2 fundamental of CH_3 (Stancu et al. 2005), the Bézard et al. (1998) CH_3 column abundance should be increased by a factor of $\sim$2. Similarly, updates in our understanding of Saturn's thermal structure at the time of the ISO observations (e.g., Lellouch et al. 2001) and of spectral line parameters (e.g., Vander Auwera et al. 2007; Jacquinet-Husson et al. 2005) necessitate some updates to the above quoted abundances from Moses et al. (2000a).

Finally, ground-based thermal infrared observations with the high-resolution TEXES spectrometer have allowed Bézard et al. (2001b) and Greathouse et al. (2006) to detect ethylene and propane, respectively. The inferred C_2H_4 column abundance lies between 2–3×10^{15} molecules.cm^{-2}. For the C_3H_8 abundance, Greathouse et al. (2006) reported 5-mbar volume mixing ratios of $(2.7 \pm 0.8) \times 10^{-8}$ at 20°S and $(2.5^{+1.7}_{-0.8}) \times 10^{-8}$ at 80°S latitude, suggesting that propane, like ethane, does not present any meridional variations of its abundance. However, from Cassini/CIRS limb soundings, Guerlet et al. (2009) find that the propane volume mixing ratio does increase from the northern winter hemisphere towards the southern summer hemisphere. This contradiction remains to be resolved by future ground-based and Cassini/CIRS observations.

5.2.4 External Oxygen Flux

An external oxygen influx to Saturn has long been postulated, originating either from micrometeoroid precipitation, large cometary impacts or from Saturn's rings and satellites. A flux of oxygen in the reducing atmosphere of Saturn can have a conspicuous effect on the stratospheric chemistry by forming new molecules, attenuating the UV flux or providing condensation nuclei. The first observational basis for such an oxygen source was presented by Winkelstein et al. (1983). These authors had to introduce a 1.6×10^{17} cm^{-2} column density of water in the stratosphere in order to fit their IUE spectrum in the 150–300 nm range. Using HST data covering a similar spectral range, Prangé et al. (2006) showed that the large water abundance inferred by Winkelstein et al. (1983) resulted from their assumption of a vertically uniform acetylene profile. Using a C_2H_2 vertical profile compatible with infrared observations and photochemical models (see Sections 5.2.3.2 and 5.3.1.2), Prangé et al. (2006) showed that only upper limits on the H_2O abundance can be derived from UV spectra. Water was first firmly detected by Feuchtgruber et al. (1997) using ISO observations. Rotational lines of H_2O were clearly detected at 30.90, 35.94, 39.37, 40.34, 43.89, 44.19, and 45.11 μm, yielding a stratospheric column density of $(1.4 \pm 0.4) \times 10^{15}$ cm^{-2} (Moses et al. 2000b). ISO observations also allowed de Graauw et al. (1997) to detect the stratospheric emission of the CO_2 ν_2 band at 14.98 μm for the first time. This emission is modeled with a CO_2 column abundance of $(6.3 \pm 1.0) \times 10^{14}$ cm^{-2} (Moses et al. 2000b).

Carbon monoxide can also act as a tracer of an external oxygen flux. Indeed, oxygen-bearing molecules are photochemically converted to CO in the stratosphere, and the CO photochemical lifetime is long compared with a saturnian year or other hydrocarbon lifetimes. However, CO can also originate from Saturn's interior as do PH_3, GeH_4, and AsH_3. To discriminate between both possible sources, observers need to retrieve the CO vertical profile: a stratospheric abundance larger than that in the troposphere constitutes the signature of an external flux. CO was first detected by Noll et al. (1986a) in the (1–0) vibration band near 4.7 μm. Noll and Larson (1991) observations of the 4.5- to 5-μm region at the IRTF superseded that of Noll et al. (1986a). About 10 CO-lines from the (1-0) band can be clearly identified in the former spectrum. As demonstrated by Moses et al. (2000b), line intensities are mostly sensitive to the CO column abundance between 100 and 400 mbar, yielding a tropospheric volume mixing ratio of $(1 \pm 0.3) \times 10^{-9}$, i. e. a column abundance of $(0.7–1.5) \times 10^{17}$ cm^{-2}. However, the observations could also be accommodated by a stratospheric volume mixing ratio of 2.5×10^{-8}; i. e. a column abundance of $\sim 6 \times 10^{17}$ cm^{-2}. Both distributions fit equally well the observations, with the exception of the P14 line at 2086.3 cm^{-1}, which is best matched by a tropospheric distribution. Unfortunately, this line is entangled with phosphine absorption and is not well reproduced by the model spectra in any case, hence undermining the conclusion.

Due to their high spectral resolution, millimetric and submillimetric heterodyne observations are well suited to discriminate between the narrow stratospheric emission and the broad tropospheric absorption. After several fruitless attempts (Rosenqvist et al. 1992; Cavalié et al. 2008), Cavalié et al. (2009) obtain the first detection of the CO(3-2) line in Saturn using the James Clerk Maxwell Telescope. They find that the CO line is best matched with a stratospheric mole fraction of 2.5×10^{-8} at pressures smaller than 15 mbar, and of 10^{-10}–10^{-9} at higher pressures. This profile favors an external origin for carbon monoxide, due to a cometary impact about 200–300 years ago. However, the water observed in Saturn's stratosphere still needs to be provided by a continuous source, so that both a continuous source and a sporadic source of oxygen seem to be active in Saturn.

Herschel and the Atacama Large Millimeter Array (ALMA) will observe stronger CO and H_2O lines at other frequencies, providing further constraints on the vertical profiles of these species. Unfortunately Cassini can contribute very marginally to the observations of oxygenated species. Water rotational lines are difficult to detect at the spectral resolution of CIRS. Hence, many spectra will have to be averaged, at the expense of meridional resolution. For CO_2, it should be possible to extract a meridional profile from CIRS limb observations at high spectral resolution, but only a few observations were performed, and have not yet been analyzed.

5.3 Chemistry

5.3.1 Tropospheric Chemistry

5.3.1.1 Thermochemistry

In the deep troposphere of Saturn, well below the regions that can be probed by remote-sensing techniques, temperatures become high enough (i.e., $\gtrsim 1000$ K) that energy barriers to chemical kinetic reactions can be overcome. Chemical reactions in these high-temperature regions therefore proceed rapidly, and the atmosphere is expected to attain thermochemical equilibrium. The thermochemical-equilibrium composition is strongly dependent on temperature and weakly dependent on pressure, as well as being dependent on the relative abundance of the different elements. Fegley and Prinn (1985) presented the first comprehensive thermochemical-equilibrium calculations for Saturn. In this model, the large background atmospheric H_2 abundance allows CH_4 to be the dominant carbon-bearing molecule throughout Saturn's atmosphere, with all other carbon species having abundances much less than CH_4. Carbon monoxide is expected to have an exceedingly low abundance in the observable regions of the upper troposphere but becomes more and more abundant with increasing temperature and depth. Similarly, H_2O, NH_3, and H_2S are the dominant forms of oxygen, nitrogen, and sulfur throughout the pressure-temperature regimes found in Saturn's troposphere. Updates to the major-element thermochemical equilibrium calculations for Saturn can be found in Fegley and Lodders (1994), Lodders and Fegley (2002), Visscher and Fegley (2005), and Visscher et al. (2006), although the qualitative conclusions of Fegley and Prinn (1985) still hold.

Fegley and Prinn (1985) expect P_4O_6 vapor (formed from the oxidation of PH_3 by water) to be the thermodynamically stable form of phosphorus in the $\sim$400–900 K region on Saturn, with PH_3 becoming dominant only at deeper levels for temperatures in excess of $\sim$1000 K. At temperatures below $\sim$400 K (i.e., still well below the water cloud), the P_4O_6 vapor is expected to be converted to a solid $NH_4H_2PO_4$ condensate, and gas-phase phosphorous-bearing species are expected to have low abundances in the upper troposphere of Saturn under thermochemical-equilibrium conditions. The phosphorous results are somewhat uncertain due to uncertain thermodynamic parameters and elemental ratios: Borunov et al. (1995) suggest that PH_3 could be the dominant equilibrium form throughout Jupiter's atmosphere at all temperatures greater than $\sim$500 K, at which point a $NH_4H_2PO_4$ condensate forms; they expect Saturn's thermochemistry to be similar. The latest phosphorus equilibrium calculations for Saturn (Visscher and Fegley 2005) produce results qualitatively similar to that of Fegley and Prinn (1985) in terms of the phosphorus speciation, with some small quantitative differences in the temperature levels at which the different major species take over.

Based on thermochemical equilibrium modeling, both GeH_4 and AsH_3 are expected to be relatively abundant at depth on Saturn (Fegley and Prinn 1985; Fegley and Lodders 1994) but fall to negligible quantities in the upper troposphere due to condensate formation (e.g., Ge, GeS, GeTe, and As solid phases). Gas-phase HF is expected to be the dominant form of fluorine until the temperature drops below $\sim$300 K, at which point NH_4F condenses. Similarly, HCl, HBr, and HI are expected to be major gas-phase species at temperatures below $\sim$1000 K, until condensation of NH_4Cl, NH_4Br, and NH_4I occurs in the 350-450 K region (Fegley and Lodders 1994).

The observed abundance of CO, PH_3, GeH_4, and AsH_3 (see Section 5.2) greatly in excess of thermochemical-equilibrium predictions provides clear evidence that thermochemical equilibrium cannot be maintained in the colder regions of Saturn's troposphere. Based on arguments first presented by Prinn and Barshay (1977), the observed disequilibrium tropospheric composition is likely caused by strong convective motions in giant-planet tropospheres that allow gas species to be mixed vertically at a rate more rapid than the chemical kinetic destruction can keep pace. As a gas parcel is transported upwards, the equilibrium composition is "quenched" or "frozen in" at the level at which the vertical mixing time scale drops below the chemical kinetic time scale of conversion between different forms of the element. The observed upper tropospheric abundance then represents equilibrium conditions at a much deeper level, with that level being dependent on the strength of vertical mixing and the kinetic rate coefficients. Because the quench level is not well constrained from theoretical arguments, the relative partitioning between different molecules containing P, Ge, and As at the quench level is uncertain, and the observed abundances of PH_3, GeH_4, and AsH_3 represent lower limits to the bulk abundances of these elements.

If one assumes the elemental abundances on Saturn are well known and the kinetics well understood, the observed abundances of these disequilibrium species can be used to back out transport time scales and one-dimensional eddy diffusion coefficients in Saturn's troposphere (e.g., see Fegley and Prinn 1985; Fegley and Lodders 1994, who derive eddy diffusion coefficients of order 10^8–10^9 cm^2 s^{-1} at the tropospheric quench levels based on model comparisons with CO and PH_3 abundances). Conversely, if one makes assumptions about the convective time constants from mixing-length theory (e.g., Stone 1976; see also Smith 1998) and the chemical time constants from the rate-limiting kinetic reactions for the conversion mechanisms of a particular molecule (e.g., Prinn and Barshay 1977; Yung et al. 1988; Bézard et al. 2002), the observed disequilibrium abundances can be used to back out the deep-atmospheric abundance of the element in question. Visscher and Fegley (2005; see also the Bézard et al. 2002 calculations for Jupiter) attempt to get the deep oxygen abundance from this method, as the deep O/H ratio on Saturn cannot be directly measured by any of the Cassini or Voyager instruments. Because the bulk elemental abundances have important implications for giant-planet formation scenarios (e.g., Lunine et al. 2004), such calculations are of great interest for understanding the origin of the solar system. In practice, however, current uncertainties in the tropospheric CO abundance on Saturn and in the kinetics and rate coefficients of the CO-CH_4 interconversion mechanisms are preventing reliable estimates of the deep oxygen abundance on Saturn from being obtained using this method (cf. the Jupiter study of Bézard et al. 2002). That situation could change in the near future with tighter observational constraints on tropospheric CO from high-resolution ground-based observations and with expanded laboratory or theoretical investigations of appropriate rate coefficients and thermodynamic parameters.

5.3.1.2 Photochemistry in the Troposphere

Rapid atmospheric mixing is not the only disequilibrium process that affects the composition of Saturn. Photochemistry, or the chemistry that is initiated when an atmospheric constituent absorbs a photon, plays a major role in controlling the abundances of trace species. For details of tropospheric photochemistry on Saturn and the other giant planets, see Atreya et al. (1984), Prinn et al. (1984), and Strobel (1975, 1983, 2005). Although Rayleigh scattering, molecular absorption, and aerosol and cloud opacity limit the depth to which solar ultraviolet photons can penetrate in Saturn's atmosphere (e.g., Tomasko et al. 1984; Pérez-Hoyos and Sánchez-Lavega 2006; see also Chapter 7 in this volume), photochemical processes can occur in the few-hundred mbar region and above. In that region, water has already been removed through condensation at depth, and NH_4SH and ammonia clouds have formed, but the vapor pressure of NH_3 may be high enough at the cloud tops (and the overlying haze opacity sufficiently low) that NH_3 may interact with ultraviolet photons with wavelengths $\lesssim$220 nm. Ammonia photochemistry may therefore occur in Saturn's upper troposphere (e.g., Strobel 1975, 1983; Atreya et al. 1980), with ammonia being photolyzed at the cloud tops to form predominantly H atoms and NH_2 radicals. The NH_2 radicals can recombine with each other to form hydrazine (N_2H_4) or with H to reform NH_3. The hydrazine will condense to form an aerosol, but some may be photolyzed or react with H to form N_2H_3 and eventually produce N_2.

Phosphine does not condense at Saturnian temperatures, and it will also participate–and likely dominate–tropospheric photochemistry. The upper tropospheric haze is relatively optically thick on Saturn (see Chapter 7), which helps to shield the ammonia from photolysis in its condensation region. Phosphine, on the other hand, can diffuse upward until haze opacity is low enough for PH_3 photolysis to occur, and phosphine photochemistry is therefore guaranteed at some atmospheric level, no matter what the haze opacity. The coupled photochemistry of phosphine and ammonia on Saturn has been studied by Kaye and Strobel (1983c, 1984). Phosphine is photodissociated by photons with wavelengths less than $\sim$235 nm. The dominant photolysis products are PH_2 and H. Analogous with NH_3 photochemistry, two PH_2 radicals can combine to form P_2H_4, which can condense at the cold Saturnian temperatures, or the PH_2 can combine with H to reform PH_3. The ammonia photolysis products NH_2 and H can also react with PH_3 to provide addition mechanisms for PH_2 production and PH_3 destruction. The yield of N_2H_4, N_2, and other nitrogen-bearing products is therefore reduced when PH_3 is present. Dominant products of the coupled PH_3-NH_3 photochemistry are expected to be P_2H_4, condensed red phosphorous, NH_2PH_2, N_2H_4, and N_2. Diphosphine (P_2H_4) is a leading candidate for the upper tropospheric haze. Appropriately designed laboratory experiments investigating the photolysis of PH_3 or NH_3-PH_3 mixtures (e.g., Ferris and Morimoto 1981; Ferris et al. 1984; Ferris and Khwaja 1985) have been valuable in elucidating the nature of the coupled phosphine-ammonia chemistry of the giant planets.

The possibility of chemical coupling between hydrocarbons and NH_3 or PH_3 is a more controversial topic. Methane is photolyzed at high altitudes on Saturn, well removed from the ammonia and phosphine photolysis regions. Numerous laboratory experiments over the years on "simulated" giant-planet atmospheres (e.g., Sagan et al. 1967; Woeller and Ponnamperuma 1969; Ferris and Chen 1975; Raulin et al. 1979; Ferris and Morimoto 1981; Bossard et al. 1986; Ferris and Ishikawa 1988; Ferris et al. 1992; Guillemin et al. 1995a) have addressed the possible formation of HCN and other nitriles and organo-nitrogen compounds or organo-phosphorus compounds, but the relative gas abundances or other conditions in those experiments have not always been good analogs for giant-planet tropospheres. Kaye and Strobel (1983a, b) have used a photochemical model to study several possible mechanisms for coupled carbon-nitrogen chemistry on Jupiter. One pathway involves CH_3 radicals produced from the reaction of methane with hot hydrogen atoms that were produced from PH_3 and NH_2 photolysis; the CH_3 combines with NH_2 to form methylamine (CH_3NH_2), which can photolyze to form a small amount of HCN. A second pathway involves reaction of NH_2 with C_2H_3 to form aziridine (or other C_2H_5N isomers), which can also be photolyzed to form HCN and other interesting compounds. In the first scenario, the hot hydrogen atoms can be quenched by H_2, effectively cutting off production of CH_3 and nitrogen-bearing organics (see also Raulin et al. 1979; Ferris and Morimoto 1981).

The second, more promising, of these scenarios has been studied in more detail by Ferris and Ishikawa (1987, 1988) and Keane et al. (1996), who have used laboratory simulations to study the detailed mechanisms and product quantum yields of coupled ammonia-acetylene photochemistry. They favor somewhat different pathways and products for the coupled photochemistry, but the reaction sequence still starts with the $NH_2 + C_2H_3$ reaction. The relevance both of these experiments and of the models Kaye and Strobel (1983a, b) to the real situation on Saturn is not completely clear because the tropospheric C_2H_2 abundance is likely much smaller than was assumed in those experiments and models (e.g., Moses et al. 2000a, 2005). Similarly, coupled PH_3-C_2H_2 photochemistry (e.g., Guillemin et al. 1995a), or coupled chemistry of NH_3 and/or PH_3 with other unsaturated hydrocarbons (e.g., Ferris et al. 1992; Guillemin et al. 1997) may be inhibited on Saturn due to the likely low tropospheric abundance of C_2H_2 and other unsaturated hydrocarbons. However, if the tropospheric haze opacity is large enough to allow PH_3 to reach the tropopause (cf. the CIRS observations of Fletcher et al. 2007b), PH_3-hydrocarbon coupling is more likely to occur. Observations that detect or put upper limits on the HCN and HCP abundance on Saturn (e.g., Weisstein and Serabyn 1996) could help constrain the coupled NH_3-C_2H_2 and PH_3-C_2H_2 photochemistry.

Because differences in the abundances of NH_3, PH_3, and (in particular) aerosols lead to differences in the shielding of the photochemically active molecules on Saturn as compared with Jupiter and because NH_3 condenses deeper in Saturn's colder atmosphere, some differences in tropospheric photochemistry are expected to exist between the two planets. PH_3 photochemistry may be more important, PH_3-hydrocarbon coupling is more likely, and NH_3-PH_3 coupling may be suppressed on Saturn. The extent and consequence of these differences remain to be quantified in up-to-date photochemical models that include accurate aerosol extinction.

Some H_2S could be present in and below the putative NH_4SH cloud on Saturn. Because H_2S has a substantial photoabsorption cross section out to $\sim$260 nm (and a weak H-S bond), some sulfur photochemistry could possibly occur on Saturn, but the deep levels to which the UV radiation must penetrate suggest that sulfur photochemistry is not very important on Saturn, except perhaps in localized areas with reduced cloud cover. The details of the possible H_2S photochemistry are highly uncertain (e.g., Lewis and Prinn 1984; Prinn and Owen 1976). Similarly, GeH_4 and AsH_3 photochemistry is poorly understood (e.g., Fegley and Prinn 1985; Nava et al. 1993; Guillemin et al. 1995b; Morton and Kaiser 2003), but more likely to be occurring. In all, tropospheric photochemistry on Saturn is less well understood than stratospheric photochemistry, due to a critical lack of reaction rate coefficients for the appropriate reactions. It is hoped that new Cassini and Earth-based observations that have helped derive the altitude profiles of some of the key tropospheric species (NH_3, PH_3, GeH_4, AsH_3, see Section 5.2) will help generate a resurgence in interest in tropospheric photochemistry on Saturn.

5.3.2 Stratospheric Hydrocarbon Chemistry

Stratospheric hydrocarbon photochemistry on Saturn and the other giant planets has been reviewed recently by Moses et al. (2005, 2004, 2000a), Strobel (2005), and Yung and DeMore (1999), and the interested reader is referred to those works for more details. Because most of the major equilibrium hydride species condense in the cold upper troposphere of Saturn, stratospheric photochemistry is dominated by photolysis of the relatively volatile molecule methane. Methane is transported up from the deep interior past the tropopause cold trap to the top of the middle atmosphere (the mesopause region), where its abundance eventually falls off due to molecular diffusion of the relatively heavy CH_4 molecules in the lighter background H_2 gas. A common misconception is that photolysis limits the CH_4 abundance at high altitudes–it is actually molecular diffusion, rather than photochemistry, that is responsible for the sharp drop off in CH_4 abundance with altitude that is observed in spacecraft ultraviolet occultation data (e.g., Festou and Atreya 1982; Smith et al. 1983; Vervack and Moses 2009; Shemansky and Liu 2009; see also Chapter 12). Methane can be photolyzed in the upper atmosphere by ultraviolet radiation, and the photolysis products react to form the plethora of complex hydrocarbons that have been observed in Saturn's stratosphere (see Section 5.2). These complex hydrocarbons diffuse down into the troposphere, where they eventually encounter high temperatures and can be thermochemically converted back to CH_4, thus completing a "methane cycle". Strobel (1969) was the first to work out the concept of this methane cycle for the giant planets.

5.3.2.1 Chemical Reactions

Hydrocarbon photochemistry on Saturn is initiated by methane photolysis in the upper atmosphere. Because the photoabsorption cross section of methane is negligible at wavelengths longer than ∼145 nm, and because the solar Lyman alpha line (121.6 nm) provides the largest flux source at wavelengths less than 145 nm, Lyman α photodissociation of CH_4 controls much of the subsequent hydrocarbon photochemistry in Saturn's stratosphere. Although the methane photolysis branching ratios (i.e., the likelihood of forming a particular set of reaction products during photolysis) at 121.6 nm are not completely understood, the major photolysis products are CH_3, excited $CH_2(a\,^1A_1)$, and CH (e.g., Cook et al. 2001; Wang et al. 2000; Smith and Raulin 1999; Brownsword et al. 1997; Heck et al. 1996; Mordaunt et al. 1993). The CH_3 radicals can recombine with another CH_3 radical to form C_2H_6, or can combine with H to recycle methane. The $CH_2(a\,^1A_1)$ radicals can react with H_2 to form CH_3 ($\lesssim$90% of the time) or can be quenched to ground-state CH_2 ($\gtrsim$10% of the time) and eventually react with H to form CH. The CH radicals can insert into CH_4 to form C_2H_4. Photolysis of both C_2H_6 and C_2H_4 produces C_2H_2. Atomic hydrogen is a major product of hydrocarbon chemistry on Saturn, and reaction of atomic H with hydrocarbons helps define the relative abundance of the observable species.

More complex hydrocarbons are produced through radical-radical combination reactions (e.g., $CH_3 + C_2H_3 + M \rightarrow C_3H_6 + M$, where M represents any third molecule or atom, $CH_3 + C_3H_5 + M \rightarrow C_4H_8 + M$, $CH_3 + C_2H_5 + M \rightarrow C_3H_8 + M$, $2\,C_2H_3 + M \rightarrow C_4H_6 + M$), through CH insertion reactions (e.g., $CH + C_2H_6 \rightarrow C_3H_6 + H$, $CH + C_2H_2 \rightarrow C_3H_2 + H$, $CH + C_2H_4 \rightarrow C_3H_4 + H$), and through other radical insertion reactions (e.g., $C_2H + C_2H_2 = C_4H_2 + H$, $CH_3 + C_2H_3 \rightarrow C_3H_5 + H$, $C_2H + C_2H_4 \rightarrow C_4H_4 + H$). Hydrocarbons are destroyed through photolysis, reaction with hydrogen atoms, and various other radical addition, insertion, and disproportionation reactions (e.g., $C_2H_3 + C_2H_3 \rightarrow C_2H_4 + C_2H_2$). Stratospheric hydrocarbon photochemistry on Saturn is rich and complex. Rather than going through the major production and loss mechanisms for each of the observed stratospheric constituents, we will direct interested readers to the reviews of Moses et al. (2000a, 2004, 2005), for which the hydrocarbon reactions and pathways thought to be most important on Jupiter and Saturn are discussed in gory detail. Note that the dominant reactions are expected to be the same on both Jupiter and Saturn, with quantitative differences caused by different stratospheric temperatures, heliocentric distances, atmospheric transport, auroral processes, condensation, and influx of external material.

Although the production and loss mechanisms for the major hydrocarbons are qualitatively understood, a full quantitative understanding has been hindered by a lack of relevant chemical kinetics data at low temperatures and low pressures. Moses et al. (2005) provide a table (their Table 6) that lists some critical data needed to improve our current understanding of hydrocarbon photochemistry. Of particular importance is information on the photochemistry of benzene and C_3H_x molecules under conditions relevant to giant-planet stratospheres, details of the photolysis branching ratios for CH_4 at Lyman alpha wavelengths (especially the quantum yield of CH formation), and information on the low-pressure limit and falloff behavior of the rate coefficients for relevant hydrocarbon termolecular reactions at temperatures of 80–170 K (e.g., Huestis et al. 2008).

5.3.2.2 One-Dimensional Models

Several one-dimensional (1-D; in the vertical direction) models for hydrocarbon photochemistry in Saturn's stratosphere have been developed in the past several decades (e.g., Strobel 1975, 1978, 1983; Atreya 1982; Festou and Atreya 1982; Lee et al. 2000; Ollivier et al. 2000; Moses et al. 2000a, b, 2005; Dobrijevic et al. 2003; Cody et al. 2003; Sada et al. 2005). In these models, the continuity equation is typically solved to predict the concentration of atmospheric species as a function of pressure. Photochemical production and loss rates for the hydrocarbons as a function of pressure are calculated after providing such inputs as a chemical reaction list, photoabsorption and photolysis cross sections, photolysis branching ratios, and temperature-dependent reaction rate coefficients. Transport of the atmospheric constituents is usually achieved by eddy and molecular diffusion. Molecular diffusion coefficients for various hydrocarbons in H_2, He, and other hydrocarbon gases have been studied theoretically and in the laboratory (e.g., Marrero and Mason 1972), and empirical formulas that depend on atmospheric temperature and gas densities are typically adopted.

Eddy diffusion, on the other hand, is a parameterization in these 1-D models that is introduced to account for macroscopic transport processes that act to keep the atmosphere well mixed (i.e., dissipative waves, small-scale and large-scale eddies, diabatic circulation, and other macroscopic mass motions). Eddy diffusion coefficients cannot be calculated from first principles without knowing the full three-dimensional wind fields from diabatic circulation, waves, etc., which are certainly not known for Saturn. The eddy diffusion coefficient K_{zz} is therefore an important free parameter in the photochemical models. Most modelers assume that the same K_{zz} profile can be used to determine the vertical transport rates for all constituents in the atmosphere, despite the warning of Strobel (1981, 2005) and West et al. (1986) that the chemical loss rate of a constituent can affect its inferred eddy diffusion coefficient K_{zz}. Constituents like C_2H_6 and CO that are long-lived chemically are typically used as tracers to help constrain K_{zz} in the lower stratosphere of the giant planets. Spacecraft ultraviolet occultations of hydrocarbons in the mesopause region of Saturn (e.g., Festou and Atreya 1982; Smith et al. 1983; Vervack and Moses 2009; Shemansky and Liu 2009) or ultraviolet emission from He 58.4 nm or H Lyman α airglow (e.g., Sandel et al. 1982; Atreya 1982; Ben-Jaffel et al. 1995; Parkinson et al. 1998) are used to help constrain the location of the methane homopause–the altitude level at which K_{zz} equals the methane molecular diffusion coefficient–and the magnitude of K_{zz} in the homopause region.

Accurate constraints on the location of the methane homopause on Saturn are important for the photochemical models because methane is photolyzed just below its homopause region, and the pressure at which methane is photolyzed affects the subsequent chemistry through termolecular and other pressure-dependent reactions. One interesting result that has arisen from the analysis of the few Cassini/UVIS stellar occultations that have been analyzed to date (Shemansky and Liu 2009), the recent analysis of all the Voyager/UVS solar and stellar occultations (Vervack and Moses 2009), the analysis of the Cassini/VIMS α CMi stellar occultation at 55° N latitude (Nicholson et al. 2006), and analysis of auroral images and spectra from Cassini/UVIS (e.g., Gérard et al. 2009; Gustin et al. 2009) is that the homopause pressure level on Saturn appears to vary *significantly* with latitude and/or time, a fact that was not realized during the Voyager era. If so, the appropriateness of 1-D models in representing the entire Saturnian atmosphere is called into question. Chapter 8 in this volume has more details about the ultraviolet occultations from Cassini.

The Cassini/UVIS stellar occultation at −43° latitude described by Shemansky and Liu (2009) implies a homopause pressure level that is nearly two orders of magnitude greater (i.e., at a lower altitude) than the Voyager 2 solar ingress occultation at 29.5° latitude analyzed by Vervack and Moses (2009). Figure 5.7 illustrates these differences. The Voyager 2 stellar egress occultation (3.8° latitude) analyzed by Smith et al. (1983) and Festou and Atreya (1982) suggests a homopause at very low pressures (high altitudes), similar to the results of the Voyager 2 solar ingress profile, whereas the Voyager 1 solar egress profile (−27° latitude) (Vervack and Moses 2009) suggests a high-pressure homopause more similar to the Shemansky and Liu (2009) Cassini/UVIS −43° latitude results (see Fig. 5.7). Although differences in analysis techniques and assumptions can play a role in these results, studies that use a consistent technique to analyze multiple occultations (i.e., Shemansky and Liu 2009; Vervack and Moses 2009) derive different homopause levels for different occultation latitudes, suggesting real variation with location and/or time. As suggested by Vervack and Moses (2009), the likely cause of the variable homopause levels is small vertical winds that vary with latitude and time. Existing and future Cassini/UVIS solar and stellar occultations, as well as Cassini/VIMS occultations, may help map out the latitude dependence of the homopause level and perhaps allow winds and circulation in the mesopause region to be derived.

Note that we try to put all quantities on a pressure scale on the giant planets because the altitude or radius scale varies significantly with latitude on these rapidly rotating, oblate planets, due to variations in the gravitational acceleration with latitude. However, occultation data are inherently a function of planetary radius, and converting to a pressure scale introduces a model dependency in the results (the greatest assumption being the temperature profile), and the reader should keep that fact in mind with the above discussion. The simultaneous or near-simultaneous measurement

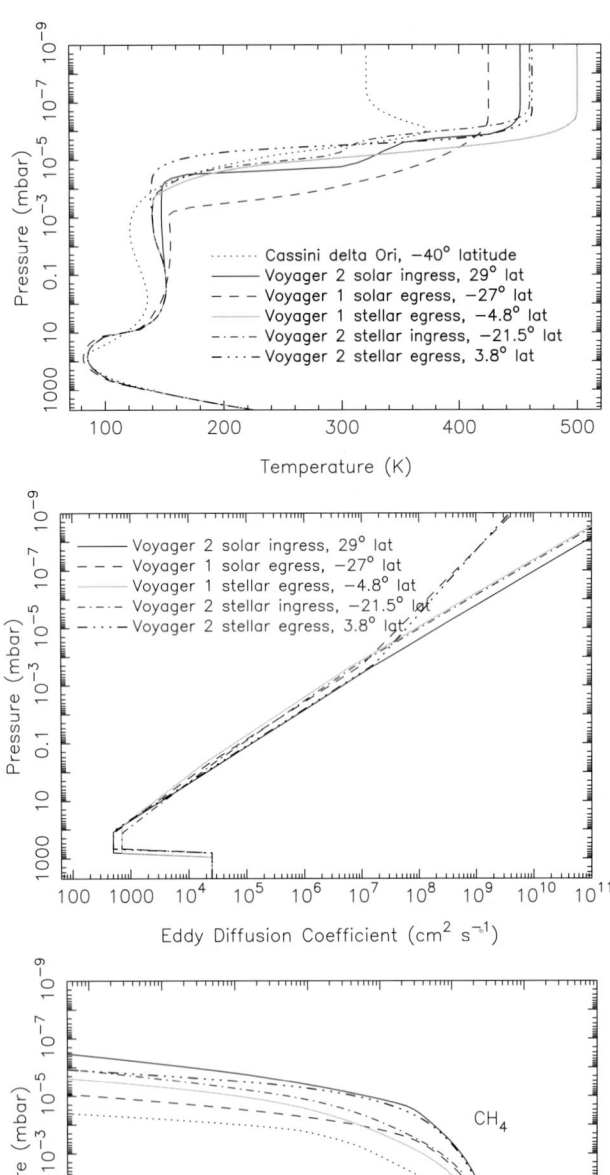

Fig. 5.7 Results from forward models developed to reproduce the ultraviolet occultation data from Cassini/UVIS (Shemansky and Liu 2009) and Voyager/UVS (Vervack and Moses 2009): model temperatures (*top*), eddy diffusion coefficients (*middle*), and methane mole fraction profiles (*bottom*). Although occultation data constrain model parameters only at altitudes above $\sim 10^{-4}$ mbar (for the Voyager data) or above 0.01 mbar (for the Cassini data), the full profiles are shown because model assumptions affect the results. Note the large derived variation in the methane homopause pressure level with latitude and/or time for Saturn

of atmospheric temperatures from 1 bar on up to ~ 0.01 mbar from Cassini/CIRS limb or other measurements at the UVIS occultation latitudes would greatly aid the occultation analyses, as would more accurate determinations of the planetary shape (i.e., determinations of the radius levels of isobaric surfaces, which are not well represented by oblate spheroids). Cassini/VIMS occultations (e.g., Nicholson et al. 2006) may be particularly useful in helping place the homopause location in pressure space.

Photochemical models for Saturn that have been published since the Voyager encounters all assume a high-altitude methane homopause most relevant to 3.8° latitude or 29° N latitude during the Voyager flybys. Figure 5.5 shows the results from one such 1-D photochemical model for Saturn (Moses et al. 2005). The eddy diffusion coefficient in the lower stratosphere for this model is constrained from the global-average ISO/SWS observations of C_2H_6 (e.g., Moses et al. 2000a, de Graauw et al. 1997). The model is designed to represent global-average conditions, with an assumed temperature profile that was derived from the globally averaged ISO observations (Lellouch et al. 2001), an assumed solar UV flux that represents low-to-average conditions within the solar cycle, and fixed vernal equinox solar insolation conditions at 30° latitude. An external influx of H_2O, CO, and CO_2 is introduced at high altitudes, and the oxygen photochemistry slightly affects the abundances of the unsaturated hydrocarbons. Diacetylene, benzene, and water are found to condense in the lower stratosphere and contribute to a stratospheric haze layer (C_4H_2 and C_6H_6 condense near ~ 10 mbar in this model, whereas water condenses closer to 1 mbar, but these values depend on temperature profiles that need to be updated based on the Cassini CIRS results, e.g., Guerlet et al. 2009).

Consistent with the infrared and ultraviolet observations discussed in Section 5.2, the Moses et al. (2005) model–and all other photochemical models–predict that ethane is the second-most abundant hydrocarbon in Saturn's stratosphere (behind methane), followed by acetylene, and propane. Ethylene is abundant at high altitudes but its mole fraction falls off in the middle and lower stratosphere due to reaction with atomic H (to eventually form C_2H_6) and photolysis (to eventually form C_2H_2). The Moses et al. model accurately predicts the global-average abundances of C_2H_6, C_2H_2, CH_3C_2H, C_4H_2, and CH_3 (not shown), but underpredicts the column abundance of C_2H_4 and C_3H_8, and greatly overpredicts the column abundance of C_6H_6. Moreover, when the same reaction list is applied to Jupiter, Uranus, and Neptune, the model has difficulty in reproducing the observed $C_2H_2/C_2H_4/C_2H_6$ ratio on all the giant planets, suggesting systematic problems with the adopted chemical inputs.

Although some of the model-data mismatch may be caused by the attempt to represent a three-dimensional atmosphere with a one-dimensional, steady-state model, the chemical reaction list used by Moses et al. (2005) clearly needs improvement. Models developed by other groups also fail to reproduce all available observations (see below), indicating that we do not have a complete quantitative understanding of hydrocarbon photochemistry under Saturnian stratospheric conditions. Sensitivity and "uncertainty" studies (e.g., Dobrijevic et al. 2003; Smith and Nash 2006) can help identify the key reactions that affect species abundances at different pressures and can help evaluate the overall robustness of the models. Laboratory measurements and theoretical calculations are needed to help fill in current uncertainties in the critical reaction schemes (e.g., see Table 6 of Moses et al. 2005), and further observations that constrain the vertical profiles of the observed constituents will provide important model constraints.

Other recent 1-D photochemical models that have proven useful in defining the hydrocarbon photochemistry on Saturn include those presented by Ollivier et al. (2000), Dobrijevic et al. (2003), Cody et al. (2003), and Sada et al. (2005). The model of Ollivier et al. (2000), like those of Moses et al. (2000a 2005), considers the chemistry of a long list of hydrocarbon and oxygen species and examines the sensitivity of the results to variations in several key input parameters. The model succeeds in reproducing the observed CH_4 and C_2H_2 abundances but greatly underestimates the C_2H_6 abundance and overestimates the CH_3C_2H, C_4H_2, and CH_3 abundances. The CH_3 problem may partly be remedied by the updated line strengths provided by Stancu et al. (2005). The C_2H_6 underestimation appears to have been caused by the adoption of a rapid reaction between C_2H_2 and C_2H_6 to form C_2H_3 and C_2H_5. This reaction between acetylene and ethane would be exceedingly slow at Saturnian temperatures; Ollivier et al. (2000) have apparently mistaken the reaction of vinylidene (H_2CC) radicals plus ethane with the acetylene plus ethane reaction in their model. Despite this reaction-rate problem, Ollivier et al. (2000) have provided sensitivity studies for such parameters as solar irradiance, latitude, H influx from the thermosphere, external oxygen influx, and absorption only vs. multiple scattering that are useful in understanding model sensitivities and limitations.

The Ollivier et al. (2000) $C_2H_2 + C_2H_6$ reaction-rate problem has been corrected in a follow-up model by Dobrijevic et al. (2003). The updated Dobrijevic et al. (2003) results compare well with the C_2H_2 and C_2H_6 observations from the ISO satellite (Moses et al. 2000a), but the model tends to underestimate CH_3C_2H and overestimates the abundance of C_4H_2, although Dobrijevic et al. (2003) emphasize that the model results are acceptable given the uncertainties in the rate coefficients and other model parameters. In fact, the main advance of the Dobrijevic et al. (2003) study is a quantification of uncertainties in product abundances in photochemical models based on uncertainties in reaction rate coefficients and other model inputs.

The model of Cody et al. (2003) is designed to specifically study the influence of new experimentally derived rate coefficients for the $CH_3 + CH_3$ self reaction on the predicted abundance of CH_3 on Saturn and Neptune. Cody et al. (2003) emphasize the relative importance of the ratio of low-pressure limiting vs. high-pressure limiting rate coefficient for the termolecular $CH_3 + CH_3$ recombination reaction in affecting the methyl radical abundance in photochemical models.

Sada et al. (2005) have developed a photochemical model for Saturn in order to better define the vertical profiles of C_2H_6 and C_2H_2 to more accurately retrieve mole fractions from ground-based infrared observations and Voyager IRIS data. The Sada et al. (2005) model simultaneously reproduces the emission in the observed C_2H_2 and C_2H_6 infrared bands, but the model profiles were shifted by some unspecified amount in mole-fraction space to obtain this good fit. Interestingly, their vertical profile for C_2H_6 differs from that of other photochemical models in the upper atmosphere such that their mole fraction continues to increase sharply from ~ 1 to 0.01 mbar. The C_2H_6 profiles in other models such as Moses et al. (2000a, 2005) and Ollivier et al. (2000) exhibit a more constant C_2H_6 mole fraction in the upper stratosphere. Limb observations acquired with Cassini/CIRS (e.g., Guerlet et al. 2009) and ultraviolet occultations with Cassini/UVIS (Shemansky and Liu 2009) may help constrain the C_2H_6 vertical profiles to help distinguish between models.

5.3.2.3 Seasonal Variation in 1-D Models

Most photochemical models to date have been concerned with steady-state conditions for either global-average atmospheres or for single specific latitudes and seasons. The time-variable, multiple-latitude model of Moses and Greathouse (2005) is an exception in that the effects of seasonally varying insolation, ring shadowing, and the solar cycle on stratospheric chemistry are examined in a more realistic manner; however, the model still considers transport in the vertical direction only–horizontal transport is neglected. Moses and Greathouse find that at very high altitudes (i.e., pressures less than ~ 0.01 mbar), the hydrocarbon abundances responded quickly to insolation changes, and the mixing ratios of the photochemical products are largest at those latitudes and seasons where the solar insolation is the greatest. At lower altitudes, however, the increased vertical diffusion time scales are found to introduce increasing phase lags in the chemical response of the atmosphere to the changing solar insolation. Below the ~ 1 mbar level, the response times of the relatively long-lived species such as C_2H_2, C_2H_6, and C_3H_8 become larger than a Saturnian year, so that their abundances

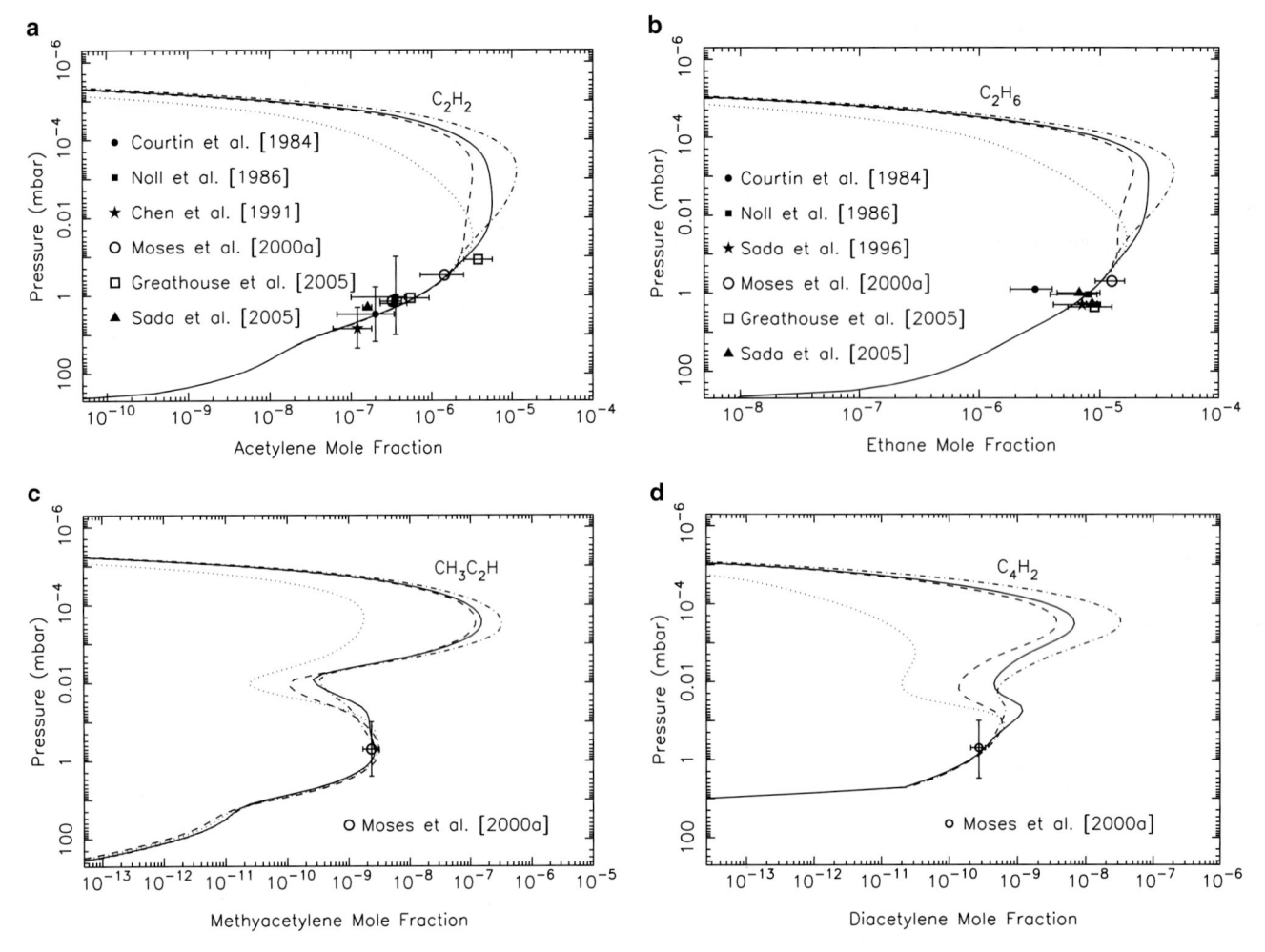

Fig. 5.8 Hydrocarbon mole fractions at $-36°$ planetocentric latitude as a function of pressure and season in Saturn's upper atmosphere, as derived from the 1-D seasonal photochemical model of Moses and Greathouse (2005). The dotted lines represent $L_s = 94.2°$ (just past southern winter solstice), the dashed lines represent $L_s = 177.4°$ (just before southern vernal equinox), the dot-dashed lines represent $L_s = 273.1°$ (just past southern summer solstice), and the solid lines represent $L_s = 360°$ (southern autumnal equinox). The data points with associated error bars represent various infrared and ultraviolet observations, as labeled (figure modified from Moses and Greathouse 2005). Note the strong seasonal variability at high altitudes and the virtual absence of seasonal variability at low altitudes

reflect annual-average insolation conditions (see Figs. 5.8 and 5.9). Because the annual-average solar insolation decreases toward the poles, Moses and Greathouse (2005) predict that the abundances of these photochemically produced species should decrease with increasing latitude at the few-mbar level (and at lower altitudes) at all seasons.

Figure 5.6 shows how the photochemical model of Moses and Greathouse (2005) compares with the ground-based, thermal-infrared observations of Greathouse et al. (2005), the Cassini/CIRS nadir data of Howett et al. (2007), the Cassini/CIRS limb data of Guerlet et al. (2009), and the HST ultraviolet observations of Prangé et al. (2006). The model roughly reproduces the observed meridional variation of C_2H_2 (albeit not as steeply in the northern hemisphere, which is likely caused by the constant-temperature-with-latitude assumption of the model) but clearly fails to reproduce the observations for the longer-lived C_2H_6 (especially in the southern hemisphere); the observations indicate that at the 2-mbar level, the C_2H_6 mixing ratio *increases* with increasing latitude in the southern summer hemisphere, whereas the model predicts a strong decrease with increasing latitude. Moses and Greathouse suggest that the long photochemical lifetime of C_2H_6 (i.e., $\sim$700 years at 2 mbar at mid-latitudes in the summer) is the root cause of this behavior–C_2H_6 is sensitive to horizontal transport, which is not included in the model (see Nixon et al. 2007 for a similar conclusion for Jupiter). Based on the failure of the model to reproduce the C_2H_6 but not the C_2H_2 meridional gradient, Moses and Greathouse suggest that the meridional transport time scale at 2 mbar on Saturn is in the $\sim$100–700 year range (i.e., somewhere between the photochemical lifetimes of C_2H_2 and C_2H_6). Although it is possible that K_{zz} profiles that vary with latitude might be developed that could reproduce some of the observed behavior, it is more likely that

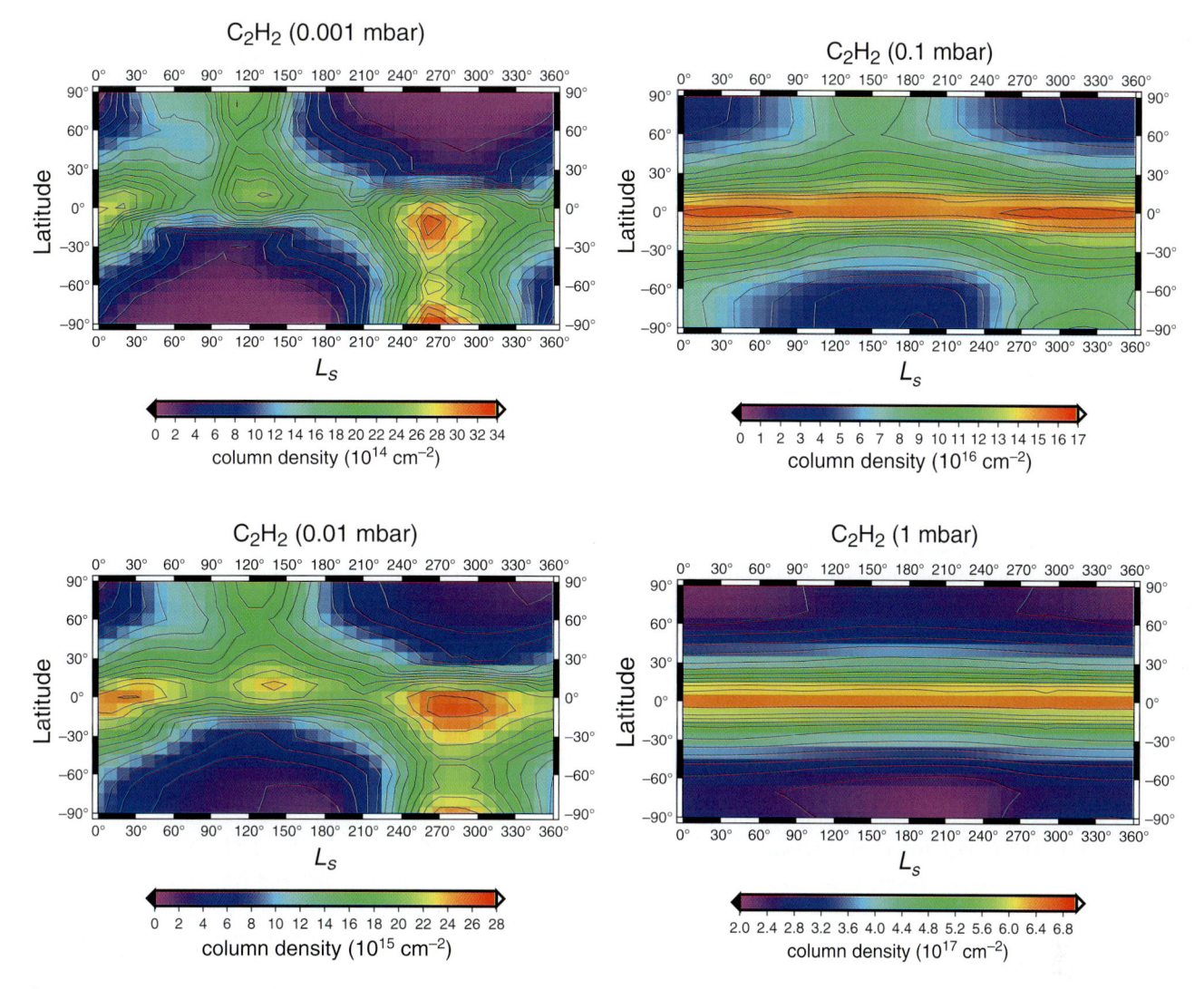

Fig. 5.9 The column density of acetylene (C_2H_2) above different pressure levels as a function of planetocentric latitude and season (L_s) from the Moses and Greathouse (2005) model: above 0.001 mbar (*top left*), above 0.01 mbar (*bottom left*), above 0.1 mbar (*top right*), above 1 mbar (*bottom right*). The contour intervals and overall scale change for each figure. Note the strong seasonal dependence at low pressures; the lack of seasonal dependence in the column abundance above 1 mbar is due to large photochemical and vertical transport time scales at pressures $\gtrsim 1$ mbar (figure modified from Moses and Greathouse 2005)

meridional transport is occurring in Saturn's stratosphere, and multi-dimensional photochemical models that include horizontal transport will be needed to make sense of these observations.

5.3.2.4 Two-Dimensional Models

The observations and modeling discussed above and shown in Figs. 5.5–5.6 make it clear that the abundances of the photochemically produced species change with altitude, latitude, and time on Saturn. Because ethane and acetylene, along with methane, are the dominant coolants in giant-planet stratospheres (e.g., Yelle et al. 2001; Bézard and Gautier 1985), variations in the abundances of these species affect the radiative properties, thermal structure, and energy balance in the atmosphere, which in turn affect atmospheric dynamics, which in turn redistributes constituents that feed back on the chemistry and temperatures. A general circulation model (GCM) will be required to fully explore this complex coupling between chemical, dynamical, and radiative processes. Unfortunately, GCMs for Saturn's stratosphere do not yet exist, and little is known about stratospheric circulation on Saturn due to a lack of clouds or other wind tracers. The two-dimensional "diabatic" or "residual mean" circulation can in principle be derived from modeling the momentum and energy budgets from seasonal radiative forcing, but the radiative-dynamical models that have been developed to date (e.g., Conrath et al. 1990; Barnet et al. 1992; see also the purely radiative seasonal climate model of Bézard and Gautier 1985) do not go high enough in the stratosphere to encompass the photochemical production regions and do not take variable hydrocarbon abundances into account. The latter variation and its effect on temperatures has been studied by Strong et al. (2007), who find that radiative time constants vary significantly with pressure such that high-altitude temperatures respond quickly to changes in insolation.

In the absence of direct information on stratospheric circulation, the observed meridional distribution of the hydrocarbons or other tracers can be used to learn something about the general nature of the stratospheric transport (see the Jupiter modeling of Friedson et al. 1999; Liang et al. 2005; Lellouch et al. 2006). Temperature maps, such as were produced from Cassini CIRS (Flasar et al. 2005; Simon-Miller et al. 2006; Fletcher et al. 2007a; Fouchet et al. 2008; Guerlet et al. 2009), also provide clues. For example, the elevated south polar temperatures (Fletcher et al. 2007a; Flasar et al. 2005) and the observed increase in the C_2H_6 abundance toward high southern latitudes (e.g., Howett et al. 2007; Greathouse et al. 2005) led Flasar et al. (2005) to suggest subsidence in the high-latitude southern stratosphere. Complex wave-like structure in the low-latitude regions suggests an effect similar to the Earth's semi-annual oscillation but on a ~15-year period (Fouchet et al. 2008; Orton et al. 2008; see also Chapter 6). This oscillatory feature seems to be influencing the C_2H_2, C_2H_6, and C_3H_8 vertical distributions in the low-latitude regions (Fouchet et al. 2008; Guerlet et al. 2009). Similarly, fine- and broad-scale features in the meridional profiles of temperatures and abundances have led Guerlet et al. (2009) to suggest vertical winds are influencing behavior at certain latitudes.

Although no 2-D photochemistry-transport models have yet been reported in the refereed literature, some preliminary models have been presented by Moses et al. (2007). In these models, the Caltech/JPL two-dimensional chemistry transport model (2-D GCM, see Morgan et al. 2004; Shia et al. 2006) is used to study the meridional and vertical distribution of stratospheric species. Both advection and/or horizontal and vertical diffusion are considered in the transport terms. Various *ad hoc* scenarios involving large-scale circulation cells or latitude- and altitude-dependent eddy diffusivities are examined, and the results are compared to the observed meridional distribution of constituents. The key conclusion of this modeling is that the abundances of C_2H_6 and C_2H_2 are highly linked by both photochemistry and vertical transport on time scales shorter than the estimated meridional transport time scales of Moses and Greathouse (2005), such that the meridional distribution of C_2H_2 in the models closely tracks that of C_2H_6. This model result is in conflict with observations (see Figs. 5.6). Moses et al. (2007) came up with some highly contrived scenarios involving latitude or altitude-dependent diffusivities that could roughly reproduce the observed C_2H_2 and C_2H_6 distributions, but the overall conclusion from this modeling is that the cause of the observed uncoupled distributions remains a mystery. Aerosol and gas-constituent shielding, multiple Rayleigh scattering, auroral chemistry, and incorrect model branching ratios for C_2H_6 have all been tentatively ruled out as possibilities. Further modeling, including the development of GCMs, is needed.

5.3.3 Oxygen Chemistry

The detection of CO_2 and H_2O that is unambiguously in the stratosphere of Saturn points to an external supply of oxygen to the planet (e.g., Feuchtgruber et al. 1997, 1999; de Graauw et al. 1997; Bergin et al. 2000; Moses et al. 2000b; Ollivier et al. 2000; Simon-Miller et al. 2005; see also the discussion in the Jupiter papers of Bézard et al. 2002; Lellouch et al. 2002, 2006). Water from the deep interior would condense in the troposphere and would not make it up into the stratosphere. Carbon dioxide from the interior, on the other hand, might not condense and some could diffuse up into the stratosphere; however, the observed stratospheric abundance is much larger than the quenched equilibrium abundance predicted in thermochemical models (e.g., Fegley and Prinn 1985; Fegley and Lodders 1994; Visscher and Fegley 2005), indicating an external source. Possible sources of the exogenic oxygen-bearing material include comets, micrometeoroids, or ring/satellite debris.

The chemistry that is initiated from this exogenic oxygen influx is discussed by Moses et al. (2000b) and Ollivier et al. (2000) and is illustrated in Fig. 5.10. Although water is readily photodissociated at wavelengths less than 185 nm into predominantly OH + H, shielding by hydrocarbons lengthens its photochemical lifetime in Saturn's stratosphere; moreover, the OH is efficiently recycled back to H_2 through the reaction $OH + H_2 \rightarrow H_2O + H$. A small portion of the H_2O can be converted to CO through addition reactions of OH with unsaturated hydrocarbons (Moses et al. 2000b):

$$H_2O + h\nu \rightarrow H + OH$$

$$OH + C_2H_2 + M \rightarrow C_2H_2OH + M$$

$$C_2H_2OH \rightarrow CH_3CO$$

$$CH_3CO + H \rightarrow HCO + CH_3$$

$$HCO + H \rightarrow CO + H_2$$

$$\text{Net}: H_2O + C_2H_2 + H \rightarrow CO + CH_3 + H_2$$

5 Saturn: Composition and Chemistry

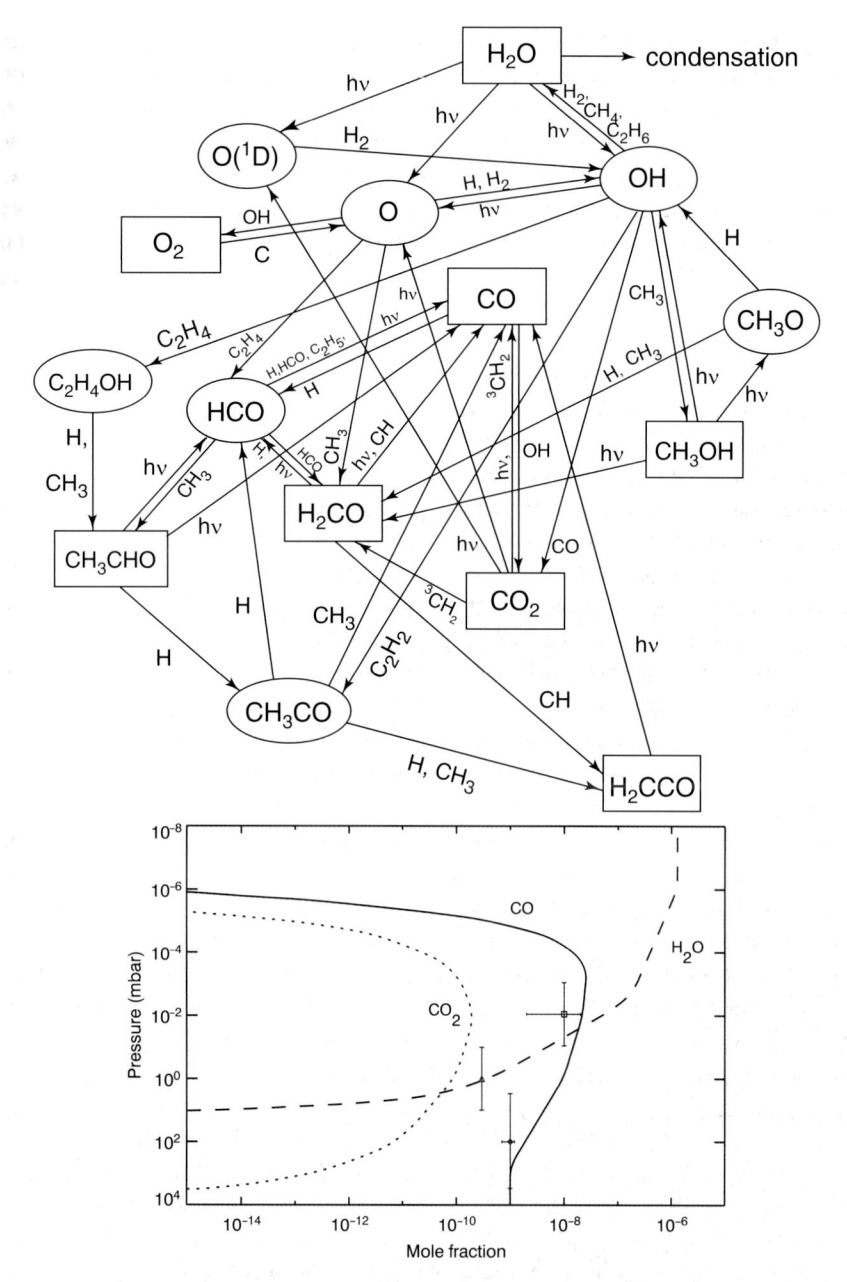

Fig. 5.10 (*top*) The important reaction pathways for oxygen photochemistry in Saturn's stratosphere. The symbol $h\nu$ corresponds to an ultraviolet photon. Figure is from Moses et al. (2000b). (*bottom*) H_2O, CO, CO_2 mole fraction profiles from Ollivier et al. (2000). These profiles were obtained with an external flux of water alone (10^7 molecules cm^{-2} s^{-1}) and tropospheric CO (with a mole fraction equal to 10^{-9}). The square and the triangle correspond, respectively, to the water and carbon dioxide mole fraction found by Feuchtgruber et al. (1997), and the diamond-shaped point corresponds to distribution CO inferred by Noll and Larson (1991)

and

$$H_2O + h\nu \rightarrow H + OH$$
$$OH + C_2H_2 + M \rightarrow C_2H_2OH + M$$
$$C_2H_2OH \rightarrow CH_3CO$$
$$CH_3CO + H \rightarrow H_2CCO + H_2$$
$$H_2CCO + h\nu \rightarrow {}^1CH_2 + CO$$
$$\text{Net}: H_2O + C_2H_2 \rightarrow CO + {}^1CH_2 + H_2.$$

Carbon monoxide can in turn be converted to CO_2 through the reaction of $CO + OH \rightarrow CO_2 + H$. The CO_2 photolyzes back to $CO + O(^1D)$ or $CO + O$. The $O(^1D)$ atoms react with H_2 to form OH and eventually water, whereas the O can react with CH_3 or unsaturated hydrocarbons to eventually reform CO. Other hydrocarbons like H_2CO, CH_3OH, and CH_3CHO are formed in the models, but not in currently observable quantities. Water can diffuse down through Saturn's stratosphere and eventually condense in the lower stratosphere. The Ollivier et al. (2000) model differs from that of Moses et al. (2000b) in that the interesting oxygen photochemistry proceeds through reaction of O + CH$_3$ or O + unsaturated hydrocarbons rather than through addition reactions of OH with unsaturated hydrocarbons. These differences have some impact on the final product abundances and the inferred influx rates from the two models.

As discussed in Section 5.2.4, most observations of CO to date have not allowed the discrimination between an internal or external source of CO on Saturn. Ollivier et al. (2000) favor an internal source based on the failure of their external-source models to reproduce the CO abundance derived by Noll and Larson (1991). Moses et al. (2000b), on the other hand, were able to produce the observed CO column abundance from external-source models under certain conditions, although they favor a situation in which CO has both and internal and external source. Both Ollivier et al. (2000) and Moses et al. (2000b) show how the models are sensitive to the assumed flux of the external oxygen and use model-data comparisons to constrain the external flux; they also discuss implications regarding the source of the external oxygen (see also Feuchtgruber et al. 1997, 1999). Ollivier et al. (2000) do not weigh in with a favored source, whereas Moses et al. (2000b) favor interplanetary dust or ring-particle diffusion as the source of the external oxygen. However, this conclusion came into effect before active venting on Enceladus was discovered by Cassini (e.g., Porco et al. 2006)–Enceladus may be a major supplier of water and oxygen species to Saturn–and before the recent 345 HZ observations of Cavalié et al. (2009) that suggest an external source of CO from a large cometary impact 200–300 years ago.

5.3.4 Auroral Chemistry

As on Jupiter (e.g., Waite et al. 1983; Perry et al. 1999; Wong et al. 2000, 2003; Friedson et al. 2002), auroral ion chemistry and subsequent neutral chemistry has the potential for affecting the global distribution and abundance of stratospheric constituents on Saturn. Models of auroral chemistry have not yet been presented for Saturn, but the Jupiter studies discussed above, as well as the obvious importance of ion chemistry in producing C$_6$H$_6$ on Titan (e.g., Waite et al. 2005, 2007; Vuitton 2007, 2008; Imanaka and Smith 2007), suggest that auroral ion chemistry may be important for forming benzene and polycyclic aromatic hydrocarbons (PAHs) on Saturn. Observations such as those presented by Gérard et al. (1995) demonstrate the possible link between the aurora and polar haze, most likely due to the formation of complex hydrocarbons. The overall effect of auroral chemistry on other molecules like C$_2$H$_2$ and C$_2$H$_6$ is unknown, although the Jupiter models of Wong et al. (2000, 2003) show an increase in the production rate of both molecules, which could affect the observed meridional distribution of C$_2$H$_2$ and C$_2$H$_6$, but which is not likely the culprit in explaining the disparate observed meridional distribution of the two species. Given the potential importance of auroral chemistry on Saturn, this topic deserves more study, especially given the high-quality observational constraints that can likely be supplied by the Cassini/CIRS and UVIS instruments.

5.4 Summary and Conclusions

Our understanding of the composition and chemistry of Saturn's atmosphere has improved widely since the *Voyager* flybys. The ever-increasing sensitivity and resolution (in the spatial and spectral dimension) of Earth-based observations have provided a wealth of new information on the Kronian atmosphere. The Cassini spacecraft is now participating in this ongoing effort, but its results are still in their infancy.

In terms of global composition, Cassini measured the carbon abundance and the carbon isotopic ratio with a precision that will not be challenged for several years. Saturn's carbon enrichment relative to the solar composition will vary more certainly in response to changes in the solar composition than in the Kronian composition. In contrast, the abundance of helium is still a subject of continuing analyses. The inversion of the collision-induced hydrogen continuum observed by Cassini in the far infrared yields different [He]/[H$_2$] ratios depending on whether the inversion is carried out including the radio occultation profiles or whether both the temperature and the composition are inferred from the far-infrared spectrum only. This discrepancy suggests that unknown systematic uncertainties still affect the data analysis. Of particular concern is the sensitivity of the inversion to the wind field. The elemental composition of nitrogen and sulfur has not been constrained from Cassini data yet. Our knowledge of these two elements still relies on ground-based centimeter observations that are difficult to calibrate and to analyse. However, there is some hope that Cassini radio science can provide new estimates in the future. Pre-Cassini observations yielded low D/H ratios in comparison with the Jupiter deuterium abundance or with the early solar abundance. The first Cassini observations have yielded even smaller values, which do not fit easily in the Solar System picture, although the error bars are still large.

Below the ammonia cloud level, at 1–5 bars, several molecular species have been measured to tremendously exceed their thermodynamic equilibrium abundances. These molecules, phosphine, germane and arsine, which are thermodynamically stable in the warm interior of the planet, are transported more rapidly to the shallow atmosphere than they are destroyed by chemical reactions. Their measurement gives clues to the vertical mixing timescales and to the elemental composition of the deep interior. Cassini/VIMS is in position to measure the meridional variation in the abundance of these elements that would reveal differential vertical mixing, a constraint of great importance for dynamical models of Saturn. However, the kinetics and rate coefficients of the interconversion mechanisms are still preventing reliable estimates of the deep abundance on Saturn, and/or the mixing timescale, from being obtained using this method. Experimental data are greatly needed here. Simultaneous measurements of aerosol opacity at ultraviolet wavelengths and species profiles would also help separate out the effects due to mixing versus the effects due to aerosol shielding.

In the upper troposphere, above the cloud tops, Cassini/CIRS has measured for the first time the meridional variation in phosphine abundance. In this atmospheric region, PH_3 is photolyzed along with NH_3, and any variation in abundance can trace the intensity of vertical and horizontal mixing. This measurement will hence help to constrain the dynamics above the cloud level. Such variations could also be linked with variations in the visible cloud colors to understand whether PH_3 and NH_3 photolysis does produce some of the chromophores that determine Saturn's visible aspect. However, to obtain quantitative constraints on the dynamics, the tropospheric photochemistry on Saturn would need to be better understood than it is presently, due to a critical lack of reaction rate coefficients for the appropriate phosphorus and nitrogen reactions under reducing conditions. It is hoped that new Cassini and Earth-based observations will generate a resurgence in interest in tropospheric photochemistry on Saturn.

In the stratosphere, several hydrocarbons have been detected (CH_3, C_2H_2, C_2H_4, C_2H_6, CH_3C_2H, C_3H_8, C_4H_2, C_6H_6) from Earth-based spectroscopy. These species are critical in controlling the stratospheric temperature. One-dimensional photochemical models are able to reproduce the observed abundance of ethane, acetylene, and several of the other observed stratospheric constituents, but have more troubles in predicting the abundance of some of the heavier hydrocarbons. Moreover, models that best reproduce Saturn's hydrocarbon photochemistry are often unable to reproduce the observed hydrocarbon abundances on the three other Giant Planets. The problem lies partly in a lack of appropriate laboratory data at the conditions (temperature, pressures) relevant to Saturn's stratosphere. Data on photolysis quantum yields, branching ratios, chemical pathways and kinetics are greatly needed. It must also be understood how dynamics interplays with the photochemistry. Ground-based and Cassini measurements have started revealing the meridional profiles in hydrocarbon abundance, while one-dimensional seasonal models and two-dimensional models are being developed. Currently the models do not succeed in reproducing the observed meridional profiles. These models now need to reproduce more accurately the dynamics, especially the meridional transport and large-scale circulation. The models would be better constrained by the measurements of the vertical and meridional distributions of large variety of hydrocarbon species. Hopefully, the Cassini equinox and solstice missions will provide insights into the seasonal variations of the hydrocarbons abundances. Observations that monitor the location of the methane homopause, and its temporal and spatial variations would also provide critical clues to understanding hydrocarbon photochemistry. Cassini CIRS limb observations and UVIS and VIMS occultations are still expected to deliver much new information on this problem.

In the upper stratosphere, Saturn is receiving an external flux of oxygenated species. A recent study supports a dual source, both from a cometary impact and from a continuous flux, conceivably from the oxygen-rich bodies in Saturn's system. The magnitude of both sources should be better constrained in the near future by new sub-millimetric observations.

If great results have been already obtained, the Cassini spacecraft still promises to deliver much more. The unprecedented spatial resolution and possible duration over almost half a Kronian year should provide a wealth of useful information to constrain the chemistry and its relation with the dynamics in Saturn's atmosphere. The supporting ground-based observations, with their unchallenged sensitivity and spectral resolution, will also contribute to our study of Saturn. Cassini, with its spectral, spatial and temporal coverage, is a unique endeavor to study in depth a Giant Planet.

References

Atreya, S.K., 1982. Eddy mixing coefficient on Saturn. *Planet. Space Sci.* **30**, 849–854.

Atreya, S.K., Kuhn, W.R., Donahue, T.M., 1980. Saturn: Tropospheric ammonia and nitrogen. *Geophyte. Res. Lett.* **7**, 474–476.

Atreya, S.K., Waite, J.H., Jr., Donahue, T.M., Nagy, A.F., McConnel, J.C. 1984. Theory, measurements, and models of the upper atmosphere and ionosphere of Saturn. In *Saturn* (Gehrels, T., Matthews, M.S., Eds.), University of Arizona Press, Arizona, Tucson, 239–277.

Baines, K.H., Drossart, P., Momary, T.W., Formisano, V., Griffith, C., bellucci, G., Bibring, J.P., Brown, R.H., Buratti, B.J., Capaccioni, F., Cerroni, P., Clark, R.N., Coradini, A., Combes, M., Cruikshank, D.P., Jaumann, R., Langevin, Y., Matson, D.L., McCord, T.B., Mennella, V., Nelson, R.M., Nicholson, P.D., Sicardy, B., Sotin, C. 2005. The atmospheres of Saturn and Titan in the near-infrared first results of Cassini/vims. *Earth Moon Planets* **96**, 119–147.

Barnet, C.D., Beebe, R.F., Conrath, B.J., 1992. A seasonal radiative-dynamic model of Saturn's troposphere. *Icarus* **98**, 94–107.

Ben-Jaffel, L., Prangé, R., Sandel, B.R., Yelle, R.V., Emerich, C., Feng, D., Hall, D.T., 1995. New analysis of the Voyager UVS H Lyman-α emission of Saturn. *Icarus* **113**, 91–102.

Bergin, E.A., 21 co-authors, 2000. Submillimeter Wave astronomy satellite observations of Jupiter and Saturn: Detection of 557 GHz

water emission from the upper atmosphere. *Astrophys. J.* **539**, L147–L150.
Bézard, B., Gautier, G. 1985. A seasonal climate model of the atmospheres of the giant planets at the Voyager encounter time. I. Saturn's stratosphere. *Icarus* **60**, 296–310.
Bézard, B., Drossart, P., Lellouch, E., Tarrago, G., Maillard, J. P. 1989. Detection of arsine in Saturn. *Astrophys. J.* **346**, 509–513.
Bézard, B., Feuchtgruber, H., Moses, J.I., Encrenaz T., 1998. Detection of methyl radicals (CH$_3$) on Saturn. *Astron. Astrophys.* **334**, L41–L44.
Bézard, B., P. Drossart, T. Encrenaz, and H. Feuchgruber 2001a. Benzene on giant planets. *Icarus*, **154**, 495–500.
Bézard, B., Moses, J. I., Lacy, J., Greathouse, T., Richter, M., Griffith, C. 2001b. Detection of Ethylene (C$_2$H$_4$) on Jupiter and Saturn in Non–Auroral Regions. *Bull. Am. Astron. Soc.* **33**, 1079.
Bézard, B., Lellouch, E., Strobel, D., Maillard, J.-P., Drossart, P. 2002. Carbon monoxide on Jupiter: Evidence for both internal and external sources. *Icarus* **159**, 95–111.
Bjoraker, G. L., Larson, H. P., Fink, U. 1981. A study of ethane on Saturn in the 3 micron region. *Astrophys. J.* **248**, 856–862.
Borunov, S., Dorofeeva, V., Khodakovsky, I., Drossart, P., Lellouch, E., Encrenaz, T., 1995. Phosphorus chemistry in the atmosphere of Jupiter: A reassessment. *Icarus* **113**, 460–464.
Bossard, A.R., Kamga, R., Raulin, F., 1986. Gas phase synthesis of organophosphorus compounds and the atmosphere of the giant planets. *Icarus* **67**, 305–324.
Brault, J. W., Fox, K., Jennings, D. E., Margolis, J. S. 1981. Anomalous ^{12}CH$_4$:^{13}CH$_4$ strengths in 3ν_3. *Astrophys. J.* **247**, L101–L104.
Briggs, F. H., Sackett, P. D., 1989. Radio observations of Saturn as a probe of its atmosphere and cloud structure. *Icarus* **80**, 77–103.
Broadfoot, A.L., Sandel, B.R., Shemansky, D.E., Holberg, J.B., Smith, G.R., Strobel, D.F., McConnell, J.C., Kumar, S., Hunten, D.M., Atreya, S.K., Donahue, T.M., Moos, H.W., Bertaux, J.L., Blamont, J.E., Pomphrey, R.B., Linick, S., 1981. Extreme ultraviolet observations from Voyager 1 encounter with Saturn. *Science* **212**, 206–211.
Brownsword, R.A., Hillenkamp, M., Laurent, T., Vasta, R.K., Volpp, H.-R., Wolfrum, J., 1997. Quantum yield for H atom formation in the methane dissociation after photoexcitation at the Lyman-alpha (121.6 nm) wavelength. *Chem. Phys. Lett.* **266**, 259–266.
Burgdorf, M. J., Orton, G. S., Encrenaz, T., Davis, G. R., Sidher, S. D., Lellouch, E., Swinyard, B. M. 2004. The far-infrared spectra of Jupiter and Saturn. *Planet. Space Sci.* **52**, 379–383.
Cavalié, T., Billebaud, F., Fouchet, T., Lellouch, E., Brillet, J., Dobrijevic, M., 2008. Observations of CO on Saturn and Uranus at millimeter wavelengths: New upper limit determinations. *Astron. Astrophys.* **484**, 555–561.
Cavalié, T., Billebaud, F., Dobrijevic, M., Fouchet, T., Lellouch, E., Encrenaz, T., Brillet, J., Moriarty-Schieven, G. H., Wouterloot, J., Hartogh, P., 2009. First observations of CO at 345 GHz in the atmosphere of Saturn with the JCMT. New constraints on its origin. *Icarus*, in press.
Cody, R.J., Romani, P.N., Nesbitt, F.L., Iannone, M.A., Tardy, D.C., Stief, L.J., 2003. Rate constant for the reaction CH$_3$ + CH$_3$ → C$_2$H$_6$ at T = 155 K and model calculation of the CH$_3$ abundance in the atmospheres of Saturn and Neptune. *J. Geophys. Res.* **108**, 5119, doi:10.1029/2002JE002037.
Combes, M., Maillard, J. P., de Bergh, C., 1977. Evidence for a telluric value of the ^{12}C/^{13}C ratio in the atmospheres of Jupiter and Saturn. *Astron. Astrophys.* **61**, 531–537.
Conrath, B.J., Gautier, D., 2000. Saturn helium abundance: A reanalysis of Voyager measurements. *Icarus* **144**, 124–134.
Conrath, B.J., Gautier, D., Hanel, R.A., Hornstein, J.S., 1984. The helium abundance of Saturn from Voyager measurements. *Astrophys. J.* **282**, 807–815.
Conrath, B.J., Gierasch, P.J., Leroy, S.S., 1990. Temperature and circulation in the stratosphere of the outer planets. *Icarus* **83**, 255–281.

Conrath, B.J., Gierasch, P.J., Ustinov, E.A., 1998. Thermal structure and para hydrogen fraction on the outer planets from Voyager IRIS measurements. *Icarus* **135**, 501–517.
Cook, P.A., Ashfold, M.N.R., Jee, Y.J., Jung, K.H., Harich, S., Yang, X.M., 2001. Vacuum ultraviolet photochemistry of methane, silane and germane. *Phys. Chem. Chem. Phys.* **3**, 1848–1860.
Courtin, R., Gautier, D., Marten, A., Bézard, B., Hanel, R., 1984. The composition of Saturn's atmosphere at northern temperate latitudes from Voyager IRIS spectra: NH$_3$, PH$_3$, C$_2$H$_2$, C$_2$H$_6$, CH$_3$D, CH$_4$, and the Saturnian D/H isotopic ratio. *Astrophys. J.* **287**, 899–916.
de Bergh, C., Lutz, B.L., Owen, T., Brault, J., Chauville, J., 1986. Monodeuterated methane in the outer solar system. II. Its detection on Uranus at 1.6 microns. *Astrophys. J.* **311**, 501–510.
de Graauw, T., Feuchtgruber, H., Bézard, B., Drossart, P., Encrenaz, T., Beintema, D.A., Griffin, M., Heras, A., Kessler, M., Leech, K., Lellouch, E., Morris, P., Roelfsema, P.R., Roos-Serote, M., Salama, A., Vandenbussche, B., Valentijn, E.A., Davis, G.R., Naylor, D.A., 1997. First results of ISO-SWS observations of Saturn: Detection of CO$_2$, CH$_3$C$_2$H, C$_4$H$_2$ and tropospheric H$_2$O. *Astron. Astrophys.* **321**, L13–L16.
Dobrijevic, M., Ollivier, J.L., Billebaud, F., Brillet, J., Parisot, J.P., 2003. Effect of chemical kinetic uncertainties on photochemical modeling results: Application to Saturn's atmosphere. *Astron. Astrophys.* **398**, 335–344.
Fegley, B., Jr., Lodders, K. 1994. Chemical models of the deep atmospheres of Jupiter and Saturn. *Icarus* **110**, 117–154.
Fegley, B., Jr., Prinn, R. G. 1985. Equilibrium and nonequilibrium chemistry of Saturn's atmosphere: Implications for the observability of PH$_3$, N$_2$, CO, and GeH$_4$. *Astrophys. J.* **299**, 1067–1078.
Ferris, J.P., Benson, R., 1981. An investigation of the mechanism of phosphine photolysis. *J. Am. Chem. Soc.* **103**, 1922–1927.
Ferris, J.P., Chen, C.T., 1975. Photosynthesis of organic compounds in the atmosphere of Jupiter. *Nature* **258**, 587–588.
Ferris, J.P., Ishikawa, Y., 1987. HCN and chromophore formation on Jupiter. *Nature* **326**, 777–778.
Ferris, J.P., Ishikawa, Y., 1988. Formation of HCN and acetylene oligomers by photolysis of ammonia in the presence of acetylene: Applications to the atmospheric chemistry of Jupiter. *J. Am. Chem. Soc.* **110**, 4306–4312.
Ferris, J.P., Khwaja, H., 1985. Laboratory simulations of PH3 photolysis in the atmospheres of Jupiter and Saturn. *Icarus* **62**, 415–424.
Ferris, J.P., Morimoto, J.Y., 1981. Irradiation of NH$_3$-CH$_4$ mixtures as a model of photochemical processes in the Jovian planets and Titan. *Icarus* **48**, 118–126.
Ferris, J.P., Bossard, A., Khwaja, 1984. Mechanism of phosphine photolysis. Application to Jovian atmospheric photochemistry. *J. Am. Chem. Soc.* **106**, 318–324.
Ferris, J.P., Jacobson, R.R., Guillemin, J.C., 1992. The photolysis of NH3 in the presence of substituted acetylenes: A possible source of oligomers and HCN on Jupiter. *Icarus* **95**, 54–59.
Festou, M. C., Atreya, S. K., 1982. Voyager ultraviolet stellar occultation measurements of the composition and thermal profiles of the Saturnian upper atmosphere. *Geophys. Res. Lett.* **9**, 1147–1150.
Feuchtgruber, H., E. Lellouch, T. de Graauw, B. Bézard, T. Encrenaz and M. Griffin 1997. External supply of oxygen to the atmospheres of the giant planets. *Nature* **389**, 159–162.
Feuchtgruber, H., Lellouch, E., Encrenaz, T., Bézard, B., Coustenis, A., Salama, A., de Graauw, T., Davis, G.R., 1999. Oxygen in the stratospheres of the giant planets and Titan. In *Proc. of The Universe as Seen by ISO*, ESA-SP 427, UNESCO, Paris.
Fink, U., Larson, H. P., 1978. Deuterated methane observed on Saturn. *Science* **201**, 343–345.
Flasar, F. M., Achterberg, R. K., Conrath, B. J., Pearl, J. C., Bjoraker, G. L., Jennings, D. E., Romani, P. N., Simon-Miller, A. A., Kunde, V. G., Nixon, C. A., Bézard, B., Orton, G. S., Spilker, L. J., Spencer, J. R., Irwin, P. G. J., Teanby, N. A., Owen, T. C., Brasunas, J.,

Segura, M. E., Carlson, R. C., Mamoutkine, A., Gierasch, P. J., Schinder, P. J., Showalter, M. R., Ferrari, C., Barucci, A., Courtin, R., Coustenis, A., Fouchet, T., Gautier, D., Lellouch, E., Marten, A., Prangé, R., Strobel, D. F., Calcutt, S. B., Read, P. L., Taylor, F. W., Bowles, N., Samuelson, R. E., Abbas, M. M., Raulin, F., Ade, P., Edgington, S., Pilorz, S., Wallis, B., and Wishnow, E. H., 2005. Temperatures, Winds, and Composition in the Saturnian System. *Science* **307**, 1247–1251.

Flasar, F. M., Schinder, P. J., Achterberg, R. K., Conrath, B. J., 2008. On combining thermal-infrared and radio-occultation data of Saturn's and Titan's atmospheres. *Bull. Am. Astron. Soc.* **40**, 495.

Fletcher , L. N., Irwin, P. G. J., Teanby, N. A., Orton, G. S., Parrish, P. D., de Kok, R., Howett, C., Calcutt, S. B., Bowles, N., Taylor, F. W., 2007a. Characterising Saturn's vertical temperature structure from Cassini/CIRS. *Icarus* **189**, 457–478.

Fletcher, L. N., Irwin, P. G. J., Teanby, N. A., Orton, G. S., Parrish, P. D., Calcutt, S. B., Bowles, N., de Kok, R., Howett, C., Taylor, F. W., 2007b. The meridional phosphine distribution in Saturn's upper troposphere from Cassini/CIRS observations. *Icarus* **188**, 72–88.

Fletcher, L.N., Irwin, P.G.J., Orton, G.S., Teanby, N.A., Achterberg, R.K., Bjoraker, G.L., Read, P.L., Simon-Miller, A.A., Howett, C., de Kok, R., Bowles, N., Calcutt, S.B., Hesman, B., Flasar, F.M., 2008. Temperature and composition of Saturn's polar hot spots and hexagon. *Science* **319**, 79–81.

Fletcher, L. N., Orton, G. S., Teanby, N. A., Irwin, P. G. J., Bjoraker, G. L., 2009a. Methane and its isotopologues on Saturn from Cassini/CIRS observations. *Icarus* **199**, 351–367.

Fletcher, L. N., Orton, G. S., Teanby, N. A., Irwin, P. G. J. 2009b. Phosphine on Jupiter and Saturn from Cassini/CIRS. *Icarus*, **202**, 543–564.

Fouchet, T., Guerlet, S., Strobel, D.F., Simon-Miller, A.A., Bézard, B., Flasar, F.M., 2008. An equatorial oscillation in Saturn's middle atmosphere. *Nature* **453**, 200–2002.

Friedson, A.J., West, R.A., Hronek, A.K., Larsen, N.A., Dalal, N., 1999. Transport and mixing in Jupiter's stratosphere inferred from Comet S-L9 dust migration. *Icarus* **138**, 141–156.

Friedson, A.J., Wong, A.-S., Yung, Y.L., 2002. Models for polar haze formation in Jupiter's stratosphere. *Icarus* **158**, 389–400.

Gautier, D., Conrath, B., Hanel, R., Kunde, V., Chedin, A., Scott, N., 1981. The helium abundance of Jupiter from Voyager. *J. Geophys. Res.* **86**, 8713–8720.

Gautier, D., Conrath, B., Flasar, M., Achterberg, R., Schinder, P., Kliore, A., The Cassini Cirs, Radio Science Teams 2006. The helium to hydrogen ratio in Saturn's atmosphere from Cassini CIRS and radio science measurement. 36th COSPAR Scientific Assembly 36, 867.

Gérard, J.C., Dols, V., Grodent, D., Waite, J.H., Gladstone, G.R., Prangé, 1995. Simultaneous observations of the Saturnian aurora and polar haze with the HST/FOC. *Geophys. Res. Lett.* **22**, 2685–2688.

Gérard, J.C., Bonfond, B., Gustin, J., Grodent, D., Clarke, J.T., Bisikalo, D., Shematovich, V., 2009. Altitude of Saturn's aurora and its implications for the characteristic energy of precipitated electrons. *Geophys. Res. Lett.* **36**, L02202, doi: 10.1029/ 2008GL036554.

Greathouse, T. K., Lacy, J. H., Bézard, B., Moses, J. I., Griffith, C. A., Richter, M. J., 2005. Meridional variations of temperature, C_2H_2 and C_2H_6 abundances in Saturn's stratosphere at southern summer solstice. *Icarus* **177**, 18–31.

Greathouse, T. K., Lacy, J. H., Bézard, B., Moses, J. I., Richter, M. J., Knez, C., 2006. The first detection of propane on Saturn. *Icarus* **181**, 266–271.

Grevesse, N., Asplund, M., Sauval, A. J., 2007. The solar chemical composition. *Space Sci. Rev.* **130**, 105–114.

Griffin, M.J., Naylor, D.A., Davis, G.R., Ade, P.A.R., Oldham, P.G., Swinyard, B.M., Gautier, D., Lellouch, E., Orton, G.S., Encrenaz, T., de Graauw, T., Furniss,I., Smith, H., Armand, C., Burgdorf, M., Di Giorgio, A., Ewart, D., Gry, C., King, K.J., Lim, T., Molinari, S., Price, M., Sidher, S., Smith, A., Texier, D., Trams, N., Unger, S.J., Salama, A., 1996. First detection of the 56 μm rotational line of HD in Saturn's atmosphere. *Astron. Astrophys.* **315**, L389–L392.

Guerlet, S., Fouchet, T., Bézard B., Simon-Miller, A.A., Flasar, F. M., 2009. Vertical and meridional distribution of ethane, acetylene and propane in Saturn's stratosphere from CIRS/Cassini limb observations. *Icarus*, in press.

Guillemin, J.-C., Janati, T., Lassalle, L., 1995a. Photolysis of phosphine in the presence of acetylene and propyne. *Adv. Space Res.* **16**, 85–92.

Guillemin, J.-C., Lassalle, L., Janati, T., 1995b. Germane photochemistry. Photolysis of gas mixtures of planetary interest. *Planet. Space Sci.* **43**, 75–81.

Guillemin, J.-C., Le Serre, S., Lassalle, L., 1997. Regioselectivity of the photochemical addition of phosphine to unsaturated hydrocarbons in the atmosphere of Jupiter and Saturn. *Adv. Space Res.* **19**, 1093–1102.

Guillot, T., 1999. A comparison of the interiors of Jupiter and Saturn. *Planet. Space Sci.* **47**, 1183–1200.

Gustin, J., Gérard, J.C., Pryor, W., Feldman, P.D., Grodent, D., Holsclaw, G., 2009. Characteristics of Saturn's polar atmosphere and auroral electrons derived from HST/STIS, FUSE, and Cassini/UVIS spectra. *Icarus* **200**, 176–187.

Heck, A.J.R., Zare, R.N., Chandler, D.W., 1996. Photofragment imaging of methane. *J. Chem. Phys.* **104**, 4019–4030.

Howett, C. J. A., Irwin, P. G. J., Teanby, N. A., Simon-Miller, A., Calcutt, S. B., Fletcher, L. N., de Kok, R., 2007. Meridional variations in stratospheric acetylene and ethane in the southern hemisphere of the saturnian atmosphere as determined from Cassini/CIRS measurements. *Icarus* **190**, 556–572.

Hubbard, W.B., Guillot, T., Marley, M.S., Burrows, A., Lunine, J.I., Saumon, S., 1999. Comparative evolution of Jupiter and Saturn. *Planet. Space Sci.* **47**, 1175–1182.

Huestis, D.L., Bougher, S.W., Fox, J.L., Galand, M., Johnson, R.E., Moses, J.I., Pickering, J.C., 2008. Cross sections and reaction rates for comparative aeronomy. *Space Sci. Rev.* **29**, doi:10.1007/11241-008-9383-7.

Imanaka, H., Smith, M.A., 2007. Role of photoionization in the formation of complex organic molecules in Titan's upper atmosphere. *Geophys. Res. Lett.* **34**, L02204, doi:10.1029/2006GL028317.

Karkoschka, E., Tomasko, M. G., 1992. Saturn's upper troposphere 1986–1989. *Icarus* **97**, 161–181.

Kaye, J.A., Strobel, D.F., 1983a. Formation and photochemistry of methylamine in Jupiter's atmosphere. *Icarus* **55**, 399–419.

Kaye, J.A., Strobel, D.F., 1983b. HCN formation on Jupiter: The coupled photochemistry of ammonia and acetylene. *Icarus* **54**, 417–433.

Kaye, J.A., Strobel, D.F., 1983c. Phosphine photochemistry in Saturn's atmosphere. *Geophys. Res. Lett.* **10**, 957–960.

Kaye, J.A., Strobel, D.F., 1984. Phosphine photochemistry in the atmosphere of Saturn. *Icarus* **59**, 314–335.

Keane, T.C., Yuan, F., Ferris, J.P., 1996. Potential Jupiter atmospheric constituents: Candidates for the mass spectrometer in the Galileo probe. *Icarus* **122**, 205–207.

Kerola, D. X., Larson, H. P., Tomasko, M. G. 1997. Analysis of the Near-IR Spectrum of Saturn: A Comprehensive Radiative Transfer Model of Its Middle and Upper Troposphere. *Icarus* **127**, 190-212.

Killen, R. M., 1988. Longitudinal variations in the Saturnian atmosphere. I–Equatorial region. *Icarus* **73**, 227–247.

Killen, R. M., Flasar, F. M. 1996. Microwave sounding of the Giant Planets. *Icarus* **119**, 67–89.

Kim, J. H., Kim, S. J., Geballe, T. R., Kim, S. S., Brown, L. R., 2006. High-resolution spectroscopy of Saturn at 3 microns: CH_4, CH_3D, C_2H_2, C_2H_6, PH_3, clouds, and haze. *Icarus* **185**, 476–486.

Larson, H.P., Fink, U., Smith, H.A., Davis, D.S., 1980. The middle-infrared spectrum of Saturn: Evidence for phosphine and upper limits to other trace atmospheric constituents. *Astrophys. J.* **240**, 327–337.

Lecluse, C., Robert, F., Gautier, D., Guiraud, M., 1996. Deuterium enrichment in giant planets. *Planet. Space Sci.* **44**, 1579–1592.

Lee, A.Y.T., Yung, Y.L., Moses, J.I., 2000. Photochemical modeling of CH_3 abundances in the outer solar system. *J. Geophys. Res.* **105**, 20207–20225.

Lellouch, E., Bézard, B., Fouchet, T., Encrenaz, T., Feuchtgruber, H., de Graauw, T., 2001. The deuterium abundance in Jupiter and Saturn from ISO-SWS observations. *Astron. Astrophys.* **370**, 610–622.

Lellouch, E., Bézard, B., Moses, J.I., Davis, G.R., Drossart, P., Feuchtgruber, H., Bergin, E.A., Moreno, R., Encrenaz, T., 2002. The origin of water vapor and carbon dioxide in Jupiter's stratosphere. *Icarus* **159**, 112–131.

Lellouch, E., Bézard, B., Strobel, D.F., Bjoraker, G.L., Flasar, F.M., Romani, P.N., 2006. On the HCN and CO_2 abundance and distribution in Jupiter's stratosphere. *Icarus* **184**, 478–497.

Lewis, J.S., Prinn, R.G., 1984. *Planets and Their Atmospheres: Origin and Evolution*. 480 pp. Academic Press, Orlando.

Liang, M.-C., Shia, R.-L., Lee, A.Y.-T., Allen, M., Friedson, A.J., Yung, Y.L., 2005. Meridional transport in the stratosphere of Jupiter. *Astrophys. J.* **635**, L177–L180.

Lindal, G.F., 1992. The atmosphere of Neptune: An analysis of radio occultation data acquired with Voyager 2. *Astron. J.* **103**, 967–982.

Lodders, K., Fegley, B., Jr. 2002. Atmospheric chemistry in giant planets, brown dwarfs, and low-mass dwarf stars. I. Carbon, nitrogen, and oxygen. *Icarus* **155**, 393-424.

Lunine, J.I., Coradini, A., Gautier, D., Owen, T.C., Wuchterl, G., The origin of Jupiter. In *Jupiter: The Planet, Satellites, and Magnetosphere* (Bagenal, F., Dowling, T.E., McKinnon, W.B., Eds.),. Cambridge University Press, Cambridge, UK, 2004, pp. 19–34.

Marrero, T.R., Mason, E.A., 1972. Gaseous diffusion coefficients. *J. Phys. Chem. Ref. Data* **1**, 3–118.

Moos, H. W., Clarke, J. T. 1979. Detection of acetylene in the Saturnian atmosphere, using the IUE satellite. *Astrophys. J.* **229**, L107.

Mordaunt, D.H., Lambert, I.R., Morley, G.P., Ashford, M.N.R., Dixon, R.N., Western, C.M., 1993. Primary product channels in the photodissociation of methane at 121.6 nm. *J. Chem. Phys.* **98**, 2054–2065.

Morgan, C.G., Allen, M., Liang, M.C., Shia, R.L., Blake, G.A., Yung, Y.L., 2004. Isotopic fractionation of nitrous oxide in the stratosphere: Comparison between model and observations. *J. Geophys. Res.* **109**, D04305, doi:10.1029/2003JD003402.

Morton, R.J., Kaiser, R.I., 2003. Kinetics of suprathermal hydrogen atom reactions with saturated hydrides in planetary and satellite atmospheres. *Planet. Space Sci.* **51**, 365–373.

Moses, J. I., Greathouse, T., 2005. Latitudinal and seasonal models of stratospheric photochemistry on Saturn: Comparison with infrared data from IRTF/TEXES. *J. Geophys. Res.* **110**, doi:10.1029/2005JE002450.

Moses, J.I., Bézard, B., Lellouch, E., Gladstone, G.R., Feuchtgruber, H., Allen, M., 2000a. Photochemistry of Saturn's atmosphere. I. Hydrocarbon chemistry and comparison with ISO observations. *Icarus*, **143**, 244–298.

Moses, J.I., Lellouch, E., Bézard, B., Gladstone, G.R., Feuchtgruber, H., Allen, M., 2000b. Photochemistry of Saturn's atmosphere. II. Effects of an influx of external oxygen. *Icarus*, **145**, 166–202.

Moses, J.I., Fouchet, T., Yelle, R.V., Friedson, A.J., Orton, G.S., Bézard, B., Drossart, P., Gladstone, G.R., Kostiuk, T., Livengood, T.A., 2004. The stratosphere of Jupiter. In *Jupiter: Planet, Satellites & Magnetosphere* (Bagenal, F., McKinnon, W., Dowling, T., Eds.), pp. 129–157, Cambridge University Press, Cambridge, UK.

Moses, J.I., Fouchet, T., Bézard, B., Gladstone, G.R., Lellouch, E., Feuchtgruber, H., 2005. Photochemistry and diffusion in Jupiter's stratosphere: Constraints from ISO observations and comparisons with other giant planets. *J. Geophys. Res.* **110**, E08001, doi:10.1029/2005JE002411.

Moses, J.I., Liang, M.-C., Yung, Y.L., Shia, R.L., 2007. Meridional distribution of hydrocarbons on Saturn: Implications for stratospheric transport. Abstract presented at the Workshop on Planetary Atmospheres, held November 6–7, 2007 in Greenbelt, Maryland, p. 85–86.

Nava, D.F., Payne, W.A., Marston, G., Stief, L.J., 1993. The reaction of atomic hydrogen with germane: Temperature dependence of the rate constant and implications for germane chemistry in the atmospheres of Jupiter and Saturn. *J. Geophys. Res.* **98**, 5531–5537.

Nicholson, P.D., Hedman, M.M., Gierasch, P.J., Cassini VIMS team, 2006. Probing Saturn's atmosphere with Procyon. *Bull. Am. Astron. Soc.* 38, 555.

Niemann, H.B., Atreya, S.K., Carignan, G.R., Donahue, T.M., Haberman, J.A., Harpold, D.N., Hartle, R.E., Hunten, D.M., Kasprzak, W.T., Mahaffy, P.R., Owen, T.C., Way, S.H., 1998. The composition of the jovian atmosphere as determined by the Galileo probe mass spectrometer. *J. Geophys. Res.* **103**, 22,831–22,846.

Niemann, H.B., Atreya, S.K., Bauer, S.J., Carignan, G.R., Demick, J.E., Frost, R.L., Gautier, D., Haberman, J.A., Harpold, D.N., Hunten, D.M., Israel, G., Lunine, J.I., Kasprzak, W.T., Owen, T.C., Paulkovich, M., Raulin, F., Raaen, E., Way, S.H. 2005. The abundances of constituents of Titan's atmosphere from the GCMs instrument on the Huygens probe. *Nature* **438**, 779–784.

Nixon, C.A., Achterberg, R.K., Conrath, B.J., Irwin, P.G.J., Teanby, N.A., Fouchet, T., Parrish, P.D., Romani, P.N., Abbas, M., LeClair, A., Strobel, D., Simon-Miller, A.A., Jennings, D.J., Flasar, F.M., Kunde, V.G., 2007. Meridional variations of C_2H_2 and C_2H_6 in Jupiter's atmosphere from Cassini CIRS infrared spectra. *Icarus* **188**, 47–71.

Noll, K. S., Larson, H.P., 1991. The spectrum of Saturn from 1990 to 2230 cm^{-1}: Abundances of AsH_3, CH_3D, CO, GeH_4, NH_3 and PH_3. *Icarus* **89**, 168–189.

Noll, K. S., Knacke, R.F., Geballe, R.R., Tokunaga, A.T., 1986a. Detection of carbon monoxyde in Saturn. *Astrophys. J.* **309**, L91–L94.

Noll, K.S., Knacke, R.F., Tokunaga, A.T., Lacy, J.H., Beck, S., Serabyn, E., 1986b. The abundances of ethane and acetylene in the atmospheres of Jupiter and Saturn. *Icarus* **65**, 257–263.

Noll, K. S., Knacke, R. F., Geballe, T. R., Tokunaga, A. T. 1988. Evidence for germane in Saturn. *Icarus* **75**, 409–422.

Noll, K. S., Geballe, T. R., Knacke, R. F. 1989. Arsine in Saturn and Jupiter. *Astrophys. J.* **338**, L71-L74.

Ollivier, J.L., Dobrijévic, M., Parisot, J.P., 2000. New photochemical model of Saturn's atmosphere. *Planet. Space Sci.* **48**, 699–716.

Orton, G.S., et al., 2008. Semi-annual oscillations in Saturn's low-latitude stratospheric temperatures. *Nature* **453**, 196–199.

Orton, G. S., Serabyn, E., Lee, Y. T., 2001. Erratum, Volume 146, Number 1, pages 48-59 (2000), in the article "Vertical Distribution of PH_3 in Saturn from Observations of Its 1–0 and 3–2 Rotational Lines,". *Icarus* **149**, 489–490.

Orton, G. S., Serabyn, E., Lee, Y. T., 2000. Vertical Distribution of PH_3 in Saturn from Observations of Its 1–0 and 3–2 Rotational Lines. *Icarus* **146**, 48–59. Owen, T., Lutz, B. L., de Bergh, C. 1986. Deuterium in the outer solar system - Evidence for two distinct reservoirs. *Nature* **320**, 244–246.

Parkinson, C.D., Griffioen, E., McConnell, J.C., Gladstone, G.R., Sandel, B.R., 1998. He 584 Å dayglow at Saturn: A reassessment. *Icarus* **133**, 210–220.

Pérez-Hoyos, S., Sánchez-Lavega, A., 2006. Solar flux in Saturn's atmosphere: Penetration and heating rates in the aerosol and cloud layers. *Icarus* **180**, 368–378.

Perry, J.J., Kim, Y.H., Fox, J.L., Porter, H.S., 1999. Chemistry of the jovian auroral ionosphere. *J. Geophys. Res.* **104**, 16541–16565.

Porco, C.C., et al., 2006. Cassini observes the active south pole of Enceladus. *Science* **311**, 1393–1401.

Prangé, R., Fouchet, T., Courtin, R., Connerney, J. E. P., McConnell, J. C., 2006. Latitudinal variation of Saturn photochemistry deduced from spatially-resolved ultraviolet spectra. *Icarus* **180**, 379–392.

Prinn, R.G., Barshay, S.S., 1977. Carbon monoxide on Jupiter and implications for atmospheric convection. *Science* **198**, 1031–1034.

Prinn, R.G., Owen, T., 1976. Chemistry and spectroscopy of the Jovian atmosphere. In *Jupiter* (Gehrels, T., Matthews, M.S., Eds.), University of Arizona Press, Tucson, Arizona, 319–371.

Prinn, R.G., Larson, H.P., Caldwell, J.J., Gautier, D., 1984. Composition and chemistry of Saturn's atmosphere. In *Saturn* (Gehrels, T., Matthews, M.S., Eds.), University of Arizona Press, Tucson, 88–149.

Profitt, C.R., 1994. Effects on heavy-element settling on solar neutrino fluxes and interior structure. *Astrophys. J.* **425**, 849–855.

Raulin, F., Bossard, A., Toupance, G., Ponnamperuma, C., 1979. Abundance of organic compounds photochemically produced in the atmospheres of the outer planets. *Icarus* **38**, 358–366.

Rosenqvist, J., E. Lellouch, P. N. Romani, G. Paubert, T. Encrenaz 1992. Millimeter-wave observations of Saturn, Uranus, Neptune: CO and HCN on Neptune. *Astrophys. J.* **392**, L99–L102.

Sandel, B.R., Shemansky, D.E., Broadfoot, A.L., Holberg, J.B., Smith, G.R., McConnell, J.C., Strobel, D.F., Atreya, S.K., Donahue, T.M., Moos, H.W., Hunten, D.M., Pomphrey, R.B., Linick, S. 1982. Extreme ultraviolet observations from the Voyager 2 encounter with Saturn. *Science* **215**, 548–553.

Sada P. V., McCabe, G.H., Bjoraker, G.L., Jennings, D.E., Reuter, D.C., 1996. ^{13}C-ethane in the atmospheres of Jupiter and Saturn. *Astrophys. J.* **472**, 903–907.

Sada, P. V., Bjoraker, G. L., Jennings, D. E., Romani, P. N., McCabe, G. H. 2005. Observations of C_2H_6 and C_2H_2 in the stratosphere of Saturn. *Icarus* **173**, 499–507.

Sagan, C., Lippincott, M.O., Dayhoff, M.O., Eck. R.V., 1967. Organic molecules and the coloration of Jupiter. *Nature* **213**, 273–274.

Sandel, B.R., McConnell, J.C., Strobel, D.F., 1982. Eddy diffusion at Saturn's homopause. *Geophys. Res. Lett.* **9**, 1077–1080.

Shemansky, D.E., Liu, X., 2009. Saturn upper atmospheric structure from Cassini EUV/FUV occultations. *Planet. Space Sci.*, submitted.

Shia, R.-L., Ha, Y.L., Wen, J.-S., Yung, Y.L., 1990. Two-dimensional atmospheric transport and chemistry model: Numerical experiments with a new advection algorithm. *J. Geophys. Res.* **95**, 7467–7483.

Simon-Miller, A.A., Bjoraker, G.L., Jennings, D.E., Achterberg, R.K., Conrath, B.J., Nixon, C.A., Cassini CIRS Team, 2005. Cassini CIRS measurements of benzene, propane, and carbon dioxide on Saturn. *Bull. Am. Astron. Soc.* **37**, 682.

Simon-Miller, A.A., Conrath, B.J., Gierasch, P.J., Orton, G.S., Achterberg, R.K., Flasar, F.M., Fisher, B.M., 2006. Jupiter's atmospheric temperatures: From Voyager IRIS to Cassini CIRS. *Icarus* **180**, 98–112.

Smith, G.P., Nash, D., 2006. Local sensitivity analysis for observed hydrocarbons in a Jupiter photochemistry model. *Icarus* **182**, 181–201.

Smith, M.D., 1998. Estimation of a length scale to use with the quench level approximation for obtaining chemical abundances. *Icarus* **132**, 176–184.

Smith, N., Raulin, F., 1999. Modeling of methane photolysis in the reducing atmospheres of the outer solar system. *J. Geophys. Res.* **104**, 1873–1876.

Smith, G. R., Shemansky, D. E., Holberg, J. B., Broadfoot, A. L., Sandel, B. R., McConnell, J. C., 1983. Saturn's upper atmosphere from the Voyager 2 EUV solar and stellar occultations. *J. Geophys. Res.* **88**, 8667–8678.

Smith, M. D., Conrath, B. J., Gautier, D., 1996. Dynamical Influence on the Isotopic Enrichment of CH_3D in the Outer Planets. *Icarus* **124**, 598–607.

Stancu, G. D., Röpcke, J., Davies, P. B., 2005. Line strengths and transition dipole moment of the ν_2 fundamental band of the methyl radical. *J. Chem. Phys.* **122**, 014306, doi:10.1063/1.1812755.

Stone, P.H., 1976. The meteorology of the Jovian atmosphere. In *Jupiter* (Gehrels, T., Matthews, M.S., Eds.), University of Arizona Press, Tucson, Arizona, pp. 586–618.

Strobel, D.F., 1969. The photochemistry of methane in the jovian atmosphere. *J. Atmos. Sci.* **26**, 906–911.

Strobel, D.F., 1975. Aeronomy of the major planets: Photochemistry of ammonia and hydrocarbons. *Rev. Geophys. Space Phys.* **13**, 372–382.

Strobel, D.F., 1978. Aeronomy of Saturn and Titan. In *JPL The Saturn System*, 185–194.

Strobel, D.F., 1981. Parameterization of linear wave chemical transport in planetary atmospheres by eddy diffusion. *J. Geophys. Res.* **86**, 9806.

Strobel, D.F., 1983. Photochemistry of the reducing atmospheres of Jupiter, Saturn, and Titan. *Int. Rev. Phys. Chem.* **3**, 145–176.

Strobel, D.F., 2005. Photochemistry in outer solar system atmospheres. *Space Sci. Rev.* **116**, 171–184.

Strong, S.B., Greathouse, T.K., Moses, J.I., Lacy, J.H., 2007. An updated radiative seasonal climate model for the Saturnian atmosphere. Abstract presented at the Workshop on Planetary Atmospheres, held November 6–7, 2007 in Greenbelt, Maryland, p. 119.

Teanby, N. A., Fletcher, L. N., Irwin, P. G. J., Fouchet, T., Orton, G. S. 2006. New upper limits for hydrogen halides on Saturn derived from Cassini-CIRS data. *Icarus* **185**, 466–475.

Tokunaga, A., Knacke, R.F., Owen, T., 1975. The detection of ethane on Saturn. *Astrophys. J.* **197**, L77–L78.

Tomasko, M.G., West, R.A., Orton, G.S., Teifel, V.G. 1984. Clouds and aerosols in Saturn's atmosphere. In *Jupiter* (Gehrels, T., Matthews, M.S., Eds.), Univ. Arizona Press, Tucson, Arizona, pp. 150–194.

Trafton, L., 1985. Long-term changes in Saturn's troposphere. *Icarus* **63**, 374–405.

van der Tak, F., de Pater, I., Silva, A., Millan, R. 1999. Time Variability in the Radio Brightness Distribution of Saturn. *Icarus* **142**, 125–147.

Vervack, R.J., Jr., Moses, J.I., 2009. Saturn's upper atmosphere during the Voyager era: Reanalysis and modeling of the UVS occultations. *Icarus*, submitted.

Visscher, C., Fegley, B. J., 2005. Chemical constraints on the water and total oxygen abundances in the deep atmosphere of Saturn. *Astrophys. J.* **623**, 1221–1227.

Visscher, C., Lodders, K., Fegley, B., Jr. 2006. Atmospheric chemistry in giant planets, brown dwarfs, and low-mass stars. II. Sulfur and phosphorus. *Astrophys. J.* **648**, 1181–1195.

von Zahn, U., Hunten, D.M., Lehmacher, G., 1998. Helium in Jupiter's atmosphere: Results from the Galileo probe helium interferometer experiment. *J. Geophys. Res.* **103**, 22,815–22,829.

Vuitton, V., et al. (2007), Ion chemistry and N-containing molecules in Titan's upper atmosphere. *Icarus* **191**, 722–742.

Vuitton, V., Yelle, R.V., Cui, J., 2008. Formation and distribution of benzene on Titan. *J. Geophys. Res.* **113**, E05007, doi:10.1029/2007JE002997.

Waite, J.H., Jr., Cravens, T.E., Kozyra, J., Nagy, A.F., Atreya, S.K., Chen, R.H., 1983. Electron precipitation and related aeronomy of the jovian thermosphere and ionosphere. *J. Geophys. Res.* **88**, 6143–6163.

Waite, J.H., Jr., et al. (2005). Ion Neutral Mass Spectrometer results from the first flyby of Titan. *Science* **308**, 982–986.

Waite, J.H., Jr., et al. (2007). The process of tholin formation in Titan's upper atmosphere. *Science* **316**, 870–875.

Wang, J.-H., Liu, K., Min, Z., Su, H., Bersohn, R., Preses, J., Larese, J., 2000. Vacuum ultraviolet photochemistry of CH_4 and isotopomers.

II. Product channel fields and absorption spectra. *J. Chem. Phys.* **113**, 4146–4152.

Weisstein, E. W., Serabyn, E., 1994. Detection of the 267 GHz $J = 1 - 0$ rotational transition of PH_3 in Saturn with a new fourier transfer spectrometer. *Icarus* **109**, 367–381.

Weisstein, E. W., Serabyn, E. 1996. Submillimeter Line Search in Jupiter and Saturn. *Icarus* **123**, 23–36.

West, R.A., Strobel, D.F., Tomasko, M.G., 1986. Clouds, aerosols, and photochemistry in the Jovian atmosphere. *Icarus* **65**, 161–217.

Winkelstein, P., J. Caldwell, S. J. Kim, M. Combes, G. E. Hunt, and V. Moore 1983. A determination of the composition of the saturnian stratosphere using the IUE. *Icarus* **54**, 309–318.

Woeller, F., Ponnamperuma, C., 1969. Organic synthesis in a simulated Jovian atmosphere. *Icarus* **10**, 386–392.

Wong, A.-S., Lee, A.Y.T., Yung, Y.L., Ajello, J.M., 2000. Jupiter: Aerosol chemistry in the polar atmosphere. *Astrophys. J.* **534**, L215–L217.

Wong, A.-S., Yung, Y.L., Friedson, A.J., 2003. Benzene and haze formation in the polar atmosphere of Jupiter. *Geophys. Res. Lett.* **30**, 1447, doi:10.1029/2002GL016661.

Yelle, R.V., Griffith, C.A., Young, L.A., 2001. Structure of the jovian stratosphere at the Galileo probe entry site. *Icarus* **152**, 331–346.

Yung, Y.L., DeMore, W.B., 1999. *Photochemistry of Planetary Atmospheres*. Oxford University Press, New York.

Yung, Y.L., Drew, W.A., Pinto, J.P., Friedl, R.R., 1988. Estimation of the reaction rate for the formation of CH_3O from $H + H_2CO$: Implications for chemistry in the solar system. *Icarus* **73**, 516–526.

Chapter 6
Saturn Atmospheric Structure and Dynamics

Anthony D. Del Genio, Richard K. Achterberg, Kevin H. Baines, F. Michael Flasar, Peter L. Read, Agustín Sánchez-Lavega, and Adam P. Showman

Abstract Saturn inhabits a dynamical regime of rapidly rotating, internally heated atmospheres similar to Jupiter. Zonal winds have remained fairly steady since the time of Voyager except in the equatorial zone and slightly stronger winds occur at deeper levels. Eddies supply energy to the jets at a rate somewhat less than on Jupiter and mix potential vorticity near westward jets. Convective clouds exist preferentially in cyclonic shear regions as on Jupiter but also near jets, including major outbreaks near 35°S associated with Saturn electrostatic discharges, and in sporadic giant equatorial storms perhaps generated from frequent events at depth. The implied meridional circulation at and below the visible cloud tops consists of upwelling (downwelling) at cyclonic (anti-cyclonic) shear latitudes. Thermal winds decay upward above the clouds, implying a reversal of the circulation there. Warm-core vortices with associated cyclonic circulations exist at both poles, including surrounding thick high clouds at the south pole. Disequilibrium gas concentrations in the tropical upper troposphere imply rising motion there. The radiative-convective boundary and tropopause occur at higher pressure in the southern (summer) hemisphere due to greater penetration of solar heating there. A temperature "knee" of warm air below the tropopause, perhaps due to haze heating, is stronger in the summer hemisphere as well. Saturn's south polar stratosphere is warmer than predicted by radiative models and enhanced in ethane, suggesting subsidence-driven adiabatic warming there. Recent modeling advances suggest that shallow weather layer theories of jet pumping may be viable if water condensation is the source of energy input driving the flow, and that deep convective cylinder models with a sufficiently large tangent cylinder radius can reproduce observed flow features as well.

6.1 Introduction

6.1.1 Saturn's Place Among Planetary Atmospheres

Surrounded by a family of exotic icy satellites, partially obscured by a complex set of rings, and pale compared to its sister giant planet Jupiter, Saturn's atmosphere has received less attention than most other celestial objects visited by the Cassini spacecraft. Nonetheless, over its 4-year nominal mission, the Cassini Orbiter has acquired a wealth of information about Saturn's atmosphere that is unprecedented among the planets other than Earth. Cassini observations have confirmed some expected similarities between Saturn and the previously much better-observed Jupiter, and provided insights into some important differences as well. The goal of this chapter is to place our understanding of the thermal structure and dynamics of Saturn on the same footing as that of other planetary atmospheres, and to consider which aspects of that science are informed by our existing knowledge of Jupiter and Earth and which are unique to Saturn.

Atmospheric circulations generally exist primarily to transport heat from regions of excess to regions of deficit. Thus, the thermal structure and circulation are intertwined

---

A.D. Del Genio
NASA Goddard Institute for Space Studies, 2880 Broadway, New York, NY 10025, USA

R.K. Achterberg
Department of Astronomy, University of Maryland, College Park, MD 20742, USA

K.H. Baines
Jet Propulsion Laboratory, 4800 Oak Grove Drive, Pasadena, CA 91109, USA

F.M. Flasar
NASA Goddard Space Flight Center, Greenbelt, MD 20771, USA

P.L. Read
Atmospheric, Oceanic & Planetary Physics, Clarendon Laboratory, Parks Road, Oxford OX1 3PU, UK

A. Sánchez-Lavega
Departamento de Física Aplicada, Universidad del País Vasco, Bilbao 48013, Spain

A.P. Showman
Department of Planetary Sciences and Lunar and Planetary Laboratory, University of Arizona, Tucson, AZ 85721, USA

– temperature gradients created by diabatic heating drive specific types of circulations, and the resulting dynamical transport modifies the temperature field and sometimes induces secondary mechanically driven circulations that may be thermodynamically indirect. Planet rotation and the vertical stratification imposed by the heating determine the "stiffness" of an atmosphere against latitudinal and vertical motions, respectively, and thereby control the extent to which poleward vs. upward dynamical heat transport dominates. To a first approximation this can be diagnosed from dimensionless parameters such as the Rossby and Richardson numbers, or alternatively, from the relative potential temperature contrasts in the vertical and horizontal over the characteristic scales of motion (Gierasch 1976; Allison et al. 1995).

By this reckoning, planets fall into three general classes. Terrestrial planets (Earth, Mars) have comparable vertical and horizontal contrasts and are dominated by baroclinic waves that transport heat both poleward and upward. Slowly rotating planets (Venus, Titan) have strong stratification but weak horizontal temperature gradients and are characterized by a planet-wide Hadley cell that efficiently transports heat poleward. Saturn's atmosphere is roughly similar to that of Jupiter in radius, depth, composition, rotation rate, albedo, and ratio of internal to external heating, and they are consequently thought to inhabit the same general dynamical regime (perhaps along with Neptune and Uranus), but one unique to the giant planets in several respects. The rapid rotation of the giant planets and the weak stratification imposed by their internal heat sources create vertical potential temperature contrasts that are less than horizontal contrasts, implying that the general circulation acts primarily to transport efficiently the internal heat upward to levels where it can easily be radiated to space. Furthermore, the giant planets' deep atmospheres allow for two very different possibilities for the dynamics. One picture assumes that large-scale convective organization of the flow driven solely by the internal heating extends through much of the depth of the atmosphere, while the other invokes a shallower upper "weather layer" that is largely decoupled from the deep convective interior by an intermediate stable layer and may respond to both solar heating from above and the convective heat flux at its lower boundary.

One major difference between Saturn and Jupiter is in their obliquities (26.7° vs. 3.1°, respectively). The greater seasonality of solar heating on Saturn thus suggests a more temporally variable atmospheric structure and dynamics in Saturn's upper troposphere and stratosphere. Observations of changes over Saturn's 29.5 year orbital period may therefore help distinguish aspects of the structure and circulation that are affected by insolation from those tied directly to the deep internal heating.

Compared to Earth, Saturn presents tremendous challenges for remote sensing. Saturn's surface area is two orders of magnitude greater than Earth's, making it impossible to observe the entire planet at once at spatial resolutions at which the relevant dynamical processes operate. Cassini remote sensing of Saturn thus combines global low-resolution mapping at long wavelengths by instruments with large fields of view when the orbiter is close to periapsis, and regional high-resolution image mosaics at shorter wavelengths by instruments with small fields of view at greater distances. Earth is partly cloud-covered and remote sensing thus sees the atmosphere from top to bottom, whereas Saturn is nearly overcast, precluding direct observations of the deep atmosphere that might differentiate deep from shallow circulation mechanisms. Earth is close to the Sun and thus strongly forced, and its solid surface dissipates atmospheric kinetic energy. As a result the dynamical phenomena of interest are frequent and grow and decay on short time scales, so observations over a limited time period suffice to diagnose the processes that control its general circulation. Saturn, however, is far from the Sun, weakly forced, and has no solid surface to dissipate energy, so much of the circulation is invariant for long intervals, and the more transient features of the dynamics are very sporadic. Thus Cassini must observe Saturn for many years to detect changes and to sample the dynamics adequately. Unlike Earth, for which routine weather balloon launches provide "ground truth" for the interpretation of remote sensing, Saturn's troposphere and stratosphere have never been sampled in situ by probes. Even compared to Jupiter, Saturn's clouds exhibit much weaker feature contrasts in reflected sunlight and its colder temperatures imply less emission of thermal radiation. Thus, Saturn must be observed for longer time intervals than Jupiter for remote sensing to acquire the same amount of information.

Several previous review articles provide an excellent foundation for our discussion of the current understanding of Saturn's atmosphere. Ingersoll et al. (1984) summarize the knowledge of Saturn gleaned from ground-based observations and the Voyager flyby encounters of the early 1980s. Ingersoll et al.'s (2004) review focuses instead on the dynamics of Jupiter as understood at the conclusion of the Galileo mission, addressing many issues relevant to our discussion of Saturn as well. Finally, Sánchez-Lavega et al. (2004a) and Vasavada and Showman (2005) review dynamical theories for all the giant planets. A synthesis of the record of temporal changes in Saturn's atmosphere obtained by ground-based observers and the Hubble Space Telescope in the interim between Voyager and Cassini is presented in Chapter 2. We briefly summarize this recent history here to set the stage for our subsequent discussion of Cassini's contribution to our evolving view of Saturn.

6.1.2 Saturn Science Between the Voyager and Cassini Epochs

Following the Voyager 1 and 2 Saturn encounters, new dynamical data were obtained by employing ground-based telescopes and instrumentation of a new generation, in particular the Hubble Space Telescope (HST), which has operated since 1990. HST images improved on ground-based resolution of Saturn by a factor of ten, first using from 1990–1994 the Wide Field Planetary Camera 1 (WFPC1) and afterwards using the WFPC2 and optical correction system that improved considerably the resolution and image quality. The introduction in the mid-80s of the Charged Coupled Device (CCD) with its high sensitivity and image digitization allowed the use of a variety of narrow spectral filters (in particular those centered in methane absorption bands) that permitted probing of the clouds at different altitudes. Detection of discrete features from 1981 to 2005 was restricted to those with a high contrast and sizes >1,000 km with ground-based telescopes or to sizes >300 km with HST.

The most conspicuous dynamical phenomenon that occurred during this period was the rarely observed "Great White Spot" (GWS). Based on the apparent recurrence of the few available recorded cases during the last century, a prediction was made of the possibility that a GWS could erupt in the 1990s (Sánchez-Lavega and Battaner 1986; Sánchez-Lavega 1989). In September 1990 the outburst of a huge bright spot occurred at Saturn's equator, being the last and best studied GWS to date (Sánchez-Lavega et al. 1991; Beebe et al. 1992). The use of the CCD and the timing of the event a few months after the HST was placed in orbit, enabled HST to obtain the first high-resolution multi-wavelength images of the phenomenon (Barnet et al. 1992; Westphal et al. 1992). This allowed the cloud vertical structure to be determined from radiative transfer modeling (Sánchez-Lavega et al. 1994; Acarretta and Sánchez-Lavega 1999), revealing the GWS to be a huge moist convective storm formed by cumulus-like cell clusters elevated above surrounding clouds (Sánchez-Lavega and Battaner 1986; Hueso and Sánchez-Lavega 2004; Section 6.4.3). Bright cloud patches were observed in 1991–1992 as residuals of the 1990 GWS (Sánchez-Lavega et al. 1993a) until a second large equatorial storm erupted in 1994 (Sánchez-Lavega et al. 1996). The second half of the 1990s was dominated by the continuous formation and disappearance of bright spots in the equatorial region perhaps as residuals of deeper processes that produced the activity in 1990 and 1994 (Sánchez-Lavega et al. 1999).

A major advance in the study of Saturn's atmosphere was the determination at high spatial resolution of the meridional structure of the alternating pattern of zonal winds and in particular of the presence of a broad intense equatorial jet (peak velocity $\sim$450 ms^{-1}) in the Voyager epoch (Ingersoll et al. 1984). This wind profile was re-measured and extended by Sánchez-Lavega et al. (2000). Another major discovery of the Voyagers was the observation of planetary scale waves in the upper cloud deck. A reanalysis of the images and the use of appropriate polar projections led to the discovery of the hexagonal wave, an unusual feature embedded in a strong eastward jet (peak velocity 100 ms^{-1}) centered at 80°N latitude (Godfrey 1988). The phenomenon was later modelled as a Rossby wave forced by a huge anticyclone, the North Polar Spot (NPS) located in one of the equatorward edges of the hexagon (Allison et al. 1990). The NPS and Hexagon were re-observed 10 years later using ground-based CCD and HST imaging (Caldwell et al. 1993; Sánchez-Lavega et al. 1993b, 1997), confirming that both features are long-lived. However no hexagon-wave counterpart was observed around the South Pole in HST images (Sánchez-Lavega et al. 2002b). Another wave discovered by the Voyagers was called the "ribbon", a thin undulating band centered at latitude 47°N in the peak of an eastward jet with velocity of 140 ms^{-1} (Sromovsky et al. 1983). The phenomenon was interpreted as a baroclinically unstable wave with cyclonic and anticyclonic circulations on either side by Godfrey and Moore (1986). It was re-observed 14 years later in HST images (Sánchez-Lavega 2002a), suggesting again that it is a long lived feature.

The detection of planetary scale thermal waves in the upper troposphere (130–270 mbar) was reported by Achterberg and Flasar (1996) from the analysis of Voyager IRIS infrared spectra and interpreted as an equatorial Rossby wave (Section 6.4.3). Long-term ground-based infrared observations of 7.8-μm and 12.2-μm stratospheric emission revealed a semiannual oscillation (14.8 years) in Saturn's low latitude stratospheric temperatures, suggesting a link to seasonal forcing (Orton et al. 2008; Section 6.3.7.2). The response of Saturn's temperature to seasonally varying radiative forcing was calculated for the upper troposphere by Bézard et al. (1984) and for the stratosphere by Bézard and Gautier (1985). Comparison with the Voyager temperature measurements gave reasonable agreement in the altitude range from $\sim$10 to 350 mbar. More advanced radiative-dynamical models were presented by Conrath et al. (1990), and including the seasonal variability in ring shadowing and ring thermal emission, by Barnet et al. (1992). Confirmation before the arrival of Cassini of the seasonal effects in the thermal structure of the stratosphere was obtained using ground-based mid-infrared imaging by Ollivier et al. (2000). Long-term changes, probably coupled to the seasonal insolation cycle, were also detected in the haze optical depth above the ammonia cloud deck from a 1967 to 1984 spectroscopic survey performed by Trafton (1985). This variability study was extended later from 1986 to 1989 by Karkoschka and

Tomasko (1992) using ground-based data and from 1994 to 2003 by Pérez-Hoyos et al. 2005 and Karkoschka and Tomasko (2005) using HST images.

That Saturn possesses an active cloud structure below the ammonia cloud level was inferred from the detection of dark spots at different latitudes at a wavelength of 5 μm (Yanamandra-Fisher et al. 2001). The spots result from the effect of variability of the cloud opacity at the 2–4 bar pressure level on radiation coming from deeper levels. At visible wavelengths, a long-term HST-imaging survey of Saturn from 1994–2002 allowed the tracking of discrete spots and the detection of an apparent strong decrease in the equatorial jet speed while the jet profile at other latitudes remained constant (Sánchez-Lavega et al. 2003, 2004b). Similar results were obtained by the first multi-spectral Cassini wind measurements (Porco et al. 2005; Sánchez-Lavega et al. 2007). The extent to which this variability represents real temporal dynamical changes versus variations in the level of seeing in the presence of vertical wind shear (Pérez-Hoyos et al. 2006) is discussed in Section 6.3.5.

6.1.3 Questions About Saturn Entering the Cassini Era

As Cassini approached the Saturn system in early 2004, a number of fundamental questions about Saturn's circulation and structure had been raised. Many of these concerned the series of alternating eastward and westward jets that Saturn has in common with Jupiter: What dynamical processes provide energy to the jets, and what is the nature of associated vertical overturning circulations? Is Saturn's apparent preference for prograde flow real or an artifact of uncertainty in its rotation rate? Given the stability of the jets at most latitudes over many decades, what is the first-order momentum balance between sources and sinks? To what depth do the jets extend, and does this tell us the relative importance of deep vs. shallow circulation mechanisms? Is Saturn's internal heating the control on its circulation, or does differential insolation also play a role? Does water moist convection and latent heat release create a stable layer at intermediate depths that decouples convective turbulence at depth from the dynamics of the cloud layer, or are convective storms merely a tracer of the large-scale dynamics?

Other questions concerned the role of the clouds and upper level hazes in both the dynamics and our ability to observe it and the transition from the flow at depth to the upper troposphere and stratosphere: Why are cloud albedo contrasts on Saturn more loosely related to the jets than on Jupiter? To what pressure do remote sensing observations "see" at levels where cloud and haze opacity are non-negligible? Has Saturn's equatorial jet actually weakened? Are the observed equatorial variations, whether in wind speed or cloud altitude, a signature of seasonal insolation variations? How important is Saturn's tropospheric haze to the local heating, static stability, and upper troposphere dynamics? What is the upper tropospheric circulation associated with the decay of the jets with height, what processes explain it, and how does it connect to the deeper circulation? How does Saturn's stratospheric temperature distribution vary seasonally, and to what degree is it modified by the dynamics?

The Cassini Orbiter remote sensing instruments together cover a broad range of wavelengths and thus sense a wide range of altitudes to address these questions (Sections 6.2.3 and 6.3.1). Most of the information presented in this chapter is the result of observations from the following instruments. The Composite Infrared Spectrometer (CIRS; Flasar et al. 2004b) senses the stratosphere and upper troposphere down to the visible cloud level at thermal infrared wavelengths. The Imaging Science Subsystem (ISS; Porco et al. 2004) provides high-resolution images of several levels within the ammonia clouds and tropospheric haze from the near ultraviolet to the near-infrared. The Cassini RADAR passive radiometry mode (Elachi et al. 2004) measures microwave emission from near the ammonia saturation level. The Visual and Infrared Mapping Spectrometer (VIMS; Brown et al. 2004) operates in both visible and infrared wavelength ranges to observe reflected sunlight and thermal emission, sometimes from relatively deep within the ammonium hydrosulfide cloud layer.

6.1.4 Scope and Organization of the Chapter

This chapter considers all aspects of the thermal structure and dynamics of Saturn's atmosphere. Our lower boundary is the ∼1–2 Mbar pressure of the molecular-metallic hydrogen transition (Guillot 1999; Chapter 4), below which Saturn's rotation rate is determined; this is the lower boundary for deep convective cylinder models. Our upper boundary is at a few μbar, above which non-neutral species and magnetospheric interactions become important (Chapters 8 and 10). Saturn's gases, aerosols and clouds determine the level we observe and act as tracers of vertical motions that cannot be directly observed; these are discussed in more detail in Chapters 5 and 7. Our discussion is organized as follows. Observational inferences about the atmosphere below the visible cloud tops and its relevance to processes at higher levels are described in Section 6.2. Global aspects of the more extensively observed visible cloud level, upper troposphere and stratosphere are considered in Section 6.3. Section 6.4

describes discrete and regional phenomena, with an emphasis on what they imply about physical processes and unobserved aspects of the atmosphere. In Section 6.5 we assess competing ideas about the maintenance of Saturn's general circulation, inferences about these mechanisms from recent modeling studies, and the extent to which observations constrain these theories. Finally, in Section 6.6 we consider the major remaining outstanding questions about Saturn structure and dynamics, how some of these may be addressed by observations taken during the Cassini Equinox and Solstice mission phases, potential avenues for modeling advances in the next few years, and long-term observational needs that might help define a future follow-on Saturn mission.

6.2 Observational Inferences About the Deep Atmosphere

6.2.1 Saturn's Rotation Period

On gas giant planets, the actual rotation period of the deep interior cannot be determined directly. Jupiter's magnetic field is tilted and thus modulates radio emissions with a periodicity that is interpreted as being representative of the interior rotation. On Saturn, however, the tilt is very small and the magnetic field nearly symmetric (see Chapter 10). Nonetheless, Voyager radio measurements detected Saturn Kilometric Radiation (SKR) that was variable but modulated at a period of 10 h, 39 min, 24 s (Desch and Kaiser 1981). This has been used ever since as the reference frame for measuring winds on Saturn.

Cloud-tracked winds based on the SKR period behave quite unlike those observed for Jupiter (Smith et al. 1982). Both planets have a strong prograde equatorial jet and a series of weaker alternating prograde and retrograde jets at higher latitudes. On Saturn, though, the equatorial jet is extremely strong and the westward jets either very weak or merely eastward minima (Section 6.3.3). The strong equatorial jet, if confined to a shallow weather layer, would require equator-pole geopotential height differences of 40% relative to the thickness of the weather layer, which is inconsistent with the small observed equator-pole temperature difference at cloud top. This led Smith et al. (1982) to conclude that the level of no motion must be quite deep, of order 10^4 bars, although they noted that if the true rotation period were ~8 min shorter the requirement for deep flow would disappear. Allison and Stone (1983) argued that thin weather layer configurations would still be viable if a significant latitudinal temperature gradient, associated with a latitudinally varying level of no motion somewhat analogous to Earth's oceanic thermocline, existed below cloud level. They suggested that the buoyancy contrasts required by such a structure might be provided by latitudinally varying latent heat release if the water abundance on Saturn were sufficiently in excess of solar.

Ulysses and Cassini observations, however, indicate that the SKR modulation period varies by ~1% on time scales of years and is now ~8 min *longer* than the Voyager SKR-based reference period (Galopeau and Lecacheux 2000; Gurnett et al. 2005; Kurth et al. 2007, 2008). These variations are considerably greater than any possible real variation in the rotation period of Saturn's massive interior. Instead the SKR period is now thought to reflect magnetospheric slippage from a centrifugally driven convective instability, associated with mass loading from the Enceladus neutral gas torus (Gurnett et al. 2007).

This leaves the actual Saturn deep interior rotation period unknown – a considerable obstacle to interpretations of its atmospheric dynamics. Anderson and Schubert (2007) returned to the original idea of a shorter period. They calculated a reference geoid from Saturn gravitational data and then estimated the rotation period that would minimize the dynamical height deviations of the 100 mbar surface from that geoid; the resulting rotation period is 10 h, 32 min, 35 s. Figure 6.1 shows that use of this period as a reference reduces Saturn's equatorial jet by ~100 m s^{-1} and produces alternating eastward and westward jets of comparable magnitude at higher latitudes. Both features make the Saturn wind profile much more like that of Jupiter.

The difficulty with the Anderson and Schubert (2007) result is that there is no physical requirement for the flow to minimize dynamic height deviations from the geoid. Physical arguments can be marshaled both for and against this being the true rotation period of Saturn. Although the Voyager SKR-based period creates a seemingly unrealistically

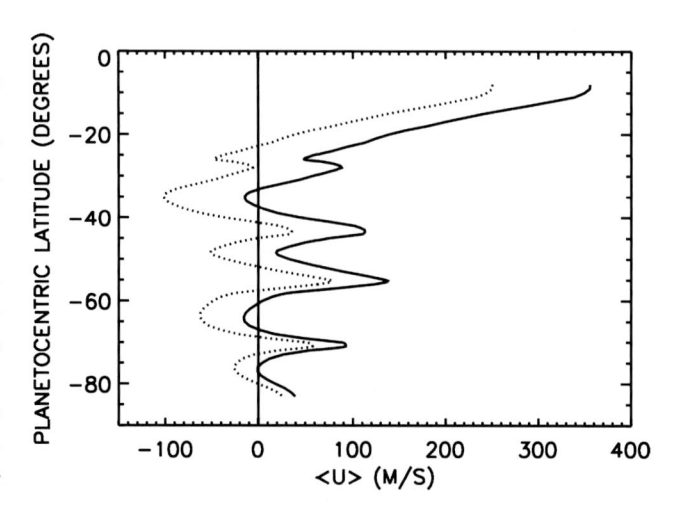

Fig. 6.1 Saturn Southern Hemisphere zonal wind profiles from the Cassini ISS data of Vasavada et al. (2006) based on the Voyager SKR rotation period (*solid*) and the Anderson and Schubert (2007) rotation period (*dotted*)

strong equatorial jet ($\sim$450 m s^{-1}), this must be placed in the context of Saturn's size and (apparently) rapid rotation. The ratio of this wind speed to the tangential speed of a point at Saturn's cloud level rotating with the planet (in effect an equatorial Rossby number) is only $\sim$.04. This is somewhat larger than the terrestrial value of $\sim$.01, but considerably smaller than the value of $\sim$50 for slowly rotating Venus. Furthermore, there is no fundamental reason to rule out global superrotation at the Saturn visible cloud level. Aurnou and Heimpel (2004), for example, show that a no-slip lower boundary condition applied in a deep convective cylinder simulation of Saturn produces a wind profile quite like that obtained from the data using the Voyager SKR period.

On the other hand, if such a lower boundary condition is relevant to Saturn, the conundrum simply shifts to deeper levels: What would the physical source of friction at the lower boundary be on Saturn, and why would a similar behavior at depth not be applicable to Jupiter, which does not show nearly as strong a preference for prograde flow? Use of something closer to the shorter Anderson and Schubert (2007) period removes this discrepancy and leaves shallow weather layer models of the cloud level dynamics in play. Possibly more compelling are two recent pieces of observational evidence. Kurth et al. (2008) report a second shorter SKR modulation period of 10.59 h, tantalizingly close to the various alternative rotation periods considered by Smith et al. (1982), Allison and Stone (1983), and Anderson and Schubert (2007), although this period too appears to be variable. Another intriguing result comes from potential vorticity analysis of Saturn's upper troposphere winds and temperatures (Read et al. 2009; Section 6.3.8), which suggests that the wind field measured in a reference frame based on a shorter rotation period might be close to neutrally stable, a desirable result given the constancy of most of Saturn's jets over long time intervals. Since the issue is not settled, we continue to report winds based on the Voyager SKR period in this chapter and await a possible resolution if highly inclined orbits in Cassini's extended mission allow the magnetic field of Saturn to be documented with greater accuracy (Bagenal 2007).

6.2.2 Convective Heat Flux and Condensation Levels

Saturn is estimated to emit 1.78 $\pm$ 0.09 times more longwave radiation than the shortwave radiation that it absorbs (Hanel et al. 1983). This implies the existence of a 2.01 $\pm$ 0.14 W m^{-2} internal heat source due to both gravitational contraction and helium differentiation (see Chapter 4). The molecular envelope is poorly conducting, and opacity due to hydrogen, helium, and several minor constituents becomes substantial at depth. Thus it has long been anticipated that Saturn's interior is convective and its lapse rate close to dry adiabatic. Ingersoll and Porco (1978) concluded that convection could transport sufficient heat poleward along sloping surfaces with very small temperature gradients to offset the latitudinal gradient of insolation and explain the small equator-pole temperature contrasts observed on the giant planets (Ingersoll 1976).

A possible difficulty with this picture was raised by Guillot et al. (1994), who pointed out that at temperatures of $\sim$1,200–1,500 K, photon emission shifts to shorter wavelengths as temperature increases, and that the resulting opacity due to hydrogen and helium absorption might decrease, creating a stable radiative zone. Gierasch (1999) showed that the existence of a radiative zone could have significant consequences for latitudinal temperature gradients at higher altitudes and might explain Saturn's preference for prograde winds in the Voyager SKR reference frame (Section 6.2.1). However, the presence of wide sodium and potassium absorption lines in spectra of brown dwarfs has led Guillot (2005) to conclude that transparent regions of the spectrum are most likely not present and therefore that a deep radiative zone is unlikely on the gas giants. Thus gases are assumed to be well-mixed in the molecular envelope until reaching their condensation levels.

Thermo-chemical equilibrium models predict that Jupiter and Saturn have three main cloud decks in the upper troposphere. For Saturn they are: an upper ammonia (NH$_3$) ice crystal layer centered at $\sim$1 bar, an ammonium hydrosulfide layer (NH$_4$SH) centered at $\sim$3–4 bar, and a water layer centered at $\sim$8–10 bar (Weidenschilling and Lewis 1973). The actual locations are uncertain because of the unknown abundances of the condensing constituents (see Chapter 5). These results have often been interpreted as implying the existence of extensive cloud decks of each species, but the terrestrial experience of a partly cloud covered planet hints at more complexity. Carlson et al. (1988) examined cloud microphysics on the giant planets. They concluded that the NH$_3$ and NH$_4$SH decks on Saturn were relatively thin, only lightly precipitating, and thus the equivalent of cirrus clouds on Earth. The water cloud is more massive, however, even for solar abundance, and is more likely to be convective and sporadic with small areal coverage. Del Genio and McGrattan (1990) used a cumulus parameterization to show that water-based moist convection would occur on the giant planets and create a stable layer over a depth of about a scale height above the water condensation level via the combined effects of latent heat release, compensating subsidence, and molecular weight gradients. This putative stable layer is too deep to have been observed but is crucial to many shallow weather layer theories of giant planet circulations (Section 6.5). Beyond the first scale height above the water condensation level, the lapse rate is assumed to once again become dry adiabatic because the other condensate species release too little latent heat to have a significant effect. Above the 1 bar level dry adiabatic lapse rates have been retrieved from Voyager occultations (Lindal et al. 1985) and Cassini CIRS (Fletcher et al. 2007b), with a transition to a stable temperature gradient occurring at ∼400–500 mbar.

6.2.3 Cassini Probing of the Atmosphere Below Cloud Top

Cassini does not directly sense the Saturn atmosphere at or below the water cloud level. However, Cassini VIMS and RADAR provide the first detailed global views of the depths of Saturn underneath the upper-level hazes and ammonia cloud tops. Each does this in modes somewhat atypical to the usual near-infrared and radar means of exploring planets.

VIMS uses wavelengths beyond 4.5 μm to image the planet bathed from within by its own thermal emission, as has been done for Venus (Baines et al. 2006; Drossart et al. 2007; Piccioni et al. 2007). In the 5 μm spectral region, the primary sources of extinction are spectrally localized molecular absorption by trace gases (e.g., phosphine, germane, ammonia) and the extinction of deep clouds comprised of large particles with radii near or larger than 5-μm. The extensive upper level haze (Section 6.3.2) does produce significant scattering of sunlight. This, together with the weak solar flux there – just 0.2% of the visible near 0.5 μm – inhibits sunlit views of these deep large-particle clouds. However, a second source of radiation is available to probe the ammonia-cloud region and below: Saturn's own indigenous heat radiation. Thus deep clouds and their motions are detected by observing their silhouettes against the background glow. For Saturn, the 5-μm signal is produced near the 6.5-bar, 245-K level and attenuated primarily by overlying large-particle clouds. Thus, relatively dark regions depict thick large-particle clouds, while relatively bright regions depict relatively cloud-free areas. The spectrum near 4.65 μm and 5.1 μm can be used to determine the cloud base pressure level for clouds located deeper than 1.8 bar, below the ammonia condensation level (Baines et al. 2009a). This technique reveals that a deep cloud layer exists with base pressure varying between ∼2.5 and 4.5 bars, depending on location, corresponding to that thermochemically predicted for a cloud comprised of NH_4SH.

RADAR, in its passive "listening" mode, views the thermal emission of Saturn as well, but at an effective wavelength of 2.2 cm, which is unaffected by haze and cloud particles but is sensitive to ammonia gas absorption, an important cloud-forming condensable. This longer wavelength has one particular drawback: the instantaneous field-of-view is an order of magnitude larger than for VIMS (∼6 mrad vs. 0.5 mrad), despite the larger telescope (in this case, the spacecraft's 4-m telecommunications antenna). Thus, useful maps (IFOV better than 2,000 km) are only acquired within ∼330,000 km above the planet, which occurs only during the 12 h around periapse on the closest orbits.

VIMS and RADAR have revealed an unexpectedly different world at levels underneath the hazes of Saturn. Figure 6.2 compares the VIMS view of Saturn in reflected sunlight at 0.9 μm, a methane absorption band that senses the upper troposphere haze, to that at 5 μm (Baines et al. 2009c). At 0.9 μm several broad bands of hazes circle the planet, with little longitudinal variability and relatively little contrast between adjacent bands. The equatorial region sports a broad band of unusually high and thick haze reaching into the stratosphere. In contrast, the lower tropospheric view at 5 μm reveals a dense array of narrow bands – cloudy "zones" and less cloudy "belts". Many of the zonal bands are broken up by discrete cloud features.

Figure 6.3 compares RADAR (Janssen et al. 2008) and VIMS global cylindrical mosaics of Saturn acquired within 1 day of each other. Two common features of dynamical significance are evident. Outside the equatorial region, both images exhibit a tendency for cloud bands to organize on a scale of ∼1.8° latitude, a behavior not observed at cloud top in reflected sunlight. The source of this fine-scale structure is unknown. The primary difference between the NH_4SH and NH_3 cloud level atmosphere sensed by VIMS and RADAR, and the visible cloud top level sensed by ISS (Section 6.3.2), is static stability. ISS images view the transition region between the stable upper troposphere and near-adiabatic lower troposphere (Fletcher et al. 2007b), while VIMS and RADAR sense the near-adiabatic region, above any stable layer below associated with water condensation. The Rossby deformation radius $L_d = NH/f$, where N is the Brunt-Väisälä frequency, H the scale height, and f the Coriolis parameter, is a characteristic scale at which Coriolis forces effectively counter pressure gradient forces and at which instabilities inject energy into the flow. Since L_d decreases as static stability decreases, these views of the deeper atmosphere might reveal something more directly about the forcing of the cloud-top dynamics, which is almost certainly driven by processes operating at or below the water cloud level. Near the equator, exceptionally large discrete clouds exist near 6°N and 6°S latitude. These clouds are likely associated with enhancements of ammonia gas observed by RADAR, and may be ammonia plumes delivering aerosols to the upper atmosphere, thereby producing the band of enhanced upper-level aerosols observed there. Since these features appear to correspond to similar scale cloudy features in the VIMS image, which senses deeper than RADAR, both features may be produced by water convection triggered at deeper levels.

Fig. 6.2 VIMS views of Saturn's upper level hazes at 0.9 μm (*left*) and lower tropospheric clouds in thermal emission at 5.1 μm (*right*). Here, the original thermal image is shown photometrically reversed so the backlit silhouetted clouds are bright and holes in the clouds are dark, thus giving a more typical black and white view of clouds as they might be observed in sunlight if the overlying hazes were to disappear (Baines et al. 2009c)

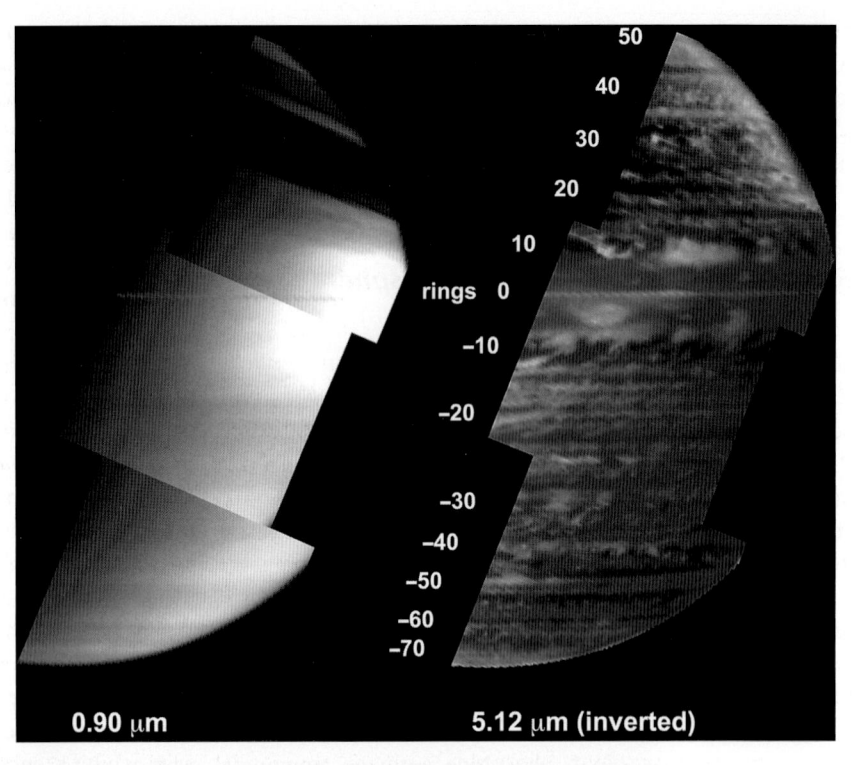

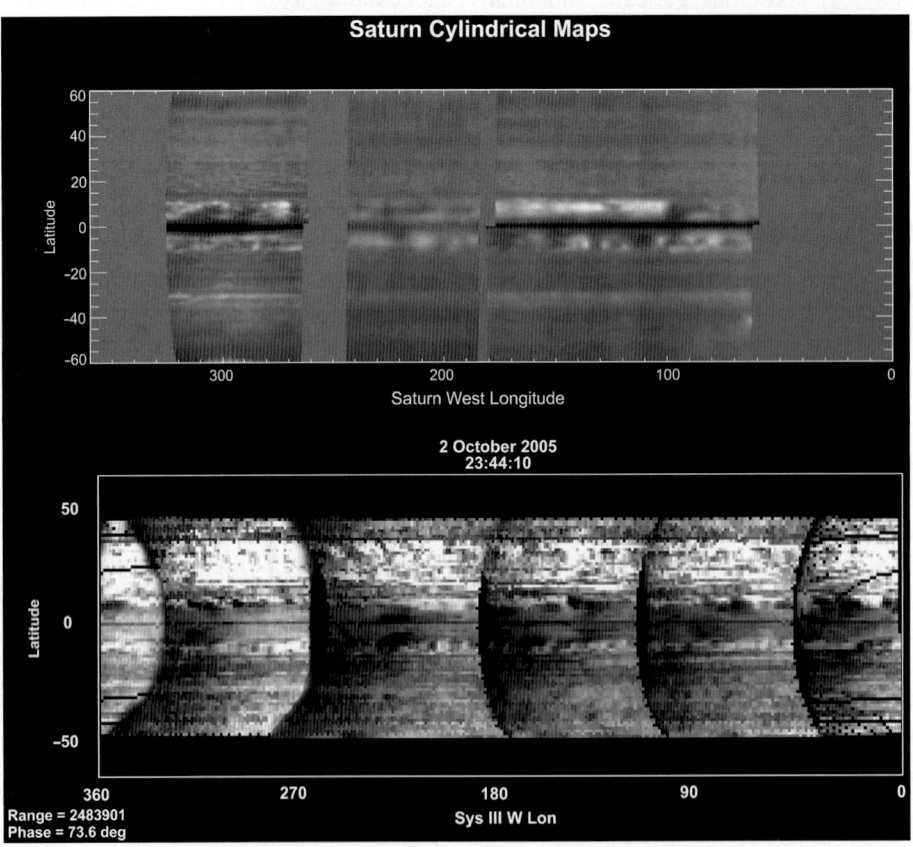

Fig. 6.3 RADAR (*upper*; Janssen et al. 2008) vs. VIMS (*lower*; Baines et al. 2009c) cylindrical mosaics of Saturn. In the RADAR view, dark discrete features near 6°N and S latitude represent enhanced ammonia gas absorption at the 1.4–2.0 bar level underneath and at the ammonia condensation level near 1.4 bar. The latitudes, size, and shape of these gaseous features correlate well with the discrete cloud plumes seen in the equatorial region in 5-μm imagery, suggesting that the discrete clouds observed underneath the thick upper-level haze of Saturn are formed from ammonia gas condensation near the 1.4-bar level

6.3 Observations at and Above the Visible Cloud Top

6.3.1 Levels Sensed by Cassini Instruments from Cloud Top to the Stratosphere

Cassini remote sensing exploits the wavelength-dependent scattering and absorption properties of gases and particulates to sense temperatures and dynamics at different pressure levels (see Chapter 7). In reflected sunlight we "see" down to a level not much greater than that for which the two-way extinction optical thickness (τ) is $\sim$1–2. Rayleigh scattering by the gaseous H_2-He Saturn atmosphere varies approximately as λ^{-4}, where λ is the wavelength of the observation. Particulates (aerosols and clouds) scatter more efficiently than gases and with a weaker ($\sim\lambda^{-1}$) dependence. Thus, in the presence of aerosols or clouds, we generally see down to the particulate scattering $\tau = 1$ level at non-absorbing (continuum) wavelengths, which is usually close to the highest cloud top. Clouds are almost conservative scatterers in the shortwave band and appear bright in continuum images, more so the more optically thick they are. Aerosol hazes (Section 6.3.2) can be either absorbing or scattering at different wavelengths and thus may appear either dark or bright in different filters. At wavelengths centered on gaseous absorption bands (usually of methane), we see no deeper than the absorption two-way $\tau = 1$ level in clear regions. In cloudy or optically thick aerosol regions, if the cloud/aerosol top is above the gas $\tau = 1$ level, absorption by the overlying gas is reduced and scattering from near the cloud/aerosol top takes place. This allows us to discriminate cloud/aerosol top heights – the brighter the feature, the higher the cloud top.

In the ultraviolet (UV), ISS would only see to $\sim$500–1,000 mbar if the atmosphere were clear, and in practice, ubiquitous stratospheric aerosols (Section 6.3.2) are sensed at these wavelengths. Most ISS imaging is done at near-infrared wavelengths (Fig. 6.4), where scattering by small aerosol particles is reduced and deeper cloud layers are easier to see. However, unlike Jupiter, Saturn has a relatively thick upper tropospheric haze (Section 6.3.2) that limits our ability to see to depth, and thus feature contrasts are more muted on Saturn than on Jupiter. Nonetheless, in the continuum filter CB2 (750 nm) we can see features attributed to reflection by optically thick clouds with a variety of cloud top altitudes within and below the haze. ISS contains three methane band filters, at wavelengths of weak (MT1, 619 nm), moderate (MT2, 727 nm), and strong (MT3, 889 nm) absorption. The clear-sky $\tau = 1$ level for MT3 is $\sim$330 mb (Tomasko et al. 1984), above the ammonia cloud top, and it thus sees only variations in haze. MT2 penetrates to $\sim$1.2 bars in clear sky and thus sees haze, some ammonia cloud tops and occasional high-penetrating convective water clouds. MT1 is sensitive to even deeper levels. Figure 6.4 (upper left) shows that CB2 gives the most detailed view of the ammonia cloud tops. MT1 (upper right) sees most but not all of the same clouds as are visible in the continuum filter, but the sizes of discrete features are smaller, suggesting either that the weak methane absorption prevents seeing of optically thinner cloud edges or that the clouds decrease in size upward. The MT2 image (lower right) is very different from the others – no small discrete cloud features are visible, though there are numerous striated features that appear to be correlated with those visible at the other wavelengths. Much of what we see in MT2 appears to be due to variations in the tropospheric haze instead. During the Cassini epoch, CB2 images of the Southern Hemisphere (lower left) generally have much less contrast than those of the Northern Hemisphere, suggesting that the haze optical thickness is lower in the winter hemisphere (see Chapter 7). This is consistent with inferences about local haze heating from the upper troposphere thermal structure (Section 6.3.6) and greater Northern Hemisphere emission variability at 5 μm in VIMS images (Fig. 6.2).

In the mid- and far-infrared ($\lambda \geq 7$ μm), reflected sunlight can be ignored, and the observed radiation is due entirely to thermal emission. If the vertical distribution of

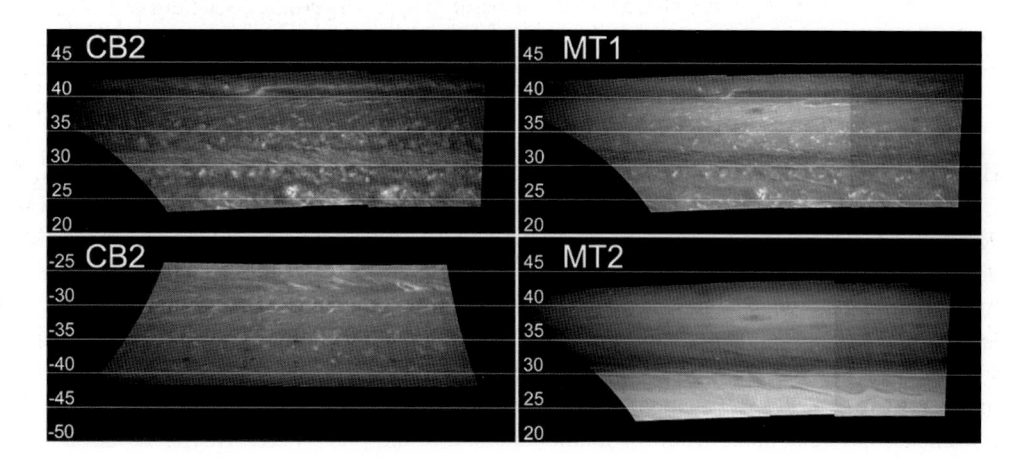

Fig. 6.4 Cassini ISS image mosaics of northern (*upper left*) and southern (*lower left*) midlatitudes in a continuum filter, and of the same northern latitudes in weak methane band (*upper right*) and moderate strength methane band (*lower right*) filters. Latitudes are planetocentric

absorption is known, spectra can be inverted to determine a vertical temperature profile. Conversely, if the temperature is known, abundances of absorbers can be determined. In practice, temperatures can be retrieved from two regions of the spectrum. The ν_4 rovibrational band of methane, at $\sim$1,200–1,400 cm^{-1}, provides information on mid-stratospheric temperatures between $\sim$0.5 and 4 mbar in nadir-viewing mode and up to a few µbar in limb-viewing mode. The S(1) and S(0) collision-induced rotational lines, plus part of the translational continuum, from 220 to 670 cm^{-1}, provide information on temperatures in the upper troposphere and tropopause regions, between $\sim$50 and 800 mbar. The collision-induced hydrogen absorption is also sensitive to the ratio of *ortho*-hydrogen to *para*-hydrogen. At temperatures less than 300 K, the equilibrium *para* to *ortho* ratio varies with temperature, and the expected equilibration timescales are comparable to or longer than the dynamical timescales (Massie and Hunten 1982; Conrath and Gierasch 1984). The *para*-hydrogen abundance can thus be used as a diagnostic of vertical motions in the troposphere. Thermal infrared spectra can also be used to retrieve abundances of several trace gaseous species that are useful tracers of the circulation. Two regions of the spectra, from about 20–200 cm^{-1} and from about 900–1,200 cm^{-1}, can be used to measure the tropospheric abundances of ammonia (NH$_3$) and phosphine (PH$_3$) at pressures around 500 mbar. Both species are transported upwards from reservoirs in the interior, and destroyed in the upper troposphere by photodissociation (NH$_3$ is also lost through condensation); their abundances are thus affected by vertical mixing and transport by the meridional circulation. Between 650 and 850 cm^{-1}, the spectrum contains stratospheric emission lines of several hydrocarbons, in particular acetylene and ethane, which can be used to measure their abundances in the middle stratosphere. These hydrocarbons have photochemical lifetimes comparable to or longer than the dynamical timescales of the meridional circulation. Thus, their distributions can be used to infer information about the meridional circulation in the stratosphere.

6.3.2 Albedo Patterns vs. Jets on Saturn vs. Jupiter

HST and Cassini reflectivity measurements from the near-UV (200 nm) to the near-infrared (3 µm), including the 619, 725 and 890 nm and 2.3 µm methane absorption bands, indicate that at least two haze layers of aerosol particles permanently exist above the ammonia cloud (see Chapter 7). An upper thin haze in the stratosphere (between pressure levels 1–50 mbar) is formed by small particles strongly absorbent in the UV (radii $\sim$ 0.2 µm, optical depth $\tau \sim$ 1). Beneath it a thicker haze extending from the ammonia cloud to the tropopause level shows a temporally variable optical depth (typically $\tau \sim$ 10 at 814 nm with a variability of a factor $\sim$ 2). Saturn's banded appearance at visible wavelengths, extending along parallels, results mostly from the combined effect of scattering and absorption of solar photons (350–800 nm) by these hazes (Karkoschka and Tomasko 1991, 2005). In the UV ISS filters the planet appears more homogeneous due to absorption and scattering of solar photons by the upper haze and gaseous atmosphere, but the latitudinal contrast between dark and bright bands is slightly greater (Vasavada et al. 2006). In the near IR filters banding is less pronounced in the continuum but becomes more evident in the deep methane absorption bands MT2 and MT3 (Vasavada et al. 2006) and at 2.3 µm where bright bands, a signature of high altitude hazes, detach at the equator and in temperate latitudes. In the thermal infrared window (4–5 µm), radiation escaping from the interior allows us to see distinct features produced by opacity variability in the NH$_3$ – NH$_4$SH cloud system and perhaps in water clouds as well, and zonally aligned features occur (Yanamandra-Fisher et al. 2001; Baines et al. 2005; Section 6.2.3).

The meridional distribution of Saturn's cloud/haze bands is related to the zonal wind profile of alternating eastward and westward jets (Karkoschka and Tomasko 1991, 2005) but is quite different from that on Jupiter in several ways:

(1) In visible and near-infrared continuum images, Jupiter's bright bands (zones) reside in the anti-cyclonic shear regions that lie equatorward of the eastward jets and its darker bands (belts) in the cyclonic shear regions poleward of these jets (Smith et al. 1979). The belts and zones are of comparable ($\sim$5°) width. A different relationship occurs on Saturn (Smith et al. 1981; Vasavada et al. 2006): Eastward jets occupy narrow dark regions, often bisected by a thin bright band, while the much broader region between successive eastward jets and including the westward jets tends to be brighter (Fig. 6.5). The eastward jet morphology resembles that seen in VIMS nighttime 5 µm thermal emission (Fig. 6.2) but that elsewhere does not.

(2) The morphology of the bands also differs on the two planets. Bright zones on Jupiter tend to have lower feature contrast than the darker belts. On Saturn, the dark bands associated with the eastward jets are generally low in contrast with mostly linear features, while the broad brighter regions show considerable evidence of turbulent, cellular, discrete features.

(3) While the jet profile is stable in its latitudinal location on Saturn (Smith et al. 1981; Sánchez-Lavega et al. 2000, 2003; Vasavada et al. 2006), the cloud/haze bands change their brightness and edge positions on time scales of both years (Sánchez-Lavega et al. 1993a; Pérez-Hoyos et al. 2005; Karkoschka and Tomasko

6 Saturn Atmospheric Structure and Dynamics

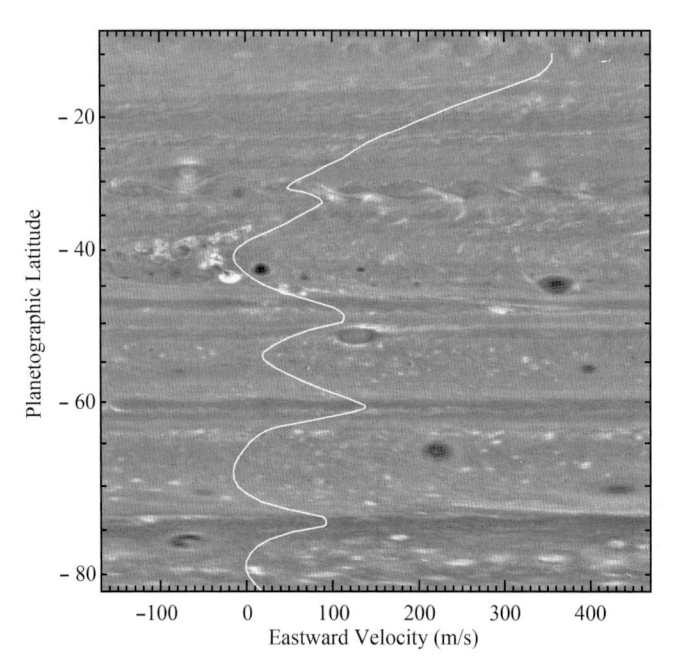

Fig. 6.5 Southern Hemisphere ISS continuum image with superimposed mean zonal wind profile (Vasavada et al. 2006)

top level, where the inferred pattern of rising and sinking motions is just the opposite (Del Genio et al. 2007a) and not obviously related to the albedo pattern (see Section 6.4.2).

6.3.3 Cloud Level Winds and Dynamical Fluxes

The early Voyager image measurements (Smith et al. 1981, 1982, later reworked by Ingersoll et al. 1984) obtained up to 1,000 velocity vectors from cloud tracers and showed a roughly symmetric pattern of eastward zonal jets about the equator, with strong eastward motion at speeds of more than $450 \, \text{m s}^{-1}$ (relative to System III) in an equatorial jet, extending to nearly 30° latitude north and south of the equator. At higher latitudes, 4–5 additional eastward jets are found (including a 'shoulder' on the flanks of the equatorial jet at ±30°) between 30° and 80° latitude, with velocities ranging from 60–$160 \, \text{m s}^{-1}$. As mentioned earlier, the cloud-tracked zonal velocity profile on Saturn exhibits a near-absence of westward jets, with the flow between the eastward jets staying close to zero in System III (Fig. 6.1). Such a pattern is quite unlike the corresponding profile on Jupiter, for which mid-latitude eastward and westward jets are of comparable strength relative to System III. This has raised doubts over the robustness of Saturn's System III, as derived from measurements of SKR emissions, which have been further reinforced by the indication from Cassini that the rotation of the SKR reference frame is not constant in time (Section 6.2.1).

More recent measurements have included a detailed study of the high latitude jets around 64°–84°N by Godfrey (1988) and a more extensive analysis of Voyager images by Sánchez-Lavega et al. (2000), which included ∼2,000 vectors covering latitudes 81°N–71°S with improved error analysis, together with comparisons with measurements taken during the 1990s using HST images. Processing of Cassini ISS images is continuing at the time of writing, and so far has led to published profiles in the southern hemisphere by Porco et al. (2005), Vasavada et al. (2006) and Sánchez-Lavega et al. (2006). A combined profile from all of these data is shown in Figure 6.6a. The new data from Cassini show few changes from the earlier Voyager measurements, though do confirm the existence of additional eastward jets at 74.5°S and 88°S. The latter represents the edge of the polar vortex over the south pole, with zonal wind speeds of more than $160 \, \text{m s}^{-1}$, in association with the polar hot-spot found in both ground-based infrared images and from Cassini CIRS (Fletcher et al. 2007b).

Zonal winds at the ammonium hydrosulfide cloud level have also been estimated by tracking features in VIMS $5 \, \mu\text{m}$ images (Baines et al. 2005, 2009a; Choi et al. 2009). These profiles (Fig. 6.7) are quite similar to those at the visible cloud level during the Voyager encounter, but suggest

2005) and months (Sánchez-Lavega et al. 1993a; Pérez-Hoyos et al. 2006). The brightness changes are wavelength dependent. A band can remain stable in brightness in one wavelength range but brighten or darken with time at others.

A long-term comparison between Voyager, Hubble and Cassini images in haze-sensitive methane band images suggests that on average, the albedo banding takes on distinct characters in the equatorial region (20°N to 20°S), the temperate region (20° to 55° in each hemisphere) and the polar area (55° to the pole in each hemisphere). The broad and intense equatorial jet correlates with an Equatorial Zone that is bright at 890 nm (due to high dense hazes) but dark in the UV (due to UV-blue absorption by this haze). The thick and vertically extended tropospheric haze contains discrete features moving at different altitude levels between 50 and 700 mbar (Pérez-Hoyos et al. 2006). Outside the equatorial region, the albedo pattern at 890 nm is well correlated with the eastward jets and the temperature field at 500 mbar (Fletcher et al. 2007a, b): equatorward of the eastward jet peak the haze is bright (high and dense) and the temperatures low, with the contrary occurring on the poleward side (dark band and high temperatures). This suggests that in the upper troposphere and lower stratosphere, ascending motions occur on the equatorward side of the jet peak whereas descending motions occur on the poleward side accompanied by adiabatic cooling and warming, in good agreement with results obtained from the Voyagers (Conrath and Pirraglia 1983). This stands in contrast to the tropospheric continuum cloud

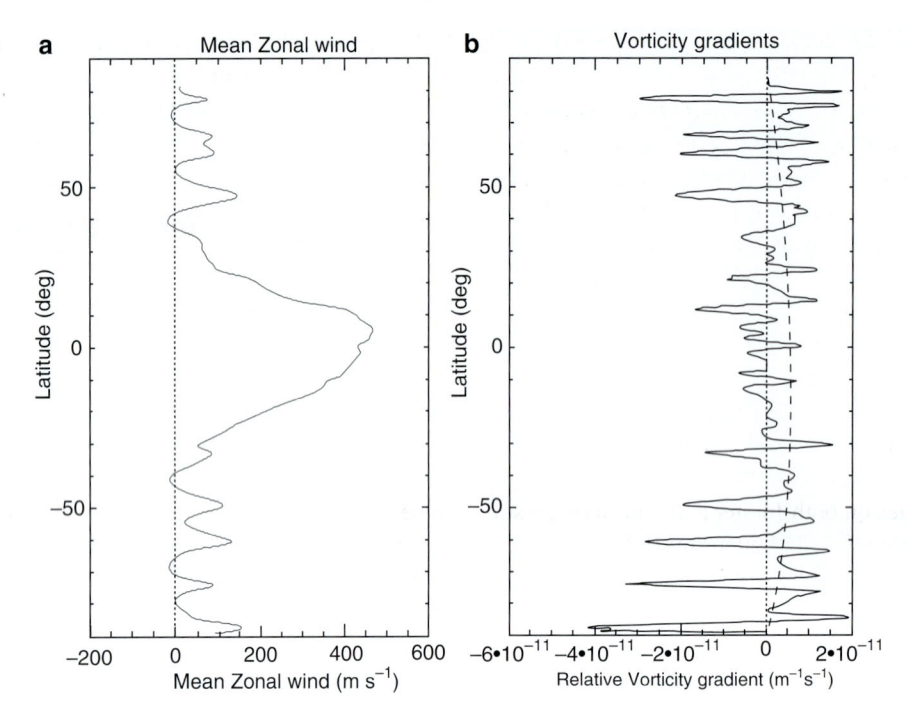

Fig. 6.6 (a) Zonally averaged zonal flow on Saturn, and (b) northward gradient of ξ (*solid line*) and β = df/dy (*dashed line*). Northern Hemisphere profiles were derived from Voyager images by Sánchez-Lavega et al. (2000) and Godfrey (1988), and Southern Hemisphere profiles from Cassini ISS images by Vasavada et al. (2006) and Sánchez-Lavega et al. (2006)

Fig. 6.7 Latitudinal profile of VIMS 5 μm cloud-tracked winds (*black dots*). Shown for comparison are Voyager (*blue*) and Cassini ISS continuum (*red*) wind profiles (Choi et al. 2009)

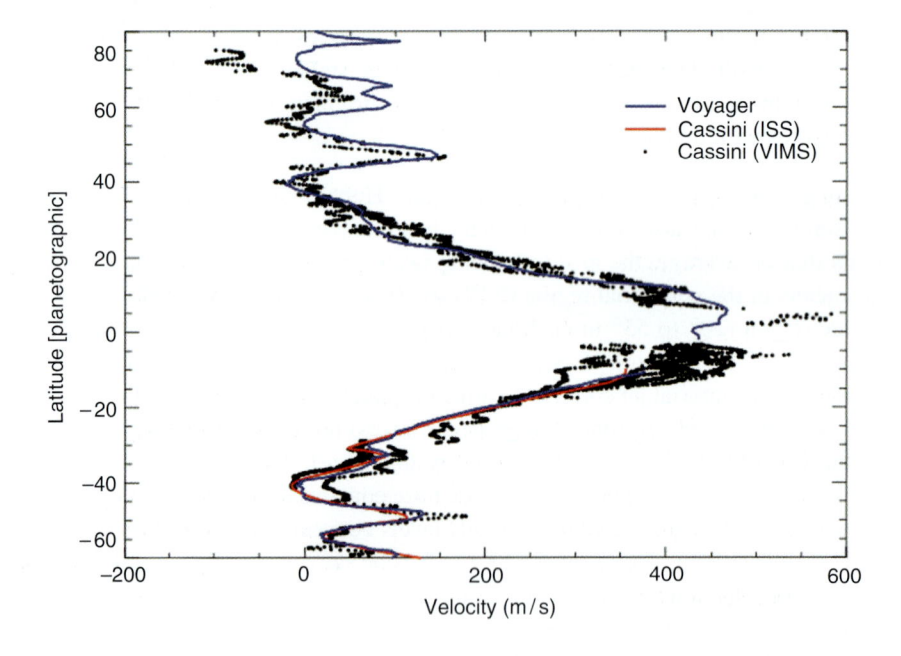

slightly higher jet speeds at the deeper level sensed by VIMS. Above the visible cloud level CIRS thermal winds indicate a further decay of the jets with height (Section 6.3.4), although tracking of features in methane band images suggests that the shear in eastward jets exceeds that in westward jets (García-Melendo et al. 2009).

It is notable that the shapes of the various eastward and westward jets take on particular characteristics that may have some dynamical significance. The equatorial jet appears to have a complex shape with either a relatively flat peak or even some tendency towards a double-peaked form with a local minimum of u on the equator itself. At higher latitudes, the eastward jets are typically much sharper and narrower than the corresponding westward jets or minima in $\bar{u}$. Such a trend is consistent with a tendency for the flow to adopt a relatively weak latitudinal gradient of absolute vorticity $\zeta = f + \xi$ (where ξ is the relative vorticity) in westward flow, while the gradient is enhanced in association with eastward jets. In practice, however, like on Jupiter, the northward gradient, $d\zeta/dy$ (Fig. 6.6b), is found to reverse in sign around many of the westward jets, such that $\beta - d\xi/dy$ changes sign several times between equator and pole in each hemisphere (where β is the planetary vorticity gradient df/dy). Such reversals of $d\zeta/dy$ would, at face value, suggest the possibility of barotropic instability, though the role of baroclinic effects should also be taken into account (see Section 6.3.8).

The processes maintaining this pattern of jets continues to excite controversy, though some indication of the possible role of eddy momentum transports is starting to become apparent from measurements of the non-zonal components of velocity from more detailed cloud-tracking. Early attempts to measure the horizontal transport of momentum via 'eddy correlation' techniques on both Jupiter and Saturn (Ingersoll et al. 1981, 1984; Mitchell and Maxworthy 1985) claimed to find quite strong correlations between $\overline{u'v'}$ and $d\overline{u}/dy$, suggesting a significant conversion of eddy into zonal mean kinetic energy. Sromovsky et al. (1982, 1983), however, suggested that these initial results may have been affected by possible selection effects, which tended to over-emphasize contributions to $\overline{u'v'}$ from bright or active cloud features. The use of relatively small numbers of vectors (<1,000) in the statistics may also have prejudiced the ability to detect a statistically significant correlation.

Recent attempts to measure this conversion from eddy to zonal mean flow (Salyk et al. 2006; Del Genio et al. 2007a) have made use of automated image correlation techniques from Cassini ISS images. Such methods enable many more cloud vectors to be measured, and appear to be much less affected by visual selection effects of cloud targets, thereby achieving a more uniform and representative sampling of velocity structures across the planet. Results indicate a reasonably clear and robust positive global correlation between $\overline{u'v'}$ and $d\overline{u}/dy$, accelerating the jet with implied mean energy conversion rates from eddies to the zonal flow of $\sim$7.1–12.3 $\times 10^{-5}$ W kg^{-1} for Jupiter (Salyk et al. 2006) and $\sim$3.3 $\times 10^{-5}$ W kg^{-1} for Saturn (Del Genio et al. 2007a). Such a correlation is broadly consistent with the notion of an anisotropic upscale turbulent cascade in a geostrophically turbulent flow (e.g., Vasavada and Showman 2005; Galperin et al. 2006; Section 6.5). However, this phenomenon needs more detailed analysis, particularly with regard to the roles of different scales. Mitchell and Maxworthy (1985), for example, noted that the correlation between $\overline{u'v'}$ and $d\overline{u}/dy$ near Jupiter's Great Red Spot was significantly *negative*, implying a conversion of kinetic energy *from* the zonal flow *into* the GRS. Such a result might indicate different mechanisms for the production and maintenance of different types of eddy, with small-scale features emerging from active upwelling convection or baroclinic instability while larger eddies are sustained (at least in part) by barotropic exchanges with the zonal jets.

6.3.4 Thermal Structure and Circulation Above Cloud Level

Information on temperatures in Saturn's upper troposphere and stratosphere comes from two primary sources. Radio occultations provide vertical temperature profiles with high vertical resolution, but at a limited number of locations. Thermal infrared sounding provides extensive spatial coverage, but the vertical resolution is limited to roughly one scale height.

The earliest spatially resolved thermal observations of Saturn, using north-south scans at 12 μm in the ν_9 band of ethane during southern summer, showed an increase in emission from north to south with a strong peak in the emission at the south pole (Gillet and Orton 1975; Rieke 1975). However, these observations did not have the information needed to determine if the spatial variations in emission were caused by variations in the temperature or ethane profiles. Later observations by Tokunaga et al. (1978) found similar north to south variations in emission from the ν_4 band of methane at 7.8 μm, indicating that there were stratospheric temperature variations, with the south (summer) pole warmer than the equator. Tokunaga et al. (1978) also observed near 20 μm, in the upper tropospheric hydrogen band, and found that the meridional emission variations were weaker than in the stratospheric emission bands, indicating that the equator-to-pole temperature gradients were smaller near the tropopause than in the stratosphere.

The first spacecraft observations of Saturn's thermal structure came from Pioneer 11 in 1979. Measurements at 20 and 45 μm by the infrared radiometer were inverted by Orton and Ingersoll (1980) to give temperatures from $\sim$60 to 500 mbar from 30°S to 10°N. They found a temperature minimum at the equator, which became weaker with increasing pressure, with no meridional gradient between 10°S and 30°S. More extensive thermal data were obtained by the infrared spectrometer IRIS on Voyagers 1 and 2. Conrath and Pirraglia (1983) used Voyager IRIS data to obtain meridional temperature profiles at three pressure levels in the upper troposphere just after northern spring equinox. At 150 mbar, they found a hemispheric asymmetry, with the north pole $\sim$5 K colder than the equator, but no large-scale gradient in the south. Superimposed on the large-scale gradients were smaller variations of $\sim$2 K which are correlated with the zonal jets. The temperature variations become weaker with depth, disappearing by 700 mbar where the temperature is uniform with latitude except for a warm region near 30°N.

Ground-based maps of emission from stratospheric ethane and methane bands during early northern summer by Gezari et al. (1989) showed emission increasing from the equator to the north (summer) pole, as well as a narrow band of enhanced emission at the equator, but they did not attempt to retrieve temperatures from their data. Later,

Ollivier et al. (2000) imaged Saturn in several stratospheric and tropospheric bands during late northern summer. They also saw enhanced emission at north polar latitudes in the stratosphere, but not in the troposphere. Greathouse et al. (2005) used high spectral resolution observations in the methane ν_4 band to retrieve stratospheric temperatures at southern summer solstice. They found a general equator-to-south-pole temperature gradient, with the south pole ~ 10 K warmer than the equator and the gradient slightly stronger at lower pressures. Orton and Yanamandra-Fisher (2005) made high spatial resolution thermal infrared images of Saturn in the stratospheric methane and upper tropospheric hydrogen bands during early southern summer, allowing them to determine temperatures at 3 mbar and 100 mbar. The 3 mbar stratospheric temperatures show a ~ 15 K temperature gradient from the equator to the south pole, with a sharp temperature increase at 70°S. The tropospheric temperatures show variations of ~ 3 K on the scale of the zonal jets, with no clear equator to pole gradient. They also found a hot spot at the pole, with a sharp 2.5 K temperature increase between 87°S and the pole (Section 6.4.5).

Cross-sections of southern hemispheric temperatures from Cassini CIRS data were presented by Flasar et al. (2005). Temperatures from CIRS mapping observations were retrieved by Fletcher et al. (2007b), giving a latitude-pressure cross section of temperatures in the upper troposphere and middle stratosphere during southern mid-summer (Figure 6.8). There is a well-defined tropopause at ~ 80 mbar, separating a strongly statically stable stratosphere with temperatures increasing with altitude, from a troposphere with temperatures increasing with depth. In the upper troposphere, the temperature gradient increases with depth down to approximately 400–500 mbar, where the gradient becomes nearly dry adiabatic (Lindal et al. 1985; Fletcher et al. 2007b). This transition to the adiabat likely indicates the radiative-convective boundary, with temperatures and dynamics at higher pressures determined primarily by convection, and the temperatures at lower pressures determined by solar heating and the solar driven circulation. Stratospheric temperatures at 1 mbar show a strong pole to pole gradient, with the south (summer) pole almost 40 K warmer than the north (winter) pole. The hemispheric temperature asymmetry weakens with increasing pressure, with the winter hemisphere ~ 10 K colder than the summer hemisphere in the upper troposphere, and disappears at pressures greater than about 500 mbar where the temperature becomes nearly uniform with latitude. In addition to the large scale equator-to-pole temperature gradients, between ~ 2–300 mbar there are temperature variations of 2–3 K on the scale of the zonal jets. Outside the equatorial region, the temperature gradients are correlated with the mean zonal winds, with warmer temperatures where the winds are cyclonic, and colder temperatures where the winds are anticyclonic (Conrath and Pirraglia 1983; Fletcher et al. 2007b, 2008). Temperatures in the equatorial region have been observed to oscillate with a period of ~ 15 Earth years (Orton et al. 2008; Section 6.3.7).

The observed hemispheric-scale temperatures and their temporal variations are qualitatively consistent with seasonal radiative-dynamic models of Saturn (Cess and Caldwell 1979; Bézard et al. 1984; Bézard and Gautier 1985; Barnet et al. 1992; Section 6.3.7). Seasonal variations are largest in the stratosphere, where the timescale for response to radiative forcing is about one Saturnian season. The radiative timescale increases at higher pressures, reaching a Saturn

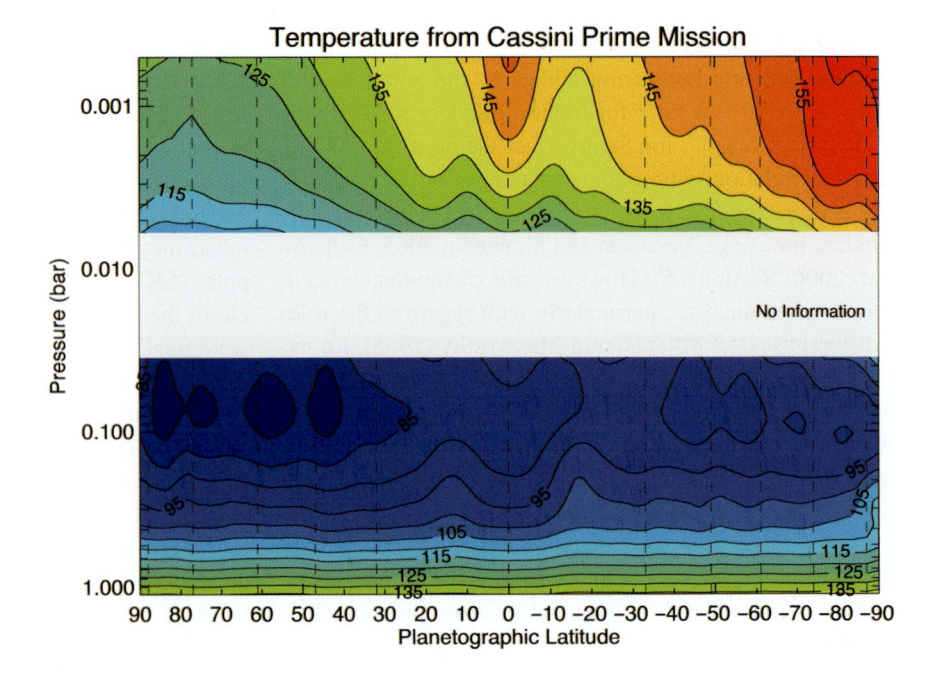

Fig. 6.8 Cross-section of temperature retrieved from Cassini CIRS nadir observations during the prime mission. The region between 5 and 50 mbar is not shown since the spectra are not sensitive to temperatures in that pressure range. Vertical dashed lines indicate the latitudes of the prograde (*eastward*) jets. (after Fletcher et al. 2007b)

year in the troposphere, resulting in a decrease in the seasonal temperature variations and an increasing lag between the solar forcing and the temperature response. The magnitude of the hemispheric differences appears to be larger than can be fully explained radiatively, however; this has implications for the stratospheric circulation.

For an atmosphere in geostrophic balance, where the Coriolis acceleration is balanced by pressure gradients, the meridional temperature gradient is related to the vertical wind gradient through the thermal wind equation:

$$\frac{du}{d \ln p} = \frac{R}{f} \frac{dT}{dy}$$

where T is temperature and R is the gas constant. Conrath and Pirraglia (1983) found that outside the equatorial region, the temperature gradient at 150 mbar in the Voyager IRIS data is anti-correlated with the measured cloud-top zonal wind velocities. The anti-correlation has also been seen in temperatures from Cassini CIRS (Flasar et al. 2005, Fletcher et al. 2007c). Application of the thermal wind equation then indicates that the zonal winds decay with altitude over about 5 scale heights.

To explain the decay of the zonal winds with altitude, Conrath and Pirraglia (1983) proposed a simple model of a meridional circulation that is mechanically forced by the zonal jets; the model was later expanded upon by Gierasch et al. (1986) and Conrath et al. (1990). In this model, turbulent or large-scale eddy processes acting on the mean zonal flow, parameterized by Rayleigh friction, are balanced by Coriolis acceleration acting on the meridional velocities. Mass continuity requires a corresponding vertical velocity. Adiabatic heating or cooling from the vertical motion is balanced by radiative damping parameterized as Newtonian cooling to an equilibrium temperature profile. The resulting meridional circulation has rising motion, and cold temperatures, on the equatorward side of eastward jets, and subsidence and warm temperatures on the poleward side of eastward jets, as is observed. Conrath et al. (1990) also found that the amplitude of the temperature variations is consistent with the temperature observation when the radiative and frictional timescales are approximately equal. Pirraglia (1989) and Orsolini and Leovy (1993a, b) have proposed a model in which the frictional damping corresponds to eddy fluxes induced by large-scale instabilities of the zonal jets. This has been recently extended to a semi-geostrophic theory by Zuchowski et al. (2009) which allows for the attribution of meridional transport circulations respectively either to frictional damping corresponding to eddy momentum fluxes or to radiative imbalances around or above the visible clouds.

Information on the tropospheric circulation may also be obtained from the distribution of disequilibrium chemical species. Meridional cross-sections of the para-hydrogen fraction f_p in the upper troposphere have been obtained from Voyager IRIS (Conrath et al. 1998) and Cassini CIRS (Fletcher et al. 2007c) data. The IRIS data did not have the spatial resolution to fully resolve the jet structure, but Conrath et al. (1998) did find a minimum of f_p near 60°S and a maximum near 15°S, roughly coincident with a local maximum and minimum in the temperatures, suggestive of upward motion near 60°S and downward motion at 15°S. The CIRS data, with higher spatial resolution, show a local minimum in f_p in a narrow band at the equator, with local maxima at ±15° but stronger in the south. This is suggestive of upwelling at the equator and subsidence at ±15°, which is consistent with the circulation implied by the temperature field. Fletcher et al. (2007a, c) used CIRS data to retrieve the upper tropospheric phosphine abundance in Saturn's southern hemisphere. At equatorial and mid-latitudes phosphine around 250 mbar is approximately anti-correlated with the temperature, consistent with the jet-scale variations in both being driven by vertical motions, except that there is no minimum in phosphine corresponding to the temperature maximum near 15°S. They also found a strong enhancement in the phosphine abundance between 60°S and 80°S for pressures less than 500 mbar. Studies of the upper tropospheric and stratospheric haze also show a change in the properties of the haze particles poleward of ∼60°S (Karkoschka and Tomasko 2005; Pérez-Hoyos et al. 2005), suggesting that the phosphine enhancement in the polar region may be produced by changes in the chemistry as well as the dynamics.

6.3.5 Temporal Variation of the Equatorial Jet

At most latitudes Saturn's zonal winds have been remarkably stable over time. The one major exception is the prograde equatorial jet (Fig. 6.9). Voyager green filter images yielded a peak zonal wind of ∼450 m s^{-1} in the System III reference frame in 1980–1981 (Sánchez-Lavega et al. 2000). However, a major outbreak of convective storms was triggered in 1990 (Section 6.4.2). The results of Sánchez-Lavega et al. (2004b), Temma et al. (2005), and Pérez-Hoyos et al. (2006) suggest that the level of tracked equatorial cloud/haze features rose from ∼200–360 mb at the time of Voyager to ∼45–70 mb after 1990. Barnet et al. (1992) found that winds measured by different HST filters sensitive to different altitudes at this time had different speeds, suggesting the presence of vertical shear in the zonal wind. HST measurements after the 1990 event and over the period 1996–2004 gave systematically weaker equatorial wind speeds of ∼275 m s^{-1}, independent of wavelength (Sánchez-Lavega et al. 2003, 2004b). Cassini ISS wind speeds in 2004 CB2 continuum images yielded equatorial wind speeds intermediate between the Voyager and HST results (∼325–400 m s^{-1}), but moderate strength

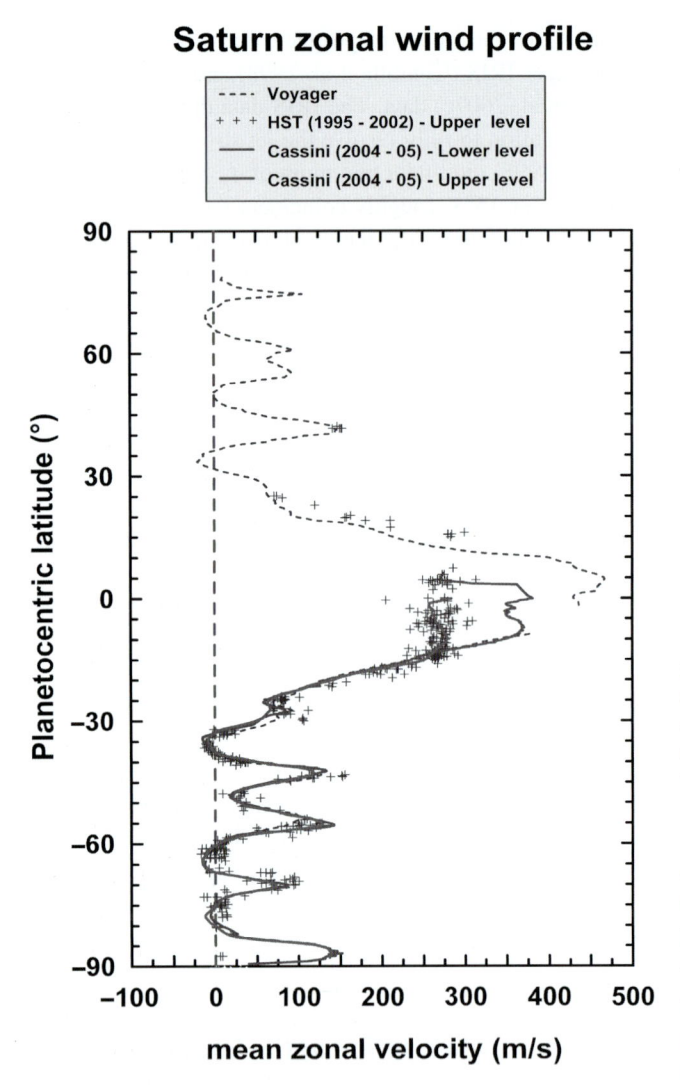

Fig. 6.9 Comparison of Voyager, HST, and Cassini ISS wind profiles derived from images in filters sensing different vertical levels at different times (Sánchez-Lavega et al. 2000, 2003, 2007)

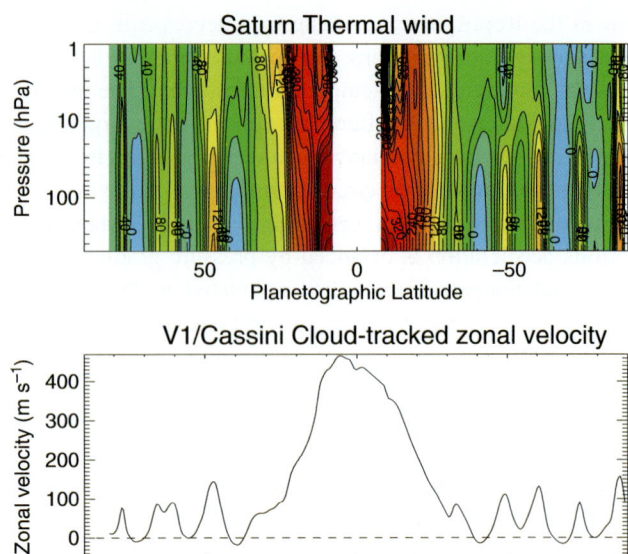

Fig. 6.10 Thermal winds derived from CIRS retrieved temperatures in the upper troposphere and stratosphere (*upper panel*) and their relationship to cloud-tracked winds at the visible cloud top derived from the results of Godfrey (1988), Vasavada et al. (2006), and Sánchez-Lavega et al. (2000, 2006) (*lower panel*) (Read et al. 2009)

methane band filter (MT2) images, which sense higher altitudes, gave speeds similar to those determined from HST (Porco et al. 2005; Sánchez-Lavega et al. 2007).

To what extent can these differences be attributed to real temporal wind changes, as opposed to changes in the altitude at which features are tracked in the presence of a vertically varying, but constant in time, zonal wind? One constraint comes from CIRS thermal winds (Flasar et al. 2005; Section 6.3.4; Fig. 6.10). If the Voyager images are assumed to sense cloud features near 500 mb, then the thermal winds imply that Cassini ISS MT2 winds of 275 m s^{-1} would need to be characteristic of levels near ∼3–30 mb. This is probably outside the uncertainties in modeling estimates of the haze altitude, suggesting that although a large portion of the apparent change may be due to an upward shift in seeing level after 1990, an additional component of real temporal wind change may be present as well. The VIMS zonal wind profile (Choi et al. 2009; Fig. 6.7), which senses the ammonium hydrosulfide cloud level, suggests that at deeper levels equatorial zonal winds are as strong as or slightly stronger than those retrieved in the visible during the Voyager era. Thus, any real temporal changes in the wind must be restricted to the upper troposphere and lower stratosphere.

Several possible effects of the 1990 convective storm development on the equatorial jet were investigated using a numerical model by Sayanagi and Showman (2007). Convection was represented in this model by the injection of mass at specified detrainment levels. Geostrophic adjustment to the mass anomaly produced an anticyclonic circulation which, in the presence of a strong β effect, spread the energy longitudinally by wave radiation. The resulting westward-propagating Rossby waves generated by the storm decelerated the jet at their level of critical layer absorption, sometimes reaching the 10 mbar level. In some of their experiments, latitudinal mixing of potential vorticity produced a local minimum in the zonal wind at the equator, a feature common to the jovian planets (Allison et al. 1995). The net result, however, was a decrease in the equatorial jet by only tens of m s^{-1} despite the excessive magnitude of the mass injection assumed.

Other mechanisms not explored by Sayanagi and Showman might further contribute to jet changes. For example, mesoscale gravity waves generated at the tops of thunderstorms are important to the momentum budget of Earth's stratosphere and mesosphere (Section 6.3.7). Convective clusters also transport momentum directly, often but not always downgradient; since Saturn's zonal winds decrease upward from the ammonium hydrosulfide cloud level, this effect would actually accelerate upper level winds instead. Thus, while the enhanced convection that occurred after 1990 is likely to have had some decelerating effect on upper level equatorial winds, it is likely to be too small an effect to explain by itself the observed cloud tracked wind differences from the Voyager to the HST to the Cassini eras. Additional long-term changes in cloud altitude, combined with the known thermal wind shear, must probably be invoked to explain the detected changes (Pérez-Hoyos et al. 2006).

Regardless of the contributions to the cloud tracking results by wind changes and cloud/haze height changes, we are still left with the mystery of why Saturn's equatorial region is more variable than the rest of the planet. The 1990 onset of the GWS disturbance is consistent with long-term behaviour that suggests a connection to the seasonal cycle (Section 6.4.2). Saturn's large obliquity gives it greater seasonality than Jupiter, and shadowing by the rings can enhance the seasonality at low latitudes. However, sunlight penetrates not much deeper than ∼1 bar on Saturn, while GWS-type storms must be driven by water convection originating at ∼10–15 bars (Section 6.4.2). Furthermore, Bézard et al. (1984) estimate that seasonal effects cause temperature to vary by 1.5 K or less in Saturn's upper troposphere. This modest upper-level destabilization ($<1\,\mathrm{K\,km^{-1}}$ in the model of Barnet et al. 1992) would seem to be incapable of causing convective penetration depths to vary by more than a few kilometres. Finally, if seasonal ring shading is the forcing mechanism for the observed variability, we are left with the question of why the time scale appears to be annual rather than the semi-annual time scale on which one or the other half of the equatorial region should be destabilized. If a longer record of observations eventually suggests semi-annual variability instead, then perhaps the tropospheric convection variations seen near the equator are related to the possible semi-annual oscillation observed in the stratosphere (Orton et al. 2008; Fouchet et al. 2008; Section 6.3.7).

6.3.6 Upper Troposphere Temperature Knee: Structure and Seasonality of Solar Heating

An interesting feature of Saturn's temperature structure is a "knee" in the vertical temperature profile in the upper troposphere, first noted in temperature retrievals by Voyager IRIS at near-equatorial and southern latitudes (Hanel et al. 1981). The knee was also seen in temperature profiles from Voyager radio occultations at 3°S and 74°S (Lindal et al. 1985).

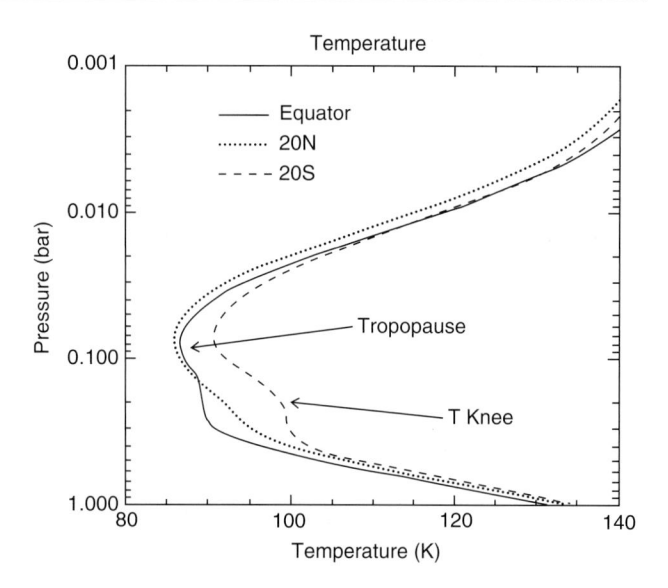

Fig. 6.11 Vertical temperature profiles in the tropopause region at 20°N (*dotted line*), equator (*solid line*) and 20°S (*dashed line*) from Cassini CIRS data, showing a knee in the upper tropospheric temperature profile (Fletcher et al. 2007b)

Fletcher et al. (2007b) used Cassini CIRS data from 2004 to 2006, during southern mid-summer, to map the strength and pressure of the knee as a function of latitude (Figure 6.11). The structure of the knee shows strong hemispheric variations, being very weak or absent in the northern (winter) hemisphere north of 20°N. The knee is higher, smaller, and weaker at the equator and south (summer) pole than at mid-latitudes, and has local maxima in strength at ±15°.

Pérez-Hoyos et al. (2006) calculated heating rates from absorption of sunlight in the visible (0.25–1 μm) for a range of plausible haze models for different latitudes and seasons. For all of their models, tropospheric heating was confined to a narrow pressure region just below the tropopause. Along with the hemispheric asymmetry in the knee, this led Fletcher et al. (2007b) to conclude that the knee is primarily a radiative effect. The latitudinal variations of the knee are then the result of latitudinal and seasonal variations in insolation, and the distribution and properties of the absorbing aerosols.

The large-scale variations in the atmospheric heating, and thus the structure of the temperature knee, can be partially explained by the latitudinal variation in insolation. In addition, Karkoschka and Tomasko (2005) found that the tropospheric aerosols are larger in the summer hemisphere than the winter hemisphere. This is consistent with the greater northern hemisphere visibility seen in ISS and VIMS images (Figs. 6.2–6.4). Furthermore, Barnet et al. (1992) found that their seasonal radiative-dynamical model could match the upper temperature structure observed by Voyager IRIS without any heating from aerosols in the northern hemisphere, but aerosol heating was required in the southern hemisphere to match the data.

6.3.7 Stratospheric Circulation

6.3.7.1 Meridional Circulation

The mean meridional circulation in Saturn's stratosphere has primarily been inferred from observations of the temperature field and the distribution of gaseous constituents. The heat equation is:

$$\frac{\partial T}{\partial t} + v\frac{\partial T}{\partial y} + w\left(\frac{\partial T}{\partial z} + \frac{g}{C_p}\right) = \frac{J}{C_p} \approx \frac{T_e - T}{\tau_r},$$

where v, w are the meridional and vertical velocities, respectively, y is the northward coordinate, C_p is the specific heat, and J is the net radiative heating. J is often parameterized in terms of a Newtonian cooling rate specified by an equilibrium temperature T_e and a radiative relaxation time, τ_r. In the (stably stratified) stratosphere the (second) horizontal advective term on the left-hand side is much smaller than the (third) vertical advective term. The thermodynamic equation does not contain divergences of eddy heat fluxes, because v, w are not Eulerian velocities discussed earlier; instead they correspond to the residual transformed Eulerian-mean circulation (see, e.g., Andrews et al. 1987). The use of these variables, common in middle-atmospheric studies, is based on the existence of so-called "non-interaction theorems," which note that for steady, inviscid flow the eddy and wave fluxes of heat and momentum induce mean meridional circulations that cancel these transports (Charney and Drazin 1961; Eliassen and Palm 1961; and Andrews et al. 1987). Some transience and dissipation exist in all real atmospheres, but these are often weak enough that the residual circulation provides a more meaningful picture of momentum and heat transports. The thermodynamic equation above is relatively simple and provides a means of probing zonal-mean vertical velocities from the observed temperature field. Dunkerton (1978) has shown that the residual mean circulation in Earth's middle atmosphere is equivalent to the Lagrangian circulation, which describes the transport of quasi-conserved constituents.

Unlike the deep troposphere, the radiative time constant in the stratosphere is sufficiently short ($\sim$9.5 yr) that large seasonal modulation of temperatures is expected (Conrath et al. 1990) and observed (Conrath and Pirraglia 1983). However, the thermal contrast between low and high southern latitudes during early southern summer (Flasar et al. 2005; Fletcher et al. 2007b), is much greater than expected from simulations of the atmospheric radiative response assuming a uniform distribution of opacity (Bézard and Gautier 1985). The CIRS observations indicated that 1-mbar temperatures near the south pole were 15 K higher than at low latitudes (Fig. 6.8), whereas the simulations predicted only 5 K in early southern summer. While the effects of the variation in opacity with latitude from hazes and gaseous coolants have yet to be modeled, Flasar et al. (2005) suggested that weak subsidence over the south polar region could also adiabatically heat the stratosphere there. Assuming a balance between the first and third terms in the heat equation, above, they estimated a subsidence velocity $\sim$0.01 cm s^{-1}. Later observations of the south-polar region at higher spatial resolution by CIRS, VIMS, and ISS showed a very compact region with a hot spot extending down to the troposphere, which was also consistent with subsidence over the south pole (Section 6.4.5).

So far, only the meridional distributions of ethane (C$_2$H$_6$) and acetylene (C$_2$H$_2$) have been mapped in detail in Saturn's stratosphere, at southern latitudes from CIRS nadir-viewing observations (Howett et al. 2007). Perhaps the most striking feature of the meridional profiles is the twofold enhancement in the C$_2$H$_6$ abundance at 2 mbar as one moves from mid to high southern latitudes (Fig. 6.12). C$_2$H$_2$ actually decreases over this range. Ground-based observations at the Infrared Telescope Facility (Hesman et al. 2009) confirm this general pattern but also indicate a localized south pole acetylene peak. Since these hydrocarbons are formed at higher altitudes than probed by CIRS, their mixing ratios should increase with altitude. Thus, one might ascribe the enhancement in C$_2$H$_6$ to subsidence in the south polar regions, as already suggested by the temperature field. However, the potential problem with this is that one would expect C$_2$H$_2$ to show a similar increase toward the south pole, because of the short time scales that characterize the photochemical link between the two species (see Chapter 5). This tight coupling in fact occurs in a 2D chemistry-climate model of Saturn's stratosphere (Moses et al. 2007). Hence the derived distribution of C$_2$H$_6$ and C$_2$H$_2$ is currently not well understood.

6.3.7.2 Equatorial Oscillations

Oscillations in middle-atmospheric zonal-mean temperatures and zonal winds at low latitudes are common on rapidly rotating planets. These have been best studied on Earth, where two major types have been identified. The quasi-biennial oscillation (QBO), dominant in the lower stratosphere, exhibits downward propagating layers of eastward and westward winds (and, from the thermal wind equation, associated warm and cold temperatures at the equator). The period is variable, but averages to $\sim$28 months. The source of this structure is from momentum stresses associated with vertically propagating waves. The wave stresses induce meridional circulations in the equatorial region, and the adiabatic heating and cooling associated with the vertical motions produce the warm and cold temperature anomalies observed to vary with altitude and time. For this oscillation to exist, a set of atmospheric waves including modes with both easterly and westerly phase velocities is needed. Because

Fig. 6.12 Latitudinal profiles of 2 mbar ethane (*upper*) and acetylene (*lower*) abundance derived from CIRS data (from Howett et al. 2007)

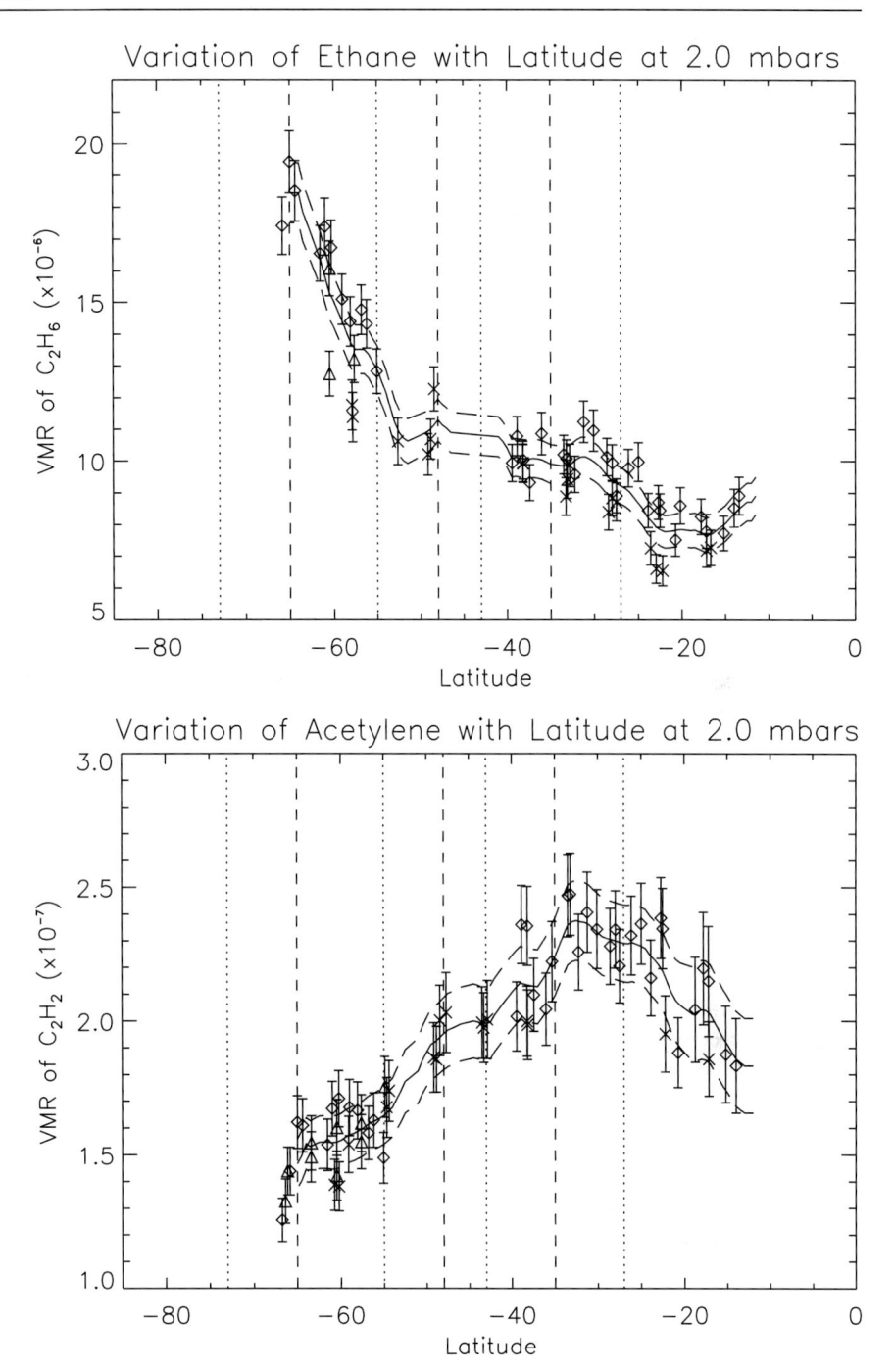

these waves can transfer zonal momentum, respectively, in the eastward and westward directions, this is a fundamental requirement. Which waves are the key players is a complicated story (see, e.g., Hamilton 1998; Baldwin et al. 2001). The candidates range from planetary-scale to mesoscale: eastward-propagating Kelvin and westward-propagating Rossby-gravity waves have been studied, as well as inertial-gravity waves and gravity waves propagating in both directions. All may well contribute to the QBO on Earth, directly through the zonal momentum deposited in the mean flow, and also by the mean meridional circulations they induce, which transport angular momentum and heat. There is evidence that the QBO influences the Earth's atmosphere at extratropical latitudes. One reason is simply that the induced vertical motions near the equator are associated with meridional cells that close at higher latitudes. Beyond this, the altered pattern of eastward and westward winds at low latitudes modifies the propagation of atmospheric waves. For example, topographically forced (i.e., stationary) planetary Rossby waves cannot propagate in regions where the zonal winds are westward.

Waves propagating at low-to-mid latitudes in the winter hemisphere would be refracted poleward, where they can deposit their zonal momentum into the mean flow.

The other major equatorial oscillation in the Earth's stratosphere and mesosphere is the semi-annual oscillation (SAO), and it has a well defined period. It dominates in the terrestrial mesosphere. Forcing by wave-generated momentum stresses is still a critical ingredient of the oscillation, but the cross-equatorial meridional circulation from the summer pole to the winter pole also appears to be important. This may account for the more regular nature of the SAO, effectively synchronizing the natural wave-driven oscillation to a simple fraction of the seasonal cycle.

Ground-based observations of Jupiter identified an oscillation in stratospheric equatorial temperatures, relative to those at adjacent latitudes, with a cycle of 4–5 years (Orton et al. 1991). By analogy with Earth, studies were undertaken to explain this as a quasi-quadrennial oscillation (QQO), based on wave-forcing mechanisms analogous to those for the QBO (Leovy et al. 1991; Friedson 1999; Li and Read 2000). During the Cassini swingby of Jupiter in 2000, CIRS nadir-viewing observations (Flasar et al. 2004a) allowed the retrieval of the meridional cross section of stratospheric (from ~1 mbar) and tropospheric (down to 400 mbar) temperatures. The temperatures showed a colder equator at the tropopause near 100 mbar, warmer near 10 mbar, and colder higher up near 1 mbar. Associated with this, the zonal wind first decayed with altitude at the tropopause, and then higher up increased again, leading to a 140 m s^{-1} eastward jet near 4 mbar that had not been previously identified. This vertical structure was broadly consistent with a wave-forced equatorial oscillation, like the terrestrial QBO.

More recently, more than two decades of ground-based observations have identified an equatorial oscillation on Saturn, with a period of ~15 years (Fig. 6.13; Orton et al. 2008). Because Cassini is currently in orbit about Saturn, CIRS observations near periapsis could be undertaken in the limb-viewing mode, providing middle atmospheric temperatures with relatively high vertical resolution and extending quite high into the atmosphere, ~ a few μbar (Fouchet et al. 2008) Fig. 6.14 shows a meridional cross section of temperatures and the zonal winds, relative to those at 10 mbar, below which the mid-infrared limb observations did not probe. One sees the characteristic reversal of relative temperatures and winds with altitude. The limb observations do not constitute a zonal average, but more complete coverage of the stratosphere at 1–3 mbar from nadir-viewing observations indicates that zonal variations at this level are small (Flasar et al. 2005; Fletcher et al. 2007c). Note that the illustrated winds overlay the much larger winds inferred at the cloud tops from feature tracking (Porco et al. 2005), so the wind field does not reverse, as it does on Earth. Nonetheless, it is plausible that some wave-forcing mechanism analogous to the terrestrial SAO or QBO is at play. One 3-D primitive equation model with prescribed equatorial forcing, for example, produces upward propagating Kelvin waves that might be relevant to the Saturn SAO (Yamazaki et al. 2005). However, the Saturn oscillation needs to be better characterized. It is hoped that the expected descent of the temperature and wind field can be measured, which may be possible if Cassini can observe long enough.

Features associated with these zonal flow oscillations may also be visible in compositional anomalies produced by changes in vertical and latitudinal transport. The meridional cross section of C_2H_6 (Fig. 6.15), for example, shows an enhanced tongue extending downward at 20–30°N. Because C_2H_6 is expected to form higher up in the atmosphere and because it is relatively long-lived, it should be a good

Fig. 6.13 Time series of brightness temperatures at 7.8 μm (CH_4) and 12.2 μm (C_2H_6) obtained from ground-based observations. The temperatures are the difference of those at latitudes 15.5° and 3.6°. A sinusoidal best fit of the individual data (*black curve*) has a period of 15.6 years; the corresponding best fit of the annual average (*dashed curve*) yields a period of 15.0 years. After Orton et al. (2008)

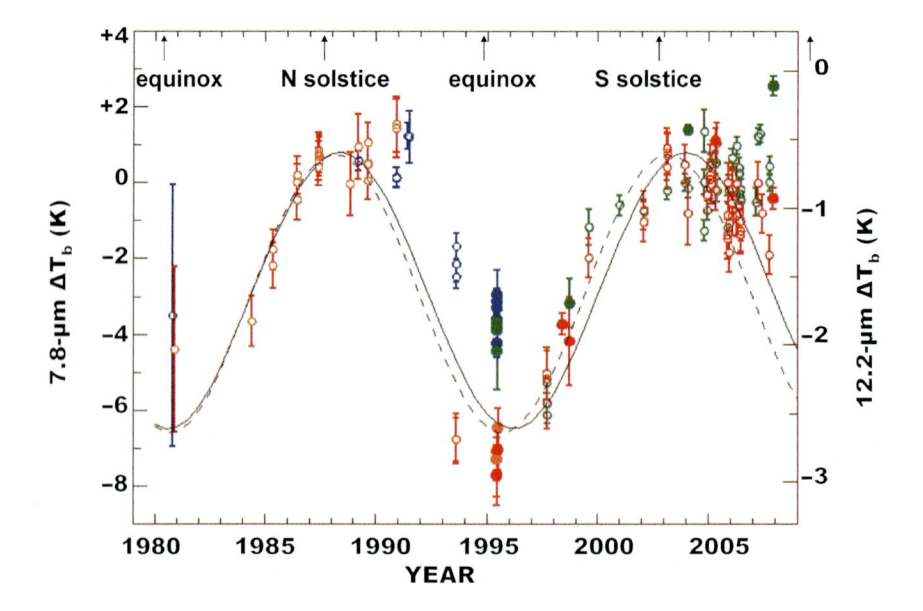

Fig. 6.14 Upper: Meridional cross section of temperatures (K) obtained from CIRS limb sounding in the mid infrared. Note that the equator is alternatively colder and warmer with altitude relative to adjacent latitudes. Lower: Zonal winds, computed from the thermal wind equation along cylinders concentric about Saturn's rotation axis. The winds are relative to the winds at the 20-mbar (hPa) level. Positive winds are eastward. The dashed parabola encloses a zone of exclusion not accessible from the thermal wind equation. Winds within this region have been interpolated along isobars. After Fouchet et al. (2008)

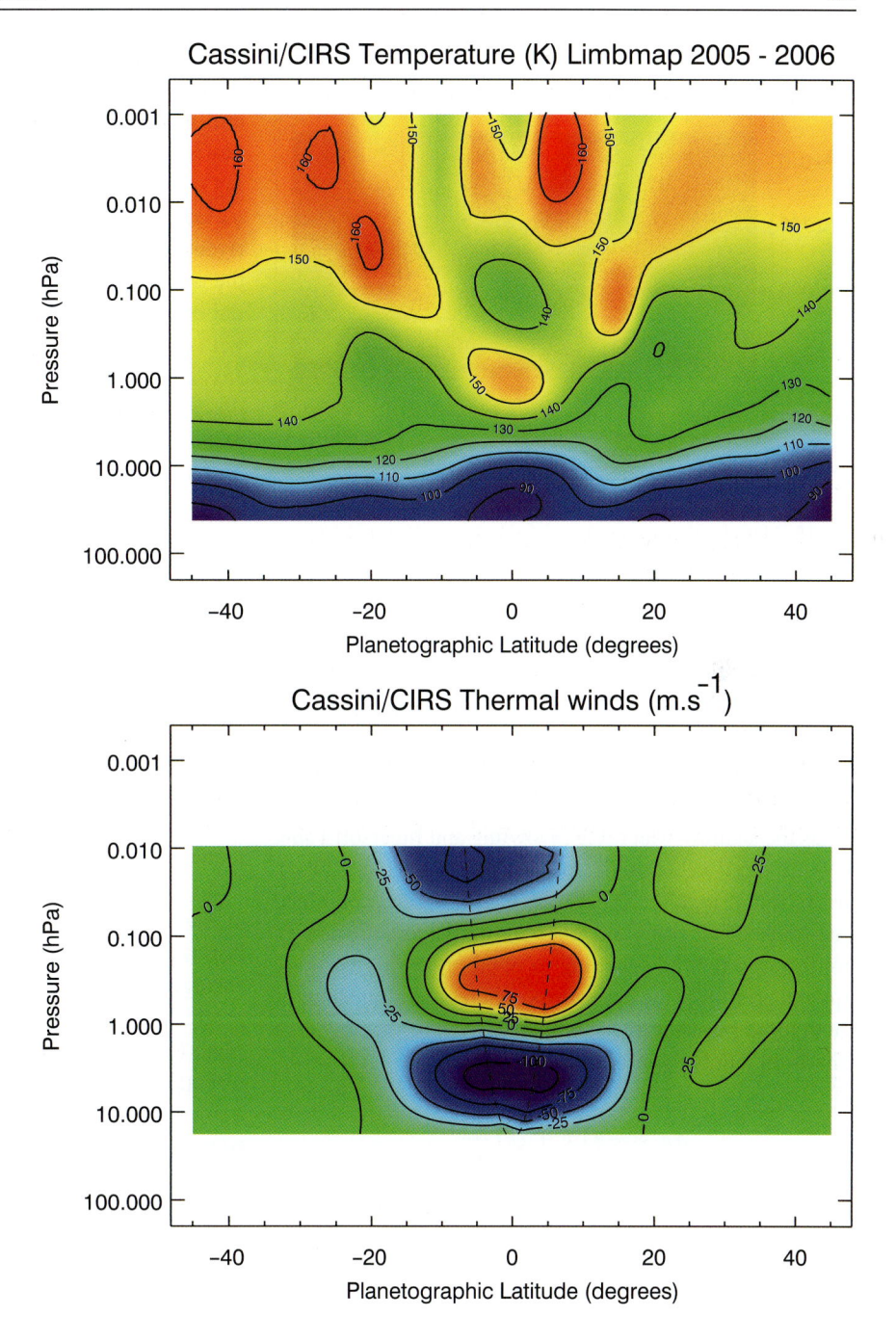

tracer of motion. This suggests sinking motion in this region. This may be associated with the circulations induced by the equatorial oscillation. That happens on Earth, where sinking motion is often observed at mid-latitudes in the winter hemisphere (Baldwin et al. 2001).

6.3.8 Potential Vorticity Diagnosis

Potential vorticity (PV) is one of the most fundamental variables in dynamical meteorology and oceanography, whose behaviour is crucial to understanding a wide variety of fluid dynamical phenomena. The Ertel form is the most fundamental, defined in terms of potential temperature θ as

$$q_E = \frac{(2\Omega + \xi) \cdot \nabla \theta}{\rho},$$

where ρ is density, and is formally conserved in frictionless, adiabatic flow *provided the composition of the air is uniform* (Gierasch et al. 2004). Unfortunately the latter is not strictly the case in the upper tropospheres of the gas giant planets, because the ortho-para ratio of molecular hydrogen

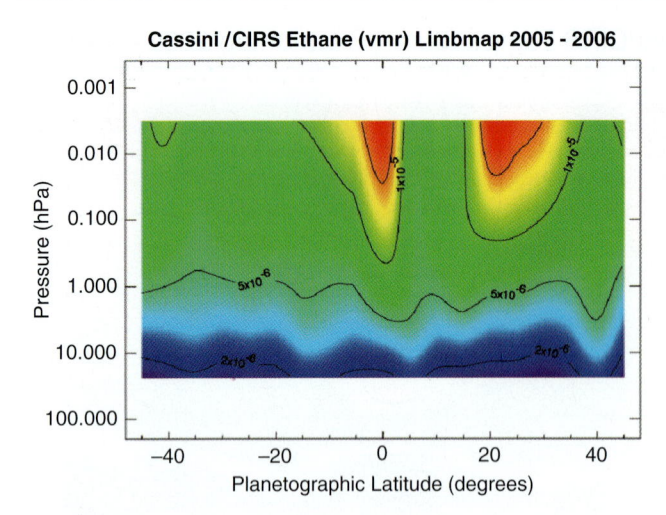

Fig. 6.15 Meridional cross section of C_2H_6 from CIRS mid-infrared limb sounding

varies across the planet. In principle, this can have significant consequences where the temperature leads to significant differences in physical properties between ortho- and para-hydrogen.

In practice, various approximations to q may be used as the basis for measuring potential vorticity from observations, such as the shallow-water (e.g., Dowling and Ingersoll 1988, 1989) or quasi-geostrophic forms (Gierasch et al. 2004). Most recently, Read et al. (2006a, b, 2009) have combined cloud-tracked velocities and retrievals of temperature and f_p from infrared remote sounding to obtain potential vorticity maps and profiles in the upper troposphere and stratospheres of both Jupiter and Saturn. This approach uses either the large Richardson number approximation to q_E,

$$q_E \approx -g\,(f + \xi_\theta)\,\frac{\partial \theta}{\partial p}$$

(where ξ_θ and $\partial\theta/\partial p$ are evaluated on surfaces of constant θ) or the quasi-geostrophic form (Gierasch et al. 2004)

$$q_G \approx f + \xi_p - \left(\frac{\partial}{\partial p}\left[\frac{pT_d(x,y,p)}{s(p)T_s(p)}\right]\right),$$

where

$$s = -\frac{pc_{p0}}{\langle c_p(p)\rangle}\frac{\partial \langle \ln\theta \rangle}{\partial \ln p}$$

is the stability parameter, defined in terms of potential temperature, $T_s(p)$ is the horizontal mean reference temperature profile, $T_d(x,y,p) = T - T_s$ and c_p is the specific heat capacity of the air. Angle brackets denote horizontal means. A possible advantage of q_G is that it is formally conserved (to O(Ro)) in the absence of diabatic effects or friction, even if f_p varies in space and time.

Typical profiles of q_E on Saturn with latitude largely reflect the behavior of zonal mean absolute vorticity, and generally show a large-scale increase from large negative values in the south to large positive values in the north, crossing zero near the equator. Figure 6.16 shows a profile of q_E at $\theta = 160\,\text{K}$, just below the Saturn tropopause, derived from Cassini data (Read et al. 2009). Superimposed on the northward increasing global trend, however, are substantial undulations, with steep positive gradients of q_E aligned with eastward jets and much weaker, or even negative, gradients aligned with each westward jet. Such local reversals in $\partial q_E/\partial y$ become most prominent at high latitudes, where the planetary vorticity gradient (β) is weakest.

The implications of these reversals for the stability of the flow are the subject of controversy, since current models disagree over whether such persistent reversals are to be expected in geostrophically turbulent flow. But the recent results using the best available velocity and thermal measurements for both Jupiter and Saturn do suggest that these reversals in $\partial q_E/\partial y$ are robust features of the circulations of both planets, at least in their troposphere. Taken at face value, however, this would seem to imply the possibility of barotropic instability around those latitudes where the reversals occur, on the flanks of the eastward jets, while Rossby wave propagation and inhibited meridional transport is anticipated around the eastward jet peaks.

In the north, reversals or near-reversals of $\partial q_E/\partial y$ occur at the latitudes of Saturn's north polar hexagon and "ribbon wave" at 76°N and 47°N, for example. But similar (or stronger) reversals are also found in the south where no such prominent wave-like disturbances occur (although Vasavada et al. (2006) do comment on more localised polygonal perturbations to the jets at 47°S and 62°S, which might indicate a role for barotropic instability, suggested by reversals in the sign of $\partial q_E/\partial y$).

Although an absence of reversals in the sign of $\partial q/\partial y$ would imply stability, the existence of a sign reversal does not necessarily imply instability. Dowling (1995) has drawn attention to the potential relevance of the stability theorems of Arnol'd (1966), the second of which (hereafter, Arnol'd II) emphasizes a flow in which $\partial q/\partial y$ and $\bar{u}$ are linearly related as representing a state of neutral stability. Such a neutrally stable flow may exhibit reversals in the sign of $\partial q/\partial y$ provided these are accompanied by corresponding reversals in $\bar{u}$ *in an appropriate reference frame*. The latter condition is necessary to ensure that Rossby wave propagation relative to the local zonal flow precludes sustained, coherent interactions between adjacent latitude bands. Dowling (1995) presented evidence to suggest that Jupiter's atmosphere approached this state of neutral stability with respect to Arnol'd II quite closely. This was subsequently confirmed using fully stratified PV by Read et al. (2006a).

Fig. 6.16 Profiles of Ertel potential vorticity (q_E, *solid line*) and mean zonal wind (u, *dashed line*) at $\theta = 160$ K in the upper troposphere, obtained by Read et al. (2009) from Cassini imaging and CIRS data. q_E is normalised by the horizontal mean value of $g\partial\theta/\partial p$ to give a potential vorticity in units of s^{-1}

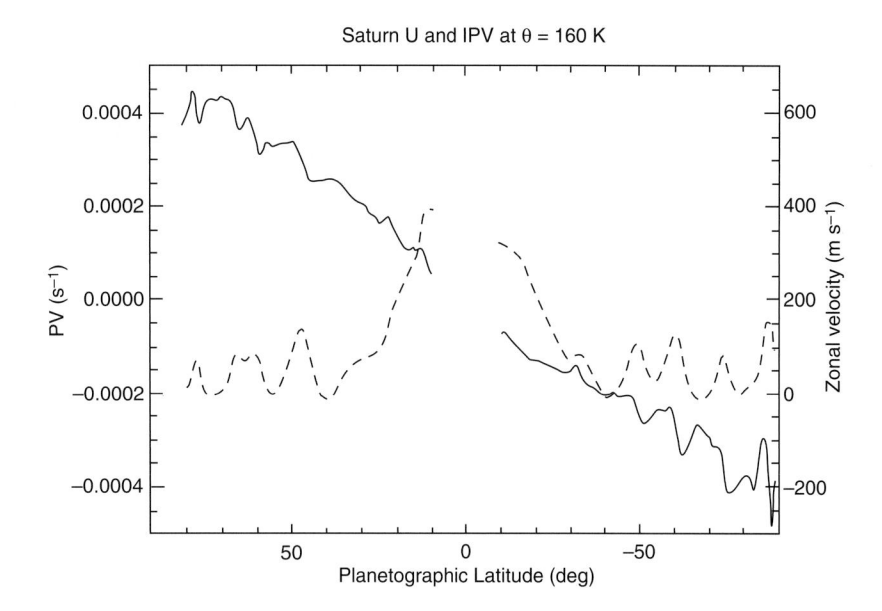

For Saturn, the PV distributions derived by Read et al. (2009) indicate that its atmosphere seems to approach this neutrally stable state even more closely than on Jupiter, with clear evidence of a locally linear relationship between $\bar{u}$ and $\partial q_G/\partial y$,

$$\bar{u} - \alpha \approx L_d^2 \frac{\partial \bar{q}}{\partial y},$$

(where α is a constant; see e.g., Dowling 1995) at almost all latitudes except close to the equator. The gradient of this relationship provides a measure of the (square of the) local baroclinic Rossby deformation radius which, in Saturn's upper troposphere, appears to range between around 1,500 km at high latitudes and more than 6,000 km in the sub-tropics. An intriguing additional result of this analysis is the suggestion from the $\bar{u}$-intercept that the reference frame for Rossby wave dynamics on Saturn at most latitudes may be quite different from the System III determined by the Voyager missions, and may even approach the interior rotation period determined by Anderson and Schubert (2007) of $10^h\ 32^m\ 35^s$, a frame that moves prograde relative to System III at more than 100 m s^{-1} at the equator (Section 6.2.1). Use of such a 'true' reference frame would render Saturn's zonal velocity profile somewhat less anomalous compared with Jupiter, with mid- and high-latitude jets then alternating in sign with latitude with roughly comparable westward and eastward amplitudes. The effect of this proposed alternative interior reference frame on the dynamics of wave propagation and instability in the upper troposphere of the outer planets has been compared by Dowling (1995) with the story of the 'princess and the pea', which seems an apt analogy with the notion that such waves and instabilities can 'sense' the circulation at much deeper levels inside the planet.

Whether such an alternative reference frame will be confirmed by more direct methods of determining the interior circulation remains to be seen. But an appealing consequence of the possible approach of both Jupiter and Saturn to a neutrally stable condition with respect to Arnol'd II is that this may explain why most zonal jets in these atmospheres remain coherent and relatively undisturbed for long periods of time. The jets are then sporadically disturbed by occasional perturbations when such jets go locally unstable to form large-scale eddies, such as the compact ovals or polygonal waves.

6.4 Discrete Features as Constraints on Processes and Structure

The zonally averaged behavior of Saturn's atmosphere defines its global circulation but does not provide all the information needed to diagnose the physical processes responsible for the observed dynamical configuration. Furthermore, many important parts of the atmosphere remain unobserved or poorly observed. The morphology, size, and temporal variation of localized meteorological features helps isolate specific physical mechanisms that may either be manifestations of the important transport processes, or at least diagnostic of physical conditions at levels that are difficult to observe.

6.4.1 Anti-Cyclonic and Cyclonic Vortices

Compared to Jupiter, Saturn has fewer long-lived vortices. Cassini images have confirmed previous Voyager results (Ingersoll et al. 1984) that anticyclones dominate in number over cyclones (Porco et al. 2005; Vasavada et al. 2006). Only four features survived the 9 months period that elapsed between the Voyager encounters (Sánchez-Lavega et al. 2000), among them the largest anticyclone, the North Polar Spot (NPS) at 75.2°N latitude, that was also observed during the years 1992–1995 using ground-based telescopes (Sánchez-Lavega et al. 1993b) and HST (Caldwell et al. 1993).

Prior to Cassini, the 1996–2004 HST imaging survey of the southern hemisphere detected the formation of vortices (seen as dark spots at red continuum wavelengths) in the polar area (65°S to 75°S) and in southern mid-latitudes (40°S to 44°S), in this last case associated with storm activity seen as bright spots in blue – red wavelengths (Sánchez-Lavega et al. 2003, 2004b). The highest concentration (10–15 spots) occurs at mid-latitudes in the equatorward flank of the eastward jets at 47° (northern and southern hemispheres). The vortex production in southern mid-latitudes was confirmed by the first Cassini images from 2004 to 2005 (Porco et al. 2005; Vasavada et al. 2006) suggesting that activity similar to that seen by the Voyagers in the north is now occurring in the south, in a jet similar in latitude location, shape and peak intensity.

Anticyclones appear as oval spots, typically with east-west length up to 5,000 km (most ∼1,000 km), dark albedo at 500–900 nm continuum wavelengths, and contrasted dark in the UV and bright in the 890-nm methane band. They drift slowly in either direction relative to the mean zonal flow at 0–10 m s^{-1}. In between, convective activity occurs with some regularity, interacting with the vortex circulation (Sromovsky et al. 1983; Porco et al. 2005; Vasavada et al. 2006; Dyudina et al. 2007). Vasavada et al. (2006) report observed episodes of formation and disappearances of vortices suggestive of particular physical mechanisms. Particularly interesting are those cases in which vortices (anticyclones and cyclones) form following the decay of a bright convective storm (Porco et al. 2005; Vasavada et al. 2006). A common interaction between close vortices is mutual orbiting and merger, occurring in particular around latitudes 41°–44° where the maximum concentration of anticyclones occurs.

Vorticity measurements are available only for the anticyclonic "Brown Spot" (BS-1) observed by Voyager with a maximum value of $4.0 \pm 1.5 \times 10^{-5}$ s^{-1}, corresponding to a tangential velocity of 45 m s^{-1} (Ingersoll et al. 1984; García-Melendo et al. 2007). Model simulations of this anticyclone (Dowling et al. 1988) from 10 mbar to 10 bar were used to constrain the vertical wind shear and static stability of Saturn's upper troposphere at this latitude (García-Melendo et al. 2007). Comparison between the observed and simulated vortex properties indicates that the Brunt-Väisälä frequency is nearly constant from the ammonia to the water cloud levels with a value above 3×10^{-3} s^{-1}. The 1-year stability of this vortex requires that the wind speed slightly decays below the visible cloud deck at a rate no larger than $\partial u/\partial z \sim$ 2–6 m s^{-1} per scale height.

The "North Polar Spot" (NPS) at 75°N was the largest and longest-lived anticyclone observed on Saturn with east-west size of 11,000 km (Sánchez-Lavega et al. 1997). This vortex drifted slowly from 1980 to 1995 at a zonal velocity of just 0.084 m s^{-1} in the Voyager System III reference frame. Calculations of the seasonal insolation at the North Pole together with a simple linear radiative response of the atmosphere to the heating at different altitudes showed a temperature variability of few Kelvins at cloud tops. Because of its long lifetime, and because its motions did not vary appreciably during the 16-year observing period, it seems that the main properties and dynamics of the NPS were insensitive to the external solar forcing.

As stated above, large (∼2,000–3,000 km) stable and coherent cyclones are rarer than anticyclones. The two cases reported in mid-latitudes, the UV spot in the northern hemisphere in the Voyager era (Smith et al. 1981, 1982) and the oval "e" in the southern hemisphere in the Cassini era (Vasavada et al. 2006), are bright in the UV but dark in the 725-nm and 890-nm methane bands, and moderately dark in the continuum. This indicates thinner hazes at their tops relative to anticyclones.

Two remarkable long-lived (>1.5 year) cloud features observed by VIMS below the 1-bar level are annular rings of clouds at 48.8°N and 57.5°N latitude planetocentric (Baines et al. 2009c). These features span ∼7,000 km in diameter, with a central cloud clearing of ∼2,800 km in diameter. The 57.5°N feature, discovered in 2005, began to dissipate after 3 years of observation, and could not be clearly discerned against the zonal background clouds a few months later. The 48.8°N feature has been observed continuously since 2006. It is located in a westward jet, with relative retrograde motion of 2.6 ± 0.1 ms^{-1}.

6.4.2 Convective Clouds and Lightning

Moist convection is the primary mechanism by which Earth's tropical atmosphere transports energy from the radiatively heated surface to high altitudes where it is radiated to space. The same is likely to be true on the jovian planets, except that the energy transported is due to the internal heat source. Different varieties of convection exist on Earth, each diagnostic of the thermodynamic structure and circulation in which it occurs. Shallow cumulus occur in the presence of large-scale subsidence, and transition to midlevel congestus clouds when moisture convergence sets in at low levels but the atmosphere is too dry above to support deep convection (Derbyshire et al. 2004). Deep cumulonimbus that account for most latent heat release require a moist atmosphere and large-scale upwelling; they may be individual cells or mesoscale clusters depending on the sustainability of the cloud base moisture source and the wind shear environment (Schumacher and Houze 2006). They have weak updrafts if the lapse rate is nearly moist adiabatic, or strong updrafts that produce lightning if instability builds before convection is triggered (Zipser et al. 2006; Del Genio et al. 2007b). Diagnosis of convection provides indirect information about conditions near and just above the water condensation level at 10–15 bars, a level not sensed directly by Cassini instruments.

Convective clouds are the most conspicuously visible features on Saturn. They are identified by their high brightness at continuum wavelengths (indicating large optical thickness) and large contrast relative to neighboring clouds and hazes, and by their rapid temporal evolution. Second, methane band filters allow for height discrimination – the brighter the feature in a methane band and the progressively stronger bands in which it is visible, the higher its cloud top. These characteristics imply features that are both high and deep. This detection method has been quite successful for Jupiter, where storms are visible even in a strong methane band (Gierasch et al. 2000; Porco et al. 2003). On Saturn, however, the thick tropospheric haze limits seeing, and no convective features have been observed in the ISS MT3 filter (Del Genio et al. 2007a).

There are two types of convective features on Saturn. The most prevalent are fields of puffy discrete clouds that primarily occupy broad bands centered on the westward jets but are also common in both polar regions (Godfrey 1988; Sánchez-Lavega et al. 1997, 2007; Vasavada et al. 2006; Dyudina et al. 2008, 2009). They are visible in the weak methane band filter MT1 but not the moderate strength band MT2 (Fig. 6.4), suggesting that their tops are only modestly higher than those of other clouds. These may be the equivalent of the midlevel congestus clouds or isolated deep convective cells common over Earth's oceans. Their frequent occurrence adjusts the lapse rate to nearly moist adiabatic, but as a result rising air parcels are only weakly buoyant and sometimes unable to penetrate to great height because of dilution by mixing with drier surrounding air.

More remarkable are the single events that appear suddenly and grow as irregular structures that are dispersed by the meridional wind shear in a few weeks. There are two types of features within this family: occasional mesoscale storm clusters and the rare "Great White Storms" (GWS) (Fig. 6.17). Outbursts of bright spots with sizes $\sim$1,000–3,000 km have been observed in the westward jets at latitude 40°N in the Voyager era (Hunt et al. 1982; Sromovsky et al. 1983; Ingersoll et al. 1984) and at 35°S in 2004 (dubbed the "dragon storm; Fig. 6.17) by Cassini (Porco et al. 2005; Vasavada et al. 2006; Del Genio et al. 2007a; Dyudina et al. 2007). Another outbreak at this latitude in 2008 was associated with the first tentative spectroscopic detection of ammonia ice on Saturn, presumably formed when the water cloud penetrated the ammonia saturation level (Baines et al. 2009b). Other less dramatic examples are found preferentially in the cyclonic shear regions and at times in the vicinity of the eastward jets, but almost never in regions of anti-cyclonic shear (Del Genio et al. 2007a). This suggests that anti-cyclonic shear regions might be analogous to Earth's subtropics, dominated by large-scale subsidence with possibly an inversion layer of stable temperature and dry air at depth that suppresses vertical development of storms (e.g., Del Genio and McGrattan 1990; Showman and de Pater 2005), while the cyclonic shear regions are areas of net large-scale upwelling at depth.

Cassini images showed dark anticyclones forming after the decay of the dragon storm (Porco et al. 2005; Vasavada et al. 2006), raising the question of whether such storms might be involved in maintenance of the jets (Section 6.5.3). The expanding area that occurs after onset indicates divergence, implying upward vertical velocities $\sim$1 m s^{-1} over several scale heights (Hunt et al. 1982; Del Genio et al. 2007a). The dragon storm is one of only a few convective features detected in the ISS MT2 filter, indicating that it is deeper than other convective storms. Interestingly, this feature is associated with short high frequency (1–15 MHz) radio emission outbursts (of typical duration 49 ms) known as "Saturn Electrostatic Discharges" (SEDs) (Porco et al. 2005; Fischer et al. 2006, 2007). There is a consensus that SED episodes are due to lightning originating within the storm clouds, in agreement with previous Voyager findings at low latitudes (Kaiser et al. 1983). However, contrary to Jupiter, visible lightning flashes have not been detected. This is probably due to the fact that lightning originates at deeper levels on Saturn ($\sim$10–20 bars) than on Jupiter, and thus may suffer more extinction by overlying cloud and the denser Saturn tropospheric haze layer as well as being obscured by ringshine (Dyudina et al. 2007). The dragon storm is not visible in the UV3 (338 nm) and MT3 filters (Dyudina et al. 2007). Using HST multi-wavelength images and a radiative transfer model, Pérez-Hoyos et al. (2005) put the convective cloud tops at $\sim$200 mbar. If the coincident appearance of SEDs and the dragon storm is evidence of lightning, it implies that the westward jets in which the Voyager and Cassini events occurred have lapse rates significantly in excess of the moist adiabatic lapse rate at depth, as is the case for lightning-producing convection on Earth.

The observed distributions of convective clouds and eddy momentum fluxes on Saturn (Section 6.3.3) allows us to infer the nature of the mean meridional overturning circulation, which is too weak to observe directly, at and below the cloud tops. Early ideas about the jovian planets, especially Jupiter, interpreted the bright anti-cyclonic zones and darker cyclonic belts to be the result of rising motion (which forms clouds) in the former and sinking motion (which evaporates clouds) in the latter (Hess and Panofsky 1951; Ingersoll and Cuzzi 1969), analogous to Earth's tropical Hadley

Fig. 6.17 Outbreaks of intense moist convective storms on Saturn. *Left upper*: False color composite of Cassini ISS images of the 2004 dragon storm at 35°S in near-infrared continuum (CB2) and moderate/strong methane band (MT2/MT3) filters. Redder shades indicate lower cloud tops, bluer shades higher cloud tops, and whiter shades high optically thick clouds and hazes. *Right upper*: HST Wide Field Planetary Camera 2 true color image of the 1994 equatorial Great White Spot. Lower: VIMS false color composite of a convective storm (*yellow*) near 35°S in 2008 (Baines et al. 2009b)

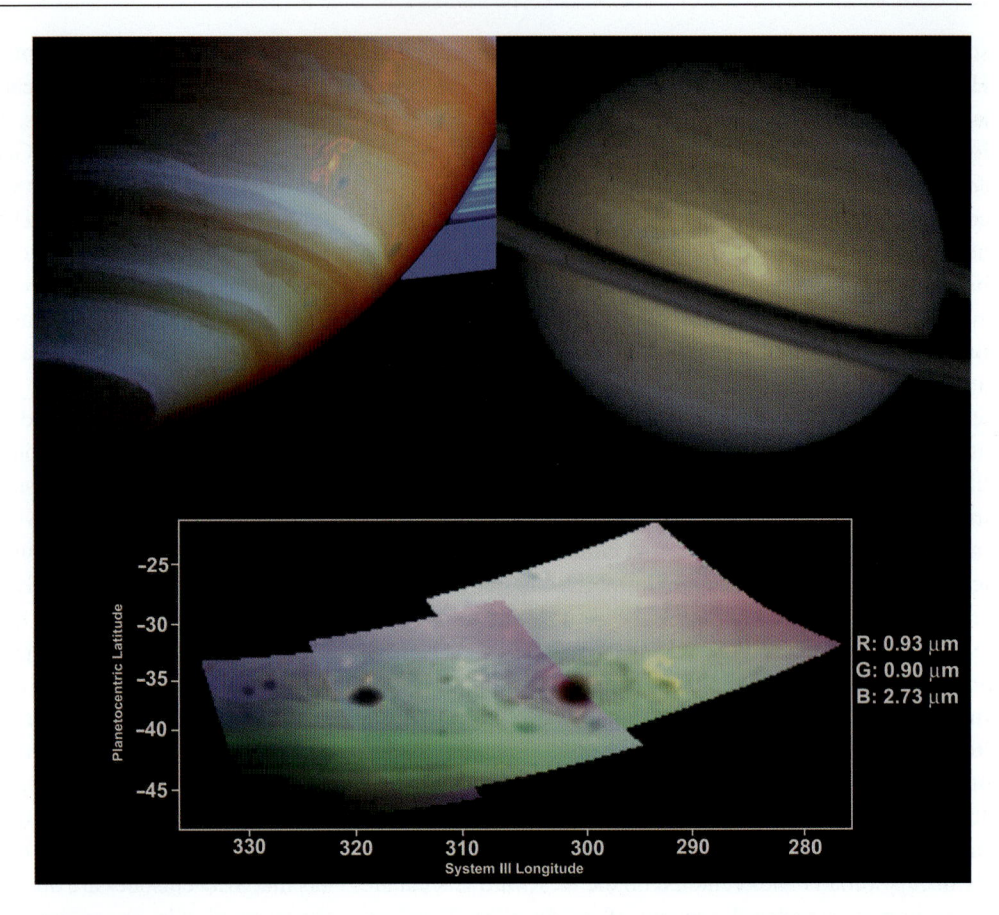

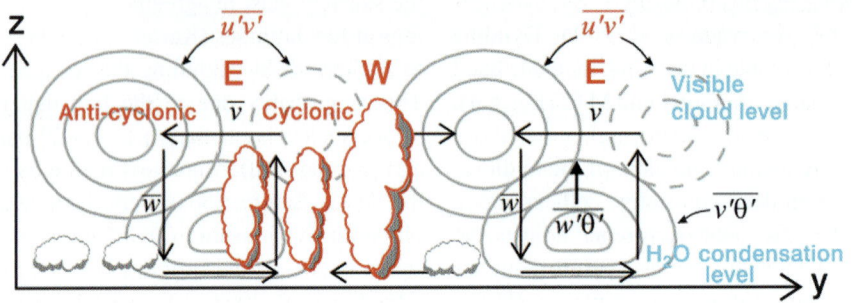

Fig. 6.18 Schematic vertical-latitudinal cross-section of the region between the visible cloud level and water condensation level indicating the eastward (E) and westward (W) jet locations and associated eddy momentum flux, observed preferred locations of convective clouds, inferred meridional overturning circulation, and eddy heat fluxes for a possible baroclinic instability source of eddies. Observed aspects of the circulation are shown in red and inferred or hypothesized features in black. Adapted from Hartmann (2007)

circulation. Poleward flow aloft from the rising to the sinking branch would experience a positive zonal acceleration due to the Coriolis force, thus maintaining the eastward jets. However, deep moist convection, which occurs overwhelmingly in the presence of low-level convergence and large-scale mean rising motion, was observed by Galileo and Cassini (during its Jupiter flyby) to be restricted to the belts (Gierasch et al. 2000; Porco et al. 2003), leading Ingersoll et al. (2000) to suggest that the belts are actually the seat of net rising motion. This picture is supported by Jupiter's belt-zone ammonia distribution at 1–5 bars (Showman and de Pater 2005). On the other hand, if Saturn had been the only giant planet the Hadley cell analogy might never have been considered, since the albedo-wind shear correlation breaks down there (Section 6.3.2). The observed distribution of convection on Saturn is generally consistent with that on Jupiter, although on Saturn convection is observed near the jet latitudes as well as in the cyclonic shear regions. Figure 6.18 illustrates a working hypothesis for the meridional circulation between the water and ammonia cloud levels on Saturn, which is exactly the opposite of the Hadley cell model. Deep moist convection, and thus net rising motion, occurs in cyclonic shear regions, and sinking in anti-cyclonic regions. At the visible cloud tops the meridional flow is then equatorward across eastward jets and poleward across westward jets (Fig. 6.18). The Coriolis force acting on this flow would decelerate both jets. Acceleration of the flow in this picture is instead provided by the observed convergence of the eddy momentum flux into the eastward jets and its divergence from the westward jets (Del Genio et al. 2007a; Showman 2007; Lian and Showman 2008). The remarkable stability of the jets outside the equatorial region suggests that these competing influences are approximately in balance, and thus $fv \sim -\partial <u'v'>/\partial y$. A reversed direction meridional return flow near the base of the water cloud would then provide moisture convergence in cyclonic shear regions needed to sustain convection there.

The dynamical source of the eddy momentum fluxes observed by Cassini at cloud top is unknown, but several possibilities exist (Section 6.5.3). Figure 6.18 illustrates how one possible mechanism, baroclinic instability, might be consistent with the observed patterns of convection. The instability would produce a poleward and upward eddy heat flux at depth beneath the eastward jets. This in turn would generate upward-propagating waves that lead to eddy momentum fluxes at cloud top. If the jet scale is broader than L_d the fluxes tend to converge on the jet (Held and Andrews 1983). The eddy heat flux convergence at the poleward edge requires rising motion and adiabatic cooling there to maintain hydrostatic equilibrium, while the eddy momentum flux convergence aloft requires the meridional flow described above to remain geostrophically balanced. This is similar to the processes responsible for the midlatitude Ferrel cell on Earth (Hartmann 2007). The deep-seated meridional temperature gradient that drives the baroclinic instability in this picture comes from the moist convection itself (Lian and Showman 2009). Assuming a near-moist adiabatic lapse rate in the cyclonic regions, subsidence along a steeper dry adiabat in the anti-cyclonic regions would suppress deep convection and lead to a warm temperature inversion just above the water condensation level (Showman and de Pater 2005). Temperatures just above cloud base would be lower, but moisture greater, in the convergent cyclonic regions, creating baroclinically unstable conditions between the two. The base-level meridional flow from the anti-cyclonic to cyclonic regions would be initially depleted in moisture but could gradually humidify via upward turbulent mixing from the convective interior, eventually allowing rising parcels to punch through the overlying inversion into the highly unstable air above. This might explain the association of the most vigorous temperate latitude storms on Saturn with the westward jets. A similar flow would occur in this scenario beneath the eastward jets, but there the upward eddy heat flux from baroclinic instability would partly stabilize the lapse rate and obviate the need for moist convection to be as intense. In fact, "slantwise" moist convection on Earth is common within the frontal regions of baroclinic eddies. We might then speculate that isentropic slopes on Saturn in the putative statically stable layer above the base of the water cloud are the combined result of baroclinic and latent heating processes and follow the approximate terrestrial relationship $\Delta_z \theta_e \sim \Delta_y \theta_e$, where θ_e is the equivalent potential temperature, a conserved quantity for saturated adiabatic ascent, and Δ_z and Δ_y are differences over characteristic vertical and horizontal length scales (Frierson 2008). This would differ from the nearly adiabatic layer at higher altitudes, for which the ratio of vertical to horizontal equivalent potential temperature contrasts is small (Allison et al. 1995).

The terrestrial midlatitude analogy may or may not be relevant, though. Baroclinic instability produces jets in numerical simulations of giant planet atmospheres (Section 6.5) but has not yet been observed on either Jupiter or Saturn, although one eastward jet feature has been proposed to be the result of such an instability (Section 6.4.6). If it does exist, the required heat fluxes should occur primarily below the visible cloud tops, making them a challenge to detect. Coordinated CIRS-VIMS-ISS feature tracks acquired at various times throughout the mission might be examined for correlations between temperature and meridional wind perturbations diagnostic of baroclinic eddies. Furthermore, the width of the terrestrial jet stream is only marginally greater than L_d, the scale at which baroclinic instability injects energy into the flow, while on Saturn the two scales may or may not be more widely separated depending on the static stability used to estimate L_d. A near-adiabatic value (small L_d) would imply different characteristics of the inverse energy cascade leading from the forcing to the jet scale on Saturn (Section 6.5). On Saturn, the inferred overturning circulation in Fig. 6.18 cannot penetrate to the tropopause, where latitudinal temperature contrasts retrieved by CIRS are of the opposite sense (Section 6.3.4), and haze top height and albedo variations suggest rising/sinking motion in the anti-cyclonic/cyclonic regions instead (Section 6.3.2). Finally, other mechanisms may be capable of providing the observed $u'v'$ to accelerate the jets. In at least one model, for example, idealized thunderstorms directly pump the jets over a wide range of assumed L_d values (Li et al. 2006; Section 6.5).

Convection in the equatorial region on Saturn differs from that elsewhere. In this region the most dramatic convective events, the "Great White Spots" (GWS), are periodically observed. GWS outbreaks start with the eruption of a very bright spot that grows in days to a size of 10,000–20,000 km (Section 6.1.2). Meridional wind shear ultimately disperses the clouds and forms a wake of regularly spaced bright clouds moving zonally in opposite directions at different latitudes dragged by the prevailing winds. The white spots suggest an associated wave phenomenon. Only five such GWS events have been recorded during the last century (in 1876, 1903, 1933, 1960 and 1990) with the three most prominent occurring at the equator in 1876, 1933 and 1990

(Sánchez-Lavega et al. 1991). The 1990 event erupted in Saturn's northern Equatorial Zone at 5°N, disturbing the whole equatorial band from 8°S to 22°N for several months (Sánchez-Lavega et al. 1991; Beebe et al. 1992; Barnet et al. 1992; Westphal et al. 1992). Radiative transfer calculations indicate that the GWS cloud tops reached the tropopause with probable overshooting into the stable region up to 60 mbar (Acarretta and Sánchez-Lavega 1999). Following this last giant disturbance the equatorial region became active in synoptic-scale cloud formation, in particular during 1991 and 1992 (Sánchez-Lavega et al. 1993a). This activity was followed by a second large-scale storm in 1994 at 10°N (Sánchez-Lavega et al. 1996) reaching a longitudinal size ∼27,000 km and a latitudinal size ∼12,000 km, with a lifetime >1 year. It moved with a zonal velocity of 274 m s^{-1}, 150 m s^{-1} slower than that of the Voyager mean wind at this latitude (Section 6.3.5). White cloud activity continued in the southern Equatorial Zone (13.5°N) during 1996, declining 1 year later (Sánchez-Lavega et al. 1999).

VIMS imagery of the deep troposphere shows discrete, convective cloud structures occupying ∼50% of the area near 6°N and S latitude (Baines et al. 2009c; Fig. 6.2). These features span 7,000–9,000 km in longitude and 2,500–4,000 km in latitude, similar to the shape, size and latitudinal position of enhanced ammonia gas absorption mapped by RADAR (Janssen et al. 2008). These results suggest that the equatorial region is typically convectively unstable at depth, perhaps continually supplying aerosols that form the thick upper tropospheric haze observed at these latitudes. The GWS and other visually observed outbursts are likely then just the manifestation of unusually energetic convective motions by these regularly occurring clouds.

Our inferences of convective clouds in Saturn images are indirect – unlike Earth, we have no measurements of precipitation or static stability beneath the visible cloud tops, nor do we even know the composition of the clouds we identify as convective features. In recent years, however, numerical models have been employed to support our interpretations. These models indicate that convection starting in the water clouds is the only plausible candidate mechanism to explain the more obvious organized, deeply penetrating features such as the GWS and dragon storm. A standard 1D ascending parcel model driven by the latent heat released upon condensation of water and ammonia, can explain the high altitude of the cloud tops (Sánchez-Lavega and Battaner 1987). The details of the development, growing stage and evolution of the storm (including microphysics and interaction with surrounding air) have been studied in detail with a 3D anelastic model by Hueso and Sánchez-Lavega (2004). Buoyant parcels can develop from an initial heating of 1–2 K at the water cloud level at 10 bar when the environment becomes saturated and the water abundance is assumed to be supersolar. In the core of the updrafts the vertical velocities can reach 150 m s^{-1}. Less vigorous moist convective storms can also develop at the ammonia cloud level, in particular under very large abundances of ammonia (10 times solar). These might explain some of the ubiquitous puffy clouds in the broad latitude bands between successive eastward jets and in the polar regions. However, deeper isolated water-based convective cells are equally likely to explain these features if the lapse rate is near-neutral and the relative humidity above cloud base subsaturated.

6.4.3 Upper Troposphere Thermal Features and Rossby Waves

Diagnosis of the physical mechanisms responsible for observed wavelike and eddy mixing behavior seen within and above the cloud layer is aided by knowledge of the vertical extent of such disturbances. Temperature retrievals in the upper troposphere with sufficient spatial and temporal resolution provide one useful constraint.

Achterberg and Flasar (1996) used a set of global mapping observations by Voyager IRIS to search for waves in Saturn's temperature field in the upper troposphere at 130 and 270 mbar, although the relatively large IRIS field of view, approximately 8° of latitude, meant they were only sensitive to planetary scale waves. The majority of statistically significant features in the data were at wavenumbers k ≤ 4 at 130 mbar at low to mid-northern latitudes. By far the most prominent wave was a quasi-stationary zonal wavenumber 2 feature extending from 10°N to 40°N at 130 mbar. The meridional structure of this wave is shown in Figure 6.19. The wave has a maximum amplitude of ∼0.5 K at 35°N–40°N, at the northern edge of the equatorial jet, with an amplitude that drops rapidly to the north, and more slowly to the south. The zonal phase of the wave is constant between 10°N and 45°N. The amplitude of the wave at 270 mbar was roughly half of the amplitude at 130 mbar.

Ray-tracing calculations by Achterberg and Flasar (1996) showed that the observed structure of the wavenumber 2 wave was consistent with a quasi-stationary Rossby wave, confined meridionally by the structure of the zonal winds, and to pressures less than ∼200 mbar by the static stability structure. Furthermore, for wave velocities near zero, the structure of the waveguide is consistent with the generation of barotropic instabilities through wave overreflection at a critical layer where the wave velocity equals the zonal wind velocity (Lindzen et al. 1980). Also, the meridional gradient of potential vorticity (Section 6.3.8), calculated from Voyager imaging and thermal data, changes sign at approximately the latitude where the mean zonal wind is zero. This led Achterberg and Flasar (1996) to propose that the observed quasi-stationary wavenumber 2 wave is forced by a

6 Saturn Atmospheric Structure and Dynamics

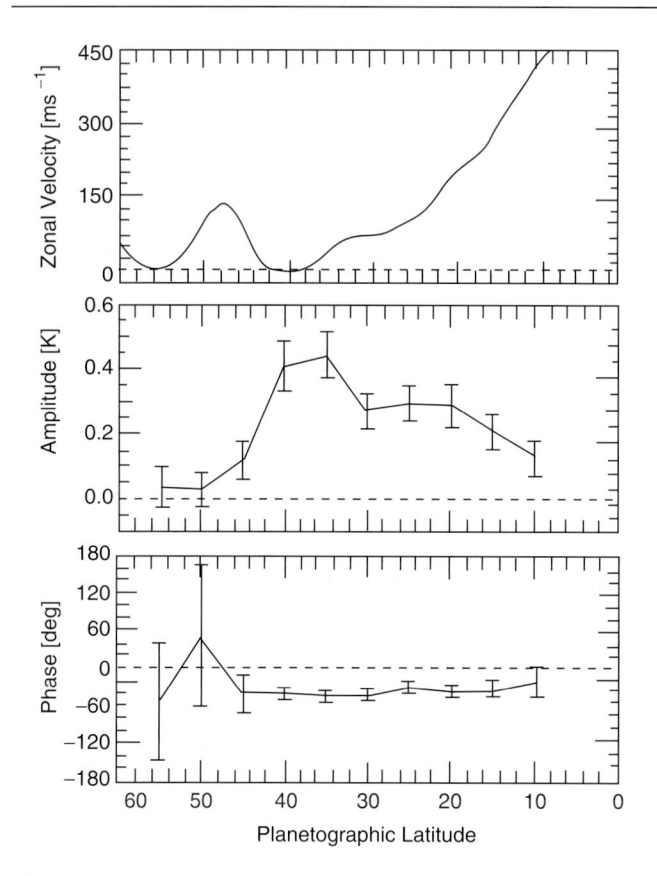

Fig. 6.19 Top: Zonal mean wind velocity from Voyager imaging data. Middle and bottom: Observed amplitude and phase of the zonal wavenumber 2 thermal wave at 130 mbar from Voyager 1 IRIS data. From Achterberg and Flasar (1996)

barotropic instability of the zonal mean flow. The wave is quasi-stationary because the instability occurs at a latitude of weak zonal wind velocity, and not because the wave is linked to Saturn's interior, as has been proposed for quasi-stationary waves on Jupiter (Magalhaes et al. 1990) and for Saturn's north polar hexagon (Godfrey 1990; Section 6.4.4).

Periodic temperature anomalies have also been detected in CIRS data at 160 mb, at the northern latitude of the ribbon wave (Fletcher et al. 2007c; Section 6.4.6). These periodic structures thus may be related to a deep-seated disturbance, perhaps associated with baroclinic instability (Godfrey and Moore 1986). Orton and Yanamandra-Fisher (2005) also observe temperature oscillations near 100 mb at 32°S, where a weaker version of the ribbon morphology is seen (Section 6.4.6).

6.4.4 North Polar Hexagon

Polar projection maps of Voyager 2 images of Saturn's north polar region by Godfrey (1988) revealed a striking hexagonal cloud structure surrounding the pole at about 76°N planetocentric latitude, with a prominent anticyclonic oval, the North Polar Spot (NPS), centered on the southern side of one of the hexagon edges. Wind measurements (Godfrey 1988) showed that the hexagon coincides with a strong eastward jet (100 m s^{-1}), with the velocity streamlines also following the hexagonal pattern. The hexagonal pattern, however, was stationary, to within measurement error, in the System III reference frame assumed at the time to indicate the rotation rate of Saturn's interior. By adding in an observation from Voyager 1, Godfrey (1990) measured a small eastward velocity for the NPS, and presumably the hexagon, of 0.108 ± 0.007 m s^{-1}.

A decade after the Voyager encounter, the hexagon and the NPS were observed in HST (Caldwell et al. 1993) and ground based (Sánchez-Lavega et al. 1993b, 1997) images. Measurements of the longitude of the NPS in these datasets give long-term averaged drift rates between 0.11 and 0.23 m s^{-1} depending upon the time span used, indicating long-term changes in the rotation period of the hexagon. There were also short term variations in the drift rate of the NPS of about ± 4 m s^{-1}.

Starting in 1996, the latitude of the hexagon was unobservable from Earth and in shadow during the Cassini mission until 2009. Prior to 2009, however, the hexagon was detected in thermal emission by both CIRS and VIMS. Maps of upper tropospheric temperatures (100–800 mbar) by CIRS (Fletcher et al. 2008) show a hexagonal warm band at 79°N, and a hexagonal cold band at 76°N (Fig. 6.20 left). The temperatures are consistent with the standard upper troposphere pattern on both Jupiter and Saturn of warm temperatures in regions of cyclonic shear on the poleward side of eastward jets, and cold temperatures in regions of anticyclonic shear equatorward of eastward jets, suggesting that the hexagon is a typical jet except for its hexagonal shape. The hexagon is also clearly visible in 5 μm thermal images from VIMS (Baines et al. 2009a), which are sensitive to large-particle cloud opacity at 1.5–4 bars. Figure 6.20 (right) shows Saturn's emission through clearings between the clouds, which are silhouetted against the background glow. Hundreds of discrete clouds populate the scene inside the polar hexagon. The hexagon, located near 78°N (planetocentric) is comprised of several nested and alternating hexagonally shaped clouds and clearings. Near the pole, alternating concentric bands of thick and somewhat optically thinner clouds extend out to ∼87°N. As observed 4 times over 6 h, the pole itself has a near-stationary cloud feature that is offset from the pole by 0.5 degrees of latitude (∼500 km), surrounded by a narrow ring of clearings in the clouds about 0.5 degrees wide. This behavior is different from that in the Southern Hemisphere, where cloud clearing occurs at the pole (Section 6.4.5). The combination of CIRS and VIMS observations indicates that the hexagon exists over an extended

Fig. 6.20 The north polar hexagon as seen in CIRS 100 mb temperatures (*left*; Fletcher et al. 2008) and in VIMS 5 μm emission (Baines et al. 2009a)

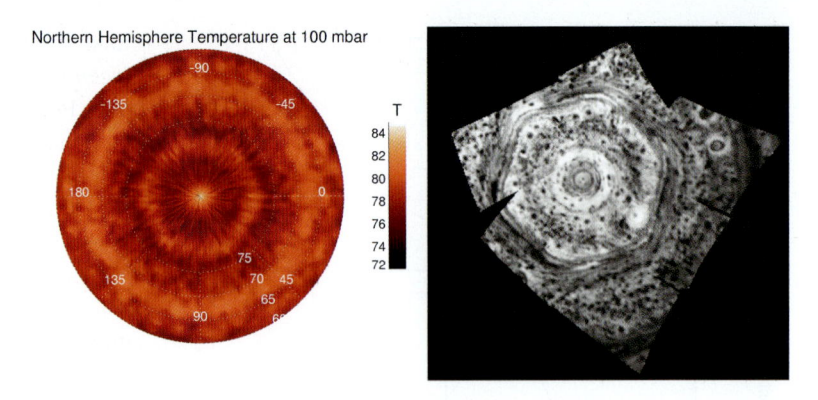

pressure range from at least the tropopause region (50 mbar) down to the ∼3 bar pressure level, consistent with features in other eastward jet regions.

The VIMS hexagon cloudy and clear latitudes correlate with the CIRS cool and warm latitudes, suggesting rising and sinking in the eastward jets are correlated over great vertical distances. This is a counter-example to the tendency elsewhere for inferred vertical motions at depth to differ from those in the upper troposphere. It is interesting to note that in baroclinic instability models of gas giant planets (Williams 2003; Lian and Showman 2008), vertical and especially horizontal eddy heat fluxes at eastward jet latitudes are coherent over great depths, a behavior that may be relevant to this question.

VIMS detections of the hexagon in 2006 and 2008 indicate that it is still stationary in the Voyager reference frame (Baines et al. 2009c). Small clouds embedded within the clear regions move at about 125 m s^{-1}, similar to the zonal speeds observed by Voyager. Flows outside the hexagon are significantly less – about 30 m s^{-1} (Baines et al. 2009a). Thus, over a quarter century, the basic structure and dynamical characteristics of the north polar hexagon appear to have changed very little.

The persistence of the hexagon over nearly a full Saturnian year suggests that it is insensitive to solar forcing. The small drift rate in System III, less than the uncertainty in the Voyager radio period, led to suggestions that the hexagon and the NPS are rooted in the deep interior (Godfrey 1988, 1990; Caldwell et al. 1993). However, the radio period is now known to be variable (see Chapter 10) and likely not representative of Saturn's interior rotation rate, making any direct relation between the hexagon and Saturn's interior unlikely. Allison et al. (1990) proposed that the hexagon is a stationary Rossby wave imbedded in and meridionally trapped by the eastward jet, and forced by the interaction of the jet with the NPS. Matching the observed phase velocity requires that the Rossby wave be vertically trapped, which may be consistent with the confinement of the observed hexagon to the troposphere (Fletcher et al. 2008).

An alternative view has been put forward by Aguiar et al. (2009), based on a laboratory analog of the barotropic instability of zonal shear layers and jets. In the experiment, a zonally symmetric shear layer or jet is maintained in a cylindrical tank by differentially rotating sections of the horizontal boundaries. The whole container is then rotated about the vertical axis of symmetry, so that a Rossby number for the flow can be defined as

$$Ro = \frac{R\omega}{2\Omega H},$$

where R is the radius of the jet or shear layer, H the depth of the tank, ω the angular velocity of differential rotation and Ω that of the background rotation. Both jets and shear layers in such a system exhibit inflection points such that the absolute vorticity gradient changes sign with radius, and can therefore exhibit fully developed barotropic instability if the effective Reynolds number is sufficiently large.

When barotropic instability occurs, the fully developed form equilibrates to a large amplitude Rossby wave-like pattern, whose manifestation may resemble polygonal meanders of the original jet or shear layer. Under conditions favoring wavenumber $m = 6$, the flow appears as a hexagonal jet (with rounded corners), which may be accompanied by cyclonic or anticyclonic vortices both inside and outside the hexagon whose strength depends on the velocity profile of the jet or shear layer. A typical example of such a flow is illustrated in Fig. 6.21 below. In this case the equilibrated hexagonal flow maintains a constant amplitude (so long as the basic jet or shear layer forcing is maintained), and moves around the tank at a rate determined by the strength of the jet or shear layer and any β-effect produced, e.g., by a sloping bottom. The scale selection process in these experiments, favoring $m = 6$ in this case, is not well understood, but these (and similar) experiments (e.g., Niino and Misawa 1984; Sommeria et al. 1991; Fruh and Read 1999) suggest a dependence of equilibrated wavenumber primarily on the ratio of the width of the jet to its radius of curvature (cf. Howard and Drazln 1964). In practice this appears as a

6 Saturn Atmospheric Structure and Dynamics

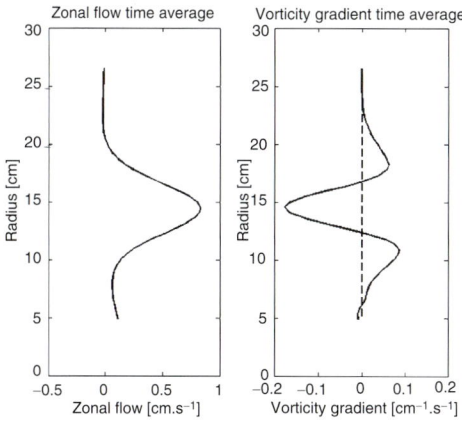

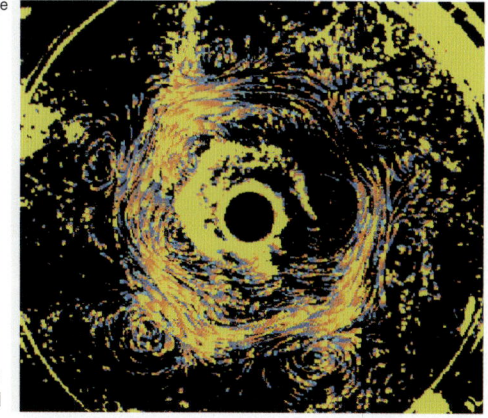

Fig. 6.21 *Left*: Typical profiles with radius of zonal flow and radial vorticity gradient in the barotropic instability laboratory experiments of Aguiar et al. (2009). *Right*: Streak image showing the fully developed hexagonal jet and weak anticyclonic peripheral vortices in the barotropically unstable jet flows

dependence primarily on the imposed Rossby number, with relatively little dependence on the viscosity of the fluid.

These experiments suggest the possibility that Saturn's hexagon owes its origin to the predominantly barotropic instability of the strong jet at 76°N. The nonlinear equilibration of this instability would lead to a steady amplitude Rossby wave which, in a spherical domain, propagates retrograde relative to the eastward jet. According to this scenario, however, there is no fundamental reason for the hexagonal wave to be stationary in any particular System III, although its retrograde propagation relative to the eastward jet might happen to render it very slowly moving in Voyager's System III. The origin of the NPS (and other, weaker and more ephemeral vortices noted by Godfrey [1988]) is then seen to be from the same instability as the hexagon itself, in the same way as vortices appear on either side of the hexagonal jet in the laboratory experiments.

6.4.5 South Polar Vortex

Ground-based measurements prior to Cassini revealed a dark disk-shaped feature at visible wavelengths centered on the South Pole and extending to 88.5°S with warm temperatures relative to its surroundings (Orton and Yanamandra-Fisher 2005). ISS imaged the feature in 2004 and at higher resolution in 2006 in continuum and methane band filters (? ?; Fig. 6.22 left). These observations revealed a dark central "eye" with either clear air or lower cloud tops relative to two surrounding "eyewalls" of thicker, higher clouds at 88 and 89°S planetocentric. The height of the inner and outer "eyewalls" was estimated to be 70 ± 30 and 40 ± 20 km relative to the surrounding clouds based on the shadows they cast. High-resolution images acquired in 2008 show a third innermost ring of somewhat elevated clouds punctuated by occasional deep convective clouds, some of which appear to be creating vortices at their divergent outflow levels. Dozens of small (<1,000 km in diameter) discrete puffy clouds occupy the south polar region farther away from the pole.

The use of the terms "eye" and "eyewall" for the cloud morphology of the south polar feature is obviously evocative of terrestrial hurricanes, which also have clear central eyes and surrounding eyewalls of towering convective clouds. It is instructive to explore what the Saturn feature does and does not have in common with terrestrial hurricanes. The structure and physics of hurricanes is reviewed by Emanuel (2003). Hurricanes are cyclonic vortices, with tangential wind speeds increasing toward the center, peaking at the eyewall, and then decreasing within the eye. The Saturn south polar feature is also a cyclonic vortex, with peak prograde winds of 160 ± 10 m s^{-1} at 88°S (the outer "eyewall") in ISS and VIMS images, decreasing poleward and more slowly equatorward to <10 m s^{-1} at ~80°S (Fig. 6.22 right). Hurricanes are warm-core features; the Saturn south polar vortex has enhanced VIMS 5 μm emission at the pole, indicating clearing at depth, and anomalously warm central temperatures in CIRS data just beneath the tropopause (by 5 K) and also in the stratosphere (by 3–4 K) (Fletcher et al. 2008). The warm central core implies a decay of the vortex wind speed with altitude of ~10 m s^{-1} per scale height (26 km at the pole), which is also consistent with the slightly stronger wind speeds measured by VIMS at deeper levels (Dyudina et al. 2009).

Hurricanes have maximum angular momentum at their outer periphery, decreasing inward due to friction experienced by the near-surface radial inflow, and upper level divergent outflow; observed latitudinal near-surface wind profiles often exhibit approximately constant vorticity. The Saturn south polar vortex wind profiles also decrease in angular momentum toward the pole and exhibit almost constant

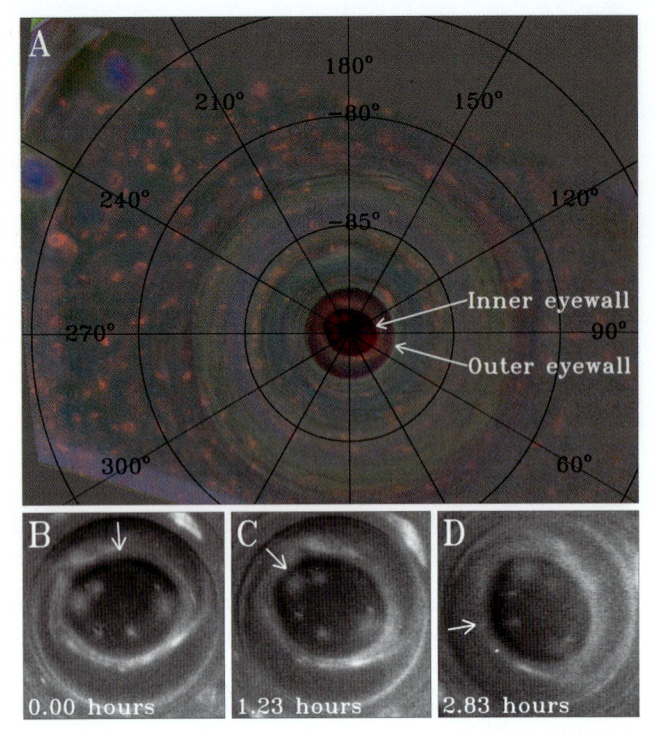

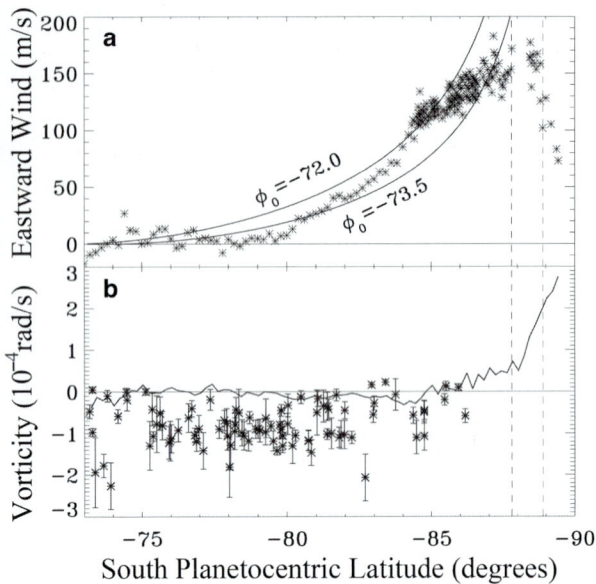

Fig. 6.22 *Left*: False-color polar stereographic image of the south polar vortex. The image is made up of MT3, MT2, and CB2 images projected into the blue, green, and red color planes, respectively. High thin clouds therefore appear blue or green, low level clouds red, and thick clouds with high tops bright pink. The lower panels show a time sequence over 2.83 h showing the shadows cast by the high clouds of the inner eyewall; the position of the Sun is indicated by the arrows. *Right*: ISS cloud-tracked zonal winds vs. latitude (*panel a*) and corresponding relative vorticity values (*panel b*). The solid curves in A represent constant absolute vorticity for a parcel moving poleward from latitude φ_o. The solid curve in B is the vorticity corresponding to the wind profile in A; the points in B are vorticities for the bright puffy clouds in the left panel. The vertical dashed lines in A and B indicate the locations of the inner and outer "eyewalls." (From Dyudina et al. 2008)

absolute vorticity (Fig. 6.22 right). However, no evidence for systematic meridional flow toward or away from the pole has been found; in the absence of such evidence a more reasonable interpretation might be that eddy mixing smoothes the vorticity, as it appears to do at lower latitudes (Section 6.3.8).

Thus, obvious morphological similarities between the Saturn south polar vortex and terrestrial hurricanes exist. However, this may be the extent to which the analogy can be carried. Hurricanes are transient propagating features, often originating in the troughs of tropical easterly waves that propagate from the continents over ocean. The warm tropical ocean surface is crucial to the genesis of hurricanes, being the source of moist entropy via surface evaporation and sensible heat fluxes during the near-surface radial inflow that provides the energy to intensify an existing convective vortex (Emanuel 2003). Hurricanes weaken quickly once they reach land and are cut off from the surface moisture source. The Saturn vortex, however, is a polar feature (itself not a problem since polar and equatorial temperatures are comparable on Saturn), fixed in location, and apparently long-lived. Hurricanes can have multiple eyewalls, but these usually migrate inward and are replaced by eyewalls farther from the center, whereas the Saturn "eyewalls" are fixed. Most importantly, there is no real equivalent to the terrestrial ocean surface source of moist entropy. Turbulent mixing of water vapor up to the condensation level may provide the condensation source that drives the convection in the Saturn vortex "eyewalls," but this may be no different than the situation occurring in eastward jets and cyclonic shear regions at other latitudes, where low-level convergence organizes moist convection (Fig. 6.18). Thus, the Saturn south polar vortex is more appropriately viewed as a feature of the mean meridional circulation.

We also note that Saturn's north and south poles are more similar than different. VIMS detects a north polar vortex with winds peaking at 88°N, but with a higher velocity (200 m s^{-1}) than that at the south pole. While the south pole does not have a hexagonal structure (Sánchez-Lavega et al. 2002b), Vasavada et al. (2006) do report polygonal segments of cloud there.

6.4.6 Other Wavelike Features

The "ribbon wave" at 47°N planetographic latitude (Section 6.1.2) was discovered by Voyager (Smith et al. 1981, 1982; Sromovsky et al. 1983), and detected again in 1994–

1995 HST images (Sánchez-Lavega 2002a). It appears to still be present, though its morphology is somewhat different. It is visible in 2007 ISS images (Fig. 6.4) near 41°N planetocentric as a more zonally oriented bright cloud band at some longitudes, interrupted by an isolated kink in its structure but more wavy in appearance to the east. VIMS viewed the ribbon in 2008 in 2.7 μm reflected sunlight, finding that it circles the planet at this latitude with mean longitudinal wavelengths of ∼4,000 km, somewhat shorter than the 5,710 ± 260 km wavelength derived by Godfrey and Moore (1986) from Voyager images. The ribbon wave is embedded in and moves with a strong (140 m s^{-1}) eastward jet, with cyclonic and anticyclonic vortices nested to its north and south. UV and methane band images indicate that the brightness reverses between the northern and southern parts of the ribbon. Radiative transfer modeling suggests that the haze top is 14 km higher to the south than to the north, consistent with upper level upwelling/downwelling in anticyclonic/cyclonic shear regions (Godfrey and Moore 1986; Sánchez-Lavega 2002a). The pattern resembles those seen in laboratory rotating annulus experiments of sloping convection (Hide et al. 1994). In view of the strong thermal gradients measured across the ribbon (Conrath and Pirraglia 1983), Sromovsky et al. (1983) and Godfrey and Moore (1986) suggested that it might be the product of baroclinic instability. The presence of a flattening or possible reversal in the sign of the latitudinal gradient of upper troposphere potential vorticity at this latitude (Section 6.3.8; Fig. 6.16) is consistent with such an interpretation. Godfrey and Moore (1986) detected the ribbon in UV images and concluded that it extends upward from the cloud level to at least 170 mb. CIRS in fact detects periodic temperature anomalies at 160 mb at this latitude (Fletcher et al. 2007c; Fig. 6.23 upper) that are likely to be associated with the ribbon. Thin undulating bright bands are also seen in VIMS 5 μm images (Fig. 6.3 lower); this might be consistent with a baroclinic instability origin if the temperature gradients driving the perturbation were due to latent heat release at the water cloud level. Saturn's southern hemisphere also exhibits subtle ribbon-like features with similar morphology in eastward jets at 48 and 32°S, but these do not fully encircle the planet (Sánchez-Lavega et al. 2000). It is interesting to note that no ribbon waves have been detected in Jupiter's midlatitude eastward jets.

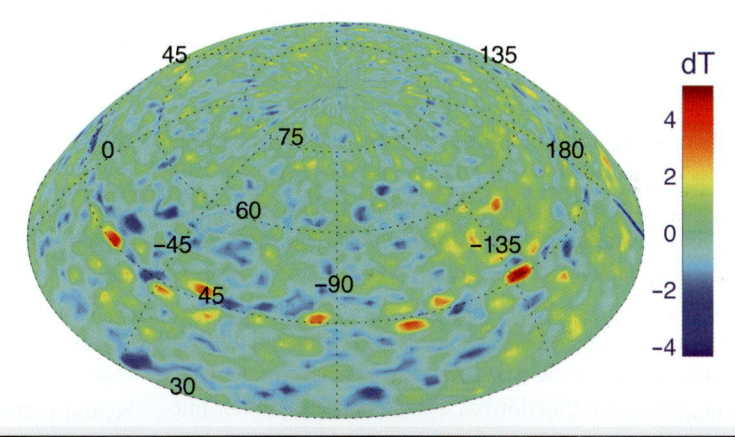

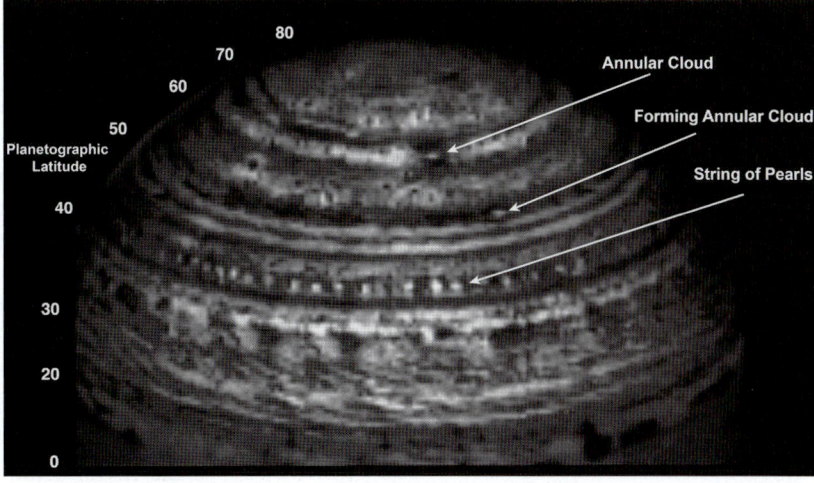

Fig. 6.23 *Upper*: CIRS temperature anomalies at 160 mbar at the ribbon wave latitude (Fletcher et al. 2007c). *Lower*: 5 μm VIMS image illustrating the string-of-pearls and annular cloud features (Baines et al. 2009c)

The "string-of-pearls" discovered in VIMS 5-μm imagery at 33.5°N planetocentric (Baines et al. 2009c; Fig. 6.23 lower) is located at pressures greater than 1 bar. It is not seen either in ISS images of the ammonia cloud and upper tropospheric haze (Fig. 6.4) or in CIRS upper troposphere temperature maps (Fig. 6.23 upper), suggesting that it is strictly a deep-seated feature with little upward propagation. It has been observed consistently for over 2.8 years. The holes of enhanced emission are spaced ∼3,500 km apart, span ∼800 km each, and are confined to a latitude range of ± 2°. The string-of-pearls resides near a westward jet and exhibits a retrograde mean zonal motion of 22.65 ± 0.04 m s^{-1} in the Voyager reference frame.

6.5 Theories and Models of the General Circulation

6.5.1 Deep Cylinders vs. Shallow Weather Layers

Two endpoint scenarios exist for the jet structure below the clouds on giant planets. In the "shallow structure" scenario, the jets are confined to a layer within several scale heights below the clouds (e.g., Hess and Panofsky 1951, Ingersoll and Cuzzi 1969). Through the thermal-wind relationship, this would require the anticyclonic regions (zones) to be warm and cyclonic regions (belts) to be cold. Such temperature contrasts might result from latent heating associated with condensation of water vapor, latent heating from ortho-para conversion, or latitudinal variations in absorbed sunlight. In the "'deep structure" scenario, the observed cloud-level jets are hypothesized to extend downward through the ∼10^4-km-thick molecular envelope. This scenario is motivated by the likelihood that Jupiter and Saturn transport their internal heat flux via convection (Section 6.2.2), and hence that their interiors must be close to a barotropic state. Under such conditions, the Taylor-Proudman theorem implies that the jets would be constant on cylinders parallel to the rotation axis (so-called "Taylor columns").

As emphasized by Vasavada and Showman (2005), hybrid scenarios are also possible with Taylor columns in the deep interior and a baroclinic thermal-wind region within the few scale heights below the clouds. The argument for near-barotropic conditions in the deep molecular envelope is strong, because convection homogenizes the entropy and plausible convective density anomalies are extremely small. However, all four giant planets exhibit distinctly *non*-barotropic conditions at the cloud level, with ∼5 K variations of temperature with latitude on constant-pressure surfaces. Such a baroclinic region could easily penetrate several scale heights or more below the clouds (Allison 2000). In this scenario, the cloud-level winds would not represent the speeds of the Taylor columns, but rather the speeds of the Taylor columns plus the height-integrated thermal-wind shear within this outermost baroclinic zone.

Early work suggested a possible problem with the shallow-structure model for Saturn (Ingersoll et al. 1984), but recent updates may have dissolved this problem. Jupiter's approximate jet speeds can be explained using thermal-shear from a level of no motion at ∼10 bars if the belt-zone temperature differences are only ∼5–10 K (Ingersoll and Cuzzi 1969). The required belt-zone temperature differences and depths could plausibly result, for example, from latent heating due to condensation of water vapor. In the case of Saturn, however, the Voyager-era winds were thought to be predominantly eastward (Fig. 6.1). If the level of no motion were at tens of bars, the required latitudinal temperature differences would be a factor of two, which are implausibly large. The required latitudinal temperature contrasts could be decreased by moving the level of no motion deeper (Ingersoll et al. 1984). However, it is now clear that Saturn's rotation rate is poorly known (see Chapter 10), and proposed revisions to the rotation rate (Anderson and Schubert 2007) imply that the jet profile consists of alternating eastward *and* westward jets. This revision largely mutes the difficulty outlined by Ingersoll et al. (1984) and has other potential advantages (Section 6.3.8), but there is as yet no definitive *physical* basis for concluding that the faster rotation rate estimate is indeed the true internal rotation rate (Section 6.2.1).

Most authors have long assumed that the strong zonal jets observed at cloud level do not penetrate into Jupiter and Saturn's metallic regions (at pressures exceeding ∼1–2 Mbar and depths of perhaps 10^4 km) because the Lorentz force would brake strong zonal flows there (Kirk and Stevenson 1987, Grote et al. 2000, Busse 2002). Recently, Liu et al. (2008) extended these arguments by suggesting that the zonal flows cannot penetrate even to the base of the molecular region (substantially above the dynamo-generating region). Motivating their work is the experimental inference (e.g., Nellis et al. 1992, 1999; Weir et al. 1996) that the transition from molecular to metallic is smooth, implying the existence of a substantial layer at the base of the molecular region that, while not metallic, exhibits significant electrical conductivity. Based on an inward extrapolation of the axisymmetric part of the observed magnetic field, and assuming the cloud-level winds extend downward on cylinders, Liu et al. (2008) calculate that the Ohmic dissipation would exceed the planet's observed luminosity if the Taylor columns penetrate deeper into the interior than 0.96 and 0.86 of the radius for Jupiter and Saturn, respectively. The implication is that the Taylor columns do not penetrate deeper than this level. As Liu et al. (2008) point out, it is difficult to envision how the Taylor columns would terminate immediately above this level (or indeed anywhere within the near-isentropic molecular envelope) because plausible deviations from barotropicity are very small within the molecular envelope. Liu et al. (2008) therefore favor a scenario with mid- and high-latitude winds confined primarily to a weather layer near the clouds. The equatorial jet may be an exception, since even if it were barotropic it does not penetrate deeper than the cutoff radius calculated by Liu et al. (2008).

Glatzmaier (2008) challenged the Liu et al. (2008) scenario on the grounds that they simply extrapolated inward the magnetic field and zonal flows using plausibility arguments (rather than, for example, a self-consistent 3D magnetohydrodynamics calculation) and that other plausible configurations for the deep zonal flows could imply much lower Ohmic dissipation, thereby circumventing the Liu et al. constraint. Non-zero Ohmic dissipation requires the poloidal component of magnetic field to have a component perpendicular to the lines of constant angular velocity of fluid motion. This is the case for the assumptions of Liu et al. Glatzmaier pointed out, however, that if one extrapolates the zonal jets inward not along cylinders parallel to the rotation axis but instead on lines parallel to the internal poloidal magnetic field, then the Ohmic dissipation would be zero. However, this configuration would require the Taylor-Proudman theorem to be significantly violated for the zonal jets in the molecular envelope, which is not an easy task. Glatzmaier (2008) proposed that this could occur if, for example, the convective Reynolds stress plays an important role in the force balance (thereby causing a violation of geostrophy, on which the Taylor-Proudman theorem is based). But, given Jupiter and Saturn's small heat fluxes, and hence the small power available for causing Reynolds stresses, the accelerations due to Reynolds stresses must likewise be weak compared with the Coriolis and pressure-gradient forces associated with the basic zonal-jet structure.

6.5.2 Distinguishing Deep-or-Shallow Structure from Deep-or-Shallow Forcing

Vasavada and Showman (2005) and Showman et al. (2006) emphasized that one must distinguish deep versus shallow models for the *structure* of the jets from deep versus shallow models for the *forcing* that maintains the jets. Most authors have implicitly assumed that shallow forcing (e.g., thunderstorms or baroclinic instabilities within the cloud layer) can only produce shallow jets (and thereby that the forcing must be deep if the jets are deep). Showman et al. (2006) and Lian and Showman (2008) constructed a model based on the primitive equations that explicitly showed, however, that deep jets can easily result from shallow forcing as well as from deep forcing. In their models, they applied a jet pumping confined to pressures less than a few bars, intended to represent the convergences of eddy momentum flux associated with weather-layer processes such as thunderstorms or baroclinic instabilities. These eddy accelerations were assumed zero at pressures exceeding a few bars. Their solutions, however, showed that the shallow eddy accelerations induce a meridional flow that (in the case of a giant planet with an almost neutrally stratified interior) extends to essentially arbitrary depths. The east-west Coriolis acceleration on these meridional motions then pumps the zonal jets at all depths, not just those within the weather layer. The result is a jet structure consisting of a baroclinic thermal-wind region in the weather layer (where the jet pumping occurs in their model) underlain by barotropic zonal jets, extending to the bottom of the domain, whose speeds can attain values similar to those at the top.[1] An implication is that the fast winds observed by the Galileo probe down to 20 bars on Jupiter (Atkinson et al. 1998) cannot be interpreted as implying that the zonal jets have a deep-convective origin, as has often been done in the literature.

It is also conceivable that *deep* forcing could drive *shallow* jets. For example, convection in the interior could generate waves that propagate upward and induce zonal winds at a low pressure detached from the generation region of these waves. Such a process is responsible, for example, for generating stratospheric zonal winds in the Quasi-Biennial Oscillation on Earth (Baldwin et al. 2001) and in the Quasi-Quadrennial Oscillations on Jupiter and perhaps the Semi-Annual Oscillation on Saturn (Leovy et al. 1991; Friedson 1999; Li and Read 2000; Fouchet et al. 2008; Orton et al. 2008).

6.5.3 Models for Jet Pumping

Vasavada and Showman (2005) provided an extensive review of the theories and models for pumping the zonal jets on Jupiter and Saturn up through early 2005. We refer the reader there for an introduction; here, we emphasize only the most recent developments. Jet-formation theories can loosely be categorized into the "shallow forcing" scenario, where jet pumping occurs by baroclinic instabilities, thunderstorms, or other weather-layer processes within the cloud layer, and the "deep forcing" scenario, where the jet pumping results from Reynolds stresses associated with convection within the molecular envelope. The past few years have seen significant development in both approaches.

As emphasized by Vasavada and Showman (2005), most modern theories for jet formation on giant planets assume

---

[1] Because Showman et al. (2006) and Lian and Showman (2008) used a shallow-atmosphere model, the jets penetrated downward vertically, but in reality these jets would penetrate downward along cylinders parallel to the rotation axis.

that the jets result from an inverse energy cascade modified by the β effect. By allowing propagating, dispersive Rossby waves, the β effect introduces anisotropy in the turbulent interactions that often lead to the production of jets with a latitudinal length scale of $L_R \sim \pi (U/\beta)^{1/2}$, called the Rhines scale, where U is a characteristic zonal wind speed (Rhines 1975). This mechanism was first explored for Jupiter by Williams (1978), and since that time, numerous numerical and laboratory studies have been performed that investigate jet formation by this process.

The traditional interpretation is that the Rhines scale is a transition wavenumber between turbulence (which acts at scales smaller than L_R) and Rossby waves (which act at scales larger than L_R). The idea is that small-scale turbulence experiences an inverse cascade that drives the energy to larger and larger length scales, but once the turbulence reaches scales of L_R, further transfer of the energy to larger scales would require wave-turbulence interactions, which are inefficient. Energy would thus pile up at scales comparable to L_R. In the older 2D-turbulence literature, the Rhines scale was thus often interpreted as an "arrest" or "halting" scale, as though the Rhines scale acted as a barrier to the inverse cascade. This picture was developed in the context of decaying turbulence studies (where turbulence is initialized in an initial state which then freely evolves without any further forcing or large-scale damping).

However, the above interpretation of L_R does not carry over to studies with small-scale forcing. Indeed, recent work has clarified that the β effect cannot provide a source of energy dissipation and thus does not act to "arrest" the inverse cascade in any meaningful sense (Galperin et al. 2006; Sukoriansky et al. 2007). Furthermore, in some contexts (known as the "zonostrophic regime"; Galperin et al. 2006, 2008), the wave-turbulence transition actually occurs at a wavelength $L_\beta \sim (\varepsilon/\beta^3)^{1/5}$ (where ε is the small-scale isotropic energy injection rate) which may be significantly smaller than L_R (Huang et al. 2001; Sukoriansky et al. 2007). In a system that lacks large-scale dissipation but that is forced at small scales, the kinetic energy increases continually over time, so the Rhines scale increases too because of its dependence on U. In a system with small-scale forcing and large-scale friction (e.g., Rayleigh drag), the drag removes the injected energy as it cascades upscale, so a statistical balance is reached with an approximately steady mean wind speed and hence L_R. In both cases (assuming the friction is not overly strong), L_R roughly represents the scale containing the most energy, which corresponds typically to the jet width.

The zonostrophic regime (Galperin et al. 2006, 2008) represents a particular situation found when L_R and L_β are widely separated (by at least a factor of 2), so that a significant inertial range develops between these scales. In this case, the turbulent energy transfers on scales smaller than L_β become highly anisotropic and tend to focus energy strongly into zonally symmetric flow components. The result is that more and more energy gets concentrated into the zonal jets until a dynamic equilibrium is reached between jet pumping and energy removal by large-scale friction processes and the barotropic instability of the jets themselves. Under these circumstances, the zonal mean jet structure may acquire a steeply sloped kinetic energy spectrum $E(k) \sim C_Z \beta^2 k^{-5}$ (where $C_Z \sim 0.5$; Huang et al. 2001, Sukoriansky et al. 2002), which contrasts with the isotropic spectrum $E(k) \sim C_K \varepsilon^{2/3} k^{-5/3}$ ($C_K \sim 4-6$) in the normal upscale energy cascade. Although ε is not well established for any of the gas giant planets, rough estimates can be obtained from the eddy-zonal flow conversion rates, $C(K_E, K_Z)$, found by Salyk et al. (2006) and Del Genio et al. (2007a) for Jupiter and Saturn. However, these conversion rates should not be confused with ε itself, which represents the isotropic spectral energy transfer rate between nearby 'shells' of similar $|k|$. Laboratory experiments in regimes approaching zonostrophic conditions (Read et al. 2004, 2007) suggest that $\varepsilon \leq 0.2 C(K_E, K_Z)$, indicating minimum values for k_β (non-dimensionalised by planetary radius) of ~ 70 for Jupiter and ~ 80 for Saturn. These may be compared with estimates of dimensionless k_R for these planets of ~ 10–20, indicating a substantial scale separation between k_β and that of the jets. This confirms that both Jupiter and Saturn lie well within the zonostrophic regime (Galperin et al. 2006, 2008), with implications for the anisotropic energy transfers within the circulation and the accumulation of kinetic energy into near-barotropic jets.

One further intriguing aspect of the zonostrophic regime is the occurrence of nonlinear Rossby-like traveling waves that do not obey the usual linear Rossby wave dispersion relation (Sukoriansky et al. 2008). These appear to have the character of forced modes, whose phase speeds are controlled by nonlinear interactions with larger amplitude (usually low wavenumber) waves. Their manifestation as unusually slowly moving Rossby-like waves is reminiscent of the slowly moving thermal waves identified on Saturn by Achterberg and Flasar (1996) (Section 6.4.3), although confirmation of this analogy will need much more detailed diagnostics of the wave flows on Saturn than are available currently.

Several 1-layer models have recently shed light on the weather-layer dynamics of Jupiter and Saturn in the presence of static stability (i.e., finite Rossby deformation radius L_d). Showman (2007) and Scott and Polvani (2007) performed forced-dissipative studies using the 1-layer shallow-water equations, which govern the evolution of a thin layer of constant-density fluid on a sphere (intended to represent the weather-layer in the 1.5 layer formulation; see Dowling and Ingersoll 1989). Showman (2007) forced the flow by episodically injecting isolated mass pulses, intended to represent the effects of episodic thunderstorms that transport mass into the weather layer. Damping, intended to mimic radiation, consisted of Newtonian relaxation that removed available potential energy by flattening the shallow-water layer. Scott and Polvani (2007), following earlier 2D studies, forced the flow by injecting turbulence randomly everywhere simultaneously; for damping they explored both frictional drag and radiative relaxation of layer thickness. Despite the differences in the formulation, both studies find that small L_d (less than a few thousand km) suppresses the β effect, leading to the production of numerous vortices rather than jets. This is consistent with 1-layer studies using the simpler quasigeostrophic (QG) equations (Okuno and Masuda 2003; Smith 2004; Theiss 2004). Because L_d decreases with increasing latitude (an effect of increasing Coriolis parameter), a critical latitude typically exists below which the flow is jet-dominated and above which the flow is vortex dominated.

In contrast, Li et al. (2006) performed a 1-layer QG study on a β plane, and found that multiple jets formed over a wide range of deformation radii and other parameters. The reasons for the differences are unclear. As in Showman (2007), forcing by moist convection in this study is represented by the injection of mass pulses that create local negative vorticity sources; these are balanced by a positive vorticity source associated with uniform radiative cooling. Perhaps the major difference is that mass is injected randomly in space in the Showman (2007) simulations, while in Li et al. (2006) mass pulses are triggered only when the thickness of the weather layer decreases below a threshold value. This restricts the convective source to cyclonic shear regions, as is observed on Jupiter and to some extent on Saturn (Section 6.4.2), and in some sense mimics the effect that organization of low-level convergence and divergence by the mean meridional circulation would have in a 3-D model. Anticyclonic vortices have been observed in the vicinity of convective outflow on Saturn (Porco et al. 2005), although imaging sequences do not yet show evidence that such features give up energy to the mean flow (Del Genio et al. 2007a).

In the forced shallow-water studies of Showman (2007) and Scott and Polvani (2007), like that of previous freely evolving shallow-water studies (Cho and Polvani 1996; Iacono et al. 1999; Peltier and Stuhne 2002), the equatorial jet is westward in the Jovian/Saturnian regime of small deformation radius and Rossby number. This behavior has long been considered a failing of the shallow-water model as applied to Jupiter and Saturn (Vasavada and Showman 2005). Recently, however, Scott and Polvani (2008) presented shallow-water simulations, forced by injection of small-scale turbulence and damped by large-scale radiative relaxation, that develop a superrotation analogous to that observed on Jupiter and Saturn. Within the context of their simulations, this result is robust and occurred over a wide range of deformation radii. Scott and Polvani (2008) suggest that the emergence of superrotation may be favored when the adopted damping is radiative relaxation rather than frictional drag. Nevertheless, the result is puzzling, because the simulations of Showman (2007) always developed equatorial subrotation despite the usage of a radiative damping. Further work is needed to resolve the issue.

Sayanagi et al. (2008) performed an idealized study, using the primitive equations, of 3D decaying stably stratified turbulence on a β plane. The motivation was to determine whether Jupiter- and Saturn-like jets form in 3D and investigate whether a small deformation radius can suppress jet formation in 3D as it can in the one-layer QG and shallow-water systems. The simulations showed, under conditions relevant to Jupiter and Saturn, that jets can indeed form. If the first-baroclinic deformation radius was substantially smaller than the barotropic Rhines scale $\pi (U/\beta)^{1/2}$, jet formation was inhibited, consistent with the one-layer studies. Interestingly, the simulations also showed that the jets became vertically coherent over a vertical length scale Lf/N, where L is the latitudinal width of the jets. This result is consistent with the scaling proposed by Charney (1971) that the inverse energy cascade is isotropic in three dimensions when the vertical coordinate is multiplied by N/f.

Motivated by the possibility that Jupiter and Saturn's jets result from baroclinic instabilities in the cloud layer, Kaspi and Flierl (2007) performed a two-layer QG study driven by imposed temperature contrasts on a β plane. The novel aspect here was the use of a shallow-atmosphere positive value of β in the top layer and a negative value, relevant for the deep interior, in the bottom layer. The top layer thus represents the shallow weather layer and the bottom layer represents the deep interior (Ingersoll and Pollard [1982] showed that if fluid columns act as Taylor columns that span the entire planet, the spherical geometry leads to a negative value of β). Kaspi and Flierl (2007) found that the two-beta model promotes baroclinic instability relative to a model with equal beta in both layers; their simulations developed multiple robust zonal jets.

Lian and Showman (2008) presented 3D simulations driven by shallow weather-layer forcing that develop numerous zonal jets similar to those on Jupiter and Saturn. The forcing consisted of latitudinal temperature gradients imposed via Newtonian heating at pressures less than $\sim$10 bars. The shallow temperature contrasts led to baroclinic instabilities that pumped the jets in the weather layer. Intriguingly, as described above, the jets developed deep barotropic components. In the simulations, the weather-layer eddy convergences induce meridional circulation cells that penetrate deeply into the interior, and the east-west Coriolis force acting on this motion pumps the deep jets. In the Jupiter-like cases, typically $\sim$25 jets occurred, similar to observations. In some cases, in agreement with observations, the simulated jets violate the barotropic and Charney-Stern stability criteria, achieving curvatures of the zonal wind with northward distance, $\partial^2 u/\partial y^2$, up to 2β, and a hyperstaircase potential vorticity profile similar to that observed is produced (Section 6.3.8). Some simulations also developed a Jupiter-like equatorial superrotation, but only if the forcing was chosen to induce a latitudinal temperature gradient near the equator. A similar response to imposed equatorial forcing had earlier been obtained in the 3-D primitive equation model of Yamazaki et al. (2005).

In Lian and Showman (2008), the Newtonian heating was intended to crudely represent possible latitudinal variations of latent heating, solar-energy absorption, or other cloud-layer heating mechanisms. The latitudinal variation of this heating function was a free parameter, and superrotation only resulted for particular choices of this function. Lian and Showman (2009) extended this work by self-consistently including a simple representation of moist convection and removing arbitrary heating functions. In addition to the momentum, heat, and continuity equations, they solved an equation for the transport of water vapor in the domain. Wherever water vapor was supersaturated, they reduced its abundance to saturation and applied the appropriate latent heating to the temperature equation. No microphysics was included; cloud particles were assumed to rain downward to the bottom of the domain. Radiative cooling was applied evenly throughout. In the simulations, ascending motion leads to supersaturation, which induces latent heating, causing a warming that enhances the ascending motion. In this way, a self-sustaining circulation with horizontal temperature contrasts develops; baroclinic instabilities then induce eddy accelerations that pump multiple east-west jet streams. Intriguingly, for a Jupiter-like simulation with three times solar water, an equatorial superrotation spontaneously develops (Fig. 6.24, left). When Uranus/Neptune cases were explored with 30 times solar water, equatorial *subrotation* occurred, reminiscent of the observed equatorial subrotation on Uranus and Neptune.

Schneider and Liu (2009) performed 3D weather-layer simulations of Jupiter using the dry primitive equations. Their simulations extended to $\sim$3 bars pressure and contained a linear drag poleward of 33° at the base of the model to represent Ohmic dissipation in the deep interior (Liu et al. 2008). Their approach is generally similar to that of Lian and Showman (2009); however, they adopted a simple grey radiative-transfer scheme, imposed an intrinsic flux at the base of the model, and included a convective parameterization that transported thermal energy whenever conditions approached neutrally stable. Schneider and Liu's (2009) simulations developed multiple east-west midlatitude jets (whose speed depends on the strength of the imposed drag) and a Jupiter-like equatorial superrotation. They suggest that the superrotation seen in their simulations results from wave/mean-flow interactions associated with Rossby waves generated by convective events near the equator.

Several advances have occurred over the past several years in 3D simulations of convection in giant planet interiors. Previous 3D convection models adopted relatively thick shells with inner/outer radius ratios typically less than $\sim$0.7, and generally produced only a few very broad jets (Aurnou and Olson 2001; Christensen 2001, 2002). Although these early studies always produced a superrotating equatorial jet, a positive feature, there was a discouraging lack of resemblance between the simulated jet profiles and those observed on Jupiter and Saturn. Recently, however, Heimpel et al. (2005) and Heimpel and Aurnou (2007) performed 3D simulations with much thinner shells (inner/outer radius ratios of 0.85 and 0.9). Like the earlier studies, these simulations are Boussinesq, meaning that they assume a basic-state density that is independent of radius, and have free-slip momentum boundary conditions at both the top and bottom. Interestingly, when the control parameters (Rayleigh and Ekman number) are tuned to obtain a Jupiter-like wind speed, these thin-shell simulations develop a surprisingly Jupiter- and Saturn-like wind profile with a superrotating jet and numerous weaker, midlatitude eastward and westward jets (Fig. 6.24, right). The convective heat fluxes predicted by

Fig. 6.24 Examples of the zonal wind field in 3-D numerical models of gas giant atmospheres that produce both equatorial superrotation and multiple higher latitude alternating jets. *Left*: The 3-D shallow atmosphere model of Lian and Showman (2009) with 3× solar water abundance. *Right*: The 3-D convective cylinder model of Heimpel and Aurnou (2007) with inner/outer shell radius = 0.85 (warm/cool colors indicate eastward/westward flow)

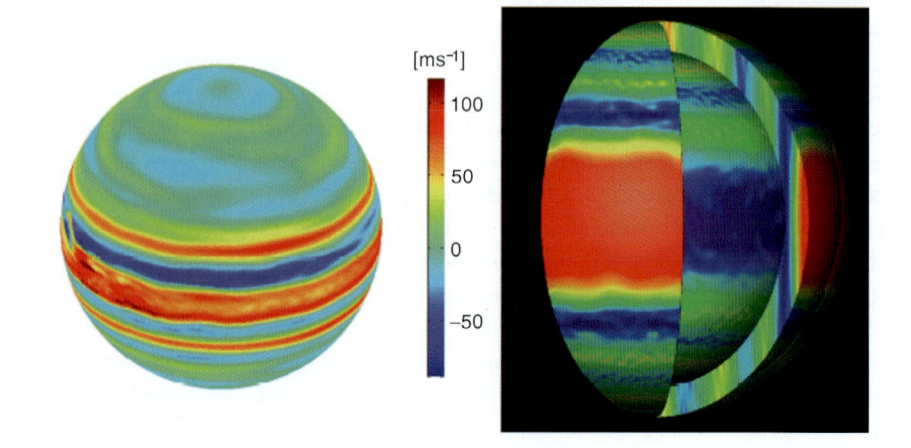

these models exhibit a minimum at the equator and maxima at the poles, consistent with the observationally inferred latitude variation of internal heat loss (Ingersoll 1976).

These are promising results. Nevertheless, the meaning of the lower boundary in these simulations is unclear. The strong barotropic jets imply the existence of strong horizontal pressure contrasts, which in these models are supported by the impermeable lower boundary. On Jupiter and Saturn, however, there is no internal boundary that could support such pressure variations. Instead, the jets would need to decay with depth before reaching the metallic region, implying the existence in the molecular envelope of thermal-wind shear that would require large lateral density contrasts to support. It is not obvious whether there exists any mechanism to produce such density contrasts (Liu et al. 2008). An alternate study that adopts a no-slip boundary condition at the bottom, perhaps more relevant to the magnetohydrodynamic drag expected there, develops a rapid equatorial jet but very weak mid- and high-latitude jets (Aurnou and Heimpel 2004). This suggests that explaining Jupiter and Saturn's mid- and high-latitude jets in the deep-convection models may be problematic.

The superrotating equatorial jet in the 3D Boussinesq simulations of Aurnou and Olson (2001), Christensen (2001, 2002), Heimpel et al. (2005), and Heimpel and Aurnou (2007) results from the interaction of columnar convective rolls with the curving spherical planetary surface (Busse 2002; Vasavada and Showman 2005). This mechanism, which can be viewed as a "topographic" β effect, requires convective rolls that act as coherent Taylor columns (i.e., the convective rolls must remain coherent along their entire length from the northern to the southern boundary). As Glatzmaier et al. (2009) point out, current 3D convection studies are relatively laminar (a result of the low resolution necessary in a 3D study), which – combined with the fact that the simulated convective structures are relatively large-scale – promotes such columnar coherence. In reality, convective plumes are probably much narrower than current models can resolve; because of their small scales, such plumes are less likely to be geostrophic and thus may not exhibit the columnar structure necessary for the topographic β effect to occur.[2]

Instead, Glatzmaier et al. (2009) propose that equatorial superrotation on Jupiter and Saturn results from the interaction of convection with the radial density gradient (an effect that is ignored in the Boussinesq models). On Jupiter and Saturn, the density varies by a factor of ~1,000 from the cloud layer to the deep interior, and most of this density variation

occurs near the surface. As a rising plume expands, it spins up anticyclonically, leading to a wavelike behavior that propagates eastward, with faster eastward propagation for fluid elements farther from the rotation axis. This radial dependence of the eastward propagation speed causes an eastward tilt in ascending convective plumes and a westward tilt in descending convective plumes (Glatzmaier et al. 2009). The Reynolds stresses acting on such tilted convective columns transport eastward momentum radially outward and westward momentum radially inward (Busse 2002; Vasavada and Showman 2005; Glatzmaier et al. 2009), leading to differential rotation with a superrotating jet at the equatorial surface. The advantage of this mechanism over the topographic β effect is that it is *local* and does not require convective columns to span the entire planet as a Taylor column. Evonuk and Glatzmaier (2006, 2007), Evonuk (2008), and Glatzmaier et al. (2009) presented 2D numerical simulations in the equatorial plane to confirm these qualitative arguments. However, a possible stumbling block for this mechanism is that it would also predict equatorial superrotation on Uranus and Neptune, where equatorial subrotation is observed instead.

6.5.4 Do the Observations Constrain Our Models?

The most basic observational constraints to be met by any model of jet formation are the observed profiles of zonal wind and temperature with latitude in the upper troposphere; specifically, the existence of a strong superrotating equatorial jet, numerous weaker high-latitude jets, and meridional temperature variations of only a few K. Recent numerical models have made significant advances in matching these constraints. Three independent shallow-atmosphere models now exist that, under Jupiter- and Saturn-like conditions, spontaneously produce multiple midlatitude jets and a strong superrotating equatorial jet similar to those on Jupiter or Saturn - the one-layer shallow-water model of Scott and Polvani (2008) and the 3D primitive-equation models of Lian and Showman (2009) and Schneider and Liu (2009). In contrast to some previous shallow-atmosphere studies, these models produce equatorial superrotation naturally, without ad-hoc forcing functions. Likewise, in the deep-convection arena, Boussinesq models in a thin shell (Heimpel et al. 2005; Heimpel and Aurnou 2007) show promise in explaining not only the equatorial superrotation (long a strength of the deep-convective models) but also the alternating mid-latitude jets (something that thick-shell convection models had previously failed to do). Nevertheless, many areas for improvement remain in both classes of models, and the approaches must ultimately be merged if we are to obtain a full understanding of the general circulation within giant planets.

---

[2] The *jets*, however, are large scale and should still exhibit a relatively columnar structure, since the Taylor-Proudman theorem applies to a geostrophically balanced, barotropic fluid regardless of whether the density varies with radius.

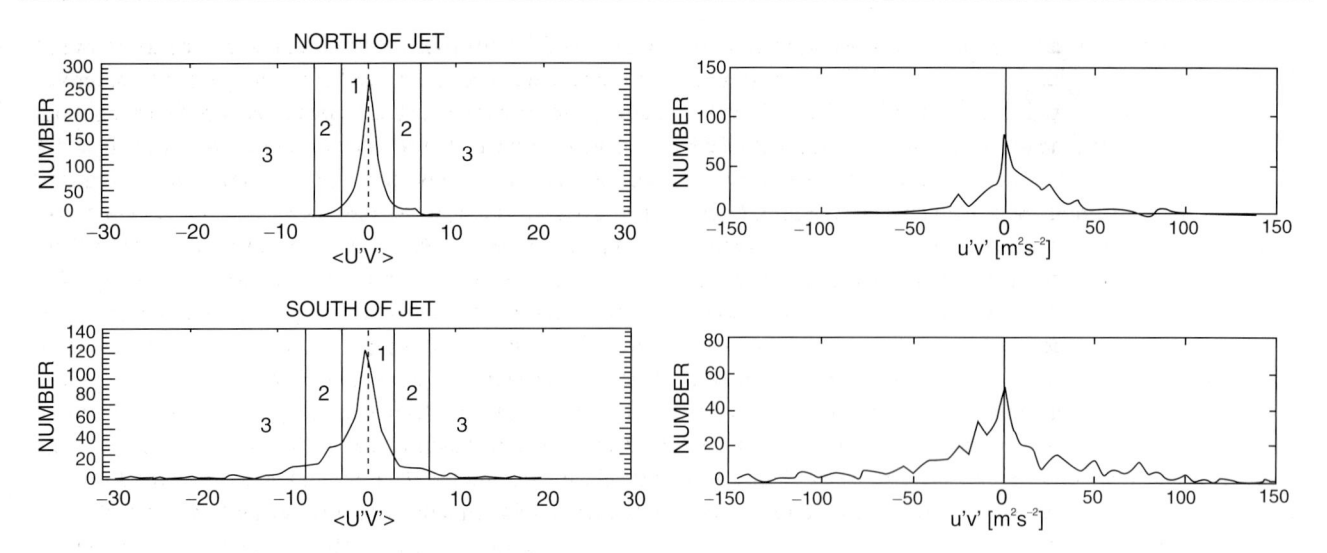

Fig. 6.25 Histograms of eddy momentum flux north and south of a westward jet (*left*) observed in Cassini ISS images (Del Genio et al. 2007a) and (*right*) produced by baroclinic instability in the simulations of Lian and Showman (2008)

Other observational constraints also exist. Cloud tracking has allowed estimation of the horizontal eddy-momentum fluxes $u'v'$; these results imply that eddies pump momentum up-gradient into the jets on both Saturn and Jupiter (Salyk et al. 2006; Del Genio et al. 2007a; Section 6.3.3). Given the observed $u'v'$ values, the power per unit area associated with the conversion of eddy to zonal mean kinetic energy would exceed Saturn's luminosity if the fluxes extended beyond ∼2 bars; thus, these eddies (or at least the accelerations they induce) appear to be truly shallow. Even then, the implied kinetic energy fluxes approach ∼1 W m^{-2}, which is a substantial fraction of the total energy available for doing work. These results imply that the baroclinic atmosphere cannot be neglected when modeling jet pumping on the giant planets.

What is the process responsible for producing these jet-pumping eddies? Histograms of the $u'v'$ values at particular latitudes on jet flanks show that the distribution peaks around zero, with long tails at both positive and negative $u'v'$, but that the distribution is skewed such that the positive tail dominates on the south flanks of eastward jets and the negative tail dominates on the north flanks of eastward jets (Fig. 6.25, left). Such a skewed distribution implies that, *on average*, eddies pump momentum up-gradient into jet cores. Interestingly, shallow-atmosphere models driven by baroclinic instabilities in the weather layer explain these distributions surprisingly well (Lian and Showman 2008; Fig. 6.25, right). This agreement supports the possibility that baroclinic instabilities pump the jets at cloud level. However, other cloud-level sources of turbulence, such as thunderstorms, might be capable of generating the observed fluxes. Indeed laboratory studies of geostrophic turbulence on a β-plane, dominated either by baroclinic instabilities (Bastin and Read 1998; Wordsworth et al. 2008) *or* unstably-stratified convection (Condie and Rhines 1994; Read et al. 2004, 2007), show a strong tendency to develop eddy-driven barotropic jets, so the question as to which process dominates on the gas giant planets must remain open until further evidence on the respective characteristics of baroclinic instability and deep convection becomes available.

6.6 Discussion

Our current tentative understanding of Saturn's atmospheric thermal structure and circulation, based on direct measurements and inferences from Cassini data, can be summarized as follows. Moist adiabatic ascent, radiative cooling aloft, and compensating dry adiabatic descent produce a reversal of the meridional temperature gradient between the water cloud level and the upper troposphere, such that cyclonic regions are cool and anti-cyclonic regions warm near the water cloud base while the opposite is true near and above the visible clouds (Showman and de Pater 2005). Between the water condensation level and the level of reversal of dT/dy, zonal winds should strengthen with height according to the thermal wind equation. Above this level, the thermal wind decays with height as observed by CIRS down to at least 800 mb. Since VIMS-derived winds are slightly stronger than ISS-derived winds, the reversal of dT/dy probably occurs below the ammonium hydrosulfide condensation level. The meridional circulation reverses sign at a somewhat higher level, near the altitude sensed in ISS images.

Despite this more comprehensive emerging picture, many questions still remain. Whether the zonal winds strengthen, decay, or remain constant down to and below the (still unknown) water cloud level is still unknown. How the inferred circulation relates to near-infrared continuum and

5 μm cloud patterns on Saturn, and why the wind-albedo correlation differs from that on Jupiter, is still uncertain. Which class of candidate mechanisms (or what combination of them) drives the observed cloud level circulation at different latitudes is yet to be determined. The eddy process that forces the mechanically driven upper troposphere circulation is also not certain. Seasonal variability is already evident near the tropopause and in the stratosphere, but the Cassini dataset itself only covers a small fraction of a Saturn year. The actual strength of Saturn's prograde equatorial jet, the reason it exists, and its more temporally variable nature relative to other latitudes, remain a mystery. Below we describe advances that we anticipate taking place in the modeling arena to more fully exploit the Cassini dataset and suggest observational strategies that could address many of the remaining questions.

6.6.1 Future Modeling Directions

Modeling of the dynamics of the gas giant planets has progressed tremendously in a short time, due both to increases in computing power and an emerging understanding of the required physics. To continue moving forward, the major weaknesses of each class of models must be addressed. Idealized models have contributed to understanding of the inverse cascade, the significance of the Rhines scale, and the conditions required for a prograde equatorial jet. The next step must be to identify and simulate the specific forcing and dissipation mechanisms responsible for the behavior seen in these models. Shallow weather layer models, which depend on moist convection as either the direct source of eddies that maintain the flow or the generator of temperature gradients that drive baroclinic instability, must begin to implement state-of-the-art cumulus parameterizations to become credible. Cumulus parameterization is itself a topic of active research, and the adaptation of such models for Saturn would benefit from the further use of cloud-resolving models (Hueso and Sánchez-Lavega 2004) to explore different assumptions about the cloud microphysics and background thermodynamic structure. Current shallow weather layer models are hydrostatic, which limits their ability to realistically portray the dynamical coupling between the visible cloud level and water condensation level, which are separated by a thick near-adiabatic layer. Global non-hydrostatic fine resolution models are imminent in Earth atmospheric science and ultimately should be pursued for the gas giants as computing power increases.

The deep convective cylinder class of models is even more computationally challenging because they simulate a much greater depth of the atmosphere. They are also limited by our lack of knowledge of the interiors of gas giant planets. The central questions that have emerged for these models are the depth to which the cylindrical configuration and the associated zonal flow extends and the physics that determines this lower boundary. Two research directions must be pursued to address these issues. First, most cylinder models are Boussinesq for computational convenience. 2-D simulations that relax this assumption produce distinctly different behavior of the convective plumes and suggest that some of the theoretical difficulties of the cylinder approach may be overcome in the presence of realistic stratification (Glatzmaier et al. 2009), but eventually it must become possible to conduct such simulations in 3-D to determine whether these ideas are truly viable. Second, fully 3-D magnetohydrodynamic simulations will be required to settle the issue of Ohmic dissipation constraints on the inner cylinder radius.

To date, simulations of Saturn's stratosphere have been largely restricted to 1-D chemistry-diffusion models, from which zeroth-order inferences about the dynamics could be made (Moses and Greathouse 2005). These are beginning to give way to 2-D interactive chemistry-dynamics models, which will be better positioned to provide insights into relationships seen in the CIRS and IRTF datasets. Already such models have raised questions about the processes that produce the differing latitudinal profiles of C_2H_6 and C_2H_2 on Saturn (Moses et al. 2007; see Chapter 5). However, even 2D models must be utilized with care, since 3D eddies can substantially influence the strength and direction of tracer transport in the zonal mean. Although 2D transport models can include a parameterization of such eddy transports, such as through the Transformed Lagrangian Mean approximation (e.g., Andrews et al. 1987), the accurate representation of fully 3D transport needs a fully 3D model. So a long term goal should be to adapt fully 3D general circulation models to incorporate interactive chemistry, much as is being developed by the terrestrial atmosphere modeling community.

The Cassini mission has provided a series of observational constraints that any model must satisfy. In addition to zonal wind profiles and temperature patterns, any viable model must also simulate other indices of the flow such as the distribution of convective clouds and disequilibrium gas concentrations, meridional and vertical velocity patterns, and process-level diagnostics such as potential vorticity and eddy momentum fluxes. Many published giant planet model simulations are limited in their diagnosis of the structure of the circulation and the energy and momentum balances that produce it, thus making it difficult to evaluate them. Future modeling studies should propose observables that would be indicative of the mechanisms at work on the gas giant planets as input for the remote sensing community to develop an effective strategy for future exploration.

6.6.2 Long-Term Observational Needs

Many of our questions about Saturn's upper troposphere and stratosphere center on the role of seasonal forcing and will thus benefit greatly by observing at different points in the seasonal cycle. The Cassini Equinox mission should be sufficient to detect hemispheric shifts in the tropospheric haze and the response of the near-tropopause temperature structure to such changes. A potential Cassini Solstice mission extension would allow us to view seasonal shifts in the stratospheric meridional circulation and to detect downward propagation of temperature and wind anomalies that are the signature of the semi-annual and quasi-biennial oscillations on Earth, if such behavior exists on Saturn.

At lower altitudes, most questions concern the structure and dynamics of the atmosphere below the visible cloud level. For Jupiter, NASA's Juno mission, scheduled to launch in 2011 and enter Jupiter orbit in 2016, will provide key constraints on the depth to which the jets extend below the clouds. For Saturn, the Cassini orbital configuration precludes such measurements because the orbit is quasi-equatorial and much too far from Saturn. However, end-of-mission scenarios under consideration would place Cassini in a highly inclined orbit with a periapse close to Saturn (below the rings), allowing the final mission phase to provide Juno-like constraints on the depth of Saturn's jets as well. Although caution will be needed in interpreting the results in terms of deep-or-shallow jet forcing (Vasavada and Showman 2005; Showman et al. 2006), such observations will nevertheless provide invaluable constraints for understanding the general circulation and the mechanisms of jet maintenance. Detailed measurements of the magnetic field at this time may also help constrain Saturn's actual deep interior rotation period.

The key feature of Juno for atmospheric dynamics is its multi-channel microwave radiometer, which will sound the water vapor profile down to about 100 bars. These observations will not only determine Jupiter's water abundance (by approximately locating the condensation level) but will also distinguish water vapor amounts in belts and zones and thereby constrain the moisture convergence/divergence at depth that we postulate to exist on the gas giants (Fig. 6.18). This information for Jupiter will obviously have ramifications for Saturn as well. However, intriguing differences in the pattern of moist convection on the two planets exist: On Jupiter convection is restricted to cyclonic shear regions (Porco et al. 2003), while on Saturn convective clouds (including some of the most vigorous storms) also appear in westward and eastward jets. Thus, in the long term, microwave radiometer water profiling of Saturn would be desirable. An alternative is a Saturn Multi-Probes Mission (Atreya 2003), which could sample both cyclonic and anti-cyclonic temperate regions on Saturn, with a possible third probe entering the apparently distinct equatorial region. Probes are less desirable than orbiters because they provide a sparse and potentially non-representative sample, as was the case for Galileo. Their advantage, however, is high vertical resolution, including the chance to detect and measure the strength of the hypothesized stable layer at depth that water condensation would create, which is crucial to shallow weather layer theories of driving the cloud level flow.

Eventually, active remote sensing must become viable for the giant planets. On Earth, lidars and millimeter cloud radars allow the thinnest hazes to be detected and precisely located in altitude and the thickest cloud systems to be profiled from top to bottom (Stephens et al. 2002). On Saturn, such instruments could accurately characterize the stratospheric and tropospheric hazes, determine the ammonia cloud top, definitively detect the ammonium hydrosulfide layer that to this point exists only in theory, and most importantly, characterize the nature of cloud structure at the water cloud level. This would tell us whether there is an extensive water cloud layer at all, as the early thermo-chemical models hypothesized, or whether water condensation is limited to scattered shallow cumulus and occasional deep penetrating cumulonimbus clouds, as is true of Earth's tropics. Active remote sensing is a challenge for the giant planets because of the mass of the instruments and the need for orbits to pass very close to the cloud tops to enhance the spatial resolution. Saturn in turn is more challenging than Jupiter because its water condensation level is likely to be much deeper. Until remote sensing down to the water condensation level and below becomes a reality, however, it will be difficult for the numerous theories of Saturn's general circulation to be regarded as anything more than simply plausible ideas.

Acknowledgments This work was supported by NASA and ESA through the Cassini-Huygens Mission and the NASA Planetary Atmospheres Program. We thank John Barbara and Lilly del Valle for assistance with several figures and two anonymous reviewers for constructive comments.

References

Acarretta, J.R., Sánchez-Lavega, A.: Vertical cloud structure in Saturn's 1990 equatorial storm. Icarus **137**, 24–33 (1999).

Achterberg, R.K., Flasar, F.M.: Planetary-scale thermal waves in Saturn's upper troposphere. Icarus **119**, 350–369 (1996).

Aguiar, A.C.B., et al.: A laboratory model of Saturn's north polar hexagon. Icarus, submitted (2009)

Allison, M.: A similarity model for the windy jovian thermocline. Planet. Space Sci. **48**, 753–774 (2000).

Allison, M., Stone, P.H.: Saturn meteorology: A diagnostic assessment of thin-layer configurations for the zonal flow. Icarus **54**, 296–308 (1983).

Allison, M., Godfrey, D.A., Beebe, R.F.: A wave dynamical interpretation of Saturn's polar hexagon. Science **247**, 1061–1063 (1990).

Allison, M., Del Genio, A.D., Zhou, W.: Richardson number constraints for the Jupiter and outer planet wind regime. Geophys. Res. Lett. **22**, 2957–2960 (1995).

Anderson, J.D., Schubert, G.: Saturn's gravitational field, internal rotation, and interior structure, Science **317**, 1384–1387 (2007).

Andrews, D. G., Holton, J. R., Leovy, C. B.: *Middle Atmosphere Dynamics*. Academic Press, New York (1987).

Arnol'd, V.I.: On an a priori estimate in the theory of hydrodynamic stability. Am. Math. Soc. Transl. Ser. 2 **79**, 267–269 (1966).

Atkinson, D.H., Pollack, J.B., Seiff, A.: The Galileo probe Doppler wind experiment: Measurement of the deep zonal winds on Jupiter. J. Geophys. Res. **103**, 22,911–22,928 (1998).

Atreya, S.K.: Composition, clouds, and origin of Jupiter's atmosphere – A case for deep multiprobes into giant planets. In Wilson, A. (ed.), *Proceeding of the International Workshop 'Planetary Probe Atmospheric Entry and Descent Trajectory Analysis and Science'*, pp. 57–62, ESA SP-544, Noordwijk, Netherlands (2003).

Aurnou, J.M., Olson, P.L.: Strong zonal winds from thermal convection in a rotating spherical shell. Geophys. Res. Lett. **28**, 2557–2559 (2001).

Aurnou, J.M., Heimpel, M.H.: Zonal jets in rotating convection with mixed mechanical boundary conditions. Icarus **169**, 492–498 (2004).

Bagenal, F.: A new spin on Saturn's rotation. Science **316**, 380–381 (2007).

Baines, K.H., et al.: The atmospheres of Saturn and Titan in the near-infrared: First results of Cassini/VIMS. Earth, Moon, Plan. **96**, 119–147 (2005).

Baines, K.H., et al.: To the depths of Venus: Exploring the deep atmosphere and surface of our sister world with Venus Express. Planet. Space Sci. **54**, 1263–1278 (2006).

Baines, K.H., et al.: Saturn's north polar cyclone and hexagon at depth revealed by Cassini/VIMS. Planet. Space Sci., in press (2009a).

Baines, K.H., et al.: Storm clouds on Saturn: Lightning-induced chemistry and associated materials consistent with Cassini/VIMS spectra. Planet. Space Sci., in press (2009b).

Baines, K.H., et al.: The deep clouds of Saturn: Morphology, spatial distribution, and dynamical implications as revealed by Cassini/VIMS. Icarus, submitted (2009c).

Baldwin, M.P., et al.: The quasi-biennial oscillation. Rev. Geophys. **39**, 179–229 (2001).

Barnet, C.D., Beebe, R.F., Conrath, B.J: A seasonal radiative-dynamic model of Saturn's troposphere. Icarus **98**, 94–107 (1992).

Bastin, M.E., Read, P.L.: Experiments on the structure of baroclinic waves and zonal jets in an internally heated rotating cylinder of fluid. Phys. Fluids **10**, 374–389 (1998).

Beebe, R.F., et al.: The onset and growth of the 1990 equatorial disturbance on Saturn. Icarus **95**, 163–17 (1992).

Bézard, B., Gautier, D., Conrath, B: A seasonal model of the Saturnian upper troposphere: Comparison with Voyager infrared measurements. Icarus **60**, 274–288 (1984).

Bézard, B., Gautier, D.: A seasonal climate model of the atmospheres of the giant planets at the Voyager encounter time. Icarus **61**, 296–310 (1985).

Brown, R.H., et al.: The Cassini visual and infrared mapping spectrometer (VIMS) investigation. Space Sci. Rev. **115**, 111–168 (2004).

Busse, F.H.: Convective flows in rapidly rotating spheres and their dynamo action. Phys. Fluids **14**, 1301–1314 (2002).

Caldwell, J., et al.: The drift of Saturn's north polar spot observed by the Hubble Space Telescope. Science **260**, 326–329 (1993).

Carlson, B.E., Rossow, W.B., Orton, G.S.: Cloud microphysics of the giant planets. J. Atmos. Sci. **45**, 2066–2081 (1988).

Cess, R.D., Caldwell, J.: A Saturnian stratospheric seasonal climate model. Icarus **38**, 349–35, (1979).

Charney, J.G.: Geostrophic turbulence. J. Atmos. Sci. **28**, 1087–1095 (1971).

Charney, J. G., Drazin, P. G.: Propagation of planetary-scale disturbances from the lower into the upper atmosphere, J. Geophys. Res. **66**, 83–109 (1961).

Cho, J.Y.-K., Polvani, L.M.: The emergence of jets and vortices in freely evolving, shallow-water turbulence on a sphere. Phys. Fluids **8**, 1531–1552 (1996).

Choi, D.S., Showman, A.P., Brown, R.H.: Cloud features and zonal wind measurements of Saturn's atmosphere as observed by Cassini/VIMS. J. Geophys. Res. **114**, E04007, doi:10.1029/2008JE003254 (2009).

Christensen, U.R.: Zonal flow driven by deep convection in the major planets. Geophys. Res. Lett. **28**, 2553–2556 (2001).

Christensen, U.R.: Zonal flow driven by strongly supercritical convection in rotating spherical shells. J. Fluid Mech. **470**, 115–133 (2002).

Condie, S. A., Rhines, P. B.: A convective model for the zonal jets in the atmospheres of Jupiter and Saturn. Nature **367**, 711–713 (1994).

Conrath, B.J., Gierasch, P.J.: Global variation of the *para* hydrogen fraction in Jupiter's atmosphere and implications for dynamics on the outer planets. Icarus **57**, 184–204 (1984).

Conrath, B.J., Pirraglia, J.A.: Thermal structure of Saturn from Voyager infrared measurements: Implications for atmospheric dynamics. Icarus **53**, 286–292 (1983).

Conrath, B.J., Gierasch, P.J., Leroy, S.S.: Temperature and circulation in the stratosphere of the outer planets. Icarus **83**, 255–281 (1990).

Conrath, B.J., Gierasch, P.J., Ustinov, E.A.: Thermal structure and para hydrogen fraction on the outer planets from Voyager IRIS measurements. Icarus **135**, 501–517 (1998).

Del Genio, A.D., McGrattan, K.B.: Moist convection and the vertical structure and water abundance of Jupiter's atmosphere. Icarus **84**, 29–53 (1990).

Del Genio, A.D., et al.: Saturn eddy momentum fluxes and convection: First estimates from Cassini images, Icarus **189**, 479–492 (2007a).

Del Genio, A.D., Yao, M.-S., Jonas, J.: Will moist convection be stronger in a warmer climate? Geophys. Res. Lett. **34**, L16703, doi:10.1029/2007GL030525 (2007b).

Derbyshire, S.H., et al.: Sensitivity of moist convection to environmental humidity. Quart. J. Roy. Meteor. Soc. **130**, 3055–3079 (2004).

Desch, M.D., Kaiser, M.L.: Voyager measurement of the rotation period of Saturn's magnetic field. Geophys. Res. Lett. **8**, 253–256 (1981).

Dowling, T. E.: Dynamics of Jovian atmospheres, Ann. Rev. Fluid Mech., **27**, 293–334 (1995).

Dowling, T. E., Ingersoll, A.P.: Potential vorticity and layer thickness variations in the flow around Jupiter's Great Red Spot and White Oval BC. J. Atmos. Sci. **45**, 1380–1396 (1988).

Dowling, T. E., Ingersoll, A.P.: Jupiter's Great Red Spot as a shallow water system. J. Atmos. Sci. **46**, 3256–3278 (1989).

Drossart, P.D.: Scientific goals for the observation of Venus by VIRTIS on ESA/Venus Express mission. Planet. Space Sci. **55**, 1653–1672 (2007).

Dunkerton, T.: On the mean meridional mass motions of the stratosphere and mesosphere. J. Atmos. Sci. **35**, 2325–2333 (1978).

Dyudina, U.A., et al.: Lightning storms on Saturn observed by Cassini ISS and RPWS during 2004–2006. Icarus **190**, 545–555 (2007).

Dyudina, U.A., et al.: Dynamics of Saturn's south polar vortex. Science **319**, 1081 (2008).

Dyudina, U.A., et al.: Saturn's south polar vortex compared to other large vortices in the solar system. Icarus **202**, 240–248 (2009).

Elachi, C., et al.: RADAR: The Cassini Titan radar mapper. Space Sci. Rev. **115**, 71–110 (2004).

Eliassen, A., Palm, E.: On the transfer of energy in stationary mountain waves, Geophys. Publ. **22**(3), 1–23 (1961).

Emanuel, K.: Tropical cyclones. Annu. Rev. Earth Planet. Sci. **31**, 75–104 (2003).

Evonuk, M., Glatzmaier, G.A.: A 2D study of the effects of the size of a solid core on the equatorial flow in giant planets. Icarus **181**, 458–464 (2006).

Evonuk, M., Glatzmaier, G.A.: The effects of rotation rate on deep convection in giant planets with small solid cores. Planet. Space Sci. **55**, 407–412 (2007).

Evonuk, M.: The role of density stratification in generating zonal flow structures in a rotating fluid. Astrophys. J. **673**, 1154–1159 (2008).

Fischer, G., et al.: Saturn lightning recorded by Cassini/RPWS in 2004. Icarus **183**, 135–152 (2006).

Fischer, G., et al.: Analysis of a giant lightning storm on Saturn. Icarus **190**, 528–544 (2007).

Flasar, F.M., et al.: An intense stratospheric jet on Jupiter. Nature **427**, 132–135 (2004a).

Flasar, F.M., et al.: Exploring the Saturn system in the thermal infrared: The composite infrared spectrometer. Space Sci. Rev. **115**, 169–297 (2004b).

Flasar, F.M., et al.: Temperatures, winds and composition in the Saturnian system. Science **307**, 1247–1251 (2005).

Fletcher, L.N., et al.: The meridional phosphine distribution in Saturn's upper troposphere from Cassini/CIRS observations. Icarus **188**, 72–88 (2007a).

Fletcher, L.N., et al.: Characterising Saturn's vertical temperature structure from Cassini/CIRS. Icarus **189**, 457–478 (2007b).

Fletcher, L.N., et al.: Saturn's atmosphere: Structure and composition from Cassini/CIRS. PhD dissertation, University of Oxford, UK (2007c).

Fletcher, L.N., et al.: Temperature and composition of Saturn's polar hot spots and hexagon. Science **319**, 79–81 (2008).

Fouchet, T., et al.: An equatorial oscillation in Saturn's middle atmosphere. Nature **453**, 200–202 (2008).

Friedson, A.J.: New observations and modeling of a QBO-like oscillation in Jupiter's stratosphere. Icarus **137**, 34–55 (1999).

Frierson, D.M.W.: Midlatitude static stability in simple and comprehensive general circulation models. J. Atmos. Sci. **65**, 1049–1062 (2008).

Fruh, W.-G., Read, P. L.: Experiments on a barotropic rotating shear layer. Part 1. Instability and steady vortices, J. Fluid Mech. **383**, 143–173 (1999).

Galopeau, P.H.M., Lecacheux, A.: Variations of Saturn's radio period measured at kilometer wavelengths. J. Geophys. Res. **105**, 13,089–13,102 (2000).

Galperin, B., et al.: Anisotropic turbulence and zonal jets in rotating flows with a β-effect. Nonlin. Processes Geophys. **13**, 83–98 (2006).

Galperin, B., Sukoriansky, S., Dikovskaya, N.: Zonostrophic turbulence. Physica Scripta **T132** 014034, doi:10.1088/0031-8949/2008/T132/014034n(2008).

García-Melendo, E., Sánchez-Lavega, A., Hueso, R.: Numerical models of Saturn's long-lived anticyclones, Icarus **191**, 665–677 (2007).

García-Melendo, E., et al.: Vertical shears in Saturn's eastward jets at cloud level. Icarus **201**, 818–820 (2009).

Gezari, D.Y., et al.: New features in Saturn's atmosphere revealed by high-resolution thermal images. Nature **342**, 777–780 (1989).

Gierasch, P.J.: Jovian meteorology: Large-scale moist convection. Icarus **29**, 445–454 (1976).

Gierasch, P.J.: Radiative-convective latitudinal gradients for Jupiter and Saturn models with a radiative zone. Icarus **142**, 148–154 (1999).

Gierasch, P.J., Conrath, B.J., Magalhães, J.A.: Zonal mean properties of Jupiter's upper troposphere from Voyager infrared observations. Icarus **67**, 456–483 (1986).

Gierasch, P.J., et al.: Observation of moist convection in Jupiter's atmosphere. Nature **403**, 628–630 (2000).

Gierasch, P.J., Conrath, B.J., Read, P.L.: Nonconservation of Ertel potential vorticity in hydrogen atmospheres. J. Atmos. Sci. **61**, 1953–1965 (2004).

Gillet, F. C., Orton, G. S.: Center-to-limb observations of Saturn in the thermal infrared. Astrophys. J. **195**, L47-L49 (1975).

Glatzmaier, G.A.: A note on "Constraints on deep-seated zonal winds inside Jupiter and Saturn." Icarus **196**, 665–666 (2008).

Glatzmaier, G.A., Evonuk, M., Rogers, T.M.: Differential rotation in giant planets maintained by density-stratified turbulent convection. Geophys. Astrophys. Fluid Dyn **103**, 31–51 (2009).

Godfrey, D.A.: A hexagonal feature around Saturn's north pole. Icarus **76**, 335–356 (1988).

Godfrey, D.A.: The rotation period of Saturn's polar hexagon. Science **247**, 1206–1207 (1990).

Godfrey, D.A., Moore, V.: The Saturnian ribbon feature – A baroclinically unstable model. Icarus **68**, 313–343 (1986).

Greathouse, T.K., et al.: Meridional variations of temperature, C_2H_2 and C_2H_6 abundances in Saturn's stratosphere at southern summer solstice. Icarus **177**, 18–31 (2005).

Grote, E., Busse, F.H., Tilgner, A.: Regular and chaotic spherical dynamos. Phys. Earth Planet Int. **117**, 259–272 (2000).

Guillot, T.: Interiors of giant planets inside and outside the solar system. Science **286**, 72–77 (1999).

Guillot, T.: The interiors of giant planets: Models and outstanding questions. Ann. Rev. Earth Planet. Sci. **33**, 493–530 (2005).

Guillot, T., et al.: Are the giant planets fully convective? Icarus **112**, 354–367 (1994).

Gurnett, D.A., et al.: Radio and plasma wave observations at Saturn from Cassini's approach and first orbit. Science **307**, 1255–1259 (2005).

Gurnett, D.A., et al.: The variable rotation period of the inner region of Saturn's plasma disk. Science **316**, 442–445 (2007).

Hamilton, K.: Dynamics of the tropical middle atmosphere: A tutorial review. Atmos.-Ocean **36**, 319–354 (1998).

Hanel, R., et al.: Infrared observations of the Saturnian system from Voyager 1. Science **212**, 192–200 (1981).

Hanel, R.A., et al.: Albedo, internal heat flux, and energy balance of Saturn. Icarus **53**, 262–285 (1983).

Hartmann, D. L.: The general circulation of the atmosphere and its variability. J. Meteor. Soc. Japan **85B**, 123–143 (2007).

Heimpel, M., Aurnou, J., Wicht, J.: Simulation of equatorial and high-latitude jets on Jupiter in a deep convection model. Nature **438**, 193–196 (2005).

Heimpel, M., Aurnou, J.: Turbulent convection in rapidly rotating spherical shells: A model for equatorial and high latitude jets on Jupiter and Saturn. Icarus **187**, 540–557 (2007).

Held, I.M., Andrews, D.G.: On the direction of the eddy momentum flux in baroclinic instability. J. Atmos. Sci. **40**, 2220–2231 (1983).

Hesman, B.E., et al.: Saturn's latitudinal C_2H_2 and C_2H_6 abundance profiles from Cassini/CIRS and ground-based observations. Icarus **202**, 249–259 (2009).

Hess, S.L., Panofsky, H.A.: The atmospheres of the other planets. In Malone, T.F. (ed.), *Compendium of Meteorology*, pp. 391–400, Amer. Meteor. Soc., Boston (1951).

Hide, R., Lewis, S.R., Read, P.L.: Sloping convection: A paradigm for large-scale waves and eddies in planetary atmospheres? Chaos **4**, 135–162 (1994).

Howard, L. N., Drazln, P. G.: On instability of parallel flow of inviscid fluid in a rotating system with variable Coriolis parameter. J. Math. Phys. N. **43**, 83–99 (1964).

Howett, C.J.A., et al.: Meridional variations in stratospheric acetylene and ethane in the southern hemisphere of the saturnian atmosphere as determined from Cassini/CIRS measurements. Icarus **190**, 556–572 (2007).

Huang, H.-P., Galperin, B., Sukoriansky, S.: Anisotropic spectra in two-dimensional turbulence on the surface of a rotating sphere. Phys. Fluids **13**, 225–240 (2001).

Hueso, R., Sánchez-Lavega, A.: A three-dimensional model of moist convection for the giant planets. II. Saturn's water and ammonia moist convective storms. Icarus **172**, 255–271 (2004).

Hunt, G.E., et al.: Dynamical features in the northern hemisphere of Saturn from Voyager 1 images. Nature **297**, 132–134 (1982).

Iacono, R., Struglia, M.V., Roncji, C.: Spontaneous formation of equatorial jets in freely decaying shallow water turbulence. Phys. Fluids **11**, 1272–1274 (1999).

Ingersoll, A.P.: Pioneer 10 and 11 observations and the dynamics of Jupiter's atmosphere. Icarus **29**, 245–253 (1976).

Ingersoll, A.P., Cuzzi, J.N.: Dynamics of Jupiter's cloud bands. J. Atmos. Sci. **26**, 981–985 (1969).

Ingersoll, A.P., et al.: Interaction of eddies and zonal flow on Jupiter as inferred from Voyager 1 and 2 images. J. Geophys. Res. **86**, 8733–8743 (1981).

Ingersoll, A.P., Pollard, D.: Motion in the interiors and atmospheres of Jupiter and Saturn: Scale analysis, anelastic equations, barotropic-stability criterion. Icarus **52**, 62–80 (1982).

Ingersoll, A.P., Porco, C.C.: Solar heating and internal heat flow on Jupiter. Icarus **35**, 27–43 (1978).

Ingersoll, A.P., et al.: Structure and dynamics of Saturn's atmosphere. In Gehrels, T., Matthews, M.S. (eds.), *Saturn*, pp. 195–238. Univ. Arizona Press, Tucson (1984).

Ingersoll, A.P., et al.: Moist convection as the energy source for the large-scale motions in Jupiter's atmosphere. Nature **403**, 630–631 (2000).

Ingersoll, A.P., et al.: Dynamics of Jupiter's atmosphere. In Bagenal, F., et al. (eds.), *Jupiter: The Planet, Satellites, and Magnetosphere*, pp. 105–128. Cambridge Univ. Press, New York (2004).

Janssen, M.A., Allison, M., Lorenz, R.D.: Saturn's thermal emission at 2-cm wavelength and implications for atmospheric composition and dynamics. Poster, Saturn After Cassini-Huygens Symposium, London, UK (2008).

Kaiser, M.L., Connerney, J.E.P., Desch, M.D.: Atmospheric storm explanation of saturnian electrostatic discharges. Nature **303**, 50–53 (1983).

Karkoschka, E., Tomasko, M.G.: Saturn's upper troposphere 1986–1989. Icarus **97**, 161–181 (1992).

Karkoschka, E., Tomasko, M.: Saturn's vertical and latitudinal cloud structure 1991–2004 from HST imaging in 30 filters. Icarus **179**, 195–221 (2005).

Kaspi, Y., Flierl, G.R.: Formation of jets by baroclinic instability on gas planet atmospheres. J. Atmos. Sci. **64**, 3177–3194 (2007).

Kirk, R.L., Stevenson, D.J.: Hydromagnetic implications of zonal flows in the giant planets. Astrophys. J. **316**, 836–846 (1987).

Kurth, W.S., et al.: A Saturnian longitude system based on a variable kilometric radiation period. Geophys. Res. Lett. **34**, L02201, doi:10.1029/2006GL028336 (2007).

Kurth, W.S., et al.: An update to a Saturnian longitude system based on kilometric radio emissions. J. Geophys. Res. **113**, A05222, doi:10.1029/2007JA012861 (2008).

Leovy, C.B., Friedson, A.J., Orton, G.S.: The quasiquadrennial oscillation of Jupiter's equatorial stratosphere. Nature **354**, 380–382 (1991).

Li, L., Ingersoll, A.P., Huang, X.: Interaction of moist convection with zonal jets on Jupiter and Saturn. Icarus **180**, 113–123 (2006).

Li, X., Read, P.L.: A mechanistic model of the quasi-quadrennial oscillation in Jupiter's stratosphere. Planet. Space Sci. **48**, 637–669 (2000).

Lian, Y., Showman, A.P.: Deep jets on gas giant planets. Icarus **194**, 597–615 (2008).

Lian, Y., Showman, A.P.: Generation of zonal jets by moist convection on the giant planets. Icarus, submitted (2009).

Lindal, G.F., Sweetnam, D.N., Eshleman, V.R.: The atmosphere of Saturn: an analysis of the Voyager radio occultation measurements. Astron. J. **90**, 1136–1146 (1985).

Lindzen, R.S., Farrell, B., Tung, K.-K.: The concept of wave overreflection and its application to barotropic instability. J. Atmos. Sci. **37**, 44–63 (1980).

Liu, J, Goldreich, P.M., Stevenson, D.J.: Constraints on deep-seated zonal winds inside Jupiter and Saturn. Icarus **196**, 653–664 (2008).

Magalhaes, J.A., et al.: Slowly moving thermal features on Jupiter. Icarus **88**, 39–72 (1990).

Massie, S.T., Hunten, D.M.: Conversion of *para* and *ortho* hydrogen in the Jovian planets. Icarus **49**, 213–226 (1982).

Mitchell, J. L., Maxworthy, T.: Large-scale turbulence in the Jovian atmosphere. In Ghil, M., Benzi, R., Parisi, G. (eds.), *Turbulence and Predictability in Geophysical Fluid Dynamics and Climate Dynamics*, pp. 226–240, Societa Italiana di Fisica, Bologna, Italy (1985).

Moses, J. I., Greathouse, T.K.: Latitudinal and seasonal models of stratospheric photochemistry on Saturn: Comparison with infrared data from IRTF/TEXES. J. Geophys. Res. **110**, E09007, doi:10.1029/2005JE002450 (2005).

Moses, J.I., et al.: Meridional distribution of hydrocarbons on Saturn: Implications for stratospheric transport. Workshop on Planetary Atmospheres, Paper #9061, Greenbelt, Maryland (2007).

Nellis, W.J., et al.: Electronic energy gap of molecular hydrogen from electrical conductivity measurements at high shock pressures. Phys. Rev. Lett. **68**, 2937–2940 (1992).

Nellis, W.J., Weir, S.T., Mitchell, A.C.: Minimum metallic conductivity of fluid hydrogen at 140 GPa (1.4 Mbar). Phys. Rev. B **59**, 3434–3449 (1999).

Niino, H, Misawa, N.: An experimental and theoretical study of barotropic instability, J. Atmos. Sci. **41**, 1992–2011 (1984).

Okuno, A., Masuda, A.: Effect of horizontal turbulence on the geostrophic turbulence on a beta-plane: Suppression of the Rhines effect. Phys. Fluids **15**, 56–65 (2003).

Ollivier, J.L., et al.: Seasonal effects in the thermal structure of Saturn's stratosphere from infrared imaging at 10 microns. Astron. Astrophys. **356**, 347–356 (2000).

Orsolini, Y., Leovy, C.B.: A model of large-scale instabilities in the Jovian troposphere. 1. Linear model. Icarus **106**, 392–405 (1993a).

Orsolini, Y., Leovy, C.B.: A model of large-scale instabilities in the Jovian troposphere. 2. Quasi-linear model. Icarus **106**, 406–418 (1993b).

Orton, G. S., Ingersoll, A. P.: Saturn's atmospheric temperature structure and heat budget. J. Geopyhs. Res. **85**, 5871–5881 (1980).

Orton, G.S., et al.: Thermal maps of Jupiter: Spatial organization and time dependence of stratospheric temperatures, 1980 to 1990. Science **252**, 537–542 (1991).

Orton, G., Yanamandra-Fisher, P.A.: Saturn's temperature field from high-resolution middle-infrared imaging. Science **307**, 696–698 (2005).

Orton, G. A., et al.: Semi-annual oscillations in Saturn's low-latitude stratospheric temperatures. Nature **453**, 196–199 (2008).

Peltier, W.R., Stuhne, G.R.: The upscale turbulent cascade: shear layers, cyclones, and gas giant bands. In Pearce, R.P. (ed.), *Meteorology at the Millenium*, pp. 43–61, Academic Press, New York (2002).

Pérez-Hoyos, S., et al.: Saturn's cloud structure and temporal evolution from ten years of Hubble Space Telescope images (1994–2003). Icarus **176**, 155–174 (2005).

Pérez-Hoyos, S., Sánchez-Lavega, A., French, R. G.: Short-term changes in the belt/zone structure of Saturn's Southern Hemisphere (1996–2004). Astronomy and Astrophysics **460**, 641–645 (2006).

Piccioni, G., et al.: South-polar features on Venus similar to those near the north. Nature **450**, 637–640 (2007).

Pirraglia, J.A.: Dissipationless decay of Jovian jets. Icarus **79**, 196–207 (1989).

Porco, C.C., et al.: Cassini imaging of Jupiter's atmosphere, satellites, and rings. Science **299**, 1541–1547 (2003).

Porco, C.C., et al.: Cassini imaging science: Instrument characteristics and capabilities and anticipated scientific investigations at Saturn. Space Sci. Rev. **115**, 363–497 (2004).

Porco, C.C., et al.: Cassini imaging science: Initial results on Saturn's atmosphere. Science **307**, 1243–1247 (2005).

Read, P.L., et al.: Jupiter's and Saturn's convectively driven banded jets in the laboratory. Geophys. Res. Lett. **31**, L22701 (2004).

Read, P.L., et al.: Dynamics of convectively driven banded jets in the laboratory. J. Atmos. Sci. **64**, 4031–4052 (2007).

Read, P.L., et al.: Mapping potential vorticity dynamics on Jupiter: 1. Zonal mean circulation from Cassini and Voyager 1 data. Quart. J. Roy. Meteor. Soc. **132**, 1577–1603 (2006a).

Read, P.L., Gierasch, P.J., Conrath, B.J.: Mapping potential-vorticity dynamics on Jupiter: II: the Great Red Spot from Voyager 1 and 2 data. Quart. J. Roy. Meteor. Soc. **132**, 1605–1625 (2006b).

Read, P.L., et al.: Mapping potential vorticity dynamics on Saturn: Zonal mean circulation from Cassini and Voyager data. Planet. Space Sci., in press (2009).

Rhines, P.B.: Waves and turbulence on a beta-plane. J. Fluid Mech. **69**, 417–443 (1975).

Rieke, G. H.: The thermal radiation of Saturn and its rings. Icarus **26**, 37–44 (1975).

Salyk, C., et al.: Interaction between eddies and mean flow in Jupiter's atmosphere: Analysis of Cassini imaging data. Icarus **185**, 430–442 (2006).

Sánchez-Lavega, A.: Motions in Saturn's atmosphere: Before Voyager encounters. Icarus **49**, 1–16 (1982).

Sánchez-Lavega, A.: Saturn's Great White Spots. Sky and Telescope **78**, 141–143 (1989).

Sánchez-Lavega, A., Battaner, E.: Long-term changes in Saturn's atmospheric belts and zones. Astr. Astrophys. Suppl. Ser. **64**, 287–301 (1986).

Sánchez-Lavega, A., Battaner, E.: The nature of Saturn's atmospheric Great White Spots. Astron. Astrophys. **185**, 315–326 (1987).

Sánchez-Lavega, A., et al.: The great white spot and disturbances in Saturn's equatorial atmosphere during 1990. Nature **353**, 397–401 (1991).

Sánchez-Lavega, A. et al.: Temporal behavior of cloud morphologies and motions in Saturn's atmosphere. Journal of Geophysical Research **98** (E10), 18857–18872 (1993a).

Sánchez-Lavega, A., et al.: Ground-based observations of Saturn's north polar spot and hexagon. Science **260**, 329–332 (1993b).

Sánchez-Lavega, A. et al.: Photometry of Saturn's 1990 equatorial disturbance. Icarus **108**, 158–168 (1994).

Sánchez-Lavega, A., et al.: Large-scale storms in Saturn's atmosphere during 1994. Science **271**, 631–634 (1996).

Sánchez-Lavega, A., et al.: New observations and studies of Saturn's long-lived north polar spot. Icarus **128**, 322–334 (1997).

Sánchez-Lavega, A., et al.: Discrete cloud activity in Saturn's equator during 1995, 1996, and 1997. Planet. Space Sci. **47**, 1277–1283 (1999).

Sánchez-Lavega, A., Rojas, J.F., Sada, P.V.: Saturn's zonal winds at cloud level. Icarus **147**, 405–420 (2000).

Sánchez-Lavega. A.: Observations of Saturn's ribbon wave 14 years after its discovery. Icarus **158**, 272–275 (2002a).

Sánchez-Lavega. A., et al.: No hexagonal wave around Saturn's Southern Pole. Icarus, **160**, 216–219 (2002b).

Sánchez-Lavega, A., et al.: A strong decrease in Saturn's equatorial jet at cloud level. Nature **423**, 623–625 (2003).

Sánchez-Lavega, A., et al.: Observations and models of the general circulation of Jupiter and Saturn. In Ulla, A., Manteiga, M. (eds.), *Lecture Notes and Essays in Astrophysics I*, pp. 63–85. Real Sociedad Española de Física, Universidad de Vigo (2004a).

Sánchez-Lavega, A., et al.: Saturn's cloud morphology and zonal winds before the Cassini encounter. Icarus **170**, 519–523 (2004b).

Sánchez-Lavega, A., et al.: A strong vortex in Saturn's South Pole, Icarus **184**, 524–531 (2006).

Sánchez-Lavega, A., Hueso, R., Pérez-Hoyos, S.: The three-dimensional structure of Saturn's equatorial jet at cloud level. Icarus **187**, 510–519 (2007).

Sayanagi, K.M., Showman, A.P.: Effects of a large convective storm on Saturn's equatorial jet. Icarus **187**, 520–539 (2007).

Sayanagi, K.M., Showman, A.P., Dowling, T.E.: The emergence of multiple robust zonal jets from freely-evolving, three-dimensional stratified geostrophic turbulence with applications to Jupiter. J. Atmos. Sci. **65**, 3947–3962 (2008).

Schneider, T., Liu, J.J.: Formation of jets and equatorial superrotation on Jupiter. J. Atmos. Sci. **66**, 579–601 (2009).

Schumacher, C., Houze, R.A. Jr.: Stratiform precipitation production over sub-Saharan Africa and the tropical east Atlantic as observed by TRMM. Quart. J. Roy. Meteor. Soc, **132**, 2235–2255 (2006).

Scott, R.K., Polvani, L.M.: Forced-dissipative shallow-water turbulence on the sphere and the atmospheric circulation of the giant planets. J. Atmos. Sci. **64**, 3158–3176 (2007).

Scott, R.K., Polvani, L.M.: Equatorial superrotation in shallow atmospheres. Geophys. Res. Letters **35**, L24202, doi:10.1029/2008GL036060 (2008).

Showman, A.P.: Numerical simulations of forced shallow-water turbulence: effects of moist convection on the large-scale circulation of Jupiter and Saturn. J. Atmos. Sci. **64**, 3132–3157 (2007).

Showman, A.P., de Pater, I.: Dynamical implications of Jupiter's tropospheric ammonia abundance. Icarus **174**, 192–204 (2005).

Showman, A.P., Gierasch, P.J., Lian, Y.: Deep zonal winds can result from shallow driving in a giant-planet atmosphere. Icarus **182**, 513–526 (2006).

Smith, B.A., et al.: The Jupiter system through the eyes of Voyager 1. Science **204**, 951–972 (1979).

Smith, B.A., et al.: Encounter with Saturn: Voyager 1 imaging science results. Science **212**, 163–190 (1981).

Smith, B. A., et al.: A new look at the Saturn system: The Voyager 2 images. Science **215**, 504–537 (1982).

Smith, K.S.: A local model for planetary atmospheres forced by small-scale convection. J. Atmos. Sci. **61**, 1420–1433 (2004).

Sommeria, J., Meyers, S.D., Swinney, H.: Experiments on vortices and Rossby waves in eastward and westward jets. In A. R. Osborne (ed.), *Nonlinear Topics in Ocean Physics*, pp. 227–269, North-Holland, Amsterdam (1991).

Sromovsky, L.A., et al.: Jovian winds from Voyager 2. Part II: Analysis of eddy transports. J. Atmos. Sci. **39**, 1413–1432 (1982).

Sromovsky, L.A., et al.: Voyager 2 observations of Saturn's northern mid-latitude cloud features: Morphology, motions, and evolution. J. Geophys. Res. **88**, 8650–8666 (1983).

Stephens, G.L., et al.: The CloudSat mission and the A-Train: A new dimension to space-based observations of clouds and precipitation. Bull. Amer. Meteor. Soc. **83**, 1771–1790 (2002).

Sukoriansky, S., Galperin, B., Dikovskaya, N.: Universal spectrum of two-dimensional turbulence on a rotating sphere and some basic features of atmospheric circulation on giant planets. Phys. Rev. Lett. **89**, 124501-1–124501-4 (2002).

Sukoriansky, S., Dikovskaya, N., Galperin, B.: On the arrest of inverse energy cascade and the Rhines scale. J. Atmos. Sci. **64**, 3312–3327 (2007).

Sukoriansky, S., Dikovskaya, N., Galperin, B.: Nonlinear waves in zonostrophic turbulence. Phys. Rev. Lett. **101**, 178501 (2008).

Temma, T., et al.: Vertical structure modeling of Saturn's equatorial region using high spectral resolution imaging. Icarus **175**, 464–489 (2005).

Theiss, J.: Equatorward energy cascade, critical latitude, and the predominance of cyclonic vortices in geostrophic turbulence. J. Phys. Ocean. **34**, 1663–1678 (2004).

Tokunaga, A. T., et al.: Spatially resolved infrared observations of Saturn: II. The temperature enhancement at the south pole of Saturn. Icarus **36**, 216–222 (1978).

Tomasko, M.G., West, R.A., Orton, G.S., Teifel, V.G.: Cloud and aerosols in Saturn's atmosphere. In Gehrels, T., Matthews, M.S., (eds.), *Saturn*, pp. 150–194, Univ. Arizona Press, Tucson (1984).

Trafton, L.: Long-term changes in Saturn's troposphere. Icarus **63**, 374–405 (1985).

Vasavada, A.R., Showman, A.P.: Jovian atmospheric dynamics: An update after Galileo and Cassini. Rep. Prog. Phys. **68**, 1935–1996 (2005).

Vasavada, A., et al.: Cassini imaging of Saturn: southern hemisphere winds and vortices. J. Geophys. Res. **111**, E05004, doi:10.1029/2005JE002563 (2006).

Weidenschilling, S.J., Lewis, J.S.: Atmospheric and cloud structures of the Jovian planets. Icarus **20**, 465–476 (1973).

Weir, S.T., Mitchell, A.C., Nellis, W.J.: Metallization of fluid molecular hydrogen at 140 GPa. Phys. Rev. Lett. **76**, 1860–1863 (1996).

Westphal, J.A., et al.: Hubble Space Telescope observations of the 1990 equatorial disturbance on Saturn – images, albedos, and limb darkening. Icarus **100**, 485–498 (1992).

Williams, G.P.: Planetary circulations 1: Barotropic representations of Jovian and terrestrial turbulence. J. Atmos. Sci. **35**, 1399–1426 (1978).

Williams, G.P.: Jovian dynamics. Part III: Multiple, migrating, and equatorial jets. J. Atmos. Sci. **60**, 1270–1296 (2003).

Wordsworth, R.D., Read, P.L., Yamazaki, Y.H.: Turbulence, waves and jets in a differentially heated rotating annulus experiment, Phys. Fluids **20**, 126602 (2008).

Yamazaki, Y.H., Read, P.L., Skeet, D.R.: Hadley circulations and Kelvin wave-driven equatorial jets in the atmospheres of Jupiter and Saturn. Planet. Space Sci. **53**, 508–525 (2005).

Yanamandra-Fisher, P.A., et al.: Saturn's 5.2-μm cold spots: Unexpected cloud variability. Icarus **150**, 189–193 (2001).

Zipser, E.J., et al.: Where are the most intense thunderstorms on Earth? Bull. Amer. Meteor. Soc. **87**, 1057–1071 (2006).

Zuchowski, L.C., Read, P.L., Yamazaki, Y.H.: Modeling Jupiter's cloud bands and decks: 1. Jet scale meridional circulations on Jupiter. Icarus **200**, document 548–562 (2009).

Chapter 7
Clouds and Aerosols in Saturn's Atmosphere

R.A. West, K.H. Baines, E. Karkoschka, and A. Sánchez-Lavega

Abstract In this chapter we review the photochemical and thermochemical equilibrium theories for the formation of condensate clouds and photochemical haze in Saturn's upper troposphere and stratosphere and show the relevant observations from ground-based and spacecraft instruments. Based on thermochemical equilibrium models we expect ammonia ice crystals to dominate in the high troposphere. There is very little spectral evidence to confirm this idea. Thanks to a stellar occultation observed by the Cassini VIMS instrument we now have spectral evidence for a hydrocarbon stratospheric haze component, and we still seek evidence for an expected diphosphine stratospheric haze component. The vertical distributions of stratospheric and upper tropospheric hazes have been mapped well with ground-based and Hubble Space telescope data, and Cassini data are beginning to add to this picture. Polar stratospheric aerosols are dark at UV wavelengths and exhibit strong Rayleigh-like polarization which suggests that auroral processes are important for their formation as is the case for the jovian polar stratospheric haze. The cloud and haze structure exhibits a variety of temporal variation, including seasonal change, long-term secular change near the equator, and short-term changes with a complicated latitudinal structure, and still not understood. Cassini instruments, especially the VIMS instrument, show an abundance of small-scale structure (convective clouds) at a pressure near 2 bar.

7.1 Introduction

Clouds and aerosols in planetary atmospheres are valuable for what they can tell us about important atmospheric chemical and physical processes. Clouds and aerosols play a role in the radiative and latent-heat terms of the energy budget, but to understand their radiative role one must have information on particle optical properties and the spatial distribution of opacity. Clouds can be used as tracers of motion to assess winds (see Chapter 6) and for that we would like to know pressure coordinates. In this chapter we start with a description of the expected vertical distributions and compositions of tropospheric clouds from thermochemical equilibrium theory and stratospheric haze from photochemistry. We examine observations made over the last three decades from the ground and from spacecraft which constrain particle optical and physical properties and vertical and horizontal distributions, and time variations of these quantities.

Images reveal three major latitude regimes and two major vertical regimes for clouds and aerosols. The troposphere and stratosphere constitute the two major vertical regimes, with the interface at the tropopause near 100 mbar. Different processes (cloud condensation/sublimation in a convecting atmosphere, and photochemical haze formation in a stably stratified atmosphere) dominate in the two locations. In terms of latitudinal distribution, the equatorial zone (generally between about $\pm 18°$) is a region of consistently high clouds and thick haze with little seasonal variation but with occasional major cloud outbursts. The polar regions north of about 60° contain a stratospheric haze that is optically thicker and darker than stratospheric haze at other latitudes, indicative of chemistry driven by auroral processes. Mid-latitude tropospheric clouds display substantial seasonal and nonseasonal variations. At middle and high latitudes small deep clouds which probably form in free convection are abundant in images of greatest atmospheric transparency. High spatial resolution limb images reveal distinct hazes layers at some latitudes. We will discuss all of these topics in more detail in the remainder of the chapter.

R.A. West and K.H. Baines
Jet Propulsion Laboratory, California Institute of Technology, CA, USA

E. Karkoschka
Lunar and Planetary Laboratory, University of Arizona, AZ, USA

A. Sánchez-Lavega
Departamento de Física Aplicada I, E.T.S. Ingenieros, Universidad del País Vasco, Spain

7.2 Expectations from Thermochemical Equilibrium Theory and Photochemistry

Oxygen, nitrogen and sulfur are the most abundant elements in solar composition which have the potential to combine with hydrogen to form clouds in the high troposphere. Lewis (1969) and Weidenschilling and Lewis (1973) were the first to offer quantitative models of the locations of condensate clouds based on thermochemical equilibrium theory. The guiding principal of these models is that solid or liquid phase condensation occurs at altitudes above the altitude where the temperature and partial pressure thermodynamically favors condensation. The models are complicated by solution chemistry in the ammonia–water cloud region and by uncertainties in the mixing ratios. Ammonia and hydrogen sulfide combine to form solid ammonium hydrosulfide (NH_4SH). Depending on the relative mixing ratios of NH_3 and H_2S, either the NH_3 or H_2S could be entirely depleted above the haze layer by this process. It is believed that H_2S is depleted and the remaining NH_3 is available to form an ammonia ice cloud at higher altitude. On Saturn the temperatures are cold enough to severely deplete ammonia gas at the level where visible and near-infrared photons sense and so gas absorption by ammonia is very weak.

The most recent thermochemical models were published by S. K. Atreya and colleagues. Figure 7.1 by Atreya and Wong (2005) shows cloud configurations for enhanced concentrations (by a factor of five over solar composition) for O, N and S. Clouds extend over a large vertical range. The base of the water–ammonia solution cloud is near 20 bar; that of the ammonium hydrosulfide cloud near 6 bar and the ammonia ice cloud base is near 2 bar. When the larger scale height, lower gravity and smaller H_2/He ratio relative to Jupiter are taken into account the clouds extend over a much larger vertical range relative to jovian clouds, and there is considerably more gas above the cloud bases than is the case for Jupiter. As a rule remote sensing does not probe as deeply as for Jupiter, but as discussed later images in window regions longer than about $0.7\,\mu m$ detect clouds as deep as about 2 bars at locations where the diffuse haze particle number density is low.

Thermochemical model results such as those shown in Fig. 7.1 provide guidance to the possible locations of condensate clouds. However, local meteorology has a large influence on the locations of clouds. Experience with water clouds in the terrestrial atmosphere shows that cloud formation can be inhibited or that multiple layers can form depending on local temperature and water content. Experience with the Galileo probe (and ground-based observation dating to 1975 – see West et al. 1986) shows that regions of descending air suppress cloud formation, although layers can still form at altitudes above those predicted for a quiescent atmosphere with enhanced abundance of volatiles. Nor do these models take into account cloud microphysical processes. Because the detailed information needed for such models is lacking the best we can do is outline the range of possibilities. West et al. (2004) reviewed cloud microphysics models for Jupiter.

Stratospheric aerosols owe their origins primarily to photochemical processes and bombardment by energetic particles, especially in the auroral zones. Cosmic dust particles impact the upper atmosphere, although their contribution is probably small relative to the other sources. We do not have a detailed model to account for all of the processes, and the roles of energetic particles and ion chemistry are not well understood. Yet a collection of ideas has emerged regarding stratospheric aerosol formation. In Chapter 5, Fouchet et al. give a comprehensive review of photochemical processes. Here we mention the main ideas via Fig. 7.2 from Atreya and Wong (2005). Figure 7.2 was generated for Jupiter but the chemical paths (with the exchange of PH_3 and P_2H_4 for NH_3 and N_2H_4) carry over to Saturn. On Saturn the ammonia mixing ratio is too low to consider the production of N_2H_4 and instead phosphine photochemistry leading to diphosphine is important. Two types of hydrocarbons might be produced – a linear chain shown as the polymerization of acetylene (C_2H_2) and ring molecules starting with benzine (C_6H_6) and leading to more complex polycyclic aromatic hydrocarbons (PAHs). Friedson et al. (2002) produced a detailed model of hydrocarbon haze formation for Jupiter. The concept should apply to Saturn.

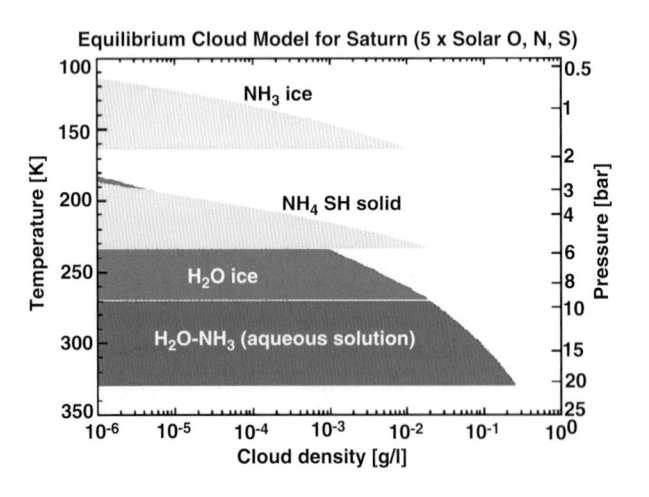

Fig. 7.1 This graphic schematically illustrates the vertical locations and compositions of condensate clouds in Saturn's atmosphere based on thermochemical equilibrium models assuming O, N, and S enrichments by a factor of five over solar composition. Cloud density expressed on the ordinate is the result of a model which neglects sedimentation or vertical mixing or advection (from Atreya and Wong (2005))

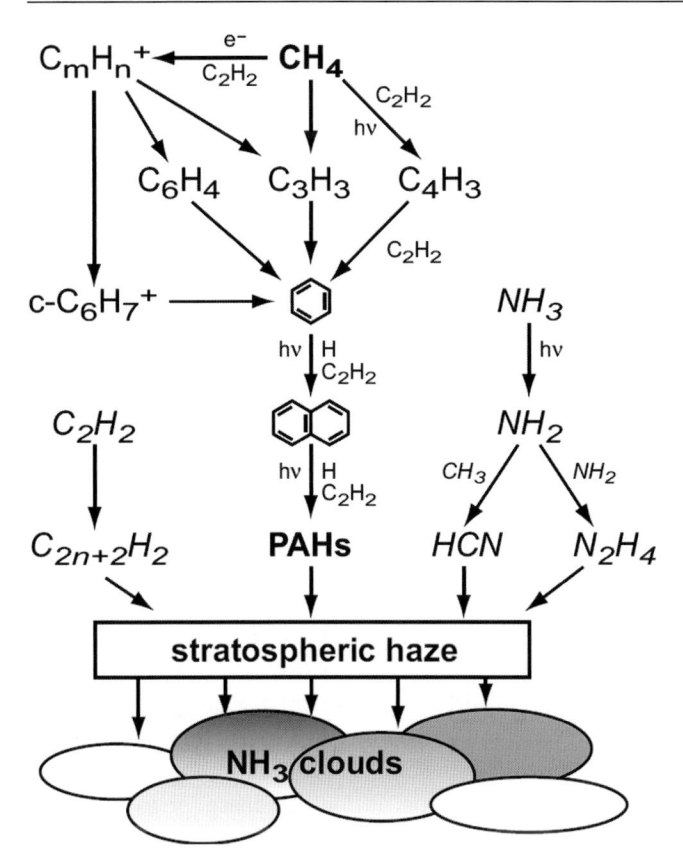

Fig. 7.2 A simplified rendering of the chemical schemes leading to the formation of stratospheric aerosols, this graphic was created for the Jupiter atmosphere. It can also be applied to Saturn if NH_3 and N_2H_4 map to PH_3 and P_2H_4. Note the two paths for hydrocarbons leading to linear chains (polyacetylenes) or polycyclic aromatic hydrocarbons (PAHs). Energetic photons and particles acting on methane or phosphine may initiate the process (from Atreya and Wong (2005); see also Wong et al. (2003))

7.3 Observational Constraints on Particle Composition and Chromophores

From thermochemical equilibrium models we expect that ammonia ice particles dominate clouds near the top of the troposphere. Spectral signatures for ammonia ice are therefore expected. Laboratory measurements (Martonchik et al. 1984) show ammonia ice spectral features at several wavelengths in the near-infrared and at 9.4 and 26 µm in the thermal infrared. Although methane and hydrogen gas absorptions interfere with detection it is remarkable that no features have been observed. Some were observed for Jupiter as reported by Brooke et al. (1998), Baines et al. (2002) and Wong et al. (2004). Spatially-resolved observations show these features to be limited to a small fraction of the cloud coverage, typically in what are thought to be newly-forming cloud regions. For both Jupiter and Saturn and especially for Saturn it is a major puzzle that we do not see more widespread spectral signatures of ammonia ice.

Several factors might conspire to mask ammonia ice spectral signatures. If the particles are large (of order 10 µm radius or larger) so much light is absorbed in the wings of the band that it is hard to detect because the feature is very broad and overlaps atmospheric absorption features. Brooke et al. (1998) believe this is the case for jovian cloud particles. If a layer of smaller particles overlays the larger particles, and if the optical depth of this layer is sufficiently large we can expect to see an ammonia ice signature. Spectroscopic evidence discussed later indicates that particles near the top of the troposphere have mean radius closer to 1 µm.

Particle non-spherical shape is another factor which might diminish an ammonia ice spectral signature, at least for those features that are sensitive to particle shape. Huffman and Bohren (1980) called attention to the shape effect for spectral features formed when surface modes dominate electro-magnetic resonances. This effect is important even for particles smaller than the wavelength (in the Rayleigh approximation). Spectral features can be broadened further by a distribution of particle shapes. West et al. (1989) explored this idea for ammonia features at 9.4 and 26 µm. The peak of the absorption dropped by a factor of 2 while the width increased for the 26 µm-feature. Whether or not the shape effect is able to thwart detection remains to be seen, but it seems unlikely that this will suffice. The peak at 9.4 µm is shifted for tetrahedral particles relative to spheres but its magnitude is not diminished. It therefore seems that shape effects alone do not account for the non-detection.

Spectral features can be masked by a foreign contaminant, either from below (water or ammonium hydrosulfide brought up from depth) or from above (photochemical products from the stratosphere). West et al. (1989) considered this possibility and found that the bulk composition of the particle must be dominated by the foreign contaminant if the resonant feature of the ammonia ice is to be significantly suppressed. Recently Atreya et al. (2005) and Kalogerakis et al. (2008) examined photochemical production rates relevant to the jovian atmosphere. Although they were able to estimate photochemical production rates they did not estimate formation rates for ammonia ice. In order for the photochemical component to be comparable to or dominate the ammonia ice component the formation rates for photochemical components must be greater than that for ammonia ice. On Saturn hydrocarbon photochemistry is operative as it is for Jupiter (although with a solar flux down by a factor of almost four but with a different methane mixing ratio in the high atmosphere, depending on the details of diffusive separation above the homopause and the depth of penetration of photolyzing radiation. Hydrocarbon haze may not be the most important contributor. Diphosphine (P_2H_4) is perhaps more important (Fouchet et al., this volume). No spectral signatures of solid P_2H_4 have been reported. If a coating or condensation nucleus is responsible for the suppression of ammonia ice spectral features it means

Fig. 7.3 The black curves are observed signal ratios (occulted signal divided by unocculted signal) at three tangent-point pressures in the atmosphere. The green curves are model calculations (from Nicholson et al. (2006))

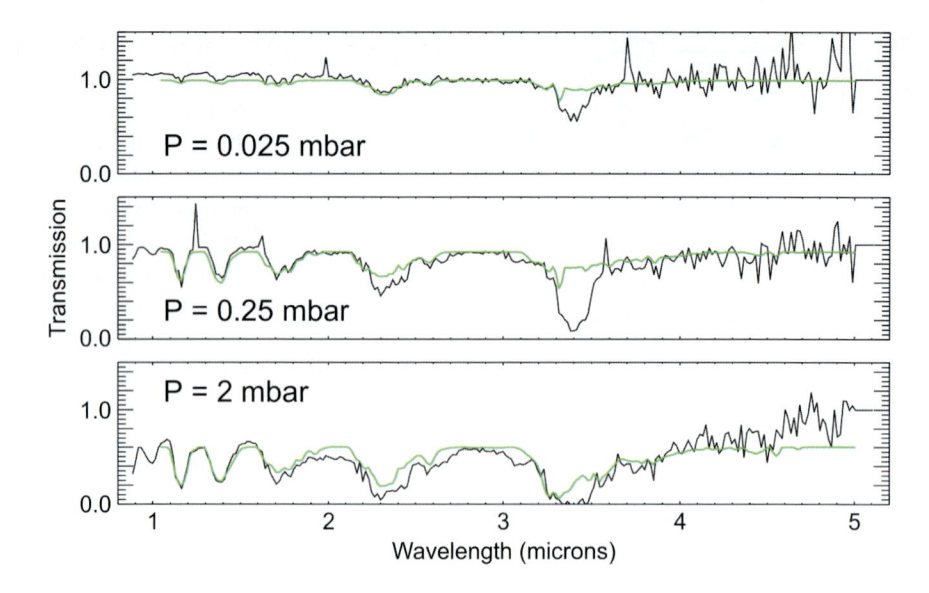

that we are seeing a photochemical haze, not an ammonia cloud, in images of Saturn.

Spectroscopic evidence for a hydrocarbon haze in Saturn's stratosphere was observed in a stellar occultation experiment performed by the Cassini VIMS instrument (Nicholson et al. 2006). The signal from the starlight passing through the atmosphere was divided by the signal taken a few moments earlier without any attenuation by the atmosphere. Instrument sensitivity drops out of the ratio making the experiment insensitive to calibration issues. Figure 7.3 shows the ratio for three values of the pressure level of the ray tangent point. Models of the methane absorption account for nearly all of the absorption features. A feature at 3.4 μm reveals a hydrocarbon haze, probably a linear chain (polyacetylene) rather than a ring (benzene or a polycyclic aromatic; P. Nicholson, private communication 2008). The occultation sampled a middle latitude. It is possible that benzene and PAHs are more abundant in the auroral zone/polar vortex regions where energy input from the auroras is substantial.

If one accepts the notion that a coating of stratospheric haze material is responsible for the suppression of ammonia ice spectral features it is tempting to think that such haze material might also be responsible for the dusky yellow/light-brown/red hue of most of Saturn at visible wavelengths. Identification of the chromophore material is ambiguous because there are no narrow spectral signatures, only a broad darkening from the red to blue wavelengths. Hydrocarbon chromophores which absorb at visible wavelengths might be derived from polyacetylene or PAH molecules that have undergone some modification due to solid-state photochemistry, driving off H and increasing the C/H ratio. Materials with higher C/H ratios absorb light more efficiently. Black carbon is an end member. A table of candidate materials (including inorganics) and references to the original proposals and laboratory measurements can be found in table V of West et al. (1986).

Compounds involving sulfur, hydrogen, and nitrogen have also been proposed as coloring agents for Jupiter and Saturn. These materials would be brought up from several bars pressure where they exist in the gas phase, or as NH_4SH solid. Solid-state photochemistry would be required to modify the composition because NH_4SH solid is white. Perhaps the best observational support for this hypothesis comes from images which show localized dark spots associated with vortices (see Fig. 7.4). Upwelling of deeper S- and N-rich gas or NH_4SH cloud particles with exposure to UV light at the top of the cloud seems the simplest mechanism to account for such localized dark albedo associated with vortices. These features are relatively bright in strong methane absorption bands which indicates that the clouds are higher and thicker than their surroundings as expected for upwelling.

Images like those in Fig. 7.4 underscore the fact that we do not have a good understanding of the relationship between atmospheric dynamics, particle microphysics, and chromophore genesis, distribution or exposure. For Jupiter West et al. (1986) proposed that chromophores are ubiquitous and that marginally stable ammonia condensation/sublimation cycles involving coating chromophore cloud condensation nuclei could account for rapid changes in color and albedo with unobservably small change in temperatures or wind. The idea advanced by Atreya et al. (2005) takes the opposite approach – it is the coating of ammonia by a photochemical material that dominates. That idea seems more attractive for Saturn if the formation rate of the photochemical product is larger than the formation rate of ammonia ice particles, but still lacking is an understanding of small-scale color and albedo contrasts associated with vortices and other features.

7 Clouds and Aerosols in Saturn's Atmosphere

Fig. 7.4 These images from the Cassini VIMS experiment show both dark and bright features at two window regions (between methane bands) in the near-infrared. At least some of the features (most notably the bright cloud in the shape of a mirror-reversed question mark) are associated with a lightning storm. Most spots are dark at both wavelengths. The mirror-reversed question mark is bright in the 0.93-μm window but not in the 2.73-μm window. Its relatively high reflectivity at 0.93 μm suggests that it is a region where an ammonia ice cloud formed recently and the ratio of ammonia cloud to chromophore material is higher than in the surrounding region (adapted from Baines et al. (2009))

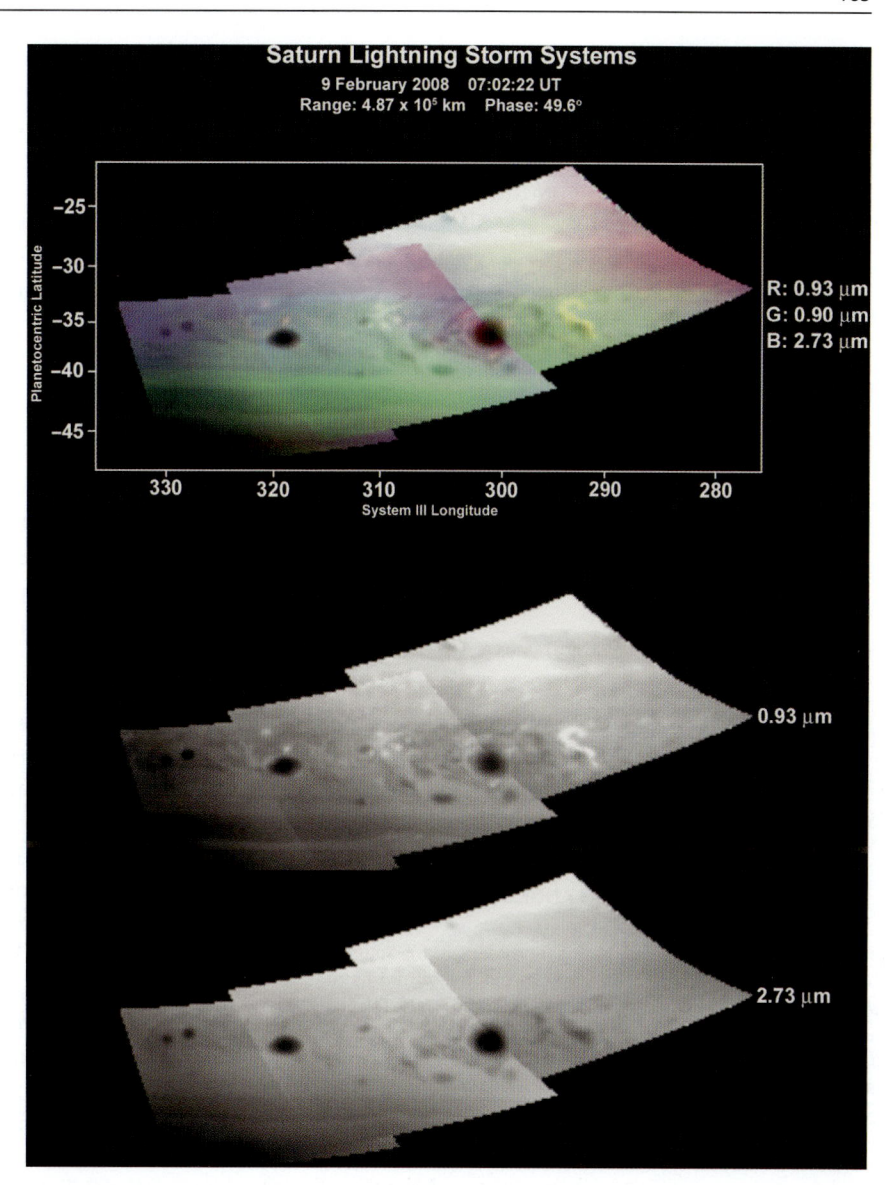

7.4 Aerosol Optical and Physical Properties

Information on particle optical and physical properties comes from the wavelength dependence of the extinction and scattering cross sections, the scattering phase function and possibly the polarization. The accumulated spacecraft and ground-based data offer the potential to retrieve this information on fine spatial scales, but thus far particle phase functions and polarization curves have been retrieved for only two latitude bands using data from the Pioneer 11 spacecraft. Analyses of optical properties at low phase angles has been performed using data from the Hubble Space Telescope (HST) for many latitudes over a time span of several years and these provide clues to time variations over short and long time scales.

Knowledge of particle scattering phase functions is useful not only for particle size studies. Knowledge of the phase function is as important as the optical depth profile for estimates of solar energy deposition and in line formation used to infer chemical abundances from reflected sunlight. Figure 7.5 shows phase functions derived from data for the Equatorial Zone and for a mid-latitude belt from Pioneer 11 data (Tomasko and Doose 1984). The curves shown in Fig. 7.5 are double Henyey–Greenstein functions with forward and backward peaks. The best-fit Henyey–Greenstein parameters are listed in Table 7.1. Note that the fits are based on measurements at scattering angles greater than 30°. The diffraction peak for particles much larger than the wavelength is confined to smaller angles, so it is difficult to gauge particle size if the mean radius is more than a few times the wavelength. It

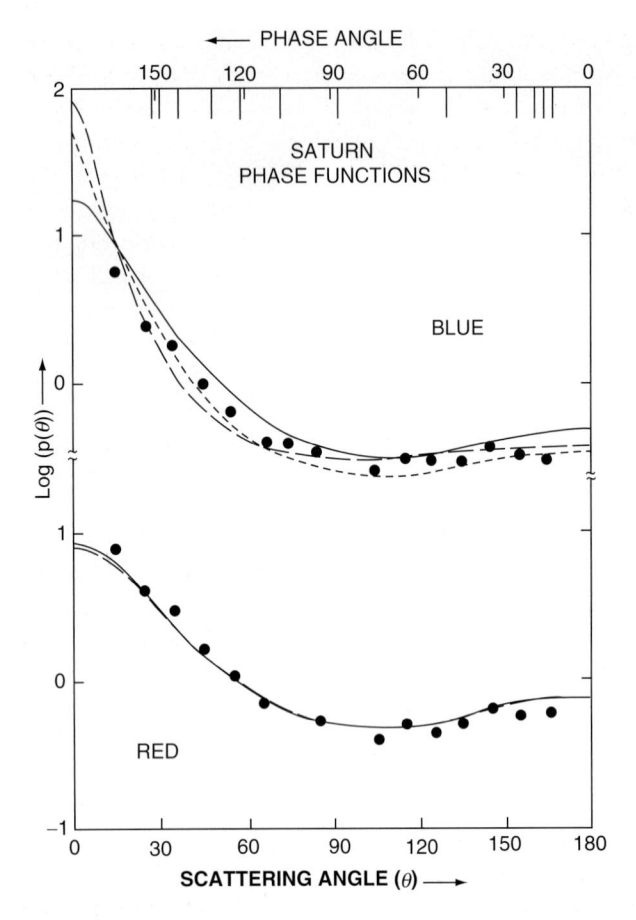

Fig. 7.5 Particle scattering phase functions derived for Saturn aerosols for the Equatorial Zone (*solid curves*) and for a mid-latitude belt (*dashed curves* with cloud-top pressures at 150 and 250 mbar). The dots are from laboratory measurements of 10-μm ammonia ice particles grown in a laboratory chamber (Pope et al. 1992) (from Tomasko and Doose (1984))

Table 7.1 Double Henyey–Greenstein (Henyey and Greenstein, 1941) phase function parameters for a Belt and Zone from Tomasko and Doose (1984). Eqs. 7.2 and 7.3 of the text give the functional form. Two sets of parameters were derived for the belt at 15–17° S for two possible values of the pressure at the top of the haze layer (P_0). The aerosol albedo for single scattering is given by ϖ. The parameter p is the geometric albedo for a homogeneous spherical body with the derived cloud properties and q is the phase integral. F_0 is a parameter for the absolute calibration of the photopolarimeter instrument

Parameters of Best-fitting Phase Functions

	7°S–11°S		15°S–17°S			
	Red	Blue	Red		Blue	
P_0	150 mb	150 mb	150 mb	250 mb	150 mb	250 mb
g_1	0.620	0.710	0.603	0.633	0.870	0.824
g_2	−0.294	−0.317	−0.302	−0.286	−0.116	−0.231
f	0.763	0.860	0.768	0.776	0.764	0.862
ϖ	0.986	0.920	0.986	0.986	0.920	0.920
F_0	85.6	50.8	80.1	80.9	51.2	50.3
p	0.501	0.287	0.502	0.496	0.264	0.276
q	1.41	1.43	1.41	1.41	1.45	1.41

is also important to recognize that the derived phase function at small scattering angles is limited to aerosols near the top of the haze layer. Rayleigh scattering from gas above the haze can introduce ambiguity in the retrieved phase function if the vertical location of the haze is uncertain and this accounts for the two models for the blue belt in Table 7.1. The use of the double Henyey–Greenstein phase function is justified because the particles are expected to be solid and therefore not spheres. Spherical particles have a deep minimum at middle scattering angles if the particle radius is larger than the wavelength and this behavior is not present for nonspherical particles which have a shallow minimum. Spherical particles larger than the wavelength also have rainbow and glory features which are either not present or weak for nonspheres. Phase functions for both spherical and nonspherical particles have a similar shape in the forward diffraction region which is controlled by the particle projected area.

Comparison with laboratory measurements for ammonia ice crystals (Pope et al. 1992) shown in Fig. 7.5 suggests that the particles which form the main haze layer have effective radii near 10 μm. While this result may be correct it seems likely that a broad range of particle size is allowed by the observations provided the mean radius is larger than visible wavelengths.

Particle polarization also supports the idea that the particles are not small compared to the wavelength. Polarization as measured by the Pioneer 11 Imaging Photopolarimeter (IPP) has a negative branch (see Fig. 7.6) at red wavelengths. Here again there is some ambiguity due to uncertainty in the vertical structure. Pope et al. (1992) infer from laboratory measurements that the particles in the main haze layer are large (size parameter 2π·radius/wavelength ∼10–50).

The Cassini ISS obtained images using polarizers at wavelengths from the near-UV to the near-IR. Some of the earliest images (Fig. 7.7) show positive polarization at high southern latitudes, similar to what was seen for Jupiter. At middle and low latitudes the polarization is negative at the phase angle of the images (60.7°). Polarization is enhanced and the angle becomes positive near the limb and terminator indicating an optically thin layer of positively-polarizing particles and/or gas above the main negatively polarizing haze layer. This behavior was also noted by Tomasko and Doose (1984).

West and Smith (1991) noted that the jovian polar aerosols have strong positive polarization and strong forward scattering, like the haze in Titan's stratosphere. This combination of optical properties cannot be matched by spheres. West and Smith proposed that the particles are aggregates of small (∼tens of nm radius) monomers. The observation shown in Fig. 7.7 suggests that particles of this type are also abundant at high latitudes in Saturn's atmosphere, although it is not yet known if they have strong forward scattering. The presence of these particles in polar regions for both Jupiter and Saturn suggests that auroral energy deposition is important in their formation. Pryor and Hord (1991) advocated this idea based on low UV reflectivity which is consistent with a hydrocarbon composition for these particles.

Observations from the Hubble Space Telescope have also shed some light on the nature of the polar stratospheric aerosol. The ultraviolet reflectivities are low at large angles of reflection, when observing near the limb of Saturn, these dark aerosols must extend up to pressures of a few millibar, because the Rayleigh scattering from a clear layer of some 10–20 mbar would make Saturn brighter than observed.

These aerosols can be seen in deep methane bands such as the 890 nm methane band, but their inferred optical depths at these near-infrared wavelengths are many times lower than at ultraviolet wavelengths (Fig. 7.8). Assuming spherical particles, Karkoschka and Tomasko (1993) found a mean aerosol radius of $0.15\,\mu m$ for the aerosols near the North Pole, and Karkoschka and Tomasko (2005) and Pérez-Hoyos et al. (2005) gave values of 0.15 and $0.1\,\mu m$ for the aerosols near the South Pole, respectively assuming the particles are spheres.

The imaginary refractive index of the polar stratospheric aerosols is about 0.1 at 300 nm wavelength with a steep wavelength dependence, approximately following a power law with wavelength with an exponent of −6.5 (Karkoschka and Tomasko 2005). This yields a single scattering albedo near 0.9 at 300 nm wavelength, increasing to 0.99 at 450 nm.

Between −65° and 65° latitude, many observations require some aerosol component in the stratosphere. However, the optical depth is down by a factor of about 4 compared to aerosols at the polar latitudes. This makes it much harder to determine the aerosol properties. Assuming that both types of stratospheric aerosol have the same properties has yielded reasonable fits to observations (Karkoschka and Tomasko 2005). However, the constraints are not tight. These aerosols may originate by solar ultraviolet radiation. Since the formation processes of both kinds of stratospheric aerosols are different, the aerosol properties could be somewhat different too. The strongest observational constraint on the low-latitude aerosols indicate that their size must be quite small too, about 0.1–0.2 μm radius assuming spherical shape (Muñoz et al. 2004; Pérez-Hoyos et al. 2005).

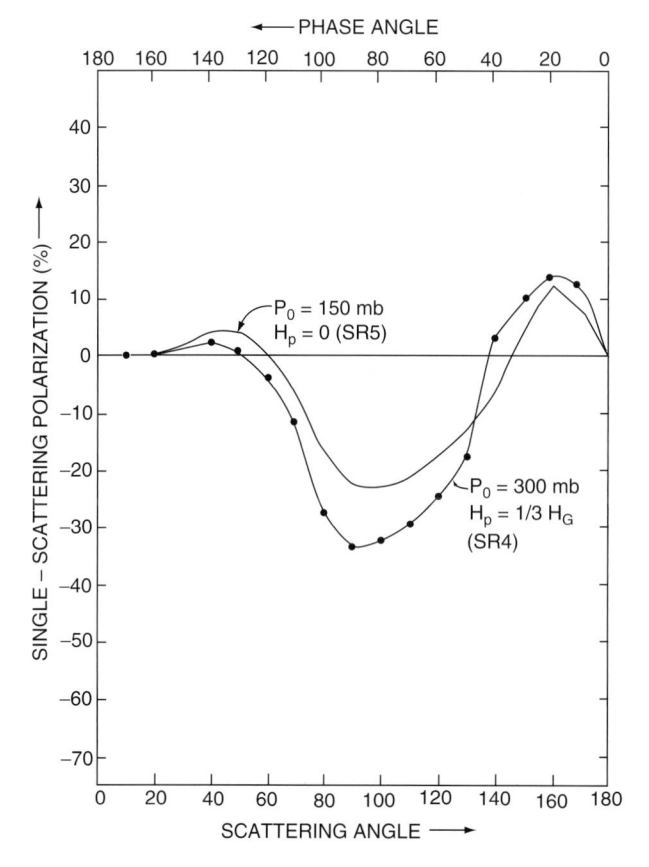

Fig. 7.6 Polarizing function for the Equatorial Zone in red light for two values of the pressure at the top of the main haze layer (P_0) and for two values of the particle scale height (H_p) (from Tomasko and Doose (1984))

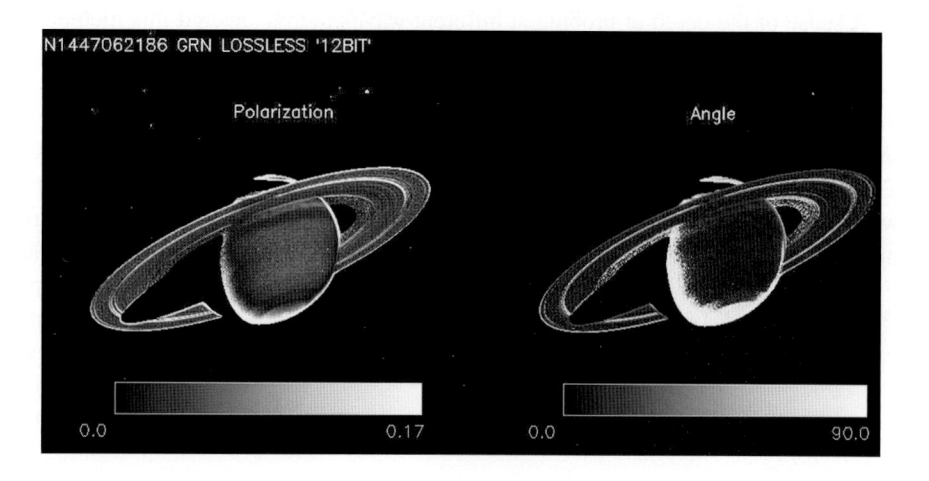

Fig. 7.7 Polarization images in the Cassini ISS Narrow-angle camera Green filter obtained in 2003 show enhanced positive polarization (angle of the electric vector near 90° with respect to the scattering plane) at high southern latitudes

Fig. 7.8 Extinction efficiency as function of wavelength for four sample aerosol parameters. The tropospheric aerosol sizes of 1.0 and 2.3 μm are typical for mid-latitudes in 1995 and 2003, respectively. The stratospheric imaginary refractive indices of 0.015 and 0.08 at 300 nm wavelength are typical for low latitudes and polar latitudes, respectively (from Karkoschka and Tomasko (2005))

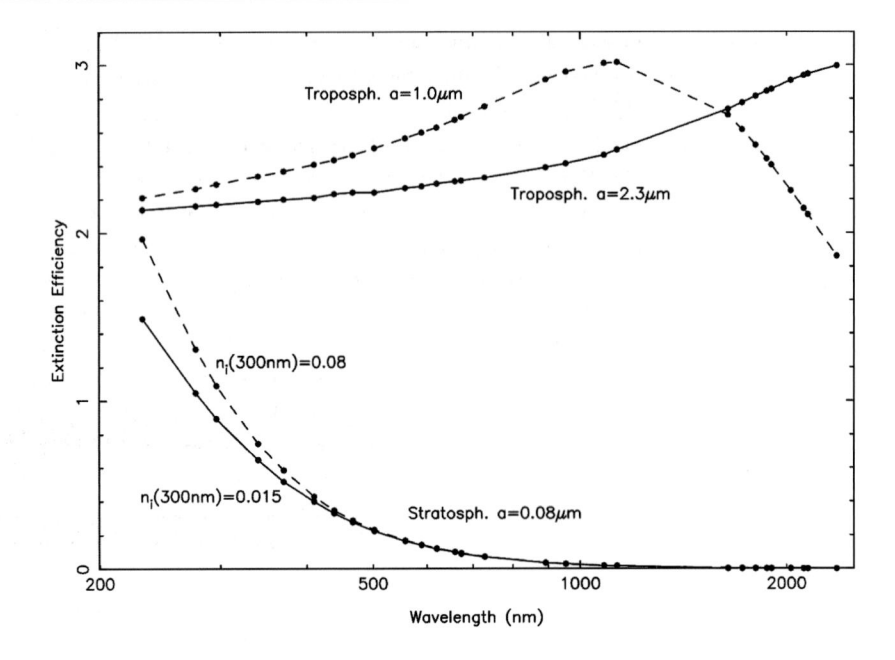

7.5 Aerosol Vertical Structure

7.5.1 Radiative Transfer Models and Haze Structure

Several techniques provide information on cloud and haze vertical structure. At visible and near-infrared wavelengths vertical sounding is possible via sampling in methane bands of various strengths. Although some have computed 'weighting functions' or 'contribution functions' these actually depend on the aerosol distribution and so are not well defined as they are for thermal temperature sensing although in the limit of single scattering one can use this technique as Stam et al. (2001) have done. In the general case which involves multiple scattering the vertical resolution of these techniques is typically of order one scale height. Other techniques such as stellar occultation (e.g. Fig. 7.3) or limb imaging provide much finer resolution in the vertical.

A sense of the depth of probing at different wavelengths can be obtained from Fig. 7.9 which shows optical depth 1/2 levels for Rayleigh scattering and methane absorption throughout the visible and near-infrared. At short wavelengths Rayleigh scattering is an important opacity source. The relation between Rayleigh optical depth and pressure was derived for Jupiter by West et al. (2004). A scaling argument leads to the following expression for Saturn.

$$\tau_R = P \cdot (24.4/g) \cdot (2.27 \cdot 10^{-3}/\mu) \cdot ([H_2]/0.86)$$
$$\cdot 0.0083 \left(1 + 0.014\lambda^{-2} + 0.00027\lambda^{-4}\right) \cdot \lambda^{-4}$$

(7.1)

where λ is the wavelength in μm, P is the pressure in bar, g (m s^{-2}; a function of latitude) is the local effective acceleration accounting for both gravitation and rotation (see Chapter 4), μ is the mean molecular weight (kg mole^{-1}) and $[H_2]$ is the H$_2$ mole fraction. The dependence on mean molecular weight and H$_2$ mole fraction is made explicit so that the equation can be used as the helium mole fraction becomes better defined (see the Chapter 5). We neglected terms of order $[H_2]^2$ which is probably acceptable in view of the uncertainty in the He mole fraction.

The method employed by Stam et al. retrieves a product of the aerosol density times the optical cross section times the particle single-scatter albedo times the phase function at the scattering angle of the observation (near 180° for the Stam et al. paper). It is impossible to know the value of each term in the product and so the method provides a measure of aerosol variations with latitude and altitude if the optical properties do not change with wavelength, or if their wavelength dependence can be estimated. Banfield et al. (1998) pioneered this method for Jupiter. The method does not make assumptions about how many layers are present or at what levels but it requires that multiple scattering be small relative to single scattering, which means that it is limited to strong absorption bands. Aerosol number density or optical depth can be inferred if the particles are assumed to be spheres. Derived aerosol distributions are shown in Fig. 7.10 which shows hemispheric asymmetry in 1995 near equinox for both stratosphere and troposphere haze layers.

The strategy employed by most investigators to retrieve the optical properties and vertical structure of the aerosols is to fit the center-to-limb variation (CTLV) curves at selected wavelengths to those obtained from a radiative transfer

7 Clouds and Aerosols in Saturn's Atmosphere

Fig. 7.9 This figure gives an indication of the sounding level for Rayleigh scattering and for methane imaging and spectra assuming photons sound to optical depth $1/2$ (for a two-way path optical depth of 1), that the methane mole fraction is 0.0045, and for an instrumental resolution for the Cassini VIMS instrument. Expected levels of the condensate clouds are also indicated. The 5-μm thermal contribution function is also shown in the second panel defined by collision-induced hydrogen opacity (adapted from Baines et al. (2005))

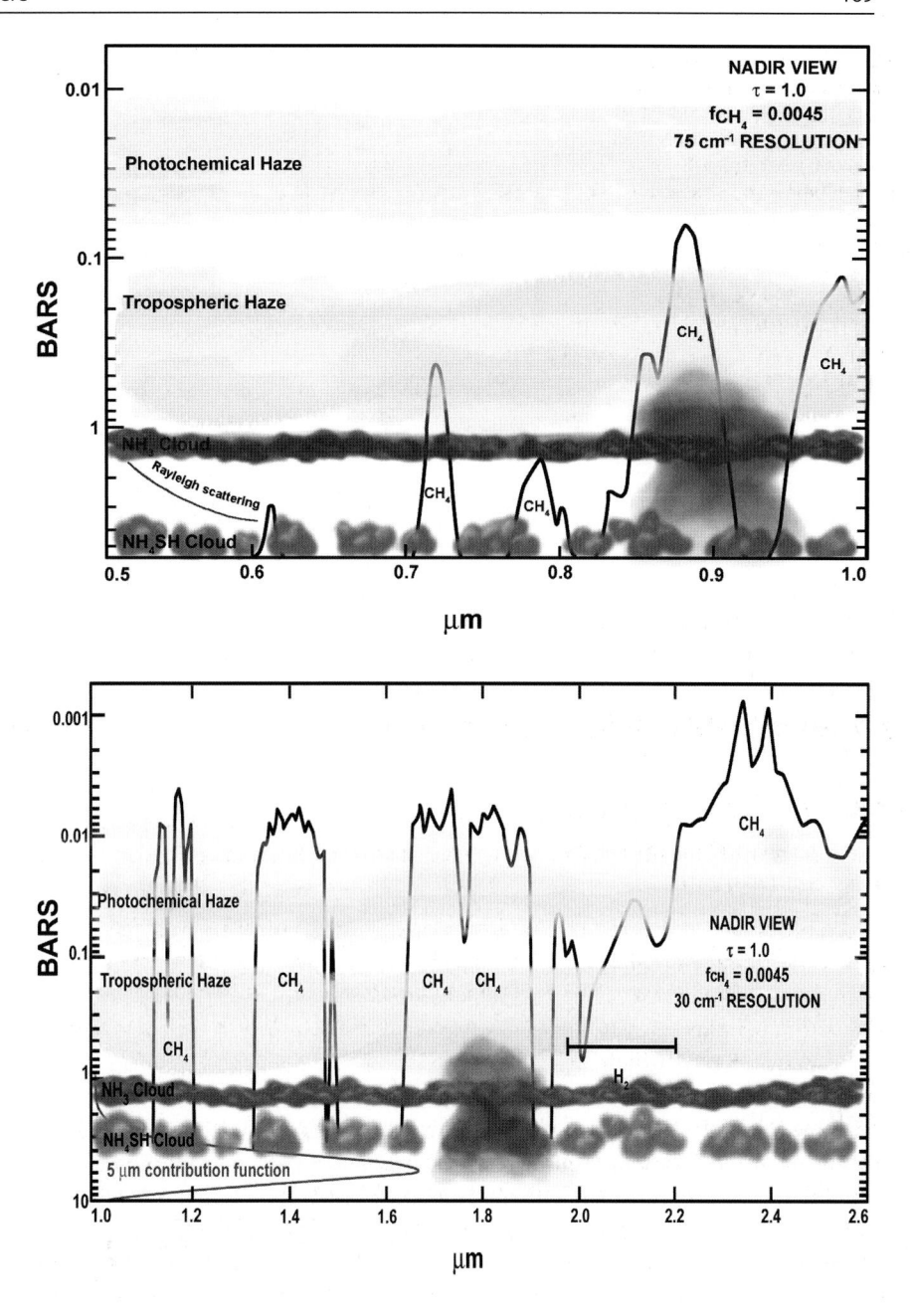

model atmosphere. Many observations are obtained with CCD sensors and for these the model takes into account the Rayleigh scattering by the major gases in the atmosphere (H_2 and He) and the weak, moderate and strong absorption by CH_4 bands at 619, 727 and 890 nm, whose absorption coefficients are known (Karkoschka 1994, 1998). Layers of particles concentrated or mixed with the atmospheric gas are incorporated by specifying their phase functions (Mie or synthetic double Henyey–Greenstein function) and optical depths. For spherical Mie particles, the size parameter, particle sizes distributions and refractive indexes are included as free parameters. The problem with this method is that we are confronted with a multi-parametric fit and usually one must assume that some of the model parameters are reasonably constrained from other observations (e.g. polarimetry and infrared radiometry).

7.5.2 Mean Vertical Structure

In Table 7.2 we summarize the results on the vertical aerosol structure of Saturn from studies performed since 1970. The most recent analysis comes from the following sources: (1) Karkoschka and Tomasko (1992) who studied in detail low and high-resolution ground-based spectra (460–940 nm) obtained from 1986 to 1989. Karkoschka

Fig. 7.10 Optical depth profiles near equinox in 1995 (soon after northern summer) are shown for a variety of pressure levels from 20 to 600 mbar (from Stam et al. (2001))

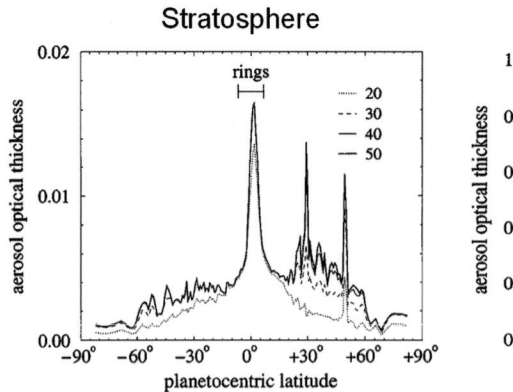

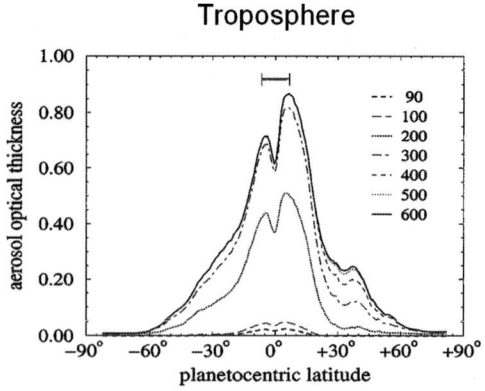

Table 7.2 Saturn's upper cloud and haze studies

References	Observations	Epoch	Number of layers	Latitude range	Spectral coverage (μm)
Macy (1977)	Ground-based	~1970	3	EZ–STZ	0.3–1.1
West (1983)	Ground-based	1979	1	34°S–34°N	0.619–0.937
Tomasko and Doose (1984)	Pioneer 11	1979	2	25°S–55°N	0.44/0.64
West et al. (1983)	Voyager 2	1981	1–2	NH	0.264/0.75
Karkoschka and Tomasko (1992)	Ground-based	1986–1989	2	NH	0.46–0.94
Karkoschka and Tomasko (1993)	HST	1991	2	NH	0.3–0.89
Ortiz et al. (1996)	Ground-based	1991–1993	3	NH	0.6–0.96
Kerola et al. (1997)	KAO	1978	1–2	Full Disk	1.7–3.3
Acarreta and Sánchez-Lavega (1999)	Ground-based	1990	2	EZ-NEB	0.336–0.89
Stam et al. (2001)	Ground-based	1995	3	NH/SH	1.45–2.5
Muñoz et al. (2004)	HST	1997	3	NH/SH	0.23–0.89
Temma et al. (2005)	Ground-based	2002	2	10°S	0.5–0.95
Pérez-Hoyos et al. (2005, 2006a)	HST	1994–2003	3	SH	0.255–1.042
Karkoschka and Tomasko (2005)	HST	1991–2004	3	NH/SH	0.23–2.37

NOTE: EZ = Equatorial Zone, STZ = South Temperate Zone, NEB = North Equatorial Belt, NH = Northern Hemisphere, SH = Southern Hemisphere

and Tomasko (1993) analyzed the structure of the stratospheric and tropospheric hazes using images obtained in the 300–889 nm wavelength range with the Hubble Space Telescope on July 1991. (2) Ortiz et al. (1996) studied the evolution of Saturn's hazes and clouds during the 1991–1993 period using ground-based images in the red methane bands and in their adjacent continuums (619–948 nm). (3) Kerola et al. (1997) presented a model of the clouds and hazes based on the near-infrared spectrum (1.7–3.3 μm wavelength range) obtained in 1978 with the Kuiper Airborne Observatory. (4) Temma et al. (2005) analyzed center-to-limb variations at many wavelengths for the Equatorial Zone. (5) Pérez-Hoyos et al. (2005) presented a study of the vertical structure of clouds and hazes in the upper atmosphere of Saturn's southern hemisphere during the 1994–2003 period, about one third of a Saturn's year, based on Hubble Space Telescope images in the spectral range between the near-UV (218–255 nm) and the near-IR (953–1042 nm), including the 890-nm methane band.

These studies envision an upper stratospheric haze and a tropospheric haze variable both with latitude and time (see Fig. 7.11). The tropospheric haze extends from the level P_2 close to the tropopause (~90–100 mbar) to a level P_1 above the ammonia cloud (~1.5–1.8 bar) and has an optical thickness τ_1. Usually the haze phase function is modeled using a double Henyey–Greenstein function (Henyey and Greenstein 1941),

$$P(\theta) = f \cdot p(g_1, \theta) + (1 - f) \cdot p(g_2, \theta) \quad (7.2)$$

where

$$p(g, \theta) = \frac{1 - g^2}{(1 + g^2 - 2g \cos(\theta))^{3/2}} \quad (7.3)$$

Values to fit for the parameters that define this phase function are the single scattering albedo ϖ_0, and the three parameters f, g_1 and g_2 that determine the contributions to the forward and backward scattering are usually taken from

7 Clouds and Aerosols in Saturn's Atmosphere

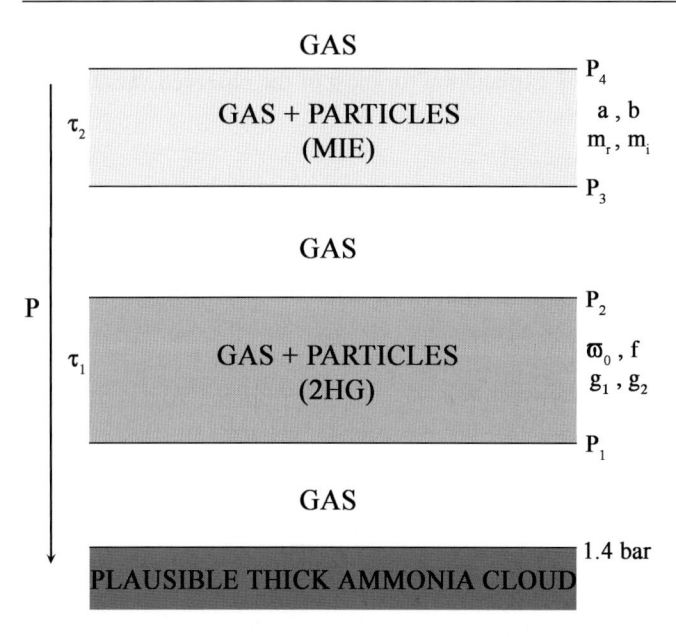

Fig. 7.11 This schematic shows haze and cloud layers and lists parameters used in radiative transfer models employed in the analysis of ground-based and spacecraft images and spectra. For a definition of the parameters see the text (from Pérez-Hoyos et al. (2005))

Tomasko and Doose (1984) as shown in Table 7.1. Alternatively a Mie phase function is also used being characterized by the real and imaginary refractive indices m_r and m_i, and the particle size distribution (with mean size a and dispersion b, from Hansen 1971).

The stratospheric haze is located between pressure levels P_3 (~10–90 mbar) and P_4 (~1 mbar) and has an optical thickness τ_2 that varies strongly with wavelength. All studies indicate that it is formed by small particles and a Mie phase function has been used to describe its behavior. The size distribution function used by Hansen and Travis (1974)

$$n(r) = r^{(1/b-3)} e^{-r/ab} \qquad (7.4)$$

is found to be appropriate. Here r is the particle's radius, $a \sim 0.15$–$0.2\,\mu\mathrm{m}$ is the effective (average) radius and $b = 0.1$ represents the variance in the size distribution. The small particle size implies that the optical depth τ_2 is dependent of wavelength. The strong UV absorption indicates that at 300 nm $\varpi_o = 0.6 \pm 0.1$ (Karkoschka and Tomasko 1993; Pérez-Hoyos et al. 2005).

7.5.3 Latitudinal Structure

The banded visual appearance of Saturn and the alternating pattern of jet streams with latitude, suggest a regional classification of the haze content in three main latitude bands: the equator (between latitudes $\pm 20°$), the middle-latitudes (from 20° to 60°) and the polar region (from 60–70° to the pole). The stratospheric particles are smaller ($a \sim 0.1\,\mu\mathrm{m}$) and more absorbent at UV wavelengths poleward of $\sim 70°$, than at other latitudes. At other latitudes, particles are pretty similar with a mean particle radii near $0.2\,\mu\mathrm{m}$.

The most important latitudinal variations take place in the tropospheric haze. There is a clear tendency to find darker particles at higher latitudes. This effect is most noticeable at 439 nm, a sensitive wavelength to single scattering. The pressure top level and especially the optical depth variations with latitude of this haze are the most important factor in generating the visual appearance of the planet. In the Equatorial Zone the haze is thicker and higher than elsewhere on the planet. At other latitudes the haze abundance is sharply reduced. In the polar region, the tropospheric haze diminished to only one tenth of its equatorial abundance (Pérez-Hoyos et al. 2005). These variations in haze altitude and thickness can be seen qualitatively in the strong 890-nm methane image (Fig. 7.12).

Karkoschka and Tomasko (2005) performed a principal-components study of 134 HST images obtained over the period 1991–2004. Their analysis included 18,000 center-to-limb curves in 30 filters from the near-UV to the near-IR. A small sample of the images is shown in Fig. 7.13. From these data four statistically-meaningful principal components emerged. The first principal variation is a strong mid-latitude variation of the aerosol optical depth in the upper troposphere. This structure shifts with Saturn's seasons, but the structure on small scales of latitude stays constant. This is what is most apparent in a casual comparison of images taken in different seasons. The second principal variation is a variable optical depth of stratospheric aerosols. The optical depth is large at the poles and small at mid- and low latitudes with a steep gradient between. This structure remains essentially constant in time. The third principal variation is a variation in the tropospheric aerosol size, which has only shallow gradients with latitude, but large seasonal variations. Aerosols are largest in the summer and smallest in the winter. The fourth principal variation is a feature of the tropospheric aerosols with irregular latitudinal structure and fast variability, on the time scale of months. This component is perhaps the most intriguing.

7.5.4 Short-term Changes

Smith et al. (1982) reported changes observed in the belt/zone pattern of Saturn between the two Voyagers encounters. Rapid changes affecting the brightness or color over an entire latitude band were observed in HST images between 1996 and 2003 (Pérez-Hoyos et al. 2006). These changes, in general, seem

Fig. 7.12 This Cassini ISS image taken with filter MT3 centered on the strong 890-nm methane absorption band shows qualitatively how cloud top altitude and optical thickness vary with latitude. The Equatorial Zone is a region of maximum vertical extent of the upper tropospheric haze and is artificially saturated in this image in order to make visible features in the darker mid- and high-latitude regions. Ring shadows in the northern hemisphere low latitudes should not be interpreted as methane absorption

not to be preceded by any individual atmospheric localized (in latitude or longitude) disturbance, evolving instead gradually and simultaneously in the whole band. Their origin is unclear, but the spectral behavior of the changes places the variability at the tropospheric haze level. They most frequently occur at equatorial and polar latitudes, and consist in the brightening or darkening by 5% to 10% of a latitude band with a width of 3° or less. About 80% of the observed phenomena produce a reflectivity variation less than 20% of the initial value. The vertical cloud structure modelling suggests variation of the single scattering albedo of the particles (probably induced by small changes in the size or composition of the particles) as the most likely explanation for most of the observed variations.

7.5.5 Seasonal Changes

Figure 7.14 shows that Saturn has gone through a full seasonal cycle starting with the 1979 measurements by the Pioneer IPP instrument (Tomasko and Doose 1984) and quantitative CCD imaging from the ground (West 1983). From those measurements and also from the Voyager color images and near-UV reflectivity obtained in 1981 (after equinox) it is clear that clouds and haze are higher and thicker in the northern mid-latitudes relative to the southern hemisphere near the end of southern summer and shortly after equinox (see Fig. 7.16 of Tomasko et al. 1984). The opposite asymmetry has prevailed thus far into the Cassini mission. During the early part of the Cassini tour beginning in 2004 Saturn's northern latitudes appeared blue, whereas the southern latitudes were yellow–red–brown. Methane absorption was stronger in the north. Both of these point to thinner, deeper haze and clouds in the north relative to the south. If the haze cycle is going to repeat in step with the seasonal phase a rapid shift in hemispheric asymmetry needs to occur very soon. Recent color changes are seen in Cassini images which indicate asymmetry reversal is underway in 2007–2008. A detailed cloud/haze microphysical model for the seasonal changes has not yet been developed but it seems likely that photochemical processes might play a role with seasonal modulation enhanced by ring shadowing. Other important processes might include seasonal changes in meridional circulation, sublimation of dirty ice grains exposed to sunlight and the atmospheric energy budget below the cloud tops.

7.5.6 Regional Structure: The Equatorial Jet

Figure 7.15 is a schematic diagram of the long-term structural changes in the tropospheric haze (from 70 mbar to 1.5 bar) of Saturn's Equatorial Zone (from 10°N to 10°S) as obtained from many studies over the period 1979–2004 (Pérez-Hoyos et al. 2005). A major change occurred following the giant storm in 1990 (GWS). In 1980–1981 the haze top was at a pressure $P_2 \sim 200$ mbar with optical thickness $\tau_1 \sim 10$ but in 2004 $P_2 \sim 40$ mbar and $\tau_1 \sim 15$. In general the haze is dense in the Equator (3.5°N and 0°) where it can reach an optical thickness of ~ 30 decreasing to ~ 7 at latitude 20°S. Observations at different wavelengths allow the detection of individual cloud tracers at different altitudes within this haze in the Equator (3° ± 2°N). For example Pérez-Hoyos and Sánchez-Lavega (2006a) placed

7 Clouds and Aerosols in Saturn's Atmosphere

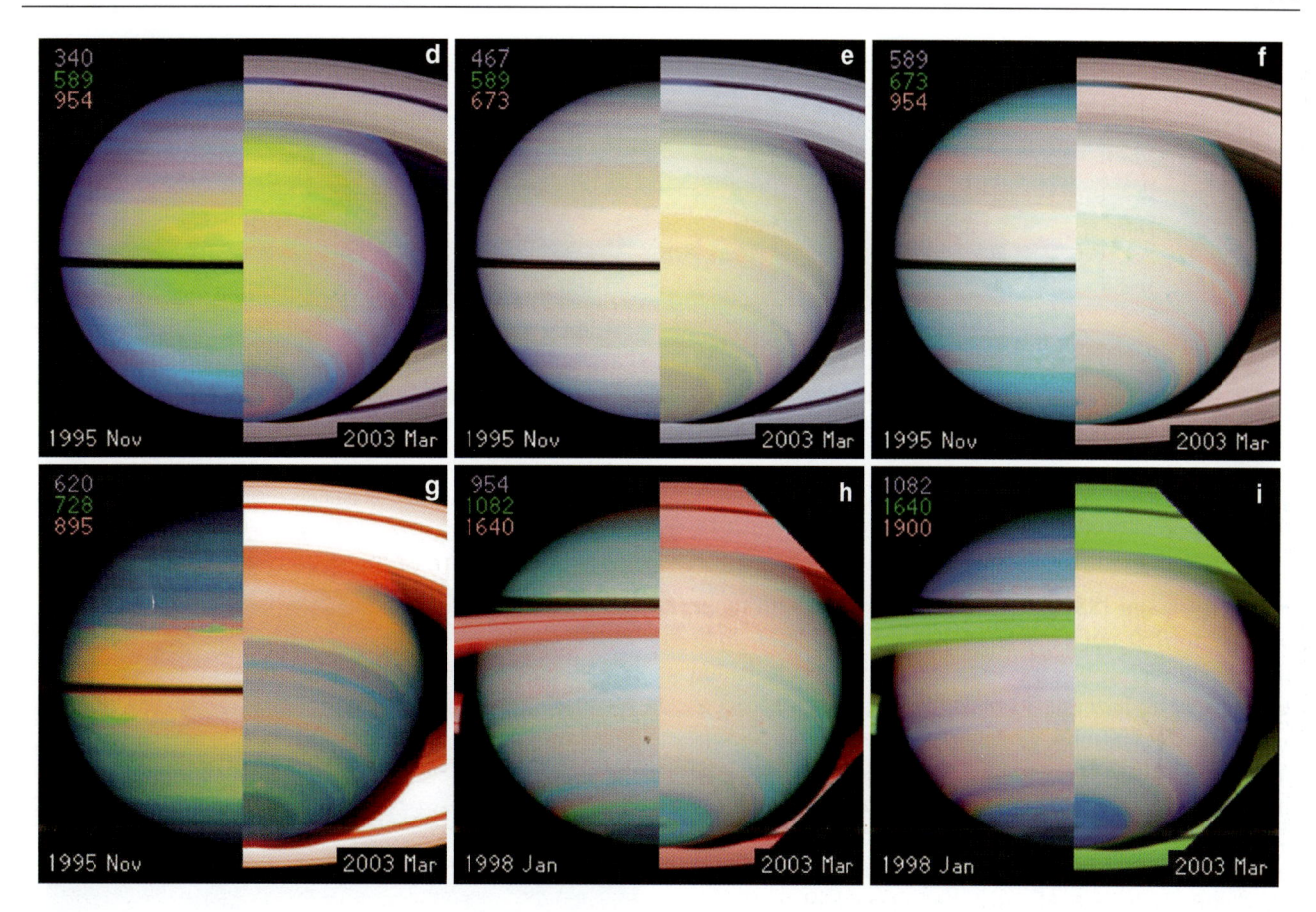

Fig. 7.13 This figure shows a small sample of Hubble Space Telescope images used in the principal-components analysis of Karkoschka and Tomasko (2005). Images at a wide variety of wavelengths were inserted into the red, green, and blue color planes as indicated by the filter effective wavelength (nm) in the notation for each color composite, and images near equinox and near solstice were split and merged to show seasonal differences

Fig. 7.14 This figure shows the coverage of modern observations of Saturn over one seasonal cycle. The Cassini observations continue to the present and planning is underway to extend the mission to northern summer solstice in 2017 (adapted from Pérez-Hoyos et al. (2005))

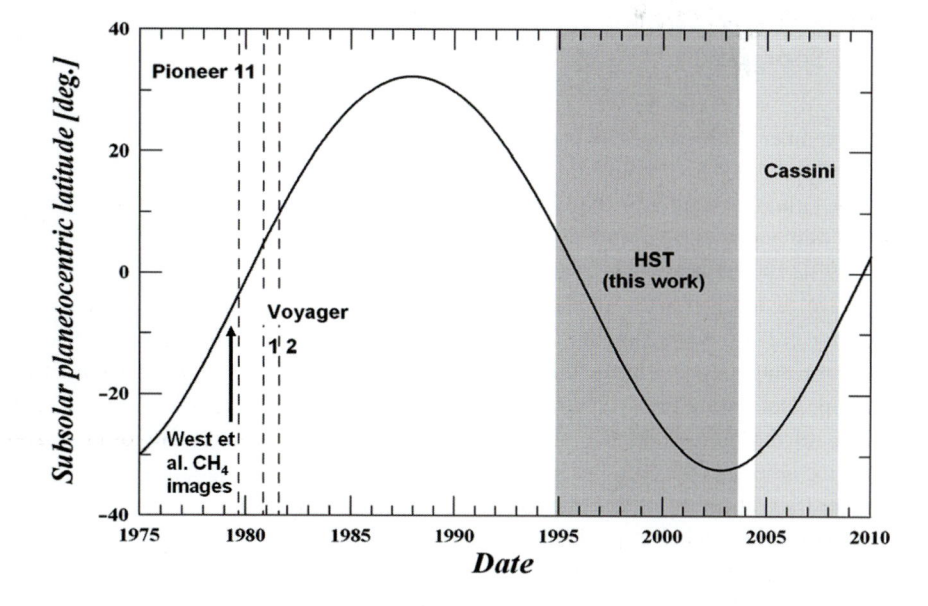

them at a pressure level of 360 ± 140 mbar in 1980–1981 where the tracers moved with zonal velocities of 455 to 465 m/s. The 2004 analysis of HST images indicates that tracers were placed high in the atmosphere at 50 ± 10 mbar moving with zonal wind speeds of 280 ± 10 m/s (Pérez-Hoyos and Sánchez-Lavega 2006a). These results were confirmed by the first reflectivity measurements and models of the equatorial zone (from 8°N to 20°S) performed in 2004 and early

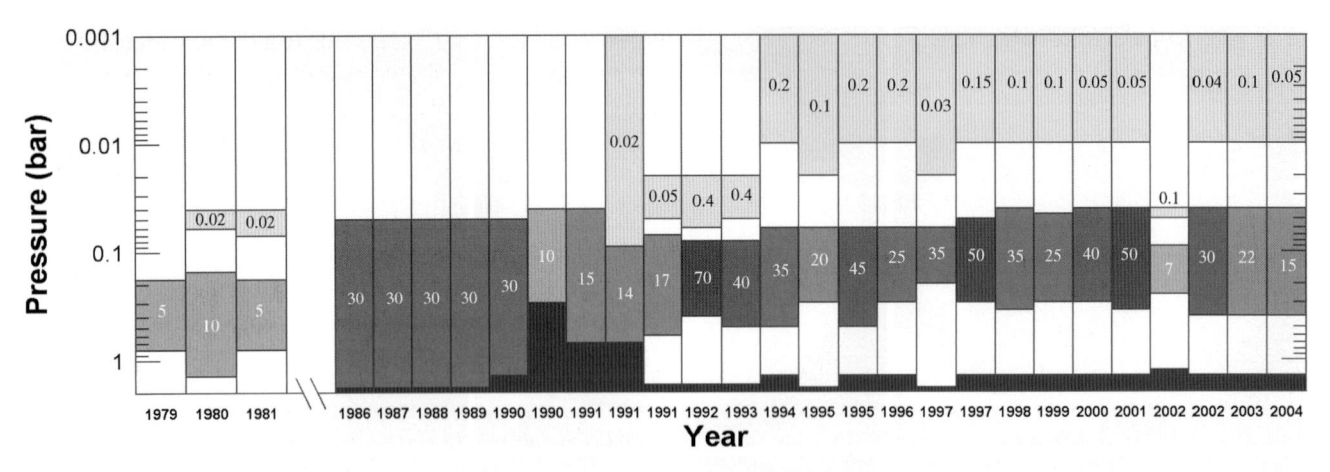

Fig. 7.15 Derived optical depths and altitudes of upper tropospheric haze and clouds are shown as a function of observation epoch from 1979 to 2004 for a variety of cloud studies (from Pérez-Hoyos and Sánchez-Lavega (2006a))

2005 with Cassini-ISS instrument in the wavelength range from 250 to 950 nm (Sánchez-Lavega et al. 2007). Individual cloud elements were detected at two levels within the tropospheric haze, at 50 mbar moving with a speed of 263 ms^{-1} and at 700 mbar moving at 364 ms^{-1} (altitudes separated by 142 km), representing a vertical shear of the zonal wind of 40 ms^{-1} per scale height. A comparison with the previous analysis indicates that the equatorial jet has undergone a significant intensity change between 1980–1981 and 1996–2005, most probably as a consequence of the large-scale equatorial storms that occurred in 1990 and 1994–1996.

7.5.7 Regional Structure: The Great White Spot (GWS)

In September 1990 (Fig. 7.16) a giant storm (the "Great White Spot", GWS) erupted in Saturn's Equatorial Zone at 5°N latitude, disturbing the dynamics and cloud structure of this region during more than a year (Sánchez-Lavega et al. 1991, 1993, 1994; Beebe et al. 1992; Barnet et al. 1992). Radiative transfer models indicated that cloud tops elevated by 1.2–1.5 scale heights during the onset and mature storm stage relative to the undisturbed atmosphere, descending later during the evolved phase by 0.7 scale heights relative to the altitude of the mature clouds. The single-scattering albedo of the particles in the UV-blue wavelengths increased significantly during the onset and mature stages of the storm (Westphal et al. 1992; Barnet et al. 1992). In 1994 another large cloud system erupted but not as large as one in 1990. The cumulative effects of these events produced elevated optical depths in the equatorial zone for years afterward.

Fig. 7.16 These panels show some of the ground-based images of the Great White Spot eruption of 1990. From top to bottom in red, green, blue and ultraviolet wavelengths (Pic-du-Midi Observatory, France). Blue-filter and UV images show the highest contrast between the bright spot and the dark ambient haze structure (from Sánchez-Lavega et al. (1991))

7.5.8 Regional Structure: The South Polar Vortex

A warm cap in the South Pole of Saturn (Orton and Yanamandra-Fisher 2005) is the site of a large vortex that surrounds the pole at latitude 87°S (Sánchez-Lavega et al. 2006). High-resolution images of the cap obtained with Cassini ISS and VIMS instruments (Fig. 7.17) show that it is a hole in the cloud layer with altitude differences in the eye wall of 40 km (inner wall) and 70 km (outer wall) as measured from the projected shadows (Dyudina et al. 2008). Radiative transfer models for latitudes poleward of 80°S show that the polar stratospheric haze is formed by ultraviolet–violet absorbing particles with a mean radius of 0.15 μm extending between the pressure levels ∼1 and 30 mbar (Sánchez-Lavega et al. 2006a). A lower tropospheric haze layer formed by particles with a radius of ∼1 μm resides between 70 and 150–300 mbar but it is thin compared to other areas of Saturn with optical thickness of ∼1 being neutral in its wavelength dependence.

7.5.9 Convective Clouds

The 5-μm panel in Fig. 7.17 probes the deepest levels (see the contribution function in Fig. 7.9) and cloud opacity is revealed by its ability to block upwelling thermal radiation. Sounding depth at other wavelengths sensitive to reflected sunlight can also be judged from Fig. 7.9. Many small-scale features can be seen in the 0.75-μm continuum image indicating that photons are penetrating deeply at that wavelength. Some of the same features can be seen even in the 0.46-μm image. Note that some small-scale features are brighter than the atmosphere they are embedded in (consistent with the idea that they are relatively dense clouds composed of white ammonia ice) but some are darker than their surroundings. The oval near the lower-left corner of each panel is bright in the blue image but dark in the near-infrared images. This is opposite the color behavior of most chromophore material. The absence of these features in the methane absorption bands shows that they reside in the deeper atmosphere, probably deeper than the 2-bar pressure level. The morphology of these features resembles terrestrial cumulus and is suggestive of formation in a freely convecting atmosphere, hence we call these convective clouds.

The Cassini VIMS ability to probe to deep levels at all latitudes and in polar night has opened up a new realm for investigation of deeper cloud features. Information on winds revealed by these images is discussed in Chapter 6. These studies are in their early phases and we give here some samples of what can be seen. Global views of Saturn in the 5-μm band are seen in Fig. 7.18. Hundreds of small cloud features can be seen. The overlying clouds are slightly thicker in the southern hemisphere in the 2005 image (panel b) as noted earlier in the section on seasonal asymmetry. Few convective clouds can be seen in the Equatorial Zone where a more zonally-uniform cloud prevails. The northern-hemisphere view in the left panel shows a series of cloud-free 'holes' in the retrograde jet at 33.5° latitude (planetocentric). Each is about 800 km in diameter. Images in the wavelength range 3.1–5.1 μm are sensitive to the slope in the particle extinction cross section. Fig. 7.19 reveals distinct bands of increasing upper troposphere cloud opacity with the lowest opacity at the highest latitudes.

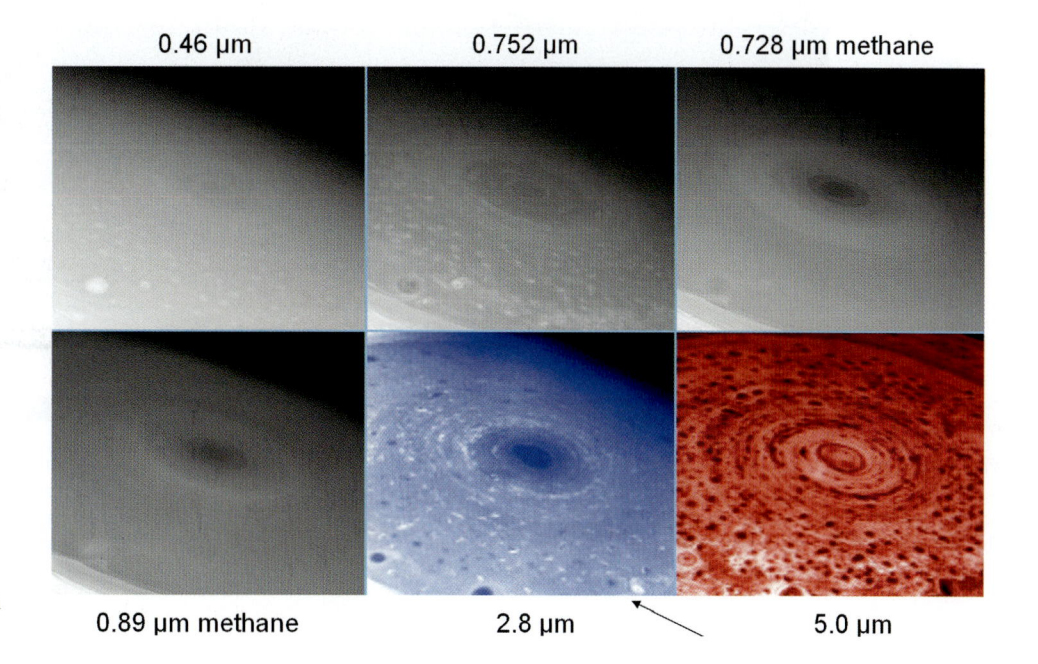

Fig. 7.17 Six panels from the Cassini VIMS experiment show views of the southern polar vortex and surrounding region from visible wavelengths to 5 μm (from Baines et al. (2008))

Fig. 7.18 Global views of Saturn at 5 μm show many details of the convective clouds in the 2–4 bar region. In the left panel Saturn's "String of Pearls" observed over 2.8 years. Observed in Saturn's thermal glow at 5 μm, a train of nearly regularly spaced holes in the clouds are repeatedly observed throughout the Cassini mission thus far, for more than 2.8 years, near 33.5° latitude (planetocentric). The holes are spaced about 3500 km (4.4° in longitude) apart, span about 1° (800 km) each, and are confined to a latitude range of <2° of latitude, 110° longitude. The right panel shows Saturn's hemispheric asymmetry (on average the intensities are slightly lower and convective clouds slightly less visible due to overlying haze). This effect is masked by scattered sunlight which elevates the intensities mostly in the southern hemisphere. The ring shadow darkens part of the northern hemisphere, right side (from Baines et al. (2008))

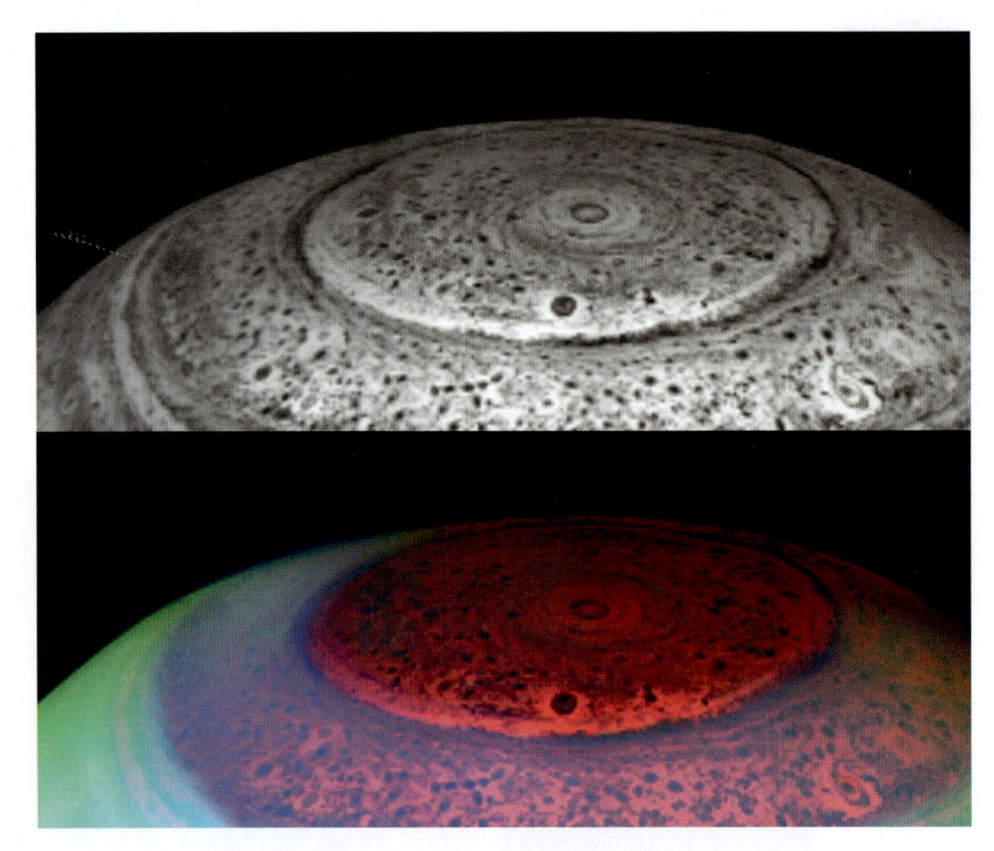

Fig. 7.19 Near-infrared views of the south pole of Saturn. Detailed views are shown as acquired on May 11, 2007 by Cassini/VIMS, from a vantage point 0.36 million km above the cloud tops. Saturn's upwelling thermal radiation at 5.1 μm illuminates the atmosphere from below, revealing deep cloud features in silhouette (*top panel*). Thermal radiation at 5.1 μm populates the red color plane in the bottom panel. Overlying haze reflects sunlight, shown as blue (3.1 μm) and green (4.1 μm), especially equatorward of the cloudy band circling the planet near 75° south latitude (from Baines et al. (2008))

7.6 Solar Radiation Penetration and Deposition

Radiative heating/cooling is a fundamental process for the general circulation of Saturn and for other dynamical phenomena. Tomasko and Doose (1984) estimated geometric albedo (the ratio of flux reflected in the back-scatter direction to that for a flat surface of equal area and having the Lambert reflection law) and phase integral for the Equatorial Zone and a mid-latitude belt. For global radiation budget the spherical albedo is a key parameter. It is the ratio of power reflected in all directions to incident power for a spherical body. It is the product of the geometric albedo and the phase integral and these are reported in Table 7.1.

The insolation at the top of Saturn's atmosphere at a given time and latitude including the effect of Saturn's oblateness, ring shadowing and the solar radiation scattered from the rings to the atmosphere, was first addressed by Brinkman and McGregor (1979) and later by Barnet et al. (1992). Pérez-Hoyos and Sánchez-Lavega (2006b) recalculated the distribution map of daily averaged insolation along a Saturn's year for a solar constant of $S_0 = 15\,\text{Wm}^{-2}$ to study the solar flux deposition in the optical wavelength range from 0.25 to 1.0 μm, a spectral band that covers about 70% of the solar radiation power. The solar flux deposition is controlled by gaseous absorption in the methane-bands and by Rayleigh scattering in the continuum (much less important are the absorptions due to ammonia bands and H_2 quadrupole lines), and more importantly by scattering and absorption by the aerosols. Radiative transfer calculations that include the internal radiation fields shows that the maximum heating level is ∼250 mbar for these wavelengths, substantially higher than previously expected because of the huge optical thickness of the tropospheric haze described in all vertical cloud structure models. Given that our spectral range accounts for about the 70% of the total solar flux, and using previous estimates for the penetration levels of infrared radiation in Saturn's atmosphere, Pérez-Hoyos and Sánchez-Lavega (2006b) concluded that almost no solar radiation will heat the levels deeper than 600 mbar. This conclusion may need to be revised in light of the recent Cassini data which show contrasts from convective clouds that are probably deeper than 600 mbar.

Once the solar flux daily mean is calculated as a function of altitude $F(P)$, the heating rate is obtained from

$$\frac{dT}{dt} = \frac{g}{C_p} \frac{dF}{dP} \qquad (7.5)$$

being g the local acceleration of gravity and $C_p(P)$ the specific heat at constant pressure. Typical values of tropospheric heating for equatorial latitudes are 0.2 K/day, whereas for mid-latitudes they decrease to 0.04–0.08 K/day. Polar tropospheric heating rates are small, not greater than 0.02 K/day.

7.7 Summary and Future Work

Observations and analyses of ground-based and spacecraft data obtained during the past three decades has led to a picture of Saturn's cloud and haze with many components covering a wide range in spatial and temporal scales. The deepest clouds trace small-scale convective activity and are seen at some latitudes all the way into the blue but are most easily seen in thermal emission at 5 μm. Examination of Figs. 7.17 through 7.19 reveals morphological forms which vary with latitude. These morphological differences may provide good clues regarding instability modes, wind shear and other attributes of dynamics on the scale of zonal jets and smaller. This information has not yet been exploited.

Observations at many wavelengths and over a significant part of Saturn's seasonal cycle reveal seasonal and nonseasonal variations. Some variations have short time scales. Recent Cassini images show that Saturn's hemispheric asymmetry is changing, in expectation with observations taken in 1979 which show an asymmetry opposite to what was observed by Cassini instruments until 2008. Over the next year the asymmetry reversal is expected to be complete with the hope that frequent and detailed Cassini observations will allow us to construct a more detailed picture of how the process unfolds.

On regional scales three latitude bands are distinct. The Equatorial Zone is consistently a region of higher and thicker clouds. It is also a region which experiences occasional major cloud eruptions which have long-term influences on cloud opacity. Images at 5 μm do not show this region to be populated with convective clouds. Rather, the haze is more uniform, possibly because of strong zonal wind shear. Middle latitudes experience strong seasonal variations. High latitudes are regions where the upper tropospheric opacity is low, convective clouds and small vortices are numerous, and where a polar vortex produces an extreme low in aerosol opacity right at the pole.

In the vertical domain the atmosphere exhibits two distinct regions – upper troposphere and stratosphere. Upper tropospheric cloud particles are larger than the wavelength, although how much larger is still not settled. Above that lies an optically thin layer of particles which are significantly smaller than visible wavelengths. Within the polar vortex region these particles are dark at UV wavelengths, more abundant than at lower latitudes, and exhibit strong positive polarization similar to jovian polar aerosols which West and Smith (1991) proposed to be aggregates of small monomers.

Their origin is probably tied to auroral energy deposition and their composition may be different than that of the main stratospheric haze at lower latitudes. A first positive identification of hydrocarbon composition was made with a stellar occultation measurement (Nicholson et al. 2006). Diphosphine haze should be abundant in the lower stratosphere based on photochemical considerations and the observation that the phosphine mixing ratio decreases with altitude.

In spite of these advances some questions remain essentially unanswered, or at least the proposed solutions are still very much open to debate. One of the major puzzles continues to be the lack of spectral evidence for ammonia ice. Atreya et al. (2005) and Kalogerakis et al. (2008) propose coating of ammonia ice particles by photochemical products (for Jupiter, with extension to Saturn). This is the inverse of the traditional picture that photochemical aerosol sedimenting from high altitude might serve as condensation nuclei for ammonia ice particles (West et al. 1986, for Jupiter, with extension to Saturn). If this mechanism can account for the masking of ammonia ice signatures the production rate of the photochemical product must outstrip the production rate of ammonia ice. It is not obvious that this can be the case. A detailed microphysical model which includes ammonia condensation might help assess this proposal. It might also help us understand why some small-scale cloud features are dark (i.e. abundant in chromophore material).

Instruments on Cassini continue to acquire data and much of the existing Cassini data has yet to be comprehended. In the coming years we can expect to gain a much better understanding of the details of seasonal change. Observations at many phase angles and many latitudes (most notably low latitudes which are obscured by the rings or ring shadow except for short times) will allow us to improve on models of the radiation budget which require observations at many phase angles. We are still using values of the phase function and phase integral based on the 1979 Pioneer data. Images of Saturn's limb at high spatial resolution will provide detailed vertical profiles of the stratospheric haze. Additional stellar and solar occultations will yield detailed profiles and compositional information on the stratospheric haze. Ultimately we can expect these new data and analyses to illuminate the important chemical and physical processes which take place over many scales in space and time.

References

Acarreta, J.R., Sánchez-Lavega, A. Vertical cloud structure in Saturn's 1990 Equatorial Storm, Icarus 137, 24–33 (1999).

Atreya, S.K., Wong, A.-S. Coupled clouds and chemistry of the giant planets – a case for multiprobes. Space Sci. Rev. 116, 121–136 (2005).

Atreya, S.K., Wong, A.-S., Baines, K.H., Wong, M.H., Owen, T.C. Jupiter's ammonia clouds – localized or ubiquitous? Planet. Space Sci. 53, 498–507 (2005).

Baines, K.H., Carlson, R.W., Kamp, L.W. Fresh ammonia ice clouds in Jupiter: I. Spectroscopic identification, spatial distribution, and dynamical implications. Icarus 159, 74–94 (2002).

Baines, K.H., Drossart, P., Momary, T.W., Formisano, V., Griffith, C., Bellucci, G., Bibring, J.-P., Brown, R.H., Buratti, B.J., Capaccioni, F., Cerroni, P., Clark, R.N., Coradini, A., Cruikshank, D.P., Jaumann, R., Langevin, Y., Matson, D.L., McCord, T.B., Mennella, V., Nelson, R.M., Nicholson, P.D., Sicardy, B., Sotin, C. The atmospheres of Saturn and Titan in the near-infrared: First results of Cassini/VIMS. Earth Moon Planets 96, 119–147 (2005).

Baines, K.H., Momary, T.W., Kim J.H., Ross-Serote, M., Showman, A.P., Atreya, S.K., Brown, R.H., Buratti, B.J., Clark, R.N., Nicholson, P.D. Saturn's dynamic atmosphere at depth: Physical characteristics, winds, and spatial constraints on trace gas variability near the 3-bar level and their dynamical implications from Cassini-Huygens/VIMS. Poster Presented at Saturn after Cassini-Huygens, Conference, Imperial College, London, United Kingdom, July 28–August 1 (2008).

Baines, K.H., Delitsky, M.L., Momary, T.W., Brown, R.H., Buratti, B.J., Clark, R.N., Nicholson, P.D. Storm clouds on Saturn: Lightning-induced chemistry and associated materials consistent with Cassini/VIMS spectra. Planetary and Space Sci. in press (2009).

Banfield, D., Conrath, B.J., Gierasch, P.J., Nicholson, P.D., Mathiews, K. Near-IR spectrophotometry of jovian aerosols: Meridional and vertical distributions. Icarus 134, 11–23 (1998).

Barnet, C.D., Westphal, J.A., Beebe, R.F., Huber, L.F. Hubble Space Telescope observations of the 1990 equatorial disturbance on Saturn: Zonal winds and central meridian albedos. Icarus 100, 499–511 (1992).

Beebe, R.F., Barnet, C., Sada, P.V., Murrell, A.S. The onset and growth of the 1990 equatorial disturbance on Saturn. Icarus 95, 163–172 (1992).

Brinkman, A.W., McGregor, J. The effect of the ring system on the solar radiation reaching the top of Saturn's atmosphere: Direct radiation. Icarus 38, 479–482 (1979).

Brooke, T.Y., Knacke, R.F., Encrenaz, T., Drossart, P., Crisp, D., Feuchtgruber, H. Models of the ISO 3-μm reflection spectrum of Jupiter. Icarus 136, 1–13 (1998).

Dyudina, U.A. Ingersoll, A.P., Ewald, S.P., Vasavada, A.R., West, R.A., Del Genio, A.D., Barbara, J.M., Porco, C.C., Achterberg, R.K., Flasar, F.M., Simon-Miller, A.A., Fletcher, L.N. Dynamics of Saturn's south polar vortex. Science 319, 1801 (2008).

Friedson, A.J., Wong, A.S., Yung, Y.L., Models for polar haze formation in Jupiter's stratosphere. Icarus 158, 389–400 (2002).

Hansen, J.E. Multiple scattering of polarized light in planetary atmospheres. Part I. The doubling method. J. Atmos. Sci. 28, 120–125 (1971).

Hansen, J.E., Travis, L.D. Light scattering in planetary atmospheres. Space Sci. Rev. 16, 527–610 (1974).

Henyey, L.G., Greenstein, J.L. Diffuse radiation in the galaxy. Annales d'Astrophysique 3, 117–137 (1941).

Huffman, D.R., Bohren, C.F. Infrared absorption spectra of nonspherical particles treated in the Rayleigh-ellipsoid approximation. In: D. Schuerman (ed) Light Scattering by Irregularly Shaped Particles, Plenum, New York, pp. 103–111 (1980).

Kalogerakis, K.S., Marschall, J., Oza, A.U., Engel, P.A., Meharchand, R.T., Wong, M.H., The coating hypothesis for ammonia ice particles in Jupiter: Laboratory experiments and optical modeling. Icarus 196, 202–215 (2008).

Karkoschka, E. Spectrophotometry of the jovian planets and Titan at 300- to 1000-nm wavelength: The methane spectrum. Icarus 111, 174–192 (1994).

Karkoschka, E. Methane, ammonia, and temperature measurements of the jovian planets and Titan from CCD-spectrophotometry. Icarus 133, 134–146 (1998).

Karkoschka, E., Tomasko, M.G. Saturn's upper troposphere 1986–1989. Icarus 97, 161–181 (1992).

Karkoschka, E., Tomasko, M.G. Saturn's upper atmospheric hazes observed by the Hubble Space Telescope. Icarus 106, 421–441 (1993).

Karkoschka, E., Tomasko, M. Saturn's vertical and latitudinal cloud structure 1991–2004 from HST imaging in 30 filters. Icarus 179, 195–221 (2005).

Kerola, D.X., Larson, H.P., Tomasko, M.G. Analysis of the near-IR spectrum of Saturn: A comprehensive radiative transfer model of its middle and upper troposphere. Icarus 127, 190–212 (1997).

Lewis, J.S. The clouds of Jupiter and the $NH_3 - H_2O$ and $NH_3 - H_2S$ systems. Icarus. 10, 365–378 (1969).

Macy, W. Inhomogeneous models of the atmosphere of Saturn, Icarus 32, 328–347 (1977).

Martonchik, J.V., Orton, G.S., Appleby, J.F. Optical properties of NH_3 ice from the far infrared to the near ultraviolet. Appl. Opt. 23, 541–547 (1984).

Muñoz, O., Moreno, F., Molina, A., Grodent, D., Gerard, J.C., Dols, V. Study of the vertical structure of Saturn's atmosphere using HST/WFPC2 images. Icarus 169, 413–428 (2004).

Nicholson, P., Hedman, M.M., Gierasch, P.J., the Cassini VIMS Team, Probing Saturn's Atmosphere with Procyon, Bull. Amer. Astron. Soc. 38, 555 (2006).

Ortiz, J.L., Moreno, F., Molina, A. Saturn 1991–1993: Clouds and hazes. Icarus 119, 53–66 (1996).

Orton, G.S., Yanamandra-Fisher, P. Saturn's temperature field from high-resolution middle-infrared imaging. Science 307, 696–698 (2005).

Pérez-Hoyos, S., Sánchez-Lavega, A., French, R.G., Rojas, J.F. Saturn's cloud structure and temporal evolution from ten years of Hubble Space Telescope images (1994–2003). Icarus 176, 155–174 (2005).

Pérez-Hoyos, S., Sánchez-Lavega, A. On the vertical wind shear of Saturn's equatorial jet at cloud level. Icarus 180, 161–175 (2006a).

Pérez-Hoyos, S., Sánchez-Lavega, A. Solar flux in Saturn's atmosphere: maximum penetration and heating rates in the aerosol and cloud layers", Icarus, 180, 368–378 (2006b).

Pérez-Hoyos, S., Sánchez-Lavega, A., French, R.G. Short-term changes in the belt/zone structure of Saturn's Southern Hemisphere (1996–2004). Astron. Astroph. 460, 641–645 (2006).

Pope, S.K., Tomasko, M.G., Williams, M.S., Perry, M.L., Doose, L.R., Smith, P.H. Clouds of ammonia ice: Laboratory measurements of the single-scattering properties. Icarus 100, 203–220 (1992).

Pryor, W.R., Hord, C.W. A study of photopolarimeter system UV absorption data on Jupiter, Saturn, Uranus, and Neptune: Implications for Auroral Haze formation. Icarus 91, 161–172 (1991).

Sánchez-Lavega, A. Saturn's great white spots, Chaos 4, 341–353 (1994).

Sánchez-Lavega, A., Colas, F., Lecacheux, J., Laques, P., Miyazaki, I., Parker, D. The Great white Spot and disturbances in Saturn's equatorial atmosphere during 1990. Nature 353, 397–401 (1991).

Sánchez-Lavega, A., Lecacheux, J., Colas, F., Laques, P. Temporal behavior of cloud morphologies and motions in Saturn's atmosphere, J. Geophys. Res. E10, 18857–18872 (1993).

Sánchez-Lavega, A., Lecacheux, J., Colas, F., Laques, P. Photometry of Saturn's 1990 equatorial disturbance, Icarus 108, 158–168 (1994).

Sánchez-Lavega, A., Hueso, R., Pérez-Hoyos, S., Rojas, J. F., A strong vortex in Saturn's South Pole, Icarus 184, 524–531 (2006).

Sánchez-Lavega, A., Hueso, R., Pérez-Hoyos, S. The three-dimensional structure of Saturn's equatorial jet at cloud level, Icarus 187, 510–519 (2007).

Smith, B.A., Soderblom, L., Batson, R., Bridges, P., Inge, J., Masursky, H., Shoemaker, E., Beebe, R., Boyce, J., Briggs, G., Bunker, A., Collins, S.A., Hansen, C.J., Johnson, T.V., Mitchell, J.L., Terrile, R.J., Cook, A.F., Cuzzi, J., Pollack, J.B., Danielson, G.E., Ingersoll, A.P., Davies, M.E., Hunt, G.E., Morrison, D., Owen, T., Sagan, C., Veverka, J., Strom, R., Suomi, V.E. A New Look at the Saturn system: The Voyager 2 images. Science 215, 504–537 (1982).

Stam, D.M., Banfield, D., Gierasch, P.J., Nicholson, P.D., Matthews, K., Near-IR Spectrophotometry of Saturnian Aerosols—Meridional and Vertical Distribution, Icarus 152, 407–422 (2001).

Temma, T., Chanover, N.J., Simon-Miller, A.A., Glenar, D.A., Hillman, J.J., Khuen, D.M. Vertical structure modeling of Saturn's equatorial region using high spectral resolution imaging, Icarus 175, 464–489 (2005).

Tomasko, M.G., Doose, L.R., Polarimetry and photometry of Saturn from Pioneer 11: observations and constraints on the distribution and properties of cloud and aerosol particles. Icarus 58, 1–34 (1984).

Tomasko, M.G., West, R.A., Orton, G.S., Tejfel, V.G. Clouds and aerosols in Saturn's atmosphere, in T. Gehrels, M.S. Matthews (eds) Saturn, University of Arizona Press, Tucson, pp. 150–194 (1984).

Weidenschilling, S.J., Lewis, J.S. Atmospheric and cloud structures on the jovian planets. Icarus 20, 465–476 (1973).

West, R.A. Spatially resolved methane band photometry of Saturn II. Cloud structure models at four latitudes. Icarus 53, 301–309 (1983).

West, R.A., Smith, P.H. Evidence for aggregate particles in the atmospheres of Titan and Jupiter, Icarus 90, 330–333 (1991).

West, R.A., Sato, M., Hart, H., Lane, L.A., Hord, C.W., Simmons, K.E., Esposito, L.W., Coffeen, D.L., Pomphrey, R.B. Photometry and polarimetry of Saturn at 2640 and 7500 Å. J. Geophys. Rev. 88, 8679–8697 (1983).

West, R.A., Strobel, D.F., Tomasko, M.G. Clouds, aerosols, and photochemistry in the jovian atmosphere. Icarus 65, 161–217 (1986).

West, R.A., Orton, G.S., Draine, B.T., Hubbell, E.A., Infrared absorption features for tetrahedral ammonia ice crystals. Icarus 80, 220–224 (1989).

West, R.A., Baines, K.H., Friedson, A.J., Banfield, D., Ragent, B., Taylor, F.W. Jovian clouds and haze. In: F. Bagenal, T.E. Dowling, W.B. McKinnon (eds) Jupiter: The Planet, Satellites and Magnetosphere, Cambridge University Press, Cambridge (2004).

Westphal, J.A., Baum, W.A., Ingersoll, A.P., Barnet, C.D., De Jong, E.M., Danielson, G.E. Hubble Space Telescope observations of the 1990 equatorial disturbance on Saturn: Images, albedos, and limb darkening, Icarus 100, 485–498 (1992).

Wong, A.-S., Yung, Y.L., Friedson, A.J. Benzene and haze formation in the polar atmosphere of Jupiter. Geophys. Res. Lett. 30 (2003).

Wong, M., Bjoraker, G., Smith, M. Flasar, D., Nixon, C. Identification of the 10-μm ammonia ice feature on Jupiter. Planet. Space Sci. 52, 385–395 (2004).

Chapter 8
Upper Atmosphere and Ionosphere of Saturn

Andrew F. Nagy, Arvydas J. Kliore, Michael Mendillo, Steve Miller, Luke Moore, Julianne I. Moses, Ingo Müller-Wodarg, and Don Shemansky

Abstract This chapter summarizes our current understanding of the upper atmosphere and ionosphere of Saturn. We summarize the available observations and the various relevant models associated with these regions. We describe what is currently known, outline any controversies and indicate how future observations can help in advancing our understanding of the various controlling physical and chemical processes.

8.1 Introduction

The direct exploration of the upper atmosphere and ionosphere of Saturn began nearly 30 years ago with the flyby of the Pioneer 11 spacecraft (September 11, 1979), followed shortly by Voyagers 1 and 2 (November 12, 1980 and August 26, 1981, respectively). These flybys offered us a glimpse of Saturn, which, combined with earlier ground based and remote measurements from Earth orbit, did provide some basic ideas on the temperature and composition of the upper atmosphere and on ionospheric electron densities. The information on the thermosphere and ionosphere during the Pioneer and Voyager flybys came from two sources: the UV spectrometer and radio occultation observations. Since the insertion of Cassini into orbit around Saturn, the amount of data has very significantly increased, although still only the same two observation techniques provide most of the information on the upper atmosphere and ionosphere. However the quality of these measurements, as well as the spatial and temporal coverage, are significantly enhanced with Cassini. In this chapter we summarize our current understanding of the upper atmosphere and ionosphere of Saturn based on spacecraft and ground based measurements and modeling activities. Some of the other chapters (e.g., Chapter 12) are somewhat associated with the topics to be discussed here.

8.2 Structure and Composition of the Neutral Upper Atmosphere

In this chapter, we are defining the upper atmosphere to be the region above the "homopause" level, which is the altitude level at which molecular diffusion begins to dominate over eddy mixing. Below the homopause, atmospheric motions act to keep the atmosphere well mixed such that the mole fractions of chemically inert species do not vary significantly with altitude. Above the homopause, vertical diffusive separation of the species occurs, and the density of each neutral species drops off with altitude with its own scale height determined by its molecular mass (assuming no sources or sinks). Species much more massive than H and H_2 drop off precipitously with altitude above the homopause region, allowing for a convenient division between the middle and upper atmosphere. Saturn's upper atmosphere above the homopause is dominated by the least massive species H_2, H, and He.

Rate processes in the photochemistry of the thermosphere depend critically on the extent of the departure of the H_2 ground state, H_2 X(v:J), from local thermodynamic equilibrium (LTE). In earlier works (see Yelle and Miller 2004;

A.F. Nagy
Department of Atmospheric Ocean and Space Sciences, University of Michigan, Ann Arbor, MI, 48109, USA

A.J. Kliore
Jet Propulsion Laboratory, California Institute of Technology, Pasadena, CA, 91109, USA

M. Mendillo and L. Moore
Center for Space Physics, Boston University, Boston, MA, 02215, USA

S. Miller
Department of Physics and Astronomy, University College, London, WC1E 6BT, UK

J.I. Moses
Lunar and Planetary Institute, Houston, TX, 77058, USA

I. Müller-Wodarg
Space and Atmospheric Physics, Imperial College London, London, SW7 2AZ, UK

D. Shemansky
Planetary and Space Science Division, Space Environment Technologies, Pasadena, CA, 91107, USA

Strobel 2005 for reviews) the H_2 X(v:J) energy states have not been explicitly calculated, leaving rate processes highly uncertain. A competent approach requires calculation at the discrete rotational level using a solar emission model at resolutions of order 500,000 ($\lambda/\Delta\lambda$). Such a calculation has been done by Hallett et al. (2005a,b) (see Killen et al. 2009 for details of the solar model), but some critical rate processes are not definitively established at this time, and thus even the more complete recent calculations have significantly underestimated the excited vibrational populations in H_2 X(v:J) compared to observation (Hallett et al. 2005b). Ionospheric processes depend on the state of H_2 X(v:J), and whether or not H_2O is assumed to be present as an inflowing component of significance in the thermosphere (see Section 8.6).

The Cassini Ultraviolet Imaging Spectrometer (UVIS) experiment (Esposito et al. 2004) provides the most accurate platform to date for extracting information on the neutral upper-atmospheric structure and composition of a giant planet, primarily because of the higher spectral resolution, signal rates, and dynamic range as compared with previous instruments. The Cassini UVIS experiment has supplied data on the atmospheric physical properties of Saturn through three observational programs: (1) solar and stellar occultations in the EUV/FUV range that allow extraction of vertical profiles of H_2 and hydrocarbon abundances from the top of the atmosphere to about 300 km above the 1 bar pressure level, (2) dayglow spectral images, which together with the ultraviolet occultation results and with constraints from the Cassini radio science measurements of ionospheric and atmospheric structure, constrain model calculations and provide atmospheric properties, and (3) images of the magnetosphere that show the escape profile of atomic hydrogen from the top of the Saturn atmosphere.

8.2.1 Determination of Atmospheric Properties from Ultraviolet Occultations

Absorptive occultations have provided much information on the structure and composition of the upper atmosphere of Saturn (e.g., Atreya et al. 1984; Smith and Hunten 1990). In such observations, the Sun or a UV-bright star provides a source of ultraviolet light that is monitored as the source passes behind the planet as viewed from a detector. Vertical profiles of temperature and the concentration of atmospheric constituents can be obtained from analysis of the observed attenuation of the light as a function of wavelength and radial distance from the planet's center. Six ultraviolet occultation experiments were performed during the encounters of the Voyager 1 and 2 spacecraft with Saturn; the results of those occultations are described in Broadfoot et al. (1981), Sandel et al. (1982), Festou and Atreya (1982), Smith et al. (1983), and Vervack and Moses (2009). To date 15 stellar occultations have been obtained by the Cassini UVIS instrument over a range of latitudes from about 43° to −50°, and 7 solar occultations have been obtained over a range of latitudes from about 66° to −60°. The results from the UVIS occultation of the star δ-Ori on day-of-year (DOY) 103 of 2005 obtained at a latitude of −42.7° are presented here and in more detail in Shemansky et al. (2009) and Shemansky and Liu (2009). The temperature profile obtained from the occultation of ζ-Ori on DOY 141 in 2006, corresponding to a latitude of 15.2° is also presented in this section.

For the Cassini UVIS occultations, the H_2 component is obtained in the EUV Channel through forward modeling of the transmission spectrum using accurate temperature dependent cross sections (Hallett et al. 2005a, b; Shemansky and Liu 2009). The transmission spectra are fitted using separate vibrational vectors of the ground state H_2 X(v:J) structure. Details of the H_2 physical properties are described by Hallett et al. (2005a, b) for the non-LTE environment that develops in the excited atmosphere. The atmospheric temperature is derived through both iterative determination of rotational temperature and through the shape of the vertical H_2 density distribution in the hydrostatic model calculations (Shemansky and Liu 2009). At lower altitudes the kinetic temperature is also constrained by the measurements of the absorption structure of the C_2H_2 diffuse temperature sensitive $(\tilde{C} - \tilde{X})$ bands (Shemansky and Liu 2009; Wu et al. 2001).

Figure 8.1 shows the preliminary forward modeled hydrostatic vertical density distributions from the Shemansky and Liu (2009) analysis of the UVIS δ-Ori stellar occultation on 2005 DOY 103 at a dayside latitude of −42.7°; the model is anchored in the 0–400 km region using the Lindal et al. (1985) radio-occultation results. Shemansky and Liu (2009) have also reanalyzed the Voyager 2 (V2) UVS δ-Sco stellar egress occultation using the H_2 model described above, and the resulting vertical H_2 profile is included in Fig. 8.1. The Voyager 2 δ-Sco egress occultation (Smith et al. 1983; Vervack and Moses 2009) occurred on the darkside at a latitude of 3.8°. The differences in H_2 density at a given altitude evident in Fig. 8.1 are mainly the consequence of the different gravitation scales adopted at the different latitudes of the two occultations (Shemansky and Liu 2009). Figure 8.1 also shows the modeled helium distribution anchored at a [He]/[H_2] = 0.12 mixing ratio at 1 bar (Shemansky and Liu 2009). The [He]/[H_2] mixing ratio affects the modeled temperature structure in the vicinity of the mesopause, which is in turn constrained by the temperature dependence of the C_2H_2 $(\tilde{C} - \tilde{X})$ band cross section, such that an upper limit to the [He]/[H_2] mixing ratio can be obtained (Shemansky and Liu 2009). In the upper

8 Upper Atmosphere and Ionosphere of Saturn

Fig. 8.1 Plot of density versus altitude obtained from forward modeling of the Cassini UVIS δ-Ori stellar occultation on 2005 DOY 103 at a latitude of −42.7° (Shemansky and Liu 2009). Density values derived from the Voyager 2 UVS δ-Sco stellar egress occultation observations at 3.8° are also shown for comparison. The overplotted light lines indicate the altitude range over which meaningful constraints can be obtained from the measured data. The magenta curve is total density from the CIRS results (Fletcher et al. 2007), after converting their pressures to densities using a hydrostatic equilibrium model

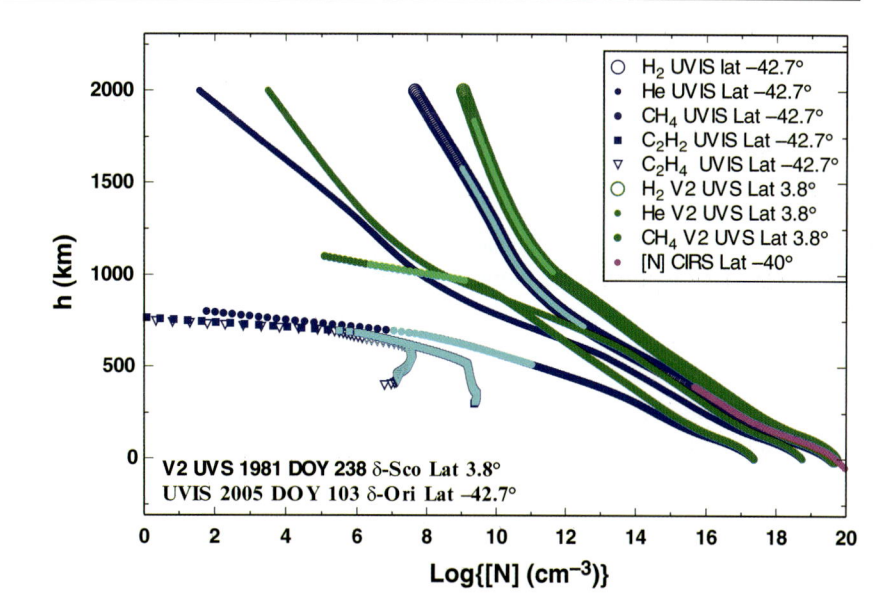

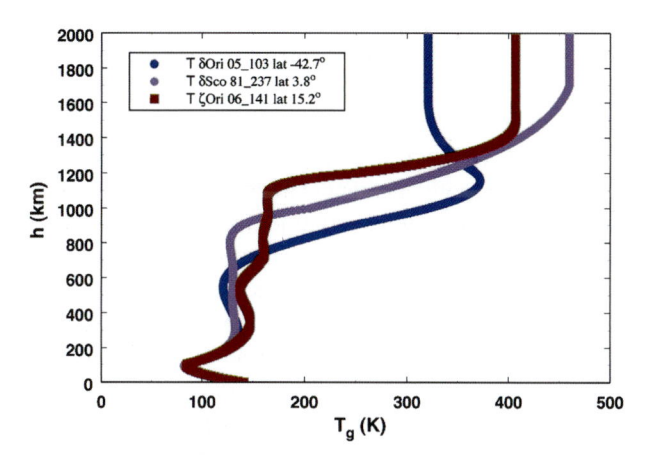

Fig. 8.2 Derived vertical temperature profiles from the Cassini UVIS occultations of δ-Ori (latitude −42.7°, 05 DOY 103) and ζ–Ori (latitude 15.2°, 06 DOY 141) and from the Voyager occultation of δ-Sco (latitude 3.8°, 81 DOY 237). These results are obtained using a common H_2 physical model (Shemansky and Liu 2009)

thermosphere, density and temperature vertical profiles evidently have a significant dependence on latitude (Shemansky and Liu 2009) (see Fig. 8.2), and the UVIS 2006 ζ-Ori occultation at 15.2° latitude has a similar vertical profile to the Voyager δ-Sco result (at 3.8° latitude) above 1,000 km (Shemansky and Liu 2009).

The UVIS FUV spectrograph stellar occultation data also lead to information on the hydrocarbon concentrations of CH_4, C_2H_2, and C_2H_4, as shown in Fig. 8.1. The evidence for other species in the transmission spectra is discussed by Shemansky and Liu (2009). In contrast, the only hydrocarbon profile that can be reliably extracted from the Voyager 2 δ-Sco stellar egress occultation is that of CH_4 (Shemansky and Liu 2009; Vervack and Moses 2009) (see Fig. 8.1). The methane homopause is just above 600 km in

the UVIS occultation and at ∼900 km in the Voyager 2 occultation, assuming the same oblate-spheroid model as used in the Cassini navigation package to define the zero altitude level at the 1-bar radius as a function of latitude (Shemansky and Liu 2009). Although the vertical displacement of the hydrocarbon homopause levels in the two cases is partially explained by the vertical displacement of the H_2 densities (see Fig. 8.1), a real difference in the location of the homopause in H_2-density space or pressure space does appear to exist, with the Cassini data implying a methane homopause located at higher pressures or H_2 densities than was the case for any of the six Voyager occultations (see Section 8.3 and Vervack and Moses 2009).

The derived temperature profiles from the UVIS 2005 δ-Ori, the UVIS 2006 ζ-Ori occultation, and the Voyager 2 1981 δ-Sco egress occultation (Shemansky and Liu 2009) are shown in Fig. 8.2. The preliminary UVIS results at −42.7° latitude shows a distinct mesopause at 545 km at a temperature of 121 K. The mesopause temperature is limited by the measured structure of the C_2H_2 ($\tilde{C} - \tilde{X}$) bands. The hydrostatic model calculation of the structure confined by the measured H_2 profile at higher altitudes, and the Voyager radio occultation results at altitudes below 400 km, is dependent on the [He]/[H_2] mixing ratio. The uncertainty in temperature above 300 km is estimated to be ±10°K for the UVIS derivation (Shemansky and Liu 2009). The UVIS δ-Ori result is one of only two analyzed occultations from the sunlit atmosphere, the other being the Voyager 2 δ-Sco stellar ingress occultation (Vervack and Moses 2009). The ζ-Ori occultation results also show a distinct mesopause, but the mesopause temperature is warmer than that derived for the δ-Ori occultation, and the overall shape of the profile is more similar to that of the Voyager 2 δ-Sco egress occultation. The derived thermospheric temperature from the ζ-Ori occultation is ~407 K. These derived temperatures, plus the 460–500 K thermospheric temperature obtained from the six Voyager ultraviolet occultations from 1980 to 1981 (Vervack and Moses 2009) suggest temperature variability in Saturn's thermosphere as a function of location and/or time.

The above figures were all plotted as a function of altitude above the 1-bar pressure level, with an oblated spheroid model used to approximate the 1-bar pressure surface. Given that the radius of the 1-bar pressure level is not accurately known for Saturn (e.g., Lindal et al. 1985) and because planetary properties such as gravity vary strongly with latitude on this unusually shaped, extended-equatorial-bulge planet, such altitude or radius profiles do not provide a good, meaningful common scale for comparison of occultations from different latitudes. In fact, finding a good common scale for comparisons is problematic. Some investigators have tried to resolve this problem by converting their radial profiles obtained at a specific latitude to an "equivalent equatorial radius" by assuming an oblate-spheroid shape for the planet (e.g., Smith et al. 1983). However, Saturn's zonal winds, which are not uniform with latitude, perturb the planet's shape significantly such that an oblate spheroid is an unacceptable approximation that can introduce errors of more than 100 km in altitude. Converting to a pressure scale for direct comparisons would be ideal, but that method also introduces uncertainties. In Fig. 8.3 we plot the derived temperature and concentration profiles as a function of pressure from most ultraviolet occultations analyzed to date – the preliminary UVIS 2005 δ-Ori DOY 103 occultation (Shemansky and Liu 2009), the Earth-based 28 Sgr stellar occultations (Hubbard et al. 1997), the original Smith et al. (1983) Voyager 2 δ-Sco egress occultation analysis, and the Vervack and Moses (2009) reanalysis of all the Voyager UVS occultations.

To get the pressures shown in Fig. 8.3, the H_2-density-radius profiles obtained from the Voyager occultations have been integrated from the top down to infer the pressure at each radius (see Vervack and Moses 2009), whereas the pressures for the other profiles were determined from either the hydrostatic equilibrium forward models (Shemansky et al. 2009) or from the H_2 densities and temperatures assuming that H_2 is the main constituent and that the ideal gas law applies. This conversion from H_2 densities and temperatures to pressures is only reliable above the homopause level, where hydrocarbons and CH_4 have already diffused out, but not so high up that H begins to compete with H_2. Note from Fig. 8.3 the general consistency in the derived H_2 and H profiles for most of the occultations; most also merge smoothly with the Hubbard et al. (1997) ground-based stellar occultation H_2 density results. This consistency suggests that Saturn's thermospheric density structure is relatively uniform across latitudes (i.e., to within a factor of ~2) on constant-pressure surfaces. On the linear temperature scale on the right-hand side of Fig. 8.3, differences between the different occultations are more apparent. The lower thermosphere of Saturn exhibits temperature variations of more than 100 K as a function of location or time. Even more striking is the much colder thermosphere derived from the Cassini UVIS results for the δ-Ori occultation in comparison with the other Voyager occultations, and that difference also shows up in the H_2 density profiles. As previously mentioned, the comparison of the Voyager and Cassini observations suggests latitudinal or temporal variations in thermospheric temperatures exist on

Fig. 8.3 The H_2 and H densities (*left*) and temperatures (*right*) as a function of pressure determined from the Cassini UVIS δ-Ori stellar occultation from 2005 at a dayside latitude of −42.7° (Shemansky and Liu 2009) are compared with various Voyager 2 UVS occultation retrievals (Vervack and Moses 2009) that have been smoothed to eliminate density scatter. The brown dot represents the Voyager 1 solar ingress occultation for which a full temperature profile could not be obtained

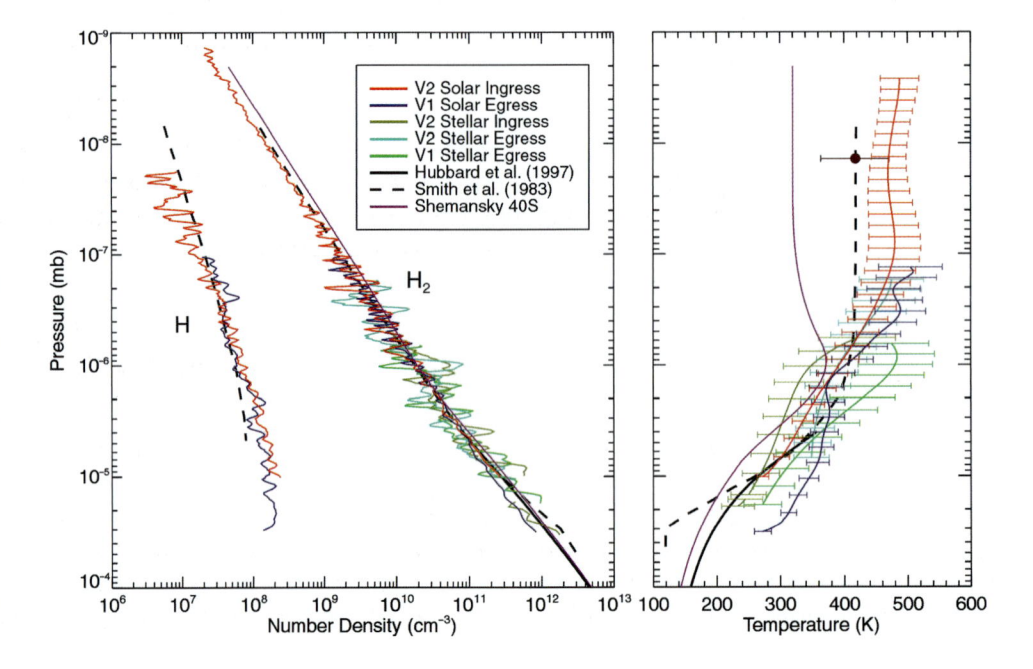

Saturn. Note, however, that the "wiggles" or "bulges" in the derived temperature profiles for several of the occultations should be interpreted with caution. If not dynamically supported, such structures would be smoothed out by conduction on very short time scales (see Section 8.4).

8.2.2 Determination of Atmospheric Properties from UVIS Spectra and Emission Maps

EUV/FUV spectra of the Saturn dayglow have been obtained with the Cassini UVIS. The UVIS spectra are the first observations of the excited atmosphere at solar minimum. The spectrum has been modeled in one dimension and pure hydrogen, constrained by ionospheric measurements (Nagy et al. 2006), and atmospheric structure, as a non-LTE system at the rotational level (Shemansky et al. 2009) with a purely solar-forced system. The model calculation establishes testable state populations, and all emission transitions in the system are predicted from radar frequencies to the EUV (Hallett et al. 2005a, b; Shemansky et al. 2009). Figure 8.4 shows an observed spectrum compared with the model calculation. The observed band intensities are a factor of 2.5 below those obtained at the Voyager encounter, and the spectra are qualitatively different, as is discussed by (Shemansky et al. 2009). Unlike the case for the Voyager observations (e.g., Shemansky and Ajello 2003), the Cassini UVIS dayglow spectra can be entirely explained (in terms of both spectral content and absolute brightness) by solar radiation deposition alone, with no excited electron source required (Shemansky et al. 2009). Note also that the non-LTE model calculations shown in Fig. 8.4 (Shemansky et al. 2009) predict a short-lived ($\sim$3,000 s) plasma population dominated by H_3^+ below about 2,000 km,
rather than H^+, and that invoking H_2O to act as a quenching agent for H^+ – a process that has been introduced by several ionospheric modelers to help explain the observed electron-density profiles (see Section 8.6) – may not be necessary at these lower altitudes.

Cassini UVIS maps of the Saturn magnetosphere have revealed distinct atomic hydrogen distributions in the region inside 4 R_S of planet center, showing the gas escaping the top of the thermosphere (Shemansky et al. 2009). The observed H Lyα brightness of the peak emission is about 1,000 R. The measurements in the sunlit southern latitudes show atomic hydrogen escaping at all latitudes below the auroral regions. The anti-solar side of the planet shows an emission distribution consistent with a combination of an orbiting and ballistic hydrogen source in the subsolar thermosphere. The hydrogen atoms in this sub-orbital portion of the corona re-enter the thermosphere within about 5 hours. A larger more broadly distributed hydrogen corona fills the magnetosphere to beyond 45 R_S in the orbital plane. This distribution is asymmetric in local time and similar to an image obtained with Voyager 1 in a different observational geometry (Shemansky and Hall 1992). The escape of atomic hydrogen from the top of the atmosphere requires a translational energy ranging from 5.5 eV at the equator to 7.2 eV at the poles and thus provides an indication of the total energy needed to create these hydrogen atoms and in turn the energy deposited in the upper atmosphere. However, there are clear problems associated with this energy estimate. It is about ten times the solar input and it cannot come from particle precipitation, because of the upper limit set by H_2 optical emissions. Shemansky et al. (2009) suggest that the hot H is the result of a high temperature ($\sim$20,000 K) electron population; this is within the constraints set by the observed H_2 UV emissions, but the basic source of this energy still remains to be identified.

8.3 Theoretical and Empirical Models of the Neutral Upper Atmosphere: Chemistry and Atmospheric Transport in the Homopause Region

Methane is photolyzed just below its homopause level, and the pressure at which photolysis occurs can affect the subsequent production and loss of complex hydrocarbons in Saturn's atmosphere (e.g., Moses et al. 2000). Therefore, the homopause-region observations and corresponding theoretical implications are discussed in some detail here, although a full discussion of hydrocarbon photochemistry is deferred to Chapter 5 by Fouchet et al. (2009) in this book. The variation of the methane abundance with altitude is controlled by molecular diffusion and/or transport – photolysis and subsequent photochemistry represent a much smaller

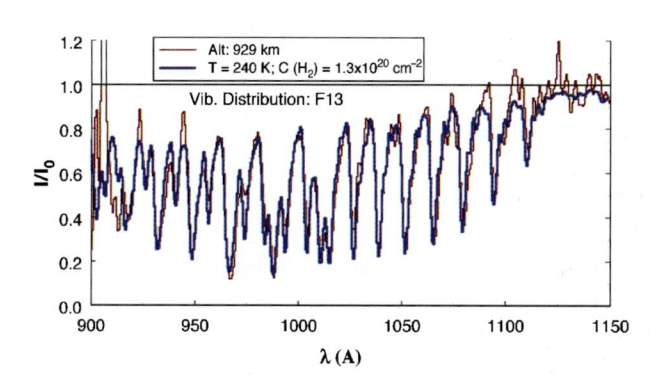

Fig. 8.4 UVIS EUV stellar occultation transmission spectrum obtained 2005 DOY 103 at an effective impact parameter of 929 km. The rotational temperature is iteratively determined assuming LTE. The vibrational population distribution is non-LTE, determined iteratively by fitting separate vibrational vectors into the model for optimal match to spectrum (Shemansky and Liu 2009)

perturbation of the CH$_4$ concentration profile. The solar and stellar ultraviolet occultation results described in Section 8.2.1 therefore provide important information needed for inferring vertical transport properties in Saturn's atmosphere (e.g., Atreya et al. 1984).

One convenient means of parameterizing atmospheric mixing in one-dimensional atmospheric models has been the use of a vertical eddy diffusion coefficient K_{zz} (e.g., Atreya et al. 1984; West et al. 1986; Strobel 2005). Different investigators have derived different K_{zz} values from the same Voyager 2 UVS δ-Sco stellar egress occultation (cf. Festou and Atreya 1982; Smith et al. 1983), illustrating the possible complexities and model dependencies of occultation analyses. Moses et al. (2000) have demonstrated that much of the difference in the quoted K_{zz} values from these two investigations results from different assumptions about the shape of the K_{zz} profile rather than true differences in the derived radius level of the methane homopause on Saturn. Both Festou and Atreya (1982) and Smith et al. (1983) agree that atmospheric mixing is relatively vigorous on Saturn compared with Jupiter and the other giant planets (cf., Atreya et al. 1984; Yung and DeMore 1999; Moses et al. 2004, 2005) such that the methane homopause is located at a relatively high altitude on Saturn.

This Voyager result was contradicted in part by the recent Cassini observations described above. Figure 8.1 demonstrates that at the time (April, 2005) of the Cassini UVIS stellar occultation at −42.7° latitude (Shemansky and Liu 2009), the methane homopause was found to reside at a significantly lower altitude – and several pressure scale heights below – the homopause level determined for the Voyager 2 δ-Sco egress occultation (August, 1981) at 3.8° latitude (see Smith et al. 1983; Shemansky and Liu 2009; Vervack and Moses 2009). Although the altitude scales at the different latitudes differ significantly due to the unusual shape and gravity variation with latitude/altitude on this rapidly rotating and high-zonal wind planet, a real difference in homopause levels of the two occultation sites does exist when the methane profiles are compared in pressure or H$_2$-density space. A reanalysis of all the Voyager solar and stellar UVS occultations (Vervack and Moses 2009) confirms that the methane homopause pressure level varies significantly with latitude and/or time on Saturn due to latitudinal and/or temporal variations in eddy mixing or vertical winds. This variation is aptly demonstrated in Fig. 8.5, which shows that the implied methane homopause pressure level is a full two orders of magnitude different between the Voyager 2 solar ingress occultation results at 29.5° latitude (Vervack and Moses 2009; methane homopause located near 10^{-6} mbar) and the Cassini UVIS stellar occultation results at −42.7° latitude (Shemansky and Liu 2009; methane homopause located near 10^{-4} mbar).

By comparing photochemical model results with the concentration profiles derived from the occultations, certain chemical and dynamical properties of Saturn's atmosphere can be constrained. In Fig. 8.5, the Voyager 2 solar ingress occultation results at 29.5° latitude (Vervack and Moses 2009) and the Cassini UVIS stellar occultation results at −42.7° latitude (Shemansky and Liu 2009) are compared with three photochemical models. The green profile, which represents a model that fits the Voyager 2 solar ingress UVS occultation light curves at methane-sensitive wavelengths (Moses and Vervack 2006), uses the hydrocarbon photochemistry from "Model C" of Moses et al. (2005), and has strong eddy mixing, with $K_{zz} \sim 2 \times 10^8$ cm^2 s^{-1} at $\sim 1 \times 10^{-5}$ mbar, dropping with decreasing altitude to $K_{zz} \sim 3 \times 10^7$ cm^2 s^{-1} at $\sim 10^{-4}$ mbar, down to $K_{zz} \sim 1 \times 10^5$ cm^2 s^{-1} at ~ 0.1 mbar. Although this model fits the methane concentration versus radius profile derived from the Vervack and Moses (2009) reanalysis of the Voyager 2 solar ingress UVS occultation quite well, the C$_2$H$_2$ and C$_2$H$_4$ model-data comparisons are much worse, suggesting problems with the chemistry and/or transport parameters in the models. The reaction rate coefficients adopted in the models are often not measured at the low pressures (and temperatures) typical of the homopause region of Saturn, and occultation observations such as these might be very useful for further constraining the chemistry and for identifying the key low-pressure reactions.

In order to fit the methane profile derived from the −42.7° Cassini UVIS δ-Ori occultation, eddy mixing must either be much less vigorous than for the Voyager case or downward winds must come into play. The red curve in Fig. 8.5 represents a model that also uses the "Model C" photochemistry of Moses et al. (2005) but has much weaker eddy mixing than the green curve described above, such that $K_{zz} \sim 2 \times 10^6$ cm^2 s^{-1} at 1×10^{-4} mbar. Even with this low value of K_{zz}, the model overpredicts the CH$_4$ concentration near 10^{-3} mbar. One interesting point to note is that both occultation profiles shown in Fig. 8.5 have much "sharper" C$_2$H$_2$ profiles than the models: the models underpredict C$_2$H$_2$ mixing ratios at the level of the peak mixing ratio and overpredict the C$_2$H$_2$ mixing ratio at lower altitudes. This failure of the models remains to be explained, but may provide useful constraints on the chemistry. For example, the chemistry in "Model A" of Moses et al. (2005) provides sharper C$_2$H$_2$ profiles and may better represent the situation on Saturn (see Table 8.1 of Moses et al. 2005 for a discussion of the differences between the chemistry in Models A and C).

The blue curve in Fig. 8.5 represents a model that uses "Model A" chemistry (to better reproduce the "sharpness" in the C$_2$H$_2$ profiles), assumes $K_{zz} \sim 2 \times 10^6$ at 10^{-4} mbar, (varying with the inverse of the square root of the pressure between 0.5 and 2×10^{-4} mbar), and adds a

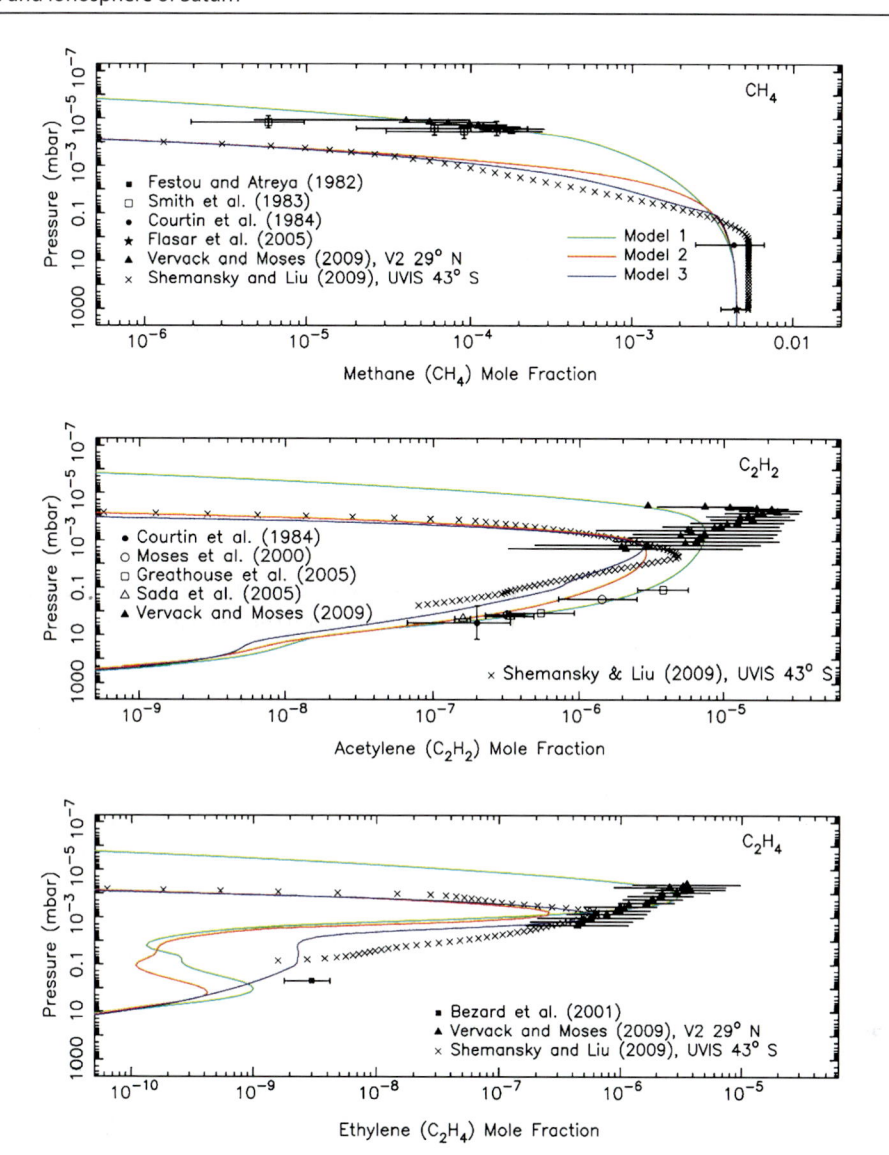

Fig. 8.5 Comparisons between photochemical models (colored solid lines) and the Voyager 2 UVS solar ingress occultation results (solid triangles, Vervack and Moses 2009), the Cassini UVIS δ-Ori −42.7° latitude stellar occultation results (×'s, Shemansky and Liu 2009), and various other infrared and ultraviolet data sets. The main point here is that the methane homopause was apparently located much higher in Saturn's atmosphere (i.e., at a pressure two orders of magnitude lower) at 29° latitude at the time of the Voyager 2 solar ingress occultation in 1981 than it was at −42.7° latitude at the time of the Cassini UVIS δ-Ori occultation in 2005, which indicates differences in vertical transport characteristics with latitude and/or time on Saturn. A secondary point is that current photochemical models do not accurately reproduce the vertical profiles derived for the C_2H_x hydrocarbons

downward wind of -0.1 mm s^{-1} above 0.1 mbar to better fit the CH_4 profile from the Cassini UVIS stellar occultation (Shemansky and Liu 2009). Such vertical wind velocities are not unreasonable in the mesopause region (Müller-Wodarg et al. 2006) and would imply vertical transport time scales of order 10 years in the middle atmosphere. This model does a better job of reproducing the derived Cassini UVIS methane and acetylene profiles, but the fit is by no means perfect.

Given the overall variation in hydrocarbon profiles from the different Voyager UVS occultations (Vervack and Moses 2009) and the Cassini UVIS δ-Ori occultation (Shemansky and Liu 2009), it appears that the standard view of vigorous atmospheric mixing in Saturn's middle atmosphere is not valid for all latitudes and times. Vertical winds and/or atmospheric mixing appear to be highly variable on Saturn, and atmospheric dynamics may play the dominant role in controlling species abundances in the upper regions of Saturn's middle atmosphere. The time scales involved are comparable to the lifetime of the Cassini mission, and the numerous solar and stellar occultations acquired by Cassini UVIS to date, along with the future planned occultations, may be invaluable in constraining stratospheric circulation on Saturn.

Caution must be exercised in interpreting the ultraviolet occultations, however. The occultations provide reliable descriptions of the concentration variations as a function of radius, albeit with some embedded assumptions about ultraviolet absorption cross sections, spectral behavior, spatial homogeneities and appropriate data processing procedures built into the analyses. Given uncertainties in Saturn's 1-bar radius as a function of latitude, in Saturn's temperature profile in the entire region from 1 bar to the high-altitude occultation levels, in Saturn's mean molecular mass variation with altitude, and in Saturn's zonal wind variation with altitude, any attempts to convert concentration-radius profiles to concentration-altitude profiles (as in Fig. 8.1 above), concentration-pressure profiles (as in Figs. 8.3 and 8.5 above), or any other such scenarios will necessarily require hydrostatic equilibrium solutions that will be highly model dependent. Similarly, during the occultation, the transmission from the H_2-sensitive wavelengths drops to zero by the radius level at which CH_4 absorption starts to be significant, which can complicate derivations of the mixing ratio (as opposed to concentration) profiles as a function of radius. All these complications and model dependencies can explain some of the very different results that have been obtained from different analyses of the same occultation data sets (e.g., Yelle et al. 1996 vs. Festou et al. 1981 for Jupiter, Festou and Atreya 1982 vs. Smith et al. 1983 for Saturn). Near-simultaneous observations that record temperature structure in the upper troposphere and middle atmosphere (e.g., from Cassini CIRS) in the regions probed by the ultraviolet occultations will greatly aid the occultation analyses. Analyses of the UVIS solar occultations, which unlike stellar occultations contain a signature atmospheric absorption in the H_2 continuum below 91.2 nm, will also be very helpful in defining thermospheric temperatures on Saturn.

8.4 Theoretical and Empirical Models of the Upper Atmosphere: Temperature Structure, Energy Balance, and Dynamics

8.4.1 Thermal Structure

The two Voyager radio occultation observations provided temperature values for the troposphere and stratosphere (Lindal et al. 1985; Lindal 1992). No information about the mesosphere could be inferred from the Voyager data, but results from ground based observations of the stellar occultation of 28 Sgr in 1986 suggested virtually constant temperatures there with a value of 141 K between 1 mbar and 0.3 μbar (Hubbard et al. 1997). Their measurements also extended into the lower thermosphere, near 0.05 μbar.

Thermospheric temperatures could first be inferred from the Voyager solar and stellar occultation experiments with the ultraviolet spectrometer (UVS) (Festou and Atreya 1982; Smith et al. 1983). These measurements placed the base of Saturn's thermosphere near the 0.1 μbar level. Initially, non-auroral exospheric temperatures on Saturn were a subject of debate since two very different values were derived from the Voyager data. The solar occultation experiment obtained a value of 420 ± 30 K near 30°N latitude (Smith et al. 1983) while the stellar occultation experiment yielded a value of 800 (+150/−120) K near 4°N latitude (Festou and Atreya 1982). These differences are not the result of the different occultation approaches and such extreme differences are not likely to be present in the atmosphere at such small latitudinal distances; the 30°N occultation result is now generally accepted as being more realistic (e.g., Vervack and Moses 2009). A comprehensive discussion of these two measurements was presented by Smith and Hunten (1990). Recently, the Voyager UVS data have been reanalyzed by Vervack and Moses (2009) and Shemansky and Liu (2009) and are shown in Figs. 8.2 and 8.3 alongside the recent Cassini UVIS occultation data discussed above (Shemansky and Liu 2009).

The thermal profile of Saturn' thermosphere, as those of the other gas giants, is poorly understood (Strobel and Smith 1973; Yelle and Miller 2004). As shown recently by the calculations of Müller-Wodarg et al. (2006) with the Saturn Thermosphere Ionosphere General Circulation Model (STIM), solar EUV heating produces exospheric temperatures on Saturn ranging from 153 K at solar minimum to 160 K at solar maximum. Those simulations assumed heating efficiencies of 50%, but even increasing this value to an unrealistic 100% raised solar driven exospheric temperatures only by 13 K (25 K) at solar minimum (maximum). It is therefore energetically not possible to heat Saturn's upper atmosphere with solar EUV radiation alone to the observed temperatures. These simple experiments also showed the solar cycle variability of exospheric temperature expected from solar EUV heating alone to be in the order of tens of degrees or less.

The main constraints on Saturn's upper atmosphere from ground-based observations have been obtained with measurements of H_3^+ emissions. The initial detection of these emissions from Saturn was made by Geballe et al. (1993) in 1992, using the CGS4 spectrometer on the United Kingdom InfraRed Telescope (UKIRT Mauna Kea, Hawaii). They found that spectra at the northern and southern limbs had roughly the same intensities, and that they were not able to measure H_3^+ emission at the equator. However, their initial measurements indicated that, for Saturn, the line intensity fell off more slowly from the limbs to the equator than was the case for Jupiter, for which auroral emission had been first detected in 1988 (Drossart et al. 1989). This indicated that the morphology was somewhat intermediate between

Jupiter, with its emission strongly concentrated around the auroral/polar regions (Baron et al. 1991), and Uranus, for which a planetwide H_3^+ glow seemed more likely (Trafton et al. 1993). Geballe et al. (1993) determined the best-fit temperature for their polar spectra on Saturn to be around 800 K, lower than for Jupiter, for which temperatures between 900 K and 1,100 K were found in the auroral regions (Drossart et al. 1989; Lam et al. 1997), and the column density of H_3^+ on Saturn to be around 1.0×10^{15} m^{-2}, about 10–50 times less than for Jupiter's auroral zones. The auroral temperature value derived by Geballe et al. (1993) is around twice the equatorial value proposed by Smith et al. (1983). Stallard et al. (1999) calculated that the total H_3^+ emission from Saturn might be as high as $1.5 \pm 0.3 \times 10^{11}$ Watts, if the temperature (~800 K) derived by Geballe et al. (1993) was correct. This figure was ~50 times less than the Jovian H_3^+ emission, but still high compared with UV emission from Saturn.

The first challenge to the "high" temperature for Saturn's auroral exosphere came from Miller et al. (2000), whose analysis of later (1999) UKIRT data indicated that 600 K was more likely, but this, too, was in error. It is now clear that the best fit temperature for Saturn's upper atmosphere, at the level of the peak H_3^+ emission in the auroral/polar zones is much closer to 400 K than 800 K. Melin et al. (2007) reanalysed the UKIRT spectra taken in 1999, and others obtained in 2004 and 2005. They found a best-fit temperature of 380 ± 70 K for the 1999 data, and 420 ± 70 K for 2004. Averaging gave 400 ± 50 K for the best fit Saturn polar thermospheric temperature. What emerged clearly was the variability in the ion column densities – assuming a constant temperature of 400 K gave 2.1×10^{16} m^{-2} and 2.9×10^{16} m^{-2} for the 1999 and 2005 data, respectively, but 20.0×10^{16} m^{-2} for 2004. The implications for the energy balance in Saturn's upper atmosphere of this large variation in H_3^+ column density are discussed below.

8.4.2 Energy Balance and Dynamics

Over the past few years, there has been considerable progress in understanding the upper atmosphere energetics and dynamics of Saturn's polar regions as a result of H_3^+ observations and modeling. Based on HST UV images, Cowley et al. (2004) first proposed that the main auroral oval resulted from the interaction between the solar wind and Saturn's magnetosphere, and, although other mechanisms have not been entirely ruled out, this seems to be a promising proposal. The Cowley et al. (2004) model of plasma dynamics in the ionosphere included flows from the Dungey (1961) and Vasyliunas (1983) cycles, and a general lag to corotation with the planet across the entire auroral/polar region due to interactions with the solar wind, a mechanism first put forward by Isbell et al. (1984). Measurements of the Doppler shifting of the H_3^+ emission line at 3.953 mm generally supported the picture of a general lag to corotation across the polar regions (Stallard et al. 2004). Cowley et al. (2004) derived a value for the angular velocity of the auroral/polar ionosphere of $\Omega_{ion} = 0.24\,\Omega_{Saturn}$, assuming a solar wind velocity of 500 km/s and an effective ionospheric Pedersen conductivity of 0.5 mho. Stallard et al. (2004) measured a value of $\Omega_{ion}/\Omega_{Saturn} = 0.34$, which suggested a value for the effective (height-integrated) Pedersen conductivity of the ionosphere of $\Sigma_P^* = 0.82$ mho, if the solar wind velocity remained at 500 km/s.

The consequences of this lag to corotation for energy balance in the Saturnian polar upper atmosphere are considerable. The auroral/polar ionosphere is produced mainly by charged particle precipitation. On Jupiter, particle precipitation may deposit in the order of 10^{12} Watts planetwide, but this energy input is largely balanced by emission from H_3^+ (Miller et al. 1994, 2006) – the so-called H_3^+ thermostat (~50–80%). Much more energetically important, however, is the energy generated by Joule heating, resulting from equatorward currents across the auroral oval, and the westward winds produced by Hall drift (see Smith et al. 2005; Miller et al. 2006). On Saturn, Melin et al. (2007) have shown that the H_3^+ thermostat is much less effective (~1%) than it is for Jupiter (Melin et al. 2006) (and probably for Uranus). Particle precipitation into the auroral/polar regions is thought to deposit around 10^{11} W, but Joule heating (ion drag) is thought to be a factor of 10 larger (Cowley et al. 2004; Miller et al. 2006).

In an attempt to understand the role of magnetospheric energy input globally, Müller-Wodarg et al. (2006) and Smith et al. (2005) carried out simulations in which Joule heating was applied at polar latitudes, depositing 8.8 TW in Saturn's lower polar thermosphere, a factor of 35–65 more than provided by planet-wide solar EUV heating. Their calculations did not include H_3^+ cooling, so polar temperatures reached around 1,000 K. Despite these unrealistically high values, equatorial temperatures in their calculations did not exceed around 250 K, considerably lower values than observed. In these simulations Saturn's fast rotation via Coriolis accelerations generated a primarily zonal flow in the atmosphere, which in turn prevented meridional transport of energy from pole to equator. In fact, calculations by Smith et al. (2007), which included the effects of ion drag that generates westward flows, suggested that polar heating on Saturn could lead to slight cooling of the low latitude thermosphere via adiabatic cooling. These calculations have assumed steady-state conditions and the effects of highly (short-term) variable inputs into Saturn's auroral/polar regions witnessed in the VIMS images (Stallard et al. 2008) on the horizontal distribution of magnetospheric energy have yet to be examined. Furthermore, future studies need to calculate the magnetosphere-ionosphere-thermosphere coupling self-consistently to allow two-way coupling between these systems, an aspect that will critically affect the effective Pedersen conductivity. Despite these shortcomings of the calculations, they have illustrated the sensitivity of the energy distribution in Saturn's thermosphere to global circulation via energy redistribution processes such as advection and adiabatic heating and cooling. The dynamics of Saturn's thermosphere are as yet unconstrained except for measured ion velocities in the auroral regions, and it should be mentioned that the calculations by Müller-Wodarg et al. (2006) and Smith et al. (2005) ignored effects of ion drag.

To "fill in" the low latitude "energy hole" and obtain thermospheric temperatures consistent with observations, Müller-Wodarg et al. (2006) proposed the presence of an additional unidentified energy source at low latitudes, an aspect we explore further in this section. Figure 8.6 shows thermospheric temperatures as derived from the Voyager 2 UVS solar ingress and stellar egress occultations by Smith et al. (1983) and from the reanalyzed UVS solar ingress occultation by Vervack and Moses (2009) (blue symbols). Also shown (light blue curve) are temperatures from the Cassini UVIS δ-Ori occultation presented in this chapter (Fig. 8.2). Super-imposed in Fig. 8.6 (red lines) are zonally averaged temperatures from three simulations carried out with the STIM General Circulation Model of Müller-Wodarg et al. (2006) for different empirical energy sources. Simulations A and C considered heating near the exobase and mesopause, respectively, and simulation B represents an in-between case. In Simulation A 3.8 TW is deposited planet-wide, with the volume heating rate peaking at the 7.3×10^{-10} mbar level, having a Gaussian shape with a mean half width of 2.5 scale heights. In Simulation B 5.2 TW is centered at the 2×10^{-7} mbar level and in Simulation C the volume heating rate peaks at the lower boundary pressure (1.6×10^{-5} mbar) with a total energy of 15 TW. As expected, the energy necessary to reach the desired exospheric temperatures becomes smaller when deposited higher up in the thermosphere since molecular conduction, a key energy loss process, becomes less effective at lower densities in the atmosphere. The profiles in Fig. 8.6 are for a mid-latitude location of 30°N consistent with the location of the Voyager 2 UVS solar ingress observations. While the observations by Cassini UVIS were made at 42.7°S and 15.2N, little difference was found in the simulations between these locations.

A key difference between the three simulations presented in Fig. 8.6 is the slope of the lower thermosphere temperature which is crucially affected by the altitude/pressure of peak volume heating deposition. Both the Voyager and Cassini derived temperature profiles are best fit by simulations B and C, where energy is deposited near and below the 10^{-7} mbar level, whereas heating in the exosphere (Simulation A) produces a temperature shape that is less consistent with these observations. It should be noted, though, as described in Section 8.2.2, that recent Cassini UVIS observations of atomic hydrogen escape have suggested the possibility of significant energy deposition at the top of the thermosphere, consistent with Simulation A, a topic which needs further investigation beyond the scope of this chapter. Depending on the exact altitude of energy deposition, the calculations suggest a total additional energy of between 5 and 13 TW being necessary to raise thermospheric temperatures at low latitudes to the observed levels. Locally, the height integrated heating rates which were applied in simulations B and C are 0.08 and 0.30 mW/m² , respectively. The model cannot currently reproduce and explain the negative temperature gradient observed by the Cassini UVIS stellar occultation curve at −42.7° above the 1×10^{-6} mbar level. While the response of the thermosphere to three possible cases of heating functions was examined, no attempt was made to explain the origin of this empirical heating, if it exists in reality. The heating function of Simulation C may suggest energy propagating upward from below, possibly via waves, but at present there are too few constraints to characterize the waves in Saturn's upper atmosphere. The required energy may also result from energy redistribution by global dynamics other than those simulated by Müller-Wodarg et al. (2006) and Smith et al. (2005).

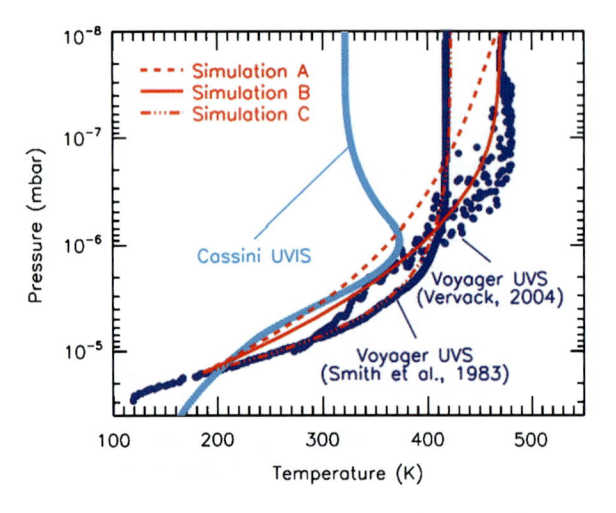

Fig. 8.6 Temperatures in Saturn's upper atmosphere as inferred from observations of the Voyager 2 solar ingress (29°N) and stellar egress (4°N) occultations (*blue curve*) by Smith et al. (1983) and from a re-analysis of the Voyager 2 UVS solar ingress occultation at 29°N (blue symbols) by Moses and Vervack (2006). Also shown are temperatures inferred from Cassini UVIS observations in 2005 (*light blue curve*) near 42.7°S (Shemansky and Liu 2009). Red lines denote temperatures at 30°N from calculations with the model by Müller-Wodarg et al. (2006), assuming three different forms of empirical heating functions. Simulation A (dashed) assumes 3.8 TW deposited near the 7.3×10^{-10} mbar level, Simulation B (solid) assumes 5.2 TW deposited near the 2×10^{-7} mbar level and Simulation C (dashed-triple-dotted) assumes 15 TW deposited near 1.6×10^{-5} mbar

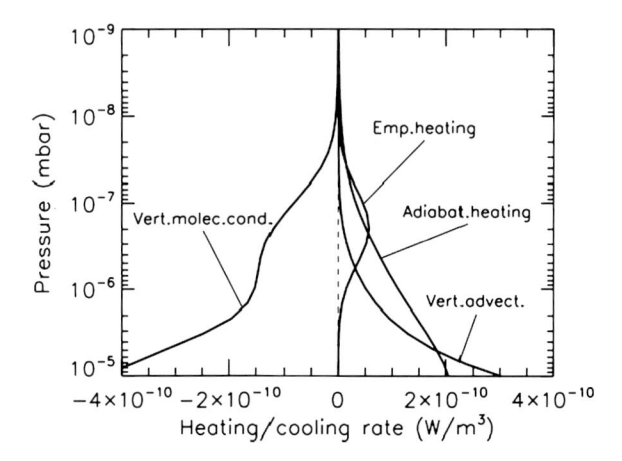

Fig. 8.7 Terms of the energy equation in Simulation B of Fig. 8.6, extracted at 30°N. The empirical heating source is balanced primarily by vertical molecular conduction. Vertical winds play an important role in depositing energy at those latitudes via advection and adiabatic heating

To analyze in more detail the energy balance in these simulations, Fig. 8.7 shows zonally averaged volume heating and cooling rates from simulation B at 30°N. Apart from the imposed empirical heating function, the main energy sources at mid latitudes are adiabatic heating and vertical advection. The combined energy sources are balanced by cooling from vertical molecular conduction. While other processes such as solar EUV heating and horizontal advection are included in the calculations, their importance is negligible compared with the above heating and cooling processes. The figure illustrates the importance of dynamical terms for the energy balance in Saturn's thermosphere, emphasizing the need to study Saturn's energetics with global dynamical models. Polar Joule heating generates strong upwelling poleward of around 60° with vertical velocities of possibly several meters per second (Müller-Wodarg et al. 2006). The high latitude heating generates a global circulation cell that leads to downwelling equatorward of around 50° with vertical velocities of up to around 1 m/s. This downwelling leads to adiabatic heating and downward transport of energy from above, where the thermosphere is hotter. While Fig. 8.7 refers to one particular case of empirical heating, namely simulation B, the same energy redistribution processes are important in the other heating cases we calculated.

8.5 Observations of the Ionosphere

8.5.1 Radio Occultation Observations of Electron Densities

Virtually all the observational evidence on the structure of the ionospheres of outer planets and their satellites has been obtained by the method of radio occultation (Lindal 1992; Kliore et al. 2004). This technique yields the vertical structure of the total electron density, $N_e(h)$, from which the plasma scale height can be derived, but neither the plasma temperature nor the ion composition is directly revealed. Pioneer 11 and the two Voyagers provided the first six $N_e(h)$ profiles at Saturn (e.g., Atreya et al. 1984), and the initial phase of the Cassini mission yielded twelve near equatorial profiles (Nagy et al. 2006). More recently, as Cassini moved away from an equatorial orbit, mid and high latitude density profiles have also been obtained (Kliore et al. 2009). The average low, middle and high latitude density profiles from Cassini radio occultations are shown in Fig. 8.8a. An increase in the averaged electron densities with latitude is clearly visible in this figure and is discussed in Section 8.6. A decrease in the mean peak density and increase in the corresponding height from dusk to dawn was seen in the average low-latitude observations, consistent with the presence of molecular ions at lower altitudes, which decay rapidly at sunset. Thus, this preferential decrease at lower altitudes leads to a decrease in the peak density and an increase in the altitude of the peak during the night. No such dusk to dawn changes could be ascertained in the mid-latitude data. (Dawn and dusk do not have the same meaning at high-latitude). Significant variations were seen in the Cassini observations, which could not be attributed to latitude and/or local time changes. A possible explanation for this variability is changing chemistry rates (e.g., water inflow that can hasten recombination, Moore and Mendillo 2007), dynamics (e.g., gravity-wave interaction with the plasma, Matcheva et al. 2001), electrodynamic effects, or variability driven by changing ionizing particle influxes. Given the uncertainty in the topside ion composition, a great deal of uncertainty is associated in deducing the topside plasma temperature from the measured scale heights. Using some simplifying assumptions, the low-latitude scale heights lead to estimated temperatures in the range of about 600–800°K, but it is important to remember that very large uncertainties are associated with these values.

Figure 8.8b also shows the electron densities deduced from the S44 orbit entry occultation, which has a very sharp "bite-out" around 2,500 km, possibly the result of some wave activity (Matcheva et al. 2001) or a surge of water influx as discussed in Section 8.6.

8.5.2 Electron Density Variations Inferred from SEDs

Impulsive short-duration bursts of broadband radio emissions were detected by the planetary radio astronomy experiment aboard both Voyagers for a few days on either side of their closest approaches to Saturn (e.g., Warwick

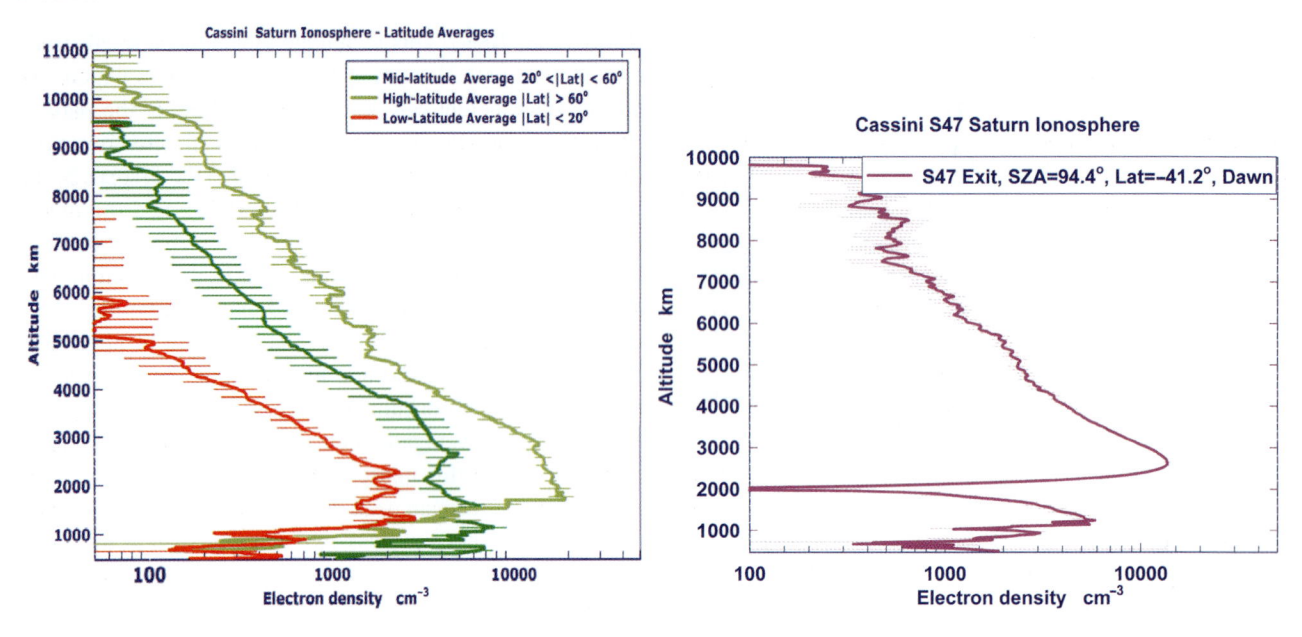

Fig. 8.8 (a) Averaged near-equatorial dusk and dawn electron density profiles (Nagy et al. 2006). (b) Electron density profile from the S47 exit observations showing a severe density "bite-out" (Kliore et al. 2009)

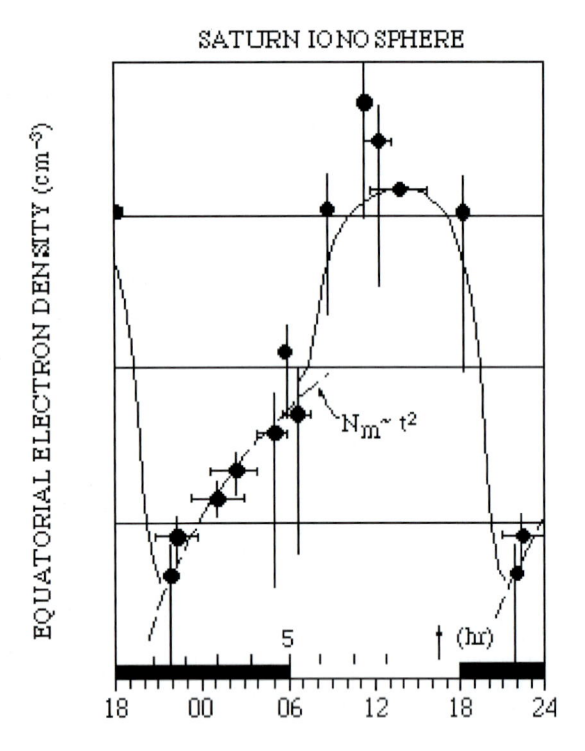

Fig. 8.9 SED inferred electron densities at Saturn from Kaiser et al. (1984)

et al. 1982). Dubbed Saturn Electrostatic Discharges (SEDs), they are now thought to be the result of low-altitude lightning storms. While there was some initial controversy over whether SEDs originated in Saturn's rings or atmosphere, Kaiser et al. (1983) demonstrated clearly that an atmospheric source was more likely. As final confirmation, Cassini's Imaging Science Subsystem instrument has imaged bright clouds whose periods matched contemporaneous SED storm detections. Four such correlated visible and radio storms have been observed as of March 2007 (Dyudina et al. 2007).

SED emission from a lower atmospheric discharge source would have to pass through Saturn's ionosphere in order to be observed by a spacecraft, and therefore the lower frequency cutoff of the emission could contain information regarding the plasma frequency of the intervening plasma. Kaiser et al. (1984) estimated the local time dependence of Saturn ionospheric density based on the low frequency SED cutoffs observed. Their results are presented in Fig. 8.9, which implies a diurnal variation in electron density of two orders of magnitude – from 10^3 cm^{-3} at midnight to 10^5 cm^{-3} at noon. The SED-inferred electron densities at dawn and dusk were of order 10^4 cm^{-3}, in agreement with the Voyager radio occultation data (Atreya et al. 1984). Thus far the observations of SEDs by Cassini have been highly sporadic, in contrast to the near constant occurrence during Voyager fly-bys, yet the inferred electron densities reproduce essentially the same diurnal behavior (Fischer et al. 2008 and references therein).

8.5.3 Ground Based Observations of H_3^+ Emission

Observations of optical signatures of ionospheric plasma at Saturn present a considerable challenge. The terrestrial detection of H_3^+ from Saturn, at infrared wavelengths, was discussed in Section 8.4.1. These observations found roughly the same brightness at the northern and southern limbs, with no detections achieved at middle and equatorial latitudes. This argued clearly for an auroral source of the emission, and hence of high-latitude ionospheric plasma produced by incoming magnetospheric charged particles that far exceeded ionization produced by sunlight. As with the radio occultation and SED results, the dominant message from the H_3^+ observations is one of extreme variability. For example, Melin et al. (2007) analyzed the spectra taken in 1999, 2004 and 2005 at the United Kingdom InfraRed Telescope (UKIRT) on Mauna Kea, Hawaii UKIRT and found that the column contents varied considerably, ranging from ~ 2 to $\sim 20 \times 10^{16}$ m^{-2}. These are high values of H_3^+ column content for a giant planet's ionosphere, with the upper value comparable to the TEC levels measured at Earth. For comparison, with solar flux being the only ionizing source considered, model calculations by Moore et al. (2004) predicted the auroral H_3^+ column content to be $\sim 0.5 \times 10^{16}$ m^{-2} and the TEC in the summer hemisphere to be $\sim 2 \times 10^{16}$ m^{-2}. As noted in Fig. 8.8b, Cassini electron density values are higher at high latitudes in comparison to the equator, and thus Saturn's ionosphere is likely the result of a blend of photo-production and auroral-production of a hydrogen plasma system, which provides a significant challenge to understand.

8.6 Models of Ionospheric Structure, Composition and Temperatures

8.6.1 Background Theory and Early Models

The first theoretical attempt at modeling Saturn's ionosphere came from McElroy's (1973) review of the four giant planets' ionospheres. He outlined most of the important photochemical reactions that all future work would draw from, and highlighted some issues that remain relevant today. The major reactions, which were usually considered, are as follows (for the sake of brevity only photoionization/photodissociation is indicated, but the same processes can result from electron impact):

$$H_2 + h\nu \rightarrow H_2^+ + e \quad (8.1a)$$
$$\rightarrow H^+ + H + e \quad (8.1b)$$
$$\rightarrow H + H \quad (8.1c)$$
$$H + h\nu \rightarrow H^+ + e \quad (8.2)$$
$$H_2^+ + H_2 \rightarrow H_3^+ + H \quad (8.3)$$
$$H_3^+ + e \rightarrow H_2 X^1 \Sigma_g^+ (v:J) + H^* \quad (8.4a)$$
$$\rightarrow H^* + H^* + H^* \quad (8.4b)$$

Reactions between ions and hydrocarbons (e.g., CH_4) were predicted to result in a pronounced shoulder on the bottomside ionosphere (Atreya and Donahue 1975). Galactic cosmic-ray induced ionization in Saturn's ionosphere was also evaluated, finding that it is likely to lead to the creation of a low-lying ledge of plasma of $\sim 7{,}000$ cm^{-3} at ~ 0.5 bar (Capone et al. 1977).

Radio occultation measurements of Saturn's electron density by Pioneer 11, Voyager 1 and Voyager 2 found an ionosphere of roughly 10^4 cm^{-3}, an order of magnitude smaller than early theoretical predictions. Therefore, a mechanism for reducing the modeled electron density was required. The chemical loss included for H^+ in early models was radiative recombination, a very slow process. In order to reduce the modeled electron density two processes were suggested (McElroy 1973; Connerney and Waite 1984), both of which act by converting long-lived atomic H^+ ions into short-lived molecular ions. The first process is the charge exchange reaction between H^+ and vibrationally excited H_2 in its ground state, as indicated by Eq. (8.5).

$$H^+ + H_2 (v \geq 4) \rightarrow H_2^+ + H \quad (8.5)$$

The main sources of vibrationally excited H_2 are via collisions with electrons (Hallett et al. 2005a):

$$e + H_2 X^1 \Sigma_g^+ (v_i : J_i) \leftrightarrow e + H_2 X^1 \Sigma_g^+ (v_j : J_j) \quad (8.6)$$

and H_3^+ recombination (Eq. 8.4a), while quenching of the excited H_2 can take place via collisions with electrons, both thermal and more energetic photoelectrons, H, H_2 and H^+. The rates for reaction (8.5) were not known for a long time and early attempts of estimating the vibrational population carried numerous uncertainties (Cravens 1987; Majeed et al. 1991). The reaction rate for (8.5) was established by Ichihara et al. (2000), and Shemansky and co-workers (e.g., Hallett et al. 2005a; Shemansky and Liu 2009) carried out comprehensive model calculations of the hydrogen system, as discussed later in this section.

The second suggestion (e.g., Connerney and Waite 1984) was that water introduced into the atmosphere from the rings and/or icy moons can reduce the H^+ density via a multistep process which converts it to H_3O^+, as shown below in Eqs. (8.7–8.9).

$$H^+ + H_2O \rightarrow H_2O^+ + H \quad (8.7)$$
$$H_2O^+ + H_2 \rightarrow H_3O^+ + H \quad (8.8)$$
$$H_2O^+ + H_2O \rightarrow H_3O^+ + OH \quad (8.9)$$
$$H_3O^+ + e \rightarrow H_2O + H \quad (8.10)$$

A discrepancy between the modeled and observed electron density had also arisen in the case of Jupiter earlier (Atreya et al. 1979). In the Jovian case, modelers had invoked a distribution of vibrationally excited H$_2$ (see Eq. 8.5) elevated above LTE, along with postulated vertical plasma drifts, in order to reproduce Jovian electron density measurements (e.g., McConnell et al. 1982; Majeed and McConnell 1991). Still, as evidenced by the model results of Waite et al. (1983) that incorporated calculations of H$_2$ vibrational levels, the disagreement was more severe for Saturn. The determination of the vibrational distribution of H$_2$ remains an important topic today at both Jupiter and Saturn, as reviewed by Yelle and Miller (2004), and most recently outlined by Shemansky et al. (2009).

8.6.2 Modern Theory and Time Dependent Models

Due to Sun–Saturn–Earth geometry, radio occultations of Saturn all occur very near the dawn or dusk terminator –a period of rapid change for any solar produced ionospheric plasma (e.g., Schunk and Nagy 2009). While all previous modeling had been steady state, the large calculated electron densities and the dominance of H$^+$ over other ion species – whose chemical lifetime is large relative to a 10-h Saturn day – meant that little diurnal variation in electron density would be expected. Thus, the disagreement between steady-state models of Saturn's ionosphere and the spacecraft observations was the first indication that a time-dependent solution was required. In addition to the peak electron density discrepancy, the observations of Saturn Electrostatic Discharges (SEDs) implied a noon-to-midnight diurnal variation in electron density of two orders of magnitude, as indicated in Section 8.5.2 (Kaiser et al. 1984).

The first time-dependent solution to Saturn's ionosphere was a 1D chemical diffusive model that attempted to directly address the SED-inferred diurnal variation in electron density (Majeed and McConnell 1996). In order to reduce the implied noontime density of 10^5 cm^{-3}, and also to induce more dramatic diurnal variation, Majeed and McConnell (1996) tested a wide range of H$_2$ vibrational temperature profiles and topside H$_2$O influxes. However, despite being able to find combinations of the above two parameters that yielded good matches to radio occultation measurements, Majeed and McConnell (1996) could not reproduce the two orders of magnitude variation in electron density inferred from SEDs. Their strongest calculated noon-to-midnight variation was ∼7, though model simulations that yielded this strong diurnal variation were not able to simultaneously reproduce Voyager electron density altitude profile observations.

The next time-dependent ionospheric model developed for Saturn (Moses and Bass 2000) combined new Saturn observations (e.g., Feuchtgruber et al. 1997; Hubbard et al. 1997; Moses et al. 2000) and new reaction rates to derive a more accurate neutral atmosphere (e.g., Moses et al. 2000). Moses and Bass (2000) solved the coupled 1D continuity equations as a function of time for a comprehensive set of 63 neutral and 46 ionized species in Saturn's atmosphere. They addressed the ionospheric effects at Saturn for water, oxygen and magnesium influx, neutral winds, electric fields, and interplanetary dust. While Moses and Bass (2000) did not comment on the issue of SED-inferred diurnal variations in electron density directly, their standard model calculations utilized an empirically derived population of vibrationally excited H$_2$ along with a planet-wide water influx of 1.5×10^6 cm^{-2} s^{-1}, and like the models of Majeed and McConnell (1996), Moses and Bass could not reproduce the large diurnal variations in electron density inferred from SEDs. The new global-average value of water influx, constrained by ISO observations and model calculations, was smaller than previously adopted values for Saturn (e.g., Connerney and Waite 1984). However, the planetary-averaged ISO observations could not explicitly exclude the possibility of strong latitudinal variations in water influx (e.g., Connerney 1986).

A recent model developed for Saturn's ionosphere is the Saturn-Thermosphere-Ionosphere-Model (STIM) (Müller-Wodarg et al. 2006). Whereas Moses and Bass (2000) provided a thorough analysis of Saturn's entire ionosphere with a focus on the lower hydrocarbon and metallic layers, STIM has concentrated on the major ions in the upper ionosphere, where the N_e peak lies. STIM is a global circulation model of Saturn's upper atmosphere (see Section 8.4), yet during its development, a series of 1D ionospheric studies were performed using the 3D thermosphere as a background. First, Moore et al. (2004) used a 1D time-dependent model that considered chemistry and plasma diffusion to investigate global ionospheric behavior, regimes of photochemical equilibrium within Saturn's ionosphere, ionospheric response to a wide range of water influxes and H$_2$ vibrational temperatures, and ionospheric conductivities (see Fig. 8.10a). Moore et al. (2004) extended the Majeed and McConnell (1996) parameter space results by demonstrating that no matter what combination of production and loss processes were included, chemistry alone could not reproduce a two order of magnitude diurnal variation in N_e during the short Saturn day. In addition, they modeled the ionospheric effects resulting from the pattern of shadows cast by Saturn's rings, finding sharp gradients and strong reductions in electron density. Mendillo et al. (2005) studied the seasonal variability of the patterns induced by ring shadowing in more detail. They argued that the electron density troughs produced by ring shadows (equinox for Voyager, solstice for Cassini) may lead to ionospheric "windows" through which SEDs could more easily escape, an interpretation that raises questions

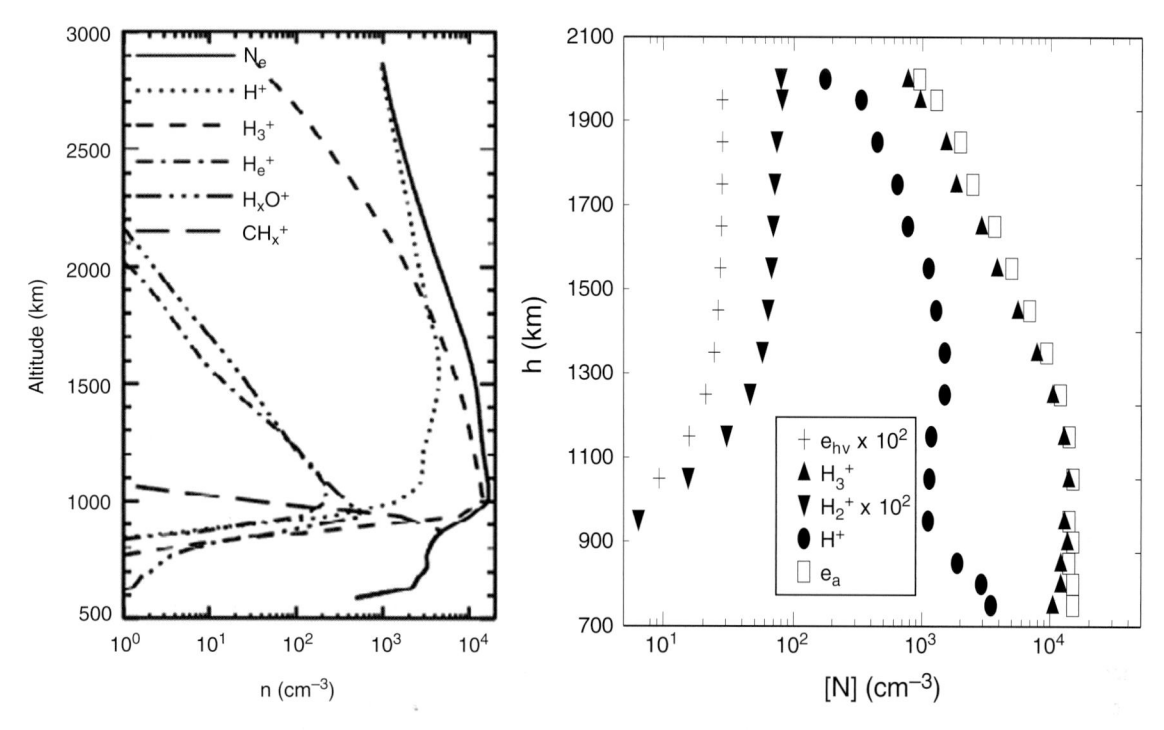

Fig. 8.10 (a) Noon ion density profiles from Moore et al. (2004). (b) Ion density profiles from Shemansky and Liu (2009)

regarding the use of SED detections as a type of ionospheric sounder for peak N_e. Second, Moore and Mendillo (2005) extended the STIM ionosphere into Saturn's inner plasmasphere based on Liousville's theorem and the method of Pierrard and Lemaire (1996, 1998), predicting an electron density of order $100\,\mathrm{cm^{-3}}$ for Cassini's closest approach ($\sim 1.3 R_S$), within the domain of extrapolation of the pre-SOI data from Gurnett et al. (2005). Third, with the twelve new radio occultation profiles of Saturn's equatorial ionosphere taken by Cassini indicating a dawn-dusk asymmetry, Moore et al. (2006) found that a topside water influx of $\sim 5 \times 10^6\,\mathrm{H_2O\,cm^{-2}\,s^{-1}}$ provided the best fit to the averaged dawn and dusk Cassini profiles, thereby reducing the importance of vibrationally excited H_2 as an atomic-to-molecular-ion catalyst. Finally, Moore and Mendillo (2007) included time-dependent neutral water diffusion calculations in which the topside water flux is increased for a short period of time leading to a bulge of water density that reduces the local electron density as it diffuses downward. They obtained a "temporary bite-out" similar to the observed one, as shown in Fig. 8.11. However, such a large water influx (their particular simulation used an augmentation of the background water flux ($5 \times 10^6\,\mathrm{cm^{-2}\,s^{-1}}$) by a factor of 50, that persisted for ~ 27 min) has not yet been observed, so one cannot draw definitive conclusion at this time. On the other hand there were only two such large bite-outs observed among the 27 occultations presented here, so one needs such large fluxes only "intermittently".

Matcheva et al. (2001) examined the potential role of gravity waves as the mechanism responsible for large and sharp electron density layering, as well as for the low peak electron densities observed by Galileo at Jupiter. They pointed out that at higher altitudes where long lived H^+ dominates diffusion is likely to dominate and act to limit large deviations from diffusive equilibrium. However, they also found that a downward electron flux at high altitudes produced from the long-term effects of gravity waves can reduce the electron densities throughout the middle and upper ionosphere, in a manner similar to that proposed from a water influx or vibrationally excited H_2. Moreover, Matcheva et al. (2001) showed that at Jupiter, in the altitude range of between about 600 to 900 km, gravity waves are likely to be important in creating the observed sharp, multiple density peaks. At this point without a more detailed quantitative calculation for conditions at Saturn, it is impossible to come to a definitive conclusion regarding the observed bite-out observed by Cassini (Fig. 8.8b). As indicated in Fig. 8.8b the very large bite-out is at 2,000 km where H^+ is likely to be a major ion and thus diffusion lifetimes are significant. Therefore, such a large perturbation is less likely to be caused by gravity waves; on the other hand the smaller bite-outs seen in many of the observed electron density profiles around 1,000 km could certainly be caused by the mechanism proposed by Matcheva et al. (2001).

The models mentioned so far predict that the dominant ions in Saturn's ionosphere are H^+ and H_3^+. H_2^+ is the ion with the greatest rate of photo-production, yet it is rapidly converted to H_3^+ (Eq. 8.3), and therefore no appreciable density of H_2^+ remains. The fast dissociative recombination of H_3^+ means that it has a strong diurnal variation, with a minimum just

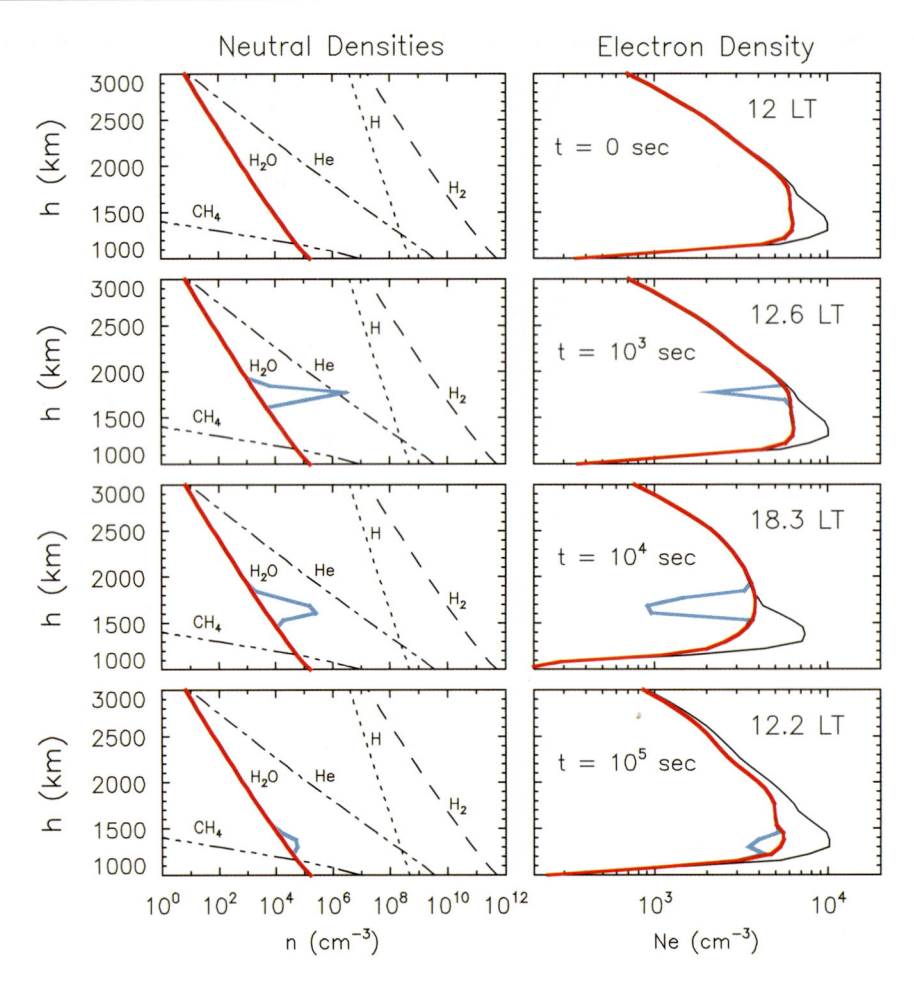

Fig. 8.11 Time evolution of a water surge simulation at Saturn. Black curves represent the nominal atmosphere for which [H$_2$O] = 0, red curves correspond to an atmosphere for which there is a constant topside water influx of 5×10^6 H$_2$O molecules s^{-1}, and blue curves present the result of a surge in the topside water influx equivalent to an augmentation in the background by a factor of 50 persisting for 1,600 s. Similar ionospheric "bite-outs" can result from lesser augmentations over longer durations. For radio occultation observations of such structure refer to Section 8.5 (from Moore and Mendillo 2007)

before sunrise. These models also indicated that the losses for H$^+$ are relatively slow, and thus it has a much milder diurnal variation that is dependent on the local populations of H$_2$O and vibrationally excited H$_2$. Finally, they all predict that H$_3^+$ is the dominant ion near and below the electron density peak during the day, while H$^+$ is the dominant ion for all local times above the peak, and certainly in the topside ionosphere and at night. Moore et al. (2004) show that the expected relative distribution of H$^+$ and H$_3^+$ vary with latitude and season. The lower ionosphere is dominated by a complicated assortment of hydrocarbon ions, of which C$_3$H$_5^+$ is the most numerous in the model of Moses and Bass (2000).

More recently Hallett et al. (2005a) and Shemansky et al. (2009) describe comprehensive calculations in which they track the physical state of the non-LTE environment and weakly ionized plasma that would develop in a pure H$_2$ atmosphere under conditions relevant to the upper atmosphere of Saturn. They are able to obtain electron densities consistent with observations without including H$_2$O, as a loss mechanism for H$^+$. Their 1D chemical/diffusive model uses a modified Monte Carlo approach and predicts that H$_3^+$ is the dominant ion throughout the Saturn ionosphere up to about 2,000 km, in agreement with previous models (see Fig. 8.10b). The conclusions of Shemansky et al. (2009) are partially the result of a more realistic calculation of the vibration/rotation state of the H$_2$ in a pure hydrogen system that includes electron forcing (Eq. 8.6) and of the new rates that they have obtained and used for the reaction of H$^+$ with H$_2$ (Eq. 8.5). Yet, as water has been observed at Saturn by ISO (Feuchtgruber et al. 1997), Hubble (Prangé et al. 2006), and now Cassini (Bjoraker et al. 2008), it is clear that the balance between the water and non-water H$^+$ loss pathways remains to be determined.

Shemansky et al. (2009) highlight the importance of the electron populations in controlling the activation of ground-state H$_2$. Most Saturn ionospheric density models to date have ignored photoelectrons, electron impact processes, electron scattering and vibrational excitation/relaxation, and have focused on solar photon processes alone. Neglect of the electron population may also be one cause of the model-data mismatches discussed above. Note that both Shemansky et al. (2009) and Huestis (2008) point out that reaction (8.5) can lead to vibrational relaxation of H_2 as well as charge-exchange, with Huestis (2008) indicating that it is likely to be the dominant process. On the other hand Shemansky et al. (2009) maintain that the energy gained by H^+, will deliver its energy back to H_2 and into vibrational excitation. Furthermore, Shemansky et al. (2009) also indicate that in their model they do not include the critically important momentum transfer and vibrational excitation reaction of the hot atomic hydrogen product with H_2 (X) (Hallett et al. 2005b), because of the lack of the appropriate collision matrix, and thus their vibrational excitation values are lower limits. The topic of the non-LTE state of the H_2 is clearly one with important and unresolved implications for upper atmospheric research at Saturn, Jupiter, and the hydrogen-rich atmospheres of other giant planets.

Galand et al. (2009) have shown that electron impact ionization by photoelectrons and secondaries is a significant ionization process in the bottomside sunlit ionosphere, increasing the electron density by a factor of ∼2–10 over calculations that ignore secondary production below about 1,100 km. Based on the work of Galand et al. (2009), a parameterization of the secondary ionization production rate at Saturn has been developed by Moore et al. (2009).

8.6.3 Plasma Temperatures in Saturn's Ionosphere

As in the case of other ionospheric processes, such as chemistry and dynamics, plasma temperature calculations at Saturn draw heritage from terrestrial and Jovian studies. Prior to 2007, there was only one published theoretical determination of ion and electron temperatures in Saturn's ionosphere (Waite 1981). However, as those calculations were performed using a now known to be unrealistic neutral temperature profile with an exospheric temperature of nearly 1,000 K (see discussion in Section 8.2), a new derivation based on more recent spacecraft data and laboratory rates was warranted. Waite (1981) predicted ion and electron temperatures ranging from 1,000 K to 100,000 K, depending on the values of various assumed parameters, such as ion-neutral differential velocities (leading to Joule heating) and downward heat fluxes at the upper boundary. Two new studies of plasma temperatures in Saturn's ionosphere were published recently. The first focused on high latitudes and used a one-dimensional multi-fluid model to study the polar wind at Saturn (Glocer et al. 2007). Glocer et al. modeled the ionosphere from below the peak to an altitude of one Saturn radius, yielding densities, fluxes and temperatures for H_3^+ and H^+. They calculated densities of a few time $10^4 \, cm^{-3}$ which are consistent with the Voyager and Cassini results. Calculated fluxes of outflowing ions over the polar cap were estimated to be in the range from 2.1×10^{26} to $7.5 \times 10^{27} \, s^{-1}$, making Saturn's auroral ionosphere an intermediate source of magnetospheric plasma, larger than the Titan Torus source (Smith et al. 2004), and smaller than the icy satellite source (e.g., Jurac and Richardson 2005). Finally, the calculated peak ion temperatures varied from about 1,500 to 3,000 K.

The second recent study of plasma temperatures in Saturn's ionosphere used STIM as a basis for the calculations (Moore et al. 2008), and focused on the low- and mid-latitude ionosphere, neglecting auroral energy inputs as well as potential energy storage at high altitudes in the field tubes. Three codes were coupled sequentially in order to derive self-consistent time-dependent ion and electron temperatures: (1) the STIM thermosphere discussed above (Müller-Wodarg et al. 2006), (2) the 1D STIM ionospheric module (Moore et al. 2004), and (3) a suprathermal electron transport code adapted to Saturn (Galand et al. 1999, 2006). Moore et al. (2008) predicted topside electron temperatures to range between 500–560 K (∼80–140 K above the neutral temperature). After sunset, plasma-neutral interactions quench the electron gas within two Saturn hours. Ion temperatures, calculated for only the major ions H^+ and H_3^+, were somewhat smaller, reaching ∼480 K during the day at the topside while remaining nearly equal to the neutral temperature at altitudes near and below the N_e peak. For easy reference, Table 8.1 provides a partial timeline of papers describing models of Saturn's ionosphere.

8.7 Summary

If the Cassini observations with respect to the upper atmosphere and ionosphere of Saturn were to be summarized with one main theme, that theme would be *variability*. The Cassini UVIS observations of the neutral upper atmosphere and the radio-science observations of the ionosphere emphasize that Saturn's upper atmosphere is much more temporally and spatially variable than has generally been realized from the Voyager observations.

The preliminary Cassini UVIS δ-Ori stellar occultation results (Shemansky and Liu 2009) imply a thermospheric temperature that is ∼140–180 K colder than temperatures inferred from a reanalysis of all six Voyager UVS occultations (Vervack and Moses 2009). Both the new Cassini data and the Voyager reanalyses suggest that thermospheric neutral temperatures on Saturn are of order 300–500 K and not as large as 800 K. However, even these 300–500 K temperatures

Table 8.1 Models of Saturn's ionosphere

Author(s)	Transport	N_{MAX} (cm^{-3})	h_{MAX} (km)	Comments (k = H$^+$ + H$_2$ ($v \geq 4$) reaction rate)
McElroy (1973)	No	30	120	Altitude referenced to the homopause. Does not include H$^+$ + H$_2$ ($v \geq 4$) reaction.
Atreya and Donahue (1975)	No	10	300	Introduced hydrocarbon chemistry to lower ionosphere. Altitudes referenced to the [H$_2$] = 10^{16} cm^{-3} level.
Capone et al. (1977)	No	30	680	Lower electron peak of 7 × 10^3 cm^{-3} at 50 km due to cosmic-ray ionization.
Waite et al. (1979)	Yes	10	2,000	Altitudes referenced to the [n] = 10^{19} cm^{-3} level. Assumed an exospheric temperature of 1,300 K and an eddy diffusion coefficient of 1.3 ×10^6 cm^2 s^{-1}.
Waite (1981)	Yes	10	1,200	For an eddy diffusion coefficient of 10^6 cm^2 s^{-1}. Also calculated plasma temperatures for a 1,000 K neutral exospheric temperature.
Majeed and McConnell (1991)	Yes	1	2,200	Nominal model N_{MAX} of 3 × 10^5 cm^{-3} at 1,200 km. Quoted values use either Φ_{H_2O} = 2.2 × 10^7 cm^{-2} s^{-1} and k = 0, or Φ_{H_2O} = 0 and k = 1.38 × 10^{-14} cm^3 s^{-1}
Majeed and McConnell (1996)	Yes	2	1,800	For Φ_{H_2O} = 10^7 cm^{-2} s^{-1} and k = 0. Also explores more values of Φ_{H_2O} and k in a time-dependent ionosphere.
Moses and Bass (2000)	Yes	1	1,400	Time-dependent solution for 46 ion species. Explores water and metal influx from ring or meteoric sources. Neutral wind and electric field induced plasma motions studied.
Moore et al. (2004)	Yes	1	1,400	Global solution using 3D thermosphere. Investigation of ring shadowing. Ionospheric conductivities.
Moore et al. (2006); Moore and Mendillo (2007)	Yes	0.1 dawn/0.4 dusk	2,500 dawn/ 1,900 dusk	Parameter space exploration of H$_2$O and k in order to reproduce average Cassini behavior. Time-dependent neutral water diffusion calculations leading to ionospheric "bite-outs".
Glocer et al. (2007)	Yes	0.3–2	1,400–3,000	Polar wind steady-state study (1 R$_S$ upper boundary). Includes plasma temperatures. Wide range of T$_{exo}$ explored.
Moore et al. (2008)	Yes	0.4 dawn/1 dusk	1,600 dawn/ 1,300 dusk	Plasma temperature calculations. Water influx. Voyager 2 era rather than Cassini (as in Moore et al., 2006).
Shemansky and Liu (2009)	Yes	2	1,100	Complete non-LTE calculations of the pure H$_2$ environment of Saturn; major ion is H$_3^+$ up to 2,000 km.

are higher than can be explained by the absorption of solar extreme ultraviolet radiation alone, and the thermospheric heat sources on Saturn have yet to be explained. Modeling, such as is described in Section 8.4, can help provide clues to the dominant heating mechanisms.

In retrospect, the older Voyager data also suggest variability in upper atmospheric structure on Saturn, especially with regard to the hydrocarbon abundance profiles and the location of the methane homopause (e.g., Vervack and Moses 2009), but because the Voyager ultraviolet occultation data have only been analyzed in a full and consistent manner recently, this variability was not recognized. Unlike the earlier Voyager view of vigorous atmospheric mixing in Saturn's stratosphere, the Cassini UVIS δ-Ori stellar occultation results (Shemansky and Liu 2009) imply a very low-altitude methane homopause at −42.7 latitude, suggesting that atmospheric mixing is relatively weak or downward vertical winds are affecting the methane profile in Saturn's middle atmosphere at this location and time. The overall variability in all the ultraviolet occultations to date suggests that vertical winds play a major role in controlling the methane profile in the homopause region on Saturn, and it is hoped that analysis of the numerous as-yet-to-be examined Cassini UVIS occultations may help constrain middle-atmospheric circulation on Saturn. Current photochemical models based on the neutral chemistry described in Moses et al. (2005) do not accurately reproduce the overall shape of the hydrocarbon vertical profiles derived from the ultraviolet occultations, and further investigation into possible reasons for the model-data mismatch are warranted. Constraints on both vertical transport and upper-atmospheric chemistry are likely to be derived from such model-data comparisons.

Although ultraviolet occultations remain one of the few techniques for which we can obtain information on thermospheric temperatures and neutral species abundances, one must keep in mind that the derived results are almost always very model dependent. Analysis and modeling techniques have improved tremendously since the early Voyager analyses, but uncertainties and poorly constrained model parameters remain that can complicate the derivations.

The Cassini radio occultation electron density profiles have significantly increased our data base. We now have good latitudinal coverage of electron densities. The diurnal variations in the peak electron densities inferred by the new Cassini SED observations are large and similar to the ones from Voyager. Significant progress has also been made in modeling the ionosphere. These new observations and models are welcome; however, there is still a lot we do not know nor understand about Saturn's ionosphere. Ion composition is one important issue without a definitive resolution. Electron density measurements can provide some indirect clues on the ion composition, but no definitive information. There appears to be agreement that H_3^+ is the dominant ion at the lower altitudes and H^+ is the major ion at the higher altitudes. The relative importance of vibrationally excited H_2 compared to water inflow or gravity waves in removing H^+ is still being debated. The scale height derived sub-auroral plasma temperatures have very large uncertainties and appear to be higher than the corresponding model values. Unfortunately, as long as the only data we have to work with are scale heights from topside electron density profiles, we will have to live with these uncertainties. Improved models will hopefully help to elucidate some of the issues associated with these parameters that are not being measured directly. The large variability in the electron densities observed from radio occultation profiles at similar latitudes and times suggests that dynamical and/or electrodynamical processes play a major role in controlling ionospheric structure on Saturn; such processes should be investigated in future models to the extent that is possible.

Note added in proof: The UVIS occultation results presented in this chapter were based on a preliminary analysis; some of the results have changed. See Shemansky and Liu (2009) for the updated results.

References

Atreya, S. K. and T. M. Donahue, The role of hydrocarbons in the ionospheres of the outer planets, *Icarus, 25*, 335–338, 1975.

Atreya, S. K., T. M. Donahue, and J. H. Waite Jr., An interpretation of the Voyager measurement of Jovian electron density profiles, *Nature, 280*, 795–796, 1979.

Atreya, S. K., J. H. Waite, Jr., T. M. Donahue, A. F. Nagy, J. C. McConnell, Theory, measurements, and models of the upper atmosphere and ionosphere of Saturn. In: Gehrels, T., Matthews, M.S. (Eds.), Saturn, Univ. Arizona Press, Tucson, pp. 239–277, 1984.

Baron R. L., R. D. Joseph, T. Owen, J. Tennyson, S. Miller, and G. E. Ballester, Infrared imaging of H_3^+ in the ionosphere of Jupiter, *Nature, 353*, 539–542, 1991.

Bezard, B., J. J. Moses, J. Lacy, T. Greathouse, M. Richter, and C. Griffith, Detection of ethylene C_2H_4 on Jupiter and Saturn in non-auroral regions, *Bull. Am. Astron. Soc., 33*, 1079–1080, 2001.

Bjoraker, G.L., R.K. Achterberg, A.A. Simon-Miller, R.C. Carlson, and D.E. Jennings, Cassini/CIRS observations of water vapor in Saturn's stratosphere, *Saturn After Cassini-Huygens*, Imperial College, London, 28 July–1 August, 2008.

Broadfoot, A.L., and 15 co-authors, Extreme ultraviolet observations from Voyager 1 encounter with Saturn, *Science, 212*, 206–211, 1981.

Capone, L. A., R. C. Whitten, S. S. Prasad, and J. Dubach, The ionospheres of Saturn, Uranus, and Neptune, *Astrophys. J., 215*, 977–983, 1977.

Connerney, J. E. P. and J.H. Waite Jr., New model of Saturn's ionosphere with an influx of water from the rings, *Nature, 312*, 136–138, 1984.

Connerney, J. E. P., Magnetic connection for Saturn's rings and atmosphere, *Geophys. Res. Lett., 13*, 773, 1986.

Courtin, R., D. Gautier, A. Marten, B. Bezard, and R. Hanel, The composition of Saturn's atmosphere at northern temperate latitudes from Voyager IRIS spectra: NH_3, PH_3, C_2H_2, C_2H_6, CH_3D, CH_4, and the Saturnian D/H isotopic ratio, *Astrophys. J., 287*, 899–916, 1984.

Cowley S. W. H., E.J. Bunce, and R. Prange, Saturn's polar ionospheric flows and their relation to the main auroral oval. *Ann. Geophys., 22*, 1379–1394, 2004.

Cravens, T. E., Vibrationally excited molecular hydrogen in the upper atmosphere of Jupiter, *J. Geophys. Res., 92*, 11083, 1987.

Drossart P., J-P. Maillard, J. Caldwell, S. J. Kim, J. K. G. Watson, W. A. Majewski, J. Tennyson, S. Miller, S. K. Atreya, J. T. Clarke, J. H. Waite Jr, and R. Wagener, Detection of H_3^+ on Jupiter, *Nature, 240*, 539–541, 1989.

Dungey J. W., The interplanetary magnetic field and auroral zones. *Phys. Rev. Lett., 6*, 47, 1961.

Dyudina, U. A., A. P. Ingersoll, P. E. Shawn, C. C. Porco, G. Fischer, W. S. Kurth, M. D. Desch, A. Del Genio, J. Barbara, and J. Ferrier, Lightning storms on Saturn observed by Cassini ISS and RPWS in the years 2004–2006, *Icarus, 190*, 545–555, 2007.

Esposito, L. W., et al, The Cassini ultraviolet imaging spectrograph investigation, *Space Sci. Rev., 115*, 299–361, 2004.

Festou M. C. and S. K. Atreya, Voyager ultraviolet stellar occulation measurements of the composition and thermal profiles of the saturnian upper atmosphere, *Geophys. Res. Lett.*, 9, 1147–1152, 1982.

Festou, M. C., S. K. Atreya, T. M. Donahue, B. R. Sandel, D. E. Shemansky, and A. L. Broadfoot, Composition and thermal profiles of the Jovian upper atmosphere determined by the Voyager ultraviolet stellar occultation experiment, *J. Geophys. Res., 86*, 5715–5725, 1981.

Feuchtgruber, H., E. Lellouch, T. de Graauw, B. Bézard, T. Encrenaz, and M. Griffin, External supply of oxygen to the atmospheres of the giant planets, *Nature, 389*, 159–162, 1997.

Flasar, F. M., et al., Temperatures, winds and composition in the Saturnian system, *Science*, 307, 1247–1251, doi: 10.1126/science.1105806, 2005.

Fischer, G. et al., Atmospheric electricity at Saturn, *Space Sci. Rev., 137*, 271–285, doi:10.1007/s11214–008–9370-z, 2008.

Fletcher, L. N., et al., Characterizing Saturn's vertical temperature structure from Cassini/CIRS, *Icarus, 189*, 457–478, 2007.

Fouchet, T., et al.: Saturn composition and chemistry. Chapter 5, this book, 2009.

Geballe T.R., M.-F. Jagod, and T. Oka, Detection of H_3^+ infrared emission lines in Saturn, *Astrophys. J.*, 408, L109–L112, 1993.

Galand, M., J. Lilensten, D. Toublanc, and S. Maurice, The ionosphere of Titan: Ideal diurnal and nocturnal cases, *Icarus*, 140, 92–105, 1999.

Galand, M., R.V. Yelle, A. J. Coates, H. Backes, and J.-E. Wahlund, Electron temperature of Titan's sunlit ionosphere, *Geophys. Res. Lett.*, 33, L21101, doi:10.1029/2006GL0247488, 2006.

Galand, M., L. Moore, B. Charney, I. C. F. Müller-Wodarg, and M. Mendillo, Solar primary and secondary ionization at Saturn, *J. Geophys Res.*, 114, A06313, doi:10.1029/2008JA013981, 2009.

Glocer, A., T. I. Gombosi, G. Toth, K. C. Hansen, A.J. Ridley, and A. Nagy, Polar wind outflow model: Saturn results, *J. Geophys. Res.*, 112, doi:10.1029/2006JA011755, 2007.

Greathouse, T. K., J. H. Lacy, B. Bezard, J. I. Moses, C. A. Griffith, and M. J. Richter Meridional variations of temperature, C_2H_2 and C_2H_6 abundances in Saturn's stratosphere at southern summer solstice, *Icarus*, 177, 18–31, 2005.

Gurnett, D. A., et al., Radio and plasma wave observations at Saturn from Cassini's approach and first orbit, *Science*, 307, 1255–1259, 2005.

Hallett, J. T., D. E. Shemansky, and X. Liu, A rotational-level hydrogen physical chemistry model for general astrophysical application. *Astrophys. J.*, 624, 448–461, 2005a.

Hallett, J. T., D. E. Shemansky, and X. Liu, Fine-structure physical chemistry modeling of Uranus H2 X quadrupole emission, *Geophys. Res. Lett.*, 32, L02204, doi:10.1029/2004GL021327, 2005b.

Hubbard, W. B. and 27 colleagues, Structure of Saturn's mesosphere from the 28 Sgr occultations. *Icarus*, 130, 404–425, 1997.

Huestis, D. L., Hydrogen collisions in planetary atmospheres, ionospheres, and magnetospheres, *Planet. Space Sci.*, 56, 1733, 2008.

Ichihara, A., O. Iwamoto, and R. K. Janev, 2000, Cross sections for the reaction $H^+ + H_2 (v = 0 - 14) -> H + H_2^+$, *J. Phys. B*, 33, 4747, 2000.

Isbell J., A. J. Dessler, and J. H. Waite, Magnetospheric energisation by interaction between planetary spin and the solar wind, *J. Geophys. Res.*, 89, 10716–10722, 1984.

Kaiser, M. L., J. E. P. Connerney, and M. D. Desch, Atmospheric storm explanation of saturnian electrostatic discharges, *Nature*, 303, 50–53, 1983.

Jurac, S. and J.D. Richardson, A self-consistent model of plasma and neutrals at Saturn: Neutral cloud morphology, *J. Geophys. Res.*, 110, A09220, doi:10.1029/2004JA010635, 2005.

Kaiser, M. L., M. D. Desch, and J. E. P. Connerney, Saturn's ionosphere – Inferred electron densities, *J. Geophys. Res.*, 89, 2371–2376, 1984.

Killen, R., D.E. Shemansky, and N. Mouawad, Expected emission from Mercury's exospheric species, and their UV-Visible signatures, *Ap. J. Suppl.*, 181, 351, 2009.

Kliore, A. J., J. D. Anderson, J. W. Armstrong, S. W. Asmar, C. L. Hamilton, N. J. Rappaport, H. D. Wahlquist, R. Ambrosini, F. M. Flasar, R. G. French, L. Iess, E. A. Marouf, and A. F. Nagy, Cassini Radio Science, *Space Sci. Rev.*, 115, 1–70, 2004.

Kliore, A. J., et al., Midlatitude and high-latitude electron density profiles in the ionosphere of Saturn obtained by Cassini radio occultation observations. *J. Geophys. Res.*, 114, A04315, doi:10.1029/2008JA0139000, 2009.

Lam H. A., N. Achilleos, S. Miller, J. Tennyson, L. M. Trafton, T. R. Geballe, and G. Ballester, A baseline spectroscopic study of the infrared aurorae of Jupiter, *Icarus*, 127, 379–393, 1997.

Lindal, G. F., D. N. Sweetnam, and V. R. Eshleman, The Atmosphere of Saturn: An Analysis of the Voyager Radio Occultation Measurements, *Astron. J.*, 90, 1136–1146, 1985.

Lindal, G. F., The atmosphere of Neptune: An analysis of radio occultation data acquired with Voyager 2, *Astron. J.*, 103, 967–982, 1992.

Majeed, T. and J. C. McConnell, The upper ionospheres of Jupiter and Saturn, *Planet. Space Sci.*, 39, 1715–1732, 1991.

Majeed, T., R. V. Yelle, and J. C. McConnell, Vibrationally excited H_2 in the outer planet thermospheres – Fluorescence in the Lyman and Werner bands, *Planet. Space Sci.*, 39, 1591–1606, 1991.

Majeed, T. and J. C. McConnell, Voyager electron density measurements on Saturn: Analysis with a time dependent ionospheric model, *J. Geophys. Res.*, 101, 7589–7598, 1996.

Matcheva, K.I., D.F. Strobel, and F.M. Flasar, Interaction of gravity waves with ionospheric plasma: Implications for Jupiter's ionosphere, *Icarus*, 152, 347–365, doi:10.1006/Icar.2001.6631, 2001.

McConnell, J. C., J. B. Holberg, G. R. Smith, B. R. Sandel, D. E. Shemansky, and A. L. Broadfoot, A new look at the ionosphere of Jupiter in light of the UVS occultation results, *Planet. Space Sci.*, 30, 151–167, 1982.

McElroy, M. B, The ionospheres of the major planets, *Space Sci. Rev.*, 14, 460–473, 1973.

Melin, H., S. Miller, T. Stallard, C. Smith, and D. Grodent, Estimated energy balance in the jovian upper atmosphere during an auroral heating event, *Icarus*, 181, 256–265, doi:10.1016/j.icarus.2005.11.004, 2006.

Melin, H., S. Miller, T. Stallard, L. M. Trafton, and T. R. Geballe, Variability in the H_3^+ emission from Saturn: consequences for ionisation rates and temperature, *Icarus*, 186, 234–241, 2007.

Melin, H., D. E. Shemansky, and X. Liu, The distribution of atomic hydrogen and oxygen in the magnetosphere of Saturn, *Planet. Space Sci.*, doi: 101016/j.pss2009.04.014, 2009.

Mendillo, M., L. Moore, J. Clarke, I. Müller-Wodarg, W.S. Kurth, and M.L. Kaiser, Effects of ring shadowing on the detection of electrostatic discharges at Saturn, *Geophys. Res. Lett.*, 32, L05107, doi:10.1029/2004GL021934, 2005.

Miller S., H.A. Lam, and J. Tennyson, What astronomy has learned from observations of H_3^+, *Can. J. Phys.*, 72, 760–771, 1994.

Miller S. and 10 others, The role of H3+ in planetary atmospheres, *Phil. Trans. Roy. Soc.* 358, 2485–2502, 2000.

Miller S., A. Aylward, and G. Millward, Giant planet ionospheres and thermospheres: The importance of ion-neutral coupling, *Space Sci. Rev.*, 116, 319–343, 2005.

Miller S., T. Stallard, C. Smith, G. Millward, H. Melin, M. Lystrup, and A. Aylward, H_3^+: the driver of giant planet atmospheres, *Phil. Trans. Roy. Soc.*, 364, 3121–3138, 2006.

Moore, L. and M. Mendillo, Ionospheric contribution to Saturn's inner plasmasphere, *J. Geophys. Res.*, 110, A05310, doi:10.1029/2004JA010889, 2005.

Moore, L. and M. Mendillo, Are plasma depletions in Saturn's ionosphere a signature of time-dependent water input?, *Geophys. Res. Lett.*, 34, L12202, doi:10.1029/2007GL029381, 2007.

Moore, L. E., M. Mendillo, I. C. F. Müller-Wodarg and D. L. Murr, Modeling of global variations and ring shadowing in Saturn's ionosphere, *Icarus*, 172, 503–520, 2004.

Moore, L., A. F. Nagy, A. J. Kliore, I. Müller-Wodarg, J. D. Richardson and M. Mendillo, Cassini radio occultations of Saturn's ionosphere: Model comparisons using a constant water flux, *Geophys. Res. Lett.*, 33, L22202, 2006.

Moore, L., M. Galand, I. Müller-Wodarg, R. Yelle, and M. Mendillo, Plasma temperatures in Saturn's ionosphere, *J. Geophys. Res.*, 113, A10, A10306, 10.1029/2008JA013373, 2008.

Moore, L., M. Galand, I. Müller-Wodarg, and M. Mendillo, Response of Saturn's ionosphere to solar radiation: Testing parameterizations for thermal electron heating and secondary ionization processes, *Planet. Space Sci.*, doi:10.1016/j.pss.2009.05.001, 2009.

Moses, J.I. and S.F. Bass, The effects of external material on the chemistry and structure of Saturn's ionosphere, *J. Geophys. Res.*, 105, 7013–7052, 2000.

Moses, J. I., B'ezard, B., Lellouch, E., Gladstone, G. R., Feuchtgruber, H., Allen, M., Photochemistry of Saturn's atmosphere. I. Hydrocarbon chemistry and comparisons with ISO observations. *Icarus*, 143, 244–298, 2000.

Moses, J. I., T. Fouchet, R. V. F. Yelle, A. J. Friedson, G. S. Orton, B. B'ezard, P. Drossart, G. R. Gladstone, T. Kostiuk, T. A. Livengood, The stratosphere of Jupiter. In: Bagenal, F., Dowling, T.E., McKinnon, W.B. (Eds.), *Jupiter: The Planet, Satellites and Magnetosphere*, Cambridge University Press, Cambridge, UK, pp. 129–157, 2004.

Moses, J. I., T. Fouchet, B. B'ezard, G. R. Gladstone, E. Lellouch, H. Feuchtgruber, Photochemistry and diffusion in Jupiter's stratosphere: Constraints from ISO observations and comparisons with other giant planets, *J. Geophys. Res.*, 110, E08001, doi:10.1029/2005JE002411 2005.

Moses, J. I. and R. J. Vervack, Jr., The structure of the upper atmosphere of Saturn, *Lunar Planet. Sci. Conf.*, 37, #1803, 2006.

Müller-Wodarg, I. C. F., M. Mendillo, R. V. Yelle, and A. D. Aylward, A global circulation model of Saturn's thermosphere, *Icarus*, 180, 147–160, 2006.

Nagy, A. F., A. J. Kliore, E. Marouf, R. French, M. Flasar, N. Rappaport, A. Anabtawi, S. W. Asmar, D. Johnston, E. Barbinis, and 2 coauthors, First results from the ionospheric radio occultations of Saturn by the Cassini spacecraft, *J. Geophys. Res.*, 111, A06310, 2006.

Pierrard, V. and J. Lemaire, Lorentzian ion exosphere model, *J. Geophys. Res.*, 101, 7923–7934, 1996.

Pierrard, V. and J. Lemaire, Erratum: Lorentzian ion exosphere model, *J. Geophys. Res.*, 101, 7923–7934, 1996; 103, 4117–4118, 1998.

Prangé, R., T. Fouchet, R. Courtin, J.E.P. Connerney, and J.C. McConnell, Latitudinal variation of Saturn photochemistry deduced from spatially-resolved ultraviolet spectra, *Icarus*, 180, 379–392, 2006.

Sada, P. V., G. L. Bjoraker, D. E. Jennings, P. N. Romani, and G. H. McCabe, Observations of C2H6 and C2H2 in the stratosphere of Saturn, *Icarus* 173, 499–507, 2005.

Sandel, B.R., and 12 co-authors, Extreme ultraviolet observations from the Voyager 2 encounter with Saturn, *Science*, 215, 548–553, 1982.

Schunk, R. W. and A. F. Nagy, *Ionospheres*, 2nd edn, Cambridge University Press, 2009.

Shemansky, D. E. and D. T. Hall, The distribution of atomic hydrogen in the magnetosphere of Saturn, *J. Geophys. Res.*, 97, 4143–4161, 1992.

Shemansky, D. E., A. I. F. Stewart, R. A. West, L. W. Esposito, J. T. Hallett, and X. M. Liu, The Cassini UVIS Stellar probe of the Titan atmosphere, *Science*, 308, 978, 2005.

Shemansky, D. E., X. Liu, and H. Melin, The Saturn hydrogen plume, *Planet. Space Sci.*, doi:101016/j.pss2009.05.002, 2009.

Shemansky, D. E. and X. Liu, Saturn upper atmospheric structure from Cassini EUV/FUV occultations, *Planet. Space Sci.*, under review, 2009.

Smith, G. R., D. E. Shemansky, J. B. Holberg, A. L. Broadfoot, B. R. Sandel, Saturn's upper atmosphere from the Voyager 2 EUV solar and stellar occultations, *J. Geophys. Res.* 88, 8667–8678, 1983.

Smith, G. R. and D. M. Hunten, Study of planetary atmospheres by absorptive occultations, *Rev. Geophys.*, 28, 117, 1990.

Smith, H.T., R.E. Johnson, and V.I. Shematovich, Titan's atomic and molecular nitrogen tori, *Geophys. Res. Lett.*, 31 L16804, doi:10.1029/2004GL020580, 2004.

Smith, C. G. A., Aylward, A. D., Miller, S., Müller-Wodarg, I. C. F., Polar heating in the Saturn's thermosphere, *Ann. Geophys.*, 23, 2465–2477, 2005.

Smith C. G. A., A. D. Aylward, G. H. Millwrad, S. Miller and L. Moore, An unexpected cooling effect in Saturn's upper atmosphere, *Nature*, 445, 399–401, 2007.

Stallard T., S. Miller, G. E. Ballester, D. Rego, R. D. Joseph, and L. M. Trafton, The H_3^+ latitudinal profile of Saturn, Astrophys. *J. Lett.*, 521, L149–L152, 1999.

Stallard, T., S. Miller, L. M. Trafton, T. R. Geballe, R. D. Joseph, Ion winds in Saturn's southern auroral/polar region, *Icarus*, 167, 204–211, 2004.

Stallard T., et al., Complex structure within Saturn's infrared aurora, *Nature*, 456, 214, 2008.

Strobel, D. F. and G. R. Smith, On the temperature of the Jovian thermosphere, *J. Atm. Sci.*, 30, 718, 1973.

Strobel, D. F., Photochemistry in outer solar system atmospheres, *Space Sci. Rev.*, 116, 155–170, 2005.

Trafton L. M., T. R. Geballe, S. Miller, J. Tennyson, and G.E. Ballester, Detection of H_3^+ from Uranus, *Astrophys. J.*, 405, 761–766, 1993.

Vasyliunas V. M., Plasma distribution and flow. In: Dessler A.J. (ed.), *Physics of the Jovian Magnetosphere*, Cambridge University Press, pp. 395–453, 1983.

Vervack, R. J., Jr., and J. I. Moses, Saturn's upper atmosphere during the Voyager era: Reanalysis and modeling of the UVS occultations, *Icarus*, under review, 2009.

Waite, J. H., S. K. Atreya, and A. F. Nagy, The ionosphere of Saturn – Predictions for Pioneer 11, *Geophys. Res. Lett.*, 6, 723–726, 1979.

Waite, J. H., T. E. Cravens, J. U. Kozyra, A. F. Nagy, S. K. Atreya, and R. H. Chen, Electron precipitation and related aeronomy of the Jovian thermosphere and ionosphere, *J. Geophys. Res.*, 88, 6143–6163, 1983.

Waite, J.H., *The ionosphere of Saturn*, PhD Thesis, Univ. of Mich., Ann Arbor, 1981.

Warwick, J. W., D. R. Evans, J. H. Romig, J. K. Alexander, M. D. Desch, M. L. Kaiser, M. G. Aubier, Y. Leblanc, A. Lecacheux, and B. M. Pedersen, Planetary radio astronomy observations from Voyager 2 near Saturn, *Science*, 215, 582–587, 1982.

West, R. A., D. F. Strobel, M. G. Tomasko, Clouds, aerosols, and photochemistry in the Jovian atmosphere, *Icarus*, 65, 161–217, 1986.

Wu, C. Y. R., F. Z. Chen, and D. L. Judge, Measurement of temperature-dependent absorption cross sections of C_2H_2 in the VUV-UV region, *J. Geophys. Res*, 106, 7629, 2001.

Yelle, R. V. and S. Miller, Jupiter's thermosphere and ionosphere. In: Bagenal, F., et al. (Ed.), *Jupiter: The Planet, Satellites, and Magnetosphere*, Cambridge Univ. Press, New York, pp. 185–218, 2004.

Yelle, R. V., L. A. Young, R. J. Vervack, Jr., R. Young, L. Pfister, and B. R. Sandel, Structure of Jupiter's upper atmosphere: Predictions for Galileo, *J. Geophys. Res.* 101, 2149–2161, 1996.

Yung, Y. L. and W. B. DeMore, *Photochemistry of Planetary Atmospheres*, Oxford University Press, Oxford, 1999.

Chapter 9
Saturn's Magnetospheric Configuration

Tamas I. Gombosi, Thomas P. Armstrong, Christopher S. Arridge, Krishan K. Khurana, Stamatios M. Krimigis, Norbert Krupp, Ann M. Persoon, and Michelle F. Thomsen

Abstract This Chapter summarizes our current understanding of Saturn's magnetospheric configuration. Observations from the Cassini Prime and Equinox Missions have answered some questions and opened new ones. One of the fundamental questions of magnetospheric physics is what are the sources of the plasma that populate the magnetosphere. At Saturn, there is a rich set of possible plasma sources: the solar wind, Saturn's ionosphere, Titan, the rings, and the icy satellites. One of the most significant discoveries of the Cassini mission was Enceladus' role as a source. Saturn's magnetospheric convection pattern falls somewhere between that of Earth and Jupiter. Earth is a slow rotator with a relatively small internal mass source and its magnetosphere is primarily dominated by the solar wind. At Jupiter the solar wind only plays a minor role since Jupiter is a fast rotator with a strong surface magnetic field and a significant plasma source (Io) deep inside the magnetosphere. As a consequence, internal processes dominate the magnetosphere and solar wind interaction is only marginally important. Saturn falls somewhere between Earth and Jupiter thus Saturn's magnetosphere exhibits both solar wind and internal controls at the same time. This fact makes the Kronian magnetosphere even more fascinating and complex than the magnetospheres of Earth and Jupiter.

T.I. Gombosi
Center for Space Environment Modeling, Department of Atmospheric, Oceanic and Space Sciences, The University of Michigan, Ann Arbor, MI, USA
e-mail: tamas@umich.edu

T.P. Armstrong
Fundamental Technologies, LLC, Lawrence, KS, USA
e-mail: armstrong@ftecs.com

C.S. Arridge
Mullard Space Science Laboratory, University College London, London, UK
and
Centre for Planetary Sciences, University College London, London, UK
e-mail: chris.arridge@physics.org

K.K. Khurana
Institute of Geophysics and Planetary Physics, University of California at Los Angeles, Los Angeles, CA, USA
e-mail: kkhurana@igpp.ucla.edu

S.M. Krimigis
Applied Physics Laboratory, Johns Hopkins University, Laurel, MD, USA
and
Center of Space Research and Technology, Academy of Athens, Athens, Greece
e-mail: tom.krimigis@jhuapl.edu

N. Krupp
Max-Planck Institute for Solar System Research, Katlenburg-Lindau, Germany
e-mail: krupp@mps.mpg.de

A.M. Persoon
University of Iowa, Iowa City, IA, USA
e-mail: ann-persoon@uiowa.edu

M.F. Thomsen
Los Alamos National Laboratory, Los Alamos, NM, USA
e-mail: mthomsen@lanl.gov

9.1 Introduction

Twenty five years is a human generation. Children are born, raised, educated and reach maturity in a quarter of a century. In space exploration, however, 25 years is a very long time. Twenty five years after the launch of the first Sputnik, Saturn was visited by three very successful spacecraft: Pioneer 11 (1979), Voyager 1 (1980) and Voyager 2 (1981). The flyby trajectories are shown in Fig. 9.1. Twenty five years after the first Saturn flyby by Pioneer 11 the most advanced planetary probe ever built started orbiting Saturn and deployed a lander to reveal the mysteries of Titan. One can only guess the exciting future missions we may have 25 years after Cassini.

9.1.1 Pre-Cassini Understanding

The three pre-Cassini flybys provided a very good overall understanding of the large-scale magnetospheric configuration. The main controlling factors were the following:

Saturn has a dipole-like intrinsic magnetic field with the dipole axis closely aligned with the axis of planetary rotation

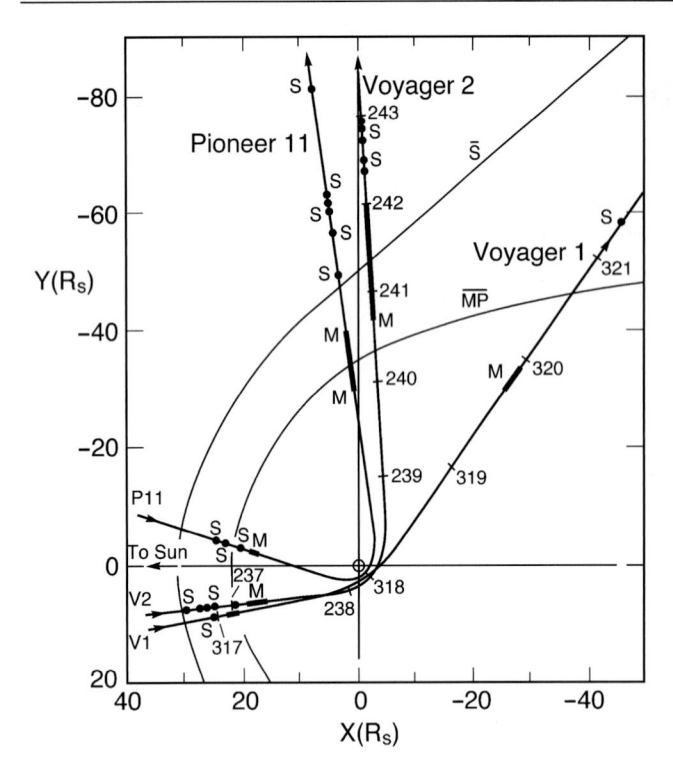

Fig. 9.1 Trajectories of the Pioneer 11, Voyager 1, and Voyager 2 spacecraft projected onto Saturn's equatorial plane. Observed bow shock (S) and magnetopause (MP) crossing are also shown (from Sittler et al. 1983)

(within about 1°). The equatorial magnetic field ($B_S \approx 2 \times 10^{-5}$ T) is very close to the equatorial magnetic field of Earth ($B_E \approx 3 \times 10^{-5}$ T) and about an order of magnitude smaller than that of Jupiter ($B_J \approx 4 \times 10^{-4}$ T).

The equatorial radius of Saturn ($R_S \approx 6.0 \times 10^4$ km) is comparable to the radius of Jupiter ($R_J \approx 7.1 \times 10^4$ km) and is about an order of magnitude larger that the radius of Earth ($R_E \approx 6.4 \times 10^3$ km). The rotation periods for Jupiter ($T_J \approx 10^h$) and Saturn ($T_S \approx 10.5^h$) are quite close, and they rotate about 2.5 times faster than Earth ($T_E \approx 24^h$). Centrifugal accelerations at the planetary equator, consequently, are comparable at Jupiter and Saturn, and they exceed that of Earth by about 1.5 orders of magnitude.

Most of the plasma in Saturn's magnetosphere is of internal origin, with the rings, the icy satellites and Titan identified as the primary plasma sources. It was recognized that the magnetospheric plasma is dominated by heavy ions (in the mass range of 10–20 amu), but due to the limited mass resolution of the plasma instrument the dominant heavy ion was not identified (both nitrogen and oxygen were suspected with the debate settling on oxygen during the 90s (Richardson et al. 1998)). The total internal mass source was estimated as $10^{27} - 10^{28.5}$ molecules per second (or $\sim 10^{1.5} - 10^{2.5}$ kg/s), a large uncertainty due to orbital and instrument limitations. In this respect the Saturnian system is again similar to Jupiter, where the Iogenic internal plasma source is about 10^{28} ($\sim 10^3$ kg/s) heavy molecules per second. This is a big difference from Earth, where under quiet conditions the ionospheric plasma source plays a relatively unimportant role in controlling the overall magnetospheric configuration.

Magnetic field lines are increasingly stretched as more and more plasma is added in the equatorial mass loading region. The increasing plasma content per unit flux tube area requires the transfer of more and more momentum from the rotating ionosphere to the increasingly stretched field line. However, the finite ionospheric conductivity limits this momentum transfer, and consequently, the angular velocity of the 'heavy' equatorial part of the magnetic field line exhibits an increasing corotation lag (Gleeson and Axford 1976; Hill 1979). Finally, the field line becomes so heavy that the ionosphere is unable to drag it around any more and corotation "breaks down" (Gleeson and Axford 1976; Hill 1979) near a radial distance of

$$r_0 = \left(\frac{4 M_S}{\mu_0 \eta_s \Omega_S^2} \right)^{1/4} \quad (9.1)$$

where M_S is the magnetic moment of Saturn, μ_0 is the magnetic permeability of vacuum, Ω_S is the angular velocity of the neutral thermosphere and η_s is the flux tube plasma content (per unit flux tube area) in the mass addition region. Expression (9.1) yields a corotation breakdown radius of $\sim 15 R_S$ assuming a flux tube content of $\sim 10^{-3}$ g/T/cm².

The solar wind encompassing Saturn's magnetosphere is two orders of magnitude more tenuous than it is at Earth orbit. The interplanetary magnetic field at Saturn is about an order of magnitude weaker than at Earth and its nominal direction is nearly azimuthal. Under these conditions the solar wind was thought to have a relatively weak control over the magnetospheric configuration.

A summary of the post-Voyager understanding of Saturn's magnetospheric configuration is shown in Fig. 9.2. We note that no direct plasma observations were available from the distant magnetotail and therefore the schematics focuses on the region within about 30 R_S.

9.1.2 Major Cassini Discoveries

The Cassini mission resulted in a number of exciting new discoveries. Some of these discoveries are noted in Fig. 9.3:

- During its approach to Saturn the Cassini spacecraft detected energetic (~ 100 keV/charge) magnetospheric ions (such as He$^+$ and O$^+$) upstream of the bow shock whenever the spacecraft was magnetically connected to Saturn's magnetosphere (Krimigis et al. 2009).

9 Saturn's Magnetospheric Configuration

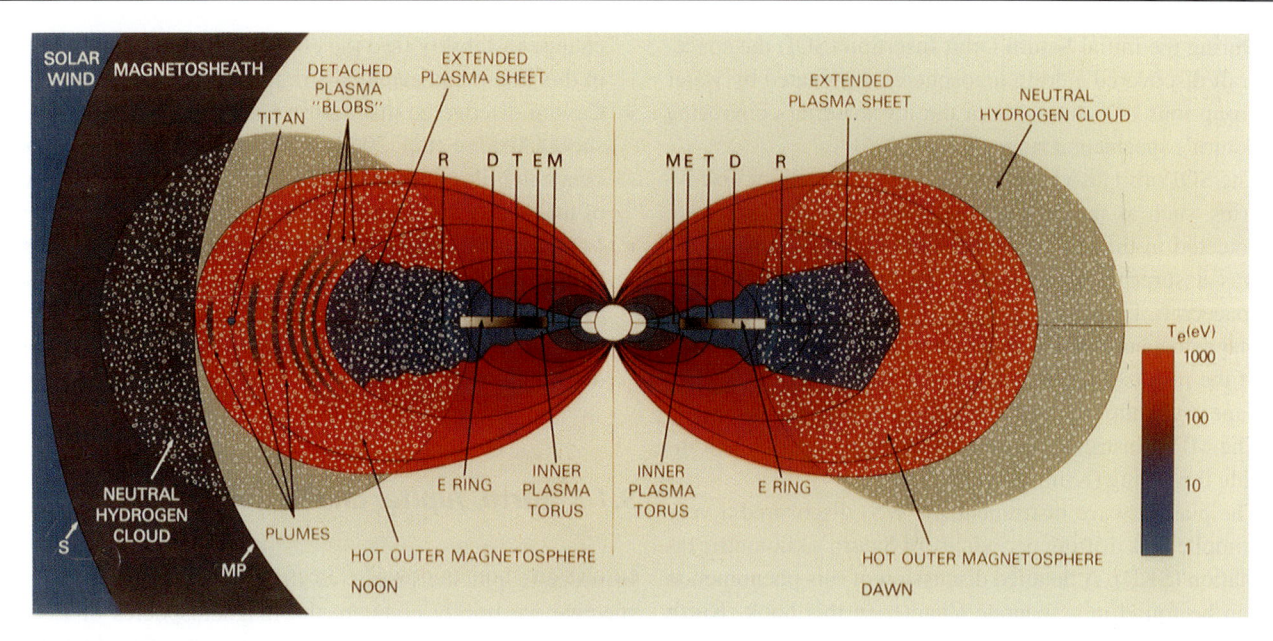

Fig. 9.2 Post-Voyager illustration of Saturn's magnetosphere. Cold regions are colored blue, regions of intermediate temperature are purple (*blue plus red*), and the hot regions are red. The satellite positions (M, E, T, D, and R for Mimas, Enceladus, Tethys, Dione, and Rhea, respectively), E ring (*gray shaped rectangular region*), neutral hydrogen cloud (*circular region with white dots*), and magnetopause boundary (MP) are displayed (from Sittler et al. 1983)

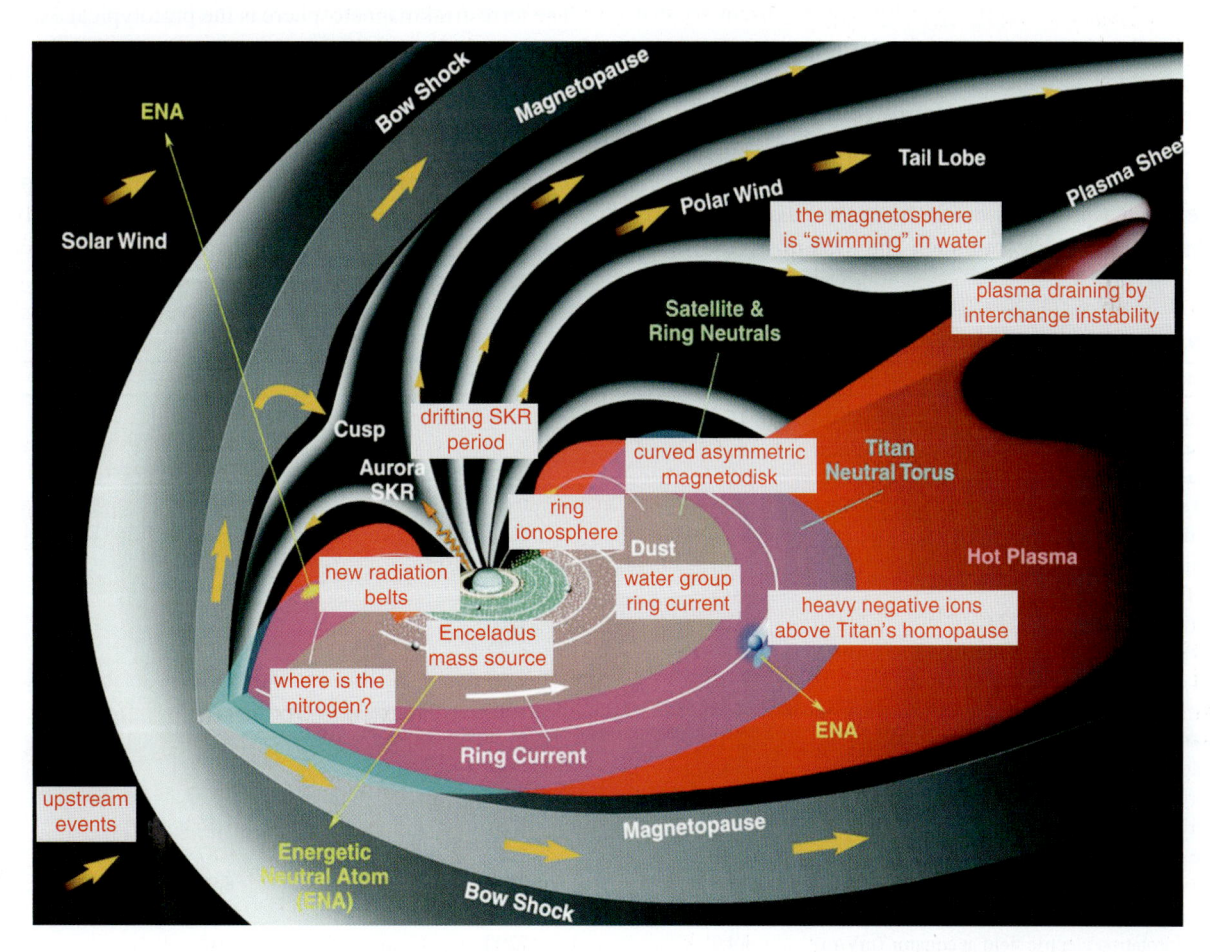

Fig. 9.3 Some of the exciting new discoveries made by the Cassini science teams during the Cassini Prime Mission. These discoveries are discussed in detail either in this Chapter or in the other magnetospheric Chapters of this book (Carbary et al. 2009, Kurth et al. 2009; Mauk et al. 2009) (background figure courtesy of the MIMI team)

- During the initial Saturn Orbit Insertion (SOI) the spacecraft discovered a "ring ionosphere" dominated by water group ions originating from the icy material composing Saturn's spectacular rings (Tokar 2006).
- The SOI orbit also brought some other important discoveries, such as the fact that very few nitrogen ions were detected in the magnetosphere (Smith et al. 2005). This was a surprise, since before the arrival of the Cassini spacecraft Titan was thought to a be a major source of magnetospheric nitrogen. For a more complete summary of the results obtained during SOI we refer to the review paper by André et al. (2008).
- The MIMI instrument discovered a new radiation belt inside the D ring (Krimigis et al. 2005).
- The plasma wave instrument (RPWS) discovered a very complicated, drifting periodicity of Saturn's kilometric radiation (SKR). A detailed discussion of this phenomenon can be found in a separate Chapter in this book (Kurth et al. 2009).
- Cassini discovered that Saturn's ring current is primarily composed of accelerated water group ions (Sittler et al. 2005, 2006, 2007; Sergis et al. 2007; Young et al. 2005). This is not surprising in light of the discovery that nearly all major magnetospheric plasma sources are water dominated. These include the rings, the icy satellites and especially Enceladus, which turned out to be the dominant mass source for Saturn's magnetospheric plasma (Waite et al. 2006).
- The plasma produced inside Saturn's magnetosphere is primarily transported through the system by the interchange instability (see the chapter by Mauk et al. (2009) in this book) and eventually drained by reconnection.
- Cassini discovered that Saturn has a bowl-shaped current sheet (Arridge et al. 2008) that is due to the interplay between the tilt of Saturn's rotational axis and the direction of the solar wind flow at Saturn's orbit.
- Very heavy negatively charged particles (most likely aerosols) were discovered above the homopause of Titan (Waite et al. 2007; Coates et al. 2007). This was a big surprise and the interpretation of this discovery is still under way.

9.1.3 Earth, Jupiter and Saturn

In order to put the results of the Cassini mission in perspective we briefly compare the magnetospheres of Earth, Jupiter and Saturn. Before Cassini the conventional wisdom was that strongly magnetized planets can have two distinct types of magnetospheres: solar wind controlled and internally controlled.

The terrestrial magnetosphere is the prototypical example of a solar wind controlled magnetosphere. As can be seen from the information presented in Table 9.1, Earth is a slow rotator (the corotational electric field is only dominant in the immediate vicinity of the planet) with a relatively small internal mass source originating from the high latitude ionosphere ($\sim$1 kg/s). The resulting solar wind interaction is usually referred to as the "Dungey cycle" (Dungey 1961). This process

Table 9.1 Comparison of physical parameters of Earth, Jupiter and Saturn

Parameter	Earth	Jupiter	Saturn
Heliocentric distance (AU)	1	5.2	9.5
Average IMF magnitude (nT)	4	1	0.5
Nominal Parker spiral angle from radial direction	45°	80°	85°
Typical solar wind ram pressure (nPa)	1.7	0.07	0.015
Equatorial radius	1 R_E = 6,371 km	1 R_J = 71,492 km = 11.2 R_E	1 R_S = 60,268 km = 9.45 R_E
Magnetic moment (T/m^3)	7.75×10^{15}	1.55×10^{20}	4.6×10^{18}
Dipole tilt	10.5°	10°	<1°
Equatorial magnetic field (μT)	31	420	20
Typical subsolar bow shock distance	$\sim$13 R_E	$\sim$70 R_J	$\sim$27 R_S
Typical subsolar magnetopause distance	$\sim$10 R_E	$\sim$50 R_J	$\sim$22 R_S
Solar wind transport time from subsolar bow shock to terminator	$\sim$3.5^m	$\sim$3.5^h	$\sim$1.1^h
Magnetospheric plasma source (kg/s)	$\sim$1	$\sim 10^3$	$\sim$300
Equatorial rotation period (hours)	23.934	9.925	10.53
Equatorial angular velocity (/s)	7.29×10^{-5}	1.76×10^{-4}	1.66×10^{-4}
Surface equatorial rotation velocity (km/s)	0.465	12.6	9.87
Surface equatorial gravitational acceleration (m/s^2)	9.78	24.8	8.96
Surface centrifugal acceleration at the equator (m/s^2)	0.034	2.22	1.62
Surface corotation electric field at equator (mV/m)	14.2	5290	197
Solar wind motional electric field (mV/m)	1.6	0.4	0.2
Radius where $E_{corot} = 0.1\, E_{sw}$	3 R_E	115 R_J	31 R_S

starts at the dayside magnetopause where reconnection between the planetary magnetic field lines and the southward component of the IMF can take place along an extended region. This reconnection creates a set of open magnetic field lines originating from the high-latitude ionosphere and extending to the free flowing solar wind. The open (interplanetary) ends of these field lines are moving with the ambient solar wind speed, while the ionospheric ends are convecting antisunward. These open field lines form the open flux magnetospheric tail lobes. Eventually, the open flux tubes close again by reconnecting in the plasma sheet. This process forms stretched closed flux tubes on the Earthward side of the tail reconnection line, which contract back toward the Earth and eventually flow to the dayside where the process can repeat. On the other side, "disconnected" field lines accelerate the tail plasma downstream and eventually "accommodate" it into the solar wind. The key feature of the "Dungey cycle" is therefore the magnetospheric convection controlled by reconnection.

Jupiter represents a prime example of a magnetosphere where the solar wind only plays a minor role. Jupiter is a fast rotator with a strong surface magnetic field (see Table 9.1). As a consequence, internal processes dominate the magnetosphere out to about $100 R_J$ and the solar wind interaction (the Dungey cycle) is only marginally important. The corotational electric field far exceeds the motional electric field in the solar wind, and Io produces about 10^3 kg/s plasma deep inside the magnetosphere. This plasma source adds considerable "new mass" to the corotating magnetic flux tubes and eventually stretches them outward.

When a mass-loaded heavy magnetic flux tube stretches beyond the corotation breakdown distance, the plasma will continue moving outward unless there is some process that inhibits this motion. On the dayside the magnetopause acts as a barrier and forces the magnetic field line to move along the inner boundary of the magnetosphere, thus forcing a corotation-like motion around the planet. On the nightside, however, the plasma can move without much resistance into the low pressure magnetotail. The magnetic field lines remain attached to the corotating ionosphere at one end and to the outward moving heavy equatorial plasma on the other end. Eventually the field line becomes so stretched and thin that a magnetic 'O' line is formed and a plasmoid is formed that can now freely move down the magnetotail. On the planetary side the newly shortened field line is 'shed' of its plasma content, and the magnetic stress pulls the equatorial part of the field line towards the planet, restoring the flow toward corotation. This process is called the "Vasyliunas cycle" (Vasyliūnas 1983) and it is shown in Fig. 9.4.

As can be seen from Table 9.1, Saturn falls somewhere between Earth and Jupiter. It is a fast rotator, but the equatorial magnetic field is comparable to that of Earth. The magnetospheric mass source is a factor of 3 smaller than that of Jupiter, and the corotation electric field is dominant inside a few tens of R_S. As a result of this intermediate parameter range Saturn's magnetosphere exhibits both a Dungey cycle and a Vasyliunas cycle at the same time (Badman and Cowley 2006, 2007). This fact makes the kronian magnetosphere even more fascinating and complex than the magnetospheres of Earth and Jupiter.

Badman and Cowley (Badman and Cowley 2006, 2007) pointed out that in the outer magnetosphere the rotational flux transport and the Dungey cycle are of comparable importance. Regions driven by planetary rotation should be dominated by heavy-ion plasmas originating from internal moon sources. The Dungey cycle layers should principally contain hot light ions originating from either the planet's ionosphere or the solar wind.

A conceptual drawing (based on Cowley et al. 2004) of Saturn's ionosphere-magnetosphere coupling and plasma

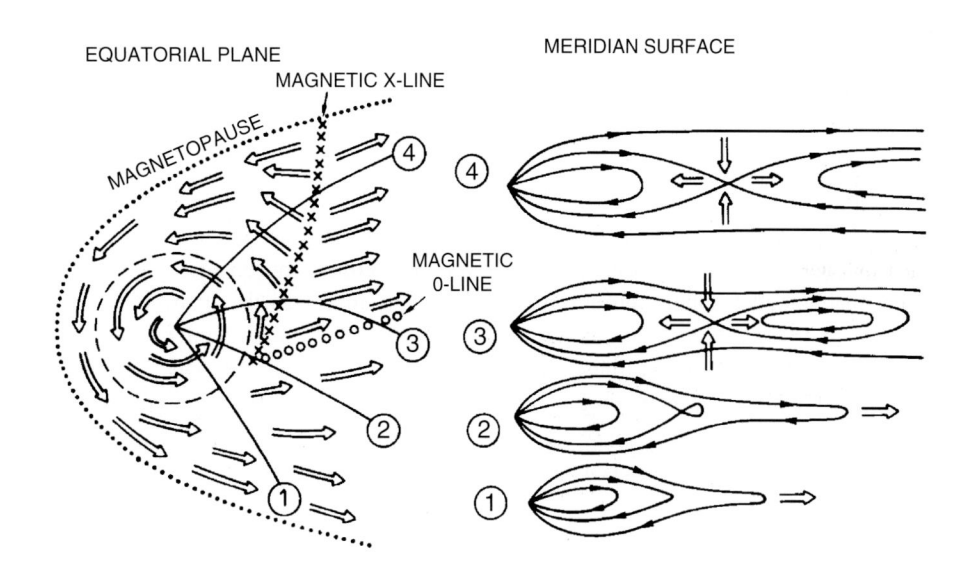

Fig. 9.4 Schematic representation of plasma flow in the equatorial plane (*left panel*) and of the associated magnetic field and plasma flow in a sequence of meridional cuts (*right panel*) (Vasyliūnas 1983)

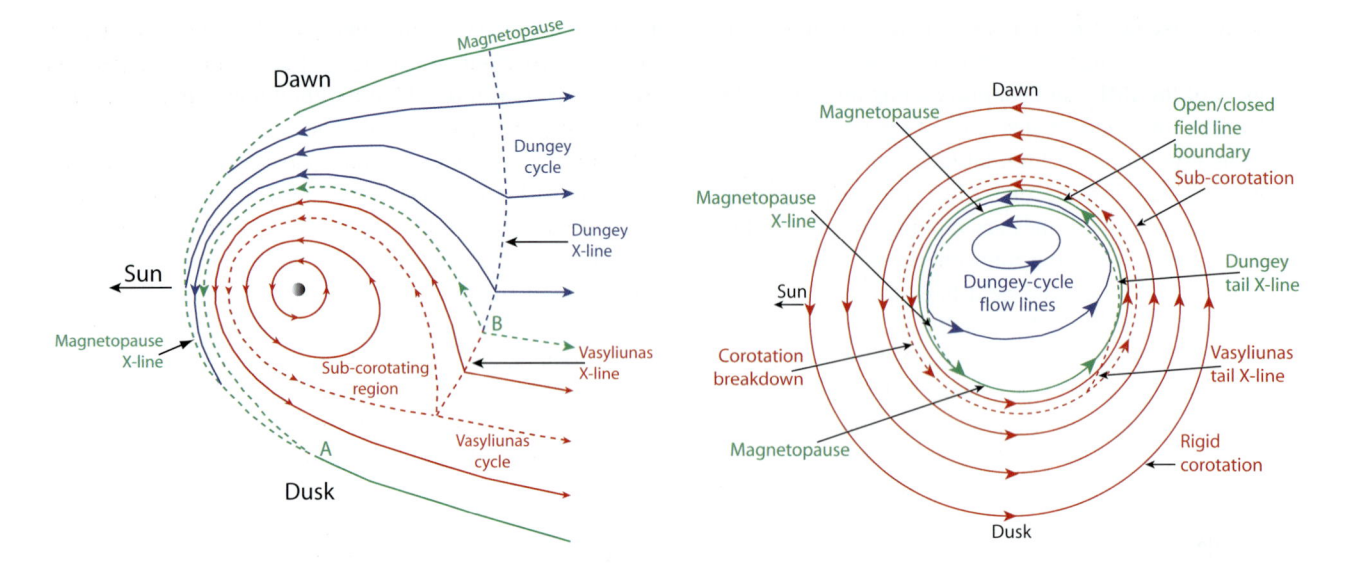

Fig. 9.5 Schematic representation of ionosphere-magnetosphere coupling at Saturn (Cowley et al. 2004). The left panel shows plasma convection in the equatorial plane of the magnetosphere and the right panel shows ionospheric convection in the northern high-latitude ionosphere. Three distinct convection regions can be distinguished: corotation, Vasyliunas-type plasmoid formation (Vasyliūnas 1983) and Dungey-type reconnection driven circulation (Dungey 1961)

circulation is shown in Fig. 9.5. There are three distinct plasma convection regions in the ionosphere and magnetosphere. Closest to the planet ($r \leq 3R_S$) the plasma corotates with the upper atmosphere. This region is shown by arrowed solid red circles in the ionosphere and arrowed solid red lines in the equatorial plane of the magnetosphere. Sub-corotation develops at larger radial distances as the magnetic field lines become increasingly mass loaded. The final breakdown of corotation is shown by the arrowed dashed red line both in the magnetosphere and in the ionosphere.

The second convection region starts at the corotation breakdown and it includes the Vasyliunas cycle. Beyond the corotation breakdown plasmoid formation "drains" magnetospheric plasma as suggested by Vasyliūnas (1983) (see Fig. 9.4). The associated X-line is shown by the dashed red line. In reality this is a highly intermittent process and numerical simulations indicate that on the dusk side small plasmoids containing plasma "blobs" are released from the stretched and tailward moving magnetic field lines. Eventually these plasma blobs evolve into a pinching of the field line and the Vasyliunas X-line is formed. Flux tubes that shed their plasma load in this way are "buoyant" in the centrifugal force and return to the inner magnetosphere through a process of flux-tube interchange described more fully in Mauk et al. (2009), thus completing the magnetic flux transport cycle.

The third distinct convection region is where the Dungey cycle (Dungey 1961) takes place. This process starts at the dayside magnetopause where reconnection between the planetary magnetic field lines and the northward component of the IMF can take place along an extended region (note that Saturn's magnetic moment is oppositely oriented as the magnetic moment of Earth). This reconnection creates a set of open magnetic field lines originating from the high-latitude ionosphere and extending to the free flowing solar wind. The open (interplanetary) ends of these field lines are moving with the ambient solar wind speed, while the ionospheric ends are convecting antisunward as shown in the right panel of Fig. 9.5 (solid blue lines with arrows). On the dusk side these open field lines form the open flux tail lobes and they stay above the equatorial plane at all times. Point 'A' is at the dusk-side magnetopause and it represents the end of the dayside reconnection line. At point 'B' the lobe field lines start reconnecting and this point marks the duskward end of the Dungey-type tail X-line. This X-line is marked with a dashed blue line in the left panel of Fig. 9.5. The Dungey cycle return flow takes place on the dawn side of the magnetosphere and it is compressed on the dayside due to the narrowing channel between the Vasyliunas cycle region and the magnetopause. The boundary between the Vasyliunas and Dungey cycle regions is marked by the arrowed green dashed line in the left panel of Fig. 9.5. Such Dungey cycle flux tubes, containing dominantly solar wind plasma, can also potentially mingle with flux tubes carrying inner magnetospheric plasma and may enter the inner magnetosphere via flux-tube interchange, as mentioned above.

9.2 Magnetic Field

9.2.1 Intrinsic Magnetic Field

The first in-situ observations of Saturn's intrinsic magnetic field were done during the Pioneer 11 flyby in 1979 (Smith et al. 1980; Acuña et al. 1980). Combined with the magnetic field observations by the two Voyager spacecraft (Connerney et al. 1982) the early measurements provided a surprisingly accurate picture of Saturn's intrinsic magnetic field. It can be described by a slightly displaced magnetic dipole that is very closely aligned with the planetary rotation axis (Smith et al. 1980, Connerney et al. 1982). Detailed studies of the Cassini observations basically confirmed this conclusion and added some additional details.

The modeling of intrinsic planetary magnetic fields has a rich heritage stretching back into the 19th century with pioneering work by Gauss. Within planetary magnetospheres there can usually be found a region where the magnetospheric volume currents are weak and the field can be considered to be approximately curl-free (this is the region where the effects of external current systems can be neglected). Under this approximation the field in this region can be written as the gradient of some magnetic scalar potential ($\mathbf{B} = -\nabla\psi$), usually using spherical harmonics, which satisfies Laplace's equation ($\nabla^2 \psi = 0$):

$$\psi(r,\theta,\phi) = a\sum_{n=0}^{\infty}\left(\frac{a}{r}\right)^{n+1}\sum_{m=0}^{n}P_n^m(\cos\theta)[g_n^m\cos(m\phi) + h_n^m\sin(m\phi)] \quad (9.2)$$

where a is the radius of the planet and r, θ and ϕ are spherical planetocentric coordinates. In expansion (9.2) the values g_n^m and h_n^m are referred to as the Gauss coefficients of the field model and indicate how strongly each 'mode' contributes to producing the total field. The values of n are described as the degree and m the order of the expansion. The choice of a Schmidt quasi-normalization allows one to directly compare the contributions from each term and assess how important each contribution is. To determine the Gauss coefficients, least squares methods are typically used to minimize the squared deviation between a model field described by Eq. (9.2) and a set of magnetometer observations.

For such a model, one can only be sure of a unique solution for ψ if measurements are available which completely cover a closed surface about the origin. Spacecraft trajectories constitute single curves in space and because of this the values of Gauss coefficients from such models are not unique. Cross-coupling and mutual dependence between the coefficients can occur. Whilst the model may provide an excellent fit to the data along a given trajectory, it might be very wrong away from that trajectory. Furthermore, terms of degree n scale as $r^{-(n+1)}$ and, consequently, a large number of spacecraft trajectories at various radial distances, longitudes and latitudes are required in order to obtain a good estimate of the most important Gauss coefficients.

The first three Gauss coefficients from $n = 1$ can be considered as three components of the magnetic dipole moment vector, one each along the X, Y and Z axes. The dipole magnetic moment is:

$$M_0 = \frac{4\pi a^3}{\mu_0}\sqrt{\left(g_1^0\right)^2 + \left(g_1^1\right)^2 + \left(h_1^1\right)^2} \quad (9.3)$$

where μ_0 is the magnetic permeability of vacuum. In addition, the colatitude of the magnetic pole can be expressed as:

$$\lambda_0 = \arccos\left(\frac{g_1^0}{\sqrt{\left(g_1^0\right)^2 + \left(g_1^1\right)^2 + \left(h_1^1\right)^2}}\right) \quad (9.4)$$

A particular property of intrinsic magnetic fields with large quadrupolar ($n = 2$) components is that the effective dipole is shifted vertically out of the equatorial plane by a distance of $z_0 = a\left(g_2^0\right)/\left(2g_1^0\right)$. This offset has been observed near the equator close to Saturn and it must be accounted for in modeling.

In order to calculate the longitudes used in Eq. (9.2) one needs an accurate rotation rate for the planet. The rotation rates of terrestrial planets can readily be obtained by tracking surface features as the planet rotates. By their very nature, the giant planets do not have any such observable solid surfaces and so other methods must be used to measure or infer the rotation rate of the interior. At Jupiter it was found that decametric radio emissions from the auroral regions were strongly modulated by the rotation of its tilted magnetic field and hence are strongly tied to the rotation rate of the deep interior of the planet (Carr et al. 1983). Thus the periodicity of decametric radio emissions provides a measure of Jupiter's internal rotation rate. Saturn's kilometric radiation (SKR) was also observed to have a periodicity close to that of cloud features and it was suggested that this also represented the rotation rate of the deep interior (Carr et al. 1981). The SKR period was used to construct a longitude system (Seidelmann et al. 2002 and Desch and Kaiser 1981) based on the (Carr et al. 1981) period of $10^h39^m22.4^s \pm 7^s$.

The accuracy of this rotation period does not pose a problem for the modeling of internal fields from a single spacecraft pass over a few days, or several spacecraft passes separated by a few months. The longitudinal error, or smear, produced by this error only amounts to several tens of degrees. But when attempting to combine datasets covering periods of several years the error becomes unacceptably large. In such cases one must ignore the longitudinal dependence and seek solutions to a zonal model. In this approximation we integrate Eq. (9.2) over the azimuth angle (ϕ) and only derive fits for the g_n^0 coefficients, effectively forcing the magnetic energy in the observations into purely zonal ($m = 0$) coefficients and producing an axisymmetric magnetic field model.

Two studies have produced non-axisymmetric field models using Pioneer and Voyager observations. Connerney et al. (1982) combined magnetometer data from Voyagers 1 and 2, the coefficients of which are presented in Table 9.2. Giampieri and Dougherty (2004) took a different approach. The Pioneer and Voyager datasets were combined and fitted with a non-axisymmetric field model but the rotation rate (and hence the longitude for each sample) was left as a free parameter. They systematically varied the rotation rate to identify the rotation rate with the smallest RMS deviation between the model and the data. This analysis yielded a rotation period consistent with the radio measurements, but the error was reduced to ± 2.4s from the ± 7s obtained from SKR. Comparing the models of Connerney et al. (1982) and Giampieri and Dougherty (2004) in Table 9.3 one can see that the dipole moments agree to within about 1%, the northward offsets to within 20%, and the dipole tilts are all less than one degree. One can easily see that the dipole moments and northward offsets from the non-axisymmetric models are in approximate agreement with all of the axisymmetric (zonal) models developed from the Pioneer and Voyager data (see Table 9.2).

Dougherty et al. (2005) fitted a zonal spherical harmonic model to the Cassini SOI magnetometer data, accounting for the external magnetospheric field with a simple symmetric disc model (Giampieri and Dougherty 2004). The results of their inversion are presented in the second column of Table 9.3 and show very similar values

Table 9.2 Gauss coefficients and internal field characteristics for zonal magnetic field models

	Z_3 (Connerney et al. 1982)	$P_{11}84$ (Davis and Smith 1986)	SPV (Davis and Smith 1990)	ZMP (Beard and Gast 1987)	Cassini SOI (Dougherty et al. 2005)	Cassini Zonal
g_1^0 (nT)	21535	21140	21160	21431	21084	21162
g_2^0 (nT)	1642	1600	1560	2403	1544	1514
g_3^0 (nT)	2743	2260	2320	2173	2150	2283
M_0 (10^{34} A m^2)	4.714	4.628	4.632	4.691	4.615	4.633
z_0 (km)	2298	2280	2220	3379	2207	2156

Table 9.3 Gauss coefficients and internal field characteristics for non-axisymmetric magnetic field models

	C_{82} (Connerney et al. 1982)	GD_{04} (Giampieri and Dougherty 2004)	B_1[a]	B_2[b]	B_3[c]	B_4[d]
g_1^0 (nT)	21439	21232	21171	21268	21278	21246
g_1^1 (nT)	−143	23	2	7	−41	−12
h_1^1 (nT)	143	60	−2	1	−43	−49
g_2^0 (nT)	1882	1563	1584	1585	1606	1492
g_2^1 (nT)	−515	−132	−37	−42	−70	−204
g_2^2 (nT)	500	5	−17	−17	37	57
h_2^1 (nT)	−433	51	111	−69	5	−27
h_2^2 (nT)	−36	−112	5	−14	16	−26
g_3^0 (nT)		2821	2240	2178	2245	2651
g_3^1 (nT)		−209	−72	−29	−205	269
g_3^2 (nT)		282	28	3	49	−296
g_3^3 (nT)		−156	5	0	22	−2
h_3^1 (nT)		1365	33	−42	52	−401
h_3^2 (nT)		−80	−75	29	−92	−99
h_3^3 (nT)		192	2	6	33	−67
RMS (%)	−		2.54	2.61	1.95	1.75
M_0 (10^{34} A m^2)	4.634	4.693	4.656	4.658	4.651	4.648
Dipole tilt (°)	0.5	0.2	0.008	0.02	0.2	0.1
z_0 (km)	2645	2218	2255	2246	2274	2116

[a] used the rotation rate given in Giampieri et al. (2006); [b] used the rotation rate given in Anderson and Schubert (2007); [c] used the rotation rate given in Kurth et al. (2007); [d] used the rotation rate given in Andrews et al. (2008).

compared to the Pioneer and Voyager inversions. This suggests that no significant secular change occurred between the Pioneer/Voyager and Cassini SOI epochs.

The most recent work on Saturn's internal magnetic field uses all the Cassini data to date and hence covers a much wider period of time than previously published studies. The axisymmetric model is in good agreement with the existing models. To treat the rotation rate issue, four different fits were produced using four different longitude systems which are based on different estimates of the rotation rate. Two of these rotation rates are based on a constant period determined from an analysis of magnetometer data (Giampieri et al. 2006) and from a study of Saturn's gravity field (Anderson and Schubert 2007). The other two are variable period systems, designed to reflect the observations of a drift in the SKR period (Kurth et al. 2007) and a drift in the period of magnetic fields in the inner and middle magnetospheres (Andrews et al. 2008). It was found that the use of these updated models greatly improve the residuals from the fit. The dipole moment, northward offset and the dipole tilt of $\sim 0.1°$ are all in good agreement with Voyager and Pioneer models. The formal errors of the non-axisymmetric coefficients, as calculated from the inversion, are smaller than the coefficients themselves, suggesting that they are well-determined. However the significance of these terms is questionable given the uncertainties in the rotation period.

9.2.2 The Magnetodisk

Planetary magnetodisks are formed when significant ring current is present over an extended region, and the dipole field becomes too weak to maintain stress balance and the current system needs to intensify in order to balance the mechanical stresses.

At Jupiter and Saturn, internal plasma sources play a significant role in this process. At Jupiter the synchronous orbit is at $2.3R_J$, while at Saturn it is even closer to the planet at $1.8R_S$. The main magnetospheric plasma sources at Jupiter and Saturn are Io (orbiting at $5.9R_J$ and producing $\sim 10^3$ kg/s new plasma) and Enceladus (orbiting at $3.9R_S$ and producing $\sim 10^{2.5}$ kg/s new plasma). The centrifugal force acting on the newly produced plasma at Io and Enceladus exceeds the gravitational force by a large factor; therefore only magnetic forces can confine the plasma. If the equatorial quasi-dipolar field cannot maintain stress balance with the plasma stresses, the field will become more and more stretched. Most of the plasma is confined to the equatorial region due to the centrifugal forces; therefore the magnetic field lines will be more and more stretched near the rotational equator as more mass is added. This magnetically confined, centrifugally outward driven plasma and the corresponding highly stretched closed magnetic field lines form the magnetodisk. We note that pressure gradient and anisotropy forces can also generate a magnetodisk. Detailed discussion of these processes can be found for instance in Russell (2004) and Vasyliūnas (2008).

The dayside magnetic field configuration of the terrestrial and Jovian and Saturnian magnetospheres are radically different. On the dayside at Earth the field has a quasi-dipolar form, not that different from the dipolar field produced by a dynamo in Earth's interior. The rather modest distortion of this field away from a dipole is produced by the Chapman-Ferraro current at the dayside magnetopause and the azimuthal electric current called the ring current. In contrast, the dayside configuration in Jupiter's magnetosphere is highly distorted. The field is quasi-dipolar out to a distance of approximately 20 Jovian radii, beyond which the field stretches out into a disk-like configuration called the magnetodisk.

Voyager and Pioneer observations of Saturn's dayside magnetic field configuration revealed a quasi-dipolar magnetosphere not too dissimilar from the terrestrial magnetosphere (Smith et al. 1980; Connerney et al. 1983). Connerney et al. (1983) concluded that Saturn's magnetosphere did not possess a magnetodisk despite the magnetosphere rotating rapidly and having internal sources of plasma (all ingredients suspected of generating the magnetodisk at Jupiter).

Using Cassini observations Arridge et al. (2007) examined magnetometer data from Saturn's dawn flank magnetosphere and found a field configuration similar to that on Jupiter's dawn flank. They suggested that this was actually Saturn's magnetodisk where the dayside quasi-dipolar configuration was a consequence of the smaller size of the magnetosphere and thus suppressed the disk on the dayside. In a later paper (Arridge et al. 2008) they surveyed the magnetometer data to determine where this magnetodisk-like field configuration was observed. They found evidence for a magnetodisk not only on the nightside and dawn flanks where it was expected to exist, but also on the dayside.

The survey showed that the dayside magnetodisk only forms during intervals when the solar wind pressure is low and hence when the magnetosphere is expanded. Specifically, Arridge et al. (2008) found that this highly stretched and distorted magnetic field configuration was only observed on the dayside when the subsolar standoff distance of the magnetopause was larger than $23R_S$. They noted that the dayside magnetosphere was compressed to less than $23R_S$ during all of the Pioneer 11 and Voyager dayside flybys, thus producing a quasi-dipolar dayside. The left panel of Fig. 9.6 illustrates this solar wind pressure-dependent distortion of Saturn's dayside magnetosphere.

The Voyager and Pioneer flybys occurred during near-equinox conditions and so were not well-placed to investigate

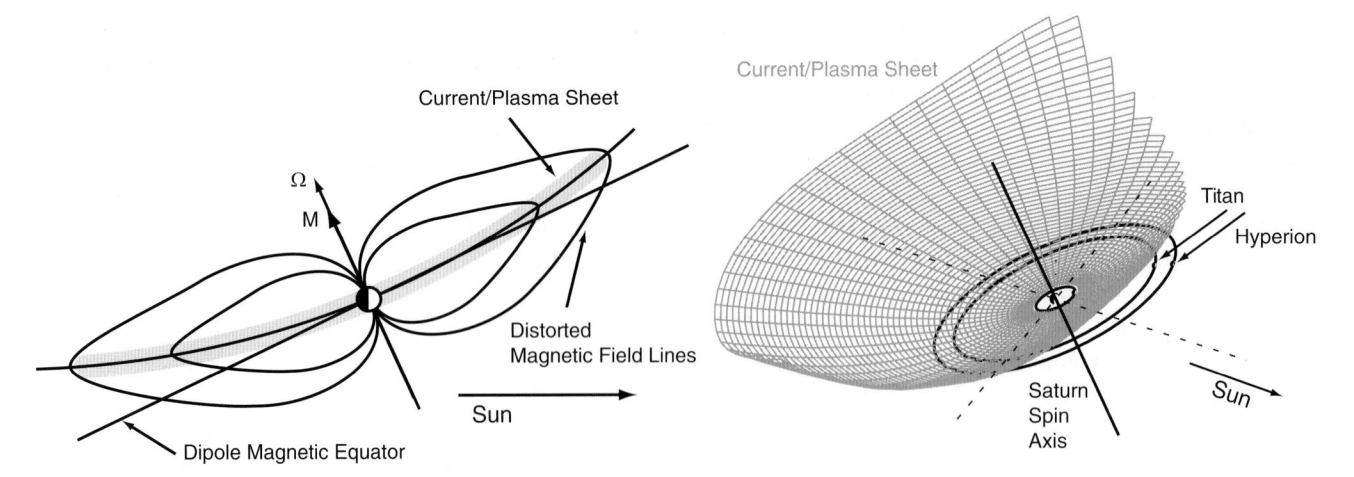

Fig. 9.6 Schematics illustrating the distortion of Saturn's magnetosphere. (*left*) The distorted plasma/current sheet and magnetic field lines in the noon-midnight meridian. (*right*) A three-dimensional view of this distortion and the resulting bowl-shaped current sheet. The orbits of Titan and Hyperion are included showing that they are underneath the sheet (from Arridge et al. 2008)

seasonal magnetospheric effects. However, Cassini's arrival shortly after solstice provided an excellent opportunity. An unexpected seasonal effect that has been first observed by Cassini (Arridge et al. 2008), is illustrated in the right panel of Fig. 9.6. The dayside magnetospheric magnetic field and the plasma sheet surface are distorted so they lie to the north of the equatorial plane, i.e., completely the opposite of the behavior inferred at Jupiter. This northward warping was observed over the noon, dawn, and midnight sectors that were surveyed (Arridge et al. 2008). The dusk sector has yet to be examined for this effect. These observations led Arridge et al. (2008) to describe the shape of the magnetosphere in terms of a bowl-shape, where the whole magnetosphere is bent above the equator beyond a characteristic distance called the hinging distance. The hinging distance was estimated to be between $15R_S$ and $30R_S$.

9.2.3 Empirical Magnetic Field Models

Planetary scientists frequently need quantitative description of the strength and the orientation of the magnetic field to understand the processes involved in magnetospheric and particle dynamics. As the spatial coverage of in situ magnetic field measurements in regions of interest is seldom complete, field models are constructed to provide the values of magnetic field globally. These models serve many purposes. For example, magnetic field models are frequently used in studies of single particle dynamics (drift and bounce of particles on stretched field lines, radial and pitch angle diffusion of particles in the magnetosphere, acceleration processes in the inner magnetosphere, etc). The models also find application in understanding the magnetospheric reconnection geometry in the magnetotail to assess the location and frequency of magnetic substorms and storms in the magnetosphere. Another area where empirical models are vital is in field line mapping studies where ionospheric phenomena such as aurorae and satellite footprints need to be related to the source populations in the equatorial region of the magnetosphere.

Empirical models are constructed by specifying scalar, vector or Euler potentials in various regions of the magnetosphere. The models are constrained by ensuring that they satisfy fundamental conservation laws such as Maxwell's equations and stress balance in the form of a momentum equation. The first global model of Saturn's magnetosphere was constructed by Maurice et al. (1996) who followed the approach of Beard (1960) where stress balance between the solar wind dynamic pressure and the magnetospheric magnetic field pressure determines the shape of the magnetopause. Maurice et al. (1996) used the GSFC Z3 model of the internal field (Connerney et al. 1984) to prescribe the internal field of Saturn and the Connerney et al. (1983) current sheet model to prescribe the contribution of the Saturnian plasma sheet. The field from the Saturnian magnetopause currents was first computed by constructing a wire frame model of the magnetopause current system and then using Biot-Savart integration. The computed field was next fitted to a spherical harmonic series for faster computation. Because of the choice of the harmonic series to express the field at the magnetopause (spherical rather than Cartesian or cylindrical harmonics which are better suited for currents arising from a paraboloidal surface), the Maurice et al. (1996) model is not applicable to the field of the stretched magnetotail. The model also lacks a magnetotail current system, does not include the hinging of the current sheet caused by the solar wind forcing and lacks bend-back of the field caused by corotation enforcement currents.

9 Saturn's Magnetospheric Configuration

Recently, Alexeev et al. (2006) have used the internal field model of Dougherty et al. (2005) and the Connerney et al. (1983) formulation of a thin current sheet with finite inner and outer boundaries to construct a new global magnetic field model. They confined the resulting field by a paraboloid magnetopause current. The model was fitted to data obtained from the SOI period. This model is more sophisticated and accurate than earlier models, but it still cannot account for the warping of the field resulting from the bowl shape of the current sheet, and it lacks radial currents that enforce corotation on the magnetospheric plasma.

New models of Saturn's magnetospheric field were constructed by using the general deformation technique successfully used by Tsyganenko (1998, 2002a, 2002b) to model the Earth's magnetosphere. First, new modules were constructed to describe the ring current and the magnetotail fields. Next, they shielded the field interior to the magnetopause by using harmonic series to represent the magnetopause field. The model includes the bowl shape of the current sheet by using a description of a stretch deformation. The models (illustrated in Fig. 9.7) are based on data from Voyager and the first 25 Cassini orbits. A comparison of the calculated field with observations shows a fairly good agreement (see Fig. 9.8) but also reveals systematic differences. The models, however, do not include the tilt of the current sheet (which produces spin periodicities in the data) and the sweep-back of the field lines.

Empirical magnetic field models have achieved a high level of sophistication. In spite of their limitations, they are a useful tool for investigating Saturn's complex

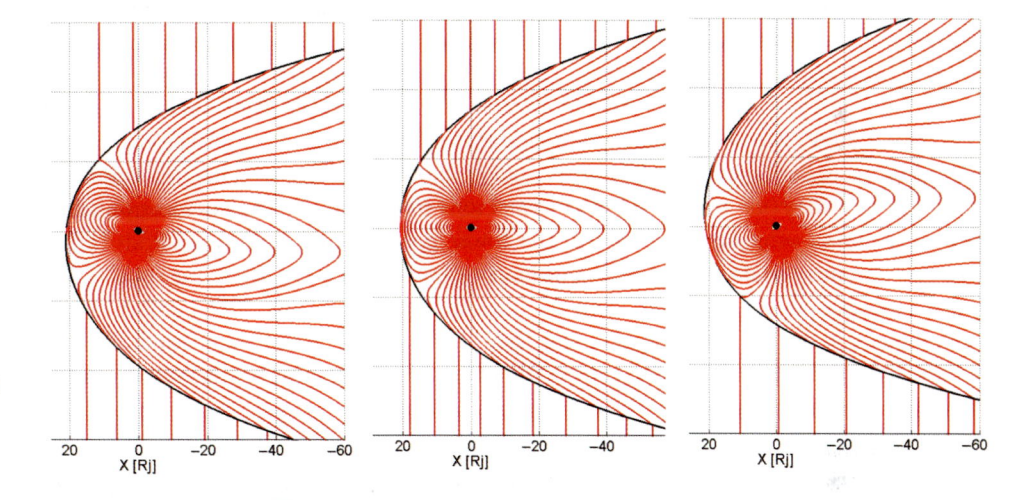

Fig. 9.7 Representative field lines from the latest field model of Saturn's magnetospheric field constructed from Voyager and Cassini data sets. Shown are field lines in the noon-midnight meridian for three situations of the dipole tilt $-26°$, $0°$ and $26°$

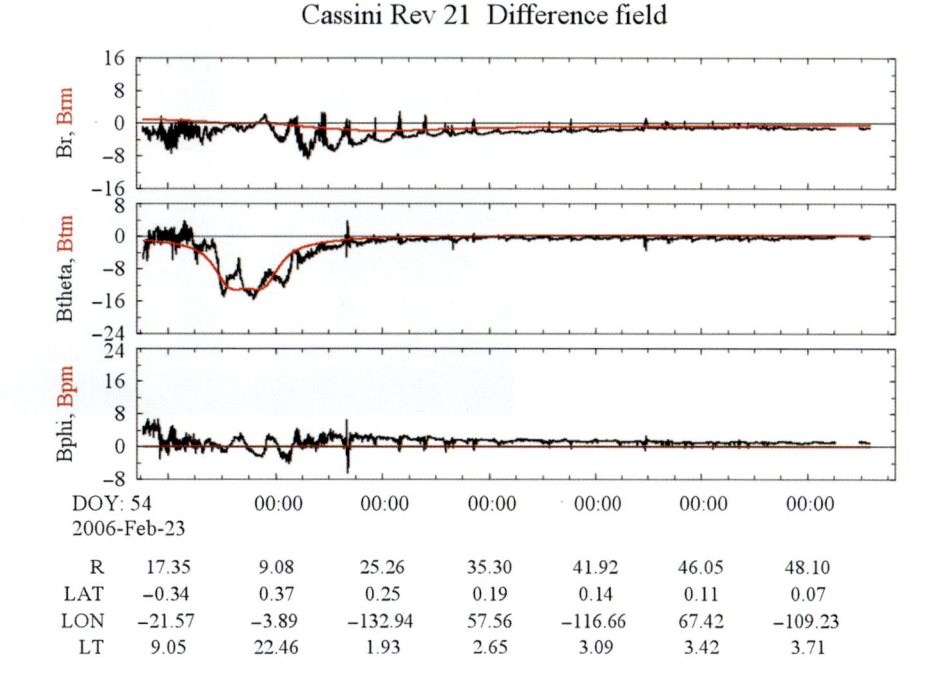

Fig. 9.8 The difference field from Rev 21 of Cassini (*black traces*) and the new model field (*red traces*). The new model predicts the average field quite accurately but does not explain the 10 h periodicity

magnetosphere. In combination with observations and numerical simulation they significantly contribute to our improving understanding of the Saturnian system.

9.3 Plasma Sources and Sinks

One of the fundamental questions of magnetospheric physics is the sources of the plasma that populates a magnetosphere. At Earth, many years of observational and theoretical work have demonstrated that there are two significant sources: the solar wind and the Earth's upper atmosphere. The mechanisms for entry and the relative importance of each source are still hotly debated, but there are no additional contenders of any significance. At Saturn, by contrast, there is a rich set of possible plasma sources: the solar wind, Saturn's ionosphere, Titan, the rings, and the icy satellites. One of the prime objectives of the Cassini mission was to assess the evidence for these various sources and their relative importance.

Figure 9.9 shows two energy-time spectrograms from the Cassini Plasma Spectrometer (CAPS) of the energy flux of the thermal electrons (*top panel*) and the thermal ions (*bottom panel*) during orbit insertion (SOI). The lower panel shows two dominant peaks in the E/q spectrum, corresponding to co-moving populations of H^+ and W^+ ions (defined as a combination of O^+, OH^+, H_2O^+, and H_3O^+). There is a clear inward gradient in the ion composition beginning near 9 R_S inbound and 8 R_S outbound, where the ratio of the water group ions to the hydrogen ions increases substantially Young et al. 2005. The region between $L \sim 5$ and $L \sim 9$ is sometimes called the inner plasmasphere (Sittler et al. 2005; Young et al. 2005).

Figure 9.10 shows the CAPS plasma parameters for the inbound portion of SOI (Sittler et al. 2005). Within the inner magnetosphere the speed of the plasma increases (panel 3 of Fig. 9.10) to near corotation, and there is a sharp increase in the densities inside $L \sim 9$ (panel 1 of Fig. 9.10). Young et al. (2005) and Rymer et al. (2007) confirm the Voyager observations (Sittler et al. 1983) that there are two electron populations in the inner plasma source region (5 to 9 R_S) (panel 4 of Fig. 9.10). The density ratio of the cold electrons (<20 keV) to the hot electrons (>100 keV) is more than one order of magnitude throughout most of this region (Young et al. 2005). The energy-time spectrogram in the upper panel of Fig. 9.9 shows that the energy of the cold electrons approximately tracks the proton corotation energy (Young et al. 2005). The average energy of the hot electron component (100 eV to >10 keV) increases with decreasing L value, consistent with the near-adiabatic inward transport (Rymer et al. 2007). The hot electrons drop out inside $L \sim 5$, due to collisions with the neutrals in Saturn's neutral cloud or losses to the E-ring (Rymer et al. 2007; Young et al. 2005).

There are a number of observational clues to the origin of a plasma population: the spatial distribution, the mass composition, the energy distribution, and the angular distribution.

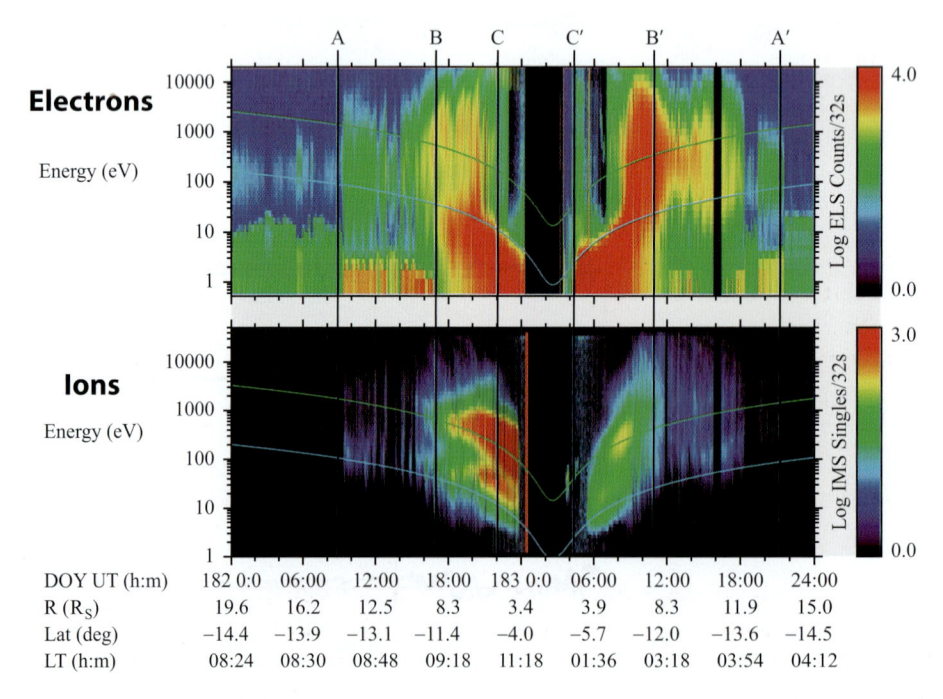

Fig. 9.9 Color-coded electron and ion count rates (proportional to energy flux) from the ELS (*top*) and the IMS/Singles (*bottom*), respectively, during Cassini's first pass through Saturn's magnetosphere (from Young et al. 2005). The lower panel shows two dominant peaks in the E/q spectrum, corresponding to co-moving populations of H^+ and W^+ ions. The curved lines superimposed on the plots give the energy corresponding to the full corotation velocity for O^+ (*upper curves in both panels*) and H^+ (*lower curves*)

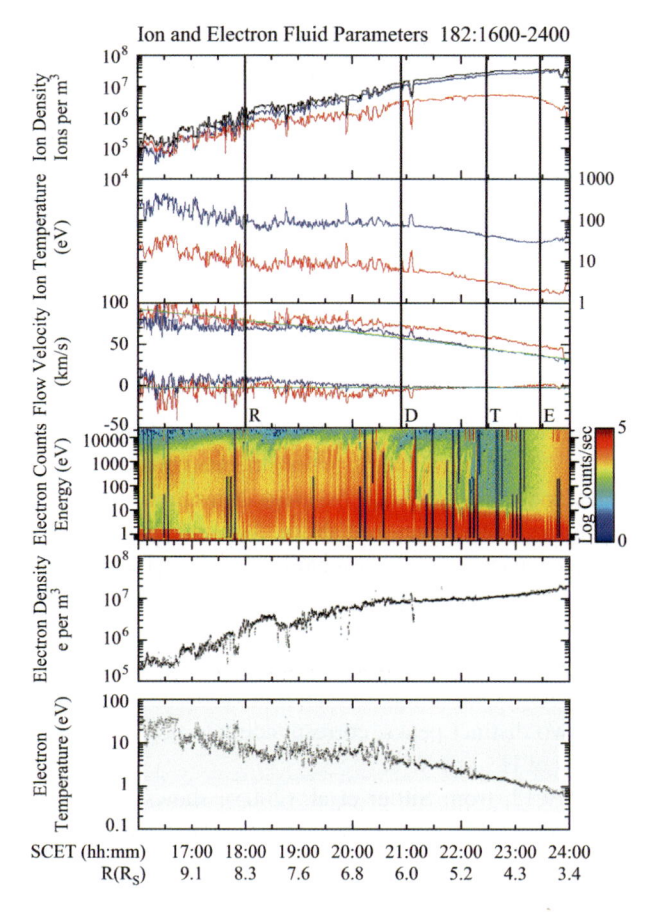

Fig. 9.10 An overview of the plasma parameters during Cassini orbit insertion on June 30, 2004: CAPS ion fluid parameters (*top three panels*) for protons (*red*) and water group ions (*blue*), electron energy spectrogram (*panel 4*), and electron densities (*panel 5*) and electron temperatures (*panel 6*) derived from CAPS/ELS using a Maxwellian fit to flux versus energy below 100 eV. Vertical lines mark the times when Cassini crossed the L-shells of Rhea, Dione, Tethys and Enceladus (from Sittler et al. 2005)

However, telltale signatures of origin can be modified or obscured by physical processes such as collisions with neutral material (gas and dust). At Saturn there is strong evidence that neutral material significantly interacts with the magnetospheric plasma, so we need to view the plasma as not completely collisionless (cf. Delamere et al. 2007). In addition, when molecular species are involved, as they are at Saturn, chemistry (e.g., gas-phase chemistry, photodissociation, etc.) can also introduce complications, and these effects need to be borne in mind when assessing the evidence for various sources.

Magnetospheric plasma primarily originates from a complex region in the inner magnetosphere where plasma is in constant motion, continually being created from sources in the atmosphere/ionosphere of the rings, the inner icy satellites, even the planet itself. Inside $L = 10$, there are plasma boundaries characterized by changes in the ion composition and in the bulk plasma properties. The most prominent ion components in Saturn's inner magnetosphere are the hydrogen ions (H^+) and the water group ions (W^+) (Krimigis et al. 2005; Sittler et al. 2005; Young et al. 2005).

We now consider Cassini evidence regarding the relative importance of the various plasma sources.

9.3.1 Rings (<3R_S)

Deep inside the inner plasma source region, ion measurements by INMS (Waite et al. 2005) and CAPS (Young et al. 2005) and high electron densities measured by the RPWS (Gurnett et al. 2005) revealed the existence of a tenuous plasma layer in the vicinity of Saturn's main rings. The electron density varies spatially in this region by more than an order of magnitude, and the electron temperature is only a few eV (Wahlund et al. 2005). The RPWS electron densities reach a peak of >100/cm^3 near the outer edge of the A ring and then decrease rapidly inside 2.2 R_S (Gurnett et al. 2005). At the same time, the electron parameters derived from the Langmuir probe show an order-of-magnitude drop in the electron temperature that strongly correlates with the order-of-magnitude electron density increase observed by RPWS (Wahlund et al. 2005). Wahlund et al. (2005) propose that the low plasma densities observed inside 2.2 R_S (Gurnett et al. 2005) are the result of absorption of the plasma by the ring particles and invoke the density increase inside the Cassini Division where ring particles have a lower density as a supporting argument. Gurnett et al. (2005) suggest that the deep electron density minimum (0.03/cm^3) at 1.7 R_S occurs because this is the location of synchronous orbit. Over the rings, Moncuquet et al. (2005) found that the cold electron temperature is ~1.5 eV over the G-ring at 2.8 R_S, dropping to ~0.5 eV at ~0.5 R_S when Cassini passes through the ring plane.

The ion composition of the tenuous plasma layer located directly over the A and B rings consists of O^+ and O_2^+ (Tokar et al. 2005; Waite et al. 2005; Young et al. 2005). The temperatures of these heavy ions drop to a minimum near synchronous orbit (~0.5 eV for O^+ and ~0.1 eV for O_2^+) and increase with increasing radial distance from Saturn (Tokar et al. 2006). Significant O_2^+ is also detected outside the main rings near the F ring (Tokar et al. 2006). There is a sharp increase in the ion and electron densities at ~1.85 R_S with the ion densities peaking at ~4/cm^3 over the B-ring (Tokar et al. 2006). Figure 9.11, from Tokar et al. (2005), shows enhanced ion fluxes consistent with the presence of O^+ and O_2^+, likely produced by UV photosputtering of the icy rings, with subsequent photoionization of the O_2. This process has been modeled by Johnson et al. (2006), who also showed that scattering and dissociation can populate the magnetosphere with O_2^+ beyond the main rings.

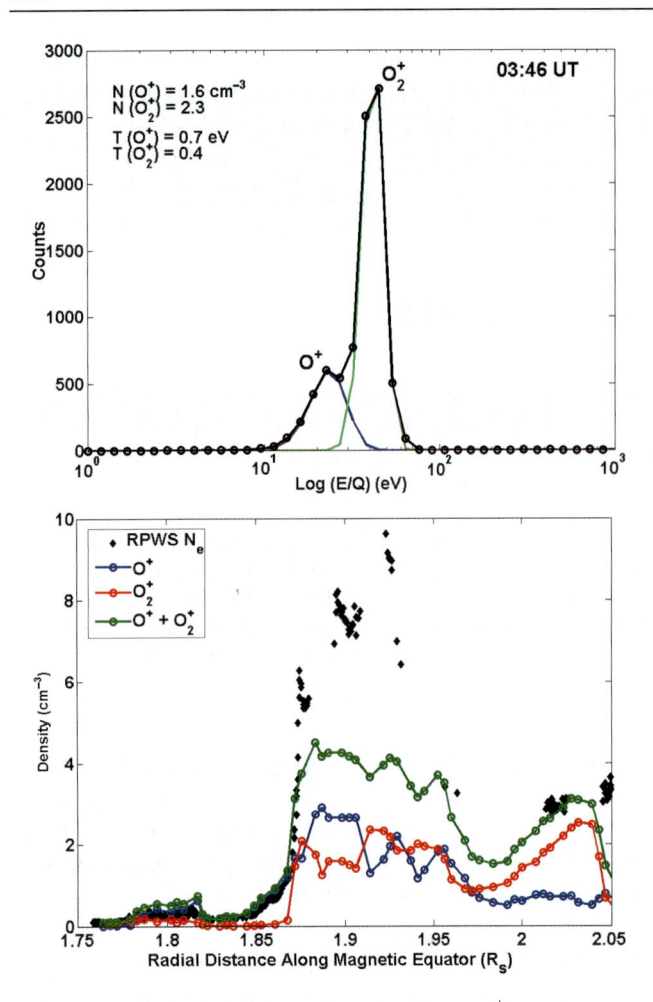

Fig. 9.11 *Top*: CAPS/IMS observations of O^+ and O_2^+ over the main rings during SOI. The corotation flow velocity was assumed. *Bottom*: Derived O^+ and O_2^+ densities over the main rings, compared with the RPWS-derived electron density (from Tokar et al. 2005)

Beyond the A ring, photo-sputtering is supplemented, indeed dominated, by charged-particle sputtering by the magnetospheric plasma. While sputtering can produce ions directly, the most likely outcome of an ion striking an icy surface is the liberation of neutral water molecules, with a small fraction of dissociation products (e.g., O_2^+, H_2^+). At one time, sputtering from icy satellites and E-ring particles was thought to be the source of the extended neutral atmosphere of water molecules and their dissociation products observed by the Hubble Space Telescope (e.g., Shemansky et al. 1993) and by Cassini (e.g., Esposito et al. 2005). Ionization of this material, by solar photons or by electron impact is indeed the primary source of plasma in Saturn's magnetosphere. However, calculations of the sputter flux gave rates significantly too low to account for the observed neutral cloud (e.g., Shi et al. 1995; Jurac et al. 2001). Recently, Johnson et al. 2008 have recalculated sputtering rates on ice grains and icy satellites, using plasma ion properties derived from Cassini CAPS observations (Sittler et al. 2006, 2007, 2008). While they still find sputtering to be a small contribution to the total neutral gas supply rate compared to the supply attributed to the gas and ice plumes observed to be emanating from Enceladus' southern polar region, it nonetheless has a significant effect on the lifetime of the small E-ring grains.

9.3.2 Icy Satellites (3R$_S$ to 6R$_S$)

The overwhelming evidence from Cassini observations is that the plasma of the inner magnetosphere was actually produced by local ionization of gases liberated from the icy satellites and the rings, with the rings determined to be of lesser importance as discussed above. As previously seen by the Voyager spacecraft (e.g., Richardson (1986)), Cassini has found inner magnetospheric plasma to consist of two dominant ion components: H^+ and water-group ions (e.g., Young et al. 2005). Figure 9.9, from Young et al. (2005), shows that the energy-per-charge distribution of the bulk plasma exhibits two distinct peaks, corresponding to nearly corotational flow of H^+ and W^+.

Figure 9.12, from Sittler et al. (2005), shows definitive evidence from the CAPS IMS instrument that the heavier component is indeed W^+. Figure 9.12 was obtained by summing observations over 6 h covering the radial range from 3.4 to 8.3 R$_S$. The left panel shows counts as a function of E/q and time-of-flight as recorded by the ST element of the IMS, and the right panel shows the corresponding measurements from the higher-mass-resolution LEF. In this representation, particular ion species should occupy specific locations, as indicated by the various labels. Noteworthy aspects of the figure include: (1) an energy-dependent background that extends across the entire range of time-of-flight, which is caused by accidental coincidences, especially with penetrating radiation; (2) the signatures of H^+ and H_2^+ extending from a few eV to several 10's of keV; (3) two peaks attributable to W^+ (in the left panel: that labeled "water groups neutrals" and that labeled "O^- Peak," the latter referring to oxygen from water group ions that enter the instrument with a positive charge, but emerges from the foil with a negative charge), extending from 10's of eV to >10 keV; (4) a significant population of N^+ clearly distinguished from the O^+ peak in the LEF (right panel); and (5) evidence for molecular O_2^+.

9.3.2.1 Enceladus

The dominant compositional signature illustrated in Fig. 9.12 (H^+, H_2^+, and W^+) points directly to water ice on rings and satellites as the primary source. This evidence is

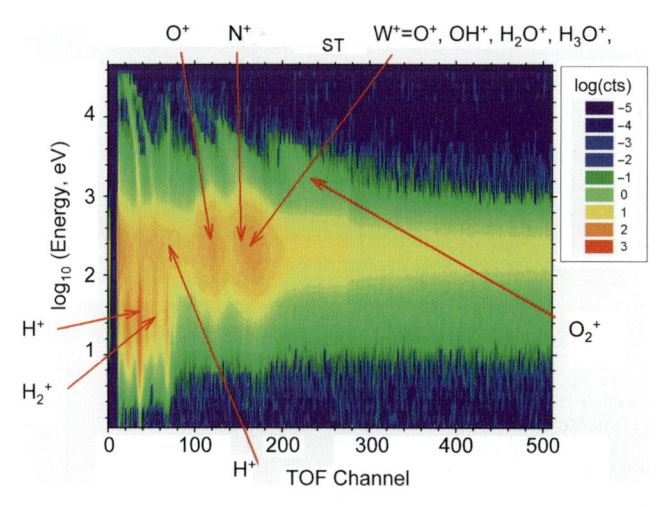

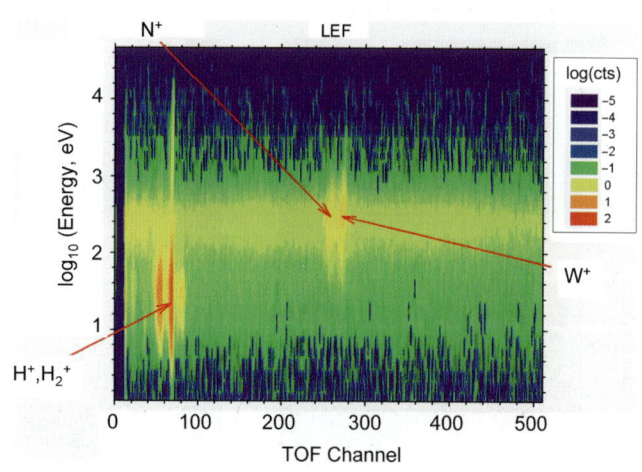

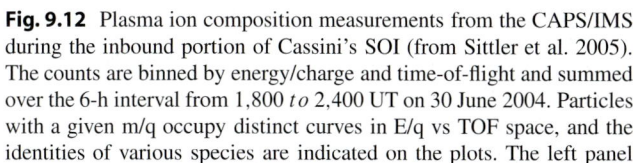

9 Saturn's Magnetospheric Configuration

Fig. 9.12 Plasma ion composition measurements from the CAPS/IMS during the inbound portion of Cassini's SOI (from Sittler et al. 2005). The counts are binned by energy/charge and time-of-flight and summed over the 6-h interval from 1,800 *to* 2,400 UT on 30 June 2004. Particles with a given m/q occupy distinct curves in E/q vs TOF space, and the identities of various species are indicated on the plots. The left panel shows measurements from the more sensitive straight-through (ST) detector, and the right panel shows the corresponding measurements from the linear-electric-field (LEF) section, which has lower sensitivity but higher resolution. The bands of counts that extend across the full TOF range are due to accidental coincidences caused mostly by penetrating radiation

consistent with the conclusion that the primary source of plasma is ionization of the cloud of neutral water-group molecules observed through remote sensing measurements (e.g., Shemansky et al. 1993), a conclusion reached previously on the basis of Voyager observations (e.g., Richardson et al. 1998).

The puzzle mentioned above as to the source of so much neutral gas, peaking near the orbit of Enceladus, was solved in 2005. Observations from a close Cassini flyby revealed the surprising fact that Enceladus is actively venting gas and ice grains (Dougherty et al. 2006; Hansen et al. 2006; Porco et al. 2006; Spahn et al. 2006; Waite et al. 2006; Sittler et al. 2008). These observations also showed that H_3O^+ ions were dominant around Enceladus which was not originally expected. The presence of molecular ions meant that the dissociative recombination loss rates of ions was more important than originally thought and boosted the source strength from Enceladus well above that originally estimated by Voyager (Richardson et al. 1998) when atomic ions were thought to be more dominant.

The primary gas emitted is water vapor, potentially accounting for the observed vast cloud of water vapor and water products. Johnson et al. (2006) have modeled the distribution of neutrals emitted from the vents and found that the emitted gas should form only a narrow, nearly uniform torus centered on Enceladus' orbit. However, when they included the effects of subsequent charge-exchange and reactive collisions with ambient corotating plasma, they found a much more extended neutral cloud, very similar to that determined from HST observations. These secondary interactions with the original narrow torus molecules also explain the large population of H_3O^+ ions that peak near Enceladus' orbit (Tokar et al. 2006; Sittler et al. 2008) since H_3O^+ is only formed in reactive collisions (e.g., $H_2O^+ + H_2O \rightarrow H_3O^+ + OH$) at low speeds (e.g., Gombosi et al. 1983; Johnson et al. 2006).

Using in-situ Cassini observations the total mass production from Enceladus was first estimated to be >10^2 kg/s (Tokar et al. 2006). This mass addition rate was consistent with values deduced from earlier HST observations (Jurac et al. 2002). UVIS observations and a more detailed analysis of the in-situ measurements yielded a total plasma production rate of ∼300 kg/s (Hansen et al. 2006; Johnson et al. 2006). While this production rate is somewhat variable, most observations are consistent with it.

9.3.2.2 H^+ and W^+

While the core ion population is dominated by <100 eV water group ions in the inner magnetosphere, CAPS detected a second non-thermal water group ion component from the vicinity of the Enceladus orbit out to about the Tethys orbit (Tokar et al. 2008). CAPS measurements of phase space velocity distributions of the water group ions in this region show the distinctive signature of pick-up ions, produced locally by local ionization and by charge-exchange collisions between the thermal water group ions and the water group neutral atoms and molecules.

Ionization of neutral atoms or molecules in the presence of a flowing plasma creates a telltale ring-type ion velocity-space distribution (for a detailed discussion of this process

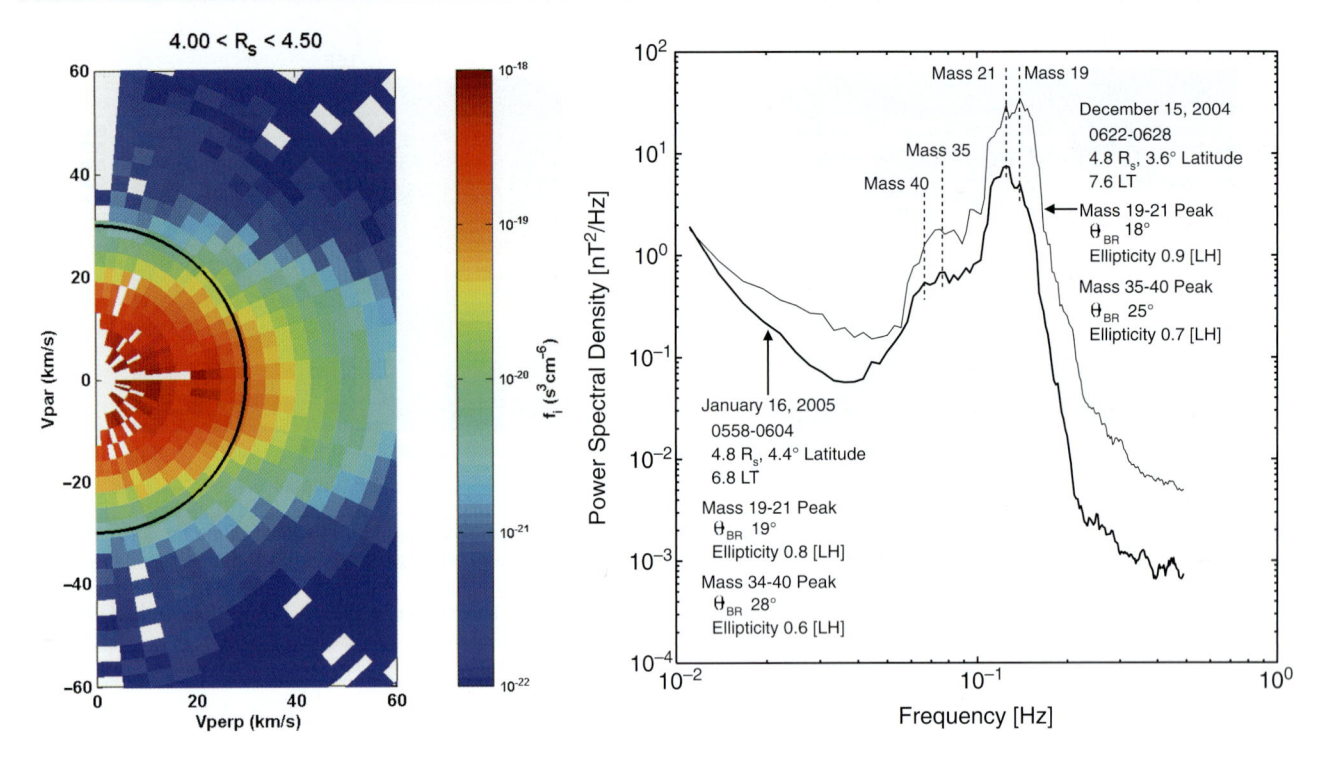

Fig. 9.13 Observational evidence for ion pick-up near the orbit of Enceladus. *Left*: The phase space density of W^+ ions as a function of $v_\perp$ and $v_\parallel$ in the frame of the bulk flow, showing the clear ring-type distribution at $(v_\perp, v_\parallel) \sim (25, 0)$ km/s, superimposed on the bulk plasma centered at (0,0) (from Tokar et al. 2008). *Right*: Magnetic field fluctuation spectra showing distinct peaks in spectral power near the gyrofrequency of W^+ and a heavier ion. Such waves are produced by an electromagnetic ion-cyclotron instability driven by an ion ring such as that shown in the left panel (from Leisner et al. 2006)

we refer to a later Chapter of this book (Mauk et al. 2009)). Such so-called "pick-up" ion distributions are shown in the left panel of Fig. 9.13. Additional evidence for the pick-up of fresh ions in this region comes from observations of electromagnetic ion cyclotron (EMIC) waves by the Cassini MAG (Leisner et al. 2006). Such waves, illustrated in the right panel of Fig. 9.13, are produced by a plasma instability driven by the ring-type velocity distributions and typically have frequencies near the gyrofrequency of the picked-up ions. The density of the non-thermal pick-up ions (see Mauk et al. 2009) is estimated to be $\sim 8\%$ of the thermal water group ion population (Tokar et al. 2008).

Sittler et al. (2006) note that the pick-up process is consistent with the observed correlation between the ion temperatures and the bulk flow speed and is likely to be the dominant energy source for the plasma in this region. As noted by Sittler et al. (2006), the CAPS observations during SOI show that both the H^+ and W^+ temperatures have similar positive radial gradients, consistent with pick-up ion energization in a plasma whose near-corotational flow speed increases linearly with r (panel 3 of Fig. 9.10). The temperature of the hydrogen ions increases steadily from ~ 2 eV just outside $3.4R_S$ to ~ 10 eV just outside $8R_S$. The water group temperature also increases steadily from ~ 40 eV to ~ 100 eV over the same radial distance (Sittler et al. 2005). The thermal electron temperature (panel 6 in Fig. 9.10) tracks the hydrogen ion temperature (Sittler et al. 2005; Rymer et al. 2007) and is found to vary as L^2 (Persoon et al. 2009; Sittler et al. 2006) over this radial distance. Electron temperatures derived from RPWS Langmuir probe measurements are qualitatively similar to, but lower than, the CAPS temperatures (Wahlund et al. 2005).

The near equality of the electron and proton temperatures has been attributed to collisional heating of the electrons (Rymer et al. 2007). However, it has been argued that Coulomb collisional heating can account for the observed electron energies outside $\sim 5R_S$, but the local heating and cooling effects seen near the E-ring and the orbit of Enceladus are probably due to cooling by dust particles in the E-ring or by the concentration of water neutrals in the vicinity of Enceladus.

The hotter electrons (>100 eV) visible in Fig. 9.9 have their source in the middle or outer magnetosphere and are transported to the inner magnetosphere by interchange-like injections of hot plasma, disappearing inside $L \sim 6$ (Rymer et al. 2007) (see Section 9.5.1).

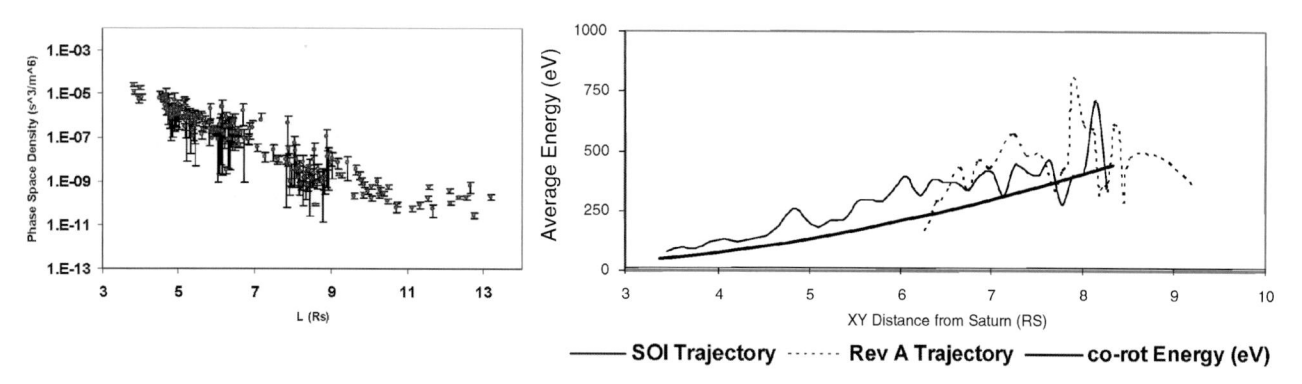

Fig. 9.14 *Left panel*: Nitrogen ion phase space density in the inner magnetosphere derived from CAPS/IMS time-of-flight measurements within 0.25 R$_S$ of Saturn's equatorial plane (from Smith et al. 2007). The negative radial gradient strongly suggests a source in the inner magnetosphere. *Right panel*: Average energy of the nitrogen ions on two passes through the inner magnetosphere (from Smith et al. 2005). The heavy solid curve is the energy of a nitrogen ion exactly corotating with the planet

9.3.2.3 N$^+$

While the N$^+$ evident in Fig. 9.12 could in principle indicate a Titan source, the radial dependence of the N$^+$ phase space density and the energy of this population point to an inner magnetosphere source (Smith et al. 2005, 2007). This is illustrated in Fig. 9.14, which shows the phase space density derived by Smith et al. (2007) as a function of radial distance (left panel) and the corresponding profile of the average energy of the population determined by Smith et al. (2005) (right panel). The inward gradient of the phase space density clearly indicates that the source is in the inner magnetosphere. There is a peak at L ∼ 4, near the orbital radius of Enceladus, but no measurements inside Enceladus' orbit were available due to penetrating radiation background. The average energy of the N$^+$ population (panel b) is consistent with pick-up in the local corotational electric field (solid line), and it is inconsistent with transport from a Titan source, which would be expected to produce an adiabatic energy increase with decreasing r. Smith et al. (2007) also found a peak in the ratio of N$^+$/W$^+$ phase space densities in the narrow Enceladus torus region (Johnson et al. 2006) near the orbital distance of that moon. Based on a comparison with simulations, they concluded that Enceladus is the principal source of N$^+$ in the inner magnetosphere. A more detailed subsequent analysis (Smith et al. 2008) sought to identify the source molecules for the observed N$^+$: N$_2^+$, consistent with INMS observations of mass 28 molecules near Enceladus (Waite et al. 2005); or NH$_3^+$, which would have potentially important implications for the physical properties of the ice on Enceladus. This study found definitive evidence for NH$_x^+$, presumably from NH$_3$, comprising a few percent of the inner magnetospheric heavy ions. While no similarly definitive evidence for N$_2^+$ was found, a best fit to CAPS data included small amounts of N$_2^+$, with upper limits near the INMS fraction, leading to the conclusion that both molecular nitrogen and ammonia are emitted from Enceladus. One possible local source might be the ionization of NH$_3$ from the surface ice of Saturn's inner satellites (Delitsky and Lane 2002).

9.3.3 Minor Sources

Titan has definitely been found to be a source of magnetospheric plasma, picked up from its upper atmosphere (e.g., Krimigis et al. 2005), but surprisingly little nitrogen, which is the signature ion for a Titan source, has been found in the outer magnetosphere (e.g., Young et al. 2005; Smith et al. 2005, 2007). Cassini has thus found no real evidence for the Titan-sourced nitrogen plasma plumes inferred from Voyager observations (Eviatar et al. 1982, 1983). This lack of nitrogen is likely attributable to the inability of flux tubes at Titan's orbital distance to execute complete drift orbits around Saturn, such that Titan-originating plasma cannot build up to substantial densities (e.g., Young et al. 2005; Smith et al. 2007). Whatever the reason, there is no evidence in the inner magnetosphere for significant amounts of plasma of Titan origin.

The solar wind likewise does not appear to be a dominant source of plasma for Saturn's inner magnetosphere, although there is some evidence for solar wind entry into the outer magnetosphere via magnetopause reconnection (e.g., McAndrews et al. 2008), as well as for the existence of open polar cap magnetic flux, which implies the existence of a Dungey-like reconnection-driven circulation that delivers solar wind plasma to the outer magnetosphere (e.g., Bunce et al. 2005b; Cowley et al. 2005; Badman et al. 2005; Krupp et al. 2005 and see Section 9.4). There is as yet, however, no definitive assessment of the relative importance of solar-wind-driven transport in populating the outer magnetosphere. With respect to the inner magnetosphere, it appears that the solar wind is at best responsible for the hot, tenuous material that accompanies low-content magnetic flux tubes in exchange for filled ones containing colder inner-magnetospheric plasma (see Section 9.1.3). And even for that hot, tenuous population, there is some compositional evidence that much of it may have originated in the inner magnetosphere as well (e.g., Sittler et al. 2005; Krimigis et al. 2005; Sergis et al. 2007).

While the icy satellite and Enceladus sources of plasma are quite significant in the inner and outer magnetosphere, Saturn's ionosphere may also play a role through the polar wind. The polar wind refers to the supersonic outflow of particles along open magnetic field lines at high latitude. While the Saturnian polar wind has not been observed yet, Glocer et al. (2007) carried out model calculations and predicted that Saturn's high latitude ionospheres might add a few kg/s of H^+, H_2^+ and H_3^+ ions to the magnetosphere. While this process is only a minor plasma source for Saturn's magnetosphere, the polar wind ions might serve as important tracers to better understand the magnetospheric configuration and dynamics.

9.3.4 Loss Processes

To maintain an approximate steady state of the magnetospheric plasma population, the production of plasma by these various sources must be balanced by roughly equivalent losses. Candidate loss processes include recombination, loss to the atmosphere through pitch-angle scattering, absorption by neutral material (satellites and ring particles), and radial transport.

The recombination timescale depends on the ion species, the electron density, and the electron temperature. As noted by Sittler et al. (2008), recombination is much slower for atomic ions (O^+, H^+) than for molecular ions (e.g., H_2O^+, OH^+, H_3O^+). Sittler et al. (2008) have estimated the effective recombination timescale for the inner magnetospheric plasma by taking a composition-weighted average of the timescale for each relevant species (H^+, O^+, OH^+, H_2O^+, H_3O^+). The relative composition of the various species was determined from CAPS time-of-flight measurements Young et al. 2005; Sittler et al. 2008. The calculated recombination lifetime is $\sim 4 \times 10^5$ s inside of L $\sim$ 5 and climbs rapidly outside of that distance.

By contrast, the timescale for radial transport is large at low and small at high L values. Using a form of the radial diffusion coefficient derived for Io by Siscoe and Summers (1981), Richardson et al. (1998) found the radial transport time to be $\sim 5 \times 10^5$ s at L = 6, decreasing as L^{-3} beyond that. If L $\sim$ 6 marks the onset of the interchange instability (see Mauk et al. 2009), radial diffusion must decrease substantially inward of that location, resulting in very long transport times inside the orbit of Dione. This break in transport properties is consistent with the flattening of the radial profiles of the electron phase space densities beyond L $\sim 6 - 8$ Rymer et al. (2007).

Evidence for plasma losses due to precipitation is indirect and primarily stems from the observations of EMIC waves (e.g., Leisner et al. 2006; see Fig. 9.13). The minimum lifetime τ_{SD} for particles pitch-angle scattering in these waves is given by the strong diffusion limit (e.g., Kennel and Petschek 1966):

$$\frac{1}{\tau_{SD}} = \frac{\Omega_B}{\pi}(1 - \cos\alpha_L) \quad (9.5)$$

where Ω_B is the particle bounce period and α_L is the equatorial loss cone angle. Taking the characteristic particle speed to be the pick-up velocity,

$$v = LR_S\Omega_S \quad (9.6)$$

where Ω_S is Saturn's rotational angular speed, Eq. (9.5) gives $\tau_{SD} \sim 2.7 \times 10^6$ s at L = 4, increasing as L^3 beyond that. Thus, in spite of the robust activity of the EMIC instability, precipitation is not likely to be a significant factor in the loss of inner magnetosphere ions.

The last potential loss mechanism is absorption by E-ring material. For low-energy particles whose range in ring material is less than the average grain size, the lifetime against loss by grain impact is Thomsen and Van Allen (1979):

$$\tau_{absorp} = \frac{T_B}{2\eta}\cos\alpha_{eq} \quad (9.7)$$

where η = optical opacity of the ring ($\eta = Nd\pi r_0^2$), N = number of ring particles per unit volume, d = ring thickness, r_0 = ring particle radius and α_{eq} = equatorial pitch angle.

From Showalter et al. (1991), the peak optical depth of the E-ring is at the orbit of Enceladus, where the geometric cross-section per unit area, i.e., the optical opacity, is $5.3 \pm 1.3 \times 10^{-5}$. Again taking $v = LR_S\Omega_S$ as the characteristic ion speed, for an equatorial pitch angle of 80°, the lifetime against absorption is 3.7×10^7 s, much longer than other loss timescales.

In summary, estimation of lifetimes against various plasma loss mechanisms shows that inside of L $\sim 5 - 6$ recombination is likely to be the most important loss process, whereas beyond L $\sim$ 6, rapid radial transport dominates. In particular, the operation of the interchange instability (cf. Section 9.5.1 and Mauk et al. 2009) beyond the peak flux-tube content at L $\sim$ 6 very efficiently removes plasma produced in the inner magnetosphere, delivering it well into the plasma sheet of the outer magnetosphere. Indeed, the cool

Fig. 9.15 Contour plots of the ion densities in the plasma source region where z is the distance above/below the equatorial plane and ρ is the perpendicular distance from Saturn's spin axis. The density contours are constructed from the comparison of the diffusive equilibrium model to measured electron densities from the RPWS instrument, assuming symmetry about the spin axis and mirror symmetry about the equator (from Persoon et al. 2009)

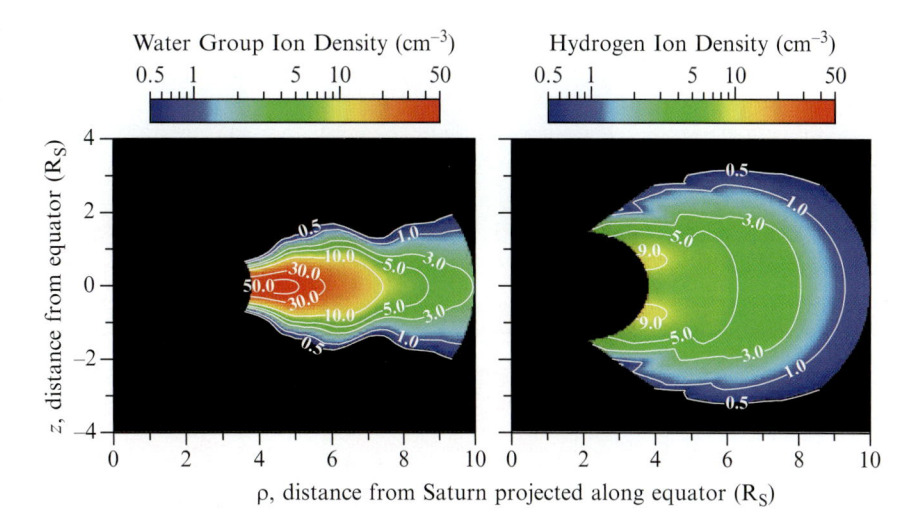

water-group plasma originating in ionization of the near-Saturn neutral cloud extends to the dayside magnetopause and well down the tail (e.g., McNutt 1983; McAndrews et al. 2009).

9.3.5 Plasma Density Models

The first density models describing the distribution of electrons and ions in Saturn's magnetosphere were developed using in situ plasma measurements from the Pioneer 11 and Voyagers 1 and 2 spacecraft, along with remote sensing observations from the Hubble Space Telescope (Richardson and Sittler 1990; Richardson 1995, Richardson 1998; Richardson and Jurac 2004). Several density models have been developed to take advantage of the wealth of in situ plasma and plasma wave measurements obtained during years of Cassini orbits around Saturn. Using plasma wave measurements of the upper hybrid resonance emission band for several of the early equatorial passes, Persoon et al. (2005) showed that the plasma diffuses radially outward from Saturn and the equatorial electron density varies inversely with increasing radial distance as $R^{-3.7}$. When sufficient high-latitude electron densities became available, Persoon et al. (2006) developed a simple scale height model, based on a dominant centrifugal force acting on the plasma. The scale height model identifies the heavy ion component of the plasma inside L = 10 and shows that the equatorial density of this ion component varies as $L^{-4.1}$ and the plasma scale height varies as $L^{1.8}$ (Persoon et al. 2006).

Sittler et al. (2008) used the CAPS ion and electron fluid parameters as boundary conditions and solved the full set of first-order differential equations which give the balance of forces acting on the plasma along Saturn's dipole field lines inside L = 10. The resulting 2D density contour plots show the strong equatorial confinement of the heavy water group ions and a butterfly distribution of the proton density, indicating that the protons peak off the equatorial plane due to the strong influence of the ambipolar electric field acting on the light ions.

Persoon et al. (2009) derived a simplified analytic solution to the same field-aligned force balance equation and developed a diffusive equilibrium model for a two-species plasma. The analytical model was compared to the RPWS electron density measurements for latitudes up to 35°. The fit of the diffusive equilibrium model to the measured densities yields the ion equatorial densities and scale heights for both dominant ion species, which are used to construct ion density contour plots in the meridian plane. Figure 9.15 shows the density contour plots for the water group ions (left panel) and the hydrogen ions (right panel) derived from the fit of the diffusive equilibrium model to the measured RPWS densities. The contour plot for the water group ions clearly shows that these ions are strongly confined to the equatorial plane at all L-values. The contour plot for the hydrogen ions (right panel) shows that the density of the lighter ions peaks off the equator at low L-values under the influence of a strong ambipolar force. However, the growing strength of the mirror force acts to cancel the ambipolar effect at larger L-values.

9.4 Magnetospheric Regions

9.4.1 Trapped Radiation

Our knowledge about Saturn's radiation belts before Cassini is based on the in-situ particles and fields measurements of the flyby missions Pioneer 11, Voyager 1, and 2 summarized in Van Allen (1983, 1984). The charged particles of the radiation belts in the Saturnian magnetosphere with energies of

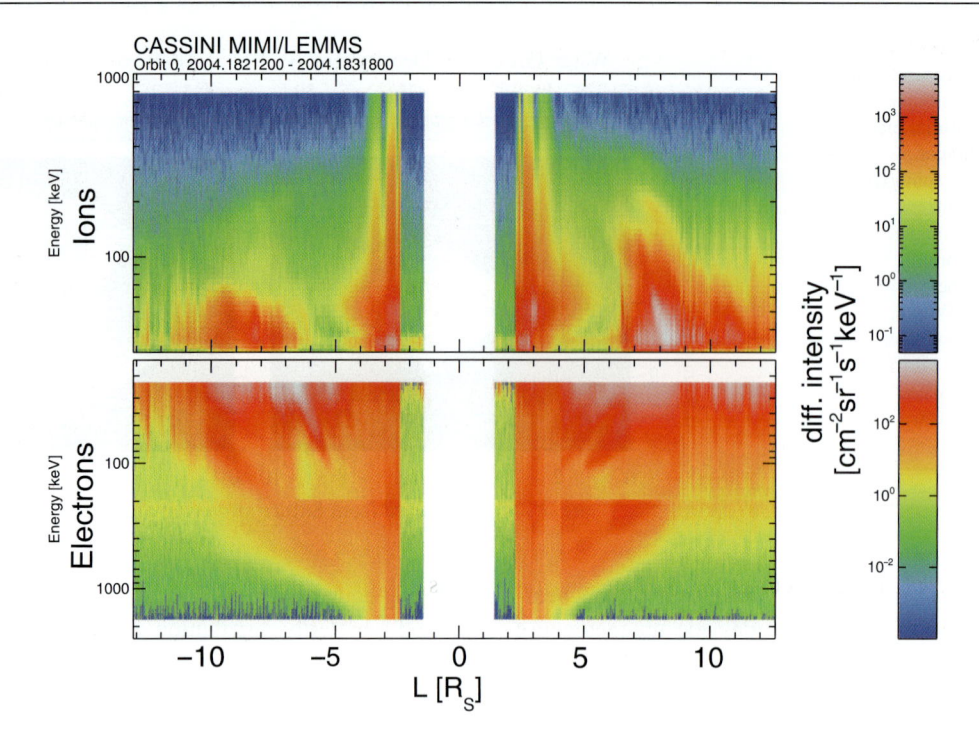

Fig. 9.16 Dynamic spectrograms (energy versus L) of energetic ion (*top panel*) and electron intensities (*bottom panel*) inside of 10 R_S for Cassini's Saturn insertion orbit in July 2004 measured with the MIMI/LEMMS instrument aboard Cassini. Electron energy is plotted increasing downward for comparison with the features in the ion population. The radiation belt outside the main rings (L > 2.3) is transitioning to an extended, highly dynamic plasma sheet outward to the magnetopause. (modified from Krimigis et al. 2005)

hundreds of keV to tens of MeV can be found mainly inside 6 R_S. Typical charged particle intensities measured in the middle and inner magnetosphere as a function of dipole L are shown in Fig. 9.16.

The top panel shows energetic ions and the bottom panel energetic electrons measured by the MIMI/LEMMS experiment aboard Cassini during SOI around Saturn in July 2004. Energetic ions are abundant in the middle magnetosphere (L > 7–8) but between 3.5–7 R_S they are absent, because of charge-exchange processes between these hot ions and the cold neutral gas that transforms these populations into hot energetic neutral atoms and cold ions. Energetic (<100 keV) electron fluxes also drop inside 7 R_S with respect to the values in the middle magnetosphere, but in the region where energetic ions are not present, hot electrons are much more abundant. Inside 3.5 R_S energetic particles in the range of 100 – 200 keV appear (typical for radiation belts), while MeV particles are also present. In the center of each panel (L < 2.3), a plasma void region is seen, where energetic particles have been depleted by Saturn's main rings. In both panels, ion and electron intensities show dispersed features as a function of energy. These are called "injections" and are thought to be the result of instabilities occurring between the middle and outer magnetosphere (where hot plasma is present) and the inner magnetosphere (where energetic ions are absent and hot electrons are less abundant) (Mauk et al. 2005). Those injection events play a major role in magnetospheric particle transport (Carbary et al. 2009). At this point the relation between the injection events and the interchange process is unclear.

The radiation belts are to some extent transient, given the continuous depletion of energetic particles by the icy moons, the rings and the neutral gas that are present in that region. All charged particles pass through the orbital plane of Saturn's moons and rings while executing one of the fundamental motions of trapped radiation: the bounce motion along the magnetic field lines. Unlike at other planets, even equatorial particles are continuously absorbed by an extended ring system and by a number of moons with almost circular and equatorial orbits within the radiation belts. As a consequence, losses of particles to the icy moon surfaces and ring particles are expected to be higher compared to the losses in other planetary magnetospheres. Carbary et al. (1983) give a good summary of the Voyager 1 and 2 findings. Among the important results were the calculation of the diffusion coefficient D_{LL} at the distance of Dione and the evaluation of various magnetic field models. Van Allen et al. (1980) extracted D_{LL} values from a Mimas absorption signature and suggested that a filtering effect to radially diffusing electrons is taking place at Enceladus that results in a monoenergetic electron spectrum in the innermost Saturnian radiation belts.

9 Saturn's Magnetospheric Configuration

The variability of the absorption signatures has been extensively studied by Roussos et al. (2008) using Cassini energetic particle data. Inside 2.3 R_S, and in the regions magnetically connected to the main rings, energetic particles are completely absent. Many of Saturn's moons are continuously immersed in the planet's radiation belts and are exposed to its trapped energetic particle population. The energetic particles are absorbed by the moons forming evacuated regions within the magnetosphere, the lifetime of which depends on the effectiveness of particle diffusion processes (and most importantly, radial diffusion) (see also Carbary et al. 1983; Mauk et al. 2009). In addition these absorption regions continuously drift in the magnetosphere. Monitoring the properties of the depleted flux tubes (depth, shape, magnetospheric coordinates, longitudinal separation from the absorbing moon), provides important information about the dynamics of the magnetosphere or even the absorbing object. The absorption regions are classified in two main categories: macrosignatures and microsignatures (Van Allen et al. 1980). Macrosignatures are the permanent and azimuthally averaged decreases of the count rates in the radial distribution of energetic particles. Microsignatures are count rate decreases that are strongly dependent on the longitudinal distance between the signature's location and the absorbing body. Satellites, rings or dust concentrations can be the source of both macrosignatures and microsignatures. Figure 9.17 gives an example of ion macrosignatures in the radiation belts of Saturn (Roussos 2008) caused by the moons Janus, Mimas, Enceladus, Tethys, and Dione. The depletion in the intensity of 10 MeV protons is seen in all latitudes and local times as a function of L. Those macrosignatures in ion fluxes with E > 10 MeV are not initially sharp or deep, as for energetic electrons. A large percentage of such ions can escape absorption by the icy moons due to gyration and bounce motion; however, these shallow absorption regions reencounter the moon before diffusion processes have any significant effect on them and therefore they become deeper, until an equilibrium is reached between ion diffusion and ion depletion (near steady-state situation). In the Cassini data from 2004–2007 it is observed that for all moons this equilibrium is not reached: the ion depletion is almost 100% around the moon orbits. This means that ion absorption rates are always faster than diffusion. For this reason, no ions of E > 10 MeV exist along the L-shells of the moons Janus, Epimetheus, Mimas, Enceladus, Tethys and Dione. At these locations, ions are absent in almost all magnetospheric local times and latitudes, independent of each moon's location.

Figure 9.18 shows two examples of many microsignatures observed in MIMI/LEMMS electron data on Cassini recorded on day 229 in 2006 in the inner magnetosphere, caused by the moons Dione (at 03:51–03:55 UT) and Helene at around 03:58 UT (Roussos et al. 2008). These microsignatures are only seconds to minutes long. From the depth and the shape of the signature as a function of longitude difference between the object and the observer, it can be

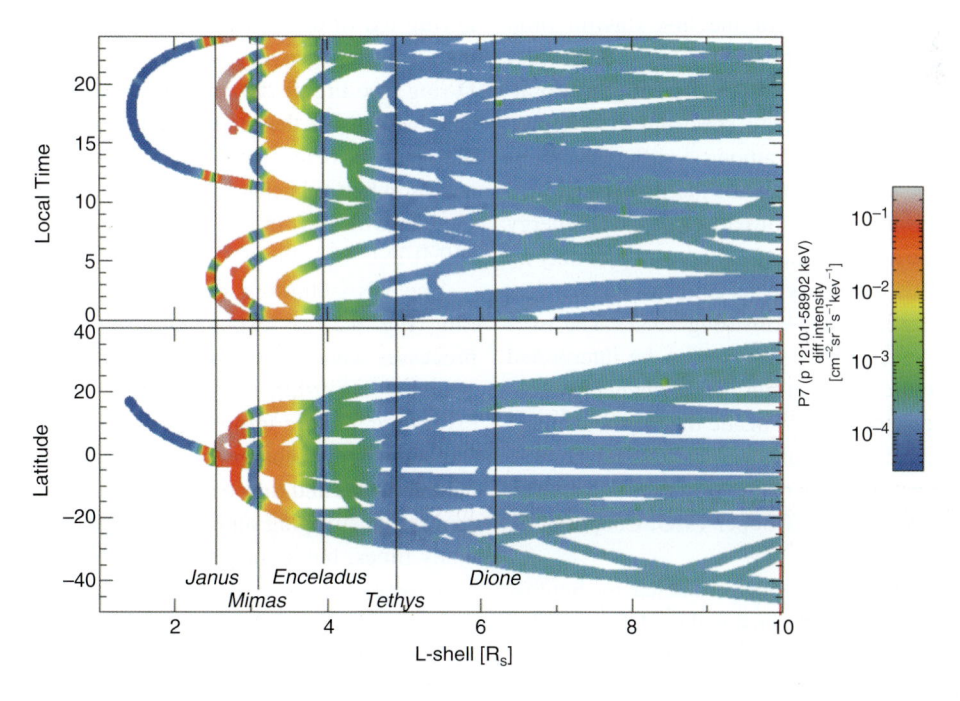

Fig. 9.17 Ion macrosignatures as a function of L-shell and local time (*top*) and latitude (*bottom*). The color denotes the ion differential intensities of the Cassini MIMI/LEMMS P7 channel, that detects ions with energy greater than about 10 MeV/nucleon. The macrosignature locations show an excellent correlation with the L-shells of the moons Janus, Epimetheus, Mimas and Enceladus. Data shown correspond to the period July 2004–July 2007 and are averaged every 86 s (from Roussos 2008)

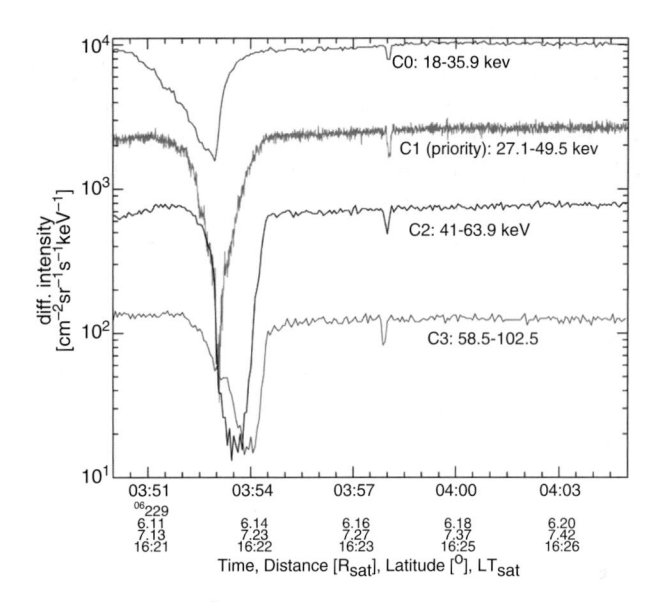

Fig. 9.18 Microsignatures of the Saturnian moons Dione (03:51–03:55 UT) and Helene (03:58 UT) in energetic electron intensities (channels C0–C3 of the MIMI/LEMMS instrument onboard the Cassini spacecraft) measured on day 229 in 2006 (from Roussos et al. 2008)

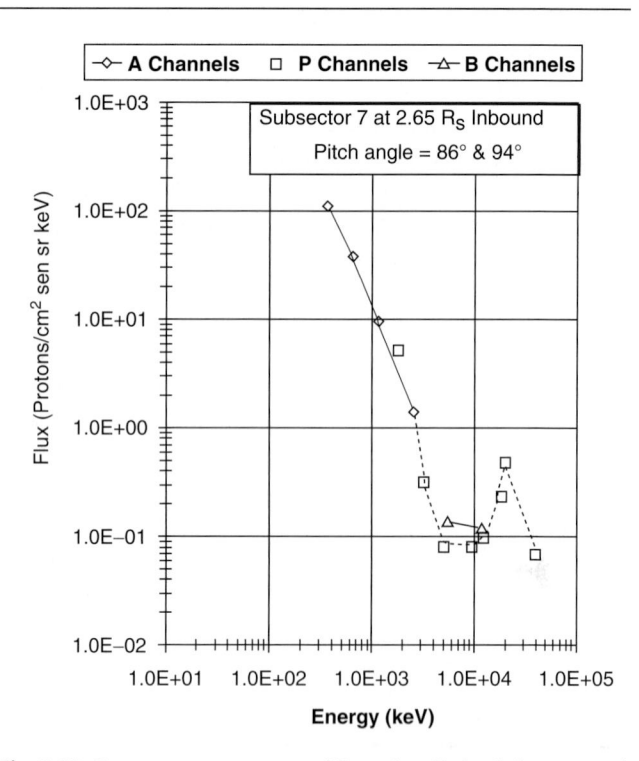

Fig. 9.19 Proton energy spectrum of Saturn's radiation belts measured with the three different sets of energy channels (A, P, and B) of the MIMI/LEMMS instrument aboard Cassini in July 2004 at 2.65 R_S (from Armstrong et al. 2009)

determined what object caused the depletion on one hand, and on the other hand diffusion processes responsible for the refilling can be studied.

A representative energy spectrum of trapped energetic protons at 2.65 R_S and at nearly perpendicular pitch angle during the inbound portion of the first Cassini orbit around Saturn in July 2004 is shown in Fig. 9.19. Besides a power-law energy dependence of the spectrum, a prominent secondary peak in the energy spectrum is observed around 20 MeV. The agreement among 3 different sensor arrangements (A, P and B-channels) of the MIMI/LEMMS instrument with measured values of particle flux and energy shows that systematic errors of measurement are probably no greater than 10%. This secondary peak confirms the Voyager observations by (Krimigis and Armstrong 1982). The two-component energy spectrum of protons can be interpreted with two different source populations: the lower energy protons originate in the solar wind or inside the magnetosphere by violating the adiabatic invariants and the secondary peak most probably originates from Cosmic Ray Albedo Neutron Decay (CRAND) described by (Cooper 1983; Cooper et al. 1985).

Saturn's radiation belts have been modeled by Santos-Costa et al. (2003). This three-dimensional model shows that absorption by dust plays the major role in the innermost part (1–2.3 R_S), while local losses from interactions with satellites are more important in the 2.3–6 R_S region, consistent with observations by the Pioneer and Voyager spacecraft and by Cassini as described above.

The uniqueness of the first Cassini orbit that provided data very close to the planet inside and above the innermost D-ring led to the discovery of a new trapped particle population. The discovery of this new radiation belt inside the D-ring (see Fig. 9.20) was only possible because the spacecraft was inside the main radiation belts and therefore only the energetic neutral atoms (ENAs) originating from the new radiation belt (via charge-exchange processes) that normally interact with ions in the main belts could be measured with MIMI/INCA (Krimigis et al. 2005).

The ENA-emitting region inward of the innermost D ring (Fig. 9.20) is explainable by double charge-exchange processes where planet-directed ENAs from the main radiation belt are stripped of electrons when they enter Saturn's exosphere and are trapped as ions which will subsequently undergo another charge-exchange collision with exospheric neutral atoms and thus be transformed back into ENAs. This process of stripping and charge exchange may be repeated many times, but some of these particles will eventually escape the exosphere as ENAs. Thus, this double (or multiple) charge-exchange process forms a low-altitude ENA emission region inside the D-ring. A similar trapped radiation belt was identified and explained by charge exchange in Earth's radiation belts by in situ measurements at low altitudes (Moritz 1972). The first direct measurements of the particle population inside the D-ring will be possible when Cassini will

9 Saturn's Magnetospheric Configuration

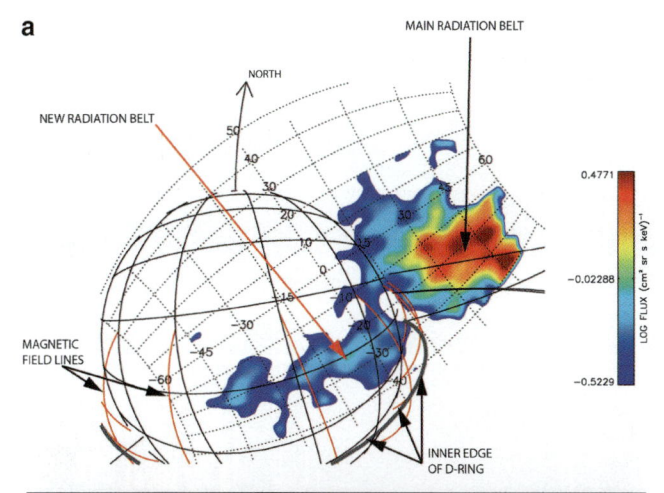

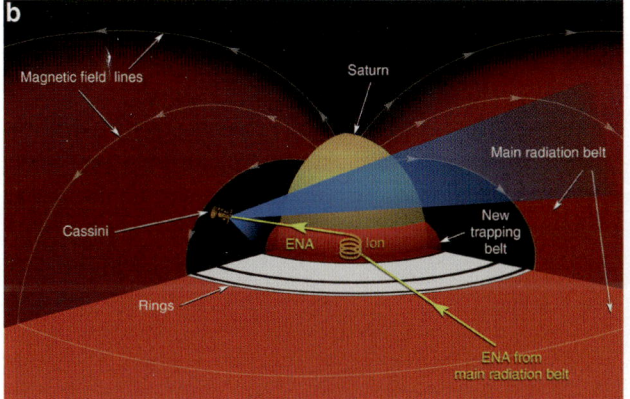

Fig. 9.20 (**a**) INCA image in 20 to 50 keV/nucl ENAs. The bright region above Saturn's limb is caused by ENAs produced by charge-exchange collisions between main radiation belt ions and near equatorial gas distributions. The band of emission above the equator is produced by the same ENAs from the main belt being stripped in Saturn's exosphere between the inner edge of the D ring and the cloud tops, trapped there temporarily as energetic ions, and then reemitted as ENAs. (**b**) Schematics of the charge exchange/stripping process that begins as ENA emission from the main belt and produces ENA emission from Saturn's exosphere (from Krimigis et al. 2005)

traverse multiple times through that region on polar orbits during the final phase of its mission.

Energetic particles (tens to thousands of keV) are found throughout the plasma source region, but the intensity is found to vary with radial distance and is often correlated with the orbits of the inner satellites. Krimigis et al. (2005) observed a depletion in the more-energetic ions between the L-shells of Dione and Enceladus, indicating energetic particle loss through charge-exchange with neutral gas in this region (Esposito et al. 2005). Just outside the orbit of Rhea $(8.7\,R_S)$, there is a rapid increase in the energy of both the energetic ion and the non-thermal electron populations (Krimigis et al. 2005).

The variability of the radiation belts in response to solar wind disturbances has been recently demonstrated by Roussos et al. (2008) who analyzed Cassini MIMI/LEMMS data. Figure 9.21 shows MeV fluxes in the inner magnetosphere. Profiles are plotted for three different periapsis passes. The 2004 profile (white curve) is the most common, identified in 27 out of the 36 orbits considered in this study. Flux peaks are clearly separated at several icy satellite L-shells, as indicated. A dropout is also seen at the G-ring L-shell. The lowest background is measured above the main rings that absorb all magnetospheric energetic ions, while Saturn's volume and the strong dipole field "shadow" the instrument from penetrating, galactic cosmic rays. Note that fluxes do not reach background at the depletion region of Janus and Epimetheus, meaning that some flux can be transported across their shared orbit. The two profiles of 2005 (yellow and turquoise curves), that correspond to an orbit with a periapsis at L = 3.5, reveal a flux enhancement centered close to Dione's L-shell (Dione belt). The enhancement is isolated only outside Tethys's L-shell. Similar enhancements are seen in all LEMMS channels between 30 keV/nuc and 10 MeV/nuc. In this study a whole series of Cassini orbits through the main radiation belts have been studied. It has been recognized that for some of the periapsis passes MeV ion fluxes increased in a region between the orbit of the moons Dione and Tethys (shown as blue curves in Fig. 9.21). Those increases correlate very well with interplanetary disturbances arriving at Saturn. After a few weeks or months the so called "Dione belt" increases disappear and the "normal" radiation belt fluxes were registered again (white and yellow curves in Fig. 9.21). It is therefore assumed that this increase is due to interaction processes between the particles in the interplanetary medium and the magnetosphere. However, inside the orbit of Tethys no change or increase could be detected during those events, from which the authors concluded that the source population of the innermost radiation belts must be different. The most probable source of those particles is the CRAND process as mentioned above.

9.4.2 Ring Current

The combination of gradient and curvature drifts in a non-uniform magnetic field generates a ring current that opposes the background field inside the ring current (cf. Gombosi 1998). The concept of an electrical current encircling the Earth at high altitudes was first proposed in the early 1900s to explain the depression of the horizontal component of the Earth's magnetic field during geomagnetic storms. Away from Earth, ring currents of a different nature and size were observed at Jupiter and Saturn. In the latter case, a ring current was inferred from magnetic field measurements during the Voyager 1 and 2 flybys (Connerney et al. 1982) and confirmed from particle measurements made by the

Fig. 9.21 MeV ion fluxes in the innermost magnetosphere. Differential ion fluxes from the ion channel P2 of LEMMS (2.28–4.49 MeV/nuc) are plotted as a function of the dipole L-shell. Negative (positive) L denotes the inbound (outbound) part of the orbit (from Roussos et al. 2008)

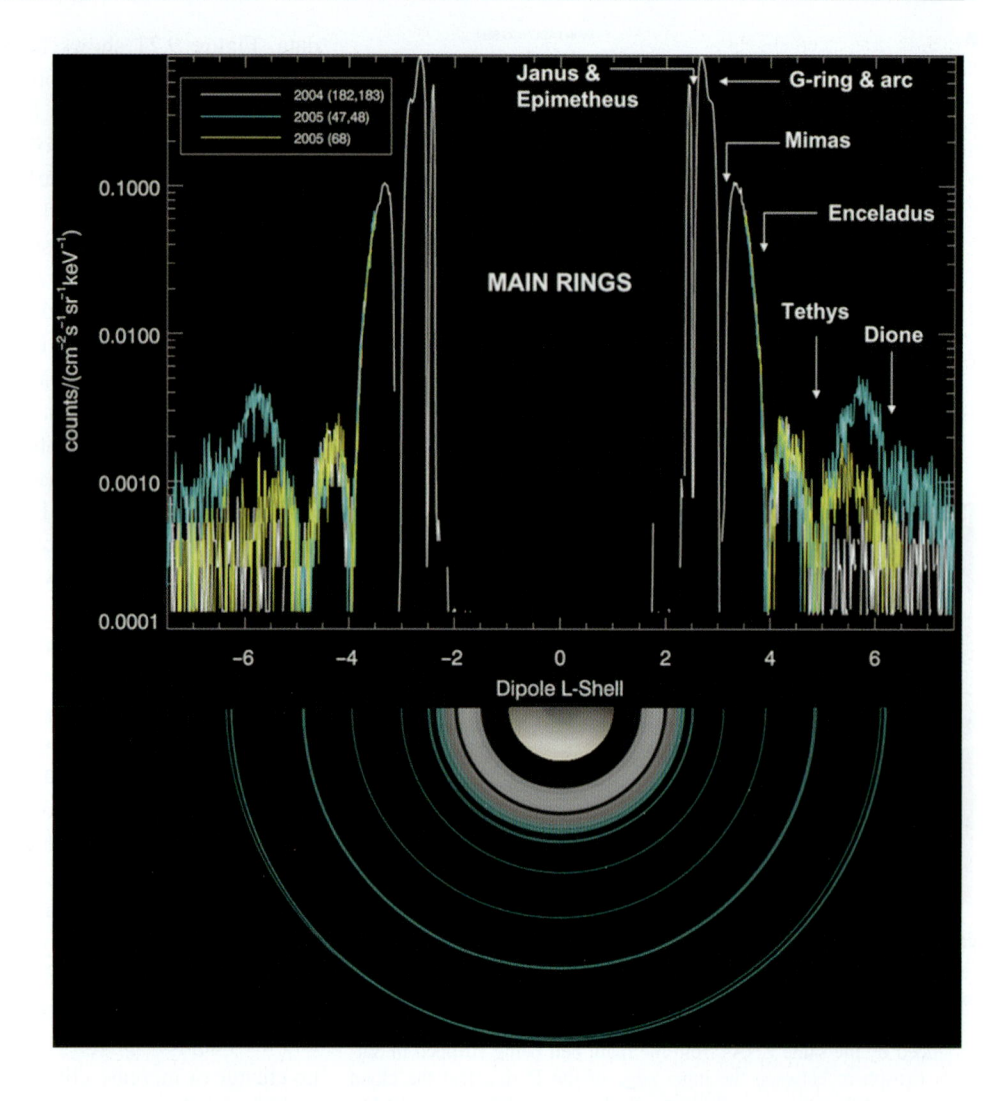

Low Energy Charged Particle (LECP) and plasma (PLS) instruments (McNutt 1983; Mauk et al. 1985; Krimigis et al. 1981, 1983).

At Saturn, the field generated by the ring current is northward and depresses the strength of the planetary field locally where the plasma energy maximizes (inside the ring current). The total current observed flowing through the ring current region is estimated to be between 8 and 17 MA (Connerney et al. 1983, Bunce et al. 2007). As the field generated by a ring current is roughly uniform inside the ring, the field remains depressed even inside the ring current region. Figure 9.22 (reproduced from Khurana et al. 2009) shows the perturbation field (observed − internal field) measured in the magnetosphere during Rev 20 in a spherical coordinate system. The perturbation field in the B_θ component is northward (negative B_θ) inside of $\sim 12 R_S$ which is a manifestation of the ring current plasma. Detailed modeling shows that a ring current starting at a radial distance of $\sim 6 R_S$ and peaking near the radial distance of $10 R_S$ is required to correctly model this perturbation field.

The inner edge of the ring current lies between 6 and 8 R_S but the outer edge of the ring current is strongly controlled by the magnetopause location on the dayside and lies anywhere between 12 and 22 R_S. In addition, the strength of the ring current is also directly related to the magnetospheric size. The lowest values of the ring current strength (8 MA) are observed when the dayside magnetopause is near its minimum location ($\sim 19 R_S$) whereas the strongest currents (17 MA) are observed when the dayside magnetopause is located near 31 R_S.

Bunce et al. (2007) used the Connerney et al. (1982) model to examine the variation of the model ring current parameters with the subsolar magnetopause distance. In this work the thickness of the current sheet was fixed at 2.5 R_S. For each pass of Cassini through the dayside, the residual (observed − model internal field) magnetic field vectors were fitted by eye to the model. The location of the last inbound magnetopause crossing and the magnetopause model of Arridge et al. (2006) were used to establish the subsolar magnetopause distance for that pass. Using their sets of ring

Fig. 9.22 The external field in Saturn's magnetosphere ($\mathbf{B}_{dif} = \mathbf{B}_{obs} - \mathbf{B}_{int}$) observed during Rev. 20. The depression in the B_θ component near the planet ($r < 14\,R_S$) arises mainly from the ring current. The perturbation field in this component can be seen to oscillate by a factor of 2. Two envelopes (*dashed lines*) representing the minimum (*upper envelope*) and the maximum value (*lower envelope*) of the ring current field have been drawn to aid the eye (from Khurana et al. 2009)

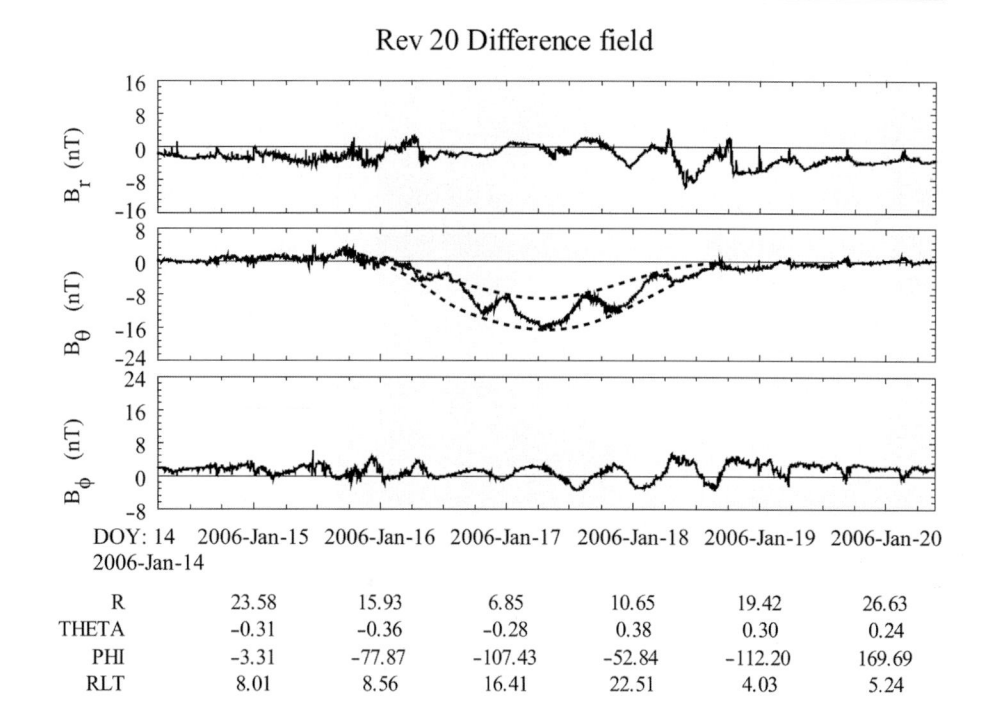

current parameters, they also calculated Saturn's total magnetic moment (dipole + ring current) and axial ring current field (essentially Saturn's Dst).

As the total magnetic flux from the planet is conserved, the field is enhanced elsewhere, namely in regions above and below the ring current (called the lobe regions) both locally and in the distant magnetosphere (cf. Bunce and Cowley 2003). The effect of the ring current on the field can be visualized as a physical pulling out of the planetary field lines from the inner low-latitude magnetosphere into the high latitude regions of the magnetosphere. The percent reduction of field strength from this "field diversion" is quite modest in the ring current regions of planets (<10% in Saturn's inner magnetosphere) because of the strong dipolar field close to the planet. However, in the outermost parts of the magnetosphere, the ring current and its extension – the current sheet – can make the lobe field many times stronger than that expected from a dipole field.

Figure 9.22 also contains evidence for the presence of an azimuthal asymmetry of the ring current, visible as a rotational modulation of the B_θ component. The azimuthally symmetric part of the ring current produces a depression of ~8 nT near the closest approach (see the upper envelope indicated by a dashed curve in Fig. 9.22), whereas the asymmetric part increases the depression by another ~8 nT (the lower envelope). A survey of such data from Cassini reveals that the ring current region anomalies are always present in the magnetic field observations obtained from the inner/middle magnetosphere. Gurnett et al. (2007), Southwood and Kivelson (2007) and Andrews et al. (2008) have identified other features of the middle magnetosphere that are fixed in a frame that rotates at the SKR period and tracks its changing value. The strength of the asymmetric ring current is modulated similarly. The amplitude of the asymmetric ring current is seen to vary from orbit to orbit over a range of 25–200% of the value of the symmetric ring current. The field and plasma data clearly establish that the energetic particle azimuthal anomalies and their associated partial ring current are a semi permanent feature of Saturn's magnetosphere.

Together with its clear presence in the in-situ particle and magnetic field measurements, the energetic particle contribution to the asymmetric Saturnian ring current is visible through the energetic neutral atom (ENA) images that the MIMI/INCA sensor obtained, once Cassini switched to high latitude orbits in July 2006 (Krimigis et al. 2007). Briefly, the ENA technique relies on charge exchange between trapped ions and a residual neutral gas that results in fast atoms escaping the system and being sensed as if they were photons. ENA images offer a complete picture of the instantaneous energetic particle distribution, for almost every region of interest within the magnetosphere. The ENA distribution provides direct information on the dynamical features of the energetic part of the ring current, which cannot possibly be revealed through the in-situ measurements from individual equatorial passes of the spacecraft. One such image is shown in Fig. 9.23. The ring current maximum intensity is generally outside the orbit of Rhea; observable intensities may extend beyond the orbit of Titan. Overall, the image in Fig. 9.23 illustrates that although this interval was chosen specifically as an example with minimal local time/ longitudinal structure, the ring current, unsurprisingly, is not the

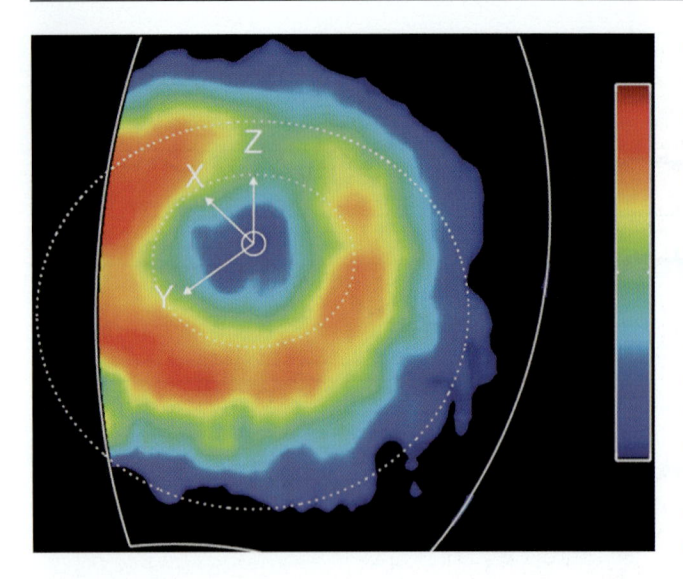

Fig. 9.23 ENA image of the ring current as viewed from above the northern hemisphere. This image, in the range 20–50 keV, was obtained on 19 March 2007, with MIMI/INCA, at a latitude of 54.5° and radial distance 24.5 R_S. Saturn is at the center, and the dotted circles represent the orbits of Rhea and Titan. The Z axis points parallel to Saturn's spin axis, the X axis points roughly sunward in the Sun-spin-axis plane, and the Y axis completes the system, pointing roughly toward dusk. The INCA field of view is marked by the white line and accounts for the cut-off of the image on the left (from Krimigis et al. 2007)

uniform, symmetric construct postulated in early modeling of Saturn's magnetic field.

The ring current geometry does not resemble the neatly-modeled, symmetrical current sheet extending from ∼8 to ∼15 R_S that fit the magnetic field data from the limited Voyager and Pioneer 11 coverage (Connerney et al. 1981). Both in-situ measurements and ENA images have shown that the ring current can at times be highly variable, possibly dominated by a series of injections (Mauk et al. 2005), with strong longitudinal asymmetries that corotate nearly rigidly with the planet, contrary to the Earth's ring current, where no rotational modulation is seen and initial asymmetries are primarily organized by local time effects. One such injection event is shown in Fig. 9.24 (Krimigis et al. 2007), a sequence of six INCA images covering a Saturn rotation. The top left panel shows a large, factor of 10, intensity increase between dawn and local midnight that moves anticlockwise through dawn, then day side, then local evening (middle right panel), then local midnight, and then returns to its original location some 11^h later (bottom right panel).

Since the Pioneer/Voyager epoch, the field due to Saturn's ring current has been modeled using a simple azimuthal symmetric disc model, originally developed to model the Jovian magnetodisk (Connerney et al. 1983 and references therein). The current density in this model is zero in the region outside of the inner and outer edges, and outside of $z \pm D$ (where D is the sheet half-thickness). Inside this region the current density is uniform in z and has a $1/r$ radial dependence. This particular radial dependence was chosen for reasons of mathematical convenience and has been criticized by studies that compare the magnetic stresses measured in situ in Saturn's magnetosphere (Mauk et al. 1985) and theoretical stress balance calculations (Vasyliūnas 1983). Beard and Gast (1987) developed a ring current model which had a more flexible current density profile, and compared more favorably to that measured from Voyager data by Mauk et al. (1985). Nevertheless, the model produces reasonably good fits to the magnetometer data and has been also applied to Cassini magnetometer data. Arridge et al. (2008) showed that fits of the Connerney et al. (1983) model in the outer magnetosphere could be radically improved by considering the bowl-shaped hinging of the current sheet.

Figure 9.25 presents these results as a function of subsolar magnetopause standoff distance. With the exception of the inner edge, each parameter increases with system size - principally due to the increase in the outer edge of the ring current. Bunce et al. (2007) showed that the ring current region occurs on a fixed band of field lines and therefore expands and contracts as the magnetopause position varies. The ring current region therefore maps to a fixed co-latitude range in the ionosphere (14° − 20°N, 16° − 22°S). This lies just equatorward of the observed southern hemisphere aurora, indicating that the aurora are not associated with the mass-loading processes occurring inside the ring current region, but with processes in the outermost layers of the magnetosphere (Bunce et al. 2008). This finding is in approximate agreement with the considerations of Arridge et al. (2007) who qualitatively observed that the magnetodisk appeared to extend right to the magnetopause.

The linear fits in Fig. 9.25 represent an empirical model for the properties of the ring current with system size. Bunce et al. (2008) used this to confirm that the total magnetospheric field on the dayside stretched out into a magnetodisk for low solar wind dynamic pressures, confirming the results of Arridge et al. (2008).

The Dst index is used as a measure of the strength of the ring current and stress state of the terrestrial magnetosphere. Leisner et al. (2007) subtracted a model of Saturn's magnetosphere from magnetometer data in the inner magnetosphere to produce a similar index for Saturn's magnetosphere. The important difference between Saturn and Earth is that both internal and external stresses may affect this kronian Dst index, producing compression or stretching of the magnetic field lines.

Figure 9.26 presents the stress indices calculated by Leisner et al. (2007) where positive (negative) stress indices indicate a compressed (expanded) magnetosphere relative to a ground state. The calculations indicate several large-scale deviations from the ground state and numerous small-scale

Fig. 9.24 Sequence of six ENA images in neutral hydrogen, taken by INCA in the range 20–50 keV on February 24, 2007, covering a full Saturn rotation. Cassini was located at 32° latitude and 26 R_S from Saturn at local time 15:12. Saturn is at the center, the X axis is pointing approximately in the solar direction, Y is pointing towards dusk, and Z is pointing along Saturn's spin vector. Dotted lines show the orbits of Dione, Rhea, and Titan in proper perspective. The images are spaced at roughly 2^h intervals (from Krimigis et al. 2007)

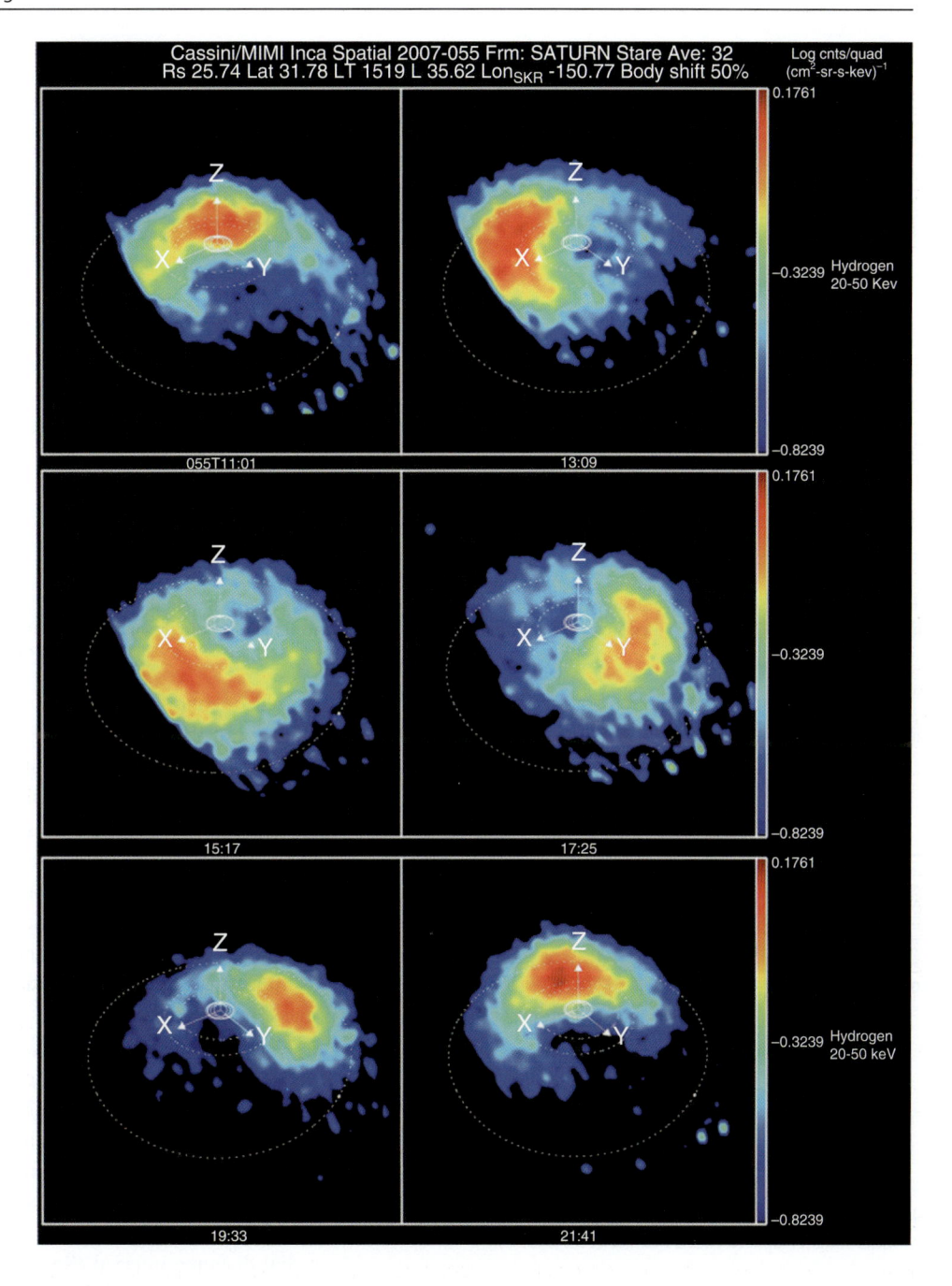

deviations near the noise level of the analysis that are consistent with variations in solar wind dynamic pressure.

The radial, steady-state form of the force balance equation for a collisionless, single-ion plasma in the equatorial plane (assuming isotropic pressure), can be written as:

$$\rho \Omega^2 r - \frac{\partial P}{\partial r} = j_\phi B_z \quad (9.8)$$

where r is radial distance, P is the particle pressure, j_ϕ is the azimuthal current density, ρ is the plasma mass density and Ω is the angular velocity of the plasma. Here we made the assumption that $B \approx B_z$ in the nominal magnetic equatorial plane (ideal dipole field). The non-trivial challenge for theory and spacecraft observations is to establish which force dominates in balancing the magnetic stress: centrifugal stresses or pressure gradients. Observations and modeling from Voyager (McNutt 1983; Mauk et al. 1985) showed that energetic particle pressure gradients (thermal current) were more important inside $\sim 14\,R_S$, but corotation centrifugal forces (inertial current) became more important at larger distances, a result consistent with the work of Arridge et al. (2007). This result is also supported by the work of Bunce et al. (2007) who showed that in the compressed state, the thermal and inertial currents were comparable in size but when the mag-

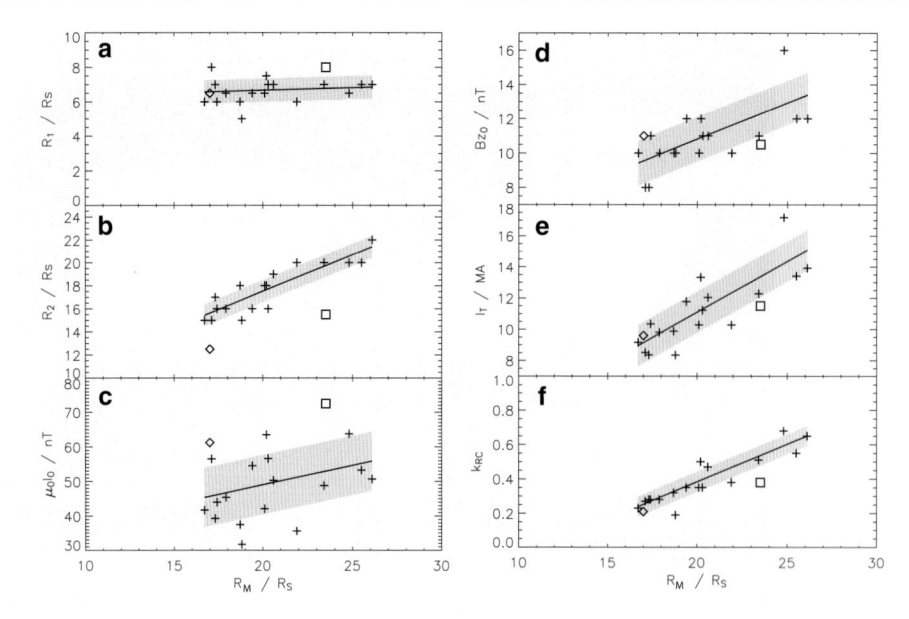

Fig. 9.25 Fitted ring current parameters as a function of subsolar magnetopause standoff distance. Panels (**a**) inner edge, (**b**) outer edge, (**c**) intensity, (**d**) axial (Dst) field, (**e**) total current, (**f**) ratio between the ring current and dipole magnetic moments (from Bunce et al. 2007)

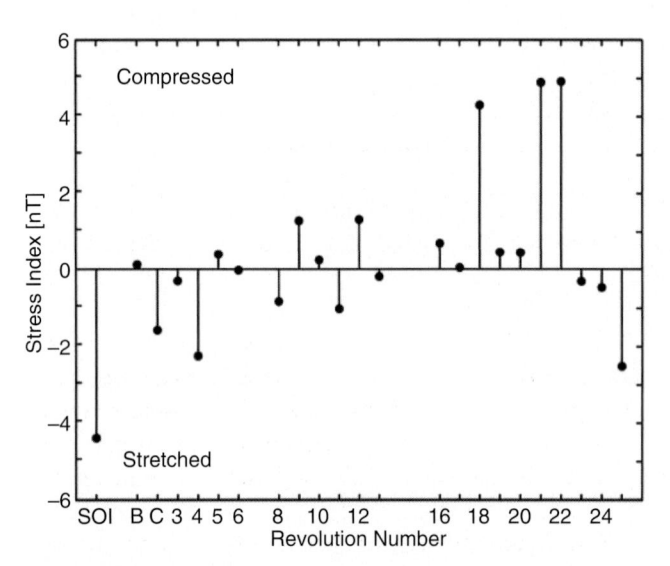

Fig. 9.26 Stress indices constructed for each pass of Cassini through the inner magnetosphere (from Leisner et al. 2007)

netosphere is in its most expanded state, the inertial currents are several times stronger than the thermal currents.

In parts of the magnetosphere where the particle pressure gradient is comparable to or greater than the inertial term, the azimuthal ring current density is decisively modified by the behavior of the suprathermal pressure. Furthermore, the pressure gradient force is expected to dominate over the centrifugal force during injection events, when the energetic particle pressure is significantly increased and highly variable.

Arridge et al. (2007) re-examined the contribution of these two terms to the force balance in the ring current using Cassini data. Transient current sheet crossings ob-served beyond $\sim 18\,R_S$ were used to estimate the magnetic stress, and the lobe magnetic field strength to infer the radial plasma pressure gradient assuming tangential pressure balance across the plasma sheet. They found that beyond 25–30 R_S centrifugal stresses dominated in this force balance, with pressure gradients more important inside this distance. Additionally, the stress balance calculations also yielded an estimate of 10^6 kg for the total mass in the magnetodisk. For mass-loading rates of 10–100 kg/s this requires 3–30 h (0.3–3 planetary rotations) to completely replace the mass in the magnetodisk.

Sergis et al. (2007) investigated the energetic plasma pressure in the middle magnetosphere between ~ 8 and 18 R_S using the MIMI/CHEMS and LEMMS instruments. Figure 9.27 shows the suprathermal equatorial pressure profile that corresponds to the first 3 years of Cassini in orbit, based on equatorial plane data that have been statistically weighted according to the time that Cassini spent within each radial distance interval (Sergis et al. 2007). The color scale is the pressure probability of occurrence. The suprathermal pressure maximizes inside the ring current, with typical values of $\sim 10^{-9}$ dyne/cm^2 between ~ 8 and ~ 10 R_S, gradually decreasing further out. The measured plasma β is kept close to or above unity outside of 8 R_S, with indications that the energetic particle population, rather than the denser cold plasma, controls most plasma pressure in this region of the magnetosphere, in agreement with the limited measurements during Voyager (Mauk et al. 1985). Given the fact that MIMI does not measure the cold plasma pressure, these values should be viewed as a lower limit of both pressure and plasma β. Inside of 8 R_S, the ring current ions are lost through charge exchange with the neutral cloud and

Fig. 9.27 Radial energetic particle (E > 3 keV) pressure profile for the nominal equatorial plane, as observed from all equatorial passes of Cassini, during 3 years in orbit around Saturn. The upper triangles correspond to the pressure medians, while the black solid line is a polynomial fit to the medians (from Sergis et al. 2009)

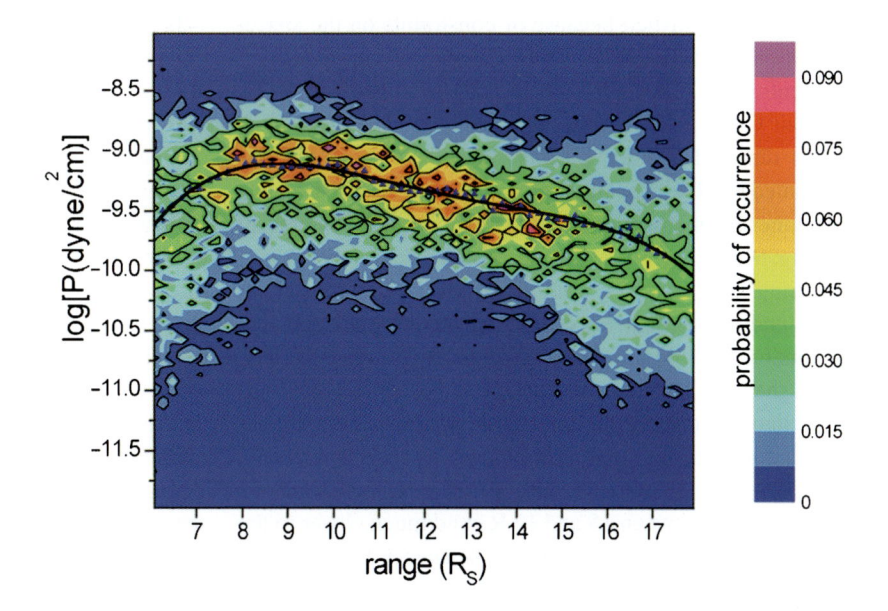

absorption by the icy satellites as described above. Some energetic electrons are lost from satellite interactions and others lose their energy by collisions with the neutral cloud and with dust from the E-ring of Saturn.

The analysis of MIMI/CHEMS compositional data obtained during the first 3 years of Cassini in orbit (July 2004–July 2007), indicates that the O$^+$ ions play a very important role in the energetic particle pressure, often being the dominant component. The O$^+$ partial pressure on average represents more than 50% of the total suprathermal pressure, and, at times, clearly dominates. This becomes more evident in the equatorial plane, in agreement with the density distribution of the neutral water products, and the several O$^+$-rich injection events often seen in the equatorial plane.

One of the direct implications of the high-β discovery in a large part of the Saturnian magnetosphere is that a realistic magnetopause pressure balance equation should necessarily include the suprathermal pressure term, in addition to the magnetic pressure supplied by the planetary field.

9.4.3 Plasma Sheet

The properties of electrons between 0.5 eV and 26 keV in Saturn's magnetotail plasma sheet have been investigated and it was found that the electron temperature is approximately 200 eV beyond 20 R$_S$ and was approximately constant with radial distance. The electron distributions are approximately Maxwellian (there is some evidence for kappa distributions) and occasionally bi-Maxwellian distributions are observed with temperatures of 20 eV and 200 eV. The density at the center of the plasma sheet varies between 10^{-1}/cm^3 at 20 R$_S$ and 10^{-3}/cm^3 at 60 R$_S$ and can be represented by a power-law: $n_e \sim r^{-1.75}$. The electron β was found to maximize at values of 1.0 to 10.0 at the center of the current sheet – the exclusion of ions and energetic particles necessarily mean that this is a lower limit to the tail plasma β.

The plasma density in the lobe was near or at the noise level for the CAPS/ELS instrument and implied that the number density was ≤10^{-4}/cm^3. No evidence was found for a variation of electron temperature between the lobe and the central plasma sheet.

Arridge et al. (2008) organized CAPS/ELS and magnetic field observations by SLS3 longitude (Kurth et al. 2008) and studied periodicities near Titan's orbit. They showed that the electron density varied by more than an order of magnitude between a minimum near 170° ± 20° and a maximum near 350° ± 20°. The electron temperature was essentially constant with longitude but with some weak evidence for an increase in temperature towards 170°. An analysis of the magnetometer data showed that the effective density modulation was primarily produced by a periodic motion of the plasma sheet and not by a density asymmetry in the frame of the plasma sheet (although one might exist). Interestingly the plasma sheet periodicity was either absent or at a lower amplitude when the dayside field structure was quasi-dipolar.

Observation of plasma-sheet ions is complicated by the fact that corotational flow dominates essentially out to the magnetopause and the fact that the ion thermal speed is typically comparable to the flow speed throughout the magnetosphere. This means that there are strong flow anisotropies, so the instrument viewing is crucial. For CAPS to see plasma-sheet ions, it needs to be looking into the flow, a condition frequently not satisfied because of constraints on the spacecraft orientation to enable measurements by the optical instruments.

From CAPS ion observations during SOI, Young et al. (2005) reported that the plasma sheet beyond $\sim 9 R_S$ was dominated by H$^+$ ions, although water-group ions indicative of an inner-magnetospheric source did continue to be present (see Fig. 9.9).

During SOI, however, Cassini was at a latitude of about $-13°$, and subsequent lower-latitude passes indicated that the apparent H$^+$ dominance was primarily a latitude effect. Due to the strong centrifugal force on the plasma, heavy ions are particularly closely confined to the equatorial plane. Near the equator, water-group ions are indeed as abundant as H$^+$, as seen in Fig. 9.28, obtained just before noon local time at a radial distance between 15 and 17 R_S and much closer to the equatorial plane than the SOI pass. The energy-time spectrogram for the ions in the upper panel of Fig. 9.28 shows the characteristic strong periodicity associated with the actuation of the CAPS field of view back and forth across the transonic flow. As in the inner magnetosphere (e.g., Fig. 9.9), there are two bands of ions, the higher-energy one corresponding to flowing W$^+$, and the lower-energy one corresponding to flowing H$^+$. The actual fluxes are variable, but it is clear that the W$^+$ constitute a significant portion of the plasma.

Figure 9.28 illustrates another aspect of the plasma sheet in the outer magnetosphere, namely that the region beyond $\sim 11 R_S$ is often characterized by significant variations in the plasma properties over fairly short time scales. Figure 9.28 shows regions of relatively cool, dense plasma alternating with regions of more tenuous, hotter plasma. This variation can be seen in both the ions (upper panel) and the electrons (lower panel). Figure 9.29 shows the ion moments derived from a numerical integration of the CAPS SNG observations, confirming the anti-correlation of the density and temperature of the plasma during this interval, especially for the W$^+$ ions. In the hot, tenuous regions, the W$^+$ is more strongly depleted than the H$^+$. The bottom panel of the figure confirms what was mentioned above, namely that the ion thermal speeds are comparable to the bulk flow velocity (dominantly azimuthal), particularly for the W$^+$.

It is possible that the alternations in the plasma properties seen in Figs. 9.28 and 9.29 are simply due to rapid variations in the effective latitude of the spacecraft, due to flapping of the plasma sheet or azimuthal variations in its thickness. It is also possible that these different regimes represent the intermingling of flux tubes with different plasma content (Sittler

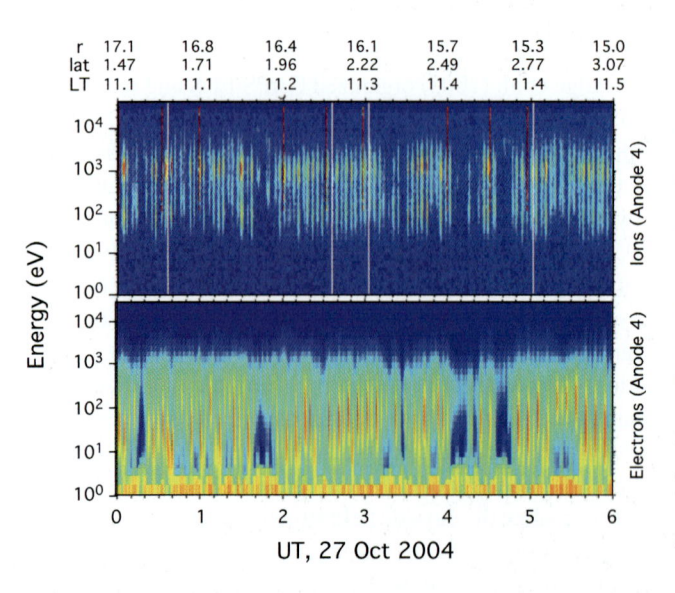

Fig. 9.28 Energy-time spectrogram of ion (*top*) and electron (*bottom*) energy flux observed by CAPS in the near-equatorial dayside plasma sheet. The two peaks in the ion energy spectrum correspond to flowing W$^+$ (*higher energy*) and H$^+$ (*lower energy*). There is strong modulation at the CAPS actuation period (~ 7 min) as the field of view is swept in and out of the flow. In addition to this modulation, there are significant variations in the plasma properties, as seen in both ions and electrons, with regions of cool, dense plasma alternating with regions of hot, tenuous plasma

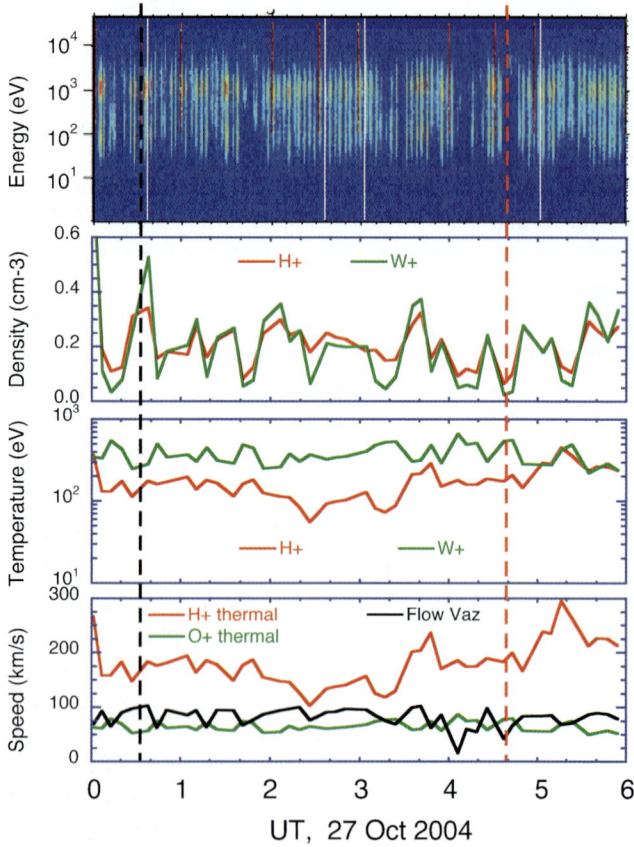

Fig. 9.29 Bulk moments derived from CAPS measurements for the same interval as Fig. 9.28, showing the general anti-correlation of the density and temperature, particularly for the W$^+$. The *bottom* panel shows that the ion thermal speed is comparable to the flow speed, accounting for the strong modulation as the instrument field of view is swept in and out of the flow

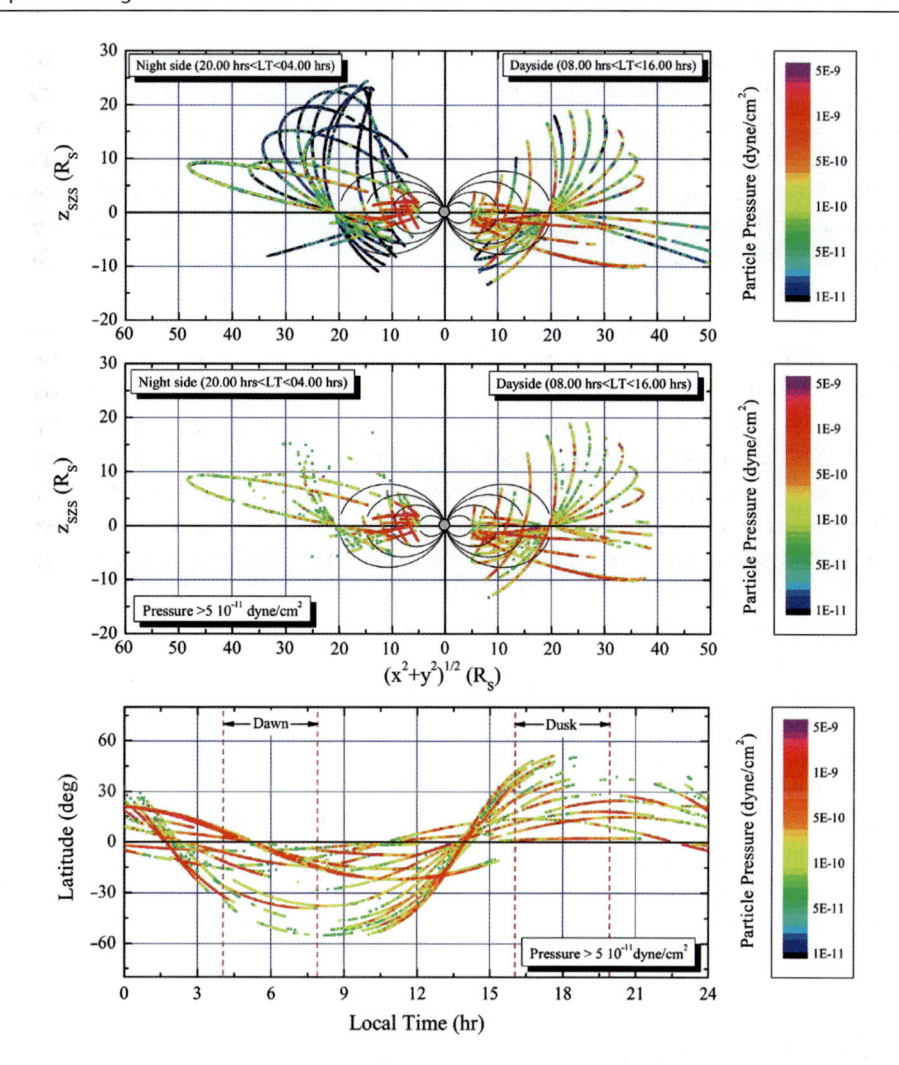

Fig. 9.30 Pressure profiles ($\sim$3–4,000 keV) obtained for the July 2004 to June 2007 period and projected onto the ($\sqrt{X^2 + Y^2}$,Z) plane. *Top*: Pressure profile on the dayside (*right*) and nightside (*left*) over the full dynamic range measured by the CHEMS and LEMMS sensors ($5 \times 10^{-13} - 5 \times 10^{-9}$ dynes/cm^2), clearly illustrating the orbital coverage. *Middle*: The same data but for a threshold $>5 \times 10^{-11}$ dynes/cm^2; the day-night asymmetry at $R > 20\,R_S$ is striking. *Bottom*: Pressure coverage in local time and along the Z axis for all data, but also including the dawn-dusk coverage not shown in the other two panels (from Krimigis et al. 2007)

et al. 2006). McAndrews et al. (2009) suggest that similar low-density, hotter regions seen in the tail plasma sheet are the remnants of previously distended flux tubes that have been broken open by centrifugal stresses on the nightside, releasing the bulk of the cool, heavy plasma near the equator. Such emptied flux tubes would therefore allow the return of magnetic flux to the inner magnetosphere, to replace the heavily loaded flux tubes there with more buoyant ones containing the hotter, tenuous remnant plasma. As discussed above in Section 9.1.3 and in Section 9.5.1, there is good evidence that this interchange process is an important means by which Saturn sheds the plasma produced in the inner magnetosphere. Discrimination between these two possible explanations of the observed plasma sheet structure awaits further analysis of the Cassini data.

The spatial extent and pressure structure of the plasma sheet has been investigated by mapping the partial ion pressure (>3 keV) over a period of $\sim$3 years, from July 1, 2004 to June 30, 2007 (McAndrews et al. 2009). The results are shown in Fig. 9.30 plotted in the ρ, Z plane and separated into the dayside and night side parts. The top panel includes all measured off-equatorial values but excludes the dawn-dusk portion so as to obtain a clear separation of day-night effects. Although the orbital coverage in Z is not uniform, the higher pressures on the dayside appear to extend to much higher latitudes than the night side, certainly at $<20\,R_S$. This fact is clearly evident in the middle panel, where pressures $<5 \times 10^{-11}$ dynes/cm^2 have been omitted. Not only is the day-night asymmetry striking, but also the shape of the night side plasma sheet beyond $\sim$20 R_S is outlined and is seen

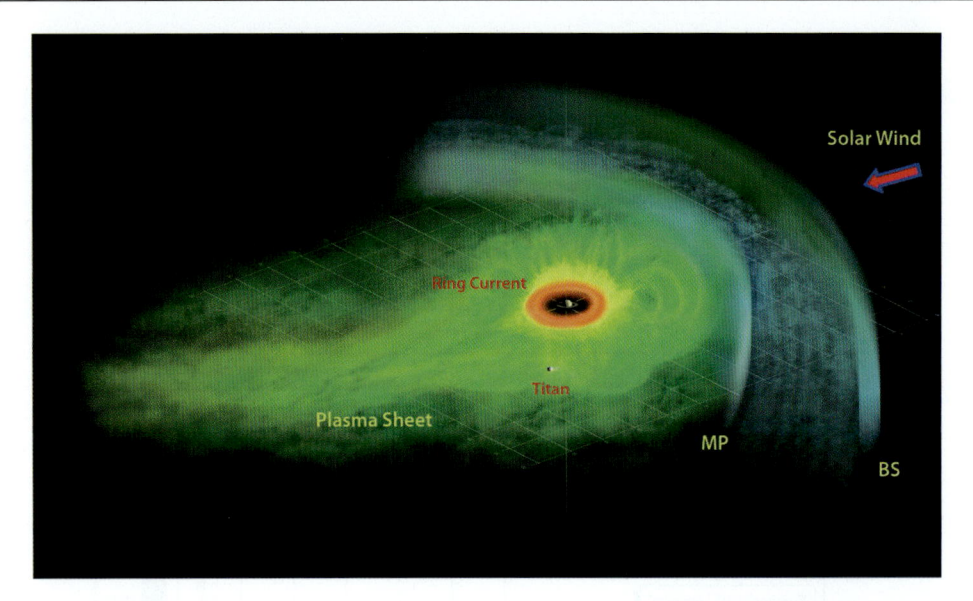

Fig. 9.31 Schematic representation of Saturn's plasma and neutral environment. Saturn is at the centre, with the red 'doughnut' representing the distribution of dense neutral gas (H, O, O_2 and OH) outside the rings. Beyond this region, energetic ions populate the plasma sheet to the day side magnetopause, filling the faintly sketched magnetic flux tubes to higher latitudes and contributing to the ring current. The plasma sheet thins gradually towards the night side (from Krimigis et al. 2007)

to be tilted northward at an angle $\sim 10°$, although the orbital coverage in this region is not extensive. Examination of each Cassini orbit at all available local times suggests that the dayside plasma sheet extends to the magnetopause at local noon but thins gradually toward the night side, even though the detailed distribution with local time is not fully determined because of incomplete latitudinal/local time coverage.

Our current understanding that has evolved from the pressure distribution so far is shown in Fig. 9.31. This view from above Saturn's equatorial plane illustrates the compressed dayside plasma sheet and indicates its expansion to northern and southern latitudes. It is expected that the sheet gradually thins on the dusk side but is drawn tail-ward at midnight and again inflates at dawn. Whether there is loss of plasma on the night side is not clear, because this sketch represents an average picture of all orbits over a nearly 3-year period. Acceleration events, however, have been observed repeatedly both in the magnetotail (McAndrews et al. 2009) and in parts of the magnetosphere, where the injected plasma cloud clearly corotates with the planet, as illustrated in Fig. 9.24.

9.4.4 Magnetotail

The interaction of the solar wind with Saturn's internal and external fields produces a long magnetotail on the night side in which the magnetic field decreases away from the planet at a much slower rate than a dipole field. Both drag applied by the solar wind on the magnetosphere, and the reconnection of the IMF field lines to those of Saturn play a role in transferring magnetic flux from the day side magnetosphere to the night side magnetotail.

At the center of the magnetotail, a current sheet maintained by the hot plasma of Saturn's magnetosphere keeps the magnetic field at a very low value. Above and below the current sheet, the magnetic field is much stronger, points nearly radially outward ($B_r > 0$) above the current sheet, and radially inward ($B_r < 0$) below the current sheet. These regions outside the current sheet are called the lobes and house the open flux of Saturn's magnetosphere, which is connected to the solar wind IMF field at one end and to Saturn's ionosphere on the other. Figure 9.32 shows the strength of the magnetic field (directed mainly in the radial direction) observed during the inbound leg of Rev. 26 and that expected from Saturn's dipole in the lobe region (dashed line).

In the low-latitude regions of the magnetotail the signs of the radial and azimuthal components of the field are observed to be opposite to each other outside a radial distance of $\sim 12\,R_S$. Thus, the field lines appear to have a bent-back configuration in the outer magnetosphere. This swept-back appearance arises from the presence of radial currents flowing in the magnetosphere, which accelerate the plasma towards corotation.

A striking feature of magnetic field observations from the magnetotail region is the presence of periodicities close to the rotation period of Saturn (Andrews et al. 2008). Figure 9.33 shows an example of such periodicities. The two reversals of B_r in each spin period illustrate that the spacecraft travels across the current sheet in a periodic fashion, which is possible only if the current sheet is tilted with re-

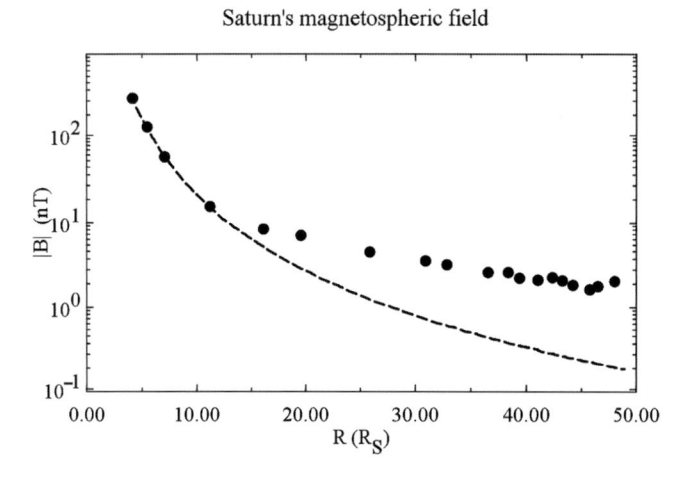

Fig. 9.32 Magnetic field strength in the low latitude lobe region (*dots*) observed during the inbound leg of Rev. 26 and the dipole field in the same region (*dashed line*)

spect to Saturn's equatorial plane and is rotating with a period close to that of Saturn. A model put forward by Khurana et al. (2009) postulates that the tilt of the current sheet arises naturally from the action of solar wind lift on an azimuthally asymmetric magnetosphere (see Fig. 9.6).

Southwood and Kivelson (2007) also propose a tilted current-sheet interpretation for the periodicities in the magnetotail. In their model they suggest that the tilt is produced by an effective dipole moment rotating in the equatorial plane in the middle magnetosphere. They show that a system of field-aligned currents flowing on magnetic shells between 12 and 15 R_S can produce a tilted dipole beyond these shells but a uniform "cam-shaft" field inside, consistent with the periodicities in the azimuthal and radial magnetic field components. Carbary et al. (2007) proposed an alternative interpretation and suggested that compressional motions in the inner and middle magnetosphere, produced by an azimuthally asymmetric density profile, would translate to transverse motions in the outer magnetosphere when Saturn's dipole was tilted with respect to the solar wind.

Plasma flow observations from Saturn's magnetotail shed light on the global convection pattern discussed in Section 9.1.3. The main issues are the extent to which corotation is enforced on the night side, evidence for the "shedding" of material originating in the inner plasma source region, and evidence for the return of magnetic flux to the day side. Figure 9.34 shows the results of a recent study by McAndrews et al. (2009) of ion flows in Saturn's magnetotail. For plasma dense enough to be detectable by the CAPS IMS, the flows are dominantly in the corotational direction, even out to very large distances ($\sim 40 R_S$), although the flow speed is well below full corotation. However, beyond $\sim 20 R_S$, there is no evidence of an inward component of these flows, and it appears that the pull from the ionosphere may be inadequate to drag the more distant flux tubes around to the day side to complete their convective circuit.

For the intervals studied by McAndrews et al. (2009), the plasma was relatively cool and dense, with a water-group/hydrogen composition similar to that seen in the dayside plasma sheet (Section 9.4.3), indicative of plasma originating in the inner magnetosphere. An estimate of the mass per unit flux for these intervals shows it to be generally at or below the critical value above which the magnetic field should be unable to confine the plasma (Goertz 1983). This evidence confirms the conclusion that these flux tubes are still closed and still loaded with inner magnetospheric plasma, although they may not be able to complete the circulation onto the dayside intact.

Another aspect of the tail plasma studied by McAndrews et al. (2009) was that between the intervals with a detectable ion population were numerous intervals of much lower density, as indicated by the more sensitive electron measurements. While some of these low-density intervals may reflect excursions to higher latitudes in the flattened tail plasma sheet, at least some of them are attributable to flux tubes depleted of their plasma content through down-tail loss to a planetary wind (McAndrews et al. 2009). They thus presumably represent the magnetic flux that must return to the dayside outer magnetosphere (cf., Vasyliūnas 1983).

The down-tail loss of plasma may occur by way of magnetic reconnection of flux-tubes strongly distended by centrifugal forces. Evidence for magnetic reconnection and the subsequent expulsion of a plasmoid has been seen in both magnetic field (Jackman et al. 2007) and plasma measurements (Hill et al. 2008). The locations of the two plasmoid events reported by (Jackman et al. 2007) for which Hill et al. (2008) could measure plasma properties, are shown in Fig. 9.34, with their corresponding flow vectors, which have a strong radial component. However, to date very few such plasmoid events have been identified in Cassini tail observations, suggesting that large-scale reconnection of this type may be relatively infrequent, and smaller-scale processes may dominate the normal mass loss down-tail. One clue to the frequency of occurrence of large-scale plasmoid release events is that just before the in situ detection of one of the events, the MIMI/INCA instrument observed a burst of energetic neutral atoms emanating from a location midway between Saturn and Cassini, probably a signature of the reconnection event that produced the plasmoid (Hill et al. 2008). Such ENA brightenings have been attributed to a substorm-like process taking place in the near-tail region (Mitchell et al. 2005), and the brightenings are associated with bursts (Kurth et al. 2005) of Saturn Kilometric Radiation (SKR) and possibly with injections of energetic particles deeper into the inner magnetosphere (Mauk et al. 2005). The occurrence rate of ENA brightenings, SKR enhancements, and energetic particle injections may therefore suggest the frequency of plasmoid releases into the tail, in-

Fig. 9.33 The difference field observed during Rev. 26. Notice, the out of phase periodicities observed in the B_r and the B_ϕ components at distances beyond about 15 RS. In each rotation of Saturn, the B_r component reverses twice

Fig. 9.34 Equatorial plane projection of plasma flow velocities derived from CAPS IMS data for various intervals during Cassini passes through Saturn's magnetotail region. The vectors originate at the spacecraft location and point in the direction of the flow, with a length proportional to the flow speed and a color determined by the density. Velocities derived by Hill et al. (2008) for two plasmoid events are also indicated (*colored black*) (from McAndrews et al. 2009)

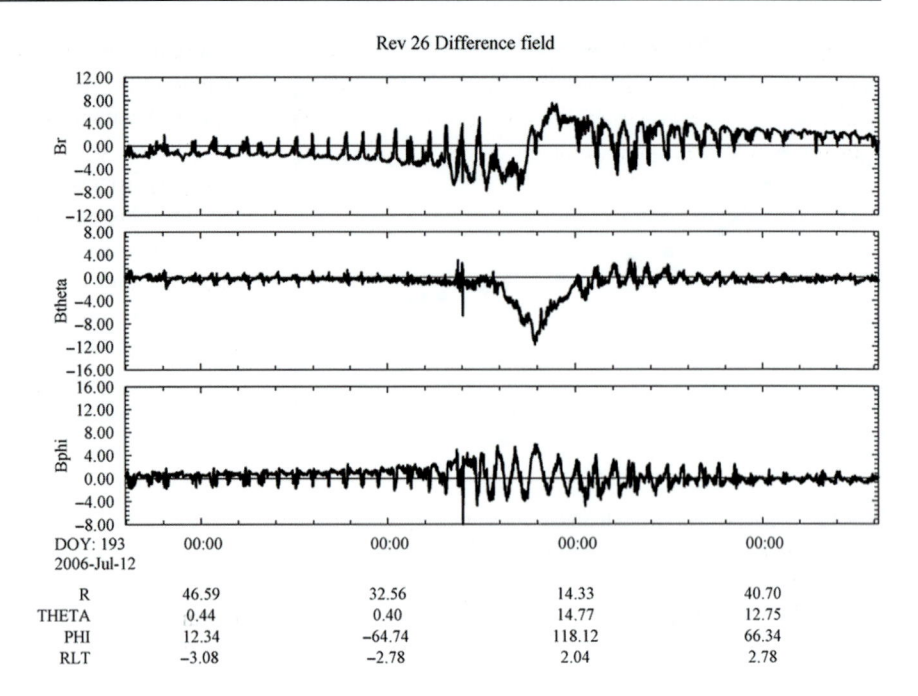

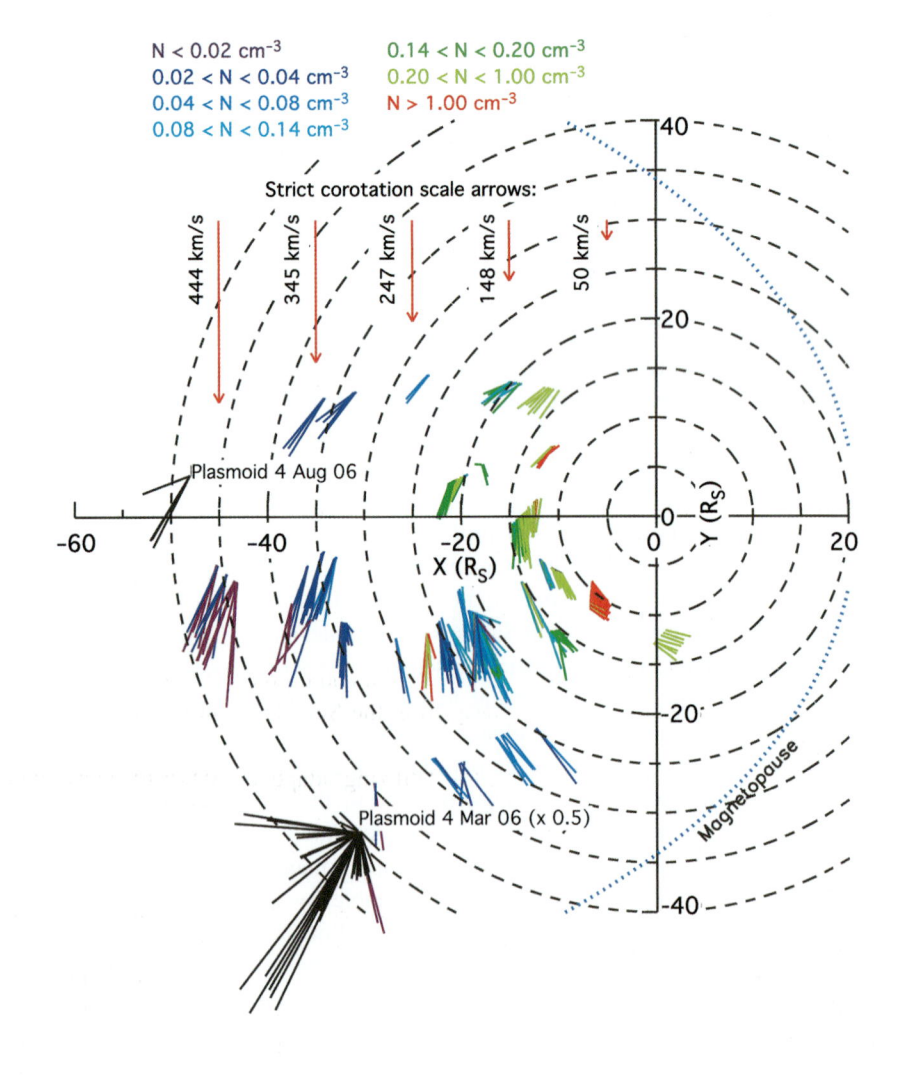

dicating whether or not this mechanism is sufficient to shed the plasma produced in the inner magnetosphere or whether other, smaller-scale processes might also be needed. This analysis has not been done to date.

9.4.5 Global MHD Models

Hansen et al. (2000) were the first to develop a global MHD model to study the large-scale interplay between the solar wind, Saturn's fast rotating embedded magnetic dipole and magnetospheric plasma sources. They modeled Saturn's magnetosphere using a version of the global, 3D MHD model BATS-R-US (Powell et al. 1999). Their present model (Hansen et al. 2005) includes several important improvements, such as a substantially modified prescription of the mass loading distribution and the use of observed conditions to prescribe the upstream solar wind. Other improvements include the use of the semi-relativistic form of the MHD equations (Gombosi et al. 2001), an implicit time stepping algorithm (Tóth et al. 2006), and better resolution.

Because the addition of mass to the kronian system is significant, Hansen et al. (2005) include this process in their MHD model through appropriate source terms for ionization, pickup, recombination and ion-neutral drag. The inner source due to dust and icy satellites is modeled as an axisymmetric torus confined near the equatorial plane using data from Richardson and Sittler (1990) and Richardson et al. (1998). They use an average mass of 16.6 amu which assumes a nearly equal mix of OH and O. A total mass loading rate of $\sim 10^{27}$/s is used in the Hansen et al. (2005) simulations (but this is an adjustable parameter that can be scaled as necessary). Mass loading due to Titan is modeled as an axisymmetric torus. The torus is centered on Titan's orbit and has a much lower mass addition rate than the inner source. The Titan related mass loading rate is $\sim 5 \times 10^{25}$/s and it was obtained by using a peak neutral density of ~ 10/cm^3 combined with a neutral lifetime ($\sim 3 \times 10^7$ s Barbosa 1987) and average mass (14 amu).

Hansen et al. (2005) assume that the rotation and dipole axes are aligned and tilted 24.48° away from the Sun. The simulation domain covers the region of $96\,R_S > X > -576\,R_S$, $-192\,RS < Y, Z < 192\,R_S$. Utilizing adaptive blocks, they are able to highly resolve the inner equatorial plane while also resolving the bow shock, magnetopause and tail regions appropriately. The smallest computational cells near the icy satellite mass loading region in the equatorial plane are 3/16 R_S across while the largest cells (6 R_S) are located far downtail. The inner boundary is at 3 R_S.

Global MHD simulations have been successfully used to put the Cassini observations in a global perspective and to predict new phenomena. Hansen et al. (2005) applied the model to describe Saturn's magnetosphere under conditions appropriate for the period just before the Cassini orbit insertion (26–29 July 2004). They successfully simulated the bow shock and magnetopause crossings (Achilleos et al. 2006; Gombosi and Hansen 2005) and investigated the compressibility of the dayside magnetopause (Hansen et al. 2005). The compressibility derived from the simulations ($\alpha \approx 5.2$) is in excellent agreement with the observed value (see Section 9.6.4.2). They also successfully simulated the bowl shaped magnetodisk, the plasma 'drizzle' along the dusk magnetotail and the global magnetospheric convection pattern (Hansen et al. 2005). These features can be seen in Fig. 9.35.

Fukazawa et al. (2007a,b) used a different global MHD model to investigate the influence of IMF B_z on Saturn's magnetosphere. They found that the subsolar magnetopause and bow shock positions are sensitive to the solar wind dynamic pressure, but they are insensitive to the IMF B_z and the amount of reconnected magnetic flux. They also found that vortices were formed in the magnetosphere that increased magnetic reconnection along the flanks of the magnetopause.

9.5 Ionosphere-Magnetosphere Coupling

9.5.1 Radial Transport

Saturn's magnetospheric plasma is primarily produced deep in the magnetosphere, partly by photoionization but predominantly by electron impact ionization of the water-group neutral gas cloud liberated by the Enceladus geysers and redistributed by charge-exchange with corotating plasma. Because the electron temperature is regulated by collisional equilibration with the pick-up H^+ ions, little electron impact ionization can occur inside $L \sim 6$, where the H^+ corotational energy finally exceeds the 10–20 eV needed for ionization. Thus, even though the neutral cloud is most dense near Enceladus itself, the peak plasma production is further out.

The peak in plasma production at $L \sim 6$ leads to a negative radial gradient in the flux-tube plasma content beyond that distance, which is unstable to flux-tube interchange. The onset of interchange rapidly transports the plasma outward into the outer magnetosphere. Inside of $L \sim 6$, the transport is much slower (see also Carbary et al. 2009), and the primary loss mechanism is recombination. Some authors have hypothesized a global circulation pattern in the inner magnetosphere (e.g., Goldreich and Farmer 2007; Gurnett et al. 2007), which would shorten the inward transport time, but so far analysis of CAPS ion data shows radial flows of less than 1–2 km/s (Sittler et al. 2006, Wilson et al. 2008), although a comprehensive study has not yet been done.

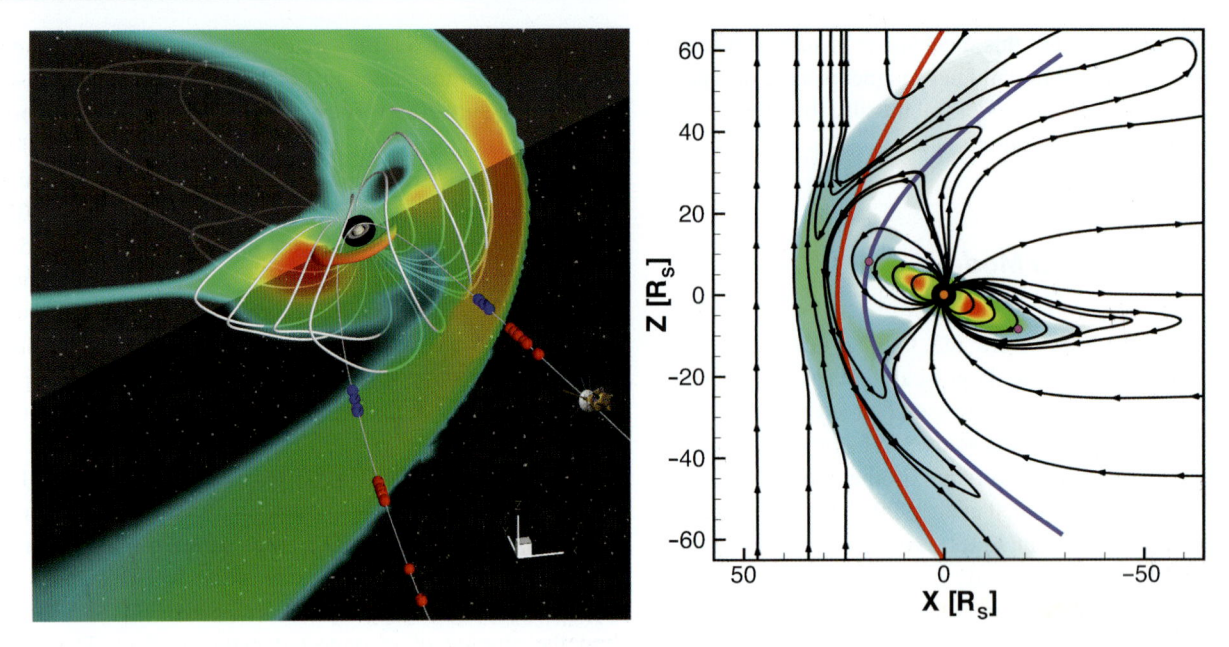

Fig. 9.35 *Left panel*: A snapshot of Saturn's simulated magnetosphere during the SOI period. The thick white lines are a 3D representation of the last closed magnetic field lines. On the dayside they represent the magnetopause while on the nightside they represent the inner magnetosphere. The crescent at 6–7 RS is a density isosurface showing that the plasma is concentrated near the rotational equator. The color code represents pressure in the equatorial and noon-midnight planes. The solid grey trajectory of Cassini is shown with observed bow shock (*red*) and magnetopause (*blue*) crossings (from Gombosi and Hansen 2005). *Right panel*: Noon-midnight meridional plane of Saturn's simulated magnetosphere at the time of the first bow shock crossing. Color contour is the thermal plasma pressure with black lines representing magnetic field lines. The solid red and blue lines are respectively the Slavin et al. (1985) average magnetopause and bow shock models. The intersection of Titan's orbit with this plane is shown as two small pink dots (from Hansen et al. 2005)

The situation is different with the electrons. Phase space density analysis of the CAPS/ELS electron measurements of both the hot and cold electron populations located the source of the cold electron population (<100 eV) in the inner magnetosphere and the source of the hot electron component (100 eV to >10 keV) in the outer magnetosphere at L $\sim$ 11 (Rymer et al. 2007). However, the simple plasma transport cycle of outward-convecting flux tubes containing cold, dense plasma and inward-convecting flux tubes containing hot, tenuous plasma required further investigation when Burch et al. (2007) found evidence of pitch angle distributions consistent with the outward transport of hot electrons. Subsequently, Rymer et al. (2008) presented a circulation model for electrons which originate in the inner magnetosphere and circulate in a cycle of outward and inward convection, driven by the centrifugal interchange instability. Heated plasma drifts out of the inward-convecting flux tubes through gradient and curvature drifts, disappearing by L $\sim$ 6 when, caught up in the corotating plasma, the heated plasma subsequently convects radially outward along with the colder and denser background plasma (Rymer et al. 2008).

The first observational evidence that the interchange process is occurring at Saturn was obtained during Cassini's orbit insertion in 2004 (André et al. 2005; Burch et al. 2005; Hill et al. 2005; Leisner et al. 2005, Mauk et al. 2005). Centrifugal interchange events have magnetic and plasma signatures consistent with distant magnetic flux tubes containing hot tenuous plasma convecting inward to replace tubes containing cold, dense plasma. Such injection events occur throughout the plasma source region in the range 5 < L < 11 (Bunce et al. 2005b; Hill et al. 2005; Rymer et al. 2007, Menietti et al. 2008). Due to adiabatic gradient and curvature drifts, the injected plasma exhibits significant longitudinal drift dispersion, which is evident as V-shaped dispersion signatures on linear energy-time plots (Hill et al. 2005). By estimating the slope and thickness of the V-shaped signature, Hill et al. (2005) were able to deduce the age and longitudinal width of the injection events.

Remote events have a dispersed energy signature with decreasing ion energy and increasing electron energy occurring sequentially within the boundaries of the event (Young et al. 2005). Similar energy dispersion curves can be seen for the more energetic ions (Mauk et al. 2005). Characteristically, local events have deep density gradients and high plasma temperatures coincident with the abrupt boundaries of the injection event (André et al. 2007), with higher plasma temperatures exceeding the background temperatures by a factor of thirty (Rymer et al. 2007). Figure 9.36 is a multi-instrument display depicting magnetic and plasma parameters during a series of local injection events taken during Cassini's first full orbit around Saturn on October 28, 2004. The top panel shows enhancements in the magnetic pressure that correlate

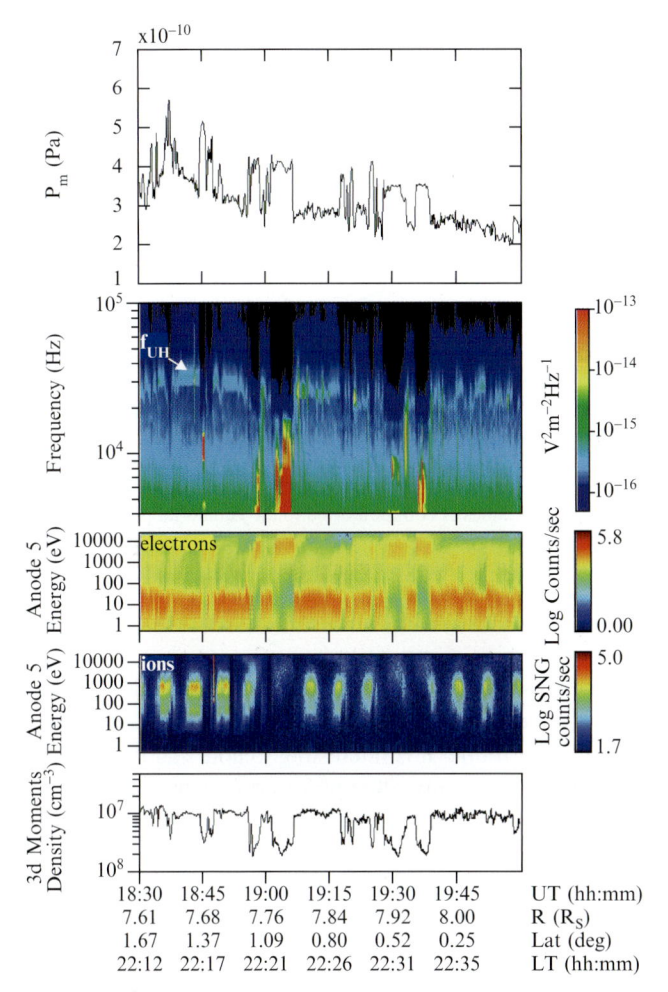

Fig. 9.36 A multi-instrument analysis of a series of injection events on October 28, 2004, showing correlations between enhancements in the magnetic pressure (*panel 1*), the disappearance of the upper hybrid emission band (*panel 2*) coinciding with significant density drop-outs (*panel 5*), and the depletion of the low-energy plasma population coinciding with the appearance of a hot plasma component (*panels 3 and 4*) inside the well-defined boundaries of the interchanging magnetic flux tubes (from André et al. 2007)

with the intermittent disappearance of the upper hybrid resonance emission band (panel 2) and significant drop-outs in the plasma density (panel 5). Another characteristic signature of these injection events is the depletion of the low-energy plasma population coinciding with the appearance of a hot plasma component (panels 3 and 4) inside the well-defined boundaries of the interchanging magnetic flux tubes (André et al. 2007). The local events show evidence of decreasing ion energies and increasing electron energies occurring simultaneously within the boundaries of the event with little dispersion (Burch et al. 2005). The low plasma densities and hotter temperatures found inside the local injections are characteristic of the magnetospheric plasma found at larger radial distances (Burch et al. 2005).

Injection events also have characteristic signatures in the magnetic field and plasma wave data. Plasma waves associated with density-depleted flux tubes include intense, narrow-banded electrostatic emissions at the electron cyclotron harmonics (Menietti et al. 2008) and whistler-mode electromagnetic chorus emissions (Hospodarsky et al. 2008). The onset of these plasma waves is clearly associated with the well-defined boundaries of the inward-convecting magnetic flux tubes. In addition to the plasma wave signatures, there are magnetic field signatures associated with these events. Typically the dispersion signatures on the energy-time CAPS spectrograms correlate with diamagnetic depressions in the magnetic field magnitude (André et al. 2007; Hill et al. 2005), although there are injection signatures which correlate with diamagnetic enhancements as well (André et al. 2007). Although both magnetic signatures correspond to inward-propagating magnetic flux tubes containing hotter, low-density plasma, the deepest density cavities and largest temperature increases within the boundaries of these injection events correlate with magnetic field enhancements (André et al. 2007).

The rapid outward transport of inner magnetospheric plasma has two major implications for the magnetospheric configuration. First, as discussed above, the presence of this plasma gives rise to a ring current that significantly distorts the magnetic field. Second, the magnetosphere has to find a way to drain it to maintain an approximate steady state. The relatively low densities in the outer magnetosphere make recombination an unlikely solution, and the tiny size of the loss cone likewise rules out significant losses from precipitation. Thus, the material must be transported out of the magnetosphere. This problem was addressed by Goertz (1983), who suggested that the plasma is lost to a planetary wind (see Vasyliūnas 1983) when loaded flux tubes convect into the night side and break open. As discussed above, this would occur when the magnetic tension is unable to enforce azimuthal motion, i.e., when the mass per unit flux exceeds a critical value related to the magnetic field strength (Goertz 1983).

A recent study of plasma in the magnetotail (McAndrews et al. 2009) has shown that, even to downtail distances $\sim 40 - 50$ R$_S$, regions of cool, dense, partially-corotating water-group plasma are commonly encountered, but the great majority of these have an estimated mass per unit flux at or below the critical value. Between these apparently intact flux tubes bearing inner magnetospheric plasma, there are regions of significantly lower density (typically below the CAPS threshold for ion detection, but still detectable in the electrons). The interpretation of this finding is that flux tubes that were sufficiently loaded to exceed the critical limit have broken open earlier in their passage through the tail, losing the bulk of their contents down-tail, and returning to the day side with a much more tenuous, hotter, and heavy-ion-depleted residual plasma. These emptied flux tubes subsequently form the low-density regions seen in the dayside outer magnetosphere, interspersed with mass-loaded flux tubes that have not yet exceeded their critical limit in their passage through the night side.

9.5.2 Corotation, Subcorotation and Corotation Breakdown

The highly conducting layer in the ionosphere is thin compared to the radius of the planet. Therefore, the height-integrated ionospheric current density can be written as Vasyliūnas 1983:

$$\mathbf{j}_{iono} = \Sigma \left(\mathbf{u}_n - \mathbf{u}_{mag} \right) \times \mathbf{B} \quad (9.9)$$

where Σ is the height integrated ionospheric conductivity, $\mathbf{u}_n$ is the velocity of the neutral atmosphere at ionospheric altitudes, $\mathbf{u}_{mag}$ is the velocity of magnetospheric plasma just above the ionosphere. The frozen-in MHD condition ($\mathbf{E} = -\mathbf{u} \times \mathbf{B}$) implies that plasma elements originally on a magnetic field line remain on the field line at any later time. This means that the plasma velocity just above the ionosphere, $\mathbf{u}_{mag}$, determines the plasma velocity perpendicular to this field line (the convection velocity) everywhere along this field line (e.g., Gombosi 1998).

If the height-integrated ionospheric conductivity is very large ($\Sigma \to \infty$) the upper atmosphere and the magnetosphere above must move with the same velocity, $\mathbf{u}_n = \mathbf{u}_{mag}$ (Vasyliūnas 1983). Since the upper neutral atmosphere corotates with the planet with an angular velocity Ω_u (note that Ω_u is not the internal rotation rate of Saturn), the magnetospheric footpoint of the field line moves with a velocity of $\mathbf{u}_{mag} = \Omega_u \times \mathbf{r}$. In other words, the footpoint of the magnetospheric field line corotates with the upper atmosphere.

As it has been discussed earlier (see Section 9.3) the rings, Enceladus and the icy satellites produce $\sim 10^2$ kg of water group ions per second near the equator between $3\,R_S$ and $5\,R_S$ radial distance. This plasma is "loaded" to rotating closed magnetic field lines that are increasingly stretched by the balance between magnetic stresses and centripetal forces (Gleeson and Axford 1976; Hill and Carbary 1978).

In the MHD limit the ionosphere rotates with Ω_S and it "drags along" closed magnetic field lines that are increasingly "mass loaded" in the equatorial region. This process generates a coupling current system that is depicted in Figure 9.37.

Subcorotation is primarily caused by addition of newly ionized particles to the local plasma population (see Gombosi 1988, 1998, Jackman et al. 2006; Mauk et al. 2009). The process of ion pick-up results in a slowing of the plasma flow

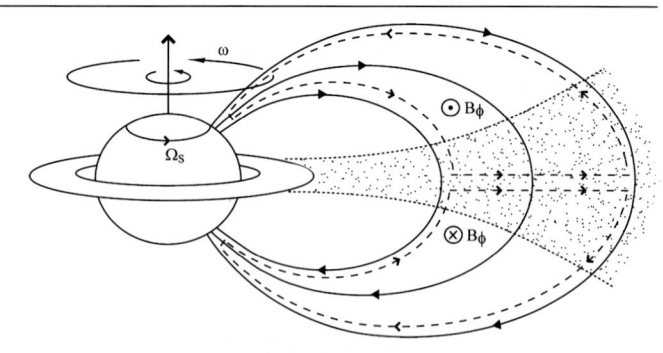

Fig. 9.37 A meridional cross section through Saturn's magnetosphere extending to distances of ~ 15–$20\,R_S$. The arrowed solid lines are closed magnetic field lines. The rotating plasma is shown by the dotted region. The angular velocity of a particular shell of field lines, Ω, and the angular velocity of the neutral upper atmosphere in the Pedersen layer of the ionosphere, Ω_S, are also shown. The frictional torque on the magnetospheric flux tubes is communicated to the equatorial plasma by the current system shown by the arrowed dashed lines. This current system bends the field lines out of meridian planes into a 'lagging' configuration (Cowley and Bunce 2003)

since newly-created ions enter the plasma essentially at rest in the Keplerian frame, thereby adding negative momentum in the plasma frame. It is interesting to note that the thermal velocity of a uniformly filled pickup ring is the same as the transport velocity of the magnetic field lines. This means that even a small percentage of heavy pickup ions (such as W^+) have a large contribution to the plasma pressure. The evidence for the pick-up of new ions in the inner magnetosphere is thus consistent with the observations of subcorotation in this region (e.g., Richardson 1986; Eviatar and Richardson 1986; Mauk et al. 2005, Wilson et al. 2008). The left panel of Fig. 9.38, from Wilson et al. (2008), shows the azimuthal plasma flow speeds derived from CAPS observations from several inner magnetosphere passes, with the flow curve inferred by Mauk et al. (2005) shown as a dashed curve.

Local ion pick-up, however, is not the only process that could produce the observed sub-corotation. In fact, there are three candidate causes:

1. Local ionization within the inner magnetosphere, via photoionization or electron-impact ionization, as discussed above (which is a plasma density source as well as a momentum-loading process)
2. Momentum exchange between existing plasma and neutrals, which could be either ion/neutral collisions or charge exchange, but which is dominated by charge exchange (e.g., Saur et al. 2004) (momentum-loading with no change in density)
3. Outward transport of plasma, with low ionospheric conductivity preventing its acceleration up to full corotation as discussed by Hill (1979)

The relative importance of these processes is determined by a complex mix of factors, including the height-integrated

9 Saturn's Magnetospheric Configuration

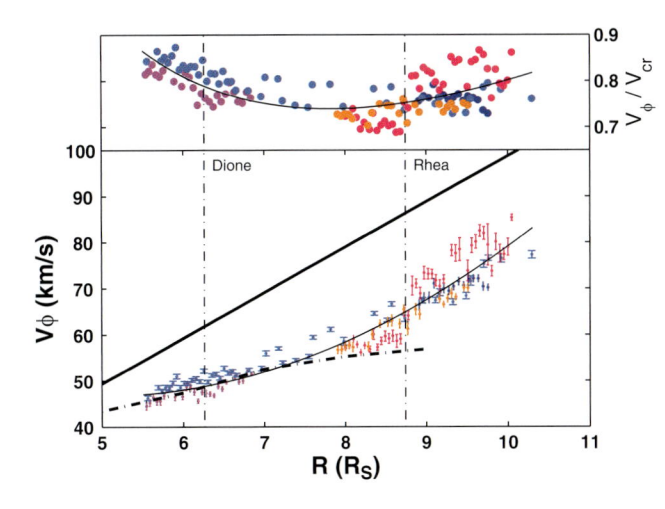

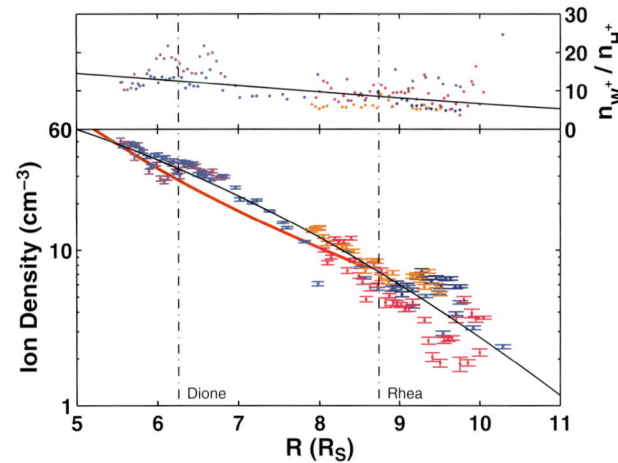

Fig. 9.38 *Left panel*: Azimuthal flow speed derived from CAPS IMS measurements made on several passes through the inner magnetosphere (different colors represent different passes) (from Wilson et al. 2008). The heavy solid line shows the value of azimuthal flow for perfect corotation, and the top panel shows the ratio of the measured speed to perfect corotation. The heavy dashed curve is the azimuthal flow speed inferred by Mauk et al. (2005) from SOI observations of energetic particle injections, and the thin curve is an analytical function fit to the CAPS measurements. *Right panel*: Total ion density derived from the same measurements as shown in the left panel. The top panel shows the ratio of the derived water-group density to the derived H^+ density. The red curve in the lower panel is the analytical function from a fit to the RPWS total electron density (Persoon et al. 2006), and the thin curve is a analytical function fit to the CAPS measurements

Pederson conductivity of Saturn's ionosphere, the radial dependence of the density and composition of neutrals, the energy distribution of magnetospheric electrons, and the ion density and composition. Using plasma parameters derived from the Voyager encounters Richardson (1986) and Saur et al. (2004) modeled these three processes for Saturn's inner magnetosphere and were able to approximately reproduce the Voyager-observed subcorotation, including some notable dips in the azimuthal velocity profile, with a very small effective Pedersen conductivity and substantial ionization and charge-exchange (at roughly equal rates throughout the modeled region).

Analysis of Cassini data can likewise help identify the relative contributions of the various processes producing subcorotation and thereby pinpoint the primary plasma source region. RPWS measurements (Persoon et al. 2005, 2006) showed that the equatorial electron density in the inner magnetosphere ($3.6 < L < 8.6$) falls off with distance as $\sim L^{-4.1}$. Conservation of flux-tube content during radial transport would produce a radial density dependence of L^{-4} (for uniform density within the flux tube; or L^{-m} where $m < 4$ if the plasma is equatorially confined due to temperature anisotropy or centrifugal effects). The fact that m is found to be greater than 4 suggests diffusive transport of plasma from a source in the inner region to a sink beyond $L \sim 8.6$.

A subsequent analysis of CAPS ion observations for a number of equatorial passes through the inner magnetosphere produced the density profile shown in the right panel of Fig. 9.38 (Wilson et al. 2008). Comparison with the Persoon et al. (2006) profile, indicated by the red curve, shows generally good correspondence, but a slightly different radial dependence. The best fit relationship found by Wilson et al. (2008) was not L^{-m}, but rather $n \sim A \times \exp(-BL^2)$. When multiplied by L^4 to approximate flux-tube content, this function actually reveals a peak in the estimated flux-tube content at $L \sim 7$. If a multiplier of L^3 is used to qualitatively account for equatorial confinement, the peak is at $L \sim 6$. This is consistent with the more sophisticated analysis of flux-tube content performed by Sittler et al. (2008) using CAPS observations from SOI and inferred temperature anisotropies. Their derived flux-tube content profile is shown in Fig. 9.39. There are two important conclusions to be drawn from Fig. 9.39: First, the presence of the peak in flux-tube content at $L \sim 6$ suggests that the net plasma transport is inwards inside of that point. Since inward transport would produce supercorotation via the Hill mechanism (process 3 above), the observation of subcorotation in this region requires significant ionization and/or mass-loading there, which agrees with the conclusions of Saur et al. (2004).

The second important point in Fig. 9.39 is that beyond the peak in flux-tube content there is substantial variability in the derived values. This variability is due to the action of the interchange instability (described above), which exchanges fully loaded flux tubes from the inner magnetosphere with flux tubes from the outer magnetosphere that contain only hot tenuous plasma (e.g., Hill et al. 2005; Burch et al. 2005; André et al. 2005, 2007). As discussed by Southwood and Kivelson (1987, 1989), plasma should be interchange-unstable when the flux-tube content decreases with radial distance, as indeed observed.

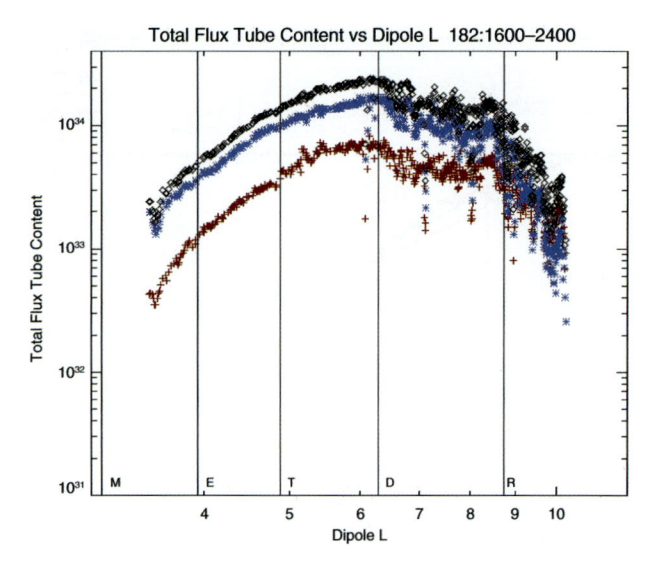

Fig. 9.39 Total flux tube content for protons, $n_p L^2$ (*red*), and for water group ions (*blue*), $n_p L^2$, and the sum of protons and water group ions (*black*) plotted as a function of dipole L. Icy satellite L shells are indicated. (from Sittler et al. 2008)

Using the neutral torus model of Johnson et al. (2006) and the electron temperature observed by CAPS, Sittler et al. (2008) calculated the ion production rate profile corresponding to the SOI observations represented in Fig. 9.39. They found a broad peak in ion production per unit L between L ∼ 4.5 and 6.5. The peak is a convolution of the neutral density, which peaks inside of L ∼ 5, and the electron energy flux, which falls off rapidly inside L ∼6. The decline in electron energy flux is due to loss of hot electrons delivered from the outer magnetosphere via interchange events, combined with the low local electron temperature produced by collisional coupling to the cool, corotating H^+ (Rymer et al. 2007). Significant electron impact ionization can only occur for electron energies greater than ∼20 eV, and the electron temperature falls below this inside of L ∼ 6 due to the small pick-up energy of H^+ (cf. Fig. 9.9). However, the non-thermal tail represented by the hot outer-magnetospheric population diffusing inwards makes a significant contribution to the ionization, even at the low flux levels near Enceladus' orbit (Delamere and Bagenal 2008).

9.6 Upstream and Solar Wind Boundaries

Saturn orbits the Sun at a heliocentric distance of approximately 9.5 AU. At this distance the nominal Parker spiral is nearly in the azimuthal direction (85° from the radial direction). The out-of-the-ecliptic component of the interplanetary magnetic field (IMF) is typically quite small, well below 1 nT. The solar wind dynamic pressure is approximately two orders of magnitude weaker at Saturn than it is at Earth (it scales with d^{-2}, where d is the heliocentric distance). As discussed earlier in this Chapter, Saturn's intrinsic magnetic field at the planetary equator is comparable to that of Earth. Combined with the fact that there are significant plasma sources deep inside Saturn's inner magnetosphere, it is expected that the solar wind interaction plays an important role in forming the magnetospheric boundaries (bow shock and magnetopause) but it does not have a major influence on the inner magnetosphere.

9.6.1 Upstream Conditions

The approach of Cassini to the Saturn system in late 2003 and early 2004 provided an opportunity to study the properties of the upstream solar wind. Jackman et al. (2004, 2005) examined the behavior of the IMF upstream of Saturn during January–June 2004 and found a structure dominated by corotating interaction regions (CIRs). These CIRs produced repeating patterns of compressions and rarefactions in the solar wind occurring at approximately the solar rotation period. Each compression region was typically associated with a crossing of the heliospheric current sheet. We note that if the heliospheric current sheet is slightly tilted, which is usually the case especially in the late declining phase of the solar cycle, we observe a two-sector structure in the ecliptic. This means that there are two current sheet crossings per solar rotation. The repeatability of these compression regions was sufficient for the authors to predict the arrival of a compression region during Cassini's SOI orbit at Saturn. The arrival of this compression was confirmed using in-situ observations, which amongst other findings, showed that the magnetopause was expanded during the inbound pass of Cassini (indicating a rarefaction region) and that the magnetopause was compressed during the outbound pass (Dougherty et al. 2005). More recently, Jackman et al. (2008) have shown that the typical spiral angle of the IMF near Saturn's orbit is in close agreement with the Parker spiral angle at Saturn's location in the heliosphere. Jackman et al. (2008) also studied the meridional angle of the field and found a distribution of angles which peaked near zero degrees. Typical field strengths in the solar wind at Saturn were $0.1 nT$ and $1 - 2 nT$ in rarefaction and compression regions, respectively.

Solar wind plasma measurements at Saturn have not been obtained routinely due to viewing limitations, but the January 2004 joint Cassini – Hubble Space Telescope (HST) campaign provided an opportunity to study the upstream IMF and solar wind parameters at the same time as HST was observing the ultraviolet aurora. Crary et al. (2005) used

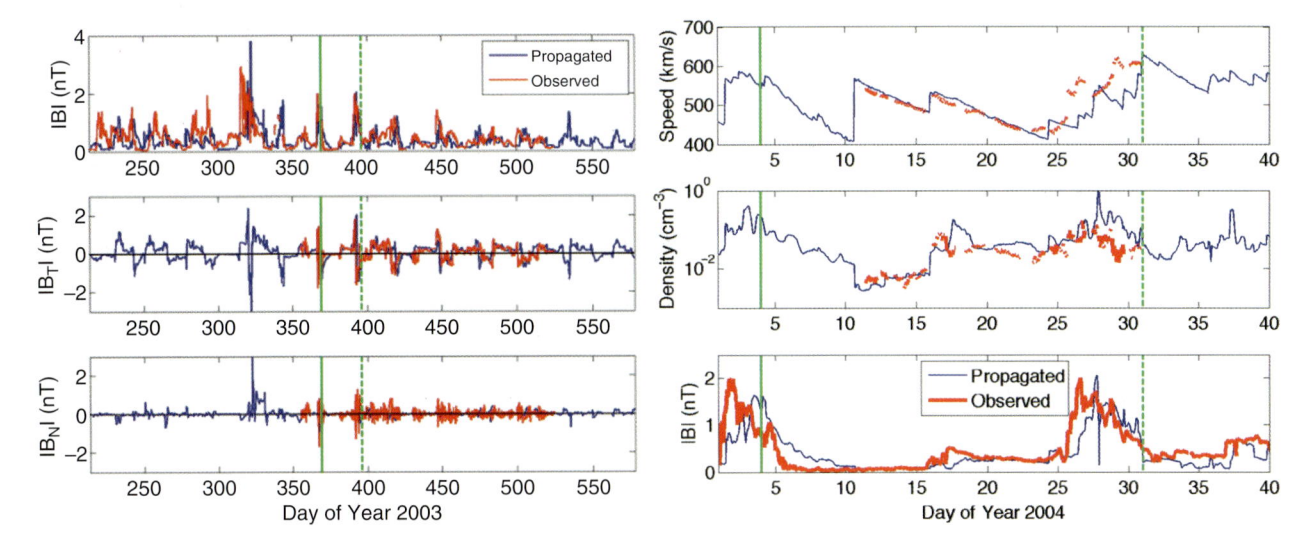

Fig. 9.40 Propagated and observed solar wind parameters for the 2003–2004 time period. *Left panel*: Propagated magnetic field values using the Zieger and Hansen (2008) method compared to Cassini MAG observations (Crary et al. 2005). *Right panel*: Propagated solar wind parameters obtained by the Zieger and Hansen (2008) method compared to the Cassini CAPS observations (Crary et al. 2005)

this dataset to show that the solar wind speed varies by approximately 25% about an average of ∼525 km/s whilst the density varies by more than two orders of magnitude. The density was found to be elevated in compression regions and reduced to less than 2×10^{-3}/cm in rarefaction regions. The range of solar wind dynamic pressures varied from 10^{-3} nPa to about 0.2 nPa.

Zieger and Hansen (2008) have introduced a 1D MHD model of solar wind propagation that is able to provide solar wind predictions at any location in the ecliptic plane between 1 and 10 AU. The boundary conditions at 1 AU are estimated from near-Earth solar wind observations assuming that the solar corona is in a quasi-steady state on the time scale of half a solar rotation. The time dependent MHD solution can be mapped to the location of any moving spacecraft, planet or other celestial body. Zieger and Hansen (2008) have validated the solar wind propagation model with 12 years of heliospheric observations from the Voyager, Pioneer and Ulysses spacecraft, quantifying the variations of the prediction efficiency both in space and time on a statistical basis. The solar wind predictions at the apparent opposition were found to be highly reliable even at solar maximum. The solar wind speed can be predicted with the highest accuracy, the second best predicted variable is the IMF magnitude, followed by the density. Their statistical analysis of time lags between predicted and observed shocks at 10 AU revealed that the error of shock arrival times is as small as 10–15 h within 75 days from apparent opposition during solar minimum. The results indicate that predicted shock arrival times tend to be late by at least 10 h during solar maximum, introducing a systematic error in the shock arrival times that can be incorporated into forecast models.

Figure 9.40 shows a comparison between predicted and observed solar wind parameters for the 2003–2004 period when Cassini was approaching Saturn.

9.6.2 Foreshock Region

A principal characteristic of interplanetary space near planetary bow shocks is the presence of energetic ions and electrons (Krimigis 1992). The Voyager 1 and 2 flybys in 1980 and 1981, respectively, observed such transient particle enhancements (Krimigis et al. 1983) whenever interplanetary magnetic field lines connected the spacecraft to the planetary bow shock. The origin of such bursts of ions cannot be established unambiguously, unless species that are characteristic of a particular source can be identified. For example, S^+ ions upstream of Jupiter are clearly of magnetospheric origin, given the Io plasma source, while He^{++} is a principal constituent of the solar wind and is likely accelerated outside the shock.

The absence of detailed composition and charge state measurements on the Voyagers at lower energies ($\leq$0.3 MeV where intensities are highest) made identification of the source plasma difficult. The Cassini instrument complement, however, includes such sensors (Krimigis et al. 2004) so it has been possible to characterize these events in detail. Figure 9.41 shows data from the MIMI and MAG investigations (see (Krimigis et al. 2004; Dougherty et al. 2004) for instrument descriptions) during the initial approach of Cassini to Saturn on day 176, 2004. Energetic ions are present throughout this period, from ∼92 R_S to the first bow shock crossing at ∼49 R_S. Superimposed are discrete, high

Fig. 9.41 Time intensity profile of energetic protons (*top panel*) upstream of Saturn's bow shock at local time ∼0740. The spikes on day 176 are shown in detail in Fig. 9.42. The *lower panels* show the direction and magnitude of the IMF (from Krimigis et al. 2009)

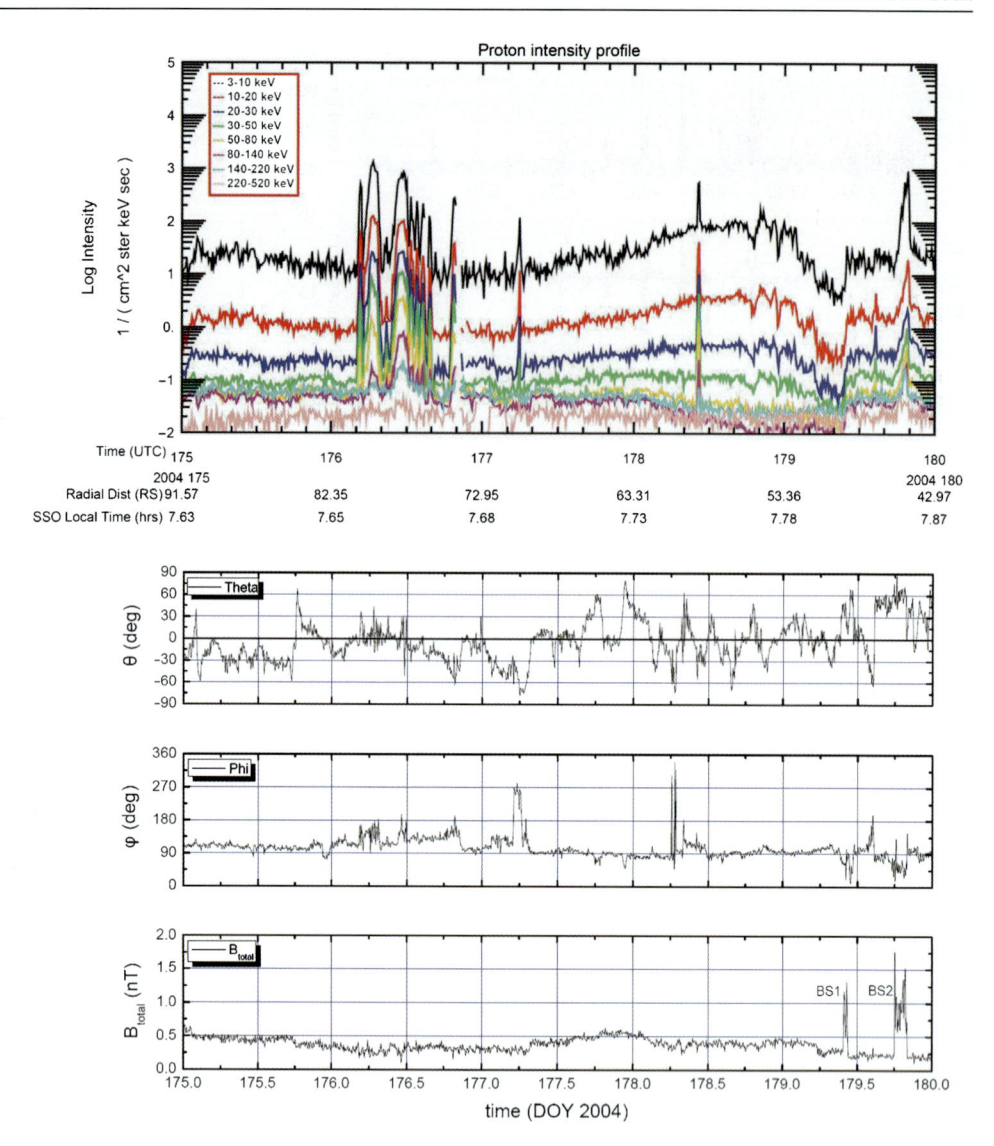

intensity bursts of varying duration, extending in energy to >220 keV. Similarly, the IMF (bottom panel) shows notable fluctuations in direction as well as magnitude that correlate well with the particle increases.

The time interval 00:00–15:00 UT of day 176 is shown in more detail in Fig. 9.42. The top left panel displays an energy spectrogram of protons while the bottom left panels show the IMF direction. Intense increases are well correlated with excursions of the IMF azimuth toward the radial direction, which would connect the spacecraft location (at ∼0740 local time, ∼80 R_S) to the dawn bow shock. The right panels show that the particle pressure exceeds that of the IMF during each of the bursts, the ratio (plasma β) being as high as ≥10. Detailed angular distribution measurements show that the onset of the event was generally field aligned, but then became isotropic, a characteristic that is common in upstream events at other planets (Krimigis et al. 2009).

Beginning with the onset of the events at ∼0400 UT there are waves in the IMF (bottom trace in Fig. 9.41) that seemingly correspond to periods of ∼45 and ∼22 min. A wavelet analysis of the IMF revealed that these were magnetic signatures consistent with the presence of O^+, and also O^{++}. The analysis of composition and charge state using the MIMI/CHEMS sensor (Fig. 9.43) shows the presence of H^+, He^{++}, He^+, and O^+ at the respective mass per charge locations.

The measurements in Fig. 9.43 suggest that the source of upstream bursts is the energetic particle population in the magnetosphere of Saturn, since singly charged oxygen is not a constituent of the upstream solar wind, but does represent a key constituent of the magnetospheric population (Krimigis et al. 2005). Protons and helium are present in both, in the upstream solar wind as plasma and in the magnetosphere as an energetic particle population. Thus the most

9 Saturn's Magnetospheric Configuration

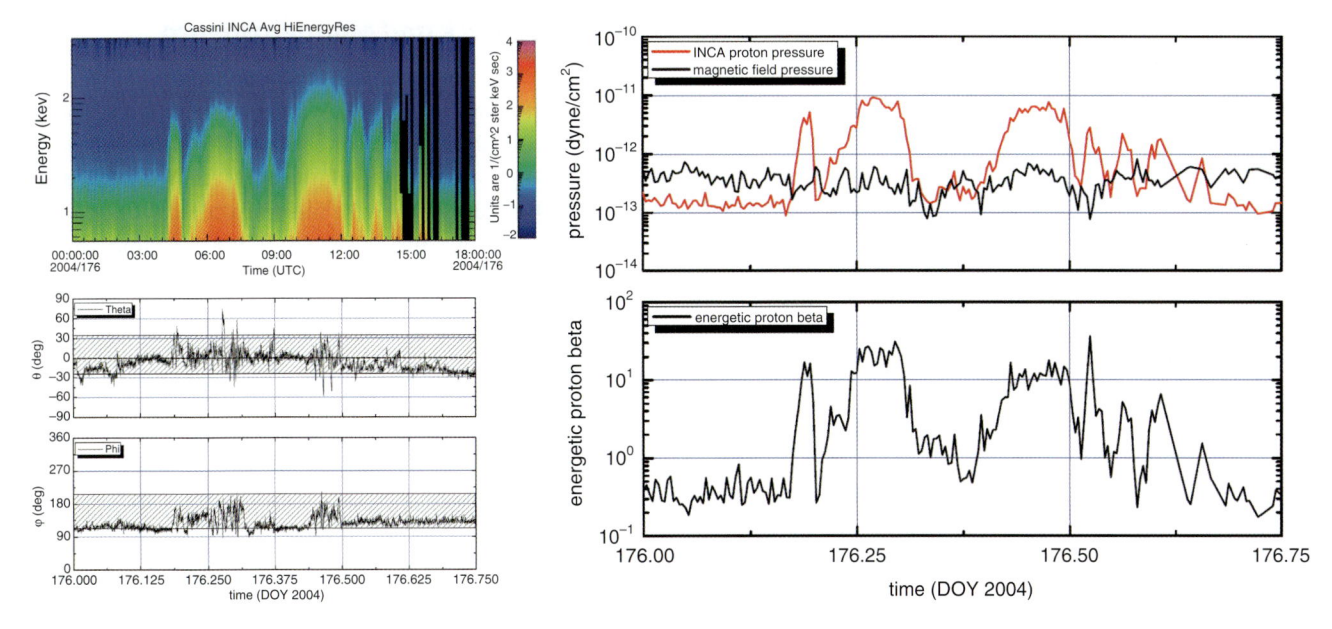

Fig. 9.42 Energy spectrogram of protons (*top left panel*) for day 176. The IMF direction and the respective particle and field pressures are shown as line-plots in the *left* and *right panels* (from Krimigis et al. 2009)

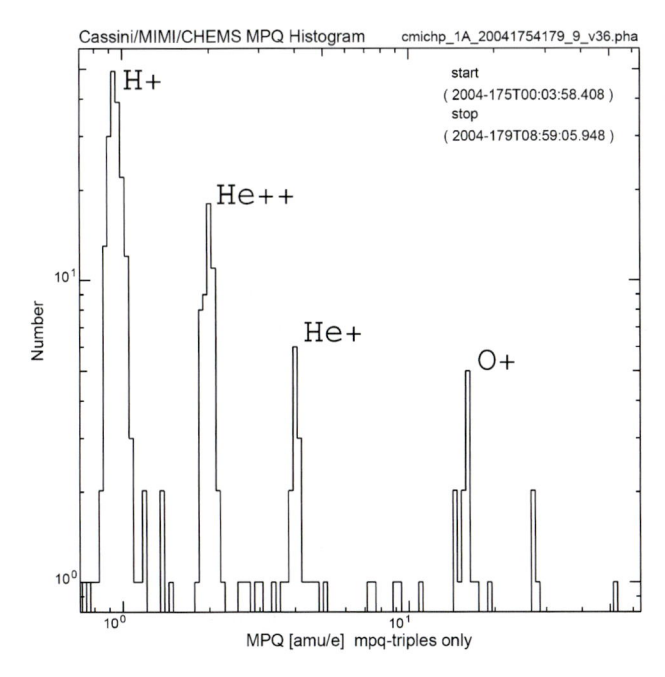

Fig. 9.43 Composition and charge state measurements (36 < E < 220 keV/e) with the MIMI/CHEMS sensor suite. A histogram from the standard display of mass vs mass per charge shows all points in the indicated time intervals during the events shown in Figs. 9.41 and Fig. 9.42. The histogram shows the clear presence of O^+ at 16 amu/e (from Krimigis et al. 2009)

plausible scenario for upstream bursts is one whereby there is nearly continuous leakage of energized particles into the upstream region that are observed whenever the IMF direction is such that it connects the spacecraft to the bow shock. It is not possible to estimate the loss rate with confidence at this time because of the lack of comprehensive statistical studies that can establish the local time dependence and the continuity or intermittency of such escape. It is likely that additional acceleration may possibly occur in the upstream region, considering the turbulence in the IMF, the high beta plasma conditions, and the isotropization of the particle distributions after the initial anisotropies during onset. These and other issues will have to await further data analyses and modeling studies.

At lower energies (<50 keV), CAPS observations show no evidence for water-group ions in the foreshock region, although there is clearly a suprathermal population of H^+ and He^{++}, as seen upstream from other bow shocks and interplanetary shocks (Thomsen et al. 2007). The implication of this observation is that the energetic water-group ions observed by MIMI were not accelerated in the foreshock region out of a leaked suprathermal population, but rather are due to direct leakage of already energized magnetospheric particles. The observed suprathermal population with solar-wind composition indicates that shock acceleration of solar wind particles occurs at Saturn's bow shock as at other shocks throughout the solar system.

Low frequency electromagnetic waves have also been detected upstream of Saturn's bow shock using the magnetometer (Bertucci et al. 2007). An analysis of waves below the local proton cyclotron frequency suggested that they were sunward propagating ion-ion resonant right-hand (fast magnetosonic) mode waves that steepen and emit whistler precursors to prevent the steepening process in the non-linear regime. These waves were observed deep into the ion foreshock and Bertucci et al. (2007) identified them as kronian counterparts of shocklets and discrete wave packets observed in the foreshock of Earth. These waves were found to be typical of the deep ion foreshock and therefore associated with diffuse back-streaming ion distributions. These ions have been observed with Cassini (Thomsen et al. 2007) but no direct correlation between superthermal ions and foreshock waves has yet been done.

Waves above the local proton cyclotron frequency were quasi-monochromatic and steepened with a right-handed polarization found closer to the bow shock than the first group. Typically the waves propagated at small angles to the magnetic field but waves with a higher amplitude exhibited a degree of compressibility and propagated obliquely to the field. The presence of apparent sunward propagating whistler precursors attached to the steepened waves led to the suggestion that the waves were produced by the ion-ion resonant left-hand instability and propagated sunward in the plasma frame. Again the hot ion distributions associated with these waves have yet to be clearly associated with specific wave events. Bertucci et al. (2007) presented direct evidence that these waves participate in the reformation of the quasi-parallel bow shock.

Narrow regions of enhanced particle and magnetic pressure called Hot Flow Anomalies (HFA) are a common observation upstream of the terrestrial bow shock. These anomalies are the result of the interaction of an interplanetary current sheet with the bow shock. Masters et al. (2008) presented Cassini observations from upstream of Saturn's bow shock which had characteristics similar to terrestrial HFAs. The analysis confirmed that the anomalies were the result of the interaction between a current sheet and the bow shock which focused energetic particles into the current sheet and formed the anomaly. In contrast with terrestrial HFAs the kronian HFAs were found to be associated with density enhancements (rather than depressions) and more modest electron heating. Estimates of the total pressure inside the HFAs suggested that their central regions were expanding. Subsequent analysis of at least one additional HFA event has revealed a profile much more similar to those seen in the terrestrial foreshock, including the strongly reduced internal density and strong flow deflection.

During the approach of Cassini to Saturn the CDA instrument observed high speed streams of nanometer-sized dust originating from the inner kronian system (Kempf et al. 2005). The dust grains were observed more often in corotating interaction regions, showing that once in the interplanetary medium the grains interact electromagnetically with the interplanetary magnetic field. Modelling and observations showed that the dust grains have radii between 2 and 25 nm, move with a speed in excess of 100 km/s, and are composed primarily of silicon, suggesting that they are impurities in the ices in Saturn's inner magnetosphere and not of an icy composition.

9.6.3 Bow Shock and Magnetosheath

The most commonly used pre-Cassini model of Saturn's bow shock was published by Slavin et al. (1985) based on bow shock crossings observed by Pioneer 11 and Voyagers 1 and 2. The shock crossings were fitted to a conic section with a focus shifted along the aberrated Saturn-Sun line. Since the size of the bow shock is a function of the upstream solar wind dynamic pressure, the standoff distance of the bow shock was assumed to be a power-law function of the solar wind dynamic pressure, $R_{BS} \propto P_{sw}^{-1/\alpha}$. Slavin et al. (1985) found that the compressibility was about $\alpha = 5.1$ with the average subsolar shock distance of $27.7\,R_S$.

In the era of Cassini-Huygens empirical models have been developed to describe Saturn's bow shock (Masters et al. 2008). An additional study used MHD simulations (Hansen et al. 2000) to investigate the pressure-size relation of the bow shock. For the latter study, Hansen et al. (2005) tuned the free parameters of the MHD model to match the locations of the boundary crossings identified in the Cassini magnetometer data. The subsolar locations of the bow shock (and magnetopause) were extracted from the simulation and plotted as a function of the solar wind dynamic pressure. Hansen et al. (2005) found that the bow shock responded to changes in solar wind dynamic pressure with $\alpha \approx 5.9$, a somewhat different power-law than published by Slavin et al. (1985). The bow shock was also found to be less flared than that identified by Slavin et al. 1985.

Two empirical models of Saturn's bow shock (Masters et al. 2008) use essentially the same dataset of shock crossings observed in Voyager data and Cassini magnetometer and electron spectrometer data, but differ in the way that they determine the upstream dynamic pressure at each crossing. Masters et al. (2008) assumed a constant solar wind speed at each aberrated shock crossing location and used Langmuir plasma wave observations to estimate the plasma density, hence providing an estimate of the solar wind dynamic pressure. Masters et al. (2008) fitted the observations to a conic section using a method similar to that employed by Slavin et al. (1985) where the crossing locations were fitted to a conic section without pressure-correction, then the pressure estimates were used to scale all the crossing locations to a common dynamic pressure and the model was refit. The resulting model is marginally hyperbolic, and thus less flared that the Slavin et al. (1985) model, and had an average size of $25 \pm 1 R_S$ in good agreement with that found by Slavin et al. (1985). The power-law exponent was found to be $\alpha = 6 \pm 2$ and hence consistent with Slavin et al. (1985) but with very large error bars. Masters et al. (2008) also investigated the effect of the IMF orientation on the shock location but found no clear relationship. Using the model to estimate the shock stand-off distance at each shock crossing suggests that the stand-off distance lies between $18\,R_S$ and $46\,R_S$.

9.6.4 Magnetopause

Jackman et al. (2004) presented an empirical voltage formula adapted from studies at Earth, to calculate the reconnection voltage across Saturn's dayside magnetopause, and thus to estimate the amount of open flux added to the system through dayside reconnection. This study indicated that Saturn's magnetosphere is immersed in highly structured solar wind, and reconnection can be driving a Dungey-type convection pattern Dungey 1961 at least in the outer magnetosphere. More recently McAndrews et al. (2008) presented in-situ evidence for dayside reconnection at Saturn's magnetopause using CAPS data, and they derived a value for the reconnection voltage from one example to be ∼50 kV, in line with the estimates of Jackman et al. (2004). Badman and Cowley (2007) considered the contribution of the solar wind driven Dungey cycle to flux transport in Saturn's magnetosphere, and found that under conditions of strong solar wind driving, the Dungey cycle return flow will make a significant contribution to the flux transport in the outer magnetosphere. The contribution of the solar wind driving to magnetospheric convection is discussed in another section of this Chapter (see Section 9.1.3).

There is also preliminary evidence suggesting the presence of surface waves on Saturn's magnetopause Masters et al. (2009). Two categories of boundary waves were identified: the first with a period of the order of hours, and the second with a period of 45 ± 9 min. Based on the propagation direction and a comparison of magnetospheric and magnetosheath magnetic fields, Masters et al. (2009) concluded that both types of waves were driven by the Kelvin-Helmholtz instability. The observed boundary perturbations were consistent with a superposition of different types of surface wave activity.

9.6.4.1 Standoff Distance

Before Cassini-Huygens the average location of Saturn's magnetopause was thought to be ∼20 R_S. Recently Achilleos et al. (2008) used an established statistical technique Joy 2002 with Cassini observations and the Arridge et al. (2006) magnetopause model to test this pre-Cassini understanding and establish the long-term statistical behavior of the magnetopause. They obtained the best fit with a bimodal distribution, with peaks at 22 R_S and 27 R_S with uncertainties in the range of 1–2 R_S. A similar bimodality of the standoff distance of the Jovian magnetopause was found by Joy et al. (2002).

To identify the physical origin of this bimodality, Achilleos et al. (2008) considered both external (i.e., a solar wind origin), and internal (mass-loading/rotational effects) mechanisms. The solar wind at the outer planets has been shown to be dominated by alternating patterns of compressions and rarefactions (Jackman et al. 2004) and so one might expect compression regions to contain enhanced solar wind dynamic pressures. Achilleos et al. (2008) compared the solar wind dynamic pressure distribution, inferred from the stand-off distance distribution, with solar wind pressures measured by the CAPS instrument (Crary et al. 2005). The solar wind pressure distribution did not exhibit the bimodality found in the magnetopause standoff distance. Furthermore, there was no ordering of the standoff distance with heliocentric longitude, as might be expected for a solar wind driver.

Processes internal to the magnetosphere, such as mass loading/release mechanisms and rotational effects can also plausibly account for the observed bimodality. Clarke et al. (2006) studied a set of magnetopause crossings on the dawn flank that were modulated at a period close to the ∼10.5 h periodicity seen in Saturn Kilometric Radiation (SKR) and interpreted these multiple crossings as evidence of an internal rotational anomaly (the camshaft Southwood and Kivelson 2007). Clarke et al. 2006 found that the standoff distance was modulated with a peak-to-trough amplitude of 4 R_S and interpreted it in terms of the effect of the camshaft. Achilleos et al. 2008 examined the magnetopause standoff distance as a function of SKR phase Kurth et al. (2007) and found no evidence that this rotational anomaly was producing the observed bimodality and concluded that it probably only contributed to producing the observed scatter about the peaks. Thus, it was concluded that internal mass-loading and mass release processes could account for the observed bimodality (Achilleos et al. 2008). Such an effect has also been observed at Jupiter (Joy et al. 2002).

9.6.4.2 Compressibility

Pioneer and Voyager studies of the Jovian magnetosphere found that the size of the magnetosphere was unusually responsive to changes in upstream dynamic pressure, compared to the rather stiff terrestrial magnetosphere (Slavin et al. 1985). The suggestion was made that this reflected the effects of internal plasma stresses in the magnetosphere that were either not present or weaker at Earth (e.g., Smith et al. 1974). Thus, examining and modeling the pressure-dependent size of the kronian magnetopause is a useful diagnostic of the stresses inside the magnetosphere. In modeling studies the relationship between the size of the magnetosphere (typically represented by the location of the sub-solar point on the magnetopause, R_0) and the upstream dynamic pressure is expressed as a power-law:

$$R_{MP} \propto P_{sw}^{-1/\alpha} \qquad (9.10)$$

The radial dependence of dipole field strength implies that the size of the terrestrial magnetopause should scale with pressure $\alpha = 6$. This expectation is consistent with the observed value of $\alpha = 6.6 \pm 0.8$ (Shue et al. 1997). The unusually compressive Jovian magnetosphere has a magnetopause that follows a power-law of $\alpha \approx 4.5$ (Huddleston et al. 1998), hence R_0 changes more rapidly with P_{SW} than $\alpha = 6$. The physical explanation is that the more stretched the magnetic field lines are on the dayside (due to a combined effect of fast rotation and mass loading of magnetic field lines) the more compressible the magnetic field lines are. Dipole field lines in vacuum are quite "rigid" and this is why the terrestrial magnetopause standoff distance scales approximately as $\alpha = 6$. This is a similar functional form as that for the size of the shock discussed above, but compressibility of the shock is different than the compressibility of the magnetopause.

Examining data from the first part of the Cassini mission, Arridge et al. (2006) obtained a compressibility of $\alpha = 4.3 \pm 0.4$. This result implied that the kronian magnetosphere was even "softer" than the Jovian one. Recently, Achilleos et al. (2008) examined a large multi-instrumental data set and obtained a compressibility parameter of $\alpha = 5.17 \pm 0.3$, implying that Saturn's dayside magnetosphere was actually "stiffer" than Jupiter's but still much "softer" than Earth's. This result is in very good agreement with the compressibility value of $\alpha = 5.2$ deduced from a series of numerical simulations early in the Cassini mission (Hansen et al. 2005). While these values show some scatter the emerging picture clearly indicates that Saturn's dayside magnetopause is somewhat less compressive than the magnetopause of Jupiter and much more compressive than the magnetopause of Earth. For comparison, we think that the best estimates for the compressibilities of Jupiter, Saturn and Earth are $\alpha \approx 4.5$, 5 and 6, respectively.

As discussed earlier, a more compressible magnetosphere is interpreted as being due to the effect of internal plasma stresses in the magnetosphere. These results suggest that the internal plasma configuration of Saturn's magnetosphere is quite similar to the Jovian magnetosphere. Bunce et al. (2007) developed an empirical model for the magnetic moment of the ring current as a function of magnetospheric size and were able to show that the total magnetic moment (intrinsic magnetic field plus ring current) varied strongly with the size of the magnetosphere. They showed that this effect could be produced if a substantial portion of the ring current is generated by centrifugal forces. They went on to argue that this naturally led to the power-law that was consistent with the observed compressibilities of Earth, Jupiter and Saturn.

9.6.4.3 Shape

Two empirical models of Saturn's magnetopause were available prior to the arrival of Cassini-Huygens at Saturn. Slavin et al. (1985) fitted conic sections to observed crossings from Pioneer 11 and Voyagers 1 & 2. The fit was highly flared (the conic section had hyperbolic geometry) and the size of the magnetopause followed an $\alpha = 6.1$ power-law. In a separate modeling study Behannon et al. (1983) described modeling by Ness et al. (1982) who used Voyager magnetopause crossings to produce a parabolic model which did not have the large flaring of the Slavin et al. (1985) model.

Using Cassini and Voyager magnetopause crossings, Arridge et al. (2006) developed an axisymmetric model of Saturn's magnetopause. Axial symmetry about the Saturn-Sun line was assumed, A Newtonian pressure balance was used to estimate the upstream dynamic pressure, and the functional form of the magnetopause was assumed to be

$$R_{MP}(\theta, P_{sw}) = A P_{sw}^{-1/\alpha} \left(\frac{2}{1 + \cos\theta} \right)^{\beta - \kappa P_{sw}} \quad (9.11)$$

Arridge et al. (2006) derived the following values for the model parameters: $\alpha = 4.3 \pm 0.4$, $A = 10\,R_S \pm 1\,R_S$, $\beta = 0.77 \pm 0.03$ and $\kappa = 1.5 \pm 0.3$.

9.7 Some Open Questions

After the completion of the Cassini mission (it is expected to plunge into Saturn's atmosphere in 2017) Saturn's magnetosphere will be the most studied planetary magnetosphere other than Earth. However, even at Earth, after many magnetospheric missions, there are many phenomena we do not fully understand. In addition, there are many individual phenomena that are intricately interrelated but we still need to understand the complex, nonlinear relationships between them. Obviously, the even more complex kronian magnetosphere cannot be explored and understood by a few flybys and a single orbiter mission, even if it is as successful as the Cassini mission is.

Here we list some open questions regarding the basic magnetospheric configuration that remain at the end of the Cassini Prime Mission. Hopefully some of these questions will be answered by the Cassini Equinox and Solstice Missions, but some of them will probably wait for the next mission to the Saturn system. Obviously, the questions reflect on the interests and priorities of the authors, and the list by no means is intended to be complete.

With these caveats, here is a list of some open questions:

- *Does the magnetosphere exhibit solar cycle or seasonal changes?*
  - What controls the interplay between the Dungey and Vasyliunas cycles?
  - Do the IMF B_y or B_z have an impact on the magnetospheric configuration?
  - What are the physical processes controlling radial mass transport in the magnetosphere?
  - What controls the periodic mass release to the magnetotail?

- *What is the temporal variability of Enceladus' plumes?* Enceladus is the dominant magnetospheric mass source ejecting ∼300 kg/s of water into the inner magnetosphere. The plume disperses and becomes ionized, adding about 300 kg/s plasma to the magnetosphere. Cassini observations show close to a factor of two change in the plume output.
  - Is the Enceladus gas production changing with solar cycle or seasons?
  - How does the magnetosphere react to the changes of gas production rate?
  - Is the neutral gas composition different in different plumes? If it changes from plume to plume it would suggest different reservoirs, and conversely, if the composition of the plume is essentially identical in all plumes it would imply the presence of a global ocean below the surface.

- *How is the ionosphere coupled to the magnetosphere?*
  - What are the coupling currents between the ring current and the ionosphere?
  - Are the rings electrodynamically coupled to the ionosphere?
  - How are auroral particles accelerated?
  - How are magnetospheric periodicities coupled to the ionosphere?
  - How is the ionospheric rotation rate coupled to the rotation of the planetary interior?
  - What process imposes a slowly varying periodicity on SKR and on other magnetospheric phenomena?
  - Is there a solar cycle or seasonal variation of magnetospheric rotation periods?
  - Is the ionosphere and/or thermosphere differentially rotating?
  - Are there UV satellite footprints on Saturn? (like at Jupiter)
  - Is there a significant polar outflow from Saturn's high latitude ionosphere? Does this outflow exhibit seasonal or solar cycle variation?

Needless to say, there are many more questions we need to answer if we want to understand Saturn's complex magnetosphere. We are also certain that as we gain a better understanding of the Saturnian system there will be even more questions to answer. We are in an exciting stage of the exploration of Saturn. The veil starts to lift, but there is so much more to see.

Acknowledgements This work was sponsored by the international Cassini/Huygens Project. Authors were supported by NASA under the following contracts: JPL-1279089 (TIG), JPL-1279973 (AMP) and NASA NNX07AE74G (KKK). Work at JHU/APL (SMK) was supported by NASA under contract NAS5-97271 and NNX07AJ69G and by subcontracts at the University of Maryland and the Office for Space Research and Technology of the Academy of Athens while work at Fundamental Technologies (TPA) was supported by APL subcontract 946749. Work at Los Alamos National Laboratory was conducted under the auspices of the U.S. Department of Energy, with support from NASA grant W-19289. CSA was supported by the STFC rolling grant to MSSL-UCL. NK was supported by the German Bundesministerium für Bildung und Forschung through the Deutsches Zentrum für Luft- und Raumfahrt e.V. under contracts No. 50 OH 0801 and 50 OH 0802, and by the Max Planck Gesellschaft. The authors are indebted to Fran Bagenal for her valuable comments and suggestions.

References

Achilleos, N., C. Bertucci, C. T. Russell, G. B. Hospodarsky, A. M. Rymer, C. S. Arridge, M. E. Burton, M. K. Dougherty, S. Hendricks, E. J. Smith and B. T. Tsurutani: Orientation, location, and velocity of Saturn's bow shock: Initial results from the Cassini spacecraft. J. Geophys. Res. **111**, A03201 (2006) doi:10.1029/2005JA011297

Achilleos, N. C. S. Arridge, C. Bertucci, C. M. Jackman, M. K. Dougherty, K. K. Khurana and C. T. Russell: Large-scale dynamics of Saturn's magnetopause: Observations by Cassini. J. Geophys. Res. **113**, A11209 (2008) doi:10.1029/2008JA013265

Acuña, M. H., N. F. Ness and J. E. P. Connerney: The magnetic field of Saturn: Further studies of the Pioneer 11 observations. J. Geophys. Res. **85**, 5675–5678 (1980).

Alexeev, I. I., V. V. Kalegaev, E. S. Belenkaya, S. Y. Bobrovnikov, E. J. Bunce, S. W. H. Cowley, and J. D. Nichols: A global magnetic model of Saturn's magnetosphere and a comparison with Cassini SOI data. Geophys. Res. Lett. **33**, L08101 (2006) doi:10.1029/2006GL025896

Anderson, J. D. and G. Schubert: Saturn's gravitational field, internal rotation, and interior structure. Science **317**, 1384–1387 (2007).

André, N., M. K. Dougherty, C. T. Russell, J. S. Leisner, and K. K. Khurana: Dynamics of the Saturnian inner magnetosphere: First inferences from the Cassini magnetometers about small-scale plasma transport in the magnetosphere. Geophys. Res. Lett. **32** L14S06 (2005) doi:10.1029/2005GL022643

André, N., A. M. Persoon, J. Goldstein, J. L. Burch, P. Louarn, G. R. Lewis, A. M. Rymer, A. J. Coates, W. S. Kurth, E. C. Sittler Jr., M. F. Thomsen, F. J. Crary, M. K. Dougherty, D. A. Gurnett and D. T. Young: Magnetic signatures of plasma-depleted flux tubes in the Saturnian inner magnetosphere. Geophys. Res. Lett. **34**, L14108 (2007) doi:10.1029/2007GL030374

André, N., M. Blanc, S. Maurice, P. Schippers, E. Pallier, T. I. Gombosi, K. C. Hansen, D. T. Young, F. J. Crary, S. Bolton, E. C. Sittler, H. T. Smith, R.E. Johnson, R. A. Baragiola, A. J. Coates, A. M. Rymer, M. K. Dougherty, N. Achilleos, C. S. Arridge, S. M. Krimigis, D.

G. Mitchell, N. Krupp, D. C. Hamilton, I. Dandouras, D.A. Gurnett, W. S. Kurth, P. Louarn, R. Srama, S. Kempf, H. J. Waite, L. W. Esposito, and J. T. Clarke: Identification of Saturn's magnetospheric regions and associated plasma processes: Synopsis of Cassini observations during orbit insertion, Rev. Geophys. **46**, RG4008 (2008) doi:10.1029/2007RG000238

Andrews, D. J., E. J. Bunce, S. W. H. Cowley, M. K. Dougherty, G. Provan, and D. J. Southwood: Planetary period oscillations in Saturn's magnetosphere: Phase relation of equatorial magnetic field oscillations and SKR modulation. J. Geophys. Res. **113**, A09205 (2008) doi:10.1029/2007JA012937

Armstrong, T. P., S. Taherion, J. W. Manweiler, S. M. Krimigis, C. Paranicas, D. G. Mitchell, and N. Krupp: Energetic ions trapped in Saturn's inner magnetosphere. Planet. Space Sci. (in press) (2009) doi:10.1016/j.pss.2009.03.008

Arridge, C. S., N. Achilleos, M. K. Dougherty, K. K. Khurana and C. T. Russell: Modeling the size and shape of Saturn's magnetopause with variable dynamic pressure. J. Geophys. Res. **111**, A11227 (2006) doi:10.1029/2005JA011574

Arridge, C. S., C. T. Russell, K. K. Khurana, N. Achilleos, N. André, A. M. Rymer, M. K. Dougherty, and A. J. Coates: Mass of Saturn's magnetodisc: Cassini observations. Geophys. Res. Lett. **34**, L09108 (2007) doi:10.1029/2006GL028921

Arridge, C. S., C. T. Russell, K. K. Khurana, N. Achilleos, S. W. H. Cowley, M. K. Dougherty, D. J. Southwood, and E. J. Bunce: Saturn's magnetodisc current sheet. J. Geophys. Res. **113**, A03214 (2008) doi:10.1029/2007JA012540

Arridge, C. S., K. K. Khurana, C. T. Russell, D. J. Southwood, N. Achilleos, M. K. Dougherty, A. J. Coates, and H. K. Leinweber: Warping of Saturn's magnetospheric and magnetotail current sheets. J. Geophys. Res. **113**, A08217 (2008) doi:10.1029/2007JA012963

Arridge, C. S., N. André, N. Achilleos, K. K. Khurana, C. L. Bertucci, L. K. Gilbert, G. R. Lewis, A. J. Coates, and M. K. Dougherty: Thermal electron periodicities at $20R_S$ in Saturn's magnetosphere. Geophys. Res. Lett. **35**, L15107 (2008) doi:10.1029/2008GL034132

Badman, S. V., E. J. Bunce, J. T. Clarke, S. W. H. Cowley, J.-C. Gérard, D. Grodent, and S. E. Milan: Open flux estimates in Saturn's magnetosphere during the January 2004 Cassini-HST campaign, and implications for reconnection rates. J. Geophys. Res. **110**, A11216 (2005) doi:10.1029/2005JA011240

Badman, S. V. and S. W. H. Cowley, J.-C. Gérard, and D. Grodent: A statistical analysis of the location and width of Saturn's southern auroras. Ann. Geophys. **24**, 3533–3545 (2006).

Badman, S. V. and S. W. H. Cowley: Significance of Dungey-cycle flows in Jupiter's and Saturn's magnetospheres, and their identification on closed equatorial field lines, Annales Geophysicae **25**, 941–951 (2007).

Barbosa, D.: Titan's atomic nitrogen torus: Inferred properties and consequences for the Saturnian magnetosphere. Icarus **72**, 53–61 (1987).

Beard, D. B.: The interaction of the terrestrial magnetic field with the solar corpuscular radiation. J. Geophys. Res. **65**, 3559–3568 (1960).

Beard, D. B. and M. A. Gast: The magnetosphere of Saturn. J. Geophys. Res. **92**, 5763–5767 (1987).

Behannon, K. W., R. P. Lepping, and N. F. Ness: Structure and dynamics of Saturn's outer magnetosphere and boundary regions. J. Geophys. Res. **88**, 8791–8800 (1983).

Bertucci, C., N. Achilleos, C. Mazelle, G. B. Hospodarsky, M. Thomsen, M. K. Dougherty, and W. Kurth: Low-frequency waves in the foreshock of Saturn: First results from Cassini. J. Geophys. Res. **112**, A09219 (2007) doi:10.1029/2006JA012098

Bunce, E. J. and S. W. H. Cowley: A note on the ring current in Saturns magnetosphere: Comparison of magnetic data obtained during the Pioneer-11 and Voyager-1 and -2 fly-bys, Ann. Geophys. **21**, 661–669 (2003).

Bunce, E. J., S. W. H. Cowley, D. M. Wright, A. J. Coates, M. K. Dougherty, N. Krupp, W. S. Kurth, and A. M. Rymer: In-situ observations of a solar wind compression-induced hot plasma injection in Saturn's tail. Geophys. Res. Lett. **32**, L20S04 (2005a) doi:10.1029/2005GL022888

Bunce, E. J., S. W. H. Cowley, and S. E. Milan: Interplanetary magnetic field control of Saturn's polar cusp. Annales. Geophys. **23**, 1405–1431 (2005b).

Bunce, E. J., S. W. H. Cowley, I. I. Alexeev, C. S. Arridge, M. K. Dougherty, J. D. Nichols, and C. T. Russell: Cassini observations of the variation of Saturn's ring current parameters with system size. J. Geophys. Res. **112**, A10202 (2007) doi:10.1029/2007JA012275

Bunce, E. J., C. S. Arridge, S. W. H. Cowley, and M. K. Dougherty: Magnetic field structure of Saturn's dayside magnetosphere and its mapping to the ionosphere: Results from ring current modeling. J. Geophys. Res. **113**, A02207 (2008) doi:10.1029/2007JA012538

Bunce, E. J., C. S. Arridge, J. T. Clarke, A. J. Coates, S. Cowley, M. K. Dougherty, J.-C. Gérard, D. Grodent, K. C. Hansen, J. D. Nichols, D. J. Southwood, and D. Talboys: Origin of Saturn's aurora: Simultaneous observations by Cassini and the Hubble Space Telescope. J. Geophys. Res. **113**, A09209 (2008) doi:10.1029/2008JA013257

Burch, J. L., J. Goldstein, T. W. Hill, D. T. Young, F. J. Crary, A. J. Coates, N. André, W. S. Kurth, and E. C. Sittler Jr.: Properties of local plasma injections in Saturn's magnetosphere. Geophys. Res. Lett. **32**, L14S02 (2005) doi:10.1029/2005GL022611

Burch, J. L., J. Goldstein, W. S. Lewis, D. T. Young, A. J. Coates, M. Dougherty, and N. André: Tethys and Dione as sources of outward flowing plasma in Saturn's magnetosphere. Nature **447**, 833–835 (2007) doi:10.1038/nature05906

Carbary J. F., S. M. Krimigis, and W.-H. Ip: Energetic particle microsignatures of Saturn's satellites. J. Geophys. Res. **88**, 8947–8958 (1983).

Carbary, J. F., D. G. Mitchell, S. M. Krimigis, and D. C. Hamilton: Spin-period effects in magnetospheres with no axial tilt. Geophys. Res. Lett. **34**, L18107 (2007) doi:10.1029/2007GL030483

Carbary, J. F., D. G. Mitchell, C. Paranicas, E. C. Roelof and S. M. Krimigis: Direct observation of warping in the plasma sheet of Saturn. Geophys. Res. Lett. **35**, L24201 (2008) doi:10.1029/2008GL035970

Carbary, J. F., D.G. Michell, S.W.H. Cowley, T.W. Hill, and P. Zarka: The Dynamics of Saturn's Magnetosphere. **Chapter 11 in this book** (2009).

Carr, T. D., J. J. Schauble, and C. C. Schauble: Pre-encounter distributions of Saturn's low frequency radio emission, Nature **292**, 745–747 (1981).

Carr, T. D., M. D. Desch, and J. K. Alexander: Phenomenology of magnetospheric radio emissions. In *Physics of the Jovian Magnetosphere*, edited by A. J. Dessler (Cambridge University Press, New York 1983), pp. 226–284.

Clarke, K. E., N. André, D. J. Andrews, A. J. Coates, S .W. H. Cowley, M. K. Dougherty, G. R. Lewis, H. J. McAndrews, J. D. Nichols, T. R. Robinson, and D. M. Wright: Cassini observations of planetary-period oscillations of Saturn's magnetopause. Geophys. Res. Lett. **33**, L23104 (2006) doi:10.1029/2006GL027821

Coates A. J., F. J. Crary, D. T. Young, K. Szego, C. S. Arridge, Z. Bebesi, E. C. Sittler, R. E. Hartle, and T. W. Hill: Ionospheric electrons in Titan's tail: Plasma structure during the Cassini T9 encounter. Geophys. Res. Lett. **34**, L24S05 (2007) doi:10.1029/2007GL030919

Connerney, J. E. P., M. H. Acuña, and N. F. Ness: Saturn's ring current and inner magnetosphere. Nature **292**, 724–726 (1981).

Connerney, J. E. P., N. F. Ness, and M. H. Acuña: Zonal harmonic model of Saturn's magnetic field from Voyager 1 and 2 observations. Nature **298**, 44–46 (1982).

Connerney, J. E. P., M. H. Acuña, and N. F. Ness: Currents in Saturn's magnetosphere. J. Geophys. Res. **88**, 8779–8789 (1983).

Connerney, J. E. P., M.H. Acuña, and N.F. Ness: The Z3 model of Saturn's magnetic field and the Pioneer 11 vector helium magnetometer observations. J. Geophys. Res. **89**, 7541–7544 (1984).

Cooper, J. F.: Nuclear cascades in Saturn's rings: Cosmic ray Albedo neutron decay and origins of trapped protons in the inner magnetosphere. J. Geophys. Res. **88**, 3945–3954 (1983).

Cooper, J. F., J. H. Eraker, and J.A. Simpson: The secondary radiation under Saturn's A-B-C rings produced by cosmic ray interactions, J. Geophys. Res. **90**, 3415–3427 (1985).

Cowley, S. W. H. and E. J. Bunce: Corotation-driven magnetosphere-ionosphere coupling currents in Saturn's magnetosphere and their relation to the auroras. Ann. Geophys. **21**, 1691–1707 (2003).

Cowley, S. W. H., E. J. Bunce, and R. Prangé: Saturn's polar ionospheric flows and their relation to the main auroral oval. Annales Geophys. **22**, 1379–1394 (2004).

Cowley, S. W. H., E. J. Bunce, and J. M. O'Rourke: A simple quantitative model of plasma flows and currents in Saturn's polar ionosphere. J. Geophys. Res. **109**, A05212 (2004) doi:10.1029/2003JA010375

Cowley, S. W. H., S. V. Badman, E. J. Bunce, J. T. Clarke, J.-C. Gérard, D. Grodent, C. M. Jackman, S. E. Milan, and T. K. Yeoman: Reconnection in a rotation-dominated magnetosphere and its relation to Saturn's auroral dynamics, J. Geophys. Res. **110**, A02201 (2005) doi:10.1029/2004JA010796

Crary, F. J., J. T. Clarke, M. K. Dougherty, P. G. Hanlon, K. C. Hansen, J. T. Steinberg, B. L. Barraclough, A. J. Coates, J.-C. Gérard, D. Grodent, W. S. Kurth, D. G. Mitchell, A. M. Rymer, and D. T. Young: Solar wind dynamic pressure and electric field as the main factors controlling Saturn's aurorae. Nature **433**, 720–722 (2005).

Davis, L. Jr. and E. J. Smith: New models of Saturn's magnetic field using Pioneer 11 vector helium magnetometer data. J. Geophys. Res. **91**, 1373–1380 (1986).

Davis, L. Jr. and E. J. Smith: A model of Saturn's magnetic field based on all available data. J. Geophys. Res. **95**, 15257–15261 (1990).

Delamere, P. A., F. Bagenal, V. Dols, and L. C. Ray: Saturn's neutral torus versus Jupiter's plasma torus. Geophys. Res. Lett. **34**, L09105 (2007) doi:10.1029/2007GL029437

Delamere, P. A. and F. Bagenal: Longitudinal plasma density variations at Saturn caused by hot electrons. Geophys. Res. Lett. **35**, L03107 (2008) doi:10.1029/2007GL031095

Delitsky, M. L. and A. L. Lane: Saturn's inner satellites: Ice chemistry and magnetosphere effects, J. Geophys. Res. **107**, 5093 (2002) doi:10.1029/2002JE001855

Desch, M. D. and M. L. Kaiser: Voyager measurements of the rotation period of Saturn's magnetic field. Geophys. Res. Lett. **8**, 253–256 (1981).

Dougherty, M. K., S. Kellock, D. J. Southwood, A. Balogh, E. J. Smith, B. T. Tsurutani, B. Gerlach, K.-H. Glassmeier, F. Gleim, C. T. Russell, G. Erdos, F. M. Neubauer, and S. W. H. Cowley: The Cassini magnetic field investigation. Spce Sci. Rev. **114**, 331–383 (2004).

Dougherty, M. K., N. Achilleos, N. André, C. S. Arridge, A. Balogh, C. Bertucci, M. E. Burton, S. W. H. Cowley, G. Erdös, G. Giampieri, K.-H. Glassmeier, K. K. Khurana, J. S. Leisner, F. M. Neubauer, C. T. Russell, E. J. Smith, D. J. Southwood, and B. T. Tsurutani: Cassini magnetometer observations during Saturn orbit insertion. Science **307**, 1266–1270 (2005).

Dougherty, M. K., K. K. Khurana, F. M. Neubauer, C. T. Russell, J. Saur, J. S. Leisner, and M. E. Burton: Identification of a dynamic atmosphere at Enceladus with the Cassini magnetometer. Science **311**, 1406–1409 (2006) doi:10.1126/science.1120985

Dungey, J. W.: Interplanetary field and the auroral zones, Phys. Rev. Lett. **6**, 47–48 (1961).

Esposito, L. W., J. E. Colwell, K. Larsen, W. E. McClintock, A. I. F. Stewart, J. T. Hallett, D. E. Shemansky, J. M. Ajello, C. J. Hansen, A. R. Hendrix, R. A. West, H.-U. Keller, A. Korth, W. R. Pryor, R. Reulke, and Y. L. Yung: Ultraviolet imaging spectroscopy shows an active Saturnian system, Science **307**, 1251–1255 (2005).

Eviatar, A., G. L. Siscoe, J. D. Scudder, E. C. Sittler, Jr., and J. D. Sullivan: The plumes of Titan. J. Geophys. Res. **87**, 8091–8103 (1982).

Eviatar, A., R. L. McNutt, Jr., G. L. Siscoe, and J. D. Sullivan: Heavy ions in the outer Kronian magnetosphere. J. Geophys. Res. **88**, 823–831 (1983).

Eviatar, A., and J. Richardson: Corotation of the Kronian magnetosphere. J. Geophys. Res. **91**, 3299–3303 (1986).

Fukazawa, K., S. Ogi, T. Ogino, and R. J. Walker: Magnetospheric convection at Saturn as a function of IMF B_Z, Geophys. Res. Lett. **34**, L01105 (2007a) doi:10.1029/2006GL028373

Fukazawa, K., T. Ogino, and R. J. Walker: Vortex-associated reconnection for northward IMF in the Kronian magnetosphere. Geophys. Res. Lett. **34**, L23201 (2007b) doi:10.1029/2007GL031784

Giampieri, G. and M. K. Dougherty: Modelling of the ring current in Saturn's magnetosphere. Annales Geophys. **22**, 653–659 (2004).

Giampieri, G. and M. K. Dougherty: Rotation rate of Saturn's interior from magnetic field observations. Geophys. Res. Lett. **31**, L16701 (2004) doi:10.1029/2004GL020194

Giampieri, G., M. K. Dougherty, E. J. Smith, and C. T. Russell: A regular period for Saturn's magnetic field that may track its internal rotation. Nature **441**, 62–64 (2006).

Gleeson, L. J. and W. I. Axford: An analytical model illustrating the effects of rotation on a magnetosphere containing low energy plasma. J. Geophys. Res. **81**, 3403–3406 (1976).

Glocer, A., T. I. Gombosi, G. Tóth, K. C. Hansen, A. J. Ridley, and A. Nagy: Polar wind outflow model: Saturn results. J. Geophys. Res. **112**, A01304 (2007) doi:10.1029/2006JA011755

Goertz, C. K.: Detached plasma in Saturn's front side magnetosphere. Geophys. Res. Lett. **10**, 455–458 (1983).

Goldreich, P. and A. J. Farmer: Spontaneous axisymmetry breaking of the external magnetic field at Saturn. J. Geophys. Res. **112**, A05225 (2007) doi:10.1029/2006JA012163

Gombosi, T. I., M. Horanyi, K. Kecskemty, T. E. Cravens, and A. F. Nagy: Charge exchange in solar wind – Cometary interactions. Astrophys. J. **268**, 889–898 (1983).

Gombosi, T. I.: Preshock region acceleration of implanted cometary H^+ and O^+. J. Geophys. Res. **93**, 35–47 (1988).

Gombosi, T. I.: *Physics of the Space Environment*. (Cambridge, New York, 1998).

Gombosi, T. I., G. Tóth, D. L. De Zeeuw, K. C. Hansen, K. Kabin, and K. G. Powell: Semi-relativistic magnetohydrodynamics and physics-based convergence acceleration. J. Comput. Phys. **177**, 176–205 (2001).

Gombosi, T. I. and K. C. Hansen: Saturn's variable magnetosphere. Science **307**, 1224–1226 (2005).

Gurnett, D. A., W. S. Kurth, G. B. Hospodarsky, A. M. Persoon, T. F. Averkamp, B. Cecconi, A. Lecacheux, P. Zarka, P. Canu, N. Cornilleau-Wehrlin, P. Galopeau, A. Roux, C. Harvey, P. Louarn, R. Bostrom, G. Gustafsson, J.-E. Wahlund, M. D. Desch, W. M. Farrell, M. L. Kaiser, K. Goetz, P. J. Kellogg, G. Fischer, H.-P. Ladreiter, H. Rucker, H. Alleyne, and A. Pedersen: Radio and plasma wave observations at Saturn from Cassini's approach and first orbit. Science, **307**, 1255–1259 (2005).

Gurnett, D. A., A. M. Persoon, W. S. Kurth, J. B. Groene, T. F. Averkamp, M. K. Dougherty, and D. J. Southwood: The variable rotation period of the inner region of Saturn's plasma disk. Science, **316**, 442–445 (2007).

Hansen, K. C., T. I. Gombosi, D. L. Dezeeuw, C. P. T. Groth, and K. G. Powell: A 3D global MHD simulation of Saturn's magnetosphere. Adv. Space. Res. **26(10)**, 1681–1690 (2000).

Hansen, K. C., A. J. Ridley, G. B. Hospodarsky, N. Achilleos, M. K. Dougherty, T. I. Gombosi, and G. Tóth: Global MHD simulations of

Saturn's magnetosphere at the time of Cassini approach. Geophys. Res. Lett. **32**, L20S06 (2005) doi:10.1029/2005GL022835

Hansen, C. J., L. Esposito, A. I. F. Stewart, J. Colwell, A. Hendrix, W. Pryor, D. Shemansky, and R. West: Enceladus' water vapor plume. Science **311**, 1422–1425 (2006).

Hill, T. W. and J. F. Carbary: Centrifugal distortion of the Jovian magnetosphere by an equatorially confined current sheet. J. Geophys. Res. **83**, 5745–5749 (1978).

Hill, T. W.: Inertial limit on corotation. J. Geophys. Res. **84**, 6554–6558 (1979).

Hill, T. W., A. M. Rymer, J. L. Burch, F. J. Crary, D. T. Young, M. F. Thomsen, D. Delapp, N. Andre, A. J. Coates, and G. R. Lewis: Evidence for rotationally-driven plasma transport in Saturn's magnetosphere. Geophys. Res. Lett. **32**, L41S10 (2005) doi:10.1029/2005GL022620

Hill, T. W., M. F. Thomsen, M. G. Henderson, R. L. Tokar, A. J. Coates, H. J. McAndrews, G. R. Lewis, D. G. Mitchell, C. M. Jackman, C. T. Russell, M. K. Dougherty, F. J. Crary, and D. T. Young: Plasmoids in Saturn's magnetotail. J. Geophys. Res. **113**, A01214 (2008) doi:10.1029/2007JA012626

Hospodarsky, G. B., T. F. Averkamp, W. S. Kurth, D. A. Gurnett, J. D. Menietti, O. Santolik, and M. K. Dougherty: Observations of chorus at Saturn using the Cassini Radio and Plasma Wave Science instrument. J. Geophys. Res. **113**, A12206 (2008) doi:10.1029/2008JA013237

Huddleston, D. E., C. T. Russell, M. G. Kivelson, K. K. Khurana, and L. Bennett: Location and shape of the Jovian magnetopause and bow shock. J. Geophys. Res. **103**, 20,075–20,082 (1998).

Jackman, C. M., N. Achilleos, E. J. Bunce, S. W. H. Cowley, M. K. Dougherty, G. H. Jones, S. E. Milan, and E. J. Smith: Interplanetary magnetic field at 9 AU during the declining phase of the solar cycle and its implications for Saturn's magnetospheric dynamics. J. Geophys. Res. **109**, A11203 (2004) doi:10.1029/2004JA010614

Jackman, C. M., N. Achilleos, E. J. Bunce, B. Cecconi, J. T. Clarke, S. W. H. Cowley, W. S. Kurth, and P. Zarka: Interplanetary conditions and magnetospheric dynamics during the Cassini orbit insertion fly-through of Saturn's magnetosphere. J. Geophys. Res. **110**, A10212 (2005) doi:10.1029/2005JA011054

Jackman, C. M. and Cowley, S. W. H.: A model of the plasma flow and current in Saturn's polar ionosphere under conditions of strong Dungey-cycle driving, Ann. Geophys. **24**, 1029–1055 (2006)

Jackman, C. M., C. T. Russell, D. J. Southwood, C. S. Arridge, N. Achilleos, and M. K. Dougherty: Strong rapid dipolarizations in Saturn's magnetotail: In situ evidence of reconnection. Geophys. Res. Lett. **34**, L11203 (2007) doi:10.1029/2007GL029764

Jackman, C. M., R. J. Forsyth, and M. K. Dougherty: The overall configuration of the interplanetary magnetic field upstream of Saturn, as revealed by Cassini observations. J. Geophys. Res. **113**, A08114 (2008) doi:10.1029/2008JA013083

Jackman, C. M., C. S. Arridge, N. Krupp, E. J. Bunce, D. G. Mitchell, H. J. McAndrews, M. K. Dougherty, C. T. Russell, N. Achilloes, G. H. Jones, and A. J. Coates: A multi-instrument view of tail reconnection at Saturn. J. Geophys. Res. **113**, A11213 (2008) doi:10.1029/2008JA013592

Johnson, R. E., H. T. Smith, O. J. Tucker, M. Liu, M. H. Burger, E. C. Sittler, and R. L. Tokar: The Enceladus and OH tori at Saturn, Astrophys. J. **644**, L137–L139 (2006).

Johnson, R. E., J. G. Luhmann, R. L. Tokar, M. Bouhram, J. J. Berthelier, E. C. Sittler, J. F. Cooper, T. W. Hill, H. T. Smith, M. Michael, M. Liu, F. J. Crary, and D. T. Young: Production, ionization, and redistribution of O_2 in Saturn's ring atmosphere. Icarus **180**, 393–402 (2006).

Johnson, R. E., M. Fam, M. Liu, R. A. Baragiola, E. C. Sittler, Jr., and H. T. Smith: Sputtering of ice grains and icy satellites in Saturn's inner magnetosphere. Planet. Space Sci. **56**, 1238–1243 (2008).

Joy, S. P., M. G. Kivelson, R. J. Walker, K. K. Khurana, C. T. Russell, and T. Ogino: Probabilistic models of the jovian magnetopause and bow shock. J. Geophys. Res. **107**, 1309 (2002) doi:10.1029/2001JA009146

Jurac, S., R. E. Johnson, J. D. Richardson, and C. Paranicas: Satellite sputtering in Saturn's magnetosphere. Plan. Space Sci. **49**, 319–326 (2001).

Jurac, S., M. A. McGrath, R. E. Johnson, J. D. Richardson, V. M. Vasyliunas, and A. Eviatar : Saturn: Search for a missing water source. Geophys Res. Lett. **29**, 2172 (2002) doi:10.1029/2002GL015855

Kempf, S., R. Srama, M. Horányi, M. Burton, S. Helfert, G. Moragas-Klostermeyer, M. Roy, and E. Grün: High-velocity streams of dust originating from Saturn. Nature **433**, 289–291 (2005).

Kennel, C. F. and H. E. Petschek: Limit on stably trapped fluxes. J. Geophys. Res. **71**, 1–28 (1966).

Khurana, K. K., D. G. Mitchell, C. S. Arridge, M. K. Dougherty, C. T. Russell, C. P. Paranicas, N. Krupp, and A. J. Coates: Sources of rotational signals in Saturn's magnetosphere. J. Geophys. Res. **114**, A02211 (2009) doi:10.1029/2008JA013312

Krimigis, S. M. and T. P. Armstrong, Two-component proton spectra in the inner Saturnian magnetosphere Geophys. Res. Lett. **9**, 1143–1146 (1982).

Krimigis, S. M. et al. (1981), Low-energy charged particles in Saturn's magnetosphere. Results from Voyager 1. Science **212**, 225–231 (1981).

Krimigis, S. M., J. F. Carbary, E. P. Keath, T. P. Armstrong, L. J. Lanzerotti, and G. Gloeckler: General characteristics of hot plasma and energetic particles in the Saturnian magnetosphere: Results from the Voyager spacecraft. J. Geophys. Res. **88**, 8871–8892 (1983).

Krimigis, S. M.: Voyager energetic particle observations at interplanetary shocks and upstream of planetary bow shocks: 1977–1990, Space Sci. Rev. **59**, 167–201 (1992).

Krimigis, S. M., D. G. Mitchell, D. C. Hamilton, S. Livi, J. Dandouras, S. Jaskulek, T. P. Armstrong, A. F. Cheng, G. Gloeckler, K. C. Hsieh, W.-H. Ip, E. P. Keath, E. Kirsch, N. Krupp, L. J. Lanzerotti, B. H. Mauk, R. W. McEntire, E. C. Roelof, B. E. Tossman, B. Wilken, and D. J. Williams: Magnetosphere Imaging Instrument (MIMI) on the Cassini Mission to Saturn/Titan. Space Sci. Rev. **114**(1–4), 233–329 (2004) doi: 10.1007/s11214-004-1410-8

Krimigis, S. M., D. G. Mitchell, D. C. Hamilton, N. Krupp, S. Livi, E. C. Roelof, J. Dandouras, T. P. Armstrong, B. H. Mauk, C. Paranicas, P. C. Brandt, S. Bolton, A. F. Cheng, T. Choo, G. Gloeckler, J. Hayes, K. C. Hsieh, W.-H. Ip, S. Jaskulek, E. P. Keath, E. Kirsch, M. Kusterer, A. Lagg, L. J. Lanzerotti, D. LaVallee, J. Manweiler, R. W. McEntire, W. Rasmuss, J. Saur, F. S. Turner, D. J. Williams, and J. Wocj: Dynamics of Saturn's magnetosphere from MIMI during Cassini's orbital insertion. Science **307**, 1270–1273 (2005).

Krimigis, S. M., N. Sergis, D. G. Mitchell, D. C. Hamilton, and N. Krupp: A dynamic, rotating ring current around Saturn. Nature **450**, 1050–1053 (2007) doi:10.1038/nature06425

Krimigis, S. M., N. Sergis, K. Dialynas, D. G. Mitchell, D. C. Hamilton, N. Krupp, M. Dougherty, and E. T. Sarris: Analysis of a sequence of energetic ion and magnetic field events upstream from the Saturnian magnetosphere. Planet. Space Sci. (2009) doi:10.1016/j.pss.2009.02.013

Krupp, N., A. Lagg, J. Woch, S. M. Krimigis, S. Livi, D. G. Mitchell, E. C. Roelof, C. Paranicas, B. H. Mauk, D. C. Hamilton, T.P. Armstrong, and M. K. Dougherty (2005). The Saturnian plasma sheet as revealed by energetic particle measurements. Geophys. Res. Lett. **32**, L20S03 (2005) doi:10.1029/2005GL022829

Kurth, W. S., D. A. Gurnett, J. T. Clarke, P. Zarka, M. D. Desch, M. L. Kaiser, B. Cecconi, A. Lecacheux, W. M. Farrell, P. Galopeau, J.-C. Gérard, D. Grodent, R. Prangé, M. K. Dougherty, and F. J.

Crary: An Earth-like correspondence between Saturn's auroral features and radio emission. Nature **433**, 722–725 (2005).

Kurth, W. S., A. Lecacheux, T. F. Averkamp, J. B. Groene, and D. A. Gurnett: A Saturnian longitude system based on a variable kilometric radiation period. Geophys. Res. Lett. **34**, L02201 (2007) doi:10.1029/2006GL028336

Kurth, W. S., T. F. Averkamp, D. A. Gurnett, and J. B. Groene: An update to a Saturnian longitude system based on kilometric radio emissions. J. Geophys. Res. **113**, A05222 (2008) doi:10.1029/2007JA012861

Kurth, W. S., E. J. Bunce, J. T. Clarke, F. J. Crary, D. C. Grodent, A. P. Ingersoll, U. A. Dyudina, L. Lamy, D. G. Mitchell, A. M. Persoon, W. R. Pryor, J. Saur, and T. Stallard: Auroral Processes. **Chapter 12 in this book** (2009).

Leisner, J. S., C. T. Russell, K. K. Khurana, M. K. Dougherty, and N. André: Warm flux tubes in the E-ring plasma and torus: Initial Cassini magnetometer observations. Geophys. Res. Lett. **32**, L14S08 (2005) doi:10.1029/2005GL022652

Leisner, J. S., C. T. Russell, M. K. Dougherty, X. Blanco-Cano, R. J. Strangeway, and C. Bertucci: Ion cyclotron waves in Saturn's E ring: Initial Cassini observations. Geophys. Res. Lett. **33**, L11101 (2006) doi:10.1029/2005GL024875

Leisner, J. S., C. T. Russell, K. K. Khurana, and M. K. Dougherty: Measuring the stress state of the Saturnian magnetosphere. Geophys. Res. Lett. **34**, L12103 (2007) doi:10.1029/2007GL029315

Masters, A., C. S. Arridge, M. K. Dougherty, C. Bertucci, L. Billingham, S. J. Schwartz, C. M. Jackman, Z. Bebesi, A. J. Coates, and M. F. Thomsen: Cassini encounters with hot flow anomaly-like phenomena at Saturn's bow shock. Geophys. Res. Lett. **35**, L02202 (2008) doi:10.1029/2007GL032371

Masters, A., N. Achilleos, M. K. Dougherty, J. A. Slavin, G. B. Hospodarsky, C. S. Arridge, and A. J. Coates: An empirical model of Saturn's bow shock: Cassini observations of shock location and shape. J. Geophys. Res. **113**, A10210 (2008) doi:10.1029/2008JA013276

Masters, A., N. Achilleos, C. Bertucci, M. K. Dougherty, S. J. Kanani, C. S. Arridge, H. J. McAndrews, and A. J. Coates: Surface waves on Saturn's dawn flank magnetopause driven by the Kelvin-Helmholtz instability. Planet. Space Sci. (2009), doi:10.1016/j.pss.2009.02.010

Mauk, B. H., S. M. Krimigis, and R. P. Lepping: Particle and field stress balance within a planetary magnetosphere. J. Geophys. Res. **90**, 8253–8264 (1985).

Mauk, B. H., J. Saur, D. G. Mitchell, E. C. Roelof, P. C. Brandt, T. P. Armstrong, D. C. Hamilton, S. M. Krimigis, N. Nrupp, S. A. Livi, J. W. Manweiler, and C. P. Paranicas: Energetic particle injections in Saturn's magnetosphere. Geophys. Res. Lett. **32**, L14S05 (2005) doi:10.1029/2005GL022485

Mauk, B. H., D. Hamilton, T. Hill, G. Hospodarsky, R. Johnson, C. Paranicas, E. Roussos, C. T. Russell, D. Shemansky, E. Sittler, and R. Thorne: Fundamental Plasma Processes in Saturn's Magnetosphere. **Chapter 11 in this book** (2009).

Maurice, S., I. Engle, M. Blanc, and M. Skubis: Geometry of Saturn's magnetopause model. J. Geophys. Res. **101**, 27053–27059 (1996).

McAndrews, H. J., C. J. Owen, M. F. Thomsen, B. Lavraud, A. J. Coates, M. K. Dougherty, and D. T. Young: Evidence for reconnection at Saturn's magnetopause, J. Geophys. Res. **113**, A04210 (2008) doi:10.1029/2007JA012581

McAndrews, H. J., M. F. Thomsen, C. S. Arridge, C. M. Jackman, R. J. Wilson, M. G. Henderson, R. L. Tokar, K. K. Khurana, E. C. Sittler, A. J. Coates, and M. K. Dougherty: Plasma in Saturn's nightside magnetosphere and the implications for global circulation. Planet. Space Sci. **in press** (2009) doi:10.1016/j.pss.2009.03.003

McNutt, R. L.: Force balance in outer planet magnetospheres. In *Physics of Space Plasmas, Proceedings of the 1982-4 MIT Symposia, number 5 in SPI Conference Proceedings and Reprint Series*, edited by J. Belcher, H. Bridge, T. Change, B. Coppi, and J. Jasperse (Scientific Publishers, Inc., 1983), pp. 179–210.

Menietti, J. D., O. Santolik, A. M. Rymer, G. B. Hospodarsky, A. M. Persoon, D. A. Gurnett, A. J. Coates, and D. T. Young: Analysis of plasma waves observed within local plasma injections seen in Saturn's magnetosphere. J. Geophys. Res. **113**, A05213 (2008) doi:10.1029/2007JA012856

Mitchell, D. G., P. C. Brandt, E. C. Roelof, J. Dandouras, S. M. Krimigis, B. M. Mauk, C. P. Paranicas, N. Krupp, D. C. Hamilton, W. S. Kurth, P. Zarka, M. K. Dougherty, E. J. Bunce, and D. E. Shemansky: Energetic ion acceleration in Saturn's magnetotail: Substorms at Saturn? Geophys. Res. Lett. **32**, L20S01 (2005) doi:10.1029/2005GL022647

Mitchell, D. G., J. F. Carbary, S. W. H. Cowley, T. W. Hill and P. Zarka: The dynamics of Saturn's Magnetosphere, **Chapter 10 in the book** (2009).

Moncuquet, M., A. Lecacheux, N. Meyer-Vernet, B. Cecconi, and W. S. Kurth: Quasi thermal noise spectroscopy in the inner magnetosphere of Saturn with Cassini/RPWS: Electron temperatures and density. Geophys. Res. Lett. **32**, L20S02 (2005) doi:10.1029/2005GL022508

Moritz, J.: Energetic protons at low equatorial altitudes. Z. Geophys. **38**, 701–717 (1972).

Ness, N. F., M. H. Acuña, K. W. Behannon, L. F. Burlaga, J. E. P. Connerney, R. P. Lepping, and F. M. Neubauer: Magnetic field studies by Voyager-2 – Preliminary results at Saturn. Science **215**, 558–563 (1982).

Paranicas C. and C. F. Cheng: A model for satellite microsignatures for Saturn. Icarus **125**, 380–396 (1997).

Persoon, A. M., D. A. Gurnett, W. S. Kurth, G. B. Hospodarsky, J. B. Groene, P. Canu, and M. K. Dougherty: Equatorial electron density measurements in Saturn's inner magnetosphere. Geophys. Res. Lett. **32**, L23105 (2005) doi:10.1029/2005GL024294

Persoon, A. M., D. A. Gurnett, W. S. Kurth, and J. B. Groene: A simple scale height model of the electron density in Saturn's plasma disk. Geophys. Res. Lett. **33**, L18106 (2006) doi:10.1029/2006GL027090

Persoon, A. M., D. A. Gurnett, O. Santolik, W. S. Kurth, J. B. Faden, J. B. Groene, G. R. Lewis, A. J. Coates, R. J. Wilson, R. L. Tokar, J.-E. Wahlund, and M. Moncuquet: A diffusive equilibrium model for the plasma density in Saturn's magnetosphere. J. Geophys. Res. **114**, A04211 (2009) doi:10.1029/2008JA013912

Porco, C. C., P. Helfenstein, P. C. Thomas, A. P. Ingersoll, J. Wisdom, R. West, G. Neukum, T. Denk, R. Wagner, T. Roatsch, S. Kieffer, E. Turtle, A. McEwen, T. V. Johnson, J. Rathbun, J. Veverka, D. Wilson, J. Perry, J. Spitale, A. Brahic, J. A. Burns, A. D. DelGenio, L. Dones, C. D. Murray, and S. Squyres: Cassini observes the active south pole of Enceladus, Science **311**, 1393–1401 (2006).

Powell, K. G., P. L. Roe, T. J. Linde, T. I. Gombosi, and D. L. De Zeeuw: A solution-adaptive upwind scheme for ideal magnetohydrodynamics. J. Comput. Phys. **154**, 284–309 (1999).

Richardson, J. D.: Thermal ions at Saturn: Plasma parameters and implications. J. Geophys. Res. **91**, 1381–1390 (1986).

Richardson, J. D. and E. C. Sittler: A plasma density model for Saturn based on Voyager observations. J. Geophys. Res. **95**, 12019–12031 (1990).

Richardson, J. D.: An extended plasma model for Saturn. Geophys. Res. Lett. **22**, 1177–1180 (1995).

Richardson, J. D., A. Eviatar, M. A. McGrath, and V. M. Vasyliunas: OH in Saturn's magnetosphere: Observations and implications. J. Geophys. Res. **103**, 20,245–20,255 (1998).

Richardson, J. D.: Thermal plasma and neutral gas in Saturn's magnetosphere. Geophys. Res. Lett. **36**, 501–524 (1998).

Richardson, J. D. and S. Jurac: A self-consistent model of plasma and neutrals at Saturn: The ion tori. Geophys. Res. Lett. **31**, L24803 (2004) doi:10.1029/2004GL020959

Roussos, E.: Interactions of weakly or non-magnetized bodies with solar system plasmas: Mars and the moons of Saturn. (PhD thesis, Technical University Braunschweig, Germany) (2008).

Roussos, E., N. Krupp, T. P. Armstrong, C. Paranicas, D. G. Mitchell, S. M. Krimigis, G. H. Jones, K. Dialynas, N. Sergis, and D. C. Hamilton: Discovery of a transient radiation belt at Saturn. Geophys. Res. Lett. **35**, L22106 (2008) doi:10.1029/2008GL035767

Roussos, E., G. H. Jones, N. Krupp, C. Paranicas, D. G. Mitchell, S. M. Krimigis, J. Woch, A. Lagg, and K. Khurana: Energetic electron signatures of Saturn's smaller moons: Evidence of an arc of material at Methone. Icarus **193**, 455–464 (2008) doi:10.1016/j.icarus.2007.03.034

Russell, C. T.: Outer planet magnetospheres: a tutorial. Advances in Space Research **33(11)**, 2004–2020 (2004).

Rymer, A. M., B. H. Mauk, T. W. Hill, C. Paranicas, N. André, E. C. Sittler, Jr., D. G. Mitchell, H. T. Smith, R. E. Johnson, A. J. Coates, D. T. Young, S. J. Bolton, M. F. Thomsen, and M. K. Dougherty: Electron sources in Saturn's magnetosphere, J. Geophys. Res. **112**, A02201 (2007) doi:10.1029/2006JA012017

Rymer, A. M., B. H. Mauk, T. W. Hill, C. Paranicas, D. G. Mitchell, A. J. Coates, and D. T. Young: Electron circulation in Saturn's magnetosphere. J. Geophys. Res. **113**, A01201 (2008) doi:10.1029/2007JA-12589

Santos-Costa, D., M. Blanc, S. Maurice, and S. J. Bolton: Modeling the electron and proton radiation belts of Saturn. Geophys. Res. Lett **30**, 2059 (2003) doi:10.1029/2003GL017972

Saur, J., B. H. Mauk, A. Kaner, and F. M. Neubauer: A model for the azimuthal plama velocity in Saturn's magnetosphere. J. Geophys. Res. **109** A05217 (2004) doi:10.1029/2003JA010207

Seidelmann, P. K., V. K. Abalakin, M. Bursa, M. E. Davies, C. De Bergh, J. H. Lieske, J. Oberst, J. L. Simon, E. M. Standish, P. Stooke, and P. C. Thomas: Report of the IAU/IAG working group on cartographic coordinates and rotational elements of the planets and satellites: 2000. Celes. Mech. Dyn. Astron. **82**, 83–110 (2002).

Sergis, N., S. M. Krimigis, D. G. Mitchell, D. C. Hamilton, N. Krupp, B. H. Mauk, E. C. Roelof, and M. K. Dougherty: Ring current at Saturn: Energetic particle pressure in Saturn's equatorial magnetosphere measured with Cassini/MIMI. Geophys. Res. Lett. **34**, L09102 (2007) doi:10.1029/2006GL029223

Sergis N., S. M. Krimigis, D. G. Mitchell, D. C. Hamilton, N. Krupp, B. H. Mauk, E. C. Roelof, and M. K. Dougherty: Energetic particle pressure in Saturn's magnetosphere measured with the Magnetospheric Imaging Instrument on Cassini. J. Geophys. Res. **114**, A02214 (2009) doi:10.1029/2008JA013774

Shemansky, D. E., P. Matheson, D. T. Hall, H.-Y. Hu, and T. M. Tripp: Detection of the hydroxyl radical in the Saturn magnetosphere. Nature **363**, 329–331 (1993).

Shi, M., R. A. Baragiola, D. E. Grosjean, R. E. Johnson, S. Jurac, and J. Schou: Sputtering of water ice surfaces and the production of extended neutral atmospheres. J. Geophys. Res. **100**, 26,387–26,395 (1995).

Showalter, M. R., J. N. Cuzzi, and S. M. Larson: Structure and particle properties of Saturn's E Ring. Icarus **94**, 451–473 (1991).

Shue, J.-H., J. K. Chao, H. C. Fu, C. T. Russell, P. Song, K. K. Khurana, and H. J. Singer: A new functional form to study the solar wind control of the magnetopause size and shape. J. Geophys. Res. **102**, 9497–9511 (1997).

Siscoe, G. L. and Summers, D.: Centrifugally-driven diffusion of iogenic plasma. J. Geophys. Res. **86**, 8471–8479 (1981).

Sittler, E. C., K. W. Ogilvie, and J. D. Scudder: Survey of low energy plasma electrons in Saturn's magnetosphere: Voyagers 1 and 2. J. Geophys. Res. **88**, 8847–8870 (1983).

Sittler, E. C., Jr., M. Thomsen, D. Chornay, M. D. Shappirio, D. Simpson, R. E. Johnson, H. T. Smith, A. J. Coates, A. M. Rymer, F. Crary, D. J. McComas, D. T. Young, D. Reisenfeld, M. Dougherty, and N. André: Preliminary results on Saturn's inner plasmasphere as observed by Cassini: Comparison with Voyager. Geophys. Res. Lett. **32**, L14S07 (2005) doi:10.1029/2005GL022653

Sittler Jr., E. C., M. Thomsen, R. E. Johnson, R. E. Hartle, M. Burger, D. Chornay, M. D. Shappirio, D. Simpson, H. T. Smith, A. J. Coates, A. M. Rymer, D. J. McComas, D. T. Young, D. Reisenfeld, M. Dougherty, and N. André: Cassini observations of Saturn's inner plasmasphere: Saturn orbit insertion results. Planet. Space Sci. **54**, 1197–1210 (2006).

Sittler Jr., E. C., M. Thomsen, R. E. Johnson, R. E. Hartle, M. Burger, D. Chornay, M. D. Shappirio, D. Simpson, H. T. Smith, A. J. Coates, A. M. Rymer, D. J. McComas, D. T. Young, D. Reisenfeld, M. Dougherty, and N. André: Erratum to "Cassini observations of Saturn's inner plasmasphere: Saturn orbit insertion results". Planet. Space Sci. **55**, 2218–2220 (2007).

Sittler Jr., E. C., N. André, M. Blanc, M. Burger, R. E. Johnson, A. Coates, A. Rymer, D. Reisenfeld, M. F. Thomsen, A. Persoon, M. Dougherty, H. T. Smith, R. A. Baragiola, R. E. Hartle, D. Chornay, M. D. Shappirio, D. Simpson, D. J. McComas, and D. T. Young: Ion and neutral sources and sinks within Saturn's inner magnetosphere: Cassini results. Planet. Space Sci. **56**, 3–18 (2008).

Slavin, J. A., E. J. Smith, J. R. Spreiter, and S. S. Stahara: Solar wind flow about the outer planets: Gas dynamic modelling of the Jupiter and Saturn bow shocks. J. Geophys. Res. **90**, 6275–6286 (1985).

Smith, E. J., L. Davis Jr., D. E. Jones, P. J. Coleman, D. S. Colburn, P. Dyal, C. P. Sonett, and A. M. A. Frandsen: The planetary magnetic field and magnetosphere of Jupiter: Pioneer 10. J. Geophys. Res. **79**, 3501–3513 (1974).

Smith, E. J., L. Davis Jr., D. E. Jones, P. J. Coleman Jr., D. S. Colburn, P. Dyal, and C. P. Sonett: Saturn's magnetosphere and its interaction with the solar wind. J. Geophys. Res. **85**, 5655–5674 (1980).

Smith, H. T., M. Shappirio, E. C. Sittler, D. Reisenfeld, R. E. Johnson, R. A. Baragiola, F. J. Crary, D. J. McComas, and D. T. Young: Discovery of nitrogen in Saturn's inner magnetosphere, Geophys. Res. Lett. **32**, L14S03 (2005) doi:10.1029/2005GL022654

Smith, H. T., R. E. Johnson, E. C. Sittler, M. Shappirio, D. Reisenfeld, O. J. Tucker, M. Burger, F. J. Crary, D. J. McComas, and D. T. Young: Enceladus: The likely dominant nitrogen source in Saturn's magnetosphere. Icarus **188**, 356–66 (2007).

Smith, H. T., M. Shappirio, R. E. Johnson, D. Reisenfeld, E. C. Sittler, Jr., F. J. Crary, D. J. McComas, and D. T. Young: Enceladus: A potential source of ammonia products and molecular nitrogen for Saturn's magnetosphere. J. Geophys. Res. **113**, A11206 (2008) doi:10.1029/2008JA013352

Southwood, D. J. and M. G. Kivelson: Magnetospheric interchange instability. J. Geophys. Res. **92**, 109–116 (1987).

Southwood, D. J. and M. G. Kivelson: Magnetospheric interchange motion. J. Geophys. Res. **94**, 299–308 (1989).

Southwood, D. J. and M. G. Kivelson: Saturnian magnetospheric dynamics: Elucidation of a camshaft model. J. Geophys. Res. **112**, A12222 (2007) doi: 10.1029/2007JA012254

Spahn, F., J. Schmidt, N. Albers, M. Horning, M. Makuch, M. Seiss, S. Kempf, R. Srama, V. Dikarev, S. Helfert, G. Moragas-Klostermeyer, A. V. Krivov, M. Sremcevic, A. J. Tuzzolino, T. Economou, and E. Grün: Cassini dust measurements at Enceladus and implications for the origin of the E ring. Science **311**, 1416–1418 (2006).

Thomsen M. F. and J. A. Van Allen: On the inference of properties of Saturn's ring E from energetic charged particle observations. Geophys. Res. Lett. **6**, 893–896 (1979).

Thomsen, M. F., J. P. DiLorenzo, D. J. McComas, D. T. Young, F. J. Crary, D. Delapp, D. B. Reisenfeld, and N. André: Assessment of the magnetospheric contribution to the suprathermal ions in Saturn's foreshock region, J. Geophys. Res. **112**, A05220 (2007) doi:10.1029/2006JA012084

Tokar, R. L., R. E. Johnson, M. F. Thomsen, D. M. Delapp, R. A. Baragiola, M. F. Francis, D. B. Reisenfeld, B. A. Fish, D. T.

Young, F. J. Crary, A. J. Coates, D. A. Gurnett, and W. S. Kurth: Cassini observations of the thermal plasma in the vicinity of Saturn's main rings and the F and G rings. Geophys. Res. Lett. **32**, L14S04 (2005) doi:10.1029/2005GL022690

Tokar, R. L., R. E. Johnson, T. W. Hill, D. H. Pontius, W. S. Kurth, F. J. Crary, D. T. Young, M. F. Thomsen, D. B. Reisenfeld, A. J. Coates, G. R. Lewis, E. C. Sittler, and D. A. Gurnett.: The interaction of the atmosphere of Enceladus with Saturn's plasma. Science, **311**, 1409–1412 (2006) doi:10.1126/science.1121061

Tokar, R. L., R. J. Wilson, R. E. Johnson, M. G. Henderson, M. F. Thomsen, M. M. Cowee, E. C. Sittler, Jr., D. T. Young, H. J. McAndrews, and H. T. Smith: Cassini detection of water group pickup ions in Saturn's toroidal atmosphere. Geophys. Res. Lett. **35**, L14202 (2008) doi:10.1029/2008GL034749

Tóth, G., D. L. De Zeeuw, T. I. Gombosi, and K. G. Powell: A parallel explicit/implicit time stepping scheme on block-adaptive grids. J. Comput. Phys. **217**, 722–758 (2006).

Tsyganenko, N. A.: Modeling of twisted/warped magnetospheric configurations using the general deformation method. J. Geophys. Res. **103**, 23,551–23,653 (1998).

Tsyganenko N. A.: A model of the near magnetosphere with a dawn-dusk asymmetry – 1. Mathematical structure. J. Geophys. Res. **107**, 1179, doi:10.1029/2001JA000219 (2002a).

Tsyganenko N. A.: A model of the near magnetosphere with a dawn-dusk asymmetry – 2. Parameterization and fitting to observations. J. Geophys. Res. **107**, 1176, doi:10.1029/2001JA0002205 (2002b).

Van Allen, J. A., B. A. Randall, and M. F. Thomsen: Sources and sinks of energetic electrons and protons in Saturn's magnetosphere. J. Geophys. Res. **85**, 5679–5694 (1980).

Van Allen, J. A.: Absorption of energetic protons by Saturn's ring G. J. Geophys. Res. **88**, 6911–6918 (1983).

Van Allen, J. A.: Energetic particles in the inner magnetosphere of Saturn. In *Saturn*, edited by T. Gehrels and M. S. Matthews (University of Arizona Press, Tucson, 1984), pp. 281–317.

Vasyliūnas, V. M.: Plasma distribution and flow. In *Physics of the Jovian Magnetosphere*, edited by A. J. Dessler (Cambridge University Press, New York 1983), pp. 395–453.

Vasyliūnas, V. M.: Comparing Jupiter and Saturn: dimensionless input rates from plasma sources within the magnetosphere. Ann. Geophys. **26**, 1341–1343 (2008).

Wahlund, J.-E., R. Bostrom, G. Gustafsson, D. A. Gurnett, W. S. Kurth, T. Averkamp, G. B. Hospodarsky, A. M. Persoon, P. Canu, A. Pedersen, M. D. Desch, A. I. Eriksson, R. Gill, M. W. Morooka, and M. André: The inner magnetosphere of Saturn: Cassini RPWS cold plasma results from the first encounter. Geophys. Res. Lett. **32**, L20S09 (2005) doi:10.1029/2005GL022699

Waite, J. H., Jr., T. E. Cravens, W.-H. Ip, W. T. Kasprzak, J. G. Luhmann, R. L. McNutt, H. B. Niemann, R. V. Yelle, I. Müller-Wodarg, S. A. Ledvina, and S. Scherer: Oxygen ions observed near Saturn's A ring. Science **307**, 1260–1262 (2005).

Waite J. H., M. R. Combi, W.-H. Ip, T. E. Cravens, R. L. McNutt, W. Kasprzak, R. Yelle, J. Luhmann, H. Niemann, D. Gell, B. Magee, G. Fletcher, J. Lunine, and W. L. Tseng: Cassini ion and neutral mass spectrometer: Enceladus plume composition and structure. Science **311**, 1419–1422 (2006).

Waite J. H., D. T. Young, T. E. Cravens, A. J. Coates, F. J. Crary, B. Magee, and J. Westlake: The Process of Tholin Formation in Titan's Upper Atmosphere. Science **316**, 870–875 (2007).

Wilson, R. J., R. L. Tokar, M. G. Henderson, T. W. Hill, M. F. Thomsen, and D. H. Pontius Jr.: Cassini plasma spectrometer thermal ion measurements in Saturn's inner magnetoshere. J. Geophys. Res. **113**, A12218 (2008) doi:10.1029/2008JA013486

Young D. T., J. -J. Berthelier, M. Blanc, J. L. Burch, S. Bolton, A. J. Coates, F. J. Crary, R. Goldstein, M. Grande, T. W. Hill, R. E. Johnson, R. A. Baragiola, V. Kelha, D. J. McComas, K. Mursula, E. C. Sittler, K. R. Svenes, K. Szeg, P. Tanskanen, M. F. Thomsen, S. Bakshi, B. L. Barraclough, Z. Bebesi, D. Delapp, M. W. Dunlop, J. T. Gosling, J. D. Furman, L. K. Gilbert, D. Glenn, C. Holmlund, J.-M. Illiano, G. R. Lewis, D. R. Linder, S. Maurice, H. J. McAndrews, B. T. Narheim, E. Pallier, D. Reisenfeld, A. M. Rymer, H. T. Smith, R. L. Tokar, J. Vilppola, and C. Zinsmeyer: Composition and dynamics of plasma in Saturn's magnetosphere, Science **307**, 1262–1266 (2005).

Zieger, B. and K. C. Hansen: Statistical validation of a solar wind propagation model from 1 to 10 AU. J. Geophys. Res. **113**, A08107 (2008) doi:10.1029/2008JA013046

Chapter 10
The Dynamics of Saturn's Magnetosphere

D.G. Mitchell, J.F. Carbary, S.W.H. Cowley, T.W. Hill, and P. Zarka

Abstract The dynamics of Saturn's magnetosphere differs considerably from that at the Earth. Saturn's magnetosphere responds to both external and internal drivers. The solar wind ram pressure, rather than the solar wind speed and interplanetary field orientation, provides the primary external driver at Saturn, while the planet's rotation provides the main internal driver. Saturn's magnetosphere generally moves in the corotation sense all the way to the magnetopause, although at speeds less than rigid corotation. Little evidence for classic substorm phenomena exists, although substorm-like processes such as plasmoid formation have been detected. Brief, narrow injections of hot plasma from the outer to inner magnetosphere play an important role in the dynamics at Saturn, as do energetic ion acceleration events in the outer magnetosphere as revealed by energetic neutral atom bursts resulting from charge exchange. Internal variations of the magnetosphere exhibit strong modulations at ~10.8 hours and ~10.6 hours: this periodicity is manifest in Saturn kilometric radiation, energetic ions and electrons, low energy plasma, magnetic fields, energetic neutral atoms, and the motions of the plasma sheet and magnetopause. Slower, long term variations (~year) in the periodicities occur, and faster (~weeks) variations are linked to changes in the solar wind speed. The mechanisms driving the periodicities are an active subject of inquiry at this writing.

D.G. Mitchell and J.F. Carbary
Johns Hopkins University Applied Physics Laboratory, Laurel, MD, USA

S.W.H. Cowley
Department of Physics and Astronomy, University of Leicester, Leicester, UK

T.W. Hill
Physics and Astronomy Department, Rice University, Houston, TX, USA

P. Zarka
Laboratoire d'Etudes Spatiales et d'Instrumentation en Astrophysique, Centre National de la Recherche Scientifique, Observatoire de Paris, Meudon, France

10.1 Introduction

The term magnetospheric dynamics will refer to those processes within a magnetosphere that vary on spatial scales larger than about one planetary radius and on time scales longer than about 1 h (chosen arbitrarily as ~1/10 the planetary period, and long relative to typical field line resonances). Thus, magnetospheric dynamics includes the global flow of plasma mass and energy, whether co-rotational (in the azimuthal direction) or non-co-rotational (in the radial direction), external (solar wind) or internal (planetary spin) drivers, magnetic reconnection at the magnetopause or in the magnetotail, solar wind compression of the dayside magnetosphere, and the effects of the centrifugal interchange process. In particular, this discussion of magnetospheric dynamics will consider the pronounced approximately 10.8-h periodicities observed in the charged particles, magnetic fields, and radio emissions of Saturn, and how these periodicities may be connected to the rotation of Saturn's atmosphere/ionosphere. Each topic will be introduced by summarizing its pre-Cassini state of understanding, then discussed in terms of discoveries made by the Cassini mission, and concluded by considering outstanding problems and unanswered questions.

10.2 Transport of Mass and Energy, and Plasma Flow

Rapidly rotating planets with strong magnetic fields can enforce plasma flow in the corotation direction (azimuthal in the same sense as the planet's spin) essentially all the way to the magnetopause (e.g., Brice and Ioannidis 1970). Coupling between the ionosphere and the magnetosphere drives the corotation of the magnetosphere, so the rotational velocity associated with corotation reflects the motion of the ionosphere, and not necessarily the rotation of the planetary interior. Any drag forces in the magnetosphere can cause the magnetospheric motion to deviate from and lag behind the ionospheric motion (referred to as subcorotation). Even before Pioneer and Voyager, investigators understood that this rapid rotation of the magnetosphere makes the centrifugal force and interchange instabilities important (Ioannidis and Brice 1971). These rotational effects play a major role in the mass and energy transport at Saturn in addition to the usual processes known to occur in all magnetospheres.

A plasma convection diagram conventionally summarizes the global plasma flow and attendant transport of mass and energy in a magnetosphere. At Earth, magnetospheric convection during periods of southward directed interplanetary magnetic field (IMF) is conceived in terms of the "Dungey cycle" wherein the magnetic field from the solar wind connects with the terrestrial field at the front of the magnetosphere, plasma carrying the field flows around the magnetosphere, and the field then reconnects in the plasma sheet of the magnetotail (Dungey 1961). Magnetic field merging forms an "X" line across the tail as a result of solar wind driving. The $\mathbf{V} \times \mathbf{B}$ electric field of the solar wind adds to the normal corotation electric field of the inner magnetosphere, and the superposition of these electric fields determines the global plasma flow pattern of the magnetosphere. Variations in the solar wind such as magnetic field turnings and traveling shocks, through the Dungey process, strongly perturb the terrestrial magnetosphere in the form of storms and substorms, which involve charged particle energization ultimately derived from the solar wind. One can say, then, that Earth's magnetospheric dynamics are externally driven by the solar wind.

In rapidly rotating magnetospheres with internal plasma sources, an "X" line is expected to form across a portion of the magnetotail, independent of solar wind effects. In this "Vasyliūnas cycle," the solar wind electric field does not significantly modify that of the magnetosphere except rather far down the tail, so that the dynamics of the magnetosphere may be almost completely controlled by the internally-driven (sub) corotation with the planet (Vasyliūnas 1983). This mechanism is driven by the build up of mass in the magnetospheric plasma from an internal source (Io at Jupiter, Enceladus at Saturn). As the centrifugal forces of the rotating plasma overwhelm magnetic field tension confining the plasma, the field configuration stretches until instabilities allow a topological change, forming an "X" line and releasing a fraction of the plasma along with the magnetic field threading it. No longer connected with the main planetary magnetic field, the disconnected plasma and field (called a plasmoid) escapes down the magnetotail. The plasma in the planetward side of the reconnected magnetic field snaps back toward the planet, accompanied by heating and particle acceleration (e.g., see Kivelson and Southwood, 2005). One can say of Jupiter that its dynamics are primarily internally driven, in accordance with this model. At Saturn, solar wind driven and internally-driven convection may coexist in a pattern that combines elements of the Dungey cycle and the Vasyliūnas cycle (e.g., Cowley et al. 2004). Figure 10.1 depicts the Vasyliūnas flow pattern. A magnetic X line would extend only partially across the magnetotail, and the release of plasma downtail as well as planetward would be strongly influenced, by the rotation of the inner magnetosphere. Dungey-type convection may occur, but only in rather thin layers in the outer magnetosphere, and then only during active solar wind conditions such as compressions from corotating interaction regions and from coronal mass ejections (e.g., Badman and Cowley 2007). In the Cowley et al. (2004) picture, the Vasyliūnas cycle is restricted to the dusk side of the tail while the dawn side is dominated by the Dungey cycle.

Before Cassini, the Pioneer and Voyager encounters provided an incomplete but suggestive picture of flow in Saturn's magnetosphere. The magnetospheric plasma flow definitely lags behind that of rigid corotation (as defined by the periodicity measured in Saturn kilometric radiation – more on

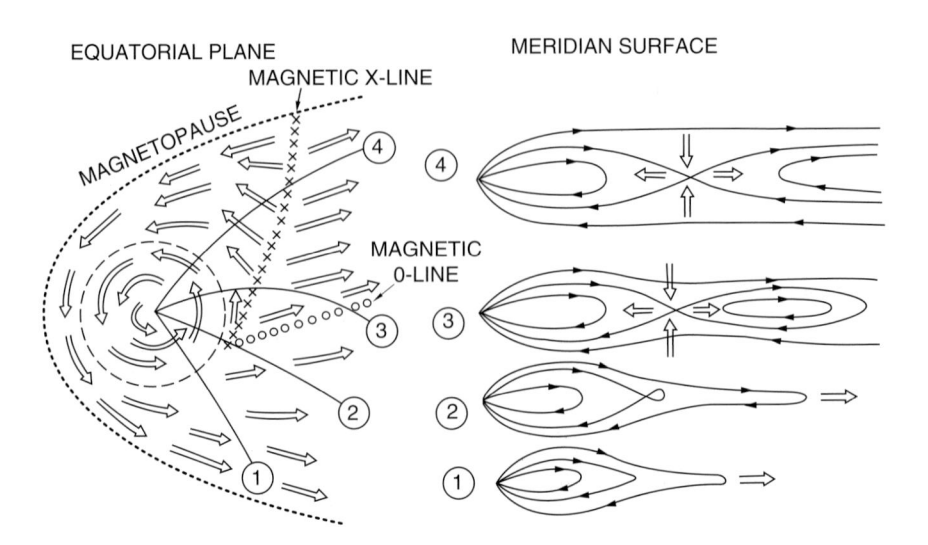

Fig. 10.1 Flow pattern (*left*) and field configuration (*right*) expected for a steady-state planetary wind, first proposed for Jupiter by Vasyliūnas (1983) but now thought to occur in a time-dependent manner at Saturn

this later), and neutral particles definitely dominate ionized particles in Saturn's inner magnetosphere (Richardson 1998). Sputtering caused by collisions of micrometeoroids and energetic charged particles with Saturn's moons and rings don't seem to provide an adequate source of these abundant neutrals (Paranicas et al. 2004), and investigators strongly suspected a large source of neutral water must exist near the orbital distance of Enceladus (e.g., Shemansky et al. 1993; Jurac et al. 2002; Jurac and Richardson 2005). The observation of subcorotational flows indicated that plasma is being added to the magnetosphere faster than the planet's rotation can accelerate it. Suggestions of outward and inward plasma flow are apparently related to effects of solar wind pressure and the centrifugal force of the plasma rotation (Richardson 1998).

Cassini discovered that Enceladus is the principal source of neutrals for Saturn's magnetosphere (Dougherty et al. 2006). The small moon emits jets of fine, icy particles and water vapor from its south polar region (Porco et al. 2006), supplying the magnetosphere with ∼100 kg/s or $\sim 3 \times 10^{27}$ water molecules/s (Tokar et al. 2006). The water jets do not directly supply the extended Enceladus neutral torus; charge exchange breaks down the ejected neutrals and accounts for a much more extended and less localized neutral torus (Johnson et al. 2006; Chapter 11 of this book). Thus, the major particle (and mass) source for Saturn's magnetosphere has been directly identified.

Present models of mass flow in Saturn's magnetosphere focus on the Enceladus source. The basic mechanism for transfer of mass from Enceladus involves the ejection of water molecules from the moon's south pole, followed by ionization of the water primarily by charge exchange with ambient charged particles, which produces a "pickup current" that accelerates the new plasma to corotational energies (e.g., Pontius and Hill 2006; Delamere et al. 2007). Unlike the similar process at Jupiter's satellite Io, ionization at Saturn near the location of the neutral source ($L \approx 4$) is limited by the smaller flux of electrons of sufficient energy to cause ionization. The Enceladus water source thus produces an abundant neutral cloud without as dense an associated plasma torus (Delamere et al. 2007; Chapter 11 of this book). The hot electron component at Saturn is largely restricted to $L > 6$ (e.g., Rymer et al. 2007), and the ionization rate thus peaks there. Any azimuthal asymmetry in the hot electron flux would also produce an asymmetry in the ionization rate (Delamere and Bagenal 2008).

Evidence from multiple Cassini instruments exists for some sort of azimuthal asymmetry moving in the corotational sense, thus causing asymmetric mass and energy flow around the planet. Such an asymmetry has been reported in the electron density determined by the RPWS instrument in the inner magnetosphere (Gurnett et al. 2007). One proposed explanation is a corotating convection pattern (Gurnett et al.

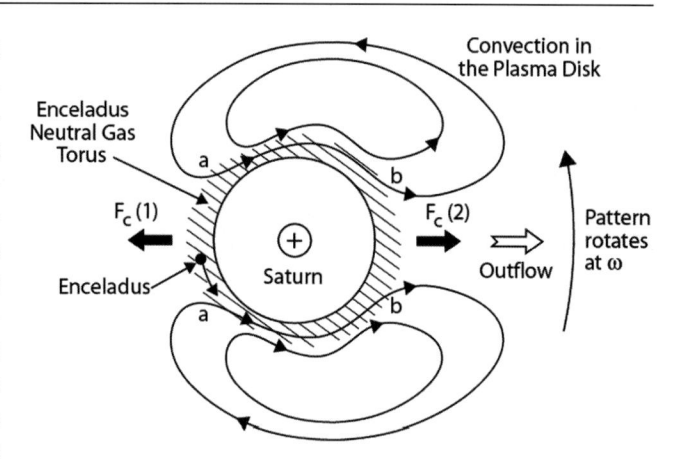

Fig. 10.2 A proposed corotating convection pattern with outflow in one particular longitude sector, based on the observation by RPWS of an electron density asymmetry. Outflow is proposed to occur at a particular longitude where a centrifugal instability is enhanced (from Gurnett et al. 2007)

2007; Goldreich and Farmer 2007). This type of pattern was initially conceived in connection with Jupiter (Vasyliūnas 1978; Hill et al. 1981). It has been proposed that the pattern need not rely on an anomaly fixed to Saturn, but instead may be generated spontaneously and maintained indefinitely by the instability itself. In effect, the centrifugal force acting on the plasma produces a single "tongue" of plasma moving outward and bending backward under the Coriolis force (Goldreich and Farmer 2007). In this case, the periodicities measured in the magnetosphere might not reflect the true rotation rate of Saturn, but provide only an upper limit. Figure 10.2 illustrates the corotating convection pattern inferred from plasma measurements in Saturn's inner magnetosphere (Gurnett et al. 2007).

Other potential mass flows include polar wind outflow from Saturn and from the atmosphere of Titan. Multifluid numerical simulations suggest that the polar wind may contribute between $\sim 10^{26}$ and 10^{28} particles/s, which could be comparable to the Enceladus source (Glocer et al. 2007). The polar outflow particles, however, may be restricted to open field lines or tail lobe field lines and could also be sporadic. There is little evidence to suggest that Titan contributes significant amounts of plasma to Saturn's magnetosphere. Measurements from the Cassini Langmuir probe suggest Titan contributes only $\sim 10^{25}$ ions/s to the magnetosphere, which is much smaller than estimates of other plasma sources (Wahlund et al. 2005). Other satellites, such as Tethys and Dione, may also represent secondary sources of plasma for the magnetosphere (Burch et al. 2007b).

Evidence for centrifugal interchange instability has been reported by several Cassini investigators. An analysis of hot plasma injections observed by the CAssini Plasma Spectrometer (CAPS; Young et al. 2004) supports the concept of drift dispersion within an interchanging flux tube as it

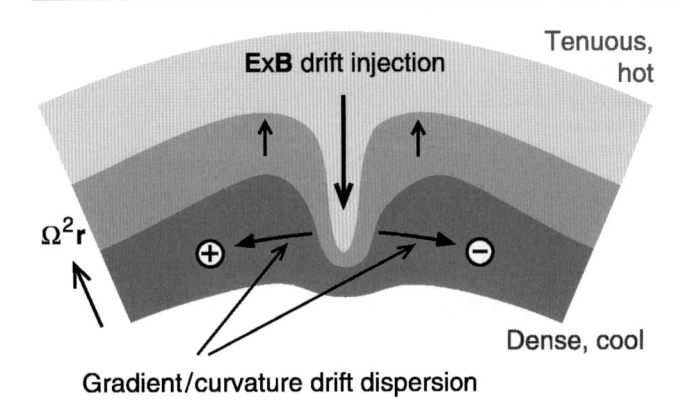

Fig. 10.3 Schematic of injection of a narrow finger of hot plasma into the cold plasma of the inner magnetosphere (from Hill et al. 2005)

convects inward through the inner magnetosphere (L < 12) (André et al. 2005; Burch et al. 2005; Hill et al. 2005; Leisner et al. 2005). In this picture, the injection itself is a narrow "finger" of hot plasma injected radially inward from the outer magnetosphere into the relatively cool plasma of the inner magnetosphere (see Fig. 10.3). Similar injections appear in Saturn's energetic ions and electrons (E > ~10 keV). The particles in these injections move according to the convective electric field (reflecting the local plasma velocity), as well as through their energy- and charge-dependent gradient and curvature azimuthal drifts. Therefore, at a given radial position as the distance between the injection longitude and the measurement longitude increases, the time of arrival of the various energy ions and electrons at the spacecraft varies (energy dispersion). By estimating their energy and charge dependent drift speeds in the nearly dipolar magnetic field, analysis of the energy dispersion signatures of these injections allows calculation of their azimuthal convective flow speeds (Mauk et al. 2005), which are comparable to the subcorotational flows deduced from Voyager (Richardson and Sittler 1990). The injections may be divided into two types: one type involves particles with energies of up to tens of keV, is intense, occurs more than once a day, and is not confined to particular local times (Hill et al. 2005; Chen and Hill 2008), while a second type extends to particle energies of hundreds of keV and occurs once every few days (Paranicas et al. 2007). An analysis of 429 plasma injection events (of the first type) in the CAPS energy range, which assumed a localized, fast injection followed by passive dispersive drift, revealed a typical width of less than ~1R_S, typical ages less than ~10.8 h (the nominal corotation period), and a random distribution in local time and Saturn longitude (Chen and Hill 2008).

A numerical model using a longitudinally narrow injection along a radial line has demonstrated that a spiral pattern will develop in the distribution of energetic charged particles as a consequence of their combined corotational and gradient/curvature drift (Brandt et al. 2008).

Theoretically, the implications of a corotating convection pattern have yet to be quantified, partly because of the intractability of the governing equations (e.g., Hill et al. 1981; Goldreich and Farmer 2007) and partly because of the lack of knowledge of the longitudinal structure of the plasma source. The corotating pattern would be expected to launch waves into the outer magnetosphere. A model of the wave phenomena thought to cause outer magnetospheric periodicities awaits development. Such a model would serve as the basis for interpreting phenomena in the entire magnetosphere.

Cassini measurements of plasma flow generally indicate magnetospheric flow is in the direction of corotation (also referred to as "prograde"), but generally below the rigid corotation speed beyond a certain radial distance. Inside ~12R_S, azimuthal speeds measured from CAPS plasma data fall ~20% below the speed of rigid corotation, although they tend to increase outside the orbit of Rhea (Wilson et al. 2008). Fits of observations of energetic protons and oxygen ions by the Cassini ion and neutral camera (INCA; Krimigis et al. 2004, 2005) to convected κ distributions reveals prograde azimuthal flow speeds close to corotation at radial distances between ~13 and ~20R_S, but which fall behind corotation outside this distance, with the disparity increasing with radial distance (Kane et al. 2008). The magnetopause occurs near 20R_S on the dayside, but predominantly prograde flows well below rigid corotation speeds can be measured at distances much exceeding this in the dawn, dusk, and midnight sectors. Figure 10.4 summarizes plasma and energetic particle measurements of flow speeds as functions of radial distance, comparing the Voyager plasma speeds (PLS) with the Cassini (INCA) speeds. The Voyager results indicate subcorotation speeds between 13 and 18R_S, so the INCA-derived speeds may be systematically high.

Other flow measurements obtained in the inner magnetosphere inside ~20R_S also indicate subcorotational flow. Tracks of "blobs" of energetic neutral atom (ENA) emissions, formed by charge exchange between gradient and curvature drifting energetic ions and cold neutrals, appear to move at speeds close to or even faster than rigid corotation for radial distances within the orbit of Rhea at ~9R_S, but fall to speeds of ~1/3 corotation outside that distance (Carbary et al. 2008b). Curiously, the ENA blobs, which reflect energetic ion motions, tend to have a small component of outward radial motion in the pre-noon sector and inward radial motion in the pre-midnight sector. Plasma measurements from the CAPS instrument indicate generally prograde flows below corotation speeds for all radial distances outside ~5R_S (McAndrews et al. 2009; Wilson et al. 2008); moreover, the plasma is also generally flowing radially outward beyond ~10R_S (McAndrews et al. 2007, 2009). Any compensating inward flow (which would be expected for magnetic flux conservation) likely occurs in narrow, fast, low-density channels where it is difficult to detect. These chan-

Fig. 10.4 Azimuthal convection speeds obtained from a moment analysis of CAPS plasma data within ∼10R$_S$ (*top panels*) (from Wilson et al. 2008) and speeds obtained by fitting a convected kappa distribution to INCA ion observations (*blue squares*) with speeds obtained from Voyager 2 plasma measurements (*open diamonds*) (I). Rigid corotation appears as a red dashed line (from Kane et al. 2008)

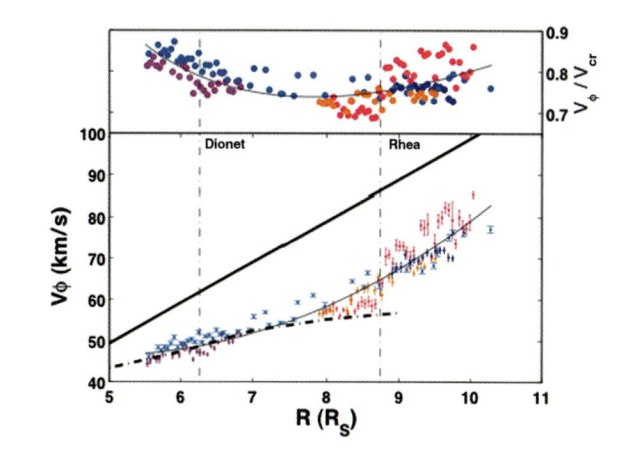

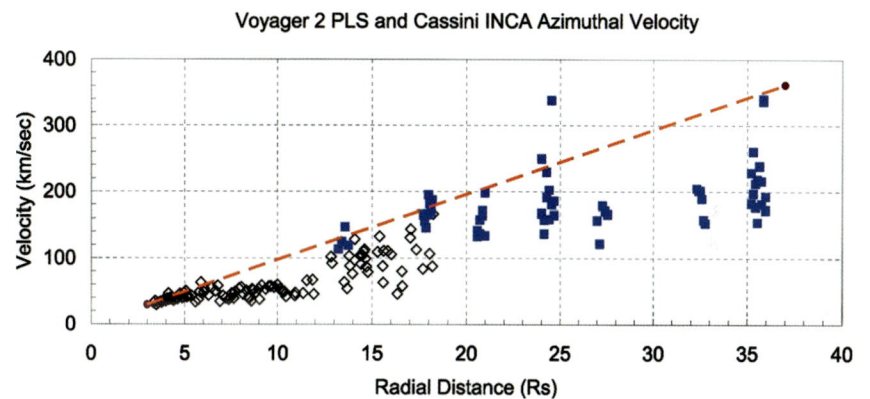

nels are presumably the same as the hot plasma injections described above. Figure 10.5 shows a compendium of the ion flow vectors measured by the plasma instrument. CAPS measurements indicate that as the azimuthal speed increases with radial distance more slowly than corotation speeds, the density decreases, as would be expected for expanding flux tubes. Whether this decrease is consistent with flux tube expansion quantitatively remains an open question.

The differences between the various data sources may in some cases reflect measurement error, but some differences may also reflect temporal variability. The Voyager results represent one cut through the system decades ago, so differences from those results could reflect either changes in the Saturn system (for example, different mass loading rates from Enceladus plumes) or systematic measurement errors. The degree to which Voyager and the CAPS measurements do agree, however, suggests limited changes between those two eras.

10.3 Rotational Modulation

Periodicities at or near a common rotation period of ∼10.8 h are ubiquitous in the magnetosphere of Saturn. When first discovered, this finding led to the notion that the observed period reflected the rotation rate of the planet. Underlying this notion was an implicit assumption that Saturn's magnetic field, like essentially every other planetary magnetic field measured, had some (not-yet-detected) nonpoloidal component that rotated with the interior, and ultimately drove the observed periodicities. To date, no such nonpoloidal component has been discovered, and this in combination with the now-chronicled variability in the periodicity renders any suggestion that these observations bear directly on the planetary interior rotation period moot. This makes the ubiquitous modulation particularly mysterious, and has generated a variety of suggestions for its causes.

Observations from the Voyager spacecraft indicated periodicities in the power of the kilometric radiation of the planet (Desch and Kaiser 1981), the spectral indices of charged particles (Carbary and Krimigis 1982), and the magnetic field itself (Espinosa and Dougherty 2000). Even the spokes in Saturn's rings seem to have a period associated with the planet's rotation (Porco and Danielson 1982). Indeed, the regularity of the Voyager-era radio period prompted the establishment of a longitude system, which was considered a definitive representation of the rotation period of the planet (Davies et al. 1996).

However, use of the radio period as the true rotation period proved premature because the precise radio period was observed to slowly change during the years after the Voyager encounters. Observations of Saturn kilometric radiation

Fig. 10.5 Plasma flow in Saturn's magnetosphere observed by the CAPS ion mass spectrometer. The arrows represent plasma flow speed and direction, while colors indicate density, all derived from a moment analysis (from McAndrews et al. 2009)

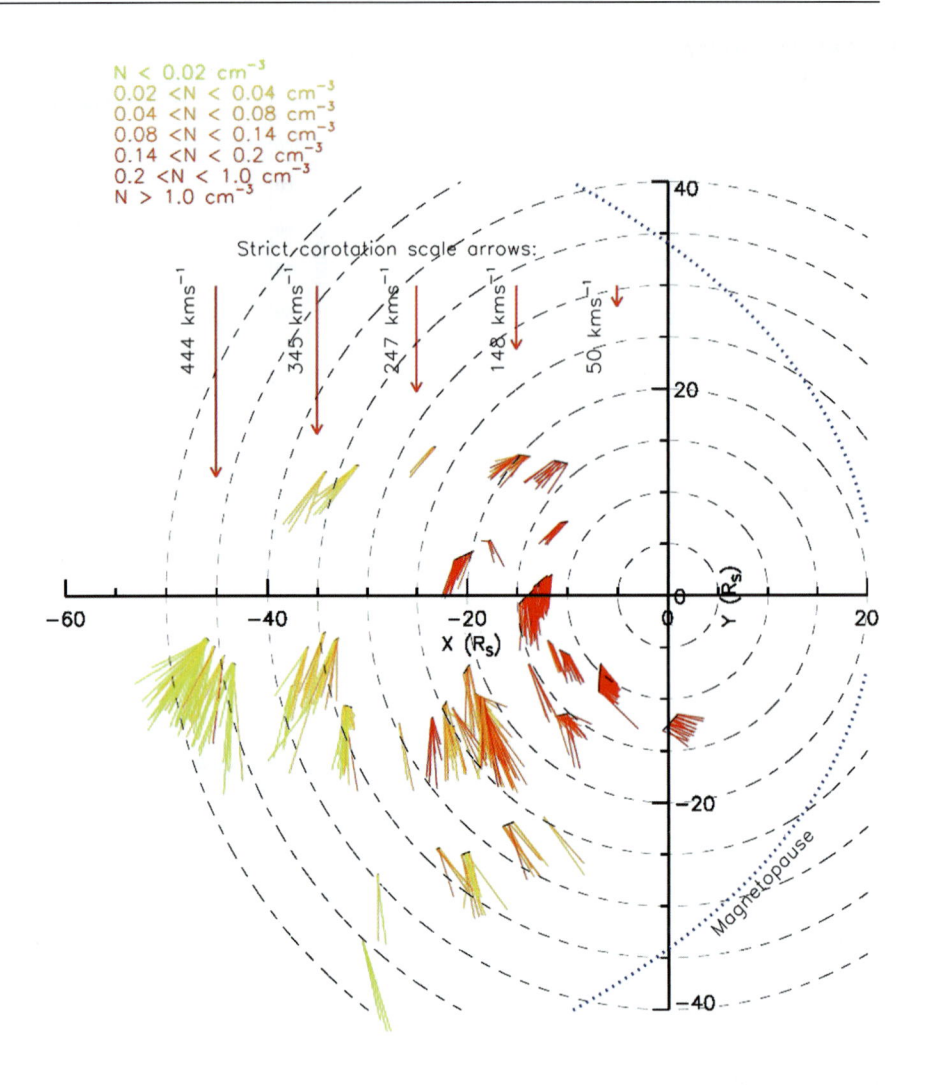

(SKR) from the Ulysses spacecraft indicated small but significant changes in the radio period over long time intervals (Galopeau and Lecacheux 2000). Over 2.5 years, the change measured by Ulysses amounted to about 1% of the nominal (Voyager-era) period of 10 h 39 min 22.4 s (Davies et al. 1996). During the several months leading up to the orbital insertion of Cassini, SKR observations of the radio and plasma wave science (RPWS) instrument confirmed that the radio period had indeed "drifted" to a period of 10 h 45 min 45 s (Gurnett et al. 2005).

The change of the radio period certainly does not reflect a change in the actual rotation period of Saturn because even a 1% change cannot be accommodated by the internal structure of the planet (e.g., Stevenson 2006). The drift in period is usually attributed to "slippage" of the magnetosphere of Saturn relative to its interior (Gombosi and Hansen 2005), although the details of this slippage may have more to do with gas dynamics and atmospheric physics and coupling to the ionosphere than with magnetospheric physics. Furthermore, the variable radio period therefore makes any relationship to the rotation period of the planet very indirect, so that a fundamental dynamical variable of Saturn is currently unknown – or at least very imprecisely known.

Given that the radio period is changing, the next question is how fast is it changing and can the change be quantified? SKR observations from early 2004 through early 2006 revealed an essentially smooth change of ∼160 s in the radio period, which could be further quantified by fitting the ΔT from a "base" period of 10.793 h to a third-order polynomial in time relative to the epoch of 1 January 2004 (Kurth et al. 2007). This initial fit proved reliable until about 1 September 2006 (day 242). Subsequent observations, especially those performed when the spacecraft orbit moved to higher latitudes, allowed an extension of the fitting to 10 August 2007 (day 222) and an improvement of the fit with new polynomials (Kurth et al. 2008). Figure 10.6 indicates the change in SKR period over the longer time sample. Significantly, as the sampling increased, a second SKR source became noticeable with a shorter (faster) period near 10.6 h. A second periodicity similar to this is also observed in electron fluxes in the same time period (Carbary et al. 2008c). The implications of this second periodicity have yet to be fully

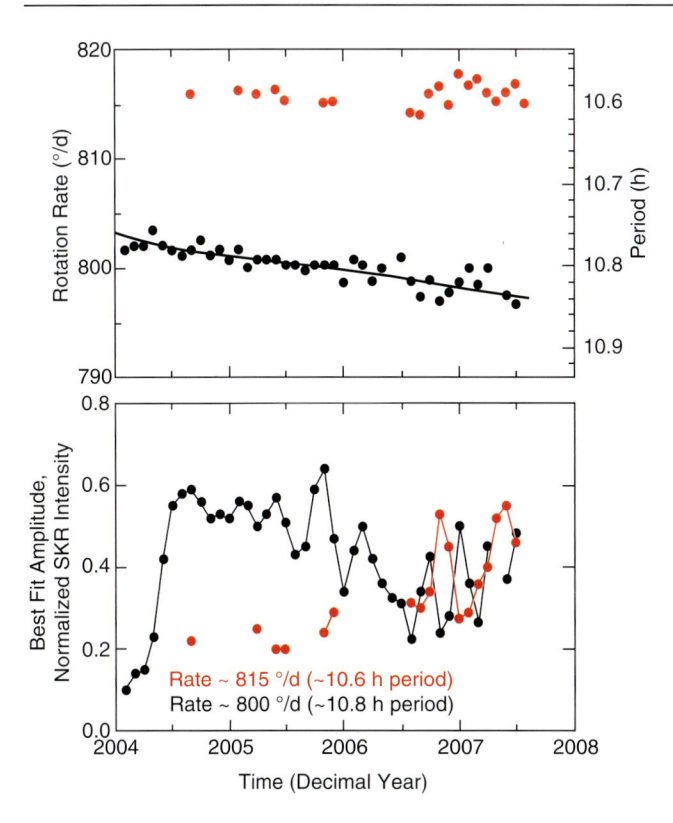

Fig. 10.6 Change in the SKR radio period measured over 3.5 years of Cassini observations (from Kurth et al. 2008). The top panel shows the change in period while the bottom panel shows the normalized amplitude of the SKR signal. The black and red dots separate two distinct SKR sources, one of which becomes prominent at a faster period later in the sampling interval

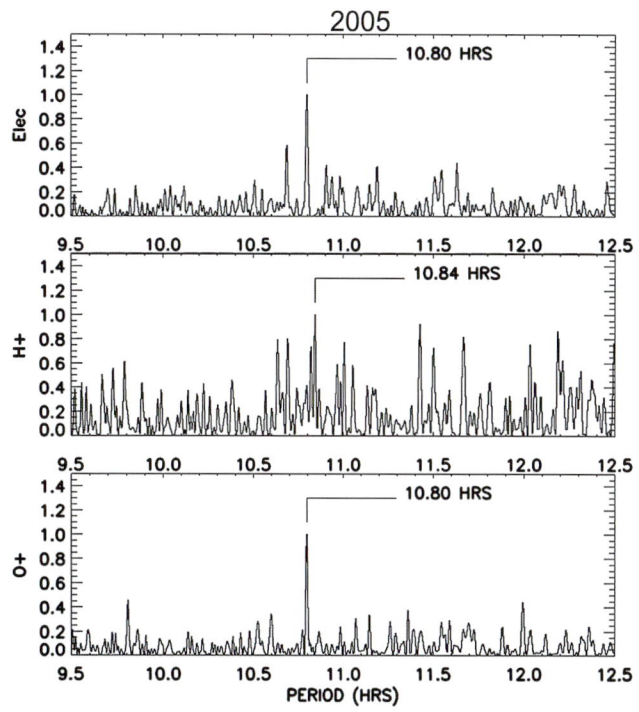

Fig. 10.7 Lomb periodogram analysis of periodic signals (in this case, count rates) from energetic electrons (*top*), protons (*middle*), and oxygen ions (*bottom*) for a 1-year sample of data. The spectral peaks are all close to the 10.793 h "base" period of the SKR (from Carbary et al. 2007b)

explored, although it may possibly be related to the true rotation rate of Saturn's interior (Anderson and Schubert 2007). At any rate, the smooth variation in SKR period has led to a variable period longitude system that can serve to organize magnetospheric phenomena even if it does not reflect the actual planetary rotation rate (Kurth et al. 2007, 2008).

Strong rotational modulation also appears in the magnetic field data. After subtraction of the contribution of the (steady) internal magnetic and magnetodisk fields, a strong periodic signal in the residual emerges. The residual displays such strong periodicity that it can be analyzed using any of the standard techniques of signal analysis. A Lomb-Scargle periodogram analysis of data from July 2004 to August 2005 revealed a period of 10 h 47 min 6 s, (10.785 h), which was distinctly different from what was then considered to be the rotation period of the planet (Giampieri et al. 2006). This signal has the same period as that expected from a tilted dipole precessing about the spin axis, but the magnetic field components are much different than such a model would predict. The polarization of the field oscillations indicates the presence of a rotating quasi-uniform equatorial field within the inner magnetosphere (Southwood and Kivelson 2007).

An investigation of 23 near-equatorial passes from October 2004 to July 2006 revealed that magnetic field oscillations generally maintain a consistent phase relative to the SKR oscillations, with the SKR power peaking when the magnetic field points radially outward and southward at 2 ± 2 h local time (Andrews et al. 2008). This study of the magnetic field phasing also indicated a slow phase drift ($\sim 0.12°$/day) of the magnetic oscillations relative to the SKR oscillations, which was interpreted as a refinement of a period common to both phenomena.

Periodicities have long been recognized in the charged particles of Saturn's magnetosphere. Indeed, the charged particles observed by Cassini display a much stronger periodic signal than those observed by the Voyagers during their passes in the 1980s (Krupp et al. 2005). The energetic charged particles ($E > 20$ keV) in particular display such strong signals that they can be analyzed using wavelet or periodogram techniques (Carbary et al. 2007a, b, 2008c). As shown in Fig 10.7, a Lomb periodogram analysis of these signals reveals strong periodicities very close to the SKR period whenever the signal-to-noise ratio is sufficiently high.

The regularity of these signals and their relation to the SKR period has prompted several investigators to organize their magnetospheric data in the new Saturn longitude system (SLS). For example, when the energetic electrons are averaged into range and SLS longitude bins, a clear spiral pattern emerges (Carbary et al. 2007c). The spiral remained essentially fixed in this revised longitude system for several months of 2006. The base of the spiral appears close to the "outflow" longitude of the corotating convection system postulated by Gurnett et al. (2007). Similar spiral patterns are observed in the perturbation magnetic field (Cowley et al. 2006; Arridge et al. 2007b).

Strong rotational modulation also appears in the neutral particle data. The ENA are spawned from charge-exchange collisions between neutrals and energetic ions in the magnetosphere, so modulations in the energetic ions should be reflected as modulations in the ENA fluxes. This is indeed the case, and the preliminary observations of energetic H and O evidenced clear modulations near the SKR frequency (Krimigis et al. 2007). These ENA modulations also show effects of relative motion between the observer and the source populations of ions as a consequence of the Compton–Getting effect (Compton and Getting 1935; Gleeson and Axford 1968), which has the effect of increasing apparent particle intensities when looking into a convective flow, and decreasing the intensity looking away from the flow (as pointed out for ENA by Paranicas et al. 2005). Observation at angles perpendicular to the source motion removes this effect and allows the true periodicity of the ENA to be measured. Such perpendicular observations reveal that the hydrogen ENA flux has only a weakly periodic behavior not consistent with the radio period. In contrast, the oxygen ENA flux evidences a very strongly periodic behavior at the radio period of 10.8 h. This disparity may suggest that the H modulations have a different driver (non-planetary) than the O modulations (planetary) (Carbary et al. 2008a). Notably, the periodicities in the ENA fluxes, as well as those in the energetic ions, seem to have disappeared in late 2007 and 2008, an effect possibly related to Saturn's movement toward its equinox or to some change in magnetospheric configuration (Carbary et al. 2008c).

Observations of energetic neutrals suggest a preferred location for particle energization. A dispersion analysis of rotationally-modulated ENA emissions suggests that under some conditions, protons and oxygen ions are accelerated once per Saturn rotation at a local time between midnight and dawn, from which a reconnection location between 15 and 20R_S in Saturn's magnetotail has been inferred (Mitchell et al. 2009). Figure 10.8 shows the statistical results of this ENA dispersion analysis. The figure shows the difference in hours between the observation time of peak oxygen ENA intensity at 250 and at 50 keV, as a function of observer local time (observations restricted to spacecraft locations near the equatorial plane). Since peak ENA intensities occur as the azimuthally rotating source ion population is convected toward the observer, the local time of the emission site lies roughly 6 h earlier than that of the observer. The zero lag dispersion (indicative of a new population which has not had time to separate in energy by gradient and curvature drifts) is observed from the late morning to noon quadrant, placing the initial acceleration site 6 h earlier in local time, between midnight and dawn. Notably, high latitude images of similar acceleration events can be tightly correlated with the onset of recurrent SKR power enhancements, and in rare cases where coincident data exist, with azimuthally rotating auroral brightenings at the same planetary longitudes as the ring current ENA enhancements (Mitchell et al. 2009).

The rotational modulations discussed so far have generally applied to regions within the magnetosphere near the equatorial plane of Saturn. However, some interesting plasma and magnetic field observations from one Cassini orbit suggest that oscillations of Saturn's magnetopause also occur near the 10.8-h rotation period (Clarke et al. 2006). These magnetopause oscillations have an amplitude of ∼2R_S, or about 10% of the noon radius of the magnetopause. Even Saturn's aurora exhibits a periodicity. Ultraviolet observations from the Hubble Space Telescope have recently revealed a 10.76 h periodicity in movement of the center of the auroral oval, with an amplitude of ∼1° (Nichols et al. 2008).

Analysis of short-term (∼10 days) variations in the SKR period from June 2003 to September 2006 revealed oscillations of ±1% (±6 min) amplitude at the time scale of 20–

Fig. 10.8 Energy dispersion (hours) in peak oxygen ENA emission vs observer local time. Red points are running boxcar averages over local time of measured (blue) points. Dispersion resets to zero in the late morning, indicating an early morning initiation site for the source ion acceleration (see text for details)

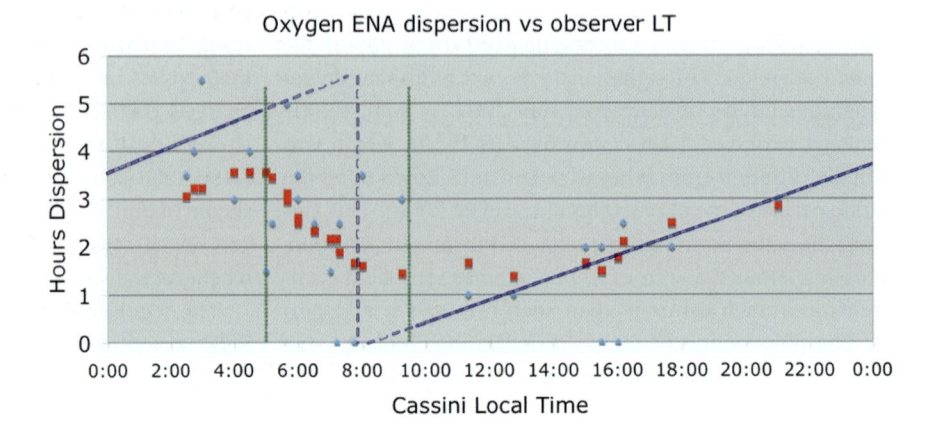

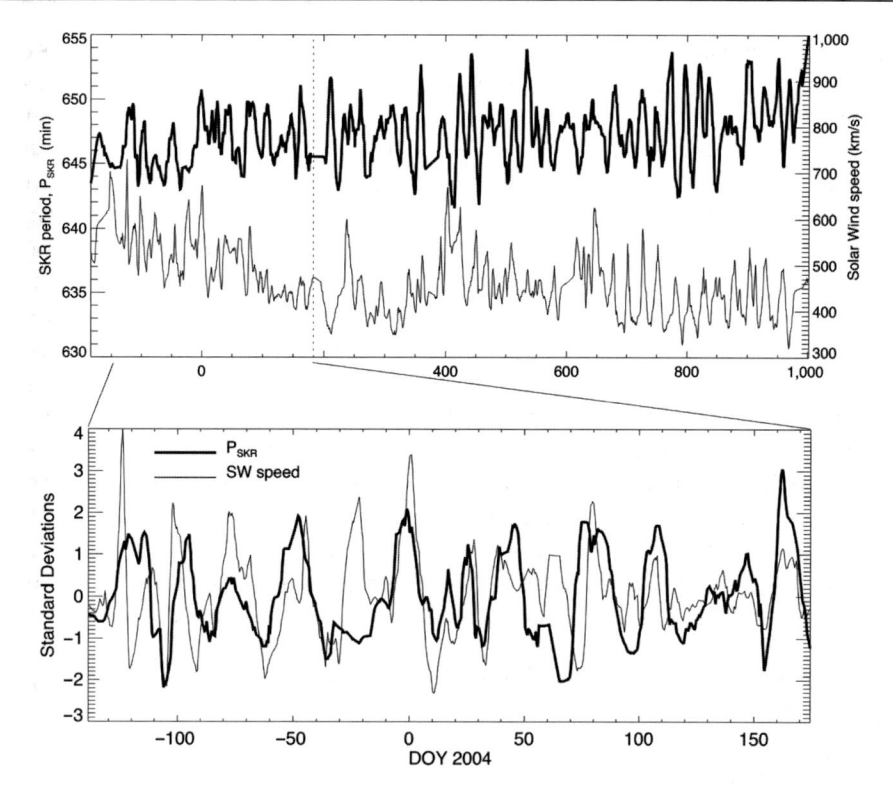

Fig. 10.9 Short-term variations of SKR period over 3.25 years (*boldface, left scale*) compared to solar wind speed at Saturn (*lightface, right scale*) measured by ACE/WIND spacecraft near Earth orbit and projected to Saturn (*top panel*). Dotted vertical line indicates Cassini's Saturn Orbit Insertion (SOI). Comparison of rapid fluctuations of SKR period and solar wind speed pre-SOI, when orbital parameters did not influence SKR visibility (*bottom panel*). Variations longer than ∼2 months have been filtered out, and both quantities normalized by their respective standard deviations. Correlation is apparent except for a few data gaps and possible coronal mass ejections (missing either ACE/WIND or Saturn). By contrast, the correlation of SKR period with solar wind pressure is very low (adapted from Zarka et al. 2007)

30 days (Zarka et al. 2007), superimposed on the long-term variation analyzed by Kurth et al. (2007). When decoupled from variations caused by Cassini's orbital motion, these short-term oscillations of the radio period exhibit a clear tendency to vary with the fluctuations of the solar wind flow speed past Saturn (Zarka et al. 2007), pointing to a cause external to the magnetosphere (see Fig 10.9). This correlation of SKR period variations with solar wind speed can be explained as the result of Kelvin–Helmholtz instabilities occurring on the dawn flank of the magnetopause where the shocked solar wind flow encounters the subcorotating magnetospheric plasma (Galopeau et al. 1995). External speed variations could induce SKR source motions in local time, modifying in turn the SKR period (Cecconi and Zarka 2005). Although the long-term averaging of these short-term variations of the SKR period result in the long-term trend of Kurth et al. (2007), it is not yet clear if this long-term trend can also be related to solar wind variations.

Table 10.1 lists the various periodicities measured by particles and fields instruments. In the case of SKR measurements, the variable period is given by a "base" or fixed period plus a polynomial in time. The assorted periodicities have strong implications for the dynamics of Saturn's magnetosphere, and several models have been proposed solely on the basis of the periodicities. At Jupiter (and Earth), the magnetic axis is tilted relative to the spin axis and rotates with the rotation of the planet, which generates outwardly traveling waves that give rise to periodic effects in the magnetosphere (Eviatar and Ershkovitch 1976). At Saturn, the magnetic axis is very closely aligned to the spin axis, so the magnetic wobble mechanism cannot operate at Saturn. Periodicities at Saturn are usually explained on the basis of some sort of rotating anomaly such as was first proposed to explain Jovian periodicities (Dessler and Hill 1975). Various incarnations of the anomaly model have been offered for Saturn. These include an "active hemisphere" on Saturn to explain particle periodicities (Carbary and Krimigis 1982), an "anomaly-swept" Kelvin–Helmholtz instability at the magnetopause to explain SKR periodicities as well as their variations (Galopeau et al. 1995; Galopeau and Lecacheux 2000; Cecconi and Zarka 2005; Zarka et al. 2007), and a rotating "camshaft" or plasma bulge to explain the magnetic field periodicities (Espinosa et al. 2003).

Cassini-era explanations are simply variations on this rotating anomaly concept. The charged particle periodicities clearly resemble a sort of rotating "searchlight" and can be

Table 10.1 Summary of Saturn periodicity measurements

Period (h:min:s)	Period (h)	Dates of measurement	Method of measurement	References
$10:39:24 \pm 7$ s	10.657 ± 0.002	Jan 1980–Sept 1981	SKR harmonic analysis (Vgr)	Kaiser and Desch (1981)
$10:21:00 \pm 48$ min	10.35 ± 0.80	Nov 1980–Aug 1981	Electron spectral index peaks (Vgr)	Carbary and Krimigis (1982)
10:39:22	10.656	1980–1981	Radio periods (IAU standard)	Davies et al. (1996)
10:45:45	10.763	Nov 1994–May 1996	SKR harmonic analysis (Ulysses)	Galopeau and Lecacheux (2000)
$10:45:45 \pm 36$ s	10.760 ± 0.010	Apr 2003–June 2004	SKR harmonic analysis (Cassini)	Gurnett et al. (2005)
$10:47:06 \pm 40$ s	10.785 ± 0.011	Jul 2004–Aug 2005	Lomb periodogram of perturbation Br	Giampieri et al. (2006)
$10:47:34 + A_0 + A_1 t + A_2 t^2 + A_3 t^3$	$10.793 + A_0 + A_1 t + A_2 t^2 + A_3 t^3$	Jan 2004–Aug 2006	SKR phase analysis	Kurth et al. (2007)
$10:53 \pm 31$ min	10.88 ± 0.52	Jul 2004–Jul 1006	Wavelet analysis of 20–300 keV electrons	Carbary et al. (2007a)
10:48:36	10.80	Jan 2005–Dec 2006	Lomb periodogram of charged particles	Carbary et al. (2007b)
$10:51 \pm 14$ min	10.85 ± 0.23	Aug 2004–Jun 2007	Lomb periodogram of oxygen ENA	Carbary et al. (2008)
$10:47:34 + B_0 B_1 t + B_2 t^2 + B_3 t^3$	$10.793 + B_0 + B_1 t + B_2 t^2 + B_3 t^3$	Jan 2004–Aug 2007	SKR phase analysis	Kurth et al. (2008)
SKR polynomial + linear phase	SKR polynomial + linear phase	Oct 2004–July 2006	Magnetic field phase analysis	Andrews et al. (2008)
$10:46 \pm 9$ min	10.76 ± 0.15	Jan. 2007 Feb. 2008	Analysis of aurora images	Nichols et al. (2008)

modeled very accurately assuming such simple dynamics (Carbary et al. 2007a). However, they can be just as easily interpreted in terms of a "wavy magnetodisk" similar to that at Jupiter. In the case of Saturn, which has no tilted dipole, a rotating longitudinal plasma or pressure anomaly would unbalance the magnetospheric rotation in the manner of a cam which can give rise to the periodic modulations without requiring any effective dipole tilt (Carbary et al. 2007d). Much the same wave motion could be induced by a "warped bowl" model in which the anomaly preferentially weights a preferred longitude against solar wind flow, which again gives rise to a warped magnetodisk (Arridge et al. 2008b; Khurana et al. 2009).

An alternative model based on the observation of rotating quasi-uniform fields in the inner equatorial magnetosphere proposes the existence of a rotating system of field-aligned currents that flow between the northern and southern hemispheres (Southwood and Kivelson 2007). Outside the current region the associated field perturbations would mimic those of a tilted dipole, with an estimated tilt of ~ 12–$15°$. Such a current system could result from a rotating convection pattern such as that proposed by Gurnett et al. (2007) and Goldreich and Farmer (2007) in the presence of an inter-hemispheric asymmetry of ionospheric conductivity, in which case the outflow sector would correspond to that in which the magnetic field perturbation points radially inward. The rotating convection pattern would produce a node in the azimuthal field at the equator, however, and such a node is not observed.

A number of outstanding problems exist in relation to understanding periodicities in Saturn's magnetosphere. Observationally, measurements of the various periodicities should be compared at different latitudes and local times to determine effects of possible differential rotation. Periodicity measurements should also be compared for different epochs to determine if solar wind and/or seasonal effects are correlated with changes in the periodicities, and this comparison requires continuing observations of the magnetosphere over many years spanning at least two solstice cycles of Saturn.

Additional information can be obtained by finding phase differences (or lack thereof) between those quantities exhibiting periodic modulation. Observations of the magnetic field in the near-equatorial quasi-dipolar magnetosphere, within radial distances of ~ 12–$15 R_S$, have shown that the magnetic field components oscillate near the ubiquitous period of ~ 10.8 h with amplitudes of a few nT (Espinosa and Dougherty 2000). The phasing of the oscillations is such that the azimuthal field is in lagging quadrature with the radial field, opposite to that of a rotating transverse dipole, indicating instead the presence of a rotating quasi-uniform equatorial field which is found to have no phase shift across the equator (Espinosa et al. 2003; Southwood and Kivelson 2007; Andrews et al. 2008). The rotating nature of the field is manifest in the Doppler shifts of the observed oscillation period due to the motion of the spacecraft around the planet (Cowley et al. 2006; Giampieri et al. 2006). An oscillating co-latitudinal magnetic field component is present as well, which is in phase with the radial component, such that overall the perturbation field lines form rotating loops with apices in the northern hemisphere (Andrews et al. 2008). The sidereal rotation period of the field has been shown to follow closely the variable period of the SKR modulations, the periods determined over the interval from October 2004 to July 2006 differing by at most ∼6 s i.e. by at most ∼0.02% (Andrews et al. 2008). Put another way, the phase drift of the two oscillations over this period was found to be at most ∼75° after a total of ∼1400 periods, this difference lying within the scatter of the SKR phase determinations. SKR maxima occur when the equatorial field is directed radially outward at a local time of ∼2± 2 h. At the same time, a phase 'jitter' of RMS amplitude ∼20° is also observed in the magnetic oscillations about this long-term behavior, which is highly correlated in the equatorial field components, but not in the co-latitudinal component (Andrews et al. 2008), which may relate to the short-term SKR phase variations reported by Zarka et al. (2007).

Southwood and Kivelson (2007) suggest that these field oscillations are associated with a rotating system of field-aligned currents that flow between Saturn's northern and southern ionospheres in opposite directions on either side of the planet, mapping to radial distances of ∼12–15R_S in the equatorial plane. Such a current system would produce a quasi-uniform equatorial field in the interior region that does not reverse about the equatorial plane as observed, while producing effects similar to that of a rotating transverse dipole in the exterior region, the effective dipole being formed by the field-aligned current loops. Such a scenario would then be consistent e.g. with planetary period oscillations of the plasma sheet in the magnetospheric tail (Arridge et al. 2007b). Planetary period field oscillations are indeed found on polar field lines in the southern hemisphere whose polarization and phase characteristics relative to the equatorial oscillations are consistent with this picture (Provan et al. 2009). However, similar oscillations in the northern hemisphere, when present, are found to have the opposite phase to this expectation, leading Provan et al. (2009) to suggest that the relevant current system is instead that of a partial ring current with stronger closure field-aligned currents in the southern hemisphere during southern summer solstice conditions. With regard to the physical origin of the rotating field and current system, however, in either case considerations based on a corotating interchange system of the same nature as that discussed e.g. by Hill et al. (1981), Gurnett et al. (2007) and Goldreich and Farmer (2007) would lead one to suppose that under these conditions the mass-loaded outflow sector would correspond to that in which the equatorial field points toward the planet (Andrews et al. 2008; Provan et al. 2009). However, Gurnett et al.'s (2007) results suggest that it instead corresponds to the sector where the azimuthal field is positive, shifted by a quarter cycle. The origin of this discrepancy thus requires further consideration.

Investigating the relation of periodicities to Saturn's moons has also recently begun and may have interesting consequences, especially in the cases of Titan and Enceladus (Menietti et al. 2007). Finally, the relation of Saturn's periodicities to solar wind driving effects is not yet known, although some interesting correlations may be possible when Cassini is outside the magnetosphere and can observe both the solar wind (in situ) and the magnetosphere (by remote ENA imagery or by SKR).

10.4 Magnetic Field Structure and Dynamics

At Earth, the principal process governing the magnetic field topology is magnetic field reconnection, which occurs primarily at the dayside magnetopause and in the magnetotail. At the magnetopause the solar wind field merges with the planetary field, producing the open field lines of the polar cap and tail lobes. These field lines are swept tailward by the solar wind flow and merge again at the center of the magnetotail, producing reconnected field lines that dipolarize and move back to the dayside again, thus completing the so-called "Dungey" cycle (Dungey 1961). Rapidly rotating planets such as Saturn have an additional (and probably more dominant) process affecting the magnetic field configuration, namely mass loading of magnetospheric plasma that corotates, either fully or partially. The cold plasma from the inner magnetosphere centrifugally stresses the field lines, causing them to migrate radially outward and interchange with "unloaded" field lines harboring hot plasma from the outer magnetosphere. This process is referred to as flux tube interchange (e.g., Siscoe and Summers 1981; Pontius and Hill 1989). Both reconnection and interchange motion of magnetic flux tubes have been proposed and detected in Saturn's magnetosphere, and both excite distinctive current systems. Finally, the magnetic field may be affected by wave motions of the magnetotail in response to a driver such as a magnetic axial tilt or a rotating plasma anomaly, complicating the interpretation of the effects of plasma transport. (e.g., Carbary et al. 2007d).

Prior to Cassini, a single Voyager magnetopause encounter suggested a reconnection signature, but its importance was discounted because the reconnection electric field (∼0.2 mV/m) was much smaller than the corotation electric field (∼1.4 mV/m) (Huddleston et al. 1997). Such a low value diminishes the importance of Dungey-type reconnection at Saturn's magnetopause relative to that at Earth's (Scurry and Russell 1991).

The Cassini mission greatly boosted the number of magnetopause observations at Saturn. At least two magnetopause encounters show signatures in plasma and magnetic field data of magnetic reconnection (McAndrews et al. 2008). Imposed across $10R_S$ along the magnetopause, the estimated reconnection voltage of 48 kV still fell short of the corotation voltage of >100 kV, even for 30% partial corotation. The electric field comparison for these two Cassini crossings supports the same conclusion as the earlier Voyager crossing, although the Dungey-cycle reconnection may have a non-negligible effect in the general motion of magnetic fields at Saturn, particularly during compression events from corotating interaction regions and from coronal mass ejections (e.g., Cowley et al. 2004; Badman and Cowley 2007).

A dearth of magnetic reconnection signatures similarly exists in the Cassini data from Saturn's magnetotail, although the Cassini trajectories to date (August 2009) have not been optimal for observing such reconnection. A survey of the component of the magnetic field normal to the current sheet when Cassini was in the magnetotail revealed only three clear instances of reconnection (Jackman et al. 2007). These classic reconnection signatures involved strong, sustained reversals of the normal magnetic field component near the current sheet between 40 and $50R_S$ in Saturn's magnetotail. The three events have been described as movements of plasmoids and their entrained magnetic fields past the spacecraft, an interpretation supported by plasma flow measurements in the cases where available (Hill et al. 2008). A more recent investigation claims to have found many more plasmoid signatures, although these have no field reversals (Burch et al. 2008). Evidence for tail reconnection has been inferred from brightening and filling of the polar cap as observed remotely by HST near the time of SOI when a corotating interaction region compressed the magnetosphere (Cowley et al. 2005; Bunce et al. 2005).

Evidence for corotational motion is considerably more prevalent at Saturn. Even before Cassini, plasma measurements indicated Saturn's at least partial corotation (Eviatar and Richardson 1986; Richardson 1998). As discussed above, Cassini data indicate partial corotation outside of about 5.5 R_S.

Corotation (or partial corotation) requires field-aligned currents (FAC), also called Birkeland currents, linking the ionosphere and magnetosphere. The enforcement of corotational flow requires FAC that close latitudinally in the polar ionospheres and radially in the equatorial plane (Hill 1979). At Saturn, an additional system has been proposed that involves FACs flowing from the northern to southern ionosphere along field lines, not closing in the equatorial plane, and having a sinusoidal variation in longitude (Southwood and Kivelson 2007). As discussed above, such a system could in principle explain the magnetic field periodicities observed by Voyager and Cassini (Espinosa et al. 2003; Giampieri et al. 2006). A comparison of the polarization and phase of magnetic field oscillations at high latitude to those at low latitudes suggests an asymmetric eastward-directed ring current together with field-aligned currents distributed in azimuth to maintain current continuity (Provan et al. 2009). This current system is not in principle different from that of Goldreich and Farmer (2007), but is distributed over a broad longitude sector rather than being restricted to the narrowing outward-moving plasma tongue discussed by the latter authors.

Cassini observations have clearly demonstrated the time-variable nature of Saturn's magnetic field and have led to at least three global models of its configuration. According to a survey of magnetometer measurements, the overall structure of Saturn's equatorial magnetic field during the Cassini prime mission (southern summer at Saturn) may be represented as a "bowl" warped northwards on the dayside in response to solar wind ram pressure and also northward on the nightside to follow the direction of solar wind flow, which asymptotically determines the magnetotail axis (Arridge et al. 2007b, 2008b). These northward deflections would be expected to disappear near the August 2009 equinox and change to southward deflections thereafter. ENA observations made during a particularly favorable viewing in late 2004, when Saturn was near its southern solstice, indicate a linear plasma sheet within $\sim 20R_S$, warped northward outside this distance on the nightside, corroborating the nightside portion of the bowl configuration (Carbary et al. 2008d). Figure 10.10 compares these two versions of the tilting of Saturn's magnetosphere.

Two suggestions have appeared explaining how the magnetotail magnetic field responds to internal and external forces given the large tilt of Saturn's magnetic axis to the planet's orbital plane ($\sim 26°$). A longitudinal anomaly in plasma loading might force the magnetic field to bulge, inducing an out-of-balance cam effect (similar to that proposed by Espinosa et al. 2003) and, in the presence of solar wind flow, causing Alfvén waves to propagate down the magnetotail (Carbary et al. 2007d). Alternatively, the corotating anomaly might cause the plasma sheet to rise and sink against the solar wind ram pressure, which would also generate periodic signals (Khurana et al. 2009). In this asymmetric lift model, the plasma sheet would tilt up and down with the rotation of the "heavy" vs "light" plasma sheet sector.

Future work will likely involve surveys of the magnetopause and magnetotail for more reconnection signatures, insofar as the Cassini trajectories allow. In particular, signatures should be sought at multiple local times across the magnetotail. ENA observations from a perspective looking edge-on to the magnetic equatorial plane might reveal the plasma sheet geometry and possibly better determine the exact bowl shape and whether the sheet responds to solar wind pressure as predicted (Carbary et al. 2008d).

Fig. 10.10 The bowl model of Saturn's equatorial current sheet (*top*) compared to the plasma sheet observed by the projection of ENA images into the noon midnight plane (*bottom*). The plasma sheet is warped northward of the equator on the nightside toward parallelism with the Sun-Saturn direction (Arridge et al. 2008b), but does not appear to be warped northward on the dayside, inside $20R_S$ (Carbary et al. 2008d)

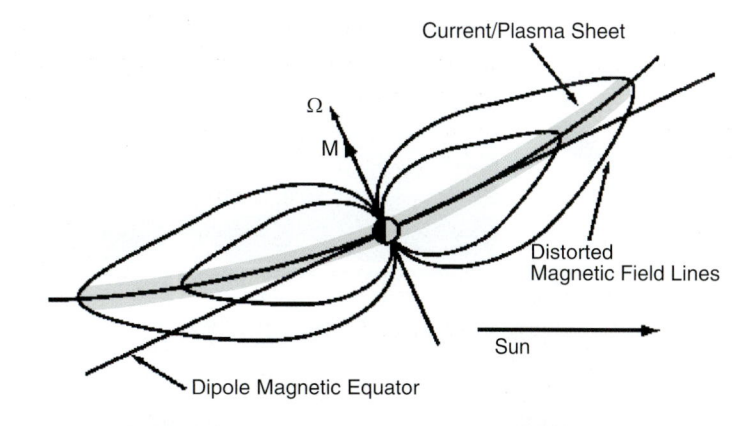

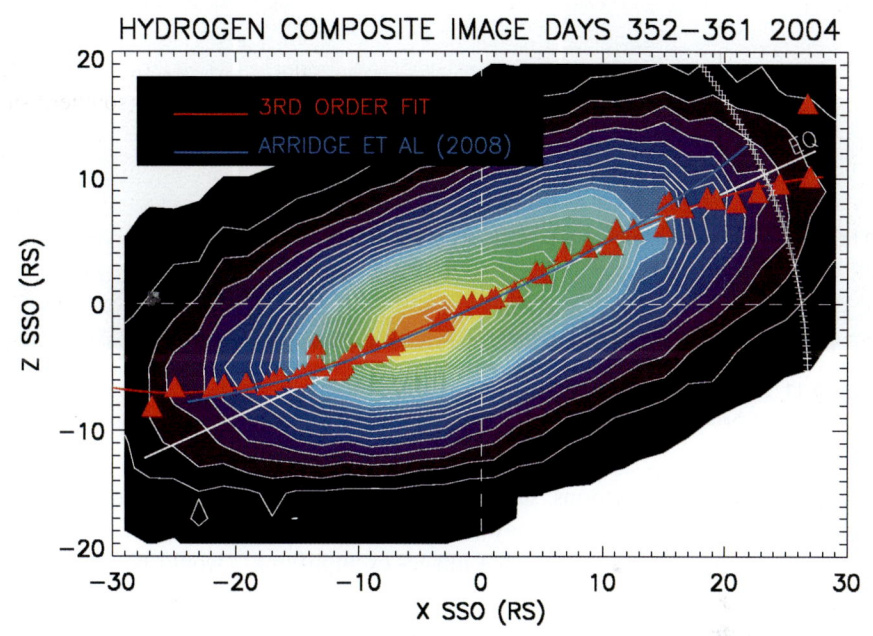

10.5 Magnetotail Dynamics

Saturn's magnetotail will be defined here as the region of the magnetosphere at radial distances greater than $\sim 20R_S$, which is the typical dayside magnetopause radius and coincidentally also the orbit of Titan, and at local times between 1800 and 0600 h.

Before the Cassini mission Saturn's magnetotail had scarcely been explored. The Pioneer 11 and Voyager 1 trajectories carried the spacecraft inbound around noon and outbound at dawn. The Voyager 2 trajectory also came in around noon but exited around 0300 h local time. Voyager 2 outbound remained at high inclination (latitude of $-40°$) and mostly sampled the magnetospheric lobe. However, these flybys allowed some general inferences about the magnetotail. First, the charged particle population of the magnetotail consists of protons and singly-charged oxygen (Lazarus and McNutt 1983; Krimigis et al. 1983), which reflects sources in the inner magnetosphere (O^+) and possibly also in the solar wind (H^+). The plasma generally flowed in the corotation sense in the outer magnetosphere, although at speeds much below corotation and with some indications of occasional radial movement (Carbary et al. 1983; Richardson 1998). The outer magnetosphere, if not the tail, also exhibited periodicities in the charged particles and magnetic field (Carbary and Krimigis 1982; Espinosa and Dougherty 2000).

Since entering orbit around Saturn, Cassini has completed several orbits through the magnetotail region and provided new information about its dynamics. Evidence has accumulated that some sort of magnetic reconnection (and subsequent particle acceleration) occurs in Saturn's magnetotail as it does at Earth and Jupiter. Images from INCA revealed bursts of energetic neutral atoms coming from Saturn's magnetotail, which have been interpreted as indicating particle acceleration (Mitchell et al. 2005). These ENA bursts were correlated with SKR activity in the same way that particle bursts are correlated with auroral kilometric

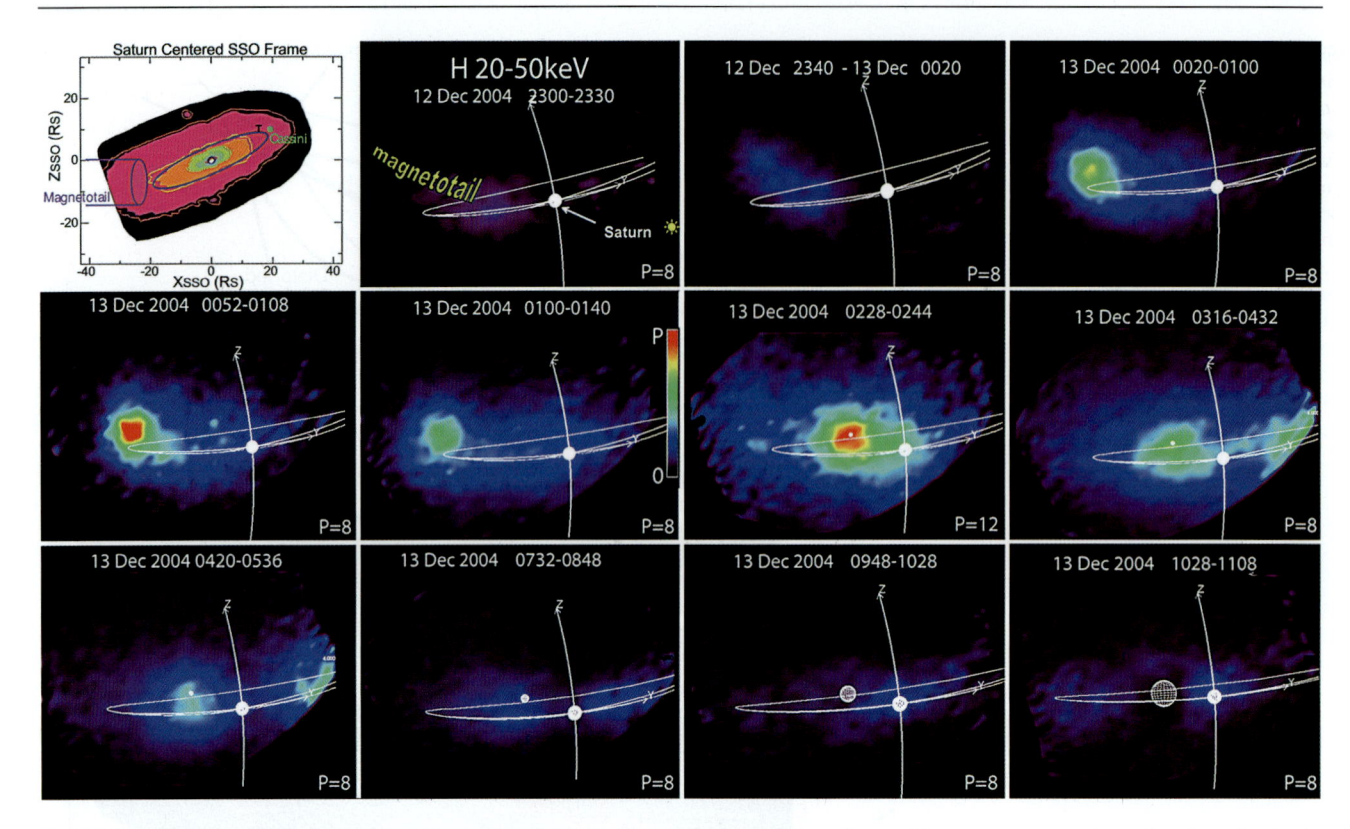

Fig. 10.11 Successive INCA images showing the initiation, growth, and decay of a burst of neutral hydrogen atoms apparently originating in the magnetotail of Saturn. The curved planar shape indicates the orbit of Titan (from Mitchell et al. 2005)

activity in terrestrial substorms, suggesting to some investigators that substorm-like activity may be occurring at Saturn. Figure 10.11 shows a sequence of ENA images exemplifying such an event.

As described in Section 10.4 above, the Cassini magnetometer observed three strong plasmoid signatures between 40 and 50R_S down the magnetotail (Jackman et al. 2007). These plasmoids were characterized by an oxygen-rich composition similar to that of the inner magnetosphere and a strongly tailward, subcorotational flow (Hill et al. 2008). Figure 10.12 displays a stacked plot of an in situ plasmoid signature seen by the Cassini magnetometer and plasma instruments. These features imply a tailward moving plasmoid originating in the inner magnetosphere between Cassini and Saturn. Indeed, INCA apparently captured the birth of this plasmoid in the ENA image sequence shown in Fig. 10.13. Careful analysis of ENA arrival times as functions of mass and energy allowed a precise determination of their common source location and time, which turned out to be midway between Saturn and Cassini and about 1/2 h before the in situ detection. Figure 10.14 indicates this source location. The space and time separations are consistent with plasma velocity measurements made by CAPS.

As noted in Section 10.4 above, plasmoid signatures have been infrequently seen in Cassini magnetotail orbits to date, probably owing to orbital geometry combined with the seasonal tilt of Saturn's rotation/magnetic axis. A further survey of the magnetotail near the August 2009 equinox would be expected to produce more signatures. The association of plasmoids with IMF reconnection and substorms at Earth suggests that magnetotail dynamics at Saturn may be internally driven by the rotation of the planet itself rather than by solar wind interactions as at Earth.

The strong periodicities of Saturn's magnetotail also imply that it is driven internally by planetary rotation rather than externally by solar wind interaction. The magnetic fields and the charged particles of the plasma sheet clearly display variations at or very close to the planetary spin period as described in Section 10.3 above. Similar spin periodicities at Jupiter are caused by the tilt of the magnetic axis relative to the spin axis (e.g., Eviatar and Ershkovitch 1976; Khurana and Kivelson 1989). Saturn's magnetic axis is not tilted, however, so the magnetotail periodicities must arise by some other mechanism. This mechanism could be some sort of outwardly-propagating wave (Espinosa et al. 2003; Cowley et al. 2006), although the source of this wave has not yet been elucidated. A longitudinal anomaly in the ionosphere or inner magnetosphere may be responsible for generating the wave. Such an anomaly was first postulated to explain the periodicities of charged particles in and upstream of Jupiter's magnetosphere (Vasyliūnas 1975; Dessler and Hill 1975, 1979; Dessler and Vasyliūnas 1979). The anomaly may be

Fig. 10.12 Magnetometer (*top panel*) and plasma (*bottom 5 panels*) data showing the signature of a plasmoid passing the Cassini spacecraft. In particular, the fourth panel shows plasma moving strongly tailward (from Hill et al. 2008)

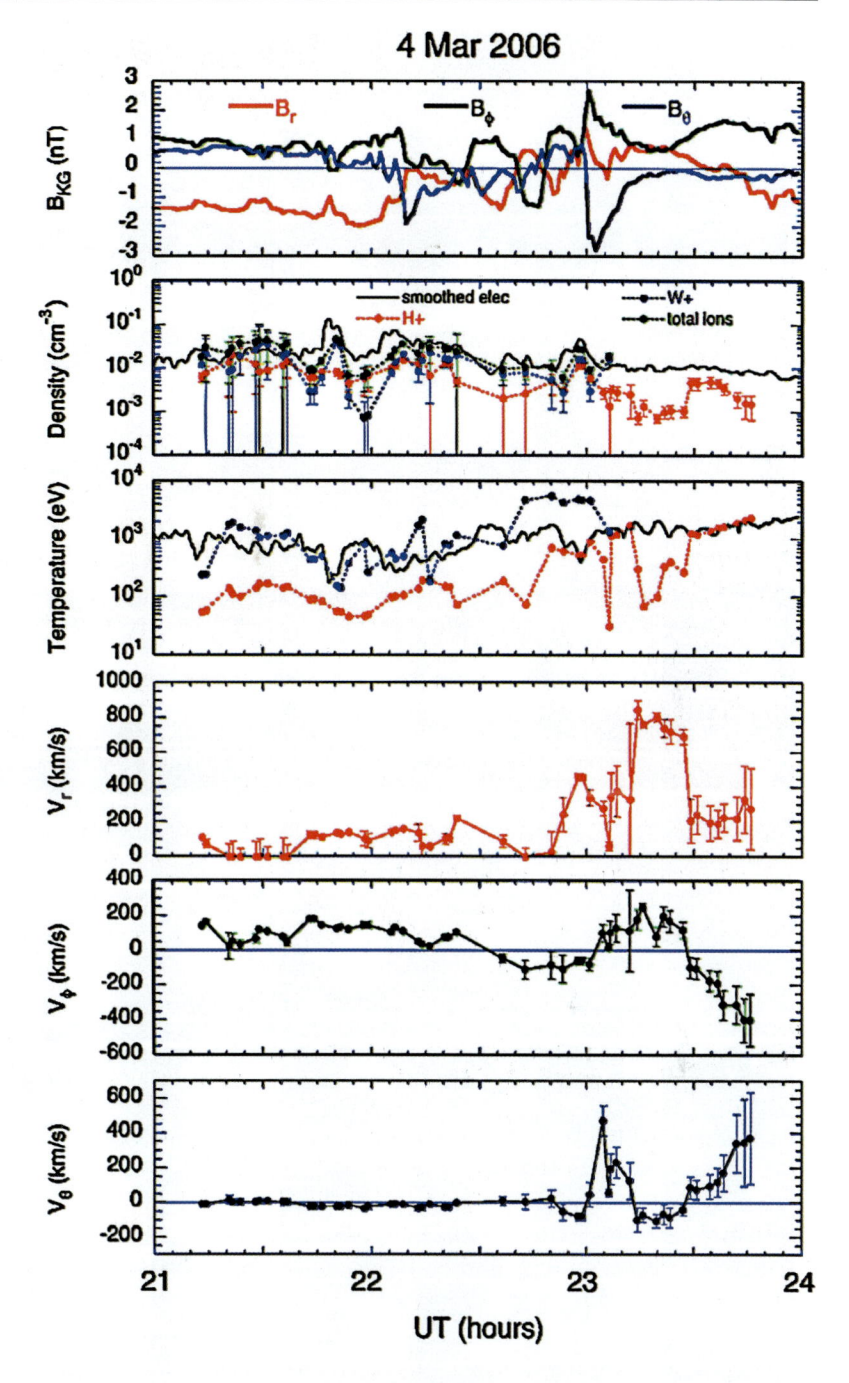

a conductance anomaly in Saturn's ionosphere (Dessler and Hill 1979), or may be an intrinsic property of the centrifugal instability of the plasma disk (Gurnett et al. 2007; Goldreich and Farmer 2007). The magnetospheric current system, including the ionosphere to ionosphere current discussed in Southwood and Kivelson (2007), may also derive from a longitudinal asymmetry in the rotating magnetic field, and those authors point out that outside the region in which these currents flow (i.e., beyond 12 to 15Rs), the effect would be to introduce an apparent dipole tilt. This would presumably produce flapping of the distant current sheet similar to that observed at Jupiter, where the dipole is intrinsically tilted.

A rotating plasma anomaly in Saturn's inner magnetosphere can generate an outwardly moving wave in the magnetotail in any of four ways. As the anomaly rotates, it may cause alternating high and low pressure and generate a compressional wave (Espinosa et al. 2003). Alternately, the anomaly may cause a sliding in and out of the "hinge" distance at which the plasma sheet bends over to parallel the solar wind flow (Carbary et al. 2007d). This motion generates a transverse wave similar to that observed at Jupiter. The rotating anomaly might also cause a variation in the plasma sheet's ability to resist the compressive force of the solar wind, which could also be manifest as a transverse

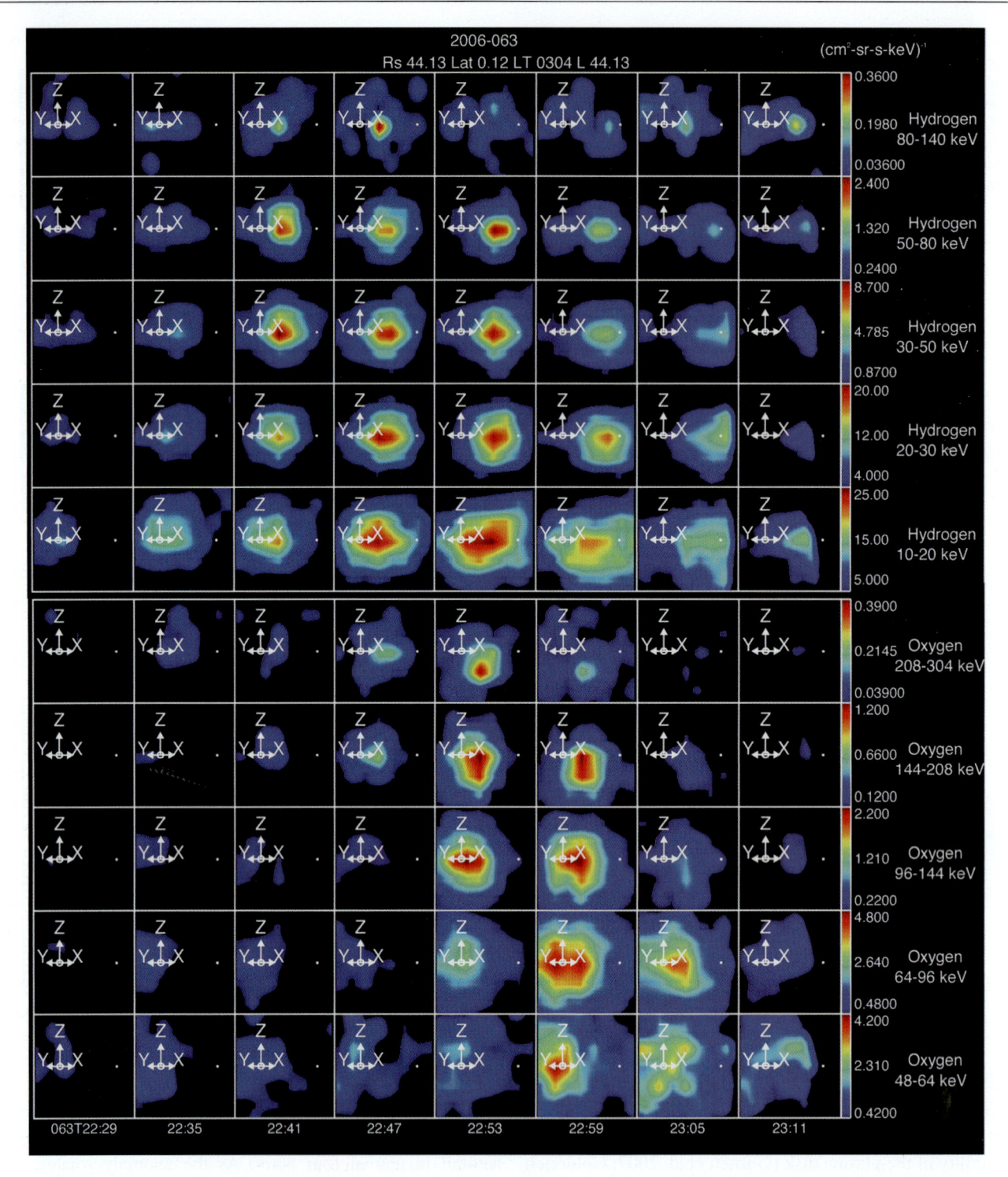

Fig. 10.13 Sequence of ENA images preceding the in situ plasmoid detection shown in the previous figure. Species appear in order of decreasing particle velocity from top to bottom. The later arrival of the slower species is treated as time dispersion to derive the distance from the sudden onset region producing the emissions and the Cassini spacecraft (from Hill et al. 2008)

wave moving down the magnetotail (Khurana et al. 2009). Finally, an interhemispheric current system fixed in longitude inside 12 to 15Rs could induce a tilt in the dipole field beyond that distance, causing flapping of the tail current sheet (Southwood and Kivelson 2007).

Considerable evidence exists for the formation of a magnetodisk-like structure in Saturn's magnetotail as well as on the dayside. Simple consideration of the balance of magnetic, pressure gradient, and centrifugal stresses in the radial direction using available data suggests that such a structure

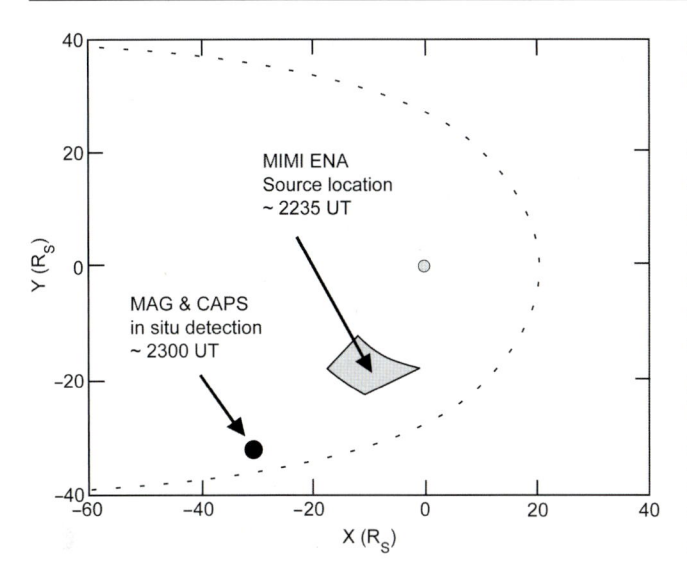

Fig. 10.14 Inferred source location (*gray box*) and time of the ENA burst shown in the preceding figure (from Hill et al. 2008)

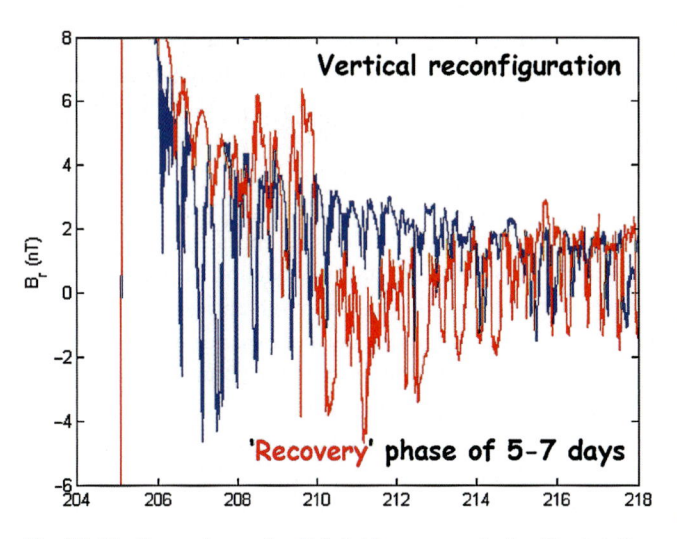

Fig. 10.15 Comparison of radial field components for Cassini Rev 26 (*blue*) and Rev 27 (*red*), which were nearly identical trajectories. The magnetic fields differ greatly, suggesting a reconfiguration of the magnetosphere caused by solar wind effects. In this case, a corotating interaction region in the solar wind seems responsible (from André et al. 2008)

forms around $16.5 R_S$ (Arridge et al. 2007a, 2008a; Bunce et al. 2008). The magnetodisk dynamics can support waves and "flapping" motions and is highly responsive to solar wind variations, which can disrupt the magnetodisk and cause a reconfiguration of the magnetosphere. Indication of such a disruption can be found by comparing the magnetic field on two identical passes of Cassini through the dawn sector (Fig. 10.15).

Additional evidence for rotational rather than solar wind control of Saturn's magnetotail is provided by the CAPS composition data shown in Fig. 10.16. Water group ions, with a mass/charge ratio of $\sim$16–18 amu/e (e.g., OH^+) dominate the protons in most but not all the data intervals out to distances of $\sim 50 R_S$ in the tail (McAndrews et al. 2009).

A great deal of research remains to be done on the dynamics of Saturn's magnetotail, and further acquisition and analysis of Cassini data will facilitate this research. First, a more complete survey of the plasma sheet in Saturn's magnetotail should be undertaken. Such a survey should include data from the several available instruments (plasma spectrometer, magnetometer, and energetic particle detectors) and should include ion and electron temperature, composition, and flow velocities, insofar as these quantities can be obtained. Second, the plasma sheet thickness and scale height should be estimated as functions of radial distance and local time. Third, magnetotail variations, especially those associated with plasmoids, ought to be characterized by a statistical analysis and correlated with changes in the inner magnetosphere (as monitored by the RPWS or INCA) or with changes in the solar wind (as propagated from the Earth or measured in situ by Cassini itself). Fourth, the two principal models for flow in Saturn's magnetotail, namely the Dungey cycle of Earth and the Vasyliūnas cycle of Jupiter must be critically evaluated in light of the new observations, especially those of flow speeds.

10.6 Magnetospheric Compression

Magnetospheric compression refers to the dynamical effects on Saturn's magnetosphere of increased solar wind pressure. At Saturn's orbit (9.6 AU), the interplanetary medium is highly variable on solar rotation time scales ($\sim$25 days) because of the regular variations in solar wind density and speed within corotating interactions regions (CIRs) and the intermittent bursts of high-speed outflow in coronal mass ejections (CMEs). The solar wind dynamic pressure at Saturn varies typically from $\sim$0.01 nPa during rarefactions lasting several days to $\sim$0.1 nPa during compressions lasting a few days, with corresponding variations of the interplanetary magnetic field (IMF) strength from $\sim$0.1 nT during rarefactions to $\sim$1 nT during compressions. Variations of similar or greater magnitude can also occur within CME disturbances lasting a few days, with field strengths up to $\sim$3 nT being possible (Jackman et al. 2004).

Prior to Cassini, observations from the Pioneer 11 and Voyager flybys demonstrated various responses of Saturn's magnetosphere to solar wind effects, most notably ram pressure. A cross-correlation analysis during the approaches of Voyager 1 and 2 to Saturn in 1980 and 1981 clearly revealed a strong connection between SKR power and solar wind ram pressure (Desch 1982; Desch and Rucker 1983). If modeled as a rectangular toroidal slab (e.g., Connerney et al. 1983), Saturn's ring current can apparently change location

Fig. 10.16 CAPS measurements of the OH^+/H^+ ion ratio as a function of radial distance (from McAndrews et al. 2007)

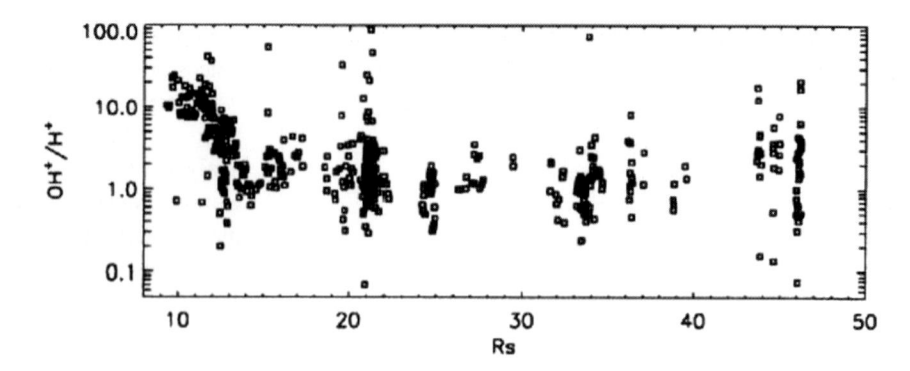

and size, moving inward and shrinking during magnetospherically compressed times seen by Pioneer 11 when compared to un-compressed times observed by Voyager (Bunce and Cowley 2003).

Considerable knowledge of the response to magnetospheric compression has derived from a series of campaigns in which Hubble Space Telescope (HST) auroral observations were combined with Earth-based or Cassini-based measurements of the solar wind. Long before the arrival of Cassini, the HST could image and monitor the aurora of Saturn (Gérard et al. 2004, 1995). Eventually, HST observed a strong, transient auroral emission correlated with the collision of an interplanetary shock with Saturn, the shock arrival being computed by propagating solar wind parameters from Earth to Saturn using a 1D model (Prangé et al. 2004). As Cassini approached Saturn in 2004, the spacecraft itself more directly monitored solar wind conditions. Two CIRs fortuitously materialized during the HST campaign at this time, and observations confirmed that Saturn's auroral oval brightens and contracts poleward in response to increases in solar wind pressure as well as the impressed electric field strength, rather than to changes in the direction of the interplanetary magnetic field (Clarke et al. 2005; Crary et al. 2005). The compression also stimulated enhanced SKR emissions (Kurth et al. 2005). A very good correlation existed between solar wind dynamic pressure (as well as convection electric field) and the auroral power (Clarke et al. 2005).

Comparisons of Saturn auroral activity have been made with estimates of the reconnection rate (or reconnection voltage) based on a Wygant coupling formula for cross-cap potential at Earth (e.g., Wygant et al. 1983), where Cassini provided the upstream values of solar wind speed and magnetic field during the CIRs discussed above (Jackman et al. 2004, 2005). Reconnection voltages as low as a few tens of kilovolts were estimated during the rarefactions, while voltages as high as 100–300 kV were inferred during the compressions, although the latter were highly variable. During the compressions, this inferred potential approaches the fractional corotational potential across the polar cap of ∼500 kV (Belenkaya et al. 2006), so the solar wind may partially or sporadically drive Dungey-like magnetospheric dynamics during the compression events. At other, more typical times, the Dungey cycle might slowly inflate the tail with open flux over intervals of days before dynamics are initiated e.g. by a subsequent compression event (e.g., Cowley et al. 2005; Bunce et al. 2005).

Cassini instruments have observed magnetospheric events apparently associated with solar wind compression. The magnetopause and bow shock responses to solar wind ram pressure represent the most direct effect of this compression. Magnetometer measurements from the first six orbits of Cassini (June 2004 to March 2005), as well as Voyager data, were used to calculate internal magnetic pressure, which was balanced against inferred solar wind dynamic pressure, to derive an empirical formula relating solar wind pressure to magnetopause distance (Arridge et al. 2006). A similar effort has empirically determined the size and shape of Saturn's bow shock (Masters et al. 2008). Modeling of Saturn's ring current using two years of Cassini observations demonstrated that the outer edge of the ring current expands or contracts in response to solar wind dynamic pressure (Bunce et al., 2007). Investigations of Saturn's ring current and current sheet show that the middle magnetosphere is quasi-dipolar during compressions and becomes a magnetodisk when the magnetosphere expands (Bunce et al. 2008; Arridge et al. 2008a). Near-equatorial magnetometer data have also been used to examine the perturbation field of the ring current, from which a Dst-like index can be constructed as a measure of the stress state of Saturn's magnetosphere. The Saturn system seems to spend most of its time near a quiet, low-index state, although some Cassini orbits show large departures from the "ground" level, apparently associated with solar-wind pressure enhancements (Leisner et al. 2007). Observations made during the Cassini approach to Saturn in 2004 reveal a correlation between peak power of SKR bursts and an increase in the magnitude of the interplanetary magnetic field (Badman et al. 2008).

Theoretical explanations of solar wind effects on Saturn's magnetosphere have been offered in a series of papers by the Leicester group in England. Assuming that Saturn's auroral oval lies at the boundary between open and closed field lines (e.g., Cowley and Bunce 2003), a four-region pattern of field aligned currents is expected in Saturn's polar ionosphere with the oval appearing near 13° colatitude and electron precipitation in Birkeland currents exciting UV aurora of a few tens of kilorayleighs (Cowley et al. 2004). In this view, the auroral effects seen in the early 2004 HST campaign represent compression-induced episodes of strong reconnection in the magnetic tail that results in the closure of a substantial fraction of the pre-existing open flux mapping within the auroral oval (Cowley et al. 2005). In this scenario, a strong correlation is expected between the solar wind dynamic pressure and Saturn's magnetospheric activity similar to that sometimes observed at Earth, but without a significant IMF B_z-dependence because the typical time scale for inflating the tail with open flux (days) is very much longer than the typical times scales for IMF B_z variations (tens of minutes to an hour or two).

Unfortunately, most of the correlations between the solar wind and Saturn's magnetosphere rely on a relatively few direct HST and Cassini observations (namely, those during 2004) or on solar wind conditions inferred by propagation from Earth. Computational propagation to Saturn of solar wind conditions observed at Earth may provide a route to future correlative studies (Zieger and Hansen 2008), as will continued use of the Hubble Space Telescope and the Cassini UVIS to monitor Saturn's aurora. Further insight into these phenomena must also await more detailed examination of the Cassini data under such compression-region conditions.

10.7 Conclusion

The years of data already accumulated by Cassini at Saturn provides a rich resource for research by the Cassini scientists, who have reached a level of confidence in and understanding of their instruments that removes many of the impediments to understanding their measurements, as well as for the entire scientific community, which has access to all of the data from the Prime Mission as well as ongoing access to extended mission measurements through the data in Planetary Data System, under the support of such programs as the Cassini Data Analysis Program and the Outer Planets Research Program.

At this writing, the Cassini mission is about half way through its first extended mission, called the Saturn Equinox Mission (SEM) because Saturn reaches equinox in August 2009. Some of the models for phenomena discussed in this chapter depend on the seasonal tilt of Saturn's rotation axis and dipole (e.g., the Khurana et al. 2009 model and the Carbary 2007d model both predict the disappearance of periodic magnetotail current sheet encounters at equinox, while the Southwood and Kivelson 2007 model predicts the persistence of such encounters through equinox). Hemispheric ionosphere-to-ionosphere current systems may change or disappear near equinox, when the hemispheric solar illumination equalizes and ionospheric conductance differences between north and south disappear. Energetic neutral atom imaging may reveal more structure in the magnetotail, as the tail current sheet comes into alignment with the equatorially enhanced cold gas cloud. New regions of the magnetosphere will receive coverage not provided during the Prime Mission, in particular the under-sampled dusk quadrant. New observing strategies are being employed to fill inadequacies of past campaigns.

Beyond the SEM, plans are maturing for a further mission extension, called the Solstice Mission (SM) as it is designed to take the Cassini tour of Saturn through to the northern summer solstice. This mission extension has the exciting attributes of carrying the observations of the magnetosphere through a full Saturn season, as well as a full solar cycle. Unique information on the effects of solar maximum conditions on the magnetosphere and on ionosphere-magnetosphere coupling will be possible with the addition of this mission.

The end of the Cassini mission as conceived in the SM mission design places Cassini into a Juno-like orbit, with periapsis skimming the dayside ionosphere inside the rings. This unique orbital geometry will allow determination of higher order moments of the internal magnetic field with high precision, perhaps revealing azimuthal asymmetries in the internal field. It should also allow direct measurement of the ionosphere in situ, and characterization of the radiation belt that almost surely exists between the inner edge of the D-ring and the upper atmosphere.

References

Anderson, J.D., and G. Schubert (2007), Saturn's gravitational field, internal rotation, and interior structure, *Science, 317*, 1384–1387, doi: 10.1126/science.1144835.

André, N., M.K. Dougherty, C.T. Russell, J.S. Leisner, and K.K. Khurana (2005), Dynamics of the Saturnian inner magnetosphere: First inferences from the Cassini magnetometers about small-scale plasma transport in the magnetosphere, *Geophys. Res. Lett., 32*, L14S06, doi:10.1029/2005GL022643.

André, N., P. Louarn, C.S. Arridge, L. Lamy, K.C. Hansen, G.R. Lewis, C. Jackman, E.J. Bunce, J.L. Burch, A.J. Coates, B. Cecconi, S.W.H. Cowley, M.K. Dougherty, K.K. Khurana, C.T. Russell, E.C. Sittler, and P. Zarka (2008), Effects of a corotating interaction region on the structure and dynamics of the Saturnian magnetosphere, Saturn after Cassini-Huygens Symposium, 27 July–1 August, London, UK.

Andrews, D.J., E.J. Bunce, S.W.H. Cowley, M.K. Dougherty, G. Provan, and D.J. Southwood (2008), Planetary spin oscillations in Saturn's magnetosphere: phase relation of equatorial magnetic field oscillations and SKR modulation, *J. Geophys. Res., 113*, A09205, doi:10.1029/2007JA012937.

Arridge, C.S., N. Achilleos, M.K. Dougherty, K.K. Khurana, and C.T. Russell (2006), Modeling the size and shape of Saturn's magnetopause with variable dynamic pressure, *J. Geophys. Res., 111*, A11227, doi:10.1029/2005JA011574.

Arridge, C.S., C.T. Russell, K.K. Khurana, N. Achileos, N. Andre, A.M. Rymer, M.K. Dougherty, and A.J. Coates (2007a), Mass of Saturn's magnetodisc: Cassini observations, *Geophys. Res. Lett., 34*, L09108, doi:10.1029/2007GL028921.

Arridge, C.S., K.K. Khurana, C.T. Russell, E.C. Sittler, N. Andre, H.J. McAndrews, A.J. Coates, and M.K. Dougherty (2007b), Periodic crossings of Saturn's magnetospheric current/plasma sheet, *EOS Trans. AGU, 88*, Fall Meeting Suppl., abstract P31A-0194.

Arridge, C.S., C.T. Russell, K.K. Khurana, N. Achileos, S.W.H. Cowley, M.K. Dougherty, D.J. Southwood, and E.J. Bunce (2008a), Saturn's magnetodisc current sheet, *J. Geophys. Res., 113*, A04214, doi:10.1029/2007JA012540.

Arridge, C.S., K.K. Khurana, C.T. Russell, D.J. Southwood, N. Achileos, M.K. Dougherty, A.J. Coates, and H.K. Leinweber (2008b), Warping of Saturn's magnetospheric and magnetotail current sheets, *J. Geophys. Res., 113*, A08217, doi:10.1029/2007JA012963.

Badman, S.V., and S.W. H. Cowley (2007), Significance of Dungey-cycle flows in Jupiter's and Saturn's magnetospheres, and their identification on closed equatorial field lines, *Ann. Geophys., 25*, 941–951.

Badman, S.V., S.W.H. Cowley, L. Lamy, B. Cecconi, and P. Zarka (2008), How do solar wind compressions affect the pulsing and intensity of Saturn kilometric radiation?, Saturn after Cassini-Huygens Symposium, 27 July–1 August, London, UK.

Belenkaya, E.S., S.W.H. Cowley, and I.I. Alexeev (2006), Saturn's aurora in the January 2004 events, *Ann. Geophys., 24*, 1649–1663.

Brandt, P., C.P. Paranicas, J.F. Carbary, D.G. Mitchell, B.H. Mauk, and S.M. Krimigis (2008), Understanding the global evolution of Saturn's ring current, *Geophys. Res. Lett., 35*, L17101, doi:10.1029/2008GL034969.

Brice, N.M., and G.A. Ioannidis (1970), The magnetospheres of Jupiter and Earth, *Icarus, 13*, 173–183.

Bunce, E.J., and S.W.H. Cowley (2003), A note on the ring current in Saturn's magnetosphere: Comparison of magnetic field data obtained during the Pioneer-11 and Voyager-1 and -2 flybys, *Ann. Geophys., 21*, 661–669.

Bunce, E.J., S.W.H. Cowley, D.M. Wrightm A.J. Coates, M.K. Dougherty, N. Krupp, W.S. Kurth, and A.M. Rymer (2005), In-situ observations of a solar wind compression-induced hot plasma injection in Saturn's tail, *Geophys. Res. Lett., 32*, L20S04, doi:10.1029/2005GL022888.

Bunce, E.J., S.W.H. Cowley, I.I. Alexeev, C.S. Arridge, M.K. Dougherty, J.D. Nichols, and C.T. Russell (2007), Cassini observations of the variation of Saturn's ring current parameters with system size, *J. Geophys. Res., 112*, A10202, doi:10.1029/2007JA012275.

Bunce, E.J., C.S. Arridge, S.W.H. Cowley, and M.K. Dougherty (2008), Magnetic field structure of Saturn's dayside magnetosphere and its mapping to the ionosphere: Results from ring current modeling, *J. Geophys. Res., 113*, A02207, doi:10.1029/2007JA012538.

Burch, J.L., J. Goldstein, T.W. Hill, D.T. Young, F.J. Crary, A.J. Coates, N. Andre, W.S. Kurth, and E.C. Sittler (2005), Properties of local plasma injections in Saturn's magnetosphere, *Geophys. Res. Lett., 32*, L14S02, doi:10.1029/2005GL022611.

Burch, J.L., J. Goldstein, W.S. Lewis, D.T. Young, A.J. Coates, M. Dougherty, and N. Andre (2007), Tethys and Dione as sources of outward moving plasma in Saturn's magnetosphere, *Nature, 447*, 833–835, doi:10.1038/nature05906.

Burch, J.L., J. Goldstein, P. Mokasji, W.S. Lewis, C. Paty, D.T. Young, A.J. Coates, M.K. Dougherty, and N. André (2008), On the cause of Saturn's plasma periodicity, *Geophys. Res. Lett., 35*, L14105, doi:10.1029/2008GL034951.

Carbary, J.F., and S.M. Krimigis (1982), Charged particle periodicity in the Saturnian magnetosphere, *Geophys. Res. Lett., 9*, 1073–1076.

Carbary, J.F., B.H. Mauk, and S.M. Krimigis (1983), Corotation anisotropies in Saturn's magnetosphere, *J. Geophys. Res., 88*, 8937–8946.

Carbary, J.F., D.G. Mitchell, S.M. Krimigis, and N. Krupp (2007a), Electron periodicities in Saturn's outer magnetosphere, *J. Geophys. Res., 112*, A03206, doi: 10.1029/ 2006JA012077.

Carbary, J.F., D.G. Mitchell, S.M. Krimigis, D.C. Hamilton, and N. Krupp (2007b), Charged particle periodicities in Saturn's outer magnetosphere, *J. Geophys. Res., 112*, A06246, doi:10.1029/2007JA012351.

Carbary, J.F., D.G. Mitchell, S.M. Krimigis, and N. Krupp (2007c), Evidence for spiral pattern in Saturn's magnetosphere using the new SLR longitudes, *Geophys. Res. Lett., 34*, L13105, doi:10.1029/2007GL030167.

Carbary, J.F., D.G. Mitchell, S.M. Krimigis, D.C. Hamilton, and N. Krupp (2007d), Spin-period effects in magnetospheres with no axial tilt, *Geophys. Res. Lett., 34*, L18107, doi:10.1029/2007GL030483.

Carbary, J.F., D.G. Mitchell, P. Brandt, C. Paranicas, and S.M. Krimigis (2008a), ENA periodicities at Saturn, *Geophys. Res. Lett., 35*, L07102, doi:10.1029/2008GL033230.

Carbary, J.F., D.G. Mitchell, P. Brandt, E.C. Roelof, and S.M. Krimigis (2008b), Track analysis of energetic neutral atom blobs at Saturn, *J. Geophys. Res., 113*, A01209, doi:10.1029/ 2007JA012708.

Carbary, J.F., S. M. Krimigis, D.G. Mitchell, C. Paranicas, and P. Brandt (2008c), Energetic neutral atom (ENA) and charged particle periodicities in Saturn's magnetosphere, submitted to *Adv. Space Res.*, July 2008.

Carbary, J.F., D.G. Mitchell, C. Paranicas, E.C. Roelof, and S.M. Krimigis (2008d), Direct Observation of warping in the plasma sheet of Saturn, *Geophys. Res. Lett., 35*, L24201, doi:10.1029/2008GL035970.

Cecconi, B., and P. Zarka (2005), Model of a variable radio period for Saturn, *J. Geophys. Res., 110*, A12203, doi:10.1029/2005JA011085.

Chen, Y., and T.W. Hill (2008), Statistical analysis of injection/dispersion events in Saturn's inner magnetosphere, *J. Geophys. Res., 113*, A07215, doi:10.1029/2008JA013166.

Clarke. J.T., J.-C. Gérard, D. Grodent, S. Wannawichian, J. Gustin, J. Connerney, F. Crary, M. Dougherty, W. Kurth, S.W.H. Cowley, E.J. Bunce, T. Hill, and J. Kim (2005), Morphological differences between Saturn's ultraviolet aurora and those of Earth and Jupiter, *Nature, 433*, 717–719, doi:10.1038/nature03331.

Clarke, K.E., N. André, D.J. Andrews, A.J. Coates, S.W.H. Cowley, M.K. Dougherty, G.R. Lewis, H.J. McAndrews, J.D. Nichols, T.R. Robinson, and D.M. Wright (2006), Cassini observations of planetary-period oscillations in Saturn's magnetopause, *Geophys. Res. Lett., 33*, L23104, doi:10.1029/2006GL027821.

Compton, A. H. and I. A. Getting (1935), An apparent effect of galactic rotation on the intensity of cosmic rays, *Phys. Rev., 47*, 817.

Connerney, J.E.P., M.H. Acuña, and N.F. Ness (1983), Currents in Saturn's magnetosphere, *J. Geophys. Res., 88*, 8779–8789.

Cowley, S.W.H., and Bunce, E.J. (2003), Corotation-driven magnetosphere-ionosphere coupling currents in Saturn's magnetosphere and their relation to the auroras, *Annales Geophysicae, 21*, 1691–1707, 2003AnGeo..21.1691C.

Cowley, S.W.H., E.J. Bunce, and J.M. O'Rourke (2004), A simple quantitative model of plasma flows and currents in Saturn's polar ionosphere, *J. Geophys. Res., 109*, A05212, doi:10.1029/2004JA010375.

Cowley, S.W.H., S.V. Badman, E.J. Bunce, J.T. Clarke, J.-C. Gerard, D. Grodent, C.M. Jackman, S.E. Milan, and T.K. Yeoman (2005), Reconnection in a rotation-dominated magnetosphere and its relation to Saturn's auroral dynamics, *J. Geophys. Res., 110*, A02201, doi:10.1029/2004JA010796.

Cowley, S.W.H., D.M. Wright, E.J. Bunce, A.C. Carter, M.K. Dougherty, G. Giampieri, J.D. Nichols, and T.R. Robinson (2006), Cassini observations of planetary-period magnetic field oscillations in Saturn's magnetosphere: Doppler shifts and phase motion, *Geophys. Res. Lett., 33*, L07104, doi:2005GL025522.

Crary, F.J., J.T. Clarke, M.K. Dougherty, P.G. Hanlon, K.C. Hansen, J.T. Steinberg, B.L. Barraclough, A.J. Coates, J.-C. Gérard, D. Grodent, W.S. Kurth, D.G. Mitchell, A.M. Rymer, and D.T. Young (2005), Solar wind dynamic pressure and electric field as the main factors controlling Saturn's aurorae, *Nature, 433*, 720–722, doi:10.1038/nature03333.

Davies, M.E., V.K. Abalakin, M. Bursa, J.H. Lieske, B. Morando, D. Morrison, P.K. Seidelmann, A.T. Sinclair, B. Yallop, and Y.S. Tjuflin (1996), Report for the IAU/IAG/COSPAR working group on cartographic coordinates and rotational elements of the planets and satellites: 1994, *Celestial Mech. Dyn. Astron., 63*, 127–148.

Delamere, P.A., and F. Bagenal (2008), Longitudinal plasma density variations at Saturn caused by hot electrons, *Geophys. Res. Lett., 35*, L03197, doi:10.1029/2007GL031095.

Delamere, P.A., F. Bagenal, V. Dols, and L.C. Ray (2007), Saturn's neutral torus versus Jupiter's plasma torus, *Geophys. Res. Lett., 34*, L09105, doi:10.1029/2007GL029437.

Desch, M.D. (1982), Evidence for solar wind control of Saturn radio emission, *J. Geophys. Res., 87*, 4549–4554.

Desch, M.D., and M.L. Kaiser (1981), Voyager measurements of the rotation period of Saturn's magnetic field, *Geophys. Res. Lett., 8*, 253–256.

Desch, M.D., and H.O. Rucker (1983), The relationship between Saturn kilometric radiation and the solar wind, *J. Geophys. Res., 88*, 8999–9006.

Dessler, A.J., and T.W. Hill (1975), High-order magnetic multipoles as a source of gross asymmetry in the distant Jovian magnetotail, *Geophys. Res. Lett., 2*, 567–570.

Dessler, A.J., and T.W. Hill (1979), Jovian longitudinal control of Io-related radio emissions, *Astrophys. J., 227*, 664.

Dessler, A.J., and V.M. Vasyliūnas (1979), The magnetic anomaly model of the Jovian magnetosphere: Predictions for Voyager, *Geophys. Res. Lett., 6*, 37–40.

Dougherty, M.K., K.K. Khurana, F.M. Neubauer, C.T. Russell, J. Saur, J.S. Leisner, M.E. Burton (2006), Identification of a dynamic atmosphere at Enceladus with the Cassini magnetometer, *Science, 311*, 1406–1409.

Dungey, J.W. (1961), Interplanetary field and the auroral zones, *Phys. Rev. Lett, 6*, 47–48.

Espinosa, S. A., and M.K. Dougherty (2000), Periodic perturbations in Saturn's magnetic field, *Geophys. Res. Lett., 27*, 2785–2788.

Espinosa, S.A., D.J. Southwood, and M.K. Dougherty (2003), How can Saturn impose its rotation period in a noncorotating magnetosphere? *J. Geophys. Res., 108*(A2), 1086, doi:10.1029/2001JA005084, 2003.

Eviatar, A., and A.I. Ershkovitch (1976), Plasma density in the outer Jovian magnetosphere, *J. Geophys. Res., 81*, 4027–4028.

Eviatar, A., and J. Richardson (1986), Corotation of the Kronian magnetosphere, *J. Geophys. Res., 91*, 3299–3303.

Galopeau, P.H.M., and A. Lecacheux (2000), Variations of Saturn's radio period measured at kilometer wavelengths, *J. Geophys. Res., 105*, 13089–13101.

Galopeau, P., P. Zarka, and D. Le Quéau (1995) Source location of SKR: the Kelvin–Helmholtz instability hypothesis, *J. Geophys. Res. Planets, 100*, 26397–26410.

Gérard, J.-C., V. Dols, D. Grodent, J.H. Waite, G.R. Gladstone, and R. Prangé (1995), Simulaneous observations of the Saturnian aurora and polar haze with the HST/FOC, *Geophys. Res. Lett., 22*, 2685–2688.

Gérard, J.-C., D. Grodent, J. Gustin, A. Saglam, J.T. Clarke, and J.C. Trauger (2004), Characteristics of Saturn's FUV aurora observed with the Space Telescope Imaging Specrograph, *J. Geophys. Res., 109*, A09207, doi:10.1029/2004JA010513.

Giampieri, G., M.K. Dougherty, E.J. Smith, and C.T. Russell (2006), A regular period for Saturn's magnetic field that may track its internal rotation, *Nature, 441*, 62–64, doi: 10.1038/nature04750.

Gleeson, I. A., and W. I Axford (1968), The Compton-Getting effect, *Astrophy. Space Sci., 2*, 431–437.

Glocer, A., T.I. Gombosi, G. Toth, K.C. Hansen, A.J. Ridley, and A. Nagy (2007), Polar wind outflow model: Saturn results, *J. Geophys. Res., 112*, A01304, doi:10.1029/ 2006JA011755.

Goldreich, P., and A.J. Farmer (2007), Spontaneous symmetry breaking of the external field of Saturn, *J. Geophys. Res., 112*, A05225, doi:10.1029/2006JA012163.

Gombosi, T.I., and K.C. Hansen (2005), Saturn's variable magnetosphere, *Science, 307*, 1224–1226, doi:10.1126/science.1108226.

Gurnett, D.A., et al. (2005), Radio and plasma wave observations at Saturn from Cassini's approach and first orbit, *Science, 307*, 1255–1259, doi:10.1026/science.1105356.

Gurnett, D.A., A.M. Persoon, W.S. Kurth, J.B. Groene, T.F. Averkamp, M.K. Dougherty, and D.J. Southwood (2007), The variable rotation period of the inner region of Saturn's plasma disk, *Science, 316*, 442–445, doi:10.1126/science.1138562.

Hill, T.W. (1979), Inertial limit on corotation, *J. Geophys. Res., 84*, 6554–6558.

Hill, T.W., A.J. Dessler, and L.J. Maher (1981), Corotating magnetospheric convection, *J. Geophys. Res., 86*, 9020–9028.

Hill, T.W., A.M. Rymer, J.L. Burch, F.J. Crary, D.T. Young, M.F. Thomsen, D. Delapp, N. Andre, A.J. Coates, and G.R. Lewis (2005), Evidence for rotationally-driven plasma transport in Saturn's magnetosphere, *Geophys. Res. Lett., 32*, L14S10, doi:10.1029/2005GL022620.

Hill, T.W., M.F. Thomsen, M.G. Henderson, R.L. Tokar, A.J. Coates, H.J. McAndrews, G.R. Lewis, D.G. Mitchell, C.M. Jackman, C.T. Russell, M.K. Dugherty, F.J. Crary, and D.T. Young (2008), Plasmoids in Saturn's magnetotail, *J. Geophys. Res., 113*, A01214, doi: 10.1029/2007JA012626.

Huddleston, D.E., C.T. Russell, G. Le, A. Szabo (1997), Magnetopause structure and the role of reconnection at the outer planets, *J. Geophys. Res., 102*, 24289–24004.

Ioannidis, G., and N. Brice (1971), Plasma densities in the Jovian magnetosphere: plasma slingshot or Maxwell demon? *Icarus, 14*, 360–373.

Jackman, C.M., N. Achilleos, E.J. Bunce, S.W.H. Cowley, M.K. Dougherty, G.H. Jones, S.E. Milan, and E.J. Smith (2004), Interplanetary magnetic field at ∼9 AU during the declining phase of the solar cycle and its implications for Saturn's magnetospheric dynamics, *J. Geophys. Res., 109*, A11203, doi:10.1029/2004JA010614.

Jackman, C.M., N. Achilleos, E.J. Bunce, B. Cecconi, J.T. Clarke, S.W.H. Cowley, W.S. Kurth, and P. Zarkaa (2005), Interplanetary conditions and magnetospheric dynamics during the Cassini orbit insertion fly-through of Saturn's magnetosphere, *J. Geophys. Res., 110*, A10212, doi:10.1029/2005JA011054.

Jackman, C.M., C.T. Russell, D.J. Southwood, C.S. Arridge, N. Achilleos, and M.K. Dougherty (2007), Strong rapid dipolarizations in Saturn's magnetotail: In situ evidence of reconnection, *Geophys. Res. Lett., 34*, L11203, doi:10.1029/2007GL029764.

Johnson, R.E., H.T. Smith, O.J. Tucker, M. Liu, M.H. Burger, E.C. Sittler, and R.L. Tokar (2006) The Enceladus and OH torii at Saturn, *Astrophys. J., 644*:L137–139.

Jurac, S., and J.D. Richardson (2005), A self-consistent model of plasma and neutrals at Saturn: neutral cloud morphology, *J. Geophys. Res., 110*, A09220, doi:10.1029/2004JA010635.

Jurac, S., M.A. McGrath, R.E. Johnson, J.D. Richardson, V.M. Vasyliūnas, and A. Eviatar (2002), Saturn: search for a missing water source, *Geophys. Res. Lett., 29*, 2172, doi: 10.1029/2002GL015855.

Kane, M., D.G. Mitchell, J.F. Carbary, S.M. Krimigis, and F.J. Crary (2008), Plasma convection in Saturn's outer magnetosphere determined from ions detected by the Cassini INCA experiment, *Geophys. Res. Lett., 35*, L04102, doi:10.1029/2007GL032342.

Khurana, K.K., and M.G. Kivelson (1989), On Jovian plasma sheet structure, *J. Geophys. Res., 94*, 11791–11803.

Khurana, K.K., D.G. Mitchell, C.S. Arridge, M.K. Dougherty, C.T. Russell, C. Paranicas, N. Krupp, and A.J. Coates (2009), Sources of rotational signals in Saturn's magnetosphere, *J. Geophys. Res., 114*, A02211, doi:10.1029/2008JA013312.

Kivelson, M.G., and D.J. Southwood (2005), Dynamical consequences of two modes of centrifugal instability in Jupiter's outer magnetosphere, *J. Geophys. Res., 110*, A12209, doi:10.1029/2005JA011176.

Krimigis, S.M., J.F. Carbary, E.P. Keath, T.P. Armstrong, L.J. Lanzerotti, and G. Gloeckler (1983), General characteristics of hot plasma and energetic particles in the Saturnian magnetosphere, *J. Geophys. Res., 88*, 8871–8892.

Krimigis, S.M., et al. (2004), Magnetospheric Imaging Instrument (MIMI) on the Cassini Mission to Saturn/Titan, *Space Sci. Rev., 114*, 233–329.

Krimigis, S.M., et al. (2005), Dynamics of Saturn's magnetosphere from MIMI during Cassini's orbital insertion, *Science, 307*, 1270–1273, doi: 10.1126/science.1105978.

Krimigis, S.M., N. Sergis, D.G. Mitchell, D.C. Hamilton, and N. Krupp (2007), A dynamic, rotating ring current around Saturn, *Nature, 450*, 1050–1053, doi:10.1038/nature06425.

Krupp, N., A. Lagg, J. Woch, S.M. Krimigis, S. Livi, D.G. Mitchell, E.C. Roelof, C. Paranicas, B.H. Mauk, D.C. Hamilton, T.P. Armstrong, and M.K. Dougherty (2005), The Saturnian plasma sheet as revealed by energetic particle measurements, *Geophys. Res. Lett., 32*, L20S01, doi:10;1029/2005GL022829.

Kurth, W.S., D.A. Gurnett, J.T. Clarke, P. Zarka, M.D. Desch, M.L. Kaiser, B. Cecconi, A. Lecacheux, W.M. Farrell, P. Galopeau, J.-C. Gérard, R. Pangé, M.K. Dougherty, and F.J. Crary (2005), An Earth-like correspondence between Saturn's auroral features and radio emission, *Nature, 433*, 722–725, doi:10.1038/nature03334.

Kurth, W.S., A. Lecacheux, T.F. Averkamp, J.B. Groene, and D.A. Gurnett (2007), A Saturnian longitude system based on a variable kilometric radiation period, *Geophys. Res. Lett., 34*, L02201, doi:10.1029/2006GL028336.

Kurth, W.S., T.F. Averkamp, D.A. Gurnett, J.B. Groene, and A. Lecacheux (2008), An update to a Saturnian longitude system based on kilometric radio emissions, *J. Geophys. Res., 113*, A05222, doi:10.1029/2007JA012861.

Lazarus, A.J., and R.L. McNutt, Jr. (1983), Low-energy plasma ion observations in Saturn's magnetosphere, *J. Geophys. Res., 88*, 8831–8846.

Leisner, J.S., C.T. Russell, K.K. Khurana, M.K. Dougherty, and N. Andre' (2005), Warm flux tubes in the E-ring plasma torus: Initial Cassini magnetometer observations, *Geophys. Res. Lett., 32*, L14S08, doi:10.1029/ 2005GL022652.

Leisner, J.S., C.T. Russell, K.K. Khurana, and M.K. Dougherty (2007), Measuring the stress state of the Saturnian magnetosphere, *Geophys. Res. Lett., 34*, L12103, doi:10.1029/ 2007GL029315.

Masters A., N. Achilleos, M.K. Dougherty, J.A. Slavin, G.B. Hospodarsky, C.S. Arridge, and A.J. Coates (2008), An empirical model of Saturn's bow shock: Cassini observations of shock location and shape, *J. Geophys. Res., 113*, A10210, doi:10.1029/2008JA013276.

Mauk, B.H., J. Saur, D.G. Mitchell, E.C. Roelof, P.C. Brandt, T.P. Armstrong, D.C. Hamilton, S.M. Krimigis, N. Krupp, S.A. Livi, J.W. Manweiler, and C.P. Paranicas (2005), Energetic particle injections in Saturn's magnetosphere, *Geophys. Res. Lett., 32*, L14S05, doi:10.1029/2005GL022485.

McAndrews, H.J., M.F. Thomsen, R.L. Tokar, E.C. Sittler, M.G. Henderson, R.J. Wilson, and A.J. Coates (2007), Ion flows in Saturn's nightside magnetosphere, *EOS Trans. AGU, 88*(52), Fall Meet. Suppl., abstract SM53A-1085.

McAndrews, H.J., C.J. Owens, M.F. Thomsen, B. Lavraud, A.J. Coates, M.K. Dougherty, and D.T. Young (2008), Evidence for reconnection at Saturn's magnetopause, *J. Geophys. Res., 113*, A04210, doi:10.1029/2007JA012581.

McAndrews, H.J., M.F. Thomsen, C.S. Arridge, C.M. Jackman, R.J. Wilson, M.G. Henderson, R.L. Tokar, K.K. Khurana, E.C. Sittler, A.J. Coates, and M.K. Dougherty (2009), Plasma in Saturn's nightside magnetosphere and the implications for global circulation, *Planet. Space Sci.*, doi:10.1016/j.pss.2009.03.003.

Menietti, J.D., J.B. Groene, T.F. Averkamp, G.B. Hospidarsky, W.S. Kurth, D.A. Gurnett, and P. Zarka (2007), Influence of Saturnian moons on Saturn kilometric radiation, *J. Geophys. Res., 112*, A08211, doi:10.1029/2007JA012331.

Mitchell, D.G., P.C. Brandt, E.C. Roelof, J. Dandouras, S.M. Krimigis, B.H. Mauk, C.P. Paranicas, N. Krupp, D.C. Hamilton, W.S. Kurth, P. Zarka, M.K. Dougherty, E.J. Bunce, and D.E. Shemansky (2005), Energetic ion acceleration in Saturn's magnetotail: Substorms at Saturn? *Geophys. Res. Lett., 32*, L20S01, doi:10.1029/2005GL022647.

Mitchell, D.G., S.M. Krimigis, C. Paranicas, P.C. Brandt, J.F. Carbary, E.C. Roelof, W.S. Kurth, D.A. Gurnett, J.T. Clarke, J.D. Nichols, J.-C. Gérard, D.C. Grodent, M.K. Dougherty, W.R. Pryor (2009), Recurrent energization of plasma in the midnight-to-dawn quadrant of Saturn's magnetosphere, and its relationship to Auroral UV and radio emissions, *Planet. Space Sci.*, doi:10.1016/j.pss.2009.04.002.

Nichols, J.D., J.T. Clarke, S.W.H. Cowley, J. Duval, A.J. Farmer, J.-C. Gérard, D. Grodent, and S. Wannawichian (2008), Oscillation of Saturn's southern auroral oval, *J. Geophys. Res., 113*, A11205, doi:10.1029/2008JA013444.

Paranicas, C., R.B. Decker, B.H. Mauk, S.M. Krimigis, T.A. Armstrong, and S. Jurac (2004), Energetic ion composition in Saturn's magnetosphere revisited, *Geophys. Res. Lett., 31*, L04810, doi:10.1029/2003GL018899.

Paranicas, C., D.G. Mitchell, E.C. Roelof, P.C. Brandt, D.J. Williams, S.M. Krimigis, and B.H. Mauk (2005), Periodic intensity variations in global ENA images of Saturn, *Geophys. Res. Lett., 32*, L21101, doi:10.1029/2005GL023656.

Paranicas, C., D.G. Mitchell, E.C. Roelof, B.H. Mauk, S.M. Krimigis, P.C. Brandt, M. Kusterer, F.S. Turner, J. Vandegriff, and N. Krupp (2007), Energetic electrons injected into Saturn's neutral gas cloud, *Geophys. Res. Lett., 34*, L02109, doi:10.1029/2006GL028676.

Pontius, D.H., Jr., and T.W. Hill (1989), Rotation driven plasma transport: the coupling of macroscopic motion and microdiffusion, *J. Geophys. Res., 94*, 15041–15053.

Pontius, D.H., Jr., and T.W. Hill (2006), Enceladus: A significant plasma source for Saturn's magnetosphere, *J. Geophys. Res., 111*, A09214, doi:10.1029/2006JA011674.

Porco, C.C., and G.E. Danielson (1982), The periodic variations of spokes in Saturn's rings, *Astron. J., 87*, 826–829.

Porco, C.C., et al. (2006), Cassini observes the active south pole of Enceladus, *Science, 311*, 1393–1401, doi:10.1126/science.1123013.

Prangé, R., L. Pallier, K.C. Hansen, R. Howard, A. Vourlidas, R. Courtin, and C. Parkinson (2004), An interplanetary shock traced by planetary auroral storms from the Sun to Saturn, *Nature, 432*, 78–81, doi:10.1038/nature02986.

Provan, G., D.J. Andrews, C.S. Arridge, S.W.H. Cowley, S.E. Milan, M.K. Dougherty, and D.M. Wright (2009) Polarization and phase of planetary-period magnetic field oscillations on high latitude field lines in Saturn's magnetosphere, *J. Geophys. Res., 114*, A02225, doi: 10.1029/2008JA013782.

Richardson, J.D. (1998), Thermal plasma and neutral gas in Saturn's magnetosphere, *Rev. Geophys., 36*, 501–524.

Richardson J.D., and E.C. Sittler, Jr. (1990), A plasma density model for Saturn based on Voyager observations, *J. Geophys. Res., 95*, 12019.

Rymer, A.M., B.H. Mauk, T.W. Hill, C. Paranicas, N. André, E.C. Sittler, D.G. Mitchell, H.T. Smith, R.E. Johnson, A.J. Coates, D.T. Young, S.J. Bolton, M.F. Thomsen, and M.K. Dougherty (2007), Electron sources in Saturn's magnetosphere, *J. Geophys. Res., 112*, A02201, doi:10.1029/2006JA012017.

Scurry, L., and C.T. Russell (1991), Proxy studies of energy transfer to the magnetosphere, *J. Geophys. Res., 96*, 9541–9548.

Shemansky, D.E., P. Matherson, D.T. Hall, and T.M. Tripp (1993), Detection of the hydroxyl radical in Saturn's magnetosphere, *Nature, 363*, 329–332.

Siscoe, G. L., and D. Summers (1981), Centrifugally driven diffusion of Iogenic plasma. *J. Geophys. Res., 86*, 8471–8479.

Southwood, D.J., and M.J. Kivelson (2007), Saturnian magnetospheric dynamics: elucidation of a camshaft model, *J. Geophys. Res., 112*, A12222, doi:10.1029/2007JA012254.

Stevenson, D.J. (2006), A new spin on Saturn, *Nature, 441*, 34–35, doi: 10.1038/441034a.

Tokar, R.L., R.E. Johnson, T.W. Hill, D.H. Pontius, W.S. Kurth, F.J. Crary, D.T. Young, M.F. Thomsen, D.B. Reisenfeld, A.J. Coates, G.R. Lewis, E.C. Sittler, and D.A. Gurnett (2006), The interaction of the atmosphere of Enceladus with Saturn's Plasma, *Science, 311*, 1409–1412, doi:10.1126/science.1121061.

Vasyliūnas, V.M. (1975), Modulation of Jovian interplanetary electrons and the longitude variation of decametric emissions, *Geophys. Res. Lett., 2*, 87–88.

Vasyliūnas, V.M. (1978), A mechanism for plasma convection in the inner Jovian magnetosphere, COSPAR Program/Abstracts, p. 66, Innsbruck, Austria, 29 May–10 June, 1978.

Vasyliūnas, V.M. (1983), Plasma distribution and flow, in *Physics of the Jovian Magnetosphere*, edited by A.J. Dessler, chap. 11, p. 395, Cambridge University Press, New York.

Wahlund, J.-E., et al. (2005), Cassini measurements of cold plasma in the ionosphere of Titan, *Science, 308*, 986–989, doi:10.1126/science.1109807.

Wilson, R.J., R.L. Tokar, M.G. Henderson, T.W. Hill, M.F. Thomsen, and D.H. Pontius, Jr., Cassini Plasma Spectrometer thermal ion measurements in Saturn's inner magnetosphere (2008), *J. Geophys. Res., 113*, A12218, doi:10.1029/2008JA013486.

Wygant, J.R., R.B. Tolbert, and F.S. Mozer (1983), Comparison of S3–3 polar cap potential drops with the interplanetary magnetic field models of magnetopause reconnection, *J. Geophys. Res., 88*, 5727–5735.

Young, et al. (2004) Cassini plasma spectrometer investigation, *Space Sci. Rev. 114*, 1–112.

Zarka, P., L. Lamy, B. Cecconi, R. Prange, and H.O. Rucker (2007), Modulation of Saturn's radio clock by solar wind speed, *Nature, 450*, 265–267, doi:10.1038/nature06237.

Zieger, B., and K.C. Hansen (2008), Statistical validation of a solar wind propagation model from 1 to 10 AU, *J. Geophys. Res., 113*, A08107, doi:10.1029/2008JA013046.

Chapter 11
Fundamental Plasma Processes in Saturn's Magnetosphere

B.H. Mauk, D.C. Hamilton, T.W. Hill, G.B. Hospodarsky, R.E. Johnson, C. Paranicas, E. Roussos, C.T. Russell, D.E. Shemansky, E.C. Sittler Jr., and R.M. Thorne

Abstract In this chapter, we review selected fundamental plasma processes that control the extensive space environment, or magnetosphere, of Saturn (see Chapter 9, for the global context). This writing occurs at a point in time when some measure of maturity has been achieved in our understanding of the operations of Saturn's magnetosphere and its relationship to those of Earth and Jupiter. Our understanding of planetary magnetospheres has exploded in the past decade or so partly because of the presence of orbiting spacecraft (Galileo and Cassini) as well as remote sensing assets (e.g., Hubble Space Telescope). This book and chapter are intended to take stock of where we are in our understanding of Saturn's magnetosphere following the successful return and analysis of extensive sets of Cassini data. The end of the prime mission provides us with an opportunity to consolidate older and newer work to provide guidance for continuing investigations.

11.1 Introduction

Our objective is to take a "fundamental processes" approach to achieve predictive understanding of Saturn's space environment rather than a descriptive or explanatory approach. Our aspiration is to address the following questions:

- What are the fundamental processes that are in play within Saturn's magnetosphere that we have learned mostly by studying other environments?
- Are those processes predictive with respect to what we are observing within Saturn's magnetosphere?
- What can we learn about the fundamental processes by applying them to this novel environment that has both similarities and differences with others?

In this chapter, where it is appropriate and useful, we will compare our understanding of Saturn to related plasma processes at Earth and Jupiter. In doing this we assume these bodies are unified by a set of fundamental processes that operate coherently across a broad spectrum of varying parametric states. By providing us with examples of the "same" processes as those that occur at Earth and Jupiter but under very different conditions, a comparative approach allows us to broaden our horizons in how we understand processes, to generalize them, and to appreciate the underlying physics more deeply.

We focus here on several categories of processes occurring in planetary magnetospheres: (1) plasma–material interactions, (2) plasma and particle transport, and (3) energy conversion processes. These topics are not all-encompassing to space plasmas but represent the broad areas of recent studies of Saturn's system. The first of these topics focuses on the

---

B.H. Mauk and C. Paranicas
The Johns Hopkins University Applied Physics Laboratory, Laurel, MD, USA
e-mail: Barry.mauk@jhuapl.edu

D.C. Hamilton
Department of Physics, University of Maryland, College Park, MD, USA

T.W. Hill
Physics and Astronomy Department, Rice University, Houston, TX, USA

G.B. Hospodarsky
Department of Physics and Astronomy, University of Iowa, Iowa City, IA, USA

R.E. Johnson
Department of Materials Science and Engineering, University of Virginia, Charlottesville, VA, USA

E. Roussos
Max-Planck-Institut für Sonnensystemforschung, Katlenburg-Lindau, Germany

C.T. Russell
Institute of Geophysics and Planetary Physics, University of California, Los Angeles, CA, USA

D.E. Shemansky
Planetary and Space Science Division, Space Environment Technologies, Pasadena, CA, USA

E.C. Sittler Jr.
NASA Goddard Space Flight Center, Greenbelt, MD, USA

R.M. Thorne
Department of Atmospheric and Oceanic Sciences, University of California, Los Angeles, CA, USA

interactions of plasmas and energetic particles with planetary surfaces, gases, and atmospheres. Plasma interactions with particulates and dust are addressed in Chapter 16. From the transport perspective, Saturn has been viewed, prior to the Cassini epoch, as residing somewhere between the predominantly solar-wind–driven magnetosphere of Earth and the rotationally driven magnetosphere of Jupiter. Now we know that Saturn's magnetosphere supports dynamical activity in the form of small-scale and large transient features, which was completely unanticipated by the Pioneer and Voyager encounters. Has our view of Saturn changed with respect to the ordering of Saturn in comparison with Jupiter and Earth? Are these dynamical features Jupiter-like, Earth-like, or unique to Saturn? What are the relative roles of the solar wind and rotational drivers in the dynamics and in the transport in general? Our final major section addresses energy conversion processes. This topic groups together several relatively disparate but important subtopics. They include magnetic reconnection, current generation processes, particle acceleration processes, and more general wave-particle interaction processes.

11.2 Plasma–Material Interaction Processes

Here we will review present understandings of the mechanisms that provide material sources of gases and plasmas within Saturn's space environment, and charged particle and plasma interactions with surfaces and with neutral gases. We also address in this section the magnetospheric plasma–satellite interactions. We focus most strongly here on magnetosphere/icy-moon interactions since the interactions with Titan are addressed in the companion book on Titan (Sittler et al. 2009; Johnson et al. 2009), a companion to this book on Saturn.

11.2.1 Signatures of Gas and Plasma Sources

This section provides a brief observational context for the discussions of the fundamental plasma–material interaction processes in subsequent sections. A more extensive presentation is provided in Chapter 9.

11.2.1.1 Neutral Gas Components

Significantly, Saturn's unique magnetosphere is dominated by neutral gas to a much larger extent than is Jupiter's. Specifically, Saturn has neutral-to-ion ratios ranging from a minimum mean of 60 in Saturn's magnetic equatorial

Table 11.1 Neutral gas populations at Saturn*

Species	Density (cm^{-3}) 3–4R_S	Total system population	Loss rate (s^{-1})
OI	500	3×10^{34}	$\sim 10^{29}$
OH	700	$\sim 4\times10^{34}$	$\sim 10^{29}$
HI	450	2×10^{35}	3×10^{30}
H$_2$O	~ 200**	–	–
NI	minor	–	–

**Shemansky et al. 2004; Shemansky et al. 2009; Melin et al. 2009.
* Theoretical.

plane in the 3 to 5R_S region and increasing inward and outward to much higher values (Melin et al. 2009; Shemansky et al. 2009). Neutral H, H$_2$O, and their dissociative products of are the dominant particle populations in the magnetosphere (Table 11.1). An OH cloud was observed by the Hubble Space Telescope (HST) (Shemansky and Hall 1992; Shemansky et al. 1993) and is inferred to be a long-term resident (see also Richardson et al. 1998), but neutral OH is not directly observable by Cassini experiments.

Cassini Ultraviolet Imaging Spectrograph (UVIS) imaging (e.g., Fig. 11.1) shows OI population spatial asymmetries and temporal effects that are substantially stronger and faster than accommodated by known oxygen source rates, seriously complicating identification of the controlling factors in this system (Shemansky et al. 2004; Melin et al. 2009; Esposito et al. 2005). On a larger scale, hydrogen distributions derived from Cassini observations show a distinct local time asymmetry, with a peak in the dusk region, similar to Voyager measurements (Shemansky and Hall 1992). The H distribution extends from planet center to more than the 45R_S limit of the observations in the orbital plane and more than 10R_S out of the plane (Figure 10.11).

Modeling (Section 11.2.3) indicates that Enceladus, largely by way of its plumes, is the largest source of heavy neutrals in Saturn's magnetosphere, supplying about $\sim 1 \times 10^{28}$ H$_2$O molecules/s (Jurac and Richardson 2005; Hansen et al. 2006; Burger et al. 2007; Sittler et al. 2008); however, other spatial imaging structures (Shemansky et al. 2004; Melin et al. 2009) suggest that other sources are also active (see Table 11.2). Cassini UVIS scans have confirmed Voyager era arguments (Shemansky and Hall 1992) that the primary source of HI is escaping dissociation products of physical chemistry at the top of the Saturn sunlit atmosphere by revealing a complex sub-solar HI ejection structure (Shemansky et al. 2009; Fig. 11.2). The density of HI at the rings is of the order 10^4 cm^{-3}; however, there is no observational evidence for a component of the HI sourced from the rings. At the orbit of Titan the only detectable neutral component is HI at densities ranging from 50 to 150 cm^{-3} depending on local time (see Shemansky and Hall 1992; Shemansky et al. 2009), with no measurable torus at 20R_S (Figure 10.11).

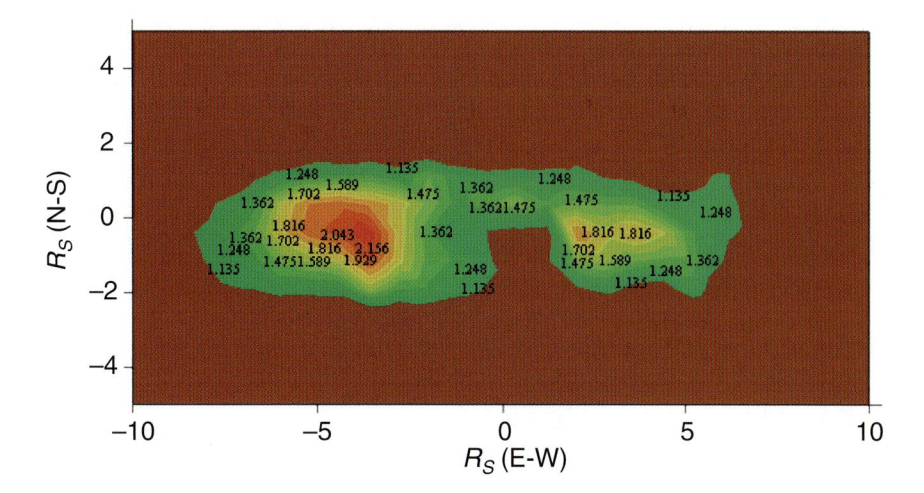

Fig. 11.1 Cassini UVIS image of OI 130.4-nm emission from the Saturn magnetosphere obtained 2004 DOY 51–72. Sub-spacecraft latitude −13.5°. Sub-solar latitude −23°. Solar flux impacts the system from the right. Brightness is indicated on the contour lines in Rayleighs. The OI emission is entirely forced by fluorescence of the solar OI flux. From Melin et al. 2009. Copyright: Elsevier, 2009 (Planetary and Space Science Journal)

Table 11.2 Neutral sources for the magnetosphere

Source	Species	Rate (10^{28}/s)	Reference
Enceladus & other icy bodies	H_2O	1.0	Jurac et al. 2002
			Hansen et al. 2006
			Burger et al. 2007
Main rings	O_2	0.2	Johnson et al. 2006a
Tenuous rings	H_2	0.4	
Enceladus	C, N	0.2	Waite et al. 2006
			HT Smith et al. 2008
Titan	CH_4	0.3	Yelle et al. 2008
			Strobel 2008
Titan	H_2	1.0	Cui et al. 2008
Saturn	H	300	Shemansky et al. 2009

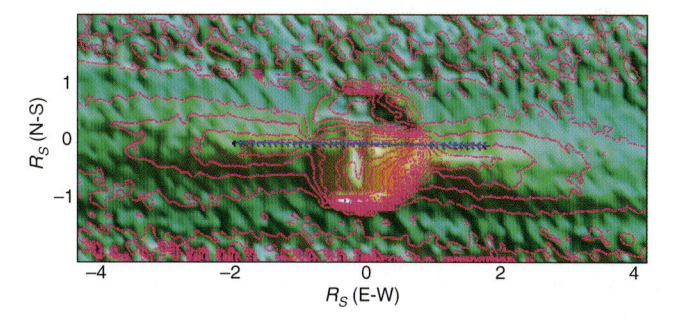

Fig. 11.2 Cassini UVIS image of Saturn and inner magnetosphere in H Ly-α emission showing structured outflow of atomic hydrogen from the top of the atmosphere. The emission is stimulated by solar flux from the right side of the image, at latitude −17°. The rings (blue markings) are viewed edge-on from the spacecraft. A ridge of emission at −8° latitude forms a propeller-shaped feature from both sides of the planet, measurable to beyond $4R_S$ as indicated in the shape of the image contours. Outflow from the sub-solar atmosphere is evident over a broad range of latitudes, apparently affected by ring shadow. Auroral emission is evident at the poles. From Shemansky et al. 2009. Copyright: Elsevier, 2009 (Planetary and Space Science Journal)

Charged Particle Components

Charged particles in Saturn's magnetosphere can be produced by ionization of the neutral clouds, discussed above, and by direct injection of ions from the solar wind, Saturn's polar wind, and auroral outflow (Sittler et al. 2006b; Glocer et al. 2007; Mitchell et al. 2009). Composition measurements provide information on the relative strengths of the various sources, but this information can be sometimes confused or masked by transport and energization processes. On Cassini, low-energy plasma composition (starting from eV levels) is measured by the Cassini Plasma Spectrometer (CAPS)/Ion Mass Spectrometer (IMS) instrument and the INMS investigation. At suprathermal (kiloelectron-volt to tens of kiloelectron-volt) and high energies (>100 keV) the Magnetospheric Imaging Instrument (MIMI)/Charge, Energy, Mass Spectrometer (CHEMS) makes composition measurements.

The relative importance of various charged species is shown in Fig. 11.3 (for highly energized components, >tens of kiloelectron-volt) and Table 11.3. The most abundant plasma species are H^+ and water group ions labeled W^+ (O^+, OH^+, H_2O^+, H_3O^+) (Young et al. 2005; Krimigis et al. 2005). In the suprathermal (>1 keV) ions the H^+ abundance is about half that of the water group (Hamilton et al. 2009a). In the thermal (<1 keV) plasma, that ratio is considerably less inside of $8R_S$ (Young et al. 2005; Sittler et al. 2008; Wilson et al. 2008).

The primary source of water group ions is electron impact or photo-ionization of the E-ring material and the neutral

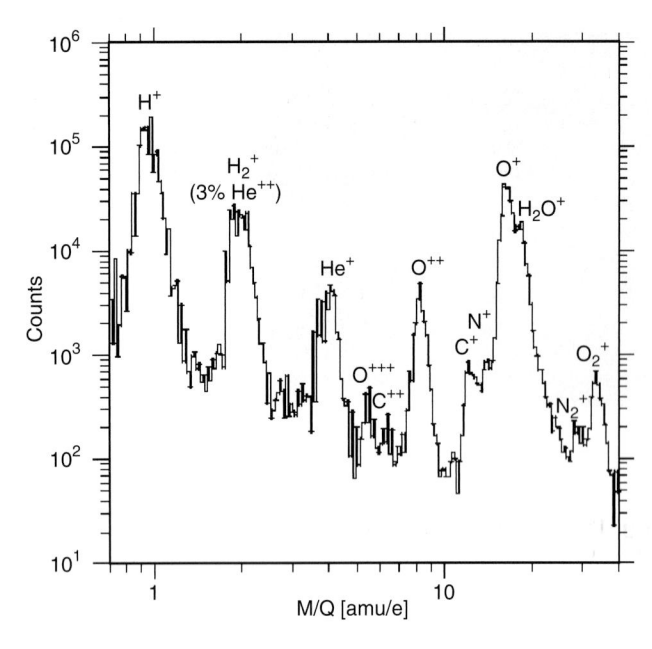

Fig. 11.3 Two-year sum of suprathermal ion composition in Saturn's magnetosphere from the MIMI CHEMS sensor (Krimigis et al. 2005; Hamilton et al. 2009a, b, c). Minimum energy increases with mass. Approximate energy ranges are 25–220 keV for H^+ and 55–220 keV for O^+

Table 11.3 Suprathermal ion abundances*

Species	Relative to H^+	Relative to O^+
H^+	$=1$	1.13
W^+	1.9	2.2
H_2^+	0.15	0.17
H_3^+	0.001	0.001
He^+	0.025	0.028
He^{++}	0.0043	0.0048
C^+	0.012	0.013
N^+	0.018	0.021
O^+	0.88	$=1$
OH^+	0.46	0.52
H_2O^+	0.47	0.54
H_3O^+	0.10	0.12
N_2^+ (or CO^+)	0.013	0.015
O_2^+	0.056	0.064

* Krimigis et al. 2005; Hamilton et al. 2009a, b, c.

cloud produced by the Enceladus plumes. Sittler et al. (2008) estimate the local (near Enceladus) ion source rate to be $\sim 1 \times 10^{27}$/s, and a more extended rate of $\sim 3 \times 10^{27}$/s, $\sim 10\%$ of the neutral production rate. Pontius and Hill (2006), however, estimate a considerably larger local ionization rate of 3–8 $\times 10^{27}$ to explain the observed mass loading at Enceladus. The need for further work is indicated.

The H^+ is also a product of water dissociation and ionization, and ionization of Saturn's neutral atmosphere (Fig. 11.2, discussed above), but the solar wind, Saturn's ionosphere, auroral upflow, and possibly Titan are additional sources.

The solar wind source of H^+ can be estimated at about 4×10^{26}/s (Hamilton et al. 1983; Sittler et al. 2006a). The average He^{++}/H^+ ratio in the solar wind is 0.04, about ten times greater than found in Saturn's magnetosphere (Table 11.3), indicating that 90% of the H^+ is from local sources ($\sim 4 \times 10^{27}$/s). Direct evidence of auroral ion upflows is provided by Mitchell et al. (2009).

Glocer et al. (2007) have modeled the outflow rate of Saturn's polar wind, a mixture of H_3^+ and H^+, to be between 2×10^{26} and 8×10^{27}/s, but the very low abundance of H_3^+ (Hamilton et al. 2009c; Mitchell et al. 2009), which is expected to be nearly equal to or greater than H^+ in the polar wind, indicates the polar wind source is probably much lower ($< 2 \times 10^{25}$/s). The quite abundant H_2^+ can arise from ionization of water dissociation products, but H_2 can also be ejected from Titan's atmosphere (Cui et al. 2008).

Trace species such as C^+ and N^+ are present at $\sim$1–2% of O^+, and currently Enceladus is thought to be the stronger source (Smith et al., 2007, 2008), with Titan also making a contribution (De La Haye et al. 2007), including observations of pickup methane near Titan (Hartle et al. 2006). The molecular ion O_2^+ is an interesting tracer, as it is primarily derived from radiation-induced decomposition of icy surfaces, as discussed in more detail in the next section. The N_2^+ cannot be distinguished in the CHEMS data from CO^+, another mass 28 species, for which Enceladus may be a source (Waite et al. 2006)

11.2.2 Charged Particle and Plasma Interactions with Surfaces

11.2.2.1 Introduction and Background

Surfaces in space are weathered by impacting photons, charged and neutral particles, and dust grains. In this section, we describe sputtering and radiolysis of surfaces in Saturn's system (Johnson 1990). Radiolysis and photolysis describe the chemical alteration of materials by incident charged particles and UV photons, respectively. Such alterations can also lead to the ejection of new volatile species such as H_2 and O_2 from ice (e.g., Johnson and Quickenden 1997; Loeffler et al. 2006). In both sputtering and radiolysis, ejected neutrals contribute to the ambient atmosphere and either escape to space or eventually return to the surface of a body with significant gravity. The surface alterations can often be seen in reflectance, and the ejected atoms and molecules can often be detected by their line emissions. In addition, once ionized they can be detected in the space plasma remotely or by an orbiting spacecraft. By these means clues can be obtained as to the composition and weathering of the surface material.

11.2.2.2 Application to Saturn

Surface Radiolysis

Unlike the Jovian plasma, which contains sulfur from Io, the Saturnian plasma is primarily composed of hydrogen from Saturn's atmosphere and Titan and water products from the icy satellites and ring particles (see Section 11.2.1), along with small amounts of light and trace ions (Table 11.3). Prior to Cassini's arrival at Saturn it was known that radiolytic products were present in the icy surfaces. This was shown by the detection of an ozone-like feature on Dione and Rhea, and a spectral slope in the UV is suggestive of the presence of peroxide produced by radiolysis in the ice matrix (Noll et al. 1997). Such features have been seen on Europa (Spencer and Calvin 2002) and Ganymede (Calvin et al. 1996; Noll et al. 1996) and are associated with the radiation processing of the surface ice (e.g., Johnson and Quickenden 1997).

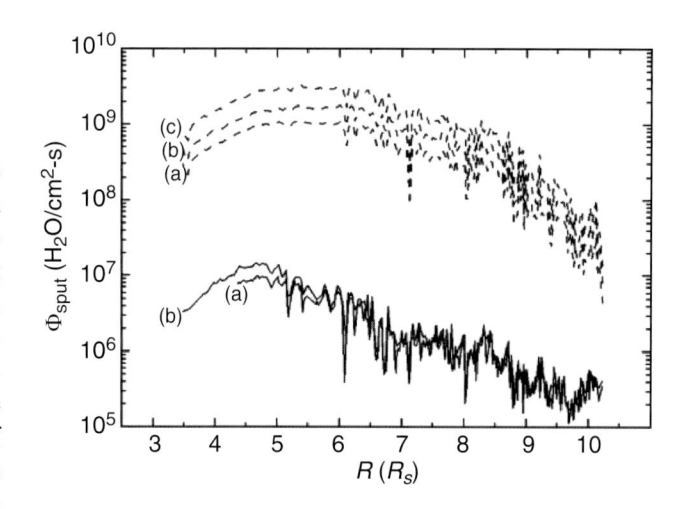

Fig. 11.4 Sputtering rate of water molecules according to Johnson et al. (2008). Solid H^+, dashed $W+$: (**a**) ion temperatures ignored ($W^+ \to O^+$): note threshold effect for H^+; (**b**) includes ion temperatures ($W^+ \diamond O^+$); (**c**) includes ion temperatures ($W^+ \to H_2O^+$). Satellites referenced in the text, Enceladus, Dione, and Rhea, reside at 4.0, 6.3, and $8.7 R_S$, respectively. Copyright: Elsevier, 2008 (Planetary and Space Science Journal)

Surface Irradiation

The radiation effects are more subtle for the icy surfaces in the Saturnian system than at Jupiter, due not only to the lack of a readily identifiable trace species like sulfur but also to the very different character of the radiation flux. Based on Cassini measurements it is now known that the flux is considerably diminished at Saturn. Near the inner icy satellites, singly charged ions are lost to space by charge exchange collisions with the prominent water product gases, and even at 10's of Saturn radii there are losses due to the extensive H cloud (e.g., Paranicas et al. 2008; Section 11.2.1). The million electron-volts ions and electrons are transient in the region around Dione's orbit (Roussos et al. 2008a; Sections 11.2.4 and 11.3.4) and continuously weather the inner satellites, putting energy into the embedded surfaces. As in the Jovian magnetosphere, thermal plasma is present throughout Saturn's inner magnetosphere (Chapter 9) that impinges on the surfaces of the embedded satellites and ring particles.

Surface Sputtering in Saturn's Magnetosphere

Although the net energy flux into the icy surfaces in the Saturnian magnetosphere is smaller than that at Europa and Ganymede, these surfaces are eroded by the incident plasma. This can limit the lifetime of the E-ring grains and produce neutrals for the magnetosphere. When the surfaces are predominantly water ice, the sputtering by the thermal plasma primarily ejects H_2O molecules. Recently Johnson et al. (2008) re-calculated the surface sputtering rate by the thermal plasma using the Cassini CAPS measurements (Sittler et al. 2008). In Fig. 11.4 is given the average sputter flux for an icy surface within a plasma scale height of the magnetic equator. One calculates from this information that the lifetime of a micrometer-size E-ring grain is of the order of 50 years in the vicinity of Enceladus and Dione and decreases rapidly with increasing distance from Dione, becoming more than an order of magnitude smaller at Rhea. These rates are surprisingly close to earlier estimates using a model flux that was very different from that seen by Cassini (Jurac et al. 2001). Surface sputtering has been suggested to account for the non-plume component of the Enceladus atmosphere (Burger et al. 2007), though that is not at all certain. This source directly populates a region around the orbit of Enceladus and is scattered by ion–molecule collisions throughout the inner magnetosphere (Johnson et al. 2006b).

Radiolytic Production of Molecular Oxygen

O_2^+ in Saturn's magnetosphere is primarily derived from O_2 that has been formed by radiation-induced decomposition of the many icy surfaces. At very low relative collision speeds such as those in the Enceladus plume, O_2^+ can be formed by ion molecule reactions involving water products (e.g., $O^+ + OH \to O_2^+ + H$), but this is not a likely process over most of the magnetosphere. Therefore, O_2^+ in Saturn's magnetosphere can be a marker for the occurrence of radiation-induced decomposition of ice.

An important source of molecular oxygen is Saturn's ring atmosphere. Because the density of icy particles in the main ring is large, there is a dearth of trapped plasma. Therefore, UV photolysis is the dominant radiation effect. Because of their low temperatures, H_2O molecules ejected from the ring particles by UV radiation, meteoroid impacts, or collisions rapidly re-condenses on the ring particle surfaces. Although UV-induced decomposition leading to H_2 and O_2 is a much smaller source, these volatiles do not stick at the suggested surface temperatures (Chapter 15). Therefore, Saturn's ring atmosphere is dominated by hydrogen and oxygen as shown by the modeling of the CAPS (Tokar et al. 2005) and INMS (Waite et al. 2006) data over the rings (Johnson et al. 2006a).

The ejected H_2 and O_2 molecules interact with the ring particles often until they are eventually lost by dissociation and ionization. The ions so formed rapidly interact, scattering neutrals into the magnetosphere and into Saturn's atmosphere (Johnson et al. 2006a; Luhmann et al. 2006; Bouhram et al. 2006; Tseng et al. 2009). Therefore, the rings are a source of molecular oxygen for Saturn's magnetosphere. Additional sources of O_2 are the tenuous G and F rings and their companion icy satellites due to the CRAND processes (Cooper 1983) dominating inside the orbit of Mimas (Paranicas et al. 2008). Jones et al. (2008) claim, based on the size of the absorption signature in energetic charged particles, that material must be present in orbit around Rhea (Section 11.2.4). This absorption also leads to the production of O_2 and hence appears to be a local source of O_2^+ in the magnetosphere near Rhea's orbit (Martens et al. 2008). In addition, the gradual absorption of energetic electrons by the E-ring particles implies that radiation energy is deposited in the grains, in which case they contribute molecular oxygen to the magnetosphere. A model for the O_2^+ production throughout the inner magnetosphere from all sources is needed.

11.2.3 Charged Particle and Plasma Interactions with Neutral Gas

11.2.3.1 Introduction and Background

Knowledge of ion–neutral and electron–neutral collisions is critical for understanding the plasma processes occurring in a planetary magnetosphere. Charge exchange involves the collision of a singly charged ion with a neutral. For high ion speeds, the original ion becomes an energetic neutral atom (ENA) with nearly the same energy of the parent ion, and a cold ion is left behind. This process results in a net loss from Saturn's inner magnetosphere of energetic ions (Paranicas et al. 2008). At lower energies, the above also occurs, but the amount of energy involved in the charge exchange is a more significant fraction of the initial ion speed, and this leads to deflection of both the initial ion and neutral. The ion–neutral collision can be linked to several other important processes. For example, the freshly formed ion, as stated above, is typically colder than the incident ion that it replaces so that it must be re-accelerated by the electric field associated with the planet's rotating magnetic field. When cold plasma is imbedded in a cloud of neutrals, as is the case at Saturn, there are many such reactions, with the net result a momentum exchange between the neutrals and the magnetized plasma. This leads to a drag on the corotating plasma (Saur et al. 2004). Likewise, ion–neutral collisions, by replacing a slow neutral with a faster neutral, effectively transfer energy to the neutral population. This has the effect of spreading the neutral distribution in a number of ways (Johnson et al. 2005, 2006a, b).

Electrons can interact with molecular ions by dissociative recombination, causing the loss of an ion and the production of hot neutrals. In addition, both electron and photons ionize the ambient neutrals. Unlike charge exchange, photoionization and electron impact ionization create an additional ion that is typically colder that the ambient thermal plasma. The electron–ion recombination and electron impact ionization rates depend on the plasma (electron) temperature and density, which vary as functions of position in Saturn's magnetosphere, meaning there are regions where these processes are most effective (e.g., Sittler et al. 2008). A consequence of the creation of ions out of neutrals by any of the processes described here is the "pickup energization" of ions and the "mass loading" of the flow system, as described in Section 11.4.2.

11.2.3.2 Application to Saturn

Neutral Sources and Predictions

Neutrals are ejected into Saturn's magnetosphere from a number of sources as discussed above (Section 11.2.1 and Table 11.2). Following Voyager, heavy neutrals in Saturn's magnetosphere were assumed to be principally from Titan with the heavy species predominantly nitrogen (Barbosa 1987). Although sputtering of the icy satellites surfaces and the E-ring grains was shown to dominate the Titan source in the inner magnetosphere (Johnson et al. 1989a, b), Shemansky and co-workers (e.g., Shemansky and Hall 1992) showed that a much more robust source of neutrals must be present in the inner magnetosphere in order to describe the OH cloud that they observed using HST. By Monte Carlo modeling of the ejecta, Jurac et al. (2001, 2002) confirmed that the sputter source was too small and the primary source of neutrals was at the orbit of Enceladus with source strength

$\sim 10^{28}$ H$_2$O/s (updated by Jurac and Richardson 2005). Remarkably, both the orbital location and the magnitude of the source rate were confirmed by Cassini observations, validating the theoretical approach. Cassini showed that this small satellite is in fact outgassing from its south polar region and was the neutral source invoked by Jurac and co-workers. Enceladus is by far the principal non-Saturn and principal heavy atom source of neutrals for the inner magnetosphere. Recent modeling of the INMS data for Titan's atmospheric corona has suggested that the escape rate of methane from Titan is much larger than pre-Cassini estimates (De La Haye et al. 2007; Yelle et al. 2008; Strobel 2008), but the largest rates proposed have been challenged (Johnson 2009; Johnson et al. 2009).

Neutral Loading at Saturn and Jupiter

Neutrals ejected from Enceladus have energies much smaller than their orbital energy. They therefore form a narrow neutral torus about the orbit of Enceladus (Johnson et al. 2006b) called the Enceladus torus to distinguish it from the OH torus initially reported by Shemansky and co-workers. Like Io at Jupiter, Enceladus is the dominant source of heavy, mass-loading neutrals for Saturn's inner magnetosphere. There is a major difference between the Jovian and Saturnian inner magnetospheres. For Saturn, neutrals dominate the ion density over a broad radial range about the orbit of Enceladus, whereas the ions dominate neutrals in the vicinity of Io's orbit, a region referred to as the Io plasma torus. This difference appears to be due to the dearth of hot electrons in the plasma near the orbit of Enceladus, reducing the ionization rate.

The difference in electron temperatures has been suggested to be due to the difference in size of the pickup ion energies and subsequent electron heating, respectively, at Enceladus and Io (Delamere et al. 2007). However, this view has been moderated (Delamere, private communication 2008). In traditional neutral cloud theory as applied, for example, to Jupiter's Io torus region, the thermal energy of pickup ions powers the torus UV emissions through the intermediary of thermal (~ 5 eV) electrons, which acquire thermal energy from ions by Coulomb collisions and return that energy to the ions at the same average rate through electronic excitation collisions. Even with the assumption of ion pickup at the full corotation speed, this theory suffered an "energy crisis" (Shemansky and Sandel 1982; Smith et al. 1988; Shemansky 1988; Barbosa 1994; Herbert and Sandel 1995), and that energy crisis becomes more acute as the pickup energy per ion is reduced. While a number explanations have been proposed, the most attractive candidate to us for the electron energy source is Birkeland (magnetic-field-aligned) currents, which are ubiquitous in the torus due to satellite-magnetosphere interactions (Piddington and Drake 1968; Goldreich and Lynden-Bell 1969; Hill and Vasyliūnas 2002; Delamere et al. 2003) and the coupling between the ionosphere and regions of the magnetosphere that are undergoing radial transport by the flux-tube interchange process (Section 11.3.4). The energy associated with the current should scale with rotational speed, which may partially explain differences between Jupiter and Saturn.

There are other important differences between the Jupiter and Saturn systems. The lower relative speeds between the plasma and the neutrals suggest that, unlike at Io, a significant fraction of the neutrals are not effectively removed by charge exchange, as discussed below. The resulting extended neutral cloud quenches the inwardly diffusing electrons and ions, removing a potential heat source. Also, because of its larger gravity, the Io source results from the interaction of the magnetosphere with the atmosphere, which is a complicated feedback process. This results, for instance, in ionospheric stripping, which directly adds ions, and atmospheric sputtering, which only gradually adds neutrals as the ion density builds up. At Enceladus, on the other hand, neutrals are directly dumped into the magnetosphere at a rate that is *independent of the local magnetosphere properties*, possibly swamping the electron heating process.

Charge Exchange

Because of the high neutral densities, charge exchange is especially important in Saturn's inner magnetosphere. As stated earlier, energetic ions that diffuse inward from beyond the orbit of Rhea are lost by charge exchange and replaced by fresh pickup ions that add to the local thermal plasma. In addition, charge exchange is also critical in redistributing neutrals in a magnetosphere and acting to clear out the gas (Johnson et al. 2005). Charge exchange is typically thought to simply replace the fast ion by a fast, undeflected neutral, the ENAs discussed above, leaving a new thermal ion. Because the nearly resonant charge exchange (e.g., $O^+ + O \to O + O^+$) cross section is primarily determined by collisions that occur with large impact parameters (e.g., Johnson 1990), such a simplification works at collisions speeds greater than about a few $\times 10^5$ m/s. In this way the ENAs are typically dispersed either out of Saturn's gravitational potential or far from their source region. At lower relative collision speeds deflections can be important or even dominant. For instance, ions in Saturn's ring atmosphere interact with the neutrals at very low average collision speeds. At these speeds the ion–neutral interactions are described by an attractive potential and the ions and neutrals are both scattered nearly isotropically in the center of mass system (Johnson et al. 2006b). In fact, it is this process that scatters O_2^+ formed in the narrowly confined equatorial ring atmosphere to altitudes well above the ring plane where such molecular ions were detected by Cassini during the Saturn Orbit Insertion (SOI) trajectory (Tokar et al. 2005; Johnson et al. 2006a). Inside about $1.86 R_S$, where the plasma corotation speed becomes *less than* the speed of a neutral in a circular orbit, pickup ions precipitate into Saturn's atmosphere (Luhmann et al. 2006; Bouhram et al. 2006).

Neutral Scattering

At Enceladus' orbital position, a neutral formed by charge exchange *in the absence of any deflection* would be typically lost from the system. However, the ion–neutral collisions have a broad distribution of speeds as determined by the ion temperature or gyro-motion, and they can experience significant deflections during charge exchange. Therefore, for a large fraction of the ion–neutral collisions near the orbit of Enceladus, deflections are important (e.g., Jurac et al. 2002; Johnson et al. 2006b; Smith et al. 2007). Although a significant fraction of the charge exchange reactions lead to scattering of neutrals well beyond the orbit of Rhea, the charge exchange scattering of neutrals from the Enceladus torus is also likely the process that forms the neutral OH torus seen by HST (Johnson et al. 2006b) as shown in Fig. 11.5. Such a process also occurs in the Io neutral torus, but the larger relative speeds tends to populate regions beyond the orbit of Callisto (Wang et al. 2001). In addition, Farmer (2008) has proposed that the Enceladus torus is sufficiently dense that neutral–neutral collisions are important in broadening the narrow Enceladus torus shown in Fig. 11.5. Due to the dominance of neutral molecules in the magnetosphere near Enceladus, electron temperatures are low and charge exchange dominates. Therefore, although the principal source of neutrals in the inner magnetosphere, as seen in Table 11.2, is Enceladus, the *dominant region of ion formation* is at a larger distance from Saturn, closer to the orbit of Dione (Sittler et al. 2008).

Electron-Induced Processes

The low electron temperatures in the Enceladus torus have a number of ramifications. In the Enceladus plume the electron temperatures are sufficiently low that electron attachment occurs, forming negative ions in the plume (Farrell et al. 2008; Coates et al. 2009). In addition, in the Enceladus torus electron impact ionization is inefficient and dissociative recombination may be the dominant ion loss process (Sittler et al. 2008). This process acts to heat the neutrals locally (Johnson et al. 2006b; Fleshman et al. 2008) contributing to the expansion of the narrow Enceladus torus shown in Fig. 11.5. Although charge exchange between the thermal plasma ions and the neutrals remains important throughout

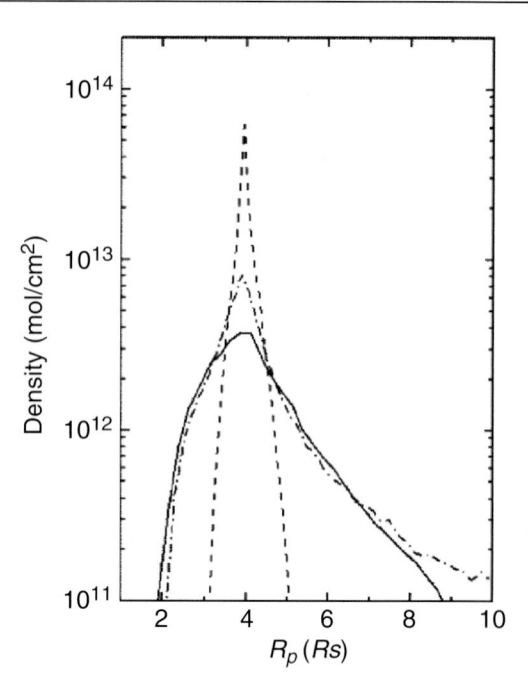

Fig. 11.5 Toroidal column density vs. radial distance from Saturn along the equatorial plane. *Solid curve*, OH component of Saturn's giant toroidal atmosphere based on observations using HST for over a decade (Jurac et al. 2002; Jurac and Richardson 2005); *dashed curve*, torus of primarily water molecules venting from south polar region of Enceladus (ignoring neutral collisions and heating by dissociation); *dot-dashed curve*, neutrals scattered from the Enceladus torus by charge-exchange and reactive collisions. An Enceladus source rate of 5×10^{27} H$_2$O s^{-1}, an average lifetime of 6×10^6 s, and a model average cross section were used. Copyright: American Astronomical Society, 2006 (Astrophysical Journal Letters)

the inner magnetosphere, with increasing distance from the Enceladus torus, the electron temperature increases so that impact ionization increases as a source of plasma and outward diffusion rapidly dominates the ion loss (Sittler et al. 2008). This dependence was recently shown to be reflected in the changing character of the pickup ion velocity distributions at Enceladus and Dione (Tokar et al. 2008). In Fig. 11.6 we give the volume ionizations rates, obtained using the data in Sittler et al. (2008), and the neutral cloud model in Fig. 11.5. Also in Fig. 11.6 is given these rates multiplied by L^3 to estimate of the ionization rate in a flux tube. It is seen that, ignoring any widening of the Enceladus torus, there is a small sharp peak in the ionization rate near Enceladus, but, unlike in the Jovian magnetosphere, ionization is primarily produced at radii much larger than the principal neutral source.

Comet-Like Interactions

Very close to Enceladus, in the plume extending from the south polar region, the neutral plasma interactions resemble

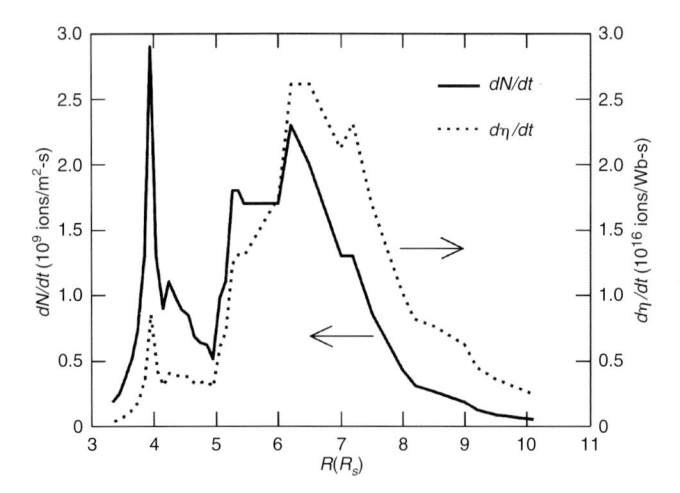

Fig. 11.6 Source rate of newly ionized plasma (excluding charge exchange) versus equatorial radial distance in Saturn's magnetosphere using model in Fig. 11.5. The solid line (see left scale) is the source rate of ion column density N per unit equatorial area, derived from a model that assumes a single neutral-gas source at $4R_S$ (the orbit of Enceladus). The dotted line (right scale) is the source rate of ion flux-tube content η per unit magnetic flux. Thus the dotted line is the solid line divided by the equatorial magnetic-field strength, assuming a dipole magnetic field

those for a comet (see Section 11.2.4 for further details). Because the plasma is slowed and the electrons are cooled further by interaction with the molecular neutrals, plasma chemistry can occur. For instance, both negative and positive ion clusters have been detected (e.g., Coates et al. 2009; Tokar et al. 2009). These might be formed either as neutral clusters in the plume, due to explosive evaporation, or on interaction with the walls.

11.2.4 Moon–Magnetosphere Interactions

11.2.4.1 Introduction and Background

Among the 61 named moons of Saturn, 8 of the largest (diameter >100 km) usually orbit the planet inside the planet's magnetopause (~$20R_S$ subsolar; Chapter 9), in circular, prograde orbits nearly in the planet's equatorial plane; including Ephimetheus ($2.5R_S$), Janus (2.5), Mimas (3.1), Enceladus (3.9), Tethys (4.9), Dione (6.3), Rhea (8.7), and Titan (~20.3). Because of their various radial positions, the space environments and exospheres are quite different. Here we consider the space environment interactions with the icy moons of Saturn. The special case of Titan is addressed in the companion book on Titan (Sittler et al. 2009; Johnson et al. 2009).

In establishing the kinds of interactions that a moon can have with its magnetospheric environment, there are a number of factors that must be considered. A useful ordering can be presented in terms of the magnetohydrodynamic (MHD) waves involved in the plasma interaction with the moon: the slow, intermediate and fast modes (Kivelson et al. 2004). In addition, if the moon has internal magnetic fields, either intrinsic like Ganymede at Jupiter (Kivelson et al., 1997a, 2004), or induced by the externally imposed magnetic fields and their variations like Europa at Jupiter (Kivelson et al. 1997b, 2004), then these fields can play a significant role in the interaction geometry. If, like Titan (Sittler et al. 2009; companion book on Titan), the moon has an atmosphere with an ionosphere (either generated by solar UV or by local, self-consistent interactions with the environment), then ionospheric currents induced by the interaction also play a role The plasma flow may be slowed and deflected, and the field-lines drape over the moon with a chevron shape, as observed near Io at Jupiter (Kivelson et al. 2004).

One important process that slows and deflects the plasma flow is mass loading (Section 11.4.2) by means of ionization of neutral gas that emanates from the moon, either by the sputtering of surfaces or atmospheres (Section 11.2.2) or by its geologic activity (e.g. Io's volcanism or Enceladus' south polar geysers). The ions generated are "picked up" by the local plasma (Section 11.4.2) and the flow slows down where mass is added to the plasma. Since the plasma is magnetically connected to faster flowing plasma, the field lines become bent. If the impinging flow is fast enough (faster than the local Alfvén and sound speeds), the draping field lines pinch together to form a magnetotail, with opposing magnetic field polarities separated by a narrow current sheet, as we find at Titan (Sittler et al. 2009). For slower impinging flows, the draping is less severe and the configuration may be described by Alfvén waves that propagate away from the moon roughly along, but not exactly aligned with, the draping field lines, forming the so-called "Alfvén wings" (Neubauer et al. 1998). If a moon is non-conducting and has no source of neutral gas that is ionized close to the moon, it means that distortion currents are absent and plasma flows undisturbed against the moon's surface forming an extended plasma cavity at the opposite hemisphere. Such an interaction is described as "plasma absorbing."

Internal Magnetic Fields

It is not currently known if induced magnetic fields are generated at any of the inner Saturnian moons as they are at Jupiter's Europa (Kivelson et al. 1997b; see discussions by Khurana et al. 2007; Anderson and Schubert 2007). The near alignment of the magnetic dipole axis of Saturn with the spin axis (Smith et al. 1980; Connerney et al. 1981) means there is no strong periodic variation of Saturn's field to diagnose induction, as there is at Jupiter. There are time-varying external fields at Titan, but Titan's ionosphere makes it difficult to detect the weak signal expected. The strongest time-varying fields suitable for sounding Saturnian moons exist around those moons that enter the solar wind such as Iapetus.

Interaction Environments and Features

Six of Saturn's large moons orbit within the strong dipolar magnetosphere. The innermost of these, Janus, Epimetheus, Mimas and Enceladus, are mainly exposed to dense and cold plasma, as well as kiloelectron-volt to million electron-volts electrons and million electron-volts ions. Below a few hundred kiloelectron-volt energy, stably trapped energetic ions are nearly absent, as they are removed by charge exchange in the neutral gas cloud (Paranicas et al. 2008). Their absence suggests that surface sputtering might not be an important atmosphere generation process for the moons of the innermost magnetosphere, contrary to what was inferred from Voyager observations (Saur and Strobel 2005). Recent analysis, however, shows that the more abundant, cold plasma could also be a significant sputtering agent (Johnson et al. 2008). Further out, Tethys and Dione still orbit within the dense neutral gas cloud, but at a distance where flux tubes with energetic plasma from injections can temporarily survive (Section 11.3.4; see Roussos et al. 2008b)

All moons considered here are overtaken by the magnetospheric plasma at a wide range of velocities; the interaction takes place in the moon's upstream hemisphere with respect to corotation (the trailing hemisphere with respect to the moon's Keplerian motion) with the interaction effects extending downstream. Plasma velocities relative to the moon's Keplerian motion (∼8–60 km/s) are typically higher or comparable to ion thermal speeds for both heavy ions and protons. Therefore, the interaction can be submagnetosonic or transmagnetosonic and a bow shock cannot form upstream of any of the moons.

Energetic electrons (around ∼1 MeV) gradient and curvature drift opposite to the direction of the corotating plasma, and thus can encounter the icy moons from their leading hemispheres. The transition energy where this occurs is termed the Keplerian resonant energy and its value depends on the L-shell, the electron pitch angle, and the bulk velocity of the ambient plasma (Thomsen and Van Allen 1980).

Physical Processes

In what follows, we will first discuss the moons that act as absorbers of charged particles and later discuss mass-loading interactions. We define the $+x$ axis along the plasma bulk velocity vector, $+z$ antiparallel to the magnetic field direction (which is southward at the orbital plane of Saturn's moons), and $+y$ toward Saturn (opposite to the corotation electric field $\mathbf{E}$).

11.2.4.2 Plasma Absorbing Interactions at Saturn

At first sight, plasma absorption seems a very simple interaction: charged particles that overtake the moon in its Keplerian motion are absorbed at its surface, as currents that could deflect the flow cannot develop from such a non-conducting obstacle. The refilling of the cavity that forms downstream could be driven by several fundamental processes in plasmas, such as two-stream instability, ambipolar diffusion, acceleration etc. Saturn's moons, with their varying space environments, represent excellent laboratories to study this problem.

Cold Plasma Response

The plasma wake presents a disturbance that the magnetosphere tends to refill with surrounding plasma. Simple analytical descriptions of the fill-in process can be extracted from a one-dimensional fluid approach of the problem, which demands quasi-neutrality between the momentum-carrying ions and the essentially massless electrons. Specifically, Samir et al. (1983), solving the momentum and continuity equations and assuming a Maxwellian distribution for the ions, found that for a given vertical cut through the wake (along the field) the plasma density increases exponentially into the wake and drops in a similar fashion from its initial value outside the cavity. Figure 11.7 illustrates this solution; note that the horizontal axis represents the vertical cut direction. This density drop propagates outwards (towards the left in the figure) as a rarefaction wave that expands at the ion sound speed. The sonic Mach number M_s defines how much and how fast the wake expands along the magnetic field. Due to the relatively low M_s (or equivalently because individual charged particles, most importantly electrons and light ions, have considerable parallel velocities and can encounter an icy moon at a large angle with respect to the bulk velocity direction) Saturnian moon wakes extend significantly north and south of the equatorial plane (Khurana et al. 2007a). For the solar wind–Moon interaction (high M_s) the opening angle of the cavity is only few degrees, while for Saturn's moons Tethys and Rhea it is between 40° and 50° (Roussos et al. 2008a; Simon et al. 2009). This angle will be reduced if there is a plasma temperature anisotropy, as indicated by Sittler et al. (2008) and Wilson et al. (2008).

This treatment agrees with results from numerical simulations and can give a sense of how fast the wake of an insulating moon can refill. For instance, Rhea's wake refills with cold plasma only 7–8 radii downstream, while the lunar wake

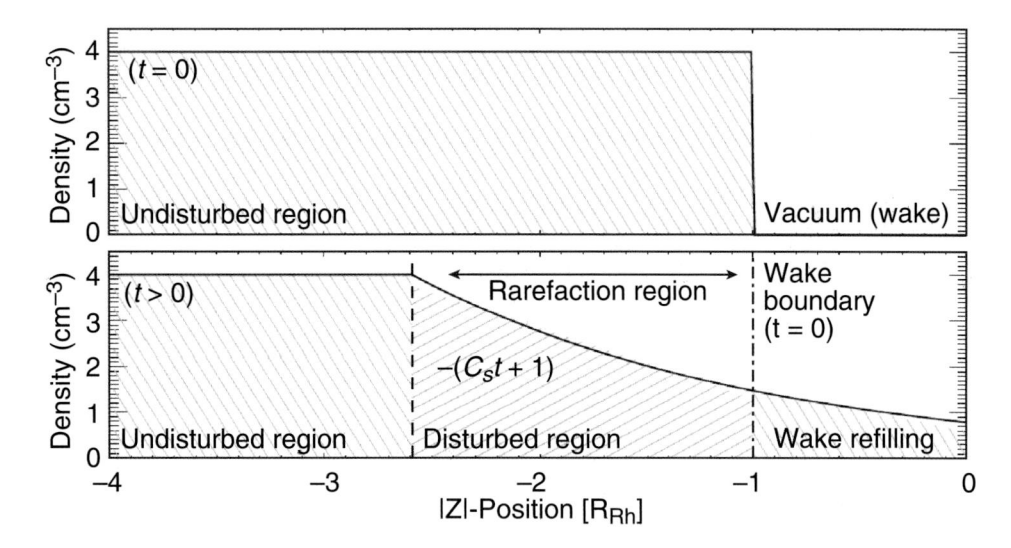

Fig. 11.7 Two cuts through a simple model of the thermal plasma wake for a plasma-absorbing magnetosphere-moon interaction, along the magnetic field direction, for a plasma with isotropic temperature. The rarefaction wave has propagated to $z = -(C_{s}t + 1)$. The dependence of the solution on the sound speed indicates that the wake structure in the plane containing the bulk velocity and the magnetic field is determined by the plasma temperature (Roussos et al. 2008a). Copyright: European Geosciences Union, 2008 (Annales Geophysicae)

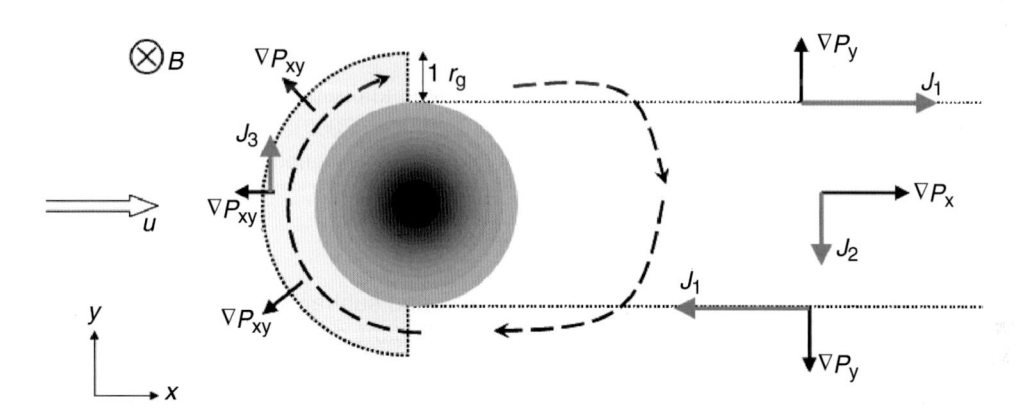

Fig. 11.8 Distribution of currents in the equatorial plane in a plasma-absorbing interaction. The wake boundaries are marked by the thin, dotted lines. These currents (J_1, J_2, J_3) are driven by pressure gradients, present in various directions mainly in the moon's wake, as indicated. Current closure partly occurs in an upstream depletion region that exists within one ion gyroradius from the moon's surface, marked by light gray. Current flow is shown with a thick, dashed line. This picture is valid in a fluid, quasi-neutral description of the plasma

in the solar wind can be observed at much larger distances, >20 lunar radii (Roussos et al. 2008a; Kallio et al. 2005).

Perpendicular to the magnetic field the plasma cavity is restricted to a region that has a width that scales with the moon's diameter. Cross-field motion generally requires the presence of an electric field within the fluid approximation. With such an approximation, electric fields are the net consequence of the divergence of electric currents generated by pressure gradients in the plasma depletion regions (diamagnetic currents Fig. 11.8) and the impedances associated with the closure of those currents, often assumed to flow along magnetic field lines and through Saturn's ionosphere.

In reality, more complex processes are taking place, as the electrons can enter the cavity faster than the sound speed and produce a polarization (or ambipolar) electric field $\mathbf{E} = -1/n_e e \nabla_{\parallel} P_e$, which can modify the fill-in process (Kalio et al. 2005; Maurice et al. 1997). This effect is considered important for the Earth's Moon wake (Birch and Chapman 2001) and can be equally important for Saturn's insulating moons, given the very high ratio of thermal electron to bulk plasma velocity. Finite ion gyro-radius effects also enhance the refilling.

Magnetic Field Response

The response of the magnetic field in the wake of an insulating body has been extensively studied for the case of the Moon's interaction with the solar wind. A common feature identified in such an interaction is the magnetic field enhancement in the plasma depletion region in order to maintain the total (plasma plus magnetic) pressure. For the Saturnian satellites, pressure balance calculations are consistent with the few-nT field enhancements observed by the Cassini magnetometer in the wakes of Tethys and Rhea (Khurana et al. 2007a). On the other hand, Simon et al. (2009) attribute the magnetic field response partly to an enhanced ion current in the wake driven by the ambipolar electric field, as mentioned in the previous paragraph.

The magnetic field enhancement is not limited to the geometrical wake of the moon, but expands north and south along the magnetic field direction over the entire plasma depletion region. This effect was observed by the Cassini magnetometer when Cassini flew by Tethys (Fig. 11.9) slightly south of the moon's geometrical wake (Khurana et al. 2007) and has been successfully reproduced by numerical simulations (Fig. 11.10; Simon et al. 2009; Roussos et al. 2008a).

On either side of the field enhancement region, perpendicular to the magnetic field direction, the magnetic field intensity drops to conserve $\nabla \bullet \mathbf{B} = 0$ (Khurana et al. 2007a; Whang 1969; Whang and Ness 1970). These regions, identified by Khurana et al. (2007a) as "expansion fans," are commonly observed in the lunar wake and have also been detected by Cassini at Tethys and Rhea (Fig. 11.9).

To better understand the magnetic field perturbations, it is useful to consider the current systems that develop in the plasma depletion regions. For instance, the J_1 currents in Fig. 11.8 along the side of the wake lead to a "double loop" perturbation of B_y and B_z (perpendicular to the xy plane), as seen in Fig. 11.10. This perturbation, added to the background field, reproduces the observed magnetic field enhancement and the expansion fan features discussed earlier. In the same way, the J_2 currents in Fig. 11.8 could lead to perturbations of the B_x and B_z components in the wake. A notable B_x perturbation has been detected at Rhea's wake (Khurana et al. 2007a) but it is uncertain if it can be attributed to such currents or to other, unidentified modes of that moon's interaction.

Energetic Particle Response

Most icy moons behave as perfect absorbers of energetic particles. A striking difference between the thermal plasma and energetic particle cavities is that the latter can be identified at large longitudinal separations from the absorbing body, close to the body's L-shell, while thermal plasma wakes tend to disappear only a few moon radii downstream (Roussos et al. 2005). The high energy of the energetic particles (mainly contained in their gyration around magnetic

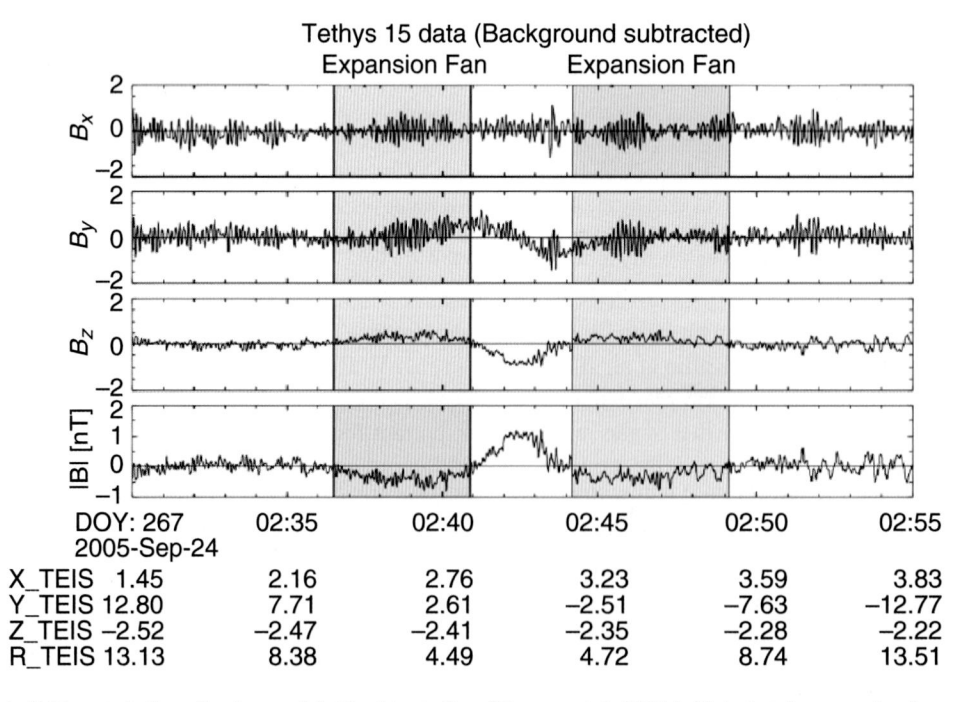

Fig. 11.9 Magnetic field perturbations (background field subtracted) observed during the Tethys flyby by Cassini. The locations of expansion fans (rarefaction regions) that bring field and plasma into the wake are marked. The total $|B|$ enhancement in the central wake is also visible (Khurana et al. 2007a). Note that the expansion fans are also referred to as "rarefaction regions," but they should not be confused with the definition of the rarefaction region by Samir et al. (1983), as it is illustrated in Fig. 11.8. Copyright: Elsevier, 2007 (Icarus)

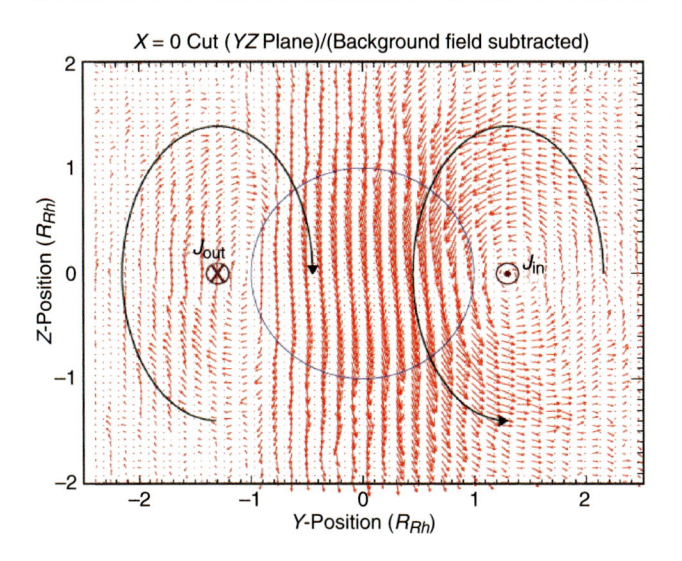

Fig. 11.10 Simulated magnetic field perturbations in a plane perpendicular to the plasma cavity of Rhea (yz plane), just behind Rhea. The direction of the plasma flow (and the wake) points out of the page. The background field has been subtracted to enhance the visibility of the "double loop" perturbation in the magnetic field induced by the pressure gradient driven currents in the wake. The blue circle indicates the location of Rhea's geometrical shadow. The peak amplitude of the residual B_y in this simulation is about 1.5 nT. The closed magnetic field loops and the inferred current flow vectors are also sketched (Roussos et al. 2008a). Copyright: European Geosciences Union, 2008 (Annales Geophysicae)

field lines) largely shapes their drift trajectory in the magnetosphere, by the action of the magnetic gradient and curvature drifts.

The local energetic ion depletions are barely observable in the Cassini data because of large gyro-radii and slow bounce effects that cause large fractions of the ions to escape absorption. Energetic electron depletions on the other hand are sharper and deeper, since these electrons have small gyro-radii (less than 1% of a moon's radius), move rapidly along the field lines and can gradient and curvature drift opposite to corotation. As a result, they cross the equatorial plane at least once while they slowly convect across a moon's diameter and cannot avoid moon encounters. As explained in Section 11.2.4.1, electrons above the Keplerian resonant energy form wakes in the upstream region that have already been observed by Cassini (e.g., Jones et al. 2006a). Energetic particle wake refilling, and the effects of dynamic injections on wake structures is discussed in Section 11.3.5).

Special Cases

Multi-instrument observations during two close Cassini flybys of Saturn's moon Rhea have shown that it is a plasma-absorbing, electromagnetically insulating moon (Jones et al. 2008; Khurana et al. 2007a; Anderson and Schubert 2007).

A deviation from the expected plasma-absorbing profile was identified in energetic electrons. On either side of the expected electron wake, an additional depletion region exists with the width of the moon's Hill sphere (the region in which Rhea's gravity dominates that of Saturn). It has been inferred that Rhea is surrounded by a tenuous disk that contains large grains (millimeter to centimeter size; Jones et al. 2008). The enhancement of the O_2^+ around Rhea's L-shell could also be attributed to the irradiation of this disk by energetic plasma (Martens et al. 2008).

11.2.4.3 Mass Loading Interactions at Saturn

Data from the Cassini spacecraft led to the discovery of plumes near the south pole of Enceladus, the only known moon apart from Titan that definitely mass loads Saturn's magnetosphere. These plumes lift gas and dust high above the surface of the satellite (Dougherty et al. 2006; Porco et al. 2006; Waite et al. 2006). Neutral gas may either fall back to the surface, become bound by Enceladus' gravity, or escape the influence of the satellite completely. The macroscopic interactions of this gas with the magnetospheric environment, including how it is distributed, ionized, and energized are also addressed in Sections 11.2.3 and 11.4.2.

Thermal Plasma and Magnetic Field Response

Mass loading interactions within a magnetospheric environment have been mainly studied in the Jovian system (e.g., Pontius and Hill 1982; Bagenal 1997; Hill and Pontius 1998; Kivelson et al. 1997b, 2004). A principal difference between plasma absorbing and mass loading interactions results from the fact that both upstream and downstream from the moon, mass addition dominates mass loss, and current directions are reversed compared to what is shown in Fig. 11.8. In the plane perpendicular to the magnetic field (the plane used in Fig. 11.8) current generation and closure is thought to occur across the mass loading region where ion pickup occurs, along Alfvén wings that are nearly but not exactly parallel to the magnetic field lines, and then through the ionosphere. Depending on how conductive and extended these two regions are, the impinging plasma flow will slow down and will be diverted around the obstacle. The resulting magnetic field perturbations lead to a magnetic field pile-up upstream of the obstacle and a field intensity dropout downstream. The field configuration for a symmetric obstacle around the moon, in two planes containing the magnetic field, is illustrated in Fig. 11.11.

Such draping signatures were observed during the first three close flybys of Enceladus by Cassini. However, it was noted in two of the three flybys that south of Enceladus, the

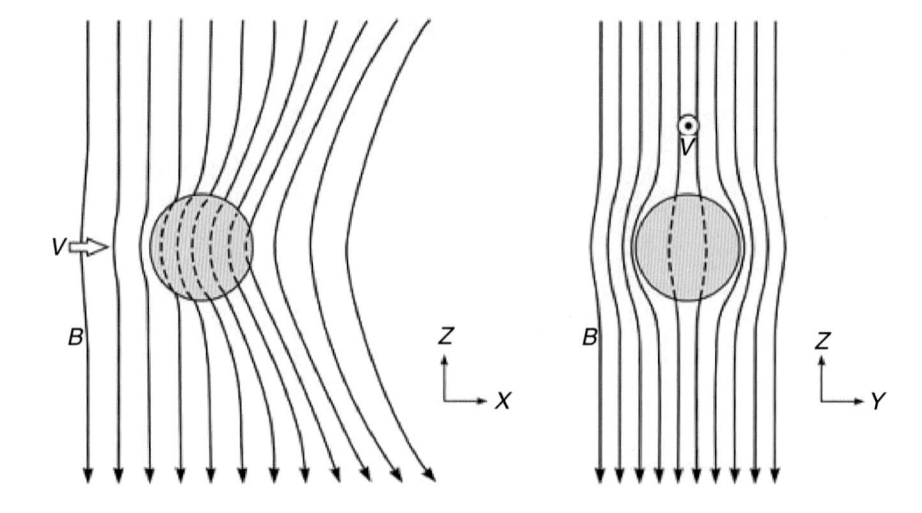

Fig. 11.11 The draping pattern of the magnetic field around a conducting obstacle in the xz and yz planes (Dougherty et al. 2006). Copyright: American Association for the Advancement of Science, 2006 (Science Magazine)

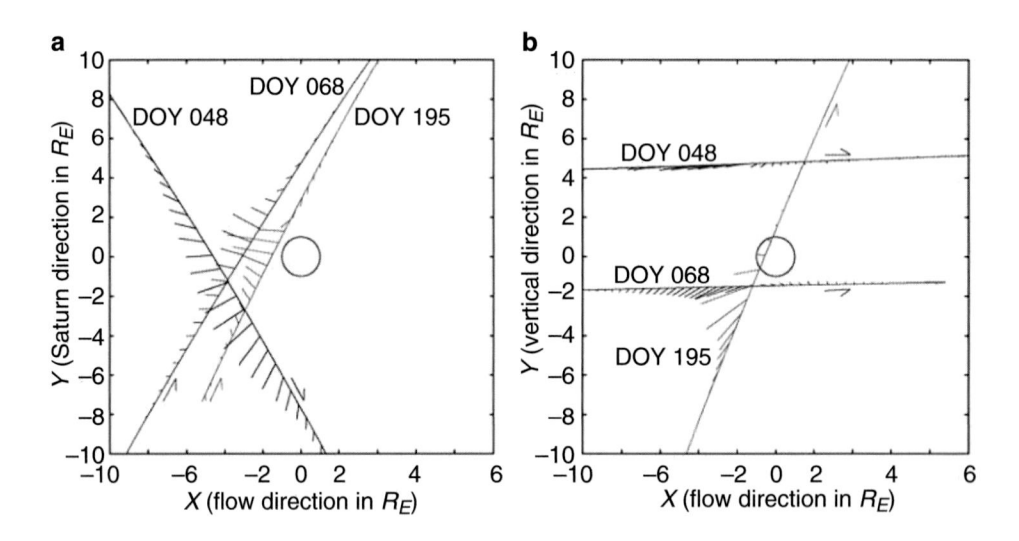

Fig. 11.12 Magnetic field residual vectors observed by the Cassini magnetometer in the first three flybys of Enceladus in the xy and xz planes (Dougherty et al. 2006). Copyright: American Association for the Advancement of Science, 2006 (Science Magazine)

B_x perturbation had the opposite sign to that expected for an Alfvén wing structure centered at the moon; this indicated that the effective obstacle's center is south of Enceladus, within its cryovolcanic plume (Fig. 11.12). The spatial extent of the magnetic field perturbations also indicated a variable and extended interaction region, with a diameter of at least $6R_E$, centered $2R_E$ below and $1R_E$ downstream of Enceladus (Khurana et al. 2007b; Dougherty et al. 2006).

A more extended interaction region was observed by CAPS, the cold plasma instrument of Cassini (Tokar et al. 2006). The slowing and/or the deflection of the bulk plasma flow were observed even 30 moon radii away from Enceladus, as shown in Fig. 11.13. Based on assumptions of the ionospheric conductance at Saturn, modeling of the flow deflection results in an estimation of $\sim$100 k/s of water molecules (Tokar et al. 2006; Pontius and Hill 2006). These neutrals, which are then ionized by charge exchange and to smaller degree by electron impact, apparently mass load the magnetosphere and result in the observed interaction features. The very different scale sizes for the magnetic field and plasma-flow perturbations requires explanation. Overall, data from the Cassini magnetometer and from CAPS indicate some mass loading of the plasma flow near the moon itself, suggesting that some fraction of the neutrals escaping Enceladus are ionized locally (Burger et al. 2007). However, it is likely the case that most neutrals travel a wide range of distances before being ionized, as described Section 11.2.3.

More detailed results can be obtained by three-dimensional (3D) MHD or hybrid simulations of the interaction. For instance, using a 3D hybrid code Kriegel et al. (2008) reproduced several of the features discussed earlier, such as the extended region of the pickup ions, as well as the pile-up, the draping and the dropout of the magnetic field upstream, around and downstream of the plume's center, respectively. They have also found that plasma absorption at the surface of Enceladus does not leave a noteworthy imprint in the magnetic field topology. Saur et al. (2008), comparing MHD simulation results with the Cassini magnetometer data,

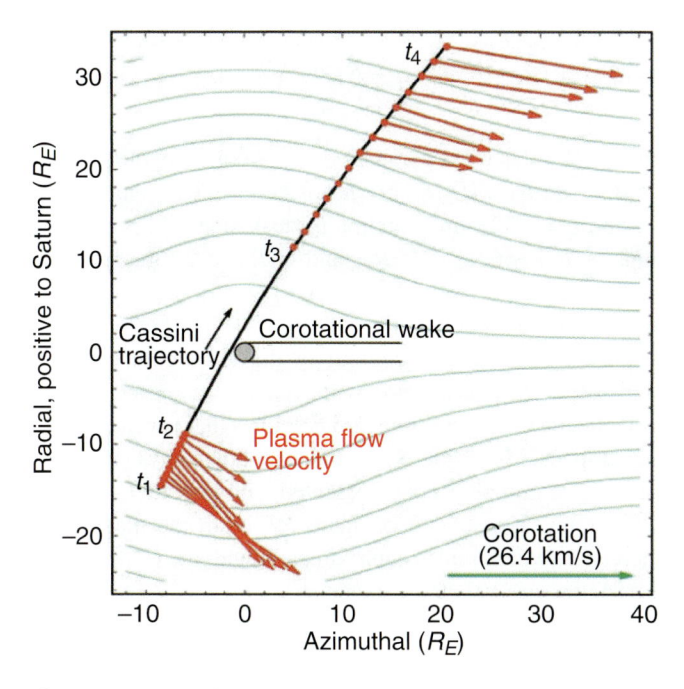

Fig. 11.13 Plasma flow vectors in the vicinity of Enceladus observed by the Cassini CAPS instrument during a close flyby on day 195 of 2005, by the CAPS instrument. Model streamlines are shown in green (Tokar et al. 2006). Copyright: American Association for the Advancement of Science, 2006 (Science Magazine)

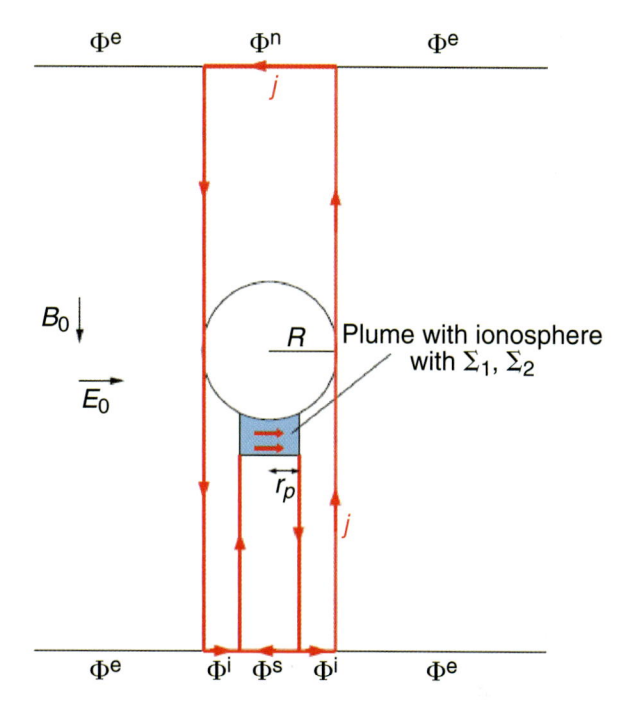

Fig. 11.14 Sketch of the electric current distribution for an analytic solution of an asymmetric plasma interaction due to excess neutral and ionized gas in the south polar region of Enceladus (Saur et al. 2007). Copyright: American Geophysical Union, 2007

have demonstrated that the total plume content of Enceladus varied up to a factor of 8, between the first three close flybys by Cassini in 2005.

The asymmetric distribution of neutral and ionized gas in the effective obstacle formed by Enceladus raises the interesting question of how this moon's hemispheres are electromagnetically linked to each other as well as to Saturn's ionosphere. An analytical approach to this problem indicates that the two hemispheres are linked through currents flowing in the ionized material in the moon's south polar region (Saur et al. 2007). Figure 11.14 shows a schematic illustration for an analytic solution of this problem. This model predicts sharp magnetic field discontinuities along the flux tube that touches Enceladus' effective obstacle (Saur et al. 2007).

Energetic Particle Response

The key question here is to what extent is the simple particle absorbing model for energetic particle interactions, as applied to a non-conducting moon, modified by the observed mass loading and other activity at Enceladus? Signatures of the Enceladus interaction with energetic particles are visible in kiloelectron-volt and million electron-volts electrons, as well as in million electron-volts ions (Jones et al. 2006a). Apart from the typical energetic electron cavities formed by the moon itself, observed as microsignatures (Sections 11.2.4.2 and 11.3.5), extended depletion regions sometimes appear in the kiloelectron-volt range. The interaction features show variable profiles from orbit to orbit and deviate from the predicted evolution that could be attributed radial diffusion alone (Jones et al. 2006a). The first three close flybys have also revealed slightly broader cavities in million electron-volts electrons, in addition to small displacements of these depletions from the moon's L-shell (Jones et al. 2006a; Paranicas et al. 2005a).

Candidate processes that have been proposed for all these variable responses, foreshadowed by studies at Jupiter (Paranicas et al. 2000; Goldstein and Ip 1983), are absorption by dust, pitch angle scattering or complex energetic electron drifts due to the disturbed plasma flow and the highly distorted magnetic field lines near the moon (Jones et al. 2006a; Paranicas et al. 2005a). None of these processes, however, have been quantitatively assessed yet.

11.3 Transport

Here we address the rotational versus solar-wind drivers of transport within the Saturnian magnetosphere, magnetosphere–ionosphere coupling as the moderator of plasma transport, the centrifugal interchange instability as a source of plasma dynamical features and associated transport, sources and processes of rotational lag within the magnetosphere, and the radial diffusion formalism for characterizing transport.

11.3.1 Rotational versus Solar-Wind Drivers

Brice and Ioannidis (1970) elucidated a simple but quantitative way to distinguish between a solar-wind–driven magnetosphere and a rotationally driven one. Close to the planet the plasma tends to corotate with azimuthal speed

$$v_\phi \sim \Omega r \qquad (11.1)$$

in the equatorial plane, where Ω is the planetary rotation rate and r is the planet-centered distance. Farther out, the plasma tends to flow Sunward (also in the equatorial plane) with a speed

$$v_x \sim \frac{\varepsilon v_{sw} B_{sw}}{B_p} \left(\frac{r}{R_P}\right)^3 \qquad (11.2)$$

where v_{sw} and B_{sw} are the solar-wind speed and magnetic field strength, respectively, B_P is the equatorial magnetic field strength at the planetary surface, R_P is the planetary radius, and ε is the efficiency of magnetic reconnection at the dayside magnetopause (Section 11.4.1 below), which is difficult to predict but has an empirically determined value at Earth of $\varepsilon \sim 0.1$–0.2 (e.g., Boyle et al. 1997). (A spin-aligned dipole magnetic field is assumed for simplicity.) These two velocities cancel on the dusk meridian at a distance R_{PP} (the "plasmapause radius"), where

$$\frac{R_{PP}}{R_P} \sim \left(\frac{\Omega R_P B_P}{\varepsilon v_{sw} B_{sw}}\right)^{1/2}. \qquad (11.3)$$

For typical solar-wind parameters ($v_{sw} \sim$ constant ~ 400 km/s and $B_{sw} \sim (5$ nT$)/R_H$, where R_H is heliocentric distance in AU, and $\varepsilon = 0.2$), this maximum (dusk side) plasmapause radius is $R_{PP} \sim 6 R_E$, $260 R_J$, and $68 R_S$ at Earth, Jupiter, and Saturn, respectively. Figure 11.15 shows flowlines obtained from the superposition of Eq. 11.1 and 11.2. Inside the plasmapause (heavy contour in the left figure panel), magnetospheric plasma circles the planet on closed flowlines; outside, it follows Sunward flowlines that approach the magnetopause. Near the magnetopause, other physical processes come into play, not represented with Eqs. 11.1–11.3.

The relevant test is to compare R_{PP} with the magnetosphere size, which scales approximately with the Chapman–Ferraro pressure-balance distance R_{MP},

$$\frac{R_{MP}}{R_P} \sim \left(\frac{2 B_P^2}{\mu_0 \rho_{sw} v_{sw}^2}\right)^{1/6}. \qquad (11.4)$$

For typical solar-wind parameters ($\rho_{sw} \sim (5/\text{cm}^3)/R_H^2$, $v_{sw} \sim 400$ km/s), this ratio has values ~ 10, 42, and 19 for Earth, Jupiter, and Saturn, respectively. (Jupiter's magnetosphere is significantly larger than this prediction because of its massive ring current that increases the effective dipole moment of the planet, but we will ignore this complication for the purpose of this comparison.) The ratio of Eq. 11.3 to Eq. 11.4 can be written as

$$\frac{R_{PP}}{R_{MP}} \sim \left(\frac{\Omega R_{MP}}{\sqrt{2} \varepsilon v_{A,sw}}\right)^{1/2}. \qquad (11.5)$$

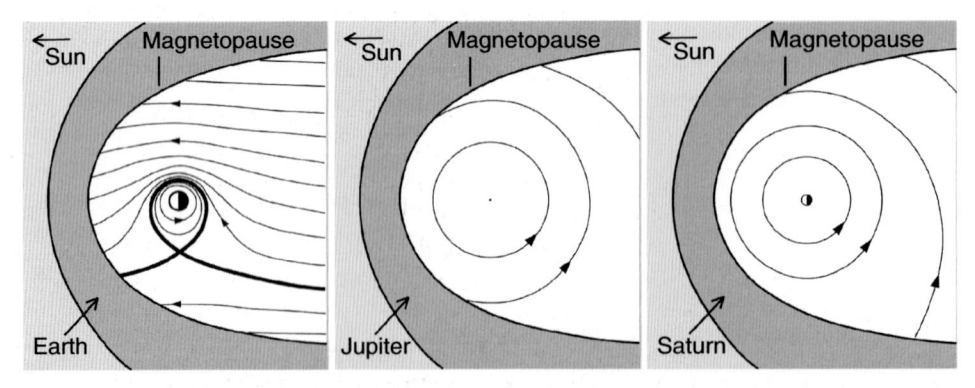

Fig. 11.15 Equatorial cross sections of three planetary magnetospheres scaled to the sizes of the respective magnetospheres. Planets are drawn to the correct scale relative to the magnetosphere, and the three flow patterns are identical, but sampled at different length scales. The flow pattern is the superposition of corotation and the Sunward flow in the equatorial plane induced by the solar-wind interaction, corresponding to a uniform dawn-dusk electric field. The concept is due to Brice and Ioannidis (1970). The flow patterns are intended to be schematic, not literally true in every detail, especially in the near vicinity of the dayside magnetopause

This is a convenient form because the solar-wind Alfvén speed $v_{A,sw}$ (typically ~ 50 km/s) is essentially independent of heliocentric distance in the steady-state Parker solar-wind model.

The ratio (Eq. 11.5) for typical solar-wind conditions and $\varepsilon = 0.2$, is $R_{PP}/R_{MP} \sim 0.6, 6$, and 4 for Earth, Jupiter, and Saturn, respectively, as illustrated in Fig. 11.15. Thus it was expected, prior to any spacecraft measurements at the giant planets, that while Earth's magnetosphere is largely solar-wind–driven ($R_{PP}/R_{MP} < 1$), those of Jupiter and Saturn should be rotationally driven ($R_{PP}/R_{MP} > 1$). Pre-Cassini spacecraft measurements (especially those of Voyagers 1 and 2) confirmed this expectation in the sense that the magnetospheric plasma flow is primarily rotational (albeit not fully corotational in magnitude) out to distances $\sim R_{MP}$ both at Jupiter (Belcher 1983 and references therein) and at Saturn (Eviatar and Richardson 1986). CAPS measurements (McAndrews et al. 2008a) show that this remains true out to distances $\sim 2.5 R_{MP}$ in Saturn's magnetotail (see Figure 10.5). Thus, while it is true that Saturn's magnetosphere is intermediate between those of Earth and Jupiter in terms of size (whether measured by R_P, R_{PP}, or R_{MP}), it is not intermediate in any meaningful dynamical sense. It is clearly rotationally driven like Jupiter's, not solar-wind–driven like Earth's.

While some theoretical estimates of the influence of the solar wind on Saturn's magnetosphere have varied (e.g., Cowley et al. 2004), Jackman et al. (2004) and Badman and Cowley (2007) have confirmed earlier expectations for rotational dominance by utilizing solar-wind statistical properties from Cassini cruise-phase observation, both for typical and for average solar-wind conditions. However, these authors suggest that solar-wind–driven ("Dungey-cycle") flow may compete with rotationally driven ("Vasyliūnas-cycle") flow in a narrow boundary layer inside the dawn-side magnetopause during strong solar-wind compression events.

Apart from the rotational nature of the observed plasma flow itself, Cassini has provided other clear evidence of a rotation-driven magnetosphere. Perhaps the most compelling evidence is the wide variety of magnetospheric phenomena that exhibit clear rotational modulation, in spite of the lack of a perceptible dipole tilt or higher-order magnetic anomaly. The spin-modulated phenomena include the Saturn Kilometric Radiation (SKR) (Kurth et al. 2007, 2008), magnetospheric field perturbations (Giampieri et al. 2006), inner magnetospheric electron density (Gurnett et al. 2007), outer-magnetospheric energetic electron flux (Carbary et al. 2007a, b), outer-magnetospheric plasma ion flux (Burch et al. 2008), and ENA bursts (Paranicas et al. 2005b). These spin modulations are discussed in detail in Chapter 10 and will not be discussed further here. In addition to the spin modulations, Cassini has also provided ample evidence of radial transport by the centrifugally driven magnetospheric interchange instability (Section 11.3.4 below). Finally, it is worth noting that Vasyliūnas (2008) has shown that the internal plasma source provided by Enceladus (Section 11.2.2), although smaller in absolute numbers than the Io plasma source for Jupiter's magnetosphere, may be more importantly dynamically in the context of Saturn's smaller magnetosphere.

Saturn's magnetosphere, like those of Earth and Jupiter, exhibits dramatic, large-scale injection events involving magnetic reconfiguration and particle acceleration in the near magnetotail (Chapter 10). At Earth, the nightside injections are clearly powered by the solar-wind interaction as part of a large-scale magnetospheric configurational instability known as a substorm (see Section 11.4.1 below and references therein). Confusion sometimes arises when such injections are described as substorms at Jupiter (e.g., Khurana et al. 2004; Krupp et al. 2004) and at Saturn (Section 11.4.1 below and references therein), although their power source is clearly internal rather than external. Compounding this confusion is the fact that nightside injections at Saturn are correlated with changes in the solar wind, and with the SKR that is, in turn, correlated with changes in the solar wind. Similar solar-wind correlations are evident at Jupiter. At Saturn, as at Jupiter, it seems likely that variations in the solar wind can trigger a nightside injection event whose energy is derived from rotation and that would have eventually occurred even in the absence of external changes.

11.3.2 Magnetosphere–Ionosphere Coupling

Although they are powered differently, the magnetospheres of Earth, Jupiter, and Saturn share many processes in common, the most obvious being magnetosphere–ionosphere (M-I) coupling. M-I coupling processes are of two general types, mass transfer and electrodynamic coupling.

Mass transfer from ionosphere to magnetosphere appears to be less important at Saturn than at Earth (see Section 11.2.1). Mass transfer in the other direction, from magnetosphere to ionosphere, involves field-aligned precipitation of magnetospheric particles into the atmosphere. This process is probably unimportant for the bulk of magnetospheric ions because of their centrifugal confinement near the equatorial plane (e.g., Hill and Michel 1976; McAndrews et al. 2008a), but it may be a significant loss process for energetic electrons in the inner magnetosphere (e.g., Rymer et al. 2007a). (For energetic particles, the centrifugal force is less important.) In any case, the mere presence of an atmospheric loss cone in the energetic electron pitch-angle distribution is an important factor in the generation of plasma waves and SKR (see Section 11.4.4 below). The resulting precipitation of energetic electrons into the atmosphere can substantially increase the ionospheric conductivity (Millward et al. 2002) and thus play a key role in the electrodynamic M-I coupling described below.

Electrodynamic M-I coupling is just as important at Saturn as it is at Earth and Jupiter. This coupling occurs whenever the motion of magnetospheric plasma perpendicular to the magnetic field **B**, when mapped along **B**, differs from that of the conducting atmosphere to which it is connected. In many (but not all) models, the neutral atmosphere at ionospheric altitudes is assumed to corotate with the planet, while the magnetosphere above the ionosphere is assumed to move in accordance with the ideal MHD approximation $\mathbf{E} + \mathbf{v} \times \mathbf{B} = 0$ (also called the "frozen-in-flux" approximation). Thus, electrodynamic M-I coupling occurs whenever the magnetospheric plasma moves in a way that differs from rigid corotation. Our discussion here assumes steady-state M-I coupling, which implies timescales longer than both the Alfvén-wave transit time along the flux tube, and the "plasma acceleration time" defined and discussed by Vasyliūnas (1994). The steady-state assumption becomes progressively worse as one moves outward in the magnetosphere. It is probably adequate in the inner magnetosphere of Saturn ($r < \sim 12 R_S$), although a quantitative assessment of this criterion remains to be done.

Steady-state electrodynamic M-I coupling involves (by assumption) a closed electric current system that includes three elements: (1) a perpendicular Pedersen conduction current in the ionosphere whose $\mathbf{J} \times \mathbf{B}$ force is balanced by the ion–neutral collisional drag force; (2) a perpendicular plasma drift current in the magnetosphere whose $\mathbf{J} \times \mathbf{B}$ force is balanced by the sum of the pressure-gradient force $-\nabla p$, the inertia of the rotating plasma itself under the influence of local plasma production and radial plasma transport, and ion–neutral collisional drag; and (3) Birkeland (magnetic-field-aligned) currents that connect the two perpendicular currents to maintain $\nabla \bullet \mathbf{J} = 0$, as required by the steady-state assumption. The Birkeland ($||\mathbf{B}$) currents are usually assumed to flow with zero resistance according to the ideal MHD assumption. A magnetic-field-aligned electric field can, in principle, reduce the strength of the coupling by introducing slippage between high and low altitudes on a given field line (Ray et al. 2008a, b), but can also increase the strength of the coupling because of the increased ionospheric conductivity resulting from the precipitation of accelerated electrons. The net effect remains to be worked out, and probably depends on geometry. The closed but complicated M-I coupling current system can be conveniently separated into its poloidal and toroidal components, as illustrated in Fig. 11.16a and b, respectively. These figures are further discussed in Sections 11.3.3 and 11.3.4 below, respectively.

11.3.3 Corotation Lag

The poloidal ionospheric Pedersen current in Fig. 11.16a extracts angular momentum from the atmosphere, and this angular momentum is delivered to the magnetosphere by the connecting Birkeland currents. The closing (perpendicular) currents in the magnetosphere accelerate plasma there toward rigid corotation with the underlying atmosphere. If the magnetospheric plasma distribution were completely static, it could persist in a state of corotation without this current system. However, azimuthal plasma acceleration is usually necessary for one or more of three reasons: (1) new plasma is being produced locally and must be accelerated up to the local background (partial) corotation speed from its former (and usually smaller) Kepler orbital speed; (2) existing plasma is moving radially outward or inward and must be accelerated or decelerated azimuthally to match the local background (partial) corotation speed, which differs

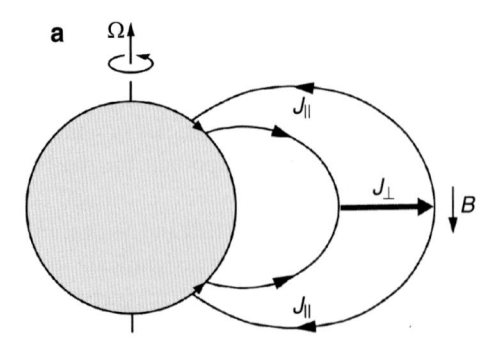

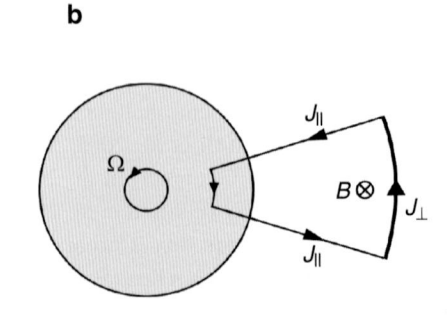

Fig. 11.16 Birkeland current systems in Saturn's magnetosphere, projected into (**a**) a magnetic meridian plane and (**b**) into the magnetic equatorial plane. The poloidal component (**a**) is associated with enforcement of (partial) corotation as described in Section 11.3.3; the toroidal component (**b**) is associated with interchange motions as described in Section 11.3.4. The size of the planet is greatly exaggerated for the sake of clarity

from the radial dependence $(\Omega \propto 1/r^2)$ that would result from angular-momentum conservation of the radially moving plasma; and/or (3) the partially corotating plasma suffers collisional drag with neutral gas orbiting at the Kepler speed, usually smaller than the partial corotation speed. As far as the angular-momentum budget is concerned, local production includes both those processes that increase the plasma density and contribute to subsequent outward transport (ionization of neutrals by UV or electron impact) and those that do not (charge exchange between ions and neutrals; see Section 11.2.3). At Saturn, the effect of ion–neutral collisional drag may become comparable to that of charge exchange (Saur et al. 2004).

These three effects (local plasma production, radial plasma transport, and ion–neutral collisional drag) are easily separated observationally at Jupiter because they occur in different places, with radial transport being the dominant effect in the region $r > \sim 10 R_J$ (Hill 1979) and the other two (mathematically equivalent) effects being dominant in the region of the Io plasma torus, $r \sim 6$–$7 R_J$ (Pontius and Hill 1982). They are not so easily separated at Saturn, where the three effects overlap significantly in space. This mathematically inconvenient fact was implicit in Voyager results (Eviatar and Richardson 1986) and is made painfully explicit in Cassini results (Pontius et al. 2007; Wilson et al. 2008; Section 11.2.3 above). In principle, the observed radial variation of the azimuthal velocity component (see Fig. 11.17, reproduced from Wilson et al. 2008) provides information about all three effects, when coupled with a physical chemistry model (Section 11.2.3 above) that distinguishes among the three. Extracting this information is a work in progress at this writing, but preliminary modeling results (Pontius et al. 2007) suggest that the observed $v_\phi(r)$ profile (Fig. 11.17) is consistent with expectations based on the physical chemistry models described in Section 11.2.3 above.

For the sake of illustration, let us neglect for the moment the effects of local plasma production and of ion–neutral collisional drag, and focus instead on the effect of outward plasma transport. If the outward transport rate is S (kg/s), the corotation lag is expected to become significant at a characteristic distance L_0 (normalized to R_S) with

$$L_0^4 = \pi \Sigma B_S^2 R_S^2 / S \tag{11.6}$$

(Hill 1979), where Σ is Saturn's ionospheric Pedersen conductance and B_S is its surface equatorial magnetic field strength. The bad news is that neither Σ nor S are known even to order-of-magnitude accuracy for Saturn; the good news is that L_0 depends only on the $1/4$ power of their ratio, and plausible values of Σ (perhaps $\sim 0.1 S$) and S (perhaps ~ 100 kg/s) are consistent with a value $L_0 \sim 8$ that would, in turn, be roughly consistent with the CAPS v_ϕ data of Fig. 11.17. However, the upturn in $v_\phi / \Omega r$ beyond $L \sim 8$ (top panel, Fig. 11.17) is not predicted by the plasma-mass-conserving radial transport theory (Hill 1979) and may reflect either an unknown loss process during the outward transport (a decrease of S with r) or, perhaps more likely, may indicate that much of the corotation lag within $L \sim 8$ results from local plasma production and/or ion–neutral collisional drag rather than from outward plasma transport. The latter option appears to be consistent with the physical chemistry models discussed in Section 11.2.3 above, but as already noted, the analysis of these data, and similar data to come in the future, is still a work in progress.

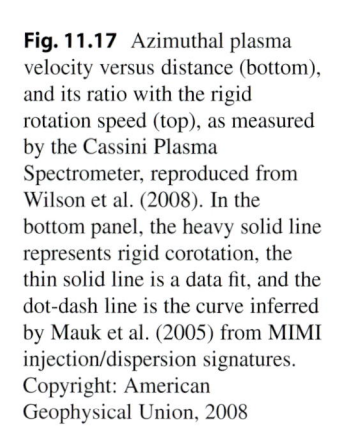

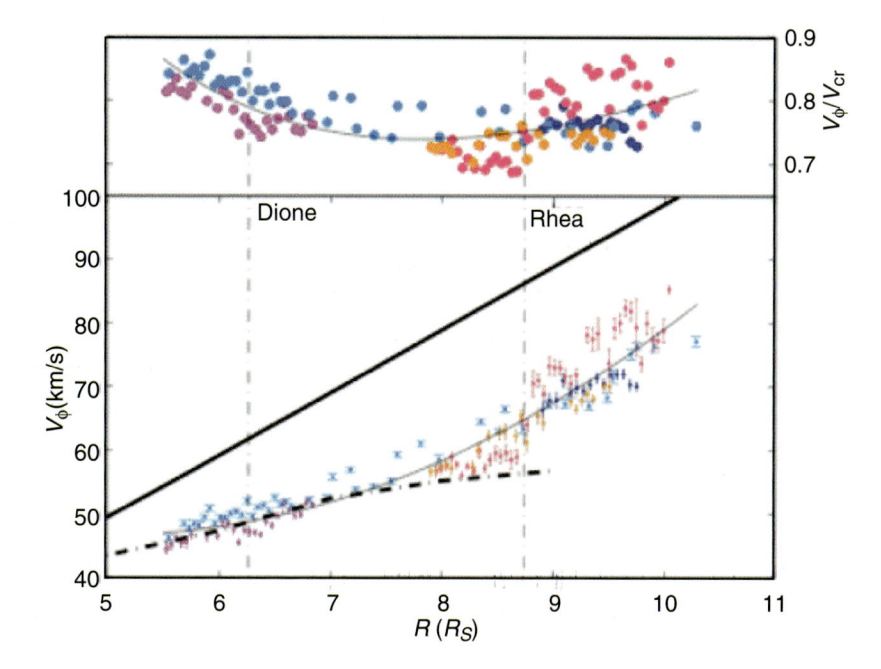

Fig. 11.17 Azimuthal plasma velocity versus distance (bottom), and its ratio with the rigid rotation speed (top), as measured by the Cassini Plasma Spectrometer, reproduced from Wilson et al. (2008). In the bottom panel, the heavy solid line represents rigid corotation, the thin solid line is a data fit, and the dot-dash line is the curve inferred by Mauk et al. (2005) from MIMI injection/dispersion signatures. Copyright: American Geophysical Union, 2008

11.3.4 Centrifugal Interchange Instability

Figure 11.16b shows the toroidal component of the M-I coupling current system that is associated with a longitude variation of the flux-tube plasma mass content and thus of the radial speed. Such longitude variations are central to the magnetospheric interchange instability that, by definition (Gold 1959), involves a radial interchange of magnetic flux tubes without significant perturbations to the background magnetic field geometry (thus implying $\beta \ll 1$ where β is the ratio of plasma to magnetic-field pressures). This condition is well met in the inner magnetospheres of Jupiter and Saturn, corresponding (perhaps coincidentally) to roughly $L < \sim 12$ for both Jupiter (Mauk et al. 2004) and Saturn (Sergis et al. 2007).

Interchange motions may be either stable or unstable, depending on whether they increase or decrease the total potential energy of the system. The potential energy has three parts. One part is the gravitational potential, which is negligible for Saturn's magnetospheric plasma outside the main rings. A second part is the centrifugal potential, equal to $-1/2\Omega^2 r^2$ per unit mass, whose gradient is the centrifugal force of corotation. With only this second contribution to the potential energy, an axisymmetric magnetosphere is centrifugally unstable (outside the spin-synchronous orbit radius, $1.87 R_S$ at Saturn, $2.11 R_J$ at Jupiter) if and only if

$$\frac{\partial \eta}{\partial L} < 0 \text{ (unstable) with } \eta = \int \frac{\rho ds}{B} \quad (11.7)$$

(e.g., Siscoe and Summers 1981; Huang and Hill 1991), where ρ is plasma mass density and the integral is along an entire field line; thus η is the mass content per unit magnetic flux. Interchange motions conserve magnetic flux, by definition, and also conserve η in the absence of plasma sources and sinks. Thus a rotationally driven magnetosphere is centrifugally unstable outside the radial location of the peak of its plasma source(s), measured in units of $d\eta/dt$. For both Jupiter and Saturn (again coincidentally) this means for $L > \sim 6$ (Fig. 11.6). At Jupiter, $L = 6$ is near the orbit of Io, the main source of both neutral gas and plasma; at Saturn, the main source of neutral gas is probably near the Enceladus orbit at $L \approx 4$, but the main plasma source is farther out, $L \sim 6$, because that is where the ambient electrons become hot enough to produce fast collisional ionization (see Section 11.2.3 above).

The third part of the potential energy is the internal (thermal) energy density of the plasma. This term gives a different instability criterion

$$\frac{\partial}{\partial L}(pV^\gamma) < 0 \text{ (unstable) with } V = \int \frac{ds}{B} \quad (11.8)$$

(e.g., Gold 1959; Wolf 1983) where p is plasma pressure, γ is the adiabatic index ($= 5/3$ for an isotropic ideal gas), and V is the flux-tube volume per unit magnetic flux. Both the centrifugal instability criterion (Eq. 11.7) and the pressure-driven instability criterion (Eq. 11.8) predict interchange instability outside the distance of the main plasma source and interchange stability inside. This criterion is, however, complicated by the presence of hot ions, with smaller density but larger temperature, that have their source in the outer magnetosphere. This hot plasma from an exterior source is stable against interchange. (The latter population is probably the remnant of the former, following loss of most of the plasma flux-tube content and heating of the remainder; Chapter 10.). Thus, a large fraction of the volume of both magnetospheres (Jupiter and Saturn) is evidently a battleground between two competing influences, the interchange instability of the internally generated plasma and the interchange stability of the externally generated plasma. This competition was first elucidated by Siscoe et al. (1981), who named it "ring current impoundment" because it offered a mechanism to impound (and hence allow for the creation of) a radially confined Io plasma torus.

The winner of this competition is evidently the interchange instability of the internally generated plasma, which must, in any case, somehow find its way out of the system at the same rate as it is produced; other loss processes such as recombination or precipitation into the atmosphere are much too slow at the observed densities. The interchange process evidently takes the form of alternating azimuthal sectors of outflowing cool, dense plasma from the internal source(s) and inflowing hot, tenuous plasma from the outer magnetosphere. Figure 11.16b illustrates the topology of the M-I coupling current associated with a single over-dense sector flowing outward through a uniform background. The perpendicular (to **B**) part of the magnetospheric current is a combination of gradient-curvature drift current and magnetization currents, dominant by definition for hot particles, and centrifugal drift current, dominant by definition for cold particles. Both current contributions are in the same (prograde) direction. The diverging part of this magnetospheric current is closed by the ionospheric Pedersen current, producing an electric field that is consistent (satisfying $\mathbf{E} + \mathbf{v} \times \mathbf{B} = 0$) with outflow of over-dense sectors and inflow of under-dense sectors. (For an under-dense sector the direction of field-aligned current flow is reversed from that of Fig. 11.16b.) Saturn's inner magnetosphere ($5 < \sim L < \sim 12$) is evidently populated by a collection of alternating inflow and outflow sectors, subject to the steady-state flux conservation requirement $\oint \mathbf{E} \cdot d\mathbf{l} = 0$.

The most dramatic Cassini evidence for this process is the multitude of "injection/dispersion" signatures that are routinely observed in the inner magnetosphere (André et al.

Fig. 11.18 CAPS data (left panel) and interpretive sketch (right panel) from Hill et al. (2005) illustrating the ever-present injection/dispersion structures in Saturn's inner magnetosphere. Copyright: American Geophysical Union, 2005

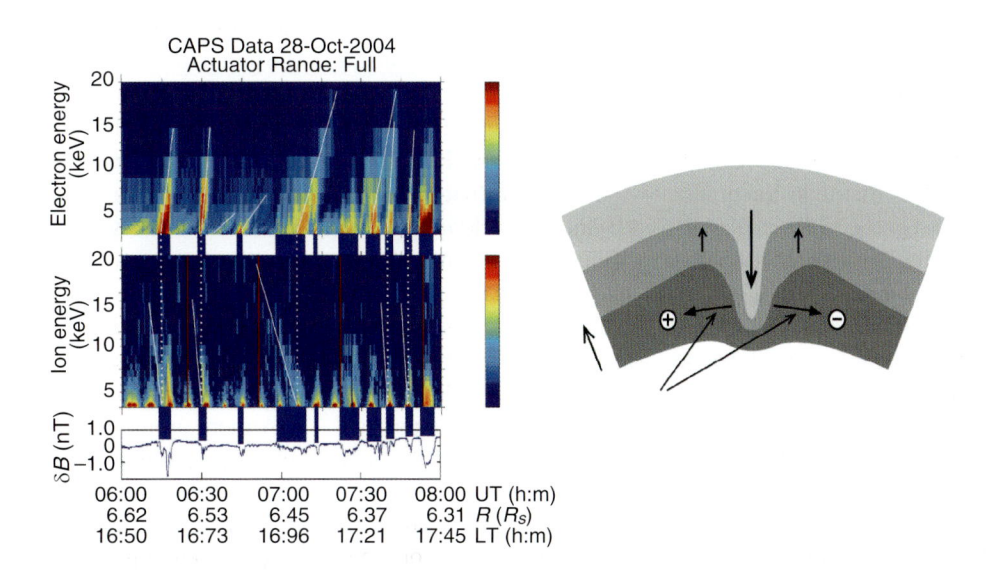

2005; Burch et al. 2005; Hill et al. 2005; Mauk et al. 2005; Chen and Hill 2008). Figure 11.18 shows a few examples in CAPS data (left panel) and an interpretive sketch (right panel), both from Hill et al. (2005). On a linear energy-time spectrogram, as shown here, an injection/dispersion event is revealed as a V-shaped structure with hot ions (electrons) forming the left (right) legs of the V. This is a simple consequence of the gradient-curvature drift speed being proportional to particle energy (non-relativistic approximation) and in opposite directions for oppositely charged particles. The electron leg of the V is more often visible than the ion leg because of the much larger electron counting rates at a given energy. The apex of the V marks the original injection longitude (assuming approximate corotation), the width of each leg indicates the width of the injection channel, and its slope is inversely proportional to the elapsed time since injection. By combining the injection longitude and the elapsed time, the local time of injection can also be determined.

The δB line plot below the ion spectrogram (Fig. 11.18) illustrates the fact that, on this (second) Cassini orbit, most injection events were accompanied by a diamagnetic reduction of B, although the diamagnetic signature was sometimes absent and occasionally even reversed in sign. This diamagnetic depression is consistent with the injection structures containing larger thermal plasma pressure than their cooler denser surroundings. Recently it has been discovered that there is a latitudinal ordering of the magnetic signatures (André et al. 2007). Near the magnetic equator, δB is positive within the injection channels, indicating a smaller plasma pressure within those channels compared to that of the surrounding cooler plasma (the density contrast is larger than the temperature contrast). Away from the equator, δB is negative within the injection channels, indicating enhanced pressure there. The latitudinal ordering probably results from the stronger equatorial centrifugal confinement of the cold plasma compared to that of the hot plasma. Energetically, the interchange instability is enhanced by the high-pressure but cool, equatorially confined plasma (Eq. 11.8).

In the fewer but very large and long-lived events observed at higher energies by the Cassini MIMI (Mauk et al. 2005), the slope of the electron leg of the V can be altered, and even occasionally reversed, by the differential rotation rate $\Omega(L)$ (Fig. 11.17 above). Because these features are so long-lived, they dominate the appearance of the inner magnetosphere at medium and high energies, showing a menagerie of different time-dispersed features produced by the combined effects of prograde (ions) and retrograde (electrons) magnetic drifts and the retrograde $\mathbf{E} \times \mathbf{B}$ drift (Paranicas et al. 2007; Brandt et al. 2008).

Similar injection/dispersion signatures have been observed at Jupiter by the Galileo spacecraft (Bolton et al. 1997; Kivelson et al. 1997c; Thorne et al. 1997), but much less frequently and much less clearly. The difference lies not with the spacecraft orbits or instrumentation, but with the magnetosphere. For a given particle energy per unit charge, and a given L value, the gradient/curvature drift is about 25 times faster at Saturn than at Jupiter, owing mostly to Saturn's much weaker magnetic field. Thus it is clear in retrospect why Saturn's magnetosphere provides the better laboratory for the study of the centrifugal interchange instability, which is presumably equally important at both planets. As at Saturn, the behavior of the higher-energy particles at Jupiter shows much larger scale injections that are relatively long-lived (Mauk et al. 1999). These injections extend into the middle magnetosphere where particle pressures compete with magnetic pressures. It remains an open question whether these middle-magnetosphere injections at Jupiter are the natural extension of interchange events in the inner magnetosphere or should be compared more with the large-scale, middle-to-outer magnetosphere injections documented at Saturn in Chapter 10.

A statistical study of 429 of the smaller-scale plasma structures from 26 Cassini orbits (Chen and Hill 2008) reveals that their longitudinal widths seldom (but occasionally) exceed $1R_S$, their ages since injection seldom (but occasionally) exceed the Saturn rotation period 10.8 h, and their distributions in both longitude and local time are essentially random. The lack of a clear longitude dependence could be a problem for the global-scale corotating convection model that has been invoked (Goldreich and Farmer 2007; Gurnett et al. 2007) to explain the clear rotational modulation of various magnetospheric properties noted above. And the lack of a clear local-time dependence distinguishes these events from otherwise similar events observed at Earth, where they occur exclusively on the nightside. Perhaps the most surprising result of this study is that the combined widths of all injection channels at a given time occupy a small fraction ($\sim$5%) of the available longitude space. The inflow channels are evidently much narrower, and hence much faster (because $\oint \mathbf{E} \cdot \mathbf{dl} = 0$), than the intervening outflow channels. A similar asymmetry was previously found to prevail at Jupiter (e.g., Krupp et al. 2004; Russell et al. 2000, 2005). This finding is at variance with theoretical models of the interchange process, in which the inflow and outflow channels are either assumed (in most cases) or derived (in a few cases) to be of roughly equal width.

An example of the latter type of model, where the roughly equal widths of inflow and outflow sectors is derived rather than assumed, is shown in Fig. 11.19 (Wu et al. 2007b). The figure shows isocontours of flux-tube content η at the same time step of two different numerical simulations utilizing the Rice Convection Model (RCM) (Wu et al. 2007a). The two simulations are identical except that the second one includes the Coriolis acceleration (right panel) while the first does not. Both simulations are initial-value problems with an initial cold plasma torus of radial width $1R_S$ centered at $L = 4$. The initial torus is azimuthally symmetric except for a miniscule level of white noise as a seed perturbation. The Coriolis acceleration has three notable effects: (1) it bends the fingers in the retrograde sense as they move outward, as expected intuitively; (2) it slows the radial growth of the fingers, as predicted theoretically (Vasyliūnas 1994; Pontius 1997); and (3) it produces a tendency for the inflow channels (dark blue) to be thinner than the ouflow channels (other colors). This last effect was not predicted, and it pushes the simulated result in the right direction to fit the Chen and Hill (2008) statistical study, although not nearly far enough in that direction.

A more promising explanation for the unequal widths of inflow and outflow channels may be provided by the presence of a continuously active plasma source in the region in question. Figure 11.20 shows the results of a more recent RCM simulation in which the initial torus was replaced by an active plasma source following closely the radial distribution of Fig. 11.6). The model magnetosphere was initially empty, and the miniscule white-noise perturbation was applied instead to the source function, otherwise azimuthally symmetric. The total source rate, integrated over L, was 100 kg/s, and the ionospheric Pedersen conductance was uniform at 0.1 S. A key difference between the two simulation results (Fig. 11.19 versus Fig. 11.20) is that, in the latter, the azimuthal width of the outflow channels tends to grow with time at the expense of the interspersed inflow channels. By the last time step the outflow channels clearly occupy most of the available longitude space in the radial range ($L \sim 5$–10) relevant to the Chen and Hill (2008) Cassini results. Grid-scale numerical diffusion is evident here as the narrowing inflow channels start to lose their intrinsically dark blue (essentially zero) value of η.

There is clearly much left to be learned about interchange transport through further acquisition and analysis of Cassini plasma data, through further, more advanced RCM simulations, and through further comparisons between the two.

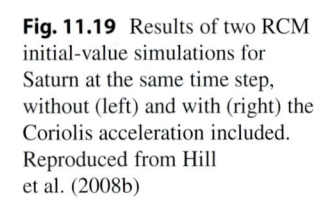

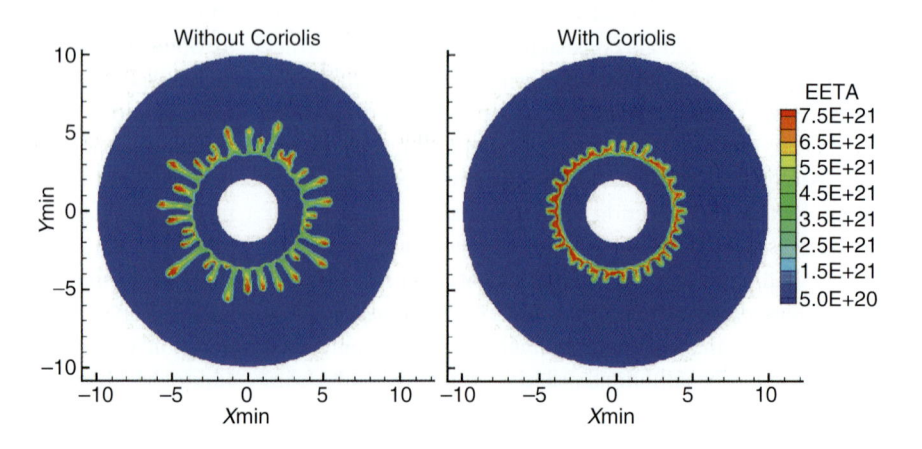

Fig. 11.19 Results of two RCM initial-value simulations for Saturn at the same time step, without (left) and with (right) the Coriolis acceleration included. Reproduced from Hill et al. (2008b)

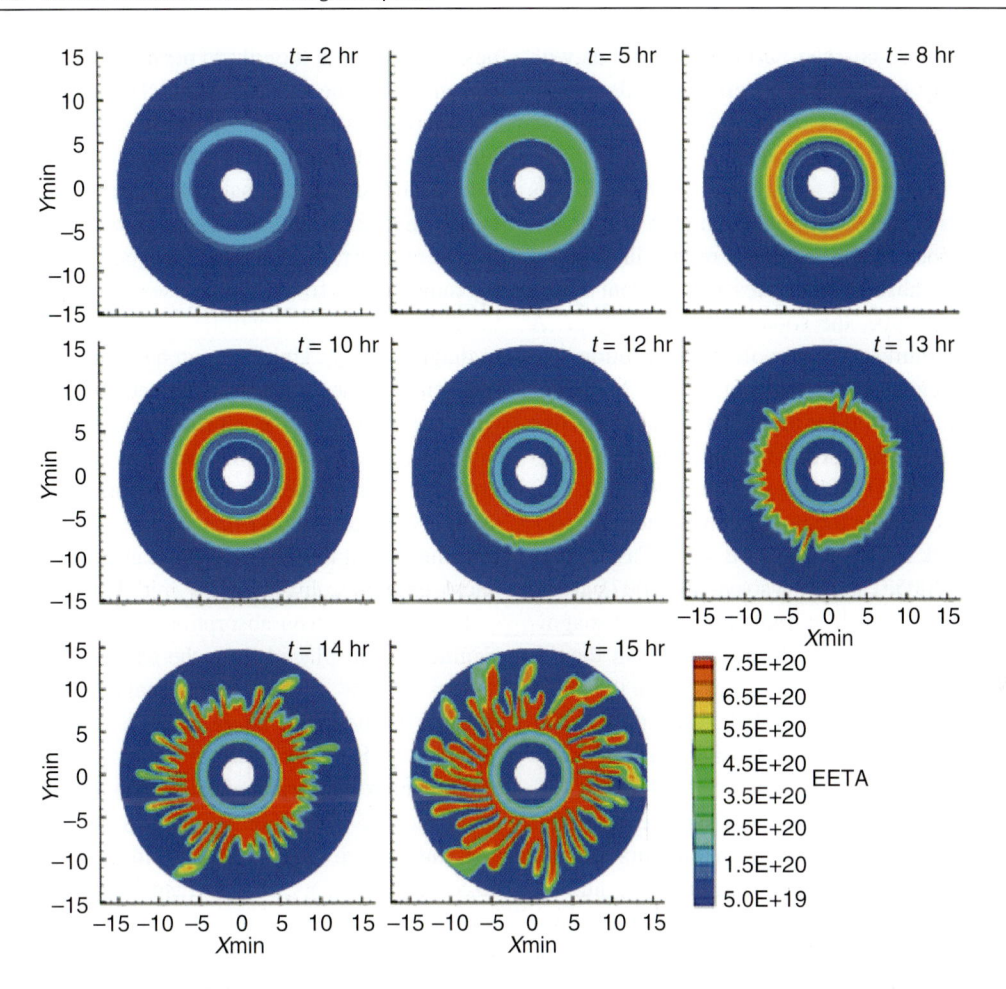

Fig. 11.20 Results of an RCM simulation for Saturn with the continuously active plasma source shown in Fig. 11.6. Reproduced from Hill et al. (2008b)

11.3.5 The Radial Diffusion Formalism

For the analysis of magnetospheric properties averaged over time and longitude, it is often useful to employ the radial diffusion formalism, which suppresses (averages over) the longitude and time dependences of the transport process and treats a one-dimensional mathematical problem in which all variables depend only on L in the spatial domain. In particular, the particle velocity distribution function (= non-relativistic phase-space density) $f(\mathbf{x}, \mathbf{v})$ is assumed to depend only on L in coordinate space, and only on the first and second adiabatic invariants in velocity space (Section 11.4.2). The quantity f in a particular volume element of velocity-invariant space is assumed to depend only on L according to the classical radial diffusion equation

$$\frac{\partial f}{\partial t} = 0 = L^2 \frac{\partial}{\partial L}\left(\frac{D_{LL}}{L^2}\frac{\partial f}{\partial L}\right) + S - \mathcal{L} \quad (11.9)$$

(e.g., Schulz and Lanzerotti 1974), where S and $\mathcal{L}$ are source and loss terms in appropriate units, averaged over longitude and time. Azimuthal transport is ignored (or averaged over) by assumption, and radial transport is described completely in this approximation by the radial diffusion coefficient $D_{LL}(L)$.

Because D_{LL} is expected to increase with increasing values of L, since radial displacements tend to increase with decreasing values of B (e.g., Walt 1994), the diffusion coefficient is usually assumed for mathematical convenience to have a power-law form, $D_{LL} \propto L^n$. Then, in the absence of sources and sinks ($S = \mathcal{L} = 0$), the solution of Eq. 11.9 is also a power law with $f(L) \propto L^{3-n}$ (or a logarithmic function in the singular case $n = 3$). This solution corresponds to a constant radial diffusion flux (the quantity in parentheses in Eq. 11.9). Deviations of the observed $f(L)$ from this smooth predictable form can then provide evidence for the locations in L of significant sources or losses (e.g., Van Allen et al. 1980a; Armstrong et al. 1983; Paranicas and Cheng 1997; Rymer et al. 2007a). Small-scale absorption signatures in $f(L)$ at the orbits of known satellites, both "macro-signatures" averaged over orbital phase and "micro-signatures" dependent on orbital phase, can provide important information on charged-particle interactions with the satellites (Carbary et al. 1983; Selesnick 1993; Paranicas and Chang 1997; Roussos et al. 2005; Jones et al. 2006a; Roussos et al. 2007).

For charged particle losses to satellite absorption, it is helpful to distinguish between macrosignatures and microsignatures. The former is a localized region in L-space where satellite or ring absorption creates a permanent reduction in $f(L)$, for instance, the so-called "Mimas gap." A microsignature, by contrast, is an evolving absorption feature that loses its integrity over time (Fig. 11.21). Microsignatures are particularly useful for diagnosing the local value of D_{LL}. For electron losses to a satellite in a region with no sources, Van Allen et al. (1980b) were able to use Eq. 11.9 on microsignatures to reduce the problem to a study of D_{LL} in isolation. This work has been extended using Cassini data by Paranicas et al. (2005a) and Roussos et al. (2007). It has been found that D_{LL} varies by almost an order of magnitude at a given L value and that the value of n in the L^n dependence of D_{LL} is not fixed in time, varying in the range $8 < n < 10$. Furthermore, the size of D_{LL} indicates that diffusive transport as diagnosed with microsignatures is a very slow process.

The high value of n in $D_{LL} \propto L^n$ is significant in that such high values are anticipated for external drivers of magnetospheric dynamics, such as solar wind buffeting and atmospheric turbulence at the feet of the magnetic field lines (e.g., Schulz and Lanzerotti 1974). Such large indices are not anticipated for internally generated interchange turbulence (e.g., Siscoe et al. 1981; Gehrels and Stone 1983) where values ∼4 have been thought to prevail. Either the expectations are wrong or the evaluation of D_{LL} on the basis of satellite microsignatures does not address the transport caused by interchange. There is good reason to accept the latter conclusion. Satellite microsignatures address stochastic transport events that have spatial scale sizes smaller than the satellite. In contrast, the interchange events that have been studied range in size from $0.1 R_S$ to several R_S (Hill et al. 2005), much larger than the satellites used by Roussos et al. (2007) to diagnose D_{LL}. Unless it can be shown that the scale size spectrum of the interchange injections extends coherently (e.g., a power law size spectrum) down to the sizes of the satellites, the microsatellite analyses cannot be used to address transport caused by the injections. There is even more direct evidence (below) that a larger-scale phenomenon is governing total radial diffusion than can be diagnosed with microsignature analyses.

Energetic electron absorption signatures are found to be significantly displaced from the expected L-shells, even in regions of low L and strong magnetic field (Fig. 11.21; Roussos et al. 2005); statistics of such displacements are provided by Roussos et al. (2007). Analysis shows that this effect cannot be attributed to a solar-wind-induced dusk to dawn electric field, nor to a dawn to dusk electric field as is inferred to exist at Jupiter. In addition to these organized displacements, chaotic displacements are also visible as "double energetic electron microsignatures" (as seen in Fig. 11.21). Their double appearance suggests that they were formed close together in time during the occurrence of dynamic events (e.g., injections). Such events may have displaced the depletion regions to slightly different L-shells (Roussos et al. 2005). These double structures are therefore indicative of localized, dynamic plasma transport processes in the inner magnetosphere. One of the significant aspects of these findings is that the radial displacements are much larger than the characteristic radial displacements associated with the filling in of the microsignatures. Therefore, microsignature analysis can address only part of the radial diffusion story, and likely not the most significant one.

The previous paragraph suggests a connection between transport and transient processes in the inner magnetosphere. It is less clear how much the injection structures contribute to the absolute value of D_{LL}, especially at higher energies, and we believe this is an important topic for future work.

11.4 Energy Conversion

Magnetospheres are large energy-storage reservoirs. Energy is stored, for example, in the particle energy in the ring current and the magnetic field energy, most prominently, in the magnetotail. Of particular interest are the mechanisms

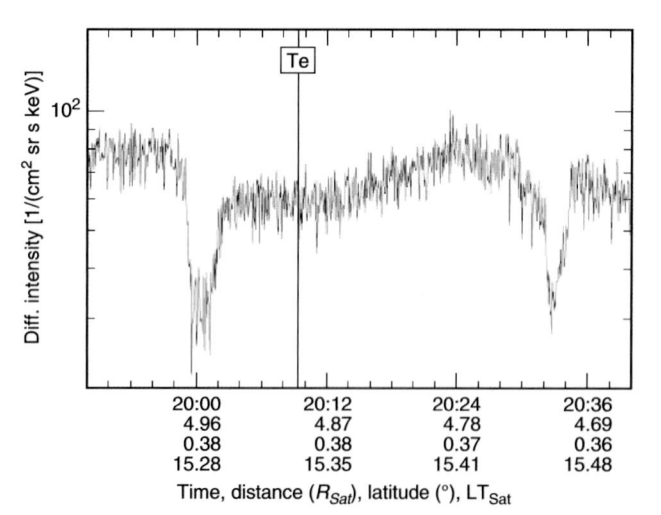

Fig. 11.21 Double signature in the 28- to 49-keV channel electrons seen during the Tethys L-shell crossing of day 47 (2005). The fact that both signatures have similar geometrical characteristics suggests a possible simultaneous formation during a variable, dynamic phase of the magnetosphere (Roussos et al. 2005; see also Roussos et al. 2007). Copyright: American Geophysical Union, 2005

leading up to both this storage and the emptying of these reservoirs, and those which convert energy from one kind of magnetospheric energy into another. The processes include magnetic reconnection, which is thought to convert solar wind flow energy across the magnetopause (Chapter 9) into magnetic field and hot plasma energy, and to convert stored magnetic field energy into plasma heating and flow in the nightside magnetotail. They include current generation that redistributes plasma flow and thermal energy into magnetic field energy. They include a menagerie of particle acceleration mechanisms that can both heat the bulk of the particle populations or generate very high energies in a small fraction of the particles, forming high-energy tails. Finally, they include various wave-particle interactions that energize particles, cause particles to be lost to the system, and generate radio waves that can be used to remotely diagnose magnetospheric structure and processes.

11.4.1 Reconnection

Reconnection changes the topology of magnetic field lines, allowing formerly distinct field lines to interconnect across a current sheet and couple kinetic energy into magnetic energy and vice versa. Such reconnection occurs trivially with magnetic fields in a vacuum, but because of the propensity of charged particles to be frozen to the magnetic field lines (in the MHD formulation), special dissipation processes must occur for reconnection to occur in a magnetized plasma. In any magnetosphere there are two very different magnetic geometries in which reconnection may occur and lead to energy conversion. These are at the dayside magnetopause and in the tail current sheet. Reconnection is most likely to occur when the magnetic fields in two adjacent regions are nearly antiparallel. At the magnetopause there is always a set of points at which the fields are antiparallel, and potentially could reconnect. In contrast, the near-tail region contains linked, almost-antiparallel fields that become more nearly antiparallel and hence more unstable to reconnection as they stretch more in the antisolar direction. In the sections below we discuss first magnetopause reconnection and then tail reconnection.

11.4.1.1 Magnetopause Reconnection

A hallmark of reconnection is accelerated flow. The discovery of accelerated flow at the Earth's magnetopause (Paschmann et al. 1979) contributed to its acceptance as the driver of the dynamics of Earth's magnetosphere, as proposed much earlier by Dungey (1961). A schematic of dayside magnetopause reconnection is shown in Fig. 11.22.

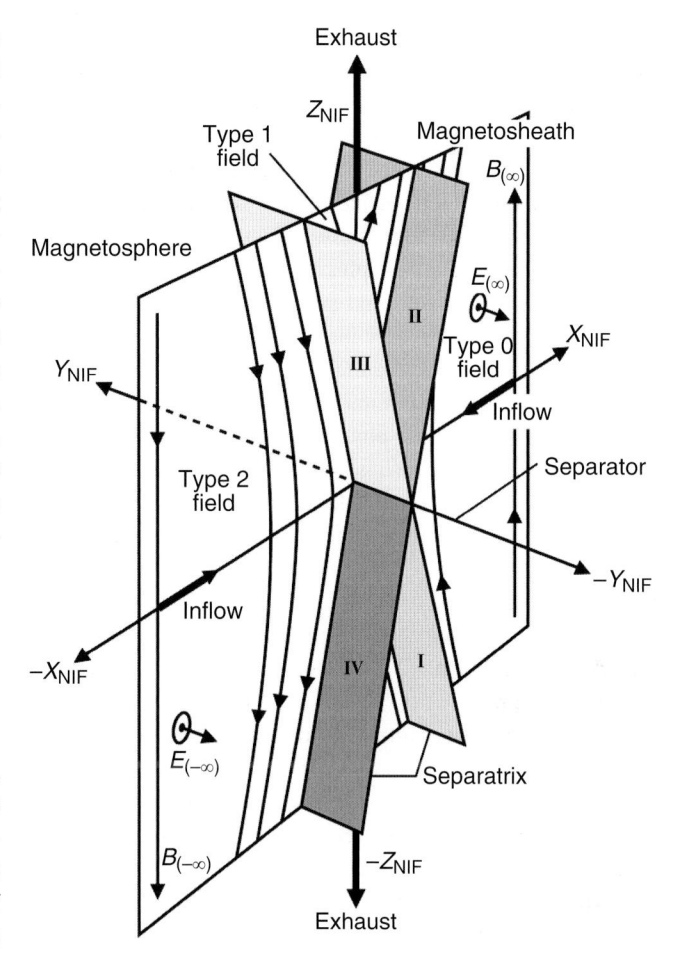

Fig. 11.22 Cartoon of reconnection on the dayside magnetopause between the magnetized shocked solar wind plasma (magnetosheath) on the right and the magnetospheric magnetic field and plasma on the left. The natural X-line field geometry leads to flow away from the X line where reconnection occurs. Field lines with 0, 1, and 2 feet on the planet are given the topological labels type 0, type 1, and type 2, respectively

Reconnection of the interplanetary magnetic field lines that originally do not intersect the surface of the Earth (type 0) with closed field lines that intersect the Earth twice (type 2) produces open field lines (type 1) that intersect the Earth's surface once. The existence of open, or type 1, field lines allows the transport of magnetic flux and plasma to the tail lobes, increasing the magnetic energy there. Ultimately, the magnetic fields in the oppositely directed tail lobes reconnect to produce closed field lines (type 2) as well as plasmoids and interplanetary field lines (both type 0) in a process called at Earth a substorm. Terrestrial substorms result from varying Interplanetary Magnetic Field (IMF) orientation, and the duration of the substorm process is controlled in part by the statistics of the IMF direction, but the reconnection process can assume a quasi-steady state, leading to a long period of energy transfer and a geomagnetic storm (Burton et al. 1975). The measure of reconnection is the potential drop along the line (Y_{NIF} in Fig. 11.22) separating the field lines of different topology (e.g., interplanetary, type 0, and closed, type 2). Assuming dissipationless flow away from the reconnection separatrix, the potential drop may be estimated with the cross product of the velocity of the plasma and the magnetic field, integrated along the separatrix, the line separating the three field topologies. This integration is difficult to obtain in practice because a spacecraft samples only one point in space and the separatrix can move around extensively, especially on the dayside magnetopause where the field lines on the two sides of the magnetopause have quite different geometries. Thus, in order to study the parametric behavior of reconnection, terrestrial magnetospheric physicists have resorted to proxy studies of reconnection using measures of geomagnetic activity as stand-ins for the dayside reconnection rate. Some of these studies are relevant to understanding reconnection at Saturn.

Proxy Studies at Earth

Using these proxy estimations, the reconnection efficiency as a function of a variety or parameters has been tested (Scurry and Russell 1991; Scurry et al. 1994). As the plasma beta increases (beta = particle-pressure/magnetic pressure), the reconnection efficiency decreases due to the relative weakening of magnetic forces to plasma thermal pressure forces. Ultimately, the magnetosonic Mach number controls the plasma conditions in the dayside magnetosheath and, at very high Mach numbers, the ion beta becomes large regardless of solar wind ion beta and the reconnection rate effectively goes to zero as shown in Fig. 11.23. This result has important implications for both Saturn and Jupiter because the Mach number of the solar wind is usually above 10 at these planets. This expected weakening of reconnection on the dayside when the Mach number is high has been confirmed with an MHD code by Fukuzawa (2007).

In Situ Studies at Earth

Two distinct modes of reconnection have been found in terrestrial data. The first is the steady-state reconnection in which persistent flows are present jetting away from the reconnection point (Sonnerup et al. 1981). A second style of reconnection is the flux transfer event (Russell and Elphic 1978) in which connected magnetic flux structures passes the spacecraft in an event lasting about a minute. These events appear to be spontaneous and not triggered by IMF changes. These structures are believed to be magnetic flux tubes that connect magnetosheath magnetic field to magnetospheric flux.

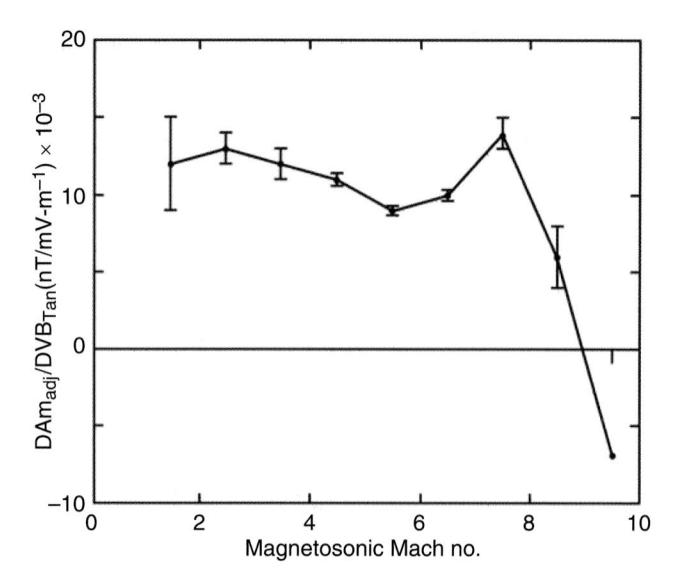

Fig. 11.23 The dayside reconnection efficiency for increasing solar-wind fast magnetosonic Mach number calculated at Earth using the Am index corrected for both dynamic pressure and velocity effects. The calculation assumes that ion beta is independent of clock angle and other reconnection controlling parameters (after Scurry et al. 1994). Copyright: American Geophysical Union, 1994

In Situ Studies at Saturn

A pre-Cassini study of the magnetic field at the Saturn magnetopause examined the joint behavior of the magnetic field and the plasma velocity across five encounters with the magnetopause. On one of the five crossings, the normal component was significant, and an accelerated velocity was observed (Huddleston et al. 1997). This crossing had a strong magnetic field in the magnetosheath, suggesting that the beta value in the magnetosheath was smaller than usual and more conducive to reconnection. A hodogram of the magnetic field variation during this crossing is shown in Fig. 11.24 (top; see that the normal component on the right is non-zero). While this reconnection rate is important for supplying open flux to the lobes, it is not large enough to overcome the corotational electric field in the equatorial plane. Thus far there are no reports of flux transfer events (FTEs) at Saturn, while FTEs were detected at the Jovian magnetopause (Walker and Russell 1985).

Cassini carries a comprehensive set of particle and field instruments with cadence comparable to those instruments used to identify reconnection at Earth. One survey for evidence of reconnection has been made (McAndrews et al. 2008b) by looking for plasma heating at crossings of the magnetopause current layer. Flow data were unavailable for study. Two heating events were seen at the magnetopause and interpreted to be reconnection events. Hodograms of the magnetic field across one of these two current layers

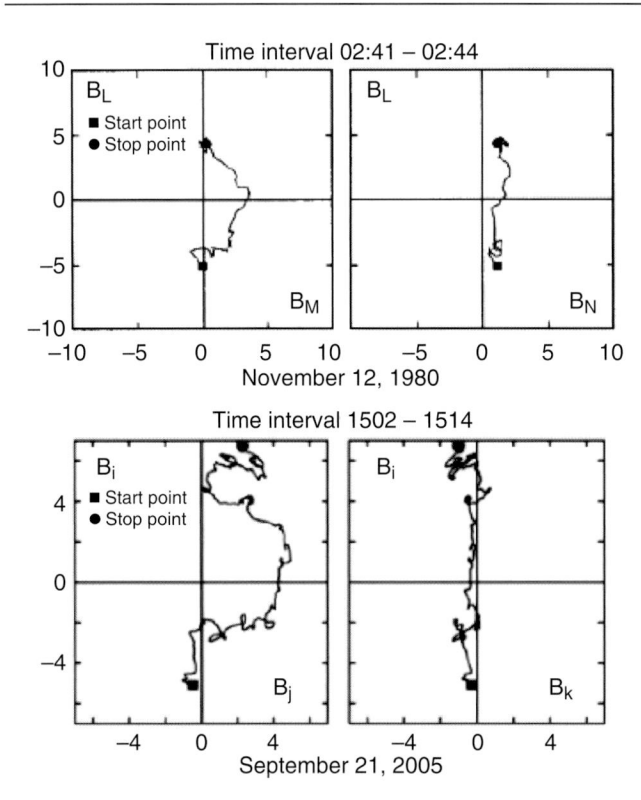

Fig. 11.24 (top) Hodogram of the magnetic field observed at the Saturnian magnetopause by Voyager on November 12, 1980 (after Huddleston et al. 1997). The hodograms show the motion of the tip of the magnetic vector projected in the plane of the magnetopause (left) and in the plane containing the major variation and the normal direction (right). Copyright: American Geophysical Union, 1997. (bottom) Hodogram of the magnetic field observed at the Saturnian magnetopause by Cassini, corresponding to one of the two crossings studied by McAndrews et al. (2008b) in which heating was observed

are shown in Fig. 11.24 (bottom). These hodograms resemble these reported by Huddleston et al. (1997) but the normal components are not statistically different from zero and there are no flow data. Thus, these crossings do not make the case for reconnection any more strongly than the Voyager data, and neither dataset informs us of the reconnection rate averaged across the dayside magnetopause.

One study of the solar wind by Jackman et al. (2004) produced an estimate of the voltage drop by assuming that the rate or efficiency of reconnection at Saturn was the same as at Earth, but as discussed above, we have no basis for making that assumption. In short, we have reason to believe there is some dayside reconnection at Saturn but have no quantitative estimate of how much. It may be very little.

Discussion/Future Directions

Terrestrial studies of the parametric dependence of dayside reconnection, and concurring simulations, suggest that such reconnection should be weak at Saturn. The studies of Huddleston et al. (1997) and McAndrews et al. (2008b) do not contradict these expectations, and the rarity of flux transfer events at Saturn tends to reinforce them. Nevertheless, both studies are suggestive of dayside reconnection, and exploration of the magnetopause is at an early stage. In particular, the possible role of the tail lobes in moderating dynamic reconnection in the tail regions (next subsection), the identification of polar cap regions, generally identified as mapping to the tail lobes, as waxing and waning in concert with auroral dynamics (Chapter 12), must ultimately be brought into concert with any conclusion that dayside reconnection is very weak.

11.4.1.2 Tail Reconnection

On the dayside, there is always some place where magnetic field lines are anti-parallel on either side of the magnetopause. In the tail, the field geometry is fixed by the solar wind interaction and by plasma circulation. In its simplest configuration, a planetary magnetotail consists of stretched quasi-dipolar field lines with the near-equatorial field lines crossing the current sheet and returning to the planet. The simplest change in the configuration that leads to reconnection is increased stretching. This stretching reduces the strength of the field component crossing the tail current sheet and allows the nearly anti-parallel field lines on either side of the current sheet to reconnect. In contrast to the dayside (Fig. 11.22), in the tail the field lines that reconnect can be those with two feet connected to the ionosphere (type 2) or one foot in the ionosphere (type 1). Both processes produce regions of magnetic field not connected to the planet (type 0).

At Earth, the varying orientation of the interplanetary field modulates the rate of transport of magnetic flux from the dayside to the nightside, and thus modulates the occurrence of magnetospheric "substorms," in which an extended period of flux transfer to the tail (the "growth phase," ~ 0.5 h) is followed by a shorter period of flux return to the closed portion of the magnetosphere (the "expansion phase," approximately a few minutes). Thus, magnetotail reconnection episodically restores the closed magnetic flux that is removed by dayside reconnection. It can also rid the magnetosphere of any mass build-up from interior plasma sources.

This latter role is particularly important in the Jovian and Saturnian magnetospheres, where Io and Enceladus, respectively, add mass to the magnetosphere that must ultimately be removed while retaining, on average, a fixed amount of closed magnetic flux in the magnetosphere. A way to accomplish this in steady state has been sketched by Vasyliūnas (1983), as reproduced in Fig. 11.25. Notice that in the extreme, this pattern can proceed in the absence of strong reconnection on the dayside.

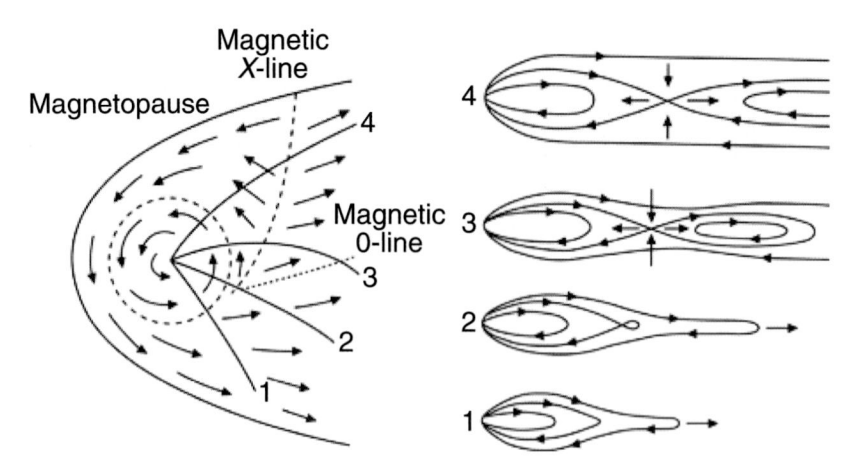

Fig. 11.25 Circulation pattern of plasma in a centrifugally driven mass-loaded magnetosphere such as those of Jupiter and Saturn (from Vasyliūnas 1983). As closed field lines are convected around to the dark side, they stretch, and the stretched type 2 (Fig. 11.22) field lines reconnect, forming shorter type 2 field lines with less plasma and a mass-loaded plasmoid of type 0 field lines. When the reconnection region cuts completely through the type 2 field lines to the type 1 field lines in the lobes, the plasmoid is ejected down the tail carrying away the mass originally added at Io (at Jupiter) or Enceladus (at Saturn). Copyright: Cambridge University Press, 1983

Tail Stretching and the Growth Phase

In the terrestrial magnetosphere, there is a clearly established growth phase in which magnetic flux is transferred from the dayside to the nightside and stored in the tail. In a rotating magnetosphere, it is not obvious whether or not an analogous growth phase should exist, because the driving mechanism is internally generated. Galileo observations provide evidence that a growth phase may exist at Jupiter (Ge et al. 2007). Jupiter displayed an increasingly stretched and strong magnetic field in its magnetotail over a 2.5-day period on two separate occasions, each of which was followed by a decreased field strength in the tail. We do not have similar data for the Saturnian tail. However, we do have evidence for rotationally controlled stretching of the tail field (Russell et al. 2008). Thus, it is possible that magnetotail stretching leads to tail reconnection at Saturn as well.

Dipolarization

At distances closer to the planet than the tail reconnection line, the magnetic field becomes more dipolar after a terrestrial substorm. A similar phenomenon was seen in the Saturnian magnetosphere (Bunce et al. 2005a; Russell et al. 2008) as the magnetic flux piles up in the nightside magnetosphere, and as at Earth, can also be associated with hot particle injections (Bunce et al. 2005a).

Near the Neutral Point

Figure 11.26 shows magnetic signatures near the reconnection line at both Jupiter (top) and Saturn (bottom; Jackman et al. 2007). The magnetic field rapidly turns southward at Jupiter because Galileo is inside the location of the reconnection line, and northward at Saturn because Cassini is outside the reconnection line. The field strength grows rapidly at both locations. With time, the field decreases and the tail returns to normal. An important point to note is that angular momentum is conserved on short timescales because it takes time to communicate ionospheric stresses to the equatorial plane. Thus, magnetized plasma pushed rapidly planetward moves around the planet more rapidly. At Jupiter, this causes the field line to bend forward so that B_r and B_{phi} have the same polarity. Plasma sent tailward slows down, and this occurs in the Cassini data because Cassini is beyond the reconnection line and the field is swept back more.

Plasmoid Formation

When reconnection occurs in the tail on closed magnetic field lines, an island is formed of field lines that close on themselves. This island, or plasmoid, contains mass but does not contain any net magnetic flux. It is therefore useful in ridding the magnetosphere of plasma from internal sources because it does not alter the amount of closed magnetic flux when it is ejected down the tail. It is anchored by surrounding closed field lines until reconnection reaches the open field lines of the tail lobe. If a spacecraft is in the tail lobe when the plasmoid is growing and moving slowly away from the planet, it sees a growing "dipolar" field followed by an anti-dipolar field as the structure begins to move down the tail. This signature, seen at Earth, has been called a Travelling Compression Region or TCR. These TCRs have been found at Saturn (Jackman et al. 2007; Hill et al. 2008a).

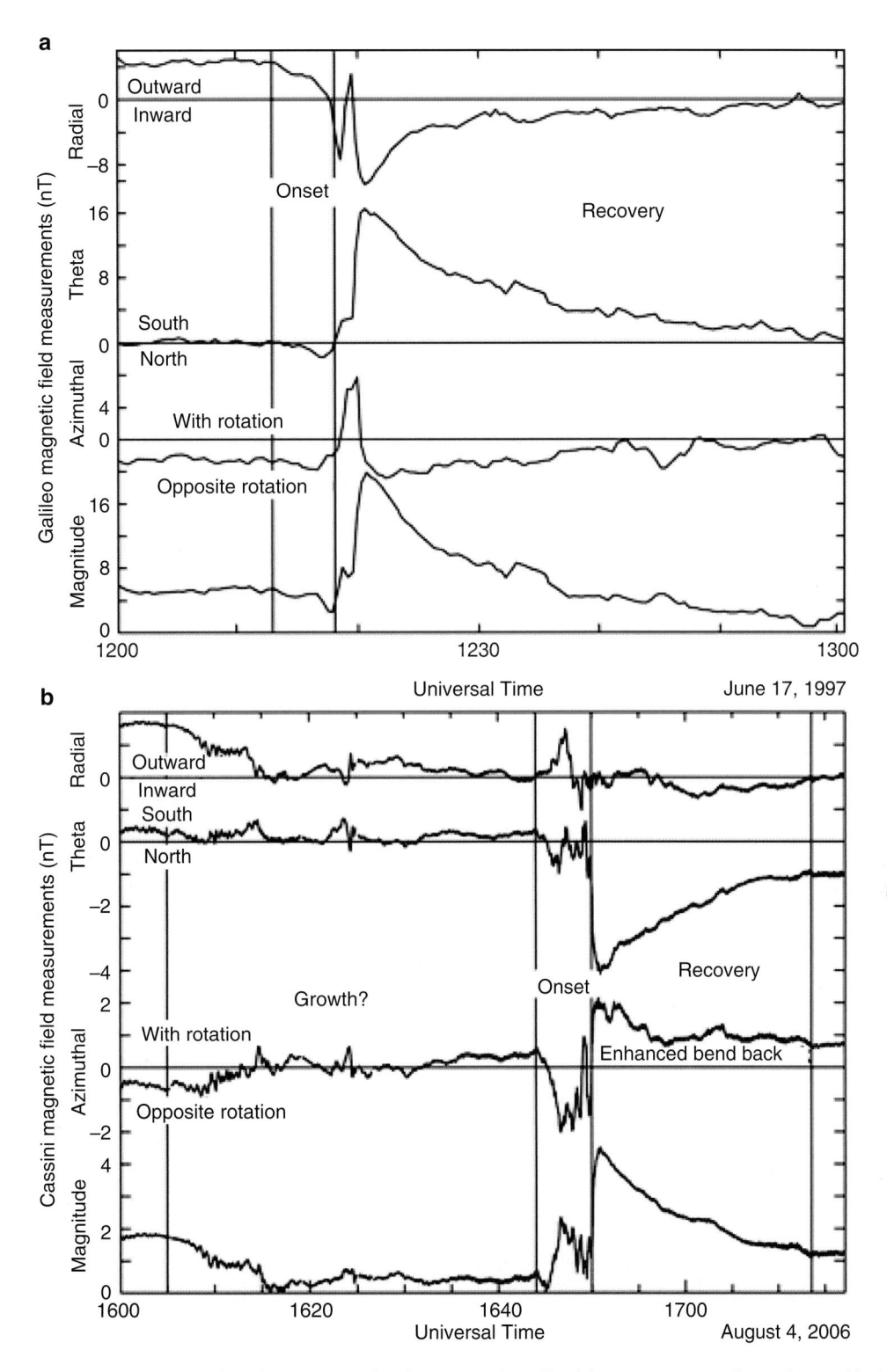

Fig. 11.26 Magnetic signatures on either side of the reconnection line. The top panel shows the magnetic field change in response to a reconnection event as observed by Galileo at Jupiter inside the location of the reconnection line. The magnetic field "dipolarizes" and becomes strong in the southward direction. Conservation of angular momentum causes the field line to sweep forward under this circumstance. In the bottom panel are Cassini measurements at Saturn but outside the location of the reconnection line at the plasmoid onset. The field turns strongly northward and becomes more swept back as the plasma rushes outward and its azimuthal velocity diminishes (modified from Jackman et al. 2007). Copyright: American Geophysical Union, 2007

Discussion/Future Directions

While both Jupiter and Saturn have rapidly rotating mass-loaded magnetospheres, they have very different internal magnetic field configurations and rotational configurations that alter the magnetosphere's behavior. At Saturn, in contrast to Jupiter, the magnetic dipole axis is aligned with the rotation axis so that the current sheet should not flap with the rotation period of the planet. Nevertheless, it does flap (Khurana et al. 2008), and that is not understood. In addition, the large obliquity, again in contrast to Jupiter, produces a bent current sheet on both the day and nightside (Arridge et al. 2008) so that a spacecraft in the equatorial plane will be below the current sheet much of the time. Thus, if we draw the sequence of substorm phases for Saturn, we obtain something like that shown in Fig. 11.27 with the current sheet curved up. This complicates observing for a near-equatorial spacecraft. There has only been a limited period of tail data for Cassini when good examples of current-sheet crossings and plasmoid formation have been seen, but we expect that situation to improve later in the mission.

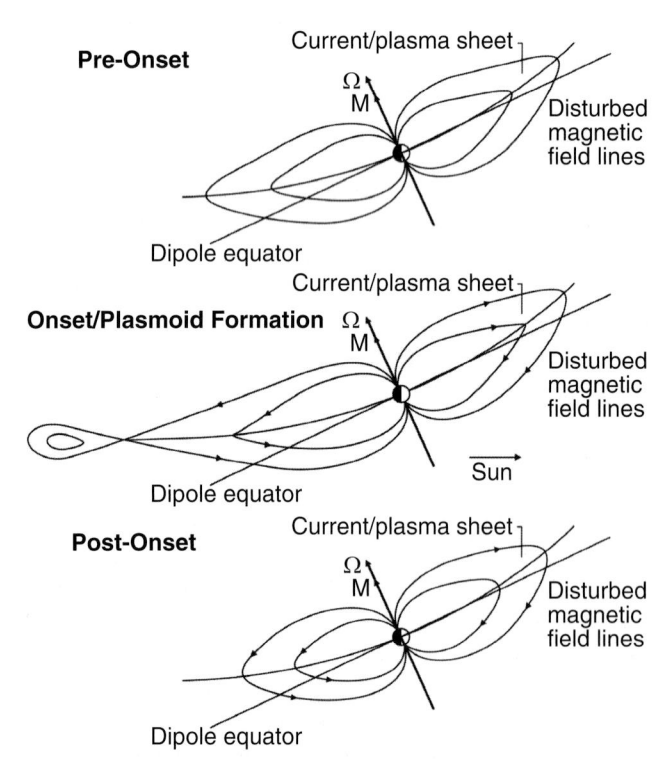

Fig. 11.27 Magnetic field configuration in the noon-midnight meridian during three phases of a Saturnian plasmoid. The bowl-shaped current sheet is due to solar wind forces on the magnetosphere when the rotation axis faces away from the Sun, on the right side of the figure (Arridge et al. 2008). This diagram is the Saturn equivalent of the three phases of a terrestrial substorm

11.4.2 Particle Acceleration

Addressed here are macroscopic processes that lead to the acceleration or energization of charged particles within Saturn's magnetosphere. There are a number of processes that are invoked to explain phenomena observed to be occurring but only several processes are addressed with specific quantitative application or tests. Here we first address in some detail the processes addressed quantitatively and then summarize those processes qualitatively invoked. Acceleration and energization primarily by wave-particle interactions is addressed in Section 11.4.4.

11.4.2.1 Adiabatic Acceleration and Related Processes

Background

The most often-invoked particle acceleration mechanism in space environment research is that which arises from transport from one region to another in a fashion that preserves the so-called single particle adiabatic invariants (e.g., Chen 1974; Parks 1991; Roederer 1970). Adiabatic invariants arise when particles undergo cyclic motions over timescales that are fast with respect to the timescales of any changes that the particle sees (from either temporal or spatial variations) in the electromagnetic fields that control the particle's motion. In magnetospheric research there are three cyclic motions that are typically invoked: (1) gyration of a particle around its central magnetic field line, (2) the bounce motion of a particle along a magnetic field line between magnetic mirror points, and (3) the guiding-center drift of a particle around its host planet. The first two of the invariants can be written:

$$\mu = \frac{p_\perp^2}{2 \cdot m_o \cdot B} \quad \text{and} \quad J = \int_{M1}^{M2} p_\| \cdot ds \qquad (11.10)$$

where p is momentum, the perpendicular and parallel direction symbols refer to the direction with respect to the magnetic field, m_o is the particle rest mass, $M1$ and $M2$ are the positions along the magnetic field line of the magnetic mirror points, and ds is an incremental distance along the controlling field line. The mirror points $M1$ and $M2$ occur at the spatial positions of the mirror magnetic field strength B_M, which can be calculated with the formula derived from $\mu =$ constant: $p_\perp/p = \sin[\alpha] = (B/B_M)^{1/2}$, where α is the particle pitch angle. The third invariant is often simplified as the total magnetic flux Φ_M contained by the orbit of the particle as it drifts all the way around the planet by means of the various guiding-center drifts that include electric field drifts, magnetic gradient drifts, centrifugal drifts associated with a planet's rapid rotation, etc.

To calculate the effects on particle distributions of adiabatic transport and acceleration one must concern oneself with the invariance of the Phase Space Density (PSD) as derived from Liouville's theorem for the condition that the particles reside in a collisionless system (Krall and Trivelpiece 1973). That is, the Phase Space Density (PSD[**q**, **p**]) that attends the position in the six-dimensional phase space of a single particle remains invariant along the trajectory of the particle in that space. In calculating the consequences of acceleration associated with adiabatic transport, we convert measured particle intensities $I(m^{-2} s^{-1} sr^{-1} J^{-1})$ into phase space densities ($f(\mathbf{p}) = I(\mathbf{p})/p^2$) and then transform the coordinate system from phase space to invariant space ($f(\mathbf{p}) \rightarrow f(p_\perp, p_\parallel) \rightarrow f(\mu, J)$, where the first transformation eliminates one dimension with the assumption that the particle distribution is gyrotropic around the field line).

Adiabatic Acceleration at Saturn

Within magnetized planet magnetospheres, the higher energy particles within the middle and inner regions are thought to have their "sources" (defined more carefully below) within the middle and outer regions of the magnetosphere (Belcher et al. 1991; Cheng et al. 1991; Fillius 1976; Mauk et al. 1995; Richardson et al. 1995; Van Allen 1976; Van Allen 1984). Qualitatively, then, one anticipates from conservation of adiabatic invariants (Eq. 11.10), based on considerations of the varying magnetic field strengths and field-line lengths, that the characteristic energies of these particles will increase dramatically with decreasing distance to the planet. Saturn follows this expectation as anticipated for the higher energy particles, but only very roughly (Fig. 11.28 after Rymer et al. 2008). Adiabatic energization is also central to detailed modeling of Saturn's radiation belts, as it is of the modeling of radiation belts at all strongly magnetized planets (Santos-Costa et al. 2003), and of the transport of other particle populations within Saturn's magnetosphere (e.g. Section 11.3.4; Hill et al. 2008b).

At Saturn, the predictive capabilities of adiabatic transport go beyond this highly qualitative picture. If we begin with an angularly isotropic particle distribution within a quasi-dipolar magnetic configuration, one finds that adiabatic planetward transport yields angular distributions that are have the so-called "pancake" or "trapped" distribution shapes, with peak intensities at equatorial pitch angle (α) values near 90° (directions perpendicular to the magnetic field **B**) (Fig. 11.29, dotted curve; Rymer et al. 2008). Similarly, beginning with the same angularly isotropic distribution, adiabatic transport away from Saturn yields field-aligned, or so-called "cigar" distributions (Fig. 11.29, dashed curve). Finally, beginning

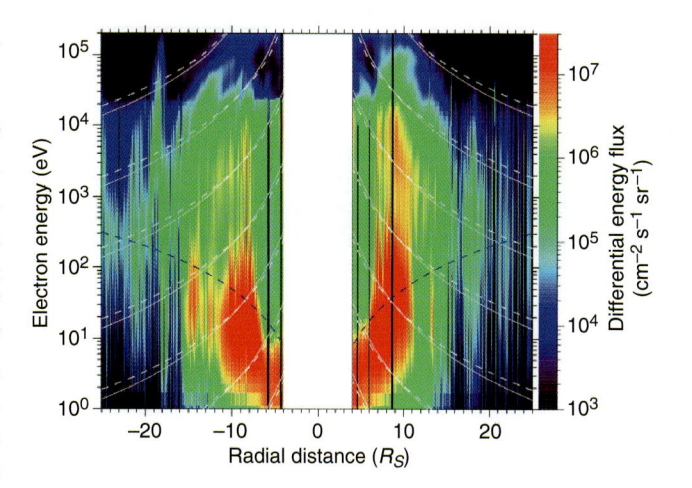

Fig. 11.28 After Rymer et al. (2008). Energy (y-axis) versus radial position (R_S; x-axis)) versus electron differential particle energy intensity (color scale) spectrogram sampled within Saturn's magnetosphere by the Cassini CAPS instrument a low energies (<20 keV) and the Cassini MIMI instrument (>20 keV). The solid white lines show contours of constant particle first adiabatic invariants; the dashed white lines show contours of constant "isotropic invariant" values, as defined by Schulz (1998); the dashed dark line is the proton rotational flow energy assuming rigid rotation

with a trapped distribution that arises naturally from the scattering loss of particles along field lines where the particles encounter Saturn's atmosphere, outward adiabatic transport yields so-called "butterfly" distributions, with the peak values intermediate between 90° and the field-aligned directions (Fig. 11.29, solid line). The discovery by Burch et al. (2007) of butterfly distributions (Fig. 11.30) in the region of radial interchange turbulence (Section 11.3.4) but in between the small-scale inward injections, provides important evidence that the inward transport provided by the injections is balanced by outward flow in the broader-scale regions in between the injections.

While the qualitative shapes of angular distributions may be explained on the basis of radial, adiabatic transport, it has been noted by Rymer et al. (2008) that it is difficult at Saturn to obtain a quantitative match between observed and adiabatically modeled angular distributions. These authors have argued that angular scattering along with the adiabatic effects plays an important role in shaping the distribution shapes, even for transport over radial distances of several Saturn radii. Strong scattering modifications of angular distributions have also been reported at Earth (Fox et al. 2006). In strong scattering environments, it is clear that neither the first nor the second adiabatic invariant of any individual particle can be viewed as being preserved.

Schulz (1998) has suggested an approach to adiabatic energization that takes advantage of strong scattering while preserving the simplicity and intuitiveness of the single particle adiabatic invariant approach. We assume that scattering is

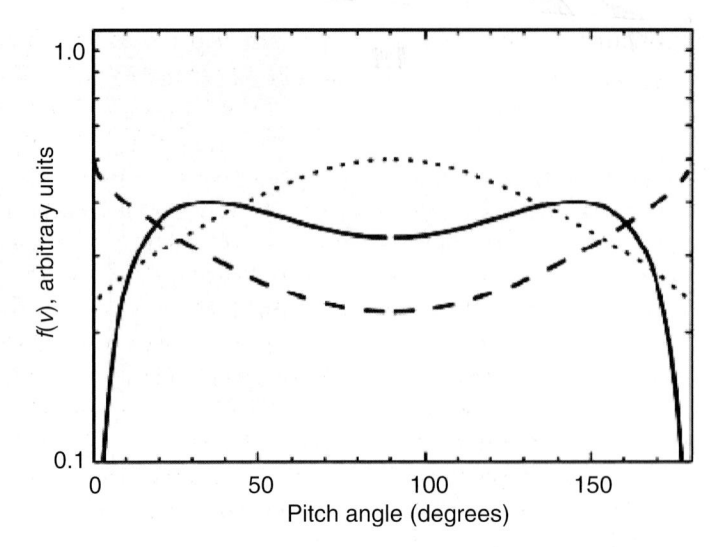

Fig. 11.29 Pitch angle distribution shapes calculated on the basis of adiabatic planetward and anti-planetward radial transport within the inner Saturnian magnetosphere. After Rymer et al. (2008). Copyright: American Geophysical Union, 2008

Fig. 11.30 Electron pitch angle distributions observed in Saturn's inner magnetosphere. After Burch et al. (2007). Copyright: Nature Publishing Group, 2007

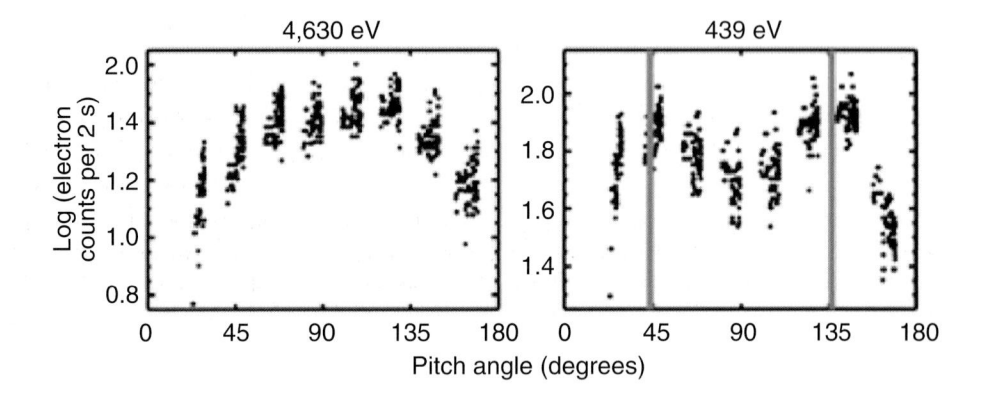

strong and approximately elastic (energy preserving). Under these assumptions we can treat sets of particles with specified energies within a distribution as constituting a perfect gas. Specifically, as summarized by Rymer et al. (2007a) after Schulz (1998), using the perfect gas law and the volume of a magnetic flux tube ($V = \int ds/B$, where ds is differential distance along the field line), one obtains the following, relativistically correct, "isotropic invariant":

$$\Lambda = \left(\frac{p^2}{2 \cdot m_o}\right) \cdot \left(\int ds/B\right)^{2/3} \cong \left(\frac{p^2}{2 \cdot m_o}\right) \cdot \left(L/B_{eq.}\right)^{2/3} \tag{11.11}$$

where the approximation on the right is for a quasi-dipolar configuration. Because the cold plasmas are centrifugally confined near the magnetic equator, L/B_{eq} may need to be replaced with $H(L)/B_{eq}$, where $H(L)$ is the L-dependent off-equatorial scale height of the plasmas (McAndrews et al. 2008a).

The isotropic invariant (Eq. 11.11) is structurally very similar to the first adiabatic invariant (Eq. 11.10), and when they are plotted side-by-side for the same Saturnian configuration they quantitatively yield very similar results (Fig. 11.28). The point is that the use of the invariant approach for examining particle transport and energization is now put onto a sounder basis in the face of strong evidence that the standard invariants are strongly violated by scattering in the magnetospheres of both Saturn and Earth.

With the confidence that this approach provides, Rymer et al. (2007a) have performed PSD analysis of Saturnian electrons, as shown in Fig. 11.31, where the PSD is plotted at constant values of Λ for two different values of Λ (top panel and bottom panel). The analysis shows that the source for the higher energy electrons, as revealed by the peaks in the PSD and interpreted on the basis of diffusion concepts, is in the middle magnetosphere, as anticipated from other planets and earlier work at Saturn (Belcher et al. 1991; Cheng et al. 1991; Fillius 1976; Mauk et al. 1995; Richardson et al. 1995; Van Allen 1976; Van Allen 1984). The lower energy electrons appear to have a source that is shifted planetward from the regions identified as the source for the higher energy electrons. The word "source" is used in the literature

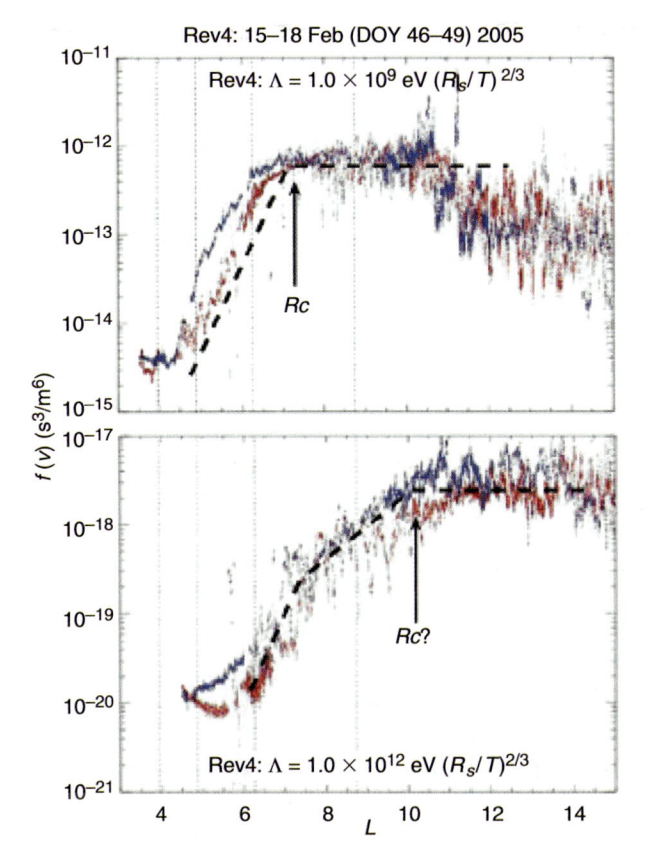

Fig. 11.31 Radial electron phase space density profiles at two selected "isotropic" (see text) adiabatic invariants for lower energy (top) and higher energy (bottom) electrons within Saturn's inner magnetosphere. After Rymer et al. (2007a). Copyright: American Geophysical Union, 2007

on PSD analysis in a somewhat ambiguous fashion. It is a region where the PSD is high in a fashion that cannot be explained on the basis of adiabatic diffusive transport from the surrounding regions. That region can be high because there is in fact a source of matter there or simply because particles are energized there by processes that violate the adiabatic invariants. Given new knowledge about the true sources of matter in the Saturnian system (Sections 11.2.1–11.2.3), the "source" regions identified in Fig. 11.31 are likely regions where non-adiabatic energization occurs, not necessarily where charged particles are born out of neutral matter. An outstanding question is: why are low-energy particles non-adiabatically energized in a region different from that where the high-energy particles are so energized?

11.4.2.2 Pickup Acceleration and Related Processes at Saturn

Background

Pickup ion energization is important and fundamental to space environments like Saturn's, where magnetized plasmas are in contact with or coexist with neutral gases. Figure 11.32, constructed by Huddleston et al. (1998) for the case of Jupiter's Io torus region, summarizes a number of key aspects of the process. In the upper left schematic there is an outward-directed electric field associated with the rotational flow of the plasma. As shown in the lower left, a neutral atom that is suddenly ionized moves initially in the direction of

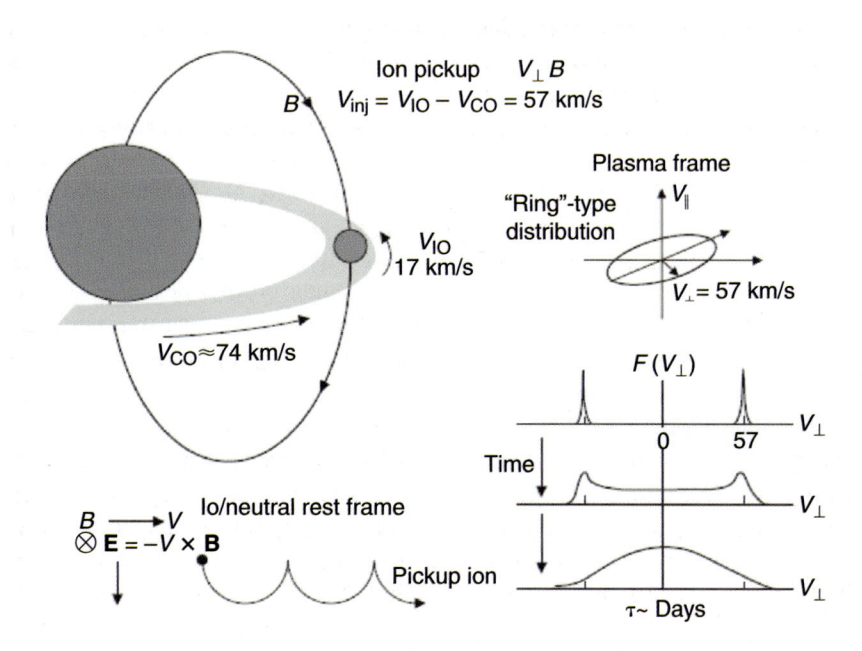

Fig. 11.32 Schematics related to the generation of pickup ions in Jupiter's inner magnetosphere. After Huddleston et al. (1998). Copyright: American Geophysical Union, 1998

the radial electric field, but then, feeling the effects of the magnetic field, executes a cycloidal motion that corresponds to a guiding center drift with the value $\mathbf{V}_D = \mathbf{E} \times \mathbf{B}/B^2$ drift. Transforming into the $\mathbf{V}_D$ motional frame of reference, one finds that the instantaneous speed of the particle is just V_D. So the particle has gained a translational or flow energy of $E_F = m \cdot V_D^2/2$ as well as a "thermal" speed of $E_T = m \cdot V_D^2/2$. As shown in the upper right of the figure, the distribution begins as ring distribution with velocity vectors strictly perpendicular to the magnetic field. In the lower right we see that the ring distribution is broadened over time into a more normal distribution by wave-particle interactions. Electromagnetic Cyclotron Waves (EMIC) play an important role in broadening the distribution (e.g., Cowee et al. 2006, 2007).

There are important secondary consequences to the pickup process. While the newly born electrons that result from the ionization, with their very small gyroradii, stay very close to the radial positions of their birth, the ions move substantial distances in the direction of the electric field. The net current J_{PU} scales as $V_D (\partial n/\partial t)/B$, where $\partial n/\partial t$ is the birth rate per volume of pickup ions, and provides the force ($\mathbf{J} \times \mathbf{B}$) to accelerate the ions to the drift speed V_D. The combined flow and thermal energy that goes into the pickup ions is provided by $q \cdot \mathbf{E} \cdot \mathbf{J}_{PU}$, extracted from the generator (the rotating planet with magnetized ionosphere) that maintains the electric field. Finally, because there are radial gradients in gas density, ionization rates, and magnetic field strength, the pickup current diverges, with closure currents thought to flow in part along the field lines and through the ionosphere (Section 11.3.2). Depending on the impedances associated with the closure currents, electric potentials build up at various radial positions with associated electric fields that partially reduce the initial rotational electric field. This change causes the rotational flow speed to lessen. This whole process is called "mass loading" of the flow.

Pickup Energization at Saturn

The importance of pickup energization is apparent from the initial reports of data from the Cassini CAPS instrument. Figure 11.33 (Young et al. 2005) shows structured ion features (bottom panel) with energy values that scale with radial position according to the rotational flow energy of protons and oxygen ions (the curved lines on the figure). This figure does not cleanly discriminate between flow and thermal energies. When proper moments are constructed within the flowing frame of reference, one finds that the thermal temperatures indeed scale roughly as expected from the pickup process with radial distance (L^2 expected) and with mass species (Sittler et al. 2006a). The large observed ion temperature anisotropy reported ($T_\perp \gg T_\parallel$; Sittler et al. 2006a) is also expected. As anticipated from our discussion of Fig. 11.32, the EMIC waves that participate in the smoothing of the predicted initial ring distributions into the more normal distributions are very active within the regions of pickup ionization (Leisner et al. 2006; Russell et al. 2006; Russell and Blanco-Cano 2007).

Vasyliūnas (2008) addresses the interesting question concerning whether Saturn's or Jupiter's magnetospheres are more mass loaded in terms of the lessening of rotational flow. He finds, based on appropriate scaling of the two systems, that Saturn is more affected despite the greater amount of mass per time released into the Jovian system. Pickup energization and mass loading is also invoked substantially in reference to the interactions between Saturn's magnetosphere and the local environments of Saturn's icy moons and Titan, as discussed in Section 11.2.4.

Pickup-Associated Electron Acceleration

While the pickup energization of ions in Saturn's magnetosphere seems very straightforward, the associated apparent energization of electrons is less so. Figure 11.33 shows that the energies of the electrons <100 eV in the inner regions also appear to scale roughly with the energies of the pickup protons (Sittler et al. 2006a). Rymer et al. (2007a) have investigated the idea that Coulomb collisions between the picked-up protons and cold electrons results in an equipartitioning of energy between the electrons and ions, based on the formulas of Spitzer (1962). Assuming that the protons have thermal energies consistent with rotational pickup, e-folding times for equipartitioning is ~15 rotations of Saturn. Rough estimates of the occupation time for the plasmas, the time available for the particles to equalize their temperatures, are at least comparable if not longer.

Other ideas for the equipartitioning of the electron and ion temperatures include (a) wave particle interactions (Sittler et al. 2006a, based on Barbosa 1986) and (b) field-aligned electron acceleration associated with the magnetosphere-ionosphere coupling associated with the mass-loading process described above and the small-scale interchange features that transport the newly created mass outward (Section 11.3.4). The energization associated with the later process is likely a result of field-aligned acceleration that occurs when there are insufficient charge carriers to carry the requisite currents, as occurs with auroral acceleration (Chapter 12).

As discussed by Delamere et al. (2007) and addressed in Section 11.2.3.2, understanding the differences between Saturn and Jupiter with regard to the scaling of electron energization to planetary rotation is important to understanding why the ratios of neutral gas to plasma contents at Jupiter and Saturn are so different.

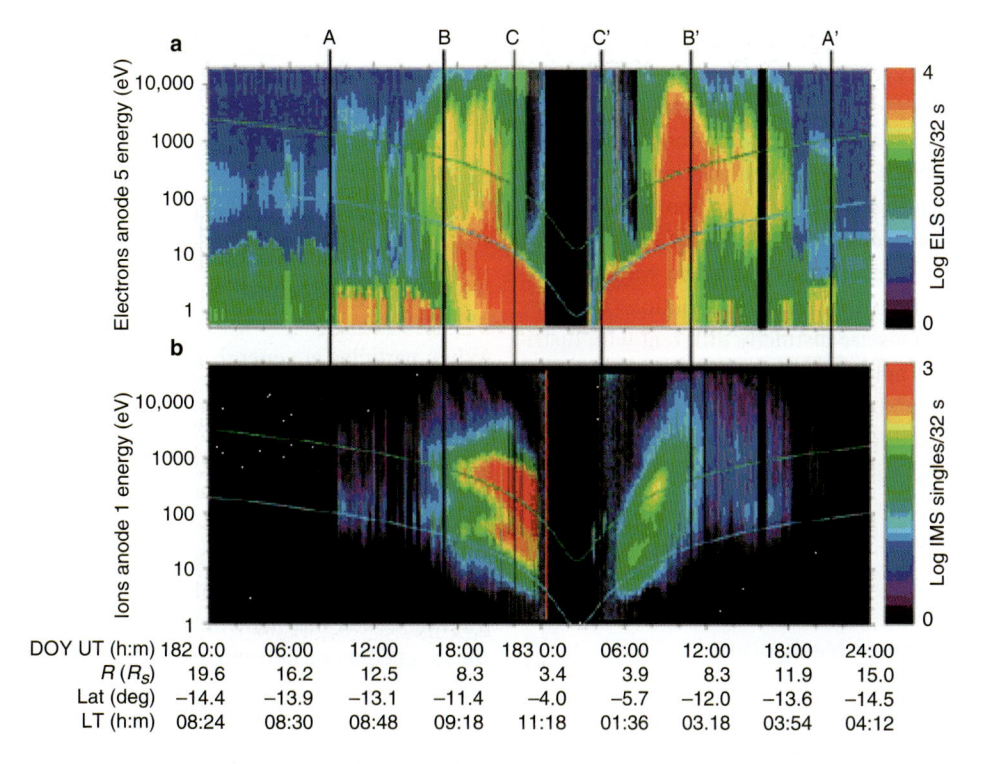

Fig. 11.33 Electron and ion plasma distributions observed in Saturn's magnetosphere. After Young et al. (2005). Copyright: American Association for the Advancement of Science, 2006 (Science Magazine)

11.4.2.3 Other Acceleration Processes

A number of other processes have been invoked for the Saturnian system on the basis of qualitative similarity between phenomena observed within the Saturnian system and those observed at Earth or at Jupiter.

Injection Energization

Mitchell et al. (2005) discovered dramatic increases in ion intensities, as diagnosed with ENA imaging, at the base of Saturn's nightside magnetotail at radial distance near and even inside $\sim 20 R_S$. These events have been likened to substorm injection phenomena at Earth (see also Bunce et al. 2005a). Significantly, the authors point out that the changes in the oxygen ion intensities is more dramatic than are the changes in the proton intensities, with the oxygen ions rising first and more steeply (see also Chapter 10). Mitchell et al. (2005) suggest that the injection process is rapid enough that the energization is sensitive to the gyroperiod of the ions, with the slow gyroperiod ions like oxygen affected more strongly by the injections than are the fast gyroperiod ions like protons, as modeled by Delcourt (1990, 2002). With this process, the electric field of the injection can "turn on" very suddenly and cause the ions to respond in much the same fashion that freshly ionized ions respond to the pickup process described in an earlier section.

However, the violation of the first adiabatic invariant is not required of the injection process to achieve large differences in the response of protons and oxygen ions. For injection timescales much longer than those required to violate the first adiabatic invariant, the so-called "centrifugal" acceleration mechanism comes into play. This process violates the second adiabatic invariant, but not the first. Specifically, for fast changes in the magnetic field configuration, the following equation is relevant for the acceleration of ions parallel to the magnetic field (Northrop 1963; see Mauk 1986, 1989):

$$\begin{aligned} m \frac{dv_\parallel}{dt} &= -\mu \cdot \frac{\partial B}{\partial s} + m \cdot \bar{V}_D \cdot \frac{d\hat{b}}{dt} \\ &= -\mu \cdot \frac{\partial B}{\partial s} + m \cdot \bar{V}_D \cdot \left(\frac{\partial \hat{b}}{\partial t} \right)_s - \frac{m \cdot V_D \cdot v_\parallel}{R_C}, \end{aligned}$$

(11.12)

where s is distance along **B**, $\mathbf{V}_D$ is the electric field convection drift speed, **b** is the unit vector of **B**, and R_C is the radius of curvature of the local magnetic field line. The last two terms of the lower row of this equation can be thought heuristically as acceleration resulting from centrifugal forces acting either against the convection velocity or the particle velocity.

Such centrifugal acceleration was proposed by Quinn and Southwood (1982) and modeled in detail by Mauk (1986) to explain acceleration observed in association with substorm injections at Earth. Oxygen ions are accelerated much more dramatically than are protons. The question is: are strong differences in the acceleration of oxygen ions and protons at Saturn explained by the centrifugal acceleration mechanism, the first-invariant-violating "pickup" process modeled by Delcourt (2002), or some combination of the two? The modeling of Sanchez et al. (1993) points the way to a resolution by showing that the angular distributions resulting from these two processes are distinctly different with distributions that favor the field-aligned direction resulting from the centrifugal terms, and distributions that favor the perpendicular directions resulting from the "pickup" term. At Earth, ion distributions in the middle magnetosphere strongly favor the centrifugal terms (Mauk 1986; Sanchez et al. 1993). We propose that such distinct signatures be sought at Saturn.

Miscellaneous Acceleration Mechanisms

Reconnection (see Section 11.4.1) has long been invoked to explain transient and/or localized energization events, although the mechanisms of acceleration by reconnection processes are highly uncertain (e.g., Hoshino 2005; Drake et al. 2005; Pritchett 2006). Invocations at Saturn include those by Bunce et al. (2005a) and Hill et al. (2008a), who both use multi-instrument observations that show likely signatures of reconnection correlated with sudden increases, perhaps injections, in the intensity of energetic particles. Connections between reconnection and auroral acceleration at Saturn have been invoked by several authors (Cowley et al. 2004, 2005; Bunce et al. 2005b; Gerard et al. 2005). Interestingly, it is acceleration that occurs at low altitudes, resulting from strong magnetic field-aligned currents generated by reconnection, that is invoked, not acceleration occurring at the localized site of the reconnection. The possible connection between auroral processes and reconnection processes is addressed in Chapter 12.

More general auroral acceleration processes operating at Saturn are also addressed by Chapter 12. There exists a tremendous literature on ion and electron acceleration processes that occur at low altitudes within the auroral zones that result from strong magnetic field-aligned current flowing in regions that are starved for charge carriers (e.g., Lysak 1993; Borovsky 1993; Bespolov et al. 2006). While specific mechanisms have not been tested at Saturn, the similarities between observed auroral particle distributions at Saturn and Earth suggest strongly that essentially the same auroral acceleration process are operating within both environments.

The most novel acceleration mechanism proposed for Saturnian electrons is that of Jones et al. (2006b) who propose that runaway electrons caused by the strong electric fields within high-altitude atmospheric lightning escape into the magnetosphere and stimulate the occurrence of Saturn's B-ring spokes.

11.4.3 Current Generation

11.4.3.1 Background

When particles are energized, a portion of the energy can be distributed into the magnetic field energy, manifesting itself into distortions in the magnetic field configuration. The distortions are causes by electric currents that are carried by the particles, and in turn, the currents are moderated by the distortions in the magnetic field configuration.

Energetic particles in a magnetic field configuration without external forces carry the currents associated with the guiding-center magnetic field drifts, inertial force drifts, and the magnetization currents associated with the gyration of the particles around the field line. When these contributions are combined (Parks 1991) one finds that for gyrotropic distributions, the currents normal to the magnetic field are

$$\bar{J}_\perp = \frac{\bar{B}}{B^2} x \bar{\nabla}_\perp p_\perp + (p_\| - p_\perp) \frac{\left[\bar{B} x \left(\hat{b}\cdot\nabla\right)\hat{b}\right]}{B^2} + m \cdot n \frac{\hat{b}}{B} x \frac{d\bar{V}_\perp}{dt}, \quad (11.13)$$

where p is total particle pressure, $\perp$ is perpendicular to **B**, $\|$ is parallel to **B**, **b** is the unit vector along **B**, **V** is the plasma velocity, and $m \cdot n$ is the mass density. The first term is the "diamagnetic current." The second term, the pressure anisotropy current, is what is left of the so-called gradient and curvature drifts when partially cancelled by magnetization currents. For an isotropic plasma $(p_\| = p_\perp)$, this term disappears. Note that in this term, $\mathbf{b} \times (\mathbf{b} \cdot \nabla) \mathbf{b}$, can be rewritten as κ, the so-called "curvature vector," that has magnitude of $1/R_C$, where R_C is the radius of curvature of the field line. Finally, under the highly restrictive assumption that the magnitude of $\mathbf{V}_\perp$ does not change with time or space along flow streamlines, the inertial term becomes the centrifugal term with $dV_\perp/dt$ becoming just $\mathbf{\Omega} \times \mathbf{V}_\perp$, where $\mathbf{\Omega}$ is the rotation vector of the plasma flow. The currents specified in Eq. 11.13 are established within a magnetic field configuration when force balance is achieved between the magnetic field and the particle flows. Specifically,

$$\bar{J}_\perp x \bar{B} = -\bar{\nabla}\left(\frac{B^2}{2\cdot\mu_o}\right) + \frac{1}{\mu_o}\left(\bar{B}\cdot\bar{\nabla}\right)\bar{B}$$
$$= \bar{\nabla}_\perp p_\perp + \bar{\kappa}\cdot(p_\| - p_\perp) + n\cdot m\cdot(\hat{b}\cdot\bar{\Omega})\cdot(\hat{b}x\bar{V}_\perp). \quad (11.14)$$

The top row is the standard decomposition of magnetic forces (Jackson 1999). The bottom row shows the particle forces that must be in balance with the field forces to establish equilibrium. There is the pressure gradient term, the pressure anisotropy term, and the centrifugal term, where again we have assumed that the magnitude of $\mathbf{V}_\perp$ does not change in time or space along the flow streamlines. Our challenge, now, is to determine which terms are important for carrying the so-called Ring Current around the planet and establishing force balance within our target planetary magnetospheres.

11.4.3.2 Force Balance at Jupiter and Saturn

Since the discovery of Jupiter's magnetodisc configuration by Pioneer (Smith et al. 1974), it has generally been believed that that configuration is supported by centrifugal forces of rotating plasmas (the last term in Eq. 11.14; Vasyliūnas 1983). However, McNutt (1984), followed by Mauk and Krimigis (1987), demonstrated that the centrifugal stresses are insufficient to balance the radial magnetic forces at just the magnetic equator by more than an order of magnitude. Subsequently, Paranicas et al. (1991) and then Frank and Paterson (2004) demonstrated that measured pressure anisotropies were sufficient to balance the equatorial radial forces within Jupiter's magnetodisc. While it continues to be believed by many in the planetary magnetosphere community that the magnetodisc configuration is synonymous with the centrifugal force, the evidence at Jupiter strongly suggests otherwise. This conclusion is not universally accepted (e.g., Russell et al. 1999); however, the evidence cited here has not been confronted in the literature.

At Saturn, problems with radial force balance were also revealed by McNutt (1984), but more detailed characterization of the magnetic forces by Mauk et al. (1985) showed that Saturn's Ring Current is configured such that centrifugal forces are the likely dominant beyond about $14R_S$, and pressure gradient forces are likely dominant in the more planetward regions. The surprise was that despite the predominance of the centrifugal forces in the middle to outer regions, a magnetodisc-like configuration was not observed.

It has now been discovered by Cassini that Saturn has a partial magnetodisc-like configuration outside of $15-20R_S$ in the dawn flank side of the magnetosphere during compressed magnetospheric conditions, and a full magnetodisc configuration during uncompressed conditions (Arridge et al. 2007, 2008; see also Dougherty et al. 2005). These authors also report that estimations of the radial magnetic forces within the magnetodisc appear to be balanced by the centrifugal forces of rotating plasmas. Other less direct findings also point to the importance of rotational centrifugal affects within Saturn's magnetosphere. Alexeev et al. (2006) followed by Bunce et al. (2007) have argued that, because of the strong sensitivity of the total magnetic moment of the ring current to variations in the size of the magnetosphere as driven by the solar wind, the centrifugal forces dominate over the pressure gradients forces overall in establishing the character of the ring current. The latter authors demonstrated that the Ring Current response to size variation is expected to be large for centrifugal contributions and relatively small for pressure gradient contributions. These authors did not test the possible sensitivity of pressure anisotropy contributions to changes in magnetospheric size.

While rotational centrifugal contributions are undoubtedly large, given the results cited above, it remains important to establish the relative contributions of the different terms in local regions. Sergis et al. (2007, 2009) and Krimigis et al. (2007) have reported high (>1) plasma beta parameters (plasma-pressure/magnetic-pressure) for the middle regions, and Sittler et al. (2006a) have reported high values for the inner magnetosphere. These findings point to an important role for plasma pressure effects, but that role is difficult to evaluate based just on the parameters provided alone. Initial efforts to use these findings to estimate the contribution of the pressure gradient term to the current generation (Kellett et al. 2008) show very small contributions within the inner magnetosphere ($<10R_S$) and low (20%) to substantial contributions (e.g., location dependent: 40%, 50%, 80%) in the middle and outer magnetosphere, depending on activity level. For the inner magnetosphere, the 3-keV lower limit on calculated contributions to the total particle pressure used here may not include important contributions from the colder plasma populations (André et al. 2007; Sittler et al. 2006a).

One simple approach to determining the relative importance of the centrifugal contributors is to test whether there are sufficient current carriers assuming centrifugal contributions alone to account for observed magnetic stresses. Figure 11.34 shows one attempt to do so. The figure, from Mauk et al. (1985), shows normalized radial magnetic forces determined with Voyager 1 magnetic field measurements (lines) compared with the centrifugal radial forces of the rotating plasmas, as determined (black dots) by Voyager 1 measurements (McNutt 1984). The red symbols are based on new Cassini measurements made just at the magnetic equator. The density and magnetic field is from Rymer et al. (2007b), and the rotational flows are from Voyager (Richardson 1986, 1998; Richardson et al. 1998). We see that as compared with the Voyager-derived centrifugal forces, the Cassini centrifugal forces are higher, perhaps because the Cassini result were obtained at just the magnetic equator, contrary to the Voyager results. None-the-less, the structure of the currents is not explained by the centrifugal forces, and the peak in the magnetic field forces is a factor of 3 greater than the forces that are offered by the centrifugal term. This exercise is clearly not definitive, but it appears likely to be the case that a mix of centrifugal forces and pressure-driven forces is needed to

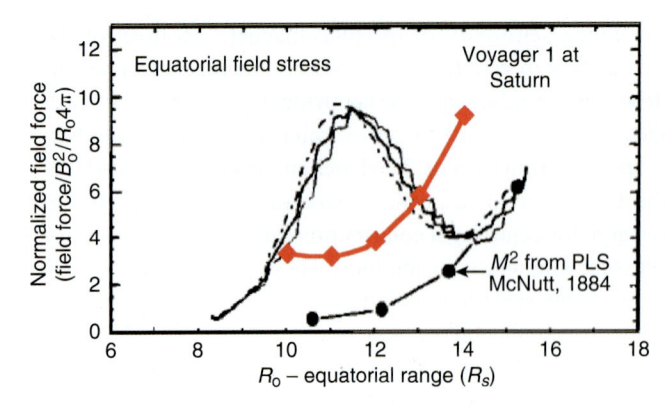

Fig. 11.34 A scoping exercise showing normalized radial magnetic density within the equatorial magnetic field configuration of Saturn as calculated using Voyager data (Mauk et al. 1985) and the normalized radial centrifugal force of ions as characterized by Voyager (dots) and by Cassini (diamonds) using plasma parameters from Rymer et al. (2007b) and flow velocities from Richardson (1986, 1998) and Richardson et al. (1998). The red points at 10, 11, 12, 13, and 14_S have densities ($1/cm^3$), field strengths (nT), and rotational speeds (km/s) of roughly (3.6, 14, 70), (2.0, 10.5, 68.5), (1.23, 8.1, 75), (0.79, 6.4, 90), and (0.58, 5.1, 110), respectively

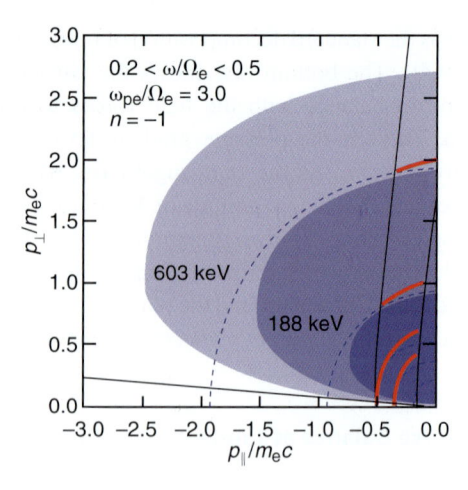

Fig. 11.35 An example of cyclotron resonant ellipses (near vertical thin solid lines) and resonant diffusion surfaces (thick solid curved line segments) in normalized momentum space for a typical band of chorus emission in the low-density region outside the Earth's plasmapause at $4.5 R_E$ (adapted from Horne and Thorne 2003). Also shown are constant energy surfaces (dashed lines) at 188, and 603 keV, which illustrate the change in resonant energy along the diffusion surfaces. The shaded regions represent contours of constant phase space density (PSD) for a typical distribution with a strong loss-cone feature. The loss cone is the straight line near the $p_\parallel$ axis

explain the structure of the ring current and the nature of the force balance between particles and the magnetic field. Much work clearly remains to be done.

11.4.4 Wave Particle Interactions

11.4.4.1 Background on Wave Particle interactions

Trapped particles in a planetary magnetosphere have three types of periodic motion: gyro motion, bounce motion, and drift motion (Section 11.4.2). Resonances between the periodic particle motion and waves can energize or scatter particles by breaking the associated adiabatic invariants and/or by scattering particles into the loss cone. For example, inward radial diffusion by ultra-low-frequency (ULF) waves can break the third invariant and accelerate particles to higher energies through conservation of the first and second invariants. Very-low-frequency (VLF) waves, such as whistler-mode chorus, can produce pitch angle diffusion, breaking the first or second invariant, resulting in either local acceleration of the particles or scattering of the particles into the loss cone.

Resonant interactions between energetic particles and magnetospheric plasma waves provide the source of energy for wave excitation, particle energization, and particle loss. As an example, Fig. 11.35 illustrates how energetic electrons in the Earth's radiation belts are affected during resonant energy exchange with whistler-mode chorus emissions (Horne and Thorne 2003). The nearly vertical thin solid lines demarcate the first-order cyclotron resonance condition in momentum space for typical terrestrial whistler-mode chorus emissions at frequencies between 0.2 and 0.5 of the local equatorial gyro-frequency. During cyclotron resonance, electrons move along the thick line segment resonant diffusion curves (Summers et al. 1998). Resonant electrons will statistically diffuse toward regions of lower PSD. The observed drop in PSD in the vicinity of the loss cone (shaded contours of constant phase space density near the bottom of the plot) ensures that resonant electrons at small pitch angles ($p_\perp < p_\parallel$) diffuse toward the loss cone; the accompanying loss of energy provides the source of free energy for wave excitation. Conversely, at larger pitch angles ($p_\perp > p_\parallel$), where the observed distributions become quasi-isotropic (again see the shaded contours of constant PSD), diffusion occurs in the direction away from the loss cone and toward higher energy where the PSD is smaller. This motion provides a local acceleration process for the high-energy trapped population. The net effect of such interactions constitutes an energy transfer from the low-energy population (10–100 keV), which is injected into planetary radiation belts by large-scale processes (convection or interchange), to the high-energy trapped radiation belts, using the waves as an intermediary.

For other wave modes discussed in this section, the diffusion paths (red traces) are different than those shown in Fig. 11.35, and the particle distribution shapes that either cause the waves to grow (transfer of energy from particles to wave) or cause particle energization (transfer of energy from waves to particles) are different, but the figure illustrates the basic idea. Such processes have been extensively studied in the terrestrial magnetosphere, and it is anticipated that similar physical processes should occur in the Saturnian system. Here we review what is known about the properties of important plasma waves in the Saturnian magnetosphere and comment on their potential affect on the particle populations.

11.4.4.2 Background on Wave Modes

The Voyager spacecraft detected a variety of plasma and radio waves during their flybys of Saturn (Gurnett et al. 1981a; Scarf et al. 1982, 1983, 1984; Warwick et al. 1981, 1982; Kaiser et al. 1984). Electromagnetic radio emissions (that freely propagate away from the system and are detected remotely) included Saturn Kilometric Radiation (SKR; Kaiser et al. 1980, 1984), lightning-produced Saturn electrostatic discharges (SEDs) (Warwick et al. 1981, 1982; Burns et al. 1983; Kaiser et al. 1983), continuum radiation emission (Kurth et al. 1982), and narrowband radio emission (Gurnett et al. 1981b). Plasma wave emissions (generally localized waves detected near their sites of generation) observed by Voyager included electron cyclotron harmonic emissions (ECH) and upper hybrid resonance emissions (UHR) (Kurth et al. 1983), whistler-mode chorus and hiss (Scarf et al. 1984), Langmuir waves observed upstream of bow shocks (Chapter 9; Gurnett et al. 1981a; Scarf et al. 1982), and electromagnetic ion cyclotron waves (Smith and Tsurutani 1983). Many of these emissions have similar characteristics to emissions detected at Earth and Jupiter where they have been shown to play important roles during wave-particle interaction.

11.4.4.3 Application to Cassini

The multiple orbits of Cassini around Saturn allow a more detailed analysis of the radio and plasma waves detected at Saturn than was possible with Voyager. Furthermore, the Radio and Plasma Science (RPWS) instrument is much more capable than earlier instruments (Gurnett et al. 2004). Typical spectrograms of the electric field intensities of the emissions measured by RPWS during a pass through Saturn's inner magnetosphere are shown for an orbit close to the magnetic equatorial orbit plane (Fig. 11.36) and for a higher inclination orbit (Fig. 11.37). The intensities are plotted as decibels (dB) above background. The white line shows the electron cyclotron frequency (f_{ce}) derived from the magnetometer instrument (Dougherty et al. 2004).

An examination of Figs. 11.36 and 11.37 shows that a variety of plasma and radio waves are usually detected in the inner magnetosphere, including SKR, ECH, and UHR bands and whistler-mode emissions, primarily chorus and hiss. At lower magnetic latitudes, a low-frequency, broadband emission is often detected, while at higher magnetic latitudes, an emission similar to auroral hiss is usually observed. Narrowband radio emissions ranging from a few kilohertz to tens of kilohertz are also often observed, especially during periods when Cassini is at higher inclination. In this chapter we discuss the ECH, UHR, narrowband radio, and whistler-mode emissions, and their importance in understanding the dynamics and processes occurring in Saturn's magnetosphere. The SKR, auroral hiss, and the SED emissions (not present in either of these two figures because of the lack of lightning-producing storms during the period of those figures) is discussed in greater detail in other chapters.

ECH, UHR, and Narrowband Radio Emissions

Electrostatic emissions at Saturn include the ECH waves and the UHR emission. ECH emission typically occurs at multiple harmonics of the electron cyclotron frequency (Ω_e), specifically $(n + 1/2)\Omega_e$, where n is an integer. ECH and UHR emissions, which are also commonly observed near the equatorial plane in the terrestrial and Jovian magnetospheres, can be excited by a loss-cone anisotropy in the electron distribution, like that illustrated in Fig. 11.35 (e.g., Kennel et al. 1970; Ashour-Abdalla and Kennel 1978). At the Earth, these electrostatic emissions play an important role in pitch-angle scattering of low-energy (hundreds of electron-volts to a few kiloelectron-volts) electrons, produce highly anisotropic electron pancake distributions (strong local maximum in intensity perpendicular to the magnetic field; Lyons 1974; Horne and Thorne 2000), and can contribute to the diffuse electron aurora (Horne et al. 2003) by scattering electrons into the loss cone. At Saturn it has been suggested that they similarly interact with electrons in the energy range of tens of kiloelectron-volts (Kurth et al. 1983; Scarf et al. 1984).

ECH emissions are highly oblique (propagation vectors nearly perpendicular to the ambient magnetic field) waves and are usually excited over a broad band of frequencies between harmonics of f_{ce}. ECH waves are one of the most intense emissions observed in planetary magnetospheres (often with amplitudes >1 mV/m), are good tracers of dynamic processes in the magnetosphere, and are believed to be the source of the narrowband radio emissions (Kurth 1982). Figure 11.38 shows a time–frequency spectrogram of high-resolution wideband receiver (WBR) data from the RPWS instrument during a period when the spacecraft encountered a plasma injection event (discussed in Section 11.3.4). As can be seen, the characteristics of the ECH waves change drastically (increase in intensity and harmonic structure) from about 17:35 to about 17:40 SCET. The change in the ECH waves matches the period of a local plasma injection event

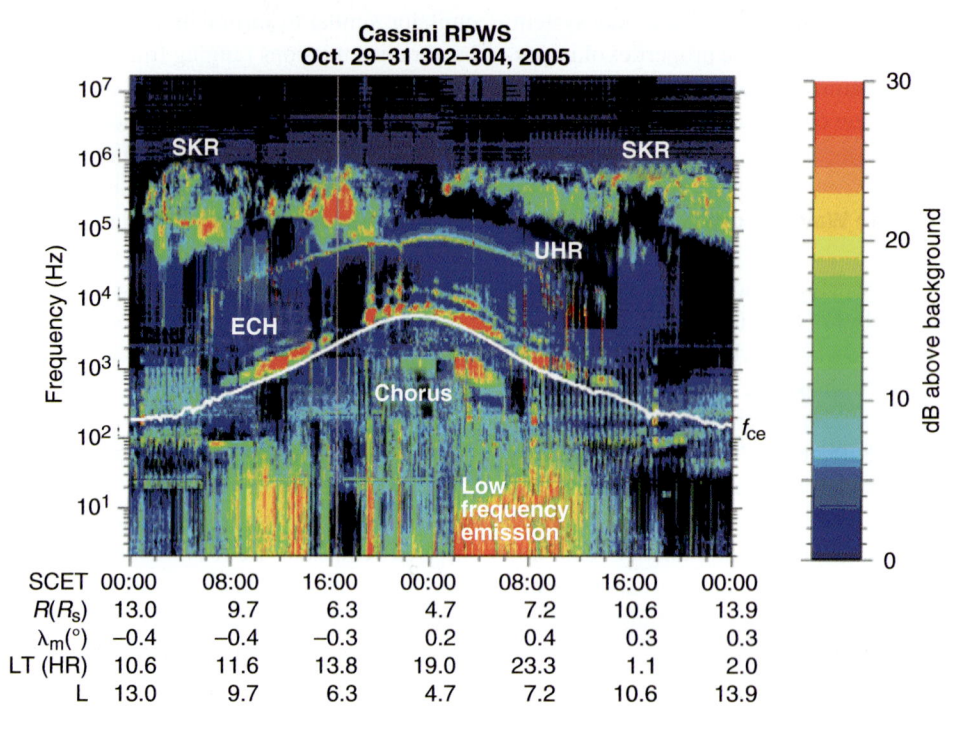

Fig. 11.36 Frequency–time spectrogram of the power spectra of the wave electric field as measured by Cassini close to Saturn's equatorial plane. (See Gurnett et al. 2005; Menietti et al. 2008a, b; and Hospodarsky et al. 2008 for discussions of the emissions shown in this figure.)

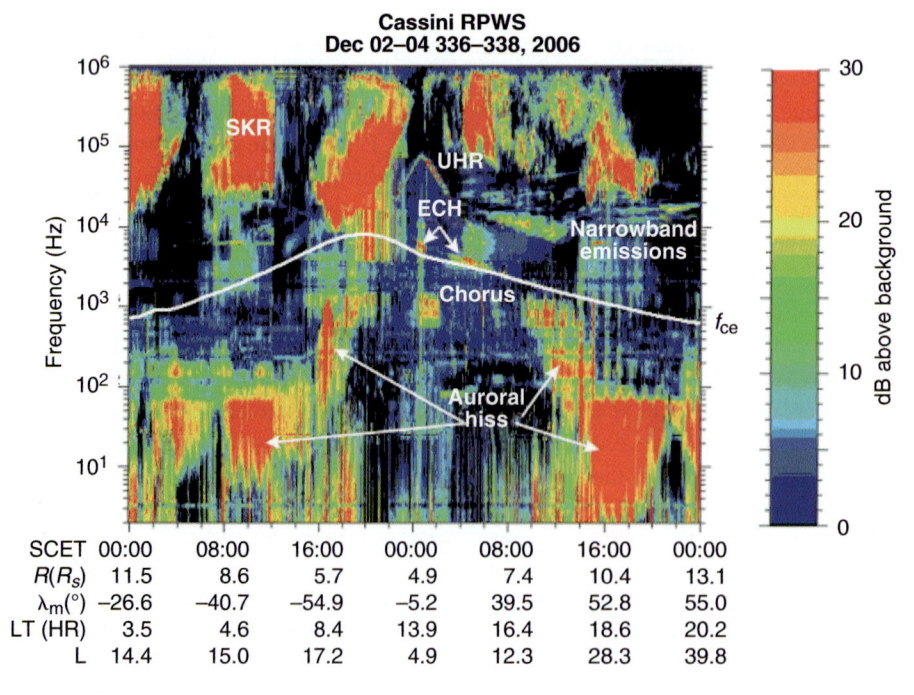

Fig. 11.37 Frequency–time spectrogram of the power spectra of the wave electric field as measured by Cassini at Saturn from high planetary latitude. (See Gurnett et al. 2005; Menietti et al. 2008a, b; Hospodarsky et al. 2208; and Mitchell et al. 2009 for discussions about the emissions shown in this figure.)

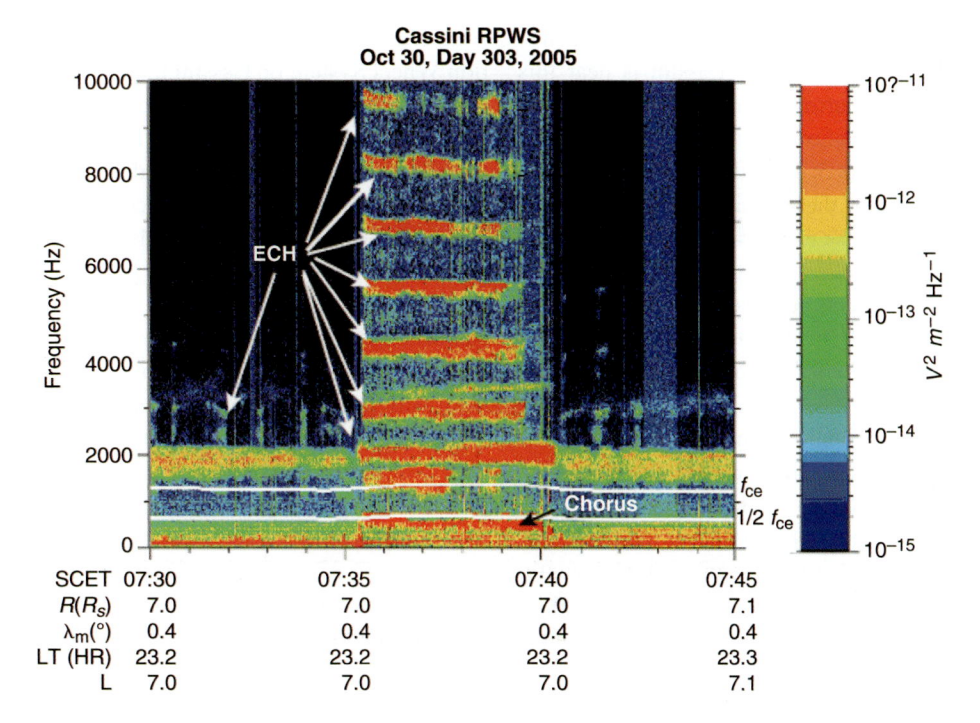

Fig. 11.38 High time resolution, frequency–time spectrogram of the power spectra of the wave electric field as measured by Cassini before, during, and after a dynamic injection event within Saturn's inner magnetosphere (from Hospodarsky et al. 2008). Copyright: American Geophysical Union, 2008

reported by Rymer et al. (2008) using the CAPS electron data. Outside of the event, the ECH waves are primarily found in the first harmonic band centered near 1.5 f_{ce}, with weaker, more sporadic bands at the higher harmonics. High time and frequency resolution observations of ECH waves at Saturn often show a more complex spectrum than simple bands at $n + 1/2$ harmonics. Inside the injection event, the intensity of the ECH emissions increases by at least an order of magnitude. Intense whistler-mode emissions are also detected during this event, and will be discussed in more detail later.

Menietti et al. (2008a, b) have modeled the electron plasma distributions observed inside (Menietti et al. 2008a) the plasma injection and for a period a few hours before (Menietti et al. 2008b) the injection event, and performed a linear dispersion analysis of the expected wave modes. For both inside and outside the injection event, the authors found that the observed ECH emissions can only be generated by the PSD gradients provided by the loss cone in the lower energy, cooler populations. In the injection event, a loss cone of about 3° (too small to be detected by the electron instrument during this period) in the cold electron population produced the observed ECH waves. Outside of the injection event, a larger loss cone of about 20° was needed to produce ECH waves with the characteristics observed (weaker fundamental and first harmonic).

ECH waves are also believed to be the source of the narrowband radio emissions (Fig. 11.37) that have been detected at every planet that has a magnetosphere: Earth (Gurnett 1975; Jones 1976; Kurth et al. 1981), Jupiter (Kaiser and Desch 1980; Gurnett et al. 1983), Saturn (Gurnett et al. 1981a; Scarf et al. 1982), Uranus (Kurth et al. 1986), Neptune (Kurth et al. 1990), and the Jovian moon Ganymede (Kurth et al. 1997). They are believed to be produced by a mode conversion of the electrostatic ECH waves when one of the ECH frequencies $(n + 1/2) f_{ce} \approx f_{UHR}$, the upper hybrid resonant frequency, usually in a region of a steep spatial density gradient (Kurth 1982). Cassini has detected this emission throughout the prime mission (Gurnett et al. 2005; Louarn et al. 2007; Wang et al. 2009).

Because of the relationship between the generation of the narrowband emissions, the ECH waves, and the density gradients, it has been suggested that the narrowband emissions can be a remote sensor of changes in the plasma conditions at the source region. Louarn et al. (2007) showed that the occurrence of the narrowband emissions is often associated with sudden intensification of the SKR emission and suggested that the events are associated with energetic "events" in the Saturn magnetosphere that produce density changes in the Saturnian plasma disk. Wang et al. (2009) argue that the narrowband emissions are related to energetic ions injected into the inner magnetosphere (Chapter 10) during periods of high magnetic activity. Ye et al. (2009) used similar techniques to suggest that the source region of the narrowband emission is located near the northern and southern edge of the main torus of high-density plasma, a suggestion supported by observations of intense electrostatic ECH and electromagnetic narrowband emissions when Cassini is near this region. A series of similar narrowband emissions were also detected over the rings of Saturn during the orbit insertion of Cassini (Gurnett et al. 2005; Farrell et al. 2005). Farrell et al. (2005) suggested that these emissions are generated by mode conversion of intense electrostatic emissions at the plasma frequency into Z-mode via whistler-mode propagation into steep density gradients near the inner edge (at about $2.3R_S$) of the plasma torus. The propagation of these emissions provides an opportunity for remote sensing of the dynamics of the plasma torus.

The identification of the UHR frequency, f_{UHR}, given by $f_{pe}^2 + f_{ce}^2 = f_{uhr}^2$ (f_{pe} is the plasma frequency and f_{ce} is the electron cyclotron frequency), can be used to determine the electron number density. Measuring the UHR frequency and determining f_{ce} from the magnetic field strength, B ($f_{ce} = 28B$ Hz), the electron number density, n_e, can be calculated from $f_{pe} = 8980\,n_e^{1/2}$. From the electron number densities derived from the UHR frequency observed by the Cassini spacecraft at Saturn, Persoon et al. (2005, 2006, 2009) have developed a plasma density model for the inner magnetosphere.

Whistler-Mode Emissions

A variety of whistler-mode emissions are detected at Saturn, including lightning-produced whistlers, auroral hiss, chorus, hiss, and a low-frequency emission detected at lower magnetic latitudes (see Figs. 11.36 and 11.37). Whistler-mode emissions, especially chorus, often play an important role in wave-particle interaction and can significantly change the distribution functions of energetic electrons, leading to both precipitation loss and energy diffusion (e.g., Horne and Thorne 2003; Bortnik and Thorne 2007).

One whistler wave packet produced by atmospheric lightning, followed by frequency-dispersed magnetic field-aligned ducting, has been detected by Cassini (Akalin 2005; Akalin et al. 2006). Using the measured dispersion of the whistler, the authors determined that the whistler must have originated from lightning in the northern hemisphere of Saturn (at about 67° latitude), and they used the dispersion properties to diagnose the composition and spatial configuration of the equatorially confined water group ions. The lack of detection of other lightning-produced whistlers by Cassini at Saturn is somewhat surprising considering that multiple whistlers were detected during the Earth flyby (Hospodarsky et al. 2001) and considering the large number of lightning-associated SED events that have been detected by Cassini (Desch et al. 2006; Fischer et al. 2006, 2007, 2008). However, Cassini has not sampled the L shells of the storms that are believed to produce the SEDs (located at about 35° southern latitude), and electromagnetic noise from a Reaction Wheel System on Cassini makes detection difficult.

Auroral hiss is detected on most orbits when Cassini is at higher inclination (Fig. 11.37). These emissions are thought to be generated by magnetic field-aligned electron distributions accelerated by auroral processes. The emissions usually have the characteristic funnel shape and can be observed out to many tens of R_S. The frequency cutoffs of the emissions are related to the local electron plasma frequency and can be used to estimate the electron number density (Gurnett et al. 2005). Propagation analysis of the auroral hiss emissions has shown that they are propagating away from the planet (up from the auroral zone) (Mitchell et al. 2009). The auroral hiss also shows a long-term periodicity with a period similar to the SKR near-planetary rotation radio period and also a short-scale periodicity of the order of one hour (Fig. 11.37 on DOY 337 from about 14:00 to 22:00 SCET). Hiss enhancements are often correlated with ion and electron beams detected by the MIMI instrument (Mitchell et al. 2009). Hiss is discussed in more detail in Chapter 12, and Chapter 10.

An emission similar to auroral hiss was also detected during the Saturn Orbit Insertion (SOI) trajectory as Cassini was passing over the rings (Gurnett et al. 2005). Using ray-tracing calculations, Xin et al. (2006), assuming propagation near the whistler-mode resonance cone, found a source near the B ring at a distance of about $1.76R_S$. Xin et al. (2006) suggest that this emission is produced by a field-aligned electron beam propagating away from the rings. The juxtaposition of this suggestion with that of Jones et al. (2006b) invoking lightning-accelerated electrons propagating toward the rings to explain B-ring spokes is interesting.

Chorus is a whistler-mode emission that has been shown to play a major role in the non-adiabatic dynamics of the electron radiation belts at the Earth (Meredith et al. 2003; Horne and Thorne 2003; Horne et al. 2005; Li et al. 2007) and at Jupiter (Coroniti et al. 1984; Horne et al. 2008). At Saturn, Scarf et al. (1983, 1984) examined the chorus detected by Voyager and found that the amplitudes were too small to play an important role as pitch-angle scatterers. However, Kurth and Gurnett (1991) noted that during the period of the Voyager flybys the magnetosphere of Saturn was not very active, and it is possible that the chorus emission would play a more significant role during more active periods.

Chorus emissions are observed by the RPWS instrument during most Cassini orbits of Saturn (Figs. 11.36 and 11.37). Unlike the Earth and Jupiter, the chorus exhibits different fine structure depending on the region of the inner magnetosphere in which it is detected (Hospodarsky et al. 2008; Fig. 11.39). The "magnetospheric" chorus occurs primarily from L shells of about 5 to 8 and shows no correlation with Saturn magnetic latitude, longitude, or local time. The "magnetospheric"

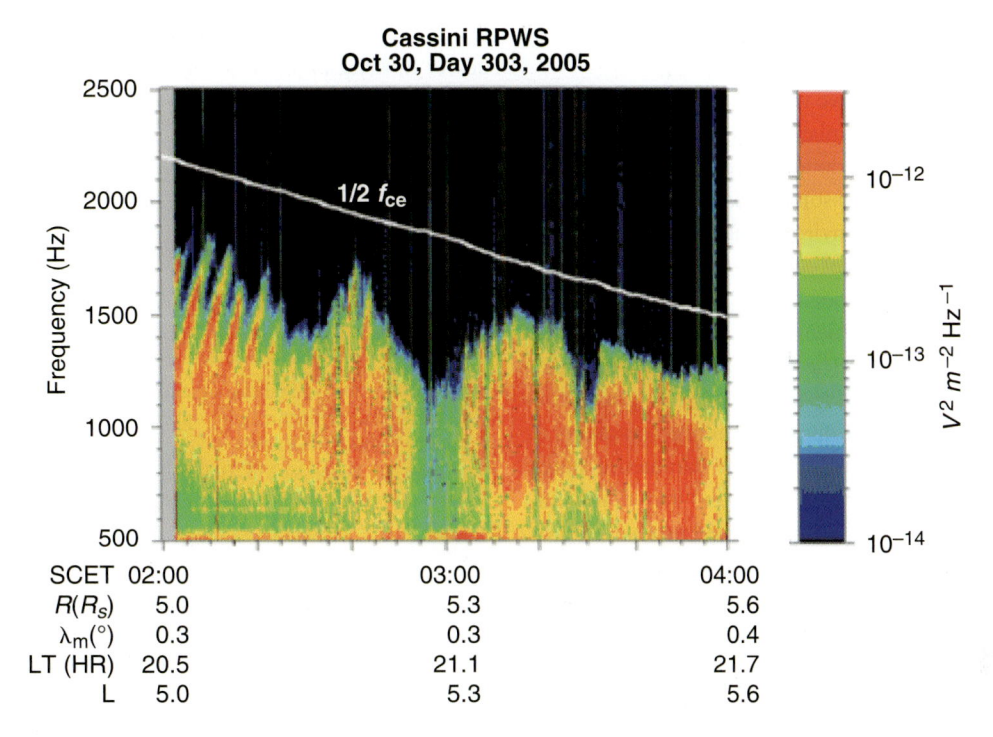

Fig. 11.39 High time resolution, frequency–time spectrogram of the power spectra of the wave electric field of "magnetospheric" chorus emissions measured by Cassini showing selected fine structure of the emissions (from Hospodarsky et al. 2008). Copyright: American Geophysical Union, 2008

chorus detected at larger L shells (between an L of 8 to 10) is usually confined to lower latitudes and occurs at local times near noon (Hospodarsky et al. 2008). The fine structure of the emissions (e.g., Fig. 11.39) often consists of rising tones, but with larger timescales (many seconds to minutes) than typically observed at the Earth or Jupiter (time scales <1 s).

An additional region of chorus detected at Saturn is in association with local plasma injections (defined as "injection event" chorus; Fig. 11.38 at low frequencies). For many plasma injection events, chorus emissions are detected both above and below half the electron cyclotron frequency, with a gap in the emission at half the cyclotron frequency. This chorus is usually detected for only a few minutes, is not observed outside of the injection event, contains fine structure (usually a series of rising tones) at a much smaller timescale (less than a second to a few seconds) than the "magnetospheric" chorus, and appears much more similar to chorus detected at the Earth and Jupiter (Hospodarsky et al. 2008).

An initial examination of the peak wave amplitudes by Hospodarsky et al. (2008) of the chorus emissions at Saturn finds amplitudes much smaller than the peak amplitudes that have been reported at the Earth, but larger by at least an order of magnitude than the amplitudes detected at Saturn by Voyager. The detection of these larger amplitudes suggests that chorus may be responsible for some pitch-angle and energy diffusion at Saturn. For comparison, recent investigations of resonant interactions with Jovian chorus indicate that wave/particle process could provide a viable source for the extremely energetic relativistic electron population in the middle Jovian magnetosphere (Horne et al. 2008; Summers and Omura 2008).

Miscellaneous Plasma Waves

Langmuir waves produced by electrons accelerated from the Saturnian bowshock were detected in the solar wind as Cassini approached Saturn and are also usually detected whenever the orbit of Cassini is large enough to leave the magnetosphere and enter the solar wind (Hospodarsky et al. 2006). Estimates of the electron density of the solar wind determined from the Langmuir wave frequency have been used to help model and predict the location of the Saturnian bowshock (Hansen et al. 2005; Achilleos et al. 2006; Bertucci et al. 2007; Masters et al. 2008).

EMIC waves are important in smoothing out ion distributions that are generated by the ion pickup process associated with the ionization of neutral gas in the inner and middle Saturnian magnetosphere. The pickup process and the associated EMIC waves are discussed in Section 11.4.2.

11.5 Closing Remarks

While Cassini observations have revolutionized our understanding of Saturn's magnetosphere, this book demonstrates that much work remains to be done before full understanding is achieved. But the greater lesson of this chapter is that the study of fundamental space environment processes has matured to a point where this novel environment is substantially understandable from the perspective of processes identified within very different environments, those of Earth, Jupiter, and other planetary objects. The plasma–surface, plasma–gas, and magnetosphere–satellite interactions are greatly informed by previous process studies of similar interactions at Jupiter, despite the vast differences in the parametric states of operant environments. The prediction and then confirmation of enhanced and originally unexplained productivity of gas in the Enceladus region highlights the understandability of Saturn's system on the basis of fundamental processes. In the arena of plasma transport, Saturn has been shown to be the best environment to test concepts that were previously explored most intensely at Jupiter. The confirmation of Saturn as a rotationally driven system again represents a success in applying fundamental processes of a novel system. Reconnection, transient particle energization, current generation, and wave-particle interactions observed at Saturn all have close analogues in phenomena observed elsewhere. The absence of direct evidence for strong reconnection activity on Saturn's dayside magnetosphere is consistent with theoretical expectations derived from the study of other systems. However, on this latter point, it remains to be seen whether observed auroral phenomena can be explained without reasonably robust dayside reconnection. This issue remains open. The apparent "understandability" of this novel Saturnian system is immensely satisfying to those of us who entered the field of planetary space environments at a time when basic exploration was the prime driver. We anticipate that the continued detailed study of processes acting within this environment will continue to test, expand, and validate our understanding of how fundamental plasmas processes behave in planetary space environments and elsewhere.

References

Achilleos N et al. (2006) Orientation, location, and velocity of Saturn's bow shock: Initial results from the Cassini spacecraft. J Geophys Res 111:A03201, doi: 10.1029/2005JA011297

Akalin F (2005) Observation of a whistler in the magnetosphere of Saturn. M.S. thesis, University of Iowa, Iowa City, Iowa

Akalin F, Gurnett DA, Averkamp TF, Persoon, AM, Santolik O, Kurth WS, Hospodarsky GB (2006) The first whistler observed in the magnetosphere of Saturn. Geophys Res Lett 33:L20107, doi: 10.1029/2006GL027019

Alexeev II, Kalegaev VV, Belenkaya ES, Bobrovnikov SY, Bunce EJ, Cowley SWH, Nichols JD (2006), A global magnetic model of Saturn's magnetosphere and a comparison with Cassini SOI data. Geophys Res Lett 33:L08101, doi: 10.1029/2006GL025896

Anderson JD, Schubert G (2007) Saturn's satellite Rhea is a homogeneous mix of rock and ice Geophys Res Lett 34:L02202, doi: 10.129/2006GL028100

André N, Dougherty MK, Russell CT, Leisner JS, Khurana KK (2005) Dynamics of the Saturnian inner magnetosphere: First inferences from the Cassini magnetometers about small-scale plasma transport in the magnetosphere. Geophys Res Lett.32(14):L14S06.1–L14S06.5, doi: 10.1029/2005GL022643

André N et al. (2007) Magnetic signatures of plasma-depleted flux tubes in the Saturnian inner magnetosphere. Geophys Res Lett 34:L14108, doi: 10.1029/2007GL030374

Armstrong TP, Paonessa MT, Bell EV II, Krimigis SM (1983) Voyager observations of Saturnian ion and electron phase space densities. J Geophys Res 88:8893–8904

Arridge CS, Russell CT, Khurana KK, Achilleos N, André N, Rymer AM, Dougherty MK, Coates AJ (2007) Mass of Saturn's magnetodisc: Cassini observations. Geophys Res Lett 34:L09108, doi: 10.1029/2006GL028921

Arridge CS, Russell CT, Khurana KK, Achilleosv, Cowley SWH, Dougherty MK, Southwood DJ, Bunce EJ (2008) Saturn's magnetodisc current sheet. J Geophys Res 113:A04214, doi: 10.1029/2007JA012540

Ashour-Abdalla M, Kennel CF (1978) Nonconvective and convective electron cyclotron harmonic instabilities. J Geophys Res 83:1531–1543

Badman SV, Cowley SWH (2007) Significance of Dungey-cycle flows in Jupiter's and Saturn's magnetospheres, and their identification on closed equatorial field lines Ann Geophys 25:941

Bagenal F (1997) The ionization source near Io from Galileo wake data Geophys Res Lett 24:2111

Barbosa DD (1986) Medium energy electrons and heavy ions in Jupiter's magnetosphere: Effects of lower hybrid wave-particle interactions. J Geophys Res 91:5605–5615

Barbosa DD (1987) Titan's atomic nitrogen torus: Inferred properties and consequences for the Saturnian aurora. Icarus 72:53–61

Barbosa DD (1994) Neutral cloud theory of the Jovian nebula: Anomalous ionization effect of superthermal electrons Astrophys J 430:376–386

Belcher JW (1983) The low-energy plasma in the Jovian magnetosphere. In: Dessler AJ (ed) Physics of the Jovian magnetosphere, pp 68–105. Cambridge, New York

Belcher JW, McNutt RL Jr, Richardson JD, Selesnick RS, Sittler EC Jr, Bagenal F (1991) The plasma environment of Uranus. In: Bergstrahl JT, Miner ED, Matthews MS (eds) Uranus, p 780. University of Arizona Press, Tucson

Bertucci C, Achilleos N, Mazelle C, Hospodarsky GB, Thomsen M, Dougherty MK, Kurth W (2007) Low-frequency waves in the foreshock of Saturn: First results from Cassini. J Geophys Res 112:A09219, doi: 10.1029/2006JA012098

Birch PC, Chapman SC (2001) Detailed structure and dynamics in particle in-cell simulations of the lunar wake Physics of Plasmas 8:4551–4559, doi: 10.1063/1.1398570

Bolton SJ, Thorne RM, Gurnett DA, Kurth WS, Williams DJ (1997) Enhanced whistler-mode emissions: Signatures of interchange motion in the Io torus. Geophys Res Lett 24:2123–2126

Borovsky JE (1993) Auroral arc thicknesses as predicted by various theories. J Geophys Res 98(A4):6101–6138

Bortnik J, Thorne RM (2007) The dual role of ELF/VLF chorus waves in the acceleration and precipitation of radiation belt electrons. J Atmos Sol Terr Phys 69:378–386

Bouhram M, Johnson RE, Berthelier J-J, Illiano J-M, Tokar RL, Young DT, Crary FJ (2006) A test-particle model of the atmosphere/ionosphere system of Saturn's main rings. Geophys Res Lett 33:L05106, doi: 10.1029/2005GL025011

Boyle CB, Reiff PH, Hairston MR (1997) Empirical polar cap potentials. J Geophys Res 102:111–125

Brandt PC, Paranicas CP, Carbary JF, Mitchell DG, Mauk BH, Krimigis SM (2008) Understanding the global evolution of Saturn's ring current. Geophys Res Lett 35:L17101, doi: 10.1029/2008GL034969

Brice NM, Ioannidis GA (1970) The magnetospheres of Jupiter and Earth. Icarus 13:173–183

Bunce EJ, Cowley SWH, Wright DM, Coates AJ, Dougherty MK, Krupp N, Kurth WS, Rymer AM (2005a) In situ observations of a solar wind compression induced hot plasma injection in Saturn's tail. Geophys Res Lett 32:L20S04, doi: 10.1029/2005GL022888

Bunce EJ, Cowley SWH, Milan SE (2005b) Interplanetary magnetic field control of Saturn's polar cusp aurora. Annal Geophys 23:1405–1431

Bunce EJ, Cowley SWH, Alexeev II, Arridge CS, Dougherty MK, Nichols JD, Russell CT (2007) Cassini observations of the variation of Saturn's ring current parameters with system size. J Geophys Res 112:A10202, doi: 10.1029/2007JA012275

Burch JL, Goldstein J, Hill TW, Young DT, Crary FJ, Coates AJ, André N, Kurth WS, Sittler EC Jr (2005) Properties of local plasma injections in Saturn's magnetosphere Geophys Res Lett 32:L14S02, doi: 10.1029/2005GL022611

Burch JL, Goldstein J, Lewis WS, Young DT, Coates AJ, Dougherty MK, André N (2007) Tethys and Dione as sources of outward-flowing plasma in Saturn's magnetosphere. Nature 447 (14 June 2007), doi: 10.1038/nature05906

Burch JL, Goldstein J, Mokashi P, Lewis WS, Paty C, Young DT, Coates AJ, Dougherty MK, André N (2008) On the cause of Saturn's plasma periodicity. Geophys Res Lett 35:L14105, doi: 10.1029/2008GL034951

Burger MH, Sittler EC Jr, Johnson RE, Smith HT, Tucker OJ, Shematovich VI (2007) Understanding the escape of water from Enceladus. J Geophys Res 112:A06219, doi: 10.1029/2006JA012086

Burns JA, Showalter MR, Cuzzi JN, Durisen RH (1983) Saturn's electrostatic discharges: Could lightning be the cause? Icarus 54:280–295

Burton RK, McPherron RL et al. (1975) An empirical relationship between interplanetary conditions and Dst. J Geophys Res 80(31):4204–4214

Calvin WM, Johnson RE, Spencer JR (1996) O_2 on Ganymede: Spectral characteristics and plasma formation mechanisms. Geophys Res Lett 23:673–676

Carbary JF, Krimigis SM, Ip W-H (1983) Energetic particle microsignatures of Saturn's satellites J Geophys Res 88:8947–8958

Carbary JF, Mitchell DG, Krimigis SM, Hamilton DC, Krupp N (2007a) Charged particle periodicities in Saturn's outer magnetosphere. J Geophys Res, doi: 10.1029/2007JA012351

Carbary JF, Mitchell DG, Krimigis SM, Krupp N (2007b) Evidence for spiral pattern in Saturn's magnetosphere using the new SKR longitudes. Geophys Res Lett doi: 10.1029/2007GL030167

Chen FF (1974) Introduction to plasma physics and controlled fusion; Volume 1: Plasma physics. Plenum, New York and London

Chen Y, Hill TW (2008) Statistical analysis of injection/dispersion events in Saturn's inner magnetosphere. J Geophys Res 113:A07215, doi: 10.1029/2008JA013166

Cheng AF, Krimigis SM, Lanzerotti LJ (1991) Energetic particles at Uranus. In: Bergstrahl JT, Miner ED, Matthews MS (eds) Uranus, p 831, University of Arizona Press, Tucson

Coates, AJ, Jones GH, Lewis GR, Wellbrock A, Young DT, Crary FJ, Johnson RE, Cassidy TA, Hill TW (2009) Negative ions in the Enceladus plume. Icarus, accepted

Connerney JEP, Acuña MH, Ness NF (1981) Saturn's ring current and inner magnetosphere. Nature 292:724–726, doi: 10.1038/292724a0

Cooper JF (1983) Nuclear cascades in Saturn's rings – Cosmic ray albedo neutron decay and origins of trapped protons in the inner magnetosphere. J Geophys Res 88:3945–3954

Coroniti FV, Scarf FL, Kennel CF, Kurth WS (1984) Analysis of chorus emissions at Jupiter. J Geophys Res 89:3801–3820

Cowee MM, Strangeway RJ, Russell CT, Winske D (2006) One-dimensional hybrid simulations of planetary ion pickup: Techniques and verification. J Geophys Res 111:A12213, doi: 10.1029/2006JA011996

Cowee MM, Russell CT, Strangeway RJ, Blanco-Cano X (2007) One-dimensional hybrid simulations of obliquely propagating ion cyclotron waves: Application to ion pickup at Io. J Geophys Res 112:A06230, doi: 10.1029/2006JA012230

Cowley SWH, Bunce EJ, O'Rourke JM (2004) A simple quantitative model of plasma flows and currents in Saturn's polar ionosphere. J Geophys Res 109:A05212, doi: 10.1029/2003JA010375

Cui J, Yelle RV, Volk K (2008) Distribution and escape of molecular hydrogen in Titan's thermosphere and exosphere J Geophys Res 113:E10004, doi: 10.1029/2007JE003032

De La Haye et al. (2007) Cassini Ion and Neutral Mass Spectrometer data in Titan's upper atmosphere and exosphere: Observation of a suprathermal corona. J Geophys Res 112:A07309, doi: 10.1029/2006JA012222

Delamere PA, Bagenal F, Ergun R, Su Y-J (2003) Momentum transfer between the Io plasma wake and Jupiter's ionosphere. J Geophys Res 108(A6):1241, doi: 10.1029/2002JA009530

Delamere PA, Bagenal F, Dols V, Ray LC (2007) Saturn's neutral torus versus Jupiter's plasma torus. Geophys Res Lett 34:L09105, doi: 10.1029/2007GL029437

Delcourt DC (2002) Particle acceleration by inductive electric fields in the inner magnetosphere. J Atmos Sol Terr Phys 64:551–559

Delcourt DC, Sauvaud JA, Pedersen A (1990) Dynamics of single-particle orbits during substorm expansion phase. J Geophys Res 95:20,853

Desch MD et al. (2006) Cassini RPWS and imaging observations of Saturn lightning. In: Rucker HO, Kurth WS, Mann G (eds) Planetary radio emissions VI, pp 103–110. Austrian Academy of Sciences, Vienna

Dougherty MK et al. (2004) The Cassini magnetic field investigation Space Sci Rev 114:331–383

Dougherty MK et al. (2005) Cassini magnetometer observation during Saturn orbit insertion Science 307:1266–1269

Dougherty MK, Khurana KK, Neubauer FM, Russell CT, Saur J, Leisner JS, Burton ME (2006) Identification of a dynamic atmosphere at Enceladus with the Cassini magnetometer Science 311:1406–1409

Drake JF, Shay MA, Thongthai W, Swisdak M (2005) Production of energetic electrons during magnetic reconnection. Phys Rev Lett 94:095001

Dungey JW (1961) Interplanetary magnetic field and the auroral zones. Phys Rev Lett 6:47–48

Esposito LW, Colwell JE, Larsen K et al. (2005) Ultraviolet imaging spectroscopy shows an active Saturnian system. Science 307:1251–1255

Eviatar A, Richardson JD (1986) Corotation of the Kronian magnetosphere. J Geophys Res 91:3299–3301

Farrell WM, Kurth WS, Kaiser ML, Desch MD, Gurnett DA, Canu P (2005) Narrowband Z-mode emissions interior to Saturn's plasma torus, J Geophys Res 110:A10204, doi: 10.1029/2005JA011102

Farrell WM, Kaiser ML, Gurnett DA, Kurth WS, Persoon AM, Wahlund JE, Canu P (2008) Mass unloading along the inner edge of the Enceladus plasma torus. Geophys Res Lett 35:L02203, doi: 10.1029/2007GL032306

Farmer A (2008) Saturn in hot water: Viscous evolution of the Enceladus torus. Saturn after Cassini Workshop, London, 28 July

Fillius W (1976) The trapped radiation belts of Jupiter. Gehrels T (ed) Jupiter p 896, University of Arizona Press, Tucson

Fischer GM et al. (2006) Saturn lightning recorded by Cassini/RPWS in 2004. Icarus 183(1):135–152, doi: 10.1016/j.icarus.2006.02.010

Fischer G, Kurth WS, Dyudina UA, Kaiser ML, Zarka P, Lecacheux A, Ingersoll AP, Gurnett DA (2007) Analysis of a giant lightning storm on Saturn. Icarus 190:528–544, doi: 10.1016/j.icarus.2007.04.002

Fischer G, Gurnett DA, Kurth WS, Akalin F, Zarka P, Dyudina UA, Farrell WM, Kaiser ML (2008) Atmospheric electricity at Saturn. Space Sci Rev 137:271–285, doi: 10.1007/s11214-008-9370-z

Fleshman BL, Delamere PA, Bagenal F (2008) A one-box chemistry model of the Enceladus torus: Preliminary results and sensitivity. Saturn after Cassini Workshop, London, July

Fox NJ, Mauk BH, Blake JB (2006) Role of non-adiabatic processes in the creation of the outer radiation belts. Geophys Res Lett 33:L18108, doi: 10.1029/2006GL026598

Frank LA, Paterson WR (2004) Plasmas observed near local noon in Jupiter's magnetosphere with the Galileo spacecraft. J Geophys Res 109:A11217, doi: 10.1029/2002JA009795

Ge YS, Jian LK, Russell CT (2007) Growth phase of Jovian substorms. Geophys Res Lett 34:L23106, doi: 10.1029/2007GL031987

Gehrels N, Stone EC (1983) Energetic oxygen and sulfur ions in the Jovian magnetosphere and their contributions to the auroral excitation. J Geophys Res 88:5537

Gerard J-C, Bunce EJ, Grodent D, Cowley SWH, Clarke JT, Badman SV (2005) Signature of Saturn's auroral cusp: Simultaneous Hubble Space Telescope FUV observations and upstream solar wind monitoring. J Geophys Res 110:A11201, doi: 10.1029/2005JA011094

Giampieri G, Dougherty MK, Smith EJ, Russell CT (2006) A regular period for Saturn's magnetic field that may track its internal rotation. Nature 441:62–64, doi: 10.1038/nature04750

Glocer A et al. (2007) Polar wind outflow model: Saturn results. J Geophys Res 112:A01304, doi: 10.1029/2006JA011755

Gold T (1959) Motions in the magnetosphere of the Earth. J Geophys Res 64:1219–1224

Goldreich P, Lynden-Bell D (1969) Io, a Jovian unipolar inductor. Astrophys J 156:59–78

Goldreich P, Farmer AJ (2007) Spontaneous axisymmetry breaking of the external magnetic field at Saturn. J Geophys Res 112:A05225, doi: 10.1029/2006JA012163

Goldstein BE, Ip W-H (1983) Magnetic drifts at Io: Depletion of 10-MeV electrons at Voyager 1 encounter due to a forbidden zone. J Geophys Res 88:6137

Gurnett D (1975) The Earth as a radio source: The nonthermal continuum. J Geophys Res 80:2751–2763

Gurnett DA, Kurth WS, Scarf FL (1981a) Plasma waves near Saturn: Initial results from Voyager 1. Science, 212:235–239

Gurnett DA, Kurth WS, Scarf FL (1981b) Narrowband electromagnetic emissions from Saturn's magnetosphere. Nature 292:733

Gurnett DA, Kurth WS, Scarf FL (1983) Narrowband electromagnetic emissions from Jupiter's magnetosphere. Nature 302:385

Gurnett DA et al. (2004) The Cassini radio and plasma wave science investigation. Space Sci Rev 114:395–463

Gurnett DA et al. (2005) Radio and plasma wave observations at Saturn from Cassini's Approach and first orbit. Science 307:1255–1259, doi: 10.1126/science.1105356

Gurnett DA, Persoon AM, Kurth WS, Groene JB, Averkamp TF, Dougherty MK, Southwood DJ (2007) The variable rotation period of the inner region of Saturn's plasma disk. Science 316(5823):442–445, doi: 10.1126/science.1138562

Hamilton DC, Brown DC, Gloeckler G, Axford WI (1983) Energetic atomic and molecular ions in Saturn's magnetosphere. J Geophys Res 88:8905–8922

Hamilton DC et al. (2009a) The major species of Saturn's ring current and their average spectra. Geophys Res Lett. submitted

Hamilton DC, DiFabio RD, Christon SP, Krimigis SM, Mitchell DG, Dandouras J (2009b) Suprathermal heavy ions in Saturn's magnetosphere. Geophys Res Lett. submitted

Hamilton DC, DiFabio RD, Mitchell DG, Krimigis SM (2009c) Suprathermal H_3^+ in Saturn's magnetosphere. Geophys Res Lett. submitted

Hansen KC, Ridley AJ, Hospodarsky GB, Dougherty MK, Gombosi TI, Toth G (2005) Global MHD simulations of Saturn's magnetosphere at the time of Cassini approach. Geophys Res Lett 32:L20S06, doi: 10.1029/2005GL022835

Hansen CJ, Esposito L, Stewart AIF, Colwell J, Hendrix A, Pryor W, Shemansky D, West R (2006) Enceladus' water vapor plume. Science 311:1422

Hartle RE et al. (2006) Preliminary interpretation of Titan plasma interaction as observed by the Cassini Plasma Spectrometer: Comparisons with Voyager 1. Geophys Res Lett 33:L08201, doi: 10.1029/2005GL024817

Herbert F, Sandel BR (1995) Radial profiles of ion density and parallel temperature in the Io plasma torus during the Voyager 1 encounter. J Geophys Res 100(a10):19,513–19,529

Hill TW (1979) Inertial limit on corotation. J Geophys Res 84:6554–6558

Hill TW, Michel FC (1976) Heavy ions from the Galilean satellites and the centrifugal distortion of the Jovian magnetosphere. J Geophys Res 81:4561–4565

Hill TW, Pontius DH Jr (1998) Plasma injection near Io. J Geophys Res 103:19,879

Hill TW, Vasyliūnas VM (2002) Jovian auroral signature of Io's corotational wake. J Geophys Res (Space Phys), 107(A12):SMP 27–1, CiteID 1464, doi 10.1029/2002JA009514

Hill TW et al (2005) Evidence for rotationally driven plasma transport in Saturn's magnetosphere Geophys Res Lett 32:L14S10, doi: 10.1029/2005GL022620

Hill TW et al. (2008a) Plasmoids in Saturn's magnetotail. J Geophys Res 113:A01214, doi: 10.1029/2007JA012626

Hill TW, Chen Y, Wu H, Johnson RE, Mauk BH (2008b) Injection structures in Saturn's inner magnetosphere (poster). Saturn After Cassini-Huygens Symposium, Imperial College, London, 28 July–1 August

Horne RB, Thorne RM (2000) Electron pitch angle diffusion by electrostatic electron cyclotron waves: the origin of pancake distributions. J Geophys Res 105:5391–5402

Horne RB, Thorne RM (2003) Relativistic electron acceleration and precipitation during resonant interactions with whistler-mode chorus. Geophys Res Lett 30(10):1527, doi: 10.1029/2003GL016973

Horne RB, Thorne RM, Meredith NP, Anderson RR (2003) Diffuse auroral electron scattering by electron cyclotron harmonic and whistler mode waves during an isolated substorm. J Geophys Res 108(A7):1290, doi: 10.1029/2002JA009736

Horne RB et al. (2005) Wave acceleration of electrons in the Van Allen radiation belts. Nature 437:227–230

Horne RB, Thorne RM, Glauert SA, Menietti JD, Shprits YY, Gurnett DA (2008) Gyro-resonant electron acceleration at Jupiter. Nature Phys 4:301

Hoshino M (2005) Electron surfing acceleration in magnetic reconnection. J Geophys Res 110:A10215, doi: 10.1029/2005JA011229

Hospodarsky GB, Averkamp TF, Kurth WS, Gurnett DA, Dougherty M, Inan U, Wood T (2001) Wave normal and Poynting vector calculations using the Cassini radio and plasma wave instrument. J Geophys Res 106:30, 253–30, 269

Hospodarsky GB, Kurth WS, Gurnett DA, Zarka P, Canu P, Dougherty M, Jones GH, Coates A, Rymer A (2006) Observations of Langmuir waves detected by the Cassini spacecraft. In: Rucker HO, Kurth WS, Mann G (eds) Planetary radio emissions VI, pp 67–79. Austrian Academy of Sciences Press, Vienna

Hospodarsky GB, Averkamp TF, Kurth WS, Gurnett DA, Santolik O, Dougherty MK (2008) Observations of chorus at Saturn using the Cassini radio and plasma wave instrument. J Geophys Res 113:A12206, doi: 10.1029/2008JA013237

Huang TS, Hill TW (1991) Drift-wave instability in the Io plasma torus. J Geophys Res 96:14,075–14,083

Huddleston DE, Russell CT et al. (1997) Magnetopause structure and the role of reconnection at the outer planets. J Geophys Res 102:24,289–24,302

Huddleston DE, Strangeway RJ, Warnecke J, Russell CT, Kivelson MG (1998) Ion cyclotron waves in the Io torus: Wave dispersion, free energy analysis, and SO_2^+ source rate estimates. J Geophys Res 103:19,887–19,889

Jackman CM, Achilleos N, Bunce EJ, Cowley SWH, Dougherty MK, Jones GH, Milan SE, Smith EJ (2004) Interplanetary magnetic field at ∼9AU during the declining phase of the solar cycle and its implications for Saturn's magnetospheric dynamics. J Geophys Res 109:A11203, doi: 10.1029/2004JA010614

Jackman CM, Russell CT, Southwood DJ, Arridge CS, Achilleos N, Dougherty MK (2007) Strong, rapid depolarization in Saturn's magnetotail: In situ evidence of reconnection. Geophys Res Lett 34:L11203, doi: 10.1029/2007GL029764

Jackson JD (1999) Classical electrodynamics, 3rd edn. Wiley, New York

Johnson RE (1990) Energetic charged particle interactions with atmospheres and surfaces. Springer, New York

Johnson RE (2009) Sputtering and heating of Titan's upper atmosphere. Proc Royal Soc (London) 367:753–771, doi: 10.1098/rsta.2008.0244

Johnson RE, Quickenden TI (1997) Photolysis and radiolysis of water ice on outer solar system bodies. J Geophys Res 102:10,985–10,996

Johnson RE, Smith HT, Tucker OJ, Liu M, Burger MH, Sittler EC, Tokar RL (1989a) The Enceladus and OH tori at Saturn. Astrophys J 644:L137–L139

Johnson RE, Pospieszalska MK, Sieveka EM, Cheng AF, Lanzerotti LJ, Sittler EC (1989b) The neutral cloud and heavy ion inner torus at Saturn. Icarus 77:311–329

Johnson RE, Liu M, Sittler EC Jr (2005) Plasma-induced clearing and redistribution of material embedded in planetary magnetospheres. Geophys Res Lett 32:L24201, doi: 10.1029/2005GL024275

Johnson RE et al. (2006a) Production, ionization and redistribution of O_2 Saturn's ring atmosphere. Icarus 180:393–402

Johnson RE, Smith HT, Tucker OJ, Liu M, Tokar R (2006b) The Enceladus and OH tori at Saturn. Astrophys J Letts 644:L137–L139

Johnson RE, Fama M, Liu M, Baragiola RA, Sittler EC Jr, Smith HT (2008) Sputtering of ice grains and icy satellites in Saturn's inner magnetosphere. Planet Space Sci 56:1238–1243

Johnson RE, Tucker OJ, Michael M, Sittler EC, Smith HT, Young DT, Waite JH Jr (2009) Mass loss processes in Titan's upper atmosphere. In: Titan after Cassini Huygens, Chapter 15, in press

Jones D (1976) Source of terrestrial nonthermal continuum radiation. Nature 260:686

Jones GH, Roussos E, Krupp N, Paranicas C, Woch J, Lagg A, Mitchell DG, Krimigis SM Dougherty MK (2006a) Enceladus' Varying imprint on the magnetosphere of Saturn Science 311:1412–1415

Jones, GH et al. (2006b) Formation of Saturn's ring spokes by lightning-induced electron beams. Geophys Res Lett 33:L21202, doi: 10.1029/2006GL028146

Jones GH et al. (2008) The dust halo of Saturn's largest icy moon, Rhea. Science. doi: 10.1126/science.1151524

Jurac S, Richardson JD (2005) A self-consistent model of plasma and neutrals at Saturn: Neutral cloud morphology. J Geophys Res 110, doi: 10.1029/2004JA010635

Jurac S, Johnson RE, Richardson JD (2001) Saturn's E ring and production of the neutral torus. Icarus 149:384–396

Jurac S, McGrath MA, Johnson RE, Richardson JD, Vasyliūnas VM, Eviatar A (2002) Saturn: Search for a missing water source. Geophys Res Lett 29(24):2172, doi: 10.1029/2002GL015855

Kaiser ML, Desch MD (1980) Narrow-band Jovian kilometric radiation: A new radio component. Geophys Res Lett 7:389–393

Kaiser ML, Desch MD, Warwick JW, Pierce JB (1980) Voyager detection of nonthermal radio emission from Saturn. Science 209:1238–1240

Kaiser ML, Connerney, JEP Desch MD (1983) Atmospheric storm explanation of saturnian electrostatic discharges. Nature 303:50–53, doi: 10.1038/303050a0

Kaiser ML, Desch MD, Kurth WS, Lecacheux A, Genova F, Pedersen BM, Evans DR (1984) Saturn as a radio source. In: Gehrels T (ed) Saturn, p 378–416. University of Arizona Press, Tucson

Kalio et al (2005) Formation of the lunar wake in quasi-neutral hybrid model. Geophys Res Lett 32:L06107, doi: 10.1029/2004GL021989

Kellett S, Bunce EJ, Cowley SWH, Dougherty MK, Krimigis SM, Sergis N (2008) Investigations into Saturn's ring current. Saturn after Cassini Workshop, London, July

Kennel CF, Scarf FL, Fredricks RW, McGehee JH, Coroniti FV (1970) VLF electric field observations in the magnetosphere. J Geophys Res 75:6136

Khurana KK et al (2004) Configuration of Jupiter's magnetosphere In: Bagenal F, Dowling T, McKinnon W (eds) Jupiter: The planet, satellites and magnetosphere, pp 513–536 Cambridge University Press, Cambridge

Khurana KK, Russell CT, Dougherty MK (2007a) Magnetic portraits of Tethys and Rhea Icarus 465–477

Khurana KK, Dougherty MK, Russell CT, Leisner JS (2007b) Mass loading of Saturn's magnetosphere near Enceladus. J Geophys Res 112(A8), doi: 10.1029/2006JA012110

Khurana KK, Russell CT, Dougherty MK (2008) Magnetic portraits of Tethys and Rhea. Icarus 193:465–474, doi: 10.1016/j.icarus2007.08.005

Kivelson MG, Khuran KK, Coronitti FV, Joy S, Russell CT, Walker RJ, Warneck J, Bennettand L, Polansk C (1997a) The magnetic field and magnetosphere of Ganymede. Geophys Res Lett 24:2155–2158

Kivelson MG, Khurana KK, Joy S, Russell CT, Southwood DJ, Walker RJ, Polanskey C (1997b) Europa's magnetic signature: Report from Galileo's pass on 19 December 1996. Science 276:1239–1241, doi: 10.1126/science.276.5316.1239

Kivelson MG, Khurana KK, Russell CT, Walker RJ (1997c) Intermittent short-duration magnetic field anomalies in the Io torus: Evidence for plasma interchange? Geophys Res Lett 24:2127–2130

Kivelson MG, Bagenal F, Kurth W, Neubauer FM, Paranicas C, Saur J (2004) Magnetospheric interactions with satellites. In: Bagenal F, Dowling T, McKinnon W (eds) Jupiter: The planet, satellites and magnetosphere, pp 513–536 Cambridge University Press, Cambridge

Krall NA, Trivelpiece AW (1973) Principles of plasma physics. McGraw-Hill Book, New York

Kriegel H, Simon S, Wiehle S, Kleindienst G, Motschmann U, Glassmeier K, Saur J, Khurana KK, Dougherty MK (2008) Hybrid Simulations of the Enceladus plasma interaction and comparison with MAG data. AGU Fall Meeting Abstracts

Krimigis SM et al. (2005) Dynamics of Saturn's magnetosphere from MIMI during Cassini's orbital insertion Science 307(5713):1270–1273

Krimigis SM, Sergis N, Mitchell DG, Hamilton DC, Krupp N (2007) A dynamic, rotating ring current around Saturn. Nature 450, doi:10.38/nature06425

Krupp N et al. (2004) Dynamics of the Jovian magnetosphere. In: Bagenal F, Dowling TE, McKinnon WB (eds) Jupiter, pp 617–638. Cambridge University Press, Cambridge

Kurt WK, Scarf FL, Sullivan JD, Gurnett DA (1982) Detection of nonthermal continuum radiation in Saturn's magnetosphere. Geophys Res Lett 9:889

Kurth WS (1982) Detailed observations of the source of terrestrial narrowband electromagnetic radiation. Geophys Res Lett 9:1341–1344

Kurth WS, Gurnett DA (1991) Plasma waves in Planetary magnetospheres. J Geophys Res 96:18,977

Kurth W, Gurnett D, Anderson R (1981) Escaping nonthermal continuum radiation. J Geophys Res 86:5519–5531

Kurth WS, Scarf FL, Gurnett DA, Barbosa DD (1983) A survey of electrostatic waves in Saturn's magnetosphere. J Geophys Res 88:8959

Kurth WS, Gurnett DA, Scarf FL (1986) Sporadic narrowband radio emissions from Uranus. J Geophys Res 91:11,958–11,964

Kurth WS, Barbosa DD, Gurnett DA, Poynter RL, Cairns IH (1990) Low-frequency radio emissions at Neptune. Geophys Res Lett 17:1649–1653

Kurth WS, Gurnett DA, Roux A, Bolton SJ (1997) Ganymede: A new radio source. Geophys Res Lett 24:2167–2170

Kurth WS, Lecacheux A, Averkamp TF, Groene JB, Gurnett DA (2007) A Saturnian longitude system based on a variable kilometric radiation period. Geophys Res Lett. doi: 10.1029/2006GL028336

Kurth WS, Averkamp TF, Gurnett DA, Groene JB, Lecacheux A (2008) An update to a Saturnian longitude system based on kilometric radio emissions. J Geophys Res. doi: 10.1029/2007JA012861

Leisner JS, Russell CT, Dougherty MK, Blanco-Cano X, Strangeway RJ, Bertucci C (2006) Ion cyclotron waves in Saturn's E ring: Initial Cassini observations. Geophys Res Lett 33:L11101, doi: 10.1029/2005GL024875

Li W, Shprits YY, Thorne RM (2007) Dynamical evolution of energetic electrons due to wave-particle interactions during storms. J Geophys Res 112:A10220, doi: 10.1029/2007JA012368

Loeffler MJ, Teolis B, Baragiola RA (2006) A model study of the thermal evolution of astrophysical ices. Astrophys J 639(2): L103–L106

Louarn P et al. (2007) Observation of similar radio signatures at Saturn and Jupiter: Implications for the magnetospheric dynamics. Geophys Res Lett 34:L20113, doi: 10.1029/2007GL030368

Luhmann JG, Johnson RE, Tokar RL, Cravens T (2006) A model of the ionosphere of Saturn's toroidal ring atmosphere. Icarus 181:465–474

Lyons LR (1974) Electron diffusion driven by magnetospheric electrostatic waves. J Geophys Res 79:557

Lysak R (1993), Auroral plasma dynamics, Geophysical monograph 80. American Geophysical Union, Washington DC

Martens HR, Reisenfeld DB, Williams JD, Johnson RE, Smith HT (2008) Observations of molecular oxygen ions in Saturn's inner magnetosphere. Geophys Res Lett, 35:L20103, doi: 10.1029/2008GL035433

Masters A, Achilleos N, Dougherty MK, Slavin JA, Hospodarsky GB, Arridge CS, Coates AJ (2008) An empirical model of Saturn's bow shock: Cassini observations of shock location and shape. J Geophys Res 113:A10210, doi: 10.1029/2008JA013276

Mauk BH (1986) Quantitative modeling of the "convection surge" mechanism of ion acceleration. J Geophys Res 91:13,423

Mauk BH (1989) Macroscopic magnetospheric particle acceleration. Solar system plasma physics, Geophysical monograph 54, p 319. American Geophysical Union, Washington DC

Mauk BH, Krimigis SM (1987) Radial force balance in Jupiter's dayside magnetosphere. J Geophys Res 92:9931

Mauk BH, Krimigis SM, Lepping RP (1985) Particle and field stress balance within a planetary magnetosphere. J Geophys Res 90:8253

Mauk BH, Krimigis SM, Cheng AF, Selesnick RS (1995) Energetic particles and hot plasmas of Neptune. In: Cruikshank DP (ed) Neptune and Triton, p 169. The University of Arizona Press, Tucson

Mauk BH, Williams DJ, McEntire RW, Khurana KK, Roederer JG (1999) Storm-like dynamics of Jupiter's inner and middle magnetosphere. J Geophys Res 104:22,759–22,778

Mauk BH, Mitchell DG, McEntire RW, Paranicas CP, Roelof EC, Williams DJ, Krimigis SM, Lagg A (2004) Energetic ion characteristics and neutral gas interactions in Jupiter's magnetosphere. J Geophys Res 109:A09S12, doi: 10.1029/2003JA010270

Mauk BH et al. (2005) Energetic particle injections in Saturn's magnetosphere. Geophys Res Lett 32(14):L14S05.1–L14S05.5, doi: 10.1029/2005GL022485

Maurice S, Blanc M, Prange R, Sittler EC Jr (1997) The magnetic-field-aligned polarization electric field and its effects on particle distribution in the magnetospheres of Jupiter and Saturn Planet Space Sci 45(11):1449–1465

McAndrews HJ et al. (2008a) Plasma in the nightside magnetosphere and the implications for global circulation (poster). Saturn After Cassini-Huygens Symposium, Imperial College, London, 28 July–1 August

McAndrews HJ, Owens CJ, Thomsen MF, Lavraud B, Coates AJ, Dougherty MK, Young DT (2008b) Evidence for reconnection at Saturn's magnetopause. J Geophys Res 113:A04210, doi: 10.1029/2007JA012581

McNutt RL Jr (1984) Force balance in the outer planet magnetospheres. In: Bridge HS et al. (eds) Proc 1982–4 Symposia on the Physics of Space Plasmas, pp 179–210. Sci. Publ., Cambridge, Massachusetts

Melin H, Shemansky DE, Liu X (2009) The distribution of hydrogen and atomic oxygen in the magnetosphere of Saturn. Planet Space Sci, in press, doi: 10.1016/j.pss.2009.04.014

Menietti JD, Santolik O, Rymer AM, Hospodarsky GB, Persoon AM, Gurnett DA, Coates AJ, Young DT (2008a) Analysis of plasma waves observed within local plasma injections seen in Saturn's magnetosphere. J Geophys Res 113:A05213, doi: 10.1029/2007JA012856

Menietti JD, Santolik O, Rymer AM, Hospodarsky GB, Gurnett DA, Coates AJ (2008b) Analysis of plasma waves observed in the inner Saturn magnetosphere. Annal Geophys 26:2631–2644

Meredith NP, Cain M, Horne RB, Thorne RM, Summers D, Anderson RR (2003) Evidence for chorus-driven electron acceleration to relativistic energies from a survey of geomagnetically disturbed periods. J Geophys Res 108(A6):1248, doi: 10.1029/2002JA009764

Millward G, Miller S, Stallard T, Aylward A, Achilleos N (2002) On the dynamics of the Jovian ionosphere and thermosphere: III. The modeling of auroral conductivity. Icarus 160:95–107

Mitchell DG et al. (2005) Energetic ion acceleration in Saturn's magnetotail: Substorms at Saturn? Geophys Res Lett 32:L20S01, doi: 10.1029/2005GL022647

Mitchell DG, Kurth WS, Hospodarsky GB, Krupp N, Saur J, Mauk BH, Carbary JF, Krimigis SM, Dougherty MK, Hamilton DC (2009) Ion conics and electron beams associated with auroral processes on Saturn. J Geophys Res 114:A02212, doi: 10.1029/2008JA013621

Neubauer et al. (1998) The sub-Alfvénic interaction of the Galilean satellites with the Jovian magnetosphere. J Geophys Res 103(E9):19,843–19,866, doi: 10.1029/97JE03370

Noll KS, Johnson RE, Lane AL, Domingue DL, Weaver HA (1996) Detection of ozone on Ganymede. Science 273:341–343

Noll KS, Roush TL, Cruikshank DP, Johnson RE, Pendleton YJ (1997) Detection of ozone on Saturn's satellites Rhea and Dione. Nature 38:45–47

Northrop TG (1963) The adiabatic motion of charged particles. Interscience, New York

Paranicas C Cheng CF (1997) A model for satellite microsignatures for Saturn, Icarus 125:380–396

Paranicas CP, Mauk BH, Krimigis SM (1991) Pressure anisotropy and radial stress balance in the Jovian neutral sheet. J Geophys Res 96:21,135

Paranicas C, McEntire RW, Cheng AF, Lagg A, Williams DJ (2000) Energetic charged particles near Europa J Geophys Res 105(A7):16005–16016

Paranicas C et al. (2005a) Evidence of Enceladus and Tethys microsignatures, *Geophys Res Lett* 32, doi: 10.1029/2005GL024072

Paranicas C, Mitchell DG, Roelof EC, Brandt PC, Williams DJ, Krimigis SM, Mauk BH (2005b) Periodic intensity variations in global ENA images of Saturn. Geophys Res Lett, doi: 10.1029/2005GL023656

Paranicas C et al. (2007) Energetic electrons injected into Saturn's neutral cloud. Geophys Res Lett 34:L02109, doi: 2006GL028676

Paranicas C et al. (2008) Sources and. losses of energetic protons in Saturn's magnetosphere. Icarus 197:519–525

Parks GK (1991) Physics of space plasmas. Addison-Wesley, New York

Paschmann G, Sonnerup BUÖ et al. (1979) Plasma acceleration at the Earth's magnetopause: Evidence for reconnection. Nature 282:243–246

Persoon AM, Gurnett DA, Kurth WS, Hospodarsky GB, Groene JB, Canu P, Dougherty, MK (2005) Equatorial electron density measurements in Saturn's inner magnetosphere Geophys Res Lett 32:L23105, doi: 10.1029/2005GL024294

Persoon AM, Gurnett DA, Kurth WS, Groene JB (2006) A simple scale height model of the electron density in Saturn's plasma disk. Geophys Res Lett 33:L18106, doi: 10.1029/2006GL027090

Persoon AM et al. (2009) A diffusive equilibrium model for the plasma density in Saturn's magnetosphere. J Geophys Res 114:A04211, doi: 10.1029/2008JA013912

Piddington JH, Drake JF (1968) Electrodynamic effects of Jupiter's satellite Io. Nature 217:935–937 (09 March), doi: 10.1038/217935a0

Pontius DH Jr (1997) Coriolis influences on the interchange instability. Geophys Res Lett 24:1961–2964

Pontius DH Jr, Hill TW (1982) Departure from corotation of the Io plasma torus: Local plasma production. Geophys Res Lett 9:1321–1324

Pontius DH, Hill TW (2006) Enceladus: A significant plasma source for Saturn's magnetosphere. J Geophys Res 111(A9), doi: 10.1029/2006JA011674

Pontius DH Jr, Hill TW, Tokar RL (2007) Inferring the radial profile of mass loading in Saturn's magnetosphere from the observed corotation lag (poster). Magnetospheres of the Outer Planets 2007 Conference, San Antonio, TX (June)

Porco CC et. al (2006) Cassini observes the active south pole of Enceladus Science 311:1393–1401

Pritchett PL (2006) Relativistic electron production during guide field magnetic reconnection. J Geophys Res 111:A10212, doi: 10.1029/2006JA011793

Quinn JM, Southwood DJ (1982) Observations of parallel ion energization in the equatorial region. J Geophys Res 87:10,536

Ray LC, Ergun RE, Delamere PA, Bagenal F, Su Y (2008a) Effect of field aligned potentials on angular momentum transfer at Jupiter, American Geophysical Union, Fall Meeting, Paper SM41B-1675, San Franscisco (18 December)

Ray LC, Ergun RE, Delamere PA, Bagenal F (2008b) Effect of field aligned potentials on magnetospheric dynamics at Saturn. Saturn after Cassini Workshop, London (28 July)

Richardson J (1986) Thermal ions at Saturn: Plasma parameters and implications J Geophys Res 91:1381–1389

Richardson J (1998) Thermal plasma and neutral gas in Saturn's magnetosphere. Rev Geophys 36:501–524

Richardson JD, Eviatar A, McGrath MA, Vasyliūnas VM (1998) OH in Saturn's magnetosphere: Observations and implications. J Geophys Res 103:20,245–20,255

Richardson JD, Belcher JW, Szabo A, McNutt RL Jr (1995) The plasma environment of Neptune. In: Cruikshank DP (ed) Neptune and Triton, p 279. The University of Arizona Press, Tucson

Roederer JD (1970) Dynamics of geomagnetically trapped radiation. In: Physics and chemistry in space, Springer, Berlin

Roussos E et al. (2005) Low energy electron microsignatures at the orbit of Tethys: Cassini MIMI/LEMMS observations Geophys Res Lett 32:L24107, doi: 10.1029/2005GL024084

Roussos E et al. (2007) Electron microdiffusion in the Saturnian radiation belts: Cassini MIMI/LEMMS observations of energetic electron absorption by the icy moons J Geophys Res 112:A06214, doi: 10.1029/2006JA012027

Roussos E, Müller J, Simon S, Bößwetter A, Motschmann U, Fränz M, Krupp N, Woch J, Khurana K, Dougherty MK (2008a) Plasma and fields in the wake of Rhea: 3D hybrid simulation and comparison with Cassini data. Ann Geophys, 26(3):619–637

Roussos E, Krupp N, Armstrong TP, Paranicas C, Mitchell DG, Krimigis SM, Jones GH, Dialynas K, Sergis N, Hamilton DC (2008b) Discovery of a transient radiation belt at Saturn. Geophys Res Lett 35(22):L22106, doi: 10.1029/2008GL035767

Russell CT, Blanco-Cano X (2007) Ion-cyclotron wave generation by planetary ion pickup. J Atmos Solar Terr Phys 69:1723–1738

Russell CT, Elphic RC (1978) Initial ISEE magnetometer results: Magnetopause observations. Space Sci Rev 22(6):681–715

Russell CT, Huddleston DE, Khurana KK, Kivelson MG (1999) Observations at the inner edged of the Jovian current sheet: Evidence for a dynamic magnetosphere. Planet Space Sci 47:521–527

Russell CT, Kivelson MG, Kurth WS, Gurnett DA (2000) Implications of depleted flux tubes in the Jovian magnetosphere. Geophys Res Lett 27(19):3133–3136

Russell CT, Kivelson MG, Khurana KK (2005) Statistics of depleted flux tubes in the Jovian magnetosphere. Planet Space Sci 53(9):937–943, doi: 10.1016/j.pss.2005.04.007

Russell CT, Leisner JS, Arridge CS, Dougherty MK, Blanco-Cano X (2006) Nature of magnetic fluctuations in Saturn's middle magnetosphere. J Geophys Res 111:A12205, doi: 10.1029/2006JA011921

Russell CT, Jackman CM, Wei, HY Bertucci C, Dougherty MK (2008) Titan's influence on Saturnian substorm occurrence. Geophys Res Lett 35:L12105, doi: 1029/GL034080

Rymer AM et al. (2007a) Electron sources in Saturn's magnetosphere. J Geophys Res, doi: 10.1029/2006JA012017

Rymer AM et al. (2007b) Plasma production and circulation in Saturn's (and Jupiter's?) magnetosphere. American Geophysical Union, Fall Meeting, San Francisco, Dec 2007, Abstract #P52B-04

Rymer AM, Mauk BH, Hill TW, Paranicas C, Mitchell DG, Coates AJ, Young DT (2008) Electron circulation in Saturn's magnetosphere. J Geophys Res 113:A01201, doi: 10.1029/2007JA012589

Samir U, Wright KH, Stone NH (1983) The expansion of plasma into a vacuum: Basic phenomena and processes and applications to space plasma physics Rev Geophys Space Phys 21:1631–1646

Sanchez ER, Mauk BH, Meng C-I (1993) Adiabatic vs. non-adiabatic particle distributions during convection surges. Geophys Res Lett 20(3):177–180

Santos-Costa D, Blanc M, Maurice S, Bolton SJ (2003) Modeling the electron and proton radiation belts of Saturn. Geophys Res Lett 30(20):2059, doi: 10.1029/2003GL017972

Saur J, Strobel D (2005) Atmospheres and plasma interactions at Saturn's largest inner icy satellites Astrophys J 620:L115-L118, doi: 10.1086/428665

Saur J, Mauk BH, Kaßner A, Neubauer FM (2004) A model for the azimuthal plasma velocity in Saturn's magnetosphere J Geophys Res 109:A05217, doi: 10.1029/2003JA010207

Saur J, Neubauer FM, Schilling N (2007) Hemisphere coupling in Enceladus' asymmetric plasma interaction. J Geophys Res 112(A11), doi: 10.1029/2007JA012479

Saur J, Schilling N, Neubauer FM, Strobel DF, Simon S, Dougherty MK, Russell CT, Pappalardo RT (2008) Evidence for temporal variability of Enceladus' gas jets: Modeling of Cassini observations. Geophys Res Lett 35(20), CiteID L20105, doi: 10.1029/2008GL03581

Scarf FL, Gurnett DA, Kurth WS, Poynter RL (1982) Voyager-2 plasma wave observations at Saturn. Science 215:587

Scarf FL, Gurnett DA, Kurth WS, Poynter RL (1983) Voyager plasma wave measurements at Saturn. J Geophys Res 88:8971

Scarf FL, Frank LA, Gurnett DA, Lanzerotti LJ, Lazarus A, Sittler EC Jr (1984) Measurements of plasma, plasma waves and suprathermal charged particles in Saturn's inner magnetosphere. In: Gehrels T (ed) Saturn, pp 318–353. University of Arizona Press, Tucson

Schulz M (1998) Particle drift and loss rates under strong pitch angle diffusion in Dungey's model magnetosphere. J Geophys Res 103:61–68

Schulz M, Lanzerotti LJ (1974) Particle diffusion in the radiation belts. Springer, New York

Scurry L, Russell CT (1991) Proxy studies of energy transfer in the magnetosphere. J Geophys Res 96:9541–9548

Scurry L, Russell CT, Gosling JT (1994) Geomagnetic activity and the beta dependence of the dayside reconnection rate. J Geophys Res 99:4,811–14,814

Selesnick RS (1993) Micro- and macro- signatures of energetic charged particles in planetary magnetospheres Adv Space Res 13(10):221–230

Sergis N, Krimigis SM, Mitchell DG, Hamilton DC, Krupp N, Mauk BM, Roelof EC, Dougherty M (2007) Ring current at Saturn: Energetic particle pressure in Saturn's equatorial magnetosphere measured with Cassini/MIMI Geophys Res Lett 34:L09102, doi: 10.1029/2006GL029223

Sergis N, Krimigis SM, Mitchell DG, Hamilton DC, Krupp N, Mauk BH, Roelof, EC, Dougherty MK (2009) Energetic particle pressure in Saturn's magnetosphere measured with the Magnetospheric Imaging Instrument on Cassini. J Geophys Res 114:A02214, doi: 10.1029/2008JA013774

Shemansky DE (1988) Energy branching in the Io plasma torus: The failure of neutral cloud theory. J Geophys Res 93(A3):1773–1784

Shemansky DE, Sandel BR (1982) The injection of energy into the Io plasma torus, J Geophys Res 87:219-229

Shemansky DE, Hall DT (1992) The distribution of atomic hydrogen in the magnetosphere of Saturn. J Geophys Res 97(A4):4143–4161

Shemansky DE, Matheson P, Hall DT, Hu H-Y, Tripp TM (1993) Detection of the hydroxyl radical in the Saturn magnetosphere. Nature 363:329–331

Shemansky DE et al. (2004) Cassini UVIS Observatory Phase Spectral, Imaging of the Saturn System COSPAR, Paris

Shemansky DE, Liu X, Melin H (2009) The Saturn hydrogen plume. Planet Space Sci, in press, doi: 10.1016/j.pss.2009.05.002

Simon S, Saur J, Neubauer FM, Motschmann U, Dougherty MK (2009) Plasma wake of Tethys: Hybrid simulations versus Cassini MAG data. Geophys Res Lett 36(4), CiteID L04108, doi: 10.1029/2008GL036943

Siscoe GL, Summers D (1981) Centrifugally driven diffusion of Iogenic plasma. J Geophys Res 86:8471–8479

Siscoe GL, Eviatar A, Thorne RM, Richardson JD, Bagenal F, Sullivan JD (1981) Ring current impoundment of the Io plasma torus. J Geophys Res 86:8480–8484

Sittler EC Jr et al. (2006a) Cassini observations of Saturn's inner plasmasphere: Saturn orbit insertion results. Planet Space Sci 54:1197–1210

Sittler EC Jr, Johnson RE, Smith HT, Richardson JD, Jurac S, Moore M, Cooper JF, Mauk BH, Michael M, Paranicus C, Armstrong TP, Tsurutani B (2006b) Energetic nitrogen ions within the inner magnetosphere of Saturn. J Geophys Res 111:A09223

Sittler EC et al. (2008) Ion and neutral sources and sinks within Saturn's inner magnetosphere: Cassini results. Planet Space Sci 56:3–18

Sittler, EC Jr, Bertucci C, Coates A, Craven T, Dandouras I, Shemansky DE (2009) Energy deposition processes in Titan's upper atmosphere. In: Titan and Cassini/Huygens, in press

Smith EJ, Tsurutani BT (1983) Saturn's magnetosphere: Observations of ion cyclotron waves near the Dione L shell. J Geophys Res 88:7831–7836

Smith EJ, Davis L Jr, Jones DE, Colburn DS, Dyal P, Sonnet CP (1974) Magnetic field of Jupiter and its interaction with the solar wind. Science 183:305–306

Smith EJ, Davis L Jr, Jones DE, Coleman PJ Jr, Colburn DS, Dyal P, Sonett CP (1980) Saturn's magnetic field and magnetosphere. Science 207:407–410 (25 January), doi: 10.1126/science.207.4429.407

Smith HT, Johnson RE, Sittler EC, Shappirio M, Tucker OJ, Burger M, Crary FJ, McComas DJ, Young DT (2007) Enceladus: The likely dominant nitrogen source in Saturn's magnetosphere. Icarus 188:356–366

Smith HT, Shappirio M, Johnson RE, Reisenfeld D, Sittler EC, Crary FJ, McComas DJ, Young DT (2008) Enceladus: A source of ammonia products and molecular nitrogen for Saturn's magnetosphere. J Geophys Res 113:A11206, doi: 10.1029/2008JA013352

Smith RA, Bagenal F, Chang AF, Strobel DF (1988) On the energy crisis in the Io plasma torus. Geophys Res Lett 15:545

Sonnerup BUÖ, Paschmann G et al. (1981) Evidence for magnetic field reconnection at the Earth's magnetopause. J Geophys Res 86:10,049–10,067

Spencer JR, Calvin WM (2002) Condensed O_2 on Europa and Callisto. Astronomical J 124(6):3400–3403

Spitzer L (1962) Physics of fully ionized gases, 2nd edn. Wiley-Interscience, Hoboken, NJ

Strobel DF (2008) Titan's hydrodynamically escaping atmosphere. Icarus 193(2):588–594

Summers D, Omura Y (2008) Ultra-relativistic acceleration of electrons in planetary magnetospheres. Geophys Res Lett 35, doi: 10.1029/2007GLos226

Summers D, Thorne RM, Xiao F (1998) Relativistic theory of wave-particle resonant diffusion with application to electron acceleration in the magnetosphere. J Geophys Res 103:20,487–20,500

Thomsen MF, Van Allen JA (1980) Motion of trapped electrons and protons in Saturn's inner magnetosphere J Geophys Res 85:5831–5834

Thorne RM, Armstrong TP, Stone S, Williams DJ, McEntire RW, Bolton SJ, Gurnett DA, Kivelson MG (1997) Galileo evidence for rapid interchange transport in the Io torus. Geophys Res Lett 24:2131–2134

Tokar RL et al. (2005) Cassini observations of the thermal plasma in the vicinity of Saturn's main rings and the F and G rings. Geophys Res Lett 32:L14S04, doi: 10.1029/2005GL022690

Tokar RL et al. (2006) The interaction of the atmosphere of Enceladus with Saturn's plasma. Science 311(5766):1409–1412, doi: 10.1126/science.1121061

Tokar RL et al. (2008) Cassini detection of water-group pick-up ions in the Enceladus torus. Geophys Res Lett 35:L14202, doi: 10.1029/2008GL034749

Tokar RL, Johnson RE, Thomsen MF, Wilson RJ, Young DT, Crary FJ, Coates AJ, Jones GH, Paty CS (2009) Cassini detection of Enceladus's cold water-group plume ionosphere. Geophys Res Lett 36, doi: 10.1029/2009GL038923

Tseng W-L, Ip W-H, Johnson RE, Cassidy TA, Elrod Bob MK (2009) The structure and time variability of the ring atmosphere and ionosphere, Icarus, in press doi: 10.1016/j.icarus.2009.05.019

Van Allen JA (1976) In: Gehrels T (ed) The high-energy particles of the Jovian magnetosphere, p 928. University of Arizona Press, Tucson

Van Allen JA (1984) Energetic particles in the inner magnetosphere of Saturn. In: Gehrels T, Matthews MS (eds) Saturn, p 281. University of Arizona Press, Tucson

Van Allen JA, Randall BA, Thomsen MF (1980a) Sources and sinks of energetic electrons and protons in Saturn's magnetosphere. J Geophys Res 85:5679–5694

Van Allen JA, Thomsen MF, Randall BA (1980b) The energetic charged particle absorption signature of Mimas J Geophys Res 85:5709–5718

Vasyliūnas VM (1983) Plasma distribution and flow. In: Dessler AJ (ed) Physics of the Jovian magnetosphere, pp 395–453. Cambridge University Press, London

Vasyliūnas VM (1994) Role of the plasma acceleration time in the dynamics of the Jovian magnetosphere. Geophys Res Lett 21:401–404

Vasyliūnas VM (2008) Comparing Jupiter and Saturn: Dimensionless input rates from plasma sources within the magnetosphere. An Geophys 26:1341–1343, doi: 2008AnGeo.26.1341V

Waite JH Jr et al. (2006) Cassini ion and neutral mass spectrometer: Enceladus plume composition and structure. Science 311:1419–1422

Walker RJ, Russell CT (1985) Flux transfer events at the Jovian magnetopause. J Geophys Res 90:7397–7404

Walt M (1994) Diffusion in L coordinate or radial diffusion In: Dessler AJ, Houghton JT Rycroft MJ (eds) Introduction to geomagnetically trapped radiation, 1st edn. pp 132–146 Cambridge University Press, Cambridge, Great Britain

Wang YL, Russell CT, Raeder J (2001) The Io mass-loading disk: Model calculations. J Geophys Res 106:26,243–26,260

Wang Z, Gurnett DA, Kurth WS, Ye S, Fischer G, Mitchell DG, Russell CT, Leisner JS (2009) Narrowband radio emissions and their relationship to rotating plasma clouds and magnetic disturbances at Saturn. J Geophys Res, submitted

Warwick JW et al. (1981) Planetary radio astronomy observations from Voyager 1 near Saturn. Science 212:239

Warwick JW et al. (1982) Planetary radio astronomy observations from Voyager 2 near Saturn. Science 215:582

Whang YC (1969) Field and plasma in the lunar wake. Phys Rev, Second Ser 186:143–150

Whang YC, Ness NF (1970) Observations and interpretation of the Lunar Mach Cone J Geophys Res 75: 6002–6009

Wilson, RJ, Tokar RL, Henderson MG, Hill TW, Thomsen MF, Pontius DH J (2008) Cassini plasma spectrometer thermal ion measurements in Saturn's inner magnetosphere. J Geophys Res 113: A12218, doi: 10.1029/2008JA013486

Wolf RA (1983) The quasi-static (slow-flow) region of the magnetosphere. In: Carovillano RL, Forbes JM (eds) Solar-terrestrial physics, pp 303–368. Reidel, Norwood, MA

Wu H, Hill TW, Wolf RA, Spiro RW (2007a) Numerical simulation of fine structure in the Io plasma torus produced by the centrifugal interchange instability. J Geophys Res 112(A2), doi: 10.1029/2006JA012032

Wu H, Hill TW, Wolf RA, Spiro RW (2007b) Numerical simulation of Coriolis effects on the interchange instability in Saturn's magnetosphere. Eos Trans. AGU 88, Fall Meet. Suppl., Abstract P43A-1006

Xin L, Gurnett DA, Santolik O, Kurth WS, Hospodarsky GB (2006) Whistler-mode auroral hiss emissions observed near Saturn's B ring. J Geophys Res 111:A06214, doi: 10.1029/2005JA011432

Ye S, Gurnett DA, Fischer G, Cecconi B, Menietti JD, Kurth WS, Wang Z, Hospodarsky GB, Zarka P, Lecacheux A (2009) Source location of narrowband radio emissions detected at Saturn. J Geophys Res 114:A06219, doi: 10.1029/2008JA013855

Yelle RV, Cui J, Müller-Wodarg ICF (2008) Methane escape from Titan's atmosphere. J Geophys Res 113:E10003, doi: 10.1029/2007JE003031

Young DT et al. (2005) Composition and dynamics of plasma in Saturn's magnetosphere. Science 307:1262–1266

Chapter 12
Auroral Processes

W.S. Kurth, E.J. Bunce, J.T. Clarke, F.J. Crary, D.C. Grodent, A.P. Ingersoll, U.A. Dyudina, L. Lamy, D.G. Mitchell, A.M. Persoon, W.R. Pryor, J. Saur, and T. Stallard

Abstract Cassini has afforded a number of unique opportunities to understand auroral processes at Saturn and to highlight both differences and similarities with auroral physics at both Earth and Jupiter. A number of campaigns were coordinated with the Hubble Space Telescope such that Cassini could provide either ground truth on the impinging solar wind or in situ measurements of magnetospheric conditions leading to qualitative and sometimes quantitative relationships between the solar wind influence on the intensity, the morphology and evolution of the auroras, and magnetospheric dynamics. The Hubble UV images are enhanced by Cassini's own remote sensing of the auroras. Cassini's in situ studies of the structure and dynamics of the magnetosphere discussed in other chapters of this book provide the context for understanding the primary drivers of Saturn's auroras and the role of magnetospheric dynamics in their variations.

W.S. Kurth (✉) and A.M. Persoon
Department Physics and Astronomy, The University of Iowa,
Iowa City, IA 52242 USA
e-mail: william-kurth@uiowa.edu

E.J. Bunce and T. Stallard
Department Physics and Astronomy, University of Leicester,
Leicester, LE1 7RH, Great Britain

J.T. Clarke
Center for Space Physics, Boston University, BO, MA 02215, USA

F.J. Crary
Southwest Research Institute, San Antonio, TX 78228, USA

D.C. Grodent
LPAP, Université de Liège, B-4000, Liège, Belgium

A.P. Ingersoll and U.A. Dyudina
Caltech, Pasadena, CA 91125, USA

L. Lamy
LESIA, Observatoire de Paris, 92195 Meudon, France

D.G. Mitchell
Applied Physics Laboratory, Johns Hopkins University,
Laurel, MD 20723, USA

W.R. Pryor
Dept. Physics, Central Arizona College, Coolidge, AZ 85228, USA

J. Saur
Institute of Geophysics and Meteorology, University of Cologne,
Cologne, Germany

Finally, Cassini's three dimensional prime mission survey of the magnetosphere culminates in high inclination orbits placing it at relatively small radial distances while on auroral field lines, providing the first such in situ observations of auroral particles and fields at a planet other than Earth. The new observations have spawned a number of efforts to model the interaction of the solar wind with the magnetosphere and understand how such dynamics influence the auroras.

12.1 Introduction

As described in Chapter 3, the study of Saturn's aurora began with Voyager (Broadfoot et al. 1981) and the IUE observatory (Clarke et al. 1981; McGrath and Clarke 1992) and continued with observations by the Hubble Space Telescope (HST), for example (Trauger et al. 1998). However, Cassini's approach to Saturn in early 2004 and the improved sensitivity of the Space Telescope Imaging Spectrograph (STIS) heralded a new age in understanding Saturn's auroral processes. In fact, the first of a number of HST-Cassini campaigns occurred in January 2004 which, for the first time, allowed the comparison of the solar wind upstream of Saturn with observations of the ultraviolet (UV) aurora. This campaign showed dramatic changes in Saturn's auroral intensity and morphology in response to solar wind compression events (in which the dynamic pressure of the solar wind and the embedded magnetic field strength are increased) as demonstrated in Fig. 12.1. This was also the first opportunity to compare the power in the UV aurora with the radiated power of the Saturn kilometric radiation (SKR), commonly thought to be an indicator of auroral activity based on experience with auroral kilometric radiation (AKR) at Earth. Additional campaigns have been carried out with Cassini in orbit at Saturn, enabling another dimension of studies, that of in situ measurements of associated particles and fields in various locations in the magnetosphere that could also be compared to the auroral activity.

Cassini also carries remote sensing instruments that can observe the auroras at UV, visual, infrared (IR), and radio

Fig. 12.1 A montage of 3 visible HST images of Saturn in pseudo-natural color with UV emissions observed by HST superimposed in blue/white from three consecutive visits during the January 2004 Cassini-HST campaign. These show an excellent example of the evolution of Saturn's main auroral oval from its quiescent state to a highly active state (from *lower left to upper right*) in response to a high pressure region in the solar wind associated with a corotating interaction region (from the cover of Nature, Vol. 433, Number 7027, 2005)

HST images of the UV aurora and some spectral observations of the IR emissions. Owing to the low intensity, no IR images of the aurora had been obtained. In both wavelength ranges, observers had heavily concentrated their time on the much brighter Jovian aurora, to the extent that relatively little was known about the properties of Saturn's aurora and the corresponding magnetospheric processes. The arrival of Cassini at Saturn has been accompanied by a surge in Earth-based observations of the auroral emissions, both UV imaging by HST and near-IR observations from ground-based telescopes.

The first observations of Saturn's aurora were carried out by the Pioneer 11 and Voyager 1 and 2 spacecraft. The spectra obtained by the Voyager UV spectrometer (UVS) revealed emissions from both polar regions near 80° latitude with no apparent emission present in the polar cap. Typical intensities were in the range 10 to 15 kR (1kiloRayleigh $= 10^9$ photons/s from a 1 cm^2 column of the atmosphere into 4π steradians) with variations up to a factor of 5 (Broadfoot et al. 1981; Sandel and Broadfoot 1981; Sandel et al. 1982). Spectra of the UV auroral emissions indicated little absorption by hydrocarbons, implying relatively low energy incident charged particles (<10 keV) (see Section 12.2.2.3). Voyager radio observations established a strong correlation between Saturn kilometric radiation and the solar wind dynamic pressure.

wavelengths. By the end of the prime mission, Cassini's orbit carried it to latitudes of ∼75° at radial distances as small as 3 Saturn radii (R$_S$). While it appears that this is yet too high in altitude to carry the spacecraft through the expected acceleration region, this geometry provides the best opportunity to study auroral processes at a planet other than Earth until Juno arrives in polar orbit at Jupiter.

In this chapter, multi-wavelength observations of Saturn's aurora both from Cassini and Earth-based platforms are reviewed along with in situ particle and field data obtained on auroral field lines. These observations form the basis of our knowledge about the morphology and variability of the auroras as well as information on their spectrum and intensity. The current ideas on the origin of the aurora are also addressed.

12.2 Observations of Auroral Emissions

12.2.1 Pre-Cassini Summary

The history of pre-Cassini observations of Saturn's aurora has been given in Chapter 3. Until the approach of Cassini to Saturn in January 2004, there had been only a few dozen

12.2.2 Earth-Based Observations Concurrent with Cassini

12.2.2.1 HST-Cassini Campaigns

The first remote observing campaign was conducted by HST using the Space Telescope Imaging Spectrograph (STIS) in January 2004, as Cassini approached Saturn and measured solar wind conditions that could be accurately extrapolated to the planet. This campaign provided evidence of the role of the solar wind in controlling the aurora at Saturn, as described in Section 12.3.1. These HST images were also the first to show the wide range of variability in the aurora as seen in Fig. 12.1.

The properties of Saturn's faint southern aurora, corresponding to quiet magnetospheric conditions, have been presented based on HST observations in 2005 (Gérard et al. 2006). In addition, in view of the recently discovered large source of plasma from Enceladus, and the known strong interaction of Io, Europa, and Ganymede leading to bright auroral footprints, a sensitive, but unsuccessful, search has been carried out for auroral emissions from the magnetic footprint of Enceladus in Saturn's atmosphere (Wannawichian et al. 2008). The low upper limits of a few kilo-Rayleighs placed on any auroral emission from the Enceladus footprint constrain the mass pickup rate and extent of the mass loading region near Enceladus, and are consistent with a model of the interaction of the extended atmosphere of Enceladus with the corotating plasma (Pontius and Hill 2006).

A large program of HST observations of UV auroral emissions at both Saturn and Jupiter was undertaken over 2007–2008, in part to take advantage of in situ measurements by Cassini at Saturn. Images were obtained using the Solar Blind Channel (SBC) of the HST Advanced Camera for Surveys (ACS), which has properties similar to the STIS instrument, although the ACS sensitivity to auroral emissions is 20–30% higher than STIS. The Saturn observations were concentrated in two campaigns of more or less daily observations, the first in January/February 2007 and the second in February 2008, both carried out when Saturn was close to opposition to optimize the accuracy of propagation of solar wind conditions from measurements obtained at Earth orbit out to Saturn. The initial results from these observations are presented in Section 12.3.1, but this large data set represents a significant resource for future studies. Movies made from images from the 2007 and 2008 HST campaigns are included on the DVD supplied with this book.

There have been several results from a detailed comparison of Cassini in situ data with the HST UV auroral images. Cassini measurements of field-aligned currents near local noon during simultaneous HST images of the auroral emission distribution have given evidence that the UV aurora is associated with these currents near the open/closed field line boundary (Bunce et al. 2008a). A detailed comparison of theoretical models of auroral current systems in Saturn's magnetosphere and Cassini and HST observations is given in Sections 12.4.2 and 12.5.

12.2.2.2 Saturn's Ultraviolet Auroral Morphology

The analysis of polar projections of HST STIS and ACS images shows that the bright ring of auroral UV emission, the main oval, actually consists of several sub-structures having different latitudinal widths and brightnesses and forming along different paths around the spin axis. Dramatic changes in morphology have been observed during the different HST campaigns. However, some order emerges from this apparent variation; the bulk of auroral emission in groups of images has been shown to fit in a set of characteristic zones (Grodent et al. 2005) fixed in local time, revealing some morphological stability.

The global morphology of the emission varies from an expanded quasi-circular distribution, almost centered on the spin axis and characterized by a mean (southern) latitude of ∼70°, to a contracted spiral shape. In most cases, the spiral starts in the midnight to dawn sector at high lati-

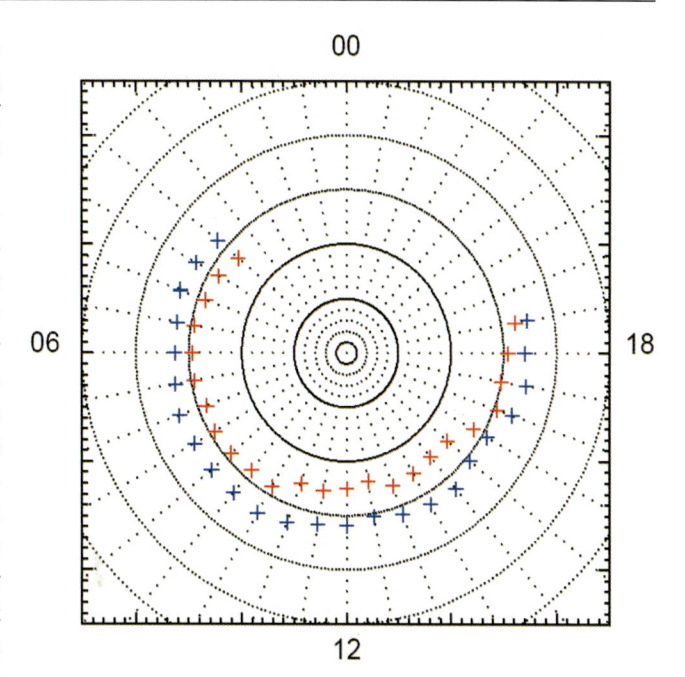

Fig. 12.2 The median equatorward (*blue*) and poleward (*red*) boundaries of the main auroral oval determined by a statistical study of HST images obtained from 1997 to 2004 by Badman et al. (2006)

tude (values can reach 80°) and then drifts equatorward with increasing local time down to values as low as 70° (Gérard et al. 2004; Clarke et al. 2005; Grodent et al. 2005). Badman et al. (2006) used a selection of HST images obtained from 1997 to 2004 to locate the median poleward and equatorward auroral boundary of the auroral oval using statistical methods. These are shown in Fig. 12.2. Near dusk it connects with the rest of the emission and starts a new, partial, revolution around the pole in the midnight to dawn sector. The spiral shape is likely a consequence of the imbalance between the processes of magnetic field line reconnection occurring at the dayside magnetosphere and in the nightside tail (Cowley et al. 2005). Extreme cases have been reported where most of the emission is concentrated in the morning side. Long term observation campaigns suggest that the global auroral morphology can remain unchanged for almost one week but can also undergo dramatic changes from one day to the next. The bright emission does not form a continuous ring but rather a string of arcs, several tens of degrees long, with variable width, ranging from a few to several thousands of kilometers (up to a few degrees in longitude), and variable brightness.

Smaller features intermittently appear poleward of the main ring of emission. They take the form of individual spots or small arcs which last several minutes and appear distinct from the rest of the auroral emission. Simultaneous remote and in-situ observations with Cassini suggest that these transient features are associated with the dynamical processes taking place in the Saturnian magnetosphere such as injection events (Radioti et al. 2009). They make no attempt to examine the local time of these features, but focus on HST images that almost overlap Cassini/MIMI (Magnetospheric Imaging Instrument) observations. Other, unpublished, images show transient auroral features at all local times.

While observations accumulated over several hours indicate that the global auroral morphology remains fixed in local time, temporal analysis of the location of individual features forming this global morphology shows that they rotate with the planet at a substantial fraction of corotation (Grodent et al. 2005). A movie showing this rotation is included on the DVD supplied with this book. This is in contrast to a structure truly fixed in local time, as in the case of the Earth. This temporal analysis indicates that the velocity is globally ~65% of corotation with rather large variations of this fraction, but there are uncertainties depending on which reference points in the images are used to determine this velocity. The rotational motion of the auroral emission may seem to contradict the local time nature of the global morphology; however, it should be stressed that the global morphology sets the poleward and equatorward boundaries of the polar region where the auroral emission is likely to occur, regardless of the identity or the previous location of the auroral features that are forming it. Accordingly, the global morphology may be seen as a frame, fixed in local time, but inside which emission (sub-)corotates with the planet's atmosphere around the pole. This may be related to the behavior of SKR in which the maximum brightness occurs when the subsolar point is near 100° SLS3 longitude (Kurth et al. 2008), yet the emission is thought to be generated on field lines threading bright auroral arcs, which appear to move in the corotational direction.

An important feature of Saturn's auroral morphology and dynamics is that longitudinal structures at ~65% of corotation have been observed during periods of very quiet solar wind activity. In addition, enhancements of the dawnside oval have been observed when Cassini did not observe any signature in the solar wind activity (Gérard et al. 2006). These observations and the onset of injection of hot plasma in the night or dawn sector in the absence of solar wind activity imply ongoing (but intermittent) dynamical phenomena associated with either the solar wind or planetary rotation through the Dungey or Vasyliunas cycle during relatively quiet intervals.

Closer inspection of a long sequence (~7 h) of images obtained in January 2004 revealed the presence of an isolated bright auroral structure for which the angular velocity constantly decreased with time as it moved across the noon to post-noon sector. The corotation factor of this auroral feature decreased from 55% to 20% and was concomitant with a rapid ~5° poleward shift of the feature. This peculiar feature is probably related to significant enhancements that are regularly observed near 1200 LT poleward of or along the main auroral oval (Gérard et al. 2005). These emissions may then be seen as the signature of the kronian cusp in the noon sector. They present two distinct states; the first is a bright arc-like feature located in the pre-noon sector that appears as an intensification of the main oval. The second is a more diffuse "spot" of aurora lying poleward of the general location of the main auroral oval. The former would be a consequence of pulsed reconnection at the low-latitude dayside magnetopause when the interplanetary magnetic field (IMF) is directed northward (antiparallel to Saturn's magnetic field lines), while the latter would correspond to the case of southward IMF and high-latitude lobe reconnection (Bunce et al. 2005).

A zone of emission commonly appears at the equatorward boundary of the nightside auroral region. In this sector, faint emission appears clearly equatorward of 70° and is detached from the brighter emission poleward of 70°. This emission is usually tenuous with an intensity only a few kiloRayleighs above the background. It is revealed by the limb brightening of the emission which, in the case of the HST/STIS 2004 dataset, increased the brightness by up to 350% near midnight. Its morphology and mean brightness remain almost constant over the 2004 HST data set, even during the most extreme events. This latter observation suggests that the lower latitude emission is produced by a different mechanism from the rest of the emission, presumably related to inwardly diffusing energetic particles deep into the corotating magnetosphere (Stallard et al. 2008c).

A general trend emerges in the brightness distribution as the dawn to noon sector is generally brighter than the noon to dusk sector, although some images show the opposite trend. The maximum brightness ranges from a few kiloRayleighs to several tens of kiloRayleighs. It usually decreases as the auroral structure rotates from dawn to dusk through noon.

12.2.2.3 H$_2$ Auroral Spectroscopy

The auroral emission at Saturn is mainly produced by collision of precipitating energetic electrons with the neutral atmosphere (Shemansky and Ajello 1983). Between 80 and 180 nm, the auroral emission is dominated by atomic H lines from the Lyman series and H$_2$ vibrational lines from the $B^1\Sigma_u^+ \to X^1\Sigma_g^+$, $C^1\Pi_u \to X^1\Sigma_g^+$, $B'^1\Sigma_u^+ \to X^1\Sigma_g^+$, $D^1\Pi_u \to X^1\Sigma_g^+$, $B'''^1\Sigma_u^+ \to X^1\Sigma_g^+$, $D'^1\Pi_u \to X^1\Sigma_g^+$ system bands (Gustin et al. 2009). The Lyman ($B \to X$) and Werner ($C \to X$) bands as well as the Lyman continuum prevail in the far UV (FUV) spectral region, and the extreme UV (EUV) bandwidth results from transitions from the B'' and higher Rydberg electronic states.

Once produced, the FUV and EUV auroral emissions interact with the atmosphere through absorption by hydrocarbons and self-absorption by H$_2$. These two mechanisms have been used to estimate the altitude of the auroral emission peak by fitting observed spectra with model spectra. A generator of H_2 spectra has been developed to simulate the effects of the impact of auroral electrons on the atmospheric H_2 molecules. It has been described and used for Jovian spectroscopy analysis by Gérard et al. (2002) and Gustin et al. (2004), and for Saturn's spectroscopy by Gérard et al. (2004) and Gustin et al. (2009).

Methane is the fourth most abundant substance in the atmosphere of giant planets, after H_2, H and He. It is produced in the troposphere and transported to the stratosphere by convection and turbulent diffusion. Methane has a large wavelength-dependent absorption cross-section in the 90–140 nm domain. As a result, it attenuates the H_2 emission below 140 nm and leaves the emission above 140 nm mostly unattenuated. Even though ethane (C_2H_6) and acetylene (C_2H_2) are known to be present within the atmosphere, only CH_4 has been previously detected from auroral UV spectra in the 115–170 nm range near local noon (Gérard et al. 2004). The level of absorption by the methane layer overlying the emitting region is characterized by the color ratio CR = I (155–162 nm) /I (123–130 nm) where I is intensity in each bandpass, and where brightness in kiloRayleighs is $4\pi I$. See Fig. 3 of Gérard et al. (2003) for an illustration of this differential absorption. The color ratio CR is 1.1 for an unattenuated spectrum, expressed in Rayleighs, and its value increases with attenuation. The color ratio is thus directly associated with the emission altitude, hence, to the penetration depth of the precipitating electrons which depends on the energy of these electrons. Analysis of six HST/STIS spectra of the southern aurora in the 115–175 nm spectral range (Gérard et al. 2004) suggests that the auroral emission is little absorbed by methane, with a nearly constant vertical CH_4 column of 6×10^{15} cm^{-2}. Using a model of Saturn's atmosphere at 30° North (Moses et al. 2000) relating the H_2 and CH_4 column density profiles, and the Grodent et al. (2001) energy degradation model that links the H_2 column density to the mean energy of the precipitated electrons, Gérard et al. (2004) derived an energy of the precipitated primary electrons in the range 12 ±3 keV. More recently, Gérard et al. (2009) used the emission altitude inferred from HST/ACS images of Saturn's southern auroral region as a constraint for a Monte Carlo electron transport code (Bisikalo et al. 1996). The peak of the auroral nightside emission is generally located 900–1300 km above the 1-bar pressure level which corresponds to a characteristic energy of the precipitating electrons between 5 and 30 keV. This energy depends on the model atmosphere; Gérard et al. (2009) considered a model in which the thermospheric temperature in the auroral region sharply increases at a higher pressure level than in the low latitude model of Moses et al. (2000).

Direct measurement of the H_2 column overlying the auroral emission can be obtained independently from EUV spectra. Below 120 nm, the photons connecting to the $v'' = 0, 1, 2$ vibrational levels may be partially or totally absorbed by the overlying column of H_2 and redistributed to the FUV portion of the spectrum. This self-absorption process provides a measure of the H_2 column overlying the UV emission peak and on the population of the ground-state vibrational levels (Gustin et al. 2004).

Gustin et al. (2009) compared synthetic H_2 spectra with UV spectra of Saturn's aurora obtained with HST/STIS, Far Ultraviolet Spectroscopic Explorer (FUSE) and with the Cassini Ultraviolet Imaging Spectrograph (UVIS) instruments. The temperature of the absorbing H_2 molecules can also be derived from self-absorption, provided the spectral resolution is sufficiently high, which is the case of the FUSE observations.

STIS and UVIS FUV spectra were found to be very little absorbed by methane, with vertical CH_4 columns less than or equal to 1.2×10^{16} cm^{-2} consistent with primary electron energy lying in the 10 to 18 keV range. This suggests that the peak of the UV auroral emission originates from above but close to the methane homopause, and primary energies near $\sim$18 keV should be considered as an upper limit.

FUSE and UVIS EUV spectra clearly exhibit self-absorbed H_2 bands. However, the low spectral resolution of the UVIS EUV spectra does not make it possible to resolve the H_2 rotational lines and cannot provide a reliable T/H_2 column pair. On the other hand, the high resolution of the FUSE EUV spectra makes it possible to determine a H_2 rotational temperature on the order of 400 K and an H_2 column of $\sim 6.0 \times 10^{19}$ cm^{-2}, corresponding to a pressure level of $\sim 0.15 \mu$bar. The H_2 columns deduced from FUSE lead to primary auroral electron energies from 10 to 15 keV. The temperature of 400 K derived from FUSE is in agreement with the 350–500 K range obtained from H_3^+ IR spectra measured at the H_3^+ peak altitude by Melin et al. (2007). These values are significantly higher than the $\sim$125 K expected from the Moses et al. (2000) equatorial model near 0.15μbar, which suggests that an equatorial temperature profile is not appropriate for polar latitudes.

12.2.2.4 Effects of Auroral Energy Input to the Atmosphere

The auroral energy input to the atmosphere of Saturn has been estimated by Gérard and Singh (1982). Their model takes into account the loss of energy of incident energetic particles from collisional processes, tracks the resulting excitation of emissions, and models the radiative transfer of the UV emissions leaving the atmosphere. From this model, the rule of thumb has emerged that the radiated UV power is approximately 10% of the total incident power. The total radiated power from Saturn's aurora is typically 10–30 GW, corresponding to 100–300 GW input power to the upper atmosphere. This input power is comparable to the solar input and, when the auroras are bright, exceeds the total input power to the global upper atmosphere from solar UV radiation. The aurora therefore has a strong and potentially dominant influence on the thermospheric dynamics of Saturn's upper atmosphere. The extent to which this input energy can be transferred from the polar regions to the global atmosphere, and contribute to the high upper atmospheric temperature, is the subject of detailed modeling (Müller-Wodarg et al. 2006). These authors found that while there is ample energy in the auroral zones to yield the global high thermospheric temperatures, the planet's large size and rapid rotation result in strong zonal winds that limit heat transport to lower latitudes.

12.2.2.5 Ground-Based Observations of Saturn's IR Aurora

Since the arrival of Cassini at Saturn, the IR aurora from H_3^+ emissions has been studied in detail by a long-term observing program based at NASA's Infrared Telescope Facility, using both high-resolution spectroscopy (Stallard et al. 2007a) and emission-line imaging (Stallard et al. 2008a). These observations have begun to categorize both the morphology and magnetospheric origin of infrared auroral emission. In broad structure, the IR emission morphology appears to be similar to that of the UV emission. Just as with the UV main auroral oval, the IR oval is roughly circular. During periods of rarefied solar wind, this appears to be low intensity on both the dawn and dusk sides of the oval, while compressions in the solar wind cause significant dawn brightening. Over the period of the Cassini mission, the overall IR intensity has varied by up to a factor of 50, from 2.56×10^{-15} W m^{-2} to 1.32×10^{-13} W m^{-2}, though the largest change seen over any individual 24-h period was a factor of two increase in intensity.

Equatorward of the main auroral oval, a secondary auroral oval has been detected. Emission from this oval appears to be relatively weak, with an intensity <25% of the main oval. As shown in Fig. 12.3, the ion wind velocities associated with the secondary oval show that it is created by the breakdown in co-rotation within the plasma disk around the planet – a weak Saturnian equivalent of Jupiter's main auroral oval (Stallard et al. 2008c). There are other significant fine scale differences between the IR and UV auroral emission. Most notable is the significantly raised level of polar emission

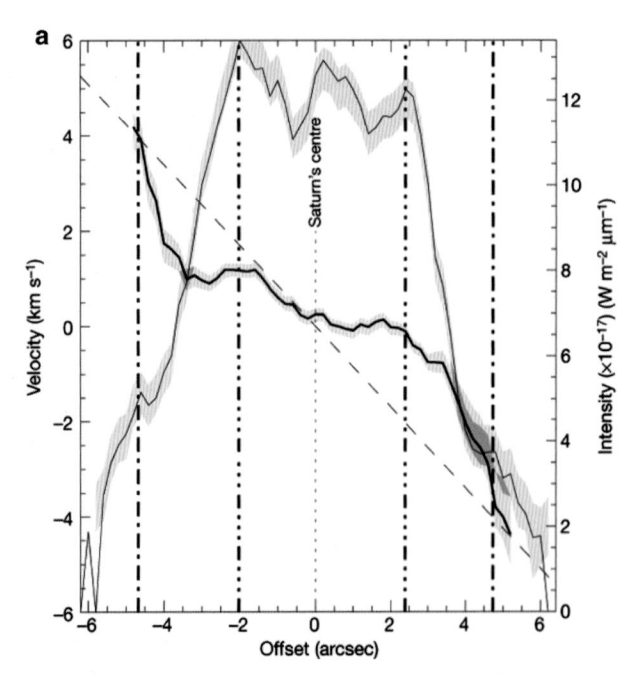

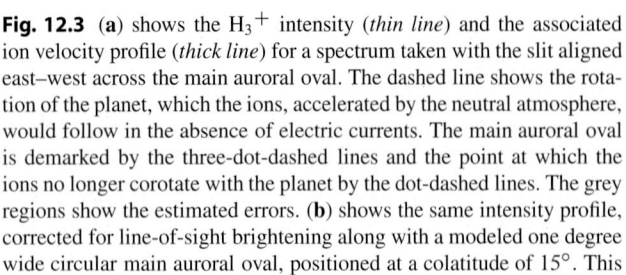

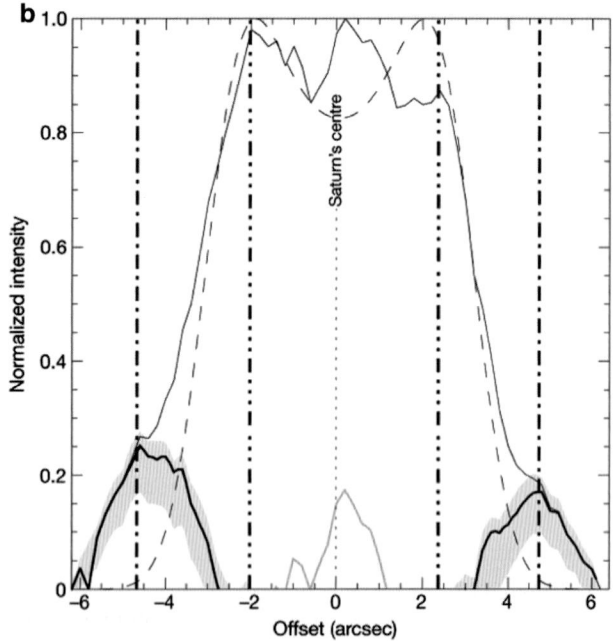

Fig. 12.3 (**a**) shows the H_3^+ intensity (*thin line*) and the associated ion velocity profile (*thick line*) for a spectrum taken with the slit aligned east–west across the main auroral oval. The dashed line shows the rotation of the planet, which the ions, accelerated by the neutral atmosphere, would follow in the absence of electric currents. The main auroral oval is demarked by the three-dot-dashed lines and the point at which the ions no longer corotate with the planet by the dot-dashed lines. The grey regions show the estimated errors. (**b**) shows the same intensity profile, corrected for line-of-sight brightening along with a modeled one degree wide circular main auroral oval, positioned at a colatitude of 15°. This model has been subtracted from the H_3^+ intensity, with the residual emission also shown. Poleward of the main oval a significant auroral brightening can be seen, which may be analogous to the polar emission seen by the Cassini-VIMS instrument. Equatorward of the main oval, two emission peaks can be seen. These coincide with the location of breakdown in corotation shown in (**a**), strongly suggesting that this is a secondary auroral oval at Saturn that is caused by the breakdown in corotation, in a similar way to the main auroral oval at Jupiter (from Stallard et al. (2008c))

seen in the majority of IR observations; the relative strength of this raised emission appears to vary with time (Stallard et al. 2007a). In addition, the main auroral oval has been seen to have an extended period of dusk brightened aurora, covering a period of up to five Saturnian days (Stallard et al. 2008b).

Ion wind velocities within the auroral region also show significant structure not seen within the emission morphology. Most notably, while the auroral region generally sub-corotates relative to the neutral atmosphere, a region poleward of the main auroral oval, generally on the dawn side, appears to consistently corotate with the planet. This is a highly unexpected result since this region is typically thought to be associated with field lines open to the solar wind. During periods of compression in the solar wind, when the auroral emission is strongly dawn brightened, this central corotating region disappears and the entire auroral region sub-corotates. It has been suggested that this region may correspond to field lines embedded in the center of the magnetotail that are shielded from the solar wind such that ions in this region are effectively controlled only by the neutral atmosphere and that a major compression in the solar wind closes these field lines (Stallard et al. 2007b).

Observations made using the United Kingdom Infrared Telescope show the overall temperature of Saturn's polar H_3^+ layer should be taken as 400 ± 50 K. The H_3^+ emission strength varied by at least an order of magnitude over the 6 year period of these observations. It was shown that this variability was driven by changes in the H_3^+ column density and that the increased level of ionization which produced this required a ~20 times variation in the particle (kiloelectron-volt electron) precipitation. However, such increases are insufficient to offset most of the heating due to the extra particle precipitation, indicating that H_3^+ does not act as a "thermostat" on Saturn, in the same way that it does on Jupiter (Melin et al. 2007).

12.2.3 Cassini Remote Sensing Observations of Auroral Emissions

Cassini can observe auroral and aurora-related emissions in wavelengths ranging from the ultraviolet through radio wavelengths. Cassini has four optically aligned instruments obtaining data relevant to the aurora. Cassini's Ultraviolet Imaging Spectrometer (UVIS) has a long-slit spectrograph with Extreme Ultraviolet (EUV, 56–118 nm) and Far Ultraviolet (FUV, 110–190 nm) channels with spectral resolution (0.275, 0.48, or 2.49 nm for EUV; 0.275, 0.48, or 1.94 nm for FUV) depending on the selected slit width (Esposito et al. 2004). Each channel has a 1024 pixel (spectral dimension) × 64 pixel (spatial dimension) microchannel plate detector, providing spectral capability and, when slewed appropriately, imaging capability. Cassini's Imaging Science Subsystem (ISS) has wide (3.5°) and narrow (0.35°) field of view 1024 × 1024 pixel CCD cameras and a set of narrow-band filters providing coverage over the range from 200–1100 nm (Porco et al. 2004). Cassini's Visual and Infrared Mapping Spectrometer (VIMS) instrument has an internal scan mirror, providing spectral and spatial coverage without additional spacecraft motions (Brown et al. 2004). Finally, Cassini's Composite Infrared Spectrometer, CIRS (Flasar et al. 2004) uses a pair of Fourier transform spectrometers to make thermal infrared observations at a wavelength range from 7 μm to 1 mm.

While not directly auroral emissions, kilometric radio emissions result from particle distributions existing on auroral field lines and can be used as indicators of auroral activity. The Radio and Plasma Wave Science (RPWS) instrument on Cassini detects waves in the frequency range from ~1 16 MHz, including the SKR spectrum extending from a few kiloHertz to 1.2 MHz and centered between 100 and 400 kHz (Gurnett et al. 2004).

12.2.3.1 UVIS Results

Esposito et al. (2005) presented an initial Saturn UV spectrum and found it indistinguishable from a Jupiter spectrum except 250 times dimmer. In both cases the observed emissions are thought to be due to electron-impact excitation of molecular and atomic hydrogen. In general, the electron energies are lower at Saturn than at Jupiter (Shemansky and Ajello 1983). Gustin et al. (2009) presented comparisons of spectra from FUSE, HST STIS, and Cassini UVIS, and found additional evidence for relatively soft electrons at Saturn.

An early campaign from Saturn orbit was undertaken during a period in which the spacecraft was above Saturn's equator near 90° phase angle and out in the solar wind on the dawn side. This period is useful for examining the response of the aurora to the solar wind. In this period the auroras are seen as brightenings at the planetary limb. UVIS observed several episodes of brightening (up to a factor of 4 changes in intensity) that took place at both poles in response to solar wind pressure increases (Esposito et al. 2005).

Figure 12.4 shows an image pair of Saturn and its rings from UVIS (2005 day 172; June 21 2005) created by slewing the spacecraft slowly across Saturn and then back. This image was prepared making use of deconvolution techniques and the estimated UVIS point-spread-function. The quality of the deconvolution was controlled by observing the effect on the appearance of the known structures in Saturn's rings. In this image blue is used to represent auroral emission wavelengths, and orange is used for reflected sunlight at longer FUV wavelengths. The two crossings of the auroral oval are

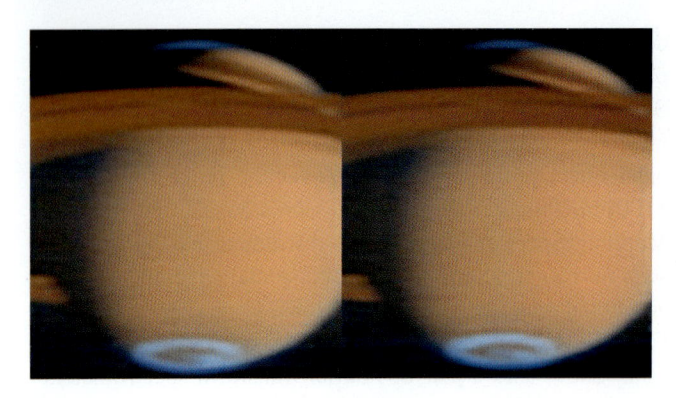

Fig. 12.4 Cassini UVIS view of Saturn from June 21, 2005

separated in time by about an hour, and show changes in the auroral brightness both on the main oval and in its interior. The auroral emission inside the oval near local noon is reminiscent of polar cusp auroras on the Earth, similar to features observed by HST (Gérard et al. 2005).

Beginning in 2007 UVIS obtained a number of movies showing Saturn's northern auroral oval with brightness features rotating on Saturn. These provide a good view of both the day and the night side of the oval, unlike most HST images. A processed version of one movie from 2008 day 129 is available on the DVD supplied with this book. This movie showed two bright auroral features on the arc moving nearly corotationally, aligned in longitude with patches of enhanced brightness in MIMI Ion and Neutral Camera (INCA) magnetospheric ring current images. The brightest part of the auroral arc separates into two distinct arcs (Mitchell et al. 2009b). Emission features on the oval in some movies brighten on the dawn side and dim on the dusk side, then rebrighten as they return to the dawn side. This enhanced emission on the dawn side is consistent with SKR observations (Galopeau et al. 1995) and theoretical considerations (Cowley et al. 2004a, b). A number of UVIS movies show a polar cusp emission near noon between the pole and the main oval, as was observed with HST (Gérard et al. 2005). The possible origin of this feature is addressed in Section 12.5. Transient bright features that show up in a single frame are common inside of the main oval. Two of the movies from July 2008 show apparent compression events where the oval contracts in size, broadens in width, and "fills in" with emission inside part of the oval.

12.2.3.2 VIMS Results

The VIMS instrument has observed Saturn's infrared auroral region using IR emission from the H_3^+ ion. Data from the instrument were combined across several separate wavebands in order to construct images shown in Fig. 12.5 of the infrared aurora at a much higher spatial resolution than had previously been possible using ground-based observations (Stallard et al. 2008a). Figure 12.5 (panels a–f) shows a sequence of recent VIMS images taken during July 2008. Figure 12.5g shows a striking image of Saturn's northern polar region, never observed in the infrared from Earth. Here the H_3^+ auroral emission coincidently overlays Saturn's polar hexagon. These observations showed that in broad structure, the main auroral oval appears to have a similar morphology as that seen in previous UV observations. In observing the planet during higher inclination orbits, it has been possible to view the entire auroral oval more clearly. This has highlighted an equatorward doubling of the main auroral oval on the midnight side, previously tentatively identified in the HST observations as a limb-brightened aurora (Grodent et al. 2005).

These observations have also highlighted the significant differences between the IR and UV auroras. Most notably, as shown in Fig. 12.5 (panels h and i), there is a high level of IR polar emission, with a much greater relative brightness relative to the UV emissions, as has previously been suggested in ground-based observations (Stallard et al. 2004). This polar emission can take the form of patchy emission across the pole, localised bright spots, extended arcs of emission and even, in some cases, an infilled region of emission, brighter than the main auroral oval, between ∼82° and the pole. This polar emission is highly variable, changing on timescales as short as 45 min, the temporal spacing between the images in panels h and i in Fig. 12.5. However, in examining the morphology of the main auroral oval during periods when such polar emission occurs, it has been concluded that this polar auroral brightening does not appear to be directly linked to compression in the solar wind.

12.2.3.3 ISS Results

Saturn's visible auroras were imaged by Cassini in 2006 and in 2007 on Saturn's night side in the north polar region near 75° latitude (Dyudina et al. 2007). These data are presented in Fig. 12.6. While the detections are much dimmer than the ultraviolet aurora and have low signal-to-noise, the spatial resolution is better than that of the other Cassini imaging instruments.

ISS observations show that Saturn's aurora is bright in a few spectral channels (in the broadband infrared channel covering 825–925 nm (∼100 R/nm), in the Hα channel covering 651–661 nm (∼200 R/nm), possibly in a UV channel covering 300–370 nm (∼250 R/nm)), and dark in the other visible channels. The average broadband visible brightness of the aurora for the 300–900 nm 2006 multi-channel observation is ∼25 R/nm. This agrees with the line emission predicted in the laboratory spectra of electron impact on molecular hydrogen and atomic hydrogen for the Saturnian aurora by Aguilar et al. (2008). This also agrees with the

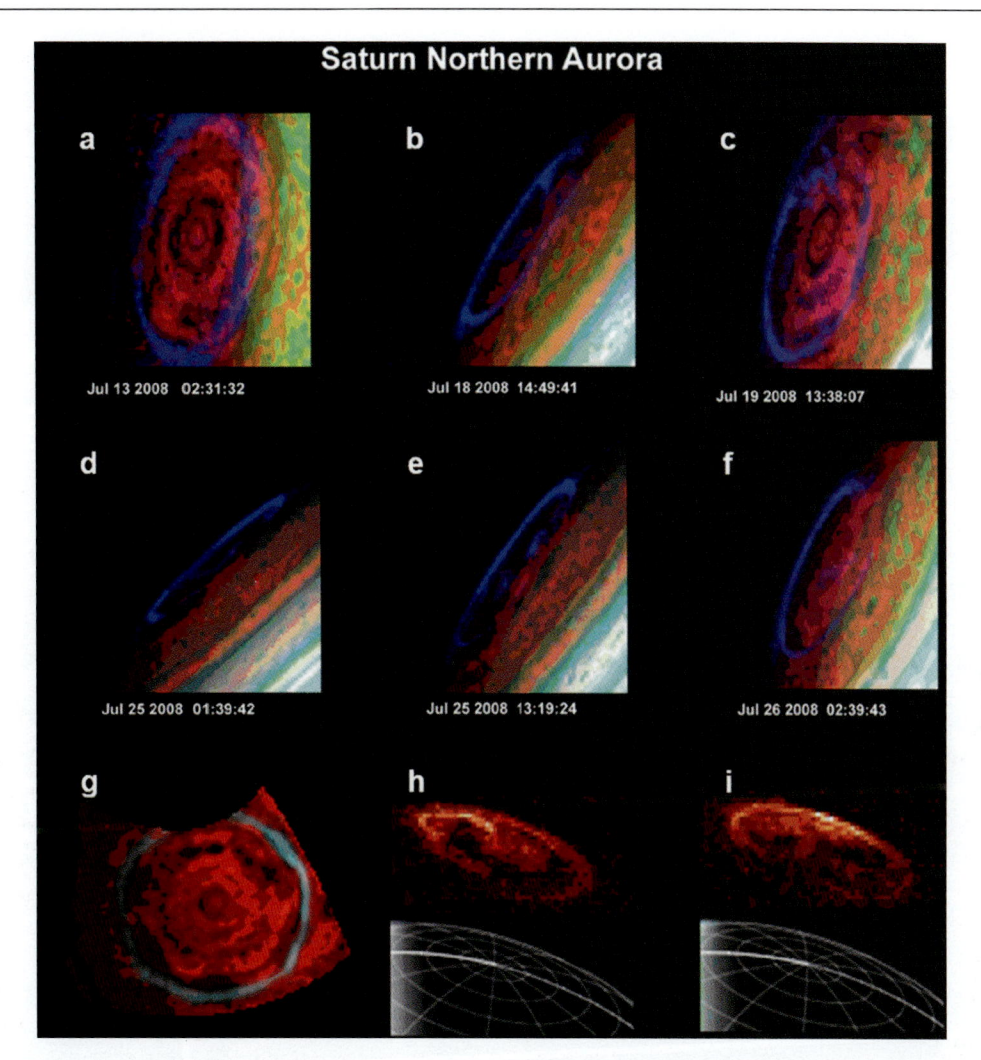

Fig. 12.5 (**a–f**) A sequence of VIMS images taken during July 2008. The (*false*) colors in these images represent the wavelengths that contributed to the images, consisting of thermal emission from the lower clouds at 5.12 μm (*red*), reflected sunlight at 41.65 μm (*green*) and emission from several lines of H_3^+ at 3.53 and 3.68 μm (*blue*). (**g**) A striking image of Saturn's northern polar region, never observed in the visible spectrum of aurora on Jupiter observed by Galileo infrared from Earth. Here the H_3^+ auroral emission overlay Saturn's polar hexagon. (**h–i**) The appearance of a bright polar auroral emission that sometimes dominates Saturn's infrared aurora between two images separated by 45 min. Below these are visible images of Saturn, with a longitude/latitude grid overlain (Stallard et al. 2008d)

visible spectrum of aurora on Jupiter observed by Galileo (Ingersoll et al. 1998; Vasavada et al. 1999). Figure 12.6 shows the auroral arcs and spots near 75° latitude (left panel), and rapid variability in auroral arc features (right panels). The two features labeled 1 and 2 in the images are seen disappearing within half an hour between the observations.

12.2.4 Auroral Radio Emissions

Like other magnetized planets, Saturn is a strong radio source at kilometric wavelengths. Zarka (1998) provides a thorough review of planetary radio emissions and serves as a basis for our understanding of Saturn's radio emissions, particularly the Saturn kilometric radiation. These radio emissions were discovered by Voyager (Kaiser et al. 1980; Warwick et al. 1981). Most interestingly, SKR displays a variable modulation period near 10.7 h, thought to be close to the internal rotation period of the planet. Decades-long studies of the period of modulation of decametric radiation from Jupiter were the basis for the IAU-adopted rotation period of the planet. However, while Jupiter has an offset and tilted magnetic moment, Saturn's magnetic field is aligned along its spin axis well within 1 degree (Davis and Smith 1990; Dougherty et al. 2005) and it is not understood why there should be a rotational modulation of the SKR. Synoptic studies using the Ulysses spacecraft detected variations in this modulation period of order 1% on time scales of years (Lecacheux et al. 1997; Galopeau and Lecacheux 2000). Cassini has verified

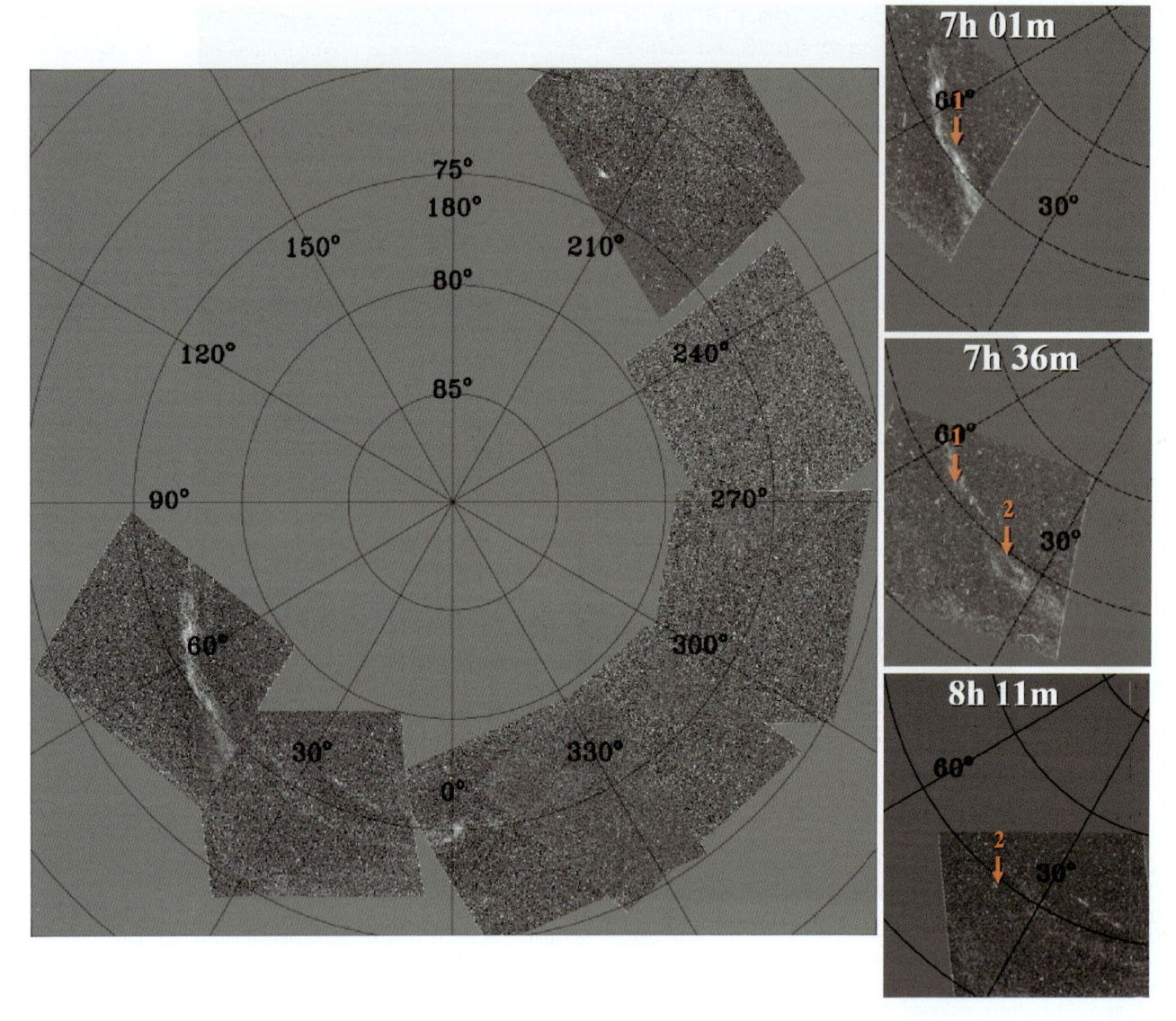

Fig. 12.6 Cassini map-projected ISS auroral images from January 2007 (*left panel*), and consecutive images of the same area showing rapid variations in auroral features (*right panels*). The times of the observations are labeled in white. Each image was taken around local midnight (from Dyudina et al. 2007)

the variation in period of SKR emissions on times scales from years to weeks (Gurnett et al. 2005 2007; Kurth et al. 2007, 2008; Zarka et al. 2007). Possible reasons for the variable SKR period are discussed in Chapter 10, but the result is that Saturn's rotation period cannot be readily deduced from the modulation of the SKR. The Cassini SKR observations are provided by the High Frequency Receiver portion of the RPWS instrument (Gurnett et al. 2004).

Given that the SKR period varies, it has been useful to define a new longitude system which takes into account the variable modulation period of the radio emissions. Kurth et al. (2007, 2008) have defined the SLS2 and SLS3 longitude systems, respectively, to provide a system that can be used to organize magnetospheric phenomena. In fact, the systems are not predictive, hence, the definition must be extended in time based on empirical models of the period. Numerous magnetospheric phenomena have been found to be organized by these systems, including the so-called 'cam' magnetic field signature (Gurnett et al. 2007; Andrews et al. 2008), the plasma density in the inner magnetosphere (Gurnett et al. 2007) and energetic neutral atom (ENA) hot spots in the middle to outer magnetosphere (e.g. Carbary et al. 2007).

Kilometric radio emissions at Earth (Benediktov et al. 1965; Gurnett 1974) were quickly associated with auroral activity by Gurnett. Wu and Lee (1979) suggested that the auroral kilometric radiation was generated by the cyclotron maser instability (CMI) and there has been consensus on this mechanism ever since. It was by analogy with these terrestrial emissions that the cyclotron maser instability was also

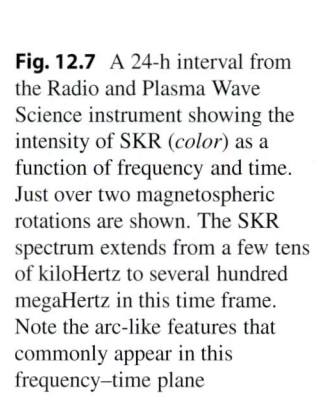

12 Auroral Processes

believed to be responsible for decametric and hectometric radiation at Jupiter and Saturn kilometric radiation. Hence, it was inferred that SKR had a close association with Saturnian aurora. Voyager observations (Desch 1982) indicated a strong correlation between SKR and the solar wind dynamic pressure; in fact, Voyager 2 found Saturn immersed in Jupiter's magnetotail around the time of its encounter and Desch (1983) showed that the lack of solar wind input to Saturn's magnetosphere greatly decreased the SKR activity.

It wasn't until the 2004 HST – Cassini campaign that evidence was obtained to indicate a direct relation between the UV aurora and SKR. HST campaigns in January and February 2007 and February 2008 have confirmed this association, but the correlation between SKR power observed by Cassini and the power in the UV aurora is not perfect (Clarke et al. 2009). The beaming of the radio emissions means that a spacecraft such as Cassini will not necessarily detect strong SKR emissions, depending on its position, even if the emissions are being generated.

Cassini has found that the general properties of the SKR spectrum are similar to those obtained by Voyager (Galopeau et al. 1989; Lamy et al. 2008b). On time scales of tens of minutes and frequency scales of a few hundred kHz, SKR demonstrates arc-like structures in the frequency–time plane, similar to, but less well defined than, Jovian decametric arcs. Examples of these can be seen in Fig. 12.7, particularly at 2300 on December 23, 2005 centered near 350 kHz; a more complex set of arcs is found between 0900 and 1100 on the same day. These have been pointed out in a number of reports, including some of the earliest Voyager studies. The arcs are discussed most recently by Lamy et al. (2008b, c).

At Jupiter, such arcs have been shown to be consistent with emission on magnetic field lines with beaming angles that vary with frequency such that an observer at a fixed location sees emissions from progressively larger or smaller distances along the field line, hence, at smaller or larger frequencies, respectively. Goldstein and Goertz (1983), for example, attempted to model the arc-like structure of Jovian decametric radiation.

The Cassini RPWS instrument (Gurnett et al. 2004) includes the capability to make high spectral and temporal resolution observations of SKR. Figure 12.8 from Kurth et al. (2005b) shows a progression of spectrograms from those used for normal survey studies to those approaching the ultimate resolution of ∼200 Hz and 125 ms (both of which are instrument mode and analysis dependent). These results showed that SKR includes both diffuse emissions, which do not appear to vary rapidly in time or frequency, and bright, narrowband tones with bandwidths of the order of 200 Hz that drift upwards or downwards in frequency. Audio files which enable one to 'hear' the complex spectrum of SKR are included on the DVD supplied with this book. Similar spectral structure is observed both in auroral kilometric radiation at Earth (Gurnett and Anderson 1981) and at Jupiter, for example in S-bursts (Carr and Reyes 1999). Since the CMI results in radio waves generated close to the local electron cyclotron frequency, such narrowband emissions imply that (a) there are small active regions emitting radio emissions and (b) these are moving up or down relative to Saturn to explain the falling or rising tones, respectively. The narrow bandwidth of the emissions essentially set an upper limit on the range of magnetic field strengths in the source region,

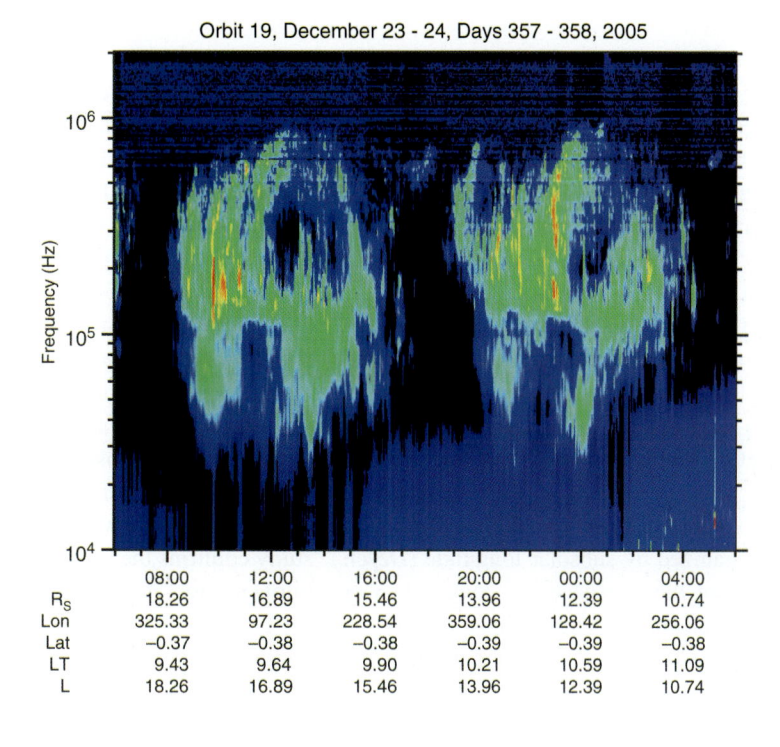

Fig. 12.7 A 24-h interval from the Radio and Plasma Wave Science instrument showing the intensity of SKR (*color*) as a function of frequency and time. Just over two magnetospheric rotations are shown. The SKR spectrum extends from a few tens of kiloHertz to several hundred megaHertz in this time frame. Note the arc-like features that commonly appear in this frequency–time plane

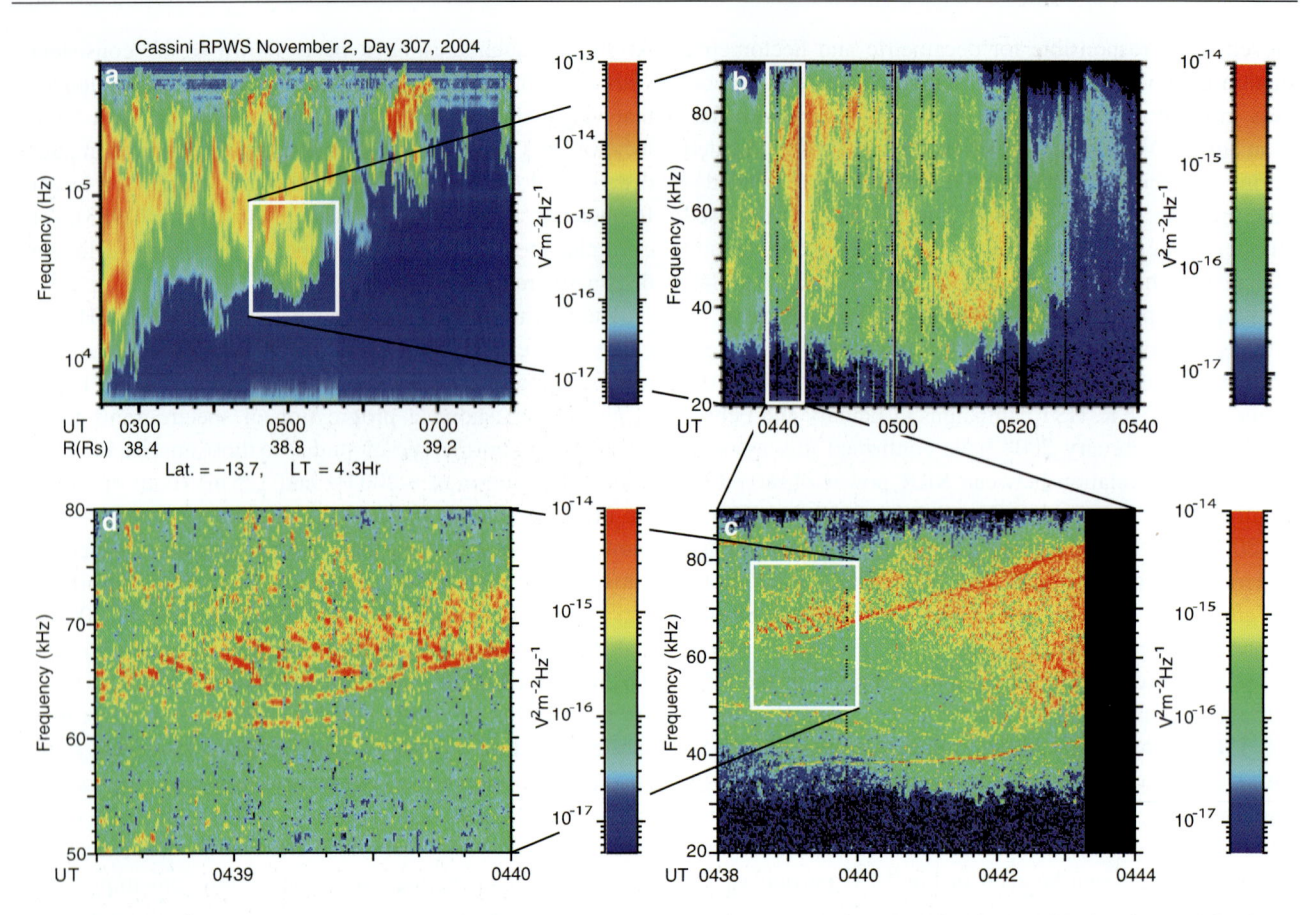

Fig. 12.8 Frequency–time spectrograms of SKR showing progressively higher resolution in both frequency and time going from panel (**a**) to (**d**) (from Kurth et al. (2005b)). Starting in panel a showing the lowest spectral and temporal resolution, a region is highlighted with a white box that is expanded in resolution in panel (**b**). Panel (**a**) comprises observations made in the RPWS survey, or low rate, mode. Panels (**b**)–(**d**) utilize the wideband receiver. Panel (**c**) shows the a higher resolution spectrogram of the data in the white box in panel (**b**). Panel (**d**) shows the observations in the white box in panel c at approximately the maximum temporal and spectral resolution of the instrument. At each level of resolution, the SKR shows an array of complex structures in the frequency–time plane, usually composed of narrowband tones drifting upwards or downwards in frequency

hence, imply small sources. And, the rate of drift of the tones implies a speed along the field line.

There is not consensus on the explanation of the narrowband tones at Earth, but one idea by Pottelette et al. (2001) with some advantages suggests that the small sources are phase space holes that are commonly observed on auroral field lines in and near the acceleration region (cf. Muschietti et al. 1999; Main et al. 2006). This idea has been further pursued by Mutel et al. (2006) suggesting that ion phase space holes might be responsible for drifting 'striated' narrowband auroral kilometric radiation features at Earth. Other work by Menietti et al. (2006) suggests that the drift rates are consistent with the propagation of ion cyclotron waves along auroral field lines.

Voyager synoptic studies showed that the peaks in SKR intensity were best organized by subsolar longitude (Desch and Kaiser 1981). When the subspacecraft longitude was used, the phase of the peak shifted with the local time of the spacecraft. This characteristic can be explained by a strobelight model where the SKR emissions are seen to peak at the same time, regardless of the observer's location. Jupiter's auroral radio emissions, on the other hand, exhibit a searchlight like pattern, where the peak rotates with the planet and the time at which the peak is observed depends on the observer's location. In addition, using various indirect methods, the average source location was deduced by Galopeau et al. (1995). This location was centered between dawn and noon at latitudes in the range of $60°$ to $>80°$. In spite of this source location, the Voyagers could observe SKR more or less without regard to their (limited) positions in local time.

Lamy et al. (2008b) have utilized approximately 2.75 years of Cassini observations to carry out a comprehensive synoptic study of the occurrence of SKR as a function of observer location with the results shown in Fig. 12.9. This study confirms the spectral range found by Voyager of a few kiloHertz to 1200 kHz. Lamy et al. (2008b) also verify that the bulk of the SKR emissions are emitted in the extraordinary (R-X) mode, that is right hand (RH) emission from the

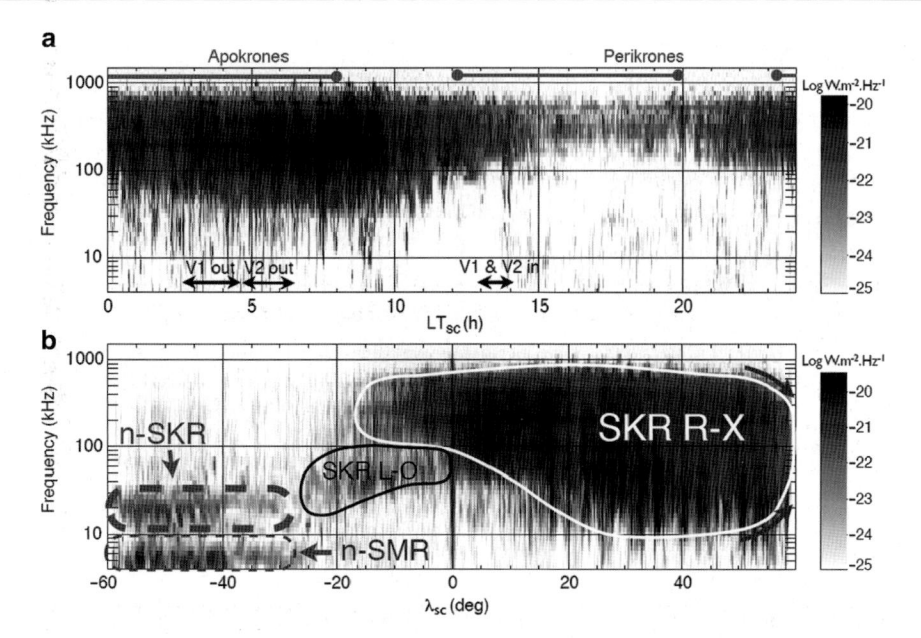

Fig. 12.9 Statistical properties of SKR (Lamy et al. 2008a). In the top panel, the average power normalized to a distance of 1 AU is given as a function of frequency and local time. The minimum in occurrence in the 15–20 h range is also where Cassini is close to Saturn. Hence, rather than a local time variation, this is indicative of an equatorial shadow zone at small radial distances. Panel (**b**) represents the average power of SKR as a function of frequency and latitude. For this analysis, north-south hemispherical symmetry was assumed so that all of the emission are assumed to come from the northern hemisphere. Labels in Panel (**b**) indicate extraordinary (R-X) mode and ordinary (L-O) mode SKR as well as two low frequency narrowband emissions that are not considered auroral radio emissions (after Lamy et al. 2008a)

northern hemisphere and left hand (LH) from the southern hemisphere. To first order, SKR can be observed at any local time. At very close planetary distances, however, there is an equatorial shadow zone which cannot be illuminated by the high latitude sources as found by Lamy et al. (2008b). Further, there is some indication that at higher latitudes emissions above 200 kHz and below 30 kHz disappear. Lamy et al. (2008b) also find general conjugacy in the emissions between the northern and southern hemispheres. Evidence for ordinary mode SKR emission is also given in Lamy et al. (2008b). Two narrowband emissions labeled n-SKR and n-SMR (narrowband Saturn myriametric radiation) are also shown in Fig. 12.9, but these are generated via mode conversion from electrostatic waves and are not related to the aurora. See Louarn et al. (2007), Wang et al. (2009), and Ye et al. (2009) for discussions of the narrowband emissions and their sources.

Lamy et al. (2008c) subsequently modeled the SKR occurrence by examining radiation from two types of electron distribution functions unstable to the cyclotron maser instability. The first of these is a loss-cone distribution which results in radio beams emitted obliquely with respect to the local magnetic field and the second, is alternately called a shell or horseshoe distribution, the result of electron trapping due to field-aligned potentials on auroral field lines. The shell electron distribution is expected to produce beaming at 90° with respect to the magnetic field. While both distributions produce radio emissions, the latter is more efficient and is thought to be the predominant source of auroral kilometric radiation at Earth (Mutel et al. 2008). Lamy et al. (2008c) modeled the frequency–time characteristics of radio emissions as observed from a virtual observer in an attempt to reproduce the arc-like structures and equatorial shadow zone observed by Cassini it its orbit. Figure 12.10 shows a comparison of a typical periapsis pass of Cassini with a clear shadow zone apparent in the SKR data that is well-modeled by Lamy et al. (2008c). They found that either distribution can reproduce the frequency–time aspects of SKR, but the loss-cone distribution provides a better fit.

A most powerful tool provided by the RPWS instrument is the ability to measure both the direction of arrival and full polarization of radio emissions. The most complete study of SKR using these capabilities was carried out by Cecconi et al. (2008), who determined the three dimensional source of SKR during one periapsis pass of Saturn by Cassini. The source locations are determined over a 24-h interval during which the spacecraft passes from the southern to the northern hemisphere and moves from dawn to dusk through local noon relative to Saturn. This study determines the direction-of-arrival, which places the source at a given frequency f in the plane of the sky, a two dimensional position. The assumption that the SKR is generated on a surface of constant electron cyclotron frequency $f_{ce} = f$, according to the theory of the cyclotron maser instability, is used to obtain the third dimension, or distance to the source. The surface of constant f_{ce} is derived from a magnetic field model, in this case the SPV

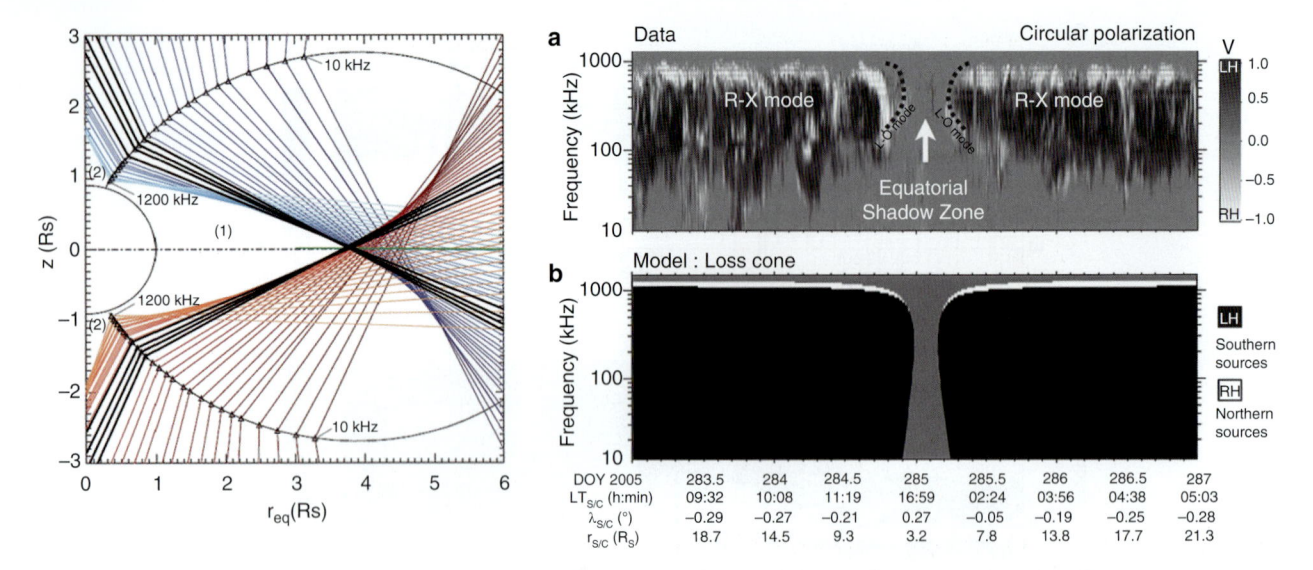

Fig. 12.10 *Upper right panel* shows the occurrence of ordinary and extraordinary mode SKR as seen by Cassini as it moves in a nearly equatorial orbit from about 19R$_S$ in to periapsis near 3R$_S$ and then back out to about 21R$_S$. The shadow zone is clearly present near periapsis. The *bottom right panel* shows the result of modeling the shadow zone; the similarity of the essential features with the data, above, is apparent. The *left-hand panel* shows how the shadow zone (labeled (1)) is formed by SKR sources distributed along auroral field lines with beaming angles that vary with frequency (from Lamy et al. 2008b)

Fig. 12.11 Source location results for a 5-min interval (Cecconi et al., 2009). In the left-hand panel, the source locations are given in a three-dimensional perspective plot as a function of frequency (*color*). In the right-hand panel, the source locations are translated to the locations of the feet of the magnetic field lines threading the sources

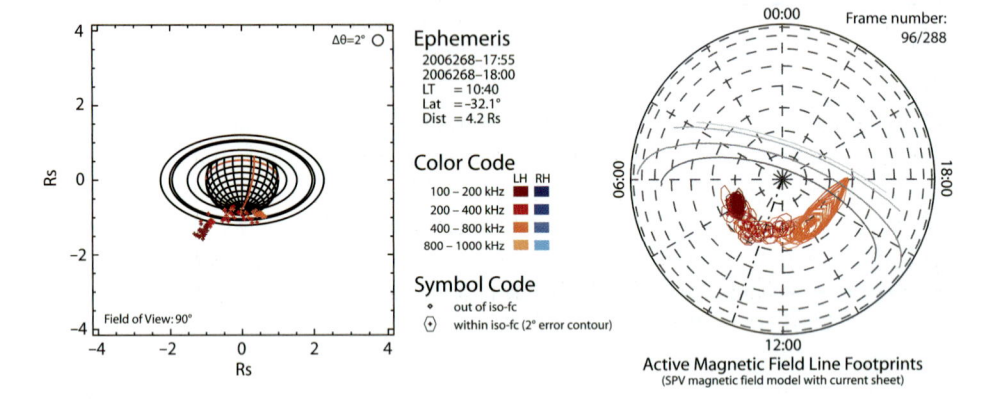

model by Davis and Smith (1990) including the current sheet model by Connerney et al. (1983). Given the source location, the field line threading the source is determined, hence, the magnetic field model can be used to trace that field line to its foot at the top of the atmosphere for comparison with the location of UV auroras.

An example set of source locations determined over a 5-min interval is given in Fig. 12.11 which shows both a perspective plot of the source positions as a function of frequency on the left and the footprints of the source field lines in the right panel. Cecconi et al. (2008) provide a movie composed of such results over the full 24-h interval of analysis. A copy of this movie is included on the DVD supplied with this book. The primary results of Cecconi et al. (2008) are that the SKR sources are located on field lines that map to the ionosphere between ∼70 and 80° latitude, consistent with the mean location of the main UV auroral oval. In the northern hemisphere the sources tend to be 2° to 5° higher in latitude than in the south. The sources are also observed over the entire 4 to 16 h local time range visible to Cassini through this particular periapsis pass. Further, the local times of the sources observed at any instant in time are strongly influenced by the local time of Cassini. That is, radio sources clustered on either side of the Cassini central meridian longitude are most often observed. This is a clear manifestation of beaming of the radio emissions.

Lamy (2008a) and Lamy et al. (2009) have, for the first time, shown the direct correspondence between field lines threading the source of SKR and those with their feet embedded in the bright dawnside auroral arc. As shown in Fig. 12.12, these authors have used direction-finding techniques to locate the source of SKR at nearly the same time as an HST image of a bright UV auroral arc on the dawn side on Jan. 17, 2007. Using a model of the magnetic field, the projection of this source along the magnetic field results in the projection of where the feet of the source field lines enter

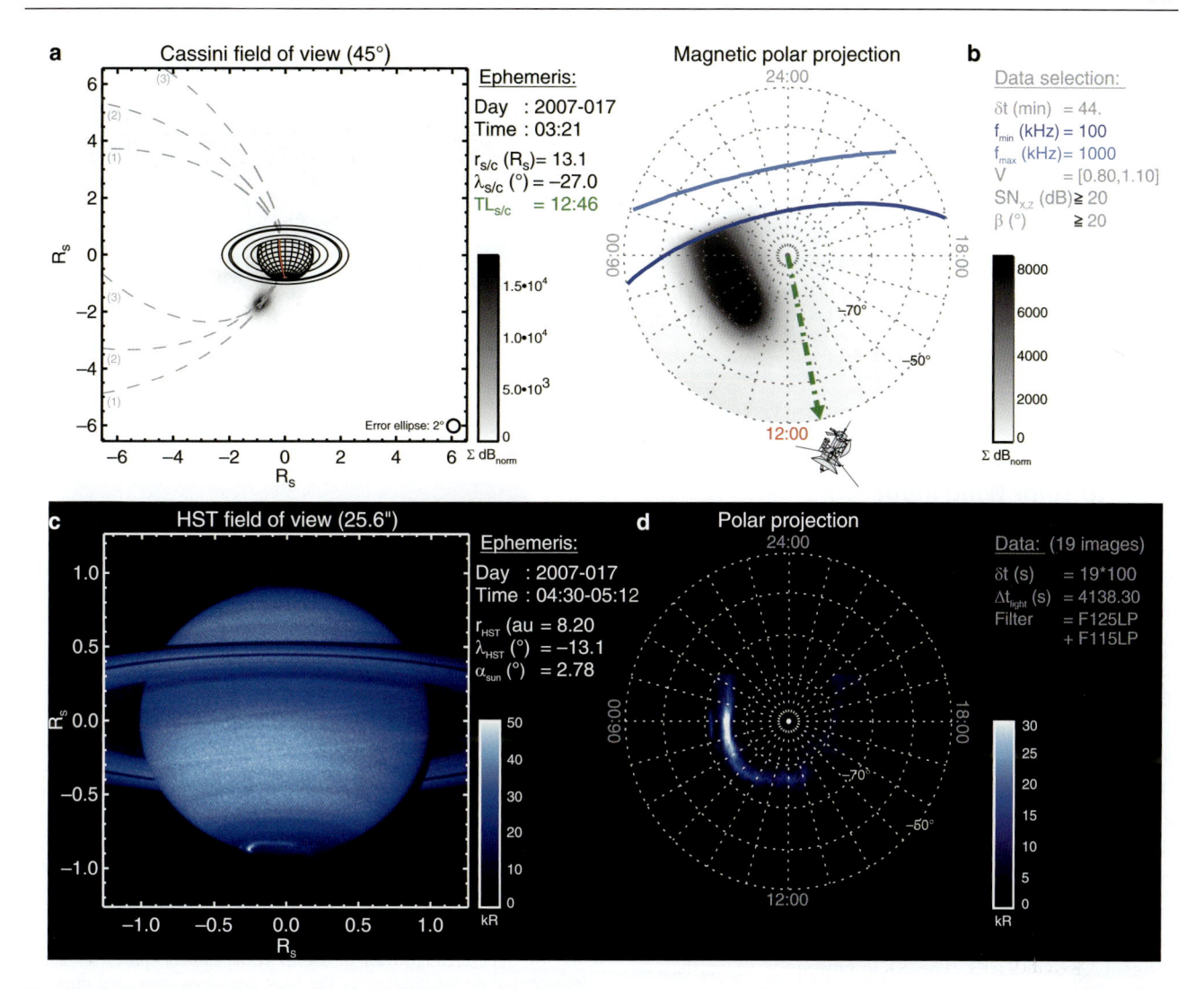

Fig. 12.12 Comparison of the distribution of SKR sources (*top*) with the location of the main UV aurora imaged by HST on the same day (Lamy et al. 2009). The top left panel shows the location of the SKR source as seen from Cassini while the top right panel shows the locations of the feet of the magnetic field lines threading the radio sources. It is clear that the field lines threading the SKR are those associated with the bright dawnside auroral arc

the pole of Saturn. A similar polar projection of the UV emissions lies in the same location. The example in Fig. 12.12 is an individual example of the correspondence between the SKR source and bright UV aurora, but Lamy (2008a) has shown a statistical correspondence, as well. A movie of the SKR radio sources is included on the DVD supplied with this book.

12.3 Magnetospheric Dynamics and the Aurora

Jupiter's and Saturn's magnetospheres are both much larger than the Earth's, and the time scales for disturbances in the solar wind to move from the bow shock past the planet are hours, compared with minutes at the Earth. A timescale of equal importance in this context is the time it takes to fill the magnetotail with open flux (which is roughly equivalent to the lifetime of an open field line) through reconnection at the dayside magnetopause, versus the rotation rate of the planet. At Earth the polar cap is filled with open flux typically in a few hours, in comparison to the 24 h rotation rate. At the outer planets, however, estimates of the dayside reconnection voltage suggest that it takes ~8 days at Saturn (Jackman et al. 2004), and ~15–25 days at Jupiter (Nichols et al. 2006) to replenish the giant magnetotails, versus the faster ~10 h rotation rate in each case. We would therefore expect the response to the solar wind to be quite different from the situation at Earth. Saturn's magnetospheric plasma content (mainly from the rings and icy moons) is much lower than Jupiter's, and the neutral content higher, but the distance to which plasma corotates with the magnetic field fills most of the magnetosphere, like Jupiter and unlike the Earth. On the basis of the mounting body of theoretical and observational evidence to be discussed in Section 12.5, Saturn's main UV auroral emission is probably not produced in a manner similar to Jupiter's, rather it seems to be associated with the flow shear between near-rigidly rotating closed field lines, and sub-corotating closed field lines. Since no emissions have been detected from magnetic footprints of the satellites, there are no direct measurements from UV images of the distance to which the auroral oval maps.

12.3.1 Response of the Aurora to Solar Wind Input

The Cassini – HST campaign in 2004, in addition to others discussed below, has contributed strong evidence of the importance of the solar wind in the intensity and morphology of Saturn's auroral emissions. This was the first chance to see how the auroral displays responded to a large shock-related compression of the dayside magnetosphere. During this campaign, the auroral emissions clearly brightened at the arrival of a large solar wind dynamic pressure increase (Clarke et al. 2005; Crary et al. 2005). For this event, the solar wind pressure, velocity, and interplanetary magnetic field (IMF) were measured by Cassini approaching Saturn. The dawn side auroral emissions brightened the most, filling the dawn side polar cap, and the main oval radius decreased in proportion to the emission brightness as shown in Figs. 12.1 and 12.13. The SKR emissions (described in Section 12.2.4), measured by the Cassini RPWS instrument, also increased in intensity during this event after correcting for the rotational modulation as shown in Fig. 12.14 (Kurth et al. 2005a). In addition to these overall brightenings, other diurnal effects were seen in specific auroral emissions. Isolated emissions appeared to move at 60–70% of the corotation speed, but some slowed to ∼20% after shifting from the morning to the afternoon sector (Grodent et al. 2005). In addition, a "comma" shaped distribution was seen at times when bright emissions from the morning sector shifted to higher latitudes in the afternoon sector. The highly-dynamic nature of Saturn's aurora revealed by the HST observations (shown in Fig. 12.13) appears to relate directly to the concurrent solar wind activity measured by Cassini. Collectively these data provide a unique insight into the solar wind driving of Saturn's magnetosphere and consequent auroral response (see Clarke et al. 2005, 2009; Bunce et al. 2006).

As shown in Fig. 12.15 the January 2004 auroral brightenings correlated well with the dynamic pressure of the solar wind, rather than the direction of the IMF which domi-

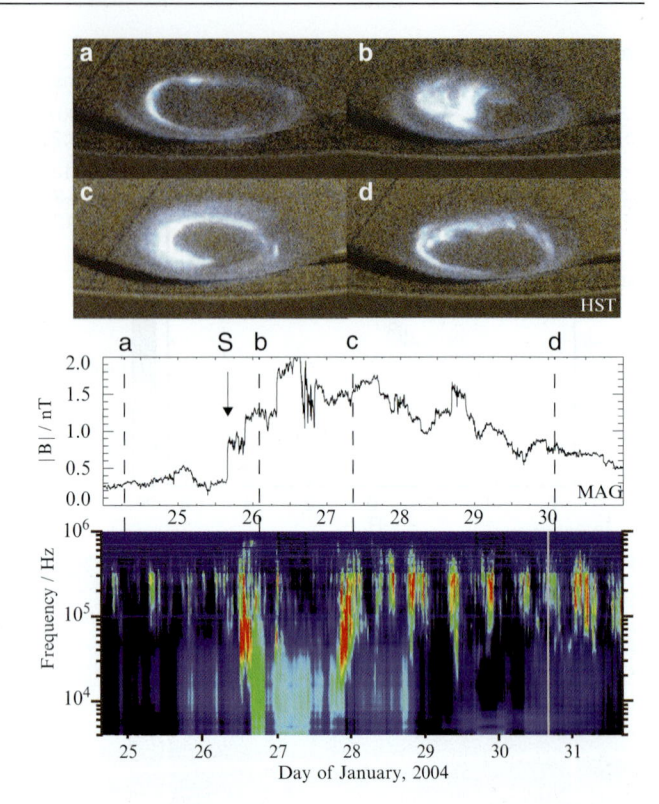

Fig. 12.13 Overview of Saturn's auroral dynamics and their relation to CIR compression regions as observed during the Cassini-HST campaign in January 2004. At the top of the figure we show four frames of Saturn's UV auroras obtained by the HST as follows: (**a**) 23:51 UT on 24 January, (**b**) 19:03 UT on 26 January, (**c**) 01:28 UT on 28 January, and (**d**) 19:02 UT on 30 January. The times given are the approximate mid-times of the imaging intervals involved. The noon-midnight meridian is near the center of each frame with noon towards the top, dawn to the left, and dusk to the right. The panels below show the strength of the IMF measured upstream of Saturn by the Cassini spacecraft, and a frequency time spectrogram of the SKR emissions measured during the interval. The vertical dashed lines in the panels marked (**a**)–(**d**) indicate the approximate corresponding times of the auroral images using a nominal 17 h lag between Cassini and Saturn (for the magnetometer data, but not for the SKR data), though these timings are uncertain to within several hours due to possible non-radial propagation effects in the solar wind and the separation of the spacecraft and the planet in heliographic longitude. The vertical arrow marked 'S' indicates the arrival of the forward shock of a CIR compression region at the spacecraft, at ∼16 UT on 25 January. The HST images are from Clarke et al. (2005), the interplanetary data are from Crary et al. (2005), and the SKR data are from Kurth et al. (2005a) (adapted from Cowley et al. 2005)

nates at the Earth, suggesting a different kind of interaction with the solar wind (Clarke et al. 2005; Crary et al. 2005). It has therefore been proposed that the brightenings are produced by shock-compressions triggering rapid reconnection and closure of open tail flux in the nightside magnetosphere (Cowley et al. 2005). This effect occurs less frequently at Earth, but due to the timescales discussed above could well be the normal mode at Saturn. The local IMF direction, largely azimuthal due to the Parker spiral configuration, was also known from Cassini measurements. As discussed this

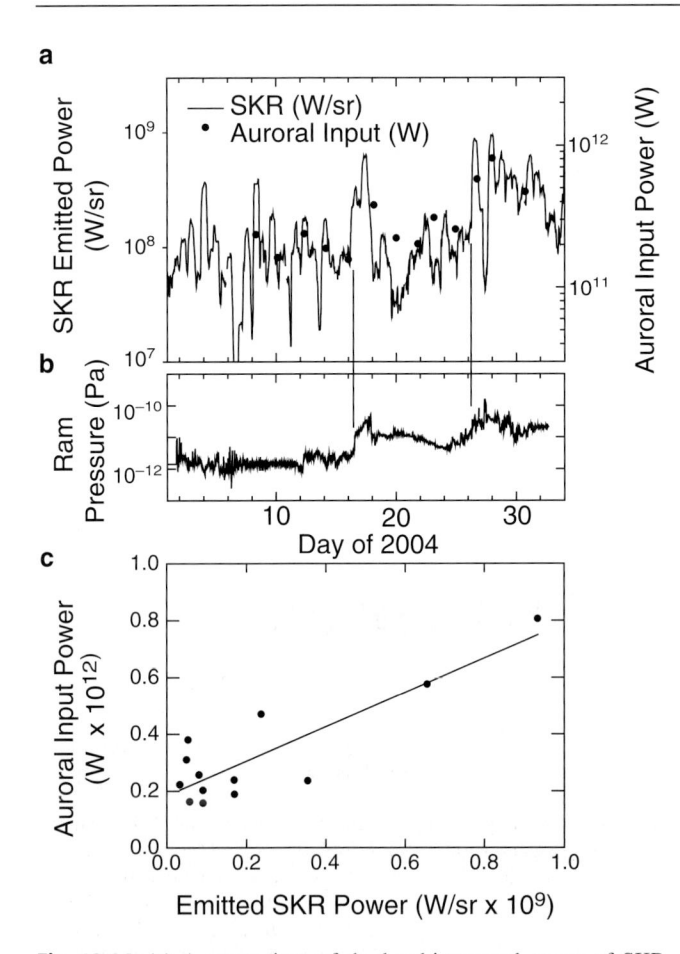

Fig. 12.14 (a) A comparison of the band-integrated power of SKR emissions (*line*) and the power input to the aurora based on the observed UV brightness. (b) The dynamic pressure in the solar wind measured by the Cassini plasma spectrometer. (c) A scatter plot comparing the auroral input power estimated from the UV brightness with the integrated SKR power (from Kurth et al. 2005a)

effect does not dominate due to the short timescales involved, however there is evidence for dayside cusp enhancements in emission at times during favourable IMF conditions (Gérard et al. 2005).

One of the principal goals of the 2007–2008 HST campaigns was to test the response of Saturn's aurora and magnetosphere to changes in the solar wind, given the evidence from the 2004 campaign. Two solar wind forward shocks (increased pressure and velocity) of varying strength arrived at Saturn during each set of observations in 2007 and 2008 (Clarke et al. 2009). Linear correlation coefficients were estimated for the nominal arrival times predicted by the MHD model and for the optimal shifted arrival time for an upper limit to the possible correlation. The correlation coefficients between auroral power and solar wind pressure were 0.22/0.51 for the 2007 data and 0.60/0.85 for the 2008 data, in both cases for the nominal and best-correlated shifted arrival times. The linear correlation coefficients between auroral power and rotation-averaged SKR power were 0.30 in 2007 and 0.02 in 2008, and in these cases there were no shifted values since the times were known. In the case of solar wind shocks, the auroral brightenings appeared to persist longer than the solar wind pressure remained elevated. The SKR emission is known to be beamed and is associated with bright auroral emissions (see Section 12.2.4). The intensity detected by Cassini would then depend on the location of the spacecraft with respect to the emission beams, and more detailed modeling is needed to estimate the observing geometry. In terms of individual events, however, there was a one to one correspondence between the arrival of solar wind forward shocks at Saturn and increases in Saturn's auroral power and SKR emission, and a decrease in the oval radius. At the times of two reverse shocks (increased velocity but decreased pressure) the SKR emission and possibly the UV emission appeared to increase, although the statistics are poor. In addition, no auroral brightenings were observed over several weeks of quiet solar wind conditions. The event data are consistent with a causal relationship between solar wind disturbances and auroral and SKR emission increases.

Both the UV auroral emissions and SKR show brightness patterns that are best organized by local time. Clearly, Saturn's magnetosphere is strongly affected by pressure variations in the solar wind and it clearly demonstrates the effects via changes in the auroral response. A later section on coupling from the solar wind through the magnetosphere to the ionosphere discusses some of the theoretical ideas of how the dynamics in the solar wind might couple into the magnetosphere and result in the patterns of auroral emissions observed in the remote sensing observations summarized here. In this regard, Saturn's magnetosphere resembles Earth's strong response to solar wind input, although there is little evidence that the orientation of the interplanetary magnetic field is as important in this interaction at Saturn as it is at Earth. The reasons for this are twofold. First, the IMF switches back and forth on timescales of tens of minutes to an hour, which for Saturn (unlike the situation at the Earth) is negligible in comparison to the open field convection times discussed above. Second, 2004 was solstice for Saturn, and therefore at this time Saturn presented an apparent lack of dipole tilt in the azimuthal (or B_y) direction. As such, IMF positive B_y is essentially the same as IMF negative B_y. Solar wind effects clearly do play a role in the auroral dynamics at Jupiter, as shown by Gurnett et al. (2002), Pryor et al. (2005), and Nichols et al. (2007), although not to the extent that they do at Saturn from two HST campaigns in 2007. The Cassini – HST campaigns in January 2004, January/February 2007, and February 2008 have contributed strong evidence of the importance of the solar wind in the intensity and morphology of Saturn's auroral emissions (Clarke et al. 2009).

Fig. 12.15 A comparison of the solar wind ram pressure (*red triangles*) and the convection electric field **v** × **B** (blue diamonds) with the UV auroral power (Crary et al. 2005)

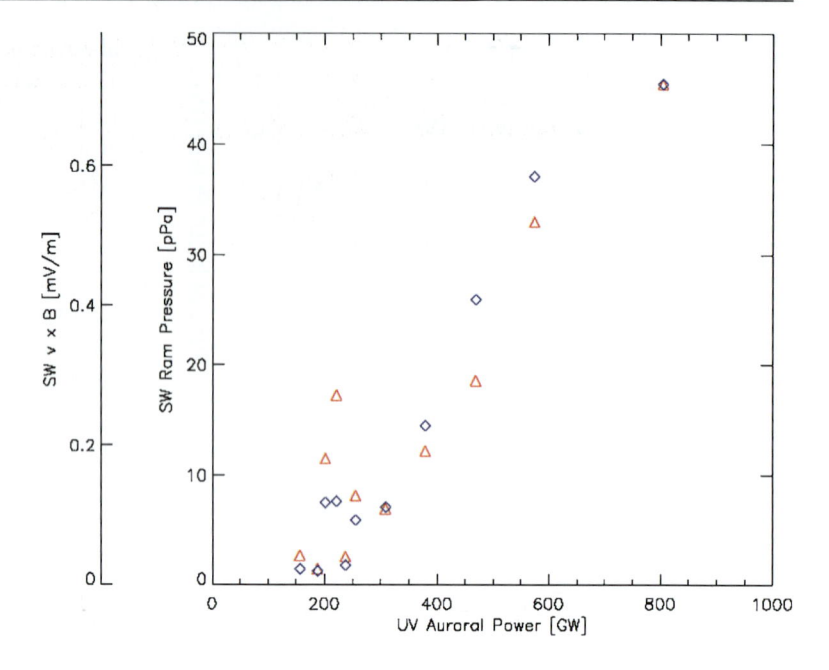

12.3.2 Response of the Aurora to Rotational Dynamics

Although there is a demonstrable correlation between the solar wind and auroral emissions, it is also clear that Saturn's rotation plays a major role in its auroral processes. Despite the importance of local time in organizing the morphology of the auroras at Saturn and the brightest SKR sources, there is clear evidence in the images of motion of the auroral bright spots in the corotational sense, albeit at only fractions of the supposed planetary rotation rate. As detailed in Chapter 10 there are numerous features in Saturn's magnetosphere that rotate at periods statistically identical to or at least similar to the SKR period. While we leave the bulk of this discussion to that chapter, we summarize here work by Mitchell et al. (2009b) which shows a particularly clear linkage between periodic injections and particle acceleration in the midnight to dawn region and SKR and UV auroral brightenings.

Mitchell et al. (2009b), using energetic neutral atom (ENA) images, show that protons and oxygen ions are periodically accelerated once per Saturn magnetosphere rotation, usually in a similar location in the midnight to dawn sector at a radial distance of 15 to 20Rs. Figure 12.16 is a sketch of this process showing the location of the simultaneous acceleration of both protons and heavy ions and the subsequent gradient drift of the particles as a function of their energy. Mitchell et al. suggest that the acceleration is related to reconnection and plasmoid formation in Saturn's magnetotail. The acceleration events, appearing as enhancements in the ENA images, correlate with periodic bursts of

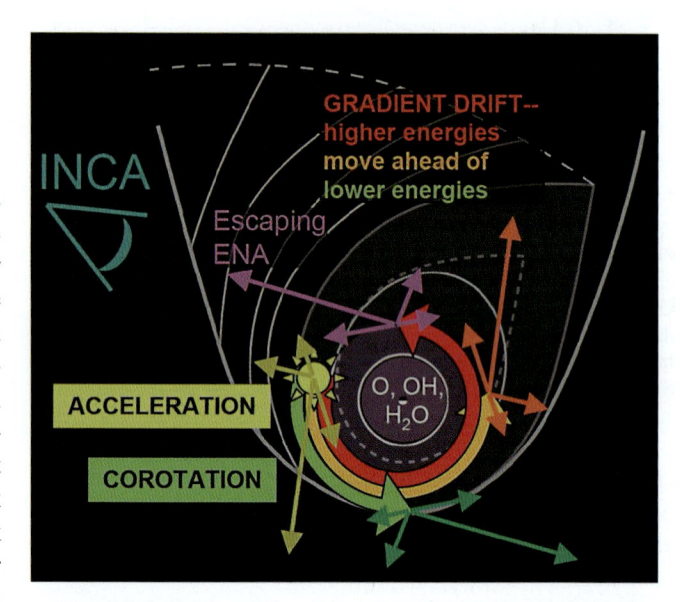

Fig. 12.16 Following an acceleration event (depicted in this schematic based on MIMI observations as the yellow burst near dawn; the sun is down in this figure) heated particles corotate along with the cold plasma. Some time later, low energies have corotated to noon, intermediate energies to dusk, and highest energies to midnight (from Mitchell et al. 2009b)

SKR emissions. A particular example of the intensification of SKR apparently correlated with the ENA enhancement is shown in Fig. 12.17. Further, Mitchell et al. (2009b) show that the enhanced regions of ENA emission rotate in the corotation direction at the same rate as UV brightenings observed by UVIS as shown in Fig. 12.18. A movie included with the book DVD shows the spatial and temporal correlations between the ENA emissions and the UV aurora, as well as the

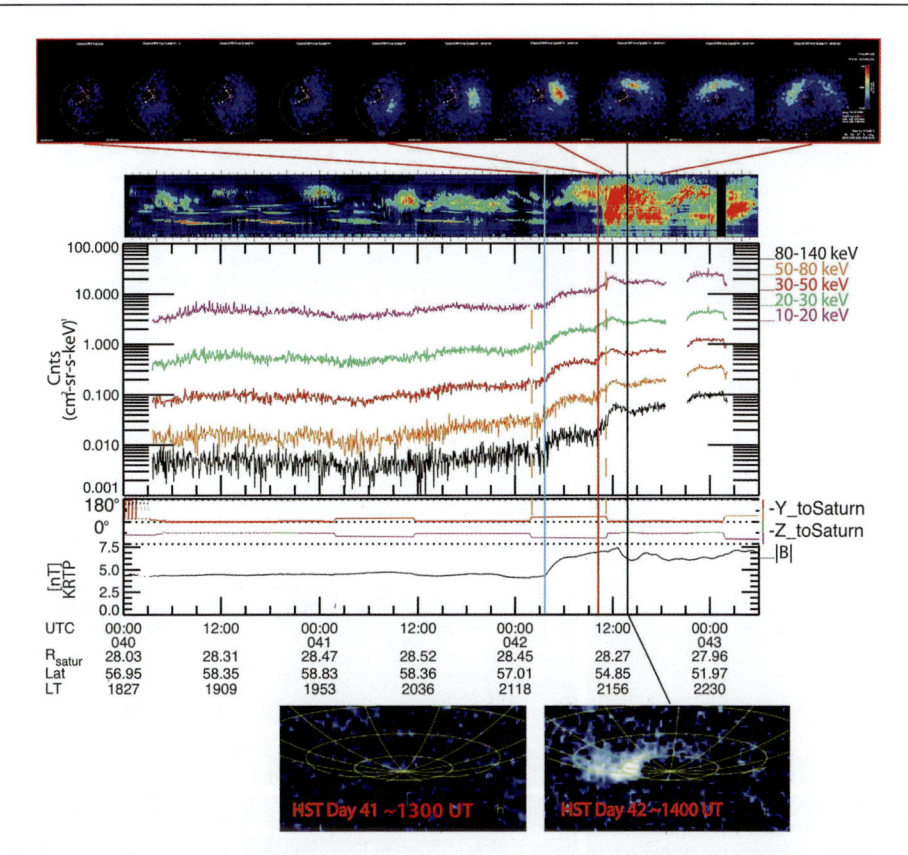

Fig. 12.17 Solar wind compression event stimulating ENA, SKR, and auroral UV emission at Saturn for days 40–43, 2007. Top panel, sequence of 50–80 keV hydrogen ENA images, with the sun to the upper left, midnight local time to the lower right in each image. Red guide lines indicate their position in the timeline organizing the line plots. Second panel is Saturn kilometric radiation between 1 kHz and 1 MHz. Third panel is ENA emission integrated over the INCA field of view for 5 hydrogen energy channels. The data in the interval between the vertical dashed lines has been moved vertically (all be the same displacement) so that they approximately match the intensities just prior and just following those times. This was done to compensate for changes in the spacecraft attitude (fourth and fifth panels). The sixth panel shows the magnitude of the magnetospheric magnetic field. The bottom images are from HST, and show that on day 41 the auroral oval was quiescent, whereas on day 42, about 1400 UT, bright auroral emission fills the dawn auroral zone (from Mitchell et al. 2009b)

temporal correlation of the brightening in these two phenomena and the enhancement in the SKR. These authors suggest that the azimuthally asymmetric ring current pressure creates a rotating field-aligned current system linking to the ionosphere through auroral field lines.

The 2007–2008 HST campaign also showed that the auroral oval was periodically shifting with time. A detailed analysis of the locations of the auroral oval in each observation was carried out to give a best fit to the oval center location. It was found that the oval center position oscillated with a period of 10.75 ± 0.15 h (2007) and 10.79 ± 0.13 h (2008), consistent with the SKR period (Nichols et al. 2008). The oval moves around highly eccentric ellipses oriented toward pre-noon and pre-midnight. Provan et al. (2009) suggest that the oval displacements are related to distortions in the magnetic field associated with rotating magnetic perturbations. This technique may in the future yield a more accurate value for the rotation rate of Saturn's magnetic field and thus its interior.

12.4 In-Situ Measurements

12.4.1 Energetic Particles

Although the Cassini spacecraft has not traversed the auroral zone at altitudes where upward field-aligned current mechanisms produce downward accelerated energetic electrons (and so such electrons may only be inferred from the remotely observed auroral emissions), upward traveling particles energized by auroral processes are frequently observed. For example, Mauk and Saur (2007) recently identified electron beams that fit such a description at Jupiter. Near-equatorial field-aligned electron beams with energies starting from ~20 keV up to several hundreds of keV were observed at Saturn by Saur et al. (2006). These electron beams were shown to be consistent with a low altitude source. Electron beams are present in Saturn's magnetosphere from the magnetopause inward to radial distances

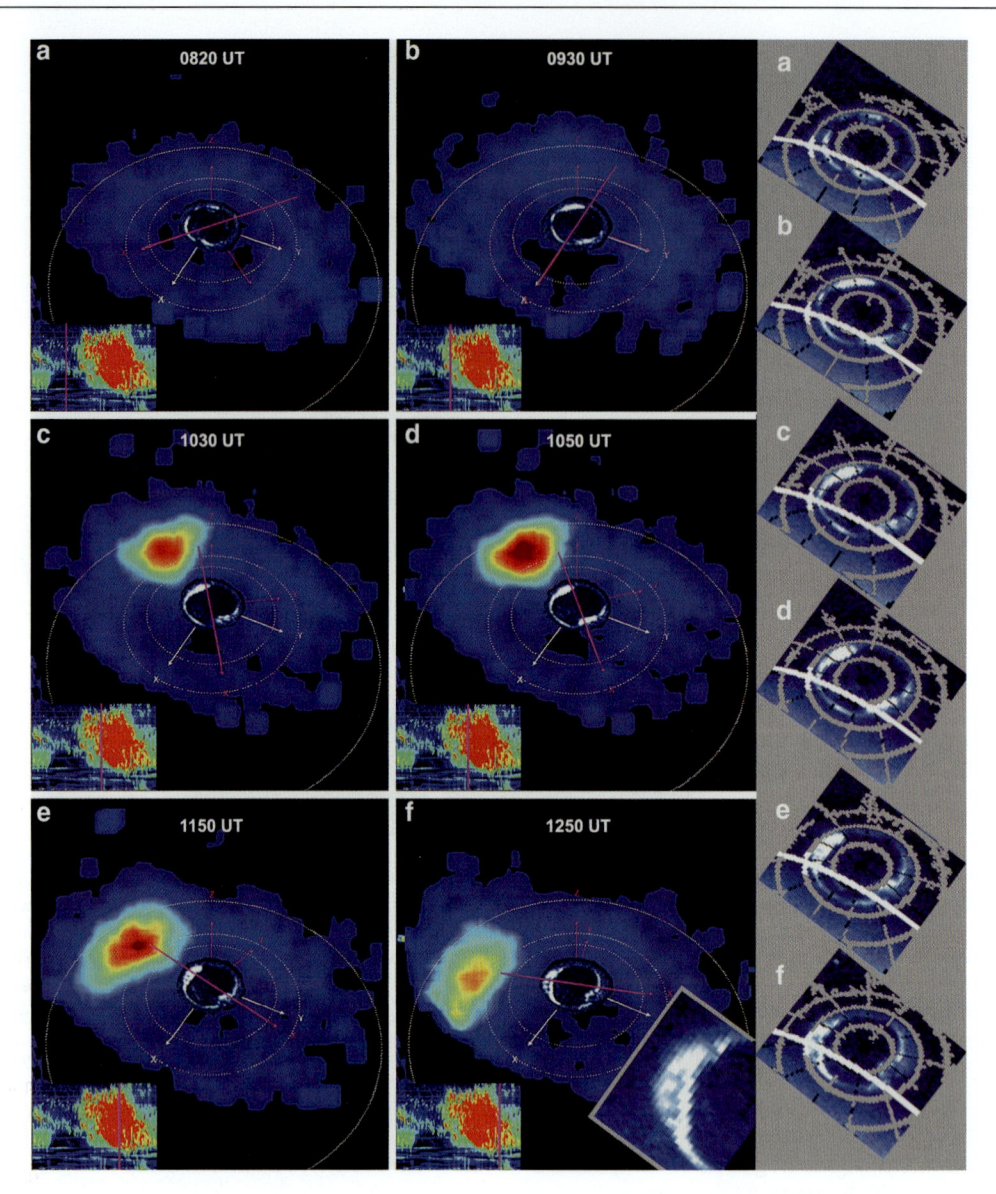

Fig. 12.18 Six images from the day 129, 2008 sequence. UVIS auroral images, each scanned over approximately 15 min, are superposed at a larger scale so they can be more easily seen. A pink line lies on top of the X axis of a coordinate system rotating about Saturn's spin axis, at the SLS3 (extrapolated) period. This is for reference, to follow the motion of the ENA enhancement and the bright auroral bulge over time. In the bottom left corner of each frame, the SKR data is reproduced, with a vertical pink line indicating the center time of that hour-long ENA accumulation (time given at *top* of each frame). To the right, each auroral image is reproduced with a superposed latitude–longitude grid (every 10° of latitude). A white line indicates the terminator, and the 70° latitude line turns to white for a segment where it crosses the terminator. An inset in Panel f shows a blow-up of the equatorward bulge at about that time (from Mitchell et al. 2009b)

as close in as $11 R_S$ (see Fig. 12.19). Even though pitch-angle coverage of the MIMI Low Energy Magnetospheric Measurement System (LEMMS) instrument did not allow Cassini to establish the existence or non-existence of electron beams along the entire first four orbits, Saur et al. (2006) still demonstrate that the electron beams are only present on parts of Cassini's orbits within the magnetosphere. Using either a dipole or a model field (Khurana et al. 2005) based on Voyager and Cassini data, the electron beams are shown to map within the general region of the UV aurora (see Fig. 12.20). Magnetospheric electron beams can thus be considered as tracers of auroral activity, assuming these are indicative of downward going return currents associated with upward auroral currents. An analogous conclusion has been reached for Jupiter (Mauk and Saur 2007).

Recently, Stallard et al. (2008c) identified a secondary auroral oval in infrared-observations. This secondary oval subcorotates, is located equatorwards of the main auroral oval, and has been argued to be a weak counterpart of Jupiter's main auroral oval. Observations of electron beams as deep inside as $11 R_S$ may be consistent with the observations of this secondary oval.

Fig. 12.19 Location of observed electron beams in Saturn's magnetosphere (*in red*) along the first four orbits of the Cassini spacecraft (from Saur et al. 2006). Sections in solid black contain no electron beams, on sections in light black the pitch-angle coverage did not suffice to establish the existence of electron beams, Blue crosses indicate the last magnetospause crossing

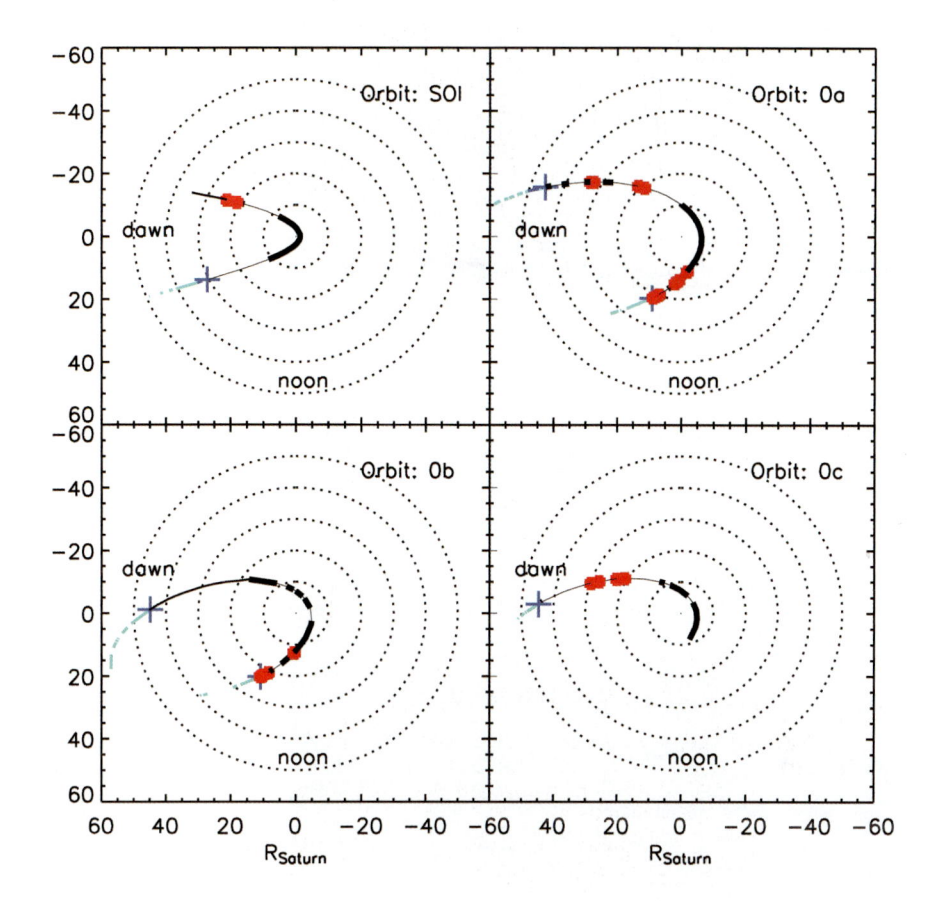

Mitchell et al. (2009a) have updated the auroral energetic particle picture at Saturn, showing that in addition to the strongly field-aligned electron beams discussed in Saur et al. (2006), upward moving field-aligned energetic ion conics and/or beams are also frequently observed, both in association with the electron beams, and at many times when simultaneous determination of the presence of electron beams could not be made.

These upward going electron beams are not peaked at a specific energy the way downward-directed auroral electrons typically are, but rather follow a power law in energy, from the lower threshold of the measurement to hundreds of kiloelectron-volts. The electron angular distributions may be either unidirectional upward, or bidirectional. Also as discussed in Saur et al. (2006), events such as these were first observed at Earth (e.g., Klumpar 1990; Carlson et al. 1998), and mechanisms for their generation were proposed by Carlson et al. (1998), Klumpar (1990), Marklund et al. (2001), and Ergun et al. (1998). The FAST satellite measurements described by Carlson et al. (1998) showed a close correspondence between upward-going energetic electron beams, upward-going ion conics, and enhanced broadband electromagnetic noise. The authors showed that all of these phenomena were located within a region of downward field-aligned current in the auroral zone Birkeland current system. At Jupiter, Mauk and Saur (2007) found that the beams mapped to regions of bright aurora, and so concluded that such bright regions, generally accepted to be associated with upward field-aligned current, must have regions of downward current intermixed.

At Saturn Mitchell et al. (2009a) show that the particle energies are higher than those typical at Earth by a factor of up to 100. They show that the composition of the ion conics indicate a likely ionospheric source for the accelerated ions. They describe two classes of field-aligned events; the first class is different from the observations at Jupiter, based on their steady appearance over hours of observation time, and are consistent with extensive, contiguous low altitude acceleration, presumably in regions of downward field-aligned current. A second class of events appears temporally pulsed, with a repetition rate on the order of 1 h.

Figure 12.21 is from an event on day 288 of 2006. Cassini was at $28.3 R_S$, 42° north latitude, dipole L ~51, and 2220 h local time. During the hours 1900 to 2400 UT the spacecraft was rolling about an axis chosen for fields and particles measurements, which for the measured field orientation allowed both the MIMI LEMMS and INCA sensors to sample nearly complete pitch angle distributions of energetic ions and electrons. In the top panel of Fig. 12.21 we display 100 keV electron counting rates from each end of the double-ended LEMMS telescope. Although this roll permits LEMMS to measure electrons over all pitch angles except those between

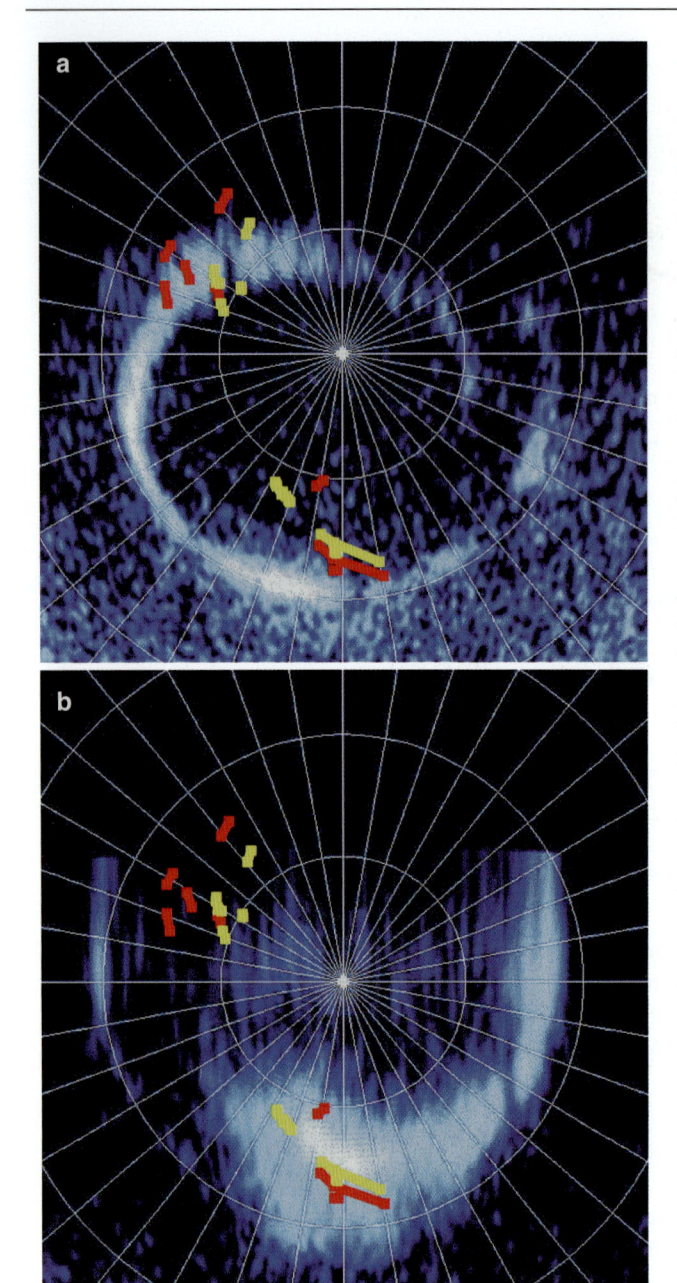

Fig. 12.20 Location of electron beams mapped into Saturn's ionosphere. Auroral images are from Gérard et al. (2004). Yellow data points are calculated with a dipole magnetic field model and red data points are calculated with a magnetic field model from Khurana et al. (2005). Note, for the time intervals when the electron beam measurements were made no simultaneous auroral images are available (from Saur et al. 2006)

$\sim$75° and 105°, the only elevated fluxes appear near 0° pitch angle. An electron spectrogram (second panel) shows that the energy of these field-aligned electrons routinely extends to >100 keV, sometimes reaching nearly 1 MeV. Note also that except when the sensor can view near 0° pitch angle, the intensities are at detector background. Mitchell et al. (2009a) show that the electron angular distribution is consistent with an acceleration altitude under $3R_S$, and likely much closer to $1R_S$. The electron energy flux in the beam is nearly flat out to $\sim$600 keV, indicating a process in the source region capable of accelerating electrons to that energy. Energetic ion data from the INCA sensor for the same event showed an upward going, field-aligned ion beam with composition consistent with an ionospheric source.

From hour 0700 to 1100 on day 269, 2006 Cassini moved from 8 to $10R_S$, and from $11 < L < 15$. In Fig. 12.22 the top panel shows electrons with energies between 200 keV and 1 MeV. LEMMS is oriented perpendicular to the magnetic field, so the pulsed enhancements are all at 90° pitch angle (not beams). Below that hydrogen intensities measured by INCA are shown. The intensities rise at 0730 as a spacecraft maneuver moves the sun out of the INCA field of view, and the INCA voltages are turned up. Four times over the next 3 h the INCA high energy proton intensities rise (lower energy protons are electrostatically excluded by the ion rejection plates). Below each proton intensification, angle–angle plots of the highest energy channel are displayed. In each case, a distinctly (upward) field-aligned intensification is evident. Below this plot, the magnetic field angles and magnitude appear. The field angles exhibit abrupt changes by $\sim$2° to 3°, then return to their pre-pulse values. This behavior is consistent with the repeated passage of field-aligned current structures. The upward direction of the ions is consistent with upward directed currents and consistent with regions where electrons could be accelerated (at lower altitudes) into the atmosphere exciting the auroral emissions even though the spacecraft is too high to observe the accelerated electrons. The bottom panel shows broad-band electromagnetic wave enhancements, again well correlated with the field-aligned currents, upward ion beams, and electron enhancements. Most of the wave energy in these broadband bursts lies below the electron cyclotron frequency f_{ce} signified by the white trace and are electromagnetic, propagating in the whistler mode. The whistler mode cannot propagate above f_{ce}, however, so the emissions above f_{ce} in this example cannot be whistler modes. In fact, no magnetic component is measured above the cyclotron frequency, suggesting an electrostatic mode.

A survey of all of the mission data to date reveals many instances where there is correspondence between electron and ion beams, in similar regions of space (high latitude, relatively low altitude, $L > 10$). In this case, we suggest that the (very weak) electron enhancements, whose timing is slightly later than the onset of the ion beams, are back-scattered from field irregularities much farther out the flux tube, from electron beams similar to those described in Fig. 12.21.

An event on day 284, 2006, unlike the one on day 269, remains steady over periods of an hour or more, and continues to reappear for over 7 h. Again, as in the earlier events, there is no evidence for heavier ions (e.g., oxygen or water

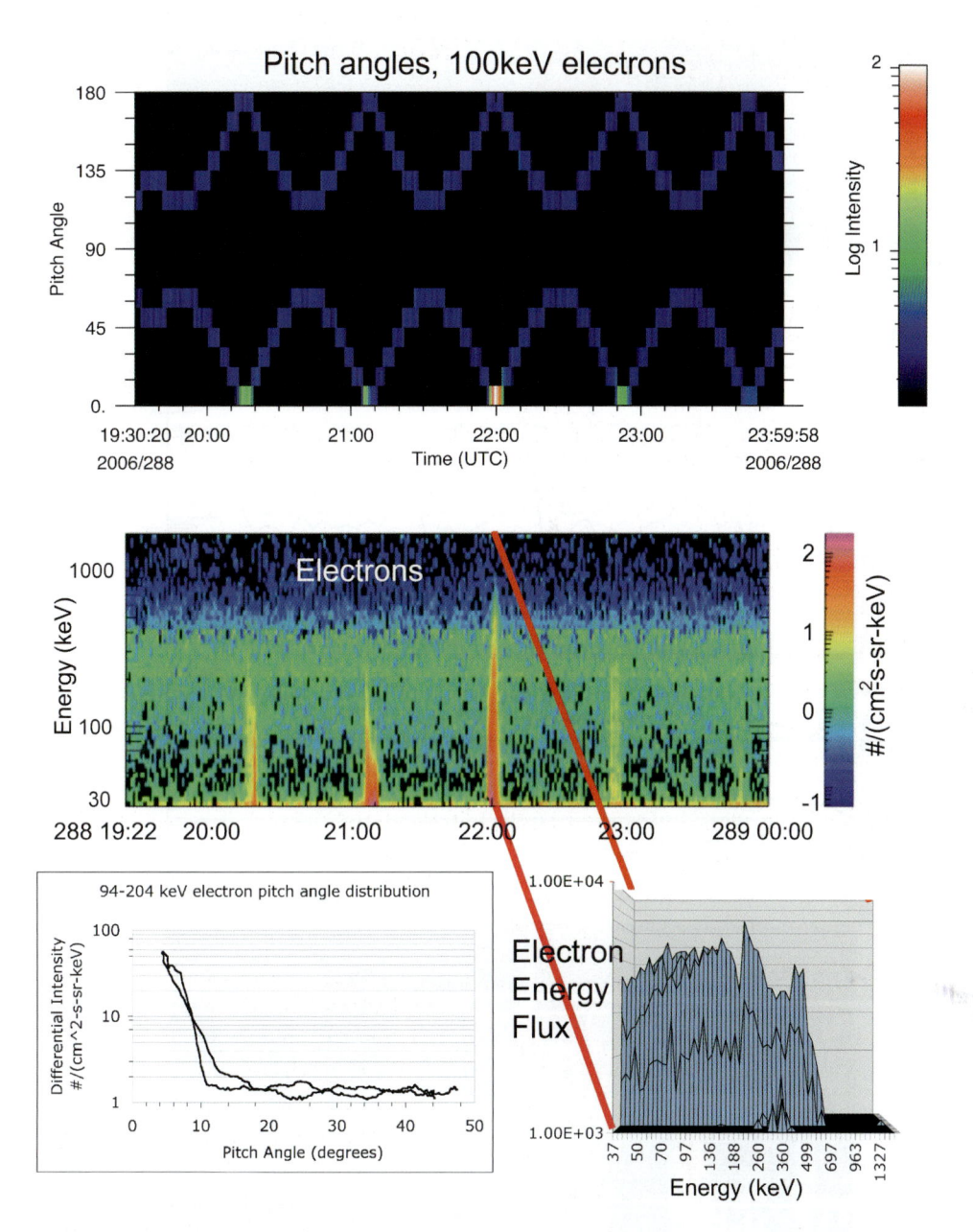

Fig. 12.21 Energetic electron beam characteristics. Top panel shows the time history of 100 keV electrons measured in opposite facing telescopes on the LEMMS sensor as the spacecraft spin sweeps the sensor through various angles with respect to the magnetic field. The sensors are counting near background except when they sample near 0° pitch angle. Second panel shows electron intensity as a function of time and energy. Spikes appear when the sensor samples 0° pitch angle. Lower left provides a quantitative plot of the electron pitch angles. Since the detector cone has a 7.5° half angle, the data are consistent with electrons confined to <3° from the magnetic field. On the lower right, the electron energy flux in the highest spike shows significant energy up to 700 keV (from Mitchell et al. 2009a)

products) in these beams. They are consistent with hydrogen, or possibly H_2 or H_3. As before, they appear as field-aligned, upward moving beams.

ENA imaging just prior to and just following the event on day 284 reveals a locus of bright hydrogen emission from the south polar region, absent in oxygen. In Fig. 12.23a and b, this emission is clear, along with the usual emission from the ring current beyond the outer edge of the E-ring (cut off by the edges of the INCA field of view). Mitchell et al. (2009a) suggest that the hydrogen emission is generated as protons that are accelerated at low altitude above the auroral zone by wave particle interactions, generating ion conics (the same mechanism invoked in Carlson et al. (1998)). Other instances of this emission have been observed, but only when Cassini is located at latitudes consistent with the pitch angles of upward going proton conics as they charge exchange in

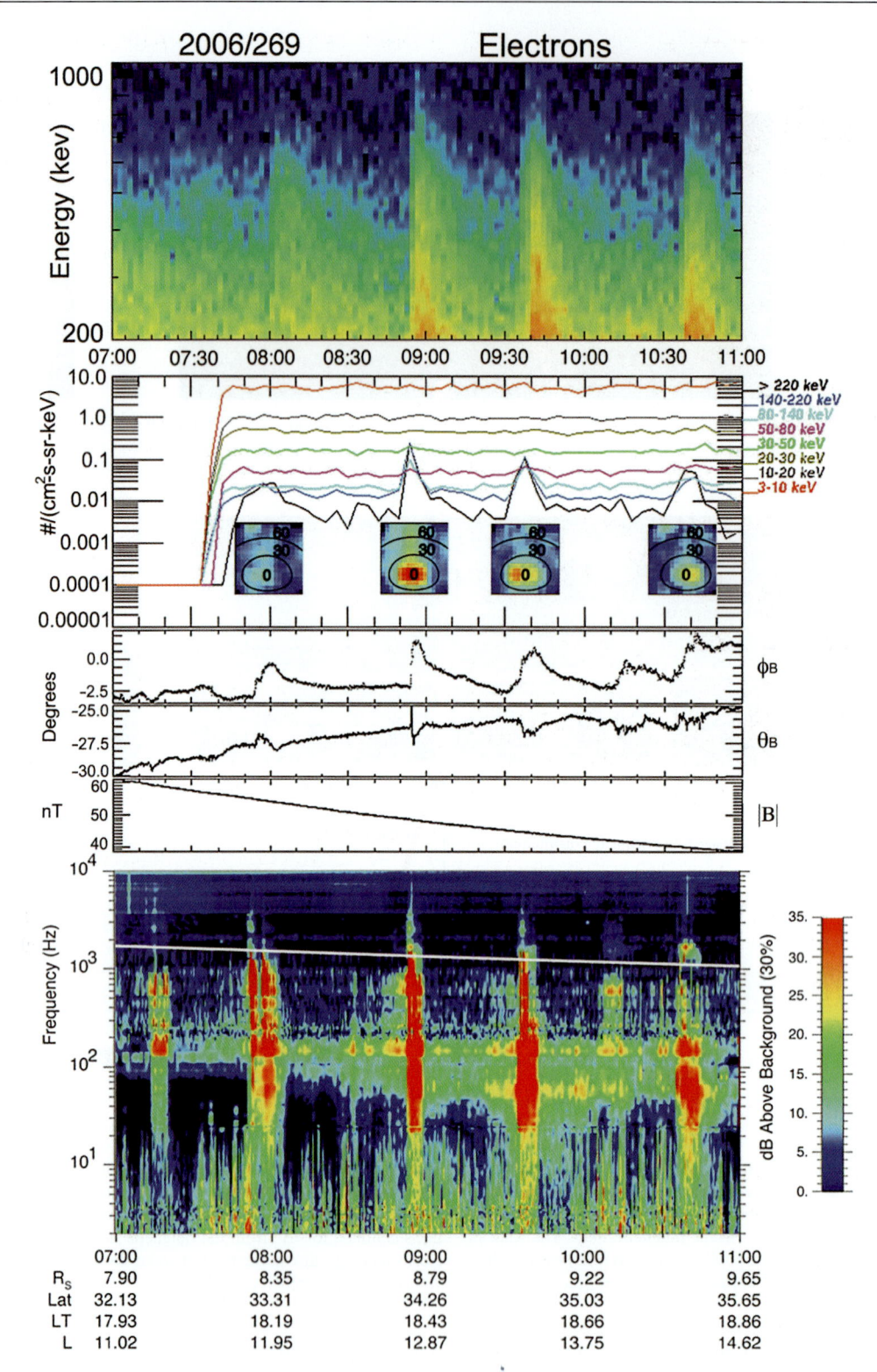

Fig. 12.22 Pulsed particle acceleration events, day 269, 2006. Top panel shows energetic electron enhancements, 200 keV to 1 MeV from MIMI LEMMS. Second panel shows time vs average intensity over the INCA FOV for eight hydrogen energy channels. Lower energies are neutrals, but sufficiently energetic ions enter through the INCA ion rejection plates. Angular plots of ion intensity with pitch angle contours appear beneath each peak in intensity. The magnetic field (*third panel*) shows sharp angular variations associated with each electron and ion enhancement. Broad band electromagnetic (below *white line*, f_{ce}) and electrostatic noise (above f_{ce}) (from RPWS, *bottom panel*) also aligns with these events (from Mitchell et al. 2009a)

12 Auroral Processes

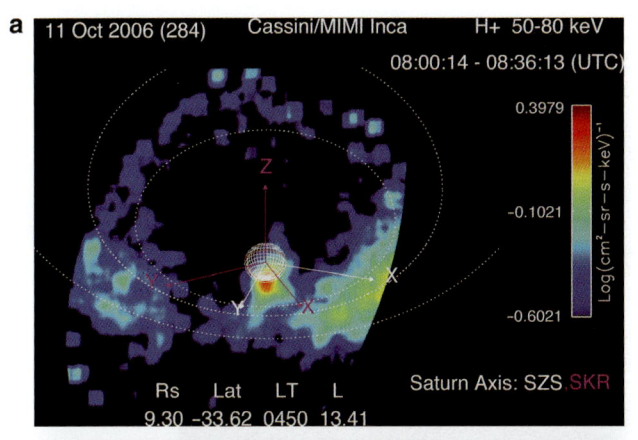

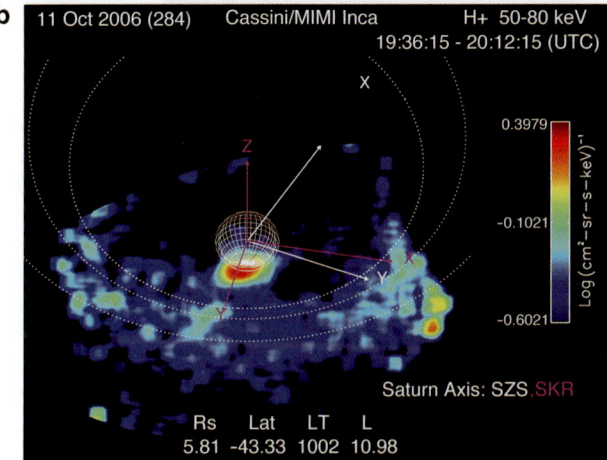

Fig. 12.23 (a) Energetic neutral atom image of intense hydrogen emission from Saturn's south polar region, just prior to an interval of a long-duration field-aligned ion beam event on day 284 of 2006. (b) ENA image of south polar hydrogen emission immediately following the end of the ion beam event. The orbits of Dione, Rhea, and Titan are included as a reference. Axes corresponding to a fixed solar frame are shown in white (Z parallel to Saturn's spin axis, X in the spin axis/sun plane, pointed generally sunward, and Y completing the system). The violet axes also have Z aligned with the spin axis, but they rotate with the SLS3 coordinate system (Kurth et al. 2008) (from Mitchell et al. 2009a)

Saturn's exospheric hydrogen and convert to neutral atoms. The emission is sufficiently weak and localized that it is only seen when Cassini is at relatively low altitudes, as in these cases.

The events presented above are all similar in most respects to the events analyzed by Carlson et al. (1998), which they attributed to perpendicular wave-particle acceleration and electrostatic confinement of ions and upward field-aligned acceleration of electrons in downward field-aligned auroral currents. The whistler mode emissions shown are consistent with VLF saucers found in the FAST observations in downward field-aligned current regimes even though the Cassini observations to date do not have sufficient resolution to show the characteristic saucer-shaped dynamic spectrum. However, Poynting flux measurements show that the waves propagate away from Saturn, consistent with the direction of the electron beams. Although the magnetic field strength in the auroral zone is similar for Earth and Saturn, the distance scale as measured by the planetary radius at Saturn is roughly 10 times that of Earth. Mitchell et al. (2009a) suggest that the magnitude of the total potential drop along an auroral field line is approximately proportional to the square of the distance scale, as the observed peak energies of the accelerated ions and electrons produced by this mechanism are roughly 100 times higher at Saturn than at Earth. At Earth, the upgoing electron beams have peak energies in the vicinity of 5 keV, at Saturn up to or above 500 keV. At Earth, the ion conics have peak energies around 100 eV to 2 keV; at Saturn the ion conics have peak energies from 30 to above 200 keV. While the differential energy flux in the events studied by Carlson et al. (1998) is an order of magnitude higher than that in the events at Saturn (consistent with higher electron densities in the events at Earth), the total energy flux integrated over energy for the electron beams at Saturn exceed those at Earth by about one order of magnitude.

Neither the ion nor the electron beams have been observed for dipole L < 9. The electron beams may be either unidirectional upward, or bidirectional (as in Saur et al. 2006), even for large L. Unidirectional downward electron beams have not been observed. The ion beams and conics are exclusively upward.

These observations have been at altitudes of ∼6R_S and higher. The altitude at which the particles are accelerated is likely to be much closer to the ionosphere, as the reason for the development of the field-aligned potentials responsible for the acceleration of the electron beams is that the magnetospheric circuit is demanding current through what would otherwise be a near-vacuum region. The ionospheric electrons are the only large reservoir of charge carriers available, but they cannot supply the current without an accompanying ion population that can maintain charge quasi-neutrality, and the ionospheric ions are cold and gravitationally bound. A few ions at sufficiently high altitude are accelerated perpendicular to the field by perpendicular stochastic electric field fluctuations and move upward along the field aided by the mirror force. A similar density of electrons can accompany those ions, and the electrostatic shock structure responsible for the field-aligned potential energizes those electrons, until their motion can supply the current required by the system. Of course, that same structure confines the ion conics, so that the ions cannot escape their acceleration region until their parallel energy exceeds the parallel potential that confines them. As discussed in Carlson et al. (1998), this results in a characteristic lower limit to the ion conic energy, and should also result in a larger cone opening angle with increasing energy. In Saturn's case, the low atmospheric and ionospheric scale heights mean that the ion acceleration process takes place in regions of lower ion density than the equivalent regions at Earth, so higher potentials are required to provide enough current from the consequently lower density of electrons permitted by quasi-neutrality, hence the much higher energies developed in these populations for downward current regions at Saturn. At the altitudes where most of the ion events have been measured, the conic opening angle is usually too narrow to resolve, and so the conic angle energy dependence has not been well identified in these events.

The ENA images of the ion conic generation region further confirm the process taking place relatively close to the auroral zone. The characteristic ENA emission from this region (Fig. 12.23a and b) cannot constrain the location precisely, because the instrument angular resolution is insufficient at the distances obtained to date. Because the neutral emission decreases with altitude due to decreasing Saturn exospheric gas density that serves as a charge exchange medium, and because as the conic angle collapses about the field with increasing altitude, the trajectories of any ENA produced would no longer intersect the spacecraft. However, it is clear from these images that the light ions reach energies of at least 100 keV in a region not more than 1R$_S$ above the surface (there is no indication of a heavy ion component, such as water products, methane or nitrogen in either the ion conics or the ENA emission).

12.4.2 Auroral Currents

A subset of the January 2007 HST campaign data discussed in Section 12.2.2.1 has been studied by Bunce et al. (2008a), and formed the basis of the subsequent modeling comparison recently discussed by Cowley et al. (2008) and Clarke et al. (2009). These observations are summarized in Fig. 12.24.

Figure 12.24 shows the HST data from two consecutive observations by the HST (Observation A at ∼05:36 UT 'Saturn time' on 16 January, and Observation B at ∼03:26 UT on 17 January). During the interval of interest, the magnetically mapped footprint of Cassini traversed magnetic field lines mapping to Saturn's southern auroral oval, from poleward to equatorward of the auroral oval in the noon sector between the two consecutive observations. Figure 12.25 shows the magnetic field and plasma electron data from Cassini during Revolution 37, for a 48 h interval from 18 UT on 15 January to 12 UT on the 17 January encompassing the HST interval above, as indicated by the vertical dashed lines in the figure. The top panel of Fig. 12.25 shows a thermal electron spectrogram from ELS in the energy range ∼0.5 eV to 26 keV color coded according to the scale on the right. The four panels beneath this then show the electron bulk parameters, the electron density N_e, the thermal energy W_{the},

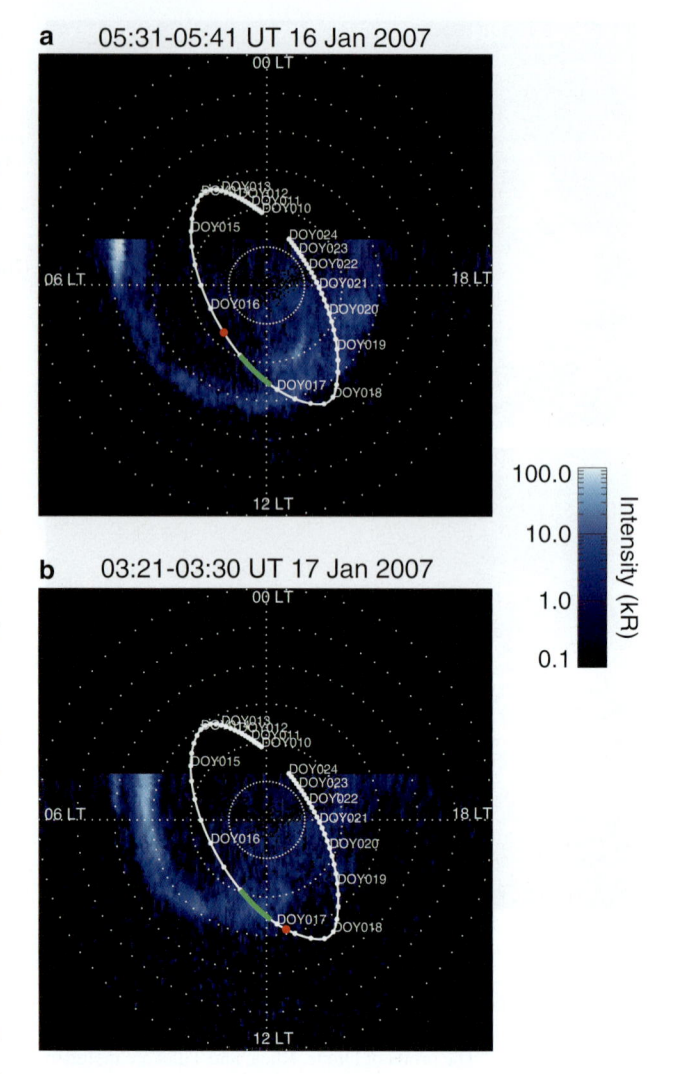

Fig. 12.24 Shows the HST data from two consecutive observations (Observation A at ∼05:36 UT 'Saturn time' on 16 January, and Observation B at ∼03:26 UT on 17 January). The white lines indicate the mapped spacecraft position, using the field model of Bunce et al. (2008b). The red dot indicates the location of the spacecraft at the time of the observation (from Cowley et al. 2008)

the field-aligned current density associated with the motion of electrons in one direction along the magnetic field lines $j_{\parallel e}$, and the corresponding field-aligned electron energy flux E_{fe}. These latter two quantities correspond to the current and energy flux delivered by the precipitating electrons to the ionosphere. The scale on the right hand side of the energy flux panel shows the estimated resulting UV emission, on the assumption that $1\,\mathrm{m\,W\,m^{-2}}$ produces ∼10 kR of UV aurora. The Cassini magnetic field data are shown beneath the CAPS-ELS data, and for comparison the red line shows the 'Cassini' planetary field model (Dougherty et al. 2005). The UV intensity at the spacecraft footprint in the southern hemisphere obtained at the time of the two observations A and B is shown at the bottom of Fig. 12.25, where the red line

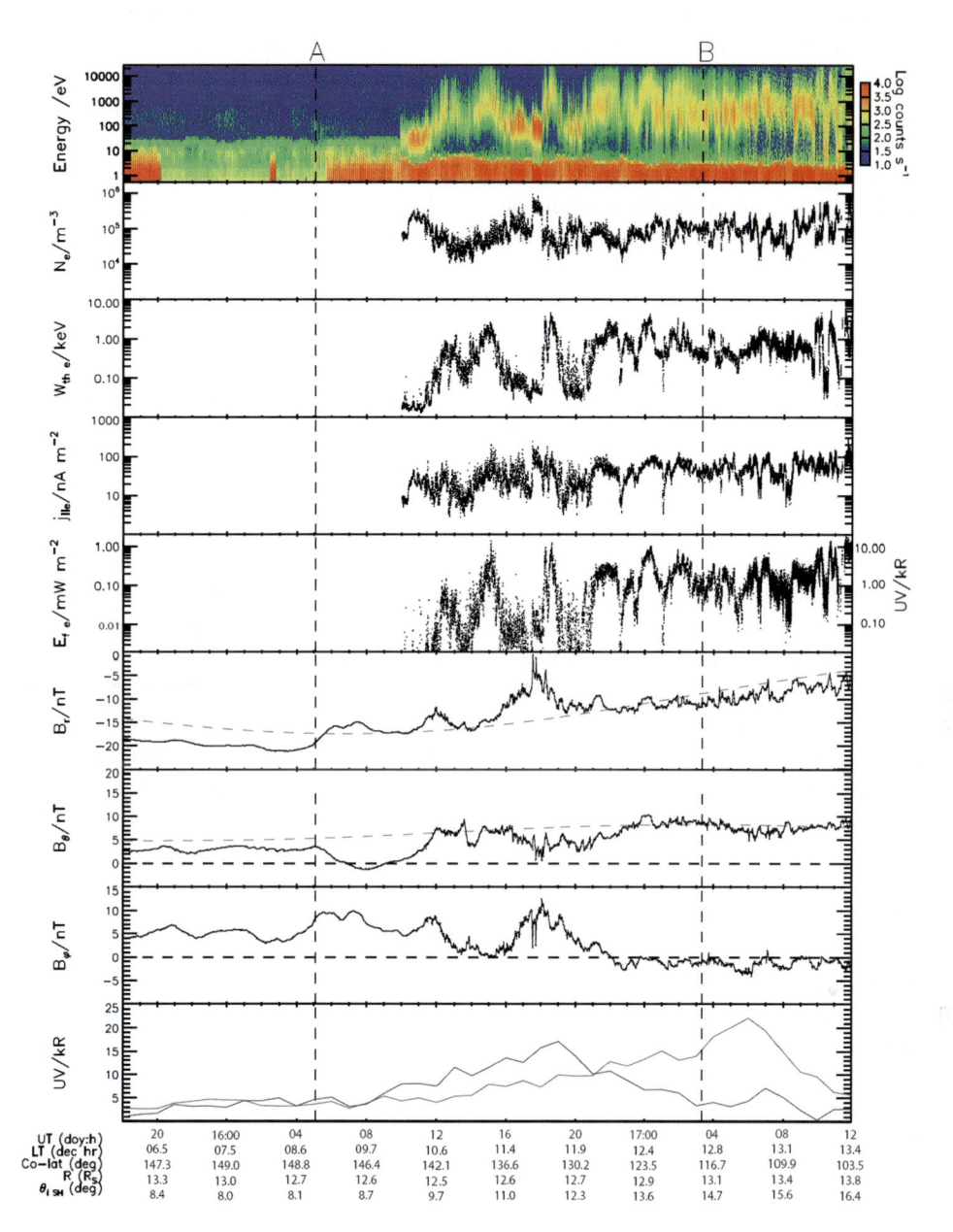

Fig. 12.25 Overview of Cassini plasma electron and magnetic field observations obtained during a 48 h interval on Rev 37 (January 2007), spanning two consecutive HST observations discussed by Bunce et al. (2008a). The top five panels show electron data obtained by the Cassini CAPS-ELS instrument, specifically an electron spectrogram from ~0.6 eV to ~26 keV color-coded according to the scale on the right (the counts at low energies are mainly spacecraft photoelectrons), followed by plots of bulk parameters obtained by numerical integration over the electron distribution assuming the distribution is isotropic. Bulk parameters values are not shown before ~10 UT on 16 January due to low electron fluxes resulting in low measurement signal-to-noise. The bulk parameters shown are the electron density, the thermal energy, the current density of electrons moving in one direction along the field lines, and the corresponding field-aligned energy flux of these electrons. The right-hand scale on the energy flux panel show the corresponding UV auroral emission expected of these electrons precipitate into the atmosphere unmodified by the field-aligned acceleration. The sixth to eighth panels show the three components of the magnetic field in spherical polar coordinates referenced to the planet's spin and magnetic axis. The red dashed lines in the B_r and B_θ panels show the internal field model of Dougherty et al. (2005). The bottom panel shows the UV auroral intensity at the ionospheric footprint of the spacecraft in the southern hemisphere obtained from the two HST image times A (*red*) and B (*blue*). Spacecraft position data are given at the foot of the plot. Taken from Cowley et al. (2008)

shows the UV intensity from observation A, and the blue line shows observation B. The peaks in the emission lines indicate where the spacecraft footprint crosses the auroral oval in each observation.

The time between the two observations (indicated by the dashed lines) labeled A and B at the top of Fig. 12.25 thus spans the crossing of the auroral oval. As discussed above, at the time of observation A the spacecraft footprint was well inside the auroral oval on the dawnside, evidenced by the low intensities in the red trace. At this time one sees a lack of measurable electron fluxes, and quiet magnetic field components, as had been the case for some $\sim$40 h previously. These conditions imply the spacecraft was on open field lines in the regions poleward of the main auroral oval, mapping to Saturn's southern tail lobe. At $\sim$10 UT on 16 January, prior to observation B, the spacecraft detected intense fluxes of warm electrons and the field components became disturbed. First a population of cool magnetosheath-like electrons is observed interspersed with hot keV electrons, followed after $\sim$21 UT by a more continuous population of hot electrons, typical of the dayside outer magnetosphere. These field and plasma data indicate that the spacecraft crossed the open closed field line boundary between the tail lobe and the outer dayside magnetosphere, in the region of the dayside cusp near noon. In addition, throughout the time on open field lines poleward of the cusp a strong positive B_φ component of the field is seen, despite the rotational symmetry of the internal planetary field. This positive B_φ component may be interpreted as sub-corotation of the open field line region as suggested by Cowley et al. (2004a), and as observed in IR data by Stallard et al. (2004). As the spacecraft moves through the cusp region, the positive azimuthal field component, with significant spatial or temporal structure, drops to near-zero values on the closed side of the boundary. The near-zero values of azimuthal field in the southern hemisphere may then be indicative of near-rigid corotation of the outer magnetosphere closed field region. This is the magnetic signature of a major layer of upward-directed field-aligned current. They show the total current flowing in the layer is $\sim$4–5 MA per radian of azimuth, over an estimated width in the ionosphere of 1.5–2° in the ionosphere, the field-aligned current density just above the ionosphere is estimated to be $\sim$275 nA m^{-2}. As can be seen in the fourth panel, this current density value exceeds that which can be provided by the flux of magnetospheric electrons alone, such that an acceleration mechanism is required.

Using Knight's (1973) kinetic theory Bunce et al. (2008a) show that the minimum accelerating voltages required to produce 275 nA m^{-2} is typically <1 kV for the cool dense sheath plasma, rising to 10 kV for the warm tenuous plasma populations. These voltages are sufficient to amplify the precipitating electron energy flux to values capable of producing emissions of 1–5 kR for cool dense populations, and 10–50 kR for the warm tenuous outer magnetosphere populations. This is excellent agreement with the peak intensities which are observed of $\sim$15–20 kR shown in the bottom panel of Fig. 12.25.

Talboys et al. (2009) have identified seven periapsis passes from the high-latitude phase of the Cassini mission between mid-2006 and mid-2007 during which the azimuthal magnetic field component exhibits similar signatures of field-aligned currents (FACs) flowing between the magnetosphere and ionosphere. In general terms, the southern hemisphere exhibits an intense layer of upward-directed FAC which occurs on closed field lines in the dawn and pre-noon sector immediately adjacent to the open-closed field line boundary as the strongly 'lagging' field consistently observed on southern open field lines first declines, and then (usually) reverses to a 'leading' configuration. 'Lagging' and 'leading' fields are generally indicative of plasma sub- and super-corotation, respectively. These 'leading' fields then decline sharply to smaller values further inside the boundary, indicative of intense downward FACs as the plasma reverts to near-corotational flow, the magnitude of the field change being dependent on the phase of the planetary-period oscillation in the interior region. Talboys et al. (2009) show that the region of upward current is co-located with the statistical UV auroral oval of Badman et al. (2006), while the downward current immediately equatorward maps to the outer ring current in the equatorial magnetosphere. In the dusk and pre-midnight northern hemisphere, however, only weak azimuthal fields are observed on open field lines, while stronger 'lagging' fields are observed immediately equatorward in the closed field region, indicative of downward current just inside the open-closed field line boundary, and upward current in the interior region where this layer interfaces with the region containing the planetary-period oscillations. A similar study of the more recent high-latitude passes is ongoing.

Recent work on Jupiter's aurora (Ray et al. 2009) uses a Vlasov code to examine the current–voltage relationship in the aurora of Jupiter, which like Saturn, is a rapidly rotating gas giant with centrifugally-confined plasma. The confinement of heavy ions to low latitudes restricts the motion of electrons due to the ambipolar electric field. For the situation at Jupiter, these authors suggest that the Knight relation does not apply and they conclude that the electron density is not a monotone function along the field, leading to a non-linear current–voltage relationship that depends on the high latitude electron density and temperature. The extension of this work to Saturn should add to the discussion of auroral currents and momentum transfer from the ionosphere to the magnetosphere.

12.5 Solar Wind-Magnetosphere–Ionosphere Coupling Currents and Their Relation to Saturn's Aurora

In this section a theoretical framework into which Saturn's auroral observations may be placed is discussed, relating directly to the global plasma flows and current systems which are expected to be present in Saturn's magnetosphere. First, the basic steady-state picture suggested by Cowley et al. (2004a, b) is introduced. Then, subsequent alterations to the model parameters are discussed based on the previous discussion of the high-latitude auroral field-aligned currents during Orbit 37. Finally, the observed highly dynamic nature of Saturn's aurora and related SKR emissions are compared to particular aspects of the solar wind-magnetosphere interaction at Saturn, and the variations of the model as a consequence of the changing upstream solar wind conditions are discussed.

12.5.1 Proposed Steady-State Theoretical Framework

A global view of a planet's auroral region provides an instantaneous picture of the state of the magnetosphere, mapped along magnetic field lines into the upper atmosphere. Typically, planetary auroras consist of two components: broad regions (in latitude) of lower intensity "diffuse" emissions produced through unmodified precipitation of magnetospheric plasma, and localized regions of more intense "discrete" auroral arcs associated with field-aligned currents and field-aligned acceleration of the current carrying particles. As such, to build a theoretical framework in which to consider the origin of the overall auroral emission region, one must incorporate a general view of the basic plasma flows and resulting field-aligned current systems that are expected to be present within the magnetosphere, and consider the requirements for field-aligned acceleration. At the giant planets, discrete auroral arcs will appear where field-aligned currents are directed upwards out of the ionosphere in regions where the plasma angular velocity is falling with increasing latitude (or equivalently increasing equatorial radial distance in the magnetosphere).

During the Voyager fly-bys some 30 years ago, it was evident from Saturn kilometric radiation observations from upstream of Saturn that the signal was modulated both by the planetary rotation rate and with solar wind dynamic pressure, and this has been studied more recently with recent joint HST/Cassini campaigns (see Section 12.3.1 and references therein). From a theoretical standpoint therefore, two major aspects of large-scale magnetospheric dynamics must be considered along with their mutual interaction. The first is the dynamics of plasma flows relating to the transfer of angular momentum from the ionosphere to the magnetosphere, associated with planetary rotation, in combination with continuous input of plasma mass to the magnetosphere from moon atmospheres and surfaces, and from ring grains. Such plasma is added into the magnetosphere in the inner regions, and is subsequently picked up by the rotational flow of the magnetosphere. This inner region is shown in the central part of Fig. 12.26a, where the plasma approximately corotates with the planet, as a result of ion–neutral collisions at the ionospheric feet of the magnetic field lines. Voyager observations showed that the flow was near-rigid to $\sim 5 R_S$, beyond which the plasma angular velocity falls with increasing radial distance to $\sim 1/2$ of rigid corotation within several R_S of the dayside magnetopause (Richardson 1986, 1995; Richardson and Sittler 1990). Wilson et al. (2008) analyses of Cassini data show similar trends. This fall in angular velocity is thought to be associated with the addition of mass from internal sources as discussed above, and the subsequent outward transport via radial diffusion and loss in the downtail regions. The plasma is subsequently lost by some release process in the outer magnetosphere. Vasyliunas (1983) suggested that the mass-loaded flux tubes in the outer part of the corotating region would be confined by the effects of the solar wind pressure compressing the dayside magnetosphere, but may stretch out into the down-tail regions of the duskside magnetosphere as they rotate around. This subcorotating region is shown in Fig. 12.26a and is bounded on the outside by an outer region (beyond the dashed line), where the variable loss of plasma downtail takes place. This process is commonly referred to as the Vasyliunas cycle, and was first suggested to be an important mass-loss process within the Jovian magnetosphere. Following reconnection of the closed stretched-out field lines in the dusk sector, the mass-reduced flux tubes contract towards the planet due to the tension force and to conserve angular momentum, and thus rotate back to the dayside with increased angular velocity. As they do so, they reload with plasma from the inner regions and subsequently slow once more. The process repeats as the flux tubes reach the dusk sector of the magnetosphere. While in this steady-state picture the Vasyliunas process has been shown as continuous, it seems likely that the reconnection and plasmoid release could be episodic, and possibly triggered by both internal and external mechanisms.

The second aspect of magnetospheric dynamics to be considered is that which is associated with the solar wind interaction with the magnetosphere, producing a large-scale cyclical flow within the magnetosphere. This process was first discussed in the context of the Earth by Dungey (1961), and is known as the Dungey cycle. The outermost region confined to the dawnside in Fig. 12.26a depicts the flows associated with the solar wind interaction, and the Dungey cycle

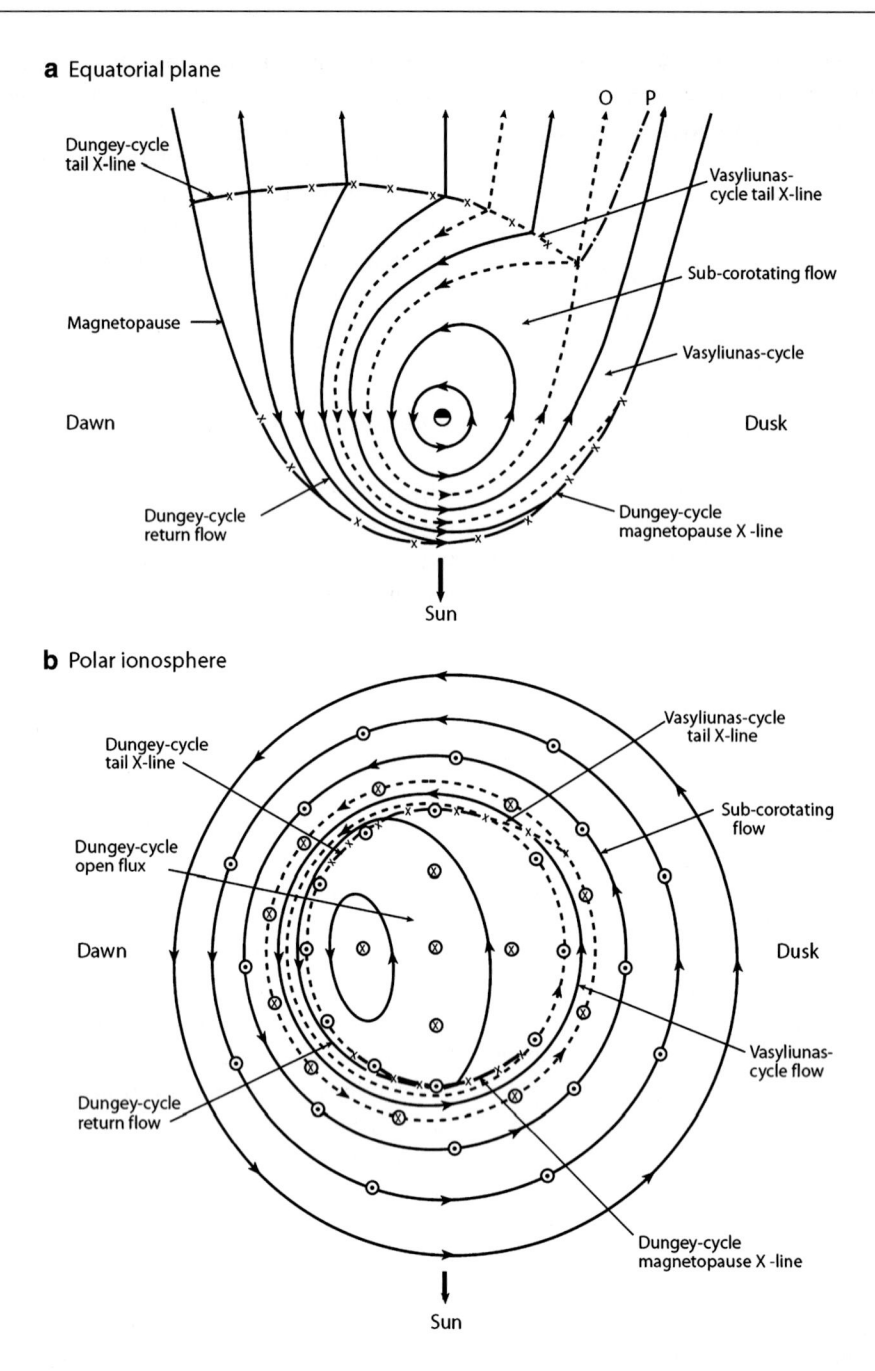

Fig. 12.26 (a) Sketch of the plasma flow in the equatorial plane of Saturn's magnetosphere, where the direction of the Sun is towards the bottom of the diagram, dawn is to the left and dusk is to the right. Arrowed solid lines indicate plasma streamlines, arrowed short-dashed lines show the boundaries between flow regimes (which are also streamlines of the plasma flow), the solid lines joined by X's the reconnection lines associated with the Dungey cycle, and the dashed lines with X's representing the reconnection associated with the Vasyliunas cycle. The two lines are shown as contiguous, but this is not necessarily the case. The lines indicated by 'O' marks the path of the plasmoid O-line in the Vasyliunas-cycle flow (also a streamline), while 'P' marks the outer limit of the plasmoid field lines, which eventually approaches the dusk magnetopause asymptotically. (b) Sketch of the plasma flow in the northern ionosphere, where the direction toward the Sun is at the bottom of the diagram, dawn is to the left, and dusk is to the right. The outermost circle corresponds to ionospheric co-latitudes of $\sim 30°$ from the magnetic pole, which maps to an equatorial radial distance of $\sim 3 R_S$. Circled dots and crosses indicate regions of upward and downward field-aligned current, respectively, as indicated by the divergence of the horizontal ionospheric current. Hall currents flow generally anti-clockwise round the pole and close within the ionosphere, while Pedersen currents flow generally equatorward and close in the field-aligned current system shown (taken from Cowley et al. 2004a)

for Saturn. Reconnection at the dayside magnetopause occurs principally when the IMF points northward, opposite to Saturn's southward directed planetary field. Following reconnection on the dayside, open field lines move anti-sunward over the polar cap (out of the plane of the diagram), and eventually sink into the magnetotail. Magnetic flux tubes are expected to remain open for up to $\sim$8 days (Jackman et al. 2004), during which time they expel the mass which was previously trapped on them. Closure of the open magnetic flux tubes takes place via reconnection in the tail, and the newly closed and emptied flux tubes then return to the dayside magnetosphere via dawn due to the presence of the Vasyliunas cycle at dusk and the action of the planetary rotational torque exerted. The emptied flux tubes may now rotate through this "corridor" in the outer magnetosphere with an increased angular velocity, and with an admixture of hot magnetospheric and solar wind plasma accelerated into the magnetosphere from the reconnection site (Badman and Cowley 2007). Once the magnetic field lines return to the dayside the process repeats.

Figure 12.26b shows the flow regimes discussed above mapped to the northern ionosphere, where the field-aligned current directions may also be depicted. Beyond a radial distance of $\sim$3R_S the angular velocity falls from near-rigid corotation with increasing radial distance, and decreasing ionospheric co-latitude, to $\sim$1/2 of rigid corotation at the boundary of the outer magnetosphere containing the Vasyliunas cycle on closed field lines. At lower co-latitudes still lies the outer magnetosphere which is bounded by the second dashed streamline, which represents the open-closed field line boundary formed through reconnection at the dayside Dungey-cycle X-line. This outer magnetosphere region contains both the Vasyliunas-cycle flow in the lower-latitude portion, and the Dungey-cycle "return" flows at higher latitudes on the dawnside, the latter flowing from the nightside to the dayside reconnection regions (merging gaps) lying on the open-closed field line boundary. In the dawn sector the flows in this region are expected to be higher than in the adjacent magnetosphere due to the lower mass-loading of these flux tubes, but will slow again to the speed of the middle magnetosphere in the dusk sector due to the subsequent refilling of the flux tubes through diffusive mass loading in the Vasyliunas cycle.

The central open flux regions in Fig. 12.26b extend from $\sim$15° co-latitude to the pole (e.g. Cowley et al. 2004b). The flows which are expected to exist within the polar cap are due to two effects; the flows associated with rotational circulation driven by ion–neutral collisions in the ionosphere twisting the open field lines (Isbell et al. 1984; Milan et al. 2005), and cyclical flows associated with the Dungey-cycle. Recent ground based Doppler observations of ionospheric IR emissions by Stallard et al. (2004) indicate plasma angular velocities of $\sim$1/3 of rigid corotation, implying a Pedersen conductance of $\sim$0.5–1 mho (compared to values of 1–2 mho derived from Voyager data by Bunce et al. 2003). When the Dungey cycle is also active, anti-sunward flow occurs between the "merging gaps" on the open-closed field line boundary, these flows add vectorially to the sub-corotational flow on open field lines discussed above giving the streamlines shown in Fig. 12.26b. Overall, the Dungey cycle flows are contained within the polar cap region, and coincide with the dashed line at dusk (the open-closed field line boundary) in Fig. 12.26b, while the return flows are confined preferentially to the dawn sector as discussed with reference to Fig. 12.26a. Finally these flow patterns are related to the aurora by considering the field-aligned current patterns which are produced and their relative importance. In Fig. 12.26b a two-ring pattern of upward field-aligned currents is driven by the implied divergence of the horizontal ionospheric current, and one would anticipate the presence of discrete auroral arcs either at the feet of the field lines connecting to the middle magnetosphere (as in the case of the main Jovian auroral oval) or at the open-closed field line boundary (more like the scenario at the Earth).

Figure 12.27 shows the quantitative results of the recently augmented theoretical model of Cowley et al. (2008), based on new Cassini and HST results to be discussed below. Full details of the model are given in the paper, and as such only an outline of the method used will be given here. The basic ingredients of a quantitative analysis of the qualitative picture represented in Fig. 12.26 are essentially simple, based on an initial assumption of axi-symmetry of the magnetic field and plasma flow. First, a model of the plasma angular velocity on magnetospheric magnetic field lines is conceived, based on a combination of results from the Voyager plasma velocity data and theoretical considerations. A simple model of the internal magnetic field (e.g. the Cassini model (Dougherty et al. 2005)) plus ring current contribution (Connerney et al. 1983), provides a mapped angular velocity profile in the ionosphere (shown in Fig. 12.27a). The profile indicates a fall from rigid corotation at lowest co-latitudes (closest to the planet) to approximately 60% of rigid corotation at the boundary between the middle and outer magnetosphere. The profile then rises in the outer magnetosphere (in the region where the flux tubes have been emptied during reconnection episodes in the tail) and peak at 80% of rigid corotation at the outermost regions of the magnetosphere. Across the open closed field line boundary, the angular velocity of the plasma sharply falls, reaching a steady value of 30% of rigid corotation across the polar cap on open field lines.

The ionospheric angular velocity profile is then combined with model ionospheric parameters (i.e. the Pedersen conductivity) to derive the horizontal ionospheric Pedersen current intensity as a function of co-latitude (Fig. 12.27b). With increasing co-latitude from the magnetic pole the Pedersen current intensity increases monotonically to a value

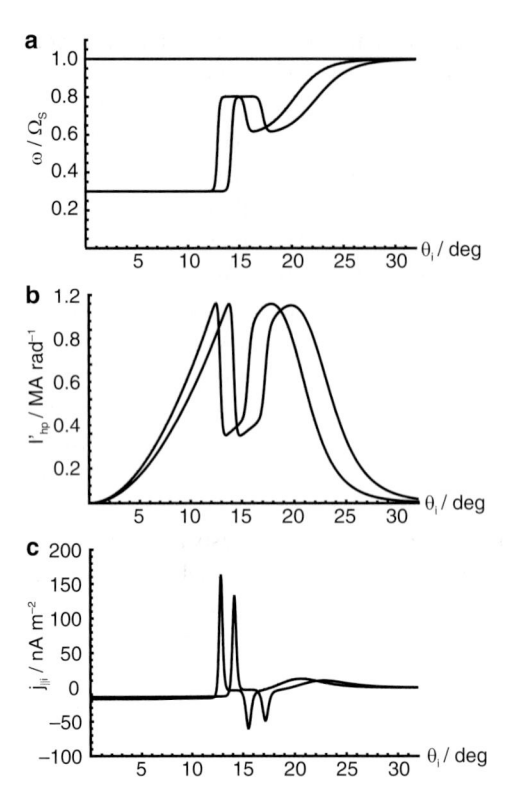

Fig. 12.27 Parameters of the Cowley et al. (2004) model of magnetosphere–ionosphere coupling at Saturn plotted versus co-latitude θi in the ionosphere for the northern (*dashed lines*) and southern hemispheres (*solid lines*), the latter relative to the southern pole. Note that the plots shown here employ updated internal magnetic field and angular velocity values compared to Cowley et al. (2004b), as described in Cowley et al. (2008), but that these result in insignificant variations at the 1% level. The panels of the figure show (**a**) the plasma angular velocity normalized to the planet's angular velocity, where the horizontal dashed line represents rigid corotation, (**b**) the equatorward-directed horizontal ionospheric Pedersen current per radian of azimuth, obtained using an effective Pedersen conductivity of 1 mho, and (**c**) the field-aligned current density just above the ionosphere required by the divergence of the horizontal Pedersen current, where positive and negative values indicate upward and downward-directed currents, respectively, in both hemispheres (adapted from Cowley et al. 2008)

of $\sim$1 MA per radian of azimuth at the open-closed field line boundary. The current intensity then falls rapidly across this boundary, where the angular velocity increases, and then grows once again with increasing co-latitude through the outer and middle magnetosphere regions. As rigid corotation is reached at largest co-latitudes the ionospheric Pedersen current falls to zero.

Taking the divergence of the Pedersen current intensity gives the field-aligned current density just above the ionosphere (Fig. 12.27c) which is required from current continuity. The profile shows that the field-aligned current is directed downward into the ionosphere (negative values) where the Pedersen current is increasing with co-latitude from the pole, and directed upward out of the ionosphere (positive) where the Pedersen current profile is falling. The model estimates show weak downward currents within the region of open field lines across the polar cap at a constant level of $\sim$15 nA m^{-2}, bounded at the open-closed field line boundary by strong upward directed field-aligned currents across a narrow layer (i.e. between $\sim$14° and 15° co-latitude in the southern hemisphere) at the level of $\sim$100 nA m^{-2}. Stronger peaks of downward currents ($\sim$50 nA m^{-2} near to $\sim$17–18° co-latitude in the southern hemisphere) are seen further equatorward in the boundary between the outer and middle magnetosphere, followed by a weaker distributed upward current throughout the middle magnetosphere (with a peak value $\sim$10 nA m^{-2} at $\sim$23° co-latitude). The auroral acceleration parameters may be calculated using Knight's (1973) kinetic theory and a model of the magnetospheric source electron parameters.

Therefore, the modeling studies have concluded that the field-aligned currents in the middle magnetosphere are too weak to require significant acceleration, and will not result in significant auroral output. They also occur at too low a co-latitude to account for the main UV emissions (Clarke et al. 2005; Grodent et al. 2005; Badman et al. 2006). However, Stallard et al. (2008c) have recently shown that at infrared wavelengths a weak oval does exist in regions which map to the middle magnetosphere, and is suggested to be the equivalent of the main auroral oval at Jupiter (see Section 12.2.2.5). Overall though, the large flow shear which occurs at the boundary between open field lines which strongly sub-corotate (Isbell et al. 1984; Stallard et al. 2003, 2004), and closed field lines moderately sub-corotating in the outer magnetosphere, is at the right co-latitude in the ionosphere to account for Saturn's main auroral oval (Badman et al. 2005), and is plausible in terms of accelerated electron and auroral parameters (Cowley et al. 2004b; Jackman and Cowley 2006). An opposing theory by Sittler et al. (2006) proposes that the auroral oval is produced by plasma heating and acceleration associated with the interchange instability at the outer edge of the plasma sheet. In this case upward field-aligned currents associated with the main oval should map to the closed field region at $\sim 15 R_S$ in the equatorial plane. However, this distance maps to $\sim 16°$ in the southern hemisphere ionosphere according to the "typical" model of Bunce et al. (2008b), and thus corresponds to the region of downward field-aligned currents shown in Fig. 12.27c.

As discussed in Section 12.2.2.2, Badman et al. (2006) have determined the average location of Saturn's main auroral oval in the southern hemisphere from multiple UV images obtained by the HST. Their results indicate that the median location of the oval in the southern hemisphere near the dawn sector is $\sim 14°$ co-latitude, and thus maps to the outer magnetosphere near the magnetopause, well beyond the middle magnetosphere corotation breakdown region (Bunce et al. 2008b). Therefore, the results of Badman et al. (2006) suggest that the auroral oval is unlikely to be associated with the magnetosphere–ionosphere coupling currents of the middle magnetosphere, but cannot clearly discriminate between the proposed positions of the aurora by Cowley et al. (2004a, b) and Sittler et al. (2006). As discussed in Section 12.4.2, above, during the initial high-inclination phase of the Cassini orbital tour in 2006, the first in situ observations of the high-latitude magnetosphere have been obtained, with coordinated HST imaging. Section 12.4.2 presented an overview of these unique observations and compared the results with the quantitative modeling of Cowley et al. (2004a, b) outlined above.

The observations presented in Section 12.4.2 are in good qualitative agreement with the model of Cowley et al. (2004b). However, Cowley et al. (2008) have compared the model predictions in a quantitative sense, and have found that while the model values are in good general agreement there are three main differences. The first is the location of the open-closed field line boundary, which occurs at $\sim 14°$ co-latitude in the model whereas the observations place the boundary closer to $12°$ co-latitude. Second, in the model the switch in the azimuthal B_φ component takes place quasi-monotonically from positive to near-zero, whilst the observations show switching back and forth indicative of multiple spatial structures, or a single oscillating layer. Finally, the magnitude of the B_φ signature in the model peaks at ~ 2 nT, while the observations show that B_φ grows to a peak value of ~ 10 nT just poleward of the open-closed field line boundary.

As a result the model gives a total upward field-aligned current of 0.8 MA rad^{-1} flowing in a layer $\sim 0.5°$ wide, producing a field-aligned current density of 150 nA m^{-2}, compared to 4–5 MA rad^{-1}, 1.5–2°, and 275 nA m^{-2} as derived from the magnetic field observations. Cowley et al. (2008) find that simple and realistic alterations to the model achieve excellent results. They relocate the open-closed field line boundary to $12°$, widen the current layer to agree with the observations, and to increase the intensity of the field-aligned current they set the angular velocity such that the open field lines do not rotate at all in the inertial frame, and increase the conductivity of the southern ionosphere by a factor of 4 (from 1 to 4 mho). Cowley et al. (2008) suggest that the main difference between the observations and the model is due to the higher conducting summer hemisphere than was previously assumed in the original model formulation. The revised model is now in excellent agreement with the Cassini-HST data, requiring downward acceleration of outer magnetosphere electrons through ~ 10 kV potential in the current layer at the open-closed field line boundary, giving an auroral oval approximately $1°$ wide, with UV emission intensities of a few tens of kiloRayleighs.

12.5.2 Time-Dependent Auroral Processes

Assuming this basic scenario, where the main auroral oval is associated with a large upward-directed field-aligned current layer at the open closed field line boundary, one would expect to see a significant modulation of the emissions as a result of changing solar wind conditions, and different levels of Dungey-cycle driving versus internal rotational effects. In particular recent coordinated HST and Cassini campaigns (e.g. January 2004) have shown that the aurora responds strongly to upstream conditions, and specifically the solar wind dynamic pressure (Clarke et al. 2005, 2009; Crary et al. 2005; Grodent et al. 2005; Bunce et al. 2006). This behavior also reflects the modulation observed in the SKR emissions (e.g. Desch 1982; Kurth et al. 2005a).

For example, Jackman et al. (2004) have investigated the reconnection-driven interaction of the solar wind with Saturn's magnetosphere with particular focus on the consequences for magnetospheric dynamics. In this study, interplanetary magnetic field (IMF) data obtained by the Cassini spacecraft en route to Saturn were collected for 8 complete solar rotations which allow the variation of the field structure to be investigated. They find that the solar wind magnetic field structure is consistent with that expected to be produced by corotating interaction regions (CIRs) during the declining phase of the solar cycle. In general the data show that the IMF structure consisted of two sectors during each rotation of the Sun, with crossings of the heliospheric current sheet generally embedded within few-day high field compression regions, surrounded by several day rarefaction regions.

During the January 2004 HST campaign (as discussed in Sections 12.2.2.1 and 12.3.1), an impact of a major CIR related solar disturbance was measured by Cassini and the effects on Saturn's magnetosphere were observed by HST. Cowley et al. (2005) have suggested that the bright auroral displays towards the end of the observation interval are triggered by the collapse of Saturn's magnetic tail in response to the impact of the compression region on Saturn's magnetosphere. Figure 12.13 shows the last four visits (a–d) from the January 2004 HST observation set. Following the start of the compression region the magnetic field strength remains high between ∼0.5–2 nT, and the solar wind velocity and density were also raised (Crary et al. 2005; Bunce et al. 2006). Image (b) occurred ∼10 h after the onset of the compression. Here, one sees that the dark polar cap which was evident in image (a) has now been filled in with bright aurora. The central dark region in image (b) is significantly contracted with respect to that in image (a), indicative of a significant amount of tail flux closure at some point between the two images. Cowley et al. (2005) have estimated that ∼15 GWb of open flux was closed during the first ∼10 h of this event, by comparison of image (a) and (b). This implies a reconnection rate of ∼400 kV over 10 h, which is considerably larger than the several tens of kV of dayside reconnection estimated by Bunce et al. (2006) during this time, implying that tail reconnection was dominant over dayside reconnection. The expansion of the oval from image (c) to image (d) is then suggestive of the cessation of the tail reconnection, as the system 'recovers' and the continuation of rapid dayside reconnection, in keeping with the raised magnetic field strength shown. In the bottom panel the SKR frequency–time spectrogram indicates that the planetary modulated signature seems to switch off after the shock-compression, and is replaced by a high-power burst of SKR which extends down to lower frequencies. However, Ulysses did not observe this switch off, and hence, the missing modulated signal is most likely due to beaming as opposed to skipping a rotation. A second burst of SKR is then eventually followed by the recommencement of the planetary modulated signatures, albeit at somewhat higher powers than prior to the compression region, although returning with the same phase.

Bunce et al. (2005a) have also shown that a similar major compression impacted Saturn during the Cassini Saturn Orbit Insertion (SOI) maneuver which took place on 1 July 2004, during the period the spacecraft was inside the magnetosphere. They witnessed the effect of this compression inside the magnetosphere, which represents the counterpart of auroral displays of January 2004. During this time a major burst of SKR was observed, indicating the approximate time that the solar disturbance impacted the magnetosphere. At the same time, Cassini measured a major injection of hot plasma into the magnetosphere from the downtail regions, and a substantial reorientation of the magnetic field structure was seen by the magnetic fields instrument. These signatures are consistent with the suggestion that major solar disturbances impacting the magnetosphere induce the magnetotail of Saturn to suddenly collapse, injecting hot plasma towards the planet. It is this hot plasma which is thought to be directly producing the auroral displays. Overall, it has been suggested that the effects of the CIR shock-compression may be one way to produce the equivalent of the terrestrial substorm at Saturn. Russell et al. (2008) have recently suggested that Titan may also play a role in the triggering of some of the substorm-like events that have been witnessed in the Cassini magnetotail data thus far (Jackman et al. 2007, 2008). Likewise, Menietti et al. (2007) have shown that the occurrence of strong SKR has some dependence on the local time of Titan.

As is well known at the Earth, transient "cusp currents" are associated with time-dependent reconnection at the dayside boundary, and are strongly modulated by the orientation of the IMF. During an interval of purely southwards IMF at the Earth, one observes an interval of transient low latitude reconnection (e.g. Milan et al. 2000). For the case of northwards IMF, high-latitude reconnection occurs between the IMF and already-open field lines in the magnetosphere. These two examples lead to two different types of impulsive flows in the ionosphere. For purely southwards IMF at Earth one would expect an impulsive twin-vortical flow localised near noon, leading to an enhanced upward field-aligned current intensity on the duskside of the main oval. During intervals of northwards IMF the impulsive vortical flows are reversed in sense and shifted poleward of the main oval, leading to an enhanced upward field-aligned current density and bright spot of aurora poleward of the main auroral oval. Milan et al. (2000) present two such examples using data from POLAR-UVI and ground-based ionospheric radars to derive the ionospheric flow directions and velocities, and show that the scenarios described above are evident in the data for different orientations of the IMF.

Pulsed dayside reconnection at Saturn then, is expected to produce similar flows across the open-closed field line boundary, as suggested to explain auroral brightening within the main auroral emission region at Jupiter (Bunce et al. 2004), but at Saturn the localised cusp emission would instead modulate the continuously upward current layer present at the open-closed field line boundary due to the flow shear between the two regimes (discussed above). Bunce et al. (2005b) have produced a variety of model scenarios for Saturn's cusp including the effects of northward and southward IMF, and varying IMF B_y orientations. An example for purely northward IMF conditions is shown on the left hand of Fig. 12.28, during strong Dungey-cycle driving which produces localised twin vortical flows straddling the open-closed field line boundary superposed on the background flows of

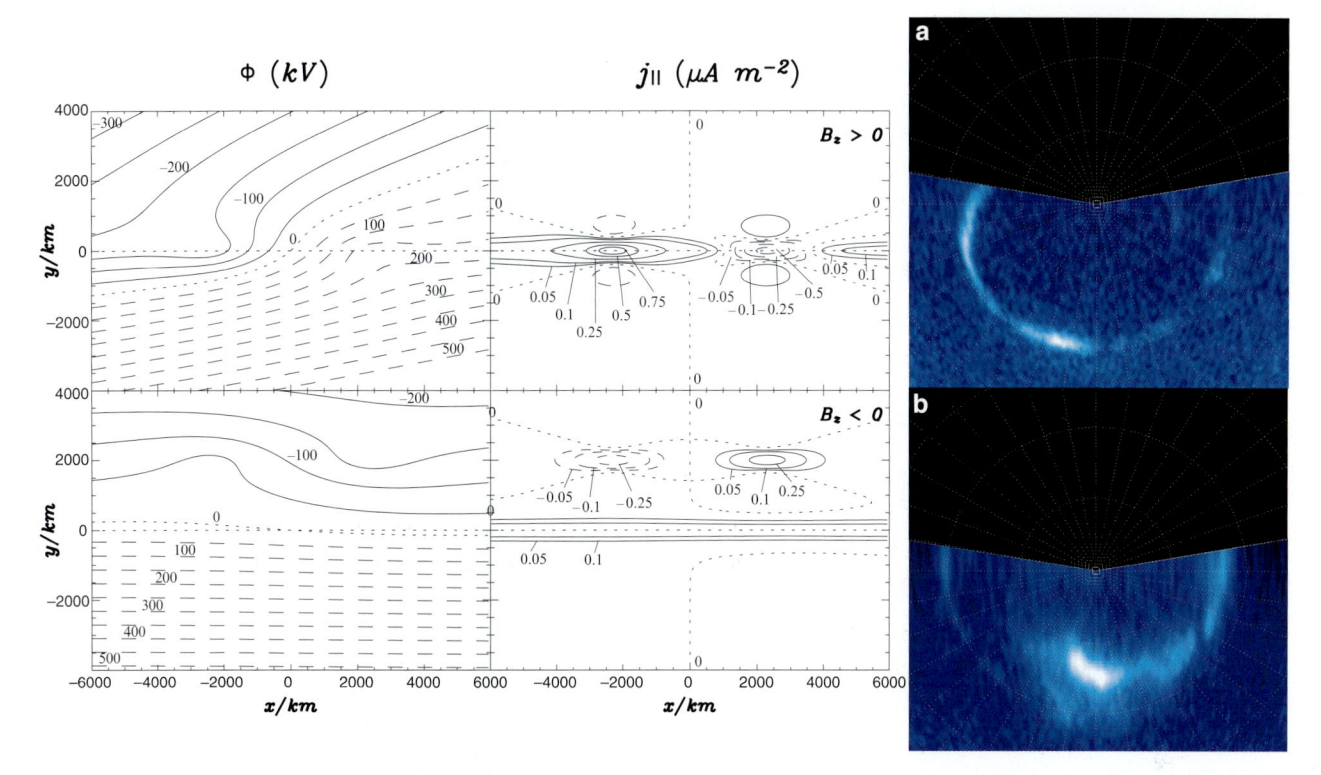

Fig. 12.28 On the left hand side we show a stack plot of electrostatic potential contours (i.e., streamlines of the plasma flow) on the left, and field-aligned current density on the right, for the case of the 'fast flow' cusp model, taken from the Bunce et al. (2005b) model. The first row in each case corresponds to the case of $B_z > 0$, and the second to the case of $B_z < 0$ as shown. Both are presented on equivalent grids of -4000–4000 km in y, and -6000–6000 km in x. The short-dashed line at $y = 0$ indicates the open-closed field line boundary. On the left, in the plasma streamlines grid the dashed lines show contours of negative electrostatic potential. Contours are labeled in steps of 50 kV. On the right, the field-aligned current grids show dotted lines indicating contours of zero field-aligned current density. Solid lines indicate the regions of upward-directed field-aligned current density while the dashed lines indicate the regions of downward-directed field-aligned current density. Contours are labeled 0.05, 0.1, 0.25, 0.5, 0.75, and 1.0, in units of $\mu A m^{-2}$. On the right hand side are two examples of the morphology of Saturn's southern aurora obtained with the HST-STIS SrF2 filter. Images are projected onto the ionosphere, where the pole is to the centre, and circles of increasing size indicate 80°, 70°, and 60° of latitude respectively. The direction to the Sun is at the bottom of the diagram, dusk is to the right and dawn to the left. Longitude meridians are shown at intervals of 10°. Image (**a**), taken on the 29th January 2001, shows an auroral oval which is brighter at dawn than at dusk, with an additional brightening in the pre-noon sector. Image (**b**) shows the high-latitude 'spot' discussed in the text (taken from Bunce et al. 2005b)

the steady-state auroral models described in this section, producing an anti-sunward burst of flow near noon (see left hand upper panel of flow streamlines). In the middle upper panel of the figure, the field-aligned current pattern produced shows that the localised cusp currents under purely northward IMF conditions appear as an enhanced 'spot' of upward field-aligned current on the morning oval, with a dimming of the main upward directed field-aligned current on the dusk side of noon. The image to the right in the upper row shows a feature along the main auroral oval which Gérard et al. (2005) describe as being the cusp, which they also show exists in two basic states suggested to be due to changing IMF orientation. The brightening on the main auroral oval, now known to be associated with the open-closed field line boundary, on the morning side of noon agrees excellently with the model prediction, along with a slight dimming of the oval to the dusk side of noon. The second example shown in the bottom row of Fig. 12.28, now shows the conditions for southward IMF.

The same basic steady-state background flows are employed but the high-latitude twin vortical flows are reversed in sense due to the reversed polarity of IMF, by simple analogy with the Earth. This produces a field-aligned current spot poleward of the open-closed field boundary and the main auroral oval currents seen in the middle panel on the bottom row. This agrees well with the second state of the 'cusp' emission seen in the HST image to the right taken from Gérard et al. (2005), which shows a significant brightening of the aurora to higher latitudes than is typically expected for the main auroral oval (e.g. Badman et al. 2005).

Badman et al. (2005, 2006) and Cowley et al. (2005) have described the overall auroral images collected during this campaign, and related them directly to the steady-state picture of the flows and currents in the ionosphere, discussed by Cowley et al. (2004a) and shown here in Fig. 12.26. Figure 12.29 shows various pictures throughout the January 2004 campaign, and the suggested modified

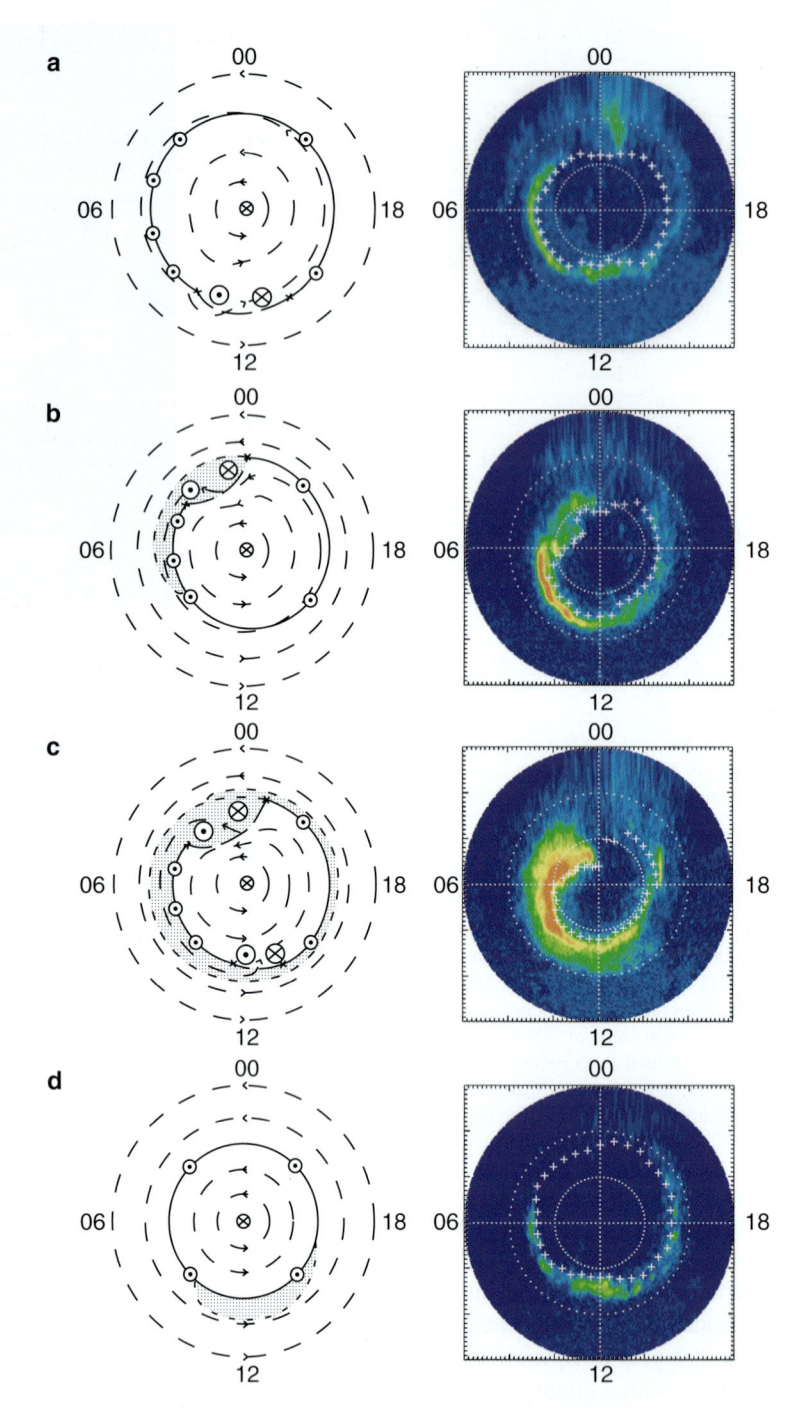

Fig. 12.29 On the left hand side, we show sketches in the ionosphere in a similar format to Fig. 12.26, where the direction towards the Sun is at the bottom of each diagram, with dawn to the left, and dusk to the right. The solid line shows the boundary between open and closed field lines, while the dashed lines with arrows show plasma streamlines. The patch of newly-closed flux is indicated by the stippled area, bounded on its equatorward side by the short-dashed line. Circled dots and crosses indicate regions of upward and downward field-aligned currents, respectively. The first two diagrams show the ionospheric consequences of intervals of steady magnetopause and tail reconnection, but where the rates are not equal to each other. In (**a**) the tail reconnection rate exceeds the dayside rate, while in (**b**) the dayside rate exceeds the tail rate. (**c**) illustrates the consequences of an interval of rapid reconnection in the tail, in which a significant fraction of the open flux in the tail lobes is closed on a time scale short compared with the typical period of plasma sub-corotation in the outer magnetosphere (after ∼20 h). Panel (**c**) show conditions in the ionosphere after ∼20 h. Dayside reconnection is also in progress in this panel at this time. Finally, panel (**d**) shows the motion of a patch of newly-opened field lines formed by a burst of reconnection at the dayside magnetopause. On the right hand side we show a series of HST images from the January 2004 campaign, which correspond well to the theoretical pictures on the left, and occur during similar conditions in the solar wind according to the in situ solar wind data measured concurrently by Cassini (adapted from Badman et al. 2005 and Cowley et al. 2005)

flows and current patterns in the ionosphere which result which successfully explain the variety of phenomena which were observed. Figure 12.29a shows the flows and currents for the situation of dayside reconnection only, with an absence of tail reconnection. This leads to an enhanced discrete aurora at dawn, and at the mapped location of the dayside "merging gap", i.e. the transient cusp currents discussed above. This scenario agrees well with the image taken on 23 January 2004. Figure 12.29b shows the flows and currents suggested by Cowley et al. (2005) to occur in response to compression-induced tail reconnection. In this picture a burst of hot plasma is produced in the tail, resulting in enhanced diffuse and discrete auroral features, as seen in the image to the right taken on the 18 January, following the "minor" compression in the middle of the campaign. Figure 12.29c shows the effects of an interval of strong dayside and nightside reconnection which in tandem produce enhanced discrete aurora at dawn and a spiral of diffuse emission on sub-corotating closed field lines. This is in keeping with the image taken on 28 January (image C in Fig. 12.13). Finally, Fig. 12.29d shows an example of the flows and currents expected in the ionosphere during an interval of intermittent tail reconnection on timescales which are small relative to the planetary rotation period. Under such conditions it is proposed that diffuse patches of aurora will sub-corotate around, lying just equatorward of the open-closed field line boundary. This idea is supported by the image to the right, taken on 14 January during the rarefaction conditions in the solar wind when dayside driving is expected to be minimal.

12.6 Summary

Cassini's prime mission, coupled with a number of joint campaigns with Earth-based remote sensing, has expanded our knowledge of auroral processes at Saturn. Observationally, Earth-based UV and IR observations have provided a consistent morphology of auroral emissions that characterize both quiet and disturbed times. This morphology is focused on the main auroral oval which is almost always present although it is often incomplete, forming a spiral that does not close upon itself. The oval resides near 70° latitude although when the spiral form is present it often begins in the midnight-dawn sector at latitudes as high as 80° and evolves equatorward as local time increases such that the end of the spiral is equatorward of the beginning, again in the region of midnight. Within this framework, there are often dawn brightenings of the oval and these bright arcs corotate at a fraction of the rotation rate of the magnetosphere, often with the bright spots moving poleward in the afternoon as they fade. Particularly in the infrared observations, there are less organized emissions poleward of the main oval. Disturbed conditions, triggered by solar wind compression regions, result in the poleward expansion of the main oval particularly on the dawn side.

Saturn kilometric radio emissions are generated by the cyclotron maser instability on field lines that thread the aurora. To first order, their power correlates with the power in the UV aurora, although the correlation is not particularly strong. Determinations of the source locations of the SKR clearly show a dependence on the observer's location, meaning beaming of the radio emissions is important. Hence, while the true correlation between SKR and UV auroral power might be very close to one, a single spacecraft cannot accurately integrate over all of the SKR emission, or may even miss the brightest emission because of its location. There is circumstantial evidence that the strongest sources of SKR are associated with auroral bright spots and move in the corotational sense. However, the brightest SKR sources are usually located between dawn and noon. Similar to the situation with UV emissions, there is evidence of rotation of the SKR sources in detail, but the primary SKR source morphology favors the dawn to noon local time quadrant. Further, the integrated intensity of SKR emissions is temporally modulated by the rotation of the magnetosphere such that the brightest SKR emissions occur, statistically, when the Sun is near 100° SLS3 longitude.

SKR can be observed at all local times, but there is evidence of at least two zones where the emission cannot be seen, that is, regions where an observer is shadowed from the SKR sources. The first of these is at low latitudes close to the planet, within a few Saturn radii. The second of these is at high latitudes to where at least the highest frequency emissions apparently cannot propagate. A more complete picture of the high latitude occlusion zone will come when occurrence studies using the high inclination orbits at the end of the prime mission and the beginning of the Cassini Equinox Mission (extended mission) are analyzed.

The ~75° inclination orbits in 2008 at the end of the prime mission and the early Cassini Equinox Mission provide excellent viewing of the aurora by the remote sensing instruments, including ENA imaging as well as increasingly better views of the northern auroral region by Earth-based instruments. Coupled with in situ measurements by the field and particle instruments on Cassini, this time period provides perhaps the best opportunity to understand extraterrestrial auroral processes until Juno arrives at Jupiter in late 2016. Nevertheless, Cassini's high latitude passes were typically at distances of 5 and above R_S and too high in altitude to fly through the expected auroral acceleration region. Thus, the electrons observed at this location by Cassini are generally upgoing, implying downward-directed field-aligned currents, which are not directly associated with auroral emissions but which may form part of a large-scale circuit, the upward-directed component of which may produce auroral emission. These electrons are in the form of beams moving away from Saturn although there are some instances of bimodal distributions moving both up and down the field lines. The electrons in these beams have energies extending to several hundred kiloelectron-volts, compared to the few to tens of kiloelectron-volts observed at Earth. The ENA images provide a glimpse of acceleration processes in the upward current region in the form of light ion (H, H^2, and He^3) conics. The ions charge exchange and reach Cassini provided they have the proper trajectory after leaving the conic as a neutral. There is currently consideration of putting Cassini into a Juno-like orbit with periapsis between the low-latitude atmosphere and the inner edge of the D ring before dropping the spacecraft into the atmosphere for planetary protection purposes. While the inclination of these orbits will not likely be higher than about 65°, there is a possibility of crossing through the acceleration region associated with the upward auroral currents during these orbits.

Considerable work has gone into placing the Cassini auroral observations into the context of a Vasyliunas cycle moderated by a Dungey cycle and with some success. Unlike Jupiter, the currents associated with the breakdown of corotation do not seem to be strong enough to require a region of parallel acceleration, although infrared observations now suggest a weak auroral oval associated with these currents. The stronger emissions, hence stronger currents, are poleward of those associated with the corotation breakdown. This has led to the suggestion, with some supporting observations, that the main auroral oval at Saturn is associated with the shear at the open-closed field line boundary. However, the situation seems to be not so simple as electron beams can be found from near the magnetopause all the way in to $\sim 11 R_S$.

Prior to the availability of Cassini observations in the Saturnian system, the natures of the auroral processes at Earth, Jupiter, and Saturn were discussed in terms of ordering by the relative importance of rotation and internal dynamics versus solar wind control. Earth is clearly at the end of the spectrum having predominantly solar wind control. Jupiter has been classified as the rotationally-dominated end-member of the set. This left Saturn as somewhere in between Earth and Jupiter. However, the situation is confused partly by the fact that the solar wind appears to play a reasonably important role in Jupiter's magnetosphere. It is perhaps more accurate to classify Saturn's auroral processes as simply unique.

References

Aguilar, A., Ajello, J.M., Mangina, R.S., James, G.K., Abgrall, H., Roueff, E.: The electron-excited middle UV to near IR spectrum of H_2: Cross-sections and transition probabilities. Astrophys. J. Supp. Ser. **177** (2008).

Andrews, D.J., Bunce, E.J., Cowley, S.W.H., Dougherty, M.K., Provan, G., Southwood, D.J.: Planetary period oscillations in Saturn's magnetosphere: Phase relation of equatorial magnetic field oscillations and Saturn kilometric radiation modulation. J. Geophys. Res. **113**, A09205, doi:10.1029/2007JA012937 (2008).

Badman, S.V., Cowley, S.W.H.: Significance of Dungey-cycle flows in Jupiter's and Saturn's magnetospheres, and their identification on closed equatorial field lines. Ann. Geophysicae **25**, 941–951 (2007).

Badman, S.V., Bunce, E., Clarke, J.T., Cowley, S., Gérard, J.C., Grodent, D., Milan, S.: Open flux estimates in Saturn's magnetosphere during the January 2004 HST-Cassini campaign, and implications for reconnection rates. J. Geophys. Res. **110**, A11216, doi:10.1029/2005JA011240 (2005).

Badman S.V., Cowley, S.W.H., Gerard, J.-C., Grodent, D.: A statistical analysis of the location and width of Saturn's southern auroras. Ann. Geophysicae **24**, 3533–3545 (2006).

Benediktov, E.A., Getmansev, G.G., Sazonov, Y.A.: Preliminary results of measurements of the intensity of distributed extra-terrestrial radio frequency emission at 725 and 1525 kc (in Russian). Kosm. Issled. **3**, 614 (1965).

Bisikalo, D.V., Shematovich, V.I., Gérard, J.-C., Gladstone, R.G., Waite J.H.: The distribution of hot hydrogen atoms produced by electron and proton precipitation in the Jovian aurora. J. Geophys. Res. **101**, 21,157–21,168 (1996).

Broadfoot, A.L., Sandel, B.R., Shemansky, D.E., Holberg, J.B., Smith, G.R., Strobel, D.F., McConnell, J.C., Kumar, S., Hunten, D.M., Atreya, S.K., Donahue, T.M., Moos, H.W., Bertaux, J.L., Blamont, J.E., Pomphrey, R.B., Linik, S.: Extreme ultraviolet observations from Voyager 1 encounter with Saturn. Science **212**, 206–211 (1981).

Brown, R.H. et al.: The Cassini Visual and Infrared Mapping Spectrometer (VIMS) investigation. Space Sci. Rev. **115**, 111–168 (2004).

Bunce, E.J., Cowley, S.W.H., Wild, J.A.: Azimuthal magnetic fields in Saturn's magnetosphere: Effects associated with plasma sub-corotation and the magnetopause-tail current system. Ann. Geophysicae **21**, 1709 (2003).

Bunce, E.J., Cowley, S.W.H., Yeoman, T.K.: Jovian cusp processes: implications for the polar aurora. J. Geophys. Res. **109**, doi:10.1029/2003JA010280 (2004).

Bunce, E.J., Cowley, S.W.H., Wright, D.M., Coates, A.J., Dougherty, M.K., Kurth, W.S., Krupp, N., Rymer, A.M.: In situ observations of a solar wind compression-induced hot plasma injection event in Saturn's tail. Geophys. Res. Lett. **32**, 20, Art No. L20S01 (2005a).

Bunce, E.J., Cowley, S.W.H., Milan, S.E: Interplanetary magnetic field control of Saturn's polar cusp aurora. Ann. Geophys. **23**, 1405 (2005b).

Bunce, E.J., Cowley, S., Jackson, C., Clarke, J.T., Crary, F., Dougherty, M.: Cassini observations of the interplanetary medium upstream of Saturn and their relation to the Hubble Space Telescope aurora data. Adv. Sp. Res. **38**, 806–814 (2006).

Bunce, E.J., Arridge, C.S., Clarke, J.T., Coates, A.J., Cowley, S.W.H., Dougherty, M.K., Gérard, J. C., Grodent, D., Hansen, K.C., Nichols, J.D., Southwood, D.J., Talboys, D.L.: Origin of Saturn's aurora: Simultaneous observations by Cassini and the Hubble Space Telescope. J. Geophys. Res. **113**, A09209, doi: 10.1029/2008JA013257 (2008a).

Bunce, E.J., Arridge, C.S., Cowley, S.W.H., Dougherty, M.K.: Magnetic field structure of Saturn's dayside magnetosphere and its mapping to the ionosphere: Results from ring current modeling. J. Geophys. Res., **113**, A02207, doi:10.1029/2007JA012538 (2008b).

Carbary, J.F., Mitchell, D.G., Krimigis, S.M., Krupp, N.: Evidence for spiral pattern in Saturn's magnetosphere using the new SKR longitudes. Geophys. Res. Lett. **34**, L13105, doi:10.1029/2007GL030167 (2007).

Carlson, C.W., McFadden, J.P., Ergun, R.E., Temerin, M., Peria, W., Mozer, F.S., Klumpar, D.M., Shelley, E.G., Peterson, W.K., Moebius, E., Elphic, R., Strangeway, R., Cattell, C., Pfaff, R.: FAST observations in the downward auroral current regions: Energetic upgoing electron beams, parallel potential drops, and ion heating. Geophys. Res. Lett. **25**, 2017–2020 (1998).

Carr, T.D, Reyes, F.: Microstructure of Jovian decametric S bursts. J. Geophys. Res. **104**, 25,127–25,141 (1999).

Cecconi, B., Lamy, L., Zarka, P., Prangé, R., Kurth, W.S., Louarn, P.: Goniopolarimetric study of the revolution 29 perikrone using the Cassini Radio and Plasma Wave Science instrument high-frequency radio receiver. J. Geophys. Res. **114**, A03215, doi:10.1029/2008JA013830 (2009).

Clarke, J.T., Moos, H.W., Atreya, S.K., Lane, A.L.: IUE detection of bursts of H Ly α emission from Saturn. Nature **290**, 226 (1981).

Clarke J.T. et al.: Morphological differences between Saturn's ultraviolet aurorae and those of Earth and Jupiter. Nature **433**, 717–719 (2005).

Clarke, J.T. et al.: The response of Jupiter's and Saturn's auroral activity to the solar wind. J. Geophys. Res. **114**, A05210, doi:10.1029/2008JA013694 (2009).

Connerney, J.E.P., Acuna, M.H., Ness, N.F.: Currents in Saturn's magnetosphere. J. Geophys. Res. **88**, 8779–8789 (1983).

Cowley, S.W.H., Bunce, E.J., Prangé, R.: Saturn's polar ionospheric flows and their relation to the main auroral oval. Ann. Geophys. **22**, 1379–1394 (2004a).

Cowley, S.W.H., Bunce, E.J., O'Rourke, J.M.: A simple quantitative model of plasma flows and currents in Saturn's polar ionosphere. J. Geophys. Res., **109**, A05212, doi:10.1029/2003JA010375 (2004b).

Cowley, S.W.H., Badman, S.V., Bunce, E.J., Clarke, J.T., Gérard, J.-C., Grodent, D., Jackman, C.M., Milan, S.E., Yeoman, T.K.: Reconnection in a rotation-dominated magnetosphere and its relation to Saturn's auroral dynamics. J. Geophys. Res. **110**, doi:10.1029/2004JA010796 (2005).

Cowley, S.W.H. et al.: Auroral current systems in Saturn's magnetosphere: Comparison of theoretical models with Cassini and HST observations. Ann Geophys. **26**, 2613–2630 (2008).

Crary, F.J. et al.: Solar wind dynamic pressure and electric field as the main factors controlling Saturn's aurorae. Nature **433**, 720–722 (2005).

Davis, L.J, Smith, E.J.: A model of Saturn's magnetic field based on all available data. J. Geophys. Res. **95**, 15,257–15,261 (1990).

Desch, M.D.: Evidence for solar wind control of Saturn radio emission. J. Geophys. Res. **87**, 4549–4554 (1982).

Desch, M.D.: Radio emission signature of Saturn immersions in Jupiter's magnetic tail. J. Geophys. Res. **88**, 6904 (1983).

Desch, M.D., Kaiser, M.L.: Voyager measurements of the rotation period of Saturn's magnetic field. Geophys. Res. Lett. **8**, 253–256 (1981).

Dougherty, M.K., Achilleos, N., Andre, N., Arridge, C.S., Balogh, A., Bertucci, C., Burton, M.E., Cowley, S.W.H., Erdos, G., Giampieri, G., Glassmeier, K.-H, Khurana, K.K., Leisner, J., Neubauer, F.M., Russell, C.T., Smith, E.J., Southwood, D.J., Tsurutani, B.T.: Cassini magnetometer observations during Saturn orbit insertion, Science, **307**, 1266–1270 (2005).

Dungey, J.W.: Interplanetary magnetic field and the auroral zones. Phys. Rev. Lett. **6**, Issue 2, 47–48 (1961).

Dyudina, U.A., Ingersoll, A.P. Ewald, S.P.: Aurora at the north pole of Saturn as seen by Cassini ISS. Presented at the Fall AGU meeting in San Francisco (2007).

Ergun, R.E. et al.: FAST satellite observations of electric field structures in the auroral zone. Geophys. Res. Lett. **25**, 2025–2028 (1998).

Esposito, L.W. et al.: The Cassini Ultraviolet Imaging Spectrograph Investigation. Space Sci. Rev. **115**, 299–361 (2004).

Esposito, L.W. et al.: Ultraviolet Imaging Spectroscopy shows an active Saturnian system. Science **307**, 1251–1255 (2005).

Flasar, F. M. et al.: Exploring the Saturn system in the thermal infrared: The composite Infrared Spectrometer. Space Sci. Rev. **115**, 169–297 (2004).

Galopeau, P., Zarka, P., Le Queau, D.: Theoretical model of Saturn's kilometric radiation spectrum. J. Geophys. Res. **94**, 8739–8755 (1989).

Galopeau, P., Zarka, P., Le Queau, D.: Source locations of SKR: The Kelvin–Helmholtz instability hypothesis. J. Geophys. Res. **100**, 26,397–26,410 (1995).

Galopeau, P.H.M., Lecacheux: A. Variations of Saturn's radio rotation period measured at kilometer wavelengths. J. Geophys. Res. **105**, A6, 13,089–13,102 (2000).

Gérard, J.-C., Singh, V.: A Model of energy deposition of energetic electrons and EUV emission in the Jovian and Saturnian atmospheres and implications. J. Geophys. Res. **87**, 4525–4532 (1982).

Gérard, J.-C., Gustin, J., Grodent, D., Delamere, P., Clarke, J.T.: The excitation of the FUV Io tail on Jupiter: Characterization of the electron precipitation. J. Geophys. Res. **107**, 1394, doi:10.1029/2002JA009410 (2002).

Gérard, J.-C., Gustin, J., Grodent, D., Clarke, J.T., Grard, A.: Spectral observations of transient features in the FUV Jovian polar aurora. J. Geophys. Res. **108**, A8, 1319, doi:10.1029/2003JA009901 (2003).

Gérard, J.-C., Grodent, D., Gustin, J., Saglam, A., Clarke, J.T., Trauger, J.T.: Characteristics of Saturn's FUV aurora observed with the Space Telescope Imaging Spectrograph. J. Geophys. Res. **109**, A09207, doi:10.1029/2004JA010513 (2004).

Gérard, J.-C., Bunce, E., Grodent, D., Cowley, S., Clarke, J.T., Badman, S.: Signature of Saturn's auroral cusp: Simultaneous Hubble Space Telescope FUV observations and upstream solar wind monitoring. J. Geophys. Res. **110**, A11201, doi:10.1029/2005JA011094 (2005).

Gérard, J.-C. et al.: Saturn's auroral morphology and activity during quiet magnetospheric conditions. J. Geophys. Res. **111**, A12210, doi:10.1029/2006JA011965 (2006).

Gérard, J.-C., Bonfond, B., Gustin, J., Grodent, D.: The altitude of Saturn's aurora and its implications for the characteristic energy of precipitated electrons. Geophys. Res. Lett. **36**, L02202, doi:10.1029/2008GL036554 (2009).

Goldstein, M.L., Goertz, C.K.: Theories of radio emissions and plasma waves. In: Dessler, A.J. (ed.): Physics of the Jovian Magnetosphere, pp. 317–352. Cambridge University Press, New York (1983).

Grodent, D., Waite, J.H., Gérard, J.C.: A self-consistent model of the jovian auroral thermal structure. J. Geophys. Res. **106**, 12,933–12,952 (2001).

Grodent, D., Gérard, J.-C., Cowley, S., Bunce, E., Clarke J.T.: The global morphology of Saturn's southern ultraviolet aurora. J. Geophys. Res. **110**, A07215, doi:10.1029/2004JA010983 (2005).

Gurnett, D.A.: The Earth as a radio source: Terrestrial kilometric radiation. J. Geophys. Res. **79**, 4227–4238 (1974).

Gurnett, D.A., Anderson, R.R.: The kilometric radio emission spectrum: Relationship to auroral acceleration processes. In Akasofu, S.-I., Kan, J.R. (eds.) Physics of Auroral Arc Formation, **25**, pp. 341–350. Geophysical Monograph Series, American Geophysical Union (1981).

Gurnett, D.A., Kurth, W.S., Hospodarsky, G.B., Persoon, A.M., Zarka, P., Lecacheux, A., Bolton, S.J., Desch, M.D., Farrell, W.M., Kaiser, M.L., Ladreiter, H.-P., Rucker, H.O., Galopeau, P., Louarn, P., Young, D.T., Pryor, W.R., Dougherty, M.K.: Control of Jupiter's radio emission and aurorae by the solar wind. Nature **415**, 985–987. (2002).

Gurnett, D.A. et al.: The Cassini Radio and Plasma Wave Science Investigation. Space Sci. Rev. **114**, 395–463 (2004).

Gurnett, D.A., Kurth, W.S., Hospodarsky, G.B., Persoon, A.M., Averkamp, T.F., Cecconi, B., Lecacheux, A., Zarka, P., Canu, P., Cornilleau-Wehrlin, N., Galopeau, P., Roux, A., Harvey, C., Louarn, P., Bostrom, R., Gustafsson, G., Wahlund, J.-E., Desch, M.D., Farrell, W.M., Kaiser, M.L., Goetz, K., Kellogg, P.J., Fischer, G.,

Ladreiter, H.-P., Rucker, H., Alleyne, H., Pedersen, A.: Radio and plasma wave observations at Saturn from Cassini's approach and first orbit. Science **307**, 1255–1259, doi: 10.1126/science.1105356 (2005).

Gurnett, D.A., Persoon, A.M., Kurth, W.S., Groene, J.B., Averkamp, T.F., Dougherty, M.K., Southwood, D.J.: The variable rotation period of the inner region of Saturn's plasma disk. Science **316**, 5823, 442–445, doi:10.1016/science.1138562 (2007).

Gustin, J., Feldman, P.D., Gérard, J.-C., Grodent, D., Vidal-Madjar, A., Ben Jaffel, L., Desert, J.-M., Moos, H.W., Sahnow, D.J., Weaver, H.A., Wolven, B.C., Ajello, J.M., Waite, J.H., Roueff, E., Abgrall, H.: Jovian auroral spectroscopy with FUSE: Analysis of self absorption and implications for electron precipitation. Icarus **171**, 336–355 (2004).

Gustin, J., Gérard, J.C., Pryor, W., Feldman, P.D., Grodent, D., Holsclaw, G.: Characteristics of Saturn's polar atmosphere and auroral electrons derived from HST/STIS, FUSE and Cassini/UVIS spectra. Icarus **200**, 176–187, doi:10.1016/j.icarus.2008.11.013 (2009).

Ingersoll, A.P., Vasavada, A.R., Little, B., Anger, C.D., Bolton, S.J., Alexander, C., Klaasen, K.P., Tobiska, W.K.: Imaging Jupiter's aurora at visible wavelengths. Icarus **135**, 251–264 (1998).

Isbell, J., Dessler, A.J., Waite, J.H., Jr.: Magnetospheric energization by interaction between planetary spin and the solar wind. J. Geophys. Res. **89**, 10,716 (1984).

Jackman, C.M, Cowley, S.W.H.: A model of the plasma flow and current in Saturn's polar ionosphere under conditions of strong Dungey cycle driving. Ann. Geophys. **24**, 1029–1055 (2006).

Jackman, C.M., Achilleos, N., Bunce, E.J., Cowley, S.W.H., Dougherty, M.K., Jones, G.H., Milan, S.E., Smith, E.J.: Interplanetary magnetic field at ∼9 AU during the declining phase of the solar cycle and its implications for Saturn's magnetospheric dynamics. J. Geophys. Res. **109**, A11203, doi:10.1029/2004JA010614 (2004).

Jackman, C.M., Russell, C.T., Southwood, D.J., Arridge, C.S., Achilleos, N., Dougherty M.K.: Strong rapid dipolarizations in Saturn's magnetotail: In situ evidence of reconnection. Geophys. Res. Lett. **34**, L11203, doi:10.1029/2007GL029764 (2007).

Jackman, C.M. et al.: A multi-instrument view of tail reconnection at Saturn. J. Geophys. Res. **113**, A11213, doi:10.1029/2008JA013592 (2008).

Kaiser, M.L., Desch, M.D., Warwick, J.W., Pearce, J.B.: Voyager detection of nonthermal radio emission from Saturn. Science **209**, 1238–1240 (1980).

Khurana, K.K., Arridge, C.S, Dougherty, M.K.: A versatile model of Saturn's magnetospheric field. Geophys. Res. Abstract, **7**, 05970 (2005).

Klumpar, D.M.: Near equatorial signatures of dynamic auroral processes. In Physics of Space Plasmas, SPI Conf. Proc. Reprint Ser. **9**, p. 265. Scientific Publishers, Inc., Cambridge, Massachusetts (1990).

Knight, S.: Parallel electric fields. Planet. Space Sci. **21**, 741–750 (1973).

Kurth, W.S. et al.: An Earth-like correspondence between Saturn's ultraviolet auroral features and radio emission. Nature **433**, 722–725 (2005a).

Kurth, W.S., Hospodarsky, G.B., Gurnett, D.A., Cecconi, B., Louarn, P., Lecacheux, A., Zarka, P., Rucker, H.O., Boudjada, M., Kaiser, M.L.: High spectral and temporal resolution observations of Saturn kilometric radiation. Geophys. Res. Lett. **32**, L20S07, doi:10.1029/2005GL022648 (2005b).

Kurth, W.S., Lecacheux, A., Averkamp, T.F., Groene, J.B., Gurnett, D.A.: A Saturnian longitude system based on a variable kilometric radiation period. Geophys. Res. Lett. **34**, L02201, doi:10.1029/2006GL028336 (2007).

Kurth, W.S., Averkamp, T.F., Gurnett, D.A., Groene, J.B., Lecacheux A.: An update to a Saturnian longitude system based on kilometric radio emissions. J. Geophys. Res. **113**, A05222, doi:10.1029/2007JA012861 (2008).

Lamy, L.: Study of Saturn auroral radio emissions, modeling and UV aurorae, PhD thesis, Université Pierre et Marie Curie (2008a).

Lamy, L., Zarka, P., Cecconi, B., Prange, R., Kurth, W.S., Gurnett, D.A.: Saturn kilometric radiation: Average and statistical properties. J. Geophys. Res. **113**, A07201, doi:10.1029/2007JA012900 (2008b).

Lamy, L., Zarka, P., Cecconi, B., Hess, S., Prangé, R.: Modeling of Saturn kilometric radiation arcs and equatorial shadow zone. J. Geophys. Res. **113**, A10213, doi:10.1029/2008JA013464 (2008c).

Lamy, L., Cecconi, B., Prangé, R., Zarka, P., Nichols, J., Clarke, J.: An auroral oval at the footprint of Saturn's kilometric radiosources, colocated with the UV aurorae. J. Geophys. Res. in press (2009).

Lecacheux, A., Galopeau, P., Aubier M.: Re-visiting Saturnian radiation with Ulysses/URAP. In Rucker, H.O., Bauer, S.J., Lecacheux, A. (eds.) Planetary Radio Emissions IV, pp. 313–325. Austrian Academy of Sciences Press, Vienna (1997).

Louarn, P., Kurth, W.S., Gurnett, D.A., Hospodarsky, G.B., Persoon, A.M., Cecconi, B., Lecacheux, A., Zarka, P., Canu, P., Roux, A., Rucker, H.O., Farrell, W.L., Kaiser, M.L., Andre, N., Harvey, C., Blanc, M.: Observation of similar radio signatures at Saturn and Jupiter: Implications for the magnetospheric dynamics. Geophys. Res. Lett. **34**, L20113, doi:10.1029/2007GL030368 (2007).

Main, D.S., Newman, D.L., Ergun, R.E.: Double layers and ion phase-space holes in the auroral upward-current region. Phys. Rev. Lett. **97**, 185001, doi: 10.1103/PhysRevLett.97.185001 (2006).

Marklund, G.T., Ivchenko, N., Karlsson, T., Fazakerley, A., Dunlop, M., Lindqvist, P.-A., Buchert, S., Owen, C., Taylor, M., Vaivalds, A., Carter, P., André, M., Balogh, A.: Temporal evolution of the electric field accelerating electrons away from the auroral ionosphere. Nature **414**, 724–727 (2001).

Mauk, B.H., Saur, J.: Equatorial electron beams and auroral structuring at Jupiter. J. Geophys. Res. **112**, A10221, doi:10.1029/2007JA012370 (2007).

McGrath, M.A., Clarke, J.T.: HI Ly-α emission from Saturn (1980–1990). J. Geophys. Res. **97**, 13,691 (1992).

Melin, H., Miller, S., Stallard, T., Trafton, L.M., Geballe, T.R.: Variability in the H_3^+ emission of Saturn: Consequences for ionisation rates and temperature. Icarus **186**, 234–241 (2007).

Menietti, J.D., Mutel, R.L., Santolik, O., Scudder, J.D., Christopher, I.W., Cook, J.M.: Striated drifting auroral kilometric radiation bursts: Possible stimulation by upward traveling EMIC waves. J. Geophys. Res. **111**, A04214, doi:10.1029/2005JA011339 (2006).

Menietti, J.D., Groene, J.B., Averkamp, T.F., Hospodarsky, G.B., Kurth, W.S., Gurnett, D.A., Zarka, P.: Influence of Saturnian moons on Saturn kilometric radiation. J. Geophys. Res. **112**, A08211, doi:10.1029/2007JA012331 (2007).

Milan, S.E., Lester, M., Cowley, S.W.H., Brittnacher, M.: Dayside convection and auroral morphology during an interval of northward interplanetary magnetic field. Ann. Geophysicae 18, 436–447 (2000).

Milan, S.E., Bunce, E.J., Cowley, S.W.H., Jackman, C.M.: Implications of rapid planetary rotation for the Dungey magnetotail of Saturn, J. Geophys. Res. **110**, A03209, doi:10.1029/2004JA010716 (2005).

Mitchell, D.G., Kurth, W.S., Hospodarsky, G.B., Krupp, N., Saur, J., Mauk, B.H., Carbary, J.F., Krimigis, S.M., Dougherty, M.K.: Ion conics and electron beams associated with auroral processes on Saturn. J. Geophys. Res. **114**, A02212, doi:10.1029/2008ja013621 (2009a).

Mitchell, D.G., Krimigis, S.M., Paranicas, C., Brandt, P.C., Carbary, J.F., Roelof, E.C., Kurth, W.S., Gurnett, D.A., Clarke, J.T., Nichols, J.D., Gerard, J.-C., Grodent, D.C., Dougherty, M.K.: Recurrent energization of plasma in the midnight-to-dawn quadrant of Saturn's magnetosphere, and its relationship to Auroral UV and radio emissions. Planet. Space Sci. in press doi: 10.1016/j.pss.2009.04.22 (2009b).

Moses, J.I., Bézard, B., Lellouch, E., Feuchtgruber, H., Gladstone, G.R., Allen, M.: Photochemistry of Saturn's atmosphere I. Hydrocarbon chemistry and comparisons with ISO observations. Icarus, **143**, 244–298 (2000).

Müller-Wodarg, I.C.F., Mendillo, M., Yelle, R.V, Aylward, A.D.: A global circulation model of Saturn's thermosphere. Icarus **180**, 147–160 (2006).

Muschietti, L., Ergun, R.E., Roth, I., Carlson, C.W.: Phase-space electron holes along magnetic field lines. Geophys. Res. Lett. **26**, 1093–1096 (1999).

Mutel, R.L., Menietti, J.D., Christopher, I.W., Gurnett, D.A., Cook, J.M.: Striated auroral kilometric radiation emission: A remote tracer of ion solitary structures. J. Geophys. Res. **111**, A10203, doi:10.1029/2006JA011660 (2006).

Mutel, R.L., Christopher, I.W., Pickett, J.S.: Cluster multi-spacecraft observations of AKR angular beaming. Geophys. Res. Lett. **35**, L07104, doi:10.1029/2008GL033377 (2008).

Nichols, J.D., Cowley, S.W.H., McComas, D.J.: Magnetopause reconnection rate estimates for Jupiter's magnetosphere based on interplanetary measurements at ∼5AU. Ann. Geophys. **24**, 393–406 (2006).

Nichols, J.D., Bunce, E.J., Clarke, J.T., Cowley, S.W.H., Crary, F.J., Dougherty, M.K., Gerard, J.-C., Grodent, D., Pryor, W.R., Rymer, A.M.: Response of Jupiter's UV auroras to interplanetary conditions as observed by the Hubble Space Telescope during the Cassini fly-by campaign. J. Geophys. Res. **112**, A02203, doi:10.1029/2006JA012005 (2007).

Nichols, J.D., Clarke, J.T., Cowley, S.W.H., Duval, J., Farmer, A.J., Gérard, J.-C., Grodent, D., Wannawichian, S.: Oscillation of Saturn's southern auroral oval, J. Geophys. Res. **113**, A11205, doi:10.1029/2008JA013444 (2008).

Pontius, D.H., Hill, T.W.: Enceladus: A significant plasma source for Saturn's magnetosphere. J. Geophys. Res. **111**, A09214, doi:10.1029/2006JA011674 (2006).

Porco, C.C. et al.: Cassini imaging science: Instrument characteristics and anticipated scientific investigations at Saturn. Space Sci. Rev. **115**, 363–497 (2004).

Pottelette, R., Treumann, R.A., Berthomier, M.: Auroral plasma turbulence and the cause of auroral kilometric radiation fine structure. J. Geophys. Res. **106**, 8465–8476 (2001).

Provan, G., Cowley, S.W.H., Nichols, J.D.: Phase relation of oscillations near the planetary period of Saturn's auroral oval and the equatorial magnetospheric magnetic field. J. Geophys. Res. **114**, A04205, doi:10.1029/2008JA013988 (2009).

Pryor, W.R., Stewart, A.I.F., Esposito, L.W., Shemansky, D.E., Ajello, J.M., West, R.A., Jouchoux, A.J., Hansen, C.J., McClintock, W.E., Colwell, J.E., Tsurutani, B.T., Krupp, N., Crary, F.J., Young, D.T., Kurth, W.S., Gurnett, D.A., Dougherty, M.K., Clarke, J.T., Waite, J.H., Grodent D.: Cassini UVIS observations of Jupiter's auroral variability. Icarus **178**, 312–326 (2005).

Radioti, A., Grodent, D., Gérard, J.C., Roussos, E., Paranicas, C., Bonfond B., Mitchell, D G., Krupp, N., Krimigis, S., Clarke J.T.: Transient auroral features at Saturn: Signatures of energetic particle injections in the magnetosphere. J. Geophys. Res. **114**, A03210, doi:10.1029/2008JA013632 (2009).

Ray, L.C., Su, Y.-J., Ergun, R.E., Delamere, P.A., Bagenal, F.: Current-voltage relation of a centrifugally confined plasma. J. Geophys. Res. **114**, A04214, doi:10.1029/2008JA013969 (2009).

Richardson, J.D.: Thermal ions at Saturn: Plasma parameters and implications. J. Geophys. Res. **91**, 1381 (1986).

Richardson, J.D.: An extended plasma model for Saturn. Geophys. Res. Lett. **22**, 1177 (1995).

Richardson, J.D., Sittler, E.C., Jr.: A plasma density model for Saturn based on Voyager observations. J. Geophys. Res. **95**, 12,019 (1990).

Russell, C.T., Jackman, C.M., Wei, H.Y., Bertucci, C., Dougherty, M.K.: Titan's influence on Saturnian substorm occurrence. Geophys. Res. Lett. **35**, L12105, doi:10.1029/2008GL034080 (2008).

Sandel, B.R., Broadfoot, A.: Morphology of Saturn's Aurora. Nature **292**, 679–682 (1981).

Sandel, B.R., Shemansky, D.E., Broadfoot, A.L., Holberg, J.B., Smith, G.R.: Extreme ultraviolet observations from the Voyager 2 encounter with Saturn. Science **215**, 548 (1982).

Saur, J., Mauk, B.H., Mitchell, D.G., Krupp, N., Khurana, K.K., Livi, S., Krimigis, S.M., Newell, P.T., Williams, D.J., Brandt, P.C., Lagg, A., Roussos, E., Dougherty, M.K.: Anti-planetward auroral electron beams at Saturn. Nature, **439**, 699–702, doi:10.1038/nature04401 (2006).

Shemansky, D.E., Ajello, J.M.: The Saturn spectrum in the EUV: Electron excited Hydrogen. J. Geophys. Res. **88**, 459 (1983).

Sittler, E.J., Blanc, M., Richardson, J.: Proposed model for Saturn's auroral response to the solar wind: Centrifugal instability model. J. Geophys. Res. **111**, A06208, doi:10.1029/2005JA011191 (2006).

Stallard, T.S., Miller, S., Cowley, S.W.H., Bunce, E.J.: Jupiter's polar ionospheric flows: Measured intensity and velocity variations poleward of the main auroral oval. Geophys. Res. Lett. **30**, 1221, doi:10.1029/2002GL016031 (2003).

Stallard, T., Miller, S., Trafton, L.M., Geballe, T.R., Joseph, R.D.: Ion winds in Saturn's southern auroral/polar region. Icarus **167**, 204–211 (2004).

Stallard, T. et al.: Saturn's auroral/polar H_3^+ infrared emission I: General morphology and ion velocity structure. Icarus **189**, 1–13 (2007a).

Stallard, T. et al.: Saturn's auroral/polar H_3^+ infrared emission II: A comparison with plasma flow models, Icarus **191**, 678–690 (2007b).

Stallard, T., Lystrup, M., Miller, S.: Emission-line imaging of Saturn's H_3^+ aurora. Astrophys. J. **675**, L117 (2008a).

Stallard, T., Miller, S., Lystrup, M., Achilleos, N., Arridge, C., Dougherty, M.: Dusk-brightening event in Saturn's H_3^+ Aurora. Astrophys. J. **673**, L203–L206 (2008b).

Stallard, T. et al.: Jovian-like aurorae on Saturn. Nature **453**, 1083–1085 (2008c).

Stallard, T. et al.: Complex structure within Saturn's infrared aurora. Nature **456**, 214–217 (2008d).

Talboys, D.L., Arridge, C.S., Bunce, E.J., Coates, A.J., Cowley, S.W.H., Dougherty, M.K.: Characterisation of auroral current systems in Saturn's magnetosphere: High-latitude Cassini observations. J. Geophys. Res. **114**, A06220, doi:10.1029/2008JA013846 (2009).

Trauger, J.T. et al.: Saturn's far-ultraviolet hydrogen aurora, imaging observations from the Hubble Space Telescope. J. Geophys. Res. **103**, E9, 20,237 (1998).

Vasavada, A.R., Bouchez, A.H., Ingersoll, A.P., Little, B., Anger, C.D.: The Galileo SSI team: Jupiter's visible aurora and Io footprint. J. Geophys. Res. **104**, 27,133–27,142 (1999).

Vasyliunas, V.M.: Plasma distribution and flow. In Dessler A.J. (ed.) Physics of the Jovian Magnetosphere, p. 395. Cambridge University Press, Cambridge (1983).

Wang, Z., Gurnett, D.A., Fischer, G., Ye, S.-Y., Kurth, W.S., Mitchell, D.G., Leisner, J.S., Russell, C.T.: Cassini observations of narrow-band radio emissions in Saturn's magnetosphere, J. Geophys. Res. submitted (2009).

Wannawichian, S., Clarke, J.T., Pontius, D.H.: Interaction evidence between Enceladus' atmosphere and Saturn's magnetosphere. J. Geophys. Res. **113**, A07217, doi:10.1029/2007JA012899 (2008).

Warwick, J. et al.: Planetary radio astronomy observations from Voyager 1 near Saturn, Science **212**, 239–243 (1981).

Wilson, R.J., Tokar, R.L., , Henderson, M.G., Hill, T.W., Thomsen, M.F., Pontius, Jr., D.H.: Cassini plasma spectrometer thermal ion measurements in Saturn's inner magnetosphere. J. Geophys. Res. **113**, A12218, doi:10.1029/2008JA013486 (2008).

Wu, C.S., Lee, L.C.: A theory of the terrestrial kilometric radiation. Astrophys. J. **230**, 621–626 (1979).

Ye, S.-Y., Gurnett, D.A., Fischer, G., Cecconi, B., Menietti, J.D., Kurth, W.S., Wang, Z., Hospodarsky, G.B., Zarka, P., Lecacheux, A.: Source locations of narrowband radio emissions detected at Saturn. J. Geophys. Res. **114**, A06219, doi: 10.1029/2008ja013855 (2009).

Zarka, P.: Auroral radio emissions at the outer planets: Observations and theories. J. Geophys. Res. **103**, 20,159–20,194 (1998).

Zarka, P., Lamy, L., Cecconi, B., Prange, R., Rucker, H.: Modulation of Saturn's radio clock by solar wind speed. Nature **450**, 265–267, doi:10.1038/nature06237 (2007).

Chapter 13
The Structure of Saturn's Rings

J.E. Colwell, P.D. Nicholson, M.S. Tiscareno, C.D. Murray, R.G. French, and E.A. Marouf

Abstract Our understanding of the structure of Saturn's rings has evolved steadily since their discovery by Galileo Galilei in 1610. With each advance in observations of the rings over the last four centuries, new structure has been revealed, starting with the recognition that the rings are a disk by Huygens in 1656 through discoveries of the broad organization of the main rings and their constituent gaps and ringlets to Cassini observations that indirectly reveal individual clumps of particles tens of meters in size. The variety of structure is as broad as the range in scales. The main rings have distinct characteristics on a spatial scale of 10^4 km that suggest dramatically different evolution and perhaps even different origins. On smaller scales, the A and C ring and Cassini Division are punctuated by gaps from tens to hundreds of kilometer across, while the B ring is littered with unexplained variations in optical depth on similar scales. Moons are intimately involved with much of the structure in the rings. The outer edges of the A and B rings are shepherded and sculpted by resonances with the Janus–Epimetheus coorbitals and Mimas, respectively. Density waves at the locations of orbital resonances with nearby and embedded moons make up the majority of large-scale features in the A ring. Moons orbiting within the Encke and Keeler gaps in the A ring create those gaps and produce wakes in the nearby ring. Other gaps and wave-like features await explanation. The largest ring particles, while not massive enough to clear a gap, produce localized propeller-shaped disturbances hundreds of meters long. Particles throughout the A and B rings cluster into strands or self-gravity wakes tens of meters across that are in equilibrium between gravitational accretion and Keplerian shear. In the peaks of strong density waves particles pile together in a cosmic traffic jam that results in kilometer-long strands that may be larger versions of self-gravity wakes. The F ring is a showcase of accretion and disruption at the edges of Saturn's Roche zone. Clumps and strands form and are disrupted as they encounter each other and are perturbed by close encounters with nearby Prometheus. The menagerie of structures in the rings reveals a system that is dynamic and evolving on timescales ranging from days to tens or hundreds of millions of years. The architecture of the rings thus provides insight to the origin as well as the long and short-term evolution of the rings.

13.1 Grand Structure of the Rings

The number of distinct rings and ring features has long surpassed the alphabetical nomenclature assigned to the rings prior to spacecraft investigations of the Saturn system. In many cases these lettered rings contain multiple ringlets within them. Nevertheless, they can be broadly grouped into two categories: dense rings (A, B, C) and tenuous rings (D, E, G). The Cassini Division, a ring region in its own right that separates the A and B rings (Fig. 13.1), resembles the C ring in many ways. The subjects of this chapter are the dense rings as well as the Cassini Division and F ring. The Roche Division, separating the A and F rings, contains tenuous material more like the D and G rings (see Chapter 16). While the dense rings covered in this chapter contain various tenuous (and likely dusty[1]) rings within them, these are covered

---

J.E. Colwell (✉)
Department of Physics, University of Central Florida, Orlando, FL 32816-2385, USA

P.D. Nicholson and M.S. Tiscareno
Department of Astronomy, Cornell University, Ithaca, NY 14853, USA

C.D. Murray
Astronomy Unit, Queen Mary, University of London, Miles End Road, London E1 4NS, UK

R.G. French
Department of Astronomy, Wellesley College, Wellesley, MA 02481, USA

E.A. Marouf
Department of Electrical Engineering, San Jose State University, San Jose, CA 95192, USA

---

[1] We use the term "dusty" to refer to particle sizes less than ∼100 μm, and not composition. All of Saturn's rings are composed primarily of water ice (Chapter 15).

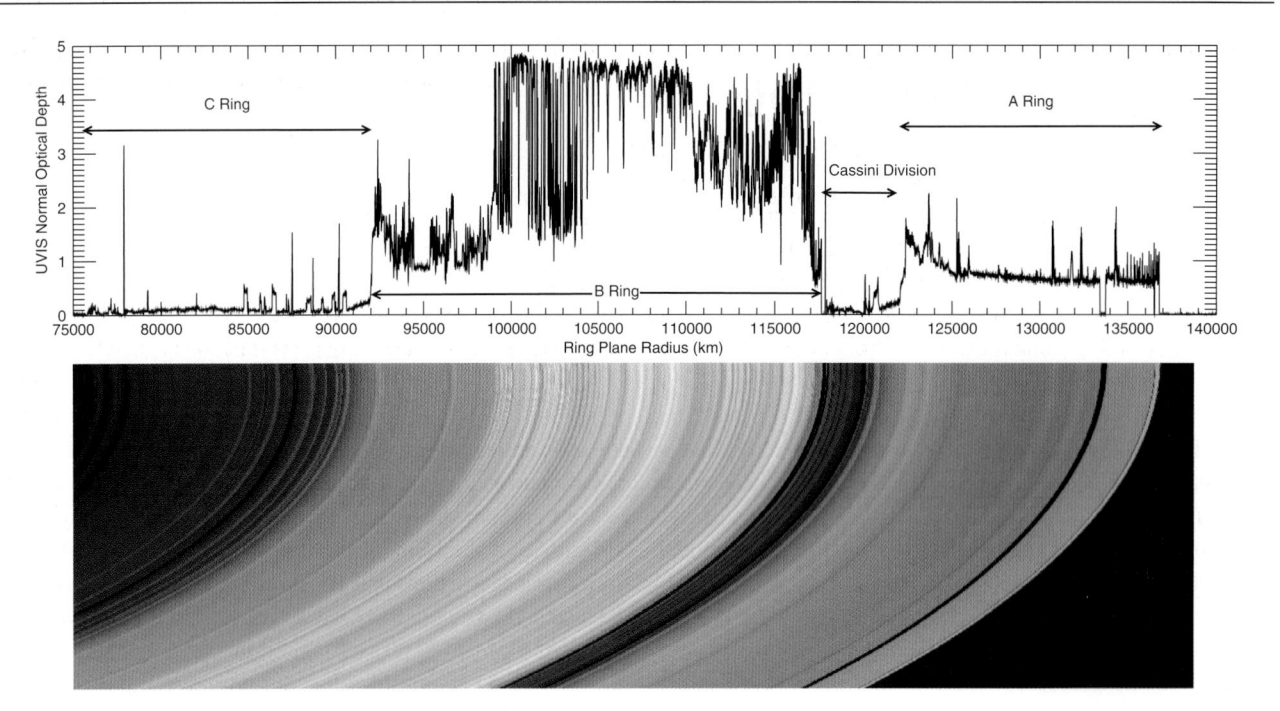

Fig. 13.1 ISS mosaic (NASA/JPL/Space Science Institute) and UVIS stellar occultation data showing the A, B, C rings and Cassini Division (CD) at approximately 10 km radial resolution. The F ring is visible in the ISS mosaic beyond the outer edge of the A ring. The images were taken from an elevation of 4° above the illuminated (southern) face of the rings, so optically thick regions appear brighter than optically thin regions

in Chapter 16. Overall the dense rings have typical optical depths greater than 0.1 and are predominantly comprised of particles larger than 1 cm while the dusty rings have optical depths of 10^{-3} and lower. The F ring, torn quite literally between the regime of moons and rings, has a dense and complex core embedded in a broad sheet of dust, and is covered in this chapter.

The study of structure within Saturn's rings originated with G. Campani, who observed in 1664 that the inner half of the disk was brighter than the outer half. More famously, G. D. Cassini discovered in 1675 a dark band between the brighter and dimmer halves, which he interpreted as a gap. Careful observations by W. Herschel confirmed in 1791 that Cassini's Division appears identical on both sides of the rings, convincing Herschel that it is in fact a gap and not just a dark marking (Alexander 1962; Van Helden 1984). In 1837, J. F. Encke observed that the middle part of the A ring appears dimmer than the inner and outer parts, but not until 1888 did J. E. Keeler clearly see what is now called the Encke Gap, narrow and sharp near the A ring's outer edge (Osterbrook and Cruikshank 1983). By that time astronomers had noted a number of other markings on the rings, though they did not agree on their locations or their nature. Meanwhile the C ring, also called the "crepe" ring, was discovered in 1850 by W. C. Bond and G. P. Bond, and independently by W. R. Dawes.

The mystery of the rings' nature was finally resolved in 1859, when J. C. Maxwell proved that they could not be solid or liquid but were instead made up of an indefinite number of small particles, each on its own orbit about Saturn (Maxwell 1859). In 1866, D. Kirkwood was the first to suggest that the Cassini Division is caused by a resonance with one of Saturn's moons (Kirkwood 1866), an idea that would later be taken up by P. Goldreich and S. Tremaine in applying the theory of Lindblad resonances in spiral galaxies to describe spiral waves in the rings (Goldreich and Tremaine 1978b). Our knowledge of ring structure was thoroughly revolutionized by the Pioneer (1979) and Voyager (1980, 1981) encounters with Saturn (Gehrels and Matthews 1984; Greenberg and Brahic 1984). Three new rings were discovered (D, F and G) as well as several new ring-shepherding moons, and detailed ring structure was revealed for the first time – by images taken at close range, by stellar occultation (observing the flickering of a star as it passes behind the rings) and by radio occultation (measuring the attenuation of the spacecraft's radio signal as it passes behind the rings as seen from Earth). Esposito et al. (1987) catalogued 216 features based on the Voyager 2 δ Sco stellar occulation. Subsequently, ring structure was probed by widespread Earth-based observation of a stellar occultation in 1989 (Hubbard et al. 1993; Nicholson et al. 2000; French and Nicholson 2000) and finally by the arrival of the Cassini orbiter in 2004.

Various structural features repeat across the ring system. Sharp-edged gaps are present in the C and A rings and Cassini Division. Waves launched by resonances with satellites are seen throughout the main ring system (Chapter 14).

The inner edges of the A and B rings are remarkably similar morphologically, a fact that has been attributed to redistribution of material at edges through ballistic transport of meteoroid ejecta (Durisen et al. 1992). Perhaps what the rings have most in common with each other, however, is the abundance of unexplained structure. Large and abrupt fluctuations in optical depth are evident in stellar and radio occultations by the rings. The most striking examples of these are in the central B ring where opaque regions ($\tau > 5$) are punctuated by valleys of moderate optical depth ($\tau < 2$). The outer C ring has numerous plateaus of $\tau = 0.4$ embedded in a background with optical depth $\tau \sim 0.1$, whose origin is still unknown. The inner A ring exhibits some unexplained structure similar to that in the B ring. Sharp gap edges in the A ring, C ring, and Cassini Division exhibit unexpectedly complex structure. A clumpy microstructure dubbed "self-gravity wakes" is beginning to be understood in the A and B rings. The A ring contains unseen embedded moonlets that reveal their presence through the "propeller"-shaped structures that form around them (Tiscareno et al. 2006a), as well as several narrow radial bands of structure that appears "ropy" and "straw"-like in images (Porco et al. 2005). Here we identify a moonlet as an individual object that opens an azimuthally limited gap but, unlike the embedded moons Pan and Daphnis, does not clear a continuous gap in the ring. It is not yet clear whether or not these moonlets simply represent the largest members of the general particle size distribution in the rings. The outer edges of the B and A rings can be understood in the context of resonant gravitational perturbations from the moons Mimas and the coorbitals Janus and Epimetheus, respectively.

We will examine the structure of the rings on all scales, organized by ring region. Although certain structural features are common to many or all of the ring regions, each main ring has distinct characteristics that showcase particular features. We begin with the A ring and its retinue of waves and wakes, followed by the opaque, massive B ring which is characterized by large-scale structure of unknown origin. We describe the C ring and Cassini Division jointly. Situated on either side of the B ring, these more tenuous regions of the main rings exhibit many structural features in common and are also similar in composition and particle size distribution (Chapter 15). The F ring, though perhaps the least massive of the main ring regions, has a variety of unique features due to its location in the outer reaches of Saturn's Roche zone and strong gravitational perturbations from the moons Prometheus and Pandora. Observations from Cassini have shown us Saturn's rings in unprecedented detail, and significant progress has been made in understanding some aspects of the large-scale and small-scale structure of the rings. Many features remain a puzzle, however. Further observations from Cassini as well as theoretical and numerical studies of particle disks should help us understand this complex and beautiful system.

13.2 A Ring

The most prominent features of Saturn's A ring (Fig. 13.2) are the multitude of density waves launched at Lindblad resonances with nearby moons (mostly Prometheus and Pandora) and the two gaps in the outer A ring cleared by the moons Pan (Encke gap) and Daphnis (Keeler gap). The outer edge of the A ring is located at the 7:6 inner Lindblad resonance (ILR) with the coorbital satellites Janus and Epimetheus. The outer region of the A ring, from the Encke gap to the outer edge, is particularly crowded with density waves as resonances with nearby Prometheus and Pandora are tightly spaced there. The central A ring, between the Encke gap and the strong Janus/Epimetheus 4:3 density wave at 125,270 km, is relatively featureless, punctuated by two strong Mimas and Janus/Epimetheus density waves, the Mimas 5:3 bending wave, and several weaker waves. It is in this region that Cassini cameras have observed the telltale propeller-shaped signatures of perturbations from 100-m sized moonlets that are too small to open a full gap in the rings (Tiscareno et al. 2006a, 2008; Sremčević et al. 2007). The inner A ring exhibits higher optical depths than the rest of the ring as well as some fluctuations in density reminiscent of the unexplained structure that permeates the B ring (Section 13.3). The strong Pandora 5:4 density wave sits at the inner edge and has the longest wavetrain of any density wave except the Janus 2:1 wave in the inner B ring. Just inside the abrupt inner edge of the A ring is a step in optical depth and a ramp that transitions to the Cassini Division.

Particles throughout the A ring are caught between the competing tendencies for gravitational clumping and Keplerian shear. As a result, at any given time most of the mass of the ring is in particles that are part of an ephemeral agglomeration of particles rather than individual isolated ring particles. This has important consequences for the interpretation of images and occultation data. These self-gravity wakes are canted 20–25° from the azimuthal direction due to Keplerian shear. This alignment of particles introduces an azimuthal dependence to the apparent optical depth measured in radio and stellar occultations (Colwell et al. 2006, 2007; Hedman et al. 2007), as well as the well-known azimuthal asymmetry in light reflected from the rings. As a result, the concept of "normal optical depth" usually used to describe the rings breaks down. The optical depth observed in an occultation depends not just on the amount of material in the

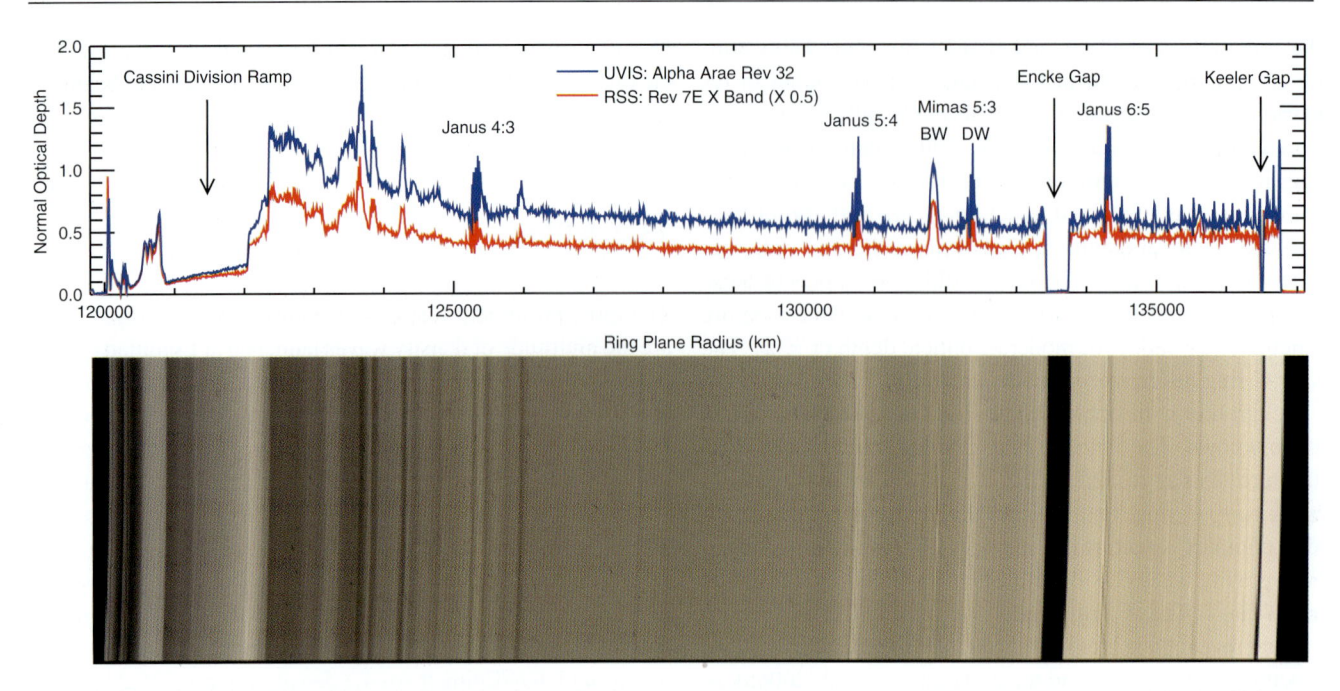

Fig. 13.2 The A ring and outer Cassini Division seen in UVIS and RSS occultation profiles and an ISS image mosaic (NASA/JPL/Space Science Institute) from above the unilluminated face of the rings with a radial resolution of ∼6 km/pixel. The occultation data are shown at 10 km resolution. Both optically thick regions (inner A Ring) and optically thin regions (inner A Ring Ramp) appear dark in the ISS image in this viewing geometry. Differences between the UVIS and RSS optical depths are due to different viewing geometries with respect to the ubiquitous self-gravity wakes

ring and the path length of the occulted beam through the ring, but also on the orientation of that beam relative to the self-gravity wakes. Nevertheless, here we report normal optical depths computed in the usual way from the slant angle, B, of the observed beam relative to the ring plane,

$$\tau_n = \mu \ln\left(\frac{I_0}{I-b}\right) \quad (13.1)$$

where I is the measured intensity of the occulted beam, I_0 is the unocculted intensity, b is any background signal, and $\mu = |\sin(B)|$. While this neglects the effect of self-gravity wakes, it provides a consistent means for presenting occultation profiles. Differences in the absolute value of optical depths between profiles can then be interpreted as due to the different aspect of the wakes with respect to the viewing geometry[2].

---

[2] There can be additional differences in optical depth at different wavelengths depending on the size distribution of the ring particles. Also, radio occultation optical depths are typically a factor of two higher than stellar occultation optical depths because the diffracted signal can be separated from the directly transmitted signal in radio occultations. See also Cuzzi (1985) and Section 13.3 for more details.

13.2.1 Satellite Resonance Features

13.2.1.1 Density and Bending Waves

Spiral density waves are densely packed throughout the A ring because of its proximity to moons and the resulting close-spacing of resonances. Spiral waves are generated in a ring at the locations of resonances with perturbing moons, and they propagate away from the resonance location in a single radial direction. Density waves are radially-propagating compressional waves that arise at locations where a ring particle's radial frequency κ is in resonance with the perturbing moon, while bending waves are transverse waves that arise at locations where a ring particle's vertical frequency ν is in resonance with the perturbing moon. The spiral structure arises because the phase of the wave depends on longitude within the ring. The multiplicity of resonance types (and thus wave types) is due to Saturn's oblateness, which causes ν and κ to differ slightly from the orbital frequency Ω.

Goldreich and Tremaine (1978a, b, 1980) were the first to point out that structure could arise in Saturn's rings due to the same physics that causes galaxies to have spiral arms (Lin and Shu 1964). This insight was proved correct by the Voyager flybys, as a plethora of waves was revealed in high-resolution stellar occultations (Lane et al. 1982; Holberg et al. 1982; Esposito et al. 1983a) and radio occultations

(Marouf et al. 1986). The general theory of resonant interactions within planetary rings was reviewed by Goldreich and Tremaine (1982) and by Shu (1984), and many aspects of the Voyager results were summarized by Cuzzi et al. (1984). Follow-up studies of spiral waves in Saturn's rings were done by Nicholson et al. (1990), who used both Voyager PPS and RSS data to pinpoint the locations of resonant wave generation and thus refine the direction of Saturn's pole, and by Rosen et al. (1991a, b), who wrote extensively on methods of density wave analysis and were the first to fit wave amplitudes in order to derive masses for the perturbing moons. Finally, Spilker et al. (2004) undertook a comprehensive survey of spiral density waves in the Voyager PPS data set.

By far the most common waves in Saturn's rings are spiral density waves driven at inner Lindblad resonances (ILR). These can be described as variations on the background surface density σ_0. Wave dynamics are described in more detail in Chapter 14; here we will discuss waves in the context of data analysis. At orbital radius r greater than the resonance location r_L, the classical linear theory (ignoring pressure effects) gives

$$\Delta\sigma(r) = Re\left\{iA_L e^{-i\phi_0}\left[1 - i\xi e^{-i\xi^2/2}\int_{-\infty}^{\xi} e^{i\eta^2}d\eta\right]\right\} e^{-(\xi/\xi_D)^3}, \tag{13.2}$$

where the dimensionless radial parameter is

$$\xi = \left(\frac{\mathcal{D}_L r_L}{2\pi G \sigma_0}\right)^{1/2} \frac{r - r_L}{r_L}, \tag{13.3}$$

and further terms are defined below. Similar equations exist for density waves generated at outer Lindblad resonances and for bending waves.

Assuming Saturn's gravity is well described as a point mass plus a J_2 harmonic, the factor $\mathcal{D}_L$ is given by Cuzzi et al. (1984) and Marley and Porco (1993)

$$\mathcal{D}_L = 3(m-1)\Omega_L^2 + J_2\left(\frac{R_S}{r_L}\right)^2\left[\frac{21}{2} - \frac{9}{2}(m-1)\right]\Omega_L^2, \tag{13.4}$$

where the second term is a small correction except for $m = 1$.

The Lindblad resonance occurs at r_L. The local mean motion is Ω_L, which must be calculated accounting for the higher-order moments of Saturn's gravity field (Murray and Dermott 1999). For the purposes of calculating spherical harmonics, Saturn's radius $R_S = 60,330$ km by convention (Kliore et al. 1980). The resonance's azimuthal parameter is m, a positive integer that gives the number of spiral arms. The amplitude A_L is related to the mass of the perturbing satellite, the phase φ_0 depends on the longitude of observation and that of the satellite, while the damping constant ξ_D describes the ring's viscous response.

The integral in Eq. 13.2 is a Fresnel integral, which significantly modulates the result near the wave's generation point, but oscillates about unity for higher values of ξ. Downstream, then, the dominant component of Eq. 13.2 has the form of a sinusoid with constantly decreasing wavelength (as well as modulating amplitude), such that the wavenumber, $k = 2\pi/\lambda$, increases linearly with distance from r_L:

$$k(r) = \frac{\mathcal{D}_L}{2\pi G \sigma_0 r_L}(r - r_L). \tag{13.5}$$

Each of the dozens of spiral waves in Saturn's rings serves as an in situ probe, by which local properties of the ring can be obtained. The ideal wave profile (Eq. 13.2) contains four tunable parameters that can be constrained by an observed wave's morphology: r_L, σ_0, A_L, ξ_D. Changing the resonance location r_L amounts to a radial translation of the entire wave. The background surface density σ_0 controls the wavelength dispersion, or the slope with which the wavenumber increases with radius (Eq. 13.5). The wave amplitude A_L is proportional to the perturbing moon's mass. The damping length ξ_D controls the point at which the wave amplitude begins to decay after its initial growth, and is a measurement of the ring's kinematic viscosity (which is dominated by interparticle collisions). A fifth tunable parameter in Eq. 13.2 is the phase φ_0, which can be calculated from the perturbing moon's position on its orbit, though this is usually so well constrained by direct observation of the moon that it is an input for spiral wave analysis rather than a result.

The easiest parameter to obtain from spiral wave data is the surface density. Values inferred by a number of investigators are plotted in Fig. 13.3. The surface mass densities measured for the three innermost waves in the Cassini Division are ~ 1 g cm^{-2} in regions where $\tau \sim 0.1$ (Porco et al. 2005; Tiscareno et al. 2007; Colwell et al. 2008). In contrast, waves in the A ring typically indicate $\sigma_0 \sim 40$ g cm^{-2} where $\tau \sim 0.5$. The larger value of the opacity $\kappa = \tau/\sigma$ in the Cassini Division suggests a different ring particle size distribution (see also Chapter 15). In addition, investigators have obtained viscosity (Esposito et al. 1983a; Tiscareno et al. 2007) and amplitudes (Rosen et al. 1991a, b) from density waves. The kinematic viscosity ν is derived from the damping length using (Tiscareno et al. 2007)

$$\nu = \frac{9}{7\kappa_L \xi_D^3}\left(\frac{r_L}{\mathcal{D}_L}\right)^{1/2}(2\pi G \sigma_0)^{3/2}, \tag{13.6}$$

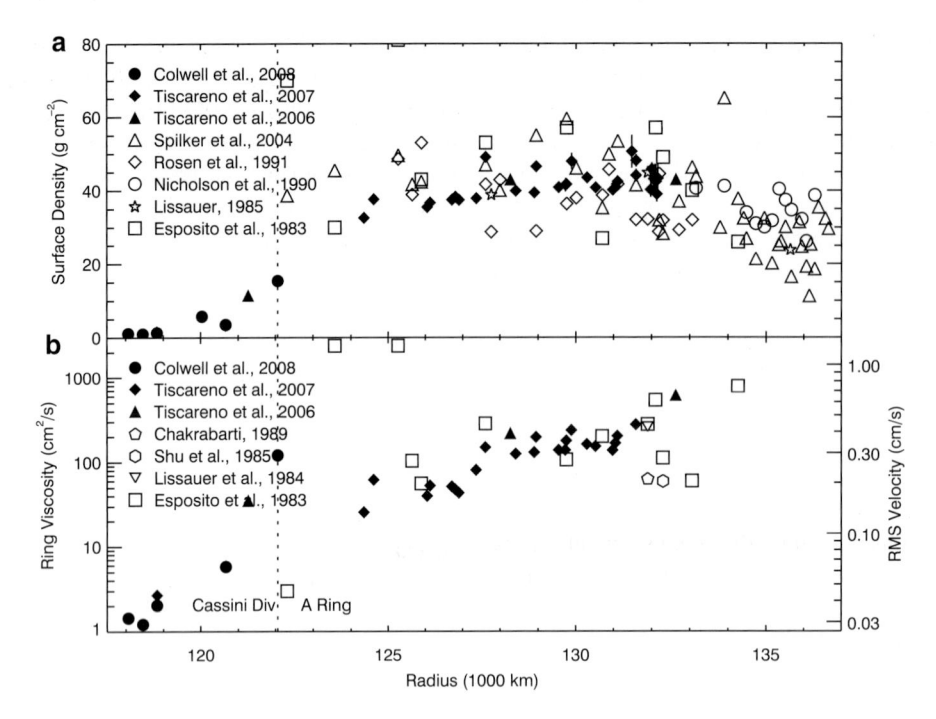

Fig. 13.3 (a) Surface density σ and (b) viscosity and RMS velocity in Saturn's A ring and Cassini Division, as inferred from density waves

where, for this purpose, the radial frequency κ_L is approximately equal to the orbital frequency Ω_L. The viscosity is directly related to the ring particles' rms random velocity c (Araki and Tremaine 1986; Wisdom and Tremaine 1988)

$$\nu = k_1 \frac{c^2}{\Omega}\left(\frac{\tau}{1+\tau^2}\right) + k_2 \Omega D^2 \tau, \quad (13.7)$$

where τ is the local optical depth, D is the particle diameter, and k_1 and k_2 are constants of order unity (see also Chapter 14). When the ring particle interactions are isolated two-particle collisions and the ring particle density is not so high that particle size becomes important, Eq. 13.7 can be approximated by the first term alone with $k_1 = 2$ (e.g. Tiscareno et al. 2007). This assumption is valid for the Cassini Division. While self-gravity wakes complicate the issue in the A ring, the expression is nonetheless useful and provides a reasonable approximation to the velocity dispersion, c.

Under the assumption that random velocities are isotropic, the ring's vertical scale height can be estimated as $H \sim c/\Omega$. However, this assumption is violated in much of the A Ring due to self-gravity wakes (see below), which cause random velocities to be larger within the equatorial plane than in the vertical direction, thus depressing the vertical scale height implied by a given magnitude of rms velocity. An approximate upper limit on the thickness or scale height of the ring, H, can be estimated from the RMS velocities shown in Fig. 13.3 by dividing those values by the value of Ω (which is $1.4 \times 10^{-4}\,\mathrm{s}^{-1}$ at $r = 124{,}500$ km). However, such an estimate is only an upper limit because it assumes that random velocities in the ring are isotropic, when in fact self-gravity wakes cause in-plane random velocities to be higher than out-of-plane velocities. Upper-limit estimates of the thickness H are 3–5 m in the Cassini Division and 10–15 m in the inner A ring (Tiscareno et al. 2007; Colwell et al. 2009).

A fundamental assumption of the classical linear theory (Eq. 13.2) is that the perturbations in density are small compared to the background density. In practice, this assumption is violated for many of the stronger waves in Saturn's rings, including probably all of the waves observed by Voyager. In non-linear waves, density peaks become strong and narrow while troughs become wide and flat, and the wavenumber increase with radius is more quadratic than linear (Shu et al. 1985). This calls into question the application of classical linear wave theory to strong waves.

Some investigators have avoided this difficulty by analyzing waves other than the prevalent non-linear density waves. Bending waves have been used to infer surface density and viscosity in a few locations (Shu et al. 1983; Lissauer et al. 1984, 1985; Chakrabarti 1989), though only five bending waves have been clearly identified in Saturn's rings (Lissauer et al. 1985; Rosen and Lissauer 1988). More recently, Tiscareno et al. (2007) used the greater sensitivity of Cassini imaging to probe weak density waves, including waves driven by the small moons Pan and Atlas and at second-order resonances with the larger moons. Yet another approach is to directly confront the challenge of constructing a comprehensive theory of non-linear density waves that has practical applications (Borderies et al. 1986; Longaretti and Borderies 1986; Rappaport et al. 2008).

The density waves driven by the co-orbital moons Janus and Epimetheus, which "swap orbits" every 4 years, have a particularly irregular morphology that changes with time. Tiscareno et al. (2006b) modeled three weak second-order Janus/Epimetheus waves (thus avoiding the problems of non-linearity) by adding together wave segments from different moon configurations, assuming that each segment ceases being generated but continues to propagate with a finite group velocity when the perturbing moons change their orbits. Their results show that density waves can be used as historical records in the case of a perturbation that changes with time.

13.2.1.2 Outer Edge of the A Ring

A Lindblad resonance will drive a spiral density wave only if sufficient dense ring material is available downstream. The original application of Goldreich and Tremaine (1978b) was that the angular momentum deposited into the ring by the resonance will entirely clear the downstream region of material. This turns out to be true not only for the outer edge of the B Ring, but for several ringlets in the C Ring and Cassini Division and for the outer edge of the A Ring.

The A Ring edge is confined by the 7:6 Lindblad resonance with the co-orbital moons Janus and Epimetheus, and to first order exhibits a 7-lobed shape consistent with streamlines perturbed by the resonance (Porco et al. 1984b). Cassini mapping of the edge is revealing more complex structure, with a time-variable component that may be linked to the changes in Janus' and Epimetheus' orbits (Spitale et al. 2008a).

13.2.2 Satellite Impulse-Driven Features

13.2.2.1 Wavy Edges of the Encke and Keeler Gaps

Unlike at resonant locations, where the effects of many distant encounters with a perturbing moon add up constructively, the orbits of ring particles that pass close by a moon are deflected with just one impulse. Because random velocities in Saturn's rings are generally quite small, one can reasonably equate a particle's semimajor axis a with the radial location at which it passes most closely to the moon, and thus call the impact parameter Δa. The outgoing ring particle (or streamline) now has an eccentricity imparted to it (Dermott 1984),

$$e = \frac{4m}{3M}\left(\frac{a}{\Delta a}\right)^2. \quad (13.8)$$

Due to Keplerian shear, ring particles inward of the moon move faster, and those outward of the moon move slower; the completion of one eccentric orbit corresponds to one sinusoid in the streamline, and the azimuthal wavelength in a frame rotating with the satellite is $3\pi\Delta a$. The canonical example of a gap of this kind is the Encke Gap in the outer A Ring (Fig. 13.4). Its wavy edges were first noticed by Cuzzi and Scargle (1985), who concluded that a moon must exist within the gap (see Section 13.2.2.2 for the continuation of that story). However, more thorough mapping of the Encke Gap edges by Cassini is revealing much more complex structure than the expected $3\pi\Delta a$ sinusoids (Tiscareno et al. 2005; Torrey et al. 2008).

Although the structure of the Encke Gap edges is enigmatic, it is relatively subdued, with amplitudes (∼1 km) a small fraction of the moon-edge distance (160 km). Not so the Keeler Gap, near the A Ring's outer edge, which is only 40 km across with an inner edge that varies by 15 km. These variations were first catalogued by Cooke (1991), who was unable to find a simple model that would fit the data. Again, mapping of the gap edges by Cassini has revealed a very complex Keeler inner edge, characterized to first order by a 32-lobe structure attributable to the Prometheus 32:31 resonance straddled by the edge, but with a great deal of other structure superimposed on top of that, some perhaps due to the nearby Pandora 18:17 resonance (Tiscareno et al. 2005; Torrey et al. 2008).

The outer edge of the Keeler Gap, at least away from Daphnis, is close to circular, punctuated periodically by feathery "wisps" (Porco et al. 2005) that have a sharp trailing edge of ∼1 km amplitude and a gentle slope on the leading side (Torrey et al. 2008). Only the region within a few degrees longitude of Daphnis shows the canonical $3\pi\Delta a$ wavy edges. Weiss et al. (2008) have shown that the usual expression for deriving Daphnis' mass from the edge amplitude does not apply for the narrow Keeler Gap. Numerical models are able to reproduce the irregular shape of the wavy edges surrounding Daphnis which arise in part from the much closer proximity of the moon to the ring edge than in the Encke Gap (Lewis and Stewart 2005, 2006; Perrine and Richardson 2006, 2007).

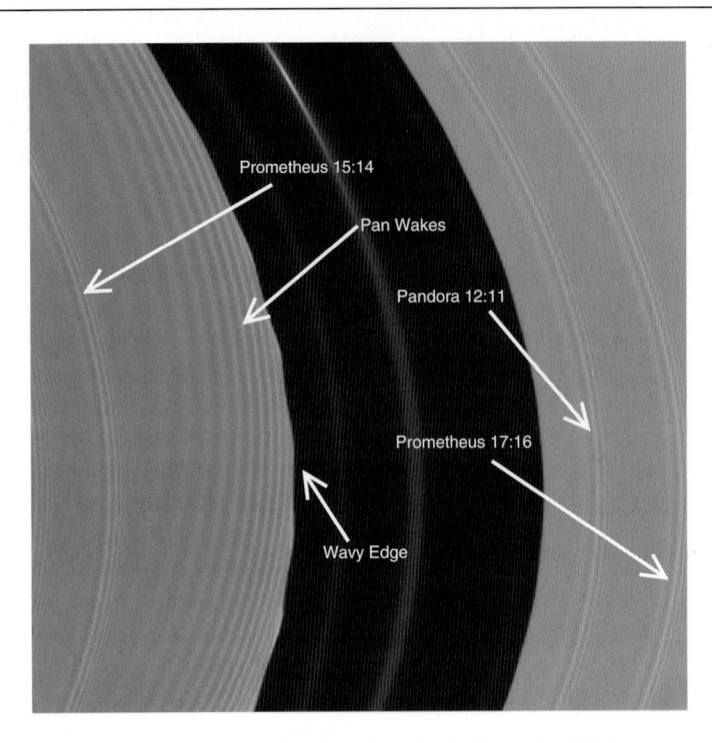

Fig. 13.4 The Encke gap (320 km wide) imaged by Cassini at Saturn Orbit Insertion showing dusty ringlets (Chapter 16), a wavy inner edge recently perturbed by the satellite Pan (roughly five image widths upstream of the inner edge, or up in this view of the south face of the rings), and satellite wakes. Density waves are also visible, indicated here by the inner Lindblad resonances that launch them. Streamlines near the edge of a gap are deflected by the embedded moon, creating a wavy edge and satellite wakes, due to the moon Pan, within the ring (see also Chapter 14) (image: NASA/JPL/Space Science Institute)

13.2.2.2 Satellite Wakes

Not only at gap edges do ring particles have their orbits deflected by close passes by a moon. Particle streamlines continue to be deflected in the interior of the ring, though with decreasing amplitude and increasing wavelength as Δa increases. This results in a pattern of alternating regions of relatively greater or lesser density which remains stationary in the moon's reference frame, thus allowing the satellite to exert a torque on the ring. First observed in Voyager data, these satellite wakes were observed in remarkable detail by Cassini at Saturn Orbit Insertion on July 1 2004 (Fig. 13.4). This pattern does not depend on mutual particle interactions and does not propagate; it is purely a kinematic effect as neighboring streamlines bunch together or spread apart.[3]

After the initial suggestion by Cuzzi and Scargle (1985) that the Encke Gap's wavy edges implied an embedded moon, Showalter et al. (1986) found the radial signature of satellite wakes in stellar and radio occultation scans. Since the radial scan of a satellite wake changes in frequency with the perturbing moon's relative longitude, Showalter et al. (1986) were able to constrain the orbit of the yet-undiscovered moon. With this guidance, Pan was discovered by Showalter (1991) in archival Voyager images.

Frequency-based analysis of radial scans of Pan's wakes (Horn et al. 1996; Tiscareno et al. 2007) shows that damping of the wakes is surprisingly inefficient. Signatures can be detected that imply a relative longitude of more than 1000°, or nearly three Pan encounters ago. By contrast, although satellite wakes are also observed in the vicinity of Daphnis, they do not extend more than a few degrees of longitude or a few tens of km radially from the moon. The proximity of Daphnis to the Keeler Gap edges results in particle streamlines (cf. Showalter 1986) crossing immediately after encountering Daphnis, and the wakes are therefore damped over a much smaller azimuthal extent than the Pan wakes.

13.2.3 Self-Gravity Wakes and Propellers

Another type of wake in the rings is the result of the competing processes of gravitational accretion of ring particles and Keplerian shear. Sometimes called gravity wakes or Julian-Toomre wakes for their similarity to wakes produced by a massive body in a galaxy (Julian and Toomre 1966; Colombo et al. 1976), these features are sheared agglomerates of ring

---

[3] The same applies to wavy gap edges, which should not strictly be called "edge waves."

particles that form due to the mutual gravitational attraction of ring particles. For a ring that is marginally unstable (Toomre's parameter Q = 1, Toomre 1964), the most unstable length scale is

$$\lambda \approx 4\pi^2 G \sigma_0 / \Omega^2 \quad (13.9)$$

where σ_0 is the ring surface mass density and Ω is the orbital frequency. In the central A ring where these wakes are most prominent, this wavelength is about 60 m. In the inner B ring, where they are also observed (see Section 13.3.6), the scale may be somewhat smaller (43 m using $\sigma_0 = 70 \, \text{g cm}^{-2}$ from the Janus 2:1 density wave (Holberg et al. 1982; Esposito et al. 1983a)). No massive seed particle is required for the formation of self-gravity wakes so that, unlike satellite wakes, they are not "wakes" in the usual sense of the word (Colwell et al. 2006). Their cant angle of ∼20–25° to the local orbital flow is a natural consequence of the gradient in orbital velocities in a Keplerian disk (e.g. Salo et al. 2004). N-body simulations produce wakes with properties that are consistent with measurements made by Cassini as well as Voyager imaging observations (e.g. Salo et al. 2004; Porco et al. 2008).

Observations of a pronounced azimuthal brightness asymmetry (Camichel 1958; Ferrin 1975; Reitsema et al. 1976; Lumme and Irvine 1976; Lumme et al. 1977; Thompson et al. 1981; Gehrels and Esposito 1981) first provided the clue to the existence of these structures in the A ring (Colombo et al. 1976; Franklin et al. 1987; Dones and Porco 1989). Direct numerical simulations by Salo (1992, 1995) and by Richardson (1994) demonstrated that self-gravity wakes do arise in a self-gravitating ring. Dones et al. (1993) measured the strength and shape of the azimuthal brightness variations in the A ring from Voyager images, and found a sharp maximum in amplitude near the center of the A ring ($R = 128,000 \, \text{km}$). In an extensive examination of Hubble Space Telescope WFPC2 images taken over a full Saturn season, French et al. (2007) systematically explored the nature of the brightness variations over a broad range of ring opening angles, and compared the observations to models based on a suite of N-body dynamical simulations of self-gravity wakes and a realistic ray-tracing code (Salo and Karjalainen 2003). They found evidence for asymmetry not only in the A ring (Fig. 13.5), but in relatively optically thin regions in the B ring as well (Fig. 13.6; see also Section 13.3.6). Porco et al. (2008) reproduced Voyager and Cassini observations of the azimuthal brightness asymmetry in the A ring, also with a combination of N-body simulations and ray-tracing calculations. Their results place constraints on the dissipation in ring particle collisions, with a smaller coefficient of restitution than previous models (e.g. Bridges et al. 1984; Supulver et al. 1995) needed to match the phase and amplitude of the asymmetry, and 40% greater energy loss in collisions than in French et al. (2007). Evidence for strong azimuthal asymmetry has also been found at radio wavelengths, from radar imaging of the rings (Nicholson et al. 2005) and from observations of Saturn's microwave thermal radiation transmitted through the rings (Dunn et al. 2004, 2005, 2007; Molnar et al. 1999).

The existence of self-gravity wakes was dramatically manifested in Cassini occultation measurements of the rings which showed significant variations in the apparent normal

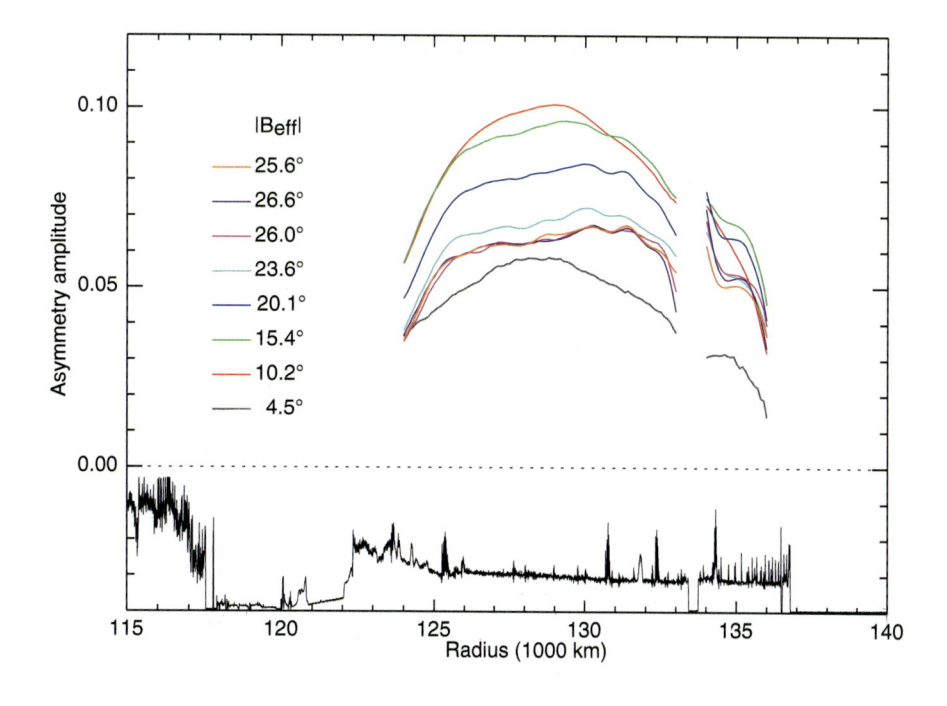

Fig. 13.5 Radial variations in the amplitude of the azimuthal brightness asymmetry in the A ring for a range of ring opening angles $|B_{eff}|$, from Hubble Space Telescope WFPC2 images obtained between 1996–2004 (French et al. 2007). The radial optical depth profile is the Cassini UVIS α Arae Rev 033 occultation

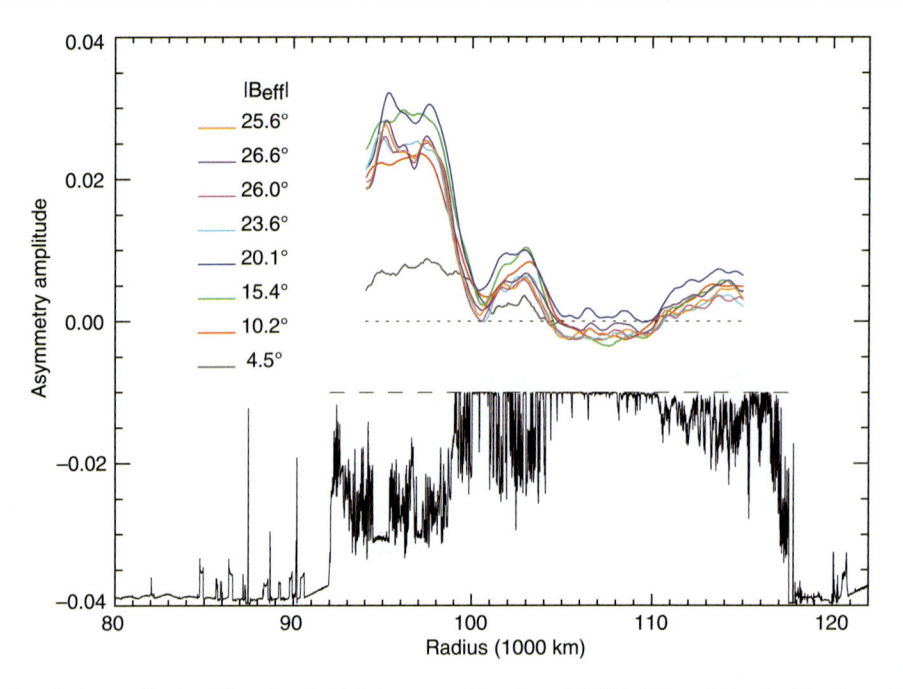

Fig. 13.6 Radial variations in the amplitude of the azimuthal brightness asymmetry in the B ring over a range of ring opening angles $|B_{eff}|$, from Hubble Space Telescope WFPC2 images obtained between 1996–2004 (French et al. 2007). The radial optical depth profile is the Cassini UVIS α Arae Rev 033 occultation

optical depth in the A ring. We use the term "apparent normal optical depth" because we have applied the normal slant path correction to the observed optical depth (Eq. 13.1), but the actual normal optical depth that would be observed from a vantage point normal to the ring plane would generally be higher than that measured by any occultation at a lower value of B. The reason for this is that the rings are not a homogeneous medium of minute absorbers but are rather a nearly planar distribution of large-scale clumps (self-gravity wakes) with a relatively sparse distribution of particles between them. The observed optical depth is determined in large part by the areal fraction of the rings sampled along the occultation line of sight that is covered by the self-gravity wakes. The same phenomenon is also seen in the B ring where the dichotomy between the self-gravity wakes and the gaps between them may be even more pronounced (Section 13.3.6).

The A ring optical depth variations are most pronounced in the central A ring (Fig. 13.7, Colwell et al. 2006; Hedman et al. 2007), also the location of the peak amplitude of the azimuthal brightness asymmetry (Fig. 13.5, Dones et al. 1993). Figure 13.7 shows stellar and radio occultation optical depths in the central A ring. The RSS optical depths have been divided by two for this comparison to account for the inherent differences in stellar and radio occultations. The occultation of a coherent, narrowband radio source (the spacecraft's X-band transmitter) can separate the directly-transmitted and diffracted signals – and thus measures the total extinction cross-section of the ring particles – whereas an occultation of a broadband source such as a star inevitably measures a combination of direct and diffracted flux. (See Cuzzi (1985) and French and Nicholson (2000) for further discussion of this topic; in their terminology, the "effective extinction efficiency", $Q_{occ} = 2$ for RSS occultations, whereas $1 < Q_{occ} < 2$ for stellar occultations, depending on the average particle size and various geometric factors.) Since the typical scattering angle for centimeter-to-meter-size ring particles at wavelengths of order 1 μm, $\theta_d \simeq \lambda/d \simeq 10^{-6}$ to 10^{-4} radians, which is less than the angular dimension of either the UVIS HSP aperture (6 mrad) or even one VIMS pixel (0.25 × 0.5 mrad), we expect that the diffracted flux is largely indistinguishable from the direct stellar signal and thus that $Q_{occ} \approx 1$ for Cassini stellar occultations (cf. Section 2.1 of French and Nicholson 2000). In this chapter when we quote an "optical depth" we will generally mean the normal optical depth measured in a stellar occultation unless specified otherwise (Eq. 13.1). This is also the quantity most relevant to the scattering of visible light.

Variations in B are not the only parameter controlling the apparent optical depth for the A ring. Because the self-gravity wakes are canted, elongated structures, the viewing angle relative to the orientation of the wakes, $\varphi - \varphi_w$, also plays a role. Initial models of the self-gravity wakes as infinitely long, periodic, opaque slabs separated by an

Fig. 13.7 Comparison of apparent normal optical depth in the central A ring from three occultations at different values of B: $B = 29.1°$ (UVIS), $B = 23.6°$ (RSS), $B = 50.9°$ (VIMS). The RSS optical depth values have been divided by 2 to account for the different measures in optical depth between radio and stellar occultations (Section 13.3)

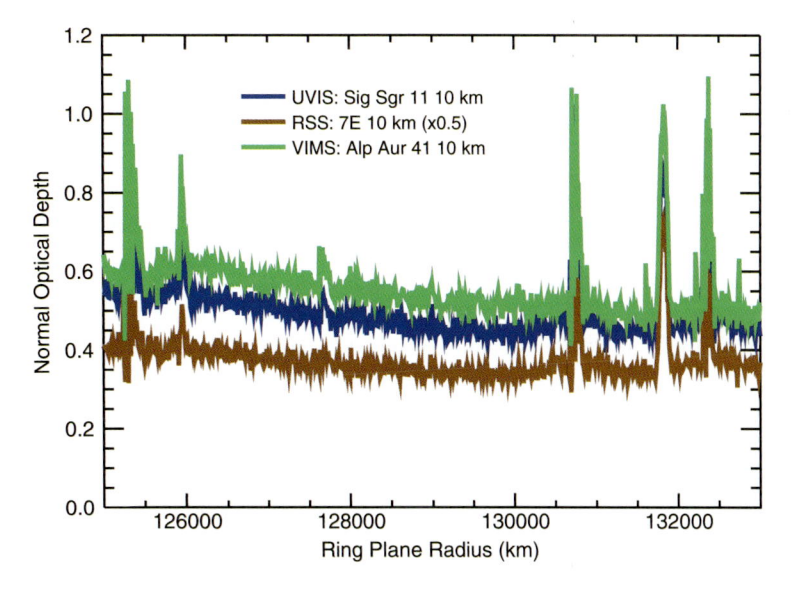

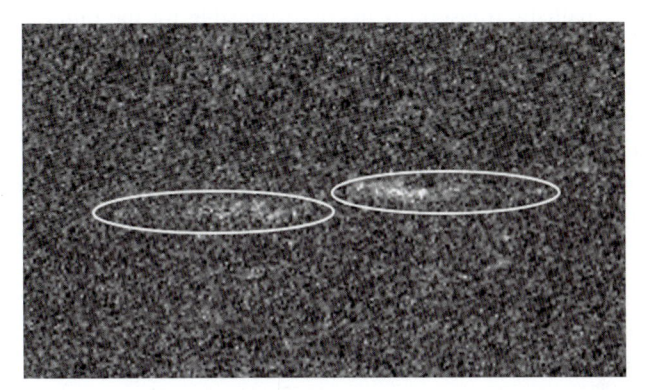

Fig. 13.8 "Propeller" observed by Cassini on 1 July 2004 (image: NASA/JPL/Space Science Institute)

optically thin cloud have successfully explained the observed variations in optical depth and provide measures of the cross-sectional aspect ratio of the wakes, the relative size and optical depth of the gaps between them, and their mean orientation (Colwell et al. 2006; Hedman et al. 2007). The A ring wakes are wider in the radial direction than they are thick by a factor of 3 to 6, with flatter wakes in the inner A ring. The spacing between the wakes is also smaller in the inner A ring with a roughly constant spacing-to-thickness ratio of about 7 for a rectangular wake cross-section (Colwell et al. 2006). Cassini thermal infrared azimuthal scans find an average height-to-width ratio of 0.4 using an ellipsoidal cross-section like that of Hedman et al. (Leyrat et al. 2008). The different wake thickness values obtained from the rectangular and ellipsoidal cross-section models appear to be reconcilable if the rectangular cross-section wakes are circumscribed by the ellipsoidal wakes (Salo et al. 2008). The rectangular thicknesses are thus more like a full-thickness at half-maximum. The gap optical depths are roughly constant across the inner two-thirds of the A ring at $\tau_g = 0.1$, with an increase to $\tau_g = 0.3$ outside the Encke gap, though the simple wake models have larger residuals in that part of the ring. This is roughly half the total optical depth of the outer A ring, suggesting that the wakes are not well-organized there.

While the fine-scale texture of the A ring is dominated by the self-gravity wakes, particularly massive ring particles can clear a local area immediately upstream and downstream of the particle, or moonlet. A large mass embedded within a ring will start to deflect the orbits of surrounding ring material, first clearing out the chaotic zones that, due to Keplerian shear, lead the embedded mass on the inward side and trail it on the outward side. Opposing this process is the ring's kinematic viscosity, which causes ring material to diffuse back into the cleared-out zones (see also Chapter 14). If the moon is large enough, the cleared-out zones will overcome viscous refilling and stretch all the way around the ring to become an annular gap, such as the Keeler or Encke Gap. Smaller moonlets, however, will be left with a smaller disturbance that moves with them through the ring. These disturbances have been dubbed "propellers," due to their morphology. They were first predicted (Spahn and Sremčević 2000; Sremčević et al. 2002; Seiß et al. 2005), then examples were found in the very highest-resolution images of the rings taken by Cassini (Fig. 13.8, Tiscareno et al. 2006a). The radial separation Δr between the two lobes of observed propellers is most commonly ~ 0.5 km, though examples with Δr larger than 1 km have been found (Tiscareno et al. 2008). There should be a direct relationship between Δr and the mass of the moonlet at the center of the propeller, except that there is controversy over whether to interpret the bright observed features as local maxima or minima in density. The propeller moonlets are probably between 40 and 500 m in diameter, placing them in a particle-size regime about which little is known (see Chapter 15). They appear to move at the local Keplerian velocity (Sremčević et al. 2007; Tiscareno et al. 2008).

Further Cassini observations determined that the occurrence of propellers is not uniform across the A Ring (Sremčević et al. 2007). In fact, 99% of known propellers occur within three narrow (∼1000 km) bands in the mid-A Ring between 126,750 km and 132,000 km from Saturn's center (Tiscareno et al. 2008). It is unclear whether these propeller-rich belts are due to specific events in which one or more parent bodies broke up into the current moonlets, or whether a larger initial moonlet population has been sculpted into bands by other ring processes. The propeller-rich belts correspond to the regions in the A Ring where self-gravity wakes are strongest (Colwell et al. 2006; Hedman et al. 2007). The propeller-poor regions that bound the propeller-rich belts correspond to the locations of spectral "halos" surrounding strong density waves, where grain sizes may be reduced (Nicholson et al. 2008a). These halos are zones which extend several hundred km on either side of the strongest density waves in the A ring within which the ring's photometric behavior is unusual (Dones et al. 1993) and where the near-IR ice bands are weaker than in surrounding regions (Nicholson et al. 2008a). The signature of self-gravity wakes in occultation data is weaker in these regions, suggesting that ring particles there are also, on average, smaller (Hedman et al. 2007).

13.2.4 The Inner A Ring

The largest optical depths in the A ring are near the inner edge where the Pandora 5:4 density wave is launched (Fig. 13.9). That wave begins in the 300 km wide "ledge" (Durisen et al. 1992) and propagates into a region of high optical depth. The weak Atlas 6:5 density wave lies near, but exterior to, the inner edge of the ledge, interior to the Pandora 5:4 resonance. Although a similar feature is seen at the inner edge of the B ring, the ballistic transport model of Durisen et al. (1992) does not produce the ledge feature. Nevertheless, ballistic transport is likely responsible for the ramp in the Cassini Division and outer C ring. Measurements of the Janus/Epimetheus 7:5 density wave at 121,260 km show that the opacity (τ/σ) in the ramp is consistent with that in the A ring (Tiscareno et al. 2006b), while waves in the rest of the Cassini Division indicate a much higher opacity (Porco et al. 2005; Colwell et al. 2009). This supports the ballistic transport model for the origin of the ramp where particles

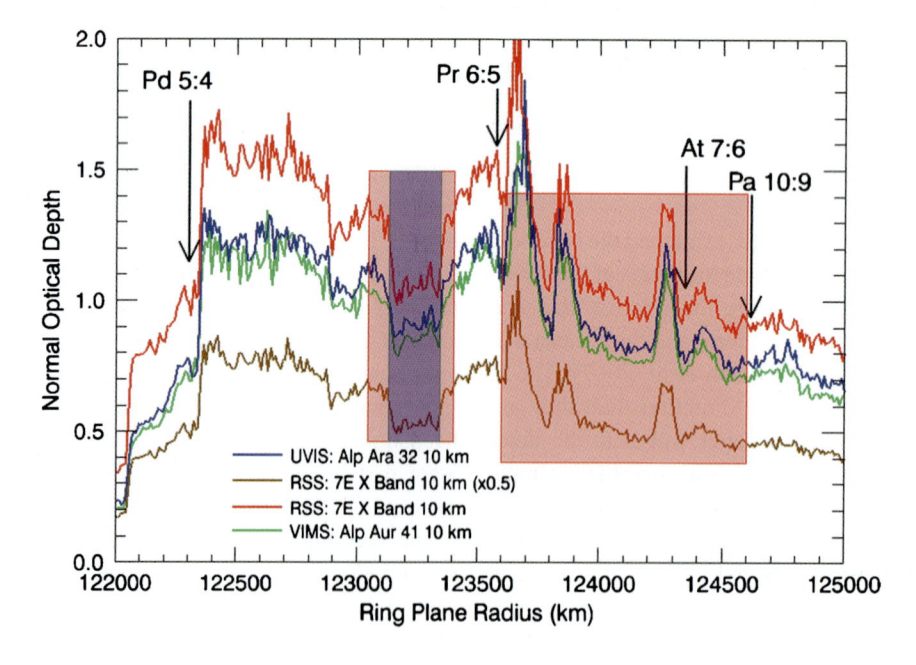

Fig. 13.9 Occultation profiles from UVIS, VIMS, and RSS of the inner A ring at 10 km resolution showing unexplained irregular structure. Major resonances are indicated. Red-shaded areas indicate the regions where RSS data show periodic microstructure (Fig. 13.10). The blue-shaded region shows the region where the wavelength of that microstructure is ∼150 m in UVIS occultation data (Sremčević et al. 2006). The wavelength is longer in the outer parts of the innermost red-shaded region. The Pandora 5:4 density wave begins in the ledge (see text) and propagates across the large increase in optical depth at 122,340 km. Individual wave peaks and troughs are not visible at the resolution of the data shown here. The humps at 123850, 124250, 124400 km, and the broad hump from 123300 to 123750 km that includes the Prometheus 6:5 density wave appear to lie on top of a smoothly varying optical depth profile that continues into the relatively featureless middle A ring (Fig. 13.7)

drift out of the A ring as they lose angular momentum due to meteoroid impacts (Durisen et al. 1989, 1992).

There are several abrupt increases in optical depth above a smoothly-varying background in the inner A ring (Fig. 13.9), mirroring, less dramatically, some of the unexplained structure seen in the B ring (Section 13.3). This structure was observed at high resolution in the Saturn Orbit Insertion images (Porco et al. 2005). Aside from these features, however, all moderate-scale fluctuations in optical depth in the A ring can be directly attributed to the perturbative effects of satellites as either density or bending waves or satellite wakes (Section 13.2.2).

13.2.5 Small-Scale Periodic Structures

With the high spatial resolution afforded by both radio and stellar occultations, the microstructure of the rings has been observed in unprecedented detail. Occultations have revealed regions throughout the rings that have a periodic structure in the radial direction with wavelengths between 150 and 250 m (Fig. 10, Thomson et al. 2007). In the RSS occultation data shown in Fig. 13.10, the structure is revealed by the presence of strong signals offset 70–100 Hz from the direct signal. These are explained by the presence of periodic structures in the rings causing the ring to act like a diffraction grating. The wavelength of the ring structure producing these diffraction lobe signals can be measured from their frequency offset in the RSS spectrograms. A similar periodic structure was observed in the transmitted light profile of a stellar occultation with unusually high radial resolution at $R = 114,150$ km (Colwell et al. 2007). Unlike the self-gravity wakes, these periodic structures have no measureable cant angle to the direction of orbital motion (Thomson et al. 2007; Colwell et al. 2007). The location of this wave structure in the A ring is restricted to two regions of relatively high optical depth shown in Fig. 13.10. It is also seen at several locations in the B ring (Thomson et al. 2007; Section 13.3). The most likely explanation for these waves is viscous overstability, predicted to occur in regions of high optical depth (see Chapter 14).

In the A ring these waves are observed between 123,100 and 124,600 km, with a 200 km break centered at 123,500 km (Thomson et al. 2007). While the optical depth is higher here than elsewhere in the A ring, it is somewhat lower than that predicted to lead to viscous overstability (Chapter 14). The wave region extends across large (unexplained) fluctuations in optical depth, and ends abruptly at 124,600 km where the optical depth of the A ring is gradually declining. On the inner edge of this wave region, the waves may be affected by the tail of the strong Pandora 5:4 density wave (Fig. 13.9).

13.3 B Ring

13.3.1 Overview

Although it is likely to contain much of the total mass in Saturn's ring system, the structure of the B ring is less well-characterized than is that of the A or C rings. This is due, in large part, to its higher optical depth, which severely limits the sensitivity of most occultation observations. There are essentially no useful Voyager radio occultation measurements of the B ring, the 3.5 cm signal being completely attenuated (i.e., $\tau > 0.85$) at 5 km resolution (Marouf et al. 1986). While the PPS and UVS stellar occultation profiles provide good ultraviolet optical depths for the inner and outer B ring at 20-km resolution, the signals in both data sets are nearly

Fig. 13.10 Spectrograms of the 3.6 cm signal in the inner A ring observed during the May 3, 2005 (**a** – ingress, **b** – egress) and August 2, 2005 (**c** – ingress, **d** – egress) RSS ring occultations. The direct signal is the central brown line in each panel. The diffraction signature of periodic microstructure appears as roughly parallel horizontal lines displaced ±70–100 Hz from the direct signal (f_{m1} and f_{m2} in a) (from Thomson et al. (2007))

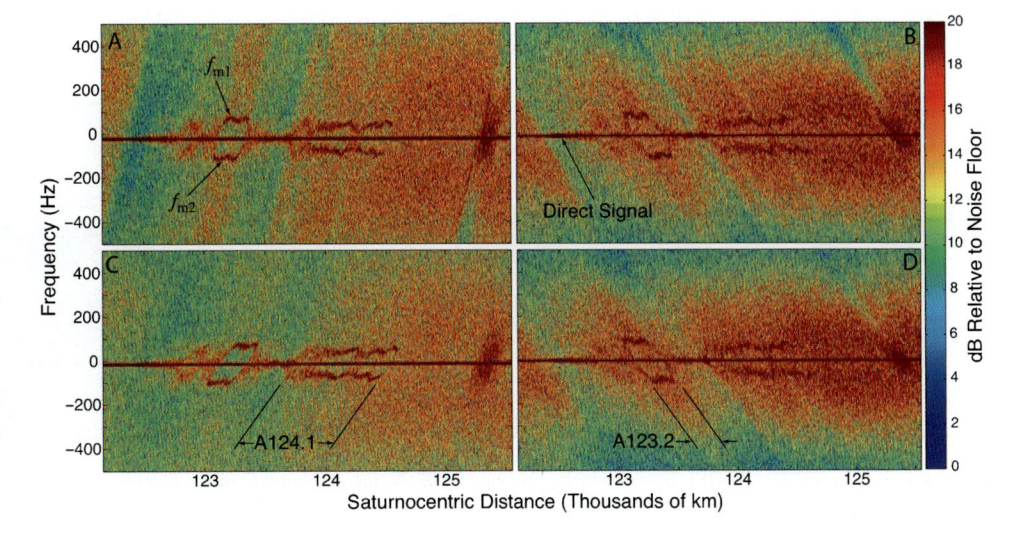

completely attenuated (i.e., $\tau > 2.6$ for PPS and $\tau > 2.8$ for UVS) in the central B ring (see also discussion and plots in Section 3.2 and 4.2 of Nicholson et al. 2000).

Ground-based observations of the occultation of the bright star 28 Sgr in 1989 provided optical depth profiles in the near-IR, including the B ring, albeit at lower radial resolution than the Voyager data. Average optical depths in the central core of the B ring (between 106,000 and 110,000 km) were found to be 2.07 ± 0.07 at 3.3 μm and 2.28 ± 0.14 at 3.9 μm (Harrington et al. 1993; Nicholson et al. 2000, Section 4.2), but the effective radial resolution of these data is reduced by scattering to ~1000 km (see French and Nicholson 2000, Section 4.1). Voyager imaging data, on the other hand, show a bewildering amount of fine-scale structure throughout the B ring (Cuzzi et al. 1984). Comparisons with the PPS occultation data show that, while the rings' brightness in diffusely-transmitted light is strongly correlated with the optical depth profile, the same is less true of images taken in reflected light. Indeed, Cuzzi et al. (1984) suggested that some of the brightness variations might reflect variations in albedo, rather than in optical depth, while Porco et al. (2008) showed that changes in ring thickness could produce brightness variations without changing the optical depth. The Voyager images also revealed subtle but real color variations across the B ring. Most prominent are the two "red bands" around 99,000 and 101,000 km and a "blue band" at 109,000 km described by Estrada and Cuzzi (1996).

In their statistical study of structure in the Voyager reflectivity (I/F) profile of the B ring, Horn and Cuzzi (1996) found that the dominant period of the radial structure varied from 80 to 200 km, with more pronounced periodicities being found in the inner third of the ring. No consensus has emerged on the origin of this structure, though many attempts have been made to identify dynamical instabilities (or "overstabilities") which might account for it (e.g. Schmit and Tscharnuter 1999; Salo et al. 2001; Schmidt et al. 2001; Tremaine 2003, see also Chapter 14).

Cassini imaging and occultation data have greatly increased our knowledge – if not yet our understanding – of structure in the B ring, though relatively little of these data has been published at the time of writing. We present below a summary of these observations, as presented at several conferences and workshops, organized by ring region rather than phenomenon. We conclude this section with a discussion of the relatively few density and bending waves which have been identified in the B ring, and of the role played by self-gravity wakes. Figure 13.11 provides an overview of the B ring, as seen by Cassini, in the same format as Figs. 13.1 and 13.2. The upper panel shows a normal optical depth profile derived from a UVIS stellar occultation while the lower panel shows a mosaic constructed from ISS images of the unilluminated side of the rings. In this geometry, and for optical depths, $\tau > \sim 0.5$, the brightness of the rings is inversely related to the optical depth. In the single-scattering

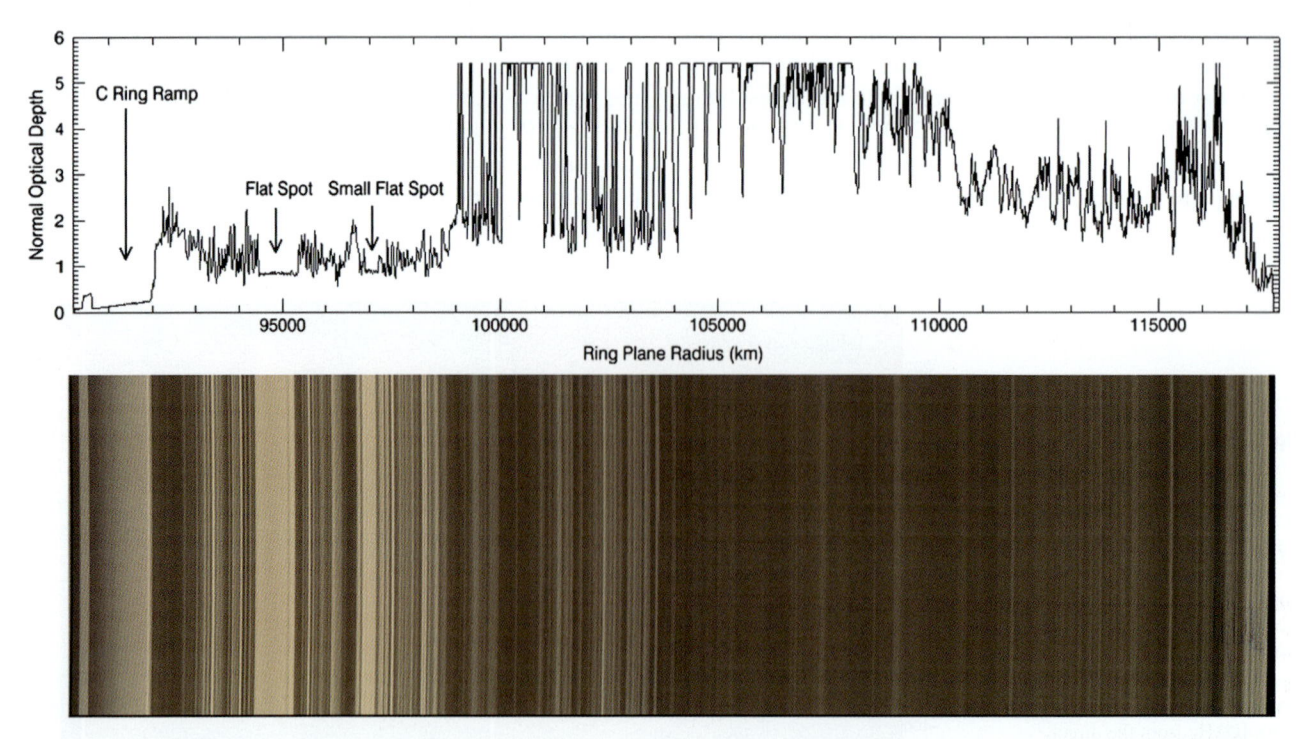

Fig. 13.11 ISS mosaic (NASA/JPL/Space Science Institute) of the B ring and outer C ring ("C Ring Ramp") and UVIS stellar occultation profile. The image is from the same mosaic shown in Fig. 13.2

limit, the diffusely-transmitted reflectivity is given by (Cuzzi et al. 1984):

$$I/F = \tfrac{1}{4}\varpi_0 P(\alpha)\frac{\mu_0}{\mu_0 - \mu}\left[e^{-\tau/\mu_0} - e^{-\tau/\mu}\right] \quad (13.10)$$

where ϖ_0 and $P(\alpha)$ are the particle single-scattering albedo and phase function, $\mu_0 = |\sin B|$ is the cosine of the incidence angle, and $\mu = |\sin B|$ is the cosine of the emission angle. For $\tau \ll \mu_e = \mu\mu_0/(\mu + \mu_0)$, the transmitted I/F increases linearly with τ, whereas for $\tau \gg \mu_e$ the I/F decreases exponentially, approaching zero as $\tau \to \infty$. Generally, the C ring falls in the former limit and the B ring in the latter. Plots of I/F vs τ for relevant Voyager and Cassini imaging geometries are given in Fig. 10 of Nicholson et al. (2008a).

At least four distinct sub-regions of the B ring can be identified in Fig. 13.11. We follow here the terminology introduced by Marouf et al. (2006) to describe the initial RSS occultation results and label these B1 through B4, in order of increasing radius. Regions B1 and B2 have low to moderate optical depths and exhibit a plethora of structure on the 100 km scale. The central zone (B3) – sometimes referred to as the "core" of the B ring – is largely opaque, while the outer B ring, intermediate in optical depth between B2 and B3, is denoted B4. The distinct nature of these regions is more apparent in a plot of the transparency of the ring rather than the optical depth (Fig. 13.12). A further distinction is useful here, with the outermost zone where significant azimuthal variations are present being denoted B5 (Table 13.1). We now discuss each of these regions in turn.

13.3.2 The Inner B Ring (Region B1)

The innermost zone, B1, is relatively transparent and extends from the inner edge of the B ring at a radius of 92,000 to 99,000 km (Fig. 13.13, Table 13.1). Apart from the first 700 km, where the structure is irregular and the optical depth mostly exceeds 1.5, this region has a median optical depth of ∼1.1 and is characterized by smooth undulations in both optical depth and I/F with radial wavelengths of ∼100 km (cf. Horn and Cuzzi 1996). Examples of both regimes are shown by Porco et al. (2005), in their Fig. 5a and b. Interrupting this pattern is the prominent "flat spot" between 94,450 and 95,350 km, where the undulations are almost completely suppressed. A second relatively featureless zone, or "small flat spot," is found between 96,900 and 97,200 km (Table 13.1). The lowest values of $\tau \approx 0.7$ are found in these two "flat spots"[4].

A comparison of Voyager stellar occultation and imaging profiles with that observed by the Cassini-VIMS instrument at Saturn orbit insertion shows that this undulating structure has remained virtually unchanged in the intervening 25 years, down to the smallest peaks and troughs (Nicholson et al. 2008a). As predicted by Eq. 13.10 above, an almost

---

[4] The word "flat" here is used to denote the appearance of these regions in plots of optical depth, and does not refer to any lack of vertical structure here compared to elsewhere in the rings.

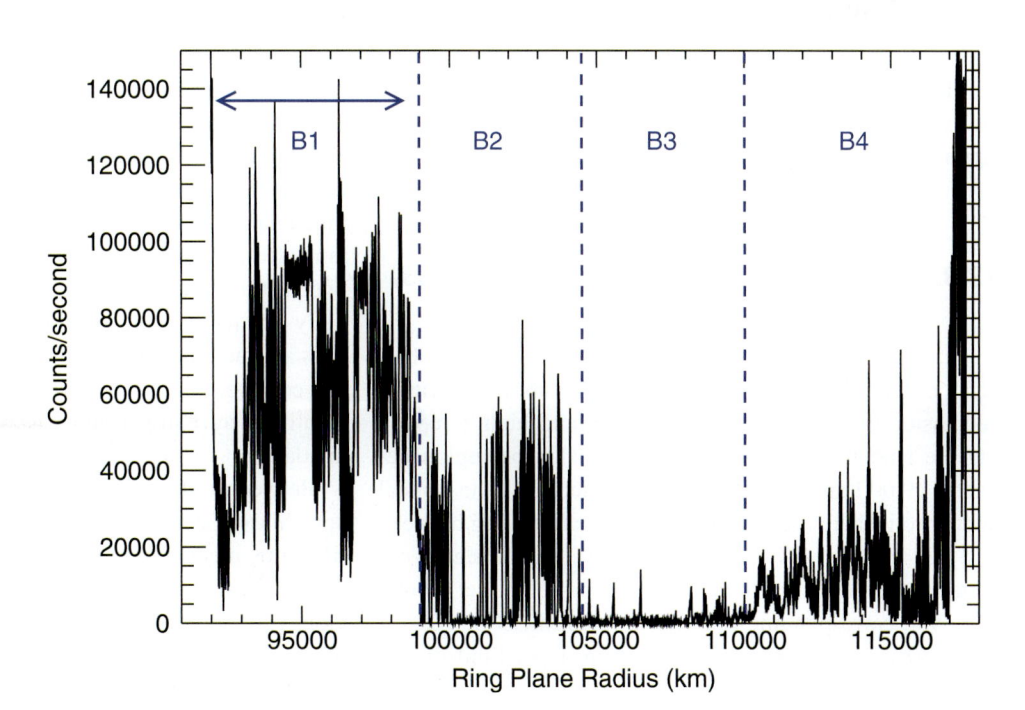

Fig. 13.12 λ Sco 029 UVIS occultation transparency showing qualitatively distinct regions of the B ring. The outermost 1000 km of B4 is a distinct region with non-axisymmetric structure dubbed B5 (see also Fig. 13.15)

Table 13.1 Major B Ring regions and features

Feature	Inner edge (km)	Outer edge (km)	Width (km)
B1 Region	92,000	99,000	7,000
Flat spot	94,450	95,350	900
Small flat spot	96,900	97,200	300
B2 region	99,000	104,500	5,500
B3 region	104,500	110,000	5,500
B4 region	110,000	116,500	6,500
B5 region[a]	116,500	117,500	1,000

[a] The outer edge of the B ring is not circular, so the width of the B5 region varies with longitude.

Fig. 13.13 The central portion of B1 (which extends to 99,000 km) illustrating the two flat spots (Table 13.1), the Janus 2:1 density wave (Fig. 13.16), and oscillatory structure with a characteristic wavelength of 80–100 km (Horn and Cuzzi 1996). The inner flat spot shows a small enhancement in optical depth ∼100 km wide near its outer edge

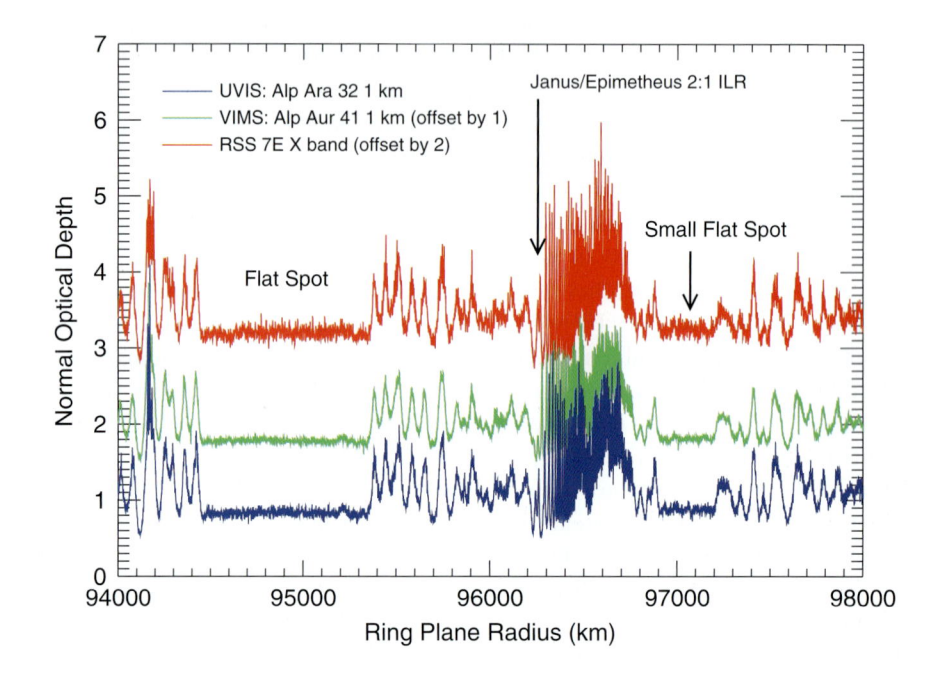

perfect anticorrelation is observed in this region between the brightness of the rings in diffusely-transmitted light and the optical depth.

The B1 region is home to the most prominent density wave in the entire B ring, driven by the strong Janus/Epimetheus 2:1 ILR, as discussed further in Section 3.4.

13.3.3 The Central B Ring (Regions B2 and B3)

At a radius of 99,000 km both the average level and the character of the optical depth profile change abruptly, with a succession of alternating zones of low and high τ replacing the smooth undulations of region B1. The region between 99,000 and 104,500 km is denoted B2. The optical depth profile here is essentially bimodal, alternating between $\tau \approx 1.5$ and $\tau \geq 4$ (Colwell et al. 2007). Cassini stellar occultation data obtained to date hit their noise floor at around $\tau = 4.6$, corresponding to a transmission of 1.0×10^{-4} at $B = 30°$, so we cannot yet set a firm upper limit on the highest optical depths in this region. Occultations planned in mid-2008, which will probe the rings at opening angles of 60–70°, should significantly improve this limit. RSS optical depths in this region alternate between 2.0 and ∼5.0, the latter also being at or close to the noise limit.

At higher resolution, the boundaries between the low-τ and high-τ zones are surprisingly abrupt, with transition zones being generally less than 10 km wide and sometimes much narrower (Fig. 13.14). The widths of the high-τ zones range from as little as 10 km to a maximum of ∼150 km. Between 100,025 and 101,010 km, however, the rings are essentially opaque except for a 40 km "window" at 100,430 km. This anomalous region, which looks more similar to region B3, corresponds closely to the outer of the two "red bands" identified by Estrada and Cuzzi (1996) in Voyager imaging data.

Between radii of 104,100 (the exact boundary is uncertain; Marouf et al. (2006) use 104,500 km) and 110,000 km (B3) the median optical depth reaches ∼3.6, its highest value anywhere in Saturn's rings. Over much of this region, $\tau > 5.0$ in the most sensitive stellar occultation data sets (Colwell et al. 2007) and $\tau \approx 5.0$ in the RSS data, when the latter are processed at 20 km resolution. Punctuating this

Fig. 13.14 The central part of the B2 region of peaks and valleys in optical depth illustrating the abrupt transitions between regions of moderately high optical depth and regions that are essentially opaque. The RSS occultation data saturate near an optical depth of 4 (2 as shown here divided by 2), so those values are a lower limit rather than a true measure of the optical depth in those optical depth peaks

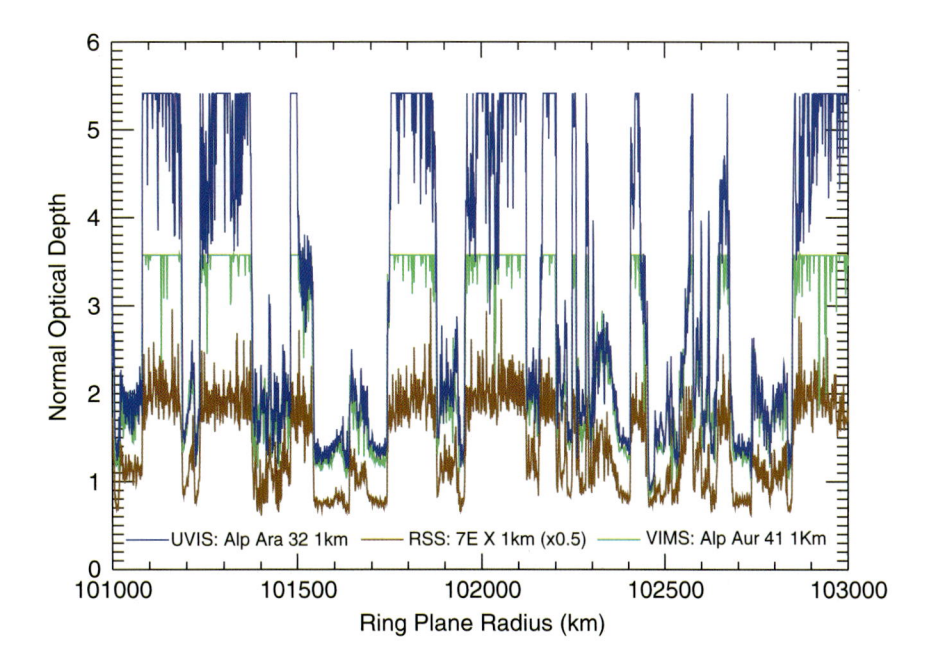

almost opaque region are at least 10 narrow windows where the optical depth drops to ∼2.0. These zones are typically ∼50 km wide and separated by distances of 300–900 km. Their profiles are smooth and often triangular, very unlike the square-wave structure seen in region B2. Three of these minima approximately coincide with strong Lindblad resonances (Prometheus 3:2, Janus 5:3 and Pandora 3:2), but others do not. The highest optical depths in the RSS profile occur in a 1600 km wide zone between 106,500 and 108,100 km, which may well be completely opaque. Nowhere in region B3 does the optical depth drop below ∼2.0 (or 3.0 in the RSS data), even at sub-km radial resolution.

It is quite possible that the peak optical depths in this region are substantially higher than the above estimates, as none of the occultations to date has clearly recorded a statistically-significant signal through the most opaque zones. Given the large lower limits on τ in these regions, they are, for all practical purposes, opaque.

13.3.4 The Outer B Ring (Regions B4 and B5)

Outside 110,300 km, the optical depth decreases gradually, but in an irregular manner with none of the periodic structure evident in B1 or the bimodal structure seen in region B2. As in regions B1 through B3, the stellar and RSS occultation profiles show essentially identical peaks and troughs. The impression left by these profiles is of a fractal structure, with power at all radial scales. Some insight into the radial structure of this region at the smallest accessible scales is provided by the UVIS α Leo occultation, whose track reached a minimum radius of 114,150 km (Colwell et al. 2007). Within 200 km of this radius, the radial sampling interval is 1.5 m or less, although the azimuthal smear relative to the ring particles is ∼90 m and Fresnel diffraction limits the true radial resolution to ∼15 m. While the innermost 2.7 km of the occultation chord shows virtually no correlation between the ingress and egress legs, both cuts show strong peaks in their power spectra at a radial wavelength of ∼160 m. This requires essentially axisymmetric structure with an azimuthal extent of several km or more.

Similar axisymmetric structures are inferred from the scattered signal seen in Cassini radio occultation data from 2005 (Thomson et al. 2007), especially in regions B2 and B4, in a narrow zone around 92,400 km in B1, and in the innermost part of the A ring where optical depths are comparable to those in the B ring (Section 13.2.5, Fig. 13.10). The inferred wavelengths are 115 ± 20 m in B1, 146 ± 14 m in B2, and ∼250 m in B4. Very fine-scale axisymmetric structure has also been glimpsed in the highest-resolution Cassini images of the innermost parts of the A and B rings (Porco et al. 2005; their Fig. 5a and f). It appears that this structure, which may reflect a viscous overstability in the denser parts of the rings (e.g., Salo et al. 2001; Schmidt et al. 2001), coexists with the canted self-gravity wakes discussed below.

Interior to 116,500 km, the median optical depth is ∼2.2 (or 3.0 in the RSS data), but at around this radius it drops rather abruptly to ∼1.3. The 1000 km wide outermost zone of the B ring is denoted B5 (Fig. 13.15). It contains a second abrupt transition at ∼117,200 km, exterior to which τ drops further to ∼0.5, similar to that in the middle A ring. Molnar and Dunn (1995) noted the coincidence of this fea-

Fig. 13.15 Optical depth profiles showing the B5 region of the B ring (from 116,500 km outward) at 1 km resolution

ture with the Mimas 2:1 corotation eccentricity resonance (CER), but no definite relationship has been established. Nicholson et al. (2000) noted that the outermost 400 km of the B ring is one of very few regions in the rings where multiple occultation profiles diverge significantly from one another (see their Fig. 11). This occurs between ingress and egress profiles from the same occultation, as well as between data sets taken at different times, so is likely to represent azimuthal rather than (or in addition to) temporal variations. Visual comparisons of Cassini occultation profiles from multiple experiments show little, if any, correlation in the small-scale structure outside ∼116,800 km.

The outer edge of the B ring, which is controlled and strongly perturbed by the Mimas 2:1 ILR at 117,555 km varies in radius by ∼160 km in a two-lobed pattern which rotates with the satellite (Porco et al. 1984b). Resonant distortions are likely to extend over a zone several hundreds of km wide, and these will obviously complicate the comparison of images and optical depth profiles obtained at different times and longitudes. Cassini images taken at Saturn orbit insertion show very complex radial structure in this region at a resolution of ∼350 m (Porco et al. 2005, Fig. 5c). A careful study taking into account these resonant distortions has yet to be made.

13.3.5 Density and Bending Waves

The most prominent density wave in the ring system is the Janus/Epimetheus 2:1 wave in the B1 region, beginning at 96,250 km and propagating over 500 km (Fig. 13.16). This wave, like the Pandora 5:4 wave and Prometheus 6:5 waves in the inner A ring, propagates across significant changes in optical depth. The change of the orbits of Janus and Epimetheus every four years modulates the density wavetrains produced by these moons. The slow propagation speed of the waves (∼0.4 mm/s) makes these transitions visible as hiccups in the dispersion of this wave at roughly 45 km intervals.

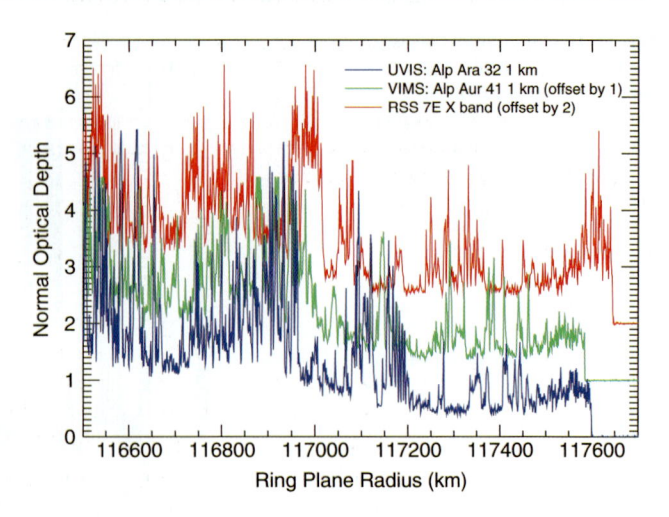

None of the other strong resonances in the B ring (Nicholson et al. 2000) produce readily recognizable density wavetrains like those in the A ring and Cassini Division. Lissauer (1985) analyzed the Mimas 4:2 bending wave in B5 and found a surface mass density of ∼54 ± 10 g cm^{-2}. With the value of 70 ± 10 g cm^{-2} from Voyager stellar occultations (Holberg et al. 1982; Esposito et al. 1983a), these are the only two direct measurements of local B ring surface mass density. Combining Cassini data from multiple observations has brought out the signature of weak waves in the Cassini Division (Colwell et al. 2009), so this technique and future observations may yet reveal the signature of waves in other regions of the B ring.

13.3.6 Self-Gravity Wakes

Comparisons of the apparent normal optical depth derived from UVIS and VIMS stellar occultations at different ring opening angles, B, show clearly that the classical expression for optical depth (Eq. 13.1) significantly underestimates the actual transmitted light through the B ring at small opening angles (Colwell et al. 2007; Nicholson et al. 2008b). Expressed another way, the apparent normal optical depth calculated from this expression for a particular ring region decreases as |B| decreases. It is likely that this non-classical behavior is due to unresolved, fine-scale structure, and in particular to self-gravity wakes similar to those seen in the A ring (see Section 2.3). Indeed, the B ring does show modest azimuthal variations in transmission and reflectivity (Fig. 13.6; French et al. 2007; Nicholson et al. 2005) similar to those in the A ring, but the variations of τ_n with B are more striking.

Fig. 13.16 Occultation profiles of the Janus/Epimetheus 2:1 density wave in the B1 region. The resonance locations for Janus and Epimetheus shift when the satellites perform their horeshoe orbit "swap" every 4 years and are between 96,210 and 96,250 km

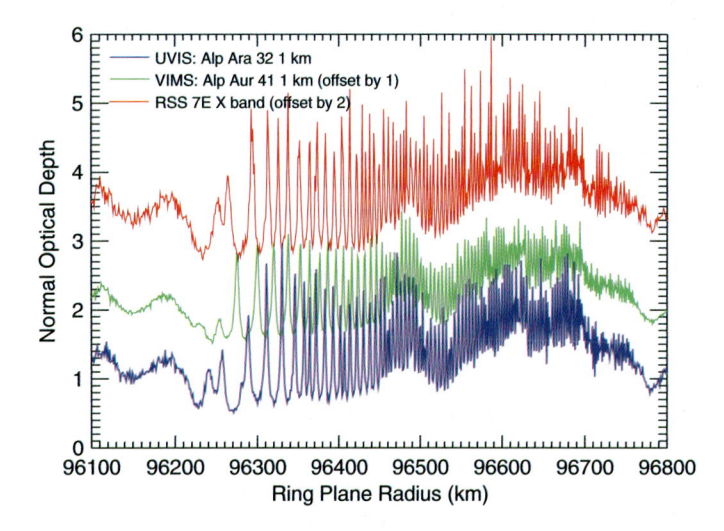

In this emerging view of the A and B rings, the average transmission is determined primarily by the width and optical depth of the narrow gaps between the wakes (Colwell et al. 2006, 2007; Hedman et al. 2007 Colwell et al. 2007), rather than by the total cross-section of particles per unit area of the rings, as has been traditionally assumed (cf. Eq. 2 in French and Nicholson 2000). In this situation, and if the "gaps" are almost transparent, the maximum transmission of the rings (i.e., that measured when the wakes are observed more-or-less end-on) can be almost independent of B, so that τ_n scales approximately as $\mu = \sin|B|$ (see Fig. 11 in Colwell et al. 2007).

In their study of a dozen stellar and solar occultation profiles of the B ring, Colwell et al. (2007) fitted both ray-tracing and approximate analytical models of self-gravity wakes to the variation of ring transmission as a function of both B and φ, the angle in the ring plane between the stellar line of sight and the local radial direction. They used data obtained by the Cassini UVIS instrument spanning ring opening angles from 9.5° to 41.7°. In the limit of opaque, infinitely-long wakes of rectangular cross-section W × H, the ring transmission is given by the analytic expression

$$I/I_0 = (G/\lambda_w - H/\lambda_w |\sin(\phi - \phi_w)| \cot B) e^{-\tau_g/\mu} \quad (13.11)$$

where G is the width of the interwake gaps, H is the wake thickness, and $\lambda_w = G + W$ is the wavelength of the self-gravity wakes. Colwell et al. find that the orientation angle of the wakes, ϕ_w, increases monotonically across the B ring from 65° to 80°, not straying far from the expected value of $\phi_w \sim 70°$ characteristic of Keplerian shear (Salo et al. 2004). Less well-determined is the optical depth in the gaps, τ_g, which averages around 0.25 but may be somewhat larger in the outer B ring. The parameter which essentially determines the average transmission of the rings is the fractional gap width, G/λ_w, or equivalently $G/W = 1/(\lambda_w/G - 1)$.

Not surprisingly, Colwell et al. find a tight negative correlation between G/W and the apparent normal optical depth, τ_n, with $G/W \simeq 3.6 \exp(-1.616\tau_n)$. The aspect ratio of the wakes, H/W is not well-constrained, but is everywhere less than 0.2 and generally decreases outwards. Colwell et al. suggest that $H/G = 0.2$ is a satisfactory assumption everywhere, with G/W given by the above exponential model. By comparison, average values for the same wake parameters in the A ring, updated from those given by Colwell et al. (2006), are $50° < \phi_w < 70°$, $\tau_g \approx 0.15$ and $0.10 < H/W < 0.25$.

Using a similar model of opaque wakes with elliptical cross-sections, again interspersed with semi-transparent zones of optical depth τ_g, but applying this only to selected occultations in which the wakes were viewed end-on, Nicholson et al. (2008b) analyzed the transmission measured in three stellar occultations observed by the VIMS instrument at opening angles between 11.4° and 50.9°. In this geometry, the transmission is given by the simplified expression

$$I/I_0 = G/\lambda_w e^{-\tau_g/\mu}. \quad (13.12)$$

This model provides a satisfactory fit to most of the B1 region and to the less-opaque parts of B2 with $0.15 < \tau_g < 0.4$ and G/λ_w ranging from 0.60 in the "flat spot" to $\sim$0.15 in B2. In region B4 the scatter is larger but on average the best-fitting parameters are $\tau_g \approx 0.35$ with G/λ_w increasing outwards from <0.05 to <0.20 (Nicholson and Hedman 2009). The same model applied to the A ring yields $0.12 < \tau_g < 0.30$, with the smallest values in the central A ring near 128,000 km, and $0.3 < G/\lambda_w < 0.65$, increasing monotonically outwards.

Perhaps the most surprising result of these models is the low optical depth in the "gaps" between the wakes; for most practical purposes these are almost empty, even when the average optical depth approaches $\tau_n = 2.0$. Numerical simulations which incorporate a range of particle sizes (e.g., Salo and Karjalainen 2003) predict that the larger, meter-sized ring particles are strongly concentrated in the wakes, whereas the smaller, centimeter-sized particles are distributed more uniformly. If the gaps do indeed comprise almost entirely small particles, then we may imagine a picture of the B ring in which broad, flat, wakes composed of densely-packed meter-sized particles are embedded in a thicker, low-density "haze" of smaller particles whose normal optical depth is only a few tenths (Nicholson et al. 2008b).

According to this picture of a wake-dominated ring, the true normal optical depth of the rings (i.e., that actually measured at normal incidence) is likely to be somewhat greater than any of the calculated values plotted here. Based on their best-fitting wake model parameters, Colwell et al. (2007) find that the true $\tau \approx 1.18\tau_n$, where τ_n is the normal optical depth computed from data obtained at $B = 41.7°$.

Even the "true" normal optical depth may not accurately reflect the amount of material in the rings, as a very large – and effectively unconstrained – density of particles might be hidden within the opaque B ring wakes (Stewart et al. 2007). This observation has called into question several key parameters previously considered known for Saturn's rings, notably their average surface density and total mass.

13.3.7 Correlations of Structure with Particle Properties

The coincidence between the 1000 km wide opaque zone within region B2 with one of the Voyager "red bands" was noted above. An examination of radial variations in the rings' reflectance spectra as measured by the Cassini-VIMS instrument shows a strong correlation between optical depths in the B ring and both the slope of the rings' visible spectrum (from 0.35 to 0.55 μm) and the depth of the water ice bands at 1.5 and 2.0 μm (Nicholson et al. 2008a; Filacchione et al. 2007). In general, the visible colors and ice band depths are almost bimodal, with stronger bands and redder colors being associated with $\tau > 2.0$ and weaker bands and more neutral (less red) colors with lower optical depths. In a profile of ice band depth or visible color vs radius, region B2 appears as a transitional zone between the more neutral B1 region and the very red B3 and B4 regions (see Chapter 15), with properties that alternate between these two extremes as the optical depth oscillates.

Although the reason for this correlation between τ and spectral properties is unkown, one possible interpretation of the curious optical depth profile of the transitional B2 region is that there exists a critical optical depth at $\sim$2.0 above which the internal structure and optical scattering properties of the rings changes, perhaps from a closely-packed mono- layer with negligible interparticle scattering to a more opaque multilayer (Nicholson et al. 2007).

13.4 The Cassini Division and C Ring

13.4.1 Overview

It has long been recognized that there are many similarities, both in particle properties and structure, between the C ring and the Cassini Division (Cuzzi et al. 1984; Esposito et al. 1984). From the structural viewpoint, the most obvious similarity is in their optical depth, which averages around 0.1 for both regions and only rarely exceeds 0.5. By comparison, the optical depth averages $\sim$0.5 in the A ring, and is larger than this almost everywhere in the B ring. A consequence of this is that good optical depth profiles for both the C ring and Cassini Division were obtained by the Voyager radio and stellar occultation experiments, in addition to many images in both reflected and transmitted light.

A second item worthy of note is the concentration of narrow gaps within these two regions: the 18,000 km wide C ring is home to five gaps, while the Cassini Division has eight within its width of only 4500 km. In contrast, the 15,000 km wide A ring has only two empty gaps and the B ring has none, despite its width of 25,500 km. Many of these gaps are not completely empty, but are inhabited by narrow ringlets, some sharp-edged and opaque and others diffuse and translucent.

Overviews of these regions are presented in Figs. 13.17 and 13.18, in the same format as for the A and B rings in Figs. 13.2 and 13.11. Note that, because of their relatively low optical depths, features in both the C ring and the Cassini Division appear in normal contrast in images taken in diffusely-transmitted light: brighter here almost always means higher optical depth.

13.4.2 Gaps and Ringlets in the Cassini Division

Table 13.2 lists all identified gaps and ringlets in the Cassini Division. Many of the sharper gap and ringlet edges in this region were used by Nicholson et al. (1990) and French et al. (1993) in their studies of the absolute radius scale of the rings, and their numerical designations are also included in the Table (e.g., the narrow Russell and Jeffreys gaps at 118,600 and 118,950 km are bounded by features 13 and 15). In Fig. 13.19 we show the optical depth of the Cassini Division from a Cassini VIMS occultation, chosen to highlight the low optical depth.

13 The Structure of Saturn's Rings

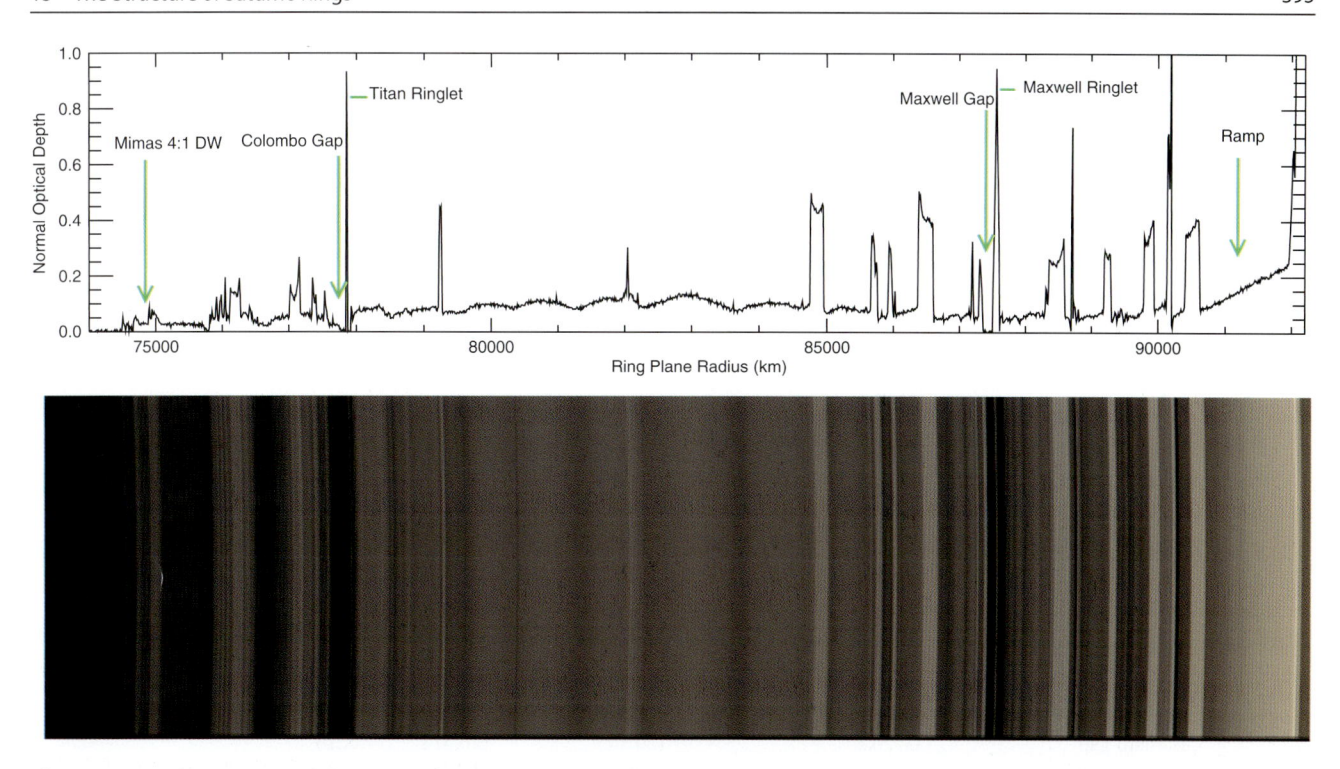

Fig. 13.17 UVIS occultation profile and ISS image (NASA/JPL/Space Science Institute) of the C ring. The image is from the same mosaic shown in Fig. 13.2

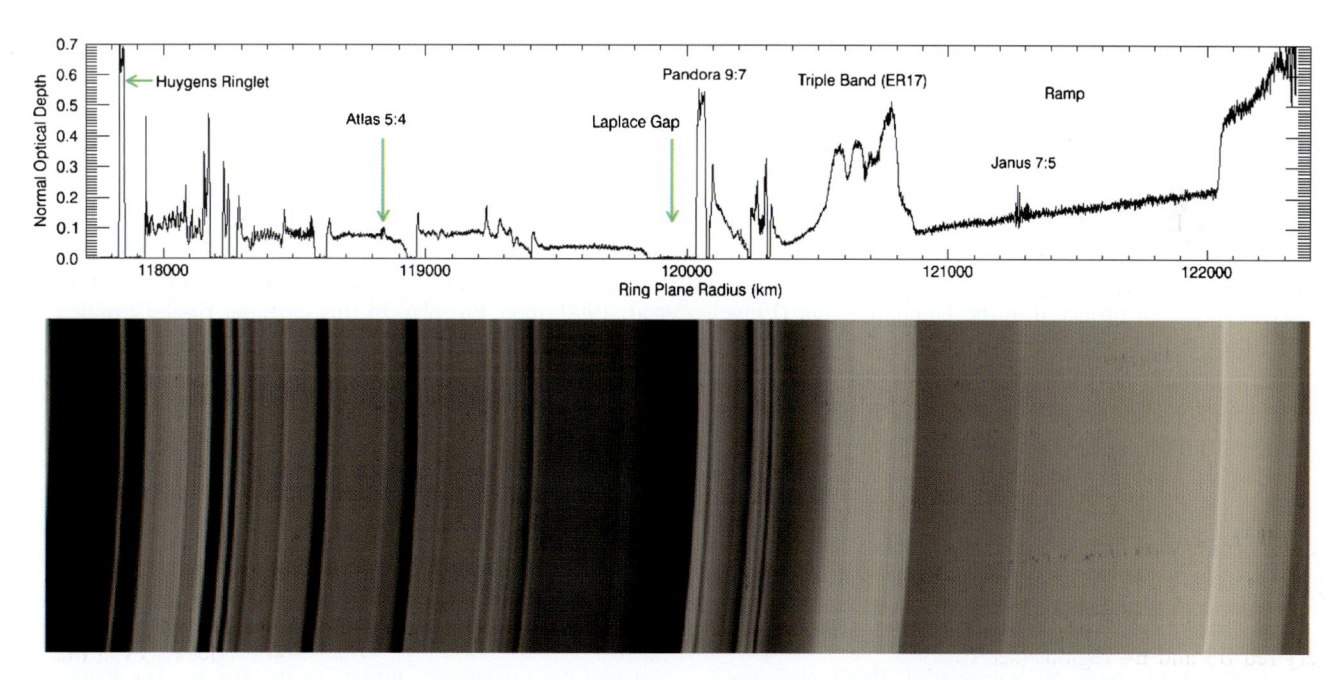

Fig. 13.18 UVIS optical depth profile (α Leo Rev 009) and ISS image (NASA/JPL/Space Science Institute) of the Cassini Division. The transition from the Cassini Division to the A ring at the outer edge of the Cassini Division ramp is much more pronounced in optical depth than in the image. The image is from the same mosaic shown in Fig. 13.2

The Huygens Gap separates the B ring, whose outer edge is defined by the strong Mimas 2:1 inner Lindblad resonance (see Sections 13.2.1 and 13.3.3), from the Cassini Division proper. It is the broadest gap in Saturn's rings, being ∼25% wider than the Encke Gap in the outer A ring. It is home to two narrow ringlets and one dusty ringlet, the latter discovered in Cassini images (Porco et al. 2005, Fig. 7b). The prominent Huygens ringlet, readily visible in Figs. 13.18 and

Table 13.2 Cassini Division ringlets and gaps. Several ringlets and pronounced features do not have designations. We denote these in numerical order from the inner edge of the C ring (Table 13.4) through the Cassini Division by type: ringlet (R, narrow ring in a gap), and embedded ringlet (ER; regions of prominent increases in optical depth over the smoothly varying background optical depth). Alternate IDs give the approximation location in 10^6 m. F93 refers to edges identified in Nicholson et al. (1990) and used by French et al. (1993). (A) Inner edge is strongly perturbed by the Mimas 2:1 resonance and is not circular. (B) Ringlet is not circular. (C) Ringlet is only seen in high phase angle images. (D) Ringlet is suspected to be inclined. (E) Pandora 9:7 density wave is located in this feature. (F) Designated the A ring "foot" by Durisen et al. (1992). Pan 7:6 density wave is located in this feature. (G) Janus/Epimetheus 7:5 density wave is located in this feature

Identification			Inner edge	Outer edge	Width	
ID (name)	Alternate	F93	(km)	(km)	(km)	Note
Huygens gap	G117.7	20, 55	117,515	117,932	417	A
R5	R117.7		17,735		∼50	B, C
Huygens ringlet	R117.8	53, 54	117,815	117,834	20	B
R6	R117.9		117,900		3	B, D
Herschel gap	G118.2	16, 19	118,183	118,285	102	
R7 (1.960 R_S ringlet)	R118.2	17, 18	118,229	118,257	28	B
Russell gap	G118.6	13	118,597	118,632	35	
Jeffreys gap	G119.0	15	118,931	118,969	38	
R8	R119.0		118,960		1	
Kuiper gap	G119.4		119,410		3	
Laplace gap (outer rift)	G119.9		119,848	120,086	238	
R9	R119.9		19,965		∼50	C
R10 (1.990 R_S ringlet)	R120.1	12, 14	120,039	120,076	37	B, E
Bessel gap	G120.2	9	120,236	120,246	10	
R11 (1.994 R_S ringlet)	R120.3	10, 11	120,246	120,305	59	B
Barnard gap	G120.3		120,305	120,318	13	
ER17 (triple band)	ER120.7		120,550	120,800	250	F
Cassini Division ramp			120,900	122,050	1,150	G

13.19, was recorded in many Voyager images and occultation data sets and is both noncircular and variable in width. A satisfactory model for its shape has, however, proven to be elusive. It appears to combine the characteristics of a freely-precessing ellipse (like the Uranian ε ring) with an $m = 2$ forced perturbation due to the nearby Mimas resonance (Porco 1983; Turtle et al. 1991). Similar multi-mode behavior has been observed in the Uranian δ ring (French et al. 1988), though in that case no external resonance is involved. The dusty ringlet mentioned above (R5, not visible in Fig. 13.18 or Fig. 13.19) is located immediately interior to the Huygens ringlet (see Chapter 16).

Less well-studied is an extremely narrow ringlet located exterior to the Huygens ringlet (R6). This ringlet, which is almost opaque in some occultations and only a few kilometer wide, is seen clearly in Cassini images (e.g., Fig. 4b of Porco et al. 2005) and in occultations (Fig. 13.19). It also appears in the Voyager PPS occultation profile (feature 131 in Esposito et al. 1987, cf. Fig. 6a in Nicholson et al. 1990), so is not a recent creation. Examination of Cassini images strongly suggests that R6 is significantly inclined with respect to the main ring plane, as it sometimes seems to leave the Huygens gap and instead appears superimposed on the region *outside* the gap (as it does in the occultation shown in Fig. 13.18, Colwell et al. 2005; Spitale et al. 2008a, b). This might also account for its apparent absence in the Voyager RSS profile, which was obtained at a very low opening angle.

The inner half of the Cassini Division, between the outer edge of the Huygens Gap at 117,932 km and the Laplace Gap at 119,848 km, is characterized by an almost uniform optical depth between 0.05 and 0.10, punctuated by four narrow, sharp-edged gaps. The innermost and widest of these gaps, the Herschel Gap, is home to a 28-km wide noncircular ringlet (R7) whose radius varies by at least 12 km (Nicholson et al. 1990, their Fig. 6b). In this case the ringlet's optical depth is comparable to that of its surroundings. Both interior and exterior to this gap are series of prominent wavelike oscillations, which were initially interpreted as satellite-driven wakes by Marouf and Tyler (1986). Later work, however, showed that these oscillations do not change with longitude as satellite wakes should (Flynn and Cuzzi 1989).

Just barely visible in Fig. 13.19 but seen clearly in several Cassini images is a narrow ringlet (R8) which inhabits the Jeffreys Gap. Although faint, this ringlet does *not* brighten at high phase angles and thus is probably not composed of fine dust like the Encke gap ringlets or the F ring. The Kuiper Gap at 119,410 km is probably the narrowest such feature in Saturn's rings; its width appears to be variable, but is generally ∼3 km.

Second only to the Huygens Gap in the Cassini Division is the Laplace Gap (or Outer Rift), with a width of ∼240 km. Like its big brother, this gap is also home to both sharp-edged and diffuse ringlets. The former, bounded by features 12 and 14 and referred to as the 1.990 R_S ringlet (R9), is located

Fig. 13.19 Occultation profile of the Cassini Division showing normal optical depth from the Cassini VIMS o Ceti Rev 008 occultation. Unlike the A and B rings, the measured normal optical depth of the Cassini Division is insensitive to viewing geometry. Prominent features are labeled (Table 13.2)

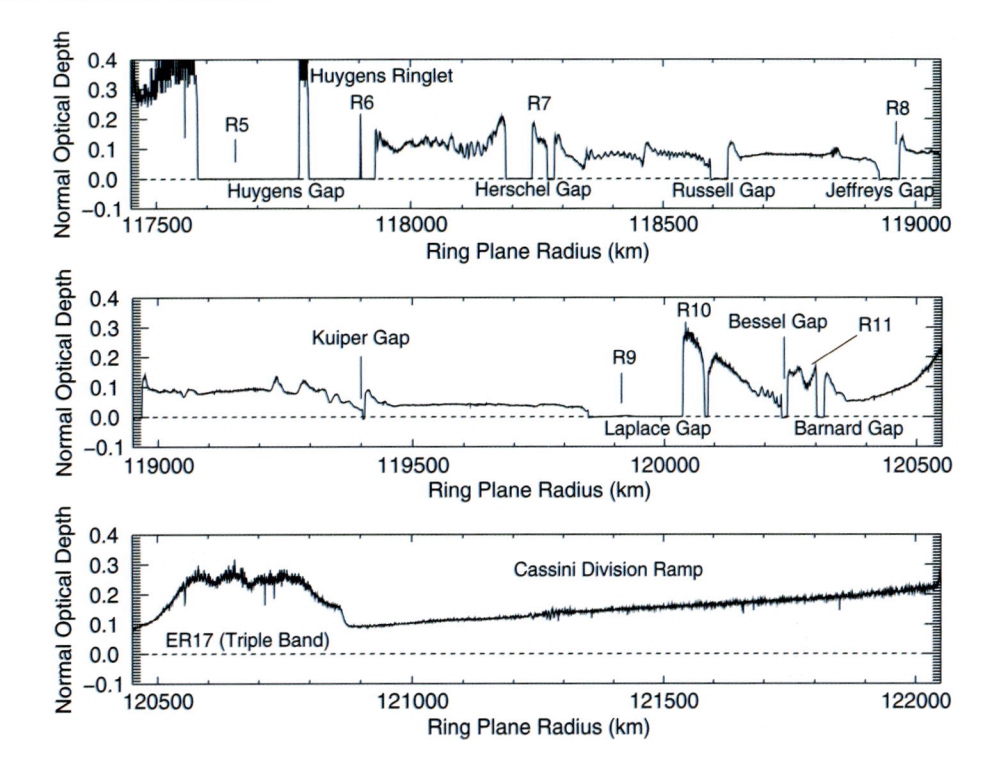

very near the outer edge of the gap and has also been suspected to be noncircular (Nicholson et al. 1990, Fig. 6c). It is 37 km wide and appreciably more opaque than its surroundings. Near the center of the Laplace Gap is a broad, diffuse ring (R9). This feature, which does appear to be dusty, is discussed in Chapter 16. Although it was first seen in the VIMS o Ceti occultation profile in Fig. 13.19, which was obtained at a very low opening angle of $B = 3.5°$, R9 is not visible in most occultation traces.

Two narrow gaps, the Bessel Gap and Barnard Gap, with widths of 10 and 13 km respectively, bracket a moderately opaque ringlet (R11). This 59-km wide ringlet, referred to as the 1.994 R_S ringlet by Nicholson et al. (1990), is also suspected to be noncircular. Its outer edge, which varies by at least 3 km in radius, falls close to the Prometheus 5:4 ILR at 120,304 km. The outer half of this ringlet also exhibits a wavelike structure of unknown origin. No models of these features, or of the ringlet at 1.990 R_S (R9), have been published.

Outside 120,400 km, the Cassini Division takes on a very different aspect, with no gaps and smoothly-varying optical depth. This region is dominated by a curious, broad feature with three distinct optical depth maxima (ER17, also referred to as the "Triple Band"). It is shown at sub-kilometer resolution in Fig. 5e of Porco et al. (2005), obtained during Saturn orbit insertion, where the absence of internal structure or any sharp edges is in striking contrast to the rest of the Cassini Division.

Completing the Cassini Division is a gradual, monotonic increase in optical depth towards the inner edge of the A ring at 122,050 km. This feature is often referred to as the "Cassini Division ramp", and with a width of ∼1150 km is very similar to the ramp at the outer edge of the C ring. As discussed in Section 13.2.4, ballistic transport processes provide a successful model for the morphology of these ramps.

13.4.3 Density Waves in the Cassini Division

Due in part to its greater distance from the satellites, there are many fewer density waves in the Cassini Division than in the A ring. Compounding the sparsity of strong resonances is the low surface density, which reduces the radial wavelength of any waves and makes them harder to identify. Nevertheless, several density waves have been identified in various Cassini data sets, and are briefly summarized here.

The first resolved spiral density wave to be reported anywhere within Saturn's rings was in the Cassini Division ramp: the Iapetus 1:0 apsidal resonance, whose long wavelength made it visible in Voyager 1 imaging data (Cuzzi et al. 1981). The reported surface density of 16 g cm^{-2} is reasonably consistent with the value of 11.5 g cm^{-2} more recently reported for the nearby Janus/Epimetheus 7:5 wave (Tiscareno et al. 2006b). Curiously, the Iapetus 1:0 wave has not been seen since, though perhaps it will appear again in 2009 when the low Sun angle (saturnian equinox) geometry of the Voyager 1 encounter is repeated.

Other density waves in the Cassini Division were first seen by Cassini (Porco et al. 2005; Tiscareno et al. 2007; Colwell et al. 2008, Table 13.3, Fig. 13.20), yielding the first

Table 13.3 Waves in the C ring and Cassini Division. Wave identification is given in 10^3 km or by the associated resonance, if known. Propagation Direction: inward means decreasing wavelength with decreasing radius; outward means decreasing wavelength with increasing radius. DW: Density wave. BW: Bending wave. "Rosen wave" identifications refer to Rosen et al. (1991b) Figs. 13.7 and 13.8. (A) On plateau P1. (B) Wavelike only in some occultations. (C) Just exterior to embedded ringlet ER10 (Table 13.4); see Fig. 13.22. (D) Only seen in low-incidence-angle occultations. (E) At 85,700 km in Voyager RSS data. On plateau P6. (F) On embedded ringlet ER13. (G) Wave launched at two locations due to resonances from the coorbital satellites. Location changes as coorbital satellite orbits change (Tiscareno et al. 2006b)

Wave	Propagation direction	Inner edge (km)	Outer edge (km)	Voyager RSS (Rosen et al. 1991b)	Note
\multicolumn{6}{c}{C ring}					
W74.67	Outward	74,666	74,669		
Mimas 4:1 DW	Outward	74,891	74,903	Rosen wave "a"	
W74.93	Outward	74,935	74,939		
W74.94	Outward	74,940	74,945	Rosen wave "b"	
W76.02	Inward	76,015	76,018		
W76.24	Outward	76,237	76,241		A
W76.44	Inward	76,432	76,436		
W76.73	Unclear	76,729	76,734		B
Titan -1:0 BW	Outward	77,520	77,560	Rosen wave "c"	
W80.98	Inward	80,975	80,985	Rosen wave "e"	
W82.00	Inward	81,998	82,006		
W82.06	Inward	82,045	82,060	Rosen wave "f"	C
W82.21	Inward	82,190	82,210	Rosen wave "g"	
W83.63	Inward	83,620	83,630	Rosen wave "h"	
W84.64	Inward	84,620	84,640	Rosen wave "i"	
W85.45	Inward	85,440	85,455	Rosen wave "j"	D
W85.48	Outward	85,480	85,495		
W85.51	Inward	85,505	85,517		
W85.52	Outward	85,520	85,550		
W85.67	Outward	85,670	85,695	Rosen wave "d"	E
W87.19	Inward	87,180	87,192		F
Atlas 2:1 DW	Outward	87,648	87,653		
\multicolumn{6}{c}{Cassini Division}					
Prometheus 9:7 DW	Outward	118,066	118,078		
Pan 6:5 DW	Outward	118,453	118,469		
Atlas 5:4 DW	Outward	118,830	118,848		
Pandora 9:7 DW	Outward	120,039	120,076		
Pan 7:6 DW	Outward	120,668	120,695		
Janus/Epimetheus 7:5 DW	Outward	121,262	121,320		G

measurements of surface density in a narrow ringlet (the Pandora 9:7 wave at 120,040 km) and in the Triple Band region (Pan 7:6 wave at 120,670 km). The measured surface mass densities for these and four other density waves in the Cassini Division are given in Fig. 13.3. There is a general trend toward higher σ at higher values of τ, but the ratio of τ/σ (the opacity or mass extinction coefficient, κ), is consistently 3–5 times higher in the Cassini Division than it is in the ramp or the A ring (Porco et al. 2005; Colwell et al. 2008). Damping of these waves places an upper limit of 5 m on the vertical thickness of the Cassini Division (Tiscareno et al. 2007; Colwell et al. 2008).

To date, no moons have been found in the Cassini Division or the C Ring, despite the many gaps that occur in those regions. However, Cassini's trajectory has not often put it into favorable geometry for detecting such moons, and searches continue.

13.4.4 C Ring Structure

The visual appearance of the C ring, as seen in Fig. 13.17, is dominated by a series of narrow, bright bands and several narrow gaps, superimposed on a faint, gently-undulating background. Comparison with occultation profiles reveals that most (or all?) of these brightness variations reflect variations in optical depth, which ranges from ~ 0.10 in

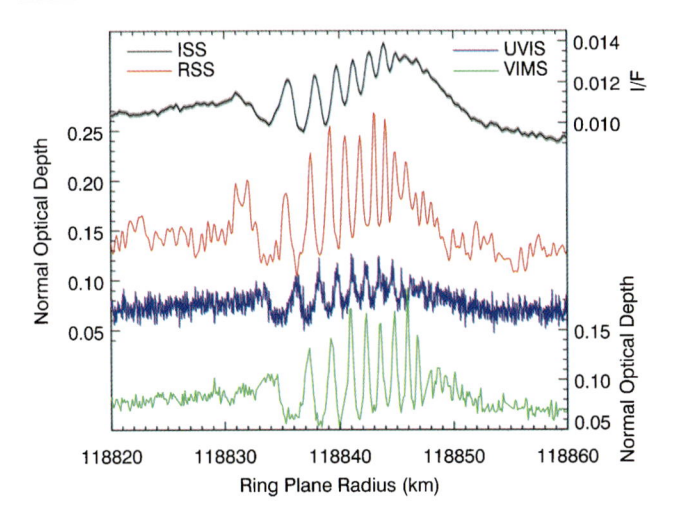

Fig. 13.20 The Atlas 5:4 density wave in the Cassini Division as seen by RSS, VIMS, and UVIS occultations and Cassini ISS images taken at Saturn Orbit Insertion. The RSS optical depths have not been divided by two in this figure

"background" regions to ∼0.4 in the bright bands, known as plateaus. (A few narrow ringlets, discussed further below, have $\tau > 1$, while the background drops to $\tau \simeq 0.05$ in the innermost C ring.) In a manner reminiscent of the Cassini Division, the C ring ends in a smooth "ramp" where the optical depth steadily increases to ∼0.25 before an abrupt transition to the B ring occurs. This ramp is ∼1340 km wide. The similar morphologies of the C ring/B ring transition and the Cassini Division/A ring transition support a common mode of origin, namely ballistic transport.

Table 13.4 lists all identified gaps, plateaus and ringlets in the C ring, including feature numbers from Nicholson et al. (1990) and French et al. (1993). As in Table 13.2, we include here various informal names which have been applied to the more prominent features, official IAU designations of gaps, and identification codes for unnamed features. In Fig. 13.21 we show the optical depth of the C ring from a Cassini RSS occultation, again chosen to highlight the low optical depth.

The plateau features are unique to the C ring, and are concentrated in the inner and outer thirds of the ring. In its archetypal form, a plateau has steep, but not perfectly sharp, edges and a U-shaped top: the optical depth is highest at the inner and outer edges and lower in the central region. In some cases, especially in the inner C ring, these edge peaks are very pronounced (e.g., P2 at 77,090 km), whereas in others the plateau is almost flat-topped (e.g., P9 at 89,250 km). Widths vary from 50 to 250 km. A total of 11 plateaus are identified in Table 13.4, 7 of which are in the outer C ring. Two features we have identified as embedded ringlets exhibit plateau-like inner edges, but gradual outer edges. In general, there is no other pattern to the distribution of plateaus in radius, and no connection with any of the relatively few strong satellite resonances in this region.

Between the inner and outer plateau regions, but extending into them for some distance as well, are optical depth background undulations with a wavelength of ∼1000 km and an amplitude of ∼30%. The average background optical depth increases from ∼0.05 around 77,000 km to a peak of ∼0.10 near 83,000 km, before declining to a minimum of ∼0.07 at 88,000 km. No specific models have been proposed for this structure, perhaps because the dominant wavelength is much longer than any known instability scale in the rings.

Sprinkled across the inner and outer C ring are five narrow gaps, similar to those in the Cassini Division, three of which contain isolated, sharp-edged, noncircular ringlets (Table 13.4). Two new ringlets (R1 and R2) were first seen in Cassini occultation data and are each separated from the inner edge of the gaps they inhabit by ∼1 km. The innermost gap is at 75,780 km, and is 80 km wide (G1). R1 is near the inner edge of this gap and has a peak optical depth of ∼0.02. Next is the Colombo Gap at 77,840 km, home to a narrow, very opaque, sharp-edged ringlet variously known as the Colombo or Titan ringlet. The latter name stems from this ringlet's association with the apsidal resonance with Titan, at ∼77,825 km, where the apsidal precession rate of ring particles matches the orbital mean motion of Titan. Analysis of Voyager observations of this ringlet showed that it is well-described by a simple ellipse precessing at this frequency, with the apoapse approximately aligned with Titan and maintained against differential precession by self-gravity (Porco et al. 1984a). Nicholson and Porco (1988) used the observed eccentricity of the Titan ringlet to infer a more accurate location for the apsidal resonance, thereby placing an important constraint on Saturn's zonal gravity harmonics, especially J_6. This gap is also home to R2 which has a similar optical depth and morphology to R1 and is similarly located near the inner edge of the gap.

In the outer C ring we find three gaps. At 260 km wide – 20% narrower than the Encke Gap – the Maxwell Gap at 87,480 km is the widest gap in the C ring. Within this gap is the Maxwell ringlet, the archetype of Saturnian eccentric ringlets (Esposito et al. 1983b; Porco et al. 1984a), plus a diffuse ringlet discovered in Cassini images (R3, Porco et al. 2005 their Fig. 7a). In many respects the Maxwell ringlet resembles the Uranian α, β, and ε rings: it is a simple Keplerian ellipse freely-precessing under the influence of Saturn's zonal gravity harmonics. Like its Uranian counterparts, its outer edge is more eccentric than the inner edge, leading to a width which is a minimum at pericenter and a maximum at apocenter (Porco et al. 1984a), a state which suggests that the ringlet's shape is preserved against differential precession by its own self-gravity (Goldreich and Tremaine 1979, 1981).

Table 13.4 C Ring ringlets, plateaus and gaps. Several ringlets and pronounced features and one gap do not have designations. We denote these in numerical order from the inner edge of the C ring through the Cassini Division (Table 13.2) by type: gap (G), ringlet (R, narrow ring in a gap), embedded ringlet (ER; regions of prominent increases in optical depth over the smoothly varying background optical depth), and plateaus (P, embedded ringlets with sharp edges and peak optical depths at the edges). Alternate IDs give the approximation location in 10^6 m. F93 refers to edges identified in Nicholson et al. (1990) and used by French et al. (1993). (A) Plateau-like outer edge. (B) Marginal plateau morphology. (C) Two peaks in optical depth. (D) The Titan and Maxwell ringlets are not circular and have variable widths (see text). Values shown here are from the UVIS α Virginis (Rev 034I) stellar occultation. (E) Adjacent to wave (see Fig. 13.22). (F) Plateau-like inner edge. (G) Porco et al. (2005). (H) Connects to P8. (I) Outer edge is inner edge of Dawes Gap, but inner edge is embedded in the C ring. This feature has abundant structure (see text). (J) This gap has variable width and a non-circular inner edge

ID (name)	Alternate	F93	Inner edge (km)	Outer edge (km)	Width (km)	Note
ER1	ER74.5	44	74,490	74,500	10	
ER2	ER75.7		75,665		<1	
G1	G75.7		75,730	75,810	80	
R1	R75.7		75,730	75,750	20	
ER3	ER75.81		75,810		6	
ER4	ER75.84		75,830	75,850	20	
ER5	ER75.91		75,895	75,925	30	
ER6	ER75.97		75,950	75,990	40	A
ER7	ER76.0		76,025	76,055	30	
P1	P76.2	40	76,110	76,265	155	
P2	P77.1	39	77,010	77,160	150	
P3	P77.4		77,345	77,410	65	B
ER8 (two peaks)	ER77.6		77,645	77,655	10	C
Colombo gap	G77.8	43	77,745	77,925	180	
R2	R77.8		77,750	77,760	10	
Titan ringlet	R77.9	62, 63	77,880	77,920	40	D
ER9	ER77.9	43	77,930	77,945	15	
P4	P79.2	37, 38	79,220	79,265	45	
ER10	ER82.0	36	82,030	82,045	15	E
P5	P84.8	34, 35	84,750	84,950	200	
P6	P85.7	33, 42	85,660	85,755	95	
ER11	ER85.9	31	85,920	85,990	70	F
ER12	ER86.0		86,015	86,030	15	
P7	P86.4	29, 30	86,375	86,605	230	
ER13	ER87.2		87,180	87,210	30	
ER14	ER87.3		87,290	87,350	60	F
Maxwell gap	G87.5		87,350	87,610	260	
R3	R87.4		87,418		~30	G
Maxwell ringlet	R87.5	60, 61	87,500	87,590	90	D
ER15	ER88.3		88,300	88,340	40	H
P8	P88.4	28	88,350	88,595	245	
Bond gap	G88.7		88,685	88,720	35	
R4 (1.470 R_S ringlet)	R88.7	58, 59	88,700	88,720	20	
P9	P89.2	27, 41	89,190	89,295	105	
P10	P89.8	25, 26	89,790	89,940	150	
ER16 (1.495 R_S ringlet)	ER90.2	56, 57	90,130	90,200	70	I
Dawes gap	G90.2		90,200	90,220	20	J
P11	P90.5	23, 24	90,405	90,610	205	
C Ring ramp			90,620	91,960	1,340	

13.4.5 Waves and Other Resonant Features in the C Ring

The strongest resonances that occur in the C Ring give rise not to waves but to gaps, some of which have eccentric ringlets embedded within them. The two outermost gaps in the C ring are the Bond Gap at 88,715 km and the Dawes Gap at 90,210 km. The Bond Gap contains a narrow, apparently noncircular ringlet, about 20 km wide (R4, Porco and Nicholson 1987). The Dawes Gap has no isolated ringlet, but is bounded on its inner edge by a 70-km-wide region featuring large fluctuations in optical depth. The outer edge of this feature (the inner edge of the Dawes Gap) also appears to be noncircular (Porco and Nicholson 1987). We denote it as an embedded ringlet (ER16). Holberg et al. (1982), Shu et al. (1983) and Porco and Nicholson (1987) all noted the coincidence between these two gap/ringlet pairs and two of the strongest satellite resonances in the C ring. Nicholson et al. (1990; see their Fig. 11) confirmed that the inner edge of R4 (the 1.470 R_S ringlet) matches the Mimas 3:1 inner vertical resonance, while the variable outer edge of ER16 (the 1.495 R_S ringlet and thus the inner edge of the Dawes Gap) matches the Mimas 3:1 ILR. The latter situation may be analogous to the outer termination of the B ring by the

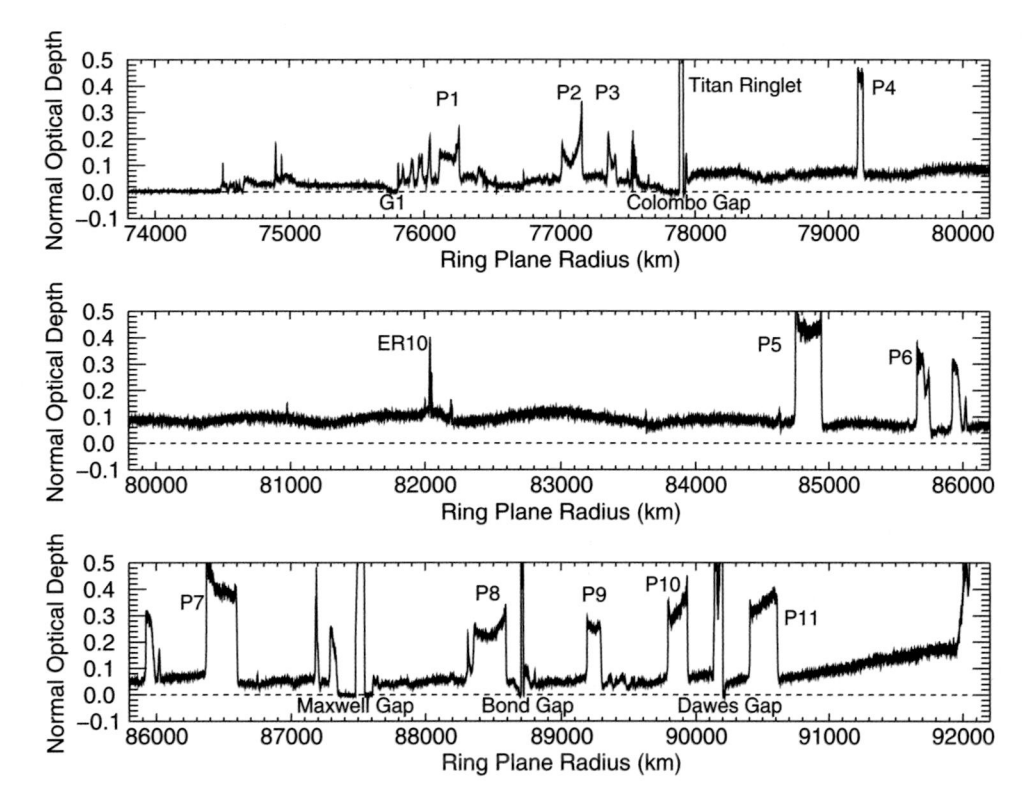

Fig. 13.21 Cassini RSS occultation profile of the C ring showing RSS optical depth divided by 2 (see text). Unlike the A and B rings, the measured normal optical depth of the C ring is insensitive to viewing geometry. Prominent features from Table 13.4 are labeled

Mimas 2:1 ILR, and of the A ring by the Janus 7:6 ILR, but involves a weaker, second order resonance. However, a study of all the available Voyager data by Porco and Nicholson (1987) failed to show conclusive evidence for resonant perturbations at either edge, though they did conclude that both ringlets are probably noncircular at the few km level. Clearly, further work is indicated here.

One of the more exotic resonances to be associated with an observed wave in the rings is the Titan – 1:0 nodal bending wave (Rosen and Lissauer 1988). This wave is excited by a resonance at which the nodal regression rate of ring particles is approximately commensurate with the orbital frequency of Titan. This is the only known bending wave that propagates outward, rather than towards Saturn, and the only observed spiral wave in which the wave pattern moves in a retrograde direction relative to individual particle orbits.

Ten other wave features in the C Ring were pointed out by Rosen et al. (1991b), though only one of them (Mimas 4:1) could be convincingly identified with a known resonance. All of these waves have also been observed by Cassini, as well as a number of new waves not previously observed (Table 13.3). Most of these waves appear to propagate *inwards* (e.g. Fig. 13.22), unlike almost all of the density waves in the A ring, and it has therefore been supposed that these are either (i) bending waves driven by an unknown inclined satellite(s), or (ii) density waves driven at Outer Lindblad resonances with some perturber which orbits interior to the C ring. Marley and Porco (1993) have suggested that this perturber is in fact Saturn itself, or more specifically its internal modes of oscillation (f-modes). Unfortunately, it is not yet possible to calculate the frequencies of such modes with sufficient accuracy to test this model.

13.5 F Ring

13.5.1 Overview

Saturn's F ring (semi-major axis 140223.7 ± 2 km, Bosh et al. 2002) is a narrow, eccentric, inclined ring with unusual, time varying structure. It was first detected by the Pioneer 11 spacecraft in 1979 (Gehrels et al. 1980) and subsequently imaged at higher resolution by the two Voyager spacecraft in 1980 and 1981 (Smith et al. 1981, 1982). The Voyager images showed a narrow ring (radial width ~50 km) with at least two additional strands. The flyby nature of the missions meant that longitudinal coverage was minimal, and it was not clear from the images if the ring's structure had changed

Fig. 13.22 A wave in the C ring adjacent to an embedded ringlet (ER10). The wave may propagate into the embedded ringlet. This is wave "f" from Rosen et al. (1991b). The UVIS and RSS occultation curves have been offset for clarity. The RSS optical depths have not been divided by two in this figure

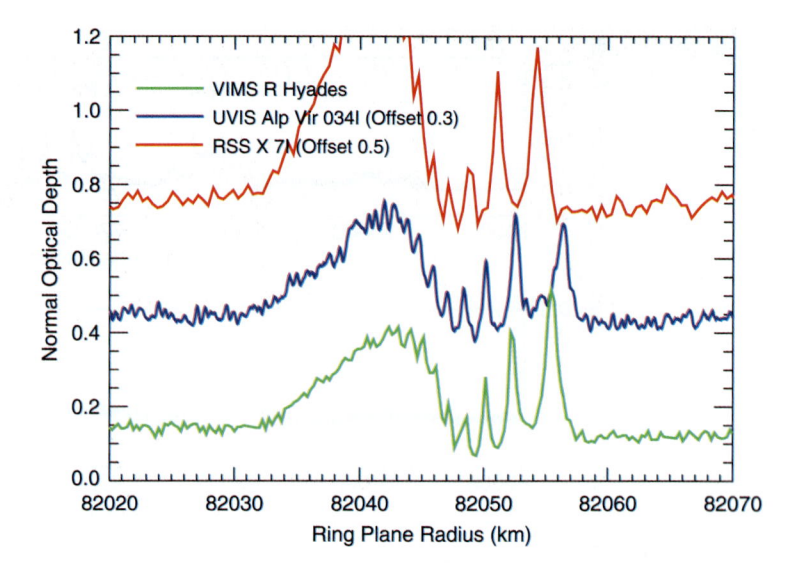

in the 9 months between the two Voyager encounters, or if the observed structure varied with longitude. The existence of the satellites Prometheus and Pandora, orbiting on either side of the ring, suggested that a shepherding mechanism prevented the ring from spreading, although this has yet to be demonstrated.

A population of moonlets in the F ring region was deduced by Cuzzi and Burns (1988) on the basis of electron absorption signatures measured by Pioneer 11. Collisions between the moonlets were seen as a source mechanism for clouds of debris that produced the electron absorptions and perhaps even, in a rarer, more energetic event, the F ring itself. Further evidence for a population of moonlets or ephemeral clumps in the F ring region was obtained in HST observations of the rings during ring plane crossing (Nicholson et al. 1996; Poulet et al. 2000; McGhee et al. 2001). The finer temporal coverage and higher spatial resolution of Cassini observations described below show the effects of these ephemeral clumps on the F ring in detail.

Using a combination of Voyager, HST and ground-based occultation data, Bosh et al. (2002) derived a freely-precessing Keplerian ellipse model with $ae = 360$ km and $a \sin i = 15$ km. Initial Cassini observations of the F ring obtained on approach and at SOI (Porco et al. 2005) confirmed its unusual, multi-stranded appearance and suggested that some of the more regular features were due to the gravitational effect of Prometheus. Additional Cassini images showed that some of the strands have a spiral structure (Charnoz et al. 2005), confirmed the key role played by Prometheus in creating regular, time varying features (Murray et al. 2005), and provided evidence for the existence of additional small satellites in the vicinity of the F ring (Porco et al. 2005; Spitale et al. 2006; Murray et al. 2008). In many respects the F ring is unique in terms of both its variety of structures and the mechanisms that determine them.

13.5.2 Perturbations by Prometheus

The proximity of Prometheus (semi-major axis 139380 km, Jacobson et al. 2008) has led several authors to undertake investigations of its perturbing effect on the F ring (see, e.g., Showalter and Burns 1982; Borderies and Goldreich 1983; Kolvoord et al. 1990). A satellite with a separation in semi-major axis Δa from a ring of material produces a feature of azimuthal wavelength $\lambda = 3\pi \Delta a$ on the ring (see Section 13.2.2). Numerical calculations by Showalter and Burns (1982) show that the relative eccentricity of the satellite and ring is important, rather than the individual eccentricities. Note that λ is a function of Δa and not Δr, as it is determined by the difference in mean motion, Δn. If the separation is sufficiently large and the orbits near circular, the effect on the ring is to produce a quasi-sinusoidal shape on the edge (see Section 13.2.2.1). In the case of the F ring-Prometheus system, $\lambda = 7952$ km, or $3.25°$ at the distance of the F ring; the equivalent value for the F ring-Pandora system is $5.72°$. Because of differential precession the closest approach between Prometheus and the F ring's core varies between 180 and 790 km on a timescale of $\sim$19 years. At the time of the Cassini SOI the closest approach distance was $\sim$500 km. In the course of one Prometheus orbital period of 14.7 h, the satellite's approach to, and retreat from, the F ring produces a characteristic "streamer-channel" phenomenon first detected in numerical simulations (see Giuliatti Winter 1994; Giuliatti Winter et al. 2000). In the course of one orbital period material is drawn out of the ring on perturbed orbits and then retreats back into the ring (Murray et al. 2005). The cycle repeats itself such that particles which have already been perturbed, continue to follow the streamer-channel behavior downstream of the encounter. The paths of ring particles involved in the streamer-channel phenomenon can be explained as primarily due to changes in the particles'

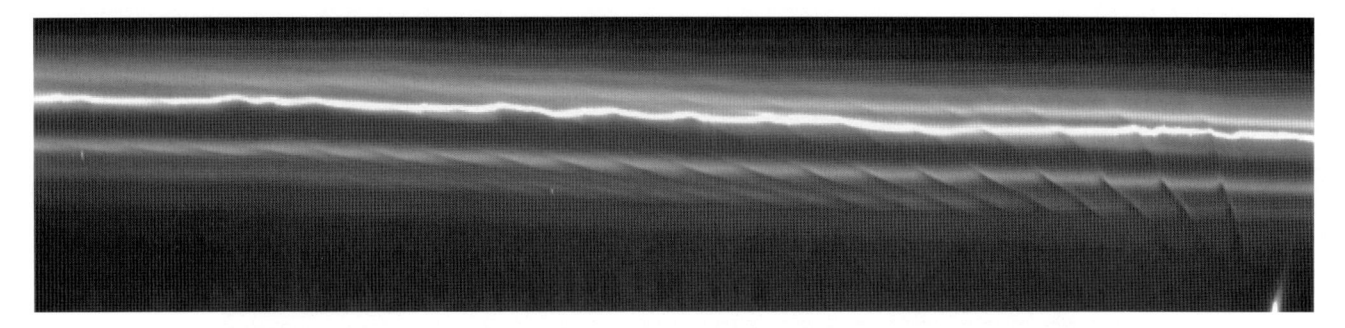

Fig. 13.23 Sheared channels in a mosaic of 15 reprojected images (NASA/JPL/Space Science Institute) of the F-ring region taken by the Cassini ISS narrow angle camera on 13 April 2005. The mosaic covers ∼60° in longitude and the radial range is 1,500 km. The slope of the ring strands is a consequence of the eccentricity of the ring becoming evident over the longitude range. The mosaic shows multiple sheared channels to the left of Prometheus (visible at the lower right corner) (see also Murray et al. (2005))

eccentricities induced by the approach to Prometheus and not to changes in a. Figure 13.23 shows a mosaic of 15 Cassini images illustrating the streamer-channel phenomenon in the F ring. The bright central part of the ring's core is visible, and Prometheus is at the lower right-hand side. Note that the system was in the channel part of the cycle at the time the images were taken, and that the channel extends across the entire region. The longitudinal spacing of the channels seen in Fig. 13.23 is $3\pi \Delta a$ and therefore smaller for the strands closer to Prometheus.

The synodic period associated with the relative motion of Prometheus and the F ring is 67.65 days whereas differential precession causes periapse alignment to be repeated every ∼19 years. Therefore the extent of the perturbations from Prometheus should vary on the same timescale. Murray et al. (2008) calculated the maximum absolute values of the perturbation in semi-major axis, a at alignment, at a periapse difference of ±90° and at anti-alignment for particles in the F ring core. These were <0.5, 4 and 19 km respectively. The corresponding change in eccentricity, e and longitude of periapse, ϖ were 2, 6 and 13×10^{-5}, and 0.5°, 1.5° and 4°, respectively. These values and accompanying simulations suggest that although Prometheus has a considerable effect on the F ring core and surrounding material, this tends to be regular in nature and cannot account for the large scale structures seen in Cassini images.

13.5.3 Jets and Kinematic Spirals

Figure 13.24 shows an annotated, 360° mosaic of the F ring derived from Cassini ISS narrow angle camera images obtained on 2005 April 13. As well as the distorted bright core and the familiar sheared channels due to Prometheus (and Pandora), the gross structure is dominated by several "jets" of material with different slopes appearing to emanate from the F ring's core. There are also several parallel lines of material covering 360° that are at an angle to the horizontal axis; these are the kinematic spirals discovered by Charnoz et al. (2005) who proposed that they are the consequences of ∼30 m/s physical collisions between the core and a nearby object, most likely S/2004 S 6 (Charnoz et al. 2005; Porco et al. 2005).

Using evidence from a series of mosaics obtained between December 2006 and May 2007; Murray et al. (2008) showed that the progress of an object with an orbit similar to that of S/2004 S 6 was associated with a succession of jets at different longitudes. These jets subsequently underwent Keplerian shear and the more prominent ones would ultimately form kinematic spirals such as those seen in Fig. 13.24. Futhermore, Murray et al. (2008) pointed out that the appearance of additional jets away from the location of S/2004 S 6 suggests that other objects in the vicinity of the F ring were also involved in collisions. Hence it is likely that the material within ∼ ± 500 km of the F ring core is fed into this region as a result of collisions between the core and a small population of ∼10 satellites (radius <2 km).

13.5.4 Embedded Objects

A satellite embedded in a ring but having a small relative eccentricity with respect to the ring produces a characteristic pattern called a "fan" (Murray et al. 2008). This consists of a sequence of sheared channels with a common point of intersection visible in any surrounding, perturbed material. The mechanism is similar to that which causes the "propeller" structures seen in the A ring (Tiscareno et al. 2006a, Section 2.5). However, whereas a propeller is essentially unchanging in its structure, a fan appears and disappears on an orbital timescale as the object completes its epicyclic motion with respect to the core. In this respect they are similar to the streamer-channel process. Two examples of fans due to embedded satellites are shown in Fig. 13.25. In this re-projected

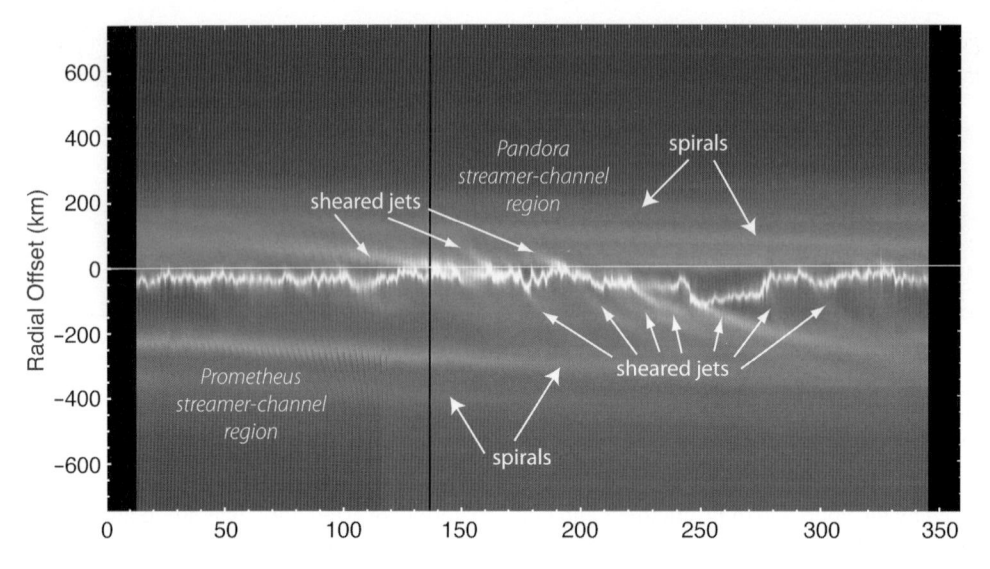

Fig. 13.24 A mosaic of re-projected images (NASA/JPL/Space Science Institute) of the F ring obtained from low phase, Cassini ISS narrow angle images on 2005 April 13 (observation ISS-006RI-LPHRLFMOV001-PRIME). The radial location of the F ring according to the Bosh et al. (2002) model is used as the vertical centre for each reprojection in which the horizontal axis is the longitude at the epoch of 12:00 UTC on 2007 January 1. The mosaic is annotated to show the more prominent jets, spirals, and the channels due to Prometheus and Pandora

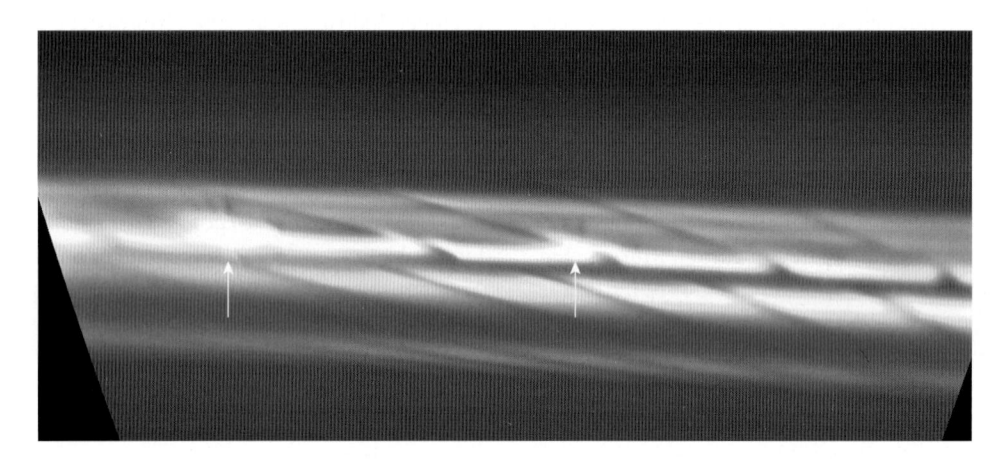

Fig. 13.25 A re-projected Cassini narrow angle camera image (NASA/JPL/Space Science Institute) of the F ring showing the presence of two "fan" structures caused by embedded satellites. The most prominent features are channels driven by encounters with Prometheus. The image, N1571964006, was obtained on 2007 October 25 and covers a longitudinal range (horizontal axis) of 18.2° and a radial range (vertical axis) of 1000∼km. The locations of two prominent fans are indicated with arrows

image each fan appears to be associated with the location of sheared channels due to the last Prometheus passage. Note that the Prometheus channels in the core and surrounding strands (i.e. sheared jets) are offset with respect to one another; this is probably due to the jets having different orbital elements with respect to the core and each other. Additional images of fans together with a direct comparison with a numerical simulation are given in Fig. 3 of Murray et al. (2008). From the simulations the observed maximum longitudinal width of a channel in a fan is ∼10 times the radius of the embedded satellite's Hill sphere. This provides a diagnostic means of calculating the size of the perturbing objects. On this basis the embedded satellites detected in the F ring core have radii that are typically 5 km or less.

It is clear that Prometheus exerts a major perturbing influence on the F ring. However, it is also clear from numerical simulations and the Cassini images of the streamer-channel phenomenon in jets (see, e.g., Fig. 13.24) that the effects of Keplerian shear would significantly reduce the detectability of the channels ∼60° downstream of Prometheus. If the ∼±50 km radial distortions seen in the core, over and above those due to the effect of collisions with objects such as S/2004 S 6, are due to Prometheus then why do they endure over the synodic period between encounters? Murray et al. (2008) suggested that the reason for the core's "memory" of earlier encounters is that the large objects in and around the core that are perturbed by Prometheus, retain their relative elements because they are less vulnerable to the effects of collisions. The only effective means of damping the relative eccentricities of the larger objects is with self-gravity or collisions with objects of similar size, and therefore the timescales are longer. Thus the memory of the core is preserved in the larger objects, with each localized radial distortion corresponding to the presence of one object. The Cassini images show numerous examples of structures in the core that oscillate on an orbital timescale (see, e.g. Fig. 8c and d of Porco et al. 2005 and Fig. 4b–d of Murray et al. 2008). More importantly these features tend to oscillate in phase, suggesting a common origin for the perturbed motion; this is entirely consistent with behavior expected from perturbations by Prometheus. There are typically several hundred localized distortions of the core at any one time implying a similar number of perturbed objects with radii <1 km.

A high resolution azimuthal scan of the F ring obtained on 2006 September 25 gives additional evidence of small objects near the core. Figure 13.26 shows one of the images from the scan, annotated to indicate the location of possible small objects and their associated structures near the core. The structures indicated by the three unfilled arrows above the core could even be associated with the three objects and associated structures seen at the lower left of the image. These may be similar to the objects detected in occultation data by the Cassini UVIS and VIMS experiments (Esposito et al. 2008).

13.5.5 A Narrow Component

Some of the images in the azimuthal scan of 2006 September 25 show a narrow (radial width ~1 km) ring component located ~15–50 km radially inward of the F ring's core. An example of one such image is shown in Fig. 13.27. This shows a narrow component located ~15 km interior to the core, as well as 2–3 objects that appear to be embedded in this component. On the basis of a limited number of detections in the imaging sequence, Murray et al. (2008) reported that segments of this component had an orbit that was closer to that derived by Bosh et al. (2002) for the F ring's occultation core than the core seen in the images. The existence of such a narrow component is more in keeping with the results of the Voyager radio science (Tyler et al. 1983; Marouf et al. 1986) and photopolarimeter occultations (Lane et al. 1982) than with the ~50 km wide feature seen in Cassini UVIS and VIMS occultation data, as well as all the Cassini ISS images. The broad Cassini-era "core" would have been detected by the Voyager PPS. Images such as that shown in Fig. 13.26 perhaps suggest that within the usual ~20–50 km

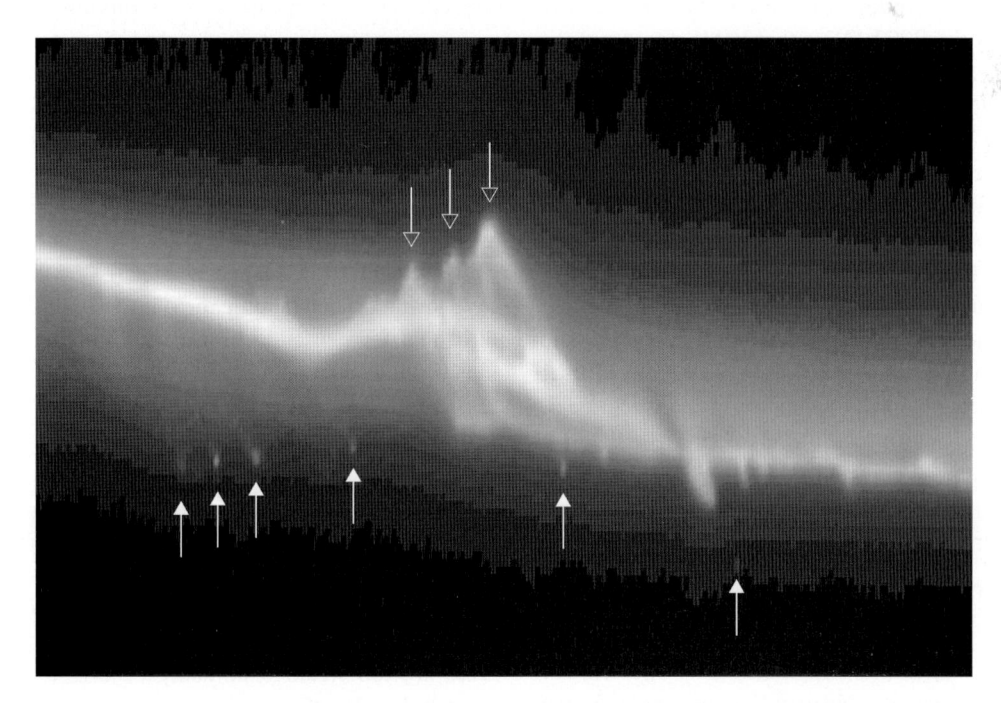

Fig. 13.26 A re-projected, stretched, high-resolution Cassini narrow angle camera image (of the F ring core (N1537898708), obtained on 2006 September 25. The longitudinal range (horizontal axis) is 0.2° while the radial range (vertical axis) is 100 km. The image shows evidence for the existence of small (<1 km) objects (indicated by filled *arrows*) traversing the core, as well as possible corresponding features (indicated by unfilled *arrows*) in the core itself (image: NASA/JPL/Space Science Institute)

Fig. 13.27 A re-projected, high-resolution Cassini narrow angle camera image (NASA/JPL/Space Science Institute) of the F ring core (N1537888059), obtained on 2006 September 25. The longitudinal range (horizontal axis) is 0.2° while the radial range (vertical axis) is 80 km. The arrow indicates the location of a narrow discontinuous ring component at a radial distance of 139,975 km; the circle indicates the location of 2–3 objects embedded in the narrow component

broad "core" is a narrower, distorted, true core. This "true core" may contain most of the F ring's mass, in the form of embedded, 1 km-sized moonlets. Collisions between these moonlets would then produce the dust in the visible core (Barbara and Esposito 2002). Note that VIMS spectra of the F ring imply typical particle sizes of 1–10 μm in the F ring's bright core and thus little mass (see Chapter 16).

Although it is difficult to account for the disparate objects and structures within the F ring region in a single theory, one possibility is that we are observing the result of an ongoing competition between accretion and fragmentation (Barbara and Esposito 2002) combined with the perturbing effect of Prometheus. This could explain the production of new objects, their physical and dynamical evolution, and their ultimate demise. Future progress on the structure of the F ring may well depend on a combination of realistic modeling of collisions and additional data derived from the Cassini Equinox Mission and beyond.

13.6 Summary

Our understanding of the structure of Saturn's rings has improved dramatically over the course of Cassini's initial 4-year mission. In large part this is due to the combination of observations from multiple instruments taken from a wide range of geometries. This has enabled, for the first time, a detailed exploration of non-axisymmetric structures in the rings as well as the vertical structure of the main rings. The temporal baseline of these observations, which may extend beyond a decade with a proposed solstice mission, also provides a unique opportunity to study the changes in ring structure due to changes in gravitational forcing as the orbits of nearby ringmoons shift on timescales of years. The macroscopic structure of Saturn's rings is virtually identical to that which was observed by Voyager. The most notable exception in the main rings covered here is the change in morphology of the F ring, which has continued to evolve underneath Cassini's multi-wavelength eyes (Section 13.5). Nevertheless, as we move toward understanding the properties of the rings on the collisional scale of clumps of particles, there are several avenues for further research.

13.6.1 Density Waves

Cassini observations provided the first measurements of a number of weak waves in the rings (Porco et al. 2005; Tiscareno et al. 2007; Colwell et al. 2009). Some of these waves are too weak to be analyzed in individual occultation profiles, but the combination of multiple datasets and azimuthal averaging makes it possible for ring properties to be extracted from these weak waves. Further observations as the mission continues should open up more of these second-order waves and weak first-order waves for detailed analysis. Waves in the C ring remain poorly understood.

In addition to having broad coverage in wave phase and observations at a variety of geometries, temporal coverage is key to understanding the propagation of waves. Preliminary work on the Janus/Epimetheus density waves has shown that ring properties can be extracted by combining observations made at different epochs relative to the orbital "swap" of these co-orbital satellites (Tiscareno et al. 2006). Other moons in near-resonance with each other also shift on timescales of several years resulting in potentially observable effects in the dispersion of density waves. These second-order effects provide a new window into ring viscosity, thickness, and surface mass density (Sremčević et al. 2008).

The larger question of angular momentum transport between rings and moons is primed for a fresh attack. The combination of multiple stellar and radio occultations at different phases of a density wave will enable mapping the two-dimensional density structure of density waves with great accuracy, perhaps enabling a direct calculation of the torque exerted on the resonant satellite (Rappaport et al. 2008).

13.6.2 Microstructure of the Rings

Many of our advances in ring structure have come from the ability of Cassini instruments to detect and measure fine-scale structures. Self-gravity wakes have been measured in unprecedented detail through the combination of high resolution stellar occultations and images at multiple geometries. From Cassini and HST observations we now know that these structures are ubiquitous in the A and B rings. If they are present in the optically thicker regions of the C ring and Cassini Division, they are less distinct and therefore have a smaller effect on observed ring properties. Further analysis of occultation data will place limits on the extent of self-gravity wakes in these regions.

Short-wavelength (~100–200 m) axisymmetric periodic structures, likely due to viscous overstability (see Chapter 14), have been observed by Cassini radio and stellar occultations. UVIS data reveal these waves directly, opening the possibility that combining observations may enable measurement of the oscillation of these structures confirming their identification as viscous overstability oscillations. The direct measurement of individual peaks and troughs of these waves in a UVIS stellar occultation show a mismatch between ingress and egress profiles, suggesting that they may have limited azimuthal extent. Indeed, they are not expected to persist uninterrupted for the full circumference of the rings (Jürgen Schmidt, personal communication, 2008). Further observations as well as a refined geometrical solution to the ring system will be needed to resolve the question of the azimuthal scale of these structures.

The highest resolution images, obtained at Saturn Orbit Insertion, revealed additional clumpiness in the rings at the sub-kilometer scale, dubbed "straw" and "rope" (Porco et al. 2005), which appears only in the troughs of strong density waves or satellite wakes. These structures may contain additional clues to the physical properties of individual ring particles and their proclivity to adhere to one another in dense regions. Extracting information about these structures will require additional images at very high resolution, perhaps possible in a proposed solstice mission, and combination of occultation measurements with an accurate geometric solution for the rings.

13.6.3 Vertical Structure

Observations in 2009 near Saturn's equinox will provide a new opportunity for unique measurements related to the vertical structure of the rings as shadows cast by bending waves will be directly observable. Thermal measurements by CIRS at this epoch will inform us about the vertical transport of ring particles as they orbit Saturn. Variations in the optical depth profiles of bending waves observed in stellar and radio occultations made at different incidence angles to the rings in effect show the shadow of these vertically extended waves, also enabling their vertical extent to be measured.

The fine spatial resolution of the UVIS stellar occultations has resolved the sharp edges of some rings for the first time. The radial profile observed at these edges is a combination of the vertical and radial structure of the ring within ~100 m of the edge. The combination of multiple occultations will enable separation of these two variations and direct measurement of the thickness at the edge.

13.6.4 Large-Scale Structure

In spite of the impressive suite of measurements that have enabled the rings to be probed at ever-smaller scales, most ring structure remains unexplained. Cassini occultations have revealed the structure of the core of the B ring for the first time, showing a nearly bimodal distribution of ring optical depth with much of the ring being essentially opaque. These abrupt transitions in optical depth may be revealing a previously unsuspected instability in the rings.

Other large-scale structure, such as the irregular structure in most of the B ring, the long-wavelength undulations in the C ring, and the C ring plateaus, remain as puzzling today as when they were first measured. Even narrow ringlets and gaps throughout the C ring and Cassini Division, long believed to be confined and cleared by embedded moonlets, remain mysterious as these moonlets, with the exceptions of Daphnis in the Keeler Gap and Pan in the Encke Gap, have evaded detection to date. Refined searches for these moons in the Cassini extended mission may yet reveal their presence.

13.6.5 Future Prospects

Many of our advances in understanding the structure of the rings were not anticipated prior to Cassini. A key aspect of these advances has been the combination of observations made at different geometries and at different times, as well as the different perspective offered by multiple instruments. We can thus safely anticipate new discoveries from the remainder of Cassini's Equinox Mission and final extended mission as the time baseline and geometric diversity of observations is extended.

Improved understanding of ring microphysics enabled by modeling in combination with Cassini observations of microstructure may reveal the source of the formation of plateaus or the bimodal structure in the B ring.

As the time baseline of measurements of the dynamic F ring grows longer, we gain a better understanding of the interaction between moon and ring and the growth and destruction of agglomerates within the F ring region.

The relative spatial precision of Cassini occultation measurements will ultimately enable highly accurate kinematic models of ring features that are linked to moons. The combination of these future developments should shed light on the question of the origin of the rings as well as how they reached their current configuration.

Acknowledgements This work was carried out with support from the NASA/ESA Cassini-Huygens project. We thank Matt Hedman for assistance with some figures. JEC and MST were also supported by the NASA Cassini Data Analysis Program. CM is grateful to the UK Science and Technology Facilities Council for financial support. The authors would also like to acknowledge the work of their colleagues on the various Cassini science and engineering teams and the dedicated science planning engineers at JPL, without whom none of the Cassini data described herein would exist.

References

Alexander, A. F. O'D. 1962. *The Planet Saturn: A History of Observation, Theory, and Discovery.* Faber and Faber, London.
Araki, S., Tremaine, S. 1986. The dynamics of dense particle disks. *Icarus* **65**, 83–109.
Barbara, J., Esposito, L. W. 2002. Moonlet collisions and the effects of tidally modified accretion in Saturn's F ring. *Icarus* **160**, 161–171.
Borderies, N., Goldreich, P. 1983. The variations in eccentricity and apse precession rate of a narrow ring perturbed by a close satellite. *Icarus* **53**, 84–89.
Borderies, N., Goldreich, P., Tremaine, S. 1986. Nonlinear density waves in planetary rings. *Icarus* **68**, 522–533.
Bosh, A. S., Olkin, C. B., French, R. G., Nicholson, P. D. 2002. Saturn's F ring: Kinematics and particle sizes from stellar occultation studies. *Icarus* **157**, 57–75.
Bridges, F., Hatzes, A., Lin, D. 1984. Structure, stability, and evolution of Saturn's Rings. *Nature* **309**, 333–335.
Camichel, H., 1958. Mesures photométriques de Saturne et de son anneau. *Ann. d'Astrophys.* **21**, 231–242.
Chakrabarti, S. K. 1989. The dynamics of particles in the bending waves of planetary rings. *Mon. Not. Roy. Astron. Soc.* **238**, 1381–1394.
Charnoz, S., Porco, C. C., Déau, E., Brahic, A., Spitale, J. N., Bacques, G., Baillié, K. 2005. Cassini discovers a kinematic spiral ring around Saturn. *Science* **310**, 1300–1304.
Colombo, G., Goldreich, P., Harris, A.W. 1976. Spiral structure as an explanation for the asymmetric brightness of Saturn's A ring. *Nature* **264**, 344–345.
Colwell, J. E., Esposito, L. W., Sremčević, M. 2005. Cassini UVIS star occultation results for Saturn's rings. American Geophysical Union, Fall Meeting 2005, abstract #P31D-03.
Colwell, J. E., Esposito, L. W., Sremčević, M. 2006. Gravitational wakes in Saturn's A ring measured by stellar occultations from Cassini. *Geophys. Res. Lett.* **33**, L07201, doi:10.1029/2005GL025163.
Colwell, J. E., Esposito, L. W., Sremčević, M., Stewart, G. R., McClintock, W. E. 2007. Self-gravity wakes and radial structure of Saturn's B ring. *Icarus* **190**, 127–144, doi:10.1016/j.icarus.2007.03.018.
Colwell, J. E., Cooney, J. H., Esposito, L. W., Sremčević, M. 2009. Density Waves in Cassini UVIS stellar occultations 1. The Cassini Division. *Icarus* **200**, 2, 574–580, doi:10.1016/j.icarus.2008.12.031.
Colwell, J. E., Esposito, L. W., Jerousek, R. G., Lissauer, J. J. 2008. Vertical structure of Saturn's rings from Cassini UVIS stellar occultations. 37th COSPAR Scientific Assembly, Montreal.
Cooke, M. L. 1991. Saturn's Rings: Radial Variation in the Keeler Gap and C Ring Photometry. Ph.D. Thesis. Cornell University, Ithaca.
Cuzzi, J. N., Lissauer, J. J., Shu, F. H. 1981. Density waves in Saturn's rings. *Nature* **292**, 703–707.
Cuzzi, J. N., 1985. Rings of Uranus: Not so thick, not so black. *Icarus* **63**, 312–316.
Cuzzi, J. N., Scargle, J. D. 1985. Wavy edges suggest moonlet in Encke's gap. *Astrophys. J.* **292**, 276–290.
Cuzzi, J. N., Burns, J. A. 1988. Charged particle depletion surrounding Saturn's F ring: evidence for a moonlet belt? *Icarus* **74**, 284–324.
Cuzzi, J. N., Lissauer, J. J., Esposito, L. W., Holberg, J. B., Marouf, E. A.,Tyler, G. L, Boischot, A. 1984. Saturn's rings: Properties and processes. In *Planetary Rings* (R. Greenberg and A. Brahic, Eds.) University of Arizona Press, Tucson, pp. 73–199.
Cuzzi, J. N., Clark, R., Filacchione, G., French, R., Johnson, R., Marouf, E., Spilker, L. 2009. Ring particle composition and size distribution. In *Saturn After Cassini-Huygens* (M. Dougherty, L. Esposito, S. Krimigis, Eds.) Springer.
Dermott, S. F. 1984. In *Planetary Rings* (R. Greenberg and A. Brahic, Eds.) University of Arizona Press, Tucson, pp. 589–637.
Dones, L., Porco, C. C. 1989. Spiral density wakes in Saturn's A ring? *Bull. Am. Astron. Soc.* **21**, 929.
Dones, L., Cuzzi, J.N., Showalter, M.R. 1993. Voyager photometry of Saturn's A ring. *Icarus* **105**, 184–215.
Dunn, D. E., Molnar, L. A., Niehof, J. T., de Pater, I., Lissauer, J. J. 2004. Microwave observations of Saturn's rings: anisotropy in directly transmitted and scattered saturnian thermal emission. *Icarus* **171**, 183–198.

Dunn, D. E., de Pater, I., Wright, M., Hogerheijde, M. R., Molnar, L. A. 2005. High-quality BIMA-OVRO images of Saturn and its rings at 1.3 and 3 millimeters. *Astron. J.* **129**, 1109–1116.

Dunn, D. E., de Pater, I., Molnar, L. A. 2007. Examining the wake structure in Saturn's rings from microwave observations over varying ring opening angles and wavelengths. *Icarus*, **192**, 56–76, doi:10.1016/j.icarus.2007.06.017.

Durisen, R. H., Cramer, N. L., Murphy, B. W., Cuzzi, J. N., Mullikin, T. L., Cederbloom, S. E. 1989. Ballistic transport in planetary ring systems due to particle erosion mechanisms. I – Theory, numerical methods, and illustrative examples. *Icarus* **80**, 136–166.

Durisen, R. H., Bode, P. W., Cuzzi, J. N., Cederbloom, S. E., Murphy, B. W. 1992. Ballistic transport in planetary ring systems due to particle erosion mechanisms. II. Theoretical models for Saturn's A- and B-Ring inner edges. *Icarus* **100**, 364–393.

Esposito, L. W., O'Callaghan, M. West, R. A. 1983a. The structure of Saturn's rings: Implications from the Voyager stellar occultation. *Icarus* **56**, 439–452.

Esposito, L. W., Harris, C. C., Simmons, K. E. 1987. Features in Saturn's rings. *Astrophys. J. Supp.*, **63**, 749–770.

Esposito, L. W., Borderies, N., Goldreich, P., Cuzzi, J. N., Holberg, J. B., Lane, A. L., Pomphrey, R. B., Terrile, R. J., Lissauer, J. J., Marouf, E. A., Tyler, G. L. 1983b. Eccentric ringlet in the Maxwell gap at 1.45 Saturn radii Multi-instrument Voyager observations. *Science* **222**, 57–60.

Esposito, L. W., Cuzzi, J. N., Holberg, J. B., Marouf, E. A., Tyler, G. L., Porco, C. C. 1984. Saturn's rings: Structure, dynamics, and particle properties. In *Saturn* (T. Gehrels and M. Matthews, Eds.) University of Arizona Press, Tucson, pp. 463–545.

Esposito, L. W., Meinke, B. K., Colwell, J. E., Nicholson, P. D., Hedman, M. M. 2008. Moonlets and clumps in Saturn's F ring. *Icarus* **194**, 278–289.

Estrada, P. R., Cuzzi, J. N. 1996. Voyager observations of the color of saturn's rings. *Icarus* **122**, 251–272.

Ferrin, I. 1975. On the structure of Saturn's rings and the "real" rotational period for the planet. *Astrophys. Space Sci.* **33**, 453–457.

Filacchione, G., 28 colleagues. 2007. Saturn's icy satellites investigated by Cassini-VIMS. I. Full-disk properties: 350–5100 nm reflectance spectra and phase curves. *Icarus* **166**, 212–222.

Flynn, B. C., Cuzzi, J. N. 1989. Regular structure in the inner Cassini Division of Saturn's rings. *Icarus* **82**, 180–199.

Franklin, F. A., Cook, A. F., Barrey, R. T. F., Roff, C. A., Hunt, G. E., de Rueda, H. B. 1987. Voyager observations of the azimuthal brightness variations in Saturn's rings. *Icarus* **69**, 280–296.

French, R. G., Elliot, J. L., French, L. M., Kangas, J. A., Meech, K. J., Ressler, M. E., Buie, M. W., Frogel, J. A., Holberg, J. B., Fuensalida, J. J., Joy, M. 1988. Uranian ring orbits from earth-based and Voyager occultation observations. *Icarus* **73**, 349–378.

French, R. G., Nicholson, P. D., Cooke, M. L., Elliot, J. L., Matthews, K., Perković, O., Tollestrup, E., Harvey, P., Chanover, N. J., Clark, M. A., Dunham, E. W., Forrest, W., Harrington, J., Pipher, J., Brahic, A., Grenier, I., Roques, F., Arndt, M. 1993. Geometry of the Saturn system from the 3 July 1989 occultation of 28 Sgr and Voyager observations. *Icarus* **103**, 163–214.

French, R. G., Nicholson, P. D. 2000. Saturn's rings II. Particle sizes inferred from stellar occultation data. *Icarus* **145**, 502–523, doi:10.1006/icar.2000.6357.

French, R. G., Salo, H., McGhee, C., Dones, L. 2007. HST observations of azimuthal asymmetry in Saturn's rings, *Icarus* (in press).

Gehrels, T., Baker, L. R., Beshore, E., Blenman, C., Burke, J. J., Castillo, N. D., Dacosta, B., Degewij, J., Doose, L. R., Fountain, J. W., Gotobed, J., Kenknight, C. E., Kingston, R., McLaughlin, G., McMillan, R., Murphy, R., Smith, P. H., Stoll, C. P., Strickland, R. N., Tomasko, M. G., Wijesinghe, M. P., Coffeen, D. L., Esposito, L. W. 1980. Imaging photopolarimeter on Pioneer Saturn. *Science* **207**, 434–439.

Gehrels, T., Esposito, L.W. 1981. Pioneer fly-by of Saturn and its rings. *Adv. Space Res.* **1**, 67–71.

Gehrels, M., Matthews, M. (Eds.). 1984. *Saturn*. University of Arizona Press, Tucson.

Giuliatti Winter, S. M. 1994. The dynamics of Saturn's F Ring. Ph.D. Thesis, Queen Mary and Westfield College, University of London.

Giuliatti Winter, S. M., Murray, C. D., Gordon, M. 2000. Perturbations to Saturn's F ring strands at their closest approach to Prometheus. *Planet. Space. Sci.* **48**, 817–827.

Goldreich, P., Tremaine, S. 1978a. The velocity dispersion in Saturn's rings. *Icarus* **34**, 227–239.

Goldreich, P., Tremaine, S. 1978b. The formation of the Cassini division in Saturn's rings. *Icarus* **34**, 240–253.

Goldreich, P., Tremaine, S. 1980. Disk–satellite interactions. *Astrophys. J.* **241**, 425–441.

Goldreich, P., Tremaine, S. 1982. The dynamics of planetary rings. *Ann. Rev. Astron. Astrophys.* **20**, 249–283.

Goldreich, P., Tremaine, S. 1979. Precession of the epsilon ring of Uranus. *Astron. J.* **84**, 1638–1641.

Goldreich, P., Tremaine, S. 1981. The origin of the eccentricities of the rings of Uranus. *Astrophys. J.* **243**, 1062–1075.

Greenberg, R., Brahic, A. (Eds.). 1984. *Planetary Rings*. University of Arizona Press, Tucson.

Harrington, J. Cooke, M. L., Forrest, W. J., Pipher, J. L., Dunham, E. W., Elliot, J. L. 1993. IRTF observations of the occultation of 28 SGR by Saturn. *Icarus* **103**, 235–252, doi:10.1006/icar.1993.1068.

Hedman, M. M., Nicholson, P. D., Salo, H., Wallis, B. D., Buratti, B. J., Baines, K. H., Brown, R. H., Clark, R. N. 2007. Self-gravity wake structures in Saturn's A ring revealed by Cassini-VIMS. *Astron. J.* **133** (6), 2624–2629.

Holberg, J. B., Forrester, W. T., Lissauer, J. J. 1982. Identification of resonance features within the rings of Saturn. *Nature* **297**, 115–120.

Horányi, M., Burns, J. A., Hedman, M. M., Jones, G. H., Kempf, S. 2009. Diffuse rings. In *Saturn After Cassini-Huygens* (M. Dougherty, L. Esposito, and S. Krimigis, Eds.) Springer.

Horn, L. J., Cuzzi, J. N. 1996. Characteristic wavelengths of irregular structure in Saturn's B ring. *Icarus* **119**, 285–310.

Hubbard, W. B., Porco, C. C., Hunten, D. M., Rieke, G. H., Rieke, M. J., McCarthy, D. W., Haemmerle, V., Clark, R., Turtle, E. P., Haller, J., McLeod, B., Lebofsky, L. A., Marcialis, R., Holberg, J. B., Landau, R., Carrasco, L., Elias, J., Buie, M. W., Persson, S. E., Boroson, T., West, S., Mink, D. J. 1993. The occultation of 28 Sgr by Saturn – Saturn pole position and astrometry. *Icarus* **103**, 215–234.

Jacobson, R. A. et al. 2008. Revised orbits of Saturn's small inner satellites. *Astron. J.* **135**, 261–263.

Julian, W. H., Toomre, A. 1966. Non-axisymmetric responses of differentially rotating disks of stars. *Astrophys. J.* **146**, 810–830.

Kirkwood, D. 1866. On the theory of meteors. *Proc. Amer. Assoc. Adv. Sci.* **15**, 8–14.

Kliore, A. J., Patel, I. R., Lindal, G. F., Sweetnam, D. N., Hotz, H. B., Waite, J. H., McDonough, T. 1980. Structure of the ionosphere and atmosphere of Saturn from Pioneer 11 Saturn radio occultation. *J. Geophys. Res.* **85**, 5857–5870.

Kolvoord, R. A., Burns, J. A., Showalter, M. R. 1990. Periodic features in Saturn's F ring. *Nature* **345**, 695–697.

Lane, A. L., Hord, C. W., West, R. A., Esposito, L. W., Coffeen, D. L., Sato, M., Simmons, K. E., Pomphrey, R. B., Morris, R. B. 1982. Photopolarimetry from Voyager 2 – Preliminary results on Saturn, Titan, and the rings. *Science* **215**, 537–543.

Lewis, M. C., Stewart, G. R. 2005. Expectations for Cassini observations of ring material with nearby moons. *Icarus* **178**, 124–143.

Lewis, M. C., Stewart, G. R. 2006. Simulating the Keeler gap in Saturn's rings: Wake and edge dynamics. 38th Meeting of the Division of Planetary Sciences of the American Astronomical Society, abstract 42.05.

Leyrat, C. et al. 2008. Poster presented at the Saturn After Cassini-Huygens Symposium, London, July 28–Aug. 1.

Lin, C. C., Shu, F. H. 1964. On the spiral structure of disk galaxies. *Astrophys. J.* **140**, 646–655.

Lissauer, J. J. 1985. Bending waves and the structure of Saturn's rings. *Icarus* **62**, 433–447.

Lissauer, J. J., Shu, F. H., Cuzzi, J. N. 1984. Viscosity in Saturn's rings. In *Planetary Rings, Proceedings of IAU Symposium No. 75*, (A. Brahic, Ed.), Toulouse, France, pp. 385–392.

Lissauer, J. J., Goldreich, P., Tremaine, S. 1985. Evolution of the Janus–Epimetheus coorbital resonance due to torques from Saturn's rings. *Icarus* **64**, 425–434.

Longaretti, P.-Y., Borderies, N. 1986. Nonlinear study of the Mimas 5:3 density wave. *Icarus* **67**, 211–223.

Lumme, K., and Irvine, W.M., 1976. Azimuthal brightness variations of Saturn's rings. *Astrophys. J.* **204**, L55–L57.

Lumme, K., Esposito, L.W., Irvine, W.M., and Baum, W.A., 1977. Azimuthal brightness variations of Saturn's rings. II. Observations at an intermediate tilt angle. *Astrophys. J.* **216**, L123–L126.

Marley, M. S., Porco, C. C. 1993. Planetary acoustic mode seismology – Saturn's rings. *Icarus* **106**, 508.

Marouf, E. A., Tyler, G. L., Rosen, P. A. 1986. Profiling Saturn's rings by radio occultation. *Icarus* **68**, 120–166.

Marouf, E. A., Tyler, G. L. 1986. Detection of two satellites in the Cassini division of Saturn's rings. *Nature* **323**, 31–35.

Marouf, E. A., French, R. G., Rappaport, N. J., McGhee, C. A., Wong, K., Thomson, F. S., Anabtawi, A. 2006. Structure and properties of Saturn's Ring B from Cassini radio occultations. *Bull. Am. Astron. Soc.* **38**, 552.

Maxwell, J. C. 1859. On the stability of Saturn's rings (London). Reprinted in *The Scientific Papers of James Clerke Maxwell*, 2 Vols. 1890. University of Cambridge Press, Cambridge, pp. 288–374.

McGhee, C. A., Nicholson, P. D., French, R. G., Hall, K. J. 2001. HST observations of Saturnian satellites during the 1995 ring plane crossings. *Icarus* **152**, 282–315, doi:10.1006/icar.2001.6635.

Molnar, L. A., Dunn, D. E. 1995. The Mimas 2:1 eccentric corotational resonance in Saturn's outer B ring. *Icarus* **116**, 397–408.

Molnar, L. A., Dunn, D. E., Niehof, J. T. 1999. Fall 1998 radio observations of Saturn's rings: New evidence of ring wakes. 31st Meeting of the Division of Planetary Sciences of the American Astronomical Society, abstract 44.05.

Murray, C. D., Dermott, S. F. 1999. *Solar System Dynamics*. Cambridge University Press, Cambridge.

Murray, C. D., Chavez, C., Beurle, K., Cooper, N., Evans, M. W., Burns, J. A., Porco, C. C. 2005. How Prometheus creates structure in Saturn's F ring. *Nature* **437**, 1326–1329.

Murray, C. D., Beurle, K., Cooper, N. J., Evans, M. W., Williams, G., Charnoz, S. 2008. The determination of the structure of Saturn's F ring by nearby moonlets. *Nature* **453**, 739–744.

Nicholson, P. D., Porco, C. C. 1988. A new constraint on Saturn's zonal gravity harmonics from Voyager observations of an eccentric ringlet. *J. Geophys. Res.* **93**, 10209–10224.

Nicholson, P. D., Cooke, M. L., Pelton, E. 1990. An absolute radius scale for Saturn's rings. *Astron. J.* **100**, 1339–1362.

Nicholson, P. D., Showalter, M. R., Dones, L., French, R. G., Larson, S. M., Lissauer, J. J., McGhee, C. A., Seitzer, P., Sicardy, B., Danielson, G. E. 1996. Observations of Saturn's ring-plane crossings in August and November 1995. *Science* **272**, 509–515.

Nicholson, P. D., French, R. G., Tollestrup, E., Cuzzi, J. N., Harrington, J., Matthews, K., Perković, O., Stover, R. J. 2000. Saturn's rings I. Optical depth profiles from the 28 Sgr occultation. *Icarus* **145**, 473–500.

Nicholson, P. D., French, R. G., Campbell, D. B., Margot, J.-L., Nolan, M. C., Black, G. J., Salo, H. J. 2005. Radar imaging of Saturn's rings. *Icarus* **177**, 32–62.

Nicholson, P. D., Hedman, M. M., Clark, R. N., Brown, R. H., Buratti, B. J., Baines, K. H., Cassini VIMS Team 2007. Through a glass darkly: Saturn's enigmatic B ring. 39th Meeting of the Division of Planetary Sciences of the American Astronomical Society, abstract 7.04, *Bull. Am. Astron. Soc.* **39**, 420.

Nicholson, P. D., Hedman, M. M., Clark, R. N., Showalter, M. R., Cruikshank, D. P., Cuzzi, J. N., Filacchione, G., Capaccioni, F., Cerroni, P., Hansen, G. B., Sicardy, B., Drossart, P., Brown, R. H., Buratti, B. J., Baines, K. H., Coradini, A. 2008a. A close look at Saturn's rings with Cassini VIMS. *Icarus* **193**, 182–212.

Nicholson, P. D., Hedman, M. M., Salo, H. J., Cassini VIMS Team 2008b. Cassini-VIMS observations of self-gravity wakes in Saturn's rings – II. 39th Meeting of Division of Dynamical Astronomy of the American Astronomical Society, abstract 18.01.

Nicholson, P. D., Hedman, M. M. 2009. *Icarus* (submitted).

Osterbrook, D. E., Cruikshank, D. P. 1983. J. E. Keeler's discovery of a gap in the outer part of the A ring. *Icarus* **53**, 165–173.

Perrine, R. P., Richardson, D. C. 2006. A computational model of moons in planetary ring gaps. 38th Meeting of the Division of Planetary Sciences of the American Astronomical Society, abstract 42.04.

Perrine, R. P., Richardson, D. C. 2007. Numerical studies of satellite-ring interactions. 39th Meeting of the Division of Planetary Sciences of the American Astronomical Society, abstract 10.01.

Porco, C. C. 1983. Voyager Observations of Saturn's Rings. 1: The Eccentric Rings at 1.29, 1.45, 1.95 and 2.27 R_S. Ph.D. Thesis. California Institute of Technology, Pasadena.

Porco, C., Nicholson, P. D., Borderies, N., Danielson, G. E., Goldreich, P., Holberg, J. B., Lane, A. L. 1984a. The eccentric Saturnian ringlets at 1.29 R_S and 1.45 R_S. *Icarus* **60**, 1–16.

Porco, C., Danielson, G. E., Goldreich, P., Holberg, J. B., Lane, A. L. 1984b. Saturn's nonaxisymmetric ring edges at 1.95 R_S and 2.27 R_S. *Icarus* **60**, 17–28.

Porco, C. C., Nicholson, P. D. 1987. Eccentric features in Saturn's outer C ring. *Icarus* **72**, 437–467.

Porco, C. C., Baker, E., Barbara, J., Beurle, K., Brahic, A., Burns, J. A., Charnoz, S., Cooper, N., Dawson, D. D., Del Genio, A. D., Denk, T., Dones, L., Dyudina, U., Evans, M. W., Giese, B., Grazier, K., Helfenstein, P., Ingersoll, A. P., Jacobson, R. A., Johnson, T. V., McEwen, A., Murray, C. D., Neukum, G., Owen, W. M., Perry, J., Roatsch, T., Spitale, J., Squyres, S., Thomas, P., Tiscareno, M., Turtle, E., Vasavada, A. R., Veverka, J., Wagner, R., West, R. 2005. Cassini imaging science: Initial results on Saturn's rings and small satellites. *Science* **307**, 1226–1236.

Porco, C. C., Weiss, J. W., Richard, D. C., Dones, L., Quinn, T., Throop, H. 2008. Simulations of the dynamical and light-scattering behavior of Saturn's rings and the derivation of ring particle and disk properties. *Astron. J.* **136**, 2172–2200.

Rappaport, N. J., Longaretti, P.-Y., French, R. G., Marouf, E. A., McGhee, C. A. 2008. A procedure to analyze nonlinear density waves in Saturn's rings using several occultation profiles, *Icarus* (in press).

Reitsema, H.J., Beebe, R.F., Smith, B.A., 1976. Azimuthal brightness variations in Saturn's rings. *Astron. J.* **81**, 209–215.

Richardson, D. 1994. Tree code simulations of planetary rings. *Mon. Not. Roy. Astron. Soc.* **269**, 493–511.

Rosen, P. A., Lissauer, J. J. 1988. The Titan – 1:0 nodal bending wave in Saturn's ring C. *Science* **241**, 690–694.

Rosen, P. A., Tyler, G. L., Marouf, E. A. 1991a. Resonance structures in Saturn's rings probed by radio occultation. I. Methods and examples. *Icarus* **93**, 3–24.

Rosen, P. A., Tyler, G. L., Marouf, E. A., Lissauer, J. J. 1991b. Resonance structures in Saturn's rings probed by radio occultation. II – Results and interpretation. *Icarus* **93**, 25–44.

Salo, H. 1992. Gravitational wakes in Saturn's rings. *Nature* **359**, 619–621.

Salo, H. 1995. Simulations of dense planetary rings. III. Self-gravitating identical particles. *Icarus* **117**, 287–312.

Salo, H., Schmidt, J., Spahn, F. 2001. Viscous overstability in Saturn's B ring: I. Direct simulations and measurement of transport coefficients. *Icarus* **153**, 295–315.

Salo, H., Karjalainen, R. 2003. Photometric modeling of Saturn's rings. I Monte Carlo method and the effect of nonzero volume filling factor. *Icarus* **164**, 428–460.

Salo, H., Karjalainen, R., French, R. G. 2004. Photometric modeling of Saturn's rings. II Azimuthal asymmetry in reflected and transmitted light. *Icarus* **170**, 70–90.

Salo, H., French, R. G., Nicholson, P. D., Hedman, M. M., Colwell, J. E., Schmidt, J. 2008. Modeling self-gravity wakes in Saturn's rings: slab models vs. N-body wakes. *Saturn After Cassini-Huygens*, Imperial College London, August 2008.

Schmidt, J., Salo, H., Spahn, F., Petzschmann, O. 2001. Viscous overstability in Saturn's B-Ring. II. Hydrodynamic theory and comparison to simulations. *Icarus* **153**, 316–331.

Schmidt, J., Ohtsuki, K., Rappaport, N., Salo, H., Spahn, F. 2009. Dynamics of Saturn's dense rings. In *Saturn After Cassini–Huygens* (M. Dougherty, L. Esposito, S. Krimigis, Eds.) Springer.

Schmit, U., Tscharnuter, W. M. 1999. On the formation of the fine-scale structure in Saturn's B ring. *Icarus* **138**, 173–187.

Seiß, M., Spahn, F., Sremčević, M., Salo, H. 2005. Structures induced by small moonlets in Saturn's rings: Implications for the Cassini mission. *Geophys. Res. Lett.* **32**, L11205.

Showalter, M. R., Burns, J. A. 1982. A numerical study of Saturn's F ring. *Icarus* **52**, 526–544.

Showalter, M. R. 1991. Visual detection of 1981S13, Saturn's eighteenth satellite, and its role in the Encke gap. *Nature* **351**, 709–713.

Showalter, M. R., Cuzzi, J. N., Marouf, E. A., Esposito, L. W. 1986. Satellite "wakes" and the orbit of the Encke Gap moonlet. *Icarus* **66**, 297–323.

Shu, F. H., Cuzzi, J. N., Lissauer, J. J. 1983. Bending waves in Saturn's rings. *Icarus* **53**, 185–206.

Shu, F. H. 1984. Waves in planetary rings. In *Planetary Rings* (R. Greenberg and A. Brahic, Eds.), University of Arizona Press, Tucson, pp. 513–561.

Shu, F. H., Dones, L., Lissaur, J. J., Yuan, C., Cuzzi, J. N. 1985. Nonlinear spiral density waves: Viscous damping. *Astrophys. J.* **299**, 542–573.

Smith, B. A., Soderblom, L., Beebe, R. F., Boyce, J. M., Briggs, G., Bunker, A., Collins, S. A., Hansen, C., Johnson, T. V., Mitchell, J. L., Terrile, R. J., Carr, M. H., Cook, A. F., Cuzzi, J. N., Pollack, J. B., Danielson, G. E., Ingersoll, A. P., Davies, M. E., Hunt, G. E., Masursky, H., Shoemaker, E. M., Morrison, D., Owen, T., Sagan, C., Veverka, J., Strom, R., Suomi, V. E. 1981. Encounter with Saturn: Voyager 1 imaging science results. *Science* **212**, 163–191.

Smith, B. A., Soderblom, L., Batson, R., Bridges, P., Inge, J., Masursky, H., Shoemaker, E., Beebe, R., Boyce, J., Briggs, G., Bunker, A., Collins, S. A., Hansen, C. J., Johnson, T. V., Mitchell, J. L., Terrile, R. J., Cook, A. F., Cuzzi, J., Pollack, J. B., Danielson, G. E., Ingersoll, A., Davies, M. E., Hunt, G. E., Morrison, D., Owen, T., Sagan, C., Veverka, J., Strom, R., Suomi, V. E. 1982. A new look at the Saturn system: The Voyager 2 images. *Science* **215**, 504–537.

Spahn, F., Sremčević, M. 2000. Density patterns induced by small moonlets in Saturn's rings? Astron. *Astrophys.* **358**, 368–372.

Spilker, L. J., Pilorz, S., Lane, A. L., Nelson, R. M., Pollard, B., Russell, C. T. 2004. Saturn A ring surface mass densities from spiral density wave dispersion behavior. *Icarus* **171**, 372–390.

Spitale, J. N., Jacobson, R. A., Porco, C. C., Owen, W. M. Jr. 2006. The orbits of Saturn's small satellites derived from combined historic and Cassini imaging observations. *Astron. J.* **132**, 692–710.

Spitale, J. N., Porco, C. C., Colwell, J. E., Hahn, J. M. 2008a. Kinematics of the outer edges of Saturn's A and B rings. AAS/Division of Dynamical Astronomy Meeting 39, #18.03.

Spitale, J. N., Porco, C. C., Colwell, J. E. 2008b. An inclined Saturnian ringlet at 1.954 R_S. 40th Meeting of the Division of Planetary Sciences of the American Astronomical Society.

Sremčević, M., Spahn, F., Duschl, W. J. 2002. Density structures in perturbed thin cold discs. *Mon. Not. Roy. Astron. Soc.* **337**, 1139–1152, doi:10.1046/j.1365-8711.2002.06011.x.

Sremčević, M., Esposito, L. W., Colwell, J. E. 2006. Size of Particles and clumps in Saturnian rings inferred from Cassini UVIS occultations. AAS/Division of Planetary Sciences Meeting 38, #42.17.

Sremčević, M., Schmidt, J., Salo, H., Seiss, M., Spahn, F., Albers, N. 2007. A belt of moonlets in Saturn's A ring. *Nature* **449**, 1019–1021, doi:10.1038/nature06224.

Sremčević, M., Stewart, G. R., Albers, N., Colwell, J. E., Esposito, L. W. 2008. Density waves in Saturn's rings: Non-linear dispersion and moon libration effects. AAS/Division of Planetary Sciences Meeting 40, #24.03.

Stewart, G. R., Robbins, S. J., Colwell, J. E. 2007. Evidence for a Primordial Origin of Saturn's Rings. 39th meeting of the Division of Planetary Sciences of the American Astronomical Society, abstract 7.06.

Supulver, K., Bridges, F., Lin, D. 1995. The coefficient of restitution of ice particles in glancing collisions: Experimental results for unfrosted surfaces. *Icarus* **113**, 188–199.

Thompson, W.T., Lumme, K., Irvine, W.M., Baum, W.A., Esposito, L.W. 1981. Saturn's rings – azimuthal variations, phase curves, and radial profiles in four colors. *Icarus* **46**, 187–200.

Thomson, F. S., Marouf, E. A., Tyler, G. L., French, R. G., Rappaport, N. J. 2007. Periodic microstructure in Saturn's rings A and B. *Geophys. Res. Lett.* **34**, L24203, doi:10.1029/2007GL032526.

Tiscareno, M. S., Burns, J. A., Hedman, M. M., Spitale, J. N., Porco, C. C., Murray, C. D., Cassini Imaging Team 2005. Wave edges and other disturbances in Saturn's Encke and Keeler gaps. 37th Meeting of the Division of Planetary Sciences of the American Astronomical Society, abstract 64.02.

Tiscareno, M. S., Burns, J. A., Hedman, M. M., Porco, C. C., Weiss, J. W., Dones, L., Richardson, D. C., Murray, C. D. 2006a. 100-metre-diameter moonlets in Saturn's A ring from observations of 'propeller' structures. *Nature* **440**, 648–650, doi:10.1038/nature04581.

Tiscareno, M. S., Nicholson, P. D., Burns, J. A., Hedman, M. M., Porco, C. C. 2006b. Unravelling temporal variability in Saturn's spiral density waves: Results and predictions. *Astrophys. J.* **651**, L65-L68.

Tiscareno, M. S., Burns, J. A., Nicholson, P. D., Hedman, M. M., Porco, C. C. 2007. Cassini imaging of Saturn's rings II. A wavelet technique for analysis of density waves and other radial structure in the rings. *Icarus* **189**, 14–34.

Tiscareno, M. S., Burns, J. A., Hedman, M. M., Porco, C. C. 2008. The population of propellers in Saturn's A ring. *Astron J.* **135**, 1083–1091.

Toomre, A., 1964. On the gravitational stability of a disk of stars. *Astrophys. J.* **139**, 1217–1238.

Torrey, P. A., Tiscareno, M. S., Burns, J. A., Porco, C. C. 2008. Mapping complexity: the wavy edges of the Encke and Keeler gaps in Saturn's rings. AAS/Division of Dynamical Astronomy Meeting Abstracts, 39, abstract 15.19.

Tremaine, S. 2003. On the origin of irregular structure in Saturn's rings. *Astron. J.* **125**, 894–901.

Turtle, E., Porco, C., Haemmerle, V., Hubbard, W., Clark, R. 1991. The kinematics of eccentric features in Saturn's Cassini Division from combined Voyager and ground-based data. *Bull. Am. Astron. Soc.* **23**, 1179.

Tyler, G. L., Marouf, E. A., Simpson, R. A., Zebker, H. A., Eshleman, V. R. 1983. The microwave opacity of Saturn's rings at wavelengths of 3.6 and 13 cm from Voyager 1 radio occultation. *Icarus* **54**, 160–188.

Van Helden, A. 1984. Saturn through the telescope: A brief historical survey. In *Saturn* (T. Gehrels and J. S. Matthews, Eds.) University of Arizona Press, Tucson, pp. 23–43.

Ward, W. R. 1981. On the radial structure of Saturn's rings. *Geophys. Res. Lett.* **8**, 641–643.

Weiss, J. W., Porco, C. C., Tiscareno, M. S. 2008. Edge-waves in ring gaps and the determination of masses of embedded satellites. AAS/Division of Dynamical Astronomy Meeting Abstracts, 39, abstract 18.02.

Wisdom, J., Tremaine, S. 1988. Local simulations of planetary rings. *Astron. J.* **95**, 925–940.

Chapter 14
Dynamics of Saturn's Dense Rings

Jürgen Schmidt, Keiji Ohtsuki, Nicole Rappaport, Heikki Salo, and Frank Spahn

Abstract The Cassini mission to Saturn opened a new era in the research of planetary rings, bringing data in unprecedented detail, monitoring the structure and properties of Saturn's ring system. The question of ring dynamics is to identify and understand underlying physical processes and to connect them to the observations in terms of mathematical models and computer simulations. For Saturn's dense rings important physical processes are dissipative collisions between ring particles, their motion in Saturn's gravity field, their mutual self-gravity, and the gravitational interaction with Saturn's moons, exterior to or embedded in the rings.

The importance of the rings' self-gravity became strikingly clear from the identification of gravitational wakes in Cassini data nearly everywhere in the A and B rings. Self-gravity wakes imply that the rings are in a flat, dynamically cold state, ring particles colliding very dissipatively, being densely packed in the ring plane, continuously forming transient gravitationally bound opaque clumps, that are disrupted again by shear on orbital timescales. Current mathematical dynamical models usually treat self-gravity in an approximate manner, which does not lead to a wake state.

The dense packing of ring particles, in turn, strongly influences the collisional dynamics, since the mean free path of the particles is then comparable to or smaller than the particle size. This leads to a strong nonlocal component of pressure and momentum transport, which determines the viscous evolution of the rings, the damping of density waves, as well as the stability properties of the ring's flow. A strong nonlocal contribution to viscosity is, for instance, favorable for viscous overstability, leading to axisymmetric waves of about 100 m wavelength. Such wavelike perturbations in the ring's opacity, consistent with overstability, are seen in Cassini stellar and radio occultations.

A classical topic of ring dynamics is the interaction of moons and rings. On the one hand, there are exterior moons with resonances in the rings, creating numerous density and bending waves. With the large sets of Cassini occultation and imaging data, improved estimates of the ring surface mass density and viscosity are obtained from fits of the observations to dynamical models. On the other hand, the embedded moons Pan and Daphnis open the Encke and Keeler gaps, respectively, and moonlets in the rings, too small to open a circumferential gap, are found to produce a characteristic propeller structure.

Comparison between theoretical studies and Cassini observations of thermal emission from the rings provides constraints on spin rates of ring particles, which are otherwise not directly observable. The size distribution of particles and small moonlets embedded in the rings, together with the observed shapes and internal densities of small moons just exterior to the rings, underline the importance of accretion and fragmentation for the dynamical evolution of Saturn's ring system.

J. Schmidt and F. Spahn
Institut f. Physik & Astronomie, Universität Potsdam, Germany

K. Ohtsuki
Lab. Atmospheric & Space Physics, University of Colorado, CO, USA;
Dept. Earth & Planet. Sci. and Center for Planet. Sci., Kobe Univ., Japan

N. Rappaport
JPL, Pasadena, CA, USA

H. Salo
Dept. Physical Sci., Div. Astronomie, University of Oulu, Finland

14.1 General Theory and Recent Advances

In this section we outline elements of the theoretical modelling of dense planetary rings. We start with a basic description of the physical processes that crucially influence the steady state properties of the rings – dissipative collisions and shear – and discuss the importance of the elastic properties of ring particles. Section 14.1.3 then gives an introduction to self-gravity in Saturn's rings.

14.1.1 Steady-State of a Dense Non-Gravitating Particle Disk

The dynamics of an unperturbed planetary ring consisting of macroscopic particles is basically determined by inelastic binary particle collisions and the shearing motion in the gravitational field of the central planet. In this section we discuss how these processes determine the energy budget of a dense planetary ring.

14.1.1.1 Dissipative Collisions

Inelastic physical collisions of the ring particles continuously dissipate energy from the particles' random motion. The dissipation can be quantified by the normal coefficient of restitution ε_n, defined as the ratio of the post-collisional to pre-collisional normal component of the relative velocity of the colliding particles[1]. Particles will collide with an impact speed on the order of the velocity dispersion c of the ring so that the kinetic energy (per unit mass) dissipated in a single collision can be estimated as $c^2 (1 - \varepsilon_n^2)$. The rate of collisional energy loss is then

$$\left.\frac{\partial c^2}{\partial t}\right|_{loss} \propto -\omega_c c^2 \left(1 - \varepsilon_n^2\right) \qquad (14.1)$$

with the collision frequency ω_c. In dilute planetary rings (where $c \gg D\Omega$, with the dominant particle diameter D), we have

$$\omega_c \approx 3\Omega\tau, \qquad (14.2)$$

with Kepler frequency $\Omega(r) = \sqrt{GM_{plan}/r^3}$ (angular velocity on a circular orbit at distance r from a planet of mass M_{plan}, where G is the gravitational constant) and dynamical optical depth τ. This expression does not depend on the velocity dispersion, although in principle $\omega_c \propto n_{space} D^2 c$, where n_{space} is the ring particles' space number density. The explicit c dependence cancels out, however, since the space density for a given τ is inversely proportional to c, due to collisional coupling between horizontal and vertical motions. For very flattened systems ($c \sim D\Omega$) the impact frequency can exceed significantly the value implied by Eq. 14.2 (Salo 1992b).

14.1.1.2 Shear Stress

In viscous shear flow, energy is transferred from the systematic motion to the random motion of the particles. There is a transport of angular momentum (related to this energy transfer), i.e. momentum in flow direction is transported perpendicular to the flow. In a system of particles with finite diameter D there are two different modes of transport, local and nonlocal, both connected to collisions and coupling to the shear rate.

In a local frame rotating at distance r_0 around the planet with Kepler frequency $\Omega(r_0)$, the Keplerian orbital velocity reads $v = [\Omega(r) - \Omega(r_0)]r$. Introducing $x = r - r_0$ this may be approximated as

$$v = -\frac{3}{2}\Omega(r_0) x + O\left[(x/r_0)^2\right] \qquad (14.3)$$

and the shear rate s reads

$$s \equiv \frac{\partial v}{\partial x} = -\frac{3}{2}\Omega(r_0). \qquad (14.4)$$

The rate of gain of kinetic energy (per unit mass) is given by the kinematic viscosity ν and shear rate s as

$$\left.\frac{\partial c^2}{\partial t}\right|_{gain} \propto \nu s^2. \qquad (14.5)$$

Local Shear Viscosity

Local shear stress is connected to the particles' radial random motions (Fig. 14.1a). This kind of momentum transport is labeled 'local' since it is not necessary to distinguish the particle positions in a collision. Let an imaginary line L separate the flow radially into an inner and outer region (Fig. 14.1). When particles cross the line between collisions (Fig. 14.1a) they carry across the line the mean momentum per particle of the ring region they come from. Since (due to shear) the mean velocity at the new position is slightly different, the particles have, on average, a small amount of excess momentum with respect to the surrounding particles, which is transferred to the flow in the subsequent collisions. In this way, neighboring ring regions are effectively coupled by a drag force: The inner (outer) part of the ring tends to accelerate (decelerate) the outer (inner) part.

For the local kinematic shear viscosity Goldreich and Tremaine (1978c) derived the formula

$$\nu_l = \frac{c^2}{\Omega} \frac{\tau}{1 + \tau^2}. \qquad (14.6)$$

For large optical depth we have $\nu_l \propto 1/\tau$, which is reminiscent of the density dependence $\nu \propto 1/\rho$ of hydrodynamics (Chapman and Cowling 1970). For low optical depth,

---

[1] In addition, kinetic energy may be lost in surface friction, reducing the tangential relative velocity between colliding particles and transferring energy of random motions into the particles' spin and vice versa. For simplicity we restrict the discussion here on normal restitution.

14 Dynamics of Saturn's Dense Rings

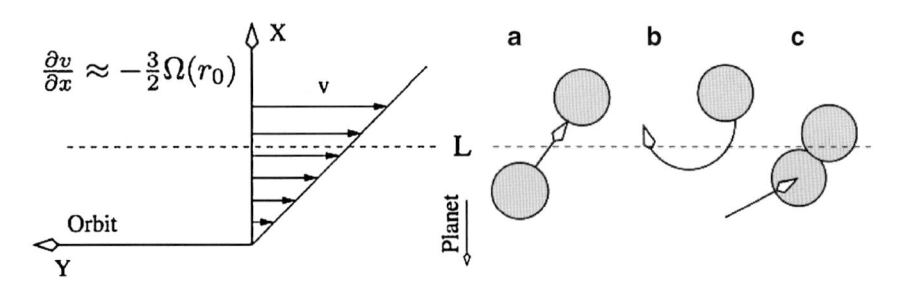

Fig. 14.1 *Left panel*: Velocity profile of the ring in a local frame co-rotating at distance r_0 from the planet. The X-axis points radially outward, the Y-axis in the direction of orbital motion. *Right panel*: Typical particle paths and collisions. (**a**) Momentum transport across the imaginary line L (*dashed*) by a crossing particle (local transport). (**b**) Particle moving on an epicycle (no transport). (**c**) Momentum transport across L in a collision (nonlocal transport)

however, $v_l \propto \tau$. This deviation from the hydrodynamic density dependence results from the motion in the central gravity field: In a dilute system the radial excursions of a particle are limited by the epicyclic length (Fig. 14.1b), which reduces the momentum transport $\propto \tau$.

Non-Local Shear Viscosity and Pressure

In a dense collisional system the mean free path is on the order of the particle dimensions and the momentum transferred by a collision (Fig. 14.1c) over the distance of one particle diameter is important. Describing this process statistically (Shukhman 1984; Araki and Tremaine 1986), the positions of particle centers in a collision must be distinguished, and hence the corresponding stress is labeled *nonlocal*.

From mean-free path arguments the nonlocal kinematic shear viscosity can be estimated to be on the order of (Shukhman 1984)

$$v_{nl} = \Omega D^2 \tau. \tag{14.7}$$

Consider the transport of momentum across the dashed line in Fig. 14.1, counting only binary collisions where particle centers are on opposite sides of the line (Fig. 14.1c). The number of particles crossing the line per unit length and unit time is roughly $J = (\sigma/m)c$, with surface mass density σ and particle mass m. To count only those particles actually intersected by the line, we multiply this particle flux density by the factor D/l, where l is the mean free path $l = c/\omega_c$. The maximal momentum transported over a radial distance of one particle diameter is $\Delta p = mDs$. Given the (vertically integrated) shear stress by $P_{xy} = J\Delta p$ and identifying it with the hydrodynamic relation $P_{xy} = \sigma v_{nl} s$, we obtain Eq. 14.7. The significance of nonlocal viscosity for planetary rings was first pointed out by Brahic (1977). Similar arguments lead to an order of magnitude estimate for the nonlocal (vertically integrated) pressure as

$$p_{nl} = \sigma \tau c \Omega D. \tag{14.8}$$

Total Shear Viscosity and Pressure

Summing up the local and non-local viscosities we obtain

$$\frac{v}{\Omega D^2} = k_1 \left(\frac{c}{\Omega D}\right)^2 \frac{\tau}{1+\tau^2} + k_2 \tau, \tag{14.9}$$

with constants k_1, k_2, and with the local pressure $p_l = \sigma c^2$ we obtain the total pressure

$$\frac{p/\sigma}{(\Omega D)^2} = \left(\frac{c}{\Omega D}\right)^2 + \frac{c}{\Omega D}\tau. \tag{14.10}$$

14.1.1.3 Steady State and Thermal Stability

Summing up the cooling (Eq. 14.1) and heating (Eq. 14.5) rates one obtains the energy budget of the ring particle ensemble

$$\frac{\partial c^2}{\partial t} = s^2 \left[k_1 \frac{c^2}{\Omega} \frac{\tau}{1+\tau^2} + k_2 \Omega D^2 \tau \right] - k_3 \Omega \tau \left(1 - \varepsilon_n^2\right) c^2 \tag{14.11}$$

where k_1, k_2, k_3 are positive dimensionless constants (e.g. Stewart et al. 1984; Morishima and Salo 2006). For a velocity independent coefficient of restitution this equation has the fixed point solution

$$\left(\frac{c^*}{\Omega D}\right)^2 = \frac{k_2 (s/\Omega)^2 (1+\tau^2)}{k_3 (1-\varepsilon_n^2)(1+\tau^2) - k_1 (s/\Omega)^2} \tag{14.12}$$

which is stable for

$$\varepsilon_n < \varepsilon_{crit} = \sqrt{1 - \frac{k_1 (s/\Omega)^2}{k_3} \frac{1}{1+\tau^2}} \tag{14.13}$$

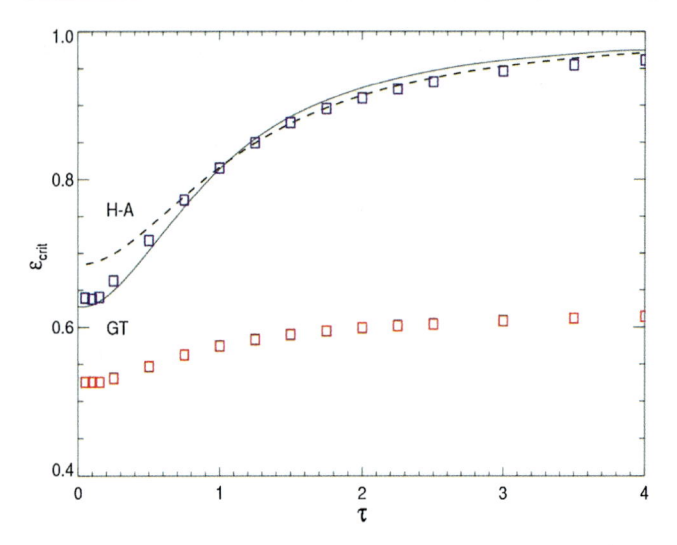

Fig. 14.2 Theoretical thermal stability boundary $\varepsilon_{crit}(\tau)$, from Hämeen-Anttila (1978) and Goldreich and Tremaine (1978c), who used different approximations in the evaluation of the collision integrals. Also shown are effective steady-state values of ε_n from two series of simulations, performed with velocity-dependent coefficient of restitution: upper points correspond to a 'hot' simulation with velocity dispersion $c \gg D\Omega$ (this dilute 'mass-point' limit approximates the assumptions behind the theoretical curves), while the lower points correspond to simulations performed with the Bridges et al. (1984) elasticity formula. The average ε_n in simulations is measured by weighting each impact with the square of the normal component of impact velocity, $<\varepsilon_n v_n^2>/<v_n^2>$. Redrawn from Salo (2001)

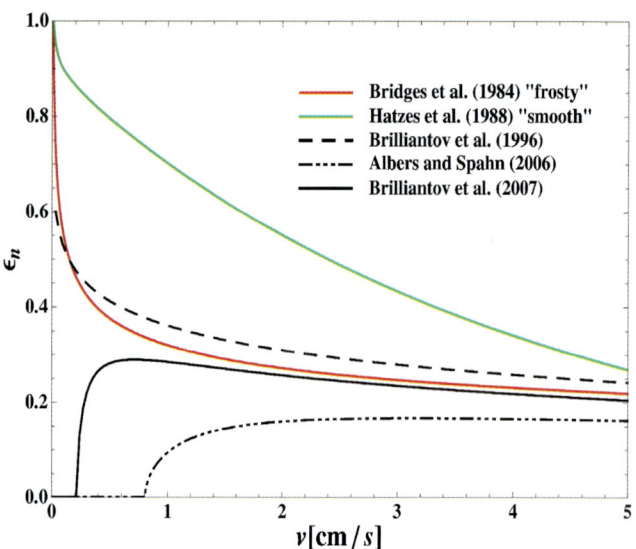

Fig. 14.3 The normal coefficient of restitution from theoretical models (black curves) and measurements for frosty ice particles (radius 2.75 cm, *red curve*) by Bridges et al. (1984) and larger ice particles with compacted frost surface-layers (radius 20 cm, *green curve*) by Hatzes et al. (1988). The triple-dotted dashed curve shows results from viscoelastic models including adhesion (Albers and Spahn, 2006) and the thick solid curve is derived from an extension of this model (Brilliantov et al., 2007). The dashed curve is from a visco-elastic model (Brilliantov et al., 1996)

and unstable otherwise. This critical value ε_{crit} was first derived by Goldreich and Tremaine (1978c) and an equivalent expression (Fig. 14.2) was given by Hämeen-Anttila (1978). In systems with constant $\varepsilon_n < \varepsilon_{crit}$ the particles establish a thermal equilibrium mainly due to the non-local viscous gain. This steady state corresponds to a flattened system with a geometric thickness of a few particle diameters only and a velocity dispersion $c \sim \Omega D$. For $\varepsilon_n > \varepsilon_{crit}$ the local viscous gain always dominates the dissipation, no energy balance can be achieved, and the system inevitably disperses via continuously growing random velocities. With increasing τ the reduced mean free path between impacts limits the local viscous gain and ε_{crit} increases (Fig. 14.2). For rough particles tangential friction adds to the dissipation and ε_{crit} is shifted closer to unity (Shukhman 1984; Araki 1988).

For realistic materials ε_n is a decreasing function of impact velocity (Fig. 14.3), or, on average, of velocity dispersion. In this case the stability behavior of Eq. 14.11 is different. Systems with $\varepsilon_n(c) > \varepsilon_{crit}$ will heat up, so that $\varepsilon_n(c)$ decreases until an equilibrium of collisional energy loss and viscous gain is established (Goldreich and Tremaine 1978c). Due to non-local shear stress, however, this equilibrium value of $\varepsilon_n(c)$ is generally smaller than ε_{crit}. Only mass point systems ($D \to 0$) can establish $\varepsilon_n(c) = \varepsilon_{crit}$. The exact value of the steady-state velocity dispersion

is determined by the dependence of ε_n on v, and, at least in principle, a ring's configuration can range from a thick multilayer of particles to a near monolayer ring. In the latter case the steady-state is in practice indistinguishable from the case of constant $\varepsilon_n < \varepsilon_{crit}$.

14.1.1.4 Mechanical Properties of Particles

The coefficient of restitution of ice particles was determined experimentally at the low temperature and pressure appropriate for the planetary ring environment. The laboratory measurements by Bridges et al. (1984) and Hatzes et al. (1988) indicated that the normal restitution coefficient ε_n decreases monotonically with the normal component of impact speed v_n, as required for thermal balance of the ring (Section 14.1.1.3). Measurements yielded the dependencies (Fig. 14.3)

$$\varepsilon_n(v_n) = (v_n/v_c)^{-0.234}, \text{ (frosty)} \quad (14.14)$$

$$\varepsilon_n(v_n) = 0.90 \exp\{-0.22\,(v_n/1\,\text{cm s}^{-1})\}$$
$$+ 0.01\,(v_n/1\,\text{cm s}^{-1})^{-0.6}, \text{(smooth)}. \quad (14.15)$$

Relation 14.14 has been widely used in computer simulations of planetary rings. It was obtained for frosty ice spheres of radius 2.75 cm (Bridges et al. 1984) with $v_c = 0.0077\,\text{cm s}^{-1}$ ($\equiv v_B$) and it corresponds to fairly inelastic impacts, or, to very flattened rings. The relation 14.15 was obtained for ice particles (radius 20 cm) covered by a compacted frost layer (Hatzes et al. 1988). We have chosen this formula from the results of Hatzes et al. (1988) to display the effect of more elastic particles in the following.

The functional dependence $\varepsilon_n(v_n)$ is sensitive to the surface properties of the particles. Higa et al. (1996, 1998) measured ε_n for a much wider range of impact velocities (1–1000 cm s^{-1}). They found that fracture of the ice ball occurs and ε_n decreases significantly when v_n is larger than a critical value. Experiments with glancing collisions showed that tangential velocities are decreased by about 10% for collisions at $v_n \geq 0.1$ cm s^{-1} (Supulver et al. 1995).

Impact experiments under micro-gravity suggest that normal and tangential rebound velocities are significantly smaller if ring particles are covered by regolith (Colwell and Taylor 1999; Colwell 2003). If ring particles are covered by a frost layer cohesive forces may even cause sticking of colliding particles at low-velocity impacts (Hatzes et al. 1991; Bridges et al. 1996; Supulver et al. 1997).

Considerable effort has been invested to derive theoretical expressions for the restitution coefficient. Dilley (1993) proposed an empirical damped oscillator model to describe dissipative collisions, which can be tuned to reproduce, for instance, the experimental results for the Bridges et al. (1984) velocity-dependence of ε_n, as well as the mass-dependence determined in later experiments (Dilley 1993; Dilley and Crawford 1996). Brilliantov et al. (1996) generalized the nonlinear elastic collision model by Hertz (1882) to the case of visco-elastic dissipation. Adjusting one single parameter, characterizing the viscous properties of the particle, the model reproduces the measurements of Bridges et al. (1984) fairly well (Fig. 14.3). Albers and Spahn (2006) included cohesive forces between elastic solids in contact (Johnson et al. 1971) in the visco-elastic model (Brilliantov et al. 1996). Their ε_n is consistent with Bridges et al. (1984) for high impact velocities, where cohesion is negligible. For low impact speeds the cohesive force becomes dominant and particles stick for impact velocities smaller than a critical value (Spahn et al. 2004; Albers and Spahn 2006; Brilliantov et al. 2007) as observed in experiments.

Although the actual mechanical properties of particles in Saturn's rings are poorly known, comparison between observations and dynamical investigations suggests that they are fairly dissipative and under-dense (Section 14.1.3).

14.1.1.5 Steady State Dynamical Properties

A difference in the elastic properties of particles is reflected in the predicted steady-state properties of the ring, and it leads to qualitative differences in the stability properties of the system. A system of fairly elastic particles, dynamically hot at low optical depth τ, exhibits a large reduction of the steady-state velocity dispersion as the optical depth increases, basically because the local viscous gain becomes less effective as the mean free path between impacts is reduced. In the left panel of Fig. 14.4 this is illustrated in terms of the effective geometric thickness: for the 'smooth' particle model (Eq. 14.15, Fig. 14.3) the thickness may drop by as

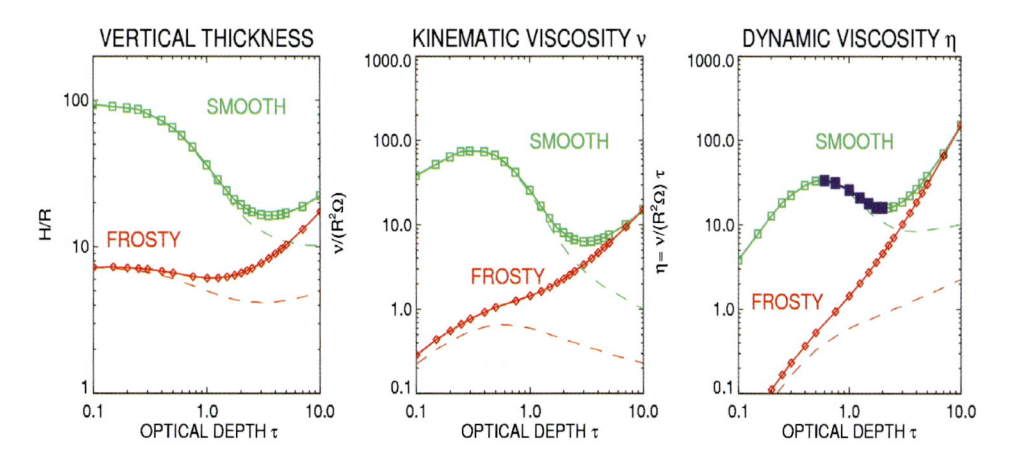

Fig. 14.4 Steady-state geometric thickness and viscosity for the 'smooth' and 'frosty' elasticity models (Eqs. 14.14 and 14.15, Fig. 14.3). *Left frame*: Symbols show the simulated effective geometric thickness $H = \sqrt{12 \langle z^2 \rangle}$, in units of particle radius R (H corresponds to the full thickness of a homogeneous layer with the same vertical dispersion as the actual distribution). The dashed line is the effective thickness estimated from the vertical velocity dispersion, $\sqrt{12}c_z/\Omega$; in the case of low filling factor these two measures for H are identical. *Middle frame*: Kinematic viscosity ν; the contribution from local viscosity is shown separately (*dashed lines*). *Right frame*: Dynamic viscosity $\tau\nu$ (τ is used instead of σ since surface density is unspecified in non self-gravitating simulations). Filled blue squares indicate the τ-regime with negative $\partial\nu\tau/\partial\tau$, susceptible to viscous instability (Section 14.2.2). A particle size $R = 1$ m and Saturnocentric distance $a = 100{,}000$ km are assumed. Adapted from Salo (2001)

much as a factor of five as optical depth increases from zero to above unity. On the other hand, the 'frosty' particle model (Eq. 14.14, Fig. 14.3) leads to a nearly constant velocity dispersion. The different steady-state velocity dispersion alters the local and, to a lesser degree, non-local contribution to viscosity. For a cool system the dynamic viscosity, given by the product $\tau \nu$, is monotonically increasing. For a hot system it decreases for some range of optical depths (marked by filled squares in Fig. 14.4), if the aforementioned reduction of ν_l with τ is strong enough. A negative $d(\tau\nu)/d\tau$ should lead to viscous instability (Section 14.2.2), whereas a strong enough positive slope may indicate overstability (Section 14.2.1).

Since the mechanical properties of ring particles are uncertain, all equilibrium states sketched above might be possible in different parts of Saturn's ring system. For ring particles covered with regolith one may expect even smaller ε_n values than those obtained in the Bridges et al. (1984) laboratory experiments. In any case, at very large optical depth the simulated steady-state properties in non-gravitating systems with viscosity are dominated by effects of finite particle size and are practically independent of the elastic model.

In systems with a particle size distribution the steady-state velocity dispersion of smaller particles generally exceeds the one of larger particles, implying a larger vertical scale height for small particles (e.g. Cuzzi et al. 1979a, b; Salo 1992b; Salo and Karjalainen 2003). For low optical depth rings this was shown by solving the evolution equation for the velocity dispersion (Ohtsuki 1999, 2006b)

$$\frac{dc_m^2}{dt} = \int n_s\left(m'\right)\{C_{VS} + \left(m'c_{m'}^2 - mc_m^2\right)C_{DF}\}dm', \quad (14.16)$$

where c_m and $n_s(m)$ are the velocity dispersion and surface number density of particles with mass m, and C_{VS} and C_{DF} are the rate coefficients related to the viscous stirring and dynamical friction due to inelastic collisions (and gravitational encounters in the case of gravitating particles). Owing to the contribution of viscous stirring and the effect of inelastic impacts, the system remains far from energy-equipartition: for example for the 'frosty' elasticity model the velocity dispersion of the smallest particles is at most a few times larger than for the largest ones, regardless of the width of the size-distribution (Salo 1992b). More elastic particles allow for a wider range of velocity dispersions, the maximum ratio still staying below about five. For a system with extended size distribution, similar to the one inferred from observations, the dynamics is governed by the largest particles. In this case the overall velocity dispersions scale proportionally to the maximum particle size, as in the case of identical particles. The size distribution and dynamical factors affecting its evolution are discussed in more detail in Section 14.4.

14.1.2 Balance Equations for Dense Rings

14.1.2.1 Kinetic Theory

Kinetic theory describes the evolution of the local velocity distribution function of an ensemble of particles in terms of the Boltzmann equation (Chapman and Cowling 1970) or a suitable generalization of it, like Enskog's theory of hard sphere gases. The kinetic equation can be derived from Liouville's theorem (Binney and Tremaine 1987; Resibois and DeLeener 1977), appearing as the leading equation in a hierarchy of equations describing n-particle distribution functions in phase space and neglecting correlations between particle pairs. Taking moments of the kinetic equation gives a set of balance equations for particle number density, momentum, and the components of the pressure tensor (see e.g. Goldreich and Tremaine (1978c); Araki and Tremaine (1986)).

Kinetic theory allows us to incorporate the full complexity of the dynamics of a planetary ring in a statistical description, such as the effects of the motion of ring particles on curved orbits between inelastic collisions, their finite size, the anisotropy of the velocity dispersion, and in principle also coagulation and fragmentation of the ring particles.

The first approach to extend kinetic theory to a system of inelastically colliding particles was undertaken by Trulsen (1972), aiming at a description of clustering and possible focusing of streams of inelastic grains in a protoplanetary disk. Goldreich and Tremaine (1978c) investigated the steady state properties of a disk consisting of point masses, assuming a (triaxial) Gaussian velocity distribution. In this limit they derived the relation between the coefficient of restitution of the particles and optical depth of the disk shown in Fig. 14.2 and the expression given by Eq. 14.6 for the local viscosity of the ring.

Shukhman (1984) included effects of the finite size of particles in the kinetic description of a steady state ring. The importance of such effects had been noted by Brahic (1977) in his collisional ring simulations. In the collision integrals Shukhman (1984) explicitly took into account the difference of the positions of particle centers in the moment of a collision. He derived an expression for the nonlocal viscosity (Eq. 14.7) and described how the disk settles into a non-local equilibrium with a vertical thickness on the order of only a few particle diameters (see Eq. 14.12). In addition, in Shukhman's work rotational degrees of freedom of ring particles and surface friction are incorporated.

Shu and Stewart (1985) used a Krook collision term to avoid the mathematical complexity of the solution of the collision integrals. In this way they obtained a closed solution for the pressure tensor of an unperturbed planetary ring. The enhancement of collision frequency in dense disks is pointed out in their work.

Araki and Tremaine (1986) extended the theory of Goldreich and Tremaine (1978c) and Shukhman (1984) for the steady state of a planetary ring in the spirit of Enskog's kinetic theory of hard spheres (Chapman and Cowling 1970). This goes beyond Shukhman's approach in that it also takes into account the finite volume occupied by particles, limiting the space available for the particles' motion, in terms of the Enskog factor in the collision integral. This effect leads in a dense system to a drastic increase of the collision frequency. Araki (1988) investigated the effect of particle spin and surface friction in a kinetic treatment of dilute rings and Araki (1991) presented a combined kinetic theory for dense and spinning particle disks.

Araki (1991) analytically reduced the collision integrals to a fourfold integration that had to be carried out numerically, which made an application to dynamical problems infeasible. Further progress was made by Latter and Ogilvie (2006) who developed a kinetic theory for dilute rings carrying out all collision integrals analytically. Their work was extended by Latter and Ogilvie (2008) to dense systems, studying stability properties of a dense ring.

An alternative statistical description of planetary rings was formulated in a series of papers by Hämeen-Anttila (Hämeen-Anttila 1978, 1981, 1982; Hämeen-Anttila and Salo 1993) improving and generalizing an earlier approach (Hämeen-Anttila 1975, 1976, 1977a, b). It uses a kinetic equation of Boltzmann type for the description of the evolution of the one-particle phase space distribution function, in a similar manner as gas-kinetics. Hämeen-Anttila gives analytical solutions for the collision integrals, where necessary in terms of appropriate approximations. The theory presented in Hämeen-Anttila and Salo (1993) can treat systems of nonidentical, rotating, inelastic particles in a Keplerian field. Arbitrary (velocity dependent) collision laws can be applied to describe translational and rotational energy losses in an impact. The effect of self-gravity is taken into account in a self-consistent manner in the local vertical gravity field of the disk and in its effect on close particle encounters. In principle the theory can treat the average effect of particle surface irregularities stochastically and it is formulated so that it can be extended to describe particle fragmentation and coagulation. The biggest advantage is that the balance equations for mass, stress, and scale hight of the ring, are given analytically as partial differential equations. Thus, the theory can be applied to investigate the dynamical evolution of a planetary ring.

14.1.2.2 Hydrodynamics

The dynamic equations resulting from kinetic theory are of great complexity. For instance the kinetic approach developed by Goldreich and Tremaine (1978c), Araki and Tremaine (1986), and Araki (1988, 1991) gives insights into steady state properties of a planetary ring, but the important question of spatial and temporal evolution is not addressed. The approaches by Latter and Ogilvie (2008) and Hämeen-Anttila and Salo (1993) allow the application to dynamical problems but the equations are so involved that even the solution of small-scale dynamical problems would require a large computational effort.

Hence, for simplicity hydrodynamic models are often employed to investigate the dynamical evolution of rings. The hydrodynamic equations, introduced below, describe the balances of mass, momentum, and energy. They can be derived from the moment equations of kinetic theory (Chapman and Cowling 1970; Stewart et al. 1984), relating the stress and the heat flow to the gradients of velocity and temperature via the concept of the transport coefficients. In dynamical studies parameterizations, suitable for a planetary ring, for the density dependence of these transport coefficients and the equation of state are often employed. These are motivated by simulations (Wisdom and Tremaine 1988; Salo et al. 2001; Schmidt and Salo 2003) or kinetic theory (Goldreich and Tremaine 1978c; Shukhman 1984).

For reference we give here vertically integrated hydrodynamic balance equations in cylindrical coordinates (r, φ) for a vertically thin planetary ring with surface density σ, planar components u_r, u_φ of velocity, and the granular temperature T (energy of particles' random motion)[2]:

$$\left(\frac{\partial}{\partial t} + u_r \frac{\partial}{\partial r} + \frac{u_\varphi}{r} \frac{\partial}{\partial \varphi}\right) \sigma = -\sigma \nabla \cdot \vec{u} \quad (14.17)$$

$$\left(\frac{\partial}{\partial t} + u_r \frac{\partial}{\partial r} + \frac{u_\varphi}{r} \frac{\partial}{\partial \varphi}\right) u_r - \frac{u_\varphi^2}{r} = -\frac{\partial \Phi_{Planet}}{\partial r} - \frac{\partial \Phi_{Disk}}{\partial r}$$
$$- \frac{1}{\sigma} \left(\nabla \cdot \hat{\mathbf{P}}\right)_r \quad (14.18)$$

$$\left(\frac{\partial}{\partial t} + u_r \frac{\partial}{\partial r} + \frac{u_\varphi}{r} \frac{\partial}{\partial \varphi} + \frac{u_r}{r}\right) u_\varphi = -\frac{1}{r} \frac{\partial \Phi_{Disk}}{\partial \varphi}$$
$$- \frac{1}{\sigma} \left(\nabla \cdot \hat{\mathbf{P}}\right)_\varphi \quad (14.19)$$

$$\frac{3}{2} \sigma \left(\frac{\partial}{\partial t} + u_r \frac{\partial}{\partial r} + \frac{u_\varphi}{r} \frac{\partial}{\partial \varphi}\right) T = -\hat{\mathbf{P}} : \nabla \vec{u}$$
$$- \nabla \cdot \vec{q} - \Gamma, \quad (14.20)$$

---

[2] The temperature $T = \mathrm{Tr}\,\hat{\mathbf{C}}/3$ is defined as the trace of the velocity dispersion tensor $\hat{\mathbf{C}} = \langle \vec{c} \cdot \vec{c} \rangle$, where $\vec{c} = \vec{v} \cdot \vec{u}$, $\vec{u}$ is the hydrodynamic mean velocity and $\vec{v}$ the instantaneous velocity of a ring particle (Chapman and Cowling, 1970). The average $\langle \cdot \rangle$ is over the particle velocity distribution function $f(\vec{v})$.

where

$$\nabla \cdot \vec{u} = \frac{1}{r} \frac{\partial (r u_r)}{\partial r} + \frac{1}{r} \frac{\partial u_\varphi}{\partial \varphi} \qquad (14.21)$$

$$(\nabla \vec{u})_{rr} = \frac{\partial u_r}{\partial r}, \quad (\nabla \vec{u})_{\varphi\varphi} = \frac{1}{r} \frac{\partial u_\varphi}{\partial \varphi} + \frac{u_r}{r} \qquad (14.22)$$

$$(\nabla \vec{u})_{r\varphi} = \frac{\partial u_\varphi}{\partial r}, \quad (\nabla \vec{u})_{\varphi r} = \frac{1}{r} \frac{\partial u_r}{\partial \varphi} - \frac{u_\varphi}{r} \qquad (14.23)$$

$$\left(\nabla \cdot \hat{\mathbf{P}}\right)_r = \frac{1}{r} \frac{\partial (r P_{rr})}{\partial r} + \frac{1}{r} \frac{\partial P_{r\varphi}}{\partial \varphi} - \frac{P_{\varphi\varphi}}{r} \qquad (14.24)$$

$$\left(\nabla \cdot \hat{\mathbf{P}}\right)_\varphi = \frac{1}{r} \frac{\partial (r P_{r\varphi})}{\partial r} + \frac{1}{r} \frac{\partial P_{\varphi\varphi}}{\partial \varphi} + \frac{P_{r\varphi}}{r} \qquad (14.25)$$

$$\nabla \cdot \vec{q} = \frac{1}{r} \frac{\partial (r q_r)}{\partial r} + \frac{1}{r} \frac{\partial q_\varphi}{\partial \varphi}. \qquad (14.26)$$

The disk's self-gravity potential Φ_{Disk} couples to the surface mass density through Poisson's equation

$$\frac{1}{r} \frac{\partial}{\partial r}\left(r \frac{\partial \Phi_{Disk}}{\partial r}\right) + \frac{1}{r^2} \frac{\partial^2 \Phi_{Disk}}{\partial \varphi^2} + \frac{\partial^2 \Phi_{Disk}}{\partial z^2} = 4\pi G \sigma \delta(z), \qquad (14.27)$$

and the gravitational potential of the planet is denoted by Φ_{Planet}. In Eqs. 14.17–14.26 $\hat{\mathbf{P}}$ is the vertically integrated symmetric pressure tensor and $\vec{q}$ the heat flux. The term Γ accounts for the cooling due to dissipative collisions (see Section 14.1.1.1, Salo et al. 2001). It does not appear in the hydrodynamic description of molecular gases, but it is common for granular flows (Brilliantov and Pöschel 2004).

If specifically $\hat{\mathbf{P}}$ is of Newtonian form then

$$P_{rr} = p - 2\eta \frac{\partial u_r}{\partial r} + \left(\frac{2}{3}\eta - \xi\right) \nabla \cdot \vec{u} \qquad (14.28)$$

$$P_{r\varphi} = -\eta \left(\frac{\partial u_\varphi}{\partial r} + \frac{1}{r}\frac{\partial u_r}{\partial \varphi} - \frac{u_\varphi}{r}\right) \qquad (14.29)$$

$$P_{\varphi\varphi} = p - 2\eta \left(\frac{1}{r}\frac{\partial u_\varphi}{\partial \varphi} + \frac{u_r}{r}\right)$$
$$+ \left(\frac{2}{3}\eta - \xi\right) \nabla \cdot \vec{u}. \qquad (14.30)$$

Fourier's law affords

$$q_r = -\kappa_D \frac{\partial T}{\partial r}, \quad q_\varphi = -\frac{\kappa_D}{r} \frac{\partial T}{\partial \varphi}. \qquad (14.31)$$

The system of Eqs. 14.17–14.20 and Eqs. 14.28–14.31 is closed by an equation of state $p = p(\sigma, T)$ and by the prescription of the dependence of the transport coefficients $\eta(\sigma, T), \xi(\sigma, T)$, and $\kappa_D(\sigma, T)$ – dynamic coefficients of shear viscosity[3], bulk viscosity, and heat conductivity – on density and granular temperature (i.e. velocity dispersion). It is important to note that for the description of a planetary ring, in general, neither the approximation of incompressibility can be adopted, nor can the transport coefficients assumed to be constant.

Further simplifying assumptions allow us to construct models that use only a part of these equations. For instance, often the approximation of an isothermal ring is applied and one deals with Eqs. 14.17–14.19 while Eq. 14.20 reduces to

$$0 = \hat{\mathbf{P}} : \nabla \vec{u} + \Gamma, \qquad (14.32)$$

i.e. to the balance of collisional cooling and viscous heating, which sets the steady state velocity dispersion of the ring (Eq. 14.12). Although thermal excitations are important (Spahn et al. 2000; Salo et al. 2001; Schmidt et al. 2001) for specific problems (Section 14.2.1), isothermal models usually provide a good qualitative description of many dynamical problems in planetary rings (Ward 1981; Lin and Bodenheimer 1981; Schmit and Tscharnuter 1995; Spahn and Sremčević 2000; Spahn et al. 2000; Schmidt et al. 2001; Schmidt and Salo 2003; Tremaine 2003).

Most instability mechanisms proposed for planetary rings (Section 14.2) aim at describing axisymmetric patterns on scales that are short compared to the dimensions of the ring. In this case, one can neglect curvature terms and restrict the analysis to radial perturbations, so that the isothermal Eqs. 14.17–14.19 and 14.27 read in a frame that co-rotates in the rings with local Kepler frequency $\Omega_0 \equiv \Omega(r_0)$ at a radial distance r_0 from the planet

$$\dot{\sigma} = -(\sigma u)' \qquad (14.33)$$

$$\dot{u} = -u u' + 2\Omega_0 v - \frac{p'}{\sigma}$$
$$+ \frac{1}{\sigma}\left(\left[\frac{4}{3}\eta + \xi\right] u'\right)' - \Phi'_{Disk} \qquad (14.34)$$

$$\dot{v} = -u v' - \frac{\Omega_0}{2} u$$
$$+ \frac{1}{\sigma}\left(\eta v'' + \eta'\left[v' - \frac{3}{2}\Omega_0\right]\right) \qquad (14.35)$$

$$\frac{\partial^2 \Phi_{Disk}}{\partial r^2} + \frac{\partial^2 \Phi_{Disk}}{\partial z^2} = 4\pi G \sigma \delta(z). \qquad (14.36)$$

---

[3] Dynamic quantities are defined as the product of the kinematic quantities and density, e.g. $\eta = \sigma \nu$, with the kinematic shear viscosity ν.

Here v is the deviation of the tangential velocity from the systematic Kepler speed (approximated by Eq. 14.3). The dots and primes denote the derivatives with respect to time and radial coordinate $x = r - r_0$, respectively. A further simplification results from the assumption that the dynamic bulk viscosity ξ has the same dependence on the ring's surface density as the dynamic shear viscosity η. If ξ_0, η_0 denote their constant unperturbed values then

$$\xi = \alpha \eta, \quad \alpha \equiv \frac{4}{3} + \frac{\xi_0}{\eta_0} = const. \quad (14.37)$$

This assumption is neither justified by kinetic theory, nor by simulations, nor is there a satisfactory understanding of the particular effect of bulk viscosity in a planetary ring at all. Generally, bulk viscosity ξ is tied to internal degrees of freedom in molecular systems (vibration and rotation). It accounts for the effect of a (small) relaxation time necessary to establish energy equipartition, say, with the translational degrees of freedom (Chapman and Cowling 1970). But in dense systems a non-zero bulk viscosity arises already due to non-local effects (Chapman and Cowling 1970). Salo et al. (2001) determined values for the bulk viscosity, fitting the theoretically expected dynamic pressure to a simulated low amplitude compression wave. Although the values of the bulk viscosity found this way[4] led to a quantitative description of the stability boundary and the growth rates of overstable modes in simulations of dense rings (Schmidt et al. 2001; Schmidt and Salo 2003), it was noted later that the same procedure failed to describe the physics of a dilute planetary ring. Namely, the values of the bulk viscosity determined from dilute simulations would lead to overstability in the theoretical model, which is not seen in the simulations. The reason for this failure is out-of-phase oscillations of the components of the pressure tensor, described in terms of kinetic theory by Latter and Ogilvie (2006). This non-Newtonian behavior of the pressure tensor they labeled as 'non-local in time'. In the limit of a dense uniform system with large collision frequency these oscillations are damped on a time scale that is faster than the orbital time (Latter and Ogilvie 2008), and one is left with a small relaxation time necessary for the adjustment of excitations between different components of the pressure tensor, i.e. a situation which is similar to the effect of bulk viscosity in a molecular system. In this case Eqs. 14.33–14.37 provide a good qualitative description of the radial viscous evolution of a uniform dense ring.

In a dense ring non-axisymmetric self-gravity wakes will form. This will generally lead to strong non-Newtonian gravitational stresses, not included in Eqs. 14.33–14.37. However, for dynamical problems on length-scales that are much larger than the wavelength of the wakes, the approximation of gravitational viscosity can then be adopted (Section 14.1.3).

14.1.3 Self-Gravity of the Ring

In Section 14.1.1 the local steady-state properties of planetary rings resulting from the balance between viscous heating and the collisional dissipation of random energy have been discussed. The inclusion of particles' mutual gravitational forces modifies the local dynamics in several, partially competing ways, depending on the mass density of the ring and the distance from the planet.

At low optical depth collective effects of self-gravity are negligible and the main effect stems from gravitational heating via close binary encounters (Cuzzi et al. 1979a, b; Hämeen-Anttila 1984; Petit and Hénon 1987; Ohtsuki 1992). For higher densities, the mean vertical self-gravity can become comparable to or even exceed the corresponding component of the central force, causing a strongly enhanced impact frequency and a reduced ring thickness (Salo and Lukkari 1982; Araki and Tremaine 1986; Wisdom and Tremaine 1988). However, in this case the ring is also susceptible to gravitational instability in the plane, which manifests in the formation of transient trailing density enhancements, called *self-gravity wakes* or *Toomre wakes* (Salo 1992a; see also Julian and Toomre 1966; Toomre and Kalnajs 1991). With increasing distance from the planet, tidal forces get weaker and eventually the direct gravitational sticking of particles becomes possible, causing the particles in the wakes to degrade into local aggregates (Salo 1995, Section 14.4); a similar clumping may take place at low optical depth via pairwise sticking of particles. In general, the inclusion of self-gravity leads to a strong enhancement of ring viscosity, due to increased impact frequency and gravitational stirring of the particle ensemble, and most importantly, due to gravitational torques exerted by the wakes, and the collective motion associated with them (Daisaka et al. 2001).

A convenient parameter characterizing the importance of self-gravity relative to the disrupting tidal force is the ratio of the mutual Hill-radius for a pair of particles to the sum of their physical radii (Lynden-Bell and Kalnajs 1972; Daisaka et al. 2001):

$$r_h \equiv \frac{R_{Hill}}{R_1 + R_2} = \left(\frac{\rho_0}{3\rho_{plan}}\right)^{1/3} \left(\frac{a}{r_{plan}}\right) \frac{(1+\mu)^{1/3}}{1+\mu^{1/3}}, \quad (14.38)$$

where a is the semi-major axis and $R_{Hill} = \{(m_1 + m_2)/3M_{plan}\}^{1/3}a$, $\mu = m_1/m_2$ with $m_i = (4\pi\rho_0/3)R_i^3$ being the mass of particle i of density ρ_0, and M_{plan}, r_{plan}, and ρ_{plan} denote mass, radius and density of the planet. In the case of

---

[4] Exceeding a dense ring's shear viscosity by a factor of $\xi_0/\eta_0 = 2\ldots 3$.

identical particles ($R_i \equiv R_0$, $m_i \equiv m_0$), r_h can be expressed in terms of physical parameters as

$$r_h = 0.82 \left(\frac{M_{plan}}{5.69 \cdot 10^{26} \text{ kg}}\right)^{-1/3} \left(\frac{\rho_0}{900 \text{ kg m}^{-3}}\right)^{1/3}$$
$$\times \left(\frac{a}{100{,}000 \text{ km}}\right), \quad (14.39)$$

and values of r_h based on Eq. 14.39 are used in the following (e.g., Figs. 14.5–14.9). With this formula the results for a given r_h can be scaled to other distances and internal densities of particles. Assuming solid ice density for the ring particles, the main rings of Saturn correspond to $r_h = 0.6 - 1.1$, from the inner C ring to the outer A ring, respectively. Similarly, Uranian rings extend between 0.65 and 0.8 if made of ice. Note that instead of r_h, $r_p \equiv r_h^{-1}$ is also often employed to parameterize gravity (Ohtsuki 1993; Salo 1995); the advantage of using r_h is that larger values correspond to stronger gravity; also the limit $r_h = 0$ corresponds to non-gravitating particles.

14.1.3.1 Gravitational Encounters

For low τ the main effect of gravity comes from close binary encounters, which act like totally elastic impacts: the kinetic energy of the encountering pair is conserved, while the deflection of mutual orbits transfers energy from the systematic velocity field to random motions. This extra heating is efficient if the velocity dispersion c is smaller than the mutual escape speed $v_{esc} = \sqrt{2Gm_0/R_0}$ of particles, but becomes inefficient for $c > v_{esc}$. Thus, encounters, if acting alone, would establish a state with $c \sim v_{esc}$ (Safronov 1969; Cuzzi et al. 1979a, b). However, if the physical impacts are able to maintain $c > v_{esc}$, then the effect of encounters is negligible. The condition for the importance of encounters can be written in terms of an upper limit for the vertical thickness, $H < H_{esc}$, where the effective geometric thickness defined by $H = \sqrt{12 \langle z^2 \rangle}$ denotes the full thickness of a uniform layer with the same vertical dispersion as the actual distribution. For low optical depths we have $H \approx 2c/\Omega$. Writing $v_{esc} = \sqrt{24}\, r_h^{3/2} R_0 \Omega$ implies

$$\frac{H_{esc}}{R_0} \approx 10\, r_h^{3/2} \quad (14.40)$$

In the case of constant coefficient of restitution $\varepsilon_n \leq 0.5$, the impacts alone maintain $H/R_0 \approx 5$, which implies that gravitational encounters dominate over physical impacts for $r_h > 0.7$.

14.1.3.2 Vertical Self-Gravity

In very dense systems (high optical depth τ and filling factor ρ/ρ_0) the collective effects of self-gravity become increasingly important. First of all, the vertical component of self-gravity, F_z, may exceed the corresponding component of the central force, $F_c = -\Omega^2 z$. For simplicity, assume an infinite homogeneous layer of identical particles with an effective geometric thickness H. Inside the layer, with given density ρ, Poisson's equation gives for the vertical self-gravity

$$F_z(z) = -2\pi G \int_{-z}^{z} \rho(z')\, dz' = -\frac{4\pi G \sigma z}{H}$$

so that

$$\frac{F_z}{F_c} = \frac{4\pi G \sigma}{H \Omega^2} = 48\tau r_h^3 \frac{R_0}{H} \quad (14.41)$$

Assuming a Gaussian vertical mass distribution, the vertical self-gravity near the equatorial plane is a factor $\sqrt{6/\pi}$ larger if F_z/F_c is parameterized in terms of H as defined above. Analogous to H_{esc} we define H_{fz} as a thickness of the system for which $F_z \sim F_c$,

$$\frac{H_{fz}}{R_0} \approx 65\tau\, r_h^3, \quad (14.42)$$

(for the case of a Gaussian mass distribution). For typical values of Saturn's B-ring, $r_h \sim 0.8$, $\tau \sim 1.5$, the vertical self-gravity exceeds the central component, unless $H/R_0 > 50$. As shown in simulations (Wisdom and Tremaine 1988; Salo 1991) the extra vertical force tends to reduce H quite markedly, both due to the increased vertical frequency itself and also indirectly via the enhanced dissipation (see Fig. 14.5). This implies a strongly enhanced viscosity for any given τ. However, there are other effects of self-gravity which will lead to an even more drastic enhancement of viscosity.

14.1.3.3 Gravitational Wakes

Intuitively, the planar components of self-gravity might be expected to have less importance than the vertical component, due to partial cancellation of forces. However, as shown by Toomre (1964), a self-gravitating differentially rotating particle disk is locally unstable against the growth of axisymmetric disturbances if its radial velocity dispersion falls below the critical value

$$c_{cr} = \frac{3.36 G \sigma}{\kappa}, \quad (14.43)$$

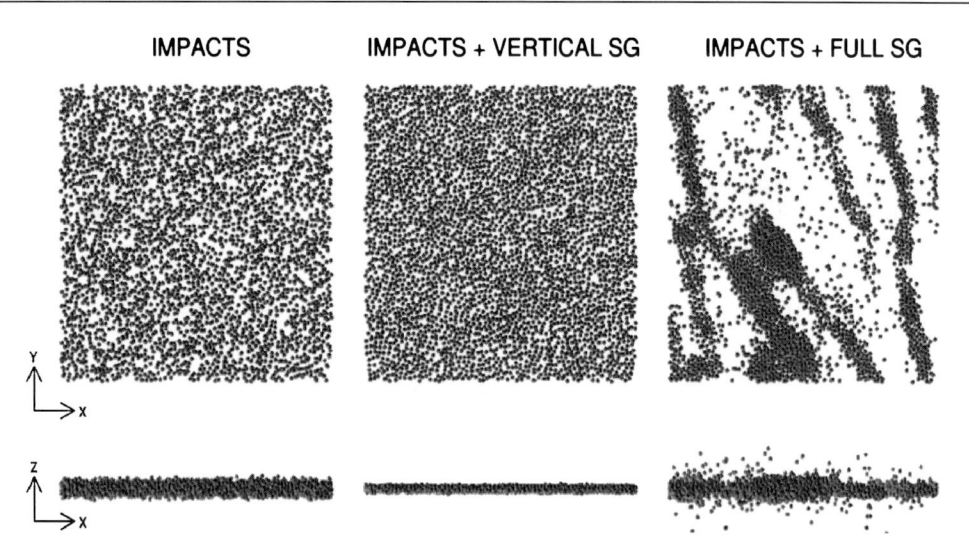

Fig. 14.5 Comparison between vertical and full self-gravity in simulations. In the left panel only physical collisions between particles are taken into account, for a simulated ring with $\varepsilon_n = 0.5$, $\tau = 0.75$, $r_h = 0.82$ (corresponding to $a = 100{,}000$ km for a solid ice particle density of 900 kg m^{-3}, or to $a = 126{,}000$ if $\rho = 450$ kg m^{-3}). In the middle panel the vertical component of self-gravity is included, calculated in a self-consistent manner from the vertical density distribution (Salo, 1991). Near the central plane the ratio of vertical self-gravity to the vertical component of central force $F_z/F_c \approx 8.8$, corresponding to an enhanced frequency of vertical oscillations by a factor $\Omega_z/\Omega = \sqrt{1 + F_z/F_c} \approx 3.1$: a very similar result would be obtained with the method of Wisdom and Tremaine (1988), who used a constant enhancement factor $\Omega_z/\Omega = 3.6$ to describe the vertical gravity. In the right panel all components of self-gravity are included, leading to the formation of self-gravity wakes. In comparison to the non-gravitating case, the inclusion of vertical gravity reduces the ring thickness in this example from $H/R_0 \approx 5 \rightarrow 3$, and increases the collision frequency by about a factor of 8. Both these effects enhance the viscosity, which in the studied example increases by a factor of two. However, when full self-gravity is included, the viscosity becomes even 30 times larger than in the non-gravitating case. A snapshot from a comoving local simulation region is displayed: the x-axis points away from the planet and the y axis in direction of orbital velocity. Note that the size of the simulation system here corresponds to $2\lambda_{cr} \times 2\lambda_{cr}$, implying that the wake structure is somewhat suppressed in comparison to what would be obtained with larger calculation regions (adapted from Salo 1995)

where κ denotes the epicyclic frequency ($\kappa = \Omega$ for the Keplerian case). This critical value offers a very convenient measure for the closeness of the system to the instability threshold in terms of the *Toomre parameter*, defined as

$$Q = \frac{c_r}{c_{cr}} = \frac{c_r \kappa}{3.36 G \sigma}, \qquad (14.44)$$

where c_r denotes the radial velocity dispersion. For identical particles,

$$\frac{c_r}{R_0 \Omega} \approx 10 Q \tau r_h^3. \qquad (14.45)$$

Comparison to Eq. 14.42 indicates that whenever vertical self-gravity is important, the system is also near the threshold of collective planar instability: $F_z/F_c > 1$ corresponds to $Q < 2.5$.

How does this gravitational near-instability manifest? The gravitational collapse is opposed by the particles' random velocities, washing out small scale agglomeration, and by differential rotation, dissolving large condensations. As long as Q exceeds at least a few times unity ($Q \gtrsim 2$–3), the collective instability is completely avoided, and the system remains practically uniform: the main effect of gravity comes via pairwise encounters stirring up the velocity dispersion. However, if the optical depth, and thus σ, increases, or alternatively, if a ring location further away from the planet is inspected, Q could fall below about 2–3. In this case, the collective gravity, together with differential rotation, leads to the formation of shearing tilted wake structures, with individual wakes forming and dissolving in a time scale of about one orbital period (Fig. 14.5). The prominence of these structures stems from the swing amplification process (interplay of self-gravity and differential rotation, Goldreich and Lynden-Bell 1965; Toomre 1981) which significantly enhances the tiny kinematic wakes triggered by small density fluctuations.

The resulting self-gravity wakes correspond to a superposition of numerous Julian–Toomre wakes excited around each individual ring particle. Although the features are transient, in contrast to the steady response around an orbiting mass enhancement in a stellar disk studied by Julian and Toomre (1966), this analogue is demonstrated by the similar autocorrelation function; this correspondence also justifies calling these features *wakes*, even in the absence of any prominent individual perturber. Self-gravity wakes have also been discussed as models of flocculent spiral structure in late type galaxies (Toomre 1991; Toomre and Kalnajs 1991), in which context, however, some form of *ad hoc* dissipation is needed to balance the gravitational heating induced by the wakes themselves, which heating otherwise would rapidly suppress the swing amplification. In planetary rings, the dissipation via partially inelastic impacts provides a natural cooling mechanism, leading to a statistical steady-state with $Q \sim 1\text{--}2$, characterized by a continuous re-generation of new wakes. The formation of wakes for plausible Saturn ring parameters was first demonstrated in simulations of Salo (1992a) and has thereafter been confirmed in several studies (Richardson 1994; Daisaka and Ida 1999; Ohtsuki and Emori 2000). Moreover, spatial auto-correlation analysis of simulated wakes (Salo 1995; Salo et al. 2004; see also Toomre and Kalnajs 1991) confirms the close correspondence to Julian–Toomre stellar wakes.

For Saturn's rings, the approximative condition for the formation of wakes, $Q < 2$, corresponds to (see Salo 1995, Ohtsuki and Emori 2000, Salo et al. 2004)

$$\tau > \tau_{\min} \approx 0.2 \left(\frac{a}{10^8 \text{m}}\right)^{-3} \left(\frac{\rho_0}{900 \text{kg m}^{-3}}\right)^{-1}, \quad (14.46)$$

implying $\tau_{\min} \approx 0.3 - 0.1$, from the inner C ring to the outer A ring, respectively, if the internal density of solid ice is assumed for ring particles. This $\tau_{\min}$ gives a conservative *lower* limit, since Eq. 14.46 is based on the assumption of fairly dissipative identical particles that in the absence of self-gravity would concentrate in a very thin ring, just a few particle diameters thick. This is the expected behavior of particles if they follow the Bridges et al. (1984) formula for the coefficient of restitution. In regions with $\tau > \tau_{\min}$, wakes *may* form, depending on the actual particle elasticity, with more elastic impacts implying an increased $\tau_{\min}$ (see Fig. 14.6, comparing the 'frosty' and 'smooth' impact models). However, $\tau > \tau_{\min}$ is not a strict boundary for wake formation: autocorrelation analysis reveals that weak wakes are always present regardless of the value of Q. Wake formation is also affected by the particle size distribution: large particles provide seeds for strong wakes. This effect is counteracted by the larger velocity dispersion achieved by small particles (Salo 1992a, b), which acts as a stabilizing factor. As a net result the wake structure is stronger among the largest particles whereas the small particles tend to have a smoother distribution (Salo et al. 2004).

The tilt angle of wakes with respect to tangential direction is determined mainly by the shear rate: for the Keplerian case the asymptotic tilt angle of the tails of the wakes is $\sim 15°$ (Julian and Toomre 1966). However, the inner portions of wakes have larger pitch angles, depending on the physical parameters. As an effective mean value $\sim 20°$ can be adopted. The typical radial spacing between wakes found in simulations (Salo 1995; Daisaka and Ida 1999) is close to Toomre's critical wavelength (Toomre 1964)

$$\lambda_{cr} = 4\pi^2 G \sigma / \kappa^2. \quad (14.47)$$

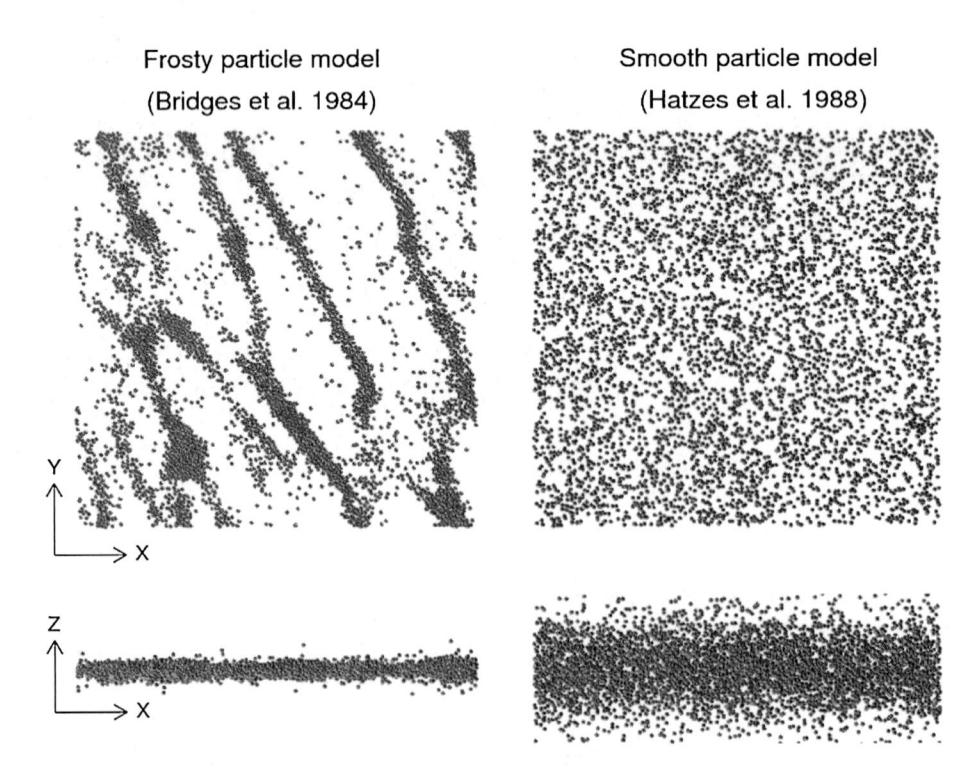

Fig. 14.6 The dependence of simulated self-gravity wakes on the assumed elasticity. In the left panel Bridges et al. (1984) formula ('frosty') is used, while in the right panel a formula from Hatzes et al. (1988) for more elastic 'smooth' particles is assumed (Fig. 14.3). In both cases a system with $\tau = 0.5$ and $r_h = 0.85$ is simulated, using a $4\lambda_{cr} \times 4\lambda_{cr}$ region. For more details of the effect of particle elasticity on the wake structure see Fig. 11 in Salo (1995)

For uni-sized particles this can be expressed in terms of the optical depth and the parameter r_h as

$$\frac{\lambda_{cr}}{R_0} = 48\pi\tau r_h^3 \qquad (14.48)$$

For Saturn's A-ring the expected spacing is $\lambda_{cr} \sim 50\text{–}100\,\text{m}$.

14.1.3.4 Survey of Self-Gravity Wake Structures

Figure 14.7 displays a simulation survey of wake structures expected for different planetocentric distances. The figure illustrates clearly the gradual increase in the strength of wakes as the assumed distance or optical depth is increased, as well as the increase in the clumpiness of the wakes, and their eventual collapse into aggregates at large distances (see Section 14.4 for detailed discussion of gravitational accretion of ring particles). Changes of the appearance of the wakes are also visible. For instance, at large r_h the simulated wakes are narrow structures separated by wide gaps, whereas for large τ and intermediate $r_h \sim 0.6$ the gaps and wake widths are more comparable. The regime of overstable oscillations ($r_h < 0.6$, $\tau > 1$) is also noticeable: apparently in this regime the self-gravity is sufficiently strong to lead to a strong increase of viscosity with density, as required for the onset of overstability (Section 14.2.1), but simultaneously the non-axisymmetric wakes are not yet too strong to suppress the coherence of axisymmetric oscillations.

14.1.3.5 Gravitational Viscosity

In the case of strong wakes the total viscosity is dominated by the angular momentum transfer related to the gravitational torques exerted by the inclined wakes (gravitational viscosity), and by the transfer associated with the large scale motion of the wakes (adds to the local viscosity, whose other

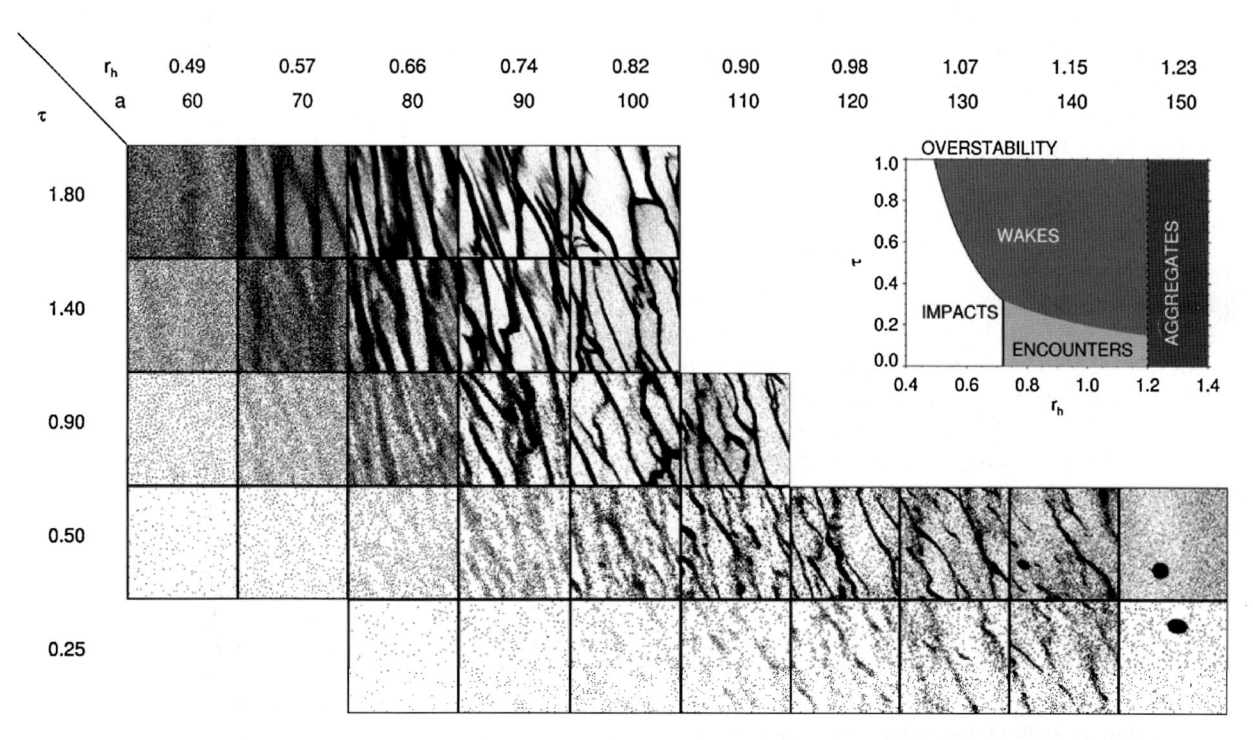

Fig. 14.7 The dependence of self-gravity wakes on optical depth τ, and the strength of gravity relative to the tidal force, quantified in terms of the r_h parameter. Also indicated are those values of Saturnocentric distance a (in units of 1,000 km) to which r_h corresponds for $\rho = 900\,\text{kg}\,\text{m}^{-3}$. Simulations use identical particles with $\varepsilon_n = 0.5$. The size of the calculation area covers $4\lambda_{cr} \times 4\lambda_{cr}$ region, thus scaling proportional to expected scale of wakes (physical size scales according to $\lambda_{cr}/R_0 \sim r_h^3\tau$, varying from 35 to 600 particle radii). Note the region in the upper left ($\tau > 1$, $r_h \sim 0.6$) where axisymmetric overstable oscillations (Section 14.2.1) coexist with the inclined gravity wakes. Also note that τ here refers to the average geometric optical depth of the system (the total area of particles divided by the calculation area). The photometric optical depth would be generally different, its value also depending on the observing direction (Salo et al., 2004; Porco et al., 2008; Robbins et al., 2009). The insert shows schematically the dynamical regimes where physical impacts, pairwise gravity, and collective gravity dominate, based on a simple estimate of which ingredient alone would maintain the largest radial velocity dispersion ($c_r = 2R\Omega$ for impacts, $c_r = v_{esc}$ for encounters, or $Q = 2$ for wakes; see Salo 1995; Ohtsuki and Emori 2000; Daisaka et al. 2001). Also indicated is the region where overstability occurs in simulations, and the boundary beyond which particles clump into local aggregates in simulations ($r_h \approx 1.2$, see Salo 1995 and Karjalainen and Salo 2004)

contribution is associated with random motions[5]). Compared to these, the nonlocal (collisional) viscosity has a minor contribution. Based on dimensional arguments, the gravitational viscosity is expected to be of the order of $L^2\Omega$, where L is the typical radial scale of momentum transport and $L\Omega$ is the associated (specific) tangential momentum. Using $L = \lambda_{cr} \propto G\sigma/\Omega^2$ and including a semi-empirical correction factor $C(r_h) = 26r_h^5$ obtained by fits to viscosities determined from simulations, Daisaka et al. (2001) derived the formula

$$\nu_{grav} \approx \frac{1}{2} C(r_h) \frac{G^2\sigma^2}{\Omega^3}. \quad (14.49)$$

Moreover, they showed that in the case of strong wakes ($r_h > 0.7$) the local viscosity associated with the streaming motion related to wakes is of the same order of magnitude as the gravitational viscosity; in this case the numerical value of total (kinematic) viscosity can be approximated as

$$\nu \approx \nu_{grav} + \nu_l \approx C(r_h) \frac{G^2\sigma^2}{\Omega^3}. \quad (14.50)$$

Similar formulae, but without the correction factor, have been discussed for example by Ward and Cameron (1978) and in the context of spiral torques in galaxy disks by Lynden-Bell and Kalnajs (1972). As discussed by Daisaka et al. (2001), the strong distance-dependence implied by the correction factor $C(r_h)$ is likely to be associated with the particulate nature of the system now in question. Namely, the maximal density of a gravitationally perturbed region is limited by the internal density of particles: this limitation is increasingly severe as r_h is decreased, since the scale of wakes is then reduced compared to the particle size (see Eq. 14.48). No such limitation is present in the analysis of a continuum fluid where the surface density perturbations can at least in principle be of the order of the surface density itself.

The values for the viscosity determined by Daisaka et al. (2001) were fully confirmed by Tanaka et al. (2003) who showed that the most convenient way of measuring the total viscosity in self-gravitating particle simulations is via the associated energy loss in impacts. When the wake amplitude has reached a statistical steady-state, the time averages of energy loss and viscous gain are equal, and ν is obtained from Eq. 14.5. A similar method was used in the nongravitating simulations of Salo et al. (2001) to check the viscosities measured with the Wisdom and Tremaine (1988) method.

---

[5] These two kinds of local viscosity formally correspond to a stress tensor which is split into two terms as $P_{xy} = \sigma \vec{v}_x \vec{v}_y + \Pi_{xy}$. Here $\vec{v}$ is the mean velocity of (a large number of) ring particles in a (still small) Lagrangian surface element of the rings. The component Π_{xy} (the pressure tensor) then corresponds to the random motion of particles *in* this Lagrangian element. With shear rate s the viscosities are defined as $\sigma s \nu = P_{xy}$.

Figure 14.8 displays the dependence of the total viscosity on τ and r_h in simulations. The dependence of various contributions are shown separately (upper panels), illustrating the dominance of ν_{grav} and the close agreement in numerical values $\nu_l \approx \nu_{grav}$ in the case of strong wake structure ($r_h > 0.7$). The lower panel displays the total viscosities (gravitational + local + nonlocal contributions), also indicating the good agreement with Eq. 14.50. Strictly speaking, this formula was obtained by fits to simulations assuming constant elasticity $\varepsilon_n = 0.5$. However, for the plausible frosty particle elasticity laws, the total viscosities are not extremely sensitive to the applied elasticity. According to Fig. 14.9, the viscosity varies less than a factor of two for a constant $0 < \varepsilon_n < 0.6$; for the Bridges et al. (1984) formula the viscosity varies at most by three if the scale factor $v_c/v_B < 10$.

14.1.3.6 Observational Signatures of Self-Gravity Wakes

Self-gravity wakes, trailing by ~20° with respect to tangential direction, offer a straightforward explanation for the long-known quadrupole azimuthal brightness variation of Saturn's rings, i.e. the fact that in Earth-bound observations the post-elongation quadrant of each ansa appears brighter than the pre-elongation quadrant. This variation results as the reflecting surface area is the smallest when the wakes are seen more or less along their long-axis (at ring longitudes ~20° before ansae), and largest when seen edge-on (see Fig. 14.10). The effect was first noted for the A ring – where it amounts to even 35% – in ground-based photographs (Camichel 1958; Lumme and Irvine 1976; Thompson et al. 1981) and was later analyzed in detail from Voyager data (Franklin et al. 1987; Dones et al. 1993).

The realization that trailing inhomogeneities would account for such variations is due to the seminal paper Colombo et al. (1976) where the probable connection to Julian and Toomre (1966) wakes was first suggested (see also Franklin et al. 1987, Dones and Porco 1989). First detailed photometric modeling of self-gravity wakes was carried out in Salo et al. (2004), demonstrating that N-body wake models (with Bridges et al. (1984) elasticity law and particle internal density one-half of solid ice density), can account in a robust manner for the overall longitude and elevation dependence of the A ring brightness variations, as seen in ground-based observations, and in Voyager observations in both reflected and transmitted light. This conclusion was confirmed by the comparison to the extensive Hubble Space Telescope data set in French et al. (2007), where the observed asymmetry amplitude and longitude of minimum brightness were characterized as a function of Saturnocentric distance for a full range of geometries accessible from Earth. It was also shown that there is a trade-off between various dynamical parameters:

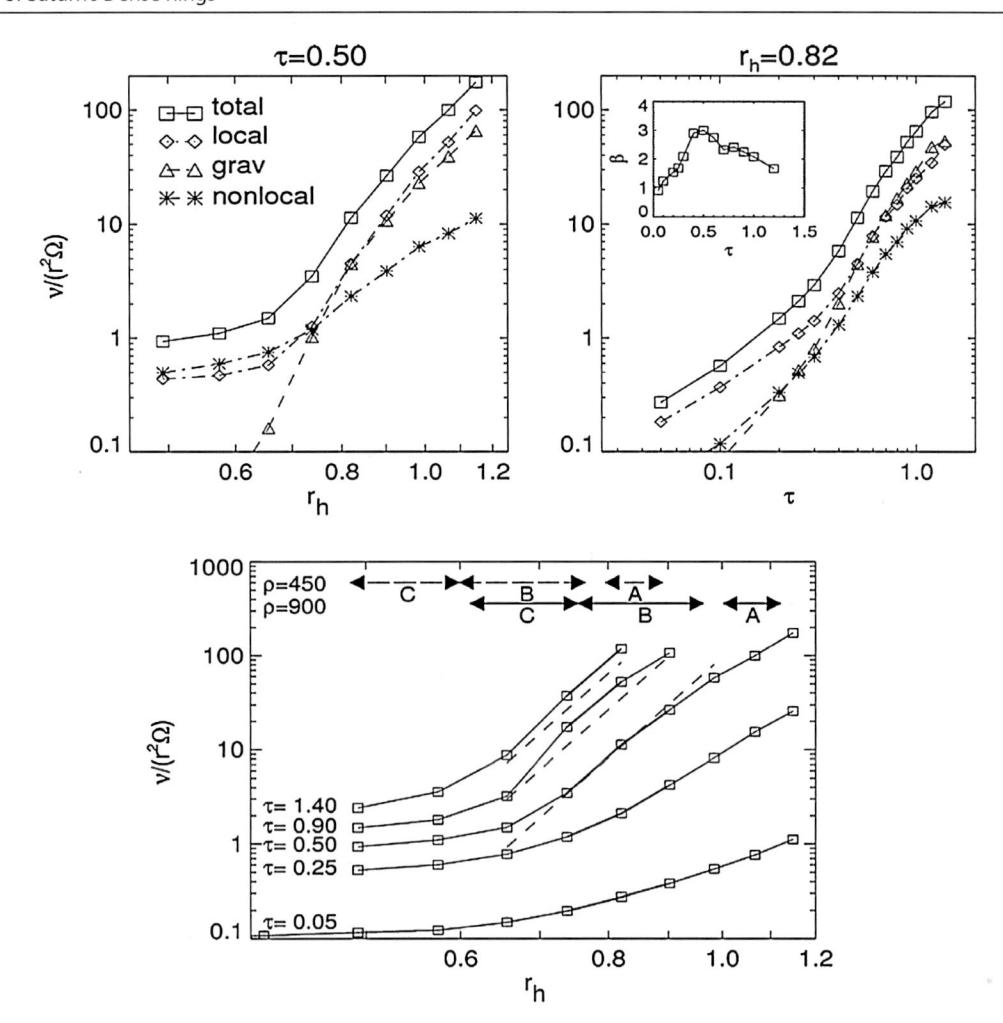

Fig. 14.8 *Upper left panel*: The dependence of various contributions to viscosity on r_h ($\tau = 0.5$ and $\varepsilon_n = 0.5$) from simulations. The gravitational viscosity ν_{grav} becomes dominant for $r_h > 0.7 - 0.8$, which increase is accompanied with an increased ν_l. This regime corresponds to the emergence of strong wake structure (Fig. 14.7). *Upper right panel*: Dependence on τ for fixed $r_h = 0.82$, $\varepsilon_n = 0.5$. The inset shows the slope $\beta = \partial \log \nu / \partial \log \tau$, rising from $\beta = 1$ at low τ to $\beta > 2$ due to the increased contribution of gravitational viscosity. *Lower panel*: Total viscosity as a function of τ and r_h (for $\varepsilon_n = 0.5$). Ring regions corresponding to the values of r_h are indicated (for internal particle densities $\rho = 900 \text{ kg m}^{-3}$ and $\rho = 450 \text{ kg m}^{-3}$). The *dashed line* indicates the Daisaka et al. (2001) fitting formula $\nu_{tot}/(R_0^2 \Omega) \approx 380 r_h^{11} \tau^2$, expected for the case of strong wakes; this formula is the same as Eq. 14.50, now written in terms of r_h and τ. In physical units, the simulated viscosity for $a = 130{,}000$ km corresponds to about $100 \text{ cm}^2\text{s}^{-1}$ (for $\sigma = 400 \text{ kg m}^{-2}$ and $\rho = 900 \text{ kg m}^{-3}$)

the modeled asymmetry is reduced for more elastic particles, lower internal density, or smaller optical depth, and vice versa (these dependencies are qualitatively consistent with the dynamical strength of wakes, measured in terms of ν_{grav} in Figs. 14.8 and 14.9); also, assuming an extended size distribution reduces the modeled asymmetry. Nevertheless, the fact that the maximum asymmetry amplitude is seen in the mid A-ring ($a \sim 129{,}000$ km) favors underdense particles: for example, for solid ice density the modeled asymmetry would peak at mid B-ring and fall sharply at the A ring region (see Fig. 23 of French et al. 2007). The HST data also showed that a weak azimuthal brightness asymmetry is present in the inner B ring, with an amplitude of roughly one-quarter of the maximum amplitude in the mid A-ring: this reduced amplitude is consistent with the weaker self-gravity wakes forming closer to the planet, even when allowing for the presumably larger surface density of the B ring. The B ring asymmetry was also inferred from the rings' radar echo (Nicholson et al. 2005); altogether the A ring asymmetry amplitude appeared about twice larger in radar signal (12.6 cm) compared to optical, while being consistent with the longitude dependence: this can be interpreted as an evidence that the wake structure is more dominant in the population of large particles as compared to smaller ones. Azimuthal variations are also present in Saturn's microwave radiation reflected and transmitted by the rings (Dunn et al. 2004, 2007).

An even more striking demonstration of gravity wakes in Saturn's rings is offered by stellar occultations observed by the Cassini Ultraviolet Imaging Spectrograph (UVIS)

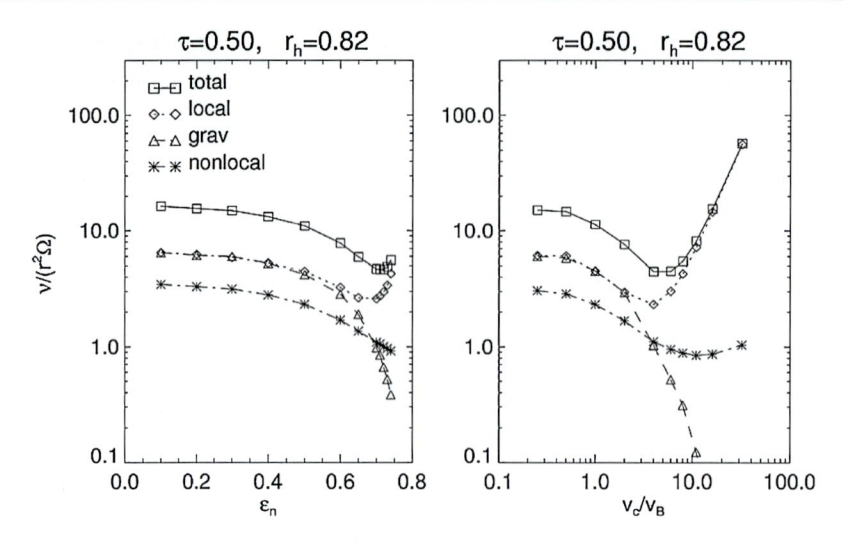

Fig. 14.9 Effect of particle elasticity on viscosity. In the left frame different values of the constant coefficient of restitution are compared, while in the right frame a Bridges et al. (1984) type velocity-dependent formula is applied with different scale factors (see Eq. 14.14). In the case of constant ε_n, the local viscosity diverges for $\varepsilon_n > 0.75$, since then no thermal equilibrium is possible in the system (see Fig. 14.2 for $\varepsilon_{crit}(\tau)$): when this limiting ε_{crit} is approached, the steady-state velocity dispersion increases and the role of self-gravity becomes insignificant. Similarly, an increase of the scale factor in the velocity-dependent $\varepsilon_n(v_n)$ leads to hotter systems, reducing all other contributions to viscosity in comparison to local viscosity (*right panel*)

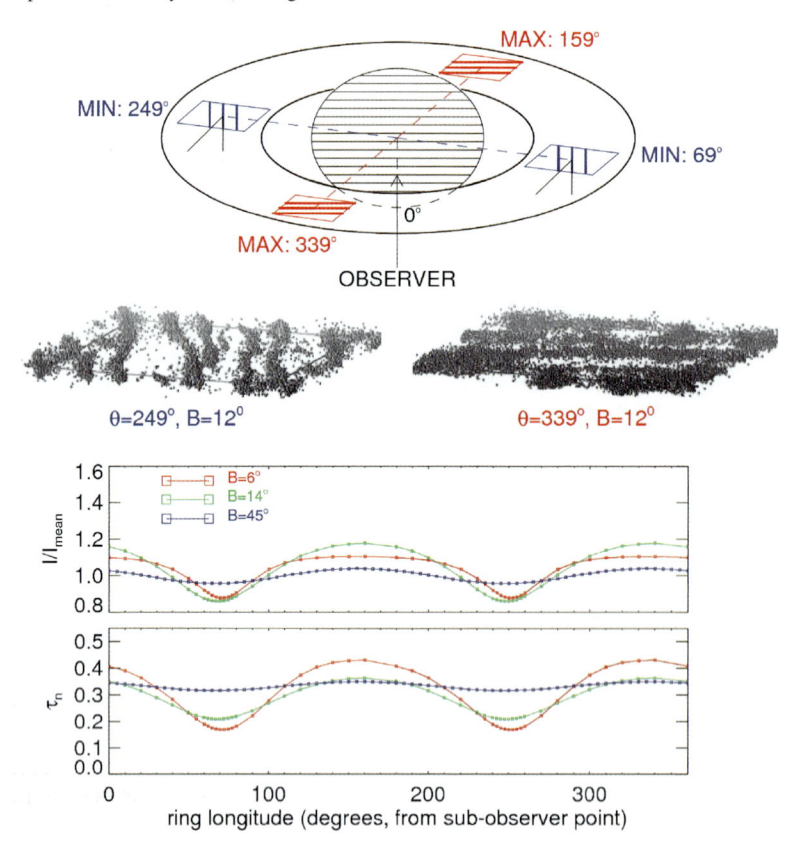

Fig. 14.10 Schematic presentation of self-gravity wakes and azimuthal brightness asymmetry (from Salo et al. 2004). At low elevation angles B the wakes, trailing here by about 21° with respect to the local tangential direction, are seen roughly along their long axis at ring longitudes of 249° and 69°, and perpendicular to their long axis at longitudes of 339° and 159°. In the former case the rarefied regions between wakes are visible, reducing the reflecting surface area: this corresponds to minimum brightness. In the latter case rarefied regions are hidden by the wakes in low tilt angle images, maximizing the reflecting area. A similar effect is seen in the apparent optical depth observed in occultations, which is smallest when viewed along the wakes. In reflected light at low phase angle the modeled asymmetry amplitude for the A ring has a maximum at $B \sim 10 - 15°$, consistent with HST observations (French et al., 2007); a similar maximum is seen also in CIRS observations of ring thermal emission (Ferrari et al., 2009). For the optical depth the expected variation increases toward smaller elevations: VIMS occultations at $B = 3.45°$ indicate a very strong longitude-dependence in the A ring transmission (Hedman et al., 2007)

(Colwell et al. 2006a, 2007), the Visual and Infrared Mapping Spectrometer (VIMS) (Hedman et al. 2007) and occultations recorded by the Radio Science Subsystem (RSS) (Marouf et al. 2006). These studies indicate a significant longitude-dependence of optical depth throughout the A and B rings, consistent with the brightness asymmetry (the non-detection of ground-based reflection asymmetry in the dense B ring is due to low brightness contrast in the case of high τ). The apparent optical depth depends also strongly on elevation (Colwell et al. 2006a, 2007), consistent with the predictions of the wake models (Salo et al. 2004). The azimuthal brightness variations based on observations with Cassini Imaging Subsystem (ISS) have been recently analyzed by Porco et al. (2008), also favoring very dissipative, under-dense particles. Azimuthal variations have also been observed in the ring's thermal radiation by the Cassini Composite and Infrared Spectrometer (CIRS) instrument (Ferrari et al. 2009), implying a striking agreement with the amplitude of reflected light variations.

Some poorly understood observations relevant to wake structure remain. For example, the A ring asymmetry drops significantly beyond the Encke gap. One possibility (French et al. 2007) is that wakes become clumpier at larger radial distances (before collapsing to aggregates) which reduces the longitude dependence of ring reflectivity or opacity. However, the observed drop is even more rapid than models predict. A more likely explanation is a wider particle size distribution in the regions beyond the Encke gap, which would reduce the contrast of wakes (Salo et al. 2004; French et al. 2007). Both factors may contribute also simultaneously: Robbins et al. (2009) suggest that clumpier wakes can stir up smaller particles, leading to apparently wider size distribution. Stronger wakes also lead to increased maximal impact velocities, which may lead to enhanced release of impact-debris, if regolith-covered particles are assumed (Salo and Schmidt 2007). This in turn would reduce the wake contrast; a similar model helps to account for the propeller brightness contrast (Section 14.3.3). Another poorly understood observation (Dones et al. 1993; French et al. 2007) is the very rapid rise in the brightness asymmetry amplitude from the inner to the mid A ring (amplitude peaks at $a \sim$ 129,000 km), in comparison to the much shallower rise in the modeled amplitude, when fixed particle properties are assumed (French et al. 2007). Again, radius-dependent size distribution could account for this, but no self-consistent models yet exist.

The extensive set of Cassini occultation profiles has made it possible to probe the gravity wake properties as a function of radial location. To facilitate this, the wake structure has been modeled as an alternating sequence of dense wakes and rarefied inter-wake regions, describing the wakes as parallel, infinitely long slabs with a rectangular (Colwell et al. 2006a, 2007, see also Dunn et al. 2004) or ellipsoidal (Hedman et al. 2007) cross section, defined by the wake's radial width W and geometric thickness H. The UVIS and VIMS occultations suggest (see also Chapter 13) that the A ring wakes have $H/L \sim 0.1$, where L is the radial separation between adjacent wakes. Identifying L with the Toomre wavelength ~ 50 m, this suggest that the geometric thickness of A ring wakes is only of the order of 5 m (Hedman et al. 2007; Colwell et al. 2006a). The UVIS observations imply $W/L \sim 0.4$ for the A ring wakes (thus the inferred gaps are wider than wakes). On the other hand, the deduced W/L approaches even unity for the optically thickest parts of the B ring (densely packed wakes with only narrow gaps); in fact Colwell et al. (2007) suggest that the photometric optical depth reflects mainly the density of packing between the wakes.

The density wave estimates of total viscosities seem consistent with the expected total viscosity in the presence of gravity wakes. In particular Tiscareno et al. (2007) estimated surface densities and viscosities of the A ring from low amplitude density waves observed in Cassini ISS images, and found a very good agreement with Eq. 14.50, for both the magnitude of the derived viscosity and its radial increase with Saturnocentric distance. However, the agreement in magnitude can be partly fortuitous, since it was obtained by assuming a solid ice density for particles, whereas all the aforementioned studies of the A ring azimuthal brightness variations favor under-dense particles. Low internal densities were deduced also in Robbins et al. (2009), based on modeling UVIS observations of ring transmission. According to Eqs. 14.39 and 14.49, $\nu_{grav} \propto \rho^{5/3}$ for a fixed distance and σ; reducing ρ by one-half would thus imply a factor of three smaller viscosity. Nevertheless, Eq. 14.49 refers to identical particles and a specific elastic law ($\varepsilon_n = 0.5$); releasing these assumptions might easily compensate for the difference.

14.2 Instabilities

A classical problem of planetary ring dynamics is the stability of the ring flow. It addresses dynamical processes on time-scales that are small compared to the secular viscous evolution of the disk. If the ring is in a steady state then small perturbations are unavoidable (for instance through the gravitational force of the satellites of the planet). The question of stability theory is whether the perturbed ring will return to the steady state, i.e. whether the ring is stable with respect to the perturbation, or, if the perturbation will grow. In the latter case the steady state is unstable. If the flow is unstable, the steady state is destroyed by perturbations of arbitrarily small magnitude and it will eventually be replaced by another configuration which is stable, and, in general more complex and structured.

The wealth of radial structure of Saturn's B ring revealed by the Voyager observations (Smith et al. 1982; Lane et al. 1982) stimulated the search for mechanisms of intrinsic instability, that could produce at least part of the observed structure. Prominent models proposed in the literature are the viscous (diffusion) instability (Lukkari 1981; Lin and Bodenheimer 1981; Ward 1981), ballistic transport models (Durisen 1995), angular momentum and mass transport related to electromagnetic effects on small grains (Goertz et al. 1986; Goertz and Morfill 1988; Shan and Goertz 1991), viscous overstability (Schmit and Tscharnuter 1995, 1999; Spahn et al. 2000; Salo et al. 2001; Schmidt et al. 2001; Latter and Ogilvie 2009), or instability of the radial shear profile due to a certain dependence of shear stress on the shear rate of the flow (Tremaine 2003).

14.2.1 Viscous Overstability

If a planetary ring is overstable, it spontaneously develops axisymmetric waves on a 100 m scale. These waves bear many similarities to density waves although overstable waves develop without external resonant perturbation. A planetary ring becomes overstable if the viscosity is increasing steeply enough with optical depth, which is expected in dense rings. In this case the shear in the compressed phase of the wave can feed back energy from the ring's flow to the oscillations. In contrast to density waves overstability is not strictly bound to self-gravity of the ring, although self-gravity generally alters the conditions necessary for the onset of overstability, and the wavelength and frequency of the oscillations. Axisymmetric 100–200 m wavelike perturbations in Saturn's A and B rings, consistent with overstability, are seen in Cassini RSS (Thomson et al. 2007) and the UVIS data (Colwell et al. 2007).

Viscous overstability was proposed by Borderies et al. (1985), expanding on earlier work (Borderies et al. 1983b), to explain the eccentricity of narrow ringlets observed around Saturn and Uranus (see also Longaretti and Rappaport (1995)). Borderies et al. (1985) also mention the possibility of axisymmetric oscillatory instability of a planetary ring, and Goldreich and Tremaine (1978b) point out that density waves may become overstable. As a spontaneous local instability of Saturn's B ring overstability was investigated in detail in terms of an isothermal hydrodynamic model by Schmit and Tscharnuter (1995, 1999). Spahn et al. (2000) included the thermal balance equation in this concept.

In local simulations of a perturbed dense planetary ring Mosqueira (1996) showed that the condition for viscous overstability derived by Borderies et al. (1985) can be fulfilled. The first direct demonstration of viscous overstability in an unperturbed dense collisional ring was given in self-gravitating local simulations by Salo et al. (2001), using parameters plausible for Saturn's B ring (see also Daisaka et al. 2001). The simulations spontaneously developed axisymmetric oscillations on a radial length of about 100 m, for an optical depth $\tau > 1$. Generally the oscillatory pattern was found to coexist with self-gravity wakes (Fig. 14.11).

In their local simulations Salo et al. (2001) also used an enhanced frequency of vertical oscillations $\Omega_z / \Omega > 1$ to approximate effects of the ring's vertical self-gravity (Wisdom and Tremaine 1988). For $\Omega_z / \Omega = 3.6$ the stability boundary was again about $\tau \sim 1$ and for $\Omega_z / \Omega = 1$ (i.e. no self-gravity) overstability developed for $\tau > 4$. This suggests that the conditions necessary for overstability are not connected to self-gravity itself but rather to a large collision frequency, which determines the viscous properties. Further, Salo et al. (2001) showed that overstability develops in simulations with a broad distribution of particle sizes. A quantitative comparison of simulations to overstability on the linear level in a hydrodynamic model was carried out by Schmidt et al. (2001), the nonlinear saturation of overstable waves was studied by Schmidt and Salo (2003), and nonlinear wavetrain solutions are derived by Latter and Ogilvie (2009).

14.2.1.1 A Linear Model

Linearizing the isothermal Eqs. 14.33–14.37 about the steady state solution $\sigma = \Sigma, u = 0, v = 0$ we obtain

$$\dot{\sigma} = -\Sigma u' \quad (14.51)$$

$$\dot{u} = 2\Omega v - \left(\frac{1}{\Sigma} \left. \frac{\partial p}{\partial \sigma} \right|_0 - \frac{2\pi G}{|k|} \right) \sigma' + \frac{\alpha}{\Sigma} \eta_0 u'' \quad (14.52)$$

$$\dot{v} = -\frac{\Omega}{2} u + \frac{1}{\Sigma} \left(\eta_0 v'' - \frac{3}{2} \Omega \left. \frac{\partial \eta}{\partial \sigma} \right|_0 \sigma' \right), \quad (14.53)$$

where η_0 denotes the dynamic shear viscosity of the unperturbed flow, α is the constant defined by Eq. 14.37, and we use

$$p(\sigma)' = \left. \frac{\partial p}{\partial \sigma} \right|_0 \sigma', \quad \eta(\sigma)' = \left. \frac{\partial \eta}{\partial \sigma} \right|_0 \sigma',$$

the subscript indicating that the derivative is taken at the steady state. Further, we inserted the solution

$$\Phi_{Disk}(r, z) = -\frac{2\pi G}{|k|} \sigma(r) \exp[-|kz|] \quad (14.54)$$

of the thin disk Poisson Eq. 14.36, valid for harmonic radial modes with wavenumber k (Binney and Tremaine 1987).

We write

$$\nu_0 = \frac{\eta_0}{\Sigma} \quad (14.55)$$

14 Dynamics of Saturn's Dense Rings

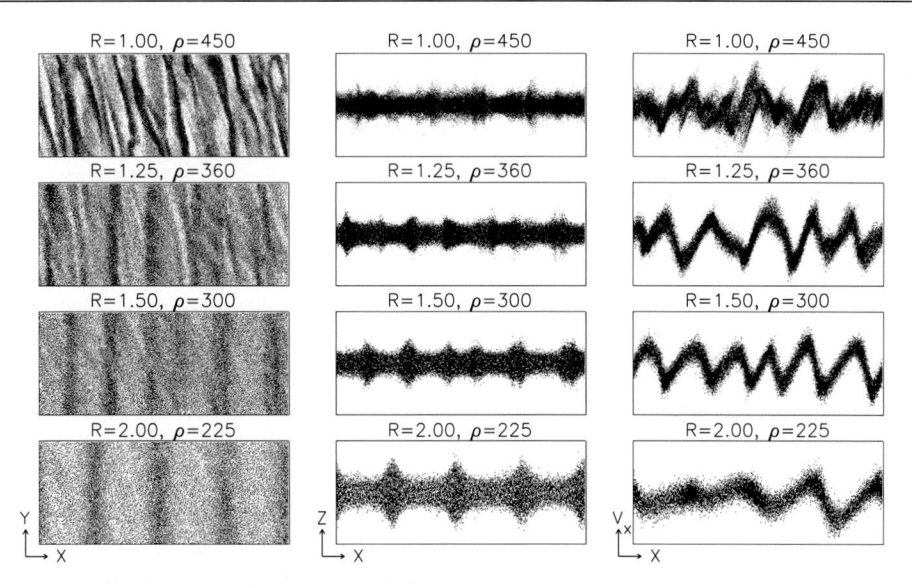

Fig. 14.11 Transition from wake dominated (upper row) to overstability dominated self-gravitating simulations. Optical depth and surface mass density are fixed ($\tau = 1.4$, $\sigma = 840\,\mathrm{kg\,m^{-2}}$) while particle radius and internal density are varied (plot labels indicate particle radius in meters and internal density in units of kg/m³). In the left panels the simulation cell is seen from above and particle centers are plotted as dots. The planet is to the left and the direction of orbital motion is upward. The size of the simulation region is $L_x \times L_y = 10\lambda_{cr} \times 4\lambda_{cr}$ where $\lambda_{cr} = 4\pi^2 G\sigma/\Omega^2$ is Toomre's critical wavelength (Toomre, 1964, see Section 14.1.3). In the case of strong wake structure the Toomre parameter Q assumes a value of about one. (The relatively small Q is due to the low internal density of the particles. In earlier simulations that used the density of solid ice $Q = 2$ was found.) On the panels in the middle the system is seen from aside. The axisymmetric oscillations are also seen in the profiles of radial velocity (right panels), as well as in tangential velocity and velocity dispersion (not shown). From Salo et al. (2001)

for the unperturbed kinematic shear viscosity and define

$$\left.\frac{\partial \eta}{\partial \sigma}\right|_0 = \nu_0 \left.\frac{\partial \ln \eta}{\partial \ln \sigma}\right|_0 \tag{14.56}$$

to display the proportionality of $\partial \eta/\partial \sigma$ to the kinematic viscosity ν_0 explicitly. For harmonic perturbations

$$\sigma, v, u \propto \exp\left[st + ikx\right] \tag{14.57}$$

we then obtain from Eqs. 14.51–14.53 the dispersion relation (Schmit and Tscharnuter 1995)

$$s^3 + s^2 \nu_0 k^2 (1+\alpha) + s\left(\Omega^2 + \left.\frac{\partial p}{\partial \sigma}\right|_0 k^2 - 2\pi G\Sigma|k|\right.$$
$$\left. + \nu_0^2 k^4 \alpha\right) + \nu_0 k^2 \left(3\Omega^2 \left.\frac{\partial \ln \eta}{\partial \ln \sigma}\right|_0 + \left.\frac{\partial p}{\partial \sigma}\right|_0 k^2\right.$$
$$\left. -2\pi G\Sigma|k|\right) = 0. \tag{14.58}$$

Equation 14.58 has, in the limit of long wavelengths, the approximate roots

$$s^{(1)} = -3\nu_0 k^2 \left(\left.\frac{\partial \ln \eta}{\partial \ln \sigma}\right|_0 - \left[\frac{1}{3} - \left.\frac{\partial \ln \eta}{\partial \ln \sigma}\right|_0\right]\frac{2\pi G\Sigma}{\Omega^2}|k|\right)$$
$$+ O\left(k^4\right) \tag{14.59}$$

$$s^{(2/3)} = \pm i \left(\Omega - \pi \frac{G\Sigma}{\Omega}|k| + \left[\frac{1}{2\Omega}\left.\frac{\partial p}{\partial \sigma}\right|_0 - \frac{\pi^2}{2}\frac{G^2\Sigma^2}{\Omega^3}\right]\right.$$
$$\times \left[1 + \frac{\pi G\Sigma}{\Omega^2}|k|\right]k^2 - \frac{3}{2}\nu_0 k^2 \left(\frac{1+\alpha}{3}\right.$$
$$\left. - \left.\frac{\partial \ln \eta}{\partial \ln \sigma}\right|_0 + \left[\frac{1}{3} - \left.\frac{\partial \ln \eta}{\partial \ln \sigma}\right|_0\right]\frac{2\pi G\Sigma}{\Omega^2}|k|\right)$$
$$+ O\left(k^4\right) \tag{14.60}$$

The first root corresponds to the viscous instability mechanism (Ward 1981; Lukkari 1981; Lin and Bodenheimer 1981), for long modes $k \to 0$ being unstable if $\left.\frac{\partial \ln \eta}{\partial \ln \sigma}\right|_0 < 0$ (see Section 14.2.2). The conjugate complex modes correspond to viscously damped or amplified wave solutions. A more rigorous analysis (valid for all k) shows that if

$$\left.\frac{\partial \ln \eta}{\partial \ln \sigma}\right|_0 > \frac{1+\alpha}{3} - \frac{\alpha}{3\Omega^2}\left(2\pi G\Sigma|k| - \left.\frac{\partial p}{\partial \sigma}\right|_0 k^2 \right.$$
$$\left. - (1+\alpha)\nu_0^2 k^4\right), \tag{14.61}$$

the amplitude of the waves grows, i.e. the system becomes overstable (i.e. it undergoes a Hopf bifurcation) if the viscosity increases steeply enough with the surface density of the ring. For the parameterization $\nu \propto \sigma^\beta$ of the surface density dependence of the shear viscosity often adopted in hydrodynamic studies (Schmit and Tscharnuter 1995; Durisen 1995; Schmit and Tscharnuter 1999; Spahn et al. 2000; Salo et al. 2001; Schmidt et al. 2001; Schmidt and Salo 2003) one recovers the criterion (Schmit and Tscharnuter 1995) for overstability ($k \to 0$)

$$\beta > \beta_{cr} := \frac{2-\alpha}{3} = \frac{1}{3}\left(\frac{\zeta}{\nu} - \frac{2}{3}\right). \quad (14.62)$$

It can be shown that the same criterion follows from the condition for overstability given by Borderies et al. (1983a), namely that their viscous coefficient t_1 is positive, at lowest order in the nonlinearity parameter q. For narrow rings Papaloizou and Lin (1988) derived an equivalent criterion, not taking into account bulk viscosity. Assuming that bulk and shear viscosities are equal ($\alpha = 7/3$) Eq. 14.62 implies $\beta_{cr} = 1/9$; inclusion of axial self-gravity (see Eq. 14.61) leads to β_{cr} even slightly less than zero (Schmit and Tscharnuter 1995). But according to direct simulation the stability limit is rather $\beta_{cr} \approx 1$. It was suggested that this discrepancy could follow from the assumption of isothermality (Spahn et al. 2000; Salo et al. 2001; Schmidt et al. 2001) made in the above analysis: in practice the velocity dispersion also adjusts locally to overstable oscillations, which acts as a stabilizing factor. The value of the bulk viscosity was determined from simulations (Salo et al. 2001; Schmidt et al. 2001) and included to the analysis, improving the agreement with linear growth rates at large optical depth. However what was interpreted as the influence of bulk viscosity is in part related to out-of-phase oscillations of the components of the velocity ellipsoid (see discussion in 14.2.2). This effect cannot be modeled by Newtonian stress (Latter and Ogilvie 2006).

Taking the time derivative of Eq. 14.52, using Eqs. 14.52 and 14.53 to eliminate v and setting

$$\sigma = -\Sigma \int_{-\infty}^{t} d\bar{t}\, u'(\bar{t})$$

we can write Eqs. 14.51–14.53 formally as a forced wave equation

$$\ddot{u} + \Omega^2 u - \left(\left.\frac{\partial p}{\partial \sigma}\right|_0 - \frac{2\pi G \Sigma}{|k|}\right) u'' = \nu_0 \left(f(r,t) + \alpha \nu_0 u''''\right) \quad (14.63)$$

where

$$f(r,t) \equiv (1+\alpha)\ddot{u}'' + \int_{-\infty}^{t} d\bar{t} \left[3\Omega^2 \left.\frac{\partial \ln \eta}{\partial \ln \sigma}\right|_0 u'' - \left(\left.\frac{\partial p}{\partial \sigma}\right|_0 - \frac{2\pi G \Sigma}{|k|}\right) u''''\right].$$

An approximate solution for the oscillatory mode is obtained by assuming that ν_0 in Eq. 14.63 is a small parameter (compared to Ω/k^2) and using a multi-scale expansion (Kevorkian and Cole 1996). In this case Eq. 14.63 indeed becomes an equation for a viscously forced inertial-acoustic wave and the dynamics splits on two well separated time scales t and $\theta = \nu_0 t$ so that

$$u(x,t,\theta) = A(\theta)\, u_0(x,t),$$

where A is a slowly variable amplitude and u_0 is the rapidly oscillating solution of the homogeneous problem. At zeroth order in ν_0 one obtains a wave with dispersion relation[6]

$$\omega = \pm\sqrt{\Omega^2 - 2\pi G \Sigma |k| + \left.\frac{\partial p}{\partial \sigma}\right|_0 k^2}. \quad (14.64)$$

At order ν_0 we get an equation for the evolution of the amplitude A

$$\frac{\partial}{\partial \theta} A = -\frac{3}{2} k^2 \left[\frac{1+\alpha}{3} - \left.\frac{\partial \ln \eta}{\partial \ln \sigma}\right|_0 + \frac{2\pi G \Sigma}{\Omega^2}|k|\right. \\ \left. \times \left(\frac{1}{3} - \left.\frac{\partial \ln \eta}{\partial \ln \sigma}\right|_0\right)\right] A + O(k^4) \quad (14.65)$$

Thus, a solution to the linear Eq. 14.63, applicable to long modes (i.e. neglecting $O(k^4)$), is

$$u = u(0)\exp\left(-\frac{3}{2}\nu_0 k^2 \left[\frac{1+\alpha}{3} - \left.\frac{\partial \ln \eta}{\partial \ln \sigma}\right|_0 \right.\right. \\ \left.\left. +\frac{2\pi G \Sigma}{\Omega^2}|k|\left(\frac{1}{3} - \left.\frac{\partial \ln \eta}{\partial \ln \sigma}\right|_0\right)\right]t\right)\cos(\omega t + kx) \quad (14.66)$$

This describes a radial wave with exponentially growing or damping amplitude, recovering the growth rate given by Eq. 14.60.

From the nonlinear Eqs. 14.33–14.35 one can derive an equation for the amplitude growth which contains a term which is cubic in the amplitude A, in addition to the linear terms in Eq. 14.65. This cubic term, arising from the nonlinearities in the system, leads to a saturation of the growth of overstable modes at a certain amplitude depending on the wavelength (Schmidt and Salo 2003; Latter and Ogilvie 2009). It has been shown analytically and in simulations that generally standing waves are unstable with respect to traveling wave solutions (Schmidt and Salo 2003).

14.2.1.2 Overstability and Self-Gravity

Self-gravity of the ring has a twofold effect on overstable oscillations. On the one hand, self-gravity promotes overstability by steepening the dependence of viscosity on density

---

[6] The same dispersion relation is obtained for tightly wound spiral density waves.

(Fig. 14.8). On the other hand, in the case of strong self-gravity, the *nonaxisymmetric* wakes tend to suppress the growth of *axisymmetric* overstable oscillations, although simulations (Salo et al. 2001) show that the co-existence of both phenomena is permitted (Figs. 14.7 and 14.11). Currently no analytical theory exists to model the formation of overstability in a ring with self-gravity wakes.

Another important effect is related to the axisymmetric component of self-gravity. Namely, in non-selfgravitating models the wavelengths of overstable oscillations are observed to grow with time (Schmit and Tscharnuter 1995, 1999; Schmidt and Salo 2003). It is not clear at present if this growth would eventually be limited by the curvature of the ring flow (which was neglected in the modeling so far) or by nonlinear effects. In contrast, the inclusion of self-gravity leads to a limited regime of wavelengths which can grow. This was noted by Schmit and Tscharnuter (1999) and is in agreement with the results from direct simulations where only axisymmetric gravity is included.

14.2.2 Viscous Instability

Viscous instability was discussed after 1980 as a promising candidate to explain the banded structure of Saturn's B ring seen in the Voyager images (Lukkari 1981; Lin and Bodenheimer 1981; Ward 1981; Hämeen-Anttila 1982; Stewart et al. 1984). It is a diffusion instability, developing in sheared collisional systems if the dynamic shear viscosity η fulfills the condition

$$\frac{\partial \eta}{\partial \sigma} < 0, \qquad (14.67)$$

which follows from the linearized Eqs. 14.51–14.53 in the non-selfgravitating or long wavelength limit (see Eq. 14.59). In this case the (viscous) collisional flux of particles is directed *away* from ring regions of depleted density, thereby amplifying perturbations in the density profile. Earlier, viscous instability was discussed for accretion disks (Lightman and Eardley 1974).

From the linearized Eqs. 14.51–14.53 one can easily derive a diffusion equation, assuming a slow viscous radial migration of material and $v = 0$ (i.e. purely Keplerian velocity). In this case Eq. 14.53 can be solved for u which gives together with Eq. 14.51

$$\dot{\sigma} = 3 \left.\frac{\partial \eta}{\partial \sigma}\right|_0 \sigma'' \qquad (14.68)$$

For harmonic perturbations (Eq. 14.57) we recover the growth-rate given by Eq. 14.59.

In Section 14.1.1.5 we demonstrated in terms of simulations that the steady state viscous properties of a planetary ring depend sensitively on the particle elasticity. If the particles are as dissipative as suggested by the Bridges et al. (1984) collision law, then the ring is in a very flat state with high filling factor. Then non-local viscosity will dominate (Wisdom and Tremaine 1988; Araki and Tremaine 1986; Araki 1991) and the condition 14.67 is not fulfilled (see the curves labeled 'frosty' in Fig. 14.4). Such a flattened and dense ring state is strongly supported by the presence of self-gravity wakes (see Fig. 14.6), which are found in Cassini data, practically all over the A and B rings (Chapter 13).

Nevertheless, assuming more elastic collision laws one can in principle expect viscous instability. In this case the ring particles maintain in thermal equilibrium a relatively large velocity dispersion, and thus, form a ring with a vertical scale height that is much larger than the particle diameter. Then the filling factors are small, local viscosity dominates at small optical depth, and the condition (14.67) can be fulfilled at intermediate optical depths, while at large optical depth the non-local viscosity becomes important. Such behavior, leading to viscous instability, is expected for example for the 'smooth' elasticity law (Fig. 14.4), for $0.6 < \tau < 2.0$. The instability can be demonstrated in direct simulation (Salo and Schmidt 2009), provided that the calculation region is large enough to cover the smallest unstable wavelengths of the order of hundred particle diameters. An example is shown in Fig. 14.12. After initial growth, the system has saturated to a bimodal quasi-steady state where collisional flux from dense dynamically cool regions is in balance with rarefied dynamically hot regions ($\eta = \nu\tau$ is radially constant). At the large wavelength limit the initial linear growth rates agree with Eq. 14.59.

The above example of viscous instability relies on the assumption of probably unrealistically elastic particles, and thus, is of limited interest. However, the studied case of identical particles with fixed elasticity is not necessarily representative of real planetary rings. Indeed, if we take into account a particle size distribution, and allow for size-dependent elastic properties, new interesting possibilities open up. In particular, if the impacts between small particles are more inelastic than those of large particles (this is at least qualitatively in accordance with laboratory experiments, Hatzes et al. 1988; Dilley and Crawford 1996), simulations indicate that it is possible to obtain selective instability, where the density fluctuations of small particles are enhanced against a more uniform background of large particles (Fig. 14.13). Qualitatively, the instability is due to the fact that an increase of the partial concentration of small particles locally reduces their velocity dispersion, since their energy balance becomes dominated by their mutual very dissipative impacts: this makes the viscosity drop with the relative concentration of small particles.

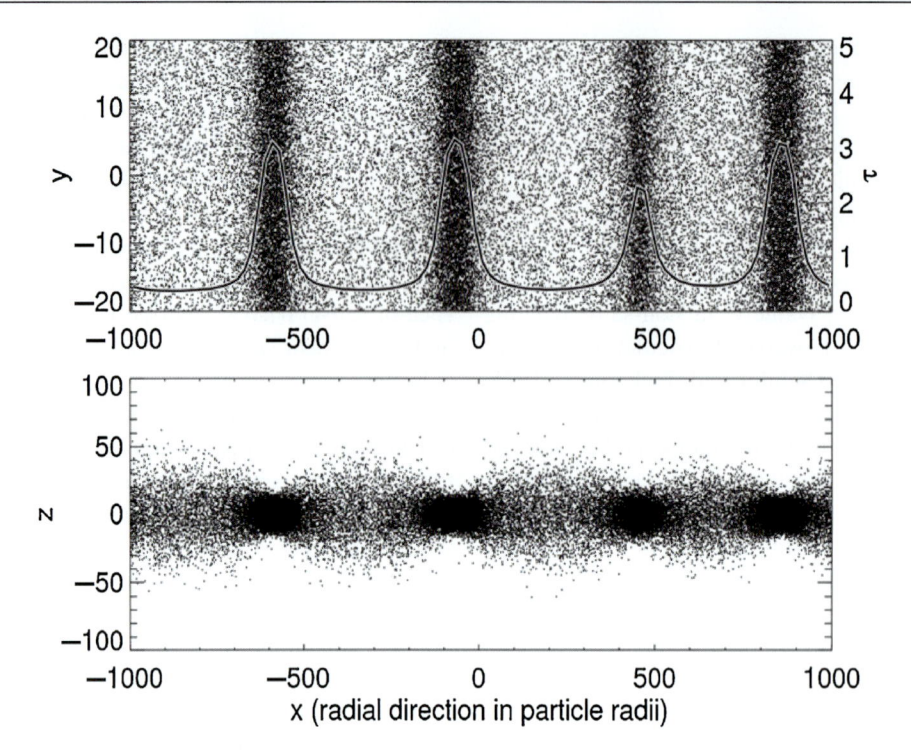

Fig. 14.12 Viscous instability in a non-gravitating simulation with $\tau = 0.94$ using the 'smooth' particle elasticity model of Fig. 14.3. During the first 100 orbital periods the system divides into several ringlets. In the plot a snapshot after 500 orbital periods is displayed. The solid line indicates the optical depth profile (upper frame). The density of the two leftmost ringlets has saturated to an almost constant level. The second ringlet from the right has started to disperse, the material being captured mainly by the rightmost ringlet which is slowly growing and migrating toward the dispersing one. It is likely that more of the ringlets will slowly disperse, and that eventually one large ringlet remains stable in the simulation box. In this way the process could generate a large-scale radial structure over long time scales. (From Salo and Schmidt 2009)

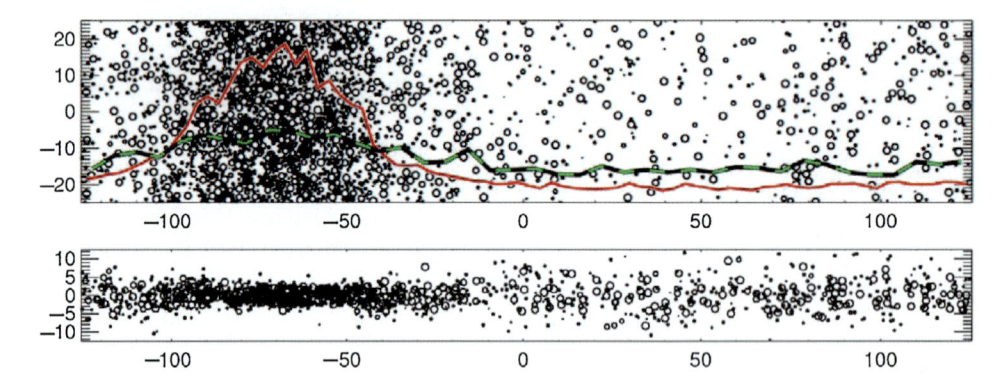

Fig. 14.13 Example of size-selective viscous instability. The non-gravitating simulation system consists of two particle sizes, with radii $R = 0.333$ m and $R = 1$ m; both components have $\tau = 0.5$. Initially both particle species were uniformly distributed; here the distribution after 700 orbital periods is displayed. The upper frame shows a slice through the equatorial plane, the lower frame shows a side view. Symbol size corresponds to particle size and 1/5th of all particles are plotted. The solid red and black-green dashed curves indicate the density profiles of small and large particles, respectively. In this simulation the Bridges et al. (1984) type elasticity model is used, with the scale parameter (see Eq. 14.14) depending on the size group of colliding particles: $v_c/v_B = 0.1, 1, 10$ for collisions of a small-small, small-large, and large-large pair of particles. Note that identical particles with $v_c/v_B = 10$ would not lead to instability. (From Salo and Schmidt 2009)

An attractive feature of this kind of selective instability is that the attained density contrast depends on the size ratio and the difference in elastic parameters between the populations: at least in principle this allows for a much richer variety of structures than the simple identical particle model leading to strictly bimodal optical depth variations. Further studies of selective instabilities, including ring self-gravity are in progress (Salo and Schmidt 2009).

14.2.3 Instabilities due to Ballistic and Electromagnetic Transport

Particles of Saturn's main rings are subject to hypervelocity impacts (10 to 50 km/s) of micro-meteoroids (100 μm to mm in size). This process leads to erosion of the ring particles by ejection of dust-sized and larger debris (Durisen 1984; Durisen et al. 1989). If the ejecta are re-absorbed at some distance by the rings it can lead to the (radial) re-distribution of mass and momentum. The latter mechanism is called ballistic transport (Ip 1983; Lissauer 1984).

It was shown by Durisen et al. (1992) that ballistic transport can produce the ramps in optical depths seen interior to the inner edges of the A and B rings (see Fig. 13.1 and discussion in Chapter 17).

As a consequence of ballistic transport, Durisen (1995) proposed an instability mechanism for rings of intermediate optical depths $0.1 < \tau < 1$. The idea is that for a given radial throw-distance of ejecta (or given distribution of throw-distances) small radial variations in the optical depth profile of certain length-scales can amplify. In principle this may happen by direct re-distribution of mass and/or by radial migration of ring material due to torques exerted by re-absorbed ejecta. Durisen (1995) investigated the linear stability of a mathematical model coupling the equations for mass and momentum conservation with appropriate gain and loss terms, using models for meteoroid bombardment and ejecta distributions developed earlier (Cuzzi and Durisen 1990).

Durisen (1995) shows that the ballistic transport instability should produce radial undulations in the inner B ring of about 100 km wavelength. In the outer B ring optical depth is too high for this instability to develop. This is consistent with analysis of pre-Cassini data (Horn and Cuzzi 1996) and structure seen in Cassini SOI images (e.g. Fig. 5b of Porco et al. 2005). For conditions plausible for the B ring the instability is driven by ballistic mass transport, momentum transport playing an insignificant role. Durisen (1995) argues that the observed structure can grow in a few million years. His derivation includes the smoothing effect of viscous diffusion but does not take into account gravitational viscosity. The qualitative conclusions, however, are not likely to change.

A conceptually similar instability mechanism was proposed by Goertz and Morfill (1988) (Shan and Goertz 1991). It appears as a consequence of the radial transport of charged sub-micron sized dust grains, lifted from the rings (Morfill et al. 1983; Goertz 1984). Due to their large charge to mass ratio such grains are affected strongly by the planetary magnetic field, which tends to force them to co-rotation. At re-accretion the angular momentum the particles gained from (or lost to) the magnetic field is transferred to the ring.

The dominant process driving this instability is angular momentum exchange of different ring segments with Saturn, mediated by the charged particles. The transported masses are insignificant. The instability should generate radial variations in the ring's surface mass density on a length of hundreds of kilometers (Goertz and Morfill 1988). Near synchronous orbit (at 1.87 planetary radii in the outer third of the B ring), however, the mechanism becomes inefficient since at that location charged grains have zero relative velocity to the corotational magnetic field. The growth rate of the instability is sufficiently large to overcome the smoothing effect of viscous diffusion only if the ring's viscosity is smaller than roughly 50 cm^2/s. The viscosities estimated from the damping of density waves in the A ring are generally larger by a factor of 2 to 4 (Tiscareno et al. 2007). However, the A ring viscosity is probably dominated by self-gravity wakes whose strength depends on the distance to Saturn and surface mass density. Extrapolation of formula 14.49 (Daisaka et al. 2001) gives for the inner B ring ($r < 100,000$ km) viscosities below 50 cm^2/s if the surface mass density is smaller than 800 kg/m^2 (assuming a particle internal density 900 kg/m^3).

14.2.4 Shear Rate Instability

The instability model proposed by Tremaine (2003) emphasizes the importance of investigating more general forms of the stress tensor in the dynamics of planetary rings than the Newtonian form (Section 14.1.2, Eqs. 14.28–14.30), demonstrating that non-Newtonian stress can lead to interesting dynamical consequences.

The basic idea is to consider shear stress as a non-monotonic function of shear rate $s = |\partial v/\partial x|$ (here x is the radial coordinate and v is circumferential velocity of the flow), such that the stress-shear curve is generally increasing with s but decreasing over some interval of shear rates. If the Keplerian shear rate $s_k = 1.5\Omega$ lies in this interval, then the flow can assume three different shear rates $s_1 < s_k < s_2$ for the same value of the stress. In this case s_k will be unstable and the flow divides into radial bands of alternating shear rates s_1 and s_2.

In principle, adhesion between particles can lead to such a non-monotonic relation between shear and stress. A solid, rigidly rotating annulus would require a certain amount of tensile stress (P_{xx}) to balance the tidal forces (Tremaine 2003). This stress can be provided by adhesive forces in a state when ring particles stick on each other. More generally, one can imagine that adhesion provides a certain amount of stress also in a non-rigidly rotating ring, so that particles stick temporarily, but are released again in a collision with a third particle, or, by tidal forces, the precise balance of sticking and release depending on the shear rate. This stress would lead to deviations from the purely Keplerian rotation. Whether adhesion can indeed produce the required non-monotonic stress-shear relation in Saturn's rings, such that the Keplerian shear rate lies somewhere in the middle of the interval with decreasing stress, is not clear at present.

For planar, incompressible, viscous shear flow, neglecting self-gravity, we have a nonlinear diffusion equation for the evolution of the shear rate (Tremaine 2003)

$$\frac{\partial s}{\partial t} = -\frac{1}{\sigma}\frac{\partial^2}{\partial x^2} P_{xy}(s) \quad (14.69)$$

with the condition that the mean shear rate is fixed at s_k. The constant surface mass density of the ring is denoted by σ.

Stability analysis of Eq. 14.69 demonstrates that any shear rate s_k in the region $\partial P_{xy}/\partial s < 0$ is unstable. Numerical solution shows that the shear profile rapidly evolves to a piecewise linear pattern that alternates between s_1 and s_2. In the long term evolution smaller domains merge, and the typical length of the pattern grows. Thus, the overall behavior of the mathematical model (Eq. 14.69) is similar to the viscous instability model (Section 14.2.2). In both cases the stress depends non-monotonically on a parameter (here shear-rate s; for viscous instability the optical depth τ) which leads to an instability such that the system develops into a state where the parameter alternates between two given stable values.

The assumption of a constant surface mass density σ, leading to Eq. 14.69, is a very strong one. Since the stress also depends on the surface density, in addition to the shear rate dependence studied here, it seems unavoidable that simultaneously perturbations in σ will develop.

14.3 Ring Moon Interactions and Narrow Rings

Saturn's rings owe a large part of their dynamical structure to the gravitational action of numerous satellites. Large satellites with orbits outside the rings generate resonant gravitational perturbations in the ring at particular radial locations, where combinations of orbital frequencies of the satellites and the ring particles are in a ratio of two integer numbers. The perturbations excite density or bending waves: spiral patterns in the ring, co-rotating with the perturbing satellite (Section 14.3.1). Such waves have been observed at numerous resonance locations mainly in Saturn's A ring. At the strongest resonances, the negative angular momentum carried by a density wave may cause an inward migration of ring particles, opening a gap in the rings.

Another class of structures are generated by small moons (moonlets or ring-moons) embedded directly in the rings. There are gaps (Section 14.3.2) cleared by the embedded satellite, density wakes at the gap edges, and also propeller-shaped features (Section 14.3.3) induced in the ring's density by skyscraper-sized moonlets. Furthermore, embedded small moons may play a crucial role in confining, sustaining and sculpting narrow ringlets (Section 14.3.4) like the F ring and the Encke-gap ringlets.

14.3.1 Spiral Waves

14.3.1.1 Background

Gravitational perturbations by satellites can excite spiral waves in planetary rings. Density waves form in the ring at radial distances where combinations of horizontal frequencies of ring particles and satellite are in the ratio of two integer numbers. They are driven by an interplay of self-gravity and inertial forces, forming a spiral pattern of compression and rarefaction of the ring. If the satellite is exterior to the rings they propagate outward with group velocity $\pi G \sigma_0/\kappa$, where σ_0 and κ are the background surface mass density and the epicyclic frequency, respectively. Such density waves transfer angular momentum from the rings to the satellite, causing the ring particles to spiral inward, while they are damped by a combination of nonlinear and viscous effects. Similarly, spiral bending waves are vertical corrugations, excited at vertical resonances with a satellite. If the satellite is external to the rings, the bending wave propagates inward from the resonance location.

Goldreich and Tremaine (1978b) predicted spiral density waves excited at locations of strong resonances with satellites such as Mimas and Titan, suggesting that such a wave (excited at the 2:1 inner Lindblad resonance with Mimas) created the Cassini Division. Many density waves associated with resonances of Prometheus, Pandora, Janus, Epimetheus, Mimas, and Iapetus (mostly in the A ring) were discovered in the Voyager data (Cuzzi et al. 1981; Holberg 1982; Holberg et al. 1982; Esposito et al. 1983; Longaretti and Borderies 1986; Rosen et al. 1991a, 1991b; Spilker et al. 2004). Bending waves were identified by Shu et al. (1983), who also developed the theory of forced bending waves, and further analyzed by Lissauer (1985) and Gresh et al. (1986). Rosen and Lissauer (1988) studied the bending wave in the C ring at the nodal resonance with Titan, where the satellite's forcing frequency equals the regression rate of the nodes of particles' orbits.

The linear theory (Goldreich and Tremaine 1978a, b, 1979b; Shu et al. 1983; Shu 1984) was often employed to analyze the waves, whereas nonlinear effects are important for most density waves found by Voyager. The nonlinear theory was developed by Shu et al. (1985a, b), and

Borderies et al. (1985, 1986), and nonlinear effects were included in the study of observed waves (e.g., Longaretti and Borderies 1986; Rosen et al. 1991a, b). The large number of high resolution occultation profiles obtained by Cassini over a longer period of time led to considerable advances in modeling of density waves (see Section 14.3.1.3).

There are several motivations for analyzing density waves. Simply speaking, in the linear regime, the surface density can be derived from the dispersion relation, which governs the decrease of wavelength with the distance from the resonance and the ring viscosity can be inferred from the damping rate, which is estimated from the decrease in the amplitude of the wave (see, e.g., Goldreich and Tremaine 1978b; Shu 1984).

A motivation to study especially nonlinear density waves is to determine the torques exerted by satellites on the rings. The measurement of the torque is relevant for the dynamical evolution of the satellites that are close to Saturn's rings, and the age of the ring-satellite system that we observe presently.

14.3.1.2 Elements of Theory

Two approaches exist to study spiral density waves in a planetary ring, the fluid dynamical approach and the streamline approach. The fluid model was introduced by Goldreich and Tremaine (1978b) based on the Euler equations with perturbations arising from the planet, a satellite, and the disk. It was further developed by Goldreich and Tremaine (1978a, 1979b), Shu (1984), and Shu et al. (1985a, b).

The streamline formalism was developed by Borderies et al. (1982) and its application to density waves is described in detail in Rappaport et al. (2009). Particles sharing the same semi-major axis are expected to follow the same m-lobe orbit in the reference frame rotating with the pattern speed. The shape of such streamlines is given by:

$$r = a\left[1 - e(a)\cos(m\phi + m\Delta(a))\right], \quad (14.70)$$

where $e(a) \ll 1$ is the eccentricity and $\Delta(a)$ is a phase angle. This description assumes a Lagrangian approach to fluid motion. An unperturbed fluid particle follows a circular orbit and has coordinates a, ϕ. Once perturbed by the satellite, this same particle follows an m-lobe orbit and has coordinates $r(a, \phi), \zeta(a, \phi)$. The degree of horizontal compression of the ring material is measured by

$$J = \left.\frac{\partial r}{\partial a}\right|_\phi = 1 - q\cos(m\phi + m\Delta + \gamma), \quad (14.71)$$

where we neglected the small term $e\cos(m\phi + m\Delta)$, and

$$q\cos\gamma = a\frac{de}{da}, \quad q\sin\gamma = mae\frac{d\Delta}{da}. \quad (14.72)$$

Streamline crossing is prevented as long as $q < 1$. We define the phase function as (Rappaport et al. 2009)

$$f(a) = m\Delta(a) + \gamma(a). \quad (14.73)$$

Conservation of the mass of a ring element between its unperturbed and perturbed state implies

$$dm \equiv \sigma(r, \zeta)\, r\, dr\, d\zeta = \sigma_0(a)\, a\, da\, d\phi,$$

so that at lowest order in e

$$\sigma = \frac{\sigma_0}{J} \quad (14.74)$$

where σ and σ_0 are the perturbed and unperturbed (or background) surface mass densities. If the ring particle distribution (with respect to size, spin, shape, etc.) is constant in time, we also have

$$\tau = \frac{\tau_0}{J}, \quad (14.75)$$

where τ_0 is the unperturbed, or background, optical depth.

Based on the above streamline formalism, the nonlinear dispersion relation for tightly wound density waves was derived by Borderies et al. (1986, see also Shu et al. 1985b; Borderies et al. 1985) as

$$K\frac{a - a_{res}}{a_{res}} = \frac{2\pi G \tau_0 |k| C(q) - 2k^2 H'(q^2) c^2 K}{3(m-1)\Omega^2}, \quad (14.76)$$

where a_{res} is the semi-major axis of the resonance, $K = \tau/\sigma$ is the opacity, and

$$k = \frac{df}{da} \quad (14.77)$$

is the wave number. The expressions

$$C(q) = \frac{4}{\pi}\int_0^{+\infty} du\, \frac{\sin^2 u}{u^2} H'\left(\frac{q^2 \sin^2 u}{u^2}\right),$$

$$H'(q^2) = \frac{1 - \sqrt{1-q^2}}{q^2\sqrt{1-q^2}}$$

result from the derivation of the self-gravitational acceleration via summation of mass-loaded streamlines of the perturbed ring. The dispersion relation, 14.76 for free density waves (Eq. 14.64 see also Shu 1984) emerges from 14.76 in the limit $q \to 0$.

14.3.1.3 Advances in Modeling

The Waves Associated with the Co-Orbital Satellites

The density waves associated with Janus and Epimetheus are especially interesting. These two co-orbital satellites swap orbits every 4 years, with the most recent reversal having occurred in January 2006. The resonance locations produced by these bodies typically shift by tens of kilometers at the times where the orbits switch around. Therefore, we expect the density waves excited at the resonances of these satellites to show temporal changes. It turns out that the group velocity is

$$v_g \approx 0.1 \left[\frac{\sigma}{100\,\mathrm{g\,cm^{-2}}} \right]\,\mathrm{cm\,s^{-1}},$$

so that the waves propagate over 100 km in a time scale of about 3 years.

Lissauer et al. (1985) first studied the long term evolution of Janus and Epimetheus under the influence of ring torques exerted on the satellites. They find that the moons should evolve from their horseshoe-type configuration to a tadpole configuration in only 20 million years. They also note the multiple wavetrain excitation at the resonance locations.

Tiscareno et al. (2007) developed a model for the complex morphology of the (linear) second order waves that result from these reversals and compared to ISS images. They assumed that the reversal occurs instantaneously while the waves generated at the previous resonance locations continue to propagate and new waves are launched at the new resonance locations. The observed morphology results from superposition of the various wavetrains.

This model shows in general good agreement with Cassini images, with the exception that there are regions where it predicts zero perturbation, while the image scans show oscillations. This may be due to the fact that the reversals are not instantaneous but take several months.

Power Spectrum Density Methods

The application of Power Density Methods to planetary rings was pioneered by Spilker et al. (2004) with the Voyager data. Porco et al. (2005) and Tiscareno et al. (2007) applied the continuous Wavelet transform to weak (i.e. nearly linear) density waves in high resolution Saturn Orbit Insertion imaging data. The wavelet technique (Burns et al. 2004) is particularly well suited for the detection of weak waves and provides the ability to isolate multiple waves superimposed on top of each other. Tiscareno et al. (2007) fitted the wave parameters of 32 density waves, most of them previously unobserved. They found a linear increase of the surface density from the inner to the mid A ring. The ring viscosity they inferred from the wave damping increases monotonically from the Cassini Division to the Encke Gap. This is likely a consequence of the increasing strength of self-gravity wakes in this ring region, which is seen in the increasing amplitude of the brightness asymmetry (Dones et al. 1993). Colwell et al. (2006b, 2009) also used spectral methods to study density waves in stellar occultation profiles recorded by the Cassini UVIS, using a linear dispersion relation. For the Atlas 5:4 density wave Colwell et al. (2006b) determined a surface density of $1.6\,\mathrm{g/cm^2}$ and a viscosity of $5 \pm 2\,\mathrm{cm^2/s}$, which are similar to the values obtained by Porco et al. (2005). For other waves which have a very short damping scale, they used Fast Fourier Transforms, Maximum Entropy Methods, and Wavelet transforms. They found results roughly consistent with previous published values, but there is considerable scatter between different occultations and different techniques.

Application of the Nonlinear Theory

Kinetic parameters $\tau_0(a)$, $q(a)$, $f(a)$, $ae(a)$, and $\gamma(a)$ are used to describe a density wave in the streamline formalism (Section 14.3.1.2). Rappaport et al. (2009) developed an inversion method to derive these parameters as a function of semi-major axis for a given density wave in the nonlinear regime, using multiple ring optical depth profiles. The method is applied to eight RSS radio occultation profiles of the Mimas 5:3 density wave. Figure 14.14 (left panel) shows the solution for the five wave functions. The dotted lines correspond to the region where the functions for at least one of the profiles were extrapolated. In this extrapolation region, the functions τ_0 and q increase. These functions are nearly constant between 50 and 120 km from the resonance radius. In this domain the phase f is very well determined. Figure 14.14 (right panel) shows the radio optical depth profiles for eight occultations (in blue) of the Mimas 5:3 density wave and the reconstructed profiles (in red). Excellent agreement is found between the observations and the model, considering the large fluctuations in optical depth in the data, notably at the peaks.

From the model the nonlinear dispersion relation (Eqn. 14.76) (with the pressure term quadratic in k) is used to compute the opacity, or equivalently, the surface density and the velocity dispersion. The background surface density σ_0 of this wave is found to be variable between 55 and $66\,\mathrm{g\,cm^{-2}}$. It is enhanced at the beginning of the wave, which can be understood by considering a uniform ring in which a strong density wave is launched. The viscous angular momentum flux is negative in the region where q is close to unity but positive inward and outward. The result is that ring material

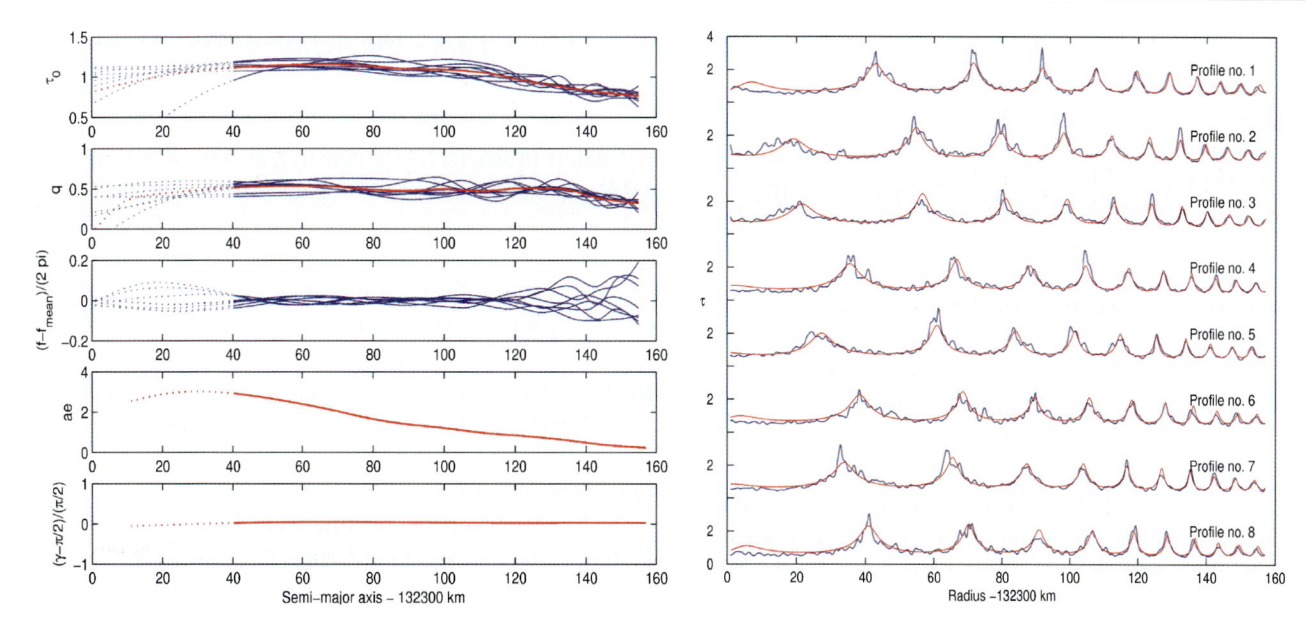

Fig. 14.14 *Left panel*: Kinematic parameters of the Mimas 5:3 density wave derived from the inversion procedure. The kinematic parameters obtained by considering the profiles independently from each other are shown in blue. The mean solution is displayed in red. The dotted lines refer to the region in which the parameters for at least one profile are simply extrapolated rather than obtained by the inversion method, because the approximation used in the method is not valid in these regions. For e and γ only mean profiles are determined (red lines in bottom panels). *Right panel*: Radio optical depth profiles for eight occultations (in blue) of the Mimas 5:3 density wave and the solution obtained by the inversion method (in red). The profiles are stacked. For each profile, the vertical scale goes from 0 to 4. Intermediate tick marks represent both the $\tau = 0$ level of the next profile, and the $\tau = 4$ level of the previous one. The resonance radius is 132,301 km (Rappaport et al., 2009)

is transported to the zone where the wave is strongly non-linear. The velocity dispersion is $\approx 0.6\,\mathrm{cm\,s^{-1}}$ in the wave region.

The determination of $\tau_0(a)$ and $q(a)$ in the above example is impaired by the considerable noise in the radio data. Improved results can be expected by applying the method to UVIS data with a better signal to noise ratio. Also an improved treatment of the first wavelength of the wave is needed for an accurate estimate of the mass of a perturbing satellite.

14.3.2 Moonlet Induced Gaps

In the previous section we discussed waves generated at resonances with satellites on orbits outside the rings. With decreasing distance to the perturbing satellite the azimuthal wavenumber m of a resonance increases and neighboring resonances and related structures become more closely spaced. This is the case for instance for the resonances of Prometheus and Pandora in the outer A ring, which, however, remain well separated.

The situation is different for a satellite directly embedded in the rings (moonlet). In the vicinity of the moonlet the resonance order diverges ($m \to \infty$) and the distance between the resonances vanishes ($|\Delta r_m| \to 0$). As a consequence, density waves cannot be excited at these resonances, since, firstly, the radial width W_m of a resonance exceeds the inter-resonant distances $W_m > |\Delta r_m|$ so that the resonance regions overlap (Wisdom 1980; Duncan et al. 1989) and phase-mixing destroys the resonant phase commensurability, leading to chaotic particle motion. Secondly, these distances do not give enough radial space for the waves to develop, i.e. $|\Delta r_m| < \lambda$, (λ – wavelength of resonant waves).

Instead, the gravitational angular momentum exchange between embedded moonlet and surrounding (perturbed) ring material tends to open a gap in the ring. Goldreich and Tremaine (1980) derived the gravitational torque density $dT_{Moonlet}/dr \propto \pm M^2_{Moonlet}/|x|^4$ exerted on a moonlet by an infinitesimally narrow ring of width dr (still containing a large number of resonances) located at radial distance $x = a - a_0$ from the moonlet at a_0, where a denotes semi-major axes. The signs correspond to ring matter inside and outside the moonlet's orbit, respectively: Angular momentum is transferred to the moonlet (positive sign) by ring material interior to its orbit and ring particles migrate towards the planet. Ring material exterior to the moonlet's orbit gains angular momentum (negative sign), particles are pushed outward. In steady state this transfer of angular momentum is counteracted by a viscous torque $T_\nu \propto 3\pi\nu\sigma r^2\Omega$, mediated by physical collisions (Lissauer et al. 1981) and gravitational interactions (Daisaka et al. 2001) between the ring particles, quantified by the viscosity ν, with the result that a gap of finite size is opened in the ring.

Integrating the torque density (summing all ringlet contributions) and balancing with the viscous torques one obtains for the width of the gap (Lissauer et al. 1981; Petit and Hénon 1988; Spahn et al. 1992a, 1993)

$$W_G \propto \begin{cases} M_{Moonlet}^{1/3} \propto R_{Moonlet}, & \text{for } M_{Moonlet} < M_0 \\ M_{Moonlet}^{2/3} \propto R_{Moonlet}^2, & \text{for } M_{Moonlet} > M_0. \end{cases} \quad (14.78)$$

where M_0 corresponds to a kilometer sized moonlet (solid ice density). A comparison of the Encke- and the Keeler gaps, created by Pan and Daphnis, respectively, confirms the scaling $W_G \propto M_{Moonlet}^{2/3}$ for large ring-moons fairly well ($M_{Moonlet}$, $R_{Moonlet}$ – moonlet mass and radius).

Density wakes form at the edges of the gap downstream from the moon (Cuzzi and Scargle 1985; Showalter et al. 1986) as a response to the perturbation by the moon. The moonlet forces an eccentricity $e \propto x^{-2}$ of ring particles and their orbital phases become temporarily synchronized, leading to the formation of a kinematic wake (Showalter et al. 1986). The streamline wavelength $\lambda_w = 3\pi x$ of a wake increases with mean radial distance x from the moon. The wakes are more pronounced for larger moons, because then diffusion ($\propto c^2$ [see Eq. 14.68]) cannot dominate the systematic wake induced velocities $\propto r_0 e(x) \Omega(r_0)$ during one synodic period

$$t_{syn} = \frac{4\pi}{3\Omega(r_0)} \left| \frac{r_0}{x} \right|. \quad (14.79)$$

Such density wakes have been found at the edges of the Encke gap (Cuzzi and Scargle 1985; Showalter et al. 1986) and the Keeler-gap (Cooke 1991) which led to the detection of the two embedded ring moons Pan (Showalter 1991) and Daphnis (Porco et al. 2005), respectively.

Cassini cameras revealed a surprising "rope" structure in the Pan wakes in the outer Encke gap edge (Porco et al. 2005, their Fig. 6c). Such features were predicted by Lewis and Stewart (2005) as a consequence of the interaction of the moonlet induced perturbation with the self-gravity of ring. Apparently the compression and stretching of self-gravitating material in the first cycles of the Pan wake leads to magnified self-gravity wakes. On the other hand, further away from the gap edges the moonlet wakes may disrupt self-gravity wakes (Lewis and Stewart 2005), which formed in the unperturbed region upstream from the moon.

The secular evolution of a satellite embedded within a gap in a ring was studied by Hahn (2008) and Hahn (2007). An eccentric satellite should excite a resonant low-amplitude density wave in the self-gravitating ring exterior to the gap (Hahn 2008), which in turn tends to damp the moon's eccentricity. In the case of Pan in the outer A ring the wavelength would be on the order of hundreds of kilometers and the amplitude $\Delta\sigma/\sigma < 4 \times 10^{-3}$. Similarly, an inclined satellite should launch a spiral bending wave in the ring exterior to the gap (Hahn 2007).

14.3.3 Propellers – the Action of Tiny Moons

If an embedded moonlet is smaller than (roughly) a few hundred meters in size (solid ice density), then the gap it induces in the ring can be closed by viscous diffusion before it extends around the complete ring circumference. This competition of gravitational scattering and viscous diffusion creates a typical structure in the rings (Fig. 14.15a) dubbed "propellers" (Spahn and Sremčević 2000; Sremčević et al. 2002). Such an S-shaped density undulation – typically induced in a disk as the local response to an embedded mass – was studied by Julian and Toomre (1966), however, not taking into account physical collisions between particles, which are essential for planetary rings.

The first four propeller features were detected in high resolution Cassini images of the A ring (Tiscareno et al. 2006b). Later, a large number of propellers were found preferentially interior to the Encke gap (Sremčević et al. 2007; Tiscareno et al. 2008) and larger propellers also outside the Encke gap (Sremčević et al. 2007; Tiscareno et al. 2008; Burns et al. 2008).

A propeller can form in the rings if the eccentricities forced by the moonlet's gravity are much larger than those implied by the quasi-steady velocity dispersion established by the balance of viscous heating and collisional cooling. The (incomplete) propeller-gaps develop roughly at radial displacement $\pm 1.5 R_{Hill}$ from the moonlet, where

$$R_{Hill} = r_0 \left(\frac{M_{Moonlet}}{3 M_{Saturn}} \right)^{\frac{1}{3}}, \quad (14.80)$$

is the Hill scale. Let the mean square radial distance $\langle \Delta r^2 \rangle \approx \nu t$ characterize viscous diffusion of a ring particle performing a random walk mediated by collisions (and gravitational encounters) in the ring. Then the distance $L(t) = \sqrt{\langle \Delta r^2 \rangle}$, arising from Gaussian solutions of the linearized diffusion (Eq. 14.68) is the radial scale particles migrate on average during time t. For moonlets with $R_{Hill} < R_{crit} := L(t_{syn}(R_{Hill}))$ all gravity-induced density undulations are smeared out during one synodic period t_{syn} so that the gaps remain incomplete (Fig. 14.15a), while a complete gap is formed when $R_{Hill} > R_{crit}$ (Fig. 14.15b).

The azimuthal extent of these density depletions can be estimated (Spahn and Sremčević 2000; Sremčević et al. 2002) by equating the radial length scales of gap formation and closing

$$R_{Hill} = L(t_E) \approx \sqrt{\nu t_E},$$

defining the time $t_E < t_{syn}$ necessary to viscously close the moonlet-induced gaps. The time t_E is related to an azimuthal longitude $\varphi_E = 3\Omega_0 t_E / 2 < 2\pi$. This simple calculation

14 Dynamics of Saturn's Dense Rings

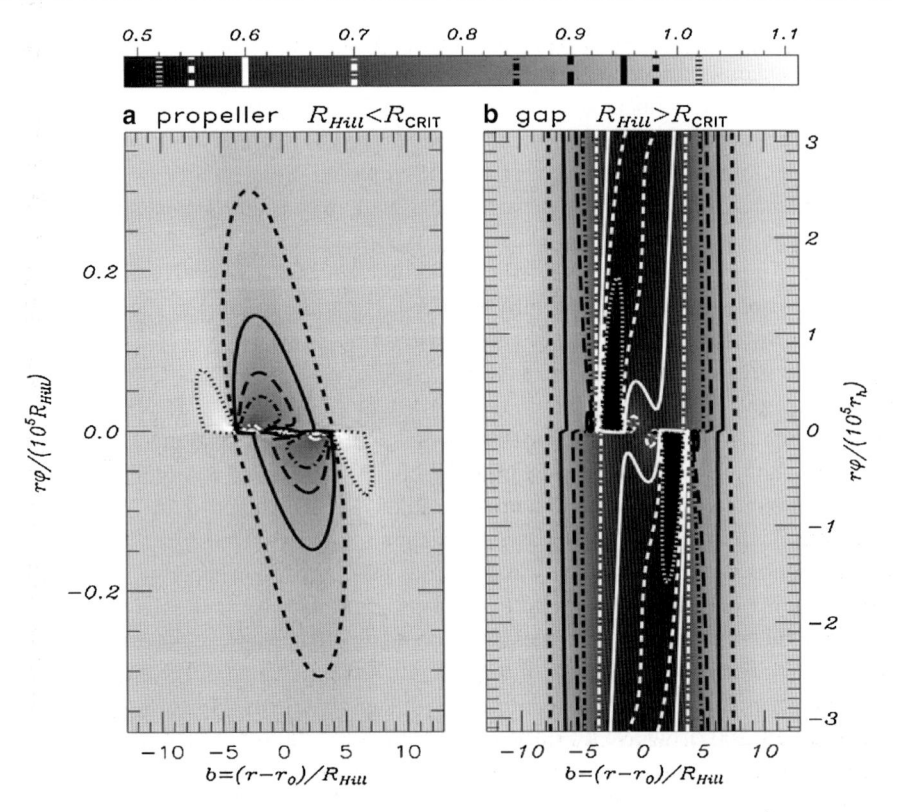

Fig. 14.15 Perturbations of the rings' surface-mass density induced by an embedded moonlet (Sremčević et al., 2002). Panel (**a**): The characteristic propeller structure induced by a small moonlet ($R = 100$ m; moonlet's density $\rho = 600$ kg m^{-3}), which does not extend around the whole circumference. Panel (**b**): A complete gap induced by a moon massive enough to create structure surviving a synodic period t_{syn}

predicts the radial and azimuthal extent of a propeller-gap as a function of moonlet mass and ring viscosity

$$R_{Hill} \propto M_{Moonlet}^{1/3} \propto R_{Moonlet}; \quad \varphi_E \propto \frac{M_{Moonlet}}{\nu} \propto \frac{R_{Moonlet}^3}{\nu}. \tag{14.81}$$

These scalings have been confirmed with N-body simulations by Seiß et al. (2005) using non-selfgravitating single-sized particles.

The comparison of the scaling (Eq. 14.81) to observations (i.e. the dependence of the azimuthal length of a propeller on the radial separation of the bright propeller streaks) is complicated by several factors. First, in images of the backlit rings, where propellers were first detected (Tiscareno et al. 2006b; Sremčević et al. 2007), the interpretation of low brightness is ambiguous since opaque or totally void regions both appear dark. Thus, the bright propeller streaks could be moonlet induced gaps or density enhancements. Second, the propeller should induce density wakes in the ring, which flank the incomplete gaps downstream, similar to the Daphnis and Pan wakes (Fig. 13.4) forming at the edges of the Keeler and Encke gaps. This wake pattern can alter the appearance of a propeller and it will scale differently from Eq. 14.81. Third, the moonlet induced structure in the ring's density is superimposed on, and may interact with, the self-gravity wakes in Saturn's A ring. As a result, the propeller induced pattern looks fuzzy (Fig. 14.16). Since the typical length of the self-gravity wakes is independent of the moonlet size this may well affect the observed scaling of the propeller dimensions, especially for small moonlets. Fourth, the perturbation of the moon may locally enhance the collision frequency and collision speeds in the ring, leading to breakup of loosely bound particle aggregates and to the release of debris residing on larger ring particles. This would increase the optical depth (Salo and Schmidt 2007), and thus, the observed pattern of brightness does not necessarily correspond directly to the pattern in the surface mass density, thus leading to differences from the scaling (Eq. 14.81). Fifth, the ring viscosity (entering the scaling as a pre-factor) might vary significantly in the propeller region (Tiscareno et al. 2007).

Analyzing twelve propeller features Sremčević et al. (2007) found a near linear dependence

$$\varphi_E \propto R_{Moonlet}^\alpha \quad \text{with} \quad \alpha \approx 1.1, \tag{14.82}$$

and not the expected cubic scaling (Eq. 14.81). Tiscareno et al. (2008) found a large scatter of the propeller dimensions in their data set (fewer than a hundred resolved propellers) and argue that a cubic scaling should not yet be ruled out. Sremčević et al. (2007) show that a near linear scaling can be understood if the bright propeller-features in the images are associated with the moonlet induced wakes and not the gaps. For 100 m-sized moonlets these wakes are not very pronounced and they are probably perturbed by the pattern of self-gravity wakes in the rings (Fig. 14.16). Nevertheless, the moonlet wakes will be destroyed near the longitude of streamline crossing $\varphi_S \propto R_{Moonlet}$, the longitude where the collision frequency in the streamline rapidly increases (Showalter et al. 1986; Spahn et al. 1994), which would result in the observed linear scaling with the moonlet size. This idea was further quantified by photometric modeling of particle configurations from self-gravitating simulations of a propeller by Sremčević et al. (2007). They showed that release of regolith and a possible destruction of the self-gravity wakes in the perturbed propeller regions in principle offers an explanation of the observed brightness of the propeller-streaks (Fig. 14.16) in backlit geometry. The re-accumulation of the debris is only completed downstream from the streamline crossing point (Sremčević et al. 2007), which would lead naturally to a slightly increased value $\alpha > 1$ in Eq. 14.82. Moreover, for this geometry the gaps are not expected to stand out relative to the ring background brightness, in agreement with observations. Earlier it has been suggested that the bright streaks seen in backlit images are actually the propeller gaps (Tiscareno et al. 2006b). However, this interpretation would require fairly large optical depths for the ring and the gap (see Fig. S12 in Sremčević et al. (2007)), and it is inconsistent with the observation of propellers on the lit side of Saturn's A ring (Tiscareno et al. 2008), since in reflected light the brightness should saturate for optical depths near unity.

Tiscareno et al. (2006b) derived a moonlet-size distribution $n(R)dR \propto R^{-q}dR$ with $q \approx 5$ including the two known ring-moons Pan and Daphnis in the fit. Sremčević et al. (2007) derived a larger slope $q > 9$, and Tiscareno et al. (2008) obtained $q \approx 6$, both considering Pan and Daphnis to be isolated from the propeller-moonlet population. This perception suggests three families of bodies in the rings: ring-particles (cm $< R <$ 10 m, roughly), propeller-moonlets (10 m $< R <$ 500 m), and ring-moons (0.5 km $< R <$ 100 km). Such a classification seems to be consistent with the scenario of ring-formation in episodic cascades (Esposito et al. 2005) where ringmoons and moonlets are continuously destroyed by meteoroid impacts (Section 14.4.1; see also Chapter 17). Self-gravitating simulations with an

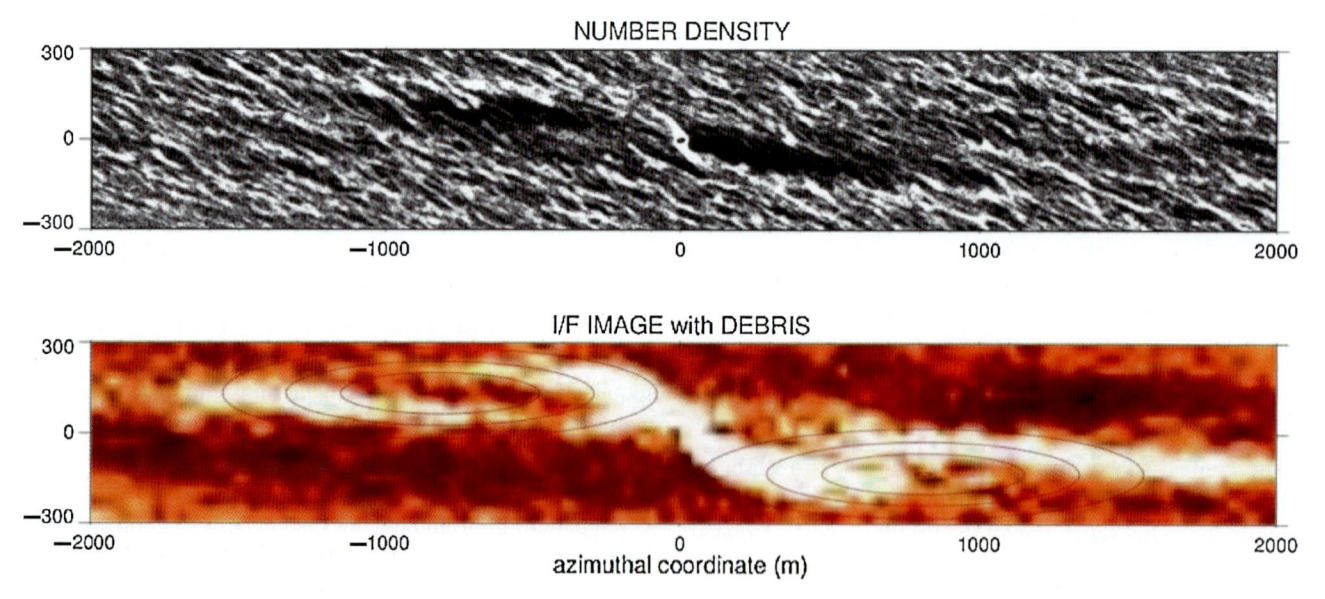

Fig. 14.16 Upper frame: Effect of a 40 m diameter moonlet in self-gravitating simulation. The frame shows a 4 km by 0.6 km azimuthally elongated simulation region co-moving with the moonlet's orbital motion (the planet is downward and orbital motion is to the right). Two symmetric density enhancements are seen downstream of the moonlet, in addition to smaller scale self-gravity wakes formed by the 1 m ring particles. The simulation also illustrates the limited-accretion mechanism pointed out by Porco et al. (2007; see Section 14.4.1.2) the moonlet with internal density 600 kg m^{-3} was able to accumulate lower density ring particles (450 kg m^{-3}) until its Hill sphere was filled. Lower frame: Synthetic image constructed for the geometry of the Cassini SOI images where the first propellers were detected (Tiscareno et al., 2006b). A fair correspondence to observed propellers (contours indicate a fit to the SOI4 propeller feature in Tiscareno et al. (2006), see Fig. 13.8) is achieved if ring particles are covered with loosely-bound regolith, released in the vicinity of the moonlet due to locally enhanced impact speeds: here it is assumed that impacts with $v_{imp} > 1$ cm s^{-1} lead to release of regolith-debris, which is re-accumulated during the subsequent impacts. Even a modest amount of such debris (here $\tau_{debris} = 0.025$ averaged over the calculation region) is able to hide the downstream gaps and enhance the brightness of the density crests. Figure modified from Sremčević et al. (2007) (supplementary online-material); photometric calculations with the method of Salo and Karjalainen (2003), Salo et al. (2004)

extended particle size-distribution (Lewis and Stewart 2009) suggest that a steep moonlet size distribution is essential for the formation of distinct propeller features, since otherwise frequent gravitational interaction with neighboring larger moonlets would tend to destroy the propeller.

With a few exceptions the propeller features appear concentrated in a belt (Sremčević et al. 2007) in the mid-A ring (interior to the Encke division) that is divided in three bands of roughly 1,000 km width (Tiscareno et al. 2008). It was suggested that the belt represents a younger ring region, formed after the destruction of a ring moon in a catastrophic collision (Sremčević et al. 2007). Tiscareno et al. (2008) argue that the perturbations induced by the strong Pandora 7:6 and Janus/Epimetheus 5:4 resonances, both lying in the propeller belt, might locally hamper the formation of the propeller structure, in this way splitting the propeller region in three sub-belts. It is intriguing that the propeller belt is found at the same ring radius as the maximum of the brightness asymmetry of Saturn's A ring (Dones et al. 1993), the latter being a measure for the strength of the self-gravity wakes. If this is a coincidence or if the two phenomena are related is unclear at present.

Exterior to the Encke gap only a few (generally larger) propellers were found so far (Sremčević et al. 2007; Tiscareno et al. 2008; Burns et al. 2008). The largest propellers show clearly the theoretically expected incomplete gaps and moonlet induced wakes. Fluctuations in the ring's surface mass density may cause (via gravitational interaction) a stochastic migration of the moonlet (Burns et al. 2008) as seen in simulations (Lewis and Stewart 2009). A very similar effect (type III migration) was suggested for growing planetary embryos embedded in pre-planetary gas-dust disks (Masset and Papaloizou 2003; Papaloizou et al. 2007), offering a unique opportunity to study this type of migration in Saturn's rings.

In Cassini images of the F ring a pattern called "fans" was observed (Fig. 13.25). These "fans" are essentially the wake pattern induced by embedded moonlets in the dusty component of the F ring.

14.3.4 Dense Narrow Rings

Prominent ringlets in the Saturn system are the Titan ringlet (Porco et al. 1984) at $1.29R_S$ ($R_S = 60,330$ km is Saturn's radius), the Maxwell ringlet at $1.45R_S$ (Esposito et al. 1983; Porco et al. 1984), the Huygens ringlet at $1.95R_S$, and the ringlets in the Encke gap. Other locations are listed in Table II of French et al. (1993) and Tables 13.2 and 13.4. The most striking narrow ring in the Saturnian system is perhaps the F ring, discussed in detail in Chapter 13.

14.3.4.1 Confinement of Narrow Rings

Goldreich and Tremaine (1979c) proposed that narrow rings are confined in radius by gravitational torques from neighboring satellites (shepherding). The confinement is due to the transport of angular momentum, induced by disturbances created in the ring by the shepherd satellites. A formula for linear torques (adequate for small perturbations) was derived by Goldreich and Tremaine (1978b) and nonlinear torques (also valid for large perturbations) were estimated by Borderies et al. (1984b).

Voyager detected the satellites Prometheus and Pandora on orbits slightly interior and exterior to the F ring. It is clear that the gravitational perturbations of Prometheus induce a wealth of pronounced structures in this ring (Kolvoord et al. 1990; Murray et al. 2008; see also Chapter 13) and both satellites are probably involved in the shepherding. This shepherding role has been questioned (Showalter and Burns 1982; Cuzzi and Burns 1988) on the ground that the F ring seems not to be in torque balance between the two satellites. However, based on the available data an accurate determination of the mass distribution in the F ring is difficult, which would be necessary for the precise evaluation of the torque balance. On the other hand, evidence for perturbations induced in the F ring by Pandora (Porco et al. 2005; Murray 2007) would lend further support to the idea that Prometheus and Pandora act as shepherds of the F ring.

Stewart and Lewis (2005) offered an alternative view of the F ring confinement, involving a weakening of the outward flux of angular momentum. In a first phase, the particles migrate to Lindblad resonances, forming ringlets while in a second phase, the ringlets collide and partially merge. This explanation is consistent with results of their N-body simulations. The dynamics of the F ring and Prometheus and Pandora was further studied in Lewis and Stewart (2007).

Hänninen and Salo (1994, 1995) found that narrow ringlets can form at isolated Lindblad resonances. Goldreich et al. (1995) explained such a single-sided shepherding as a consequence of the net negative angular momentum flux (i.e. integrated over a streamline) promoted by satellite perturbations of the streamlines of the flow of particles. Lewis and Stewart (2005) showed that a surprisingly complex and variable structure can already arise from the perturbations exerted by one nearby eccentric moon on a narrow, self-gravitating ring of uni-sized particles. Using typical scales and elements for Prometheus and the F ring, their simulations develop many features similar to those seen in the F ring.

Dermott et al. (1980) proposed a confinement of a narrow ring by an embedded ring-moon forcing the particles in the ringlet area to revolve in horseshoe and tadpole orbits. The long-term behavior and the gravitational action of multiple moonlets in a narrow ringlet was studied by Spahn and Sponholz (1989; Spahn et al. 1992b) pointing to the existence of a kilometer-sized moonlet in the F ring's core (Spahn and Wiebicke 1989). Cassini experiments (Esposito et al. 2008; Murray et al. 2008) confirm moonlets populating the F ring which can migrate chaotically (Winter et al. 2007), driven by the shepherds and higher gravity moments due to Saturn's oblateness. If the ringlet contains a large fraction of dust (Showalter et al. 1992; Encke ringlets, F ring) the continuous loss of dust due to non-gravitational perturbations must be balanced by sources in the ringlets (e.g. small moonlets; Ferrari and Brahic 1997), replenishing the dusty material.

Rappaport (1998) studied the possible confinement of a large ring by a narrow ring applied to the ringlet and gap at $1.470 R_S$. She found that the torque exerted by the ringlet is able to confine the gap's inner edge, but not the outer edge. This suggests that either the gap's edges are, as the ringlet itself, confined by reversal of the viscous flux of angular momentum, or that one or several small satellites located within the gap are responsible for the confinement of the edges.

14.3.4.2 Rigid Precession

Most narrow rings are eccentric and/or inclined. Goldreich and Tremaine (1979a) proposed that differential precession (due to the quadrupole moment of the planet) can be balanced by the effect of a ringlet's self-gravity, thus allowing a rigid precession of eccentric ringlets. The dynamics of narrow elliptical rings in the presence of eccentricity excitation by shepherd satellites, self-gravity and viscous forces was studied by Borderies et al. (1983b). They find that the mean eccentricity of the ringlet is determined by a balance between viscous damping and excitation by the shepherd satellites, and the eccentricity gradient is positive. Due to viscous forces, the line of apsides of the inner and outer edges are not aligned in their model.

Chiang and Goldreich (2000) and Mosqueira and Estrada (2002) studied the effect of particle collisions on the interplay of the effects of planetary oblateness and self-gravity. They show that collisional forces felt by the material in the last ∼100 m of a ∼10 km wide ring can increase equilibrium ring masses by a factor up to ∼100, and can lead to apse alignment of the inner and outer ring edges, in accordance with Voyager radio measurements of the Uranian ε, α, and β rings. Papaloizou and Melita (2005) used a fluid dynamics approach to derive a condition for the steady maintenance of the eccentricity of a ring, which requires the external satellite torque to balance the dissipative effects due to collisions. Melita and Papaloizou (2005) further considered the case where the pattern frequency of the eccentric ringlet is in a secular resonance with an external satellite. Applying the results to the Titan ringlet (in a 1:0 resonance with Titan), they estimated the mean surface density of the ringlet.

14.3.4.3 Excitation of Eccentricities and Inclinations

Goldreich and Tremaine (1981) showed that torques exerted at a Lindblad resonance with a satellite excite eccentricities of the ring particles, while torques exerted at a corotation resonance damp eccentricities. If the corotation resonances are not saturated, these resonances win and the eccentricity of the ring decreases. If the corotation resonances are partially or fully saturated, then the Lindblad resonances win and the eccentricity of the ring increases.

The excitation of inclinations was studied by Borderies et al. (1984a). Gravitational interaction of rings and satellites at vertical resonances produce a secular increase of their inclination until viscous dissipation leads to an equilibrium. Vertical resonances are similar to Lindblad resonances, involving inclination instead of eccentricity. However, there is no equivalent of the corotation resonances.

14.3.4.4 Viscous Overstability

Longaretti and Rappaport (1995) showed that spontaneous oscillations of viscous origin (overstabilities) can arise in a dense narrow ring (see also Papaloizou and Lin 1988). They found two possible regimes of instability, one in which the mean eccentricity of the ring decreases to a small but finite and nearly constant value, while internal modes of libration reach comparable amplitudes. In the other regime the mean eccentricity of the ring increases to a much larger asymptotic value, while internal librations are strongly reduced (though not fully damped). Wave like features in the Maxwell ringlet might be caused by this mechanism (Porco et al. 2005).

14.3.4.5 Ring Edges

Many ringlets have remarkably sharp edges. This can be explained by the reversal of angular momentum luminosity in the regions of the rings which are strongly perturbed (see Borderies et al. 1982, 1983a, 1989). Angular momentum luminosity is the momentum flux integrated over a streamline, i.e. the rate at which angular momentum crosses the streamline. The perturbed streamlines oscillate with slightly different wavelengths $\lambda = 3\pi x$ (x is the modulus of the difference in semi-major axes of satellite and streamline) so that neighboring streamlines eventually intersect. The intersection occurs at quadrature, midway between periapsis and apoapsis. If the streamlines are sufficiently perturbed, the orbital angular velocity of the particles increases outward at the point of intersection (while for a circular orbit it would decrease). As a consequence, in this region the angular momentum associated with particle collisions flows inward rather than outward. Also, energy is dissipated and the streamline eccentricities are damped. Considering the total angular momentum flux over the 360° azimuth of the streamlines, the net torque due to the collisions between particles can balance the torque exerted by the satellite for a certain distortion of the streamline. A more complete analysis indicates that the balance of the energy dissipation and angular momentum transport occurs well before the hypothetical streamline crossing for $\tau \ll 1$ and close to it for $\tau \gg 1$. A sharp edge forms because the angular momentum balance is established over a few streamlines only. In this picture, the equivalent width of one streamline is the radial width corresponding to the excursions associated with the particle orbital eccentricities.

Hahn et al. (2009) revisited the streamline approach for a ring edge located at a Lindblad resonance with a perturbing satellite. Applying their model to the outer B ring edge, they point at difficulties in achieving the torque balance, unless a fairly high value of the rings bulk viscosity is assumed (see discussion on bulk viscosity in Section 14.1.2.2), or additional friction for the particles' motion is invoked. Shepelyansky et al. (2009) have shown that synchronization of the orbital phases of ring particles near the resonance may lead to a decreased collision frequency. As a result, diffusion of particles can be drastically suppressed, which helps to maintain a sharp ring edge.

14.4 Size Distribution and Spins of Ring Particles

Radio and stellar occultations at various wavelengths provide information about the size distribution of ring particles smaller than a few tens of meters, while the size distribution of ~100 m-sized moonlets can be inferred from observation of propeller structures (Section 14.3.3). The importance of accretion and fragmentation for the evolution of the size distribution is implied by the observed shapes and internal densities of small inner satellites, which is supported by dynamical studies (Section 14.4.1).

Oblique impacts between ring particles with rough and irregular surfaces lead inevitably to rotation (Section 14.4.2). Although the rotational states of ring particles are not directly observable, particle spins have been constrained indirectly from observations of the thermal emission from Saturn's rings made by ground-based telescopes, the Pioneer 11 infrared radiometer, Voyager's infrared interferometer spectrometer, and Cassini CIRS (see e.g., Cuzzi et al. 1984; Spilker et al. 2003, 2006). In principle, constraints on the dynamical behavior and physical properties of ring particles can be derived from the comparison of those observations and model calculations including particle spins.

14.4.1 Particle Size Distribution and Its Evolution

14.4.1.1 Particle Size Distribution Derived From Observations

The results of the Voyager 1 radio occultation experiments significantly advanced our knowledge about ring particle size distribution (Marouf et al. 1983; Zebker et al. 1985). In this experiment, the radio source onboard Voyager 1 was used to measure the radio opacity at two wavelengths (3.6 and 13 cm). A power-law distribution $n(R) = n(R_0)[R/R_0]^{-q}$ with $R_{min} < 1$ cm and $R_{max} \approx 1$ m was assumed, and constraints on the values of q and $n(R_0)$ were derived from the opacities. Also, measurements of the differential scattering cross section and integral inversion were used to derive a general size distribution over the range of $1 \lesssim R \lesssim 15$ m. Marouf et al. (1983) assumed a many-particle-thick model, while Zebker et al. (1985) allowed for a near monolayer system. The results imply that there exists an upper cutoff in the size distribution at $R \sim 5$ m. For centimeter-to-meter-sized particles $q \approx 3$ was found, with a slightly larger value for the C ring ($q \approx 3.1$), and slightly smaller in the Cassini Division and the A ring (~2.7–3.0; Chapter 15). A tendency of increasing q was also found from the inner to the outer A ring, which indicates a greater abundance of small particles in the outer A ring (Tyler et al. 1983). The observation that the optical depth is similar at visible and radio wavelengths suggests that sub-centimeter particles do not dominate the surface area. The Cassini RSS occultations support these earlier results (Chapter 15).

French and Nicholson (2000) derived the particle size distribution in Saturn's rings from ground-based observations of the 3 July 1989 occultation of 28 Sgr, and the Voyager photopolarimeter (PPS) optical depth profile. They used the PPS occultation data (Esposito et al. 1983) to estimate and then remove the directly transmitted signal from the 28 Sgr observations, and derived particle size distributions from the so obtained high SNR scattered light profiles at wavelengths of 3.9, 2.1, and 0.9 μm. Using a two-dimensional forward-scattering model with the many-particle-thick-layer approximation and a single power-law size distribution for each major ring region, they obtained values for q, R_{min}, and R_{max} for each region, including the B ring, for which the Voyager radio occultation experiment was unable to derive the particle size distribution due to its high opacity. The results for q and R_{max} derived for the A and C rings were fairly consistent with those obtained by the Voyager radio occultation measurements, although the values of R_{max} obtained by French and Nicholson (2000) were somewhat larger ($R_{max} = 10$ m for the C ring, and 20 m for other locations). The values of R_{min} for the inner A ring and the B ring were as large as 30 cm, while a greater population of small particles with $R_{min} = 1$ cm was suggested for the C ring and the outer A ring.

An independent estimate of the size of the largest particles was obtained by an analysis of the statistical properties of the Voyager PPS stellar occultation data by Showalter and Nicholson (1990). Owing to the finite size of ring particles, the variance of the photon counts during stellar occultation can be significantly larger than expected from Poisson statistics. Showalter and Nicholson (1990) derived an expression for the noise in the PPS scan due to large particles, and used the measured noise to constrain their sizes. The excess variance depends on the fourth moment of the particle size distribution. Showalter and Nicholson (1990) expressed their results in terms of an effective radius defined by $R_{eff}^2 = \int_{R_{min}}^{R_{max}} R^4 n(R)\, dR / \int_{R_{min}}^{R_{max}} R^2 n(R)\, dR$, finding $R_{eff} \approx 2 - 10$ m in the main rings (Chapter 15). French and Nicholson (2000) calculated the values of R_{eff} defined above, using the size distribution derived from their observations, and found excellent agreement with the estimates by Showalter and Nicholson (1990).

The surface mass densities estimated from the size distribution by French and Nicholson (2000) are much larger than those derived from studies of density and bending waves. Their large mass densities for the A and B rings ($q < 3$) reflect their inferred value of $R_{max} \approx 20$ m. The disagreement may suggest that the largest particles with $R \approx R_{max}$ actually represent loosely bound aggregates of meter-sized objects in gravitational wakes. A smaller value of R_{max} would lead to smaller surface mass densities consistent with density and bending wave studies. Strong gravitational wakes can also affect the apparent particle size distribution, by enhancing the release of impact-debris or stirring up small particles (Sremčević et al. 2007, Robbins et al. 2009, Sections 14.1.3 and 14.3.3).

The number of particles larger than the upper cutoff of 5–20 m is much smaller than extrapolation of the $q \approx 3$ power-law would predict. The power-law index for moonlets with $R \geq 10$ m estimated from the analysis of propeller structures in the A ring (Section 14.3.3) is $q \approx 5$ (Tiscareno et al. 2006b), $q \approx 6$ (Tiscareno et al. 2008), and $q \approx 9-11$ (Sremčević et al. 2007).

14.4.1.2 Accretion of Particles in the Roche Zone

The evolution of the size distribution due to gravitational accretion and fragmentation has been extensively studied in the context of accumulation of planetesimals in the protoplanetary disk (e.g., Wetherill and Stewart 1993; Weidenschilling et al. 1997). Planetary accretion occurs far outside the Sun's Roche limit, and the effect of tidal forces on the outcome of planetesimal collisions can be neglected. However, the tidal effect is significant in the Roche zone, where planetary rings are located. The Roche limit a_R, which is defined as

$$\frac{a_R}{r_{plan}} = 2.456 \left(\frac{\rho_{plan}}{\rho}\right)^{1/3}, \qquad (14.83)$$

is the distance inside which no figure of hydrostatic equilibrium exists (e.g., Roche 1847; Chandrasekhar 1969), and is not the distance at which a planet's tidal force exceeds a satellite's gravitational attraction. Thus, gravitational accretion is possible in the Roche zone under certain conditions (e.g., Smoluchowski 1979, Dobrovolskis and Burns 1980, Davis et al. 1984, Weidenschilling et al. 1984, Chapter 17).

The condition that the attraction between two radially aligned, synchronously rotating particles in contact with mass ratio μ exceeds the tidal force due to the planet is (e.g., Weidenschilling et al. 1984)

$$\frac{a}{r_{plan}} > \left(\frac{3\rho_{plan}}{\rho}\right)^{1/3} \frac{1+\mu^{1/3}}{(1+\mu)^{1/3}}. \qquad (14.84)$$

In terms of the ratio of the sum of the physical radii of the colliding particles to their mutual Hill radius denoted by r_p (or $r_h = r_p^{-1}$; Eq. 14.38), this condition reads $r_p < 1$ (or $r_h > 1$). Figure 14.17a shows the values of r_p in Saturn's rings, as a function of the distance from Saturn and the mass ratio of colliding particles. For icy particles around Saturn this condition gives $a > 126,000$ and $80,000$ km for identical particles or for very differently sized particles, respectively. On the basis of this criterion, Weidenschilling et al. (1984) and Davis et al. (1984) argued that particle collisions should produce accretion in Saturn's rings, especially in the A ring, and that the rapid accretionary processes are counterbalanced by tidal disruption of the larger accreted aggregates, which they called "dynamic ephemeral bodies." Salo (1992a) confirmed the formation of gravitational wakes and aggregates in Saturn's rings by local N-body simulations (Section 14.1.3).

On the other hand, Ohtsuki (1993) derived accretion criteria for colliding particles from an energy perspective, considering both relative kinetic and potential energies under the Hill approximation in the three-body problem. The linearized equations of relative motion of the two particles in a local coordinate system rotating with the Keplerian angular velocity Ω are written as

$$\ddot{x} = 2\dot{y}\Omega + 3x\Omega^2 - \frac{G(m_1 + m_2)x}{r^3}$$
$$\ddot{y} = -2\dot{x}\Omega - \frac{G(m_1 + m_2)y}{r^3} \qquad (14.85)$$
$$\ddot{z} = -z\Omega^2 - \frac{G(m_1 + m_2)z}{r^3}$$

14 Dynamics of Saturn's Dense Rings

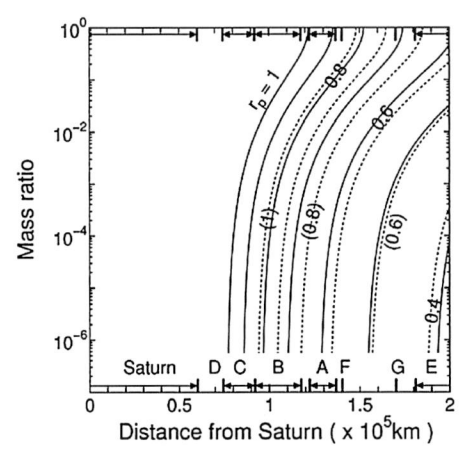

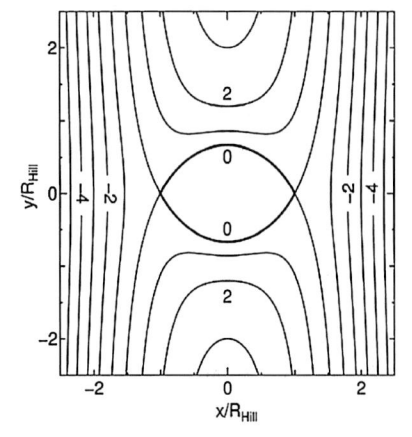

Fig. 14.17 *Left panel*: Contours of r_p as a function of the distance from Saturn and the mass ratio of colliding particles. Solid lines show the case with a particle internal density of $0.9\,\mathrm{g\,cm^{-3}}$ ($0.4 \leq r_p \leq 1$, with an increment of 0.1), while the dotted lines represent the case with $0.5\,\mathrm{g\,cm^{-3}}$ ($0.5 \leq r_p \leq 1$). The radial locations of Saturn's rings are also shown. *Right panel*: Contours of the Hill potential U for the $z = 0$ plane. Contour lines inside the Hill region are not shown

where the x-axis points radially outward, the y-axis in the direction of orbital motion, the z-axis is perpendicular to the x-y plane, and $r = \left(x^2 + y^2 + z^2\right)^{1/2}$. On the right-hand side of Eq. 14.85, the $2\dot{y}\Omega$ and $-2\dot{x}\Omega$ terms represent Coriolis forces; $3x\Omega^2$ and $-z\Omega^2$ are the tidal terms; and those proportional to $1/r^3$ represent the mutual gravity between the two particles. Equation 14.85 holds a constant of motion

$$E = \frac{1}{2}\left(\dot{x}^2 + \dot{y}^2 + \dot{z}^2\right) + U(x, y, z), \qquad (14.86)$$

where $U(x, y, z)$ is the Hill potential given as

$$U(x, y, z) = -\frac{3}{2}x^2\Omega^2 + \frac{1}{2}z^2\Omega^2 - \frac{G(m_1 + m_2)}{r} + \frac{9}{2}R_{Hill}^2\Omega^2. \qquad (14.87)$$

A constant $(9/2)R_{Hill}^2\Omega^2$ has been added so that U vanishes at the Lagrangian points $(x, y, z) = (\pm R_{Hill}, 0, 0)$. The contour lines of the Hill potential at the $z = 0$ plane are shown in Fig. 14.17b (see, e.g., Nakazawa and Ida 1988). The $U = 0$ surface defines the Hill sphere, which is actually lemon-shaped with a half-width of unity in the radial direction, $2/3$ in the azimuthal direction, and ≈ 0.638 in the vertical direction. Since the Hill sphere is identical to the $U = 0$ surface and $U < 0$ inside the sphere, only particles with positive E can enter the sphere, and they cannot escape out of the sphere if their energy is reduced to negative values by an inelastic collision. Therefore, the criteria for gravitational accretion are (i) the relative position (x, y, z) must be inside the Hill sphere, and (ii) $E < 0$ after impact.

Gravitational accretion is inhibited at the radial locations close to the planet where $r_p > 1$ (or $r_h < 1$), while colliding pairs with an arbitrary mass ratio can become gravitationally bound in the outermost part of the ring system where $r_p \ll 1$ (or $r_h \gg 1$), if there is sufficient energy dissipation in a collision. In the intermediate region, gravitational accretion is possible for particles which differ greatly in mass, while like-sized particles overflow their mutual Hill sphere and cannot remain gravitationally bound. Numerical results of three-body orbital integration show that the capture probability decreases abruptly for $r_p \geq 0.7$, because particles overflow the Hill sphere in the azimuthal direction when $r_p > 2/3$ (Ohtsuki 1993, Fig. 14.18a). The capture probability increases when surface friction is taken into account, while it decreases significantly when the relative random velocity becomes comparable to or larger than the escape velocity (Ohtsuki 1993; Morishima and Salo 2004, Fig. 14.18b). Neglecting the tidal terms in the Hill potential, Ohtsuki (1993) obtained an analytic expression for the capture probability in the case of $r_p \ll 1$ and relative random velocity larger than the escape velocity. Canup and Esposito (1995) derived a simple expression for the capture criterion by retaining the tidal terms and performing an angle-average (Chapter 17). The derived criteria are roughly consistent with the numerical results of Ohtsuki (1993) for low random velocities. Cohesive forces between Saturn's ring particles may extend the zone for particle accretion radially inward and allow small particles to be deposited on larger ones to form regolith layers (Spahn et al. 2004, Albers and Spahn 2006).

In dense rings, collective effects among more than two particles are important (Salo 1992a, 1995, Section 14.1.3). Using N-body simulation, Karjalainen and Salo (2004) examined the criterion for aggregate formation in detail. Results of their simulations with equal-sized particles in Saturn's rings with $\rho = 0.9\,\mathrm{g\,cm^{-3}}$, and a velocity-dependent normal coefficient of restitution, based on Bridges et al. (1984), showed that formation of transient aggregates should occur beyond $a \approx 137{,}000\,\mathrm{km}$ and stable aggregates form beyond $a \approx 146{,}000\,\mathrm{km}$, respectively. The critical radial

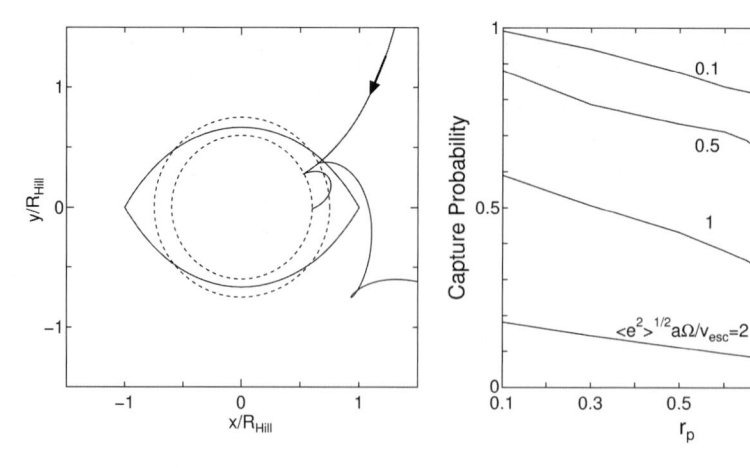

Fig. 14.18 *Left panel:* Examples of particle orbits leading to collision with targets of different values of r_p (0.6 and 0.75). In the case of $r_p = 0.6$, the orbit leads to accretion (i.e., E becomes negative) after two impacts, while the orbit in the case of $r_p = 0.75$ results in escape after the first collision (Redrawn from Ohtsuki). *Right panel:* Capture probability averaged over the Rayleigh distribution of particles' eccentricities and inclinations obtained by three-body orbital integration is shown as a function of r_p. $<e^2>^{1/2} a\Omega/v_{esc}$ represents particles' random velocity scaled by their escape velocity. The values of the normal and tangential coefficients of restitution are 0.5 and 0.9, respectively

location for the formation of stable aggregates corresponds to $r_p \sim 0.84$ ($r_h \sim 1.2$), which is slightly larger than the aforementioned $r_p \sim 0.7$ criterion based on three-body calculations. This indicates that gravitational accretion is facilitated by many-body effects, as compared to accretion between two solid bodies. Karjalainen and Salo (2004) also showed that the inclusion of the particles' surface friction and/or size distribution facilitates accretion, shifting the above accretion region by about 5,000 and 10,000 km, respectively. Karjalainen (2007) performed N-body simulations of impacts between gravitational aggregates in Saturn's rings, and found that impacts between aggregates, with mass ratios from 1 to 10, result on average in disruption, while net accretion typically should occur at $a > 145,000$ km. The shapes of the aggregates formed in simulations are well described with Roche ellipsoids, approaching spherical shapes as the distance increases (Karjalainen and Salo 2004).

Porco et al. (2007) noted that accretion is facilitated if under-dense ring particles accumulate around a high-density core, as compared to the case of accretion of low-density particles without core: under-dense particles are inferred from comparison between dynamical simulations and observational signatures of gravitational wakes (Section 14.1.3). The mean density of such an aggregate which has a high-density core decreases with increasing size due to accretion of particles. Therefore the size of the aggregate's Hill sphere grows more slowly than its physical size. As a result, accretion is halted when particles fill the aggregate's Hill sphere. Porco et al. (2007) analyzed Cassini images of Saturn's small inner satellites to derive their sizes, shapes, and mean densities. They found that the long axes of Pan, Daphnis, Atlas, and Prometheus agree within 15% with the long axis of the Hill sphere for a body of the given satellite mass and orbit.

They also confirmed the above process of particle accretion around a high-density core using N-body simulation. Moreover, Cassini images revealed that Pan and Atlas have equatorial ridges, which may have formed by preferential accretion of ring particles onto the equatorial surfaces of already-formed satellites embedded in the rings (Charnoz et al. 2007, Chapter 17).

14.4.1.3 Processes and Models for Particle Size Evolution

Multiple processes must have been involved to produce the observed size distribution of ring particles and small moonlets. Studies of the size evolution of asteroids through a collisional cascade show that a power-law $n(R)dR \propto R^{-q}dR$ with $q = 3.5$ is expected in steady state if the dependence of the critical specific energy for fragmentation on target size is neglected (e.g., Dohnanyi 1969; Williams and Wetherill 1994; Tanaka et al. 1996; Durda et al. 1998). Additional processes, including the tidally modified accretion in the Roche zone described above, are likely to play a role for the size evolution of ring particles. The role of satellite fragmentation and tidally modified accretion in the origin and evolution of ring systems is further discussed in Chapter 17.

Davis et al. (1984) and Weidenschilling et al. (1984) examined the evolution of the particle size distribution in a planetary ring, using a statistical simulation that includes accretion, rebound, as well as the collisional and tidal disruption of particles. They found that their simulation can reproduce the size distribution inferred from the Voyager observation (Marouf et al. 1983), if tidal disruption is assumed to occur at $R = 10$ m, producing large fragments and a shower of small particles with a power-law size distribution. Longaretti (1989) developed an analytic theory for the particle size distribution and argued that the upper cutoff can be explained by efficient collisional erosion of the large particles.

Esposito and Colwell (1989) and Colwell and Esposito (1992, 1993) (also Colwell et al. 2000) considered the evolution of the sizes of satellites and their fragments around Uranus and Neptune due to meteoroid impacts. Their simulations showed that once the initial disruption of a parent satellite occurs, subsequent disruptions of its fragments occur relatively quickly. Their earlier simulations, which used the critical impact energy for catastrophic fragmentation based on a scaling theory (e.g., Housen and Holsapple 1990), showed that many of the small moons of Uranus and Neptune have lifetimes against catastrophic disruption shorter than the age of the solar system (Colwell and Esposito 1992). These lifetimes are significantly longer when the criterion for catastrophic disruption based on the asteroid collisional evolution model of Durda et al. (1998) or from hydrodynamic impact simulations (Benz and Asphaug 1999) is used. Colwell et al. (2000) found that a model intermediate to that of the scaling theory and that of Durda et al. (1998) would allow the fragments from a 10–20 km radius parent satellite to produce a more steady population of rings, moonlet belts, and small moons.

Canup and Esposito (1995) considered the effect of accretion and performed numerical simulations of the accretional growth of a disrupted satellite. In the Roche zone where only bodies which differ greatly in mass can remain gravitationally bound, Canup and Esposito (1995) found that a fragmentation-produced debris distribution basically evolved into a bimodal population, with one element consisting of a swarm of small, high-velocity bodies and the other composed of a small number of large moonlets with low random velocities. Canup and Esposito applied a similar model to Saturn's G ring, and found that evolution from the disruption of a 1.5–3 km progenitor satellite can explain the G ring's particle and dust population inferred from observations (Showalter and Cuzzi 1993; Gurnett et al. 1983; Van Allen 1983). Barbara and Esposito (2002) performed simulations of collisional evolution of particles in the F ring, including accretion and fragmentation. They argued that collisions between moonlets (which themselves formed as a result of accretion) can explain the anomalous localized brightenings in the F ring observed by Voyager (Showalter 1998).

The size and spatial distributions of the 20–250 m radius moonlets, derived from the propeller structures observed in the A ring region by Cassini, also provide a clue to the evolution of the size distribution of rings and moonlets. As we mentioned before, the derived size distribution for propeller moons has a larger power-law index compared to $q \approx 3$ for ring particles with $R \leq 10$ m (Tiscareno et al. 2006b, 2008; Sremčević et al. 2007). Moreover, Sremčević et al. (2007) found that the propeller structures are concentrated in a narrow 3,000-km-wide annulus at $a \approx 130,000$ km. On the basis of these observations, they proposed an explanation that such embedded moonlets are remnants of fragments of a ring-moon of Pan size or larger disrupted by a meteoroid impact. Large fragments produced by the disruption evolve by further shattering by meteoroid bombardment, explaining the steepened size distribution of the moonlets responsible for the propeller structures. They argue that the steepness of the moonlets' size distribution and their apparent lack in the rest of the A ring represent different phases of the moonlet destruction and subsequent evolution.

As we mentioned earlier, observations suggest $R \approx 0.1$–30 cm as a minimum particle size depending on the radial location in Saturn's rings (French and Nicholson 2000). Poynting-Robertson drag (loss of orbital angular momentum by absorption and reemission of radiation) and plasma drag (angular momentum transfer due to physical collisions and/or long-range charged particle interactions between ring particles and ions and electrons tied to the planet's magnetic field) are the important mechanisms of removal of micrometer-sized ring particles (see, e.g. Burns et al. 1984).

As another possible mechanism, the removal of small ring particles due to thermal torques has been recently proposed. The effect of thermal torques produced by the absorption and re-emission of sunlight from an asteroid's surface on its orbital evolution is called the Yarkovsky effect (e.g., Bottke et al. 2002). Similar effects on small particles in Saturn's rings, i.e. thermal torques due to Saturn shine and the torques due to illumination by the sun (the Yarkovsky–Schach effect) have been recently studied by Rubincam (2006) and Vokrouhlický et al. (2007). The removal of subcentimeter-sized particles due to thermal torques may help to explain the observed paucity of such small particles in Saturn's rings. However, frequent collisions of the small particles with large ring particles will rapidly re-distribute any extra torque received by the small particles over the whole population. For this reason the evolution of the small particles, most sensitive to the Yarkovsky and Yarkovsky–Schach effects, cannot be considered independently from the large particles. Further studies including the effects of particle collisions and resonances are needed to clarify such effects on the long-term orbital evolution of the ring particle size distribution.

14.4.2 Particle Spins

14.4.2.1 Dynamical Studies

For rings consisting of non-gravitating equal-sized particles on circular orbits undergoing inelastic collisions due to Kepler shear, the particles' rotation rate ω is expected to be on the order of the orbital angular velocity Ω (e.g., Weidenschilling et al. 1984). Detailed calculations, using N-body simulation, show that spin rates follow a Gaussian distribution with mean spin rate being slow prograde rotation with $\langle\omega\rangle/\Omega \sim 0.3$–$0.5$, both in cases of non-gravitating and gravitating particles (Salo 1987a, b, 1995; Richardson 1994; Ohtsuki and Toyama 2005). However, the dispersion of spins can be much larger, scaling as $\sqrt{\langle\omega^2\rangle}/\Omega \propto c/R\Omega$, where the proportionality constant depends on the coefficient of tangential friction[7], ε_t. Besides friction, also small-scale deviations from spherical size are efficient in promoting particle spins (Salo 1987a, b).

Because of the additional energy dissipation due to surface friction, inclusion of rough surfaces and particle spins allow a large critical value (ε_{cr}) of the normal restitution coefficient for the thermal stability discussed in Section 14.1.1.3 (Salo 1987a, b; Araki 1988, 1991; Morishima and Salo 2006; Ohtsuki 2006a). The strength of surface friction also determines the ratio of the particles' rotational energy $E_{rot} = mR^2 \langle\omega^2\rangle /5$ (for a uniform sphere with mass m) and kinetic energy E_{kin}. In the non-gravitating case, this ratio in the equilibrium state is obtained analytically (Salo 1987a; Morishima and Salo 2006) as

$$\frac{E_{rot}}{E_{kin}} = \frac{2(1-\varepsilon_t)}{14 - 5(1-\varepsilon_t)},$$

which shows that equipartition between random motion and spins is not realized, unless surface friction is extremely strong (i.e. $\varepsilon_t = -1$). Compared to a system of smooth particles, the inclusion of energy dissipation due to friction slightly decreases the equilibrium velocity dispersion in the non-gravitating case as well as for gravitating particles in a low optical depth ring. In self-gravitating rings of larger optical depth, i.e. where gravitational wakes are formed, the additional energy dissipation due to surface friction facilitates the formation of gravitational wakes, and thus tends to slightly increase velocity dispersion (Salo 1995). However, such an effect is minor as compared to the major effect of self-gravity in dense rings.

As an extreme case of the rotational evolution due to collisions between particles with different sizes, the spin of a moonlet embedded in a ring of small equal-sized particles was studied in terms of three-body calculations and N-body simulations (Morishima and Salo 2004; Ohtsuki 2004a, b). Calculations show that the moonlet's rotation reaches an equilibrium state of slow prograde spin with $\langle\omega\rangle/\Omega \sim 0.3$–$0.5$ when ring particles are much smaller than the moonlet, while the moonlet can spin both in prograde and retrograde directions when the particle size is comparable to the moonlet size.

The spin state of ring particles with size distribution was examined in N-body simulations for various cases, including dense self-gravitating rings (Salo 1987b; Richardson 1994; Ohtsuki and Toyama 2005; Morishima and Salo 2006), while analytic approaches and numerical three-body calculations were used to study the size-dependence of rotation rates in low-optical depth rings (Ohtsuki 2005, 2006a, b). Results of these studies show that large particles spin slowly, with mean spin period comparable to the orbital period, and a spin dispersion which is much smaller than the mean, while small particles generally spin much faster, with a dispersion that considerably exceeds the mean.

In the case of an extended size distribution, the spin period was found to be roughly proportional to the particle size (Fig. 14.9). Rapidly spinning small particles have larger orbital inclinations than slowly spinning large particles (Salo and Karjalainen 2003; Ohtsuki 2005, 2006b; Morishima and Salo 2006); thus, the ring particles' rotational states have vertical heterogeneity. The spin axes of slowly spinning large particles tend to be aligned normally to the ring plane, while the small particles' spin axes are nearly randomly distributed (see Fig. 9 in Salo 1987b; Ohtsuki and Toyama 2005). Since in the case of a continuous size distribution the rotation frequency of the largest particles is on the order of Ω, and that of small particles is roughly inversely proportional to the particle radius, the values of the rotation rates of smaller particles depend on the size of the largest particles. The overall rotation rates also tend to increase slightly for stronger surface friction and/or a shallower size distribution of particles, but the R^{-1}-dependence of the rotation rates is insensitive to the values of these parameters (Ohtsuki 2005, 2006b, Morishima and Salo 2006, Fig. 14.19).

In the case of rings with low optical depth, the rate of change of the mean rotational energy of particles with mass m due to collisions with other particles with mass m' can be written as

$$\frac{dE_{rot}(m)}{dt} = \int n_s(m')\{C_{CS} + (E_{rot}(m') - E_{rot}(m))C_{RF}\}dm', \quad (14.88)$$

where C_{CS} and C_{RF} are the rate coefficients, which depend on the particles' relative random velocity and can be evaluated by three-body calculation (Ohtsuki 2005, 2006a). The first term in parentheses in the integrand represents energy exchange between random motion and rotation ('collisional stirring'), while the second term tends to equalize the mean rotational energy among particles with different sizes and to decrease the rotation rates of large particles

---

[7] ε_t is defined so that the tangential component of the relative velocity of the two contacting points is changed by a factor of ε_t ($-1 \leq \varepsilon_t \leq 1$) due to collision. Perfectly smooth spheres have $\varepsilon_t = 1$ while perfectly rough spheres have $\varepsilon_t = -1$ (Araki and Tremaine, 1986).

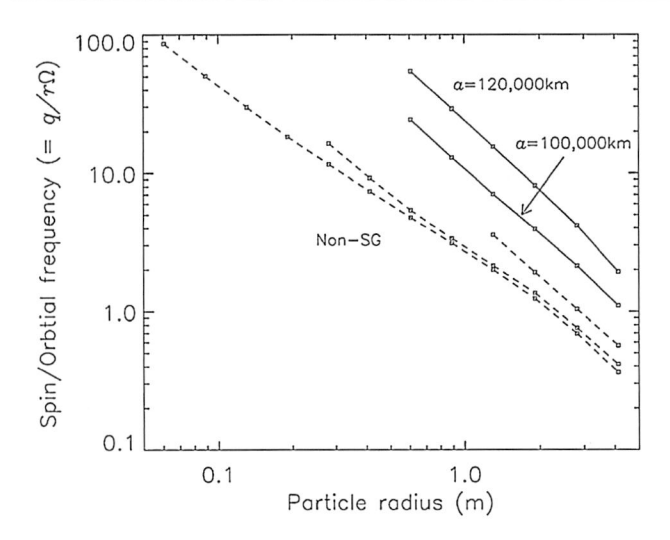

Fig. 14.19 Results of N-body simulations for the rotation rates of particles in units of orbital angular velocity (i.e., $q \equiv R\langle\omega^2\rangle^{1/2}$), as a function of particle radius. Solid lines show results for self-gravitating particles at two different radial locations around Saturn ($\varepsilon_n = \varepsilon_t = 0.5$). Dashed lines represent results for non-gravitating particles with different widths of the particle size distribution (Morishima and Salo, 2006)

('rotational friction', by analogy with the dynamical friction term in the velocity evolution equation; see Eq.14.16). Because of the former effect, equipartition of rotational energy among particles with different sizes (which would imply $\langle\omega^2\rangle^{1/2} \propto R^{-5/2}$) is not realized.

14.4.2.2 Relation to Observations of the Rings' Thermal Emission

While the primary heat source for the rings is the sunlight, thermal radiation and reflected sunlight from Saturn also contribute to the heating of the rings. Also, mutual heating between nearby particles can be significant when the ring is dense enough (Aumann and Kieffer 1973; Spilker et al. 2003). The response of ring particles to such heating depends on their physical and dynamical properties, including albedo, thermal inertia, and spin rates. Observations of the thermal response of the rings during and after eclipse suggest low thermal inertia of ring particles, with a thermal relaxation time of about one hour, i.e. about one tenth of an orbital period (e.g., Froidevaux et al. 1981; Spilker et al. 2003; Ferrari et al. 2005). In this case, particles with spin period much longer than the thermal relaxation time can be regarded as slow rotators, which radiate their thermal emission mainly from the face illuminated by the sunlight, while fast rotators with random spin orientations radiate over their whole surface area (Froidevaux 1981; Kawata 1983). Furthermore, in dilute rings, some fraction of fast rotators have their spin axes pointing nearly toward the Sun. The north and south sides of such a particle with respect to its spin axis can have a temperature contrast regardless of the rotation rate, if the interval of collisions that would change the spin orientation is longer than the thermal relaxation time (Morishima and Salo 2006).

Comparison between observations of the rings' thermal emission with model calculations provides constraints on physical properties as well as spin states and structure of the rings (e.g., Froidevaux 1981; Kawata 1983; Esposito et al. 1984; Spilker et al. 2003; Ferrari and Leyrat 2006). Cassini CIRS has acquired an extensive set of thermal measurements of Saturn's main rings for a number of different viewing geometries (Spilker et al. 2006; Leyrat et al. 2008; Altobelli et al. 2007). For example, the observed temperature decrease with increasing solar phase angle suggests that the rings include a population of slow rotators. However, the spin states of particles depend on their sizes, as mentioned above. Furthermore, the thermal emission of the rings likely depends on the particles' vertical distribution and the rate of vertical mixing as well as on the rings' fine structure, such as gravitational wakes in the A ring (Leyrat et al. 2008; Ferrari et al. 2009; Morishima et al. 2009a, b). Further studies using models that include such effects and detailed comparison with observations will provide unique constraints on physical and dynamical properties of ring particles.

14.5 Open Problems

Cassini has considerably advanced our understanding of the dynamics of Saturn's dense rings. Naturally, new questions arise and problems remain open.

An outstanding problem of ring dynamics remains the pronounced structure of the B ring, notably region B2 where the ring alternates between states that are practically opaque and states of moderate optical depth (see Fig. 13.14). It seems clear that classical viscous instability will not work here, still, the fact that this instability would produce a bimodal pattern of optical depth is compelling. In this chapter we have shown how the stability properties of the rings can drastically depend on the elasticity of the ring particles. The study of the influence of a particle-size dependence of elastic properties on the ring dynamics has only started. In particular in an interplay with adhesion and self-gravity this might well lead to new types of instabilities (or modification of old ones) applicable to Saturn's rings.

There are several Cassini observations that point at local changes in the particle size distribution in perturbed ring regions. One example are the propellers, where the brightness of the propeller streaks relative to the surrounding ring is difficult to understand. A plausible explanation is the release of small particles, otherwise sticking on the large ring-particles, in those regions that are most perturbed. Another example are the halos around strong density waves observed in the A ring (Dones et al. 1993; Nicholson et al. 2008). They can be similarly interpreted as the effect of small particles locally released from the large ones where the density waves perturb the ring (Nicholson et al. 2008). The third example is the outermost region of the A ring outside the Keeler gap. There the amplitude of the ring's brightness asymmetry is very small, if any. This means that the self-gravity wakes either do not form here, or they are hidden by some process. One possibility is that the combined perturbations of the numerous resonances with Pandora and Prometheus in that region lead to a release of small particles, decreasing the contrast between the crests and troughs of self-gravity wakes, and, in this way, the amplitude of the brightness asymmetry. Altogether this implies that ring particles are loosely bound aggregates, as predicted by Davis et al. (1984; Weidenschilling et al. 1984). Their size is determined by a balance of coagulation and fragmentation as discussed in Spahn et al. (2004), which in turn depends on the typical speed of particle collisions, so that a local increase in the velocity dispersion, induced by external perturbations, leads to a gradual breakup of the aggregates, changing the size distribution. The consequences of such a variable size distribution for the ring dynamics (e.g. properties of self-gravity wakes, density waves, or instabilities) has not been studied so far.

One challenge for future studies is a unified description of ring dynamics and self-gravity wakes. This might significantly change the conclusions drawn from uniform ring models using approximations of self-gravity. For example, the interaction self-gravity wakes with periodically expanding and contracting density waves might well lead to non-trivial effects on the dispersion relation and the damping of the wave. On the other hand the perturbed ring state in the density wave region can alter the properties of the wakes.

Another question is the relation between viscous overstability and density waves. In fact, spontaneous overstable waves and density waves obey the same dispersion relation. In the derivation of the formula for the damping of tightly wound waves, in the simple hydrodynamic treatment of Shu (1984), the density dependence of the viscosity has been neglected. Such a term would, however, strongly modify the damping behavior of the wave in that model, as noted by Goldreich and Tremaine (1978b). Therefore, the viscosities fitted from that model should be taken with care. Future study could address the question if density waves can undergo an instability similar to the viscous overstability. If this is the case, then the resonance might even lead to such a strong perturbation of the ring that no regular density wave-train is seen. For instance the Janus/Epimetheus 2:1 and 4:3 resonances lead to pronounced density waves in the inner B ring and inner A ring, respectively, while it is a puzzle that the strong 3:2 resonance in the outer B ring does not produce a wave.

Acknowledgment This work was supported by DLR and DFG (JS and FS), NASA (KO), the CASSINI project (KO and NR), and the Akademy of Finland (HS). We would like to thank the reviewers for their thorough and constructive reports and Larry Esposito for his comments on the manuscript.

References

Albers, N. and Spahn, F. (2006). The influence of particle adhesion on the stability of agglomerates in Saturn's rings. *Icarus*, 181:292–301.

Altobelli, N., Spilker, L., Pilorz, S., Brooks, S., Edgington, S., Wallis, B., and Flasar, M. (2007). C ring fine structures revealed in thermal infrared. *Icarus*, 191:691–701.

Araki, S. (1988). The dynamics of particle disks. II. Effects of spin degrees of freedom. *Icarus*, 76:182–198.

Araki, S. (1991). The dynamics of particle disks III. Dense and spinning particle disks. *Icarus*, 90:139–171.

Araki, S. and Tremaine, S. (1986). The dynamics of dense particle disks. *Icarus*, 65:83–109.

Aumann, H. H. and Kieffer, H. H. (1973). Determination of particle sizes in Saturn's rings from their eclipse cooling and heating curves. *Astrophys. J.*, 186:305–311.

Barbara, J. M. and Esposito, L. W. (2002). Moonlet collisions and the effects of tidally modified accretion in Saturn's F ring. *Icarus*, 160:161–171.

Benz, W. and Asphaug, E. (1999). Catastrophic disruptions revisited. *Icarus*, 142:5–20.

Binney, J. and Tremaine, S. (1987). *Galactic Dynamics*. Princeton, New Jersey, Princeton University Press, 1987, p 747.

Borderies, N., Goldreich, P., and Tremaine, S. (1982). Sharp edges of planetary rings. *Nature*, 299:209–211.

Borderies, N., Goldreich, P., and Tremaine, S. (1983a). Perturbed particle disks. *Icarus*, 55:124–132.

Borderies, N., Goldreich, P., and Tremaine, S. (1983b). The dynamics of elliptical rings. *Astron. J.*, 88:1560–1568.

Borderies, N., Goldreich, P., and Tremaine, S. (1984a). Excitation of inclinations in ring-satellite systems. *Astrophys. J.*, 284:429–434.

Borderies, N., Goldreich, P., and Tremaine, S. (1984b). Unsolved problems in planetary ring dynamics. In Greenberg, R. and Brahic, A., editors, *Planetary Rings*, pp. 713–734, Tucson Arizona, University of Arizona Press.

Borderies, N., Goldreich, P., and Tremaine, S. (1985). A granular flow model for dense planetary rings. *Icarus*, 63:406–420.

Borderies, N., Goldreich, P., and Tremaine, S. (1986). Nonlinear density waves in planetary rings. *Icarus*, 68:522–533.

Borderies, N., Goldreich, P., and Tremaine, S. (1989). The formation of sharp edges in planetary rings by nearby satellites. *Icarus*, 80:344–360.

Bottke, W. F., Vokrouhlický, D., Rubincam, D. P., and Broz, M. (2002). The effect of Yarkovsky thermal forces on the dynamical evolution of asteroids and meteoroids. In Bottke, W. F. J., Cellino, A., Paolicchi, P., and Binzel, R. P., editors, *Asteroids III*, pp. 395–408, Tucson Arizona. University of Arizona Press.

Brahic, A. (1977). Systems of colliding bodies in a gravitational field. I – Numerical simulation of the standard model. *Astron. Astrophys.*, 54:895–907.

Bridges, F., Hatzes, A., and Lin, D. (1984). Structure, stability and evolution of Saturn's rings. *Nature*, 309:333–338.

Bridges, F., Supulver, K. D., Lin, D., Knight, R., and Zafra, M. (1996). Energy loss and sticking mechanisms in particle aggregation in planetesimal formation. *Icarus*, 123:422–435.

Brilliantov, N. V. and Pöschel, T. (2004). *Kinetic Theory of Granular Gases*. Oxford, Oxford University Press, 2004, 340 p.

Brilliantov, N., Spahn, F., Hertzsch, J.-M., and Pöschel, T. (1996). Model for collisions in granular gases. *Phys. Rev. E*, 53:5382–5392.

Brilliantov, N. V., Albers, N., Spahn, F., and Pöschel, T. (2007). Collision dynamics of granular particles with adhesion. *Phys. Rev. E*, 76(5):051302.

Burns, J. A., Showalter, M. R., and Morfill, G. E. (1984). The ethereal rings of Jupiter and Saturn. In Greenberg, R. and Brahic, A., editors, *Planetary Rings*, pp. 200–272, Tucson Arizona. University of Arizona Press.

Burns, J.A., Tiscareno, M.S., Porco, C.C., Dones, L., Murray, C.D., and Cassini Imaging Team (2004). Weak waves and wakes in Saturn's rings: Observations by Cassini ISS. *Bull. American Astron. Soc.*, 36: p. III.

Burns, J. A., Tiscareno, M. S., Spitale, J., Porco, C. C., Cooper, N. J., and Beurle, K. (2008). Giant propellers Outside the Encke gap in Saturn's rings. In *AAS/Division for Planetary Sciences Meeting Abstracts*, 40: p. 30.07.

Camichel, H. (1958). Mesures photométriques de Saturne et de son anneau. *Annales d'Astrophysique*, 21:231–242.

Canup, R. M. and Esposito, L. W. (1995). Accretion in the Roche zone: Co-existence of rings and ringmoons. *Icarus*, 113:331–352.

Canup, R. M. and Esposito, L. W. (1997). Evolution of the G ring and the population of macroscopic ring particles. *Icarus*, 126:28–41.

Chandrasekhar, S. (1969). *Ellipsoidal Figures of Equilibrium*. New Haven, Yale University Press.

Chapman, S. and Cowling, T. G. (1970). *The Mathematical Theory of Non-Uniform Gases*. Cambridge, Cambridge, University Press, 1970, 3rd edn.

Charnoz, S., Brahic, A., Thomas, P., and Porco, C. (2007). The equatorial ridges of Pan and Atlas: Terminal accretionary ornaments? *Science*, 318:1622–1624.

Chiang, E. I. and Goldreich, P. (2000). Apse alignment of narrow eccentric planetary rings. *Astrophys. J.*, 540:1084–1090.

Colombo, G., Goldreich, P., and Harris, A. W. (1976). Spiral structure as an explanation for the asymmetric brightness of Saturn's A ring. *Nature*, 264:344–345.

Colwell, J. E. (2003). Low velocity impacts into dust: Results from the COLLIDE-2 microgravity experiment. *Icarus*, 164:188–196.

Colwell, J. E. and Esposito, L. W. (1992). Origins of the rings of Uranus and Neptune: I. Statistics of satellite disruptions. *J. Geophys. Res.*, 97:10227–10241.

Colwell, J. E. and Esposito, L. W. (1993). Origins of the rings of Uranus and Neptune: II. Initial conditions and ring moon populations. *J. Geophys. Res.*, 98:7387–7401.

Colwell, J. E. and Taylor, M. (1999). Low-velocity microgravity impact experiments into simulated regolith. *Icarus*, 138:241–248.

Colwell, J. E., Esposito, L. W., and Bundy, D. (2000). Fragmentation rates of small satellites in the outer solar system. *J. Geophys. Res.*, 105:17589–17599.

Colwell, J. E., Esposito, L. W., and Sremčević, M. (2006a). Self-gravity wakes in Saturn's A ring measured by stellar occultations from Cassini. *Geophys. Res. Lett.*, 33:L07201.1–L07201.4

Colwell, J. E., Esposito, L. W., Sremčević, M., Stewart, G. R., and McClintock, W. E. (2007). Self-gravity wakes and radial structure of Saturn's B ring. *Icarus*, 190:127–144.

Colwell, J. E., Esposito, L. W., and Stewart, G. R. (2006b). Density waves observed by Cassini stellar occultations as probes of Saturn's rings. In Mackwell, S. and Stansbery, E., editors, *37th Annual Lunar and Planetary Science Conference*, volume 37 of *Lunar and Planetary Inst. Technical Report*, p. 1221.

Colwell, J. E., Cooney, J. H., Esposito, L. W., and Sremčević, M. (2009). Density waves in Cassini UVIS stellar occultations. 1. The Cassini Division. *Icarus*, 200:574–580.

Cooke, M. L. (1991). *Saturn's Rings: Photometric Studies of the C Ring and Radial Variation in the Keeler Gap*. Ph.D. thesis, AA (Cornell University, Ithaca, NY).

Cuzzi, J. N. and Burns, J. A. (1988). Charged particle depletion surrounding Saturn's F ring – Evidence for a moonlet belt? *Icarus*, 74:284–324.

Cuzzi, J. N. and Durisen, R. H. (1990). Bombardment of planetary rings by meteoroids – General formulation and effects of Oort Cloud projectiles. *Icarus*, 84:467–501.

Cuzzi, J. N. and Scargle, J. D. (1985). Wavy edges suggest moonlet in Encke's gap. *Astrophys. J.*, 292:276–290.

Cuzzi, J. N., Burns, J. A., Durisen, R. H., and Hamill, P. M. (1979a). The vertical structure and thickness of Saturn's rings. *Nature*, 281:202–204.

Cuzzi, J. N., Durisen, R. H., Burns, J. A., and Hamill, P. (1979b). The vertical structure and thickness of Saturn's rings. *Icarus*, 38:54–68.

Cuzzi, J. N., Lissauer, J. J., and Shu, F. H. (1981). Density waves in Saturn's rings. *Nature*, 292:703–707.

Cuzzi, J. N., Lissauer, J. J., Esposito, L. W., Holberg, J. B., Marouf, E. A., Tyler, G. L., and Boischot, A. (1984). Saturn's rings: Properties and processes. In Greenberg, R. and Brahic, A., editors, *Planetary Rings*, pp. 73–199, Tucson Arizona, University of Arizona Press.

Daisaka, H. and Ida, S. (1999). Spatial structure and coherent motion in dense planetary rings induced by self-gravitational instability. *Earth Planets Space*, 51:1195–1213.

Daisaka, H., Tanaka, H., and Ida, S. (2001). Viscosity in a dense planetary ring with self-gravitating particles. *Icarus*, 154:296–312.

Davis, D. R., Weidenschilling, S. J., Chapman, C. R., and Greenberg, R. (1984). Saturn ring particles as dynamic ephemeral bodies. *Science*, 224:744–747.

Dermott, S. F., Murray, C. D., and Sinclair, A. T. (1980). The narrow rings Jupiter, Saturn and Uranus. *Nature*, 284:309–313.

Dilley, J. (1993). Energy loss in collisions of icy spheres: Loss mechanism and size-mass dependence. *Icarus*, 105:225–234.

Dilley, J. and Crawford, D. (1996). Mass dependence of energy loss in collisions of icy spheres: An experimental study. *J. Geophys. Res.*, 101:9267–9270.

Dobrovolskis, A. R. and Burns, J. A. (1980). Life near the Roche limit: Behavior of ejecta from satellites close to planets. *Icarus*, 42:422–441.

Dohnanyi, J. S. (1969). Collisional model of asteroids and their debris. *J. Geophys. Res.*, 74:2531–2554.

Dones, L., Cuzzi, J. N., and Showalter, M. R. (1993). Voyager Photometry of Saturn's A Ring. *Icarus*, 105:184–215.

Dones, L. and Porco, C. C. (1989). Spiral density wakes in Saturn's A ring? *Bull. American Astron. Soc.*, 21:929.

Duncan, M., Quinn, T., and Tremaine, S. (1989). The long-term evolution of orbits in the solar system – A mapping approach. *Icarus*, 82:402–418.

Dunn, D. E., Molnar, L. A., Niehof, J. T., de Pater, I., and Lissauer, J. J. (2004). Microwave observations of Saturn's rings: Anisotropy in directly transmitted and scattered saturnian thermal emission. *Icarus*, 171:183–198.

Dunn, D. E., de Pater, I., and Molnar, L. A. (2007). Examining the wake structure in Saturn's rings from microwave observations over varying ring opening angles and wavelengths. *Icarus*, 192:56–76.

Durda, D. D., Greenberg, R., and Jedicke, R. (1998). Collisional models and scaling laws: A new interpretation of the shape of the main-belt asteroid size distribution. *Icarus*, 135:431–440.

Durisen, R. H. (1984). Transport effects due to particle erosion mechanisms. In Greenberg, R. and Brahic, A., editors, *Planetary Rings*, pages 416–446, Tucson Arizona, University of Arizona Press.

Durisen, R. H. (1995). An instability in planetary rings due to ballistic transport. *Icarus*, 115:66–85.

Durisen, R. H., Cramer, N. L., Murphy, B. W., Cuzzi, J. N., Mullikin, T. L., and Cederbloom, S. E. (1989). Ballistic transport in planetary ring systems due to particle erosion mechanisms. I – Theory, numerical methods, and illustrative examples. *Icarus*, 80:136–166.

Durisen, R. H., Bode, P. W., Cuzzi, J. N., Cederbloom, S. E., and Murphy, B. W. (1992). Ballistic transport in planetary ring systems due to particle erosion mechanisms. II – Theoretical models for Saturn's A- and B-ring inner edges. *Icarus*, 100:364–393.

Esposito, L. W. and Colwell, J. E. (1989). Creation of the Uranus rings and dust bands. *Nature*, 339:605–607.

Esposito, L. W., Borderies, N., Goldreich, P., Cuzzi, J. N., Holberg, J. B., Lane, A. L., Pomphrey, R. B., Terrile, R. J., Lissauer, J. J., Marouf, E. A., and Tyler, G. L. (1983). Eccentric ringlet in the Maxwell gap at 1.45 Saturn radii Multi-instrument Voyager observations. *Science*, 222:57–60.

Esposito, L. W., Ocallaghan, M., Simmons, K. E., Hord, C. W., West, R. A., Lane, A. L., Pomphrey, R. B., Coffeen, D. L., and Sato, M. (1983). Voyager photopolarimeter stellar occultation of Saturn's rings. *J. Geophys. Res.*, 88:8643–8649.

Esposito, L. W., Cuzzi, J. N., Holberg, J. B., Marouf, E. A., Tyler, G. L., and Porco, C. C. (1984). Saturn's rings: Structure, dynamics, and particle properties. In Greenberg, R. and Brahic, A., editors, *Planetary Rings*, pages 463–544, Tucson Arizona, University of Arizona Press.

Esposito, L. W., Colwell, J. E., Larsen, K., McClintock, W. E., Stewart, A. I. F., Hallett, J. T., Shemansky, D. E., Ajello, J. M., Hansen, C. J., Hendrix, A. R., West, R. A., Keller, H.-U., Korth, A., Pryor, W. R., Reulke, R., and Yung, Y. L. (2005). Ultraviolet imaging spectroscopy shows an active Saturnian system. *Science*, 307:1251–1255.

Esposito, L. W., Meinke, B. K., Colwell, J. E., Nicholson, P. D., and Hedman, M. M. (2008). Moonlets and clumps in Saturn's F ring. *Icarus*, 194:278–289.

Ferrari, C. and Brahic, A. (1997). Arcs and clumps in the Encke division of Saturn's rings. *Planet. Space Sci.*, 45:1051–1067.

Ferrari, C., Galdemard, P., Lagage, P. O., Pantin, E., and Quoirin, C. (2005). Imaging Saturn's rings with CAMIRAS: Thermal inertia of B and C rings. *Astron. Astrophys.*, 441:379–389.

Ferrari, C. and Leyrat, C. (2006). Thermal emission of spherical spinning ring particles: The standard model. *Astron. Astrophys.*, 447:745–760.

Ferrari, C., Brooks, S., Edgington, S., Leyrat, C., Pilorz, S., and Spilker, L. (2009). Structure of self-gravity wakes in Saturn's A ring as measured by Cassini CIRS. *Icarus*, 199:145–153.

Franklin, F. A., Cook, A. F., Barrey, R. T. F., Roff, C. A., Hunt, G. E., and de Rueda, H. B. (1987). Voyager observations of the azimuthal brightness variations in Saturn's rings. *Icarus*, 69:280–296.

French, R. G. and Nicholson, P. D. (2000). Saturn's rings II: Particle size inferred from stellar occultation data. *Icarus*, 145:502–523.

French, R. G., Nicholson, P. D., Cooke, M. L., Elliot, J. L., Matthews, K., Perkovic, O., Tollestrup, E., Harvey, P., Chanover, N. J., Clark, M. A., Dunham, E. W., Forrest, W., Harrington, J., Pipher, J., Brahic, A., Grenier, I., Roques, F., and Arndt, M. (1993). Geometry of the Saturn system from the 3 July 1989 occultation of 28 SGR and Voyager observations. *Icarus*, 103:163–214.

French, R. G., Salo, H., McGhee, C. A., and Dones, L. (2007). HST observations of azimuthal asymmetry in Saturn's rings. *Icarus*, 189:493–522.

Froidevaux, L. (1981). Saturn's rings: Infrared brightness variation with solar elevation. *Icarus*, 46:4–17.

Froidevaux, L., Matthews, K., and Neugebauer, G. (1981). Thermal response of Saturn's ring paricles during and after eclipse. *Icarus*, 46:18–26.

Goertz, C. K. (1984). Formation of Saturn's spokes. *Adv. Space Res.*, 4:137–141.

Goertz, C. K. and Morfill, G. (1988). A new instability of Saturn's rings. *Icarus*, 74:325–330.

Goertz, C. K., Morfill, G. E., Ip, W. H., Grün, E., and Havnes, O. (1986). Electromagnetic angular momentum transport in Saturn's rings. *Nature*, 320:141–143.

Goldreich, P. and Lynden-Bell, D. (1965). II. Spiral arms as sheared gravitational instabilities. *Mon. Not. R. Astron. Soc.*, 130:125–+.

Goldreich, P. and Tremaine, S. (1978a). The excitation and evolution of density waves. *Astrophys. J.*, 222:850–858.

Goldreich, P. and Tremaine, S. (1978b). The formation of the Cassini division in Saturn's rings. *Icarus*, 34:240–253.

Goldreich, P. and Tremaine, S. (1978c). The velocity dispersion in Saturn's rings. *Icarus*, 34:227–239.

Goldreich, P. and Tremaine, S. (1979a). Precession of the epsilon ring of Uranus. *Astron. J.*, 84:1638–1641.

Goldreich, P. and Tremaine, S. (1979b). The excitation of density waves at the Lindblad and corotation resonances by an external potential. *Astrophys. J.*, 233:857–871.

Goldreich, P. and Tremaine, S. (1979c). Towards a theory for the Uranian rings. *Nature*, 277:97–99.

Goldreich, P. and Tremaine, S. (1980). Disk–satellite interactions. *Astrophys. J.*, 241:425–441.

Goldreich, P. and Tremaine, S. (1981). The origin of the eccentricities of the rings of Uranus. *Astrophys. J.*, 243:1062–1075.

Goldreich, P., Rappaport, N., and Sicardy, B. (1995). Single sided shepherding. *Icarus*, 118:414–417.

Gresh, D. L., Rosen, P. A., Tyler, G. L., and Lissauer, J. J. (1986). An analysis of bending waves in Saturn's rings using Voyager radio occultation data. *Icarus*, 68:481–502.

Gurnett, D. A., Grün, E., Gallagher, D., Kurth, W. S., and Scarf, F. L. (1983). Micron-sized particles detected near Saturn by the Voyager plasma wave instrument. *Icarus*, 53:236–254.

Hahn, J. M. (2007). The secular evolution of a close ring-satellite system: The excitation of spiral bending waves at a nearby gap edge. *Astrophys. J.*, 665:856–865.

Hahn, J. M. (2008). The secular evolution of a close ring-satellite system: The excitation of spiral density waves at a nearby gap edge. *Astrophys. J.*, 680:1569–1581.

Hahn, J. M., Spitale, J. N., and Porco, C. C. (2009). Dynamics of the sharp edges of broad planetary rings. *Astrophys. J.*, 699:686–710.

Hämeen-Anttila, K. A. (1975). Statistical mechanics of Keplerian orbits. *Astrophys. Space Sci.*, 37:309–333.

Hämeen-Anttila, K. A. (1976). Statistical mechanics of Keplerian orbits. II – Dispersion in particle size. *Astrophys. Space Sci.*, 43:145–174.

Hämeen-Anttila, K. A. (1977a). Statistical mechanics of Keplerian orbits. III – Perturbations. *Astrophys. Space Sci.*, 46:133–154.

Hämeen-Anttila, K. A. (1977b). Statistical mechanics of Keplerian orbits. IV. *Astrophys. Space Sci.*, 51:429–437.

Hämeen-Anttila, K. A. (1978). An improved and generalized theory for the collisional evolution of Keplerian systems. *Astrophys. Space Sci.*, 58:477–519.

Hämeen-Anttila, K. A. (1981). Quasi-equilibrium in collisional systems. *Earth Moon Planets*, 25:477–506.

Hämeen-Anttila, K. A. (1982). Saturn's rings and bimodality of Keplerian systems. *Earth Moon Planets*, 26:171–196.

Hämeen-Anttila, K. A. (1984). Collisional theory of non-identical particles in a gravitational field. *Earth Moon Planets*, 31:271–299.

Hämeen-Anttila, K. A. and Salo, H. (1993). Generalized theory of impacts in particulate systems. *Earth Moon Planets*, 62:47–84.

Hänninen, J. and Salo, H. (1994). Collisional simulations of satellite Lindblad resonances. 2: Formation of narrow ringlets. *Icarus*, 108:325–346.

Hänninen, J. and Salo, H. (1995). Formation of isolated narrow ringlets by a single satellite. *Icarus*, 117:435–438.

Hatzes, A., Bridges, F., Lin, D., and Sachtjen, S. (1991). Coagulation of particles in Saturn's rings: Measurements of the cohesive force of water frost. *Icarus*, 89:113–121.

Hatzes, A., Bridges, F. G., and Lin, D. N. C. (1988). Collisional properties of ice spheres at low impact velocities. *Mon. Not. R. Astron. Soc.*, 231:1091–1115.

Hedman, M. M., Nicholson, P. D., Salo, H., Wallis, B. D., Buratti, B. J., Baines, K. H., Brown, R. H., and Clark, R. N. (2007). Self-gravity wake structures in Saturn's A ring revealed by Cassini VIMS. *Astron. J.*, 133:2624–2629.

Hertz, H. (1881). Ueber die Berührung fester elastischer Körper. *J. f. Reine Angew. Math.*, 92:156.

Higa, M., Arakawa, M., and Maeno, N. (1996). Measurements of restitution coefficients of ice at low temperatures. *Planet. Space Sci.*, 44:917–925.

Higa, M., Arakawa, M., and Maeno, N. (1998). Size dependence of restitution coefficients of ice in relation to collision strength. *Icarus*, 133:310–320.

Holberg, J. B. (1982). Identification of 1980S27 and 1980S26 resonances in Saturn's A ring. *Astron. J.*, 87:1416–1422.

Holberg, J. B., Forrester, W. T., and Lissauer, J. J. (1982). Identification of resonance features within the rings of Saturn. *Nature*, 297:115–120.

Horn, L. J. and Cuzzi, J. N. (1996). Characteristic wavelengths of irregular structure in Saturn's B ring. *Icarus*, 119:285–310.

Housen, K. R. and Holsapple, K. A. (1990). On the fragmentation of asteroids and planetary satellites. *Icarus*, 84:226–253.

Ip, W.-H. (1983). Collisional interactions of ring particles – The ballistic transport process. *Icarus*, 54:253–262.

Johnson, K., Kendall, K., and Roberts, A. (1971). Surface energy and the contact of elastic solids. *Proc. R. Soc. Lond. A*, 324:301.

Julian, W. H. and Toomre, A. (1966). Non-axisymmetric responses of differentially rotating disks of stars. *Astrophys. J.*, 146:810–827.

Karjalainen, R. (2007). Aggregate impacts in Saturn's rings. *Icarus*, 189:523–537.

Karjalainen, R. and Salo, H. (2004). Gravitational accretion of particles in Saturn's rings. *Icarus*, 172:328–348.

Kawata, Y. (1983). Infrared brightness temperature of Saturn's rings based on the inhomogeneous multilayer assumption. *Icarus*, 56:453–464.

Kevorkian, J. and Cole, J. (1996). *Multiple Scale and Singular Perturbation Methods*. Springer.

Kolvoord, R. A., Burns, J. A., and Showalter, M. R. (1990). Periodic features in Saturn's F ring – Evidence for nearby moonlets. *Nature*, 345:695–697.

Lane, A., Hord, C., West, R., Esposito, L., Coffeen, D., Sato, M., Simmons, K., Pomphrey, R., and Morris, R. (1982). Photopolarimetry from Voyager 2: Preliminary results on Saturn, Titan, and the rings. *Science*, 215:537–543.

Latter, H. N. and Ogilvie, G. I. (2006). The linear stability of dilute particulate rings. *Icarus*, 184:498–516.

Latter, H. N. and Ogilvie, G. I. (2008). Dense planetary rings and the viscous overstability. *Icarus*, 195:725–751.

Latter, H. N. and Ogilvie, G. I. (2009). The viscous overstability, nonlinear wavetrains, and finescale structure in dense planetary rings. *Icarus*, 202:565–583.

Lewis, M. C. and Stewart, G. R. (2005). Expectations for Cassini observations of ring material with nearby moons. *Icarus*, 178:124–143.

Lewis, M. C. and Stewart, G. R. (2007). Collisional simulations of the F ring with Prometheus and Pandora. In *AAS/Division for Planetary Sciences Meeting Abstracts*, volume 39, 26.08.

Lewis, M. C. and Stewart, G. R. (2009). Features around embedded moonlets in Saturn's rings: The role of self-Gravity and particle size distributions. *Icarus*, 199:387–412.

Leyrat, C., Ferrari, C., Charnoz, S., Decriemb, J., Spilker, L. J., and Pilorz, S. (2008). Spinning particles in Saturn's C ring from midinfrared observations: Pre-Cassini mission results. *Icarus*, 196:625–641.

Lightman, A. P. and Eardley, D. M. (1974). Black holes in binary systems: Instability of disk accretion. *Astrophys. J. Lett.*, 187:L1–L3.

Lin, D. N. C. and Bodenheimer, P. (1981). On the stability of Saturn's rings. *Astrophys. J. Lett.*, 248:L83–L86.

Lissauer, J. J. (1984). Ballistic transport in Saturn's rings – an analytic theory. *Icarus*, 57:63–71.

Lissauer, J. J. (1985). Bending waves and the structure of Saturn's rings. *Icarus*, 62:433–447.

Lissauer, J. J., Shu, F. H., and Cuzzi, J. N. (1981). Moonlets in Saturn's rings. *Nature*, 292:707–711.

Lissauer, J. J., Goldreich, P., and Tremaine, S. (1985). Evolution of the Janus-Epimetheus coorbital resonance due to torques from Saturn's rings. *Icarus*, 64:425–434.

Longaretti, P.-Y. (1989). Saturn's main ring particle size distribution: An analytic approach. *Icarus*, 81:51–73.

Longaretti, P.-Y. and Borderies, N. (1986). Nonlinear study of the Mimas 5:3 density wave. *Icarus*, 67:211–223.

Longaretti, P.-Y. and Rappaport, N. (1995). Viscous overstabilities in dense narrow planetary rings. *Icarus*, 116:376–396.

Lukkari, J. (1981). Collisional amplification of density fluctuations in Saturn's rings. *Nature*, 292:433–435.

Lumme, K. and Irvine, W. M. (1976). Azimuthal brightness variations of Saturn's rings. *Astrophys. J. Lett.*, 204:L55–L57.

Lynden-Bell, D. and Kalnajs, A. (1972). On the generating mechanism of spiral structure. *Mon. Not. R. Astron. Soc.*, 157:1–30.

Marouf, E., Rappaport, N., French, R., McGhee, C., and Anabtawi, A. (2006). Azimuthal variability of radial structure of Saturn's rings observed by Cassini radio occultations. *36th COSPAR Scientific Assembly*, volume 36 of *COSPAR, Plenary Meeting*, pages 2806–+.

Marouf, E. A., Tyler, G. L., Zebker, H. A., Simpson, R. A., and Eshleman, V. R. (1983). Particle size distributions in Saturn's rings from Voyager 1 radio occultation. *Icarus*, 54:189–211.

Masset, F. S. and Papaloizou, J. C. B. (2003). Runaway migration and the formation of hot Jupiters. *Astrophys. J.*, 588:494–508.

Melita, M. D. and Papaloizou, J. C. B. (2005). Resonantly forced eccentric ringlets: Relationships between surface density, resonance location, eccentricity and eccentricity-gradient. *Cel. Mech. Dyn. Astron.*, 91:151–171.

Morfill, G. E., Fechtig, H., Grün, E., and Goertz, C. K. (1983). Some consequences of meteoroid impacts on Saturn's rings. *Icarus*, 55:439–447.

Morishima, R. and Salo, H. (2004). Spin rates of small moonlets embedded in planetary rings: I. Three-body calculations. *Icarus*, 167:330–346.

Morishima, R. and Salo, H. (2006). Simulations of dense planetary rings IV. Spinning self-gravitating particles with size distributions. *Icarus*, 181:272–291.

Morishima, R., Salo, H., and Ohtsuki, K. (2009a). A Multi-layer model for thermal infrared emission of Saturn's rings: Basic formulation and implications for Earth-based observations. *Icarus*, 201:634–654.

Morishima, R., Salo, H., Ohtsuki, K., Spilker, L., and Altobelli, N. (2009b). Dynamics and particle properties of Saturn's rings inferred from thermal infrared emission. Accepted to *Icarus*.

Mosqueira, I. (1996). Local simulations of perturbed dense planetary rings. *Icarus*, 122:128–152.

Mosqueira, I. and Estrada, P. R. (2002). Apse alignment of the Uranian rings. *Icarus*, 158:545–556.

Murray, C. D. (2007). F Ring objects and ring history. *AGU Fall Meeting Abstracts*.

Murray, C. D., Beurle, K., Cooper, N. J., Evans, M. W., Williams, G. A., and Charnoz, S. (2008). The determination of the structure of Saturn's F ring by nearby moonlets. *Nature*, 453:739–744.

Nakazawa, K. and Ida, S. (1988). Hill's approximation in the three-body problems. *Prog. Theor. Phys. Suppl.*, 96:167–174.

Nicholson, P. D., French, R. G., Campbell, D. B., Margot, J.-L., Nolan, M. C., Black, G. J., and Salo, H. J. (2005). Radar imaging of Saturn's rings. *Icarus*, 177:32–62.

Nicholson, P. D., Hedman, M. M., Clark, R. N., Showalter, M. R., Cruikshank, D. P., Cuzzi, J. N., Filacchione, G., Capaccioni, F., Cerroni, P., Hansen, G. B., Sicardy, B., Drossart, P., Brown, R. H., Buratti, B. J., Baines, K. H., and Coradini, A. (2008). A close look at Saturn's rings with Cassini VIMS. *Icarus*, 193:182–212.

Ohtsuki, K. (1992). Equilibrium velocities in planetary rings with low optical depth. *Icarus*, 95:265–282.

Ohtsuki, K. (1993). Capture probability of colliding planetesimals: Dynamical constraints on accretion of planets, satellites, and ring particles. *Icarus*, 106:228–246.

Ohtsuki, K. (1999). Evolution of particle velocity dispersion in a circumplanetary disk due to inelastic collisions and gravitational interactions. *Icarus*, 137:152–177.

Ohtsuki, K. (2004a). Formulation and analytic calculation for the spin angular momentum of a moonlet due to inelastic collisions of ring particles. *Earth Planets Space*, 56:909–919.

Ohtsuki, K. (2004b). On the rotation of a moonlet embedded in planetary rings. *Icarus*, 172:432–445.

Ohtsuki, K. (2005). Rotation rates of particles in Saturn's rings. *Astrophys. J.*, 626:L61–L64.

Ohtsuki, K. (2006a). Rotation rate and velocity dispersion of planetary ring particles with size distribution. I. Formulation and analytic calculation. *Icarus*, 183:373–383.

Ohtsuki, K. (2006b). Rotation rate and velocity dispersion of planetary ring particles with size distribution. II. Numerical simulation for gravitating particles. *Icarus*, 183:384–395.

Ohtsuki, K. and Emori, H. (2000). Local N-body simulations for the distribution and evolution of particle velocities in planetary rings. *Astron. J.*, 119:403–416.

Ohtsuki, K. and Toyama, D. (2005). Local N-body simulations for the rotation rates of particles in planetary rings. *Astron. J.*, 130:1302–1310.

Papaloizou, J. C. B. and Lin, D. N. C. (1988). On the pulsational overstability in narrowly confined viscous rings. *Astrophys. J.*, 331:838–860.

Papaloizou, J. C. B. and Melita, M. D. (2005). Structuring eccentric-narrow planetary rings. *Icarus*, 175:435–451.

Papaloizou, J. C. B., Nelson, R. P., Kley, W., Masset, F. S., and Artymowicz, P. (2007). Disk-planet interactions during planet formation. In Reipurth, B., Jewitt, D., and Keil, K., editors, *Protostars and Planets V*, pages 655–668.

Petit, J. M. and Hénon, M. (1987). A numerical simulation of planetary rings. I. Binary encounters. *Astron. Astrophys.*, 173:389–404.

Petit, J. M. and Hénon, M. (1988). A numerical simulation of planetary rings. III – Mass segregation, ring confinement, and gap formation. *Astron. Astrophys.*, 199:343–356.

Porco, C., Nicholson, P. D., Borderies, N., Danielson, G. E., Goldreich, P., Holberg, J. B., and Lane, A. L. (1984). The eccentric Saturnian ringlets at 1.29 R(s) and 1.45 R(s). *Icarus*, 60:1–16.

Porco, C., Thomas, P., Weiss, J. W., and Richardson, D. C. (2007). Saturn's small inner satellites: Clues to their origins. *Science*, 318:1602–1607.

Porco, C. C., Baker, E., Barbara, J., Beurle, K., Brahic, A., Burns, J. A., Charnoz, S., Cooper, N., Dawson, D. D., Del Genio, A. D., Denk, T., Dones, L., Dyudina, U., Evans, M. W., Giese, B., Grazier, K., Helfenstein, P., Ingersoll, A. P., Jacobson, R. A., Johnson, T. V., McEwen, A., Murray, C. D., Neukum, G., Owen, W. M., Perry, J., Roatsch, T., Spitale, J., Squyres, S., Thomas, P., Tiscareno, M., Turtle, E., Vasavada, A. R., Veverka, J., Wagner, R., and West, R. (2005). Cassini Imaging Science: Initial results on Saturn's rings and small satellites. *Science*, 307:1226–1236.

Porco, C. C., Weiss, J. W., Richardson, D. C., Dones, L., Quinn, T., and Throop, H. (2008). Simulations of the dynamical and light-scattering behavior of Saturn's rings and the derivation of ring particle and disk properties. *Astron. J.*, 136:2172–2200.

Rappaport, N. (1998). Ring–ringlet interactions. *Icarus*, 132:36–42.

Rappaport, N. J., Longaretti, P.-Y., French, R. G., Marouf, E. A., and McGhee, C. A. (2009). A procedure to analyze nonlinear density waves in Saturn's rings using several occultation profiles. *Icarus*, 199:154–173.

Resibois, P. and DeLeener, M. (1977). *Classical Kinetic Theory of Fluids*. Wiley, New York.

Richardson, D. (1994). Tree code simulations of planetary rings. *Mon. Not. R. Astron. Soc.*, 269:493–511.

Robbins, S. J., Stewart, G. R., Lewis, M. C., Colwell, J. E., and Sremčević, M. (2009). Estimating the masses of Saturn's A and B rings from high-optical depth N-body simulations and stellar ocuultations. Submitted to *Icarus*.

Roche, R. A. (1847). Acad. des scis et Lettres de Montpelier. *Mem. de la Section des Sciences*, 1:243–262.

Rosen, P. A. and Lissauer, J. J. (1988). The Titan-1:0 nodal bending wave in Saturn's ring C. *Science*, 241:690–694.

Rosen, P. A., Tyler, G. L., and Marouf, E. A. (1991a). Resonance structures in Saturn's rings probed by radio occultation. I – Methods and examples. *Icarus*, 93:3–24.

Rosen, P. A., Tyler, G. L., Marouf, E. A., and Lissauer, J. J. (1991b). Resonance structures in Saturn's rings probed by radio occultation. II – Results and interpretation. *Icarus*, 93:25–44.

Rubincam, D. P. (2006). Saturn's rings, the Yarkovsky effects, and the ring of fire. *Icarus*, 184:532–542.

Safronov, V. S. (1969). Evolution of the protoplanetary cloud and the formation of the earth and planets. *Nauka*, NASA TTF-677.

Salo, H. and Lukkari, J. (1982). Self-gravitation in Saturn's rings. *Moon Planets*, 27:5–12.

Salo, H. (1987a). Collisional evolution of rotating, non-identical particles. *Moon Planets*, 38:149–181.

Salo, H. (1987b). Numerical simulations of collisions between rotating particles. *Icarus*, 70:37–51.

Salo, H. (1991). Numerical simulations of dense collisional systems. *Icarus*, 92:367–368.

Salo, H. (1992a). Gravitational wakes in Saturn's rings. *Nature*, 359:619–621.

Salo, H. (1992b). Numerical simulations of dense collisional systems. II. Extended distribution of particle sizes. *Icarus*, 96:85–106.

Salo, H. (1995). Simulations of dense planetary rings. III. Self-gravitating identical particles. *Icarus*, 117:287–312.

Salo, H. (2001). Numerical simulations of the collisional dynamics of planetary rings. In Pöschel, T. and Luding, S., editors, *Granular Gases*, volume 564 of *Lecture Notes in Physics, Berlin Springer Verlag*, pages 330–349.

Salo, H., Schmidt, J., and Spahn, F. (2001). Viscous overstability in Saturn's B ring: I. Direct simulations and measurement of transport coefficients. *Icarus*, 153:295–315.

Salo, H. and Karjalainen, R. (2003). Photometric modeling of Saturn's rings. I. Monte Carlo method and the effect of nonzero volume filling factor. *Icarus*, 164:428–460.

Salo, H., Karjalainen, R., and French, R. G. (2004). Photometric modeling of Saturn's rings. II. Azimuthal asymmetry in reflected and transmitted light. *Icarus*, 170:70–90.

Salo, H. J. and Schmidt, J. (2007). Release of impact-debris in perturbed ring regions: Dynamical and photometric simulations. In *Bulletin of the American Astronomical Society*, volume 38 of *Bulletin of the American Astronomical Society*, pages 425–+.

Salo, H. and Schmidt, J. (2009). N-body simulations of viscous instability of planetary rings. Accepted by *Icarus*.

Schmidt, J. and Salo, H. (2003). A weakly nonlinear model for viscous overstability in Saturn's dense rings. *Phys. Rev. Lett.*, 90(6):061102.

Schmit, U. and Tscharnuter, W. (1995). A fluid dynamical treatment of the common action of self-gravitation, collisions, and rotation in Saturn's B-ring. *Icarus*, 115:304–319.

Schmit, U. and Tscharnuter, W. (1999). On the formation of the fine-scale structure in Saturn's B ring. *Icarus*, 138:173–187.

Schmidt, J., Salo, H., Spahn, F., and Petzschmann, O. (2001). Viscous overstability in Saturn's B ring: II. Hydrodynamic theory and comparison to simulations. *Icarus*, 153:316–331.

Seiß, M., Spahn, F., Sremčević, M., and Salo, H. (2005). Structures induced by small moonlets in Satrun's rings: Implications for the cassini mission. *Geophys. Res. Lett.*, 32:11025.

Shan, L.-H. and Goertz, C. K. (1991). On the radial structure of Saturn's B ring. *Astrophys. J.*, 367:350–360.

Shepelyansky, D. L., Pikovsky, A. S., Schmidt, J., and Spahn, F. (2009). Synchronization mechanism of sharp edges in rings of Saturn. *Mon. Not. R. Astron. Soc.*, 395:1934–1940.

Showalter, M., Pollack, J., Ockert, M., Doyle, L., and Dalton, J. (1992). A photometric study of saturn's f ring. *Icarus*, 100:394–411.

Showalter, M. R. (1991). Visual detection of 1981S13, Saturn's eighteenth satellite, and its role in the Encke gap. *Nature*, 351:709–713.

Showalter, M. R. (1998). Detection of centimeter-sized meteoroid impact events in Saturn's F ring. *Science*, 282:1099–1102.

Showalter, M. R. and Burns, J. A. (1982). A numerical study of Saturn's F-ring. *Icarus*, 52:526–544.

Showalter, M. R. and Cuzzi, J. N. (1993). Seeing ghosts: Photometry of Saturn's G ring. *Icarus*, 103:124–143.

Showalter, M. R. and Nicholson, P. D. (1990). Saturn's rings through a microscope: Particle size constraints from the Voyager PPS Scan. *Icarus*, 87:285–306.

Showalter, M. R., Cuzzi, J. N., Marouf, E. A., and Esposito, L. W. (1986). Satellite "wakes" and the orbit of the Encke Gap moonlet. *Icarus*, 66:297–323.

Shu, F. H. (1984). Waves in planetary rings. In Greenberg, R. and Brahic, A., editors, *Planetary Rings*, pages 513–561, Tucson Arizona, University of Arizona Press.

Shu, F. H., Cuzzi, J. N., and Lissauer, J. J. (1983). Bending waves in Saturn's rings. *Icarus*, 53:185–206.

Shu, F. H. and Stewart, G. R. (1985). The collisional dynamics of particulate disks. *Icarus*, 62:360–383.

Shu, F. H., Yuan, C., and Lissauer, J. J. (1985b). Nonlinear spiral density waves – An inviscid theory. *Astrophys. J.*, 291:356–376.

Shu, F. H., Dones, L., Lissauer, J. J., Yuan, C., and Cuzzi, J. N. (1985a). Nonlinear spiral density waves – Viscous damping. *Astrophys. J.*, 299:542–573.

Shukhman, I. (1984). Collisional dynamics of particles in Saturn's rings. *Sov. Astron.*, 28:574–585.

Smith, B. A., Soderblom, L., Batson, R., Bridges, P., Inge, J., Masursky, H., Shoemaker, E., Beebe, R., Boyce, J., Briggs, G., Bunker, A., Collins, S. A., Hansen, C. J., Johnson, T. V., Mitchell, J. L., Terrile, R. J., Cook, A. F., Cuzzi, J., Pollack, J. B., Danielson, G. E., Ingersoll, A., Davies, M. E., Hunt, G. E., Morrison, D., Owen, T., Sagan, C., Veverka, J., Strom, R., and Suomi, V. E. (1982). A new look at the Saturn system: The Voyager 2 images. *Science*, 215:505–537.

Smoluchowski, R. (1979). The rings of Jupiter, Saturn and Uranus. *Nature*, 280:377–378.

Spahn, F., Albers, N., Sremčević, M., and Thornton, C. (2004). Kinetic description of coagulation and fragmentation in its granular particle ensembles. *Europhys. Lett.*, 67:545–551.

Spahn, F. and Sponholz, H. (1989). Existence of moonlets in Saturn's rings inferred from the optical depth profile. *Nature*, 339:607–608.

Spahn, F. and Sremčević, M. (2000). Density patterns induced by small moonlets in Saturn's rings? *Astron. Astrophys.*, 358:368–372.

Spahn, F. and Wiebicke, H. J. (1989). Long-term gravitational influence of moonlets in planetary rings. *Icarus*, 77:124–134.

Spahn, F., Greiner, J., and Schwarz, U. (1992a). Moonlets in Saturn's rings. *Advances Space Res.*, 12:141–147.

Spahn, F., Saar, A., Schmidt, S., and Schwarz, U. (1992b). The influence of various moonlets on the optical depth profile in planetary rings. *Icarus*, 100:143–153.

Spahn, F., Petit, J. M., and Bendjoya, P. (1993). The gravitational influence of satellite Pan on the radial distribution of ring-particles in the region of the Encke-division in Saturn's a ring. *Celest. Mech. Dyn. Astron.*, 57:391–402.

Spahn, F., Scholl, H., and Hertzsch, J. M. (1994). Structures in planetary rings caused by embedded moonlets. *Icarus*, 111:514–535.

Spahn, F., Schmidt, J., Petzschmann, O., and Salo, H. (2000). Stability analysis of a Keplerian disk of granular grains: Influence of thermal diffusion. *Icarus*, 145:657–660.

Spilker, L., Ferrari, C., Cuzzi, J. N., Showalter, M., Pearl, J., and Wallis, B. (2003). Saturn's rings in the thermal infrared. *Planet. Space Sci.*, 51:929–935.

Spilker, L. J., Pilorz, S., Lane, A. L., Nelson, R. M., Pollard, B., and Russell, C. T. (2004). Saturn A ring surface mass densities from spiral density wave dispersion behavior. *Icarus*, 171:372–390.

Spilker, L. J., Pilorz, S. H., Wallis, B. D., Pearl, J. C., Cuzzi, J. N., Brooks, S. M., Altobelli, N., Edgington, S. G., Showalter, M., Flasar, F. M., Ferrari, C., and Leyrat, C. (2006). Cassini thermal observations of Saturn's main rings: Implications for particle rotation and vertical mixing. *Planet. Space Sci.*, 54:1167–1176.

Sremčević, M., Spahn, F., and Duschl, W. (2002). Density structures in perturbed thin cold disks. *Mon. Not. R. Astron. Soc.*, 337:1139–1152.

Sremčević, M., Schmidt, J., Salo, H., Seiß, M., Spahn, F., and Albers, N. (2007). A belt of moonlets in Saturn's A ring. *Nature*, 449:1019–1021.

Stewart, G. R. and Lewis, M. C. (2005). Why Saturn's F ring breaks up into multiple ringlets. In *Bulletin of the American Astronomical Society*, volume 37, page 528.

Stewart, G. R., Lin, D. N. C., and Bodenheimer, P. (1984). Collision-induced transport processes in planetary rings. In Greenberg, R. and Brahic, A., editors, *Planetary Rings*, pages 447–512, Tucson Arizona, University of Arizona Press.

Supulver, K. D., Bridges, F. G., and Lin, D. N. C. (1995). The coefficient of restitution of ice particles in glancing collisions: Experimental results for unfrosted surfaces. *Icarus*, 113:188–199.

Supulver, K. D., Bridges, F. G., Tiscareno, S., Lievore, J., and Lin, D. N. C. (1997). The sticking properties of water frost produced under various ambient conditions. *Icarus*, 129:539–554.

Tanaka, H., Inaba, S., and Nakazawa, K. (1996). Steady-state size distribution for the self-similar collision cascade. *Icarus*, 123:450–455.

Tanaka, H., Ohtsuki, K., and Daisaka, H. (2003). A new formulation of the viscosity in planetary rings. *Icarus*, 161:144–156.

Thompson, W. T., Lumme, K., Irvine, W. M., Baum, W. A., and Esposito, L. W. (1981). Saturn's rings – Azimuthal variations, phase curves, and radial profiles in four colors. *Icarus*, 46:187–200.

Thomson, F. S., Marouf, E. A., Tyler, G. L., French, R. G., and Rappoport, N. J. (2007). Periodic microstructure in Saturn's rings A and B. *Gephys. Res. Lett.*, 34:L24203.1–L24203.6

Tiscareno, M. S., Nicholson, P. D., Burns, J. A., Hedman, M. M., and Porco, C. C. (2006a). Unravelling temporal variability in Saturn's spiral density waves: Results and predictions. *Astrophys. J. Lett.*, 651:L65–L68.

Tiscareno, M. S., Burns, J. A., Hedman, M. M., Porco, C. C., Weiss, J. W., Murray, C. D., and Dones, L. (2006b). Observation of "propellers" indicates 100-metre diameter moonlets reside in Saturn's a-ring. *Nature*, 440:648–650.

Tiscareno, M. S., Burns, J. A., Nicholson, P. D., Hedman, M. M., and Porco, C. C. (2007). Cassini imaging of Saturn's rings: II. A wavelet technique for analysis of density waves and other radial structure in the rings. *Icarus*, 189:14–34.

Tiscareno, M. S., Burns, J. A., Hedman, M. M., and Porco, C. C. (2008). The population of propellers in Saturn's A ring. *Astron. J.*, 135:1083–1091.

Toomre, A. (1964). On the gravitational stability of a disk of stars. *Astrophys. J.*, 139:1217–1238.

Toomre, A. (1981). What amplifies the spirals. In Fall, S. M. and Lynden-Bell, D., editors, *Structure and Evolution of Normal Galaxies*, pages 111–136.

Toomre, A. (1991). Gas-hungry Sc spirals. In Wielen, R., editor, *Dynamics and Interactions of Galaxies*, pages 292–303. Springer, Berlin.

Toomre, A. and Kalnajs, A. J. (1991). Spiral chaos in an orbiting patch. In Sundelius, B., editor, *Dynamics of Disc Galaxies*, pages 341–358. Almquist-Wiksell, Göteborg.

Tremaine, S. (2003). On the origin of irregular structure in Saturn's rings. *Astron. J.*, 125:894–901.

Trulsen, J. (1972). Numerical simulation of jetstreams, 1: The three-dimensional case. *Astrophys. Space Sci.*, 17:241–262.

Tyler, G. L., Marouf, E. A., Simpson, R. A., Zebker, H. A., and Eshleman, V. R. (1983). The microwave opacity of Saturn's rings at wavelengths of 3.6 and 13 cm from Voyager 1 radio occultation. *Icarus*, 54:160–188.

Van Allen, J. A. (1983). Absorption of energetic protons by Saturn's ring G. *J. Geophys. Res.*, 88:6911–6918.

Vokrouhlický, D., Nesvorný, D., Dones, L., and Bottke, W. F. (2007). Thermal forces on planetary ring particles: Application to the main system of Saturn. *Astron. Astrophys.*, 471:717–730.

Ward, W. R. (1981). On the radial structure of Saturn's rings. *Geophys. Res. Lett.*, 8:641–643.

Ward, W. R. and Cameron, A. G. W. (1978). Disc evolution within the Roche limit. In *Lunar and Planetary Institute Conference Abstracts*, volume 9 of *Lunar and Planetary Inst. Technical Report*, pages 1205–1207.

Weidenschilling, S. J., Chapman, C. R., Davis, D. R., and Greenberg, R. (1984). Ring particles: Collisional interactions and physical nature. In Greenberg, R. and Brahic, A., editors, *Planetary Rings*, pages 367–415, Tucson Arizona, University of Arizona Press.

Weidenschilling, S. J., Spaute, D., Davis, D. R., Marzari, F., and Ohtsuki, K. (1997). Accretional evolution of a planetesimal swarm. 2. The terrestrial zone. *Icarus*, 128:429–455.

Wetherill, G. W. and Stewart, G. R. (1993). Formation of planetary embryos: Effects of fragmentation, low relative velocity, and independent variation of eccentricity and inclination. *Icarus*, 106:190–204.

Williams, D. and Wetherill, G. (1994). Size distribution of collisionally evolved asteroidal populations: Analytical solution for self-similar collision cascade. *Icarus*, 107:117–128.

Winter, O. C., Mourão, D. C., Giuliatti Winter, S. M., Spahn, F., and da Cruz, C. (2007). Moonlets wandering on a leash-ring. *Mon. Not. R. Astron. Soc.*, 380:L54–L57.

Wisdom, J. (1980). The resonance overlap criterion and the onset of stochastic behavior in the restricted three-body problem. *Astron. J.*, 85:1122–1133.

Wisdom, J. and Tremaine, S. (1988). Local simulations of planetary rings. *Astron. J.*, 95:925–940.

Zebker, H. A., Marouf, E. A., and Tyler, G. L. (1985). Saturn's rings: Particle size distributions for thin layer models. *Icarus*, 64:531–548.

Chapter 15
Ring Particle Composition and Size Distribution

Jeff Cuzzi, Roger Clark, Gianrico Filacchione, Richard French, Robert Johnson, Essam Marouf, and Linda Spilker

Abstract We review recent progress concerning the composition and size distribution of the particles in Saturn's main ring system, and describe how these properties vary from place to place. We discuss how the particle size distribution is measured, and how it varies radially. We note the discovery of unusually large "particles" in restricted radial bands. We discuss the properties of the grainy regoliths of the ring particles. We review advances in understanding of ring particle composition from spectrophotometry at UV, visual and near-IR wavelengths, multicolor photometry at visual wavelengths, and thermal emission. We discuss the observed ring atmosphere and its interpretation and, briefly, models of the evolution of ring composition. We connect the ring composition with what has been learned recently about the composition of other icy objects in the Saturn system and beyond. Because the rings are so thoroughly and rapidly structurally evolved, the composition of the rings may be our best clue as to their origin; however, the evolution of ring particle composition over time must first be understood.

15.1 Introduction

Cassini will revolutionize our understanding of the composition and size distribution of the particles making up Saturn's main rings. We say "will" with confidence, because only a fraction of the relevant data obtained by Cassini during its 4-year prime mission has actually been thoroughly analyzed as of this writing. At this time, only very broad regional averages of various observable properties have been looked at; characterizing the most opaque regions, or increasing radial resolution, will require more sophisticated data analysis; also, instrumental calibration remains in flux to some degree. In this chapter, we provide a sense of the direction indicated by the current sample of newly analyzed data. We will emphasize Cassini results (comparing them with Earth-based results that have not been reviewed previously). A very thorough pre-spacecraft historical review is provided by Pollack (1975). A series of extensive reviews covering ring particle composition and size from the Voyager era includes Cuzzi et al. (1984) and Esposito et al. (1984). Some post-Voyager reviews that include more recent work include Cuzzi (1995), Dones (1998) and Cuzzi et al. (2002). A short meta-review of the pre-Cassini status may be found in this volume (Chapter 2).

The composition and size distribution of the particles in Saturn's main rings are tied together from the observational standpoint, and both are key factors in any serious modeling of the origin and evolution of the main rings (Chapter 17). The fact that the ring composition evolves with time is a fairly recent insight; the particle composition can change as particles are irradiated by photons, bombarded by magnetospheric and/or ionospheric particles or primitive interplanetary meteoroids, or perhaps as they interact chemically or mineralogically with their locally produced oxygen atmosphere. For instance, it has long been known that water ice constitutes the bulk of the ring material (Chapter 2); however the rings are noticeably red at visual wavelengths, manifesting the presence of another substance. Moreover, interplanetary debris is primarily non-icy material – silicates and carbon-rich organics – so the rings become increasingly "polluted" over their lifetime. The degree to which this happens depends critically on the local particle size and surface mass density. Only by understanding the evolutionary processes that transpire in the rings can we look back from their current state to infer their primordial state, and thus get a clue

J. Cuzzi
Ames Research Center, Moffett Field, CA

R. Clark
United States Geological Survey, Denver, CO

G. Filacchione
Istituto di Astrofisica Spaziale e Fisica Cosmica, Rome

R. French
Wellesley College, Wellesley, MA

R. Johnson
University of Virginia, Charlottesville, VA

E. Marouf
San Jose State University, San Jose, CA

L. Spilker
Jet Propulsion Laboratory, Pasadena CA

as to their provenance (Chapter 17). The way in which the ring composition is observed to vary with local ring properties will provide important evidence that will allow us to understand and unravel this evolution.

Our chapter is divided into six sections. In Section 15.2, we review the size distribution of the particles in the main rings, sketching several methods by which particle size distributions are inferred. Ring particles range in size from a centimeter to meters or perhaps tens of meters, and the particle size distribution (in particular the abundance of 1–10 cm particles) changes dramatically across the rings. We comment on the distinction between "particles" and transient, dynamical entities composed of particles. In Section 15.3 we briefly describe Cassini's discovery of an entirely new class of "particles"– objects hundreds of meters across which make their presence known only by their disturbance of surrounding material, and summarize their implications for the ring mass. In Section 15.4, we discuss what we have learned about the composition of the particles in the main rings, primarily from remote sensing spectroscopy and photometry. The rings are composed almost entirely of water ice – in its crystalline phase and of unusual purity – but the puzzle of the reddening material – the so-called "UV absorber" has perhaps even deepened. A combination of laboratory studies of icy mixtures, theoretical models, and analogies with other icy objects are employed to interpret these observations. In Section 15.5, we discuss two possible evolutionary influences on ring composition. The ring atmosphere was newly characterized by Cassini to be composed not of the expected water products (OH and H), but of O atoms and O_2 molecules. Such chemically reactive molecules might play a role in the compositional evolution of the rings. In this section we also briefly describe some of the issues related to compositional evolution by meteoroid bombardment, deferring to the Chapter 17 for the details. Finally, in Section 15.6, we broaden the discussion connecting ring composition to ring provenance, comparing the properties of the main ring material to those of Saturn's icy moons, icy moons of other systems, and icy and non-icy outer solar system objects.

15.2 Ring Particle Size Distribution

Ring particle size information is captured in observations of the interaction of electromagnetic radiation with the ring material. In general, the particle sizes, shape, composition and spatial distribution (clustering, packing, and spread normal to the mean ring plane) control the manner in which the electromagnetic radiation is extinguished and scattered in all spatial directions. The size information is usually captured, along with the other physical properties, in several types of Earth-based and spacecraft observations, which we review here.

Readers interested only in results rather than methods can skip to Sections 15.2.8 and 15.2.9.

Radio and stellar ring occultations provide two especially sensitive ways to determine ring particle sizes, because of their near-forward-scattering observation geometry. The first is direct measurement of the extinction of the incident electromagnetic radiation passing through the rings and hence oblique optical depth, a parameter especially sensitive to the particle sizes relative to the radiation wavelength. The second is indirect measurement of the near-forward scattering pattern; that is, of the collective diffraction-lobe. The lobe shape and width are primarily controlled by the particle size distribution and are relatively insensitive to particle composition and shape.

15.2.1 Models and Theory

Modeling the interaction of electromagnetic radiation with the rings has been traditionally based on the so-called "classical" model. The model is rooted in radiative-transfer-like approaches to the electromagnetic interaction problem, where the particles are assumed to be uniformly distributed in a loosely-packed, extended layer many-particles-thick. No particle clustering is assumed, although individual large particles can be thought of as ephemeral aggregates of densely packed smaller particles (Marouf et al. 1982, 1983; Tyler et al. 1983).

Dynamical simulations and observations provide compelling evidence for the prevalence of "gravitational wakes", or extended transient structures which form by virtue of the self-gravity of the ring particles, nearly across the full extent of Rings A and B (Chapters 13 and 14). Particles within the wakes form chains of spatially correlated canted and elongated clusters, invalidating basic assumptions of the classical model. Extension of the electromagnetic interaction problem to include wake models is an ongoing endeavor. We base the discussion below on the classical model for the lack of a better electromagnetic interaction model at this time and to enable comparison with results of previous published work. Preliminary results regarding the impact of wakes on extinction and forward scattering observations are discussed briefly in Section 15.2.10.

For the classical ring model, the normal optical depth τ and its oblique value τ_q are related by $\tau_q = \tau/\mu_0$, $\mu_0 = \sin(|B|)$, and B is the ring opening angle (the angle between the planet-observer line and the ring plane). The optical depth τ is given by

$$\tau(\lambda) = \int_0^\infty \pi a^2 Q_e(a,\lambda) n(a) \, da \quad (15.1)$$

where $Q_e(a,\lambda)$ is the extinction efficiency of a spherical particle of radius a, λ is the radiation wavelength, and $n(a)$ is the size distribution (particles m^{-2} m^{-1}). Dependence on the particle composition is implicit in Q_e. In principle, measurements of $\tau(\lambda)$ at several λ may be used to invert the integral equation to recover $n(a)$. Alternatively, parameters of an assumed model of $n(a)$ may be constrained by matching predictions based on the integral above to the observed $\tau(\lambda)$. A commonly adopted model is the power-law model, where

$$n(a) = n_0 \, (a/a_0)^{-q}, \quad a_{\min} \leq a \leq a_{\max}, \quad (15.2)$$

and is zero otherwise. It is characterized by the minimum radius $a_{\min}$, the maximum radius $a_{\max}$, the value n_0 at an arbitrary reference radius a_0, and the power-law index q.

When τ is small, multiple scattering effects can be neglected and the single scattered near-forward signal intensity $I_1(\theta, \lambda)$ relative to the "free-space" incident power per unit area I_i can be approximated by

$$I_1(\theta,\lambda)/I_i = \frac{e^{-\tau/\mu_0}}{4\pi\mu_0} \int_{a_c}^{\infty} \pi a^2 \left[\frac{2J_1(ka \sin\theta)}{\sin\theta}\right]^2 n(a)\, da \quad (15.3)$$

where $k = 2\pi/\lambda$, J_1 is the Bessel function of first kind and order 1, a_c is a lower bound on the radius a of particles effectively contributing to the scattered signal, and θ is the scattering angle. The particles are assumed large compared to the wavelength ($ka > ka_c \gg 1$). Here too, the size distribution $n(a)$ may be recovered from the measured $I_1(\theta, \lambda)$ using integral inversion. Alternatively, parameters of an assumed power-law model of $n(a)$ may be constrained by matching computed values of the right-hand side to the observed collective diffraction pattern $I_1(\theta, \lambda)/I_i$. The approach applies equally to λ's in the ultraviolet, visible, infrared, and radio spectral regions.

For realistic optical depths of order unity, the effects of multiple scattering on the observed near-forward scattered signal $I_s(\theta, \lambda)/I_i$ must first be deconvolved to recover the single scattered component $I_1(\theta, \lambda)/I_i$. In the case of scattering by particles of optical size $ka \gg 1$, and assuming a classical ring model, it is possible to express $I_s(\theta, \lambda)/I_i$ as the sum of terms each representing a distinct order of scattering (Marouf et al. 1982, 1983)

$$I_s(\theta,\lambda)/I_i = \sum_{n=1}^{\infty} I_n(\theta,\lambda)/I_i$$

$$= \sum_{n=1}^{\infty} \left[\frac{1}{n!} \tau_q^n e^{-\tau_q}\right] \left[\frac{1}{4\pi}\varpi_0\Phi(\theta)\right]^{*n} \quad (15.4)$$

where $\Phi(\theta)$ is the particle phase function and ϖ_0 is the single scattering albedo (the ratio of the particle's scattering and extinction cross-sections). $\Phi(\theta)/4\pi$ is normalized to unity over 4π solid angle, and the symbol $[.]^{*n}$ denotes convolution of the term with itself n times. When $\tau_q = \tau/\mu_0 \ll 1$, the $n=1$ term (single scattering) dominates, hence Eqs. 15.3 and 15.4 imply

$$I_s(\theta,\lambda)/I_i \cong I_1(\theta,\lambda)/I_i = \frac{\tau e^{-\tau/\mu_0}}{4\pi\mu_0}\varpi_0\Phi(\theta) \quad (15.5)$$

which, when compared with Eqs. 15.1 and 15.3, defines $\varpi_0\Phi(\theta)$ in terms of the particle size distribution $n(a)$ to be

$$\varpi_0\Phi(\theta) = \frac{\int_{a_c}^{\infty} \pi a^2 \left[\frac{2J_1(ka\sin\theta)}{\sin\theta}\right]^2 n(a)\, da}{\int_0^{\infty} \pi a^2 Q_e(a,\lambda) n(a)\, da} \quad (15.6)$$

Terms of the infinite series in Eq. 15.4 can also be interpreted as the sum of probabilistic events. An n-th order scattering event occurs with Poisson distribution of parameter τ_q. After each interaction, the radiation is scattered (not absorbed) with probability ϖ_0. After a single interaction, the probability density function of the scattered energy emerging in any given direction θ is $\Phi(\theta)/4\pi$. After n independent interactions, the density function is the convolution of $\Phi(\theta)/4\pi$ with itself n times, which is denoted by the symbol $[.]^{*n}$ in the infinite sum above. This multiple scattering formulation leads to an infinite number of interactions, albeit with rapidly decreasing probability for $n > \tau_q$.

Rings are not many particles thick (see Chapter 14), and the number of interactions as the incident radiation crosses a ring of relatively small vertical extent around the mean ring plane is likely to be limited to some upper limit N. Replacing the Poisson distribution above by a binomial distribution of parameter p yields the alternative formulation (Zebker et al. 1985)

$$I_s(\theta,\lambda)/I_i = \sum_{n=1}^{N} \left[\binom{N}{n} p^n (1-p)^{N-n}\right] \left[\frac{1}{4\pi}\varpi_0\Phi(\theta)\right]^{*n} \quad (15.7)$$

Since p represents the probability of a single interaction, the incident radiation emerges without any interactions with probability $(1-p)^N = \exp(-\tau_q)$, hence,

$$p = (1 - e^{-\tau_q/N}) \quad (15.8)$$

For self-consistency, Eq. 15.1 for the classical optical depth now assumes the form

$$\tau(\lambda) = -2\mu_0 N \, \ln\left[1 - \frac{1}{2\mu_0 N}\int_0^{\infty} \pi a^2 Q_e(a,\lambda) n(a)\, da\right] \quad (15.9)$$

which reduces to the classical form when N is large. Although still "classical" in its basic assumptions regarding mutual particle interactions and uniform spatial distribution, the finite N model provides an additional degree offreedom to better match the observations. The model is referred to as the thin-layers model (Zebker et al. 1985). Both the classical and the thin-layer models above allow closed form summation of the order of scattering terms in the Hankel transform domain and subsequent recovery of the single scattered component $I_1(\theta, \lambda)/I_i$ from the measured $I_s(\theta, \lambda)/I_i$, an important first step for recovery of $n(a)$ from Eq. 15.3 (see Marouf et al. 1982, 1983, Zebker et al. 1985).

15.2.2 Cassini RSS Extinction Observations

The Cassini Radio Science Subsystem (RSS) ring occultations are conducted using three simultaneously transmitted microwave frequencies. The corresponding wavelengths (λ) are 0.94, 3.6, and 13.0 cm, and the corresponding microwave bands are Ka-, X-, and S-bands, respectively. The sole ring occultation before Cassini was conducted by the Voyager 1 spacecraft in 1980 using dual-frequency (X-S; Tyler et al. 1983). As of the end of 2008, Cassini had completed 28 one-sided 3-frequency ring occultations.

The nearly pure sinusoidal signals are generated from a common ultra-stable oscillator on board Cassini (Kliore et al. 2004). The phase coherence of the signals allows measurement of the amplitude, frequency, and phase of the sinusoids after they are perturbed by ring material. Although the amplitude measurements by themselves are diffraction limited, the phase information allows reconstruction of the observations to remove diffraction effects, providing optical depth profiles of ring structure with radial resolution approaching few hundred meters (Marouf et al. 1986).

Reconstructed RSS normal optical depth (τ) profiles of Rings C, B, and A are depicted in the upper panels of Figs. 15.1 a–c, respectively (Marouf et al. 2008a). The profiles are from the first Cassini diametric radio occultation (Rev 7 ingress) completed on March 3, 2005 at a ring opening angle $|B|=23.6°$. The blue, green, and red profiles correspond to 0.94, 3.6, and 13.0 cm $-\lambda$ (Ka-, X-, and S-band), respectively, averaged to a radial resolution of 10 km. The lower dashed horizontal line in each panel identifies the baseline $\tau = 0$, while the upper one (when it falls within the plotted limits) identifies the optical depth level at which the measurements are noise-limited (the so-called "threshold" optical depth τ_{TH}; Marouf et al. 1986). The different signal power at each frequency transmitted by Cassini, and the different noise temperature of the various groundbased receiving systems, combine to cause τ_{TH} to be wavelength dependent, as Fig. 15.1b shows, with S-band being the noisiest and X-band the least noisy profile for the same radial resolution (Marouf et al. 2008a).

As discussed above, reliable measurements of $\tau(\lambda)$ provide information about the particle size distribution $n(a)$.

The normalized differential profiles $[\tau(X) - \tau(S)]/\tau(X) \equiv \Delta\tau(XS)/\tau(X)$ and $[\tau(K) - \tau(X)]/\tau(X) \equiv \Delta\tau(KX)/\tau(X)$ are particularly suited to achieving this objective (Marouf et al. 1983, Zebker et al. 1985). Differential profiles corresponding to $\tau(\lambda)$ in each of the upper panels of Fig. 15.1 are shown in the corresponding lower panel. To reduce statistical scatter, the $\Delta\tau/\tau$ profiles were smoothed to a relatively coarse resolution of 80 km. In addition, the plotted points were restricted to values $0.05 < \tau(\lambda) < 3$. The lower bound ensures reliably detectable $\tau(\lambda)$ and the upper one ensures $\tau(\lambda) < \tau_{TH}$ for all three signals.

Figure 15.1 reveals remarkably rich $\Delta\tau/\tau$ variability among and within the three main ring regions: C, B, and A. A relatively large $\Delta\tau(XS)/\tau(X) = \sim 30–35\%$ characterizes almost the full extent of the tenuous background structure of Ring C, with possibly systematic decreasing trend as radius increases. Deviations from the background values within the denser "plateaus" in the outer C ring are likely to be real ($\sim$86,000–90,600 km). Different $\tau(\lambda)$ variations are evident within individual dense ringlets in the upper panel of Fig. 15.1a. Voyager X-S observations indicated similar behavior, albeit at coarser radial resolution (Tyler et al. 1983). Clearly detectable $\Delta\tau(KX)/\tau(X)$ approaching 10% characterizes the tenuous wavy region in mid Ring C ($\sim$78,000 to 86,000 km). Little mean $\Delta\tau(KX)/\tau(X)$ is evident over the plateau region (84,500–90,500 km). Although the $\tau(\lambda)$ profiles of the outer Ring C ramp ($\sim$90,600 to 92,000 km) show systematically increasing $\Delta\tau(KX)$ and $\Delta\tau(XS)$ with radius, when normalized by the also increasing $\tau(X)$ with radius, the normalized differential has nearly constant $\Delta\tau(KX)/\tau(X) = \sim$9% and $\Delta\tau(XS)/\tau(X) = \sim$30%.

Much less evident differential extinction of the three radio signals characterizes two of the four main regions of Ring B, provisionally identified as regions B1 to B4 (Marouf et al. 2008a). As Fig. 15.1b shows, the two are regions B2 ($\sim$99,000–104,100 km) and B4 ($\sim$110,000–117,500 km). Region B3 ($\sim$104,100–110,000 km; the "core" of Ring B) yields mostly noise-limited measurements at all three frequencies (except for a few narrow "lanes" of smaller optical depth). In sharp contrast, the innermost region of Ring B, region B1 ($\sim$92,000–99,000 km), exhibits clearly detectable $\Delta\tau(XS)/\tau(X)$ over most of its extent, including the two relatively "flat" features evident in the upper panel of Fig. 15.1b. An estimated $\Delta\tau(XS)/\tau(X)$ of $\sim$20% across both flat features is distinctly smaller than typical values across most of Ring C. Marginally detectable $\Delta\tau(KX)/\tau(X)$ of a few percent characterizes the wider of the two flat regions, with little or no Ka-X differential detectable elsewhere.

Differential optical depth profiles of Ring A (Fig. 15.1c) present their own distinct behavior. Like Rings B2 and B4, small or no X-S or Ka-X differential is detectable over the relatively optically thick inner region neighboring the

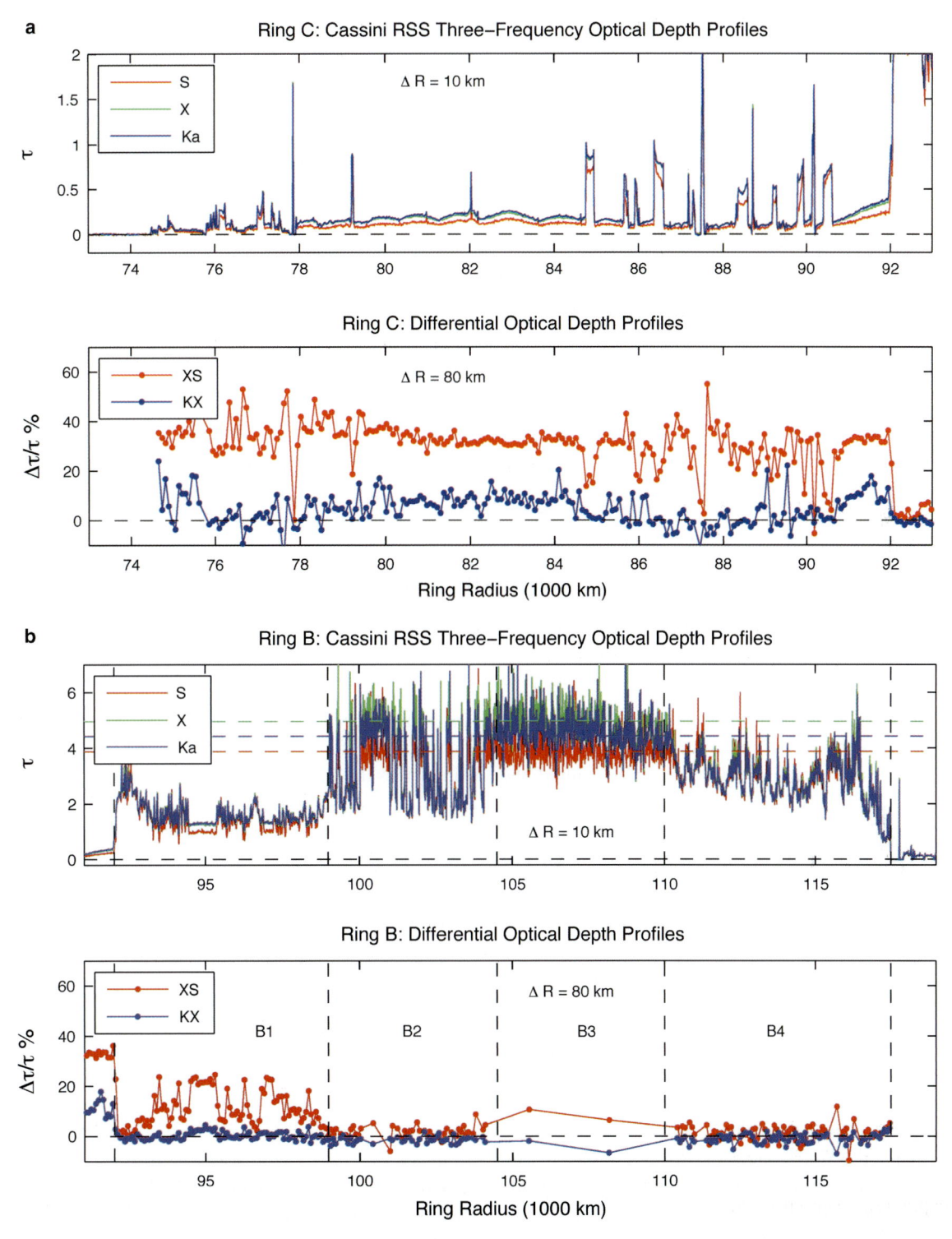

Fig. 15.1 (**a**) *Upper panel*: Cassini Radio Science Subsystem (RSS) normal optical depth profiles of Ring C observed at the three microwave wavelengths (bands) indicated. The profiles were observed during the Rev 7 ingress ring occultation on March 3, 2005, at ring opening angle $|B| = -23.6°$. The normal value is the measured oblique value scaled by $\sin(|B|)$. The radial resolution is $\Delta R = 10$ km. *Lower Panel*: The corresponding normalized X-S (*red*) and Ka-X (*blue*) differential optical depth. The radial resolution is degraded to $\Delta R = 80$ km to reduce scatter. The plotted differential is limited to regions of normal optical depth >0.05 and <3.5 to ensure reliability. (Marouf et al. 2008a). (**b**) Same as Figure 1a but for Ring B. The three dashed horizontal lines in the upper panel identify the optical depth level at which the measurement signal-to-noise ratio (SNR) drops to ∼1 (the threshold optical depth). The different threshold values reflect the different intrinsic free-space SNR for the Ka-, X-, and S-band signals (*blue, green, and red lines*, respectively). B1, B2, B3, and B4 identify four main regions of Ring B bounded by the approximate radius values 92,000, 99,000, 104,100, 110,000, 117,500 km. The measurements are noise limited in region B3 (the "core" of Ring B), except for few narrow "lanes." (Marouf et al. 2008a). (**c**) Same as Figure 1a but for the Cassini Division and Ring A. Detailed differential profiles for the Cassini Division region interior to the outer ramp feature require careful calibration of the free-space signals level and are still to be determined (Marouf et al. 2008a)

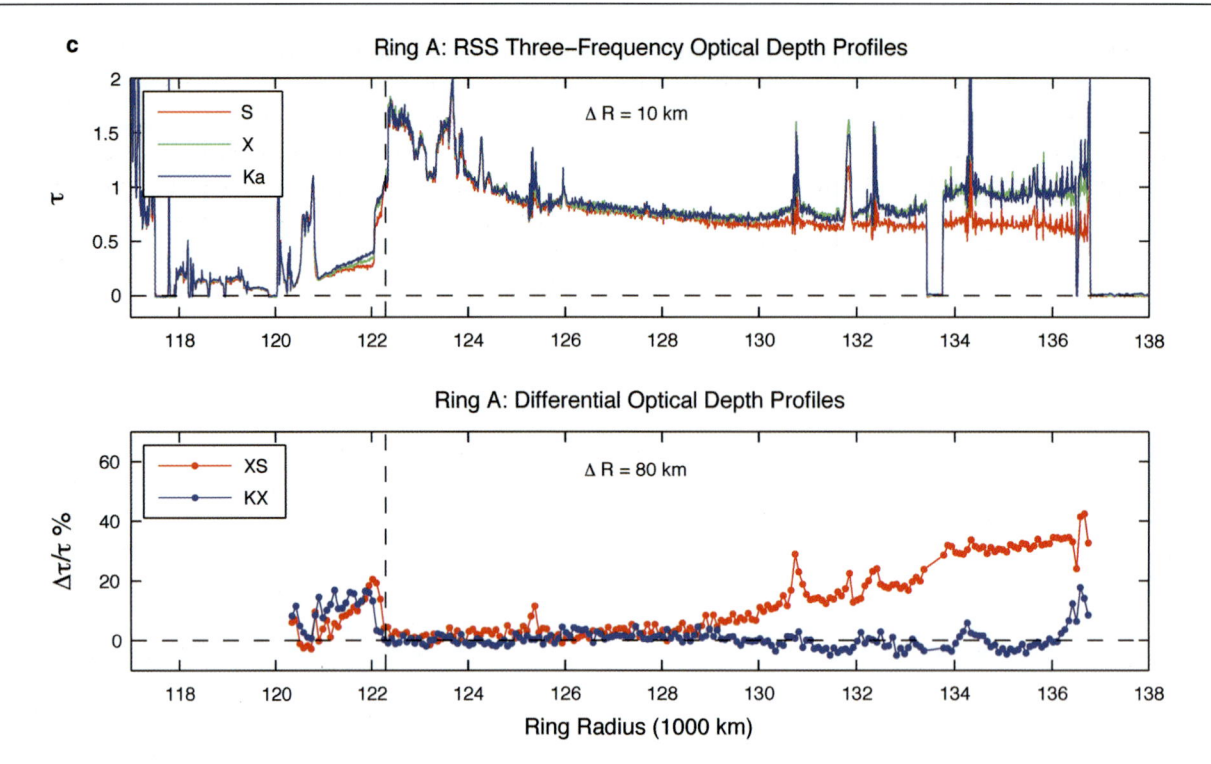

Fig. 15.1 (continued)

outer ramp of the Cassini Division. Over the ∼6,500 km wide inner region of Ring A (∼122,500–129,000), $\Delta\tau(XS)/\tau(X)$ increases by no more than a few percent, while $\Delta\tau(KX)/\tau(X)$ remains close to zero. Unlike the observed behavior in Rings B or C, a gradual, and significant, increase in the X-S differential characterizes the outer ∼7,800 km region of Ring A (∼129,000–136,800 km). The rate of increase is largest over the region inside the inner boundary of the Encke Gap (roughly at 133,500 km) where the X-S differential reaches ∼20%. As Fig. 15.1c illustrates, enhancements in $\Delta\tau(XS)/\tau(X)$ over the background level are also evident within some major wave features.

The Ka-X differential, on the other hand, remains small or absent over most of the extent of Ring A, exhibiting occasional negative, albeit small, values over the radius region outside ∼130,000 km. The exception is the narrow band between the Keeler Gap and the outer edge of the rings, where a relatively large Ka-X differential of ∼10% is observed. The band is also distinguished by exhibiting the largest X-S differential observed in Ring A (40%). The large differentials are reminiscent of values observed in Ring C.

Not discussed in comparable detail here is the Cassini Division (117,500–122,500 km roughly). Reliable characterization of the differential behavior of small optical depth features requires careful calibration of the free-space baseline. Nonetheless, we point out the remarkable similarity between the behavior of the $\tau(\lambda)$ profiles of the outer ramp of the Cassini Division (at ∼121,500 km) and that of Ring C (at ∼91,300 km; Figs. 15.1a and c). Despite the morphological similarity in $\tau(\lambda)$, the behavior of the $\Delta\tau(XS)/\tau(X)$ profiles appear different for the two features, being nearly constant across the Ring C ramp and increasing with increasing radius for the Cassini Division ramp. For both ramp features, $\Delta\tau(KX)/\tau(X)$ appears nearly constant across the feature's width.

15.2.3 Model Results

Assuming a classical ring model, the observed $\Delta\tau(XS)/\tau(X)$ and $\Delta\tau(KX)/\tau(X)$ may be used to constrain parameters of an assumed power-law particle size distribution $n(a)$, as discussed above. Starting from Eq. 15.1, the dependence of $\tau(\lambda)$ on wavelength λ is attributed to the behavior of the extinction efficiency $Q_e(a,\lambda)$. In particular, the strong dependence of $Q_e(a,\lambda)$ on wavelength for $2\pi a/\lambda < 1$ (van de Hulst 1957; Tyler et al. 1983, Fig. 6; Marouf et al. 2008a) is the physical mechanism responsible for the measured differentials depicted in Fig. 15.1. Model values of $\Delta\tau(XS)/\tau(X)$ and $\Delta\tau(KX)/\tau(X)$, computed using Eq. 15.1, depend on the differential Q_e averaged over $n(a)$, and hence on a_{min}, a_{max}, and q of the assumed power-law size distribution (see Eqs. 15.2 and 15.2). The ratio eliminates dependence on n_0 (see Marouf et al. 2008a for details).

Predicted differentials, based on model calculations of this type, may be directly compared with actual measurements for selected ring features. The comparison is carried out for seven of the eight features identified in Fig. 15.2. The feature extent is identified by the dashed vertical lines. For brevity, the features are referred to below as mid-C, C-ramp, B1-flat, in-B2, in-B4, inner-A, and outer-A. The Cassini Division ramp is included in Fig. 15.2 for profile comparison with the C-ramp but is not included in the model matching analysis below.

Figure 15.3 depicts results of the comparison for both the classical and thin layer ($N = 1, 2, 3$ and 4) ring models. The two independent measurements $\Delta\tau(KX)/\tau(X)$ and $\Delta\tau(XS)/\tau(X)$ define the two orthogonal axes, and the set of measured sample points for a given feature define a cluster in this measurements plane. The seven clusters shown correspond to seven of the features in Fig. 15.2, as labeled. Superposed are power-law model predictions spanning a range of (a_{min}, a_{max}, q). For a given q, the predicted $\Delta\tau(KX)/\tau(X)$ and $\Delta\tau(XS)/\tau(X)$ are plotted as a continuous curve parameterized by a_{min}. Results for $a_{min} = 1, 3, 5$ mm, 1, 3, 10 cm, and 1 m are explicitly labeled. For each q, results for $a_{max} = 3$ and 10 m are shown.

The model and results in Fig. 15.3 have three important implications. The first is that for $q > \sim 2.8$, a detectable $\Delta\tau(XS)/\tau(X)$ provides a direct estimate of q if $a_{min} < \sim 1$ cm. The larger the observed differential, the larger the implied q. For the mid-C, C-ramp, and outer-A features, where $\Delta\tau(XS)/\tau(X) \sim 30$–35%, the implied q is ~ 3.2–3.3, in agreement with the Voyager results for the assumed classical model (Marouf et al. 1982). The smaller $\sim 20\%$ differential for the B1-flat feature implies a smaller $q \sim 3.1$, a new result (Ring B was mostly noise limited in the Voyager case). In general, the inferred q is somewhat smaller if the thin layer model is assumed instead, dropping for the B1-flat feature to $q = \sim 3$ for $N = 2$ to 4.

The second important implication of results in Fig. 15.3 is that, for $q > \sim 2.8$, simultaneous measurement of $\Delta\tau(XS)/\tau(X)$ and $\Delta\tau(KX)/\tau(X)$ also determine or constrain a_{min}, a unique Cassini capability. In particular, for $a_{min} < \sim 1$ cm, $\Delta\tau(KX)/\tau(X)$ uniquely determines a_{min}, independently of the model assumed. In the case of the mid-C and the C-ramp features, a $\Delta\tau(KX)/\tau(X) \sim 10\%$ implies $a_{min} \sim 4$ mm; the few percent differential in the outer-A and the B-flat features imply $a_{min} \sim 5$ mm. The implied lower cutoff is sharp. Further numerical simulations suggest that the cutoff need not be an actual sudden drop in the particle number density to very small values, and can be relative flattening of the distribution to values $q < \sim 2.7$ over $a < \sim 4$ mm.

The third important implication of results in Fig. 15.3 is that, for either the classical or thin layers ring models, small or undetectable X-S and Ka-X differential, as is the case for the two Ring B features (in-B2 and in-B4) and the inner-A features, can be due to either a true absence of particles smaller than ~ 50 cm in radius or a still broad distribution with $q < \sim 2.7$. In the latter case, the relative abundance of the millimeter- to decimeter- size particles that differentially

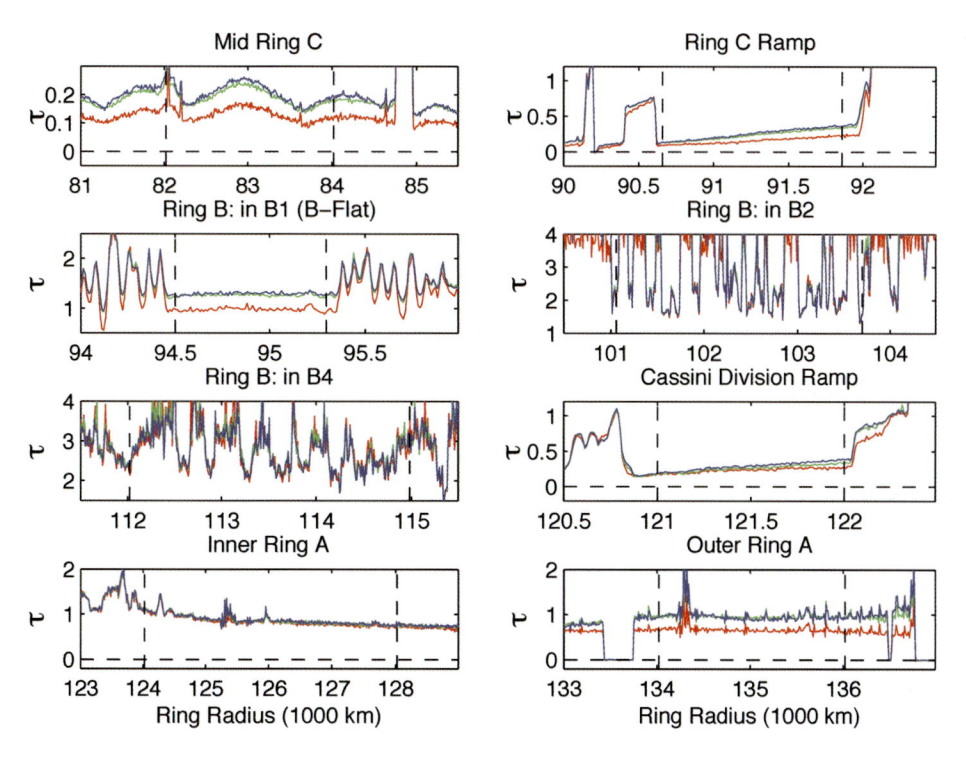

Fig. 15.2 Three-wavelength RSS optical depth profiles of eight ring features identified for examination of the constraints imposed by observed interesting differential extinction, or lack thereof, on parameters of a power-law size distribution model (? ?)

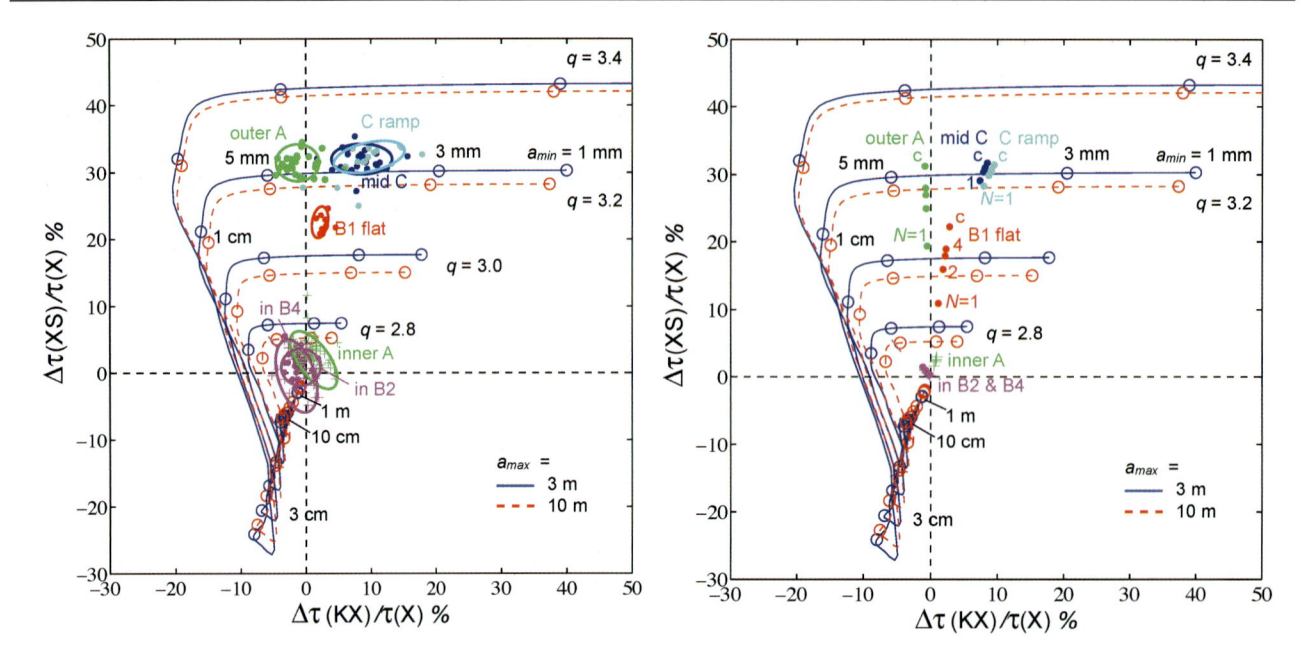

Fig. 15.3 (*Left*) Comparison of the measured X-S (*vertical axis*) and Ka-X (*horizontal axis*) differential optical depth with predictions of a power-law size distribution model of the indicated parameters. For a given power-law index q, the predicted differentials are plotted using the minimum radius a_{min} as a parameter. Points a_{min} = 0.1, 0.3, 0.5, 1, 3, 10, and 100 cm are as identified. For each case, dependence on a_{max} is illustrated for the two cases a_{axn} = 3 (*solid blue*) and 10 m (*dashed red*). The seven clusters are the values measured for seven of the ring features identified in Figure 3, as labeled. Individual points within each cluster are 80 km resolution samples. The comparison is based on the assumption of a classical (many-particle-thick) ring model. (*right*) Same figure, except that the measured values are scaled based on the thin layers ring model. The five points for each feature correspond to number of layers fours N = 1 to 4 and the classical model result 'c', as labeled. Each point is an average over the radial width of the corresponding ring feature (Marouf et al. 2008a)

affect the three radio wavelengths remains too small to cause any detectable effect.

As Fig. 15.3 also illustrates, the results described above are relatively insensitive to the exact upper bound of the size distribution a_{max}, a parameter better determined or constrained by near-forward scattering observations. We note here that limits on a_{min} can also be placed by the scattering observations at wavelengths for which $ka_{min} \gg 1$, and hence all particles are large compared with the wavelength (see Section 15.2.6). We discuss inferences of a_{max} from scattered signal observations below.

15.2.4 Near-Forward Scattered Signal Observations

In radio and stellar occultations, the signal power lost from the electromagnetic radiation passing straight through the rings (the direct signal or direct flux) is either absorbed by ring particles and/or scattered into other spatial directions. Absorption is negligible if the absorption coefficient (or imaginary refractive index) of ring particles is small at the observational wavelength. For the near-forward occultation geometry, the scattered signal component is dominated by diffraction by particles large compared to the wavelength ($ka \gg 1$). Eq. 15.3 may then be used to recover $n(a)$ over the radius range $a > a_c$. The exact size range depends on the wavelength of the observations, but is typically $a > \sim 1$ cm for stellar occultations (French and Nicholson 2000), and $a > \sim 1$ m for radio occultations (Marouf et al. 1983).

In both types of occultations, separation of the contributions of the direct and scattered signals requires special effort. In the case of radio occultations, the coherent nature of the incident radiation allows the separation based on the distinct spectral nature of each component (Marouf et al. 1982; Thomson et al. 2007). In stellar occultations, the spacecraft radio transmitter is replaced by a distant star and the Earth-based receiver is replaced by a spacecraft-based detector. Because the incident stellar flux is incoherent in nature, the direct and scattered flux components cannot be separated based on their spectral properties and other approaches are required (Section 15.2.6).

The schematic of the idealized occultation observation shown in Fig. 15.4 illustrates the conditions. Normal incidence is assumed for simplicity. A detector at distance D from the rings, modeled for example as a circular aperture of radius a_{ap}, has an acceptance or resolution angle $\theta_{ap} \sim \lambda/2a_{ap}$. Particles of radius $a > a_{ap}$ will diffract the incident signal over an angle $\theta_a = \lambda/2a < \theta_{ap}$, and the field of view of the detector on the ring plane (the circle of radius $\theta_{ap}D$ in Fig. 15.4) will encompass the diffraction

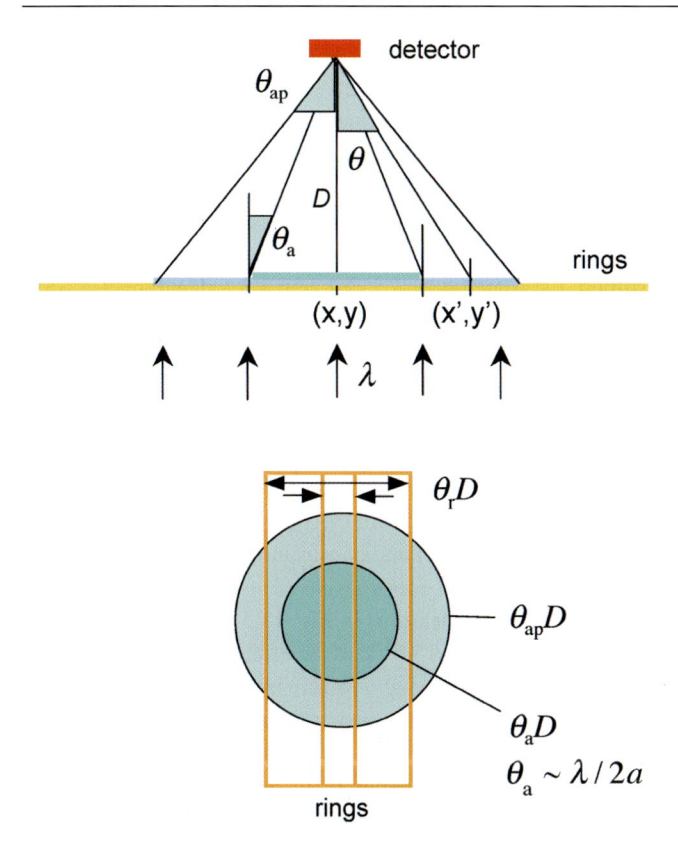

Fig. 15.4 Schematic of geometric parameters related to scattered signal observations (stellar and radio; see text). Idealized normal incidence geometry is assumed for clarity. The upper panel is a side-view; the lower panel is a top-view (not to scale). A detector of acceptance angle θ_{ap} is located a distance D away from the ring plane. Its field of view on the rings is the circle of radius $\theta_{ap}D$. The diffraction lobe of a ring particle of cone angle $\theta_a \sim \lambda/2a$ is fully captured by the detector if $\theta_a < \theta_{ap}$ and affects both the shape and strength of the scattered signal observed by the detector. The diffraction lobe footprint in that case is the circle of radius $\theta_a D$ and is fully contained within the detector's field of view. Particles for which $\theta_a > \theta_{ap}$ scatter nearly isotropically over angles $0 < \theta < \theta_{ap}$, hence do not affect the shape of the scattered signal, only its strength. In either case, the exact shape and strength of the scattered signal also depends on the radial extent of the ring region being observed, $\theta_r D$, relative to $\theta_{ap}D$ and $\theta_a D$

lobe footprint (the circle of radius $\theta_a D$). If the condition $\theta_a D < \theta_{ap} D$ holds for particles of all sizes, then the total diffracted flux will be captured by the detector as part of the measured flux (the other part being the directly transmitted flux), and is not counted by the detector as "removed" from the direct beam (Cuzzi 1985). This condition defines a critical particle minimum radius $a_c = 1/2\theta_{ap}$ (see below) Special care is required to determine the exact fraction of the diffraction lobe captured within the observing aperture when ring structure is not uniform across the aperture field of view, that is, when $\theta_r D < \theta_{ap} D$ in Fig. 15.4, where θ_r is the angle subtended by any given ring feature (Marouf et al. 1982, Cuzzi 1985, French and Nicholson 2000). A final detail enriches the radio occultations; because the coherent transmitted signal emerges after scattering from particles in the detected zone of scatterers (the zone within $\theta_a D$) with a Doppler-shifted frequency that depends on its location in the zone, spectral analysis can refine the spatial resolution to regions smaller than $\theta_a D$, especially useful if the Doppler contours are aligned with lines of constant ring radius (see Marouf et al. 1982, 1983, Zebker et al. 1985, and Marouf et al. 2008a for details).

It is well known that a particle of radius a large compared to the wavelength ($ka \gg 1$) removes from the incident wave exactly twice the amount of light it intercepts (van de Hulst 1957). In that case, the extinction cross section $C_e = 2\pi a^2$ and the extinction efficiency $Q_e = C_e/\pi a^2 = 2$. Exactly half of the power per unit area lost from the incident signal is accounted for by the total power in the diffraction pattern. If the diffraction lobe is fully captured by the detector, the apparent extinction efficiency Q_e drops from 2 to 1 (see, e.g., Cuzzi 1985). Thus, in an occultation for which the condition $ka \gg 1$ holds for all ring particles of radius $a > a_c$, the observed normal optical depth reduces to the geometric optical depth, defined as

$$\tau_g = \int_{a_c}^{\infty} \pi a^2 n(a) \, da \quad (15.10)$$

Figure 15.4 also helps illustrate the limit on the smallest particle radius that contributes to the shape of the rings' collective diffraction lobe. In the radio case, the spacecraft high-gain antenna (HGA) plays the role of the detector in the stellar case (because its illumination selects the sampled area on the rings), and the HGA beamwidth plays the role of the detector acceptance angle θ_{ap}. Particles of radius a satisfying $\theta_a D > \theta_{ap} D$ scatter nearly isotropically over θ_{ap}, hence contributing little or no information regarding the shape of the collective diffraction pattern. Only particles of size comparable to or larger than the antenna meaningfully contribute to any observed angular variations. Both Voyager and Cassini use a 2 m radius dish, setting the limit $a > \sim 1$ m in the radio occultation case (Marouf et al. 1982, 1983; Zebker et al. 1985). The limit is much smaller in the stellar occultation case, e.g., $a > \sim 4$ cm for $\lambda = 1\,\mu$m and $a > \sim 15$ cm for $\lambda = 4\,\mu$m (French and Nicholson 2000).

15.2.5 Size Distribution from the Voyager RSS Observations

The Voyager 1 radio occultation in 1980 provided the first definitive detection of near-forward scattered X-band signal in the time sequence of observed spectra (Tyler et al. 1983, Marouf et al. 1983). The small ring-opening angle at the time ($B = 5.9°$) caused the Voyager antenna beam to sample a relatively large ring area at any given observation time. The experiment geometry was optimized to closely align contours of constant Doppler-shift with contours of constant ring radius. Both the scattered signal observations and the X-S differential extinction observations were used to determine self-consistent size distributions for several main ring features, including mid Ring C, Ring C ramp, Cassini Division ramp, inner Ring A, and outer Ring A (Marouf et al. 1983, Zebker et al. 1985). The features are similar to, but not identical, to those in Fig. 15.2; see Table 15.1 for exact definitions. Voyager measurements in Ring B were largely noise-limited.

Assuming the classical model, Marouf et al. (1983) recovered the first explicit size distribution of ring particles over the radius range $1 < a < 15$ m for four main ring features (Fig. 15.5, Table 15.1). The distributions revealed a sharp upper size cutoff in the 3–5 m radius range, depending on the feature. Knowledge of $n(a)$, $a > 1$ m, allows computing the contribution of particles in this radius interval to the measured X- and S-band optical depth, constraining the adjusted optical depth due to smaller particles. Modeling the distribution over $a < 1$ m by a power-law having $a_{\min}$ small enough to contribute negligibly ($a_{\min} = 1$ cm) and $a_{\max} = 1$ m, the index q follows from the differential $\Delta\tau(XS)/\tau(X)$ and the scaling factor n_0 follows from $\tau(X)$, both computed over the limited range 1 cm $< a < 1$ m. The combined power-law model and the explicit inversion results yielded the first detailed characterization of $n(a)$ over the broad range 1 cm $< a < 15$ m, for the three optically thin features in Fig. 15.5 (Marouf et al. 1983); see also Table 15.1.

Figure 15.5 illustrates a problem with the classical model assumed. Estimated $q = 3.5$, 3.4, and 3.3 for features C1.35, C1.51, and CD2.01 (the red lines in Fig. 15.5; Table 15.1) yield $n(a)$ values that connect poorly with the explicit inversion over $a > \sim 1$ m. Overcompensating for multiple scattering effects in recovering $n(a)$ from $I_s(\theta, \lambda)$ would cause an overestimate of the absolute $n(a)$ values recovered over $a > 1$ m, hence the mismatch near $a = 1$ m. The thin-layers model (see Section 15.2.1) controls the contribution of multiple scattering to $I_s(\theta, \lambda)$ and constrains values of the number of layers N that yield self-consistent results near $a \sim 1$ m. Results for the three features in Fig. 15.5 and others in Ring A are shown in Fig. 15.6 (Zebker et al. 1985). Less steep (smaller) q's over 1 cm $< a < 1$ m are implied in this case; see Table 15.1. The more self-consistent matching of the power-law with the upper-size cutoff results suggests that the finite-thickness models are a better fit to reality, consistent with dynamical expectations (see Chapters 13 and 14).

Table 15.1 Particle Size Distribution from Radio and Stellar Occultations of Saturn's Rings

Ring region	Radius Range (km)	q	$a_{\min}$ (cm)	$a_{\max}$ (m)	$n_0(1$ cm$)$ (#/m²/m)	Q	a_{eff}(PPS) (m)	a_{eff}(RSS) (m)	a_{eff}(28 Sgr) (m)
Voyager RSS[a]						**Voyager PPS**[c]			
C1.35	78,430–84,460	3.11	0.1	4.5	2700	0.0028	1.4	0.84	
C1.51	90,640–91,970	3.05	0.1	2.4–5.3	2990	0.0086	2.3	1.22	
CD2.01	120,910–122,010	2.79	0.1	7.5	1780	0.026	3.9	2.44	
A2.10	125,490–127,900	2.70	0.1	5.4	3300	0.242	11.6	1.55	
A2.12	125,490–130,310	2.74	0.1	5.0	2870				
A2.14	127,900–130,310	2.75	0.1	6.3	3530	0.262	11.9	1.65	
A2.19	130,860–133,270	2.93	0.1	11.2	5650	0.252	11.2	1.82	
A2.24	133,930–136,350	3.03	0.1	8.9	8950	0.180	9.6	1.32	
Earth-Based 28 Sgr[b]									
Ring C	74,490–91,983	3.1	1	10		<0.002–0.012	<1.2–2.8		2.3
Ring B	91,183–117,516	2.75	30	20		0.05–0.12	5.7–8.8		8.3
Cassini Division	117,516–122,053	2.75	0.1	20		<0.002–0.035	<1.1–4.5		7.0
Inner Ring A	122,053–133,423	2.75	30	20		0.23–0.27	11.2–12.2		8.3
Outer Ring A	133,745–136,774	2.9	1	20		0.16–0.16	9–10.7		6.0

[a] Size distribution from the Voyager radio occultation observation (Zebker et al., 1985). The distribution parameters are inferred from inversion of the near-forward scattered 3.6 cm $-\lambda$ (X-band) signal over $a > 1$ m and modeling of the 3.6 and 13 cm $-\lambda$ (S-band) differential extinction as a power-law distribution over 0.1 cm $< a < 1$ m,. A minimum radius $a_{\min} = 0.1$ cm is assumed for all ring regions. The results are based on the thin-layers ring model (see Section 15.2.1).

[b] Size distribution from the Earth-based 28 Sgr stellar occultation (French and Nicholson, 2000). The distribution parameters are inferred from comparison of the strength and shape of profiles of the observed near-forward scattered stellar flux at 0.9, 2.1, and 3.9 μm wavelengths with theoretical predictions based on a power-law size distribution model. The model parameters are assumed uniform across each main ring region and are selected to provide a compromise match to data at all three wavelengths. Results for the Cassini Division are not well determined. The results are based on the classical ring model.

[c] Effective radius from the variance of the statistical fluctuations in photon count observed during the Voyager PPS stellar occultation (Showalter and Nicholson, 1990). The parameter Q characterizes the increase in variance above Poisson count statistics. It provides an estimate of the effective particle radius a_{eff}(PPS) which is controlled by the 4th moment of the size distribution. For comparison purposes, the last two columns also lists a_{eff} computed based on the inferred RSS and 28 Sgr size distributions (French and Nicholson, 2000, Showalter and Nicholson, 1990). The results are based on the classical ring model.

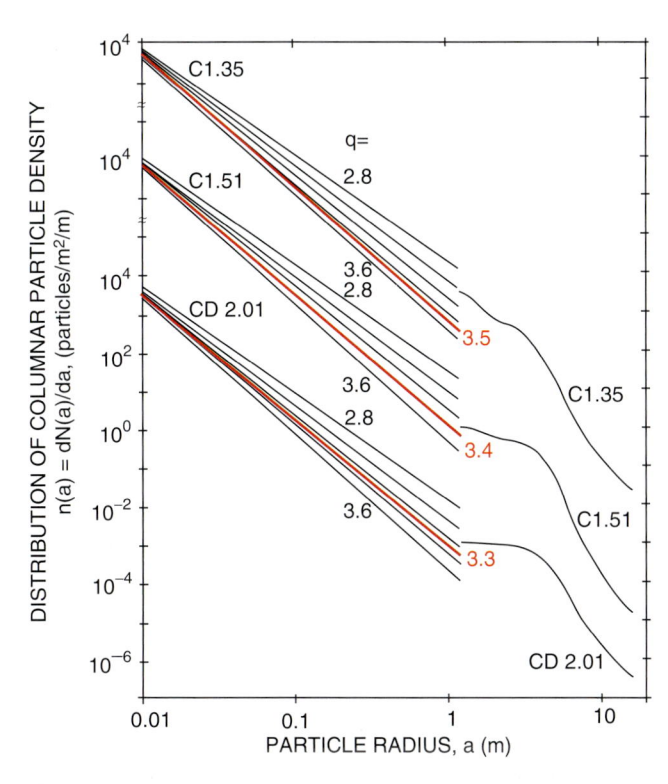

15 Ring Particle Composition and Size Distribution

Fig. 15.5 Inversion of the near-forward scattered signal observed during the Voyager 1 radio occultation for several ring features (see Table 15.1) revealed a sharp upper radius cutoff of 3–5 m. In addition, measurement of the X-S differential optical depth provided constraints on parameters of a power-law model over the radius interval 1 cm to 1 m (*the red lines*). The size distribution discontinuity at $a \sim 1$ m motivated searching for a better ring model than the classical (many-particles-thick) model used in the analysis (adapted from Marouf et al. 1983)

15.2.6 Size Distribution from 28 Sgr Stellar Occultations

The 1989 stellar occultation of 28 Sgr by the Saturn system was widely observed, and provided the first detailed post-Voyager examination of the geometry, structure and scattering properties of Saturn's rings. This occultation was unique in that the star was unusually bright, and its diminished signal could be detected even on top of sunlight reflected from the rings. The observed intensities were a complicated blend of directly attenuated starlight and starlight diffracted into the detector from other regions of the rings. French and Nicholson (2000) used ring occultation profiles from the Lick ($\lambda = 0.9\,\mu$m), McDonald ($\lambda = 2.1\,\mu$m), and Palomar ($\lambda = 2.1\,\mu$m) observatories to infer the size distribution of the ring particles. The Voyager PPS optical depth profile was used to estimate and remove the direct signal contribution to the observed total flux, and the method concentrated on interpreting the diffracted signal. For the idealized geometry of Fig. 15.4, an Earth-based detector of acceptance angle θ_{ap} looking back at the rings at distance D collects the superposition of contributions from all ring elements at angles θ within its field of view. Assuming single scattering, the intensity of the diffracted light $I_1(\theta, \lambda)$ is governed by Eq. 15.5. The total scattered flux was modeled by a two dimensional convolution, for the exact 28 Sgr observation geometry, constraining an assumed power-law size distribution parameters to achieve a good match to the measured flux. The power-law parameters were fit separately for

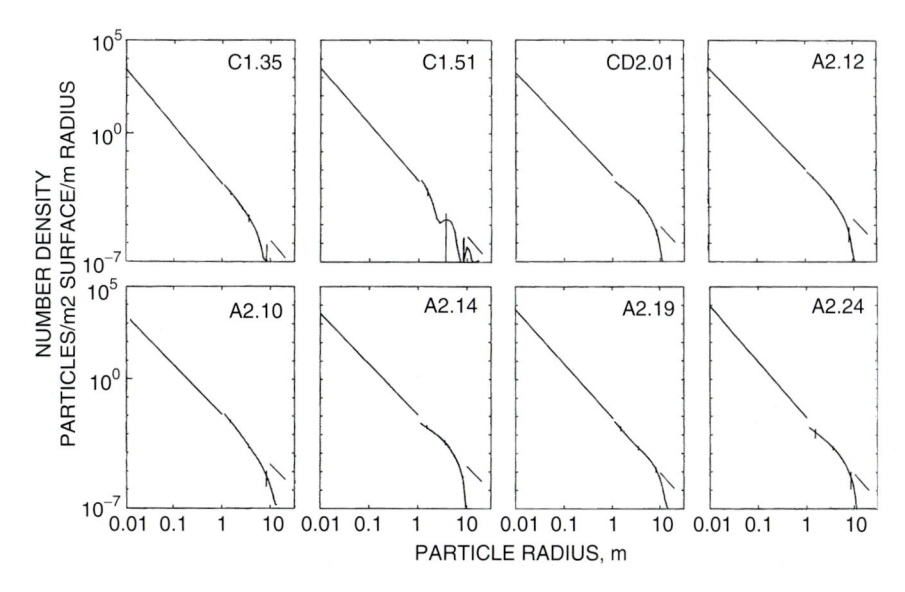

Fig. 15.6 As discussed in Section 15.2.5, a "thin-layers" ring model provided more self-consistent results for the Voyager RSS particle size distribution over the full radius range 1 cm to 15 m for eight ring features (from Zebker et al. 1985)

each main ring region. The piecewise best solutions are then used collectively to compute a predicted composite scattered signal for the entire ring system. Table 15.1 lists the compromise power-law model parameters for each ring region that gave the best overall fit to the observations at all three observation wavelengths.

15.2.7 Size Information from the Excess Variance in Stellar Occultations

This is a fundamentally different approach to constraining ring particle sizes using stellar occultation measurements (Showalter and Nicholson 1990). It stipulates that the statistical fluctuations in the photon count k (not to be confused with the wavenumber k used earlier) measured by the photodetector behind the rings are partly intrinsic and partly due to the random nature of the local ring area blocking the incident stellar flux. The intrinsic part originates in the stochastic nature of the incident stellar photon count (S) and any background contribution (B), for example from Saturnshine. Both intrinsic components are well modeled by Poisson distributions of parameters λ_S and λ_B. The expected value $E(k)$ during occultation by a ring region of oblique optical depth τ/μ_0 is

$$E(k) = \lambda_S\, e^{-\tau/\mu_0} + \lambda_B = \lambda_S P + \lambda_B \quad (15.11)$$

where $P = \exp(-\tau/\mu_0)$ is the fraction of ring area not blocked by ring particles and τ accounts for whatever role near-forward diffraction plays (Section 15.2.4).

Showalter and Nicholson (1990) argue that independent additional information about particle sizes is provided by the higher order statistical averages of the photon count k, in particular, its variance $\sigma^2(k)$. Ring particles large enough to stochastically perturb the fraction of ring area not covered (P) would introduce additional stochastic fluctuations in k, hence contribute to $\sigma^2(k)$. Treating P as a random variable of mean $\exp(-\tau/\mu_0)$ and variance $\sigma^2(k)$, they show that

$$\sigma^2(k) = E(k) + \lambda_S^2 \sigma^2(P) \quad (15.12)$$

The first term is the variance if P were deterministic, and the second is the "excess variance," that is, the additional contribution to $\sigma^2(k)$ due to "ring noise", which is a measure of the *variation* in blockage fraction of the sampled patch of local ring material. The effective area contributing to the direct signal, A_d, depends on the size of the first Fresnel zone smeared by the motion of the spacecraft. The effective area contributing to the scattered signal, A_s, is determined by the detector field of view (the circle of radius $\theta_{ap}D$ in Fig. 15.4 adjusted for oblique incidence). In addition to dependence on A_d and A_s, $\sigma^2(k)$ is strongly controlled by a dimensionless parameter Q (not to be confused with the extinction efficiency Q_e) defined as

$$Q \equiv \frac{\int (\pi a^2)^2 n(a)\, da}{\mu_0 A_d \int \pi a^2 n(a)\, da} = \frac{\pi a_{\text{eff}}^2}{\mu_0 A_d} \quad (15.13)$$

where Q can iteratively be estimated from the observed time series k measured during the Voyager PPS occultation, and

$$a_{\text{eff}} = \sqrt{\frac{\int a^4 n(a)\, da}{\int a^2 n(a)\, da}} \quad (15.14)$$

Hence, a_{eff} is strongly weighted toward the largest particle sizes and the Q-profiles provide constraints on the largest particle sizes across the main ring regions with achievable radial resolution as fine as 20 km. Q-profiles of Rings C, B, and A are shown in Fig. 15.7; see Table 15.1 for translation of these Q values into particle radii.

15.2.8 Summary of Current Knowledge and Limitations

Table 15.1 groups comparative results from the Voyager radio occultation, the 28 Sgr stellar occultations, and the Voyager PPS excess variance observations. The table is adapted from similar tables in Zebker et al. (1985), Showalter and Nicholson (1990), and French and Nicholson (2000). More recent Cassini results were presented in Figs. 15.1 and 15.3 (Marouf et al. 2008a).

The upper left side of Table 15.1 lists $(q, a_{\max}, n_0)$ of a power-law model (Eq. 15.2) consistent with the direct and near-forward scattered (diffracted) signals observed during the Voyager radio occultation. The Voyager observations did not constrain $a_{\min}$, which was assumed to be much less than the $\lambda = 3.6$ cm wavelength of the X-band signal ($a_{\min} = 0.1$ cm). Results for eight ring features and their radial extent are tabulated.

Similar results for the 28 Sgr occultation are listed in the lower left part of the table. Here, parameters $(q, a_{\min}, a_{\max})$ are constrained by the estimated near-forward scattered flux shape and strength. Before Cassini, the 28 Sgr results provided the only available direct constraints on $a_{\min}$. Cassini radio occultation observation of the Ka-X differential optical depth (Figs. 15.1 and 15.3) provide new tight constraints on $a_{\min}$. Except for Ring A, the 28 Sgr results are assumed to uniformly apply to each of Ring C, Ring B, and the Cassini Division as a whole. Results for inner (interior to the Encke Gap) and outer (between the Encke and Keeler Gaps) regions of Ring A are inferred independently (Table 15.1).

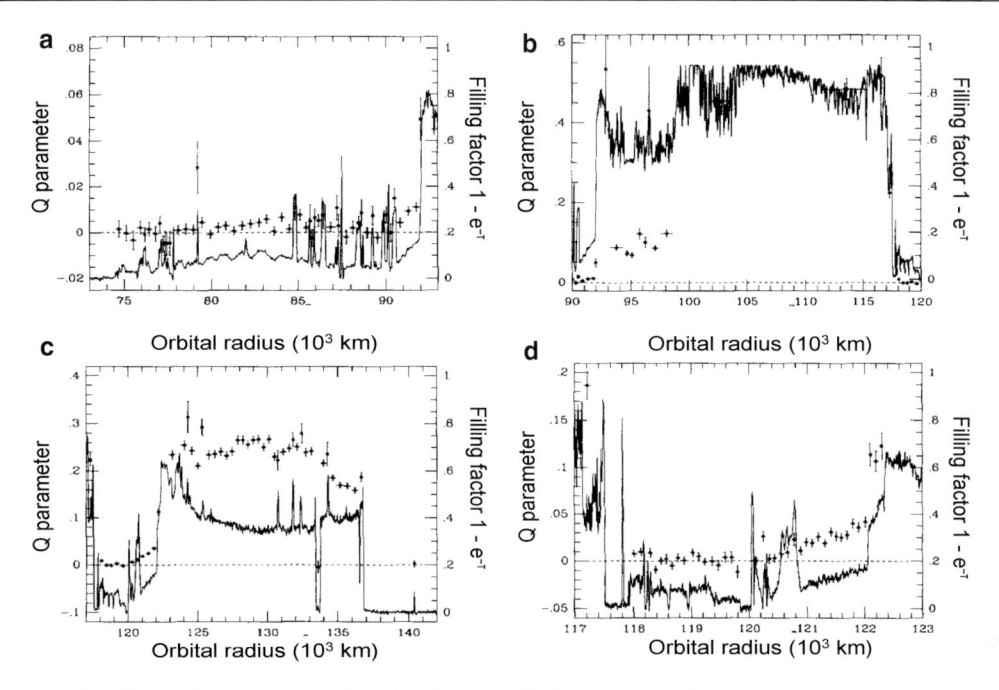

Fig. 15.7 Figure 15.10 Profiles of the Q-parameter for (**a**) Ring A, (**b**) Ring B, (**c**) The Cassini Division and Ring A. The profiles are generated from 2 km resolution estimates averaged over up to 20 points. The horizontal bars on each point indicate the radial averaging interval.

(**d**) Closer look at the behavior of Q in the Cassini Division and inner Ring A. The 2 km resolution estimates are averaged over up to 5 points. In all cases, 50 km resolution Voyager PPS optical depth profiles are shown for reference (Showalter and Nicholson 1990)

The last four columns of Table 15.1 list results based on the Q-parameter from the PPS excess variance observations. The first of these lists Q itself and the second the implied a_{eff}, denoted a_{eff}(PPS) to emphasize its inference from the PPS data. The third column lists a_{eff}(RSS) implied by Eq. (15.14) if the size distribution $n(a)$ is assumed to be a power-law of the same parameters determined by the radio occultation observations of the corresponding feature (values are in Showalter and Nicholson 1990). The last column, a_{eff}(28Sgr), is the same except that $n(a)$ is determined by the 28 Sgr occultation (values are from French and Nicholson 2000).

15.2.9 Comparison of the Four Main Ring Regions

Overall the particle sizes in the main rings seem to follow powerlaw distributions in radius of the form $n(r, r + dr) = n_0 r^{-q} dr$, where the slope q of the powerlaw, and the upper and lower radius limits, vary with location in the rings. Generally speaking $q \sim 3$, suggesting there is equal surface area per decade and most of the mass is in the larger particles, the lower radius limit is in the 1–30 cm range, and the upper radius limit is in the 2–20 m range. More detailed discussion is given below.

Ring C: Both the Voyager radio and the 28 Sgr stellar occultations suggest a relatively steep power-law index $q \sim 3.1$ in Ring C. The Cassini Ka-X differential optical depth suggests slightly steeper $q \sim 3.2$ and strongly constrains a_{min} to be ~4 mm (Fig. 15.3), in general agreement with $a_{min} = 1$ cm from the 28 Sgr observations (Table 15.1). A largest particle radius $a_{max} = 4.5$ m in mid ring C and ~2.5–4.5 m in the Ring C ramp from the Voyager radio occultation is smaller than $a_{max} = 10$ m from the stellar occultation over the full Ring C. The inferred values are still within an estimated factor of 2 to 3 uncertainty in the latter, however. Both the 28 Sgr occultation and the excess variance (Q-based) result imply similar $a_{eff} \sim 2.3$ m, and hence similar a_{max}. An estimated Q increasing with radius over the Ring C ramp (Fig. 15.7a) suggests an a_{max} increasing with radius across this feature. Cassini Ka-X and X-S differential optical profiles (Fig. 15.1a) show no evidence for significant variations in q or a_{min} across the Ring C ramp feature (Fig. 15.3).

Ring B: The Voyager radio occultation observation of Ring B was mostly noise limited, since the rings were nearly closed at the time ($B = 5.9°$); not so for Cassini at much larger B. The differential X-S and Ka-X optical depth Cassini observations suggest a size distribution for the innermost region of Ring B (region B1) that is different from the other three regions (B2, B3, and B4; Fig. 15.1b). In particular, the nearly flat feature between ~94,400 and 95,300 km is characterized by a $q \sim 3$–3.1 and $a_{min} \sim 4$ mm (Fig. 15.3;

Regions B2 and B4 show little detectable X-S or Ka-X differential, indicating either relatively large $a_{min} > \sim 50$ cm or relatively flat power law index $q < \sim 2.7$. Clearly detectable Q-values over region B1 (Fig. 15.7) imply an a_{eff}(PPS) = 5.7 m within the flat feature, and a larger a_{eff}(PPS) = 8.8 m on either side of the feature (Table 15.1). No reliable Q-based estimates of a_{eff} are available for other regions of Ring B. For the 28 Sgr case, a single uniform size distribution for regions B1, B2, and B4 has parameters $q = 2.75$, $a_{min} = 30$ cm, and $a_{max} = 20$ m (Table 15.1). Comparison with the more localized estimates above must be regarded with due care. An implied a_{eff}(28Sgr) = 8.3 m appears consistent with the Q-based estimate of 5.7–8.8 m in region B1, and an $a_{min} = 30$ cm is more or less consistent with the Cassini radio inference of $a_{min} > 50$ cm as one potential reason for the lack of observed X-S-Ka differential in regions B2 and B4. Region B3 is noise limited in all observation types.

Cassini Division: The size distribution from the Voyager radio occultation is limited to the outer Cassini Division ramp feature, where the estimated $q \sim 2.79$ and $a_{max} = 7.5$ m are also comparable to their values in the inner A Ring (Table 15.1). The Cassini X-S differential optical depth exhibits a systematic increase with radius (Fig. 15.1c), suggesting that the size distribution may be varying across the 1,100 km extent of this feature. A similar systematic increase in estimated Q with radius (Fig. 15.7c and d) suggests that a_{max} may be increasing with increasing radius. A mean a_{eff}(PPS) = 3.9 m for the ramp feature is larger by about a factor of 2 than a_{eff}(RSS) = 2.4 m from the Voyager RSS size distribution, and smaller by about the same factor than a_{eff}(28Sgr) = 7 m from the 28 Sgr size distribution. Smaller estimated Q values for the tenuous Cassini Division region interior to the ramp imply smaller a_{eff}(PPS) = 1.1 m, hence smaller a_{max}. Because of its relative narrowness ($\sim$4,500 km wide), the size distribution from the 28 Sgr occultation is not well determined in this ring region. Nonetheless, inferred $q = 2.75$ and $a_{max} = 20$ m are in general agreement with inferences in the neighboring inner Ring A (Table 15.1).

Ring A: Both Voyager and Cassini radio occultations reveal interesting X-S differential optical depth that increases with increasing ring radius (Tyler et al. 1983, Marouf et al. 2008a; Fig 15.1c). The X-S differential increase could be explained by either an increasing q or a decreasing a_{min} with increasing ring radius. A small observed Ka-X differential suggests that the increasing X-S differential over mid and outer Ring A is likely due to an increasing q. An exception perhaps is the outermost region between the Keeler Gap and the outer edge of Ring A, where the Ka-X differential is not small. Inferences from the Voyager radio occultation suggest an increase of q from about 2.7 in inner and mid Ring A to about 3 in the neighborhood of the Encke Gap. The estimated a_{max} also appears to increase from about 5–6 m to about 9–11 m in these two regions (Table 15.1).

Estimates from the 28 Sgr stellar occultation yield $q = \sim 2.75$, $a_{min} = 30$ cm, and $a_{max} = 20$ m in the inner and mid Ring A region (interior to the Encke Gap), and $q = \sim 2.9$, $a_{min} = 1$ cm, and $a_{max} = 20$ m in the outer region (between the Encke and Keeler Gaps). The q values and the trend are consistent with the Voyager radio estimates, and somewhat smaller than a Cassini radio estimate of $q = \sim 3.15$–3.2 (Fig. 15.3). A large $a_{min} = 20$ cm in the inner region is compatible with the observed small X-S-Ka differential optical depth in the innermost part of Ring A but is difficult to reconcile with regions closer to the Encke gap where a relatively large X-S differential is observed. On the other hand, an $a_{min} = 1$ cm in outer Ring A is compatible with the Cassini radio observations where $a_{min} = 4$–5 mm is inferred (Fig. 15.3). It's likely that the size distribution varies continuously across Ring A. The 28 Sgr estimate of $a_{max} = 20$ m in both inner and outer Ring A is a factor of 2 to 4 larger than the radio values (Table 15.1). A large a_{max} is also suggested by the Q-based estimates of a_{eff}(PPS) = 9.6 to 11.6 m, in general agreement of a_{eff}(28Sgr) = 6 to 8.3 m from the 28 Sgr inferred size distributions. Both estimates are much larger than a_{eff}(RSS) = 1.5 to 1.8 m implied by the size distribution from the Voyager radio occultation (Table 15.1). The significant differences may be caused, at least in part, by particle clumping due to the gravitational wakes that permeate Ring A. The Q-profile in outer Ring A exhibits an interesting systematic decrease of estimated Q with increasing ring radius suggesting systematically decreasing a_{max} with radius over that region (Fig. 15.7c). The behavior is reminiscent of the systematic X-S differential optical depth behavior in outer Ring A (Fig. 15.1c), although the latter is likely more related to variations in q and/or a_{min}.

15.2.10 Caveats Regarding Modeling "Ring Particles" vs. "Self-Gravity Wakes"

In concluding this section, we point out two important limitations of the results summarized in Table 15.1. First, objective comparison of the particle size distribution inferences must be based on the same ring model. Although the classical ring model is at the heart of all three approaches discussed, only the Voyager radio results have been adapted to the perhaps more realistic thin layers ring model. Especially in ring models of likely small vertical extent, it is also desirable to understand electromagnetic interaction with possibly close-packed ring particles.

Second, and perhaps more important, all analysis procedures need to be extended to account for the presence of gravitational wakes in Rings A and B (Chapters 13 and 14). Particle clustering in elongated and preferentially oriented formations fundamentally impacts the observed optical depth and its dependence on the ring viewing geometry. It also impacts the strength and shape of the collective near-forward scattering (diffraction) pattern as well as the higher order moments of random fluctuations in the observed signal intensity, all of which being important elements of self-consistent determination of the size distribution. The observations therefore not only provide information about the individual ring particles and their size distribution, but also the physical properties of the wake structure that hosts the individual particles. The challenge therefore is to separate and determine both.

Two idealized models have been used to infer characteristic dimensions of the wake structure in Ring A (Colwell et al. 2006, Hedman et al. 2007). For Cassini radio occultations, preliminary results have been obtained from numerical simulations of signal extinction and forward scattering by ring models that simulate gravitational wakes as clusters of ring particles that are randomly packed in the ring plane (Marouf et al. 2008a,b). The clusters can be of arbitrary width, length, vertical thickness, and packing fraction, and can be embedded in a classical layer of arbitrary thickness. All wake models predict strong dependence of the observed optical depth on wake orientation relative to the observation geometry and ring-opening angle B. The dependence invalidates the classical $\tau(\text{oblique}) = \tau(\text{normal})/\sin(|B|)$ scaling, and shows especially strong dependence on the wake orientation when B is small (when the rings are relatively closed). In principle, the measured optical depth variations with observation geometry provide constraints on the physical wake properties (Chapter 13). A corresponding self-consistent inference of the normal optical depth and its variation with wavelength should provide information about the particle size distribution – as was the case in the absence of wakes. From all indications, wakes are so much larger than the radio wavelengths that no wavelength dependence should be expected, only elevation and longitudinal dependence which can be modeled.

Additional complementary information is provided by the near-forward scattered signal measured during radio occultations. Wakes composed of long formations of spatially correlated particles diffract the incident radio signal much like cylindrical structures, with the forward lobe being much stronger and narrower than the diffraction pattern of the constituent particles. The phase coherency required to maintain the cylindrical scattering behavior is limited to very small angles close to the exact forward direction. The randomized phase of wake-diffracted signals scattered to larger angles cause their intensity to add incoherently, yielding behavior similar to that of the classical model. Numerical simulations validate this behavior (Marouf et al. 2008a,b); comparison of the predicted scattered signal spectra based on the Voyager particle size distribution with those measured by Cassini in inner Ring A reveals the clear presence of a narrower and stronger spectral component – likely due to wakes. Its angular width provides a measure of a characteristic physical dimension of the narrow dimension of the wakes, which is large compared to the individual few-to-tens-of-meter-size particles, clearly distinguishing collective wake effects from individual particle effects. Quantitative results will require careful consideration of the impact of observation geometry and multiple scattering on the diffraction pattern.

In principle, near-forward scattered signal observed during stellar occultations should also be affected by the presence of wakes and the effects on analysis procedures remain to be assessed. Because of the obvious impact on the random ring area blocked during a stellar occultation, the effect of wakes on the excess variance observations is likely to be significant and may be responsible for the differences in typical "sizes" between the Voyager radio and Q-based inferences in Table 15.1. Understanding and quantifying the impact of wakes on all particle size inference techniques will be an active area of current and future research. Hopefully, more general analysis procedures that account for the wakes will not only yield the particle size distribution but also physical properties of the wake structure itself.

15.3 "Propeller" Objects: Shards of the Ring Parent or Locally Grown?

As discussed in Section 15.2, the distribution of "ring particles" follows a powerlaw with a noticeable upper limit on particle radius in the 5–10 m range. Cassini has also discovered an entirely separate class of "particles" in, at least, the A ring, with radii that are up to 100 times larger. These objects are not seen directly, but are revealed by the very characteristic disturbances they create in passing ring material. For lack of a better name they have been dubbed "propeller objects" after the shapes of their associated disturbances (see Chapter 14 for a theoretical discussion relating the objects to their observable disturbances). Here we will summarize the observational aspect of this population and briefly discuss the implications.

The observations were made by Tiscareno et al. (2006, 2009) and Sremcevic et al. (2007). Several hundred objects have been analyzed in terms of their size and radial distribution (Figs. 15.8 and 15.9). Tiscareno et al. (2009) showed that the propeller objects are restricted to three radial bands. These locations are in the mid-A ring, in good agreement with where French et al. (2007) have observed wake-related nonaxisymmetrical brightness variations to maximize as well (Chapter 13). New observations by Cassini (Fig. 15.10) indicate visually the nonuniform distribution of these objects and the potential richness of this database.

It appears that the propeller objects lie on quite a steep size distribution, much steeper than the ring particles themselves,

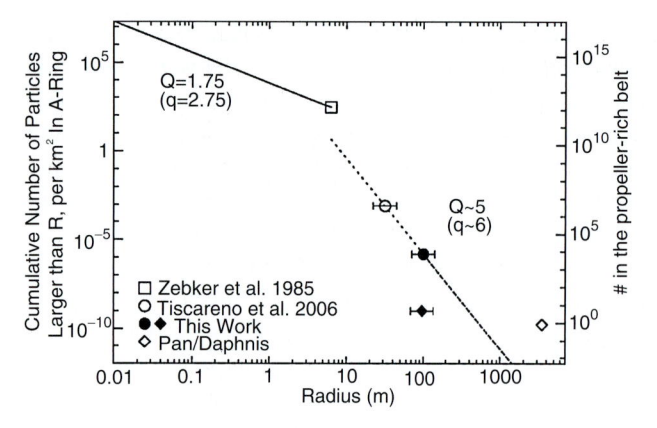

Fig. 15.8 Number density plot showing the abundance of newly discovered propeller objects in 30 m radius and 100 m radius ranges (from Tiscareno et al. 2006, 2009 respectively) in the belts where they are observed. The filled diamond indicates the density at one other radius where complete coverage was obtained, showing the strong abundance contrast between propeller belts and their surroundings. At the top left is a line indicating a ring particle powerlaw distribution $n(r) = n_0 r^{-q}$, in which for $q = 2.75$, most of the mass lies towards the upper end. Because the size distribution of propellers falls off so rapidly, they contain negligible mass compared to the ring particle population itself. Note also that the revealed ringmoons Pan and Daphnis do not fall on a line connecting the propellers to the largest ring particles, and even they contain far less mass than the ring particles

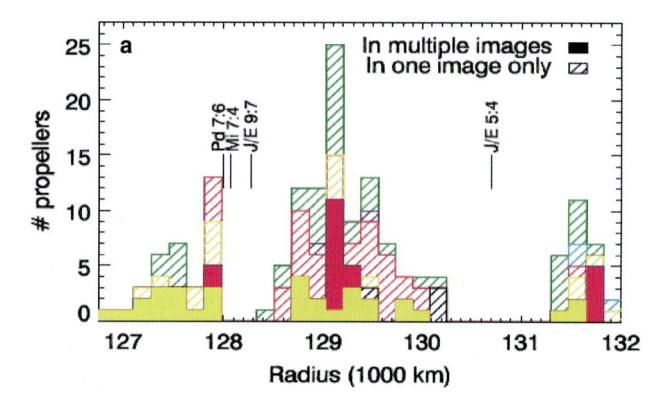

Fig. 15.9 Radial distribution of 100 m diameter propellers (from Tiscareno et al. 2009). Three distinct radial bands are seen

and that they apparently do not simply connect the largest ring particle with the few known embedded moonlets Pan and Daphnis. Converting number densities of Tiscareno et al. (2006, 2009) into surface mass densities gives 10^{-2} g cm^{-2} for the "small" 30 m radius SOI propellers and 10^{-3} g cm^{-2} for the "larger" 100 m radius propellers, insignificant relative to typical A ring surface mass densities of 40 g cm^{-2} (Tiscareno et al. 2007; Chapter 13). Because we now have ways of detecting objects in the full size range between ring particles and revealed moonlets, it seems that the mass inventory of the A ring is now complete. As yet, no propeller objects have been discovered in the B ring.

15.4 Ring Particle Composition, Its Radial Variations, and Comparison with Other Icy Objects

In this section we review and preview studies directly related to the composition of the particles of the main rings. We start with a discussion of some of typical observations, the advantages brought to bear by Cassini over prior studies, and some observational challenges. We present ring spectra through the near- and thermal-infrared, visual, and UV spectral regions. We first present large-scale radial averages at low phase angle, reaching some general qualitative conclusions about which materials are, and are not, found in the rings. We then show how the spectra of broad regions (A, B, C, Cassini Division, F ring) vary with phase angle, and discuss the significance. We next select certain key spectral properties and show how they vary with radius on finer scales. These radial spectral variations suggest radial variation of composition, although their significance remains unclear in detail.

Next, we discuss the analysis needed to obtain particle composition from spectral observations, involving models of both the ring layer as a whole, and of the grainy surfaces of the ring particles. Some model studies have attempted to extract both surface grain size and material composition from observed spectra; we discuss these and describe two interesting options for explaining ring color in terms of ring composition. Finally, we compare and contrast the spectral properties of the rings with those of a number of icy objects from the Saturn system and beyond as a prelude to the discussion of ring provenance in Section 15.6.

15.4.1 Observations

The reflected brightness of Saturn's rings varies with wavelength λ, solar incidence, phase, and ring opening angles[1], and radial location due to the scattering properties of individual ring particles (Section 15.4.6.2) as well as their collective spatial and size distributions (Sections 15.2, 15.4.6.1, and below). Denoting the incident solar flux across some spectral band as πF erg cm^{-2} s^{-1}, the observed intensity I of the rings (erg cm^{-2} s^{-1} str^{-1}) in some geometry is ratioed to the intensity of a perfect Lambert surface (incident flux/$\pi = F$), defining the normalized reflectance I/F. In the case of the rings, this I/F includes the effects of finite ring optical depth,

---
[1] The ring opening angle B is the elevation angle of the observer from the ring plane. The phase angle α is the angle between the sun, the viewed target, and the observer, or the angle between the sun and observer as seen from the target. The phase angle is zero in direct backscattering.

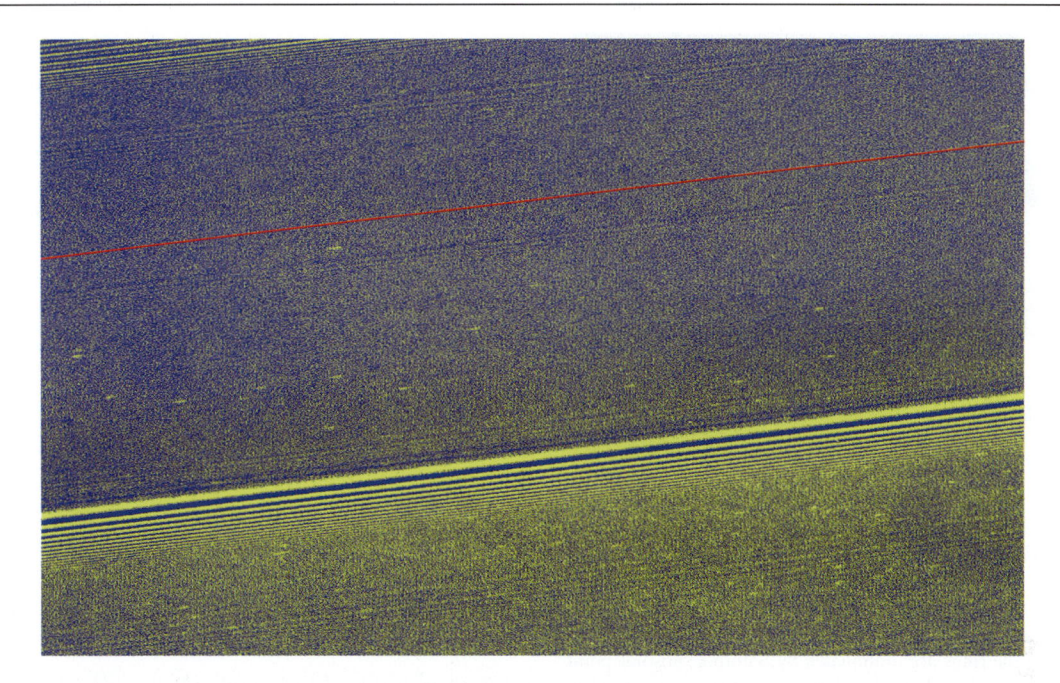

Fig. 15.10 Cassini image PIA10505 visually indicating how the abundance of propellers (each appearing as a short, bright dash along the orbit direction) is localized in distinct radial belts; the red line is at 128,600 from Saturn, at the inner edge of the central, largest belt of Tiscareno et al. (2009). Note that the propellers are apparently unaffected by the Prometheus 9:8 density wave at 128,946 km, towards the bottom of the image. For closeup views of propeller structure see Chapter 14 Image credit: JPL and NASA

the effects of multiple scattering between particles, and the properties of individual particles which can be regarded as small (but very irregular) moons. Moreover, rings generally contain some admixture of wavelength-size "dust" particles which scatter light in a much different way than macroscopic objects, so the variation of ring I/F with phase angle can become quite complicated; the reader is referred to Cuzzi et al. (1984, 2002) or Cuzzi (1985) for more detail on photometric definitions.

Observations of Saturn's rings from Earth are restricted to solar phase angles $\alpha < 6°$ and ring opening angles $B < 26°$, but in spite of these limitations a great deal of interesting variation with viewing geometry has been seen. The Voyager 1 and 2 flybys in 1980 and 1981 provided snapshots of the rings at two illumination geometries over a wide range of phase angles. Cassini observations are a quantum step forward in covering a full range of viewer geometry and solar incidence angle, and in providing numerous stellar and radio occultations of the rings to determine their optical depth at a variety of elevation angles and longitudes. Cassini also provided our first true spectroscopy at UV, near-IR, and thermal-IR wavelengths.

In most cases we do not observe particles acting alone, but as a thick slab where particles can cover and illuminate each other. Relating the overall ring reflectance to the individual particle reflectance is a complex matter (Section 15.4.6.1). Once the albedo and phase function of a typical particle are known from models of the ring layer, one then turns to a different kind of model to infer the particle's composition by modeling multiple scattering of photons in its granular regolith, complicated by facet-related shadowing effects. Several models have emerged to handle this problem. All of them are simplified and their various assumptions introduce uncertainty in the properties inferred – primarily, regolith grain size and composition. These models are discussed in Section 15.4.6.2. Also, numerous Cassini occultation studies have shown that, on a local scale of hundreds of meters, most of the rings resolve into inhomogeneous collections of dense "self-gravity wakes" which are azimuthally extended, tilted to the orbital direction, and perhaps entirely opaque, separated by much more transparent gaps (see Chapters 13 and 14). This structure greatly complicates the modeling of observed ring brightness as a function of viewing geometry.

Another complication in studies of ring particle properties is that the rings are not only illuminated by the sun, but also by reflected light from Saturn, which is not spectrally neutral, and the relative importance of this illumination varies with viewing geometry. The top panel in Fig. 15.11 is a VIMS reflected light image at a wavelength where scattered light from Saturn is very low due to strong methane absorption in Saturn's atmosphere. The bottom two panels represent ring brightness at a wavelength where Saturn's methane is known to absorb, but weakly, thus allowing us to detect where "Saturnshine" is reflected from Saturn to the rings and then back to the spacecraft. The most obvious effect is the angular brightening near the 10-o-clock position due to backscattering of light from Saturn's fully lit hemisphere by the rings. At this phase angle (135 degrees), the Saturn

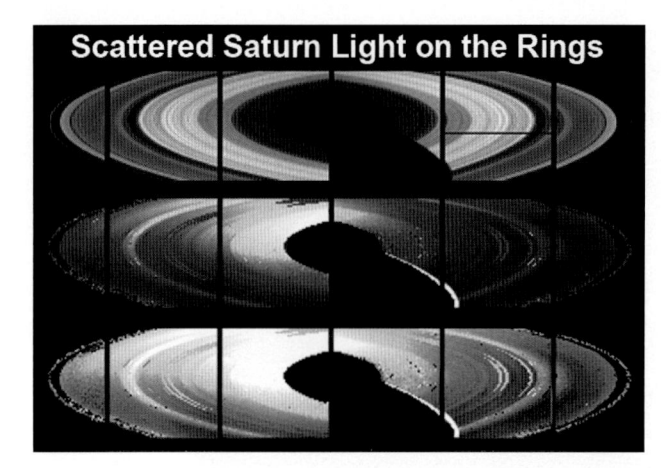

Fig. 15.11 Effect of Saturnshine on the rings as observed by VIMS. *Top panel*: monochromatic image at 2.2 μm where methane absorbs strongly and Saturnshine is minimal; *Center*: methane absorption strength image at 1.17 μm (a weaker absorption band which is detectable but where the planet remains bright) stretched between 0–50%; *Bottom*: the 1.17 μm image stretched between 0–15%. Saturnshine is a maximum where the rings are illuminated by Saturn's subsolar point (noon), but there is a bright streak along the shadow edge due to light refracted through the atmosphere

noon position on the B ring reflects so much Saturnshine that the methane band depths in ring spectra are on the order of 50% (the C ring is even more strongly affected). A less obvious but even stronger effect is the bright line tracing the shadow edge on the rings, due to Saturn's penumbra or light refracted through its high atmosphere onto the rings. There is some evidence for contamination of C ring spectra taken during Saturn Orbit Insertion (SOI), where the C ring pointing was extremely close to the edge of the planet's shadow (Nicholson et al. 2008). Other (primarily radial) variations in the lower two panels of Fig. 15.11 may not represent Saturnshine, but instead actual ring spectral variations between, specifically, the Cassini Division and C ring, and the A and B rings. At low phase angles (characterizing all Earthbased observations), models suggest that Saturnshine should be in the percent range (Dones et al. 1993; French et al. 2007) – negligible except when searching for extremely weak non-ice spectral signatures where strong atmospheric methane absorptions are also known to lie.

15.4.2 Global VIMS Ring Spectra and Overall Composition

VIMS spectra of the lit face of the rings at low phase angles, and near (but not too near) the shadow boundary avoid all the complications of Saturnshine and most of the complications of multiple interparticle scattering of dust grain forward scattering (Section 15.4.6.1), so are most easily interpreted as the spectra of individual ring particles with grainy regoliths (at least, in spectral shape). Such ring-averaged spectra are shown in Figs. 15.12–15.15 (see also Clark et al. 2008a). At low phase angles, the spectra show classic crystalline ice spectral features, except for the steep red slope at wavelengths <550 nm which is caused by some unknown UV absorber.

Note in Figs. 15.12 and 15.13, there is no evidence for any C-H stretch organic signature in the 3.3–3.5 μm region, or CO_2 signature at 4.2 μm wavelength (see also Nicholson et al. 2008), such as seen on Iapetus. Some ices, notably CH_4 in the 2.3 μm spectral region and CO_2 at 4.2 μm, are more strongly absorbing than water ice at those wavelengths and would be visible in the VIMS spectra if their mixing fractions were larger than a fraction of a percent. It may be of interest that the CO_2 feature seen in Saturn's moons is restricted to regions where dark, non-icy material is more prevalent and is generally absent in regions where water ice dominates (Buratti et al. 2005; Clark et al. 2005; Clark et al. 2008b; Cruikshank et al. 2007; Filacchione et al. 2007; Coradini et al. 2008). On the other hand, a number of tholin-like organics that have been proposed to redden the rings are much more weakly absorbing than ice in the 3.4 μm region, and might escape detection at percent and smaller abundances (see Section 15.4.8).

These broadly averaged spectra are also useful for addressing the crystalline-amorphous balance in the water ice of the rings. Poulet et al. (2003) inferred from IRTF data that the A and B rings were primarily crystalline, and Cassini confirms this (Figs. 15.14 and 15.15); however, Poulet et al. also suggested that the C ring might contain primarily amorphous ice, whereas the Cassini observations show that the C ring ice (along with the A and B ring ice) appears to be crystalline (see, e.g., the shape and location of the sharp 3.1 μm feature in Fig. 15.15 (Clark et al. 2008a).

The 1.5- and 2-μm band positions also favor crystalline ice: ring spectra generally show the 2 μm band shifted towards 2.05 μm, more compatible with the crystalline form (for amorphous ice the band is slightly shorter than 2.0 μm). Similarly, the 1.5 μm complex band position also favors crystalline ice. An analogous effect is seen on the disk-integrated spectra of the Saturnian satellites (Filacchione et al. 2008a,b).

15.4.3 Regional and Phase Angle Variations of VIMS Ring Spectra

Below we break the VIMS spectra, from Clark et al. 2008a, down further, showing how the I/F of each main region varies with phase angle (Figs. 15.16–15.20). Spectra at high phase angles are complicated by multiple scattering between

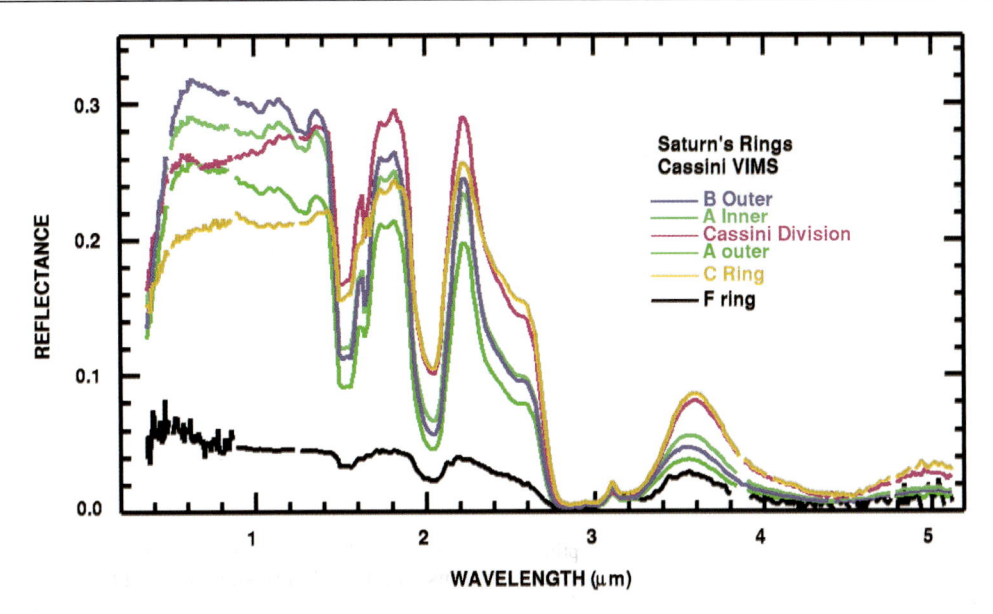

15 Ring Particle Composition and Size Distribution

Fig. 15.12 Average reflectance spectra of different regions of the main rings (A, B, CD) as measured by VIMS. The ring distances for each spectrum are: F-ring: 140,200 km, A-outer: 134,000–136,700, A-inner: 123,700–125,700, Cassini Division: 118,000–125,700, B-outer: 103,000–117,000, C-ring: 75,000–87,000 km. Data from Cassini VIMS Rev 75, LATPHASE observation at 19 degrees phase angle. The spectra are fill-factor corrected (from Clark et al. 2008b)

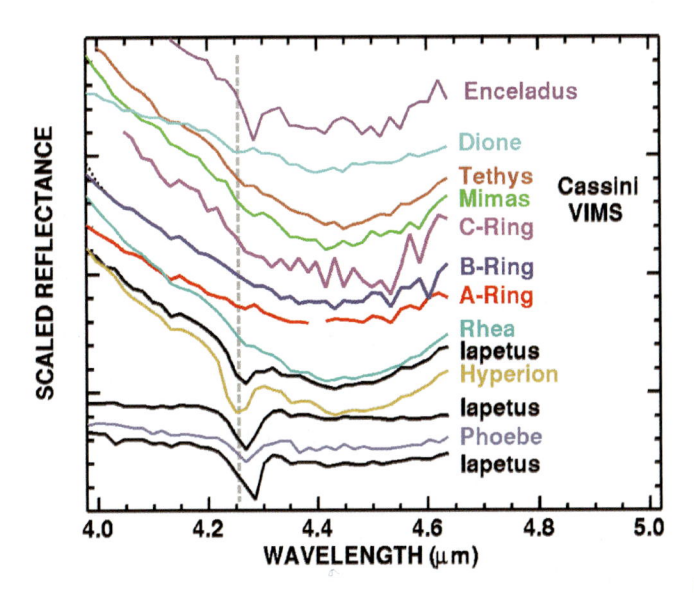

Fig. 15.13 Changing abundance of CO_2 in the icy satellites and rings. No CO_2 is seen in the rings. From Clark et al. (2008b)

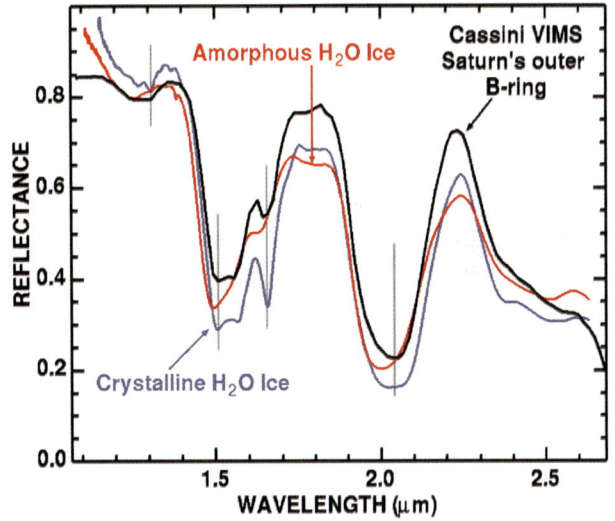

Fig. 15.14 B-ring spectrum (*black*) is compared to amorphous (*red*) and crystalline (*blue*) water ice spectral models using optical constants from Mastrapa et al. (2008, 2009). The absorption band positions and shapes indicate the B-ring spectra are dominated by crystalline water ice. The B-ring spectrum is for the 103,000 to 117,000 km region. From Clark et al. (2008b)

ring particles (which deepens spectral features) and also by primarily diffractive forward scattering by dust grains (which is spectrally featureless). The main rings show decreasing I/F with increasing phase angle, and strong ice spectral features, up to at least 135 degrees phase angle (Figs. 15.16–15.19), as expected for large particles with grainy surfaces. The F-ring I/F *increases* with phase angle from 19 to 178.5 degrees, indicating forward scattering by a predominance of small particles (Showalter et al. 1992). But the F-ring also shows the 1.5 and 2 μm ice bands and 3.1 μm Fresnel peak (Fig. 15.20), consistent with large grains of crystalline ice, showing that it also contains particles at least several millimeters in diameter (Clark et al. 2008a,b), consistent with the fact that the F ring core, at least, displays a significant (if variable) radio occultation signal (Section 15.2).

The weakening of the main 1.5-, 2-, and 3 μm ice spectral features in the 178.5 degree phase spectra of all the rings

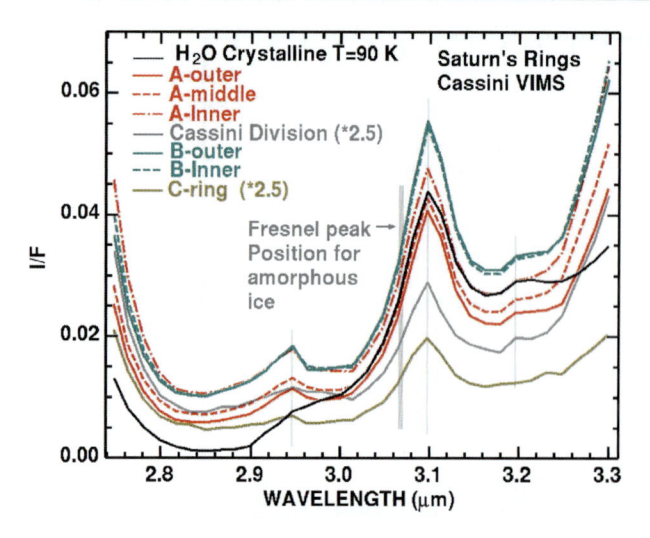

Fig. 15.15 The 3.1 μm Fresnel reflectance peak in Saturn's A-ring (*red*), B-ring (*blue*), Cassini Division (*gray*), and C ring (*gold*). VIMS spectra are compared to that of crystalline water ice (*black*). The observed peak width and position, along with the side features at 3.2 and 2.94 μm, also indicate crystalline H_2O ice. The ring distances for each spectrum are: A-outer: 134,000–136,700 km, A-middle: 126,000–134,000, A-inner: 123,700–125,700, Cassini Division: 118,000–125,700, B-outer: 103,000–117,000, B-inner: 92,000–103,000, C-ring: 75,000–87,000 km. The feature at 2.94 μm occurs at an order-sorting filter gap in the VIMS spectrometer, so has greater uncertainty. From Clark et al. (2008b)

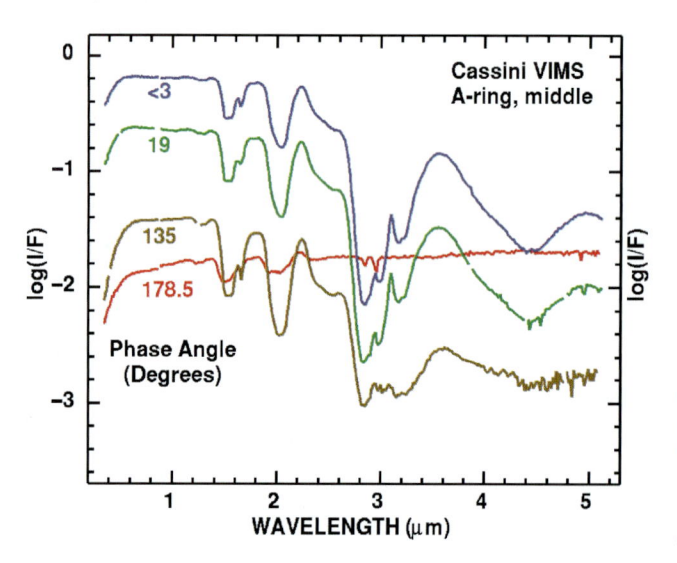

Fig. 15.16 Cassini VIMS spectra of the middle A ring (126,000–134,000 km) as a function of phase angle. The <3 degrees phase is from Rev 44, 0PHASE001, 19 degrees from Rev 75, LATPHASE001, 135 degrees from Rev A, LATPHASE001, 178.5 degrees from Rev 28, HIPHASE001, From Clark et al. (2008b)

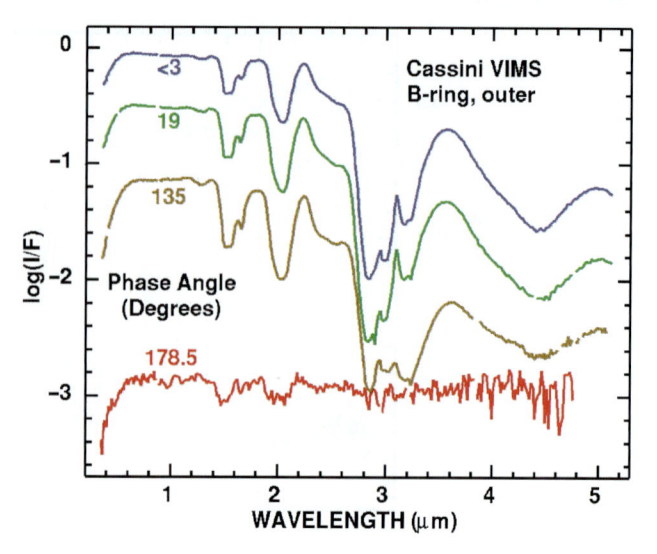

Fig. 15.17 Cassini VIMS spectra of the outer B ring (103,000–117,000 km) as a function of phase angle (observation sets are the same as in Fig. 15.16). From Clark et al. (2008b)

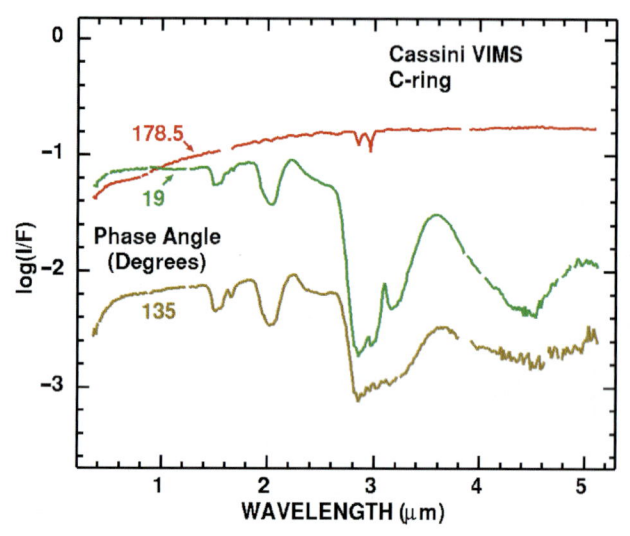

Fig. 15.18 Cassini VIMS spectra of the C ring (75,000–87,000 km) as a function of phase angle (observation sets are the same as in Fig. 15.16). From Clark et al. (2008b)

(and their complete absence in the F ring and C ring) is an indication that the observed brightness at high phase angles is primarily forward scattering by small, free floating grains. That is, the light scattered at this geometry is primarily diffraction and does not sample the material of the particle. The Cassini division appears intermediate in this regard.

The broad hump in spectral shape in the C and F-rings at very high phase angles, being dominated by diffraction, constrains the grain size distribution, and detailed modeling is in progress (Hedman et al. 2008, Vahidinia et al. 2008). The UV absorber and some weak 1.5- and 2.0 μm features are visible in all the main ring spectra to the highest observed phase angle. The simplest interpretation of this is that there is some fraction of multiply scattered light (in the particle regoliths or possibly between large ring particles) reaching the observer, perhaps mixed with a not-entirely-dominant, spectrally featureless contribution from tiny, forward-scattering grains.

In the highest phase angle spectra, two dips appear near 3 μm: the first is probably the Christiansen frequency of

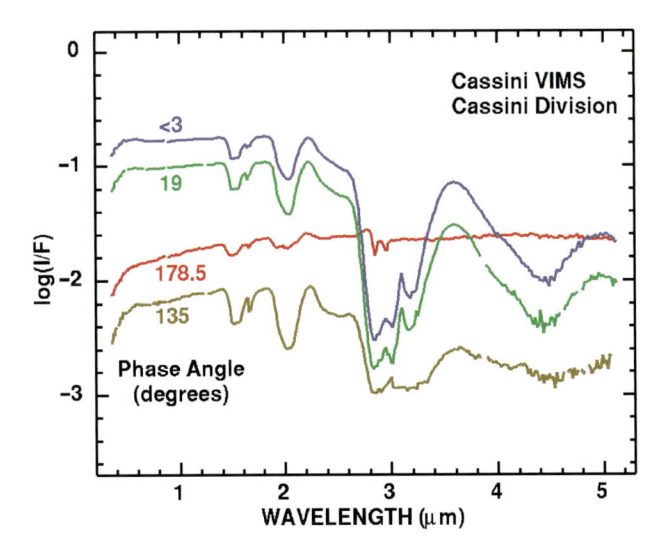

15 Ring Particle Composition and Size Distribution

Fig. 15.19 Cassini VIMS spectra of the Cassini Division (118,000–125,700 km) as a function of phase angle (observation sets are the same as in Fig. 15.16). From Clark et al. (2008b)

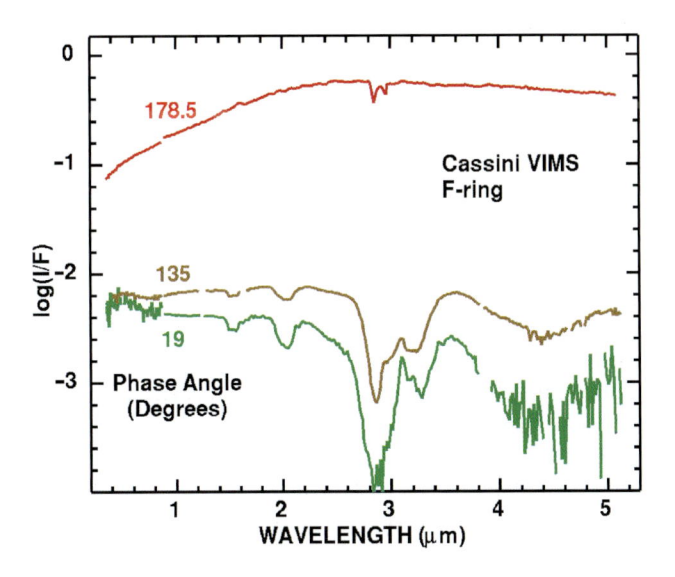

Fig. 15.20 Cassini VIMS spectra of the F ring as a function of phase angle (observation sets are the same as in Figure 15.16). From Clark et al. (2008b)

water ice (Nicholson et al. 2007, Vahidinia et al. 2008, Clark et al. 2008b); the second is at the location of an order sorting filter gap in the VIMS instrument, as well as the N-H stretch fundamental, so needs confirmation by a different instrument or observing geometry.

15.4.4 UVIS Spectra of the Main Ring Regions

UVIS I/F spectra are shown in Fig. 15.21 at four 4,000-km-wide locations in the main rings (Bradley et al. 2009). The bin centers are at 87,400 km (C ring), 111,400 km (B ring), 119,400 km (Cassini Division), and 127,400 km (A ring) from Saturn center, respectively. The sharp decrease in I/F below 175 nm is due to water ice absorption of incident solar photons. Also apparent above 175 nm is the variation in I/F for the different regions of the rings. The difference between the ring brightnesses are due at least partly to their different optical depths.

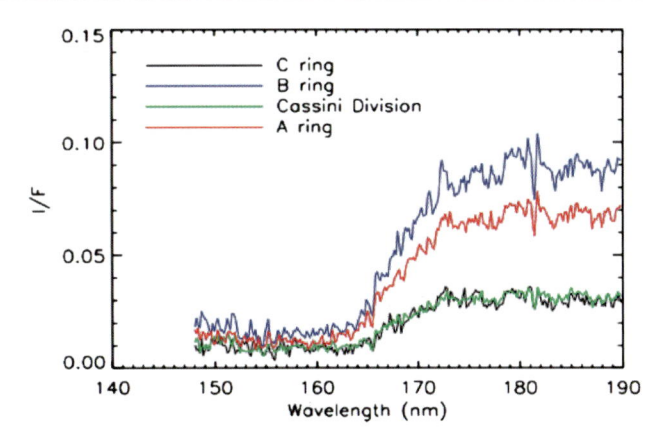

Fig. 15.21 The I/F at ultraviolet wavelengths from the Cassini UVIS spectrometer is shown at four different locations in the main ring (Bradley et al. 2009)

15.4.5 Radial Profiles of ISS and VIMS Spectral Properties

ISS, VIMS, and CIRS observations can be used to explore the radial variation of ring composition in a qualitative fashion, until detailed modeling allows ring particle albedos (and their spectral variation) to be extracted from ring brightness (see Sections 15.4.1 and 15.4.6.1).

15.4.5.1 Radial Profiles of ISS and VIMS Spectral Properties

Estrada and Cuzzi (1996) and Estrada et al. (2003) generated color ratios as a function of radius from Voyager color images. Spectral slopes are equivalent to color ratios in bright regions, and are less sensitive to uncertain backgrounds in regions of low I/F. We create normalized spectral slopes S_{ij} between wavelengths λ_i and λ_j, for the full ring system from lit face data, using both VIMS and ISS data. For VIMS data we present slopes between 350–520 nm ($S_{350-520}$) and 520–950 nm ($S_{520-950}$) (Filacchione et al. 2007, 2008a,b; Nicholson et al. 2008). The 520–950 nm slopes are new to Cassini, and were not observed by Voyager. Cuzzi et al. (2002) attempted some analysis of HST profiles at long visual wavelengths, but were unsuccessful because of a combination of lower resolution and wavelength-dependent scattered light problems. The VIMS spectral slopes are defined by the best fitting linear trend to the I/F in the spectral ranges $\Delta\lambda_{ij}$ (where $\Delta\lambda_{ij}$ is given in μm):

$$S_{ij} = \frac{I/F_j - I/F_i}{(I/F_j)\Delta\lambda_{ij}} \quad (15.15)$$

Normalization by the I/F itself removes illumination effects and decouples the ring color from its brightness. For the ISS data we simply construct slopes between the various filter bands UV3 (0.34 μm), BL1 (0.44 μm), and GRN (0.55 μm). Similarly (for the VIMS data) we can define a Band Depth metric BD_i, for a band at wavelength λ_i, typically the 1.5 and 2.0 μm water ice bands, as

$$BD_i = \frac{I/F_{cont} - I/F_i}{I/F_{cont}} \quad (15.16)$$

where I/F_{cont} is the average continuum I/F on both sides of each band (1.345 & 1.790 μm, and 1.790 & 2.234 μm respectively).

The data of Fig. 15.22 cover the lit face of the rings.[2] As found by Estrada and Cuzzi (1996), Estrada et al. (2003), and Cuzzi et al. (2002), the slopes (or color ratios) do not merely echo the radial ring brightness variations, but show uncorrelated variations of their own. The C ring and Cassini Division particles are "less red" or more neutral in color than the A and B ring particles, and the colors vary smoothly with radius across very abrupt ring boundaries. The two Cassini ISS profiles split the Voyager spectral range into two spectral ranges, and the radial behavior is different between them at several locations (note 83,000–91,000 km, 105,000–110,000, and 120,000–125,000 km) – probably indicating radial variation of composition. Notice how the optically thick central B ring is the reddest region in $S_{440-550}$, but the inner B ring is reddest in $S_{340-440}$ (has the deepest UV absorption).

Figure 15.23 shows comparable results from a VIMS radial scan taken at Saturn Orbit Insertion (SOI; Nicholson et al. 2008). The $S_{350-550}$ color ratios (blue curve) are in good qualitative agreement with those of Estrada et al. (2003) and those in Fig. 15.22, even though the VIMS observations were of the unlit face and the other observations were of the lit face. In particular, smooth radial variations are seen across abrupt boundaries between regions of different optical depth, such as the A ring inner edge denoted CD-A. The inner part of the Cassini division (118,000–120,500 km) is less red than the A ring at short visual wavelengths ($S_{350-550}$) but more red at long visual wavelengths ($S_{550-950}$). Figure 15.23 also shows that $S_{350-550}$ correlates quite well with the water ice band depths $BD_{1.5}$ and $BD_{2.0}$. This suggests that the UV absorber is localized to, and perhaps even residing within, the water ice regolith grains. The distinctly different behavior of $S_{550-950}$ indicates that some different material has a greater abundance in the Cassini Division.

Following Nicholson et al. (2008), we created new, full-ring visual slope and BD plots from VIMS lit face data at somewhat lower resolution (Fig. 15.24).[3] The VIMS visual slope profiles of Fig. 15.24 repeat the overall behavior of the ISS profiles in Fig. 15.22, but at lower resolution and with more scatter due to lower fidelity geometrical registration; they are adequate for our purposes of exploring general regional behavior (Section 15.4.9). The two water ice BD radial profiles are correlated, showing the same radial variations; the BD are largest in the A (130,000 km to Encke gap) and B rings (from 104,000 to 116,000 km). In the outer B ring (from 104,000 to 117,000 km) the BD are almost flat, with a local minimum at 109,000 km; in the central B ring (from 98,500 to 104,000 km) there are several regions with high BD, coinciding with the visual wavelength "red bands" of Estrada and Cuzzi (1996); in the inner B ring (92,000 to 98,500 km) the BD are flat to moderately decreasing towards the inner part; Nicholson et al. (2008), in their higher

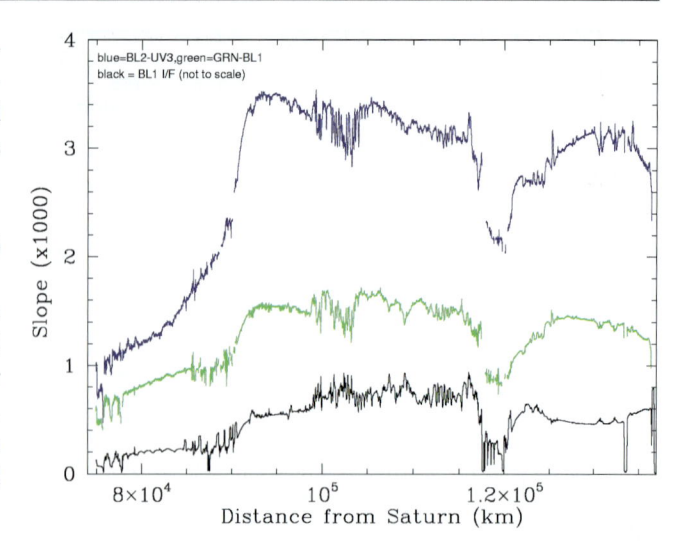

Fig. 15.22 Radial profiles of ring spectral slopes between ISS UV3-BL1(*blue line*) and BL1-GRN (*green line*) filters, compared with the ring I/F profile (*black line*) in the BL1 filter[2]

---

[2] Color images in multiple filters were obtained on 2004-day 347 as part of ISS observation RADCOLOR001_PRIME, at phase angle = 45.2°, elevation angle = 4.1°, and distance from Saturn of approximately 120000 km (7.2km/pixel). The data were calibrated using standard techniques and scanned radially with approximately 100 pixel azimuthal averaging. See Porco et al. (2005) for a description of the filter wavelengths and widths.

[3] The data come from rings mosaic S36-SUBML001, acquired by VIMS on a CIRS-prime observation, on 19–20 December 2007 with a solar phase angle of 32°, a solar elevation angle of −12° and from a mean distance of about 545000 km, giving a radial resolution of 125 km.

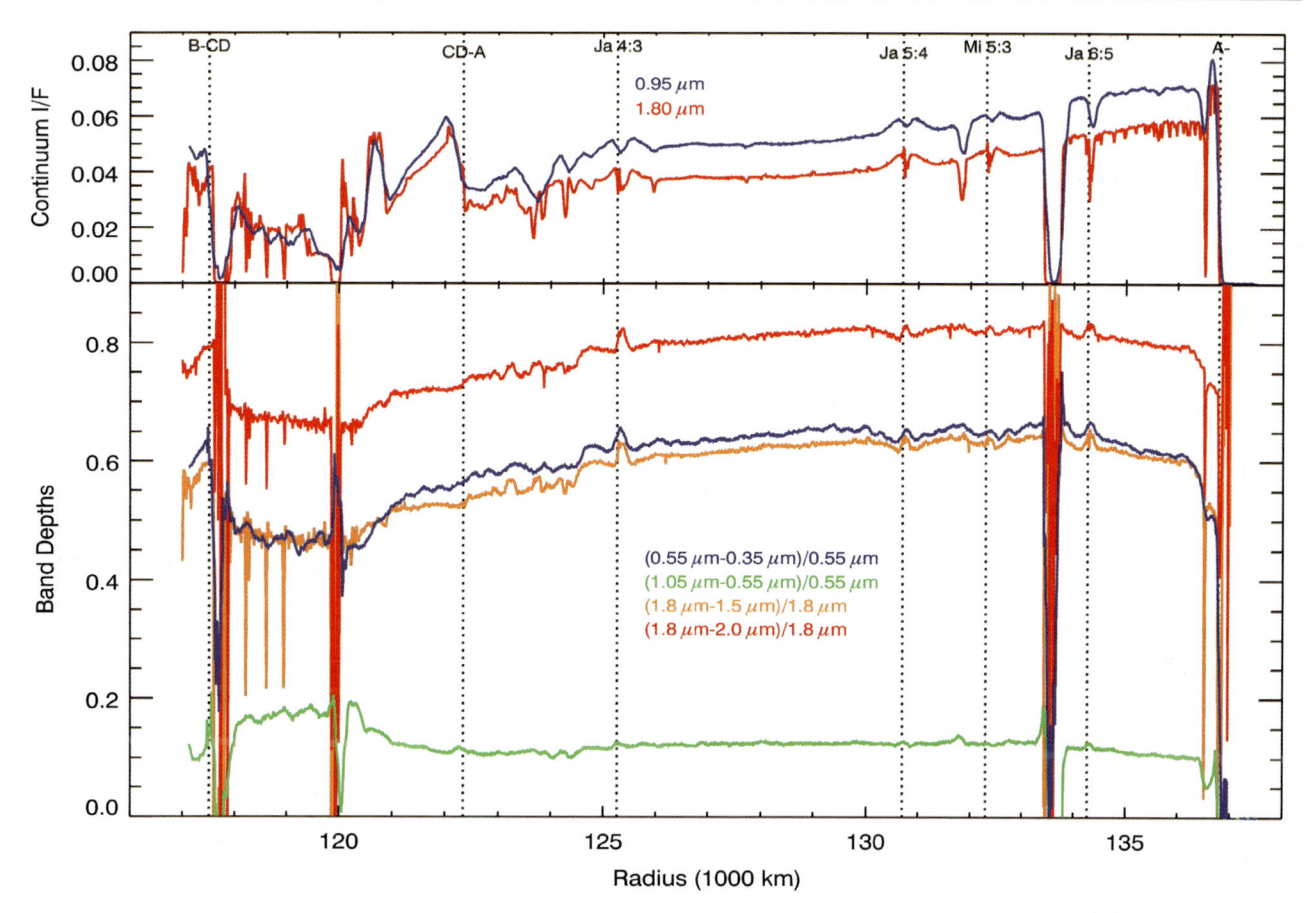

Fig. 15.23 *Top*: Radial profiles of ring I/F at wavelengths where water ice does not (*blue*) and does (*red*) absorb light, from SOI VIMS data; (*bottom*) comparison of two different visual wavelength spectral slopes ($S_{350-520}$ and $S_{520-950}$) and two different water ice band depths ($BD_{1.5}$ and $BD_{2.0}$). $S_{350-520}$, indicating the short-wavelength redness of the rings or the abundance of the UV absorber, tracks the water ice band depths very closely, while $S_{520-950}$, indicating the longer-wavelength ring color, does not (Nicholson et al. 2008)

resolution data, find some evidence of structure in the inner B ring. The middle and inner C ring is similar to the inner CD and the outer part of the C ring (outside the Maxwell gap) is similar to the outer part of the CD. As seen in greater detail in the A ring and Cassini Division by Nicholson et al. (2008; Fig. 15.23), $S_{350-520}$ correlates extremely well everywhere with $BD_{1.5}$ and $BD_{2.0}$, indicating a close spatial correlation between the UV absorber and the water ice, but $S_{520-950}$ is decorrelated from $BD_{1.5}$, $BD_{2.0}$, and $S_{350-520}$, hinting at a different constituent.

15.4.5.2 Particle Albedo Variation from CIRS Ring Temperature Profiles

It is possible to estimate the albedo of a ring particle from its physical temperature; lower albedo particles absorb more sunlight and are warmer. Cassini CIRS observations over the 10–100 µm spectral range can be fit spectrally to determine the physical temperature of the ring particles, assuming the particle emissivity is independent of wavelength (see Section 15.4.7.2 for how emissivity variation is a major concern at longer wavelengths, however). In Fig. 15.25 we compare a radial profile of ring temperature obtained by CIRS[4] with a VIMS BD profile (Fig. 15.24).

The C ring and Cassini Division particles are considerably warmer than the A and B ring particles, as was first observed by Voyager when the Sun was at a much lower elevation angle (Hanel et al. 1982; see Esposito et al. 1984 for a review of ring thermal models). This is a direct indication of lower albedos, consistent with the idea that the two lower optical depth regions are more polluted by non-icy material. Low albedos of the C ring and Cassini division particles (0.15, relative to 0.5 for the A and B ring particles) were reported by Smith et al. (1981). A more polluted particle composition can explain the smaller VIMS water ice band depths seen

---

[4] CIRS lit face scan, on the West Ansa, obtained in 2006 (day 349) near zero phase angle (∼5.9 deg) when the Sun was −14.6 deg south of the ring plane. The radial distance between each CIRS footprint was ∼100 km on the ring plane, although the radial resolution was limited by the field of view to ∼1700–1800 km. For clarity, the data have been binned every 200 km.

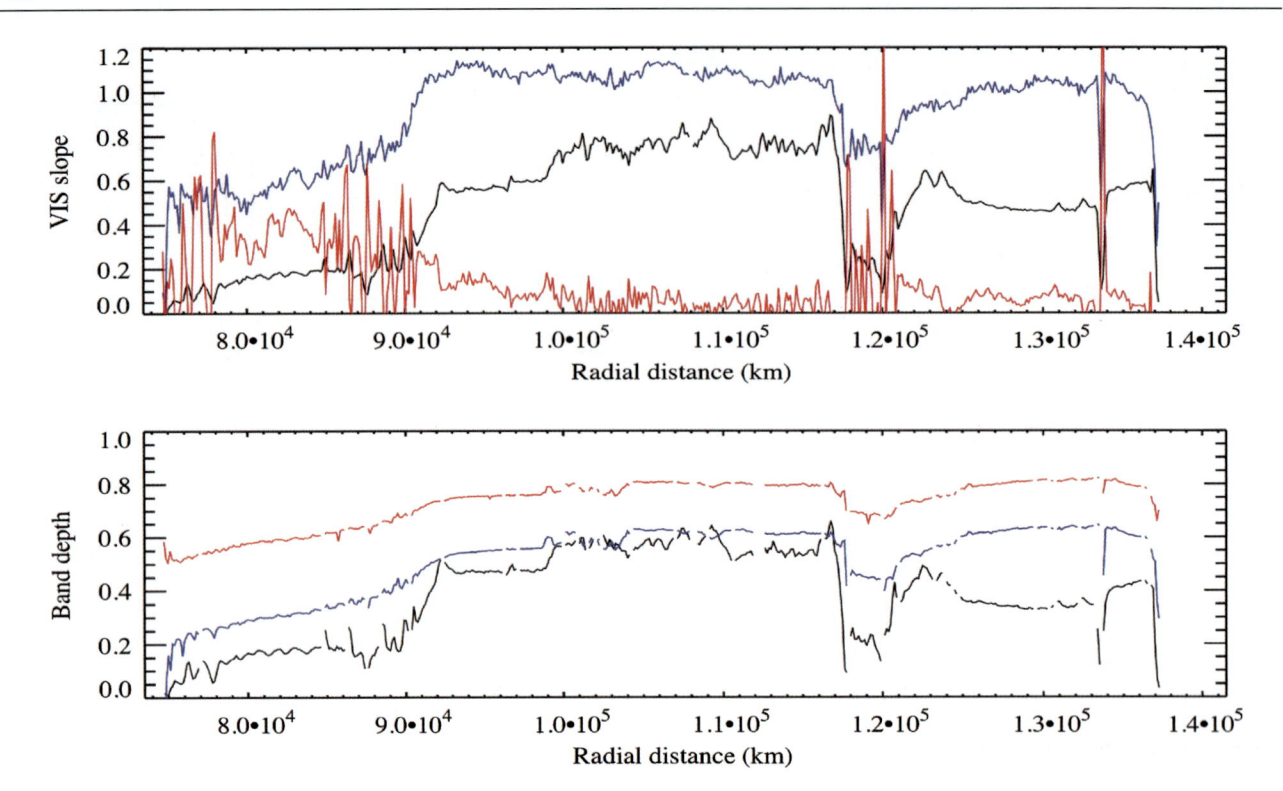

Fig. 15.24 *Top panel*: Radial profiles of the visible spectral slopes $S_{350-520}$ (*blue line*, magnification factor ×300) and $S_{520-950}$ (*red line*, ×1,000); for comparison an I/F profile measured at 550 nm is shown (*black line*, ×4). *Bottom panel*: water ice band depths at 1.5 μm (*blue line*) and 2.0 μm (*red line*); an I/F profile at 1.822 μm is shown (*black line*, ×3). These 125 km per sample profiles are retrieved by VIMS from S36-SUBML001 mosaic (32° solar phase angle)[3]

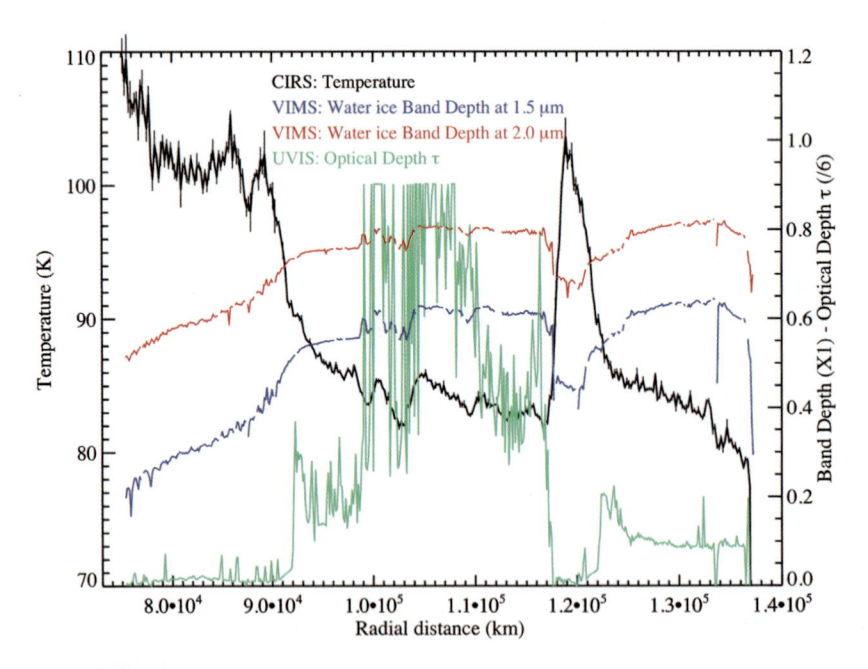

Fig. 15.25 Radial profiles of CIRS ring temperature at low phase angle (*black line*) and VIMS water ice band depth (*red* and *blue lines*). Also shown is a plot of optical depth from UVIS (J. Colwell, personal communication 2007). The ring particle temperature is higher in the C ring and Cassini Division; it also decreases outwards in the A ring and has local minima in dense, optically thick regions of the inner B ring; note that in the 101,000–104,000 km and 108,000–109,000 km regions the temperature correlation with ice band depth reverses sign from the global behavior described above

in these regions (see Section 15.4.7). However, this comparison raises some puzzles. The simple interpretation above is not consistent with the behavior in the 101,000–104,000 and 108,000–109,000 km regions, or in the very outermost parts of the A ring, where both ice band depth and particle temperature decrease in unison. Perhaps here regolith grain size itself is playing a role (Section 15.4.7.2), or perhaps, for the outer A ring, impinging E ring material or plasma play a role (Farrell et al. 2008, Jurac and Richardson 2007).

15.4.6 Modeling Individual Particle Properties from Observed Ring Reflectance

As discussed in the introduction to this section, several distinct stages of modeling are required to extract ring composition and particle properties from ring I/F observations. Here we discuss two in turn.

15.4.6.1 Modeling the Layer of Ring Particles as a Whole

In most cases we do not observe particles acting alone, but as a thick slab where particles can cover and illuminate each other. Relating the overall ring reflectance to the individual particle reflectance is a complex matter. Voyager-era radiative transfer models generally relied on classical concepts of ring structure (low volume density and a vertical structure many well-separated particles in extent). Layers of this kind can be modeled by traditional doubling-adding techniques (e.g., Dones et al. 1993; Doyle et al. 1989). The ring regions and viewing geometries where the classical models are the most successful are those where single scattering is dominant (low phase angle, low albedo particles and/or where the optical depth is low; e.g., Cuzzi et al. 1984, Cooke 1991, Cuzzi et al. 2002, Nicholson et al. 2008). However, even in the Voyager era there were observational indications that these models were unsatisfactory in the A and B rings, suggesting a dense, and indeed dynamically preferred, ring vertical structure (Dones et al. 1989, 1993).

Recent HST (Cuzzi et al. 2002, French et al. 2007) and Cassini observations (Hapke et al. 2005, 2006, Nelson et al. 2006) and their analyses (see Appendix), have called into question the classical, many-particle-thick ring that can be modeled using doubling techniques or their single-scattering limits for densely packed regions of moderate to high optical depth (Cuzzi et al. 2002, Salo and Karjalainen 2003, Porco et al. 2008, Chambers and Cuzzi 2008). Recent ring radiative transfer models using ray-tracing techniques to address layers of closely-packed particles (Salo and Karjalainen 2003, Porco et al. 2008, Chambers and Cuzzi 2008) are removing the limitations of traditional models, and can even handle the anisotropy of self-gravity wakes, but still make a number of simplifying assumptions (spherical particles with idealized surface scattering laws, for instance). Clearly, one cannot extract actual particle albedos from observed ring brightnesses until this modeling problem has been addressed carefully. However, for studying spectral variations of reflectivity rather than absolute value, and at low phase angle observations of the lit face, assuming single scattering by individual surfaces and comparing to laboratory reflectance spectra might be acceptable (see Sections 15.4.2–15.4.5).

A concern in interpreting ring spectra in terms of grain size and/or compositional makeup, is how much light comes from multiple scattering in the ring particle regoliths versus how much comes from multiple scattering between ring particles, or interparticle scattering. The amount of multiple interparticle scattering should increase with phase angle, which will increase the spectral contrast and affect compositional inferences. Cuzzi et al. (2002) found from HST data that the rings redden significantly with increasing phase angle, over the spectral range where the rings are already red. At the same time, they found no tendency for the ring color to vary with ring opening angle at a given phase angle, or where the spectrum of the rings is flat. They concluded that multiple scattering within the near-surface of individual ring particles was important, but that multiple scattering between separate ring particles in a vertically extended layer was not.

An even more sensitive measure of spectral contrast is the I/F ratio between 2.86 and 2.6 μm; 2.86 μm is in the deepest part of the water ice absorption band, and 2.6 μm is a nearby wavelength where absorption by ice is fairly weak; multiple scatterings quickly amplify this brightness difference. Examining the spectral contrast in the Cassini VIMS data, Clark et al. (2008a) found that the 2.86/2.6 μm brightness ratio in the rings stayed relatively constant at low phase angles, decreasing slightly at 135° phase and thus indicating a small increase in multiple scattering between low and moderate phase angles. The phase variation of the 2.86–2.6 μm ratio is always weaker for the rings than seen in pure ice, consistent with scattering from an icy regolith with trace contaminants but inconsistent with substantial interparticle-scattering (Clark et al. 2008b).

These results suggest that there is little interparticle scattering in the rings at low phase angles (at least), which allows us to directly compare low phase spectra of the rings with laboratory analog spectral data and standard regolith radiative transfer models for single surfaces (next section).

15.4.6.2 Modeling Ring Particle Regoliths

This step consists of modeling the reflectance of a surface element of a single ring particle in terms of the combined multiple scatterings of all the grains in its regolith. This general class of models is often generally referred to as "Hapke theory" after its most popular variant (Hapke 1981, 1993), and is more generically called a Regolith Radiative Transfer (RRT) model. The models start by deriving an individual grain albedo, assuming geometric optics where a particle is so much larger than a wavelength that Fresnel reflection coefficients can be averaged over; thus the theory is not applicable to regolith grains of wavelength size or smaller. RRT models then proceed by calculating the multiple scattering between grains, in different ways. Reflectance spectra can then be computed for pure minerals or mineral mixtures with any grain size distribution, using the known refractive indices of the various materials. Clark and Roush (1984) also showed that a reflectance spectrum can be inverted to determine quantitative information on the abundances and grain sizes of each component. The inversion of reflectance to quantitative abundance has been tested in laboratory mixtures (e.g., Johnson et al. 1988, 1992; Clark 1983, Mustard and Pieters 1987a, 1989; Shipman and Adams 1987; Sunshine and Pieters 1990, 1991; Sunshine et al. 1990; Gaffey et al. 1993, Mustard and Pieters 1987b, Li et al. 1996, Adams et al. 1993 and references therein). Generally the results are fairly good when the regolith grains are all very large compared to a wavelength.

Perhaps the oldest RRT is generically referred to as "Mie-Conel" theory (Conel 1969). Instead of assuming geometrical optics to get the grain albedo, it uses Mie theory to determine the albedo and degree of forward scattering for an isolated particle, and a transformation suggested by the two-stream approximation of radiative transfer to handle the multiple scattering component. The approach used by Hansen and McCord (2004) and Filacchione et al. (2008a,b) is similar, replacing the two-stream treatment of multiple scattering with a doubling model. Moersch and Christensen (1995) tested a generic Mie-Conel theory and Hapke theory against actual laboratory reflectance spectra where the composition (SiO_2) and particle size (tens of microns) were known; general agreement was fair, but detailed agreement in and out of absorption features was not good for any of the theories – possibly because here, the grain size was not always much larger than the wavelength. A more recent RRT has been developed by Shkuratov et al. (1999). The albedo of an individual grain is found in a way similar to Hapke theory, but the degree of forward scattering by such a grain is retained. This model does not predict particle phase functions. Poulet et al. (2003) compared results obtained using Shkuratov and Hapke theory, and found that Hapke theory underestimates spectral contrasts relative to Shkuratov theory for the same regolith grain size because of this difference. The size of the effect increases with brighter surfaces. Because of these systematic uncertainties, caution must be exercised regarding quantitative compositional inferences from models such as these.

Moreover, the actual phase function of the ring particle as a whole (which enters into layer models) must account for the role of shadowing by rough surfaces and facets. Poulet et al. (2003) derived a roughness parameter from HST observations over a small range of low phase angles, and found it to be extremely large relative to the value found to characterize icy satellites in general. This suggests that the ring particle surfaces might be extremely lumpy – not inconsistent with our mental picture of ring 'particles' as aggregates. Dones et al. (1993) showed that the wavelength-dependent single-particle phase function $P_\lambda(\alpha)$ was well-matched by a rather strongly backscattering power law: $P_\lambda(\alpha) = c_n(\pi-\alpha)^n$ where c_n is a normalization constant and n = 3.3 gives a good match to the phase function of Callisto and to Saturn's A ring particles. Hapke (1984; 1993, his Chapter 12) and Kreslavsky and Shkuratov (2003, and references therein) present theories including macroscopic shadowing, which result in phase functions having this shape. This strongly backscattering behavior leads to the general dominance of single scattering by the rings at low phase angle (Cuzzi et al. 1984, 2002; Dones et al. 1993).

15.4.7 Laboratory and Model Water Abundance and Regolith Grain Size

We first deal with the more straightforward observational geometries (ring spectra at low phase angles) and describe compositional and grain size implications we can obtain from these observations. In the Appendix, we present a discussion of modeling the zero-phase opposition effect, which combines the complications of regolith and ring layer **properties**.

15.4.7.1 Water Ice Band Depths from VIMS Data

Laboratory ice spectra at various grain sizes are shown in Fig. 15.26. The spectral contrasts from 2.6 to 2.86 μm for spectra of fine to medium sized grains provide the best match to A and B ring data (Figs. 15.16–19).

Below we interpret band depth variations in terms of grain size, but an important caveat should be kept in mind. The overall light backscattered by a regolith-covered particle is (nonlinearly) related to the energy absorbed by a single regolith grain. For grains which are large compared to a wavelength (the regime of all current RRT models), the energy absorbed by a grain is determined by the product of the

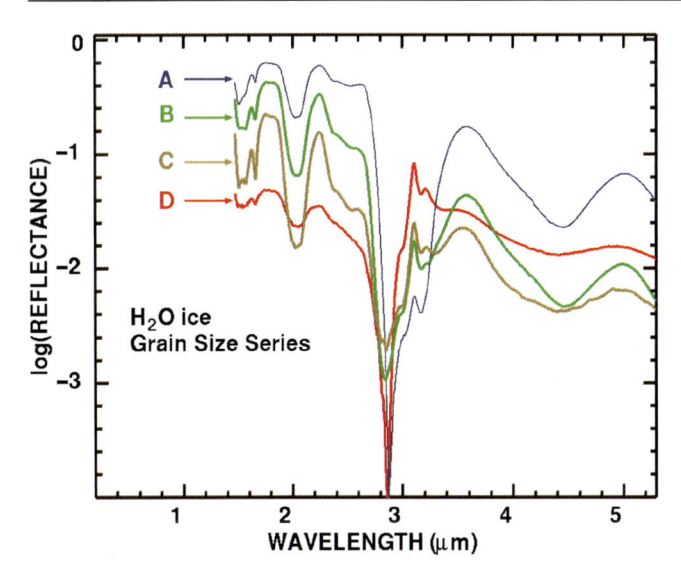

15 Ring Particle Composition and Size Distribution

Fig. 15.26 Log reflectance of pure laboratory ice as a function of ice grain size (**a**: <4 microns in diameter, **b**: fine grained, **c**: medium grained, tens of microns, and **d**: an ice block). Temperatures from 80 to 83 K, incidence angle = 20 degrees, emission = 35 degrees, phase angle = 20 degrees. From Clark et al. (2008b)

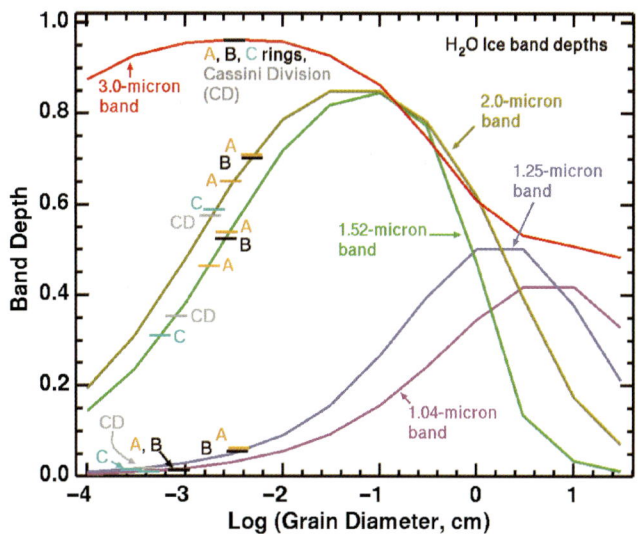

Fig. 15.27 Water ice band depth as a function of grain diameter (calculated using a Hapke model from Clark and Lucey 1984) is compared with observed ice absorption band depths for Saturn's rings. Different ice bands probe to different depths, with shorter wavelength absorptions probing deeper due to the lower absorption coefficients. The A and B-ring band depths imply a depth-independent grain size. The Cassini Division and C ring display a slightly smaller grain size, which may instead be due to a larger proportion of non-icy contaminants in grains of the same size. The two depths for the A-ring are for A-inner and A-outer, with radial ranges given in Figure 15.15. The A-inner band depths are greater than A-outer. From Clark et al. (2008b)

absorption coefficient and a typical path length, say the grain diameter (e.g., Irvine and Pollack 1968, Hapke 1981). Because we know the relative amount of water ice varies with location, we must realize that radial variations in band depth can map out either grain size variation, compositional variation, or a combination. The C ring and Cassini Division particles are more "polluted" with nonicy material, based on their lower albedoes (sect. 15.4.5.2); thus their weaker ice band depths might not represent smaller grains, but less water-rich grain composition.

Clark et al. (2008b) compared the ice band depths in Saturn's rings spectra to that from Hapke models of ice spectra (Fig. 15.27). They found the mean grain size of ice to be on the order of 30–50 μm diameter in the A and B-rings.

Another way to map the regolith grain size is through the I/F(3.6 μm)/I/F(1.822 μm) ratio. Both lab data and models for pure water ice (Clark and Lucey 1984, Hansen and McCord 2004, Jaumann et al. 2008) indicate that the reflectance at 3.6 μm depends on the water ice grain size (higher reflectances are measured for fine grains). The 3.6 μm reflectance is normalized to that at 1.822 μm to remove the effects of radially variable optical depth. Figure 15.28 (Filacchione et al. 2008b) uses a Hansen-McCord (2004) RRT model and infers particle sizes about twice as large, across the rings, as obtained from the analysis of Fig. 15.27. The size difference could be due to the different band used, or to the different regolith radiative transfer model used, illustrating the caveats expressed in Section 15.4.6.2 above. Nevertheless, the radial variation is of interest as a possible indication of compositional variations.

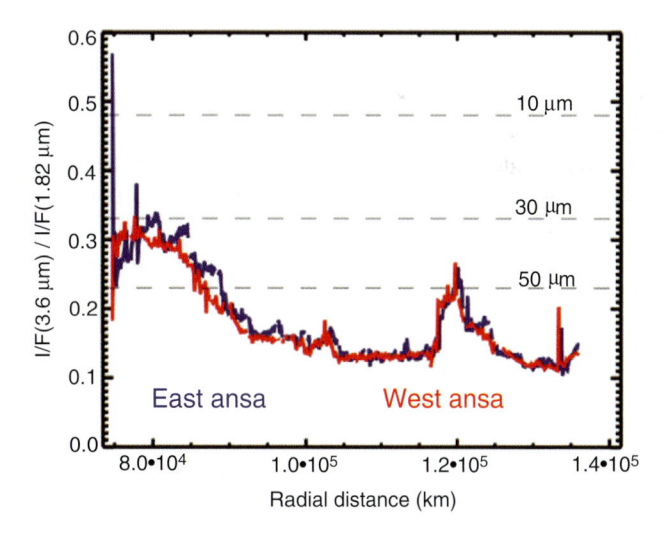

Fig. 15.28 Radial profiles for water ice regolith grain diameters, from 3.6 μm band strengths and the RRT model of Hansen and McCord (2004). Note the factor of two difference relative to the results of figure 15.27, which use a different RRT model

15.4.7.2 Regolith Properties from CIRS Spectra at Long Thermal Infrared Wavelengths

The far infrared spectra of the main rings exhibit a decrease in spectral intensity, or brightness temperature, relative to a black body at wavelengths longer than 100 μm (Fig. 15.29a).

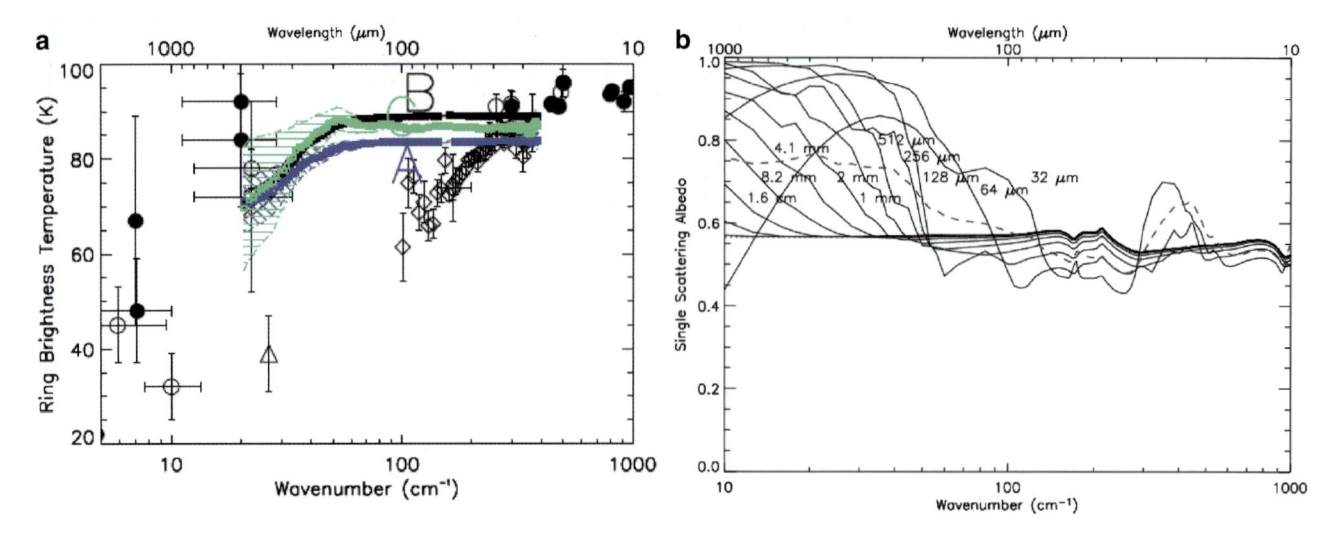

Fig. 15.29 (a) CIRS spectral observations of large regions of the A, B, and C rings (A ring: 123,100 to 128,700 km, B ring: 93,700–118,400 km, C ring: 83,700–90,800 km), converted to "adjusted" brightness temperature (Spilker et al. 2005). Black symbols are various groundbased spectral observations (see Esposito et al. 1984 for a tabulation and discussion). (b) Mie theory calculations for the spectral albedos of water ice particles of different sizes (*solid lines*) and for a powerlaw distribution (*dashed line*). The strong variation near 20–70 μm is due to a water ice absorption feature, and is not seen in ring observations. (From Spilker et al. 2005)

This rolloff in intensity continues to microwave wavelengths where the ring brightness temperatures are only a few degrees K, providing our currently strongest evidence that the rings are nearly pure water ice even below their surfaces (Chapter 2; see Esposito et al. 1984 for a discussion). Spilker et al. (2005) corrected for ring optical depth effects by fitting the shape of the ring intensity spectrum between 100–400 cm^{-1}, where the particle emissivities and ring optical depths are likely to be wavelength-independent. In this way a temperature close to the physical temperature is obtained at the shortest wavelengths, and the rolloff towards long wavelengths can be ascribed to decreasing emissivity of surface grains. The CIRS short-wavelength ring temperatures are lower than groundbased values because they are made at higher phase angles (120° for the A and C rings, 67° for the B ring), where more shadowed area is visible (Spilker et al. 2006, Altobelli et al. 2008).

A simple model was constructed by calculating the spectral albedos of different particle sizes, assuming Mie theory and water ice composition (Fig. 15.29b). High-albedo particles have the low emissivities needed to cause the ring brightness temperature to roll off as seen in Fig. 15.29a. For particles larger than a cm or so, no spectral albedo variation is seen; thus most emitting particles must be smaller in size and are most plausibly ascribed to regolith grains on actual ring particles (Section 15.2). Moreover, strong spectral variation of water ice itself at 20–70 μm wavelengths (Warren 1984; Johnson and Atreya 1996; solid lines in Fig. 15.29b) is not observed, arguing for a broad distribution of regolith grain sizes (dashed line in Fig. 15.29b). Specifically, a powerlaw of the form r^{-q} ($q = 3.4$) provided the average grain albedo as a function of wavelength. Note that other studies have found it advantageous to model a broad regolith grain size distribution (Section 15.4.8). The doubling code used to calculate emergent intensity assumes independent, well-separated particles, not in fact valid in a regolith for these combinations of wavelength and particle size (Section 15.4.6.2). New theories will have to be employed which properly account for close packing of particles which are wavelength-sized, non-spherical and clumpy. However, these results are certainly indicative that the properties of realistic ring particle regoliths can account in a general way for the observed brightness temperature rolloff in the far-infrared.

15.4.8 Global Models of Ring Composition

Poulet et al. (2003) conducted the most recent comprehensive study of the ring composition prior to Cassini arrival (see Poulet et al. (2003) or Chapter 2, for a summary of previous studies of the composition of Saturn's rings). They modeled their ring spectra (Fig. 15.30) with a radiative transfer model by Shkuratov et al. (1999), using ice, a dark colorless component (amorphous carbon) to adjust the albedo, and a UV absorber (organic tholins) to reproduce the UV-visible reddening. Their study concluded that spectra of the A and B rings indicated water ice with no evidence for other volatile ices. The depths of the ice absorptions differed for each ring, and their model called for a wide spread in grain sizes from 10 to 1,000 μm to explain all the band strengths. They concluded that the lower albedo and the less blue slope in the near-infrared reflectance of the C ring indicated a different fractional amount of dark material relative to the A and B

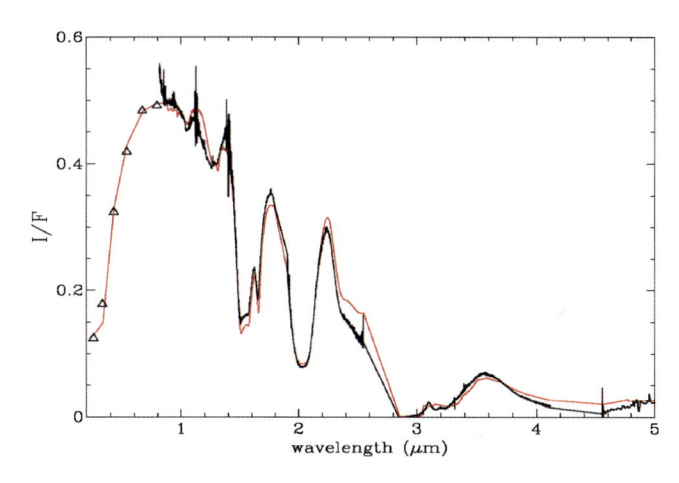

Fig. 15.30 Groundbased visual and near-IR spectra of Saturn's B ring (Poulet et al. 2003) and their best-fit model, which employed three different grain sizes to mimic a grain size distribution, and a combination of ice, carbon, and tholins. See text for discussion. Apparent dip at $\lambda > 3.3\,\mu$m might be an (unobserved) model signature of tholins, or might be an artifact of how refractive indices were handled in the region $\lambda > 2.9\,\mu$m[5]

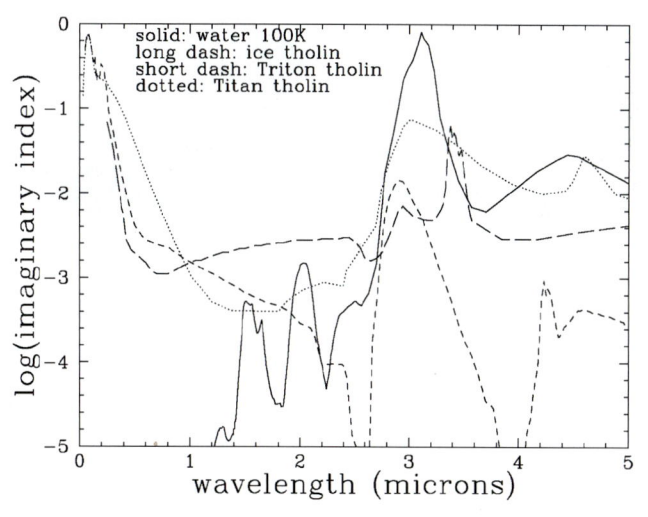

Fig. 15.31 Comparison of the imaginary refractive index n_i of water ice with that of three tholins (data from Cruikshank et al. 2005); the wavelength variation of n_i is primarily responsible for observed absorption features. Only one tholin has a noticeable spectral feature near 3.4 microns, and the relative importance of this feature would be negligible in the sub-percent mixing ratios needed to explain the reddish ring color (notice how n_i for tholins dominates at visual wavelengths, even if the tholins were present in very small abundances

rings, as did Cuzzi and Estrada (1998) and Smith et al. (1981) from Voyager data alone. Moreover, they required a different *kind* of contaminant mixing in the C ring compared to the A and B rings. Poulet et al. (2003) interpreted their C-ring spectra as evidence for C-H absorption in the rings (most easily seen in their ratio spectra). However, C-ring spectra from Cassini VIMS data, obtained near the shadow boundary (but away from the refracted light problem zone of Fig. 15.11), show no such absorptions. The C-H features seen by Poulet et al. (2003) might have been due to light scattered from Saturn directly into the detector.

Their model spectra provided good matches to the UV/Visible spectra of the rings and the ice absorptions in the 1–2.5 μm region, but close inspection of their models shows distortion in the 3.3 μm region which probably represents inappropriate water ice optical constants.[5] Figure 15.31 compares the spectral variation of the imaginary indices of water ice and several tholins. Notice that one of these materials has a spectral feature near 3.4 μm which might be visible in an ice mixture if it had abundance *comparable to the ice*; two other candidates would not be spectrally obvious even in significant proportions. However, in the models of Poulet et al. (2003), less than a percent of Tholins were needed to provide the red color of the rings (the grain refractive index is a volume-weighted average of the indices of the materials

present). Thus it seems that none of these tholins would be obvious in the ring spectra at 3.4 μm, if they were present in the proposed abundances.

"Tholins" are only one kind of organic material – created as huge macromolecules by UV irradiation and/or charged particle bombardment of various simple organics like CH_4, N_2, NH_3, and H_2O (Cruikshank et al. 2005 and references therein). Another possible alternate coloring agent is one of the many classes of "PAHs" (Polycyclic Aromatic Hydrocarbons), which are much simpler molecules typically consisting of large patches of perhaps only a few to dozens of benzene rings (see Section 15.6.3 for more detail). As shown in Fig. 15.32, many PAHs are visually reddish and indeed might be the fundamental source of the reddening caused by Tholins, which are enormous arrangements of many PAHs in different orientations.

Comparisons between the spectra of rings and reddish outer solar system objects which are generally presumed to be the *carriers* of the Tholins (Cruikshank et al. 2005, Barucci et al. 2008) also show some qualitative differences in shape (Section 15.5). Thus, one might wish to explore other options for explaining the ring redness. One such option was suggested to us by the newly-discovered oxygen (O, O_2) rich ring atmosphere (Section 15.5). If there ever were a component of the rings made of fine grained iron, such as is found in meteorites, one can speculate that this iron could be "rusted" over time in this atmosphere into the oxidized iron oxide hematite, which has attractive spectral properties in the ring context.

---

[5] Suspecting a temperature-dependent effect, Poulet et al. (2003) shifted the optical constants of ice by 0.07μm for wavelengths longer than 2.9 μm – right where a glitch is seen in the models (F. Poulet, personal communication, 2008). More modeling work needs to be done using the most up to date optical constants of water ice at the appropriate temperature.

Fig. 15.32 Samples of six different Polycyclic Aromatic Hydrocarbon ("PAH") molecules; from left to right and back to front: $C_{48}H_{20}$, $C_{22}H_{12}$, $C_{20}H_{12}$, $C_{22}H_{12}$; and $C_{36}H_{16}$, $C_{40}H_{18}$, $C_{42}H_{48}$. PAHs dominate the organic component of the interstellar medium. The visual colors are all reddish. These are much smaller molecules than the macromolecular "tholins" that are often used to mimic planetary organics; they may act alone or as subunits within tholins to provide a reddening agent (see also Section 15.6.3)

Nanohematite (very fine-grained hematite or Fe_2O_3) is a strong UV absorber that matches the spectral structure observed in spectra of Saturn's rings (Fig. 15.33) and has no strong IR absorptions. Nanohematite has muted spectral features compared to larger grained hematite, due to crystal field effects at grain surfaces and the high surface to volume ratio when the particles are less than a few tens of nanometers in diameter (Morris et al. 1985). Clark et al. (2008b, 2009) have used combinations of nanohematite and carbon or fine-grained metallic iron to model the spectra of various of Saturn's icy satellites, and showed that admixture of *un-oxidized* fine-grained iron along with the (*oxidized*) nanohematite led to improved fits across the entire visual range. Analog laboratory spectra indicate that nanohematite abundance of only 0.25 wt% is needed to match spectra of the Cassini Division and C-ring, the more heavily contaminated rings. Less nanohematite might be needed to explain spectra of the A and B rings because less dark material is contaminating those rings, allowing increased multiple scattering within the regolith. Such small abundances of

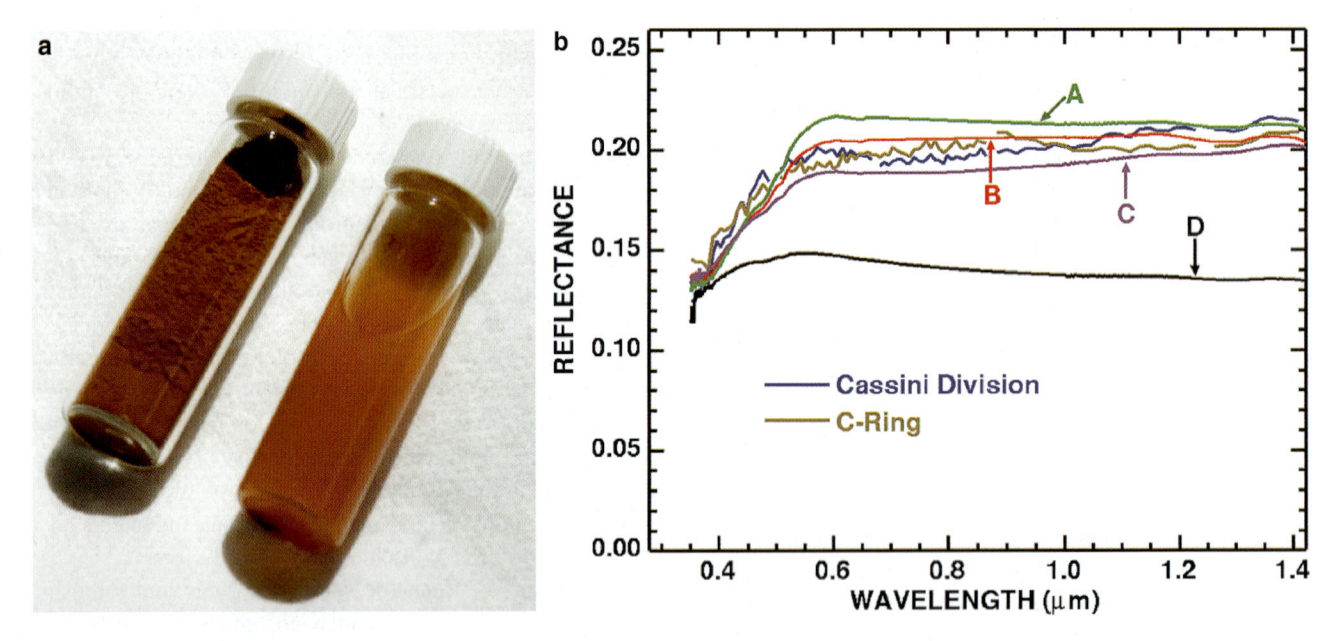

Fig. 15.33 (**a**, *left*): Photo of hematite showing its redness as a solid and suspended in water. (**b**, *right*): Laboratory analog models with metallic iron, nano-hematite, carbon black, and water ice are compared to VIMS spectra of the C-ring and Cassini Division. The Cassini Division VIMS spectrum is from S42, Rev 75 LATPHASE00, Phase Angle = 19 degrees, a 41 pixel average, and the C-Ring VIMS spectrum is from the same observation, Phase Angle = 19 degrees, a 69 pixel average. Spectra A–D are laboratory analogs all at 84–90 K at a phase angle of 20 degrees and have been scaled close to the rings spectra to illustrate varying UV absorber and NIR slopes. A) H_2O + 2% (75:25 nano-iron (metallic): nano-hematite), B) H_2O + 1% (75:25 nano-iron (metallic): nano-hematite), C) H_2O + 0.25% nano-hematite + 0.25% carbon black + 33% metallic iron <10 μm, and D) H_2O + 0.5% carbon black + 0.25% nano-hematite. The nano-metallic iron and nano-hematite samples contain approximately 100 nm diameter particles. Particles much smaller than the wavelength of visible light produce NIR blue slopes while the addition of larger grains of metallic iron produces NIR red slopes in agreement with the ring observations. Clark et al. (2008b) argue that the amount of metallic iron in the ring particles could be 10–30 times smaller than in the sample shown here (to avoid violating microwave observations; Chapter 2), if finer grained iron were used instead of the <10 μm grains shown here

nanohematite would not perceptibly add to the 3 μm ice absorption, but could slightly decrease the contrast in the 2.6–2.86 μm region, making the mixture better fit the ring data than pure ice (see Fig 15.26). This is because the ice index of refraction is so low at 2.86 μm that trace contaminants increase the reflectance, while the 2.6 μm reflectance will remain little changed. This possibility might suggest that the ring redness reflects an extrinsic, rather than an intrinsic, material (see Section 15.5.6). The pros and cons of the hematite, tholin, and PAH options are discussed further in Section 15.6.3.

15.4.9 Comparison of Ring Spectral Properties with Other Icy Objects

Alternate ring origin scenarios have included either regular Saturnian moons, or interlopers from the Kuiper belt, as possible ring parents. In this subsection we present a comparative study of the icy surfaces of the Saturnian system (including rings A, B, C and CD, regular and minor satellites) as well as of some TransNeptunian Objects (TNOs). For this study we selected about 1,500 full-disk observations of both regular satellites (Mimas, Enceladus, Tethys, Dione, Rhea, Hyperion, Iapetus) and minor satellites (Atlas, Prometheus, Pandora, Janus, Epimetheus, Telesto, Calypso and Phoebe) from a wide range of distances, hemispheric longitudes, and solar illumination angles (Filacchione et al. 2007, 2008b). For the main rings, we used an East-West mosaic taken in reflectance by VIMS in high spatial resolution[6] (Filacchione et al. 2008a). In addition, some Earth-based VIS-NIR spectra of TNOs were considered: 1996TO66 (Brown et al. 1999), 1999UG5 (Bauer et al. 2002), 2003UB313 (Licandro et al. 2006a), 2005FY9 (Licandro et al. 2006b), 90377/Sedna (Barucci et al. 2005) and Triton (Buratti et al. 1994; Quirico et al. 1999; Tryka and Bosh 1999; Grundy and Young 2004; Hicks and Buratti 2004). As in Section 15.4.5 we selected the following indicators: $S_{350-520}$, $S_{520-950}$, $BD_{1.5}$ and $BD_{2.0}$.

Figure 15.34 shows the distribution of $S_{350-520}$ vs. $S_{520-950}$. The ring region clusters are circumscribed to simplify the plot. The highest values of $S_{350-520}(>3.0)$ are observed in the B ring (cyan points); the A ring corresponds to $2.5 < S_{350-520} < 3.0$ (red points), the Cassini Division (CD; green points) contains $1.2 < S_{350-520} < 2.0$ while the C ring (blue points) shows $0.6 < S_{350-520} < 1.1$. For $S_{520-950}$ we observe values $-0.05 < S_{520-950} < 0.3$ in the A and B rings, while the CD and C ring show small but noticeable differences ($-0.15 < S_{520-950}(CD) < 0.1$ and $0.05 < S_{520-950}(C) < 0.25$). Many regular satellites (Tethys, Dione, Rhea, Mimas, Iapetus trailing) have spectral slopes similar to the CD. The C ring is not quite like any of the icy moons. Overall the rings are remarkable in having much higher $S_{350-520}$ than any of the regular satellites, as pointed out by Cuzzi and Estrada (1998). However, Hyperion has $S_{520-950} > 0.5$, redder than any ring region at these long visual wavelengths; in fact Iapetus and Hyperion form an entirely separate branch on this plot. On the other hand, Enceladus and Phoebe have very low $S_{350-520}$, and have a slightly negative $S_{520-950}$. The TNOs we considered have the highest spread in spectral slopes (Sedna and 1999UG5, not shown here, are extremely red compared to Saturnian objects (Barucci et al. 2005; Bauer et al. 2002). 2003UB313 and 2005FY9 have spectral slopes similar to the A ring and to Rhea. Charon and Eris, like Phoebe and Enceladus, are fairly neutral. Triton seems to have a spectrum that varies with time (Fig. 15.35); in its most neutral appearance it is compatible to the C ring; at its reddest, it is comparable to the A and B rings (Hicks and Buratti 2004).

The principal spectral indicators in the IR range are the water ice band depths at 1.5 and 2.0 μm (respectively $BD_{1.5}$ and $BD_{2.0}$). Figure 15.36 shows a scatter plot of these two band depths. In this case the observed points are dispersed along two well-defined diagonal branches. The upper branch contains the ring points, which reach the largest $BD_{1.5}$ (>0.6) and $BD_{2.0}$ (>0.7) in the A and B rings. On each branch, the fractional abundance of non-icy contaminants decreases from lower left to upper right. The CD and C rings are grouped towards the faintest ice band strengths. A similar distribution is found for the icy satellites, which are grouped on a second branch characterized by lower $BD_{2.0}$ with respect to the rings. Pandora and Prometheus more closely follow the ring trend than the satellite trend. The more pure water ice objects are at one extreme of this branch (Enceladus and Tethys at $BD_{1.5} > 0.5$, $BD_{2.0} > 0.65$) and the least pure at the other extreme (Phoebe and Iapetus (leading side) at $BD_{1.5} < 0.3$, $BD_{2.0} < 0.4$). 1995UG5 and Triton are compatible with Iapetus (leading side) while Sedna has very weak water ice bands. The fact that some TNOs have unobservable water ice bands does not mean that water ice is absent in them, but more plausibly that it is simply obscured by thick surface layers of more volatile material such as methane, nitrogen, etc.

Finally, in Fig. 15.37 we show the distribution of $BD_{1.5}$ vs. $S_{350-520}$. We have combined results from both east and west ansae in the S36-SUBML001 VIMS mosaic. We see several diagonal branches, with that containing the C ring and Cassini Division being the most dramatic, trending upwards from their inner portions which connect to several satellites, to their outer portions which connect to the A and B rings. The A and B rings have the highest values of both

---

[6] (LATPHASE001 in sequence S14 - VIS IFOV 166 × 166 μrad, IR IFOV 250 × 500 μrad, with exposure times of 5.12 sec (VIS) and 80 msec (IR) from a distance of about 1,400,000 km from Saturn (inclination angle = 16°, phase = 51°)

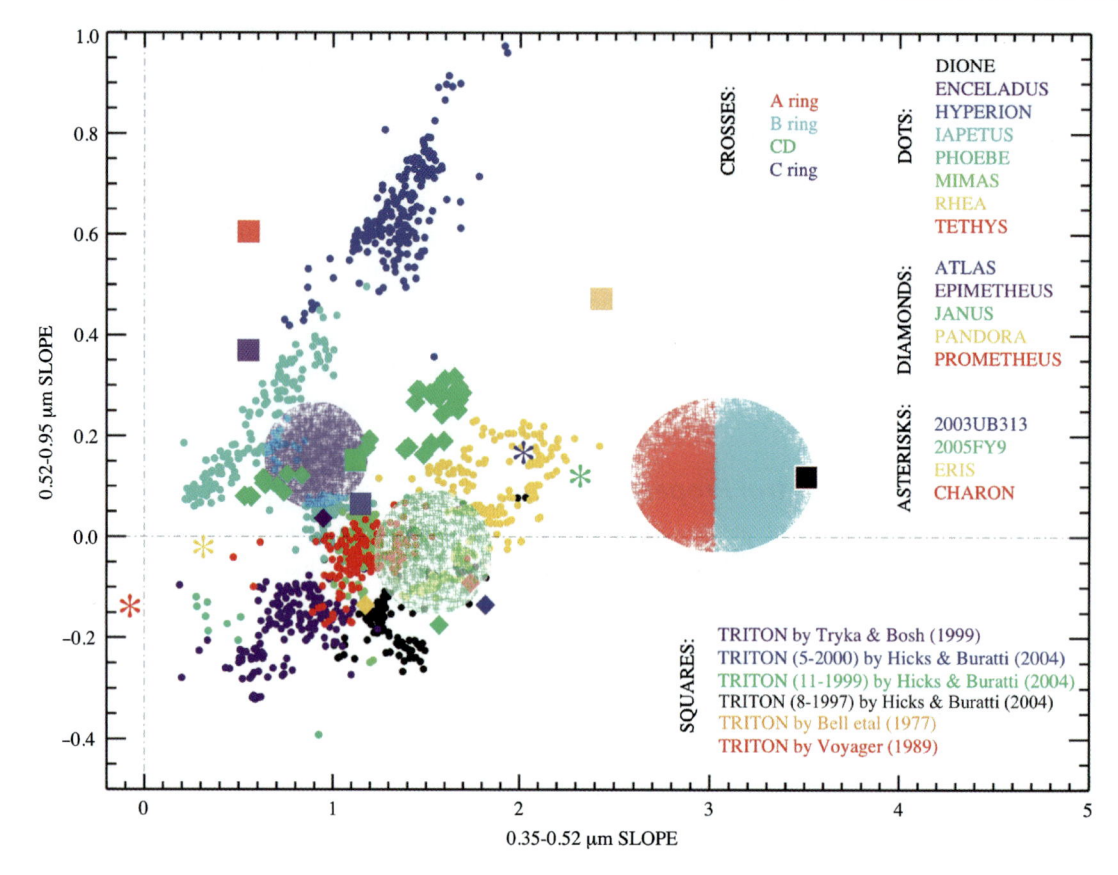

Fig. 15.34 Scatter plot of the 0.35–0.52 ($S_{350-520}$) vs. 0.52–0.95 μm ($S_{520-950}$) spectral slopes measured by VIMS on ring, Saturnian icy satellite, TNOs and Triton. The A and B rings are characterized by the highest $S_{350-520}$, denoting the reddest visible spectra at short wavelengths. The ring scatter plot is circumscribed for clarity

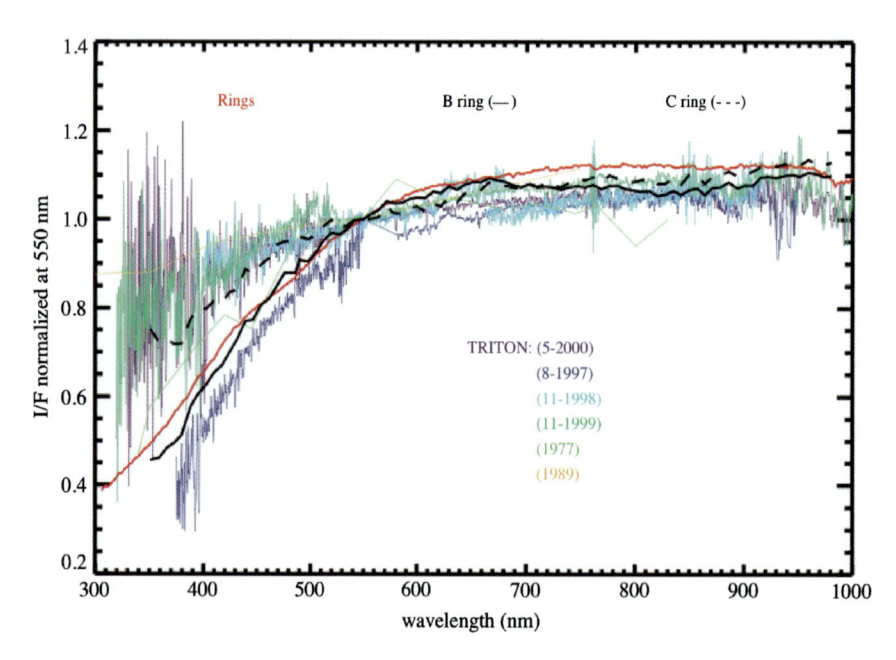

Fig. 15.35 Triton's visual wavelength spectrum is the closest outer solar system analog to that of the rings, being steep at short wavelengths and fairly flat at long wavelengths. This figure shows Triton's (variable) visible spectra at a number of epochs from 1997 to 2000 (Hicks and Buratti 2004) compared with B and C ring spectra from VIMS (*black lines*) and an HST average ring spectrum by Karkoschka (1994, *red line*) with which HST spectra by Cuzzi et al. (2002) are in good agreement). Whether the difference between the VIMS B ring spectrum and the Karkoschka spectrum in the 680–950 nm range is real, or a calibration issue, merits further study

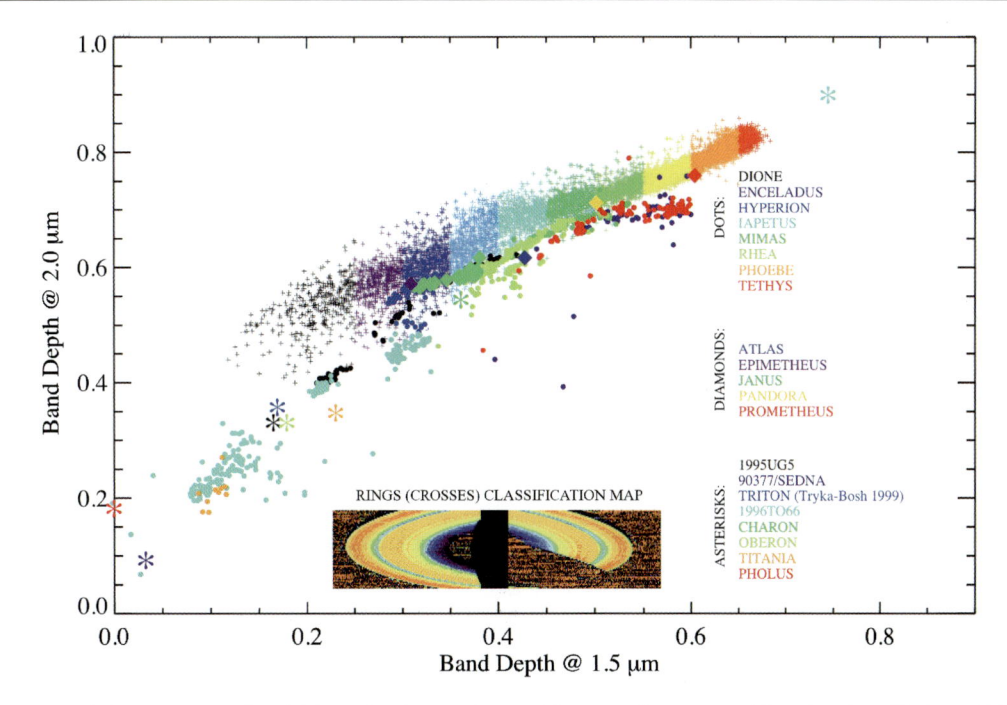

15 Ring Particle Composition and Size Distribution

Fig. 15.36 Scatterplot of the water ice 1.5 and 2.0 μm band strengths (BD$_{1.5}$ vs. BD$_{2.0}$) as measured by VIMS on rings (*plus signs*) and various positions on the Saturnian icy satellites (*dots*), along with comparable properties measured from Earth on TNOs and Triton. *Inset*: classification map of the water ice band depths across the rings

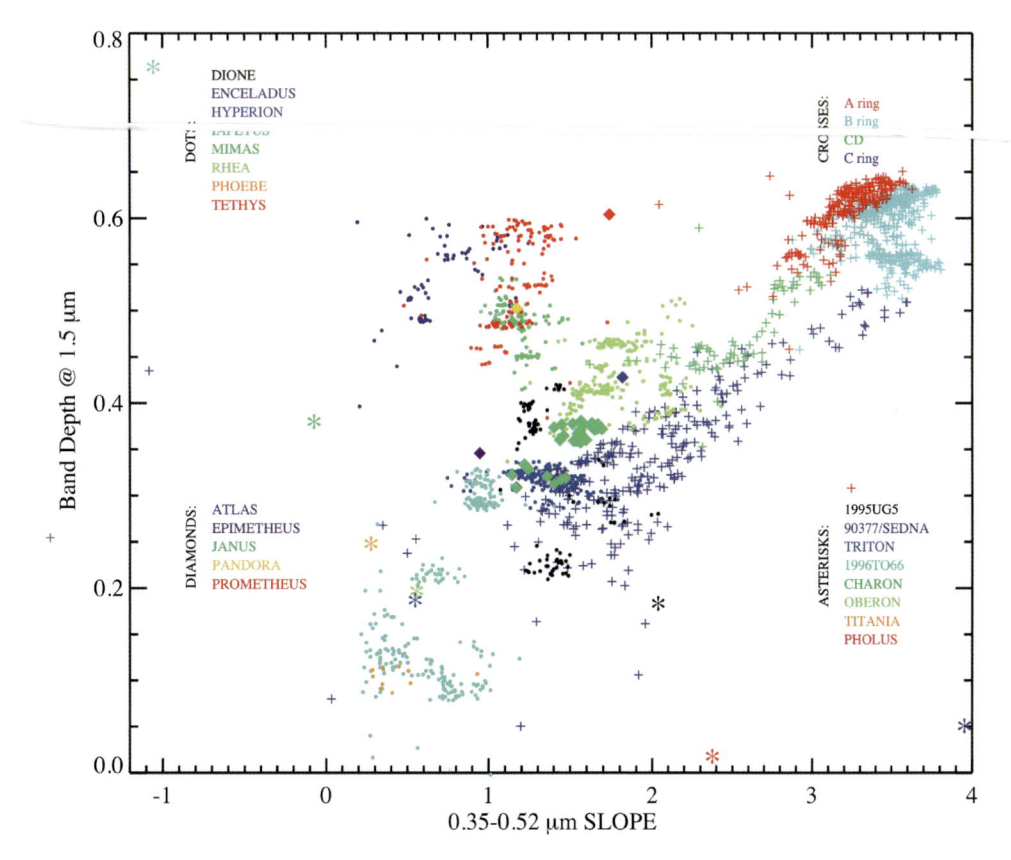

Fig. 15.37 Scatterplot of the S$_{350-520}$ spectral slope vs. the water ice band strength at 1.5 μm (BS$_{1.5}$) as measured by VIMS on rings (*plus signs*) and various locations on the Saturnian icy satellites (*dots*), along with comparable properties measured from Earth on TNOs and Triton. The A and B rings have the reddest slope while maintaining a high 1.5 μm band strength

$S_{350-520}$ and $BD_{1.5}$. The B ring separates into at least two differently sloped structures: the upper branch contains points in the outer B ring (R > 105,000 km) while the lower branch contains points in the inner B ring (R < 105,000 km). The CD has intermediate values. Epimetheus, Dione, and Hyperion overlay the innermost C ring in this plot. Mimas, Enceladus, Tethys, Pandora and Prometheus differ from the ring primarily in their higher ice BD at the same color. Iapetus, Phoebe, 1995UG5 and Triton have lower BD. Sedna has the highest $S_{350-520}$ and the smallest $BD_{1.5}$, being primarily covered with methane. In fact, it is notable that, while the TNO data are sparse, a trend can be detected which groups all the TNOs (upper left to lower right) which is nearly orthogonal to the trend of the ring properties (in which redness increases with water ice band depth).

15.5 Ring Atmosphere and Meteoroid Bombardment

15.5.1 Introduction

Saturn's extensive ring and satellite system is exposed to the ambient photon radiation field, the magnetospheric plasma, and meteoroid flux. These cause ejection of surface material, producing a toroidal gaseous envelope. Although the highest neutral densities are over the main rings, scattered atoms and molecules from the ring atmosphere extend from Saturn's atmosphere to beyond Titan's orbit (Johnson et al. 2006a; Fig. 15.38). This extended ring atmosphere is superimposed on a large toroidal atmosphere produced primarily by direct outgassing of water molecules from Enceladus (Johnson et al. 2006b) and, to a lesser extent, by the sputtering of the icy satellite surfaces and the grains in the tenuous E, F and G rings. The combined toroidal atmosphere is dominated by water products: H_2O, OH, O, H, H_2 and O_2 and their ions. This toroidal atmosphere is the principal source of plasma in Saturn's magnetosphere and, possibly, the principal source of oxygen for the upper atmospheres of both Saturn and Titan. However, the dominant molecular components from the two largest sources differ. Enceladus primarily outgases H_2O with trace amounts (~4%) of carbon and nitrogen species (Waite et al. 2006), while the atmosphere over the main rings appears to be dominated by molecular oxygen (Johnson et al. 2006a).

15.5.2 Main Rings

A plasma has been reported containing O_2^+ formed from O_2 that is produced in and ejected from the surfaces of icy ring

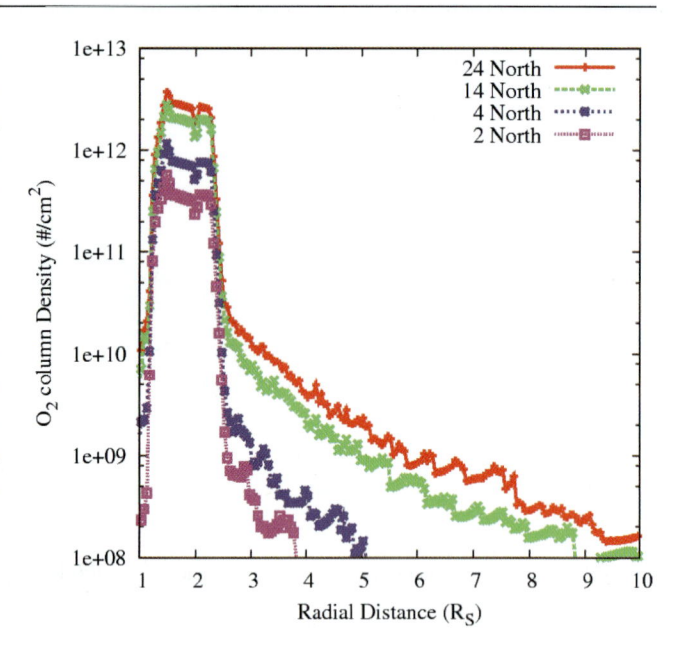

Fig. 15.38 The column density of O_2 ring atmosphere vs. distance in R_S from Saturn in the equatorial plane, for a number of solar zenith angles. The O_2 has significant density inside the B-ring and outside the A-ring, which are its principal sources. Ion-molecule collisions scatter the ring atmosphere both inwards (providing a source of oxygen for Saturn's atmosphere) and outwards (providing a source of O_2^+ ions for the magnetosphere). From Tseng et al. 2009

particles (and satellites; e.g., Johnson and Quickenden 1997). It was initially discovered by the CAPS (CAssini Plasma Spectrometer) instrument at SOI along Cassini's trajectory over the main rings from ~1.82 to 2.05 R_S (Tokar et al. 2005) and since studied from ~4 to 12 R_S (Martens et al. 2008; Tseng et al. 2009); see Figs. 15.38 and 15.39. In addition, during SOI, INMS detected H^+, O^+, and O_2^+ in proportions of 1.0:0.3:1.0 over the A ring from ~ 2.05 to 2.23 R_S. They reported an O_2^+ density at ~2.2 R_S of ~0.1–1 cm^{-3} and a very rough upper limit of ~10^5 neutrals cm^{-3} (Waite et al. 2005). INMS also detected bursts of molecular hydrogen ions over the A ring. Although H_2^+ must be present at some level, it has yet to be determined whether these observations were due to transients in the ring atmosphere, impacts of grains onto the instrument, or an artifact due to surface desorption within the instrument.

Since ions and electrons are efficiently absorbed by ring particles, the energetic particle flux is very small over the main rings and is dominated by a low flux of energetic ions produced by cosmic ray impacts (Cooper 1983). Therefore, the plasma-induced decomposition and sputtering rate are both very small. Carlson (1980) estimated the UV photodesorption of water molecules into the ring atmosphere as also being negligible. Meteoroid bombardment probably dominates the initial production of the vapor environment of the rings (Morfill 1983). An interplanetary value of 3×10^{-17} g cm^{-2} s^{-1} gives a two-sided flux at the rings

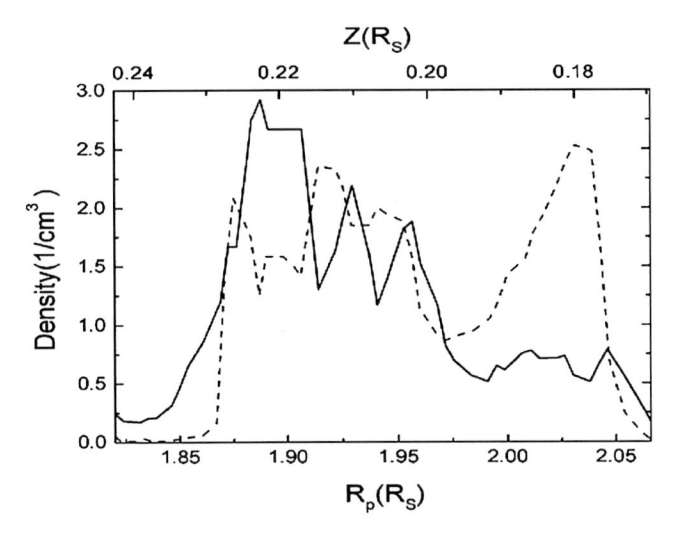

Fig. 15.39 Local volume density of O_2^+ (*dashed*) and O^+ (*solid*) by the CAPS instrument at Cassini altitudes above the ring plane, $Z(R_S)$ vs. distance from Saturn along the equatorial plane $R_p(R_S)$ where R_S is Saturn's radius. The principal peak in the O_2^+ density is related to the Cassini Division and that for O^+ occurs close to the point where the orbit speed of the neutrals is equal to the rotation speed of the magnetic field and hence the ions. The steep drop "inside of" $1.87\,R_S$ is more appropriately attributable to a steep vertical dropoff (see Fig. 15.40). Figure from Tokar et al. (2005)

of about $2.5 \times 10^{-16}\,\mathrm{g\,cm^{-2}\,s^{-1}}$, after adjusting for gravitational focussing (Cook and Franklin 1970; Ip 1984; Cuzzi and Durisen 1990) and a water vapor production rate $\sim 5 \times 10^6\,H_2O\,\mathrm{cm^{-2}\,s^{-1}}$ ($\sim 10^{27}\,H_2O\,\mathrm{s^{-1}}$ averaged over the ring system; Ip 1984; Pospieszalska and Johnson 1991). Because the emitted H_2O molecules re-condense on ring particles, the average column density ($\sim 10^{11}\,\mathrm{cm^{-2}}$) is much lower than the O_2 column densities at SOI ($> 2 \times 10^{12}\,O_2\,\mathrm{cm^{-2}}$). A meteoroid flux at the rings as large as $5 \times 10^{-14}\,\mathrm{g\,cm^{-2}\,s^{-1}}$ (Ip 2005) would result in $\sim 10^9\,H_2O\,\mathrm{cm^{-2}\,s^{-1}}$ ($\sim 10^{29}\,H_2O\,\mathrm{s^{-1}}$ averaged over the ring system), and would lead to a density of water ions comparable to that detected for O_2^+. The lack of definitive detection of water ions would suggest fluxes that are at least an order of magnitude smaller. Arguments in Cuzzi and Estrada (1998) and Cuzzi et al. (2002) allow an upper limit of $\sim 3 \times 10^{-15}\,\mathrm{g\,cm^{-2}\,s^{-1}}$ *at the rings*, even assuming a large gravitational focussing factor of 40 (an order of magnitude high for projectiles on highly inclined or eccentric orbits).

Ip (1995) suggested that photo-dissociation products from such a primary water atmosphere could react to produce O_2. Since O_2 would not condense out at the temperatures of the ring particles, it could accumulate in the ring atmosphere. Based on a surface source of $H_2O \sim 5 \times 10^{27}\,\mathrm{s^{-1}}$, he predicted a tenuous atmosphere $\sim 5 \times 10^{11}\,O_2\,\mathrm{cm^{-2}}$, about an order of magnitude smaller than inferred from SOI data. In addition however, the incident UV flux can decompose solid water ice, producing H_2 and O_2 directly (Johnson and Quickenden 1997). Therefore, although photodecomposition of ice is not a very efficient process, O_2 and H_2 are directly supplied to the ring atmosphere on the illuminated side. Since ejected water molecules and their dissociation products will stick on re-impacting the ring particles, but the H_2 and O_2 formed by decomposition do not, an atmosphere containing H_2 and O_2 can accumulate (Johnson et al. 2006a). At equinox, production from an impact-generated H_2O atmosphere may provide a low residual O_2 ring atmosphere at the level predicted by Ip (2005).

15.5.3 Modeling of the Ring Atmosphere

Laboratory experiments in which ice is exposed to a Lyman-alpha photo-flux (Westley et al. 1995) show that O_2 and H_2 are directly produced by the solar EUV/UV flux at an estimated rate $> 10^6\,O_2\,\mathrm{cm^{-2}\,s^{-1}}$ (Johnson et al. 2006a). Using the CAPS data and this source rate resulted in a number of simulations of the formation and structure of the ring atmosphere (Johnson et al. 2006a; Bouhram et al. 2006; Luhmann et al. 2006; Farmer and Goldreich 2007; Tseng et al. 2009). These simulations also predict the abundance of neutral molecules needed to produce the observed ions. The results from one set of simulations (Fig. 15.40) gives estimates of the spatial distributions of O_2, O_2^+, and O^+ above and below the ring plane. The essence of these simulations and their implications for the Saturnian system are given below.

Since the O_2 and H_2 produced by photolysis (primarily on the lit side of the rings) do not condense out, they orbit with and thermally equilibrate with the ring particle surfaces. Therefore, O_2 atmospheres exist both north and south of the ring plane with slightly different scale heights due to the different surface temperatures of the ring particles ($\sim 0.025\,R_s \sim 1{,}500\,\mathrm{km}$ at $\sim 2\,R_S$). A corresponding H_2 atmosphere is also produced in such a model, having a scale height about 4 times larger (Johnson et al. 2006a). The instantaneous O_2 and H_2 column densities are limited by their destruction rates, primarily photo-dissociation. Since the O and H produced by dissociation have excess energy, they are rapidly lost to Saturn, ionized in the magnetosphere, or re-impact and stick to ring particle surfaces. However, because of the significant mass difference, hydrogen is lost preferentially.

Ions are formed from the orbiting neutrals primarily by photo-ionization: $O_2 + h\nu \rightarrow O_2^+ + e$ or $O + O^+ + e$ in about a 4:1 ratio. These freshly produced ions are then "picked-up" (accelerated by Saturn's advective electric field). Because the ions are formed by photolysis, at the time of SOI the production rate south of the ring plane was larger than north of the ring plane. The O^+ are formed with additional energy, but the O_2^+ are not. Therefore, the molecular ions are picked-up

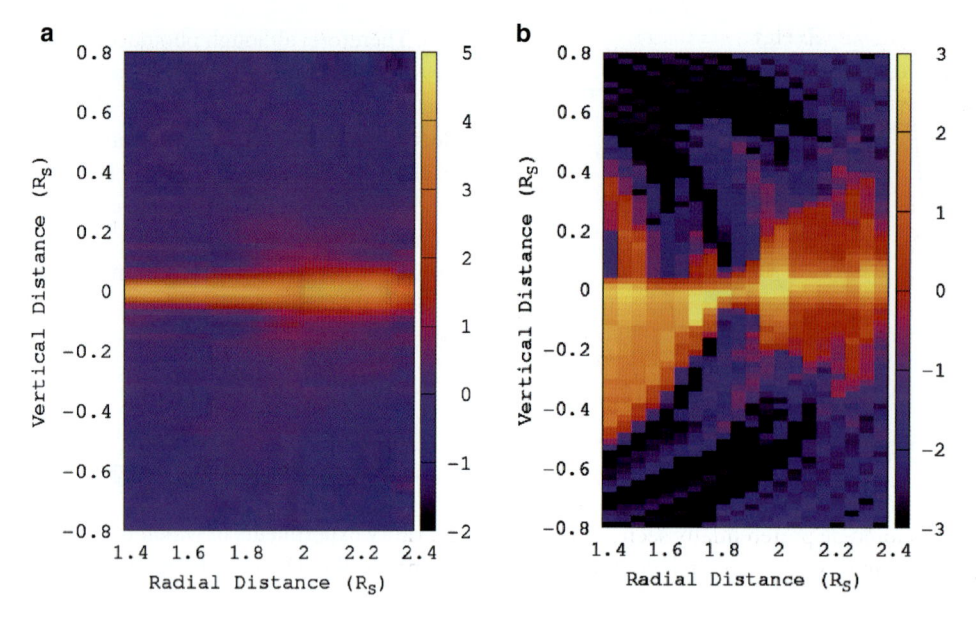

Fig. 15.40 Densities given in powers of 10 from a test-particle simulation of O_2 molecules (**a**; *left*), and O_2^+ ions (**b**; *right*) in the vicinity of the Saturnian rings. Since the magnetic equator lies north of the ring plane, ions formed south of the ring plane try to move northward and impact the rings. However, ions formed inside 1.86 Rs, especially those produced on the southward side, escape to the planet (see flux at lower left of right hand figure). Figure courtesy W. Tseng; see also Tseng et al. (2009)

with a velocity that is primarily perpendicular to the magnetic field and will oscillate about the magnetic equator. However, the magnetic equator is $\sim 0.04\,R_s$ north of the ring plane, a distance larger than the O_2 scale height ($\sim 0.025\,R_s$). Thus the for the A and B rings, a denser O_2^+ atmosphere will be found preferentially north of the ring plane, and a seasonal variation would be anticipated once the sun again illuminates the north face of the rings (Tseng et al. 2009). Newly formed ions can be absorbed by ring particles as they move along the magnetic field lines attempting to cross the ring plane. In the modeling results of Fig. 15.40, the absorption probability is determined by the local optical depth; thus, abundances are higher above locations crossing the optically thin Cassini Division.

The nature of the ion pick-up process changes closer to Saturn. For equatorial distances from Saturn $> \sim 1.86\,R_S$ (the corotation radius) the rotation speed of the magnetic field, which guides the motion of the ions, is larger than the average speeds of the neutrals as they orbit. The opposite is the case when $R < \sim 1.86\,R_S$, so that freshly ionized neutrals are typically *slowed* by the magnetic field. This slowing, combined with Saturn's gravity, can cause ions formed well within $\sim 1.86\,Rs$ to precipitate along the field lines into Saturn's southern atmosphere (Northrop and Hill 1983; Ip 1984; Luhmann et al. 2006, Tseng et al. 2009) as indicated by the 2nd panel in Fig. 15.40. Plasma loss to Saturn's atmosphere results in the net erosion of the ring particles and becomes the dominant ion loss process for the inner ring system. On ionization, neutrals ejected from particles in the low-optical depth C ring are lost with an especially high probability to Saturn's atmosphere, because of their lower probability of re-encountering a ring particle.

15.5.4 Atmosphere-Driven Chemistry on Icy Ring Particle Surfaces

One of the principal uncertainties in modeling the ring atmosphere and ionosphere is the fate of radicals and ions when they impact the ring particle surfaces. H_2 and O_2 only briefly adsorb on the surface, becoming thermally accommodated to the surface temperature before they return to the gas phase. The radicals (O, H) and ions (O_2^+, O^+, H_2^+ and H^+) either stick or react. Since hydrogen is preferentially lost from the system, the ring particle surfaces are, on average, slightly oxidizing as discussed elsewhere for Europa's surface (Johnson et al. 2004) and also likely charged. In addition, the returning O_2^+, O^+ and O are reactive. Johnson et al. (2006a) obtain agreement with CAPS ion data by *requiring* that a significant fraction of the returning oxygen reacts on the surface and returns to the atmosphere as O_2. This was also suggested by Ip (2005). This recycling resulted in roughly an order of magnitude increase in the densities and loss rates (Section 15.5.5).

Based on the above, the surface chemistry is such that non-water ice contaminants would tend, on average, to become oxidized as is the case at Europa where the principal contaminants (sulfur and carbon) are observed primarily as oxides. Therefore, near-surface, refractory carbon species, such as hydrocarbons, tholins, or PAHs, would likely experience reactions in which they would be degraded and oxidized to form volatiles such as CO and CO_2 in ice. The most volatile product (CO) would, like the O_2 formed, be desorbed. It also would not recondense, and would be eventually scattered from the ring atmosphere, removing surface carbon. Near-surface CO_2 can also be photolyzed, producing CO, which can then be lost. In laboratory experiments NH_3 in ice mixtures is rapidly destroyed by EUV photons (Wu et al. 2006) and, since it is more volatile than H_2O, its lifetime on the surface of a grain is not long. In addition, it can form N_2 under irradiation (e.g., Johnson 1998; Loeffler et al. 2006), which could be trapped in inclusions, like O_2 at Europa (Johnson et al. 2004), or it can diffuse out and be removed like O_2 (Teolis et al. 2005) With the removal of volatiles, more refractory species and/or heavy oxides (such as carbon suboxides) should be preferentially seen in the surface. Similarly the returning oxygen can interact with Fe, either as metal or some other reduced state, forming an iron oxide; Fe_2O_3 has been suggested to help explain the ring reflectance data (Section 15.4.8).

15.5.5 The Ring Atmosphere as a Magnetospheric and Atmospheric Source

Because the O^+ ions are formed with a significant, randomly oriented energy, they will have a distribution in velocity large enough for a fraction of them to reach the spacecraft altitude. On the other hand, O_2^+ is formed with no additional kinetic energy. Therefore, the O_2^+ detected at altitudes $>0.1 R_S$ must have been scattered by collisions with neutrals in the ring atmosphere (Johnson et al. 2006a). Farmer and Goldreich (2007) examined the collisional interaction between the neutral and the ion component of the ring atmosphere to constrain the estimates of density of the atmosphere and ionosphere, as suggested by models of the ring spoke phenomenon (Morfill and Thomas 2005); however, their upper limit of $\sim 2 \times 10^{15}$ O_2 cm^{-2} did not improve the estimates based on models of the Cassini ion data (Johnson et al. 2006a).

Whereas the motion of a scattered ion is restricted by the magnetic field, neutrals are subject only to gravity and their centripetal motion. Therefore, they can be scattered into Saturn's atmosphere or into the Saturnian magnetosphere beyond the edge of the main rings forming *an extended ring atmosphere* as simulated in Fig. 15.40. Ionization of the O_2 in the extended ring atmosphere contributes to O_2^+ detected outside of the ring system (Tokar et al. 2005; Young et al. 2005; Martens et al. 2008). Based on a Monte Carlo model (Tseng et al. 2009), about 5×10^{26} O s^{-1} are scattered out of the ring atmosphere for the illumination at SOI, either as O or O_2. Most of these are in large orbits which eventually re-impact the ring particles. A small fraction is ionized as they orbit in the magnetosphere, about 0.4% escape and about 6% are scattered into Saturn's atmosphere (Johnson et al. 2006a; Tseng et al. 2009). This gives an O source rate $\sim 10^5$ O cm^{-2} s^{-1} which is about an order of magnitude smaller than the required flux ($\sim 4 \times 10^6$ O cm^{-2} s^{-1}; Moses et al. 2000, Shimizu 1980, Connerney and Waite 1984; Moore et al. 2006; Moore and Mendillo 2007). Based on the numbers in Johnson et al. (2006a), the oxygen contribution by direct ion precipitation into Saturn's atmosphere is even smaller: $\sim 0.2 \times 10^5$ O cm^{-2} s^{-1} as O^+ or O_2^+. These rates can be up to an order of magnitude larger depending on how oxygen from impacting O, O^+ and O_2^+ is recycled on the grain surfaces (Johnson et al. 2006a, Ip 2005). Moses et al. (2000) suggested a direct meteoroid flux into Saturn's atmosphere of $\sim 3 \pm 2 \times 10^{-16}$ gm cm^{-2} s^{-1} could explain its atmospheric oxygen; this is consistent with estimates of direct meteoroid infall by Cuzzi and Estrada (1998), and, given the inadequacy of the indirect flux from the ring atmosphere, constitutes an independent method of estimating meteoroid flux.

Neutrals scattered from the ring atmosphere can be ionized and contribute to the magnetospheric plasma outside of the main rings (e.g., Fig. 15.38). Initially the toroidal atmosphere of water dissociation products seen by HST (Shemansky et al. 1993) was thought to be derived from the E-ring grains. However, Jurac et al. (2002) showed that the principal source region was near the orbit of Enceladus, and Cassini eventually identified this source as outgassing from Enceladus's south polar region (Waite et al. 2006; Hansen et al. 2006), which is also the source of the E-ring grains. Cassini data have also shown that inside the orbit of Rhea there is a dearth of the energetic particle radiation that is the source of molecular oxygen at Europa and Ganymede (Johnson et al. 2004). Since the Enceladus plumes and sputtering primarily supply water products to the plasma (H^+, O^+, OH^+ and H_2O^+), and O_2 is hard to create in the magnetosphere, the ionization of neutral O_2 originally produced in the ring atmosphere is the primary source of O_2^+ inside the orbit of Rhea (e.g., Fig. 15.38). Therefore, the plasma measurements of O_2^+ give a clear marker for the extent of Saturn's ring atmosphere.

15.5.6 Meteoroid Bombardment, Ring Mass, and Ring Composition

The subject of meteoroid bombardment is reviewed in detail by Chapter 17, so we will only mention several aspects relating to ring compositional properties. The rings are constantly bombarded by primitive interplanetary meteoroids, which move mass around and pollute the rings with considerable amounts of silicate and carbonaceous material, changing the reflectivity and color of the dominantly icy ring particles. In this scenario, regions with small mass density (C ring and Cassini Division) are expected to be – and are seen to be – characterized by lower particle albedos and more neutral colors (Cuzzi and Estrada 1998). Recent VIMS results showing a smooth variation of water ice band depth across the abrupt inner A and B ring boundaries (Section 15.4.5) are also consistent with these models. Extension of the models, along with refinement of their parameters, has some promise to constrain the "exposure age" of the rings; best current estimates of this scenario give a ring age on the order of several hundred million years, about one-tenth the age of the solar system. The most significant uncertainties in the inferred ring age derive from the incoming mass flux of meteoroids, and the surface mass density of the rings. Some recent suggestions have arisen that the ring "exposure age" might be much greater, if the surface mass density were much larger than currently inferred, or if the incoming mass flux were much smaller (or both).

Ring surface mass density: In order for some ring region to avoid becoming polluted over 4.5 Gyr at the currently estimated meteoroid mass flux, it must have a much larger unpolluted mass reservoir cloistered away somewhere, upon which to draw occasionally (see Chapter 17; also Esposito 2008). In the A ring, the mass density has been measured to be about 40 g cm^{-2} by dozens of spiral density waves that cover nearly its entire radial extent (see Chapter 13), consistent with the observed ring particle size distribution (Section 15.2). Compared to this, the mass of both the visible embedded moonlets Pan and Daphnis, and that of the indirectly observed 100 m radius "propeller" shards (Section 15.3), is insignificant. Thus there is no reason to believe the A ring has any unseen reservoir from which to replenish it with fresh material. Neither is the surface mass density of the *inner* B ring likely to be greatly in error; the Janus 2:1 density wave propagates across 600 km of radial extent; Holberg et al. (1982) and Esposito et al. (1983) give the mass density in this region as 70 g cm^{-2}, consistent with observed optical depths and ice particles of several meter upper radius limit. There is a single estimate of mass density in the *outer* B ring, from a bending wave in a complex region (Lissauer 1985), giving 54 ± 10 g cm^{-2}, which is consistent with canonical particle sizes and local ring optical depth. Like the A ring, these parts of the B ring seem to have no unseen mass reservoirs.

This leaves us with the dense central core of the B ring (see e.g. Fig. 15.1b, regions B2 and B3, and Chapter 13). Here, it is difficult to place an upper limit on the mass density, as no waves or wakes have been found. If this opaque and largely unexplored region is the only place where large amounts of excess mass *are* secluded, it then becomes a puzzle why its color and brightness are not more different from the color and brightness of the adjacent inner B ring, where the surface mass density is in a range which should be darkened considerably by meteoroid bombardment over the age of the solar system. That is to say, if there were a huge contrast in surface mass density between the inner/outer and central B rings, allowing only the central B ring to be primordial, there should be a strong change in particle color and brightness between these regions of such greatly differing mass density, which is not seen. Cassini will attempt dedicated observations of the ring mass (Section 15.6.4).

Meteoroid mass flux: Cuzzi and Estrada (1998) reanalyzed prior analyses of the meteoroid mass flux, and favor a value of 4.5×10^{-17} g cm^{-2} s^{-1} for the incoming, unfocussed, one-sided mass flux. Using this value they, and Durisen et al. (1992, 1996) arrived at ring exposure ages in the range of a few hundred million years. The density of the ring atmosphere (Section 15.5) was once thought to constrain the meteoroid mass flux; however, its high density and surprising O-rich composition suggests that meteoroid bombardment is not in fact the driving mechanism except perhaps at solar equinox when photo-desorption ceases. The value of the mass flux in the jovian system was addressed by the Galileo spacecraft (Sremcevic et al. 2005), using measurements of the "albedo" dust mass ejected into the Hill spheres of several of the jovian satellites. Their conclusion (cf. their Section 4.4) was that the unfocussed, one-sided mass flux at Jupiter was 3×10^{-17} g cm^{-2} s^{-1}. If the mass flux in the jovian planet region is primarily cometary and Kuiper-belt related, the value at Saturn is not likely to be significantly different, so this measurement somewhat supports the current best estimate noted above. Cassini will attempt dedicated observations (Section 15.6.4).

15.6 Summary, Discussion, and Future Directions

15.6.1 Summary of Observational Properties

Cassini observations are only in the very early stages of analysis, because many of the investigators remain deeply involved in design of ongoing and future observations; thus, this chapter represents only a progress report on what will be a decades-long study. More in-depth studies, new data, and even calibration refinements might change some results and inferences reported here.

Ring particles are likely to be chunky aggregates of smaller particles (Section 15.2), with permanence that remains unknown. They are surely obliterated frequently by incoming meteoroids of various sizes. Several lines of argument (phase function at low phase angles, radar reflectivity) suggest the observed particles, or aggregates, are highly irregular, in the nature of dense grape clusters rather than spheres. Scattering properties indicate that these particles, or aggregates, obey a rough powerlaw with a fairly sharp upper cutoff size; both lower and upper cutoff sizes vary somewhat with location (Section 15.2.9). These "particle entities" do seem to be smaller than "wakes" as observed by UVIS (Section 15.2.10). The area fraction of tiny dust grains in the main rings is generally small. The particle size distribution is not a strong constraint on the origin and evolution of the rings, rather being determined by local collisional dynamics (Chapter 14). However ring particles (and self-gravity wakes) seem to contain most of the ring mass in, at least, the A and inner B rings.

Ring composition: The primary composition of the rings is water ice; it is quite pure and predominantly crystalline, to the sensitivity level of the measurements (there is probably less than a few percent amorphous ice, if any). The reddish color of the rings at wavelengths shorter than 500 nm testifies to a non-icy component which is strongly absorbing at near-UV and blue wavelengths, which must represent less than a few percent by mass in order not to violate microwave brightness observations (Chapter 2). The detailed shape of the spectrum between 600–900 nm differs slightly between Cassini and HST observations, perhaps due to calibration uncertainties. There appears to be no CO_2 or CH_4 in the rings. There is no sign of spectral features in the 3.3–3.5 μm region that would give supporting evidence for C-H organics; however, this feature is intrinsically quite weak in many tholins and its absence does not preclude reddish organics or PAHs (Section 15.4.8). An alternative compositional interpretation for the UV absorber is nanohematite. Nanohematite is a strong UV absorber and has no other strong spectral features in the 1–5 micron spectral range, consistent with high-signal-to-noise-ratio VIMS spectra. The presence of hematite is consistent with oxidation of nanophase iron particles by highly oxidizing particle surfaces, a result of the oxygen atmosphere around the rings (Section 15.5). To date, there is no clear spectral evidence for silicates. Regolith grain sizes on the surfaces of ring particles have been inferred from near-IR and far-IR observations; different regolith radiative transfer models lead to at least factor-of-two different grain sizes (Section 15.4.7).

Radial composition variations: Ring composition varies from place to place in systematic ways that are obviously, but not directly, correlated with local surface mass density and/or optical depth. This variation is inferred from radial variations of ring color, particle albedo, and water ice band depth (Section 15.4.5). The C ring and Cassini Division are more contaminated by non-icy material than the A and B rings, but the composition of this pollution remains uncertain (see however Fig. 15.11 for intriguing behavior in the Cassini Division and C ring near 1 μm). The degree of visual redness (caused by some UV absorber) is highly correlated with ice band strengths, suggesting the UV absorber is distributed within the ice grains rather than as a distinct component. More detailed mixing models should be explored. The radial profile of 340–440 nm redness is slightly, but clearly, distinct from that of 440–550 nm redness, and the radial profile of 500–900 nm spectral slope is entirely different and uncorrelated with water ice band depth (Figs. 15.22–15.24). In fact, the most plausible extrinsic pollutant – material found in the C ring and Cassini division – is *less* red at *short* visual wavelengths – where the main rings are most dramatically red, and *more* red at *long* visual wavelengths than the A and B ring material. These distinct radial variations point to several different processes and/or contaminants with different radial distributions – some perhaps representing primordial or intrinsic properties and some representing extrinsic or evolutionary influences.

15.6.2 Origin – the Big Picture

Formation and compositional evolution: The rings are under the influence of vigorous evolutionary processes (satellite torques, meteoroid bombardment, and perhaps ring-atmospheric chemistry) that reshape their structure and change their composition on timescales which are apparently much shorter than the age of the solar system. The rings are very pure (>90%) water ice, requiring their parent to have differentiated significantly from a primitive solar mix (roughly equal proportions of ice, silicate, and refractory carbon compounds). Post-Voyager interpretation of this combination of factors suggested that the rings are the secondary product of the destruction of a differentiated body, occurring well after the time the Saturn system formed, rather than some primordial residue formed in place at the time of Saturn's origin (Section 15.5.6). Whether some or all parts of the main rings can be as old as the solar system is a question for which Cassini hopes to provide the answer, before the end of its mission (Section 15.6.4).

Where did the ring parent come from? Regardless of the formation *epoch* of the rings, one still needs to distinguish between the birth *location* of the ring parent(s). That is, the rings can be connected with two alternate formation hypotheses: disruption by impact of a *locally formed* inner regular satellite, and disruption by tides and collision of some *remotely formed*, heliocentric interloper (Chapter 17). The probability of either of these events happening significantly after the era of the "late heavy bombardment" (ca. 3.8 Gya) is only about 0.01 (Dones 1991, Chapter 17). Below we assess these two alternate scenarios in terms of known ring properties and processes, and the properties of parent bodies from these candidate source regions.

(1) *Saturn system icy bodies*: We immediately limit our view to the regular satellites; the (probably captured) irregular satellites never differentiated, have very low albedo, and are not candidate parents for Saturn's rings. By comparison, the rings of Jupiter, Uranus, and Neptune are far less massive, and seem to be composed of very dark material so are simple to explain by disruption of small, primitive bodies (Colwell and Esposito 1990, 1992, 1993). Saturn's regular satellites, on the other hand, are at least ice-mantled and many of them are largely ice in bulk (Iapetus for instance; Chapters 18 and 20). The regular satellites probably formed in some kind of circumplanetary disk (Canup and Ward 2002, 2006; Mosqueira and Estrada 2003a,b; Estrada and Mosqueira 2006; also Chapter 3). There are ways in which these disks – *if they were gas rich* (Estrada and Mosqueira 2006) – might have become enriched in water ice relative to cosmic abundances, but achieving the enormous amounts characterizing the rings remains an unmet challenge.

Saturn's regular satellite surfaces generally have spectra that are qualitatively similar, but quantitatively different, from the spectra of the rings. All have deep, crystalline ice bands in the 1–3 μm range. All have red spectra from 340–520 nm and fairly flat spectra from 520–950 nm (see Fig. 15.34). The main difference from an overall spectral standpoint is that the ring spectra are much redder from 340–550 nm than the satellite spectra. The rings lack CO_2, as do most of the regular satellites; however, see Section 15.5 regarding its local destruction. Only Phoebe (an obvious interloper), the dark regions of Iapetus (covered with possibly extrinsic material), and Hyperion (a mystery in all regards) show strong CO_2 while the signature on the inner large satellites is weak or nonexistent. Spectrally then, the rings share a number of properties with those regular moons that are most plausibly locally formed. The primary difference between the rings and the regular moons is the larger amount of "UV absorber" in the rings (which provides their much steeper 330–550 nm spectral slope).

The lack of typical cosmic abundances of silicates in most of the main rings requires a coreless parent, or a way of segregating the core of a parent and keeping it from becoming increasingly fragmented and mixed into the rubble of the rings. Could Pan and Daphnis be such primordial shards? Any primordial shards would probably be deeply buried in icy ring material today and their composition unknown (Porco et al. 2008). No moonlets even close to the size of Daphnis have been detected in any other empty gaps, in spite of dedicated searches by the Cassini ISS team. Could the enigmatic, nearly opaque, central B ring hide larger shards, overwhelming their attempts to clear gaps about themselves (Chapter 13)? Indeed there are two fairly narrow, *relatively* clear radial bands in the densest part of the B ring (Chapter 13). More careful searches for B ring "propellors" in these regions would be valuable. Formation of the rings by destruction of a local differentiated parent would be problematic unless all the silicate core remained in large fragments which have been not only hidden from our view, but also protected from subsequent disruption over subsequent aeons (e.g., Colwell and Esposito 1990, 1992, 1993) by shrouds of enveloping icy material. It seems to us that keeping core silicates out of the current rings represents a serious challenge to ring parentage by disruption of a locally formed and differentiated moon.

(2) *Icy denizens of the outer solar system*: In this group we include Centaurs, TNOs, and KBOs, which we will collectively refer to as Outer Solar System Objects or OSSOs. Many OSSOs are well known to be "reddish", with this reddish color generally ascribed to organic "tholins" (Cruikshank et al. 2005). Formation of the rings from such an object would involve dynamical disturbance into Saturn-crossing orbit and close encounter, with tidal or collisionally aided disruption (Dones 1991; Chapter 17). It is believed that Triton incurred a very close encounter with Neptune (and, probably, a collision) that led to its capture; it is also known that Jupiter has tidally disrupted numerous heliocentric passersby; this scenario could be thought of as "Shoemaker-Levy-Triton". One advantage of this scenario is that the core of the differentiated object could continue on its way, leaving only ice-rich mantle material behind to be captured (Dones 1991, Chapter 17).

Looked at more closely, this concept has its own problems. The reddish 340–520 nm wavelength spectral properties of most OSSOs persist through the 500–1,000 nm spectral range, reminiscent of the properties of Hyperion (see Barucci et al. 2008), and distinct from main ring and (most) icy satellite spectra which flatten at wavelengths longer than 550 nm. However, two of Saturn's regular moons (Hyperion and Iapetus) have spectra that are strikingly different from the others, and more qualitatively similar to TNOs and Centaurs (Section 15.4.9). On the other hand, Triton itself does apparently have a spectrum that resembles that of the rings, at least during certain observing apparitions (Hicks and Buratti 2004). In the near-IR, other differences become apparent. The most reddish OSSOs have weak (Triton) or nonexistent (Pluto, Sedna, etc.) water ice bands at 1–3 μm wavelengths, instead displaying absorption by CH_4, N_2, CO (and sometimes CO_2). Water is probably present, but presumably coated, perhaps to significant depth, by degassing and freezing of more volatile constituents. The most obviously water-ice rich OSSO (EL61) has a very flat visible wavelength spectrum with no reddening at all (Merlin et al. 2007) – proving that pure water ice actually exists in the outer solar system, even if only on fragments of catastrophic disruptions! If a reddish, methane-mantled object were to be perturbed into disruption and capture at Saturn, the most volatile material would evaporate over time and the water ice and reddish material might remain behind; however one might expect any CO_2 carried in this way to also persist at Saturnian temperatures. Perhaps CO_2 is merely a trace surface radiation byproduct on OSSOs rather than a widely mixed component of importance, or perhaps it is quickly destroyed in the ring particle surfaces (Section 15.5).

15.6.3 Candidate "UV Absorbers"

Below we summarize some of the pros and cons of the alternate suggestions that have arisen for the material that provides the steep reddish visual spectra of the rings (and perhaps, to a lesser degree, of the moons as well).

(1) *Tholins*: It has long been argued that these macromolecular organics, created by the action of diverse energy sources on simple molecules like CH_4, NH_3, N_2, and H_2O, are responsible for reddening the surfaces of OSSOs (Cruikshank et al. 2005). Note that the traditional mechanism of radiation reddening of simple ices (see Hudson et al. 2008 for a recent review) produces spectra that are uniformly red out beyond 1 μm, which is consistent with OSSO colors (Barucci et al. 2008) but not with ring or regular satellite colors (Section 15.4.9). The observed lack of a C-H spectral feature in the 3.5 μm spectral region of the rings might seem like an argument against the presence of tholin-like red material; however, this feature has never been seen on OSSOs either, and might never *be* visible in the presence of water ice, because its absorption coefficient at 3.5 μm is generally less than that of water ice and it is only required in small abundance to explain the red color (Section 15.4.8). Even some pure tholins show no sign of a 3.4–3.5 μm absorption (Bernard et al. 2006). CO_2 and CH_4, on the other hand, have relatively large absorption coefficients compared to water at their most detectable wavelengths, so are much more easily seen if present. If CO_2 is always formed and present in an environment where tholins are formed and present, then the lack of CO_2 in the rings may be an argument against the presence of tholins, whether produced in the Saturn system or carried in by a heliocentric interloper – unless it is quickly destroyed in the ring environment (Section 15.5).

(2) *PAHs (organic molecules much smaller than Tholins)*: Compared to tholins, PAHs are very simple molecules, containing a few, to a few tens, of benzene rings (Salama et al. 1996, Li 2008). Unfortunately, very little is known about their optical properties in bulk or in ice, especially at visual and near-IR wavelengths (Salama et al. 1996 and references therein), but most of them are visually reddish (Fig. 15.32).

Their color is related to their physical size, which determines the wavelength of radiation sufficiently short to sense them as conducting/absorbing particles (even if far smaller than the wavelength) (e.g., Schutte et al. 1993, Draine and Li 2001, Mattioda et al. 2005). Photons with energies less than the band gap energy in a PAH (which decreases linearly as its linear dimension increases, vanishing in the "graphite limit") are less likely to be absorbed; the transition between absorption and lack of it is fairly sharp, so the overall spectrum will depend on the PAH size and structural distribution. The flattening of ring and satellite spectra beyond 550 nm could correspond to the properties of PAHs not much larger than 4–6 rings in linear extent. However, it remains unknown how PAH-forming conditions in the circumplanetary nebula might differ from those of tholins (which seem to be adequately reproduced in lab experiments), and which lead to spectra which continue to absorb to longer wavelengths than seen in the rings, suggestive of larger PAHs. It is possible that disordered tholin structures, containing a random mixture of carbon rings in different lengths and orientations, might also provide just this type of absorption. A speculative possibility is that highly energetic micrometeoroid impacts on the rings process pre-existing graphitic and/or "tholin" material, either in the projectile or in the target particle, into much smaller fragments – the 4–6 ring PAHs of Fig. 15.32, for instance. Impacts are much more intense, and at higher speeds, in the rings than on the surfaces of icy satellites. Some PAHs, or their ionized states common in ice, can display telltale absorption features that are roughly 100 nm wide, at visual wavelengths (Salama et al. 1996 and references therein).

(3) *Nanophase iron and nanophase hematite*: A number of experimental studies have found that admixture of nanophase hematite and/or iron particles, in extremely small doses, can color icy material reddish (Fig. 15.33) and might help explain some of the ring (and even satellite) spectra. The physics behind this is due to a strong charge transfer absorption extending into the UV. A turnover to relatively neutral behavior is found at about the right wavelength (500 nm). The reddish color of Mars is due to nanophase hematite, for instance (Morris et al. 1985). Nano-hematite particles, moreover, exhibit far less absorption at 850 nm than larger grains (because of their tiny size compared to a wavelength and particle-field effects introduced at that scale), which is relevant because the rings seem to have very little excess absorption at 850 nm, in spite of initial suspicions (Clark 1980) and very careful inspection of the VIMS data by one of us (RC).

Cosmochemically speaking, one expects iron oxides and iron metal to be associated with silicates, rather than ices, and the very low abundance of silicates in the rings suggests a very low abundance of iron metal. Recent in situ observations, however, motivate some openness of mind on the subject. First, during cruise to Saturn, CDA detected six particles, and measured the composition of two – both iron or iron oxide/carbide and no silicates (Hillier et al. 2007). CDA also detected "stream" particles coming from the ring system (Srama et al. 2006), which were judged to be primarily silicates (Kempf et al. 2005), although the water abundance remains uncertain because of the unusually high speed of the impacts (F. Postberg, personal communication 2008). While in Saturn orbit, mostly in the E ring, CDA has predominantly detected ice particles with up to percent-level impurities of silicate, organics or sodium salts, and perhaps 1% by mass of pure iron/iron oxide-or-sulfide particles, which are free of water ice, unlike other non-icy impurities (Postberg et al. 2007, 2008, 2009). So the in situ sampling of particles currently in and around the Saturn system and potentially polluting the surfaces of the rings and satellites, while not yet understood, remains moot on whether iron, silicate, or organic grains are the most important non-icy pollutants. The second surprise was the (O, O_2)-rich ring atmosphere (Section 15.5). Perhaps even a tiny amount of Fe-metal in the ring material could have been oxidized in-situ over the age of the rings to create this unexpected constituent in such abundance. The greater density of the ring O-atmosphere compared to the environment in which the moons reside (Fig. 15.38) might lead to a larger Fe_2O_3 production, and a more extreme 340–520 nm redness unique to the rings.

Finally, we note that nano-grains of metal and moderate size PAHs are fairly similar from a physics standpoint – they are both "conductors" with physical sizes smaller – perhaps much smaller - than a wavelength, and might behave similarly from an optical standpoint. More studies of nanophase, conducting absorbers of different composition would be helpful.

15.6.4 Future Work Needing to Be Done

Data Analysis and calibration: First, of course, is a thorough reduction and analysis of ring data regarding size distribution (from stellar and radio occultations) and composition (from UVIS, ISS, VIMS, and CIRS spectra). Only a small fraction of these data, in only a small fraction of observing geometries, have yet been analyzed. Cassini has obtained 2 cm radiometry, with resolution better than the groundbased interferometry that still provides our strongest overall constraints on the abundance of non-icy material (Chapter 2), but only some calibration and preliminary analysis of the data has yet been done. Careful attention must be paid to calibration of all Cassini observations, using available ties to groundbased and HST observations under similar observing conditions where possible.

A new generation of radiative transfer models must be developed and deployed: Inferring composition from remote observations is a multi-stage process. Particle composition is most directly related to particle albedo (as a function of wavelength) by "Hapke"-type regolith radiative transfer models (Section 15.4.6.2). Improved models will need to account for grain size-wavelength similarity, assess the plausibility of nanophase inclusions of profoundly different refractive index than their water ice matrix, and address gross irregularity of the particle aggregate itself. The spectral behavior of contaminants seems to change in significant ways when their sizes decrease into the nano-regime; more experimental data is needed here to provide the optical constants for future modeling efforts. Moreover, different possibilities exist for the configuration and structure in the grainy regolith surfaces themselves: non-icy contaminants can be mixed on a molecular level with ice molecules, or on a grain-by-grain basis; these differences all have physical significance and they make a substantial difference in the inferences of fractional abundances which are derived from modeling – amounting to a systematic uncertainty that is usually overlooked (see, e.g., Poulet et al. 2003).

The structure of the probably very irregular aggregate ring particles (in particular how their facets shadow and illuminate each other) will determine their phase function (Hapke 1984, Shkuratov et al. 2005). The phase function enters into models of the overall ring scattering behavior and is likely to be considerably more strongly backscattering than analogues explored to date (Poulet et al. 2002), and wavelength-dependent as well; Cuzzi et al. 2002). Finally, ring layer radiative transfer models are needed to combine the individual particle albedo and phase function with the effects of multiple scattering and particle volume density to determine the reflectivity of the layer as a function of viewing geometry. It has been shown that traditional "adding-doubling" codes, which assume widely-separated particles, cannot properly match the full range of observations (because of their inability to handle high packing densities) and lead to erroneous, geometry-dependent inferences of particle albedo (Salo and Karjalainen 2003, Porco et al. 2008, Chambers and Cuzzi 2008). On top of all this, we now also know that the rings are not a homogeneous slab, but a two-phase system of gaps and dense clumps (Chapter 13), where the clumps have a preferred orientation! Finally, the ring brightness component due to reflected "Saturnshine" needs to be properly accounted for.

Chemical Evolution models: Models of circumplanetary satellite formation should be improved to include thermal and chemical evolution to track the history of CO_2 (vis-a-vis CO, CH_4, etc.). Moreover, the role of a persistent O_2 atmosphere regarding production of oxidized minerals such as Fe_2O_3 should be considered.

Measure the meteoroid mass flux and the ring mass: During Cassini's Equinox mission (2008–2010) the spacecraft will fly by Rhea closely to measure the mass fux indirectly, sampling the ejected mass filling its Hill sphere in the approach used by Sremcevic et al. (2005) at Jupiter. The geometry of the flyby will make it possible to distinguish this ejecta from whatever equatorial debris might or might not be responsible for the charged article absorptions observed by MAPS instruments (Jones et al. 2008). At the end of Cassini's mission, it is hoped that a number of orbits can be implemented with the periapse inwards of the D ring. In these close orbits, it is anticipated that a ring mass comparable to Mimas (the post-Voyager consensus; Esposito et al. 1984) can be detected to a few percent accuracy. A primordial ring compatible with current estimates of mass flux would need to be 5–10 times more massive and would be easily detected. Until the time that these fundamental measurements can be made, the question of the ring exposure age to pollution will not be resolved.

Acknowledgements All of us are very grateful to the hundreds of engineers and analysts who have worked so tirelessly over the last three decades to make Cassini such a huge success. We especially thank all our science planning colleagues at JPL, especially Brad Wallis and Kelly Perry, who have guided the integration and implementation of the many complicated ring observations made by Cassini. JC thanks James Gearhart and Kari Magee for critical early help regarding integration, Bill Owen for his star catalog, and Pauline Helfenstein, Emma Birath, Ken Bollinger, Emily Baker, Rich Achterberg, and Alain Couchoux for help with observation design. We also thank L. Allamandola, T. Bradley, M. Brown, B. Buratti, J. Colwell, D. Cruikshank, B. Draine, S. Edgington, W. Grundy, M. Hedman, M. Hicks, K. Mjaseth, R. Nelson, F. Postberg, F. Poulet, T. Roush, F. Salama, M. Tiscareno, and W. Tseng for conversations, insights, data analysis, and material in advance of, or addition to, its publication. We thank our reviewers (B. Hapke and F. Poulet) for their helpful comments. This paper was partially supported by grants from the Cassini project and from the Italian Space Agency (ASI).

Appendix 15: The Zero-Phase Opposition Effect

An entirely separate subset of scattering theory must be considered for very small phase angles (less than a degree or so), characterized by very strong brightening with the approach of true opposition. This so-called 'opposition effect' was initially interpreted in terms of shadow hiding in the regolith surface, and porosities were derived from the strength and width of the opposition surge. Early measurements of the opposition effect in Saturn's rings were obtained by Franklin and Cook (1965) and Lumme and Irvine (1976). Lumme et al. (1983) concluded that the opposition effect resulted from shadow hiding (SH) amongst different ring particles in a classical many-particle-thick layer (Irvine 1966) with a very low volume filling factor. This was at odds with dynamical studies (Brahic 1977, Goldreich and Tremaine 1978) indicating that the rings should be only a few particles thick, as shown by N-body dynamical simulations (Salo 1987, 1992; Wisdom and Tremaine 1988, Richardson 1994, Salo et al. 2004, Karjalainen and Salo 2004; see Chapters 13 and 14). A partial resolution to the apparent contradiction between the photometric observations and the simulations was work by Salo and Karjalainen (2003), who used Monte Carlo ray tracing studies in dense particle layers. Interparticle shadowing can even produce a narrow, sharp opposition brightening for broad particle size distributions (French et al. 2007, Salo et al. 2008 DPS).

In addition however, SH *within the regolith of an individual ring particle* can contribute to the opposition brightening (Hapke 1986) and *coherent backscattering* (CB), or the constructive interference of incoming and outgoing light rays (Muinonen et al. 1991; Mishchenko and Dlugach 1992; Hapke 1990; Mishchenko 1993), can also contribute. Both SH in regoliths and CB are complicated functions of the surface structure of the particles and the optical properties of the grains, and have been the subjects of extensive theoretical and laboratory studies (Nelson et al. 2000, Nelson et al. 2002, Hapke et al. 2005, 2009).

It is a challenge to separate individual-particle scattering behavior (either SH or CB) from collective SH effects. In January 2005, Saturn's rings were observed from the earth at true opposition. French et al. (2007) used HST's WFPC2 to measure the sharp brightening of the rings with the approach of zero phase. Combined with the previous decade of HST observations at each opposition (Poulet et al. 2002), the WFPC2 data represent a uniform set of photometrically precise, multiwavelength measurements of the opposition effect of Saturn's rings at ring opening angles from $|B| = 6-26°$ and phase angles from $\alpha = 0-6°$. Figure 15.41 (Fig. 4 of French et al. 2007) shows the opposition phase curve of the A ring from HST observations. Note the very strong, roughly two-fold increase in I/F at small phase angles, most noticeable at short wavelengths. For comparison, the mutual-particle SH opposition effect is plotted for a range of assumed particle size distributions. At left, the dashed curves show the mutual-particle opposition effect for a monodispersion of 5 m radius particles. The solid lines show the narrower, more intense opposition surge resulting from a broader size distribution. At right, several even broader size distributions are assumed, but none of them exceed an amplitude of 1.5, compared to the observed surge of a factor of two.

Clearly, the narrow core of the opposition surge cannot be explained by interparticle shadowing alone. French

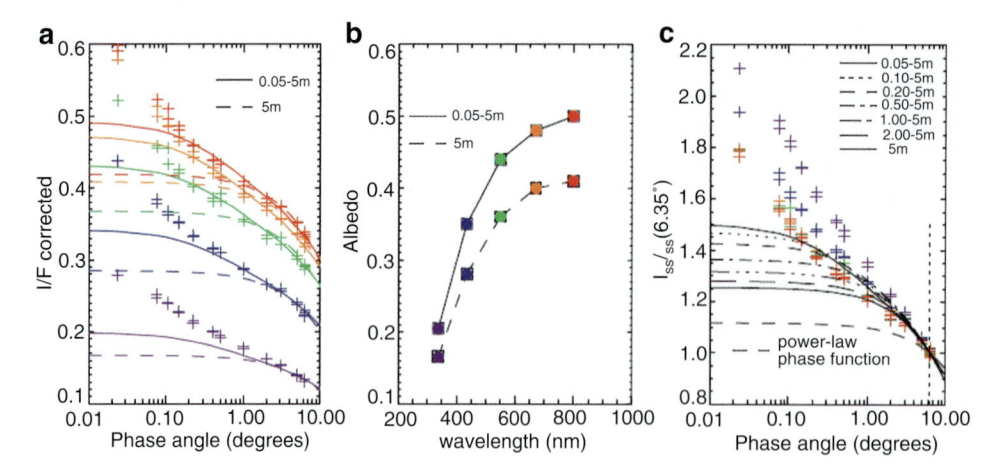

Fig. 15.41 Comparison of the observed A ring phase curves (*crosses*) to the mutual shadowing opposition effect calculated by photometric Monte Carlo simulations (*curves*). Dynamical simulations with seven different particle size distributions were conducted, ranging from q = 3 power laws for 0.05–5 m radius, to simulations with identical 5 m particles, (shown by different line types). At left, the two extreme size distribution models are compared to observations at different wavelengths. The single scattering albedos for the models, indicated in the middle panel, are chosen to fit the observed I/F at $\alpha \sim 6°$. At right, the observations and single-scattering models are normalized to $\alpha = 6.35°$. Also shown is the contribution from the adopted power-law phase function alone, (lowest dashed line) amounting to about 1.1 for the interval $\alpha = 0°$ to $6.35°$. The color code refers to the wavelength of the observation, as shown in the center panel

et al. (2007) fitted the opposition measurements to the composite model of Hapke (2002), which incorporates a wavelength-dependent CB component based on the theoretical predictions of Akkermans et al. (1988) and an explicit representation of SH by a particulate surface. The fits imply that the porosities of the ring particle regoliths are very high, ranging from 93% to 99%, and that the width of the narrow CB surge actually decreases with wavelength, rather than increasing. However, current CB models are somewhat idealized, and thus far, agreement between theory and experiments has been imperfect (Shepard and Helfenstein 2007, Hapke et al. 2009).

Regional variations in the opposition effect: The opposition effect in Saturn's rings shows strong regional variability. French et al. (2007) fitted a simple linear-exponential model to the opposition effect, and Fig. 15.42 shows the variations in the fitted amplitude and half-width with ring radius; qualitatively similar results were obtained by Poulet et al. (2003) regarding radial variation, but with different 'scale lengths' inferred.

It seems likely that most of these variations are attributable to differences in the degree of interparticle shadowing and to the relative widths of particle size distributions, rather than to strong regional variations in the intrinsic particle or regolith scattering properties. In the C ring, the detailed variations correlate strongly with the optical depth variations, which affects the amount of interparticle shadowing. The opposition effect changes markedly at the boundary between the outer C and inner B ring, while (as shown in Section 15.4.5), the particle albedo and color, and thus presumably regolith properties, do not. Over the least opaque (inner) part of the B ring, the amplitude exceeds 0.5, decreasing gradually with increasing radius and optical depth. The Cassini division resembles the C ring in optical depth, composition and color, and possibly in particle size distribution, and these similarities are also seen in the opposition effects of these two separated ring regions. The A ring and the inner B ring have comparable optical depths, and the overall characteristics of the opposition effect are similar, including significant strengthening and broadening at short wavelengths. The particle size distribution in the inner A ring is similar as well. There is a striking contrast between the inner and outer A ring opposition effect. Salo and French (2009) used the wavelength-dependence of the opposition effect, its variation with ring tilt, and numerical modeling, to disentangle the interparticle and intraparticle oppositions effects using HST observations, and concluded that there is a very narrow, wavelength-dependent CB contribution to the opposition effect.

Cassini observations: In June 2005 (B = −21°) and July 2006 (B = −21°), Cassini conducted remote sensing observations of the opposition spot traversing the rings over a range of phase angles restricted by the angular half-width of the VIMS and ISS fields of view. Only preliminary analyses are available at the time of this writing (Nelson et al. 2006, Hapke et al. 2005, 2006, Deau et al. 2006). Based on thermal infrared observations from CIRS, Altobelli et al. (2008) measured temperature phase curves of the rings. For the C ring and Cassini Division, they interpret the opposition effect as caused by regolith on the surface of individual grains, whereas for the more optically thick A and B rings, the opposition surge is attributed to interparticle shadowing.

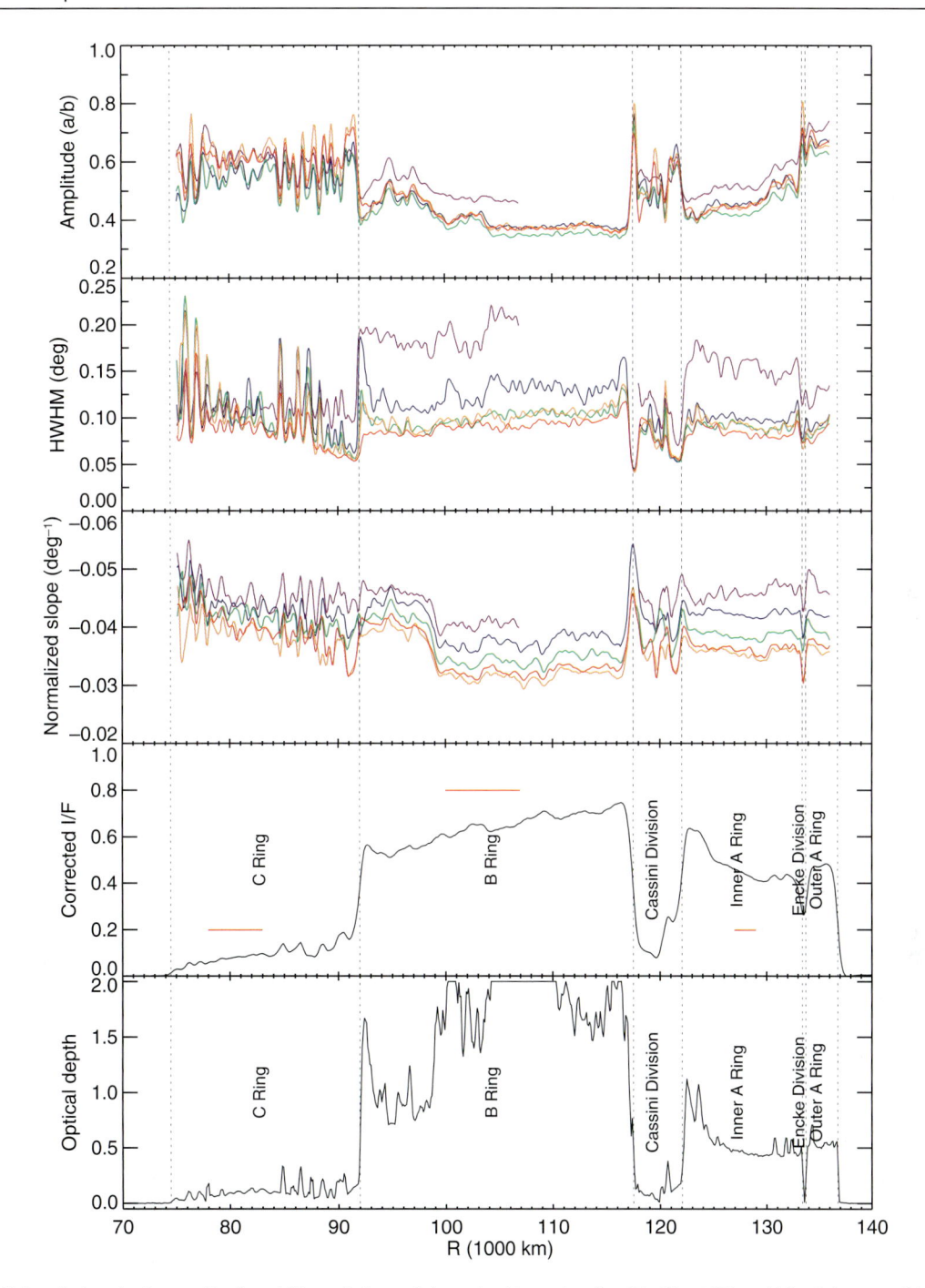

15 Ring Particle Composition and Size Distribution 503

Fig. 15.42 Radial variations in the amplitude, width, and slope of the opposition surge from linear-exponential model fits to HST WFPC2 observations of Saturn's rings at five wavelengths, taken during Cycles 10–13. The colors are the same as in Fig. 15.41 The amplitude of the opposition effect (*top*) is nearly independent of wavelength except for the F336W filter (*violet line*), especially in the A and B rings, where the amplitude increases sharply at short wavelengths. (The gap in the F336W profiles between 107,000–118,000 km results from saturation of a unique low phase angle image, making the model fits unreliable in this region for this filter.) The width of the opposition surge varies strongly with ring region at short wavelengths in the A and B rings, and shows strong correlations with optical depth in the inner and outer C ring. The normalized slope (*third panel*) is most shallow for the optically thick central B ring. A radial profile of ring brightness is shown in the *fourth panel*, taken near true opposition ($\alpha = 0.0043°$ on January 14, 2005). The *bottom panel* shows the Voyager PPS optical depth profile, truncated at optical depth $= 2$ because of limited signal to noise at high optical depths.

References

Akkermans, E., P. Wolf, R. Maynard, G. Maret (1988) Theoretical study of the coherent backscattering of light by disordered media. J. Phys. 49, 77–98.

Altobelli, N., L. J. Spilker, C. Leyrat, S. Pilorz (2008) Thermal observations of Saturn's main rings by Cassini CIRS: Phase, emission and solar elevation dependence, Planetary and Space Science, 56, 134–146.

Barucci, M. A., D. P. Cruikshank, E. Dotto, F. Merlin, F. Poulet, C. Dalle Ore, S. Fornasier, C. de Bergh (2005) Is Sedna another Triton? A&A 439, L1–L4.

Barucci, M. A., M. E. Brown, J. P. Emery, F. Merlin (2008) Composition and surface properties of transneptunian objects and centaurs; in *The Solar System Beyond Neptune*, M. A. Barucci, H. Boehnhardt, D. P. Cruikshank, and A. Morbidelli eds., University of Arizona Press, Tucson, 143–160.

Bauer, J. M., K. J. Meech, Y. R. Fernandez, T. L. Farnham, L. Ted, T. L. Roush (2002) Observations of the Centaur 1999 UG5: Evidence of a Unique Outer Solar System Surface. Publ. Astronom. Soc. Pacific 114, 1309–1321.

Bernard, J.-M. and 9 coauthors (2006) Reflectance spectra and chemical structure of Titan's tholins: application to the analysis of Cassini-Huygens observations. Icarus 185, 301–307.

Bouhram, M., R. E. Johnson, J.-J. Berthelier, J.-M. Illiano, R. L. Tokar, D. T. Young, F. J. Crary (2006) A test-particle model of the atmosphere/ionosphere system of Saturn's main rings, Geophys. Res. Lett., 33, L05106, doi:10.1029/2005GL025011.

Bradley, E. T., J. E. Colwell, L.W. Esposito, H. Tollerud, L. Bruesch-Chambers, J. N. Cuzzi (2009) Far ultraviolet reflectance spectrum of Saturn's rings from Cassini UVIS; Icarus, submitted.

Brahic, A (1977) Systems of colliding bodies in a gravitational field. I – Numerical simulation of the standard model. Astronom. Astrophys. 54, 895–907.

Brown, R. H., D. P. Cruikshank, Y. Pendleton (1999). Water ice on Kuiper belt object 1996 TO66. Astrophys. J. 519(1), L101–L104.

Buratti, B. J., D. P. Cruikshank, R. H. Brown, R. N. Clark, J. M. Bauer, R. Jaumann, T. B. McCord, D. P. Simonelli, C. A. Hibbitts, G. B. Hansen, T. C. Owen, K. H. Baines, G. Bellucci, J.-P. Bibring, F. Capaccioni, P. Cerroni, A. Coradini, P. Drossart, V. Formisano, Y. Langevin, D. L. Matson, V. Mennella, R. M. Nelson, P. D. Nicholson, B. Sicardy, C. Sotin, T. L. Roush, K. Soderlund, A. Muradyan (2005) Cassini visual and infrared mapping spectrometer observations of Iapetus: Detection of CO_2. Astrophys. J. 622, L149–L152.

Buratti, B. J., J. D. Goguen, J. Gibson, J. Mosher, (1994) Historical photometric evidence for volatile migration on Triton. Icarus 110, 303–314.

Canup, R. M., W. R. Ward (2002) Formation of the Galilean satellites: conditions of accretion. Astronom. J. 124, 3404–3423.

Canup, R. M. and W. R. Ward (2006) A common mass scaling for satellite systems of gaseous planets; Nature, 441, 834–839.

Carlson, R. W. (1980) Photosputtering of Saturn's rings. Nature 283, 461–462.

Chambers, L. S., J. N. Cuzzi (2008) Beam: A ray-tracing Monte Carlo code for photometric studies of Saturn's Rings; American Geophysical Union, Fall Meeting 2008, abstract #P13A-1293.

Clark, R. N. (1980) Ganymede, Europa, Callisto, and Saturn's rings: Compositional analysis from reflectance spectroscopy, Icarus, 44, 388–409.

Clark, R. N. (1983) Spectral properties of mixtures of montmorillonite and dark carbon grains: Implications for remote sensing minerals containing chemically and physically adsorbed water, J. Geophys. Res., 88, 10635–10644.

Clark, R. N., R. H. Brown, R. Jaumann, D. P. Cruikshank, R. M. Nelson, B. J. Buratti, T. B. McCord, J. Lunine, K. H. Baines, G. Bellucci, J.-P. Bibring, F. Capaccioni, P. Cerroni, A. Coradini, V. Formisano, Y. Langevin, D. L. Matson, V. Mennella, P. D. Nicholson, B. Sicardy, C. Sotin, T. M. Hoefen, J. M. Curchin, G. Hansen, K. Hibbits, K.-D. Matz (2005). Compositional maps of Saturn's moon Phoebe from imaging spectroscopy. Nature 435, 66–69.

Clark, R. N., D. P. Cruikshank, R. Jaumann, G. Filacchione, P. D. Nicholson, R. H. Brown, K. Stephan, M. Hedman, B. J. Buratti, J. M. Curchin, T. M. Hoefen, K. H. Baines, R. Nelson (2008b) Compositional mapping of Saturn's rings and icy satellites with Cassini VIMS, poster presented at Saturn after Cassini-Huygens, London, July, 2008a.

Clark, R. N., J. M. Curchin, R. Jaumann, D. P. Cruikshank, R. H. Brown, T. M. Hoefen, K. Stephan, J. M. Moore, B. J. Buratti, K. H. Baines, P. D. Nicholson, R. M. Nelson (2008b) Compositional mapping of Saturn's satellite Dione with Cassini VIMS and implications of dark material in the Saturn system, Icarus 193, 372–386.

Clark, R. N., P. G. Lucey (1984). Spectral properties of ice–particulate mixtures and implications for remote sensing. I. Intimate mixtures. J. Geophys. Res. 89, 6341–6348.

Clark, R.N., T. Roush (1984) Reflectance spectroscopy – Quantitative analysis techniques for remote sensing applications; J. Geophys. Res. 89, 6329–6340.

Colwell, J. E., W. Esposito (1990) A numerical model of the Uranian dust rings; Icarus 86, 530–560.

Colwell, J. E., L. W. Esposito (1992) Origins of the rings of Uranus and Neptune. I – Statistics of satellite disruptions; J. Geophys. Res. 97, 10,227–10,241.

Colwell, J. E., L. W. Esposito (1993) Origins of the rings of Uranus and Neptune. II – Initial conditions and ring moon populations. J. Geophys. Res. 98, E4, 7387–7401.

Colwell, J. E., L. W. Esposito, M. Sremcevic (2006) Gravitational wakes in Saturn's A ring measured by stellar occultations from Cassini. Geophys. Res. Lett. 33, doi:10:1029/2005GL025163. L07201.

Conel, J. (1969) Infrared emissivities of silicates: Experimental results and a cloudy atmosphere model of spectral emission from condensed particulate media. J. Geophys. Res. 74, 1614–1634.

Connerney, J. E. P., J. H. Waite (1984) New model of Saturn's ionosphere with an influx of water from the rings. Nature 312, 136–138.

Cook, A. F., F. A. Franklin (1970) The effect of meteoroidal bombardment on Saturn's tings. Astronom. J. 75, 195–205.

Cooke, M. (1991) Saturn's rings: Photometric studies of the C ring and radial variation in the Keeler Gap; unpublished PhD thesis, Cornell University, Ithaca, NY.

Cooper, J. (1983) Nuclear cascades in Saturn's rings: Cosmic ray Albedo neutron decay and origins of trapped protons in the inner magnetosphere. J. Geophys. Res. 88, 3945–3954.

Coradini, A., F. Tosi, A. Gavrishin, F. Capaccioni, P. Cerroni, G. Filacchione, A. Adriani, R. H. Brown, G. Bellucci, V. Formisano, E. D'Aversa, J. I. Lunine, K. H. Baines, J.-P. Bibring, B. J. Buratti, R. N. Clark, D. P. Cruikshank, M. Combes, P. Drossart, R. Jaumann, Y. Langevin, D. L. Matson, T. B. McCord, V. Mennella, R. M. Nelson, P. D. Nicholson, B. Sicardy, C. Sotin, M. M. Hedmann, G. B. Hansen (2008) Identification of spectral units on Phoebe. Icarus 193, 233–251.

Cruikshank, D. P., H. Imanaka, C. M. Dalle Ore (2005) Tholins as coloring agents on outer Solar System bodies. Adv. Space Res. 36, 178–183.

Cruikshank, D. P., J. B. Dalton, C. M. Dalle Ore, J. Bauer, K. Stephan, G. Filacchione, A. R. Hendrix, C. J. Hansen, A. Coradini, P. Cerroni, F. Tosi, F. Capaccioni, R. Jaumann, B. J. Buratti, R. N. Clark, R. H. Brown, R. M. Nelson, T. B. McCord, K. H. Baines, P. D. Nicholson, C. Sotin, A. W. Meyer, G. Bellucci, M. Combes, J.-P. Bibring, Y. Langevin, B. Sicardy, D. L. Matson, V. Formisano,

P. Drossart V. Mennella (2007). Surface composition of Hyperion. Nature 448, 54–56.

Cuzzi, J. N., J. J. Lissauer, L. W. Esposito, J. B. Holberg, E. A. Marouf, G. L. Tyler, A. Boischot (1984) Saturn's rings: Properties and Processes; in *Planetary Rings*, R. Greenberg and A. Brahic, eds. University of Arizona Press, Tucson.

Cuzzi, J. N. (1985) Rings of Uranus – Not so thick, not so black; Icarus 63, 312–316.

Cuzzi, J. N. (1995) Evolution of planetary ring-moon systems; in *Comparative Planetology*, Kluwer Press; also, Earth, Moon, and Planets 67; 179–208.

Cuzzi, J. N., R. H. Durisen (1990) Bombardment of planetary rings by meteoroids: General formulation and effects of Oort cloud projectiles. Icarus 84, 467.

Cuzzi, J. N., P. R. Estrada (1998) Compositional evolution of Saturn's rings due to meteoroid bombardment. Icarus 132, 1–35.

Cuzzi, J. N., J. E. Colwell, L. W. Esposito, C. C. Porco, C. E. Murray, P. D. Nicholson, L. Spilker, E. A. Marouf, R. C. French, N. Rappaport, D. Muhleman (2002) Saturn's rings: Pre-Cassini status and mission goals; Space Sci. Rev. 104, 209–251.

Cuzzi, J.N., R.G. French, L. Dones (2002) HST multicolor (255–1042 nm) photometry of Saturn's main rings I: Radial profiles, phase and opening angle variations, and regional spectra. Icarus 158, 199–223.

Deau, E., S. Charnoz, L. Dones, A. Brahic, C. Porco (2006) Sunshine on the rings: The opposition effect seen at high resolution with CASSINI-ISS; AAS/DPS meeting #38, paper #51.01; B. A. A. S. 38, 577.

Dones, L., J. N. Cuzzi, M. R. Showalter (1989) Simulations of light scattering in planetary rings; in Dynamics of astrophysical discs; Proceedings of the Conference, Manchester, England, Dec. 13–16, 1988 (A90-51451 23-90). Cambridge, England and New York, Cambridge University Press, 25–26.

Dones, L. (1991) A recent cometary origin for Saturn's rings? Icarus 92, 194–203.

Dones, L., J. N. Cuzzi, M. R. Showalter (1993) Voyager photometry of Saturn's A ring. Icarus 105, 184–215.

Dones, L. (1998) The rings of the giant planets; in *Solar System Ices*, B. Schmitt, C. de Bergh, and M. Festou, eds. Kluwer, Dordrecht, pp. 711–734.

Doyle, L. R., J. N. Cuzzi, L. Dones (1989) Radiative transfer modeling of Saturn's outer B ring. Icarus 80, 104–135.

Draine, B. T., A. Li (2001) Infrared emission from interstellar dust. I. Stochastic heating of small grains. Astrophys. J. 551, 807–824.

Durisen, R. H., P. W. Bode, J. N. Cuzzi, S. E. Cederbloom, B. W. Murphy (1992) Ballistic transport in planetary ring systems due to particle erosion mechanisms. II – Theoretical models for Saturn's A- and B-ring inner edges; Icarus 100, 364–393.

Durisen, R. H., P. W. Bode, S. G. Dyck, J. N. Cuzzi, J. D. Dull, J. C. White, II (1996) Ballistic transport in planetary ring systems Due to particle erosion mechanisms; III. Torques and mass lsoading by meteoroid impacts; Icarus 124(1), 220–236.

Esposito, L. W. (2008) *Regolith Growth and Darkening of Saturn's Ring Particles*; 21.01, AAS/DPS meeting, Ithaca, NY.

Esposito, L. W., J. N. Cuzzi, J. B. Holberg, E. A. Marouf, G. L. Tyler, C. C. Porco (1984) Saturn's rings: Structure, dynamics, and particle properties; in *Saturn*, T. Gehrels and M. Matthews, eds. University of Arizona Press, Tucson.

Esposito, L. W., M. O'Callaghan, K. E. Simmons, C. W. Hord, R. A. West, A. L. Lane, B. Pomphrey, D. L. Coffeen, M. Sato (1983) Voyager photopolarimeter stellar occultation of Saturn's rings. J. Geophys. Res. 88, 8643–8649.

Estrada, P. R., J. N. Cuzzi (1996) Voyager observations of the color of Saturn's rings; Icarus 122, 251–272.

Estrada, P. R., J. N. Cuzzi, M. R. Showalter(2003) Voyager color photometry of Saturn's main rings: a correction. Icarus 166, 212–222.

Estrada, P., I. Mosqueira (2006) A gas-poor planetesimal capture model for the formation of giant planet satellite systems; Icarus 181(2) 486–509.

Farmer, A. J., P. Goldreich (2007) How much oxygen is too much? Constraining Saturn's ring atmosphere. Icarus 188, 108–119.

Farrell, W. M., M. L. Kaiser, D. A., Gurnett, W. S. Kurth, A. M. Persoon, J. E. Wahlund, P. Canu (2008) Mass unloading along the inner edge of the Enceladus plasma torus Geophys. Res. Lett. 35, CiteID L02203; doi:10.1029/2007GL032306

Filacchione, G., F. Capaccioni, T. B. McCord, A. Coradini, P. Cerroni, G. Bellucci, F. Tosi, E. D'Aversa, V. Formisano, R. H. Brown, K. H. Baines, J. P. Bibring, B. J. Buratti, R. N. Clark, M. Combes, D. P. Cruikshank, P. Drossart, R. Jaumann, Y. Langevin, D. L. Matson, V. Mennella, R. M. Nelson, P. D. Nicholson, B. Sicardy, C. Sotin, G. Hansen, K. Hibbits, M. Showalter, S. Newman (2007) Saturn's icy satellites investigated by Cassini-VIMS. I. Full-disk properties: 350–5100 nm reflectance spectra and phase curves. Icarus 186, 259–290.

Filacchione, G., F. Capaccioni, F. Tosi, A. Coradini, P. Cerroni, R. N. Clark, P. D. Nicholson, J. N. Cuzzi, M. H. Hedman, M. R. Showalter, R. Jaumann, K. Stephan, D. P. Cruikshank, R. H. Brown, K. H. Baines, R. M. Nelson, T. B. McCord (2008a) The diversity of Saturn's main rings: A Cassini-VIMS perspective. 39th Lunar and Planetary Science Conference (Lunar and Planetary Science XXXIX), held March 10–14, 2008 in League City, Texas. LPI Contribution No. 1391, p. 1294.

Filacchione, G., F. Capaccioni, F. Tosi, A. Coradini, P. Cerroni, R. N. Clark, P. D. Nicholson, J. N. Cuzzi, M. M. Hedman, M. R. Showalter, R. Jaumann, K. Stephan, D. P. Cruikshank, R. H. Brown, K. H. Baines, R. M. Nelson, T. B. McCord (2008b) The diversity of Saturn's main rings investigated by Cassini-VIMS. Saturn After Cassini-Huygens symposium, Imperial College London, UK, 28 July–1 August 2008.

Franklin, F. A., Cook, F. A. (1965) Optical properties of Saturn's rings. II. Two-color phase curves of the two bright rings. Astronom. J. 70, 704.

French, R. G., P. D. Nicholson (2000). Saturn's rings II: Particle sizes inferred from stellar occultation data. Icarus 145, 502–523.

French, R. G., H. Salo, C. A. McGhee, L. Dones (2007) HST observations of azimuthal asymmetry in Saturn's rings. Icarus 189, 493–522.

French, R. G., A. Verbiscer, H. Salo, C. McGhee, L. Dones (2007) Saturn's rings at true opposition. Publ. Astronom. Soc. Pacific 119, 623–642.

Gaffey, M. J., T. H. Burbine, R. P. Binzel (1993) Asteroid spectroscopy – Progress and perspectives. Meteoritics 28, 161–187.

Goldreich, P. and S. D. Tremaine (1978) The velocity dispersion in Saturn's rings. Icarus 34, 227–239.

Grundy, W. M., L. A. Young (2004) Near-infrared spectral monitoring of Triton with IRTF/SpeX I: Establishing a baseline for rotational variability. Icarus 172, 455–465.

Hanel, R., B. Conrath, F. M. Flasar, V. Kunde, W. Maguire, J. C. Pearl, J. Pirraglia, R. Samuelson, D. P. Cruikshank, D. Gautier, P. J. Gierasch, L. Horn, C. Ponnamperuma (1982) Infrared observations of the Saturnian system from Voyager 2; Science 215, 544–548.

Hansen, C. J., L. Esposito, A. I. F. Stewart, J. Colwell, A. Hendrix, W. Pryor, D. Shemansky, R. West (2006) Enceladus' water vapor plume. Science 311, 1422–1425.

Hansen, G.B., T.B. McCord (2004). Amorphous and crystalline ice on the Galilean satellites: A balance between thermal and radiolytic processes. J. Geophys. Res. 109, doi:10.1029/2003JE002149. E001012.

Hapke, B. (1981) Bidirectional reflectance spectroscopy 1. Theory, J. Geophys. Res. 86, 3039–3054.

Hapke, B. (1984) Bidirectional reflectance spectroscopy. III – Correction for macroscopic roughness. Icarus 59, 41–59.

Hapke, B. (1986) Bidirectional reflectance spectroscopy. IV – The extinction coefficient and the opposition effect. Icarus 67, 264–280.

Hapke, B. (1990) Coherent backscatter and the radar characteristics of outer planet satellites. Icarus 88, 407–417.

Hapke, B. (1993) *Introduction to the Theory of Reflectance and Emittance Spectroscopy*, Cambridge University Press, New York.

Hapke, B. (2002) Bidirectional reflectance spectroscopy 5. The coherent backscatter opposition effect and anisotropic scattering. Icarus 157, 523–534.

Hapke, B., R. M. Nelson, R. H. Brown, L. J. Spilker, W. D. Smythe, L. Kamp, M. Boryta, F. Leader, D. L. Matson, S. Edgington, P. D. Nicholson, G. Filacchione, R. N. Clark, J. Bibring, K. H. Baines, B. J. Buratti, G. Bellucci, F. Capaccioni, P. Cerroni, M. Combes, A. Coradini, D. P. Cruikshank, P. Drossart, V. Formisano, R. Jaumann, Y. Langevin, T. McCord, V. Menella, B. Sicardy (2005) Physical Properties of the Saturnian ring system inferred from Cassini VIMS opposition observations. AGU Fall Meeting Abstracts 6.

Hapke, B. W., M. K. Shepard, R. M. Nelson, W. D. Smythe, J. L. Piatek (2009) A quantitative test of the ability of models based on the equation of radiative transfer to predict the bidirectional reflectance of a well-characterized medium. Icarus 199, 210–218.

Hapke, B., and 30 coauthors (2006) Cassini observations of the opposition effect of Saturn's rings 2. Interpretation: Plaster of paris as an analog of ring particles; 37th Annual Lunar and Planetary Science Conference, March 13–17, 2006, League City, Texas, abstract no. 1466.

Hedman, M. M., P. D. Nicholson, H. Salo, B. D. Wallis, B. J. Buratti, K. H. Baines, R. H. Brown, R. N. Clark (2007) Self-gravity wake structures in Saturn's A ring revealed by Cassini VIMS. Astronom. J. 133, 2624–2629.

Hedman, M. M., J. A. Burns, P. D. Nicholson, M. R. Showalter (2008) Backlit views of Saturn's dusty rings: Clues to their origins and evolution; Saturn after Cassini-Huygens symposium, Imperial College London, UK, 28 July – 1 August 2008.

Hicks, M. D., B. J. Buratti (2004). The spectral variability of Triton from 1997–2000. Icarus 171, 210–218.

Hillier, J. K., S. F. Green, N. McBride, N. Altobelli, F. Postberg, S. Kempf, J. Schwanethal, R. Srama, J. A. M. McDonnell, E. Grün (2007) Interplanetary dust detected by the Cassini CDA Chemical Analyser; Icarus 190, 643–654.

Holberg, J. B., W. T. Forrester, J. J. Lissauer (1982) Identification of resonance features within the rings of Saturn. Nature. 297, 115–120.

Hudson, R. L., M. E. Palumbo, G. Strazzulla, M. H. Moore, J. F. Cooper, S. J. Sturner (2008) Laboratory Studies of the Chemistry of Transneptunian Object Surface Materials; in *The Solar System Beyond Neptune*, M. A. Barucci, H. Boehnhardt, D. P. Cruikshank, and A. Morbidelli eds., University of Arizona Press, Tucson, 507–523.

Ip, W.-H. (1984) The ring atmosphere of Saturn: Monte Carlo simulations of the ring source model. J. Geophys. Res. 89, 8843–8849.

Ip, W.-H. (1995) Exospheric systems of Saturn's rings. Icarus 115, 295–303.

Ip, W.-H. (2005) An update on the ring exosphere and plasma disc of Saturn. Geophys. Res. Lett. 32, doi:10.1029/2004GL022217. L13204.

Irvine, W. M. (1966) The shadowing effect in diffuse reflection. J. Geophys. Res. 71, 2931–2937.

Irvine, W. M., J. B. Pollack (1968) Infrared optical properties of water and ice spheres. Icarus 8, 324–360.

Johnson, B. R., S. Atreya (1996) Feasibility of determining the composition of planetary ices by far infrared observations: Application to Martian cloud and surface ices. Icarus 119, 405–426.

Johnson, R. E., M. Nelson, T. McCord, J. Gradie (1988), 'Analysis of Voyager Images of Europa: Plasma Bombardment'. Icarus 75, 423–436.

Johnson, R. E. (1998) Sputtering and desorption from icy satellite surfaces; in *Solar System Ices*, B. Schmitt and C. beBergh eds., Kluwer Acad. Pub., Netherlands, pp. 303–334.

Johnson, R. E., R. W. Carlson, J. F. Cooper, C. Paranicas, M. H. Moore, M. C. Wong (2004) Radiation effects on the surface of the Galilean satellites; in *Jupiter-The Planet, Satellites and Magnetosphere*, F. Bagenal, T. Dowling, W. B. McKinnon, eds., Cambridge Univ. Press, Cambridge, Chapter 20, pp. 485–512.

Johnson, R. E., J. G. Luhmann, R. L. Tokar, M. Bouhram, J. J. Berthelier, E. C. Sittler, J. F. Cooper, T. W. Hill, H. T. Smith, M. Michael, M. Liu, F. J. Crary, D.T. Young (2006a) Production, ionization and redistribution of O_2 Saturn's ring atmosphere. Icarus180, 393–402.

Johnson, R. E., T. I. Quickenden (1997) Photolysis and radiolysis of water ice on outer Solar System bodies. J. Geophys. Res. 102, 10985–10996.

Johnson. R. E., H. T. Smith, O. J. Tucker, M. Liu, R. Tokar (2006b) The Enceladus and OH Tori at Saturn, Astrophys. J. Lett. 644, L137–L139.

Jones, G. H. and 34 coauthors (2008) The dust halo of Saturn's largest icy moon Rhea; Science 319, 1380–1384.

Jurac, S., M. A. McGrath, R. E. Johnson, J. D. Richardson, V. M. Vasyliûnas, A. Eviatar (2002) Saturn: Search for a missing water source. Geophys. Res. Lett. 29, 2172, 25–1–4.

Jurac, S., J. D. Richardson (2007) Neutral cloud interaction with Saturn's main rings; Geophys. Res. Lett. 34, CiteID L08102; doi:10.1029/2007GL029567

Karkoschka E. (1994) Spectrophotometry of the jovian planets and Titan at 300- to 1000-nm wavelength: The methane spectrum. Icarus 111, 174–192.

Kliore, A. J., et al. (2004). Cassini radio science. Space Sci. Rev. 115, 1–69.

Karjalainen, R., Salo, H. (2004) Gravitational accretion of particles in Saturn's rings. Icarus 172, 328–348.

Kempf, S., R. Srama, F. Postberg, M. Burton, S. F. Green, S. Helfert, J. K. Hillier, N. McBride, J. A. M. McDonnell, G. Moragas-Klostermeyer, M. Roy, E. Grün (2005) Composition of Saturnian stream particles. Science 307, 1274–1276.

Li, A. (2009) PAHs in comets: an overview; in *Deep Impact as a World Observatory Event – Synergies in Space, Time, and Wavelength*, H. U. Kaufl, C. Sterken, eds., ESO Astrophys. Symp., ISBN 978-3-540-76958-3. Springer Berlin Heidelberg, p. 161.

Licandro, J., W. M. Grundy, N. Pinilla-Alonso, P. Leisy (2006a) Visible spectroscopy of 2003 UB313: Evidence for N2 ice on the surface of the largest TNO? A&A, 458, L5–L8.

Licandro, J., N. Pinilla-Alonso, M. Pedani, E. Oliva, G. P. Tozzi, W. M. Grundy, (2006b) The methane ice rich surface of large TNO 2005 FY9: A Pluto-twin in the trans-neptunian belt? A&A 445, L35–L38.

Lissauer, J. J. (1985) Bending waves and the structure of Saturn's rings. Icarus 62, 433–447.

Loeffler, M. J., U. Raut, R. A. Baragiola (2006) Enceladus: A source of nitrogen and an explanation for the water vapor plume observed by Cassini. Astrophys. J. 649, L133–L136.

Luhmann, J. G., R. E. Johnson, R. L. Tokar, T. Cravens (2006) A model of the ionosphere of Saturn's toroidal ring atmosphere, Icarus 181, 465–474.

Lumme, K., W. M. Irvine (1976) Photometry of Saturn's rings. Astronom. J. 81, 865–893.

Lumme, K., W. M. Irvine, L. W. Esposito (1983) Theoretical interpretation of the ground-based photometry of Saturn's B ring. Icarus 53, 174–184.

Marouf, E. A., G. L. Tyler, V. R. Eshleman (1982) Theory of radio occultation by Saturn's rings, Icarus 49, 161–193.

Marouf, E. A., G. L. Tyler, H. A. Zebker, R. A. Simpson, V. R. Eshleman (1983) Particle size distribution in Saturn's rings from Voyager 1 radio occultation. Icarus 54, 189–211.

Marouf, E. A., G. L. Tyler, P. A. Rosen (1986) Profiling Saturn's rings by radio occultation. Icarus 68, 120–166.

Marouf, E., R. French, N. Rappaport, C. McGhee, K. Wong, F. Thomson, A. Anabtawi (2008a) Structure and physical properties of Saturn's rings from Cassini radio occultations. Abstracts for "Saturn after Cassini-Huygens" Symposium, Imperial College London, U.K., July 28 to August 1, p. 113.

Marouf, E. A., R. French, N. Rappaport, K. Wong, C. McGhee, A. Anabtawi (2008b) Physical properties of Saturn's rings from Cassini radio occultation (Abstract). Bull. Am. Astronom. Soc. 40, 3, 23.03.

Martens, H. R., D. B. Reisenfeld, J. D. Williams, M. F. Thomsen, H. T. Smith, A. Eviatar, R. E. Johnson, D. T. Young, E. C. Sittler, R. A. Baragiola (2008) Molecular oxygen ions in Saturn's inner magnetosphere for the first 24 Cassini orbits. Geophys. Res. Letts. 35, L20103, doi:10.1029/2008GL035433.

Mastrapa, R. M., M. P. Bernstein, S. A. Sandford, T. L. Roush, D. P. Cruikshank, C. M. Dalle Ore (2008) Optical constants of amorphous and crystalline H_2O-ice in the near infrared from 1.1 to 2.6 μm. Icarus 197, 307–320.

Mastrapa, R. M., S. A. Sandford, T. L. Roush, D. P. Cruikshank, C. M. D. Ore (2009) Optical constants of amorphous and crystalline H_2O-ice: 2.5 – 22 micrometers (4000 – 455 cm^{-1}). Astrophysical Journal, in press.

Mattioda, A., L. J. Allamandola, D. M. Hudgins (2005) The UV to far-IR optical properties of PAHs: A semiempirical model. Astrophys. J. 629, 1183–1187.

Merlin, F., A. Guilbert, C. Duma, M. A. Barucci, C. de Bergh, P. Vernazza (2007) Properties of the icy surface of the TNO 136108 (2003EL61). Astronom. Astrophys. 466, 1185–1188.

Mishchenko, M. I., Z. M. Dlugach (1992) Can weak localization of photons explain the opposition effect of Saturn's rings? Monthly Notices of the Roy. Astronom. Soc. 254, 15P–18P.

Moersch, J. E., P. R. Christensen (1995) Thermal emission from particulate surfaces: A comparison of scattering models with measured spectra. J. Geophys. Res. 100, 7465–7477.

Moore, L., A. F. Nagy, A. J. Kliore, I. Müller-Wodarg, J. D. Richardson, M. Mendillo (2006) Cassini radio occultations of Saturn's ionosphere: Model comparisons using a constant water flux. Geophys. Res. Lett. 33, L22202, doi:10.1029/2006GL027375.

Moore, L., M. Mendillo (2007) Are plasma depletions in Saturn's ionosphere a signature of time dependent water input? Geophys. Res. Lett. 34, L12202, doi:10.1029/2007GL029.

Morfill, G. E., H. M. Thomas (2005) Spoke formation under moving plasma clouds – The Goertz Morfill model revisited. Icarus 179, 539–542.

Morris, R. V., H. V. Lauer, C. A. Lawson, E. K. Jr. Gibson, G. A. Nace, C. Stewart (1985) Spectral and other physiochemical properties of submicron powders of hematite (-Fe2O3), maghemite (-Fe2O3), maghemite (Fe3O4), goethite (-FeOOH), and lepidochrosite (-FeOOH). J. Geophys. Res. 90, 3126–3144.

Moses, J. I., E. Lellouch, B. Bezard, G. R. Gladstone, H. Feuchtgrube, M. Allen (2000) Photochemistry of Saturn's Atmosphere II. Effects of an influx of external oxygen. Icarus 145, 166–202.

Mosqueira, I., P. R. Estrada (2003a) Formation of the regular satellites of giant planets in an extended gaseous nebula I: Subnebula model and accretion of satellites. Icarus 163, 198–231.

Mosqueira, I., P. R. Estrada (2003b) Formation of the regular satellites of giant planets in an extended gaseous nebula II: Satellite migration and survival. Icarus 163, 232–255.

Muinonen, K. O., A. H. Sihvola, I. V., Lindell, K. A. Lumme (1991) Scattering by a small object close to an interface. II. Study of backscattering. J. Opt. Soc. Am. A 8, 477–482.

Mustard, J. F., C. M. Pieters (1987a) Quantitative abundance estimates from bidirectional reflectance measurements, Proc. 17th Lunar Planet. Sci. Conf., J. Geophys. Res. 92, E617–E626.

Mustard, J. F., C. M. Pieters (1987b) Abundance and distribution of serpentinized ultramafic microbreccia in Moses Rock dike: Quantitative application of mapping spectrometer data, J. Geophys. Res. 92, 10376–10390.

Mustard, J. F., C. M. Pieters (1989) Photometric phase functions of common geologic minerals and applications to quantitative analysis of mineral mixture reflectance spectra, J. Geophys. Res. 94, 13619–13634.

Nelson, R. M., B. W. Hapke, W. D., Smythe, L. J., Spilker (2000) The opposition effect in simulated planetary regoliths. Reflectance and circular polarization ratio change at small phase angle. Icarus 147, 545–558.

Nelson, R. M., W. D. Smythe, B. W. Hapke, A. S. (2002) Hale low phase angle laboratory studies of the opposition effect: search for wavelength dependence. Planet. Space Sci. 50, 849–856.

Nelson, R. P. and 29 coauthors (2006) Cassini observations of the opposition effect of Saturn's rings-1; 37th Annual Lunar and Planetary Science Conference, March 13–17, 2006, League City, Texas, abstract no. 1461.

Nicholson, P. D., M. M. Hedman, B. D., Wallis, Cassini-VIMS Team (2007) Cassini-VIMS observations of stellar occultations by Saturn's rings American astronomical society, DDA meeting #38, #12.05.

Nicholson, P. D., M. M. Hedman, R. N. Clark, M. R. Showalter, D. P. Cruikshank, J. N. Cuzzi, G. Filacchione, F. Capaccioni, P. Cerroni, G. B. Hansen, B. Sicardy, P. Drossart, R. H. Brown, B. J. Buratti, K. H. Baines, A. Coradini (2008) A close look at Saturn's rings with Cassini VIMS. Icarus 193, 182–212.

Northrop, T. G., J. R. Hill (1983) The inner edge of Saturn's B ring. J. Geophys. Res. 88, 6102–6108.

Pollack, J. B. (1975) The Rings of Saturn; Space Science Reviews 18, 3–93.

Pospieszalska, M. K., R. E. Johnson (1991) Micrometeorite erosion of the main rings as a source of plasma in the inner Saturnian Plasma Torus. Icarus 93, 45–52.

Porco, C., J. Weiss, D. Richardson, L. Dones, T. Quinn, H. Throop (2008) Simulations of the dynamical and light-scattering behavior of Saturn's rings and the derivation of ring particle and disk properties. Astronom. J. 136, 2172–2200.

Postberg, F., S. Kempf, J. K. Hillier, R. Srama, U. Beckmann, S.F. Green, N. McBride, and E. Grün (2007) Composition of submicron-sized particles in the Saturnian System; EPSC 2007-A-00221.

Postberg, F., S. Kempf, J. K. Hillier, R. Srama, S. F. Green, N. McBride, E. Grün (2008) The E-ring in the vicinity of Enceladus. II. Probing the moon's interior – The composition of E-ring particles. Icarus 193, 438–454.

Postberg, F., S. Kempf, J. Schmidt, N. Brillantov, A. Beinsen, B. Abel, U. Buck, R. Srama (2009) Sodium salts in E ring ice grains from an ocean below Enceladus' Surface; Nature 459, 1098–1101.

Poulet, F., D. P. Cruikshank, J. N. Cuzzi, T. L. Roush, R. G. French (2003) Composition of Saturn's rings A, B, and C from high resolution near-infrared spectroscopic observations; Astronom. Astrophys. 412, 305–316.

Poulet, F., J. N. Cuzzi, R. G., French, L. Dones (2002) A study of Saturn's ring phase curves from HST observations. Icarus 158, 224–248.

Quirico, E., S. Doutè, B. Schmitt, C. de Bergh, D. P. Cruikshank, T. C. Owen, T. R. Geballe, T. L. Roush (1999) Composition, physical state, and distribution of ices at the surface of triton. Icarus 139, 159–178.

Richardson, D. C. (1994) Tree code simulations of planetary rings. Monthly Notices of the Roy. Astronom. Soc. 269, 493.

Salama, F., E. L. O. Bakes, L. J. Allamandola, A. G. G. M. Tielens (1996) Assessment of the polycyclic aromatic hydrocarbon–diffuse intersttellar band proposal; Astrophys. J. 458, 621–636.

Salo, H. (1987) Numerical simulations of collisions between rotating particles. Icarus 70, 37–51.

Salo, H. (1992) Gravitational wakes in Saturn's rings. Nature 359, 619–621.

Salo, H., R. G. French (2009) Photometric modeling of Saturn ring's opposition and tilt effects: Disentangling intrinsic and interparticle contributions based on HST observations. Icarus, in review, 2009.

Salo, H. and R. Karjalainen (2003) Photometric modeling of Saturn's rings I. Monte Carlo method and the effect of nonzero volume filling factor. Icarus 164, 428–460.

Salo, H., R. Karjalainen (2003) Photometric modeling of Saturn's rings I. Monte Carlo method and the effect of nonzero volume filling factor. Icarus 164, 428–460.

Salo, H., R. Karjalainen, R. G. French (2004) Photometric modeling of Saturn's rings. II. Azimuthal asymmetry in reflected and transmitted light. Icarus 170, 70–90.

Salo, H. J., J. Schmidt, M. Sremcevic, M. Sremcevic, F. Spahn (2008) N-body survey of viscous overstability in Saturn's rings. AAS/Division for Planetary Sciences Meeting Abstracts 40, #30.03.

Schutte, W., A., A. G. G. M. Tielens, L. J. Allamandola (1993) Theoretical modeling of the infrared fluorescence from interstellar polycyclic aromatic hydrocarbons. Astrophys. J. 415, 397–414.

Shemansky, D. E., P. Matheson, D. T. Hall, H.-Y. Hu, T. M. Tripp, (1993) Detection of the hydroxyl radical in the Saturn magnetosphere. Nature 363, 329.

Shepard, M. K., P. Helfenstein (2007) A test of the Hapke photometric model. J. Geophys. Res. (Planets) 112, 3001.

Shimizu, M. (1980) Strong interaction between the ring system and the ionosphere of Saturn. Moon Planets 22, 521–522.

Shipman, H., J.B. Adams (1987) Detectability of minerals on desert alluvial fans using reflectance spectra, J. Geophys. Res. 92, 10391–10402.

Shkuratov. Y. G., D. G. Stankevich, D. V. Petrov, P. C. Pinet, A, M. Cord, Y. H. Daydou, S. D. Chevrel (2005) Interpreting photometry of regolith-like surfaces with different topographies: shadowing and multiple scattering. Icarus 173, 3–15.

Shkuratov, Y., L. Starukhina, H. Hoffmann, G. Arnold (1999) A model of spectral albedo of particulate surfaces: Implications for optical properties of the Moon. Icarus 137, 235–246.

Showalter, M. R., P. D. Nicholson (1990) Saturn's rings through a microscope- Particle size distribution from the Voyager PPS scan. Icarus 87, 285–306.

Showalter, M. R., J. B. Pollack, M. E. Ockert, L. R. Doyle, J. B. Dalton (1992) A Photometric study of Saturn's F ring. Icarus 100, 394–411.

Smith, B. A. and 26 coauthors (1981) Encounter with Saturn – Voyager 1 imaging science results. Science, 212, 163–191.

Spilker, L. J., S. H. Pilorz, S. G. Edgington, B. D. Wallis, S. M. Brooks, J. C. Pearl, F. M. Flasar (2005) Cassini CIRS observations of a roll-off in the Saturn ring spectra at submillimeter wavelengths. Earth Moon Planets 96, 149–163.

Spilker, L. J., S. H. Pilorz, B. D. Wallis, J. C. Pearl, J. N. Cuzzi, S. M. Brooks, N. Altobelli, S. G. Edgington, M. Showalter, F. M. Flasar, C. Ferrari, C. Leyrat (2006) Cassini thermal observations of Saturn's main rings: Implications for particle rotation and vertical mixing, Planet. Space Sci. 54(12), 1167–1176.

Srama, R. and 41 coauthors (2006) In situ dust measurements in the inner Saturnian system. Planet. Space Sci. 54, 967–987.

Sremcevic, M., A. V. Krivov, H. Krueger, F. Spahn (2005) Impact-generated dust clouds around planetary satellites: models vs Galileo data. Planet. Space Sci. 53, 625–641.

Sremcevic, M., J. Schmidt, H. Salo, M. Seiss, F. Spahn, N. Albers (2007) A belt of moonlets in Saturn's A ring; Nature 449, 1019–1021.

Sunshine, J. M., C. M. Pieters (1990) Extraction of compositional information from olivine reflectance spectra: new capability for lunar exploration (abstract), in Lunar and Planetary Science XXI, 962–963, Lunar and Planetary Institute, Houston.

Sunshine, J. M., C. M. Pieters, S. R. Pratt (1990) Deconvolution of mineral absorption bands: an improved approach, J. Geophys. Res. 95, 6955–6966.

Sunshine, J.M., C.M. Pieters (1991) Identification of modal abundances in spectra of natural and laboratory pyroxene mixtures: a key component for remote analysis of lunar basalts (abstract), in Lunar and Planetary Science XXII, 1361–1362, Lunar and Planetary Instute, Houston.

Teolis, B. D., R. A. Vidal, J. Shi, R. A. Baragiola (2005) Mechanisms of O_2 sputtering from water ice by keV ions; Phys. Rev. B, 72, 245422 (9 pages).

Thomson, F. S., E. A. Marouf, G. L. Tyler, R. G. French, N. J. Rappoport (2007) Periodic microstructure in Saturn's rings A and B. Geophys. Res. Lett. 34, L24203, doi:10.1029/2007GL032526.

Tiscareno, M. S., J. A. Burns, M. M. Hedman, C. C. Porco, J. W. Weiss, L. Dones, D. C. Richardson, C. D. Murray (2006) 100-metre-diameter moonlets in Saturn's A ring from observations of 'propeller' structures. Nature 440, 648–650.

Tiscareno, M. S., J. A. Burns, M. M. Hedman, C. C. Porco (2009) The population of propellers in Saturn's A ring. Astronom. J. 135, 1083–1091.

Tiscareno, M. S., J. A. Burns, P. D. Nicholson, M. M. Hedman, C. C. Porco (2007) Cassini imaging of Saturn's rings II. A wavelet technique for analysis of density waves and other radial structure in the rings. Icarus 189, 14–34.

Tokar, R. L. and 12 authors (2005) Cassini observations of the thermal plasma in the vicinity of Saturn's main rings and the F and G rings, GRL 32, L14S04, doi:10.1029/2005GL022690 (2005).

Tryka, K. A., A. S. Bosh (1999). A visual spectrum of Triton from the Hubble space telescope. Icarus 142, 571–574 (1999).

Tseng, W.-L., W.-H. Ip, R. E. Johnson, T. A. Cassidy M. K. Elrod (2009) The structure and time variability of the ring atmosphere and ionosphere; Icarus, in press.

Tyler, G. L., E. A. Marouf, R. A. Simpson, H. A. Zebker, V. R. Eshleman (1983) The microwave opacity of Saturn's rings at wavelengths of 3.6 and 13 cm from Voyager 1 radio occultation. Icarus 54, 160–188.

Vahidinia, S., J. N. Cuzzi, M. Hedman, R. Clark, B. Draine, G. Filacchione, P. Nicholson (2008) Modeling the F ring's aggregates; 29.02, AAS/DPS meeting, Ithaca, NY.

van de Hulst, H. C. (1957). *Light Scattering by Small Particles*. Wiley, New York (also available as a Dover Publication, 1981).

Waite, J. H., T. E. Cravens, W.-H. Ip, W. T. Kasprzak, J. G. Luhmann, R. L. Mc-Nutt, H. B. Niemann, R. V. Yelle, I. Müller-Wodarg, S. A. Ledvina, S. Scherer (2005) Cassini ion and neutral mass spectrometer measurements of oxygen ions near Saturn's A-ring. Science 307, 1260–1262.

Waite, J. H. and 13 authors (2006) Cassini ion and neutral mass spectrometer: Enceladus plume composition and structure. Science 311, 1419–1422.

Warren, S. G. (1984) Optical constants of ice from the ultraviolet to the microwave. Appl. Opt. 23, 1206–1225.

Westley, M. S., R. A. Baragiola, R. E., Johnson, G. A., Barrata (1995). Ultraviolet photodesorption from water ice. Planet. Space Sci. 43, 1311–1315.

Wisdom, J., S. Tremaine (1988) Local simulations of planetary rings. Astronom. J. 95, 925–940.

Wu, C. Y. R, T. Nguyen, D. L. Judge, H.-C. Lu, H.-K. Chen, B.-M. Cheng (2006) Destruction yields of NH3 produced by EUV photolysis of various mixed cosmic ice analogs. Adv. Geosci. Planet. Sci. 7, 101–113.

Young, D. T., et al. (2005) Composition and dynamics of plasma in Saturn's magnetosphere, Science 307, 1262–1266.

Zebker, H. A., E. A. Marouf, G. L. Tyler (1985) Saturn's rings: Particle size distribution for thin layer models. Icarus 64, 531–548.

Chapter 16
Diffuse Rings

M. Horányi, J.A. Burns, M.M. Hedman, G.H. Jones, and S. Kempf

Abstract In order to give context to *Cassini*'s findings about Saturn's diffuse rings, this chapter first recalls the *Voyager* and telescopic observations prior to 2004. *Cassini* has investigated these faint rings composed of small particles with remote sensing (visual and infrared imaging) and in-situ detectors (charged-particle and dust detectors), for the first time allowing results obtained by the different techniques to be compared. Generally the agreement is good. The description of the observations are organized by increasing distance from Saturn, and includes (a) the faint rings in and around the main rings; (b) spokes in the B-ring; (c) the narrow outer faint rings; (d) the E-ring with emphasis on its connection to Enceladus's geysers; and (e) the Saturnian dust streams. These discussions also summarize relevant models that have been proposed to explain the behavior of charged dust grains. Except for the spokes and much of the E ring, the particles in these rings are collisional debris. Saturn's D ring has changed significantly since Voyager; part of it seems to be inclined and winding up while another portion (and the Roche Division) has periodic structures that are forced by Saturn's magnetic field. The faint rings in ring gaps are also time-variable and some have Sun-aligned elliptical orbits. The reappearance of the enigmatic spokes should allow several recent theories to be tested. Rings and arcs have been discovered to accompany *Cassini*-found small moons that are trapped in satellite resonances. The realization that Enceladus feeds the E ring and the opportunity to make in-situ measurements, including the electric charge and composition of grains, has made this a rich topic. The dust streams are composed of nanoscale particles moving at speeds of many tens to hundreds of km s^{-1}; they likely originate in the outer reaches of the E ring.

16.1 Introduction

In addition to its majestic main rings, Saturn also possesses a suite of diffuse, low optical depth rings composed primarily of particles less than 100 microns in radius. Interparticle collisions are rare in these tenuous rings, and the small sizes of the particles make them sensitive to non-gravitational forces, so the dynamics of these diffuse rings are qualitatively different from those in the main rings. Furthermore, while the main rings of Saturn can be studied only by remote sensing, the diffuse rings offer a unique opportunity to combine both remote-sensing and in-situ observations. The combination of these sets of data permits us to learn about the effects of phenomena such as radiation pressure, magnetospheric interactions, and plasma drag.

Figure 16.1 shows the entire ring system as seen by *Cassini* when it flew through Saturn's shadow on September 15, 2006. In this particular viewing geometry, small particles scatter light very efficiently, so all of the dusty rings can be detected with a relatively high signal-to-noise ratio. The D ring, the innermost component of Saturn's ring system, can be seen just interior to the main rings. Within the main rings, several narrow dusty ringlets can be detected, and even a few spokes are visible hovering over the B ring. Beyond the F ring, which is the brightest ring of all in this image, there is a series of narrow dusty rings, the brightest of which is the G ring. Furthest out, the extensive E ring fills the entire space between the orbits of Mimas and Rhea.

After summarizing the ground-based and Hubble Space Telescope (HST) observations of Saturn's diffuse rings since

---

M. Horányi (✉)
Laboratory for Atmospheric and Space Physics, and Department of Physics, University of Colorado, Boulder, CO 80309–0392, USA

J.A. Burns and M.M. Hedman
Department of Astronomy, Cornell University, Ithaca, NY 14853, USA

J.A. Burns is also at the Department of Theoretical and Applied Mechanics, Cornell University, Ithaca, NY 14853, USA

G.H. Jones
Mullard Space Science Laboratory, University College London, Holmbury St. Mary, Dorking, Surrey, RH5 6NT, UK
and
The Centre for Planetary Sciences at UCL/Birkbeck, Gower St., London WC1E 6NT, UK

S. Kempf
Max Planck Institute for Nuclear Physics, Saupfercheckweg 1, Heidelberg, 69117, Germany

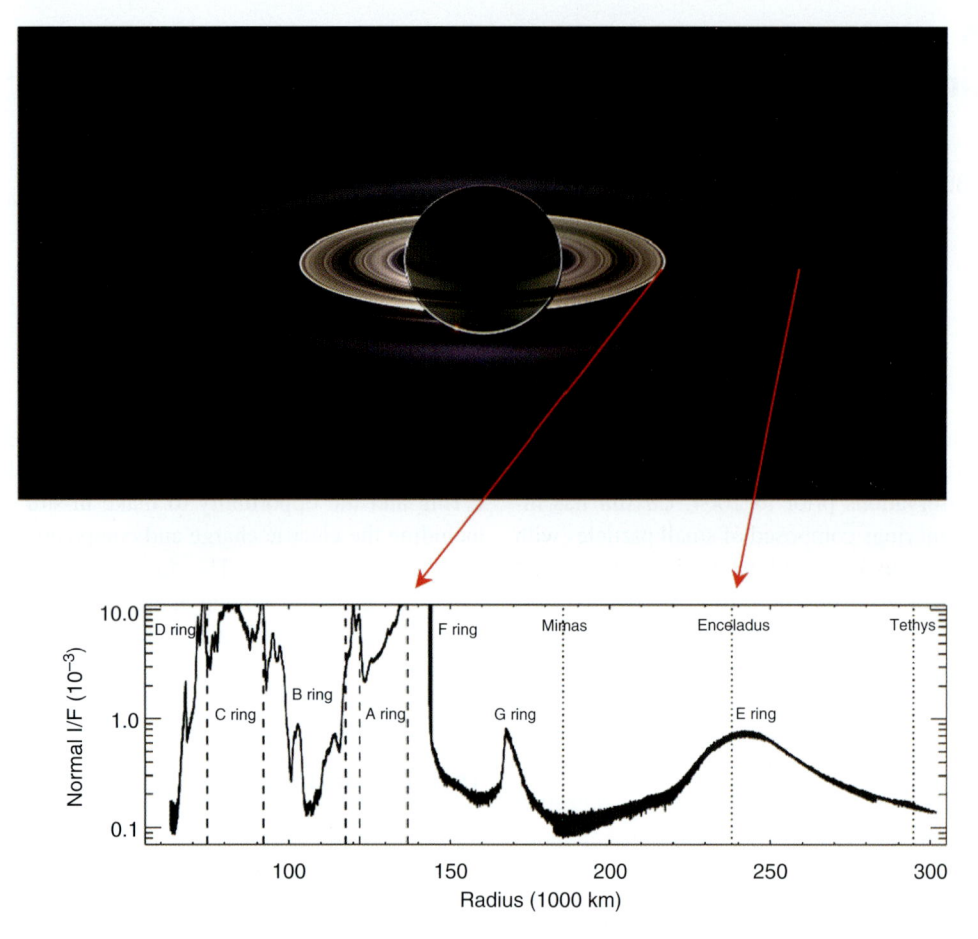

Fig. 16.1 *Top*: A mosaic of images taken on September 15, 2006 while *Cassini* was in the shadow of Saturn (image # PIA08329). The red, green and blue colors in this image are derived from images taken in the IR3, clear and VIO filters. In this geometry the small particles that mostly comprise the diffuse rings scatter light very efficiently, so these normally faint rings appear especially bright. *Bottom*: the brightness of the rings as a function of radial distance from Saturn for a constant phase angle of 178.5° as observed through the camera's clear filter (central wavelength of 635 nm). Brightness is plotted in terms of a quantity called normal I/F, which is proportional to the fraction of the incoming solar radiation scattered into the camera by the material

Voyager, we describe new *Cassini* results for the diffuse rings. We also mention the dust streams of nanoparticles because they are subject to non-gravitational forces and because they likely originate in the outer reaches of Saturn's E ring. Table 16.1 provides the locations and properties of the diffuse rings discussed in this chapter. Note that the F ring, while also composed predominantly of small particles and in most places having low optical depth, is not discussed here, but is instead described in Chapter 13.

16.2 Pre-*Cassini* Observations

Before *Cassini*, the state of knowledge about the properties of Saturn's faint rings and the processes responsible for shaping them were mainly based on *Voyager* observations, which have been discussed in various reviews (Burns et al. 1984, Grün et al. 1984, Mendis et al. 1984, Burns et al. 2001, Horányi et al. 2004). Detailed studies of the D, G and E rings (Showalter et al. 1991, Showalter and Cuzzi 1993, Showalter 1996) and the spokes in the B ring (Porco 1983) that review Voyager data are also available, and therefore do not need to be repeated here.

After the *Voyager* flybys, the next valuable opportunity to observe Saturn's faint rings came in 1995–1996, when Earth passed three times through the planet's ring plane. At this time, the line-of-sight optical depth through the faint rings was greatly enhanced, while the glare from the main rings was reduced. The G and E rings were each observed with HST as well as with large ground-based telescopes like Keck. These observations confirmed that the core of the E ring has a strong blue spectral slope in backscattered light, while the G ring has a slightly red slope between the visible and the near infrared (Nicholson et al. 1996, de Pater et al. 1996, Bauer et al. 1997, de Pater et al. 2004). These color differences provided evidence that these two rings had very different particle size distributions. The G ring's red

Table 16.1 Properties of Saturn's Diffuse Rings

Main ring	Diffuse ring	Radial location	Optical depth	Source
	D Ring Inner Edge	65,000 km		Hedman et al. (2007a)
	D68	67,600 km		
	D72	71,600 km		
	D73	73,300 km	~10^{-4} (inside 73,000 km)	
	D Ring Outer Edge	74,500 km	~10^{-3} (outside 73,000 km)	
74,500 km				
C	Dusty Ringlet in the Maxwell Gap	87,420 km	10^{-4}	Porco et al. (2005)[a]
91,980 km				
B	Spokes	100,000–117,500 km	$10^{-2} - 10^{-1}$	Smith et al. (1981, 1982)
117,500 km				
Cassini Division	Dusty Ringlet in the Huygens Gap	117,490 km	10^{-4}	Porco et al. (2005)[a]
	Charming Ringlet in the Laplace Gap	119,940 km	10^{-3}	Porco et al. (2006)[b]
122,100 km				
A	Inner Encke Gap Ringlet	133,490 km	10^{-3} (up to 0.1 in clumps)	Porco et al. (2005)[a]
	Central Encke Gap Ringlet	133,590 km	10^{-3} (up to 0.1 in clumps)	
	Fourth Encke Gap Ringlet	133,660 km	~10^{-4}	
	Outer Encke Gap Ringlet	133,720 km	10^{-3} (up to 0.1 in clumps)	
136,800 km				
	Roche Division	136,800 km–139,500 km	~10^{-4}	Burns et al. (2002), Porco et al. (2005)
	F ring Core	140,200 km	0.2	Bosh et al. (2002)
	F ring Spiral	139,500 km–141,000 km	~10^{-2}	Charnoz et al. (2005), Murray et al. (2008)
	Janus/Epimetheus Ring	151,450 km	~10^{-7}	Porco et al. (2006)[c]
	G Ring	165,000 km–175,000 km	10^{-6}	Hedman et al. (2007b)
	G Ring Arc	167,500 km	10^{-5}	
	Methone Ring Arc	194,230 km	~10^{-7}	Hedman et al. (2009a)[c]
	Anthe Ring Arc	197,650 km	~10^{-7}	Hedman et al. (2009a)[c]
	Pallene Ring	212,280 km	~10^{-7}	Hedman et al. (2009a)[c]
	E ring	180,000 km–700,000 km	10^{-5} (Peak)	Showalter et al. (1991)

[a] Optical depth assuming rings primarily dust (τ_{min} in Table 3).
[b] Optical depth based of VIMS measurements.
[c] Optical depth estimates based of brightness relative to G ring.

color, similar to those of other dusty rings, is consistent with broad size distributions like power laws and physical models of collisional debris (Showalter and Cuzzi 1993, Throop and Esposito 1998). By contrast, the blue color of the E ring suggests a very steep or narrow size distribution (Showalter et al. 1991), indicating that the particles in the E ring are generated or dispersed by different mechanisms than those active in the G ring.

Observations during this ring-plane crossing also provided improved measurements of the radial and vertical structure of these rings. The G ring was found to have a relatively sharp inner edge and a more diffuse outer boundary (Lissauer and French 2000), in agreement with *Voyager* measurements (Showalter et al. 1991), although the implications of this shape were not yet understood. The E ring (Fig. 16.2) was confirmed to have an asymmetric radial profile that peaked outside Enceladus's orbit (de Pater et al. 2004). Ground-based observations were also able to resolve the vertical structure of the E ring, showing it had a minimum vertical thickness at Enceladus's orbit and became progressively wider with increasing distance from that moon (Nicholson et al. 1996, de Pater et al. 2004). Finally, observers (Roddier et al. 1998) saw what might have been a temporary arc of material in the E ring close to the orbit of Enceladus. All this reinforced an early model that the E ring was closely linked to Enceladus, an idea that would be amply corroborated by *Cassini*.

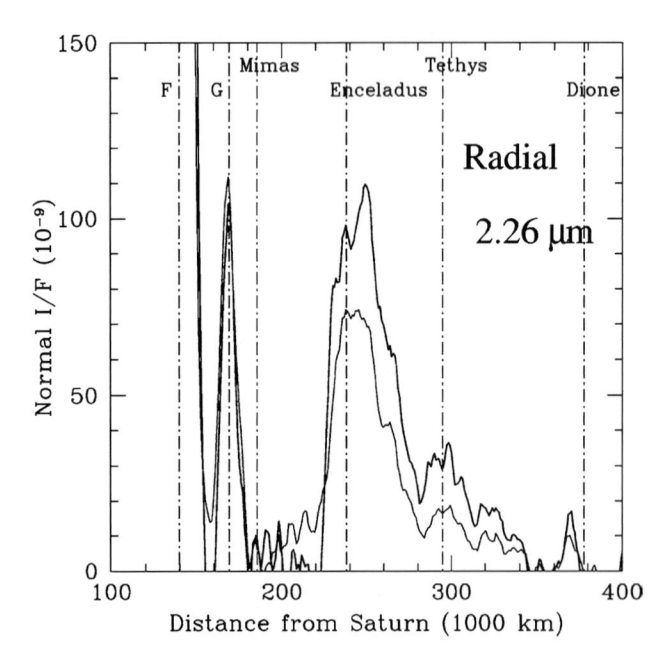

Fig. 16.2 Radial profiles of the back-lit ring derived from Keck near-infrared observations at a wavelength $\lambda = 2.26\,\mu$m. The upper (*heavy line*) profile is vertically integrated over the ring's entire height (0.5 R_S, or 30,000 km), while the lower (*thin line*) profile is integrated over 8,000 km. These thicknesses are much greater than the FWHM of the ring as measured by CDA (cf. Fig. 16.11) (from de Pater et al. 2004)

The D ring, lying inside the main rings, could not be imaged during the ring plane crossing, but was detected in an occultation by the star GSC5249–01240 observed on 21–22 November 1995 with HST (Bosh and Olkin 1996). The outer D ring was noted to have a normal optical depth of around 10^{-3}, while the inner D ring, which included the brightest feature observed by *Voyager* (Showalter 1996), had no detectable optical depth. These data were puzzling at the time, but they began to make more sense in the context of *Cassini* observations. In particular, periodic variations observed in the outer D ring would later be interpreted as the first detection of vertical corrugations in this ring (Hedman et al. 2007a).

HST monitored the activity of the spokes starting shortly before the ring-plane crossing in 1995 until October 1998, when spokes were no longer apparent (McGhee et al. 2005). The implications of these observations are outlined below.

16.3 *Cassini* Observations and Current Theories

Cassini has detected dusty material in numerous locations throughout the Saturn system. The remote-sensing instruments have observed dusty regions extending interior and exterior to the main rings, dusty ringlets within gaps in the main rings, and spokes above the B ring. Further from Saturn, both remote-sensing and in-situ measurements provided information about the G ring and about narrow faint rings and arcs associated with several small moons. Finally, the dust detectors have directly sampled the particles in the extensive E ring and those ejected into interplanetary space. The following sections will consider each of these different features in turn, summarizing both the currently available observational data and the present state of theoretical models.

16.3.1 The D Ring

Lying between Saturn and the classical main rings, the D ring is among the most complex of the faint rings. Both *Cassini* images and earlier *Voyager* observations have revealed a number of distinct structures in this region. The *Voyager* spacecraft detected three features designated as ringlets in this region, along with more subtle, quasi-periodic brightness variations (Showalter 1996). At least two of these ringlets were recovered in *Cassini* images, but these data also indicate significant changes in the structure of the D ring over the last 25 years (Hedman et al. 2007a). For example, the brightest feature in the D ring that was present during the *Voyager* observations was a narrow (<40 km) ringlet located

16 Diffuse Rings

71,710 km from Saturn's center named D72. Images taken by *Cassini* show no narrow ringlet at this location (Fig. 16.3). Instead there is a much broader feature (with a full-width at half-maximum of roughly 300 km and centered around 71,600 km) with a peak brightness that is significantly less than other structures in the D ring. The connection between this structure and the D72 ringlet is unclear, but regardless of whether the ringlet has vanished or transformed into a broader, more diffuse feature, this would be among the largest secular changes in Saturn's rings observed to date.

Noteworthy features in the D ring include not only brightness variations but also trends in the particle size distribution. Various parts of the D ring have very different photometric and spectral properties that almost certainly reflect variations in the shape of the local particle size distributions (Hedman et al. 2007a). In general, regions closer to Saturn have larger

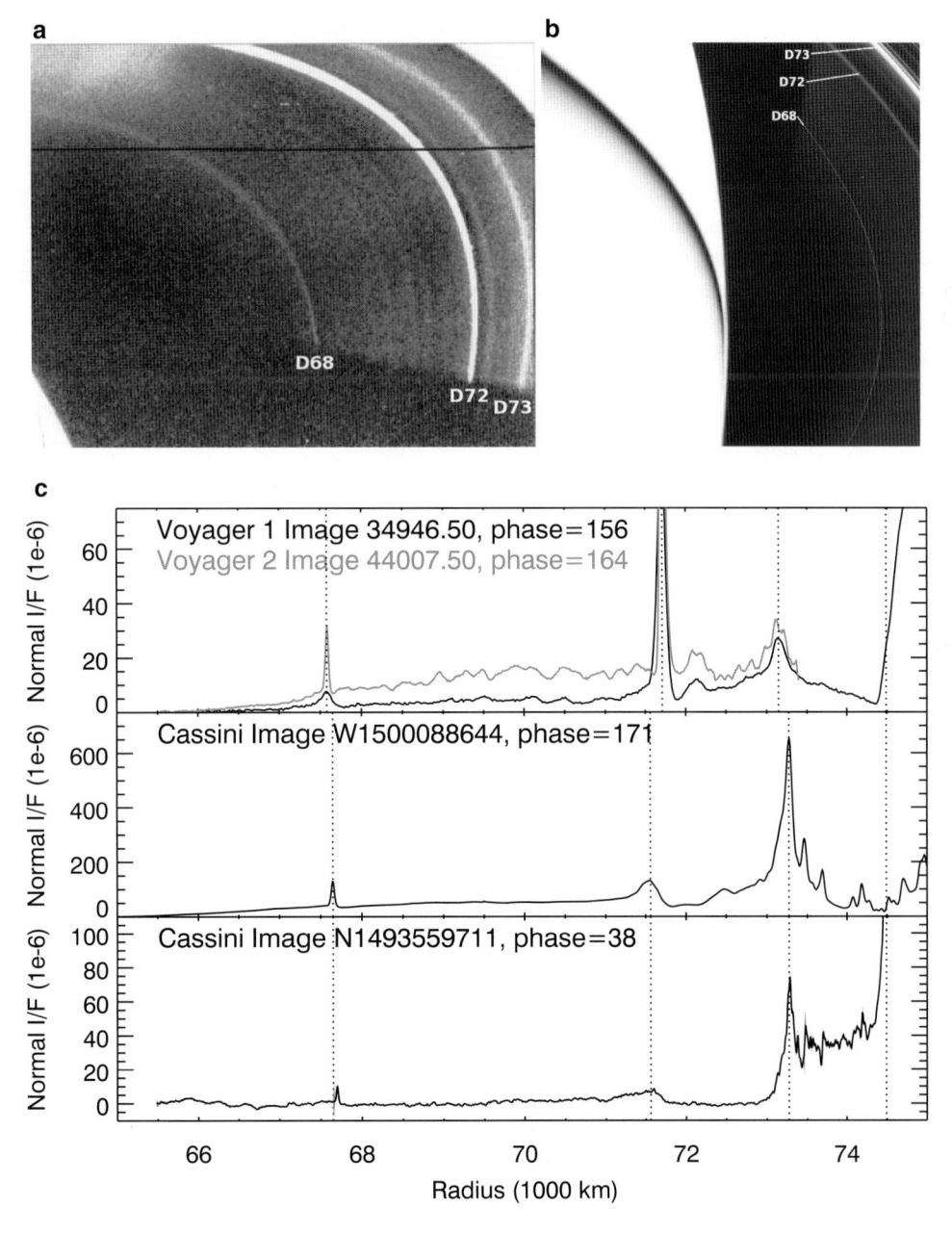

Fig. 16.3 D ring (**a**) *Voyager* image 1 (34946.50) taken at a phase angle of 156°. The three brightest bands were called D68, D72 and D73. The horizontal black line is a data dropout. (**b**) The same region imaged by *Cassini* (W1500088644), at a phase angle of 171°; an over-exposed Saturn fills the image's left half. (**c**) Radial brightness profiles derived from the two images above, a *Voyager* 2 frame (#44007.50 at 164° phase) and an additional *Cassini* image (N1493559711) taken at a phase angle of 38°. The vertical dotted lines mark the positions of D68, D72 and D73 and the C ring's inner edge from left to right (from Hedman et al. 2007a)

fractions of smaller particles (radius of 1–10 μm) relative to larger particles (10–100 μm) compared with regions further out. Moreover, the size distributions vary on more local scales as well. For example, the appearance of the outer part of the D ring (between 73,200 km and 74,500 km) is dramatically different when it was viewed at various phase angles (Fig. 16.3c). At high phase angles, this region appeared to be populated by an array of ringlets, but at low phase angles this entire region seems to be filled with a continuous sheet of material having several brightness minima that look to be correlated with some of the ringlets observed at high phase angles (Hedman et al. 2007a). Likely, non-gravitational processes operate on various scales within this ring to sort particles according to size.

Given this complexity, we consider the various features in the D ring separately, starting from its inner edge and moving outwards. An apparently empty span of about 5,000 km lies between Saturn's cloud-tops and the innermost region of the D ring where material is detectable. This clearing could potentially represent a place where spacecraft could fly close to the planet in order to measure the higher-order components of the planet's gravity and magnetic fields.

Between the inner edge of the D ring at 65,000 km and the inner edge of the D73 ringlet at 73,000 km, there are broad sheets of material that are strongly forward-scattering, implying that they are composed primarily of small particles 1–10 microns across. This may be material derived from the various ringlets that is spiraling inward towards Saturn under the influence of various drag forces.

Embedded in this sheet, roughly 67,650 km from Saturn center is a narrow ringlet called D68. This is the innermost discrete feature in the rings. High-resolution observations of this ringlet reveal that it sometimes has two components, separated by up to 20 km. Lower resolution images indicate the apparent location of this ringlet can vary by up to 50 km, implying that this ringlet is non-circular and/or inclined (Hedman et al. 2007a).

Between 71,000 km and 73,000 km from Saturn center, a region of enhanced brightness occurs around 71,500 km that could be related to the no-longer-visible D72 ringlet seen by *Voyager*, and a local minimum in the surface density appears around 72,000 km. On top of these broad radial structures, there are interesting finer-scale brightness variations that change with time and longitude (see below).

Outside 73,000 km, the character of the D ring changes dramatically: As seen in an occultation (Bosh and Olkin 1996), it has a detectable normal optical depth of 10^{-3} and shows radial brightness variations on scales as small as tens of kilometers. Particularly interesting is a structure that appears as a quasi-sinusoidal variation in the ring's brightness extending between 73,000 and 74,000 km (Fig. 16.4a, Hedman et al. 2007a). Observations of this region at sub-degree ring opening angles show "contrast reversals" similar

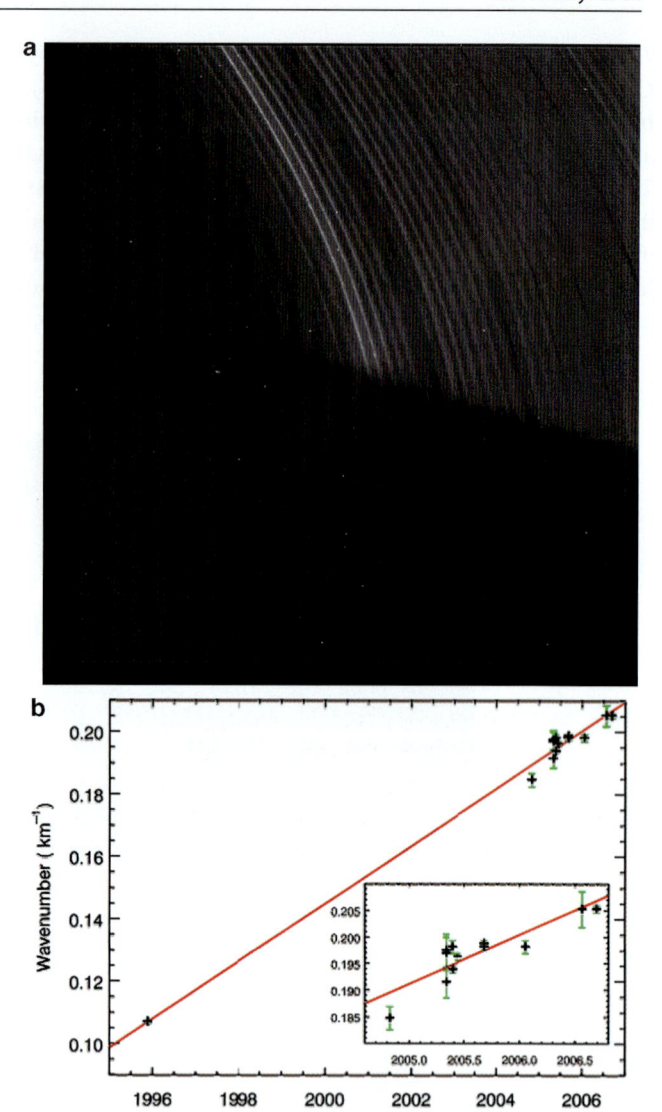

Fig. 16.4 The ∼30-km wavelength structure in the outer D ring. (**a**) One of the highest-resolution images of the outer D ring, obtained at a phase angle of 41° and a resolution of 1.6 km/pixel. The over-exposed inner edge of the C ring lies at the upper right-hand corner of the image, while the shadow cuts across the very lower half of the image. The regular brightness variations in the D ring are apparent, and the ringlet D73 corresponds to the innermost 2–3 brightest features. (**b**) A plot showing the wavenumber $k = 2\pi/\lambda$ of this pattern versus time. The line corresponds to a steady increase in the wavenumber consistent with a progressive winding of the pattern due to differential nodal regression (from Hedman et al. 2007a). This winding up has continued into 2009

to those noted in the Jovian ring (Ockert-Bell et al. 1999, Showalter et al. 2001). These brightness variations can be generated by a vertical corrugation in this part of the D ring. Such a corrugation produces periodic variations in the amount of material along the line of sight that lead to apparent brightness variations in *Cassini* images and optical depth variations in the 1995 occultation data (cf. Gresh et al. 1986).

A comparison of observations taken at different times shows that the wavelength of the corrugation has been decreasing steadily over time. Specifically, the radial wavenumber $k = 2\pi/\lambda$ has been increasing with time at roughly $2.5 \times 10^{-5}\,\text{km}^{-1}\,\text{day}^{-1}$ (Fig. 16.4b, Hedman et al. 2007a); this rate has proven to be an excellent predictor of the wavelength into 2009. This steady reduction in the pattern's wavelength can be attributed to differential nodal regression of inclined particle orbits. Extrapolating backwards in time, one can compute when $k = 0$ corresponding to a simple inclined ring, i.e., a plane. The ring would have been in such a state in March 1984. Something, perhaps an impact, may have disturbed this ring at this time.

16.3.2 The Roche Division

Material in the Roche Division was first detected in *Voyager* data (Burns et al. 1984), and *Cassini* observations now show significant substructure in this region (Fig. 16.5). Early *Cassini* observations had seemed to suggest two brightness enhancements near the orbits of the satellites Prometheus and Atlas (Porco et al. 2005). Subsequent observations have indicated that a nearly continuous ring-like structure is present around 139,000 km from Saturn center and just interior to Prometheus's orbit. However, the structure observed near the orbit of Atlas was found to be part of a more complex structure that varies with longitude and time (see below and Hedman et al. 2009b).

16.3.3 Resonant Structures in the D Ring and the Roche Division

Despite one being exterior – and the other interior – to the main rings, the Roche Division and the D ring are connected to each other by a common dynamical phenomenon. In both the inner Roche Division (~138,000 km) and in the middle D ring (70,000–73,000 km) complex patterns exist that can be decomposed into multiple series of alternating bright and dark bands tilted relative to the local radial direction (Fig. 16.5). The pattern speeds and morphology of these structures are consistent with patterns generated by multiple Lindblad resonances with periodic perturbing forces (Hedman et al. 2009b). These are reminiscent of the Lorentz resonances, driven by magnetic-field periodicities like those that are important in shaping Jupiter's faint rings (Burns et al. 1985, Hamilton 1994, Ockert-Bell et al. 1999). The forcing periods operating in both regions range between 10.5 and 10.9 h, commensurate with the rotation periods of Saturn's atmosphere (Sanchez-Lavega et al. 2000) and with periods observed in radio emissions (Kurth et al. 2007, Gurnett et al. 2007), but not with the expected periods of acoustic oscillations in Saturn's interior (Marley and Porco 1993). If the forcing in these regions was primarily gravitational, many strong additional resonances would make their presence known in the C ring, and they are not seen. The forcing therefore is more likely due to non-gravitational driving terms, so the dynamics of particles in these regions could

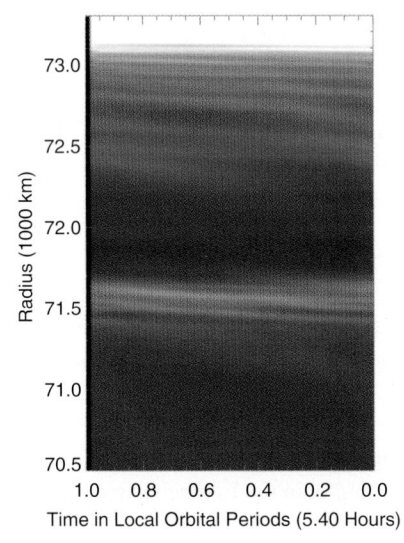

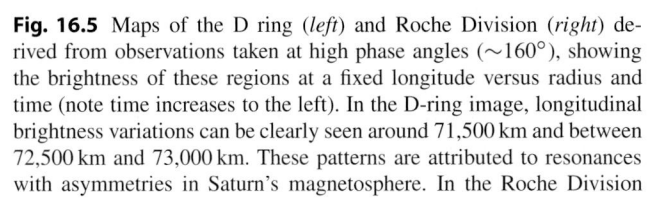

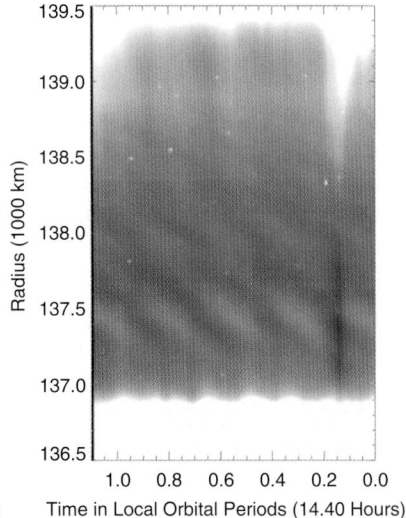

Fig. 16.5 Maps of the D ring (*left*) and Roche Division (*right*) derived from observations taken at high phase angles (~160°), showing the brightness of these regions at a fixed longitude versus radius and time (note time increases to the left). In the D-ring image, longitudinal brightness variations can be clearly seen around 71,500 km and between 72,500 km and 73,000 km. These patterns are attributed to resonances with asymmetries in Saturn's magnetosphere. In the Roche Division image, the bright streaks near the top of the image are due to bright features in the F ring. A brightness concentration can be seen around 139,000 km (just interior to the orbit of Prometheus), and a periodic structure is visible near 137,500 km (close to the orbit of Atlas). The latter also seems to be produced by asymmetries in Saturn's magnetosphere (from Hedman et al. 2009b)

potentially provide insight into the various asymmetries in the magnetosphere and their influence on small dust grains.

16.3.4 Faint Ringlets Within Main-Ring Gaps

A number of largely empty gaps can be found throughout Saturn's otherwise dense main rings (see Chapter 13). Four of these gaps contain low optical depth ringlets that are strongly forward-scattering (i.e., they are particularly bright in backlit images like Fig. 16.1) and therefore seem to be composed primarily of small ($<100\,\mu$m) particles. Despite containing very little total mass, such ringlets interest modelers because their presence in otherwise open gaps may provide hints as to the mechanism that clears those gaps. In order of increasing distance from Saturn, the ringlet-containing gaps are the Maxwell Gap in the C-ring ($\sim$87,420 km from Saturn center), the Huygens and Laplace Gaps in the Cassini Division ($\sim$117,740 km and $\sim$119,940 km from Saturn center, respectively), and the Encke Gap in the A ring ($\sim$133,590 km from Saturn center). Example Images are displayed in Fig. 16.6.

The Maxwell, Huygens and Laplace gaps each contain a single dusty ringlet in the space between the gap's inner edge and the innermost edge of the optically thick ringlet that occupies each gap (Fig. 16.6 a–c; Porco et al. 2005, 2006). These dusty ringlets do not show strong variations in brightness with longitude. By contrast, the dusty material in the Encke Gap has a much more complex structure, being organized into three narrow ringlets (referred to here as the "inner", "central" and "outer" Encke Gap Ringlets) plus another broader, fainter feature known as the "fourth" ringlet (Fig. 16.6 d, Porco et al. 2005). The inner, central and outer Encke Gap ringlets all contain "clumps", localized regions that are up to an order of magnitude brighter than the background ringlet; these are associated with "kinks" in the radial position of the ringlet (Ferrari and Brahic 1997, Burns et al. 2005). While the morphology of individual clumps may vary and the clumps in a given ringlet may drift slowly relative to each other, the overall distribution of the clumps in each ringlet has been remarkably persistent over the course of the *Cassini* mission to date. Undulations in the radial position of the inner ringlet have been noted, and these are likely due to the same gravitational perturbations that produce the curious and confounding periodicities on the edges of the Encke Gap.

Many aspects of the origin, evolution and dynamics of these dusty ringlets remain obscure. For example, it is not obvious why the Encke Gap has three or four dusty ringlets while the other gaps have only one. However, comparisons among these features reveal some interesting aspects of these ringlets' behavior. For instance, the spaces occupied by these

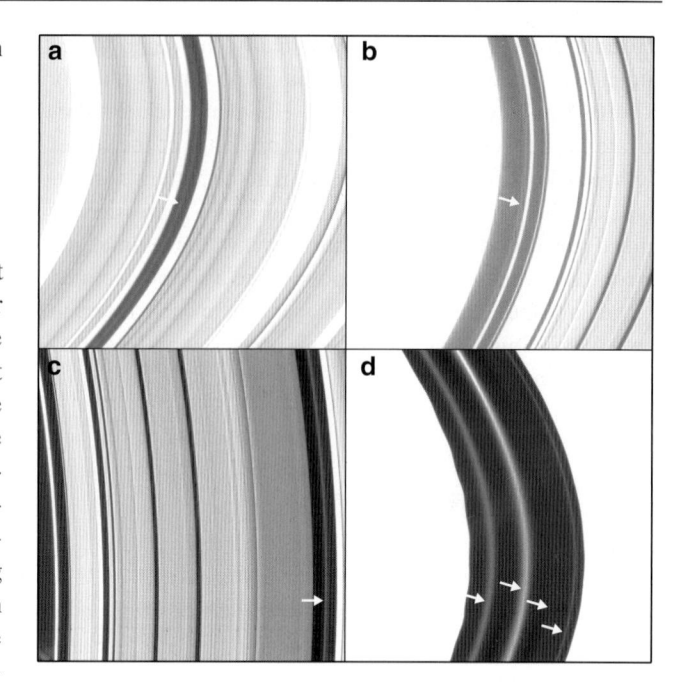

Fig. 16.6 Diffuse ringlets within Saturn's rings. Images (**a**) and (**b**) were taken with a solar incidence angle of 114.5°, an emission angle of 99°, and a phase angle of 145°. (**a**) Image centered on a radius of 84,394 km, has a radial scale of 7 km per pixel, and shows a radial region of 4,290-km extent, including the Maxwell gap in the C ring. The arrow points to a diffuse ring lying interior to the dense Maxwell ringlet. (**b**) This image is centered on a radius of 117,292 km, has a radial scale of 7 km per pixel, and covers 3,870 km, including the Huygens gap immediately outside the outer B-ring edge. The arrow points to a diffuse ring lying interior to the dense Huygens ringlet. Image (**c**) was taken at a phase angle of 111°, an emission angle of 108°, and has a resolution of 2.4 km per pixel centered on 118,907 km, covering a range of 2,571 km, including the Laplace Gap in the outer *Cassini* Division. The arrow points to a dusty ringlet that lies inside this gap. Image (**d**) is from the dayside orbit insertion sequence and is centered on a radius of 133,557 km interior to the Encke gap, has a radial scale of 1.15 km/pixel, a radial extent of 858 km, and was taken from a phase angle of 134°. The arrows point to (from the *left*) the inner, central, fourth, and outer Encke gap ringlets (Images **a**, **b** and **d** from Porco et al. 2005, Image **c** from PIA 08330)

ringlets are the widest gaps in the rings, and in fact every open space in the main rings wider than 100 km across seems to contain a dusty ringlet. While some narrower gaps contain low optical depth ringlets, open spaces with a mean width less than 100 km do not contain forward-scattering *dusty* ringlets with $\tau \geq 10^{-4}$. This may be related to the fact that most of the dusty ringlets are non-circular. In particular, the ringlet in the Laplace gap exhibits "heliotropic" behavior, where the geometric center of the ringlet is displaced towards the Sun (Hedman et al. 2007c). This ringlet also appears to be inclined and/or displaced out of the ringplane (Burt et al. 2008). This behavior can be explained in part by the perturbations to the particle orbits induced by solar radiation pressure, and similar phenomena are probably operating on some, if not all, of the other dusty ringlets. Non-gravitational forces like solar radiation pressure are therefore having a significant effect on the shape and structure of these rings. If successfully modeled, these distorted ringlets could provide otherwise unobtainable information on the rings' electrodynamic environment.

16.3.5 Spokes in the B Ring

Spokes are intermittent, approximately radial markings on Saturn's B ring, thought to consist of small charged dust particles lofted from their parent ring bodies owing to electrostatic repulsion. While they were first recognized in images taken by *Voyagers 1* and *2* (Smith et al. 1981, 1982), spokes were possibly noticed earlier in ground-based observations (Robinson 1980) (Fig. 16.7). These perplexing features have attracted great attention; following their appearance in *Voyager 1* images, *Voyager 2*, during its approach to Saturn, dedicated sequences to spoke observations, providing an invaluable dataset. The spokes' characteristics have been derived with an increasing level of sophistication since their acquisition in these images. The desire to provide an explanation for the spokes has played a large role in the emergence of the research field of 'Dusty Plasmas'.

The key physical characteristics of spokes based on the *Voyager* data (Porco 1983, Grün et al. 1983, 1992, Eplee and Smith 1984, Doyle and Grün 1990) can be summarized as follows: These features are generally most common surrounding the dawn ring ansa, and seem to form primarily in that region. Spokes develop on a timescale of minutes, and can become more intense over a period of a few hours. Appearing at radial distances that are near to, or straddle, kronosynchronous orbit, they move around the ring nearly co-rotating with the planet. During increases in spoke intensity, these features extend forwards and backwards from the kronosynchronously moving longitude inside and outside the co-rotation distance, respectively, while their central

Fig. 16.7 *Top*: Spokes in the B ring as seen by *Voyager* 2 (Smith et al. 1982). The left image was captured in back-scattered light before closest encounter, with the spokes appearing as dark radial features across the ring's center. The right image was taken in forward-scattered light after the spacecraft crossed the ring plane, and was looking back towards the Sun; the spokes now occur as bright markings. Typical dimensions of these spokes are 10,000 km in length and 2,000 km in width. The nature of the changing brightness indicates that spokes consist of small grains with radii (<1 μm), i.e., that are comparable to the wavelength of visible light. At the time these images were taken, the rings' opening angle to the sun was $B' = 8°$. *Bottom*: The initial spoke observations by taken *Cassini* on September 5, 2005 ($B' = 20.4°$), over a span of 27 min. These faint and narrow spokes were seen from the un-illuminated side of the B ring. These spokes are $\approx 3,500$ km long and ≈ 100 km wide, much smaller than the average spokes seen by *Voyager*. These images were taken with a resolution of 17 km per pixel at a phase angle of 145° when *Cassini* was 13.5° above the unlit side of the rings as the spokes were about to enter Saturn's shadow (from Mitchell et al. 2006)

regions intensify; this indicates that the spoke material follows Keplerian trajectories, in broad terms at least. Once spokes no longer intensify, they fade while traveling around the day-side of the rings. Newly formed spokes in the *Voyager* data often coincided with the positions of older spokes that seem to have survived an entire revolution around Saturn. This periodicity in spoke formation, with a period near that of the Saturn Kilometric Radiation (SKR) measured by the *Voyagers* (Porco and Danielson 1982), suggests a formation trigger that is linked to Saturn's magnetic field.

As summarized below, numerous formation theories were proposed to explain the spokes' existence, but none could be definitively tested without further observations. HST monitored spoke activity from shortly before the ring-plane crossing in 1995 until October 1998, when HST no longer detected spokes. McGhee et al. (2005) proposed that spokes were possibly always present, but only detectable when the observer lay close to the ring plane. It was therefore anticipated that *Cassini* would detect spokes on its 2004 arrival at Saturn and that its observations would finally decide which, if any, of the competing theories were correct (Horányi et al. 2004). However, contrary to predictions, *Cassini* did not observe spokes, even when close to the ring plane, until September 2005 (Fig. 16.7). The variability in spoke occurrence in HST data was therefore not an observational effect: spokes are indeed a seasonal phenomenon, and their formation can be suspended for extended periods (Fig. 16.8). This seasonal variation of spoke activity may be a consequence of the variable plasma density near the ring. The plasma density is a function of the solar elevation angle B′, measured from the ring plane, since it is generated mainly from the rings by photoelectron production and by photo-sputtering of neutrals that are subsequently ionized (Mitchell et al. 2006, Farrell et al. 2006). Although this may explain the seasonality of spoke activity after their formation, we still lack a generally accepted model for how they are triggered.

Spokes comprise dust particles in a narrow size distribution centered at about $s \sim 0.6\,\mu\text{m}$ (Doyle and Grün 1990). It is generally believed that spoke formation involves charging and thus electric felds acting on these small grains, but this process requires – as we show below – a much higher plasma density than is commonly expected near the rings (Hill and Mendis 1982, Goertz and Morfill 1983). When formed, spokes initially cover an approximately radial strip with an area of $A \sim 10^3 \times 10^4\,\text{km}^2$, with a characteristic optical depth of $\tau \sim 0.01$. The total number of elevated grains can be estimated to be on the order of $N_d \sim A\tau/(\pi s^2) \sim 10^{23}$. If the grains are released approximately at the same time and carry just a single electron when released from their parent bodies, the formation of the spoke cloud requires a minimum surface charge density (measured in units of electron charges e) $\sigma_e^* = N_d/A \sim 10^6\,\text{cm}^{-2}$, orders of magnitude higher than the charge density, σ_o, expected from the nominal plasma conditions in the B ring.

The nominal plasma environment near the optically thick B ring is set by the competing electron and ion fluxes to and from the ring due to photoelectron production from the ring (as well as the ionosphere) and the photo-ionization of the rings' neutral atmosphere that is maintained by photo-sputtering. All of these are expected to show a seasonal modulation with the ring's opening angle with respect to the Sun, B′. The characteristic energy for photo-scattered electrons is

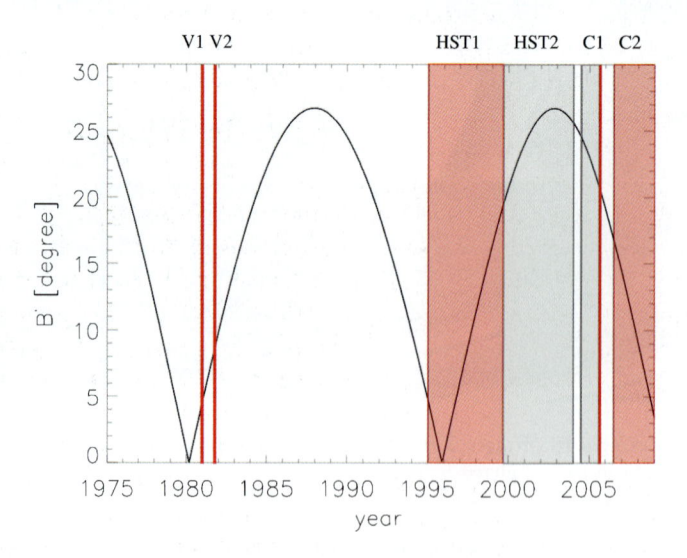

Fig. 16.8 The variation of the absolute value of solar elevation angle B′ as a function of time. Red lines and red-shaded areas identify when spokes were visible. The vertical red lines V1 and V2 indicate the *Voyager* encounters in 1980 and 1981. HST1 and HST2 mark the periods of spoke observations by HST (McGhee et al. 2005). C1 shows the period without spoke activity as reported by *Cassini* following its orbit insertion in June 2004, ending with its first spoke sighting in September 2005. C2 is the current episode when spokes are active. Due to *Cassini*'s low inclination orbit between September 2005 and July 2006, spokes could not be observed

$T_e \sim 2$ eV, and the plasma density is expected to be $n \sim 0.1-1$ cm^{-3} (Waite et al. 2005). The characteristic plasma shielding distance is $\lambda_D = 740 \, (T_e/n)^{1/2} = 1-3 \times 10^3$ cm, larger than the average distance between the cm - m sized objects in the B ring, which has a comparable vertical thickness, $h \sim 10$ m. Hence, it is reasonable to treat the B ring as a simple sheet of material (Goertz and Morfill 1983). The nominal surface potential, including its possible seasonal variations, is expected to be in the range of -5 V $< \phi_R < 5$ V. The surface charge density can be estimated from Gauss's law,

$$\sigma_0 \sim \varphi_R/(4\pi e \lambda_D) \sim 2.5 \, \phi_R \, (n/T_e)^{1/2} < 1-3 \times 10^3 \text{ cm}^{-2}.$$

Since $\sigma_0 << \sigma_e^*$, the formation of a spoke requires higher than normal plasma densities.

Several spoke formation theories, as described by McGhee et al. (2005), have been put forward. Of these, the proposed spoke formation trigger theories that arguably have been most widely accepted are those of meteoroid impacts onto the rings (Goertz and Morfill 1983) and field-aligned electron beams originating from the auroral regions of Saturn (Hill and Mendis 1982): both could transiently increase the plasma density above a critical threshold, and trigger the formation of spokes.

A meteoroid impact-produced plasma cloud was shown to expand, cool and recombine as it rapidly propagates in the radial direction, possibly explaining many of the observed spoke characteristics. However, the estimated propagation speed of such a cloud seems to have been overestimated (Farmer and Goldreich 2005, Morfill and Thomas 2005). An electron-beam mechanism has been suggested to loft small particles instantaneously along the entire radial extent of a spoke (Hill and Mendis 1982). Other spoke formation ideas include dusty plasma waves (Tagger et al. 1991, Yaroshenko et al. 2008) and impact-induced avalanches of small charged dust particles (Hamilton 2006). *Cassini* image sequences with high-temporal resolutions could easily determine the radial propagation speed of a forming spoke, but such sequences have not yet been obtained (Mitchell et al. 2008).

Recently, an alternative formation process has been proposed which suggests that, although the electrostatic charging mechanism is responsible for spoke formation as first postulated by Hill and Mendis (1981), the exact cause of the charging is linked to electrical storms in the atmosphere of Saturn itself (Jones et al. 2006). Following a terrestrial lightning discharge, strong electric fields are thought to exist above the associated thunderstorms. The ionization of atmospheric particles by incoming cosmic rays, in the presence of this electric field, can set off an electron avalanche, as has been suggested to cause gamma-ray emission from above thunderstorms. When the atmospheric density is low enough, such electron avalanches can escape into the magnetosphere (e.g., Lehtinen et al. 2000). The escaping electrons are guided by the planetary magnetic field to the thunderstorm's magnetic conjugate point in the opposite hemisphere. At Saturn, if this occurs within a certain range of latitudes, the escaping electrons will strike the rings and possibly trigger spokes. The ionospheric density varies with local time, and reaches a minimum near local dawn (Moore et al. 2004), where spoke formation is indeed most prevalent.

Although radio emissions, termed Saturn Electrostatic Discharges (SEDs), are known to be linked to Saturnian thunderstorms (Burns et al. 1983, Fischer et al. 2006), the detection of whistler radio waves in the absence of an SED (Akalin et al. 2206) suggests that SEDs may not be as reliable markers of thunderstorms as previously concluded from *Voyager* observations. *Cassini*'s MIMI instrument (Krimigis et al. 2004) has detected a pair of magnetic field-aligned electron beams (Jones et al. 2006), which in many respects are similar to those expected from thunderstorms (Lehtinen et al. 2000). No accompanying observation of SED emission or whistler events were noted during this time, so a link with thunderstorms remains unproven.

During the first four years (2004–2008) of *Cassini* observations, spokes remained a high priority. For most of this interval, spokes were much fainter and less frequent than those seen by the *Voyagers* (Fig. 16.8). By late 2008, B' had reached values similar to those during the *Voyager* encounters, and spoke activity was indeed approaching - if not matching - the activity observed by the *Voyagers* (Mitchell et al. 2008). Some observations suggest that a periodicity linked to the SKR emission period is appearing again. Based on the increase in spokes at the time of writing, it is anticipated that *Cassini* should answer key questions regarding the nature of these perplexing ring features around the equinox period of 2009–2010.

16.3.6 The G Ring

Several relatively narrow rings reside between the F ring and the core of the E ring. The brightest and best known of these is the G ring, located approximately between 165,000 and 175,000 km from Saturn center; the exact boundaries of this ring are difficult to define precisely as its edges blend smoothly into the background E ring. This ring has an asymmetrical profile, with a sharp inner edge and a diffuse outer boundary. Near the inner edge of this ring, at 167,500 km, a bright arc of material extends over roughly 60° in longitude, with a peak brightness several times that of the background G ring and a radial full-width at half-maximum of approximately 250 km (Fig. 16.9). This arc has been observed multiple times over the course of the *Cassini* mission, indicating that it is a persistent feature in the G ring. Furthermore, these observations allow us to measure the mean motion of this

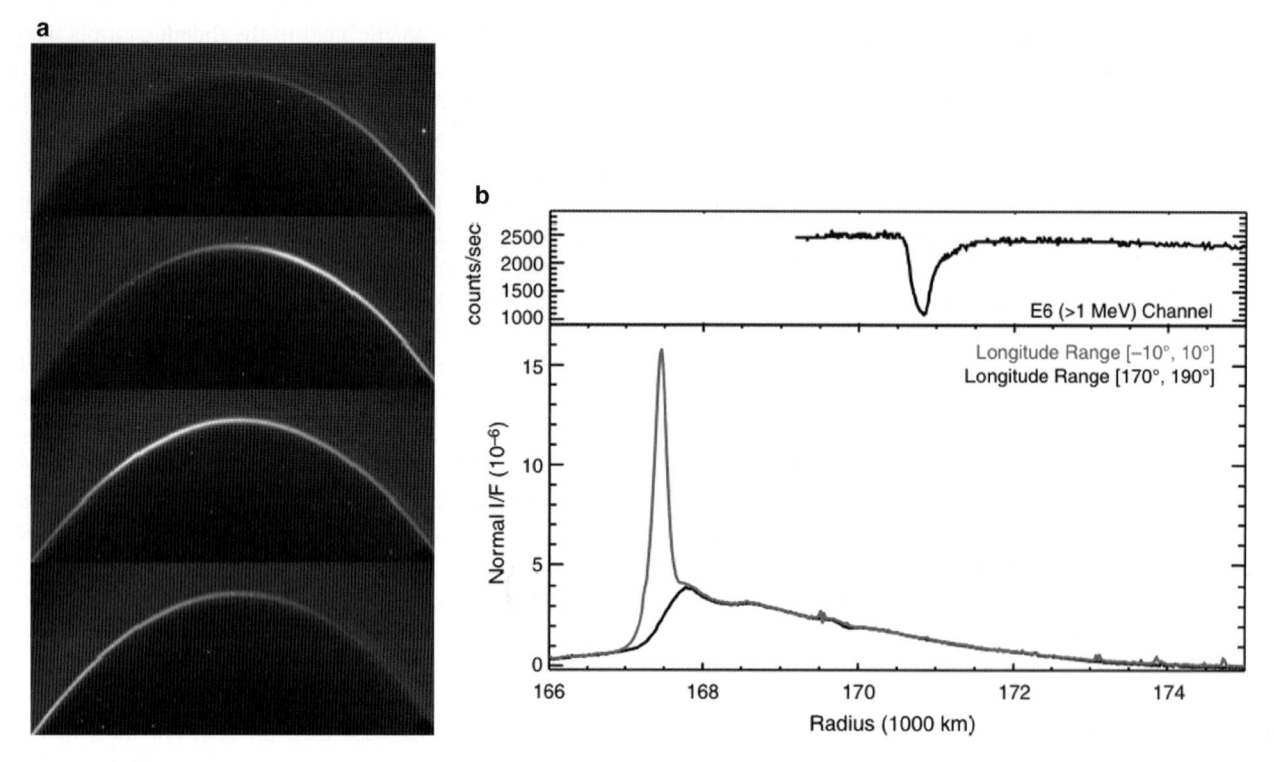

Fig. 16.9 (**a**) Images of the G ring arc obtained on September 19, 2006 at 12:37, 13:11, 13:44, and 14:18 UTC from top to bottom. The bright arc moves from right to left through the field of view. (**b**) *Top*: The drop-out in charged-particle flux detected during *Cassini*'s passage over the arc region on September 5, 2005. The radial scale corresponds to the equatorial distance of the unperturbed magnetic field lines that thread *Cassini* at the time of the observation. *Bottom*: Average (offset-subtracted) radial brightness profiles of the G ring at different longitudes relative to the arc's peak visible in **a**. The profiles through the arc (grey) and elsewhere (black) are essentially identical outside 168,000 km, whereas the arc has a sharp peak at 167,500 km. The absorption feature's radial width is comparable to that of the visible arc. While the absorption is radially displaced from the arc, this may be explained by larger-scale magnetospheric processes or through deformations in the magnetic field by the arc (from Hedman et al. 2007b).

feature. This arc lies near the 7:6 co-rotation eccentricity resonance with the moon Mimas, suggesting that this resonance is likely responsible for confining the arc in longitude. Indeed, numerical models of the motions of particles trapped in this resonance can reproduce the arc's observed radial and longitudinal extent (Hedman et al. 2007b).

On 5 September 2005, *Cassini* flew through the magnetic field lines that pierced the arc. During this passage, the MIMI experiment detected a strong, ∼50% depletion in energetic electrons (Fig. 16.9b). Although a subtle energetic proton macrosignature is always present in association with the G ring, and indeed was observed by Pioneer 11 before the ring was identified in images (Van Allen et al. 1983, and references therein), such a deep depletion was not present in previous passages near the G ring, implying that the absorption was due to material trapped in the arc. Like some other microsignatures of small satellites in this region of Saturn's magnetosphere (cf. Roussos et al. 2008), the G-ring's absorption is displaced radially (see Fig. 16.9b), possibly because of local currents (cf. Thomsen and Van Allen 1980; Roussos et al. 2007). The magnitude of the absorption indicates that the arc contains a total mass between 10^8 and 10^{10} kg, equivalent to a 100-meter-wide ice-rich moonlet (Hedman et al. 2007b). In fact, a small sub-kilometer moonlet Aegaeon was recently observed embedded in the G ring arc (Porco et al. 2009). Given the breadth of the absorption feature observed by MIMI, and the fact that the cross-section of this object is much less than the total cross section of large particles computed by Van Allen (1987), it is unlikely that Aegaeon is the only absorbing object in this region. Instead, Aegaeon probably shares the arc with a population of particles between 1 and 100 m across.

While the larger particles are likely resonantly confined to the arc by Mimas's action, the dust they produce has stronger interactions with the ambient plasma and therefore can escape to produce the rest of the visible G ring. Since the local orbital speed is lower than the speed of the plasma (which co-rotates with the magnetosphere), interactions between the dust and plasma will tend to accelerate the dust grains and cause them to drift away from Saturn (Burns et al. 2001). As the particles move outwards, they are eroded by processes like sputtering, causing the density of material to decline with distance from the arc. Such a model would explain why the bulk of the G ring lies exterior to the arc, and can even account for the presence of relatively large particles throughout the G ring, consistent with the detection of a large (>100 micron) grain around 176,700 km by the dust detector (Hedman et al. 2007b).

16.3.7 Other Narrow Outer Faint Rings

Just as the bodies in the arc can provide a source for ring material in the G ring, small moons can also potentially supply material to narrow rings (Burns et al. 1999, 2001). Thus far, diffuse rings have not been noticed in the vicinity of Daphnis in the Keeler Gap or around the Lagrangian moons of Tethys and Dione. However, diaphanous rings have been observed near several small moons that reside between the F ring and Enceladus. In extremely high-phase imaging of the region around the G ring, at least two rings were discovered. A relatively broad feature is found surrounding the orbit of the co-orbital moons Janus and Epimetheus, and another ringlet overlaps the orbit of the small moon Pallene (Porco 2006, Hedman et al. 2009a).

The highest-phase images did not show clear evidence for ringlets associated with two other small moons of Saturn, Methone and Anthe. However, in-situ measurements and subsequent images demonstrate that these moons are also embedded in tenuous arcs of material. Roussos et al. (2008) reported >0.6 MeV electron microsignatures detected by *Cassini*'s MIMI instrument on 2006 September 9, approximately centered at equatorial distances of 3.23 and 3.31 R_s. The former was detected when *Cassini* was ~12,000 km from Methone, at the exact orbital distance of the moon. Although Methone is only ~3 km wide (Spitale et al. 2006), the microsignature width measured ~1,500 km. Radial diffusion could account for a signature up to ~100 km wide; the observed signature therefore suggests the presence of an arc of material sharing an orbit with Methone. Given the 3.31 R_s microsignature's relatively close proximity to Methone, it too may be associated with that moon, but that signature would have been displaced radially due to magnetospheric currents (e.g. Hedman et al. 2007b). We note that the more distant signature's equatorial radius is close to the orbit of Anthe, a moon measuring approximately 1 km in diameter (Cooper et al. 2008). However, as the latter moon was separated from *Cassini* by ~131°, it is more likely that Methone and its putative arc of material was again the cause. Imaging data later confirmed the existence of an arc of material extending ±5° in longitude around Methone (Fig. 16.10, Hedman et al. 2009a).

These images also demonstrate that an arc of material extending over 20° in longitude surrounds Anthe (Fig. 16.10, Hedman et al. 2009a). Both these moons are trapped in co-rotation eccentricity resonances with Mimas (Spitale

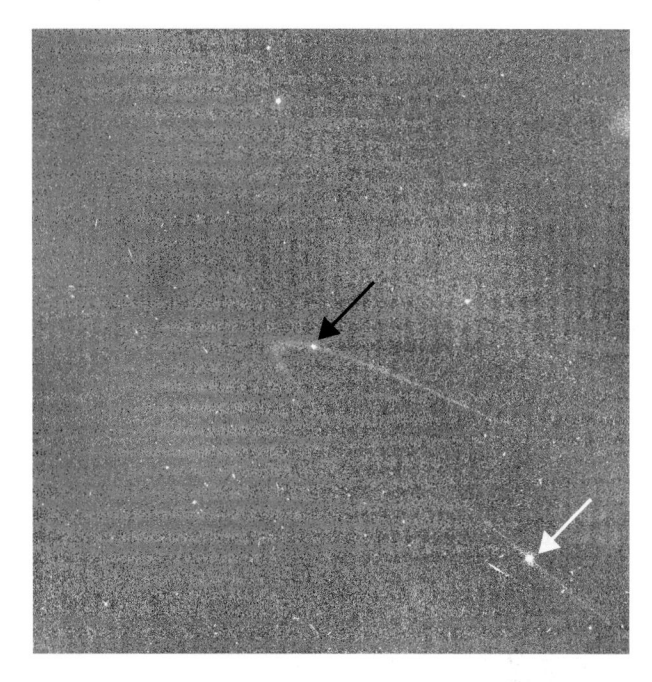

Fig. 16.10 Image showing arcs of debris associated with the small moons Anthe (*black arrow*) and Methone (*white arrow*) (from Hedman et al. 2009b)

et al. 2006, Cooper et al. 2008, Hedman et al. 2009a), and the longitudinal extents of these arcs are consistent with them being resonantly trapped populations of particles. These arcs are therefore directly analogous to the G-ring's arc. Comparisons among these different rings and their relationships with their parent bodies therefore promise to be very productive, especially since the range of parent-body sizes involved brackets the ~10 km "optimal size" for dust production derived by Burns et al. (1984, 1999).

16.3.8 The E Ring

The E ring is the most extensive planetary ring in the solar system, enveloping the icy satellites Mimas, Enceladus, Tethys, Dione, Rhea and, as *Cassini* has discovered, Titan. Since the maximum edge-on brightness occurs near Enceladus's mean orbital distance, the icy moon was early on proposed to be the dominant source of ring particles (Baum et al. 1981). In telescopic data, brightness enhancements were suggested to occur also near Tethys (de Pater et al. 1996, 2004) and Dione (Baum et al. 1981), but are not apparent in *Cassini* data. Ever since the ring's discovery (Feibelman 1967), it has been mainly investigated using ground- and space-based images. Through such analysis, a global description of the ring was achieved (Showalter et al. 1991). More recently, the ring has been extensively imaged by the remote-sensing instruments onboard the *Cassini* spacecraft (e.g., Fig. 16.1).

In-situ spacecraft measurements in the Saturnian system provide a complementary view of the E ring by measuring dust impacts during passages through the ring. *Voyager 1*'s planetary radio-astronomy (PRA) and plasma-wave (RPWS) instruments were first to identify dust impacts by their characteristic electromagnetic signatures, as the spacecraft traversed the E ring in 1980 (Aubier et al. 1983, Gurnett et al. 1983, Meyer-Vernet et al. 1996). The Cosmic Dust Analyzer (CDA) onboard *Cassini* is the first dedicated dust detector to investigate the local properties in the E ring, including the spatial and size distributions of the dust particles, their charge state, as well as their chemical composition (Srama et al. 2004). By the end of 2006, *Cassini* performed two almost equatorial and eleven steep traversals through the E ring inside the orbit of Dione in a favorable configuration for dust measurements. Steep passages through the ring plane are particularly useful for determining the ring's vertical profile where the ring-plane is pierced (Fig. 16.11). Unfortunately, little similar vertical data has been obtained since 2006; in fact, because of the primary instrument (HRD) was partially damaged on *Cassini*'s closest passage to the G ring, it is unlikely that data of this quality will be obtained in the near future.

These new data must be interpreted in the context of the discoveries that Enceladus is a geologically active moon, and that the plume of particles launched from the vents at Enceladus' south pole is likely the primary source for most of the E ring. Enceladus and its plume are described in detail in other chapters in this book, but for completeness, we briefly review the findings relevant to the E ring here. From the plume's brightness profile, Porco et al. (2006) concluded that about 1% of the particles, ejected with a mean velocity of 60 m s^{-1} escape at a rate of about 10^{13} particles s^{-1} (0.04 kg s^{-1}) to the E ring, while Spahn et al. (2006) inferred from the dust data that about $5 \cdot 10^{12}$ particles larger than $2 \, \mu\text{m}$ escape from the moon's gravity – amounting to an escaping dust mass of at least 0.2 kg s^{-1}. The latter authors also constrained the escape rate of ejecta particles created by

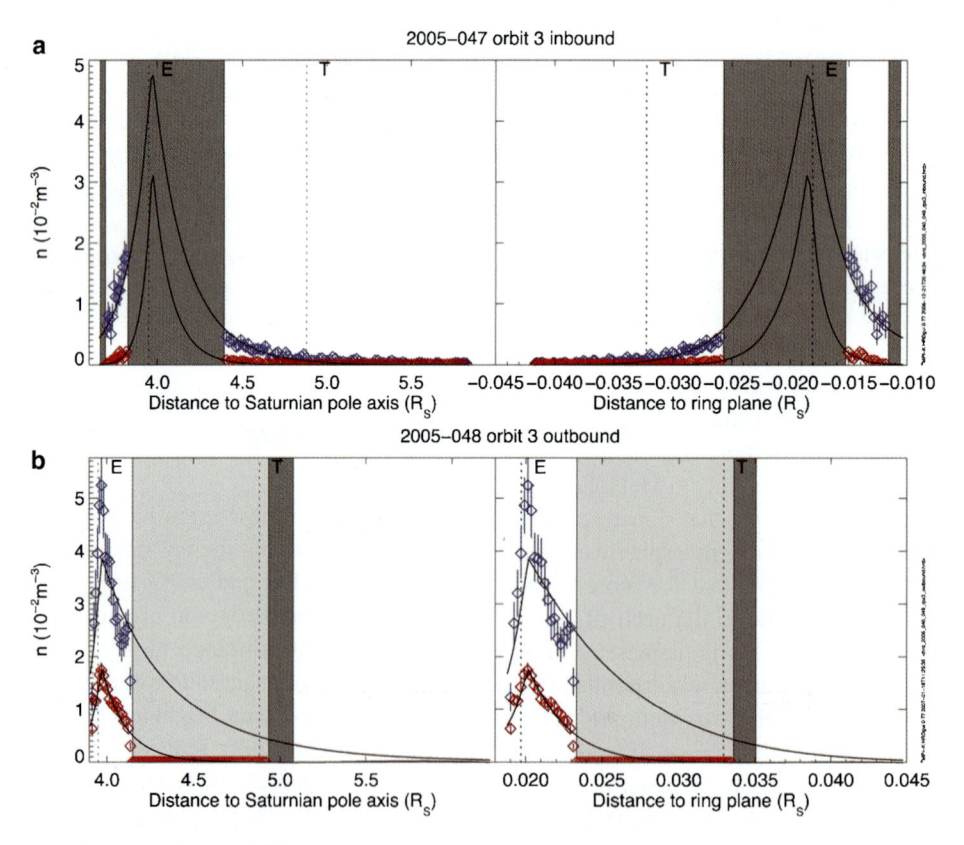

Fig. 16.11 Spatial distribution of E-ring particles with radii $s \geq 1.3 \, \mu\text{m}$ (blue) and $s \geq 2.4 \, \mu\text{m}$ (red) inferred from CDA measurements inside $6 \, R_S$ (364,000 km) during *Cassini* orbit 3. The dust number density versus distance to Saturn's rotation axis (*left*) and versus the elevation above the ring plane (*right*) for the (**a**) inbound, and (**b**) outbound segments of the trajectory. Light grey areas mark periods when the detector was either insensitive to E-ring dust particles or the data were not transmitted to Earth; areas in dark grey indicate periods when the instrument's operation interfered with data acquisition. The curves show the empirical model presented in the text. The dotted vertical lines labeled E and T show the positions of the satellites Enceladus and Tethys (from Kempf et al. 2008).

hypervelocity impacts of interplanetary meteoroids (Krivov et al. 2003) or E ring particles (Hamilton and Burns 1994) onto the moon's surface to be at most 10^{12} particles s^{-1}.

Schmidt et al. (2008) suggested that the plume particles nucleate and condense from the water vapor expanding inside fractures in the moon's surface, while the size and speed distributions of the emerging grains are established by wall collisions within the vents before the grains escape to vacuum. By adjusting their model to match the available imaging and in-situ data, they obtained a total dust production rate in the plume of about 5 kg s^{-1}. The resulting size-dependent speed distributions of the bigger plume particles are consistent with the distributions inferred from infrared spectra of the plume at altitudes ranging between 50 and 300 km (Fig. 16.12), obtained by *Cassini* VIMS (Hedman et al. 2009c). The spectral data provide clear evidence that the gradient of the particles' speed distribution increases with the grain size, implying that bigger grains are predominantly found at lower altitudes. Hedman et al. (2009c) estimate for grains of one-micron radius ejected at 120 m s^{-1} a total flux of a few times 10^{18} per m^2 (total number of such grains per second and per velocity increment per size increment). Schmidt et al. (2008) compute about the same flux. Spitale and Porco (2007) showed that the plume is composed of at least eight dust jets emerging from discrete sources localized at thermal hot spots identified by *Cassini* CIRS (Spencer et al. 2006).

The overall structure of the E ring can be described by an empirical model derived from CDA dust measurements interior to Rhea's orbit for grains with s > 0.9 micrometers. This model gives the number density n as a function of the distance r to Saturn's spin axis and the altitude z above the ring's symmetry plane. The radial distribution of ring particles with s > 0.9 µm is reasonably well described by a pair of power laws centered at the densest point within the ring plane, while the vertical ring thickness increases linearly with distance from the densest point:

$$n(\rho, z) = n_0 e^{-(z-z_0)^2/2\sigma^2} \begin{cases} (\rho/\rho_c)^{+e_i}, \rho \leq \rho_c \\ (\rho/\rho_c)^{-e_o}, \rho > \rho_c \end{cases}$$

with

$$\sigma(\rho) = \sigma_c + (\rho - \rho_c) \begin{cases} \frac{\sigma_i - \sigma_c}{\rho_i - \rho_c}, \rho \leq \rho_c \\ \frac{\sigma_o - \sigma_c}{\rho_o - \rho_c}, \rho > \rho_c \end{cases} \text{ and }$$

$$z_0(\rho) = \begin{cases} z_0(\rho_i) \frac{\rho - \rho_c}{\rho_i - \rho_c}, \rho \leq \rho_c \\ 0, \rho > \rho_c \end{cases},$$

where $e_i \sim 50$, $e_o \sim 20$, $\sigma_i = 2,293$ km, $\rho_i = 3.16$ R$_S$ (Mimas's orbital semi-major axis), $\sigma_c = 1,826$ km, $\rho_c = 3.98$ R$_S$, $\rho_o = 4.75$ R$_S$ (Tethys's semi-major axis), $\sigma_0 = 2,336$ km, and $z_0(\rho_i) = -1,220$ km (see Fig. 16.11). Note that the variables σ and z_0 give the vertical extent and the vertical offset of the ring as a function of ρ.

Densities change by about a factor of two from orbit to orbit; presumably, such changes reflect the time-variability in the plume source, the plasma properties and the magnetosphere's configuration. Also, images like Fig. 16.1 indicate that the color and brightness of the ring is modulated with longitude and/or hour angle, perhaps because particles of various sizes are responding in different ways to the dynamics, as predicted by Hamilton (1993). Thus a radially symmetric model oversimplifies the ring's actual complexity.

The peak number density, according to CDA, was found to be 0.16 to 0.21 m^{-3} for grains with s > 0.9 µm and 0.021 to 0.076 m^{-3} for s > 1.6 µm. The in-situ results roughly match the local number densities for s > 0.9 µm derived from the shadows of embedded moons in the E ring seen in *Cassini* images, which were estimated to be about 0.03 m^{-3} in the vicinity of Enceladus, and about a tenth this value at Tethys (Schmidt and Sremčević 2009). The analysis of the shadows is consistent with a local depletion in space number density of grains with radii smaller than about half a micron near the mid-plane of the E ring. This could be a consequence of the rapid orbital evolution of the population of smaller grains and their dispersal over a larger radial and vertical domain.

Remarkably, the location of the densest point does not coincide with Enceladus's orbit but is displaced outwards by at least 3,000 km (Kempf et al. 2008), which is likely due to plasma drag (Juhász et al. 2007). This displacement was noticed earlier in ground-based (Fig. 16.2, de Pater et al. 2004) and HST observations. As seen through these telescopes, the ring's brightness blends with the background at about 7.5 R$_S$ (450,000 km), whereas, according to the impacts on *Cassini*'s CDA, the ring extends to much greater radial distances, engulfing even Titan's orbit. Remarkably, outside Enceladus's orbit, the ring's radial profile decays smoothly until Titan's orbit (Srama et al. 2006). The large spatial extent of the E ring in both the radial and vertical directions was already noted by the RPWS instrument onboard Voyager *1* when it crossed the ring plane in 1980 (Gurnett et al. 1983).

The vertical profiles at Enceladus's orbit derived from the *Cassini* CDA measurements are in good agreement with those derived from RPWS and edge-on images of this ring (Kurth et al. 2006, Kempf et al. 2008). Interior to Enceladus's orbit, the vertical distribution of ring particles with radii s > 0.9 µm is well described by a Gaussian profile in CDA data, but exterior to Enceladus the vertical ring structure no longer shows a Gaussian distribution. Images of the E ring and the RPWS data also indicate that near the orbit of Enceladus the ring has a two-banded vertical structure with maxima occurring ± 1,000 km from the mid-plane (Kurth et al. 2006). The ring's full-width-half-maximum (FWHM) has its minimum of ≈4,200 km at Enceladus and rises to ≈5,400 km by

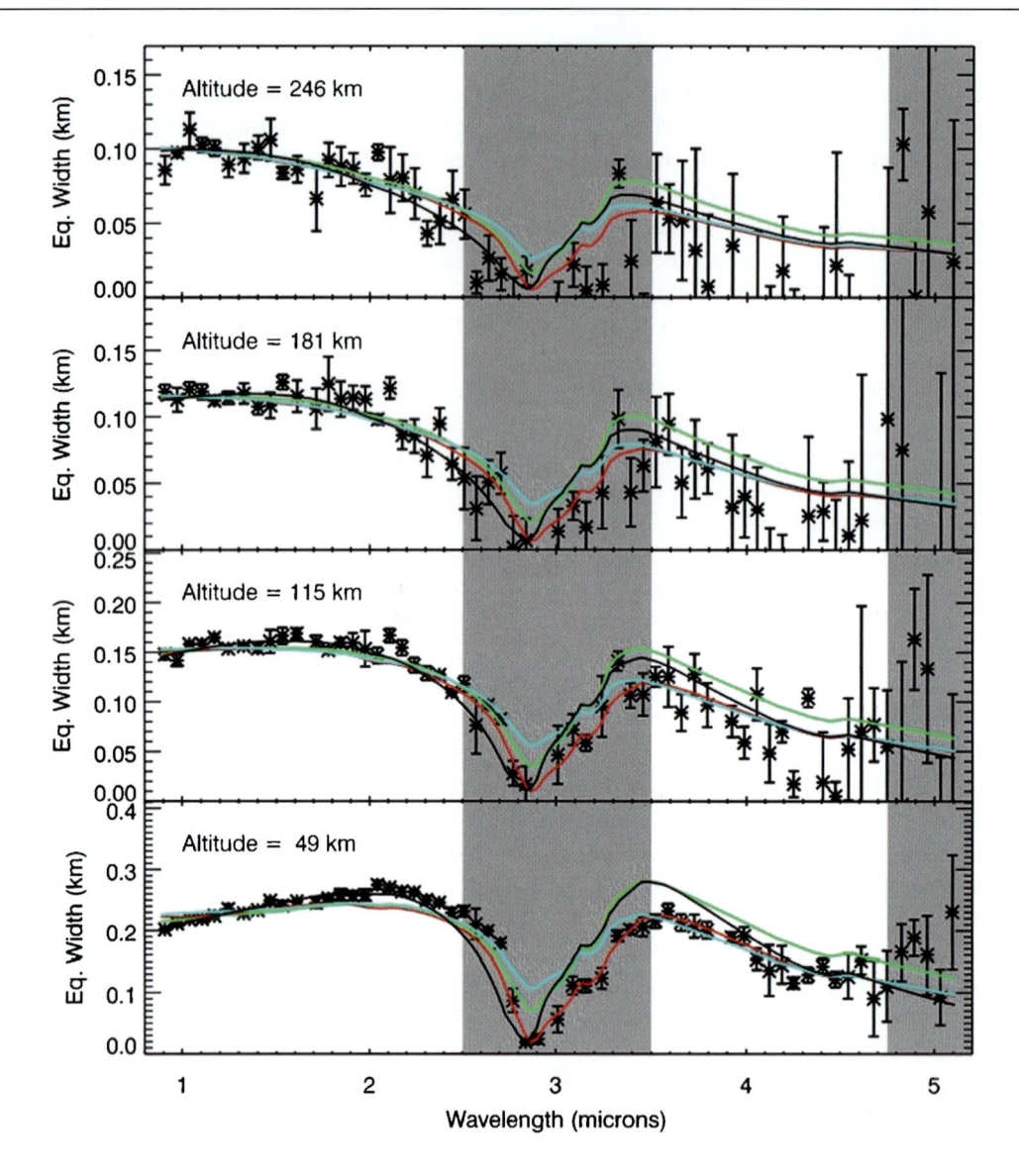

Fig. 16.12 Sample IR spectra of Enceladus's plume derived from *Cassini* VIMS data. The absorption band at 3 microns is due to water ice. Outside this band, the overall slopes of the spectra vary in ways that are sensitive to the size distribution of the ejected ice grains. The black lines are the best-model fits, while the colored lines are scaled spectra for the best-fit size distributions computed using Mie theory for spheres (red) and for irregular shape models 3 (green) and 5 (blue) from Pollack and Cuzzi (1980) (from Hedman et al. 2009c)

Mimas's orbit (Kempf et al. 2008). The ring also flares exterior to Enceladus, and its FWHM at Tethys' orbit is about 5,500 km for grains with $s \geq 1\,\mu m$, and 6,300 km at 4.97 R_S (301,000 km) for grains with $s \geq 0.6\,\mu m$. The ring thickness at Enceladus does not depend on the grain size at least for particles with $s \geq 1\,\mu m$ (Kempf et al. 2008). At Mimas, the ring is displaced southwards by $\approx 1,200$ km with respect to the planet's equatorial plane, while the CDA data provide no evidence for the ring's vertical displacement between the orbits of Enceladus and Dione (some evidence for such displacements may be seen in images like PIA07803, however). The relevance of the directional plume-particle injection for the vertical E ring structure was swiftly recognized (Porco et al. 2006, Juhász et al. 2007). Numerical simulations of particle paths within the jets (Kempf et al. 2009) indicate that the small-scale features apparent in the vertical E ring profiles measured by *Cassini* CDA are associated with specific Enceladus dust jets (Fig. 16.13).

The particle sizes in the E ring found by the CDA experiment are similar to those in the Enceladus plume inferred by the VIMS team. So far, particle size distributions in the E ring have only been obtained for grains with $s > 0.9\,\mu m$. Near Enceladus's orbit, the differential size distribution $n(s)ds \sim s^{-q}ds$ has slopes between $4.2 < q < 5.4$. The slope derived from CDA measurements is generally smaller than the slope derived from the RPWS data (Kurth

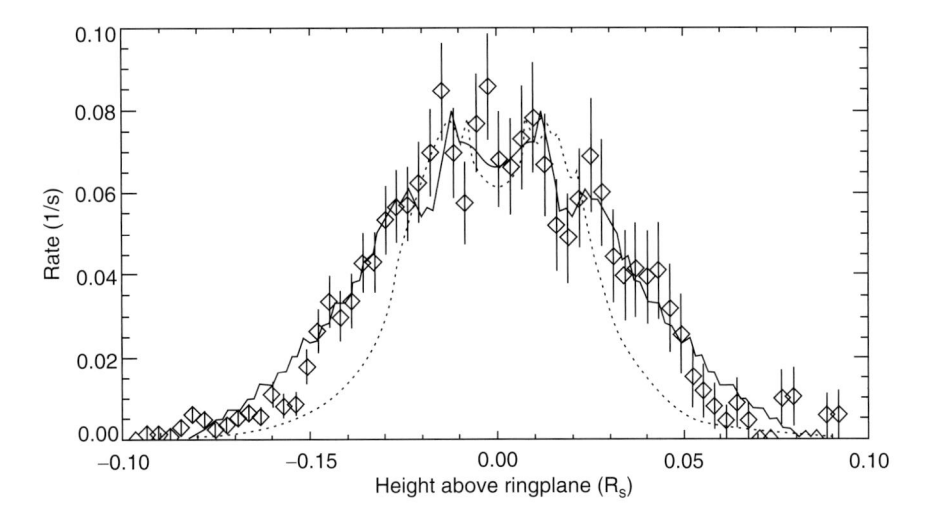

Fig. 16.13 Comparison between the dust impact rate recorded by the *Cassini* dust detector during a steep crossing of the ring plane at a radial distance from Saturn's center of 3.93 R_S (diamonds) and the vertical ring profile resulting from simulations of the Enceladus jet particle propagation (*solid line*). The dotted curve shows the solution of a simple four-parameter model for the inclination distribution for each of the individual jets (from Kempf et al. 2009)

et al. 2006), perhaps because RPWS was not designed to measure dust and, thus, its derived grain sizes may be uncertain. Different parts of the E ring could have different size distributions due to variations in the relative concentrations of freshly ejected plume particles like those measured during the close Enceladus flyby in orbit 11, versus background E-ring particles (Spahn et al. 2006, Kempf et al. 2008).

For the first time, CDA's time-of-flight mass spectrometer allowed the composition of the ring particles to be determined in situ. Measurements obtained during *Cassini*'s initial E-ring crossing in October 2004 have concluded that the particle composition in the outer E ring is dominated by water ice (Hillier et al. 2007). Based on a detailed analysis of 2,000 mass spectra, two major types of ring particles have been identified (Fig. 16.14). Type I spectra show hardly any traces of ions other than the bulk water ice material and a tiny amount of sodium; Type II spectra exhibit impurities of organic compounds and/or silicate minerals within the ice particles (Postberg et al. 2008).

It seems unlikely that water-ice particles with embedded impurities are surface ejecta produced by hypervelocity impacts of interplanetary meteoroids or E ring particles onto Enceladus's clean ice crust (Cuzzi and Durisen 1990). This suggests that Type II dust exclusively originates from Enceladus's ice geysers while Type I particles are surface ejecta (Postberg et al. 2008). The similar composition of Saturnian stream particles (discussed in the following section) and the impurities within the Type II particles suggests that the stream particles are the sputtered remnants of larger E ring particles created in the interior of Enceladus (Kempf et al. 2005b, Postberg et al. 2008). Both particle types can be found in abundance everywhere in the E ring, indicating

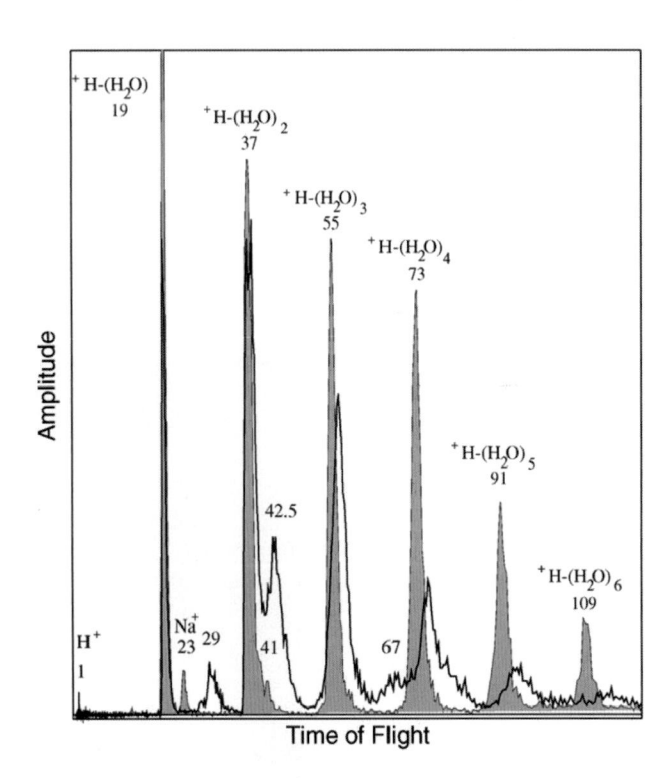

Fig. 16.14 Type I (grey) and Type II (solid line) particle mass spectra. The only non-water features observed in the Type I spectra come from known contaminants (from Postberg et al. 2008)

the rapid dispersal of Enceladus plume particles throughout the entire ring. Two additional minor dust populations consist of sodium-rich water ice (Type III) (Postberg et al. 2009) and of pure minerals (Type IV).

Model calculations for grain growth inside Enceladus's vents show that the trace sodium concentrations in Type I and Type II particles are in good agreement with the condensation of vapor emerging from liquid water with much higher salt concentrations. However, the high Na content of Type III grains is not consistent with the rapid sublimation of solid ice nor the decomposition of clathrates (Postberg et al. 2009). Consequently, evaporation of a liquid water reservoir below Enceladus's ice crust, which is – or was – in contact with the moon's rocky core, appears to be the most significant plume-producing process.

CDA has also measured the charge of particles in the E ring. Figure 16.15 shows the charges measured on 367 large E-ring particles by CDA early in *Cassini*'s tour (Kempf et al. 2006). We compare these measurements to those calculated for the expected surface potentials on dust using a model of the magnetospheric plasma environment based on *Voyager* measurements (Juhász et al. 2002). Inside 6.5 R_S (394,000 km), all grains carried negative charges, while outside 7.5 R_S (455,000 km) the detected

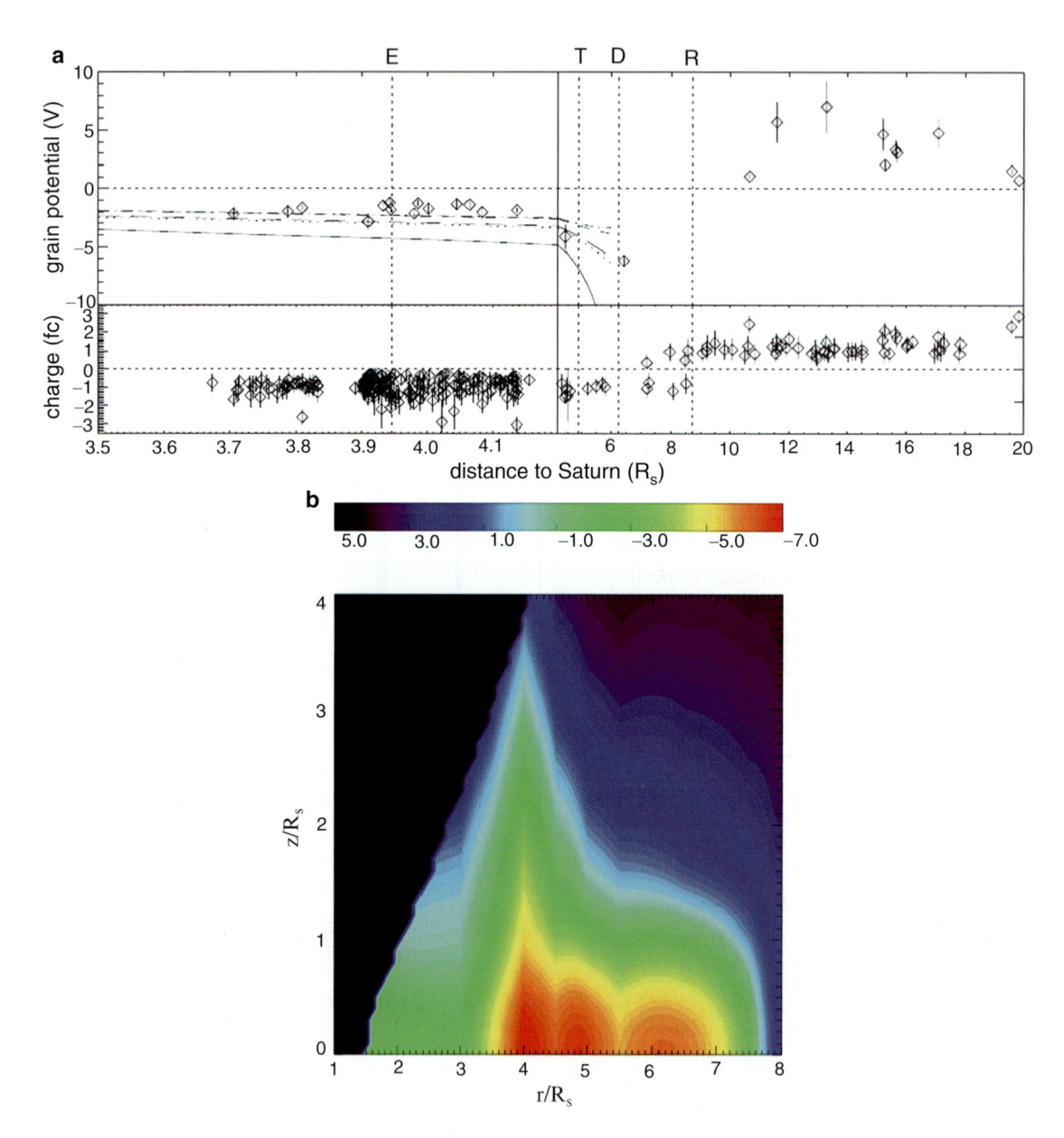

Fig. 16.15 (a) The electrostatic grain potential (*upper panel*) and grain charge (*lower panel*) measured by *Cassini*'s CDA as a function of the radial distance to Saturn. The section between 3.5 and 4.2 R_S is displayed on an expanded scale. Dotted vertical lines give the orbital distances of Saturn's moons Enceladus (E), Tethys (T), Dione (D), and Rhea (R). The electrostatic potential is only plotted for impacts having a charge feature with a signal-to-noise better than 3. CDA measures the charge and the mass of an impacting particle, assuming a composition of ice; this can be used to calculate the surface potential (from Kempf et al. 2006). (b) Model calculation of the equilibrium surface potential (measured in Volts) of circum-Saturnian grains using a magnetospheric plasma model based on *Voyager* measurements (from Juhász et al. 2002)

grains were positively charged. The electrostatic potential of the grains, detected between $3.5\,R_S$ (212,000 km) and $4.4\,R_S$ (267,000 km), was about −1.6 V, which is consistent with the spacecraft potential measured by the RPWS Langmuir probe (Wahlund et al. 2005), but inconsistent with model calculations using the Voyager model for the Saturnian plasma environment (Richardson 1995, Juhász et al. 2007) which predicts much higher potentials (−5 V). Most likely this results from *Cassini*'s ability to extend electron measurements towards lower energies than the *Voyagers* could. However, the models correctly predict the switch from negative to positive surface potentials to occur in the region between the moons Dione and Rhea; this happens mainly because secondary and photoelectron emissions become increasingly important in this region.

The motion of small grains around a planet can be surprisingly complex due to the competing effects of gravity, drag, radiation pressure, and electromagnetic forces. We now discuss Saturn's E ring as an example where relatively straightforward orbital calculations following the short-term dynamics of particles ejected from Enceladus are capable of elucidating many observations. Since *Cassini*'s discovery of the plumes leaving Enceladus's south polar regions, it seems that the dust production from these geysers is sufficient to sustain the E ring and to fill the entire magnetosphere with grains far beyond the classical boundaries of this ring (Srama et al. 2006).

Dust particles gain an electric charge in Saturn's plasma environment (Fig. 16.15). The orbital regression caused by the corresponding electromagnetic forces for particles with $s \sim 1\,\mu$m can compensate for the orbital precession due to Saturn's oblateness (see the equation immediately below). This allows the solar radiation pressure to induce large orbital eccentricities for micron-sized particles, spreading them over a broad radial span. Grains much larger or smaller than about a micron remain closer to Enceladus and re-impact the satellite in a short time (Horányi et al. 1992, Hamilton and Burns 1994). With escape speeds ≪ orbital speed of Enceladus, the predicted dust distribution in the E ring extends from ~ 2 to $\sim 7\,R_S$ in the radial direction with the vertical thickness increasing from $\sim 1{,}000$ to $\sim 5{,}000$ km. However, the calculated normal optical depth remained symmetric relative to Enceladus, contrary to observations.

The semi-major axes of dust grains originating from Enceladus will secularly increase once plasma drag is incorporated in the dynamics (Dikarev and Krivov 1998). Such evolution can explain the observed asymmetry in the dust distribution inside and outside Enceladus's orbit. In addition, particle erosion due to sputtering can be surprisingly fast, limiting the lifetime of micron-sized particles to ~ 50 years in the E ring (Jurac et al. 2001). Detailed dynamical models involve the simultaneous integration of three coupled nonlinear differential equations: (a) the equation of motion that includes gravity, radiation pressure, plasma and neutral drags, and the Lorentz force, calculated assuming a model of the magnetospheric fields and plasmas; (b) the current balance equation to follow a grain's time-dependent charge, including several currents (collection of electrons and ions, secondary and photoelectron production), assuming some material properties for the grains; and (c) the equation describing the mass-loss rate of a grain due to sputtering. These equations involve assumptions about the material properties of the grains, including their density, light-scattering efficiency, and yields of photoelectrons, secondary electrons, and sputtering.

Two *Cassini* discoveries – the active dust-producing geysers in Enceladus's south-polar regions and the E-ring's large radial span, reaching even Titan's orbit – indicate that Enceladus may be dominantly responsible for filling Saturn's entire magnetosphere with dust. Particles are transported outwards from Enceladus due to plasma drag (Morfill et al. 1983, Havnes et al. 1992, Dikarev 1999), but along the way they lose mass via sputtering (Jurac et al. 2001, Burns et al. 2001). These processes compete to eventually determine those regions that particles with a given initial size can reach (Horányi et al. 2008). Plasma drag acting alone produces a slow adiabatic increase in semi-major axis, while radiation pressure induces periodic changes in the orbital eccentricity of a particle, with a period $P = 2\pi/(\dot{\varpi})$, and amplitude $A \sim 1/(\dot{\varpi})$, where $\dot{\varpi}$ is the precession rate of the longitude of pericenter. Both the planetary oblateness and the Lorentz force acting on a charged grain cause orbital precession that, for Saturn and for small eccentricity and inclination (e, i ≪ 1), can be written as (Horányi et al. 1992)

$$\dot{\varpi} = 51.4 a^{-3.5} + 5.1 \frac{\phi^V}{s_\mu} a^{-3}, \qquad (16.1)$$

where a is the semi-major axis in units of Saturn's radius, ϕ^V is the grain's surface potential in Volts, and s_μ is the dust-particle's radius in μm. In regions where the plasma environment sets $\phi^V < 0$ (Fig. 16.15), the terms could cancel each other. The outward-drifting and eroding grains can temporarily experience 'locking' $(\dot{\varpi} \sim 0)$, at which point they swiftly develop large eccentricities. The out-of-plane component of radiation pressure will simultaneously force particles onto inclined orbits (Hamilton 1993, Burns et al. 2001). While this effect remains modest for negatively charged grains, it can be quite significant in regions of positive grain charges, which become destabilized against vertical oscillations (Howard et al. 1999). Additionally, the time-dependent charge of the grains on elliptical orbits moving along and against the co-rotational electric field can lead to swift gains or losses in orbital energy (Burns and Schaffer 1989), while the magnetic-field interaction can result in rapid changes in a particle's orbital angular momentum (Horányi 1996).

Computer simulations following the evolution of the charge, mass, and the orbit of particles escaping from Enceladus's geysers reproduce many E-ring characteristics that have been observed in-situ and remotely, including the spatial and size distributions of the particles comprising the ring, on both large scales and also near the source moon itself (Figs. 16.16 and 16.17). Particles transported from Enceladus to 20 R_S arrive there in about 300 years, but meanwhile have shrunk to only a few percent of their original radii due to sputtering losses. If grains with $s \gg \mu$m are found at this distance, they would have to be born at one of the more distant moons. However, to date, no dust-density enhancements have been noticed while crossing the ring plane near the orbits of any other satellites. Nonetheless, since small grains have been detected far from Saturn, these calculations therefore indicate that the geysers on Enceladus have been supplying the E-ring material at an approximately constant rate at least for the last 300 years.

16.3.9 Dust Streams

Figure 16.18 documents the discovery of the high-velocity streams of nanometer-sized dust particles originating from the inner Saturnian system (Kempf et al. 2005a, Hsu et al. 2009). This topic is included here because such particles likely are an end-state for E-ring material, and because the same processes that govern the lives of faint-ring material are active on stream particles. Previously, only the Jovian system was known to eject dust (Grün et al. 1993).

During *Cassini*'s approach to Saturn, the dust detector often registered impact signals, which were most likely caused by particles moving faster than 70 km s^{-1}, the fastest impact speed for which CDA was calibrated at the Heidelberg dust-accelerator facility. Small, positively charged grains within Saturn's magnetosphere can be accelerated to high speeds as a result of the outward-pointing electric field induced by Saturn's co-rotating magnetic field, which has a similar configuration to Jupiter's (Horányi et al. 1993, Hamilton and Burns 1993). Calculating the work done by the co-rotating electric field and the expected charges of the grains, a simple order-of-magnitude relationship can be derived to show that the expected escape velocity is inversely proportional to the size of the particles (Horányi 2000):

$$V_{escape} = \frac{3}{s_\mu} \quad \text{km s}^{-1} \text{ for Jupiter, and}$$

$$V_{escape} = \frac{0.6}{s_\mu} \quad \text{km s}^{-1} \text{ for Saturn,}$$

where s_μ is the grain's radius measured in μm. These expressions assume that Jovian dust particles start at Io, and that Saturn's grains are accelerated outward from Dione's distance, since that is where the grains' charge become positive. The faster speeds at Jupiter result mainly from its stronger magnetic field. Dust grains with sizes $0.001 < s_\mu < 0.01$ are expected to leave Saturn's magnetosphere with a speed in the range of 600 km s^{-1} > V_{escape} > 60 km s^{-1}, indicating that the detected particles were a few nm in size.

Both the mass and speed of Saturnian stream particles are outside the calibrated range of the *Cassini* dust detector. However, the amount of plasma generated by the impacts should scale with the particle mass and speed. Particles detected after *Cassini*'s orbit insertion generally produced little impact plasma, which is dynamically compatible with an E ring origin. Interestingly, stream particles detected during the approach phase at distances larger than 500 R_S caused large impact charges, which supports an origin from the outskirts of the A ring (Kempf et al. 2005a). Beyond Saturn's magnetosphere, the dynamics of the stream particles are governed by interactions with the interplanetary magnetic field convected by the solar wind (Hsu et al. 2009). *Cassini* observed a continuous flow of stream particles arriving from directions close to the line-of-sight to Saturn. Both the directionality and the strength of the dust streams changed almost instantaneously whenever the *Cassini* magnetometer sensed 'co-rotating interaction regions' (CIRs), which are characterized by compressed flow of high-speed solar-wind plasma and by enhanced magnetic-field strengths. This coincidence suggests that the periodic impact bursts observed by *Ulysses, Galileo, and Cassini* arise from the speed increase of the dust streams during their CIR traversals (Hamilton and Burns 1993, Kempf et al. 2005a, Krüger et al. 2006a).

Saturnian stream particles were found to be composed predominantly of oxygen, silicon and iron, suggesting that they consist of silicate materials (Kempf et al. 2005b). Because Saturn's main rings as well as the E ring are primarily made of water ice, stream particles are suspected to be the impurities embedded in the icy ring material rather than the bulk composition of the particles themselves. Numerical simulations of the stream-particle propagation showed that the majority of these grains started at distances to Saturn between 7 and 9 R_S and leave Saturn's magnetosphere closely aligned with the ring-plane (Hsu et al. 2009). In this region, ring particles are electromagnetically trapped, while their size decreases slowly due to sputtering by the ambient plasma until their charge–to-mass ratio exceeds the critical value for ejection from the ring region (Beckmann et al. 2009).

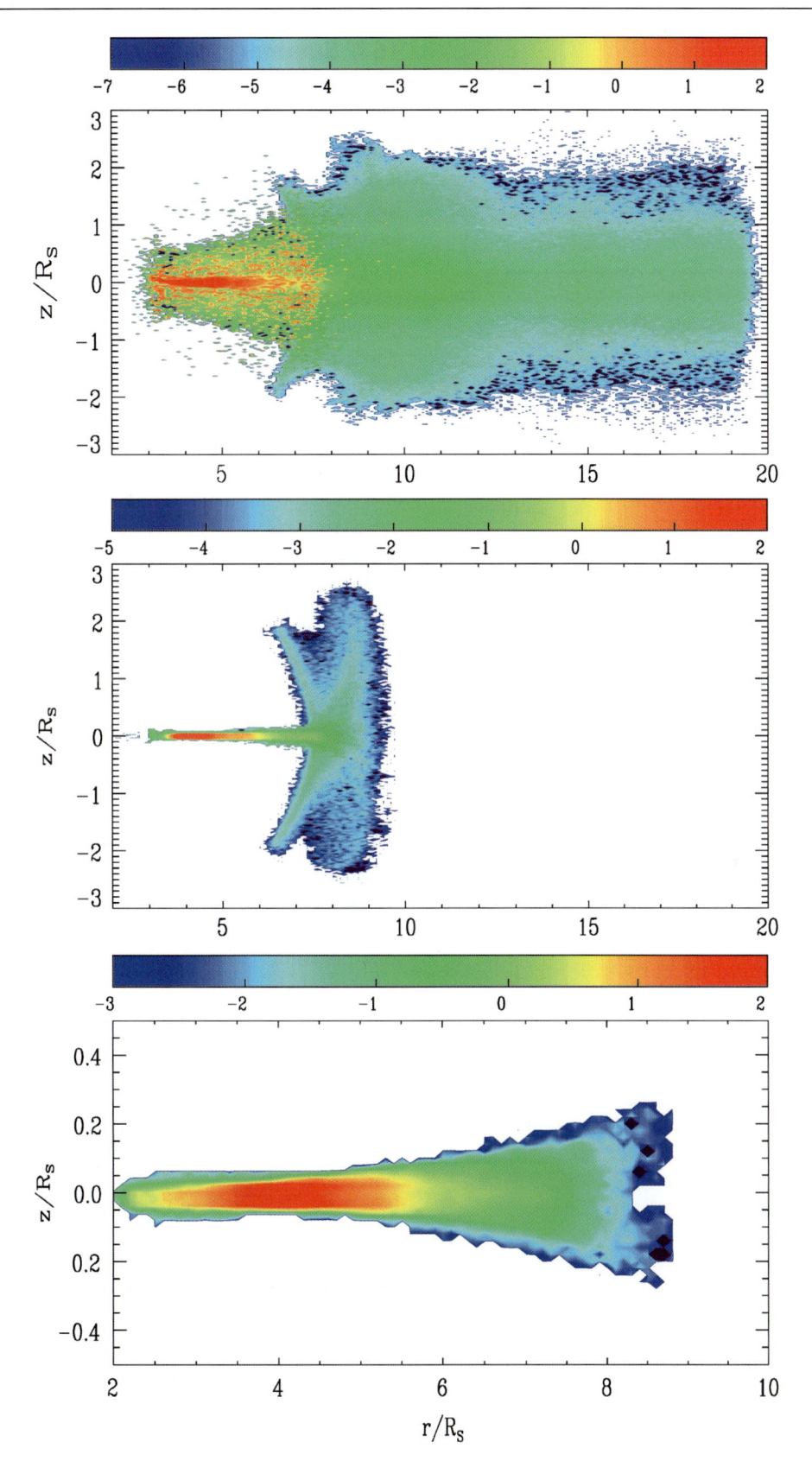

Fig. 16.16 The results of a numerical simulation for the azimuthally averaged density distribution of grains having particular radii: (*top*) $0.1 < s < 0.5\,\mu$m; (*middle*) $0.5 < s < 1\,\mu$m; and (*bottom*) $1 < s < 3\,\mu$m. The largest grains remain confined to the classical E ring, while the smallest particles can reach the orbit of Titan. These simulations started with a power-law size distribution (exponent of -2.5) between 0.1–$10\,\mu$m with all grains being released from Enceladus with an initial southward velocity of 100 m/s. The color scale is logarithmic, and in each case normalized to 100, corresponding to maximum number densities of 7; 1.2; and $0.3\,\text{m}^{-3}$ (from Horányi et al. 2008)

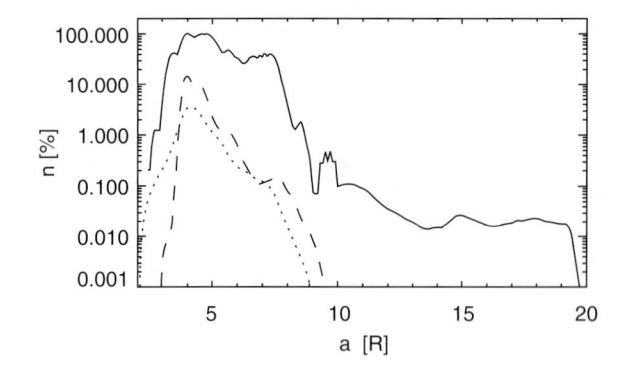

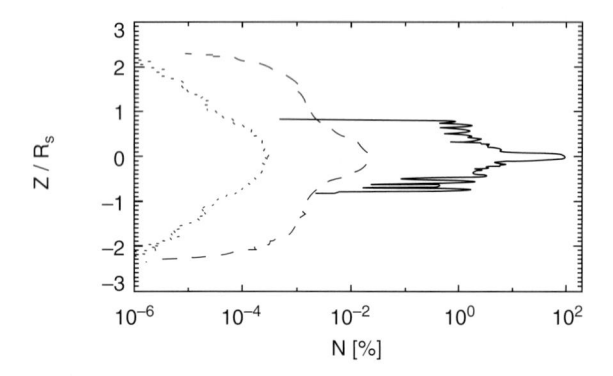

Fig. 16.17 The column densities (*top panel*) as function of distance from Saturn for the same groups of particles as in Fig. 16.16, $0.1 < s < 0.5\,\mu m$ (*continuous line*); $0.5\,\mu m < s < 1\,\mu m$ (*dashed line*); and $1 < s < 3\,\mu m$ (*dotted line*). The curves are normalized by setting the maximum density of the smallest grains to 100, corresponding to a real column density of $0.3\,m^{-2}$. The vertical distribution of the particles (*bottom*) in the size range $0.1 < s < 0.5\,\mu m$, at 5 (*continuous line*); 10 (*dashed line*); and 15 R_S (*dotted line*). The curves are normalized by setting the maximum density at 5 R_S to 100, corresponding to a real number density of $7\,m^{-3}$ (from Horányi et al., 2008)

16.4 Summary: Dynamical Connections Between Diffuse Rings

The above sections have described various characteristics of individual faint rings and dust populations around Saturn, as well as some attempts that have been made to interpret and model the observed features in specific regions. However, it is important to realize that many of the same physical processes are active in various different dusty rings. In this section we summarize some of the phenomena that appear to be active in several different rings. We maintain that the same ingredients may lead to remarkably different outcomes.

Some of the dust-sized grains observed in the Saturn system are thought to be collisonal debris produced by impacts into, and among, various source bodies close to, or within, the dusty rings (notable exceptions to this being the E ring particles generated by Enceladus's geological activity and the possibly electrostatically levitated grains in the spokes). The Type I particles in the E ring may be generated by impacts of both interplanetary meteoroids and eccentrically orbiting E-ring grains into the surface of the mid-sized satellites like Enceladus. Also, discrete rings are associated with tiny moons (Janus/Epimetheus, Pallene, Methone and Anthe), reminiscent of Jupiter's ring (Burns et al. 1999). A population of larger source bodies has also been identified in the G-ring arc. Comparing the particle properties and size distributions of the material derived from these different sources could clarify how efficiently different-sized source bodies can generate dust grains of various sizes (Burns et al. 1984, 2001).

Several dusty rings show evidence of being influenced by solar radiation. The E-ring's broad expanse could be partly due to periodic changes in the orbital parameters induced

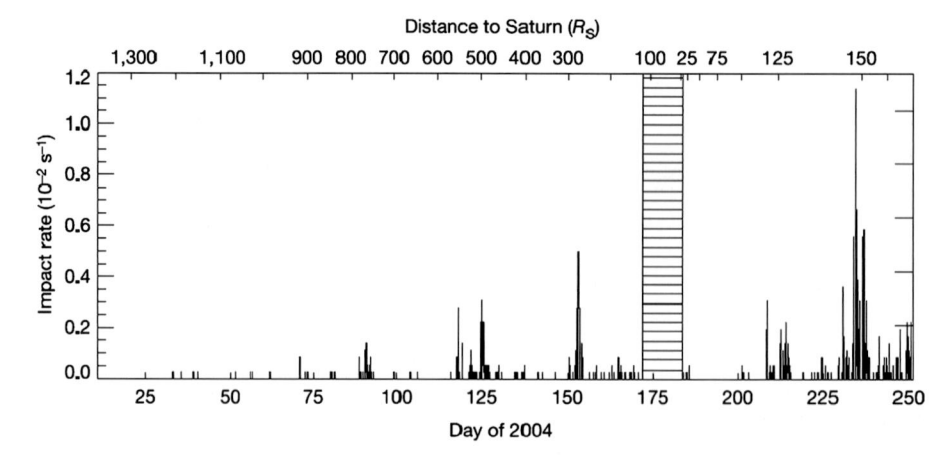

Fig. 16.18 The impact rate registered by the cosmic dust analyzer (CDA) between 10 January and 6 September 2004. During Saturn orbit insertion, marked by horizontal bars, CDA was powered off. The upper scale shows *Cassini*'s distance to Saturn. In total, 1,409 impacts were detected and exhibited the characteristic features of high-velocity impacts by tiny dust particles. Owing to *Cassini*'s changing orientation relative to the arriving particles, the observed flux does not necessarily represent the true temporal variability of the stream particles (from Kempf et al. 2005a)

either by solar radiation pressure (Horányi et al. 1992, Hamilton 1993) or by variations in the particles' charges as they pass through the planet's shadow (Horányi and Burns 1991, Hamilton and Krüger 2008). Meanwhile, some ringlets within the main rings exhibit heliotropic behavior where the entire ringlet seems to be displaced towards the sun. These two systems provide two very different environments in which to explore how solar radiation alters the dynamics of small particles.

Many dusty rings show evidence of being sculpted by resonances with various periodic perturbing forces. Arcs of material confined by co-rotation resonances with Mimas have been found at the orbits of the small moons Anthe and Methone, and in the G ring (Hedman et al. 2007b, 2009a). The outer G ring also contains a structure that appears to be driven by a Lindblad resonance with Mimas. Intriguingly, multiple dust populations in and around the main rings (the middle D ring, inner Roche Division and the spokes in the B ring) reveal periodic structures that appear to be driven by perturbations with characteristic periods between 10.5 and 11 h. This range of periods is close to the periods observed in the SKR and other magnetospheric phenomena, suggesting that all of these regions are being influenced by asymmetries in Saturn's magnetosphere (Burns et al. 1985, Hedman et al. 2009b). Since these dusty systems are qualitatively different from other environments (e.g., dense rings) where resonant phenomena have been extensively studied, detailed modeling of the particle dynamics promise to be very illuminating and may constrain local plasma properties.

Dissipative and drag forces can be observed operating in several different ring systems. The E-ring's broad span almost certainly requires plasma drag to produce secular changes in the semi-major axes of particles launched from Enceladus. The outer flank of the G ring appears to be consistent with particles drifting away from the source bodies in the arc under via interactions with the surrounding plasma. The inner parts of the D ring contain sheets of material that could consist of fine material spirally in towards the planet from various source regions (like the D68 ringlet) due to drag from the planet's upper atmosphere or ionosphere, as has been suggested for the Uranian rings (Colwell and Esposito 1990). Evidence for dissipative processes can also be found in the response of various dusty rings to resonant perturbations. The structures associated with resonances in the G ring, Roche Division and D ring all appear to require some additional physics to dissipate eccentricities on short timescales (10–100 days). Comparisons of the orbital evolution and damping timescales at these various locations should clarify the nature and strength of various sources of dissipation (Hedman et al. 2009b).

Diffuse planetary rings are an excellent laboratory to study dusty plasma processes acting on large scales. Imaging observations, when combined with in-situ measurements of the dust and of the plasma parameters, including the electric and magnetic fields, provide a rich complementary data set. The extension of *Cassini* beyond its prime mission period of 2004–2008 for four more years provides an unprecedented series of observations that cover a full period of Saturnian seasons. To understand the role of the UV radiation due to solar-cycle variability, and to enable the comparison of the diffuse rings and the spokes in the B-ring to their state during the *Voyager* encounters, *Cassini* must remain healthy into 2016 and beyond.

Acknowledgements All authors appreciate the efforts of *Cassini*'s engineers in providing an exceptional mission. MH thanks Antal Juhász for years of fruitful collaboration. JAB recognizes the contributions of numerous former graduate students on ethereal ring problems. JAB, MH and MMH have received support from NASA, the *Cassini-Huygens* Project and CDAP. GHJ holds an STFC Advanced Fellowship. SK acknowledges support from DLR.

References

Akalin, F., Gurnett, D.A., Averkamp, T.F., Persoon, A.M., Santolik, O., Kurth, W.S., Hospodarsky, G.B.: First whistler observed in the magnetosphere of Saturn. Geophys. Res. Ltrs. **33**, L20107 (2006), doi:10.1029/2006GL027019

Aubier, M.G., Meyer-Vernet, N., Pedersen, B.M.: Shot noise from grain and particle impacts in Saturn's ring plane. Geophys. Res. Ltrs. **10**, 5–8 (1983).

Bauer, J., Lissauer, J.J., Simon, M.: Edge-on observations of Saturn's E and G rings in the near-IR. Icarus **125**, 440–445 (1997).

Baum, W. A., Kreidl, T., Westphal, J. A., Danielson, G. E., Seidelmann, P. K., Pascu, D., Currie, D. G.: Saturn's E ring. Icarus **47**, 84–96 (1981).

Beckmann, U., Kempf, S., Spahn, F.: Long-term evolution of Saturn's E ring particles, Icarus, submitted (2009).

Bosh, A.S., Olkin, C.B.: Low optical depth features in Saturn's rings: The occultation of GSC5249–01240 by Saturn and its rings. Bull. Am. Astron. Soc. **28**, 1124 (1996).

Burns, J.A., Schaffer, L.: Orbital evolution of circumplanetary dust by resonant charge variations. Nature **37**, 340–343 (1989).

Burns, J.A., Showalter, M.R., Cuzzi, J.N., Durisen, R.H.: Saturn's electrostatic discharges – Could lightning be the cause? Icarus **54**, 280–295 (1983).

Burns, J.A., Showalter, M.R., Morfill, G.E.: The ethereal rings of Jupiter and Saturn. In: Greenberg, R., Brahic, A. (eds.) *Planetary Rings*, pp. 200–274. Univ. Ariz. Press, Tucson (1984).

Burns, J.A., Schaffer, L., Greenberg, R.J., Showalter, M.R.: Lorentz resonances and the structure of the Jovian ring. Nature **316**, 115–119 (1985).

Burns, J.A., Showalter, M.R., Hamilton, D.P., Nicholson, P.D, de Pater, I., Ockert-Bell, M., Thomas, P.C.: The formation of Jupiter's faint rings. Science **284**, 1146–1150 (1999).

Burns, J.A., Hamilton, D.P., Showalter, M.R.: Dusty rings and circumplanetary dust: Observations and simple physics. In: Grün, E., Gustafson, B., Dermott, S., Fectig, H. (eds.) Interplanetary Dust, pp. 641–725. Springer, Berlin (2001).

Burns, J.A., Hedman, M.M., Tiscareno, M.S., Nicholson, P.D., Streetman, B.J., Colwell, J.E., Showalter, M.R., Murray, C.D., Cuzzi, J.N., Porco, C.C., ISS Team: Morphology, movements and models of ringlets in Saturn's Encke Gap. Bull. Am. Astro. Soc. **37**, 766 (2005).

Burt, J.A., Hedman, M.M., Tiscareno, M.S., Burns, J.A.: The where and why of Saturn's inclined "Charming" ringlet. Bull. Am. Astro. Soc. **40**, 445 (2008).

Colwell, J.E., Esposito, L.W.: A numerical model of the Uranian dust rings. Icarus **86**, 530–560 (1990).

Cooper, N.J., Murray, C.D., Evans, M.W., Beurle, K., Jacobson, R.A., Porco, C.C.: Astrometry and dynamics of Anthe (S/2007S4), a new satellite of Saturn. Icarus **195**, 765–777 (2008).

Cuzzi, J.N., Durisen, R.H.: Bombardment of planetary rings by meteoroids – General formulation and effects of Oort Cloud projectiles. Icarus **84**, 467–501 (1990).

de Pater, I., Showalter, M.R., Lissauer, J.J., Graham, J.R.: Keck infrared observations of Saturn's E and G rings during Earth's 1995 ring plane crossings. Icarus **121**, 195–198 (1996).

de Pater, I., Martin, S.C., Showalter, M.R.: Keck near-infrared observations of Saturn's E and G rings during Earth's ring-plane crossing in August 1995. Icarus **172**, 446–454 (2004).

Dikarev, V.V.: Dynamics of particles in Saturn's E ring: Effects of charge variations and the plasma drag force. Astron. Astrophys. **346**, 1011–1019 (1999).

Dikarev V.V., Krivov, A.V.: Dynamics and spatial distribution of particles in Saturn's E ring. Solar Sys. Res. **32**, 128 (1998).

Doyle, L.R., Grün, E.: Radiative transfer modeling constraints on the size of the spoke particles in Saturn's rings. Icarus **85**, 168–190 (1990).

Eplee, R.E., Smith, B.A.: Spokes in Saturn's rings – Dynamical and reflectance properties. Icarus **59**, 188–198 (1984).

Farmer, A.J., Goldreich, P.: Spoke formation under moving plasma clouds. Icarus **179**, 535–538 (2005).

Farrell, W.M., Desch, M.D., Kaiser, M.L., Kurth, W.S., Gurnett, D.A.: Changing electrical nature of Saturn's rings: Implications for spoke formation, Geophys. Res. Ltrs., **33**, L07203(2006). doi:10.1029/2005GL024922

Feibelman, W.: Concerning the "D" ring of Saturn. Nature **214**, 793–794 (1967).

Ferrari, C., Brahic, A.: Arcs and clumps in the Encke division of Saturn's rings. Planet. Space Sci. **45**, 1051–1067 (1997).

Fischer, G., Desch, M.D., Zarka, P., Kaiser, M.L., Gurnett, D.A., Kurth, W.S., Macher, W., Rucker, H.O., Lecacheux, A., Farrell, W.M., Cecconi, B.: Saturn lightning recorded by *Cassini* RPWS in 2004. Icarus **183**, 135–152 (2006).

Goertz, C.K., Morfill, G.: A model for the formation of spokes in Saturn's rings. Icarus **53**, 219–229 (1983).

Gresh, D.L., Rosen, P.A., Tyler, G.L., Lissauer, J.J.: An analysis of bending waves in Saturn's rings using *Voyager* radio occultation data. Icarus **68**, 481–502 (1986).

Grün, E., Morfill, G.E., Terrile, R.J., Johnson, T.V., Schwehm, G.: The evolution of spokes in Saturn's B ring. Icarus **54**, 227–252 (1983).

Grün, E., Morfill, G.E., Mendis, D.A.: Dust-magnetosphere interactions. In: Greenberg, R., Brahic, A. (eds.) *Planetary Rings*, pp. 275–332. Univ. Ariz. Press, Tucson (1984).

Grün, E., Goertz, C.K., Morfill, G. E., Havnes, O.: Statistics of Saturn's spokes. Icarus **99**, 191–201 (1992).

Grün, E., Zook, H., Baguhl, M., Balogh, A., Bame, S., Fechtig, H., Forsyth, R., Hanner, M., Horányi, M., Kissel, J., Lindblad, B.-A., Linkert, D., Linkert, G., Mann, I., McDonnell, J.A.M., Morfill, G., Phillips, J., Polanskey, C., Schwehm, G., Siddique, N., Staubach, P., Svestka, J., Taylor, A.: Discovery of Jovian dust streams and interstellar grains by the *Ulysses* spacecraft. Nature **362**, 428–430 (1993).

Gurnett, D.A., Grün, E., Gallagher, D., Kurth, W.S., Scarf, F.L.: Micron-sized particles detected near Saturn by the *Voyager* plasma wave instrument. Icarus **53**, 236–254 (1983).

Gurnett, D.A., Kurth, W.S., Wang, Z., Persoon, A.M., Groene, J.B., Averkamp, T.F., Zarka,P., Farrell, W.M., Kaiser, M.L.: An overview of the rotational modulation of three types of Saturnian radio emissions. AGU Fall Meeting P23C-07 (2007).

Hamilton, D.P.: Motion of dust in a planetary magnetosphere: Orbit-averaged equations for oblateness, electromagnetic and radiation forces with application to Saturn's E ring. Icarus **101**, 244–264 (1993).

Hamilton, D.P.: A comparison of Lorentz, gravitational, and satellite gravitational resonances. Icarus **109**, 221–240 (1994).

Hamilton, D.P.: The collisional cascade model for Saturn's ring spokes. Bull. Am. Astron. Soc. **38**, 578 (2006).

Hamilton, D.P., Burns, J.A.: Ejection of dust from Jupiter's gossamer ring. Nature **364**, 695–699 (1993).

Hamilton, D.P., Burns, J.A.: Origin of Saturn's E ring: Self-sustained, naturally. Science **264**, 550–553 (1994).

Hamilton, D.P., Krüger, H.: The sculpting of Jupiter's gossamer rings by its shadow. Nature **453**, 72–75 (2008).

Havnes, O., Morfill, G.E., Melandso, F.: Effects of electromagnetic and plasma drag forces on the orbital evolution of dust in planetary magnetospheres. Icarus **98**, 141–150 (1992).

Hedman, M.M., Burns, J.A., Showalter, M.R., Porco, C.C., Nicholson, P.D., Bosh, A.S., Tiscareno, M.S., Brown, R.H., Buratti, B.J., Baines, K.H., Clark, R.: Saturn's dynamic D ring. Icarus **188**, 89–107 (2007a).

Hedman, M.M., Burns, J.A., Tiscareno, M.S., Porco, C.C., Jones, G.H., Roussos, E., Krupp, N., Paranicas, C., Kempf, S.: The source of Saturn's G ring. Science **317**, 653–656 (2007b).

Hedman, M.M., Burns, J.A., Tiscareno, M.S., Porco, C.C.: The heliotropic rings of Saturn. Bull. Am. Astron. Soc. **39**, 427 (2007c).

Hedman, M.M., Murray, C.D., Cooper, N.J., Tiscareno, M.S., Beurle, K., Evans, M.W., Burns, J.A.: Three tenuous rings/arcs for three tiny moons. Icarus **199**, 378–386 (2009a).

Hedman, M.M., Burns, J.A., Tiscareno, M.S., Nicholson, P.D., Porco, C.C.: Organizing some very tenuous things: Resonant structures in Saturn's faint rings. Icarus **202**, 260–279 (2009b). doi: 10.1016/j.icarus.2009.02.016

Hedman, M.M., Nicholson, P.D., Showalter, M.R., Brown, R.H., Buratti, B.J., Clark, R.N.: Spectral observations of the Enceladus plume with *Cassini* VIMS. Astrophys. J. **693**, 1749–1762 (2009c).

Hill, J.R., Mendis, D.A.: On the braids and spokes in Saturn's ring system, Moon Planets **24**, 431–436 (1981).

Hill, J.R., Mendis, D.A.: The dynamical evolution of the Saturn ring spokes. J. Geophys. Res. **87**, 7413–7420 (1982).

Hillier, J.K., Green, S.F., McBride, N., Schwanethal, J.P., Postberg, F., Srama, R., Kempf, S., Moragas-Klostermeyer, G., McDonnell, J.A.M., Grün, E.: The composition of Saturn's E ring. MNRAS **377**, 1588–1596 (2007).

Horányi, M.: Charged dust dynamics in the solar system. Ann. Rev. Astron. Astrophys. **34**, 383–418 (1996).

Horányi, M.: Dust streams from Jupiter and Saturn. Phys. Plasmas **7**, 3847–3850 (2000).

Horányi, M., Burns, J.A.: Charged-dust dynamics – Orbital resonance due to planetary shadows. J. Geophys. Res. **96**, 19,283–19,289 (1991).

Horányi, M., Burns, J.A., Hamilton, D.P.: The dynamics of Saturn's E ring particles. Icarus **97**, 248–259 (1992).

Horányi, M., Morfill, G., Grün, E.: Mechanism for the acceleration and ejection of dust grains from Jupiter's magnetosphere. Nature **363**, 144–146 (1993).

Horányi, M., Hartquist, T.W., Havnes, O., Mendis, D.A., Morfill, G.E.: Dusty plasma effects in Saturn's magnetosphere. Rev. Geophys. **42** (2004) RG4002. doi: 10:1029/2004 RG000151

Horányi, M., Juhász, A., Morfill, G.: The large scale structure of Saturn's E ring, Geophys. Res. Ltrs. **35**, L04203 (2008). doi: 10.1029/2007GL032726

Howard, J.E., Horányi, M., Stewart, G.R.: Global dynamics of charged dust grains in planetary magnetospheres. Phys. Rev. Ltrs. **83**, 3993–3996 (1999)

Hsu, H.-W., Kempf, S., Jackman, C.M.: Observations of Saturnian stream particles in the interplanetary space. Icarus, in press (2009).

Juhász, A., Horányi, M.: Saturn's E ring: A dynamical approach. Jnl. Geophys. Res. **107** (2002). doi: 10.1029/2001JA 000182

Juhász, A., Horányi, M., Morfill, G. E.: Signatures of the Enceladus plumes in Saturn's E-ring. Geophys. Res. Ltrs. **34**, L09104 (2007). doi: 10.1029/2006 GL 029120

Jones, G. H., Krupp, N., Krüger, H., Roussos, E., Ip, W.-H., Mitchell, D. G., Krimigis, S. M., Woch, J., Lagg, A., Fränz, M., Dougherty, M. K., Arridge, C. S., McAndrews, H. J.: Formation of Saturn's ring spokes by lightning-induced electron beams. Geophys. Res. Ltrs. **33**, L21202 (2006). doi:10.1029/2006GL028146

Jurac, S., Johnson, R.E., Richardson, J.D., Paranicas, C.: Satellite sputtering in Saturn's magnetosphere. Planet. Space Sci. **49**, 319–326 (2001).

Kempf, S., Srama, R., Horányi, M., Burton, M., Helfert, S., Moragas-Klostermeyer, G., Roy, M., Grün, E.: High-velocity streams of dust originating from Saturn. Nature **433**, 289–291 (2005a).

Kempf, S., Srama, R., Postberg, F., Burton, M., Green, S. F., Helfert, S., Hillier, J. K., McBride, N., McDonnell, J. A. M., Moragas-Klostermeyer, G., Roy, M., Grün, E.: Composition of Saturnian stream particles. Science **307**, 1274–1276 (2005b).

Kempf, S., Beckmann, U., Srama, R., Horányi, M., Auer, S., Grün, E.: The electrostatic potential of E ring particles. Planet. Space Sci. **54**, 999–1006 (2006).

Kempf, S., Beckmann, U., Moragas-Klostermeyer, G., Postberg, F., Srama, R., Economou, T., Schmidt, J., Spahn, F., Grün, E.: The E ring in the vicinity of Enceladus. I: Spatial distribution and properties of the ring particles. Icarus **193**, 420–437 (2008).

Kempf, S., Beckmann, U., Schmidt, J.: Icarus, submitted (2009).

Krimigis, S.M., Mitchell, D.G., Hamilton, D.C., Livi, S., Dandouras, J., Jaskulek, S., Armstrong, T.P., Boldt, J.D., Cheng, A.F., Gloeckler, G., Hayes, J.R., Hsieh, K.C., Ip, W.-H., Keath, E.P., Kirsch, E., Krupp, N., Lanzerotti, L.J., Lundgren, R., Mauk, B.H., McEntire, R.W., Roelof, E.C., Schlemm, C.E., Tossman, B.E., Wilken, B., Williams, D.J.: Magnetosphere Imaging Instrument (MIMI) on the *Cassini* Mission to Saturn/Titan. Space Sci. Rev. **114**, 223–329 (2004).

Krivov, A.V., Sremčević, M., Spahn, F., Dikarev, V.V., Kholshevnikov, K.V.: Impact-generated clouds around planetary satellites: Spherically symmetric case. Planet. Space Sci. **51**, 251–269 (2003).

Krüger, H., Graps, A., Hamilton, D. P., Flandes, A., Forsyth, R.J., Horányi, M., Grün, E.: *Ulysses* jovian latitude scan of high-velocity dust streams originating from the jovian system. Planet. Space Sci. **54**, 919–931 (2006a).

Krüger, H., Altobelli, N., Andweiler, B., Dermott, S.F., Dikarev, V.V., Graps, A.L., Grün, E., Gustafson, B.A.: Five years of *Ulysses* dust data: 2000–2004. Planet. Space Sci. **54**, 932–956 (2006b).

Kurth, W.S., Averkamp, T.F., Gurnett, D.A., Wang, Z.: *Cassini* RPWS observations of dust in Saturn's E Ring. Planet. Space Sci. **54**, 988–998 (2006).

Kurth, W.S., Lecacheux, A., Averkamp, T.F., Groene, J.B., Gurnett, D.A.: A Saturnian longitude system based on a variable kilometric radiation period. Geophys. Res. Ltrs. **34**, L02201 (2007).

Lehtinen, N.G., Inan, U.S., Bell, T.F.: Trapped energetic electron curtains produced by thunderstorm-driven relativistic runaway electrons. Geophys. Res. Ltrs. **27**, 8, 1095–1098 (2000).

Lissauer, J.J., French, R.G.: HST high-resolution backscatter image of Saturn's G ring. Icarus **146**, 12–18 (2000).

Marley, M., Porco, C.C.: Planetary acoustic mode seismology – Saturn's rings. Icarus **106**, 508–524 (1993).

McGhee, C.A., French, R.G., Dones, L., Cuzzi, J.N., Salo, H.J., Danos, R.: HST observations of spokes in Saturn's B ring. Icarus **173**, 508–521 (2005).

Mendis, D.A., Hill, J.R., Ip, W.-H., Goertz, C.K., Grün, E.: Electrodynamic processes in the ring system of Saturn. In: Gehrels, T. (ed.) *Saturn*, pp. 546–589. Univ. Ariz. Press, Tucson (1984).

Meyer-Vernet, N., Lecacheux, A., Pedersen, B. M.: Constraints on Saturn's E ring from the *Voyager* 1 radio astronomy instrument. Icarus **123**, 113–128 (1996).

Mitchell, C.J., Horányi, M., Havnes, O., Porco, C.C.: Saturn's spokes: Lost and found. Science **311**, 1587–1589 (2006).

Mitchell, C.J., Porco, C., Dones, L., Spitale, J.: Analysis of the behavior of spokes in Saturn's B ring as observed in *Cassini* ISS images. Bull. Am Astro. Soc. **40**, 429 (2008).

Moore, L. E., Mendillo, M., Müller-Wodarg, I. C. F., Murr, D. L.: Modeling of global variations and ring shadowing in Saturn's ionosphere. Icarus **172**, 503–520 (2004).

Morfill, G. E., Grün, E., Johnson, T.V.: Saturns's E, G and F rings – Modulated by the plasma sheet? Jnl. Geophys. Res. **88**, 5573–5579 (1983).

Morfill, G.E., Thomas, H.M.: Spoke formation under moving plasma clouds – The Goertz-Morfill model revisited. Icarus **179**, 539–542 (2005).

Nicholson, P.D., Showalter, M.R., Dones, L., French, R.G., Larson, S.M., Lissauer, J.J., McGhee, C.A., Seitzer, P., Sicardy, B., Danielson, G.E.: Observations of Saturn's ring-plane crossing in August and November 1995. Science **272**, 509–515 (1996).

Ockert-Bell, M.E., Burns, J.A., Daubar, I.J., Thomas, P.C., Veverka, J., Belton, M.J.S., Klaasen, K.P.: The structure of Jupiter's ring system as revealed by the *Galileo* imaging experiment. Icarus **138**, 188–213 (1999).

Porco, C. C.: *Voyager* observations of Saturn's rings. 1. The eccentric rings at 1.29, 1.45, 1.95 and 2.27 R_s. 2. The periodic variation of spokes. PhD thesis, California Institute of Technology, Pasadena, CA (1983).

Porco, C.C. on behalf of *Cassini* ISS team: Rings of Saturn (R/2006S1, R/2006S2, R/2006S3, R/2006S4). IAU Circ. **8759** 1 (2006).

Porco, C.C. on behalf of *Cassini* ISS team: S/2008 S1. IAU Circ. **9023** 1 (2009).

Porco, C.C., Danielson, G.E.: The periodic variation of spokes in Saturn's rings. Astron. J. **87**, 826–833 (1982).

Porco, C.C., Baker, E., Barbara, J., Beurle, K., Brahic, A., Burns, J.A., Charnoz, S., Cooper, N., Dawson, D.D., Del Genio, A.D., Denk, T., Dones, L., Dyudina, U., Evans, M.W., Giese, B., Grazier, K., Helfenstein, P., Ingersoll, A.P., Jacobson, R.A., Johnson, T.V., McEwen, A., Murray, C.D., Neukum, G., Owen, W.M., Perry, J., Roatsch, T., Spitale, J., Squyres, S., Thomas, P., Tiscareno, M., Turtle, E., Vasavada, A.R., Veverka, J., Wagner, R., West, R.: *Cassini* imaging science: Initial results on Saturn's rings and small satellites. Science **307**, 1226–1236 (2004).

Porco, C.C., Helfenstein, P., Thomas, P.C., Ingersoll, A.P., Wisdom, J., West, R., Neukum, G., Denk, T., Wagner, R., Roatsch, T., Kieffer, S., Turtle, E., McEwen, A., Johnson, T.V., Rathbun, J., Veverka, J., Wilson, D., Perry, J., Spitale, J., Brahic, A., Burns, J.A., Del Genio, A.D., Dones, L., Murray, C.D., Squyres, S.: *Cassini* observes the active South Pole of Enceladus. Science **311**, 1393–1401 (2005).

Postberg, F., Kempf, S., Hillier, J. K., Srama, R., Green, S. F., McBride, N., Grün, E.: The E-ring in the vicinity of Enceladus II: Signatures of Enceladus in the elemental composition of E-ring particles. Icarus **193**, 438–454 (2008).

Postberg, F., Kempf, S., Brilliantov, N., Schmidt, J., Buck, U., Srama, R.: Sodium in E ring particles implies ocean below the Enceladus' surface. Nature **459**(7250), 1098–1101 (2009).

Richardson, J.: An extended plasma model for Saturn. Geophys. Res. Lett. **22**, 1177–1180 (1995).

Robinson, L. J.: Closing in on Saturn. Sky Tel. **60**, 481 (1980).

Roddier, C., Roddier, F., Graves, J.E., Northcott, M.J.: Discovery of an arc of particles near Enceladus' orbit: A possible key to the origin of the E ring. Icarus **136**, 50–59 (1998).

Roussos, E., Jones, G.H., Krupp, N., Paranicas, C., Mitchell, D.G., Lagg, A., Woch, J., Motschmann, U., Krimigis, S.M., Dougherty, M.K.: Electron microdiffusion in the Saturnian radiation belts: *Cassini* MIMI/LEMMS observations of energetic electron absorption by the icy moons, J. Geophys. Res. **112**, A06214 (2007).

Roussos, E., Jones, G.H., Krupp, N., Paranicas, C., Mitchell, D.G., Krimigis, S.M., Woch, J., Lagg, A., Khurana, K.: Energetic electron signatures of Saturn's smaller moons: Evidence of an arc of material at Methone. Icarus **193**, 455–464 (2008).

Sanchez-Lavega, A., Rojas, J.F., Sada, P.V.: Saturn's zonal winds at cloud level. Icarus **167**, 405–420 (2000).

Schmidt. J, Sremčević, M.: The shadows of Saturn's icy satellites in the E Ring. Geophys. Res. Ltrs., submitted (2009).

Schmidt, J., Brilliantov, N., Spahn, F., Kempf, S.: Slow dust in Enceladus's plume from condensation and wall collisions in tiger stripe fractures. Nature **451**, 685–688 (2008).

Showalter, M.R.: Saturn's D ring in the *Voyager* images. Icarus **124**, 677–689 (1996).

Showalter, M.R., Cuzzi, J.N.: Seeing ghosts: Photometry of Saturn's G ring. Icarus **103**, 124–143 (1993).

Showalter, M.R., Cuzzi, J.N., Larson, S.M.: Structure and particle properties of Saturn's E ring. Icarus **94**, 451–473 (1991).

Showalter, M.R., Hamilton, D.P., Burns, J.A., de Pater, I., Simonelli, D.P.: Structure of Jupiter's main ring and halo from *Galileo* SSI and Earth-based images. In: Bagenal, F. (ed.) *Conference "Jupiter: Planet, Satellites and Magnetosphere,"* pp. 101–102. LASP, Boulder (2001).

Smith, B.A., Soderblom, L., Beebe, R.F., Boyce, J., Briggs, G., Bunker, A., Collins, S.A. Hansen, C.J., Johnson, T.V., Mitchell, J.L., Terrile, R.J., Carr, M.H., Cook, A.F., Cuzzi, J.N., Pollack, J.B., Danielson, G.E., Ingersoll, A.P., Davies, M.E., Hunt, G.E., Masursky, H., Shoemaker, E.M., Morrison, D., Owen, T., Sagan, C., Veverka, J., Strom, R., Suomi, V.E.: Encounter with Saturn – *Voyager 1* imaging science results. Science **212**, 163–191 (1981).

Smith, B.A., Soderblom, L., Batson, R., Bridges, P., Inge, J., Masursky, H., Shoemaker, E., Beebe, R., Boyce, J., Briggs, G., Bunker, A., Collins, S.A., Hansen, C.J., Johnson, T.V., Mitchell, J.L., Terrile, R.J., Cook, A.F., Cuzzi, J., Pollack, J.B., Danielson, G.E., Ingersoll, A., Davies, M.E., Hunt, G.E., Morrison, D., Owen, T., Sagan, C., Veverka, J., Strom, R., Suomi, V.E.: A new look at the Saturn system: The *Voyager 2* images. Science **215**, 505–537 (1982).

Spahn, F., Schmidt, J., Albers, N., Hörning, M., Makuch, M., Seiß, M., Kempf, S., Srama, R., Dikarev, V., Helfert, S., Moragas-Klostermeyer, G., Krivov, A.V., Sremčević, M., Tuzzolino, A.J., Economou, T., Grün, E.: *Cassini* dust measurements at Enceladus and implications for the origin of the E ring. Science **311**, 1416–1418 (2006).

Spencer, J., Pearl, J.C., Segura, M., Flasar, F.M., Manoutkine, A., Romani, P., Buratti, B.J., Hendrix, A.R., Spilker, L.J., Lopes, R.M.C.: *Cassini* encounters Enceladus: Background and the discovery of a south-polar hot spot. Science **311**, 1401–1405 (2006).

Spitale, J.N., Porco, C.C.: Association of the jets of Enceladus with the warmest regions on its south-polar fractures. Nature **449**, 695–699 (2007).

Spitale, J.N., Jacobson, R.A., Porco, C.C., Owen, W.M.: The orbits of Saturn' small satellites derived from combined historic and *Cassini* Imaging observations. Astron. J. **132**, 692–710 (2006).

Srama, R., Ahrens, T.J., Altobelli, N., Auer, S., Bradley, J.G., Burton, M., Dikarev, V.V., Economou, T., Fechtig, H., Görlich, M., Grande, M., Graps, A., Green, S.F., Grande, M., Grün, E., Havnes, O., Helfert, S., Hillier, J.K., Horányi, M., Igenbergs, E., Jessberger, E.K., Johnson, T.V., Kempf, S., Krivov, A.V., Krüger, H., Mocher-Ahlreep, A., Morgas-Klostermeyer, G., Lamy, P., Landgraf, M., Linkert, D., Linkert, G., Lura, F., McDonnell, J.A.M., Möhlmann, D., Morfill, G.E., Müller, M., Roy, M., Schafer, G., Schlotzhauer, G., Schwehm, G.H., Spahn, F., Stübig, M., Svestka, J., Tschernjawski, V., Tuzzolino, A.J., Wäsch, R., Zook, H.A.: The *Cassini* Cosmic Dust Analyzer. Space Sci. Rev. **114**, 465–518 (2004).

Srama, R., Kempf, S., Moragas-Klostermeyer, G., Helfert, S., Ahrens, T.J., Altobelli, N., Auer, S., Beckmann, U., Bradley, J.G., Burton, M., Dikarev, V.V., Economou, T., Fechtig, H., Green, S.F., Grande, M., Havnes, O., Hillier, J.K., Horányi, M., Igenbergs, E., Jessberger, E.K., Johnson, T.V., Krüger, H., Matt, G., McBride, N., Mocker, A., Lamy, P., Linkert, D., Linkert, G., Lura, F., McDonnell, J.A.M., Möhlmann, D., Morfill, G.E., Postberg, F., Roy, M., Schwehm, G.H., Spahn, F., Svestka, J., Tschernjawski, V., Tuzzolino, A.J., Wäsch, R., Grün, E.: In-situ dust measurements in the inner Saturnian system. Planet. Space Sci. **54**, 967–987 (2006).

Tagger, M., Henricksen, R.N., Pellat, R.: On the nature of the spokes in Saturn's rings. Icarus **91**, 297–314 (1991).

Thomsen, M.F., Van Allen, J.A.: Motion of trapped electrons and protons in Saturn's inner magnetosphere. J. Geophys. Res. **85**, A11, 5831–5834 (1980).

Throop, H.B., Esposito, L.W.: G ring particle sizes derived from ring plane crossing observations. Icarus **131**, 152–166 (1998).

Van Allen, J.: Absorption of energetic protons by Saturn's Ring G. J. Geophys. Res. **88**, A9, 6911–6918 (1983).

Van Allen, J.: An upper limit on the sizes of the shepherd satellites of Saturn's ring G. J. Geophys. Res. **92**, 1153–1159 (1987).

Wahlund, J.E., Boström, R., Gustafsson, G., Gurnett, D.A., Kurth, W.S., Averkamp, T., Hospodarsky, G.B., Persoon, A.M., Canu, P., Pedersen, A., Desch, M.D., Eriksson, A.I., Gill, R., Morooka, M.W., André, M.: The inner magnetosphere of Saturn: *Cassini* RPWS cold plasma results from the first encounter. Geophys. Res. Lett. **32**, L20S09 (2005).

Waite, J. H., Cravens, T. E., Ip, W.-H., Kasprzak, W. T., Luhmann, J. G., McNutt, R. L., Niemann, H. B., Yelle, R. V., Müller-Wodarg, I., Ledvina, S. A., Scherer, S.: Oxygen ions observed near Saturn's A ring. Science **307**, 1260–1262 (2005).

Yaroshenko, V., Horányi, M., Morfill, G.: The wave mechanism of spoke formation in Saturn's rings. In: Multifacets of Dusty Plasmas, Fifth Int'l. Conf. Physics of Dusty Plasmas. AIP Conf. Proc. **1041**, 215–216 (2008).

Chapter 17
Origin and Evolution of Saturn's Ring System

Sébastien Charnoz, Luke Dones, Larry W. Esposito, Paul R. Estrada, and Matthew M. Hedman

Abstract The origin and long-term evolution of Saturn's rings is still an unsolved problem in modern planetary science. In this chapter we review the current state of our knowledge on this long-standing question for the main rings (A, Cassini Division, B, C), the F Ring, and the diffuse rings (E and G). During the Voyager era, models of evolutionary processes affecting the rings on long time scales (erosion, viscous spreading, accretion, ballistic transport, etc.) had suggested that Saturn's rings are not older than 10^8 years. In addition, Saturn's large system of diffuse rings has been thought to be the result of material loss from one or more of Saturn's satellites. In the Cassini era, high spatial and spectral resolution data have allowed progress to be made on some of these questions. Discoveries such as the "propellers" in the A ring, the shape of ring-embedded moonlets, the clumps in the F Ring, and Enceladus' plume provide new constraints on evolutionary processes in Saturn's rings. At the same time, advances in numerical simulations over the last 20 years have opened the way to realistic models of the rings' fine scale structure, and progress in our understanding of the formation of the solar system provides a better-defined historical context in which to understand ring formation. All these elements have important implications for the origin and long-term evolution of Saturn's rings. They strengthen the idea that Saturn's rings are very dynamical and rapidly evolving, while new arguments suggest that the rings could be older than previously believed, provided that they are regularly renewed. Key evolutionary processes, timescales and possible scenarios for the rings' origin are reviewed in the light of these recent advances.

17.1 Introduction

17.1.1 New Results on an Old Question

Although Saturn's rings were first observed in the seventeenth century by Galileo Galilei, their origin and long-term evolution is still a matter of passionate debate. Whereas the origins of the rings of Jupiter, Uranus and Neptune, as well as of the dusty E and G Rings of Saturn, seem to be linked to the presence of nearby moonlets (via their destruction or surface erosion, see Esposito 1993; Colwell 1992; Burns et al. 2001; Hedman et al. 2007a; Porco et al. 2006), the unique characteristics of Saturn's main rings still challenge any scenario for their origin. Saturn's main rings have a mass on the order of one to several Mimas masses (Esposito et al. 1983, 2007; Stewart et al. 2007) and are mainly composed of pure water ice, with little contamination (Cuzzi and Estrada 1998; Poulet et al. 2003; Nicholson et al. 2008).

In the present chapter we detail the processes at work in the ring system, their associated timescales, and their possible implications for the origin and long term evolution of the rings. Meteoroid bombardment, viscous spreading and satellite perturbations imply a rapid evolution of the main ring system and support the idea of geologically young rings, although it seems very unlikely that the rings formed within the last billion years. Given the current state of knowledge, the destruction of a massive satellite (Pollack et al. 1976; Harris 1984) or the tidal splitting of comets grazing Saturn (Dones 1991; Dones et al. 2007) could be viable mechanisms for implanting such a large amount of ice inside Saturn's Roche zone. However, the actual cometary bombardment rate seems too low, by orders of magnitude, for these scenarios to work. Such a paradoxical situation could be resolved if the seemingly young rings are constantly renewed by some

S. Charnoz (✉)
Laboratoire AIM, Université Paris Diderot/CEA/CNRS, Gif sur Yvette, France
e-mail: charnoz@cea.fr

L. Dones
Southwest Research Institute, Boulder, CO, USA

L.W. Esposito
Laboratory for Atmospheric and Space Physics, University of Colorado, Boulder, CO, USA

P.R. Estrada
SETI Institute, 515 N. Whisman Rd., Mountain View, CA 94043

M.M. Hedman
Department of Astronomy, Cornell University, Ithaca, NY 14853, USA

material recycling mechanism (sometimes called "Cosmic Recycling"). Developments in the last 20 years have shed new light on these ideas.

- *The Cassini spacecraft* has provided invaluable data that constrain particle properties, size distributions, and processes at work in the rings, such as: the discovery of propellers (Tiscareno et al. 2006, 2008), measurements of the shape of small inner satellites (Charnoz et al. 2007; Porco et al. 2007), multiple stellar occultations with resolutions better than 10 m (Colwell et al. 2006, 2007), high resolution spectra of rings (Nicholson et al. 2008), measurements of the rings' thermal inertia (see e.g., Leyrat et al. 2008) and the discovery of Enceladus' plumes (Porco et al. 2006).
- *Numerical simulations* have brought new insights into the microphysics and dynamics of ring particles below the kilometer scale, refining the notion of the Roche Limit and the conditions of accretion in a tidally dominated environment.
- *Solar system origin and evolution* is now much better understood than 20 years ago. In particular the discovery of the Kuiper Belt has brought important constraints on the primordial dynamical evolution of the giant planets, and our knowledge of satellite formation has also improved.

In this chapter we review how these three factors have modified and improved our conceptions regarding the origin and evolution of Saturn's rings.

17.1.2 Organization of the Chapter

The challenge of writing such a chapter is that it requires a global, extensive study of ring physics and satellite dynamics, as well as planet and satellite formation. Since it is not possible to treat every aspect of these subjects in depth here, we will refer to other chapters of the present book in which several of these questions are addressed. Due to the large amount of material involved, we believe that a linear presentation is not desirable, as it would overwhelm a casual reader. Whereas the specialist will be interested in the details of some models, the graduate student will be more interested in the global picture. The present chapter was written to satisfy both readers.

Sections 17.2–17.4 are mostly dedicated to Saturn's main rings (A, Cassini Division, B, and C). Section 17.2 is a self contained overview and should satisfy graduate students: it presents the main observations and the main evolutionary processes in a somewhat simplified way, as well as the associated timescales. This includes a discussion on the putative youth of the main ring system and a brief presentation of three possible scenarios for its origin.

Sections 17.3 and 17.4 are intended for specialists. Section 17.3 deals with physical models of evolutionary processes at work in the rings. It ends with a discussion on the long term evolution of Saturn's main rings. Section 17.4 describes in detail three scenarios for the main rings' origin, making use of recent results on satellite and planet formation. Since some material in Section 17.2 is presented in more detail in Sections 17.3 and 17.4, the discussion in Section 17.2 will periodically refer to these sections.

Sections 17.5 and 17.6 describe the origins of Saturn's F Ring and the diffuse E and G Rings, respectively. While the origins and evolution of these rings are likely very different from those of the main rings, new *Cassini* data have revealed a variety of transient structures in the F Ring and have also provided new evidence connecting diffuse rings to material shed by satellites. A chapter on ring origins would not be complete without some discussion of these fainter rings.

In the conclusion we summarize what we have learned in the last 20 years. We also suggest future key observations that would help answer the long-standing question of the rings' origin and long term evolution.

17.2 Basic Observational Constraints and Theoretical Considerations

We review here the basic elements necessary to build a coherent model of the origin and evolution of Saturn's rings. The main physical characteristics of Saturn's rings that any formation and evolution scenario should account for are first recalled (Section 17.2.1) and the main driving processes are reviewed in a simple way (Section 17.2.2). Throughout this section, the reader will notice that there are still a lot of unknowns, both on the observational and theoretical side. Consequently we focus our attention on the rings' basic fundamental characteristics (material composition, mass, and spatial distribution). Only considering these three basic properties is already very challenging, as we will see. In Section 17.2.3 we depict the basic scenarios for the origin of Saturn's massive ring system. Technical aspects of Section 17.2 are described in detail in Sections 17.3 and 17.4, which are intended mainly for specialists.

17.2.1 Ring Structure

Saturn's rings are the largest and the brightest of the four ring systems surrounding each of the giant planets (for comparison, see Esposito 2006). They likely contain at least as much mass as the moon Mimas (see Section 17.2.1.3), and

Table 17.1 Saturn's rings (Burns et al. 2001; Nicholson and Dones 1991. For more detail, see Chapter 14 by Colwell et al. See Section 17.2.1.1 for details)

	Location[a] (width)	Optical depth	Dust fraction (%)	Size power-law index[b]	Notes
D ring	66,000–74,000	10^{-3}	50–100	?	Internal structure
C ring	74,490–91,983	~0.1	<3	3.1	Some isolated ringlets
B ring	91,983–117,516	1–5	<3	2.75	Abundant structure
Cassini Division	117,516–122,053	0.05–0.15	<3		Several plateaus
A ring	122,053–136,774	~0.5	<3	$a < 10$ m[c] 2.75–2.90 $a > 10$ m[c] 5–10	Many density waves Propellers[c]
F ring	140,200 (W ≅ 50 km)	0.1–0.5	>98	2–3	Narrow, broad components
G ring	166,000–173,000	10^{-6}	>99	1.5–3.5	
E ring	180,000–450,000	10^{-5}	100		Peak near Enceladus

[a] Distance units in km
[b] Power law index of the differential size distribution
[c] Tiscareno et al. (2006); Sremčević et al. (2007)
[a–c] show that there is a knee in the size distribution around a = 10 m

display all the phenomena found in the smaller ring systems of Jupiter, Uranus and Neptune. These phenomena include gaps with embedded moons and ringlets, narrow rings, broad rings, ethereal rings, waves, wakes and wiggles. Ring D lies interior to the brighter A, B and C Rings; Ring E is a broad, tenuous ring centered on the orbit of the moon Enceladus. The F Ring is a narrow ring just outside the A Ring, discovered by the Pioneer 11 flyby in 1979. The G Ring is another narrow ring outside ring F (Cassini entered the Saturn system in an apparently empty area between the F and G Rings in July 2004). Saturn's named rings are given in Table 17.1. Figure 17.1 compares the various rings, scaled to the equatorial radius of each planet. Saturn's rings include both massive and diffuse rings, and thus provide examples of phenomena occurring in all the ring systems. Because of the large mass of Saturn's rings, building a model of their origin may be the most difficult. However, any theory that explains Saturn's ring origin must be consistent with all the ring systems, and the main rings and faint rings are expected to have different origins and evolutions. Indeed, Saturn's rings are usually divided into two broad categories, main rings and diffuse rings, which display different physical properties and are driven by different physical processes:

- The *main ring system* containing rings A, Cassini Division, B, and C that are below Saturn's Roche limit for ice (about 140,000 km from Saturn's center, see Section 17.2.2.1) and are made of particles larger than ~1 cm across. They are bright and collisionally evolved, with optical depths ranging from ~0.1 (C Ring, Cassini Division) to more than 5 (B Ring). Collisions are a major driving process for the main rings. Due to their closeness to Saturn tidal affects are also strong.
- The *faint ring system* includes the E and G Rings. They are mainly made of micrometer-sized dust. The E and G Rings are very faint and in general hardly visible from Earth. Due to their low densities, mutual collisions play almost no role in their evolution, although due to the small size of their constituent particles, non-gravitational forces (e.g., radiation pressure and Poynting-Robertson effects) are important to their evolution. By contrast, non-gravitational forces are generally assumed to play almost no role in the main ring system. Tidal effects play only a minor role in the E and G Rings because they are located outside the Roche limit for ice.

The F Ring is somewhat intermediate between the main rings and the faint rings because (1) it is located near the Roche limit and (2) it seems to include a mix of big bodies (larger than 1 m, and perhaps as large as 1 km) surrounded by dusty transient structures comprised of micrometer-sized "dust". For more information about the structure of the main rings and faint rings, see the chapters by Colwell et al. and Horanyi et al., respectively.

17.2.1.1 Ring Particle Sizes

Saturn's rings are composed of myriad individual particles which continually collide. See the chapter by Schmidt et al. for a detailed discussion of ring dynamics. A growing consensus is that ring particles are actually agglomerates of smaller elements that are at least temporarily stuck together: these temporary bodies are subject to both growth and fragmentation. The balance between these competing processes yields a distribution of particle sizes and velocities (see Sections 17.2.2.2 and 17.3.2). Like the ring systems of the other giant planets, Saturn's rings overlap with numerous small moons, including Pan and Daphnis. Not only do the

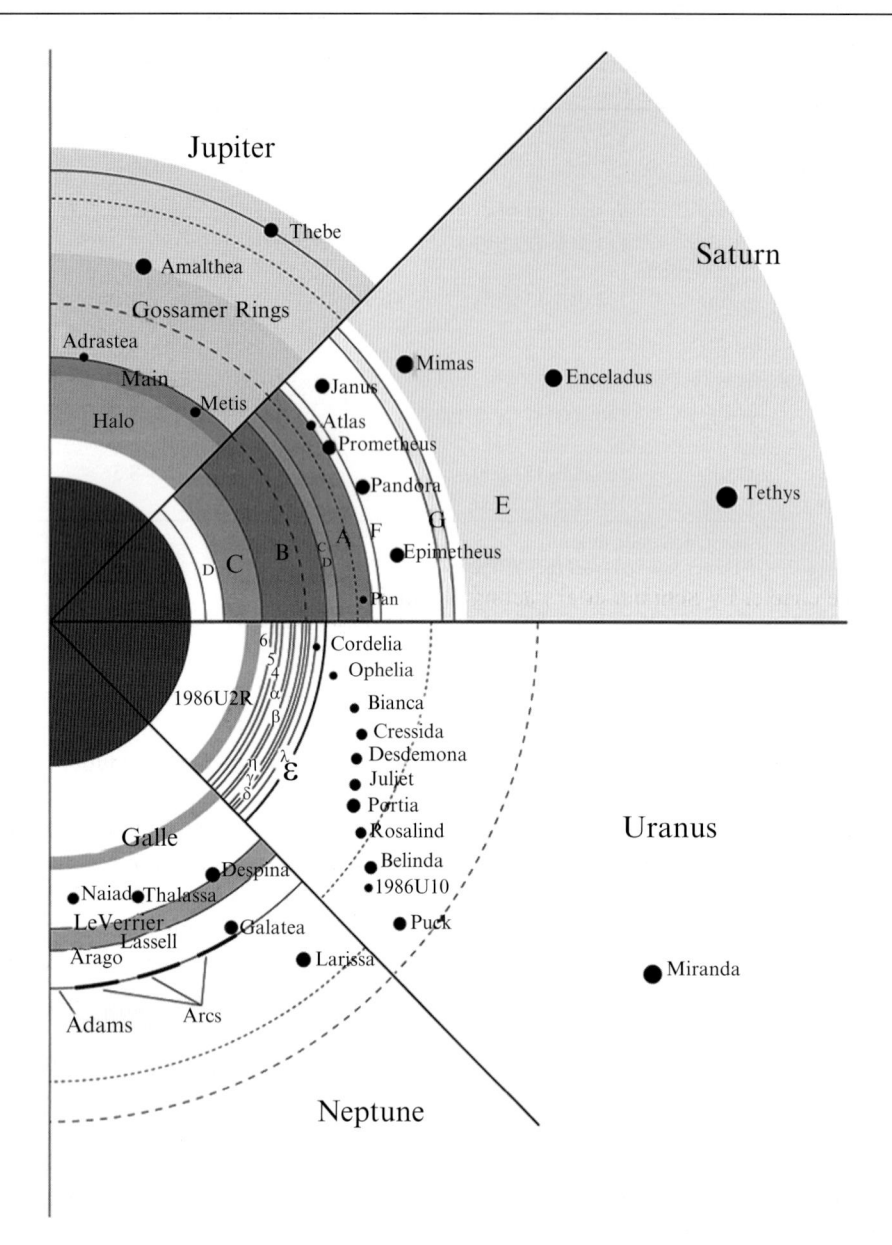

Fig. 17.1 A comparison of the four planetary ring systems, including the nearby satellites, scaled to a common planetary equatorial radius. Density of crosshatching indicates the relative optical depth of the different ring components. Synchronous orbit is indicated by a *dashed line*, the Roche limit for a density of 1 g/cm^3 by a *dot-dash line* (figure courtesy of Judith K. Burns, from Burns et al. 2001)

nearby moons affect the rings dynamically, but they can also interchange material. These "ring moons" thus provide a source of material for making the rings, and also possible sinks, affecting the size distributions of particles.

The size of Saturn's ring particles extends over many decades, from fine dust to embedded moonlets, kilometers across (see chapter by Cuzzi et al.). The observations can often be fit with a size distribution following a power law

$$N(a)\, da = C_0\, a^{-q}\, da \text{ for } a_{min} < a < a_{max}, \quad (17.1)$$

where $N(a)\, da$ is the number of particles with radius between a and $a + da$, C_0 is a constant related to the total opacity, and a_{min} and a_{max} are the radii of the smallest and largest particles in the distribution. Typical values of q are around 3 for $a < 10$ m and range from 5 to 11 for $a > 10$ m (see Table 17.1 and Tiscareno et al. 2006; Srem čević et al. 2007). Note that $q \sim 3.5$ is also characteristic of the asteroid belt and of size distributions created by shattering objects in the laboratory. Indeed, fragmentation processes at equilibrium tend to produce size distributions with $q \sim 3.5$ (assuming size independent material strength, see Dohnanyi 1969). These similarities are likely not coincidental: both the asteroids and particles in planetary rings were probably created by fragmentation of larger objects and were subject to subsequent collisional evolution. However, collisions and limited accretion should both be at work in the rings, and the current size distribution is likely to be the result of these two competitive processes (see Sections 17.2.2.4 and 17.3).

Numerical simulations show that the collisions between particles tend to equalize the kinetic energy of the various particles, a process called the *equipartition* of energy. In ideal gases, this is achieved at the state of thermal equilibrium. For rings, this state is only partly reached: the smaller bodies have only 2–20% of the kinetic energy of motion of the largest (Salo 2001). However, because of their much smaller masses, the small ring particles have significantly higher velocities relative to a purely circular orbit. These larger velocities represent larger eccentricities and inclinations and are equivalent to a higher "temperature" that causes their vertical excursions also to be larger. Particles may sometime agglomerate in the form of larger extended structures (called "gravitational wakes", see chapter by Schmidt et al.) acting like big particles that may efficiently "heat up" small particles. Simulations also show that the size distribution should vary with distance: close to Saturn, tidal forces are strong so that large aggregates are prevented from forming (Albers and Spahn 2006) whereas close to the Roche Limit (around 140,000 km, see Section 17.2.2) in the A Ring or F Ring for example, big rubble piles could exist.

The collisions of the ring particles can cause them either to grow in size or to be disrupted. The dynamic balance between these competing processes establishes an equilibrium state of aggregate bodies that resemble piles of rubble. Models show that particles tend to gather together, quickly growing to sizes that resist tidal disruption, only to be broken apart by mutual collisions (Weidenschilling et al. 1984a; Karjalainen and Salo 2004; Albers and Spahn 2006). Since relative velocities are low (about mm/s to cm/s) and collisions are inelastic, accretion is very rapid (see Section 17.3.2). Large particles can hold smaller ones on their surfaces by their mutual gravitational attraction (Canup and Esposito 1995) or by adhesion (Albers and Spahn 2006). In Saturn's rings, the time scale is only weeks for house-sized objects to accrete. After rapid growth beyond several meters, ring particles become increasingly prone to disruption, and were therefore called "Dynamical Ephemeral Bodies" (Weidenschilling et al. 1984a). These large rubble piles are indeed dynamic and are very different from the simple idea of spherical ring particles of a uniform size. In the outer regions of Saturn's rings temporary aggregations are typically elongated and sheared, as seen in numerical simulations (e.g., Salo 1995; Lewis and Stewart 2005) and in various measurements (e.g., Colwell et al. 2007) of the self-gravity wakes in the A Ring, as well as one partially transparent F Ring feature ("Pywacket") observed by the Cassini UVIS and VIMS instruments (Esposito et al. 2008a).

17.2.1.2 Ring Particle Composition

Since spacecraft have not directly sampled the particles in Saturn's main rings, we must use reflection spectra and color to get some indication of their composition. In general, ring particles are similar to the nearby moons. Saturn's rings are predominantly water ice and therefore bright (Poulet et al. 2003; Nicholson et al. 2008); by comparison, Uranus's and Jupiter's are dark. Color variations across Saturn's rings may indicate varying composition, possibly due in part to the effects of the interplanetary dust that bombards them and darkens the particles. It is likely that Saturn's ring particles have rough, irregular surfaces resembling frost more than solid ice. There is good indication that the particles are under-dense (internal densities $\rho \ll 1 \text{ g/cm}^3$), supporting the idea of ring particles as temporary rubble piles. These slowly spinning particles collide gently with collision velocities of just mm/s. For more details on the spectroscopic signature of ring particles, see the chapter by Cuzzi et al.

17.2.1.3 Mass of Saturn's Rings

The most striking characteristic of Saturn's rings is their vast extent and brightness: the rings of Jupiter, Uranus and Neptune are very faint, whereas Saturn's rings are among the most easily visible objects in the solar system. Explaining this unique characteristic is also a major challenge for any formation scenario, and it is why measuring the mass and surface mass density of the rings is of major interest. Unfortunately we have very few data constraining the total mass of Saturn's rings. Pioneer 11 set an upper limit on the ring mass of $10^{-6} M_{\text{sat}}$ ($M_{\text{sat}} \sim 5.7 \times 10^{26}$ kg stands for Saturn's mass) from the lack of any discernible perturbation on the spacecraft's trajectory (Null et al. 1981). Holberg et al. (1982) and Esposito et al. (1983) estimated the ring mass by using density waves to calculate the mass extinction coefficient κ of the ring material from the measured density wave dispersion ($\kappa = \tau/\sigma$, with τ and σ standing for the optical depth and the surface mass density, respectively). Esposito's measurements have now been confirmed by numerous measurements by Cassini in Saturn's A Ring (Tiscareno et al. 2007) and the Cassini Division (Colwell et al. 2008). Esposito et al. (1983) estimated the ring mass by assuming that this extinction coefficient measured in density wave regions applied everywhere in the rings, multiplying the optical depth measured by stellar occultation by the value $\kappa = 1.3 \pm 0.5 \text{ cm}^2/\text{g}$. Integration over the observed ring optical depth gives the ring mass $M_{\text{ring}} = 5 (\pm 3) \times 10^{-8} M_{\text{sat}}$. However, these results still do not measure the material in the densest parts of the rings, where most of the ring mass may reside.

Esposito et al. (1983) noted that 40% of the calculated ring mass came from opaque regions: since no detectable starlight was observed in these regions, this gives only a lower limit. We now know that the measured opacity in the B Ring is mostly due to gaps between self-gravity wakes (e.g., Colwell et al. 2007). Neither the stellar or radio occultations from Cassini penetrate the densest regions. Colwell et al. (2007) set a lower limit on the wake optical depth of 4.9. No upper limit can be set. This means that even if the assumption of constant extinction coefficient is true for the B Ring, the integration cannot be trusted because the optical depth in much of the B Ring is still unmeasured. Dynamical calculations by Stewart et al. (2007) and Robbins et al. (2009) were used to predict the optical depth inferred from stellar occultations, given the clumpiness evident in dynamical simulations. For optically thick regions, the occultations would underestimate the amount of ring material by factors of 4 or greater. Using the model of Cooper et al. (1985), estimating the ring density from secondaries created by galactic cosmic rays bombarding the rings, Esposito et al. (2008a) shows that a non-uniform ring fits the data consistently with Cassini occultation results reported in Colwell et al. (2007). Thus, an alternate interpretation of the Pioneer 11 results is that the majority of the mass in the B Ring is in dense wakes. In that case, the measured secondary fluxes could be consistent with a total mass more than five times larger than originally reported by Cooper et al. for the B Ring.

17.2.2 Processes in Saturn's Rings: Simple Considerations

We describe below the main physical processes that drive the evolution of Saturn's rings. We begin with a description of the notion of the Roche Limit, which seems to provide a natural answer to the simple and fundamental question: why are there rings rather than satellites close to Saturn (and the other giant planets)? Then we show that Saturn's rings are expected to evolve rapidly under the action of processes such as viscous spreading, surface darkening and material recycling. All these processes imply different evolution timescales, and raise the long-debated question: are the rings young or old? This is discussed in the last section.

17.2.2.1 Tidal Forces: the Roche Limit

Edouard Roche (1847; see Chandrasekhar 1969) calculated the orbital distance below which a purely fluid satellite would be pulled apart by tidal forces. This is the distance at which the gradient in a planet's gravitational force overcomes the gravitational attraction of the satellite's own material: its self-gravity alone is insufficient to hold it together. Of course, solid objects (and we humans, for example), can exist inside the Roche limit without being disrupted by a planet's tides due to their material strength. Even loose aggregates possess some strength. Roche's criterion can be written:

$$\frac{\alpha_r}{R} = 2.456 \left(\frac{\rho_p}{\rho}\right)^{1/3} \qquad (17.2)$$

where fluid objects would suffer tidal disruption inside the Roche limit, α_r, for a central planet with radius R and average density ρ_p. The particle's density is ρ. Thus, more dense objects could avoid tidal disruption closer to the planet (smaller α_r). For real bodies, stripping of loose material or fracture by tidal stresses occurs much closer to the planet than in the above equation. See Smoluchowski (1978, 1979) for discussion. The region surrounding the classical Roche limit, where the mutual gravitational attraction of two bodies is comparable to their mutual gravitation, is called the *Roche zone*. This is the same region where accretionary growth must compete with tidal disruption, so that formation of natural satellites around the planet would also be impeded there.

However, the definition of the Roche Limit is still ambiguous since its value depends sensitively on the underlying assumptions concerning the material strength. For example, Canup and Esposito (1995) numerically characterized accretion inside the classical Roche limit if one body is much larger than the other. In fact, if they are not rotating, a small body on the surface of a larger one will remain attached due to gravitational attraction if the distance to the planet is larger than a critical distance α such that:

$$\frac{\alpha}{R} \geq 1.26 \left(\frac{\rho_p}{\rho}\right)^{1/3} \qquad (17.3)$$

So, considering Eqs. (17.2) and (17.3), the region between α and α_r could be considered as a (large) transition region in which some limited accretion could take place despite the tidal forces. Using Saturn's parameters and material density of water ice, this region extends from 70,000 to 136,000 km from Saturn's center, or about the totality of Saturn's main ring system. This opens the way to some possible material recycling through accretion processes (see Sections 17.2.2.4 and 17.3).

17.2.2.2 Collisions: Flattening and Viscous Spreading

A planetary ring consists of small particles in nearly circular orbits, with orbital angular velocities given approximately by Kepler's law,

$$\Omega = \sqrt{\frac{GM_{sat}}{r^3}} \qquad (17.4)$$

where G is the gravitational constant, r is the distance from the planet's center, and Ω is the angular rotation rate. The optical depth of a ring of equal-sized particles is

$$\tau = \pi a^2 \sigma / m \qquad (17.5)$$

(where a and m are the radius and the mass of an individual particle). For small values of τ, the average number of collisions per orbit per particle is $\sim 2\tau$, since a particle will cross the ring plane twice. So the collision frequency is $\nu_c \sim \tau \Omega / \pi$. For an optically thick ($\tau > 1$) ring like Saturn's B Ring, collisions occur every few hours or less. This rapid collision rate explains why each ring is a nearly flat disk. Starting with a set of particle orbits on eccentric and mutually inclined orbits (say, the fragments of a small, shattered moon), collisions between particles dissipate energy: they are not perfectly elastic, but also must conserve the overall angular momentum of the ensemble. Thus, the relative velocity is damped out, and the disk flattens after only a few collisions to a set of nearly coplanar, circular orbits (Brahic 1976).

Once the disk is flattened, it spreads radially, but on what timescale? Consider the mean free path λ (average radial distance between collisions). For thick rings, this is the average random speed c multiplied by the time between collisions: $\lambda = c/(\tau \Omega)$. Cook and Franklin (1964) included both limiting values in their prescription (which was also adopted by Goldreich and Tremaine 1978a):

$$\lambda^2 = \frac{c^2}{\Omega^2} \frac{1}{1+\tau^2} \qquad (17.6)$$

The behavior of any individual particle experiencing repeated collisions can be seen as a simple random walk with the step size in radius given by λ. Let $\Delta r = n\lambda$. For a random walk, it takes on the average n^2 steps to reach a distance $n\lambda$ from the origin. Thus, the time for a typical particle to diffuse a distance Δr is n^2 steps, each of duration $\Delta t = 1/(\Omega \tau)$, giving the total time for a particle to diffuse a distance Δr:

$$T \approx \frac{n^2}{\Omega \tau} = \left(\frac{\Delta r}{\lambda}\right)^2 \frac{1}{\Omega \tau} = \left(\frac{\Delta r}{c}\right)^2 \Omega \frac{1+\tau^2}{\tau} \qquad (17.7)$$

However, the particle velocity dispersion c is not an easily measurable quantity, while a related quantity known as the viscosity ν can be inferred from the observation of spiral density and bending waves (see the chapter by Schmidt et al.). It is therefore useful to consider an expression equivalent to Eq. (17.7), but using viscosity:

$$T \approx \frac{(\Delta r)^2}{\nu} \qquad (17.8)$$

Equations (17.7) and (17.8) show that the spreading timescale depends sensitively on ν, c and τ (which depend on each other). Using Eq. (17.8) and viscosities reported in Tiscareno et al. (2007), we find that the A Ring doubles its width in 240 My to 2.4 Gy, a timescale shorter than the solar system's age (using $\Delta r = 15,000$ km, 3×10^{-3} m^2/s $< \nu < 3 \times 10^{-2}$ m^2/s). Conversely for the B Ring, using Eq. (17.7) with $\Delta r = 25,000$ km, $\tau = 3$, $c = 10^{-3}$ m/s (corresponding to a 10 m thick ring, as suggested by some numerical simulations, such as in Salo 1995), we find $T \sim 12$ Gy, longer than the age of the solar system. For the C Ring, using Eq. (17.7), with $\Delta r = 20,000$ km, $\nu = 2.7 \times 10^{-5}$ m^2/s (from the Daisaka et al. 2001 results for $\tau = 0.1$) we find $T \sim 700$ Gy, much longer than the age of the solar system. Thus there is a strong suspicion that the A Ring could be much younger than the B Ring (at least) by one or two orders of magnitude, and younger than the solar system. Some theoretical results strengthen this argument. It has been shown that Eq. (17.7) does not hold for dense rings (Daisaka et al. 2001). The finite size and the gravity of the ring particles (e.g., enhanced transport by self-gravity wakes) are responsible for a global increase of the viscosity with the optical depth. Indeed, viscosity can be split into three different contributions (see chapter by Schmidt et al.): $\nu = \nu_{\text{trans}} + \nu_{\text{coll}} + \nu_{\text{grav}}$, corresponding respectively to the effect of particle transport (ν_{trans}), the finite size of particles (ν_{coll}), and the self-gravity of particles (ν_{grav}). At low density, low optical depth and close to the planet, ν_{trans} and ν_{coll} dominate, consistent with Eq. (17.7). However, Daisaka et al. (2001) show that at higher densities and further away from the planet, random motions induced by the formation of wakes increase the values of ν_{trans} and ν_{grav}, which become of the same order and increase rapidly with τ. A fit to numerical simulations shows that $\nu(\tau) \propto \tau^\beta$ with $\beta \geq 2$ (Daisaka et al. 2001). Equation 17.8 shows that increasing the viscosity decreases the spreading timescale. As a consequence the A Ring, and perhaps the B Ring, could spread more rapidly than we have estimated here due to the presence of self-gravity wakes (Colwell et al. 2006, 2007).

17.2.2.3 Meteoroid Bombardment

Like all solar system objects, Saturn's rings suffer meteoroid bombardment. Due to their huge surface/volume ratio they are an efficient collecting area to capture the incoming meteoroid flux, which is strongly focused by the planet. The effect of meteoroid bombardment was studied in several papers (Ip 1984; Durisen et al. 1989, 1992, 1996; Cuzzi and Durisen 1990; Cuzzi and Estrada 1998) that highlighted its major role in the evolution of large-scale ring structures.

When a meteorite hits the rings, typically at an impact speed larger than 30 km/s, the impact ejecta are thrown predominantly in the prograde orbital direction (Cuzzi and Durisen 1990) and spread in radius over hundreds to thousands of kilometers. These ejecta tend to be re-incorporated into the rings because they do not reach escape velocity from the ring system. The mass ratio of the ejecta to the impactor can be quite large (possibly up to 10^6 depending on the impact geometry, material strength, etc.). All this results in a redistribution of material, as well as in a net transport of angular momentum across the system. Due to their higher surface/volume ratio, low-density rings such as the C Ring and the Cassini Division are more susceptible to being altered compositionally and dynamically. Under some assumptions on the meteoritic flux at Saturn, Cuzzi and Durisen (1990) found that the C Ring should spread into the planet in 10^7–10^8 years due to the addition of mass from the meteoroids with very little net angular momentum. Note also that the C Ring could be regenerated by spillover of material from the B Ring on the same timescale, especially if the B Ring is more massive than thought (see Section 17.3.1). The other major effect is a rapid darkening due to an accumulation of exogenic organic material. Over the age of the solar system, Saturn's rings may have collected a mass comparable to their own mass, resulting in a strong darkening of the rings. Cuzzi and Estrada (1998) showed that the A and B ring's observed brightness implies that they are much younger than the age of the solar system. They further suggest that the C to B ring transition is consistent with an exposure to meteoroid flux of around 5×10^8 years. Ballistic transport of material across the ring system may also create structures like sharp edges, wavy patterns, and optical depth ramps between regions of low to high optical depth, which recall many structures seen in the rings (Durisen et al. 1989, 1992, 1996). A more detailed discussion can be found in Section 17.3.1.

Whereas meteoroid bombardment seems to be a major mechanism at play in Saturn rings, the exact magnitude of the bombarding flux is still poorly known. Attempts to observe the flash produced by 10 cm meteoroids onto the rings with the UVIS High-Speed Photometer (HSP) yielded no detections (but this is due to the low flux of light in the HSP bandpass for such impacts, Chambers et al. 2008). Measurements by Cassini's Cosmic Dust Analyzer are dominated by dust in the Saturn system, primarily the E Ring, and its cruise measurements did not allow for a determination of the micrometeoroid flux at Saturn (Srama et al. 2006; Altobelli 2006). The New Horizons dust experiment should provide a useful measure of the heliocentric variation in micrometeoroid flux on its way to Pluto (M. Horanyi, 2006, personal communication). Note however that the meteoritic flux at Jupiter was recently evaluated on the basis of Galileo data (Sremčević et al. 2005), and appeared to be in agreement, within a factor of 3, with previous estimates.

17.2.2.4 Cosmic Recycling

As shown in Section 17.2.2.1, although the rings are located in Saturn's Roche zone, some limited accretion could be possible there, and accretion processes could substantially recycle material. Some evidence for this recycling can be found in Saturn's F Ring. Although the F Ring is clearly different from the main rings, the same processes of accretion and fragmentation occur there, and are more easily visible. If we can use the F Ring as an indicator of less obvious processes in Saturn's A and B Rings, this can provide a possible explanation of phenomena there. Cassini UVIS star occultations by the F Ring detect 13 events ranging from 27 m to 9 km in width (see Fig. 17.2). Esposito et al. 2008a interpret these structures as likely temporary aggregations of multiple smaller objects, which result from the balance between fragmentation and accretion processes. One of these features was simultaneously observed by VIMS and nicknamed "Pywacket". There is evidence that this feature is elongated in azimuth. Some features show sharp edges. At least one F Ring object is opaque, nicknamed "Mittens", and may be a "moonlet". F Ring structures and other youthful features detected by Cassini may result from ongoing destruction of small parent bodies in the rings and subsequent aggregation of the fragments. If so, the temporary aggregates are ten times more abundant than the solid objects, according to Esposito et al. (2008a).

Calculations by Canup and Esposito (1995, 1997), Throop and Esposito (1998), and Barbara and Esposito (2002) (see Fig. 17.3) show that the balance between fragmentation and

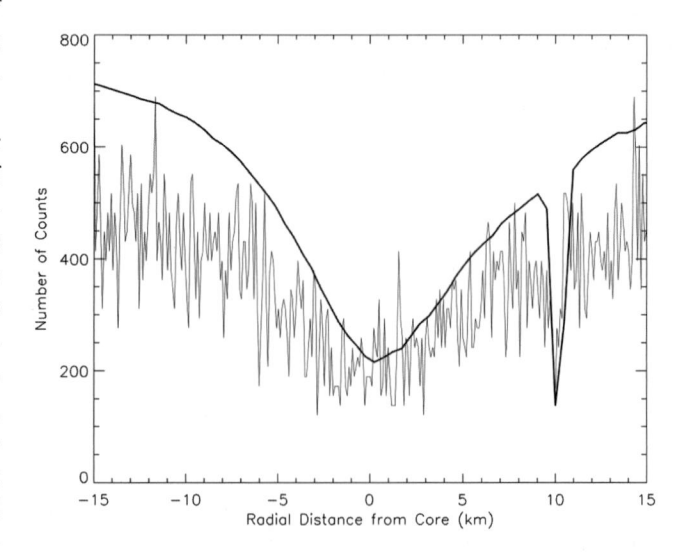

Fig. 17.2 VIMS (*solid, smooth black curve*) and UVIS (*thin, gray curve*) alpha Sco Egress occultation data overplotted. The UVIS data curve (scaled to match VIMS unocculted flux far from the event, off the figure) appears noisier, but at the center of the event has higher spatial resolution. Pywacket, the event 10 km outside the F ring core, is detected by both instruments

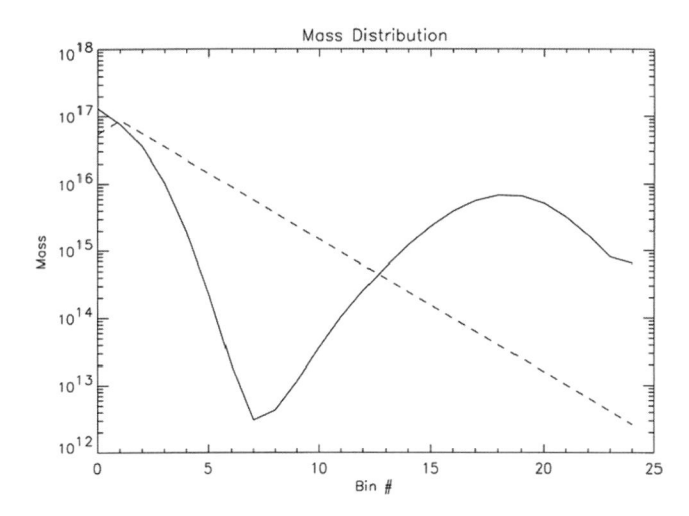

Fig. 17.3 Initial and evolved mass distribution for Saturn's F ring. Initially a shattered moon gives a power-law size distribution (*dashed line*). The system evolves to a bi-modal distribution (*solid line*) due to particle interactions and accretion. The size bins run from dust to small moons (Barbara and Esposito 2002)

accretion leads to a bi-modal size distribution in the Roche zone, yielding a small number of larger bodies that coexist with the ring particles and dust. Thus, after disruption, some significant mass fraction of a shattered moonlet would be recaptured by other moonlets and would be available for producing future rings. Calculations by Barbara and Esposito (2002) show that at equilibrium after the disruption of a small moon, about 50% of the mass of small material is collected by the larger bodies, and the other half remains in the form of small fragments. This current loss rate implies no significant loss of mass from that ring over the age of the solar system. These calculations show that recycling can extend the age of rings almost indefinitely, although this conclusion must be tempered by the results of Pöschel et al. (2007), which show that dissipation is an irreversible process and so processes such as viscous spreading cannot be reversed. If the recycling is large enough, this can significantly extend the ring lifetime (Esposito and Colwell, 2003, 2004, 2005; Esposito 2006). Larger bodies in the Roche zone can gradually grow if they can attain a roughly spherical shape (Ohtsuki et al. 2002). Those that do stick will continue to grow, perhaps until they are large enough to form a propeller structure (Spahn and Sremčević 2000; Sremčević et al. 2002). The growth process, including many starts and stops, disruptions and rare events, may be very slow: Esposito et al. (2008a) call this "creeping" growth. Although adhesion and gravity alone may be insufficient to achieve km-size objects (Albers and Spahn 2006), solid cores (Charnoz et al. 2007) may persist to re-seed new agglomerates, or compaction may convert fragments into more competent objects. However, there was some suggestion in the past that moonlets could be present in the middle of the ring system (Spahn and Sponholz 1989), or next to its outer edge (see e.g. Spahn and Wiebicke 1989). Current numerical simulations do not include all the processes or long enough timescales to constrain this "creeping" growth.

17.2.2.5 Young or Old Rings?

The question of the rings' age is central to understanding their origin. Spreading of Saturn's A Ring due to mutual collisions among the particles (Esposito 1986; see Section 17.2.2.2) and darkening of the rings due to infall and incorporation of meteoroid material (Doyle et al. 1989; Cuzzi 1995) both give ages shorter than the age of the solar system (see Section 17.3.1), from several $\times 10^5$ to several $\times 10^9$ years. Unless confined, the rings viscously spread as their particles exchange momentum via collisions and gravitational scattering (e.g. Goldreich and Tremaine 1982). Even if the rings are confined by shepherding satellites, the process only slows: the momentum is instead transferred to the shepherding moons via the resonance at the ring's edge. Because of the additional mass provided by the moon, the system's evolution is slower, but nonetheless the moon steadily moves away from the ring due to conservation of angular momentum (see Section 17.3.4). The abundant density waves in the rings also transfer momentum to the perturbing moons, again causing the moons to recede. As an example, tiny Atlas, which orbits just outside Saturn's A Ring, would have evolved to its present location in less than 10 million years if it formed at the outer edge of the A Ring. Similar short timescales are found for Prometheus and Pandora, the F ring "shepherds", which could collide or be resonantly trapped in less than 20 million years (Poulet and Sicardy 2001). Mutual collisions and meteoroid bombardment grind the ring particles while charged particles sputter molecules from their surfaces.

Estimates of the age of the rings can discriminate between possible scenarios for ring formation. If the lifetimes of some ring features are much less than the age of the solar system, those parts cannot have a primordial origin, but indicate the recent origin or possibly, renewal, of the material we observe. In Table 17.2 key observations and estimated ages are reported, and in particular the possibility that each feature could be renewed or not over time is reported. Inspecting Table 17.2 shows clearly that the rapid evolution of Saturn's rings argues for either a recent origin or renewal (or Cosmic Recycling, see Section 17.2.2.4). However, if recycling seems an appealing possibility, Saturn's A Ring and the material in the Cassini Division must have been recently emplaced. Most of the ring mass is in the B Ring, where in much of the ring the optical depth is so large we cannot directly measure the mass from density waves. The B Ring could have survived longer and be less polluted now *if* its

Table 17.2 Inferred ages of various ring features and consistency with three models for ring formation (adapted from Esposito 2006, presented at the Montana Rings Workshop)

Ring feature	Inferred/observed age	Notes	Old	Young	Renewed
Narrow ringlets in gaps	Months	Variable during Cassini mission		OK	OK
F ring clumps	Months	Sizes not a collisional distribution			OK
F ring moonlets	Tens to millions of years	Create fans and jets		OK	OK
Cassini Division density waves	100,000 years	Low mass quickly ground to dust	NO	OK	
Embedded moonlets	Millions of years	Low bulk density shows accretion		OK	OK
"Propeller" objects	Millions of years	Steep size distribution from recent disruption	NO	OK	
Pollution and color of A ring	10^7–10^8 years	Expected more polluted than B	NO	OK	NO
Pollution and color of B ring	10^8–10^9 years	Meteoroid flux not so high? More massive?	NO	OK	OK if massive
Color/spectrum varies in A	10^6–10^7 years	Ring composition not homogenized			OK
Shepherd moons	Breakup: 10^7 years			OK	OK
Radial spreading	Momentum: 10^7 years	Breakup/momentum: No contradiction in ages!	NO	OK	
Self-gravity wakes	days	Particles continually collide; self-gravity and adhesion enhance aggregation	OK	OK	OK

OK: can be accommodated; NO: serious contradiction; Blank: unclear, or deserves more study

mass has been underestimated. As we will see, large uncertainties remain on the rings' age and processes at work.

At current rates of evolution, if the rings formed with Saturn, they would now be all gone or all dark (see Table 17.2). However, it is hard to reconcile the youthful aspects of the ring system with the large mass of Saturn's rings (perhaps greater than that of the moon Mimas). Although the rings of Jupiter, Uranus and Neptune are thought to be the result of the destruction of nearby kilometer-sized moonlets due to meteoritic bombardments (Esposito 1993; Colwell 1994) forming Saturn's rings would imply the destruction of a 200 km radius moon (!). An alternate explanation is the destruction of a close-passing comet (Dones 1991; Dones et al. 2007). As stated by several authors (e.g., Lissauer et al. 1988; Ip 1988) such events are very rare and are unlikely to have occurred in the last billion years.

Ecliptic comets, which are generally thought to be the primary impactors on the giant planets and their regular satellites, are believed to originate in the so-called "Scattered Disk", a component of the Kuiper Belt. Since its discovery in 1993 (Jewitt and Luu 1993) the Kuiper Belt has been extensively observed and its dynamics has been simulated on computers. Colwell (1994) gives the number of cometary impacts (N_i) on a satellite (with radius r_s) with comets (radius r_c):

$$N_i(r_s, r_c) = N_c(r_c) \cdot P \cdot \left(1 + \frac{1}{2}\left(\frac{V_{esc}}{V_\infty}\right)^2\right)\left(\frac{r_s + r_c}{1 km}\right)^2 \quad (17.9)$$

where P is the intrinsic collision probability per year per km^2 with Saturn, V_{esc} is the escape velocity from Saturn at the location of the satellite, and V_∞ is the velocity at infinity of an ecliptic comet. $N_c(r_c)$ is the number of comets with size larger than r_c in the cometary reservoir (here the Scattered Disk). Note that the $1/2$ factor in front of the $(V_{esc}/V_\infty)^2$ is still a matter of debate (see Charnoz et al. 2009a). Levison et al. (2000) give $V_\infty \sim 3$ km/s and find $P \sim 5.6 \times 10^{-24}$ km^{-2}/year in the present-day solar system (also see Charnoz et al. 2009). Assuming a Mimas sized satellite ($r_s \sim 200$ km) is present at 100,000 km from Saturn, a ~ 20 km radius comet is necessary to destroy such an object (Harris 1984, Charnoz et al. 2009). Duncan and Levison (1997) estimate the current number of comets in the Scattered Disk with radii larger than 1 km is about 10^9. Assuming a size distribution at collisional equilibrium, so that $N_c(>r) \propto r^{-2.5}$, we get $N_c(>20 km) \sim 5 \times 10^5$. Using Eq. (17.9) with these values, we find that the number of destructive impacts on a ring progenitor located at 100,000 km from Saturn is about $N_i \sim 6.4 \times 10^{-12}$ per year (assuming the current dynamics and Kuiper Belt population). So the probability that such an impact happened in the last 10^7 or 10^8 years is only about 6×10^{-5} or 6×10^{-4}. These numbers are so small that destruction of a ring parent body by a cometary impactor is unlikely to have happened in the last 10^8 years, as pointed out by several authors (Lissauer 1988; Ip 1988).

So we are left in a paradoxical situation where, on the one hand, evolutionary processes suggest that the rings are $<10^8$ years old (or at least evolve on this timescale) and, on the other hand, cometary passages (either to break a satellite or to tidally disrupt a comet) are too rare to have happened recently in solar system history.

For the moment, this paradox has no good solution. A possibility might be recycling (see discussion above) that may make the ring material look young despite having an age as old as the solar system. Clearly, we do not really understand all the detailed processes at work in Saturn's rings and perhaps here lies the solution of this paradox.

17.2.3 An Overview of Possible Scenarios for the Origin of the Main Rings

Keeping in mind that we still do not understand how the rings could look so young if they formed shortly after Saturn's formation, we briefly summarize below several possibilities for ring formation, each of them with its pros and cons. The reader interested in details is invited to read Section 17.4.

Because of the short time scales for viscous spreading of the accretion disk of the forming planet, gas drag, particle coagulation, and transport of momentum to the forming planet, Harris (1984) argues that rings did not form at the same time as their primary planets, but were created later by disruption of satellites whose large size had made them less subject to the early destructive processes. This could happen either at the end of accretion, or well after its completion. The pieces of the disrupted satellite are within the Roche zone, where tidal forces keep them from coagulating. This explains naturally the presence of shepherd satellites and ring moons around the giant planets as the largest pieces remaining after the destruction.

Another possibility is that Saturn's rings result from the same process that created the regular satellites. Like the ring particles, the satellites' orbits are prograde, nearly equatorial, and nearly circular. A question that immediately arises is whether rings are (1) the uncoagulated remnants of satellites that failed to form, or (2) the result of a disruption of a preexisting object, either a satellite or a comet. A related question highlighted by the apparent youth of the rings is whether this latter process of ring creation by satellite destruction continues to the present time. This possibility thus mixes the origin of the rings with their subsequent evolution. Whatever their origin, the sculpted nature of the rings of Saturn, Jupiter, Uranus, and Neptune requires active processes to maintain them.

Are the main rings remnants from the Saturn's nebula? Whereas the formation of giant planets is not fully understood, simulations show that during their formation giant planet cores (about 10–30 $M_\oplus$) are surrounded by a gaseous envelope which eventually collapses to form a compact disk. As the planet approaches its final mass, it forms an extended disk component (see Estrada et al. 2009; chapter by Johnson and Estrada). It is within this subnebula that the saturnian satellites form (Mosqueira and Estrada 2003a, b; Estrada et al. 2009). Some authors (Pollack et al. 1976) have suggested that today's rings are the unaccreted remnants of this disk. The main problem here is how this material may have survived long enough for the subnebula to dissipate. Gas drag could have easily swept all the ring-material into the planet's atmosphere. Another problem is the ring composition: why only ice would have survived, whereas we know there were silicates in this disk (from the density of Saturn's satellites). Whereas this scenario seems today the most unlikely, it has never been really investigated in detail. See discussion in Section 17.4.1 for further details.

Are the main rings remnants from a destroyed satellite? Since the total ring mass is comparable to a Saturnian midsized satellite (like Mimas) it is tempting to suggest that a 200–300 km radius satellite was destroyed in Saturn's Roche Zone. The first question is: how to bring such a big object into Saturn's Roche Zone? One possibility (see Section 17.4.2.1) is that it was brought there during Saturn's formation through type 1 migration in Saturn's subnebula or gas drag (Mosqueira and Estrada 2003a, b; chapter by Johnson and Estrada), in a similar fashion to "Hot Jupiters" in extrasolar systems. Once in the Roche Zone, the satellite must be destroyed by an impactor, since tidal forces alone cannot grind a satellite to 1 cm sized particles (see Section 17.4.2 for more details). An appealing aspect of this scenario is that it has been recently shown (Charnoz et al. 2009) that a Mimas-mass moon located 10^5 km from Saturn can be destroyed during a "Late Heavy Bombardment" (LHB) type event (Tsiganis et al. 2005) about 700 My after the planet's birth. Conversely, such a mechanism cannot work for Uranus and Neptune because tidal migration should have removed the satellite from their host planet's Roche Zone before the onset of the bombardment (Charnoz et al. 2009). This encouraging property may be a partial explanation for Saturn's having massive rings. However, the LHB hypothesis sets a ring age around 3.8 or 3.9 Gy, still much too old to be reconciled with the apparent youth of the rings. In addition, a satellite should be made of ice and silicates, and silicates seem nearly absent in Saturn's rings, in apparent contradiction with this scenario. See Section 17.4.2 for further discussion of this issue.

Are the main rings remnants from a tidally split comet? Dones (1991) and Dones et al. (2007) suggested that a big (300 km radius or more) "comet" or Centaur could be the progenitor of Saturn's rings, provided it passed very close to Saturn and was disrupted by the tidal forces close to the planet, in a fashion very similar to how comet Shoemaker-Levy 9 was disrupted after a close passage with Jupiter. Again, the main problem is that the current flux of objects larger than 300 km is far too low, by orders of magnitude, for such a close passage to have been likely in the last 10^8 years. However, this scenario was revisited in Charnoz et al. (2009), who considered the possibility that this happened during the LHB. Charnoz et al. (2009) show that the cometary flux is so high during this event that several tens of Mimas masses of cometary material may have been brought into Saturn's Hill sphere, and that a fraction of it could have ended in Saturn's Roche zone. Surprisingly, the same thing may have happened to other giant planets during the LHB. So explaining why only Saturn has massive rings is a mystery in this scenario. See Section 17.4.3 for more details.

17.2.4 Beyond the Paradox?

We have presented the main observations and the main ideas concerning processes affecting the evolution of Saturn's main rings and possible scenarios for their origin. We have reached a point of giving seemingly paradoxical conclusions (rings look young, whereas they cannot have formed recently). Such a situation arises from a lack of knowledge of the processes and of the underlying assumptions, such as material recycling and total ring mass. For example, recycling of ring material can not explain the limited micrometeoroid darkening of Saturn's rings (Cuzzi and Estrada 1998). Why are the rings not darker now, if they are truly ancient? One possibility is that the total mass of the rings, mostly in Saturn's B Ring, has been underestimated. As said in Section 17.2.1.3, for example, Cooper et al. (1983) selected the smaller of two possible values for the B Ring mass consistent with Pioneer 11 results. Esposito (2008a) showed that the "granola bar" model of Colwell et al. (2007) is also consistent with Pioneer 11 results, and implies a B Ring mass about 4 times greater than estimated by Esposito et al. (1983). Because the total optical depth of the B Ring is still unmeasured and may be more than two times greater than previous estimates (Colwell et al. 2007; Stewart et al. 2007), meteoritic pollution would have a smaller effect. This is seen in the Markov chain simulations of Esposito et al. (2008b). An important consideration is that meteoritic pollution could affect mostly the exterior of objects (Esposito and Eliott 2007). If the rings (particularly the B Ring) are much more massive than we now estimate, the interior of the largest ring objects (which may encompass most of the ring mass) can remain more pristine until disrupted. However, in the thinner parts of the rings where density waves are visible, we have good ring mass estimates. The mass there would be quickly polluted by the micrometeoroid flux (Cuzzi and Estrada 1998). Esposito (1986) noted that most of the age problems involve Saturn's A Ring. Perhaps the A and F Rings are more recent? This raises the problem of how the material that formed these (possibly more recent) rings had been preserved, perhaps as large, unconsolidated objects with competent solid cores encased in rubble. If the A Ring is much younger than the B ring, we also need to find an explanation for the Cassini Division. It would not have originated simply by a density wave clearing a gap, as proposed by Goldreich and Tremaine (1978b).

From this discussion, it is clear that processes affecting the rings' evolution and scenarios for their origin must be investigated with much care in order to determine the limits of our ignorance. This is the subject of the following two sections.

17.3 Evolution of the Main Rings

This section details processes affecting the large scale evolution of the rings. It can be skipped by undergraduate students and is mainly intended for specialists. It illustrates the diversity and complexity of processes at work, whose interactions are still not really investigated. It also shows that the timescales of these processes, discussed in the previous sections, sometimes rely on some poorly constrained parameters (like the meteoritic flux, the B Ring surface mass density, etc.). Better understanding these complex processes will undoubtedly help to resolve the paradox discussed in the previous section.

Viscous spreading was treated in the previous section. Section 17.3.1 presents the effect of meteoritic bombardment. Section 17.3.2 presents the effects of accretion within the rings, which is a key element for the Cosmic Recycling process (and might be a solution for the apparent youth of the rings). Section 17.3.3 presents the effects of collisional cascades that counteract accretion processes. Section 17.3.4 presents the effects of ring–moon interactions that may give further clues on the rings' age and evolution. Section 17.3.5 is a tentative exploration of the future evolution of Saturn's ring system.

17.3.1 Meteoritic Bombardment and Ballistic Transport

Because the rings have a huge surface area-to-mass ratio, they are particularly susceptible to modification due to extrinsic meteoroid bombardment. While other mechanisms can lead to the erosion of ring particle surfaces, such as interparticle collisions, sputtering of energetic ions or photons, and the sublimation of ices, most tend to produce atomic and molecular by-products, whereas meteoroid impacts can produce a large amount of particulate ejecta in addition to significant amounts of gas and plasma (e.g., see Morfill et al. 1983). Moreover, the vast majority of the dust and debris produced from these collisions are ejected at speeds much less than the velocity needed to escape the Saturn system at the distance of the rings ($\sim$27 km/s). As a result, a copious exchange of ejecta between different ring regions can occur, which over time can lead to the structural and compositional evolution of the rings on a global scale. This process by which the rings evolve subsequent to meteoroid bombardment is referred to as "ballistic transport" of impact ejecta (Ip 1983; Lissauer 1984; Durisen 1984a, b).

17.3.1.1 Principles

The essence of the ballistic transport mechanism is shown schematically in Fig. 17.4. Impact ejecta from a given (nondisruptive) meteorite impact are thrown predominantly in the prograde orbital direction. This results naturally from consideration of impact geometries and probabilities azimuthally averaged over the rings (Cuzzi and Durisen 1990). The yield Y of a single impact, which is defined as the ratio of ejecta mass to impactor mass, can be quite large, depending on several factors. Oort cloud-type projectiles, for example, impact at speeds much greater than the sound speed in either the target ring particle or impactor. As a result, shocks vaporize part of or the entire projectile along with a small volume of the target, and shatter and excavate a large volume of the target (e.g., Durisen 1984b). The velocity (or velocity distribution) at which it is ejected depends on the hardness of the target and the angle of impact (Cuzzi and Durisen 1990). If the target is powdery, yields can be on the order of $\sim 10^5$–10^6 at normal incidence (Stöffler et al. 1975; Burns et al. 1984), while micrometer-sized particles impacting into granular surfaces have yields from 1 to $\geq 10^3$ (Vedder 1972; Koschny and Grün 2001). Ejecta velocities for the bulk of the material tend to range from $\sim$1 m/s to tens of meters/second, while escape velocities are achieved for a mass comparable to the projectile (see Durisen 1984b and references therein). This means that, in general, one need not consider the total mass added to or lost from the system unless one considers very long evolutionary times.

Impact ejecta carry not only mass, but angular momentum as well. Because most ejecta are prograde, they tend to land in the rings at larger distances where the specific angular momentum is greater, so the net resultant drift is inward. Ring structure (i.e., optical depth and surface mass density) can have an effect on the rate of material drift because the ejecta absorption probability (which determines the

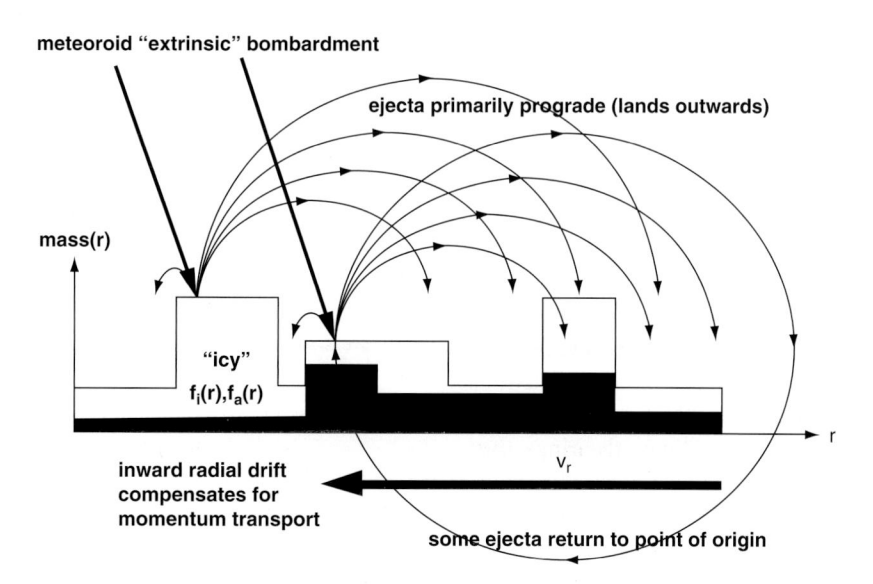

Fig. 17.4 Schematic of the ballistic and subsequent pollution transport process. Impact ejecta (which are primarily prograde) carry both mass and angular momentum to outer regions of the rings where the specific angular momentum is typically larger. This leads to a net inward radial drift of material where ejecta preferentially land which compensates for the direct momentum transport with respect to drift due to the rings' inherent viscosity. The various plateaus represent radial distribution of mass density, while shaded regions represent the variable fraction of material that is non-icy (from Cuzzi and Estrada 1998)

actual mass that hits the rings as opposed to merely passing through them) depends weakly on the local optical depth, but its angular momentum depends linearly on the surface density (Cuzzi and Durisen 1990; Cuzzi and Estrada 1998). Moreover, ring particles tend to be smaller in low optical depth regions. Thus, lower surface density regions are more quickly altered compositionally (e.g., darkened) relative to higher surface density regions simply because the mass fraction of extrinsic ("polluting") material relative to the local overall mass surface density (which is related to the optical depth) will be higher in low surface density regions, for a given micrometeoroid flux.

The value of the micrometeoroid flux at Saturn (now, and in the past when it was most likely greater) remains uncertain. Past estimates of the micrometeoroid flux at Saturn (Morfill et al. 1983; Ip 1984; Cuzzi and Durisen 1990; Cuzzi and Estrada 1998) vary slightly but all imply that the main rings would be impacted by close to their own mass over the age of the solar system (Landgraf et al. 2000). More recently, Galileo measurements of the flux at Jupiter have provided estimates of the mass flux that may at most be too low by a factor of 2–3 only (Sremčević et al. 2005) compared to previous estimates. However, the micrometeorite mass flux at Saturn has not been observed, and will not be until the Cassini Extended Mission, which is now underway, using an indirect technique similar to that described in Sremčević et al. (2005).

17.3.1.2 Dynamical Evolution

For nominal values of ring mass and meteoroid flux, micrometeorite impacts on the rings will have two different effects. First, they will lead to angular momentum loss which would lead, for example, to the loss of the C Ring to the planet in $\sim 10^7$–10^8 years (Cuzzi and Durisen 1990). A similar age for the rings more closely related to ejected material was obtained by Northrop and Connerney (1987), who suggested that water molecules generated by impacts were lost to Saturn's ionosphere, providing an explanation for Saturn's unusually low ionospheric electron density. Note that resonant interactions with nearby "ring moons" can also lead to significant angular momentum loss by the rings (Goldreich and Tremaine 1982; Poulet and Sicardy 2001). Second, in addition to angular momentum arguments, meteoroid material also darkens and pollutes the rings over time. Doyle et al. (1989) and Cuzzi and Estrada (1998) noted that the relatively high albedo of the A and B Rings was inconsistent with their having retained more a small fraction of primitive, carbonaceous material from the large mass they would have accreted over the age of the solar system (using a mass flux from Morfill et al. 1983), thereby suggesting a geologically young age for the rings.

The first efforts to model the global effects of ballistic transport on Saturn's rings indicated that ballistic transport can produce prominent edge effects, enhance pre-existing density contrasts, and produce "spillovers" from high to low optical depth regions (Ip 1983; Durisen 1984a). Since then, the influence of meteoroid bombardment and ballistic transport has been found to explain certain aspects of ring structure such as the fairly abrupt, and remarkably similar, inner edges of the A and B Rings, which include "ramp" features which connect them to the Cassini Division and C Ring, respectively (Durisen et al. 1989, 1992, 1996), given evolutionary times of ~ 100 "gross erosion" times ($\sim 10^8$ years, see Fig. 17.5). A gross erosion time is defined as the time in which a reference ring annulus would disappear due to ejected material if nothing returned. Durisen et al. (1992) concluded from their simulations that the sharpness of the A and B Ring inner edges are maintained for very long times at their observed widths by a balance between ballistic transport, which tends to sharpen low-optical-depth to high-optical-depth transitions, versus the broadening effect that results from viscous transport. In addition, Durisen and colleagues found that prograde ejecta distributions (Cuzzi and

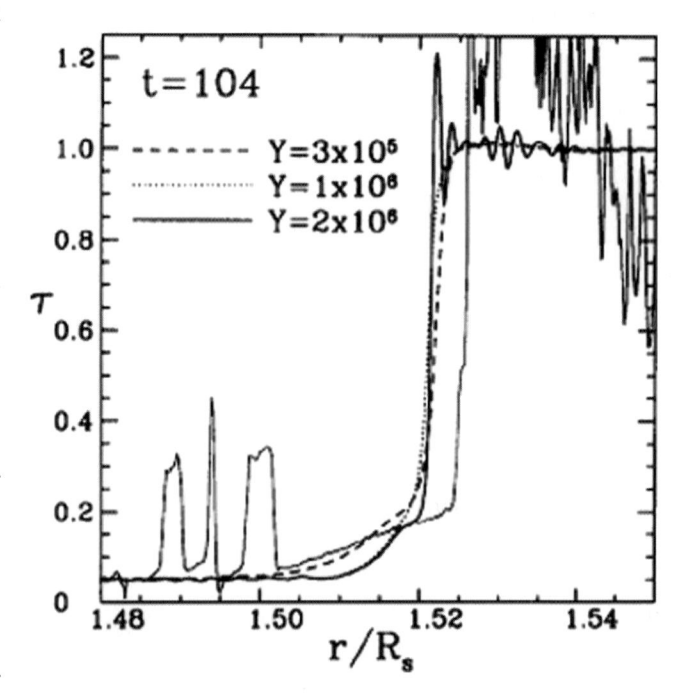

Fig. 17.5 A comparison of the Voyager ISS optical depth (*light solid curve*) with ballistic transport simulation results using a power-law distribution of ejecta velocities and different values for the total yield Y. The simulations were run for 104 "gross erosion times," which refers to the time in which a reference ring annulus would disappear due to ejected material if nothing returned. The lower bound for the ejecta distribution is varied, but is typically ~ 1 m/s. The undulatory structure discussed in the text in the inner B Ring is clearly seen. For details, see Durisen et al. (1992)

Durisen 1990) inevitably lead to the formation of a ramp on the low optical depth side of the ring edge due to advective effects.

Another effect unique to ballistic transport found by Durisen et al. (1992) is the formation of wavelike structure. Undulatory structure not associated with known resonances is observed in the C Ring, and throughout the B Ring (e.g., see Horn and Cuzzi 1996; Estrada and Cuzzi 1996). In their early simulations, Durisen and colleagues found that undulatory, wavelike structure with length scales on the order of the maximum "throw" distance of ejected material ($\sim 10^2$–10^3 km) was generated near, and propagated away from, ring edges. This phenomenon was referred to as "echoing" of edge structure (see Fig. 17.5), and it was originally thought that these structures were driven by the presence of the sharp inner edge (and to a lesser degree the ramp), with their growth dominated by either direct mass exchanges (in high optical depth regions) or advective effects (in low optical depth regions). More recently, Durisen (1995) used a local linear instability analysis to demonstrate that ballistic transport causes these undulatory structures to arise spontaneously in an otherwise uniform ring system due to a linear instability, and does not require driving by the presence of an inner edge. Durisen concluded that the $\sim$100 km structures seen in the inner B Ring (and possibly the more suppressed undulations seen in the C Ring) arise as a result of the ballistic transfer process, and perhaps viscous overstability, as suggested by some authors on the basis of stellar occultations (Colwell et al. 2007). However, in order to maintain the sharp inner edge, and produce the inner B Ring structure, a low-speed ejecta distribution component (that characterizes disruptive-type impacts) may be required in addition to the prograde distribution.

Finally, Durisen et al. (1996) considered the "direct" effects of meteoroid bombardment on the rings which include mass deposition (of the incoming projectiles), radial drifts that arise due to mass loading and/or loss of ejecta angular momentum, and radial drifts due to torques caused by the asymmetric absorption of the aberrated meteoroid flux (e.g., Cuzzi and Durisen 1990). The main result these authors find is that the effects of radial drifts due to mass loading are dominant by an order of magnitude, and that its most notable effect would be to cause parts of the Saturn ring system (i.e., the C Ring) to be lost to the planet in a timescale less than the age of the solar system (10^7 to 10^8 years, assuming the flux of Cuzzi and Estrada 1998) due to a decrease in the ring material's specific angular momentum. The result, however, is subject to uncertainties in the rate of mass deposition. Conversely, if the rings are old, there would have been significant mass loss over the age of the solar system, and we would need the C Ring to somehow be "regenerated" regularly, possibly from spillover of material from the B Ring. Nonetheless, it would still only require some small fraction of the age of the solar system for the C Ring to darken to its current state from the B Ring color (whereas the B Ring color presumably represents the effects of meteoroid bombardment over the age of the solar system, see below). Then the C Ring we see now might be some grandchild of an "original C Ring", with the timescale for recycling of the C Ring likely being on the same order as the C Ring lifetimes discussed earlier. Unfortunately, such a scenario has yet to be studied in detail.

17.3.1.3 Spectral Evolution

Subsequently, Cuzzi and Estrada (1998) developed a "pollution transport" code to model the evolution of ring composition with time under the influence of meteoroid bombardment and ballistic transport. Their goal was to see whether the meteoroid bombardment process was consistent with the regional and local color differences in Saturn's ring system given a uniform "primordial" or intrinsic composition, and if so, to attempt to constrain the intrinsic and extrinsic (bombarding) materials, and provide an independent estimate of ring age. The need for both intrinsic and extrinsic sources of material stems from the argument that if the rings started as pure water ice, then meteoritic bombardment by dark, spectrally neutral material such as carbon black (or even something slightly red and dark) would explain the C Ring color fairly well, but one could never produce the B Ring color. On the other hand, if the incoming material were spectrally red in color, then one could drive the B Ring to its present observed color, but the C Ring would be even more red than the B ring. The assumption, then, was that the red material was primarily intrinsic in origin (which might be consistent, e.g., with the breakup up of a small icy satellite that formed in the Saturnian subnebula, see Section 17.4.1) and that the meteoritic material bombarding the rings was dark and neutral in color. Their model consisted of two parts. The first was a dynamical code developed from the structural code of Durisen et al. (1989, 1992) that assumed time-invariant structure (e.g., constant optical depth, surface density), in which they calculated how the fractional amounts of non-icy constituents build up over time and how these impurities are redistributed over the rings. These authors assumed time-invariant structure because Durisen and colleagues had found in their structural evolution studies that constant optical depth regions and inner edges can remain more or less unchanged for very long timescales. The second part was a radiative transfer calculation that used the results of the ballistic transport code to see how ring particle composition was manifested in ring particle color and albedo.

Using their radiative transfer code, Cuzzi and Estrada (1998) calculated the albedo of a ring particle using reflectance values obtained from three different

Voyager filters (G, V, and UV, which corresponded to 562, 413, and 345 nm, respectively; Estrada and Cuzzi 1996; also see Estrada et al. 2003) by volumetrically mixing together icy and non-icy constituents with different mass fractions and real and imaginary indices of refraction. An implicit assumption made by these workers was that in the low-phase-angle Voyager geometry, ring brightness was dominated by single scattering, and that the particle phase function was independent of wavelength so that ring reflectivity (color) ratios were just ratios of ring particle albedo at two wavelengths (however, see below). This assumption allowed them to calculate the ring color at two extreme radial locations (which bracketed the B Ring/C Ring transition) in order to determine the "current" refractive indices of ring material at these locations. The two reference points were treated as identical initial "primordial" composition, which evolved independently to their current composition due to the deposition of meteoritic material. Primordial (i.e., initial) composition was then inferred using a linear volume mixing approximation, while the composition of the extrinsic material was determined through iteration using the results of the ballistic transport code for the fractional amounts of non-icy constituents.

The final stage of the modeling process involved fitting all three radial color profiles (G/UV, G/V, V/UV) using their mixing model (Fig. 17.6). It should be noted that the mass mixing ratios of the individual constituents in their simple two-point model can only provide a constraint on the absolute ring age if the micrometeoroid flux, retention efficiency, and impact yields are known. They resolved this ambiguity by modeling the actual shape of the B Ring/C Ring transition, because it turns out that the shape depends on a combination of all these parameters. The best fits of Fig. 17.6 correspond to an estimated timescale of ~ 3–4×10^8 years, much less than the age of the solar system. Thus, Cuzzi and Estrada (1998) found that they could simultaneously explain both the C Ring/Cassini Division versus A Ring/B Ring dichotomy and the form and shape of the B Ring/C Ring transition on a timescale similar to that found by Durisen and colleagues. Moreover, the structural evolution conclusions and timescales and the conclusions of the pollution models were obtained independently of one another, making the overall result quite robust.

The actual age of the rings derived from these models remains quite uncertain because the absolute timescale for ring erosion depends on the typical impact yields of ejecta; the efficiency with which extrinsic material retains its absorptive properties (rather than changing drastically in composition during high-speed impacts onto the rings); and the micrometeoroid flux. It should be noted that, although the historical scenario is that the rings are primordial (e.g., Pollack 1976), and the post-Voyager paradigm became that they are young, as described above (also see Section 17.2), some workers have revived the idea that the rings could be as old as the solar system. For example, Esposito et al. (2005) find variations in ring brightness over scales of 1,000–3,000 km which they assert are not clearly consistent with ballistic transport, and posit that a local event such as the breakup of a small icy moon may be responsible for resetting the "ballistic transport" clock. Stewart et al. (2007) and Robbins et al. (2009) suggest that gravitational instabilities in massive rings could easily hide most of the mass in dense, but transient, clumps and streamers, while a small fractional percentage of fairly empty gaps between these opaque structures would then provide the observed ring transparency or finite optical depth. Esposito (2006) suggested that if the ring surface mass density is an order of magnitude larger than currently inferred from optical depth and sparse spiral density wave measurements, the rings can more easily resist becoming polluted and darkened, and could perhaps indeed be as old as the solar system (see Section 17.3.2.1). A caveat to keep in mind, however, is that the micrometeorite flux was certainly higher in the early solar system. As much as an order of magnitude (or more) higher is not out of the question, which might require that an ancient B Ring have even more mass than what is currently theorized to remain as icy as observed, despite the cumulative effects of pollution over such a long period of time.

In the midst of the Cassini era, a renewed vigor has emerged regarding ballistic transport modeling. Dozens of occultations cutting all parts of the main rings at a number of longitudes and elevations are allowing for a more detailed understanding of ring properties such as opacity, surface density, optical depth, and ring viscosity – all of which are key quantities for structural evolution modeling. On the other hand, a wide range of Cassini spectral data which spans a larger range of the electromagnetic spectrum in more viewing geometries than previously obtained by Voyager is allowing for a comprehensive compositional study. For example, the assumption made by Cuzzi and Estrada (1998) that ring particle phase functions are wavelength independent turns out to be incorrect (Cuzzi et al. 2002; Poulet et al. 2002; see chapter by Cuzzi et al.). Thus modeling ring color and composition requires a more sophisticated approach which involves a combination of ring layer and regolith radiative transfer modeling. Fortunately, given the wide array of Cassini spectral data available, the problem is a tractable one.

17.3.2 Limited Accretion

As seen in the previous section, there are several processes capable of eroding and altering the bulk composition of the ring system, and thus a renewing mechanism is often invoked (see Sections 17.2 and 17.3.4). A combination of accretion

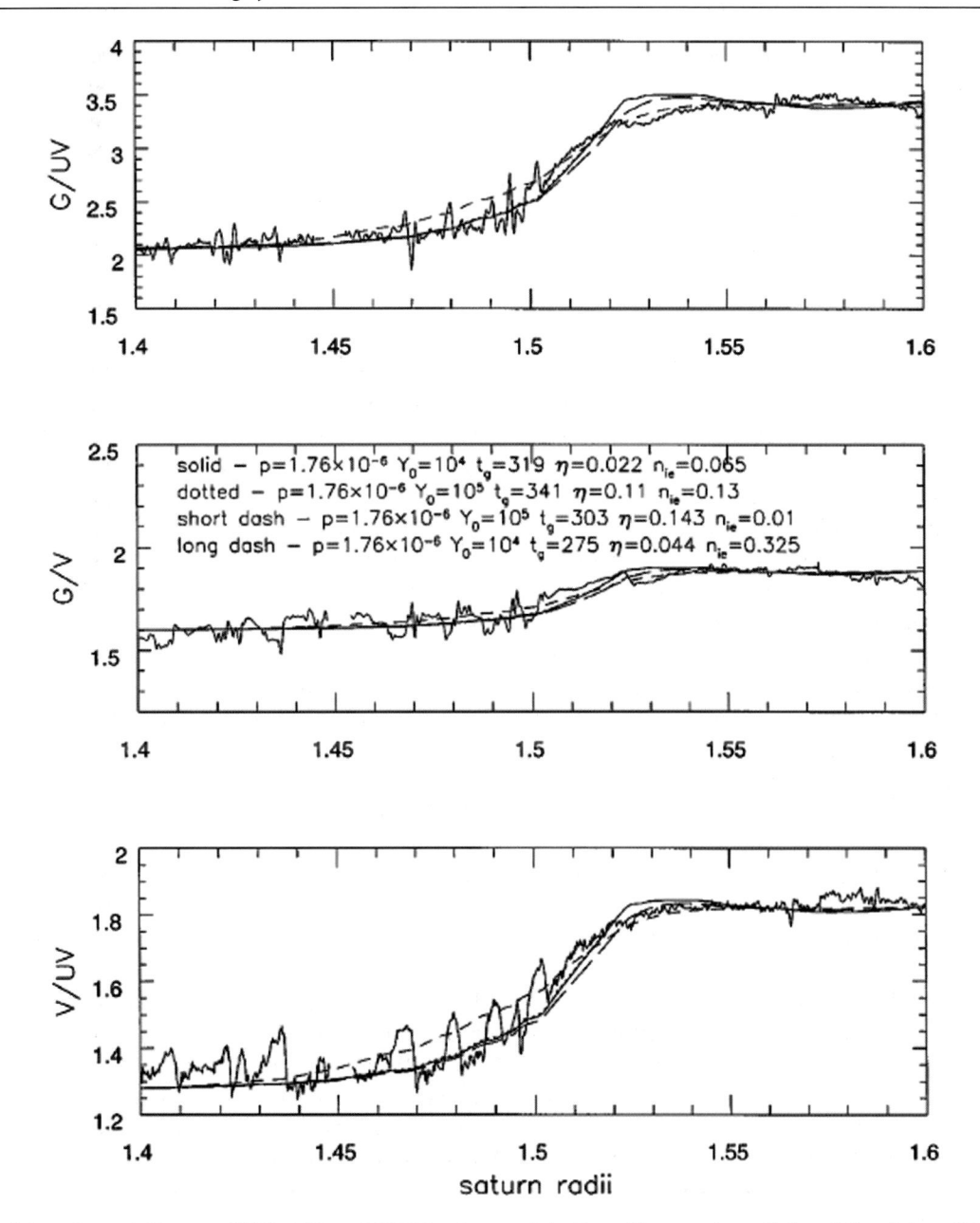

Fig. 17.6 Radial profiles of Voyager G/UV, G/V, and V/UV color ratios, compared with the results of combined ballistic transport and radiative transfer models. Although the absolute age of the rings depends on the micrometeoroid flux, the retention efficiency η, and the impact yield, the shape of the B Ring/C Ring transition depends on a combination of these parameters embodied by the shape parameter p. The figure illustrates how all three full radial profiles of color ratio can be simultaneously matched. The best case fit (*solid curve*) corresponds to $t_G \sim 300$ "gross erosion times" or a timescale of ~ 3–4×10^8 years. In the figure, Y_0 is the ejecta yield, and n_{ie} is the imaginary component of the refractive index of extrinsic material

and erosion could be such a mechanism. Here we present how a limited form of accretion could take place inside the rings, despite the tidal forces. Further discussion of accretion physics can be found in the chapter by Schmidt et al.

The concept of accretion is embodied in the convenient, but sometime confusing, notion of the "Roche Limit", which depends on several assumptions (fluid vs. rigid body, etc.). Depending on these assumptions, the Roche Limit is located somewhere between 1.26 R (where R is the planet's radius, and assuming the body has the same density as the planet, Section 17.2.2.1) and 2.456 R, corresponding, respectively, to the inner C Ring and about 8,000 km beyond the F Ring. So it is plausible that substantial accretion could occur in Saturn's rings; however, it should be severely limited by

Saturn's strong tides. Indeed, the Cassini spacecraft reveals indirect evidence for accretion. The main observations are the following:

1. The detection of self-gravity "wakes" in Saturn's A Ring (Colwell et al. 2006; Hedman et al. 2007b) thanks to multiple stellar occultations. Wakes are a gathering of particles due to a local gravitational instability (see chapter by Schmidt et al. and below).
2. The detection of propeller-shaped structures in Saturn's A Ring (Tiscareno et al. 2006, 2008; Sremčević et al. 2007), revealing the presence of chunks of material in the 50–100 m range. Propellers could be either the result of fragmentation of an ancient moon (Sremčević et al. 2007), or, the result of "creeping" accretion processes (Esposito et al. 2008b).
3. VIMS data have revealed that the regolith on the surface of Saturn's ring particles shows a general trend for increasing grain size with distance across the rings (Nicholson et al. 2008).
4. ISS images of Pan and Atlas, two small satellites embedded in, or close to, the rings have revealed prominent equatorial bulges on these moons (Porco et al. 2007). Simulations show that these bulges are likely the results of ring particles accreted at the satellites' equators (Charnoz et al. 2007)
5. ISS surveys of the F Ring have revealed the presence of embedded bodies that could not be tracked for more than a couple of orbits (Porco et al. 2005) suggesting the idea of "ephemeral moonlets". Such bodies might be produced by accretion. Stellar occultations of Saturn's F Ring (Esposito et al. 2008a) have revealed the presence of clumps embedded in the ring's core that could well be the result of some accretion process owing to their optical translucency, suggesting loose aggregates of material (Fig. 17.2).

All this indirect evidence supports the idea that accretion may indeed take place in Saturn's rings, and that an active recycling of material is possible. In the last twenty years several analytical studies and numerical models have characterized accretion processes under tidal forces (e.g., Weidenschilling et al. 1984a; Ohtsuki 1993; Canup and Esposito 1995; Barbara and Esposito 2002; Karjalainen and Salo 2004; Albers and Spahn 2006), and have shown that an exotic form of accretion may take place in the rings.

17.3.2.1 Gravitational Instability

A gravitational instability occurs when the local self-gravitational potential energy exceeds both internal energy (due to pressure) and shear kinetic energy (due to Keplerian shear, i.e., the variation of the ring particles' angular speed with distance from the planet (Eq. (17.4)). It is embodied in the "Jeans-Toomre" criterion, implying that the gravitational instability occurs for small Q, such that (Toomre 1964; Karjalainen and Salo 2004, and see chapter by Schmidt et al.):

$$Q = \frac{c\Omega}{3.36 G \sigma} \quad (17.10)$$

where c is the velocity dispersion, G is the gravitational constant, Ω is the local keplerian angular velocity and σ is the surface density. For Q < 2 the collective gravity together with keplerian shear creates shearing, tilted wake structures. Numerical simulations (Salo 1995; Karjalainen and Salo 2004) show that gravitational wakes are progenitors of gravitational aggregates: in the inner regions of the A Ring, wakes are like parallel rods with moderate density contrast with the inter-wake medium. Further out in the A Ring, the density contrast increases and the wakes become more and more clumpy. Finally, just outside of the A Ring, the wakes coalesce into clumps, recalling small satellites (see, e.g., Fig. 2 of Karjalainen and Salo 2004). For more details on wakes, the reader is invited to read the chapters by Schmidt et al. and Colwell et al.

17.3.2.2 Tidally Modified Accretion

Even in the absence of gravitational instability (i.e., when Q > 2), accretion may still be possible via binary encounters, as for planetesimal growth in the protoplanetary disk. The classic two-body criterion for accretion is simple: the rebound velocity, V_r (which is a fraction $\varepsilon < 1$ of the impact velocity V_i), must be less than the escape velocity of the two bodies. This is valid in free space, but becomes incorrect under the tidal influence of a nearby massive body (Saturn in our case), called the "primary": close to the primary, the differential tidal acceleration can overcome the two bodies' gravitational acceleration. To describe three-body modified accretion, it is necessary to properly take into account the potential energy of the colliding bodies. We follow the approaches of Ohtsuki (1993) and Canup and Esposito (1995). Consider a body with mass m and radius a, orbiting on a circular orbit at distance r from Saturn, with keplerian angular velocity Ω. The specific potential energy of a body with mass m', radius a', located at $\vec{r}'$, in the (non-inertial) frame rotating with body m is:

$$E_p = \frac{-GM_s}{r'} - \frac{-Gm}{\|\vec{r} - \vec{r}'\|} - \frac{\Omega r^2}{2} \quad (17.11)$$

After developing Eq. (17.11) to first order in (r − r') and nondimensionalizing time scales with Ω^{-1} and lengths with the Hill radius ($R_{hill} = r[(m+m')/(3M_s)]^{1/3}$), and using relative coordinates in the rotating frame x,y,z (Nakazawa and Ida 1988; Ohtsuki 1993) so that $r = \left(x^2 + y^2 + z^2\right)^{1/2}$, we get (see Section 14.4.1.2 in the chapter by Schmidt et al. for more details):

$$E_p = -\frac{3}{2}x^2 + \frac{1}{2}z^2 - \frac{3}{r} + \frac{9}{2} \quad (17.12)$$

We continue, following the simplified model of Canup and Esposito (1995), which preserves the essential physics well enough for our purposes. When the two bodies collide, the scaled rebound velocity is computed as (after averaging over all impact angles) $V_r = \varepsilon \left[V_b^2 + V_e^2 + 2r_p^{2/3}\right]^{1/2}$, with $r_p = (a + a')/R_{\text{hill}}$. In this expression V_b is the relative velocity (the quadratic sum of the random velocity plus the gradient of orbital velocity due to different semi-major axes), V_e is the scaled mutual escape velocity $\left(= (6/r_p)^{1/2}\right)$, and $2r_p^{2/3}$ stands for the finite keplerian velocity difference between the two mass centers (see Canup and Esposito (1995) for details). Finally, the requirement for gravitational sticking is that the total energy $E = 1/2 V_i^2 + E_P$ must be negative after rebound. In the limit of zero random velocities among particles (i.e., when relative velocities are simply due to keplerian shear), new behaviors appear. Since the potential energy cannot be arbitrarily negative (due to the finite radii of both bodies setting a lower bound to their distance) it turns out that even for $\varepsilon = 0$ (perfect sticking in free space), no sticking occurs for particles above a given size, so two large bodies are always separated from each other by the tidal force. More precisely, the condition $E < 0$ defines a critical coefficient of restitution $\varepsilon_{\text{crit}}$, above which accretion is impossible, in the case of zero random velocity:

$$\varepsilon_{crit} = \left[\frac{v_e^2 + 2r_p^{2/3} - 9}{v_e^2 + v_b^2 + 2r_p^{2/3}}\right]^{1/2} \quad (17.13)$$

$\varepsilon_{\text{crit}}$ is displayed in Fig. 17.7, showing a steep fall to 0 for $r_p \geq 0.691$ (the condition for zero numerator in Eq. (17.13), see Canup and Esposito 1995). Note that the case of non-zero random velocity has been treated in Ohtsuki (1993), and it is found that accretion is still be possible in this case, although severely modified by tides, provided the coefficient of restitution is low enough (see the chapter by Schmidt et al. for details).

The sticking requirement $r_p = (a + a')/R_{\text{hill}} \leq 0.691$ means that the sum of the physical radii of the particles must be smaller than the Hill sphere of the agglomerate, that is, the two bodies' mass centers must be gravitationally bound, despite the tidal forces, in order to stick. This requirement prevents accretion between equal size bodies, but allows accretion between unequal size bodies (a small particle at the surface of a big one); indeed, a small particle can be stored

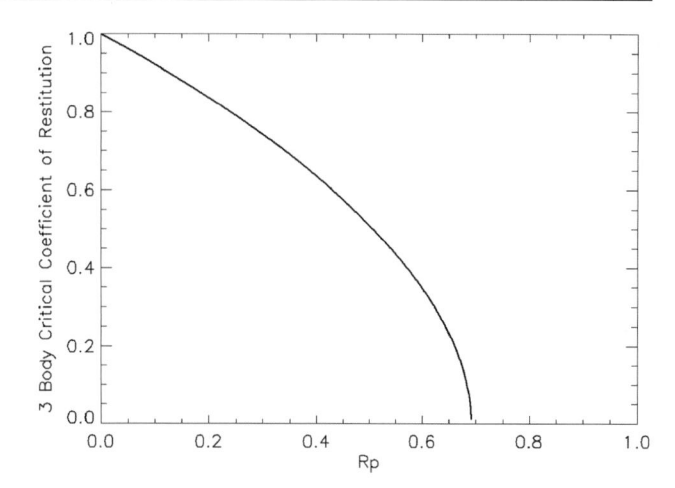

Fig. 17.7 Critical coefficient of restitution for the classical three body gravitational encounter (from Canup and Esposito 1995)

in the empty spaces of the Hill sphere of the big particle. In consequence, we would expect gravitational aggregates to naturally adopt the shape of Hill spheres. Another way of understanding this result is that for growth to occur, an aggregate must be denser than the average density of its own Hill Sphere (ρ_{HS}), given by (assuming a point mass, see Porco et al. 2007):

$$\rho_{HS} \approx \frac{3M_s}{1.59 r^3} \quad (17.14)$$

Finally, tidally modified accretion implies a critical mass ratio, so that m/m' must be larger than some factor, about 100 to 1,000 in the A Ring (Canup and Esposito 1995). In the limit of zero mass ratio, and under the assumption of zero random velocity, this also yields a new definition of the Roche Limit α:

$$\alpha = 2.09\, R\, (\rho/\rho_p)^{-(1/3)} \quad (17.15)$$

Equation 17.15 sets the Roche Limit for ice at $\sim$138,000 km, in very good agreement with the A Ring's outer edge at $\sim$136,800 km. Numerical simulations (Karjalainen and Salo 2004; Karjalainen 2007) have qualitatively confirmed all the previous considerations. Further discussion can be found in the chapter by Schmidt et al.

17.3.2.3 Surface Sticking

Up to now, only gravitational accumulation was considered. As a direct consequence, a gravitational aggregate of any size can grow inside the rings as long as its density remains high enough so that the body is geometrically contained inside its Hill sphere: numerical simulations show that it should grow without bound, as long as it is immersed in the disk (Porco et al. 2007). However, for a body with internal strength (for example, due to intrinsic material strength, surface sticking, etc.) there should be a size vs. distance dependence because the tidal stress may exceed the material strength at some locations. The taking into account of material strength was done in Davis et al. (1984) and in Weidenschilling et al. (1984a) and introduced the notion of "Dynamical Ephemeral Bodies", that is, bodies that grow to a critical size, determined by the balance between their internal strength (gravitational + material strength) and tidal stress, before being torn apart when the tidal forces exceed their internal strength. A recent work (Albers and Spahn 2006) introduced an efficient formalism to take into account adhesive forces, using a viscoelastic model of collisions. When two particles stick on each other, their contact surface is (to first order) a circle with radius a, requiring a surface energy U_s given by:

$$U_s = -\pi \gamma a^2 \quad (17.16)$$

where γ is the surface energy per surface unit. Eq. (17.16) shows that larger contact areas will result in higher surface energy. Solving the collision dynamics shows that above a certain size, smaller grains favorably stick to larger ones. This translates into an adhesion force (Albers and Spahn 2006):

$$F_{ad} = 3\gamma \pi R_{eff}/2 \quad (17.17)$$

where $R_{eff} = R_1 R_2 / (R_1 + R_2)$ is the effective radius of particles with radii R_1 and R_2. The value of γ is poorly known, but is estimated to be about 0.74 N/m for ice (Albers and Spahn 2006). By setting F_{ad} plus the mutual gravitational force equal to the tidal acceleration, a new criterion is established for the stability of a two-particle aggregate with a dependence on the particle size. Let $\mu = R_2/R_1$ and $R = R_1$. A new "Roche Limit" (R_{crit}) can be defined as the distance where a stable aggregate can exist (Eq. 18 of Albers and Spahn 2006):

$$R_{crit}^3 = \frac{24 G M_s \rho R^3 \mu^2 (1+\mu)^3 \left[(k+1)^2 + 2\right]}{(1+\mu^3)\left[27\gamma(1+\mu) + 32\pi G \rho^2 \mu^2 R^3\right]} \quad (17.18)$$

where k is the order of 1 and is proportional to the rotation frequency of the particle. We now see that (i) a dependence on the particle size has appeared and (ii) that R_{crit} also depends on the size ratio μ. R_{crit} is plotted for R = 10 m in Fig. 17.8. We see that for equal size bodies ($\mu = 1$), agglomerates up to R = 10 m can be stable in the D Ring and up to R = 100 m for the B Ring. Conversely, in the D Ring, aggregates with R > 10 m are stable only for μ much larger than 1, corresponding to aggregates of smaller grains sitting on top of much larger ones.

17.3.2.4 Accretion of Small Embedded Satellites?

At this point a natural question is: did the small satellites embedded within or just outside Saturn's main rings

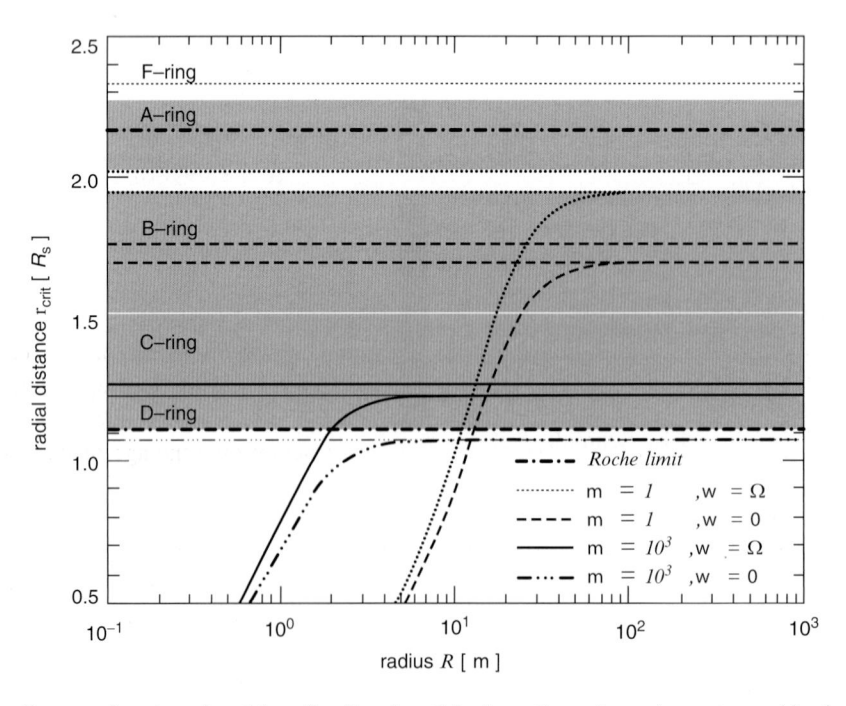

Fig. 17.8 Critical distance R_{crit} as a function of particle radius R and particle size ratio μ. An agglomerate combination is stable in the regime above/left of the corresponding line. Any combination below/right of the line is unstable (from Albers and Spahn 2006)

17 Origin and Evolution of Saturn's Ring System

(e.g. Pan, Daphnis, Atlas) accrete *inside* Saturn's rings? Recent Cassini observations have provided high resolution images and have allowed a determination of the shapes of Pan and Atlas (Porco et al. 2007). They are significantly flatter than Hill spheres, thus suggesting they are not the result of pure tidally modified accretion inside the rings (Charnoz et al. 2007; Karjalainen 2007; Porco et al. 2007). Pan and Atlas exhibit prominent equatorial ridges whose locations are well reproduced in simulations of ring particle accretion onto the surface of pre-existing bodies (Charnoz et al. 2007). This suggests that both Pan and Atlas may hide an inner core, on top of which an envelope of ring particles was accreted. Computation of the gravitational potentials at their surfaces shows that the equatorial ridges extend beyond their Hill spheres (Charnoz et al. 2007), implying that the ridges are maintained by inter-particle adhesion forces, or material strength, rather than pure gravitation.

In conclusion, there are numerous reasons to think that there is ongoing accretion in Saturn's rings; however, its characteristics seem very sensitive to the particles' size, shape and location, as well as the physics at work (gravitation, surface sticking, etc.). There should be a general trend for bigger bodies as a function of distance from the planet. Analytical models are supported, at least qualitatively, by numerical simulations. More extensive accretion may occur in the F Ring region, which in turn, may give us clues to understanding some of the peculiar aspects of this ring (Section 17.5).

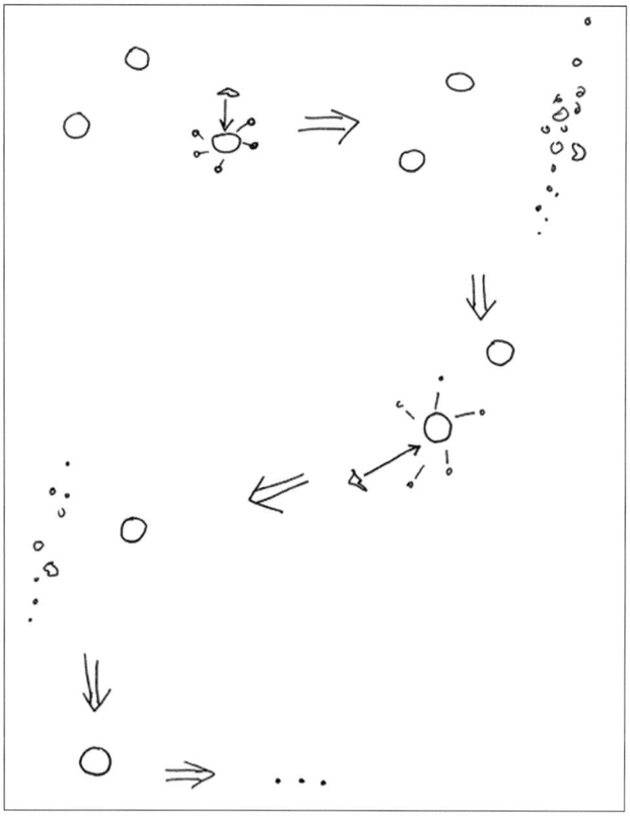

Fig. 17.9 In the collisional cascade, moons are shattered and their fragments further broken to make rings and dust. Eventually, the last moon is destroyed and the original material is completely ground to dust (from Esposito and Colwell 2003)

17.3.3 Collisional Cascade

We have seen in the previous section that under some circumstances, some limited accretion could be possible in the rings, especially near their outer edge. In order for material to recycle back into the rings, a destruction process is needed. Destruction of the biggest bodies, or moonlets, via meteoroid bombardment could lead to a cascade of smaller collisions. A moon shattered by a large impact from an interplanetary projectile would become a ring of material orbiting the planet. Big moons are the source of small moons; small moons are the source for rings. Rings are eventually ground to dust that is lost by becoming charged and carried away by the planet's rotating magnetic field or by atmospheric drag into the planetary atmosphere (where it shines briefly as a meteor), and may produce a "ring of fire" as suggested by Rubincam (2006). This process is called a "collisional cascade" (see Fig. 17.9).

Since the shattering of a moon is a random event, ring history will be stochastic and somewhat unpredictable. The differences between the various ring systems might be explained by the different random outcomes of this stochastic process. Thus, the collisional cascade can provide an explanation for the apparently different ring systems around each of the giant planets. Catastrophic events provide the tempo for creating planetary rings: new rings are episodically created by destruction of small moons near the planet (Colwell et al. 1992, 2000). This disorderly history arises from singular events.

The most serious problem with this explanation is that the collisional cascade uses the raw material (a planet's initial complement of moons) too rapidly. If we imagine we are now looking at the remnants of 4.5 billion years of successive destruction over the age of the solar system, then this process is almost at its end. The small moons that now remain as the source of future rings have a lifetime of only some few hundred million years, based on calculations by Colwell et al. (2000). This is less than 10% of the age of the solar system. Why are we humans so fortunate as to come upon the scene with robotic space exploration, just in time to see the rings' finale? A direct consequence of this process for Saturn's rings is the destruction of the population of moonlets near the outer edge of the rings, like Pan, Daphnis and Atlas.

17.3.4 Ring–Moon Interactions

Whereas ring–moon interactions do not seem directly connected to the question of ring origin and evolution, dynamical interactions between rings and satellites can result in rapid orbital evolution; indeed, this rapid evolution provides the original argument for the possible youth of Saturn's rings. Unfortunately the torque exerted by the ring on a moon is complex to compute and depends on several assumptions and on the ring surface density, which is poorly known. This topic is discussed in detail in the chapter by Schmidt et al., but we briefly outline the relevant physics here.

Consider a ring particle and a satellite orbiting exterior to it, with both bodies on low-eccentricity, low-inclination orbits around Saturn. At an arbitrary location in the rings, the frequency at which the satellite perturbs the ring particle will not be a natural oscillation frequency of the particle (such as its radial (epicyclic) frequency, κ), and the satellite will have no systematic effect on the ring particle's orbit. However, at the satellite's inner Lindblad resonances (ILRs), a more dramatic interaction can take place. The location of these resonances are given by the condition $r = r_L$, where

$$\omega = m\Omega(r_L) - \kappa(r_L). \quad (17.19)$$

In this expression, ω is the forcing frequency due to the moon; m is a positive integer; $\Omega(r_L)$ is the orbital frequency of a ring particle at r_L; and $\kappa(r_L)$ is the radial frequency of the ring particle (Shu 1984). In the simplest case, $\omega = m\Omega_M$, where Ω_M is the moon's orbital frequency. If Saturn were a sphere, its potential would be keplerian, like a point mass, and the orbital and radial frequencies of a ring particle would be equal. Substituting $\omega = m\Omega_M$ and $\Omega(r_L) = \kappa(r_L)$ in Eq. (17.19), we find

$$m\Omega_M = (m-1)\Omega(r_L), \quad (17.20)$$

which is known as an m:(m − 1) resonance because a ring particle completes m orbits for every m − 1 orbits of the satellite. For example, the outer edges of the B Ring and A Ring lie near the 2:1 and 7:6 resonances with Mimas and Janus/Epimetheus, respectively. The Prometheus 2:1 ILR lies in the C Ring, while the Prometheus 6:5, 7:6, ..., 33:32, and 34:33 ILRs lie in the A Ring (Lissauer and Cuzzi 1982; Nicholson et al. 1990; Tiscareno et al. 2007.)

By Kepler's third law, $r_L = r_M [(m-1)/m]^{2/3}$, where r_M is the semi-major axis of the moon's orbit. For large m, successive resonances are separated by $2r_M/(3m^2)$, which is less than 100 km for the ILRs of Prometheus and Atlas in the outermost part of the A Ring. (In reality, Saturn's oblateness causes κ to be slightly smaller than Ω, causing these resonances to fall slightly further from Saturn than in the keplerian case. In general, this shift does not affect the resonance dynamics or relative locations in any important way.)

At the inner Lindblad resonances, there is a torque which transfers angular momentum outward from the ring to the satellite. In the main rings, particularly in the A Ring, the rings' response to the satellite takes the form of a density wave that propagates outward from the resonance. The density wave is a nonaxisymmetric pattern which exerts a back-reaction on the forcing satellite, causing it to recede from the ring. Goldreich and Tremaine (1980, 1982) calculated the torque using a linear theory; but even for strong (nonlinear) waves, the torque appears to be similar to the value given by linear theory (Shu et al. 1985). In the case of a satellite of mass m_s at distance r_M from the center of Saturn that is close to a ring of surface mass density σ at distance r ("close" means that $r_M - r \ll r$), the torque is proportional to $m_s^2 \sigma (r_M - r)^{-3}$, and the timescale for the satellite's orbit to expand is proportional to $(r_M - r)^3 / (m_s \sigma)$. Assuming that they started at the outer edge of the A Ring, Prometheus and Atlas would have taken only some 10–100 million years to reach their current positions. (These timescales are about ten times larger than those given by Goldreich and Tremaine (1982), because the satellites are less massive and the rings' surface density in the outer A Ring is much smaller than they assumed.) Some tens of millions of years in the future, Prometheus will cross the orbit of Pandora, and the satellites are likely to collide with each other (Poulet and Sicardy 2001). We may wonder what would be the result of such a collision: will they merge into a single new satellite or will they be destroyed ? Since they are located close to the Roche Limit for ice (see Section 17.3.2) a variety of outcomes are possible that will mainly depend on their impact velocity. It is possible that a dusty ring would form after the impact, but we have no idea if it would be similar to the actual F Ring or not.

When the Cassini mission was being planned, it was expected that the expansion of the orbits of some of the ring moons due to density wave torques would be directly measurable as a lag in their longitudes that would increase over the course of the mission. However, the dynamics of the ring moons has proven to be quite chaotic; for example, overlapping 121:118 resonances between Prometheus and Pandora cause "jumps" in their orbits at ∼6-year intervals (e.g., French et al. 2003; Goldreich and Rappaport 2003a, b; Renner et al. 2005; Farmer and Goldreich 2006; Shevchenko 2008). As a result, the predicted recession of the moons from the rings has not yet been detected. In principle, the ring moons' orbital evolution could be much slower if the moons were resonantly locked to larger satellites (Borderies 1984), but no such link is known to exist.

Satellites produce many other dynamical effects on rings. These include the formation of gaps at strong isolated resonances, such as the inner edge of the Cassini Division, which is associated with the Mimas 2:1 ILR; gaps due to the overlapping resonances of embedded satellites, such as Pan and Daphnis, which produce the Encke and Keeler Gaps, respectively; "shepherding," or confinement, of narrow ringlets; and the excitation of the random velocities of ring particles in wave regions due to the energy input by the wave. This dynamical "heating" increases the viscosity, and hence spreading rate, of the ring (see Section 17.2.2.2). For details, we refer the reader to the reviews by Cuzzi et al. (1984), Esposito et al. (1984), Lissauer and Cuzzi (1985), and the chapter by Schmidt et al. in this volume.

17.3.5 The Long Term Evolution of Saturn's Main Rings

From the different processes described above, is it possible to draw a picture of the long term evolution of Saturn's rings? This is a difficult task since the interactions between these effects have never been studied. In addition local effects (like accretion) and large scale effects (like meteoroid bombardment) are difficult to couple in the same formalism or simulation. We now try to briefly imagine what the long term evolution of the rings might be like. First, viscous evolution and meteoroid bombardment would likely spread the C Ring (and perhaps the B Ring itself) closer to the planet, on timescales less than 10^9 years. The C Ring could also be replaced by material leaking from the B Ring (through meteoroid bombardment, for example). Small moonlets orbiting at the edge of the rings will migrate to larger radii, and be trapped in mutual resonances, perhaps into horseshoe or tadpole orbits (like Janus and Epimetheus) on timescales of about 10^8 years. If the resonant trapping is ineffective, collision and re-accretion could occur, creating a new generation of bigger satellites from the former generation of smaller ones. Whereas the torque exerted by the rings on Mimas and Janus/Epimetheus are not well determined because their strongest resonances lie near the outer edges of the B and A Rings, respectively, these satellites should migrate to larger radii. It is even possible to imagine a distant future in which the Cassini Division has shifted outward due to the outward movement of the 2:1 Mimas resonance (which establishes the inner edge of the Cassini Division). In addition, since Mimas is in a 4:2 mean motion resonance with Tethys, the ring torque should be transferred simultaneously to both moons. The A Ring is expected to viscously spread out beyond the Roche limit in 10^8 years (due to the recession of Janus/Epimetheus), and new moonlets should form at this location due to the sharp fall of tidal forces at the Roche Limit. The final fate of the rings is very difficult to imagine, at least because we do not really understand how the rings could

ever exist today. Viscous spreading and meteoroid bombardment could be the ultimate mechanisms, acting without end until the rings have finally fallen into the planet and spread outside the Roche Limit. However, so many mechanisms remain poorly understood that these conclusions should be taken with much care.

17.4 Scenarios for Origin of the Main Rings

The origin of the faint rings of the giant planets seems to be quite well understood: with the exception of Saturn's E Ring, they are attributed to the erosion and periodic destruction of km-sized moonlets due to meteoroid bombardment (see Section 17.6). However, due to their very high mass and putative young age, such scenarios cannot apply directly to Saturn's massive ring system. Keeping in mind that we still do not understand how the rings could have formed in the last billion years, we try here to present the details of three scenarios for the origin of Saturn's main rings, and present the pros and the cons for each of them. The first scenario (Section 17.4.1) suggests that Saturn's rings are a remnant from Saturn's primordial nebula, the second one suggests that Saturn's rings are debris from a destroyed satellite (Section 17.4.2), and the last one suggests that they are debris of one or several comets tidally disrupted by Saturn (Section 17.4.3).

17.4.1 Remnant from Saturn's Sub-nebula Disk?

The circumplanetary gas disk or subnebula is a by-product of the later stages of giant planet accretion. Its formation occurs over a period in which the giant planet transitions from runaway gas accretion to its eventual isolation from the solar nebula, at which time planetary accretion ends. This isolation occurs because the giant planet has either tidally opened a well-formed gap, effectively pushing the nebula gas away from itself, or the nebula gas dissipates. The formation of the regular satellites is expected to occur towards the tail end of giant planet accretion when the planet is approaching its final mass, and at a time when any remaining inflow of nebula gas through the giant planet gap is weak and waning. A more detailed discussion of satellite accretion in the context of the combined process of giant planet and circumplanetary disk formation may be found in Estrada et al. (2008; see also chapter by Johnson et al.). If the rings are the remnants of Saturn's subnebula, then their formation must have taken place in the same environment in which the satellites formed, which must be discussed first.

17.4.1.1 Satellite Formation

Satellite formation appears to be a natural consequence of giant planet accretion. The fact that all of the giant planets have ring systems suggests that their origin results from processes common to the giant planets' environments. The observed giant planet satellite systems are quite diverse in appearance and composition, and the rings are no less so. The Jovian rings are ethereal, diffuse, and composed of non-icy material; those of Uranus and Neptune appear to be "dusty", dark and neutral in color (e.g. Cuzzi 1985; Baines et al. 1998; Karkoschka 2001), although some Uranian rings have been found to be spectrally red or blue (e.g., de Pater et al. 2006). On the other hand, the rings of Saturn are prominent and massive, composed mostly of water ice, but distinctly reddish in color, suggesting the presence of organic material (e.g., Estrada and Cuzzi 1996; Cuzzi et al. 2002; Estrada et al. 2003), possibly tholins, in addition to some neutrally colored darkening agent (Cuzzi and Estrada 1998; Poulet et al. 2003). The darkening of the rings over time can most likely be associated with extrinsic meteoroid bombardment (see Section 17.3.3.1). However, if the rings of the giant planets did form during the time of satellite formation, then their different compositions might, in great part, reflect differences in their respective accretional environments. It would also imply that at least some parts of the rings are as old as the solar system, a conclusion that may be at odds with the various lines of evidence that suggest that the rings are geologically young.

17.4.1.2 Implanting the Ring System

The model in which the rings formed in a relatively massive subnebula from essentially the same processes that lead to the formation of the satellites has been referred to as the "condensation" model (Pollack 1975; Pollack et al. 1977; Pollack and Consolmagno 1984). In this model the ring particles form through the sticking of sub-micron sized dust grains which are dynamically coupled to the subnebula gas, and collide at low (size-dependent) relative velocities that can be caused by a variety of mechanisms (e.g., Völk et al. 1980; Weidenschilling 1984b; Nakagawa et al. 1986; Ossenkopf 1993; Ormel and Cuzzi 2007). As grains grow into agglomerates, they begin to settle to the midplane, and may continue to grow through coagulation and coalescence, and as the subnebula cools, through vapor phase deposition. In the satellite forming regions, particles may continue to grow further as they settle by sweeping up dust and rubble (Cuzzi et al. 1993; Mosqueira and Estrada 2003a), eventually growing large enough that they may "decouple" from the gas; however, close to the planet, specifically within the Roche zone where tidal forces begin to overcome gravitational sticking, the final stages of growth are stymied, so that one tends to be left with a population of smaller particles (Pollack and Consolmagno 1984). In addition, growth close to the planet, where dynamical times are quite short relative to the solar nebula, is likely further limited simply as a natural consequence of relative velocities between particles entrained in the gas being too high to allow coagulation beyond some fragmentation barrier. For example, the dynamical times at the A Ring would be comparable to those at ~ 0.01 AU in the solar nebula (a distance at which refractories may not condense due to high temperatures in any case). In spite of these difficulties, the main issue facing this picture is how these small particles can survive long enough for the subnebula gas to dissipate.

In the condensation model, the composition of the rings being primarily icy has been attributed to two things, both of which lead to the formation of ice particles late in the lifetime of the subnebula. The first is that the atmosphere of Saturn, even after envelope contraction, extended beyond the region of the rings. In the early stages of gas accretion long before satellite formation can even begin, the giant planet's atmosphere fills up a fair fraction of its Hill sphere (Bodenheimer and Pollack 1986; Pollack et al. 1996); but, once envelope contraction happens, it occurs fairly rapidly compared to the runaway gas accretion epoch, which lasts $\sim 10^4 - 10^5$ years (Hubickyj et al. 2005), and at a time when the planet is only a fraction of its final mass (Lissauer et al. 2009). The exact timing of the collapse depends on several factors (see Lissauer and Stevenson 2007).

17.4.1.3 Collapse and Cooling of the Envelope

The envelope collapse results notably in two things relevant to rings and satellites: the formation of a compact subnebula disk component due to the excess angular momentum of accreted nebula gas in the envelope prior to collapse (Stevenson et al. 1986; Mosqueira and Estrada 2003; Estrada et al. 2009); and, as implied above, the planet's radius shrinks down to a few planetary radii (e.g., Lissauer et al. 2009, and references therein). Although contraction of the giant planet's radius to its current size takes the remainder of the planet's lifetime, by the time the subnebula gas dissipates, the planet's atmospheric boundary (within which temperatures at this time might still remain too high for condensation of water ice) may lie just within or interior to the radial location of the B Ring (e.g., Pollack et al. 1977). If so, the atmospheric boundary might provide a natural mechanism for explaining an initial sharp edge for the B Ring (which is subsequently maintained by resonant and ballistic transport processes). As Saturn continued to contract to its present size, gas drag may have leeched off particles from the inner B Ring, leading to the first incarnation of the C Ring, and as suggested in Section 17.3, other processes may be able to continually reproduce the short-lived C Ring over the age of the solar system.

Second, the luminosity of Saturn had to decrease sufficiently to allow for the condensation of ice in the innermost regions of the circumsaturnian gas disk. During the beginning stages of satellite formation, when the planet is approaching its final mass and any remaining gas inflow from the nebula wanes, the planet's excess luminosity alone is enough to produce inner disk temperatures that would be initially too high to allow for the condensation of water (and depending on ambient subnebula conditions, condensates more volatile than water, for example, Pollack et al. 1976). As a result, silicates would likely condense first, and be lost to the planet due to gas drag forces and/or settling. In this way, the subnebula may be "enhanced" in water ice and depleted in rocky material, particularly in the ring and mid-sized satellites region (Pollack et al. 1976; Mosqueira and Estrada 2003a, b; also see the chapter by Johnson et al. for a discussion of water–ice enrichment). Because the cooling time for the Saturnian subnebula post-planet accretion ($\sim 10^5$ years, Pollack et al. 1977) may be significantly shorter than the circumsaturnian disk lifetime ($\sim 10^6$–10^7 years, e.g., by photoevaporation, Shu et al. 1993), this enhancement would suggest that late stage accretion would lead to ice-rich satellites and rings. Contrast this with the Jovian subnebula, in which the cooling time is an order of magnitude longer, and comparable to the disk dissipation time (Pollack and Reynolds 1974; Lunine and Stevenson 1982). In this case, water ice may never condense in the inner regions prior to disk dissipation. If the rings are formed as a result of condensation, then the Jovian ring composition would be consistent with this picture.

17.4.1.4 The Role of Turbulence

The satellites are expected to accrete in an environment in which turbulence, driven by the gas inflow, decays (Mosqueira and Estrada 2003a, b). This presumes that there is no source of intrinsic turbulence that could continue to drive subnebula evolution once the gas inflow from the solar nebula is cut off. This assumption is supported by both numerical (e.g., Hawley et al. 1999; Shen et al. 2006), and laboratory experiments (Ji et al. 2006) that question the ability of purely hydrodynamical turbulence to transport angular momentum efficiently in Keplerian disks. Yet, it may be possible that in a scenario in which turbulence decays, some mechanism capable of driving turbulence may persist very close to the planet (Mosqueira and Estrada 2003a), which could provide another reason why the condensation of water ice could be delayed or prevented altogether.

For example, turbulence due to a magneto-hydrodynamic instability (MRI, Balbus and Hawley 1991) may apply, assuming that a combination of density, ionization, and temperature conditions allows for it (e.g., Gammie 1996; McKee and Ostriker 2007), but may be rendered ineffective due to the constant production of dust as a result of the persistent fragmentation of particles. Another potential driving mechanism that may operate is convection (e.g., Lin and Papaloizou 1980). The continuous fragmentation of particles due to both high systematic and turbulence-induced relative velocities within the Roche zone may keep the dust opacity high, allowing for a strong vertical temperature gradient to be sustained. However, if convection drives turbulence, the angular momentum transport may be quite weak (e.g., Stone and Balbus 1996), and furthermore, directed inwards (e.g., Ryu and Goodman 1992; Cabot 1996). This means that a "cavity" may be created close to the planet, which would effectively terminate gas accretion onto Saturn. The consequences of such a cavity on satellite and ring formation have not been explored.

17.4.1.5 Caveats

The rings' being primarily water ice seems to suggest that if they condensed from the subnebula at their present location, then turbulence and the viscous heating associated with it, at least at the tail end of the lifetime of the subnebula, was absent even close to the planet. Moreover, if water ice were allowed to condense, ring particles would still need to be able to survive being lost to the planet by gas drag, which is responsible for clearing wide regions of the circumplanetary disk (the essentially empty regions between the satellites). This difficulty can be overcome by presuming that the condensation of icy ring particles continued up until disk dispersal (Pollack 1975), which may allow for conditions in which gas drag may not have had enough time to clear the remaining solid material before the subnebula dissipated.

Another difficulty is explaining the presence of the small moonlets embedded in the rings, such as Pan and Daphnis. Pollack and Consolmagno (1984) suggested that their growth may have been enabled at resonance locations. Recently, Charnoz et al. (2007) and Porco et al. (2007) have concluded that these moons may have been collisional shards initially one-third to one-half their present sizes that grew to their present-day sizes by accreting ring material. In order to facilitate growth, a core of sufficient density is required in order to be stable against tidal disruption. Their growth is then mostly limited by tidal truncation (i.e., they open gaps), rather than tidal shear. An alternative possibility is that such objects may have been satellitesimals that drifted in via gas drag and were left stranded as the disk dissipated, and also grew to their present sizes by the accumulation of ring material. Note that gas drag and type I migration may have also provided a means for bringing in larger objects such as embryos or small moons close in to the planet where they may have eventually been broken up (see Section 17.4.2.1).

If the rings are much younger than the age of the solar system, then their origin is almost assuredly due to some collisional process. However, if the rings (at least the B ring) turn out to be much more massive than previously thought, then it allows for the possibility that the rings are as old as the solar system (although, it should be noted that by themselves, massive rings do not imply ancient rings), and their origin is open to both collisional and condensation model hypotheses. The spectral shape of the ring material being most similar to that of the Saturnian mid-sized satellites (Barucci et al. 2008; although cf. Buratti et al. 1999 and Section 17.3.1) may tend to support the idea that the materials that compose the rings and satellites were processed under similar conditions (e.g., processed in the subnebula). Given that it may only take a very small fraction of non-icy intrinsic material (<1% if volumetrically mixed; Grossman 1990; Cuzzi and Estrada 1998; Poulet et al. 2003) to give the rings their reddish color, their composition does not appear inconsistent with the condensation scenario. Thus, although physically plausible if the rings are as old as might be allowed by a much higher mass than was previously believed, the condensation model may require that conditions and timing be "just so" to allow for the rings to survive the processes of planet formation and gas disk dispersal.

17.4.2 Debris from a Destroyed Satellite?

It was proposed some time ago (Pollack et al. 1973; Pollack 1975; Harris 1984) that Saturn's rings could be the remnant of the catastrophic disruption of one, or several, of Saturn's satellites. Similarly, Uranus's and Neptune's rings are now thought to be the result of the periodic destruction and re-accretion of small moons (see, e.g., Colwell 1994). The "destroyed satellite scenario" requires that a large satellite, already present in Saturn's Roche zone, is destroyed by some mechanism and its fragments scattered away, thus forming a disk like that of Saturn's rings we see today. Such a scenario has two critical points on which some recent results cast new light: (1) how to bring a massive satellite inside Saturn's Roche Zone; and (2) how to destroy it? We now discuss these points.

17.4.2.1 Bringing and Keeping a Satellite in the Roche Zone

Recent advances in the formation of satellites around gas giant planets suggest possible ways in which satellites may migrate to the planet's Roche zone (Mosqueira and Estrada 2003a, b; Canup and Ward 2006; Estrada et al. 2009). As satellites grow, either by sweep up of dust and rubble followed by gas-drag drift-augmented accretion of smaller satellitesimals and embryos (Cuzzi et al. 1993; Mosqueira and Estrada, 2003a, b), or via binary accretion in a fashion similar to planetary formation (see, e.g., Wetherill 1989; Spaute et al. 1991) occurring under low gas density conditions (Estrada and Mosqueira 2006), several mechanisms can trigger their migration inwards towards the planet. The dominant mechanism for their migration depends on their size, with the transition from gas-drag dominated migration to type-I migration induced by the satellite's tidal interaction with the disk (e.g., Ward 1997) occurring for significantly large bodies (∼500 km). For example, the inner icy Saturnian satellites remain within the gas-drag dominated to transitional regime (gas drag and torque are similar in importance) for the wide range of subnebula gas surface densities that bracket the models of workers mentioned above (e.g., see Fig. 6 of Mosqueira and Estrada 2003a).

For larger satellites, tidal interactions that trigger type-I migration can lead to their rapid infall towards the planet with some even being lost. In the case of a non-turbulent subnebula, a satellite can migrate into the planet if the satellite's perturbation of the disk is insufficient to stall its migration and open a gap in the gas (Mosqueira and Estrada 2003b). On the other hand, if the subnebula is persistently turbulent, stalling and gap opening are prevented and satellites continue to migrate via type-I. In either case, as the subnebula dissipates or is cleared, eventually both the gas drag and type-I migration timescales become longer than the lifetime of the disk, migration stops, and the surviving satellites remain frozen in their final orbital positions. In this context, then, it is not unlikely that a smaller moon that migrated via gas drag or a larger satellite that migrated via type-I is eventually found near or even inside the planet's Roche Zone (see e.g., Fig. 1 in the Supplementary Online Material of Canup and Ward 2006).

If a satellite is able to migrate and be left stranded in the vicinity of Saturn's Roche zone, a key difficulty then becomes one of keeping the satellite close to Saturn: because of Saturn's tides, a satellite above (resp. below) the synchronous orbit, at distance a_s from the center of Saturn, would see its semi-major axis increase (resp. decrease), at a rate given by Murray and Dermott (1999):

$$\frac{da}{dt} = sign(a - a_s)\frac{3k_{2p}m_s G^{1/2} R_p^5}{Q_p m_p^{1/2} a^{11/2}} \qquad (17.21)$$

where a, m_s, m_p, R_p, and G, stand for the satellite's semi-major axis and mass; the planet's mass and radius; and the gravitational constant, respectively, and Q_p and k_{2p} (∼0.3 for

Saturn; Dermott et al. 1988) stand for the dissipation factor and the Love number of the planet. However, the value of Q_p that describes all dissipative processes in the planet's interior is very uncertain. Dermott et al. (1988) suggest $Q_p > 1.6 \times 10^4$. In addition it is now thought that Q_p should depend on the excitation frequency; however, this dependence is still not understood or constrained for giant planets. Integrating Eq. (17.21) yields an ejection timescale of about 1 Gy for a satellite about 3 times Mimas' mass with $Q_p \sim 10^5$ and $\sim$600 My for satellite with five times Mimas' mass. So if the "destroyed satellite" scenario is correct, the destruction must have happened less than 1 Gy after the satellite was brought into Saturn's Roche Zone.

17.4.2.2 Destruction of the Satellite

Despite some common ideas, Saturn's tides alone may not be able to destroy and grind a big satellite penetrating its Roche Zone to dust unless the satellite comes close to the planet's surface. The Davidsson (1999) model of tidal fractures shows that a Mimas-sized satellite is disrupted only below 76,000 km from Saturn's center (Saturn's equatorial radius is about 60,300 km for comparison), well inside the C Ring. On the basis of similar arguments, Goldreich and Tremaine (1982) suggest that a 100 km radius satellite can survive undisrupted at 100,000 km from Saturn's center, well inside the B Ring. For these reasons it has been suggested for some time now that only a catastrophic impact with an impactor coming from outside of Saturn's system would be able to destroy such a big satellite and reduce it to a swarm of meter-sized particles (Pollack et al. 1973, Pollack 1975). Because of Saturn's tides, the swarm would not re-accumulate into a single body. Instead, it would then evolve under the effect of dissipative collisions between fragments that lead inexorably to flattening and radial spreading, as for any astrophysical disk (see, e.g., Brahic 1976).

Extrapolating results from hydrocode simulations of catastrophic impacts, we can get an idea of the impactor size necessary to break a parent body (see, e.g., the work of Benz and Asphaug 1999). Assuming an impact velocity $\sim$34 km/s ($\sim\sqrt{3}$ times the circular orbital velocity at 10^5 km; Lissauer et al. 1988, Zahnle et al. 2003), breaking an icy body with radius $r = 200$ km would require an icy impactor between 10 and 20 km radius (Charnoz et al. 2009). The flux of large bodies around Saturn is not well constrained, and extrapolating from the actual visible flux and size distribution of comets in the solar system (see, e.g., Zahnle et al. 2003) it seems that such an event would be very unlikely in the last 10^8 years of solar system history (Harris 1984). Thus both the "destroyed satellite" scenario and the "tidally split comet" scenario (discussed below) face the same problem: the current cometary flux is not sufficient, by orders of magnitude, to provide enough large bodies passing close enough to Saturn (see Section 17.2.2.4).

One possibility is that the breakup of a satellite occurred at the tail end of formation as the subnebula dissipated, or shortly thereafter by a circumsolar interloper. Models of satellite formation suggest that deep in Saturn's potential well, the mid-sized icy moons may have undergone substantial collisional evolution as a result of impacts with incoming planetesimals during their formation process due to a number of factors (Mosqueira and Estrada 2003a; Estrada et al. 2009). However, a problem with this scenario in which a moon is broken up so close to the completion of the satellite system is that it is unclear how much solid material may have been around at the tail end of planet and satellite formation. Presumably, most of the solid mass in Saturn's feeding zone would have been scattered away, but this is not quantified yet.

Another possibility recently proposed by Charnoz et al. (2009) is that a satellite trapped in Saturn's Roche Zone was destroyed much later, during a period called the "Late Heavy Bombardment" (LHB): a short (a few tens of My) global phase of intense bombardment that may have happened throughout the solar system about 700–800 My after its origin. Whereas the reality of a solar system-wide bombardment at this time is still a matter of debate, the so-called "Nice Model" (Tsiganis et al., 2005) suggests that the primordial Kuiper Belt, originally 100 to 1,000 times more massive than today, was destabilized by Saturn's crossing of the 2:1 mean-motion resonance with Jupiter, triggering a global bombardment in the solar system. Charnoz et al. (2009) find that a 200 km radius satellite could expect about two destructive impacts during the LHB; that is, the probability that the satellite would be destroyed is $1 - e^{-2} \sim 90\%$.

The impact scenario during the LHB has some interesting consequences for the uniqueness of massive rings around giant planets: it requires that the ring progenitor remain inside the planet's Roche Zone for 700–800 My in order to be destroyed during the LHB, which is a constraining requirement because of the rapid radial migration induced by the planet's tides. In particular, a satellite below the planet's synchronous orbit would fall rapidly onto the planet, because the migration rate scales as $a^{-11/2}$ (Eq. (17.21)). Thus any planet with its Roche Limit below its synchronous orbit may not be able to keep a large satellite for 800 My inside the Roche Limit. This is precisely the case of Uranus and Neptune, which do not have massive rings like Saturn. Conversely, Saturn's and Jupiter's synchronous orbits are well below their Roche Limits, so that each could maintain a ring progenitor within its Roche Limit at the time of LHB, if it was originally there at the time of the dissipation of the subnebula. Note, however, that the way the quality factor Q depends on the excitation frequency is not well understood for the moment.

Some other major problems remain with the destroyed satellite scenario. In this model, Saturn's rings would be about 3.8–3.9 to 4.5–4.6 Gy old (if formed either during the LHB or during the satellite accretion phase), apparently conflicting with the putative youth of Saturn's rings, unless some mechanism for renewal exists (see Sections 17.2 and 17.3). In addition, a 200 km moon may contain a substantial fraction of silicates, which is not visible in Saturn's rings (see, e.g., Nicholson et al. 2008). A mechanism to eliminate, or efficiently hide, the silicates still remains to be identified. Alternatively, it may be the case that a broken-up (∼200 km) moon may have been predominantly icy. Such a notion is not unfounded, given the wide range of densities observed in the Saturnian mid-sized moons. Furthermore, recent models identify and allow for the possibility that there may have been mechanisms at work in the Saturn subnebula that lead to the enhancement of water ice (see Section 17.4.1), especially in the mid-sized satellite region. This is still an open question.

17.4.3 Debris from Tidally Disrupted Comets

Dones (1991) proposed that Saturn's rings could have arisen through the tidal disruption of a large "comet," or outer solar system planetesimal. A similar "disintegrative capture" model for the origin of the Moon had been proposed by Öpik (1972) and Mitler (1975) prior to the rise of the giant impact theory for lunar origin (Fig. 17.10). The disintegrative capture scenario relies on the difference in gravitational potential across an interloper, that is, a small body that undergoes a very close encounter (well within the classical Roche limit) with a planet. Consider a small body of radius r and mass m_p, that undergoes a parabolic encounter with a planet (i.e., its velocity with respect to the planet at "infinity," v_∞, is equal to zero). Ignore energy dissipation within the body during the encounter. If the body instantaneously breaks into a large number of fragments at its closest approach distance to the planet's center, q, half the fragments (those on the hemisphere facing the planet) will have speeds less than the local escape velocity, $(2GM_{sat}/q)^{1/2}$, from Saturn, and hence will become weakly bound to the planet. For the more realistic case of $v_\infty > 0$, smoothed-particle hydrodynamics simulations, originally carried out for stars tidally disrupted by black holes, indicate that a fraction

$$f = \max\left[\frac{1}{2} - \frac{\left(\frac{1}{2}v_\infty^2 + \frac{Gm_p}{Q}\right)}{1.8\Delta E}, 0\right] \quad (17.22)$$

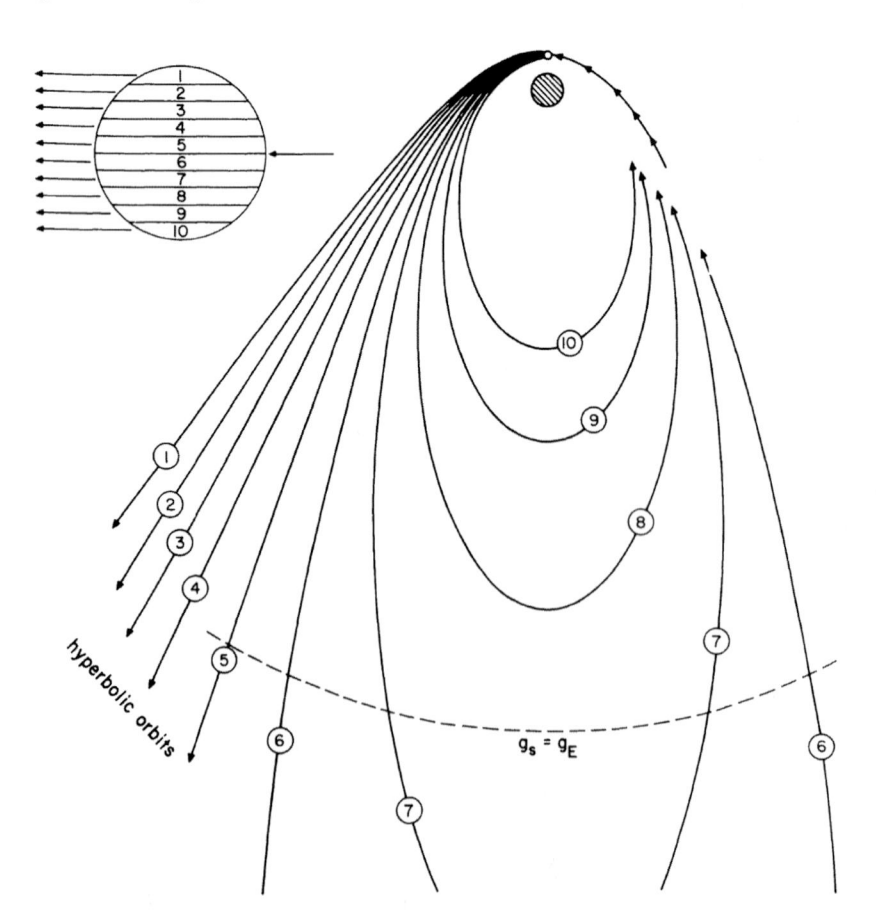

Fig. 17.10 Diagram illustrating the process of disintegrative capture (from Wood and Mitler 1974; Wood 1986)

of the mass of the interloper is captured into orbits around the planet with apocenter distances less than Q. (Immediately after capture, Q is typically hundreds of Saturn radii.) In this expression, $\Delta E = Gm_p r/q^2$ represents the difference in gravitational potential per unit mass across the interloper. The larger ΔE is, the larger the fraction of material that can be captured. Thus, encounters by large interlopers with small pericenter distances (q only slightly greater than the planet's radius) are most effective in this scenario.

Dones (1991) assumed that Chiron, which has $v_\infty = 3$ km/s with respect to Saturn, was a typical Saturn-crosser. At that time, Chiron was the only large Saturn-crosser known. Small bodies with orbits in the Saturn–Neptune region are now known as Centaurs. Even now, only three large Saturn-crossing Centaurs are known (Zahnle et al. 2003; Horner et al. 2004): Chiron, Pholus, and 1995 SN55. Specifically, Dones (1991) carried out Monte Carlo simulations of Centaurs encountering Saturn, using Eq. (17.22), along with the assumptions that (1) Chiron has a mean radius r = 90 km; (2) Centaurs have internal densities of 1 g/cm^3; (3) the distribution of v_∞ for Centaurs encountering Saturn is uniform between 0 and 6 km/s; (4) the cumulative size distribution N(r), i.e., the number of bodies with mean radius larger than r for Centaurs is a power law with an index of 2, 2.5, or 3 (i.e., $N(r) \propto r^{-2}, r^{-2.5}$, or r^{-3}); and (5) these size distributions extend up to 250, 500, or 1,000 km. Dones estimated that, on average, Centaurs with r > 90 km pass through the "Roche zone" between Saturn's surface and the classical Roche limit once every 28 My. Since the mass of the rings corresponds to (at least) a Mimas-sized body (r = 200 km) with unit density, and at most half of the interloper can be captured (Eq. (17.22)), possible ring progenitors (those capable of depositing the mass of Mimas in a single encounter) pass through the Roche zone once every 200–600 My. However, most of those bodies are moving too fast or do not come close enough to Saturn to leave mass in orbit around the planet.

Dones found that 0.1–1 Mimas masses were captured in 4 Gy at current encounter rates, with most of the capture taking place in rare events by bodies with large r, small q, and small v_∞. He argued that the captured fragments would collide and, by conservation of angular momentum, form a ring near two Saturn radii. If the ring progenitor was differentiated, disintegrative capture predicts that material from its icy shell is most easily captured, possibly explaining the icy composition of the rings. However, the rate of ring formation by disintegrative capture is too low in the last billion years to be a likely way to form "young" rings, and there is no obvious reason why Saturn, rather than another giant planet, should have massive rings.

The realization that comet Shoemaker-Levy 9 had been tidally disrupted by Jupiter prior to its fatal descent into the planet in 1994 stimulated efforts to reproduce the "string of pearls" morphology of the comet's fragments. Asphaug and Benz (1996) found that only a strengthless model for the comet, with a pre-disruption diameter of $\sim$1.5 km and a material density of $\sim$0.6 g/cm^3, could match the observations. Asphaug and Benz also pointed out that because of its low density, Saturn was the giant planet *least* effective in tidally disrupting small bodies.

Dones et al. (2007) revisited the disintegrative capture model, using a modified version of the code that Asphaug and Benz had used to model the disruption of Shoemaker-Levy 9. The main improvement over Dones (1991) was that the disruption and mass capture were modeled explicitly, rather than relying on results in the literature (Fig. 17.11). Overall, the results of Dones et al. (2007) confirm those of Dones (1991), though it is not known whether Centaurs hundreds of km in size are rubble piles as they assume.

Nonetheless, disintegrative capture remains unlikely to have produced a ring system within the last billion years. Dones et al. (2007) used the latest estimates for the population and size distribution of Centaurs, which are generally thought to arise in the Scattered Disk component of the

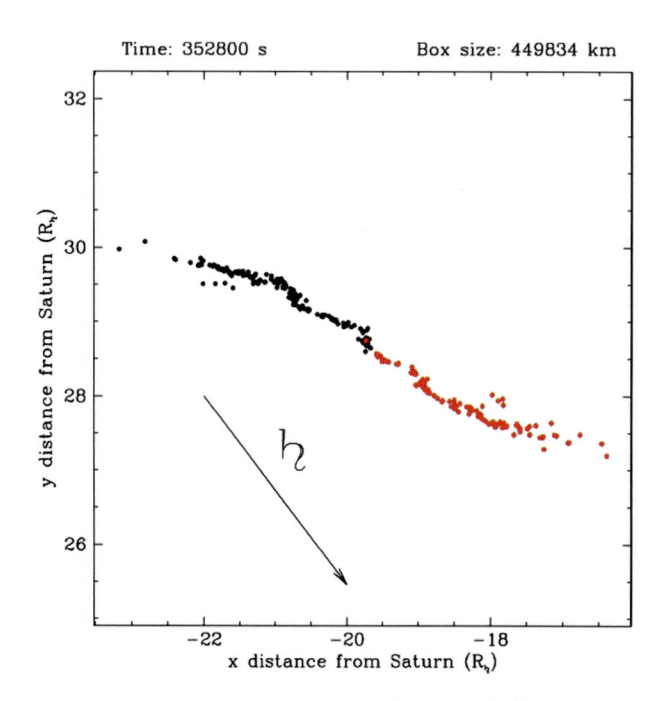

Fig. 17.11 Panel from simulations by Dones et al. (2007) of a large Centaur tidally disrupted by Saturn after it passed within 1.1 radii of the planet's center. Several days after the encounter, the fragments are beyond Titan's orbit. The black dots represent fragments that will escape the Saturn system. The red dots represent fragments bound to Saturn; these fragments were on the hemisphere of the Centaur facing Saturn at the time of disruption. About 40% of the Centaur's mass was captured. The objects on bound orbits have semi-major axes around Saturn of hundreds of planetary radii, in the region of Saturn's irregular satellites. Plausibly, collisions between the fragments will ultimately result in a ring within Saturn's Roche Zone, but this process has not yet been modeled

Kuiper Belt. As with large Kuiper Belt Objects, the size distribution of large Centaurs appears to be steep ($N(r) \propto r^{-q}$, with $q \geq 3$), implying that large Centaurs are rare. Since $q = 3$ was the largest value considered by Dones (1991), and there appear to be only a few Chiron-sized bodies on Saturn-crossing orbits, the rates calculated by Dones (1991) may have been too high. Thus making Saturn's rings recently is difficult, assuming that the current population of Centaurs is representative of the long-term average.

Charnoz et al. (2009) model the influx of planetesimals through the Hill spheres of the giant planets during the Late Heavy Bombardment in the context of the Nice model (see, e.g., Tsiganis et al. 2005). They find that are more than enough interlopers for disintegrative capture to have occurred at each of the giant planets, but find, as do Asphaug and Benz (1996), that Saturn is the planet least likely to tidally disrupt Centaurs. They argue that collisional disruption of a satellite within Saturn's Roche Zone is a better way to explain the origin of the rings, though the rings' pure ice composition is hard to understand in this scenario. This topic is discussed further in Section 17.4.2.

17.4.4 A Conclusion?

After having presented the pros and cons of the different options in detail, is it possible to draw a conclusion about the origin of Saturn's main rings? As shown in Sections 17.2 and 17.3, for the moment we are still stuck in the longstanding paradox of the apparent youth of the rings, but with no means to form them recently.

Still, the situation is not desperate: our understanding of the origin of other ring systems (those of Uranus, Neptune and Jupiter) has improved, and each time, the collisional origin was the best working scenario, and perhaps it is again the case for Saturn, whereas we still are missing some important observations. Because rings are bombarded with material from all over the solar system, it seems that, unexpectedly, better understanding the rings' origin requires a better understanding of the full evolution of the solar system. For example, the recent findings about the Late Heavy Bombardment (Tsiganis et al., 2005), while still controversial, open new possibilities for understanding the rings' origins. Another possibility is that the rings formed just after the accretion of planets (~10–100 My after the origin of the solar system), as we know the solar system was dominated by giant collisions between planetary embryos during this epoch. Unfortunately we do not have quantitative measurements about this period.

Twenty years after the Voyager era, the conclusion that rings are a rapidly evolving system still holds. Perhaps a solution lies in the possibility of recycling material, or that the meteoroid flux is much lower than we have assumed by orders of magnitude, or, finally, that some process somehow slowed down the evolution of the ring system. Since Saturn's main rings are more like a granular medium, whose physics is still poorly understood, we cannot dismiss the fact that we have still a lot to learn about the physics of particulate systems.

So in conclusion, the question of the rings' origin is still open, although formation mechanisms are now better understood.

After having extensively discussed the origin of Saturn's main rings, we now turn to diffuse rings: the F Ring and the E and G Rings. These rings are subject to different processes and their origin and evolution may be somewhat better constrained than those of the main rings.

17.5 Saturn's F Ring: Processes and Origin

Saturn's F Ring is one of the most dynamic objects in the solar system. It was first detected by the Pioneer 11 imaging experiment in 1979 (Gehrels et al., 1980). This narrow ring lies 3,400 km beyond the A Ring's outer edge, precisely at the classical Roche limit of Saturn for ice (see Section 17.3). A year later the ring appeared with much greater clarity under the scrutiny of the Voyager 1 cameras, which revealed a remarkable wealth of longitudinal structures, including clumps, kinks, and the so-called "braids" (Smith et al. 1981, 1982). Twenty-five years later, Cassini has provided high resolution images, maps and "movies" of the F Ring, confirming the abundance of delicate dynamical structures. In addition to the visible ring, sharp drops in the flux of magnetospheric electrons detected by Pioneer 11 suggested the presence of a nearby moonlet belt (Cuzzi and Burns 1988). Several moonlets that could be members of this putative belt have been found in Cassini images (Porco et al. 2005; Murray et al. 2005, 2008). The F Ring is also famous for its "shepherding moons" Pandora and Prometheus, which were believed initially to confine the ringlet radially (Goldreich and Tremaine 1982). However, things appear more complex today, and it is not clear at all whether this mechanism is really responsible for the F Ring's narrowness.

All these elements make the F Ring region a complex environment, where the coupling of various physical processes implies a rich evolution.

17.5.1 Characteristics of the F Ring Relevant for Its Origin and Evolution

A detailed description of the F Ring is provided in the chapter by Colwell et al., but here we emphasize the aspects relevant for this chapter. The F Ring is an eccentric ringlet, with about 2,000 km full radial width but radially sub-divided into a main bright central component (the "core") about 20 km wide in Cassini ISS images, which is surrounded by non-continuous strands of material (Showalter et al. 1992; Murray et al. 1997; Porco et al. 2005). Showalter et al. (1992) found that the ring was 98% fine dust. The outer strands are also composed of dust, and they may form a single spiral structure (Charnoz et al. 2005) due to keplerian shear. These structures seem to be embedded in a continuous sheet of faint material. The core is a moderately high optical depth structure (a few tenths, Showalter et al. 1992), whose orbit seems to be accurately described by the Bosh et al. (2002) model of an eccentric, inclined ringlet precessing under the influence of Saturn's oblateness (Murray et al. 2005). A model of the F Ring occultation data suggests that its opacity is dominated by centimeter-sized particles (Showalter et al. 1992), which may hide within a 1-km wide inner core seen in some recent Cassini images (Murray et al. 2008). Recent occultation data (Esposito et al. 2008a) have also revealed the presence of a population of extended (30 m to 1.5 km) bodies that could be either solid or clumpy aggregates (owing to their translucent appearance). Very high resolution images of the F Ring core (Murray et al. 2008) reveal a wealth of km-scale dynamical structures that still remain to be explained. Conversely, the envelope and strands that surround the core are made of micrometer-sized dust (which is especially bright at high phase angles) with a steep particle size distribution (index of the differential size distribution q $\sim$ 4.6, Showalter 1992). In Showalter (1998, 2004) the largest transient bright features were interpreted as dust clouds generated by meteoroid bombardment, while other authors suggest that local collisional activity involving moonlets (or clumps) could be the cause of these events (Poulet et al. 2000a; Barbara and Esposito 2002; Charnoz 2009). Recent Cassini imaging data seem to support the latter model (Charnoz et al. 2005; Murray et al. 2008; Charnoz 2009). While meteoroid bombardment must also be an active process, it is difficult to quantify for the moment.

The F Ring is much more extended vertically than Saturn's main ring system, and dominates the edge-on brightness during the Earth's crossing of the ring plane (Nicholson et al. 1996; Poulet et al. 2000a; Charnoz et al. 2001). Photometric models (Poulet et al. 2000b) suggest a vertical extent of about 21 km, thus implying an ongoing dynamical excitation.

The F Ring's mass is unknown. Showalter et al. (1992) obtains a dust mass of 2–5 $\times$ 10^9 kg from the derived size distribution of the envelope, whereas arguments concerning the survival of the core against meteoroid bombardment over the age of the solar system leads to a core mass estimate equivalent to a 15–70 km radius body, roughly the size of Prometheus or Pandora. The apparent alignment of the F Ring strands with the F Ring core (Murray et al. 1997) suggested that the F Ring core is as massive as Prometheus or Pandora in order to counter-balance the differential precession between the core and the strands. However new models of the strands and jets (Charnoz et al. 2005; Murray et al. 2008) suggest they are transient structures that still require a population of moonlets or clumps to be produced (Murray et al. 2008; Charnoz 2009), thus converging to the same conclusion as Cuzzi and Burns (1988) that suggest that the moonlet population re-accretes the dusty material, thus extending its lifetime.

In addition to the population of clumps inside the F Ring core found in the UVIS occultations, a population of moonlets, or clumps, exterior to the ring's core has been seen in several ISS images (Porco et al. 2005; Murray et al. 2005, 2008). However, their orbits were difficult to constrain, as the identity of each body is not easily determined due to multiple orbital solutions being possible and the changing appearance of these objects (J. Spitale, N. Cooper, 2009). The most famous member of this family is S/2004 S6 (Porco et al. 2005), whose orbit seems to cross the F Ring core at the precise location where the strands appear to originate (Charnoz et al. 2005).

17.5.2 Processes at Work in the F Ring

Such a complex environment should have a rapid dynamical evolution and accordingly, there are strong suspicions that Saturn's F Ring could be young unless a replenishment mechanism exists. Micron-sized particles in the F Ring envelope and strands should leave the ring in far less than 10^6 years due to Poynting-Robertson drag (Burns et al. 1984). It has been suggested (Showalter 1992) that massive moonlets could maintain the material on horseshoe orbits despite the Poynting-Robertson decay. However this does not seem to be confirmed by numerical simulations (Giuliatti Winter et al. 2004; Giuliatti Winter and Winter 2004). A major factor of dynamical evolution is surely the orbital chaos induced by Prometheus and Pandora, and possibly by the nearby population of moonlets. Since the seminal work of Goldreich and Tremaine (1982), who proposed that the F Ring/Prometheus/Pandora system could be stable due to the so-called "shepherding mechanism" relying on an exchange of angular momentum and collisional dissipation, later analytical studies and numerical simulations (see, e.g., Goldreich and Rappaport 2003a, b; Winter et al. 2007) have shown that the full system could be globally unstable and that orbital chaos is active due to complex interactions between Prometheus and Pandora. Giuliatti Winter et al. (2007) show that moonlets in the F Ring region are rapidly placed on chaotic orbits and scattered over more than 1,000 km in only 160 years, implying that 20 km wide structures like the F Ring core and strands should be young. Charnoz et al. (2005)

find that the F Ring strands should disappear in about 1,800 orbits, only ∼3 years, for the same reasons. In addition, the evolution of Pandora and Prometheus's semi-major axes due to tidal interactions with Saturn's rings should lead to an encounter of Prometheus with the F Ring in less than 10^7 years (Poulet and Sicardy 2001). Defying all these arguments, the presence of a narrow, uniformly precessing core within such a chaotic region remains unexplained (Bosh et al. 2002; Murray et al. 2008). Due to meteoroid bombardment, big bodies should erode by $\sim 3 \times 10^{-5}$ cm/year (Showalter 1992), so that bodies smaller than 100 m should be much younger than the solar system. All these destruction mechanisms are tempered by the possibility that substantial accretion could be expected in Saturn's F Ring (see Section 17.2.2.4). Barbara and Esposito (2002) show that the competition between accretion and fragmentation should result in a bimodal size distribution. While such a distribution has not been observed in UVIS data (Esposito et al. 2008a) we note that the average number of bodies larger than 1 km diameter seems to be correctly predicted by this model (Barbara and Esposito 2002).

17.5.3 Origin and Evolution of Saturn's F Ring

It is not an easy task to draw a coherent picture of the F Ring's origin with such a diversity of antagonistic mechanisms. However, we note that different models have been proposed with a similar underlying principle: the progressive erosion of a population of big bodies by a diversity of mechanisms such as meteoroid bombardment and collisional evolution.

Cuzzi and Burns (1988) proposed that the dusty component of Saturn's F Ring is a cloud of particles released by the collisional erosion of an underlying population of 0.1–1 km sized moonlets. They developed a detailed model in which moonlets spread over a region ∼2000 km wide collide and release surface material that gives birth to the F Ring itself. The material is subsequently re-accreted by the same moonlets. They computed that a population of 100 m-sized moonlets must represent a total optical depth about 10^{-4} (Cuzzi and Burns 1988). This model is very similar to the modern picture of a circumstellar debris disk like Beta Pictoris (see, e.g., Lagage and Pantin 1994), in which unseen belts of small bodies, stirred by one or several planets, produce a dusty disk that is visible thanks to its infrared excess. Showalter (1992) pointed out that the existence of an F Ring core only ∼1 km wide (recently confirmed by Murray et al. 2008) is hard to explain in the Cuzzi and Burns model. In addition, Voyager data revealed the presence of some clumps suddenly appearing and dissipating. Showalter (1992) proposed that the population of clumps and moonlets embedded in the core is ground to dust via meteoritic bombardment impacting the ring at high velocity (∼30 km/s), as in other dusty rings of the solar system. Note that the Showalter (1992) model is not incompatible with the presence of moonlets suggested by Cuzzi and Burns (1988); it merely proposes an alternative mechanism for dust production. After the Cassini high-resolution images of the F Ring and the discovery of the population of nearby moonlets (Porco et al. 2005), it was proposed (Charnoz et al. 2005; Murray et al. 2008; Charnoz 2009) that the population of moonlets exterior to the core (among which S/2004 S6 is a member) regularly collides with the population inside the core, releasing material whose orbital motion form structures described as "spirals" (Charnoz et al. 2005) or "jets" when they are still young (Murray et al. 2008; Charnoz 2009). In this view (close to the Cuzzi and Burns model), these two populations interact physically and gravitationally and create the transient and dusty structures surrounding the F Ring.

So it seems that the F Ring's origin is linked to the origin of the population of km-sized moonlets. Cuzzi and Burns (1988) proposed that one or several small moons were destroyed in the past, whose fragments are slowly eroding today. Conversely, Barbara and Esposito (2002) suggest that there is on-going accretion in the F Ring core, regularly producing clumps and moonlets, whose subsequent collisional erosion produces the F Ring. UVIS stellar occultation data (Esposito et al. 2008a) have revealed the presence of a population of clumps in the core, among which some are translucent, like loose gravitational aggregates, suggesting that accretion is active in the densest regions of the F Ring (see Section 17.2.3 and Fig. 2). However, none of these models at the moment can be reconciled with the "hot" structure of the ring (i.e., the large random velocities) induced by the strong nearby perturbations (see chapter by Colwell et al.) with the confirmed presence of a narrow and thin core that clearly must be a "cold" structure (Showalter 2004; Murray et al. 2008). Is self-gravity the solution? We have no clue for the moment.

17.6 Diffuse Rings: Processes and Origins

After having discussed the origin and evolution of Saturn's F Ring, we now turn to the origin and evolution of the E and G Rings. While they share similarities with the F Ring in terms of their optical properties, the E and G Rings may have a different origin and fate due to their erosional origins and the importance of non-gravitational forces and resonances in their dynamics. The chapter by Horanyi et al. provides a detailed discussion of these rings, while here we briefly review Cassini's contributions to our understanding of how the E and G Rings are sustained.

The E and G Rings, like the D and F Rings and a number of other structures, can be distinguished from Saturn's better known main rings not only by their low optical depths, but also by the fact that they dramatically increase in brightness at high phase angles. All these rings therefore must contain a sizable fraction of particles that are sufficiently small to efficiently scatter light by as much as several tens of degrees via diffraction. Such particles are of order 1–10 μm in size, and are therefore extremely sensitive to non-gravitational processes that can both erode and disperse material on relatively short timescales (Burns et al. 2001). For example, sputtering by energetic particles in Saturn's magnetosphere can destroy particles of this size in a few thousand years, while charge variations can potentially cause orbital evolution in even less time. The visible dust therefore almost certainly needs to be continuously re-supplied to these rings by various sources.

The source of the most extensive of the dusty rings, the E Ring, was suspected to be the satellite Enceladus long before Cassini reached Saturn. After all, Enceladus' orbit is located near the peak in the E Ring's brightness. However, prior to Cassini's arrival it was not clear how Enceladus generated the E Ring. The blue color of this ring in backscattered light observed by Earth-based telescopes, which strongly contrasts with the neutral or red color of many other dusty rings (Nicholson et al. 1996; de Pater et al. 1996, 2004; Bauer et al. 1997) suggested that this ring had an unusually narrow or steep particle size distribution (Showalter et al. 1991), such that 1-μm particles were emphasized. Such an unusual size distribution can be explained if non-gravitational forces are involved in generating and/or dispersing material in this ring. For example, Hamilton and Burns (1994) developed a model in which the E Ring was self-sustained by the impacts of small grains in the ring onto various satellites. In this scenario, solar radiation pressure and electromagnetic interactions between the particles in the ring and the surrounding plasma preferentially gives particles of a certain size large eccentricities, and so only particles of that size would be able to populate an extensive ring or to impact moons with enough speed to yield additional E Ring material. One issue with such models was that Enceladus has a relatively high escape speed, and yet it was not obvious why its particle yield was sufficiently large to support the E Ring.

Cassini images finally showed how Enceladus is able to supply the E Ring: its south pole is geologically active, generating a plume of small particles that extends thousands of kilometers above its surface (Porco et al. 2006). While the physical processes responsible for generating this plume are still being debated (Porco et al. 2006, Kieffer et al. 2006, Kieffer and Jakosky 2008), the connection between the plume and the E Ring is secure. Both contain ice-rich particles a few microns across (Kempf et al. 2008; Postberg et al. 2008; Hedman et al. 2009b; Brilliantov et al. 2008). In fact, the dynamics of the plume particles can provide a natural explanation for the E Ring's peculiar size distribution. For the plume particles to populate the E Ring, they must be launched from the surface with sufficient velocity to escape Enceladus' gravity. Physical models of the condensation and acceleration of solid matter within cracks indicate that larger grains reach the surface with lower average velocities (Schmidt et al. 2008), and so the size distribution of particles that make it into the E Ring is expected to be skewed towards smaller particles. Detailed analyses of the plume's spectral and photometric properties have recently confirmed that different sized particles are launched from Enceladus with different velocity distributions (Hedman et al. 2009b), and further studies of remote-sensing and in-situ measurements should provide additional constraints on the dynamics of how the plume supplies the E Ring. For more information about how the particles supplied by Enceladus are dispersed throughout the E Ring, see the chapter by Horanyi et al.

In contrast to the E Ring, the origins of the G Ring were obscure before Cassini. The limited in-situ measurements and the observed spectral and photometric properties of this ring were consistent with models of collisional debris (Showalter 1993; Throop and Esposito 1998; de Pater et al. 2004). The visible dust in this ring therefore appeared to be generated by collisions among and into a suite of larger (>1 m) bodies located in the region of the G Ring (Canup and Esposito 1997). However the G Ring's brightness peaks around 168,000 km from Saturn's center, over 15,000 km from the nearest known satellite. There was no clear explanation why source bodies would be concentrated in this particular location.

A combination of in-situ and remote-sensing data from the Cassini spacecraft has clarified the source of the G Ring (Hedman et al. 2007a). Images taken by the cameras reveal a localized brightness enhancement at 167,500 km from Saturn's center, near the inner edge of the G ring. This arc has a radial full-width at half-maximum of ∼250 km and extends over only 60° in longitude. Cassini has imaged this arc multiple times over the course of the nominal mission, and these data show that the arc moves around Saturn at a rate consistent with the 7:6 corotation eccentricity resonance (CER) with Mimas. Numerical simulations confirm that the gravitational perturbations due to this resonance are able to confine material into an arc.

In-situ measurements of the charged particle environment by the MIMI instrument demonstrated that this arc does not just contain the dust visible in most images. MIMI detected a ∼50% reduction in the energetic electron flux when it passed through magnetic field lines that thread through the arc. Such deep absorptions were not observed on other occasions when the spacecraft flew over longitudes in the G Ring far from the arc, and the radial width of this absorption was comparable to the width of the arc observed in images, so this absorption can reasonably be attributed to material trapped in the arc. The magnitude of the absorption indicates that this arc contains a total mass between 10^8 and 10^{10} kg. This greatly exceeds the observed amount of dust, and is therefore direct evidence that larger bodies exist in the arc. The existence of at least one such object has since been confirmed, thanks to images that appear to show a small moonlet embedded within the arc (Porco et al. 2009).

Based on the above information, Hedman et al. (2007a) have suggested that the arc contains a population of large particles that may be the remnants of a disrupted moonlet. The 7:6 Mimas CER prevents the material from dispersing and therefore helps explain why a relatively dense ring exists at this location. Collisions into and among these larger bodies produce the dust that forms the visible arc. Unlike larger objects, these small dust grains are subject to significant non-gravitational forces. In particular, interactions between the dust grains and the local plasma (which co-rotates with Saturn's magnetosphere) can accelerate these small particles, enabling them to leak out of the arc and drift outward to form the rest of the G Ring. As the dust grains drift away from Saturn, processes like sputtering and micrometeorite bombardment steadily erode them, causing the density and brightness of the ring to steadily decline with radial distance from the arc. Now that ultra-faint rings and resonantly-confined arcs of debris have been found associated with several small moons of Saturn (Hedman et al. 2009a), comparisons among these different moon-ring systems should allow such models to be more thoroughly tested and therefore yield a better understanding of the origins and evolution of the G Ring.

17.7 Conclusions

Thanks to Cassini findings, we now seem to have quite a clear picture of the origin of the E and G Rings: the E Ring is fed by Enceladus' plumes and the progressive destruction of a moonlet feeds the G Ring. The question of the F Ring's origin is still unsolved, but the discovery of a population of nearby moonlets, as well as high resolution images of the core, seem to qualitatively confirm the model of Cuzzi and Burns (1988), according to which the F Ring is the result of the collisional evolution of a moonlet belt, whose origin is still to be understood.

Conversely, the question of the origin of Saturn's main rings is still wide open more than 25 years after the flybys of Voyager 1 and 2. Since then, new observational data and theoretical results have brought new insights both into ring dynamics and into the history of the solar system, opening new possibilities for understanding the main rings' origin and long term evolution. These developments include new models of the formation of giant planets and their satellites, local numerical simulations of planetary rings, and the discovery of the "propellers" in the A Ring. The Voyager-era idea of young rings created recently by disruption of a Mimas-size moon (e.g., Esposito 1986) seems untenable, since the likelihood of such an event in the last 100 My is tiny. Still, no fully satisfactory answer has been found, and the apparent contradiction between the rings' apparent youth and the difficulty of forming them in the last billion years still holds.

One of the strongest constraints on the rings' age – the darkening by meteoroid bombardment – might be solved if the rings were more massive than previously thought or, alternatively, if the meteoroid bombardment was much smaller than previously estimated. However, as mentioned in Section 17.3.1, the meteoroid flux may also have been much higher in the distant past. The idea of recycling material, embodied in the concept of "Cosmic Recycling" still needs modeling and investigation. The evidence for ancient rings and continued recycling is indirect, but more than just intriguing. It includes the larger optical depths measured by Cassini occultations, the clumpiness seen in the rings that implies that the mass of the rings may have been underestimated, and the variety of structures visible in the outer A Ring and in the F Ring.

Concerning the mechanism responsible for the implantation of the ring system, new theoretical results on the Late Heavy Bombardment open the possibility that either the destruction of a satellite or the tidal splitting of passing comets could have taken place about 700–800 My after the origin of the solar system. If either of these is the correct explanation for the origin of the rings, the rings would be about 3.8–3.9 billion years old, still too old given the pollution and evolutionary processes at work in the rings, and a recycling mechanism would still be necessary.

Data that would help further progress on the question of rings' origins are:

1. The mass of Saturn's main rings, in particular for the B Ring.
2. The meteoroid flux at Saturn. This would give almost an absolute measurement of the rings' age by at least two independent methods (structural and photometric appearances).
3. Images and spectra of individual particles, their size distribution and how it depends on the distance to Saturn. Such data would help to better constrain evolutionary processes and in turn, the rings' origin. In addition, we still do not know whether the silicates are really absent from Saturn's rings or if they are deeply buried inside the ring particles. An answer to this question would be critical and would provide strong constraints on the scenario of ring formation, since we do not know of an obvious way to eliminate silicates from Saturn's rings.

Data #1 and #2 could be provided during the Cassini extended-extended mission after 2010, whereas datum #3 could be only obtained with an in situ mission in Saturn's rings (see the chapter by Guillot et al.). A final important source of data that may be available in the moderately near future would be the occurrence of ring systems around extra-solar planets, which would allow us to test whether massive rings are either a natural or an extraordinary outcome of planetary formation. These data would also be invaluable for building a scenario for the origins of Saturn's main rings. The Corot and Kepler missions will hopefully bring new constraints about the presence of rings around extra-solar giant planets, and in turn, give new insight into the origin of Saturn's rings.

Acknowledgements We wish to thank N. Albers, M. Showalter and K. Ohtsuki for valuable comments. We also thank our three referees, whose comments improved the quality of this chapter. LD thanks the NASA Planetary Geology and Geophysics program for support.

References

Albers N., Spahn F. The influence of particle adhesion on the stability of agglomerates in Saturn's rings. Icarus 181, 292–301 (2006)

Altobelli N., Grün R., Landgraf M. A new look into the Helios dust experiment data: presence of interstellar dust inside the Earth's orbit. A&A 448, 243–252 (2006)

Asphaug E., Benz W. Size, Density, and structure of Comet Shoemaker-Levy 9 inferred from the physics of tidal breakup. Icarus 121, 225–248 (1996)

Baines, K. H., Yanamandra-Fisher, P. A., Lebofsky, L. A., Momary, T. W., Golisch, W., Kaminski, C., Wild, W. J. Near-infrared absolute photometric imaging of the uranian system. Icarus 132, 266–284 (1998)

Balbus, S. A., Hawley, J. F. A powerful local shear instability in weakly magnetized disks. II. Nonlinear evolution. Astrophys. J. 376, 223–233 (1991)

Barbara J.M., Esposito L.W. Moonlet collisions and the effects of tidally modified accretion in Saturn's F ring. Icarus 160, 161–171 (2002)

Barucci, M. A., Brown, M. E., Emery, J. P., Merlin, F. Composition and surface properties of transneptunian objects and centaurs. In: The Solar System Beyond Neptune. (Barucci, M. A., Boehnhardt, H., Cruikshank, D. P., Morbidelli, A., eds.), University of Arizona Press, Tucson, AZ, pp. 143–160 (2008)

Bauer, J., Lissauer, J.J., Simon, M. Edge-on observations of Saturn's E and G rings in the near-IR. Icarus 125, 440–445 (1997)

Benz W., Asphaug E. Catastrophic disruptions revisited. Icarus 142, 5–20 (1999)

Bodenheimer, P., Pollack, J. B. Calculations of the accretion and evolution of the giant planets: the effects of solid cores. Icarus 67, 391–408 (1986)

Borderies N. Dynamics of ring-satellite systems around Saturn and Uranus. Cel. Mech. 34, 297–327 (1984)

Bosh A. S., Olkin C. B., French R. G., Nicholson P. D. Saturn's F ring: kinematics and particle sizes from stellar occultation studies. Icarus 157, 57–75 (2002)

Brahic A. Numerical simulation of a system of colliding bodies in a gravitational field. J. Comp. Phys. 22, 171–188 (1976)

Brilliantov, N.V. Schmidt, J., Spahn, F. Geysers of Enceladus: Quantitative analysis of qualitative models. Planet Space Sci. 56, 1596–1606 (2008)

Buratti, B. J., Hicks, M. D., Newburn, Jr., R. L. Does global warming make Triton blush? Nature 397, 219 (1999)

Burns, J. A., Showalter, M. R., Morfill, G. E. The ethereal rings of Jupiter and Saturn. In: Planetary Rings. (Greenberg, R., Brahic, A., eds.), University of Arizona Press, Tucson, AZ, pp. 200–274 (1984)

Burns, J.A., Hamilton, D.P., Showalter, M.R. Dusty rings and circumplanetary dust: observations and simple physics. In: Interplanetary Dust. (Grün, E., Gustafson, B., Dermott, S., Fechtig, H., eds.), Springer-Verlag, Berlin, pp. 641–725 (2001)

Cabot, W. Numerical simulations of circumstellar disk convection. Astrophys. J. 465, 874–886 (1996)

Canup, R. M., Esposito, L. W. Accretion in the Roche zone: coexistence of rings and ring moons. Icarus 113, 331–352 (1995)

Canup R.M., Esposito L.W. Evolution of the G Ring and the population of macroscopic ring particles. Icarus 126, 28–41 (1997)

Canup R.M., Ward W.R. Formation of the Galilean satellites: conditions of accretion. Astron. J. 124, 3404–3423 (2002)

Canup R.M., Ward W.R. A common mass scaling for satellite systems of gaseous planets. Nature 411, 834–839 (2006)

Chambers L.S., Cuzzi J.N., Asphaug E., Colwell J., Sugita S. Hydrodynamical and radiative transfer modeling of meteoroid impacts into Saturn's rings. Icarus 194, 623–635 (2008)

Chandrasekhar, S. Ellipsoidal Figures of Equilibrium. The Silliman Foundation Lectures. Yale Univ. Press, New Haven (1969)

Charnoz S., Brahic A., Ferrari C., Grenier I., Roddier F., Thébault, P. Detection of arcs in Saturn's F ring during the 1995 Sun ring-plane crossing. A&A 365, 214–221 (2001)

Charnoz, S., Porco C. C., Déau E., Brahic A., Spitale J. N., Bacques G., Baillie K. Cassini Discovers a Kinematic Spiral Ring Around Saturn. Science 310, 1300–1304 (2005)

Charnoz S., Brahic A., Thomas P.C., Porco C.C. The equatorial ridges of Pan and Atlas: terminal accretionary ornaments? Science 318, 1622–1624 (2007)

Charnoz S., Morbidelli A., Dones L., Salmon J. Did Saturn's rings form during the Late Heavy Bombardment? Icarus 199, 413–428 (2009)

Charnoz S. Physical collisions of moonlets and clumps with the Saturn's F-ring core. Icarus 201, 191–197 (2009)

Colwell, Joshua E., Esposito, Larry W. Origins of the rings of Uranus and Neptune. I – Statistics of satellite disruptions. J. Geophys. Res. 97, 10227–10241 (1992)

Colwell J.E. The disruption of planetary satellites and the creation of planetary rings. P&SS 42, 1139–1149 (1994)

Colwell J.E., Esposito L.W., Bundy D. Fragmentation rates of small satellites in the outer solar system. J. Geophys. Res. 105, 17589–17600 (2000)

Colwell, J. E.; Esposito, L. W.; Sremčević, M. Self-gravity wakes in Saturn's A ring measured by stellar occultations from Cassini. Geophys. Res. Lett. 33, L07201 (2006)

Colwell J. E., Esposito L. W., Sremčević M., Stewart G. R., McClintock W. E. Self-gravity wakes and radial structure of Saturn's B ring. Icarus 190, 127–144 (2007)

Colwell J.E., Esposito L., Sremčević, M., Jerousek R., Cooney J., Lissauer J. Mass density and ring thickness from Cassini UVIS stellar occultations. AGU meeting, San Francisco, abstract #P32A-01 (2008)

Cook, A. F., Franklin, F. A. Rediscussion of Maxwell's Adams Prize Essay on the stability of Saturn's rings. Astron. J. 69, 173–200 (1964)

Cooper J.F., Eraker J.H., Simpson, J. A. The secondary radiation under Saturn's A-B-C rings produced by cosmic ray interactions. J. Geophys. Res. 90, 3415–3427 (1985)

Cuzzi J. N., Lissauer J. J., Esposito L. W., Holberg J. B., Marouf E. A., Tyler G. L., Boishchot A. Saturn's rings - Properties and processes.

In: Planetary Rings. (Greenberg, R., Brahic, A., eds.), University of Arizona Press, Tucson, AZ, pp. 73–199 (1984)

Cuzzi, J. N. Rings of Uranus: not so thick, not so black. Icarus 63, 312–316 (1985)

Cuzzi, J. N., Burns J. A. Charged particle depletion surrounding Saturn's F ring – evidence for a moonlet belt? Icarus 74, 284–324 (1988)

Cuzzi, J. N., Durisen R.H. Meteoroid bombardment of planetary rings: general formulation and effects of Oort cloud projectiles. Icarus 84, 467–501 (1990)

Cuzzi, J. N., Dobrovolskis, A. R., Champney, J. M. Particle-gas dynamics in the midplane of the protoplanetary nebula. Icarus 106, 102–134 (1993)

Cuzzi J.H., Estrada P.R. Compositional evolution of Saturn's rings: ice, tholin, and 'Chiron'-dust. BAAS 29, 1124 (1996)

Cuzzi, J.N., Estrada, P.R. Compositional evolution of Saturn's rings due to meteoroid bombardment. Icarus 132, 1–35 (1998)

Cuzzi, J. N., French, R. C., Dones, L. HST multicolor (255–1042 nm) photometry of Saturn's main rings. I: radial profiles, phase and tilt angle variations, and regional spectra. Icarus 158, 199–223 (2002)

Daisaka H., Tanaka H., Ida S. Viscosity in a dense planetary ring with self-gravitating particles. Icarus 154, 296–312 (2001)

Davidsson B.J.R. Tidal splitting and rotational breakup of solid spheres. Icarus 142, 525–535 (1999)

Davis D.D., Weidenschilling S.J., Chapman C.R., Greenberg R. Saturn ring particles as dynamic ephemeral bodies. Science 224, 744–747 (1984)

Dermott S.F., Malhotra R., Murray C.D. Dynamics of the Uranian and Saturnian satellite systems – a chaotic route to melting Miranda? Icarus 76, 295–334 (1988)

de Pater, I., Showalter, M.R., Lissauer, J.J., Graham, J.R. Keck infrared observations of Saturn's E and G Rings during Earth's 1995 ring-plane crossings. Icarus 121, 195–198 (1996)

de Pater, I., Martin, S.C., Showalter, M.R. Keck near-infrared observations of Saturn's E and G rings during Earth's ring plane crossing in August 1995. Icarus 172, 446–454 (2004)

de Pater, I., Hammel, H. B., Gibbard, S. G., Showalter, M. R. New dust belts of Uranus: one ring, two ring, red ring, blue ring. Science 312, 92–94 (2006)

Dohnanyi, J. Collisional model of asteroids and their debris. J. Geophys. Res. 74, 2531–2554 (1969)

Dones L. A recent cometary origin for Saturn's rings? Icarus 92, 194–203 (1991)

Dones L., Agnor C. B., Asphaug E. Formation of Saturn's rings by tidal disruption of a Centaur. BAAS 38, 420 (2007)

Doyle, L. R., L. Dones, J. N. Cuzzi. Radiative transfer modeling of Saturn's outer B ring. Icarus 80, 104–135 (1989)

Duncan, M.J., Levison, H.F. A disk of scattered icy objects and the origin of Jupiter family comets. Science 276, 1670–1672 (1997)

Durisen, R. H. Particle erosion mechanisms and mass redistribution in Saturn's rings. Adv. Space Res. 4, 13–21 (1984a)

Durisen, R. H. Transport effects due to particle erosion mechanisms. In: Planetary Rings. (Greenberg, R., Brahic, A., eds.), University of Arizona Press, Tucson, AZ, pp. 416–446 (1984b)

Durisen, R. H., N. L. Cramer, B. W. Murphy, J. N. Cuzzi, T. L. Mullikin, and S. E. Cederbloom. Ballistic transport in planetary ring systems due to particle erosion mechanisms. I: Theory, numerical methods, and illustrative examples. Icarus 80, 136–166 (1989)

Durisen, R. H., P. W. Bode, J. N. Cuzzi, S. E. Cederbloom, and B. W. Murphy. Ballistic transport in planetary ring systems due to particle erosion mechanisms. II: Theoretical models for Saturn's A- and B-ring inner edges. Icarus 100, 364–393 (1992)

Durisen, R. H. An instability in planetary rings due to ballistic transport. Icarus 115, 66–85 (1995)

Durisen, R. H., P. W. Bode, S. G. Dyck, J. N. Cuzzi, J. D. Dull, and J. C. White II. Ballistic transport in planetary ring systems due to particle erosion mechanisms. Icarus 124, 220–236 (1996)

Esposito L.W., O'Callaghan M., West R.A. The structure of Saturn's rings – Implications from the Voyager stellar occultation. Icarus 56, 439–452 (1983)

Esposito L.W., Cuzzi J. N., Holberg J.B., Marouf E.A., Tyler G.L., Porco C. C. Saturn's rings – structure, dynamics, and particle properties. In Planetary Rings. (Greenberg, R., Brahic, A., eds.), University of Arizona Press, Tucson, AZ, pp. 463–545 (1984)

Esposito L.W. Structure and evolution of Saturn's rings. Icarus 67, 345–357 (1986)

Esposito. L.W. Understanding planetary rings. Ann. Rev. Earth Planet. Sci. 21, 487–523 (1993)

Esposito, L.W., Colwell, J.E. Estimating the effectiveness of cosmic recycling in the history of the planetary rings and ring moons. Presented at Fall AGU, San Francisco, CA (2003)

Esposito, L.W., Colwell, J.E. Explaining radial spectral variations. Presented at Fall AGU, San Francisco, CA (2004)

Esposito, L.W., Colwell, J.E. Cassini observations and ring history. Presented at Fall AGU, San Francisco, CA (2005)

Esposito L.W., Colwell J.E., Larsen K., McClintock W.E., Stewart A.I.F., Hallett J.T., Shemansky D.E., Ajello J.M., Hansen C.J., Hendrix A.R., West R.A., Keller H.U., Korth A., Pryor W.R., Reulke R., Yung Y.L. Ultraviolet imaging spectroscopy shows an active Saturnian system. Science 307, 1251–1255 (2005)

Esposito, L. W. Cassini observations and the history of Saturn's rings. AGU Fall Meeting, San Francisco, abstract no. P23E-0110 (2006)

Esposito L.W., Eliott J.P. 2007. Regolith Growth and Pollution on Saturn's Moons and Ring Particles. DPS meeting #39, #7.09

Esposito L.W., Meinke B.K., Colwell J.E., Nicholson P.D., Hedman M.H. Moonlets and clumps in Saturn's F ring. Icarus 194, 278–289 (2008a)

Esposito L.W., Eliott. J.P., Albers N. Regolith growth and darkening of Saturn's ring particles. AGU meeting, San Francisco, Abstract #P13A-1295 (2008b)

Estrada, P. R., J. N. Cuzzi. Voyager observations of the color of Saturn's rings. Icarus 122, 251–272; Also – Erratum, Icarus, 125, 474 (1996)

Estrada, P. R., J. N. Cuzzi, M. R. Showalter. Voyager color photometry of Saturn's main rings: a correction. Icarus 166, 212–222 (2003)

Estrada, P. R., Mosqueira, I., Lissauer, J. J., D'Angelo, G., and Cruikshank, D. P. Formation of Jupiter and conditions for the accretion of the Galilean satellites. Submitted to appear in: Europa. (McKinnon, W., Pappalardo, R., Khurana, K., eds.), University of Arizona Press, Tucson, AZ (2009)

Farmer A.J., Goldreich P. Understanding the behavior of Prometheus and Pandora. Icarus 180, 403–411 (2006)

French R.G., McGhee C.A., Dones L., Lissauer J.J. Saturn's wayward shepherds: the peregrinations of Prometheus and Pandora. Icarus 162, 143–170 (2003)

Gammie, C. F. Linear theory of magnetized, viscous, self-gravitating gas disks. Astrophys. J. 463, 725–731 (1996)

Gehrels T., Baker L. R., Beshore E., Blenman C., Burke J. J., Castillo N. D., Dacosta B., Degewij J., Doose L. R., Fountain J. W., Gotobed J., Kenknight C. E., Kingston R., McLaughlin G., McMillan R., Murphy R., Smith P. H., Stoll C. P., Strickland R. N., Tomasko M. G., Wijesinghe M. P., Coffeen D. L., Esposito L. W. Imaging photopolarimeter on Pioneer Saturn. Science 207, 434–439 (1980)

Giuliatti Winter, S. M., Winter, O. C. Some comments on the F ring-Prometheus–Pandora environment. Adv. Space Res. 33, 2298–2302 (2004)

Giuliatti Winter S. M., Winter O. C., Guimarães A. H. F. A note on the horseshoe confinement model: the Poynting-Robertson effect. Astron. Astrophys. 418, 759–764 (2004)

Giuliatti Winter S. M., Mourão D. C., Freitas, T. C. A. The strands of the F ring disturbed by its closest satellites. Adv. Space Res. 38, 781–787 (2006)

Goldreich, P., Tremaine, S. D. The formation of the Cassini Division in Saturn's rings. Icarus 34, 240–253 (1978a)

Goldreich, P., Tremaine, S. D. The velocity dispersion in Saturn's rings. Icarus 34, 227–239 (1978b)

Goldreich P., Tremaine S. Disk–satellite interactions. Astrophys. J. 241, 425–441 (1980)

Goldreich, P., S. Tremaine. Dynamics of planetary rings. Ann. Rev. Astron. Astrophys. 20, 249–283 (1982)

Goldreich P., Rappaport N. Chaotic motion of Prometheus and Pandora. Icarus 162, 391–399 (2003a)

Goldreich P., Rappaport N. Origin of the chaos in the Prometheus–Pandora system. Icarus 166, 320–327 (2003b)

Grossman, A. W. Microwave imaging of Saturn's deep atmosphere and rings. Unpublished Ph.D. thesis, California Institute of Technology, Pasadena, CA (1990)

Hamilton, D.P., Burns, J.A. Origin of Saturn's E ring: self-sustained, naturally. Science 264, 550–553 (1994)

Harris A. The origin and evolution of planetary rings. In: Planetary Rings. (Greenberg, R., Brahic, A., eds.), University of Arizona Press, Tucson, AZ, pp. 641–659 (1984)

Hawley, J. F., Balbus, S. A., Winters, W. F. Local hydrodynamic stability of accretion disks. Astrophys. J. 518, 394–404 (1999)

Hedman, M.M., Burns, J.A., Tiscareno, M.S., Porco, C.C., Jones, G.H., Roussos, E., Krupp, N., Paranicas, C., Kempf, S. The source of Saturn's G ring. Science 317, 653–656 (2007a)

Hedman M.M., Nicholson P.D., Salo H., Wallis B.D., Buratti B.J., Baines K.H., Brown R.H., Clark R.N. Self-Gravity Wake Structures in Saturn's A Ring Revealed by Cassini VIMS. Astron. J. 133, 2624–2629 (2007b)

Hedman M.M., Murray, C.D., Cooper, N.J., Tiscareno, M.S., Beurle, K., Evans, M.W., Burns, J.A. Three tenuous rings/arcs for three tiny moons. Icarus 199, 378–386 (2009a)

Hedman M.M., Nicholson, P. D., Showalter, M. R., Brown, R. H., Buratti, B. J., Clark, R. N. Spectral observations of the Enceladus Plume with Cassini-VIMS. ApJ 693, 1749–1762 (2009b)

Holberg J.B., Forrester W.T., Lissauer J.J. Identification of resonance features within the rings of Saturn. Nature 297, 115–120 (1982)

Horn, L. J., Cuzzi, J. N. Characteristic wavelengths of irregular structure in Saturn's B ring. Icarus 119, 285–310 (1996)

Horner J., Evans N. W., Bailey M. E. Simulations of the population of Centaurs – I. The bulk statistics. Mon. Not. R. Astron. Soc. 354, 798–810 (2004)

Hubickyj, O., Bodenheimer, P., Lissauer, J. J. Accretion of the gaseous envelope of Jupiter around a 5–10 Earth-mass core. Icarus 179, 415–431 (2005)

Ip, W.-H. Collisional interactions of ring particles – the ballistic transport process. Icarus 54, 253–262 (1983)

Ip, W.-H. Ring torque of Saturn from interplanetary meteoroid impact. Icarus 60, 547–552 (1984)

Ip, W.H. An evaluation of a catastrophic fragmentation origin of the Saturnian ring system. A&A 199, 340–342 (1988)

Jewitt, D.C., Luu, J.X. Discovery of the candidate Kuiper belt object 1992 QB1. Nature 362, 730–732 (1993)

Ji, H., Burin, M., Schartman, E., Goodman, J. Hydrodynamic turbulence cannot transport angular momentum effectively in astrophysical disks. Nature 444, 343–346 (2006)

Karjalainen R., Salo H. Gravitational accretion of particles in Saturn's rings. Icarus 172, 328–348 (2004)

Karjalainen R. Aggregate impacts in Saturn's rings. Icarus 189, 523–537 (2007)

Karkoschka, E. Comprehensive photometry of the rings and 16 satellites of Uranus with the Hubble Space Telescope. Icarus 151, 51–68 (2001)

Kempf, S., Beckmann, U., Moragas-Klostermeyer, G., Postberg, F., Srama, R., Economou, T., Schmidt, J., Spahn, F., Grün, E. The E-ring in the vicinity of Enceladus. Icarus 193, 420–437 (2008)

Kieffer, S.W., Lu, X., Bethke, C.M., Spencer, J.R., Marshak, S., Navrotsky, A. A clathrate reservoir hypothesis for Enceladus' south polar plume. Science 314, 1764–1767 (2006)

Kieffer, S.W., Jakosky, B.M. Enceladus—Oasis or ice ball? Science 320, 1432–1433 (2008)

Koschny D. Grün E. Impacts into ice-silicate mixtures: crater morphologies, volumes, depth-to-diameter ratios, and yield. Icarus 154, 391–401 (2001)

Lagage P. O., Pantin E. Dust depletion in the inner disk of Beta-Pictoris as a possible indicator of planets. Nature 369, 628–630 (1994)

Landgraf, M., W. J. Baggaley, E. Grün, H. Krüger, G. Linkert. Aspects of the mass distribution of interstellar dust grains in the solar system from in-situ measurements. J. Geophys. Res. 105, 10343–10352 (2000)

Levison H.F., Duncan M.J., Zahnle K., Holman M., Dones, L. Planetary impact rates from ecliptic comets. Icarus 196, 258–273 (2000)

Lewis M.C., Stewart G.R. Expectations for Cassini observations of ring material with nearby moons. Icarus 178, 124–143 (2005)

Leyrat C., Spilker L.J., Altobelli N., Pilorz S., Ferrari C. Infrared observations of Saturn's rings by Cassini CIRS: Phase angle and local time dependence. P&SS 56, 117–133 (2008)

Lin, D. N. C., Papaloizou, J. On the structure and evolution of the primordial solar nebula. MNRAS 191, 37–48 (1980)

Lissauer J.J., Cuzzi J.N. Resonances in Saturn's rings. Astron. J. 87, 1051–1058 (1982)

Lissauer J.J. Ballistic transport in Saturn's rings – an analytical theory. Icarus 57, 63–71 (1984)

Lissauer J.J., Cuzzi J.N. Rings and moons – clues to understanding the solar nebula. In Protostars and Planets II. (Black, D., Matthews, M. S., eds.), University of Arizona Press, Tucson, AZ, pp. 920–956 (1985)

Lissauer J.J, Squyres S.W., Hartmann W.K. Bombardment history of the Saturn system. J. Geophys. Res. 93, 13776–13804 (1988)

Lissauer, J.J., Stevenson, D. J. Formation of giant planets. In: Protostars and Planets V. (Reipurth, B., Jewitt, D., Keil, K., eds.), University of Arizona Press, Tucson, AZ, pp. 591–606 (2007)

Lissauer J.J., Hubickyj O., D'Angelo G., Bodenheimer P. Models of Jupiter's growth incorporating thermal and hydrodynamic constraints. Icarus 199, 338–350 (2009)

Lunine, J. I., Stevenson, D. J. Formation of the Galilean satellites in a gaseous nebula. Icarus 52, 14–39 (1982)

McKee, C. F., and Ostriker, E. C. Theory of star formation. Ann. Rev. Astron. Astrophys. 45, 565–687 (2007)

Mitler, H. E. Formation of an iron-poor moon by partial capture, or: Yet another exotic theory of lunar origin. Icarus 24, 256–268 (1975)

Morfill, G. E., H. Fechtig, E. Grün, C. Goertz. Some consequences of meteoroid bombardment of Saturn's rings. Icarus 55, 439–447 (1983)

Mosqueira, I., Estrada, P. R. Formation of the regular satellites of giant planets in an extended gaseous nebula I: Subnebula model and accretion of satellites. Icarus 163, 198–231 (2003a)

Mosqueira, I., Estrada, P. R. Formation of the regular satellites of giant planets in an extended gaseous nebula II: Satellite migration and survival. Icarus 163, 232–255 (2003b)

Murray C.D., Gordon M.K., Giuliatti Winter, Silvia M. Unraveling the strands of Saturn's F ring. Icarus 129, 304–316 (1997)

Murray C.D., Dermott S.F. Solar System Dynamics. Cambridge University Press, New York (1999)

Murray C.D., Chavez C., Beurle K., Cooper N., Evans M.W., Burns J.A., Porco C.C. How Prometheus creates structure in Saturn's F ring. Nature 437, 1326–1329 (2005)

Murray C.D., Beurle K., Cooper N.J., Evans M.W., Williams G.A., Charnoz S. The determination of the structure of Saturn's F ring by nearby moonlets. Nature 453, 739–744 (2008)

Nakagawa, Y., Sekiya, M., Hayashi, C. Settling and growth of dust particles in a laminar phase of a low-mass solar nebula. Icarus 67, 375–390 (1986)

Nakazawa K., Ida S. Hill's approximation in the three-body problem. Prog. Theo. Phys. Supp. 96, 716–174 (1988)

Nicholson P.D., Cooke M.L., Pelton E. An absolute radius scale for Saturn's rings. Astron. J. 100, 1339–1362 (1990)

Nicholson, P.D., Showalter, M.R., Dones, L., French, R.G., Larson, S.M., Lissauer, J.J., McGhee, C.A., Seitzer, P., Sicardy, B., Danielson, G.E. Observations of Saturn's ring-plane crossing in August and November 1995. Science 272, 509–515 (1996)

Nicholson P.D., Hedman M.M., Clark R.N., Showalter M.R., Cruikshank D.P., Cuzzi J.N., Filacchione G., Capaccioni F., Cerroni P., Hansen G.B., Sicardy B., Drossart P., Brown R.H., Buratti B.J., Baines K.H., Coradini A. A close look at Saturn's rings with Cassini VIMS. Icarus 193, 182–212 (2008)

Northrop, T. G., J. E. P. Connerney. A micrometeorite erosion model and the age of Saturn's rings. Icarus 70, 124–137 (1987)

Null G.W., Lau E.L., Biller E.D., Anderson J.D. Saturn gravity results obtained from Pioneer 11 tracking data and earth-based Saturn satellite data. Astron. J. 86, 456–468 (1981)

Ohtsuki, K. Capture probability of colliding planetesimals: dynamical constraints on accretion of planets, satellites, and ring particles. Icarus 106, 228–246 (1993)

Ohtsuki K., Stewart G.R., Ida S. Evolution of planetesimal velocities based on three-body orbital integrations and growth of protoplanets. Icarus 155, 436–453 (2002)

Öpik E.J. Comments on lunar origin. Ir. Astron. J. 10, 190–238 (1972)

Ormel, C. W., Cuzzi, J. N. Closed-form expressions for particle relative velocities induced by turbulence. Astron. Astrophys. 466, 413–420 (2007)

Ossenkopf, V. Dust coagulation in dense molecular clouds: the formation of fluffy aggregates. Astron. Astrophys. 280, 617–646 (1993)

Pollack, J. B., Reynolds, R. T. Implications of Jupiter's early contraction history for the composition of the Galilean satellites. Icarus 21, 248–253 (1974)

Pollack, J. B. The rings of Saturn. Space Sci. Rev. 18, 3–93 (1975)

Pollack, J. B., Grossman, A. S., Moore, R., Graboske Jr., H. C. The formation of Saturn's satellites and rings as influenced by Saturn's contraction history. Icarus 29, 35–48 (1976)

Pollack, J. B., Grossman, A. S., Moore, R., Graboske Jr., H. C. A calculation of Saturn's gravitational contraction history. Icarus 30, 111–128 (1977)

Pollack, J. B., Consolmagno, G. Origin and evolution of the Saturn system. In: Saturn. (Gehrels, T., Matthews, M. S., eds.), University of Arizona Press, Tucson, AZ, pp 811–866 (1984)

Pollack, J. B., Hubickyj, O., Bodenheimer, P., Lissauer, J. J., Podolak, M., Greenzweig, Y. Formation of the giant planets by concurrent accretion of solids and gas. Icarus 124, 62–85 (1996)

Porco C.C., Baker E., Barbara J., Beurle K., Brahic A., Burns J.A., Charnoz S., Cooper N., Dawson D.D., Del Genio A.D., Denk T., Dones L., Dyudina U., Evans M.W., Giese B., Grazier K., Helfenstein P., Ingersoll A.P., Jacobson R.A., Johnson T.V., McEwen A., Murray C.D., Neukum G., Owen W.M., Perry J., Roatsch T., Spitale J., Squyres S., Thomas P., Tiscareno M., Turtle E., Vasavada A.R., Veverka J., Wagner R., West R. Cassini imaging science: initial results on Saturn's rings and small satellites. Science 307, 1226–1236 (2005)

Porco, C.C., Helfenstein, P., Thomas, P.C., Ingersoll, A.P., Wisdom, J., West, R., Neukum, G., Denk, T., Wagner, R., Roatsch, T., Kieffer, S., Turtle, E., McEwen, A., Johnson, T.V., Rathbun, J., Veverka, J., Wilson, D., Perry, J., Spitale, J., Brahic, A., Burns, J.A., Del Genio, A.D., Dones, L., Murray, C.D., Squyres, S. Cassini observes the active south pole of Enceladus. Science 311, 1393–1401 (2006)

Porco C. C., Thomas P. C., Weiss J. W., Richardson D. C. Saturn's small inner satellites: clues to their origins. Science 318, 1602–1607 (2007)

Porco, C.C. on behalf of the ISS Team: S/2008 S 1. IAU Circular Number 9023 (2009)

Pöschel T., Brilliantov N.V., Formella A. Granular gas cooling and relaxation to the steady state in regard to the overpopulated tail of the velocity distribution. Int. J. Mod. Phys. C, 18, 701–711 (2007)

Postberg, F., Kempf, S., Hillier, J.K., Srama, R., Green, S.F., McBride, N., Grün, E. The E-ring in the vicinity of Enceladus II. Icarus 193, 438–454 (2008)

Poulet F., Sicardy B., Nicholson P.D., Karkoschka E., Caldwell J. Saturn's ring-plane crossings of August and November 1995: a model for the new F-ring objects. Icarus 144, 135–148 (2000a)

Poulet F., Sicardy B., Dumas C., Tiphène D. The crossing of Saturn ring plane by the Earth in 1995: Ring thickness. Icarus 145, 147–165 (2000b)

Poulet F., Sicardy B. Dynamical evolution of the Prometheus–Pandora system. Mon. Not. R. Astron. Soc. 322, 343–355 (2001)

Poulet, F., J. N. Cuzzi, D. Cruikshank, T. Roush, and C. Dalle Ore. Comparison between the Shkuratov and Hapke scattering theories for solid planetary surfaces. Application to the surface composition of two centaurs. Icarus 160, 313–324 (2002)

Poulet F., Cruikshank D.P., Cuzzi J.N., Roush T.L., French R.G. Compositions of Saturn's rings A, B, and C from high resolution near-infrared spectroscopic observations. A&A 412, 305–316 (2003)

Renner S., Sicardy B., French R.G. Prometheus and Pandora: masses and orbital positions during the Cassini tour. Icarus 174, 230–240 (2005)

Robbins, S.J., Stewart, G.R., Lewis, M.C., Colwell, J.E., and M. Sremčević. Estimating the masses of Saturn's A and B rings from high-optial depth N-body simulations and stellar occultations. Submitted to Icarus (2009)

Roche E.A. Académie des sciences et lettres de Montpellier. Mém. Sect. Sci. 1, 243–262 (1847)

Rubincam, D.P. Saturn's rings, the Yarkovsky effects, and the Ring of Fire. Icarus 184, 532–542 (2006)

Ryu, D., Goodman, J. Convective instability in differentially rotating disks. Astrophys. J. 308, 438–450 (1992)

Salo, H. Simulations of dense planetary rings. III. Self-gravitating identical particles. Icarus 117, 287–312 (1995)

Salo H. Numerical simulations of the collisional dynamics of planetary rings. In: Granular Gases. (Pöschel, T., Luding, T., eds.) Lecture Notes in Physics 564, 330–349 (2001)

Schmidt, J., Brilliantov, N., Spahn, F., Kempf, S. Slow dust in Enceladus' plume from condensation and wall collisions in tiger stripe fractures. Nature 451, 685–688 (2008)

Shen, Y., Stone, J. M., Gardiner, T. A. Three-dimensional compressible hydrodynamic simulations of vortices in disks. Astrophys. J. 653, 513–524 (2006)

Shevchenko I.I. Adiabatic chaos in the Prometheus–Pandora system. Mon. Not. R. Astron. Soc. 384, 1211–1220 (2008)

Showalter, M.R., Cuzzi, J.N., Larson, S.M. Structure and particle properties of Saturn's E ring. Icarus 94, 451–473 (1991)

Showalter M.R., Pollack J.B., Ockert M.E., Doyle L.R., Dalton J.B. A photometric study of Saturn's F ring. Icarus 100, 394–411 (1992)

Showalter, M.R., Cuzzi, J.N. Seeing ghosts: photometry of Saturn's G ring. Icarus 103, 124–143 (1993)

Showalter M.R. Detection of centimeter-sized meteoroid impact events in Saturn's F ring. Science 282, 1099–1102 (1998)

Showalter, M. R. Disentangling Saturn's F ring. I. Clump orbits and lifetimes. Icarus 171, 56–371 (2004)

Shu F.H. Waves in planetary rings. In: Planetary Rings. (Greenberg, R., Brahic, A., eds.), University of Arizona Press, Tucson, AZ, pp. 513–561 (1984)

Shu F.H., Dones L., Lissauer J.J., Yuan C., Cuzzi J. N. Nonlinear spiral density waves – viscous damping. Astrophys. J. 299, 542–573 (1985)

Shu F.H., Johnstone D., Hollenbach D. Photoevaporation of the solar nebula and the formation of the giant planets. Icarus 106, 92–101 (1993)

Smith B.A., and 26 co-authors. Encounter with Saturn: Voyager 1 imaging science results. Science 212, 163–191 (1981)

Smith B.A., and 28 co-authors. A new look at the Saturn system: the Voyager 2 images. Science 215, 504–537 (1982)

Smoluchowski R. Width of a planetary ring system and the C-ring of Saturn. Nature 274, 669–670 (1978)

Smoluchowski R. The ring systems of Jupiter, Saturn and Uranus. Nature 280, 377–378 (1979)

Spahn, F.; Wiebicke, H.-J. Long-term gravitational influence of moonlets in planetary rings. Icarus 77, 124–134 (1989)

Spahn F., Sponholz H. Existence of moonlets in Saturn's rings inferred from the optical depth profile. Nature 339, 607–608 (1989)

Spahn F., Sremčević M. Density patterns induced by small moons in Saturn's rings? Astron. Astrophys. 358, 368–372 (2000)

Spaute D., Weidenschilling S.J., Davis D.R., Marzari F. Accretional evolution of a planetesimal swarm. I – A new simulation. Icarus 92, 147–164 (1991)

Sremčević M., Spahn F., Duschl W. Density structures in perturbed thin cold discs. Mon. Not. R. Astron. Soc. 337, 1139–1152 (2002)

Sremčević, M., A. V. Krivov, H. Krüger, F. Spahn. Impact-generated dust clouds around planetary satellites: model versus Galileo data. Planet. Space Sci. 53, 625–641 (2005)

Sremčević M., Schmidt J., Salo H., Seiss M., Spahn F., Albers N. A belt of moonlets in Saturn's A ring. Nature 449, 1019–1021 (2007)

Srama R., and 40 co-authors. In situ dust measurements in the inner Saturnian system. P&SS 54, 967–987 (2006)

Stevenson D.J., Harris A.W., Lunine J.I. Origins of satellites. In: Satellites. (Burns, J., Matthews, M. S., eds.), University of Arizona Press, Tucson, AZ, pp. 39–88 (1986)

Stewart, G. R., S. J. Robbins, J. E. Colwell. Evidence for a primordial origin of Saturn's rings. 39th DPS meeting, abstract no. 7.06 (2007)

Stöffler, D., Gault, D. E., Wedekind, J. A., Polkowski, G. Experimental hypervelocity impact into quartz sand: distribution and shock metamorphism of ejecta. J. Geophys. Res. 80, 4062–4077 (1975)

Stone, J. M., Balbus, S. A. Angular momentum transport in accretion disks via convection. Astrophys. J. 464, 364–372 (1996)

Takeuchi, T., Lin, D. N. C. Radial flow of dust particles in accretion disks. Astrophys. J. 581, 1344–1355 (2002)

Throop, H.B., Esposito, L.W. G ring particle sizes derived from ring plane crossing observations. Icarus 131, 152–166 (1998)

Tiscareno M.S., Burns J.A., Hedman M.M., Porco C.C., Weiss J.W., Dones L., Richardson D.C., Murray C.D. 100-metre-diameter moonlets in Saturn's A ring from observations of 'propeller' structures. Nature 440, 648–650 (2006)

Tiscareno M.S., Burns J.A., Nicholson P.D., Hedman M.M., Porco C.C. Cassini imaging of Saturn's rings. II. A wavelet technique for analysis of density waves and other radial structure in the rings. Icarus 189, 14–34 (2007)

Tiscareno M.S., Burns J.A., Hedman M.M., Porco C.C. The population of propellers in Saturn's A ring. Astron. J. 135, 1083–1091 (2008)

Toomre, A. On the gravitational stability of a disk of stars. Astrophys. J. 139, 1217–1238 (1964)

Tsiganis K., Gomes R., Morbidelli A., Levison H.F. Origin of the orbital architecture of the giant planets of the solar system. Nature 435, 459–461 (2005)

Vedder, J. F. Craters formed in mineral dust by hypervelocity microparticles. J. Geophys. Res. 77, 4304–4309 (1972)

Völk H.J., Jones F.C., Morfill G.E., Roeser S. Collisions between grains in a turbulent gas. A&A 85, 316–325 (1980)

Ward W.R. Survival of planetary systems. ApJL 482, L211 (1997)

Weidenschilling, S.J., Chapman, C.R., Davis, D.R., Greenberg, R. Ring particles. In: Planetary Rings. (Greenberg, R., Brahic, A., (eds.), University of Arizona Press, Tucson, AZ, pp. 367–415 (1984a)

Weidenschilling, S. J. Evolution of grains in a turbulent solar nebula. Icarus 60, 553–567 (1984b)

Wetherill G. W., Stewart G. R. Accumulation of a swarm of small planetesimals. Icarus 77, 330–357 (1989)

Winter O. C., Mourão D. C., Giuliatti Winter S. M., Spahn F., da Cruz, C. Moonlets wandering on a leash-ring. Mon. Not. R. Astron. Soc. 380, L54-L57 (2007)

Wood, J. A., Mitler, H. E. Origin of the Moon by a modified capture mechanism, or half a loaf is better than a whole one. Abstracts of the Lunar and Planetary Science Conference 5, 851 (1974)

Wood, J. A. Moon over Mauna Loa – a review of hypotheses of formation of Earth's Moon. In: Origin of the Moon; Proceedings of the Conference, Kona, HI, October 13–16, 1984, Houston, TX, Lunar and Planetary Institute, pp. 17–55 (1986)

Zahnle K., Schenk P., Levison H., Dones L. Cratering rates in the outer solar system. Icarus 163, 263–289 (2003)

Chapter 18
The Thermal Evolution and Internal Structure of Saturn's Mid-Sized Icy Satellites

Dennis L. Matson, Julie C. Castillo-Rogez, Gerald Schubert, Christophe Sotin, and William B. McKinnon

Abstract The Cassini-Huygens mission is returning new geophysical data for the midsize, icy satellites of Saturn (i.e., satellites with radii between 100 and 1,000 km). These data have enabled a new generation of geophysical model studies for Phoebe, Iapetus, Rhea, Mimas, Tethys, Dione, as well as Enceladus (which is addressed in a separate chapter in this book). In the present chapter we consider the new model studies that have reported significant results elucidating the evolutionary histories and internal structures of these satellites. Those results have included their age, the development of their internal structures and mineralogies, which for greatest fidelity must be done concomitantly with coupled dynamical evolutions. Surface areas, volumes, bulk densities, spin rates, orbit inclinations, eccentricities, and distance from Saturn have changed as the satellites have aged. Heat is required to power the satellites' evolution, but is not overly abundant for the midsized satellites. All sources of heat must be evaluated and taken into account. This includes their intensities and when they occur and are available to facilitate evolution, both internal and dynamical. The mechanisms of heat transport must also be included. However, to model these to high fidelity the material properties of the satellite interiors must be accurately known. This is not the case. Thus, much of the chapter is devoted to discussion of what is known about these properties and how the uncertainties affect the estimation of heat sources, transport processes, and the consequential changes in composition and evolution. *Phoebe* has an oblate shape that may be in equilibrium with its spin period of ∼9.3 h. Its orbital properties suggest that it is not one of the regular satellites, but is a captured body. Its density is higher than that of the other satellites, consistent with formation in the solar nebula rather than from material around Saturn. Oblate shape and high density are unusual for objects in this size range, and may indicate that Phoebe was heated by ^{26}Al decay soon after its formation, which is consistent with some models of the origin of Kuiper-Belt objects. *Iapetus* has the shape of a hydrostatic body with a rotation period of 16 h. It subsequently despun to its current synchronous rotation state, ∼79 day period. These observations are sufficient to constrain the required heating in Iapetus' early history, suggesting that it formed several My after CAI condensation. Since Saturn had to be present for Iapetus to form, this date also constrains the age of Saturn and how long it took to form. Both shape and gravitational data are available for *Rhea*. Gravity data were obtained from the single Cassini flyby during the prime mission and within the uncertainties cannot distinguish between hydrostatic and non-hydrostatic gravitational fields. Both Dione and Tethys display evidence of smooth terrains, with Dione's appearing considerably younger. Both are conceivably linked to tidal heating in the past, but the low rock abundance within Tethys and the lack of eccentricity excitation of Tethys' orbit today make explaining this satellite's geology challenging.

18.1 Introduction

In this chapter we consider the midsize satellites of Saturn for which the Cassini-Huygens mission has returned new geophysical data. These data have provided much needed constraints for models. Existing models were not adequate for the interpretation of these data and the development of some new approaches has been required. Chief among these is the realization that thermophysical modeling and dynamical modeling cannot be carried out in the absence of the other. They must be done simultaneously. It appears that these satellites accreted early (i.e., less than ∼10 Myr after the formation of the Solar System). Thus, depending upon the date a simulation is started, the correct amount of heat from short-lived radioactive isotopes must be considered in the models. Our review is focused on these new developments that are described in the

---

D.L. Matson, J.C. Castillo-Rogez, and C. Sotin
Jet Propulsion Laboratory, California Institute of Technology, Pasadena, CA, 91109, USA

G. Schubert
University of California, Los Angeles, Los Angeles, CA, 90095, USA

W.B. McKinnon
Washington University in St. Louis, St. Louis, MO, 63139, USA

recent literature, which has focused on Iapetus, Phoebe, Rhea, and Enceladus. Thus, this is not a complete review of the scientific literature on the midsize satellites. The reader is referred to Johnson (1998) and McKinnon (1998) and relevant chapters in *Satellites* (Burns and Matthews 1986) for earlier reviews that more explicitly deal with Mimas, Tethys, and Dione. New information on Enceladus is found in the chapter by Spencer et al. in this book.

In the following we discuss some geological considerations and then address why the midsize satellites are so important for understanding the more general questions of the histories of the satellites, and the Saturnian system.

Geology: Satellite surfaces are shaped by both exogenic and endogenic processes. The most obvious of the exogenic processes is impact cratering. Interactions with the magnetosphere and its trapped particles are less obvious but may drive chemical reactions on the surface. Also, through sputtering, the satellites contribute material, e.g., ions and atoms, to the magnetosphere.

Endogenic geologic processes are driven by heat and are the main shapers of the geologic features observed. Changes in the volume of a satellite can arise from internal evolution: melting and differentiation, silicate serpentinization or other rock hydration reactions, or transformation of ice I to/from ice II (the midsize satellites are too small to stabilize higher pressure ice polymorphs). Some of these processes have been described, for example, by Ellsworth and Schubert (1983) and Squyres and Croft (1986). The volume change resulting from pore and void space compaction could also create large-scale compressive features.

More generally, if initial temperatures were as low as 70 K, and the main heat source is from long-lived radioisotope decay, then the time needed for the temperature to reach the peritectic ammonia-water melting temperature, i.e., 176 K, is so long that it might occur only beneath a very thick lithosphere (given solid-ice thermal conductivity). Under these conditions it is difficult for the interior to influence the geology of the surface. While the satellites, by and large, show ancient surfaces (i.e., as inferred from the cratering record; see Dones et al., this book), there is also plenty of evidence of activity during their geological histories.

Importance of the midsize icy satellites: The midsize icy satellites of Saturn preserve evidence of their formation and evolution, and thus, also, evidence of the history of the whole Saturnian system, including Saturn itself. By studying the satellites we can learn more about their present properties, particularly about their interiors, and how they came to this state. The desire to decipher and understand the evidence is motivating present studies and the development of increasingly sophisticated models for all elements of the Saturnian system. These better models include higher computational accuracy and more of the relevant physical and chemical processes than were previously available or needed to interpret available data. If the early results from these efforts are correct, then we will be learning much about the early solar system and how bodies evolve.

The recovery of evolutionary evidence started with the *Voyager* images that showed variations in the impact crater densities across the surfaces of some of the satellites (see Chapman and McKinnon 1986 for a review). Clearly, there had been geologic periods when resurfacing occurred. The resurfacing erased existing impact craters and this allowed crater counting to measure impacts referenced to a new starting date. The *Cassini-Huygens* mission obtained much higher precision geophysical data for important properties such as satellite shapes, motion (e.g., librations) and gravitational moments. These facts have turned out to be very important for constraining the evolutionary history of the Saturnian satellites.

There is a reason why the midsize icy satellites can make a unique contribution. They have the "right" size. Smaller satellites lose heat by thermal conductivity very fast, and are likely to have been inactive throughout their evolution. At the other extreme, large satellites, such as Titan, have evolved more and, as a result, key information about their formation and evolution has been destroyed as a result of geological activity. Thus, it is the midsize, icy satellites that were large enough to have evolved in response to environmental processes (such as tidal heating) but were small enough that their evolution reached its end before key evidence was destroyed. What was preserved depends on the evolution of each of the satellites.

Today most of the midsize icy satellites are cold and inert. One, Enceladus, is active and may have relatively high internal temperatures. Some have possibly differentiated and undergone endogenic activity in the not too distant geologic past, as expressed in their geology, such as Dione. Several of them, e.g., Iapetus, Mimas, Rhea, show primitive, heavily cratered surfaces that let us wonder whether these objects have undergone any endogenic activity. The Cassini-Huygens mission has provided important geophysical constraints on satellite evolution, for example gravity data for Rhea, high-resolution shape and topography measurements, as well as high-resolution imaging that allows the geological evolution of the satellites to be studied. These data are also crucial for assessing the hydrostatic (compensation) state of the satellites, thus their internal evolution, and, in some cases, their dynamical evolution. Over the past two decades our understanding of processes involved in the geophysical evolution of icy satellites has grown. Work on theory has enabled more accurate modeling, especially in areas such as convection, tidal dissipation, and the understanding of the thermomechanical properties of ice. As examples of advances in modeling, one can point to the most recent generation of icy satellite models that combines simultaneous dynamical and geophysical modeling.

This chapter builds on these new observational, theoretical, and laboratory developments to convey the present state of knowledge about the midsized icy satellites. First we will present the Cassini-Huygens and other observations that can be interpreted as geophysical constraints on the evolution of icy satellites (Section 18.2). Then we present and discuss the key factors involved in the evolution of these objects – sources of heat (Section 18.3) and heat transfer (Section 18.4). In particular, these involve assumptions about the initial state of these satellites, i.e., initial composition, temperature, etc. (Section 18.5). This is followed by a discussion of the main aspects of satellite modeling which leads to a comparison of model results with the observations (Sections 18.6 and 18.7). A roadmap for future studies (Section 18.9) and a summary (Section 18.10) close out the chapter.

18.2 Satellite Properties

The midsize satellites of Saturn are a diverse group. An inspection and comparison of the entries in Table 18.1 will reveal the extent of this diversity. Pay particular attention to the radii, the densities, and the silicate fractions, x_s. The differences in these parameters are significant. Later we will argue that these values are responsible for the model results showing that the satellites have different evolutionary histories, internal structures, and levels of activity today.

18.2.1 Size and Shape

Size is important. The internal pressures are relatively low for bodies with radii of less than about a hundred kilometers. As a result they retain their initial porosities because the internal pressure is not sufficient to compact their material. They also tend to have irregular shapes. They may even be "rubble piles" especially as they have likely been subject to heavy if not catastrophic bombardment early in Solar System history (see Chapter 19). Larger satellites with radii greater than several hundred kilometers have higher internal pressures and tend to be spherical in shape.

Shapes are an indication of radial mass distribution, provided that the bodies are in hydrostatic equilibrium with the tidal and rotation forces that act on them (e.g., Schubert et al. 2004). The shapes measured in *Cassini* images (Thomas et al. 2007b) indicate that most satellites deviate by less than a few kilometers from uniform density bodies in hydrostatic equilibrium. The shape data are compared in Fig. 18.1.

The trend is for the satellites to cluster about an $(a-c)/(a-c)_{hydrostatic}$ value of 1.00, indicating that they are either in hydrostatic equilibrium or close to that condition. Iapetus and the Moon are outliers. In fact until Cassini-Huygens returned data for Iapetus, the Moon was the most non-hydrostatic satellite known. Iapetus is now the extreme case. It supports a 33-km shape anomaly. This is huge compared to a 10 m bulge that is the amplitude expected for such body in hydrostatic equilibrium with the present rotation period of 79.3 days.

Why is there such a huge anomaly? Studies of satellite rotation suggest that the present periods are highly evolved from their initial values that were much shorter (see discussion by Peale, 1977). Since the shape of Iapetus corresponds to that of a hydrostatic body with a rotation period of about 16 h, this suggests a faster spin in the past, but there are few constraints on the initial rotation period. The present day spin rates of other satellites cannot be used for guidance because they also experienced despinning. For guidance we can look to the distribution of rotation periods among the asteroids (e.g., Dermott and Murray 1982) and the transneptunian objects. These distributions suggest that periods ranging from 5 to 10 h should be considered.

Unless a body is spinning particularly rapidly, the theoretical shape as a function of spin period for a uniform density, hydrostatic body is a Maclaurin spheroid. Its rotational oblateness can be computed using Chandrasekhar's (1969) formulation. The clear implication is that all satellites had faster spin rates in the past. Depending upon their degrees of hydrostaticity, their shapes also changed with time as they despun. Castillo-Rogez et al. (2007) have suggested that this was the evolution for Iapetus. Over time it became less hydrostatic as it cooled. By the time it slowed down to a spin period of about 16 h its lithosphere had become strong enough to maintain the nonhydrostatic figure. If this is the case, then the formation and long-term preservation of the 16-h figure strongly constrains models for Iapetus' geophysical evolution.

18.2.2 Density

With the exception of Hyperion, the apparent densities of all of the satellites are consistent with their being mixtures of solid rock and ice. The Cassini data produced a significant change in the density of Enceladus which is now 1,608 kg/m^3, an increase of ~60% over that previously used in geophysical modeling. It is now the most rock rich of the midsized satellites.

The Cassini spacecraft's close flyby of Iapetus' dark, leading, hemisphere on December 31, 2004, collected the data for

Table 18.1 Physical and dynamical properties of the Saturnian satellites discussed in this chapter

Properties	Ref.	Mimas	Enceladus	Tethys	Dione	Rhea	Hyperion	Iapetus	Phoebe
Physical properties									
GM (km^3/s^2)	a	2.5026 ± 0.0006	7.2027 ± 0.0125	41.2067 ± 0.0038	73.1146 ± 0.0015	153.9426 ± 0.0037	0.3727 ± 0.0012	120.5038 ± 0.0080	0.5532 ± 0.0006
Mean Radius R$_s$ (km)	a	198.20 ± 0.25	252.10 ± 0.10	533.00 ± 0.70	561.70 ± 0.45	764.30 ± 1.10	135.00 ± 4.00	735.60 ± 1.50	106.60 ± 1.00
(a−c) (km)	b	16.8 ± 0.6	8.3 ± 0.6	12.9 ± 1.9	3.5 ± 1.2	4.1 ± 2.1	–	35.0 ± 3.7	–
(b−c) (km)	b	2.7	1.1	0.5	0.2	1.6	–	–	–
Density $\bar{\rho}$ (g/cm^3)	a	1.150 ± 0.004	1.608 ± 0.003	0.973 ± 0.004	1.476 ± 0.004	1.233 ± 0.005	0.542 ± 0.048	1.083 ± 0.007	1.634 ± 0.046
Surface Gravity (m/s^2)	c	0.06	0.11	0.14	0.23	0.26	0.02	0.22	0.05
Maximum Central Pressure P$_c$ (MPa)	c	7.4	23.0	36.7	95.4	124.0	0.7	89.2	4.3
Silicate Mass Fraction x$_s$	d	0.26	0.57	0.06	0.50	0.33	–	0.20	–
Silicate Volume Fraction	d	0.09	0.26	0.02	0.21	0.12	–	0.06	–
Geometric Albedo	a	0.6	1	0.8	0.6	0.6	0.3	0.6	0.081 ± 0.002t
Dynamical Properties									
Orbital Period (d)	a	0.942	1.370	1.888	2.737	4.518	21.28	79.33	550.31
Present Eccentricity	a	0.0196	0.0047	0.0001	0.0022	0.0010	0.0274	0.0283	0.1635
Semi-Major Axis (km)	a	185540	238040	294670	377420	527070	1500880	3560840	12947780
Inclination (deg.)	a	1.572	0.009	1.091	0.028	0.331	0.630	7.489	175.986
Obliquity (deg.)	a	0.001	0.001	0.002	0.005	0.029	0.461	14.968	26.891

References are the following: (a) http://ssd.jpl.nasa.gov and references therein; (b) Thomas et al. (2007b); (c) Parameter computed using a combination of density and radius; the central pressure is computed assuming the satellite is homogenous: $P_c = \frac{2\pi}{3} G \bar{\rho}^2 R_s^2$ with G the gravitational constant; (d) Parameter computed assuming an ice density of 931 kg/m^3, a rock density of 3,510 kg/m^3, and zero porosity. The obliquity is the angle between the satellite's equator and its Laplace plane (the plane normal to the satellite's orbital precession pole).

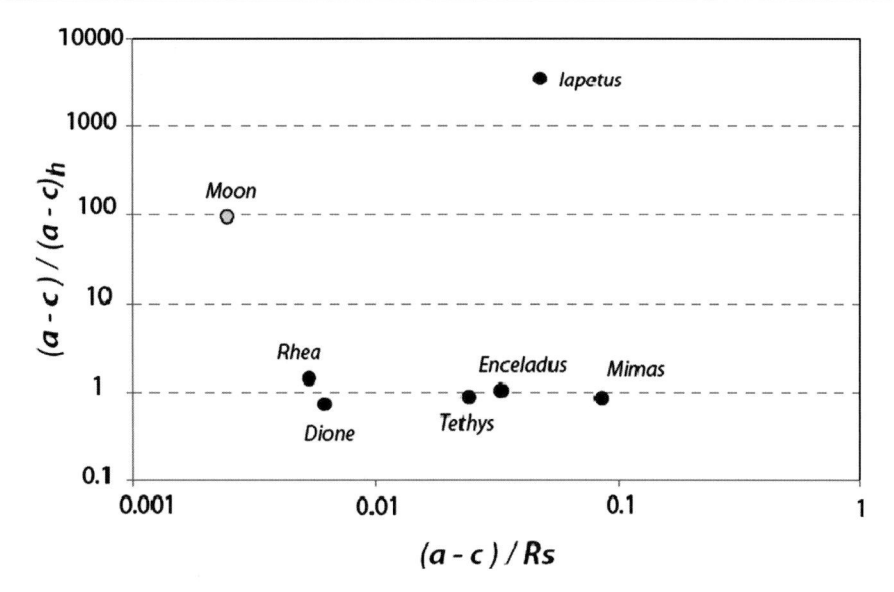

Fig. 18.1 Satellite shapes. This logarithmic plot shows the deviation from hydrostatic equilibrium (using the shape determinations by Thomas et al. (2007b)). The ordinate is the difference between the polar and an equatorial axis, a–c, scaled by the corresponding difference for a hydrostatic, uniform density body. The abscissa is a–c scaled by the mean radius of the satellite. The Moon is shown for reference

determining that satellite's density and shape (Thomas et al. 2007b). Iapetus' density of $1,090 \pm 32\,\text{kg/m}^3$ is the third lowest density, after Hyperion and Tethys (see Table 18.1). This density is ∼20% lower than the weighted mean density of the midsized Saturnian satellites. This mean density is computed from the densities of Mimas, Enceladus, Tethys, Dione, Rhea, Iapetus weighted by their actual mass (see Johnson and Lunine (2005) for details of the calculation). The lower density implies that Iapetus' material is ice rich and suggests that it accreted from solids condensed directly from the circum-Saturnian-nebula rather than accreting from relatively unprocessed heliocentric solids (as in recent satellite formation models (Canup and Ward 2006)), as the latter would have given a density some fifty percent higher (Prinn and Fegley 1989). It has been suggested that Iapetus is composed of a mixture of water ice and carbonaceous chondritic material with an enrichment in volatiles and, possibly, light hydrocarbons (Cruikshank et al. 2008). It is also significant that there is no simple, monotonic relation between satellite density and distance from Saturn (i.e., no parallel to the situation for the Galilean satellites of Jupiter). (Also see chapter on the Origin of the Saturn System).

18.2.3 Porosity

Hyperion's density is $540\,\text{kg/m}^3$, so even pure, solid water ice is too dense to account for the bulk of this satellite. Thus the presence of a significant amount of porosity must be considered in order to account for Hyperion's density. Porosity is responsible for the difference between a satellite's apparent density and the density of the material of which it is composed. Thus the porosity must be taken into account before the relative amounts of rock and ice in a satellite can be inferred. At the time of accretion the satellites were presumably relatively porous. The presumption that porosity was significant in the early history of small ice-rock bodies is supported by several studies (see McKinnon 2008 for a review). Even today a significant fraction of their volume may still be pore space.

A number of processes play a role in compacting the satellites. When hydrostatic pressure exceeds the strength of ice it produces compaction through brittle fracture and reorganization of the material (e.g., Durham et al. 2005). Several laboratory studies (Leliwa-Kopystynski and Maeno 1993; Leliwa-Kopystynski and Kossacki 1995); Durham et al. (2005) and models (Leliwa-Kopystynski and Maeno 1993) show or imply that porosity evolution is not linear with depth but is characterized by a substantial change in porosity due to brittle reorganization of the material at pressures between 1 and 10 MPa. Laboratory measurements by Durham et al. (2005) indicate that in pure water up to 20% porosity can be sustained up to pressures as large as 150 MPa when the temperature is less than 120 K. Leliwa-Kopystynski and Maeno (1993) have shown that if ammonia is a substantial fraction of the volatile component present in the ice, the porosity decreases for temperatures greater than 100 K. For pressures less than 1 MPa, porosity can be as large as 40%. Volatile migration can also play a role in decreasing porosity or modifying the local structure (e.g., McKinnon 2008). When conditions, especially temperature, become suitable, ice creep and sintering results in further structural evolution and porosity reduction.

Leliwa-Kopystynski et al. (1995) have pointed out the importance of composition for these different mechanisms. For example, the decrease of porosity is impeded by the presence of rocky material in the ice. Such an aggregation of fine-grained rock particles, well mixed with the ice, can produce a substantially stronger material.

The present day porosity of a satellite cannot be measured directly but must be inferred via an iterative procedure using thermal models to estimate how the porosity evolved. This procedure has important consequences for interpreting the observed or apparent densities. The satellite interiors can be thought of as having three components: ice, rock, and void space. In order to match the apparent density, the fractional abundance of rock must be increased to compensate for the decrease in density due to porosity. Fortunately, it is possible to model the evolution of porosity with time and iteratively obtain estimates for the amount of void space in a satellite. The procedure will be illustrated later in the chapter.

18.2.4 Initial Composition

The temperature of the accreting planetesimals is a function of the characteristics of the Saturnian subnebula that presumably existed around Saturn when the giant planet was forming. In such a subnebula, it is the temperature and pressure that determines whether or not some ambient chemical species such as CH_4 and NH_3 will be present in solid form and be available to accrete (Prinn and Fegley 1981; Mousis et al. 2002; Hersant et al. 2008). In treating the satellites' composition we use the terms "ice", "rock", and "porosity". We follow the definitions of Johnson and Lunine (2005). By "ice" we mean the volatile phase that contains all of the ices and gases. The "rock" phase contains the silicate and oxide minerals and any metals and sulfides. "Porosity", of course, refers to void space. The presently available constraints on the compositions of ice and rock phases are now discussed.

18.2.4.1 Volatile Composition

Conditions in the solar nebula near Saturn's position or in the Saturnian subnebula were such that the ice likely accreted in crystalline form (Hersant et al. 2004; Gautier and Hersant 2005; cf. Garaud and Lin 2007). Given the weighted-average density of the regular midsized satellites above, Johnson and Lunine (2005) have inferred that the Saturnian subnebula was enriched in volatiles over the composition of the Solar nebula. While the subnebula models predict the materials that should be present in the satellite interiors, the exact composition of the ice that accreted is not so obvious. For the most part, the best information we have on the internal volatile composition of the Saturnian satellites comes from in situ measurements, such as those made in Enceladus' plumes, in the E-ring, and in Titan's atmosphere. As detected by the *Cassini* Ion and Neutral Mass Spectrometer (INMS) (Waite et al. 2006), Enceladus' plumes contain substantial CO_2, either N_2 or CO, methane, some ammonia, and traces of propane and acetylene, Greatly improved signal-to-noise ratios for later encounters allow for refinement of these results, noted below. Enceladus is the second Saturnian satellite, after Titan, for which methane has been detected. It remains to be determined whether this component is primordial, i.e., trapped as clathrates (e.g., Hersant et al. 2008) and/or the result of internal thermochemical processes (e.g., Fischer-Tropsch reaction, Matson et al. 2007). However, the possible presence of N_2 Is a real puzzle. It was either trapped in clathrates (or adsorbed on grain surfaces) at very low temperatures (~ 27 K) in the Saturnian subnebula, or it was produced by the decomposition of NH_3, thus indicating temperatures of at least a few hundred degrees K (Matson et al. 2007; Glein et al. 2008). Such conditions may have also been favorable for Fischer-Tropsch and many other organic synthesis reactions. If this was the case, then it would favor the argument that the methane we see today was synthesized in Enceladus. Recent, high signal-to-noise measurements of plume composition by the INMS (Waite et al. 2009) have failed to detect primordial ^{36}Ar, which supports the synthesis interpretation.

Spectrometric measurements of the satellites' surface reflectances indicate that their ices are predominantly (if not overwhelmingly) composed of water. Also, the presence of CO_2 and HCN, as well as simple organics (Cruikshank et al. 2005) have also been detected at the surface of most midsize Saturnian satellites, Mimas and Tethys being exceptions. No CO has been detected so far, even though it is predicted to have condensed in the Saturnian subnebula models of Hersant and Gautier e.g., (Hersant et al. 2004). Under the present conditions, however, CO is not stable on the satellite surfaces, so its absence is not a surprise.

Recent Saturnian subnebula evolution models indicate that ammonia should constitute 0.5 to 11wt% of the total mass of water included in the satellites (Mousis et al. 2002; Alibert and Mousis 2007). Conclusive observational evidence of NH_3 hydrates, however, has not been reported. Ground-based observations are also contradictory (Cruikshank et al. 2005; Emery et al. 2005; Verbiscer et al. 2008). The *Cassini* Visual and Infra-red Spectrometer (VIMS) team has detected NH_3 itself and sets its concentration to be less than 2% at Enceladus' surface (Brown et al. 2006). The *Cassini* Ion and Neutral Mass Spectrometer (INMS) has conclusively detected NH_3 in Enceladus' vented plume gas at a concentration of nearly 1 mole% (Waite et al. 2006, 2009). RADAR measurements, while not diagnostic, suggest the absence of ammonia at the surface of the satellites, except for Iapetus (Ostro et al. 2006). For the latter, the presence of a small amount of ammonia appears as the best possibility for explaining the electrical properties of the returned radar signal. Ostro et al. (2006) refer to Lanzerotti et al. (1984) for an explanation of the non-detection of ammonia on the inner Saturnian satellites. They proposed that ammonia is destroyed as a result of the interactions between the surfaces and Saturn's magnetospheric plasma, which is too weak to affect Iapetus at its relatively great distance from Saturn.

The reasons for the difficulty in detecting ammonia in the Saturnian system may be due to the low concentration of ammonia in the satellites when they formed, as well as subsequent thermal and sputtering loss from their surfaces. As noted, ammonia hydrates are a crucial parameter in geophysical evolution and modeling because they depress the ice melting temperature to as low as 176 K for a multicomponent mixture (Kargel 1992), and thus can influence thermal evolution and heat transfer. However, the effects of ammonia may be mitigated if it is present at levels below 1 or 2 wt%, such as suggested for Iapetus (Ostro et al. 2006).

18.2.4.2 Rock Composition

Many geophysical models for the outer planet satellites assume or argue that anhydrous rock accreted or is representative of the rock fraction today (e.g., Kuskov and Kronrod 2001; Sohl et al. 2002, 2003). However, in the case of meteorite parent bodies, it has been suggested that some fine components of the silicate phase could have been hydrated in the Solar nebula as a result of shock waves (Ciesla et al. 2003). Moreover, bulk aqueous alteration was widespread on carbonaceous asteroids (e.g., Kargel 1991), and similar infalling planetesimals could have contributed to the accreting satellites (Canup and Ward 2006). Leliwa-Kopystynscki and Kossacki (2000) have also suggested that Mimas could have accreted amorphous silicate. Cosmochemical models for the Saturnian satellites, however, tend to favor a scenario in which the minerals were crystalline (e.g., Gautier and Hersant 2005), and evidence from the Stardust mission implies substantial mixing of crystalline silicates throughout the solar nebula (Brownlee et al. 2006). Thus the rock phase was likely composed of silicate and metallic minerals.

The constraints on the composition of the rock phase are poor. Ordinary (L) chondritic and mean CI (i.e., C1) carbonaceous chondritic compositions have been used as analogs for the rock phase. These also provide a basis for estimating the radionuclide content and thus the amount of heat that will become available from radioactive decay. Note also that the average ordinary chondrite density, $\sim$3,500 kg/m^3 (Consolmagno et al. 1998), is very close to the density of Io ($\sim$3,528 kg/m^3) often used for modeling the rock phase in the Galilean satellites (Schubert et al. 2004) (though it is important to remember that Io is very hot by comparison to meteorite samples!).

Alteration of the rock by water results in the production of the serpentine minerals, montmorillonite, as well as oxides (e.g., goethite) and silicate hydroxides (e.g., talc and brucite) (e.g., Scott et al. 2002). The mean density for this mixture of minerals is between 2,300 and 2,700 kg/m^3. This means that hydrated rock can sequester up to 15% water by volume.

18.2.4.3 Rhea's Gravitational Field

Rhea is especially important among the medium size icy satellites of Saturn because it is the only moon in this family for which we have data on the quadrupole gravitational field (Table 18.2). The data are limited, consisting of only 1 near-equatorial flyby by the Cassini spacecraft. Moreover, the inference of J_2 and C_{22} from the radio Doppler data has been controversial; Table 18.2 lists 3 separate sets of values for these gravitational coefficients. The mass of Rhea, expressed by the GM values in Table 18.2, is not in contention. The mean radius of Rhea (764.3 $\pm$ 2.2 km) and its mass yield a density of 1,233 $\pm$ 11 kg/m^3 (Thomas et al., 2007b). If Rhea is composed of ice with density 1,000 kg/m^3 and rock with density 3,527.5 kg/m^3 (2,500 kg/m^3) (the larger value of rock density is the mean density of Io) then its silicate mass fraction is 0.26 (0.31). Rhea has a larger density and silicate mass fraction than Iapetus, but a smaller density and rock mass fraction than Dione. The rock mass fraction determines the quantity of long-term radiogenic heat available to a satellite. Rhea is intermediate between Dione and Iapetus in terms of the magnitude of the satellite radiogenic heat source.

The different values of the gravitational coefficients of Rhea listed in Table 18.2 derive mainly from different

Table 18.2 Published values for Rhea's gravitational coefficients

	Anderson and Schubert (2007)	Iess et al. (2007)	Mackenzie et al. (2008)
GM (km^3 s^{-2})	153.9372 $\pm$ 0.0013	153.9395 $\pm$ 0.0018	153.9398 $\pm$ 0.0008
$J_2(10^{-6})$	889.0 $\pm$ 25.0	794.7 $\pm$ 89.2	931.0 $\pm$ 12.0
$C_{22}(10^{-6})$	266.6 $\pm$ 7.5	235.3 $\pm$ 4.8	237.2 $\pm$ 4.5
J_2/C_{22}	10/3 (assumed)	3.377	3.925
C/MR^2	0.3911 $\pm$ 0.0045	0.3721 $\pm$ 0.0036	–

approaches to the interpretation of the radio Doppler data. The gravitational coefficients J_2 and C_{22} are strongly coupled for a near equatorial flyby as can be seen by writing the terms contributing to the second-degree equatorial, gravitational potential, V_{eq}

$$V_{eq} = -\left(\frac{GM}{r}\right)\left[1 + 1/2 J_2 \left(\frac{R}{r}\right)^2 + 3 C_{22}\left(\frac{R}{r}\right)^2 \cos 2\lambda\right]$$
(18.1)

where G is the universal gravitational constant, M is the mass of the satellite, λ is longitude and r is the distance from the center of the satellite of radius R and $S_{22} \ll C_{22}$. Generally, as a practical matter, only a linear combination of J_2 and C_{22} can be inverted from a near equatorial flyby of a satellite (Schubert et al. 2004) although Mackenzie et al. (2008) have reported the values of J_2 and C_{22} listed in Table 18.2. Anderson and Schubert (2007) assumed that Rhea is in hydrostatic equilibrium, an assumption that connects J_2 and C_{22} by $J_2/C_{22} = 10/3$; they derive the values of J_2 and C_{22} given in Table 18.2. Iess et al. (2007) did not make the a priori assumption of hydrostatic equilibrium but derived values of J_2 and C_{22} consistent with it (Table 18.2). If Rhea is in hydrostatic equilibrium then the value of C_{22} can be used to infer the satellite's moment of inertia factor C/MR^2, where C is the axial moment of inertia. Anderson and Schubert (2007) find $C/MR^2 = 0.391$ (Table 18.2) consistent with an undifferentiated Rhea in which the ice component of Rhea's ice/rock interior undergoes a phase transition from ice I to ice II at depth. Iess et al. (2007) obtain a slightly smaller value of C/MR^2 (Table 18.2) implying a partial separation of ice and rock inside Rhea. Since the values of J_2 and C_{22} according to Mackenzie et al. (2008) are not consistent with hydrostatic equilibrium, nothing about Rhea's interior can be inferred from them (other than Rhea is not in hydrostatic equilibrium).

To better understand the differences in the reported values of Rhea's quadrupole gravitational coefficients, Anderson and Schubert (2009) calculated the Doppler residuals for the Rhea flyby for the three inferred gravitational fields. The results are shown in Fig. 18.2 where it is seen that the residuals are essentially indistinguishable from one another. In other words, the three Rhea gravitational fields listed in Table 18.2 fit the Doppler data from the Rhea flyby equally well. The reason for this agreement is that the fits to the Doppler flyby data are mainly determined by the values of C_{22} that are similar to about the ten percent level for all three gravitational fields. The different values of J_2 are all a priori possible. The only way to distinguish among the possible

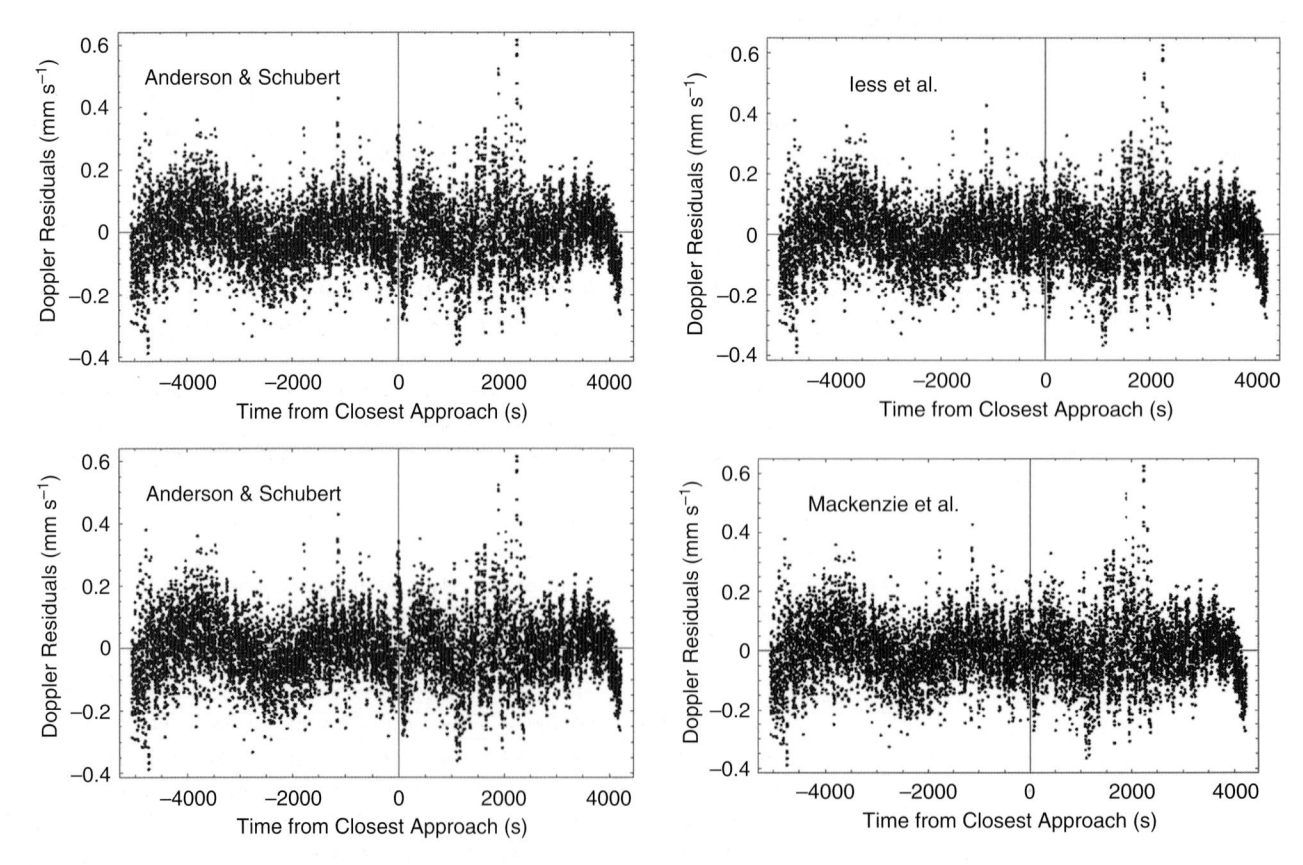

Fig. 18.2 Calculated Doppler residuals for the Rhea flyby for the three inferred gravitational fields of Table 18.2

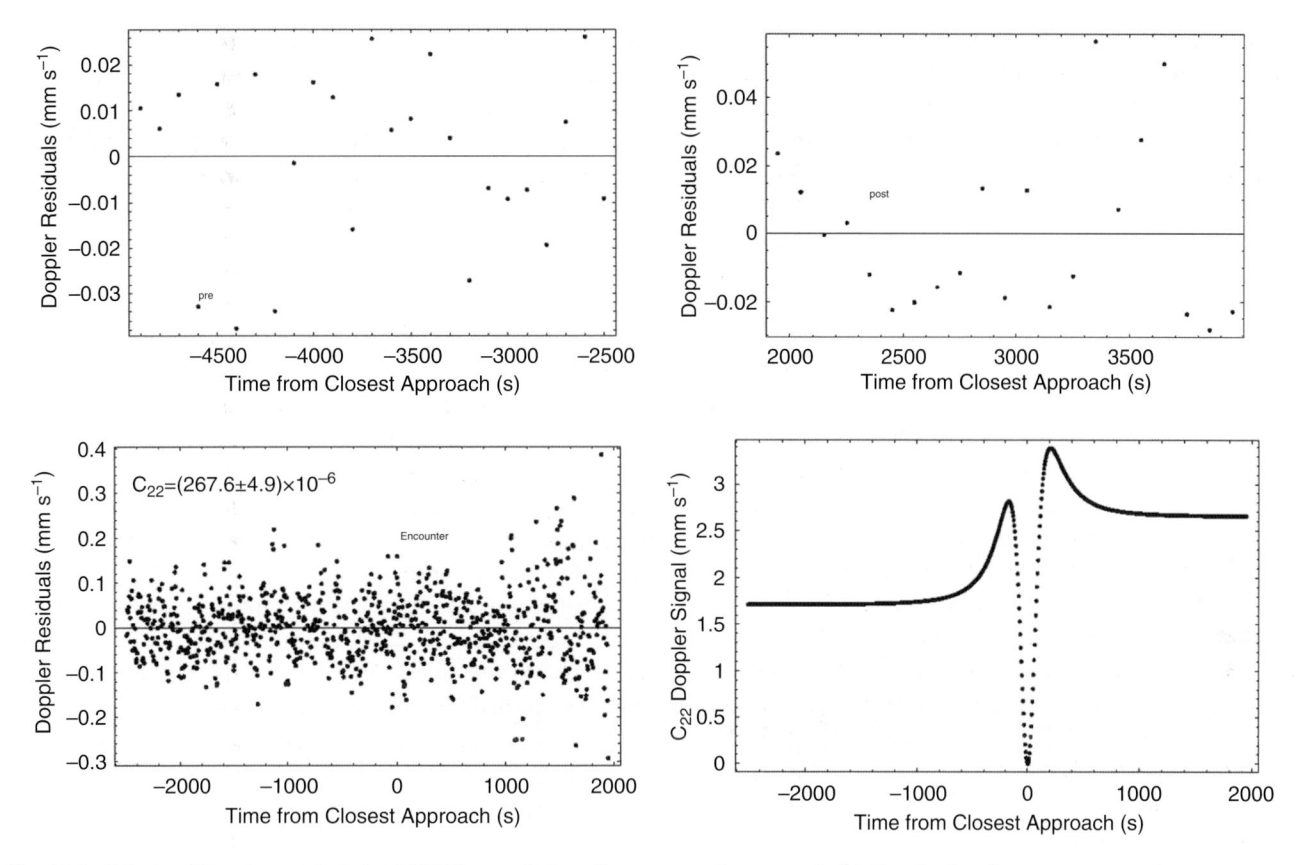

Fig. 18.3 Calculated Doppler residuals (with RTG heat radiation effects removed) compared with Doppler data for pre-encounter, close-encounter, and post-encounter spacecraft trajectory segments

J_2 values is on the basis of some physical argument. The assumption of hydrostatic equilibrium provides such a basis although hydrostaticity cannot be proven. There is no physical basis on which to select a non-hydrostatic J_2. The bottom line is that the single near equatorial flyby of Rhea cannot distinguish between hydrostatic and non-hydrostatic gravitational fields. A hydrostatic Rhea is consistent with the Doppler flyby data despite the claim of Mackenzie et al. (2008) to the contrary (Anderson and Schubert 2009). And as this discussion makes clear, until we can determine with some degree of certainty whether Rhea is truly hydrostatic, inferences about Rhea's MOI based on C_{22} alone should be treated with caution.

The Doppler residuals in Fig. 18.2 contain an unmodeled signal. Anderson and Schubert (2009) attribute this signal to the radiation of RTG heat from the Cassini spacecraft. The signal can be removed and a better estimate of Rhea's C_{22} obtained by dividing the analysis of the Doppler data into three segments, a pre-encounter segment, a close-encounter segment, and a post-encounter segment (Anderson and Schubert 2009). Figure 18.3 shows the results of separate fits to the Doppler data in each of the individual segments. The separate fits reveal no further residual signal. The fit to the close-encounter segment yields the improved determination of $C_{22} = (267.6 \pm 4.9) \times 10^{-6}$ (Anderson and Schubert 2009), in essential agreement with the value of C_{22} determined by Anderson and Schubert (2007) (see Table 18.2). Figure 18.3 also shows that the gravitational signal from Rhea's C_{22} is fully contained within the time span of the close-encounter segment (Anderson and Schubert 2009).

An undifferentiated or partially differentiated Rhea is consistent with its heavily cratered surface and its lunar-like, inert interaction with the Saturnian plasma (Khurana et al. 2008). Rhea's shape is also consistent with hydrostatic equilibrium although uncertainties in the shape data leave open the question of the satellite's homogeneity (Thomas et al. 2007b).

18.3 Sources of Heat

The relative significance of the different sources of heat depends upon the time when they occur and how long they continue to supply heat. Heating of the satellites starts off with accretion. Then other sources of heat become important. The decay of short-lived radioactive isotopes (hereafter, SLRI) provides a heat pulse during the first 10

My following accretion. By contrast, the decay of long-lived radiogenic isotopes (hereafter, LLRI) provides heat over the long-term. Saturn's luminosity plays a role, perhaps up to the first hundred million years following the subnebula dissipation (Lissauer et al. 2009). Other heat sources are provided by transient (and sometimes dramatic) events, such as the release of gravitational energy associated with internal structure evolution (porosity collapse, differentiation), and despinning. Tidal dissipation associated with orbital evolution is a more complex process, which is highly dependent on the initial dynamical situation and on the internal temperature evolution. Let us consider these heat sources in some more detail.

18.3.1 Heating by Radioactivity

During accretion, the satellites incorporate radionuclides in proportion to their rock content. The concentration of radionuclides in the rock is a function of time and is usually referenced to the time when the Calcium–Aluminum Inclusions (CAIs) accreted in the inner solar nebula.. This, in turn, was approximately the time when the Solar system formed (Wasserburg and Papanastassiou 1982). The volumetric radiogenic heating rate is given by:

$$H_R = \rho x_S \sum_{i=1}^{n} C_0 H_{0,i} e^{-\lambda_i t_{0-CAIs}} \qquad (18.2)$$

where ρ is the density of the mixture of ice and rock, x_s is the mass fraction of silicates, C_0 is the initial concentration of radiogenic elements, n is the number of radiogenic elements included in the sum, $H_{0,i}$ is the initial power produced by radiogenic decay per unit mass of radiogenic element i with decay constant λ_i. Time, t, is time since CAIs formation, labeled as t_{0-CAIs}.

The SLRI have half-lives of less than 10 My (see Cohen and Coker 2000, for a review of SLRI nuclides). The important radionuclides, in terms of abundance and heat production are ^{26}Al and ^{60}Fe. Collectively they are often referred to as "Aluminium-26" because it produces the largest amount of heat. The origin of these elements is not well constrained. Two models have been proposed: (1) the "X-wind" model by Shu et al. (1993), since then highly explored, for example by Gounelle and Russell (2005); (2) the supernova injection model proposed by Vanhala and Boss (2002).

SLRI are extensively referred to in the literature about asteroids and the formation of the inner Solar system and they are widely used in models for meteorite parent bodies. However their presence in the outer Solar system has been rarely addressed, except in the works by Prialnik and Bar-Nun (1990), Leliwa-Kopystynski and Kossacki (2000), Prialnik and Merk (2008) and related, see the review by McKinnon et al. (2008). While these works considered SLRI as a potential component of thermal models, it is only recently that it has been suggested that these nuclides are useful in explaining the Iapetus and Enceladus observations by the *Cassini* spacecraft (Castillo-Rogez et al. 2007; Matson et al. 2007; Schubert et al. 2007). The initial concentration of ^{26}Al/^{27}Al has been established as 5×10^{-5} (Wasserburg and Papanastassiou 1982). This is referred to as the "canonical" value. Recently a "supercanonical" value equal to 6.5×10^{-5} has been proposed by Young et al. (2005), but it remains to be confirmed by further studies.

With respect to ^{60}Fe, it is important to note that there has been significant progress in the last decade in instrumentation for measuring the ^{60}Fe daughter product (^{60}Ni) in meteorites. The initial concentration of ^{60}Fe/^{56}Fe is now reported to be between 0.5×10^{-6} and 1×10^{-6} (Mostefaoui et al. 2005; Tachibana et al. 2006). This is an increase of a factor of $\sim$1,000 with respect to the discovery measurements by Shukolyukov and Lugmair (1993).

The concentrations of radionuclides for the ordinary and mean CI chondritic compositions are presented in Table 18.3. These data are based on the elemental compositions from Wasson and Kalleymen (1988) convolved with isotopic abundances from Van Schmus (1995), Kita et al. (2005) and Tachibana et al. (2006). The data for LLRI are found in Table 18.4 and those for SLRI in Table 18.5. These are our preferred values and they are presented here because a variety of values have been used for some of the parameters in the recent literature, creating a confusing situation (see Castillo-Rogez et al. 2009, for a listing of the values that have been used). Note that the ordinary chondritic composition provides about 10% more specific radiogenic heat in the long term than the mean CI chondritic composition (though CI compositions are of course hydrated, oxidized, and carbonaceous by comparison).

18.3.2 Tidal Heating

Tidal heating is due to changing tides. The gravitational fields of other bodies raise tides. The tides convert power into heat as they distort the bulk of the satellite and as they work against friction when they cause motion along faults and fractures near the surface. Concurrent with tidal heating is dynamical evolution. The heat energy must have a source, such as the energy stored in the spin of the planet and the orbital energy and spin of the satellite. We now consider two of the most common situations, namely the heat produced by the despinning of the satellite and by the satellite's orbital evolution.

Table 18.3 Major radioactive isotopes: Initial compositions for the CI and ordinary chondrites (Wasson and Kalleymen 1988)

	CI chondrite	Ordinary chondrite
Density (kg/m^3)	2,756	3,510
^{26}Al (ppb)	525	600
^{60}Fe (ppb)	111–222	107.5–215
^{53}Mn (ppb)	23.2	25.7
^{40}K (ppb)	943	1,104
^{232}Th (ppb)	44.3	53.8
^{235}U (ppb)	6.27	8.2
^{238}U (ppb)	20.2	26.2

Table 18.4 Decay information for the long-lived radioisotopes. Adapted from Van Schmus (1995)

Element	Potassium	Thorium	Uranium	
Isotope	^{40}K	^{232}Th	^{235}U	^{238}U
Isotopic abundance (%wt)	0.01176	100.00	0.71	99.28
Decay constant (per year)	5.54×10^{-10}	4.95×10^{-11}	9.85×10^{-10}	1.551×10^{-10}
Half-life λ (My)	1,277	14,010–14,050	703.81	4,468
Specific heat production (W/kg of elements) today	29.17×10^{-6}	26.38×10^{-6}	568.7×10^{-6}	94.65×10^{-6}

Table 18.5 Decay data for ^{26}Al and ^{60}Fe

Parent nuclide	^{26}Al	^{60}Fe
Daughter nuclide	^{26}Mg	^{60}Ni
Initial isotopic abundance	^{26}Al/^{27}Al	^{60}Fe/^{56}Fe
	5×10^{-5}	0.1–1×10^{-6}
Half-life λ (My)	0.717	1.5
Specific heat production (W/kg) at 4.5 Ga (t$_o$ = CAI)	0.357	0.063

18.3.2.1 Despinning

Despinning is a rapid event for most satellites, as was recognized by Peale (1977). If taking place instantaneously, despinning increases the internal temperature by the amount (Burns 1976):

$$\Delta T = \frac{1}{2}\gamma \left(\Omega_o^2 - \Omega^2\right) \frac{R^2}{C_P} \quad (18.3)$$

where Ω_o and Ω are the initial and final angular rates, and γ is the dimensionless moment of inertia. C_p is the temperature-dependent specific heat of the material. For a $C_p \sim 0.5$ kJ/kg/K, appropriate to a 60/40 ice-rock mixture at 100 K (McKinnon 2002), and an initial rotation period of 5 h, despinning can contribute to a globally averaged increase in temperature of up to 20 K in satellites as large as Iapetus and Rhea.

The despinning rate (in rad/s) as a function of time, t, is given by

$$d\Omega/dt = -[3k_2(t)\,GM_p^2\,R_{eq}^5(t)]/[C(t)\,D^6(t)\,Q(t)] \quad (18.4)$$

where M_p is Saturn's mass, R_{eq} is equatorial radius of the satellite, C is the polar moment of inertia of the satellite, and D is the semi-major axis of the orbit. The dissipation factor Q and the tidal Love number k_2 are functions of the frequency- and temperature-dependent viscoelastic (and other) properties of the satellite and thus vary as a function of time. The values of k_2 and Q can be computed by numerical integration (Takeuchi and Saito 1972) (and see Tobie et al. 2005a; Wahr et al. 2009). Usually despinning occurs rapidly and D can be taken as constant. But, if the despinning takes a sufficiently long time, and the orbit evolves substantially on that time scale, then D must be updated as a function of time.

Peale (1977) estimated the times required for tidal dissipation to reduce the satellites' spins to values equal to their planetary orbital periods. The times are from Peale's Table 6.1 and are plotted in Fig. 18.4. The initial rotation period was 2.3 h. The units are in "years/Q" where Q is the specific dissipation or "quality" factor. Despinning time is proportional to Q, so a despin time/Q of 10^7 years means a despin time of 10^9 years for $Q = 100$. The red horizontal line is drawn for the age of the Solar System using the common assumption that 100 is a reasonable quality factor for icy satellites (Goldreich and Soter 1966).

Most of the satellites despin rapidly, as expected. The satellites above the red line require a time longer than the age of the solar system to despin. Hyperion is rotating chaotically and thus this analysis does not apply to it. However, this analysis does apply to Iapetus and it should not be rotating synchronously. Iapetus is relatively far from Saturn and the difficulty in despinning a distant satellite results directly

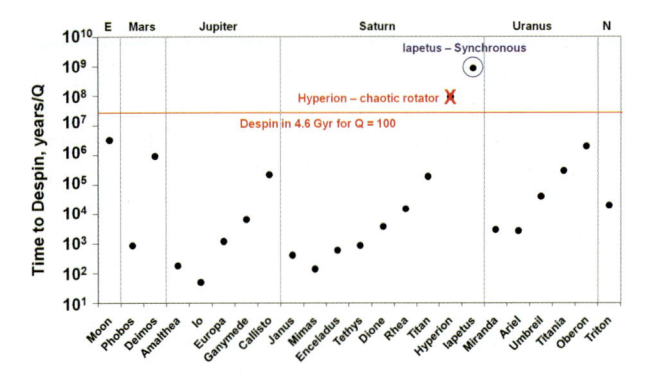

Fig. 18.4 Despinning times for satellites of the outer solar system

from the strong (inverse sixth power) dependence on distance of the despinning torque. However, the internal evolution of a satellite is also a complicating factor that was beyond the scope of the assumptions used for Peale's estimates. Castillo-Rogez et al. (2007) have shown that despinning for Iapetus proceeds non-linearly. Most of the despinning takes place when internal conditions are favorable for substantial tidal dissipation. Thus, in some cases, the heat from despinning is not available immediately but is supplied on a timescale that can stretch over several hundred My. For Saturn's inner satellites, despinning occurs in less than a few My and the heat produced by this process generally raises their internal temperatures by less than 10K.

18.3.2.2 Orbital Eccentricity

The average heating rate due to eccentricity tides is expressed as (Peale 1999):

$$\frac{dE}{dt} = \frac{21}{2} \frac{k_2}{Q} \frac{(nR_{eq})^5}{G} e^2 \qquad (18.5)$$

where n is the orbital mean motion, and e is the orbital eccentricity. A major uncertainty in the modeling of tidal dissipation is the absence of laboratory data for frequency and temperature dependent, rheological parameters. Several models for rheology have been considered in the literature. Those more frequently used are Maxwell, Cole, and Burgers. The Maxwell model is a theoretical one (see Zschau 1978) while the Cole and Burgers models are based on terrestrial analogues. The latter are especially valid for ice for temperatures greater than 230 K. Moreover, for these temperatures, the Maxwell, Cole, and Burgers models are essentially in agreement for the dissipation factor as a function of frequency (see Sotin et al. 2009 for a detailed discussion).

At lower temperatures, there is a wide range of possible values for the dissipation factor, and the discrepancy among models is further exacerbated at lower frequencies (cf. Tobie et al. 2005b). In general, the interiors of the midsize icy satellites remain cold during the larger part of their histories. Thus, better knowledge of the responses of planetary materials at orbital frequencies and low temperatures is necessary to make the modeling more accurate. For example, measurements on ice at frequencies of 10 Hz (Nakamura and Abe 1977), yielded a dissipation factor of 300 at 100 K, whereas on a theoretical basis, Showman and Malhotra (1997) have proposed that Q for icy satellites should not be greater than 10^4. For a more detailed discussion (from a planetary perspective) of dissipation in ice/rock, see Sotin et al. (2009).

18.3.3 Heat from the Gravitational Field

Heat is obtained from a gravitational field when a mass falls and potential energy is converted into kinetic energy that is subsequently dissipated as heat. This is the case during accretion as incoming material impacts the satellite's surface. Later, when the interior of the satellite experiences a phase change or differentiation and as a result mass moves downward, its potential energy is converted to kinetic energy that is viscously dissipated, producing heat. Accretion and internal evolution will now be discussed as processes that produce heat by extracting energy from the gravitational potential field.

18.3.3.1 Accretion

Accretional heating occurs at the time the satellite is formed. The temperature increase due to accretion is relatively small for the midsize satellites. It can be as high as 90 K for the largest such as Rhea and Iapetus, but is only a few tens of degrees for the smaller satellites (Ellsworth and Schubert 1983). The temperature profiles resulting from accretion can be computed using the following formula (Squyres et al. 1988)

$$T(r) = \frac{h_a}{C_p(T)} \left[\frac{4\pi}{3} \rho G r^2 + \frac{<v>^2}{2} \right] + T_i \qquad (18.6)$$

where h_a is the fraction of mechanical energy turned into material heat and varies between 0 and 1, C_p is the temperature-dependent specific heat of the material, T_i is the temperature of the planetesimals, r is the instantaneous radius, and v is the mean encounter velocity of a planetesimal with the growing satellite, but outside the satellite's gravitational sphere of influence (i.e., v_∞. For our application, $<v>^2$ is usually some small fraction (typically 1/5 to 1/3) of the square of a satellites' instantaneous circular orbital speed, $4\pi\rho G r^2/3$.

The value of h_a depends on the characteristics of accretional processes, especially the duration of accretion (Stevenson et al. 1986), and the size-frequency distribution of the accreting planetesimals. Midsized satellite accretion is considered to be a rapid process (e.g., Mosqueira and Estrada 2005; Canup and Ward 2006) that takes place in less in a few 10^5 years. Coradini et al. (1989) have proposed that h_a ranges between 0.1 and 0.5 for giant planet satellites. Recently, Barr and Canup (2008) have assumed $h_a = 0$, characteristic of a very slow accretion process where all of the accretionary heat is thermally radiated to space as it is produced. However, the precise link between h_a and the actual accretion conditions is not clear. It is also at least conceivable that a greenhouse effect developed due to the formation of an atmosphere (in a very rapid accretion case for the largest midsize satellites) and this would tend to retard the loss of heat and increase the value of h_a (see Lunine and Stevenson, 1982).

According to modeling by Squyres et al. (1988), the maximum temperature is reached at a depth of about 20 km depth. They assumed that eighty percent of the accretional energy was retained as heat. If accretion occurred very slowly, much less heat would be retained. If enough ammonia accreted to control the rheology, then early compaction and early melting could occur. Otherwise, the corresponding processes for water-ice rheology will control the evolution. In most of the satellites, the maximum temperature increase resulting from accretional energy is between 20 and 50 K. Assuming that the initial temperature is about 75 K, the ammonia creep temperature (to the extent this is known at low temperatures and stresses; Durham et al. 1993) is barely reached, and this only in a relatively small region just below the surface.

A major difference between midsized and larger icy satellites is the temperature profile immediately after accretion. A comparative plot of these temperatures is shown in Fig. 18.5. Models for Europa, Ganymede, and Titan indicate that there may have been substantial melting, if not full melting, of their interiors during accretion, for high enough h_a (e.g., Schubert et al. 1986; Deschamps and Sotin 2001; Grasset et al. 2001). In those cases the lithosphere grows by freezing of ocean water as heat is conducted to the surface and radiated to space. This cooling continues until the conditions become favorable for convection to start. On the other hand, if the ammonia concentration is significant then the midsized icy satellite interiors will reach the water-ammonia peritectic/eutectic melting point within a few hundred My after accretion, but at temperatures too low (and thus too early) for solid state convection to get started in water ice.

It is generally assumed that the satellites accreted homogeneously. This is mainly for lack of information about the accretion process. Depending on the characteristics of the accreting planetesimal (size, velocity, composition), chemical

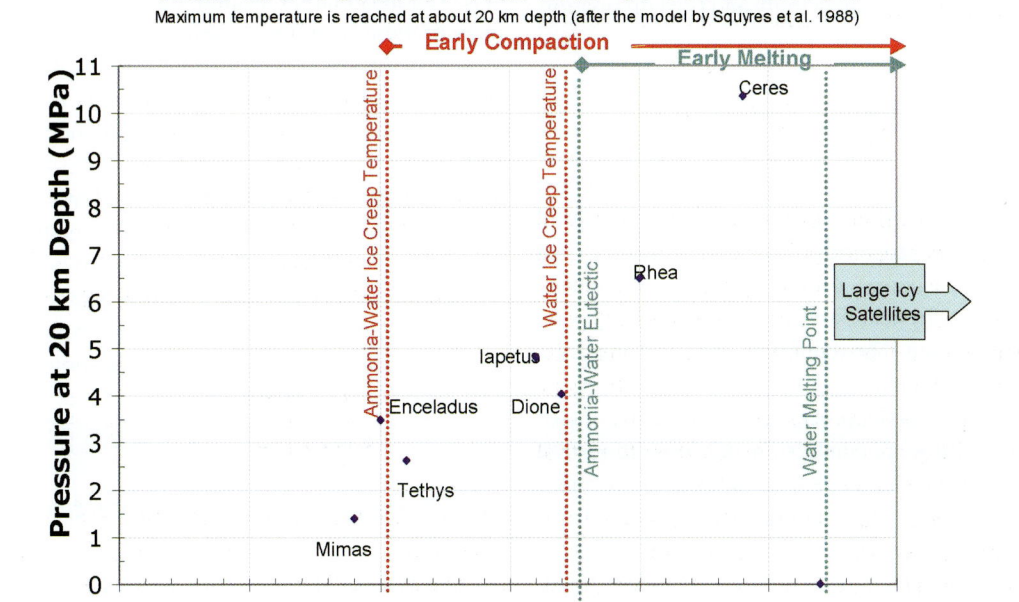

Fig. 18.5 Maximum temperatures reached by the end of accretion. The *creep temperature* is the lowest temperature at which crystals of material can glide when in contact with each other, yielding substantial deformation on a geologic time scale

and structural heterogeneities could have been present in the original satellite. Pressures generated by impacts during accretion could also have contributed to uneven compaction (e.g., Leliwa-Kopystynski and Kossacki 1994; Blum 1995). However, at the present, there is insufficient information for treating heterogeneous properties acquired during accretion in the models for midsize satellites.

18.3.3.2 Internal Conversion of Gravitational Potential into Heat

The shrinking and differentiation of a satellite releases gravitational energy in the form of heat. This phenomenon has been studied and modeled by Leliwa-Kopystynscki and Kossacki (2000).

The specific gravitational energy of self-compaction related to the closing of pores is defined by (e.g., Leliwa-Kopystynski and Kossacki 2000):

$$Eg = R^2 \left[1 - (1-\Psi)^{1/3}\right] \times 0.8\pi G\bar{\rho} \text{ (in J/kg)} \quad (18.7)$$

where $\bar{\rho}$ is the satellite's mean density, Ψ is porosity. For medium-sized satellites, the increase in temperature resulting from full compaction is a few degrees.

For phase changes and differentiation, the difference in the gravitational potential of the mass distribution before and after the event characterizes the heat produced. A quantitative assessment of how much heat is produced requires a model of the event. In general, however, this is a small source of heat compared, for example, to radioactivity and tidal dissipation.

18.4 Thermal Transfer

Modeling temperature as a function of time is important because temperature, through the temperature dependence of chemical and physical properties of ices, controls processes such as chemical reactions, orbital dynamics, tectonism, and differentiation. Physical parameters like viscosity vary greatly with temperature. The stability of chemical species is also temperature dependent.

The basic thermal transfer processes relevant to the interiors of midsize satellites are conduction and convection. The possibility of convection has obvious geological implications because convection is an important regulator of the thermal state of the interior. Determining whether or not it can occur is a key issue for understanding a satellite's evolution. Solid-state convection may also play a role in providing thermal and rheological anomalies that can change the distribution of tidal dissipation in a satellite's icy shell.

18.4.1 Heat Transfer by Conduction

Because the timescales related to heat transfer are longer than those related to accretion, one can start the models with a temperature profile described by Eq. 18.6. Assuming that lateral variations are negligible, one can use the one-dimensional, radial, conservation of energy if there is no advection of heat. Thus, in the early history, just following accretion, heat is transferred by conduction following:

$$\frac{\partial \left(k(T).\partial T(r)/\partial r\right)}{\partial r} + \frac{2}{r}\left(k(T)\frac{\partial T(r)}{\partial r}\right) + \rho(r)C_P(T)\left(\frac{dT(r)}{dt}\right) - H(r) = 0 \quad (18.8)$$

where k is the thermal conductivity, T is temperature, and H is internal heating (radiogenic, tidal dissipation) as a function of radius. Radiogenic heating is computed from the parameters presented in Tables 18.2 to 18.4.

Other things being equal, the maximum temperature that can be reached inside a satellite depends on the relative amount of the rock phase and the size of the satellite (Table 18.6). First, one can note that the maximum temperature that can be reached is equal to several hundred degrees except for Tethys whose density is so close to that of ice that very little radiogenic heating is expected. Such temperature variations suggest that ice would melt and that chemical reactions and differentiation could have happened. The second observation is that the heating rate is small (Table 18.6). For more of these satellites, it varies from 0.1 to 0.3 K/My. It implies that it takes several hundreds of million years to 1 Gy before the melting temperature of ice is reached. This calculation does not include the effect of short-live radiogenic elements, which may allow for a much faster heating rate if the accretion was completed within a few half-life times of ^{26}Al and ^{60}Fe (cf. Section 18.3.1). The melting of water ice can take place if convection does not start before the melting temperature is reached. The equations describing convective heat transfer are now described for the specific case of the mid-sized icy satellites. Then the question of the onset of convection will be addressed.

18.4.2 Heat Transfer by Convection

Considering the likelihood of convection occurring is important because both the thermal evolution and internal dynamics of the satellite depend strongly on whether or not convection occurs. For convection to operate, the viscosity of the interior must be low enough. Because these bodies are

18 The Thermal Evolution and Internal Structure of Saturn's Mid-Sized Icy Satellites

Table 18.6 Thermal characteristics of mid-sized Saturnian satellites

			Mimas	Enceladus	Tethys	Dione	Rhea	Iapetus
Radiogenic power (GW)		(a)	0.06	0.37	0.13	3.28	4.62	2.13
Radiogenic power (GW) at t = 0		(b)	0.40	2.48	0.86	21.85	30.80	14.17
Cp at 250 K		(c)	1,671.17	1,348.87	1,907.96	1,425.82	1,597.45	1,740.25
Cp at 273 K			1,789.87	1,416.64	2,064.08	1,505.75	1,704.49	1,869.87
Total temperature increase		(d)	476.24	1,286.38	55.43	1,059.67	632.72	341.63
ΔT in first 10 My		(e)	1.06	2.86	0.12	2.35	1.41	0.76
equiv heat flux (mW/m^2)		(f)	0.12	0.46	0.04	0.82	0.63	0.31
$T_{conv} = 250 K$	TBL thickness (km)	(g)	7.02	5.20	5.72	4.23	4.29	4.73
	heat flux (mW/m^2)	(h)	9.21	12.44	11.31	15.30	15.08	13.69
	Power GW	(i)	4.57	9.95	40.87	60.85	110.62	92.81
	Equivalent ΔT variations in 10 My	(j)	22.70	21.48	10.92	12.27	9.45	9.30
$T_{conv} = 273 K$	TBL thickness (km)	(g)	3.06	2.26	2.49	1.84	1.87	2.06
	heat flux (mW/m^2)	(h)	25.22	34.05	30.96	41.89	41.28	37.48
	Power GW	(i)	12.53	27.24	111.90	166.57	302.80	254.08
	Equivalent ΔT variations in 10 My	(j)	58.05	56.00	27.65	31.80	24.24	23.70
Rayleigh number at 200 K		(k)	1.74	31.7	57.9	2,970	1.51	3,280

(a) and (b) are the radiogenic heating (calculated using the values in Tables 18.3 through 18.5 and the mass fractions in Table 18.1). The values of specific heat at two temperatures are given in row (c). These values are calculated using the mass fraction of silicates and ices with the specific heat of ice being temperature dependent (McCord and Sotin 2005). The temperature increase assuming that all the radioactive heat becomes sensible heat (neglecting heat transfer) is given in row (d). The temperature increase during the first 10 My is provided in row (e) and the equivalent heat flux at present time (radiogenic heating in line "a" divided by the surface) is given in row (f). Rows (g) to (j) provide some similar information obtained with models of heat transfer by parameterized convection (cf. text). (k) gives values of the Rayleigh number at T D 200 K for the different satellites.

mainly composed of ice (Table 18.1), the viscosity of ice is the main parameter that controls the possibility of convection. Because viscosity depends strongly on temperature, the interior must become warm enough and thus the viscosity low enough before convection can start. The interior warms due to the decay of radiogenic elements contained in the rock fraction (Tables 18.3 through 18.5). At the same time the interior of the satellite is being cooled from its surface by the radiation of heat to space. There is a competition between the heating of the interior and cooling from the surface. This manifests itself in the downward propagation of a cold front.

As will be described in the following, the interiors of some of the mid-sized moons of Saturn become warm enough that convection is likely to start, eventually. The key issue is assessing when it might start and what the effect would be on the evolution of the satellite (Section 18.2.4.3). Before addressing this question, however, the use of scaling laws must be discussed. The geometry for convection within a fluid that is heated from within and cooled from above is driven by *descending cold plumes* that form at the cold boundary layer and by global upwelling (Parmentier et al. 1994). The geometry is fully three-dimensional. In addition, the curvature due to the size of these moons must be taken into account as well as the dependence of the gravitational acceleration with depth. In order to investigate a large range of parameters, scaling laws can be used to quickly assess whether convection can occur under a given set of conditions (e.g., Ellsworth and Schubert 1983; Multhaup and Spohn 2007).

Following the study of Multhaup and Spohn (2007), it is assumed that the mid-sized moons of Saturn are undifferentiated although recent studies suggest that Enceladus may be fully differentiated (Schubert et al. 2007) and that the other moons may be partly differentiated (e.g., Dione, as suggested by its surface morphology).

Because the viscosity depends very strongly on temperature, convection in these icy moons is likely to be of the "conductive lid" type (Solomatov and Moresi 2000) in which a conductive layer overlays the convective shell in which the convection takes place. In this regime, cold instabilities grow at the bottom of the conductive lid where a thermal boundary layer builds up between the lid and the convecting shell below.

After the initial onset phase when convection begins, the convection reaches a steady-state phase during which the convection process can be described by simple scaling laws. The temperature difference across the thermal boundary layer (ΔT_{TBL}) is proportional to a *viscous temperature scale* (ΔT_η) which describes how the viscosity (η) varies with temperature (T) (Davaille and Jaupart 1993; Moresi and Solomatov 1995; Grasset and Parmentier 1998). It must be noted that viscosity depends on a series of parameters including the fractional amount of rock, the deviatoric stress, and the grain size (Durham and Stern 2001). But because temperature is the parameter which most influences the viscosity, results for Newtonian fluids (i.e., strain rate proportional to deviatoric stresses) are emphasized here, and which can be justified on theoretical grounds (very low convective stresses; (Kirk and Stevenson 1987; McKinnon 2006; Barr and McKinnon 2007). In this case the viscosity follows an Arrhenius type of law:

$$\eta = A \exp\left(Q_{act}/R_{gas}T\right) \quad (18.9)$$

where A is a constant, Q_{act} is the activation energy (typically $\approx 60\,kJ/mol$ for ice) and R_{gas} is the gas constant. The constant A can be adjusted empirically to take into account the grain size and the presence of rock. If the viscosities of terrestrial glaciers and polar ices are assumed as good analogues, the constant A is chosen such that Eq. 18.9 yields a viscosity of 10^{14} Pa s at the melting point. In practice, this value may vary by a factor 10 or more and the effects of different values can be investigated. The viscous temperature scale, ΔT_η, is then defined by

$$\Delta T_\eta = \left|\frac{1}{\partial ln(\eta)/\partial T}\right|_{T=T_m} = \frac{R_{gas}T_m^2}{Q_{act}} \quad (18.10)$$

where T_m is the temperature of the convective interior. Following the experimental work by Davaille and Jaupart (1993) and the three-dimensional, numerical study by Grasset and Parmentier (1998), the temperature variations across the thermal boundary layer, ΔT_{TBL}, can be expressed by

$$\Delta T_{TBL} = T_m - T_c = 2.23\frac{R_{gas}T_m^2}{Q_{act}} \quad (18.11)$$

where T_c is the temperature at the base of the conductive lid. The constant has been empirically determined by the inversion of laboratory and numerical experiments (e.g., Grasset and Parmentier 1998; Solomatov and Moresi 2000).

It is then possible to determine the heat flux that can be transported by convection, because the thickness of the thermal boundary layer (δ) is controlled by the growth of thermal instabilities through the *thermal boundary layer Rayleigh number* (Ra_{TBL}), which is defined by

$$Ra_{TBL} = \frac{\alpha\rho g \Delta T_{TBL}\delta^3}{\kappa\eta} \quad (18.12)$$

where α, ρ, g, and κ are the volume thermal expansion coefficient, density, gravitational acceleration, and thermal diffusivity, respectively. For a volumetrically heated fluid, the value of the thermal boundary layer Rayleigh number is about 20 (e.g., Sotin and Labrosse 1999, and references therein). This value depends slightly on the global Rayleigh number itself (Ra), which is defined by

$$Ra = \frac{\alpha\rho g H' R^5}{k\kappa\eta} \quad (18.13)$$

Where H' is the volumetric heating rate in W/m³, which contains both the radiogenic heating, tidal heating if any, and the cooling or heating rate,

$$H' = H_{rad} + H_{tid} - \rho C_p\,(\partial T/\partial t) \quad (18.14)$$

Using Eqs. 18.11 through 18.13, one can determine the convective heat flux (q_{conv}):

$$q_{conv} = k\frac{\Delta T_{TBL}}{\delta} = k\frac{\Delta T_{TBL}}{R}\left(\frac{Ra\Delta T_{TBL}}{Ra_{TBL}H'R^2/k}\right)^{1/3}$$
$$= H'R\left(\frac{Ra}{Ra_{TBL}}\right)^{1/3}\left(\frac{\Delta T_{TBL}}{H'R^2/k}\right)^{4/3} \quad (18.15)$$

With this approach, one can determine how much heat can be removed by convection. If the heat that can be removed by convection is smaller than internal heat sources, then the internal temperature increases, the viscosity decreases, the Rayleigh number increases and the convective heat flux increases. Table 18.6 gives two examples, one for a temperature close to melting temperature (viscosity close to 10^{14} Pa s), and the other for a temperature 20 K lower which corresponds to a viscosity almost one order of magnitude larger. The tabulated results show that the amount of the convective heat that can be transported is much higher than the production by radiogenic heating. The temperature increase due to radiogenic heating (row) can also be compared with the temperature decrease due to convection (rows j) to show the effect (Table. 18.6). It is obvious that for viscosities smaller than 10^{16} Pa s, convection processes are sufficient to remove the radioactive heat as it is produced.

Table 18.6 shows that convection is very efficient at removing heat. In the case of Enceladus, row (i) shows that convection can support fluxes of 9–20 GW of heat to the surface. Barr and McKinnon (2007) derived somewhat lower heat flows, but they specifically considered a differentiated Enceladus. In any event, these heat flows are global values, and an important issue for Enceladus is being able to concentrate the heat flux within a relatively small geographic area at the south pole (see Chapter 21). In the case of Iapetus, such an efficient process would have quickly removed so much heat that Iapetus would not have evolved to its present day properties. The conclusion of studies that have used scaling laws to describe convection (e.g., Ellsworth and Schubert 1983; Multhaup and Spohn 2007) is that convection is very efficient at viscosities occurring at temperatures lower than the water-ice melting point. They predict that temperatures would not become high enough for ice to melt and that convection would prevent the differentiation of the satellite. A legitimate question is whether the amount of rock in each satellite, if undifferentiated, sufficiently stiffens the ice so that ice melting occurs before convection can start as internal temperatures increase (as in the classic paper by Friedson and Stevenson 1983). This question was not specifically addressed by Multhaup and Spohn (2007), although the rock volume fractions in question (Table 18.1) are generally not very high, so this is not likely to be a significant effect for most of the midsized satellites. Another way to look at this is to ask whether convection can start at all. This question has been addressed by Barr and Pappalardo (2005) and Barr and McKinnon (2007) and is discussed in the following section.

18.4.3 Onset of Convection

Linear stability analysis (Turcotte and Schubert 1982; McCord and Sotin 2005) shows that thermal anomalies grow and lead to convection once a critical value of the Rayleigh number has been reached. This critical value depends on the wavelength of the anomaly, on the heating mode (basal versus internal) and on the amplitude of the anomaly (Barr and Pappalardo 2005). These linear stability analyses suggest that typical values of the critical Rayleigh number for convection to start are around 10^3. In order to compare the different mid-sized Saturnian satellites, Rayleigh numbers at a given temperature have been calculated for each satellite using Eq. (18.13) where the radius has been replaced by an effective thickness equal to one third of the radius. This allows us to compare with analytical solutions that are obtained in Cartesian coordinates instead of spherical coordinates. The different parameters have been calculated using volumetric average of ice and silicates. Values of thermal expansion, thermal conductivity and thermal diffusivity for ice and silicates were taken from McCord and Sotin (2005). Equation 18.9 is used for viscosity with constant A being chosen such that the 10^{14} Pa s is obtained at melting temperature (T = 273 K). With these data, the Rayleigh number at T = 200 K varies from 1.5 for Rhea to 3,280 for Iapetus. Actually, Dione and Iapetus have Rayleigh numbers two orders of magnitude larger than those of Enceladus and Tethys and three orders of magnitude larger than those of Mimas and Rhea. It suggests that convection would start at much lower temperature (higher viscosity) within Iapetus and Dione. However the mid-sized Saturnian satellites have many complexities including the spherical shape, the rate of conductive cooling versus the heating rate of the interior, the strong temperature dependence of viscosity, and the depth dependence of the gravitational acceleration. Although all of these complexities have not yet been extensively explored, the following paragraphs describe some of their effects.

A density perturbation must occur before buoyant forces can start parcels of "fluid" in motion. In conductive lid convection the density perturbation might be due to a temperature variation (in the sense of being cooler) caused by irregularities in surface topography. If the size of the perturbation is sufficiently large, both in terms of its density contrast and its physical size (and depending on rheology), the onset phase of convection will be initiated. Viscosity and gravity control the velocity with which the parcels can move. The three dimensional size of the volume where convection will occur sets the distances that the parcels must traverse. These factors govern the time it takes for convection to start.

The time required to reach a temperature large enough for the Rayleigh number to exceed the critical value depends on the amount of internal heating available. Table 18.6 (line e) provides some numbers for each satellite. As discussed previously, the temperature must be large enough for the viscosity to reach a value that yields a Rayleigh number larger than the critical value. This table suggests that it takes from some tens to hundreds of millions of years up to a billion years if the long-lived radiogenic elements are the only source of internal heating. During this time, the satellite is cooled from above. Determining the onset of convection for a fluid cooled from above and heated from within has not yet been performed for materials having a strongly temperature dependent viscosity. On the other hand, the scaling rules that govern the onset of convection have been studied in the case of a hot fluid cooled from above (Davaille and Jaupart 1994; Choblet and Sotin 2000; Korenaga and Jordan 2003; Zaranek and Parmentier 2004; Solomatov and Barr 2006). These rules have been applied to convective instabilities in large icy satellites (Barr et al. 2004), instabilities of the oceanic lithosphere (Korenaga and Jordan 2003), and the onset of convection for Mars (Choblet and Sotin 2000). Although these models can be applied to instabilities at shallow depth, they are *not* the designed to describe what happens much deeper in a satellite where curvature and gravity variations must be taken into account.

The onset of convection is initiated once the internal viscosity, which is strongly temperature-dependent, reaches the critical value for the colder fluid above to become unstable. Work by Barr and McKinnon (2007) provides a detailed consideration of this for Enceladus and their results are qualitatively applicable to similar midsize satellites. Assuming that the upper layer is pure water ice Ih, they found that the driving stresses due to buoyancy were quite low, and (in the absence of strong tidal stresses) would probably be accommodated by volume diffusion creep of the ice crystals. Under these conditions they found that convection could start within

Enceladus only if the ice grain size was very small, ~0.3 mm. They noted that for this to be the case, the grain sizes have to be kept small, suggesting the presence of other processes or materials that are rigid compared to water ice, such as small rock particles, salts, hydrated sulfates, and/or clathrates, to pin the grain boundaries and limit grain growth. With regard to the larger midsize satellites, Ra is such a strong function of R (see Eq. 18.13) that the grain size (really, viscosity) limit is swiftly relaxed for satellites the size of Tethys and Dione and larger (McKinnon 1998).

18.4.4 Convective Evolution and an Example

There are two time periods during which convection might start in the medium-sized icy satellites: (a) in their early history (<10 My after formation), as a result of heat produced by the decay of SLRI, (b) later, as the decay of LLRI warms the deep interior up to temperatures at which the viscosity of ice becomes low enough for subsolidus convection to occur. As previously discussed, the Rayleigh number has to exceed some critical value. Application to the Saturnian mid-sized satellites requires including the effects of two processes which are the cooling from above and heating from within. Robuchon et al. (2009) provides some examples. Their study uses a viscous law, described by Eq. 18.9, with an activation energy of 60 kJ/mol and a viscosity of 10^{14} Pa s at melting temperature. With these conditions, the convective instabilities start (see also discussion above). For the viscosity law used by Robuchon et al. (2009), illustrated in Fig. 18.6, this condition occurs when the temperature is above 240 K. The Rayleigh number at the onset time is larger than 10^6, which is much higher than the critical value.

Now, down-welling plumes may be prevented from extending all the way to the center of an undifferentiated satellite. First, consider the spherical geometry of such a small satellite as being like the layers of an onion. With successively smaller areas (and volumes) for each lower layer, the relative amount of ascending hot material would very quickly be overwhelmed by the relatively larger amount of cold plume material coming down from above. Second, (for a homogeneous satellite) the gravity decreases with depth toward a value of zero at the center. Between radii of 400 and 200 km, for example, the gravity acceleration decreases by a factor of two, resulting in a severe decrease in the buoy-

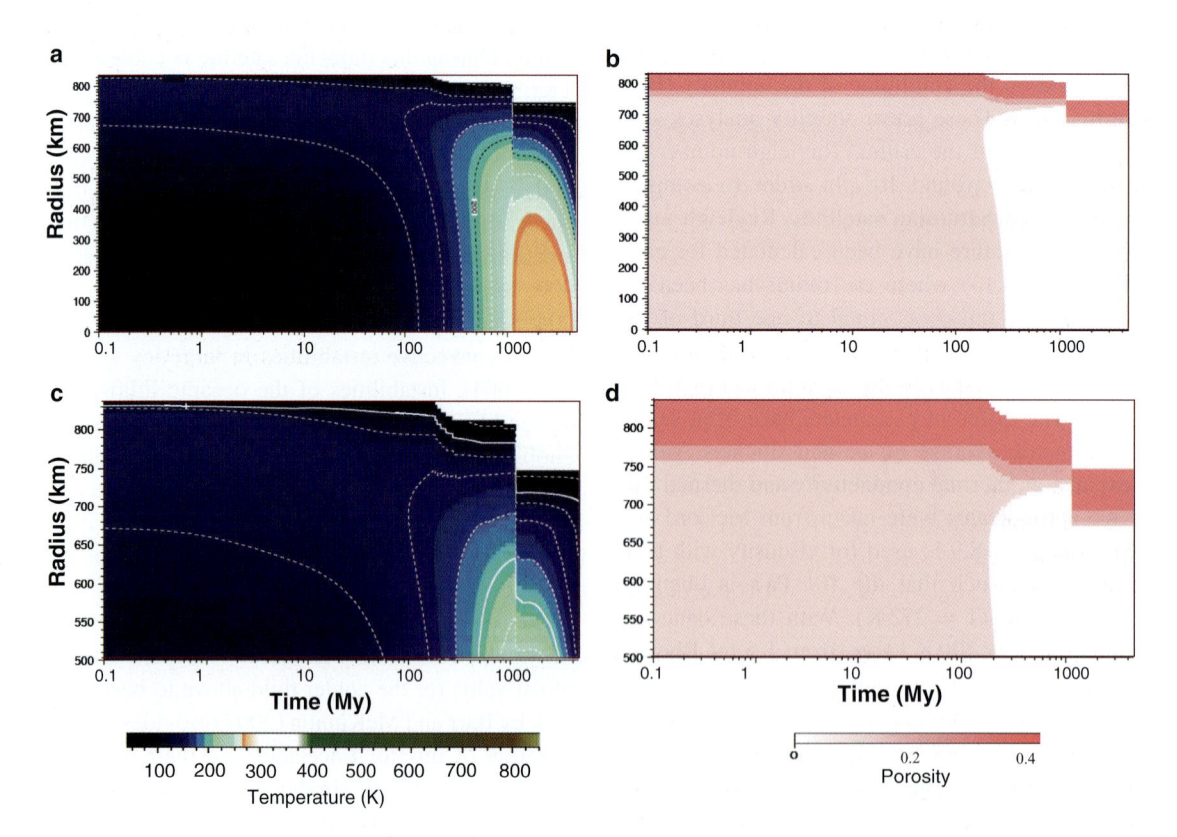

Fig. 18.6 Illustration of the evolution of porosity using a model for Iapetus. The rock/ice ratio is initially uniform, and only accretion and long-lived radiogenic heating is considered. The ordinate is equatorial radius; the abscissa is time (on a log scale). The time at the extreme left is the start of the model (time of accretion); at the extreme right is the present. The lower panels are magnified views of the upper 350 km. *Left-hand panels* show temperature; *right-hand panels* show the corresponding porosity. Temperature contour interval is 25 K. The *left-hand* color scheme highlights geophysically significant temperature regions. Adapted from Castillo-Rogez et al. (2007)

ancy force. These two effects may constrain any convection to a shell between about 200 and 400 km from the center (if a satellite is large enough), leaving the center of the satellite free of convection.

The temperature-dependence of viscosity is critical to initiating the onset of convection. As a fluid cools from above, a relatively denser, colder layer is formed at the surface, but its viscosity is too large for convective instabilities to start. Thus, a conductive "lid" forms, with the temperature inside the lid increasing with depth. When the temperature is high enough, a thermal boundary layer is formed, defining the base of the lid. Here convective instabilities can start and grow (Multhaup and Spohn 2007). A convective instability begins once the thickness of the thermal boundary layer and its Rayleigh number (Eq. 18.12) reach critical values and the viscosity contrast across the boundary layer is less than one order of magnitude (Solomatov and Barr 2006).

There are at least two additional complexities that need to be discussed. First, if dissipation is important for the satellite in question, the viscosity used for convection is not necessarily the same as that used for tidal dissipation, which is frequency dependent. Internal temperatures can be favorable for both despinning (for example) and convection. In such a case, because despinning is a very rapid process with respect to the time needed for the onset of convection, despinning finishes before convection reaches steady-state. Second, the onset of convection is further delayed by the accretional temperature profile, which is coldest at the center and warmest near the surface. This is not conducive to convection and must be overcome before convection can start. This takes 10–300 My (Ellsworth and Schubert 1983; Multhaup and Spohn 2007).

Depending on the temperature at which convection starts, it takes from 2 My (at 250 K) to 20–200 My (at 220 K) to reach steady-state (Sotin et al. 2006; Multhaup and Spohn 2007). However, the initial pulse in the heating rate due to SLRI decay, if any, lasts less than 10 My and then can no longer contribute to sustaining the development of convection. Numerical simulations suggest that convection would exist for a few million years in this case. As soon as convection starts, the amount of heat that can be removed exceeds the radiogenic heating. Convection cools the interior down to a temperature at which the viscosity becomes too high for convection to continue. With the viscosity law used in Robuchon et al. (2009) convection stops when the temperature is equal to 220 K. At that time the LLRI are the only remaining, significant heat source. The heating rate is such that high temperature can be attained only at large depths. For Iapetus, it is then possible that convection may have existed very early in its history, preventing high enough temperature (i.e., low enough viscosity) from being achieved and thus not permitting despinning during these first several millions of years. However, we note again our earlier caveat that at large depths in these satellites the effects of increasing curvature with depth and waning gravity sap the vigor of convection, making it less effective.

The time needed for LLRI to heat the deep interior up to 240 K (the approximate temperature for the onset of convection), can be several hundred million years. The arguments developed above on the effect of sphericity and the lack of numerical work on convection in an infinite Prandtl number regime, volumetrically heated, self-gravitating sphere, lead us to assume that in any case an inner core of about 200 km in radius could be free of convection and would provide a low viscosity volume where tidal dissipation could effect despinning.

We close this section with a final caveat reminding the reader that much depends upon accurate data for the rheological properties of the materials of which the satellites are actually composed. Presently much of this data does not exist, a topic that will be revisited in Section 18.9.2.

18.5 Constraints on Thermal Parameters

A major difference between the present models and "classical" models of the Saturnian satellites (e.g., Ellsworth and Schubert 1983; Schubert et al. 1986) comes from the use of temperature-dependent thermal parameters, especially thermal conductivities and specific heats. In this section we focus on the effect of the material's structural and chemical properties on the thermal conductivity. The reader seeking more information is invited to consult the review by Ross and Kargel (1998).

18.5.1 Ice Thermal Conductivity

The thermal conductivity of pure water ice ranges from $\sim 6.2\,\text{W/m/K}$ at 100 K to 2.3 W/m/K at 270 K (Petrenko and Whitworth 1999). A review of ice thermal properties has been published by Ross and Kargel (1998, Fig. 1). Data show that the thermal conductivity of some clathrates can be one order of magnitude less compared to that of water ice. The thermal conductivity of ammonia-water has been crudely measured by Lorenz and Shandera (2001). They show that the thermal conductivity of ice can be decreased by a factor of two to three by increasing its ammonia content.

18.5.2 Rock Thermal Conductivity

The thermal conductivity of the rock phase is not that well constrained, or at least not that well appreciated, for the full range of temperatures and compositions expected in icy satellite interiors. Values of 3–4 W/m/K are often used (e.g., McCord and Sotin 2005; Multhaup and Spohn 2007), which are typical STP values for igneous and metamorphic rock, although the higher values are more applicable to ultramafic (i.e., mantle) rocks (Turcotte and Schubert 2002 Appendix 2, Table E). Crystalline rocks are considerably more conductive at low temperatures, and less conductive at higher temperatures, whereas amorphous minerals (glasses) do not show pronounced temperature dependence (Hofmeister 1999). Hydrated silicate conductivities at STP range between 0.5 and 3 W/m/K (Clauser and Huenges 1995) depending on composition.

18.5.3 Effect of Porosity

Porosity can have a significant effect on thermal conductivity (see Ross and Kargel 1998, for a review). The affect of porosity is difficult to assess accurately because it is a function of the microstructure of the material. Indeed, depending on porosity the thermal contacts between the grains will be more or less efficient in transferring heat. Results presented below from Castillo-Rogez et al. (2007) follow the approach of McKinnon (2002) who defined two bounds for thermal conductivity. The upper bound is a function of the *effect of void volume fraction* on the equation for the thermal conductivity of a mixture. The lower bound is set by making a rough assessment of the effect of the *structural arrangement* and the resulting contact between grains. Based on laboratory and other constraints (e.g., Ross and Kargel 1998), McKinnon (2002) suggested (for modeling) purposes that imperfect thermal contacts in a rubble or otherwise unconsolidated layer can decreases the thermal conductivity of such a rock-ice-void mixture by up to an additional order of magnitude. Based on geological analogues, we infer that the upper layers (and regolith) of a midsize icy satellite could be as much as 40% porous (also see discussion in McKinnon et al. 2008). We estimate that the thermal conductivity should be approximately less by a factor of 10 compared to solid. This value also roughly corresponds to laboratory measurements performed on lunar regolith (Langseth et al. 1976) and the conductivity of Europa's surface inferred from telescope observations of thermal emission during eclipses (Matson and Brown 1989), and detailed (or at least complex) theoretical models of cometary conductivity (Shoshany et al. 2002). How deep a given regolith extends depends on the satellite in question and its respective thermal and bombardment history (e.g., Eluszkiewicz et al. 1998), aspects which are very poorly understood at present.

18.6 Structural Evolution

Structural evolution needs to be considered over a range of three different size scales: micro- and meso-scales (multi-scale porosity decrease), and global scale (melting, differentiation, ice phase change). These processes are driven by the material rheology as a function of temperature, pressure, composition, and structure. Volume changes and conversion of some gravitational energy into heat occur when mass is redistributed. We will now discuss changes due to the evolution of porosity, melting, differentiation, and chemical alteration.

18.6.1 Porosity Evolution

As the internal temperature increases, ice creeps and porosity decreases as a function of pressure and composition of the material. We apply the results and empirical relationships developed by Leliwa-Kopystynski and Maeno (1993) and Leliwa-Kopystynski and Kossacki (1995) noted earlier. Although these have only been defined for a limited range of compositions and ice properties compared with the range of conditions expected in the midsize Saturnian satellites, they do give some insight into the compaction kinetics over a few tens of millions of years when the ice is at what we term the creep temperature and for a few million years when the ice is close to the melting point (or solidus). We define the *creep temperature* as the lowest temperature for which ice can viscously deform and release elastic stress (and so reduce porosity). It is related to the well-known Maxwell time (e.g., Turcotte and Schubert, 2002), and is equivalent to the elastic blocking temperature mentioned in Section 18.4, or the brittle-ductile transition temperature. Now, because solid-state creep is a thermally activated process, there is no one, fixed creep temperature; rather it depends on the time scale involved, stress level, and composition. It is nonetheless a useful benchmark. For satellite modeling purposes, the creep temperature can be taken ∼176 K for pure water ice (Durham et al. 1983). If sufficient ammonia is present in the ice then the creep temperature may be depressed to as low as 100 K (Leliwa-Kopystynski and Kossacki 1995).

To give a specific example, we show how porosity evolves in a model for Iapetus (Fig. 18.6). The initial porosity profile is based on Durham et al. (2005) and Kossacki and Leliwa-Kopystynski (1993). It has a porosity of 45% at the surface, decreasing approximately exponentially with depth to about 10% at 120 MPa, following the laboratory values of Durham et al. (2005). The mechanisms for the collapse of porosity are as discussed in Section 18.2. In the model, the creep temperature is reached between 200 and 300 K; as internal temperatures increase the porosity collapses. The thermal conductivity becomes close to that for solid ice and is about an order of magnitude more conductive than the porous material. As porosity decreases, the radius of the satellite decreases from about 850 km to about 800 km. This occurs between ∼200–300 My. The gradual radius reduction is due to the creep temperature being reached, and densification achieved, at different times at different depths. A second episode of radius change occurs much later, at ∼1 Gy, when the satellite has developed conditions (solidus temperatures) favorable for tidal dissipation. As a result, despinning occurs rapidly, the model becomes more spherical, and the equatorial radius changes from about 800 km to the present day 734.5 km, but there is no global volume change. The evolution of the porosity profile is best seen in Fig. 18.6d.

We note that the timescales for porosity collapse are much shorter than the timescale for heating by LLRI decay. However, the internal heating sets the time when compaction can occur.

18.6.2 Melting and Differentiation

The mechanisms by which the rock phase separates from the ice have not been studied in detail. Clearly, if the rocky phase is agglomerated in relatively large chunks in an icy matrix, it will be easier for differentiation to occur (Friedson and Stevenson 1983; Nagel et al. 2004). De La Chapelle et al. (1999) show that when a few percent of pure water ice melts, separation of the liquid (water) from ice (by gravity) occurs quasi-instantaneously (as might be expected). Obviously, for sufficient ice melting in an icy satellite, rock particles can be released and differentiation can occur.

At temperatures below the melting point of pure ice, the separation of rock from ice is more complicated. For example, if ammonia is present in the ice, when the ammonia-water peritectic/eutectic temperature is reached (∼176 K), the remaining water-ice-and-rock matrix would be too rigid to convect, but it may be able to deform because the water ice creep temperature has been reached or because residual, interstitial melt weakens the ice. Ammonia-rich melt should be able to percolate upward under the action of gravity, but it is not clear whether downward rock separation is feasible at all (the depleted ice-rock residuum would be even stiffer than before). This is especially true if no more than a few percent ammonia is present, which may be the case for the icy satellites. Furthermore, the time scale of rock separation, by either ice melting or Stokes flow, is poorly constrained. Thus, in a fundamental way, we do not know how rapidly subsolidus differentiation (for example) progresses compared to other processes taking place in the satellite (i.e., ice-rock mixture heating from LLRI decay versus cooling).

An understanding of the above processes is crucial, because differentiation affects the long-term thermal evolution when and if the radiogenic heat sources are concentrated in a rocky core. Depending on the temperature at which differentiation occurs, conditions may or may not be favorable for tidal dissipation. In the long term, the warming of the core can heat the ice phase from below and trigger the development of hot upwelling plumes in the ice. However the conditions under which this develops in very cold ice and the timescale for it starting (once conditions are favorable) have not been studied in detail. Differentiation and its aftermath are deserving of much continued study.

Two scenarios can be envisioned for the process of differentiation when ammonia is mixed with ice: (1) the ammonia concentration is significant and the rock chunks large enough for differentiation to take place; (2) the ammonia is present in a concentration of only a few percent and the rock cannot separate from the ice when the eutectic temperature is reached. At present there is no definitive constraint on the amount of ammonia needed for favoring one process or the other.

18.6.3 Long-Term Evolution of a Rock Core

At present, Enceladus is the only midsized satellite that is generally thought to have formed a rock core (e.g., Barr and McKinnon 2007; Schubert et al. 2007; Roberts and Nimmo 2008; Tobie et al. 2008). The reader is referred to the Enceladus chapter in this book for a detailed discussion of Enceladus. Nevertheless, given the possible presence of SLRI, if early conditions differ from the present estimates, other satellites might have formed rocky cores as well. If that proves to be the case, and differentiation occurred, then depending on the temperature at which differentiation occurred, hydration of the silicate phase could have taken place. Silicate hydration kinetics are temperature dependent. Thermodynamically optimum conditions for hydration start at 300–350 K and such temperatures have been observed in hydrothermal sites on the Earth (e.g., Kelley et al. 2005). Under these conditions, the serpentinization (for example) of the bulk of the rock phase can be completed in less than 100 years, assuming that the rock phase is relatively finely divided and in contact with water. Thus, in an icy satellite it is possible that hydration of anhydrous rock (and metal) could go to completion while differentiation is still taking place, and thus before core formation has been completed. On the other hand, hydration reactions will be slow at low temperatures. For example, if differentiation could occur via ammonia-water eutectic melting, hydration kinetics might be sufficiently inhibited, and lead to the formation of an anhydrous core. The reactivity of cold ammonia-water solutions with anhydrous minerals would, however, make an interesting laboratory study.

During serpentinization, olivine and pyroxene are altered into the serpentine minerals, brucite and oxides, such as magnetite (Scott et al. 2002). These reactions are also accompanied by an abundant release of H$_2$. An example of the reaction (assuming 20% Fe and 80% Mg by mole) is (McCord and Sotin 2005):

$$2\,(Fe_{0.2}, Mg_{0.8})_2\,SiO_4 + 3H_2O = Mg_3Si_2O_5\,(OH)_4$$
$$+ (Mg_{0.2}, Fe_{0.8})\,(OH)_2$$

or

$$2\text{ olivine} + 7.4/3\,H_2O = 1\text{ chrysotile} + 0.2\text{ brucite}$$
$$+0.8/3\text{ magnetite} + 0.8/3\,H_2$$
(18.16)

Such serpentinization can produce a rock volume increase of up to 60% (e.g., McCord and Sotin 2005) and is accompanied by the release of 233 kJ per kg of rock (Grimm and McSween 1989). Also, other geochemical activity is likely to take place with reactions such as the production of methane from the Fischer-Tropsch reaction (e.g., Giggenbach 1980; Atreya et al. 2006) or amino acid synthesis (Shock and McKinnon 1993). Matson et al. (2007) have proposed that such an environment inside Enceladus was favorable to the decomposition of ammonia into the molecular nitrogen (N$_2$) that is observed in Enceladus' plumes.

This context also favors hydrothermal circulation, as has been described for terrestrial analogues (e.g., Kelley et al. 2005). Hydrothermal circulation is expected to play a significant role in cooling Enceladus' core, if the core was heated by SLRI (Matson et al. 2007; Glein et al. 2008). Many parameters are involved in modeling hydrothermal circulation, however. A major one is the permeability of the material, which governs the depth of penetration of the water flow, and thus, its capacity to cool down, and chemically interact with, the core (e.g., Travis and Schubert 2005). McKinnon and Zolensky (2003) discuss the fact that sulfur compounds released during the interaction of rocks and warm water, and bulking due to alteration (the volume change referred to above), could lead to deposits that plug and/or collapse the porosity near the surface of the core (and below) and slow down hydrothermal circulation.

18.7 Model Studies of Thermal and Dynamical Evolution

Modeling the midsize satellites requires simultaneous simulation of a satellite's thermophysical and dynamical evolution. This is necessary because each affects the other. For example, as the interior of a satellite warms up, its temperature and frequency-dependent rheology properties change and it responds differently to the external gravitational field. Consequently, its tidal response changes and this, in turn, directly affects its despinning (if not already despun) and the evolution of inclination, eccentricity, and semimajor axis. The changing dynamical parameters, in turn, affect the interior of the satellite by presenting it with a different, time-variable (i.e., spin and orbit), gravitational field. The tides and the amount of tidal dissipation will now be different and this will be reflected by a change in the internal temperature distribution that affects the interior's rheology, bringing us full cycle, back to where this example started.

At the practical level, this means that all of a simulation's parameters must be updated at appropriate computational time steps. For example, in the models reported by Castillo-Rogez et al. (2007) for Iapetus, a one-dimensional temperature field is used to calculate heat transfer and the new temperature field. All of the known sources and sinks of energy are included (e.g., Section 18.3) as well as applicable heat transfer processes (e.g., Section 18.4). The new temperatures are used to update parameters related to tides and tidal dissipation and these, in turn, feed into calculations of the changes in the rotation rate and the orbital elements. Thus, the material properties are updated at each time step based on the current temperature and dynamical state. Part of this also involves keeping track of size, shape (e.g., changes in volume and surface area) and structure (e.g., phase changes, degree of differentiation, etc.).

In the following we shall see what such models have to say about the initial composition of the satellites, and their global evolution, with Iapetus as the prime example. The latter topic encompasses issues such as differentiation, lithosphere formation, and global shape and age.

18.7.1 Effect of Initial Composition

An important uncertainty in initial composition is how much ammonia was present. Depending upon the amount, it can affect the evolutionary path of a satellite. To investigate this and other effects model studies were carried out by Castillo-Rogez et al. (2007) for three general cases: (1) the volatile phase was only pure water; (2) a small amount of ammonia was present such that it plays a role in ice creep but is rather ineffective in causing differentiation; (3) a larger amount of ammonia is present such that it plays a significant role, enabling internal differentiation (though we stress that the latter is an assumption, and generally requires large rock "particles" or batch melting).

The main result is that depending on the composition, conditions favorable for ice creep compaction occur relatively late in the history of a satellite. The presence of any volatiles that depress the water ice melting point helps internal differentiation (at least to some degree) take place.

The heat from SLRI decay appears as a heat pulse during the first few million years following accretion. This results in a decrease in porosity early in the histories of the satellites (see Section 18.6). The ammonia-water eutectic temperature is reached just below the surface, but the ice that is ammonia free remains solid.

18.7.1.1 Rock-Rich Models

Some satellites, such as Enceladus or Dione, have more rock than ice by mass. Models for these bodies that include ^{26}Al show quick melting that enables differentiation that then proceeds rapidly, with the rock phase descending to form a core. This occurs whether or not ammonia is present, because the water ice melting point can be reached in a few million years, and in this case convection cannot transfer heat as efficiently as it is supplied by ^{26}Al decay (see Section 18.4). Furthermore, conditions may become favorable for mafic silicate alteration (e.g., serpentinization) under optimum kinetics conditions (see Section 18.6). In these cases, serpentinization of the whole silicate phase can be achieved on a rapid timescale (Section 18.6.3).

The decay of ^{60}Fe provides a second, smaller but broader, pulse of heat during the five million years following accretion. As a result the core temperature rapidly reaches several hundred K, depending on when the model started with respect to the formation time of CAIs. For extreme models with intervals as short as 2 My or less, the boiling point of water and even critical point can be reached inside cores with a large rock fraction. In the long-term an outer icy shell develops, made of relatively pure water with a potentially salt-and/or-sulfate water solution at its base (e.g., Kargel et al. 2000; McCarthy et al. 2006). The shell has little porosity except for that caused by impact bombardment and large-scale faulting and tectonic activity. Soluble volatiles and salts released during hydrothermal circulation concentrate in a deep, ocean-like layer, with volatiles such as H_2 probably escaping (Zolotov and Shock 2003; Zolotov 2007).

18.7.1.2 Rock-Poor Models

The effect of SLRI decay is less dramatic in the case of rock-poor bodies such as Tethys. However, the heat pulse can be significant enough to trigger early compaction of the deep interior. This increases the thermal conductivity, accelerates cooling, and leads to a colder interior. We will return to this point below when we consider some models for Iapetus (Section 18.7.2, Fig. 18.7).

18.7.2 Global Evolution

18.7.2.1 Assessing the State of Differentiation of an Icy Satellite

The long-term evolution of a satellite's core temperature is a function of: (1) the nature of the rock phase, hydrated or not, and its thermal properties, especially conductivity; (2) the initial amount of SLRI that can increase the temperature by a few hundred degrees. Depending on the core pressure and temperature profiles, different processes can take place. Most probably, if the largest satellites are differentiated, their cores could be stratified into an internal dehydrated central core and a hydrated outer core.

The final thermal state and internal structure depend on whether major temperature thresholds are achieved: (1) the temperature at which the viscous creep of ice becomes important which is a function of composition; (2) the melting temperature, also a function of composition; (3) the onset of convection. The depth at which these temperatures are achieved strongly influences the geological evolution of the satellites.

We identify two chief evolutionary paths and they depend on the initial conditions, especially the time of formation with respect to CAIs. Short-lived radiogenic isotopes always bring about an early porosity decrease. In the long term this results in the development of a lithosphere that is more elastically rigid than for models without SLRI. Other consequences include earlier differentiation and the formation of a rocky core, and related processes, such as hydrothermal activity, that can give rise to interesting geochemical environments. On the other hand, models without SLRI will be cooler and are likely to become, at most, only partially differentiated in the long term.

Now we will compare some models with and without SLRI and times of formation with respect to CAIs of 2 to 5 My (i.e., time when the model was started) (Fig. 18.7). The choice of formation time sets the amount of SLRIs present. In all cases the composition of the model is pure water ice and rock (as defined in Section 18.2). The heat sources include accretion (assuming several heat retention values, h_a) and the decay of long-lived radioactive isotopes (LLRI), and different amounts of SLRI as indicated by the time after CAI at which the model was started (t_0-CAIs). The lower bound on the initial ^{60}Fe/^{56}Fe ratio is 0.5×10^{-6} and the upper bound is 1.0×10^{-6} (see Section 18.3.1). The amount of ^{60}Fe, even at the upper bound, does not influence the models very much.

The model "(a)" is the hottest of the four models shown in Fig. 18.7. It started at 2.0 My and has the highest concentration of SLRIs. The accretion heat-retention coefficient is moderate, 0.5. It has a minimal amount of ^{60}Fe, with

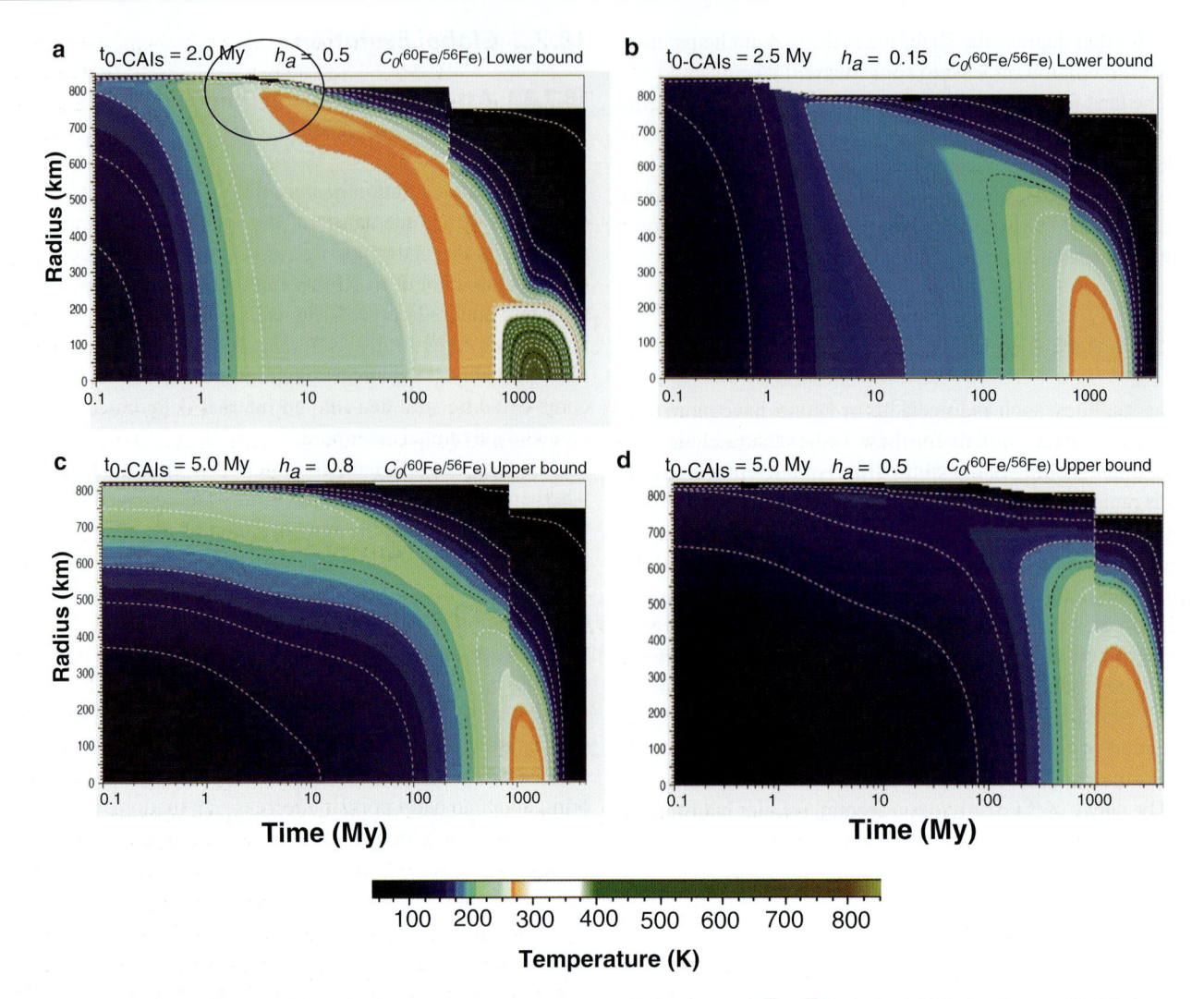

Fig. 18.7 Sensitivity of thermal evolution to parameter values in several models for Iapetus. Temperature plots are as in Fig. 18.6. Pure water ice is assumed for the ice phase. (t0-CAIs) is the time when the model started in My after CAIs condensed. h_a is the accretion-heat-retention coefficient, and $C_0(^{60}Fe/^{56}Fe)$ is the initial abundance ratio. Their typical radial grid size was 5 km and the time step was $\sim$100 years, for a three billion year evolution. Adapted from Castillo-Rogez et al. (2007)

$C_0{}^{60}Fe/{}^{56}Fe$ at the lower bound. Within 10 My a significant collapse of porosity has started and some melting occurs near the surface. The big change in radius at $\sim$250 My is due to change in shape as the model despins. By $\sim$300 My much of the interior ice has melted (orange tone) and the rock has sunk to the center. By $\sim$1.5 Gy temperatures in excess of 600 K have been reached at the center. After that, the model cools down. (The method of handling tidal dissipation is the same in all of these models and is discussed in Section 18.3.2).

Model "(b)" has significantly less initial SLRIs, having started at 2.5 My. Its value of $h_a = 0.15$ means that 85 percent of the heat of accretion escapes. This corresponds to a slower rate of accretion than model "(a)". There is ample time for the heat deposited by infalling material to be conducted to the surface and radiated to space. From the center outward $\sim$200 km, ice melting occurs at $\sim$1 Gy, and then the model cools.

Models "(c)" and "(d)" start at 5.0 My. This is late enough that most of the SLRIs have decayed and there is no significant heating from this source. These two models show the effect of differing levels of retention of accretion heat. Even though "(d)" is the model that has the least amount of total heat available to it, it has the longest lasting major region of melted ice, because its internal porosity was maintained for more than 100 Myr, promoting thermal insulation.

18.7.2.2 Evolution of the Lithosphere

For icy satellites the bottom of the lithosphere, or more precisely, the *mechanical lithosphere* is conventionally defined by the isotherm at which the ice responds to stress by viscous creep rather than by elastic deformation (e.g., Deschamps and Sotin 2001). For the midsize satellites the evolution of the lithosphere is primarily a function of the initial conditions that determine the evolution of the internal temperature gradient and mechanical properties. The mechanical properties are dependent upon temperature, viscosity, elastic moduli, and the rock and ice composition. In models without SLRIs (such as Figs. 18.7c and 18.7d) the warming of the interior is primarily by the decay of the long-lived radionuclides and the time scale for this is longer than the thermal conduction time scale for the satellite as a whole. As a consequence the satellites have a relatively thick, porous, outer shell during the first hundred million years or so following accretion. If sufficient ammonia is present, conditions will become favorable for dihydrate-ice creep sooner. Temperatures near the surface never get very high and the lithosphere remains structurally weak, in the sense of being porous and possessing a low ridigity, although it might also be thought of as strong in the sense of being thick. In models with significant SLRI heating, porosity collapse and compaction take place relatively early in the history, especially if ammonia is present. Full melting of the ice can also take place relatively early. The outer, icy shell that grows afterwards will be relatively stronger due to the absence of porosity in most of the lithosphere. At the surface there will still be some porosity that will be maintained by impact gardening, fracturing and other processes.

18.7.2.3 Shape Evolution

The data for the shapes of the midsize, icy satellites were presented in Section 18.2. There was a short discussion about the satellites spinning faster in the past and that perhaps some shapes reflect back to a time when they were in hydrostatic equilibrium with the earlier spin rate. Now we will consider the evolution of the satellite shape from the time of accretion until the present when all but chaotically rotating Hyperion are spinning synchronously with their orbital periods about Saturn. While our discussion is perfectly general and applies to all satellites, we will focus as before on Iapetus because the details of the models we need to consider have been published (Castillo-Rogez et al. 2007).

Cassini ISS images show that Iapetus is an oblate spheroid with a 33 km difference between the equatorial and polar radii (Thomas et al. 2007b). This difference does not include the prominent equatorial ridge. The shape is inconsistent with the present slow 79.33 day period. If the moon were in hydrostatic equilibrium with this period, it would have an axial difference of only 10 m. The measured shape requires a spin period of 16 ± 1 h to attain such an equilibrium hydrostatic shape.

Among the satellites in the Solar System with synchronously locked rotation, Iapetus is by far the most distant from its planet. Given the very strong inverse dependence on the orbital semi-major axis of the tidal despinning torque, it is somewhat surprising that Iapetus is in a synchronous state at all (see Fig. 18.4). The time scale and dissipation levels required for despinning Iapetus have been the subjects of several studies. In a general study of the despinning of giant planet satellites, Peale (1977) showed that due to its orbit's large semi-major axis, despinning Iapetus requires a dissipation factor of $Q < 50$, corresponding to a moderately dissipative interior. If we model Iapetus with a Maxwell rheology (Zschau 1978) and the best values for the temperature dependent material properties, this constraint on Q corresponds to an internal viscosity lower than 5×10^{15} Pa s. Models of Iapetus' thermal evolution that assume radioactive heating only by long-lived isotopes lack a significant, early, warming phase that is needed to match the data. Furthermore, tidal dissipation in such relatively cold models is not a significant source of heat for a body in Iapetus' distant orbit. For example, Ellsworth and Schubert's (1983) model never reaches viscosities lower than 10^{17} Pa s throughout Iapetus' history whereas a viscosity of less than 10^{15} Pa s is needed for significant tidal dissipation by Maxwell dissipation. Thus, using the current best knowledge of rheology and a low density, rock poor composition, we find that satellites such as Iapetus would not have despun to a synchronous spin state in the age of the Solar System. One solution to this problem, summarized by Fig. 18.8, invokes a source of heat that temporarily made the interior relatively warm, and thus dissipative enough for despinning to occur. However, the preservation of the 16-h shape requires that the lithosphere become thick and strong enough to retain this shape before despinning has been completed.

Heating by short-lived radionuclides provides a temporary source of heat that makes the interior relatively warm and dissipative enough for despinning to occur. When they are added to the model, it can produce the required evolution.

The idea of SLRI in satellites and small bodies is not new. For example, McCord and Sotin (2005) invoked them in comparing possible models for Ceres' interior as a function of its formation time. However, Iapetus is the first case where other sources of heat appear not to be sufficient to explain its evolution, and heat from SLRIs may be required.

18.7.2.4 Age of Iapetus

Castillo-Rogez et al. (2007, 2009), found models that accounted for Iapetus' shape and its despinning if the models were started between ∼3.4 and 5.4 My after the CAIs were formed. If we accept the Pb–Pb age of CAIs measured by Amelin et al. (2002) of $4{,}567.2 \pm 0.6$ My, for

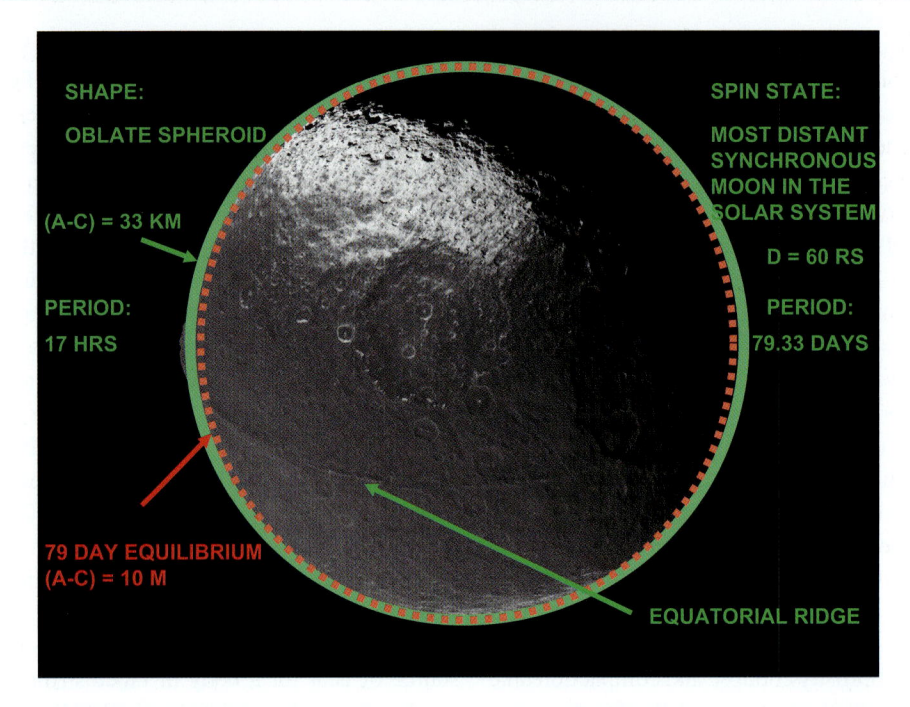

Fig. 18.8 Iapetus' enigmatic geophysical properties. The nearly circular, dashed red curve shows the shape Iapetus should have if it were in equilibrium with its spin period. The solid green curve traces Iapetus' actual shape which corresponds to a hydrostatic body rotating with a period of 16 h. In addition to its broad, non-hydrostatic bulge of 33 km, it has a relatively narrow equatorial ridge that reaches heights of up to 20-km in some places

example, then the age of Iapetus is between 4.5638 and 4.5618 Gy ± 0.0006 Gy.

The situation can be succinctly summarized with the aid of Fig. 18.9. The figure plots the amount of heat that can be obtained from SLRI versus the time after CAIs when the model started. The horizontal band represents the additional amount of heat that must be added to the models for them to produce the shape of a 16-h period of rotation, hydrostatic body and to despin to the present synchronous value. The needed heat is available in the time interval of ∼3.4–5.4 My.

A different model for Iapetus has recently been reported by Robuchon et al. (2009). They find a suggested time for Iapetus' formation of ≤4 My after CAI condensation.

Since tidal despinning of Iapetus requires the existence of Saturn, and since Iapetus is almost certainly a regular satellite of Saturn, rather than a captured object (Canup and Ward 2006), the 5.4 My upper bound on its age also bounds the time available for the formation of Saturn itself. This sort of timescale for giant planet formation is supported by recent astronomical observations of dust clearing in circum-stellar disks suggest that giant planets form on timescales of two to five million years (Najita and Williams 2005).

18.7.2.5 Iapetus' Equatorial Ridge

Another surprising feature of Iapetus is its prominent equatorial ridge. It extends along most of the equator and is centered

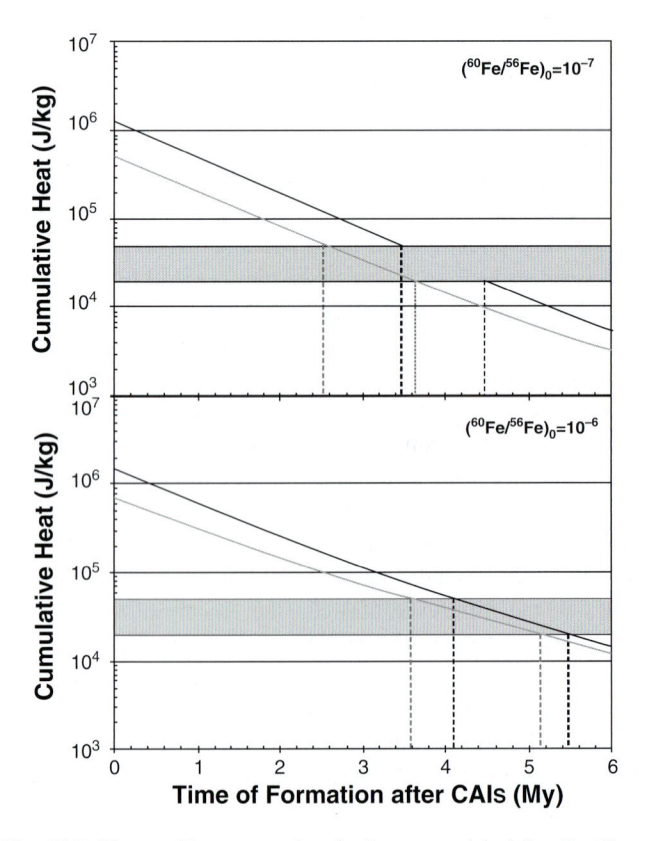

Fig. 18.9 Time and heat constraints for Iapetus models (after Castillo-Rogez et al. 2007)

upon it. It is well developed over a length of ∼1,600 km. This feature also sits right on top of the equatorial bulge. The ridge appears to be cratered as densely as the surrounding terrain, indicating that it is comparable in age. The ridge is clearly older than the large basins that overlap it.

There is no consensus among researchers as to the mechanism that created the ridge. Porco et al. (2005) and Castillo-Rogez et al. (2007) have suggested that it is related to despinning. Denk et al. (2005) proposed that it resulted from volcanic activity. Giese et al. (2005, 2008) suggest that the morphology of the ridge indicates upwarping of the surface due to a tectonic (rather than a volcanic) event. Ip (2006) suggested that it was due to the collapse of a ring in orbit about Iapetus. Czechowski and Leliwa-Kopystynski (2008) and Roberts and Nimmo (2009) argue for it being a result of convection within Iapetus. Melosh and Nimmo (2009) suggested that dykes may be involved in raising the ridge.

18.8 Other Satellites

18.8.1 Phoebe

Despite its small size and obvious lack of activity at its surface, Phoebe is a fascinating object that may hold the key to better understanding the early history of the outer Solar system. Phoebe differs from other small icy objects in many ways, as demonstrated by the analysis of data from the *Cassini-Huygens* June 11, 2004 flyby.

Imaging by the *Cassini-Huygens* remote sensing instruments found that Phoebe has one of the most complex surface compositions in the outer Solar system. At the surface it has patches of water ice. CO_2 ice, a variety of organic compounds, and signs of hydrated minerals (Clark et al. 2005). Overall, its global spectral properties (albedo, spectral slope) appear similar to C-type asteroids. Clark et al. (2005) have suggested that Phoebe's surface may sample a variety of primitive materials in the outer Solar system. They also raise the possibility that, similarly to Iapetus, Phoebe is coated with dark outer Solar System materials, and that its interior may be much less ice-rich. If Phoebe is a C-type asteroid, it would be the best-observed object of its class, and as such it can help better understand the origin of that population of small bodies (Hartmann 1987). C-type asteroids are characteristic of the main asteroid belt, however, so a dynamical and chemical link between Phoebe and this asteroid class remains to be demonstrated. Regardless, Phoebe could have been part of the population of outer Solar System planetesimals, the building blocks that led to the formation of large icy objects in the outer Solar System.

Phoebe's density of ∼1,630 kg/m³ (cf. Table 18.1) is significantly higher than the average density for the regular, midsize Saturnian satellites, which is about 1200 kg/m³. This led Johnson and Lunine (2005) to suggest that Phoebe's density could be closer to the density for Kuiper-Belt Objects, offset by 15–20% due to bulk porosity (see also discussion in Wong et al. 2008). This is further evidence that Phoebe was captured at Saturn, as also indicated by its peculiar dynamical properties: a retrograde orbit with an inclination of 176 deg. and eccentricity of 0.164. An origin in the Kuiper Belt (i.e., sharing a common genetic pool with ecliptic comets, Centaurs, and possibly D-type asteroids (e.g., Gomes et al. 2005; McKinnon 2008), is also consistent with the variety of organic compounds and possibly hydrated silicates identified on Phoebe's surface. Most interestingly, Buratti et al. (2008) drew a comparison between Phoebe and the nucleus of Comet 19P/Borrelly, which both show variations in albedo that Buratti et al. (2004) linked to areas of significant degassing. Whether or not Phoebe could have spent some time of its life as a comet or Centaur is difficult to constrain with the available observations and dynamical modeling (Turrini et al. 2008).

Another outstanding characteristic of Phoebe is its subspherical shape, which makes it the only roughly equidimensional and low-porosity object in the 100-km size range visited by a spacecraft (Johnson et al. 2009). Other objects that are sub-spherical and low-porosity are, for example, the asteroids of the binary 90-Antiope. However, most icy bodies in that size category, e.g., Amalthea, Janus, and Himalia, have very different physical properties: irregular shapes analogous to rubble-piles, and (where determined) densities less than that of water ice, between 0.4 and 900 kg/m³ (e.g., Anderson et al. 2005). These data indicate a porosity of about 5–30%, assuming these objects are composed of pure water, and up to "rubble pile" values of 30–50% if we assume that their bulk densities match the average density for Saturn's midsize regular satellites, equal to 1240 kg/m³ (Johnson and Lunine 2005). The most notable object in that category is probably Hyperion, whose density is 544 ± 50 kg/m³ despite its 135 km mean radius (Thomas et al. 2007a). In comparison, the density of the 6-km diameter Comet 9P/Tempel 1 is about 400 kg/m³ (Richardson et al. 2007). This indicates that these objects (in their present physical form, i.e., subsequent to any catastrophic impact) never had internal temperatures high enough for them to relax to a more spherical shape.

From compaction tests on pure water ice at 77–120 K (Durham et al. 2005) demonstrated that compaction is a pressure-limited process for temperatures below which either solid state creep or vapor-induced sintering is important. For the maximum pressure inside Phoebe, about 4.5 MPa, and assuming the same porosity for the whole ice and rock assemblage, compaction as a result of ice crushing is limited and the total remnant porosity expected in the center could at least be 25% and possibly up to 35% assuming that the internal temperature always remained low. Impact-induced compaction may decrease micro-porosity, but also can create porosity on larger size scales. Durham et al. (2004) suggested that, in the absence of creep-induced densification, Phoebe's intermediate porosity could have resulted from accretion of a "wide-size spectrum of ice-rock fragments," i.e., a well-graded sample with efficient filling of voids. That concept, however, does not address how such an assemblage of fragments would have relaxed to a more-or-less spherical shape.

Crater shape analysis hints, from the number of relatively deep, conical craters, that Phoebe's outer "layer" is substantially porous (Giese et al. 2006); from numerous landslides on crater walls, Phoebe's surface certainly appears unconsolidated. As modeled by Johnson et al. (2009), a 10- to 15-km thick outer porous layer (45% porosity) is sufficient to account for most of the porosity in the framework of Johnson and Lunine (2005) and is a natural result from some thermal models that assume that the satellite formed within a few My after the beginning of the Solar system (i.e., a few million years after CAI formation) (Johnson et al. 2009). Such low-porosity layers are thermally insulating, and Phoebe may have warmed up quite a bit in that case, and even LLRI may prove important for such a relatively rock-rich body if thermal conductivity remains low enough (Fig. 18.10). Future models should explore the evolution of Phoebe for different assumed formation times and compositions (especially considering suggestions of the accretion of primordial amorphous ice, as has been suggested to be the case for KBOs and comets (e.g., Prialnik and Bar-Nun 1992; Prialnik and Merk 2008).

18.8.2 Mimas, Tethys and Dione

Only one targeted flyby was dedicated to each of Mimas, Dione and Tethys during Cassini's primary mission. High-resolution imaging led to better mapping the cratering record for these satellites (Chapter 19). However, the geophysical constraints on their internal structures are few. From their physical and geological properties, it is envisioned that Dione and Tethys have undergone some endogenic activity (e.g., Schenk and Moore 1998; Schenk and Moore 2009). Signs have been reported that Dione might be slightly active based on its interaction with Saturn's magnetosphere (Khurana et al. 2007), but the interpretation of the Cassini Magnetometer data in terms of geophysical and geological constraints is not definitive at this time and remains to be investigated further.

In comparison, Tethys' very low density indicates a small silicate mass fraction and possibly some non-negligible internal porosity. The most striking feature observed at Tethys is Ithaca Chasma, a rift that runs over 2,000 km in length and up to 100-km across (Giese et al. 2007). Interpreting the uplifted flanks of the chasma in terms of elastic flexure, Giese et al. (2007) inferred constraints on the thickness of the lithosphere and thus the corresponding heat flux at the time the ridge formed. They inferred a significant heat flux value of 18–30 mW/m^2, a value dependent upon their assumed thermodynamic properties of the surface material and its porosity. This value is surprisingly large for such a small satellite with a small silicate mass fraction. About an order of magnitude less would be expected, even 4 billion years ago. Nevertheless, this flux led Chen and Nimmo (2008) to infer that at the time Ithaca Chasma formed Tethys must have been subject to significant tidal heating. These

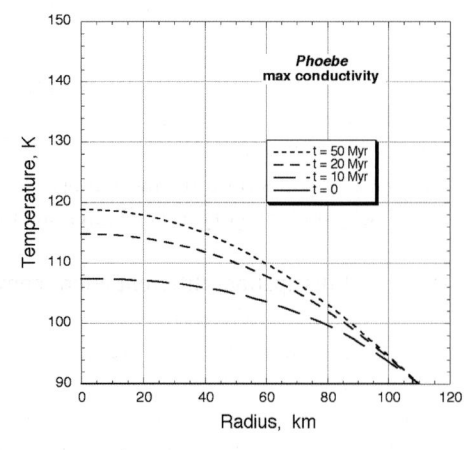

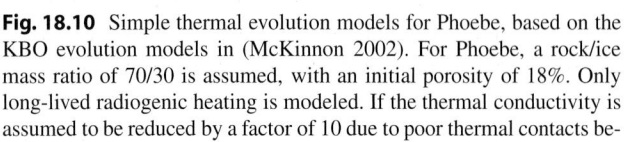

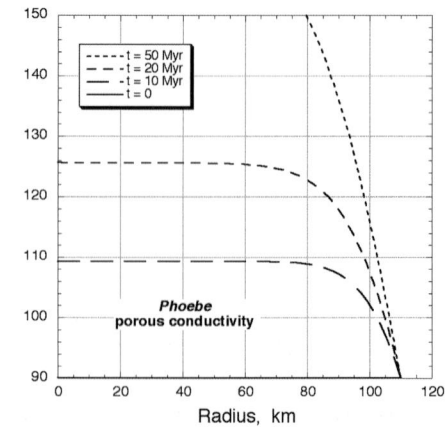

Fig. 18.10 Simple thermal evolution models for Phoebe, based on the KBO evolution models in (McKinnon 2002). For Phoebe, a rock/ice mass ratio of 70/30 is assumed, with an initial porosity of 18%. Only long-lived radiogenic heating is modeled. If the thermal conductivity is assumed to be reduced by a factor of 10 due to poor thermal contacts between (crystalline) ice grains, Phoebe's interior may have gotten warm enough for creep densification, sintering, and perhaps shape change, although only the inner 50% of the satellite (by volume) may have been affected (from Durham et al. 2004)

authors suggested the presence of a warm interior, possibly associated with a deep ocean, as a likely context for substantial tidal heating during Tethys' passage through an orbital resonance with Dione (the present inclination resonance between Mimas and Tethys does not excite the eccentricity of either body). The conditions leading to internal melting and the conjectured creation of a deep ocean in Tethys remain to be investigated.

Mimas was much studied after Voyager flyby by Eluszkiewicz et al. (1998) and Leliwa-Kopystynscki and Kossacki (2000). The latter authors inferred that the satellite has preserved some porosity although the exact amount is model dependent. Recent modeling by Charnoz et al. (2008) indicates the high probability that Mimas could have been destroyed during the last period of intense bombardment in the outer Solar system. A better understanding of this satellite is important because it theoretically undergoes at least 35 times as much tidal stress as Enceladus, due to its proximity to Saturn, and assuming similar internal structures. However, Mimas does not show obvious signs of recent or even past activity, and even its eccentricity is not fully tidally evolved as illustrated by its relatively high, relict eccentricity of 0.0196. This demonstrates that the satellite is barely affected by tidal heating, no doubt due to a cold, non-dissipative interior, which would be quite different compared with Enceladus (McKinnon and Barr 2007).

18.9 At the Frontiers: Space, Laboratory, Processes, and Modeling

In this section we discuss ways to advance our understanding of the midsize satellites of Saturn. The areas prime for advancement are further exploration by spacecraft, laboratory data for the properties of candidate materials for the composition of the midsize satellites, research on the operative processes like convection, and better modeling and computational techniques.

18.9.1 Space

With the Cassini-Huygens mission expected to be active until about 2017, there will be more data becoming available on the properties of midsize icy satellites. The shapes and masses of the satellites are now accurately known. Observations that could potentially greatly enhance our knowledge of satellite structure and evolution are geophysical measurements, especially higher-order gravity fields (presently only available for Rhea). Such provide constraints on the density structure and, if hydrostaticity prevails, the moment-of-inertia. However, most of the satellites' shapes have been identified as potentially non-hydrostatic to some degree (Thomas et al. 2007b). In these cases more flybys are desirable in order to constrain (1) non-hydrostatic anomalies due to large scale heterogeneities in porosity distributions (e.g., in the case of Mimas between the sub- and anti-Saturnian hemispheres), (2) non-hydrostatic rock cores, (3) other types of mass concentrations (e.g., bodies of subsurface water). The last are particularly important for Enceladus as potential ways to explain its non-hydrostatic shape (Collins and Goodman 2007; Thomas et al. 2007b; Schenk and McKinnon 2009).

From image analysis it is possible to infer constraints on the geological evolution of the satellites, especially the counting of craters and the analysis of their sizes, distributions, and morphologies. The identification of surface composition and tectonic features can also provide additional constraints for models of the interior processes and structure (e.g., Moore et al. 2004). Furthermore, there will be important geological consequences for models of internal evolution when good evidence can be developed for cryovolcanism, resurfacing, and recent tectonic activity. Other types of remote sensing of potentially great utility are heat flow measurements (as now done at Enceladus) and, on a future mission to the Saturn system, radar sounding. But perhaps the greatest advance, in a comparative sense, would come from a more detailed reconnaissance of the Uranian midsize icy satellites. While a flyby with modern instruments would itself be of great value, we would argue that what is most needed is an orbital mission, with its opportunities for multiple encounters with the bodies most analogous to those discussed in this chapter.

18.9.2 Laboratory Data Needed

Data on the properties of the materials from which the satellites are made are uneven – plentiful for water ice, for example, but sparse for all other ice formers – and do not provide adequate support for the modeling of satellite evolution. Also in the same category are mixtures such as ice combined with rock particles of various sizes. Impacted are the modeling of tidal dissipation, differentiation, convection, as well as geological and orbital evolution and the understanding of their timescales. Such measurements are not necessarily easy to achieve. Many of the properties are known to be temperature and frequency dependent. Data are needed down to temperatures as low as ∼80 K (lower if one is to model conditions at even greater heliocentric distance). This must also be done for frequencies characteristic of satellite spin rates and orbital periods over the whole course of their evolution. Particularly urgent are the measurement of low-temperature properties and the characterization of icy materials near the visco-elastic transition. Such data are essential for accurate simulations of how a satellite evolves (geophysically and dynamically) from the time of accretion until the present.

18.9.3 Processes

The processes of accretion and convection in satellites are not well understood. This is especially true of accretion, for which laboratory simulation or field observation are not yet possible. Improving our knowledge of these, and other, processes would allow their inclusion in more detail in models for the satellites. At the present we must make many assumptions about some of these processes and as a result it is very difficult to assess the accuracy of the results that we obtain. Improved knowledge will make the process and evolution simulations more accurate.

18.9.4 Modeling

Dynamical evolution of the satellites can no longer be treated in isolation. Similarly, the thermophysical histories cannot be calculated in isolation of the ongoing dynamical situation. Each affects the other. Satellite modeling must simultaneously treat thermophysical and dynamical evolution. Likewise, models with early starting dates must take into account heat from short-lived radioisotopes.

Since the medium-sized satellites are sensitive to initial and early heating and have preserved evidence regarding that history, it is important to better model the processes expected to take place in their interiors. The proper interpretation of the preserved evidence is one way that these satellites can be used to provide a window into conditions in the early Saturnian system.

Because the midsize satellites are sensitive to heat, latitudinal variations in properties can create lateral variations in temperature and porosity. In some cases two- or three-dimensional modeling is necessary to better characterize heat transfer and the evolution of the satellite. Another process for future models is the evolution of a hydrated-rock core at low pressure and high temperature and the resulting hydrothermal transfer of heat.

18.10 Concluding Remarks

The return of data by Cassini-Huygens has reinvigorated icy satellite research. While the previous section emphasized that there is much work to do to learn more about these satellites, we should not lose sight of the major advances that have been made. While researchers are by in large exploring different assumption and parameter spaces, there are emerging some common threads in their results. As a result we are learning much about the early solar system. For example, a strong case is being made that these satellites are very old objects. The studies of Iapetus by Castillo-Rogez et al. (2007, 2009) suggest that Iapetus formed some 3.4–5.4 My after the nominal beginning of the Solar System. Robuchon et al. (2009), using a somewhat different approach, find 4 My or earlier. Meanwhile, the study by Barr and Canup (2008) suggests that Rhea formed at a time no earlier than 4 My after the CAI condensation, assuming Rhea is indeed undifferentiated. Taking this last date as a point of reference, that means that these satellites formed some 4.563 Gy ago, with the uncertainties being discussed by investigators being variations in the last decimal point, the several million year level!

Since Saturn had to be present before the satellites, these dates also constrain the age of Saturn and how long it took to form. Interestingly, a formation time of a few million years for a giant planet is supported by recent astronomical observations of dust clearing in circum-stellar disks that suggest giant planets can form on timescales of two to five million years (e.g., Najita and Williams 2005).

With these and other results, we have demonstrated our initial thesis. At the beginning of the chapter it was stated that the midsized icy satellites are unique in that they have preserved important geophysical evidence of their early history and evolution. Thus, the future of icy satellite studies is bright. There will be more data returned from spacecraft. Laboratory programs are obtaining data at the proper temperatures, pressures, and frequencies needed for assessing the behavior of icy materials inside the satellites over their entire histories. Improvements in modeling techniques are occurring and many more are on the horizon. Numerical, computational techniques as well as computer speeds are all improving at impressive rates. And, for the first time, we can treat the icy satellites as a system. What we learn from one satellite, like Iapetus, we can apply to the others to unlock additional information. Thus, we can expect that in the not too distant future there will continue to be impressive advances in understanding the evolution and history of the midsize icy satellites and indeed, Saturn itself, and, for that matter, the whole outer solar system.

Acknowledgements This work has been conducted at the Jet Propulsion Laboratory, California Institute of Technology, Under a contract with the National Aeronautics and Space Administration. Copyright 2008 California Institute of Technology. Government sponsorship acknowledged. W.B.M. thanks the Cassini Data Analysis Program.

Appendix: Glossary of Symbols

A = constant (Eq. 18.9)
C = polar moment of inertia of the satellite (Eq. 18.4)
C_0 = initial concentration of radiogenic elements. (Eq. 18.2)
C_p = temperature-dependent specific heat (Eqs. 18.3, 18.6, 18.8)
C_{22} = second degree gravitational harmonic (i.e., ellipticity of the equator) (Eq. 18.1)
D = the semi-major axis of the orbit (Eq. 18.4)
e = eccentricity (Eq. 18.5)
dE/dt = average tidal heat produced during one orbit (Eq. 18.5)
g = gravity (Eq. 18.12)
G = universal gravitational constant (Eqs. 18.1, 18.4 through 18.7)
h_a = fraction of mechanical energy retained as heat (Eq. 18.6)
H_R = volumetric radiogenic heating rate (Eq. 18.2)
H = internal heating rate (radiogenic, tidal dissipation) (Eq. 18.8)
$H_{0,i}$ = initial power produced by radiogenic decay per unit mass of element i (Eq. 18.2)
J_2 = second degree gravitational harmonic ($= -C_{20}$) (i.e., oblateness) (Eq. 18.1)
k = thermal conductivity (Eqs. 18.8 and 18.13)
k_2 = the periodic, potential, tidal Love number (Eqs. 18.4 and 18.5)
M = satellite mass (Eq. 18.1)
M_p = Saturn's mass (Eq. 18.4)
n = number of radiogenic elements included in the sum (Eq. 18.2)
n = mean orbital motion (Eq. 18.5)
q_{conv} = convective heat flux (Eq. 18.13)
Q = the dissipation factor (Eqs. 18.4, 18.5, 18.10, 18.11)
Q_{act} = the activation energy (typically 60 kJ/mol for ice) (Eq. 18.9)
r = distance from the center of the satellite (Eqs. 18.1, 18.6, and 18.8)
R = satellite radius (Eqs. 18.1, 18.3, and 18.7)
R_{eq} = equatorial radius of the satellite (Eqs. 18.4 and 18.5)
R_{gas} = the perfect gas constant (Eqs. 18.9 and 18.11)
Ra = Rayleigh number (Eq. 18.13)
Ra_{TBL} = thermal boundary layer Rayleigh number (Eqs. 18.12 and 18.13)
t = time (Eq. 18.8)
t_{0-CAIs} = time since CAIs formation (Eq. 18.2)
T = temperature (Eq. 18.8)
T_i = temperature of the planetesimals (Eq. 18.6)
ΔT = increase in the internal temperature (Eq. 18.3)
ΔT_η = viscous temperature scale is then defined by (Eq. 18.10)
ΔT_{TBL} = temperature variation across thermal boundary layer (Eqs. 18.11 and 18.13)
T_m = temperature of convective interior (Eq. 18.11)
T_c = temperature at the base of the conductive lid (Eq. 18.11)
T_m = temperature of the convective interior (Eq. 18.10)
$T(r)$ = temperature profile resulting from accretion (Eq. 18.6)
V_{eq} = second degree equatorial gravitational potential (Eq. 18.1)
x_s = mass fraction of silicates (Eq. 18.2)
α = thermal expansion coefficient (Eq. 18.12)
δ = thickness of thermal boundary layer (Eqs. 18.12 and 18.13)
γ = moment of inertia (Eq. 18.3)
η = viscosity (Eqs. 18.9 and 18.12)
κ = thermal diffusivity (Eq. 18.12)
λ = longitude (Eq. 18.1)
λ_i = decay constant of radiogenic element i (Eq. 18.2)
$<v>$ = mean satellitesimal encounter velocity
Ω = the initial angular rate (Eq. 18.3)
ρ = the density (Eqs. 18.2, 18.6, 18.8, 18.12)
$\bar{\rho}$ = the satellite's mean density (Eq. 18.7)
Ψ = porosity (Eq. 18.7)

References

Alibert, Y., Mousis, O., 2007. Formation of Titan in Saturn's subnebula: Constraints from Huygens probe measurements. Astronomy and Astrophysics. 465, 1051–1060.

Amelin, Y., Krot, A. N., Hutcheon, I. D., Ulyanov, A. A., 2002. Lead isotopic ages of chondrules and calcium-aluminum-rich inclusions. Science. 297, 1678–1683.

Anderson, J. D., Johnson, T. V., Schubert, G., Asmar, S., Jacobson, R. A., Johnston, D., Lau, E. L., Lewis, G., Moore, W. B., Taylor, A., Thomas, P. C., Weinwurm, G., 2005. Amalthea's density is less than that of water. Science. 308, 1291–1293.

Anderson, J. D., Schubert, G., 2007. Saturn's satellite Rhea is a homogeneous mix of rock and ice. Geophysical Research Letters. 34, L02202–L02202.

Anderson, J. D., Schubert, G., 2009. Rhea's gravitational field and interior structure inferred from archival data files of the 2005 Cassini Flyby. Physics of the Earth and Planetary Interiors. submitted.

Atreya, S. K., Adams, E. Y., Niemann, H. B., Demick-Montelara, J. E., Owen, T. C., Fulchignoni, M., Ferri, F., Wilson, E. H., 2006. Titan's methane cycle. Planetary and Space Science. 54, 1177–87.

Barr, A. C., Pappalardo, R. T., 2005. Onset of convection in the icy Galilean satellites: Influence of rheology. Journal of Geophysical Research. 110, doi: 10.1029/2004 JE002371.

Barr, A. C., McKinnon, W. B., 2007. Convection in Enceladus' ice shell: Conditions for initiation. Geophysical Research Letters. 34, doi:10.1029/2006GL028799. L09202.

Barr, A. C., Canup, R. M., 2008. Constraints on gas giant satellite formation from the interior states of partially differentiated satellites. Icarus. 198, 163–177.

Barr, A. C., Pappalardo, R. T., Zhong, S., 2004. Convective instability in ice I with non-Newtonian rheology: Application to the icy Galilean satellites. Journal of Geophysical Research 109, E12008.

Blum, J., 1995. Laboratory and space experiments to study preplanetary growth. Advances in Space Research. 15, 39–54.

Brown, R. H., Clark, R. N., Buratti, B. J., Cruikshank, D. P., Barnes, J. W., Mastrapa, R. M. E., Bauer, J., Newman, S., Momary, T., Baines, K. H., Bellucci, G., Capaccioni, F., Cerroni, P., Combes, M., Coradini, A., Drossart, P., Formisano, V., Jaumann, R., Langevin, Y., Matson, D. L., McCord, T. B., Nelson, R. M., Nicholson, P. D., Sicardy, B., Sotin, C., 2006. Composition and physical properties of Enceladus' surface. Science. 311, 1425–8.

Brownlee, D., Tsou, P., Aleon, J., Alexander, C., Araki, T., Bajt, S., Baratta, G. A., Bastien, R., Bland, P., Bleuet, P., 2006. Comet 81P/Wild 2 under a microscope. Science. 314, 1711.

Buratti, B. J., Hicks, M. D., Soderblom, L. A., Britt, D., Oberst, J., Hillier, J. K., 2004. Deep Space 1 photometry of the nucleus of Comet 19P/Borrelly. Icarus. 167, 16–29.

Buratti, B. J., Soderlund, K., Bauer, J., Mosher, J. A., Hicks, M. D., Simonelli, D. P., Jaumann, R., Clark, R. N., Brown, R. H., Cruikshank, D. P., Momary, T., 2008. Infrared (0.83–5.1 μm) photometry of Phoebe from the Cassini Visual Infrared Mapping Spectrometer. Icarus. 193, 309–322.

Burns, J. A., 1976. Consequences of the tidal slowing of Mercury. Icarus. 28, 453–458.

Burns, J. A., Matthews, M. S. (Eds.), 1986. *Satellites*. Univ. Arizona Press, Tucson.

Canup, R. M., Ward, W. R., 2006. A common mass scaling for satellite systems of gaseous planets. Nature. 441, 834–839.

Castillo-Rogez, J., Johnson, T., Lee, M. H., Turner, N. J., Matson, D., Lunine, J., 2009. 26Al Decay: Heat production and a revised age for Iapetus. Icarus. in press.

Castillo-Rogez, J. C., Matson, D. L., Sotin, C., Johnson, T. V., Lunine, J. I., Thomas, P. C., 2007. Iapetus' geophysics: Rotation rate, shape, and equatorial ridge. Icarus. 190, 179–202.

Chandrasekhar, S., 1969. *Ellipsoidal Figures of Equilibrium*. Yale Univ. Press, New Haven, CT.

Chapman, C. R., McKinnon, W. B., Cratering of planetary satellites. In: J. A. Burns, M. S. Matthews (Eds.), *Satellites*. Univ. Arizona Press, Tucson, 1986, pp. 492–580.

Charnoz, S., Morbidelli, A., Dones, L., Salmon, J., 2009. Did Saturn's rings form during the Late Heavy Bombardment? Icarus. 199, 413–428.

Chen, E. M. A., Nimmo, F., 2008. Implications from Ithaca Chasma for the thermal and orbital history of Tethys. Geophysical Research Letters. 35, L19203–L19203.

Choblet, C., Sotin, C., 2000. 3-D thermal convection with variable viscosity: Can transient cooling be described by a quasi-static scaling law? Physics Of the Earth and Planetary Interiors. 119, 321–336.

Ciesla, F. J., Lauretta, D. S., Cohen, B. A., Hood, L. L., 2003. A nebular origin for chondritic fine-grained phyllosilicates. Science. 299, 549–552.

Clark, R. N., Brown, R. H., Jaumann, R., Cruikshank, D. P., Nelson, R. M., Buratti, B. J., McCord, T. B., Lunine, J., Baines, K. H., Bellucci, G., Bibring, J. P., Capaccioni, F., Cerroni, P., Coradini, A., Formisano, V., Langevin, Y., Matson, D. L., Mennella, V., Nicholson, P. D., Sicardy, B., Sotin, C., Hoefen, T. M., Curchin, J. M., Hansen, G., Hibbits, K., Matz, K. D., 2005. Compositional maps of Saturn's moon Phoebe from imaging spectroscopy. Nature. 435, 66–69.

Clauser, C., Huenges, E., Thermal conductivity of rock and minerals. *Rock Physics and Phase Relations American Geophysical Union*, Washington, DC, 1995, pp. 105–125.

Cohen, B. A., Coker, R. F., 2000. Modeling of liquid water on CM meteorite parent bodies and implications for amino acid racemization. Icarus. 145, 369–381.

Collins, G. C., Goodman, J. C., 2007. Enceladus' south polar sea. Icarus. 189, 72–82.

Consolmagno, G., Britt, D. T., Stoll, C. P., 1998. The porosities of ordinary chondrites: Models and interpretation. Meteoritics and Planetary Science. 33, 1221–1230.

Coradini, A., Cerroni, P., Magni, G., Federico, C., 1989. Formation of the satellites of the outer solar system-Sources of their atmospheres. IN: Origin and evolution of planetary and satellite atmospheres (A89-43776 19–90). University of Arizona Press, Tucson, AZ, 1989, pp. 723–762.

Cruikshank, D. P., Owen, T. C., Ore, C. D., Geballe, T. R., Roush, T. L., de Bergh, C., Sandford, S. A., Poulet, F., Benedix, G. K., Emery, J. P., 2005. A spectroscopic study of the surfaces of Saturn's large satellites: H$_2$O ice, tholins, and minor constituents. Icarus. 175, 268–283.

Cruikshank, D. P., Wegryn, E., Ore, C. M. D., Brown, R. H., Bibring, J. P., Buratti, B. J., Clark, R. N., McCord, T. B., Nicholson, P. D., Pendleton, Y. J., Owen, T. C., Filacchione, G., Coradini, A., Cerroni, P., Capaccioni, F., Jaumann, R., Nelson, R. M., Baines, K. H., Sotin, C., Bellucci, G., Combes, M., Langevin, Y., Sicardy, B., Matson, D. L., Formisano, V., Drossart, P., Mennella, V., 2008. Hydrocarbons on Saturn's satellites Iapetus and Phoebe. Icarus. 193, 334–343.

Czechowski, L., Leliwa-Kopystynski, J., 2008. The Iapetus's ridge: Possible explanations of its origin. Advances in Space Research. 42, 61–69.

Davaille, A., Jaupart, C., 1993. Transient high Rayleigh number thermal convection with large viscosity variations. Journal of Fluid Mechanics 253, 141–166.

Davaille, A., Jaupart, C., 1994. Onset of thermal convection in fluids with temperature-dependent viscosity: Application to the oceanic mantle. J. Geophys. Res. 99, 19853–19866.

De La Chapelle, S., Milsch, H., Castelnau, O., Duval, P., 1999. Compressive creep of ice containing a liquid intergranular phase: Rate-controlling processes in the dislocation creep regime. Geophysical Research Letters 26, 251–254.

Denk, T., Neukum, G., Helfenstein, P., Thomas, P. C., Turtle, E. P., McEwen, A. S., Roatsch, T., Veverka, J., Johnson, T. V., Perry, J. E., Owen, W. M., Wagner, R. J., Porco, C. C. 2005. The Cassini ISS Team, 2005. The first six months of Iapetus observations by the Cassini ISS Camera. Lunar and Planetary Science. 36, 2262–2263.

Dermott, S. F., Murray, C. D., 1982. Asteroid rotation rates depend on diameter and type. Nature. 296, 418–421.

Deschamps, F., Sotin, C., 2001. Thermal convection in the outer shell of large icy satellites (Paper 2000JE001253). Journal of Geophysical Research-Part E-Planets. 106, 5107–5121.

Durham, W. B., Heard, H. C., Kirby, S. H., 1983. Rheology of Ice ih at High Pressure and Low Temperature. Lunar and Planetary Science. 14, 169–170.

Durham, W. B., Stern, L. A., 2001. Rheological properties of water ice – applications to satellites of the outer planets Annual Review of Earth and Planetary Sciences. 29, 295–330.

Durham, W. B., McKinnon, W. B., Stern, L. A., 2004. Cold compaction of porous ice and the density of Phoebe. EOS Transactions AGU. 85, P43B-07.

Durham, W. B., McKinnon, W. B., Stern, L. A., 2005. Cold compaction of water ice. Geophysical Research Letters 32, 1–5.

Ellsworth, K., Schubert, G., 1983. Saturns icy satellites – Thermal and structural models. Icarus. 54, 490–510.

Eluszkiewicz, J., Leliwa-Kopystynski, J., Kossacki, K. J., Metamorphism of solar system ices. In: B. Schmitt, et al. (Eds.), *Solar System Ices*. Kluwer, Dordrecht Netherlands. 1998, 119–138.

Emery, J. P., Burr, D. M., Cruikshank, D. P., Brown, R. H., Dalton, J. B., 2005. Near-infrared (0.8–4.0 micron) spectroscopy of Mimas, Enceladus, Tethys, and Rhea. Astronomy & Astrophysics. 435, 353–62.

Friedson, A. J., Stevenson, D. J., 1983. Viscosity of rock-ice mixtures and applications to the evolution of icy satellites. Icarus. 56, 1–14.

Garaud, P., Lin, D. N. C., 2007. The effect of internal dissipation and surface irradiation on the structure of disks and the location of the snow line around Sun-like stars. The Astrophysical Journal. 654, 606–624.

Gautier, D., Hersant, F., 2005. Formation and composition of planetesimals. Space Science Reviews. 116, 25–52.

Giese, B., Neukum, G., Roatsch, T., Denk, T., Porco, C. C., 2006. Topographic modeling of Phoebe using Cassini images. Planetary and Space Science. 54, 1156–66.

Giese, B., Wagner, R., Neukum, G., Helfenstein, P., Thomas, P. C., 2007. Tethys: Lithospheric thickness and heat flux from flexurally supported topography at Ithaca Chasma. Geophysical Research Letters. 34, L21203–L21203.

Giese, B., Denk, T., Neukum, G., Roatsch, T., Helfenstein, P., Thomas, P. C., Turtle, E. P., McEwen, A., Porco, C. C., 2008. The topography of Iapetus' leading side. Icarus. 193, 359–371.

Giese, B., Denk, T., Neukum, G., Porco, C. C., Roatsch, T., Wagner, R., 2005. The topography of Iapetus' leading side. Bulletin of the American Astronomical Society. 37, 3.

Giggenbach, W. F., 1980. Geothermal gas equilibria. Geochimica et Cosmochimica Acta. 44, 2021–2032.

Glein, C. R., Zolotov, M. Y., Shock, E. L., 2008. The oxidation state of hydrothermal systems on early Enceladus. Icarus. 197, 157–63.

Goldreich, P., Soter, S., 1966. Q in the solar system. Icarus. 5, 375–389.

Gomes, R., Levison, H. F., Tsiganis, K., Morbidelli, A., 2005. Origin of the cataclysmic Late Heavy Bombardment period of the terrestrial planets. Nature. 435, 466–469.

Gounelle, M., Russell, S. S., 2005. On early solar system chronology: Implications of an heterogeneous spatial distribution of ^{26}Al and ^{53}Mn. Geochimica et Cosmochimica Acta. 69, 3129–3144.

Grasset, O., Parmentier, E. M., 1998. Thermal convection in a volumetrically heated, infinite Prandtl number fluid with strongly temperature dependent viscosity: implications for planetary evolution. Journal of Geophysical Research. 103, 18171–18181.

Grasset, O., Mevel, L., Mousis, O., Sotin, C., 2001. The pressure dependence of the eutectic composition in the system MgSO$_4$-H$_2$O: Implications for the deep liquid layer of icy satellites. Lunar Planetary Science. 32, 1524.pdf.

Grimm, R. E., McSween, H. Y., 1989. Water and the thermal evolution of carbonaceous chondrite parent bodies. Meteoritics. 24, 273–274.

Hartmann, W. K., 1987. A satellite-asteroid mystery and a possible early flux of scattered C-class asteroids. Icarus. 71, 57–68.

Hersant, F., Gautier, D., Lunine, J. I., 2004. Enrichment in volatiles in the giant planets of the Solar System. Planetary and Space Science. 52, 623–641.

Hersant, F., Gautier, D., Tobie, G., Lunine, J. I., 2008. Interpretation of the carbon abundance in Saturn measured by Cassini. Planetary and Space Science. 56, 1103–1111.

Hofmeister, A. M., 1999. Mantle values of thermal conductivity and the geotherm from phonon lifetimes. Science. 283, 1699–1706.

Iess, L., Rappaport, N. J., Tortora, P., Lunine, J., Armstrong, J. W., Asmar, S. W., Somenzi, L., Zingoni, F., 2007. Gravity field and interior of Rhea from Cassini data analysis. Icarus. 190, 585–593.

Ip, W. H., 2006. On a ring origin of the equatorial ridge of Iapetus. Geophysical Research Letters. 33, 16203.

Johnson, T. J., Introduction to icy satellite geology. In: B. Schmitt, et al. (Eds.), *Solar System Ices*. Kluwer, Dordrecht, Netherlands, 1998, pp. 511–524.

Johnson, T. V., Lunine, J. I., 2005. Saturn's moon Phoebe as a captured body from the outer Solar System. Nature. 435, 69–71.

Johnson, T. V., Castillo-Rogez, J. C., Matson, D. L., Thomas, P. C., 2009. Phobe's Shape: Possible constraints on internal structure and origin. Lunar Planetary Science. 40, 2334.pdf.

Johnson, T. V., Lunine, J., 2005. Saturn satellite densities and the C/O chemistry of the solar nebula. Lunar Planetary and Science 36, 1410–1411.

Kargel, J. S., 1991. Brine volcanism and the interior structures of asteroids and icy satellites. Icarus. 94, 368–390.

Kargel, J. S., 1992. Ammonia-water volcanism on icy satellites: Phase relationships at one atmosphere. Icarus. 100, 556–574.

Kargel, J. S., Kaye, J. Z., Head, J. W., Marion, G. M., Sassen, R., Crowley, J. K., Ballesteros, O. P., Grant, S. A., Hogenboom, D. L., 2000. Europa's crust and ocean: Origin, composition, and the prospects for life. Icarus. 148, 226–265.

Kelley, K. A., Plank, T., Farr, L., Ludden, J., Staudigel, H., 2005. Subduction cycling of U, Th, and Pb. Earth and Planetary Science Letters. 234, 369–383.

Khurana, K. K., Burger, M. H., Leisner, J. S., Dougherty, M. K., Russell, C. T., 2007. Does Dione have a tenuous atmosphere? Eos Transactions AGU. 88, P43A-03.

Khurana, K. K., Russell, C. T., Dougherty, M. K., 2008. Magnetic portraits of Tethys and Rhea. Icarus. 193, 465–474.

Kirk, R. L., Stevenson, D. J., 1987. Thermal evolution of a differentiated Ganymede and implications for surface features. Icarus. 69, 91–134.

Kita, N. T., Huss, G. R., Tachibana, S., Amelin, Y., Nyquist, L. E., Hutcheon, I. D., Constraints on the origin of chondrules and CAIs from short-lived and long-lived radionuclides. In: A. N. Krot, et al. (Eds.), *Chondrites and the Protoplanetary Disk*, Vol. 341. Astronomical Society of the Pacific, Kaua'i, Hawaii, 2005, pp. 558.

Korenaga, J., Jordan, T. H., 2003. Onset of convection with temperature- and depth-dependent viscosity. Geophysical Research Letters. 29, 29.

Kossacki, K. J., Leliwa-Kopystynski, J., 1993. Medium-sized icy satellites: Thermal and structural evolution during Accretion. Planetary and Space Science. 41, 729–741.

Kuskov, O. L., Kronrod, V. A., 2001. Core sizes and internal structure of Earth's and Jupiter's satellites. Icarus. 151, 204–227.

Langseth, M. G., Keihm, S., Peters, K., 1976. The revised lunar heat flow values. Lunar and Planetary Science 7, 3143.

Lanzerotti, L. J., Brown, W. L., Marcantonio, K. J., Johnson, R. E., 1984. Production of ammonia-depleted surface layers on the Saturnian satellites by ion sputtering. Nature. 312, 139.

Leliwa-Kopystynski, J., Maeno, N., 1993. Ice/rock porous mixtures: compaction experiments and interpretation. Journal of Glaciology. 39, 643–655.

Leliwa-Kopystynski, J., Kossacki, K. J., 1994. Evolution of small icy satellites – the role of ammonia admixture. Bulletin of the American Astronomical Society 26, 1160.

Leliwa-Kopystynski, J., Kossacki, K. J., 1995. Kinetics of compaction of granular ices H$_2$O, CO$_2$ and NH$_3$ × H$_2$O$_{1-x}$ at pressures of 2–20 MPa and in temperatures of 100–270 K. Application to the physics of the icy satellites. Planetary and Space Science. 43, 851–861.

Leliwa-Kopystynski, J., Kossacki, K. J., 2000. Evolution of porosity in small icy bodies. Planetary and Space Science. 48, 727–745.

Lissauer, J. J., Hubickyj, O., D'Angelo, G., Bodenheimer, P., 2009. Models of Jupiter's growth incorporating thermal and hydrodynamic constraints. Icarus. 199, 338–350.

Lorenz, R. D., Shandera, S. E., 2001. Physical properties of ammonia-rich ice: Application to Titan. Geophysical Research Letters. 28, 215–218.

Lunine, J. I., Stevenson, D. J., 1982. Formation of the Galilean satellites in a gaseous nebula. Icarus. 52, 14–39.

Mackenzie, R. A., Iess, L., Tortora, P., Rappaport, N. J., 2008. A non-hydrostatic Rhea. Geophysical Research Letters. 35, L05204–L05204.

Matson, D. L., Brown, R. H., 1989. Solid-state greenhouses and their implications for icy satellites. Icarus. 77, 67–81.

Matson, D. L., Castillo, J. C., Lunine, J., Johnson, T. V., 2007. Enceladus' plume: Compositional evidence for a hot interior. Icarus. 187, 569–573.

McCarthy, C., Cooper, R. F., Kirby, S. H., Durham, W. B., 2006. Ice/hydrate eutectics: The implications of microstructure and rheology on a multiphase Europan Crust. Lunar and Planetary Science. 37, 2467.

McCord, T. B., Sotin, C., 2005. Ceres: Evolution and current state. Journal of Geophysical Research. 110, E05009.

McKinnon, W., Geodynamics of icy satellites. In: B. Schmitt, et al. (Eds.), *Solar System Ices*. Kluwer, Dordrecht, Netherlands, 1998, pp. 525–550.

McKinnon, W. B., On the initial thermal evolution of Kuiper Belt objects. In: B. Warmbein (Ed.), *Asteroids, Comets, Meteors – ACM 2002*, Vol. ESA SP-500. ESA Publications Division, Noordwijk, the Netherlands, Berlin, Germany, 29 July–2 August 2002, 2002, pp. 29–38.

McKinnon, W. B., Zolensky, M. E., 2003. Sulfate content of Europa's ocean and shell: Evolutionary considerations and some geological and astrobiological implications. Astrobiology. 3, 879–897.

McKinnon, W. B., 2006. On convective instability in the ice I shells of outer solar system bodies, with detailed application to Callisto. Icarus. 183, 435–450.

McKinnon, W. B., Barr, A. C., The Mimas Paradox revisited plus crustal spreading on Enceladus?, Workshop on Ices, Oceans, and Fire: Satellites of the Outer Solar System, Vol. 1357. Lunar and Planetary Institute, Boulder, Colorado, 2007, pp. id.6083.

McKinnon, W. B., Could Ceres be a Refugee from the Kuiper Belt?, Asteroids, Comets, Meteors 2008 LPI Contribution, Vol. LPI Contribution No. 1405, paper id. 8389. Lunar and Planetary Institute, Baltimore, Maryland, 2008.

McKinnon, W. B., Prialnik, D., Stern, S. A., Coradini, A., Structure and evolution of Kuiper belt objects and dwarf planets. In: M. A. Barucci, et al. (Eds.), *The Solar System Beyond Neptune*. University of Arizona Press, Tucson, 2008, pp. 213–241.

Melosh, H. J., Nimmo, F., 2009. An intrusive dike origin for Iapetus' enigmatic ridge? Lunar and Planetary Science. 40, #2478.

Moore, J. M., Schenk, P. M., Bruesch, L. S., Asphaug, E., McKinnon, W. B., 2004. Large impact features on middle-sized icy satellites. Icarus. 171, 421–43.

Moresi, L. N., Solomatov, V. S., 1995. Numerical investigation of 2D convection with extremely large viscosity variations. Physics of Fluids. 7, 2154–2162.

Mosqueira, I., Estrada, P. R., 2005. On the origin of the saturnian satellite system: Did Iapetus form in-situ? Lunar and Planetary Science. 36, 1951–1952.

Mostefaoui, S., Lugmair, G. W., Hoppe, P., 2005. 60Fe: A heat source for planetary differentiation from a nearby supernova explosion. The Astrophysical Journal. 625, 271–277.

Mousis, O., Gautier, D., Bockelee-Morvan, D., 2002. An evolutionary turbulent model of Saturn's subnebula: Implications for the origin of the atmosphere of Titan. Icarus. 156, 162–175.

Multhaup, K., Spohn, T., 2007. Stagnant lid convection in the mid-sized icy satellites of Saturn. Icarus. 186, 420–435.

Nagel, K., Breuer, D., Spohn, T., 2004. A model for the interior structure, evolution, and differentiation of Callisto. Icarus. 169, 402–412.

Najita, J., Williams, J. P., 2005. An 850 µm survey for dust around solar-mass stars. The Astrophysical Journal. 635, 625–635.

Nakamura, T., Abe, O., Internal friction of snow and ice at low frequency. *Proceedings of the Sixth International Conference on Internal Friction and Ultrasonic Attenuation in Solids*. University of Tokyo Press, Tokyo, 1977, p. 285.

Ostro, S. J., West, R. D., Janssen, M. A., Lorenz, R. D., Zebker, H. A., Black, G. J., Lunine, J. I., Wye, L. C., Lopes, R. M., Wall, S. D., Elachi, C., Roth, L., Hensley, S., Kelleher, K., Hamilton, G. A., Gim, Y., Anderson, Y. Z., Boehmer, R. A., Johnson, W. T. K., 2006. Cassini RADAR observations of Enceladus, Tethys, Dione, Rhea, Iapetus, Hyperion, and Phoebe. Icarus. 183, 479–490.

Parmentier, E. M., Sotin, C., Travis, B. J., 1994. Turbulent 3D thermal convection in an infinite Prandtl number, volumetrically heated fluid: Implications for mantle dynamics. Geophysical Journal International. 116, 241–251.

Peale, S. J., Rotational histories of the natural satellites. In: J. A. Burns (Ed.), *Planetary Satellites*. University of Arizona Press, Tucson, AZ, 1977, pp. 87–112.

Peale, S. J., 1999. Origin and evolution of the natural satellites. Annual Review of Astronomy and Astrophysics. 37, 533–602.

Petrenko, V. F., Whitworth, R. W., 1999. *Physics of Ice*. Oxford Univ. Press, Oxford.

Porco, C. C., Baker, E., Barbara, J., Beurle, K., Brahic, A., Burns, J. A., Charnoz, S., Cooper, N., Dawson, D. D., Del Genio, A. D., Denk, T., Dones, L., Dyudina, U., Evans, M. W., Giese, B., Grazier, K., Helfenstein, P., Ingersoll, A. P., Jacobson, R. A., Johnson, T. V., McEwen, A., Murray, C. D., Neukum, G., Owen, W. M., Perry, J., Roatsch, T., Spitale, J., Squyres, S., Thomas, P. C., Tiscareno, M., Turtle, E., Vasavada, A. R., Veverka, J., Wagner, R., West, R., 2005. Cassini imaging science: Initial results on Phoebe and Iapetus. Science. 307, 1237–1242.

Prialnik, D., Bar-Nun, A., 1990. Heating and melting of small icy satellites by the decay of Al-26. Astrophysical Journal. 355, 281–286.

Prialnik, D., Bar-Nun, A., 1992. Crystallization of amorphous ice as the cause of Comet P/Halley's outburst at 14 AU. Astronomy and Astrophysics (ISSN 0004–6361). 258.

Prialnik, D., Merk, R., 2008. Growth and evolution of small porous icy bodies with an adaptive-grid thermal evolution code. I. Application to Kuiper belt objects and Enceladus. Icarus. 197, 211–20.

Prinn, R. G., Fegley, B., 1981. Kinetic inhibition of CO and N_2 reduction in circumplanetary nebulae – Implications for satellite composition. Astrophysical Journal. 249, 308–317.

Prinn, R. G., Fegley, B., Solar nebula chemistry: Origin of planetary, satellite, and cometary volatiles. In: S. Atreya, et al. (Eds.), *Origin and Evolution of Planetary and Satellite Atmospheres*. University of Arizona Press, Tucson, Arizona, 1989, pp. 78–136.

Richardson, J. E., Melosh, H. J., Lisse, C. M., Carcich, B., 2007. A ballistics analysis of the deep impact ejecta plume: Determining comet tempel 1's gravity, mass, and density. Icarus. 190, 357–390.

Roberts, J. H., Nimmo, F., 2008. Tidal heating and the long-term stability of a subsurface ocean on Enceladus. Icarus. 194, 675–689.

Roberts, J. H., Nimmo, F., 2009. Tidal dissipation due to despinning and the equatorial ridge on Iapetus. Lunar and Planetary Science. 40, #1927.

Robuchon, G., C, C., Tobie, G., Cadek, O., Sotin, C., Grasset, O., 2009. Coupling of thermal evolution and despinning of early Iapetus. Icarus. Submitted.

Ross, R. G., Kargel, J. S., Thermal conductivity of solar system ices, with special reference to Martian Polar Caps. *Solar System Ices*. Kluwer, Dordrecht, 1998.

Schenk, P. M., Moore, J. M., Geologic landforms and processes on icy satellites. In: B. Schmitt, et al. (Eds.), *Solar System Ices*. Kluwer, Dordrecht, Netherlands. 1998, 551–578.

Schenk, P. M., McKinnon, W. B., 2009. One-hundred-km-scale basins on Enceladus: Evidence for an active ice shell. Geophys. Res. Lett. 36, in press.

Schenk, P. M., Moore, J. M., 2009. Eruptive volcanism on saturn's icy moon Dione. Lunar and Planetary Science. 40, id.2465.

Schubert, G., Spohn, T., Reynolds, R. T., Thermal histories, compositions, and internal structures of the moons of the solar system. In: J. A. Burns, M. S. Matthews (Eds.), *Satellites*. University of Arizona Press, Tucson, 1986, pp. 224–292.

Schubert, G., Anderson, J. D., Spohn, T., McKinnon, W. B., Interior composition, structure and dynamics of the Galilean satellites. In: F. Bagenal, et al. (Eds.), *Jupiter: The Planet, Satellites and Magnetosphere*. Cambridge University Press, Cambridge, U.K. 2004, pp. 281–306.

Schubert, G., Anderson, J. D., Travis, B. J., Palguta, J., 2007. Enceladus: Present internal structure and differentiation by early and long-term radiogenic heating. Icarus. 188, 345–355.

Scott, H. P., Williams, Q., Ryerson, F. J., 2002. Experimental constraints on the chemical evolution of large icy satellites. Earth and Planetary Science Letters. 203, 399–412.

Shock, E. L., McKinnon, W. B., 1993. Hydrothermal processing of cometary volatiles – Applications to Triton. Icarus. 106, 464–477.

Shoshany, Y., Prialnik, D., Podolak, M., 2002. Monte Carlo modeling of the thermal conductivity of porous cometary ice. Icarus. 157, 219–227.

Showman, A. P., Malhotra, R., 1997. Tidal evolution into the Laplace resonance and the resurfacing of Ganymede. Icarus. 127, 93–111.

Shu, F. H., Johnstone, D., Hollenbach, D., 1993. Photoevaporation of the solar nebula and the formation of the giant planets. Icarus. 106, 92–101.

Shukolyukov, A., Lugmair, G. W., 1993. Fe-60 in eucrites. Earth and Planetary Science Letters (ISSN 0012–821X). 119, 159–166.

Sohl, F., Spohn, T., Breuer, D., Nagel, K., 2002. Implications from Galileo observations on the interior structure and chemistry of the Galilean satellites. Icarus. 157, 104–119.

Sohl, F., Hussmann, H., Schwentker, B., Spohn, T., Lorenz, R. D., 2003. Interior structure models and tidal Love numbers of Titan. Journal of Geophysical Research. E. Planets. 108, 5130.

Solomatov, V. S., Moresi, L. N., 2000. Scaling of time-dependent stagnant lid convection: Application to small-scale convection on Earth and other terrestrial planets. Journal of Geophysical Research. 105, 21795–21818.

Solomatov, V. S., Barr, A. C., 2006. Onset of convection in fluids with strong temperature-dependent, power-law viscosity. Physics of the Earth and Planetary Interiors. 155, 140–145.

Sotin, C., Labrosse, S., 1999. Three-dimensional thermal convection in an iso-viscous, infinite Prandtl number fluid heated from within and from below: Applications to the transfer of heat through planetary mantles. Physics of the Earth and Planetary Interiors. 112, 171–190.

Sotin, C., Castillo, J. C., Tobie, G., Matson, D. L., 2006. Onset of convection in mid-sized icy satellites. EGU. EGU06-A-08124.

Sotin, C., Tobie, G., Wahr, J., McKinnon, W. B., Tides and tidal heating on Europa. In: R. T. Pappalardo, et al. (Eds.), *Europa*. University of Arizona Press, Tucson, 2009.

Squyres, S. W., Croft, S. K., The tectonics of icy satellites. In: J. A. Burnes, M. S. Matthews (Eds.), *Satellites*. University of Arizona Press, Tucson, AZ, 1986, pp. 293–341.

Squyres, S. W., Reynolds, R. T., Summers, A. L., Shung, F., 1988. Accretional heating of the satellites of Saturn and Uranus. Journal of Geophysical Research. 93, 8779–8794.

Stevenson, D. J., Harris, A. W., Lunine, J. I., Origins of satellites. In: J. A. Burns, M. S. Matthews (Eds.), *Satellites*. University of Arizona Press, Tucson, 1986, pp. 39–88.

Tachibana, S., Huss, G. R., Kita, N. T., Shimoda, G., Morishita, Y., 2006. 60Fe in Chondrites: Debris from a nearby supernova in the early solar system? The Astrophysical Journal. 639, L87–L90.

Takeuchi, H., Saito, M., Seismic surface waves. *Methods in Computational Physics*. Academic Press, New York, 1972, pp. 217–295.

Thomas, P. C., Armstrong, J. W., Asmar, S. W., Burns, J. A., Denk, T., Giese, B., Helfenstein, P., Iess, L., Johnson, T. V., McEwen, A., Nicolaisen, L., Porco, C., Rappaport, N., Richardson, J., Somenzi, L., Tortora, P., Turtle, E. P., Veverka, J., 2007a. Hyperion's sponge-like appearance. Nature. 448, 50–53.

Thomas, P. C., Burns, J. A., Helfenstein, R., Squyres, S., Veverka, J., Porco, C., Turtle, E. P., McEwen, A., Denk, T., Giese, B., Roatsch, T., Johnson, T. V., Jacobson, R. A., 2007b. Shapes of the saturnian icy satellites and their significance. Icarus. 190, 573–584.

Tobie, G., Grasset, O., Lunine, J. I., Mocquet, A., Sotin, C., 2005a. Titan's internal structure inferred from a coupled thermal-orbital model. Icarus. 175, 496–502.

Tobie, G., Mocquet, A., Sotin, C., 2005b. Tidal dissipation within large icy satellites: Applications to Europa and Titan. Icarus. 177, 534–549.

Tobie, G., Cadek, O., Sotin, C., 2008. Solid tidal friction above a liquid water reservoir as the origin of the south pole hotspot on Enceladus. Icarus. 196, 642–652.

Travis, B. J., Schubert, G., 2005. Hydrothermal convection in carbonaceous chondrite parent bodies. Earth and Planetary Science Letters. 240, 234–250.

Turcotte, D. L., Schubert, G., 1982. *Geodynamics*. John Wiley & Sons, New York.

Turcotte, D. L., Schubert, G., 2002. Geodynamics, 2nd edn. Cambridge University Press, Cambridge.

Turrini, D., Marzari, F., Beust, H., 2008. A new perspective on the irregular satellites of Saturn-I. Dynamical and collisional history. Monthly Notices of the Royal Astronomical Society. 391, 1029–1051.

Van Schmus, W. R., Natural radioactivity of the crust and mantle. In: T. J. Ahrens (Ed.), *Global Earth physics: A Handbook of Physical Constants*. American Geophysical Union., Washington, DC, 1995, pp. 283–291.

Vanhala, H. A. T., Boss, A. P., 2002. Injection of radioactivities into the forming solar system. Astrophysical Journal. 575, 1144–1150.

Verbiscer, J., Peterson, D. E., Skrutskie, M. F., Cushing, M., Helfenstein, P., Nelson, M. J., Smith, J. D., Wilson, J. C., Ammonia Hydrate on Tethys' Trailing Hemisphere. Vol. 1406. Lunar and Planetary Institute, 3600 Bay Area Boulevard, Houston, TX, 77058–1113, USA, 2008, pp. 156–157.

Wahr, J., Selvans, Z. A., Mullen, M. E., Barr, A. C., Collins, G. C., Selvans, M. M., Pappalardo, R. T., 2009. Modeling stresses on satellites due to non-synchronous rotation and orbital eccentricity using gravitational potential theory. Icarus. 200, 188–206.

Waite, J. H., Lewis, W. S., Magee, B. A., Lunine, J. I., McKinnon, W. B., Glein, C. R., Mousis, O., Young, D. T., Brockwell, T., Westlake, J., Nguyen, M.-J., Teolis, B., Niemann, H., McNutt, R. L., Perry, M., Ip, W. H., 2009. Liquid water on Enceladus from observations of ammonia and ^{40}Ar in the plume. Nature. 460, 487–490.

Waite, J. H., Jr., Combi, M. R., Wing-Huen, I., Cravens, T. E., McNutt, R. L., Jr., Kasprzak, W., Yelle, R., Luhmann, J., Niemann, H. B., Gell, D., Magee, B., Fletcher, G., Lunine, J., Wei-Ling, T., 2006. Cassini Ion and Neutral Mass Spectrometer: Enceladus plume composition and structure. Science. 311, 1419–22.

Wasserburg, G. J., Papanastassiou, D. A., Some short-lived nuclides in the early Solar System. In: C. A. Barnes, et al. (Eds.), *Essays in Nuclear Astrophysics*, ed. CA Barnes, DD Clayton, and DN Schramm. Cambridge University Press, New York, 1982, p. 77.

Wasson, J. T., Kalleymen, G. W., 1988. Composition of chondrites. Philosophical Transactions Royal Society of London A. 325, 535–544.

Wong, M. H., Lunine, J. I., Atreya, S. K., Johnson, T., Mahaffy, P. C., Owen, T. C., Encrenaz, T., Oxygen and other volatiles in the giant planets and their satellites. In: G. MacPherson, W. Huebner (Eds.), *Oxygen in Earliest Solar System Materials and Processes Mineralogical Society of America*, Chantilly, VA, 2008, pp. 219–246.

Young, E. D., Simon, J. I., Galy, A., Russell, S. S., Tonui, E., Lovera, O., 2005. Supra-Canonical 26Al/27Al and the residence time of CAIs in the Solar protoplanetary disk. Science. 308, 223–227.

Zaranek, S. E., Parmentier, E. M., 2004. The onset of convection in fluids with strongly temperature-dependent viscosity cooled from above with implications for planetary lithospheres. Earth and Planetary Science Letters. 224, 371–386.

Zolotov, M. Y., Shock, E. L., 2003. Energy for biologic sulfate reduction in a hydrothermally formed ocean on Europa. Journal of Geophysical Research. 108.

Zolotov, M. Y., 2007. An oceanic composition on early and today's Enceladus. Geophysical Research Letters. 34, L23203.

Zschau, J., Tidal friction in the solid earth: loading tides versus body tides. In: P. Brosche, J. Sundermann (Eds.), *Tidal Friction and the Earth's Rotation*. Springer, Berlin, 1978.

Chapter 19
Icy Satellites of Saturn: Impact Cratering and Age Determination

Luke Dones, Clark R. Chapman, William B. McKinnon, H. Jay Melosh, Michelle R. Kirchoff, Gerhard Neukum, and Kevin J. Zahnle

Abstract Saturn is the first giant planet to be visited by an orbiting spacecraft that can transmit large amounts of data to Earth. Crater counts on satellites from Phoebe inward to the regular satellites and ring moons are providing unprecedented insights into the origin and time histories of the impacting populations. Many Voyager-era scientists concluded that the satellites had been struck by at least two populations of impactors. In this view, the Population I impactors, which were generally judged to be "comets" orbiting the Sun, formed most of the larger and older craters, while Population II impactors, interpreted as Saturn-orbiting ejecta from impacts on satellites, produced most of the smaller and younger craters. Voyager data also implied that all of the "ring moons," and probably some of the mid-sized classical moons, had been catastrophically disrupted and reaccreted since they formed. We examine models of the primary impactor populations in the Saturn system. At the present time, "ecliptic comets," which likely originate in the Kuiper Belt/Scattered Disk, are predicted to dominate impacts on the regular satellites and ring moons, but the models require extrapolations in size (from the observed Kuiper Belt Objects to the much smaller bodies that produce the craters) or in distance (from the known active Jupiter family comets to 9.5 AU). Phoebe, Iapetus, and perhaps even moons closer to Saturn have been struck by irregular satellites as well. We describe the Nice model, which provides a plausible mechanism by which the entire Solar System might have experienced an era of heavy bombardment long after the planets formed. We then discuss the three cratering chronologies, including one based upon the Nice model, that have been used to infer surface ages from crater densities on the saturnian satellites. After reviewing scaling relations between the properties of impactors and the craters they produce, we provide model estimates of the present-day rate at which comets impact, and catastrophically disrupt, the saturnian moons. Finally, we present crater counts on the satellites from two different groups. Many of the heavily cratered terrains appear to be nearly saturated, so it is difficult to infer the provenance of the impactors from crater counts alone. More large craters have been found on Iapetus than on any other satellite. Enceladus displays an enormous range of surface ages, ranging from the old mid-latitude plains to the extremely young South Polar Terrain. Cassini images provide some evidence for the reality of "Population II". Most of the observed craters may have formed in one or more "cataclysms," but more work is needed to determine the roles of heliocentric and planetocentric bodies in creating the craters.

19.1 Introduction: Understanding of Saturnian Impact Crater Populations Through the Voyager Era

During the three centuries following the discovery of Titan by Christiaan Huygens and Iapetus, Rhea, Tethys, and Dione by Gio Domenico Cassini, the nature of these moons' surfaces was unknown. Ostro and Pettengill (1978) had concluded that the "anomalous" radar behavior of Europa, Ganymede, and Callisto could be explained if these moons had icy, cratered surfaces, Nonetheless, the existence of macroscopic craters on icy satellites remained uncertain because of doubts about whether ice could support topography over geological timescales. Torrence Johnson wrote (Johnson 1978):

> We do not know what the satellite surfaces look like, whether there are craters, mountains, rifts, or new and unexpected tectonic features.

L. Dones and C.R. Chapman
Southwest Research Institute, Boulder, CO 80302, USA

W.B. McKinnon
Washington University in Saint Louis, St. Louis, MO 63130, USA

H.J. Melosh
University of Arizona, Tucson, AZ 85721, USA
Now at Purdue University, West Lafayette, IN 47907

M.R. Kirchoff
Lunar and Planetary Institute, Houston, TX 77058, USA
Now at Southwest Research Institute, Boulder CO 80302, USA

G. Neukum
Freie Universität, 12249 Berlin, Germany

K.J. Zahnle
NASA Ames Research Center, Moffett Field, CA 94035, USA

The Voyagers, of course, discovered craters on every satellite they imaged at sufficient resolution (with the singular exception of Io). Early Voyager studies of the saturnian satellites in particular (Smith et al. 1981, 1982) proposed that there were two sources of bodies responsible for the visible craters in that system. The earlier impactors, "Population I," most clearly expressed on Rhea, were "characterized by a relatively high abundance of craters larger than ∼20 km" (Smith et al. 1981), with a size-frequency distribution (SFD) crudely similar to the SFDs on the Moon and terrestrial planets. However, Population I was thought to have been formed by "debris in heliocentric orbits," i.e., "comets".[1] "Population II" appeared on the younger (i.e., less cratered) terrains on Dione and Tethys, dominated Mimas and (apparently) Enceladus (where craters could be seen at all), and featured "abundant craters smaller than ∼20 km and a general absence of larger craters" (Smith et al. 1981). Population II craters were generally taken to have been formed by circum-Saturnian debris from the cratering or breakup of adjacent satellites, although Plescia and Boyce (1985) took them to be late-arriving bodies from heliocentric orbit. These attributes of Saturn satellite SFDs were further documented by crater statistics published by Strom (1981) and others and were summarized by Chapman and McKinnon (1986). The reality of the two populations was questioned by Lissauer et al. (1988), who argued instead that all of the craters could be explained by saturation equilibrium by a single population of heliocentric comets rich in small bodies, although their crater statistics show significant differences between Rhea and Mimas. Cratering of Mimas was later modeled by Chapman (1990) using Population II characteristics very different from Population I.

Voyager-era estimates of cratering rates at Saturn were given by Smith et al. (1981, 1982) and by Shoemaker and Wolfe (1981). These authors estimated that "Saturn-family" comets (a population akin to what we now call ecliptic comets; see Section 19.2) dominated cratering on Rhea and moons interior to its orbit, while Oort Cloud comets dominated on moons further from Saturn. This calculation was based on the only Saturn-family "comet" known at that time, the 140-km-diameter Chiron, which had been discovered in 1977 (and is now known as the first Centaur to be found). According to these rate estimates, cratering at the present-day rate could have not produced the observed number of craters on most of the terrains observed by Voyager in the age of the Solar System. Presumably, then, there was an intense early bombardment of the satellites, which might or might not be related to the lunar Late Heavy Bombardment (LHB), i.e., the formation of large basins such as Imbrium and Orientale some 3.8–3.9 billion years ago (Hartmann et al. 2000; Chapman et al. 2007).

Besides these a priori estimates, cratering rates can be scaled from moon to moon, allowing for the higher spatial densities and velocities of (heliocentric) impactors closer to Saturn due to gravitational focusing. Scaling from Iapetus' cratering record, Smith et al. (1982) inferred that all the moons interior to Dione's orbit should have been catastrophically disrupted one or more times. Here catastrophic disruption was equated with the formation of a crater whose diameter equaled the diameter of the satellite. However, the use of Iapetus as a template can be questioned, because Voyager had found, surprisingly, that the density of large craters on Iapetus was slightly higher than on Rhea and Mimas. Models, by contrast, predict much larger cratering rates per unit area on Rhea, Mimas, and the other inner moons than on Iapetus because of gravitational focusing by Saturn (e.g., Zahnle et al. 2003). The Voyager results seem to imply one or more of the following: (1) Iapetus and Rhea were cratered by different populations of impactors; (2) the most heavily cratered terrains on all the moons are saturated; (3) more crater relaxation has taken place on Rhea than on Iapetus; or (4) Rhea's surface is, on average, considerably younger than that of Iapetus (Lissauer et al. 1988).

The key goal of crater counting was expressed by Shoemaker et al. (1963) in the classic paper "Interplanetary Correlation of Geologic Time":

> ...if the frequency of meteoroid impact and its variation with time on the different planets can be established, the age of rock bodies exposed on their surfaces can be estimated from the distribution of superimposed impact craters.

Hartmann (1965, 1966) used crater counts on the Canadian shield to correctly deduce the age of the lunar maria, and concluded that the average early impact rate on the lunar highlands must have been at least 100 times the present-day rate (Hartmann et al. 2000). Radiometric dating of lunar samples returned by the Apollo astronauts provided an absolute chronology for the Moon, and the lunar chronology was applied to Mars (Neukum and Hiller 1981). Asteroids are the primary impactors on the terrestrial planets, so it is possible to relate their chronologies to each other (Hartmann and Neukum 2001; Strom et al. 2005). However, comets are widely (Smith et al. 1981, 1982; Zahnle et al. 1998, 2003; Schenk et al. 2004), though not universally (Neukum et al. 1998, 1999, 2005, 2006; Schmedemann et al. 2008, 2009) thought to be the main impactors on the giant planets and their satellites. Since no radiometric dates exist for surfaces in the outer Solar System, and the size-frequency distribution of comets is less well understood than that of asteroids, the chronology of the moons of the giant planets remains in a primitive state. Nevertheless, we may rejoice that there are copious impact craters to count, their survival

---

[1] In this chapter, we use the term "comet" as a generic term for a planetesimal in the region of the giant planets or beyond. By our definition, comets need not have comae.

over geologic time against viscous relaxation now understood (Dombard and McKinnon 2000).

In Section 19.2 we discuss impactor populations; in Section 19.3, recent ideas about the impact flux in the first billion years of Solar System history; in Section 19.4, cratering chronologies; in Section 19.5, impact physics and scaling laws; in Section 19.6, predicted cratering rates; in Section 19.7, observed crater statistics; and in Section 19.8, conclusions.

19.2 Impactor Populations

Potential impactors on the saturnian moons can be divided into six categories.

First, most *main-belt asteroids* follow orbits with average distances from the Sun in the range 2.1–3.3 Astronomical Units (AU) from the Sun. The main belt is estimated to contain of order 10^6 objects with diameters $d > 1$ km (Bottke et al. 2005).

Second, both Jupiter and Neptune have *Trojans*, small bodies that librate around the Lagrangian points some 60° ahead of and behind each planet in its orbit. The number of km-sized Trojans of Jupiter is comparable to the number of main-belt asteroids (Shoemaker et al. 1989; Jewitt et al. 2000). The number of km-sized Neptune Trojans is unknown; it may be far greater than the population of Jupiter Trojans. Because of Neptune's great distance from the Sun, only six of its Trojans, all larger than ∼100 km, have been detected thus far (Sheppard and Trujillo 2006). Although Trojans are sometimes called "Trojan asteroids," these bodies appear to be more closely related to comets than to most main-belt asteroids.[2]

Third, *Centaurs and ecliptic comets*, which are believed to originate in the Kuiper Belt/Scattered Disk[3] (Gladman et al. 2008; Morbidelli et al. 2008; Kenyon et al. 2008; and other chapters in *The Solar System Beyond Neptune* [Barucci et al. 2008]; Jewitt 2008) beyond Neptune's orbit, are generally thought to be the primary impactors on the regular satellites of the giant planets at the present epoch. The Kuiper Belt is estimated to contain nearly 10^5 bodies with diameters, d, larger than 100 km. The Kuiper Belt size distribution flattens markedly at smaller sizes, with a "knee" in the distribution at $d = 90 \pm 30$ km (Fuentes et al. 2009; see Fig. 19.1), where the sizes are derived assuming a representative albedo of 0.04. The smallest body detected in the Kuiper Belt thus far has $d \sim 25$ km (Bernstein et al. 2004). Since impactors from heliocentric orbit typically produce craters larger than the impactors themselves by a factor of 10 or more, our current census of the belt provides little direct information on the size range relevant to cratering the satellites.

Bodies slowly leak from the Kuiper Belt to become "ecliptic comets," which include Scattered Disk Objects with eccentric orbits outside that of Neptune and Centaurs with orbits in the Jupiter-Neptune region (Levison and Duncan 1997; Duncan and Levison 1997; Tiscareno and Malhotra 2003; Duncan et al. 2004; Di Sisto and Brunini 2007; Volk and Malhotra 2008). The steady-state population of Saturn-crossing Centaurs is ∼10^{-4} times the population of the Kuiper Belt (Irwin et al. 1995; Fig. 19.2)[4]. However, the population of km-sized Kuiper Belt Objects (KBOs) [and hence Centaurs] is highly uncertain; we defer discussion of the absolute number of Saturn-crossing Centaurs to Section 19.6.

More than 300 Jupiter-family comets (JFCs) have been discovered (Fernández 2009). JFCs, which have small orbital inclinations, are thought to be Centaurs that have come close enough to the Sun (typically, perihelion distances <3 AU) that they release gas and dust. Most *known* JFCs have nuclei with diameters in the range 1–10 km; the true size-frequency distribution of JFC nuclei can be approximated by a power law with a possible flattening or cutoff near 1 km (Lamy et al. 2004). Comet Shoemaker-Levy 9 (SL9) was a JFC that was tidally disrupted by Jupiter in 1992; the fragments struck the planet in 1994. Prior to its tidal disruption, the diameter of SL9's nucleus is estimated to have been 1.5–1.8 km (Asphaug and Benz 1996). Zahnle et al. (2003) use the historical rate of close passages of comets such as SL9

---

[2] We do not break out the Hilda asteroids, a relatively populous group occupying the 3:2 mean-motion resonance with Jupiter (Brunini et al. 2003) separately. For the purposes of this chapter, the Hildas can be considered as a source population combined with either the Trojan asteroids or the ecliptic comets.

[3] For the remainder of this article, we define the Kuiper Belt as the group of small bodies in heliocentric orbit with perihelion distances between 30 and 50 AU, not including Neptune's Trojans. By this definition, the Kuiper Belt includes the "Scattered Disk". The Kuiper Belt appears to be a mix of at least two populations with different dynamical and physical properties, and distinct size-frequency distributions. Duncan et al. (2004) have stated that the Scattered Disk is the primary source of the Centaurs and ecliptic comets (and therefore the cometary impactors on the moons of the giant planets), but the size distributions of the different populations in the Kuiper Belt are not known at most of the sizes relevant for cratering, so this conclusion is premature (Volk and Malhotra 2008).

[4] The number of Jupiter- and Saturn-crossing Centaurs is small because the giant planets quickly eject most bodies that cross their orbits from the Solar System. If N_{Saturn} is the number of Saturn-crossing bodies larger than a given size, N_{KB} is the number of Kuiper Belt Objects larger than that size, and assume no physical disruption of Centaurs, we have $N_{Saturn}/N_{KB} \sim f_{escape} \, f_{Saturn} \, t_{Saturn}/t_{KB}$, where f_{escape} is the fraction that escapes the Kuiper Belt in the age of the Solar System, f_{Saturn} is the fraction of escapees that become planet-crossing, t_{Saturn} is the typical lifetime of Saturn crossers found from numerical orbit integrations, and t_{KB} is the typical lifetime of a Kuiper Belt Object. Taking $f_{escape} = 0.4$, $f_{Saturn} = 0.5$, $t_{Saturn} = 3 \times 10^6$ years, $t_{KB} = 4 \times 10^9$ years, we have $N_{Saturn}/N_{KB} \sim 10^{-4}$ (Irwin et al. 1995).

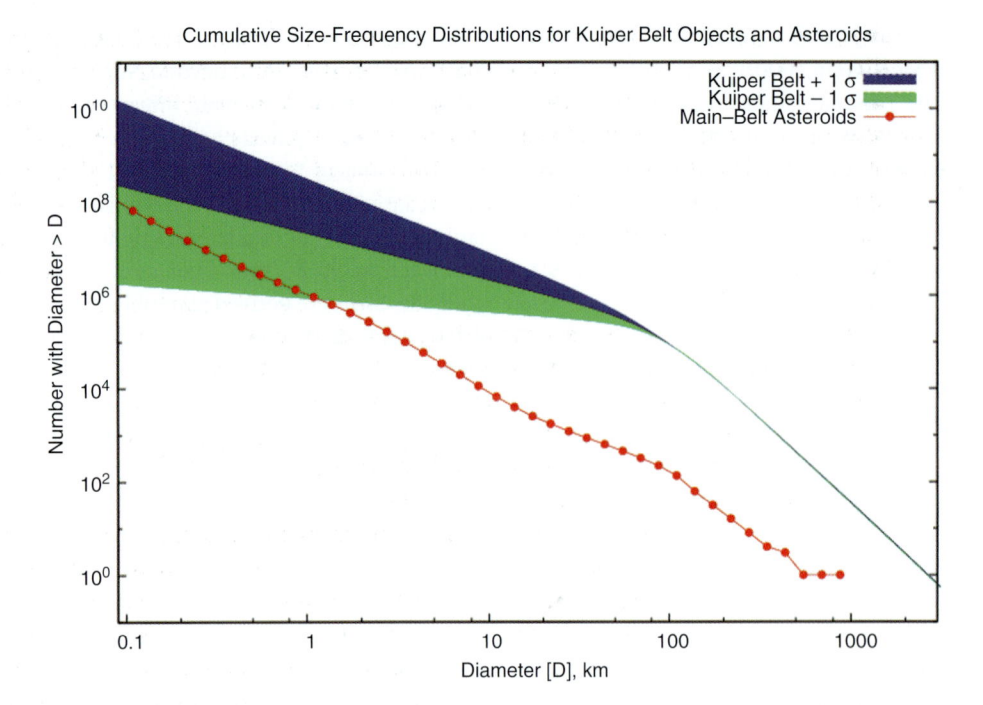

Fig. 19.1 Cumulative size-frequency distribution of Kuiper Belt Objects (*blue and green bands*) from Fuentes et al. (2009) and asteroids (*red dots*) from Bottke et al. (2005). Representative albedos of 0.04 and 0.092 were used to convert the respective absolute magnitude distributions to physical size distributions. The Kuiper Belt size distribution is well constrained only for diameters larger than tens of km, while the asteroidal size distribution is known down to sizes a bit smaller than 1 km (Gladman et al. 2009). The known number of Jupiter-family comets exceeds the range of values predicted by the Fuentes et al. (2009) distribution. The cause of this discrepancy is unknown

to Jupiter, other observations, and the model Centaur orbital distribution from Levison and Duncan (1997) to estimate that 0.003–0.009 ecliptic comets with $d > 1$ km strike Saturn itself per year.

In Fig. 19.1 we show the cumulative size-frequency distributions (SFDs) of Kuiper Belt Objects (Fuentes et al. 2009) and main-belt asteroids (Bottke et al. 2005). The blue and green bands for the Kuiper Belt Objects represent $\pm 1\sigma$ uncertainties. To lowest order, the SFD of Centaurs should be the same as that of KBOs, since Centaurs rarely collide with each other (Durda and Stern 2000). However, some comets split (Boehnhardt 2004) or undergo outbursts (West et al. 1991) at heliocentric distances beyond 10 AU, thereby perhaps changing the SFD of Saturn-crossing bodies. Even ignoring this complication, the uncertainties in Fig. 19.1 are enormous by the time one reaches the km-sized bodies that produce craters with diameters of tens of km in the Saturn system. Kuiper Belt Objects are more numerous than asteroids, at least for bodies larger than about 1 km. The asteroidal size-frequency distribution is well-determined down to 1 km, and uncertain at smaller sizes.

In Fig. 19.2 we show the heliocentric distance distribution, dN/dr, of Centaurs in simulations by Di Sisto and Brunini (2007). The distribution can be approximated by power laws, dN/dr $\propto r^4$ for r < 30 AU and dN/dr $\propto r^{-1.5}$ for r > 30 AU, in general agreement with Levison and Duncan (1997) and Tiscareno and Malhotra (2003). If

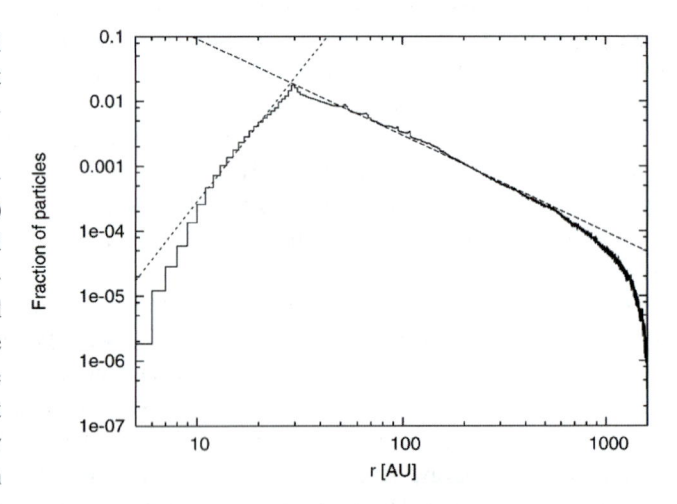

Fig. 19.2 Theoretical radial distribution, dN/dr, of Centaurs and Scattered Disk Objects, from Di Sisto and Brunini (2007). In the region of the giant planets, $r < 30$ AU, the numbers falls steeply inward because gravitational scattering by the giant planets, particularly Jupiter and Saturn, rapidly ejects most planet-crossing small bodies from the Solar System

Centaurs remain intact, the SFD of Centaurs should have the same shape as that in Fig. 19.1, but scaled to each planet by factors given in Fig. 19.2.

Fourth, Saturn's *irregular satellites* follow distant, highly inclined (both prograde and retrograde), eccentric orbits around the planet. The irregular satellites of the giant planets are thought to have been captured from heliocentric orbit through some form of dissipation such as gas drag, collisions with other small bodies within the Hill sphere of the parent planet (Turrini et al. 2009), or various dynamical "freeze-in" processes (Morbidelli et al. 2005; Vokrouhlický et al. 2008). Until the year 2000, Phoebe, which has an average diameter of 213 km (Thomas et al. 2007b), was the only known saturnian irregular, but a total of 37 are now known (JPL Solar System Dynamics 2009). The newly discovered irregulars have diameters ranging between 4 and 40 km. Many of the irregulars cross each others' orbits, and so can collide. The present-day cratering rate on Phoebe by other irregulars is estimated to be larger than the rate due to ecliptic comets (Nesvorný et al. 2003). Some of Saturn's irregular satellites may belong to collisional families (Turrini et al. 2008) like those in the main asteroid belt (Nesvorný et al. 2002; Parker et al. 2008), Trojan swarms of Jupiter (Shoemaker et al. 1989; Roig et al. 2008), and Kuiper Belt (Brown et al. 2007; Ragozzine and Brown 2007). When the irregulars were captured, their orbits may have extended closer to Saturn (Nesvorný et al. 2007), and these early irregulars might have produced many of the craters on Iapetus, or perhaps even moons closer to Saturn (Bottke et al. 2009; W. F. Bottke and D. Nesvorný, personal communication, 2009). Irregular satellites appear to have a shallow size-frequency distribution, even when observational biases are properly taken into account (Nicholson et al. 2008).

Fifth, *planetocentric bodies* can arise as (a) fragments, launched from cratered or disrupted satellites, which then orbit Saturn for some time, typically decades or hundreds of years, before usually colliding with the satellite from which they were launched (Alvarellos et al. 2005), excepting ejecta from Hyperion (Farinella et al. 1990; Dobrovolskis and Lissauer 2004); (b) comets or asteroids temporarily captured by Saturn, akin to the capture of SL9 by Jupiter (Kary and Dones 1996); (c) bodies captured permanently after tidal disruption (Dones 1991); or (d) asteroids permanently captured by Saturn by some unspecified process (Schmedemann et al. 2008, 2009). Alvarellos et al. (2005) estimate that satellite ejecta fragments typically produce craters only a few km in size. However, larger and/or faster fragments, such as might be produced by catastrophic disruption of a moon, might create bigger craters.

Finally, *Nearly Isotropic Comets* (NICs), which encompass long-period and Halley-type comets, are thought to originate in the Oort Cloud (Dones et al. 2004). The classical "outer" Oort Cloud is estimated to contain between 1×10^{11} and 3×10^{11} comets with d > 1 km (Francis 2005). Of these, roughly 10^8 comets are likely to be on Saturn-crossing orbits, assuming a cometary perihelion distribution which is flat or gradually increasing with distance from the Sun (Mazeeva 2007). However, because of their long orbital periods around the Sun and the limited degree of gravitational focusing by Saturn they undergo, impact rates due to NICs on Saturn's moons interior to Titan's orbit are probably one to two orders of magnitude smaller than the rate due to ecliptic comets (Zahnle et al. 1998, 2003). Because NICs encounter Saturn at higher speeds and therefore undergo less gravitational focusing, they become relatively less important for the moons closest to Saturn. By the same logic, cratering by NICs could be important for Iapetus and the irregular satellites, however.

19.3 Heavy Bombardments

Even before the Apollo landings, it was clear that crater densities on the lunar highlands were larger than could have been produced at present-day impact rates (Hartmann et al. 2000). Dating of lunar samples appeared to show that some basins, such as Imbrium, formed late, 3.8–3.9 Ga, i.e., some 700 Myr after the planets had formed. In a classic paper, Wetherill (1975) considered possible reservoirs of long-lived impactors, including accretional leftovers in the region of the terrestrial planets and Uranus-Neptune planetesimals (cf. Wetherill 1981). Bottke et al. (2007) dealt a blow to the idea of long-lived leftovers of terrestrial planet accretion. They showed that, even if an implausibly large mass of small bodies had been present soon after planet formation, collisional and dynamical evolution would have depleted the small bodies so severely that too few were left to make even a single basin as recently as 3.9 billion years ago.

In the past decade, Wetherill's ideas about long-lived Uranus-Neptune planetesimals have been combined with insights about dynamical chaos acting on long timescales (Sussman and Wisdom 1992; Laskar 1996; Thommes et al. 1999, 2002, 2008; Levison et al. 1998, 2001, 2004), culminating in the "Nice model".

The Nice model proposes that the orbits of the giant planets changed significantly long after the planets had formed (Tsiganis et al. 2005; Morbidelli et al. 2005; Gomes et al. 2005). In this picture, the four giant planets formed close together, between ∼5.5 and ∼14–17 AU of the Sun, compared with the present range of 5.2 to 30.1 AU. A massive (∼35 Earth mass) disk of planetesimals, i.e., a weighty ancestor of today's Kuiper Belt, is assumed to lie outside the planetary orbits. Some planetesimals near the inner edge of the disk, initially around 15 AU, evolve onto planet-crossing orbits, resulting in slow migration of the planets (Fernández and Ip 1984; Malhotra 1993; Murray et al. 1998; Hahn and Malhotra 1999; Gomes et al. 2004). In the simulations of the Nice model, slow migration continues for times ranging from 350 Myr to 1.1 Gyr, during which the disk loses about one-third of its mass. At this point, Jupiter and Saturn cross the strong 1:2 mean-motion resonance. This event excites the eccentricities of Jupiter and Saturn's orbits, triggering an instability in the orbits of all the giant planets. The orbits of Uranus and Neptune become chaotic, causing them to suffer close encounters with each other and, in many cases, with Saturn. The ice giants are scattered onto eccentric orbits that traverse the planetesimal disk, thereby scattering bodies in the massive "Kuiper Belt" all over the Solar System. The planets' orbits are eventually circularized by dynamical friction exerted by the planetesimals (e.g., Del Popolo 2001; O'Brien et al. 2006; Leinhardt et al. 2009). In a few Myr, half the mass left in the disk prior to the resonance crossing is eliminated, and after some tens of Myr, 99% of the mass has been scattered out of the Solar System. During the chaotic phase, most main-belt asteroids would have become unstable as well due to sweeping resonances as the outer planets moved about. In the baseline model, the mass striking the Moon in comets would have been roughly comparable to the amount in asteroids (Gomes et al. 2005; Strom et al. 2005).

Depending upon the initial conditions, Jupiter and Saturn might have passed through a strong resonance other than the 1:2 as they migrated (Morbidelli et al. 2007). For our purposes, the key point is that it is quite plausible that the giant planets underwent a dynamical instability long after they formed. Independent of the details of what actually happened, it is plausible that there were profound implications for the saturnian satellites that were present at that time.

Whether or not the Nice model scenario actually happened remains unproven, although the model explains many features of the outer planets, most remarkably their semi-major axes, eccentricities, and mutual inclinations (Tsiganis et al. 2005), as well as other features such as the orbital distribution and mass of the jovian Trojans (Morbidelli et al. 2005) and the inclinations of neptunian Trojans (Lykawka et al. 2009). Other models for the LHB have been proposed, some of which would predominantly operate in the inner Solar System and have lesser effects on the saturnian satellites. While a bombardment of saturnian satellites roughly simultaneous with the inner Solar System LHB has been widely assumed, it has not been established. As we noted earlier and will see in Section 19.6, the current impact rate in the Saturn system by heliocentric bodies is insufficient to explain the apparently saturated populations of large craters on some of Saturn's moons, so the impact rate must have been higher at some point in the past. But whether it was the declining phase of early satellite accretion, due to one or more Solar System-wide or local impact spikes, or largely due to collisional evolution of satellites within the Saturn system cannot be securely assumed, despite the early post-Voyager assumptions. Nevertheless, at this time, the Nice model provides the most detailed Solar System-wide cratering chronology that is available.

19.4 Cratering Chronologies

A cratering chronology is a function that specifies the number of craters on a surface as a function of age. Three chronologies have been used in recent studies of cratering in the outer Solar System.

The *Neukum lunar chronology* has been used by Neukum and colleagues (Neukum et al. 2001). In Neukum's view, the same impactors (asteroids) have produced most of the primary craters on both the Moon and the satellites of Jupiter and Saturn, and so the rate at which bodies of diameter d strike the saturnian moons is taken to be proportional to the rate at which these bodies strike the Moon. The chronology is an updated version of that derived by the Basaltic Volcanism Study Project (1981 [Table 8.4.2, p. 1072]). Neukum (1983) fit the lunar data with a linear term, representing a constant cratering rate, plus an exponential term with an e-folding time of 144 Myr which dominates at early times. Marchi et al. (2009) use the same functional form, $N_1(t) = a(\exp(t/\tau) - 1) + ct$, where $N_1(t)$ represents the number of craters with diameter $D > 1$ km per km^2, t is the age of the terrain in Gyr, and a, τ, c are empirically derived constants. Marchi et al. (2009) find numerical values of $a = 1.23 \times 10^{-15}$, $\tau = 0.127$, and $c = 1.30 \times 10^{-3}$. The rate of crater formation per unit area, dN_1/dt, equals $a \exp(t/\tau)/\tau + c$. The exponential term exceeds the constant term for ages larger than 3.5 Gyr.

The Neukum model assumes that the flux of asteroids on any surface in the Solar System occurred with the same time dependence as measured for the Moon, and that the cratering record in the saturnian system stems from asteroids captured into planetocentric orbits about Saturn, not simply asteroids moving on heliocentric, Saturn-crossing orbits. The planetocentric aspect of this chronology is strongly, but not entirely, driven by a perceived need to explain the lack of strong apex-antapex cratering asymmetries in the outer Solar System. We discuss the apex-antapex issue in Section 19.6, and return to the issue of asteroid capture in Section 19.8. To calibrate the lunar chronology at Saturn, Neukum assumes that the oldest surface on Iapetus is 4.4 Ga (Castillo-Rogez et al. 2007). Chronologies for other saturnian satellites are derived by scaling from Iapetus' record, assuming that the impactors orbit Saturn with eccentricities of 0.6 and inclinations of 15° (Horedt and Neukum 1984b). The Neukum model assumes captured asteroids as the main source of impactors, but allows comets to play some role during recent epochs.

The *cometary chronology* uses estimates of the numbers of "comets" of different sizes crossing Saturn's orbit at the present epoch to calculate cratering rates (Smith et al. 1981, 1982; Shoemaker and Wolfe 1981, 1982). The cratering rate dN/dt is assumed either to be constant with time (a reasonable approximation for at least the last three billion years) or to have monotonically decreased with time. For example, based on the work of Holman and Wisdom (1993), Zahnle et al. (1998, 2003) assumed that the population of bodies crossing the orbits of the giant planets has declined inversely with time since the Solar System formed, i.e., dN/dt $\propto (t_0-t)^{-1}$, where $t_0 = 4.567$ Gyr. With this assumption, the crater density on a surface of age t is proportional to ln $[t_0/(t_0-t)]$. The cometary chronology is primarily intended for application to young surfaces. It takes the view that N(t) is unknown for the outer Solar System, since the impacting population (comets) differs from that of the Moon (asteroids), and there are no radiometrically dated surfaces on the satellites of the giant planets. Zahnle et al. (2003) computed the cometary chronology for satellites of the giant planets under two assumptions about the impactors' size distribution: "Case A," based upon the paucity of small primary impact craters on the galilean satellites (Bierhaus et al. 2005), and thus poor in small comets; and "Case B," based on high-resolution crater counts on Triton (McKinnon et al. 2000) and richer in small comets. Because of the much larger abundance of small comets in Case B, a given crater density translates into a younger age in Case B.

The *Nice model chronology* assumes that most of the impacts on the regular satellites of the giant planets took place around the time of the dynamical instability in the model. The Nice model cannot predict exactly when the resonance crossing took place, but 700 Myr after the formation of the giant planets (i.e., 3.9 billion years ago) lies within the range found in the simulations. Gomes et al. (2005) assumed that the lunar LHB 3.9 Ga was caused by the flood of outer Solar System planetesimals and asteroids released by the instability. If that assumption is correct, the Nice model provides the basis for a Solar System-wide chronology, although "comets" (which, on average, bombard planetary surfaces somewhat earlier than asteroids) produce almost all the impacts on bodies in the outer Solar System, while both asteroids and comets contribute on the terrestrial planets and their satellites. Charnoz et al. (2009) applied the Nice model chronology to the origin of Saturn's rings, calibrating the model by the observed density of basins and 10-km craters on Iapetus, and found that the rings could have formed at that time by disruption of a satellite at the current location of the rings.

19.5 Impact Physics and Scaling Laws

The size-frequency distribution of craters on the surface of a planetary body is not a faithful reflection of the size-frequency distribution of the hypervelocity impactors that create them. Because the final craters are typically at least 10 times larger in diameter than the object that strikes the surface, the final crater diameter distribution may present a distorted picture of the impactor population. Factors such as the porosity of the surface (an extreme factor in the case of Hyperion), varying impact velocity, and impact angle all contribute to differences between the populations of craters and impactors. The same parent population of projectiles may produce radically different crater size-frequency distributions on different satellites, especially when considered in an absolute sense. It is thus critical to understand the relation between the diameter of the final crater and the impacting projectile and its dependence on the various conditions of impact.

Because we are as yet unable to perform direct experiments with kilometer-size projectiles impacting full-scale planetary targets, we must rely on so-called scaling relations that relate the results of small-scale experiments and numerical simulations to actual craters in the Solar System. Refined by many investigators over many decades of work, such scaling relations now give reasonably reliable results. They are no longer wholly empirical: physics-based numerical simulations have recently achieved great success in both reproducing the results of small-scale laboratory and field experiments[5] and justifying the empirical extrapolations of the past.

Employing the highly successful methods long established in fluid mechanics, crater scaling is based on relations between dimensionless combinations of variables describing the properties of the projectile and target, often expressed in the language of the Buckingham Π theorem (Buckingham 1914). The most important groups are the cratering efficiency, described as the ratio of either the crater diameter to the projectile diameter or the ratio of crater volume to projectile volume, and the inverse Froude number that expresses the ratio between gravitational forces and inertial forces. In all cases the crater diameter or volume pertains to the "transient crater", which is formed immediately after the impact. In nearly all planetary circumstances the transient crater subsequently collapses and expands in diameter (while it shrinks in volume) to produce the final, observed, crater. We thus also need a relation between the transient crater and final crater before we can link the impactor population to that of the observed craters.

The cratering efficiency is defined as either π_V or π_D, depending on whether the transient crater volume or diameter is more convenient. These definitions also traditionally incorporate the ratio between the projectile and target densities in the following forms:

$$\pi_V = \frac{\rho V}{m} \qquad (19.1a)$$

---

[5] Field experiments here refer to explosion cratering tests, and to laboratory impact experiments performed at elevated gravity in a centrifuge, which simulate similar scales (e.g., Schmidt and Housen 1987).

and

$$\pi_D = D_t \left(\frac{\rho}{m}\right)^{1/3} \quad (19.1b)$$

where V is the excavated volume of the crater, D_t is the transient crater diameter, ρ is the density of the target, and m is the mass of the projectile. The inverse Froude number, or "gravity-scaled size," is usually expressed as:

$$\pi_2 = \frac{3.22 g a}{U^2} \quad (19.2)$$

where U is the speed of the impact, g is the surface gravity of the target and a is the mean radius of the projectile.[6] The relation between the cratering efficiency and the gravity scaled size closely approximates a power law (Holsapple 1993, 2009). The exponents of this power law depend on whether the crater size is more dependent upon the momentum or the energy of the impactor, a dependence that is a function of the dynamics of the impact and which cannot be determined from the scaling relations alone.

Thus, we can write

$$\pi_V \propto \pi_2^{-3\mu/(2+\mu)} \left(\frac{\rho}{\delta}\right)^{(2+\mu-6\nu)/(2+\mu)} \quad (19.3)$$

where δ is the density of the impactor and μ and ν are experimentally determined constants (e.g., Melosh 1989; Holsapple 1993, 2009). The case when $\mu = 1/3$ corresponds to "momentum scaling" (i.e., the crater's volume depends on the momentum of the impactor, independent of its velocity), while $\mu = 2/3$ gives "energy scaling" (the crater's volume depends on the energy of the impactor, independent of its velocity) (Holsapple and Schmidt 1982).

These relations are frequently expressed by the simpler relations for practical use:

$$\pi_V = C_V \pi_2^{-\alpha} \quad (19.4a)$$

$$\pi_D = C_D \pi_2^{-\beta} \quad (19.4b)$$

where C_V, C_D, α, and β are dimensionless constants that are either determined empirically or from computer simulations. For impacts into competent rock (or ice) or water-saturated soil these constants are given by $C_V = 0.2$, $C_D = 1.6$, $\alpha = 0.65$, and $\beta = 0.22$.

There may be other dimensionless constants in the problem. For example, it is well known that the cratering efficiency for sand is substantially smaller than that for competent rock; for sand $C_V = 0.24$, $C_D = 1.68$, $\alpha = 0.51$, and $\beta = 0.17$ may be used (cf. Table 1 in Holsapple [1993] and

---

[6] The factor $3.22 = 2\sqrt[3]{4\pi/3}$ reflects historical inheritance from explosion cratering studies, and is not of any fundamental physical significance.

Table 7.1 in Melosh [1989]). Craters on the highly porous satellite Hyperion, therefore, should not follow the same scaling relations as craters on more densely compacted icy satellites. (Cratering in porous bodies has been studied experimentally [Housen and Holsapple 2003], but impact simulations involving porosity are only now becoming possible [Jutzi et al. 2008, 2009; Leinhardt and Stewart 2009]). So-called sand or dry soil scaling probably also applies to saturnian satellites of more modest porosity (such as Phoebe; Giese et al. 2006) and to craters formed in the regoliths and megaregoliths of highly battered midsize satellites such as Rhea and Iapetus. Until recently, it was not possible to disentangle the independent effects of porosity and internal friction on cratering efficiency because these two factors are closely linked for all materials on which experiments have been performed to date. However, recent numerical work (Wünneman et al. 2006, 2008) using a newly formulated model for introducing porosity into hydrocode computations has evaluated these factors separately and finds both to be about equally important.

The angle of impact is also an important factor whose effect on cratering efficiency has been poorly understood. A commonly used *ansatz* has been to replace U in the above scaling relations with the vertical component of velocity, U sin θ, where θ is the angle between the trajectory of the impactor and the plane of the target surface (e.g., Chapman and McKinnon 1986). This *ansatz* has recently received strong support from 3-dimensional hydrocode simulations (Collins et al. 2009) for dense targets, but it remains to be verified for porous targets.

The diameter of a simple crater, D_s, is somewhat larger than the diameter of the transient crater, D_t, because even simple craters suffer some degree of rim collapse after they form (Melosh 1989). For the purposes of this paper, we relate the simple crater diameter D_s, measured in km, to the impactor diameter d through

$$D_s = 11.9 \left(U^2/g\right)^{0.217} (\delta/\rho)^{0.333} (d/\text{km})^{0.783} \quad (19.5)$$

with U in km/s and g in cm/s^2, and "final" diameter D in km given by

$$D = D_s \text{ for } D_s < D_c \quad (19.6a)$$

$$D = D_s (D_s/D_c)^\xi \text{ for } D_s > D_c \quad (19.6b)$$

where D_c is the transition diameter from the simple to collapsed, complex morphology. Equation 19.5 is derived from Eq. 19.4b with competent rock parameters (see Zahnle et al. 2003); Holsapple (2009) states that these parameters are also appropriate for cold ice. In this paper we take $D_c = 2.5$ km for Titan and $D_c = 15$ km for Saturn's other satellites (Schenk 1989; Schenk et al. 2004). We take $\xi = 0.13$

(McKinnon et al. 1991). We assume an impactor density of $\delta = 0.6\,\text{g/cm}^3$, consistent with the density of the nucleus of Comet Shoemaker-Levy 9 (Asphaug and Benz 1996, and see Sosa and Fernández 2009 for density estimates for ten periodic comets). Eqs. 19.5 and 19.6 represent a compromise between unwarranted precision and practicality, though we recognize that accurate simple-to-complex transition diameters can now be derived for the mid-sized saturnian satellites from Cassini images, and on theoretical grounds are expected to vary according to satellite gravity (e.g., McKinnon 2007)[7]. The reader can experiment with various scaling relations using the interactive calculators provided by Melosh and Beyer (2009) and Holsapple (2009).

Typical impact velocities are given by

$$U_{mean} = \left(3v_{orb}^2 + v_\infty^2 + v_{esc}^2\right)^{1/2} \quad (19.7)$$

where v_{orb} is the orbital speed of the satellite around Saturn, v_∞ is the velocity at "infinity" of the body encountering the Saturn system, and v_{esc} is the escape velocity from the surface of the satellite (Lissauer et al. 1988; Zahnle et al. 2003). The final term is negligible for our purposes, so we will neglect it henceforth. The minimum and maximum impact speeds are approximately

$$U_{slow} = \left(2v_{orb}^2 + v_\infty^2\right)^{1/2} - v_{orb} \quad (19.8a)$$

and

$$U_{fast} = \left(2v_{orb}^2 + v_\infty^2\right)^{1/2} + v_{orb} \quad (19.8b)$$

respectively.

In Fig. 19.3, we use Eqs. 19.5 and 19.6 to plot the diameter (d) of the cometary impactor (vertical axis) needed to make a crater of diameter D on Mimas, Rhea, and Iapetus. The assumed impact speeds on Mimas are 6.2 ("slow"), 25.0 ("mean"), and 34.8 km s^{-1} ("fast"); for Rhea, 3.9, 15.0, and 20.9 km s^{-1}; and for Iapetus, 2.2, 6.4, and 8.8 km s^{-1}. Ten-km craters on Mimas are made by fast impactors with diameters of only 0.2 km, 0.3 km at the mean impact speed, and 0.6 km for slow impactors. For Rhea, the corresponding impactor sizes are 0.4, 0.5, and 1.1 km, respectively; for Iapetus, 0.7, 0.8, and 1.4 km. Three-hundred-km craters, i.e., basins, are made by 11-, 13-, and 26-km-diameter impactors at Mimas; 22-, 26-, and 55-km-diameter impactors at Rhea; and 33-, 40-, and 71-km-diameter impactors at Iapetus, again, all at the impact speeds given above, respectively. If we wish to tie the cratering record of the satellites to the

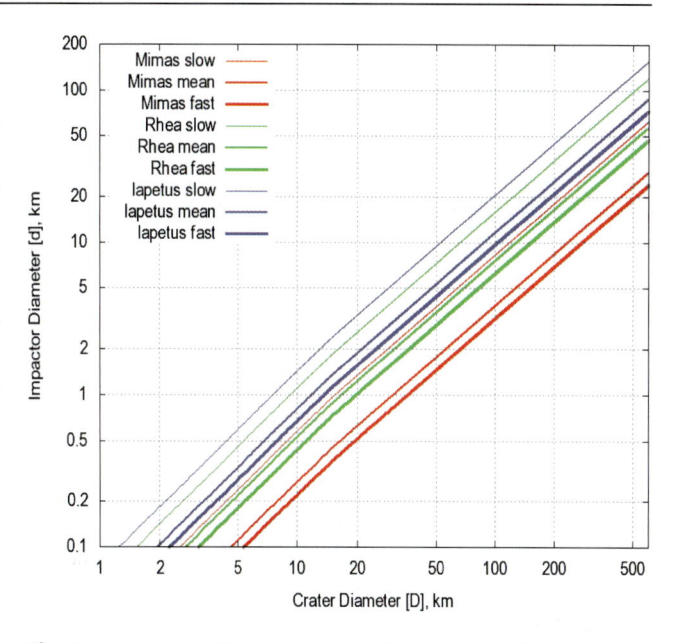

Fig. 19.3 Impactor diameter d (*vertical axis*) required to produce a crater of diameter D (*horizontal axis*) on Mimas, Rhea, and Iapetus. We use Eqs. 19.2 and 19.3 with the following parameters: $\rho_i = 0.6$; $\rho_t = 0.9$ (all three moons are assumed to have icy shells); U given by Eqs. 19.4–19.6 with v_{orb} = 14.32, 8.49, and 3.27 for Mimas, Rhea, and Iapetus, respectively; v_∞ = 3; g = 6.36, 26.4, and 22.3; Dc = 15; and ξ = 0.13

putative impactors from the swarm of ecliptic comets, basins on Iapetus, Titan, and Rhea are our best bets, because the required impactor sizes overlap the size range for which we know a bit about the Kuiper Belt size-frequency distribution, and for which we have some constraint on absolute numbers and rates (e.g., Fuentes et al. 2009).

19.6 Predicted Cratering Rates by Comets

Heliocentric (Sun-orbiting) "comets" on Saturn-crossing orbits, which roam the outer Solar System in numbers too big to ignore, are the only present-day *known* population of potential impactors for Saturn's inner moons. The cometary population can be characterized by astronomical observations, both current and historical, the observed orbital distribution can be corrected for discovery biases, and their statistical impact probabilities can be computed. Such astronomically based impact rates have been used to estimate the ages of young and middle-aged surfaces on the satellites of the outer planets (Zahnle et al. 2003). We will review this line of reasoning here.

On the other hand, there is evidence that many or perhaps most of the craters in the Saturn system may be of planetocentric (Saturn-orbiting) origin. The key issues are the global uniformity of cratering and the large abundance of small craters. Heliocentric origin strongly favors cratering of

---
[7] The scaling exponent $\xi = 0.13$ was derived from the depth-diameter relations for lunar craters assuming volume conservation during collapse. The value of this exponent should be reassessed for the mid-sized icy satellites.

the leading hemisphere of synchronously rotating satellites (Shoemaker and Wolfe 1982; Horedt and Neukum 1984a; Zahnle et al. 2001), while planetocentric cratering weakly favors the leading or trailing side, depending on whether the debris fall from outside or inside the satellite's orbit, respectively (Horedt and Neukum 1984a, Alvarellos et al. 2005). Cratering asymmetries due to captured irregular satellites would likewise depend on whether the irregulars were prograde or retrograde. Strong apex-antapex asymmetries have not been detected on any satellite in the outer Solar System, except for the perverse case of Neptune's moon Triton (Schenk and Zahnle 2007). The origin of voluminous planetocentric debris is often attributed to a cataclysm or cataclysms, possibly linked to the destruction of a moon or the formation of the rings or both (Smith et al. 1981, 1982; Chapman and McKinnon 1986; Lissauer et al. 1988). If planetocentric cratering has been dominant, the surfaces are younger than we have inferred by assuming that the impactors come from heliocentric orbit. If there has been a geologically *recent* cataclysm, such as the post-Voyager idea that Saturn's rings formed within the last 100 Myr (Esposito 1986; cf. Lissauer et al. 1988; Dones 1991; Charnoz et al. 2009; and Chapter 17), we can use heliocentric impact rates to estimate the probability of such an event, but not precisely when it occurred.

As we discussed in the Introduction, Population I and Population II craters were generally taken by Voyager-era scientists to be made by heliocentric and planetocentric bodies, respectively. The craters themselves can be called "primary" and "sesquinary" (formerly "poltorary"; Dobrovolskis and Lissauer 2004), respectively, to distinguish them from conventional secondaries that fall back to the cratered satellite immediately (Zahnle et al. 2008). In one picture Population I is (or was) responsible for the larger craters, while Population II is responsible for most of the small craters.

More recent observations support the importance of both heliocentric and planetocentric impactors on the satellites of the giant planets, but unequally. The Galileo spacecraft returned images of Ganymede and Europa that show remarkably few small (1-to-20 km diameter) *primary* impact craters on young surfaces (Bierhaus et al. 2001; Schenk et al. 2004; Bierhaus et al. 2005; McEwen and Bierhaus 2006), but numerous secondaries. On the other hand, Triton has a young surface with a "typical" size-frequency distribution of small craters. But on retrograde Triton, the leading-trailing asymmetry is so extreme that it is not clear whether we are seeing the results of heliocentric comets or head-on collisions with prograde planetocentric debris (Croft et al. 1995; Schenk and Sobieszczyk 1999; Stern and McKinnon 2000; Schenk and Zahnle 2007).

Here we focus on heliocentric comets and primary craters. We take the chief source of primary craters in the outer Solar System to be the ecliptic comets [ECs] (Shoemaker and Wolfe 1982; Smith et al. 1982, 1986, 1989; Zahnle et al. 1998, 2003; Levison et al. 2000; Schenk et al. 2004). ECs refer to low inclination, prograde comets that interact strongly with planets: Jupiter-family comets, Centaurs, and Scattered Disk Objects. Long-period comets (also known as Oort Cloud comets) and Halley-type comets, even when lumped together as "Nearly Isotropic Comets" (NICs; Levison 1996), are important at Saturn (at the $\sim$10%–30% level) only for Iapetus and the irregular satellites (see Section 19.2). The distinct inclination distributions of the ecliptic comets and the nearly isotropic comets suggest that they arise from different reservoirs, with the former evolving inward from the Kuiper Belt/Scattered Disk, while the latter fall from the Oort Cloud (Duncan et al. 1988; Levison and Duncan 1997; Gladman et al. 2001; Dones et al. 2004).

To compute cratering rates, we have two types of information. The modern cratering rate can be estimated by characterizing the current populations of impacting bodies (comets), calculating impact probabilities, and using scaling rules to estimate the sizes of the resulting craters (see Section 19.5). A time-averaged ("historical") cratering rate can be estimated from the numbers of craters and basins observed on unsaturated satellite terrains (Zahnle et al. 2003; Kirchoff and Schenk 2009b). The major uncertainties are the age of the surface and the constancy of the impact rate. In practice the only usable date is the 3.9 Gyr age of the LHB, which is sensibly explained in the Nice model as a Solar System-spanning feature caused by a rearrangement of planetary orbits (Section 19.3; Morbidelli et al. 2007; Levison et al. 2008). We can take the cratering rate as constant, or we can impose the inverse-time decay relation suggested by numerical integrations (Holman and Wisdom 1993; see Dobrovolskis et al. 2007 for a more general discussion of impact rate decay curves). During the era of heavy bombardment effects on satellite surfaces may have been especially severe; such effects include saturation cratering, complete disruption, and surface melting (Chapman and McKinnon 1986; Levison et al. 2001; Barr and Canup 2008). Whatever the source or sources of the heavy bombardment on the saturnian satellites, the historical rates refer to the time since the heavy bombardment ended.

Arguably, the best source for historical crater counts in the outer Solar System is Ganymede's younger (bright) terrains. These terrains have been geologically reworked since the LHB. The upper bound age of $\sim$3.9 Gyr and a constant impact rate imply that Jupiter has been struck 2×10^{-3} per year by cometary nuclei with diameters d $\geq$ 2 km, which corresponds to 8×10^{-4} per year striking Saturn, using the ratio of 42 Saturn impacts to 100 Jupiter impacts calculated by Levison and Duncan (1997). The characteristic time scale set by the orbital rearrangement in the Nice model is $\sim$10 Myr (Gomes et al. 2005). The dN/dt $\propto$ $(t_0-t)^{-1}$ decay law then implies that the current impact rate is smaller than the

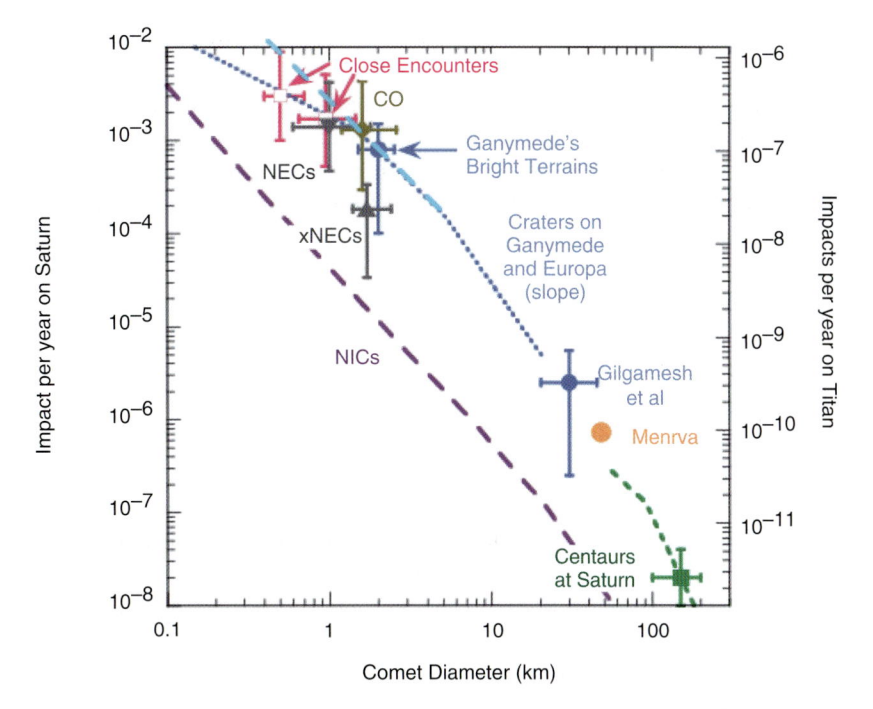

Fig. 19.4 Impact rates at Saturn and Titan. "Close Encounters" is deduced from the historical record of comets known to have passed near Jupiter. "CO" refers to the impact rate that would be deduced if excess carbon monoxide in Jupiter's stratosphere had been caused by comet impact. Points labeled "NECs" and "xNECs" refer to impact rates computed from the roster of known Jupiter-family comets by Levison et al. (2000) and Bottke et al. (2002), respectively. Points labeled "Ganymede" and "Gilgamesh" are impact rates inferred from impact craters on Ganymede and Callisto. Titan's 440-km basin Menrva (Lorenz et al. 2007) is conservatively taken to represent a once in 10 Gyr event. The associated lines are deduced from the size-frequency distributions of craters on Europa and Ganymede. The point labeled "Centaurs" refers to the three big Centaurs currently known to be on Saturn-crossing orbits. The line through "Centaurs" is based on the observed populations of Kuiper Belt Objects. The dashed line refers to impacts by long period and Halley-type comets ("NICs"). Case A (blue dots) refers to the relative abundance of small comets at Europa and Ganymede, while Case B (blue dashes) refers to the relative abundance of small impacting objects at Triton

long term average by a factor $\ln(3.9 \text{ Gyr}/10 \text{ Myr}) \sim 6$.[8] The corresponding current impact rate on Saturn of $d \geq 2 \text{ km}$ comets would be 1×10^{-4} per year. Since the bright terrains might be considerably younger than 4 Gyr, we plot this point (in light blue) in Fig. 19.4 at $8 (+16, -7) \times 10^{-4} \text{ year}^{-1}$ at $d = 2.0 \pm 0.5 \text{ km}$. If geological reworking of Ganymede took place independently of the great rearrangement in the Nice model (or other heavy bombardment), a relatively high impact rate is appropriate.

There are four (possibly five) big, young impact basins on Ganymede and Callisto, with Gilgamesh and Valhalla being the best-known (e.g., Schenk and McKinnon 2008). These basins were made by comets with $d \geq 30 \pm 10 \text{ km}$. Spread uniformly over 4 Gyr, these four impacts imply an impact rate of 30 km comets on Saturn of dN/dt $(d \geq 30 \pm 10 \text{ km}) = 2.5 (+3, -2) \times 10^{-6} \text{ year}^{-1}$. As above, the $(t_0-t)^{-1}$ decay implies current impact rates that are 6 times smaller.

The current population of ecliptic comets can be inferred from any of (a) the number of small comets observed near the Earth; (b) the historical record of small comets known to have closely encountered Jupiter[9]; or (c) the observed number of Centaurs in the more distant Solar System.

Levison and Duncan (1997) modeled the evolution of test particles from sources in the Kuiper Belt. In modeling the migration of ecliptic comets, Levison and Duncan assumed a dynamically cold (low inclination, low eccentricity) classical Kuiper Belt source. Levison and Duncan calibrated their model to the number and orbital properties of JFCs that reach deep into the inner Solar System. Levison et al. (2000) developed this to estimate an impact rate at Saturn dN/dt $(d > 1 \text{ km}) = 1.4 \times 10^{-3} \text{ year}^{-1}$, with a quoted uncertainty of "at least an order of magnitude." This point is plotted on Fig. 19.4 as "NECs," which

---

[8] t_0 here refers to initiation of the LHB, e.g., when Jupiter and Saturn cross their 1:2 mean-motion resonance.

[9] Because of the steep decline in reflected brightness with distance from the Sun for small bodies, the census of Saturn-crossing comets is vastly inferior to that of JFCs. Lagerkvist et al. (2000) and Hahn et al. (2006) have identified three ecliptic comets that have passed within 0.01 and 0.03 AU of Saturn, i.e., roughly at Iapetus' distance from the planet. Nevertheless, this represents a vast improvement with respect to the Voyager era, when Shoemaker and Wolfe (1982) of necessity based their bombardment rates on the observations of single bodies (e.g., asteroid/extinct comet Hidalgo, or Chiron).

stands for "Near-Earth Comets". Bottke et al. (2002) updated the arguments of Levison et al. (2000) by using observations of inactive comets obtained from their analysis of discovery rates in the Spacewatch Near-Earth Object survey. From six objects Bottke et al. deduce that there are currently 61 ± 50 dormant JFCs with d > 1.7 (+0.7, −0.3) km, for which the corresponding impact rate on Saturn would be dN/dt (d > 1.7 [+0.7, −0.3] km) = (1.8 ± 1.4) × 10^{-4} year^{-1}. This point is plotted on Fig. 19.4 as "xNECs".

An independent approach exploits the historical record of comets observed to have made close approaches to Jupiter. As reviewed by Schenk and Zahnle (2007), observers have recorded four close approaches within 3 Jupiter radii and two direct hits since the dawn of the telescopic era. A lower bound on the sizes of these six is d > 1.0 (+0.5, −0.3) km. Six encounters inside of 4 Jupiter radii over 350 years implies an impact rate on Jupiter of at least 4×10^{-3} year^{-1}, corresponding to 1.7×10^{-3} year^{-1} on Saturn[10]. Another source of information is more recent comets making more distant encounters. Observations and orbital simulations indicate that at least 9 comets crossed Callisto's orbit between 1950 and 1999 (a sample that we have some hope might be complete). The smallest of these has a diameter of d ∼ 0.5 km (Lamy et al. 2004). The corresponding impact rate on Saturn would be $3 \cdot 10^{-3}$ year^{-1} for d > 0.5 km. Another independent argument uses excess CO present in Jupiter's stratosphere to infer the recent impact history (Bézard et al. 2002; also see Lellouch et al. 2005 for a possible similar event on Neptune). Excess CO is removed by mixing into the troposphere over ∼300 years. The impact of Shoemaker-Levy 9 (SL9) produced $∼7 \times 10^{14}$ g of CO (Lellouch et al. 1997) from a $∼10^{15}$ g comet (Harrington et al. 2004), but this falls well short both in quantity and location of being able to account for all the excess CO in the jovian stratosphere; the old CO suggests that another comet as big or bigger than SL9 has hit Jupiter in the past 300 years. The point is plotted on Fig. 19.4 as "CO".

We would like to estimate the impact rate on Saturn from the numbers and characteristics of Saturn-crossing ecliptic comets. For very big Saturn-crossing Centaurs, discovery may be nearly complete. The problem is that they require a huge extrapolation of the size-frequency curve to scale them to the desired quantity, the number of small Centaurs, because counts of smaller Centaurs and Saturn-crossing comets are severely incomplete. Here we will use the three known large Centaurs on Saturn-crossing orbits: Chiron, Pholus, and 1995 SN$_{55}$ (the latter unnamed and unnumbered because it is lost). All appear to be roughly 150 km diameter, with absolute magnitudes, H, in the range 6–7 (Fernández et al. 2002; Groussin et al. 2004; JPL Solar System Dynamics Small-Body Database Browser 2009). Their annual statistical ("Öpik") impact probabilities with Saturn are 1.5×10^{-8}, 1.0×10^{-9}, and 3.5×10^{-9}, respectively (though, naturally, the orbital elements upon which such probabilities are based vary strongly on million-year time scales for these and all Saturn-crossing Centaurs, and are especially uncertain for 1995 SN$_{55}$; Horner et al. 2004). Added together, they imply that d = 150 ± 50 km objects hit Saturn at a rate of 2 (+2, −1) × 10^{-8} year^{-1}.

To extend this datum to smaller sizes requires a size distribution. Size-frequency distributions have been determined for Kuiper Belt Objects, but these are not yet trustworthy for diameters smaller than 50 km (see the review of Petit et al. 2008), while the sizes of the bodies that crater satellites range from <1 km (for 20 km craters on the moons interior to Dione; see Table 1 of Zahnle et al. 2003) to ∼70 km for Iapetus' ∼580-km basin Turgis (Giese et al. 2008; Kirchoff and Schenk 2009b; USGS Astrogeology Gazetteer of Planetary Nomenclature 2009; see Section 19.5). An SFD with a power-law index a = 4–5 now seems well established for KBOs larger than a few tens of km in diameter (Petit et al. 2008)[11]. More recently, Fuentes et al. (2009) report *differential* SFDs with power-law indices of 4.65 (+0.4, −0.45) for d > 90 ± 30 km, with a sharp knee, with an index of 2.0 (+0.6, −0.7) for d < 90 km (see Fig. 19.1). Because the smallest bodies that have been detected in the shallow, small-size end of the distribution are tens of km in diameter, there is little information for bodies with d < 50 km.

From crater counts on Ganymede, Zahnle et al. (2003) inferred effective slopes of the cumulative impactor size distribution of b = 1.7 for 2 km < d < 5 km and b = 2.5 for 5 km < d < 20 km. Jupiter's Trojan asteroids provide another sample of quasi-local debris at sizes that can bridge the gap, assuming no collisional evolution within each cloud. (In the Nice model, Jupiter's Trojans are captured from the same outer Solar System population that gave rise to today's Kuiper Belt [Morbidelli et al. 2005, 2009], so the link is genetic.) Szabó et al. (2007) find b = 2.2 ± 0.25 for absolute magnitudes between 9 and 13.5, which respectively correspond to diameters of 110 and 13 km, respectively, assuming a 4% albedo. For specificity we will use here a single b = 2.5 cumulative slope for 5 km < d < 120 km and b = 1.7 for 1.5 km < d < 5 km.

For small craters we use populations on Europa and those superposed on the young large basins Gilgamesh (on Ganymede) and Lofn (on Callisto). We invert the crater

---

[10] The extrapolated impact rate of $(3 \pm 2) \times 10^{-2}$ year^{-1} for impacts on Jupiter for d > 0.3 km is consistent with the July 2009 impact on Jupiter, which appears to have had an energy comparable to that of the smaller fragments of SL9 such as W.

[11] A differential distribution $dN/dD \propto D^{-a}$ is said to have a power-law index of a; the corresponding cumulative distribution would have a power-law index $b = a - 1$.

counts using Eqs. 19.5 and 19.6 to obtain the size-frequency distribution (SFD) of small comets at Jupiter. The inversions are shown in Fig. 19.4. Based on 150 estimated Europan craters with D > 1 km (Schenk et al. 2004), we infer a cumulative cometary SFD $N(>d) \propto d^{-b}$ with $b = 0.9$ for d < 1 km. From craters on Gilgamesh, the inferred slope of the power law distribution is b∼1.2 for d < 2 km.

The shallow size distribution we adopt for cometary nuclei with diameters between about 1 and 10 km is similar to size distributions of JFCs inferred from (some) observational studies, which are reviewed by Lamy et al. (2004). For example, Fernández and Morbidelli (2006) derive a value of b = 1.25 for faint JFCs with perihelion distances <1.3 AU. Size determinations of cometary nuclei are not easy, because the nucleus is unresolved in Earth-based telescopic images, and thus requires either careful subtraction of comae, or dubious empirical relationships between total cometary brightness and the size of the nucleus (the former technique is now much preferred, and is reasonably well-calibrated for HST images). In any case, the apparent progressive steepening of the cometary SFD at larger sizes has been suggested before (e.g., Weissman and Levison 1997). However, models of the formation and collisional evolution of Kuiper Belt Objects (Kenyon et al. 2008) do not yet match the size distribution inferred from the variety of observations we have described (Petit et al. 2008).

With these admittedly provisional, but constrained, limits on the cometary SFD and impact rates (Fig. 19.4), we consider the implications for Saturn's satellites. We consider two cases. In Case A, the mass distribution of small comets is consistent with what we find at Jupiter and on its satellites. Case A requires an additional source of (presumably) planetocentric debris to account for most small craters at Saturn (in that the relative flatness of the predicted crater SFD at smaller sizes is hard to reconcile with observations; see next section). In Case B, we use a mass distribution of small comets that would be consistent with small craters on Triton, which is steeper at smaller sizes. Case B may or may not require planetocentric debris to account for small craters. The results are given in Table 19.1.

For comparison, cratering rates at Saturn were estimated by Smith et al. (1982) and Lissauer et al. 1988). These studies predate the discovery of the Kuiper Belt, and so underestimate heliocentric impact rates and correspondingly overestimate cratering time scales and crater retention ages, typically by roughly an order of magnitude.

Finally, we find the Nearly Isotropic Comets (NICs) to be relatively unimportant for cratering at Saturn, given the number of such comets on Saturn-crossing orbits estimated earlier. these. We use the scalings discussed by Zahnle et al. (2003) to estimate the total impact rate by NICs (Table 19.1). Because NICs tend to strike at higher velocity than ECs, especially for the more distant moons, they contribute somewhat more to the cratering record there, but are relatively less important as one moves inward toward Saturn due to their limited gravitational focusing. Among the moons of Saturn listed in Table 19.1, NICs are most important for Iapetus, where they contribute 10% of the craters, and for Phoebe, where they contribute 30% of the impact craters due to comets. They only contribute at the 1% level at Dione.

19.6.1 Implications for Catastrophic Disruption

It has been suggested that a satellite breaks up if the predicted transient crater diameter exceeds the satellite's diameter

Table 19.1 Present-day cratering rates at Saturn

	Cratering rates				Cratering times	
	$\dot{C}_A(>10)$	$\dot{C}_B(>10)$	$\dot{C}_S(>10)$	$\dot{C}_{NIC}(>10)$	τ_A	τ_B
Mimas	$5.6 \cdot 10^{-14}$	$5.0 \cdot 10^{-13}$	$1.6 \cdot 10^{-14}$	$5.6 \cdot 10^{-16}$	80	17
Enceladus	$3.7 \cdot 10^{-14}$	$2.8 \cdot 10^{-13}$	$1.0 \cdot 10^{-14}$	$4.2 \cdot 10^{-16}$	80	19
Tethys	$2.6 \cdot 10^{-14}$	$1.8 \cdot 10^{-13}$	$4.3 \cdot 10^{-15}$	$2.8 \cdot 10^{-16}$	25	6.5
Dione	$1.7 \cdot 10^{-14}$	$1.0 \cdot 10^{-13}$	$2.7 \cdot 10^{-15}$	$2.2 \cdot 10^{-16}$	34	10
Rhea	$1.1 \cdot 10^{-14}$	$6.2 \cdot 10^{-13}$	$1.5 \cdot 10^{-15}$	$1.6 \cdot 10^{-16}$	29	9
Titan	$3.4 \cdot 10^{-15}$	$1.4 \cdot 10^{-14}$	$1.3 \cdot 10^{-15}$	$9.0 \cdot 10^{-17}$	9	4
Hyperion	$7.0 \cdot 10^{-14}$	$6.2 \cdot 10^{-14}$	$1.8 \cdot 10^{-15}$	$2.3 \cdot 10^{-16}$	1,400	300
Iapetus	$1.1 \cdot 10^{-15}$	$4.2 \cdot 10^{-15}$	$7.9 \cdot 10^{-16}$	$1.1 \cdot 10^{-16}$	380	180
Phoebe	$3.4 \cdot 10^{-16}$	$1.4 \cdot 10^{-15}$	$1.3 \cdot 10^{-15}$	$1.2 \cdot 10^{-16}$	54,000	2,400

These are calibrated to a Saturn impact rate of 0.0012 cometary nuclei with d > 1.5 km per year. Quoted cratering rates should be regarded as uncertain to a factor of 4. The rates given here are typically 0.6 times the rates given in Table 19.4 of Zahnle et al. (2003), primarily because we have adopted a slightly smaller cometary impact rate with Saturn than in Zahnle et al. (2003). The rates given for Titan apply to a body with Titan's mass and size, but without an atmosphere.
$\dot{C}_A(>10)$: Case A cratering rate D > 10 km[km^{-2} year^{-1}].
$\dot{C}_B(>10)$: Case B cratering rate D > 10 km[km^{-2} year^{-1}].
$\dot{C}_S(>10)$: Smith et al. (1982) cratering rates, D > 10 km[km^{-2} year^{-1}].
$\dot{C}_{NIC}(>10)$: NIC (=Halley-type + Long-period comet) cratering rate D > 10 km[km^{-2} year^{-1}].
τ_A: Case A timescale for D > 20 km craters [Myr].
τ_B: Case B timescale for D > 20 km craters [Myr].

(e.g., Smith et al. 1981, 1982), when calculated as if the satellite were a planar target at fixed gravity. Simulations by Melosh and Ryan (1997) and Benz and Asphaug (1999) indicate that catastrophic disruption is more difficult than this criterion would predict, but that a better match is found if the expected transient crater *volume* is set equal to the satellite volume. Leinhardt and Stewart (2009), in contrast, find that disruption may be easier than found by Benz and Asphaug (1999) if the target is weak. In general, we might expect somewhat different disruption thresholds depending on whether the target is solid and coherent or weak and porous, even in the gravity regime (somewhat analogous to the difference between rock/ice and sand crater scaling discussed in Section 19.5).

Equating the transient crater diameter to satellite diameter well exceeds the threshold for icy satellite *shattering*, as determined by the scaling analysis in McKinnon et al. (1991) for the mid-sized uranian satellites. We will therefore continue to use the crater-size criterion above to compute crude (but illustrative) estimates of the characteristic impact disruption times for small satellites in the Saturn system, despite some uncertainty in the literature as to how representative such a calculation is for small bodies. Table 19.2 lists disruption time scales at Saturn at current impact rates. We expect that a 400 km crater on Mimas requires a 20 km diameter impactor releasing about 10^{24} J, which exceeds the gravitational binding energy of Mimas by a factor of five. (The same energy could be dispersed by raising the temperature of a cold Mimas by a modest 50 K, which implies that disruption is not the only option[12].) *At current impact rates* Mimas experiences such events on a time scale of 50 Gyr. In any case, satellites well outside the Roche limit, such as Mimas, should be able to reassemble themselves quickly (Dobrovolskis et al. 2007). Moreover, impacts that shatter and modestly rearrange the fragments, as opposed to catastrophically dispersing more than 50% of the satellite's mass into (very) temporary saturnian orbit are equivalent from a geological perspective – the cratering record is wiped clean.

The mass of Saturn's rings is estimated to equal or exceed that of Mimas (Robbins et al. 2009), so it is interesting to ask how long a Mimas-sized satellite would last at the distance of the rings. A 20-Gyr lifetime against disruption by a ~ 16-km-diameter comet seems like a reasonable requirement for an event that could not be commonplace (Lissauer et al. 1988; Dones 1993). But Table 19.2 makes it plain that the smaller number of small impactors in case A, when extrapolated to bodies tens of meters in size, leads to surprisingly long lifetimes for the very small moons such as Anthe (diameter $d \sim 2$ km). On the other hand, Pan ($d \sim 28$ km) and Daphnis ($d \sim 8$ km) live within gaps in the rings, and probably would not have been able to reassemble themselves if they had been catastrophically disrupted. It is therefore comforting that their disruption timescales are not extremely short.

19.7 Observed Crater Statistics and Interpretation

The number of craters of different sizes on planetary surfaces is usually presented in the literature in two different ways (Crater Analysis Techniques Working Group 1979).

---

[12] It is well known that impacts partition their initial kinetic energy into both internal energy and kinetic energy of the target (Melosh 1989).

Table 19.2 Present-day catastrophic disruption time scales at Saturn

Satellite	Semi-major axis (R_s)	Diameter of catastrophic disruptor (km)	Catastrophic disruption timescales (Gyr) Case A	Case B	Smith et al. (1982)
Ring parent	2.17	16	20	20	32
Pan	2.20	0.09	16	0.1	
Daphnis	2.26	0.04	3.3	1.1	
Prometheus	2.31	1.5	2.1	0.8	120
Pandora	2.35	1.2	2.8	0.9	130
Janus	2.51	4.1	4.4	3.8	200
Epimetheus	2.51	2.2	3.9	2.0	190
Mimas	3.12	20	40	40	360
Anthe	3.28	0.005	10	0.07	
Enceladus	3.94	37	180	180	360
Telesto, Calypso	4.88	0.25	16	1.8	
Helene	6.26	0.43	18	2.8	
Polydeuces	6.26	0.02	17	0.3	
Hyperion	24.6	12	230	230	2,000

The quantities given in the table include the semi-major axis of the moon in Saturn radii (R_s); the diameter of a cometary impactor, with an assumed density of $0.6\,\mathrm{g\,cm^{-3}}$, striking the satellite at the mean impact velocity for ecliptic comets (Eq. 19.7), that would make a transient crater on a flat surface with a diameter equal to the diameter of the satellite; and the expected present-day timescales for catastrophic disruption using the Case A and Case B size distributions (Zahnle et al. 2003), and based upon rates given by Smith et al. (1982). The "ring parent" is a moon the mass of Mimas placed at the current location of Saturn's A Ring.

The first is the cumulative plot, in which the number of craters larger than diameter D per unit area, N(D), is plotted as a function of D on a log-log scale. Frequently N(D) can be approximated by a power law, $N(D) \propto D^{\gamma}$, or by a series of power laws for different size ranges. The "canonical" value of the exponent γ for primary craters is near –2 (corresponding to an index of the differential size distribution, dN/dD, of –3.) The second approach is to give a relative plot ("R-plot"), again on a log-log scale, which displays the differential size distribution of the craters. However, to enhance structure, the R-plot divides dN/dD by a power law with an index of –3. Common crater size distributions thus plot as roughly horizontal lines in R-plots. The expression plotted is $R = D_{ab}^3 n/[A(D_b - D_a)]$, where n is the number of craters with diameters between D_b and D_a; A is the surface area of the region counted; and $D_{ab} = (D_a D_b)^{1/2}$ (see e.g., Barlow 2008). Frequently D_b is taken to be $\sqrt{2} D_a$; in that case, $R = 2^{3/4}/(\sqrt{2}-1) D_a^2 n/A$, or $R \sim 4.06 D_a^2 n/A$. In this review we will primarily use R-plots to present crater size-frequency distributions.

An R-plot giving an overview of crater statistics for saturnian satellites imaged by Cassini, the Moon, and Callisto is shown in Fig. 19.5 (Kirchoff and Schenk 2008, 2009b). The SFD curves for these heavily cratered terrains are generally *convex upwards* (except for an apparent concave upwards dip between ~75 and 250 km diameter, depending on satellite), reaching maximum values of $R \sim 0.1 - 0.3$, i.e., slightly below or above the "empirical saturation" line of Hartmann (1984). As Hartmann (1984) showed, many heavily cratered surfaces in the Solar System reach a maximum density near $R \approx 0.2$, about a factor of five below "geometric saturation" ($R = 1$) due to effects, like ejecta blanketing, that degrade and erase smaller craters well beyond crater rim crests (also see Melosh 1989, Chap. 10). Chapman and McKinnon 1986; their Fig. 19) showed, and more recently Richardson (2008) confirmed, that narrow SFDs may extend significantly above the average empirical saturation line, as is apparently true for Mimas and Dione in Fig. 19.5.

As was true for the Voyager results, Cassini images show that Rhea's SFD may depart somewhat from those of the other satellites, in being straighter with less convex curvature, more nearly characteristic of the putative Population I. In these new counts, based on far more extensive and better imagery than before, Rhea appears to have lost the transition to a somewhat steeper slope at greater than ~64-km diameter (or at least its statistical significance; Kirchoff and Schenk 2009b). Strom (1987) had taken this to be characteristic of the production population at Saturn, and Lissauer et al. (1988) argued that it indicated saturation at smaller crater sizes. A bump at ~10-km

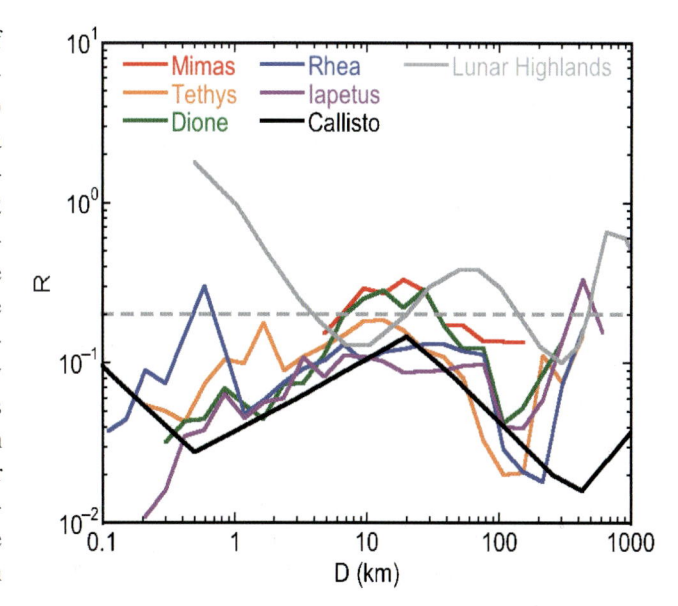

Fig. 19.5 Relative plot (R-plot) of the spatial densities of craters versus diameter for heavily cratered terrains on five satellites of Saturn (Kirchoff and Schenk 2009b), the lunar highlands (Ivanov et al. 2002), and Callisto (Schenk et al. 2004). Error bars have been omitted for clarity (see Fig. 19.6). The horizontal dashed line represents the approximate spatial density at which empirical saturation is reached according to Hartmann (1984). (Note the dip in crater densities [R values] between about 75- and 250-km diameter for these satellites and Mimas is not inconsistent, for the ≈145-km-diameter Herschel is the single crater within the "gap," and statistically speaking is consistent with a count of 1 ± 1.) The dip is remarkable because cometary impact speeds are generally much higher on the inner satellites, so there is essentially no overlap between the size of the impactors on, say, Tethys and Iapetus that would otherwise produce craters within the "dip" (see Figs. 19.3). This plot has been constructed by combining image sequences of small areas at high resolution and global mosaics at coarser resolution, the latter typically at $0.4 \text{ km pixel}^{-1}$. In all cases, the minimum crater size that is tabulated is about 10 pixels wide

diameter, which was interpreted as the signature of population II on Rhea's high northern latitudes (see McKinnon 1990; McKinnon et al. 1991, Fig. 17), does not appear either (though we note that Rhea's north pole was not in sunlight during the Cassini prime mission). Interestingly, Iapetus' SFD bears some degree of resemblance to that of Rhea (Kirchoff and Schenk 2009b). In general, all of the observed SFDs in Fig. 19.5 may be taken to represent, at least approximately, the true production functions for cratering in the Saturn system. As demonstrated by Chapman and McKinnon (1986), even if crater populations are "saturated," they may achieve a quasi-equilibrium that expresses the production function, provided that the production function is "shallow" (relatively depleted in small craters; i.e., R is constant or increases with increasing crater diameter) and is unchanging in time.

The SFDs for the saturnian satellites appear to be broadly similar to that of Callisto, and unlike that of the Moon (although see below). Assuming a cometary chronology applies to the data shown in Fig. 19.5, Kirchoff and Schenk (2008, 2009a, b) derive *crater-retention* ages for cumulative counts of craters greater than 5- and 10-km diameter. Unless the surface is undersaturated at the diameters in question, such "ages" do not refer to the true surface age, as many more craters could have formed earlier and been destroyed. Also, these ages assume the cometary flux decreases as $1/(t - t_0)$, where $t_0 = 4.56$ Gyr (Zahnle et al. 1998), and if this is not adequate to account for the observed crater density, an age of 4.56 Gyr (the age of the Solar System) is given. For the case A cometary mass distribution described in the last section, all of the heavily cratered surfaces in Fig. 19.5 are ancient, ≈ 4.4–4.6 Gyr; for, case B, with more small cometary impactors, there is a range of ages, from ≈ 0.8 to 4.6 Gyr, with a steady increase in crater-retention age from ≈ 0.8 Gyr (Mimas) to 4.6 Gyr (Iapetus). In Table 19.3 we list surface ages for terrains on Mimas, Tethys, Dione, and Rhea from Kirchoff and Schenk (2008). The younger ages for the moons closer to Saturn are a consequence of the much higher predicted impact flux due to gravitational focusing of comets in the inner saturnian system. Intriguingly, the crater density on the ejecta blanket of Mimas' large central-peak "Death Star" crater Herschel is low; Kirchoff and Schenk (2009b) derive a formation age of ≈ 0.9 Gyr for Herschel in case A and ≈ 0.1 Gyr in case B. All the ages use the cometary fluxes in Zahnle et al. (2003), not the revised rates in Table 19.1; the reader can make the corresponding correction for the latter and any future revisions to the estimated cometary flux (which in any event remains highly uncertain).

Using the Neukum lunar chronology, Neukum et al. (2006) derived model ages for Iapetus from 4.2 to 4.4 Gyr; for Dione, from 3.8_2 Gyr for the "youthful" Evander basin in the southern polar region to 4.2_5 Gyr on its densely cratered plains; for Tethys, from 3.8_7 Gyr for the interior of the Odysseus basin to 4.1_2 Gyr for cratered plains east of Ithaca Chasma; and on Enceladus, from less than 4 Myr for the South Polar Terrain ("tiger stripes") to 4.1_4 Gyr for its cratered plains. Because there is no absolute calibration in the Neukum chronology, whose cratering flux is presumed to be due to captured (i.e., planetocentric) asteroids, the time scale is set by assuming that Iapetus began to retain craters at 4.4 Gyr (this, in turn, is based on theoretical models of Iapetus' internal evolution [Castillo-Rogez et al. 2007; see chapter by Matson et al.]), and no gravitational focusing is assumed. The third significant figure in the ages illustrates the formal precision of this model, but not necessarily its accuracy. In general, ages derived using the cometary chronology (Section 19.6) are younger than Neukum's (*much* younger for "case B"). However, Zahnle et al. (2003) and Kirchoff and Schenk (2008, 2009b) allow craters to be formed and retained as far back as 4.56 Gyr, compared with Neukum's upper limit of 4.4 Gyr.

Figure 19.5 shows no clear relationship between a satellite's distance from Saturn and its crater density. This may indicate that all of the heavily cratered surfaces illustrated may have reached "saturation equilibrium". In particular, at the largest sizes, the crater density is highest on Iapetus, the most distant of the regular satellites (the number of multi-hundred-km impact features is remarkable). To look into this issue in more detail, in Table 19.4 we give the number of confirmed large craters (>100-km diameter) and basins (defined as craters in excess of 300-km in diameter)[13] on the five satellites shown in Fig. 19.5 (Kirchoff and Schenk 2008,

Table 19.3 Terrain ages in Gyr

Satellite	D ≥ 5 km A	D ≥ 5 km B	D ≥ 10 km A	D ≥ 10 km B
Mimas	4.39	0.75	4.35	1.33
Tethys	4.56	1.66	4.44	2.10
Odysseus basin	–	–	3.76	1.06
Dione – cp	4.56	2.60	4.56	3.22
Dione – sp	4.55	1.97	4.43	1.96
Evander basin	3.62	0.60	3.61	1.00
Rhea – cp	4.56	3.05	4.56	3.67
Rhea – sp	–	–	4.48	2.47

Surface ages inferred by Kirchoff and Schenk (2008), using the chronology of Zahnle et al. (2003) for Mimas, Tethys, the Odysseus basin on Tethys, cratered plains ('cp') on Dione, smooth plains ('sp') on Dione, the Evander basin on Dione, cratered plains on Rhea, and smooth plains on Rhea.

Table 19.4 Observed and predicted numbers of large craters and basins

Satellite	Satellite diameter (km)	Number of basins with D > 300 km Observed	Number of basins with D > 300 km Predicted	Number of large craters and basins with D > 100 km
Mimas	396	0	0.42	1
Tethys	1,066	2	0.69	8
Dione	1,123	2	0.45	10
Rhea	1,529	3	0.41	13
Iapetus	1,471	5	0.02	18

[13] Basins on the Moon are defined as those impact structures with two or more rings (see Schultz and Merrill 1981), such as the 320-km-diameter Schrödinger. On the midsized saturnian satellites the largest craters may or may not have rings, but in general their central floors appear more complex than a simple central peak, so we use the lunar terminology ("basins") for these magnificent structures.

2009b). We also list the number of 300-km or greater basins predicted to form in 4 Gyr at current impact rates from Zahnle et al. (2003).

If we add up the known >300-km-diameter basins on Mimas, Tethys, Dione, and Rhea, we have 7 in total, compared with 2 expected at current rates in 4 Gyr, so the results might be consistent with an early era that brought a few times more impactors in total than have fallen in the last several eons. However, Iapetus has more than 100 times the predicted number of basins in 4 Gyr at current rates. Possible explanations for the discrepancy include cratering of Iapetus by a population such as early irregular satellites that did not crater the inner moons (Nesvorný et al. 2007), or viscous relaxation on Rhea, which has a thinner lithosphere than Iapetus (Iapetus' unrelaxed shape certainly indicates a long-term, thick lithosphere; e.g., Giese et al. 2007), or simply, substantial oversaturation at these large scales on the inner satellites (or a combination of all three). Neukum et al. (2006) find a crater SFD for large craters and basins on Iapetus with a differential index of −2.2, corresponding to an impactor SFD with a power-law index of ∼−2. This impactor SFD is compatible with the known SFDs of both Kuiper Belt Objects and irregular satellites in the relevant size range of tens of km (Fuentes et al. 2009; Nicholson et al. 2008; see Fig. 19.1).

In Fig. 19.6, we show individual R-plots for Enceladus, Tethys, Dione, and Rhea from Kirchoff and Schenk (2009a, b). Enceladus shows huge variations in crater density (by a factor of at least 1000 when the South Polar Terrain [Porco et al. 2006], not shown here, is counted). Kirchoff and Schenk (2007, 2009b) note that Enceladus shows steeper dropoffs for craters with both $D < 2$ km and $D > 6$ km than do the other three satellites. This might be a signature of Population II impactors (such as are usually taken to characterize Mimas' crater population; see Fig. 19.5). Alternatively, it could be a result of endogenic modification, e.g., burial of the smallest craters by plume and E Ring material and viscous relaxation, followed by burial by plume fallout, of larger craters (Kirchoff and Schenk 2007, 2009b; Smith et al. 2007). Evidence for relaxed crater topography is provided by central peak craters that rise above the local mean elevation on Mimas, Enceladus, Tethys, and Dione, but not on Rhea and Iapetus (Schenk and Moore 2007; Dombard et al. 2007).

Tethys and Rhea are geologically inactive, at least in comparison with Enceladus and Dione. It might therefore be seen as somewhat disquieting that counts from different image sequences on Rhea in Fig. 19.6 do not always agree with each other. These discrepancies are most noticeable between the high-resolution ("hr") and global counts on Rhea. Such differences in R-values illustrate the effects of global averaging versus higher-resolution counts from very limited areas (as is the case here); the latter can be affected by unrecognized secondaries, lighting geometry, and local geological history.

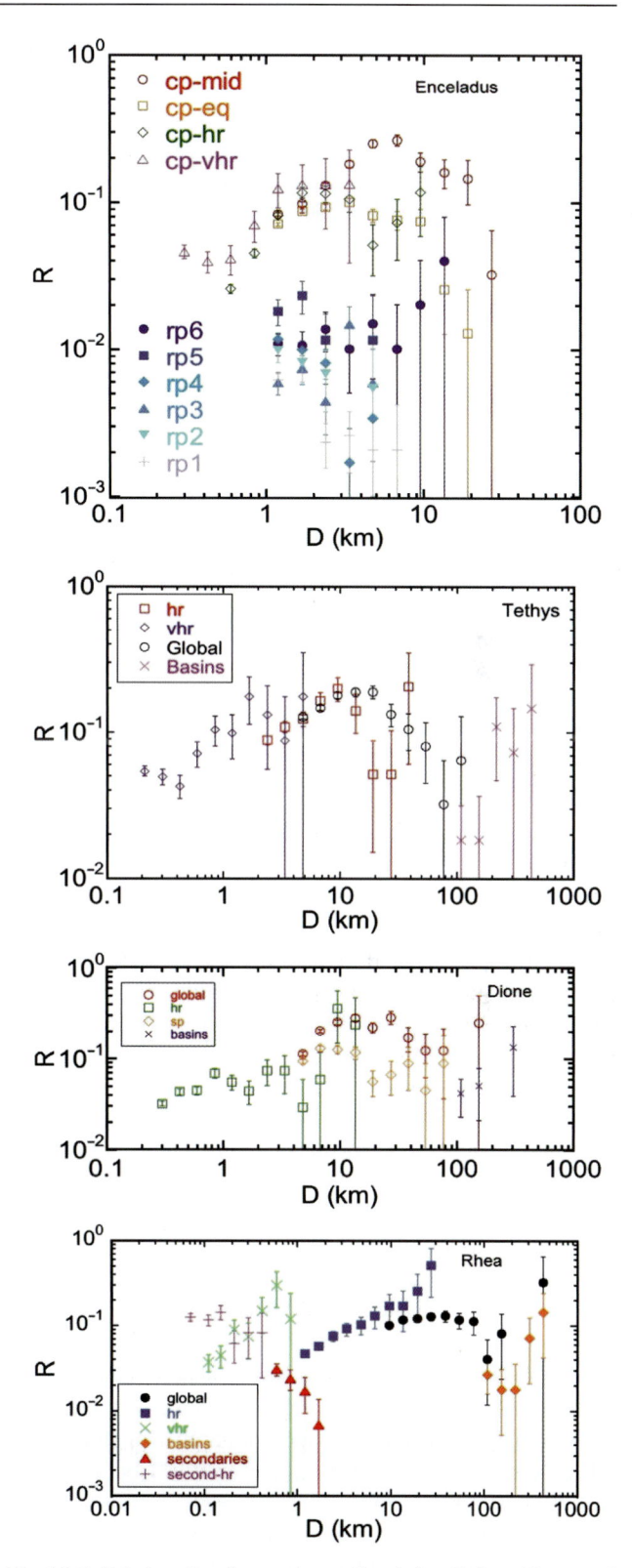

Fig. 19.6 Relative plots for terrains on Enceladus, Tethys, Dione, and Rhea (from Kirchoff and Schenk 2009b). The terrains include the heavily cratered plains ("cp") and lightly cratered ridged plains ("rp") on Enceladus (as defined by Kargel and Pozio 1996); and global counts on Tethys, Dione, and Rhea, supplemented with high-resolution ("hr") or very high-resolution ("vhr") coverage of limited areas

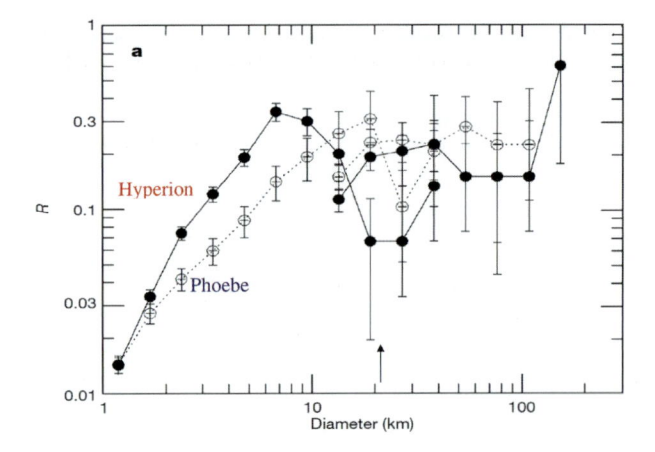

Fig. 19.7 Relative plots for Hyperion and Phoebe, from Thomas et al. (2007a). The rapid increase of R with increasing crater diameter is interpreted by Thomas et al. and by Richardson and Thomas (2007) as being due to the combined effects of a shallow impactor size distribution and erasure by larger craters. The red arrow near 20-km diameter represents the typical size of a crater made by a 1-km ecliptic comet on Hyperion (assuming solid-ice scaling, which is an approximation at best for this highly porous world)

For completeness we briefly discuss crater counts on the distant irregular satellite Phoebe, as well as those (such as they are) on Titan. Crater counts for Phoebe were presented by Porco (2005), and for Hyperion and Phoebe by Thomas et al. (2007a). In Fig. 19.7, we show the R-plots from Thomas et al., who find shallow size distributions (R rapidly increasing with crater diameter, D) for values of D between 1 and 10 km, and near empirical saturation (R ~ 0.2) for $D > 10$ km. Richardson and Thomas (2007) used a numerical stochastic cratering model to infer that Hyperion and Phoebe have been impacted by a population whose differential SFD has an index between -2.0 and -2.6. Extending the stochastic model of Chapman and McKinnon (1986) and Bottke (2006), Richardson and Thomas showed that an impactor population with a shallow size-frequency distribution produces a "quasi-equilibrium" paucity of small craters.

Craters on Titan are best seen by radar. Lorenz et al. (2007) list three confirmed craters: Menrva (~ 440 km diameter), Sinlap ($D = 80$ km) and Ksa (29 km). The USGS Astrogeology Gazetteer of Planetary Nomenclature (2009) adds two more: Afekan (115 km) and Selk (29 km). The lack of craters with $D < 20$ km was expected because of shielding by Titan's thick atmosphere (e.g., Lorenz 1997), but the rarity of 20-to-100-km features seems to imply erosion and/or burial of craters; simple observational incompleteness may also be involved. Menrva likely formed billions of years ago, but models suggest crater retention ages of 100 Myr to 1 Gyr for the mid-sized craters (Lorenz et al. 2007; Jaumann and Neukum 2009). Wood et al. (2008) find 70 possible degraded craters on the 22% of Titan's surface seen by radar as of early 2008.

19.8 Conclusions

We are in the early days of interpreting the cratering record of the saturnian satellites, but at least we now have the proper data in hand. In preparing this chapter, we have examined crater counts carried out independently by Neukum's group (Porco 2005, 2006; Thomas et al. 2007a) and by Kirchoff and Schenk (2009a, b), and the counts generally appear to be consistent in an overall cumulative sense. The devil is always in the details, though, and until all counts are published and available, it will not be possible to fully intercompare the structures of the various crater SFDs (i.e., their R-values). Moreover, convergence on the interpretation of the counts will probably remain elusive for some time to come.

Kirchoff and Schenk (2008, 2009a, b) consider the crater size-frequency distributions (SFDs) to fall into two groups, each different from the lunar highlands SFD, and attribute the satellite cratering to an outer Solar System heliocentric source. Rhea's and Iapetus' SFDs appear similar, but different from the others (Dione, Tethys, Enceladus' cratered terrain, and especially Mimas). The difference between Rhea and Mimas, in particular, was originally found by Voyager analyses and has been confirmed by all subsequent studies. These two different classes of crater SFD parallel the original Voyager-era definitions of populations I and II, but whether population II (now seen clearly only on satellites *interior* to Rhea's orbit) is truly planetocentric cannot be determined from its SFD alone.

Ecliptic comets from the Kuiper Belt are far more numerous than asteroids, at least at sizes greater than ~ 1 km, and the direct evidence that such comets may dominate the present-day impact rate is provided by (1) the collisions of Shoemaker-Levy 9's fragments with Jupiter and very close passages of several other comets by Jupiter, (2) observations of Saturn-crossing ecliptic comets and Centaurs (a total of 6 bodies as of this writing), and (3) the detection of the apex-antapex crater density asymmetry for younger (bright terrain) craters on Ganymede (Zahnle et al. 2001; Schenk et al. 2004). At Saturn, however, neither the strong apex-antapex asymmetries nor the steep increase in crater density as one approaches Saturn that are predicted for cometary impactors is observed. Both of these unmet expectations can be explained by crater saturation equilibrium, in that there are limits to how heavily cratered any surface, or hemisphere, can get. The lack of apex-antapex asymmetries could also result from satellite reorientation following large impacts (Lissauer 1985; Nimmo and Matsuyama 2007), and in the case of Enceladus, polar wander or other reorientation due to internal activity (Nimmo and Pappalardo 2006).

Neukum et al. (2006) take main-belt asteroids to be the primary impactors on the terrestrial planets and the satellites of Jupiter and Saturn, at least in early Solar System history, based upon common features in these bodies' crater SFDs.

For example, the maximum at 60-km diameter in the R-plot for the Moon is identified with maxima at about 12 km for Mimas, Dione, and Tethys (Fig. 19.5). The smaller crater sizes for the saturnian moons imply considerably smaller impact velocities, given by Horedt and Neukum (1984b) as ranging from $1.6\,\mathrm{km\,s^{-1}}$ for Iapetus to $5.9\,\mathrm{km\,s^{-1}}$ for Enceladus. These speeds are so small that they imply planetocentric impactors. However, no mechanism has been suggested for populating Saturn's vicinity with more bodies from the main asteroid belt than from the cometary reservoirs. Moreover, there is no known mechanism for capturing any such near-Saturn asteroids into the required low-eccentricity orbits around Saturn. We do believe, however, that planetocentric populations such as a large early population of irregular satellites and the debris of catastrophically disrupted satellites may have played important roles in the history of the Saturn system. Perhaps an earlier generation of satellites underwent an instability like that in the Nice model.

In the absence of independent radiometric dates, crater counts cannot support one proposed history of the Saturn system over another. However, continued work on crater counting, dynamical modeling, impact simulation, and crater modification and saturation processes will prepare us well for the day when we return the first sample from a moon of our Solar System's most iconic planet.

Acknowledgments We thank Nico Schmedemann, Amy Barr, Beau Bierhaus, Bill Bottke, Hal Levison, Alessandro Morbidelli, David Nesvorný, and David Vokrouhlický for discussions, and the editors for their patience. This research was supported by grants from the Cassini Data Analysis Program to LD, MRK, and WBM.

References

Alvarellos, J. L., Zahnle, K. J., Dobrovolskis, A. R., Hamill, P. 2005. Fates of satellite ejecta in the Saturn system. Icarus 178, 104–123.

Asphaug, E., Benz, W. 1996. Size, density, and structure of comet Shoemaker-Levy 9 inferred from the physics of tidal breakup. Icarus 121, 225–248.

Barlow, N. 2008. *Mars: An Introduction to its Interior, Surface and Atmosphere.* Cambridge Univ. Press, Cambridge, UK, 276 pp.

Barr, A. C., Canup, R. M. 2008. Constraints on gas giant satellite formation from the interior states of partially differentiated satellites. Icarus 198, 163–177.

Barucci, M. A., Boehnhardt, H., Cruikshank, D. P., Morbidelli, A., Eds. 2008. *The Solar System Beyond Neptune.* Univ. Arizona Press, Tucson, 632 pp.

Basaltic Volcanism Study Project 1981. *Basaltic Volcanism on the Terrestrial Planets.* Pergamon, New York. 1286 pp.

Benz, W., Asphaug, E. 1999. Catastrophic disruptions revisited. Icarus 142, 5–20.

Bernstein, G. M., Trilling, D. E., Allen, R. L., Brown, M. E., Holman, M., Malhotra, R. 2004. The size distribution of trans-Neptunian bodies. Astron. J. 128, 1364–1390. Erratum: Astron. J. 131, 2364.

Bézard, B., Lellouch, E., Strobel, D., Maillard, J.-P., Drossart, P. 2002. Carbon monoxide on Jupiter: Evidence for both internal and external sources. Icarus 159, 95–111.

Bierhaus, E. B., Chapman, C. R., Merline, W. J. 2005. Secondary craters on Europa and implications for cratered surfaces. Nature 437, 1125–1127.

Bierhaus, E. B., Chapman, C. R., Merline, W. J., Brooks, S. M, Asphaug E. 2001. Pwyll secondaries and other small craters on Europa. Icarus 153, 264–76.

Boehnhardt, H. 2004. Split comets. In *Comets II* (M. C. Festou, H. U. Keller, and H. A. Weaver, Eds.), pp. 301–316. Univ. Arizona Press, Tucson.

Bottke, W. F., Chapman, C. R. 2006. Determining the main belt size distribution using asteroid cratering records and crater saturation models. Lunar and Planetary Institute Science Conference Abstracts 37, #1349.

Bottke, W. F., Durda, D. D., Nesvorný, D., Jedicke, R., Morbidelli, A., Vokrouhlický, D., Levison, H. F. 2005. The fossilized size distribution of the main asteroid belt. Icarus 175, 111–140.

Bottke, W. F., Levison, H. F., Nesvorný, D., Dones, L. 2007. Can planetesimals left over from terrestrial planet formation produce the lunar Late Heavy Bombardment? Icarus 190, 203–223.

Bottke, W. F., Morbidelli, A., Jedicke, R., Petit, J.-M., Levison, H. F., Michel, P., Metcalfe, T. S. 2002. Debiased orbital and absolute magnitude distribution of the Near-Earth Objects.

Bottke, W. F., Nesvorný, D., Vokrouhlický, D., Morbidelli, A. 2009. The irregular satellites: The most collisionally evolved populations in the solar system. Astron. J., submitted.

Brown, M. E., Barkume, K. M., Ragozzine, D., Schaller, E. L. 2007. A collisional family of icy objects in the Kuiper Belt. Nature 446, 294–296.

Brunini, A. di Sisto, R. P., and Orellana, R. B. 2003. Cratering rate on the jovian system: The contribution from Hilda asteroids. Icarus 165, 371–378.

Buckingham, E. 1914. On physically similar systems; illustrations of the use of dimensional equations. Phys. Rev. 4, 345–376.

Castillo-Rogez, J. C., Matson, D. L., Sotin, C., Johnson, T. V., Lunine, J. I., Thomas, P. C. 2007. Iapetus' geophysics: Rotation rate, shape, and equatorial ridge. Icarus 190, 179–202.

Chapman, C. R. 1990. Crater saturation simulation. Bull. Amer. Astron. Soc. 22, 1057.

Chapman, C. R., Cohen, B. A., Grinspoon, D. H. 2007. What are the real constraints on the existence and magnitude of the late heavy bombardment? Icarus 189, 233–245.

Chapman, C. R., McKinnon, W. B. 1986. Cratering of planetary satellites. In *Satellites* (J. A. Burns and M. S. Matthews, Eds.), pp. 492–580. Univ. Arizona Press, Tucson.

Charnoz, S., Morbidelli, A., Dones, L., Salmon, J. 2009. Did Saturn's rings form during the Late Heavy Bombardment? Icarus 199, 413–428.

Collins, G. S., Davison, T., Elbeshausen, D., Wünnemann, K. 2009. Numerical simulations of oblique impacts: The effect of impact angle and target strength on crater shape. LPSC 40th, Abstract #1620.

Crater Analysis Techniques Working Group 1979. Standard techniques for presentation and analysis of crater size-frequency data. Icarus 37, 467–474.

Croft, S. K., Kargel, J. S., Kirk, R. L., Moore, J. M., Schenk, P. M., Strom, R. G. 1995. The geology of Triton. In *Neptune and Triton* (D. P. Cruikshank, Ed.), pp. 879–947. Univ. Arizona Press, Tucson.

Del Popolo, A., Gambera, M., and Ercan, N. 2001. Migration of giant planets in planetesimal discs. Mon. Not. R. Astron. Soc. 325, 1402–1410.

Di Sisto, R. P., Brunini, A. 2007. The origin and distribution of the Centaur population. Icarus 190, 224–235.

Dobrovolskis, A. R., Alvarellos, J. L., Lissauer, J. J. 2007. Lifetimes of small bodies in planetocentric (or heliocentric) orbits. Icarus 188, 481–505.

Dobrovolskis, A. R., Lissauer, J. J. 2004. The fate of ejecta from Hyperion. Icarus 169, 462–473.

Dombard, A. J., Bray, V. J., Collins, G. S., Schenk, P. M., Turtle, E. P. 2007. Relaxation and the formation of prominent central peaks in large craters on the icy satellites of Saturn. Bull. Amer. Astron. Soc. 38, 429.

Dombard, A.J., McKinnon, W. B. 2000. Long-term retention of impact crater topography on Ganymede. Geophys. Res. Lett. 27, 3663–3666.

Dones, L. 1991. A recent cometary origin for Saturn's rings? Icarus 92, 194–203.

Dones, L., Weissman, P. R., Levison, H. F., Duncan, M. J. 2004. Oort cloud formation and dynamics. In *Comets II* (M. C. Festou, H. U. Keller, and H. A. Weaver, Eds.), pp. 153–174. Univ. Arizona Press, Tucson.

Duncan, M. J., Levison, H. F. 1997. A scattered comet disk and the origin of Jupiter family comets. Science 276, 1670–1672.

Duncan, M., Levison, H., Dones, L. 2004. Dynamical evolution of ecliptic comets. In *Comets II* (M. C. Festou, H. U. Keller, and H. A. Weaver, Eds.), pp. 193–204. Univ. Arizona Press, Tucson.

Duncan, M., Quinn, T., Tremaine, S. 1988. The origin of short-period comets. Astrophys. J. Lett. 328, L69–L73.

Durda, D. D., Stern, S. A. 2000. Collision rates in the present-day Kuiper Belt and Centaur regions: Applications to surface activation and modification on comets, Kuiper Belt Objects, Centaurs, and Pluto-Charon. Icarus 145, 220–229.

Esposito, L. W. 1986. Structure and evolution of Saturn's rings. Icarus 67, 345–357.

Farinella, P., Paolicchi, P., Strom, R. G., Kargel, J. S., Zappalà, V. 1990. The fate of Hyperion's fragments. Icarus 83, 186–204.

Fernández, J. A., Ip, W.-H. 1984. Some dynamical aspects of the accretion of Uranus and Neptune – The exchange of orbital angular momentum with planetesimals. Icarus 58, 109–120.

Fernández, J. A., Morbidelli, A. 2006. The population of faint Jupiter family comets near the Earth. Icarus 185, 211–222.

Fernández, Y. 2009. Web page at http://www.physics.ucf.edu/~yfernandez/cometlist.html#jf. Accessed April 25, 2009.

Fernández, Y. R., Jewitt, D. C., Sheppard, S. S. 2002. Thermal properties of Centaurs Asbolus and Chiron. Astron. J. 123, 1050–1055.

Francis, P. J. 2005. The demographics of long-period comets. Astrophys. J. 635, 1348–1361.

Fuentes, C. I., George, M. R., Holman, M. J. 2009. A Subaru pencil-beam search for $m_R \sim 27$ trans-Neptunian bodies. Astrophys. J. 696, 91–95.

Giese, B., Denk, T., Neukum, G., Roatsch, T., Helfenstein, P., Thomas, P. C., Turtle, E. P., McEwen, A., Porco, C. C. 2008. The topography of Iapetus' leading side. Icarus 193, 359–371.

Giese, B., Neukum, G., Roatsch, T., Denk, T., and Porco, C. C. 2006. Topographic modeling of Phoebe using Cassini images. Planet Space Sci. 54, 1156–1166.

Giese, B., Wagner, R., Roatsch, T., Denk, T., Neukum, G. 2007. The topographies of Rhea and Iapetus in comparison. American Geophysical Union, Fall Meeting 2007, abstract #P12B-07AGU.

Gladman, B. J., Davis, D. R., Neese, C., Jedicke, R., Williams, G., Kavelaars, J. J., Petit, J.-M., Scholl, H., Holman, M., Warrington, B., Esquerdo, G., Tricarico, P. 2009. On the asteroid belt's orbital and size distribution. Icarus 202, 104–118.

Gladman, B., Kavelaars, J. J., Petit, J.-M., Morbidelli, A., Holman, M. J., Loredo, T. 2001. The structure of the Kuiper Belt: Size distribution and radial extent. Astron. J. 122, 1051–1066.

Gladman, B., Marsden, B. G., VanLaerhoven, C. 2008. Nomenclature in the outer Solar System. In *The Solar System Beyond Neptune* (M. A. Barucci, H. Boehnhardt, D. P. Cruikshank, and A. Morbidelli, Eds.), pp. 43–57. Univ. Arizona Press, Tucson.

Gomes, R., Levison, H. F., Tsiganis, K., Morbidelli, A. 2005. Origin of the cataclysmic Late Heavy Bombardment period of the terrestrial planets. Nature 435, 466–469.

Gomes, R. S., Morbidelli, A., Levison, H. F. 2004. Planetary migration in a planetesimal disk: Why did Neptune stop at 30 AU? Icarus 170, 492–507.

Groussin, O., Lamy, P., Jorda, L. 2004. Properties of the nuclei of Centaurs Chiron and Chariklo. Astron. Astrophys. 413, 1163–1175.

Hahn, G., Lagerkvist, C.-I., Karlsson, O., Oja, T., Stoss, R. M. 2006. P/2004 A1 (Loneos) – A comet under transition from Saturn to Jupiter. Astronomische Nachrichten 327, 17–20.

Hahn, J.M., Malhotra, R. 1999. Orbital evolution of planets embedded in a planetesimal disk. Astron. J. 117, 3041–3053.

Harrington, J., de Pater, I., Brecht, S. H., Deming, D., Meadows, V., Zahnle, K., Nicholson, P. D. 2004. Lessons from Shoemaker-Levy 9 about Jupiter and planetary impacts. In *Jupiter – The Planet, Satellites and Magnetosphere* (Fran Bagenal, Timothy E. Dowling, William B. McKinnon, Eds.), pp. 159–184. Cambridge Univ. Press, Cambridge, UK.

Hartmann, W. K. 1965. Terrestrial and lunar flux of large meteorites in the last two billion years. Icarus 4, 157–165.

Hartmann, W. K. 1966. Early lunar cratering. Icarus 5, 406–418.

Hartmann, W. K. 1984. Does crater "saturation equilibrium" occur in the Solar System? Icarus 60, 56–74.

Hartmann, W. K., Neukum, G. 2001. Cratering chronology and the evolution of Mars. Space Sci. Rev. 96, 165–194.

Hartmann, W. K., Ryder, G., Dones, L., Grinspoon, D. 2000. The time-dependent intense bombardment of the primordial Earth/Moon system. In *Origin of the Earth and Moon* (R. M. Canup and K. Righter, Eds.), pp. 493–512. Univ. Arizona Press, Tucson.

Holman, M. J., Wisdom, J. 1993. Dynamical stability in the outer Solar System and the delivery of short period comets. Astron. J. 105, 1987–1999.

Holsapple, K. A. 1993. The scaling of impact processes in planetary sciences. Ann. Rev. Earth Planet. Sci. 21, 333–373.

Holsapple, K. A. 2009. Web pages at http://keith.aa.washington.edu/craterdata/scaling/index.htm and http://keith.aa.washington.edu/craterdata/scaling/theory.pdf. Accessed April 25, 2009.

Holsapple, K. A., Schmidt, R. M. 1982. On the scaling of crater dimensions. II – Impact processes. J. Geophys. Res. 87, 1849–1870.

Horedt, G. P., Neukum, G. 1984a. Cratering rate over the surface of a synchronous satellite. Icarus 60, 710–717.

Horedt, G. P., Neukum, G. 1984b. Planenetocentric versus heliocentric impacts in the jovian and saturnian satellite system. J. Geophys. Res. 89, 10,405–10,410.

Horner, J., N.W. Evans, and M.E. Bailey 2004. Simulations of the population of Centaurs II: Individual objects. Mon. Not. R. Astron. Soc. 355, 321–329.

Housen, K. R., Holsapple, K. A. 2003. Impact cratering on porous asteroids. Icarus 163, 102–119.

Irwin, M., Tremaine, S., Żytkow, A. N. 1995. A search for slow-moving objects and the luminosity function of the Kuiper Belt. Astron. J. 110, 3082–3092.

Ivanov, B. A., Neukum, G., Bottke, W. F., Jr., Hartmann, W. K. 2002. The comparison of size-frequency distributions of impact craters and asteroids and the planetary cratering Rate. In *Asteroids III* (W. F. Bottke Jr., A. Cellino, P. Paolicchi, and R. P. Binzel, Eds.), pp. 89–101. Univ. Arizona Press, Tucson.

Jaumann, R., Neukum, G. 2009. The surface age of Titan. Lunar and Planetary Institute Science Conference Abstracts 40, abstract #1641.

Jewitt, D. 2008. Kuiper Belt and comets: An observational perspective. *Saas-Fee Advanced Course 35: Trans-Neptunian Objects and Comets* (K. Altwegg, W. Benz, and N. Thomas, Eds.), pp. 1–78. Springer, New York.

Jewitt, D. C., Trujillo, C. A., Luu, J. X. 2000. Population and size distribution of small jovian Trojan asteroids. Astron. J. 120, 1140–1147.

Johnson, T. V. 1978. The galilean satellites of Jupiter – Four worlds. Ann. Rev. Earth Planet. Sci. 6, 93–125.

JPL Solar System Dynamics 2009. Web page at http://ssd.jpl.nasa.gov/?satellites. Accessed April 25, 2009.

JPL Solar System Dynamics Small-Body Data Browser 2009. Web page at http://ssd.jpl.nasa.gov/sbdb.cgi. Accessed April 25, 2009.

Jutzi, M., Benz, W., Michel, P. 2008. Numerical simulations of impacts involving porous bodies. I. Implementing sub-resolution porosity in a 3D SPH hydrocode. Icarus 198, 242–255.

Jutzi, M., Michel, P., Hiraoka, K., Nakamura, A. M., Benz, W. 2009. Numerical simulations of impacts involving porous bodies: II. Comparison with laboratory experiments. Icarus 201, 802–813.

Kargel, J. S., Pozio, S. 1996. The volcanic and tectonic history of Enceladus. Icarus 119, 385–404.

Kary, D. M., Dones, L. 1996. Capture statistics of short-period comets: Implications for comet D/Shoemaker-Levy 9. Icarus 121, 207–224.

Kenyon, S. J., Bromley, B. C., O'Brien, D. P., Davis, D. R. 2008. Formation and collisional evolution of Kuiper Belt objects. In *The Solar System Beyond Neptune* (M. A. Barucci, H. Boehnhardt, D. P. Cruikshank, and A. Morbidelli, Eds.), pp. 293–313. Univ. Arizona Press, Tucson.

Kirchoff, M. R., Schenk, P. M. 2007. Impact crater distributions on the saturnian satellites from Cassini ISS imaging – Implications for geologic Histories and ages. Amer. Geophys. Union Fall Meeting Abstracts 545.

Kirchoff, M. R., Schenk, P. M. 2008. Bombardment history of the saturnian satellites. Workshop on Early Solar System Impact Bombardment, LPI Contrib. 1439, abstract #3023.

Kirchoff, M. R., Schenk, P. M. 2009a. Crater modification and geologic activity in Enceladus' heavily cratered plains: Evidence from the impact crater distribution. Icarus 202, 656–668.

Kirchoff, M. R., Schenk, P. M. 2009b. Impact cratering records of the mid-sized, icy saturnian satellites. Submitted to Icarus.

Lagerkvist, C.-I., Hahn, G., Karlsson, O., Carsenty, U. 2000. The orbital history of two periodic comets encountering Saturn. Astron. Astrophys. 362, 406–409.

Lamy, P. L., Toth, I., Fernández, Y. R., Weaver, H. A. 2004. The sizes, shapes, albedos, and colors of cometary nuclei. *Comets II* (M. C. Festou, H. U. Keller, and H. A. Weaver, Eds.), pp. 223–264. Univ. Arizona Press, Tucson.

Laskar, J. 1996. Large scale chaos and marginal stability in the Solar System. Cel. Mech. Dyn. Astron. 64, 115–162.

Leinhardt, Z. M., Richardson, D. C., Lufkin, G., Haseltine, J. 2009. Planetesimals to protoplanets II: Effect of debris on terrestrial planet formation. Mon. Not. R. Astron. Soc. 396, 718–728.

Leinhardt, Z. M., Stewart, S. T. 2009. Full numerical simulations of catastrophic small body collisions. Icarus 199, 542–559.

Lellouch, E., Bézard, B., Moreno, R., Bockelée-Morvan, D., Colum, P., Crovisier, J., Festou, M., Gautier, D., Marten, A., Paubert, G. 1997. Carbon monoxide in Jupiter after the impact of comet Shoemaker-Levy 9. Planet. Space Sci. 45, 1203–1212.

Lellouch, E., Moreno, R., Paubert, G. 2005. A dual origin for Neptune's carbon monoxide? Astron. Astrophys. 430, L37–L40.

Levison, H. F. 1996. Comet taxonomy. In *Completing the Inventory of the Solar System* (T.W. Rettig and J.M. Hahn, Eds.), pp. 173–191. Astron. Soc. Pac. Conf. Proc., vol. 107, Astronomical Society of the Pacific, San Francisco.

Levison, H. F., Dones, L., Chapman, C. R., Stern, S. A., Duncan, M. J., Zahnle, K. 2001. Could the lunar "Late Heavy Bombardment" have been triggered by the formation of Uranus and Neptune? Icarus 151, 286–306.

Levison, H. F., Duncan, M. J. 1997. From the Kuiper Belt to Jupiter-family comets: The spatial distribution of ecliptic comets. Icarus 127, 13–32.

Levison, H. F., Duncan, M. J., Zahnle, K., Holman, M., Dones, L. 2000. Note: Planetary impact rates from ecliptic comets. Icarus 143, 415–420.

Levison, H. F., Lissauer, J. J., Duncan, M. J. 1998. Modeling the diversity of outer planetary systems. Astron. J. 116, 1998–2014.

Levison, H. F., Morbidelli, A., VanLaerhoven, C., Gomes, R., Tsiganis, K. 2008. Origin of the structure of the Kuiper Belt during a dynamical instability in the orbits of Uranus and Neptune. Icarus 196, 258–273.

Levison, H. F., Thommes, E., Duncan, M. J., Dones, L. 2004. A fairy tale about the formation of Uranus and Neptune and the lunar Late Heavy Bombardment. In *Debris Disks and the Formation of Planets* (L. Caroff, L. J. Moon, D. Backman, and E. Praton, Eds.), ASP Conference Series 324, pp. 152–167. San Francisco, Astronomical Society of the Pacific.

Lissauer, J. J. 1985. Can cometary bombardment disrupt synchronous rotation of planetary satellites? J. Geophys. Res. 90, 11289–11293.

Lissauer, J. J., Squyres, S. W., Hartmann, W. K. 1988. Bombardment history of the Saturn system. J. Geophys. Res. 93, 13776–13804.

Lorenz, R.D. 1997. Impacts and cratering on Titan: A pre-Cassini view. Planet. Space Sci. 45, 1009–1019.

Lorenz, R. D., and 11 colleagues 2007. Titan's young surface: Initial impact crater survey by Cassini RADAR and model comparison. Geophys. Res. Lett. 34, L07204.

Lykawka, P. S., Horner, J. A., Jones, B. W., Mukai, T. 2009. Origin and dynamical evolution of Neptune Trojans – I: Formation and planetary migration. Submitted to Mon. Not. R. Astron. Soc.

Malhotra, R., 1993. The origin of Pluto's peculiar orbit. Nature 365, 819–821.

Marchi, S., Mottola, S., Cremonese, G., Massironi, M., Martellato, E. 2009. A new chronology for the Moon and Mercury. Astron. J. 137, 4936–4948.

Mazeeva, O. A. 2007. Long-period comet flux in the planetary region: Dynamical evolution from the Oort cloud. Solar System Research 41, 118–128.

McEwen, A. S., Bierhaus, E. B. 2006. The importance of secondary cratering to age constraints on planetary surfaces. Ann. Rev. Earth Planet. Sci. 34, 535–567.

McKinnon, W. B. 1990. Planetary evolution: The Hyperion hypothesis. Nature 346, 414–415.

McKinnon, W. B. 2007. The mechanics of complex crater and ringed basin formation: Constraints from 30 years of planetary observations. Bridging the Gap II: Effect of Target Properties on the Impact Cratering Process, LPI Contrib. 1360, abstract #8072.

McKinnon, W. B., Chapman, C. R., Housen, K. R. 1991. Cratering of the Uranian satellites. In *Uranus* (J.T. Bergstralh, E. D. Miner, and M. S. Matthews, Eds.), pp. 629–692. Univ. Arizona Press, Tucson.

McKinnon, W. B., Schenk, P. M., Stern, S. A. 2000. New constraints on the small Kuiper Belt object population from high-resolution images of Triton. *The Transneptunian Population*, IAU 24, Joint Discussion 4, Manchester, England.

Melosh, H. J. 1989. *Impact Cratering: A Geologic Process*. Oxford University Press (Oxford Monographs on Geology and Geophysics, No. 11), 1989, 253 pp.

Melosh, H. J., Beyer, R. A. 2009. Web page at http://www.lpl.arizona.edu/tekton/crater.html. Accessed April 25, 2009.

Melosh, H. J., Ryan, E. V. 1997. Asteroids: Shattered but not dispersed. Icarus 129, 562–564.

Morbidelli, A., Levison, H. F., Gomes, R. 2008. The dynamical structure of the Kuiper belt and its primordial origin. In *The Solar System Beyond Neptune* (M. A. Barucci, H. Boehnhardt, D. P. Cruikshank, and A. Morbidelli, Eds.), pp. 275–292. Univ. Arizona Press, Tucson.

Morbidelli, A., Levison, H. F., Tsiganis, K., Gomes, R. 2005. Chaotic capture of Jupiter's Trojan asteroids in the early Solar System. Nature 435, 462–465.

Morbidelli, A., Levison, H. F., Bottke, W. F., Dones, L., Nesvorný, D. 2009. Considerations on the magnitude distributions of the Kuiper Belt and of the Jupiter Trojans. Icarus 202, 310–315.

Morbidelli, A., Tsiganis, K., Crida, A., Levison, H. F., Gomes, R. 2007. Dynamics of the giant planets of the Solar System in the gaseous protoplanetary disk and their relationship to the current orbital architecture. Astron. J. 134, 1790–1798.

Murray, N., Hansen, B., Holman, M., Tremaine, S. 1998. Migrating planets. Science 279, 69–72.

Nesvorný, D., Alvarellos, J. L. A., Dones, L., Levison, H. F. 2003. Orbital and collisional evolution of the irregular satellites. Astron. J. 126, 398–429.

Nesvorný, D., Bottke, W.F., Levison, H., Dones, L., 2002. A recent asteroid breakup in the main belt. Nature 417, 720–722.

Nesvorný, D., Vokrouhlický, D., Morbidelli, A. 2007. Capture of irregular satellites during planetary encounters. Astron. J. 133, 1962–1976.

Neukum G. 1983. *Meteoritenbombardement und Datierung planetarer Oberflächen*. Habilitation Dissertation, Ludwig-Maximilians Univ. München, Germany, 186 pp.

Neukum, G., Hiller, K. 1981. Martian ages. J. Geophys. Res. 86, 3097–3121.

Neukum, G., Ivanov, B. A., Hartmann, W. K. 2001. Cratering records in the inner Solar System in relation to the lunar reference system. Space Sci. Rev. 96, 55–86.

Neukum, G., Wagner, R. J., Denk, T., Porco, C. C., the Cassini ISS Team 2005. The cratering record of the saturnian satellites Phoebe, Tethys, Dione and Iapetus in comparison: First results from analysis of the Cassini ISS imaging data. Lunar and Planetary Institute Conference Abstracts 36, abstract #2034.

Neukum, G., Wagner, R., Wolf, U., Denk, T. 2006. The cratering record and cratering chronologies of the Saturnian satellites and the origin of impactors: results from Cassini ISS data. Euro. Planet. Sci. Conf., 610.

Neukum, G., Wagner, R., Wolf, U., Ivanov, B. A., Head, J. W. III, Pappalardo, R. T., Klemaszewski, J. E., Greeley, R., Belton, M. J. S., Galileo SSI Team 1998. Cratering chronology in the jovian system and derivation of absolute ages. Lunar and Planetary Institute Conference Abstracts 29, abstract #1742.

Neukum, G., Wagner, R., Wolf, U., the Galileo SSI Team 1999. Cratering record of Europa and implications for time-scale and crustal development. Lunar and Planetary Institute Conference Abstracts 30, abstract #1992.

Nicholson, P. D., Ćuk, M., Sheppard, S. S., Nesvorný, D., Johnson, T. V. 2008. Irregular satellites of the giant planets. In *The Solar System Beyond Neptune* (M. A. Barucci, H. Boehnhardt, D. P. Cruikshank, and A. Morbidelli, Eds.), pp. 411–424. Univ. Arizona Press, Tucson.

Nimmo, F., Matsuyama, I. 2007. Reorientation of icy satellites by impact basins. Geophysical Research Letters 34, L19203.

Nimmo, F., Pappalardo, R.T., 2006. Diapir-induced reorientation of Saturn's moon Enceladus. Nature 441, 614–616.

O'Brien, D. P., Morbidelli, A., Levison, H. F. 2006. Terrestrial planet formation with strong dynamical friction. Icarus 184, 39–58.

Ostro, S. J., Pettengill, G. H. 1978. Icy craters on the Galilean satellites. Icarus 34, 268–279.

Parker, A., Ivezić, Ž., Jurić, M., Lupton, R., Sekora, M. D., Kowalski, A. 2008. The size distributions of asteroid families in the SDSS Moving Object Catalog 4. Icarus 198, 138–155.

Petit, J-M., Kavelaars, J. J., Gladman, B., Loredo, T. 2008. Structure and evolution of Kuiper Belt Objects and dwarf planets. In *The Solar System Beyond Neptune* (M.A. Barucci, H. Boehnhardt, D. Cruikshank, and A. Morbidelli, Eds.), pp. 71–87. Univ. Arizona Press, Tucson.

Plescia, J. B., Boyce, J. M. 1985. Impact cratering history of the Saturnian satellites. J. Geophys. Res. 90, 2029–2037.

Porco, C. C., and 34 colleagues 2005. Cassini Imaging Science: Initial results on Phoebe and Iapetus. Science 307, 1237–1242.

Porco, C. C., and 24 colleagues 2006. Cassini observes the active South Pole of Enceladus. Science 311, 1393–1401.

Ragozzine, D., Brown, M. E. 2007. Candidate members and age estimate of the family of Kuiper Belt object 2003 EL61. Astron. J. 134, 2160–2167.

Richardson, J. E. 2008. Interpreting cratered terrains: A new model investigation of crater saturation conditions. AAS/Division for Planetary Sciences Meeting Abstracts 40, #09.05.

Richardson, J. E., Thomas, P. C. 2007. Modeling the cratering records of Hyperion and Phoebe: Indications of a shallow-sloped impactor population. Bull. Amer. Astron. Soc. 38, 430.

Robbins, S.J., Stewart, G.R., Lewis, M.C., Colwell, J.E., Sremčević, M. 2009. Estimating the masses of Saturn's A and B Rings from high-optical depth N-Body simulations and stellar occultations. Icarus, in press.

Roig, F., Ribeiro, A. O., Gil-Hutton, R. 2008. Taxonomy of asteroid families among the Jupiter Trojans: Comparison between spectroscopic data and the Sloan Digital Sky Survey colors. Astron. Astrophys. 483, 911–931.

Schenk, P. M., 1989. Crater formation and modification on the icy satellites of Uranus and Saturn: Depth/diameter and central peak occurrence. J. Geophys. Res. 94, 3813–3832.

Schenk, P. M., Chapman, C. R., Zahnle, K., Moore, J. M. 2004. Ages and interiors: The cratering record of the Galilean satellites. In *Jupiter. The Planet, Satellites and Magnetosphere* (Fran Bagenal, Timothy E. Dowling, William B. McKinnon, Eds.), pp. 427–456. Cambridge Univ. Press, Cambridge, UK.

Schenk, P. M., McKinnon, W. B. 2008. A gallery of multiring basins on Europa, Ganymede, and Callisto. *Large Meteorite Impacts and Planetary Evolution IV*, LPI Contrib. 1423, abstract #3107.

Schenk, P. M., Moore, J. M. 2007. Impact crater topography and morphology on saturnian mid-sized satellites. Lunar and Planetary Institute Conference Abstracts 38, abstract #2305.

Schenk, P. M., Sobieszczyk, S. 1999. Cratering asymmetries on Ganymede and Triton: From the sublime to the ridiculous. Bull. Amer. Astron. Soc. 31, 1182.

Schenk, P. M., Zahnle, K. J. 2007. On the negligible surface age of Triton. Icarus 192, 135–149.

Schmedemann, N., Neukum, G., Denk, T., Wagner, R. 2009. Impact crater size-frequency distribution (SFD) on Saturnian satellites and comparison with other Solar-System bodies. Lunar and Planetary Institute Science Conference Abstracts 40, abstract #1941.

Schmedemann, N., Neukum, G., Denk, T., Wagner, R., Hartmann, O., Michael, G. 2008. Comparison of the production size-frequency distribution (SFD) of craters on saturnian satellites with the lunar crater SFD and asteroid diameter SFD. AAS/Division for Planetary Sciences Meeting Abstracts 40, abstract #61.09.

Schmidt, R.M., Housen, K.R., 1987. Some recent advances in the scaling of impact and explosive cratering. Int. J. Impact Eng. 5, 543–560.

Schultz, P.H., Merrill, R. B., Eds., 1981. *Multi-ring Basins, Proc. Lunar Planet. Sci. 12A*. Pergamon Press, New York.

Sheppard, S. S., Trujillo, C. A. 2006. A thick cloud of Neptune Trojans and their colors. Science 313, 511–514.

Shoemaker, E. M., Hackman, R. J., Eggleton, R. E. 1963. Interplanetary correlation of geologic time. Adv. Astronaut. Sci. 88, 70–89.

Shoemaker, E. M., Shoemaker, C. S., Wolfe, R. F. 1989. Trojan asteroids – Populations, dynamical structure and origin of the L4 and L5 swarms. In *Asteroids II* (R. P. Binzel, T. Gehrels, and M. S. Matthews, Eds.), pp. 487–523. Univ. Arizona Press, Tucson.

Shoemaker, E. M., Wolfe, R. F. 1981. Evolution of the saturnian satellites: The role of impact. Lunar Planet Sci. XII, A1–A3. LPI Contribution 428, Houston, Texas.

Shoemaker, E. M., Wolfe, R. F. 1982. Cratering time scales for the galilean satellites. In *Satellites of Jupiter* (Ed. D. Morrison, Ed.), pp. 277–339. Univ. Arizona Press, Tucson.

Smith, B. A., and 26 colleagues 1981. Encounter with Saturn – Voyager 1 imaging science results. Science 212, 163–191.

Smith, B. A., and 28 colleagues 1982. A new look at the Saturn system – The Voyager 2 images. Science 215, 504–537.

Smith, B. A., and many colleagues 1986. Voyager 2 in the Uranian system: Imaging science results. Science 233, 43–64.

Smith, B. A., and many colleagues 1989. Voyager 2 in the Neptunian system: Imaging science results. Science 246, 1422–1449.

Smith, D. E., Bray, V. J., Turtle, E. P., Melosh, H. J., Perry, J. E. 2007. Studies of viscous relaxation of craters on Enceladus. LPI Contribution 1357, 127–128.

Sosa, A., Fernández, J. A. 2009. Cometary masses derived from non-gravitational forces. Mon. Not. R. Astron. Soc. 393, 192–214.

Stern, S. A., McKinnon, W. B. 2000. Triton's surface age and impactor population revisited in light of Kuiper Belt fluxes: Evidence for small Kuiper Belt Objects and recent geological activity. Astron. J. 119, 945–952.

Strom, R. G. 1981. Crater populations on Mimas, Dione and Rhea. Lunar Planet Sci. XII, A7–A9. LPI Contribution 428, Houston, Texas.

Strom, R. G. 1987. The Solar System cratering record: Voyager 2 results at Uranus and implications for the origin of impacting objects. Icarus 70, 517–535.

Strom, R. G., Malhotra, R., Ito, T., Yoshida, F., Kring, D. A. 2005. The origin of planetary impactors in the inner Solar System. Science 309, 1847–1850.

Sussman, G. J., Wisdom, J. 1992. Chaotic evolution of the Solar System. Science 257, 56–62.

Szabó, Gy. M., Ivezić, Ž., Jurić, M., Lupton, R. 2007. The properties of Jovian Trojan asteroids listed in SDSS Moving Object Catalogue 3. Mon. Not. R. Astron. Soc. 377, 1393–1406.

Thomas, P. C., and 17 colleagues 2007a. Hyperion's sponge-like appearance. Nature 448, 50–56.

Thomas, P. C., Burns, J. A., Helfenstein, P., Squyres, S., Veverka, J., Porco, C., Turtle, E. P., McEwen, A., Denk, T., Giese, B., Roatsch, T., Johnson, T. V., Jacobson, R. A. 2007b. Shapes of the saturnian icy satellites and their significance. Icarus 190, 573–584.

Thommes, E. W., Bryden, G., Wu, Y., Rasio, F. A. 2008. From mean motion resonances to scattered planets: Producing the Solar System, eccentric exoplanets, and late heavy bombardments. Astrophys. J. 675, 1538–1548.

Thommes, E. W., Duncan, M. J., Levison, H. F. 1999. The formation of Uranus and Neptune in the Jupiter-Saturn region of the Solar System. Nature 402, 635–638.

Thommes, E. W., Duncan, M. J., Levison, H. F. 2002. The formation of Uranus and Neptune among Jupiter and Saturn. Astron. J. 123, 2862–2883.

Tiscareno, M. S., Malhotra, R. 2003. The dynamics of known Centaurs. Astron. J. 126, 3122–3131.

Tsiganis, K., Gomes, R., Morbidelli, A., Levison, H. F. 2005. Origin of the orbital architecture of the giant planets of the Solar System. Nature 435, 459–461.

Turrini, D., Marzari, F., Beust, H. 2008. A new perspective on the irregular satellites of Saturn – I. Dynamical and collisional history. Mon. Not. R. Astron. Soc. 391, 1029–1051.

Turrini, D., Marzari, F., Tosi, F. 2009. A new perspective on the irregular satellites of Saturn – II. Dynamical and physical origin. Mon. Not. R. Astron. Soc. 392, 455–474.

USGS Astrogeology Gazetteer of Planetary Nomenclature 2009. Web page at http://planetarynames.wr.usgs.gov/jsp/SystemSearch2.jsp?System=Saturn. Accessed April 25, 2009.

Vokrouhlický, D., Nesvorný, D., Levison, H. F. 2008. Irregular satellite capture by exchange reactions. Astron. J. 136, 1463–1476.

Volk, K., Malhotra, R. 2008. The Scattered Disk as the source of the Jupiter family comets. Astrophys. J. 687, 714–725.

Weissman, P. R., Levison, H. 1997. The population of the trans-Neptunian region: The Pluto-Charon environment. In *Pluto and Charon* (S. A. Stern and D. J. Tholen, Eds.), pp. 559–604. Univ. Arizona Press, Tucson.

West, R. M., Hainaut, O., Smette, A. 1991. Post-perihelion observations of P/Halley. III – an outburst at R = 14.3 AU. Astron. Astrophys. 246, L77–L80.

Wetherill, G. W. 1975. Late heavy bombardment of the moon and terrestrial planets. Lunar and Planetary Science Conference 6, 1539–1561.

Wetherill, G.W. 1981. Nature and origin of basin-forming projectiles. In *Multi-ring Basins, Proc. Lunar Planet. Sci. 12A* (P. H. Schultz and R. B. Merrill, Eds.), 1–18.

Wood, C. A., Lunine, J. I., Stofan, E. R., Lorenz, R. D., Lopes, R. M. C., Radebaugh, J., Wall, S. D., Paillou, P., Farr, T. 2008. Degraded impact craters on Titan. Lunar Planet. Sci. Conf. 39, abstract #1990.

Wünnemann K., Collins, G. S., Elbeshausen D. 2008. Limitations of point-source analogy for meteorite impact and implications to crater-scaling. Large Meteorite Impacts and Planetary Evolution IV, LPI Contrib. 1423, abstract #3076.

Wünnemann K., Collins, G. S., Melosh, H. J. 2006. A strain-based porosity model for use in hydrocode simulations of impacts and implications for transient crater growthin porous targets. Icarus 180, 514–527.

Zahnle, K. J., Alvarellos, J. L., Dobrovolskis, A. R., and Hamill, P. 2008. Secondary and sesquinary impact craters on Europa. Icarus 194, 660–674.

Zahnle, K., Dones, L., Levison, H. F. 1998. Cratering rates on the galilean satellites. Icarus 136, 202–222.

Zahnle, K., Schenk, P., Levison, H., Dones, L. 2003. Cratering rates in the outer Solar System. Icarus 163, 263–289.

Zahnle, K., Schenk, P., Sobieszczyk, S., Dones, L., Levison, H. F. 2001. Differential cratering of synchronously rotating satellites by ecliptic comets. Icarus 153, 111–129.

Chapter 20
Icy Satellites: Geological Evolution and Surface Processes

Ralf Jaumann, Roger N. Clark, Francis Nimmo, Amanda R. Hendrix, Bonnie J. Buratti, Tilmann Denk, Jeffrey M. Moore, Paul M. Schenk, Steve J. Ostro[†], and Ralf Srama

Abstract The sizes of the Saturnian icy satellites range from ~1,500 km in diameter (Rhea) to ~20 km (Calypso), and even smaller 'rocks' of only a kilometer in diameter are common in the system. All these bodies exhibit remarkable, unique features and unexpected diversity. In this chapter, we will mostly focus on the 'medium-sized icy objects' Mimas, Tethys, Dione, Rhea, Iapetus, Phoebe and Hyperion, and consider small objects only where appropriate, whereas Titan and Enceladus will be described in separate chapters. Mimas and Tethys show impact craters caused by bodies that were almost large enough to break them apart. Iapetus is unique in the Saturnian system because of its extreme global brightness dichotomy. Tectonic activity varies widely – from inactive Mimas through extensional terrains on Rhea and Dione to the current cryovolcanic eruptions on Enceladus – and is not necessarily correlated with predicted tidal stresses. Likely sources of stress include impacts, despinning, reorientation and volume changes. Accretion of dark material originating from outside the Saturnian system may explain the surface contamination that prevails in the whole satellite system, while coating by Saturn's E-ring particles brightens the inner satellites.

So far, among the surprising Cassini discoveries are the volcanic activity on Enceladus, the sponge-like appearance of Hyperion and the equatorial ridge on Iapetus – unique features in the solar system. The bright-ray system on Rhea was caused by a relatively recent medium impact which formed a ~40 km crater at 12°S latitude, 112°W longitude, while the wispy streaks on Dione and Rhea are of tectonic origin. Compositional mapping shows that the dark material on Iapetus is composed of organics, CO_2 mixed with H_2O ice, and metallic iron, and also exhibits possible signatures of ammonia, bound water, H_2 or OH-bearing minerals, and a number of as-yet unidentified substances. The spatial pattern, Rayleigh scattering effect, and spectral properties argue that the dark material on Iapetus is only a thin coating on its surface. Radar data indicate that the thickness of the dark layers can be no more than a few decimeters; this is also consistent with the discovery of small bright-ray and bright-floor craters within the dark terrain. Moreover, several spectral features of the dark material match those seen on Phoebe, Iapetus, Hyperion, Dione and Epimetheus as well as in the F-ring and the Cassini Division, implying that throughout the Saturnian system. All dark material appears to have a high content of metallic iron and a small content of nano-phase hematite. However, the complete composition of the dark material is still unresolved, and additional laboratory work is required. As previously concluded for Phoebe, the dark material appears to have originated external to the Saturnian system.

The icy satellites of Saturn offer an unrivalled natural laboratory for understanding the geological diversity of different-sized icy satellites and their interactions within a complex planetary system.

20.1 Introduction

Covering a wide range of diameters, the satellites of Saturn can be subdivided into three major classes: 'icy rocks,'

R. Jaumann
DLR, Institute of Planetary Research, Rutherfordstrasse 2, 12489 Berlin, Germany
and
Freie Universität Berlin, Institute of Geological Sciences, Malteserstr. 74-100, 12249 Berlin, Germany

T. Denk
Freie Universität Berlin, Institute of Geological Sciences, Malteserstr. 74–100, 12249 Berlin, Germany

R.N. Clark
U.S. Geological Survey, Denver Federal Center, Denver CO 80225, USA

F. Nimmo
Department of Earth and Planetary Sciences, University of California Santa Cruz, Santa Cruz, CA 95064, USA

A.R. Hendrix, B.J. Buratti, and S.J. Ostro[†]
Jet Propulsion Laboratory, California Institute of Technology, Pasadena, CA 91109, USA
[†]Deceased

J.M. Moore
NASA Ames Research Center, MS 245-3 Moffett Field, CA 94035-1000, USA

P.M. Schenk
Lunar and Planetary Institute, Houston, TX 77058, USA

R. Srama
Max Planck Institut für Kernphysik, 69117, Heidelberg, Germany

with radii <10 km, 'small satellites' with radii <100 km, such as Prometheus, Pandora, Janus, Epimetheus, Telesto, Calypso, and Helene, and 'medium-sized satellites' with radii <800 km: Mimas, Enceladus, Tethys, Dione, Rhea, Hyperion, Iapetus and Phoebe. The largest satellite, Titan, has a radius of 2,575 km. We focus here on the 'medium-sized satellites,' considering the 'small satellites' where appropriate. Enceladus is discussed in a separate chapter (Spencer et al. 2009) and Titan in a dedicated book (Brown et al. 2009; Jaumann et al. 2009); therefore, these two satellites will not be discussed in detail here. The main orbital and physical characteristics of the medium and small satellites of Saturn are shown in Table 20.1.

Common properties of the Saturnian moons include a very large amount of water ice on the surface and in the interior, resulting in low mean densities between $\sim$1.0 g/cm^3 and $\sim$1.6 g/cm^3, and numerous impact craters scattered over the surface. Tectonic features and less heavily cratered plains are restricted to only a few satellites and indicate endogenic geological activity. The Voyager missions (e.g., Smith et al. 1981, 1982) changed our perception of the medium-sized Saturnian moons from small dots in large telescopes into real worlds with alien landscapes. Impact craters were recognised as the dominant landforms, except for parts of Enceladus that appeared to be crater-free. It was found that instead of cratering widespread tectonics seems to have shaped the surface of Enceladus, and most craters showed evidence of viscous relaxation. A causal connection with Saturn's E-ring was also suspected because the E-ring is densest at Enceladus' orbit (Pang et al. 1984). However, similarly sized Mimas was found to look very different. Its trademark is a relatively large, 110 km crater, Herschel. Besides Herschel, there are numerous other impact-related but few tectonic structures. On Tethys, a large valley (Ithaca Chasma), spanning about 3/4 of Tethys' globe, and a giant impact basin (Odysseus, $\sim$400 km in diameter) were found to dominate the surface. Dione and Rhea both showed wispy structures in early views of their trailing sides, but the resolution of the Voyager images was insufficient to determine the origin of these surface features. Although it is the largest medium-sized moon of Saturn, Rhea shows numerous craters and few endogenic geological structures. On Rhea's trailing side, an extended bright-ray pattern was detected in very low-resolution Voyager images (Smith et al. 1981), possibly caused by a recent impact. On other outer solar system moons, only Pwyll on Europa is known to have a similar structure, with

Table 20.1 Physical and orbital characteristics of the small (<100 km radius) and medium-sized moons (<800 km radius) of Saturn (from JPL at http://ssd.jpl.nasa.gov/?satellites)

Satellite	a (10^3 m)	P (days)	e	i (deg)	Mean radius (km)	Mean density	GM	V$_0$ or R	P
Pan	134	0.575	0.0000	0.001	12.8	0.56	0.00033 ± 0.00005	19.4	0.5
Atlas	138	0.602	0.0012	0.003	10.	0.5	0.00014	19.0	0.4
Prometheus	139	0.613	0.0022	0.008	46.8 ± 5.6	0.435 ± 0.159	0.01246 ± 0.00083	15.8	0.6
Pandora	142	0.629	0.0042	0.050	40.6 ± 4.5	0.530 ± 0.194	0.00995 ± 0.00155	16.4	0.5
Janus	151	0.695	0.0068	0.163	90.4 ± 3.0	0.612 ± 0.062	0.1266 ± 0.0017	14.4	0.6
Epimetheus	151	0.694	0.0098	0.351	58.3 ± 3.1	0.634 ± 0.102	0.0351 ± 0.0004	15.6	0.5
Mimas	186	0.942	0.0202	0.001	198.2	1.150	2.530 ± 0.012	12.8	0.6
Enceladus	238	1.370	0.0045	0.001	252.1	1.608	7.210 ± 0.011	11.8	1.0
Tethys	295	1.888	0.0000	0.002	533	0.973	41.210 ± 0.007	10.2	0.8
Telesto	295	1.888	0.0002	1.180	12. ± 3.0	1.0	0.00048	18.5	1.0
Calypso	295	1.888	0.0005	1.499	9.5 ± 1.5	1.0	0.00024	18.7	0.7
Dione	377	2.737	0.0022	0.005	561.7	1.476	73.113 ± 0.003	10.4	0.6
Helene	377	2.737	0.0071	0.213	16. ± 4.0	1.5	0.0017	18.4	0.6
Rhea	527	4.518	0.0010	0.029	764.3	1.233	154.07 ± 0.16	9.6	0.6
Hyperion	1501	21.28	0.0274	0.461	133.0 ± 8.0	0.569 ± 0.108	0.37 ± 0.02	14.4	0.3
Iapetus	3561	79.33	0.0283	14.968	735.6	1.083	120.50 ± 0.03	11	0.6
Phoebe	12948	550.31	0.1635	26.891	106.6 ± 1.1	1.633 ± 0.049	0.5531 ± 0.0006	16.4	0.081 ± 0.002
Paaliaq	15200	686.95	0.3630	45.084	11.0	2.3	0.00055	21.3R	0.06
Albiorix	16182	783.45	0.4770	34.208	16	2.3	0.0014	20.5R	0.06
Siarnaq	17531	895.53	0.2960	46.002	20	2.3	0.0026	20.1R	0.06

a – mean value of the semi-major axis.
P – sidereal period P.
e – mean eccentricity.
i = mean inclination with respect to the reference plane ecliptic, ICRF, or local Laplace.
V$_0$ = mean opposition magnitude or R = red magnitude.
p = geometric albedo (geometrical albedo is the ratio of a body's brightness at zero phase angle to the brightness of a perfectly diffusing disk of the same position and apparent size as the body). (After Thomas et al. 1983; Morrison et al. 1984; Thomas 1989; Showalter 1991; Jacobson and French 2004; Jacobson et al. 2004; Spitale et al. 2006; Jacobson 2006; Porco et al. 2005b; Jacobson et al. 2005; Jacobson 2007)

rays extending over thousands of kilometers away from the crater. During the Voyager era, the leading sides of Tethys, Dione and Rhea were found to be brighter than the trailing sides. Later, Mimas and Enceladus were found to have slightly brighter trailing hemispheres; this general pattern of albedo dichotomies is consistent with alteration by Saturn's E-ring (Hamilton and Burns 1994, Buratti et al. 1998). On Iapetus, the elliptical shape of the dark terrain on the leading hemisphere was discovered along with giant mountains on the anti-Saturnian side. Hyperion was found by Voyager to be very irregularly shaped and by ground-based telescopic observations to be without a regular rotational period (Klavetter 1989).

Most of our post-Voyager knowledge has been summarized by Gehrels and Matthews (1984) and Morrison et al. (1986). However, important questions remained unanswered before Cassini: the retrograde orbit and dark surface of Phoebe threw its Saturnian origin into doubt (e.g., Cruikshank et al. 1983; Burns et al. 1979). Hyperion is darker than the inner regular satellites (e.g., Cruikshank et al. 1984), exhibits a surface dominated by cratering and spallation (e.g., Thomas and Veverka 1985) and rotates chaotically (e.g., Wisdom et al. 1984), indicating a unique geological evolution. Almost all the small inner satellites move in remarkable coorbital, shepherd or Langrangian orbits, which suggests that they originated from the break-up of larger bodies (e.g., Morrison et al. 1986). The highly symmetrical distribution of Iapetus' dark material along the direction of orbital motion called for some external control, whereas the topographical relationship of dark material on the floors of bright-side craters indicated endogenic control, so that the origin of the dark material was problematic (Morrison et al. 1986). During the Voyager era the origin, nature and distribution mechanism of dark material in the Saturnian system was not understood. The origin of bright features like the 'wispy streaks' on Rhea and Dione could not be determined by Voyager images due to their low resolution. Differences in crater density between some of the medium-sized satellites, such as Dione, Rhea and Enceladus, as well as extended grabens, faults and valleys suggested endogenic activities. However, the nature of these activities was not well understood. For a more detailed discussion of our knowledge before Cassini, see Dougherty et al. (2009) and Orton et al. (2009). In terms of geology, the Cassini mission was designed to provide significantly better area coverage, spatial resolution, and spectral range, resulting in numerous new discoveries. A detailed description of the Cassini mission can be found in Dougherty et al. (2009) and Seal et al. (2009). Maps of the Saturnian satellites are included in Roatsch et al. (2009).

20.2 Cassini's Exploration of Saturn's Icy Satellites

The Cassini spacecraft is equipped with instruments tailored to investigate the surfaces, environments and interiors of icy satellites. The optical remote sensing (ORS) instrument suite includes cameras and spectrometers designed for high spatial and spectral resolution covering wavelengths between $0.06\,\mu m$ and $1,000\,\mu m$. With the imaging subsystem (ISS) (Porco et al. 2004), morphologic, stratigraphic and other geological surface properties can be observed at spatial resolutions down to a few meters (locally) or a few hundred meters (globally), depending on flyby distances. Moreover, as many as 33 color and polarization filter combinations permit mapping geologically diverse terrain spatially at wavelengths between $\sim 0.3\,\mu m$ and $\sim 1.1\,\mu m$. The visible and infrared mapping spectrometer (VIMS; $0.4\,\mu m$ to $5.1\,\mu m$) provides chemical and compositional spectral information with spatial resolutions from a few kilometers (globally) to better than one hundred meters (locally) (Brown et al. 2004; Jaumann et al. 2006). The composite infrared spectrometer (CIRS) determines global and regional surface temperatures and thermal properties on a kilometer scale (Flasar et al. 2004). The ultraviolet imaging spectrograph (UVIS) provides information about thin atmospheres and volcanic plume structures as well as about water ice and other minor constituents on the surface (Esposito et al. 2004), operating in the 60 nm – 190 nm wavelength range. A suite of magnetosphere and plasma science (MAPS) instruments characterizes the satellites' environments by in-situ methods. Micron-sized dust grains and neutral molecules released from the surface by active or passive processes (volcanoes/sputtering) carry surface composition information to distances as far as hundreds of kilometers above the surface. Released material affects the plasma surrounding the satellite and is registered by variations in the magnetic field, ion density, and neutral gas and dust density as well as gas and dusty composition.

The primary task of the radio science subsystem (RSS) is to determine the mass of the moons (Table 20.1) by tracking deviations in Cassini's trajectory. RADAR data can provide unique information about the upper sub-surface (Elachi et al. 2004), though few close-up RADAR SAR observations were performed during satellite flybys. The lack of a scan platform for the remote sensing instruments prevents the radio science, RADAR and remote sensing systems from operating simultaneously during a flyby because the antenna and remote sensing instruments are oriented $90°$ apart on the spacecraft. Even joint surface scans by the ORS instruments require significant compromises between individual observations due to data rate and integration time constraints. While the ISS instrument needs short 'dwell' times for mosaicking, the VIMS depend on long 'dwells' for improved signal-to-noise ratios. On the other hand, the CIRS and UVIS instruments contain line-scanning devices, which depend on either slow or fast slews to scan the surfaces. Thus, reaching satisfactory compromises is a major challenge in the planning process.

Since the Cassini spacecraft orbits the planet rather than individual satellites, the moons can only be observed at various distances and illumination conditions, ideally during very close-targeted flybys. Depending on the distance of a moon to Saturn, the flybys occur at very different velocities. In March 2008, for instance, Enceladus was passed at ∼14 km/s, while the Iapetus flyby in September 2007 took place at a leisurely 2.4 km/s. The flyby geometry can also vary significantly. The closest-ever approach was during the Enceladus flyby on October 9, 2008, when the spacecraft skimmed as low as 25 km over Enceladus' surface. Other, more typical targeted flyby altitudes occur between 100 km and 2,000 km. A flyby can be polar, equatorial, or in between, and at the closest approach, the sub-spacecraft point over a moon can be located either over its illuminated or over its unlit side.

The MAPS instruments measure densities or field gradients over time. Sputtering processes lead to high dust and neutral-gas densities above the surface, so that environmental in-situ instruments rely on measurements as close to the surface as possible. Further plasma density variations caused by satellites moving in the magnetosphere are empty flux tubes (wakes) and the drop of plasma along the related L-shell. Crossing those regions is of high value for in-situ plasma investigations that provide indirect information about the satellite surface and the density of its neutral and plasma environment.

During the nominal mission, Cassini performed nine targeted as well as numerous close flybys of icy satellites (Table 20.2).

Three of the targeted flybys were dedicated to Enceladus, two in 2005 and one in 2008. In addition, a very close but non-targeted Enceladus flyby happened in February 2005. As the data provided by the MAPS instruments in March 2005 had raised suspicions about potential internal activity (Spahn et al. 2006; Waite et al. 2006), the July 2005 flyby was lowered from 1,000 km to 173 km to allow for increased sensitivity in MAPS measurements. This flyby supplied overwhelming evidence from multiple instruments of current activity on Enceladus, eventually confirmed by direct images of plumes taken during a distant flyby in November 2005. The locations of the plumes were found to coincide with linear tectonic features near the south pole, dubbed 'tiger stripes'. The CIRS instrument also identified these 'tiger stripes' as unusually warm linear features (Spencer et al. 2006); the tiger stripes contain relatively large ice particles identified by VIMS (Brown et al. 2006; Jaumann et al. 2008).

All other moons had only one targeted flyby during the nominal mission, if any. The Tethys, Hyperion, and Dione flybys in the late summer/early fall of 2005 were performed within 3 weeks of each other during the same orbit of Cassini around Saturn. Due to Hyperion's chaotic rotation, it was impossible to predict which part of its surface would be visible and how the surface might look. By coincidence, the viewing perspective during the approach was almost identical with the best Voyager 1 views. It showed a giant crater, dark material within numerous smaller craters and a very low mean density of only $0.54 \, g/cm^3$. The 'wispy streaks' of Dione were observed at very oblique viewing angles during the October 2005 flyby, and were confirmed to be tectonic in origin (Wagner et al. 2006). The images from this targeted Dione flyby are among the most spectacular pictures of the mission. The Rhea flyby in November 2005 was dedicated to radio science, and the results from that flyby are reported in Matson et al. (2009). However, while the spacecraft itself was inertially pointed with its antenna towards Earth, the cameras slewed across the surface, obtaining the highest-resolved images of Rhea to date.

In 2006, no targeted flybys took place, primarily because the spacecraft spent most of its time in highly inclined orbits. The very close but non-targeted Rhea flyby in August 2007 was dedicated to remote sensing and revealed small details of the prominent bright-ray crater and its ejecta environment. This might be the youngest large crater in the Saturnian system. The targeted Iapetus flyby in September 2007 over the anti-Saturnian and trailing hemispheres yielded very high-resolution observations of the highest parts of the equatorial ridge, the transition zone between the dark and the bright terrain, and the bright trailing side. Combined with data from a more distant, non-targeted flyby on December 31, 2004, and other distant observations, these data permitted developing a model of the formation of the global and unique brightness dichotomy of this unusual moon (Denk et al. 2009; Denk and Spencer 2009). For more detailed descriptions of the Cassini mission, see Dougherty et al. (2009), Orton et al. (2009) and Seal et al. (2009).

20.3 Morphology, Geology and Topography

Voyager's partial global mapping at 1 km resolution suggested that the morphology of the Saturnian satellites was dominated by impact craters (Smith et al. 1981). Indeed, Voyager's best images, 500 m resolution views of Rhea, showed a landscape saturated with craters. Voyager also found evidence of limited volcanic and tectonic activity in

Table 20.2 Mission characteristics of the Cassini icy satellites flybys; C/A... closest approach

Target	Orbit	Flyby date	Altitude (km)	Solar Phase Angle 2 h before, at, and 2 h after C/A (deg)	Target (T)/ non-targeted (nT)
Phoebe	0	June 11, 2004	1,997	85/25/90	T
Dione	B	December 15, 2004	72,100	108/85/62	nT
Iapetus	B/C	December 31, 2004	123,399	87/94/100	nT
Enceladus	3	February 17, 2005	1,176	25/114/158	nT
Enceladus	4	March 9, 2005	500	47/42/130	T
Hyperion	9	June 10, 2005	168,000	25/35/47	nT
Enceladus	11	July 14, 2005	170	47/63/132	T
Mimas	12	August 2, 2005	62,700	41/57/91	nT
Tethys	15	September 24, 2005	1,503	21/68/155	T
Hyperion	15	September 26, 2005	505	51/50/127	T
Dione	16	October 11, 2005	500	23/66/156	T
Iapetus	17	November 12, 2005	415,400	89/91/93	nT
Rhea	18	November 26, 2005	1,264	19/87/161	T
Rhea	22	March 21, 2006	82,200	113/137/157	nT
Iapetus	23	April 11, 2006	602,400	123/124/125	nT
Enceladus	27	September 9, 2006	39,900	105/118/105	nT
Dione	33	November 21, 2006	72,300	125/144/119	nT
Hyperion	39	February 16, 2007	175,000	66/64/64	nT
Tethys	47	June 27, 2007	18,400	148/116/49	nT
Tethys	49	August 29, 2007	55,500	135/102/72	nT
Rhea	49	August 30, 2007	5,800	125/46/41	nT
Iapetus	49	September 10, 2007	1,620	139/59/27	T
Dione	50	September 30, 2007	43,400	89/49/19	nT
Hyperion	51	October 21, 2007	122,000	115/116/115	nT
Mimas	53	December 3, 2007	84,200	161/139/82	nT
Enceladus	61	March 12, 2008	52	115/135/65	T
Enceladus	80	August 11, 2008	50	108/110/73	T
Enceladus	88	October 9, 2008	25	108/112/73	T
Enceladus	91	October 31, 2008	200	108/113/73	T
Tethys	93	November 16, 2008	57,100	79/41/58	nT
Tethys	94	November 24, 2008	24,200	123/159/84	nT
Tethys	115	July 26, 2009	68,400	105/89/79	nT
Rhea	119	October 13, 2009	40,400	140/82/25	nT
Mimas	119	October 14, 2009	44,200	152/102/44	nT
Hyperion	119	October 16, 2009	127,000	155/140/126	nT
Enceladus	120	November 2, 2009	100	177/90/5	T
Enceladus	121	November 21, 2009	1,600	146/87/36	T
Rhea	121	November 21, 2009	24,400	127/58/10	nT
Tethys	123	December 26, 2009	52,900	136/77/17	nT
Dione	125	January 27, 2010	45,100	158/106/53	nT
Mimas	126	February 13, 2010	9,500	173/99/21	nT
Rhea	127	March 2, 2010	100	176/87/3	T
Dione	129	April 7, 2010	500	167/79/11	T
Enceladus	130	April 28, 2010	100	171/93/9	T
Enceladus	131	May 18, 2010	200	153/108/29	T
Tethys	132	June 3, 2010	52,600	154/99/45	nT

the form of relatively smooth plains and linear features (Smith et al. 1981, 1982; Plescia and Boyce 1982, 1983; Moore et al. 1985; Moore and Ahern 1983), especially on Enceladus. The Enceladus discoveries aside, Cassini's considerably improved resolution and global mapping coverage confirmed the abundance of impact craters on all satellites (Dones et al. 2009) and also revealed surprisingly varied degrees of past endogenic activity on Tethys, Rhea and especially Dione, indicating that these satellites were more dynamic than originally thought. Observed deformations range from linear grooves (Mimas) and nearly global-scale grabens (Tethys, Rhea) to repeated episodes of tectonics and resurfacing (Dione, Enceladus). To date, little or no direct evidence of past or present endogenic activity has been identified on Hyperion, Phoebe or Iapetus (other than its potentially endogenic equatorial ridge). This diversity is in itself a puzzle that has not yet been explained. In the following sections, we examine the features observed, their morphology, and implications for internal evolution.

20.3.1 Craters

Comparison with craters on larger icy satellites (e.g., Ganymede) is illuminating because surface gravities differ by a factor of 5–10. Central-peak craters are common on Saturnian satellites. Central peaks in larger craters can be as high as 10 km, many of them rising above the local mean elevation. Cassini has revealed little evidence of terrace formation, and floor uplift dominates clearly over rim collapse in the formation of complex craters (Schenk and Moore 2007). Pancake (formerly called pedestal) ejecta have been observed in smooth plains on Dione. This kind of ejecta deposit, once only known as a feature of the Galilean satellites, now appears to be common on icy satellites in general. The fact that it is not obvious on other satellites is most likely due to the difficulty of identifying it on these rugged, heavily cratered surfaces.

A transition to larger, more complex crater landforms (e.g., central pit craters, multi-ring basins) on these satellites is predicted to occur at much larger diameters (>150 km) than those seen on Ganymede and Callisto (McKinnon and Melosh 1980; Schenk 1993), due to their weaker gravity. True central-pit craters like those on Ganymede or Callisto (Schenk 1993) have not been observed, although the central complex of Odysseus features a central depression and closely resembles its counterparts on large icy moons. The dominant morphology at larger diameters (>150 km) is peak-ring or similar, with a prominent central peak surrounded by a ring or an elevated circular plateau. The number of such basins discovered by Cassini on Iapetus and Rhea was greater than expected. Indeed, at least 10 basins larger than 300 km across have been seen on Iapetus (e.g., Giese et al. 2008).

The most ancient large basins on Rhea and Tethys have been modified by relaxation and long-term impact bombardment (e.g., Schenk and Moore 2007). The younger basins are generally deep and unrelaxed. However, the floor of Evander, is the largest basin identified on Dione, has been relaxed and uplifted to the level of the original surface, and although its rim and peak remain prominent (e.g., Schenk and Moore 2007). Relaxation of this basin is consistent with elevated heat flows on Dione, as indicated by the relaxation of smaller (diameters of 25–50 km) craters within the extensive smooth plains to the north of Evander. Relaxation in this basin, which has a low superposed crater density, is consistent with elevated heat flows, as indicated by the smaller relaxed craters nearby and the extensive smooth plains to the north. The relaxation state of impact basins can be determined by comparing their depth to that of unmodified craters and examining their shape. The most ancient large basins on Rhea (e.g., Tirawa) and Tethys are roughly 10–20% shallower than predicted and feature prominent domical or conical central uplifts that approach or exceed the local elevation of the satellite (Schenk and Moore 2007), which suggests they have been modified to some degree by relaxation. The youngest basins appear to be deeper and possibly unrelaxed. The most prominent examples include Herschel on Mimas (diameter ∼120 km, depth ∼11 km) and Odysseus (diameter ∼450 km, depth >8 km). In neither case do the central uplifts rise above the local topography. A prominent exception is the relatively young 400 km wide Evander basin on Dione, which may be either ∼3 Gyr (Wagner et al. 2006) or less than 2 Gyr old (Kirchoff and Schenk 2008; 2009). For a more detailed discussion of surface ages, see Dones et al. (2009).

Iapetus' leading side shows more relief than the other medium-sized Saturnian satellites (Thomas et al. 2007b; Giese et al. 2008). The lack of basin relaxation on Iapetus is consistent with the presence of a thick (50–100 km) lithosphere in Iapetus' early history. Such a lithosphere might also support Iapetus' prominent equatorial ridge. The ridge shows a complex morphology, probably the result of impact erosion in some places and of post-formation tectonic modifications in others.

The abundance of large impact basins, even in Voyager's limited view, suggested that such basins could have disruptive effects on these satellites (Smith et al. 1981). Several attempts to discover evidence of these effects (e.g., seismic shaking, global fracturing) have met with limited success (e.g., Moore et al. 2004), but Cassini-based studies are now in progress.

20.3.2 Tectonics

The Voyager 1 and 2 spacecraft revealed that the medium-sized icy Saturnian satellites have undergone varying degrees of tectonic deformation (Smith et al. 1981; 1982; Plescia and Boyce 1982, 1983). Although such deformation can provide constraints on the interior structure and evolution of a satellite, the imaging resolution was often insufficient to permit sustainable conclusions. Post-Voyager advances include a vastly better understanding of tectonic deformation and thermal evolution in the Jovian system thanks to *Galileo*, and greatly improved experimental measurements of relevant ice properties (Schmitt et al. 1998). In this section, we combine this improved understanding with the results of the ongoing Cassini mission to discuss the current state of knowledge of icy-satellite tectonics in the Saturnian system. We also examine the extent to which cryovolcanic processes may be relevant. Most of the topics covered in this section are treated in greater detail in Collins et al. (2009).

Tectonics on icy satellites comprises three major aspects: (1) the various mechanisms by which ice deforms when subjected to tectonic stress; (2) the mechanisms by which tectonic stress may arise; and (3) the observational constraints on these deformation mechanisms.

From a tectonics point of view, ice is rather similar to silicate materials: under surface conditions its behavior is rigid or brittle, while at greater depths it is ductile. The rheology of ice is complicated and cannot be discussed in depth here; useful discourses on its elastic, brittle and viscous properties may be found in Petrenko and Whitworth (1999), Beeman et al. (1988) and Durham and Stern (2001), respectively. Although a real ice shell will exhibit all three kinds of behavior, it can be modeled as a simple elastic layer (e.g., McNutt 1984). This effective elastic thickness depends on the temperature gradient within the ice shell during deformation and can be inferred from topographical measurements (e.g., Giese et al. 2007). Thus, measuring the elastic thickness places a useful constraint on satellite thermal evolution (e.g., Nimmo et al. 2002). Pressures within Saturnian satellites are generally low (see Table 20.3).

Thus, although ice transmutes into high-pressure forms at pressures greater than about 0.2 GPa (e.g., Sotin et al. 1998), this effect is relevant mainly on Titan.

Tectonic features may be used to infer the direction and (sometimes) the magnitude of the driving stresses. Observations can thus place constraints on the mechanisms which gave rise to these tectonic features. We briefly discuss several mechanisms which are potential candidates for producing tectonic features.

Satellites experience both rotational and tidal deformation. If the rotation rate or the magnitude of the tide changes, the shape of the satellite will change as well, generating global-scale patterns of stress. A recent compilation of the present-day shapes of the Saturnian satellites is given by Thomas et al. (2007b); a good general description of tidal and rotational satellite deformation is given in Murray and Dermott (1999).

If a synchronously rotating satellite's orbital eccentricity is zero, the tidal bulge will be at a fixed geographic point, and of constant amplitude. The maximum amplitude H of this static (or permanent) tidal bulge depends on the mass and radius of the satellite, the mass of the primary, the distance between them, and the rigidity of the satellite. The latter is described by the h_2 Love number, where h_2 has a value of 5/2 for a fluid body of uniform density which, however, declines as the rigidity (or shear modulus) μ of the satellite increases (e.g., Murray and Dermott 1999).

In a synchronous satellite with non-zero orbital eccentricity, the amplitude and direction of the static tidal bulge will change slightly, resulting in a time-varying diurnal tide. Causing diurnal stresses, this tide has an amplitude expressed by $3eH$, where e is the eccentricity and H is the equilibrium tidal bulge (e.g., Greenberg and Geissler 2002). Obliquity librations are potentially an additional source of tidal stress (Bills and Ray 2000); they can be large if their period is a small multiple of the orbital period (Wisdom 2004).

Table 20.3 shows the diurnal tidal stresses of the icy Saturnian satellites at zero rigidity, demonstrating the general decrease in stress as the semi-major axis lengthens. For comparison, the diurnal stresses on Europa thought to be responsible for cycloidal features are a few tens of kPa (Hoppa et al. 1999a). Because the principal stresses rotate during each orbit, contraction, extension and mixed-mode horizontal principal stresses which promote strike-slip motions are all possible as a satellite orbits (Hoppa et al. 1999b). This motion can lead to both shear heating (Nimmo and Gaidos 2002) and the monotonic accumulation of strike-slip offsets (Hoppa et al. 1999b).

Table 20.3 Data from Yoder (1995) and Thomas et al. (2007b). R, M_s, a and e are the radius, mass, semi-major axis, period and eccentricity, respectively. H is the permanent tidal bulge assuming $h_2 = 2.5$. $3eH$ is the approximate magnitude of the diurnal tidal bulge – note that this is likely an overestimate for small satellites because their rigidity will reduce h_2. "Tidal" is the tidal heat production assuming a homogeneous body with $k_2 = 1.5$ (again, a likely overestimate for small satellites) and $Q = 100$. "Rad" is the radiogenic heat production assuming a chondritic rate of 3.5×10^{-12} W/kg. "Stress" is the approximate diurnal tidal stress given by EeH/R where E is Young's modulus (assumed 9 GPa). P_{cen} is the central pressure, assuming uniform density. "MG" is the "metamorphic grade", modified after Johnson (1998), which is a qualitative assessment of the degree of deformation: I = unmodified, II = intermediate, III = heavily modified

	R (km)	M_s (10^{20} kg)	A (10^3 km)	e	H (m)	$3eH$ (m)	Tidal 10^{10} (W)	Rad. 10^{10} (W)	Stress (kPa)	P_{cen} (MPa)	MG
Mimas	198	0.37	186	0.0202	9,180	556	78	0.013	8,400	7.1	I
Enceladus	252	1.08	238	0.0045	3,940	53	2	0.038	630	23	III
Tethys	533	6.15	295	0.	7,270	0	0	0.22	0	37	II
Dione	562	11	377	0.0022	2,410	16	0.86	0.39	85	97	II
Rhea	764	23	527	0.001	1,440	4.3	0.067	0.81	17	124	II
Titan	2,575	1,346	1,222	0.0292	254	22	45	47	26	3,280	-
Iapetus	736	18	3,561	0.0283	5	0.4	2.6×10^{-5}	0.63	1.7	88	I/II

A satellite surface may move longitudinally with respect to the static tidal bulge due to tidal (Greenberg and Weidenschilling 1984) or atmospheric torques (Lorenz et al. 2008) or, in the case of floating ice shells, if the thermally equilibrated ice shell thickness variations are not compatible with rotational equilibrium (Ojakangas and Stevenson 1989a). This non-synchronous rotation leads to surface stresses (Helfenstein and Parmentier 1985). Stresses from non-synchronous rotation increase with the angular speed of rotation but will dissipate with time owing to viscous relaxation (Greenberg and Weidenschilling 1984; Wahr et al. 2009). Maximum tensile stress depends on the amount of reorientation and the degree of flattening, f, which depends in turn on the satellite rotation rate (Leith and McKinnon 1996). In non-rigid satellites, the stresses due to non-synchronous rotation will exceed the diurnal tidal stresses for a reorientation typically in excess of roughly one degree. In certain circumstances, the surface of a satellite may reorient relative to its axis of rotation (Willemann 1984; Matsuyama and Nimmo 2008) and give rise to global fracture patterns (e.g., Schenk et al. 2008). This process of true polar wander is conceptually very similar to non-synchronous rotation (Melosh 1980; Leith and McKinnon 1996), and 'reorientation' will in the following be understood as implying either non-synchronous rotation or true polar wander. In general, rotation axis motion occurs roughly perpendicular to the tidal axis because this path is energetically favored (Matsuyama and Nimmo 2007). True polar wander can occur because of ice shell thickness variations (Ojakangas and Stevenson 1989b), long-wavelength density anomalies (Janes and Melosh 1988; Nimmo and Pappalardo 2006), volatile redistribution (Rubincam 2003) or impacts, either directly (Chapman and McKinnon 1986) or due to the creation of a new impact basin (Melosh 1975; Murchie and Head 1986; Nimmo and Matsuyama 2007). Like non-synchronous rotation, the maximum stress developed by true polar wander depends on the amount of reorientation and the effective flattening f, with the value of f depending on whether any reorientation of the tidal axis has occurred (Matsuyama and Nimmo 2008).

Satellites that are initially rotating at a rate faster than synchronous will despin (i.e., reduce their rotation rate) to synchronous rotation in timescales that are generally very short compared to the age of the solar system (e.g., Murray and Dermott 1999). Spin rate reductions result in shape changes and stresses – compressive at the equator as the equatorial bulge collapses, and tensile at the poles as the poles elongate. The maximum differential stress caused by despinning in a hydrostatic body depends mainly on the satellite radius and the difference between the initial and the final angular rotation velocities (Melosh 1977). Spin-up may occur in some circumstances (e.g., if the satellite undergoes differentiation), and in this case the signs of all the stresses are reversed. Stresses along lines of longitude are always larger than the stresses along lines of latitude. Thus, irrespective of whether a satellite is spinning down or spinning up, equatorial tectonic features are expected to be oriented in a north-south direction. This result is important when considering the origin of the global equatorial ridge on Iapetus.

Synchronous satellites, which are evolving outwards due to tidal torques from the primary, will undergo both a reduction in their spin rate and a parallel reduction in the tidal bulge amplitude. This combination of despinning and tidal bulge reduction causes a stress pattern in which a region around the sub-planet point experiences compressive stress, middle latitudes experience horizontal shear stress, and the poles undergo tension (Melosh 1980; Helfenstein and Parmentier 1983). The maximum principal stress difference is similar to that produced by despinning alone (Melosh 1980).

A large variety of mechanisms can lead to volume changes within a satellite, and thus to extensional or contractional features on the surface (Squyres and Croft 1986; Kirk and Stevenson 1987; Mueller and McKinnon 1988). Volume changes generate isotropic stress fields on the surface. A potentially important source of expansion or contraction is the large density contrast between ice I and water. As water freezes to ice I, large surface extensional stresses may result (Cassen et al. 1979; Smith et al. 1981; Nimmo 2004a). The reason for this effect is that, on a spherical body, the increased volume of ice relative to water drives the ice shell outwards; and this increase in radius results in extension. A fractional change in radius of 0.1% gives rise to stresses of about 10 MPa. This effect is much reduced if there are high-pressure ice phases freezing simultaneously (Squyres 1980; Showman et al. 1997).

Similar but smaller effects are caused by thermal expansion or contraction (Ellsworth and Schubert 1983; Hillier and Squyres 1991; Showman et al. 1997; Nimmo 2004a). A global temperature change of 100 K will cause stresses on the order of 100 MPa for a thermal expansivity of $10^{-4}/K$. Warming may also lead to silicate dehydration (e.g., Squyres and Croft 1986), which in turn causes expansion.

Satellites are quite likely to have non-axisymmetric structures, in which case some of the above-mentioned global mechanisms may lead to local deformation. For instance, local zones of weakness or pre-existing structures can result in enhanced tidal dissipation and deformation (e.g., Sotin et al. 2002; Nimmo 2004b) and/or the alteration of local stress trajectories. Convection (thermal or compositional) and buoyancy forces due to lateral shell thickness variations can generate local deformation. They are discussed at greater length in Collins et al. (2009). We will not discuss these processes further here, as they do not tend to produce globally organized tectonic structures.

Large impact events can also potentially create global fracture networks. This mechanism is implicit in the model presented by E. Shoemaker in Smith et al. (1981) for the catastrophic breakup and re-accretion of the medium-sized icy satellites of Saturn. Presumably, impacts not quite large enough to completely disrupt a satellite could do considerable damage in the form of global fracturing, and post-Voyager studies of these satellites focused in part on attempts to find evidence of an incipient break-up of this kind. For instance, the surface of Mimas is crossed by a number of linear-to-arcuate, sub-parallel troughs (Fig. 20.1) that are thought to be the consequence of global-scale fractures during the Herschel impact event (McKinnon 1985; Schenk 1989a).

Voyager and Cassini mapping indicates that these features, if related to Herschel, do not form a simple radial or concentric pattern. Nor has it been demonstrated that any of these features form on a plane intersecting Herschel, or whether they should do so. It thus remains possible that these are random fractures formed during freeze expansion of the Mimas interior. Detailed mapping remains to be done to test these hypotheses. Mimas does not otherwise exhibit endogenic landforms and will, therefore, not be discussed further.

Enceladus shows the greatest tectonic deformation of any of the Saturnian satellites, and very significant spatial variations in surface age (Smith et al. 1982; Kargel and Pozio 1996; Porco et al. 2006; Jaumann et al. 2008). Spencer et al. (2009) discusses the tectonic behavior of Enceladus in much greater detail; only a very brief summary is given here.

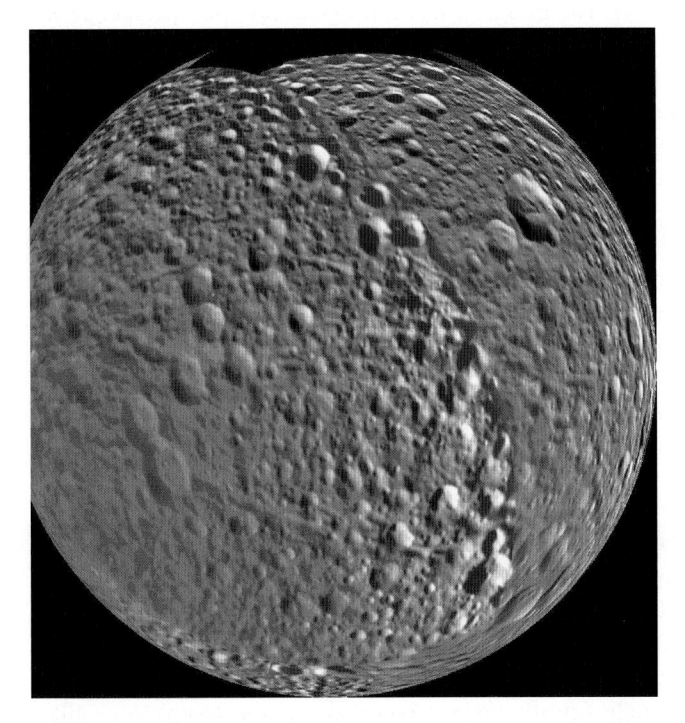

Fig. 20.1 Linear troughs on Mimas, extending east to west. These troughs are interpreted as fractures, possibly related to the Herschel crater. Orthographic map projection at 400 m/pixel is centered on the antipode to Herschel, the largest impact basin on Mimas

Three different terrain types exist on Enceladus. Ancient cratered terrains cover a broad band from 0 to 180° longitude. Centred on 90° and 270° longitude there are younger, deformed terrains roughly 90° wide at the equator. The southern polar region south of 55 °S is heavily deformed and almost uncratered. The most prominent tectonic features of this region are the linear depressions called 'tiger stripes' (Porco et al. 2006). These tiger stripes are typically ∼500 m deep, ∼2 km wide, up to ∼130 km long and spaced ∼35 km apart. Crosscutting fractures and ridges with almost no superimposed impact craters characterize the area between the stripes. The tiger stripes are apparently the source of the geysers observed to emanate from the south polar region (Spitale and Porco 2007). The energy source of these geysers is unknown, but it may be shear heating generated by strike-slip motion at the tiger stripes (Nimmo et al. 2007). The coherent (but latitudinally asymmetrical) global pattern of deformation on Enceladus is currently unexplained, but it is most probably due to shape changes, perhaps related to a hypothetical episode of true polar wander (Nimmo and Pappalardo 2006).

Nearly encircling the globe, Ithaca Chasma is the most prominent feature on Tethys (Fig. 20.2) (Chen and Nimmo 2008).

It extends approximately 270° around Tethys and is not a complete circle. It is narrowly confined to a zone which lies along a large circle whose pole is only ∼20° from the center of Odysseus, the relatively fresh largest basin on the satellite which is 450 km wide (Smith et al. 1982; Moore and Ahern 1983). Smith et al. (1981) suggested that Ithaca Chasma was formed by freeze-expansion of Tethys's interior. They noted that if Tethys was once a sphere of liquid water covered by a thin solid ice crust, freezing in the interior would have produced an expansion of the surface comparable to the area of the chasm (∼10% of the total satellite surface area). This hypothesis fails to explain why the chasm occurs only within a narrow zone; besides, it is difficult to account for the geophysical energy needed for the presence of a molten interior. Expansion of the satellite's interior should have caused fracturing over the entire surface in order to effectively relieve stresses in a rigid crust (Moore and Ahern 1983).

The nearly concentric geometrical relationship between Odysseus and Ithaca Chasma prompted Moore and Ahern (1983) to suggest that the trough system was an immediate manifestation of the impact event, perhaps caused by a damped, whole-body oscillation of the satellite. Based on the deep topography of Odysseus, Schenk (1989b) also suggested that it had not undergone substantial relaxation, and that Ithaca Chasma was the equivalent of a ring graben formed during the impact event by a prompt collapse of the floor involving a large portion of the interior. In a Cassini-era study, Giese et al. (2007) doubted the Odysseus-related

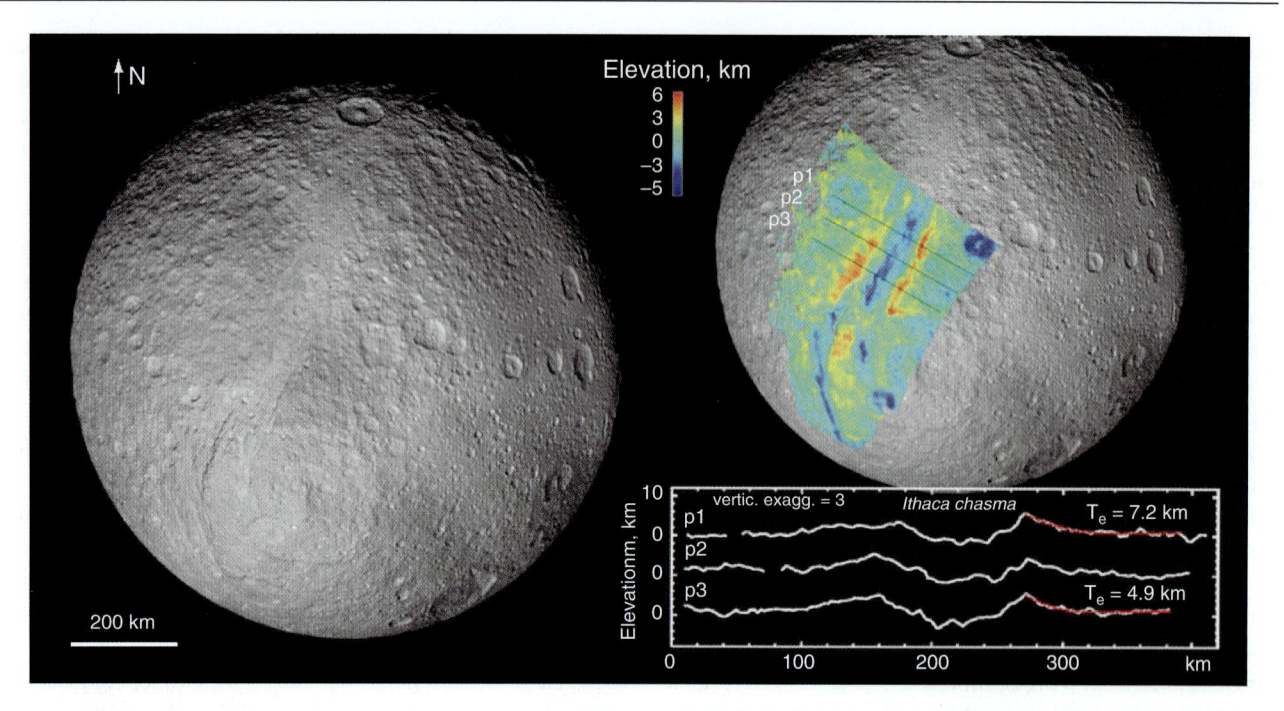

Fig. 20.2 Tethys and Ithaca Chasma. Topography was derived from Cassini stereo images and has a horizontal resolution of 2–5 km and a vertical accuracy of 150–800 m. Profiles have superimposed model flexural profiles (*in red*). T_e is elastic thickness (reproduced from Giese et al. (2007))

hypothesis (Moore and Ahern 1983; Schenk 1989b) on the basis of crater densities within Ithaca Chasma relative to Odysseus, from which they concluded that Ithaca Chasma is older than Odysseus and therefore could not have influenced its formation. They went on to measure photogrammetrically the raised flanks of Ithaca Chasma, which they attributed to flexural uplift. Assuming that these raised flanks were indeed due to flexural uplift, they derived a mechanical lithospheric thickness of ∼20 km and surface heat fluxes of 18–30 mW/m² at assumed strain rates of 10^{-17} to 10/s. They concluded that Ithaca Chasma is an endogenic tectonic feature but could not explain why it is confined to a narrow zone approximating a large circle.

Several Voyager-era studies of Dione (Plescia 1983; Moore 1984) noted a near-global network of tectonic troughs and the presence of a smooth plain crossed by both ridges and troughs located near the 90°W longitude and extending eastward. Cassini coverage revealed the westward extension of this plain (Wagner et al. 2006) and demonstrated that the fractured and cratered dark plains are spectrally distinct (Stephan et al. 2008). Like the eastern portion observed by Voyager, the western plain also exhibits both ridges and troughs (Fig. 20.3).

However, the ridge system is much more pronounced. The western ridges are planimetrically narrower and morphologically better expressed (e.g., their relative lack of degradation). These western ridges also conform much more closely to the boundary between the plain and the older cratered rolling terrain to the west. The linear to arcuate troughs interpreted to have been created by tectonic extension form a complex network that divides Dione's surface into large polygons (Smith et al. 1981, 1982; Plescia 1983; Moore 1984). The troughs often form parallel to sub-parallel sets, somewhat reminiscent of the grooved terrain on Ganymede. The global orientation of Dione's tectonic fabric is non-random (Moore 1984) and exhibits patterns consistent with a decline in oblateness due to either despinning or orbital recession (Melosh 1977, 1980).

The topography derived from Cassini images of Dione shows one clear case of tectonic orientation regionally influenced by a preexisting but very degraded large impact basin centered on the smaller crater Amata (Schenk and Moore 2007 (Figs. 20.3 and 20.4)).

Moore et al. (2004) identified another example of tectonic lineament orientation associated with an unambiguously identified impact feature, 90 km Turnus. Such crater-focused tectonics has been identified on Ganymede from Galileo data (e.g., Pappalardo and Collins 2005). This crater-lineament relationship on Dione is the most pronounced among any of the middle-sized icy satellites except Enceladus (Spencer et al. 2009). Moore (1984) proposed a tectonic history of extension followed by (at least regional) compression, the latter being responsible for the ridges. Cassini-era images of the tectonic troughs (mostly grabens) show them to be morphologically fresh, implying that extension continued into geologically recent times (Wagner et al. 2006).

After Voyager, Rhea was held to be the least endogenically evolved satellite larger than 1,000 km in diameter.

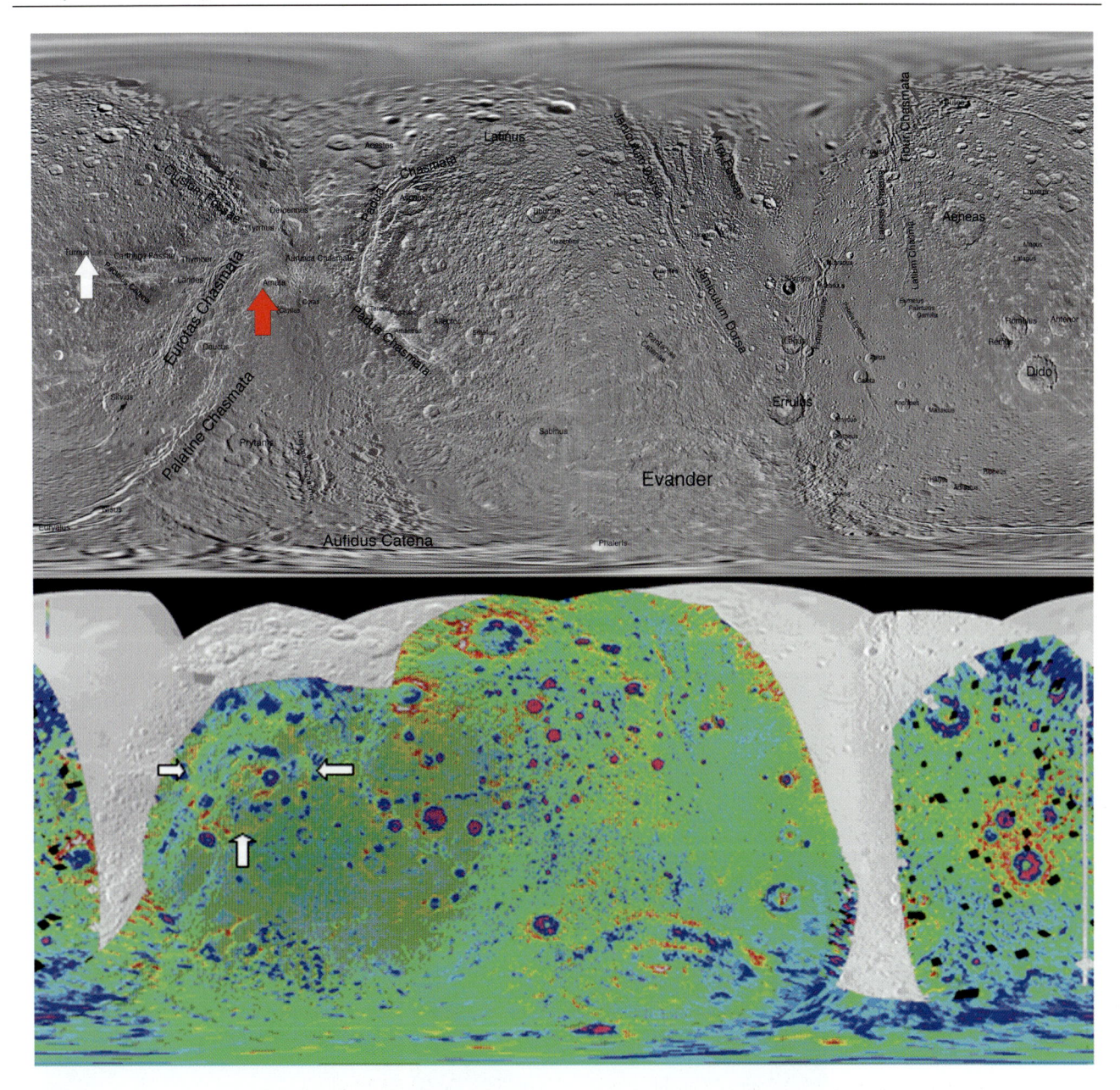

Fig. 20.3 (a) Image mosaic of Dione, centered on 180° longitude. Red and white arrows point to Amata and Turnus impact structures, respectively. (b) Topography of Dione, from Schenk and Moore (2007). An ancient degraded impact basin centered just west of Amata can be identified by the two concentric rings in the DEM (*white arrows*). DEM has a dynamic range of 8 km (red is highest, magenta is lowest)

Moore et al. (1985) pointed out a few parallel lineaments and noted several large ridges or scarps (which they termed 'megascarps') and speculated that these features were formed by a period of extension followed by compression. Cassini images, and digital terrain models derived from them, of the trailing hemisphere of Rhea seen from both middle northern and far southern latitudes clearly show that the wispy arcuate albedo markings seen in Voyager images are associated with a graben/extensional fault system dominantly trending north-south (Schenk and Moore 2007; Wagner et al. 2007) (Fig. 20.5a).

This is the first identification of unambiguous regional endogenic activity on Rhea. The digital terrain models of Rhea derived from Cassini images also show a set of roughly north-south trending ridges (Figs. 20.5b and 20.6) that correspond to the 'megascarps' of Moore et al. (1985).

If these ridges are in fact due to compressional tectonics, their geometrical relation to the graben/scarp system may also indicate a genetic relationship. However, the poor morphological presentation of these ridges implies that they are old, while the 'wispy terrain' grabens are fresh and presumably much more recently formed.

While the troughs and ridges suggest mainly extensional and (minor) compressional tectonics (Thomas 1988), the en-echelon pattern of the scarps and troughs on the trailing hemisphere suggests shear stress (Wagner et al. 2007). This pattern may have been influenced by the possible presence of a large, degraded impact basin (basin C of Moore et al. 1985).

The equatorial ridge of Iapetus (Fig. 20.7) is the most enigmatic feature on any of the medium-sized icy satellites (Porco et al. 2005a; Giese et al. 2008); it extends across more than half of Iapetus' circumference (Giese et al. 2008; Denk et al. 2008). Different parts show very different morphologies, from trapezoidal to triangular cross-sections with steep walls and a sharp rim to isolated mountains with moderate slopes and rounded tops (Giese et al. 2008). Although the ridge is variable in height, it rises up to 10 km above the local mean but also has a gently sloping flank. The ridge is abundantly cratered, indicating that it is comparable in age to other terrains on Iapetus (Schmedemann et al. 2008). Although the equatorial location of the ridge might be due to despinning (Castillo-Rogez et al. 2007), it represents such a large mass that, if it had formed in another location, it would

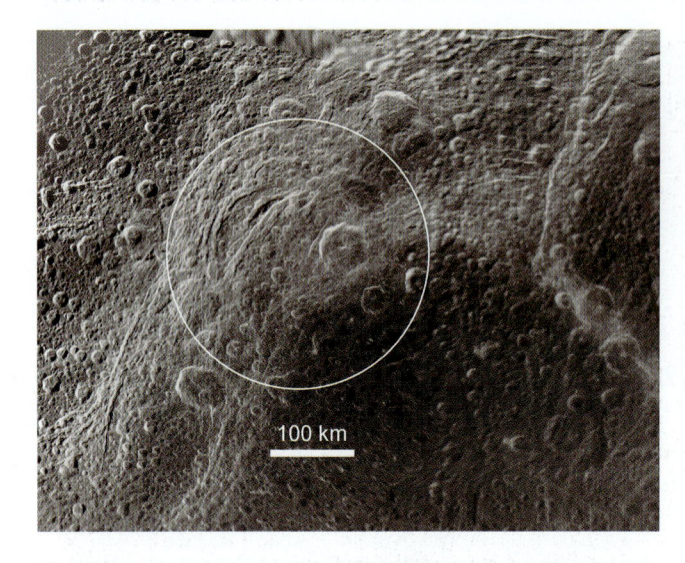

Fig. 20.4 Fracture zones on Dione. Global base map of Cassini imaging has a resolution of 400 m. The circle represents the outer ridge corresponding to the rim of the degraded basin near Amata shown in Fig. 20.3

Fig. 20.5 (**a**) Cassini image of the northern trailing hemisphere of Rhea showing the bright scarps of the northern extension of the graben system and their relationship to 'wispy' terrain. (**b**) Mosaic of Cassini images showing curvilinear grabens (*black arrows*) and an orthogonal set of ridges and troughs (*white arrow*). Base mosaic resolution is 1 km/pixel

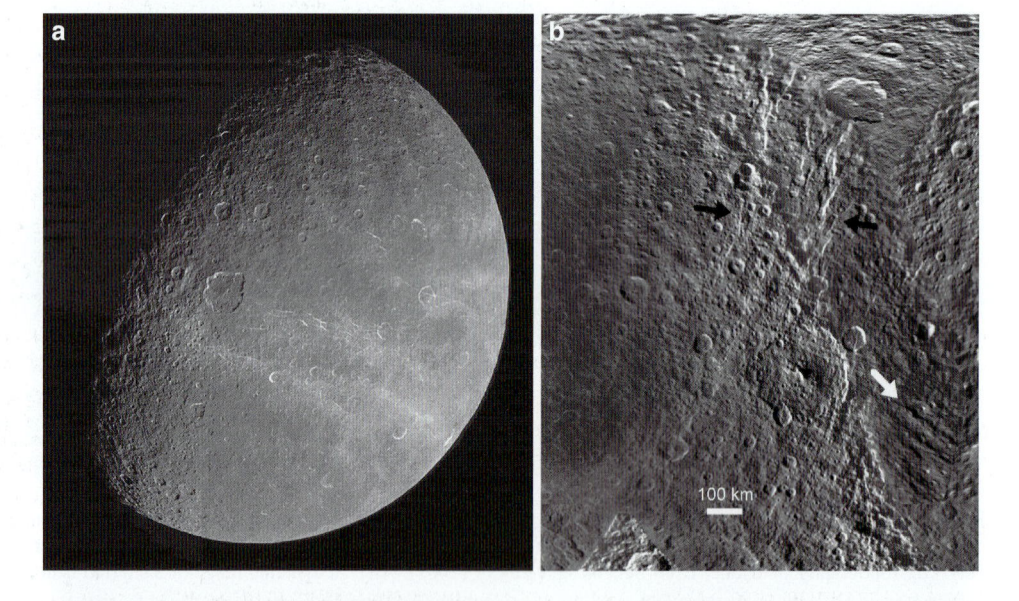

Fig. 20.6 Preliminary stereo topography map, from Schenk and Moore (2007). White arrows indicate graben systems. The DEM has a dynamic range of 10 km (red is highest, magenta is lowest)

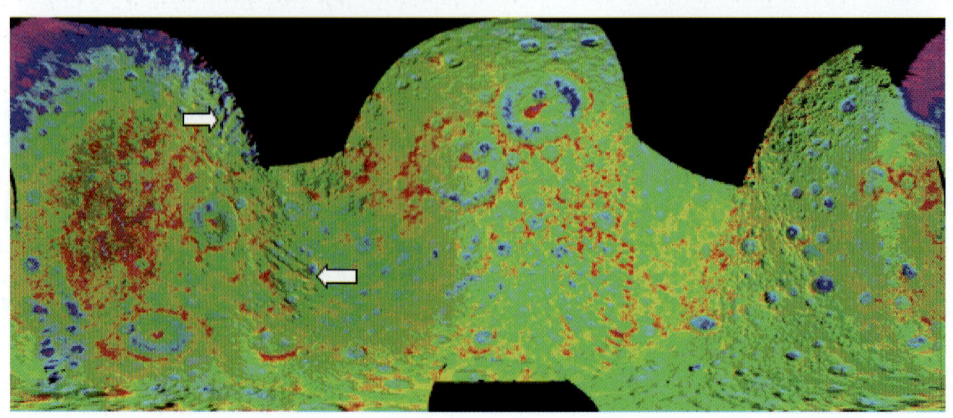

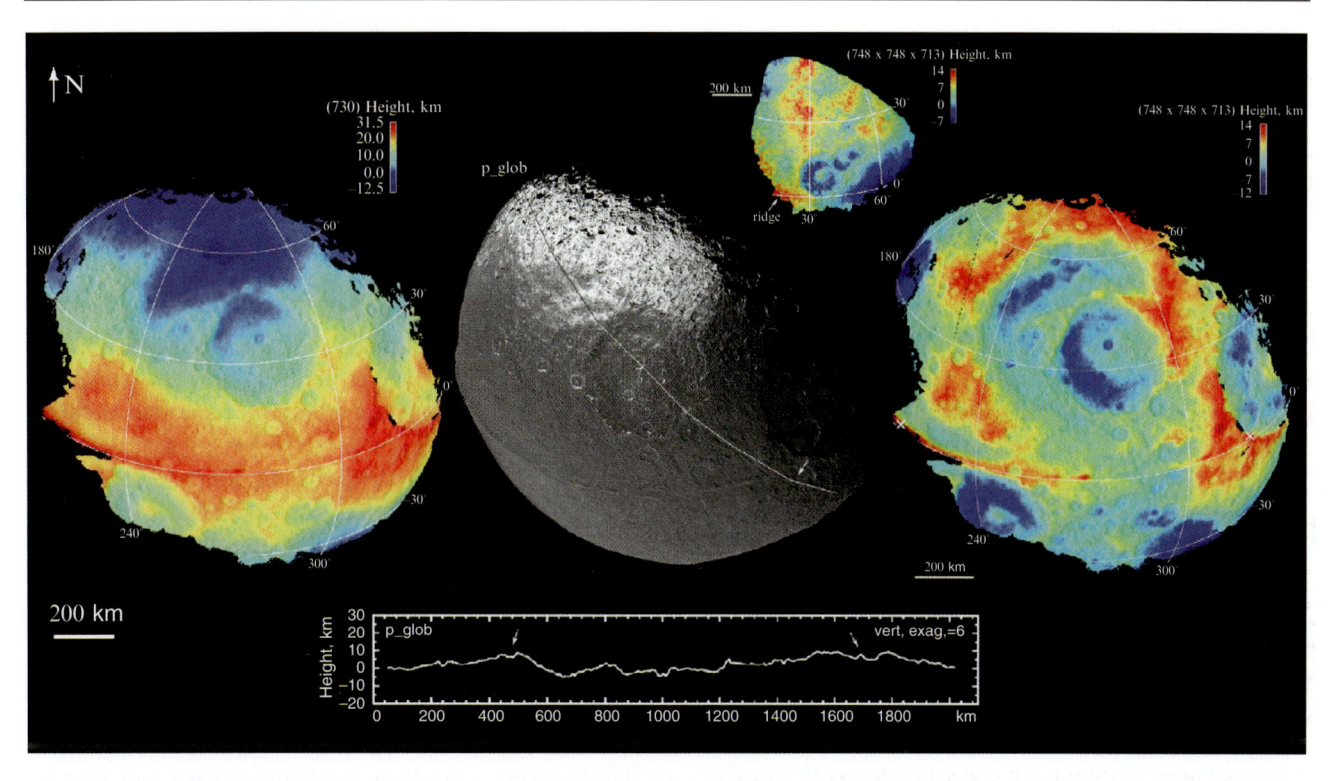

Fig. 20.7 Topography of Iapetus as derived from stereo Cassini imaging. Note the sharply defined equatorial ridge. The DEM at left is referenced to a sphere, the DEM at right to a biaxial ellipsoid (adapted from Giese et al. 2008)

almost certainly have reoriented to the equator anyway. However, as already mentioned above, irrespective of whether a satellite is spinning down or spinning up, equatorial tectonic features can be expected to be oriented north-south, making it difficult to explain the ridge with despinning. One possibility is that despinning stresses combined with lithospheric thickness variations due to convection may have resulted in equatorial extension and diking (Roberts and Nimmo 2009; Melosh and Nimmo 2009). Other proposals suggest that the morphology of the ridge indicates that the surface was warped up by tectonic faulting (Giese et al. 2008), that it is a result of extensional forces acting above an ascending current of solid-state convection from a two-cell convection pattern (Czechowski and Leliwa-Kopystyński 2008), or that it was formed by the impact of an ancient debris disk on Iapetus (Ip 2006). The great apparent elastic thickness suggested by the survival of the ridge may be used to place constraints on the thermal evolution of Iapetus (Castillo-Rogez et al. 2007).

Although the global shape of Iapetus does suggest an earlier, more rapid spin rate (Castillo-Rogez et al. 2007), despinning cannot explain the equatorial ridge (see above): stresses in the north-south direction are always smaller than those in the east-west direction, and will result in N-S oriented features (Melosh 1977). Thus, the origin of the equatorial ridge is currently a mystery.

The deformation of Saturnian satellites varies wildly, from currently active and heavily deformed Enceladus to nearby Mimas, which is tectonically uninteresting. Nor does the degree of deformation correlate in any straightforward way with predicted tidal stresses. However, there are at least two global yet somewhat contradictory aspects, which may be identified. On the one hand, coherent global tectonic patterns are evident on Dione, Tethys and Enceladus and to a lesser extent on Rhea and Iapetus. On the other, several satellites, especially Tethys and Enceladus, show very pronounced spatial variations in the extent of deformation. These two aspects suggest that global sources of stress are common, but also that lateral variations in the mechanical properties of the satellites' ice shells play an important role.

Regarding this second point, localization of deformation (e.g., at the south pole of Enceladus or at Ithaca Chasma) may arise naturally in ice shells (e.g., Sotin et al. 2002; Nimmo 2004b) but is currently poorly understood and makes modeling tectonic deformation much harder. As far as global deformation patterns are concerned, several satellites (Enceladus, Iapetus, perhaps Dione) show patterns with simple symmetries about the rotational and tidal axes. This symmetry may be due either to feature reorientation or the operation of mechanisms (tides, despinning etc.), which have a natural symmetry. On Dione, Rhea and Iapetus at least, diurnal tidal stresses are rather small (Table 20.3), and most of the tectonic features are apparently ancient, which suggests that despinning, volume changes or reorientation are more likely to be responsible for the observed deformation than present-day tidal stresses.

As in the Galilean satellites, extensional features are generally more common than compressional features on the other Saturnian satellites. This observation is readily explained if we assume that the satellites once contained subsurface oceans which progressively froze (e.g., Smith et al. 1981; Nimmo 2004a); alternative explanations are that compressional features such as folds (Prockter and Pappalardo 2000) are less easy to recognize than extensional features, or that larger stresses are required to form compressional features (see Pappalardo and Davis 2007).

20.3.3 Cryovolcanism

Cryovolcanism can be defined as 'the eruption of liquid or vapor phases (with or without entrained solids) of water or other volatiles that would be frozen solid at the normal temperature of the icy satellite's surface' (Geissler 2000). Cryovolcanism therefore includes effusive eruptions of icy slurries and explosive eruptions consisting primarily of vapor (Fagents 2003). Voyager-era data suggested that such activity might be common in the outer solar system, resulting in a flurry of theoretical papers (e.g., Stevenson 1982; Allison and Clifford 1987; Crawford and Stevenson 1988; Kargel 1991; Kargel et al. 1991). Higher-resolution images from Galileo generally revealed tectonic rather than cryovolcanic resurfacing processes, although some features may be due to cryovolcanic activity (e.g., Schenk et al. 2001; Fagents et al. 2000; Miyamoto et al. 2005). In the following, we discuss the current state of knowledge about cryovolcanism in the Saturnian satellites. Enceladus, which is definitely cryovolcanically active (Porco et al. 2006; Dougherty et al. 2006), and Titan, which may be (Lopes et al. 2007; Nelson et al. 2009a,b), are discussed by Spencer et al. (2009), Jaumann et al. (2009) and Sotin et al. (2009).

The principal difference between cryovolcanism and silicate volcanism is that in the former, the melt is denser than the solid (by ~10%). This means that special circumstances must exist for an eruption of non-vapor cryovolcanic materials to occur (though of course on Earth it is not uncommon for denser basaltic lavas to erupt through lighter granitic material). Such circumstances include one or more of the following: (1) the melt includes dissolved volatiles, which exsolve and reduce the bulk melt density (Crawford and Stevenson 1988); (2) the melt is driven upwards by pressure differences caused by tidal effects (Greenberg et al. 1998), surface topography (Showman et al. 2004) or freezing (Fagents 2003; Manga and Wang 2007); (3) compositional differences render the ice shell locally dense or the melt less dense (e.g., by the addition of ammonia) (Kargel 1992, 1995). The latter mechanism can be nothing more than a local effect, since if the bulk ice shell is denser than the underlying material; it will sink if the ice shell is disrupted.

Erupted material consisting primarily of water plus solid ice will simultaneously freeze and boil, the latter process ceasing once a sufficiently thick ice cover is formed (Allison and Clifford 1987). The resulting flow velocity will depend on the viscosity of the ice-water slurry (Kargel et al. 1991) as well as on flow thickness and local slope (Wilson et al. 1997; Miyamoto et al. 2005). If the erupted material is dominated by the vapor phase, the vapor plume will expand outwards (Tian et al. 2007) and escape entirely if the molecular thermal velocities exceed the escape velocity. Any associated solid material will follow ballistic trajectories, ultimately forming deposits centered on the eruption vents (Fagents et al. 2000).

The plume properties measured on Enceladus by Cassini imply considerably smaller velocities for icy dust grains than for vapor. On hypothesis is that the dust grains condense and grow in vertical vents of variable width. Repeated wall collisions and re-acceleration by the gas in the vents lead to effective friction, which explains the observed plume properties (Schmidt et al. 2008).). Most dust grains are growing by direct condensation of the ascending water vapor within the ice channels and contain almost pure water ice; they are free of atomic sodium. In 2008 the CDA instrument while analyzing the composition of E-ring particles discovered a second type of icy dust particles. These grains are rich in sodium salts (0.5–2% in mass) and there origin is explained by the existence of an ocean of liquid water below the surface of Enceladus and which is in direct contact with its rock core (Postberg et al. 2009). This shows that cryovolcanism can provide valuable information about geological conditions of the interior.

The plains of Dione are another candidate for cryovolcanic resurfacing (Fig. 20.8). Several Voyager-era studies concluded that these plains were formed in this way (e.g., Plescia and Boyce 1982; Plescia 1983; Moore 1984). Moore (1984) suggested that several branching broad-rimmed troughs north of the equator at 60°W may have been the source vents. Surprisingly, the plains unit may be topographically higher than the cratered terrain (Schenk and Moore 2007). This observation seems a little difficult to reconcile with the plains being emplaced as a low-viscosity cryovolcanic flow. One possibility is that the plains material was emplaced as a deformable plastic unit (like terrestrial glaciers), but this, or any other, explanation needs both further evidence and thought. The ridges discussed above bound (or at least occur near the edges) the plains unit in the west and southeast and may be related to their formation.

A plains unit on Tethys is located antipodal to the very large Odysseus impact feature. Moore and Ahern (1983) proposed that the plains were formed by cryovolcanism. Alternatively, Moore et al. (2004) considered the possibility that it

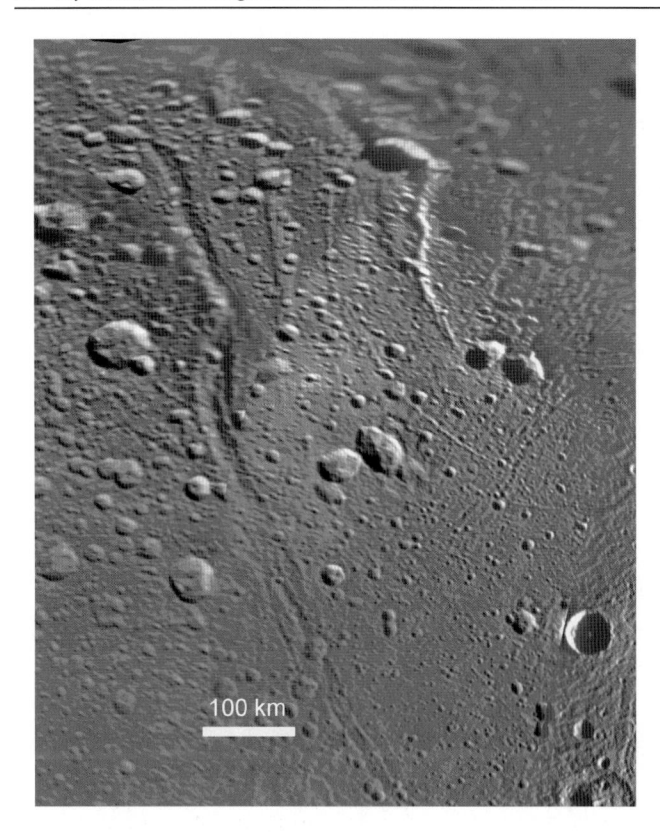

Fig. 20.8 Cassini view of smooth plains on Dione. Older heavily cratered terrains lie to the west. Note the prominent north-south ridge and the subtle linear features to the east. Mosaic resolution is 400 m/pixel

was extreme seismic shaking caused by energy focused there by the Odysseus impact that formed the antipodal plains. If seismic shaking was responsible, there would probably be a significant transition zone between that unit and the adjacent hilly and cratered terrain. Digital terrain models derived from Cassini images indicate that the transition between the cratered terrain and the plains is abrupt and the plains themselves are level, supporting the hypothesis that they are endogenic (Schenk and Moore 2007).

Both Tethys and Dione show evidence for weak plasma tori (Burch et al. 2007) somewhat analogous to the much more marked feature at Enceladus, which is caused by geysers (Porco et al. 2006). It is not yet clear whether surface activity or some other process is generating these tori. Similarly, Clark et al. (2008a) report a possible, but currently unconfirmed, detection of material around Dione, which may also indicate some surface activity. The Cassini magnetometer has also measured small amounts of mass loading near Dione (Burch et al. 2007).

So far, none of the digital terrain models of Rhea developed in the Cassini era shows any regions that would qualify as plains, supporting the perception that Rhea shows no record of cryovolcanic resurfacing. However, unlike Dione, the diffuse nature of the wisps (the 'wispy terrain') on Rhea

has yet to be explained as numerous smaller bright scarps associated with fracture and faulting, and may yet be shown to be an expression of cryo-pyroclastic mantling (Stevenson 1982). Furthermore, Rhea, surprisingly, exhibits evidence for faint particle ring (Jones et al. 2008), the origin of which is unclear but may be related to endogenic or impact processes. On the other hand, Cassini VIMS observations found no evidence of a water vapor column around Rhea (Pitman et al. 2008). On Iapetus, no evidence of cryovolcanism has been detected thus far. If volcanism ever operated on these satellites, it occurred in ancient times and its evidence is now unseen.

20.4 Composition and Alteration of Surface Materials

The composition of the satellites of Saturn is determined by remote spectroscopy, by sampling materials that were sputtered from the satellite surfaces or, in the case of Enceladus, injected into space where they can be directly sampled by spacecraft instruments.

It has long been known that the surfaces of Saturn's major satellites, Mimas, Enceladus, Tethys, Dione, Rhea, Hyperion, Iapetus and Phoebe, are predominantly icy objects (e.g., Fink and Larson 1975; Clark et al. 1984, 1986; Roush et al. 1995; Cruikshank et al. 1998a; Owen et al. 2001; Cruikshank et al. 2005; Clark et al. 2005, 2008a; Filacchione et al. 2007, 2008). While the reflectance spectra of these objects in the visible range indicate that a coloring agent is present on all surfaces except Mimas, Enceladus and Tethys (Buratti 1984). Only Phoebe and the dark hemisphere of Iapetus display near-IR spectra markedly different from very pure water ice (e.g., Cruikshank et al. 2005; Clark et al. 2005).

The major absorptions in icy satellite spectra shown in Figs. 20.9 a–e are due to water ice and can be observed at $1.5\,\mu m$, $2.0\,\mu m$, $3.0\,\mu m$, and $4.5\,\mu m$. Weaker ice absorptions appear at $1.04\,\mu m$ and $1.25\,\mu m$ in the spectra of bright regions where the ice is purer. Trapped CO_2 was first detected in Galileo near-infrared mapping spectrometer (NIMS) spectra of the Galilean satellites (Carlson et al. 1996) and in the Saturnian system on Iapetus (Buratti et al. 2005), Phoebe (Clark et al. 2005), Hyperion (Cruikshank et al. 2007), and Enceladus (Brown et al. 2006). Weak CO_2 absorptions were also detected in VIMS data on Dione, Tethys, Mimas and Rhea (Clark et al. 2008a).

Organic compounds were detected directly in the Enceladus plume by the INMS (Waite et al. 2006) and in trace amounts by the VIMS on Phoebe (Clark et al. 2005), Hyperion (Cruikshank et al. 2007) and Iapetus (Cruikshank et al. 2008), and possibly Enceladus (Brown et al. 2006). While the spectral signatures indicate aromatic hydrocarbon molecules,

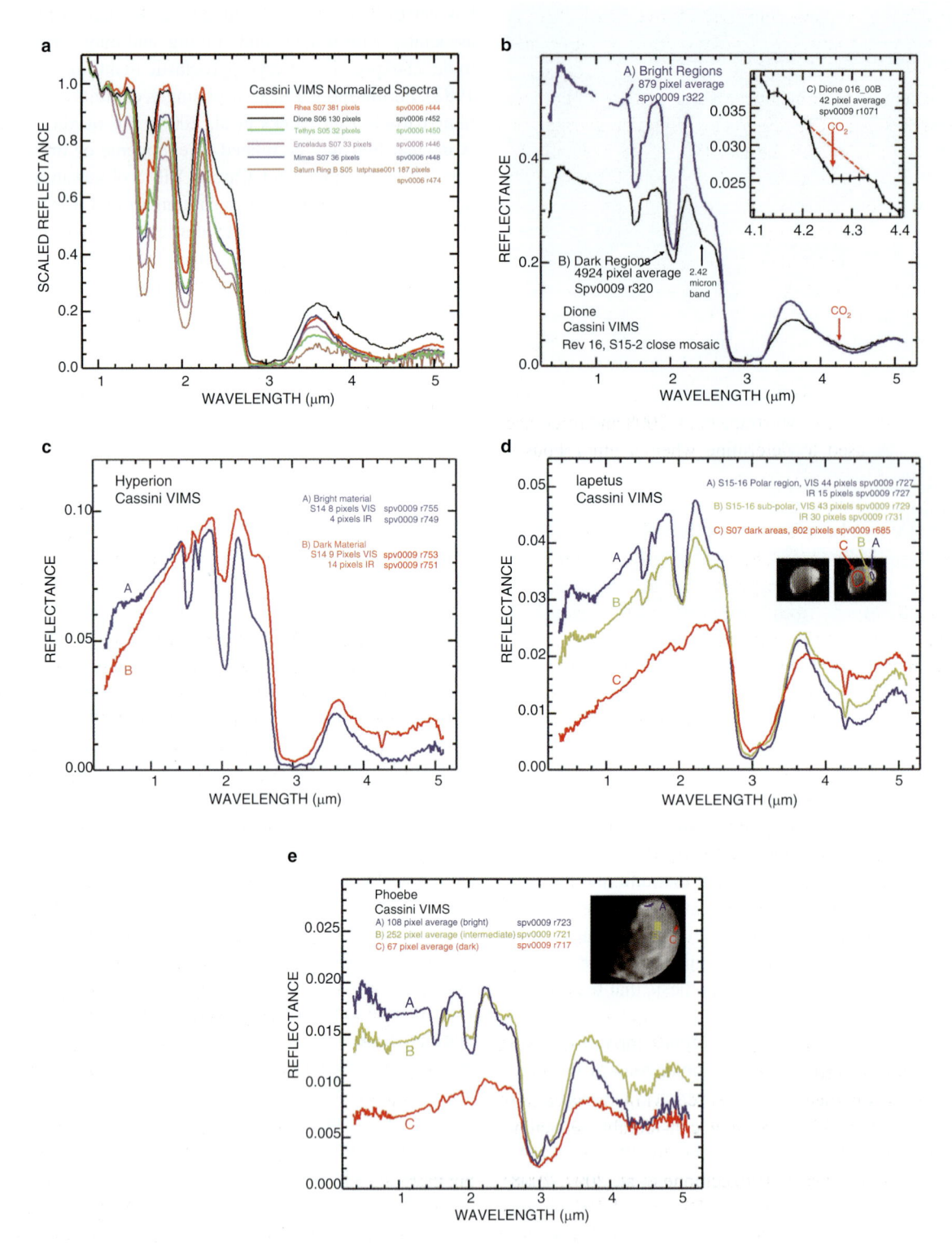

Fig. 20.9 VIMS spectra of the Saturnian icy satellites: (**a**) The major absorptions in icy satellite spectra are due to water ice and appear at 1.5, 2.0, 3.0, and 4.5 μm; (**b**) Representative (A) bright and (B) dark region Dione spectra (the gaps in the bright region spectra (A) near 1 μm are due to sensor saturation); the inset shows an area of high concentration of CO_2 (after Clark et al. 2008a); (**c**) Spectra of Hyperion showing an increase in blue reflectivity and absorptions of water ice and CO_2 (after Clark et al. 2008a); (**d**) Spectra of Iapetus showing a range of blue peak intensities, absorptions of different intensities by water ice and CO_2 absorption bands (after Clark et al. 2008a); (**e**) Spectra of Phoebe showing a range of blue peak intensities and absorptions of different intensities by water ice and CO_2 absorption bands (after Clark et al. 2008a). However, no CO_2 was found on the small satellites Atlas, Pandora, Janus, Epimetheus, Telesto, and Calypso (Buratti et al. 2009a)

exact identifications have remained elusive, partly because the detections have low signal-to-noise ratios. Another spectral feature observed in VIMS data of dark material on Phoebe, Iapetus and Dione appears to be an absorption centered near 2.97μm (Fig. 20.9b,d,e). The detection of this feature is complicated by the fact that VIMS has an order-sorting filter change at this wavelength. Clark et al. (2005) presented evidence that this feature is real and maps to geological patterns on Dione and Iapetus (Clark et al. 2008a, 2009). However, a better understanding of the instrument response is needed to increase confidence in the identification, including confirmation by a different instrument. If correct, the feature strength indicates approximately 1% ammonia in the dark material.

The position shift of amorphous ice absorptions from their crystalline counterparts (Mastrapa et al. 2008 and references therein) can be used to determine whether amorphous or crystalline ice is present on a surface (Newman et al. 2008). Traditionally, amorphous ice displays almost no 1.65 μm feature and the 3.1μm Fresnel peak shifts, broadens and weakens. Newman et al. (2008) used variations in the 3.1μm peak and the 1.65 μm absorption to argue for amorphous ice around the 'tiger stripes' on Enceladus. However, Clark et al. (2009) showed that scattering and sub-micron grain sizes also influence the intensities of the 3.1-μm peak and the 1.65-μm absorption, and can be confused with temperature and crystallinity effects. Further examination of the ice feature positions in spectra of Enceladus and other satellites in the Saturnian system showed that positions and strengths are consistent with those of fine-grained crystalline ice (Clark et al. 2009). Newman et al. (2009) showed that water ice on Dione exists exclusively in crystalline form.

The spectra of the satellites in the visible wavelength range are smooth with no sharp spectral features, but all show an ultraviolet absorber affecting the blue end of the visible-wavelength spectral slope. Visible spectra can be classified by their spectral slope: from bluish Enceladus and Phoebe to redder Iapetus, Hyperion and Epimetheus. In the 1μm to 1.3μm range, the spectra of Enceladus, Tethys, Mimas and Rhea are characterized by negative slope, consistent with a surface mainly dominated by water ice, while the spectra of Iapetus, Hyperion and Phoebe show considerable reddening, pointing out the relevant role played by the darkening materials present on the surface (Filacchione et al. 2007, 2008). In between these two classes are Dione and Epimetheus, which have a relatively flat global spectrum in this range. At wavelengths less than about 0.5 μm, all satellite spectra display a UV absorber, which might be due to charge transfer in oxides such as nano-phase hematite (Clark et al. 2008a), with the apparent strength of the absorber generally indicating the amount of contaminant. Trends in the UV absorber are complicated by ice dust contributions to satellite surfaces transmitted from Enceladus via the E-ring. Satellites close

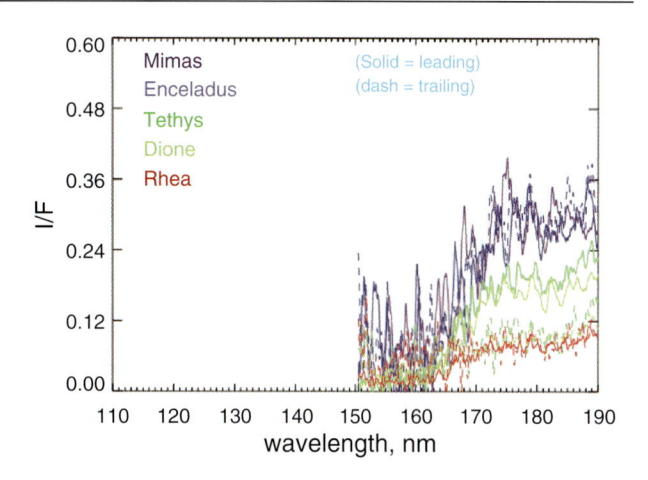

Fig. 20.10 Cassini UVIS disk-integrated spectra of the icy moons, all at phase angles between 10° and 13.6

to the rings, such as Janus, Atlas and Prometheus, show an intermediate red tint between that of the rings and Enceladus. Mimas and Tethys show a red tint between that of Janus, Atlas and Prometheus and that of Enceladus on the other (e.g., Cuzzi et al. 2009; Buratti et al. 2009a).

Disk-integrated ultraviolet observations by the Cassini UVIS (Esposito et al. 2004; Hendrix and Buratti 2009) (Fig. 20.10) show that Tethys and Dione exhibit marked UV leading-trailing differences whereas Mimas, Enceladus and Rhea do not, suggesting compositional variations among the satellites as well as spectrally active components in the ultraviolet. UVIS spectra of all the icy moons of Saturn are dominated by a pronounced water-ice absorption edge at ∼0.165 μm. Its strength varies with the abundance of water ice, and its wavelength depends on grain size. Nearly all UVIS spectra display a small absorption feature at ∼0.185 μm (Hendrix and Hansen 2008b) whose origin is not clear at present. It is possible that this feature is due to water ice as it is seen in published reflectance spectra of water ice (Pipes et al. 1974; Hapke 1981). Published optical constants (Warren 1982) do not exhibit the feature, but the constants have not been well measured in the ∼0.17 μm to 0.4 μm range.

Cassini spectral coverage of the icy moons is complete in the 0.4 μm to 5.2 μm and in the 0.1 μm to 0.19 μm region, but there is a gap in the 0.19 μm to 0.4 μm region (except for the broad band spectral measurements of ISS). This is an important wavelength region for non-water ice absorptions, and the Cassini UVIS data in the far UV provide a critical tie-point between the far UV and the visible band that is helpful in understanding the nature of the non-ice species present in the surfaces of the moons. The icy moons are dark in the far UV compared to their spectra in the Vis-NIR. For instance, the disk-integrated albedo of Enceladus at 0.19 μm is ∼0.4, while at 0.439 μm (Verbiscer et al. 2005) it is ∼1.2. Water ice is thought to have a relatively flat spectral signature in this wavelength range, which suggests that some other species darkens Enceladus in the FUV. Ammonia, which exhibits an abrupt absorption edge at 0.2 μm (Pipes et al. 1974), is a candidate (and ammonia hydrate has been tentatively detected at Enceladus and Tethys in Earth-based spectra (Verbiscer et al. 2005; Verbiscer et al. 2008)), though at this time other species cannot be ruled out other materials.

20.4.1 Dark Material

Cassini (1672) was the first to infer the presence of dark material in the Saturnian system, which was verified by Murphy et al. (1972) and Zellner (1972). Dark material appears throughout the system, most predominantly on Dione, Rhea, Hyperion, Iapetus and Phoebe. The nature of the dark material especially on Iapetus has been studied by numerous authors, often with conflicting conclusions, including Cruikshank et al. (1983), Wilson and Sagan (1995, 1996) Vilas et al. (1996), Jarvis et al. (2000), Owen et al. (2001), Buratti et al. (2002) and Vilas et al. (2004).

The dark material tends to darken and redden the satellites' spectra, particularly in the visible. In dark-region spectra, additional absorptions are seen at 2.42 μm and 4.26 μm, with the latter due to CO_2, as well as stronger water ice absorption near 3μm (Fig. 21.5.1d). The 2.42 μm feature was first tentatively identified as a CN overtone band by Clark et al. (2005) in spectra of Phoebe, but evidence of CN fundamentals in the 4μm to 5μm region has not been confirmed. At present, the 2.42 μm feature is believed to be due to OH or hydrogen (Clark et al. 2008a, b, 2009), but complications with the VIMS calibration have limited any detailed definition of the absorption.

Diverse spectral features of the dark material on Dione match those seen on Phoebe, Iapetus, Hyperion and Epimetheus as well as in the F-ring and the Cassini Division, implying that its composition is the same throughout the Saturnian system (Clark et al. 2008a, 2009). Models of the dark material on Iapetus have used tholins to explain the reddish visible spectral slope (e.g., Cruikshank et al. 2005). However, VIMS data, which better isolate the spectral signature of the dark material from that of icy regions thanks to the high spatial resolution afforded by close fly-bys, show that the dark material has a remarkably linear spectral increase in the near infrared (Clark et al. 2008a, 2009) (Fig. 20.12) not explained by tholins which have a non-linear spectral response in this wavelength region. Tholins also have strong CH absorbers in the 3–4-μm region as well as other marked absorptions at wavelengths longer than 2 μm that are not seen in, for example, spectra of the dark regions of Iapetus. The problem of matching one spectrum might be accomplished with a specific mixture of many compounds, but with spatially resolved spectra on the satellites we now see a large range of mixtures, so models must show consistency across that range of observed mixtures, and many models do not.

The source of the dark material in the Saturnian system is unclear. Several researchers (e.g., Clark et al. 2005) suggest a source outside the system, perhaps the Kuiper belt or comets. Owen et al. (2001) suggested that an impact on Titan created a spray of dark, organic-rich material, part of which was swept up by Iapetus' leading hemisphere, another part being transported to the inner system to accrete onto the surfaces of the inner icy moons. The Dione VIMS spectra provide a key to solving some of the mysteries of the dark material. Clark et al. (2008a) showed that a pattern of bombardment by fine particles below 0.5 μm impacted Dione from the trailing-side direction. Several lines of evidence point to an external origin of the dark material on Dione, including its global spatial pattern, local patterns including crater and cliff walls shielding implantations on slopes facing away from the trailing side, exposing clean ice, and slopes facing the trailing direction that show higher abundances of dark material.

A blue scattering peaks is seen in the spectrum of the dark material and was initially observed in the spectra of Tethys, Dione and Rhea by Noland et al. (1974). Clark et al. (2008a) suggest that the blue peak is caused by particles less than 0.5 μm in diameter embedded in the ice. The VIMS resolved areas on Phoebe, Iapetus and Hyperion also show blue peaks (Fig. 20.9 c–e), indicating that this is a common property of satellites in the Saturnian system (Clark et al. 2008a, 2009). The blue scattering peak with a strong UV/visible absorption was modelled with ice plus 0.2 μm diameter carbon grains (to provide the Rayleigh scattering effect) plus nano-phase hematite (Fe_2O_3) to produce the UV absorber (Fig. 20.11). But carbon is inconsistent with the linear-red slope seen in the purest dark material spectra from 0.4 μm to 2.5 μm. Metallic iron is the only known compound to display this spectral property and nano-phase metallic iron plus nano-phase hemtatite mixed with ice show consistency with the mixture spectra observed on the icy satellites and in Saturn's rings (Clark et al. 2009).

Previous models of the dark material on Iapetus typically used tholins (e.g., Cruikshank et al. 2005). However, VIMS data, which better isolate the spectral signature of the dark material from that of icy regions thanks to the high spatial resolution afforded by close fly-bys, show that the dark material has a remarkably linear spectrum (Clark et al. 2008a, 2009) (Fig. 20.12).

Tholins have strong CH absorbers in the 3–4-μm region as well as other marked absorptions at wavelengths longer than 2 μm that are not seen in, for example, spectra of the dark regions of Iapetus.

Using VIMS data, Clark et al. (2008a) found the general composition of the dark material in the Saturnian system

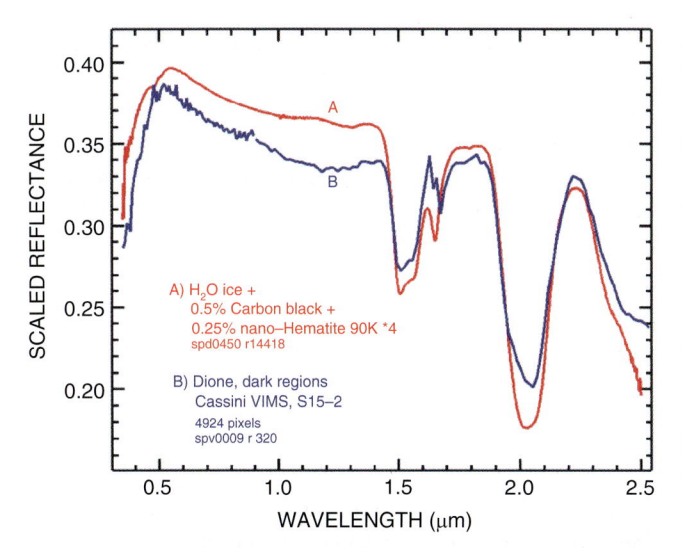

Fig. 20.11 (a) Spectrum of Dione compared to a spectrum of silicon carbide. (b) Spectrum of Dione compared to the spectrum of a laboratory mixture of 99.25 wt% H_2O ice, 0.5 wt% carbon black, and 0.25 wt% nano-crystalline hematite (reagent by Sigma-Aldrich). The carbon black grains are 0.2 μm in diameter. Both the carbon black and hematite grains create Rayleigh scattering in the sample, but at wavelengths shorter than about 0.5 μm, absorption by hematite causes a downturn similar to that seen in the Dione spectra (after Clark et al. 2008a)

likely contains bound water, CO_2 and, tentatively, ammonia. More recently, Clark et al. (2008b, 2009) argued that the signature of iron identifies the major coloring agent on the icy surfaces in the Saturnian system. Metallic iron does not exhibit sharp diagnostic spectral features, but it does have unique spectral properties in that its absorption increases linearly with decreasing wavelength (Clark et al. 2009).

The theory of Clark et al. (2008b, 2009) envisages small-particle meteoroid dust contaminating the system as well as meteoroids with a high metallic-iron content. As one moves closer to the rings, the dark material diminishes (e.g., Mimas and Enceladus have little dark material). However, the competing process of accretion by Saturn's E-ring, which originates on Enceladus, also contributes to the greater abundance of high-albedo material on Mimas and Enceladus (Buratti et al. 1990, Verbiscer et al. 2007), The main rings do not show any strong signature of metallic iron particles, but the Cassini Division displays spectra that closely match those of Iapetus in Iapetus' transition zone from dark to light material, implying mixtures of the same materials. The ring spectra are much redder in the 0.3 μm to 0.6 μm region than the spectra of dark material on Phoebe, Iapetus and Hyperion. Cosmic rays impacting the huge surface area of the rings break up water molecules, thus creating an atmosphere of hydrogen and oxygen. The hydrogen escapes, leaving an atmosphere of oxygen around the rings (Johnson et al. 2006). Oxygen ions

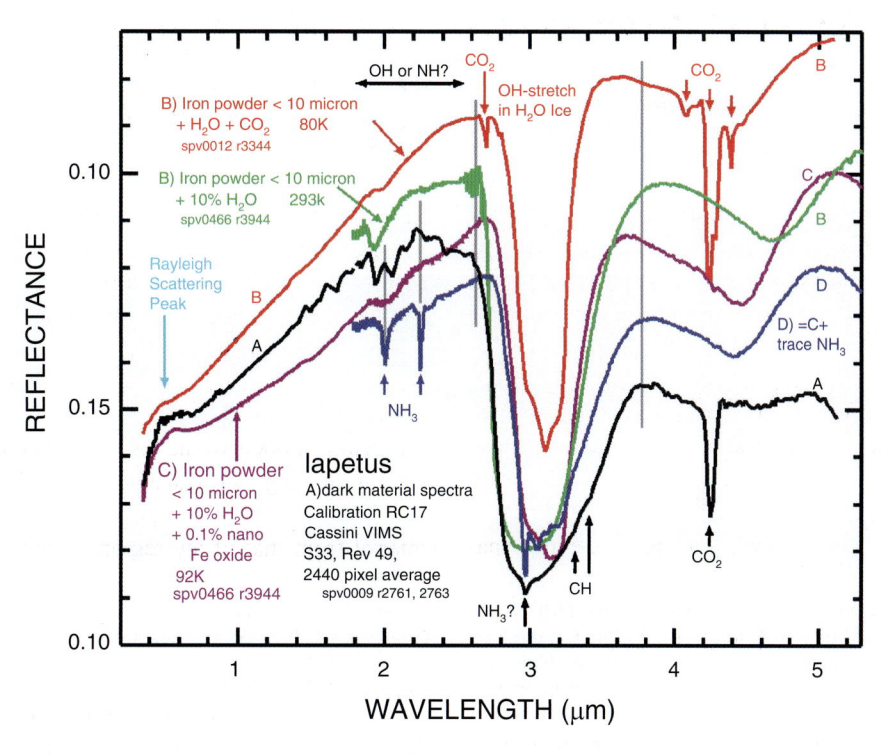

Fig. 20.12 Spectrum of the dark material of Iapetus compared to laboratory spectra of mixtures of iron powder, H_2O, CO_2 and NH_3. The small Rayleigh scattering peak and UV absorber can be explained by traces of iron oxide nano-powder. The 3-micron absorption is best explained by a combination of bound water, ice and trace ammonia. Finer-grained iron would require higher abundances of water, ammonia and carbon dioxide to match the spectrum of Iapetus

are very reactive, and when they encounter iron particles, they oxidize them and turn them into nano-phase hematite (Fe_2O_3). The spectra of nano-phase hematite mixed with ice and dark particles closely match the UV red slope seen in the spectra of Dione (Clark et al. 2008a) (Fig. 20.11). To explain the dark material in the Saturnian system, Clark et al. (2008b, 2009) suggest that oxidized sub-micron iron particles from meteorites, mixed with ice, might explain all the colors and spectral shapes observed in the system. A small amount of carbon might also be identified in the mixture, which would change the slope of iron and further lower the albedo. Sub-micron particles also explain the origin of the observed Rayleigh scattering. Trace amounts of CO_2, ammonia and organics would not significantly influence the visible and UV spectra down to 0.3 µm. Finally, the inner satellites may be exposed to as much iron as the outer satellites, but E-ring ice particles mix with and cover up the iron particles. Species found by INMS, CAPS and MIMI in the magnetosphere are candidates for surface components as the magnetospheric material originates either directly from satellite surfaces or from ring particles. Compositional analyses of Enceladus' plume are described in Spencer et al. (2009).

The idea of iron in the Saturn system is supported by the Cassini CDA instrument, which has discovered small (<20 nm) particles escaping the Saturnian system, which are composed predominantly of oxygen, silicon and iron, with some evidence of H_2O ice, ammonium, and possibly carbon (Kempf et al. 2005). The escaping iron is not in a silicate and is not metallic; iron oxide is most consistent with the data (Kempf et al. 2008, personal communication). Iron has also been detected in the E-ring by CDA (Kempf et al. 2008). The origin of the iron may be meteoric dust falling into the Saturnian system, and may indicate that space weathering processes operate there as in other parts of the solar system (e.g., Chapman 2004).

20.4.2 Iapetus' Hemispheric Dichotomy

Iapetus has intrigued planetary scientists for centuries, primarily due to its striking hemispheric albedo dichotomy. The leading hemisphere (centered exactly on the apex of motion at 90 °W) is very dark, reflecting just ~5% of the visible light that hits it (at 3° phase angle), while the trailing hemisphere (centered on 270 °W), is relatively bright and has a visible albedo of ~50% to ~60% (Fig. 20.13). A long-standing question has been how the global albedo dichotomy of

Iapetus originated. Has the leading hemisphere's dark terrain been created by exogenic processes, as most scientists hypothesized (e.g., Cook and Franklin 1970; Soter 1974; Cruikshank et al. 1983; Squyres and Sagan 1983; Bell et al. 1985; Tabak and Young 1989; Matthews 1992; Buratti and Mosher 1995; Wilson and Sagan 1996; Denk and Neukum 2000; Owen et al. 2001; Buratti et al. 2002, 2005a), or

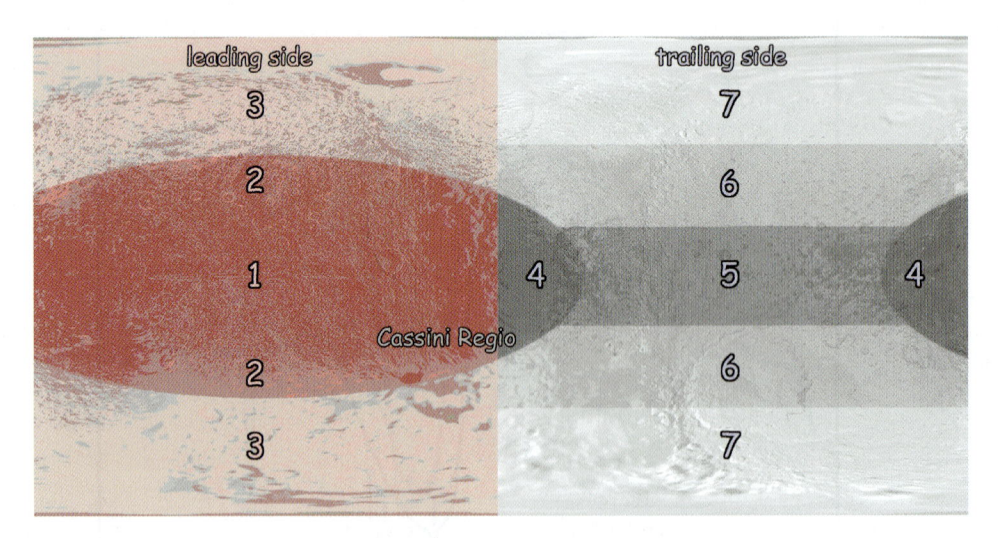

Fig. 20.13 Patterns of the global dichotomies on Iapetus (Denk et al. 2009). 1 ... Uniformly dark, reddish terrain 'deep within' Cassini Regio; 2 ... Dark, reddish terrain similar to (1) but showing bright slopes (esp. crater walls) facing poleward; 3 ... Bright, reddish terrain on the leading side, with dark spots located on equatorward-facing slopes; 4 ... Dark terrain, but less reddish (at visible to IR wavelengths) than (1); 5 ... Bright equatorial trailing-side terrain containing numerous dark spots colored similarly to (4); 6 ... Low- to mid-latitude bright terrain with multiple dark equatorward-facing slopes; 7 ... Bright 'polar caps' on the trailing side, no dark spots, flattest spectra (near UV to near IR) of all Iapetus terrains. These might be the sinks for the water ice in the thermal segregation process. The global color dichotomy separates areas (1)–(3) (leading side; redder color in visible to IR wavelengths) from (4) to (7) (trailing side; less reddish). The global brightness dichotomy separates areas (1), (2) and (4) (mainly dark terrain; named Cassini Regio) from (3), (5), (6) and (7) (mainly bright terrain, named Roncevaux Terra and Saragossa Terra). Latitudinal inspection: the differences between areas (1), (2) and (3) result from the thermal segregation of water ice, as do the differences between areas (5), (6) and (7)

did geological activity emplace dark material from within Iapetus (Smith et al. 1981, 1982)? Voyager images of dark-floored craters within the bright terrain point to an endogenic source; they also suggest that the bright-dark boundary is too irregular to be consistent with infalling dust (Smith et al. 1981, 1982; Denk and Spencer 2008). However, albedo patterns observed by Cassini cameras in late 2004 suggest external material emplacement (e.g., dark material on ram-facing crater walls at high latitudes) (Porco et al. 2005a), but this was later found to be due to photometric effects. The initial theory of an exogenically-created dark pattern (Cook and Franklin 1970) suggested that pre-existing dark material was uncovered by meteoritic bombardment; this idea was extrapolated by Wilson and Sagan (1996). Researchers theorized (Soter 1974) that the dark material is exogenically emplaced on Iapetus' leading hemisphere as material is lost from the moon Phoebe (Burns et al. 1979). Retrograde Phoebe dust from a distance of 215 Saturn radii would travel inward from the effects of Poynting-Robertson drag and impact the leading hemisphere of Iapetus orbiting at 59 Saturn radii. However, Phoebe is spectrally neutral at visible wavelengths, while the Iapetus dark material is reddish (Cruikshank et al. 1983; Squyres et al. 1984). If the material does come from Phoebe, then some sort of chemistry or impact volatilization must occur to change the color and darken the material (Cruikshank et al. 1983; Buratti and Mosher 1995). Another possibility is that the exogenic source of the dark material is Hyperion (Matthews 1992; Marchi et al. 2002), Titan (Wilson and Sagan 1995; Owen et al. 2001), Iapetus itself (Tabak and Young 1989) or the debris remnants of a collision between a former outer moon and a planetocentric object (Denk and Neukum 2000). Matthews (1992) suggested that the hypothetical impact that disrupted Hyperion created a debris cloud that subsequently hit Iapetus. Both Hyperion and Titan dark materials are spectrally reddish (Thomas and Veverka 1985; McDonald et al. 1994; Owen et al. 2001), though not as dark as those of Iapetus. Buratti et al. (2002, 2005a) suggested that both Hyperion's and Iapetus' leading hemispheres are impacted by dark, reddish dust from recently discovered retrograde satellites exterior to Phoebe. Clark et al. (2008a) offer a solution for the color discrepancy. Red, fine-grained dust (less than one-half micron in diameter) when mixed in small amounts with water ice produces a Rayleigh scattering effect, enhancing the blue response. This blue enhancement appears just enough on Phoebe to create an approximately neutral spectral response. An examination of the spectral signatures of Dione, Hyperion, Iapetus and Phoebe by Clark et al. (2008a) showed a continuous mixing space on all these satellites, demonstrating variable mixtures of fine-grained dark material in the ice. Thus, the colors observed are simply variable amounts of dark material, with the redder slopes indicating greater abundance.

Ground-based radar observations at 13 cm (Black et al. 2004) and Cassini RADAR data at 2.2 cm (Ostro et al. 2006; 2009) indicate that the dark terrain on Iapetus must be quite thin (one to several decimeters); an ammonia-water ice mixture may be present at a depth of several decimeters below the surface on both the leading and trailing hemispheres of Iapetus. The RADAR results appear to rule out any theories of a thick dark material layer (Matthews 1992; Wilson and Sagan 1996) and are consistent with the spectral coating indicated by VIMS data as described by Clark et al. (2008a). Cassini radio science results (Rappaport et al. 2005) indicate a bulk density for Iapetus of 1.1 g/cm^3, from which it can be inferred that the moon is composed primarily of water ice.

Because the large craters within the dark terrain appear evenly colored by the dark material in Voyager data, no craters significantly break up the dark material to expose bright underlying terrain (Denk et al. 2000, 2009). This suggests that the emplacement of dark material by whatever mechanism is relatively new or ongoing.

With the arrival of Cassini at Saturn and the first Iapetus flyby (Dec. 2004), the idea of thermal segregation developed (Spencer et al. 2005; Hendrix and Hansen 2008a; Spencer and Denk 2009). This idea was originally mentioned as one option by Mendis and Axford (1974). The September 2007 close flyby of Iapetus (Fig. 20.14) permitted testing of this hypothesis, resulting in the following theory. A small amount of exogenic dust from an unknown source created a 'global color dichotomy': the leading side became slightly redder and darker than the trailing with fuzzy boundaries close to the sub-Saturn and anti-Saturn meridians (Denk et al. 2009). A runaway thermal segregation process might have been initiated primarily at the low and middle latitudes of the leading side (Spencer and Denk 2009) as well as in some local areas at low latitudes on the trailing side, especially on equatorward-facing slopes (Denk et al. 2009). Indeed, poleward-facing slopes even within the dark terrain (at middle latitudes) are bright. On these slopes the temperature never raises enough to allow the thermal process to act faster than the destruction process by micrometeorite gardening, which transports bright sub-surface material to the surface. In those locations that are now dark, the originally bright surface material becomes warm enough for volatiles (mainly water ice) on the top part of the surface to grow unstable and migrate towards cold traps, leaving behind residual dark material that was originally intimately mixed with the water ice as a minor constituent. The slow 79.3 day rotation is a crucial boundary condition for this process to work in the Saturnian system. ISS images suggest that significant cold traps might be located at high latitudes on the trailing side, where lower-resolution images show bright white polar caps (Denk et al. 2009).

Compositional data from the UVIS and the VIMS (Hendrix and Hansen, 2009a, b; Clark et al. 2009) indicate variable amounts of ice and volatiles in the dark material. The water-ice absorption edge in the UV at 0.165 μm was shown to increase with latitude, consistent with decreasing temperatures and increasing numbers of visibly bright regions (Hendrix and Hansen 2009a). This is consistent with thermal segregation as a possible process explaining the overall appearance of Iapetus and, more particularly, the observed correlations of the dark/bright boundary with geographical location and topography (Fig. 20.15) and the small-scale dramatic brightness variations (Fig. 20.15).

The 3μm absorption observed in VIMS data (Clark et al. 2009) might be due to ice, ammonia or bound water. As the 3μm absorption is not been well matched by those due to ice, bound water and ammonia, a combination seems to be indicated, but a model showing a good match has not yet been developed (Clark et al. 2009). The amount of ice contained in the dark materials from VIMS data could be on the order of 10%, but is estimated to be <5% from UVIS data. The presence of H_2O- or NH_3-ice in the dark material is inconsistent with sublimation rates of ∼1 μm/10 years. Therefore, further spectral observations of Iapetus and laboratory investigations are needed to finally solve the problem.

20.4.3 Surface Alterations and Photometry

Modifications to the surfaces of the icy Saturnian satellites include those due to charged-particle bombardment and sputtering, E-ring grain bombardment or coating, thermal processing, UV photolysis, and micrometeoroid bombardment, many of which can lead to leading-trailing asymmetries. Photometry is a particularly useful observation method for detecting surface alterations, and it was used in the first

Fig. 20.14 Cassini ISS images from the September 2007 flyby, demonstrating the large- and small-scale albedo effects of thermal segregation. (*Left*) The central longitude of the trailing hemisphere is 24° to the left of the mosaic's center. (*Right*) The mosaic consists of two image footprints across the surface of Iapetus. The view is centered on terrain near 42° southern latitude and 209.3° western longitude, on the anti-Saturn facing hemisphere. Image scale is approximately 32 m/pixel; the crater in the center is about 8 km across

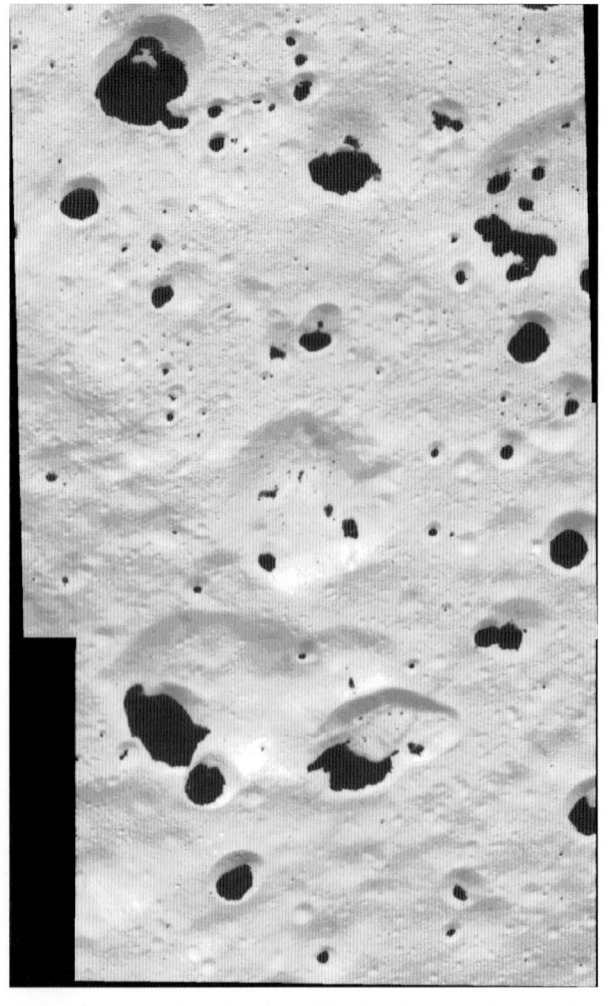

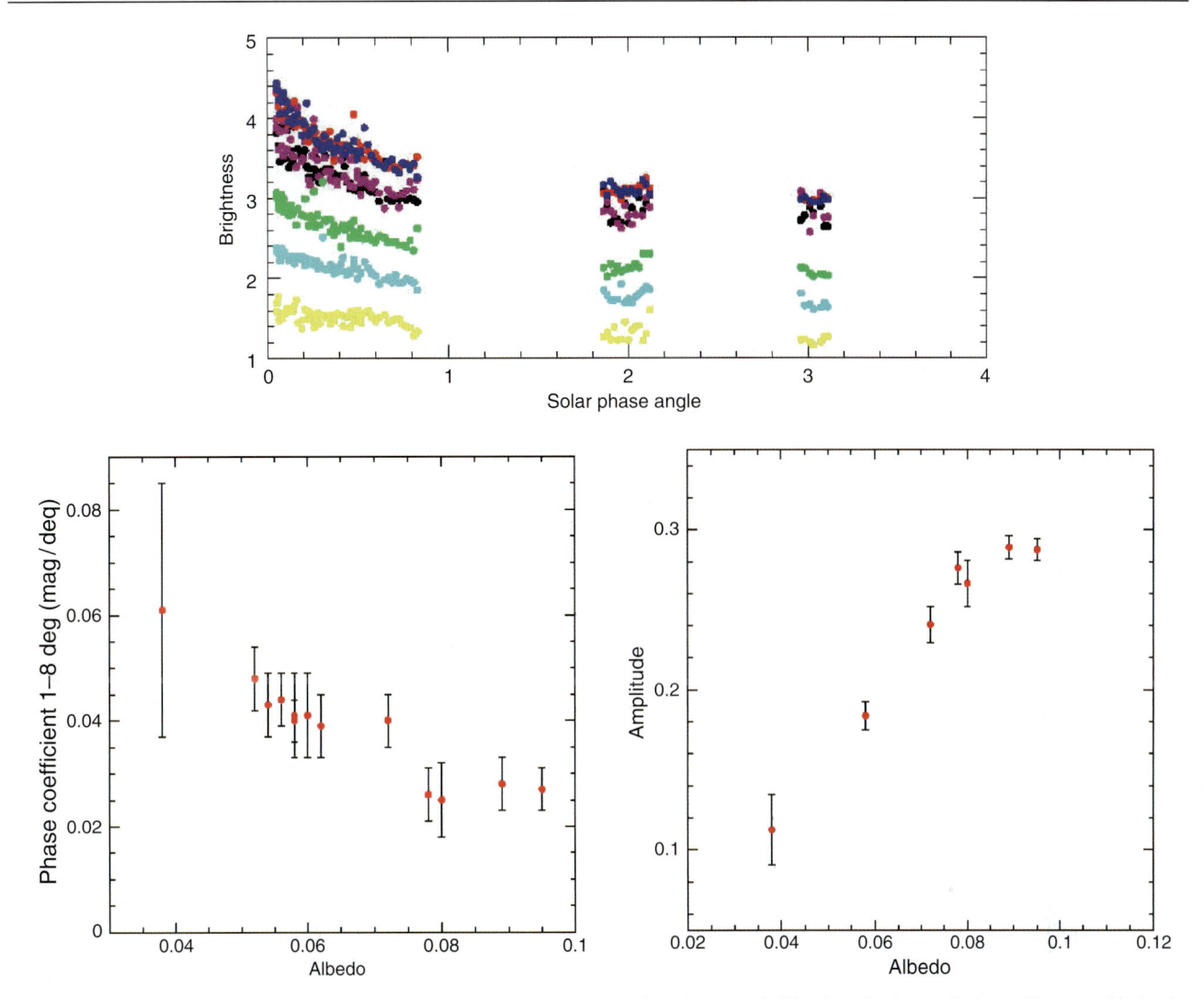

Fig. 20.15 (*Above*) Cassini visual infrared mapping spectrometer (VIMS) disk-integrated observations of the opposition surge on Iapetus between 0.35 and 3.6 μm. (*Lower left*) The slope of the solar phase curve (the 'phase coefficient') between 1° and 8° decreases as the albedo increases. This effect is expected for shadow hiding, since bright surfaces have partly illuminated primary shadows. (*Lower right*) On the other hand, the amplitude of the 'spike' under 1° (shown as a fractional increase above the value of the phase curve at 1°) increases with the albedo, suggesting that it is a multiple scattering effect, such as coherent backscatter

detection of large-scale surface alterations that are the result of exogenic processes.

All the major Saturnian satellites (with the exception of Hyperion) are synchronously locked, so that one hemisphere faces Saturn at all times. The Saturn-facing hemisphere is approximately centered on 0°W longitude, with the leading hemisphere (facing the direction of orbital motion) centered on 90°W and the trailing hemisphere on 270°W. It is expected that gardening of the regolith by micrometeorites is especially important on the leading hemisphere, which 'scoops up' incoming dust and exogenic material throughout an orbit. In contrast, charged-particle bombardment affects primarily (but not exclusively) the trailing hemisphere. Because Saturn's magnetosphere co-rotates at a rate faster than the orbital speed of these moons, the satellites' trailing hemispheres are preferentially affected by magnetospheric particle bombardment (cold ions and keV electrons). Meteoritic bombardment of icy bodies acts in two major ways to alter the optical characteristics of the surface. First, the impacts excavate and expose fresh material. Second, impact volatilization and the subsequent escape of volatiles results in a lag deposit enriched in opaque, dark materials. The relative importance of the two processes depends on the flux, size distribution and composition of the impacting particles as well as on the composition, surface temperature and mass of the satellite (Buratti et al. 1990). Impact gardening and excavation tend to brighten an icy surface at visible wavelength (Smith et al. 1981), while impact volatilization tends to darken it. Buratti (1988) suggested that E-ring coating dominates any micrometeorite gardening.

Coating by E-ring grains affects the reflectance spectra of the satellites and could mask the signatures of other species. The E-ring is a region of fine-grained ($\sim$1 μm) particles primarily consisting of H_2O ice that extends from the orbit of Mimas ($\sim$3 Saturn radii) outwards past the orbit of Rhea ($\sim$15 Saturn radii). The ultimate source of the E-ring is Enceladus and its geyser-like plumes (e.g., Porco et al. 2006; Spahn et al. 2006). Compositional measurements of dust particles in the ring directly reveal the composition of their sources. E-ring particles are of two types: (1) H_2O ice and (2) organics and/or silicates in icy particles (Postberg et al. 2008). E-ring particles coat Rhea and other satellites closer to Saturn like Mimas, Enceladus, Tethys and Dione as well as contributing to the G-, F-, and A-rings at least (Buratti et al. 1990; Hamilton and Burns 1994; Buratti et al. 1998; Verbiscer et al. 2007). The geometric albedos of these embedded icy moons were measured in the visible wavelength by the Hubble space telescope (Verbiscer et al. 2007). Their brightness changes with their distance from the E-ring, which is consistent with the assumption that their albedo results from E-ring coating. Hamilton and Burns (1994) predicted that the relative velocities between the E-ring particles and the icy satellites might explain the large-scale longitudinal albedo patterns on the icy satellites. Because of their higher relative velocity, the satellites exterior to the densest point in the E-ring have brighter leading hemispheres, while those interior to it have brighter trailing hemispheres because of the higher velocity of the particles. Indeed, Mimas and Enceladus are both slightly darker on their leading than on their trailing hemispheres, while the leading hemispheres of Tethys, Dione and Rhea are brighter than their trailing hemispheres (Buratti et al. 1998). The visible orbital phase curve of Enceladus shows that the leading hemisphere is $\sim$1.2 times darker than its trailing hemisphere. At visible wavelengths, the leading hemisphere of Tethys is $\sim$1.1 times brighter than its trailing hemisphere, Dione's leading hemisphere is up to $\sim$1.8 times brighter and Rhea's leading hemisphere is $\sim$1.2 times brighter than the trailing hemisphere (Buratti and Veverka 1984).

The bombardment of icy surfaces by charged particles causes both structural and chemical changes in the ice (Johnson et al. 2004). Effects include the implantation of magnetospheric ions, sputtering and grain size alteration. Low-energy electrons can cause chemical reactions on the surface, while high-energy electrons and ions tend to cause radiation effects at depth. The latter can result in bulk chemical changes in composition as well as structural damage. A primary effect of energetic particle bombardment on ice is to cause defects in it (Johnson 1997; Johnson and Quickenden 1997; Johnson et al. 1998). Extended defects, in the form of voids and bubbles, affect the light-scattering properties of the surface. They also act as reaction and trapping sites for gases, which can produce spectral absorption features. Incident ions deposit energy over a small region around their path through the ice, causing local damage. Depending on excitation density and cooling rate, this can result in either local crystallization or amorphization (Strazzulla 1998).

Energetic ions and electrons have been seen to brighten laboratory surfaces in the visible by producing voids and bubbles (Sack et al. 1992). Depending on the size and density of these features, a clear surface can become brighter due to enhanced scattering in the visible. The plasma-induced brightening competes with annealing, which operates efficiently in regions where the ice temperature exceeds $\sim$100 K and can make equatorial regions less light-scattering despite the plasma bombardment.

A significant effect of the ion bombardment of ice is the creation of new species through radiolysis. Bombarding particles can decompose H_2O ice, creating molecular hydrogen and oxygen. Mini-atmospheres of trapped volatiles have been suggested to form in voids in the ice (Johnson and Jesser 1997). As a result, species such as O_3, H_2O_2, O_2 and oxygen-rich trace species have been detected as gases trapped in the regoliths of the icy Galilean satellites (Nelson et al. 1987; Noll et al. 1996; Hendrix et al. 1998, 1999; Spencer et al. 1995; Carlson et al. 1996, 1999; Hibbitts et al. 2000, 2003). On these satellites, the hydrogen and oxygen produced by radiation decomposition form tenuous atmospheres; in contrast, atmospheres have not been detected on satellites in the Saturnian system (with the exception of the gaseous plume of Enceladus). There are several possible reasons for this. Lower gravities might support the loss of volatiles; temperatures are lower, which might slow down the breakup of H_2O and the loss of H; the particle environment is different (discussed below); the electron density and temperature is not effective in exciting and 'lighting up' any gases around the moons; and sputtering is not as intense as in the Jupiter system due to differences in the net charge and energy of the particle environment. On Saturn, the primary surface alterations found thus far are due to the E-ring, dust (e.g., on Dione) and thermal processing (especially apparent on Iapetus, as previously discussed).

Neutrals, rather than ions as on Jupiter, dominate the Saturnian system. The energetic heavy ions ($\sim$10 keV to MeV) that are critical at Europa and Ganymede are lost by charge exchange to the neutral cloud of the Saturnian system as they diffuse inwards from beyond $\sim$10 Saturn radii (Paranicas et al. 2008). However, although the energy fluxes due to trapped plasma and the solar UV are relatively low on the icy Saturnian satellites (Paranicas et al. 2008), radiation processing appears to be occurring. Radiation fluxes are known to cause decomposition in ice, possibly producing ozone in the icy surfaces of Dione and Rhea (Noll et al. 1997) and O_2^+ in Saturn's plasma (Martens et al. 2007). In addition, the role of low energy ions has been recently re-examined using Cassini CAPS data (Sittler et al. 2008), indicating that sputtering by this component is not negligible (Johnson et al. 2008). Spencer (1998) searched for but did not find O_2 inclusions on Rhea and Dione. Such inclusions have been seen on Ganymede in the low latitudes of the trailing hemisphere, and it was suggested that they might be related to charged-particle bombardment or UV photolysis. The lack of O_2 on Rhea and Dione suggests that low temperatures preclude the formation of O_2 bubbles in the ice. Hydrogen peroxide (H_2O_2) has been tentatively VIMS data (Newman et al. 2007) of the south pole tiger-stripes region of Enceladus, although it is not clear whether exogenic processing is needed to create peroxide.

It is important to distinguish between hemispheric dichotomies in color, albedo, texture and macroscopic structure because each exogenic alteration mechanism leaves a different signature on the physical and optical properties of the surface. Photometric methods have been employed over the past two decades to characterize the roughness, compaction state, particle size and phase function, and single scattering albedo of planetary and satellite surfaces, including the Saturnian satellites (Buratti 1985; Verbiscer and Veverka 1989, 1992, 1994; Domingue et al. 1995). An indication of the scattering properties of the satellites' surfaces can be obtained by extracting photometric scans across their disks at small solar phase angles. Dark surfaces such as that of the Moon, in which single scattering is dominant, are not significantly limb-darkened, whereas bright surfaces in which multiple scattering is important could be expected to show limb darkening. Physical properties such as roughness, compaction state and particle size give clues as to the past evolution of the satellites' surfaces and reveal the relative importance of exogenic alteration processes. Accurate photometric modeling is also necessary to define the intrinsic changes in albedo and color on a planetary surface (Noland et al. 1974); much, and in many cases most of the change in intensity is due to variations in the radiance viewing geometry (incident, emission, and solar phase angle) and is not intrinsic to the surface.

Planetary surface scattering models have been developed (e.g., Horak 1950; Goguen 1981; Lumme and Bowell 1981a, b; Hapke 1981, 1984, 1986, 1990; Shkuratov et al. 1999, 2005) which express the radiation reflected from the surface in terms of the following physical parameters: single scattering albedo, single particle phase function, the compaction properties of the optically active portion of the regolith, and the scale and extent of macroscopically rough surface features. Observations at small solar phase angles (0–10°) are particularly important for understanding the compaction state of the upper regolith: this is the region of the well-known opposition effect, in which the rapid disappearance of shadowing among surficial particles causes a nonlinear surge in brightness as the object becomes fully illuminated. At the smallest phase angles (<1°), coherent backscatter dominates the surge (Hapke 1990). Larger phase angles (>40°) are necessary for investigating the macroscopically rough nature of surfaces, ranging in size from clumps of several particles to mountains, craters and ridges. Such features cast shadows and alter the local incidence and emission angles. Two formalisms have been developed to describe roughness for planetary surfaces: a mean-slope model in which the surface is covered by a Gaussian distribution of slope angles with a mean angle of θ (Hapke 1984; Helfenstein and Shepard 1998), and a crater model in which the surface is covered by craters with a defined depth-to-radius ratio (Buratti and Veverka 1985). The single particle phase function, which expresses the directional scattering properties of a planetary surface, is an indicator of the physical character of individual particles in the upper regolith, including their size, size distribution, shape, and optical constants. Small or transparent particles tend to be more isotropically scattering because photons survive to be multiply scattered; forward scattering can exist if photons exit in the direction away from the observer. The single particle phase function is usually described by a 1 or 2-term Henyey-Greenstein phase function defined by the asymmetry parameter g, where g $=$ 1 is purely forward-scattering, g $=$ -1 is purely backscattering, and g $=$ 0 is isotropic. The opposition surge of airless surfaces has been described by parameters that define the compaction state of the surface and the degree of coherent backscatter of multiply scattered photons (Irvine 1966; Hapke 1986, 1990). One final physical photometric parameter is the single scattering albedo ($\tilde{\omega}$), which is the probability that a photon will be scattered into 4π steradians after one scattering.

Three other important physical parameters are the geometrical albedo (p), the phase integral (q) and the Bond albedo (A_B). The geometrical albedo is the flux received from a planet at a solar phase angle of 0° compared with that of a perfectly diffusing disk of the same size, also at 0°. The phase integral (q) is the flux, normalized to unity at a solar phase angle of 0°, integrated over all solar phase angles. The Bond albedo is $p \cdot q$, and when integrated over all wavelengths it yields the bolometric Bond albedo, which is a measure of the energy balance on the planet or satellite. Except for the bolometric Bond albedo, all these parameters are wavelength-dependent.

Buratti (1984) and Buratti and Veverka (1984) performed photometric analyses of visible-wavelength Voyager data of the icy Saturnian moons based on Voyager disk-resolved data sets, and disk-integrated Voyager observations at large phase angles to supplement existing ground-based observations at smaller phase angles. These studies investigated limb darkening with implications for multiple scattering (significant limb darkening suggests that multiple scattering is important). Physical photometric modeling of Saturnian moons based on Voyager and ground-based data was done by Buratti (1985), Verbiscer and Veverka (1989, 1992, 1994), Domingue et al. (1995) and Simonelli et al. (1999). In general, the surfaces of the Saturnian satellites are backscattering and exhibit roughness comparable to that of the Moon, indicating that impact processes produce similar morphological effects on rocky and icy bodies at low temperatures. The photometric and physical properties of the satellites determined so far from both Voyager and Cassini data are summarized in Table 20.4.

Opposition surge data from 0.2 to 5.1 μm were obtained for Enceladus, Tethys, Dione, Rhea and Iapetus. Figure 20.16 shows the opposition curve of Iapetus in between 0.35 μm and 3.6 μm (Buratti et al. 2009b; Hendrix and Buratti 2009). The curve is divided into two components: at phase angles greater than 1°, brightness increases linearly with decreasing phase angle, while at phase angles smaller than one degree the increase is exponential. Moreover, the albedo-dependence of the two types of surge is different. For the first type, the magnitude of the surge decreases as the albedo increases, while for the second type of surge the reverse is true. This result suggests that different phenomena are responsible for the two types of surge. At angles greater than 1°, the surge is caused by mutual shadowing, in which shadows cast among particles of the regolith rapidly disappear as the face of the satellite becomes fully illuminated to the observer. Since multiply scattered photons, which partly illuminate primary shadows, become more numerous as the albedo increases, this effect becomes less pronounced for bright surfaces. On the other hand, the huge increase at very small solar phase angles can be explained by coherent backscatter, a phenomenon in which photons following

Table 20.4 Photometric and physical parameters of the icy Saturnian satellites (references: (1) Buratti and Veverka 1984; (2) Buratti 1985; (3) Verbiscer and Veverka 1994; (4) Verbiscer and Veverka 1989; (5) Verbiscer and Veverka 1992; (6) Simonelli et al. 1999; (7) Buratti et al. 2008; (8) Verbiscer et al. 2005)

Parameter	Mimas	Enceladus	Tethys	Dione	Rhea	Phoebe
Mean slope angle θ (°)	30 (2) 0 ± 1 (5)	6 ± 1 (3)			13 ± 5 (4)	31 ± 4 (6) 3 ± 3 (7)
Asymmetry parameter g	−0.3 ± 0.05 (2) 0.213 ± 0.005 (5)	−0.35 ± 0.03 (2) 0.399 ± 0.005 (3)			−0.287 ± 0.007 (4)	−0.24(6) −0.3 ± 0.1 (plus a forward component) (7)
Single scattering albedo $\acute{\Omega}$	0.93 ± 0.03 (2) 0.951 ± 0.002 (5)	0.99 ± 0.02 (2) 0.998 ± 0.001 (3)			0.861 ± 0.008 (4)	0.068 (6) 0.07 ± 0.01 (7)
Phase integral q	0.82 ± 0.05 (1) 0.78 ± 0.05 (5)	0.86 ± 0.1 (1) 0.92 ± 0.05 (3)	0.7 ± 0.1 (1)	0.8 ± 0.1 (1)	0.70 ± 0.06 (1) 0.78 ± 0.05 (4)	0.24 ± 0.05 (6) 0.29 ± 0.03 (7)
Visual geometric albedo p_v	0.77 ± 0.15 (1)	1.04 ± 0.15 (1) 1.41 ± 0.03 (8);	0.80 ± 0.15 (1)	0.55 ± 0.15 (1)	0.65 (1)	0.081 ± 0.002 (6); 0.082 ± 0.002 (7);
Bond albedo A_B	0.6 ± 0.1 (1) 0.5 ± 0.1 (5)	0.9 ± 0.1 (1) 0.91 ± 0.1 (3)	0.6 ± 0.1 (1)	0.45 ± 0.1 (1)	0.45 ± 0.1 (1) 0.49 ± 0.06 (4)	0.020 ± 0.004 (6) 0.023 ± 0.007 (7)

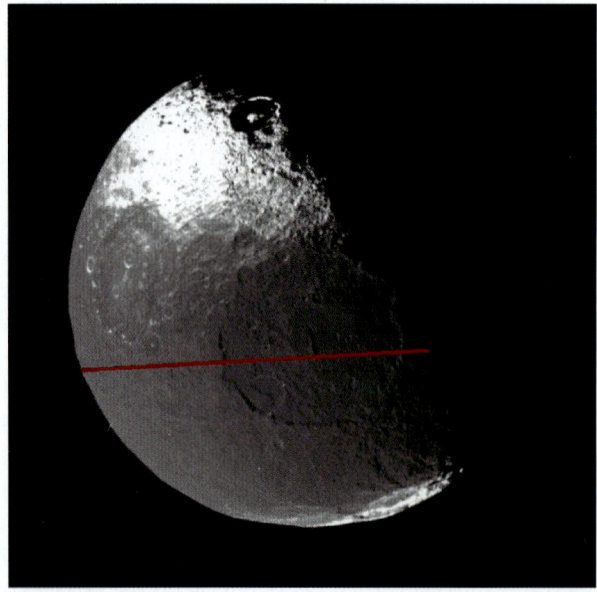

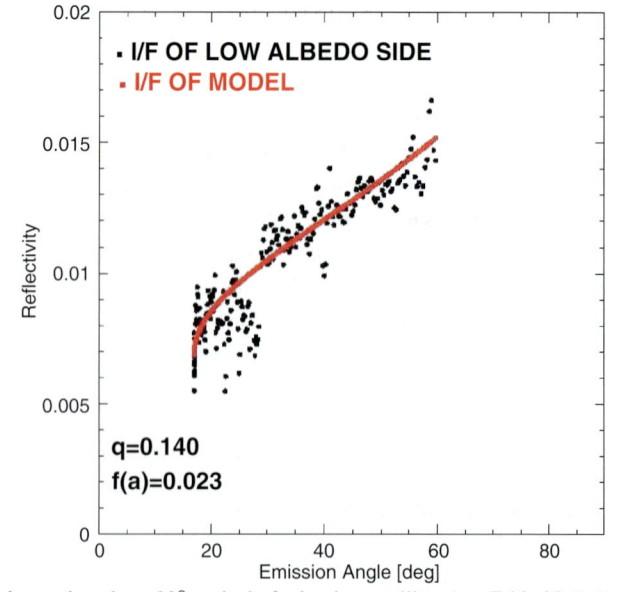

Fig. 20.16 A macroscopic roughness model for the low-albedo hemisphere of Iapetus, based on the crater roughness model by Buratti and Veverka (1985). The best-fit roughness function corresponds to a depth-to-radius of 0.14, or a mean slope angle of 11°, which is much lower than the ∼30° typical of other icy satellites (see Table 20.4). For these fits, a surface phase function $f(\alpha)$ of 0.023 was derived. From Lee et al. 2009

identical but reversed paths in a surface interfere constructively in exactly the backscattering direction, causing brightness to increase by up to a factor of two (Hapke 1990).

The Cassini flyby in June 2004 prior to Saturn orbit insertion afforded views at large phase angles – important for modeling roughness - and over a full excursion in geographical longitudes. Phoebe exhibits a substantial forward-scattering component to its single particle phase function. This result is consistent with a substantial amount of fine-grained dust on the surface of the satellite generated by particle infall, as suggested by Clark et al. (2005, 2008a). As observed in Cassini images (Porco et al. 2005a), Phoebe also shows extensive macroscopic roughness, hinting at the violent collisional history predicted by Nesvorny et al. (2003).

ISS images of the low-albedo hemisphere of Iapetus were fitted to the crater roughness model (Buratti and Veverka 1985), yielding a markedly smooth value for the degree of macroscopic roughness on this side with a mean slope angle of only a few degrees (Lee et al. 2009). This result suggests that small-scale rough features on the dark side have been filled in (features which probably dominate the photometric effects; see Helfenstein and Shepard 1998), which is consistent with the idea of Spencer and Denk (2009) that thermal migration of water ice leaves behind a fluffy residue of dark material. The low-albedo side of Iapetus shows a roughness much smaller than that of the other satellites and comparable within a factor of 2 with Enceladus, which is coated with micron-sized particles from its plume and the E-ring (Verbiscer and Veverka 1994; see Table 20.4). Model-fits for the low albedo hemisphere are shown in Fig. 20.16.

Finally, observations at very large solar phase angles (155°–165°) have been used to search for cryovolcanic activity on satellites. By comparing the brightness of Rhea to Enceladus in this phase angle range at 2.02 μm, Pitman et al. (2008) were able to place an upper limit on water vapor column density based on a possible plume of 1.52×10^{14} to $1.91 \times 10^{15}/cm^2$, two orders of magnitude below the observed plume density of Enceladus (Fig. 20.17).

The microphysical structure of the E-ring grain coatings of the inner icy moons embedded in the E-ring can be probed by investigating the opposition surge. Verbiscer et al. (2007) found that the amplitude and angular width of the opposition effect on the inner icy moons are correlated with the moons' position relative to the E-ring.

20.5 Constraints on the Top-Meter Structure and Composition by Radar

The wavelength of Cassini's 13.8 GHz RADAR instrument, 2.2 cm, is about six times shorter than the only ground-based radar wavelength available to study the satellites

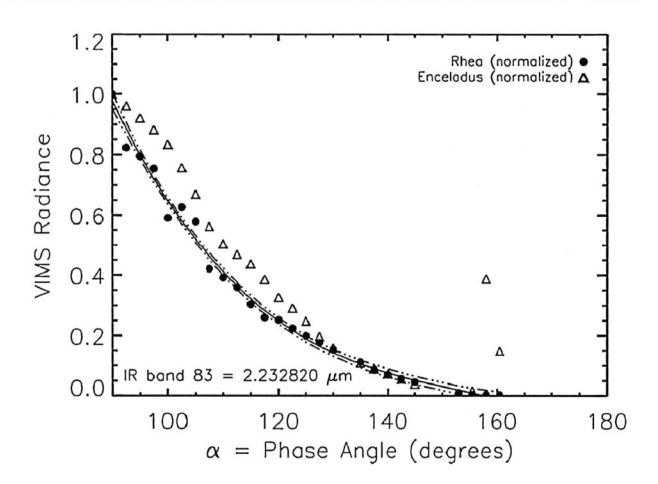

Fig. 20.17 Cassini VIMS disk-integrated brightness as a function of solar phase angle at 2.23 μ. Symbols: Normalized Rhea and Enceladus VIMS brightness (every fifth data point plotted). Solid line: Third-order polynomial fit to Rhea data. Dashed lines: (polynomial fit to Rhea data). Enceladus's plume can be seen as a peak at a solar phase angle of 159°. Adapted from Pitman et al. (2008)

(13 cm, at Arecibo) and 22 times longer than the millimeter wavelengths at the limit of the CIRS.

The echoes result from volume scattering, and their strength is sensitive to ice purity. Therefore, they provide unique information about near-surface structural complexity as well as about contamination with non-ice material. This section summarizes the most basic aspects of Cassini- and ground-based radar observations of the satellites, the theoretical context for interpreting the echoes, and the inferences drawn about subsurface structure and composition.

Cassini measures echoes in the same linear polarization as transmitted, whereas most ground-based radar astronomy consists of Arecibo 13 cm or 70 cm observations (Black et al. 2001a) and Goldstone 3.5 cm observations that almost always use transmission of a circularly polarized signal and simultaneous reception of echoes in same-circular (SC) and opposite-circular (OC) polarizations. A radar target's radar albedo is defined as its radar cross section divided by its projected area. The radar cross section is the projected area of a perfectly reflective isotropic scatterer which, if observed at the target's distance from the radar and using the same transmitted and received polarizations, would return the observed echo power. The total-power radar albedo is the sum of the albedos in two orthogonal polarizations, TP = SL + OL = SC + OC. Here, we will use 'SL-2' to denote the 2 cm radar albedo in the SL polarization, 'TP-13' to denote the 13 cm total-power albedo, etc.

For a smooth, homogeneous target, single back reflections would be entirely OC when the transmission is circularly polarized, or entirely SL when the transmission is linearly polarized, so the polarization ratios SC/OC and OL/SL would equal zero. Saturn's icy satellites and the icy Galilean satellites not only have radar albedos an order of magnitude greater than those of the Moon and most other solar system targets but also unusually large polarization ratios.

These strange radar signatures were discovered in observations of Europa, Ganymede and Callisto in the mid-1970s. It took a decade and a half to realize that the signatures are the outcome of 'coherent backscattering,' which is phase-coherent, multiple scattering within a dielectric medium that is both heterogeneous and non-absorbing (e.g., MacKintosh and John 1989; Hapke 1990).

During the next decade, Peters (1992) produced a vector formulation of coherent backscattering that led to predictions of radar albedos and polarization ratios, and Black (1997) and Black et al. (2001b) built on Peters' work with models of scatterer properties.

Extremely clean ice is insufficient for anomalously large radar albedos and circular polarization ratios, which arise from multiple scattering within a random, disordered medium whose intrinsic microwave absorption is very low. Thus, perfectly clean ice that is structurally homogeneous (constant density at millimeter scales) would not give anomalous echoes. The regoliths on solar system objects are structurally very heterogeneous, with complex density variations from the distribution of particle sizes and shapes produced by impacts of projectiles of different sizes and impact energies.

It is to be expected that the uppermost few meters of Saturn's icy satellite regolith are basically similar to those of the icy Galilean satellites or the Moon's in terms of particle size distribution, porosity and density as functions of depth, and lateral/vertical heterogeneity. It is the combination of naturally complex density variations and the extreme transparency of pure water ice that gives icy regolith its exotic radar properties.

The icy Galilean satellites were observed in dual-circular-polarization experiments from the ground at 3.5, 13, and 70 cm; experiments measuring SL-13 and OL-13 were done only for Europa, Ganymede and Callisto (Ostro et al. 1992). Arecibo obtained estimates of SC-13 and OC-13 for Enceladus, Tethys, Dione, Rhea and Iapetus. The Cassini radar measured SL-2 for Mimas, Hyperion and Phoebe. The satellites' albedos show different styles and are location- and wavelength-dependent, as discussed below. The results described in this chapter make use of 73 Cassini SL-2 measurements, more than twice the number reported by Ostro et al. (2006).

If a radar target is illuminated by a uniform antenna beam (that is, if the antenna gain is constant across the disk), then it is a simple matter to use the appropriate 'radar equation' to convert the measured echo power spectrum (power as a function of Doppler frequency) into a radar cross section (e.g., Ostro 1993). For Cassini measurements, except for the SAR imaging of Iapetus discussed below, the antenna beam width is within a factor of several times the target's angular width, so the gain varies across the target's disk, usually - significantly. Data reduction yields an estimate of echo power, which is the sum of contributions from different parts of the surface, with each contribution weighted by the two-way gain. However, the distribution of echo power as a function of location on the target disk is not known a priori. Thus, we do not know how much the echo power from any given surface element has been 'amplified' by the antenna gain. To estimate a radar albedo from our data we must make an assumption about the homogeneity and angular dependence of the surface's radar scattering. For the purpose of obtaining a radar albedo estimate from an echo power spectrum, one assumes uniform, azimuthally isotropic, cosine scattering laws, and uses least squares to estimate the value of a single parameter that defines the echo's incidence-angle dependence. In general, only modest departures from Lambert limb darkening are found.

20.5.1 Radar-Optical Correlations

Figure 20.18 shows all measurements of the satellites' individual SL-2 albedos along with their average optical geometrical albedos, plotted vs. distance from Saturn. A very pronounced radar-optical correlation means that variations in optical and radar albedos involve the same or related contaminant(s) and/or surface processes. Absolutely pure water ice at satellite surface temperatures is extremely transparent to radar waves. Contamination of water ice with no more than trace concentrations of virtually any other substance (including earth rocks, chondritic meteorites, lunar soil, ammonia, metallic iron, iron oxides and tholins) can decrease its radar transparency (and hence the regolith's radar albedo) by one to two orders of magnitude. Phenomena that depend on the distance from Saturn include primordial composition, meteoroid flux and the influx of dark material (perhaps from several sources), the balance between the influx of ammonia-containing particles and the removal of ammonia by a combination of ion erosion and micrometeoroid gardening, variation in E-ring fluxes, and the systematics of thermal volatiles redistribution.

Modeling of icy Galilean satellite echoes (Black et al. 2001b) suggests that penetration to about half a meter should be adequate to produce the range of measured values of SL-2.

The explanation by Clark et al. (2008b, 2009) for dark Saturnian system material, which suggests that it consists of sub-micron iron particles from meteorites which oxidized and became mixed with ice, would also suffice to explain, at least qualitatively, the overall radar-optical albedo correlation in Fig. 20.18 if the contamination of the ice extends some one to several decimeters below the optically visible

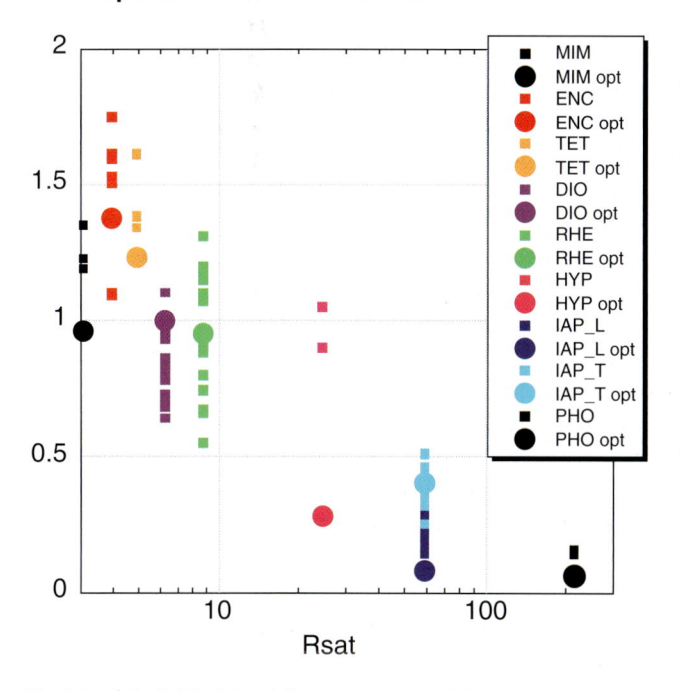

Fig. 20.18 Individual Cassini measurements of SL-2 (small symbols) and average optical geometrical albedos, plotted for the satellites vs. their distances from Saturn. Optical albedos of Mimas, Enceladus, Tethys, Dione and Rhea are from Verbiscer et al. (2007); those of Iapetus and Phoebe are from Morrison et al. (1986) and involve a V filter; Hyperion's is from Cruikshank and Brown (1982). Note the similar distance dependences. Hyperion's radar albedo is large and appears to break the pattern, perhaps (Ostro et al. 2006) because Hyperion's optical coloring may be due to a thin layer of exogenously derived material that, in contrast to the other satellites, has never been worked into the icy regolith, which remains very clean and radar-bright

surface, as expected by Clark et al. (2008b, 2009). The tentative detection of ammonia in VIMS spectra by Clark et al. (2008a) opens up the possibility that contamination by this material may play an important role in inter- and intra-satellite radar albedo variations. If so, this contamination would affect the radar but not the optical albedo.

Almost all Cassini radar data on the icy satellites consist of real-aperture scatterometry, which means that the beam size determines the resolution. Cassini has no real-aperture resolution of Mimas, Enceladus, Hyperion or Phoebe, but does have abundant measurements for Tethys, Dione, Rhea and Iapetus with beams smaller than, and in many cases less than half, the size of the disk (Ostro et al. 2006).

The SL-2 distributions of the individual satellites can be summarized briefly as follows. As regards Mimas, Herschel may or may not contribute to the highest value of the disc-integrated albedo. Enceladus shows no dramatic leading/trailing asymmetry and no noteworthy geographical variations except for a low albedo in an observation of the Saturn-facing side. Tethys' leading side is brighter both optically and in SL-2. Dione's highest-absolute-latitude views give the highest radar albedos. Rhea's leading side has a lower SL-2 in the north, where there is much cratering, and higher in the south, where there are optically bright rays. Its trailing and Saturn-facing sides, which look younger, show a relatively high SL-2. Iapetus displays an extremely marked correlation between optical brightness and SL-2.

In general, a first-order understanding of the SL-2 distribution is that higher SL-2 means cleaner near-surface ice, but the nature of the contaminant(s) and the depth of the contamination are open questions. The latter can be constrained by measuring radar albedos at more than one wavelength. Fortunately, there are Arecibo 13-cm observations of Enceladus, Tethys, Dione, Rhea (Black et al. 2004) and Iapetus (Black et al. 2007).

Also fortunately, the wavelength dependence observed for Europa, Ganymede and Callisto (EGC) gives us a context for discussing the wavelength dependence observed for Saturn's satellites.

20.5.2 Wavelength Dependence

EGC total-power radar albedos and circular polarization ratios at 3.5 cm and 13 cm are indistinguishable. This means that their regolith structure (distribution of density variations) and ice cleanliness look identical at the two wavelengths, for each object. That is, near-surface structures are 'self-similar' at 3.5 cm and 13 cm scales; there are no depth-dependent phenomena, like contamination levels or structural character, perceptible in EGC's 3.5 cm and 13 cm radar signatures.

At 70 cm, moreover, EGC SC/OC ratios show little wavelength dependence, so subsurface multiple scattering remains the primary source of the echo even on this longer scale. However, the radar albedos (TP-70) plummet for all three objects. The implication is that at 70 cm, scattering heterogeneities are relatively sparse, and the self-similarity of the regolith structure disappears on the larger scale. On Europa at least, with the most striking albedo drop from TP-13 to TP-70, the regolith may simply be too young for it to be heterogeneous at 70 cm.

Within the framework of the Black et al. (2001b) model, this implies that at 70 cm 'the scattering layer is disappearing in the sense that there are too few scatterer at this scale and they are confined to a layer too shallow to permit numerous multiple scatterings. At even longer wavelengths, of the order of several meters, the reflectivity from the surfaces of these moons, and Europa in particular, may drop precipitously or even vanish as the layers responsible for the centimeter-wavelength radar properties become essentially transparent.'

Now, given that the 3.5 cm and 13 cm polarization properties of EGC show no significant wavelength dependence, and that the 13 cm OL/SL ratios of those objects are very similar, let us assume that the linear polarization properties of the Saturnian satellites are wavelength-invariant. Let us then use EGC's mean OL/SL value to estimate total-power 2 cm albedos: TP-2 = (1.52 ± 0.13) SL-2 and compare them to the Arecibo 13 cm total-power albedos (TP-13).

Figure. 20.19 shows relevant Cassini and Arecibo information.

Moving outward from Saturn, the magnitude of 2 cm to 13 cm wavelength dependence is unremarkable for Enceladus' leading side, extreme for Enceladus' trailing side, very large for Tethys, large for Dione, unremarkable for Rhea, large for Iapetus' leading side, and extreme for Iapetus' trailing side.

Whereas compositional variation (ice cleanliness) almost certainly does not contribute to the wavelength dependence seen in EGC, it may well play a key role in the wavelength dependence seen in Saturn's satellites. In this complicated system, the tapestry of satellite radar properties may involve many factors, including inter- and intra-satellite variations in E-ring flux, variations in the meteoroid or ionizing flux, variations in the concentration of radar absorbing materials (including optically dark material and possibly ammonia) as a function of depth, the systematics of thermal volatile redistribution (especially for Iapetus), and/or constituents with wavelength- or temperature-dependent electrical properties. Here, we will present what we consider reasonable hypotheses about what the most important factors are.

The unremarkable 2- to-13-cm wavelength dependence seen on Enceladus' leading side and also on Rhea is reminiscent of Europa's 3.5- to-13-cm wavelength invariance. The simplest interpretation is that at least the top few decimeters are uniformly clean and old enough for meteoroid bombardment to have created density heterogeneities that are able to produce coherent backscattering and extreme radar brightness.

Enceladus' leading-side 2 cm and 13 cm TP albedos resemble Europa's 3.5 cm and 13 cm TP albedos, whereas Rhea's 2 cm and 13 cm TP albedos resemble Ganymede's 3.5 cm and 13 cm TP albedos, reflecting much greater contamination of Rhea's upper regolith.

Dione's and Rhea's TP-2 albedos are similar, but Dione is dimmer at 13 cm. Dione VIMS data show evidence of a bombardment of apparently dark particles on the trailing side (Clark et al. 2008a). If the dark material contains ammonia, and the ejecta systematics of meteoroid bombardment efficiently distribute the dark material over the surface of the object, multiple scattering will be cut off at a depth that may be too shallow for TP-13 to be enhanced.

The striking wavelength dependence of Enceladus' trailing side (TP-2 >> TP-13) mirrors Europa's (TP-13 >> TP–70). It is easily understood if we assume that the top meter or so of ice was deposited so recently that there has not been time for the regolith to be matured by meteoroid bombardment. That is, there are small-scale heterogeneities that produce coherent backscattering at 2 cm, but none of the larger-scale heterogeneities needed to produce coherent backscattering at 13 cm.

This idea is consistent with the fact that Enceladus' trailing side is optically brighter than its leading side, apparently because of the greater E-ring flux it receives (Buratti 1988; Showalter et al. 1991; Buratti et al. 1998). So all of Enceladus is very clean, but the uppermost layer of one or several meters on the trailing side is also extremely young, resulting in a striking hemispheric dichotomy at 13 cm that is the most intense seen at that wavelength in the icy-satellite systems of either Jupiter or Saturn.

At least some of the decrease in average SL-2 (Fig. 20.18) and the decrease in average wavelength dependence outward from Enceladus (Fig. 20.19) is probably due to the outward decrease in E-ring flux (Verbiscer et al. 2007).

Finally, we get to Iapetus, where TP-2 is strongly correlated with the optical albedo: low for the optically dark leading-side material and high for the optically bright trailing-side material. However, Iapetus' TP-13 values show

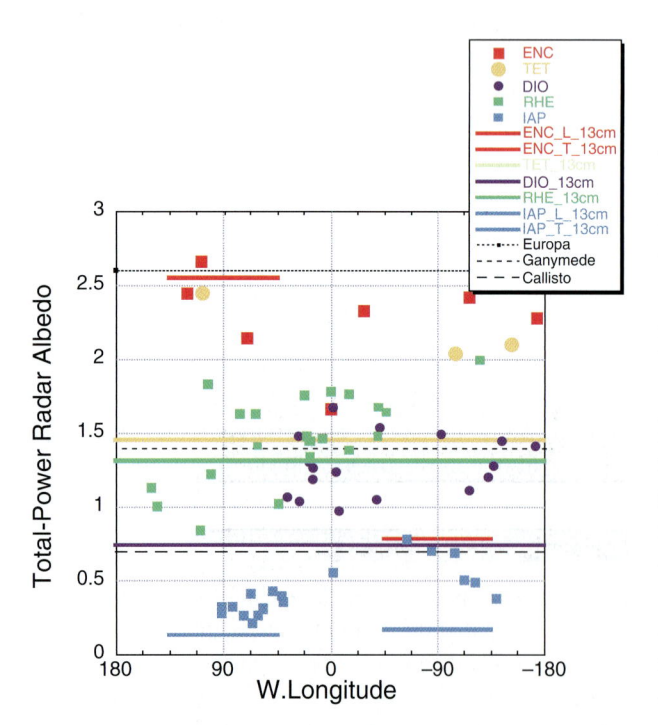

Fig. 20.19 Wavelength dependence of total-power radar albedos. Individual symbols are estimates of TP-2 obtained from individual measurements of SL-2 using TP-2 = 1.52 SL-2 ± 0.13 SL. Arecibo 13-cm albedos are horizontal lines. Dashed lines represent EGC

much less of a dichotomy and are several times lower than TP-2, being indistinguishable from the weighted mean of TP-13 for main-belt asteroids, 0.15 ± 0.10 (Table 20.5).

Therefore, whereas Iapetus is anomalously bright at 2 cm (similar to Callisto) and mimics the optical hemispheric dichotomy, at 13 cm it looks like a typical main-belt asteroid, with minimal evidence of mimicking the optical hemispheric dichotomy.

The Iapetus results are understandable if (1) the leading side's optically dark contaminant is present to depths of at least one to several decimeters, so that SL-2 can see the optical dichotomy, and (2) ammonia and/or some other radar-absorbing contaminant is globally much less abundant within the upper one to several decimeters than at greater depths, which would explain the wavelength dependence, that is, the virtual disappearance of the object's 2-cm signature at 13 cm. The first conclusion is consistent with the dark material being a shallow phenomenon, as predicted from the thermal migration hypothesis discussed elsewhere in this chapter.

Moreover, ammonia contamination may be uniform within the top few meters, so that the wavelength dependence arises because ammonia's electrical loss is much greater at 13 cm (2.38 GHz) than at 2 cm (13.8 GHz). Unfortunately, the likelihood of this possibility cannot yet be judged because measuring the complex dielectric constant of ammonia-contaminated water ice as a function of concentration, frequency and temperature is so difficult that, although needed, it has not been done yet. This is a pity, because the lack of this information undermines the interpretation of the 2 cm to 13 cm wavelength dependence seen in other Saturnian satellites as well as of the possible influence of ammonia and its 'modulation' by magnetospheric bombardment (which varies with distance from Saturn) on the dependence of SL-2 on the distance from Saturn.

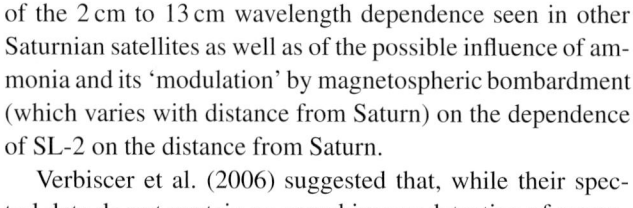

Verbiscer et al. (2006) suggested that, while their spectral data do not contain an unambiguous detection of ammonia hydrate on Enceladus, their spectral models do not rule out the presence of a modest amount of it on both hemispheres. However, they note that the trailing hemisphere is also exposed to increased magnetospheric particle bombardment, which would preferentially destroy the ammonia. In any case, ammonia does not appear to be a factor in Enceladus' radar albedos.

20.5.3 Iapetus Radar Image

During the Iapetus 49 flyby, RADAR obtained an SAR image at 2 km to 12 km surface resolution that covers much of the object's leading (optically darker) side. From a comparison with optical imaging (Fig. 20.20), we see that, as expected, the SAR reliably reveals the surface features seen in ISS images.

The pattern of radar brightness contrast is basically similar to that seen optically, except for the large basin at 30°N, 75°W, which has more SL-2-bright than optically bright highlights. This suggests that the subsurface of these highlights at decimeter scales may be cleaner than in other parts of the optically dark terrain, which is consistent with the idea that the contamination by dark material might be shallow, at least in this basin. Conversely, many optically bright crater rims are SL-2-dark, so the exposure of clean ice cannot be very deep.

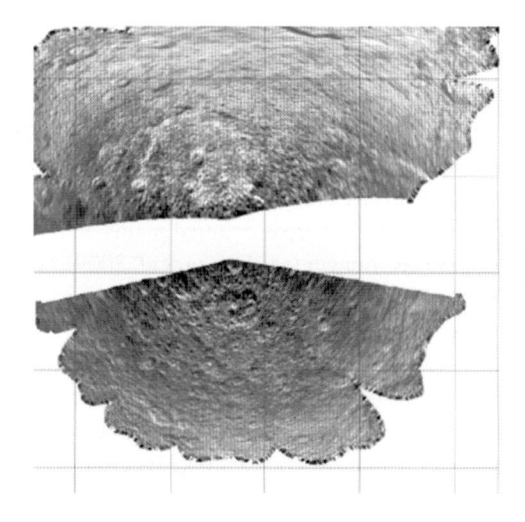

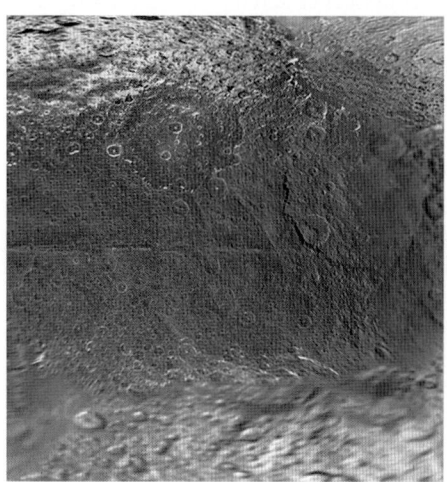

Table 20.5 Total-power radar albedos

	Dark/Leading	Bright/Trailing
2.2-cm	0.32 ± 0.09	0.58 ± 0.18
13-cm	0.13 ± 0.04	0.17 ± 0.04

Fig. 20.20 RADAR SAR image of the leading side of Iapetus (Sept. 2007, *left*) and the ISS mosaic PIA 08406 (Jan. 2008), using similar projections and similar scales (the large crater on the right in the ISS image is about 550 km in diameter)

20.6 Geological Evolution

The densities of the Saturnian satellites constrain their composition to be about 2/3 ice and 1/3 rock. Accretional energy, tidal despinning and radioactive decay might have heated and melted the moons. Due to the very low melting temperatures of various water clathrates, it is likely that the larger bodies differentiated early in their evolution while smaller satellites may still be composed and structured in their primitive state. For a more detailed discussion of the aspects of origin and thermal evolution, see Johnson et al. (2009) and Matson et al. (2009).

The heavily cratered solid surfaces in the Saturnian system indicate the role of accretional and post-accretional bombardment in the geological evolution of the satellites. Observed crater densities are summarized in Dones et al. (2009). However, the source as well as the flux of bombarding bodies is still under discussion, and Dones et al. (2009) came to the conclusion that, since no radiometric sample data exist from surfaces in the outer solar system and the size-frequency distribution of comets is less well understood than that of asteroids, the chronology of the moons of the giant planets remains in a primitive state. Therefore, we refer the reader to Dones et al. (2009) for the crater chronology discussion. In interpreting the geological evolution of the Saturnian satellites, we will rely on relative stratigraphic relations. Nevertheless, absolute ages are given in Dones et al. (2009).

In a stratigraphic sense, three major surface unit categories can be distinguished on all icy satellites:

(1) Cratered plains, which are characterized by dense crater populations that are superimposed by all other geological features and, thus, are the relatively oldest units.
(2) Tectonized zones of narrow bands and mostly subparallel fractures, faults, troughs and ridges of global extension indicating impact-induced or endogenic crustal stress. These units dissect cratered plains and partly dissect each other, indicating different relative ages.
(3) Resurfaced units consisting of terrain that has undergone secondary exogenic and endogenic processes such as mass wasting, surface alterations (due to micrometeorite gardening, sputtering, thermal segregation and crater ray formation by recent impacts), contamination by dark material and cryovolcanic deposition mostly resulting in smoothed terrain. Superimposed on cratered plains and tectonized regions, these are the youngest surface units.

Heavily cratered surfaces are found on all the icy rocks and small satellites (Fig. 20.21) as well as on the medium-sized moons Mimas, Enceladus, Tethys, Dione, Rhea, Titan, Hyperion, Iapetus and Phoebe, indicating an intense post-accretional collision process relatively early in the geological history of the satellites. Cassini observations added to the number of large basins and ring structures mainly on Rhea,

Fig. 20.21 Phoebe, 213 km in average diameter, was the first one of the nine classic moons of Saturn to be captured by the ISS cameras aboard Cassini in a close flyby about three weeks prior to Cassini's Saturn Orbit Insertion (SOI) in July 2004. The mosaic shows a densely cratered surface. Higher-resolution data revealed that the high frequency of craters extends with a steep slope down to craters only tens of meters in diameter (Porco et al. 2005a). Image source: http://photojournal.jpl.nasa.gov, image identification PIA06064

Dione and Iapetus. Differences in the topographical expression of these impact structures suggest post-impact processes such as relaxation, which may reflect differences in the thermal history and/or crustal thickness of individual satellites (e.g., Schenk and Moore 2007; Giese et al. 2008). Although age models are based on different impactor populations, the cratered surfaces on Iapetus seem to be the oldest in the Saturnian system, whereas cratered plains on the other satellites seem to be younger (see Dones et al. 2009). One large (48 km in diameter) stratigraphically young ray crater was found on Rhea (Wagner et al. 2007) (Fig. 20.22).

While the small satellites lack tectonic features, the medium-sized bodies experienced rotation-, tide- and impact-induced deformation and volume changes, which created extensional or contractional tectonic landforms. Parallel lineaments on Mimas seem to be related to the Herschel impact event (McKinnon 1985; Schenk 1989a), although freezing expansion cannot be ruled out as the cause (Fig. 20.23).

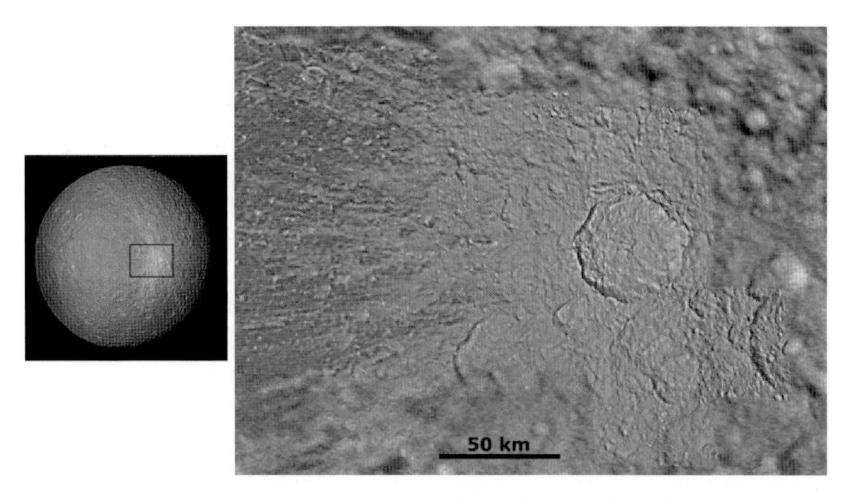

Fig. 20.22 A prominent bright ray crater with a diameter of 48 km is located approximately at 12.5 °S, 112 °W on Saturn's second-largest satellite Rhea. The crater is stratigraphically young, indicated by the low superimposed crater frequency on its floor and continuous ejecta (Wagner et al. 2008). The high-resolution mosaic (*right*) of NAC images, shown in context of a WAC frame, was obtained by Cassini ISS during orbit 049 on Aug. 30, 2007. The global view to the left showing the location of the detailed view is a single NAC image from orbit 056

Fig. 20.23 The anti-Saturnian hemisphere of Mimas, centered approximately at lat. 22 °S, long. 185 °W, is shown in a single image (NAC frame N1501638509) at a resolution of 620 m/pixel, taken during a non-targeted encounter in August 2005. The densely cratered landscape implies a high surface age. Grooves and lineaments infer weak tectonic on this geologically unevolved body

Enceladus exhibits the greatest geological diversity in a complex stress and strain history that is predominantly tectonic and cryovolcanic in origin, with both processes stratigraphically correlated in space and time, although the energy source is so far not completely understood (e.g., Smith et al. 1982; Kargel and Pozio 1996; Porco et al. 2006; Nimmo and Pappalardo 2006; Nimmo et al. 2007; Jaumann et al. 2008). Enceladus is discussed in detail in Spencer et al. (2009).

Located in densely cratered terrain, the nearly globe-encircling Ithaca Chasma rift system on Tethys (Fig. 20.2) is 100 km wide, consists of at least two narrower branches towards the south and was caused either by expansion when the liquid-water interior froze (Smith et al. 1981) or by deformation by the large impact which formed Odysseus (Moore and Ahern 1983). However, crater densities derived from Cassini images suggest Ithaca Chasma to be older than Odysseus (Giese et al. 2007), so that the rift system may be an endogenic feature.

Dione exhibits a nearly global network of tectonic features such as troughs, scarps, ridges and lineaments partly superimposed on each other (Fig. 20.24), *which* correlate with smooth plains (e.g., Plescia 1983; Moore 1984; Stephan et al. 2008). The global orientation of Dione's tectonic pattern is non-random, suggesting a complex endogenic history of extension followed by compression. Cassini images indicate that some of the grabens are fresh, which implies that tectonic activity may have persisted into geologically recent times (Wagner et al. 2006). Cassini also revealed a crater-lineament relationship on Dione (Moore et al. 2004; Schenk and Moore 2007), indicating exogenic-induced tectonic activity in the early geological history of the moon. Thus, Dione's crust not only experienced stress caused by different exogenic and endogenic processes, such as ancient impacts, despinning and volume change over a long period of time, but also more recent tidal stress. As on Dione, wispy accurate albedo markings on Rhea are associated with an extensional fault system (Schenk and Moore 2007; Wagner et al. 2007), although Rhea seems less endogenically evolved than the other moons (Fig. 20.25).

Parallel north-south trending lineaments are thought to be the result of extensional stress followed by compression

Fig. 20.24 This image of Tethys, taken by the Cassini ISS NA camera in May 2008 in visible light, shows the old, densely cratered surface of Tethys and a part of the extensive graben system of Ithaca Chasma. The graben is up to 100 km wide and is divided into two branches in the region shown in this image. Using digital elevation models infers flexural uplift of the graben flanks and a total topography ranging several kilometers from the flanks to the graben floor (Giese et al. 2007). Image details: ISS frame number N1589080288, scale 1 km/pixel, centered at lat. 42 °S, long. 50 °W in orthographic projection

(Moore et al. 1985) and appear in Cassini observations as poorly developed ridges. This suggests that they are older than the grabens identified in the wispy terrain that appear to have formed more recently (Wagner et al. 2006).

Iapetus' equatorial ridge (Porco et al. 2005a; Giese et al. 2008) (Fig. 20.7) is densely cratered, making it comparable in age to any other terrains on this moon (Schmedemann et al. 2008). Models explain the origin of the ridge either by endogenic shape and/or spin state changes (e.g., Porco et al. 2005a; Giese et al. 2008) or exogenic causes, with the ridge accumulating from ring material left over from the formation of proto-Iapetus (Ip 2006). On Iapetus, diurnal tidal stress is rather low, indicating that the ridge is old, and its preservation suggests the formation of a thick crust early in the moon's history (Castillo-Rogez et al. 2007).

Enceladus, where cryovolcanism is definitely active (Porco et al. 2006), is discussed in Spencer et al. (2009). Titan, where some evidence of cryovolcanism exists (Lopes et al. 2007; Nelson et al. 2009a, b) is discussed in Jaumann et al. (2009) and Sotin et al. (2009).

There is no direct evidence of cryovolcanic processes (Geissler 2000) on the other Saturnian satellites. Although it has been suggested in pre-Cassini discussions that the plains on Dione (Plescia and Boyce 1982; Plescia 1983; Moore 1984) and Tethys (Moore and Ahern 1983) are endogenic in origin, Cassini has not directly observed any cryovolcanic activity on these satellites. The plains correlate with tectonic features and exhibit distinct boundaries with more densely cratered units (Schenk and Moore 2007), which indicate

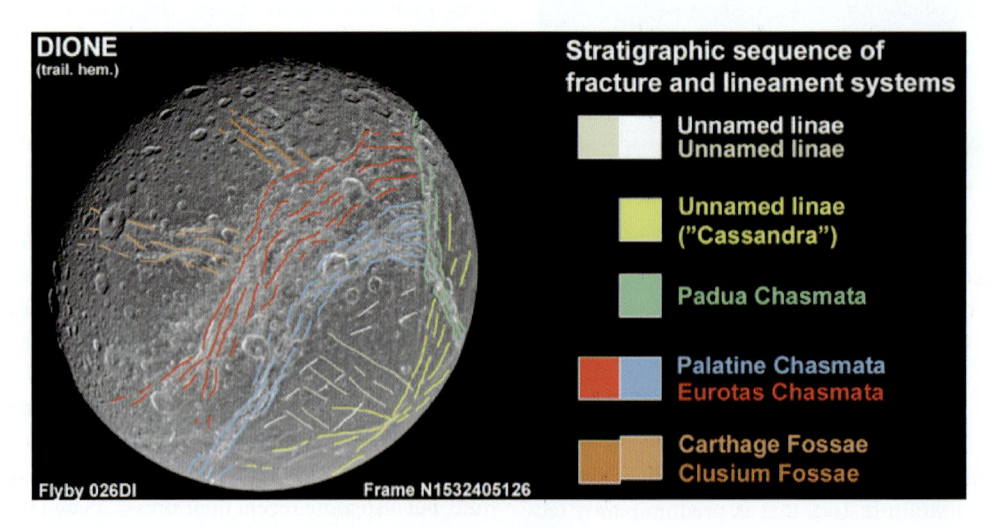

Fig. 20.25 Map of Dione's trailing hemisphere showing an age sequence of troughs, graben and lineaments, inferred from various cross-cutting relationships (Wagner et al. 2009). Three age groups of troughs and graben can be distinguished: Carthage and Clusium Fossae are oldest, cut by the younger systems of Eurotas and Palatine Chasmata, which in turn are truncated by Padua Chasmata which are the youngest. Several systems of lineaments with various trends, either parallel or radial (the latter previously assumed to be a bright ray crater named Cassandra) can be mapped which appear to be younger than the troughs and graben

that the plains are a stratigraphically younger formation. In addition, there is evidence of material around Dione, Tethys and even Rhea (Burch et al. 2007; Jones et al. 2008; Clark et al. 2008a), but its origin is unclear. There is no evidence of cryovolcanism on Iapetus. Although the plains on Dione and Tethys are younger than the surrounding terrain, cryovolcanic activity must have occurred in ancient times, assuming that such processes caused these surface units.

All medium-sized icy satellites appear to have a debris layer, as may be inferred from (1) the non-pristine morphology of most craters; (2) the presence of groups of large boulders and debris lobes seen in the highest-resolution ISS images; and (3) derived thermal-physical properties consistent with loose particulate materials. With the exception of Enceladus, which is addressed in Spencer et al. (2009), these debris layers were generated by impacts and their ejecta on the other satellites. The upper limiting case for a given satellite is crustal disruption from extreme global impact-induced seismic events. A way to evaluate the depth of fracture in such events is to assume that these events were energetic enough to open fractures in the brittle ice (brittle on the timescale of these events) composing the interiors of the target satellite, and that, during the impact event, the lithostatic pressure everywhere was less than the brittle tensile strength of cold ice. Brittle fracture caused by global seismicity, all else being equal, extends deeper into smaller satellites, given that larger satellites are sheltered from dynamic rupture by their interior pressure. These being transient events, a simple estimate of the realm where brittle fracture may be expected to be induced, and visibly sustained, can be obtained from global seismicity by applying an expression (e.g., Turcott and Schubert 2002) of the lithostatic pressure at radius R_p, and comparing this to the brittle tensile strength of water ice at timescales in which extreme global impact-induced seismic events produce energy sufficient to break ice mechanically. The determination of the fracture depth is based on laboratory-derived dynamic values of $\sim$1 MPa (Lange and Ahrens 1987; Hawkes and Mellor 1972). With ρ the (bulk) density of the body (approximated as $1,000 \text{ kg/m}^3$), G the gravitational constant, and R the radius of the satellite, the depth of the fracture can be constrained. Based on this, the value of $R - R_p$ for Mimas is $\sim$20 km ($\sim$10% of its radius), $\sim$10 km ($\sim$2% of radius) for Tethys, and $\sim$5 km ($\sim$0.5% of radius) for Rhea. This is a measure of the depth to which it is reasonable to expect the surface layer (i.e., the megaregolith) to be intensely fractured. Of course, a body may be fractured to greater depths by large impacts, and the resulting pore space may not easily be eliminated. Deep porosity is most easily maintained on smaller and presumably colder bodies such as Mimas (Eluszkiewicz 1990). Thus, for Mimas we expect that relatively recent major impact events, such as Herschel, might exploit pre-existing fracture patterns in the 'megaregolith' which might penetrate deep into the body rather than focus energy, as has been suggested for the formation of the plains of Tethys (e.g., Bruesch and Asphaug 2004; Moore et al. 2004). Consequently, the lack of antipodal focusing or axial symmetry about a radius through Herschel in the lineament pattern on Mimas may be unsurprising, if that pattern is indeed an expression of tectonic movements initiated by the Herschel impact.

The impact-gardened regolith of most middle-sized icy satellites can be assumed to be many meters thick if the regolith in the lunar highlands is any guide. Calculations of impact-generated regolith, based on crater densities observed in Voyager images, yielded mean depths of 500 m for Mimas, 740 m for Dione, 1,600 m for Tethys and 1,900 m for Rhea (Veverka et al. 1986).

This regolith will be further distributed by gravity-driven mass wasting (non-linear creep, i.e., disturbance-driven sediment transport that increases nonlinearly with the slope due to granular creep (e.g., Roering et al. 2001; Forsberg-Taylor et al. 2004)) which reduces high-standing relief on craters (e.g., lowering of crater rims and central peaks) and tectonic features (e.g., the brinks of scarps) as well as filling depressions (e.g., Howard 2007). The process of non-linear creep, when not competing with other processes, produces a landscape, which has a repetitive consistent appearance. Impact-induced shaking and perhaps diurnal and seasonal thermal expansion facilitate such mass wasting. Note that this common expression of mass wasting on regolith-covered slopes assumes no topography-maintaining precipitation of frosts on local topographical highs, as discussed below.

Bedrock erosion might be caused by the sublimation of mechanically cohesive interstitial volatiles. The icy bedrock would crumble, move down slope through nonlinear creep and form mantling slopes, perhaps leaving only topographical highs (crests) composed of bedrock exposed. This situation would occur if the initial bedrock landscape disaggregates to form debris at a finite rate, so that exposure of bedrock gradually diminishes, lasting longest at local crests. Such a situation might explain the landscape of Hyperion. If there is re-precipitation of volatiles in the form of frosts on local topographical highs, these frosts would armor the topographical highs against further desegregation. It has been proposed that landscapes evolved in this fashion on Callisto (Moore et al. 1999; Howard and Moore 2008). Thermally driven global and local frost migration and redistribution may occur on Iapetus (Spencer et al. 2005; Giese et al. 2008; Spencer and Denk 2009), as previously discussed. This surficial frost redistribution may occur regardless of whether volatiles in the bedrock of Iapetus sublimate and contribute to bedrock erosion.

20.7 Conclusions

The icy satellites of Saturn show great and unexpected diversity. In terms of size, they cover a range from ~1,500 km (Rhea) to ~270 km (Hyperion) and even smaller 'icy rocks' of less than a kilometer in diameter. The icy satellites of Saturn offer an unrivalled natural laboratory for understanding the geology of diverse satellites and their interaction with a complex planetary system.

Cassini has executed several targeted flybys over icy satellites, and more are planned for the future. In addition, there are numerous other opportunities for observation during non-targeted encounters, usually at greater distances. Cassini used these opportunities to perform multi-instrument observations of the satellites so as to obtain the physical, chemical and structural information needed to understand the geological processes that formed these bodies and governed their evolution.

All icy satellites exhibit densely cratered surfaces. Although there have been some attempts to correlate craters and potential impactor populations, no consistent interpretation of the crater chronology in the Saturnian system has been developed so far (Dones et al. 2009). However, most of the surfaces are old in a stratigraphical sense. The number of large basins discovered by Cassini was greater than expected (e.g., Giese et al. 2008). Differences in their relaxation state provide some information about crustal thickness and internal heat flows (Moore and Schenk 2007). The absence of basin relaxation on Iapetus is consistent with a thick lithosphere (Giese et al. 2008), whereas the relaxed Evander basin on Dione exhibits floor uplift to the level of the original surface (Moore and Schenk 2007), suggesting higher heat flow.

The wide variety in the extent and timing of tectonic activity on the icy satellites defies any simple explanation, and our understanding of Saturnian satellite tectonics and cryovolcanism still is at an early stage. The total extent of deformation varies wildly, from Iapetus and Mimas (barely deformed) to Enceladus (heavily deformed and currently active), and there is no straightforward relationship to predicted tidal stresses (e.g., Nimmo and Pappalardo 2006; Matsuyama and Nimmo 2007, 2008; Schenk et al. 2008; Spencer et al. 2009). Extensional deformation is common and often forms globally coherent patterns, while compressional deformation appears rare (e.g., Nimmo and Pappalardo 2006; Matsuyama and Nimmo 2007, 2008; Schenk et al. 2008; Spencer et al. 2009). These global patterns are suggestive of global mechanisms, such as despinning or reorientation, which probably occurred early in the satellites' histories (e.g., Thomas et al. 2007b); present-day diurnal tidal stresses are unlikely to be important, except on Enceladus (e.g., Spencer et al. 2009) and (perhaps) Mimas and Dione (e.g., Schenk and Moore 2007). Ocean freezing gives rise to isotropic extensional stresses; modulation of these stresses by pre-existing weaknesses (e.g., impact basins) may explain some of the observed long-wavelength tectonic patterns (e.g., Schenk and Moore 2007). Except for Enceladus, there is as yet no irrefutable evidence of cryovolcanic activity in the Saturnian system, either from surface images and topography or from remote sensing of putative satellite atmospheres. The next few years will hopefully see a continuation of the flood of Cassini data, and will certainly see a concerted effort to characterize and map in detail the tectonic structures discussed here. Understanding the origins of these structures and the histories of the satellites will require both geological and geophysical investigations and probably provide a challenge for many years to come.

Although the surfaces of the Saturnian satellites are predominantly composed of water ice, reflectance spectra indicate coloring agents (dark material) on all surfaces (e.g., Fink and Larson 1975; Clark et al. 1984, 1986; Roush et al. 1995; Cruikshank et al. 1998a; Owen et al. 2001; Cruikshank et al. 2005; Clark et al. 2005, 2008a, 2009; Filacchione et al. 2007, 2008). Recent Cassini data provide a greater range in reflected solar radiation at greater precision, show new absorption features not previously seen in these bodies, and allow new insights into the nature of the icy-satellite surfaces (Buratti et al. 2005b; Clark et al. 2005; Brown et al. 2006; Jaumann et al. 2006; Cruikshank et al. 2005, 2007, 2008; Clark et al. 2008a, b, 2009). Besides H_2O ice, CO_2 absorptions are found on all medium-sized satellites (Buratti et al. 2005b; Clark et al. 2005; Cruikshank et al. 2007; Brown et al. 2006; Clark et al. 2008a), which make CO_2 a common molecule on icy surfaces as well. Various spectral features of the dark material match those seen on Phoebe, Iapetus, Hyperion, Dione and Epimetheus as well as in the F-ring and the Cassini Division, implying that the material has a common composition throughout the Saturnian system (Clark et al. 2008a, 2009). Dark material diminishes closer to Saturn, which might be due to surface contamination of the inner moons by E-ring particles and some chemical alteration processes (Clark et al. 2008a). Although the general composition of the dark material in the Saturnian system has not yet been determined, it is known that it at least contains CO_2, bound water, organics and, tentatively, ammonia (Clark et al. 2008a). In addition, a blue scattering peak that dominates the dark spectra (Clark et al. 2008a, 2009) can be modelled by nano-phase hematite mixed with ice and dark particles. This led Clark et al. (2008b, 2009) to infer that sub-micron iron particles from meteorites, which oxidize (explaining spectral reddening) and become mixed with ice could best explain all the colors and spectral shapes observed in the system. Iron has also been detected in E-ring particles and in particles escaping the Saturnian system (Kempf et al. 2005). Previous explanations for the dark material were based on tholins (e.g., Cruikshank et al. 2005), but this is inconsistent with the lack of marked absorptions due to CH in the 3 μm to 4 μm region as well as other major absorptions at wavelengths longer than 2 μm. Iron particles are consistent with previous Cassini studies that suggested an external origin for the dark material on Phoebe, Iapetus and Dione (Clark et al. 2005, 2008a, b, 2009). Another remarkable issue relating to dark material is the albedo dichotomy on Iapetus, whose leading hemisphere is the darkest surface in the Saturnian system. In contrast to the outermost satellite, Phoebe, also covered by dark material, Iapetus' dark hemisphere is spectrally reddish, making it doubtful whether Phoebe is the source of the dark material. Clark et al. (2008a, 2009) explained the color discrepancy by red fine-grained dust (<0.5 μm) that, mixed in small amounts into water ice, produces a Rayleigh scattering effect enhancing the blue response, so that the colors observed simply depend on the amount of dark material.

Radar observations constrain the dark material on Iapetus to a thickness of only a few decimeters (Ostro et al. 2006), which is consistent with the thin dark coating derived from spectral data (Clark et al. 2008a, 2009) as well as with the low density of $1.1\,\mathrm{g/cm^3}$ of Iapetus (Rappaport et al. 2005). In addition, the coated surface in the dark terrain is nowhere broken up by large impacts, suggesting that the emplacement of dark material occurred relatively recently (Denk et al. 2000). In addition to its brightness dichotomy, Iapetus also shows a hemispheric color with the leading side dark material being redder than the trailing side bright material (Bell et al. 1985; Buratti and Mosher 1995). This also seems to correlate with the low degree of macroscopic roughness on the dark side (Lee et al. 2009), indicating the infilling of rough surface structures. Moreover, there are also latitude-dependent spectral variations in the UV (Hendrix and Hansen 2009b) that indicate temperature decreases and increases in the bright regions, particularly at the dark/bright boundary. These observations are consistent with a model process of thermal segregation in which water ice leaves behind a residue of dark material (Denk and Spencer 2009).

Alteration mechanisms cause not only compositional variations on the satellites but also structural differences. The satellites serve as sources and sinks of dust grains within the Saturnian system. Plume activity on Enceladus, meteoritic bombardment and subsequent escape of particles, capture of interplanetary dust by Saturn's gravitational field, and ring particles all serve to provide material that is subsequently accreted by the satellites. The fate of these dust grains in the system depends on a complex set of interactions determined by competing processes that move particles inward or outward. Forces at work include gravitational perturbations, plasma drag, radiation pressure and Poynting-Robertson drag (Burns 1979). These mechanisms are sensitive to particle composition and size; furthermore, the latter diminishes over time as particles are subjected to collisions.

Radar echos result from volume scattering, and as their strength is sensitive to ice purity they provide unique information about near-surface structural complexity as well as contamination with non-ice material. The observed radar-optical albedo correlation (Ostro et al. 2006) is consistent with the explanation developed for the dark material in the Saturnian system in which Clark et al. (2008b, 2009) propose that sub-micron iron particles from meteorites might oxidize and become mixed with ice, assuming that the contamination of the ice extends between one and several decimeters below the optically visible surface. Inter- and intra-satellite radar albedo variations might be due to contamination by ammonia, which was tentatively detected in VIMS spectra (Clark et al. (2008a) and would affect the radar but not the optical albedo.

The Saturnian icy satellites exhibit stratigraphically old surfaces showing craters and early-formed tectonic structures induced by despinning, reorientation and volume changes. However, they also show evidence of younger geological processes such as tide-induced crustal stress and plains formation. Recent and current cryovolcanism exists on Enceladus, and ancient cryovolcanism cannot be ruled out for Dione, Tethys and Rhea. Younger events, such as the formation of ray craters, can be observed on Rhea. Surface alteration by charged-particle bombardment, sputtering, micrometeorite bombardment and thermal processing are ongoing geological activities. Surface coating by E-ring particles is also a recent process. Coating of the satellites' surfaces by dark material has been observed throughout the system, and there is some evidence that this contamination might also be a relatively recent process.

Acknowledgements We gratefully acknowledge the long years of work by the entire Cassini team that allowed these data of the Saturnian satellites to be obtained. We also acknowledge NASA, ESA, ASI, DLR, CNES and JPL that provide support for the international Cassini team. We thank K. Stephan, R. Wagner and M. Langhans for data processing. The discussions of reviewers C. Chapman and D. Domingue are highly appreciated. Part of this work was performed at the DLR Institute of Planetary Research, with support provided by the Helmholtz Alliance 'Planetary Evolution and Life' and the Jet Propulsion Laboratory under contract to the National Aeronautics and Space Administration.

We dedicate this work to Steve Ostro, who provided outstanding scientific contributions to the exploration of the Saturnian System.

References

Allison, M.L. and Clifford, S.M., 1987, Ice-covered water volcanism on Ganymede, J. Geophys. Res. 92, 7865–7876.
Beeman, M., Durham, W.B., Kirby, S.B., 1988, Friction of ice, J. Geophys. Res. 93, 7625–7633.
Bell, J.F., Cruikshank, D.P., and Gaffey, M.J., 1985, The composition and origin of the Iapetus dark material, Icarus 61, 192–207.
Bills, B.G. and Ray, R.D., 2000, Energy dissipation by tides and librations in synchronous satellites, *31st Lunar and Planetary Lunar and Planetary Science Conference*, Houston, abstract #1709.

Black, G.J., 1997, Chaotic rotation of Hyperion and modeling the radar properties of the icy Galilean satellites as a coherent backscatter effect. PhD thesis, Cornell University.

Black, G.J., Campbell, D.B., and Ostro, S.J., 2001a, Icy Galilean satellites: 70 cm radar results from Arecibo, Icarus 151, 160–166.

Black, G.J., Campbell, D.B., and Nicholson, P.D., 2001b, Icy Galilean satellites: Modeling radar reflectivities as a coherent backscatter effect, Icarus 151, 167–180.

Black, G.J., Campbell, D.B., Carter, L.M., and Ostro, S.J., 2004, Radar detection of Iapetus, Science 304, 553.

Black, G.J., Campbell, D.B., and Carter, 2007, Arecibo radar observations of Rhea, Dione, Tethys, and Enceladus, Icarus 191, 702–711.

Brown, R.H., Baines, K.H., Bellucci, G., Bibring, J.-P., Buratti, B.J., Bussoletti, E., Capaccioni, F., Cerroni, P., Clark, R.N., Coradini, A., Cruikshank, D.P., Drossart, P., Formisano, V., Jaumann, R., Langevin, Y., Matson, D.L., McCord, T.B., Miller, E., Nelson, R.M., Nicholson, P.D., Sicardy, B., and Sotin, C., 2004, The Cassini visual and infrared mapping spectrometer investigation, Space Sci. Rev. 115, 111–168.

Brown, R.H., Clark, R.N., Buratti, B.J., Cruikshank, D.P., Barnes, J.W., Mastrapa, R.M.E., Bauer, J., Newman, S., Momary, T., Baines, K.H., Bellucci, G., Capaccioni, F., Cerroni, P., Combes, M., Coradini, A., Drossart, P., Formisano, V., Jaumann, R., Langevin, Y., Matson, D.L., McCord, T.B., Nelson, R.M., Nicholson, P.D., Sicardy, B., and Sotin, C., 2006, Composition and physical properties of Enceladus surface, Science, 311, 1425–1428.

Brown, R.H., Lebreton, J.-P., Waite, H., (eds.), 2009, *Titan from Cassini-Huygens*, Springer NY, 2009.

Bruesch, L.S. and Asphaug, E., 2004, Modeling global impact effects on middle-sized icy bodies: Applications to Saturn's moons, Icarus 168, 457–466.

Buratti, B.J., 1984, Voyager disk resolved photometry of the Saturnian satellites, Icarus 59, 426–435.

Buratti, B.J. and Veverka, J., 1984, Voyager photometry of Rhea, Dione, Tethys, Enceladus, and Mimas, Icarus 58, 254–264.

Buratti, B.J., 1985, Application of a radiative transfer model to bright icy satellites, Icarus 61, 208–217.

Buratti, B.J. and Veverka, J. 1985, Photometry of rough planetary surfaces: The role of multiple scattering, Icarus 64, 320–328.

Buratti, B.J. 1988, Enceladus: Implications of its unusual photometric properties, Icarus 75, 113–126.

Buratti, B.J., Mosher, J., and Johnson, T.V. 1990, Albedo and color maps of the Saturnian satellites, Icarus 87, 339–357.

Buratti, B.J. and Mosher, J., 1995, The dark side of Iapetus: New evidence for an exogenous origin, Icarus 115, 219–227.

Buratti, B.J., Mosher, J.A., Nicholson, P.D., McGhee, C., and French, R., 1998, Photometry of the Saturnian satellites during Ring Plane Crossing, Icarus 136, 223–231.

Buratti, B.J., Hicks, M.D., Tryka, K.A., Sittig, M.S. and Newburn, R.L., 2002, High-resolution 0.33–0.92 μm spectra of Iapetus, Hyperion, Phoebe, Rhea, Dione, and D-type asteroids: How are they related? Icarus 155, 375–381.

Buratti, B., Hicks, M., and Davies, A., 2005a, Spectrophotometry of the small satellites of Saturn and their relationship to Iapetus, Phoebe, and Hyperion, Icarus 175, 490–495.

Buratti, B.J., Cruikshank, D.P., Brown, R.H., Clark, R.N., Bauer, J.M., Jaumann, R., McCord, T.B., Simonelli, D.P., Hibbitts, C.A., Hansen, G.A., Owen, T.C., Baines, K.H., Bellucci, G., Bibring, J.-P., Capaccioni, F., Cerroni, P., Coradini, A., Drossart, P., Formisano, V., Langevin, Y., Matson, D.L., Mennella, V., Nelson, R.M., Nicholson, P.D., Sicardy, B., Sotin, C., Roush, T.L., Soderlund, K., and Muradyan, A., 2005b, Cassini visual and infrared mapping spectrometer observations of Iapetus: Detection of CO_2, Astrophys. J. 622, L149–L152.

Buratti, B.J., Soderlund, K., Bauer, J., Mosher, J.A., Hicks, M.D., Simonelli, D.P., Jaumann, R., Clark, R.N., Brown, R.H., Cruikshank, D.P., and Momary, T., 2008, Infrared (0.83–5.1 μm) Photometry of Phoebe from the Cassini visual infrared mapping spectrometer, Icarus 193, 309–322.

Buratti, B.J., Bauer, J., Hicks, M.D., Mosher, J.A., Filacchione, G., Momary, T.R., Baines, K. Brown, R.H., Clark, R.N., and Nicholson, P.D. 2009a, Cassini spectra and photometry 0.25–5.1 μm of the small inner satellites of Saturn, Icarus, in revision.

Buratti, B.J., Hillier, J., Mosher, J., DeWet, S., Abramson, L., Akhter, N., Clark, R.N., Brown, R.H., Baines, K., and Nicholson, P., 2009b, Opposition surges on the icy satellites of Saturn as seen by Cassini, Icarus, submitted.

Burch, J.L., Goldstein, J., Lewis, W.S., Young, D.T., Coates, A.J., Dougherty, M.K., and André, N., 2007, Tethys and Dione as sources of outward-flowing plasma in Saturn's magnetosphere, Nature 447, 833–835. doi:10.1038/nature05906.

Burns, J.A., Lamy, P.L., and Soter, S., 1979, Radiation forces on small particles in the solar system, Icarus 40, 1–48.

Carlson, R., Smythe, W., Baines, K., Barbinas, E., Burns, R., Calcutt, S., Calvin, W., Clark, R., Danielson, G., Davies, A., Drossart, P., Encrenaz, T., Fanale, F., Granahan, J., Hansen, G., Hererra, P., Hibbitts, C., Hui, J., Irwin, P., Johnson, T., Kamp, L., Kieffer, H., Leader, F., Lopes-Gautier, R., Matson, D., McCord, T., Mehlman, D., Ocampo, A., Orton, G., Roos-Serote, M., Segura, M., Shirley, J., Soderblom, L., Stevenson, A., Taylor, F., Weir, A., and Weissman, P., 1996, Near-infrared spectroscopy and spectral mapping of Jupiter and the Galilean satellites: First results from Galileo's initial orbit, Science, 274, 385–388.

Carlson, R.W., Anderson, M.S., Johnson, R.E., Smythe, W.D., Hendrix, A.R., Barth, C.A., Soderblom, L.A., Hansen, G.B., McCord, T.B., Dalton, J.B., Clark, R.N., Shirley, H.J., Ocampo, A.C., and Matson, D.L., 1999, Hydrogen peroxide on the surface of Europa, Science, 283, 2062–2064.

Cassen, P., Reynolds, R.T., and Peale, S.J., 1979, Is there liquid water on Europa? Geophys. Res. Lett. 6, 731–734.

Cassini, G.D., 1672, Phil. Trans. 12, 831. Quoted in Alexander, A.F. O'D., 1962, *The Planet Saturn*, McMillan, New York, 474 p.

Castillo-Rogez, J.C., Matson, D.L., Sotin, C., Johnson, T.V., Lunine, J.I., and Thomas, P.C., 2007, Iapetus' geophysics: Rotation rate, shape and equatorial ridge, Icarus 190, 179–202.

Chapman C.R. and McKinnon, W.B., 1986, Cratering of planetary satellites. In: *Satellites* (Burns, J.A., Matthews, M.S. (eds.)), Univ. Ariz. Press, Tucson, pp. 492–580.

Chapman, C.R., 2004, Space weathering of asteroid surfaces. Annual. Rev. Earth Planet Sci. 32, 539–567.

Chen, E.M.A. and Nimmo, F., 2008, Implications from Ithaca Chasma on the thermal and orbital history of Tethys, Geophys. Res. Lett. 35, L19203.

Clark, R.N., Brown, R.H., Owensby, P.D., and Steele, A., 1984, Saturn's satellites: near-infrared spectrophotometry (0.65–2.5 μm) of the leading and trailing sides and compositional implications, Icarus 58, 265–281.

Clark, R.N., Fanale, F.P., and Gaffey, M.J., 1986, Surface composition of natural satellites. In: *Satellites* (Burns, J.A., Matthews, M.S. (eds.)), Univ. Ariz. Press, Tucson, pp. 437–491.

Clark, R.N., Brown, R.H., Jaumann, R., Cruikshank, D.P., Nelson, R.M., Buratti, B.J., McCord, T.B., Lunine, J., Baines, K.H., Bellucci, G., Bibring, J.-P., Capaccioni, F., Cerroni, P., Coradini, A., Formisano, F., Langevin, Y., Matson, D.L., Mennella, V., Nicholson, P.D., Sicardy, B., Sotin, C., Hoefen, T.M., Curchin, J.M., Hansen, G., Hibbits, K., and Matz, K.-D., 2005, Compositional maps of Saturn's moon Phoebe from imaging spectroscopy, Nature 435, 66–69, doi:10.1038/nature03558.

Clark, R.N., Curchin, J.M., Jaumann, R., Cruikshank, D.P., Brown, R.H., Hoefen, T.M., Stephan, K., Moore, J.M., Buratti, B.J., Baines, K.H., Nicholson, P.D., and Nelson, R.M., 2008a, Compositional

Mapping of Saturn's Satellite Dione with Cassini VIMS and Implications of Dark Material in the Saturn System, Icarus 193, 372–386.

Clark, R.N., Cruikshank, D.P., Jaumann, R., Filacchione, G., Nicholson, P.D., Brown, R.H., Stephan, K., Hedman, M., Buratti, B.J., Curchin, J.M., T.M., Hoefen, Baines, K.H., and Nelson, R., 2008b, Compositional Mapping of Saturn's Rings and Icy Satellites with Cassini VIMS, Saturn After Cassini-Huygens, London, July, 2008.

Clark R.N., Cruikshank, D.P., Jaumann, R., Brown, R.H., Cruchin, J.M., Hoefen, T.M., Stephan, K., Dalle Ore, C., Buratti, J.B., Filacchione, G., Baines, K.H., and Nicholson, P.D., 2009, The Composition of Iapetus: Mapping Results from Cassini VIMS, Icarus, in review.

Collins, G.C., McKinnon, W.B., Moore, J.M., Nimmo, F., Pappalardo, R.T., Prockter, L.M., and Schenk, P.M., 2009, Tectonics of the outer planet satellites, in Planetary Tectonics, (R.A. Schultz and T.R. Watters (eds.)), Cambridge Univ. Press, in press.

Cook, A.F. and Franklin, F., 1970, An explanation for the light curve of Iapetus, Icarus 13, 282–291.

Crawford, G.D. and Stevenson, D.J., 1988, Gas-driven water volcanism and the resurfacing of Europa, Icarus 73, 66–79.

Cruikshank, D.P. and Brown, R.H., 1982, Surface composition and radius of Hyperion, Icarus 50, 82–87.

Cruikshank, D.P., Bell, J.F., Gaffey, M.J., Brown, R.H., Howell, R., Beerman, C., and Rognstad, M., 1983, The Dark Side of Iapetus, Icarus 53, 90–104.

Cruikshank, D.P., Veverka, J., Lebofsky, L.A., 1984, Satellites of Saturn: The optical properties, In Saturn, T. Gehrels and M.S. Matthews (eds.), Univ. Ariz. Press, 640–667.

Cruikshank, D.P., Brown, R.H., Calvin, W., Roush, T.L., and Bartholomew, M.J., 1998a, Ices on the satellites of Jupiter, Saturn, and Uranus. In: *Solar System Ices* (Schmitt, B., de Bergh, C., Festou, M. (eds.)), Kluwer Academic, Dordrecht, pp. 579–606.

Cruikshank, D.P., Owen, T.C., Ore, C.D., Geballe, T.R., Roush, T.L., de Bergh, C., Sandford, S.A., Poulet, F., Benedix, G.K., and Emery, J.P., 2005, A spectroscopic study of the surfaces of Saturn's large satellites: H$_2$O ice, tholins, and minor constituents, Icarus 175, 268–283.

Cruikshank, D.P., Dalton, J.B., Dalle Ore, C.M., Bauer, J., Stephan, K., Filacchione, G., Hendrix, C.J., Hansen, C.J., Coradini, A., Cerroni, P., Tosi, F., Capaccioni, F., Jaumann, R., Buratti, B.J., Clark, R.N., Brown, R.H., Nelson, R.M., McCord, T.B., Baines, K.H., Nicholson, P.D., Sotin, C., Meyer, A.W., Bellucci, G., Combes, M., Bibring, J.-P., Langevin, Y., Sicardy, B., Matson, D.L., Formisano, V., Drossart, P., and Mennella, V., 2007, Surface composition of Hyperion, Nature 448, 54–57, doi:101038/nature05948.

Cruikshank, D.P., Wegryn, E., Dalle Ore, C.M., Brown, R.H., Baines, K.H., Bibring, J.-P., Buratti, B.J., Clark, R.N., McCord, T.B., Nicholson, P.D., Pendleton, Y.J., Owen, T.C., Filacchione, G., and the VIMS Team, 2008, Hydrocarbons on Saturn's Satellites Iapetus and Phoebe, Icarus 193, 334–343, doi:10.1016/j.icarus.2007.04.036.

Cuzzi, J.N., Dones, L., Clark, R.N., French, R.G., Johnson, T.V., Marouf, E.A., Spilker, L.J., 2009, Ring composition. In: *Saturn after Cassini-Huygens, Saturn from Cassini-Huygens* (M. Dougherty, L. Esposito and S. Krimigis, (eds.)), Springer NY, 2009.

Czechowski, L. and Leliwa-Kopystyński, J., 2008, The Iapetus's ridge: Possible explanations of its origin, Adv. Space Res. 42, 61–69.

Denk, T., Matz, K.-D., Roatsch, T., Wolf, U., Wagner, R.J., Neukum, G., and Jaumann, R., 2000, Iapetus (1): size, topography, surface structures, craters. *31th Lunar and Planetary Science Conference*, Houston, abstract #1596.

Denk, T. and Neukum, G., 2000, Iapetus (2): Dark-side origin. *31th Lunar and Planetary Science Conference*, Houston, abstract #1660.

Denk, T., Neukum, G., Roatsch, T., Giese, B., Wagner, R.J., Helfenstein, P., Burns, J.A., Turtle, E.P., Johnson, T.V., Porco, C.C., Cassini Imaging Team, 2006, Iapetus: Two Years of Observations by the Cassini ISS Camera, 1st European Planetary Science Congress, EPSC2006-A-00571.

Denk, T., Neukum, G., Schmedemann, N., Roatsch, T., Wagner, R.J., Giese, B., Perry, J.E., Helfenstein, P., Turtle, E.P., and Porco, C.C., 2008, Iapetus imaging during the targeted flyby of the Cassini spacecraft, *39th Lunar and Planetary Science Conference, Houston*, abstract #2533.

Denk, T. and Spencer, J.R., 2008, Iapetus: A two-step explanation for its unique global appearance. BAAS 40 (3), 61.04. http://www.abstractsonline.com/viewer/viewAbstract.asp?CKey={105D4425-F206-476D-BDF1-D155267B27BD}&MKey={35A8F7D5-A145-4C52-8514-0B0340308E94}&AKey={AAF9AABA-B0FF-4235-8AEC-74F22FC76386}&SKey={2B16A2FD-1D5E-4F2A-B234-3EDCA2413097} http://www.abstractsonline.com/viewer/viewAbstract.asp?CKey = {105D4425-F206–476D-BDF1-D155267B27BD}& MKey = {35A8F7D5-A145–4C52–8514–0B0340308E94} & AKey = {AAF9AABA-B0FF-4235–8AEC-74F22FC76386}& SKey = {2B16A2FD-1D5E-4F2A-B234–3EDCA2413097}.

Denk, T., Neukum, G., Roatsch, Th., Porco, C.C., Burns, J.A., Galuba, G.G., Schmedemann, N., Helfenstein, P., Thomas, P.C., Wagner, R.J., West, R.A., 2009, Iapetus: Unique Surface Properties and a Global Color Dichotomy from Cassini Imaging. Science under revision.

Domingue, D.L., Lockwood, G.W., and Thompson, D.T., 1995, Surface textural properties of icy satellites: A comparison between Europa and Rhea, Icarus 115, 228–249.

Dones, L., Chapman, C., Melosh, H.J., McKinnon, W.B., Neukum, G., and Zahnle, K., 2009, Impact cratering and age determination. In *Saturn after Cassini Huygens, Saturn from Cassini-Huygens* (M. Dougherty, L. Esposito and S. Krimigis, (eds.)), Springer NY, 2009.

Dougherty, M.K., Khurana, K.K., Neubauer, F.M., Russell, C.T., Saur, J., Leisner, J.S., and Burton, M.E., 2006, Identification of a dynamic atmosphere at Enceladus with the Cassini magnetometer, Science 311, 1406–1409.

Dougherty, M.K., Esposito, L.W., Krimigis, S.M., 2009, *Overview, Saturn from Cassini-Huygens*, (M. Dougherty, L. Esposito and S. Krimigis, (eds.)), Springer NY, 2009.

Durham, W.B. and Stern, L.A., 2001, Rheological properties of water ice – applications to satellites of the outer planets, Ann. Rev. Earth Planet. Sci. 29, 295–330.

Elachi, C., Allison, M., Borgarelli, L., Encrenaz, P., Im, E., Janssen, M.A., Johnson, W.T.K., Kirk, R.L., Lorenz, R.D., Lunine, J.I., Muhleman, D.O., Ostro, S.J., Picardi, G., Posa, F., Roth, L.E., Seu, R., Soderblom, L.A., Vetrella, S., Wall, S.D., Wood, C.A., and Zebker, H.A., 2004, RADAR: The Cassini Titan radar mapper, Space Sci. Rev. 115, 71–110.

Ellsworth, K. and Schubert, G., 1983, Saturn's icy satellites – thermal and structural models, Icarus 54, 490–510, 1983.

Eluszkiewicz, J.,1990, Compaction and internal structure of Mimas, Icarus 84, 215–225.

Esposito, L.W., Barth, C.A., Colwell, J.E., Lawrence, G.M., McClintock, W.E., Stewart, A.I.F., Keller, H.U., Korth, A., Lauche, H., Festou, M.C., Lane, A.L., Hansen, C.J., Maki, J.N., West, R.A., Jahn, H., Reulke, R., Warlich, K., Shemansky, D.E., Yung, Y.L., 2004, The Cassini ultraviolet imaging spectrograph investigation, Space Sci. Rev. 115, 299–361, doi:10.1007/s11214–004–1455–8.

Fagents, S.A., Greeley, R., Sullivan, R.J., Pappalardo, R.T., Prockter, L.M., The Galileo SSI Team, 2000, Cryomagmatic mechanisms for the formation of Rhadamanthys Linea, triple band margins, and other low-albedo features on Europa, Icarus 144, 54–88, doi:10.1006/icar.1999.6254.

Fagents, S.A., 2003, Considerations for effusive cryovolcanism on Europa: the post-Galileo perspective, J. Geophys. Res. 108, 5139.

Filacchione, G., Capaccioni, F., McCord, T.B., Coradini, A., Cerroni, P., Bellucci, G., Tosi, F., D'Aversa, E., Formisano, V., Brown, R.H., Baines, K.H., Bibring, J.-P., Buratti, B.J., Clark, R.N., Combes, M., Cruikshank, D.P., Drossart, P., Jaumann, R., Langevin, Y., Matson,

D.L., Mennella, V., Nelson, R.M., Nicholson, P.D., Sicardy, B., Sotin, C., Hansen, G., Hibbitts, K., Showalter, M., and Newmann, S., 2007, Saturn's icy satellites investigated by Cassini VIMS. I. Full-disk properties: 0.35–5.1 μm reflectance spectra and phase curves, Icarus 186, 259–290.

Filacchione, G., Capaccioni, F., Tosi, F., Cerroni, P., McCord, T.B., Baines, K.H., Bellucci, G., Brown, R.H., Buratti, B.J., Clark, R.N., Cruikshank, D.P., Cuzzi, J.N., Jaumann, R., Stephan, K., Matson, D.L., Nelson, R.M., and Nicholson, P.D., 2008, Analysis of the Saturnian icy satellites full-disk spectra by Cassini-VIMS, Saturn After Cassini-Huygens, London, July, 2008.

Fink, U. and Larson, H.P., 1975, Temperature dependence of the water-ice spectrum between 1 and 4 microns: application to Europa, Ganymede, and Saturn's rings, Icarus 24, 411–420.

Flasar, F.M., Kunde, V.G., Abbas, M.M., Achterberg, R.K., Ade, P., Barucci, A., Bézard, B., Bjoraker, G.L., Brasunas, J.C., Calcutt, S., Carlson, R., Césarsky, C.J., Conrath, B.J., Coradini, A., Courtin, R., Coustenis, A., Edberg, S., Edgington, S., Ferrari, C., Fouchet, T., Gautier, D., Gierasch, P.J., Grossman, K., Irwin, P., Jennings, D.E., Lellouch, E., Mamoutkine, A.A., Marten, A., Meyer, J.P., Nixon, C.A., Orton, G.S., Owen, T.C., Pearl, J.C., Prangé, R., Raulin, F., Read, P.L., Romani, P.N. Samuelson, R.E., Segura, M.E., Showalter, M.R., Simon-Miller, A.A., Smith, M.D., Spencer, J.R., Spilker, L.J., and Taylor, F.W., Exploring the Saturn system in the thermal infrared: The composite infrared spectrometer, 2004, Space Sci. Rev. 115, 169–297, doi 10.1007/s11214-004-1454-9.

Forsberg-Taylor, N.K., Howard, A.D., and Craddock, R.A. 2004, Crater degradation in the Martian highlands: Morphometric analysis in the Sinus Sabaeus region and simulation modeling suggest fluvial processes, J. Geophys. Res. 109, doi:10.1029/2004JE002242.

Gehrels, T. and Matthews, M.S., 1984, *Saturn*, Univ. Ariz. Press, Tucson.

Geissler, P.E., 2000, Cryovolcanism in the outer solar system. In *Encyclopedia of Volcanoes* (Sigurdsson, H. (ed.)), Academic, San Diego, pp. 785–800.

Giese, B., Wagner, R., Neukum, G., Helfenstein, P., and Thomas, P.C., 2007, Tethys: Lithospheric thickness and heat flux from flexurally supported topography at Ithaca Chasma, Geophys. Res. Lett. 34, L21203.

Giese, B., Denk, T., Neukum, G., Roatsch, T., Helfenstein, P., Thomas, P.C., Turtle, E.P., McEwen, A., and Porco, C.C., 2008, The topography of Iapetus' leading side, Icarus 193, 359–371.

Goguen, J. 1981, A theoretical and experimental investigation of the photometric functions of particulate surfaces, PhD Thesis, Cornell Univ., Ithaca, NY.

Greenberg, R. and Weidenschilling, S.J., 1984, How fast do Galilean satellites spin? Icarus 58, 186–196.

Greenberg, R. Geissler, P., Hoppa, G., Tufts, B.R., Durda, D.D., Pappalardo, R., Head, J.W., Greeley, R., Sullivan, R., Carr, M.H., 1998, Tectonic processes on Europa: Tidal stresses, mechanical response, and visible features, Icarus 135, 64–78, doi:10.1006/icar.1998.5986.

Greenberg, R. and Geissler, P., 2002, Europa's dynamic icy crust, Meteorit. Planet. Sci. 37, 1685–1710.

Hamilton, D.P. and Burns, J.A., 1994, Origin of Saturn's E-ring: Self sustained, naturally, Science 264, 550–553.

Hapke, B. 1981, Bidirectional reflectance spectroscopy, 1. Theory, J. Geophys. Res. 86, 4571–4586.

Hapke, B. 1984, Bidirectional reflectance spectroscopy, 3. Correction for macroscopic roughness, Icarus 59, 41–59.

Hapke, B. 1986, Bidirectional reflectance spectroscopy, 4. The extinction coefficient and the opposition effect, Icarus 67, 264–280.

Hapke, B. 1990, Coherent backscatter and the radar characteristics of outer planet satellites, Icarus 88, 407–417.

Hawkes, I. and Mellor, M., 1972, Deformation and fracture of ice under uniaxial stress, J. Glaciol. 11, 61, 103–131.

Helfenstein P. and Parmentier, E.M., 1983, Patterns of fracture and tidal stresses on Europa, Icarus 53, 415–430.

Helfenstein, P. and Parmentier, E.M., 1985, Patterns of fracture and tidal stresses due to nonsynchronous rotation – implications for fracturing on Europa, Icarus 61, 175–184.

Helfenstein, P. and Shepard, M.K., 1998, Submillimeter-scale topography of the lunar Regolith, Icarus 124, 262–267.

Hendrix, A.R. Barth, C.A., Hord, C.W., and Lane, A.L., 1998, Europa: Disk-resolved ultraviolet measurements using the Galileo ultraviolet spectrometer, Icarus 135, 79–94.

Hendrix, A.R., Barth, C.A., Hord, C.W., 1999, Ganymede's ozone-like absorber: Observations by the Galileo Ultraviolet Spectrometer, J. Geophys. Res. 104, 14169–14178.

Hendrix, A.R. and C.J. Hansen, 2008a, The albedo dichotomy of Iapetus measured at UV wavelengths, Icarus 193, 344–351.

Hendrix, A.R. and C.J. Hansen, 2008b, Ultraviolet observations of Phoebe from the Cassini UVIS, Icarus 193, 323–333.

Hendrix, A.R., and C.J. Hansen, 2009a, The UV albedo of Enceladus: Implications for surface composition, Icarus under revision.

Hendrix, A.R. and Hansen, C.J., 2009b, Iapetus: New results from Cassini UVIS, *40th Lunar and Planetary Lunar and Planetary Science Conference*, Houston, abstract #2200.

Hendrix, A.R. and Buratti, B.J., 2009, Multi-wavelength photometry of icy Saturnian satellites: A fist look, *40th Lunar and Planetary Lunar and Planetary Science Conference*, Houston, abstract 2438#.

Hibbitts, C.A., McCord, T.B., and Hansen, G.B., 2000, Distributions of CO_2 and SO_2 on the surface of Callisto, J. Geophys. Res. 105, 22541–22557.

Hibbitts, C.A., Pappalardo, R.T., Hansen, G.B., and McCord, T.B., 2003, Carbondioxide on Ganymede, J. Geophys. Res. 108, doi:10.1029/2002JE001956.5036.

Hillier, J. and Squyres, S.W., 1991, Thermal stress tectonics on the satellites of Saturn and Uranus, J. Geophys. Res. 96, 15665–15674.

Hoppa, G.V., Tufts, B.R., Greenberg, R., and Geissler, P.E., 1999a, Formation of cycloidal features on Europa, Science 285, 1899–1902.

Hoppa, G., Tufts, B.R., Greenberg, R., and Geissler, P., 1999b, Strike-slip faults on Europa: Global shear patterns driven by tidal stress, Icarus 141, 287–298.

Horak, H.G., 1950, Diffuse reflection by planetary atmospheres, Ap. J. 112, 445–463.

Horányi, M., Juhász, A., and Morfill, E.G., 2008, Large-scale structure of Saturn's E-ring, Geophys. Res. Lett., 35, L04203, doi:10.1029/2007GL032726.

Howard, A.D., 2007, Simulating the development of Martian highland landscapes through the interaction of impact cratering, fluvial erosion, and variable hydrologic forcing, Geomorphology, 9, 332–363.

Howard, A.D. and Moore, J.M., 2008, Sublimation-driven erosion on Callisto: A landform simulation model test, Geophys. Res. Let. 35, L03203, doi:101029/2007GL032618.

Ip, W.-H., 2006, On a ring origin of the equatorial ridge of Iapetus. Geophys. Res. Lett. 33, doi:10.1029/2005GL025386. L16203.

Irvine, W. 1966, The shadowing effect in diffuse radiation, J. Geophys. Res. 71, 2931–2937.

Jacobson, R.A. and French, R.G., 2004, Orbits and masses of Saturn's coorbital and F-ring shepherding satellites, Icarus 172, 382–387.

Jacobson, R.A. Antreasian, P.G., Bordi, J.J., Criddle, K.E., Ionasescu, R., Jones, J.B., Meek, M.C., Owen, W.M., Jr., Roth, D.C., Roundhill, I.M., and Stauch, J.R., Cassini Navigation, 2004, The orbits of the major Saturnian satellites and the gravity field of the Saturnian system, Bull. Am. Astronom. Soc. 36(3), 1097.

Jacobson, R.A., Antreasian, P.G., Bordi, J.J., Criddle, K.E., Ionasescu, R., Jones, J.B., Mackenzie, R.A., Pelletier, F.J., Owen Jr., W.M., Roth, D.C., and Stauch, J.R., 2005, The gravity field of the Saturnian system and the orbits of the major Saturnian satellites, BAAS 37, 729.

Jacobson, R.A., 2006, SAT252 – JPL satellite ephemeris.

Jacobson, R.A., 2007, SAT270, SAT271 – JPL satellite ephemerides.

Janes, D.M. and Melosh, H.J., 1988, Sinker tectonics – An approach to the surface of Miranda, J. Geophys. Res. 93, 3217–3143.

Jarvis, K.S., Vilas, F., Larson, S.M., and Gaffey, M.J., 2000, Are Hyperion and Phoebe linked to Iapetus? Icarus 146, 125–132.

Jaumann, R., Stephan, K., Brown, R.H., Buratti, B.J., Clark, R.N., McCord, T.B., Coradini, A., Capaccioni, P., Filacchione, G., Cerroni, P., Baines, K.H., Bellucci, G., Bibring, J.P., Combes, M., Cruikshank, D.P., Drossart, P., Formisano, V., Langevin, Y., Matson, D.L., Nelson, R.M., Nicholson, P.D., Sicardy, B., Sotin, C., Soderblom, L.A., Griffith, C., Matz, K.D., Roatsch T., Scholten, F., and Porco, C.C., 2006, High-resolution Cassini-VIMS mosaics of Titan and the icy Saturnian satellites, Planet Space Sci. 54, 1146–1155.

Jaumann, R., Stephan, K., Hansen, G.B., Clark, R.N., Buratti, B.J., Brown, R.H., Baines, K.H., Bellucci, G., Coradini, A., Cruikshank, D.P., Griffith, C.A., Hibbits, C.A., McCord, T.B., Nelson, R.M., Nicholson, P.D., Sotin, C., and Wagner, R., 2008, Distribution of icy particles across Enceladus' surface as derived from Cassini-VIMS measurements, Icarus 193, 407–419, doi:10.1016/j.icarus.2007.09.013.

Jaumann, R., Kirk, R.L., Lorenz, R.D., Lopez, R.M.G., Stofan, E.R., Turtle, E.P., Keller, H.U., Wood, C.A., Sotin, C., Soderblom, L.A., and Tomasko, M., 2009, Geology and surface processes on Titan, Titan from Cassini-Huygens (Brown, R. Lebreton, J.-P., and Waite, H. (eds.)), Springer NY, 2009.

Johnson, R.E., 1997, Polar "caps" on Ganymede and Io revisited, Icarus 128, 469–471.

Johnson, R.E. and Jesser, W.A., 1997, O_2/O_3 microatmospheres in the surface of Ganymede, Astrophys. J. 480, L79–L82.

Johnson, R.E. and Quickenden, T.I., 1997, Photolysis and radiolysis of water ice on outer solar system bodies, J. Geophys. Res. 102, 10985–10996.

Johnson, R.E., Killen, R.M., Waite, J.H., and Lewis, W.S., 1998, Europa's surface composition and sputter-produced ionosphere, Geophys. Res. Lett. 25, 3257–3260.

Johnson, R.E., Carlson, R.W., Cooper, J.F., Paranicas, C., Moore, M.H., and Wong, M.C., 2004, Radiation effects on the surfaces of the Galilean satellites. In: *Jupiter – the Planet, Satellites and Magnetosphere* (Bagenal, F., Dowling, T., and McKinnon, W. B. (eds.)), Cambridge University, Cambridge.

Johnson, R.E., Fama, M., Liu, M., Baragiola, R.A., Sittler, E.C., and Smith, H.T., 2008, Sputtering of ice grains and icy satellites in Saturn's inner magnetosphere, Planet. Space Sci. 56, 1238–1243.

Johnson, T.V., 1998, Introduction to icy satellite geology. In *Solar System Ices* (Schmitt, B., De Bergh, C., and Festou, M. (eds.)), Kluwer, Dordrecht, pp. 511–524.

Johnson, T.V., Estrada, E., and Lissauer, J., 2009, *Origin of the Saturnian System, Saturn from Cassini-Huygens* (Dougherty, M. Esposito, L. and Krimigis, S. (eds.)), Springer NY, 2009.

Jones, G.H., Roussos, E., Krupp, N., Paranicas, C., Woch, J., Lagg, A., Mitchell, D.G., Krimigis, S.M., and Dougherty, M.K., 2006, Enceladus' varying imprint on the magnetosphere of Saturn, Science 311, 1412–1415.

Jones, G.H., Roussos, E., Krupp, N., Beckmann, U., Coates, A.J., Crary, F., Dandouras, I., Dikarev, V., Dougherty, M.K., Garnier, P., Hansen, C.J., Hendrix, A.R., Hospodarsky, G.B., Johnson, R.E., Kempf, S., Khurana, K.K., Krimigis, S.M., Krüger, H., Kurth, W.S., Lagg, A., McAndrews, H.J., Mitchell, D.G., Paranicas, C., Postberg, F., Russell, C.T., Saur, J., Seiß, M., Spahn, F., Srama, R., Strobel, D.F., Tokar, R., Wahlund, J.-E., Wilson, R.J., Woch, J., and Young, D., 2008, The dust halo of Saturn's largest icy moon, Rhea, Science 319, 1380–1384. doi:10.1126/science.1151524.

Kargel, J.S., 1991, Brine volcanism and the interior structures of asteroids and icy satellites, Icarus 95, 368–390.

Kargel, J.S., Croft, S.K., Lunine, J.I., and Lewis, J.S., 1991, Rheological properties of ammonia-water liquids and crystal-liquid slurries: Planetological implications, Icarus 89, 93–112.

Kargel, J.S., 1992, Ammonia-water volcanism in icy satellites: Phase relations at 1 atmosphere, Icarus 100, 556–574.

Kargel, J.S., 1995, Cryovolcanism on the icy satellites, Earth Moon Planets 67, 101–113.

Kargel, J.S. and Pozio, S., 1996, The volcanic and tectonic history of Enceladus, Icarus 119, 385–404.

Kempf, S., Srama, R., Postberg, F., Burton, M., Green, S.F., Helfert, S. Hillier, J.K., McBride, N., McDonnell, J.A.M., Moragas-Klostermeyer, G., Roy, M., and Grüen, E., 2005, Composition of Saturnian stream particles, Science, 307, 1274–1276.

Kempf, S., Srama, R., Beckmann, U., and Schmidt, J., Saturn's E ring as seen by the Cassini dust detector, 2008, AGU Fall Meeting, http://www.agu.org/meetings/fm08/fm08-sessions/fm08_P32A.html.

Kirchoff, M. and Schenk, P., 2008, Cratering records of Saturnian satellites, *39th Lunar and Planetary Science Conference*, Houston, abstract # 2234.

Kirchoff, M. and Schenk, P., 2009, Cratering records of Saturnian satellites, Icarus under revision.

Kirk, R.L. and Stevenson, D.J., 1987, Thermal evolution of a differentiated Ganymede and implications for surface features, Icarus 69, 91–134.

Klavetter, J.J., 1989 Rotation of Hyperion, I: Observations Astron. J. 97, 570–579.

Lange, M.A. and Ahrens, T.J., 1987, The dynamic tensile strength of ice and ice-silicate mixtures, J. Geophys. Res. 88, 1197–1208.

Lee, J., Buratti, B.J., Mosher, J.A., and Hicks, M.D., 2009, The relative surface roughness of the dark sides of Iapetus from the 2005 flyby, Icarus under revision.

Leith, A.C. and McKinnon, W.B., 1996, Is there evidence for polar wander on Europa? Icarus 120, 387–398.

Lopes, R.M.C., Mitchell, K.L., Stofan, E.R., Lunine, J.I., Lorenz, R., Paganelli, F., Kirk, R.L., Wood, C.A., Wall, S.D., Robshaw, L.E., Fortes, A.D., Neish, C.D., Radebaugh, J., Reffet, E., Ostro, S.J., Elachi, C., Allison, M.D., Anderson, Y., Boehmer, R., Boubin, G., Callahan, P., Encrenaz, P., Flamini, E., Francescetti, G., Gim, Y., Hamilton, G., Hensley, S., Janssen, M.A., Johnson, W.T.K., Kelleher, K., Muhleman, D.O., Ori, G., Orosei, R., Picardi, G., Posa, F., Roth, L.E., Seu, R., Shaffer, S., Soderblom, L.A., Stiles, B., Vetrella, S., West, R.D., Wye, L., and Zebker, H.A., 2007, Cryovolcanic features on Titan's surface as revealed by the Cassini Titan Radar mapper, Icarus 187, 395–412, doi:10.1016/j.icarus.2006.09.006.

Lorenz, R.D., Stiles, B., Kirk, R.L., Allison, M., Persi del Marmo, P., Iess, L. Lunine, J.I., Ostro, S.J., and Hensley, S., 2008, Titan's rotation reveals an internal ocean and changing zonal winds, Science 319, 1649–1651.

Lumme, K. and Bowell, E., 1981a, Radiative transfer in the surfaces of atmosphereless bodies, I. Theory. Astron. J. 86, 1694–1704.

Lumme, K. and Bowell, E., 1981b, Radiative transfer in the surfaces of atmosphereless bodies, II. Interpretation of Phase Curves. Astron. J. 86, 1705–1721.

MacKintosh, F.C. and John, S., 1989, Diffusing-wave spectroscopy and multiple scattering of light in correlated random media, Phys. Rev. B 40, 2383–2406.

Manga, M. and Wang, C.Y., 2007, Pressurized oceans and the eruption of liquid water on Europa and Enceladus, Geophys. Res. Lett. 34, L07202.

Marchi, S., Barbieri, C., Dell'Oro, A., and Paolicci, P., 2002, Hyperion-Iapetus: Collisional relationships, Astron. Astrophys. 381, 1059–1065.

Martens, H.R., Reisenfeld, D.B., Williams, J.D., Johnson, R.E., Smith, H.T., Baragiola, R.A., Thomsen, M.F., Young, D.T., and Sittler, E.C.

2007, Molecular oxygen ions in Saturn's inner magnetosphere for the first 24 Cassini orbits, AGU abstract #P43A-1032.

Mastrapa, R.M., Bernstein, M.P., Sandford, S.A., Roush, T.L., Cruikshank, D.P., and Dalle Ore, C.M., 2008, Optical constants of amorphous and crystalline H$_2$O-ice in the near infrared from 1.1 to 2.6 μm, Icarus 197, 307–320, DOI:10.1016/j.icarus.2008.04.008, 307–320.

Matson, D.L., Castillo-Roget, J.C., McKinnon, Schubert, G., and Sotin, C., 2009, Thermal evolution and internal structure. In *Saturn from Cassini-Huygens* (Dougherty, M., Esposito, L., and Krimigis, S. (eds.)), Springer NY, 2009.

Matsuyama, I. and Nimmo, F., 2007, Rotational stability of tidally deformed planetary bodies, J. Geophys. Res. 112, E11003.

Matsuyama, I. and Nimmo, F., 2008, Tectonic patterns on reoriented and despun planetary bodies, Icarus 195, 459–473.

Matthews, R.A.J., 1992, The darkening of Iapetus and the origin of Hyperion, Quart. J. Roy. Astron. Soc. 33, 253–258.

McDonald, G.D. Thompson, W.R., Heinrich, M., Khare, B.N., and Sagan, C., 1994, Chemical investigation of Titan and Triton tholins, Icarus 108, 137–145.

McKinnon, W.B. and Melosh, H.J., 1980, Evolution of planetary lithospheres – Evidence from multi-ringed structures on Ganymede and Callisto, Icarus 44, 454–471.

McKinnon, W.B., 1985, Geology of icy satellites. In: *Ices in the Solar System* (Klinger, J., Benest, D.,Dollfus, A., Smoluchowski, R. (eds.)), Reidel, Dordrecht, pp. 820–856.

McNutt, M.K., 1984, Lithospheric flexure and thermal anomalies, J. Geophys. Res. 89, 1180–1194.

Melosh H.J., 1975, Large impact basins and the Moon's orientation, Earth Planet Sci. Lett. 26, 353–360.

Melosh, H.J., 1977, Global tectonics of a despun planet, Icarus 31, 221–243.

Melosh, H.J., 1980, Tectonic patterns on a tidally distorted planet, Icarus 43, 334–337.

Melosh, H.J. and F. Nimmo, 2009, An intrusive dike origin for Iapetus' enigmatic ridge? *40th Lunar and Planetary Science Conference*, Houston, abstract #2478.

Mendis, D.A. and Axford, W.I., 1974, Satellites and magnetospheres of the outer planets, Ann. Rev. Earth Plant. Sci. 2, 419–475.

Miyamoto, H., Mitri, G., Showman, A.P., and Dohm, J.M., 2005, Putative ice flows on Europa: Geometric patterns and relation to topography collectively constrain material properties and effusion rates, Icarus 177, 413–424.

Moore, J.M. and Ahern, J.L., 1983, The geology of Tethys, JGR, 83, A577–A584.

Moore, J.M., 1984, The tectonic and volcanic history of Dione, Icarus 59, 205–220.

Moore, J.M., Horner, V.M., and Greeley, R., 1985, The geomorphology of Rhea: Implications for geologic history and surface processes, J. Geophys. Res. 90, C785–C795.

Moore, J.M., Asphaug, E., Morrison, D., Klemaszewski, J.E., Sullivan, R.J., Chuang, F., Greeley, R., Bender, K.C., Geissler, P.E., Chapman, C.R., Helfenstein, P., Pilcher, C.B., Kirk, R.L., Giese, B., and Spencer, J.R., 1999, Mass movement and landform degradation on the icy Galilean satellites: Results from the Galileo nominal mission, Icarus 140, 294–312.

Moore, J.M., Schenk, P.M., Bruesch, L.S., Asphaug, E., and McKinnon, W.B., 2004, Large impact features on middle-sized icy satellites, Icarus 171, 421–443.

Morrison, D., Johnson, T.V., Shoemaker, E.M., Soderbloom, L.A., Thomas, P., Veverka, J., and Smith, B.A. 1984, Satellites of Saturn: Geological perspective. In: *Saturn* (Gehrels, T. and Matthews, M. S. (eds.)), Univ. Ariz. Press, Tucson, pp. 609–63.

Morrison, D., Owen, T., and Soderblom, L.A., 1986, The satellites of Saturn. In: *Satellites* (Eds. Burns, J.A., Matthews, M.S. (eds.)) Univ. Ariz. Press, Tucson, pp. 764–801.

Mueller, S. and McKinnon, W.B., 1988, 3-layered models of Ganymede and Callisto – compositions, structures and aspects of evolution, Icarus 76, 437–464.

Murchie, S.L. and Head, J.W., 1986, Global reorientation and its effect on tectonic patterns on Ganymede, Geophys. Res. Lett. 13, 345–348.

Murphy, R.E., Cruikshank, D.P., Morrison, D., 1972, Radii, albedos, and 20-micron brightness temperatures of Iapetus and Rhea, Astrophys. J. Lett. 177, L93–L96.

Murray, C.D. and Dermott, S.F., 1999, *Solar System Dynamics*, Cambridge Univ. Press, Cambridge, NY.

Nelson, R.M., Lane, A.L., Matson, D.L., Veeder, G.J., Buratti, B.J., and Tedesco, E.F., 1987, Spectral geometric albedos of the Galilean satellites from 0.24 to 0.34 micrometers: Observations with the International Ultrviolett Explorer, Icarus 72, 358–380.

Nelson, R.M., Kamp, L.W., Lopes, R.M.C., Matson, D.L., Kirk, R.L., Hapke, B.W., Wall, S.D., Boryta, M.D., Leader, F.E., Smythe, W.D.,. Mitchell, K.L., Baines, K.H., Jaumann, R., Sotin, C., Clark, C.N., Cruikshank, D.P., Drossart, P., Lunine, J.I., Combes, M., Bellucci, G., Bibring, J.-P., Capaccioni, F., Cerroni, P., Coradini, A., Formisano, V., Filacchione, G., Langevin, Y., McCord, T.B., Mennella, V., Nicholson, P.D., Sicardy, B., Irwin, P.G.J., and Pearl, J.C., 2009a, Photometric changes on Saturn's Titan: Evidence for active Cryovolcanism, Geophys. Res. Lett. 36, L04202, doi:10.1029/2008GL036206.

Nelson, R.M., Kamp, L.W., Matson, D.L., Irwin, P.G.J., Baines, K.H., Boryta, M.D., Leader, F.E., Jaumann, R., Smythe, W.D., Sotin, C., Clark, C.N., Cruikshank, D.P., Drossart, P., Pearl, J.C., Habke, B.W., Lunine, J.I., Combes, M., Bellucci, G., Bibring, J.-P., Capaccioni, F., Cerroni, P., Coradini, A., Formisano, V., Filacchione, G., Langevin, R.Y., McCord, T.B., Mennella, V., Nicholson, P.D., and Sicardy, B., 2009b, Saturn's Titan: Surface change, ammonia, and implications for atmospheric and tectonic activity, Icarus 199, 429–441.

Nesvorny, D., Alverellos, J.L., Dones, L., and Levison, H.F., 2003, Orbital and collisional evolution of the irregular satellites, Astron. J. 126, 398–429.

Newman, S.F., Buratti, B.J., Jaumann, R., Bauer, J.M., and Momary, T.M., 2007, Hydrogen peroxide on Enceladus, Ap. J. 670, L143–L146.

Newman, S.F., Buratti, B.J., Brown, R.H., Jaumann, R., Bauer, J., and Momary, T., 2008, Photometric and spectral analysis of the distribution of crystalline and amorphous ices on Enceladus as seen by Cassini, Icarus 193, 397–406.

Newman, S.F., Buratti, B.J., Brown, R.H., Jaumann, R., Bauer, J., and Momary, T., 2009, Water ice crystallinity and grain sizes on Dione Icarus, in press.

Nimmo, F. and Gaidos, E., 2002, Thermal consequences of strike-slip motion on Europa, J. Geophys. Res. 107, 5021, 10.1029/2000JE001476.

Nimmo, F., Pappalardo, R.T., and Giese, B., 2002, Elastic thickness and heat flux estimates on Ganymede, Geophys. Res. Lett. 29, 1158.

Nimmo, F., 2004a, Stresses generated in cooling viscoelastic ice shells: Application to Europa, J. Geophys. Res. 109, E12001, doi:10.1029/2004JE002347.

Nimmo, F., 2004b, Dynamics of rifting and modes of extension on icy satellites, J. Geophys. Res.109, E01003, doi:10.1029/2004JE002347.

Nimmo, F. and Pappalardo, R.T., 2006, Diapir-induced reorientation of Saturn's moon Enceladus, Nature 441, 614–616.

Nimmo, F., Spencer, R.J., Pappalardo, R.T., Mullen, and M.E., 2007, Shear heating as the origin of the plumes and heat flux on Enceladus, Nature 447, 289–291.

Nimmo, F. and Matsuyama, I., 2007, Reorientation of icy satellites by impact basins, Geophys. Res. Lett. 34, L19203.

Noland, M., Veverka, J., Morrison, D., Cruikshank, D.P., Lazarewicz, A.R., Morrison, N.D., Elliot, J.L., Goguen, J., and Burns J.A., 1974,

Six-color photometry of Iapetus, Titan, Rhea, Dione, and Tethys, Icarus 23, 334–354.

Noll, K.S., Johnson, R.E., Lane, A.L., Domingue, D.L., and Weaver, H.A., 1996, Detection of ozone on Ganymede, Science 273, 341–343.

Noll, K.S., Roush, T.L., Cruikshank, D.P., Johnson, R.E., and Pendleton, Y.J., 1997, Detection of ozone on Saturn's satellites Rhea and Dione, Nature 388, 45–47.

Orton, G.S., Beines, K.H., Cruikshank, D.P., Krimigis, S.M., and Miller, E., 2009, Knowledge of the Saturn system prior to Cassini. In *Saturn from Cassini-Huygens* (Dougherty, M., Esposito, L., and Krimigis, S. (eds.)), Springer NY, 2009.

Ojakangas G.W. and Stevenson D.J., 1989a, Thermal state of an ice shell on Europa, Icarus 81, 220–241.

Ojakangas G.W. and Stevenson D.J., 1989b, Polar wander of an ice shell on Europa, Icarus 81, 242–270.

Ostro, S.J., Campbell, D.B., Simpson, R.A., Hudson, R.S., Chandler, J.F., Rosema, K.D., Shapiro, I.I., Standish, E.M., Winkler, R., and Yeomans, D.K., 1992, Europa, Ganymede, and Callisto – New radar results from Arecibo and Goldstone, J. Geophys. Res. 97, 18227–18244.

Ostro, S.J., 1993, Planetary radar astronomy, Rev. Mod. Phys. 65, 1235–1279.

Ostro, S.J., West, R.D., Janssen, M.A., Lorenz, R.D., Zebker, H.A., Black, G.J., Lunine, J.I., Wye, L.C., Lopes, R.M., Wall, S.D., Elachi, C., Roth, L., Hensley, S.T., Kelleher, K., Hamilton, G.A., Gim, Y., Anderson, Y.Z., Boehmer, R.A., Johnson, W.T.K., the Cassini RADAR Team, 2006, Cassini radar observations of Enceladus, Tethys, Dione, Rhea, Iapetus, Hyperion, and Phoebe, Icarus 183, 479–490. doi:10.1016/j.icarus.2006.02.019.

Owen, T.C., Cruikshank, D.P., Dalle Ore, C.M., Geballe, T.R., Roush, T.L., de Bergh, C., Pendleton, Y.J., and Khare, B.N., 2001, Decoding the domino: The dark side of Iapetus, Icarus 149, 160–172.

Pang, K.D., Voge, C.C., Rhoads J.W., and Ajello, J.M., 1984, The E-ring of Saturn and satellite Enceladus. J. Geophys. Res. 89, 9459–9470.

Pappalardo, R.T. and Collins, G.C., 2005, Strained craters on Ganymede, J. Struct. Geol. 27, 827–838.

Pappalardo, R.T. and Davis, D.M., 2007, Where's the compression? Explaining the lack of contractional structures on icy satellites, Ice, Oceans, Fire Workshop August 2007, Boulder, # 6080.

Paranicas, C., Mitchell, D.G., Krimigis, S.M., Hamilton, D.C., Roussos, E., Krupp, N., Jones, G.H., Johnson, R.E., Cooper, J.F., and Armstrong T.P., 2008, Sources and losses of energetic protons in Saturn's magnetosphere, Icarus 197, 519–525.

Peters, K.J., 1992, The coherent backscatter effect: A vector formulation accounting for polarization and absorption effects and small or large scatterers, Phys. Rev. B. 46, 801–812.

Petrenko V.F. and Whitworth, R.W., 1999, *Physics of Ice*. Oxford Univ. Press, Oxford.

Pipes, J.G., Browell, E.V., and Anderson, R.C., 1974, Reflectance of amorphous-cubic NH_3 frosts and amorphous-hexagonal H_2O frosts at 77 K from 1,400 to 3,000 A. Icarus 21, 283–291.

Pitman, K.M., Buratti, B.J., Mosher, J.A., Bauer, J.M., Momary, T.W., Brown, R.H., Nicholson, P.D., and Hedman, M.M., 2008, First high solar phase angle observations of Rhea using Cassini VIMS: upper limits on water vapour and geologic activity, Astrophys. J. 160, L65, doi:10.1086/589745.

Plescia, J.B. and Boyce, J.M., 1982, Crater densities and geological histories of Rhea, Dione, Mimas and Tethys, Nature 295, 285–290.

Plescia, J.B. and Boyce, J.M., 1983, Crater numbers and geological histories of Iapetus, Enceladus, Tethys and Hyperion, Nature 301, 666–670.

Plescia, J.B., 1983, The geology of Dione, Icarus 56, 255–277.

Porco, C.C., West, R.A., Squyres, S., McEwen, A., Thomas, P., Murray, C.D., Del Genio, A., Ingersoll, A.P., Johnson, T.V., Neukum, G., Veverka, J., Dones, L., Brahic, A., Burns, J.A., Haemmerle, V., Knowles, B., Dawson, D., Roatsch, T., Beurle, K., and Owen, W., 2004, Cassini imaging science: Instrument characteristics and capabilities and anticipated scientific investigations at Saturn, Space Sci. Rev., 115, 363–497, doi:10.1007/s11214–004–1456–7, 2004.

Porco, C.C., Baker, E., Barbara, J., Beurle, K., Brahic, A., Burns, J.A., Charnoz, S., Cooper, N., Dawson, D.D., Del Genio, A.D., Denk, T., Dones, L., Dyudina, U., Evans, M.W., Giese, B., Grazier, K., Helfenstein, P., Ingersoll, A.P., Jacobson, R.A., Johnson, T.V., McEwen, A., Murray, C.D., Neukum, G., Owen, W.M., Perry, J., Roatsch, T., Spitale, J., Squyres, S., Thomas, P.C., Tiscareno, M., Turtle, E., Vasavada, A.R., Veverka, J., Wagner, R., and West, R., 2005a, Cassini imaging science: initial results on Phoebe and Iapetus, Science 307, 1237–1242, doi:10.1126/science.1107981.

Porco, C.C., Baker, E., Barbara, J., Beurle, K., Brahic, A., Burns, J.A., Charnoz, S., Cooper, N., Dawson, D.D., Del Genio, A.D., Denk, T., Dones, L., Dyudina, U., Evans, M.V., Giese, B., Grazier, K., Helfenstein, P., Ingersoll, A.P., Jacobson, R.A., Johnson, T.V., McEwen, A., Murray, C.D., Neukum, G., Owen, W.M., Perry, J., Roatsch, T., Spitale, J., Squyres, S., Thomas, P., Tiscareno, M., Turtle, E., Vasavada, A.R., Veverka, J., Wagner, R., West, R. et al., 2005b, Cassini imaging science: Initial results on Saturn's rings and small satellites, Science 307, 1226–1236.

Porco, C.C. Helfenstein, P., Thomas, P.C., Ingersoll, A.P., Wisdom, J., West, R., Neukum, G., Denk, T., Wagner, R., Roatsch, T., Kieffer, S., Turtle, E., McEwen, A., Johnson, T.V., Rathbun, J., Veverka, J., Wilson, D., Perry, J., Spitale, J., Brahic, A., Burns, J.A., DelGenio, A.D., Dones, L., Murray, C.D., and Squyres, S., 2006, Cassini observes the active south pole of Enceladus, Science 311, 1393–1401.

Postberg, F., Kempf, S., Hillier, J.K., Srama, R., Green, S.F., McBride, N., Grün, E., 2008, The E-ring in the vicinity of Enceladus II, Probing the moon's interior – The composition of E-ring particles, Icarus 193, 438–454.

Prockter, L.M. and Pappalardo, R.T., 2000, Folds on Europa: Implications for crustal cycling and accommodation of extension, Science 289, 941–943.

Rappaport, N.J., Iess, L., Tortora, P., Asmar, S.W., Somenzi, L., Anabtawi, A., Barbinis, E., Fleischman, D.U., and Goltz, G.L., 2005, Gravity Science in the Saturnian system: The masses of Phoebe, Iapetus, Dione and Enceladus, DPS abstract #39.02.

Roatsch, T., Jaumann, R., Stephan, K., and Thomas, P.C., 2009, Cartographic mapping of the icy satellites using ISS and VIMS data. In: *Saturn from Cassini-Huygens* (Dougherty, M., Esposito, L., and Krimigis, S. (eds.)), Springer NY, 2009.

Roberts, J.H. and F. Nimmo, 2009, Tidal dissipation due to despinning and the equatorial ridge on Iapetus, *40th Lunar and Planetary Science Conference*, Houston, abstract #1927.

Roering, J.J., Kirchner, J.W., Sklar, L.S., and Dietrich, W.E., 2001, Hillslope evolution by nonlinear creep and landsliding: An experimental study, Geology 29, 143–146.

Roush, T.L., Cruikshank, D.P., and Owen, T.C., 1995, Surface ices in the outer solar system. In: *Volatiles in the Earth and Solar System* (Farley, K.A. (ed.)), Am. Inst. Phys. Conf. Proc. 341, 143–153.

Rubincam, D.P., 2003, Polar wander on Triton and Pluto due to volatile migration, Icarus 163, 469–478.

Sack, N.J., Johnson, R.E., Boring, J.W., and Baragiola, R.A., 1992, The effect of magnetospheric ion bombardment on the reflectance of Europa's surface, Icarus 100, 534–540.

Schenk, P.M., 1989a, Mimas grooves, the Herschel impact, and tidal stresses, *20th Lunar and Planetary Science Conference*, Houston, abstract #960.

Schenk, P.M., 1989b, Crater morphology and modification on the icy satellites of Uranus and Saturn: Depth/diameter and central peak occurrence, J. Geophys. Res. 94, 3815–3832.

Schenk, P.M., 1993, Central pit and dome craters – Exposing the interiors of Ganymede and Callisto, J. Geophys. Res. 98, 7475–7498.

Schenk, P.M., McKinnon, W.B., Gwynn, D., and Moore, J.M., 2001, Flooding of Ganymede's bright terrains by low-viscosity water-ice lavas, Nature 410, 57–60.

Schenk P.M. and Moore, J.M., 2007, Impact crater topography and morphology on Saturnian mid-sized satellites, *38th Lunar and Planetary Science Conference*, Houston, abstract #2305.

Schenk, P.M., Matsuyama, I., and Nimmo, F., 2008, True polar wander on Europa from global-scale small-circle depressions, Nature 453, 368–371.

Schmedemannn, N., Neukum, G., Denk, T., and Wagner, R., 2008, Stratigraphy and surface ages on Iapetus and other Saturnian satellites, *39th Lunar and Planetary Science Conference*, Houston, abstract #2070.

Schmidt, J., Brilliantov, N., Spahn, F., and Kempf, S., 2008, Slow dust in Enceladus' plume from condensation and wall collisions in tiger stripe fractures, Nature 451, 685–688, doi:10.1038/nature06491.

Schmitt, B., C. De Bergh M. Festou (eds.), 1998, *Solar System Ices*, Kluwer, Dordrecht.

Seal, D., Brown, R.H., Buffington, B., Dougherty, M.K., and Esposito, L.W., 2009, *Cassini Huygens Extended Mission, Saturn from Cassini-Huygens* (Dougherty, M., Esposito, L., and Krimigis, S. (eds.)), Springer, 2009.

Shkuratov, Yu., Starukhina, L., Hoffmann, H., Arnold, G., 1999, A model of spectral albedo of particulate surfaces: implication to optical properties of the Moon, Icarus 137, 235–246.

Shkuratov, Y.G., Stankevich, D.G., Petrov, D.V., Pinet, P.C, Cord, A.M., Daydou, Y.H., and Chevrel, S.D., 2005, Interpreting photometry of regolith-like surfaces with different topographies: Shadowing and multiple scattering, Icarus 173, 3–15.

Showman A.P., Stevenson D.J., and Malhotra, R., 1997, Coupled orbital and thermal evolution of Ganymede, Icarus 129, 367–383.

Showman A.P., Mosqueira I., and Head J.W., 2004, On the resurfacing of Ganymede by liquid-water volcanism, Icarus 172, 625–640.

Showalter, M.R., 1991, Visual detection of 1981 S13, Saturn's eighteenth satellite, and its role in the Encke Gap, Nature 351, 709–713.

Showalter, M.R., Cuzzi, J.N., and Larson, S.M., 1991, Structure and particle properties of Saturn's E Ring, Icarus 94, 451–473.

Simonelli, D.P., Kay, J., Adinolfi D., Veverka, J., Thomas, P.C., and Helfenstein, P., 1999, Phoebe: Albedo map and photometric properties, Icarus 138, 249–258.

Sittler, E.C., Andre, N., Blanc, M.; Burger, M., Johnson, R.E., Coates, A., Rymer, A., Reisenfeld, D., Thomsen, M.F.; Persoon, A., Dougherty, M., Smith, H.T., Baragiola, R.A., Hartle, R.E., Chornay, D., Shappirio, M.D., Simpson, D., McComas, D.J., and Young, D.T., 2008, Ion and neutal sources and sinks within Saturn's inner magnetosphere: Cassini results, Planet. Space Sci. 56, 3–18, doi:10.1016/j.pss.2007.06.006.

Smith, B.A., Soderblom, L., Beebe, R., Boyce, J., Briggs, G., Bunker, A., Collins, S.A., Hansen, C.J., Johnson, T.V., Mitchell, J.L., Terrile, R.J., Carr, M., Cook, II, A.F., Cuzzi, J., Pollack, J.B., Danielson, G.E., Ingersoll, A., Davies, M.E., Hunt, G.E., Masursky, H., Shoemaker, E., Morrison, D., Owen, T., Sagan, C., Veverka, J., Strom, R., and Suomi, V.E., 1981, Encounter with Saturn: Voyager 1 imaging science results, Science 212, 163–191, doi:10.1126/science.212.4491.163.

Smith, B.A. Soderblom, L., Batson, R., Bridges, P., Inge, J., Masursky, H., Shoemaker, E., Beebe, R., Boyce, J., Briggs, G., Bunker, A., Collins, S.A., Hansen, C.J., Johnson, T.V., Mitchell, J.L., Terrile, R.J., Cook, II, A.F., Cuzzi, J., Pollack, J.B., Danielson, G., Ingersoll, A.P., Davies, M.E., Hunt, G.E., Morrison, D., Owen, T., Sagan, C., Veverka, J., Strom, R., and Suomi, V.E., 1982, A new look at the Saturn system: The Voyager 2 images, Science 215, 504–536, doi:10.1126/science.215.4532.504.

Soter, S., 1974, *Brightness of Iapetus, Presented at IAU Colloq. 28*, Cornell Univ, Ithaca.

Sotin C., Grasset O., and Beauchesne, S., 1998, Thermodynamical properties of high pressure ices. Implications for the dynamics and internal structure of large icy satellites. In: *Solar System Ices* (Schmitt, B., De Bergh, C., and Festou, M. (eds.)), Kluwer, Dordrecht, The Netherlands, pp. 79–96.

Sotin, C., Head, J.W., and Tobie, G., 2002, Europa: Tidal heating of upwelling thermal plumes and the origin of lenticulae and chaos melting, Geophys. Res. Lett., 29, doi:10.1029/2001GL013844.

Sotin, C., Stevenson, D.J., Rappaport, N.J., Mitri, G., and Schubert, G., 2009, Titan's interior structure. In: *Titan from Cassini-Huygens* (Brown, R., Lebreton, J.-P., and Waite, H. (eds.)), Springer NY, 2009.

Spahn, F., Schmidt, J., Albers, N., Hörning, M., Makuch, M., Seiß, M., Kempf, S., Srama, R., Dikarev, V., Helfert, S., Moragas-Klostermeyer, G., Krivov, A.V., Sremcevic, M., Tuzzolino, A.J., Economou, T., and Grün, E., 2006, Cassini dust measurements at Enceladus and implications for the origin of the E ring, Science 311, 1416–1418.

Spencer, J.R., Calvin, W.M., and Person, M.J., 1995, CCD spectra of the Galilean satellites: Molecular oxygen on Ganymede. J. Geophys. Res. 100, 19049–19056.

Spencer, J.R., 1998, Upper limits for condensed O_2 on Saturn's icy satellites and rings, Icarus 136, 349–352.

Spencer, J.R., Pearl, J.C., Segura, M., and The Cassini CIRS Team, 2005, Cassini CIRS observations of Iapetus' thermal emission, *36th Lunar and Planetary Science Conference*, Houston, abstract #2305.

Spencer, J.R., Pearl, J.C., Segura, M., Flasar, F.M., Mamoutkine, A., Romani, P., Buratti, B.J., Hendrix, A.R., Spilker, L.J., and Lopes, R.M.C., 2006, Cassini encounters Enceladus: Background and the discovery of a south polar hot spot, Science 311, 1401–1405.

Spencer, J., Barr, A., Esposito, L., Helfenstein, P., Ingersoll, A., Jaumann, R., Kieffer, S., McKay, C., Nimmo, F., Porco, C.C., and Waite, W., 2009, Enceladus: a cryovolcanic active satellite. In: *Saturn from Cassini-Huygens* (Dougherty, M., Esposito, L., and Krimigis, S. (eds.)), Springer NY, 2009.

Spencer, J.R. and Denk, T., 2009, Formation of Iapetus's Extreme Albedo Dichotomy by Exogenically-Triggered Thermal Migration of Water Ice, Science, in revision.

Spitale, J.N., Jacobson, R.A., Porco, C.C., and Owen, Jr., W.M., 2006, The orbits of saturn's small satellites derived from combined historic and Cassini imaging observations, Astron. J. 132, 692.

Spitale, J.N. and Porco, C.C., 2007, Association of the jets of Enceladus with the warmest regions on its south-polar fractures, Nature 449, 695–697.

Spahn, F., Schmidt, J., Albers, N., Hörning, M., Makuch, M., Seiß, M., Kempf, S., Srama, R., Dikarev, V., Helfert, S., Moragas-Klostermeyer, G., Krivov, A.V., Sremevi, M., Tuzzolino, A.J., Economou, T., and Grün, E., 2006, Cassini dust measurements at Enceladus and implications for the origin of the E ring, Science, 311, 1416–1418, doi:10.1126/science.1121375.

Squyres, S.W., 1980, Volume changes in Ganymede and Callisto and the origin of grooved terrain, Geophys. Res. Lett. 7, 593–596.

Squyres, S.W. and Sagan, C., 1983, Albedo asymmetry of Iapetus, Nature 303, 782–785.

Squyres, S.W. Buratti, B., Veverka, J., and Sagan, C., 1984, Voyager photometry of Iapetus, Icarus 59, 426–435.

Squyres, S.W. and Croft, S.K., 1986, The tectonics of icy satellites. In: *Satellites* (Matthews, M.S. (ed.)), Univ. Ariz. Press, Tucson, pp. 293–341.

Stephan, K., Jaumann, R., Wagner, R., Roatsch, Th., Clark., R.N., Cruikshank, D.P., Hibbitts, C.A., Hansen, G.B., Buratti, B.J., Filiacchione, G., Baines, K.H., Nicholson, P.D., McCord, T.B., and Brown, R.H, 2008, Relationship of Dione's spectral properties to geological surface units, *39th Lunar and Planetary Science Conference*, Houston, abstract #1717.

Stevenson, D., 1982, Volcanism and igneous processes in small icy satellites, Nature 298, 142–144.

Strazzulla, G., 1998, Chemistry of ice induced by bombardment with energetic charged particles. In: *Solar System Ices* (B. Schmitt et al. (ed.)), Kluwer Academic Publ., Dordrecht, The Netherlands, pp. 281–302.

Tabak, R.G. and Young, W.M., 1989, Cometary collisions and the dark material on Iapetus. Earth, Moon, Planets 44, 251–264.

Tian, F., Stewart, A.I.F., Toon, O.B., Larsen, K.W., Esposito, L.W., 2007, Monte Carlo simulations of the water vapor plumes on Enceladus, Icarus 188, 154–161.

Thomas, P.C., Veverka, J., Morrison, D., Davies, M., and Johnson, T.V., 1983, Saturn's small satellites voyager imaging results, J. Geophys. Res. 88, 8743–8754.

Thomas, P. and Veverka, J., 1985, Hyperion: Analysis of voyager observations, Icarus 64: 414–424.

Thomas P.G., 1988, The tectonic and volcanic history of Rhea as inferred from studies of scarps, ridges, troughs and other lineaments, Icarus 74, 554–567.

Thomas, P.C., 1989, Shapes of Small Satellites, Icarus 77, 248–274.

Thomas, P.C., Armstrong, J., Asmar, S.W., Burns, J.A., Denk, T., Giese, B., Helfenstein, P., Iess, L., Johnson, T.V., McEwen, A.S., Nicolaisen, L., Porco, C.C., Rappaport, N., Richardson, J., Somenzi, L., Tortora, P., Turtle. E.P., Veverka, J., 2007a, Hyperion's sponge-like appearance, Nature 448, 50–53, doi: 10.1038/nature05779.

Thomas, P.C., Burns, J.A., Helfenstein, P., Squyres, S., Veverka, J., Porco, C., Turtle, E.P., McEwen, A., Denk, T., Giese, B., Roatsch, T., Johnson, T.V., Jacobson, R.A., 2007b, Shapes of the Saturnian icy satellites and their significance, Icarus 190, 573–584. doi:10.1016/j.icarus.2007.03.012.

Turcott, D.L. and Schubert, G., 2002, *Geodynamics*, 2nd ed. Cambridge Univ. Press, Cambridge.

Verbiscer, A.J. and Veverka, J., 1989, Albedo dichotomy of Rhea: Hapke analysis of Voyager photometry, Icarus 82, 336–353.

Verbiscer, A.J. and Veverka, J., 1992, Mimas: Photometric roughness and albedo map, Icarus 99, 63–69.

Verbiscer, A.J. and Veverka, J., 1994, A photometric study of Enceladus, Icarus 110, 155–164.

Verbiscer, A.J., French, R.G., and McGhee, C.A., 2005, The opposition surge of Enceladus: HST observations 0.338–1.022 μm, Icarus 173, 66–83.

Verbiscer, A.J., Peterson, D.E., Skrutskie, M.F., Cushing, M., Helfenstein, P., Nelson, M.J., Smith, J.D., and Wilson, J.C., 2006, Near-infrared spectra of the leading and trailing hemispheres of Enceladus, Icarus 182, 211–223.

Verbiscer, A., French, R., Showalter, M., and Helfenstein, P., 2007, Enceladus: Cosmic graffiti artist caught in the act, Science 315, 815.

Verbiscer, A.J., Skrutskie, M., Wilson, J., Nelson, M., Helfenstein, P., 2008, Ammonia hydrate in the Saturnian System, Bulletin of the American Astronomical Society, 40, 510.

Veverka, J., Thomas, P.C., Johnson, T.V., Matson, D., and Housen, K.R., 1986, The physical characteristics of satellites surfaces. In *Satellites* (Burns, J.A., Matthews, M.S. (eds.)), Univ. Ariz. Press, Tucson, pp. 242–402.

Vilas, F., Larson, S.M., Stockstill, K.R., and Gaffey, M.J., 1996, Unraveling the zebra: Clues to the Iapetus dark material composition, Icarus 124, 262–267.

Vilas, F., Jarvis, K.S., Barker, E.S., Lederer, S.M., Kelly, M.S., and Owen, T., 2004, Iapetus dark and bright material: giving compositional interpretation some latitude, Icarus 170, 125–130.

Wahr, J., Selvans, Z.A., Mullen, M.E., Barr, A.C., Collins, G.C., Selvans, M.M., Pappalardo, and R.T., 2009, Modeling stresses on satellites due to non-synchronous rotation and orbital eccentricity using gravitational potential theory, Icarus 200, 188–206.

Wagner R. et al., Neukum, G., Giese, B., Roatsch, T., Wolf, U., Denk, T., and the Cassini ISS Team, 2006, Geology, ages and topography of Saturn's satellite Dione observed by the Cassini ISS camera, *37th Lunar and Planetary Science Conference*, Houston, abstract #1805.

Wagner, R.J., Neukum, G., Giese, B., Roatsch, T., and Wolf, U., 2007, The global geology of Rhea: Preliminary implications from the Cassini ISS data, *38th Lunar and Planetary Science Conference*, Houston, abstract #1958.

Wagner, R.J., Neukum, G., Giese, B., Roatsch, T., Denk, T., Wolf, U., and Porco, C.C., 2008, Geology of Saturn's satellite Rhea on the basis of the high-resolution images from the targeted flyby 049 on Aug. 30, 2007, *39th Lunar and Planetary Science Conference*, Houston, abstract #1930.

Wagner, R.J., Neukum, G., Stephan, K., Roatsch, T., Wolf, U., and Porco, C.C., 2009, Stratigraphy of tectonic features on Saturn's satellite Dione derived from Cassini ISS camera data, *40th Lunar and Planetary Science Conference*, Houston, abstract #2142.

Waite, J.H., Jr., Combi, M.R., Ip, W.H., Cravens, T.E., McNutt, R.L., Jr., Kasprzak, W., Yelle, R., Luhmann, J., Niemann, H., Gell, D., Magee, B., Fletcher, G., Lunine, J., and Tseng, W.L., 2006, Cassini Ion and Neutral Mass Spectrometer: Enceladus Plume Composition and Structure, Science 311, 1419–1422.

Warren, S.G., 1982. Optical properties of snow, Rev. Geophys. Space Phys. 20, 67–89.

Willemann, R.J., 1984, Reorientation of planets with elastic lithospheres, Icarus 60, 701–709.

Wilson, P.D. and Sagan, C., 1995, Spectrophotometry and organic matter on Iapetus, 1. Composition models, J. Geophys. Res. 100, 7531–7537.

Wilson, P.D. and Sagan, C., 1996, Spectrophotometry and organic matter on Iapetus, 2: Models of interhemispheric asymmetry, Icarus 122, 92–106.

Wilson L., Head J.W., and Pappalardo R.T., 1997, Eruption of lava flows on Europa: Theory and application to Thrace Macula, J. Geophys. Res. 102, 9263–9272.

Wisdom, J., Peale, S.J., and Mignard, F., 1984, The chaotic rotation of Hyperion, Icarus 58, 137–152.

Wisdom, J., 2004, Spin-orbit secondary resonance dynamics of Enceladus, Astron. J. 128, 484–491.

Yoder, C.F., 1995, Astrometric and geodetic properties of Earth and the Solar System. In: *Global Earth Physics* (Ahrens, T.J. (ed.)), Amer. Geophys. Union Washington, pp. 1–31.

Zellner, B., 1972, On the nature of Iapetus, Astrophys. J. Lett. 174, L107–L109.

Chapter 21
Enceladus: An Active Cryovolcanic Satellite

John R. Spencer, Amy C. Barr, Larry W. Esposito, Paul Helfenstein, Andrew P. Ingersoll, Ralf Jaumann, Christopher P. McKay, Francis Nimmo, and J. Hunter Waite

Abstract Enceladus is one of the most remarkable satellites in the solar system, as revealed by Cassini's detection of active plumes erupting from warm fractures near its south pole. This discovery makes Enceladus the only icy satellite known to exhibit ongoing internally driven geological activity. The activity is presumably powered by tidal heating maintained by Enceladus' 2:1 mean-motion resonance with Dione, but many questions remain. For instance, it appears difficult or impossible to maintain the currently observed radiated power (probably at least 6 GW) in steady state. It is also not clear how Enceladus first entered its current self–maintaining warm and dissipative state- initial heating from non-tidal sources is probably required. There are also many unanswered questions about Enceladus' interior. The silicate fraction inferred from its density of 1.68 g cm^{-2} is probably differentiated into a core, though we have not direct evidence for differentiation. Above the core there is probably a global or regional liquid water layer, inferred from several models of tidal heating, and an ice shell thick enough to support the ∼1 km amplitude topography seen on Enceladus. It is possible that dissipation is largely localized beneath the south polar region. Enceladus' surface geology, ranging from moderately cratered terrain to the virtually crater-free active south polar region, is highly diverse, tectonically complex, and remarkably symmetrical about the rotation axis and the direction to Saturn. South polar activity is concentrated along the four "tiger stripe" fractures, which radiate heat at temperatures up to at least 167 K and are the source of multiple plumes ejecting ∼200 kg s^{-1} of H$_2$O vapor along with significant N$_2$ (or C$_2$H$_4$), CO$_2$, CH$_4$, NH$_3$, and higher-mass hydrocarbons. The escaping gas maintains Saturn's neutral gas torus, and the plumes also eject a large number of micron-sized H$_2$O ice grains that populate Saturn's E-ring. The mechanism that powers the plumes is not well understood, and whether liquid water is involved is a subject of active debate. Enceladus provides perhaps the most promising potential habitat for life in the outer solar system, and the active plumes allow the unique opportunity for direct sampling of that zone. Enceladus is thus a prime target for Cassini's continued exploration of the Saturn system, and will be a tempting target for future missions.

21.1 Cassini's Exploration of Enceladus

Enceladus (Fig. 21.1) has been known to be an unusual satellite since the early 1980s. The Voyager encounters in 1980 and 1981 revealed Enceladus' extraordinarily high albedo, the highest of any solid body in the solar system, and its heterogeneous cratered and fractured surface terrains (Smith et al. 1982). At around the same time, ground-based observations revealed that Saturn's tenuous E-ring was concentrated at the orbit of Enceladus (Baum et al. 1981). The inferred short lifetimes of the micron-sized E-ring particles implied a continuous source of supply at Enceladus, and geyser-like activity on Enceladus was suggested as a possible source by several authors (Haff et al. 1983, Pang et al. 1984). The neutral OH cloud discovered in the 1990s by the Hubble Space Telescope (Shemansky et al. 1993) also appeared to require a strong source near Enceladus (Jurac et al. 2001).

J.R. Spencer, A.C. Barr
Southwest Research Institute, Boulder, USA

J.H. Waite
Southwest Research Institute, San Antonio, USA

L.W. Esposito
LASP, University of Colorado, Boulder, USA

P. Helfenstein
Center for Radiophysics and Space Research, Cornell University, USA

A.P. Ingersoll
Division of Geological and Planetary Sciences, California Institute of Technology, Pasadena, USA

R. Jaumann
Institute of Planetary Research, DLR, Berlin, Germany

C.P. McKay
NASA Ames Research Center, Mountain View, USA

F. Nimmo
Dept. Earth and Planetary Sciences, University of California Santa Cruz, Santa Cruz, USA

Fig. 21.1 Global false-color view of Enceladus taken by ISS on July 14 2005. The location of the south pole is shown by the red dot, and the four active "tiger stripe" fractures appear as prominent blue-green lines surrounding the pole. From Cassini Press Release PIA06254, NASA/JPL/SSI

Enceladus was thus a high-priority target of the Cassini mission since the program's inception: of the ten targeted flybys of Saturn's mid-sized satellites included in the prime mission tour, three were of Enceladus. In addition, the tour included a fourth close non-targeted Enceladus flyby in February 2005. Flybys in prime mission and subsequent mission extensions are listed in Table 21.1, and the geometry of the closest flybys in the prime and Equinox missions is illustrated in Fig. 21.2.

Cassini's first important Enceladus-related discovery was made even before Saturn orbit insertion. Scans of the Saturn system with the UVIS instrument in early 2004 revealed neutral O emission (Esposito et al. 2005) concentrated near the orbit of Enceladus, complementing the neutral OH emission previously detected. The first close encounters, on orbits 3 and 4, viewed the surface from low latitudes. The ISS images revealed spectacular details of the intense tectonic disruption of the surface and the modification of older cratered terrain (Porco et al. 2006). A stellar occultation observed by the UVIS instrument during the Orbit 3 flyby detected no trace of an atmosphere at the latitudes probed (15 N and 31 S) (Hansen et al. 2006). However, a large perturbation of the ambient magnetic field was seen, indicating a conducting barrier near Enceladus but larger than the moon itself (Dougherty et al. 2006). As a result of a presentation made to the Cassini Project by the Magnetometer Team in April of 2005, in which they noted that their data indicated that Enceladus has a significant atmosphere composed of water-group molecules and ions being continuously replenished (Dougherty et al. 2006), the upcoming July 2005 flyby was modified to reduce Cassini's closest approach distance from a planned 1,000 km to 173 km, to more closely investigate.

It was the July 2005 flyby that revealed with certainty the active nature of Enceladus. The southerly approach provided a good view of the south polar region, and the CIRS instrument detected anomalous thermal emission from the south pole (Spencer et al. 2006) which correlated with four prominent fresh fractures informally dubbed "tiger stripes" seen in detail by the ISS cameras (Porco et al. 2006, Fig. 21.1). Near closest approach, UVIS observed another stellar occultation, and this time detected water vapor over the south pole (Hansen et al. 2006), but not at equatorial latitudes. The south polar water vapor cloud was also detected by INMS, which also found a few percent of CO_2, CH_4, and CO and/or N_2 (Waite et al. 2006). Simultaneously, CDA detected a large concentration of dust grains above the south pole (Spahn et al. 2006a). Combining these data, it became clear that Enceladus' south pole was actively erupting plumes of ice grains, water vapor, and other gases, probably from the tiger stripe fractures, to supply both the E-ring and the neutral torus. Near-IR spectra from VIMS added to the picture, showing concentrations of CO_2 and C-H stretch features near the tiger stripes, consistent with the gas composition (Brown et al. 2006). The dust plumes were then seen directly at high resolution in high-phase ISS images taken on November 27 2005, which revealed multiple individual jets (Porco et al. 2006). Lower-resolution ISS images taken in January and February 2005 also showed the dust plume (Porco et al. 2006), but the plume was not recognized with any certainty at that time, and was not announced.

It was over 2 years till the next close Enceladus flyby, but in the meantime ISS obtained numerous high-phase-angle images of the plume, allowing triangulation of the major plume sources and confirming their location along the tiger stripes, near the warmest regions seen by CIRS (Spitale and Porco 2007), and CIRS observed the south polar hot spot again in November 2006, showing no change in activity since the previous year (Abramov and Spencer 2009).

The very close March 2008 flyby provided an opportunity to penetrate the plume much more deeply than in July 2005, resulting in high quality INMS measurements of the gas composition, and CIRS obtained a high resolution view of the thermal emission along the tiger stripes, revealing heat radiation along the full length of each the fractures imaged. However, because Enceladus was in Saturn's shadow

Table 21.1 Cassini's past and planned close Enceladus flybys

Orbit, encounter name	Date and time (UT)	Speed (km s^{-1})	Close approach Altitude (km)	Close approach Sub-spacecraft point	Geometry 1 h before encounter Sub-spacecraft point	Geometry 1 h before encounter Solar phase	Geometry 1 h after encounter Sub-spacecraft point	Geometry 1 h after encounter Solar phase
3, E0	2005/02/17 03:30	6.7	1,260	239 W, 51 N	318 W, 1 N	27	163 W, 3 N	159
4, E1	2005/03/09 09:08	6.6	497	304 W, 30 S	202 W, 1 S	45	41 W, 0 S	130
11, E2	2005/07/14 19:55	8.2	166	327 W, 22 S	197 W, 48 S	46	36 W, 47 N	132
61, E3	2008/03/12 19:06	14.4	47	135 W, 21 S	125 W, 70 N	114	327 W, 70 S	65
80, E4	2008/08/11 21:06	17.7	50	97 W, 28 S	87 W, 62 N	107	289 W, 63 S	72
88, E5	2008/10/09 19:06	17.7	25	97 W, 31 S	86 W, 62 N	107	289 W, 63 S	72
91, E6	2008/10/31 17:14	17.7	197	97 W, 26 S	86 W, 62 N	107	288 W, 63 S	72
120, E7	2009/11/02 07:41	7.7	99	159 W, 88 S	329 W, 1 S	174	170 W, 0 N	5
121, E8	2009/11/21 02:09	7.7	1,602	112 W, 82 S	189 W, 3 S	145	31 W, 3 S	33
130, E9	2010/04/28 00:10	6.5	99	147 W, 89 S	317 W, 1 S	172	159 W, 0 S	7
131, E10	2010/05/18 06:04	6.5	434	304 W, 34 S	201 W, 0 S	154	42 W, 1 S	27
136, E11	2010/08/13 22:31	6.9	2,550	29 W, 78 S	198 W, 17 S	153	39 W, 4 N	23
141, E12	2010/11/30 11:53	6.3	47	53 W, 62 S	315 W, 0 S	166	156 W, 0 S	14
142, E13	2010/12/21 01:08	6.3	47	232 W, 62 S	314 W, 0 S	166	156 W, 0 S	15
154, E14	2011/10/01 13:52	7.4	99	204 W, 89 S	192 W, 0 S	151	34 W, 1 S	28
155, E15	2011/10/19 09:23	7.5	1,229	114 W, 14 S	189 W, 0 S	154	36 W, 1 S	31
156, E16	2011/11/06 04:59	7.4	495	295 W, 30 S	193 W, 0 S	151	34 W, 0 S	29
163, E17	2012/03/27 18:29	7.2	82	191 W, 76 S	191 W, 0 N	160	32 W, 1 S	20
164, E18	2012/04/14 14:00	6.5	175	206 W, 49 S	191 W, 0 N	160	32 W, 1 S	20
165, E19	2012/05/02 09:31	7.6	73	291 W, 72 S	191 W, 0 N	159	32 W, 1 S	21
223, E20	2015/10/14 10:41	8.4	1,838	106 W, 78 N	185 W, 4 N	23	24 W, 4 N	148
224, E21	2015/10/28 15:22	8.5	48	66 W, 84 S	336 W, 0 N	41	177 W, 1 S	139
228, E22	2015/12/19 17:48	9.5	4,999	353 W, 88 S	342 W, 6 S	46	183 W, 10 S	144

while the south pole was in view, the highest-resolution remote sensing with other instruments covered the northern hemisphere.

Cassini's first extended mission, the "Equinox Mission", from mid-2008 to -2010, includes seven additional close Enceladus flybys designed to follow up on the discoveries of the prime mission (Table 21.1). The first three were high-inclination passes with good sunlit views of the south pole on departure, and included deep passages through the plume in orbits 80 and 88 and very high resolution remote sensing on orbits 80 and 91. The four flybys in orbits 120 – 131 are low inclination passes which can only observe the south pole near closest approach, but will provide good coverage of low latitudes. These flybys include the deepest plume penetrations yet planned, on orbits 120 and 130, and a mixture of occultations, gravity observations, and remote sensing including high-resolution observations of the plume jets.

At the time of writing, a second extended mission, the "Solstice Mission", has been designed but is not yet approved for funding. This extension includes 12 additional Enceladus flybys through 2015 (Table 21.1). The south pole will be largely in darkness for most of this mission phase, and remote sensing of the south pole will be largely restricted to thermal and Saturn-light observations, and observations of the upper parts of the jets and plume that extend thousands of kilometers above the pole. There will also be several more close south polar flybys to sample the plume and investigate its longer-term variability.

21.2 Interior and Tidal Heating

21.2.1 Introduction

Cassini observations of Enceladus' active south polar region confirm that this tiny moon ranks alongside the Earth, Io, and possibly Triton as one of the few currently volcanically or cryovolcanically active worlds in the solar system. Much of the endogenic power output is concentrated within linear features referred to as "tiger stripes" (Spencer et al. 2006, Porco et al. 2006), which are also the source of the plume materials (Spitale and Porco 2007). Characterization of tectonic features elsewhere on Enceladus indicates that heat flows comparable to values presently observed in the south polar terrain (SPT) may have been present in those regions at the time of deformation (Bland et al. 2007, Giese et al. 2008).

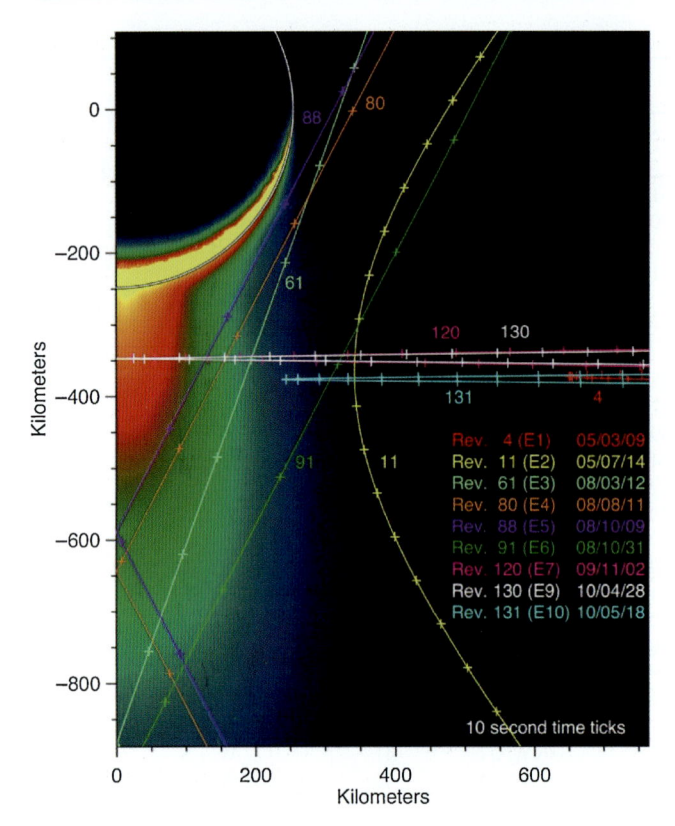

Fig. 21.2 The geometry of Cassini's closest Enceladus flybys during the prime and Equinox missions (Table 21.1) is shown relative to the south polar plume, assuming rotational symmetry about the spin axis. The plume image, taken on April 24 2007, is stretched logarithmically to highlight the faint outermost regions of the plume. Tick marks are shown at 10 s intervals

This suggests that the strong heating occurring presently in the SPT may have occurred at other locations in Enceladus in the past (Section 21.3.2). In contrast, other regions of the surface, particularly in the northern hemisphere, are quite heavily cratered and may be several billion years old (Kargel and Pozio 1996, Porco et al. 2006). Such profound variations in interior structure and surface morphology across the satellite suggest that Voyager-era models, all of which assume a radially symmetric interior structure, probably do not capture key processes driving its past and present-day activity.

Observations of Enceladus offer a chance to understand the origin of cryovolcanism, and the observed heat flux from the SPT provides a unique opportunity to constrain geodynamical models. The main source of Enceladus' energy is undoubtedly tidal dissipation, maintained over long time scales by Enceladus' 2:1 mean motion resonance with neighbor satellite Dione (e.g., Murray and Dermott 1999). Thus, Enceladus provides a unique natural laboratory for the study of how the interaction between tidal flexing, interior thermal and compositional heterogeneity, and heat transport gives rise to cryovolcanism and resurfacing on icy bodies. In this section we will focus on the interior of Enceladus, addressing its state of differentiation, heat sources, heat transport mechanisms, and the time evolution of its interior structure. The question of whether or not Enceladus has a subsurface ocean is of particular interest, not only because the presence or absence of an ocean strongly affects the generation of tidal heat, but also because its ocean is potentially habitable.

21.2.2 Bulk Structure

21.2.2.1 Observational Constraints

Three types of observations can be made by Cassini to infer the internal structure of Enceladus: bulk density, shape, and gravity moments. Based on Cassini observations, the mean radius of Enceladus is 252 km, $GM = (7.207 \pm 0.011) \times 10^9 \, \text{m}^3 \, \text{s}^{-2}$, and its bulk density is $1,608 \pm 5 \, \text{kg m}^{-3}$ (Jacobson et al. 2006, Thomas et al. 2007, Rappaport et al. 2007), much higher than Voyager density estimates of roughly $1,200 \, \text{kg m}^{-3}$ (Smith et al. 1982). Given its mean density, Enceladus is likely composed of rock and ice, with a rock mass fraction of roughly 50–60%. A fully differentiated Enceladus would have an H_2O layer 60–100 km thick, which may be partly liquid in the form of a global ocean (Nimmo and Pappalardo 2006, Barr and McKinnon 2007a, Schubert et al. 2007). Pressures reach 1 MPa at depths of ~10 km on Enceladus, so near-surface porosity is unlikely to significantly affect its bulk density.

Enceladus' measured principal axes indicate that its interior is not in hydrostatic equilibrium. If Enceladus were in hydrostatic equilibrium, its shape could be used to infer its polar moment of inertia (e.g., Dermott and Thomas 1988) and state of differentiation. The measured radii a, b and c are 256.6, 251.4 and 248.3 km (Thomas et al. 2007), respectively, with uncertainties of 0.6, 0.2 and 0.2 km, respectively. The derived quantity $(a-c)/(b-c) = 2.7 \pm 0.5$, much lower than 4.0, the expected value for a hydrostatic body. Porco et al. (2006) argue that the observed value of $a - c$ is also smaller than that expected for a hydrostatic, differentiated body with the present-day rotation rate, and thus that Enceladus is either undifferentiated (or slightly differentiated), or was rotating more rapidly (and perhaps closer to Saturn) when the shape was acquired. The former possibility is unlikely (see Schubert et al. 2007 and below); the latter possibility requires that Enceladus have a finite rigidity (in order to maintain its shape) and is thus non-hydrostatic at the present day, though it is unlikely that Enceladus has undergone much orbital evolution (Porco et al. 2006). If Enceladus is fully differentiated and in hydrostatic equilibrium, its shape indicates an asymmetric interior distribution of mass (Porco et al. 2006). Collins and Goodman (2007) point out that the original value of c might have been affected by melting and subsidence at the south pole, and that a differentiated Enceladus may be consistent with the shape observations. Note that this approach again implicitly assumes a non-hydrostatic Enceladus at the present day.

Degree-2 gravitational coefficients determined from spacecraft radio tracking are commonly used to infer a satellite's interior density structure, but this inversion can only be performed if the satellite's interior is in hydrostatic equilibrium (e.g., Rappaport et al. 2007). At present, the gravity coefficients of Enceladus are unknown, but given the apparently non-hydrostatic nature of this satellite, measuring these coefficients will at best provide only a limited improvement in our understanding of its internal structure. One method of determining the moment of inertia without requiring the hydrostatic assumption is to measure both the degree-2 gravity coefficients and the obliquity (Bills and Nimmo 2008). However, this technique requires an alternative assumption: that the obliquity is tidally damped, such that the spin axis, orbit pole and pole of the ecliptic are coplanar (i.e., a Cassini state, Bills et al. 2009). At present we do not know Enceladus' rotational state well enough to test this assumption.

21.2.2.2 Indirect Arguments

In addition to direct observational constraints, theoretical arguments based on heat sources and thermal evolution models can be used in combination with observations to give likely interior structures (see e.g., Schubert et al. 2007 for discussion). Existing arguments are based on thermal evolution models of Enceladus, or on its observed present-day heat flux. The present-day observed heat flux at the south pole of Enceladus is at least 5.8 ± 1.9 GW (Spencer et al. 2006). This power is much larger than the current radiogenic heat production, and is likely a result of tidal heating (Spencer et al. 2006, see Section 21.2.4 below). Even in the absence of tidal heating, a conductive Enceladus is marginally capable of differentiating in the presence of long-lived radionuclides, and is certain to differentiate if significant amounts of the short-lived nuclide ^{26}Al are present at the time of its formation (Schubert et al. 2007).

Because the tidal heating experienced by a differentiated Enceladus depends strongly on its interior structure, estimates of Enceladus' tidal heat flux may provide clues about its interior structure. Ross and Schubert (1989) showed that a homogeneous undifferentiated (uniform-density ice-rock) Enceladus could generate as much as 920 GW, but only if the viscosity were as low as 10^{12} Pa s. This value is too low even for pure ice (Goldsby and Kohlstedt 2001), and certainly too low once the likely effects of silicate particulates on ice viscosity (Friedson and Stevenson 1983, Durham et al. 1992) are included. For a more reasonable viscosity of 10^{14} Pa s (Durham and Stern 2001), assuming a plausible grain size

near 1 mm (Barr and McKinnon 2007a) and rigidity of 3 GPa, the heat production for a uniform Enceladus is 2 GW, less than but more comparable to the observed value. The presence of a rigid silicate core will reduce the amplitude of the tidal deformation and produce a lower heat flow. However, the presence of an ocean decoupling the ice shell from the silicate interior enhances the total volumetric tidal heat production by a factor of ~ 10 (Roberts and Nimmo 2008a). A differentiated Enceladus without an ocean does not generate enough heat to account for the observations, but in cases with a subsurface ocean, the modeled *global* heat production is comparable to that observed (Roberts and Nimmo 2008a). Tobie et al. (2008) examined heating with lateral variations in viscosity, and concluded that to reproduce the amplitude and pattern of heating observed, either a global ocean or a liquid layer spanning a major part of the southern hemisphere is required. Nimmo et al. (2007) examined a different mechanism of tidal heating, wherein friction due to strike-slip motion along faults generates heat in the shallow subsurface of the ice shell, but also concluded that a subsurface ocean was required to generate the observed heat flow.

An alternative to a global ocean model is a model wherein the south polar hot spot sits above a local "sea" (Collins and Goodman 2007). Melting of the ice shell to generate the sea could explain the low topography at the south pole, but requires either strong heating from Enceladus' silicate interior or the downwards percolation of melt from within the ice shell (Tobie et al. 2008). As discussed in Section 21.2.4 below, it is difficult to generate enough tidal heating within the silicates to melt the ice shell in the manner envisioned by Collins and Goodman. Melt generation within the ice shell is possible, but has not yet been quantitatively demonstrated.

The high regional heat fluxes in the south polar region, averaging at least ~ 100 mW m^{-2} over the region S of 55° S, could imply the presence of water at a depth of only 6 km. However, this calculation assumes that the heat is transported by conduction, which may not be correct: Spencer et al. (2006) and Nimmo et al. (2007) argue that the bulk of the heat is advected (by flow of vapor) and thus that the subsurface thermal gradient cannot be inferred from the heat flow measurements. See Section 21.5.3.1.

21.2.2.3 Discussion and Summary

Most existing models assume that Enceladus is presently differentiated, even though the direct evidence supporting this hypothesis is weak. The strongest argument for a differentiated Enceladus is its large heat flux, which seems to require a differentiated structure and a global subsurface ocean (see below). Shape measurements suggesting a non-hydrostatic interior, and the significant north-to-south variations in Enceladus' geology, both indicate that deductions

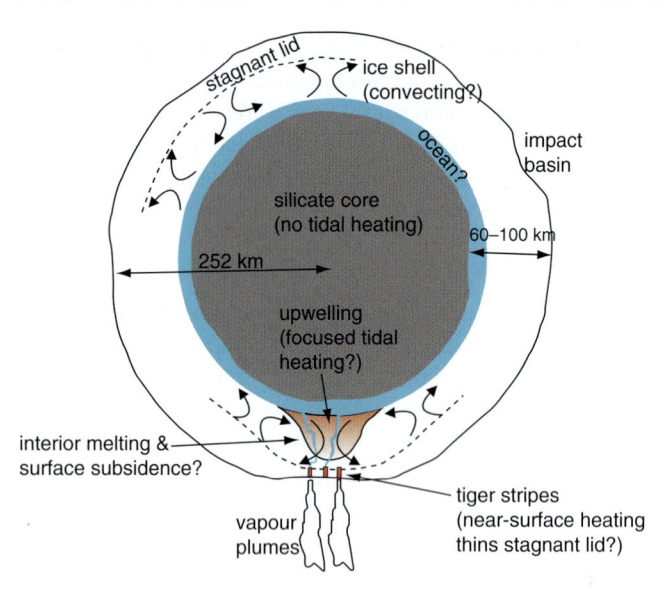

Fig. 21.3 A sketch of the potential internal structure of Enceladus, after Nimmo and Pappalardo (2006), Schubert et al. (2007) and Tobie et al. (2008). The amplitude of the surface topography is exaggerated. The arguments for a convective interior and thinning of the lithosphere near the south pole are presented in Section 21.2.5

about the interior structure cannot be made from the static gravitational field by assuming hydrostatic equilibrium and a radially symmetric structure. More complex models will be required, which will be less likely to yield unique solutions.

Figure 21.3 is a sketch illustrating our current understanding of Enceladus, aspects of which are discussed through the remainder of this section. The presence of question marks indicates the uncertainty of many of the inferences made.

21.2.3 Interior Composition and Chemistry

The plume of water vapor and other materials erupting from Enceladus' south pole provides the first opportunity to directly sample material from the interior of an icy satellite. Cassini INMS data (Section 21.5.2.4) suggest that Enceladus' gas plume is composed of roughly 90–94% water vapor by mass, plus 5% CO_2, 0.9% CH_4, and 0.8% NH_3, and small amounts of organics, H_2S, and ^{40}Ar. There may also be significant N_2, though this is uncertain because INMS cannot distinguish N_2 from C_2H_4 (Waite et al. 2009). CAPS measurements of magnetospheric nitrogen ions detect nitrogen in the plume: the data imply a source of nitrogen and perhaps NH_3 at Enceladus (Smith et al. 2008a, b). Most E-ring particles measured by CDA appear to contain a small amount of sodium, and some are rich in both NaCl and $NaHCO_3$ (Postberg et al. 2009 and Section 21.5.2.5), suggesting an oceanic provenance. However, the abundances of CO_2, CH_4, and perhaps other gas species are too high to originate from gas dissolved in liquid (Waite et al. 2009), and subsurface clathrates have been suggested as a source for these species (Kieffer et al. 2006).

The compositional measurements from Enceladus have so far been interpreted in two contexts: as a constraint on its present-day ocean chemistry, and as a constraint on Enceladus' early thermal evolution. Calculations of the water-rock chemistry occurring over Enceladus' lifetime suggest that its present ocean should include significant Na-Cl-HCO_3 (Zolotov 2007), consistent with the CDA measurements of E-ring particle composition.

Because nitrogen is likely delivered to Enceladus from the solar nebula in the form of NH_3 clathrate within ice grains during its accretion (Mousis et al. 2002, Lunine and Gautier 2004), the conversion from ammonia to molecular nitrogen (if present) likely occurs within Enceladus itself (Matson et al. 2007, Glein et al. 2008). Matson et al. (2007) suggest that N_2 is formed from NH_3 by thermal decomposition, which requires high temperatures (~500–800 K) within the silicate core, and further suggest that the satellite's interior had been heated to these temperatures by a combination of short-lived radioisotopes (SLRI's, such as ^{26}Al and ^{60}Fe) and tidal heat early in its history. Glein et al. (2008) favor a different evolutionary pathway for nitrogen, in which the NH_3 to N_2 conversion occurs within hydrothermal systems created during an early phase of core volcanism driven by SLRI's.

The apparent presence in the plume of ^{40}Ar, which originates from radioactive decay in Enceladus' silicate fraction, is also likely to constrain the interior evolution of Enceladus (Waite et al. 2009).

It should also be noted that if the current plume gas production rate of about $200 \, kg \, s^{-1}$ has been maintained over the age of the solar system, Enceladus would have lost about 20% of its mass and 40% of its water, with drastic consequences for its chemical and physical evolution (Kargel 2006), including a loss of volatile elements and a major reduction in radius.

21.2.4 Heat Production and Tides

Observations of the south polar heat flow anomaly place direct constraints on heat production within Enceladus. The observed heat flow is at least $5.8 \pm 1.9 \, GW$ (Spencer et al. 2006), and is likely to be higher, because published observations are insensitive to heat radiated at temperatures below about 85 K. The observed power significantly exceeds the expected radiogenic contribution of ~0.3 GW (Porco et al. 2006, Schubert et al. 2007), suggesting that the bulk of the energy is derived from tidal dissipation (cf. Squyres et al. 1983,

Ross and Schubert 1989). Although this conclusion appears robust, many details of the heat production are still poorly understood.

Tidal heating arises because a satellite in an eccentric orbit experiences a tidal bulge that varies in amplitude and location over the course of the orbit. These time-dependent strains will dissipate energy if the deforming material is not perfectly elastic (i.e., if there is a phase lag). For a synchronous satellite the dissipation rate can be written as (e.g., Peale and Cassen 1978)

$$\dot{E} = -\frac{21}{2}\frac{k_2}{Q}\frac{(\Omega R)^2}{G}e^2$$

where Ω, e and R are the orbital angular frequency and eccentricity and the satellite radius, respectively, and G is the gravitational constant. The quantity k_2 is the magnitude of the tidal Love number, which depends on density and rigidity of the satellite and determines the amplitude of the tidal bulge. The quantity Q is the quality factor: $Q/(2\pi)$ is the ratio between the total energy of an oscillation and the energy dissipated per cycle, and is proportional to the phase lag of the tidal bulge. Q can be calculated if the specific dissipation mechanism is known- smaller values imply a more dissipative interior. The tidal Q is generally poorly known and on the Earth is observably frequency-dependent (e.g., Anderson 1989, Benjamin et al. 2006).

Geological materials exhibit both a brittle/elastic behavior described by a rigidity, μ, and ductile behavior, described by a viscosity, η. The most commonly used viscoelastic model, which best approximates the behavior of real geological materials, is the Maxwell model (Ross and Schubert 1986). For Maxwellian materials the phase lag depends on the ratio of the forcing period to the characteristic relaxation or Maxwell time ($= \eta/\mu$) (Ross and Schubert 1986). For a Maxwell body the Love number is complex and k_2/Q is given by $Im(k_2)$ (Zschau 1978). Assuming a Maxwell rheology, the total tidal dissipation, and the spatial distribution of this heating, may be calculated for an arbitrary satellite structure (Segatz et al. 1988, Tobie et al. 2005).

An early application of this method to Enceladus was performed by Ross and Schubert (1989), who demonstrated that significant (up to $\sim$1 TW) tidal heating could occur in a viscoelastic Enceladus, depending on material properties, though the properties required for the highest heat flows were not necessarily realistic (Section 21.2.2.2). More recently, Roberts and Nimmo (2008a) investigated viscoelastic tidal dissipation in Enceladus in some detail. They assumed a radially symmetric body with a uniform silicate layer, an optional ocean and an ice shell in which the temperature and viscosity structure were appropriate to either a conductive or a convective setting. Tidal dissipation in the silicates was found to be negligible for any likely viscosity. This conclusion is important because Collins and Goodman (2007) require significant silicate heating to explain the south polar topographic depression. Dissipation in the ice shell increases by a factor of $\sim$10 if the shell is decoupled from the interior by an ocean. For models with a subsurface ocean, the global heat flow is comparable to or exceeds the measured heat flow of the south polar thermal anomaly. However, the spatial pattern of the modeled heating has maxima at both poles, in contrast to the observations.

This last point illustrates a basic problem with almost all published tidal heating models: they are symmetrical about the equator, which is not the case for the heat flow on Enceladus. Enceladus may have large lateral variations in mechanical properties, which can in turn focus tidal heating. A recent study by Tobie et al. (2008) calculates the tidal dissipation in a non-radially symmetric body and confirms that tidal heating can indeed be focused in a warm upwelling, as suggested by previous authors (Sotin et al. 2002, Mitri and Showman 2008). The focusing of tidal heating may explain the high heat flux at the south pole, if this region is indeed the site of an upwelling diapir (Nimmo and Pappalardo 2006).

Most tidal heating models focus only on eccentricity tides. However, tidal heating can also occur if Enceladus has a non-zero obliquity, or if it is undergoing forced rotational librations (Wisdom 2004, see his Equation 45). Forced librations can occur if the libration and spin frequencies are in a simple integer ratio (Wisdom 2004), however, no librations have been detected within uncertainty (± 0.026 radian), implying a current librational heating that is at most comparable to radiogenic heating (Porco et al. 2006). The obliquity of Enceladus is currently unknown, but based on theoretical investigations of the Galilean satellites (Bills 2005) it is likely to be small but non-zero, and could contribute to the total tidal heating.

It is also not yet clear that the Maxwellian rheology is appropriate to real planetary materials, including water ice. For instance, both observations and laboratory measurements of silicates suggest that Q has a significantly weaker dependence on frequency than the Maxwell model implies (e.g., Anderson 1989, Gribb and Cooper 1998, Karato 1998, Jackson et al. 2004, Benjamin et al. 2006). Another non-Maxwellian process that may be important is friction on shallow faults. Tidal stresses can cause oscillatory strike-slip motion (Hoppa et al. 1999), which can in turn cause shear heating (Nimmo and Gaidos 2002). Shear heating is a possible way of explaining the high temperatures of the tiger stripes relative to the surrounding terrain (Nimmo et al. 2007), and exhibits fundamentally different dependencies compared with equation (1). A third non-Maxwellian process of possible interest is turbulent dissipation in the subsurface ocean (Nash et al. 1986, Ross and Schubert 1989, Tyler 2008). Shallow seas on Earth are the dominant location of tidal dissipation: on Enceladus, dissipation in a thin ocean could generate several GW of heat. Developing dissipation models using more realistic rheologies is likely to be another focus of future work.

21.2.5 Thermal Structure

The spatial localization of Enceladus' activity at its south pole strongly suggests that tidal dissipation has become concentrated there. As described in Section 21.2.4, most tidal dissipation models (e.g., Ojakangas and Stevenson 1989, Tobie et al. 2005, Roberts and Nimmo 2008a) imply that dissipation should be concentrated in warm ice in Enceladus' ice shell, and should be broadly distributed as a function of latitude and longitude. Tobie et al. (2008), however, showed that a low viscosity region can focus tidal deformation and heating, and suggested that episodic behavior might result. Nimmo and Pappalardo (2006) suggest that the activity is concentrated in the south polar region because a large density perturbation in its ice shell and/or rocky interior caused the satellite to "roll over" to place this active region on the satellite's rotation axis. Solid state convection would be one way of creating a thermal and density anomaly in the ice shell, although it is currently unclear why only one such convective plume should have developed (cf. Grott et al. 2007), or whether a purely thermal plume could generate a sufficiently large mass anomaly. Localization of tidal heating probably requires some thermal, compositional, or mechanical anomaly.

Convection may also play a role in Enceladus' interior evolution by regulating the temperature in its interior. Barr and McKinnon (2007a) find that initiating convection in a floating conductive ice I shell on Enceladus may be difficult, in part because of the satellite's low gravity, which implies that thermal buoyancy stresses created by local temperature highs and lows (which are required as initial perturbations for the onset of convection) are very low. As a result, convection can only begin in a the thickest possible ice shell ($D \sim 100$ km) with a low melting point viscosity, implying that grain sizes in the warm ice at the base of its shell must be less than 0.3 mm, close to the lower limit on grain size in natural terrestrial systems (i.e., glaciers) (McKinnon 2006, Barr and McKinnon 2007b). Tidal dissipation may play a role in helping to trigger ice shell convection if tidal heat can create pockets of warm ice on length scales comparable to the thickness of Enceladus' ice shell, a length scale comparable to the size of temperature fluctuations maximally efficient at triggering convection (Solomatov and Barr 2007). Other modes of convection have also been proposed, including "whole-Enceladus" degree-1 convection in an undifferentiated or partially differentiated satellite (Grott et al. 2007). Such a pattern would provide a natural means of creating a large thermal anomaly at one location in Enceladus, which could be further amplified by tidal dissipation to create the north/south dichotomy.

Spread over the $\sim 70,000$ km^2 region of the south polar terrain (Porco et al. 2006), the total power output of at least 5.8 ± 1.9 GW (Spencer et al. 2006) corresponds to a heat flux $F = 55$ to 110 mW m^{-2}. Because the viscosity of water ice is strongly temperature dependent, most existing models of convective heat transfer in Enceladus' ice shell assume that convection occurs beneath a thick "stagnant lid" of highly viscous, immobile ice at the satellite's surface. The stagnant lid inhibits convective heat transport by preventing convective plumes from reaching close to the surface of the satellite. On Enceladus, the upper limit on the heat flux from stagnant lid convection is $F \sim 20$ mW m^{-2} (Barr and McKinnon 2007a, Barr 2008, Roberts and Nimmo 2008a), a factor of ~ 3–4 less than the regional heat flux within the SPT.

However, if the near-surface ice on Enceladus is very weak, convective plumes can reach closer to the surface of the satellite, permitting the observed SPT heat flux to be carried by convection in a thick ice shell rather than by conduction in a thin ice shell. If the yield strength of Enceladus' lithosphere within the SPT region is less than ~ 10 kPa, convection occurs in the "sluggish" or "mobile-lid" regime, characterized by larger heat fluxes comparable to the observed value, and horizontal velocities ~ 1–10 m year^{-1} (Barr 2008). Lithospheric spreading associated with convection in a weak ice shell may provide a natural explanation for the region's <0.5 Myr surface age (Porco et al. 2006) and the compressional tectonics at its flanks (Helfenstein et al. 2009, and Section 21.3). The ice shell of Enceladus within the SPT may become mechanically weakened by cyclical tidal deformation (Barr 2008), or thermally weakened by tidal dissipation in the shallow subsurface (Roberts and Nimmo 2008b).

The effects of the recent detection of NH$_3$ in the plume (Waite et al. 2009, and Section 21.5.2.4), and therefore in the interior of Enceladus, on the internal rheology and thermal state of the ice shell may be significant, but have yet to be considered in detail.

The surface topography of Enceladus provides an important additional constraint. Limb profiles (Thomas et al. 2007), suggest topographic variations with amplitudes of hundreds of meters on wavelengths of ~ 100 km. These wavelengths are comparable to those predicted by theoretical convection models (see Fig. 21.4), but the amplitudes are much larger than predicted for stagnant lid convection. The amplitudes will be larger if there is lithospheric yielding (Showman and Han 2005) or compositional buoyancy contrasts. The latter effect may arise if tidal heating is focused within upwelling regions (Sotin et al. 2002, Mitri and Showman 2008), leading to melting and removal of dense contaminants (cf. Pappalardo and Barr 2004).

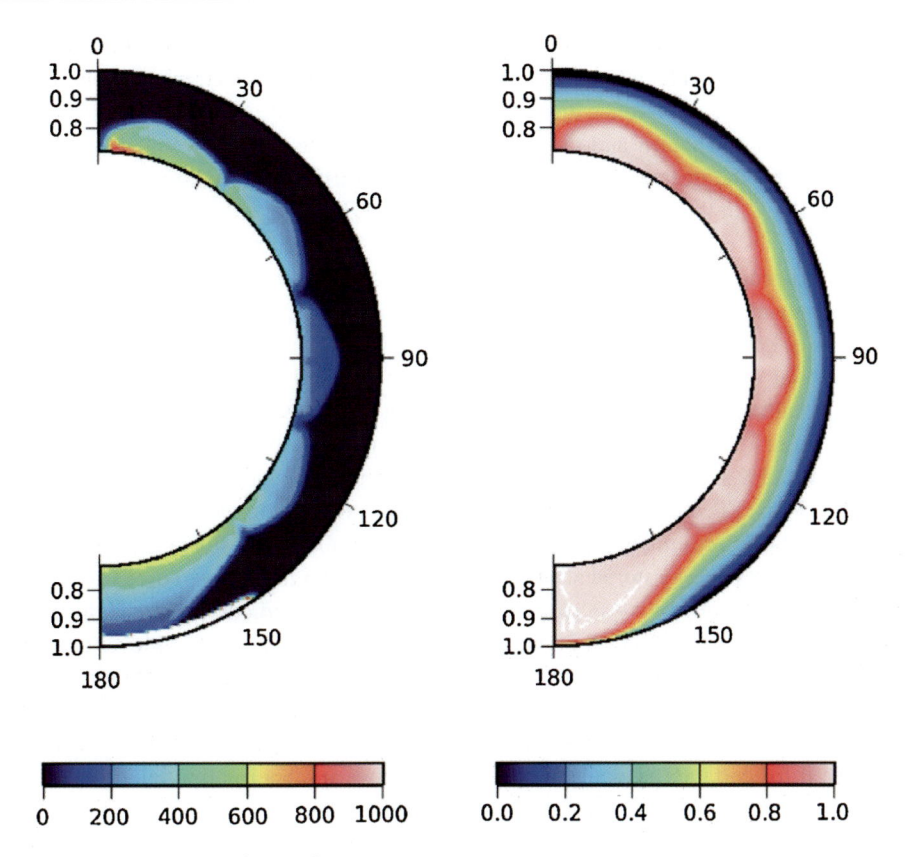

Fig. 21.4 Plot of heating rate in units of 10^{-9} W m^{-3} (*left*) and dimensionless internal temperature (*right*: 0.0 and 1.0 correspond to 75 K and 273 K respectively) for a convecting axisymmetric Enceladus, from Roberts and Nimmo 2008b). Maxwellian tidal heating is greatest at the poles and at the base of the ice shell; heating in the stagnant lid is negligible owing to the high viscosity. An additional (frictional) shallow heat source has been imposed polewards of 55°S; note that this generates an abnormally large upwelling and thinned stagnant lid at the south pole

A number of hypotheses have been proposed for the relationship between tidal heat generation, heat transport, and the tiger stripes. The high brightness temperatures observed close to the tiger stripes (Spencer et al. 2006 and Fig. 21.14) led Nimmo et al. (2007) to suggest that shear heating due to cyclical strike-slip motion along fault zones within the stripes could be the dominant mode of tidal heat generation in this region. Because the SPT sits topographically lower than the surrounding terrain, Collins and Goodman (2007) suggest that the SPT is a region where the ice shell is locally thin. Tobie et al. (2008) suggest that subsurface tidal dissipation focused within the SPT causes melting within the ice shell, leading to subsidence. Conversely, in the Nimmo et al. (2007) scenario, frictional heating on faults within the tiger stripes is currently the dominant mode of tidal dissipation in the region. If shear heating occurs in an ice shell that is convecting, convective upwellings are "drawn to" the fault zones, resulting in locally enhanced heat fluxes and a thinned stagnant lid (Roberts and Nimmo 2008b, see Fig. 21.4). The interaction between shear heating and convection is an active area of research in the context of Europa's double ridges (Han and Showman 2008): similar efforts for Enceladus could provide insight into the modes of heat generation and transport within the SPT.

21.2.6 Evolution Through Time

The thermal evolution of satellites is much more complicated than that of their planetary cousins because tidal heating, which can show large temporal variability, provides the main satellite heat source. Dissipation in a satellite depends on the mechanical (and thus thermal) structure of the satellite, and causes the orbital eccentricity to decrease, while dissipation in the primary increases the satellite semi-major axis (e.g., Dermott et al. 1988). Changing the satellite eccentricity changes the tidal heating and thus the thermal structure; as the thermal structure changes, so does the dissipation, and so on. Orbital and thermal evolution are thus intimately coupled (Ojakangas and Stevenson 1986, Showman et al. 1997, Hussman and Spohn 2004). In the case of Enceladus the eccentricity is prevented from damping out by the Enceladus:Dione 2:1 mean motion resonance.

Meyer and Wisdom (2007) recently pointed out that the present-day eccentricity of Enceladus cannot be a steady-state value. A constant eccentricity, e, implies that the increase in e driven by dissipation in Saturn and the resonance with Dione is balanced by the decrease in e driven by dissipation in the two satellites. In this steady-state case, the total dissipation in the satellites can be calculated based on the Q of Saturn, without needing to know Q or k_2 of either satellite (cf. Lissauer et al. 1984). Assuming it is frequency-independent, a lower bound on Q_{Saturn} can be derived by considering the orbital evolution of Mimas (e.g., Dermott et al. 1988). The resulting upper bound on dissipation within Enceladus is about 1.1 GW for the minimum Q_{Saturn} of 18,000, less than the $>5.8 \pm 1.9$ GW currently observed (Spencer et al. 2006). Thus, Enceladus has an eccentricity that is not in steady state but is presumably decreasing with time, while the satellite releases energy stored at an earlier epoch.

The orbital evolution of Enceladus turns out to be complicated (Meyer and Wisdom 2008). Capture into its present-day 2:1 "e-Enceladus" resonance was probably preceded by temporary capture into several other nearby 2:1 resonances, and could only have occurred for certain values of Dione's eccentricity and k_2/Q. None of these resonances generate an eccentricity in excess of 0.007, which limits the amount of tidal heating that could have occurred in the recent past. Passage through the Mimas:Enceladus 3:2 resonance does not appear to generate eccentricities above the equilibrium value (resulting in a maximum heat flow of 0.48 GW); furthermore, there is no obvious way of escaping this resonance (Meyer and Wisdom 2008). An alternative way to generate a transient, high-eccentricity episode in the recent past might be for Janus to have passed through a 2:1 resonance with Enceladus, which was then disrupted by some unspecified process (Lissauer et al. 1984), but recent estimates of the resulting tidal heating (Meyer and Wisdom 2007) are much smaller than is observed on Enceladus.

As outlined above, it seems likely that Enceladus possesses a subsurface ocean at the present day. The lifetime of the ocean is therefore a question of considerable interest from both an astrobiological and a geophysical point of view. In particular, because tidal heating is much reduced in cases lacking an ocean, it is unclear how an initially solid Enceladus could be tidally heated enough to allow an ocean to form, unless it did so early in its history when radioactive heating was important (Schubert et al. 2007). The long-term stability of an ocean beneath tidally heated Maxwellian conductive and convective ice shells was examined by Roberts and Nimmo (2008a). These authors concluded that in both cases the removal of heat from the top of the ocean significantly exceeded the likely (radioactive) heat flow into the base of the ocean, and thus that the ocean would freeze over timescales of a few tens of Myr. Tobie et al. (2008) reached a similar conclusion, but also suggested that a local subsurface "sea" might be sustained indefinitely by melting in the overlying ice shell. The presence of antifreezing agents (e.g., ammonia) in a global ocean would delay but not halt the freezing process. Turbulent dissipation in a thin ocean might prevent it from freezing altogether (cf. Ross and Schubert 1989). Alternatively, an eccentricity three times higher than the present-day value, if it could be maintained, would allow an ocean to persist indefinitely, again strongly pointing towards an earlier epoch when Enceladus had a higher eccentricity.

The coupled thermal-orbital evolution of Enceladus has not yet been investigated in any detail. The arguments of Meyer and Wisdom (2007) suggest that either heat production within Enceladus has varied with time, or heat production has stayed constant but the heat output has varied. Ojakangas and Stevenson (1986) pointed out that Enceladus' eccentricity might undergo oscillatory behavior in a manner similar to that predicted for the Galilean satellites (cf. Hussmann and Spohn 2004), but Meyer and Wisdom (2008) have shown that such a model can't work for Enceladus because it is too small. Tobie et al. (2008) argue that tidal heating itself might result in oscillatory behavior, with Enceladus currently generating an abnormally high level of heat. These authors also argue that the putative ocean must be ancient, because in the absence of an ocean it is impossible to initiate sufficient tidal heating, as noted above. Coupling the thermal and orbital evolution of icy satellites is a challenging task that will undoubtedly form the focus of much future work.

In addition to explaining the current activity on Enceladus, any self-consistent theory has to also explain the fact that Mimas appears to be geologically inactive. If Mimas and Enceladus had the same value of Q, tidal heating in Mimas should be about 25 times stronger than in Enceladus (Squyres et al. 1983). There are two possible solutions to this problem: either Enceladus has an additional source of heat not available to Mimas; or small variations in initial conditions lead to different evolutionary paths. In the first category, Wisdom (2004) proposed that Enceladus was in a spin-orbit resonance which causes enhanced (librational) tidal heating, though as discussed above this heat source is probably not important at present. In the second category, it is well known that tidal heating can lead to softening of the interior, a decrease in Q and thus enhanced tidal heating, leading to a runaway effect: the tidal heating becomes self-sustaining if there is sufficient energy in the orbital resonance to maintain the heating. Thus, if Enceladus were initially more strongly heated than Mimas, they might have experienced different evolutionary paths (e.g., Czechowski 2006). One possibility is that Enceladus, which is presumed to have a higher rock fraction than Mimas (Thomas et al. 2007), experienced more heating due to the decay of short-lived radioisotopes such as ^{26}Al (e.g., Matson et al. 2007). While reasonable, this hypothesis is probably not unique, given that the orbital evolution of the Saturn system was complex and likely involved various satellite resonant encounters. For short-lived radioisotopes to have initiated the currently observed tidal heating, the tidal heating must have been nearly continuous since the origin of the Saturn system, which is difficult to arrange (Meyer and Wisdom 2007).

21.2.7 Summary and Future Work

After nearly 5 years of observing Enceladus from Cassini, we are only beginning to unravel its interior structure. Given the amount of tidal and radiogenic heating Enceladus has experienced, and the present-day tidal heating rate (see Section 21.2.2.2), it seems likely that Enceladus is differentiated. Its shape indicates that its interior is probably not in hydrostatic equilibrium, which means that direct observational constraints on its internal structure from static gravity or shape data will prove difficult (Section 21.2.2.1).

The interior structure and material properties of Enceladus are very likely laterally heterogeneous, and probably include a warm upwelling region at the south pole, in which tidal heating is focused (Section 21.2.5). This upwelling may have been promoted by near-surface shear heating along the tiger stripes, and may have led to reorientation of the satellite. The magnitude of the observed heating strongly suggests the presence of either a global subsurface ocean or a regional "sea" (Sections 21.2.2.2, 21.2.4). Whether such an ocean could have survived over geological timescales depends on the history and amount of tidal heat production and is currently very much an open question.

The amount of heating observed at the south polar terrain exceeds that which can be produced by standard tidal heating models if Enceladus' eccentricity is in steady state (Section 21.2.6). Because there is no obvious alternative to the Maxwell viscoelastic tidal heating model, this result strongly suggests that the orbital and thermal evolution of Enceladus has been complex and non-monotonic.

The long-term thermal evolution of Enceladus is a major unsolved question, and will benefit from coupled thermal-orbital models in a manner similar to those developed for the Jovian satellites by Ojakangas and Stevenson (1986), Showman et al. (1997) and Hussmann and Spohn (2004). The spatial heterogeneity of Enceladus' interior structure and mechanical/rheological properties pose a challenge to current radially symmetric geodynamical and orbital models. As geodynamical models become more complex, and involve more realistic and detailed models of Enceladus' interior and material properties, more laboratory data about ice behavior and observations of Enceladus will be needed to constrain them.

The complexity of Enceladus' interior and the observations of its behavior from Cassini have sparked renewed interest in icy satellite interiors, and have suggested new areas of theoretical development and new laboratory experiments. We have been granted a unique opportunity to characterize an icy satellite during the time of its cryovolcanic activity with a well-instrumented spacecraft. Understanding the processes at work in Enceladus' interior will no doubt lead to substantial progress in our understanding of tidal heating, icy satellite resurfacing, and coupled thermal/orbital evolution.

Several arguments suggest that Enceladus is currently in an atypically warm and active state. Its current energy output is difficult to sustain over the long term. In addition, there is no geological evidence for the large change in radius implied if current mass loss rates have been maintained through Enceladus' lifetime, and ^{40}Ar is apparently being lost through the plume much faster than it is being generated by radioactive decay (Waite et al. 2009 and Section 21.5.2.4). Against these arguments must be weighed the probability of our appearing on the scene at just the right epoch to witness this unusual activity. The apparent difficulty in re-starting tidal heating in a cold and frozen Enceladus, if heating were ever to cease completely, may also place limits on how much the level of activity has varied.

21.3 Geology

21.3.1 Introduction

Voyager's highest resolution coverage ($\sim$1 km per pixel) of Enceladus provided views only of the Saturn facing hemisphere and its north polar region. That coverage was sufficient to show that Enceladus' surface morphology and geology is characterized by distinct provinces that show abrupt changes in crater densities and which are confined by major tectonic contacts. The major geological provinces identified in Voyager images are heavily cratered terrain and ridged and fractured plains. Each of these provinces can further be subdivided in terms of specific regional crater densities and complexities as well as their tectonic characteristics (Smith et al. 1982, Squyres et al. 1983, Passey 1983, Plescia and Boyce 1983, Lissauer et al. 1988, Kargel and Pozio 1996).

Cassini has now provided near-global coverage at Voyager resolution or better (Roatsch et al. 2008, Helfenstein et al. 2009). Globally, Enceladus exhibits large-scale geological and tectonic patterns that have remarkable symmetry relative to the spin axis and the Saturn/anti-Saturn direction (Porco et al. 2006, Helfenstein et al. 2009). Four broad terrain regions, with subdivisions that encompass and extend the earlier classes, are systematically arranged relative to these directions. These are cratered plains, western hemisphere fractured plains, eastern hemisphere fractured plains, and the South Polar Terrain (SPT) province (Fig. 21.5). For detailed image maps and feature names referred to below, see Chapter 24, this book.

21.3.2 Major Terrains: Nature and Global Distribution

21.3.2.1 Cratered Plains:

A nearly continuous N-S band of heavily cratered materials wraps from the Saturn-facing side of Enceladus, over the north pole, and down the anti-Saturn side. It is interrupted only near the south pole by the SPT province. The band of cratered plains separates the tectonically resurfaced terrains on the leading and the trailing hemispheres. Kilometer and smaller scale craters in the cratered plains often exhibit bowl shapes, however almost all larger craters show varying stages of viscous relaxation (Smith et al. 1982, Passey 1983, Moore et al. 2004) with updomed floors, subdued topographic relief, and lobate or ramparted crater rims (Fig. 21.6A1, A2). The craters are all less than 40 km in diameter with topographic relief that varies from less than 100 m to about 1,000 m (Kargel and Pozio 1996). Large craters in the cratered plains are also heavily dissected by numerous fine fractures with spacing downward from kilometers to hundreds of meters (Miller et al. 2007). The degradation state of craters increases toward southern latitudes. The most extreme examples of viscously relaxed impact features occur especially near the outer boundaries of the cratered plains, such as at the northern margin of Sarandib and Dyar Planitias (near 35°N, 290°W), and adjacent to the South Polar Terrain province. Within the cratered plains, prominent orthogonal sets of N-S and E-W trending rifts occur near the sub-Saturn (0°N, 0°W) and anti-Saturn (0°N, 180°W) points. Evidence of strike-slip motion is seen on the sub-Saturn example (Kargel and Pozio 1996). Their symmetric placement relative to the tidal axis of Enceladus suggests that tidal interactions played a role in their formation (cf. Helfenstein et al. 2009).

21.3.2.2 Eastern (Trailing) Hemisphere Fractured Plains

The eastern hemisphere (centered at 270°W) is heavily resurfaced by tectonism. From equatorial to mid-northern latitudes, it is dominated by two major fractured plains regions, Sarandib Planitia (5°N, 305°W) and Diyar Planitia (5°N, 250°W). The fractured plains exhibit both negative and positive tectonic relief forms. Troughs, scarps, grooves, and pit chains reflect mainly extensional stress while ridges may indicate compressional forces (Passey 1983). Networks of kilometer-scale troughs, fractures, and ridges further subdivide these plains into crudely polygonal sections, each of which is several tens of kilometers across. These polygonal subsections exhibit varied fracturing styles, although commonly the fractures and ridges within a given polygon form a parallel fabric that may differ in orientation from those in adjacent polygons. In the northern parts of Sarandib and Diyar, terrains are mostly made up of striated & ridged plains (Kargel and Pozio 1996) and relatively smooth plains (cf. Passey 1983). The striated & ridged plains are bounded against the cratered plains by ridged & grooved materials; Harran Sulcus in the North and East, and Samarkand Sulcus on the West side. Also within the fractured terrains are flow-like features resembling glacier surfaces or pasty cryovolcanic deposits (Smith et al. 1982, Squyres et al. 1983, Kargel and Pozio 1996, Kargel 2006, Porco et al. 2006).

Sarandib Planitia and Diyar Planitia are separated in the north by a peculiar polygonal system of rounded narrow ridges called dorsa, and elongate domes, all of which appear to have been extruded from pre-existing fractures. The most prominent examples Ebony Dorsum and Cufa Dorsa (Fig. 21.6, C1), were visible in Voyager images (Smith et al. 1982). At the highest Cassini ISS resolution of this area (65 m per pixel), the features are crosscut by sub-kilometer scale parallel cracks that seem to cover the entire surface of Enceladus (Miller et al. 2007). The only other Eastern hemisphere ridge structure of comparable relief is a 120 km long, N-S trending ridge at the southern end of Samarkand Sulcus along the western edge of Sarandib Planitia. The Samarkand Sulcus ridge is morphologically distinct from the dorsa: It appears to be a compressional feature composed of angular, tilted icy slabs that are stacked en echelon as chevron-shaped cuestas that dip southward at shallow angles. Their morphology suggests that the Sarandib ridge is a compressional feature, most likely a thrust (Kargel and Pozio 1996, Helfenstein et al. 2009).

At about 15°S, a distinct south-facing scarp marks an abrupt southward transition from the equatorial and northern fractured plains units to the eastern hemisphere reticulated plains (Helfenstein et al. 2009). Reticulated plains are characterized by a complex network of crosscutting fractures that contain at least two recognizable superposed fracture patterns of generally different stratigraphic age. The base tectonic fabric is a stratigraphically old pattern of curvilinear fractures that generally intersect at acute angles and divide the terrain into lenticular mesas and hills. Superposed on this pattern is a system of quasi-parallel bands that cut the pre-existing terrain into lanes along which lateral shear displacements are sometimes evident. In the eastern hemisphere, the reticulated terrain is overprinted with sparse impact craters

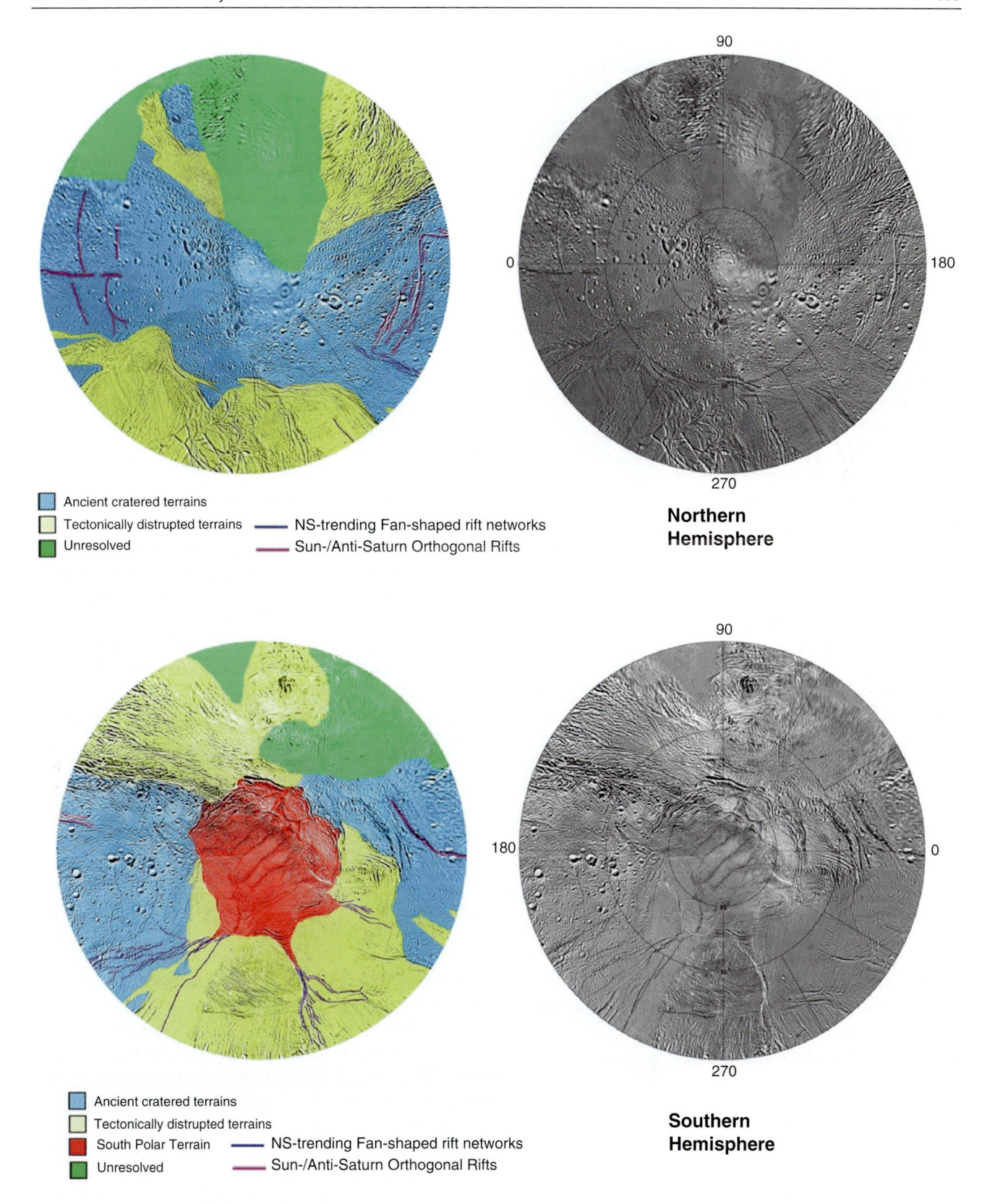

21 Enceladus: An Active Cryovolcanic Satellite

Fig. 21.5 Simplified geological unit map of the southern hemisphere (*bottom*) and northern hemisphere (*top*), respectively, showing principal terrain subdivisions and prominent tectonic patterns. Colored unit maps at left are keyed to the corresponding grayscale basemaps at right. Longitude 0 points towards Saturn. See text for discussion

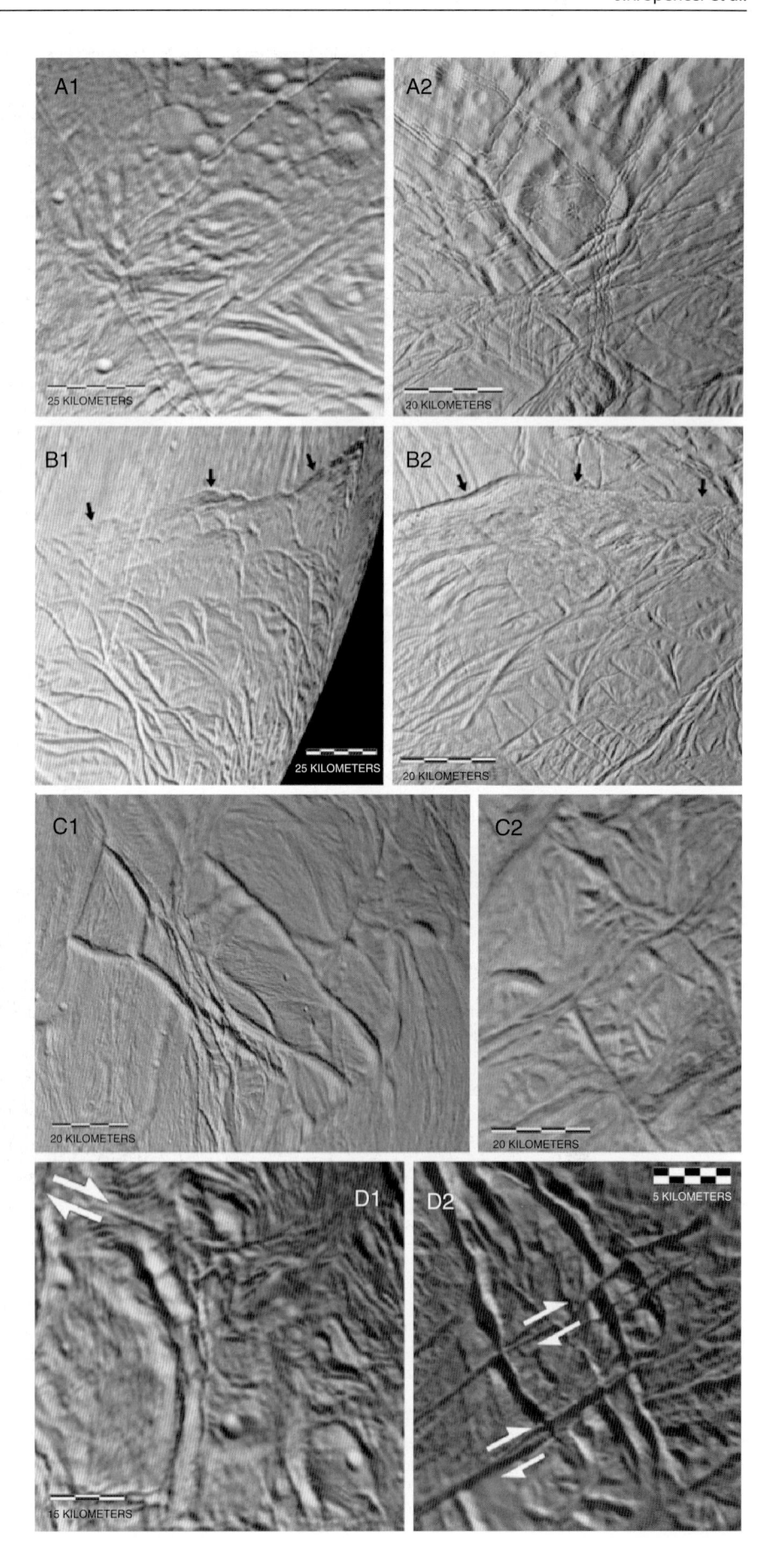

Fig. 21.6 Comparison of proposed relict SPT-style features in Enceladus' eastern hemisphere with SPT examples (Helfenstein et al. 2009). A1-A2 shows highly relaxed, laterally deformed craters that suggest that regions bounding Sarandib Planitia (A1 at 38°N, 316°W) and the SPT (A2 at 50°S, 177°W) have undergone very similar thermal histories and margin deformation. B1-B2 compares (B1) the south facing scarp and enclosed reticulated plains (near 255°W, 15°S) in the eastern hemisphere with (B2) a similar bounding scarp and enclosed reticulated plains in the SPT (55°S, 200°W). C1-C2 compares the eastern hemisphere dorsum network (C1 at 5°N, 285°W) containing Ebony Dorsum and Cufa Dorsa with (C2 at 73°S, 320°W) a similar unnamed dorsa network within the south polar reticulated plains. D1-D2 compares (D2 at 76°S, 0°W) morphologically degraded relict tiger stripe features at the end of Damascus Sulcus with (D1 at 20°N, 277°W) a morphologically similar example in the Eastern hemisphere within Diyar Planitia. Arrows in D1-D2 identify right-lateral strike-slip offsets

that have diameters up to about 10 km. The eastern hemisphere reticulated terrain is bounded in the south by the South Polar Terrain margin (see below and Fig. 21.5).

21.3.2.3 Western (Leading) Hemisphere Fractured Plains

While the eastern hemisphere of Enceladus has been imaged at kilometer-scale resolution or better by both Voyager and Cassini spacecraft, at present only about 30% of the western hemisphere (centered at 90°W) has been covered at this resolution. Coverage is sufficient to show that, as in the eastern hemisphere, large areas of the western hemisphere have been modified by fracturing and tectonism. However, unlike in the eastern hemisphere, tectonic resurfacing of the western hemisphere may not have eradicated all evidence of early large impact craters. Figure 21.5 shows that, in the western hemisphere, the edge of the near-global N-S band of cratered plains borders against fractured plains. To the extent that they can currently be identified, the fractured plains are ridged & grooved materials, generally trending N-S where they are closest to cratered plains, transitioning to E-W near the boundary with north polar cratered plains and the south polar Terrain boundary.

In spite of widespread fracturing, the remnants of the early geological history of the western hemisphere may possibly remain. Figure 21.7A shows possible highly relaxed ancient impact features (palimpsests). While there is some evidence for overprinting by fractures, the features, if they are indeed palimpsests, are relatively undistorted by widespread tectonism. Other large, possible relict impact craters are shown Fig. 21.7B, where arcuate patterns of narrow ridges outline a cluster of large quasi-circular features. These are perhaps the buried or thermally eroded rims of large ancient impact craters. If these are impact features, they are larger in diameter by a factor of two or more than any unambiguously identified impact crater on Enceladus. In that case the western hemisphere would preserve some of the oldest terrain on Enceladus. The high state of thermal degradation of the possible palimpsests in the western hemisphere is evidence that this region, like the eastern hemisphere and SPT, may have evolved through a period of elevated thermal flux (Bland et al. 2007, Giese et al. 2008).

21.3.2.4 South Polar Terrain (SPT)

The SPT province is a geologically active region centered on the south pole (Porco et al. 2006). It is bounded at about 55°S latitude by a prominent circumpolar chain of south-facing arcuate scarps and confined mountain chains, from which a series of N-S trending fractures and rifts extend toward the equator (Figs. 21.1, 21.5, and 21.8, see also Porco et al. 2006, Fig. 2A). These fractures crosscut all other terrains in their path and are thus stratigraphically youthful

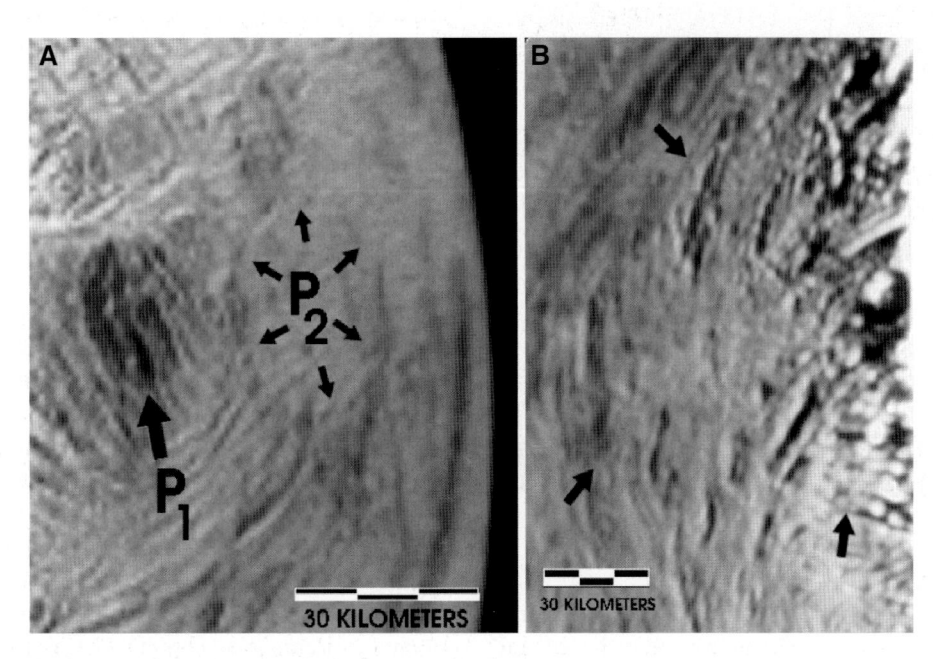

Fig. 21.7 Possible palimpsests in the western hemisphere of Enceladus. (**A**) Portion of ISS NAC image N1489087056 that shows a possible dark palimpsest (P$_1$, 11°N, 79°W) and adjacent bright palimpsest (P$_2$, 15°S, 70°W). The dark palimpsest possibly the darkest feature on the surface of Enceladus. Its albedo contrast with the surrounding brighter terrain is about 40%. North is up. (**B**) Portion of Cassini ISS NAC image N1485423252 that shows an array of arcuate ridges that appear to outline relict crater rims (*arrows*) near 18°N, 140°W. The phase angle is 105° and north is 15° counter clockwise from up (Helfenstein et al. 2009)

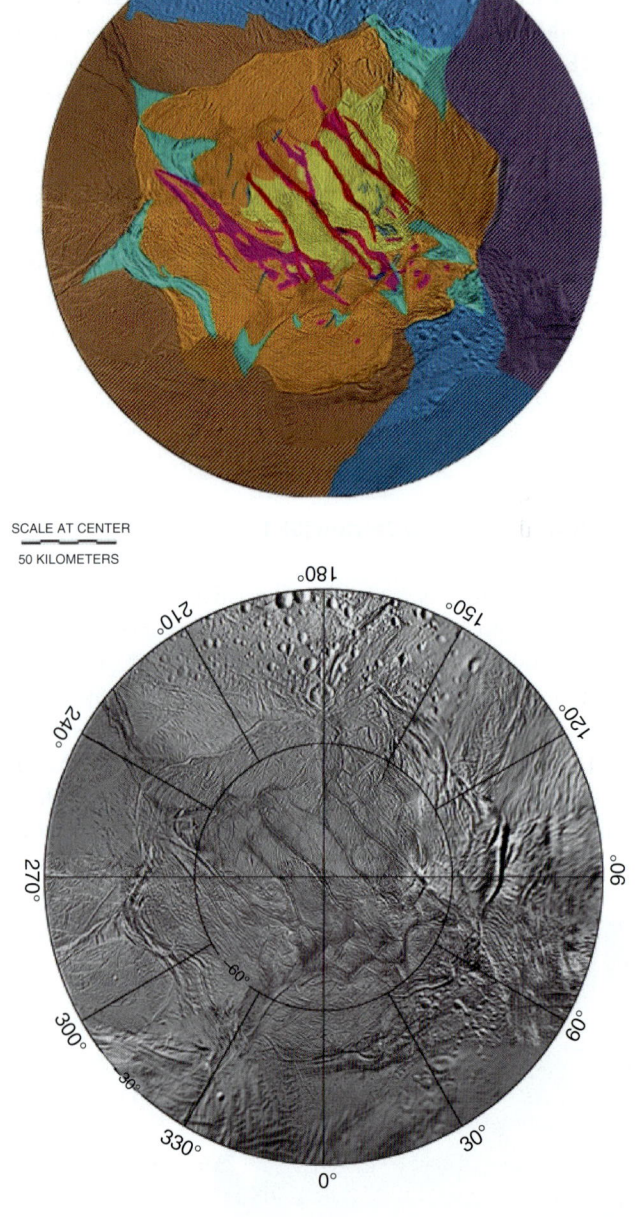

Table 21.2 Enceladus South Polar Terrain Units (Fig. 21.8)

Color	Unit
■ (red)	Tiger stripe smooth flank formation
■ (pink)	Tiger stripe platy flank formation
■ (dark blue)	Tiger stripe medial dorsum structure
■ (magenta)	Relict tiger stripe structure
■ (yellow)	South polar funiscular plains
■ (teal)	South polar lateral fold-and-wedge formation
■ (blue-gray)	Dorsum
■ (tan)	South polar reticulated plains
■ (dark brown)	Eastern hemisphere reticulated plains
■ (orange)	Shear zone
■ (purple)	Western hemisphere fractured plains
■ (light blue)	Cratered plains
▨	Uncertain due to inadequate imaging

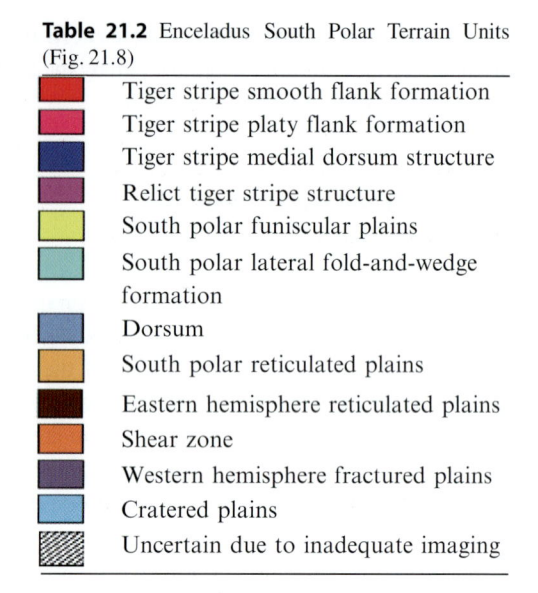

The terrain features within the SPT province are complex and varied (Fig. 21.8). In addition to the marginal arcuate scarps and mountain ranges, they include a series of four prominent parallel rifts, informally called tiger stripes (Porco et al. 2006) from which plumes of water vapor (Hansen et al. 2008) and icy particles (Spitale and Porco 2007) are erupting; a network of quasi-parallel ropy ridges, or funiscular plains, between the tiger stripes; and surrounding the tiger stripes, an outer circumpolar ring of chaotic, highly fractured south polar reticulated plains.

Along more than half the perimeter of the SPT margin, between longitudes 110°W to 320°W, the arcuate scarps that bound the SPT exhibit convex faces that uniformly point southward, and at longitudes where the arcs border cratered plains to the north, they intersect to form cusp-shaped "kinks" that crudely resemble the dihedral cusps of intersecting terrestrial island arc systems (Porco et al. 2006). Along much of the remainder of the SPT boundary, the arcuate scarps are concave southward and the dihedral cusps point toward rather than away from the SPT region (Helfenstein et al. 2009). The circumpolar boundary has been interpreted as a convergent tectonic boundary that formed in response to compressive stresses oriented N-S (Porco et al. 2006, Helfenstein et al. 2009). At longitudes where convex, south facing arcuate scarps bound pre-existing eastern hemisphere reticulated terrains to the north, they do not intersect in a sharp cusp. Instead, they bend northward through about 90° in azimuth to form the parallel, tapered ends of Y-shaped discontinuities (Porco et al. 2006) that interrupt the otherwise continuous circumpolar outline. The Y-shaped discontinuities transition northward into sub-parallel N-S trending rifts and cracks (Figs. 21.1, 21.5, and 21.8). The cracks are most closely spaced at the tapered end of the Y-shaped discontinuities from which they emerge and progressively fan-out as

Fig. 21.8 Preliminary South Polar Terrain unit map (*top*) and the corresponding ISS clear-filter photomosaic map (*bottom*) (Helfenstein et al. 2009). The legend is given in Table 21.2. The polar stereographic projection is centered on the south pole with latitude range in 30° increments from 30°S to 90°S with west longitudes given. Diagonally hashed sections delineate areas where the classification is tentative due to inadequate imaging

(Porco et al. 2006). Topographic measurements from Cassini images (Helfenstein et al. 2009) indicate that the central portion of the SPT region is topographically depressed by 500 m or more relative to the mean elevation, consistent with recent determinations of Enceladus' global figure (Thomas et al. 2007).

they propagate toward equatorial latitudes and often merge with major N-S rift systems, e.g., Labtayt Sulci and Lahej Sulci that, near the equator, are each parallel to a stratigraphically old system of ridges and grooves. A south-trending rift system from Anbar Sulcus merges with cracks that extend north from another Y-shaped discontinuity.

Prominent icy mountain ranges with hundreds of meters of relief occur just within the SPT margin (Porco et al. 2006, Gioia et al. 2007, Helfenstein et al. 2009). The mountain ranges consist of curved, parallel ridges and grooves that are nested and arc together so that in plan-form they have a laterally folded appearance with the fold axis running approximately N-S. A triangular wedge-shaped prism, often topographically raised relative to flanking terrain, is usually present at the crest of the fold shape. On the eastern hemisphere side of the SPT, these Lateral Fold-and-Wedge structures penetrate the SPT margin scarps into the bordering eastern hemisphere reticulated terrains. The mountain belts are particularly well developed on the eastern hemisphere side of the SPT region, and especially when they are adjacent to or confined by the southward-facing openings of Y-shaped discontinuities. There, in plan form the lateral folds generally bend northward such that the horizontal fold axis is in line with the discontinuity's axis of symmetry. The Lateral Fold-and-Wedge structures are interpreted as thrust belts (Helfenstein et al. 2009). The icy thrust wedges evidently focus stresses at the tapered end of the Y-shaped discontinuities and appear to cleave apart the already highly cracked reticulated terrain and initiate a fan of north-going tension cracks from the wedge tip, where crack-tip propagation stresses are most focused (Gioia et al. 2007, Helfenstein et al. 2009). The Y-shaped discontinuities appear to be more evolved examples of the dihedral margin cusps. Helfenstein et al. (2009) propose that the Y-shaped discontinuities can penetrate through reticulated plains, but not cratered plains because the highly fractured reticulated terrains are mechanically weaker than the cratered plains lithosphere and that reactivation of the many reticulated terrain fractures allow the lithosphere to deform on a large scale in a pseudo-ductile behavior.

The thinnest region of Enceladus' lithosphere may be right at the South Pole where the tiger stripes could extend to the depth of a volatile reservoir of liquid water (Collins and Goodman 2007, Manga and Wang 2007) or clathrate (Kieffer et al. 2006). Currently published estimates of this thickness vary: Roberts and Nimmo (2008b) give a minimum thickness of 5 km based on modeling of shear heating along tiger stripes, while Gioia et al. (2007) postulate that tiger stripe fractures penetrate a 35 km thick lithosphere equivalent to the mean spacing between them, and Rudolph and Manga (2009) calculate that complete cracking is of the crust is possible if it is less than 25 km thick.

21.3.2.5 Tiger Stripes

As described by Porco et al. (2006), the tiger stripes are linear depressions or cracks, typically about 500 m deep, 2 km wide and 130 km in length. They are flanked on both sides by 100-m-high ridges. The four major tiger stripes (Alexandria, Cairo, Baghdad, and Damascus Sulci) are sequentially arranged in a right lateral en echelon placement across the south polar region (Helfenstein et al. 2009), suggesting that the tiger stripes and south polar region have evolved through a regional system of right-lateral shear parallel to the trends of the tiger stripes. The tiger stripes are known sources of thermal emission (Spencer et al. 2006 and Fig. 21.14) and plume jets (Spitale and Porco 2007). Cassini obtained high-resolution (as fine as 7 m per pixel) images of most of the known jet vents during two close flybys of Enceladus in August and October 2008. These have been mapped by Helfenstein et al. 2009 (e.g., Fig. 21.9).

All of the known active portions of the tiger stripes are covered with smooth deposits identified as Smooth Flank Formations. These smooth deposits also extend along most of the lengths of the tiger stripes. These deposits appear to drape over the raised flanks and valley walls of tiger stripes and extend laterally outward to surrounding terrains, where they gradually thin out, especially toward the ends of the tiger stripes. The summits and outer limbs of raised flanks have a smoothly undulating texture, suggesting that they are mantled with a particulate blanket. They are overlain by numerous ice blocks up to tens of meters in size. Smooth Flank Formations along the walls of the tiger stripe rifts are often striated, perhaps from the down slope movement of icy blocks or from scouring by plume jets.

The appearance, shape, and surface texture of long inactive tiger stripe flanks are subtly different from the active regions. Platy Flank Formations are noticeably less covered by smooth mantling material than Smooth Flank Formations. They can be distinguished by the presence of visible small cracks and the dissection of the exposed surface into small slabs or plates that are absent on active tiger stripes. On multi-kilometer scales, Platy Tiger Stripe Formations develop gradual lengthwise topographic undulations along tiger stripe flanks, with lobate margins, and partial segmentation of the flanks into parallel chains of rounded connected hills.

Relict Tiger Stripe materials represent more evolved degradational states of the aforementioned tiger stripe formations. They cover a range of rift structures with raised smooth or segmented parallel flanks and they have a similar scale, branching structure, width dimensions, and sometimes trend parallel to or extend from other Tiger Stripe formations. They can occur as long smooth-appearing parallel ridges that are similar in appearance, but often wider with more muted

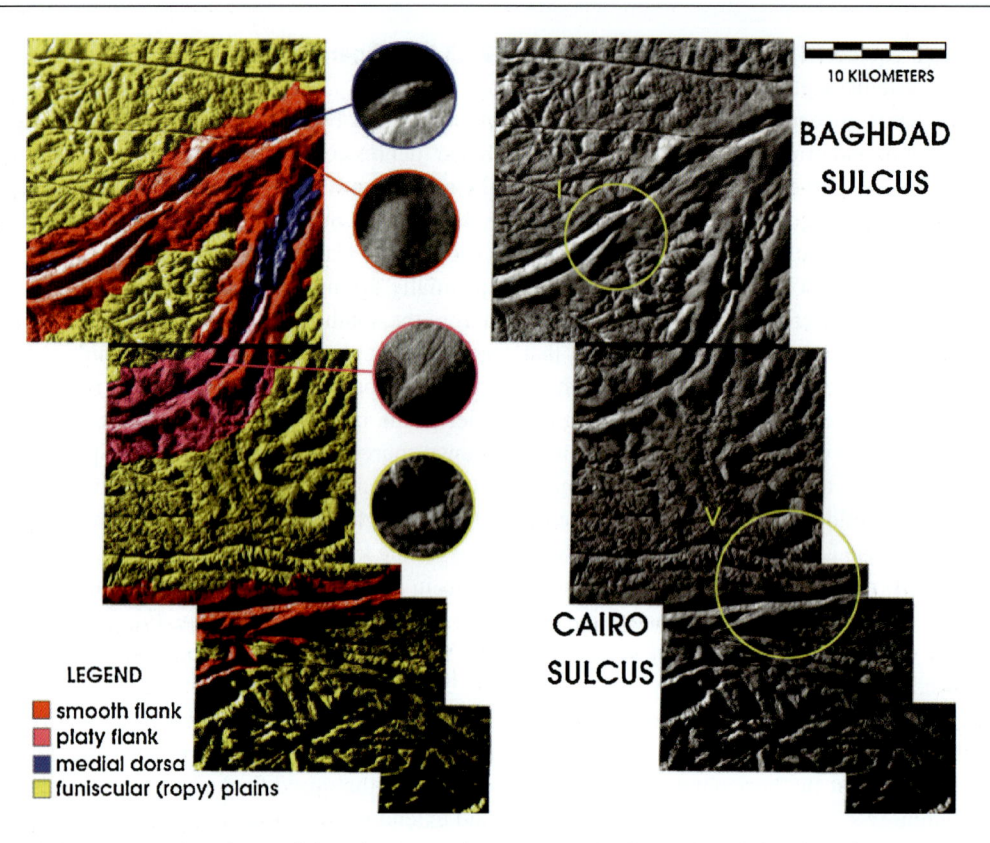

Fig. 21.9 High-resolution local terrain unit map (*left*) and corresponding photomosaic (*right*) of parts of Cairo and Baghdad Sulci, from Cassini's August 11, 2008 flyby (Helfenstein et al. 2009). Terrain unit maps are keyed to the legend in Table 21.2. Insets show textural and structure details that help define the units (see text). Yellow circles labeled with Roman numerals I and V correspond to erupting jet sources (Spitale and Porco 2007)

topography and higher albedos than the smooth tiger stripe formations. They appear to be extinct or dormant tiger stripes that have accumulated a coating of icy, probably particulate, regolith material over time.

The high-resolution ISS images reveal narrow ridges and elongated domes within the tiger stripe valleys. Helfenstein et al. (2009) class these features as Tiger Stripe Medial Dorsum Structures: They are narrow (∼100 m wide) lenticular or sinuous ridges, or clustered collinear or parallel formations of such ridges, that rise tens to hundreds of meters from within a tiger stripe valley. Solitary examples have a distinct free-standing dorsal fin like shape, and are informally called "shark fins". They may also appear as irregular masses of oblate, smoothly rounded hills. They are interpreted as transpressional ridges that are extruded from the tiger stripe rifts, perhaps as a result of cyclic tidal flexing (cf. Hurford et al. 2007, Helfenstein et al. 2009). As such, their presence lends support for recent theories that invoke shear-heating along tiger stripes as a mechanism for generating plume jets (Nimmo et al. 2007, Smith-Konter and Pappalardo 2008). Free-standing dorsa occur throughout the SPT region, as polygonal networks (Fig. 21.6D) and also along strike-slip faults, restraining bends, and shear zones (Fig. 21.8). Their frequent association with shear zones implies that, like tiger stripe medial dorsa, they are transpressional features that have been extruded from faults.

Between the tiger stripes are curved, quasi-parallel knobby ridges and clustered arrangements of elongated barrel-shaped knobs (Fig. 21.9), classified as funiscular (i.e., ropy) plains by Helfenstein et al. (2009). They were first recognized as "grooved bands" by Porco et al. (2006) and trend approximately parallel to tiger stripes. Their lengths and kilometer-scale widths are comparable to those nearby tiger stripe flanks. On scales of tens to hundreds of meters, they have a broken, platy surface texture. They are often finely dissected by thin parallel cracks spaced tens of meters apart and less. Near their distal ends where funiscular plains intersperse and grade into south polar reticulated terrain (see below) they appear to break into a more chaotic arrangement of knobs and cylindrical segments. Color ISS and VIMS data show that the funiscular plains are generally covered with more fine-grained ice than the tiger stripes themselves (Brown et al. 2006, Porco et al. 2006, Jaumann et al. 2008).

Surrounding the tiger stripes and funiscular plains is an approximately 50 km wide annulus of South Polar Reticulated Plains (see also Gioia et al. 2007, Helfenstein et al. 2009). This terrain unit is broadly similar to eastern hemisphere reticulated plains material and likely represents a stratigraphically younger South Polar extension of that terrain. It is characterized by at least two superposed fracture patterns of generally different age, as determined by cross-cutting relationships. The base unit is a tectonic fabric of old curvilinear fractures that usually intersect at acute angles and divide the terrain into lenticular mesas and hills. The terrain is further divided by a system of quasi-parallel bands that cut the pre-existing terrain into lanes along which lateral shear displacements sometimes are evident. This unit is morphologically distinguished from eastern hemisphere reticulated plains by its near-absence of recognizable kilometer sized or larger impact craters and by the frequent occurrence of isolated polygonal islands of recognizable Relict Tiger Stripe material within it. Among the most significant mechanical roles that this annulus of reticulated terrain may play in SPT tectonism is that, like its eastern hemisphere counterpart, it is highly fractured, apparently mechanically weak and can likely deform easily in response to regional stresses. Because it surrounds the tiger stripes and funiscular terrain, it may act to partially decouple deformations of the central polar cap from those near the SPT margin (Helfenstein et al. 2009).

21.3.3 Surface Age Distribution Through Crater Counting

Crater counts on Enceladus confirm the wide range of surface ages (Plescia and Boyce 1983, Lissauer et al. 1988, Kargel and Pozio 1996, Porco et al. 2006). Depending on the flux model used, which can be either lunar-like and therefore asteroidal (Neukum 1985) or, more plausibly, based on models of cometary impact rates (Zahnle et al. 2003), cratered plains materials have absolute ages as old as 4.2 billion or 1.7 billion years, respectively. In general, the eastern fractured terrain has formed either between 3.7 and 1.0 billion years ago in the lunar-like scenario or it is as young as 200–10 million years based on the Zahnle et al. model. The south polar terrain shows crater ages younger than 100 million to few hundreds of thousands years, respectively, consistent with the ongoing activity at the tiger stripes.

21.3.4 Global Tectonics and Possible Rotational Changes

The newly recognized, remarkably simple global symmetry and systematic placement of terrain provinces on Enceladus have inspired different theories about the large scale mechanisms that have produced them. Among the first hypotheses to emerge (Porco et al. 2006) was that a simple uniaxial global deformation mechanism might be responsible for the circumpolar SPT boundary at about 55°S with attendant N-S tension cracks that extend to the equator. Increased flattening of the rotational axis was proposed, as would be predicted from an increased spin rate (perhaps as a result of a large, suitably oriented impact) or some global deformation mechanism that mimics the hemispheric deformation effects of spin-up (like the melting of a subsurface sea at the South Pole (Collins and Goodman 2007)). To first order, this single mechanism appears to account for the circumpolar location and orientation of the SPT boundary if it is interpreted entirely as a convergent tectonic margin (evidenced by convex-southward arc-and-cusp scarp and thrust-faults associated with N-S lateral fold-and-wedge features). However, this simple hypothesis fails to explain the tiger stripes as a system of parallel tension cracks at the South Pole and it does not account for the absence of an SPT type province at the North Pole. Separate mechanisms for the origin of tiger stripes and the hemispheric north–south asymmetry are thus needed to amend this hypothesis. The model also ignored changes in the tidal figure of Enceladus.

Matsuyama and Nimmo (2008) present a more complete global deformation model that considers changes in the triaxial tidal figure that would accompany increased spin rate. In their refined model, the zone of polar thrust faulting begins poleward of about ±70° latitude, 15° closer to the South Pole than the SPT margin. To extend the thrust zone to SPT boundary latitudes, a decrease in internal volume of Enceladus would need to accompany the increase in spin rate. Matsuyama and Nimmo also show that, if Enceladus fractures always form by shear failure, strike-slips faults that are inclined relative to lines of latitude and longitude are predicted to occur closer to the equator, contrary to observation, except for regions that surround the sub-Saturn and anti-Saturn points on the surface where N-S trending normal faults are predicted. Consequently, simple uniaxial deformation about the spin axis is insufficient to account for the observed arrangement of SPT tectonic features. And of course, stress patterns from changes in spin rate and orientation are always symmetric between the north and south hemispheres, so additional mechanisms are required to explain the hemispheric asymmetry.

Gioia et al. (2007) propose a model in which thermotectonic stresses arise from a postulated deep-seated hot thermal anomaly under south polar cap. The thermal expansion of warm materials under their "Tiger Stripe Terrain" or TST (corresponding to tiger stripes, funiscular plains, and adjoining portions of the south polar reticulated plains in Fig. 21.9) causes lateral tension throughout the central south polar cap occupied by tiger stripes. Toward the SPT boundary, horizontal radial stresses transition to compression, while hoop stresses parallel to lines of latitude remain tensile. Gioia et al. note that the TST region is not circular, but somewhat rectangular in shape. If a 210 × 148 km rectangular shape is chosen for the thermal anomaly, with the longest side approximately matching the length and orientation of the tiger stripes, then the resulting azimuthal variations in horizontal TST tensile stresses predict the formation of tiger stripes as tension cracks or normal faults with the observed orientation. At the same time, horizontal radial stresses near the SPT become preferentially more compressive parallel to the tiger stripes. Finally, if the depth of the lithosphere and thickness of the thermal anomaly is chosen to be about 35 km, then the model predicts to first order a reasonable match for the spacing of dihedral cusps along the SPT bounding scarps and the mean spacing of the tiger stripes.

The Gioia et al. model is interesting especially because it invokes a single mechanism to predict the placement and orientation of the SPT boundary, the fold-and-wedge structures, and the tiger stripes. However, the model requires additional development: It treats the SPT region as a flat plate rather than realistically as a curved thick shell, it ignores temperature-dependent rheological behaviour of deep seated warm materials, and it postulates post hoc the existence and orientation of a rectangular-shaped thermal anomaly on the basis of the current shape and orientation of the tiger stripes and adjacent terrains.

As described in earlier sections, Nimmo and Pappalardo (2006) propose that the current placement of South Polar volcanism and active tectonism is a result of global reorientation of a warm diapir from a different southern latitude closer to the equator. Collins and Goodman (2007) further suggest that such a reorientation could also be caused by the melting of a subsurface sea. Matsuyama and Nimmo (2008) find that the global deformation model which most easily explains the observed placement and orientations of the SPT boundary involves approximately 90° of true polar wander about an almost stationary tidal axis. That is, the paleo-South Pole and paleo-tidal axis would have been located at (2°N, 270°W) and (4°S, 0°W), respectively. An additional small amount of global dilatation is needed to place circumpolar thrust faults at their currently observed latitude of about 55°S, but the model otherwise accounts for the SPT boundary and the N-S tension cracks closer to the equator. As in the rotational spin-up model, polar thrust faults are predicted to occur in the SPT region and thus a separate mechanism such as diurnal tidal flexure (Porco et al. 2006, Hurford et al. 2007) must be invoked to account for the presence and orientation of tiger stripes as tension cracks.

Two independent lines of geological evidence from the analysis of Cassini ISS images reinforce a polar wander mechanism for the origin of the SPT and the evolution of fractured plains on the eastern hemisphere. Helfenstein et al. (2009) identified a suite of tectonic features in the eastern hemisphere fractured plains that are possible analogs to SPT tectonic features, some examples of which are compared in Fig. 21.6. Their placement in Sarandib Planitia and Diyar Planita, and the spatial patterns of tectonic forces that they imply, suggest that either polar wander has occurred following a path from eastern hemisphere equatorial latitudes toward the south pole, or at least that SPT-style tectonism was once more widespread, extending to these eastern hemisphere locations and progressively dissipating over time toward the only remaining active region at the South Pole.

The second line of evidence involves the possibility that tiger stripes and the funiscular terrain are surface manifestations of lithospheric spreading (Helfenstein et al. 2009). Helfenstein et al. (2009) have recently identified transform faults perpendicular to tiger stripes and apparent lateral spreading across the funsicular terrain. Preliminary reconstructions of lithospheric spreading across the tiger stripe region indicates that at least 70 km of lateral spreading has occurred from the western hemisphere side of the SPT region toward the eastern hemisphere side. This would be most easily accomplished by the presence of a buried warm diapir or subsurface sea, or both.

21.4 The Surface: Composition and Processes

21.4.1 Surface Composition

21.4.1.1 Water Ice, and Its Spatial Variability

Near infrared spectroscopy reveals that Enceladus' surface is almost completely dominated by water ice, which is mainly in a crystalline state (Cruikshank 1980, Grundy et al. 1999, Cruikshank et al. 2005, Emery et al. 2005, Verbiscer et al. 2006, Brown et al. 2006, Jaumann et al. 2008, Newman et al. 2008, Fig. 21.10).

H_2O band depths are sensitive to the abundance of ice, degree of contamination by other materials, and the particle sizes of the H_2O-ice grains (Clark 1981, Clark and Lucey 1984). Ice temperature is also important (Grundy et al. 1999). Given that water ice dominates the surface composition (Brown et al. 2006, Newman et al. 2008) spectral variations are mainly due to differences in particle sizes, with increasing absorption band depth indicating increasing particle size according to a simple power law (Jaumann et al. 2008). Global grain size variations are seen in ground-based near-IR spectra, which find somewhat stronger absorptions, implying larger grain sizes, on the leading hemisphere than on the trailing hemisphere (Verbiscer et al. 2006). VIMS data provide more detail on spatial variability, and

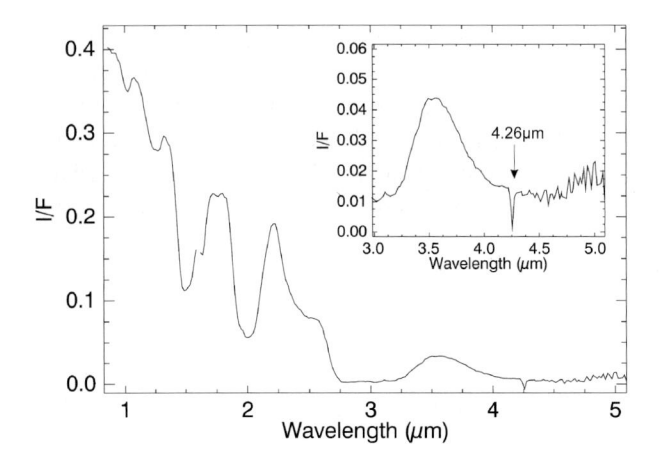

Fig. 21.10 VIMS spectrum of Enceladus' south polar region. The absorptions at 1.04 µm, 1.25 µm, 1.5 µm, 1.65 µm, 2.0 µm, and the Fresnel peak at 3.1 µm (see small *inset*) indicate almost pure water ice, with an additional strong CO_2 absorption at 4.26 µm (Brown et al. 2006)

show significant spectral variations across the surface of Enceladus, particularly in the H_2O ice absorption bands (Jaumann et al. 2008, Newman et al. 2008).

The spatial resolution of VIMS image cubes is sufficient to distinguish the three major geologic units on Enceladus: heavily cratered terrain, fractured and ridged terrain and active tiger stripes at the south pole region. Variations in the main water ice absorptions at 1.04, 1.25, 1.5, and 2.0 µm indicate that on Enceladus particle size is directly correlated with geological units (Jaumann et al. 2008, and Fig. 21.11). The heavily cratered terrains exhibit a mean particle size of 15 ± 5 µm within a VIMS pixel. Local increases of particle diameters up to 30 ± 10 µm seems to be correlated either with impact craters that expose fresh material from underneath or have locally melted the water crust, or with ridges and fractures marking the transition zone to the lightly cratered plains. Icy particles in the fractured and ridged terrain peak at a size of about 30 ± 10 µm. The tectonically deformed regions exhibit particles ranging from 40 µm to about 120 µm with the largest ones (about 2% reaching values of about 0.2 mm) located close to the center of the south polar tiger stripes.

Similar patterns are seen in the ISS multispectral images of the south polar region, which show enhanced near-infrared absorption near the tiger stripes and other tectonically disrupted regions, consistent with larger grain sizes (up to 100 µm) near the tiger stripes (Porco et al. 2006). These regions of larger grain size appear blue/green in false-color ISS images (Fig. 21.1). Possible causes for the large grain sizes near the tiger stripes include preferential fallout of larger, slower, plume grains near their sources (Jaumann et al. 2008, Hedman et al. 2009, Section 21.5.2.6), or sintering and growth of grains resulting from deposition of plume gases or from the enhanced temperatures near the tiger stripes.

Disk-integrated spectra of Enceladus indicate that the water ice in the equatorial and mid- to high-latitude regions is primarily crystalline (Grundy et al. 1999, Emery et al. 2005). However, VIMS has detected locally abundant amorphous ice in the south polar region between the tiger strips, while the stripes themselves are dominated by crystalline ice (Brown et al. 2006, Newman et al. 2008). Amorphous ice can form when water condenses from a vapor at temperatures below 100 K, and it is thus possible the south polar amorphous ice results from surface condensation of plume gases. Away from the tiger stripes, south polar surface temperatures are well below 100 K (Spencer et al. 2006), so amorphous ice should be able to form and remain stable for long periods there.

21.4.1.2 Minor Constituents

Other constituents have minor abundance and are restricted to certain areas. Some telescopic spectra show a weak 2.2–2.4 µm feature tentatively ascribed to NH_3 absorption (Grundy et al. 1999, Emery et al. 2005, Fig. 21.12), though the match is not perfect, and other spectra do not show this feature (Cruikshank et al. 2005). Spectral modeling of earth-based observations by Verbiscer et al. (2006) suggest the possibility of modest concentrations (~1%), of ammonia hydrate ($NH_3 \cdot 2 H_2O$) on both hemispheres, though no specific ammonia hydrate spectral features were identified. No features compatible with NH_3 absorptions have been detected VIMS data of Enceladus, indicating an upper limit to NH_3 content on the surface of no more than 2% (Brown et al. 2006). H_2O_2 has been tentatively identified in VIMS spectra, via a weak 3.5 µm absorption feature (Newman et al. 2007). The wavelength of the feature varies slightly with terrain type, perhaps due to varying crystallinity of the H_2O_2 or the surrounding H_2O.

Another constituent found in VIMS spectra at 4.26 µm is CO_2-ice, seen in small amounts on a global scale and in higher concentrations near the south pole (Fig. 21.10, Brown et al. 2006). Near the tiger stripes, CO_2 is highly abundant but the band position indicates that it is not in the form of free CO_2 ice. Instead, it is complexed, probably with water ice. Small amounts of free CO_2 ice are seen elsewhere on the satellite. The lack of free CO_2 ice near the tiger stripes may be due to the higher temperatures in this area (Spencer et al. 2006), which would cause free CO_2-ice to rapidly migrate northward to colder temperatures (Brown et al. 2006). The high concentration of complexed CO_2 in the tiger stripes

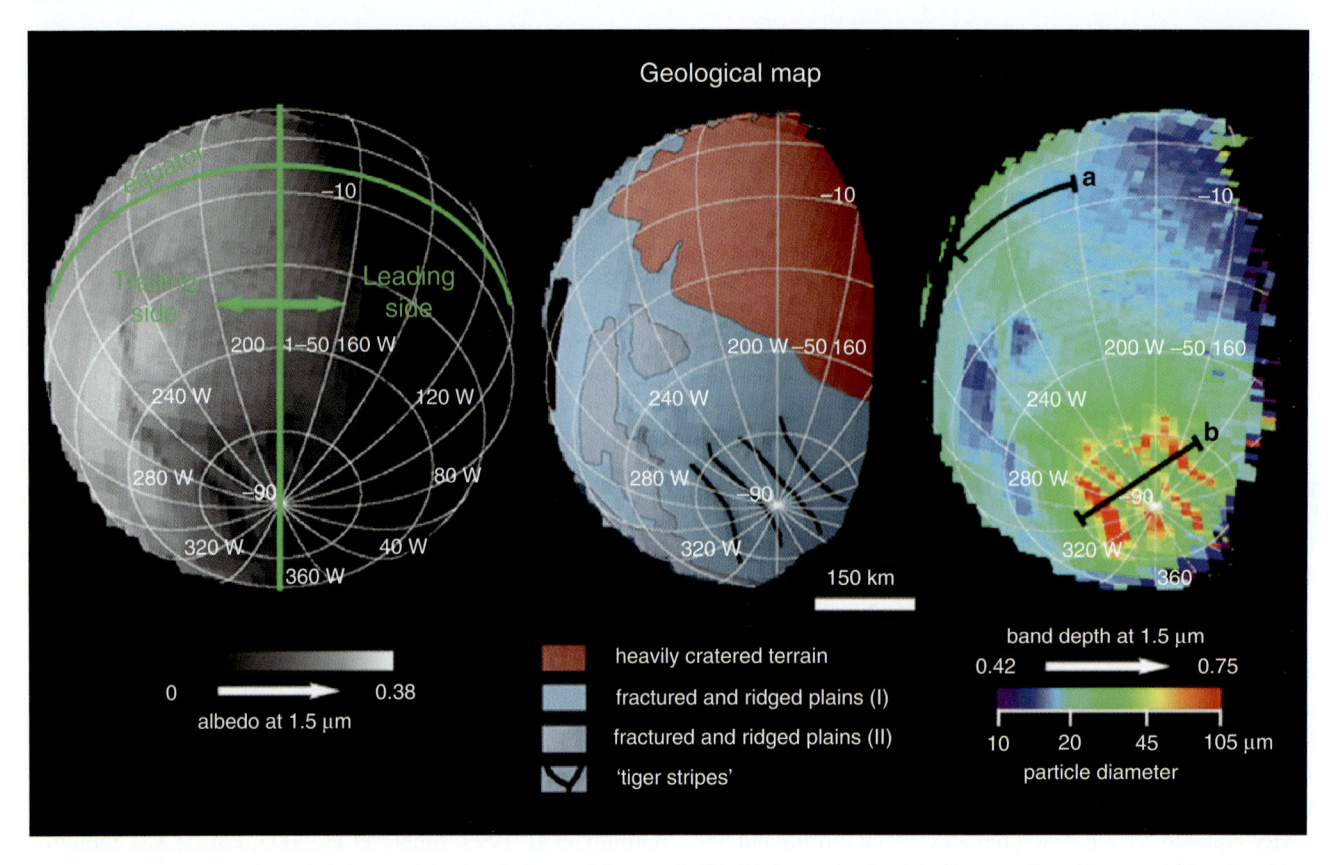

Fig. 21.11 VIMS cube image of the southern hemisphere of Enceladus, from orbit 11, with resolution <30 km per pixel. Left: albedo at 1.5 μm; middle: geologic map; right: absorption band depth at 1.5 μm indicating increasing particle diameters from blue to red. From Jaumann et al. (2008)

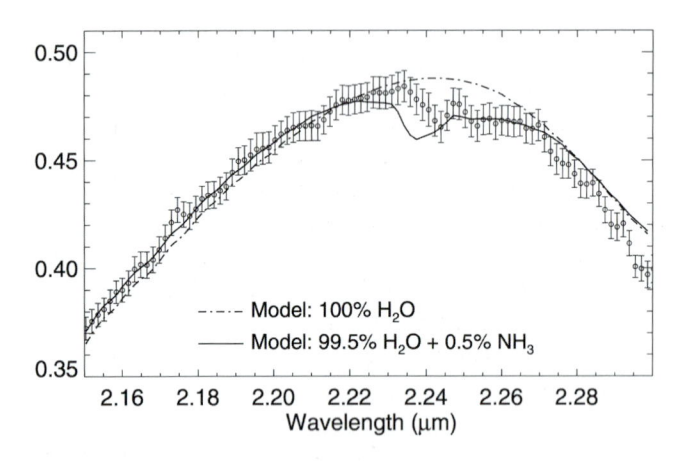

Fig. 21.12 A portion of a disk-integrated ground-based spectrum of Enceladus' trailing side, with central longitude 320 W, from Emery et al. (2005), showing a feature at 2.25 μm possibly due to NH$_3$ ice or ammonia hydrate. Emery et al. suggest that the wavelength shift of the band relative to the model might be due to complexing of NH$_3$ with water ice

may result from active replenishment by cryovolcanic activity in this area (Brown et al. 2006). No signature of CO in ice, gas, clathrate, or any other physical form is detected on the surface (Brown et al. 2006). However, VIMS spectra show 3.44 and 3.53 μm C-H stretch features near the tiger stripes, indicating simple organics in the south polar tectonic features (Brown et al. 2006). Hydrated salts or acids, such as those found on Europa (McCord et al. 1998), which might be indicative of surface materials derived from a salt-rich ocean in contact with the silicate core, have not been detected on the surface so far.

The surface composition of Enceladus, dominated by H$_2$O with small quantities of CO$_2$, and simple organics near the tiger stripes, and possible NH$_3$, is quite similar to that of the plume gases (Section 21.5.2.4), consistent with the possibility that much of the surface is coated by plume fallout.

21.4.2 Surface Processes

Surface modification processes on the icy Saturnian satellites include charged particle bombardment (implantation or sputtering), E-ring grain bombardment or coating, thermal processing, UV photolysis, and micrometeoroid bombardment. Such processes result in chemical as well as structural changes.

Models of Enceladus' surface must explain its extraordinarily high albedo, the highest of any known solar system body. Geometric albedo, including the strong opposition effect, is 1.41 at 0.55 μm (Verbiscer et al. 2005) and bolometric albedo is 0.80 (Spencer et al. 2006). The high albedo is probably due to a coating of plume particles, derived both directly from plume fallout and from re-impact of plume particles stored temporarily in the E-ring. Support for the importance of direct or indirect plume fallout comes from the observation that the geometric albedo of the inner Saturnian satellites correlates closely with the local density of the E-ring (Verbiscer et al. 2007), with both quantities peaking at Enceladus. This coating affects the reflectance spectra of the satellites, and could mask the signatures of other species: Buratti (1988) found no relationship between crater density and albedo in Voyager images of Enceladus and suggested that E-ring coating dominates any micrometeorite gardening. The visible orbital phase curve of Enceladus shows that the trailing hemisphere is ∼1.2 times brighter than the leading hemisphere (Buratti and Veverka 1984, Buratti et al. 1998), and ground-based spectra show smaller ice particles on the trailing hemisphere (Verbiscer et al. 2006): these differences may reflect hemispheric differences in Enceladus' interaction with the E-ring.

The Cassini CDA team is performing extensive modeling of Enceladus' interaction with the E-ring, constrained by CDA's measurements of the dust population (e.g., Spahn et al. 2006b, Kempf et al. 2008). Bombarding E-ring particles and interplanetary dust particles will both eject more dust into the E-ring, with E-ring particles greatly dominating the dust production rate (Spahn et al. 2006b). These impacts should produce an isotropic dust source for the E-ring that is in addition to, but much smaller than, the south polar plume source (Spahn et al. 2006a).

Plasma sputtering rates have been calculated recently using Cassini plasma data (Johnson et al. 2008). Water ions are the dominant sputtering agents, and rates are estimated to be about 0.1 μm year^{-1}, or 10 cm per million years, which is not insignificant.

The VIMS spectra reveal a consistent pattern of increasing grain size with decreasing surface age (Jaumann et al. 2008), as discussed above. This might be due to gradual comminution of particles over time by impact gardening or sputtering. In the south polar region, patterns of plume fallout are probably also important, with larger particles falling closer to the tiger stripe plume sources. However, it is puzzling that older surfaces, such as the lightly cratered fractured and ridged terrain and the heavily cratered terrain, have distinctive grain sizes, as their optical surfaces might be expected to be dominated by E-ring and plume fallout that is younger than either surface and would not reflect the underlying geology. Perhaps enhanced mass wasting on the rougher slopes of the younger tectonically disrupted terrain exposes fresher, coarser-grained material there.

Surface temperatures away from the active south pole probe the physical structure of the regolith by constraining the thermal inertia, and are also important for understanding the stability of volatiles and for modeling temperature-dependent spectral features. Observations by CIRS show that Enceladus has a very low average thermal inertia of 15 W m^{-2} s$^{-1/2}$ K^{-1}, which is 3–4 times lower than the Galilean satellites, though similar to the other inner Saturnian satellites, with increasing thermal inertia towards the active south pole (Spencer et al. 2006, Howett et al. 2009). A highly unconsolidated particulate surface at least a few centimeters deep is implied. Equatorial temperatures reach a peak of about 75 K at noon and drop to 50–60 K at night. At the north pole, which had been in darkness since 1995, CIRS measured an upper-limit effective temperature of 33 K during the July 2005 flyby, implying a thermal inertia of <100 W m^{-2} s$^{-1/2}$ K^{-1}, and thus an unconsolidated surface to a depth of at least ∼1 m (Spencer et al. 2006).

21.5 Plumes and System Interaction

21.5.1 Introduction

The Enceladus plumes are both a probe of subsurface conditions and a remarkable phenomenon in themselves. Some of the important questions regarding the plumes are: Where (at what depth) and how is the heat generated? How does it get to the surface? Is liquid water involved? What is the composition of the ice? How deep do the cracks extend? How long does each plume remain active? Potentially useful observations include: the rate of vapor and particle mass loss in the plumes; the composition of the gas; the speed of the gas; the speed, size, and composition of the particles; the total power in the plumes; the time dependence of the plumes; and the association between the plumes and surface features. We review the most significant observational evidence from Cassini instruments and ground-based sources, and then consider how modeling can help us to interpret these observations.

21.5.2 Observations

21.5.2.1 Imaging Science Subsystem (ISS)

The ISS measures the brightness of the ice grains in the plumes in spatially resolved images at wavelengths from

340 to 930 nm. The ISS images reveal the brightness of the solid particle portion of the plumes, the multiplicity of plume sources, the degree of collimation, the rate of fall-off of brightness with altitude, and the plumes source locations on the surface.

Because the plumes are optically thin, brightness at a fixed wavelength and phase angle is directly proportional to the column density $\int n ds$, where n is the number of particles per unit volume and s is distance along the line of sight. The constant of proportionality depends on the cross sectional area of the particles, their albedo, and their scattering phase function. The latter governs the fraction of scattered light that goes into each direction and is a strong function of the phase angle (the sun-plume-spacecraft angle). Forward scattering is when the phase angle is close to 180°. Particles whose radius a is larger than the wavelength of light λ have a strong forward scattering lobe that narrows as λ/a decreases.

One of the highest resolution ISS images of the plume jets is shown in Fig. 21.13. The phase angle is just outside the forward scattering lobe for particles whose radius is 1.0 μm. To determine plume particle abundance, the ISS team assumed a particle size distribution with an effective radius $a = 1$ μm. This is the dominant radius of particles in the E ring (e.g., Nicholson et al. 1996), which Cassini has revealed to be supplied by the Enceladus plumes. Brightness is measured in units of I/F, where I is the observed radiance in a given filter and πF is the solar irradiance in the same filter. The I/F values for the plume in this image are typically in the range 10^{-3} to 10^{-2}. With these assumptions about particle size, the ISS team estimated a column number density of 6×10^8 m^{-2}, corresponding to mass of ice in the column of 3×10^{-6} kg m^{-2}. This is for a line of slight whose closest approach to Enceladus is at an altitude of 15 km and a latitude of 76°S. Comparison to the column density of vapor of 7×10^{-6} kg m^{-2} reported by the UVIS instrument (Hansen et al. 2006, see Section 21.5.2.3 below) gave an ice/gas mass ratio of ~ 0.4. This ratio is quite uncertain because the particle size distribution is uncertain, and smaller mass ratios (ice/gas <0.1) have been derived from a recent re-analysis of the original data (Kieffer et al. 2009). The ice/gas ratio is a significant constraint on the processes that form the plumes (Section 21.5.3.2).

Plume brightness falls off with altitude more or less exponentially. The scale height within 50 km of the surface is ~ 30 km, which is much less than the radius of the planet, and implies that most of the particles close to the ground are falling back. The scale height is larger at higher altitudes, which means that many particles are escaping as well. The vertical fall-off fits a model in which the distribution of speeds of the particles leaving the vent is a Gaussian of the form $\exp[-(v/v_0)^2]$. The best fit gives a mean velocity $v_0/\pi^{1/2} = 60$ m s^{-1}. This is much less than the escape speed (235 m s^{-1}) and much less than the velocity of the gas molecules, $\sim(2RT)^{1/2}$, where R is the specific gas constant (the ideal gas constant divided by the gas molecular weight), of 406 m s^{-1} for $T = 180$ K (Spencer et al. 2006), or the $\sim$300–500 m s^{-1} speed inferred from UVIS observations (Tian et al. 2007). The inference again is that most of the particles are falling back to the surface.

Spitale and Porco (2007) use images taken over a 2-year period to triangulate the source locations for the most prominent jets and compare them with the observed hotspot locations (Spencer et al. 2006) and surface features (Fig. 21.14). The jets are located on or very close to the tiger stripes, and tend to occur near the warmest regions of the tiger stripes. These observations also show that at least some of the plume jets are persistent over the 2-year period.

Further analysis of the ISS images should yield an estimate of the phase function, and therefore the particle size, though possible temporal variability is a potential complication, since the observations at different phase angles necessarily occur at different times. The wavelength dependence of the plume brightness provides a further particle size constraint.

Fig. 21.13 Image of the Enceladus plumes take on November 27 2005 at a phase angle of 161.4° with a spatial resolution of 0.9 km per pixel. Individual plume jets are visible in image **A**, and image **B** has been enhanced using pseudocolor to bring out the fainter portions of the plume. From Cassini Press Releases PIA07758 and PIA07759, NASA/JPL/SSI, and Porco et al. (2006)

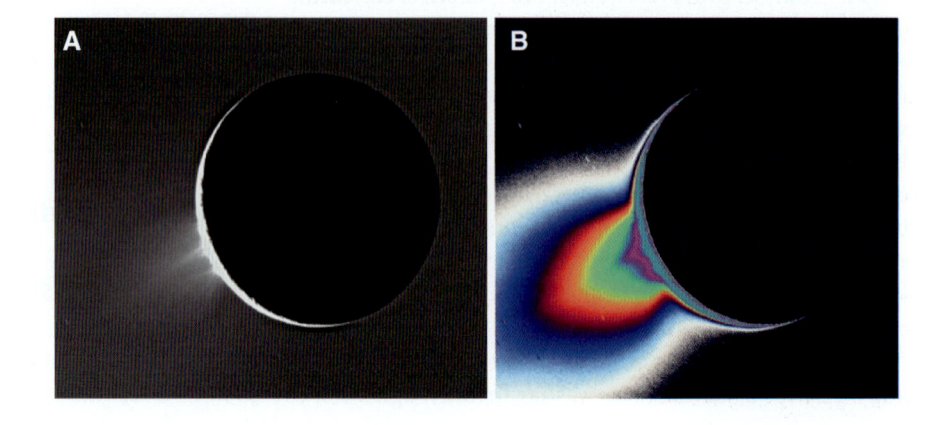

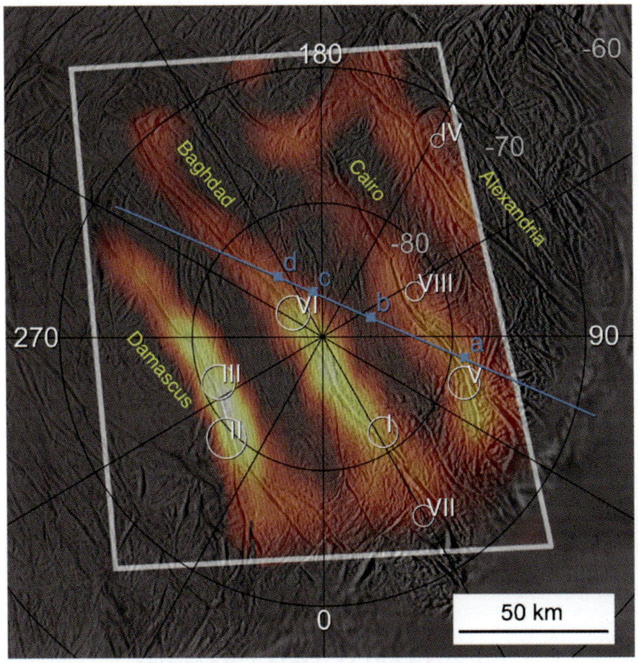

21 Enceladus: An Active Cryovolcanic Satellite

Fig. 21.14 Locations of thermal emission, plume sources, and jets seen in stellar occultations relative to the geological features of Enceladus' south polar region. The warm colors superposed on the base map show the distribution of 9–17 μm thermal emission seen by CIRS on March 12th 2008 (Cassini press release image PIA10361 NASA/JPL/GSFC/SwRI/SSI), with a spatial resolution of about 15 km. Only the region within the white trapezoidal outline was mapped by CIRS. The white circles show the locations (with uncertainties) of dust plume sources seen by ISS, from Spitale and Porco 2007). The blue line shows the surface trace of a stellar occultation observed by UVIS (Hansen et al. 2008), with letters identifying the locations, projected onto the occultation trace, of discrete gas jets seen in the occultation (Fig. 21.17)

21.5.2.2 Composite Infrared Spectrometer (CIRS)

The thermal emission from Enceladus' warm south pole is observed by the CIRS instrument, which obtains spatially resolved spectra with resolution down to a few kilometers, sufficient to map the warm areas in relation to surface features such as the tiger stripes. CIRS uses three detectors covering the 10–1,500 cm^{-1} range, and CIRS' most useful observations prior to 2008 were taken with the 600–1,100 cm^{-1} (9.1–16.6 μm) detector. CIRS observations constrain the total radiated power from the south polar region, and the spatial distribution of temperatures.

Figure 21.15 shows mid-IR maps, superposed on an ISS base map, and spectra of the south polar region from the 14 July 2005 flyby and from a distant observation on November 9th 2006 (Abramov and Spencer 2009). The prominent south polar hot spot seen in both observations corresponds to the location of the tiger stripes, and higher spatial resolution CIRS observations in March 2008 (Fig. 21.14) show that nearly all the thermal emission comes from the tiger stripes and associated fractures. The observed emission is much brighter than the expected passive thermal emission from the sun-heated surface (blue curves in Fig. 21.15): the elevated temperatures are therefore maintained by heat from within, escaping along the tiger stripes.

The shape of the average CIRS south polar spectra in Fig. 21.15 require relatively high temperatures occupying a small fraction of the CIRS field of view. After subtracting the expected passive thermal radiation, the average spectrum south of 65°S from July 2005 can be fit by a blackbody at about 133 K occupying an area of about 345 km^2 (Fig. 21.15), which is a small fraction of the 37,000 km^2 area of the region and is equivalent to a strip with an average width of 700 m along the approximately 500 km length of the tiger stripe fractures. This width is a small fraction of the 2 km distance between the ridge crests of the fractures. 1,100–1,500 cm^{-1} spectra of Damascus Sulcus from the August 2008 flyby, at higher spatial resolution, indicate places on the surface as warm as 167 K (JPL news release 2008–185). The power radiated by the tiger stripes, determined from the July 2005 blackbody fits, is 5.8 ± 1.9 GW, with an uncertainty that is dominated by the assumed temperature. This is a lower limit to Enceladus' total endogenic power, because additional power is likely to be radiated at temperatures too low to be detected in the CIRS 600–1,100 cm^{-1} bandpass. Assuming the total length of the tiger stripes is 500 km, the average power per unit length is thus at least ∼12 kW m^{-1}.

Spencer et al. (2006) note that if the plume gases seen by UVIS were in vapor pressure equilibrium with surface ice that was visible to CIRS, a temperature of at least 180 K could be inferred for the gas source: lower temperatures would require a larger area of exposed ice to produce the H$_2$O flux observed by UVIS, and would thus produce a brighter thermal infrared signature than CIRS observed. The corresponding upper limit to the width of the plume source is 50 m, averaged along the length of the tiger stripes. Because of the strong dependence of vapor pressure and sublimation rate on temperature, the source width decreases rapidly for higher temperatures: ∼10 cm for 225 K, and ∼1 mm for 273 K.

Additional CIRS observations of the south pole in November 2006 (Fig. 21.15) show that the total radiated power and its spatial distribution was essentially unchanged since July 2005, with a <15% change in total 15 μm radiance (Abramov and Spencer 2009).

21.5.2.3 Ultraviolet Imaging Spectrometer (UVIS)

The Cassini UVIS instrument has probed the structure and chemistry of the Enceladus plume gases by means of two stellar occultations: a vertical cut through the plume on July

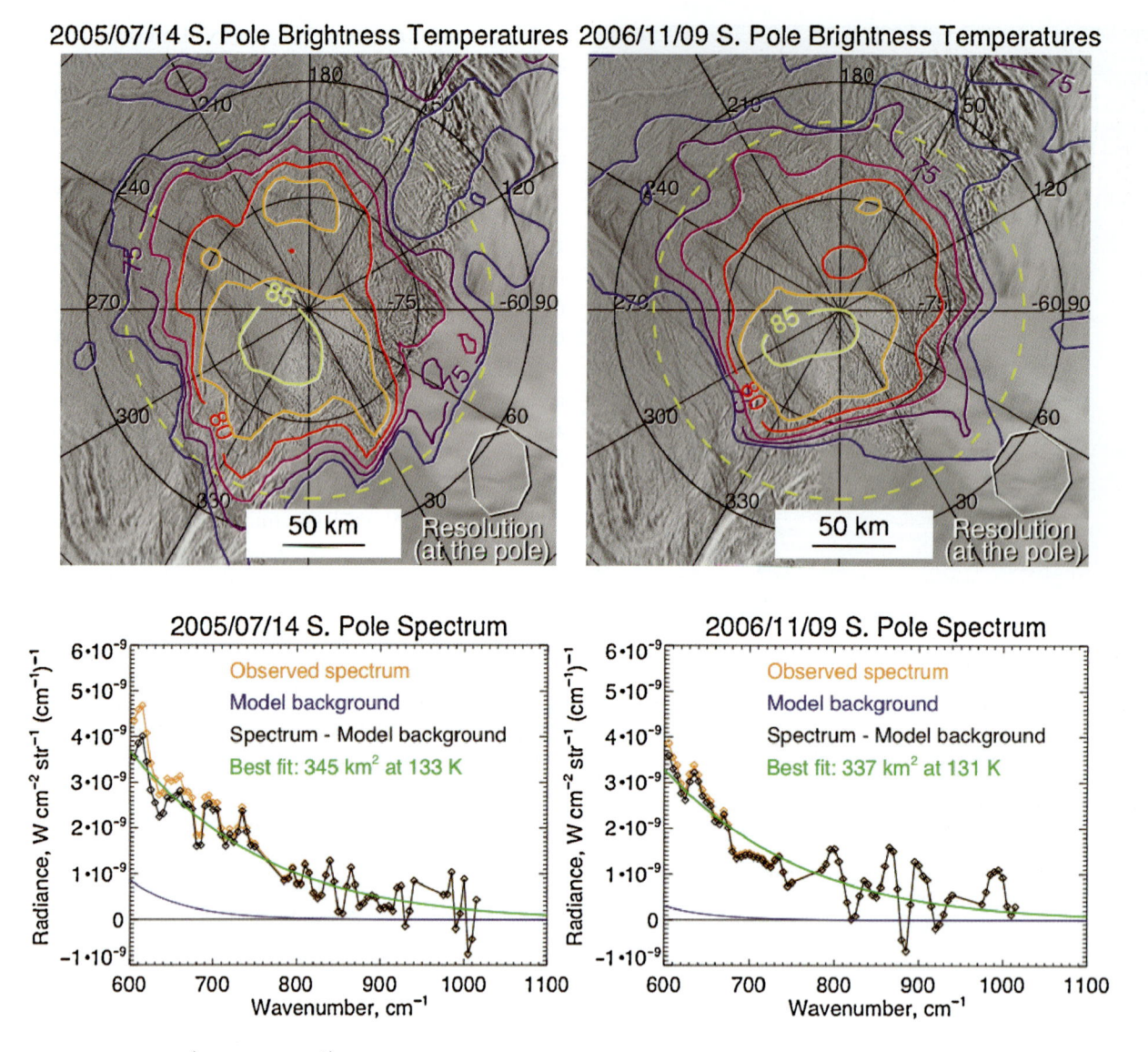

Fig. 21.15 9.1–16.6 μm (600–1,100 cm^{-1}) maps and spectra of the south pole of Enceladus, seen by CIRS during the July 14th 2005 flyby and during more distant observations on Nov. 9th 2006 (Abramov and Spencer 2009). The spectra show the average thermal emission from within the dashed yellow circle. Contours show brightness temperature, which is the temperature of a pixel-filling blackbody radiating the observed flux. See the text for details

14th 2005 (Hansen et al. 2006), extending down to the surface, and a horizontal cut on October 24th 2007 (Hansen et al. 2008), with a minimum altitude of 16 km. Two components of UVIS are used: the High Speed Photometer (HSP), which provides high time resolution over a 1,100–1,900 Å bandpass, and the Far Ultraviolet Spectrograph (FUV), which generates 1,115–1,914 Å spectra with lower time resolution. Both instruments are sensitive to absorption of starlight by water vapor and potentially by other gases.

The absorption spectrum of the plume gases from both occultations clearly shows the signature of water vapor (Fig. 21.16) and allows its abundance to be mapped as a function of position in the plume, as discussed below. Absorption due to CO is not observed in the UVIS FUV spectrum in either 2005 or 2007, with a formal 2-σ upper limit of 3.6×10^{14} cm^{-2}, corresponding to a mixing ratio with H_2O of <3% (Hansen et al. 2008). This may be consistent with the presence of CO in the INMS spectra if, as is likely (Waite et al. 2009, Section 21.5.2.4), the CO seen by INMS is primarily a breakdown product of CO_2 within the instrument.

The two occultation cuts through the plume reveal much about the plume's structure and dynamics. The attenuation of the star signal during the 2005 occultation is best fit by an exponential decline with altitude, with scale height of 80 km, and the horizontal cut through the plume in the 2007 occultation (Hansen et al. 2008) reveals additional details (Fig. 21.17). There are four statistically significant narrow jets superposed on the broader plume background. In

Fig. 21.16 The spectrum of the occultation star shortly before the July 14, 2005 occultation, divided by the average unocculted star spectrum (I_0), showing absorption due to H$_2$O vapor in the Enceladus plume. A water spectrum with column density $n = 1.5 \times 10^{16}$ cm^{-2} is compared (*thick line*) (Hansen et al. 2006)

Fig. 21.17 HSP data from the 2007 UVIS plume stellar occultation is shown binned to 200 ms intervals. The large absorption is due to the water vapor in the large background plume. The narrower numbered absorption features a, b, c, and d indicate individual jets in the plume (Hansen et al. 2008)

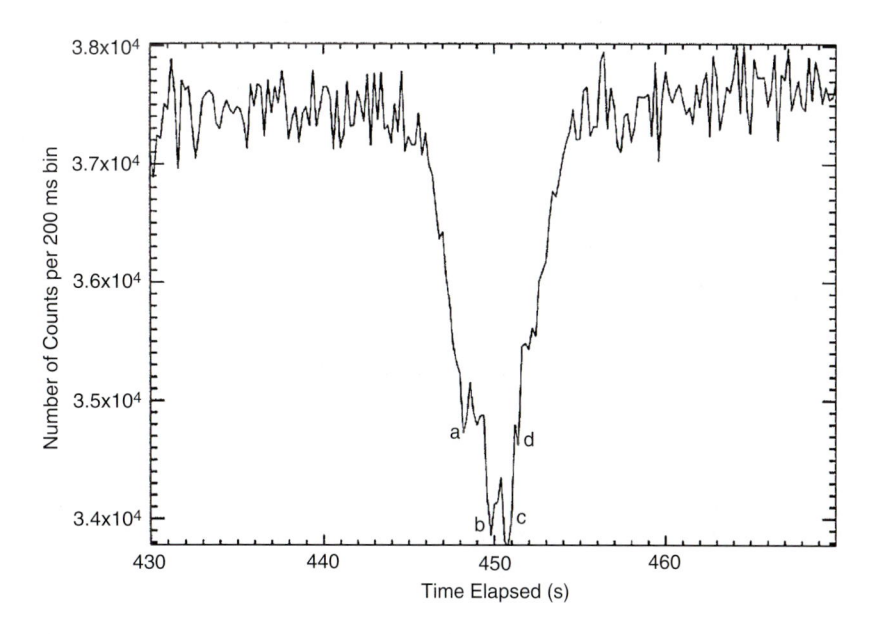

Fig. 21.14 the blue line is the path of the point on Enceladus closest to the ray as the 2007 occultation progressed-these jets correspond roughly, but not precisely, to some of the Spitale and Porco (2007) jets. The points labeled a, b, c, d correspond to the absorption events labeled a, b, c, d in Fig. 21.17. The prominent jets 'b' and 'c' are spaced 20 km apart, and the jet width combined with the height above the surface constrains the jet opening angle, which in turn constrains the ratio v_B/v_T where v_B is the vertical bulk velocity and v_T is the thermal velocity (see Tian et al. 2007). The idea is that the gas expands horizontally at speed v_T and travels vertically at speed v_B. From these data, $v_B/v_T \sim 1.5$. The whole occultation profile can be well represented with 8 equally spaced jets whose strengths are not equal but which all share this ratio. ISS images (Spitale and Porco 2007) reveal a much more complex situation, but a model with more closely spaced jets would require a higher value of this ratio. Assuming a gas temperature of $T = 180$ K, the lower limit inferred by CIRS (Spencer et al. 2006), gives a thermal velocity $v_T = 406$ m s^{-1}, and thus a vertical velocity of $406 \times 1.5 = 609$ m s^{-1}, well above the Enceladus escape velocity of 235 m s^{-1}.

The 2007 HSP data in Fig. 21.17 give a peak H$_2$O column density of 2.6×10^{16} cm^{-2} along a horizontal path passing 15 km above the surface of Enceladus. The UVIS FUV spectra, which have lower time resolution and thus provide densities more representative of the average plume density, give water column abundances $1.26 \pm 0.14 \times 10^{16}$ cm^{-2}, and $1.36 \pm 0.14 \times 10^{16}$ cm^{-2} for the two spectra closest to Enceladus, comparable to near-surface densities seen in the 2005 occultation (Fig. 21.16). The escape rate of water molecules from Enceladus with these column densities and the 609 m s^{-1} vertical velocity, and a plume horizontal dimension of 80 km (assumed similar to the scale height) is $\sim 6.5 \times 10^{27}$ molecule s^{-1} or ~ 200 kg s^{-1} using the methodology of Hansen et al. (2006). Modeling of the 2005 occultation data with similar assumptions also gives a 200 kg s^{-1} escape flux. Tian et al. (2007), using a more detailed model of the plume geometry, calculate a source flux of $4-6 \times 10^{27}$ molecules s^{-1} and a vertical velocity of 300–500 m s^{-1} in order to match the UVIS vertical profile, consistent with these results.

If the density at the source n_0 is approximately the vapor pressure of the surface ice, this limits the area of a single source. For $T = 180$K, $(n_0 = 2 \times 10^{12}$ cm$^{-3})$, with 8 sources and the total water flux of $\sim 6.5 \times 10^{27}$ molecule s^{-1}, a single source has area 0.6 km^2. This is highly temperature-dependent: for $T = 200$K, the area would be 20 $\times$ smaller.

Comparison of the 2007 to the 2005 plume column density at ~ 15 km altitude from the FUV spectra gives a value 1.7 times higher in 2007. HSP values at the same altitudes, 15.6–22 km, give a ratio of the attenuation in 2007–2005 of 1.4 ± 0.4, which is consistent. An increased plume density in 2007 is inferred (Hansen et al. 2008), though comparison is complicated by the differing geometry of the two occultations.

21.5.2.4 Ion and Neutral Mass Spectrometer (INMS)

INMS has sampled the structure and chemistry of the plume on several flybys. The July 2005 flyby (Waite et al. 2006) indicated that the plume gas composition (Table 21.3) was dominantly H_2O with minor CO_2, CH_4, and a mass 28 peak which could be due to CO and/or N_2. The structure of the water vapor profile during the encounter suggested the existence of both a plume source for the gas and a more distributed source due to sputtering of the surface of Enceladus.

In March and October 2008 there were two much deeper plume penetrations that were optimized for plume sampling. Cassini came within 25–50 km of the surface of Enceladus and flew through the center of the plume at altitudes of a few 100 km on its outbound path (Figure 21.2). The proximity to the plume source resulted in a two order of magnitude increase in signal to noise relative to the July 2005 flyby. The mass spectrum obtained from the October 9th 2008 flyby (Waite et al. 2009) is shown in Fig. 21.18. The complexity of the organic mixture leads to many overlapping mass

Table 21.3 Composition of the Enceladus plume as measured by INMS during the October 9th 2008 flyby (Waite et al. 2009)

Species	Molar mixing ratio
H_2O[a]	0.90 ± 0.01
CO_2[a]	0.053 ± 0.001
H_2CO	$(3.1 \pm 1) \times 10^{-3}$
CH_3OH	$(1.5 \pm 0.6) \times 10^{-4}$
C_2H_4O	$<7.0 \times 10^{-4}$
C_2H_6O	$<3.0 \times 10^{-4}$
H_2S	$(2.1 \pm 1) \times 10^{-5}$
^{40}Ar	$(3.1 \pm 0.3) \times 10^{-4}$
$^{36}Ar + {}^{38}Ar$	$<1 \times 10^{-5}$
NH_3	$(8.2 \pm 0.2) \times 10^{-3}$
N_2[b]	<0.011
HCN[b]	$<7.4 \times 10^{-3}$
CH_4	$(9.1 \pm 0.5) \times 10^{-3}$
C_2H_2	$(3.3 \pm 2) \times 10^{-3}$
C_2H_4[b]	<0.012
C_2H_6	$<1.7 \times 10^{-3}$
C_3H_4	$<1.1 \times 10^{-4}$
C_3H_6	$(1.4 \pm 0.3) \times 10^{-3}$
C_3H_8	$<1.4 \times 10^{-3}$
C_4H_2	$(3.7 \pm 0.8) \times 10^{-5}$
C_4H_4	$(1.5 \pm 0.6) \times 10^{-5}$
C_4H_6	$(5.7 \pm 3) \times 10^{-5}$
C_4H_8	$(2.3 \pm 0.3) \times 10^{-4}$
C_4H_{10}	$<7.2 \times 10^{-4}$
C_5H_6	$<2.7 \times 10^{-6}$
C_5H_{12}	$<6.2 \times 10^{-5}$
C_6H_6	$(8.1 \pm 1) \times 10^{-5}$

[a] Includes CO and H_2 which are assumed to be derived from dissociation of H_2O and CO_2 in the instrument.
[b] INMS cannot distinguish between (N_2 + HCN) and C_2H_4. N_2 and HCN abundances equal the given upper limit if C_2H_4 is absent, and C_2H_4 abundance equals the given upper limit if N_2 and HCN are absent. The N_2/HCN ratio, however, is well constrained by the data.

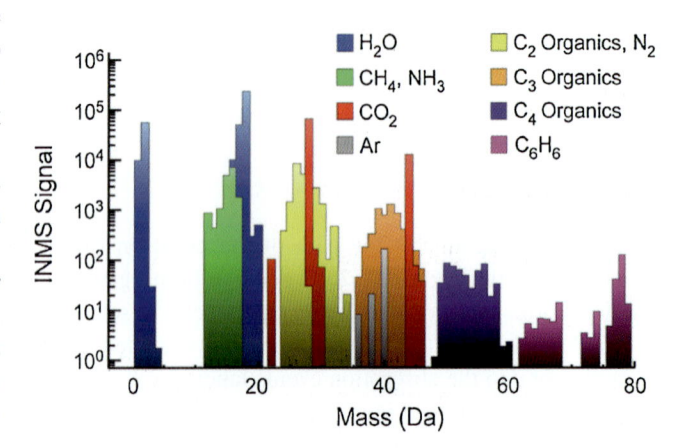

Fig. 21.18 Mass spectrum of the Enceladus plume from the October 9th 2008 flyby. The colors show contributions from various species and their breakdown products using the composition shown in Table 21.3. Reprinted by permission from Macmillan Publishers Ltd: Nature (Waite et al. 2009), copyright 2009

peaks that cannot be uniquely separated with the mass resolution of INMS, and many molecules are dissociated on impact with the instrument, especially at the high speeds of the 2008 flybys (14.4 and 17.7 km s^{-1}), resulting in detection of molecular fragments. Another complication is the tendency for water vapor to stick to the interior wall of the antechamber. This has two consequences: (1) it may displace gases already absorbed onto the surface and complicate the interpretation of minor species, and (2) the stickiness produces a shift in time and broadening of the water vapor time dependent structure. Mass deconvolution and modeling of impact dissociation leads to the estimated composition shown in Table 21.3. Water is again the dominant constituent, and may be even more abundant than indicated here, due to instrumental effects. The mass 28 species appears to be a mixture of CO and N_2 and/or C_2H_4, with the CO derived primarily from CO_2 dissociation in the instrument. The fact that the observed CO may be largely created within INMS is consistent with the lack of CO in the UVIS (Hansen et al. 2006) or VIMS (Brown et al. 2006) data. Methane (CH_4) and CO_2 are seen as in July 2005, and there is also, for the first time, a firm detection of NH_3, which is important because of its potential role as an antifreeze inside Enceladus. There are also many heavier organics, including benzene. Radiogenic ^{40}Ar is also apparently detected, at surprisingly high abundance, but primordial, non-radiogenic, ^{36}Ar and ^{38}Ar are not seen. The data also yield a D/H ratio of 2.9 (+1.5/−0.7) × 10^{-4}, which is close the cometary value, twice the terrestrial value, and ten times the protosolar value (Waite et al. 2009).

The water density in the March and October 2008 flybys was strongly peaked over the south pole, indicating a dominant plume source for the gas. Unlike the July 2005 flyby, there was no evidence of more widely distributed source of gas at other latitudes on Enceladus. The difference may be due to the different flyby geometry, though instrumental effects may also be involved.

21.5.2.5 Cosmic Dust Analyzer (CDA)

The CDA makes direct measurements of the particles encountered by the spacecraft. ("Dust" includes particles of any composition, including ice). Spahn et al. (2006a) show data collected during the July 2005 flyby with a CDA sensor that is sensitive to particles with a radius larger than 2 μm. The sensor recorded a maximum impact rate of ∼4 particles s^{-1} at 1 min before the closest approach to Enceladus, during passage through the plume. Modeling these data, including the dynamics of particles that escape from Enceladus and go into orbit around Saturn, Spahn et al. inferred a rate of particles larger than 2 μm emitted by the south pole source and escaping the moon's gravity of 5 × 10^{12} particles s^{-1}. They also inferred an isotropic source, perhaps impact ejecta, with about 5× smaller strength., similar to the isotropic gas source seen by INMS in July 2005 (but not in March 2008). These dust numbers correspond to an escaping dust mass of at least 0.2 kg s^{-1}, assuming a 2 μm radius for all grains. The authors concluded that these processes probably provide the dominant source of Saturn's E ring. Kempf et al. (2008) show that the ejected particles must have a speed greater than 222 m s^{-1} to avoid re-collision with Enceladus. This is significantly larger than the 3-body escape speed from the South pole.

CDA has also measured the compositions of dust particles in the E-ring, which are derived from the Enceladus plume. Postberg et al. (2008) report several classes of particles, including pure water ice particles, and water ice particles contaminated by organic compounds and/or silicate minerals. They speculate that the pure ice particles might be impact ejecta from Enceladus' surface, while the impure particles are from the Enceladus plumes, though it is also plausible that both particle types come from the plumes. Most particles also contain a small amount of sodium. However, a few percent of measured particles are enriched in the sodium salts NaCl, NaHCO$_3$, and/or Na$_2$CO$_3$ by several orders of magnitude, with potassium salts also present (Postberg et al. 2009).

More detailed dust measurements were made by CDA during the deeper plume passages on August 11th 2008 and October 9th 2008, but these results were not published at the time of writing.

21.5.2.6 Visual and Infrared Mapping Spectrometer (VIMS)

VIMS has obtained spectra of both the plume itself and the surface deposits around the plume sources, constraining both particle size and particle composition. Spectra of surface plume deposits are roughly consistent with the plume composition, as discussed in Section 21.4.1. However alteration of the particle properties – radiation damage, sintering, loss of volatiles, and change in crystal size – complicates the interpretation since the time scale and degree of alteration are hard to determine.

Spectra of the surface seen through the plume provide an upper limit on CO in the plume gas of 10^{14} molecules cm^{-2} (Brown et al. 2006), which is less than 1% of the water vapor column density detected by UVIS, and the likely global upper limit to NH_3 on the surface is 2%. Both of these limits are consistent with recent interpretation of INMS results (Section 21.5.2.4). VIMS has not yet seen thermal emission from the plume sources: the upper limit on the temperature of filled pixels 5–10 km on a side from 2005 observations is 140 K (Brown et al. 2006), which is consistent with CIRS results that show much smaller areas at these temperatures.

Hedman et al. (2009) use November 2005 VIMS spectra of the plume itself, seen at high phase angle, to infer the size of particles in the plumes at altitudes between 50 and 300 km. Smaller particles have weaker water ice spectral signatures, because a greater fraction of the light is scattered off the particle surfaces without penetrating into the material. They find typical particle radii of order 1 μm, consistent with the size of E ring particles. They also find a falloff in particle size with altitude, and used this to model the distribution of ejection speeds, in the 80–180 m s^{-1} range, as a function of particle radius. They inferred a uniform distribution of ejection speeds in this range for 1 μm particles, but a strong preference for speeds at the slow end of the range for 3 μm particles.

21.5.2.7 Magnetometer (MAG) and Cassini Plasma Spectrometer (CAPS)

During the February and March 2005 Enceladus flybys, the Cassini magnetometer detected the magnetic signature of the deflection of the Saturn magnetospheric plasma around a conductive barrier at Enceladus, and also detected cyclotron waves at the water group gyrofrequency which indicated a major ion source at Enceladus (Dougherty et al. 2006). These discoveries prompted the reduction in the altitude of the July 2005 flyby and thus enabled the first direct sampling of the plume. The July 2005 flyby also provided additional details of the magnetic signature, showing that the source of the magnetic perturbation was centered south of Enceladus, consistent with other observations of the south polar plume. Differences between the magnetic signatures on the three 2005 flybys suggested temporal variations in plume activity. Saur et al. (2008) modeled the magnetic signature and for the March and July 2005 flyby data obtained densities of H$_2$O molecules in the plume that are in good agreement with those inferred from the UVIS stellar occultations (Hansen et al. 2006). The February 2005 data, in contrast, indicated a 8× larger gas density during that flyby, suggesting large temporal variability on several-week timescales, though again the different geometry of the various observations complicates comparison.

During the March 2008 flyby, the Cassini plasma spectrometer (CAPS) saw signatures apparently due to the impact of nanometer-sized ice grains with the instrument (Jones et al. 2009), with flux spike locations consistent with the jet locations determined by Spitale and Porco (2007).

21.5.2.8 Groundbased Observations

Although almost all recent data on the Enceladus plumes has come from Cassini, recent ground-based work provides additional constraints on the plume composition. Schneider et al. (2009) use Earth-based spectroscopy to establish a strong upper bound on neutral sodium from Enceladus (Na/H$_2$O $< 4 \times 10^{-7}$ in the plume and $<7 \times 10^{-6}$ in the E-ring). This constraint is consistent with the CDA sodium detection in the E-ring particles discussed above even if (as is likely) the particles release their sodium as they are sputtered away, because the particles comprise a small fraction of the total plume mass, and only a small fraction of the particles contain significant quantities of sodium.

21.5.3 Plume Models

The plumes offer a unique window into the interior of Enceladus, and there has already been much work to infer the nature of the plume source from the available data. A critical question is whether liquid water is required to generate the plumes, and if so, how close to the surface the water must be.

21.5.3.1 Transfer of Heat to the Surface

To use the surface temperatures and radiated power to infer subsurface conditions, it is necessary to consider the mechanisms that bring the observed heat to the surface where it is detected. The heat may be transferred to the surface from the source by conduction through the ice, or by the plume gases, either as sensible or latent heat, or by flow of liquid water. The radiated power per unit length of the tiger stripes, assuming the >6 GW total is distributed evenly along the 500 km cumulative length of the tiger stripes is at least 12 kW m^{-1}. An additional ~0.5 GW (the vapor loss rate of ~200 kg s^{-1} multiplied by the latent heat, L, of the water vapor, ~2.8×10^6 J kg^{-1}), is carried away in the form of latent heat in the plume.

Spencer et al. (2006) note that liquid water would have to be very close to the surface if the warm surface temperatures were maintained only by thermal conduction from below. For instance a 180 K blackbody has a heat flux $F = 60$ W m^{-2}, and ice near its melting point has a thermal conductivity $k = 2.4$ W (m K)$^{-1}$, which means that the temperature gradient dT/dz into the ice would be 25 K m^{-1}. With a surface temperature of 180 K, the ice would melt at 4 m depth if thermal conduction were the only heat loss mechanism.

Spencer et al. (2006), and Nimmo et al. (2007) point out that the vapor provides another way to transport heat to the surface. The simplest picture is that vertical fractures run the length of the tiger stripes, and plume vapor traveling up these fractures delivers heat to the fracture walls. The shape of the CIRS thermal emission spectra is roughly consistent with this model. Abramov and Spencer (2009) use a 2-D thermal model to show that both the total power and the wavelength distribution of the thermal emission spectra can be matched quite well by assuming that most of the internal heat is radiated from the near-fracture surfaces, which are warmed by conduction through the ice from the fractures themselves. Assuming that the conduction is through solid ice, fracture temperatures near 225 K match the CIRS spectra well, though higher fracture temperatures cannot be ruled out if there is a thin insulating layer on the surface. However, multiple fractures are required along at least some portions of the tiger stripes to match the total south polar radiated power.

Ingersoll and Pankine (2009) argue that in a gas-filled fracture, condensation and evaporation are so rapid that the vapor pressure of the gas is fully determined by the saturation vapor pressure of the walls. Tiny differences between the two pressures drive large mass fluxes of water between the walls and the gas. These fluxes maintain the gas at the saturation vapor pressure of the walls. Condensation and evaporation occur, but they occur under quasi-equilibrium, i.e., reversible conditions.

Assuming the gas moves upward at thermal speeds, and is in vapor pressure equilibrium with the walls, the rate of upward heat transport is given by $wL\,e(T)\,(RT)^{-1/2}$, where $e(T)$ is the ice vapor pressure and w is the fracture width (Ingersoll and Pankine 2009). The observed average heat flow of $12\,\text{kW}\,\text{m}^{-1}$ can be matched by a single fracture with temperatures of ~ 215 K and ~ 276 K for fracture widths of 1 m and 2 mm, respectively. Heat can thus be transported fast enough to prevent near-surface melting if the fractures are wider than a few millimeters. If there are several cracks per tiger stripe, each crack has to supply only a fraction of the $12\,\text{kW}\,\text{m}^{-1}$, and the temperatures will be lower.

Schmidt et al. (2008) note that condensation of vapor on the fracture walls may close the fractures on a timescale of months. Ingersoll and Pankine (2009) reach similar conclusions. Very near the surface, where the depth is comparable to the width of the crack, they note that the heat flux into the walls is comparable to the heat flux that is radiated into space. This means that the crack closes at a rate $F/(L\rho_i) = 0.74$ m year^{-1} using F appropriate for a surface at 180 K, where $\rho_i = 917\,\text{kg}\,\text{m}^{-3}$ is the density of ice and $F = 60\,\text{W}\,\text{m}^{-2}$. A crack of width 1 m will close in a little more than a year, while crack of width 2 mm will close in less than a day. New cracks would therefore have to be forming constantly. If the surface is warmer, then F is greater and the crack will close even faster. Narrow cracks might perhaps be kept open by erosion of the walls by the vapor and entrained particulates, or by mechanical stresses associated with the tidal flexing.

The efficiency of vapor transport also affects conditions at depth, where the tidal dissipation is occurring. The Nimmo et al. (2007) shear heating model assumes efficient transport of heat from the shear zone to the surface by water vapor – the heat transport reduces temperatures in the shear zone, thus increasing viscosity and enhancing shear heating, and allowing generation of the observed heat and vapor flux without subsurface melting. This model is able to reproduce the observed heat flow and the observed vapor production rate. However, the model overestimates the efficiency of vapor transport because it neglects condensation of the vapor onto the surrounding ice matrix, as noted by Ingersoll and Pankine (2009). The water vapor pressures assumed by the Nimmo et al. model (Fig. D of their supplemental material) are several orders of magnitude higher than the vapor pressure of ice at the assumed temperature, so the vapor should condense and would not be able to transport the heat to the surface, as the model requires. Sufficient heat transport can be maintained at the lower equilibrium vapor pressures only if the diffusion coefficient of the vapor in the subsurface ice is much higher than assumed by Nimmo et al.

Ingersoll and Pankine (2009) use a hydrodynamic model to show that if the heat is generated on meter-sized cracks in the near-surface brittle zone, and if the cracks open and close with position in the orbit (Hurford et al. 2007), the upward flow of vapor can carry the heat away. However removal of any frictional heat generated in the underlying ductile zone is more difficult, because open fractures are less likely, and conduction or convection are likely to be the dominant heat transport methods. If the ductile zone warms to a temperature where viscosity is low enough that frictional heating is negligible, melting in the ductile zone might be avoided, but detailed self-consistent models have not yet been developed.

Kieffer et al. (2006) and Gioia et al. (2007) postulate a low-temperature plume source, below 200 K, driven by clathrate decomposition at depth (see the discussion of plume composition below). In this model open cracks extend all the way down to the source, at depths of 35 km or greater, and again, advection of plume vapor up these fractures is sufficient to carry the heat away. Accordingly, they call this the "Frigid Faithful" model. In this model, thermal expansion in the source region accounts for the tiger stripes and other surface features seen in the south polar region. The cracks must have a certain minimum width, on the order of 10 cm. Gioia et al. argue that the crust is so cold and is strengthened by the presence of clathrate hydrates, that it behaves like a brittle elastic solid, so the cracks do not close up.

In summary, vapor transport along the tiger stripe fractures may be able to bring heat to the surface of Enceladus efficiently enough to explain the observed thermal signature without requiring liquid water near the surface or at depth. Whether temperatures reach the melting point depends on the width of the fractures and the details of the heat generation mechanism, and new observations and models may provide a clearer answer.

Liquid water would presumably transport heat to the surface more efficiently than gas. However, Postberg et al.

(2009) show (in the paper's supplementary information) that for evaporation from a water surface in a fracture of constant width, heat cannot be transported upward through the water, by conduction or convection, quickly enough to replace the latent heat lost by evaporation at the water surface. Such a water column would therefore rapidly freeze at the surface. For heat transport to be sustainable, the cross-sectional area of the water column must be several orders of magnitude greater than the area venting to space. Postberg et al. therefore suggest that the plumes are fed by narrow fractures in the roof of vapor-filled chambers overlying relatively large bodies of water.

21.5.3.2 Plume Dynamics and Particle Formation

A high-pressure gas erupting into a low-pressure environment is a classic fluid dynamics problem, with applications ranging from rocket nozzles to volcanoes and geysers (e.g., Kieffer 1982). In ideal cases the gas can convert all of its initial enthalpy $c_p T$ into kinetic energy. The specific heat c_p of water vapor is 1,850 J (kg K)$^{-1}$, which is almost exactly four times the gas constant R for water vapor. Thus the plumes could theoretically reach speeds of $(2C_p T)^{1/2} = 816$ and 1,005 m s^{-1} for $T = 180$ and 273 K, respectively. Actual gas speeds inferred from the plume shape seen by UVIS ($\sim$600 m s^{-1}, see Section 21.5.2.3 above), are lower than this maximum but higher than the thermal speed.

However, the fact that the temperature of the walls completely controls the temperature and pressure of the vapor means that the dynamics of the vapor is uncertain. If the temperature gradient dT/dz is small, then the pressure gradient dP/dz is small, and the force that accelerates the fluid is small. The velocity will be small as well. The fact that the observed velocity is not small, i.e., is larger than the thermal speed $(RT)^{1/2}$ implies that dT/dz is not small. Clearly the length, width, and temperature gradient in the crack are important parameters for the dynamics, but they depend on the thermal and mechanical history of the crack, which is uncertain.

The ice particles in the plume are the most readily observed aspect of Enceladus' activity. Their number density, speed distribution, size, and composition provide valuable constraints on the nature of the plumes and their source regions. Schmidt et al. (2008, Fig. 21.19) use models of the plume behavior in subsurface conduits to infer that the plume particle speeds and densities are consistent with particle formation by direct condensation of the water vapor in the plume before it reaches the surface. Their detailed hydrodynamic model includes homogeneous nucleation and particle growth in the gas as it expands toward the surface. The reservoir is at a given temperature T. The flow does not interact with the walls after it leaves the reservoir. Schmidt et al. derive particle speed and size distributions that match the observed and inferred properties of the plumes. The cracks have variable width, and Schmidt et al. show that repeated collisions of the grains, with re-acceleration by the gas, induce an effective friction that explains the reduced velocity of the grains when they leave the vent. Large grains emerge at slower speeds, consistent with the VIMS results (Hedman et al. 2009). The distance between collisions L_{coll} is assumed to be similar to the width of the crack. Thus $L_{coll} = 0.1$–1.0 m. They then adjust the density of the gas to get the right velocity and cutoff radius for the

Fig. 21.19 Water vapor and ice grain interaction with the walls of a model plume vent, reprinted by permission from McMillan Publishers Ltd: Nature Schmidt et al. (2008), copyright 2008. (**a**) Model schematic. Gas flow upward from a liquid reservoir is modified by constrictions in the vent, and ice particles collide with the vent walls, slowing the particles. (**b**) Typical solution for gas density and condensed grain mass fraction as a function of depth z. Most grains nucleate at the narrowest part of the vent. (**c**) Profiles of gas speed and temperature. The gas becomes supersonic near the narrowest part of the vent. (**d**) Example distribution of condensed particle radii R

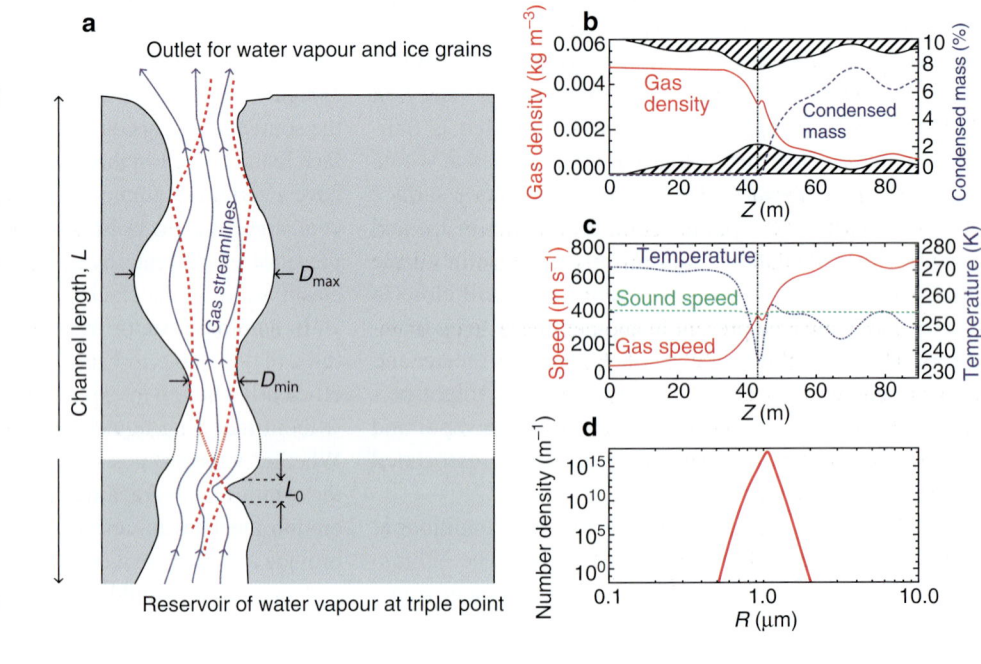

particles that escape to supply the E ring, given the gas speed inferred from the UVIS observations (Section 21.5.2.3). Since density depends on temperature when the gas is saturated, Schmidt et al. are able to estimate the temperature of the source. They find it is above 260 K, suggesting liquid water at the plume sources. Note that the Schmidt et al. model does not consider the effects on the gas and particle dynamics of vapor condensation on the walls, which is probably important especially near the surface (Ingersoll and Pankine 2009), though whether inclusion of these effects would change the conclusions remains to be seen.

The high ice/gas ratio in the plumes inferred from comparison of ISS images and UVIS measurements of the plume (Section 21.5.2.1) led Porco et al. (2006) to suggest that the observed plume particles might form directly from near-surface boiling of liquid water, which could provide an efficient way to generate particles. However, modeling by Brilliantov et al. (2008) implies that such particles would have speeds much lower than is inferred for the actual plume particles. Also, recent estimates of the ice/gas mass ratio are considerably lower (Kieffer et al. 2009), and ice/gas *production* ratios are lower still because of the average speed of the ice particles is much lower than that of the gas. The Kieffer et al. (2006) clathrate dissociation model may also have trouble explaining the speed of ice particles, because it is difficult for the low H_2O vapor density, resulting from the low temperatures of that model, to accelerate particles to the speeds observed (Brilliantov et al. 2008).

21.5.3.3 Plume Composition

While the plume is dominated by water vapor and water ice particles, the presence of additional species, including numerous gas species and salts in the plume particles, provides additional valuable clues to the plume origin.

Kieffer et al. (2006) point out that the other gases seen by INMS in the plume, notably N_2 (if present), CO_2 and CH_4, are insufficiently soluble in liquid water to have originated in their observed abundances from a liquid aqueous phase. Improved INMS 2008 plume composition measurements and solubility calculations confirm this conclusion (Waite et al. 2009). If the plume source is dominantly liquid water, the other gases must be present in another phase, such as gas bubbles. However, the solubility of these gases in clathrate hydrates (ices with a cage-like structure in which water ice traps other volatile components) is enormous compared with their solubility in liquid water. The observed molar ratio of H_2O vapor to non-condensable gases in the plume is $\sim 10:1$. The similarity between this ratio and the ratio of water to guest molecules in a clathrate (hydration number is 6:1 to 8:1) suggests that the reservoir could consist of clathrates or clathrates plus water ice. Kieffer et al. suggest that the water and other gases can be released by depressurization of the clathrate to form the plumes at relatively low temperatures, below 200 K.

A problem with the Kieffer et al. model is that the sudden de-pressurization of clathrate-hydrate does not necessarily lead to vaporization of all the constituents. Only the more volatile constituents come off at low temperatures. For the Enceladus plumes, they come off in the order N_2, CH_4, and then CO_2 followed by water, which vaporizes at relatively high temperatures only. The plume should therefore be depleted in water vapor relative to the clathrate, and it is thus not clear whether clathrate dissociation can explain the abundance of water in the plume. Further laboratory work is needed. Halevy and Stewart (2008) elaborate on the clathrate dissociation model by considering non-equilibrium dissociation, modulated by the tidal opening and closing of the fractures described by Hurford et al. (2007). They predict that variations in both plume composition and density are possible as function of Enceladus' position in its orbit.

The ultimate source of the complex organics seen by INMS is an intriguing question. The detailed compositional information provided by INMS is likely to provide a strong constraint on possible formation mechanisms for the organics and other species, including interactions between the presumed ocean and the silicate core (e.g., Matson et al. 2007).

The detection of ice particles rich in NaCl and $NaHCO_3$ in the E-ring by CDA (Postberg et al. 2009) provides perhaps the strongest evidence yet for a liquid water plume source. The composition of these particles strongly suggests that they originate by direct freezing of saline water at the plume source, while the more common sodium-poor particles may be vapor condensates. The salt-rich particle compositions are similar to those expected for an Enceladus ocean that has been in contact with warm rock (Zolotov 2007), suggesting direct communication of the plume source with such an ocean. However, it is possible that preferential evaporation of H_2O has increased the salinity in the plume source reservoir, and that the ocean itself is less saline (Schneider et al. 2009). The extremely low sodium content of the neutral gases near Enceladus' orbit determined by Schneider et al. rules out plume production by wholesale boiling into space, without fractionation, of water derived from a salty ocean.

The apparent presence of abundant radiogenic ^{40}Ar in the plume provides additional indirect evidence for liquid water within Enceladus, because aqueous processes provide an efficient way to leach ^{40}Ar (or its parent species, ^{40}K) from the core and to concentrate the ^{40}Ar in the plume (Waite et al. 2009).

21.5.4 System Interaction

Saturn's system is filled with neutral products from the electron-impact dissociation and photodissociation of H_2O molecules. From measured O and OH abundances and theoretical estimates of the loss processes of all water products from the system, various investigators have estimated H_2O supply rates necessary to maintain a steady state as $> 2 \times 10^{27}$ H_2O molecules s^{-1} (Shemansky et al. 1993), possibly as high as 3.75×10^{27} (Jurac et al. 2002) or 10^{28} H_2O molecules s^{-1} (Jurac et al. 2005). These numbers are very similar to the inferred UVIS source rate of 6.5×10^{27} H_2O molecules s^{-1} (Hansen et al. 2008). It is thus very probable that Enceladus is the dominant source of the observed O and OH in the Saturn system. Johnson et al. (2006) suggest that the extended OH torus seen by Hubble is derived from charge exchange with a denser, colder, neutral gas torus that is populated directly by the Enceladus gas plumes. The neutrals are ionized to produce much of the magnetospheric plasma, though most of the mass loading occurs at some distance from Enceladus (Khurana et al. 2007). It is likely that nitrogen or ammonia from Enceladus' plumes is also the dominant source of the nitrogen found in the magnetosphere (Smith et al. 2008a), and products of ammonia have also been tentatively identified in the magnetosphere (Smith et al. 2008b).

Similarly, Enceladus' particle plumes are almost certainly the dominant source of Saturn's E-ring (Fig. 21.20), though a significant contribution from impact ejecta from the surface has also been suggested (Spahn et al. 2006a, Postberg et al. 2008). The ring, composed of water ice grains that are primarily 0.3 to 3 μm in size (Showalter et al. 1991, Nicholson et al. 1996), extends from about three to at least eight Saturn radii, with a peak density at or very close to Enceladus' orbit (Baum et al. 1981, Showalter et al. 1991). The lifetimes of 1-μm grains are <50 years (Jurac et al. 2002, Johnson et al. 2008), because water molecules are sputtered from the grains' surfaces by the plasma trapped in Saturn's magnetosphere. If Enceladus' plume has a comet-like particle-to-gas production ratio of 10%, then the mass of ice coming from Enceladus is roughly $20 \, kg \, s^{-1}$, which is more than sufficient to compensate for the estimated particle loss rate of the E-ring of $1 \, kg \, s^{-1}$ (Juhász and Horányi 2002). The Cassini CDA instrument has extensively sampled the E-ring, and the measured vertical structure can be used to constrain the plume dynamics. The ring thickness appears to be controlled by particles with ejection speeds near $230 \, m \, s^{-1}$ (Kempf et al. 2008). Plasma drag effects are also important in determining the morphology of the ring (Juhász et al. 2007).

The E-ring particles in turn modify the surfaces and albedos of the other mid-sized Saturnian moons, as implied by the decrease in satellite albedo with distance from Enceladus (Verbiscer et al. 2007). In addition, the fact that Rhea, Dione, and Tethys, orbiting outside Enceladus, have trailing hemispheres that are darker than their leading hemispheres, while the reverse is true on Enceladus itself and Mimas (Buratti et al. 1998), is consistent with brightening of their surfaces by E-ring particles (Hamilton and Burns 1994).

21.5.5 Variability of the Activity

The E-ring has been in existence since at least 1966, when it was discovered (Feibelman 1967). Because the lifetime of the E-ring particles against sputtering is about 50 years (Jurac et al. 2002, Johnson et al. 2008), this implies that the current dust plumes have been feeding the E-ring for many

Fig. 21.20 Enceladus' complex interaction with the E ring, as seen by ISS at a phase angle of 175° on September 15, 2006. Enceladus is the dark dot in the center of the image (only its night side is visible at this phase angle), and the plume is the bright patch immediately below it. In contrast, no enhancement in ring density is visible near Tethys, which is on the far left of the image. Cassini press release image PIA08321, NASA/JPL/SSI

decades, but we have no direct information on activity earlier than about 100 years ago. We have direct evidence for gas plume activity only since the early 1990s, when the OH torus was discovered (Shemansky et al. 1993). OH lifetimes are tens of days, so gas torus variability should reveal variations in the plume activity on short timescales that would not be reflected in variations in the E-ring. Significant variability in neutral oxygen emission near Enceladus was indeed seen by UVIS during Cassini's approach to Saturn in 2004 (Esposito et al. 2005).

There is some evidence for variable features in the E-ring itself. An unusual discrete arc of particles in the E-ring, near Enceladus' orbit was seen in August 1995 using adaptive optics techniques by Roddier et al. (1998). However, they point out that its orbital longitude was distant from that of Enceladus and its lifetime was only a few hours, so it was probably not directly ejected from Enceladus.

Hurford et al. (2007) calculate the tidal stresses in an icy shell, focusing on the timing of the opening and closing of the cracks vs. position of Enceladus in its eccentric orbit, and provide maps of the stress state on each fracture as a function of orbital position. They predict greater plume activity at orbital longitudes when the cracks are open. These predictions have not yet been tested in detail, though the 1.7-fold difference between the plume gas densities inferred from the two UVIS stellar occultations (Hansen et al. 2008 and Section 21.5.2.3 above), which occurred at true anomalies of 98° and 254° respectively, are in the opposite sense to that predicted by Hurford et al. As noted above, the Cassini magnetometer data seem to indicate eightfold variations in gas plume density in the 20 days between the February and March 2005 Enceladus encounters, and comparison of E-ring particle models with CDA dust observations during close encounters also suggests variability in the particle plume output (Kempf et al. 2008). All these inferences of plume variability are somewhat model-dependent, as the geometry of the observations being compared differs significantly. As noted above, CIRS saw negligible change in the thermal emission from the south pole between July 2005 and November 2006 (Abramov and Spencer 2009).

On the longer term, several lines of evidence suggest that Enceladus is unusually active in the current epoch. See Section 21.2.7.

21.5.6 Conclusions: The Nature of the Plume Source

While many questions remain about the nature of the plume source, and the possible involvement of liquid water, progress has been made. Upward transfer of latent heat by water vapor is an effective means of bringing heat to the surface along the tiger stripes, though liquid water transport is also possible if surface freezing can be avoided. It seems likely that at least most of the observed plume particles can be produced by subsurface condensation of water vapor, though this is most easily accomplished at high subsurface gas pressures which themselves suggest near-melting temperatures. The presence of a small fraction of salt-rich particles in the E-ring is strong evidence that there is liquid water at the plume source, and that those particles are derived directly from that water. Possible alternative origins for these intriguing particles have not yet been explored, however. If the source is liquid water, it is not clear how the several percent of non-H_2O gases are introduced into the plume, and whether clathrates have a role to play. Continued analysis of the rich harvest of data from the current series of Cassini Enceladus flybys should provide a clearer picture in the near future.

21.6 Biological Potential of Enceladus

The observed plume and the associated thermal activity on Enceladus may indicate the presence of liquid water below the surface. From an astrobiological perspective, the possibility of liquid water opens up the question of life.

Discussions of life in subsurface water on Enceladus must logically begin with a consideration of the possible origin of life on that world. Unfortunately, we do not know how life originated on Earth nor have we been able to reproduce it in the laboratory. Indeed, there are many theories for how life originated on Earth (Davis and McKay 1996) and we can determine which of these would apply to Enceladus (McKay et al. 2008). Two of the theories for the origin of life on Earth would seem to apply to Enceladus, these are the organic "soup" theory and the deep sea vent theory.

The theory that the origin of life occurred in a soup of abiotically produced organic material is arguably the original theory for the origin of life and traces back to Darwin. It was given an experimental basis in the Miller–Urey experiment (Miller 1953) and has been extended since that time. On Earth the organics that comprised the soup could have been produced on the Earth or may have come in with comets and interplanetary dust. Applied to Enceladus the organic soup model might involve the formation of Enceladus from organic rich ices similar to cometary materials, consistent with INMS plume composition results. The organics would be in the mix from the start. As tidal heading forms a subsurface aquifer the solution would therefore be rich in organics comprising a suitable prebiotic soup.

The alternative theory for the origin of life on Earth that might apply well to Enceladus is the deep sea vent theory. In this scenario life began at the interface where chemically rich fluids heated by tidal dissipation emerge from the silicate core below the sea bottom. Energy is derived from reduced gases such as H_2S and H_2 coming out from the vent in contact with a suitable oxidant, which can even be CO_2, to form a source of chemical energy (e.g., Corliss et al. 1981, Shock 1990, Wächtershäuser 1990). In this model for the origin of life the first organisms would be chemotrophs – consuming inorganics - whereas in the Miller–Urey model they would be hetereotrophs – eating organics.

The time required for the origin of life is unknown and one possible objection for the origin of life on Enceladus is that there might not be enough time. Would liquid water exist for long enough for life to start? There is evidence from cratering ages that the activity on Enceladus has persisted for at least hundreds of millions of years (Kargel and Pozio 1996, Porco et al. 2006), so long periods of time might be available, though we do not know how continuous the activity has been. The origin of life appears to have happened relatively quickly on Earth, before 3.5 Gyr ago (e.g., Tice and Lowe 2004). In a paper addressing the question of the time required for the origin of life, Lazcano and Miller (1994) suggest that, "in spite of the many uncertainties involved in the estimates of time for life to arise and evolve to cyanobacteria, we see no compelling reason to assume that this process, from the beginning of the primitive soup to cyanobacteria, took more than 10 million years." However, Orgel (1998) criticized this result stating that we do not understand the steps leading to life and consequently, we cannot estimate the time required. "Attempts to circumvent this essential difficulty are based on misunderstandings of the nature of the problem." The resolution of this question must await new data on Earth, the discovery of other life forms, or the ability to synthesize life in the laboratory.

A possibility that must also be considered is that life is present on Enceladus but did not originate there. This may also be true of the Earth. The idea that life was carried to the Earth from elsewhere is known as panspermia. The evidence that rocks from Mars have reached the Earth without experiencing temperatures high enough to sterilize their interiors has heightened interest in panspermia between the planets (Melosh 1988, Weiss et al. 2000, Gladman et al. 2005). McKay et al. (2008) point out that it is important to consider two possibilities for Enceladus: (1) life carried there from the inner solar system and (2) life brought to Enceladus as part of the material from which the Solar System formed. The former is unlikely given the distance to Enceladus and there is no data to support the latter model in which life in the Solar System comes from the pre-solar nebulae. These hypotheses are testable. If life is detected on Enceladus and found to be related to the phylogenetic tree of life on Earth this would argue for shared origin via panspermia.

For life to persist requires an environment of liquid water, the essential elements and nutrients, and a source of energy. The composition of the plume from the South Pole of Enceladus indicates that C and N, and probably S, are present in the source regions of the ice jet. If this source region is a subsurface liquid water aquifer then if phosphorous is present, all the biogenic elements (C,H,N,O,P,S) are present. Phosphorous is expected based on its abundance in rocky material. Thus, the remaining question to establish the possibility of an ecosystem is that of an energy source.

On Earth the dominant energy source for all ecosystems is sunlight. Even most subsurface ecosystems on Earth derive their energy from sunlight produced organic material from the surface. The deep sea vent communities derive energy from the reaction of H_2S from the vent with O_2 from the ambient seawater. However, this O_2 comes from surface photosynthesis. So these ecosystems are also dependent on surface photosynthesis.

To be the basis for a subsurface ecosystem on Enceladus we must turn to ecosystems on Earth that are completely independent of O_2 or organic material produced by surface photosynthesis (because of the cold vacuum conditions at the surface, we do not consider photosynthetic communities). Three such ecosystems have been reported.. The first ecosystem discovered that does not require oxygen, organic material, or sunlight is based on methanogens using H_2 derived from rock-water reactions in the Columbia River Basalts (Stevens and McKinley 1995). Chapelle et al. (2002) reported a similar system in the basalts of Idaho. A different ecosystem, also without requirements for O_2, organic material, or sunlight was discovered deep below the surface in South Africa (Lin et al. 2006) There the primary productivity of the system is based on sulfur-reducing bacteria using redox couples produced ultimately by radioactive decay.

These three microbial ecosystems are models for life that might be present inside Enceladus today – or in liquid water deep below the surface of other worlds such as Europa and Mars. Following this line of thought, McCollom (1999) considered methanogen-based biota for Europa as well as one based on sulfur-reducing bacteria. Most, if not all, of his analysis would apply to Enceladus. Figure 21.21 shows a proposed methane cycle that could support methanogens in a subsurface aquifer on Enceladus (or Europa) following the approach of McCollom (1999). Hydrogen and CO_2 present in the water provides a basis for methanogens which form CH_4 and biomass, acting as the primary producers in the community.

Gaidos et al. (1999) concluded that life trapped under thick ice covers in subsurface oceans eventually dies out due to depletion of redox pairs. In particular the availability of oxidants would eventually limit life in such an isolated ocean. A possible source of oxidants which would obviate this problem has been proposed for Europa based on oxidants produced by radiation on the surface of the ice and could be

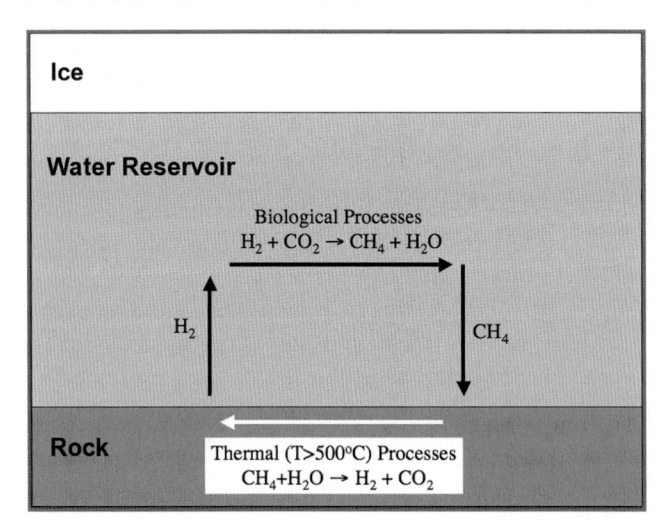

Fig. 21.21 Possible methane cycle on Enceladus (McKay et al. 2008)

carried to subsurface liquid water reservoirs that may contain reductants (Chyba and Phillips 2001, Hand et al. 2007). Parkinson et al. (2007) consider Enceladus and argue for a similar and strong source of surface oxidants there. There is no similar system on Earth but there is a rough analogy with the oxidants produced by radioactivity as observed by Lin et al. (2006).

However, it is worth noting that the three isolated microbial ecosystems discussed above provide a basis for considering how a long term biological cycle can persist in an ice-covered ocean even without a source of oxidants from the surface (McKay et al. 2008). If similar ecosystems existed on Enceladus, the chemical energy that supports life would come from long-lived non-biological energy sources (geothermal heat and radioactive decay). As shown above, these energy sources create redox pairs that recombine very slowly. Life catalyzes the recombination reaction harnessing the energy. The cycle can persist as long as the geochemical or radioactive energy is present – even if the system is isolated from outside sources of energy.

Parkinson et al. (2008) consider the geochemical cycles that might operate on Enceladus, and conclude that these may be sufficient to sustain life. Analyzing the organic compounds in the plume of Enceladus (e.g., Waite et al. 2006, 2009) is clearly the best way to detect any life in the subsurface aquifer. Life might be implicated by the presence of extreme chemical disequilibrium, for instance. The Cassini spacecraft is flying progressively more deeply through the plume as part of its extended mission, providing a steadily improving inventory of the plume's organic and inorganic chemistry, and a clearer picture of the possibility of life beneath the surface.

21.7 Summary and Future Exploration

By discovering ongoing geological activity at Enceladus' south pole, Cassini has confirmed, with a wealth of detail, earlier hints that this moon is one of the most remarkable bodies in the solar system. With the possible exception of Triton, which has active plumes that may result from either internally driven geology or sunlight-driven volatile processes (Kirk et al. 1995), Enceladus is the only icy body in the solar system known to be geologically active.

Enceladus provides living examples of phenomena that have been important at some time throughout the outer solar system. These processes can be studied as they happen on Enceladus, leading to understanding that can be applied to less currently active worlds. Already, Enceladus is testing models of shear heating and ice tectonics that were first developed for Europa, is challenging our understanding of tidal heating and orbital evolution of icy satellites, and the composition of its plume is shedding light on the internal chemistry of icy satellites and their possible oceans. Enceladus is also important because it is the source of the E-ring and the extensive torus of neutral and ionized species that fill the middle Saturnian magnetosphere, thus playing a pivotal role in the Saturnian system similar in some ways to Io's role in the Jovian system.

Evidence is accumulating that the plume source may involve liquid water, possibly in communication with a global or regional ocean that is likely to be in contact with the silicate core. Coupled with the rich chemistry apparent from the plume composition, and the energy supplied by tidal heating, the plume source provides a potential habitat for life. The plumes themselves allow the study of fresh samples from this environment, allowing relatively easy investigation of Enceladus' interior and its biological potential.

Many questions remain unanswered, including the following:

- Is the currently high level of activity on Enceladus atypical, and if not, how is the tidal heating and mass loss sustained over the long term?
- What accounts for the extreme variations in surface age, and why is the geology so symmetrical about the spin axis and the direction to Saturn?
- What mechanisms produce Enceladus' abundant and varied tectonic features?
- What generates the south polar plumes, and what are the mass loss rates?
- Is liquid water present in the interior, as a global or regional ocean, or locally at the source of the plumes? What is the detailed chemistry of these water bodies?
- Finally, can subsurface conditions sustain life, and is life actually present there?

Cassini will continue to address these questions with four more flybys in 2009 and 2010 (Table 21.1, Fig. 21.2), and (if approved) twelve additional flybys through 2015. Highlights in 2009 and 2010 will include deeper passages through the plume for in situ sampling, high resolution imaging of the plumes and additional very high resolution remote sensing of the surface, a plume solar occultation to map nitrogen abundance, and a gravity pass to search for south polar gravity anomalies.

Cassini still has much to tell us about Enceladus, but its orbit and instrumentation do not allow us to answer all the above questions. For instance, analysis of plume composition is limited by the mass range and resolution of the mass spectrometer, and the high flyby speeds prevent analysis of intact dust grains and may destroy complex molecules. The remote sensing instruments are not optimized for wide-area high-resolution coverage during the very limited time close to Enceladus afforded by the spacecraft's relatively high-speed flybys, and the configuration of the spacecraft does not permit all instruments to gather optimal data during all flybys. Cassini also lacks some instruments that would be of great value at Enceladus, such as ice-penetrating radar to map subsurface structure. Finally Cassini's Saturn-centric orbit prevents homogeneous global mapping and precludes many geophysical observations, including comprehensive measurements of Enceladus' tidal response, static gravity field, and magnetic properties, which would greatly expand our current limited information on its interior structure.

Accordingly, several recent studies have considered return missions to Enceladus. A 2006 JPL study (Reh et al. 2007) looked at missions that might be possible with a <$1 B New Frontiers budget, including a Stardust-like plume sample return mission, and a Saturn orbiter with multiple Enceladus flybys, like Cassini but with its mission and instrumentation optimized for Enceladus. The study concluded that while such missions would be scientifically worthwhile, expected costs of approximately $1.5 B each would not fit within the New Frontiers program, and the Enceladus sample return in particular would require significant technology development for high-speed plume sample capture. A NASA study of several more ambitious Flagship-class Enceladus missions was conducted by Goddard Spaceflight Center in 2007 (Razzaghi et al. 2007). Mission concepts included a Saturn orbiter with multiple Enceladus flybys and a soft lander, an Enceladus orbiter, and an Enceladus orbiter with soft lander, with costs in the $2.1 B–$3.3 B range. The Enceladus orbiter missions were the most challenging but most scientifically valuable, and made use of large numbers of gravity-assist flybys of Rhea and Dione to reduce delta-V for Enceladus orbit insertion. Orbiting Enceladus is challenging due to large orbital perturbations from Saturn, with polar orbits, which are preferred for global mapping and plume sampling, having stable lifetimes of only a few days. On the other hand, the mass of Enceladus is low enough that landing on the surface, once in orbit, would be relatively easy. The European TANDEM (Titan and Enceladus Mission) mission concept was proposed to the ESA Cosmic Vision program in 2007 (Coustenis et al. 2008). This was primarily a Titan mission, with an orbiter and landers, but included substantial Enceladus science from Saturn orbit, including a surface penetrator. The goals of this mission concept were incorporated into the 2008 joint NASA/ESA study of a Flagship-class Titan/Saturn System Mission, which would conduct Enceladus science with multiple flybys before entering Titan orbit. Such a mission may follow the Europa Jupiter Orbiter as the next-but-one outer solar system Flagship mission.

While the nature of the next mission after Cassini to visit Enceladus is still undecided, it remains a compelling target for future exploration, and Cassini itself will provide many new discoveries in the next few years. We look forward to the continuing exploration of this remarkable moon.

Acknowledgements We wish to thank everyone who has made these discoveries possible, by contributing to the success of the Cassini/Huygens project. This work was funded by the Cassini Project and by NASA.

List of Acronyms

CAPS	Cassini Plasma Spectrometer (Cassini instrument)
CDA	Cosmic Dust Analyzer (Cassini instrument)
CIRS	Composite Infrared Spectrometer (Cassini instrument)
FUV	Far Ultraviolet Spectrograph (part of UVIS)
HSP	High Speed Photometer (part of UVIS)
INMS	Ion and Neutral Mass Spectrometer (Cassini instrument)
IR	Infrared
ISS	Imaging Science Subsystem (Cassini instrument)
MAG	Magnetometer (Cassini instrument)
SLRI	Short-lived radioactive isotope
SPT	South Polar Terrain
TANDEM	Titan and Enceladus Mission
TST	Tiger Stripe Terrain
UV	Ultraviolet
UVIS	Ultraviolet Imaging Spectrograph (Cassini instrument)
VIMS	Visible and Infrared Mapping Spectrometer (Cassini instrument)

References

Abramov, O., Spencer J.R.: Endogenic heat from Enceladus' south polar fractures: New observations, and models of conductive surface heating. Icarus **199**, 189–196 (2009).

Anderson, D.L.: *Theory of the Earth*. Blackwell, London (1989).

Barr, A.C., McKinnon, W.B.: Convection in Enceladus' ice shell: Conditions for initiation. Geophys. Res. Lett. **34**, L09202. (2007a).

Barr, A.C., McKinnon, W.B.: Convection in ice I shells and mantles with self-consistent grain size, J. Geophys. Res. **112**, E02012 (2007b). doi: 10.1029/2006JE002781.

Barr, A.C.: Mobile lid convection beneath Enceladus' south polar terrain, J. Geophys. Res. **113**, E07009 (2008). doi: 10.1029/2008JE003114.

Baum, W.A., Kreidl, T., Westphal, J.A., Danielson, G.E., Seidelmann, P.K., Pascu, D., Currie, D.G.: Saturn's E ring. Icarus **47**, 84–96 (1981).

Benjamin, D., Wahr, J., Ray, R.D., Egbert, G.D. Desai, S.D.: Constraints on mantle anelasticity from geodetic observations, and implications for the J(2) anomaly, Geophys. J. Int. **165**, 3–16 (2006).

Bills, B.G.: Free and forced obliquities of the Galilean satellites of Jupiter. Icarus **175**, 233–247 (2005).

Bills, B.G. Nimmo, F.: Forced obliquity and moments of inertia of Titan. Icarus **196**, 293–297 (2008).

Bills, B.G., Nimmo F., Karatekin O., Van Hoolst, T., Rambaux, N, Levrard B., Laskar, J.: Rotational dynamics of Europa. In: Pappalardo, R., McKinnon, W. Khurana, K (eds.) *Europa*, Univ. Ariz. Press., Tucson, in press (2009).

Bland, M.T., Beyer R.A., Showman A.P.: Unstable extension of Enceladus' lithosphere. Icarus **192**, 92–105 (2007).

Brilliantov, N.V., Schmidt, J., Spahn, F.: Geysers of Enceladus: Quantitative analysis of qualitative models. Planet. Space Sci. **56**, 1596–1606 (2008).

Brown, R.H. and 24 co-authors: Composition and physical properties of Enceladus' surface. Science **311**, 1425–1428 (2006).

Buratti, B. J.: Enceladus – Implications of its unusual photometric properties. Icarus **75**, 113–126 (1988).

Buratti, B., Veverka, J.: Voyager photometry of Rhea, Dione, Tethys, Enceladus and Mimas. Icarus **58**, 254–264 (1984).

Buratti, B.J., Mosher, J.A., Nicholson, P.D., McGhee, C.A., French, R.G.: Near-infrared photometry of the Saturnian satellites during ring plane crossing. Icarus **136**, 223–231 (1998).

Chapelle, F.H., O'Neill K., Bradley P.M., Methe B.A., Ciufo S.A., Knobel L.L., Lovley D.R.: A hydrogen-based subsurface microbial community dominated by methanogens. Nature **415**, 312–315 (2002).

Chyba, C.F., Phillips, C.B.: Possible ecosystems and the search for life on Europa. Proc. Natl Acad. Sci. **98**, 801–804 (2001).

Clark, R.N.: Water frost and ice: The near-infrared spectral reflectance 0.65–2.5 µm. J. Geophys. Res. **86**, 3087–3096 (1981).

Clark, R.N., Lucey, P.G.: Spectral properties of ice–particulate mixtures and implications for remote sensing. I. Intimate mixtures. J. Geophys. Res. **89**, 6341–6348 (1984).

Collins, G.C., Goodman, J.C.: Enceladus' south polar sea Icarus **189**, 72–82 (2007).

Corliss, J.B., Baross, J.A., Hoffman, S.E.: An hypothesis concerning the relationship between submarine hot springs and the origin of life. Oceanologica Acta **4**. suppl. C4, 59–69 (1981).

Coustenis, A. and 155 co-authors: TandEM: Titan and Enceladus mission. Exper. Astron. **23** (2008). doi: 10.1007/s10686-008-9103-z

Cruikshank, D.P.: Near-infrared studies of the satellites of Saturn and Uranus. Icarus 41, 246–**258** (1980).

Cruikshank, D.P., Owen, T.B., Dalle Ore, C., Geballe, T.R., Roush, T.L., de Bergh, C., Sandford, S.A., Poulet, F., Benedix, G.K., Emery, J.P.: A spectroscopic study of the surface of Saturn's large satellites: H_2O ice, tholins, and minor constituents. Icarus **175**, 268–283 (2005).

Czechowski, L.: Parameterized model of convection driven by tidal and radiogenic heating. Adv. Space Res. **38**, 788–793 (2006).

Davis, W.L., McKay, C.P.: Origins of life: A comparison of theories and application to Mars. Origins Life Evol. Biosph. **26**, 61–73 (1996).

Dermott, S.F., Thomas P.C.: The shape and internal structure of Mimas, Icarus **73**, 25–65 (1988).

Dermott, S.F., Malhotra, R., Murray, C.D.: Dynamics of the Uranian and Saturnian satellite systems: A chaotic route to melting Miranda? Icarus **76**, 295–334 (1988).

Dougherty, M.K., Khurana, K.K., Neubauer, F.M., Russell, C.T., Saur, J., Leisner, J.S., Burton, M.E.: Identification of a dynamic atmosphere at Enceladus with the Cassini magnetometer. Science **311**, 1406–1409 (2006).

Durham, W. B., Kirby S.H., Stern, L.A.: Effect of dispersed particulates on the rheology of water ice at planetary conditions, J. Geophys. Res. **97**, 20,883–20,897 (1992).

Durham, W.B., Stern, L.A.: Rheological properties of water ice-Applications to satellites of the outer planets. Ann. Rev. Earth Planet. Sci. **29**, 295–330 (2001).

Emery, J.P., Burr, D.M., Cruikshank, D.P., Brown, R.H., Dalton, J.B.: Near-infrared (0.8–4.0 µm) spectroscopy of Mimas, Enceladus, Tethys, and Rhea. Astron. Astrophys. **435**, 353–362 (2005).

Esposito, L.W., and 15 co-authors: Ultraviolet imaging spectroscopy shows an active Saturnian system. Science **307**, 1251–1255 (2005).

Feibelman, W. A.: Concerning the "D" ring of Saturn. Nature **214**, 793–794 (1967).

Friedson, A.J., Stevenson, D.J.: Viscosity of rock ice mixtures and applications to the evolution of icy satellites, Icarus **56**, 1–14 (1983).

Gaidos, E.J., Nealson, K.H., Kirschvink, J.L.: Life in ice-covered oceans. Science **284**, 1631–1633 (1999).

Giese, B., Wagner, R., Hussmann, H., Neukum, G., Perry, J., Helfenstein, P, Thomas, P.C: Enceladus: An estimate of heat flux and lithospheric thickness from flexurally supported topography. Geophys. Res. Lett. **35**, L24204 (2008).

Gioia, G., Chakraborty, P., Marshak, S., Kieffer, S.W.: Unified model of tectonics and heat transport in a frigid Enceladus. Proc. Natl Acad. Sci. **104**, 13578–13581 (2007).

Gladman, B., Dones, L., Levison, H.F., Burns, J.A.: Impact seeding and reseeding. in the inner solar system. Astrobiology **5**, 483–496 (2005).

Glein, C.R., Zolotov, M.Y., Shock, E.: The oxidation state of hydrothermal systems on early Enceladus, Icarus **197**, 157–163 (2008).

Goldsby, D.L., Kohlstedt, D.L.: Superplastic deformation in ice: Experimental observations. J. Geophys. Res. **106**, 11,017–11,030 (2001).

Gribb, T.T., Cooper, R.F.: Low frequency shear attenuation in polycrystalline olivine: Grain boundary diffusion and the physical significance of the Andrade model for viscoelastic rheology J. Geophys. Res. **103**, 27,267–27,279 (1998).

Grott, M., Sohl, F., Hussmann, H.: Degree-one convection and the origin of Enceladus' dichotomy, Icarus **191**, 203–210 (2007). doi:10.1016/j.icarus.2007.05.001.

Grundy, W.M., Buie, M.W., Stansberry, J.A., Spencer, J.R., Schmitt, B.: Near-infrared spectra of icy outer Solar System surfaces: Remote determination of H_2O ice temperatures. Icarus **142**, 536–549 (1999).

Haff, P.K., Siscoe, G.L., Eviatar, A.: Ring and plasma – The enigmae of Enceladus. Icarus **56**, 426–438 (1983).

Halevy, I., and Stewart S.T.: Is Enceladus' plume tidally controlled?, Geophys. Res. Lett. **35**, L12203 (2008). doi: 10.1029/2008GL034349.

Hamilton, D.P., Burns, J.A.: Origin of Saturn's E-ring: Self-sustained, naturally. Science **264**, 5158 (1994).

Han, L., Showman, A.P.: Implications of shear heating and fracture zones for ridge formation on Europa, Geophys. Res. Lett. **35**, L03202 (2008). doi: 10.1029/2007GL031957.

Hand, K.P., Carlson, R.W., Chyba, C.F.: Energy, chemical disequilibrium, and geological constraints on Europa. Astrobiology 7, 1006–1022 (2007).

Hansen, C. J., Esposito L., Stewart, A.I.F., Colwell, J., Hendrix, A., Pryor, W., Shemansky, D., West R.: Enceladus' water vapor plume. Science **311**, 1422–1425 (2006).

Hansen, C.J., Esposito, L.W., Stewart, A.I.F., Meinke, B., Wallis, B., Colwell, J.E., Hendrix, A.R., Larsen, K., Pryor, W., Tian, F.: Water vapor jets inside the plume of gas leaving Enceladus. Nature **456**, 477–479 (2008).

Hedman, M.M., Nicholson, P.D., Showalter, M.R., Brown, R.H., Buratti, B.J., Clark R.N.: Spectral observations of the Enceladus plume with Cassini-VIMS. Astron. J. **693**, 1749–1762 (2009).

Helfenstein, P., Thomas, P.C., Veverka, J., Burns, J.A., Roatsch, T., Giese, B., Wagner, R., Denk, T., Neukum, G., Turtle, E.P., Perry, J., Bray, V., Rathbun, J., Porco, C.C.: Tectonism and terrain evolution on Enceladus,. Icarus, in press (2009).

Hoppa, G.V., Tufts, R., Greenberg, R., Geissler, P.E.: Formation of cycloidal features on Europa, Science **285**, 1899–1902 (1999).

Howett, C.J.A., Spencer, J.R., Pearl, J.C., Segura, M.: Thermal inertia and bolometric albedo for Mimas, Enceladus, Dione, Rhea and Iapetus as derived from Cassini/CIRS measurements. Icarus: doi: 10.1016/j.icarus.2009.07.016 (2009).

Hurford, T.A., Helfenstein, P., Hoppa, B.V., Greenberg, R., Bills, B.: Eruptions arising from tidally controlled periodic openings of rifts on Enceladus. Nature **447**, 292–294 (2007).

Hussmann, H., Spohn, T.: Thermal-orbital evolution of Io and Europa. Icarus **171**, 391–410 (2004).

Ingersoll, A.P., Pankine, A.A.: Models of the Enceladus plumes: Criteria for melting. Icarus, submitted (2009).

Jackson I., Faul, U.H., Gerald, J.D.F., Tan B.H.: Shear wave attenuation and dispersion in melt-bearing olivine polycrystals 1. Specimen fabrication and mechanical testing, J. Geophys. Res. **109**, B06201 (2004).

Jacobson, R.A., and 13 co-authors: The gravity field of the Saturnian system from satellite observations and spacecraft tracking data. Astron. J. **132**, 2520–2526 (2006).

Jaumann, R., and 18 co-authors: Distribution of icy particles across Enceladus' surface as derived from Cassini-VIMS measurements. Icarus **193**, 407–419 (2008).

Johnson, R.E., Famá, M., Liu, M., Baragiola, R.A., Sittler, E.C., Smith, H.T.: Sputtering of ice grains and icy satellites in Saturn's inner magnetosphere. Planet. Space Sci. **56**, 1238–1243 (2008).

Johnson, R.E., Smith, H.T., Tucker, O.J., Liu, M., Burger, M.H., Sittler, E.C., Tokar R. L.: The Enceladus and OH tori at Saturn. Astrophys. J. **644**, L137–L139 (2006).

Jones, G.H. and 20 co-authors: Fine jet structure of electrically charged grains in Enceladus's plume. Geophys. Res. Lett., in press (2009).

Juhász, A., Horányi, M.: Saturn's E ring: A dynamical approach. J. Geophys. Res. **107**, 1066 (2002).

Juhász, A., Horányi, M., Morfill, G.E.: Signatures of Enceladus in Saturn's E ring. Geophys. Res. Lett. **34**, L09104 (2007).

Jurac, S., Johnson, R.E., Richardson, J.D.: Saturn's E Ring and production of the neutral torus. Icarus **149**, 384–396 (2001).

Jurac, S., Richardson, J.D.: A self-consistent model of plasma and neutrals at Saturn: Neutral cloud morphology. J. Geophys. Res. **110**, A09220 (2005).

Jurac, S., McGrath, M.A., Johnson, R.E., Richardson, J.D., Vasyliunas, V.M., Eviatar, A.: Saturn: Search for a missing water source. Geophys. Res. Lett. **29**, 25–1 (2002).

Karato, S.: A dislocation model of seismic wave attenuation and microcreep in the Earth: Harold Jeffreys and the rheology of the solid Earth. Pure Appl. Geophys. **153**, 239–256 (1998).

Kargel J. and Pozio, S. (1996) The volcanic and tectonic history of Enceladus. Icarus 119, 385–404.

Kargel, J.S.: Enceladus: Cosmic gymnast, volatile miniworld. Science **311**, 1389–1391 (2006).

Kempf, S., Beckmann, U., Moragas-Klostermeyer, G., Postberg, F., Srama, R., Economou, T., Schmidt, J., Spahn, F., Grün, E.: The E ring in the vicinity of Enceladus. I. Spatial distribution and properties of the ring particles. Icarus **193**, 420–437 (2008).

Khurana, K.K., Dougherty, M.K., Russell, C.T., Leisner, J.S.: Mass loading of Saturn's magnetosphere near Enceladus. J. Geophys. Res. **112**, A08203 (2007).

Kieffer, S. W.: Dynamics and thermodynamics of volcanic eruptions: Implications for the plumes of Io. In: Morrison, D. (ed.) *Satellites of Jupiter*, pp. 647–723. Univ. Ariz. Press, Tucson (1982).

Kieffer, S., Lu, X., Bethke, C.M., Spencer, J.R., Marshak, S., Navrotsky, A.: A clathrate reservoir hypothesis for Enceladus' south polar plume. Science **314**, 1764–1766 (2006).

Kieffer, S.W., Lu, X., McFarquhar, G., Wohletz, K.H.: A redetermination of the ice/vapor ratio of Enceladus' plumes: Implications for sublimation and the lack of a liquid water reservoir. Icarus, submitted (2009).

Kirk, R.L., Soderblom, L.A., Brown, R.H., Kieffer, S.W., Kargel, J.S.: Triton's plumes: Discovery, characteristics, and models. In: Cruikshank, D.P. (ed.) *Neptune and Triton*, pp. 949–989. Univ. Ariz. Press, Tucson (1995).

Lazcano, A., Miller, S.L.: How long did it take for life to begin and evolve to cyanobacteria? Journal of Molecular Evolution **39**, 546–554 (1994).

Lin, L.-H., and 13 co-authors: Long-term sustainability of a high-energy, low-diversity crustal biome. Science **314**, 479–482 (2006).

Lissauer, J.J., Peale, S.J., Cuzzi J.N.: Ring torque on Janus and the melting of Enceladus, Icarus **58**, 159–168 (1984).

Lissauer, J.J., Squyres, S.W., Hartmann, W.K.: Bombardment history of the Saturn system. J. Geophys. Res. **93**, 13776–13804 (1988).

Lunine, J.I., Gautier, D.: Coupled physical and chemical evolution of volatiles in the protoplanetary disk: A tale of three elements. In: Festou, M.C., Keller, H.U., Weaver, H.A. (eds.), *Comets II*, pp. 105–113, Univ. Ariz. Press, Tucson (2004).

Manga, M., Wang, C.-Y.: Pressurized oceans and the eruption of liquid water on Europa and Enceladus. Geophys. Res. Lett. **34**, L07202 (2007).

Matson, D.L., Castillo, J.C., Lunine, J., Johnson, T.V.: Enceladus' plume: Compositional evidence for a hot interior. Icarus **187** 569–573 (2007).

Matsuyama, I., Nimmo, F.: Tectonic patterns on reoriented and despun planetary bodies. Icarus **195**, 459–473 (2008).

McCollom, T.M.: Methanogenesis as a potential source of chemical energy for primary biomass production by autotrophic organisms in hydrothermal systems on Europa. J. Geophys. Res. **104**, 30,729–30,742 (1999).

McCord, T.B., and 11 co-authors: Salts on Europa's surface detected by Galileo's Near Infrared Mapping Spectrometer. Science **280**, 1242 (1998).

McKay, C.P., Porco, C.C., Altheide, T. Davis, W.L., Kral, T.A.: The possible origin and persistence of life on Enceladus and detection of biomarkers in the plume. Astrobiology **8**, 909–919 (2008).

McKinnon, W. B.: On convection in ice I shells of outer solar system bodies, with specific application to Callisto. Icarus **183**, 435–450 (2006).

Melosh, H.J.: The rocky road to panspermia. Nature **332**, 687–688 (1988).

Meyer, J., Wisdom, J.: Tidal heating in Enceladus. Icarus **188**, 535–539 (2007).

Meyer, J., Wisdom, J.: Tidal evolution of Mimas, Enceladus, and Dione. Icarus **193**, 213–223 (2008).

Miller, D.J., Barnash, A.N., Bray, V.J., Turtle, E. P., Helfenstein, P., Squyres, S. W., Rathbun, J. A.: Interactions between impact craters and tectonic fractures on Enceladus and Dione. Workshop on Ices,

Oceans, and Fire: Satellites of the Outer Solar System, held August 13–15, 2007. Boulder, Colorado, LPI Contribution No. 1357, pp. 95–96 (2007).

Miller, S.L.: A production of amino acids under possible primitive Earth conditions. Science **117**, 528–529 (1953).

Mitri, G., Showman, A.P.: A model for the temperature-dependence of tidal dissipation in convective plumes on icy satellites: Implications for Europa and Enceladus, Icarus **195**, 758–764 (2008).

Moore, J.M., Schenk, P.M., Bruesch, L.S., Asphaug, E., McKinnon, W.B.: Large impact features on middle-sized icy satellites. Icarus **171**, 421–443 (2004).

Mousis, O, Gautier, D., Bockelee-Morvan, D.: An evolutionary turbulent model of Saturn's subnebula: Implications for the origin of the atmosphere of Titan. Icarus **156**, 162–175 (2002). doi:10.1016/icar.2001.6782.

Murray, C., Dermott, S.: *Solar System Dynamics*. Cambridge Univ. Press, Cambridge (1999).

Nash, D.B., Carr, M.H., Gradie, J., Hunten, D.M., Yoder, C.F.: Io. In: Burns J.A., Matthews, M.S. (eds.) *Satellites*, pp. 629–688. Univ. Ariz. Press, Tucson (1986).

Neukum, G.: Cratering records of the satellites of Jupiter and Saturn. Adv. Space Res. **5**, 107–116 (1985).

Newman, S.F., Buratti, B.J., Jaumann, R., Bauer, J.M., Momary, T.W.: Hydrogen peroxide on Enceladus. Astrophys. J. **670**, L143–L146 (2007).

Newman, S.F., Buratti, B.J., Brown, R.H., Jaumann, R., Bauer, J., Momary, T.: Photometric and spectral analysis of the distribution of crystalline and amorphous ices on Enceladus as seen by Cassini. Icarus **193**, 397–406 (2008).

Nicholson, P.D., Showalter, M.R., Dones, L., French, R.G., Larson, S.M., Lissauer, J.J., McGhee, C.A., Sicardy, B., Seitzer, P., Danielson, G.E.: Observations of Saturn's ring plane crossings in August and November 1995. Science **272**, 509–515 (1996).

Nimmo, F., Pappalardo, R.T.: Diapir-induced reorientation of Saturn's moon Enceladus. Nature **441**, 614–616 (2006).

Nimmo, F., Gaidos E.: Thermal consequences of strike-slip motion on Europa. J. Geophys. Res. **107**, 5021 (2002).

Nimmo, F., Spencer, J.R., Pappalardo, R.T., Mullen, M.E.: Shear heating as the origin of the plumes and heat flux on Enceladus. Nature **447**, 289–291 (2007).

Ojakangas, G.W., Stevenson, D.J.: Thermal state of an ice shell on Europa. Icarus **81**, 220–241 (1989).

Ojakangas, G.W., Stevenson, D.J.: Episodic volcanism of tidally-heated satellites with application to Io. Icarus **66**, 341–358 (1986).

Orgel, L.E.: The origin of life – how long did it take? Origins Life Evol. Biosphere. **28**, 91–96 (1998).

Pang, K.D., Voge, C.C., Rhoads, J.W., Ajello, J.M.: The E ring of Saturn and satellite Enceladus. J. Geophys. Res. **89**, 9459–9470 (1984).

Pappalardo, R.T., Barr A.C., The origin of domes on Europa: The role of thermally induced compositional diapirism. Geophys. Res. Lett. **31**, L01701 (2004).

Parkinson, C.D., Liang, M.-C., Hartman, H., Hansen, C.J., Tinetti, G., Meadows, V., Kirschvink, J.L., Yung Y.L.: Enceladus: Cassini observations and implications for the search for life. Astron. Astrophys. **463**, 353–357 (2007).

Parkinson, C.D., Liang, M.-C., Yung, Y.L., Kirschvink, J.L.: Habitability of Enceladus: Planetary conditions for life. Origins Life Evol. Biospheres (2008). doi:10.1007/s11084-008-9135-4.

Passey, Q.: Viscosity of the lithosphere of Enceladus. Icarus **53**, 105–120 (1983).

Peale S.J., Cassen P.: Contribution of tidal dissipation to lunar tidal history, Icarus **36**, 245–269 (1978).

Plescia, J.B., Boyce, J.M.: Crater numbers and geological histories of Iapetus, Enceladus, Tethys and Hyperion. Nature **301**, 666–670 (1983).

Porco, C. C., and 24 co-authors: Cassini observes the active south pole of Enceladus. Science **311**, 1393–1401 (2006).

Postberg, F., Kempf, S., Hillier, J.K., Srama, R., Green, S.F., McBride, N., Grün, E.: The E-ring in the vicinity of Enceladus II. Probing the moon's interior—The composition of E-ring particles. Icarus **193**, 438–454 (2008).

Postberg, F., Kempf, S., Schmidt, J., Brillantov, N., Beinsen, A., Abel, B., Buck, U., Srama, R.: Sodium salts in E Ring ice grains from an ocean below Enceladus' surface. Nature **459**, 1098–1101 (2009).

Rappaport, N.J., Iess, L., Tortora, P., Anabtawi, A., Asmar, S., Somenzi, L., Zingoni, F.: Mass and interior of Enceladus from Cassini data analysis, Icarus **190**, 175–178 (2007).

Razzaghi, A., Di Pietro, D., Simon-Miller, A., Spencer, J.: Enceladus Flagship Mission Concept Study. NASA-Goddard Spaceflight Center, http://www.lpi.usra.edu/opag/Enceladus_Public_Report.pdf (2007). Accessed 9 April 2009.

Reh, K., Elliot, J., Spilker, T., Jorgensen, E., Spencer, J., Lorenz, R.: Titan and Enceladus $1B mission feasibility study report. JPL D-37401B (2007).

Roatsch, T., Wählisch, M., Giese, B., Hoffmeister, A., Matz, K.-D., Scholten, F., Kuhn, A., Wagner, R., Neukum, G., Helfenstein, P., Porco, C.: High-resolution Enceladus atlas derived from Cassini-ISS images. Planet. Space Sci. **56**, 109–116 (2008).

Roberts, J.H., Nimmo F.: Tidal heating and the long-term stability of a subsurface ocean on Enceladus. Icarus **194**, 675–689 (2008a).

Roberts, J.H., Nimmo F.: Near-surface heating on Enceladus and the south polar thermal anomaly. Geophys. Res. Lett. **35** L09201 (2008b).

Roddier, C., Roddier, F., Graves, J.E., Northcott M.: Discovery of an arc of particles near Enceladus' orbit: A possible key to the origin of the E ring. Icarus **136**, 50–59 (1998).

Ross, M., Schubert, G.: Tidal dissipation in a viscoelastic planet. Proc. 16th Lunar Planet Sci. Conf. Pt 2, J. Geophys Res. **91**, D447–D452 (1986).

Ross, M.N., Schubert, G.: Viscoelastic models of tidal heating in Enceladus, Icarus **78**, 90–101 (1989).

Rudolph, M.L., Manga, M.: Fracture penetration in planetary ice shells. Icarus **199**, 536–541 (2009).

Saur, J., Schilling, N., Neubauer, F.M., Strobel, D.F., Simon, S., Dougherty, M.K., Russell, C.T., Pappalardo, R.T.: Evidence for temporal variability of Enceladus' gas jets: Modeling of Cassini observations. Geophys. Res. Lett. **35**, 20105 (2008).

Schmidt, J., Brilliantov, N., Spahn, F., Kempf, S.: Slow dust in Enceladus' plume from condensation and wall collisions in tiger stripe fractures. Nature **451**, 685–688 (2008).

Schneider, N.M., Burger, M.H., Schaller, E.L., Brown, M.E., Johnson, R.E., Kargel, J.S., Dougherty, M., Achilleos N.: No sodium in Enceladus' vapor plumes. Nature **459**, 1102–1104 (2009).

Schubert, G., Anderson, J.D., Travis, B.J., Palguta, J.: Enceladus: Present internal structure and differentiation by early and long-term radiogenic heating. Icarus **188**, 345–355 (2007).

Segatz, M.T., Spohn, T., Ross, M.N., Schubert G.: Tidal dissipation, surface heat flow, and figure of viscoelastic models of Io. Icarus **75**, 187–206 (1988).

Shemansky, D.E., Matheson, P., Hall, D.T., Hu, H.-Y., Tripp, T.M.: Detection of the hydroxyl radical in the Saturn magnetosphere. Nature **363**, 329–331 (1993).

Shock, E.L.: Geochemical constraints on the origin of organic compounds in hydrothermal systems. Origins Life Evol Biosphere **20**, 331–367 (1990).

Showalter, M.R., Cuzzi, J.N., Larson, S.M.: Structure and particle properties of Saturn's E Ring. Icarus **94**, 451–473 (1991).

Showman A.P., Stevenson, D.J., Malhotra, R.: Coupled orbital and thermal evolution of Ganymede. Icarus **129**, 367–383 (1997).

Showman, A.P., Han, L.J.: Effects of plasticity on convection in an ice shell: Implications for Europa. Icarus **177**, 425–437 (2005).

Smith, B.A., and 28 co-authors: A new look at the Saturn system: The Voyager 2 images, Science **215**, 504–537 (1982).

Smith, H.T., Shappirio, M., Johnson, R.E., Reisenfeld, D., Sittler, E.C., Crary, F.J., McComas, D.J., Young, D.T.: Enceladus: The likely dominant nitrogen source in Saturn's magnetosphere. Icarus **188**, 356–366 (2008a).

Smith, H.T., Shappirio, M., Johnson, R.E., Reisenfeld, D., Sittler, E.C., Crary, F.J., McComas, D.J., Young, D.T.: Enceladus: A potential source of ammonia products and molecular nitrogen for Saturn's magnetosphere. J. Geophys. Res. **113**, A11206 (2008b).

Smith-Konter, B., Pappalardo R.T.: Tidally driven stress accumulation and shear failure of Enceladus's tiger stripes. Icarus **198**, 435–451 (2008).

Solomatov, V.S., Barr A.C.: Onset of convection in fluids with strongly temperature-dependent, power-law viscosity 2. Dependence on the initial perturbation. Phys. Earth and Planet. Int. **165**, 1–13 (2007).

Sotin, C., Head, J.W., Tobie G.: Europa: Tidal heating of upwelling thermal plumes and the origin of lenticulae and chaos melting. Geophys. Res. Lett. **29**, 1233 (2002).

Spahn, F. and 15 co-authors: Cassini dust measurements at Enceladus and implications for the origin of the E ring. Science **311**, 1416–1418 (2006a).

Spahn, F. Albers, N., Hörning, M., Kempf, S., Krivov, A.V., Makuch, M., Schmidt, J., Seiß, M., Miodrag, S.: E ring dust sources: Implications from Cassini's dust measurements. Planet. Space Sci. **54**, 1024–1032 (2006b).

Spencer, J.R., Pearl, J.C., Segura, M., Flasar, F.M., Mamoutkine, A., Romani, P., Buratti, B.J., Hendrix, A.R., Spilker, L.J., Lopes, R.M.C.: Cassini Encounters Enceladus: Background and the discovery of a south polar hot spot, Science **311**, 1401–1405 (2006).

Spitale, J.N., Porco, C.C.: Association of the jets of Enceladus with the warmest regions on its south-polar fractures. Nature **449**, 695–697 (2007).

Squyres, S.W., Reynolds, R.T., Cassen P.M., Peale, S.J.: The evolution of Enceladus, Icarus **53**, 319–331 (1983).

Stevens, T.O., McKinley J.P.: Lithoautotrophic microbial ecosystems in deep basalt aquifers. Science **270**, 450–454 (1995).

Thomas, P.C. Burns, J.A., Helfenstein, P., Squyres, S., Veverka, J., Porco, C., Turtle, E.P., McEwen, A., Denk, T., Giese, B., Roatsch, T., Johnson, T.V., Jacobson, R.A.: Shapes of the Saturnian icy satellites and their significance. Icarus **190**, 573–584 (2007).

Tian, F., Stewart, A.I.F., Toon, O.B., Larsen, K.W., Esposito, L.W.: Monte Carlo simulations of the water vapor plumes on Enceladus. Icarus **188**, 154–161 (2007).

Tice, M.M., Lowe, D.R.: Photosynthetic microbial mats in the 3,416-Myr-old ocean. Nature **431**, 549–552 (2004).

Tobie, G., Mocquet, A., Sotin, C.: Tidal dissipation within large icy satellites: Applications to Europa and Titan. Icarus **177**, 534–549 (2005).

Tobie, G., Cadek, O., Sotin, C.: Solid tidal friction above a liquid water reservoir as the origin of the south pole hotspot on Enceladus, Icarus **196**, 642–652 (2008). doi:10.1016/j.icarus.2008.03.008.

Tyler, R. H.: Strong ocean tidal flow and heating on moons of the outer planets. Nature **456**, 770–772 (2008).

Verbiscer, A.J., French, R.G., McGhee, C.A.: The opposition surge of Enceladus: HST observations 338–1022nm. Icarus **173**, 66–83 (2005).

Verbiscer, A.J., Peterson, D.E., Skrutskie, M.F., Cushing, M., Helfenstein, P., Nelson, M.J. Smith, J.D., Wilson J. C.: Near-infrared spectra of the leading and trailing hemispheres of Enceladus. Icarus **182**, 211–223 (2006).

Verbiscer, A., French, R., Showalter, M., Helfenstein, P.: Enceladus: Cosmic graffiti artist caught in the act. Science **315**, 815 (2007).

Wächtershäuser, G.: The case for the chemoautotrophic origin of life in an iron-sulfur world. Origins Life Evol. Biosphere **20**,173–176 (1990).

Waite, J. H., and 16 co-authors: Ammonia, radiogenic Ar, organics, and deuterium measured in the plume of Saturn's icy moon Enceladus. Nature **460**, 487–490 (2009).

Waite, J. H., and 13 co-authors: Cassini Ion and Neutral Mass Spectrometer: Enceladus plume composition and structure. Science **311**, 1419–1422 (2006).

Weiss, B.P., Kirschvink, J.L., Baudenbacher, F.J., Vali, H., Peters, N.T., Macdonald, F.A., Wikswo J.P.: A low temperature transfer of ALH84001 from Mars to Earth. Science **290**, 791–795 (2000).

Wisdom, J.: Spin-orbit secondary resonance dynamics of Enceladus. Astron. J. **128**, 484–491 (2004).

Zahnle, K., Schenk, P., Levison, H., Dones L.: Cratering rates in the outer Solar System. Icarus **163**, 263–289 (2003).

Zolotov, M. Y.: An oceanic composition on early and today's Enceladus, Geophys. Res. Lett. **34**, L23203 (2007). doi:10.1029/2007GL031234.

Zschau, J. Tidal friction in the solid Earth: Loading tides versus body tides. In: Brosche P., Sundermann, J. (eds.) Tidal Friction and the Earth's Rotation. Springer, Berlin (1978).

Chapter 22
The Cassini Extended Mission

David A. Seal and Brent B. Buffington

Abstract Based on the overwhelming success of the Cassini/Huygens 4-year tour of Saturn from July 2004 to June 2008, NASA Headquarters approved at least two years of extended mission for continued study of the target-rich Saturnian system. After a rigorous phase of science objective definition and trajectory design and analysis, the Cassini project initiated an efficient, scientifically intense and operationally challenging mission phase, including 60 orbits around Saturn, 26 close Titan flybys, and 10 close icy satellite flybys – including seven more flybys of Enceladus. At the conclusion of the 2-year extended mission, substantial operating margins should be present with some fascinating options for further extensions.

22.1 Introduction

The Cassini-Huygens program, a joint mission between the National Aeronautics and Space Administration (NASA), the European Space Agency (ESA), and the Italian Space Agency (ASI), completed an outstanding prime mission in June of 2008. After a highly successful Saturn Orbit Insertion (SOI) in July of 2004 and Huygens Probe descent to the surface of Titan in January of 2005, Cassini went on to complete a 4-year tour of Saturn, its rings, satellites, and magnetosphere via 75 orbits around Saturn, 45 close Titan flybys, 10 close icy satellite encounters, and dozens of radio, stellar, and solar occultations by Saturn, the rings, and Titan (Cassini Extended Mission Navigation Plan 2008; Buffington et al. 2008).

Saturn is arguably the most target-rich environment in the Solar System, with 61 known moons, an active Saturn atmosphere, magnetosphere, and network of aurorae, a highly dynamic and diverse ring system, a prominent system of faint rings and ringlets, active water ice geysers spouting from the south polar region of Enceladus, and a very large satellite – Titan – with the only appreciable atmosphere of any moon and more Earth-like features than any other extraterrestrial body in the Solar System. Nowhere else beyond Earth can one find in one place so many intriguing phenomena with dynamics on the time scale of a human lifetime.

By the end of its prime mission, Cassini had demonstrated a sustained and staggering pace of scientific discovery, with nearly one thousand independent publications in prominent science journals. This was made possible by highly successful spacecraft operations and subsystem maintenance, sequencing, instrument monitoring, and risk management. In particular, the Cassini Navigation Team has excelled in negotiating the most complex gravity-assist tour in the history of unmanned space flight, affording the many geometric opportunities present in its 4-year tour. By the end of the prime mission, including all of the encounters, Saturn periapses, maneuvers, occultations, sequence uploads, and flight software updates, Cassini averaged two key events per week, a pace of operations unmatched by any other deep space mission.

Through all of these activities, all spacecraft subsystems and instruments remained healthy, and substantial margins existed (and had been projected) in both bi- and mono-propellant propulsion systems used for orbital maneuvering and attitude control. Lastly, power levels from Cassini's Radioisotope Thermal Generators (RTGs) allowed for continued science operations for many more years.

In 2006, NASA Headquarters was considering possible continued operations for Cassini and debating four key questions common to any mission extension:

- Is the spacecraft likely to remain operational?
- Is the project capable of continued operations and science sequencing?
- Will the remaining consumables permit further operations?
- Does the potential science return justify the required expenditures of resources?

D.A. Seal and B.B. Buffington
Jet Propulsion Laboratory, California Institute of Technology, Pasadena, CA

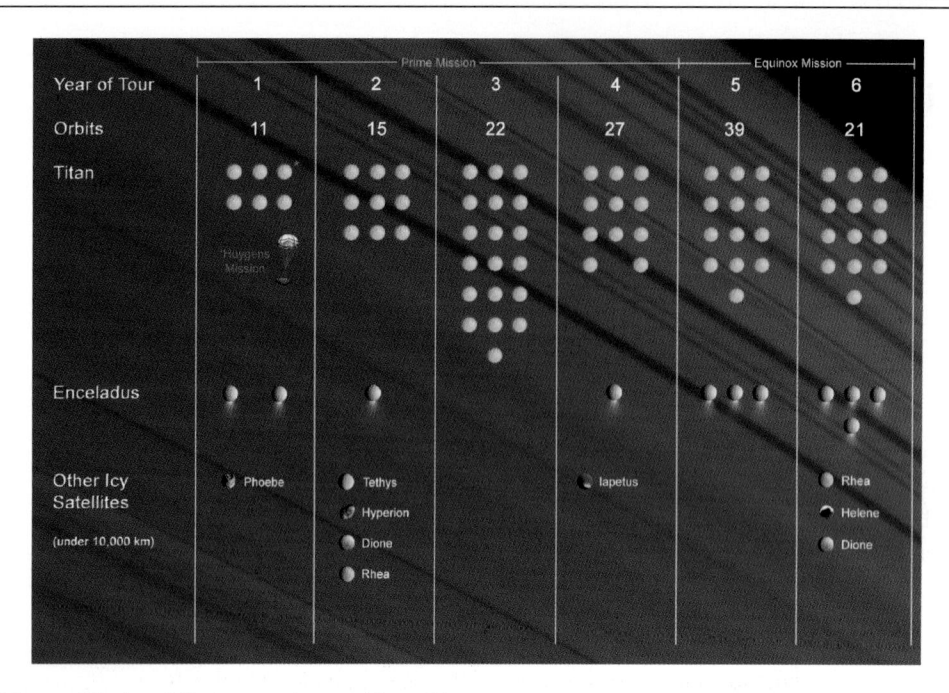

Fig. 22.1 Cassini Prime and Equinox Mission overview – orbits and key encounters

To answer these questions, in early 2006 NASA Headquarters funded the investigation of a possible 2-year extension to Cassini's prime mission. At the conclusion of these studies, and more specifically through reviews held in 2007 and 2008, the Cassini program answered all four of the above questions in the affirmative. Therefore, NASA officially approved Cassini's 2-year Equinox Mission (EM), to be conducted from July 2008 through June 2010, whose purpose was to continue scientific investigations at the same or greater pace as the prime mission. The mission contains 60 orbits around Saturn, dozens of additional radio, stellar, and solar occultations of Saturn, the rings, and Titan, 26 close Titan flybys, and 10 close icy satellite flybys – including seven more flybys of Enceladus. (See Fig. 22.1.)

The mission is so named because it carries Cassini through the Saturn north hemispheric vernal equinox in August of 2009, when the Sun passes through Saturn's ring plane. This is a key geometric event with unique scientific merit to a variety of studies, and a symbol for Cassini's continued study of Saturn through its nearly 30-year seasonal cycle.

Even though this mission is technically an extension, it is far more similar in scope and funding profile to the prime mission, unlike many programs who pass beyond their official end date. The pace of science observation, and as a result, the pace of navigation and operations continue at the same or arguably greater level as the prime mission, with only minor modifications to sequencing strategies. The EM could therefore be thought of as a direct extension to the prime mission and is not a radically different phase of the project.

22.2 Equinox Mission Design Overview

The Cassini science community adheres to a discipline-driven model, composed of five teams representing the study of Saturn, Titan, the icy satellites, the magnetosphere, and the rings. This organization is designed to focus science teams on how best to study phenomena within their domain, rather than merely on what specific measurements one instrument can make, resulting in an efficient strategic and balanced approach to scientific observation. Each discipline is given equal weight, and decisions are made by consensus of the Project Science Group (PSG) wherever possible, or by the Project Scientist or Project Manager where necessary.

Since NASA HQ had directed that science continue at a nearly even pace, with a commensurate funding profile, few fundamental trajectory design or operational changes were required. Furthermore, NASA indicated no particular preference of scientific discipline or target. Therefore, the Equinox Mission was derived exclusively from the science objectives, operations and safety constraints developed within the project.

Starting in the spring of 2006, the PSG – organized in terms of the five disciplines – was solicited for science objectives for the EM. At the same time, a handful of engineering team reviews were held to study the operational and safety (i.e., environmental hazard) constraints and update them based on lessons learned during the prime mission to date. These data were given to the tour designers and an iterative process was begun, where trajectories were developed and rated versus the science objectives and constraints, resulting in an ever-increasing maturation of tour design.

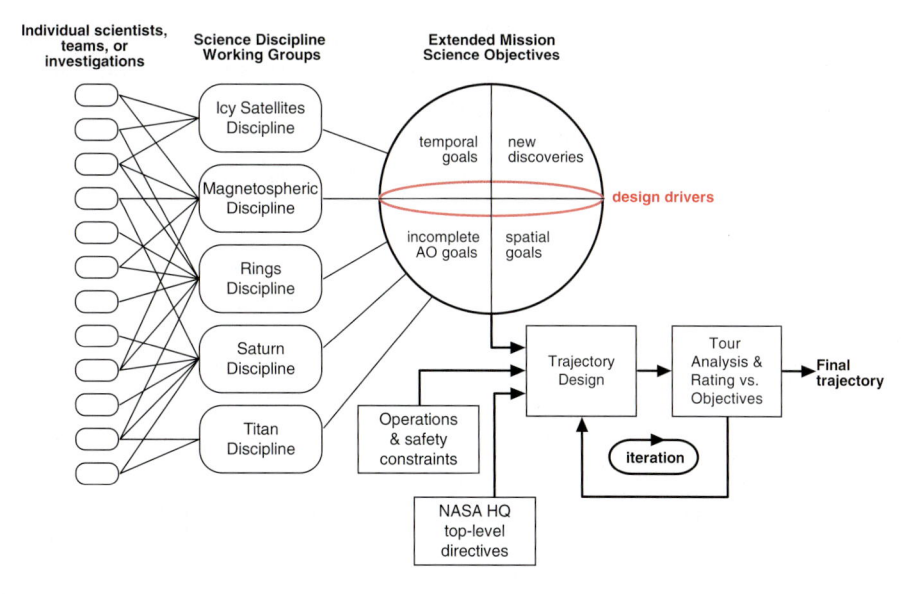

Fig. 22.2 Equinox Mission development process

Each major iteration of candidate tours was tied to a Cassini Project Science Group (PSG) meeting which took place every three to four months during the design period. Prior to each of these meetings, about a half dozen candidate tours were released to the science community for evaluation and discussion at the subsequent meeting. Twelve classes of tours were identified which encompassed the likely range of possible designs, and during the 1.5 year design period, 29 candidate tours spanning nine of these classes were designed. Each iteration of tours significantly improved the science return as evaluated by the Cassini science discipline working groups, and the feedback obtained from each meeting heavily influenced the subsequent crop of tours. At each stage, every tour was rated based on its ability to address each science objective, and the results were reviewed by the full PSG group together with engineering representatives across all project elements. Five full iterative loops were conducted, beginning with the most primitive "tour classes" and ending with a polished trajectory complete with encounter-specific targeting designed to optimize all science. The development of the Equinox mission followed the process illustrated in Fig. 22.2.

The value of direct interaction between the trajectory designers and the PSG cannot be overstated. No written record can ever fully convey the nuances inherent in any one science objective – whether it be the criticality of one aspect of the required geometry or timing, the flexibility of another, or how the objective could improve our basic understanding about the Saturnian system (i.e., "the story" behind the objective).

22.3 The Equinox Mission Science Objectives

In total, 75 objectives were developed and delivered to the project in detail to dictate the equinox mission design (Cassini Extended Mission Science Objectives 2007). In addition to being discipline-specific, these objectives also fall within four distinct categories, also shown in Fig. 22.2: temporal campaigns – intended to continue the time baseline of existing studies, or determine the effects of seasonal change on the system; geometric campaigns – designed to cover geometrical ranges un- or poorly sampled during the prime mission; prime mission goal completion – to round out the AO goals not fully completed by the first 4 years of tour; and objectives developed in response to new discoveries made by Cassini. In addition, many of the science objectives, with particular emphasis on high-resolution Titan and Enceladus coverage, lend themselves to providing essential data necessary for future missions to the Saturnian system. As is typical, some objectives cross disciplinary boundaries, and/or fall into more than one of the above categories.

Due to the high number and intricacy of science objectives and the complexity of the Saturnian system, a subset of scientific objectives referred to as design drivers were identified to guide the scope of the trade space and interactions between objectives (Buffington et al. 2008). Design drivers were defined as scientific objectives that required observational geometries that narrowed the tour design trade space; not all science objectives were design drivers. This set of design drivers was further separated into two categories: time critical and geometric. Time critical design drivers consisted of observations that could be carried out only during a specific time period. Geometric design drivers consisted of scientific objectives that could be carried out at any time during the EM, as long as the geometric constraint(s) were met. Identifying these characteristic differences allowed the tour designers to strategically incorporate specific drivers, discern which drivers were mutually exclusive, and which drivers were compatible.

The science objectives that were considered to be of the highest priority and/or design drivers for the EM are described below. Each objective is labeled according to the discipline(s) it falls within, as well as the applicable category or categories.

22.3.1 Icy Satellite Objectives

- Complete the global mapping of Enceladus (500 m/pixel or better) at small to moderate phase angles (10–60°), specifically of the leading hemisphere and north pole, to complete the understanding of global tectonic and geological patterns, and their stratigraphy. [*Spatial coverage*].
- Observe Enceladus' south polar terrain at very high resolution (<20 m/pixel), tiger stripes in particular, to investigate detailed morphology. Requires flybys under the south pole which are close, but far enough so target motion compensation is possible. [*Spatial coverage, new discoveries*].
- Explain the endogenic activity of Enceladus and morphology via remote sensing of hot spots, in the infrared in particular, at higher resolution (6 km at spacecraft range of 20,000 km). Also, search for temporal changes and compositional variations along the tiger stripes, important for landing site selection of future missions. [*Spatial, temporal coverage, new discoveries*].
- Monitor temporal changes in Enceladus' plumes via remote sensing during multiple, temporally spaced encounters and at high phase angles. Also, determine the plumes' connection to time variability in the O_2 environment. Stellar occultations of the plume are important. [*Temporal coverage, new discoveries*].
- Measure detailed composition of the plume and other icy satellites, Dione in particular, via in-situ collection of dust and gas from the surface at very close range (within one body radius, preferably downstream of the magnetospheric wake). [*Spatial coverage, new discoveries*].
- Explore the differentiation of Enceladus via a close gravity pass, to determine the basic composition of Enceladus, the timing of its formation and the rapidity of accretion. Requires a close (250 km or closer) dedicated pass. [*Spatial coverage*].

22.3.2 Magnetospheric Objectives

- Pass through the auroral and electro-static discharge regions, located at latitudes between 68–80° and several Saturn radii (Rs[1]), multiple times to increase the probability that Cassini will be close during the highly time variable active periods. [*Temporal coverage*].

---
[1] Rs = 60,330 km (Saturn equatorial radius at 100 mbar)

- Study Saturn's magnetosphere in the poorly sampled dusk sector (looking down on Saturn, with the sun at top, this would be to the left), to study the flow pattern of plasma and shed light on the relative importance of co-rotation and solar wind driven convection in the magnetosphere. [*Spatial coverage*].
- Conduct continuous, dedicated plasma and particle measurements across inner magnetosphere (inclined and equatorial orbits with periapsis <5 Rs). [*Incomplete AO objective*].
- Obtain a continuous, dedicated dust and mapping campaign across full extent of E ring (radially and vertically, at low inclination). [*Incomplete AO objective*].
- Measure the dust flux during close flybys of Enceladus (within one body radius, away from plume) and Rhea to determine the flux of interplanetary dust into the Saturnian system. [*Spatial coverage*].
- Determine the nature of the Enceladus–plume–magnetospheric interaction via in-situ passages through/near plume and Enceladus + plume wake. This would benefit follow-on missions designed to orbit or land on Enceladus. [*Spatial coverage, new discovery*].
- Resolve the nature of Dione–magnetospheric and Dione–E ring interactions via wake mapping with a downstream, close (<1100 km) flyby of Dione. [*Spatial coverage*].
- Complete the understanding of the interaction of Saturn's magnetosphere with Titan, i.e., Titan's induced magnetotail and wake, via a relatively close (5,100–7,600 km) encounter in the wake. [*Spatial coverage, incomplete AO objective*].
- Sound Titan's lower ionosphere and interior via a flyby as close as can be done safely. [*Spatial coverage*].

22.3.3 Rings Objectives

- During equinox, observe new and unexpected vertical structure, ring and moonlet shadowing, thermal behavior and relative trends of the various rings, and other effects. This event occurs only once every 15 years. [*Temporal coverage*].
- Conduct low opening angle radio occultations of Saturn's diffuse rings near equinox within 10 Rs for superior signal to noise measurements not achievable in other geometries. This may result in detection of diffuse material in gaps, the D ring, the F ring region, and beyond. [*Spatial coverage*].
- Observe changes in ring density waves from the Janus-Epimetheus orbit swap in Jan 2010. The resulting altered forcing of density waves will change the locations of existing waves and allow the observation of a density wave from scratch. [*Temporal coverage*].

- Observe Prometheus' invasion of the F ring in Dec. 2009 and its gravitational interactions. The extent to which Prometheus will invade the F ring recurs only once every 17 years. *[Temporal coverage]*.
- Determine the mass flux of interplanetary meteoroids into the Saturnian system via close flybys (<1,000 km) of larger moons. Rhea and Tethys are the most attractive targets as they lie further from the center of the E ring. *[Spatial coverage]*.
- Implement multiple highly inclined orbits to supplement the infrequent and poor quality ring imaging and stellar occultations to date. *[Spatial coverage]*.
- Conduct low opening angle stellar occultations of diffuse rings to supplement the infrequent and poor quality stellar occultations to date. *[Spatial coverage]*.
- Attain high resolution (<10 Rs), low phase angle (<45°) observations for embedded ringmoon searches and observations of newly discovered features in the rings. *[Spatial coverage]*.

22.3.4 Saturn Objectives

- Conduct continuous mapping of two full Saturn rotations in infrared and radar wavelengths at low inclination and near periapsis (within 6 Rs) as well as low to moderate inclination at 20 Rs to map the deep atmosphere, stratosphere and troposphere. *[Spatial coverage]*.
- Obtain close inclined orbits of Saturn (2 within 4 Rs between 20–70° inclination, at different locations, unocculted) to fully determine the interior gravity and magnetic field. *[Spatial coverage]*.
- Complete a latitudinally comprehensive set of radio occultations of Saturn atmosphere to fully map temperature, pressure, electron density, and microwave absorption. *[Spatial coverage]*.
- Complete a latitudinally comprehensive set of stellar and solar occultations of Saturn atmosphere for temperature and compositional mapping of the thermosphere. *[Spatial coverage]*.
- Maintain spacecraft and consumables beyond equinox to extend the time baseline of Saturn observations towards northern summer solstice. *[Temporal coverage]*.

22.3.5 Titan Objectives

- Repeat radar mapping of Titan lakes and fluvial channels for temporal changes, to study the precipitation cycle and seasonal atmospheric influence. *[Temporal coverage, new discoveries]*.
- Conduct a Titan encounter at the lowest altitude possible to study Titan's ionosphere and search for internal magnetic field signatures which may only be detectable significantly below the ionosphere (≤1,000 km). *[Spacial coverage]*.
- Observe the evolution of Titan's polar vortex, including changes in atmospheric chemistry. Compare and contrast the vortex with winter polar vortices in Earth's stratosphere. *[Temporal coverage]*.
- Complete the understanding of Titan's induced magnetotail and wake, building on earlier incomplete observations. Wake crossings are very diagnostic of Titan's interaction with the magnetosphere. *[Spatial coverage]*.
- Expand the radar surface coverage of Titan for age, cratering, topographical and cryovolcanism studies, as well as global inventory of precipitation and fluvial activity. *[Spatial coverage]*.
- Complete latitudinally comprehensive set of radio and stellar occultations, and thermal and in situ observations of Titan's atmosphere (required for confidence in future aerobraking/aerocapture) *[Spatial coverage]*.
- Obtain further close flybys of Titan to determine higher-order internal structure to understand Titan's origin. *[Spatial coverage]*.
- Determine the global properties of Titan's atmosphere via a distribution of low-, mid- and high-latitude flybys to fill in gaps from the prime mission. *[Spatial coverage]*.
- Complement the four Titan bistatic surface scattering flybys in the prime mission via a bistatic (50–70° incidence angle) scattering observation to cover likely surface compositions. *[Spatial coverage]*.

22.4 Operational and Safety Constraints

In addition to the challenges pertaining to balancing and assessing the quantity of complex, high-level science objectives levied on the tour design, operational and safety constraints also played an important role. Some constraints were a straightforward consequence of the orbital mechanics, while others pertain to spacecraft and/or ground system operational limitations (Wolf 2002).

Ring Plane Crossings: Saturn's rings are a fascinating plexus of moonlets, dust, water-ice, and other constituents with distinct structures varying temporally and spatially in size and abundance. While the solar system's most massive and diverse ring system is one of intrigue from a scientific investigation point of view, much of the system – the main rings, clearly – is too hazardous to navigate. Specifically, traversing the ring plane inside 2.35 Rs (i.e. inside the outer edge of the F ring) and between 2.7 and 2.9 Rs (i.e. the G ring) must be avoided. In addition, due to the nonzero thickness of the

G ring (several hundred km), all low inclination ring plane crossings near the G ring were evaluated to ensure that the spacecraft did not penetrate too deep into the G ring. Lastly, the total probability of loss of mission due to particle impact must not exceed 1.25% per year, per project requirement (Cassini Project Policies and Requirements 2001).

Minimum Flyby Altitudes: Due to torques imparted to the spacecraft by the dense atmosphere that completely enshrouds Titan, a minimum Titan altitude was imposed in order to maintain control authority throughout each flyby. During the design of all EM candidate tours, the minimum Titan flyby altitude was set at 1,000 km based on the Titan Atmospheric Modeling Working Group's (TAMWG) best estimate of Titan's atmosphere and the predicted spacecraft control capability over the EM. Once the final tour was selected, the Attitude and Articulation Control Subsystem (AACS) team then determined the actual flyby-specific altitudes based on the TAMWG atmospheric model, spacecraft attitude profiles, and thruster control authority.

For Enceladus, the minimum flyby altitude was set to 100 km for flybys with closet approach latitudes $\geq 70°S$ (i.e., flybys that penetrate deep into Enceladus' plume), and 25 km for all other flybys. The lower bound for all other icy satellite flyby altitudes was set at 100 km. This altitude was derived in part from navigation delivery accuracies as well as a hazard study conducted by Cassini's Enceladus Plume Working Group, which concluded that no hazardous large material should be present in the plume at such altitudes.

Solar Conjunction: Coherent two-way Doppler measurements (~6–8 hrs. in duration) at regular intervals (~1 day) are required to accurately determine the orbit(s) of the spacecraft to the precision necessary to navigate a tour of similar intensity and aggressiveness as the prime mission (Cassini Extended Mission Navigation Plan 2008). However, during solar conjunction (when the Sun-Earth-Probe angle (SEP) $\leq 7.5°$), the radio signal must propagate through the solar plasma, making the data noisy and of limited (if any) use. This phenomenon puts restrictions on certain mission events, namely maneuvers and flybys. Specifically: (1) prior to solar conjunction, at least two good days of tracking (without flybys or other maneuvers) must be obtained at SEP $\geq 10°$ when a maneuver is to be performed; (2) no maneuver can be performed below 5° SEP (telemetry data cannot be downlinked and scheduling backup maneuver windows are troublesome); and (3) after solar conjunction, maneuvers must be at least 3 days after 10° SEP to allow for 2 days of tracking plus an additional day to design the maneuver. In addition, the lower the SEP angle at a flyby near conjunction, the higher the flyby altitude must be due to the growth in error from the tracking at 10° SEP used for the approach maneuver.

Time of Flight between Flybys: A sufficient amount of time is needed between successive targeted flybys to accurately determine the spacecraft's orbit and design, uplink, and perform two deterministic[2] and one statistical orbital trim maneuvers (OTMs). This requirement puts a lower limit on the time-of-flight (*tof*) between encounters. As with the prime mission, this limit was set to 16 days for the EM (since staffing levels would remain the same). An exception to this rule is a scenario referred to as a "double flyby" in which two flybys occur sufficiently close (~2–3 days) to one another that no OTMs can be performed between the two encounters. Instead, only one body is targeted (Buffington et al. 2005 and Cassini Extended Mission Navigation Plan 2008). The other flyby, which is not directly targeted, must have a higher flyby altitude. Double flybys were proven navigationally feasible and highly valuable from a science return point of view during the prime mission.

Available Propellant: The predicted Δv from the bi-propellant propulsion system available at the beginning of the EM was approximately 351 m/s.

Mission Duration: As previously stated, NASA HQ directed a 2-year duration for the Cassini EM (01-Jul-2008–30-Jun-2010).

Initial conditions: The prime mission ended with a highly inclined short period (7.1 days) resonant Titan-to-Titan outbound transfer, having a vacant ring plane crossing (i.e., the node crossing not at Titan's orbital radius) between the F and G rings, and a apoapsis orientation of approximately 11:00 local Saturn time (LST) (refer to Fig. 22.3). The first Titan flyby (T45) in the extended mission is an outbound flyby at the ascending node, with a v_∞ of 5.879 km/s.

22.5 Tour Design and Development Process

For Saturn-centered gravity-assist tours with a high v_∞ such as Cassini, Titan is the "tour-engine". Titan is over 58 times more massive than Saturn's second largest moon, Rhea, and is the only Saturnian satellite sufficient in mass to power the Cassini tour. A single 1,000 km altitude Titan flyby provides the Cassini spacecraft a gravity-assist Δv (or change in velocity) of approximately 800 m/s, compared with typical Cassini maneuvers that typically span the comparatively diminutive range of 0.2–20 m/s. As a result, gravity-assist tours are built as a sequence of Titan-to-Titan transfers: even if another moon is encountered, the leg is designed principally to return to Titan. If Cassini can't return to Titan, it has "fallen off the tour" and must either execute a poten-

---

[2] Special cases do exist where only one deterministic maneuver is utilized between targeted encounters.

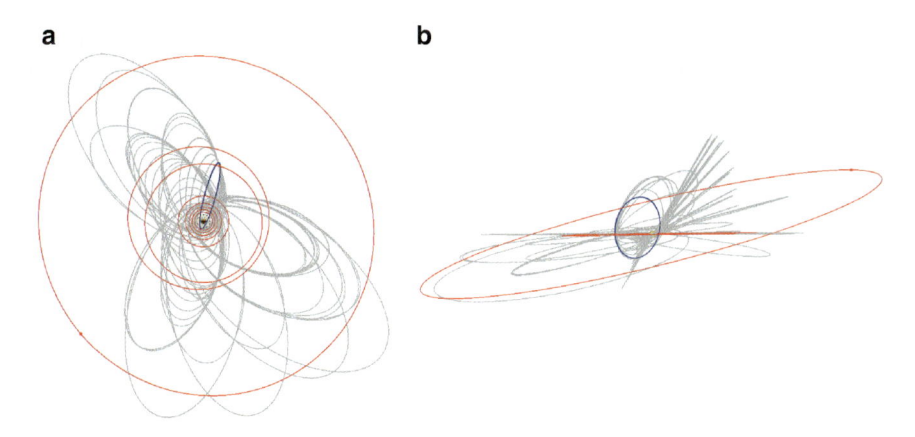

Fig. 22.3 4-year Cassini prime mission (Grey – Cassini, Red – Eight inner major satellites, Mimas through Iapetus) with the last transfer highlighted in blue; (**a**) Saturn north-pole view (sun direction towards *top* of page), and (**b**) Saturn equatorial-plane view

tially large maneuver to return to Titan or remain in the same Saturn-relative orbit.

The requirement to always return to the same moon results in three[3] basic types of same-body transfers: resonant, non-resonant, and pi-transfers (Uphoff et al. 1974; Strange and Sims 2001).

A *resonant transfer* has a time-of-flight that is an integer multiple of the gravity-assist body's period and the flybys at either end of the transfer occur at the same place in the gravity-assist body's orbit. Because of this, the encounters' longitude occurs on a fixed line from the central body and the resonant transfer may be inclined. A resonant transfer is typically labeled as *m:n*, where *m* is the number of gravity-assist body revs and *n* is the number of spacecraft revs.

A *non-resonant transfer*'s time-of-flight is not an integer multiple of the gravity-assist body's orbit. Other than the special case of a pi-transfer, the flybys of a non-resonant transfer do not occur at the same longitude in the gravity-assist body's orbit and, therefore, constrain the spacecraft's orbit plane to be the same as the gravity-assists body's orbit plane.

A *pi-transfer* is a special case of a non-resonant transfer where the time-of-flight of the transfer is *m* plus one-half times the gravity-assist body's period. The flybys of a pi-transfer occur on a line passing through the central body, hence these transfers can be inclined. In fact, they typically must be inclined with a specific inclination determined by the v_∞ magnitude (Strange and Sims 2001). A pi-transfer changes the longitude of the encounter by 180° (i.e., ΔLST = 12 h).

The development of gravity-assist satellite tours consisted of three major steps: (1) Initial tour design using patch conics or patch-integrated techniques (Buffington and Strange 2007), (2) numerically integrating and converging the gravity-assist sequence in a high-fidelity optimizer, and (3) further refinement of the trajectory attempting to maximize science return while minimizing Δv costs.

The first step was carried out using a computer program that patches together central body integrated trajectories via zero sphere-of-influence patch-conics modeling each gravity assist (i.e., Titan flybys) (Buffington and Strange 2007). The program is highly interactive, allowing the user to specify, from a finite range of possibilities (constrained by a limited bending angle), flyby conditions that target different subsequent flybys and modify the post-flyby central body orbit(s). In this way, tours can be built sequentially, one transfer as time, in a very quick and fairly precise manner, to evaluate the efficiency and scientific quality of different routes. However, the complexity in choosing a sequence of gravity-assist flybys and associated orbital transfers between the flybys (i.e., the pathfinding problem (Strange, N., and Sims 2001)) grows geometrically as the number of flybys increases (Strange et al. 2007). The combinatorics of the problem quickly overwhelmed attempts to computationally solve the problem. While finding an optimal solution to this problem is extraordinarily difficult, a graphical method for designing transfers between the same gravity-assist body was developed (out of necessity for the Cassini EM) which collapsed the large and complex space of possible trajectories to a map (Strange et al. 2007). This "tour map," coupled with the feedback and visualization available from the patched-integrated gravity-assist program, allowed the tour designers to efficiently develop the intuition necessary to reduce the vast design-space and pursue only the most auspicious satellite tours (i.e., tours dense in scientific return and acceptable Δv costs).

Once a promising tour has been built, the next step is to converge the conic approximation into a fully numerically integrated trajectory. This is accomplished through a multi-mission optimization program that numerically integrates the equations of motion of a point-mass spacecraft subject to

---
[3] With large maneuvers or large third-body perturbations, leveraging transfers or other tricks are possible.

gravitational accelerations due to any combination of Sun, planets, and satellites of the central body, including their oblateness (Hatfield and Rinderle 2001). As with all existing trajectory optimization techniques/programs, the program requires the sequence of gravity assists and other necessary trajectory quantity inputs to be determined a priori. Hence, the program only determines the optimal solution for the specified sequence of gravity assists, where "optimal" in this case refers to the minimization of the total deterministic Δv. Since the initial guess only includes central body motion with J2 perturbations, velocity and position discontinuities will be present when first implemented into the full force model. In order to achieve convergence, the optimizer must drive the position discontinuities to zero. Velocity discontinuities at user specified points/epochs between flybys represent the deterministic maneuvers needed to fly the trajectory.

The third phase of the design process involves fine-tuning various aspects of the tour using different combinations of user specified constraints to further maximize the number of science objectives met while minimizing additional Δv expenditures and adhering to operational constraints. Typical examples included, adjusting central body inclination profiles, lowering and converting distant non-targeted satellite flybys into targeted or double flybys (Buffington et al. 2005; Buffington and Strange 2006), finding near-optimal maneuver locations, and reducing spacecraft safety risks during potentially hazardous ring plane crossings.

22.6 The Equinox Mission Trajectory

In June of 2007, after one and a half years of development and five iteration cycles between the tour designers and the PSG, a tour was chosen which addressed the majority of the top priority objectives and used significantly less than the total available propellant. Tour refinement had progressed greatly from each stage to the next, and there was general agreement amongst the science community that little further optimization was possible even with additional time and resources.

The Cassini prime mission was the most complex gravity-assist tour ever flown. The equinox mission maintains this level of design and navigational complexity in order to meet and balance the myriad of disparate scientific objectives requested by the Cassini discipline working groups within the 2-year time frame. As a result, the equinox mission trajectory, when compared to the prime mission, clearly meets or exceeds the intensity of scientific opportunities. This conclusion is apparent with a brief glance at the pace of encounters and orbits as shown in Fig. 22.1. The 2-year extension consists of 26 close Titan flybys, 11 close icy satellite flybys (of which seven are of Enceladus), 60 orbits around Saturn, and dozens more radio, solar and stellar occultations of Saturn, its rings, and Titan. The EM trajectory is shown in Fig. 22.4, and close encounters are listed in Table 22.1.

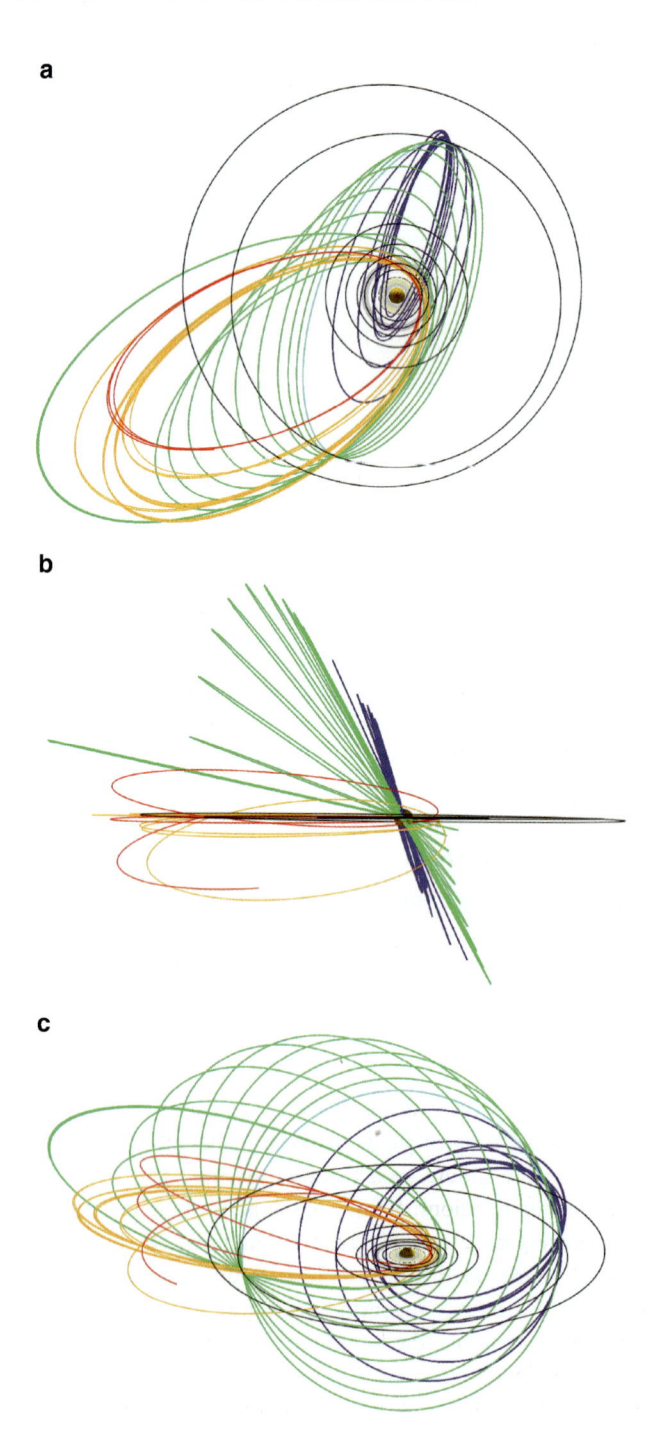

Fig. 22.4 Two-year Cassini extended mission trajectory; (**a**) Saturn north-pole view (sun direction towards *top* of page), (**b**) Saturn equatorial-plane view, and (**c**) oblique view. Key to phases: blue – high inclination, cyan – pi-transfer, green – equinox viewing, orange – icy satellite flybys and ansa-to-ansa occultations, and red – high northern Titan groundtracks

Table 22.1 Cassini extended mission encounters (Buffington 2008)

Encounter	Body	Date[ET]	Altitude (km)	B-Plane Angle (deg)	In/Out	V-inf (Km/s)	Period (days)	Inc. (deg)	Periapsis (RS)
T45/78Ti	Titan	31 Jul 2008 02:14:16	1,613	126.8	O	5.9	7.4	74.4	3.94
E4/80En	Enceladus	11 Aug 2008 21:07:24	50	90.0	I	17.7	7.4	74.5	3.95
E5/88En	Enceladus	09 Oct 2008 19:07:45	25	90.0	I	17.7	7.3	74.5	3.94
E6/91En	Enceladus	31 Oct 2008 17:15:56	200	90.0	I	17.7	7.3	74.5	3.95
T46/91Ti	Titan	03 Nov 2008 17:36:28	1,100	6.6	O	5.9	8.0	70.5	4.27
T47/93Ti	Titan	19 Nov 2008 15:57:33	1,023	151.1	O	5.9	8.0	72.2	5.21
T48/95Ti	Titan	05 Dec 2008 14:26:50	960	164.1	O	5.9	8.0	72.9	5.76
T49/97Ti	Titan	21 Dec 2008 13:00:57	970	110.2	O	5.9	9.6	69.0	9.09
T50/103Ti	Titan	07 Feb 2009 08:51:57	960	59.9	O	5.9	11.9	65.3	13.57
T51/106Ti	Titan	27 Mar 2009 04:44:42	960	114.0	O	5.9	16.0	61.7	19.67
T52/108Ti	Titan	04 Apr 2009 01:48:53	4,150	178.5	I	5.5	16.0	61.7	18.69
T53/109Ti	Titan	20 Apr 2009 00:21:51	3,600	169.3	I	5.5	16.0	61.4	16.94
T54/110Ti	Titan	05 May 2009 21:27:47	3,244	158.4	I	5.5	16.0	60.6	15.07
T55/111Ti	Titan	21 May 2009 21:27:47	965	146.8	I	5.5	16.0	58.8	12.17
T56/112Ti	Titan	06 Jun 2009 20:01:06	965	132.8	I	5.5	15.9	55.7	9.52
T57/113Ti	Titan	22 Jun 2009 18:33:41	955	121.9	I	5.5	15.9	51.2	7.21
T58/114Ti	Titan	08 Jul 2009 17:05:09	965	113.2	I	5.5	15.9	44.5	5.30
T59/115Ti	Titan	24 Jul 2009 15:35:09	955	106.2	I	5.5	15.9	34.8	3.85
T60/116Ti	Titan	09 Aug 2009 14:04:59	970	100.0	I	5.5	16.0	21.1	2.93
T61/117Ti	Titan	25 Aug 2009 12:52:44	970	161.1	I	5.5	24.0	12.4	4.29
T62/119Ti	Titan	12 Oct 2009 08:37:30	1,300	61.4	I	5.6	19.0	0.6	3.20
E7/120En	Enceladus	02 Nov 2009 07:43:04	100	90.0	O	7.7	19.0	0.5	3.21
E8/121En	Enceladus	21 Nov 2009 02:10:56	1,604	82.0	I	7.8	19.0	0.5	3.20
T63/122Ti	Titan	12 Dec 2009 01:04:20	4,850	−146.7	O	5.5	16.0	4.9	2.60
T64/123Ti	Titan	28 Dec 2009 00:18:05	955	−94.5	O	5.5	16.0	21.6	3.03
T65/124Ti	Titan	12 Jan 2010 23:11:42	1,073	86.1	O	5.5	16.0	5.2	2.59
T66/125Ti	Titan	28 Jan 2010 22:29:55	7,490	52.7	O	8.6	17.5	0.3	2.89
R2/127Rh	Rhea	02 Mar 2010 17:41:42	100	−98.8	I	9.1	17.6	0.4	2.91
He1/127He	Helene	03 Mar 2010 13:41:28	1,816	134.1	O	9.1	17.6	0.4	2.91
T67/129Ti	Titan	05 Apr 2010 15:51:44	7,462	180.0	I	5.5	20.4	0.3	3.53
D2/129Di	Dione	07 Apr 2010 05:17:16	504	0.0	I	8.4	20.4	0.4	3.53
E9/130En	Enceladus	28 Apr 2010 00:11:23	100	90.0	O	6.5	20.5	0.3	3.54
E10/131En	Enceladus	18 May 2010 06:05:11	200	122.6	I	6.5	20.5	0.3	3.52
T68/131Ti	Titan	20 May 2010 03:25:26	1,400	130.7	O	5.5	16.0	12.1	2.76
T69/132Ti	Titan	05 Jan 2010 02:27:33	2,044	−89.2	O	5.5	16.0	2.0	2.64
T70/133Ti	Titan	21 Jun 2010 01:28:23	955	−93.6	O	5.5	16.0	18.8	2.97

[a]B-plane angle relative to the satellite's mean equator of epoch, In/Out = flyby inbound (I) or outbound (O), Period = Spacecraft period after the encounter, Inc. = Inclination after the encounter, Periapsis = Spacecraft periapsis distance relative to Saturn after the encounter.

During each phase of the tour, different science objectives dominate the tour design. Some science objectives are time critical (optical studies while Enceladus' south pole remains lit, Rings observations during equinox, etc.) while other objectives are geometrically driven (orbit apoapsis orientation, occultation intercept latitudes, inclination profiles, etc.), all of which can require several Titan flybys to achieve. The different phases of the EM are delineated in Fig. 22.5 in a plot of inclination versus time, and they will be covered in detail in the subsequent section. Figures 22.6–22.9 describe the flybys and coverage of Titan and Enceladus, and the periapsis, apoapsis and ring-plane crossing (node) locations during the EM.

The EM begins at high inclination, where the prime mission left off, with the apoapsis of its orbit oriented nearly sunward. This geometry is maintained for 9 months for continued in-situ study of Saturn's auroral region, complementing incomplete observations made at the end of the prime mission. Also, these high inclinations also allow for high priority stellar occultations of the main rings, particularly the very dense B ring. Since one of the ring-plane crossings of this orbit geometry was near Enceladus, it was also possible to insert three low Enceladus flybys during this phase while meeting the other objectives. Lastly, the Titan flybys during this *high inclination phase* (see Fig. 22.5) were designed to obtain a mid-northern latitude Titan radio occultation and a high quality groundtrack over the Huygens landing site.

This first phase completes with a short Titan-Titan *"Pi-transfer" phase*, with the first encounter on one side of Titan's orbit and the second encounter one-half of one orbit later, on the opposite side of Titan's orbit. This creative

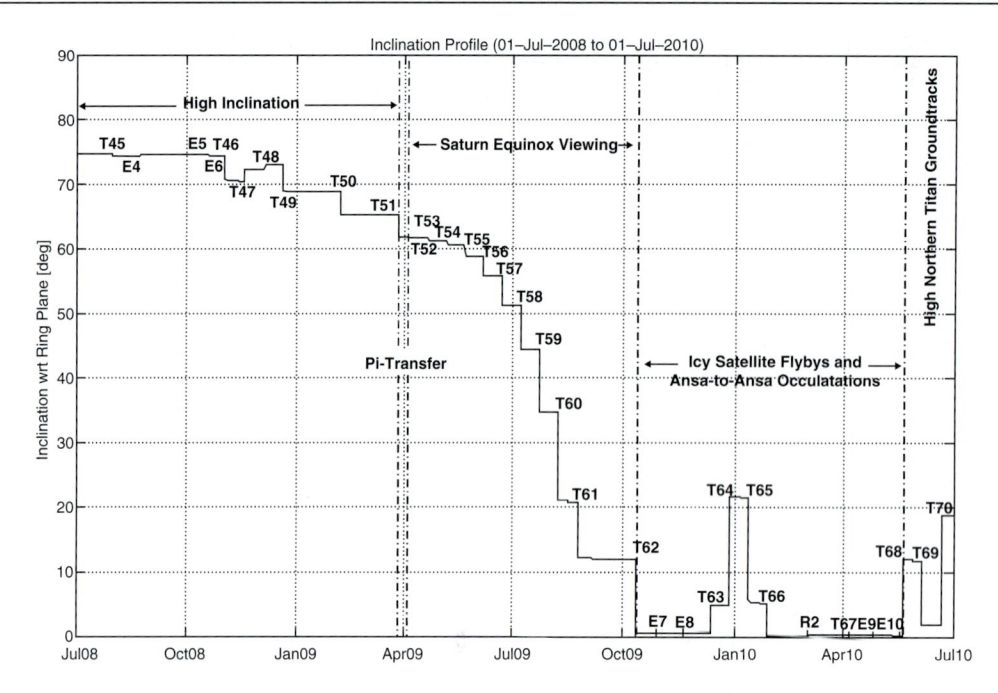

Fig. 22.5 Cassini extended mission inclination profile (Saturn mean equator), including the major phases of the tour

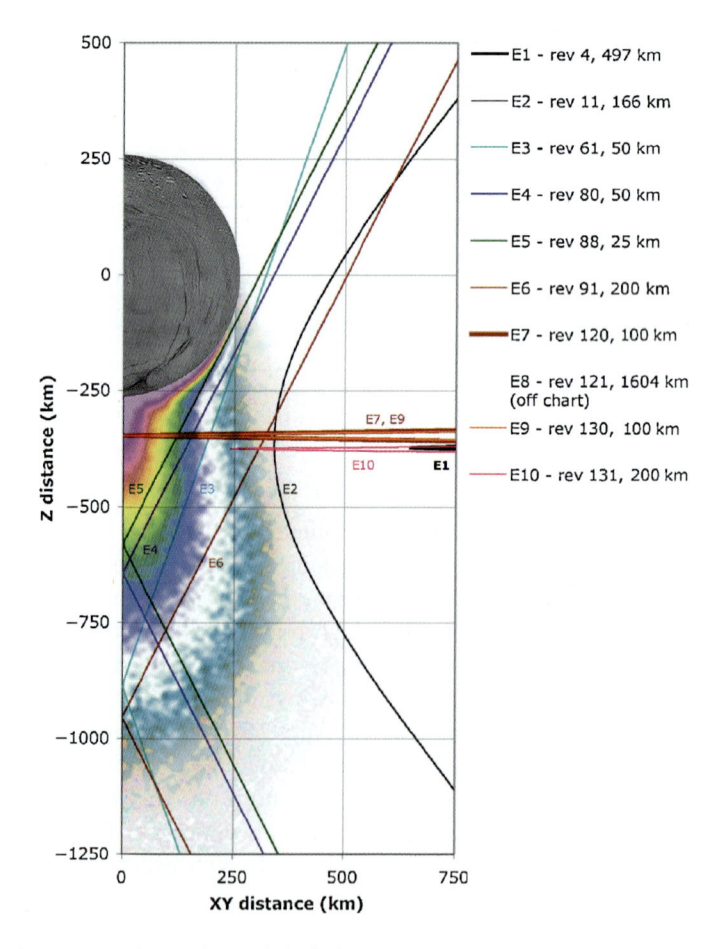

Fig. 22.6 Cassini plume penetration passes (prime and extended mission)

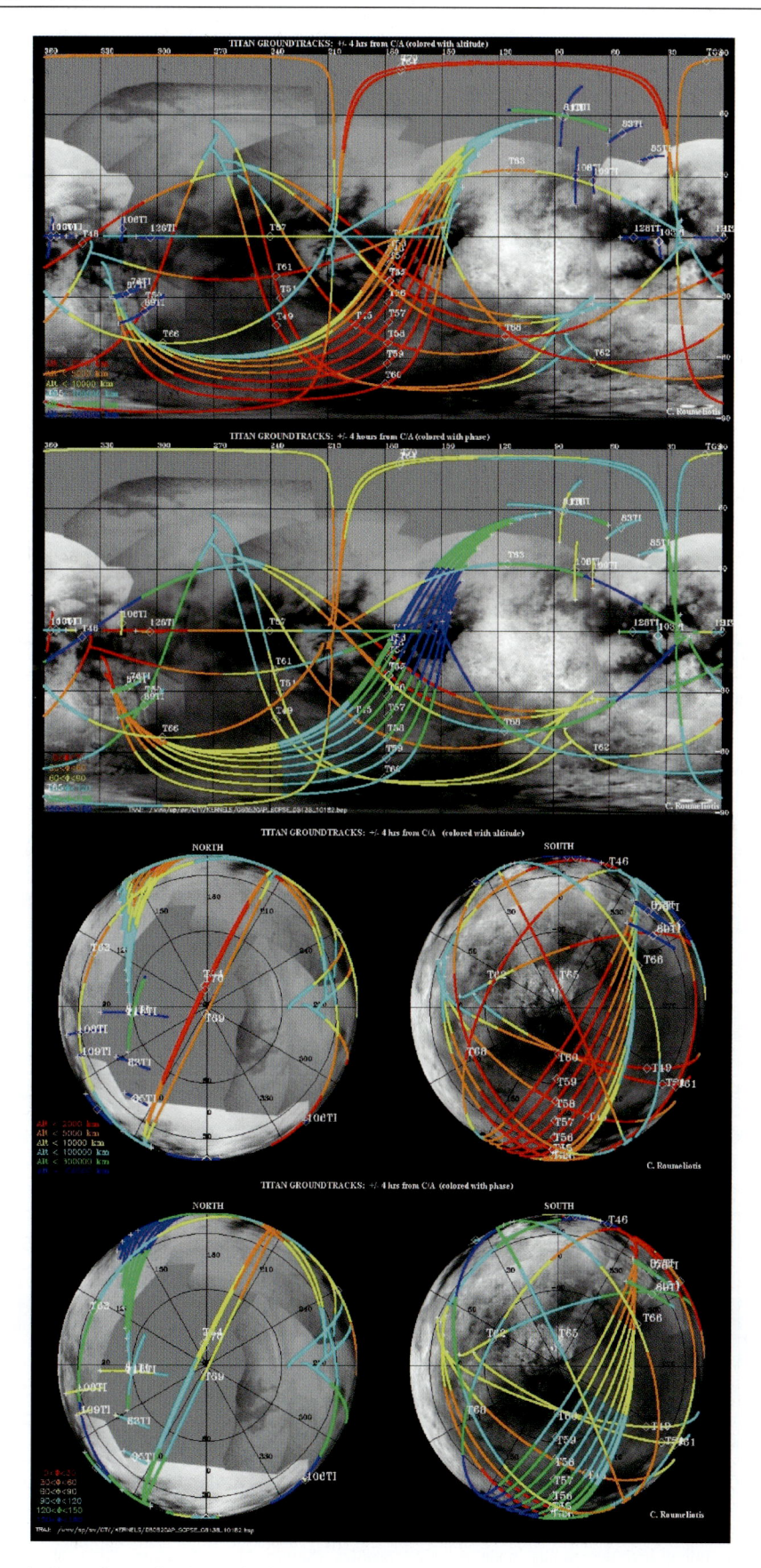

Fig. 22.7 Titan groundtracks (cartesian and polar projections) with color contoured altitude and phase profiles (courtesy of Chris Roumeliotis; 2007 ISS base-map)

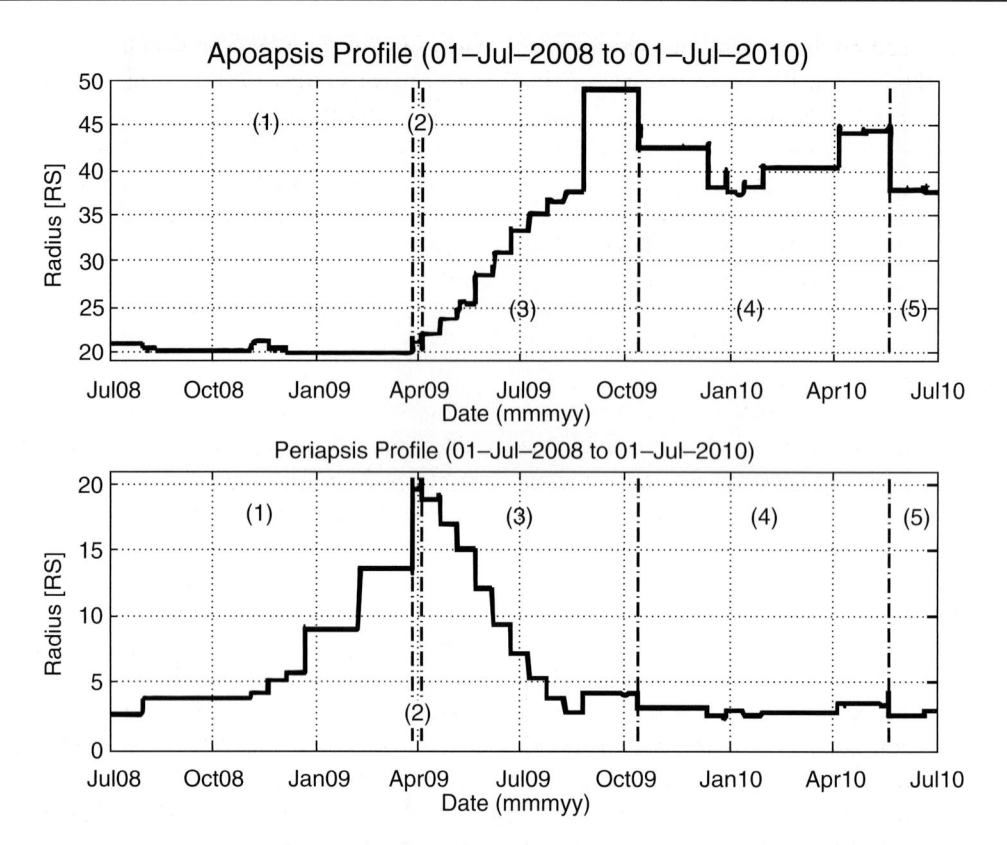

Fig. 22.8 Saturn-centered apoapsis and periapsis profiles (Saturn Mean Equator). Tour phases: (**1**) High inclination, (**2**) Pi-transfer, (**3**) Saturn equinox viewing, (**4**) Icy satellite flybys and ansa-to-ansa occultations, and (**5**) High northern groundtracks

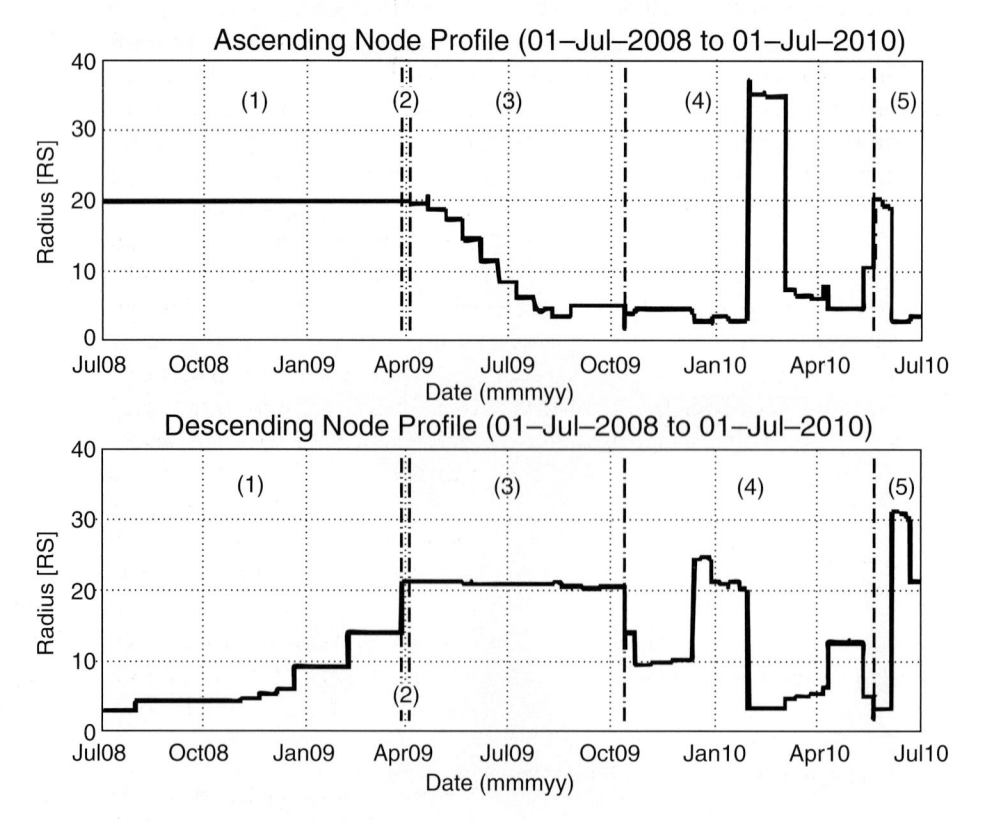

Fig. 22.9 Saturn-centered ascending and descending node crossing profiles (Saturn Mean Equator). Tour phases: (**1**) High inclination, (**2**) Pi-transfer, (**3**) Saturn equinox viewing, (**4**) Icy satellite flybys and ansa-to-ansa occultations, and (**5**) High northern groundtracks.

strategy was implemented by the Cassini tour designers as a highly efficient method to transfer the orbit from one side of Saturn to another, speeding up the exploration of the environment compared to a more conventional, encounter-by-encounter rotation.

The second major phase of the EM takes place over 6 months in the poorly sampled "dusk sector" of Saturn, i.e., the side outward from Saturn's dusk terminator. This phase contains the Saturnian equinox in mid-August of 2009, and is Cassini's only chance to study this infrequent geometric event, as it occurs only every 15 years. During this *equinox viewing phase*, Cassini's orbit inclination is slowly reduced, offering a generous variety of observation geometries for Saturn's rings, where the bulk of the equinox-related objectives are focused. In addition, Titan flyby geometries allow for key high-resolution radar mapping and Titan occultations.

The third major phase is dominated by equatorial orbits and icy satellite encounters. The fourth through seventh Enceladus encounters occur here, as well as close flybys of Mimas, Dione, Rhea, and Helene (a Dione coorbital). Two of the Enceladus encounters, E7 and E9, are designed to pass deep within the plume emanating from the southern polar region (see Fig. 22.7) – should continuing safety analyses permit. In addition, two high quality horizontal (ansa-to-ansa[4]) ring occultations, long Saturn observation opportunities, a Titan wake crossing, and mid-northern latitude Saturn occultations are included in this icy satellite and ansa-to-ansa occultation phase.

The final phase of the Equinox Mission is less than 2 months long, and contains a series of short Titan transfers designed for gravity measurements, more high northern Titan observations, and a third horizontal ring occultation. It is generally referred to as the *high northern Titan groundtrack phase*.

In total, this Equinox Mission requires 202 m/s, approximately two-thirds of the estimated propellant budget available as of the end of the prime mission. It is a very aggressive and scientifically intense design worthy of a flagship-class NASA mission, but clearly achievable given the project's capabilities and successes, and spacecraft health to date.

22.6.1 Detailed Description of EM Trajectory Phases

22.6.1.1 High Inclination Phase (T45–T51)

The first phase of the extended mission was designed for two main purposes. The first was to maintain a high inclination with respect to Saturn's equatorial plane ($i \geq 65°$), which would be beneficial to both the Magnetospheric and Plasma Science Group (MAPS) and the Rings Working Group (RWG). For MAPS, one of the highest priority extended mission objectives was additional in situ measurements of Saturn's auroral region (Cassini Extended Mission Science Objectives 2007) to complement a set of five auroral observations taken near the end of the prime mission (between 19-Apr-2007 and 31-May-2007). Furthermore, the observations needed to be spaced sufficiently forward in time (relative to the first set) to give the MAPS team the ability to change observation strategy arising from any discoveries made during the prime mission. These auroral observations are best accomplished by using highly inclined short-period orbits to minimize the spacecraft's altitude relative to Saturn when passing through the $L = 15$ L-shell[5] (with the assumption the auroral region is located at, or near, 75°N latitude). Based on the EM initial conditions (i.e., highly inclined) and other high priority science objectives required low inclination orbits, fulfilling these objectives at the beginning of the extended mission was the most efficient use of time. The transfers between T45 and T48 provide 15 additional $L = 15$ L-shell crossings below 5 RS.

For the RWG, the additional high inclination orbits provide favorable geometry (the inclination must be greater than approximately 64° due to the declination of bright ultraviolet and infrared stars) for a high number of stellar occultations that penetrate the densest parts of the B-ring (Cassini Extended Mission Science Objectives 2007).

The second main design driver for this phase of the EM was to increase the vacant node distance near 20 RS to set up the geometry needed for the 8-day pi-transfer.

While achieving a particular set of observations may drive the general design of any given tour phase, it was the intent to fulfill as many Discipline Working Group's scientific objectives as possible over a finite amount of time (i.e., a discrete set of transfers). For the Icy Satellites Working Group, a high number of low altitude targeted Enceladus flybys was a high priority due to the discovery (by the Cassini spacecraft in 2005) of thermally anomalous fissures emanating water-ice in excess of 250 km from Enceladus' south polar region (lat $\geq$ 75°S). By implementing a 6:13 resonant transfer at a 74.4° inclination from T45 to T46, three low altitude Enceladus flybys (E4, E5, and E6) were achieved. All three have similar geometries to E3 (last Enceladus flyby in prime mission), occurring at descending nodes with high northern latitude inbound asymptotes and high southern latitude outbound asymptotes. E4 and E5, both targeted flybys,

---

[4] Ansa: The outermost point of a ring that is visible from a particular point in space.

[5] Parameter used to describe the set of magnetic field lines that cross a planet's magnetic equator at a number of planet-radii. Each L-shell intersects/connects to specific latitudes on a planet, and hence, are often used to locate/describe phenomenon.

have flyby altitudes of 50 and 25 km respectively. However, since E6 is part of a double flyby and not directly targeted, the flyby was limited to a 200 km closest approach (c/a) altitude. Both E4 and E6 have B-plane angles of 90° to minimize the spacecraft's flyby altitude over Enceladus' south pole (Buffington, B., and Strange 2006). The E5 B-plane angle was initially set to 70° in order to render a 90°S egress intercept latitude for the planned Ultraviolet Imaging Spectrograph (UVIS) stellar occultation occurring near closest approach. However, it was later decided that allocating this flyby to MAPS would net the best overall science return, so the E5 B-plane angle was later changed to 90° to minimize the spacecraft's flyby altitude over Enceladus' south pole just as was done with the design of E4 and E6 (Buffington 2008).

Lastly, the T45 to T51 sequence was chosen to obtain both a mid northern latitude near-diametric Titan radio science (RSS) occultation (T46), and a low solar phase groundtrack over the Huygens landing site for the Visual and Infrared Mapping Spectrometer (VIMS) instrument team (T47).

22.6.1.2 8-Day Pi-Transfer (T51–T52)

The Titan Working Group (TWG) requested several Titan flybys in the dusk sector (LST between 15:00 and 21:00) due to the sparse distribution of them in the prime mission (only two). In addition, MAPS requested low inclination orbits with apoapsis in the dusk sector to map Saturn's magnetosphere. Given the high inclination at the beginning of the extended mission, a pi-transfer is the most time-efficient way to change the LST of the Titan encounters. The T51-T52 transfer implements an outbound-to-inbound, 8-day (transfer time) pi-transfer at 61.7° inclination, which changes the LST of the Titan encounter from 10:00 to 22:00, and the corresponding apoapsis LST from 11:00 to 19:30 LST.

22.6.1.3 Saturn Equinox (T52–T62)

Saturn's equinox[6] will occur near the midpoint of the EM (11-Aug-2009), a geometry that will not recur for another 15 years. The Rings Working Group's highest ranked EM objective (and one of the highest priority EM objectives) was to observe the ring system at equinox from an inclination greater than 15°, observe it during the surrounding time (± 1–2 months) from an inclination greater than or equal to 10°, and place apoapses over the northern hemisphere to enable thermal observations of the rings heating up.

The first three flybys (T52–T54), were designed for quality high northern latitude and equatorial RSS and UVIS Titan occultations. In addition, by forcing these flyby altitudes to be higher, an acceptable inclination profile for rings equinox viewing ($i = 21°$ at equinox, and $i = 12°$ for 2 months post-equinox) was attained while simultaneously minimizing the altitudes of the T55 – T61 flybys (for in situ atmospheric measurements and high resolution Radar mapping). The T61-T62 transfer implements a 3:2 resonant transfer to avoid flybys during the 2009 solar conjunction period. Lastly, the T62 flyby was set to 1,300 km for a UVIS Titan occultation, and was used to place the spacecraft's orbit in the same plane as Titan's ($i = 0.4°$).

22.6.1.4 Icy Satellites and Ansa-to-Ansa Occultations (T62–T68)

This phase of the tour is dominated (in time) by equatorial orbits via non-resonant transfers designed to obtain six of the nine close icy satellite flybys, two of the three RSS ansa-to-ansa ring occultations[7] (see Fig. 22.10), long un-obstructed (by the rings) observations of Saturn, mid-northern latitude Saturn occultations, and numerous Titan flybys in the dusk sector.

The fourth and fifth Enceladus flybys of the extended mission, E7 and E8, occur on the T62-T63 non-resonant transfer. Unlike the first three Enceladus flybys (E4-E6), E7 and E8 (as well as E9 and E10) occur on near-zero inclination orbits and have inbound/outbound asymptotes near the equator which allows closest approach to be placed[8] at any latitude (the corresponding c/a longitude will be offset by $\pm 90°$ from the asymptote longitude). This freedom enables flybys directly through Enceladus' plume at an arbitrarily low altitude – one of the highest priority objectives in the EM. Both E7 and E9 (occurring between T67 and T68), were optimized

Fig. 22.10 First RSS ansa-to-ansa ring occultation (25-Dec-2009). Spacecraft path as seen from Earth is shown in red

---

[6] Moment when the Sun crosses Saturn's ring plane.

[7] Occultation in which the spacecraft passes behind one side of the farthest part of the rings from Saturn, referred to as the ansa, to the other ansa as seen from Earth.

[8] For a given icy satellite flyby, a Δv-optimal set of flyby parameters exists. The "freedom" to vary the flyby parameters to increase scientific return is associated with a Δv cost.

for in situ measurements of Enceladus' plume by selecting a 90° B-plane angle, to place c/a directly over 90° S latitude, and by setting the flyby altitude to 100 km. The E8 flyby was optimized for optical remote sensing; the B-plane angle was set to 82°, placing the groundtrack directly along one of the illuminated thermally anomalous fissures (referred to as "tiger stripes") believed to be the source of Enceladus' plumes and the E ring. The corresponding closest approach altitude was set to 1,604 km as this altitude minimized the Δv required for the desired groundtrack placement. In addition, the T62–63 non-resonant transfer rotates the tail petal clockwise (decreases LST) to set up the correct petal orientation for the first RSS ansa-to-ansa ring/Saturn occultation.

The T63 flyby was designed to obtain an acceptable MAPS mid-tail Titan wake crossing and set up the inclination needed for the first ansa-to-ansa ring occultation. The latter involved adjusting the inclination to maximize and balance the free-space baseline (FSB), which is the time interval between the top of the ionosphere (68,000 km) and the inner edge of Ring C (74,500 km), on each side of Saturn (Marouf 2007). The ingress and egress FSB for the first ansa-to-ansa occultation are 7.7 and 4.9 min, respectively.

The T64 and T65 flybys were designed in unison to achieve a high northern Titan groundtrack and a second RSS ansa-to-ansa ring occultation. The T63 aimpoint was chosen to change the inclination while holding the orbit period constant in order to achieve a high northern Titan groundtrack for optical remote sensing (82°N c/a latitude). Using a similar approach, cranking in the opposite direction (i.e., a T65 aimpoint on approximately the opposite side of Titan as T64) negates the majority of the inclination change to set up the correct inclination for the second ansa-to-ansa occultation on the T65-T66 transfer. However, unlike the first ansa-to-ansa occultation, the second ansa-to-ansa egress FSB was compromised in order to fine-tune the inclination (on the order of 0.0025°) to obtain a distant RSS Enceladus plume occultation with a polar crossing altitude of 50 km (see Fig. 22.11). The plume occultation will occur approximately 2.18 h before the start of the second ansa-to-ansa occultation. The ingress and egress FSB for the second ansa-to-ansa occultation are 4.6 and 0.2 min, respectively.

The T66 flyby lowers the inclination back down into Titan's orbit plane, and initiates a non-resonant Titan-to-Titan transfer that includes a 100 km targeted Rhea flyby. In addition, the two closest non-targeted flybys of the EM occur on the T66-T67 transfer: Mimas (9,532 km) and Helene (1,806 km). This transfer was also designed to get numerous empty (i.e., no icy satellite flybys with altitudes ≤10,000 km) periapsis passages for optical remote sensing observations of Saturn requested by the Saturn Working Group (SWG).

The last transfer in this phase, T67-T68, is a non-resonant transfer designed to obtain three close icy satellite flybys; one of Dione and two of Enceladus. The Dione flyby, D2, is a 500 km targeted flyby, and has been optimized for MAPS by constraining the B-plane angle to 0° in order to fly through the center of the downstream plasma wake (closest approach: 72°W, 0°N). Of the two Enceladus flybys, E9 is a targeted, 100 km flyby optimized for in situ measurements (previously discussed), and E10, like E6, is part of a double flyby (occurring only 2 days prior to a T68), and has a 200 km flyby altitude.

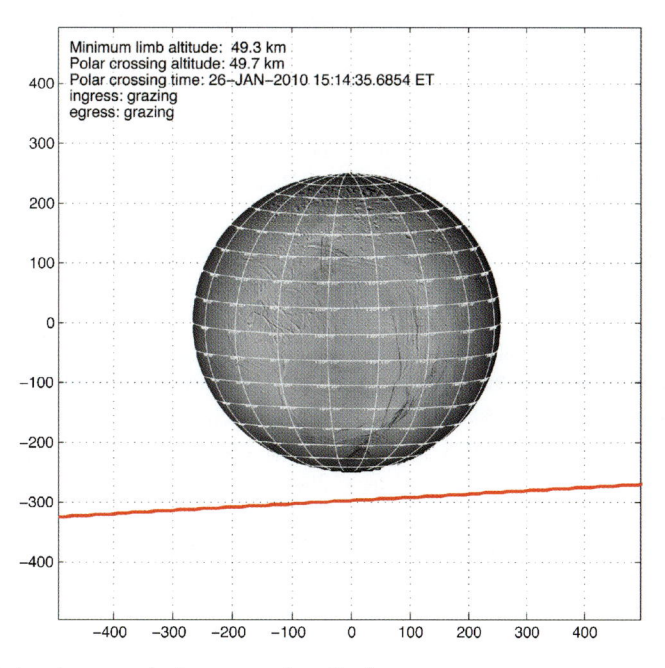

Fig. 22.11 26-Jan-2010 RSS Enceladus plume occultation as seen from Earth

22.6.1.5 High Northern Titan Groundtracks (T68–T70)

The final phase of the Cassini extended mission consists of a series of 16-day Titan-to-Titan transfers designed for Titan gravity measurements, two additional high northern Titan groundtracks, and a third RSS ansa-to-ansa ring occultation.

The T68 flyby, designed for an RSS gravity pass, has a flyby altitude of 1,400 km, an SEP angle of 112°, and a Titan-relative inclination of approximately 49°. Similar to T65, the 2,100 km T69 flyby reverses most of the inclination change induced by the T68 flyby, setting up the correct inclination (1.96°) for the third RSS ansa-to-ansa ring occultation and attaining a high northern groundtrack with inbound asymptotic and closest approach phase between 60° and 90° for VIMS observations. The last Titan flyby in the Cassini extended mission, T70, continues increasing the inclination in the same direction using a 1:1 resonant transfer (16-day), obtaining a third high northern Titan groundtrack. Although the T70 altitude is currently 955 km, the flyby altitude will be lowered at a future date to maximize the science return from the Dual Technique Magnetometer (MAG) instrument. The T70 altitude will be lowered as much as possible while maintaining the health and safety of the Cassini spacecraft pending analysis to be performed by the AACS and Thermal teams.

22.7 Beyond the Equinox Mission

Even during the middle stages of development of the Equinox Mission, JPL and NASA HQ revisited the same basic questions described in this chapter's introduction. It became clear very quickly, in answer to those four queries, that spacecraft health could easily continue long-term – for years, if not decades; further project operations and sequencing were feasible; some level of consumables would remain at the end of the EM; and further scientific study justifying continued expenditures was possible – though not yet proven.

As the end of the prime mission neared, the limiting question of those four seemed to be the third: that of consumables. Propellant margins and power levels appeared to be present, but at what levels? To address this and other post-EM questions, JPL, with concurrence from NASA headquarters, began studying further mission extension options shortly after EM development was complete.

Power levels, though decreasing, do so slowly at a pace of slightly less than 1 W per month. At an end of prime mission power estimate of just under 700 W, to an end-of-equinox-mission power level of 675 W, this profile supports all operational modes and suites of science instruments through the end of the Equinox Mission, except one – the three-frequency Radio Science mode. This mode required unique support where other instruments and components are put in sleep or turned off to support each observation. This limitation was known for some time and does not appreciably degrade any priority one objectives of the EM.

As far as the post-EM period is concerned, the project quickly concluded that all core spacecraft systems can be sustained for decades. Through a combination of power transient analysis, instrument power usage updates, and reaction wheel mode management, most of the workhorse science modes can also be sustained until at least 2017, throughout the drop of about 60 W past the end of the EM. The remaining modes also remain workable, mostly via turning different combinations of instruments to sleep and managing the reaction wheel power modes. No unpleasant instrument off cycles would be required, even through 2017. In short, disciplinary-specific observations should still be fully viable for at least 9 years after the EM, with only slight complications and additional work to operations.

The other major consumable of concern for post-EM activities is propellant. Cassini has two propellant systems. The bipropellant system (mono-methyl hydrazine and nitrogen tetraoxide) is used exclusively for orbital maneuvers (translational). The monopropellant system (hydrazine) is used for attitude control (when the reaction wheels are not operating), reaction wheel biasing, and small orbital maneuvers.

Figure 22.12 illustrates the propellant usage profile for both systems during the prime and extended mission tours, along with the surplus expected to be present at the end of the Equinox Mission in June of 2010. As observed on the chart, the bipropellant system, used exclusively for significant orbital trim maneuvers, should have about 110 kg of usable propellant available. The monopropellant system has even more significant margin, with over 60 kg remaining. (Propellant usages during arrival and the Huygens mission have not been included in these charts, since they were unique events not to be repeated.)

Based on the power and propellant margins, some level of extension beyond June of 2010 is feasible. None of the other spacecraft consumables constitute a measurable risk to further mission operations.

Scientifically, strong cases may be made for continued study of the Saturn system. Saturn has shown a variety of dynamic phenomena, which no doubt carry some response to seasonal change. Determining that response demands study over a sufficient fraction of a season on Saturn, which takes 29.5 years to complete – or at least, nearly one half of a seasonal cycle, from northern winter to summer, for example. The flybys of Pioneer and Voyager from 1979–1981 took place nearly one full season before Cassini's arrival. Even now Cassini is approaching the near-equinox geometry observed by the earlier spacecraft, so a minority of even one Saturnian half-season (e.g., Northern winter to

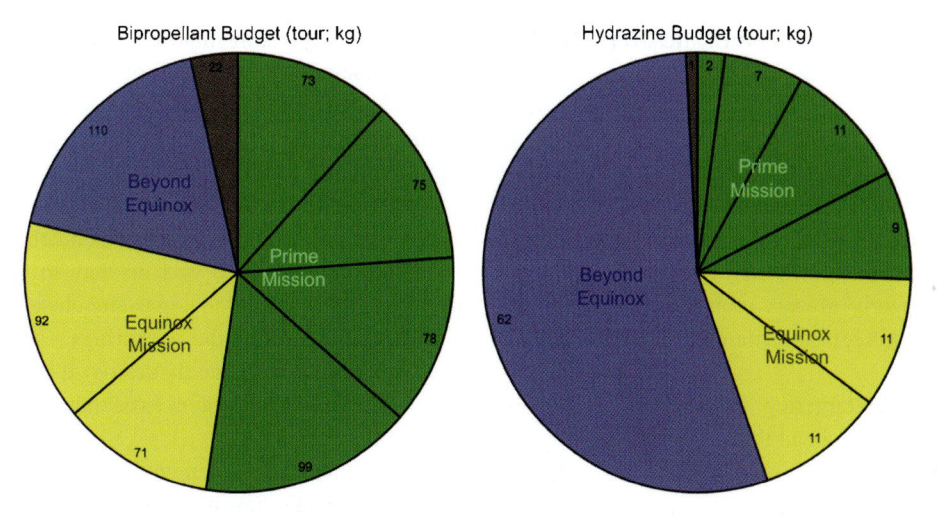

Fig. 22.12 Bi- and mono-propellant (hydrazine) usage for Prime and Equinox Missions tours; subdivisions in pie chart show usage per year of tour

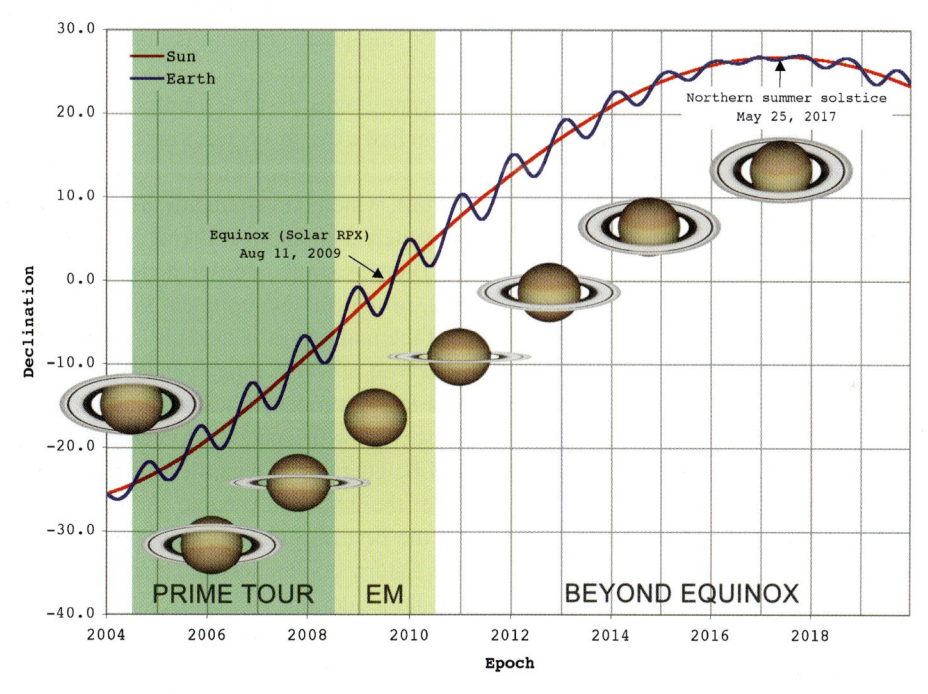

Fig. 22.13 Seasonal declination of the Sun and Earth as seen at Saturn

summer solstice) will have been observed by the end of the Equinox Mission, equivalent to less than the full months of January through March on Earth. If scientists were to study one place on Earth, or even the entire planet itself, over those few months alone, they likely would achieve only limited understanding of the processes and phenomena of the environment.

This concept is illustrated in Fig. 22.13, which shows the declination of the Sun and the Earth with respect to Saturn from Cassini's arrival through one full half season – just after northern winter solstice to summer solstice, which occurs in May of 2017. Views of Saturn from the Sun (or Earth) are added to illustrate the varying geometry over time.

Initial thoughts and discussions on a possible extension beyond equinox reasoned that no more than a year or two would be feasible. The propellant usage illustrated in the upper panel of Fig. 22.12 seems to imply, with even a cursory examination, that the basic cost of being in Saturn orbit is between 70–100 kg of maneuver bipropellant per year. With only 110 kg of bipropellant projected at the end of the Equinox Mission, Cassini would deplete the system no later than the fall of 2011. However, this simplistic conclusion is in fact far from the truth.

Many project members were surprised to learn that the majority of propellant had not, in fact, been expended merely to remain in orbit. Three key elements of the prime and

Equinox missions are responsible for the majority of the expenditures shown in Fig. 22.12:

- Dense interleaving of science objectives – often with incompatible geometries – required to maximize the science in the prime mission.
- Highly inclined orbits with close periapses (3–5 Saturn radii). These orbits are significantly perturbed by Saturn's oblateness, and nearly always require large maneuvers to counteract this effect, and maintain a Titan-crossing trajectory.
- Optimized encounter targeting, particularly for the icy satellite flybys, resulting in additional costs for aimpoints inconvenient for tour design.

Relaxing some or all of these elements – even slightly – results in trajectories with far lower propellant expenditures per unit time. The density of scientific opportunities is somewhat less in some disciplines, but is not proportional to the propellant used. Only a subset of possible scientific objectives are affected by such a strategy – specifically, those which require close inclined periapses and inconvenient satellite targeting. Such geometries are still feasible, but should be paired with intensive discipline-specific observation during periods brief enough to have a reasonable cost.

With this strategy, the remaining propellant could be made to last for more than 1 or 2 years – perhaps far enough to reach the Northern summer solstice. Overall, there would be some reduction in scientific intensity, but paired with a moderate reduction in operations funding, the potential scientific advancement from full exposure to the seasonal cycle could be a judicious use of resources.

22.7.1 Cassini End-of-Mission

Eventually there will come a time when the orbiter is nearly out of propellant. The Pioneer and Voyager missions have been free to roam into the recesses of deep space unchecked and with very slowly depleting consumables. The Cassini spacecraft must similarly leave the Saturnian system, or dispose of the spacecraft in a manner consistent with Planetary Protection requirements. A variety of options of both types have been studied in detail (Davis et al. 2007; Patterson et al. 2007; Okutsu et al., 2007; Yam et al. 2009). However, one very intriguing option has captured the attention of the project and the outer planet community.

It is possible to hop the spacecraft orbit's inner ring-plane crossing all the way over Saturn's entire main ring system, from just beyond the F ring, to the cleared gap between Saturn's upper atmosphere and inner portion of the main rings, using one single Titan flyby.

So far, it has been impossible to get very close to Saturn itself due to the presence of the main rings out to over 140,000 km from the center of Saturn. During Saturn Orbit Insertion (SOI), the spacecraft passed about 20,000 km from the cloud tops, and that is the closest the project ever expected to get to the planet and rings, provided by the hyperbolic approach trajectory. For much of the mission, it was believed that there was no way to construct an orbit with reasonable period that gets any closer to Saturn and yet does not pass through the main rings – the implementation of which would undoubtedly destroy the spacecraft. No safe gaps in the main ring system exist that could be employed as intermediate steps.

This strategy is implemented by raising the orbit inclination to near the "critical inclination" (which minimized perturbations due to Saturn's oblateness), and placing the inner ring-plane crossing just outside the main rings. The other ring-plane crossing is located at Titan, as usual, since Cassini must continually encounter Titan to employ its significant gravity assist resources. Then, with the last targeted Titan flyby, the spacecraft's orbit is altered such that the inner ring-plane crossings are between 62,000 and 65,000 km from Saturn, which correspond to the upper extent of Saturn's atmosphere and the inner D ring, respectfully.

There is some additional risk associated with this strategy, in that the environment in the gap between the planet and the main rings is not fully understood and has not been observed at close range (via remote sensing or in-situ) – this itself is a partial justification for exploring this region. However, with specific orbital geometries, Saturn impact can be guaranteed regardless of whether the spacecraft survives the first orbit (i.e., no s/c maneuvers are necessary after the first ring-plane crossing) by exploiting perturbations due Saturn's oblateness and distant non-targeted flybys. This should render this spacecraft disposal option more attractive to Planetary Protection representatives.

If this technique is successful, and the gap proves to be safe, this "proximal orbit" geometry could be maintained for some time, perhaps for dozens of orbits (Fig. 22.14), and science comparable to that of the $700 million Juno Mission at Jupiter (scheduled for launch in 2011, arrival in 2016) can be done by Cassini at Saturn.

With the possibility of multiple orbits for continued science observation, this possible end-of-mission option offers a significant wealth of scientific opportunities heretofore unanticipated for Cassini. Whole new avenues of study could be opened for the Saturn, magnetospheric, and rings disciplines. In short, these orbits create the possibility of a final mission phase at Saturn wholly beyond the initial scope of the Cassini project.

In conclusion, Cassini's Equinox Mission should advance our understanding of the Saturnian system, prepare for future missions to Saturn, and continue to dazzle the public. As to

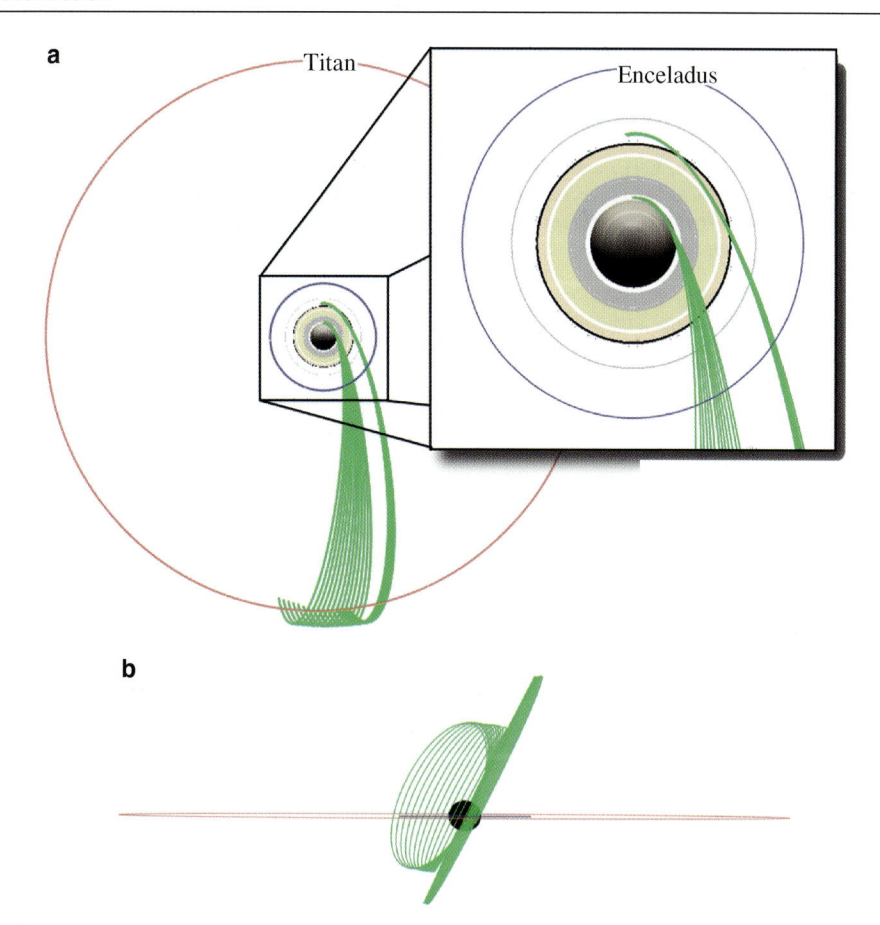

Fig. 22.14 Potential end-of-mission proximal orbit scenario. (**a**) Saturn north-polar view (sun direction towards *top* of page) – only the s/c trajectory (green) above the ring plane is shown to illustrate the evolution of the node distances, (**b**) Saturn equatorial view – exhibits the change in longitude of the ascending node due to J2 perturbations

what comes after the Equinox Mission, NASA will determine the best use of Cassini's remaining resources and complete the program in the best interests of the NASA Vision, the public's investment and the advancement of science.

Acknowledgments The Equinox Mission owes its successful design to the heroic and sustained efforts over 2 years of the Cassini tour designers – Brent Buffington, John Smith and Nathan Strange – and to the dedication of the Cassini science community.

This research was carried out at the Jet Propulsion Laboratory, California Institute of Technology, under a contract with the National Aeronautics and Space Administration. Reference to any specific commercial product, process, or service by trade name, trademark, manufacturer or otherwise, does not constitute or imply its endorsement by the United States Government or the Jet Propulsion Laboratory, California Institute of Technology.

References

Buffington, B., 080520 Reference Trajectory Update, JPL IOM 343M-08-003, 2008.

Buffington, B. and Strange, N., Science-driven Design of enceladus flyby geometry, *57th International Astronautical Congress*, Paper IAC-06-C1.6.05, Valencia, Spain, October 2–6, 2006.

Buffington, B. and Strange, N., Patched-integrated gravity-assist trajectory design, *AAS/AIAA Astrodynamics Specialist Conference*, Mackinac Island, MI, AAS Paper 07–276, August 19–23, 2007.

Buffington, B., Ionasescu, R., and Strange N., Addition of a low altitude Tethys flyby to the nominal Cassini tour, *AAS/AIAA Astrodynamics Specialists Conference*, Lake Tahoe, CA, Aug. 2005, AAS Paper 05–270, August 2005.

Buffington B., Strange, N., and Smith, J., Overview of the Cassini extended mission trajectory, *26th AIAA/AAS Astrodynamics Specialist Conference*, Paper AIAA-2008-6752, Honolulu, Hawaii, August 2008.

Cassini Project Policies and Requirements, Rev. E, May 2001.

Cassini Extended Mission Science Objectives, Revision 1: Final Tour Selection, February 2007.

Cassini Extended Mission Navigation Plan, Tech. rep., 699–101 Update, JPL D-11621, March 2008.

Davis, D.C., Patterson, C., and Howell, K., Solar gravity perturbations to facilitate long-term orbits: Application to Cassini, *AAS/AIAA Astrodynamics Conference*, AAS Paper 07–255. Mackinac Island, MI, Aug. 2007.

Hatfield, J. and Rinderle, E., User's Guide for CATO 4.3, internal JPL documentation, April 2001.

Initial concept design for Cassini's post-equinox mission, Cassini internal presentation to the Project Science Group, January 2008.

Jones, G.H., et al., The Dust Halo of Saturns largest icy moon: Evidence of rings at Rhea? Science, 319, 2008, 1380.

Marouf, E., Personal Communications, May 9, 2007.

Okutsu, M., Yam, C.H., Longuski, J.M., and Strange, N.J., Cassini end-of-life trajectories to the outer planets, *AAS/AIAA Astrodynamics Conference*, AAS Paper 07–258. Mackinac Island, MI, Aug. 2007.

Patterson, C., Kakoi, M., Howell, K.C., Yam, C.H., and Longuski, J.M., 500-Year eccentric oribts for the Cassini spacecraft within the Saturnian system, *AAS/AIAA Astrodynamics Conference*, AAS Paper 07–256. Mackinac Island, MI, Aug. 2007.

Strange, N. and Sims, J., Methods for the design of V-infinity leveraging maneuvers, *AAS /AIAA Astrodynamics Specialist Conference*, Quebec, Quebec, August 2001, AAS Paper 01–437, August 2001.

Strange, N., Russell, R., and Buffington, B., Mapping the V-infinity globe, *AAS/AIAA Astrodynamics Specialists Conference*, AAS Paper 07–277, Mackinac Island, MI, August 19–23, 2007.

Uphoff, C., Roberts, P. H., and Friedman, L. D., Orbit design concepts for Jupiter orbiter missions, *AIAA Mechanics and Control Conference*, Anaheim, CA, Aug. 1974, AIAA Paper 74–781, August 1974.

Wolf, A. A., Touring the Saturnian system, Space Science Review, 104, 2002, 101–128.

Yam, C.H., Davis, D.C., Longuski, J.M., Howell, K.C., and Buffington, B., Saturn impact trajectories for Cassini end-of-mission, Journal of Spacecraft and Rockets, 46(2), 2009, 353–364.

Chapter 23
Saturn's Exploration Beyond Cassini-Huygens

Tristan Guillot, Sushil Atreya, Sébastien Charnoz, Michele K. Dougherty, and Peter Read

Abstract For its beautiful rings, active atmosphere and mysterious magnetic field, Saturn is a fascinating planet. It also holds some of the keys to understanding the formation of our Solar System and the evolution of giant planets in general. While the exploration by the Cassini-Huygens mission has led to great advances in our understanding of the planet and its moons, it has left us with puzzling questions: What is the bulk composition of the planet? Does it have a helium core? Is it enriched in noble gases like Jupiter? What powers and controls its gigantic storms? We have learned that we can measure an outer magnetic field that is filtered from its non-axisymmetric components, but what is Saturn's *inner* magnetic field? What are the rings made of and when were they formed?

These questions are crucial in several ways: a detailed comparison of the compositions of Jupiter and Saturn is necessary to understand processes at work during the formation of these two planets and of the Solar System: was the protosolar disk progressively photoevaporated of its hydrogen and helium while forming its planets? Did Jupiter and Saturn form at the same time from cores of similar masses? Saturn is also a unique laboratory for studying the meteorology of a planet in which, in contrast to the Earth, the vapor of any condensing species (in particular water) is *heavier* than the surrounding air. A precise measurement of its magnetic field is needed to constrain dynamo theories and apply it to other contexts, from our Earth to extrasolar planets. Finally,

T. Guillot
Observatoire de la Côte d'Azur, CNRS, UMR 6202, BP4229, 06304 Nice, France

S. Atreya
Department of Atmospheric, Oceanic, and Space Sciences, University of Michigan, Ann Arbor, MI 48109-2143, USA

S. Charnoz
Equipe AIM, CEA/SAp, Université Paris Diderot, 1191 Gif-Sur-Yvette Cedex, France

M.K. Dougherty
Imperial College London, SW7 2AZ, UK

P. Read
University of Oxford, Clarendon Laboratory, Parks Road, OX1 3PU Oxford, UK

the theory behind the existence of its rings is still to be confirmed, and has consequences for a variety of subjects from theories of accretion of grains to the study of physical mechanisms at work in protoplanetary systems.

All in all, this calls for the continued exploration of the second largest planet in our Solar System, with a variety of means including remote observations and space missions. Measurements of gravity and magnetic fields very close to the planet's cloud tops would be extremely valuable. Very high spatial resolution images of the rings would provide details on their structure and the material that form them. Last but not least, one or several probes sent into the atmosphere of the planet would provide the critical measurements that would allow a detailed comparison with the same measurements at Jupiter.

23.1 Introduction

Saturn was probably first observed with a telescope by Galileo in 1610 but it was not until 1655 that Christian Huygens discovered its largest moon, Titan. Four years later, he correctly inferred that the planet has rings. Then, between 1671 and 1684, Jean-Dominique Cassini discovered Saturn's moons Japetus, Rhea, Thethys and Dione, as well as the now so-called Cassini division. Although the planet fascinated many, the following major milestones in the discovery of this world had to await the first space missions, Pioneer 11 in 1977, Voyager 1 in 1980 and Voyager 2 in 1981. Among many results, the missions measured the planet's atmospheric composition, discovered Saturn's magnetic field, measured Saturn's wind patterns. They gave evidence of the amazing thinness of the rings and of their structure. Then in 2004 came the Cassini-Huygens spacecraft which, to list but a few results, further extended our knowledge of Saturn's system of moons, composition of the rings, and unveiled Saturn's meteorology in all its complexity.

Saturn is a truly major planet in the Solar System: with 95 times the mass of the Earth, it is the second largest planet. It contains a large fraction of hydrogen and helium, gases that were most abundant when the Solar System was formed. As such, it is a witness of events that occurred very early during the formation of the system, for which the study of its formation provides us with invaluable information. About 800 million years later, it was probably responsible for a reorganization of the system of outer planets and Kuiper-Edgeworth belt, which led to the so-called "late heavy bombardment" in the inner Solar System. Saturn also has a complex atmosphere, both in terms of chemistry and dynamics and, as such, it is a fantastic laboratory for understanding planetary atmospheres in general. Among the planets in the Solar System, its magnetic field is second only to Jupiter in intensity, and it has an unusual, completely axisymmetric form that is still unexplained. Its rings are a laboratory for understanding disks and can be sought as miniature protoplanetary systems. Finally, in the era of the discovery of planets, and particularly giant planets, around other stars than our Sun, understanding Saturn's thermal evolution is crucial.

In spite of all the efforts and progress made in the past 30 years or so, Saturn remains mysterious. We will review the many questions, some of them unexpected, that the Cassini-Huygens mission has left us with and that call for a continued exploration of this planet. First, we will discuss how the interior of the planet remains uncertain. We will then see how understanding the evolution, and hence the composition, of giant exoplanets is tied to understanding the evolution of Saturn. The next sections will discuss Saturn's atmospheric composition, atmospheric dynamics, and rings respectively. We will then see how a better understanding of the planet would help us in deciphering the mystery of the origin of our Solar System. The means to explore Saturn further during the next several decades will be discussed.

23.2 Saturn's Interior

Saturn is known to be mostly made of hydrogen and helium but to contain other elements ("heavy elements") in a proportion that is significantly larger than is the case in the Sun. It is also known to contain a central dense core of 10–25 Earth masses that was probably the seed of the formation of the planet, before it accreted its hydrogen–helium envelope from the protosolar disk (e.g., Guillot 2005). *Qualitatively*, these conclusions have stood almost unchanged for more than 25 years (e.g., Stevenson 1982), although *quantitative* improvements are due to better equations of state, and improvements in computing power. These inferences have rested on the calculation of interior models fitted to several key measurements: the planet's mass, radius, atmospheric temperature and pressure, atmospheric helium abundance, interior rotation rate and gravity field (gravitational moments J_2, J_4 and J_6). An important astrophysical constrain is the primordial (protosolar) helium to hydrogen ratio, with the hypothesis that missing helium in the planet's atmosphere is hidden deeper into the planet. A final, crucial ingredient is an equation of state, most importantly for hydrogen, the most abundant element, but ideally for all species to be considered.

While Cassini-Huygens has provided a much better measurement of the gravity field of the planet by an order of magnitude, it has also demonstrated that the inner rotation rate is much less constrained than it was thought to be. Furthermore, uncertainties have remained as to whether Saturn's atmosphere is very poor, or moderately poor in helium, with a mass fraction in helium still ranging between 7% and 17% (28% being the protosolar value). In parallel, while progresses have been made with high-pressure experiments in the Mbar regime, both in the lab and numerically, hydrogen has proven to be a surprisingly difficult substance to comprehend. As a result, several theories exist to explain several experiments, but they generally do not agree with each other! Naturally, the case of the hydrogen–helium mixture has remained even more difficult and no reliable prediction exist as to the temperature–pressure regime in which hydrogen and helium may separate out into two mixtures of differing compositions in the giant planets. Yet, this is crucial for explaining the missing atmospheric helium, as such a phase separation would yield the rapid formation of helium-rich droplets that would fall towards the deeper regions (Stevenson and Salpeter 1977) with consequences for the planet's interior structure.

Figure 23.1 shows two possible structures for Saturn's interior: in the traditional view, the helium phase separation occurs close to the molecular-metallic transition. The planet is then thought to consist of a helium poor molecular hydrogen envelope, a helium-rich metallic hydrogen envelope and a central dense core of unknown composition (e.g., Saumon and Guillot 2004). Alternatively, helium sedimentation may proceed all the way to the core in which case a helium shell may be present on top of the central core in which case most of the envelope should be helium-poor (Fortney and Hubbard 2003). It is also important to note that differential rotation in the interior has been traditionally neglected, but may play an important role (Hubbard 1999; see Chapter 4.).

Altogether, progressing in our knowledge of Saturn's interior will require:

– To determine accurately the abundance of helium in Saturn's atmosphere.
– To obtain a reliable hydrogen–helium equation of state and phase diagram in the relevant pressure-temperature range (from 0.1 to 30 Mbar, and 1,000–20,000 K) and for mixtures of hydrogen, helium and heavy elements (most importantly water).
– To measure the deep water abundance (water is extremely important both because it forms almost half of the mass of a solar composition mixture of heavy elements, and because it directly impacts the planet's meteorology).

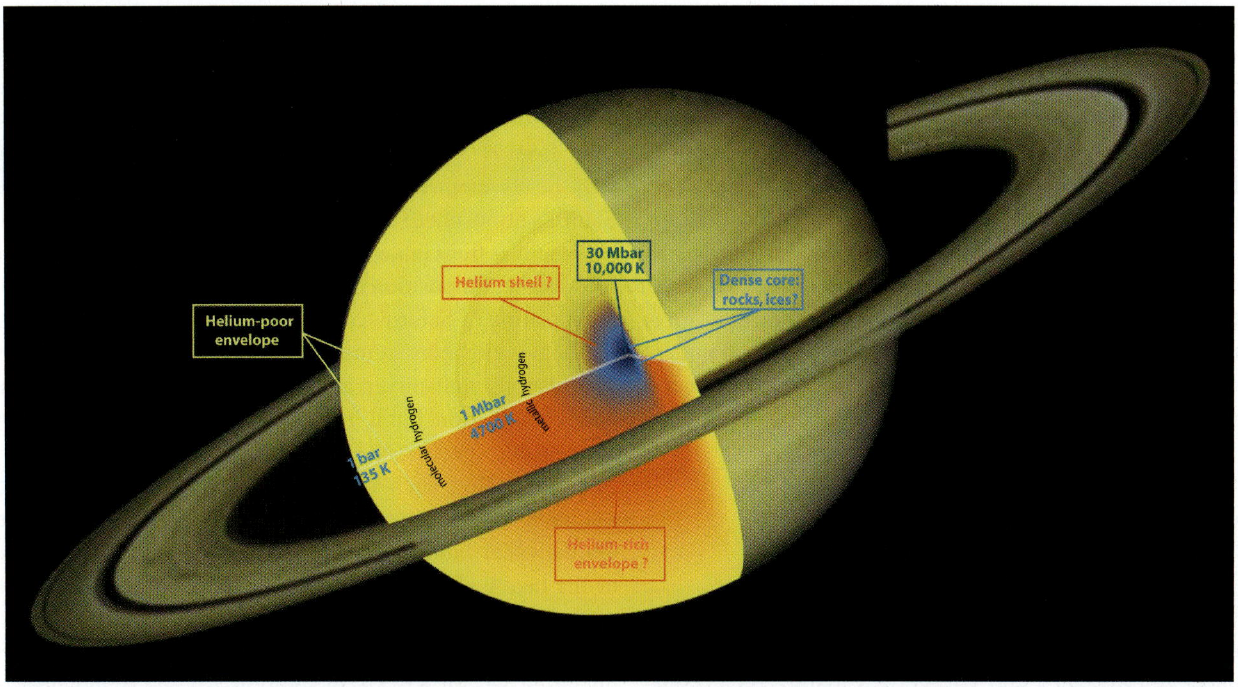

Fig. 23.1 Sketch of the interior of Saturn showing two possible classes of models: (1) *Bottom*: The envelope is split in a helium-poor region in which hydrogen is in molecular form and a helium-rich region with metallic hydrogen. (2) *Top*: Helium sedimentation is supposed to be complete down to the core so that a helium shell is formed at the bottom of a helium-poor envelope. In both cases, a central core of made of dense material (most probably a mixture of iron, rocks, ices in unknown proportion and state) is supposed to be present

- To measure accurately Saturn's interior rotation rate.
- To obtain accurate measurements of Saturn's gravity field including high order gravitational moments.
- With numerical experiments, to model Saturn's interior rotation to determine how atmospheric zonal winds and interior rotation are linked, with the aim of being able to invert gravitational field measurements for accurate constraints on the density profile.
- To model the extent to which elements can be mixed upward or inversely settle towards central regions as a function of their behaviors at high pressures.

Progresses in our knowledge of Saturn's interior should parallel that on Jupiter, the comparison between the two planets being extremely instructive in terms of constraints on their interior structures and formation.

23.3 Structure and Evolution of Low-Density Giant Planets

The discovery that extrasolar giant planets abound in the Galaxy and the possibility to directly characterize them especially when they transit in front of their star is a great opportunity to know more about these planets and the formation of planetary systems (e.g., Charbonneau et al. 2007). Different Saturn-mass extrasolar planets have been found: HD149026b (Sato et al. 2005) has a mass of $1.2\,M_{Saturn}$ ($114\,M_\oplus$) for a radius of $0.9\,R_{Saturn}$, implying that it contains about $70\,M_\oplus$ of heavy elements (e.g., Ikoma et al. 2006)! On the other side of the spectrum, HAT-P-1b has a mass of $1.8\,M_{Saturn}$ ($170\,M_\oplus$) for a radius of $1.4\,R_{Saturn}$ (Bakos et al. 2007), implying that it contains little heavy elements (Guillot 2008). However, most of the knowledge that can be gained on these objects that lie many tens of light years away rests on what we have learned on the structure and evolution of giant planets in our Solar System. Details such as the influence of atmospheric winds on the structure and cooling, the interior rotation of the planets, the presence of a dynamo, the extent of a phase separation of elements in the planet. etc., all must rest on models tuned to reproduce the structure and evolution of giant planets in our system for which minute details are known.

It is important in this respect to stress that the known 4.56 Ga age of giant planets in our Solar System is unfortunately not yet precisely accounted for. While Jupiter models traditionally reach that value to within 10% (which in itself should be improved), the situation is dire for Saturn for which traditional models fall short of this value by 2–2.5 Ga (e.g., Hubbard et al. 1999). This is most probably due to the presence of the helium-hydrogen phase separation at high pressures and to the subsequent release of gravitational energy as the helium-rich droplets fall towards the planetary interior. A model age of $\sim$4.5 Ga can be obtained by accounting for this extra energy source and properly tuning the H/He phase diagram. However there is yet no model that properly accounts for the ages of both Jupiter and Saturn at the same time. The application to extrasolar planets is to be made with caution (see Fortney and Hubbard 2004).

It is therefore most important that the evolution of the two giant planets closest to us is better understood, in order to apply this knowledge to the ensemble of extrasolar giant planets found so far. This requires using better equations of state including phase diagrams of the major species and their mixtures, updated atmospheric models, and an improved understanding of differential rotation and mixing in the planet's interior.

23.4 Saturn's Atmospheric Composition

Saturn's atmosphere has been investigated by remote sensing from the ground, earth orbit (HST, ISO) and spacecraft (Pioneer 11, Voyager 1 and Voyager 2, Cassini). As a result, much is known about the composition of Saturn's stratosphere. Column abundances of various hydrocarbons, including CH_4, CH_3, C_2H_2, C_2H_4, C_2H_6, C_3H_8, C_4H_2, C_6H_6, oxygen-bearing molecules, including H_2O, CO, CO_2, disequilibrium species, including PH_3, GeH_4, AsH_3 (and CO), and certain other minor constituents such as NH_3 and HCl, have been determined (Atreya et al. 1999). However, their vertical density distributions have not been measured, with the exception of CH_4 and C_2H_2 whose profiles were derived over a couple of scales heights in the vicinity of Saturn's homopause using the solar occultation technique in the ultraviolet from Voyager. No composition or temperature information is available for the middle atmosphere, the "ignorosphere", extending roughly from 1 mb to 1 nanobar. Furthermore, composition in the troposphere has not been determined except for methane. Being the principal reservoir of carbon at Saturn, methane provides the carbon elemental abundance. The stratospheric hydrocarbons are photochemical products of methane, and thus inappropriate for deriving the C/H.

Both Voyager and Cassini spacecraft were able measure the abundance of methane in the well-mixed troposphere. The high precision data from the Cassini Composite Infrared Spectrometer (CIRS) have yielded a CH_4 mole fraction of $4.5 \pm 0.9 \times 10^{-3}$, or $CH_4/H_2 = 5.1 \pm 1.0 \times 10^{-3}$ (Flasar et al. 2005). This implies the carbon elemental ratio, $C/H = 9.3 \pm 1.8 \times$ solar, using the Grevesse et al. (2005) solar elemental abundances. CIRS also determined $P/H = 10 \times$ solar from PH_3 in the upper troposphere (Fletcher et al. 2007). Although similar to C/H, it may not represent the true deep, well-mixed atmosphere elemental abundance of phosphorus, as PH_3 is a disequilibrium species which is in thermochemical equilibrium at several thousand bars where the temperatures are several thousand degrees Kelvin. Finally, ground-based VLA observations in the microwave have yielded rather uncertain, model-dependent results for N/H = 2–4x solar from NH_3 and $S/H = 12 \times$ solar from H_2S (Briggs and Sackett 1989). In summary, the only heavy element with reliable data on its abundance in Saturn's atmosphere is carbon.

Whereas the Galileo entry probe measured the bulk atmospheric composition of Jupiter except for water, the bulk composition of Saturn's atmosphere, hence the abundance of the heavy elements, remains mysterious for the most part. The planet's well-mixed atmosphere is representative of the bulk composition, but it lies below the clouds, especially for condensible species (NH_3, H_2S, H_2O in Saturn and Jupiter). Remote sensing, such as that from the Cassini orbiter, is not a suitable technique for sampling this part of the atmosphere. On the other hand, elemental abundances, especially those of the heavy elements are required to constrain the models of the formation of Saturn and the origin of its atmosphere. The most critical heavy elements are O, C, N, S and the noble gases Ne, Ar, Kr, Xe. In addition, isotope abundances of the noble gases, D/H, $^3He/^4He$, $^{14}N/^{15}N$ are also important for gaining an insight into the origin and evolution of the atmosphere. Finally, a precise determination of the He/H ratio in the atmosphere provides a window into interior processes such as the conversion of helium into liquid metallic form in the 3–5 megabar region (as predicted from the equation of state and laboratory experiments), release of gravitational potential energy, etc. The revised Voyager analysis yielded $He/H_2 = 0.11$–17 (Conrath and Gautier 2000), which has too large an uncertainty to be of value in discriminating between various models of the interior of Saturn. Direct in situ measurements of the helium abundance are needed, as was done at Jupiter by the Galileo probe. Moreover, determination of the elemental and isotope abundances requires accessing and measuring the well-mixed part of the atmosphere for a large number of key constituents. The pressure levels at which the atmosphere can be considered as well-mixed depend on the constituent, as illustrated in Fig. 23.2.

Ammonia, hydrogen sulfide and water vapor are all condensible species in Saturn's atmosphere. Their well-mixed atmosphere abundances yield, respectively, the elemental ratios N/H, S/H and O/H. On Saturn, NH_3 clouds are the topmost clouds, forming in the 1–2 bar region, depending on the enrichment of N, S, and O as NH_3 dissolves in water and forms a cloud of ammonium hydrosulfide upon combining with H_2S (Fig. 23.2). Thus, it would seem that ammonia is well-mixed below the 2 bar level, if thermodynamic equilibrium prevails and dynamics play no role in the atmosphere. It was evident from the Galileo probe measurements that the atmosphere of Jupiter is far from being ideal. Saturn is no different. The Cassini VIMS observations (Baines et al. 2009) show that a great degree of convective activity

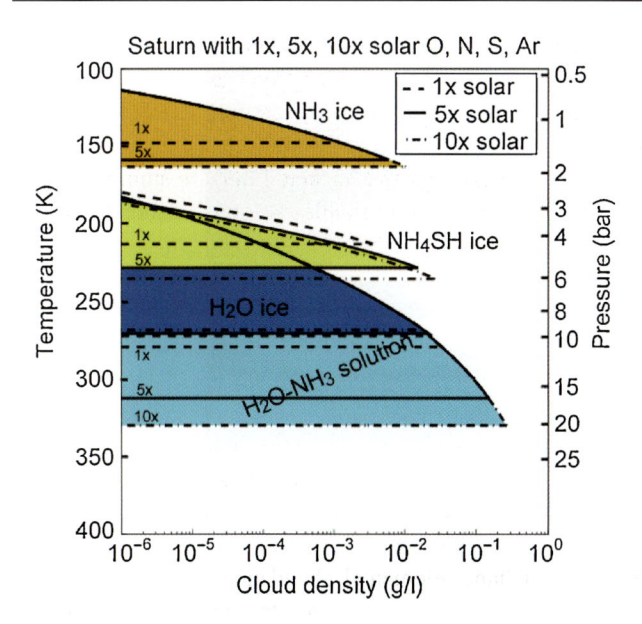

Fig. 23.2 Mean vertical distribution of cloud layers on Saturn, deduced from a simple thermochemical model. The cloud concentrations (in g/L) represent upper limits that are likely to be depleted by factors of 10–1,000 due to precipitation and dynamics as on Earth (updated from Atreya and Wong (2005), using the Grevesse et al. (2005) solar elemental abundances)

extends to at least several bars in Saturn's atmosphere as well. As a consequence, well mixed ammonia may only lie much deeper. In fact a true measure of well-mixed NH_3 could come only from the atmosphere below the water cloud since the NH_3 above these clouds may already be depleted due to solution in water. The well-mixed part for H_2S is below the base of the NH_4SH cloud or >6 bar level. Water forms the deepest cloud. Its base could range from 10 bars (ice cloud only, with $1 \times$ solar O/H) to >20 bars (an aqueous ammonia droplet cloud, with $10 \times$ solar O/H). Again, because of convective processes, the well-mixed water may not be found until 50–100 bars. Thus water places the most severe constraint on the well mixed region of the atmosphere. Water is critical to determine, as it was presumably the original carrier of the heavy elements that formed the core of the planet and comprised 50–70% of its core mass according to formation models. The noble gases and all the isotopes listed above are accessible at pressures less than 10 bars, however, and are therefore an important design consideration for future probe missions to Saturn.

23.5 Saturn's Atmospheric Dynamics

As earlier chapters have abundantly demonstrated, the Cassini orbiter has provided a wealth of new data relating to the dynamics and circulation of Saturn's atmosphere, from the high stratosphere down to the deep troposphere, at least as far as the NH_4SH cloud deck around 2–4 bars (Fig. 23.3). This has led to major new insights into the chemistry and transport processes in the stratosphere, including the discovery of a substantial semi-annual oscillation of the zonal flow in the tropics. This is akin to the Earth's Quasi-Biennial and Semi-Annual Oscillations and is presumably driven by upward-propagating waves from the troposphere that break and dissipate in the middle stratosphere. The structure of the stratosphere is also seen to be strongly affected by seasonal variations, with substantial differences in temperature and composition between winter and summer hemispheres around solstice.

In the troposphere itself, new measurements from Cassini have discovered a surprisingly intense and compact vortex at the south pole, and have begun to measure how waves and eddies in Saturn's atmosphere interact to produce the well known banded pattern of zonal winds, clouds and hazes. Both Saturn and Jupiter appear to be in a special ('zonostrophic' – see Chapter 6) dynamical regime, in which energetic stirring of the atmosphere takes place on relatively small scales, dominated perhaps by baroclinic instabilities or intense but highly intermittent convective storms that may be driven, at least partly, by moist condensation effects. An upscale turbulent cascade then passes energy to ever larger scales, but becomes highly anisotropic due to dispersive wave propagation, the end-point of which is the energizing of intense but remarkably persistent zonal jets. In this regime, however, the effects of mechanical friction are very weak except at the largest scales, so jets may grow to sufficient intensity to become unstable on large scales, at least locally, leading to the production of large-scale waves and eddies that also often appear to be highly persistent. Such wave systems may include Saturn's unique and mysterious North Polar Hexagon and 'Ribbon wave' patterns, which seem to be remarkably stable and long-lived. However, the mechanisms by which these persistent circulation patterns sustain themselves remain somewhat mysterious.

Despite the considerable progress that has been, and continues to be made, by the Cassini-Huygens enterprise, however, it is already clear that a number of key questions confronting theoreticians and modelers will not be solved by the range of measurements enabled by the current mission. Such questions are likely to require new developments within the modeling community and, even more challenging, new kinds of measurements beyond those offered by Cassini, even given further mission extensions.

As discussed in Chapter 5 by Fouchet et al., measurements of composition in Saturn's stratosphere reveal anomalies that cannot be explained or modeled by local chemical changes or conventional 'eddy diffusion' parameterizations of transport. Large-scale transport across and between hemispheres is evidently more important than hitherto realized, and will

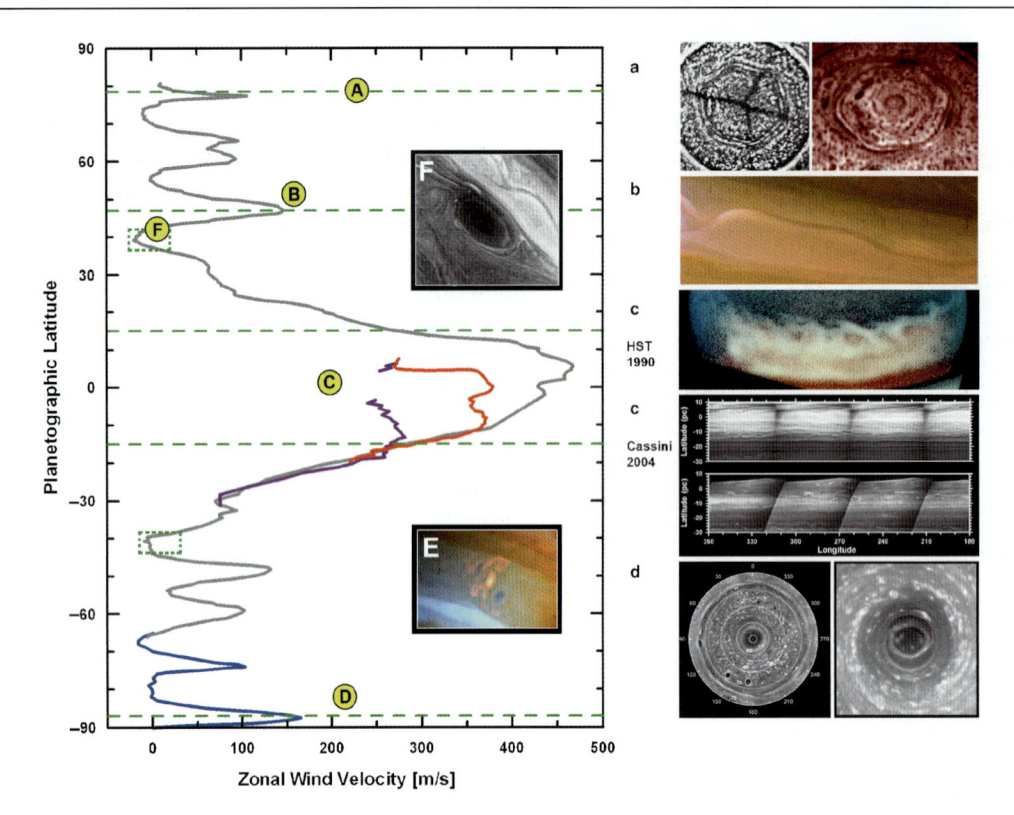

Fig. 23.3 General circulation of Saturn and relevant atmospheric features on its atmosphere (from Marty et al. 2009). Winds at cloud level, relative to the interior reference frame measured by Voyager, traced by the Voyagers (grey line) and Cassini data of the Southernmost latitudes (blue) and equatorial region in different filters (red and violet). Relevant meteorological structures appear on the insets: (A) North polar hexagon in visible (Voyager 1) and infrared light (Cassini); (B) The Ribbon; (C) Saturn Great White Spot in the Equatorial Region in 1990 and the state of the Equator as seen by Cassini in the methane absorption band and continuum filters; (D) The South Polar jet and the inner polar vortex; (E) Convective storms seen by Cassini; (F) Anticyclones from Voyager 1. The location of most convective storms appear marked with green dashed boxes

require new approaches to chemical modeling in Saturn's stratosphere, most likely involving a full representation of chemical transport in 2D or even 3D models.

Cassini measurements of clouds and aerosols in Saturn's atmosphere (see Chapter 7) have partly confirmed earlier ideas concerning the composition, nature and distribution of NH_3 and NH_4SH, although the lack of direct spectroscopic evidence for ammonia ice or NH_4SH indicates that our knowledge of cloud microphysics on Saturn has some way to go. The interaction between microphysics, thermodynamics and atmospheric transport and circulation on a variety of scales should be a major objective of future work, both in terms of modeling and new observations post-Cassini. The large-scale distribution of clouds is critically dependent on the strength and form of the circulation in the upper troposphere on the scale of the banded zonal jets, and this continues to be quite uncertain. In this respect, also, the deep water abundance on Saturn is a key parameter that remains relatively poorly constrained on a global level.

On convective scales (~100 km or less), high resolution imagery from Cassini has shown a wealth of structure that is only just beginning to be explored – especially in the vicinity of the intense polar vortex. Current discussion of the polar vortex has tended to centre around supposed analogies with the core of terrestrial tropical cyclones, but this may be premature. The interaction between individual moist convection cells and a larger scale balanced vortex circulation is critical to the dynamics of terrestrial hurricanes, but whether this interaction operates in the same way in Saturn's polar vortex remains to be confirmed.

Much uncertainty in our understanding of the structure, dynamics and composition of Saturn's atmosphere stems from huge uncertainties in our knowledge of the nature and form of the atmospheric circulation beneath the ubiquitous upper cloud decks of NH_3, NH_4SH and H_2O condensates. Visible and near-infrared imagery only effectively senses the motions and structure of the uppermost NH_4 clouds in the main, although occasional glimpses of deeper seated structures are occasionally possible in relatively clear regions. Imaging in the thermal infrared around 5 μm wavelength by the VIMS instrument on Cassini has shown a wealth of structure on clouds at intermediate depths, most likely around 1–3 bars, largely representing the classical NH_4SH cloud decks (see Fig. 23.4 below for an example). These structures appear

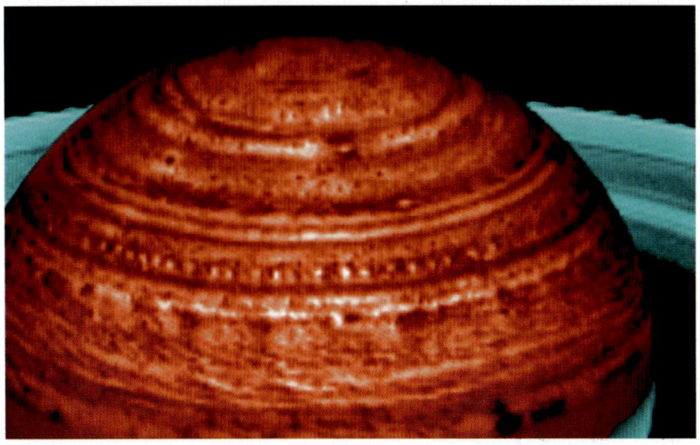

Fig. 23.4 Infrared image taken by the Cassini VIMS instrument in the 5 μm band of the northern hemisphere of Saturn, showing zonal bands and various waves and eddies (such as the 'string of pearls' around 40°N), representing features located in the vertical around pressure levels of 1–3 bars

quite differently and on a smaller scale than those apparent in visible images, suggesting a quite complex, baroclinic character for certain types of wave and eddies. Cassini has also begun to supplement the 5 μm images from VIMS with microwave remote sensing around 2 cm wavelength using the microwave RADAR instrument in passive mode. Microwave observations can penetrate significantly below the visible clouds, from a few bars at wavelengths of ∼1 cm to several tens of bars at ∼50 cm (Briggs and Sackett 1989; de Pater and Dickel 1991). This approach, utilizing long wavelength microwave measurements for remote sensing, is being developed for the JUNO mission to Jupiter (Janssen et al. 2005) and could be readily adapted for Saturn in due course.

But the issue remains that no remote sensing method, either from orbit or from the ground, is able to penetrate to levels deeper than a few tens of bars into Saturn's interior. This means that we are highly dependent on models to infer what may be happening beneath the clouds. Recent models (reviewed in Chapter 6), in particular, have highlighted a need to distinguish between the pattern of motions that may (or may not) be present to great depths below the visible clouds and the processes that may be driving them. It now seems clear (Lian and Showman 2008) that a deeply-penetrating meridional circulation pattern could be driven by energetic processes that are relatively shallow (such as moisture-enhanced convection in the water cloud layer), so a deep-seated circulation may not necessarily imply deep forcing. Moreover, where a circulation pattern penetrates very deeply into Saturn's troposphere, the electrical conductivity of the deep atmosphere may become a factor affecting the forces acting on fluid elements, in relation to Saturn's magnetic field. Theoretical and model studies of this magnetohydrodynamic aspect of Saturn's hypothetical deep atmospheric circulation are still in their infancy, but it is clear that Cassini alone will not be able to unravel them. Not least amongst these issues is the problem of Saturn's interior bulk rotation, whose period remains frustratingly illusive although some hints of a way forward have begun to appear (see Chapters 4, 6 and 8). This is because the radio/SKR measurements from Voyager and Cassini have proved to provide only an ambiguous and uncertain estimate of magnetic field rotation. Anderson and Schubert (2007), for example, have recently obtained a very different value for the interior rotation rate, based on a combination of gravity and radio-occultation measurements, with a period of 10 h, 32 min, and 35 ± 13 s, though this is still regarded as somewhat controversial. Within a reference frame with such a rotation period, however, Saturn's zonal winds would appear more symmetrical about zero, with eastward and westward jets (other than the equatorial jet) of more or less equal strength, much as found on Jupiter.

For the future it will be necessary to focus on a set of key questions that should guide the design and objectives of both future spacecraft missions and theoretical and numerical modelling activities. In the dynamics and circulation arena these questions should include the following:

- What is the role of dynamical transport in determining the distribution of relatively short-lived chemical tracers in the troposphere and lower stratosphere?
- How do dynamics and microphysics interact to produce the observed clouds on Saturn (and Jupiter)?
- How is the cloud-level circulation maintained? What is the nature (and location) of both energetic forcing (moist convection, baroclinic instability...?) and large-scale energy dissipation?
- What is the nature and role of observed convective storm systems, coherent waves and stable vortices in the general circulation?
- How deep do the observed cloud-level circulation systems penetrate? Do they reach levels where electrical conductivity becomes important? If so, how does this affect the nature of the deep circulation?
- What is the nature of the polar vortices on Saturn (and the other gas giant planets)? How are they sustained? What determines their stability and what is their role in the planet-scale circulation?

This is not intended to be an exhaustive list, and doubtless others working in the field will come up with additional issues that are felt to be important and compelling. But these are a core of recurring questions that have challenged the planetary atmospheres community for decades now, and will continue to hold up further progress in understanding more detailed aspects of Saturn until they are resolved. This will require observations of Saturn's atmosphere, in particular at deep levels, but it also will require ambitious hydrodynamic modeling of the Saturn's atmosphere and interior including all relevant physical processes (see Chapter 6).

23.6 Saturn's Magnetosphere

An initial understanding of Saturn's magnetosphere was obtained by three spacecraft flybys in the late seventies and early eighties, that of Pioneer 11 and Voyagers 1 and 2 (Fig. 23.5). These limited in-situ observations confirmed via plasma, fields and particles data that Saturn's magnetosphere resulted from the interaction of the solar wind plasma with Saturn's internal planetary magnetic field trapping plasma and energetic charged particles. The magnetosphere seemed to share many of the characteristics of the Earth's solar wind dominated magnetosphere and Jupiter's rotation-dominated one (Dougherty et al. 2004). However it was clear that this magnetosphere is unique as a result of the extremely diverse nature of the coupling mechanisms which exist between the numerous components which made up the Saturn system, including the solar wind, the ionosphere of Saturn; Titan, the icy satellites, the rings and dust and neutral clouds. Addressing the importance and complexity of these various mechanisms was one of the main science objectives of the Cassini-Huygens mission (Blanc et al. 2004) with a major strength of an orbiting mission such as Cassini being the potential to be able to resolve temporal from spatial processes via extended coverage of radial distance, latitude, longitude and Saturn local time.

The primary magnetospheric science objectives for the first 4 years of the Saturn tour by Cassini can be described as follows:

- To characterize the magnetic field, plasma and energetic particle population within the magnetosphere as a function of time and position
- To determine the relationship of the magnetic field orientation to Saturn's kilometric radiation
- Investigate Titan-magnetosphere interactions (this topic is described in the companion Titan book)
- Study the interaction between the magnetosphere and the icy satellites
- Investigate the interaction of the rings and the magnetosphere

On completion of the primary Cassini mission some major discoveries have resulted which will be briefly described (further details are given in Chapters 9–12). Resolution of some of the objectives is still incomplete and requires further observations during the Cassini Extended Mission (XM) as well as during the planned extension onto 2017, known as the Solstice Mission. Following on from the discoveries at Titan and Enceladus in particular there is the possibility of a future Titan orbiter with numerous Enceladus flybys which will allow further study of the Saturn environment. Recent

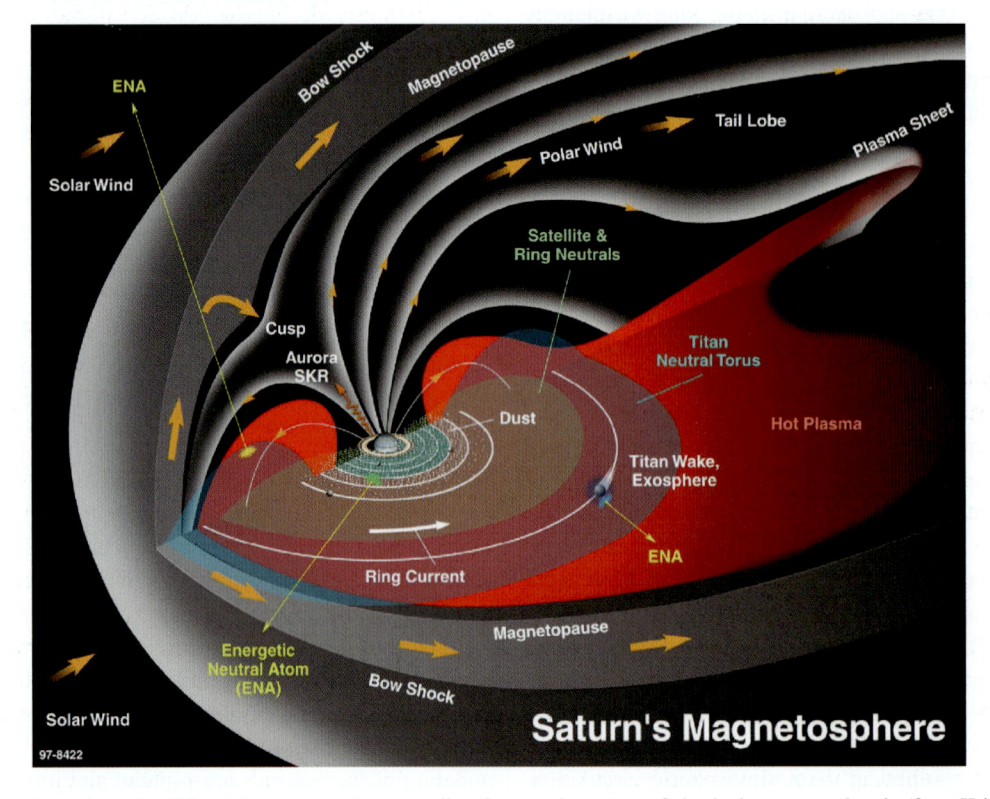

Fig. 23.5 An overview schematic of Saturn's magnetosphere revealing the complex nature of physical processes therein (from Krimigis et al. 2004)

studies of a Saturn probe mission have also been carried out between US and European scientists (Marty et al. 2009), with the primary goal of determining the abundance of the heavy elements, as was done at Jupiter with the Galileo probe in 1995. Complementary remote sensing observations from such a mission have the potential of yielding other valuable pieces of information including that on the core and the magnetospheric environment of the planet.

Characterization of the magnetosphere as a function of time and position has been carried out by a survey of the inner magnetosphere via the MAPS instruments onboard Cassini. Some major discoveries include: (i) That solar wind control is relatively weak compared to the rotational and mass-loading effects; (ii) A new radiation belt has been found inside of the D-ring; (iii) Imaging of the rotating dynamic ring current; (iv) The magnetosphere has a time-varying rotation period. However complete characterization is incomplete since the orbit of Cassini did not reach the magnetotail reconnection region and the neutral sheet during the prime mission. Observations during the XM and SM will fill this gap in local time and allow a more complete characterization of the magnetospheric processes. In addition the first in-situ observations of Saturn's auroral zones were only achieved late in the prime mission (see Chapter 12 and PSS special issue articles reference therein) with further orbits covering this region in the XM and SM. This data is critical in order for a better understanding to be gained of the physical processes driving the aurora. In order to best understand a complex three-dimensional system such as the magnetosphere as much temporal and spatial coverage as possible is desirable in order to separate the importance of the different plasma processes arising. We have at the end of the prime mission a 4-year timeline allowing a much better understanding of the effect of temporal variations; however extended mission observations will further refine this understanding as well as take account of the seasonal effects which arise at Saturn. Higher order moments of Saturn's internal magnetic field have finally been resolved during the prime mission (Burton et al. 2009) however in order to accurately determine the planetary field and better resolve the effect of the ring ionosphere on observations inside of the D-ring we require much more complete spatial coverage at a wide range of latitude, longitude and radial distances.

Our understanding of the relationship between Saturn's magnetic field orientation and the SKR is much advanced following the prime mission observations. It has been determined that the SKR period does not represent the internal rotation period of the planet, that the SKR period is variable and continues to evolve and that many magnetospheric phenomena have a period similar to the SKR period (Chapter 10) However there are still unresolved questions which require further measurements during the XM; the prime mission orbit in fact only resulted in very limited observation in the 3–5 Rs region, the 4-year mission to date has not allowed for a long enough observation in order to be able to distinguish between competing theories of why the SKR period varies and why it does not reflect the internal rotation period of the planet; and we are still yet to determine what this internal rotation period is.

Many major icy satellite discoveries have been made (Chapters 18–21) here we will focus on the interactions of the icy satellites and the magnetosphere. Some major discoveries include the discovery of a dynamic and exotic atmosphere at Enceladus, which led to the discovery of the plumes of Enceladus and confirmation that this moon is the source of Saturn's E-ring (Chapter 21). In addition a unique charge particle interaction with Rhea may be due to cloud and dust particles trapped within the Hill sphere (Chapter 11). There are as yet many unexplored areas concerning satellite/magnetosphere interaction and the XM and SM will yield additional close flybys (both upstream and downstream) of Mimas, Dione, Thetys and Rhea; and further Enceladus flybys will allow details of the plumes to be further studied as well as it's interaction with the magnetosphere.

The magnetospheric interaction with Saturn's rings has become more complex than originally though due the discovery of ring ionospheres (Chapter 13) and this requires further study. Saturn Orbit Insertion (SOI) demonstrated the unique nature of the region just above the main ring system. However due to the critical nature of the SOI burn, this orbit was not configured for prime science observations and additional close periapses during the XM and in particular the end of mission scenario linked to the SM which will consist of numerous polar orbits inside of the D ring will enable a clear determination of the ring ionosphere and its properties to be made as well as help constrain the internal planetary magnetic field.

23.7 Saturn's Rings

Saturns rings are a perfect, and the closest, example of an astrophysical disk. Unlike protoplanetary disks, material cannot accrete inside rings because of the strong planet's tides. However, since tidal forces decrease with the distance from Saturn, at the ring's outer edge (A and F ring) accretion and erosion processes are expected to occur and are manifested by the formation of Jeans Toomre waves in the A ring ("wakes') or clumps in the F ring (Esposito et al. 2008). Taking advantage of a multi-scale approach, exploring rings would help to understand fundamental astrophysical processes (gravitational instabilities, gap-opening in disks, accretion, viscous evolution, etc..) via remote sensing and, ideally, direct observation. Of course rings are also interesting for themselves, and are still poorly understood. Whereas the Cassini spacecraft has brought outstanding new data, some fundamental questions remain unanswered about its origin and evolution, and the milestone of future exploration would be in-situ images and spectra, with spatial resolution better than 10 cm. On the basis of theoretical arguments and numerical simulations (see Chapters 14 and 17) a lot of large scales processes and evolutionary processes (viscous spreading, gap opening, evolution of particle size distribution, accretion, meteoritic bombardment, brightness asymmetries) depends on the micro-scale structures (at scales <10 m) and 3-dimensional organization of the particle, which have been never observed. Stellar occultation data (Colwell et al. 2006, 2007, Hedman et al. 2007, Esposito et al. 2008) allowed probing the rings with resolution ∼10 m in 1 dimension. But the detected structures could be aggregates rather than individual particle. Among the outstanding question is the rings' mass. It is believed to be of the order of Mimas' mass, but could be much larger (Esposito et al. 2008, Charnoz et al. 2009, see also the Chapter 17). The mass is a determinant parameter for understanding the long term evolution of the rings. Another point is the composition of Saturn's rings: They seem to be made of almost pure ice with some contaminants (see, e.g., Nicholson et al. 2008), defying any formation scenario. We no review these points in more details.

23.7.1 Direct Imaging of Particle Size Distribution

Are Saturn's rings young or old? What is their origin? Theses two questions are deeply linked, but a long-standing paradox arises when one try to answer them jointly: numerous arguments suggest that the rings are young ($<10^8$ years) because of fast evolutionary processes (erosion due to meteoroid bombardment, surface darkening, viscous spreading, see Chapter 17). However, in the current state of our knowledge, it seems very improbable that they originated less than 1Ga ago, mainly because of the too low cometary flux (Harris 1984, Charnoz et al. 2009, and Chapter 17). To solve this paradox, it is proposed that the ring material is constantly reprocessed due to self-gravity and collisions, and thus, may appear much younger. In order to constrain this "cosmic-recycling" process, a detailed knowledge of the size distribution would be an invaluable data. Indeed, different material strength, different surface sticking properties and different accretion regimes (gravity vs. surface-sticking) would lead to different size distributions. Voyager and Cassini radio occultation experiments allowed a rough estimate of the size distribution, in the 1cm-10m range (e.g., Marouf et al. 1983), but it is somewhat model dependant. The intermediate-size range (∼100 m) has been probed indirectly by the detection of 100 m–1 km moonlets or aggregates (Tiscareno et al. 2006, Esposito et al. 2008). However, when put together, these measurements still do not draw a coherent picture because (1) we do not know if we are observing individual particles or aggregates and (2) these measurements are taken at different radial locations whereas we know that the size distribution should change with distance (Nicholson et al. 2008, see also Chapters 14 and 17). Only direct imaging at different radial locations in the rings would allow to unveil all these degeneracies and thus provide new and strong constrains on the particle size's distribution and evolution.

Indeed accretion and erosion within the rings may strongly alter the particle's size distribution. Despite the strong tidal field of Saturn, limited accretion is theoretically possible (e.g., Canup and Esposito 1995, see Chapters 14 and 17). Ring particles have weak cohesive forces, and therefore can assemble into transient structures much larger than an individual ring particle. Thus, an exotic accretion physics takes place, resulting in the formation of temporary aggregates, either called "wakes" in the A rings, where material assemble into elongated structures (see Chapter 14) or further away, close to the A-ring's edge or in the F ring, the aggregates can accrete into moonlets called "ringmoons" (Colwell and Esposito 1993). The presence of 100 m objects in the A ring has been revealed by the tidal arm they imprint in their surrounding (Tiscareno et al. 2008). Are theses objects primordial, fragments of larger bodies, or are they the natural product or local accretion? Direct imaging of "propellers" would be invaluable to understand the ring evolution and origin. It is also possible that moonlets embedded in the rings are made of two components: an outer shell of ring-particles accreted at the surface of a dense core that could be anterior to the formation of the ring (as was suggested by the odd shape of Pan and Atlas, see Porco et al. 2007, Charnoz et al. 2007).

23.7.2 Microstructuration

Only a handful of ring structures are explained, and it is thought that a substantial number of observable phenomena (brightness asymmetry in the A ring, fractal appearance, ring texture, sharp edges..., see Chapter 13) could be the large scale manifestation of "microstructuration": collective structuration processes occurring below the kilometre scale. For example, this could be either gravitational wakes in the A ring, viscous overstabilities in the B ring (see Chapter 14), self-organization of particles nearby a satellite resonance (Shepelyansky et al. 2009). Gravitational wakes in the A ring have been predicted for long (Toomre et al. 1964) and their wavelength is about 100 m. Direct observation of wakes would allow to constrain their morphology (size, orientation, separation, etc.) the ring's surface density, viscosity and

Fig. 23.6 Saturn rings observed in transmitted light by the Cassini/ISS camera system. In this point-of-view, denser rings appear darker because they block the light coming from the Sun. The B ring appears as a wide dark lane in the middle of the ring system because of its high surface density.

the Toomre Q parameter (although some of these parameters could be measured by indirect means, see e.g., Hedman et al. 2007). In addition, Daisaka et al. (2001) have shown that the presence of gravitational aggregates in Saturn rings result in a larger macroscopic viscosity than for a swarm of individual particles, which would have strong consequences on the rings lifetime. The B ring is maybe the most mysterious one due to its high density leading to non-linear effects. Overstability in the B ring produces parallels transient structures, with ∼50 m wavelength, that could explain the "microsillon" appearance of the B ring. Some stellar occultation data (Colwell et al. 2007) have provided further evidences that over-stability is still active in the B ring, but densest regions of the rings could not be investigated (Fig. 23.6).

23.7.3 Rings Thickness

The main-ring thickness has never been measured because the edge-on brightness of Saturn's rings is dominated by the dusty F ring, which is dynamically excited. We hence have no direct measurement of the A,B,C,D ring thickness, which is a critical dynamical parameter, like in any astrophysical disk. In particular, the age of the rings is directly linked to their thickness since the timescale of viscous spreading depends closely on the local particle velocity dispersion. Photomet-

ric arguments suggest $H < 1$ km (Nicholson et al. 1996), but dynamical arguments suggest values of the order of, or less than, 5m in the B ring (Salo 1995).

23.7.4 Chemical Composition: the Mystery of Silicates

Saturn's rings are composed mainly of water ice (see, e.g., Cuzzi and Estrada 1998, Poulet et al. 2003, Nicholson et al. 2008), and a few contaminant brought by the meteoroid bombardment. The abundance of silicates must be lower than 1% if uniformly distributed within the rings particles (Grossman 1990). Such purity is difficult to explain in any cosmogonic scenario of Saturn's rings, and there is no obvious mechanism that could remove silicates from Saturn's rings (see for example Harris 1984, Charnoz et al. 2009). However, when probing the material composition by remote-sensing, the penetration depth of light is about of the order of the wavelength, so that only the particle surface is probed and the bulk of the dense B ring remains maybe unobserved. Maybe some silicates may hide in the core of big ring particles, but we have no clue about this.

23.7.5 Rings' Mass

The total rings' mass is still unknown. It is thought to be of the order of Mimas' mass (Esposito et al. 1983). However, there are strong suspicions that it could be much larger (see, e.g., Stewart et al. 2007), which in turn would help solve some aspects of the question of the ring's lifetime and origin (see Chapter 17). A lot of mass could be stored in the B ring. Unfortunately, estimating its mass through remote sensing is difficult because it is opaque even to radio emissions (normal optical depth can be larger than 5) and its surface density shows strong spatial and temporal variability (Colwell et al. 2007). A direct measurement of the mass of Saturn's B ring may be obtained from a gravity field inversion after flybys around the entire ring system.

23.8 Saturn and the Formation of the Solar System

Saturn formed about 4.56 Ga ago, probably within a few Ma of the formation of the Sun, and at an epoch when a circumstellar disk, mainly made of hydrogen and helium, still surrounded the infant Sun (e.g., Pollack et al. 1996,

Wuchterl et al. 2000, Cassen 2006). On the other hand, the terrestrial planets were almost certainly fully formed later, in the 10–30 Ma or so after the gaseous disk had disappeared (e.g., Chambers 2004). Based on various observations like the structure of the Kuiper belt, the present orbital parameters of the planets in the Solar System and the traces of a late heavy bombardment in the inner system, it appears that Saturn (and the whole system of outer planets) was initially closer to the Sun, i.e., around 8 AU, instead of 9.54 AU now (Tsiganis et al. 2005). According to this scenario, known as the "Nice model", it is Saturn's crossing of the 1:2 resonance that has led to the so-called late heavy bombardment responsible for the formation of the lunar basins 4 Ga ago, or about 500 Ma after the formation of the first solids in the system (Gomes et al. 2005).

However, not much is known concerning the early phase (few Ma) when the gaseous protoplanetary disk was still present and the giant planets were forming. Models of planet formation that attempt to explain the observed ensemble of planets orbiting other stars (Ida and Lin 2004, Alibert et al. 2004, Thommes et al. 2008) indicate that giant planets such as Jupiter and Saturn grew from planetary embryos that had reached by accretion masses of a few times the mass of the Earth. However, it is not clear whether these protoplanets had migrated significantly (e.g., Alibert et al. 2005, see also Cresswell and Nelson 2008), or whether a nearly joint formation of Jupiter and Saturn locked them into a resonance with little migration (e.g., Masset and Snellgrove 2001, Morbidelli et al. 2008). Also we don't know whether Jupiter, Saturn, Uranus and Neptune formed at the same time or in sequence, and whether they formed in a protoplanetary disk still massive or relatively light. There are arguments that indicate that planetary embryos grew while the disk was being evaporated (see Ida et al. 2008), so that it seems plausible that Jupiter would have formed first and thereby acquired the largest mass in hydrogen and helium while Saturn would have formed a little after. In that scenario, Uranus and Neptune would have formed near the end of the lifetime of the protosolar disk. This, however, remains a conjecture given the scarce amount of evidence and constraints. We need to better constrain the structure and precise composition of the giant planets in order to truly understand how the Solar System was formed.

In this mysterious early phase of the formation of the Solar System, two planets stand out, because they were probably the first to appear and had a tremendous impact on the structure of the planetary system that formed around the Sun: namely, Jupiter and Saturn. It is therefore crucial that their characteristics be compared with a similar level of detail. Jupiter's atmospheric composition has been measured by the Galileo probe, and Jupiter's interior will be further examined by the Juno mission. We presently lack similar observations for Saturn. As discussed in Section 23.2, we would want to compare Jupiter and Saturn's central core masses as well as their total mass of heavy elements to know how they formed, and possibly where. A striking example of this need for a comparison is tied to the abundance of noble gases measured in Jupiter by the Galileo probe, but not in Saturn.

In Jupiter, noble gases (Ar, Kr, Xe) are enriched compared to their abundance in the Sun by a factor ∼2. This is puzzling and still unexplained, mostly because noble gases (particularly argon) are very difficult to trap into solids and deliver into the planet's atmosphere. Several explanations have been put forward: (i) Jupiter was formed at very low temperatures, at a place where even argon would be able to condense onto grains (Owen et al. 1999); (ii) Jupiter was impacted with planetesimals made of crystalline ice in which an efficient clathration process occurred (Gautier and Hersant 2005; Alibert et al. 2005); (iii) Jupiter's formation occurred late in a photoevaporating disk in which the midplane had been progressively enriched in all elements capable of sticking to grains in the disk's very cool outer regions, including noble gases (Guillot and Hueso 2006). These scenarios explain Jupiter's case, but yield very different answers at Saturn, as shown by Fig. 23.7. Measuring Saturn's abundance in noble gases would allow a decision between these very different possibilities.

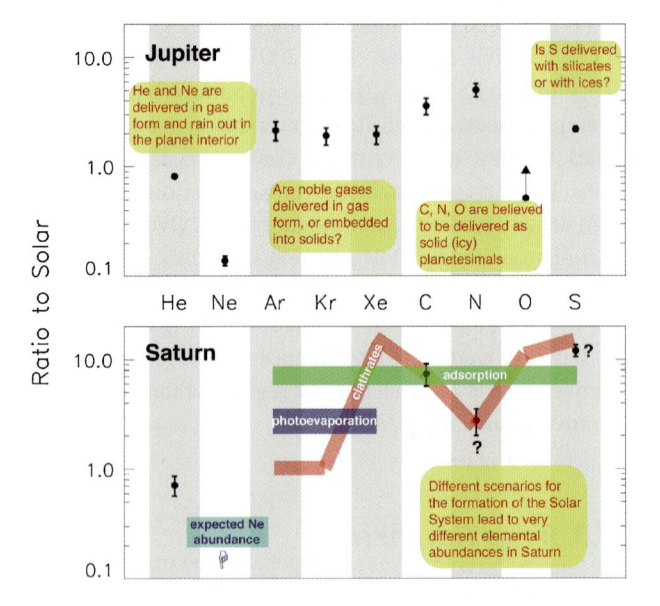

Fig. 23.7 Elemental abundances measured in the tropospheres of Jupiter (*top*) and Saturn (*bottom*) in units of their abundances in the protosolar disk. The elemental abundances for Jupiter are derived from the in situ measurements of the Galileo probe. The abundances for Saturn are spectroscopic determinations from Cassini for He/H and C/H, and model-dependant ground based measurements for N/H and S/H. A determination of the abundance of noble gases in Saturn would allow distinguishing between different formation scenarios whose predictions are shown as green, blue and pink curves, respectively (see text) (adapted from Marty et al. (2009))

23.9 The Means of Saturn's Future Exploration

In this chapter we have attempted to demonstrate that to satisfactorily address the questions of the formation of the outer solar system, at the very minimum a comparative study of the two gas giant planets, Jupiter and Saturn, is essential. While Jupiter has been explored with flybys (Pioneer, Voyager, Cassini), an orbiter (Galileo), a probe (Galileo), and will be further investigated with the Juno orbiter, the complete exploration of Saturn still has long way to go despite the highly successful Cassini orbiter mission. The bulk composition, particularly the abundance of heavy elements, existence and magnitude of the core, deep meteorology and dynamics and internal rotation period are some pieces of the Saturn's puzzle that need to be pursued vigorously in any future exploration of Saturn. Some questions can be addressed with a flyby spacecraft while others may require an orbiter. In either case, entry probes are key. Complementary observations from 1 AU also have an important role to play.

23.9.1 Flyby

A flyby of Saturn is among the simplest solution among space missions. Its scientific return from Saturn would be highly valuable with a properly chosen flyby geometry. With a very close flyby of a few thousand kilometers above Saturn's cloud tops at pericenter, it is possible to obtain an extremely accurate measurement of the planet's gravity field. This would allow to determine whether Saturn's winds extend deep into the interior. This determination and a precise measurement of the gravitational moments of the planet would allow to much better constrain the planet's interior structure. Furthermore, the flyby would yield Saturn's true magnetic field, unfiltered by the rings. As with Juno at Jupiter (Janssen et al. 2005), microwave radiometry from the flyby spacecraft can allow the determination of NH_3 and H_2O in the well-mixed atmosphere, hence the oxygen and nitrogen elemental abundances, that are crucial components of Saturn's bulk composition since water was presumably the original carrier of heavy elements. The determination of remaining heavy elements and isotopes would still require entry probes, however.

23.9.2 Orbiter

To date, most infrared remote sounding has been carried out using nadir methods which, although sensitive to trace constituents, are very limited in vertical resolution. Moreover, a significant part of Saturn's (and Jupiter's) middle atmosphere is still unexplored. This gap, the "ignorosphere", is important to fill, however, for a complete understanding of Saturn's physico-chemical workings. A valuable extension of presently used methods to include systematic limb-sounding could allow much more detailed coverage of the stratosphere, both for dynamics and composition, especially if carried out systematically from a stable, near-circular polar orbiting platform for a significant fraction of a Saturnian year. This extension of the record of horizontal velocity fields from cloud tracking in the visible and infrared would also make a valuable contribution to our understanding of the turbulent nature of Saturn's atmosphere. Again, this would be particularly valuable if carried out systematically and globally, over a long period of time from a suitable orbiting platform, and at higher spatial resolution even than achieved by Cassini in order to resolve individual convective circulations. Even though such measurements would be restricted in altitude coverage to pressure levels less than 1–2 bars, this is probably the only feasible means in the foreseeable future to obtain the kind of detailed, global information on winds and transport of energy, momentum and vorticity necessary for comparison with the most sophisticated numerical models of atmospheric circulation.

At deeper levels, however, the only means of recovering information on bulk motions in the planetary interior would have to come from highly detailed and accurate measurements of the gravity and magnetic fields, as suggested above for a flyby. Following the example of Juno, this ideally requires the placing of a suitably instrumented spacecraft into a very low altitude polar orbit with the aim of measuring the high order gravitational moments $>J_{10}$. As noted by Hubbard (1999), were Jupiter's (or Saturn's) cloud-level banded zonal flow to penetrate barotropically into the deep interior, the need for the pressure field to accommodate a deep geostrophic flow would require adjustments in the interior mass distribution that would manifest themselves in the fine structure of the external gravity field. Because such high order moments decay rapidly in amplitude with distance from the centre of the planet, it is necessary to measure them at very low altitude – in Saturn's case within the inner radius of its ring system. Under current plans, NASA's Juno mission will achieve this objective for Jupiter, and a similar approach will also be necessary to constrain Saturn's deep zonal flow. To complete the picture, if Saturn's banded zonal flows were to penetrate to the deep interior, it is presumably likely that they would perturb the magnetic field on scales similar to the gravity field. Thus, high precision measurements of the high order moments of the magnetic field would be expected to provide important constraints on patterns of deep-seated circulations.

Microwave radiometric measurements done from an orbiter, as opposed to a flyby, will also have the advantage of providing a map of deep NH$_3$ and H$_2$O over Saturn, as at Jupiter with Juno. This would be particularly useful for understanding the convective processes in the deep atmosphere that could prevent these species from being well-mixed in localized regions even at very deep levels of Saturn's atmosphere.

23.9.3 Probes

Entry probes offer the potential advantages of high vertical resolution in the measurement of vertical profiles of temperature, winds, clouds and composition during their descent. They are also capable of measuring the much needed bulk composition and isotopes, including the noble gases, He, Ne, Ar, Kr and Xe, together with their isotope abundances, condensable species, H$_2$O, NH$_3$ and H$_2$S, CH$_4$, isotopic ratios of 14N/15N, 3He/4He, D/H. The well-mixed abundances of only CH$_4$ (infrared) and NH$_3$ and H$_2$O (microwave radiometry from flyby or orbiter) can be measured by remote sensing. Probes are essential for the other measurements in the list. However, as the Galileo probe demonstrated, the condensable species (H$_2$O, NH$_3$ and H$_2$S) may be vulnerable to being unrepresentative of the planet as a whole unless a significant number of locations are sampled. The practicalities of communication with such probes also places limits on the depth from which they can recover information, unless they can be made sufficiently sophisticated to adjust their buoyancy autonomously and resurface to relay the results of their measurements at a later time. But even relatively modest probes could provide much-needed information on the profiles of static stability and humidity that would significantly constrain current models of moist convection on Saturn, especially if this can be recovered from depths greater than 10 bars. Profiles of the concentrations of condensable species would be particularly valuable, especially if a representative range of meteorological phenomena (e.g. belts, zones, plumes and hot-spots) could be sampled. Doppler wind measurements at a number of locations would also allow extension of the cloud-top wind patterns to greater depths, which may be diagnostic of processes maintaining the banded system of jets and clouds.

A probe-orbiter or probe-flyby combination would provide the most crucial information necessary to unravel Saturn's mysteries, as the noble gases and the isotopes can be sampled at relatively shallow depths whereas deep water and ammonia can be mapped with microwave radiometry from an orbiter. The orbiter also has the potential of providing data for discerning the presence of a core.

23.9.4 Microprobe in Saturn's Rings

In order to achieve the rings science objectives listed above, the most valuable project would be an in-situ mission into Saturn's rings. Ideally, a big probe with multiple sensors should be dropped into the rings. However, the technical difficulties to inject the spacecraft into the rings (bring it close to Saturn, brake it to a nearly circular orbit, keep the instruments safe, etc.) and to navigate it (requiring an efficient propulsion system to keep the spacecraft above the ring plane) imply that this is a project for the long-term future. Fortunately, a large fraction of the science objectives could be achieved with a couple of instruments (a camera and/or a spectro-imager) and a communication system onboard very simple probes, called "microprobes". A spacecraft would drop microprobes at different locations into the rings (A,B,C by order of priority) on impact trajectories. On close approach this would provide opportunity for very high resolution images of ring particles. A difficulty would be to put the microprobes on very low inclination trajectories to maximize the time for science return before impacting into the ring material. In addition, if a gap is targeted, and if a probe survives the ring plane crossing, this would allow direct measurement of ring's thickness. Data would have to be collected, compressed and transmitted in real time, either to a spacecraft or directly transmitted to Earth, as was shown to be possible by the Huygens probe from Titan's surface.

23.9.5 Observations from 1AU

Earth-based planetary astronomy has always played an important supporting role in the exploration of the planets and Saturn is no exception. A long-term database is important to discern regular and irregular temporal changes, and is practical only with observations carried out from the earth. Observations in the infrared with ground-based telescopes can provide valuable data on certain stratospheric molecules and the tropospheric meteorology particularly above the ammonia cloud tops. High-resolution submillimeter measurements with the ALMA telescope have the potential of extending the coverage. Space-borne telescopes such as Herschel and the James Webb Space Telescope will undoubtedly open new vistas into Saturn's atmosphere not accessible to previous ground-based or Earth-space observations. It should be stressed, however, the observations made from Earth are only complementary to, not a replacement for, the above probe-flyby/orbiter observations.

23.10 Conclusions

Saturn is a beautiful, intriguing, complex planet. Years of research, observations, experiments, theories, several flybys, measurements in orbit by the Cassini-Huygens spacecraft, all have greatly expanded our knowledge of this giant planet. They have at the same time highlighted or raised many questions that remain unanswered: What is Saturn's interior composition? How does the planet evolve? Why can't we predict its storms? What maintains its strange magnetic field? What is the origin of its rings? How was the planet formed?

A continued exploration of Saturn is essential to understand our Solar System and progress in areas of science as diverse as the study of atmospheric dynamics to dynamo theory and the formation of planetary systems. Future exploration of Saturn will benefit from an appropriately parallel approach to Jupiter's in the past, i.e., with a complement of similar instruments and techniques including in situ measurements. The comparison of similarities and differences will yield a gain of knowledge much greater than the sum of the independent pieces of information obtained for either planet. Even very different theories of the formation of the Solar System could not be tested by measurements in only one planet; they require comparison of key data at least the two gas giant planets.

Saturn's exploration should proceed with a variety of methods. Sending a probe into its atmosphere, even at moderate pressures of a few bars will allow for a unique comparison with the Galileo measurements at Jupiter, including data on the heavy elements, clouds and winds. Measurements of the planet's deep water abundance, interior rotation, magnetic field inside the D ring are extremely important both for understanding the planet's interior, evolution, origin, meteorology and magnetic field. They require either a flyby or a very close orbiter. Similarly, important gains in our understanding of planetary meteorology and dynamics can be made by monitoring Saturn's clouds and weather systems at very high spatial resolution and, ideally, over relatively long timescales. Continued monitoring of the planet remotely with ground- and space-based instruments will be highly valuable, as they will provide complementary and synoptic information. Finally, the study of the rings itself also requires high spatial resolution images to detail its components, something that may be most easily obtained with microprobes.

References

Alibert, Y., Mordasini, C., Benz, W. Migration and giant planet formation. A&A 417, L25–L28 (2004).
Alibert, Y., Mousis, O., Benz, W.: On the volatile enrichments and composition of Jupiter. Astrophys. J. 622, L145–L148 (2005).
Anderson, J. D., Schubert, G.: Saturn's gravitational field, internal rotation, and interior structure, Science, 317, 1384–1387 (2007).
Atreya, S.K., Wong, A.S.: Coupled chemistry and clouds of the giant planets – A case for multiprobes. Space Sci. Rev. 116 (1–2), 121–136 (2005).
Atreya, S.K., Wong, M.H., Owen, T.C., Mahaffy, P.R., Niemann, H.B., de Pater, I., Drossart, P., Encrenaz, Th.: A comparison of the atmospheres of Jupiter and Saturn: Deep atmospheric composition, cloud structure, vertical mixing, and origin. Planet. Space Sci. 47, 1243–1262 (1999).
Baines, K. H., Momary, T. W., Fletcher, L. N., Showman, A. P., Roos-Serote, M., Brown, R. H., Buratti, B. J., Clark, R. N.: Saturn's north polar cyclone and hexagon at depth revealed by Cassini/VIMS. Planet. Space Sci. submitted (2009).
Bakos, G.A., et al.: HAT-P-1b: A large-radius, low-density exoplanet transiting one member of a stellar binary. Astrophys. J. 656, 552–559 (2007).
Blanc M., Bolton, S. Bradley, J. Burton, M. Cravens, T. E. Dandouras, I. Dougherty, M. K. Festou, M. C. Feynman, J. Johnson, R. E. Gombosi, T. G. Kurth, W. S. Liewer, P. C. Mauk, B. H. Maurice, S. Mitchell, D. Neubauer, F. M. Richardson, J. D. Shemansky, D. E. Sittler, E. C. Tsurutani, B. T. Zarka, Ph. Esposito, L. W. Gruen, E. Gurnett, D. A. Kliore, A. J. Krimigis, S. M. Southwood, D. Waite, J. H., Young, D. T. Magnetospheric and plasma science with Cassini-Huygens. Space Sci. Rev. 104, 253–346 (2004).
Briggs, F. H.: Sackett, P. D.: Radio observations of Saturn as a probe of its atmosphere and cloud structure. Icarus 80, 77–103 (1989).
Burton, M. E., Dougherty, M. K., Russell, C. T.: Model of Saturn's internal planetary magnetic field based on Cassini observations, Planet. Space Sci., doi:10.1016 (2009).
Canup R. M., Esposito L. W.: Accretion in the Roche zone: Coexistence of rings and ring moons, Icarus 113, 331–352 (1995).
Cassen, P.: Protostellar disks and planet formation. In *Extrasolar Planets*, M. Mayor, et al. eds., Saas-Fee adv courses, vol. 31, pp. 369–448 (2006), Springer, Berlin.
Chambers, J. E.: Planetary accretion in the inner Solar System. Earth Planet. Sci. Lett. 223, 241–252 (2004).
Charbonneau, D., Brown, T.M., Burrows, A., Laughlin, G.: When extrasolar planets transit their parent stars. Protostars Planets V, 701–716 (2007).
Charnoz, S., Brahic, A., Thomas, P.C., Porco, C.C. The equatorial ridges of Pan & Atlas: Terminal accretionary ornaments? Science 318, 1622 (2007).
Charnoz, S., Morbidelli, A., Dones, L., Salmon, J.: Did Saturn rings form during the Late Heavy Bombardment ? Icarus 199, 413–428 (2009).
Colwell, J.E., Esposito, L.W.: Origins of the rings of Uranus and Neptune. II – Initial conditions and ring moon populations, JGR 98, 7387–7401 (1993).
Colwell, J. E., Esposito, L. W., Sremčević, M.: Self-gravity wakes in Saturn's A ring measured by stellar occultations from Cassini. GeoRL 33, L07201 (2006).
Colwell, J. E., Esposito, L. W., Sremčević, M., Stewart, G. R., McClintock, W. E.: Self-gravity wakes and radial structure of Saturn's B ring. Icarus 190,127–144 (2007).
Conrath, B., Gautier, D.: Saturn helium abundance: A reanalysis of the Voyager measurements. Icarus 144, 124–134 (2000).
Cresswell, P., Nelson, R.P.: Three-dimensional simulations of multiple protoplanets embedded in a protostellar disc. A&A 482, 677–690 (2008).
Cuzzi, J.N., Estrada, P.R.: Compositional evolution of Saturn's rings due to meteoroid bombardment. Icarus 132, 1–35. (1998).
de Pater, I., Dickel, R.: Multifrequency radio observations of Saturn at ring inclination angles between 5 and 26 degrees. Icarus 94, 474–492 (1991).

Daisaka H., Tanaka H., Ida S.: Viscosity in a dense planetary ring with self-gravitating particles. Icarus 154, 296–312 (2001).

Dougherty, M. K., Kellock, S., Southwood, D. J., Balogh, A., Smith, E. J., Tsurutani, B. T., Gerlach, B., Glassmeier, K.-H., Gleim, F., Russell, C. T., Erdos, G, Neubauer, F. M. and Cowley, S. W. H., The Cassini magnetic field investigation. Space Sci. Rev. 114, 331–383 (2004).

Esposito, L.W., O'Callaghan, M., West, R.A. The structure of Saturn's rings: Implications from the Voyager stellar occultation. Icarus 56, 439–452 (1983).

Esposito, L.W., Meinke, B.K., Colwell, J.E., Nicholson, P.D., Hedman, M.H.: Moonlets and clumps in Saturn's F ring. Icarus 194, 278–289 (2008).

Flasar, M., and the CIRS Team: Temperature, winds, and composition in the Saturnian system. Science 307, 1247–1251 (2005).

Fletcher, L. N., Irwin, P. G. J., Teanby, N. A., Orton, G. S., Parrish, P. D., Calcutt, S. B., Bowles, N., Kok, R. de, Howett, C., Taylor, F. W.: The meridional phosphine distribution in Saturn's upper troposphere from Cassini/CIRS observations. Icarus 188, 72–88 (2007).

Fortney, J.J., Hubbard, W.B.: Phase separation in giant planets: inhomogeneous evolution of Saturn. Icarus 164, 228–243 (2003).

Fortney, J.J., Hubbard, W.B.: Effects of helium phase separation on the evolution of extrasolar giant planets. Astrophys. J. 608, 1039–1049 (2004).

Gautier, D., Hersant, F.: Formation and composition of planetesimals – Trapping volatiles by clathration. Space Sci. Rev. Space Sci. Rev. 116, 25–52 (2005).

Gomes, R., Levison, H. F., Tsiganis, K., Morbidelli, A.: Origin of the cataclysmic Late Heavy Bombardment period of the terrestrial planets. Nature 435, 466–469 (2005).

Grevesse, N., Asplund, M., Sauval, J.: The new solar composition. In *Element Stratification in Stars: 40 Years of Atomic Diffusion*, Alecian, G., Richard, O., Vauclair, S. eds., pp. 21–30, EAS Publications Series (2005).

Grossman, A.W.: Microwave imaging of Saturns deep atmosphere and rings. PhD Thesis, California Institute of Technology (1990).

Guillot, T.: The interiors of giant planets: models and outstanding questions. Ann. Rev. Earth Planet. Sci. 33, 493–530 (2005).

Guillot, T.: The composition of transiting giant extrasolar planets. Physica Scripta T 130, 014023 (2008).

Guillot, T., Hueso, R.: The composition of Jupiter: Sign of a (relatively) late formation in a chemically evolved protosolar disc. Mon. Not. R. Astron. Soc. 367, L47–L51 (2006).

Harris, A.: The origin and evolution of planetary rings. In Planetary Rings, Greenberg, R., Brahic, A. eds., University of Arizona Press, Tucson, pp. 641–659 (1984).

Hedman, M.M., Burns, J.A., Tiscareno, M.S., Porco, C.C., Jones, G.H., Roussos, E., Krupp, N., Paranicas, C., Kempf, S.: The source of Saturn's G ring. Science 317, 653–656 (2007).

Hubbard, W.B.: Gravitational signature of Jupiter's deep zonal flows. Icarus 137, 357–359 (1999).

Hubbard, W. B., Guillot, T., Marley, M. S., Burrows, A., Lunine, J. I., Saumon, D. S.: Comparative evolution of Jupiter and Saturn. Plan. Space Sci. 47, 1175–1182 (1999).

Ida, S., Lin, D.N.C.: Toward a deterministic model of planetary formation. I. A desert in the mass and semimajor axis distributions of extrasolar planets. ApJ 604, 388–413 (2004).

Ida, S., Guillot, T., Morbidelli, A.: Accretion and destruction of planetesimals in turbulent disks. ApJ 686, 1292–1301 (2008).

Ikoma, M., Guillot, T., Genda, H., Tanigawa, T., Ida, S.: On the Origin of HD 149026b. Astrophys. J. 650, 1150–1159 (2006).

Janssen, M. A., Hofstadter, M.D., Gulkis, S., Ingersoll, A. P., Allison, M., Bolton, S. J., Levin, S.M., Kamp, L.W. Microwave remote sensing of Jupiter's atmosphere from an orbiting spacecraft, Icarus, 173, 447–453 (2005).

Krimigis, S. M., Mitchell, D. G., Hamilton, D. C., Livi, S. Dandouras, J. Jaskulek, S. Armstrong, T. P. Cheng, A. F. Gloeckler, G. Hsieh, K. C. Ip, W.-H. Keath, E. P. Kirsch, E. Krupp, N. Lanzerotti, L. J. Mauk, B. H. McEntire, R. W. Roelof, E. C. Tossman, B. E. Wilken, B., Williams, D. J.: Magnetosphere Imaging Instrument (MIMI) on the Cassini Mission to Saturn/Titan. Space Sci. Rev. 114(1–4), 233–329 (2004).

Lian, Y., Showman, A. P. Deep jets on gas-giant planets. Icarus 194, 597–615 (2007).

Marouf, E. A., Tyler, G. L., Zebker, H. A., Simpson, R. A., Eshleman, V. R.: Particle size distributions in Saturn's rings from Voyager 1 radio occultation. Icarus 54, 189–211 (1983).

Marty, B., Guillot, T., Coustenis, A., Kronos consortium: Achilleos, N., Alibert, Y., Asmar, S., Atkinson, D., Atreya, S., Babasides, G., Baines, K., Balint, T., Banfield, D., Barber, S., Bézard, B., Bjoraker, G. L., Blanc, M., Bolton, S., Chanover, N., Charnoz, S., Chassefière, E., Colwell, J. E., Deangelis, E., Dougherty, M., Drossart, P., Flasar, F. M., Fouchet, T., Frampton, R., Franchi, I., Gautier, D., Gurvits, L., Hueso, R., Kazeminejad, B., Krimigis, T., Jambon, A., Jones, G., Langevin, Y., Leese, M., Lellouch, E., Lunine, J., Milillo, A., Mahaffy, P., Mauk, B., Morse, A., Moreira, M., Moussas, X., Murray, C., Müller-Wodarg, I., Owen, T. C., Pogrebenko, S., Prangé, R., Read, P., Sanchez-Lavega, A., Sarda, P., Stam, D., Tinetti, G., Zarka, P., Zarnecki J.: Kronos: exploring the depths of Saturn with probes and remote sensing through an international mission. Exp. Astron. 23, 977–980 (2009).

Masset, F., Snellgrove, M.: Reversing type II migration: Resonance trapping of a lighter giant protoplanet. MNRAS 320, L55-L59 (2001).

Morbidelli, A., Crida, A., Masset, F., Nelson, R.P.: Building giant-planet cores at a planet trap. A&A 478, 929–937 (2008).

Nicholson, P.D., Showalter, M.R., Dones, L., French, R.G., Larson, S.M., Lissauer, J.J., McGhee, C.A., Seitzer, P., Sicardy, B., Danielson, G.E.: Observations of Saturn's ring-plane crossing in August and November 1995. Science 272, 509–515 (1996).

Nicholson, P.D., Hedman, M.M., Clark, R.N., Showalter, M.R., Cruikshank, D.P., Cuzzi, J.N., Filacchione, G., Capaccioni, F., Cerroni, P., Hansen, G.B., Sicardy, B., Drossart, P., Brown, R.H., Buratti, B.J., Baines, K.H., Coradini A.: A close look at Saturn's rings with Cassini VIMS. Icarus 193, 182–212 (2008).

Owen, T., et al.: A low-temperature origin for the planetesimals that formed Jupiter. Nature 402, 269–270 (1999).

Pollack, J.B., Hubickyj, O., Bodenheimer, P., Lissauer, J.J., Podolak, M., Greenzweig, Y.: Formation of the giant planets by concurrent accretion of solids and gas. Icarus 124, 62–85 (1996).

Porco, C. C., Thomas, P. C., Weiss, J. W., Richardson, D. C.: Saturn's small inner satellites: Clues to their origins. Science 318, 1602 (2007).

Poulet, F., Cruikshank, D.P., Cuzzi, J.N., Roush, T.L., French, R.G.: Composition of Saturn's rings A, B, and C from high resolution near-infrared spectroscopic observations. Astron. Astrophys. 412, 305–316 (2003).

Salo, H.: Simulations of dense planetary rings. III. Self-gravitating identical particles. Icarus 117, 287–312 (1995).

Sato, B., et al.: The N2K consortium. II. A transiting hot Saturn around HD 149026 with a large dense core. Astrophys. J. 633, 465–473 (2005).

Saumon, D., Guillot, T.: Shock Compression of deuterium and the interiors of Jupiter and Saturn. Astrophys. J. 609, 1170–1180 (2004).

Shepelyansky, D.L., Pikovsky, A.S., Schmidt, J., Spahn, F. Synchronization mechanism of sharp edges in rings of Saturn. MNRAS 395, 1934–1940 (2009).

Stevenson, D.J.: Interiors of the giant planets. Ann. Rev. Earth Planet. Sci. 10, 257–295 (1982).

Stevenson, D.J., Salpeter, E.E.: The phase diagram and transport properties for hydrogen-helium fluid planets. Astrophys. J. Suppl. Ser. 35, 221–237 (1977).

Stewart, G. R., Robbins, S. J., Colwell, J. E.: Evidence for a primordial origin of Saturn's rings. 39th DPS meeting, abstract no. 7.06 (2007).

Thommes, E.W., Matsumura, S., Rasio, F.A.: Gas disks to gas giants: Simulating the birth of planetary systems. Science 321, 814 (2008).

Tiscareno, M.S., Burns, J.A., Hedman, M.M., Porco, C.C.: The population of propellers in Saturn's A ring. Astron. J. 135, 1083–1091 (2008).

Toomre, A.: On the gravitational stability of a disk of stars. ApJ 139, 1217–1238 (1964).

Tsiganis, K., Gomes, R., Morbidelli, A., Levison, H. F.: Origin of the orbital architecture of the giant planets of the Solar System. Nature 435, 459–461 (2005).

Wuchterl, G., Guillot, T., Lissauer, J.J.: The formation of giant planets. In *Protostars & Planets IV*, Mannings, V., Boss, A.P., Russell, S. S., eds., p. 1081, Tucson: University of Arizona Press (2000).

Chapter 24
Cartographic Mapping of the Icy Satellites Using ISS and VIMS Data

Th. Roatsch, R. Jaumann, K. Stephan, and P.C. Thomas

Abstract The sizes and shapes of six icy Saturnian satellites have been measured from Cassini Imaging Science Subsystem (ISS) data, employing limb coordinates and stereogrammetric control points. Mimas, Enceladus, Tethys, Dione and Rhea are well described by triaxial ellipsoids; Iapetus is best represented by an oblate spheroid. The ISS acquired many high-resolution images (<1 km/pixel) during close flybys of the medium-sized icy Saturnian satellites (Mimas, Enceladus, Tethys, Dione, Rhea, Iapetus, and Phoebe). We combined these images with lower-resolution coverage and a few images taken by Voyager cameras to produce high-resolution mosaics of these satellites. The global mosaics are the baseline for high-resolution atlases. The atlases consist of 15 tiles each for Enceladus, Dione, and Tethys, whereas the Iapetus, Mimas, and Phoebe atlases consist of 3, 1, and 1 tile, respectively. The nomenclature used in these atlases was suggested by the Cassini-ISS team and approved by the International Astronomical Union (IAU). The whole atlases are available to the public through the Imaging Team's website (http://ciclops.org/maps/) and from the Planetary Data System (PDS, http://pds-imaging.jpl.nasa.gov/). Additionally to ISS, the Visual and Infrared Mapping Spectrometer (VIMS) onboard the Cassini spacecraft detected the chemical and physical surface properties of the Saturnian satellites. Multiple VIMS observations were combined into global VIMS maps representing the VIMS coverage achieved during the nominal Cassini mission. Progressed mapping has been done for the satellites Dione, Rhea, and Enceladus.

24.1 Introduction

The Cassini Imaging Science Subsystem (ISS) consists of two framing cameras. The narrow angle camera is a reflecting telescope with a focal length of 2,000 mm and a field of view of 0.35°. The wide angle camera is a refractor with a focal length of 200 mm and a field of view of 3.5°. Each camera is outfitted with a large number of spectral filters which, taken together, span the electromagnetic spectrum from 0.2 to 1.1 μm. At the heart of each camera is a charged coupled device (CCD) detector consisting of a 1,024 square array of pixels, each 12 microns on a side. The data system allows many options for data collection, including choices for on-chip summing and data compression. The stated objective of the ISS is to obtain global coverage for all medium-sized icy satellites with a resolution better than 1 km/pixel and high-resolution images of selected areas (Porco et al. 2004). This goal was achieved with image sequences obtained during close flybys supplemented by images from greater distances to complete the coverage. Close flybys of all medium sized satellites except Mimas were executed during the nominal mission of the Cassini spacecraft. The first flybys during the mission were those of Phoebe in June 2004 and Iapetus in December 2004 followed by three flybys of Enceladus in February, March, and July 2005 (Porco et al. 2005, 2006).

As an imaging spectrometer the Visual Infrared Mapping Spectrometer (VIMS) onboard the Cassini spacecraft combines the characteristics of both a spectrometer and an imaging instrument. This makes it possible to analyze the spectrum of each pixel separately and to map the spectral characteristics spatially, which is important to study the relationships between the spectral information and geological and/or geomorphological surface features. VIMS operates in a spectral range from 0.35 to 5.2 μm generating spectral image cubes in which each pixel represents a spectrum of 352 contiguous wavebands (Brown et al. 2004). It is the first imaging spectrometer operating in the Saturnian system and providing spectral data with sufficient pixel ground resolution up to 2 km/pixel. This allows the mapping of spectral variations related to physical and chemical surface properties across the satellites surfaces that could be combined into composition maps of a specific satellite.

In this work we first review in Section 24.2 the method of shape determination and evaluation for the icy satellites, followed by reporting the measurement results for the satellites. We then examine how close each moon is to an equilibrium shape. Then we view the results in total, and finally evaluate the significance of departures from equilibrium forms. Details of the image processing that is necessary to generate the global mosaics of the medium-sized satellites are described in Section 24.3. Section 24.4 summarizes the cartographic work that produced our high-resolution atlas. During Cassini's nominal tour VIMS was able to observe especially the Saturnian satellites Dione and Enceladus with sufficient pixel ground resolution and signal-to-noise ratio (SNR) to combine the VIMS results into regional and global maps as described in Section 24.5. The final section describes our plans for the extended mission.

24.2 Shapes and Sizes of the Saturnian Satellites

The shapes and sizes of the icy Saturnian satellites, required for mapping and global geophysical interpretations, are derived primarily by limb profile analysis for the ellipsoidal satellites (Mimas, Enceladus, Tethys, Dione, Rhea, and Iapetus) and by stereo control points and limb matching for the smaller, irregularly-shaped ones. Limb analysis is described in Thomas et al. (1998); small satellite shape modeling in Simonelli et al. (1993).

Images suitable for limb analysis are full-disk with phase angles <85°, plus higher phase images in which part of the solar-shadowed region has sufficient Saturn shine for measurement. Transits of a satellite across the illuminated disk of Saturn provide the best measurements, essentially full disk limb data. Limb measurement can be made with precisions of order 0.1 pixels, but in practical terms the roughness of the limb limits center-fixing accuracy, hence the accuracy of the calculated shape. Greater geographical distribution of the limb tracks enhances the accuracy of the solution. A rough object such as Tethys can yield very different triaxial solutions with only a few ground tracks, then the results of the fit usually stabilize as more profiles are added.

The progressive refinement of the shape solution of Tethys with increasing image coverage is shown in Fig. 24.1. Portions of limbs can provide some information about local topography, such as large crater depth profiles, rim profiles, and relief on suitably aligned scarps.

The Cassini results greatly improved the Voyager data. Dione and Rhea's mean radii were uncertain to several km, Iapetus' shape and mean radius have improved drastically (more than 10 km) and the shapes of Enceladus and Mimas are sufficiently refined to change interpretations. The small satellites were only approximately measured prior to Cassini data.

Shapes are useful in scientific interpretation as well as for cartographic work. Among the ellipsoidal satellites, the most revealing shape is that of Iapetus. As noted by Castillo et al. (2007), the oblate spheroid shape of this object indicates a frozen figure from a time of more rapid spin (∼16 h period). Subsequent limb data have only strengthened the conclusion that this oblate form represents a fossil bulge from rapid rotation.

The shapes of the other satellites all come close to those of equilibrium, effectively fluid, ellipsoids. In Tables 24.1 and 24.2 we list the derived global properties, one of which, $F = (b - c) / (a - c)$ can be an indicator of how close to an equilibrium shape (here assumed fluid equilibrium) the object is. The observed values of F and the predicted ones are also listed in Table 24.2. The predicted values of F decrease slightly for the very fast rotating satellites (Chandrasekhar 1969; Dermott and Thomas 1988). While many of the observed F values are close to the ideal 0.25–0.21, only Tethys and Dione actually satisfy the relationship at the measurement uncertainty. Yet the data, and the smoothness of limb topography (Thomas et al. 2007) suggest all these objects have approached equilibrium forms.

The chief scientific interest is in the modest departures from the idealized shapes. For instance, while Enceladus' F parameter is fractionally quite distinct from the theoretical value, it in fact represents less than 2 km deviation from that ideal. The (a–c) value for Enceladus is appropriate for

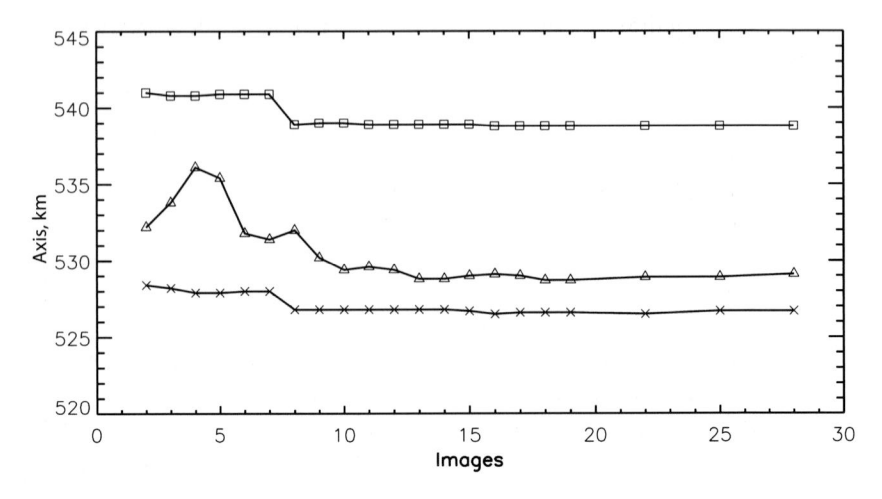

Fig. 24.1 Tethys axial fits with increasing number of images

Table 24.1 Satellite shapes from limb data

Satellite	a	b	c	Mean radius	a–c	im	Data
Mimas	207.8 ± 0.5	196.7 ± 0.5	190.6 ± 0.3	198.2 ± 0.4	17.2 ± 0.5	36	21,527
Enceladus	256.6 ± 0.6	251.4 ± 0.2	248.3 ± 0.2	252.1 ± 0.2	8.3 ± 0.6	34	23,226
Tethys	538.4 ± 0.3	528.7 ± 1.7	526.3 ± 0.6	531.1 ± 0.6	12.1 ± 0.9	28	18,128
Dione	563.4 ± 0.6	561.3 ± 0.5	559.6 ± 0.4	561.4 ± 0.4	3.8 ± 0.7	43	33,229
Rhea	766.2 ± 0.6	762.8 ± 1.0	762.4 ± 1.2	763.8 ± 1.0	3.8 ± 1.1	38	29,606
Iapetus	746.0 ± 2.9	746.0 ± 2.9	712.0 ± 1.6	734.5 ± 2.8	34.0 ± 3.6	51	14,607

Here we use a–c (all in km) to denote the Saturn-facing, orbit-facing, and polar radii. im: number of images used; data: number of data points.

Table 24.2 Satellite shape parameters

Satellite	Mean radius (km)	ρ (kg/m^3)	a–c (km)	(a–c)h (km)	F_{obs}	F_{pred}
Mimas	198.2 ± 0.4	1,149 ± 7	17.2 ± 0.6	19.7	0.35 ± .02	0.21
Enceladus	252.1 ± 0.2	1,609 ± 5	8.3 ± 0.6	8.0	0.37 ± 0.04	0.23
Tethys	531.1 ± 0.6	984 ± 3	12.1 ± 0.9	14.7	0.28 ± 0.20	0.24
Dione	561.4 ± 0.4	1,478 ± 3	3.8 ± 1.2	4.9	0.20 ± 0.37	0.25
Rhea	763.8 ± 1.0	1,236 ± 5	3.8 ± 1.5	2.9	0.05 ± 0.20	0.25
Iapetus	734.5 ± 2.8	1,088 ± 13	34.0 ± 3.6	0.01	n/a	0.25

ρ, mean density, determined from masses reported in Jacobson et al. (2006).
(a–c) h is predicted (a–c) for homogeneous model.

Table 24.3 Sizes of irregularly-shaped Saturnian satellites

Satellite	A	B	C	Mean radius	Mass	ρ
Pa	17.4 ± 2.0	15.8 ± 1.3	10.4 ± 0.84	14.2 ± 1.3	0.495 ± .075	410 ± 150
Daphni	4.5 ± 0.8	4.3 ± 0.8	3.1 ± 0.9	3.9 ± 0.8	0.0077 ± 0.0015	310 ± 200
Atla	20.9 ± 1.4	18.1 ± 2.5	8.9 ± 0.8	15.1 ± 1.4	0.66 ± 0.06	440 ± 190
Prometheus	67.1 ± 3.0	39.6 ± 3.0	30.0 ± 1.8	43.1 ± 2.6	15.68 ± 0.20	470 ± 90
Pandora	52.1 ± 1.8	40.6 ± 2.0	32.0 ± 0.9	40.7 ± 1.5	13.58 ± 0.23	480 ± 60
Epimetheus	62.6 ± 1.5	54.9 ± 2.4	51.4 ± 0.7	56.8 ± 1.6	53.10 ± 0.14	690 ± 130
Janus	100.3 ± 1.7	94.7 ± 1.0	76.1 ± 1.1	89.8 ± 1.4	191.37 ± 0.005	640 ± 64
Methone	1.6 ± 0.6	1.6 ± 0.6	1.6 ± 0.6	1.6 ± 0.6		
Pallene	2.6 ± 0.4	2.2 ± 0.3	1.8 ± 0.2	2.2 ± 0.3		
Telesto	15.8 ± 0.6	11.7 ± 0.3	10.2 ± 0.3	12.3 ± 0.4	–	
Calypso	15.0 ± 0.3	11.5 ± 2.3	7 ± 0.6	10.6 ± 0.7	–	
Polydeuces	1.5 ± 0.6	1.2 ± 0.4	1.0 ± 0.2	1.3 ± 0.4		
Helene	19.4 ± 0.2	18.5 ± 1.0	12.3 ± 1.0	16.5 ± 0.6		
Hyperion	–	–	–	135 ± 4	561.99 ± 5	544 ± 50
Phoebe	107.7 ± 1.1	108.6 ± 2.3	101.5 ± 0.4	106.6 ± 1	829.2 ± 1	1,629 ± 50

Radii a–c are given in km, Mass in 10^{19}g and density in kg/m^3.
Size data have been updated from Porco et al. (2007) for Prometheus, Pandora, Epimetheus, and Janus. Mean radii are from shape models and may differ slightly from calculations using the ellipsoidal axis fits. Masses, in units of 10^{19} g, are those reported in Porco et al. (2007) from different sources.

a homogeneous equilibrium body. However, the overwhelming expectation (Porco et al. 2006) is that the interior has had enough heating to allow substantial segregation of materials and thus a simple onion layer model will leave the observed surface departing from an equipotential. The solution presumably involves lateral changes in crustal density, topography on any core, or lateral changes in mantle properties.

Rhea's shape also suggests it is not a highly differentiated, relaxed object.

The more irregularly shaped, smaller bodies may represent effects of accretion or effects of overlapping large impacts or even complete disruption. Apart from Atlas and Pan, none show strong signs of accretionary remnants. Atlas and Pan have smoother, equatorial bulges, which may reflect interactions with ring particles after the ring system collapsed to its present small thickness (Porco et al. 2007; Charnoz et al. 2007). It has been found that complex figures can be supported by unconsolidated bodies with characteristics of a "sand pile", which may play a role in some small object shapes (Minton 2008). Telesto has considerable regolith filling of craters, but it does not affect the overall shape and its regolith appears to be undergoing active downslope motion. Phoebe's shape is more equidimensional than most irregular satellites, but the range of topography relative to estimated equipotentials (28 km) and the obvious topographic effects of large craters indicates this object does not show significant effects of global viscous relaxation (Porco et al. 2005) (Table 24.3).

24.3 Global Basemaps Derived from Cassini-ISS Images

24.3.1 Data Processing

Though the Cassini-ISS camera takes images using many different filters (Porco et al. 2004), we used only images taken with the filters CL1, CL2 or GRN, as these images show similar contrast. The processing of the Cassini images follows the typical processing chain for framing cameras: radiometric correction, geometric correction and map projection, and mosaicking (Roatsch et al. 2006). For the Cassini mission, spacecraft position and camera pointing data are available in the form of SPICE kernels (http://naif.jpl.nasa.gov). While the orbit information is sufficiently accurate to be used directly for mapping purposes, the pointing information must be corrected using limb fits (Roatsch et al. 2006) or least-squares adjustment methods if enough stereo data are available as for Enceladus (Roatsch et al. 2008a). High-resolution images that do not contain the limb were registered to limb images to improve the pointing.

24.3.2 Coordinate System

The coordinate system adopted by the Cassini mission for satellite mapping is the IAU "planetographic" system, consisting of planetographic latitude and positive West longitude. The surface position of the prime meridians are defined by Davies and Katayama (1983a–c, 1984) and adopted by the IAU cartography working group as standard (Seidelmann et al. 2007) are defined by small craters. Our mosaics that were calculated using rotational parameters derived from Voyager data and the limb-fitted attitude data or the attitude data that were improved using a least-squares adjustment had a slight offset to this definition. Therefore we decided to shift the whole mosaics to be consistent with the IAU longitude definition. This requires also an update of the rotational parameters, which are used for the calculation of the prime meridian location, the so called rotational parameter W_0 (Seidelmann et al. 2007). Table 24.4 summarizes the reference craters, the necessary shift and the new values for W_0.

24.3.3 Basemaps

Digital global mosaics that are also called basemaps were prepared in simple cylindrical projection, a special case of equirectangular projection. The mapping cylinder is tangent to the equator of the sphere, the longitude range is 0°–360° W and latitude range –90° to 90° (Kirk et al. 1998). The prime meridian is in the center of the map. All basemaps were calculated using Cassini images. Voyager images were used to fill remaining gaps. The only exception is the Rhea basemap that is still based on Voyager data (Roatsch et al. 2006) overlayed by higher-resolution Cassini images. The calculation of a Cassini based Rhea basemap will be part of our future work.

Figures 24.2–24.8 show the basemaps of the medium sized satellites. We marked the most prominent features with their names. Some features were already named based on the Voyager images. Many new feature names were suggested by the Cassini imaging time and approved by the IAU. New Cassini based feature names are not yet available for Rhea. In naming the features on the Saturnian satellites, the Working Group for Planetary System Nomenclature within the International Astronomical Union (IAU) has expanded the mythological theme first used on the Jovian satellites. On the Saturnian satellites, however, the features bear names derived from the great epics and legends of the world (Batson 1984; http://planetarynames.wr.usgs.gov/). A summary of the epics used for feature names in the Saturnian system is given in Table 24.5.

24.4 High-Resolution Atlases

High-resolution atlases were produced to conform to the design and standards of the USGS airbrush maps and photomosaics, established by Greeley (1990), widely used in planetary cartography. The selection of the atlas format depends on the resolution of the mosaics and the size of the satellites. Three different formats were used for the generation of the atlases (Figs. 24.9–24.11):

– Synoptic format for making planetwide maps on a single sheet, used for Phoebe and Mimas (Roatsch et al. 2006, 2009)
– Subdivision of the synoptic format for making planetwide maps with four quadrangles on three sheets, used for Iapetus (Roatsch et al. 2009)

Table 24.4 New values of rotational parameter W_0

Satellite	Reference crater	Longitude shift (West)	W_0
Mimas	Palomides	−4.0	333.46
Enceladus	Salih	3.5	6.32
Tethys	Arete	−1.5	8.95
Dione	Palimurus	0.6	357.6
Rhea	Tore	0	235.16
Iapetus	Almeric	5.0	355.2

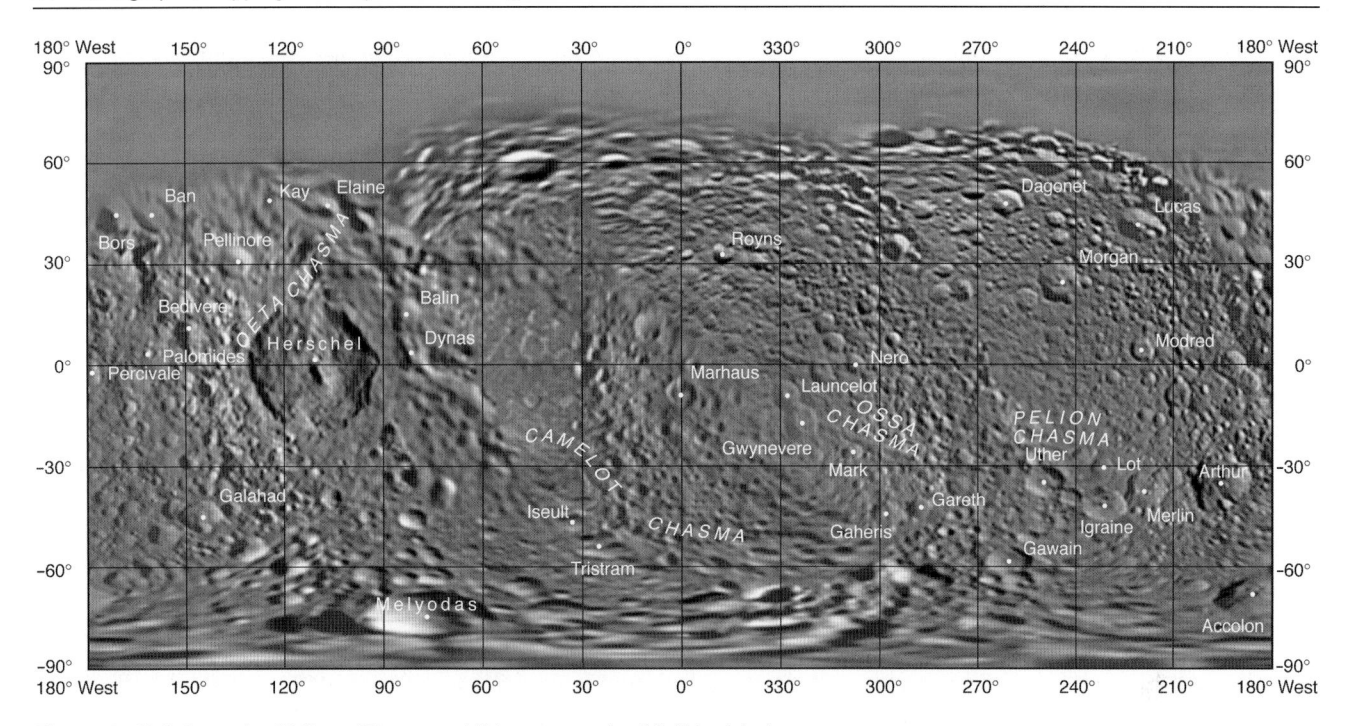

Fig. 24.2 Global mosaic of Mimas. The map exhibits a map scale of 0.43 km/pixel

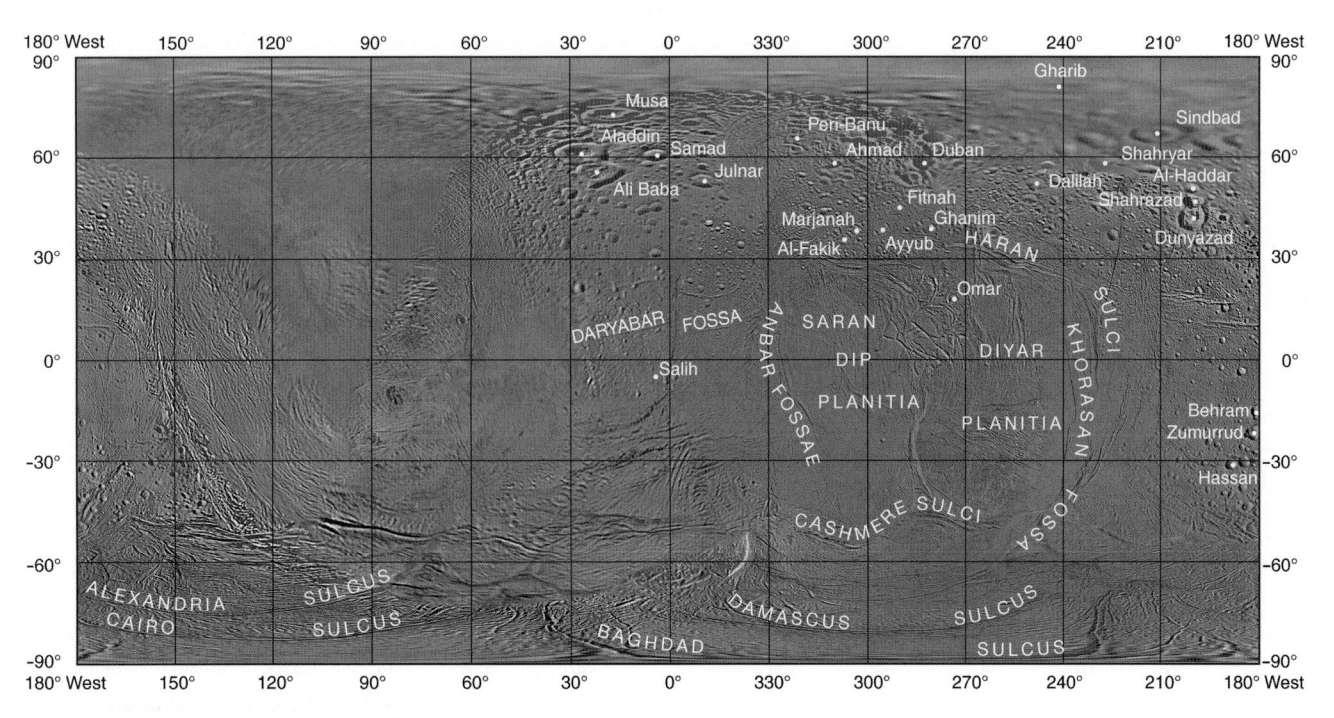

Fig. 24.3 Global mosaic of Enceladus. The map exhibits a map scale of 0.11 km/pixel

- 15 quadrangles format for medium sized bodies and high-resolution imaging, used for Enceladus, Tethys, and Dione (Roatsch et al. 2008a, b)

Three map sheets of Enceladus, Tethys, and Dione are shown in Figs. 24.12–24.14 as examples of the atlases. We did not generate a high-resolution atlas of Rhea so far since a new Cassini basemap has yet to be produced. Resolution and scale of the atlases are given in Table 24.6.

The Enceladus atlas was generated using a basemap, which was calculated using the flyby data from 2004 to 2005, no longitude shift as described in Section 24.3.2 was applied for

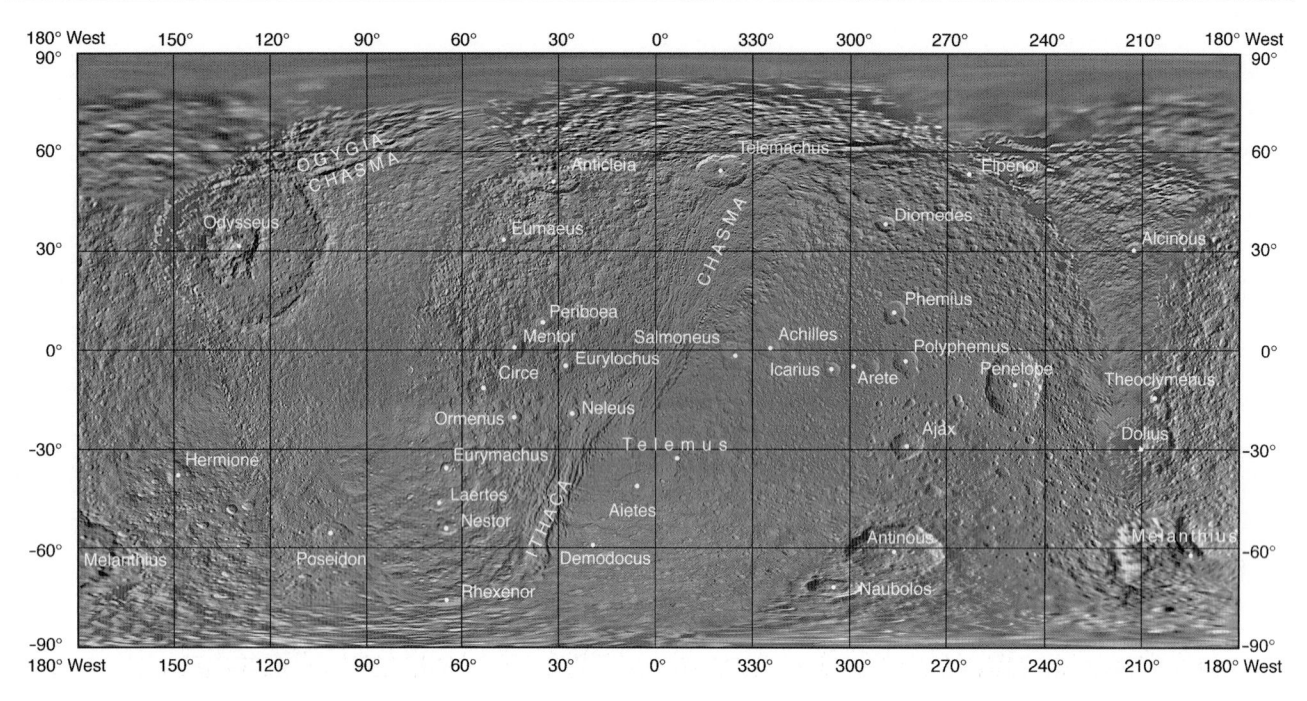

Fig. 24.4 Global mosaic of Tethys. The map exhibits a map scale of 0.29 km/pixel

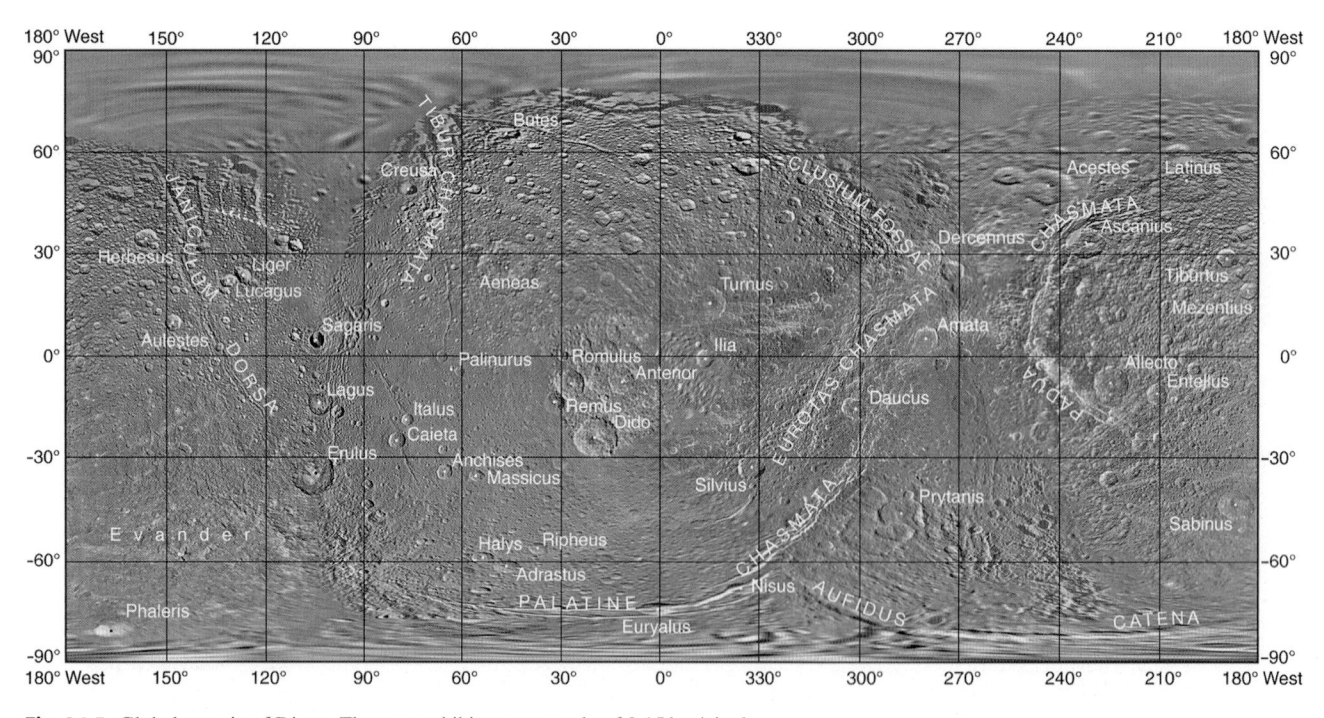

Fig. 24.5 Global mosaic of Dione. The map exhibits a map scale of 0.15 km/pixel

this basemap and the current atlas (Roatsch et al. 2008a). Crater Salih, which determines the longitude system on Enceladus, was imaged with high resolution during the flyby in March 2008, which allowed us to determine the necessary longitude shift also for Enceladus. The basemap shown in Fig. 24.3 and the south pole map sheet shown in Fig. 24.12 were shifted accordingly. A new version of the complete atlas is planned for 2010 after the end of the Cassini extended mission.

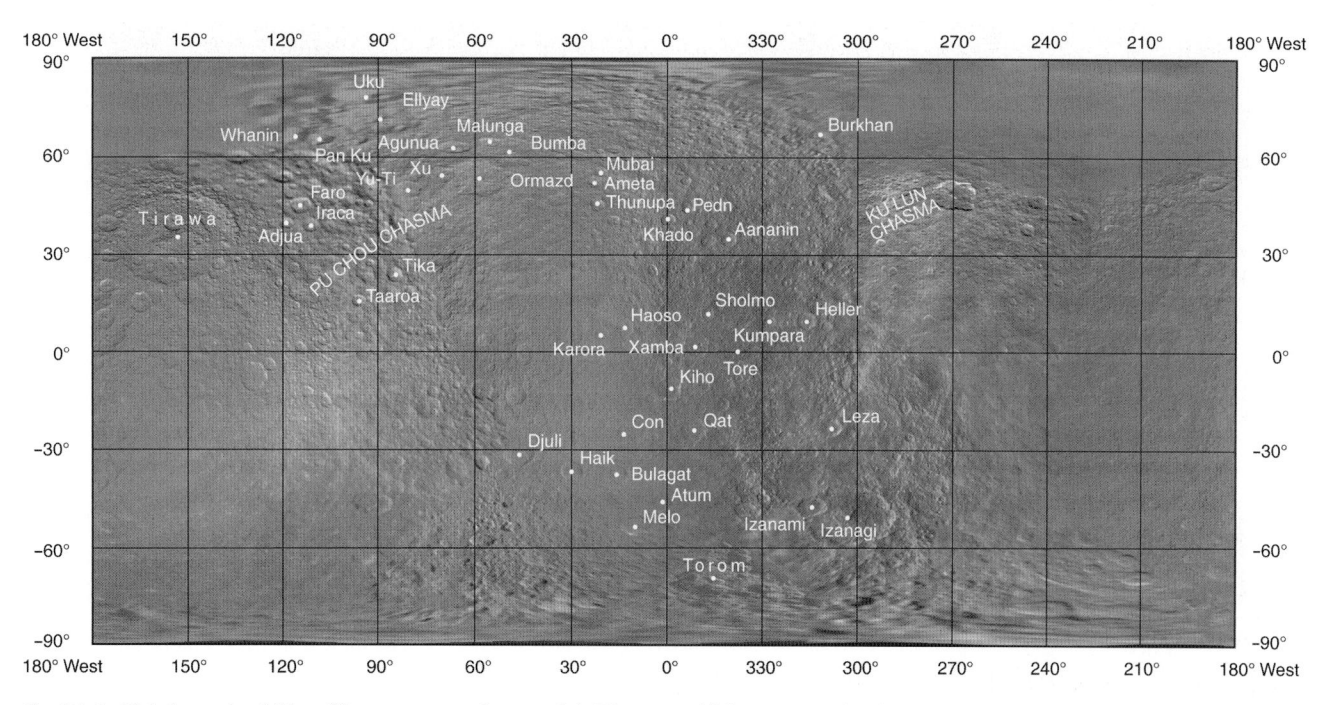

Fig. 24.6 Global mosaic of Rhea (Voyager nomenclature only). The map exhibits a map scale of 0.67 km/pixel

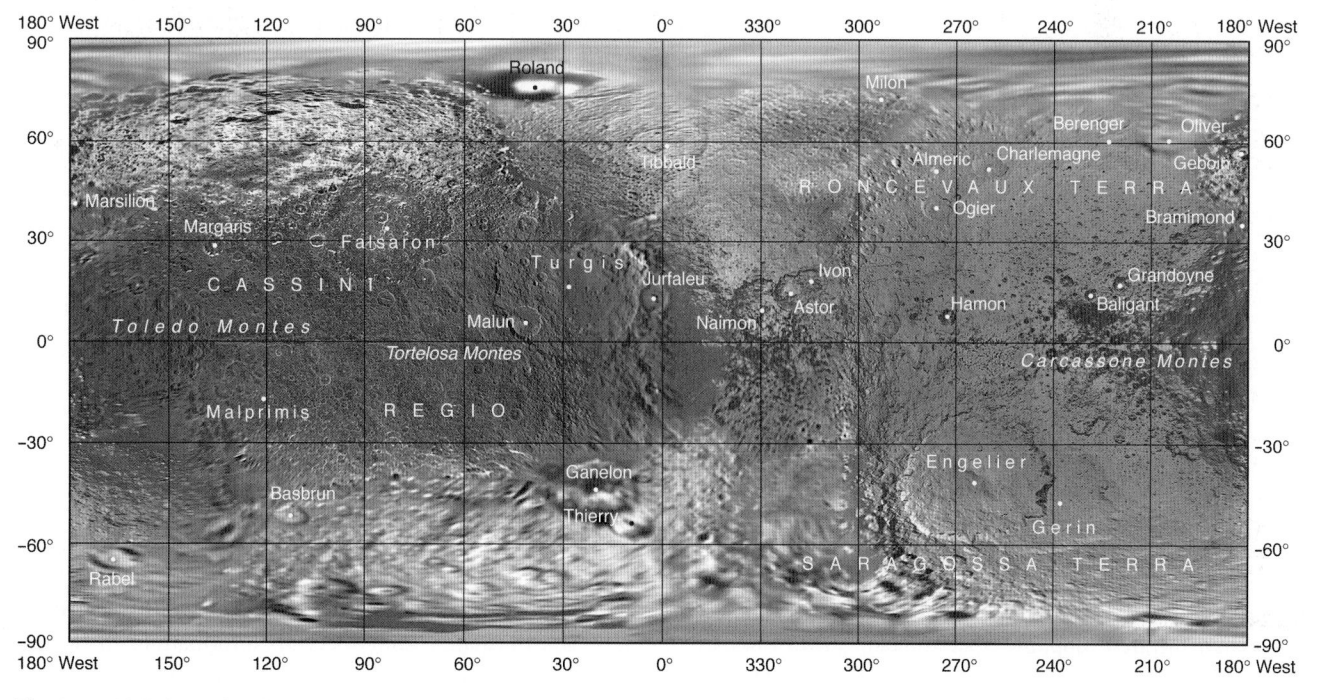

Fig. 24.7 Global mosaic of Iapetus. The map exhibits a map scale of 0.8 km/pixel

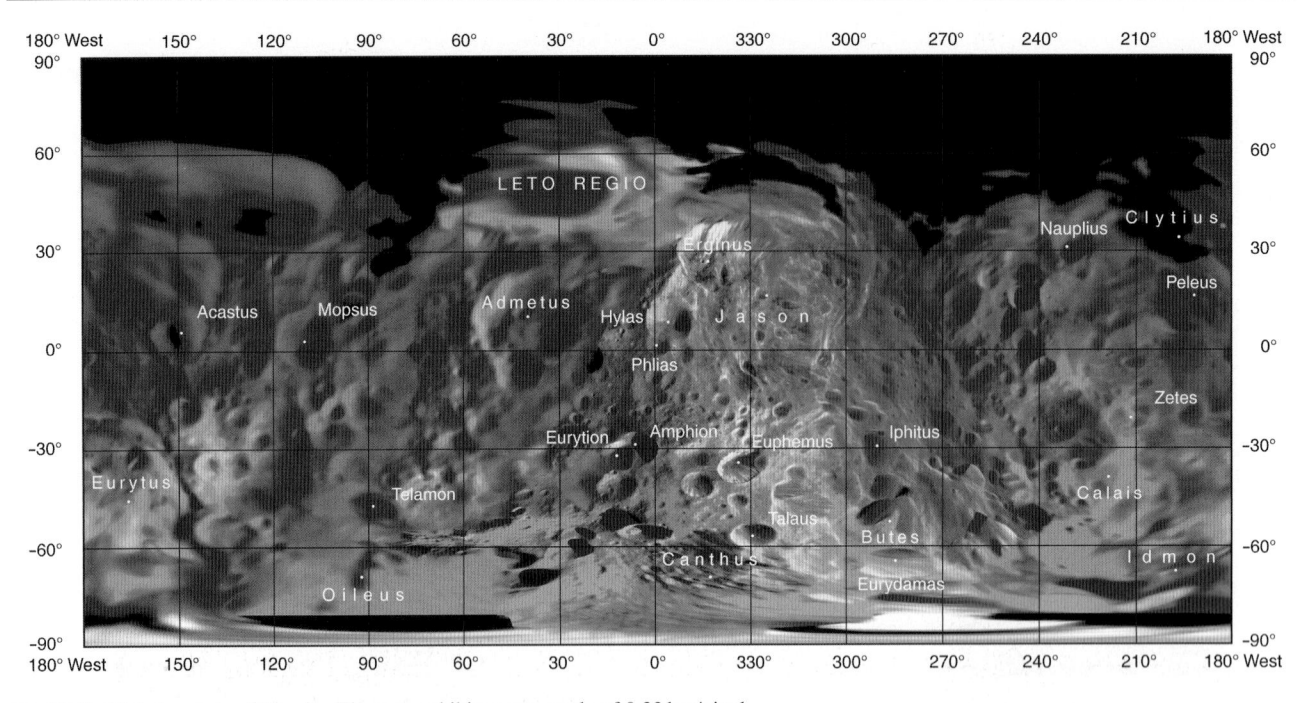

Fig. 24.8 Global mosaic of Phoebe. The map exhibits a map scale of 0.23 km/pixel

Table 24.5 Epics used for feature names on the Saturnian satellites

Satellite	Epic	References
Mimas	Le Morte d' Arthur	Baines (1962)
Enceladus	The Thousand Nights and a Night	Burton (1900)
Tethys	The Odyssey of Homer	Bates (1929)
Dione	The Aeneid of Virgil	Mandelbaum (1972)
Rhea	Names are characters and places from creation myths selected from various cultures around the world; Asian names were emphasized.	
Iapetus	The Song of Roland	Sayers (1967)
Phoebe	The Argonautica	Mozley (1934)

24.5 Compositional Maps Derived from Cassini-VIMS Data

24.5.1 Data Processing

The Visual Infrared Mapping Spectrometer VIMS onboard the CASSINI spacecraft obtained new spectral data of the icy Saturnian satellites after its arrival at Saturn in June 2004. VIMS operates in a spectral range from 0.35 to 5.2 μm, generating image cubes in which each pixel represents a spectrum consisting of 352 contiguous wavebands (Brown et al. 2004). As an imaging spectrometer VIMS combines the characteristics of both a spectrometer and an imaging instrument. This makes it possible to analyze the spectrum of each pixel separately and to map the spectral characteristics spatially, which is important to study the relationships between spectral information and geological and geomorphologic surface features.

VIMS maps presented here show the global coverage achieved by VIMS during Cassini's nominal mission. The re-projection and mosaicking procedures that were applied to the VIMS data are described in detail by Jaumann et al. (2006) and include the following steps:

1. The spatial analysis of the spectral data requires the determination of the exact geographic position of each pixel on the specific surface and that all 352 spectral elements of each pixel show the same region of the target. Therefore the position of each pixel was geometrically corrected and the spectral data converted into map projected image cubes. A nearest-neighbor algorithm was used in order to not modify the original spectra and the map resolution of the map-projected VIMS cubes oversamples the original pixel ground resolution at least by a factor of two to avoid any loss of spatial (and spectral) information. In order to guarantee the accuracy of the VIMS maps and mosaics their quality have been checked by comparison with maps of Voyager and Cassini ISS imaging data (Roatsch et al. 2006). Usually the accuracy of maps based on lower spatial resolution VIMS data is within the limit of one VIMS pixel. If the inaccuracy exceeds one pixel, an additional registration of the VIMS data to ISS basemaps has been applied.

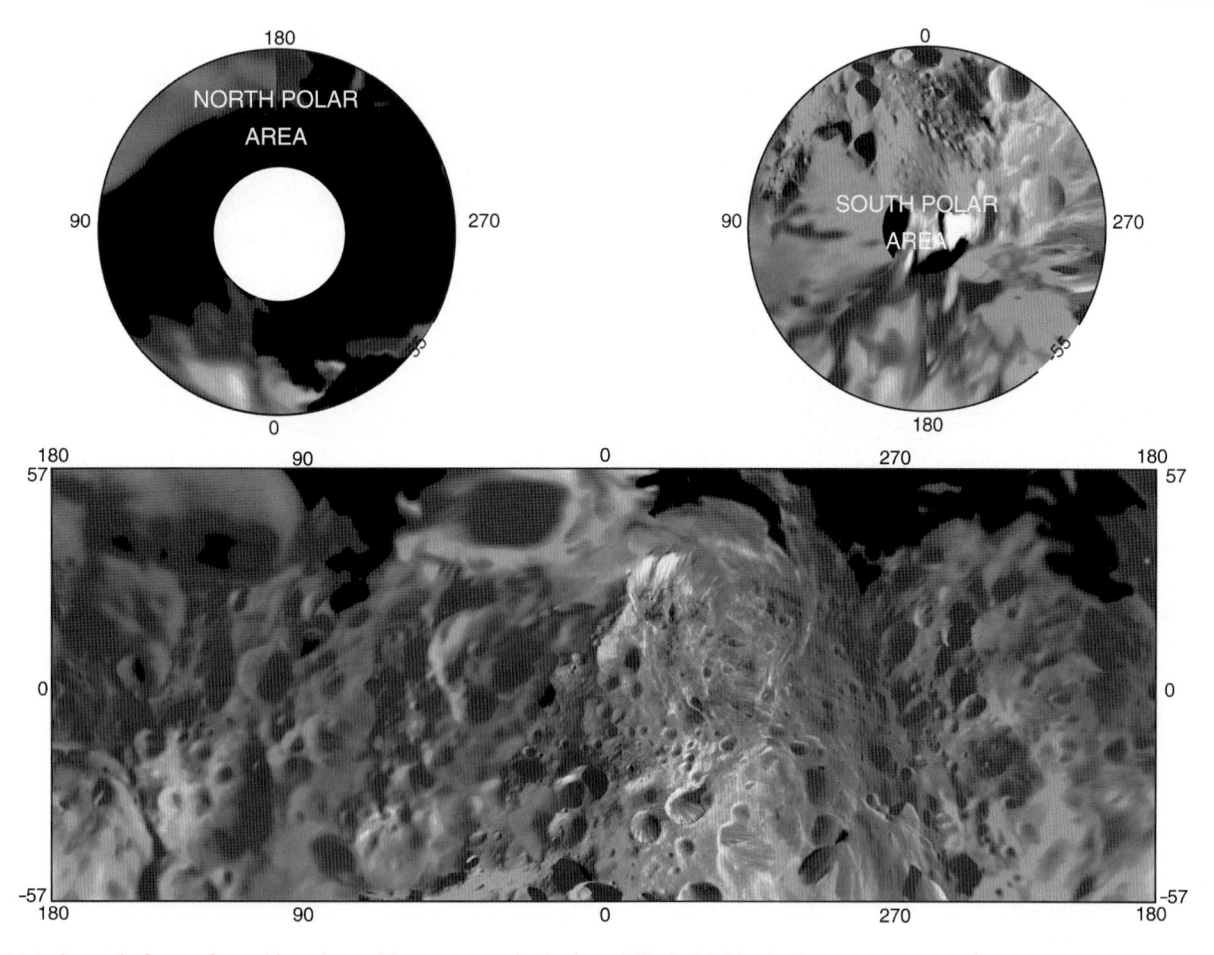

Fig. 24.9 Synoptic format for making planetwide maps on a single sheet. Filled with Phoebe data

2. In order to reduce the influences of the viewing geometry (i.e. incidence, emission and phase angle), that greatly varies between the VIMS observations acquired during Cassini's numerous flybys, in the resulting global VIMS maps, key spectral parameters like the band depths of water ice absorptions were derived according to Clark and Roush (1984) and Clark et al. (2003) for each single map-projected VIMS cube. These band depths are known to indicate changes in the abundance of main surface compounds i.e. water ice and visually dark non-ice materials in the surface material of icy satellites (Clark et al. 2008; Stephan et al. 2009b) but also the sizes and/or crystallinity of water ice particles (Jaumann et al. 2008).

3. Finally, multiple VIMS band depth maps were combined into global VIMS mosaics. Individual projected VIMS maps that were incorporated into a VIMS mosaic were selected according to the following criteria: 1. pixel ground resolution, 2. signal-to-noise ratio, and 3. illumination conditions. Only VIMS cubes with an original pixel ground resolution of at least 150 km/pixel were included into the final mosaics. The individual maps were sorted by pixel ground resolution and signal-to-noise ratio, and the image cube with the highest resolution is located on top of the mosaic. A map resolution of 1 km/pixel was chosen for the final mosaics defined by the VIMS cube with the highest spatial resolution of the specific set. No limb observations were used in the mosaicking process. All resulting VIMS maps were overlaid onto Cassini basemaps in order to relate spectral properties to the surface geology and morphology seen in ISS data.

Progressed mapping by the VIMS instrument during Cassini's nominal mission has been done for the Saturnian satellites Dione, Rhea and Enceladus and will be presented in detail below.

24.5.2 VIMS Composition Map of Dione

Dione was observed during two targeted flybys (16, 50) in October 2005 and September 2007 and also during the

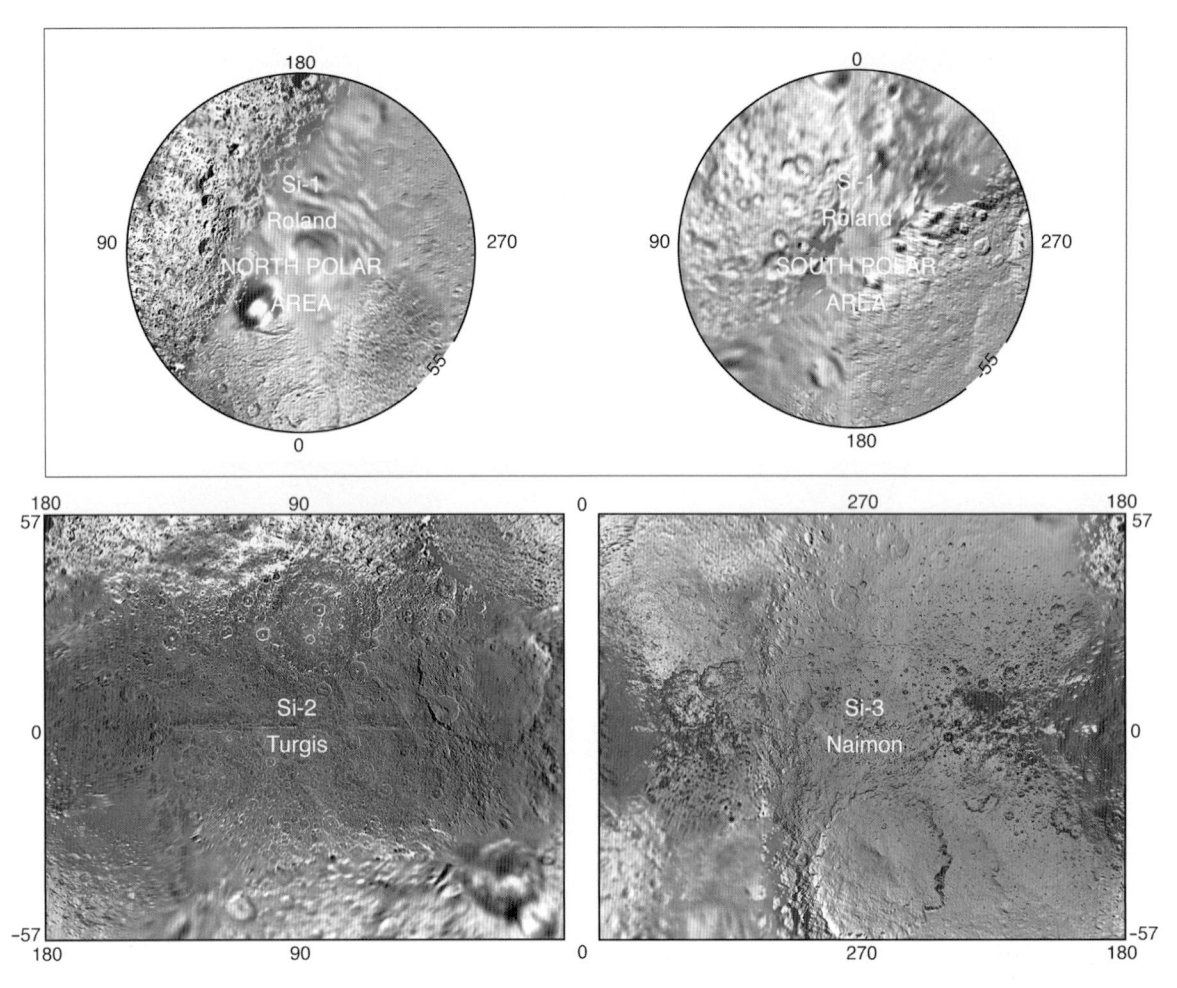

Fig. 24.10 Subdivision of the synoptic format for making planetwide maps on three sheets. Filled with Iapetus data

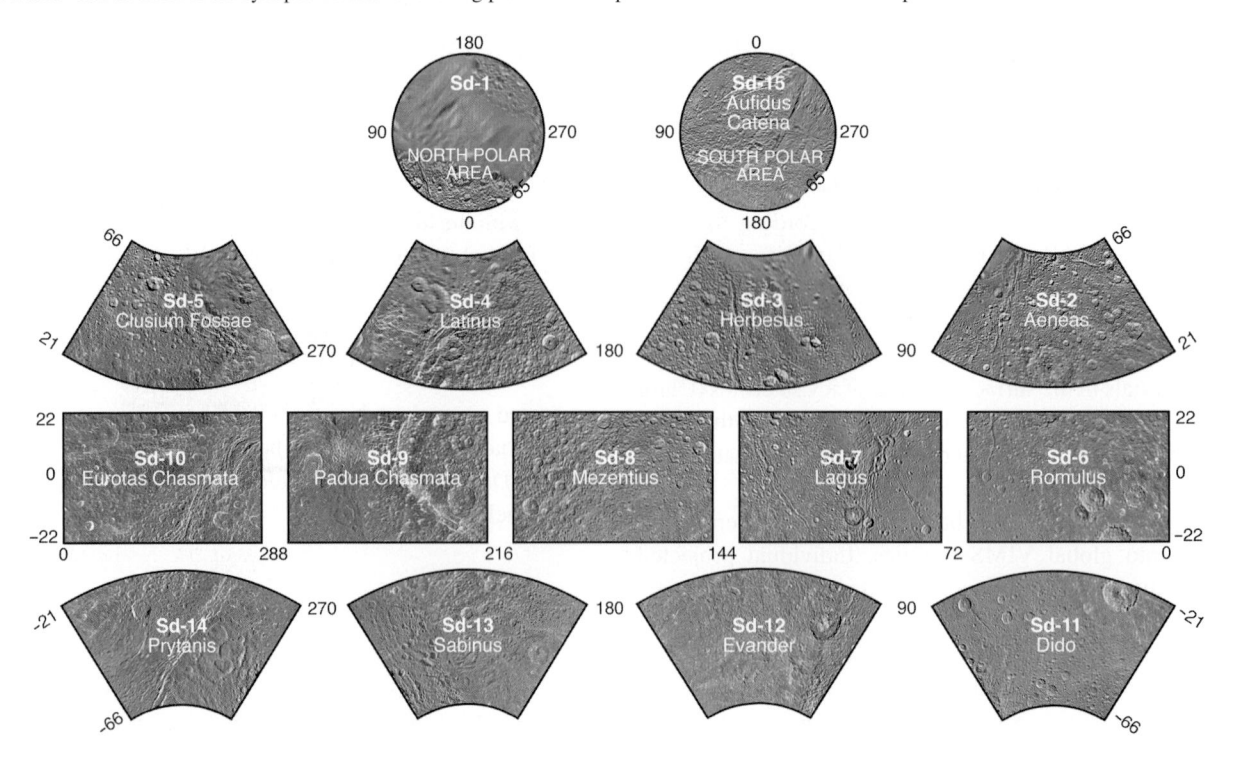

Fig. 24.11 The schema for medium sized bodies and high-resolution imaging with 15 quadrangles. Filled with Dione data

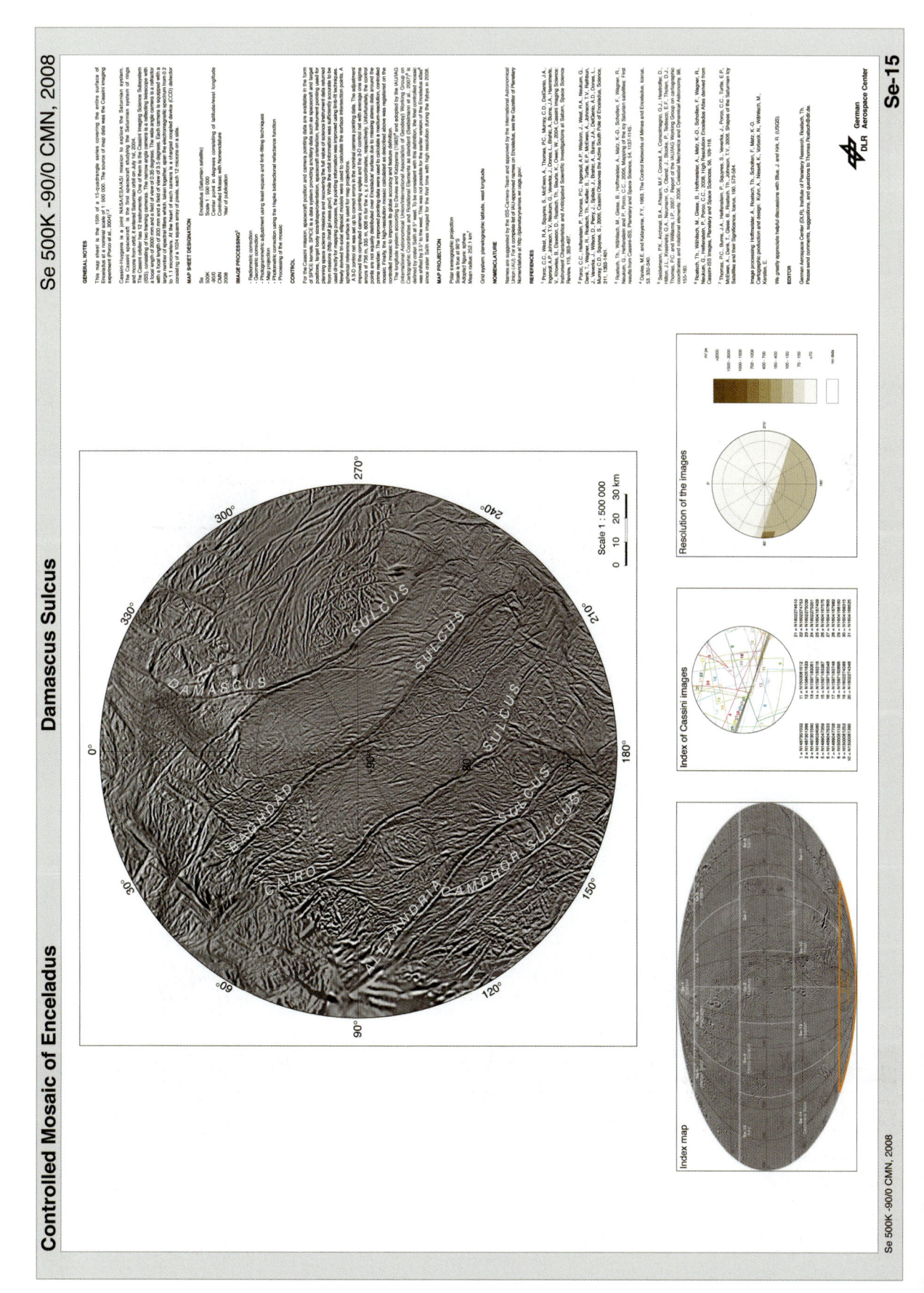

Fig. 24.12 Enceladus map sheet 15: Damascus Sulcus

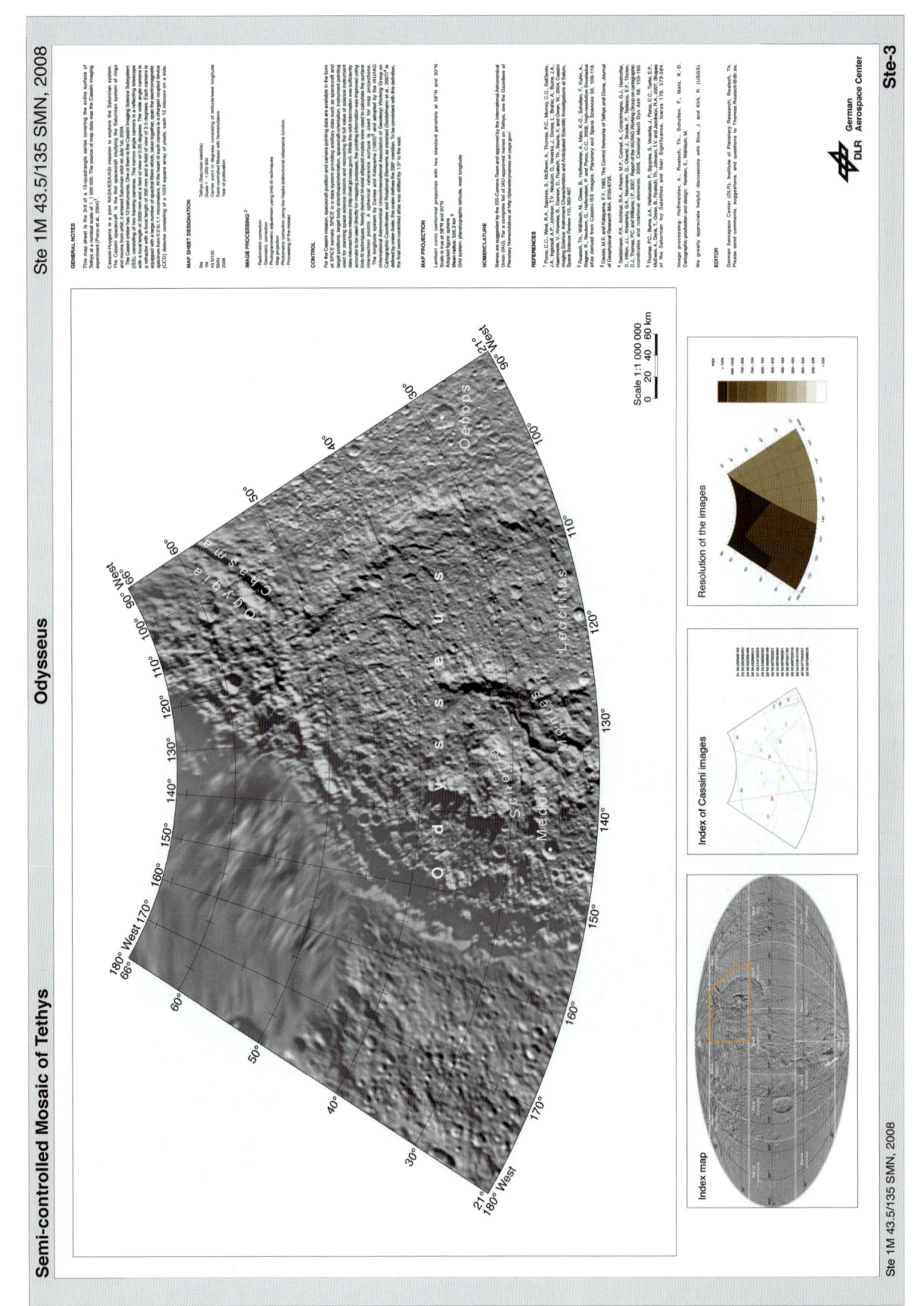

Fig. 24.13 Tethys map sheet 3: Odysseus

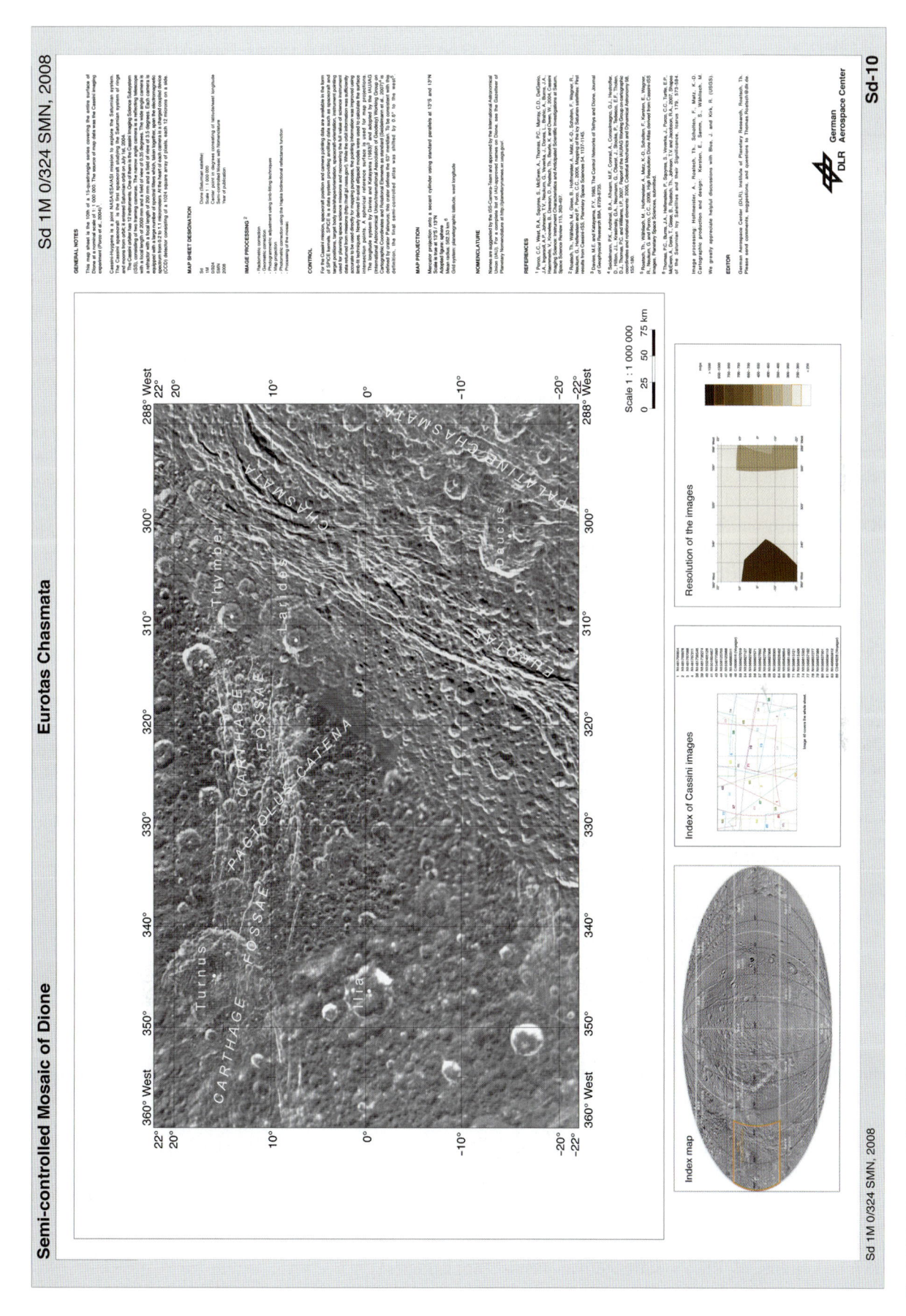

Fig. 24.14 Dione map sheet 10: Eurotas Chasmata

Table 24.6 Resolution and scale of the mosaics and atlases

Satellite	Mean radius used for the map projection (km)	Resolution of the map digmap (pixel/°)	Map scale of the atlas
Mimas	198.8	8	1:1,500,000
Enceladus	252.1	40	1:500,000
Tethys	536.3	32	1:1,000,000
Dione	562.53	64	1:1,000,000
Rhea	764.0	20	–
Iapetus	736.0	16	1:3,000,000
Phoebe	106.8	8	1:1,000,000

non-targeted flybys B, 9, 26, 27 and 43 with sufficient pixel ground resolution of better than 150 km/pixel (Table 24.7). This resulted into a complete VIMS coverage of this satellite. Especially the trailing hemisphere is best covered by high-resolution observations. Best pixel ground resolutions (<5 km/pixel) were achieved during the observation sequence DIONE205 in orbit 16 with the closest approach to Dione's surface of 498 km (Stephan et al. 2009b). During orbit B and 50 the satellite was still observed with 10 to 40 km/pixel. However, during orbits 9, 26, and 27 pixel ground resolutions only occasionally are better than 40 km/pixel. Nevertheless the acquired VIMS data provide the context information to the highly resolved VIMS data. This is also the case for Dione's leading hemisphere which was observed mainly during orbit 43 (Table 24.7).

Figure 24.15 shows a global VIMS map of Dione that illustrates the color coded spatial variations of the water ice absorption depth at 1.5 μm. Especially, the increasing amount of the dark non-ice material in the surface material of Dione, that is concentrated on its trailing hemisphere, is known to reduce the depth of this water ice absorption significantly. For scientific discussion the reader is referred to the work of Clark et al. (2008) and Stephan et al. (2009b).

24.5.3 VIMS Composition Map of Rhea

Since July 2004 numerous flybys were performed at Saturn's satellite Rhea (Table 24.8). VIMS data acquired during 11 flybys (up to Cassini's 54th orbit in December 2007) exhibiting pixel ground resolutions of at least 150 km/pixel were incorporated into the global VIMS map in Fig. 24.16 (Stephan et al. 2009a). Despite of the polar regions, Rhea's surface has been observed by VIMS almost completely. However, most observations exhibit relatively low pixel ground resolutions of less than 40 km/pixel. Only VIMS observations of Rhea's anti-Saturnian hemisphere acquired during one non-targeted flyby (Cassini orbit 49 on Aug. 30, 2007) reach pixel ground resolutions of about 1 km/pixel.

Like presented for Dione the resulting map shows the variations in band depth of the water ice absorption at 1.5 μm dominated by the varying amount of water ice and visually dark rocky material (Stephan et al. 2009a).

24.5.4 VIMS Composition Map of Enceladus

The Saturnian satellite Enceladus was observed during the targeted flybys 3, 4, 11 and 61 in March and July 2005 and in March 2008, respectively, as well as the quasi-targeted flyby 3 in February 2005 with sufficient pixel ground resolution and signal-to-noise ratio. Best pixel ground resolutions achieved during orbit 11 reached 1 km/pixel (Table 24.9) and covers especially the trailing hemisphere between 180° and 360° W including Enceladus' South Polar Region. In contrast, the coverage of the leading hemisphere from 0° and 180° W is still very rare and remains to be a completed during Cassini's extended (equinox) mission through the Saturnian system.

Figure 24.17 shows the global VIMS map produced for Enceladus. Like shown for Dione and Rhea this color coded map also illustrates the variations in depth of the water ice absorption at 1.5 μm. However, the surface material on Enceladus consists of pure water ice (Brown et al. 2006; Jaumann et al. 2008). Therefore, variations in band depth are here indicative to the varying sizes of the water ice particles as described in the work of Jaumann et al. (2008). Figure 24.18 shows the particle size map separately for the South Polar Region of Enceladus including the so called 'tiger stripes' (Porco et al. 2006).

24.6 Future Work

The Cassini spacecraft will continue its imaging campaign through the Saturnian system through at least 2010. The upcoming flybys will help to replace the low-resolution parts of the mosaics and atlases with higher resolution image data derived from ISS images. This will be also an important step for the VIMS coverage of Saturn's satellites in addition to the completion of VIMS coverage of Enceladus and the remaining mid-sized satellites Iapetus, Tethys and Mimas that were not or only preliminary mapped so far. The northern part of the Saturnian satellites will be illuminated during the extended (equinox) mission providing a unique opportunity to obtain high-resolution Cassini coverage of high northern latitudes.

24 Cartographic Mapping of the Icy Satellites Using ISS and VIMS Data

Table 24.7 VIMS observations of Dione combined into the global map in Fig. 24.15

Cassini orbit	Observation time	VIMS Observations sequences	Number of cubes	Geographic location	Average pixel ground resolution of the input cubes (km/pixel)
00B	Dec. 2004	REGMAP001	3	47°N–52°S/276–350°W	25–17.4
		GLOCOL001	2	60°N–55°S/140–280°W	~38
009	June 2005	DIONE001	13	35°N–70°S/164–275°W	79.3–40.8
016	Oct. 2005	DIONE205	32	60°N–85°S/96–260°W	44.5–2.9
026	July 2006	GLOCOL001	4	60°N–60°S/245–340°W	63.8–64.8
027	Aug. 2006	REGEODB001	30	67°N–37°S/240°–320°W	66.5–97.4
		REGEODC001	24	60°N–17°S/220°–330°W	102–150
043	Apr. 2007	REGMAPB001	5	40°N–90°S/100–360°W	29.5–60.6
050	Sep. 2007	DIONE001	4	60°N–75°S/350–250°W	27.2–23.3
		DIONE002	6	70°N–60°S/276–67°W	13.3–29.1
		REGMAPE001	17	60°N–75°S/250–320°W	16.2–32.6
		REGMAPF001	31	76°N–70°S/270–50°W	10.0–21.5
		REGMAPG001	13	67°N–70°S/85–310°W	24.3–87.4

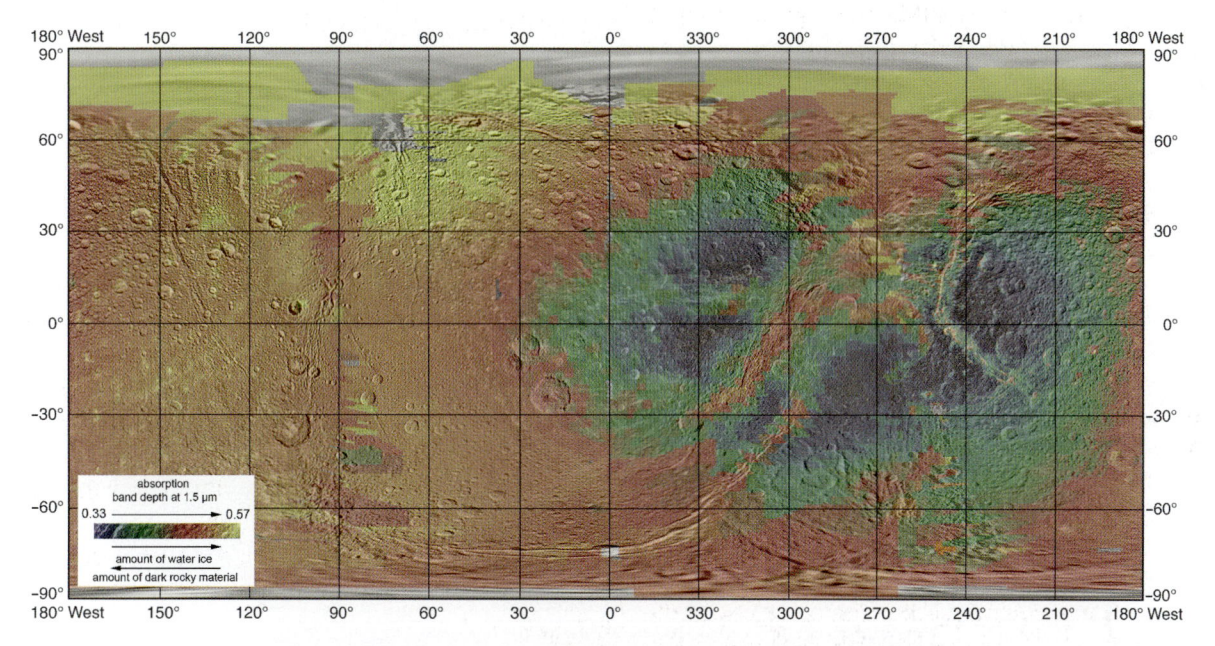

Fig. 24.15 Global VIMS composition map of Dione based on VIMS observations summarized in Table 24.7 in a simple cylindrical map projection overlaid onto the corresponding ISS basemap. The map exhibits a map scale of 1 km/pixel (Stephan et al. 2009b)

Table 24.8 VIMS observations of Rhea combined into the global map in Fig. 24.16

Cassini orbit	Observation time	VIMS Observations sequences	Number of cubes	Geographic location	Average pixel ground resolution of the input cubes (km/pixel)
00C	Jan. 2005	RHEA003	4	38°N–87°S/178–334°W	38.4–36.3
		RHEA004	4	44°N–80°S/346–174°W	40.8–39.6
		RHEA005	3	42°N–76°S/48–186°W	43.0–41.7
		RHEA006	4	55°N–77°S/25–174°W	46.5–45.2
		RHEA007	4	52°N–76°S/44–182°W	54.3–50.7
		RHEA102	2	88°N–68°S/250–36°W	120.3–119.7
003	Feb. 2005	RHEA101	5	64°N–90°S/10–170°W	84.8–81.0
005	Mar. 2005	RHEA001	12	60°N–75°S/15–160°W	66.3–33.1
006	Apr. 2005	RHEA001	8	60°N–88°S/10–165°W	59.3–57.3

(continued)

Table 24.8 (continued)

Cassini orbit	Observation time	VIMS Observations sequences	Number of cubes	Geographic location	Average pixel ground resolution of the input cubes (km/pixel)
011	July 2005	RHEA001	2	20°N–90°S/0°–360°W	60.5–60.0
		RHEA002	8	10°N–88°S/0°–360°W	56.7–50.5
		RHEA003	6	30°N–90°S/330°–270°W	82.4–79.5
		RHEA004	4	6°S–90°S/0–360°W	43.6–43.1
012	Aug. 2005	RHEA001	16	26°N–90°S/0–360°W	64.6–61.6
		RHEA003	6	34°N–90°S/0–360°W	52.7–52.0
		RHEA004	3	40°N–90°S/100–360°W	55.7–55.5
018	Nov. 2005	RHEA002	8	88°N–64°S/90°–248°W	48.0–29.1
		RHEA111	16	60°N–75°S/350–250°W	42.5–32.9
		RHEA112	14	40°N–40°S/110°–180°W	19.7–16.2
019	Dec. 2005	RHEA101	14	70°N–64°S/148°–270°W	81.5–74.5
020	Jan. 2006	RHEA002	6	74°N–54°S/230°–15°W	57.2–54.7
022	Mar. 2006	RHEA002	4	63°N–70°S/268°–28°W	34.9–34.45
		RHEA003	9	32°N–45°S/168°–10°W	33.8–28.7
		RHEA004	3	62°N–72°S/285°–46°W	26.8–26.4
		RHEA006	3	58°N–75°S/310°–48°W	23.1–22.6
049	Aug. 2007	RHEA001	6	68°N–66°S/350°–80°W	39.5–32.2
		RHEA010	34	58°N–72°S/96°–340°W	6.28–1.32
		RHEA011	19	72°N–48°N/−90°–240°W	8.48–6.87

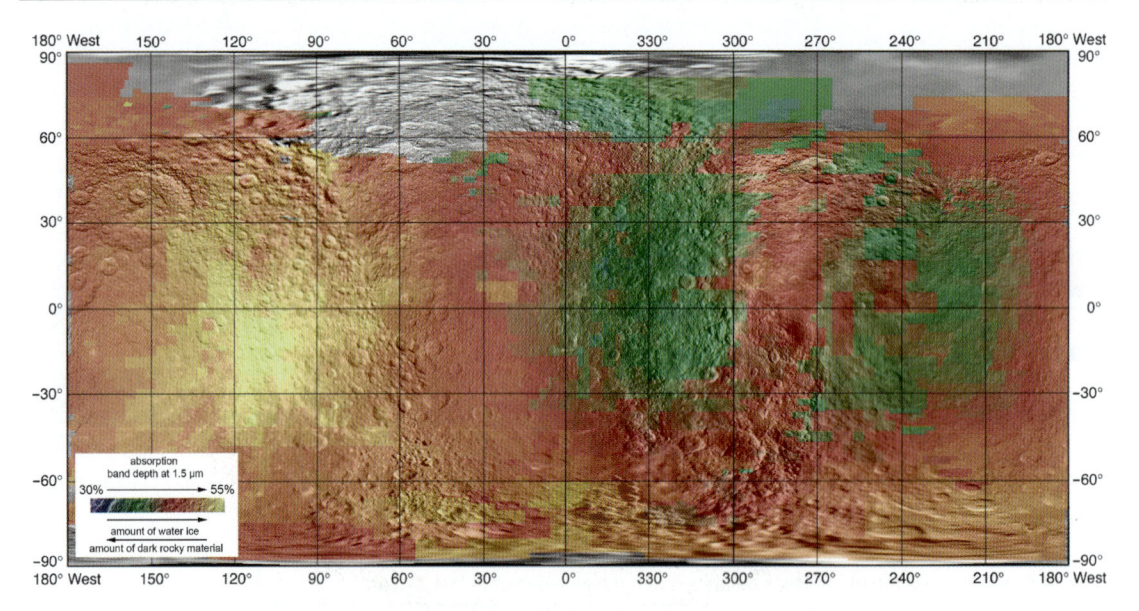

Fig. 24.16 Global VIMS composition map of Rhea based on VIMS observations summarized in Table 24.8 in a simple cylindrical map projection pixel overlaid onto the corresponding ISS basemap. The map exhibits a map scale of 1 km/pixel (Stephan et al. 2009a)

Table 24.9 VIMS observations of Enceladus that were combined into the global map in Fig. 24.17

Cassini orbit	Observation time	VIMS Observations sequences	Number of cubes	Geographic location	Average pixel ground resolution of the input cubes (km/pixel)
003		ENCELADUS007	13	68°N–77°S/2°–229°W	7.7–2.6
004	Mar 2005	ENCELADUS019	7	70°N–30°S/160–260°W	10.2–13.2
011	July 2005	ENCELADUS108	4	30°N–90°S/160–280°W	30.3–13.2
		ENCELADUS110	5		18.2–1.0
		ENCELADUS111	14		6.85–3.6
061		ENCELADUS002	14	90°–8.5°N/250°–60°W	22.5–18.5
		ENCELADUS005	2		25.8–25.4
		ENCELADUS007	3		8.2–6.8

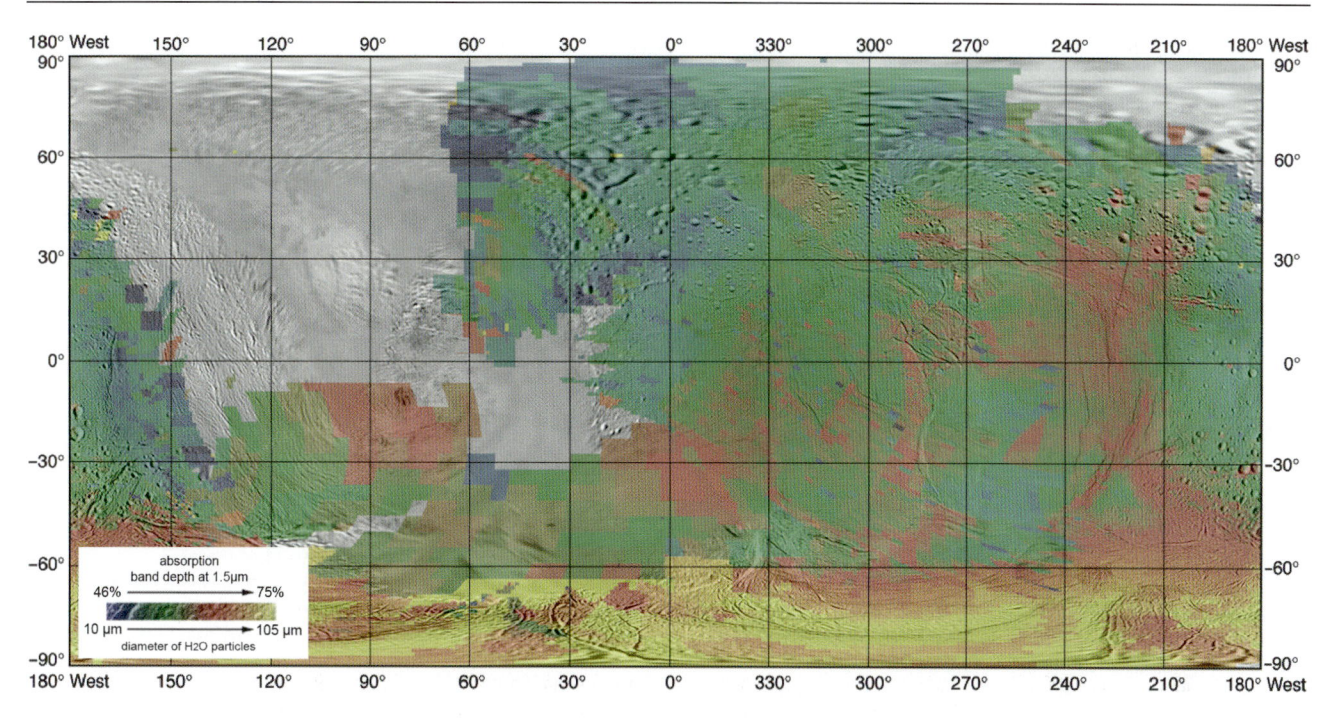

Fig. 24.17 Global VIMS composition map of Enceladus based on VIMS observations summarized in Table 24.9 in a simple cylindrical map projection pixel overlaid onto the corresponding ISS basemap. The map exhibits a map scale of 1 km/pixel (after Jaumann et al. 2008)

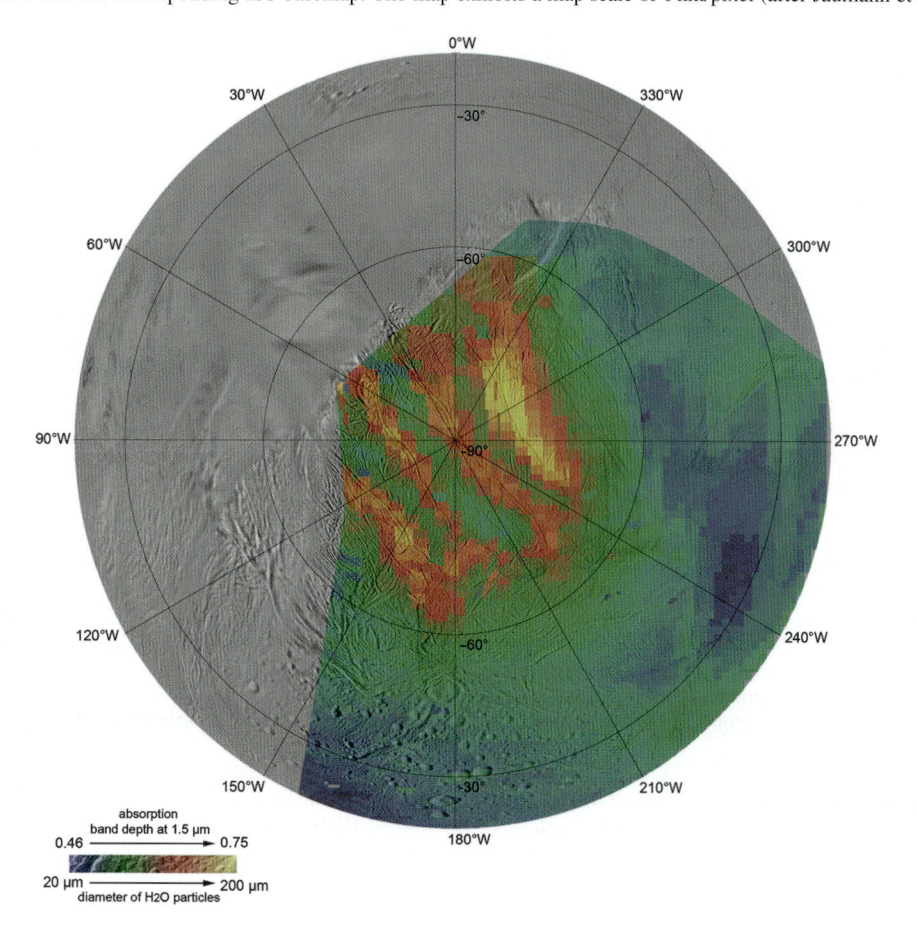

Fig. 24.18 VIMS mosaic of the South Polar Region illustrating the particle size variations derived by VIMS measurements using data of observation sequence ENCELADUS111 (Table 24.9) of orbit 11 overlaid onto the corresponding ISS map (after Jaumann et al. 2008)

Acknowledgements The authors gratefully acknowledge the work for the software development, image processing and the generation of the atlases of M. Wählisch, A. Hoffmeister, E. Kersten, K.-D. Matz, F. Scholten, R. Wagner, and B. Giese (DLR). They also acknowledge the image planning and camera operation work of their colleagues from the Cassini imaging team especially the operations team at the Cassini Imaging Central Laboratory for Operations (CICLOPS). Furthermore, they acknowledge helpful discussions with J. Blue and R. Kirk (USGS) about the proposed feature nomenclature and for reviewing the atlases. Finally, the authors acknowledge the planning and operation work of their colleagues from the Cassini VIMS team.

References

Baines, K., 1962: *Malor's Le Morte d'Arthur*. Mentor, New York.

Bates, H., 1929: *The Odyssey of Homer*. Harper and Brothers, New York.

Batson, R., 1984: Voyager 1 and 2 Atlas of Six Saturnian Satellites. Scientific and Technical Information Branch, National Aeronautics and Space Administration, Washington, D.C., http://history.nasa.gov/SP-474/sp474.htmp.

Brown, R.H., Baines, K.H., Bellucci, G., Bibring, J.-P., Burratti, B.J., Bussoletti, E., Capaccioni, F., Cerroni, P., Clark, R.N., Coradini, A., Cruikshank, D.P., Drossart, P., Formisano, V., Jaumann, R., Langevin, Y., Matson, D.L., McCord, T.B., Miller, E., Nelson, R.M., Nicholson, P.D., Sicardy, B., and Sotin, C., 2004: The Cassini Visual and Infrared Mapping Spectrometer Investigation. Space Science Reviews 115, 111–168.

Brown, R.H., Clark, R.N., Buratti, B.J., Cruikshank, D.P., Barens, J.W., Mastrapa, R.M.E., Bauer, J., Newman, S., Momary, T., Baines, K.H., Bellucci, G., Capaccioni, P., Cerroni, P., Combes, M., Coradini, A., Drossart, P., Formisano, V., Jaumann, R., Langevin, Y., Matson, D.L., McCord, T.B., Nelson, R.M., Nicholson, P.D., Sicardy, B., and Sotin, C., 2006: Composition and physical properties of Enceladus' surface. Science 311, 1425–1428.

Burton, R., 1900: Alf Laylah Wa Laylah, *The Book of the Thousand Nights and a Night*. Larsen-Harper, Colorado Press, Denver, CO.

Castillo-Rogez, J.C., Matson, D.L., Sotin, C., Johnson, T.V., Lunine, J.I., Thomas, P.C., 2007: Iapetus' geophysics: Rotation rate, shape, and equatorial ridge. Icarus 190, 179–202.

Chandrasekhar, S., 1969: *Ellipsoidal Figures of Equilibrium*, Yale University Press, New Haven.

Charnoz, S., Brahic, A., Thomas, P.C., and Porco, C.C., 2007: The equatorial ridges of Pan and Atlas: Terminal accretionary ornaments?. Science 318, 1622.

Clark, R.N. and Roush, T.L., 1984: Reflectance spectroscopy: Quantitative analysis techniques for remote sensing applications. J. Geophys. Res. 89, 6329–6340.

Clark, R.N., Swayze, G.A., Livo, K.E., Kokaly, R.F., Sutley, S.J., Dalton, J.B., McDougal, R.R., and Gent, C.A., 2003: Imaging spectroscopy: Earth and planetary remote sensing with the USGS Tetracorder and expert systems. J. Geophys. Res. 108(E12), 5131, doi:10.1029/2002JE001847.

Clark, R., Brown, R.H., Jaumann, R., Cruikshank, D.P., Buratti, B., Baines, K.H., Nelson, R.N., Nicholson P.D., Moore, J.M., Curchin, J., Hoefen, T., and Stephan, K., 2008: Compositional mapping of Saturn's Satellite Dione with Cassini VIMS and implications of dark material in the Saturnian system. Icarus 1933, 72–386.

Davies, M. E. and Katayama, F. Y., 1983a: The control networks of Mimas and Enceladus. Icarus 53, 332–340.

Davies, M.E. and Katayama, F.Y., 1983b: The control networks of Tethys and Dione. Journal of Geophysical Research 88A, 8729–8735.

Davies, M. E. and Katayama, F. Y., 1983c: The control network of Rhea. Icarus 56, 603–610.

Davies, M. E. and Katayama, F. Y., 1984: The control network of Iapetus. Icarus 59, 199–204.

Dermott, S. F. and Thomas, P.C., 1988: The shape and internal structure of Mimas. Icarus 73, 25–65.

Greeley, R. and Batson, G., 1990: *Planetary Mapping*. Cambridge University Press.

Jacobson, R.A., Antreasian, P.G., Bordi, J.J., Criddle, K.E., Ionasescu, R., Jones, J.B., Mackenzie, R.A., Pelletier, F.J., Owen, W.M. Jr., Roth, D.C., and Stauch, J.R., 2006: The gravitational field of Saturn and the masses of its major satellites. Astronomical Journal 132, 2520–2526.

Jaumann, R., Stephan, K., Brown, R.H., Buratti, B.J., Clark, R.N., McCord, T.B., Coradini, A., Capaccioni, P., Filacchione, G., Cerroni, P., Baines, K.H., Bellucci, G., Bibring, J.P., Combes, M., Cruikshank, D.P., Drossart, P., Formisano, V., Langevin, Y., Matson, D.L., Nelson, R.M., Nicholson, P.D., Sicardy, B., Sotin, C., Soderblom, L.A., Griffith, C., Matz, K.D., Roatsch, T., Scholten, F., and Porco, C.C., 2006: High-resolution Cassini-VIMS mosaics of Titan and the icy Saturnian satellites. Planetary and Space Science 54, 1146–1155.

Jaumann, R., Stephan, K., Hansen, G.B., Clark, R.N., Buratti, B.J., Brown, R.H., Baines, K.H., Newman, S.F., Bellucci, G., Filacchione, G., Coradini, A., Cruikshank, D.P., Griffith, C.A., Hibbitts, C.A., McCord, T.B., Nelson, R.M., Nicholson, P.D., Sotin, C., and Wagner, R., 2008: Distribution of icy particles across Enceladus' surface as derived from Cassini-VIMS measurements. Icarus 193, 407–419.

Kirk, R.L., Becker, T.L., Rosanova, T., Soderblom, L.A., Davies, M.E., and Colvin, T.R., 1998: Digital maps of the Saturnian satellites – First steps in cartographic support of the Cassini Mission. Jupiter after Galileo, *Saturn before Cassini Conference*, Nantes, France.

Mandelbaum, A., 1972: *The Aeneid of Virgil*. Bantam, New York.

Minton, D.A. 2008: The topographic limits of gravitationally bound, rotating sand piles. Icarus 195, 698–704.

Mozley, J. H., 1934: *The Argonautica, by Valerius Flaccus* (English translation). The Loeb Classical Library, Cambridge, MA.

Porco, C.C. and 19 co-authors, 2004: Cassini imaging science: Instrument characteristics and anticipated scientific investigations at Saturn. Space Science Review 115, 363–497.

Porco, C.C. and 34 co-authors, 2005: Cassini imaging science: Initial results on Phoebe and Iapetus. Science 307, 1237–1242.

Porco, C.C. and 24 co-authors, 2006: Cassini observes the active south pole of Enceladus. Science 311, 1393–1401.

Porco, C.C., Thomas, P.C., Weiss, J.W., and Richardson, D.C., 2007: Saturn's small inner satellites: Clues to their origins. Science 318, 1602.

Roatsch, T., Wählisch M., Scholten, F., Hoffmeister, A., Matz K.-D., Denk, T., Neukum G. Thomas, P., Helfenstein, P., and Porco, C.C., 2006: Mapping of the icy Saturnian satellites: First results from Cassini-ISS. Planetary Space Sciences 54, 1137–1145.

Roatsch, T., Wählisch, M., Giese, B., Hoffmeister, A., Matz, K.-D., Scholten, F., Kuhn, A., Wagner, R., Neukum, G., Helfenstein, P., and, Porco, C.C., 2008a: High-resolution Enceladus atlas derived from Cassini-ISS images. Planetary Space Sciences 56, 109–116.

Roatsch, T., Wählisch, M., Hoffmeister, A., Matz, K.-D., Scholten, F., Kersten, E., Wagner, R., Denk, T., Neukum, G., and Porco, C.C., 2008b: High-resolution Dione atlas derived from Cassini-ISS images. Planetary Space Sciences 56, 1499–1505.

Roatsch, T., Wählisch, M., Hoffmeister, A., Kersten, E., Matz, K.-D., Scholten, F., Wagner, R., Denk, T., Neukum, G., Helfenstein, P., and Porco, C.C., 2009: High-resolution atlases of Mimas, Tethys, and Iapetus derived from Cassini-ISS images. Planetary Space Sciences 57, 83–92.

Sayers, D.L., 1967: *The Song of Roland*. Penguin Books Inc., Baltimore.

Seidelmann, P.K., Archinal, B.A., A'hearn, M.F., Conrad, A., Consolmagno, G.J., Hestroffer, D., Hilton, J.L., Krasinsky, G.A., Neumann, G., Oberst, J., Stooke, P., Tedesco, E.F., Tholen, D.J., Thomas, P.C., and Williams, I.P., 2007: Report of the IAU/IAG Working Group on cartographic coordinates and rotational elements. Celestial Mechanics and Dynamical Astronomy 98, 155–180.

Simonelli, D.P., Thomas, P., Carcich, B.T., and Veverka, J., 1993: The generation and use of numerical shape models for irregular solar-system objects. Icarus 103, 49–61.

Stephan, K., Jaumann, R., Wagner, R., Clark, R., Cruikshank, D.P., Hibbitts, C.A., Roatsch, T., Brown, R.H., Buratti, B.J., Filiacchione, G., Hansen, G.B., McCord, T.B., Baines, K.H., and Nicholson, P.D., 2009a: VIMS coverage of Saturn's icy satellite Rhea. LPSC, 1377.

Stephan, K. Jaumann, R., Wagner, R., Clark, R.N., Cruikshank, D.P., Hibbitts, C.A., Roatsch, T., Hoffmann, H., Brown, R.H., Filiacchione, G., Buratti, B.J., Hansen, G.B., McCord, T.B., Nicholson, P.D., and Baines, K.H., 2009b: Dione's spectral and geological properties, submitted to Icarus.

Thomas, P.C., Davies, M.E., Colvin, T.R., Oberst, J., Schuster, P., Neukum, G., Carr, M.H., McEwen, A., Schubert, G., Belton, M.J.S., and the Galileo Imaging Team, 1998: The Shape of Io from Galileo Limb Measurements. Icarus 135, 175–180.

Thomas, P.C., Burns, J.A., Helfenstein, P., Squyres, S., Veverka, J., Porco, C.C., Turtle, E.P., McEwen, A., Denk, T., Giese, B., Roatsch, T., Johnson, T.V., and Jacobson, R.A., 2007: Shapes of the Saturnian icy satellites and their significance. Icarus 179, 573–584.

Appendix: The Cassini Orbiter, Behind the Scenes

In the pages that follow are pictures of Cassini orbiter operations personnel, beginning with those working at Caltech's Jet Propulsion Laboratory (JPL) (Cassini's Mission Operations Center), and then moving to pictures of operations personnel at each of the sites based outside of JPL where individual science instrument operations take place.

The people depicted herein represent a subset of all of the people who have been part of the Cassini-Huygens Mission, from developing it as a candidate planetary-exploration mission in the early 1980s, to eventually being responsible for its design, construction, launch and flight. It is hoped that the knowledge exposited within this book will represent a small token of appreciation for the prodigious efforts of all of the people that have been involved with the Cassini-Huygens Mission. It is they who have made Cassini-Huygens a shining success, and their tireless efforts will continue to bear important scientific and cultural fruit far into the future.

1 The Cassini Mission Operations Center at Caltech's Jet Propulsion Laboratory

1.1 The Cassini Program-Office Management, Resource Management, Mission Planning, and Outreach Teams

Fig. 1 Row 1, left to right: M. Pao, J. Jones, C. Martinez, R. Pappalardo, S. Chatterjee, R. Zimmerman-Brachman, K. Chan; Row 2, left to right: G. Yee, D. Matson, C. Vetter, J. Nelson, E. Manor-Chapman, S. Payan; Row 3, left to right: K. Munsell, L. Spilker, A. Wessen, R. Woodall, V. Barlow, S. McConnell; Row 4: J. Smith, D. Bradford, R. Mitchell, D. Seal

1.2 The Cassini Navigation Team

Fig. 2 Row 1, left to right: E. Gist, I. Roundhill, J. Pojman, P. Antreasian, J. Frautnick, D. Vaughan; Row 2, left to right: M. Wang, R. Ionacescu, P. Thompson, J. Costello, S. Wagner, P. Williams; Row 3, left to right: K. Criddle, J. Jones, C. Ballard, F. Pelletier, B. Buffington, S. Gillam, S. Nolet; Row 4: V. Legerton, D. Roth, R. Jacobson, T. Goodson, P. Stumpf; Row 5, left to right: N. Strange, M. Wong, B. Stavert

1.3 The Cassini Mission Support Services Office

Fig. 3 Row 1, left to right: O. Castillo, N. Patel, V. Trinh, T. Fujii, B. Wilson, M. Carranza, M. Rubio; Row 2, left to right: V. Villa, J. Ibanez, C. Wong, P. Smith, R. Aguilar, D. Coppedge, J. Kesterson; Row 3, left to right: R. Jobsky, C Lush, G. Eller, M. Weisenfelder, B. Elgin, D. Doan; Row 4, left to right: B. Mogensen, D. Doody, L. Mellinger, G. Chin

1.4 The Cassini Science Planning Team

Fig. 4 Row 1, left to right: J. Pitesky, K. Steadman, A. Aguinaldo, R. Lopes, N. Vandermey, A. Hendrix, S. Edgington; Row 2, left to right: B. Larsen, L. Cheng, R. Lange, T. Ray, K. Grazier, B. Paczkowski, M. Burton, N. Kelly

1.5 The Cassini Instrument Operations Team (1)

Fig. 5 Row 1, left to right: E. Martinez, J. Yoshimizu, C. To, P. Lee; Row 2, left to right: P. Andres, S. Linick, P. Callahan, H. Mortensen, D. Fleishman; Row 3: A. Culver, O. Harrison, C. Acton

1.6 The Cassini Instrument Operations Team (2)

Fig. 6 Row 1, left to right: L. Ly-Hollins, K. Kelleher, Y. Anderson; Row 2, left to right: A. Stevenson, B. Brooks, A. Tinio, R. Boehmer, P. Meegyeong; Row 3, left to right: C. Cordell, J. Gerhard, A. Anabtawi, E. Barbinis, M. Roy; Row 4, left to right: F. Loaiza, C. Avis, F. Leader, D. Kahan, S. Asmar, R. West

1.7 The Cassini Science and Uplink Office

Fig. 7 Row 1, left to right: K. Weld, N. Rouse, C. Chouinard, B. Landry, K. Yetter, D. Tong; Row 2, left to right: R. Espinueva, J. Boyer, S. Goo, J. Berkeley, S. Chatterjee, L. Nakamura; Row 3, left to right: K. Magee, D. Conner, S. Javidnia, W. Heventhal, J. Carter, J. Krueger

Appendix: The Cassini Orbiter, Behind the Scenes

1.8 The Cassini Spacecraft Operations Office

Fig. 8 Row 1, left to right: R. Lin, L. Burke, K. Garcia, E. Wang, S. Adamiak; Row 2, left to right: J. Webster, K. Herman, P. Meakin, R. Lim, P. Morgan, M. Luna, J. Brown, A. Lee; Row 3, left to right: A. Ging, J. Wertz-Chen, C. Lee, T. Burke, M. Pellegrin, C. Mittelsteadt, S. Sarani, J. Millard, D. Beach; Row 4, left to right: C. Kirby, D. Bates, F. Chrisney, D. Morgan, R. Jurenko

1.9 The Cassini Spacecraft Operations Team

Fig. 9 Row 1, left to right: G. Yang, R. Somawardhana, N. Grenander, C. Huynh, R. Mukai; Row 2, left to right: P. Yoder, C. Sagoian, K. Baddarudin, A. Thomas, R. Weaver; Row 3, left to right: L. Christodoulou, T. Barber, T. Zorn, S. Clark

2 The Cassini Plasma Spectrometer (CAPS) Operations Group

Fig. 10 The CAPS Operations Group. Left to right: Frank Crary, Prachet Mokashi, Greg Ferris and Judith Furman

3 The Cosmic Dust Analyzer (CDA) Operations Group

Fig. 11 The CDA Operations Group: From left to right: S. Hsu (bottom), G. Matt, S. Helfert (top), G. Linkert, S. Kempf (top), D. Linkert, R. Srama (bottom), F. Postberg, E. Grün, G. Moragas-Klostermeyer (bottom), U. Beckmann

Appendix: The Cassini Orbiter, Behind the Scenes

4 The Composite Infrared Spectrometer (CIRS) Operations Group

Fig. 12 The CIRS Operations Group. Left to right, at Goddard Spaceflight Center: N. Gorius, G. Bjoraker, M. Segura, C. Nixon, D. Jennings, S. Albright, R. Achterberg, J. Pearl, A. Simon-Miller, A. Mamoutkine, E. Guandique, C. Anderson, J. Brasunas, R. Carlson, M. Kaelberer, J. Tingley; not pictured: V. Kunde and P. Romani; at the Jet Propulsion Laboratory: S. Edgington, S. Brooks, C. Roumeliotis; at the Observatoire de Paris-Meudon: E. Lellouch; at Oxford University, United Kingdom: S. Calcutt and N. Bowles

5 The Ion and Neutral Mass Spectrometer (INMS) Operations Group

Fig. 13 The INMS Operations Group: Back row, left to right: David Gell (Analysis), Rob Thorpe (Ground System). Front row, left to right: Greg Fletcher (Operations team Lead), Aimee Cardenes (Operations Engineer), June Dunkelburger (Operations Engineer)

6 The Imaging Science Subsystem (ISS) Group

Fig. 14 The ISS Group. Left to right: Joe Spitale, Andre Brahic, Josh Riley, Robert Jacobson, Joseph Veverka, Andrew Ingersoll, Joe Ferrier, Tilmann Denk, Doug Dawson, Michael Evans, Gerhard Neukum (front), Ben Knowles (back), Thomas Roatsch (middle), Torrence Johnson (back), Henry "Luke" Dones, Carolyn Porco, Peter Thomas, Carl Murray, Sebastien Charnoz, Joseph Burns, Elizabeth "Zibi" Turtle, John Weiss, Emma Birath, Jason Perry, Preston Dyches, Daren Wilson, Paul Helfenstein, Ashwin Vasavada, Anthony DelGenio, Matthew Tiscareno, Michael Belanger

Fig. 15 Individuals not included in the photo above. Top row, left to right: Emily Baker, Kevin Beurle (deceased), Bobby DiDia, Daiana DiNino; Bottom row, left to right: Pauline Helfenstein, Nicole Martin, Joe Mason

7 The Magnetospheric Imaging Instrument (MIMI) Operations Group

Fig. 16 The MIMI Operations Group. Left to right: Jon Vandegriff, Analysis and Display Software Engineer; John Aiello, Science Planning Engineer; David LaVallee, Operations Team Lead (uplink); Martha Kusterer, Downlink Lead Engineer, INCA Display Software Engineer; Linda Burke, Operations Team Engineer (uplink and downlink); Scott Turner, INCA, SPICE, Science Planning Tool, and Command Automation Software Engineer; Stuart Nylund, Operations Team Engineer (uplink)

8 The Cassini Magnetometer (MAG) Operations Group

Fig. 17 The MAG Operations Group: Top row, left to right: Steve Kellock (Instrument Manager), Nick Achilleos (Operations Engineer), Leah Alconcel (Operations Engineer). Bottom row, left to right: Charlotte Dunford (Archive Engineer), Tim Seears (Operations Engineer), Peter Slootweg (Operations Engineer). Right column: Adrian Hitchman (Archive Engineer), Joyce Wolf (Software Developer), Louise Lee (Software Developer). All personnel except JW and LL are current and former associates of Imperial College London. JW is associated with the Jet Propulsion Laboratory and LL with University of California, Los Angeles

9 The Cassini RADAR Operations Group

Fig. 18 The RADAR Operations Group: Front row, left to right: Young Gym, Alice Le Gall, Yanhua Anderson, Kathleen Kelleher. Back row, left to right: Richard West, Mike Janssen, Bill Johnson, Phil Callahan, Bryan Stiles, Gary Hamilton

10 The Radio and Plasma Wave Science (RPWS) Operations Group

Fig. 19 The RPWS Operations Group. Front row: Robert Johnson, Terry Averkamp, Don Kirchner, George Hospodarsky; Back row: Bill Kurth, Bill Robison, Larry Granroth, Jessica Swanner, Ann Persoon, Chris Piker

11 The Radio Science Subsystem (RSS) Operations Group

Fig. 20 The RSS Operations Group. From left to right: Sami Asmar (supervisor), John Klose, Elias Barbinis, Aseel Anabtawi (technical lead), Daniel Kahan, Don Fleischman

12 The Ultraviolet Imaging Spectrometer (UVIS) Operations Group

Fig. 21 The UVIS Operations Group. From left to right: Michelle Kelly, Darren Osborne, David Judd, Alain Jouchoux, Heather Buck, Crystal Salcido, and John Donnelly

13 The Visual and Infrared Mapping Spectrometer (VIMS) Operations Group

Fig. 22 The VIMS Operations Group. From left to right: Dyer Lytle, Virginia Pasek, John Ivens, Bob Watson and Dan Moynihan

Index

A

Absortive occultation(s), 182
Abundance, 83–86, 88–98, 100–104, 106, 107
Acceleration, 344, 350, 353, 354, 356, 357, 359–361, 364, 365, 369, 370
Accretion, 425, 433, 442, 445–449, 537, 538, 541, 542, 544, 545, 547, 548, 552–557, 559–562, 564, 566, 568
Accretional heating, 589–590
Acetylene (C_2H_2), 92, 94, 97, 100, 101, 103, 107, 122, 130, 131, 153, 182, 183, 186, 337
ACS. *See* Advanced camera for surveys
Activity
  endogenic, 641, 642, 647
  geologic, 638, 657
  tectonic, 640, 669, 672
  volcanic, 640
Adhesion, 416, 435, 451
Adiabatic acceleration
  adiabatic energization, 311
  butterfly distributions, 311, 312
  invariants, 310
  isotropic invariant, 312
  pancake/trapped distribution shapes, 311
  phase space density (PSD), 311–313
  pitch angle distribution shapes, 311, 312
Advanced camera for surveys (ACS), 335, 337
Aerosol(s), 85, 86, 90, 92, 96, 97, 104, 107, 161–178
Aerosol structure
  Earth-based telescopes, 17–19
  pioneer/voyager era measurements, 16–17
Aggregate formation, 447
Aggregate particles, 460, 484, 496, 497, 500
AKR. *See* Auroral kilometric radiation
^{26}Al, 61, 70, 71
Albedo (geometric), 660, 683, 697, 700, 704–706, 716
Albedo-wind relationship, 138
Alfvén wings, 289
Alkali halide, 90
Amalthea, 63, 64
Ammonia (NH_3), 11–12, 88, 89, 92, 95–98, 106, 107, 653–658, 664–667, 672, 673, 688, 690, 692, 703, 704, 711, 716
Ammonia hydrate, 667
Ammonia ice, 162–166, 175, 178
Ammonium hydrosulfide (NH_4SH), 88, 96, 97, 162, 163
Ammonium salt, 95–97
Amorphization, 660
Amorphous carbon, 486

Amorphous ice, 476, 497
Angular momentum, 414, 425, 430, 435, 436, 438, 439, 443–445, 449
Annular rings, 136
Anthe, 523, 532, 533
^{40}Ar, 690, 695, 713, 717
A rings, 375–381, 383–387, 391–399, 401–403, 473, 474, 478–486, 489, 492, 496, 501, 502
Arsine (AsH_3), 10, 13–14, 89, 90, 92, 106
Asteroid belt, 57, 59, 64
Atlas, 381, 386, 399, 438, 448
Atmosphere and auroral emission
  infrared observations, 27–29
  ultraviolet observations, 26–27
Atmospheric dynamics, 23–25, 749–751
Aurora, 98, 99, 104, 106, 228, 242, 264, 275, 333–335, 337–341, 343, 345–353, 360–369
Auroral arc, 336, 340, 341, 346, 347, 361
Auroral chemistry, 104, 106
Auroral current systems, 335
Auroral electrons, 337, 353
Auroral energy, 337–338
Auroral field lines, 334, 344–346, 351, 357
Auroral footprint, 334
Auroral hiss detection, 322
Auroral kilometric radiation, 333
Auroral morphology, 335–336
Auroral oval, 334–336, 338, 339, 340, 346, 348, 351, 352, 358, 360, 364, 365, 367, 369, 370
Auroral/polar ionosphere, 29
Auroral power, 349, 350, 369
Azimuthal brightness variation, 426, 429

B

Ballistic transport, 430, 435
Barnard gap, 397
Beaming, 343–346, 366, 369
Bending waves, 377–380, 387, 388, 392, 398, 401, 407, 436, 446
Benzene (C_6H_6), 14, 92, 94, 98, 100, 106, 107
Bessel gap, 397
Bimodal optical depth variations, 434
Binary encounters, 421, 422
Birkeland current, 353
Birkeland current systems, 298
Boltzmann equation, 418
Bond gap, 400
Bound water, 655, 658, 672
Bow shock, 204, 237, 238, 242–246

795

Branching ratio, 98, 99, 104, 107
Brightness asymmetry, 428, 429, 438, 443, 452
B ring core, 462, 463, 496
B rings, 375, 377, 381, 383, 384, 386–397, 399–401, 407, 408, 471, 474, 476–492, 494, 496–498, 502, 503, 511, 512, 514, 518–521, 533
Broadband electromagnetic noise, 353
Bulk densities, 62–64
Bulk viscosity, 420, 421, 432, 445
Butterfly distributions, 311

C

Calcium aluminum inclusions (CAIs), 56, 64, 70, 71
Callisto, 61, 63, 64, 66, 68–70
Capture probability, 448
Carbon, 62, 65–67, 83, 86–89, 94, 95, 97, 106
Carbon dioxide (CO_2), 94, 95, 100, 104, 105, 651, 652, 654–656, 672, 684, 688, 703, 704, 708, 710, 711, 715, 718
Carbon monoxide (CO), 10, 13–14, 89, 94–96, 99, 100, 104, 105, 106
Cartographic mapping
   high-resolution atlases, 766–768, 771–776
   ISS images, 766–775
   Saturnian satellites, 764–766
   south polar region, 779
   VIMS data, 770–779
Cassini division, 375–381, 386, 392, 394–401, 407, 408, 436, 438, 463–465, 468, 470–472, 474, 476–482, 485, 488, 489, 493, 494, 496, 497, 502, 518, 654, 655, 672
Cassini dust analyzer (CDA), 246
Cassini Equinox mission, 154, 369
Cassini Extended mission
   Equinox mission design, 726–727
   Equinox mission science objectives, 727–729
   Equinox mission trajectory, 732–740
   operational and safety constraints, 729–730
   spacecraft subsystems, 725
   tour design and development process, 730–732
Cassini/Huygens, 55, 62, 68, 69
Cassini-Huygens mission
   atmospheric composition, 748–749
   atmospheric dynamics, 23–25, 749–751
   clouds and aerosols, 16–19
   composition, 10–16
   exploration, 756–757
   icy satellites, 41–46
   interior, 16, 746–747
   low-density giant planets, 747–748
   magnetosphere, 752–753
   planetary magnetic field and magnetosphere, 29–36
   ring system, 36–41, 753–755
   Solar system, 755–756
   temperature, 19–23
Cassini magnetometer (MAG), 218, 228, 243, 246
Cassini plasma spectrometer (CAPS), 214–221, 231, 232, 235–243, 245, 247, 260–262, 270, 273, 274
Cassini Solstice mission, 154
Catastrophic disruption, 625–626
Centaurs and ecliptic comets, 615
Centrifugal acceleration mechanism, 315–316
Centrifugal interchange instability
   Cassini orbits, 301, 302
   "injection/dispersion" signatures, 300–301
   magnetic equatorial plane, 298, 300
   potential energy, 300
   Rice Convection Model (RCM) simulation, 302, 303
   ring current impoundment, 300

C_2H_2. *See* Acetylene (C_2H_2)
C_2H_4. *See* Ethylene (C_2H_4)
Chaotic particle motion, 439
Chapman-Ferraro current, 211
Charged dust, 511, 519, 521
Charge exchange, 193, 197, 355, 370
Charon, 63, 66, 68
CH_3D/CH_4 ratio measurements, 14–15
Chemical lifetime, 194
Chemistry, 83–107
Chorus emissions, 322–323
Chromophore, 107
CIR. *See* Corotating interaction regions (CIRs)
Circulation, 94, 99, 103, 104, 107
Circumplanetary disk, 60–62
Circumplanetary nebula, 499
Circumstellar disk, 58, 60
CIRS. *See* Composite infrared spectrometer (CIRS)
Clathrate, 688, 699, 704, 713, 715, 717
Closed field line, 335, 348, 360, 361, 363–370
Cloud(s), 83, 84, 86–88, 91, 96, 97, 103, 106, 107, 161–178
Cloud decks
   ammonia, 115, 118
   ammonium hydrosulfide, 118
   water, 118
Cloud microphysical models, 162
Cloud particle properties, 163, 164, 166
Clouds and aerosols
   Earth-based telescopes, 17–19
   pioneer/voyager era measurements, 16–17
Cloud top level
   infrared (5 micron), 122
   near-infrared/visible, 119, 122
   ultraviolet, 121
Cloud vertical structure, 168
Clustering, 418
CMI. *See* Cyclotron maser instability (CMI)
Coagulation, 418, 419, 452
Coefficient of restitution, 414–417, 422, 424, 428, 447, 448, 450
Cohesion, 417
Cohesive force, 417, 447
Collision, 539–543, 545, 546, 549, 556, 558, 559, 561–563, 565–567, 569, 570
Collisional cascade, 548, 557
Collisional cooling, 420, 440
Collisional evolution, 67, 540, 563, 568, 570
Collision frequency, 414, 418, 423, 430, 441, 442, 445
Collision integrals, 416, 418, 419
Colombo gap, 399
Column density, 189
Cometary chronology, 618
Cometary impact, 94, 95, 106, 107
Comet-like interactions, 288–289
Comets, 58, 63, 64, 621–626
Compaction state, 661
Composite infrared spectrometer (CIRS), 85–91, 93–95, 97, 100–102, 104, 106, 107, 116, 119, 123, 124, 126–135, 139, 141–143, 145, 146, 152, 153, 339, 639, 640, 663
Composition, 83–107
   ionosphere, 193
   upper atmosphere, 181–182
Compositional maps, VIMS data
   data processing, 770–771
   Dione, 771, 776, 777
   Enceladus, 776, 778–779
   Rhea, 776–778

Compton-Getting effect, 264
Condensate, 65–68
Condensate clouds, 162, 169
Condensation, 83, 92, 94, 96–98, 560–562, 569, 650, 703, 713–715, 717
Condensation levels, 118–120, 137–139, 144, 152–154
Condensible, 83, 88
Conduction, heat transfer, 590
Contamination, 663–668, 672, 673
Continuum filter, 121
Convection, 90
   evolution of, 594–595
   heat transfer, 590–593
   onset of, 593–594
Convective clouds
   discrete, 137, 140
   dragon storm, 137, 138
   great white spot, 137–140
Cooling rates, 191
Co-orbital satellites, 438
Core nucleated accretion model, 57
Coriolis acceleration, 189, 302
Coriolis force(s), 447
Coronal mass ejection (CME), 273
Corotating interactions regions (CIRs), 273, 274, 334, 365
Corotating plasma, 335
Corotation, 189, 257–261, 267–270, 274, 336, 338, 348, 360, 361, 363–365, 370, 435
Corotation resonance(s), 444
Cosmic dust analyzer (CDA), 514, 524–528, 530, 532, 533
Cosmic rays, 655
Cosmic recycling, 538, 544–545, 548, 570
Cratering chronology, 618–619
Craters, 683, 684, 686, 693, 694, 696–699, 701, 703, 705, 718
Crater statistics and interpretation
   Enceladus, Tethys, Dione and Rhea, 629
   Gyr, 628–629
   Hyperion and Phoebe, 630
   R-plot, 626–627
   size-frequency distribution (SFD), 627–630
C rings, 375–377, 381, 386–389, 394–402, 407, 408, 462–464, 468, 470, 471, 476–482, 485–490, 492, 494, 496, 497, 502, 503
Cryovolcanism/cryovolcanic, 70, 683–720
Crystalline ice, 476, 477, 498
Crystallization, 660
Cryovolcanic eruption, 650
Current generation
   curvature vector, 316
   force balance, Jupiter and Saturn, 317–318
   gradient and curvature drifts, 316
   ring current, 317
Cusp, 336, 340, 349, 360, 366, 367, 369
Cyclotron maser instability (CMI), 342, 345

D
Damping scale, 438
Daphnis, 377, 381, 382, 408, 440–442, 448
Dark material, 639, 640, 652–659, 663, 664, 666–668, 672, 673
Dawes gap, 400
Dawn, 191, 192, 194, 195
Dayglow, 182, 185
Deep atmosphere, 114, 117–120
Deformation radius, 119, 135, 148, 149
Density(ies), 55, 57, 60–70, 639, 640, 642–644, 650, 657, 660, 663, 664, 666, 671, 673

Density profiles
   atomic hydrogen, 182, 185, 190, 197
   electron, 181, 185, 191–195, 197, 199
   ion, 189–191, 194–197, 199
   molecular hydrogen, 199
Density waves, 375, 377–383, 386, 387, 390, 392, 393, 396–399, 401, 407, 429, 430, 435–439, 452
Deuterium, 88, 106
Diacetylene (C_4H_2), 14, 92, 94, 98, 100, 101, 107
Diamagnetic current, 316
Dichotomy, 640, 656–658, 666, 667, 673
Dielectric constant, 667
Differential rotation, 423
Differentiation, 686
Diffuse aurora, 336, 369
Diffuse rings, 511–533, 538, 539, 568–570
Diffusion instability, 430, 433
Diffusive separation, 181
Dione, 4–5, 44, 69, 523, 524, 526, 528–530, 686, 691, 692, 716, 720
   ISS basemaps, 766, 768, 772, 775
   thermal evolution and internal structure, 604–605
   VIMS composition map, 771, 776, 777
Diphosphine, 162, 163, 178
Direction-finding, 346
Discrete aurora, 361, 363, 369
Dispersion relation, 431, 432, 437, 438, 452
Disruption, 446, 448, 449
Dissipative collisions, 413, 414, 417, 420
Dissociative recombination, 195
Diurnal variation, 192, 194–196, 199
D ring, 511, 514–518, 533
Dst index, 228
Dungey cycle, 205, 207, 208, 247, 258, 267, 268, 273, 274, 361–363, 365, 366, 370
Dungey process, 258
Dusk, 191, 192, 194
Dust, 684, 705, 707, 711, 716, 717, 720
Dust streams, 511, 512, 530–532
Dynamical ephemeral bodies, 446, 541, 556
Dynamical regime, 114
Dynamic pressure, 421
Dynamics, 83, 86, 88, 103, 107
Dynamic viscosity, 417, 418

E
Earth-orbiting spacecraft, 9–10
Eccentricity, 643
Eddy diffusion, 96, 99, 100, 186
Eddy momentum flux, 127, 137–139, 147, 152, 153
Ejecta, 435
Electromagnetic cyclotron waves (EMIC), 314
Electron collisions, 193
Electron cyclotron frequency, 343, 354
Electron cyclotron harmonic (ECH) emissions, 319–321
Electron-impact excitation, 339
Electron-induced processes, 288
Electron-neutral collisions, 286
ENAs. *See* Energetic neutral atoms (ENAs)
Enceladus, 42–43, 64, 65, 69–71, 511, 514, 523–533
   ISS basemaps, 766–767, 773
   VIMS composition map, 776, 778–779
Enceladus flybys, 733
Enceladus orbiter, 720
Enceladus plume, 538, 570
Enceladus torus, 287, 288

Encke gap, 376, 377, 381, 382, 385, 395, 396, 399, 408, 429, 436, 438, 440, 441, 443, 464, 470, 472, 480
Energetic neutral atoms (ENAs), 3, 36–37, 222, 224, 225, 227–229, 235, 260, 264, 267–270, 272, 273, 286, 287, 342, 350, 357
Energetic particles, 222, 223, 225, 227, 229–231, 235, 241, 244, 246
   electron counting rates, 33–34
   energy intervals, 32–33
   equatorial field stress calculations, 34–35
   phase-space density, 34–35
   radial dependences, 32–34
   spectrograms, 32–33
Energy balance, 188–191
Energy cascade, 139, 148, 149
Enskog factor, 419
Enskog's theory, 418
Epics, ISS images, 766, 770
Epicyclic frequency, 423, 436
Epicyclic length, 415
Epimetheus, 375, 377, 381, 386, 390, 392, 393, 396, 397, 407, 436, 438, 443, 452, 523, 532
Equation of state, 419, 420
Equatorial ridge, 640, 642, 644, 648, 649, 670
Equilibrium condensation, 64–68
Equinox and Solstice missions, 7
Equinox mission, 684–686
   Bi-and mono-propellant profile, 740–741
   Cassini end-of-mission, 742–743
   design overview, 726–727
   icy satellite objectives, 728
   magnetosphere, 728
   ring, 728–729
   Saturn, 729
   seasonal declination, 741
   trajectory, 732–740
Equipartition, 418, 421, 450, 451
E ring, 511, 512, 514, 521, 523–532, 683–684, 688, 704–706, 711, 712, 715–717, 719
Eris, 63, 66
Erosion, 435, 449
Escape speed, 422
Ethane (C_2H_6), 14–15, 89, 92–94, 98–104, 106, 107, 122, 125, 130–132, 134, 153, 337
Ethylene (C_2H_4), 92, 94, 98, 100, 101, 107, 183, 186
Evander, 642, 672
Evaporation, 713–715
Excess variance, 446
Extraordinary (R-X) mode, 345

F
Far Ultraviolet Spectroscopic Explorer (FUSE), 337
Fast Fourier transform, 438
Fast rotators, 451
Field-aligned current (FACs), 335, 351, 353, 354, 357, 358, 360–367
Field-aligned potentials, 345, 357
Flow shear, 348, 364, 366
Flux tube interchange, 208, 237
Flyby, Phoebe, 757
Focusing, 418
Formation, 83, 84, 88, 89, 90, 96–98, 106
Fourier's law, 420
Fragmentation, 418, 419, 445, 448, 449, 452
Frequency–time spectrogram
   dynamic injection event, 319–321
   high planetary latitude, 319, 320
   Saturn's equatorial plane, 319, 320
Friction, 416, 418, 419, 445, 447, 448, 450, 451

F ring, 375–377, 396, 401–406, 408
F-ring clumps, 554, 566–568
F-ring spiral, 568
F-ring strands, 567, 568
FUSE. See Far Ultraviolet Spectroscopic Explorer (FUSE)

G
Galilean satellites, 64, 70
Ganymede, 63, 64, 68, 69
Gap formation, 60, 62
Gas instability model, 57, 58
Gas-kinetics, 419
Geometric thickness, 416, 417, 422, 429
Germane (GeH_4), 10, 13–14, 89, 90, 92, 95–98, 106
Geyser, 645, 651, 660
Giant planets, 55–62, 64, 65, 68, 70
Global evolution
   Iapetus, 601–603
   icy satellite, 599–600
   lithosphere, 600–601
   shape, 601
Global MHD models, 237
Graben, 639, 641, 645, 647, 648, 669, 670
Gradient drift, 350
Granular flow(s), 420
Granular temperature, 419, 420
Gravitational accretion, 425, 447, 448
Gravitational encounters, 418, 422, 440
Gravitational instability, 421, 554
Gravitational potential, 420
Gravitational stirring, 421
Gravitational torques, 425, 443
Gravitational viscosity, 421, 425–427, 435
Gravitational wakes, 460, 472, 473
Gravity field and shape, 75–76
Gravity waves, 195, 199
Great White Spot (GWS), 24–25
G ring, 511–514, 521–524, 532, 533
Ground state, 181, 182, 193, 196

H
H, 181, 184, 185, 191, 193
H+, 185, 193, 195, 196
H_2, 181–186, 193–197, 199
H_2+, 195
H_3+, 188, 189, 192–193
Hapke theory, 484
Hard spheres, 419
Haze, 86, 90, 91, 96, 97, 100, 106
   particle properties, 127
   vertical structure, 168
He, 181–183, 193
Heat flow, 419
Heat flux, 420
Heating rate, 190
Heat sources
   radioactivity, 585–586
   tidal heating, 586–588
Heat transport, 686, 690, 691, 713, 714
Heavy elements, 84
Helium (He), 10–11, 83–86, 99, 106
Helium partitioning problem, 79–80
Hematite, 655, 656, 672
Herschel, 638, 642, 645, 665, 668, 671
Herschel gap, 396
Hexagon, 115, 134, 141–144

Index

Hexagonal wave structure, 24
Hill approximation, 446
Hill potential, 447
Hill radius, 421, 446
Hill scale, 440
Hill sphere, 59, 61, 404, 442, 447, 448
History, 10
H_3^+ molecular ion, 27–28
H_2O, 182, 185, 194, 196
Homopause, 83, 99, 100, 107, 181, 183–188, 198
Hopf bifurcation, 431
HST. *See* Hubble Space Telescope (HST)
Hubble Space Telescope (HST), 23–24, 114–116, 122, 123, 127–129, 136–138, 141, 145, 268, 274, 275, 333–337, 339, 340, 343, 346–349, 351, 358, 359, 361, 363, 365–368
Huygens gap, 395, 396, 518
Hydrocarbon haze, 162–164
Hydrocarbons, 14–15, 84, 92–95, 97–107, 184, 185, 187, 193, 334
Hydrodynamic equations, 419
Hydrogen (H_2), 10–12, 84–91, 95, 97–99, 104–106, 655, 660, 661
Hydrogen corona, 185
Hydrogen escape, 190
Hydrogen peroxide ($H2O2$), 703
Hydrogen (D/H) ratio, 15–16
Hydrostatic, 69, 182–184, 188
Hyperion, 44–45
Hypervelocity impacts, 435

I

Iapetus, 4, 45, 61, 62, 64, 67, 68, 70, 71, 397
    global evolution, 601–603
    ISS basemaps, 766, 769, 772
Ice-II, 63
Ice III, 63
Icy satellites, 4–5
    Ansa-to-Ansa occultations (T62–T68), 738–739
    cartography, 763–779
    comets, 621–626
    cratering chronology, 618–619
    dimensions, densities and rotational properties, 41
    Dione, 44
    dynamical evolution, 598–603
    Enceladus, 42–43
    Equinox mission science objectives, 728
    geology, 42
    heat sources, 585–590
    heat transfer, 590–595
    heavy bombardments, 617–618
    Hyperion, 44–45
    Iapetus, 45
    impactor populations, 615–617
    laboratory data, 605–606
    Mimas orbits, 42
    Mimas, Tethys and Dione, 604–605
    modeling, 606
    Phoebe, 603–604
    physics and scaling laws, 619–621
    porosity effect, 596
    processes, 606
    Rhea, 44
    rock thermal conductivity, 595–596
    satellite properties, 579–585
    small satellites, 45–46
    space, 605
    statistics and interpretation, 626–630
    structural evolution, 596–598
    surface compositions and optical properties, 41–42
    Tethys, 43–44
    thermal conductivity, 595
    voyager color image, 42–43
    voyager era, 613–615
I/F. *See* Reflectivity
Imaging Science Subsystem (ISS), 116, 117, 119, 121–125, 127–130, 137–139, 143–146, 152, 339, 639, 640, 657, 658, 663, 667–671
Imaging science subsystem (ISS), 763–779
IMF. *See* Interplanetary magnetic field (IMF)
Impact basins, 59, 70, 71
Impact bombardment, 642
Impact cratering and age determination
    comets, 621–626
    cratering chronology, 618–619
    heavy bombardments, 617–618
    impactor populations, 615–617
    physics and scaling laws, 619–621
    statistics and interpretation, 626–630
    voyager era, 613–615
Impact craters
    bright ray crater, 640, 669, 671
    central pit craters, 642
    complex craters, 642
    ejecta, 640, 642, 666, 669, 671
Impact frequency, 421
Impacts, 517, 521, 524, 525, 527–530, 532
INCA. *See* Ion and Neutral Camera (INCA)
Incompressibility, 420
Inertial-acoustic wave, 432
Infrared (IR), 333, 338–341, 352, 364, 369, 370
Infrared auroral observations, 27–29
Infrared space observatory (ISO), 87–90, 92, 94, 100, 101
Infrared telescope facility (IRTF), 338
Inhomogeneous cloud structure, 19
Injection, 259–261
Injection energization, 315–316
Injection events, 222, 228, 230, 231, 238, 239, 335
Instability
    Arnol'd II, 134, 135
    baroclinic, 125, 138, 139, 141, 142, 145, 147, 149, 150, 152, 153
    barotropic, 125, 134, 140–143, 148
    Charney-Stern, 149
    convective, 117
Interchange, 365
Internal heat source, 114, 118, 136
International ultraviolet explorer (IUE), 26
Interplanetary magnetic field (IMF), 204, 242, 243, 246, 305, 336, 348, 349, 363, 365–367
Ion and neutral camera (INCA), 260, 261, 269, 270, 273, 340, 351, 353–356
Ion and neutral mass spectrometer (INMS), 215, 219
Ion conics, 353, 355, 357, 358
Ion cyclotron waves, 344
Ion drag, 189, 190
Ion–neutral collisions, 286
Ionosphere, 257, 262, 267, 268, 270, 271, 275
    equatorial, 191, 192
    polar, 189
Ionospheric conductivity, 204, 240
Ion phase space holes, 344
Ion pickup, 237
IR. *See* Infrared (IR)
IR aurora, 338–339
Irregular satellites, 59, 66
ISO/SWS spectrum, 11

Isothermal, 420, 430
Isotopic ratios, 15–16
ISS. *See* Imaging Science Subsystem (ISS)
ISS basemaps, 766
    Dione, 766, 768, 772, 775
    Enceladus, 766–767, 773
    Epics, 766, 770
    Iapetus, 766, 769, 772
    Mimas, 766–767
    Phoebe, 766, 770, 771
    Rhea, 766, 769
    Tethys, 766, 768, 774
Ithaca Chasma, 638, 645, 646, 649, 669, 670

J
Janus, 375, 377, 381, 383, 386, 390–393, 396, 397, 401, 407, 436, 438, 443, 452, 523, 532
Jeffreys gaps, 394, 396
Jets
    eastward, 115, 122–124, 126, 127, 132, 134, 136–145, 152, 154
    equatorial, 115–118, 123, 124, 127, 128, 140, 147, 149–151, 153
    westward, 116, 117, 122–125, 134, 136, 137, 139, 146, 150, 152
Joule heating, 189, 191, 197
Julian–Toomre wakes, 423
Juno, 275
Jupiter, 79–80

K
Ka-band, 462, 463
Keeler gaps, 375, 377, 381–382, 408, 440, 452, 464, 470, 472, 523
Kepler speed, 421
KiloRayleigh, 334, 336, 337, 365
Kinematic viscosity, 414, 417, 426
Kinetic equation, 418, 419
Kinetic theory, 418–419, 421
Kuiper Belt Objects (KOBs), 63, 615–616
K_{zz}, 186

L
Lagrangian points, 447
Langrangian orbits, 639
Laplace gaps, 396, 397, 518
Lapse rate, 118, 119, 137, 139, 140
Late Heavy Bombardment (LHB), 56, 59, 70, 71, 547, 563, 566, 570
Libration, 444
Life, 717–719
Lightning, 136–140
Lindblad resonance(s), 376, 377, 379, 381, 382, 391, 395, 401, 436, 443–445
Liouville's theorem, 418
Lithosphere, 642, 672
Local thermodynamic equilibrium (LTE), 181
Local viscosity, 414, 417, 418, 425, 426, 428, 433
Loss-cone distribution, 345
L-shell, 640
Lyman continuum, 336
Lyman series, 336

M
MAG. *See* Magnetometer (MAG)
Magnetic field
    earth, 204
    equatorial, 204, 207
    Gauss coefficients, 209, 210
    induced, 289, 293
    intrinsic, 203, 209–211
    Jupiter, 207

Magnetic field perturbations, 292
Magnetic flux transport, 208
Magnetodisk, 263, 266, 272, 273, 274
Magnetometer (MAG), 348
Magnetopause, 204, 205, 207, 208, 211–213, 219, 221, 222, 226, 228, 230, 231, 234, 237, 238, 242, 247–248, 257, 260, 264, 265, 267–269, 274, 336, 347, 351, 361–363, 365, 368, 370
    compressibility, 247–248
    shape, 248
    standoff distance, 228, 230, 247
Magnetopause reconnection
    dayside magnetopause, 305
    interplanetary magnetic field (IMF), 305
    pre-Cassini study, 306–307
    proxy estimation, Earth, 306
Magnetosheath, 246
Magnetosphere, 3, 6, 333, 335, 342, 343, 347–350, 352, 353, 360–370
    corotation breakdown, 204, 207, 208
    corotation lag, 204
    current sheet, 206, 212, 213, 226
    Earth, 213
    Equinox mission science objectives, 728
    hinging distance, 212
    Jupiter, 211
    plasma environment, 31–32, 36–37
    plasma sheet, 212, 220, 231–234
Magnetosphere–ionosphere (M–I) coupling, 297–298
Magnetospheric and plasma science group (MAPS), 737
Magnetospheric dynamics, 258, 274, 347–351, 361, 365
Magnetospheric Imaging Instrument (MIMI), 206, 222–225, 227–231, 235, 243–245, 336, 340, 350, 352, 353, 356
Magnetospheric plasma
    composition, 220
    corotation, 12, 31, 33, 34, 38
    Enceladus source, 220
    losses, 19
    source rate, 3, 16, 19, 36, 38
    sources, 204, 206, 207, 211, 214, 215, 219
Magnetotail, 204, 207, 212, 213, 231, 234–237, 239, 258, 264, 267–273, 275, 343, 347, 350, 363, 366
    current, 213
    lobes, 234
Magnitude (opposition), 638
Main-belt asteroids, 615
Mass flux, 6–7
Mass loading interactions
    candidate process, 295
    cryovolcanic plume, 294
    3D hybrid code, 294
    draping pattern, magnetic field, 293, 294
    electric current distribution, 295
    plasma flow vectors, 294, 295
Mass point, 416
Mass transfer, 297
Maximum entropy methods, 438
Maxwell gap, 399, 518
Maxwellian distribution, 290
Mean free path, 415–417
Meridian line projections, 30
Meteoritic bombardment, 657, 659, 673
Meteoroid bombardment, 460, 492–497, 537, 543–545, 549–551, 557, 559, 560, 567, 568, 570
Meteoroid mass flux, 496, 501
Meteoroids, 521, 525, 527, 532
Methane (CH_4), 10–11, 85–89, 92, 95–101, 103, 107, 183, 184, 186–188, 337, 358, 688, 710, 711, 715, 718, 719

Index

Methane absorption, 164, 168, 172
Methane band filters, 121, 128, 137, 143
Methone, 523, 532, 533
Methylacetylene (CH$_3$C$_2$H), 14, 92, 94, 100, 101
Methyl radical (CH$_3$), 14, 92, 94, 97, 98, 100, 101, 105–107
Micrometeorite, 705
Micro-meteoroids, 435
Microprobe, 758
Mimas, 42, 64, 375, 377, 392, 395, 396, 400, 401, 436, 438, 439, 692, 716
   ISS basemaps, 766–767
   thermal evolution and internal structure, 604–605
Mimas gap, 304
MIMI. *See* Magnetospheric Imaging Instrument (MIMI)
Miscellaneous acceleration mechanisms, 316
Mixing (atmospheric), 186, 187, 198
Models of the general circulation
   Boussinesq, 151
   chemistry-diffusion, 153
   deep cylinders, 146–147
   3D general circulation models, 153
   2D transport models, 153
   shallow water, 149, 151
   shallow weather layers, 149
   two-layer, 149
Molecular conduction, 190, 191
Molecular oxygen, radiolytic production, 285–286
Moment equations, 419
Monolayer, 416, 445
Moonlet(s), 436, 439–445, 448–450
Moon–magnetosphere interactions
   Alfvén wings, 289
   internal magnetic fields, 289–290
   Keplerian motion, 290
   magnetohydrodynamic (MHD) waves, 289
   mass loading interactions, 293–295
   physical process, 290
   plasma absorbing interactions, 290–293
   plasma flow, 289
Morphology, 640–651, 671
Multilayer, 416
Multiple scattering, 461, 468, 473, 475, 476, 483, 484, 488, 500
Multi ring basins, 642
Multi-scale expansion, 432

N

Nanohematite, 488, 489, 497
Narrowband Saturn kilometric radiation (n-SKR), 345
Narrowband Saturn myriametric radiation (n-SMR), 345
Nearly isotropic comets (NICs), 625
Negative ions, 5
Neukum lunar chronology, 618
Neutral loadings, 287
Neutral scattering, 288
Newtonian, 420, 432, 435
Nice model, 56–59, 62, 67, 70
Nice model chronology, 619
Nitrogen (N$_2$), 684, 688, 710, 711, 715, 716, 720
Nitrogen ions, 206, 219
Non-local viscosity, 433
Non-Newtonian, 421, 435
North polar spot, 115, 136, 141–143
n-SKR. *See* Narrowband Saturn kilometric radiation (n-SKR)
n-SMR. *See* Narrowband Saturn myriametric radiation (n-SMR)

O

Oblateness, 444
Obliquity, 129
Occultation, 84–87, 93, 94, 98–101, 106, 107, 684, 685, 707–710, 712, 717, 720
Ocean, 686–689, 692, 693, 704, 715, 718, 719
Odysseus, 638, 642, 645, 646, 650, 651, 669
OH, 684, 716, 717
Open-closed field line boundary, 360, 363–367, 369, 370
Open field line, 347, 360, 363–366
Opposition effect, 484, 501–503
Optical constants, 653, 661
Optical depth (Ë), 163, 165, 167–171, 174, 414, 417, 418, 421–423, 425, 427–439, 441, 442, 445, 450, 451, 511–514, 516, 518, 520, 529
Orbital eccentricity, 588
Orbital evolution, 686, 692, 693, 719
Orbital period, 643
Orbiter, 757–758
Ordinary mode, 345, 346
Organics, 459, 476, 486–488, 497–500, 656, 660, 672
Organic tholins, 486, 498
Oxygen, 15, 83, 84, 88, 94–96, 100, 101, 104–107, 655, 656, 660

P

Pallene, 523, 532
Pan, 377, 381, 382, 396, 398, 408, 440–442, 448, 449
Pandora, 377, 381, 386, 387, 391, 392, 396, 398, 402–404, 436, 439, 443, 452
Pan wakes, 440, 441
Para fraction, 86, 91
Para-hydrogen fraction, 127
Particles
   composition, 163–165
   scattering phase function, 165, 166
   single scattering albedo, 170, 172
Particle spin, 419, 445, 449–451
Pedersen current, 362–364
Periodicity, 258, 261–264, 266
Phase-mixing, 439
Phase space density (PSD), 311–313
Phoebe, 45, 55, 63, 65, 66, 68
   ISS basemaps, 766, 770, 771
   thermal evolution and internal structure, 603–604
Phosphine (PH$_3$), 13, 89–91, 95–98, 106, 107, 122, 127
Photochemical haze, 161, 164
Photochemical model, 186, 187, 198
Photochemistry, 14–15, 83, 84, 90, 96–101, 104–107
Photodissociation, 493
Photoelectrons, 193, 196, 197
Photolysis, 90–92, 97–100, 107, 493, 658, 661
Photosputtering, 492
Physics and scaling laws, 619–621
Pickup acceleration
   electron and ion plasma distributions, 314, 315
   Jupiter's inner magnetosphere, 313
   pickup energization, 314
Pickup energization, 286
Pioneer, 55, 64, 334
Pioneer 11 observations, 209, 211, 221, 228, 246, 248
Planetary magnetic field, 29–31
Planetary magnetospheres, 296, 297
Planetary migration, 57–59
Planetesimals, 57–62, 65, 67, 446
Plasma absorbing interactions
   cold plasma response, 290–291

energetic particle response, 292–293
magnetic field response, 291–292
Rhea's gravity, 293
Plasma absorption
macrosignatures, 223
microsignatures, 223, 224
Plasma diffusion, 221
Plasma drag, 449
Plasma transport, 238, 241
Plasma waves, 37–38
Plasmoid, 207, 208, 235, 236, 258, 268, 270–273
Plume, 639, 640, 650, 651, 656, 660, 663, 673, 684–686, 688, 690, 693, 698–700, 703–720
Pluto, 63, 65, 66, 68
Poisson's equation, 420, 422
Polar cap, 334, 347, 348, 363, 364, 366
Polarization, 639, 663–666
Polar stratospheric haze, 2
Polycyclic aromatic hydrocarbons (PAHs), 487, 488, 494, 497, 499, 500
Porosity, 63, 64, 66
Potential vorticity, 118, 128, 133–135, 140, 145, 150, 153
Poynting flux, 357
Poynting-Robertson drag, 449
Precipitating electrons, 337, 358
Pressure tensor, 418, 420, 421
Probes, 758
Production rate, 197
Prograde rotation, 450
Prometheus, 375, 377, 381, 386, 391, 392, 397, 402–406, 436, 439, 443, 448, 452
Propane (C_3H_8), 92, 94, 98, 100, 101, 104, 107
Propeller(s), 375, 377, 382–386, 403, 429, 436, 440–443, 445, 446, 449, 451
Propeller belt, 443
Propeller objects, 473–474
Protoplanet, 56, 57, 60
Protoplanetary disk, 446

R
Radar, 116, 119, 120, 140, 154
Radau–Darwin relation, 75–76
Radial diffusion coefficient, 303, 304
Radial diffusion equation, 303
Radiation, 660, 661, 672, 673
Radiation belts, 221–225
Radiation pressure, 511, 518, 519, 529, 533
Radiative heating, 177
Radiative recombination, 193
Radiative time scale, 127
Radio and plasma wave science (RPWS), 339, 342–345, 348, 356
Radio and plasma wave science (RPWS) instrument, 259, 262, 273
Radio and plasma wave spectrometer (RPWS), 206, 215, 216, 218, 221, 241
Radio emissions, 339, 341–347, 369
Radiogenic heating, 69, 70, 689, 693
Radiogenic heat production, 643
Radio occultation observations, 181, 188, 191, 196
Radio occultations, 462, 466–473, 475, 477, 500
Radio science subsystem (RSS), 639
Ramps in optical depth, 435
Random walk, 440
Rayleigh scattering, 166–169, 177, 654–657, 673
Reconnection, 264, 267–270, 274, 275, 335, 336, 347, 348, 350, 361–363, 365, 366, 368, 369

magnetopause, 207, 208, 247
X-line, 208
Reflectivity, 483, 496, 497, 500
Regolith, 417, 418, 429, 442, 447, 659–662, 664–666, 671
Regolith grain size, 475, 483–487, 497
Regolith radiative transfer, 483–485, 497, 500
Regular satellite system, 55, 60, 61, 66
Remote sensing, 162
Resonance(s), 413, 436–439, 443–445, 449, 452
Restitution coefficient, 416, 417, 450
Resurfacing, 642, 650, 651
Rhea, 4, 44, 69, 70
ISS basemaps, 766, 769
VIMS composition map, 776–778
Rhines scale, 148, 149, 153
Ribbon, 115, 134, 141, 144, 145
Ring atmosphere
composition, 492, 500
oxygen, 460, 487, 492–496, 500
Ring color, 474, 480, 481, 483, 487, 497
Ring current, 206, 211, 213, 225–231, 234, 239, 248, 264, 267, 268, 273, 274, 340, 351, 355, 360, 363
Ring darkening, 542, 544, 545, 548, 560, 570
Ring ionosphere, 206
Ring many-particle-thick models, 466, 483, 501
Ring-moon interaction
Ring-moons, 436, 440, 442, 474
Ring optical depth, 460, 462, 463, 465, 467, 472, 473, 474, 475, 479–481, 486, 494, 496, 502, 503
Ring parent body, 498
Ring particle albedo, 460, 479, 481–483, 496, 497, 500, 502
Ring particle composition, 459–503
Ring particle maximum size, 461, 476
Ring particle phase function, 461, 475, 484, 497, 500, 502
Ring particle regoliths, 478, 483–484, 486, 502
Ring particle size, 459–503
Ring particle size distribution, 459–503
Ring radiative transfer models, 483, 485, 486, 497, 500
Ring "red bands," 480
Rings, 55, 62, 64, 66, 69, 71
Rings-age, 496, 500, 501, 545, 546, 548, 570
Rings and KBOs, 498
Rings and satellites, 476, 477, 484, 488–492, 496–500
Ring's density, 541, 544, 548, 552, 554, 555, 558
Ring shadows, 194
Rings-mass, 460, 474, 495–497, 501, 541–542, 545, 546, 548, 550, 567, 570
Rings-origin, 459, 489, 497–499, 538, 539, 547–548, 558, 559, 566–571
Rings-parent, 473–474, 489, 497, 498
Ring spectrum, 477, 483, 486, 490
Ring surface mass densities, 459, 474, 496, 497
Rings working group (RWG), 737
Ring system
A and B rings, 36–37
Cassini-Huygens mission, 753–755
C ring, 37–38
E, F and G rings, 38
Enke and Keeler gaps, 37
Equinox mission science objectives, 728–729
origin and evolution, 40–41
particle size distribution, 38–39
ring particle composition, 39–40
voyager image, 36–37
Ring temperature, 481–483, 486, 493
Ring thickness, 468, 473

Index

Ring UV absorber, 460, 476, 478, 480, 481, 486, 488, 497–500
Ring UVIS spectra, 479
Ring vertical structure, 461, 472, 483
Ring VIMS spectra, 476–481, 488
Roche division, 375, 511, 517–518, 533
Roche ellipsoids, 448
Roche limit, 446, 538–542, 553, 555, 556, 558, 559, 563–566
Roche zone, 375, 377, 446–449
Rock/ice fraction, 64–68
Rock-poor models, 599
Rock-rich models, 599
Rock thermal conductivity, 595–596
Rope structure, 440
Rossby number, 118, 142, 143, 149
Rossby waves, 115, 128, 131, 134, 135, 140, 142, 143, 148, 150
Rotation, 686, 690
    modulation, 261–267
    period, 117–118, 135, 141, 154
    state and equations, 76–78
Rotational degrees of freedom, 418
Rotational energy, 419
Rotational levels, 182, 185
Rotational modulation, 341, 348
Rotational period, 639
Roughness, 661–663, 673
RPWS. *See* Radio and plasma wave science (RPWS)
Rubble piles, 541, 565
Russell gaps, 394

S
Sample return, 720
Satellite accretion, 60–62
Satellite migration, 23
Satellite properties
    density, 579, 581
    gravitational field, 588–590
    physical and dynamical properties, 579–580
    porosity, 581–582
    Rhea's gravitational field, 583–585
    rock composition, 583
    size and shape, 579, 581
    volatile composition, 582–583
Satellite system, 7
Saturn
    atmosphere, 2, 5–6
    detailed evolutionary models, 79–80
    evolution, 78–79
    interior, 16, 75–80
    ring system, 3–4, 6–7, 36–41
    system prior, 1–2
Saturn electrostatic discharges (SEDs), 137, 192
Saturn Equinox (T52–T62), 738
Saturn Equinox Mission (SEM), 275
Saturnian rings (E-, F-), 638, 639, 650, 653–656, 658–660, 663, 664, 666, 668, 670, 672, 673
Saturnian satellites
    Albiorix, 638
    Atlas, 652, 653
    Calypso, 652
    Dione, 651, 665, 673
    Enceladus, 645, 649, 650
    Epimetheus, 638, 652, 653, 672
    Helene, 638
    Hyperion, 651, 652, 659
    Iapetus, 638, 642, 651, 668
    Janus, 638, 652
    Mimas, 638, 651
    Paaliaq, 638
    Pan, 638
    Pandora, 638, 652
    Phoebe, 638
    Prometheus, 638, 653
    Rhea, 638, 669
    Telesto, 638, 652
    Tethys, 638
    Titan, 638
Saturn Kilometric Radiation (SKR), 3, 26, 117, 118, 123, 206, 209, 210, 211, 227, 235, 247, 258, 261–265, 267, 269, 273, 274, 333, 334, 341, 343, 351, 361
Saturn longitude system (SLS), 263, 264
Saturn Orbit Insertion (SOI), 474, 476, 480, 481, 492, 493, 495
Saturnshine, 475, 476, 500
Saturn's ionosphere, 207, 214, 220, 234, 241
Saturn's magnetosphere, fundamental plasma processes
    adiabatic acceleration and related processes, 310–313
    Cassini ultraviolet imaging spectrograph (UVIS) imaging, 282, 283
    centrifugal interchange instability, 300–303
    corotation lag, 298–299
    current generation, 316–318
    dominant particle populations, 282
    magnetopause reconnection, 305–307
    magnetosphere–ionosphere coupling, 297–298
    moon–magnetosphere interactions, 290–295
    neutral gas, 286–289
    pickup acceleration, 313–314
    radial diffusion formalism, 303–304
    rotational *vs.* solar-wind drivers, 296–297
    suprathermal ion composition and abundances, 283, 284
    surfaces, 284–286
    tail reconnection, 307–310
    water dissociation and ionization, 284
    wave particle interactions, 318–323
Saturn's rings, 468, 469, 474–476, 485, 488, 490, 495, 498, 501–503
S-band, 462, 463, 468
Scale height, 181, 186, 190, 191, 199, 418, 433
Season, 86, 101–103
Seasonality, 114, 129
Secondary ionization, 197
SEDs. *See* Saturn electrostatic discharges
Selective instability, 433, 434
Self-gravity potential, 420
Self-gravity wakes, 375, 377, 378, 380, 383–386, 388, 391–393, 407, 413, 421, 423–430, 433, 435, 438, 440–443, 452
Semi-annual oscillation (SAO), 129, 132, 147
Semi-major axis, 421, 437, 438
Shear flow, 414, 436
Shear rate, 414, 424, 430, 435–436
Shear stress, 414–416, 430, 435
Shear viscosity, 414–415, 420, 421, 430, 431, 433
Shepherding, 443
Sidereal period, 638
Single scattering phase function, 661
Size-frequency distribution (SFD)
    comets, 624–625
    crater statistics and interpretation, 627–630
    impactor populations, 616
    voyager, 614
SKR. *See* Saturn kilometric radiation (SKR)
SKR period, 342, 350, 351
Slow rotators, 451
SLS3 longitude, 336, 342, 369
Smooth plains, 641, 642, 646, 651, 669

Snow line, 65
Sodium, 688, 711, 712, 715
Solar composition, 57, 64–68, 84, 86, 87, 88, 106
Solar composition condensate, 66
Solar nebula, 56, 57, 64–66, 68, 70, 84
Solar occultations, 188
Solar system, 755–756
Solar wind, 257–259, 265–271, 273–275, 333–336, 338–340, 343, 347–351, 361, 363, 365, 366, 368–370
    dynamic pressure, 212, 228, 229, 237, 242, 243, 246, 247
    interaction, 206, 207, 234, 242
Solar wind dynamic pressure, 334, 343, 348, 361, 365
Solar wind shock, 349
Solstice mission (SM), 275, 685
South polar region, 69, 70
South polar vortex, 143–144
Space telescope imaging spectrograph (STIS), 333–337, 339, 367
Spectrograms, 32–33
Spokes, 511, 512, 514, 519–521, 532, 533
Sputtering, 639, 640, 658, 660, 668, 673, 704, 705, 710, 716
Standing waves, 432
Stellar occultations, 182–188, 190, 197, 466, 468–470, 473
STIS. *See* Space telescope imaging spectrograph (STIS)
Stratosphere, 84, 92, 94, 95, 98–101, 103–105, 107, 114–116, 119, 121–123, 125, 126, 128–130, 132, 134, 143, 153, 154
Stratospheric composition
    hydrocarbons and photochemistry, 14–15
    isotopic ratios, 15–16
    oxygen supply, 15
Stratospheric haze, 127
Streamline(s), 437, 438, 440, 442–445
String of pearls, 145, 146
Structural evolution
    melting and differentiation, 597
    porosity, 596–597
    rock core, 597–598
Subcorotation, 240–242, 258, 259, 260, 270
Sublimation, 707
Subnebula, 60, 61, 66, 67
Substorm, 366
Sulfur, 83, 88, 95, 97, 106
Surface ages, 690, 701, 705, 719
Surface friction, 418, 419, 447, 448, 450
Surface irregularities, 419
Surface sticking, 555–557
Surface temperature, 703, 705, 712
Swing amplification, 423
Synchronization, 445
Synchronous orbit, 435
System III, 123, 127, 135, 136, 141–143

T
Tail reconnection
    circulation pattern, plasma, 307, 308
    dipolarization, 308
    field geometry, 307
    magnetic signatures, 308, 309
    plasmoid formation, 308
    Saturnian plasmoid, 310
    tail stretching and growth phase, 308
Tangential friction, 416, 450
TEC, 193
Tectonic, 70
Tectonic structures
    faults, 639, 668
    graben, 647, 669
    grooves, 641, 669
Tectonism, 694, 697, 701, 702
Temperature
    electron, 193, 194, 197
    ion, 191, 197
    ionospheric, 193–197
    mesopause, 182, 183, 187, 190
    plasma, 191, 197
    polar, 189
    thermospheric, 183–184, 188–190, 197
    upper atmosphere, 188–191
Temperature knee, 129
Temperature profile
    para/ortho-H_2 ratio, 20–21
    spatial variations, 20
    stratosphere, 22–23
    troposphere, 19–20
    voyager, 20–21
Tethys, 43–44, 64, 523–526, 528
    ISS basemaps, 766, 768, 774
    thermal evolution and internal structure, 604–605
Thermal equilibrium, 416, 428, 433
Thermal evolution, 642, 643, 649, 668
Thermal evolution and internal structure
    dynamical evolution, 598–603
    geology, 578
    heat sources, 585–590
    ice thermal conductivity, 595
    laboratory data, 605–606
    midsize icy satellites, 578
    Mimas, Tethys and Dione, 604–605
    modeling, 606
    Phoebe, 603–604
    porosity effect, 596
    processes, 606
    rock thermal conductivity, 595–596
    satellite properties, 579–585
    space, 605
    structural evolution, 596–598
    transfer, 590–595
Thermal inertia, 451, 705
Thermal profile, 188
Thermal radiation, 429, 451
Thermal relaxation time, 451
Thermal stability, 415–416, 450
Thermal torques, 449
Thermal waves, 115, 141, 148
Thermal wind, 124, 127–130, 133, 146, 147, 151, 152
Thermochemistry, 83, 95–96
Thermospheric dynamics, 338
Tholin, 476, 487–489, 494, 497–499
Three-body problem, 446
Tidal deformation, 643
Tidal force(s), 421, 425, 435, 446
Tidal heating, 685–693, 719
Tidally disrupted comet, 564–566
Tidally modified accretion, 554–555, 557
Tidal migration, 547
Tiger stripes, 69, 671, 684, 685, 689, 693, 696, 698–707, 712, 713, 717
Time dependent model, 194
Titan, 55, 63, 65–71, 357, 366, 399–401, 436, 443, 444
Titan and Enceladus Mission (TANDEM), 720
Titan groundtracks, 735, 737
Toomre factor, 554
Toomre parameter, 383, 423, 431
Toomre wakes, 421, 423

Topography, 640–651, 658, 670, 687, 688, 690, 700
Torque density, 439
Tour design and development process, 730–732
Trajectory phases
  apoapsis and periapsis profiles, 733, 736
  ascending and descending node crossing profiles, 733, 736
  cassini extended mission encounters, 732–733
  8-day Pi-transfer (T51–T52), 738
  Enceladus flybys, 733
  equinox viewing phase, 737
  high inclination phase (T45–T51), 733–734, 737–738
  high northern Titan groundtracks (T68–T70), 740
  icy satellites and Ansa-to-Ansa occultations (T62–T68), 738–739
  plume penetration passes, 733–734
  Saturn Equinox (T52–T62), 738
  Titan groundtracks, 735, 737
Transmission spectra, 182, 183
Transport coefficients, 419, 420
Traveling waves, 432
Triton, 63, 65, 67, 68
Trojans, 615
Tropopause, 90, 97, 98
Troposphere, 84, 86, 88–92, 95–98, 104, 107
Tropospheric composition
  ammonia, water and hydrogen sulfide, 11–12
  disequilibrium species, 13–14
Tropospheric haze, 116, 121, 123, 137, 140, 146, 154
Troughs, 194
Type III migration, 443

U
Ultraviolet (UV), 333–335, 337–340, 347, 349–351, 358, 364, 365, 369
Ultraviolet aurora (UV aurora), 333–335, 337, 338, 340, 343, 346–350, 352, 358, 359, 360, 369
Ultraviolet auroral observations, 26–27
Ultraviolet imaging spectrograph (UVIS), 86, 94, 99–101, 106, 107, 337, 339–340, 350, 352, 639, 640, 653, 658
United Kingdom infrared telescope (UKIRT), 339
Upper hybrid resonance (UHR) emissions, 319, 322
Upper troposphere, 114–116, 118, 119, 121–123, 125–129, 133–136, 140–142, 145, 146, 151–154
Upstream ions, 204, 243
UV. *See* Ultraviolet (UV)
UV aurora. *See* Ultraviolet aurora (UV aurora)
UVIS. *See* Ultraviolet imaging spectrograph (UVIS)
UV spectrometer (UVIS), 181–188, 190, 197–199

V
Vapor transport, 713
Vasyliunas cycle, 207, 208, 249, 336, 361–363, 370,
Velocity dispersion, 414, 416–418, 420, 422–425, 428, 431–433, 438–440, 450, 452
Velocity dispersion tensor, 419
Velocity distribution function, 418
Vertical density distributions, 182
Vibrational excitation, 196, 197
Vibrational populations, 182, 193
Vibrational temperatures, 191, 193–196, 199
VIMS. *See* Visual and infrared mapping spectrometer (VIMS)
VIMS composition map
  Dione, 771, 776, 777
  Enceladus, 776, 778–779
  Rhea, 776–778
Visco-elastic dissipation, 417
Viscosity, 413–415, 417, 418, 420–423, 425–433, 435, 437–439, 441, 445, 452, 543, 549, 552, 559
Viscous cooling, 420, 421, 440
Viscous diffusion, 435, 440
Viscous instability, 417, 418, 431, 433–434, 436, 451
Viscous overstability, 387, 391, 407, 413, 430–433, 444, 452
Viscous relaxation, 638, 644, 694
Viscous spreading, 537, 542–543, 545, 547, 548, 559
Viscous stirring, 418
Viscous torque, 439
Visual and infrared mapping spectrometer (VIMS), 90, 91, 99, 100, 107, 116, 119–124, 128–130, 136, 138–146, 152, 338–341, 639, 640, 651–654, 657–659, 661, 663, 665, 666, 673, 763–779
Volatiles, 650, 657–660, 671
Vortices, 135–136, 142, 143, 145, 149
Voyager, 55, 62, 64, 69, 84–87, 89, 92, 94, 96, 99, 100, 101, 106, 114–119, 123–125, 127–129, 135–137, 140–146, 333, 334, 341, 343, 344, 352, 361, 363, 462, 465, 467–473, 475, 479–481, 483, 487, 497, 501, 503, 511, 512, 514–517, 519–521, 524, 525, 528, 529, 533, 683, 686, 693, 694, 697, 705
  aerosol structure, 16–19
  atmosphere, 2
  icy satellites, 42–43
  impact cratering and age determination, 613–615
  magnetospheric plasma environment, 31–32
  observations, 210, 214, 217, 219, 224
  temperature profile, 20–21

W
Wakes, 539, 541–543, 554
Water, 11–12, 84, 88, 94–96, 100, 104–106, 162, 163
Water group ions, 206, 214–218, 221, 232, 240, 242, 2445
Water ice (H_2O), 638, 639, 650–658, 660, 661, 663, 664, 667, 671, 672
Water ice band depths, 480–482, 484–486, 491, 496, 497
Water ice laboratory spectra, 460, 484
Water influx, 191, 194–196
Wave activity, 191
Wavelet transform, 438
Wave particle interactions
  ECH, UHR, and narrowband radio emissions, 319–321
  Langmuir waves, 319
  miscellaneous plasma waves, 323
  resonant energy exchange, 318
  whistler-mode emissions, 322–323
Whistler mode, 354, 357
Winds, 86, 99, 103, 104, 106
  anticyclonic/cyclonic shear, 145
  meridional circulation, 127
  vertical shear, 127
  vertical velocity, 127
  zonal jets, 127

X
X-band, 462, 467, 470
"X" line, 258

Y
Yarkovsky effect, 449
Yarkovsky-Schach effect, 449

Z
Zonostrophic regime, 148